Organic Electronic Spectral Data
Volume XXVI 1984

Organic Electronic Spectral Data, Inc.

BOARD OF DIRECTORS

Organic Electronic Spectral Data

Volume XXVI 1984

JOHN P. PHILLIPS, DALLAS BATES
HENRY FEUER & B.S. THYAGARAJAN

EDITORS

CONTRIBUTORS

Dallas Bates

H. Feuer

L.D. Freedman

C.M. Martini

F.C. Nachod

J.P. Phillips

AN INTERSCIENCE ® PUBLICATION
JOHN WILEY & SONS, INC.
New York · Chichester · Brisbane · Toronto · Singapore

An Interscience ® Publication

Copyright © 1990 by John Wiley & Sons, Inc.

Library of Congress Catalog Card Number: 60-16428

ISBN 0-471-51941-3

Printed in the United States of America

10 9 8 7 6 5 4 3 2 1

INTRODUCTION TO THE SERIES

In 1956 a cooperative effort to abstract and publish in formula order all the ultraviolet-visible spectra of organic compounds presented in the journal literature was organized through the enterprise and leadership of M.J. Kamlet and H.E. Ungnade. Organic Electronic Spectral Data was incorporated in 1957 to create a formal structure for the venture, and coverage of the literature from 1946 onward was then carried out by chemists with special interests in spectrophotometry through a page by page search of the major chemical journals. After the first two volumes (covering the literature from 1946 through 1955) were produced, a regular schedule of one volume for each subsequent period of two years was introduced. In 1966 an annual schedule was inaugurated.

Altogether, more than fifty chemists have searched a group of journals totalling more than a hundred titles during the course of this sustained project. Additions and subtractions from both the lists of contributors and of journals have occurred from time to time, and it is estimated that the effort to cover all the literature containing spectra may not be more than 95% successful. However, the total collection is by far the largest ever assembled, amounting to over a half million spectra in the twenty-six volumes so far.

Volume XXVII is in preparation.

PREFACE

Processing of the data provided by the contributors to Volume XXVI as to the last several volumes was performed at the University of Louisville.

John P. Phillips
Dallas Bates
Henry Feuer
B.S. Thyagarajan

ORGANIZATION AND USE OF THE DATA

The data in this volume were abstracted from the journals listed in the reference section at the end. Although a few exceptions were made, the data generally had to satsify the following requirements: the compound had to be pure enough for satisfactory elemental analysis and for a definite empirical formula; solvent and phase had to be given; and sufficient data to calculate molar absorptivities had to be available. Later it was decided to include spectra even if solvent was not mentioned. Experience has shown that the most probable single solvent in such circumstances is ethanol.

All entries in the compilation are organized according to the molecular formula index system used by Chemical Abstracts. Most of the compound names have been made to conform with the Chemical Abstracts system of nomenclature.

Solvent or phase appears in the second column of the data lists, often abbreviated according to standard practice; there is a key to less obvious abbreviations on the next page. Anion and cation are used in this column if the spectra are run in relatively basic or acidic conditions respectively but exact specifications cannot be ascertained.

The numerical data in the third column present wavelength values in nanometers (millimicrons) for all maxima, shoulders and inflections, with the logarithms of the corresponding molar absorptivities in parentheses. Shoulders and inflections are marked with a letter s. In spectra with considerable fine structure in the bands a main maximum is listed and labelled with a letter f. Numerical values are given to the nearest nanometer for wavelength and nearest 0.01 unit for the logarithm of the molar absorptivity. Spectra that change with time or other common conditions are labelled "anom." or "changing", and temperatures are indicated if unusual.

The reference column contains the code number of the journal, the initial page number of the paper, and in the last two digits the years (1984). A letter is added for journals with more than one volume or section in a year. The complete list of all articles and authors thereof appears in the References at the end of the book.

Several journals that were abstracted in previous volumes in this series have been omitted, usually for lack of useful data, and several new ones have been added. Most Russian journals have been abstracted in the form of the English translation editions.

ABBREVIATIONS

s	shoulder or inflection
f	fine structure
n.s.g.	no solvent given in original reference
$C_6H_{11}Me$	methylcyclohexane
C_6H_{12}	cyclohexane
DMF	dimethylformamide
DMSO	dimethyl sulfoxide
THF	tetrahydrofuran

Other solvent abbreviations generally follow the practice of Chemical Abstracts.

Underlined data were estimated from graphs.

JOURNALS ABSTRACTED

Organic Electronic Spectral Data
Volume XXVI 1984

Compound	Solvent	λ_{max}(log ϵ)	Ref.
CBrF$_2$N Methanimine, N-bromo-1,1-difluoro-	gas	240(2.81),340s(--)	35-4266-84
CFN$_3$O Carbonazidic fluoride	gas	195(3.11),205(3.27)	64-1211-84B
CH$_2$Cl$_2$Si Silane, dichloromethylene-	Ar at 10°K	246(--)	24-2351-84
CH$_2$D$_2$Si Silane-d$_2$, methylene-	Ar at 10°K	258(--)	24-2351-84
CH$_3$ClOS Methanesulfinyl chloride	hexane	196(3.80),289(2.78)	23-0610-84
CH$_3$ClSi Silane, chloromethylene-	Ar at 10°K	255(--)	24-2351-84
CH$_4$Si Methylene, silyl-	Ar at 10°K	258f(--)	24-2351-84
C$_2$H$_2$N$_4$ Acetonitrile, azido-	MeCN	221(2.494)	1-0895-84
C$_2$H$_2$O$_4$ Oxalic acid	gas	277(1.40)	23-1414-84
C$_2$H$_3$Cl$_3$O$_3$S$_2$ Methanesulfinyl chloride, dichloro- (methylsulfonyl)-	hexane	226(3.54),274(2.58)	23-0610-84
C$_2$H$_4$O$_2$ Acetaldehyde, hydroxy-	H$_2$O	<u>265</u>(0.22)	110-0057-84
C$_2$H$_5$NO$_2$ Ethanol, 2-nitroso-	pH 8.8	270(4.03)	94-2609-84
C$_2$H$_6$NO$_4$P Ethanephosphonic acid, 2-(hydroxy- amino)-2-oxo- (dianion)	pH 13	219(3.95)	69-2779-84
C$_2$H$_6$NO$_5$P Ethanephosphonic acid, 2-nitro-, lithium salt	dianion	232(3.98)	69-2779-84
C$_2$H$_6$N$_2$O$_2$ Methanamine, N-methyl-N-nitro-	dioxan	244(3.66)	44-4860-84
C$_2$H$_6$N$_2$O$_3$ Methanol, (methylnitroamino)-	MeCN	231(3.80)	5-1494-84
C$_2$H$_6$Si Silane, methylmethylene- Silylene, dimethyl-	Ar at 10°K cyclopentane	260(--) 350(--)	24-2351-84 35-7267-84

Compound	Solvent	λ_{max} (log ϵ)	Ref.
C₃ClF₆NOS Methanamine, N-[(chlorocarbonyl)thio]- 1,1,1-trifluoro-N-(trifluoromethyl)-	C_6H_{12}	212(3.23),301s(2.33), 285(2.54)	155-0485-84A
C₃F₆S₂Se Carbonoselenodithioic acid, S,Se- bis(trifluoromethyl) ester	hexane	271(3.80),306(3.88), 318(3.64),499(1.34)	78-4963-84
C₃HBr₂NS Thiazole, 2,4-dibromo- Thiazole, 2,5-dibromo-	EtOH EtOH	213s(3.48),258(3.59) 222.5(3.45),260(3.84)	142-1077-84 142-1077-84
C₃HN₂ Propanedinitrile, anion	H_2O	<u>222(4.4)</u>	44-0482-84
C₃H₂BrF₃O 2-Propanone, 3-bromo-1,1,1-trifluoro-	gas	<u>222(1.5),315(0.70)</u>	46-1157-84
C₃H₂BrNS Thiazole, 2-bromo- Thiazole, 4-bromo- Thiazole, 5-bromo-	EtOH EtOH EtOH	211(3.49),248(3.73) 246.5(3.45) 218(3.49),245(3.64)	142-1077-84 142-1077-84 142-1077-84
C₃H₂N₄O Acetamide, 2-cyano-2-diazo-	H_2O	247(3.93)	103-1038-84
C₃H₃BrN₄O 1H-1,2,3-Triazole-4-carboxamide, 5-bromo-	H_2O	234(3.72)	103-1038-84
C₃H₃ClN₄O 1H-1,2,3-Triazole-4-carboxamide, 5-chloro-	H_2O	231(3.81)	103-1038-84
C₃H₃IN₄O 1H-1,2,3-Triazole-4-carboxamide, 5-iodo-	H_2O	218(3.96),242s(3.77)	103-1038-84
C₃H₃NOS 2(3H)-Oxazolethione	n.s.g.	211(3.64),268(4.17)	47-0739-84
C₃H₃NO₂S₂ 1,2,4-Dithiazolidine-3,5-dione, 4-methyl-	EtOH	235(3.53),256(3.51)	4-0241-84
C₃H₃NS Thiazole	EtOH	208(c.3.42),233(c.3.5)	142-1077-84
C₃H₃NSe Isoselenazole	isooctane	265(3.75)	78-0931-84
C₃H₃N₃ 1-Propyne, 3-azido-	MeCN	215(2.510)	1-0895-84
C₃H₃N₃O₂S₃ 1,3,5,2,4,6-Trithiatriazepine-7-carbox- ylic acid, methyl ester	EtOH	266(4.09),332(3.40)	77-0055-84
C₃H₃N₅O 2-Propanone, 1-azido-3-diazo-	MeOH	250(3.94),272(3.92)	103-0416-84

Compound	Solvent	$\lambda_{max}(\log \epsilon)$	Ref.
C_3H_4			
1,2-Propadiene	gas	171f(4.36),185f(3.51)	35-0952-84
C_3H_4BrNO			
2-Propenamide, 3-bromo-, (E)-	H_2O	225(4.10)	19-0085-84
	MeOH	222(4.10)	19-0085-84
	dioxan	220(4.11)	19-0085-84
	MeCN	219(4.13)	19-0085-84
$C_3H_4BrN_3$			
1-Propene, 3-azido-2-bromo-	MeCN	221(2.723)	1-0895-84
C_3H_4ClNO			
2-Propenamide, 3-chloro-, (E)-	H_2O	215(4.25)	19-0085-84
	MeOH	213(4.26)	19-0085-84
	dioxan	211(4.23)	19-0085-84
	MeCN	212(4.23)	19-0085-84
$C_3H_4ClN_3$			
1-Propene, 3-azido-2-chloro-	MeCN	221(2.446)	1-0895-84
C_3H_4INO			
2-Propenamide, 3-iodo-, (E)-	H_2O	245(3.91)	19-0085-84
	MeOH	242(3.95)	19-0085-84
	dioxan	240(3.92)	19-0085-84
	MeCN	240(3.94)	19-0085-84
$C_3H_4N_4$			
1H-Tetrazole, 1-ethenyl-	EtOH	224(3.79)	39-1933-84C
C_3H_4O			
2-Propenal	EtOH	210(4.00)	70-0414-84
$C_3H_5F_3OSSi$			
Ethanethioic acid, S-[(trifluorosilyl)-	heptane	233(3.54)	70-0726-84
methyl] ester	Bu_2O	234(3.62)	70-0726-84
	MeCN	237(3.73)	70-0726-84
	CH_2Cl_2	237(3.73)	70-0726-84
C_3H_5NO			
2-Propenal, 3-amino-, (Z)-	C_6H_{12}	287(4.06)	131-0019-84K
$C_3H_5N_3$			
1H-Imidazol-2-amine, hydrochloride	MeOH	245(3.97)	105-0126-84
1-Propene, 3-azido-	MeCN	213(2.790)	1-0895-84
$C_3H_6N_2O_2$			
Allylnitroamine	EtOH	232(3.80)	104-0449-84
Ethenamine, N-methyl-N-nitro-	MeCN	278(3.91)	5-1494-84
C_3H_6O			
Acetone	M KCl	265(1.27)	35-1351-84
$C_3H_6O_2$			
Acetaldehyde, methoxy-	octane	300f(1.52)	110-0057-84
$C_3H_6O_3$			
Propanal, 2,3-dihydroxy-, (±)-	H_2O	<u>285</u>(-0.16)	110-0057-84

Compound	Solvent	$\lambda_{max}(\log \epsilon)$	Ref.
C_3H_6Te			
Ethene, (methyltelluro)-	heptane	232(3.69),249s(3.58), 266(3.68)	104-1642-84
	EtOH	231(3.67),248s(3.59), 263(3.66)	104-1642-84
$C_3H_6Te_2$			
1,2-Ditellurolane	benzene	581(2.4),676(2.4)	48-0467-84
	DMSO	581(2.6)	48-0467-84
$C_3H_7NO_2$			
Ethanamine, 2-hydroxy-N-methylene-, N-oxide	pH 8.8	229(3.85)	94-2609-84
1-Propanol, 2-nitroso-	pH 8.8	270(3.97)	94-2609-84
2-Propanol, 1-nitroso-	pH 8.8	270(4.01)	94-2609-84
$C_3H_7NS_2$			
1,2-Dithiolan-4-amine, oxalate (1:1)	H_2O	333(2.00)	104-0773-84
$C_3H_8N_2O_3$			
1-Propanol, 2-(hydroxynitrosoamino)-	pH 3	224(4.14)	158-84-114
	pH 11	246(4.25)	158-84-114
inner salt	MeOH	240(3.67)	158-84-117
	MeOH-acid	225(3.59)	158-84-117
	MeOH-base	207(3.67),245(3.74)	158-84-117
$C_3H_8N_4O_2$			
Urea, [(methylnitrosoamino)methyl]-	pH 7	227(3.92),345(c.1.95)	118-0311-84

Compound	Solvent	$\lambda_{max}(\log \epsilon)$	Ref.
$C_4Cl_2F_4O_3$ Acetic acid, chlorodifluoro-, anhydride	gas	240(2.35),248(2.32)	60-2323-84
$C_4F_6O_2Se_2$ Ethanebis(selenoic) acid, Se,Se-bis(trifluoromethyl) ester	hexane	207(4.47),301(3.50), 394(1.90),412(1.88), 457(1.52)	78-4963-84
$C_4HF_9Se_3$ Methane, tris[(trifluoromethyl)seleno]-	hexane	208(3.51),262(3.00)	78-4963-84
$C_4H_2N_4O_2$ 1,2,4-Triazine-6-carbonitrile, 2,3,4,5-tetrahydro-3,5-dioxo-	EtOH	207(3.82),283(3.90)	103-0455-84
$C_4H_3BrN_2$ Pyrimidine, 5-bromo- cation	pH 2.0 H_0 -2.3	219(3.83),261(3.35) 227(3.82),273(3.49)	59-0075-84 59-0075-84
$C_4H_3BrN_2O_2$ 2,4(1H,3H)-Pyrimidinedione, 5-bromo-	neutral anion	276(3.85) 299(3.81)	41-0021-84 41-0021-84
$C_4H_3BrN_3S_2$ Pyrazolodithiazolium, 6-bromo-5-methyl-, chloride	CF_3COOH	328(3.94),500(3.07)	44-1224-84
$C_4H_3ClN_2O_5S$ 2(5H)-Thiophenone, 3-chloro-4-nitro-, oxime, 1,1-dioxide (class spectrum)	n.s.g.	233(4.4),303(4.1)	104-0809-84
$C_4H_3FN_2O_2$ 2,4(1H,3H)-Pyrimidinedione, 5-fluoro-	neutral anion	266(3.82) 268(3.60),296s(3.44)	41-0021-84 41-0021-84
$C_4H_3F_2N_3O_2S$ 6-Azauracil, 5-[(difluoromethyl)thio]-	pH 7 pH 13 EtOH	291(3.74) 303(3.86) 295(3.78)	104-2395-84 104-2395-84 104-2395-84
$C_4H_3IN_2O_2$ 2,4(1H,3H)-Pyrimidinedione, 5-iodo-	neutral anion	283(3.79) 303(3.86)	41-0021-84 41-0021-84
$C_4H_3N_2S_3$ Isothiazolodithiazolium, 6-methyl-, chloride	CF_3COOH	277(4.22),434(2.04)	44-1224-84
$C_4H_3N_3O_4$ 1,2,4-Triazine-6-carboxylic acid, 2,3,4,5-tetrahydro-3,5-dioxo-	EtOH	203(3.79),268(3.42)	103-0455-84
$C_4H_3N_5O_2$ 1,2,5-Oxadiazolo[3,4-d]pyrimidin-7(3H)-one, 5-amino-	EtOH	221(4.05),273(3.68), 325(3.60)	78-0879-84
C_4H_4 Cyclopropene, methylene-	pentane at -78°	206(c.3.5),242(--), 309(--)	35-3699-84

Compound	Solvent	$\lambda_{max}(\log \epsilon)$	Ref.
$C_4H_4Cl_2N_4O_{10}$			
2,3-Butanediol, 1,4-dichloro-1,1,4,4-tetranitro-, (±)-	H_2O	365(4.10)	104-2277-84
meso-	pH 13	230(3.76),386(4.18)	104-2277-84
	$CHCl_3$	273(2.53)	104-2277-84
$C_4H_4N_2O_2$			
3-Isoxazolecarbonitrile, 4,5-dihydro-, 2-oxide	H_2O	320(3.85)	104-0599-84
Uracil	neutral	261(3.91)	41-0021-84
	cation	275(3.85)	41-0021-84
	anion	285(3.76)	41-0021-84
$C_4H_4N_2O_5S$			
2(5H)-Thiophenone, 4-nitro-, oxime, 1,1-dioxide	n.s.g.	233(4.4),303(4.1)	104-0809-84
$C_4H_4N_3S_2$			
Imidazodithiazolium, 6-methyl-, chloride	CF_3COOH	382(3.54)	44-1224-84
Pyrazolodithiazolium, 5-methyl-, chloride	CF_3COOH	326(4.01),452(3.20)	44-1224-84
$C_4H_4N_4O$			
Acetamide, 2-cyano-2-diazo-N-methyl-	H_2O	247(3.98)	103-1038-84
$C_4H_4N_4O_3$			
1,2,4-Triazine-6-carboxamide, 2,3,4,5-tetrahydro-3,5-dioxo-	EtOH	210(3.77),277(3.84)	103-0455-84
$C_4H_4N_4O_6$			
2-Butenedial, 1,4-dinitro-, dioxime	pH 13	345(3.96),360(3.99), 412(3.53)	104-1264-84
	MeCN	283(4.01),370(3.42)	104-1264-84
di-aci-nitro form	pH 13	315(4.02)	104-1264-84
	EtOH	280(3.83)	104-1264-84
$C_4H_4N_4O_8$			
2-Butene, 1,1,4,4-tetranitro-	MeCN	245(4.00)	104-1264-84
$C_4H_4N_6$			
2-Butyne, 1,4-diazido-	MeCN	221(3.94),247(3.92)	89-0736-84
$C_4H_4O_3S$			
3(2H)-Thiophenone, 1,1-dioxide	pH 12	225(3.48),245(3.45), 330(3.22),350(3.04), 410(2.76)	104-0278-84
	EtOH	220(3.74)	104-0278-84
$C_4H_4O_4$			
2,4(3H,5H)-Furandione, 3-hydroxy-	EtOH	244(4.05)	39-1539-84C
C_4H_5Br			
1,2-Butadiene, 4-bromo-	MeCN	219(4.172),247(3.848)	89-0736-84
$C_4H_5BrN_4O$			
1H-1,2,3-Triazole-4-carboxamide, 5-bromo-N-methyl-	H_2O	232(3.96)	103-1038-84
$C_4H_5ClN_4O$			
1H-1,2,3-Triazole-4-carboxamide,	H_2O	207(3.84),237(3.91)	103-1038-84

Compound	Solvent	$\lambda_{max}(\log \epsilon)$	Ref.
5-chloro-N-methyl- (cont.)			103-1038-84
$C_4H_5IN_4O$			
1H-1,2,3-Triazole-4-carboxamide, 5-iodo-N-methyl-	H_2O	219(3.94)	103-1038-84
C_4H_5NSe			
Isoselenazole, 3-methyl-	isooctane	268(3.79)	78-0931-84
Isoselenazole, 5-methyl-	isooctane	221(3.51),265(3.74)	78-0931-84
$C_4H_5N_3O$			
Cytosine	neutral	265(3.73)	41-0021-84
	cation	275(3.95)	41-0021-84
	anion	282(3.84)	41-0021-84
$C_4H_5N_3OS$			
1,2,4-Triazin-5(4H)-one, 2,3-dihydro-4-methyl-3-thioxo-	EtOH	218(4.40),270(4.27), 310(3.55)	142-1049-84
1,2,4-Triazin-5(2H)-one, 3-mercapto-2-methyl-	EtOH	220(4.27),267(4.40), 312s(3.56)	142-1049-84
1,2,4-Triazin-5(2H)-one, 3-(methyl-thio)-	EtOH	234(4.27),300s(3.00)	142-1049-84
$C_4H_5N_5O$			
2-Butanone, 3-azido-1-diazo-	MeOH	250(4.12),268(3.97)	103-0416-84
$C_4H_5N_5O_3$			
Furazancarboxylic acid, [(aminoimino-methyl)amino]-	pH 13	235(3.88),273(3.58)	78-0879-84
C_4H_6BrNO			
2-Propenamide, 3-bromo-N-methyl-, cis	C_6H_{12}	212(3.94)	19-0085-84
	H_2O	224(3.91)	19-0085-84
	MeOH	227(3.91)	19-0085-84
	dioxan	218(3.97)	19-0085-84
	MeCN	221(3.92)	19-0085-84
trans	C_6H_{12}	223(4.20)	19-0085-84
	H_2O	227(4.21)	19-0085-84
	MeOH	224(4.19)	19-0085-84
	dioxan	222(4.19)	19-0085-84
	MeCN	223(4.18)	19-0085-84
C_4H_6ClNO			
2-Propenamide, 3-chloro-N-methyl-, cis	C_6H_{12}	206(3.97)	19-0085-84
	H_2O	215(3.97)	19-0085-84
	MeOH	213(3.95)	19-0085-84
	dioxan	209(3.91)	19-0085-84
	MeCN	211(3.91)	19-0085-84
trans	C_6H_{12}	215(4.22)	19-0085-84
	H_2O	216(4.22)	19-0085-84
	MeOH	214(4.20)	19-0085-84
	dioxan	214(4.18)	19-0085-84
	MeCN	213(4.21)	19-0085-84
C_4H_6INO			
2-Propenamide, 3-iodo-N-methyl-, cis	C_6H_{12}	232(3.81)	19-0085-84
	H_2O	240(3.79)	19-0085-84
	MeOH	243(3.86)	19-0085-84
	dioxan	244(3.82)	19-0085-84
	MeCN	243(3.88)	19-0085-84

Compound	Solvent	$\lambda_{max}(\log \epsilon)$	Ref.
2-Propenamide, 3-iodo-N-methyl-, trans	C_6H_{12}	245(4.02)	19-0085-84
	H_2O	242(4.03)	19-0085-84
	MeOH	242(4.08)	19-0085-84
	dioxan	241(4.07)	19-0085-84
	MeCN	240(4.08)	19-0085-84
$C_4H_6N_2$ 2,3-Diazabicyclo[2.1.1]hex-2-ene	hexane	316(2.09),320(2.27), 325(2.16),331(3.13)	35-4211-84
$C_4H_6N_2O_2$ 2-Propyn-1-amine, N-methyl-N-nitro-	EtOH	237(3.78)	104-0449-84
$C_4H_6N_2O_2S$ 1,2,4-Thiadiazolidine-3,5-dione, 2,4-dimethyl-	EtOH	237(3.61)	4-0241-84
$C_4H_6N_2O_3S$ 2H-1,2,6-Thiadiazin-3(6H)-one, 4-methyl-, 1,1-dioxide	50%EtOH-HCl + NaOH	283(3.65) 304(3.88)	73-0840-84 73-0840-84
$C_4H_6N_3O_3P$ Phosphonic acid, (cyanodiazomethyl)-, dimethyl ester	EtOH	207(3.71),235(4.09)	65-2359-84
$C_4H_6N_4$ 1H-Tetrazole, 1-(1-propenyl)-, (E)-	EtOH	229(3.87)	39-1933-84C
$C_4H_6N_4O_2$ 1,2,4,5-Tetrazinium, 3,4,5,6-tetrahydro-1,4-dimethyl-3,6-dioxo-, hydroxide, inner salt	dioxan	227(4.23),571(3.50), 603(3.49)	64-1790-84B
	HCOOH	524(3.56)	64-1790-84B
	70% $HClO_4$	217(4.13),464(3.47)	64-1790-84B
$C_4H_6O_2$ 2-Butenal, 4-hydroxy-, (E)-	MeOH	283(2.06)	158-0718-84
$C_4H_6O_3S$ 1,3-Butadiene-1-sulfonic acid, lithium salt, (Z)-	H_2O	231(3.74)	70-2519-84
C_4H_7NO 3-Buten-2-one, 4-amino-, (Z)- 2-Propenal, 3-(methylamino)-, (Z)-	C_6H_{12} C_6H_{12}	282(4.08) 306(4.15)	131-0019-84K 131-0019-84K
$C_4H_7N_3OS$ Thiophen, 2-azidotetrahydro-, 1-oxide	MeCN	223(2.98),286(1.80)	23-0586-84
$C_4H_7N_3O_2S$ Thiophene, 2-azidotetrahydro-, 1,1-dioxide	MeCN	229(3.69),253(3.48), 283(3.38)	23-0586-84
$C_4H_7N_3S$ Thiophene, 2-azidotetrahydro-	MeCN	222(2.95)	23-0586-84
$C_4H_7N_5O$ 1,2,4-Triazin-5(4H)-one, 3-amino-6-(aminomethyl)-, monohydrochloride	pH 1 pH 7 pH 13	255(3.78) 236(3.77),250s(3.30) 227(3.95),288(3.86)	4-0697-84 4-0697-84 4-0697-84

Compound	Solvent	$\lambda_{max}(\log \epsilon)$	Ref.
$C_4H_8N_2O_2$			
Ethenamine, N-ethyl-N-nitro-	MeCN	280(3.94)	5-1494-84
2-Propen-1-amine, N-methyl-N-nitro-	EtOH	240(3.80)	104-0449-84
$C_4H_8N_2O_4$			
Methanol, (methylnitroamino)-, acetate	MeCN	233(3.82)	5-1494-84
$C_4H_8OS_2$			
Propane(dithioic) acid, 3-hydroxy-, methyl ester	EtOH	305(4.18),454(1.18)	39-0859-84C
$C_4H_8OTe_2$			
1,4,5-Oxaditellurepane	benzene	369(2.7)	48-0467-84
C_4H_9NO			
Acetamide, N,N-dimethyl-	heptane or EtOH	214(1.94)	70-0310-84
$C_4H_9NO_2$			
1-Propanol, 2-(methyleneimino)-, N-oxide	pH 8.8	228(3.86)	94-2609-84
$C_4H_9NS_2$			
1,2-Dithiolan-4-amine, N-methyl-, oxalate	H_2O	333(2.02)	104-0773-84
$C_4H_9N_3O_2$			
1,1-Ethenediamine, N,N'-dimethyl-2-nitro-	pH 6.5	226(3.62),313(4.18)	39-1761-84B
$C_4H_{10}N_2O_3$			
Ethanol, 2,2'-(nitrosoimino)bis-	H_2O	234(3.83),345(c.2.04)	118-0311-84
$C_4H_{10}Te_2$			
Ditelluride, diethyl	benzene	394(2.8)	48-0467-84
$C_4H_{22}B_{18}Co$			
Cobaltate(1-), bis[(7,8,9,10,11-η)-undecahydro-7,8-dicarbaundeca-borato(2-)]-, cesium	MeOH	216(4.56),293(4.65), 345s(3.34),445(2.60)	73-2776-84
$C_4H_{22}B_{18}Fe$			
Ferrate(1-), bis[(7,8,9,10,11-η)-undecahydro-7,8-dicarbaundeca-borato(2-)]-, trimethylamine salt	MeCN	272(4.33),296(4.26), 444(2.77),520s(2.60)	73-2776-84
$C_4H_{22}B_{18}Ni$			
Nickelate(1-), bis[(7,8,9,10,11-η)-undecahydro-7,8-dicarbaundeca-borato(2-)]-, trimethylamine salt	MeCN	237(3.93),337(4.33), 435s(3.56)	73-2776-84

Compound	Solvent	$\lambda_{max}(\log \epsilon)$	Ref.
C_5BrN_3 Ethenetricarbonitrile, bromo-	CH_2Cl_2	270(3.85)	88-0117-84
$C_5F_{12}N_2O_2S$ Methanamine, N-[[[[bis(trifluoromethyl)amino]oxy]carbonyl]thio]-1,1,1-trifluoro-N-(trifluoromethyl)-	hexane	207.5(2.26)	155-0485-84A
$C_5F_{12}Se_4$ Methane, tetrakis[(trifluoromethyl)-seleno]-	hexane	211(3.70),267(3.30), 298s(3.11)	78-4963-84
$C_5HBr_2NO_4$ 2(5H)-Furanone, 3-bromo-5-(bromonitromethylene)-, (Z)-	$CHCl_3$	253(3.66),344(4.21)	12-0055-84
$C_5H_2BrClN_4$ 1H-Pyrazolo[3,4-d]pyrimidine, 3-bromo-4-chloro-	EtOH	262(3.62),286s(3.48)	87-1119-84
$C_5H_2N_4O$ Cyanamide, (3-oxo-1-cyclopropene-1,2-diyl)bis-, dipotassium salt	H_2O	243(4.34)	118-0752-84
$C_5H_2N_4S$ Cyanamide, (3-thioxo-1-cyclopropene-1,2-diyl)bis-, dipotassium salt	H_2O	243(4.40),253s(4.37)	118-0752-84
$C_5H_2N_4Se$ Cyanamide, (3-selenoxo-1-cyclopropene-1,2-diyl)bis-, disodium salt	H_2O	240(4.12),285s(3.90)	118-0752-84
$C_5H_3ClN_4$ 1H-Purine, 6-chloro-	neutral cation anion	266(3.93) 250(3.96) 274(3.91)	41-0021-84 41-0021-84 41-0021-84
$C_5H_3F_3N_2O_2$ 2,4(1H,3H)-Pyrimidinedione, 5-(trifluoromethyl)-	pH 12 pH 1-7	270(3.94) 256(--)	48-0985-84 48-0985-84
$C_5H_3F_4N_3O_2S$ 1,2,4-Triazin-3(2H)-one, 5-(difluoromethoxy)-6-[(difluoromethyl)thio]-	pH 7 pH 13 EtOH	287(3.70) 284s(3.58) 289(3.72)	104-2395-84 104-2395-84 104-2395-84
$C_5H_3N_3$ 5-Pyrimidinecarbonitrile cation	pH 5.0 H_0 -2.1	221(3.83),245(3.14), 250(3.17),255(3.07), 275(2.59) 275(3.77)	59-0075-84 59-0075-84
$C_5H_4Cl_2N_2O$ Pyridazinium, 4,5-dichloro-3-hydroxy-1-methyl-, hydroxide, inner salt	MeOH	229(4.44),328(3.56)	87-1613-84
$C_5H_4N_2O_2$ 5-Pyrimidinecarboxylic acid cation	pH 7.0 H_0 -1.1	245(3.30),275(2.71) 225(3.58),227(3.57)	59-0075-84 59-0075-84

Compound	Solvent	λ_{max}(log ϵ)	Ref.
C$_5$H$_4$N$_2$O$_2$S			
2(1H)-Pyridinethione, 5-nitro-	EtOH	386(4.15),470(3.00)	140-1797-84
C$_5$H$_4$N$_4$			
Purine	neutral	263(3.91)	41-0021-84
	cation	261(3.81)	41-0021-84
	anion	272(3.90)	41-0021-84
C$_5$H$_4$N$_4$O			
Hypoxanthine	neutral	250(3.94)	41-0021-84
	cation	249(3.99)	41-0021-84
	anion	260(3.95)	41-0021-84
C$_5$H$_4$N$_4$O$_2$S$_2$			
2H-Thiazolo[3,2-b][1,2,4]thiadiazol-2-imine, 6-methyl-N-nitro-	isoPrOH	279(3.15),343(4.20)	94-3483-84
C$_5$H$_4$N$_4$O$_3$			
Uric acid	neutral	283(3.90)	41-0021-84
	anion	293(3.87)	41-0021-84
C$_5$H$_4$N$_4$O$_3$S			
1H-Pyrazino[2,3-c][1,2,6]thiadiazin-4(3H)-one, 2,2-dioxide	acid	255(4.15),386(3.79)	4-0861-84
	pH 3.0	240(3.99),332(3.62)	4-0861-84
dianion	pH 8.0	257(4.16),375(3.70)	4-0861-84
C$_5$H$_4$N$_4$S			
6H-Purine-6-thione, 1,7-dihydro-	neutral	324(3.92)	41-0021-84
	anion	312(3.88)	41-0021-84
C$_5$H$_4$O$_2$			
2-Furancarboxaldehyde	C$_6$H$_{12}$	218(3.63),266(4.24), 317(1.78)	18-0844-84
	MeOH	218(3.49),270(3.97), 314(1.90)	18-0844-84
C$_5$H$_4$O$_3$			
2-Furancarboxylic acid	C$_6$H$_{12}$	228(4.04),255(4.23), 294(2.23)	18-0844-84
	MeOH	240(4.10)	18-0844-84
C$_5$H$_5$BrN$_2$O$_2$			
1H-Pyrrole, 2-bromo-1-methyl-5-nitro-	MeOH	343(4.11)	44-4065-84
C$_5$H$_5$ClN$_2$O$_2$			
1H-Pyrrole, 2-chloro-1-methyl-5-nitro-	MeOH	340(4.08)	44-4065-84
C$_5$H$_5$F$_2$N$_3$O$_2$S			
1,2,4-Triazine-3,5(2H,4H)-dione, 6-[(difluoromethyl)thio]-2-methyl-	pH 7	300(3.84)	104-2395-84
	pH 13	295(3.76)	104-2395-84
	EtOH	303(3.86)	104-2395-84
C$_5$H$_5$F$_3$O$_2$			
2,4-Pentanedione, 1,1,1-trifluoro-	EtOH	284(3.93)	39-2863-84C
C$_5$H$_5$NOS			
2(1H)-Pyridinethione, 1-hydroxy-	C$_6$H$_{11}$Me	291(4.28),372(3.63)	39-2031-84B
	MeOH-EtOH	282(4.28),347(3.81)	39-2031-84B
	EtOH-Et$_2$O	283(4.32),350(3.82)	39-2031-84B
	PrCN	282(4.32),350(3.85)	39-2031-84B

Compound	Solvent	λ_{max} (log ϵ)	Ref.
$C_5H_5NO_2$			
2(1H)-Pyridinone, 1-hydroxy-	C_6H_{12}	234(2.83),314(2.74)	40-0001-84
	MeOH	228(2.86),303(2.71)	40-0001-84
	MeCN	232(2.79),308(2.70)	40-0001-84
$C_5H_5N_3O_4$			
1,2,4-Triazine-2(3H)-acetic acid, 4,5-dihydro-3,5-dioxo-	pH 2 and 7	267(3.82)	73-2541-84
$C_5H_5N_5$			
Adenine	H_2O	207(4.36),260(4.12)	158-1284-84
	neutral	260(3.85)	41-0021-84
	cation	262(3.85)	41-0021-84
	anion	268(3.81)	41-0021-84
1H-Pyrazolo[3,4-d]pyrimidin-4-amine	pH 1	258(3.98)	103-0215-84
	pH 7	259.5s(--),272(4.00), 281s(--)	103-0215-84
	pH 11	262(4.02),282(3.82)	103-0215-84
$C_5H_5N_5O$			
4H-Imidazo[4,5-d][1,2,3]triazin-4-one, 1,5-dihydro-6-methyl-	H_2O	250(3.19),285(3.04)	4-1221-84
6H-Purin-6-one, 2-amino-1,7-dihydro-	neutral	275(3.85)	41-0021-84
	cation	248(4.02),271s(3.83)	41-0021-84
	anion	273(3.88)	41-0021-84
$C_5H_5N_5OS$			
[1,2,5]Oxadiazolo[3,4-d]pyrimidin-5-amine, 7-(methylthio)-	EtOH	210(4.31),304(3.91), 360(3.70)	78-0879-84
$C_5H_5N_5O_2S$			
1H-Pyrazino[2,3-c][1,2,6]thiadiazin-4-amine, 2,2-dioxide	pH 0	246(3.95),336(3.66), 401(3.47)	4-0861-84
anion	pH 7	259(4.16),381(3.73)	4-0861-84
$C_5H_6BrN_3O_2$			
1H-1,2,3-Triazole-4-carboxylic acid, 5-bromo-, ethyl ester	H_2O	237(3.84)	103-1038-84
$C_5H_6ClN_3O_2$			
1H-1,2,3-Triazole-4-carboxylic acid, 5-chloro-, ethyl ester	H_2O	237(4.08)	103-1038-84
$C_5H_6Cl_2Si$			
Silacyclohexa-2,4-diene, 1,1-dichloro-	pentane	264(3.74)	24-2337-84
Silacyclohexa-2,5-diene, 1,1-dichloro-	pentane	end absorption at 230 nm	24-2337-84
$C_5H_6IN_3O_2$			
1H-1,2,3-Triazole-4-carboxylic acid, 5-iodo-, ethyl ester	H_2O	241(3.75)	103-1038-84
C_5H_6N			
Pyridinium	H_2O	256(3.70)	142-0505-84
$C_5H_6N_2$			
2-Pyridinamine	heptane	290(3.63)	151-0343-84D
	PrOH	296(3.64)	151-0343-84D
	MeCN	293(3.68)	151-0343-84D
Pyrimidine, 5-methyl-	pH 4.1	252(3.43),256(3.41)	59-0075-84
cation	H_0 -1.0	255(3.53)	59-0075-84

Compound	Solvent	$\lambda_{max}(\log \epsilon)$	Ref.
$C_5H_6N_2O$			
2-Propenamide, 3-cyano-N-methyl-, (E)-	MeOH	218(4.16),231s(3.99), 252(3.68)	39-2097-84C
Pyrimidine, 5-methoxy-	pH 3.4	221(3.77),272(3.61), 266(3.24)	59-0075-84
cation	H_0 -1.1	227(3.87),287(3.66)	59-0075-84
$C_5H_6N_2O_2$			
Thymine	neutral	264(3.94)	41-0021-84
	anion	290(3.88)	41-0021-84
	n.s.g.	264.5(3.76)	156-0051-84B
$C_5H_6N_2O_5S$			
2(5H)-Thiophenone, 3-methyl-4-nitro-, oxime, 1,1-dioxide	n.s.g.	233(4.4),303(4.1)	104-0809-84
$C_5H_6N_4O_2$			
1H-Imidazole-4-carboxamide, 5-(formyl-amino)-	EtOH	230(3.76),267(4.11)	103-0204-84
2H-Pyrrol-2-one, 4-azido-1,5-dihydro-5-methoxy-	MeCN	210(3.75),239(3.84), 265(3.86)	142-1179-84
$C_5H_6N_4O_2S_2$			
Thiourea, (4-methyl-5-nitro-2-thiazo-lyl)-	isoPrOH	268s(3.78),297s(3.63), 364(4.12),440(3.00)	94-3483-84
C_5H_6O			
Bicyclo[1.1.1]pentanone	pentane	247(1.15)	44-4978-84
$C_5H_6O_2$			
2-Cyclopenten-1-one, 4-hydroxy-, (R)-	EtOH	215(3.77)	39-2089-84C
$C_5H_6O_3$			
2,4(3H,5H)-Furandione, 3-methyl-	EtOH	228(3.95),256(3.78)	39-1539-84C
	EtOH-NaOH	256(4.27)	39-1539-84C
$C_5H_6O_3S$			
2,4(3H,5H)-Furandione, 3-(methylthio)-	EtOH	228(3.91),258(3.82)	39-1539-84C
	EtOH-NaOH	252(4.14)	39-1539-84C
$C_5H_6O_4$			
2,4(3H,5H)-Furandione, 3-methoxy-	EtOH	237(4.03)	39-1539-84C
	EtOH-NaOH	260(4.31)	39-1539-84C
C_5H_6Si			
Silabenzene (low temperature)	n.s.g.	212(--),272(--), 305.0(--),313.4(--), 320.6(--)	24-2337-84
C_5H_7ClO			
3-Penten-2-one, 4-chloro-, (E)-	EtOH	239(3.86)	80-0193-84
C_5H_7ClSi			
Silacyclohexa-2,4-diene, 1-chloro-	C_6H_{12}	267(3.91)	24-2337-84
$C_5H_7NO_2$			
3(2H)-Furanone, 5-amino-4-methyl-	EtOH	268(3.95)	39-1539-84C
C_5H_7NSe			
Isoselenazole, 3,5-dimethyl-	isooctane	267(3.78)	78-0931-84
Isoselenazole, 3-ethyl-	isooctane	267(3.81)	78-0931-84

Compound	Solvent	$\lambda_{max}(\log \epsilon)$	Ref.
$C_5H_7N_3O$			
3(2H)-Pyridazinone, 6-(aminomethyl)-, hydrochloride	EtOH	224(3.79),293(3.38)	39-0229-84C
4(3H)-Pyrimidinone, 2-amino-3-methyl-	H_2O	226(3.91),284(4.04)	44-4021-84
$C_5H_7N_3OS$			
1,2,4-Triazin-5(2H)-one, 3,4-dihydro-2,6-dimethyl-3-thioxo-	pH 11	237(4.15),262(4.42)	142-1049-84
1,2,4-Triazin-5(2H)-one, 3,4-dihydro-4,6-dimethyl-3-thioxo-	pH 11	264(4.05),326(3.74)	142-1049-84
$C_5H_7N_3S_2$			
Thiourea, N-(4-methyl-2-thiazolyl)-	isoPrOH	252(3.82),293(4.29)	94-3483-84
$C_5H_7N_5O_3$			
Furazancarboxylic acid, [(aminoimino-methyl)amino]-, methyl ester, hydrochloride	EtOH	245(4.07)	78-0879-84
C_5H_8BrNO			
2-Propenamide, 3-bromo-N,N-dimethyl-, (E)-	C_6H_{12}	226(4.09)	19-0085-84
	H_2O	230(4.07)	19-0085-84
	MeOH	227(4.10)	19-0085-84
	dioxan	225(4.08)	19-0085-84
	MeCN	223(4.06)	19-0085-84
C_5H_8ClNO			
2-Propenamide, 3-chloro-N,N-dimethyl-, (E)-	C_6H_{12}	217(4.11)	19-0085-84
	H_2O	219(4.08)	19-0085-84
	MeOH	217(4.10)	19-0085-84
	dioxan	216(4.10)	19-0085-84
	MeCN	214(4.10)	19-0085-84
C_5H_8INO			
2-Propenamide, N-ethyl-3-iodo-, (E)-	C_6H_{12}	245(3.98)	19-0085-84
	H_2O	245(4.03)	19-0085-84
	MeOH	243(4.04)	19-0085-84
	dioxan	242(4.03)	19-0085-84
	MeCN	239(4.03)	19-0085-84
2-Propenamide, 3-iodo-N,N-dimethyl-, (E)-	C_6H_{12}	246(4.04)	19-0085-84
	H_2O	246(4.02)	19-0085-84
	MeOH	247(4.01)	19-0085-84
	dioxan	244(4.02)	19-0085-84
	MeCN	244(3.99)	19-0085-84
$C_5H_8N_2$			
2,3-Diazabicyclo[2.2.1]hept-2-ene	hexane	341(2.62)	24-0517-84
$C_5H_8N_4$			
1H-Tetrazole, 5-methyl-1-(1-propenyl)-, (E)-	EtOH	226(3.85)	39-1933-84C
$C_5H_8N_4O$			
1H-Imidazole-4-carboxamide, 5-amino-2-methyl-, monohydrochloride	H_2O	270(3.86)	4-1221-84
1,2,4-Triazin-5(2H)-one, 6-(aminometh-yl)-3-methyl-, dihydrochloride	EtOH	234(3.96)	39-0229-84C
$C_5H_8N_4OS$			
1H-Pyrazole-4-carboxamide, 3-amino-5-(methylthio)-	pH 1,7,11	256(3.82)	87-1119-84

Compound	Solvent	$\lambda_{max}(\log \epsilon)$	Ref.
$C_5H_8N_8O_2$			
Urea, N,N''-(6-amino-1,3,5-triazine-2,4-diyl)bis-	H_2O	216.0(4.62),243s(3.86)	18-2009-84
$C_5H_8O_2$			
2,4-Pentanedione	EtOH	274(4.00)	39-2863-84C
$C_5H_8O_3S$			
1,3-Butadiene-1-sulfonic acid, 2-methyl-, lithium salt, (Z)-	H_2O	235(3.70)	70-2519-84
1,3-Pentadiene-1-sulfonic acid, lithium salt, (Z,E)-	H_2O	238(4.04)	70-2519-84
C_5H_9NO			
2-Propenal, 3-(dimethylamino)-, (E)-	C_6H_{12}	272(4.47)	131-0019-84K
2-Propenamide, N,N-dimethyl-	C_6H_{12}	242(3.51)	19-0085-84
	H_2O	234(3.66)	19-0085-84
	MeOH	239(3.43)	19-0085-84
	dioxan	243(3.43)	19-0085-84
	MeCN	241(3.54)	19-0085-84
(Z)-	heptane	201(3.88)	131-0019-84K
$C_5H_{10}N_2O_2$			
2-Propanamine, N-ethenyl-N-nitro-	MeCN	275(3.86)	5-1494-84
$C_5H_{10}N_2O_4$			
Ethanol, 1-[(methylazoxy-NNO)oxy]-, acetate	MeCN	206(3.84)	5-1494-84
Ethanol, 1-(methylnitroamino)-, acetate	MeCN	234(3.75)	5-1494-84
Methanol, (ethylnitroamino)-, acetate	MeCN	234(3.84)	5-1494-84
Pentanediamide, N,N'-dihydroxy-	H_2O	208(3.95)	34-0345-84
$C_5H_{10}OS_3$			
Ethane(dithioic) acid, (methylsulfinyl)-, ethyl ester	EtOH	316(4.32)	39-0085-84C
$C_5H_{10}Te_2$			
1,2-Ditellurepane	benzene	398(2.8)	48-0467-84
$C_5H_{11}NOS$			
Propanal, 2-methyl-2-(methylthio)-, oxime	MeCN	235(<u>3.0</u>)	98-0873-84
$C_5H_{11}NO_2$			
1-Propanol, 2-methyl-2-(methyleneimino)-, N-oxide	pH 8.8	225(3.87)	94-2609-84
2-Propanol, 3-(ethylidenimino)-, N-oxide	pH 8.8	225(3.93)	94-2609-84

Compound	Solvent	$\lambda_{max}(\log \epsilon)$	Ref.
$C_6F_{12}SSe_3$ Ethene, tris[(trifluoromethyl)seleno]-[(trifluoromethyl)thio]-	hexane	207(3.94),301(3.68)	78-4963-84
$C_6HBr_3O_2$ 2,5-Cyclohexadiene-1,4-dione, 2,3,5-tribromo-	$CHCl_3$	300(4.03)	12-0055-84
C_6HCl_5O Phenol, pentachloro-	CCl_4	<u>297s(0.8),308(1.0)</u>	93-1038-84
$C_6H_2Br_2N_4S_2$ 1,2,5-Thiadiazole, 3,3'-(1,2-dibromo-1,2-ethenediyl)bis-, (E)-	C_6H_{12}	206(3.96),226(3.94), 246(4.01),267(4.12), 300(3.95)	4-1157-84
(Z)-	C_6H_{12}	200(4.32),230(4.27), 260(4.20),295(4.11)	4-1157-84
$C_6H_2Br_2O_2$ 2,5-Cyclohexadiene-1,4-dione, 2,6-di-bromo- ($1\lambda,2\epsilon$)	$CHCl_3$	293(4.08),?(3.93)	12-0055-84
$C_6H_2Br_4O_2$ 1,2-Benzenediol, 3,4,5,6-tetrabromo-	pH 1 borate	276(3.28) 281(3.46)	3-1935-84 3-1935-84
$C_6H_2N_4O_6$ 4-Benzofurazanol, 5,7-dinitro-	H_2O MeOH EtOH PrOH BuOH isoAmOH dioxan 80% dioxan MeCN $CHCl_3$ $C_2H_4Cl_2$	305(3.89),425(4.35) 305(3.96),425(4.37) 305(3.95),425(4.37) 305(3.95),425(4.37) 305(3.94),425(4.36) 305(3.98),425(4.36) 295(3.85),405(4.35) 305(3.93),431(4.46) 309(3.98),432(4.45) 325(3.95),370(3.53) 328(3.97)	65-0602-84 65-0602-84 65-0602-84 65-0602-84 65-0602-84 65-0602-84 65-0602-84 65-0602-84 65-0602-84 65-0602-84 65-0602-84
$C_6H_2N_4SSe$ [1,2,5]Selenadiazolo[3,4-e]-2,1,3-benzothiadiazole	MeCN	291(3.58),310s(3.24), 332s(2.87),348s(2.40)	64-0485-84B
$C_6H_3Cl_2N_3$ 1H-Pyrrolo[2,3-d]pyrimidine, 2,4-di-chloro-	MeOH MeOH	228(4.43),275s(3.60), 290(3.63) 227(4.48),290(3.62)	5-0722-84 35-6379-84
$C_6H_3Cl_3O$ Phenol, 2,4,5-trichloro-	CCl_4	<u>290(1.2),300(1.2)</u>	93-1038-84
$C_6H_3F_5N_2O_2$ 2,4(1H,3H)-Pyrimidinedione, 5-(penta-fluoroethyl)-	pH 1-7 pH 14	258.3(--) 281(3.71)	48-0985-84 48-0985-84
$C_6H_3N_3O_3$ Benzofurazan, 4-nitro- (in 15% DMF)	pH 2.7	327(3.83)	48-0385-84
$C_6H_3N_3O_4$ 4-Benzofurazanol, 7-nitro- (in 15% DMF)	pH 2.7	467(4.31)	48-0385-84

Compound	Solvent	$\lambda_{max}(\log \epsilon)$	Ref.
$C_6H_3N_5S$ 1H-Pyrazolo[3,4-d]pyrimidine-3-carbo- nitrile, 4,5-dihydro-4-thioxo-	EtOH	253(3.83),329(3.94)	103-0210-84
$C_6H_3N_7$ 1,3,4,6,7,9,9b-Heptaazaphenalene	MeCN	219(3.99),279s(3.13), 293(4.31),298s(4.28), 305(4.36),391s(1.98), 416(2.30),443(2.43)	35-2805-84
$C_6H_4BrNO_3$ Furan, 2-bromo-5-(2-nitroethenyl)-	MeOH	208(3.78),234(3.80), 355(4.29)	73-2496-84
$C_6H_4BrN_2$ Benzenediazonium, 4-bromo-, tetra- fluoroborate	DMSO	285(4.18)	123-0075-84
$C_6H_4Br_3N$ Benzenamine, 2,4,6-tribromo-	MeOH	315(3.60)	65-0289-84
$C_6H_4ClN_3O_4$ Benzenamine, 4-chloro-2,6-dinitro- anion	MeOH MeOH	441(3.92) 447(3.90)	65-0289-84 65-0289-84
$C_6H_4Cl_2N_4$ 1H-Purine, 6-chloro-8-(chloromethyl)-	pH 1 pH 13	270(4.06) 280(4.07)	39-0879-84C 39-0879-84C
$C_6H_4N_2O$ 1,2,3-Benzoxadiazole (calcd.) (equil. with following compd.) 2,4-Cyclohexadien-1-one, 6-diazo-	hexane MeOH-5% hexane	203(4.04),244(3.65), 295(3.34) 232(4.3),254(3.8), 407(3.7)	89-0509-84 89-0509-84
$C_6H_4N_2O_3S$ 2-Thiophenecarbonitrile, 5-methoxy- 4-nitro-	MeOH	278(3.94),313(3.72)	39-0317-84B
$C_6H_4N_2O_5$ Furan, 2-nitro-5-(2-nitroethenyl)-	MeOH	208(3.49),246(3.91), 350(4.28)	73-2496-84
$C_6H_4N_2S$ 1,2,3-Benzothiadiazole	hexane CH_2Cl_2	215(4.28),256(3.67), 310(3.36) 252(3.6),260(3.6), 310(3.6)	89-0509-84 24-0107-84
$C_6H_4N_3O_2$ Benzenediazonium, 4-nitro-, tetra- fluoroborate	DMSO	310s(3.24)	123-0075-84
$C_6H_4N_4$ 1H-Imidazo[1,2-b]pyrazole-7-carboni- trile	pH 1 pH 7 pH 11	240(3.79) 240(3.82) 268(3.66)	44-3534-84 44-3534-84 44-3534-84
$C_6H_4N_4O_2$ Benzene, 1-azido-3-nitro- Benzene, 1-azido-4-nitro-	2% MeOH 2% MeOH	246(4.26),330(3.18) 222(3.88),320(4.08)	149-0545-84B 149-0545-84B

Compound	Solvent	$\lambda_{max}(\log \epsilon)$	Ref.
$C_6H_4N_4O_3$ 4-Benzofurazanamine, 7-nitro- (in 15% DMF)	pH 2.7	466(4.37)	48-0385-84
$C_6H_4N_4O_6$ Benzenamine, 2,4,6-trinitro-	MeOH	410(3.85)	65-0289-84
$C_6H_4N_6$ 2,4-Hexadiyne, 1,6-diazido- 1H-Pyrazolo[3,4-d]pyrimidine-3-carbo- nitrile, 4-amino-	MeCN EtOH	223(3.86) 239(4.00),282(3.94), 287(3.93)	1-0623-84 103-0215-84
$C_6H_4O_4$ 2,5-Cyclohexadiene-1,4-dione, 2,5-di- hydroxy-	aq base EtOH EtOH-HCl	316(--),324(4.28), 497(2.30) 233(--),296(--), 493(2.34) 396(2.06)	94-2406-84 12-0611-84 12-0611-84
$C_6H_5BrN_2OS$ 4(3H)-Pyrimidinethione, 6-bromo-3-eth- enyl-5-hydroxy-	EtOH	280(3.75),365(3.96)	4-1149-84
$C_6H_5ClN_2O_4$ 1(2H)-Pyrimidineacetic acid, 5-chloro- 3,4-dihydro-2,4-dioxo-	pH 2 and 7 pH 12	276(3.93) 278(3.81)	73-2541-84 73-2541-84
C_6H_5NOS 2-Pyridinecarbothioic acid iron complex	MeOH MeOH	267(3.85),340(3.35) 267(4.25),407(3.05), 670(2.50)	64-1607-84B 64-1607-84B
$C_6H_5NO_2$ Nitrobenzene (protonated)	CF_3SO_3H	344(4.03)	88-0325-84
$C_6H_5NO_2S$ 5H-Thiopyrano[4,3-d]isoxazol-4(7H)-one	EtOH	235.5(3.80)	4-1437-84
$C_6H_5NO_3$ Acetaldehyde, (3-amino-5-oxo-2(5H)- furanylidene)- (basidalin) Furan, 2-(2-nitroethenyl)-	pH 2 pH 12 MeOH MeOH	251(4.22),308(3.98) 260(4.31) 220(3.89),277(4.20) 207(3.58),238(3.60), 346(4.23)	158-84-33 158-84-33 158-84-33 73-2496-84
$C_6H_5NO_4$ 1,2-Benzenediol, 4-nitro-	pH 1 borate	345(3.85) 385(3.93)	3-1935-84 3-1935-84
$C_6H_5N_3$ Benzene, azido-	MeCN	250(<u>4.0</u>)	35-0715-84
$C_6H_5N_3O$ Imidazo[1,2-a]pyrimidin-5(1H)-one Imidazo[1,2-c]pyrimidin-5(6H)-one 4-Pyridazinecarbonitrile, 2,3-dihydro- 2-methyl-3-oxo- 4-Pyridazinecarbonitrile, 3-methoxy- Pyridazinium, 4-cyano-2,3-dihydro-1- methyl-3-oxo-, hydroxide, inner salt	H_2O H_2O MeOH MeOH MeOH	213(4.23),296(3.93) 268(3.94) 328(3.71) 292(3.60) 222(4.31),362(3.63)	44-4021-84 44-4021-84 87-1613-84 87-1613-84 77-0422-84 +87-1613-84

Compound	Solvent	$\lambda_{max}(\log \epsilon)$	Ref.
$C_6H_5N_3O_3S$			
Benzenesulfonic acid, 4-azido-	2% MeOH	256(4.20)	149-0545-84B
$C_6H_5N_3O_4$			
Benzenamine, 2,4-dinitro-	MeOH	337(4.16)	65-0289-84
Benzenamine, 2,6-dinitro-	MeOH	423(4.00)	65-0289-84
$C_6H_5N_3O_7S$			
Benzenesulfonic acid, 4-amino-3,5-di-nitro-	MeOH	421(3.85)	65-0289-84
	anion	411(4.15)	65-0289-84
$C_6H_5N_5$			
Cyanamide, 7H-imidazo[1,5-a]imidazol-5-yl-	MeOH	240(4.20)	4-0753-84
$C_6H_5N_5O_3$			
Acetamide, N-(3,7-dihydro-7-oxo[1,2,5]-oxadiazolo[3,4-d]pyrimidin-5-yl)-	EtOH	235(4.05),260s(3.81), 319(3.60)	78-0879-84
$C_6H_6BrNO_2$			
1H-Pyrrole-2,5-dione, 3-bromo-1,4-di-methyl-	MeOH	238(4.21)	39-2097-84C
$C_6H_6BrN_5$			
1H-Purin-6-amine, 8-bromo-1-methyl-	pH 1.0	263(4.20)	39-0879-84C
	pH 11.0	273(4.21)	39-0879-84C
$C_6H_6ClN_5$			
1H-Purin-6-amine, 8-chloro-1-methyl-	pH 1.0	261(4.17)	39-0879-84C
	pH 11.0	272(4.18)	39-0879-84C
$C_6H_6Cl_2N_2O$			
Pyrimidine, 4,5-dichloro-2-methoxy-6-methyl-	MeOH	222(3.63),280(3.38)	4-1161-84
$C_6H_6Cl_2N_2S$			
Pyrimidine, 4,5-dichloro-6-methyl-2-(methylthio)-	MeOH	220(2.56),261(4.32), 300(2.44)	4-1161-84
$C_6H_6Cl_2O$			
3,5-Hexadien-2-one, 1,4-dichloro-, (E)-	MeCN	270(4.26)	104-1884-84
$C_6H_6N_2OS$			
4(3H)-Pyrimidinethione, 3-ethenyl-5-hydroxy-	EtOH	259(3.78),281(3.77), 359(3.97)	4-1149-84
$C_6H_6N_2O_2$			
Aziridine, 1-(5-isoxazolylcarbonyl)-, monohydrochloride	EtOH	208(4.28),238(4.28), 304(4.3)	120-0326-84
Benzenamine, 2-nitro-	MeOH-CCl₄	404(3.71)	12-0341-84
3-Pyridinecarboxamide, 4-hydroxy-	MeOH	251(3.94),282(2.60)	94-1355-84
5-Pyrimidinecarboxylic acid, methyl ester	pH 3.0	221(3.62),241(3.26), 245(3.25),252(3.12), 280(2.87)	59-0075-84
cation	H₀ -1.1	228(3.48),278(3.52)	59-0075-84
$C_6H_6N_2O_4S$			
2-Thiophenecarboxamide, 5-methoxy-4-nitro-	MeOH	250(4.16),282(3.99), 322(3.73)	39-0317-84B

Compound	Solvent	$\lambda_{max}(\log \epsilon)$	Ref.

$C_6H_6N_2O_4S_3$
  1,3,5,2,4-Trithia(3-S^{IV})diazepine-
  6,7-dicarboxylic acid, dimethyl ester
	EtOH	274(4.26),334(3.56)	77-0055-84

$C_6H_6N_4$
  Benzenamine, 4-azido-
  1H-Purine, 6-methyl-
	pH 2.0
	neutral
	cation
	anion
		248(4.08),262(4.15)
		261(3.93)
		264(3.88)
		271(3.95)
			149-0545-84B
			41-0021-84
			41-0021-84
			41-0021-84

1H-Pyrrolo[2,3-d]pyrimidin-2-amine	MeOH	230(4.41),252s(3.45), 316(3.66)	5-1719-84
$C_6H_6N_4O$ Imidazo[1,2-b]pyrazole-7-carboxamide	pH 1 pH 7 pH 11	237s(3.81),258(3.91) 245s(3.88),263(4.09) 222(3.97),267(3.91)	44-3534-84 44-3534-84 44-3534-84
4H-Imidazo[4,5-c]pyridin-4-one, 6-amino-1,5-dihydro-	pH 1 pH 7 pH 11	274(4.06),312(3.81) 262(4.02),298(3.91) 262(4.00),298(3.89)	87-1389-84 87-1389-84 87-1389-84

$C_6H_6N_4OS$
  6H-Purin-6-one, 1,2,3,7-tetrahydro-
  1-methyl-2-thioxo-
	pH 13	235(4.45),286(4.30)	2-0316-84

$C_6H_6N_4O_2$
  1H-Imidazole-1-carboxylic acid, 4-ami-
  no-5-cyano-, methyl ester
  1H-Imidazole-1-carboxylic acid, 5-ami-
  no-4-cyano-, methyl ester
	EtOH
	
	EtOH
		210(4.11),230(3.80),
		280(3.86)
		252(3.99),270(3.74)
			103-0204-84
			
			103-0204-84

$C_6H_6N_4O_2S_2$
  2H-Thiazolo[3,2-b][1,2,4]thiadiazol-
  2-imine, 5,6-dimethyl-N-nitro-
  2H-Thiazolo[3,2-b][1,2,4]thiadiazol-
  2-imine, 6-ethyl-N-nitro-
	isoPrOH
	
	isoPrOH
		242s(3.53),260s(3.26),
		282s(3.08),348(4.13)
		279(3.18),343(4.19)
			94-3483-84
			
			94-3483-84

$C_6H_6N_4O_4$ 2(1H)-Pyridinimine, 1-(dinitromethyl)-	MeCN	335(4.28)	39-0069-84C

$C_6H_6N_6$
  4H,9H-Bis[1,2,3]triazolo[1,5-a:1',5'-
  d]pyrazine
	MeCN	216(3.83)	1-0623-84

$C_6H_6N_6S$
  1H-Pyrazolo[3,4-d]pyrimidine-3-carbo-
  thioamide, 4-amino-
	n.s.g.	234(4.06),258(3.99),
		303(3.97)
			103-0210-84

$C_6H_6N_8$ 1H-Purin-6-amine, 8-azido-1-methyl-	pH 1.0 pH 11.0	282(--) 291(--)	39-0879-84C 39-0879-84C
$C_6H_6O_2$ 1,2-Benzenediol	pH 1 borate	274(3.35) 282(3.58)	3-1935-84 3-1935-84
Ethanone, 1-(2-furanyl)-	C_6H_{12} MeOH	261(4.24),309(1.90) 219(3.66),267(4.27), 307(1.98)	18-0844-84 18-0844-84
$C_6H_6O_3$ 1,2,3-Benzenetriol	pH 1 borate	265(2.82) 272(3.17),277s(--)	3-1935-84 3-1935-84

Compound	Solvent	$\lambda_{max}(\log \epsilon)$	Ref.
2-Butynal, 4-acetoxy-	hexane	215s(3.71),217.8(3.87), 226(3.82),258s(1.99), 277s(1.12),310s(1.27), 319(1.33),329.5(1.36), 342.3(1.31),357.5(1.12), 375.0(0.72)	44-1898-84
2H-Pyran-2-one, 4-hydroxy-6-methyl-	EtOH	276(3.86)	39-1317-84B
	dioxan	284(3.72)	39-1317-84B
	CHCl₃	284(3.77)	39-1317-84B
	80% DMSO	284(3.78)	39-1317-84B
$C_6H_6O_8S_2$			
1,3-Benzenedisulfonic acid, 4,5-dihydroxy-	pH 1	290(3.56)	3-1935-84
	borate	297(3.74)	3-1935-84
$C_6H_7BrN_2O$			
2H-Pyrrol-2-one, 3-bromo-1,5-dihydro-5-imino-1,4-dimethyl-	MeOH	249(4.29)	39-2097-84C
$C_6H_7ClN_2OS$			
4(1H)-Pyrimidinone, 5-chloro-6-methyl-2-(methylthio)-	MeOH	218(3.64),246(3.92), 289(3.49)	4-1161-84
$C_6H_7ClN_2O_2$			
4(1H)-Pyrimidinone, 5-chloro-2-methoxy-6-methyl-	MeOH	224(3.39),275(3.47)	4-1161-84
$C_6H_7F_3O_2S_2$			
Acetic acid, trifluoro-, 3-(methylthio)-3-thioxopropyl ester	ether	304.5(4.29),454(1.15)	39-0859-84C
C_6H_7NO			
Methanamine, N-(2-furanylmethylene)-	C_6H_{12}	264(4.25),360(0.88)	18-0844-84
	EtOH	264(4.39),354(1.02)	18-0844-84
4(1H)-Pyridinone, 1-methyl-	EtOH	263(4.28)	118-0485-84
$C_6H_7NO_2$			
1H-Pyrrole-2,5-dione, 1,3-dimethyl-	MeOH	222(4.18)	39-2097-84C
$C_6H_7NO_2S$			
2H-Thiopyran-3(6H)-one, 5-hydroxy-4-(iminomethyl)-	EtOH	250(4.00),292.5(4.12)	4-1437-84
$C_6H_7NO_3S$			
2H-Thiopyran-4-carbonitrile, 3,4,5,6-tetrahydro-5,5-dihydroxy-3-oxo-	EtOH	234(3.80),269.5(3.97), 295s(3.69)	4-1437-84
$C_6H_7NO_5S_2$			
Thiophene, 2-methoxy-5-(methylsulfonyl)-3-nitro-	MeOH	240(4.18),265(3.92), 310(3.71)	39-0317-84B
$C_6H_7N_3O$			
Imidazo[1,2-c]pyrimidin-5(6H)-one, 7,8-dihydro-	MeOH	231(3.33)	24-2597-84
$C_6H_7N_3O_3$			
Imidazo[2,1-b]oxazole, 2,3-dihydro-2-methyl-5-nitro-	n.s.g.	334(3.97)	2-0363-84
Imidazo[2,1-b]oxazole, 2,3-dihydro-3-methyl-6-nitro-	n.s.g.	324(3.87)	2-0363-84

Compound	Solvent	$\lambda_{max}(\log \epsilon)$	Ref.
$C_6H_7N_3S$ Imidazo[1,2-c]pyrimidine-5(6H)-thione, 7,8-dihydro-	MeOH	284(4.03)	24-2597-84
$C_6H_7N_5O_2S$ 8H-Pyrazino[2,3-c][1,2,6]thiadiazin- 4-amine, 8-methyl-, 2,2-dioxide	pH 7 MeOH	258(3.98),313(3.18), 401(3.76) 260(4.00),324s(3.23), 410(3.78)	4-0861-84 4-0861-84
$C_6H_7N_5O_4S$ Urea, N-methyl-N'-(4-methyl-5-nitro- 2-thiazolyl)-N-nitroso-	isoPrOH	235(4.04),342(4.05), 416(3.85)	94-3483-84
$C_6H_7N_5O_5S$ Urea, N-methyl-N'-(4-methyl-5-nitro- 2-thiazolyl)-N-nitro-	isoPrOH	233(4.10),345(4.11), 420s(2.95)	94-3483-84
$C_6H_7N_7O$ 1H-Pyrazolo[3,4-d]pyrimidine-3-carbox- imidamide, 4-amino-N-hydroxy-	n.s.g.	275(3.99)	103-0210-84
C_6H_8N Pyridinium, 3-methyl- Pyridinium, 4-methyl-	H_2O H_2O	262(3.70) 252(3.65)	142-0505-84 142-0505-84
$C_6H_8NO_3P$ 1-Aza-5-phosphabicyclo[4.2.0]oct-2-en- 8-one, 5-hydroxy-, 5-oxide, compd. with methyl carbamimidothioate	EtOH	243(3.72)	77-0200-84
$C_6H_8N_2$ 1-Cyclopen tene-1-carbonitrile, 2-amino-	MeOH	265(4.15)	107-0967-84
$C_6H_8N_2O$ 2-Propenamide, 3-cyano-N,2-dimethyl-, (E)- (Z)- 2H-Pyrrol-2-one, 1,5-dihydro-5-imino- 1,3-dimethyl- 2H-Pyrrol-2-one, 1,5-dihydro-5-imino- 1,4-dimethyl-	MeOH MeOH MeOH MeOH	223(4.13) 217(4.00) 236(4.17),295(3.17) 235(4.21),294(3.20)	39-2097-84C 39-2097-84C 39-2097-84C 39-2097-84C
$C_6H_8N_2O_6S$ Thiophene, 2,5-dihydro-3,4-dimethyl- 2,5-dinitro-, 1,1-dioxide, disodium salt Thiophene, 2,5-dihydro-3-methyl-2- nitro-4-(nitromethyl)-, 1,1-dioxide, disodium salt	H_2O H_2O	252(3.96),382(4.10) 227(3.63),270(3.39), 390(4.46)	104-2172-84 104-2172-84
$C_6H_8N_2S$ Pyrimidine, 5-(ethylthio)- cation	pH 3.4 H_0 -1.1	260(3.83),300s(3.17) 278(3.95),337(3.13)	59-0075-84 59-0075-84
$C_6H_8N_4OS$ 5-Pyrimidinecarboxamide, 4-amino-1,2- dihydro-N-methyl-2-thioxo-	EtOH	241(4.17),299(4.37)	142-2259-84

Compound	Solvent	$\lambda_{max}(\log \epsilon)$	Ref.
$C_6H_8N_4O_2$			
Acetamide, N-(4-amino-6-oxo-1,6-di-hydro-5-pyrimidinyl)-	pH 1	260(3.93)	39-0879-84C
	pH 13	255(3.69)	39-0879-84C
1H-Imidazole-4-carboxamide, 5-(acetyl-amino)-	EtOH	230(3.80),267(4.16)	103-0204-84
$C_6H_8N_4O_2S$			
1H-Imidazole-1-carboxylic acid, 5-ami-no-4-(aminothioxomethyl)-, methyl ester	EtOH	205(3.99),268(3.88), 330(3.95)	103-0204-84
$C_6H_8N_4O_2S_2$			
Thiourea, (3,4-dimethyl-5-nitro-2(3H)-thiazolylidene)-	isoPrOH	259(3.34),297(3.45), 386(3.62)	94-3483-84
Thiourea, (4-ethyl-5-nitro-2-thiazo-lyl)-	isoPrOH	238(4.08),270s(3.70), 295s(3.58),366(4.05), 442(3.18)	94-3483-84
Thiourea, N-methyl-N'-(4-methyl-5-nitro-2-thiazolyl)-	isoPrOH	240(4.05),270s(3.61), 297(3.48),366(4.04)	94-3483-84
$C_6H_8N_4O_3$			
Acetamide, N-(4-amino-1,6-dihydro-6-oxo-5-pyrimidinyl)-2-hydroxy-	pH 1	259(3.92)	39-0879-84C
	pH 13	256(3.71)	39-0879-84C
Carbamic acid, [5-(aminocarbonyl)-1H-imidazol-4-yl]-, methyl ester	EtOH	230(3.75),265(4.02)	103-0204-84
1H-Imidazole-1-carboxylic acid, 4-ami-no-5-(aminocarbonyl)-, methyl ester	EtOH	265(3.93)	103-0204-84
1H-Imidazole-1-carboxylic acid, 5-ami-no-4-(aminocarbonyl)-, methyl ester	EtOH	265(4.10)	103-0204-84
$C_6H_8N_4O_3S$			
Urea, N-methyl-N'-(4-methyl-5-nitro-2-thiazolyl)-	isoPrOH	237(3.81),353(4.08), 435(2.85)	94-3483-84
$C_6H_8N_6$			
1H-Purine-6,8-diamine, 1-methyl-	pH 1.0	271(4.07)	39-0879-84C
	pH 13.0	284(4.18)	39-0879-84C
$C_6H_8N_6O$			
4H-Imidazo[4,5-c]pyridin-4-one, 5-ami-no-6-hydrazino-1,5-dihydro-	pH 1	271(3.89),292s(3.73)	87-1389-84
	pH 7	262(3.84),299(3.68)	87-1389-84
	pH 11	262(3.87),298(3.58)	87-1389-84
$C_6H_8N_8$			
1H-Pyrazolo[3,4-d]pyrimidine-3-carbox-imidic acid, 4-amino-, hydrazide	n.s.g.	277(3.99)	103-0210-84
C_6H_8O			
1,3-Cyclopentadiene, 1(and 2)-methoxy-	n.s.g.	257(3.17)	104-0061-84
C_6H_8OS			
3(2H)-Thiophenone, 2,2-dimethyl-	C_6H_{12}	310(3.75),352s(2.34), 366s(1.90)	33-1402-84
$C_6H_8O_2$			
Propanedial, cyclopropyl-	pH 13	273.5(4.42)	5-0649-84
$C_6H_8O_3$			
2(5H)-Furanone, 3-hydroxy-4,5-di-methyl-, (R)-	MeOH	232(3.86)	44-2714-84
	MeOH-NaOH	274(--)	44-2714-84

Compound	Solvent	$\lambda_{max}(\log \epsilon)$	Ref.
2(5H)-Furanone, 5-(1-hydroxyethyl)-, [R-(R*,S*)]-	H_2O	203(3.83)	94-2815-84
L-Hex-2-enopyran-4-ulose, 2,3,6-trideoxy-	EtOH	215(3.84),294(1.38), 345(1.54)	94-4350-84
2H-Pyran-2-one, 5,6-dihydro-5-hydroxy-6-methyl-, (5R-trans)-	H_2O	197(4.02)	94-2815-84
4H-Pyran-4-one, tetrahydro-3-(hydroxymethylene)-	EtOH	275.5(3.90)	4-1441-84

$C_6H_8O_4$
| 2-Cyclopenten-1-one, 4,5-dihydroxy-5-(hydroxymethyl) (pentenomycin) | EtOH | 218(3.43) | 39-2089-84C |

$C_6H_8O_6$
| L-Ascorbic acid | pH 2.0 | 243(4.00) | 98-0021-84 |
| | pH 7.0 | 265(4.22) | 98-0021-84 |

C_6H_9ClO
| 3-Hexen-2-one, 4-chloro-, (E)- | EtOH | 239(3.87) | 80-0193-84 |

C_6H_9N
| 1H-Pyrrole, 2,5-dimethyl | hexane or dodecane | 209(3.9) | 142-1513-84 |
| oxygenated (approx. absorbances) | " | 209(5),224(4.4), 275(3.6),282(3.5) | 142-1513-84 |

C_6H_9NO
| 2-Cyclopenten-1-one, 3-(methylamino)- | MeOH | 278(4.57) | 44-0220-84 |
| 4(1H)-Pyridinone, 1,4-dihydro-1-methyl- | EtOH | 324(4.18) | 118-0485-84 |

$C_6H_9NO_2$
Cyclopentene, 1-methyl-2-nitro-	EtOH	276(3.80)	12-1231-84
3-Furanmethanol, 5-(aminomethyl)-	pH 2 and 7	214(4)	35-3636-84
2(5H)-Furanone, 3-amino-4,5-dimethyl-, (R)-	MeOH	206s(--),257(4.07)	44-2714-84

$C_6H_9NO_3$
3(2H)-Isoxazolone, 4-(2-hydroxyethyl)-5-methyl-	EtOH	225(3.1)	23-1940-84
2H-Pyrrole-2-carboxylic acid, 3,4-dihydro-, methyl ester, 1-oxide	EtOH	238(3.77)	78-4513-84
2H-Pyrrol-2-one, 1,5-dihydro-5,5-dimethoxy-	MeCN	204(3.57)	142-1179-84

C_6H_9NS
| 1H-Pyrrole, 1-methyl-2-(methylthio)- | MeOH | 223(4.41),249(4.47) | 4-1041-84 |

C_6H_9NSe
| Isoselenazole, 3-propyl- | isooctane | 268(3.82) | 78-0931-84 |

$C_6H_9N_3$
| 1,3,5-Triazine, 2,4,6-trimethyl- | heptane | 177(4.51),195(3.78), 225(2.40) | 70-0842-84 |

$C_6H_9N_3O$
| 3(2H)-Pyridazinone, 6-(1-aminoethyl)-, monohydrochloride | EtOH | 225(3.80),297(3.38) | 39-0229-84C |

$C_6H_9N_3O_2$
| 1H-Imidazolium, 4-(aminocarbonyl)-5-hydroxy-1,3-dimethyl-, hydroxide, | H_2O | 239(3.54),279(4.05) | 4-0529-84 |
| | M HCl | 240(3.65),280(3.89) | 4-0529-84 |

Compound	Solvent	$\lambda_{max}(\log \epsilon)$	Ref.
inner salt (cont.)	M NaOH	240s(--),279(4.00)	4-0529-84
$C_6H_9N_3S_2$			
Carbamimidothioic acid, N-methyl-N'-2-thiazolyl-, methyl ester	isoPrOH	226(3.78),305(4.26)	94-2224-84
Thiourea, 1-(4,5-dimethyl-2-thiazolyl)-	isoPrOH	253(3.68),298(4.14)	94-3483-84
Thiourea, (3,4-dimethyl-2(3H)-thiazolylidene)-	isoPrOH	279(3.82),326(4.16)	94-3483-84
Thiourea, (4-ethyl-2-thiazolyl)-	isoPrOH	253(3.79),293(4.28)	94-3483-84
Thiourea, N-methyl-N'-(4-methyl-2-thiazolyl)-	isoPrOH	255(3.95),290(4.31)	94-2224-84
$C_6H_9N_4S$			
Thiazolo[3,2-a]pyrimidin-4-ium, 5,7-diamino-2,3-dihydro-, bromide	pH 2	223(4.43),235(4.35), 278(4.02)	70-1285-84
	pH 7	233(4.43),235(4.34), 278(4.00)	70-1285-84
$C_6H_9N_5O_3$			
Furazancarboxylic acid, [(aminoiminomethyl)amino]-, ethyl ester, hydrochloride	EtOH	247(3.80)	78-0879-84
$C_6H_9N_9$			
1H-Pyrazolo[3,4-d]pyrimidine-3-carboximidic acid, 4-hydrazino-, hydrazide	EtOH	300(4.19)	103-0210-84
$C_6H_9N_9O_3$			
Urea, N,N'',N''''-1,3,5-triazine-2,4,6-triyltris-	H_2O	215.5(4.86),247s(4.08)	18-2009-84
C_6H_{10}			
1,3-Butadiene, 2,3-dimethyl-	hexane	228(4.34)	44-2981-84
$C_6H_{10}ClNO$			
2-Propenamide, 3-chloro-N-(1-methylethyl)-, (E)-	C_6H_{12}	214(4.12)	19-0085-84
	H_2O	217(4.13)	19-0085-84
	MeOH	217(4.13)	19-0085-84
	dioxan	215(4.10)	19-0085-84
	MeCN	214(4.11)	19-0085-84
$C_6H_{10}INO$			
2-Propenamide, 3-iodo-N-(1-methylethyl)-, (E)-	C_6H_{12}	247(4.10)	19-0085-84
	H_2O	244(4.04)	19-0085-84
	MeOH	244(4.13)	19-0085-84
	dioxan	242(4.12)	19-0085-84
	MeCN	240(4.12)	19-0085-84
$C_6H_{10}N_2$			
2,3-Diazabicyclo[2.1.1]hex-2-ene, 1,4-dimethyl-	hexane	325(2.43),330(2.35), 334(2.43),341(3.12)	35-4211-84
$C_6H_{10}N_2O$			
5-Isoxazolamine, 3-(1-methylethyl)-	EtOH	243(4.06)	39-1079-84C
$C_6H_{10}N_2O_2S_2$			
Thiazole, 4,5-dihydro-2-[(1-methyl-1-nitroethyl)thio]-	EtOH	275(4.00)	39-2327-84C
$C_6H_{10}N_2O_3$			
2(1H)-Pyrazinone, 5,6-dihydro-1-hydroxy-	EtOH	261(4.09)	103-0097-84

Compound	Solvent	$\lambda_{max}(\log \epsilon)$	Ref.
6,6-dimethyl-, 4-oxide (cont.)			103-0097-84
$C_6H_{10}N_3O_3P$ Phosphonic acid, (cyanodiazomethyl)-, diethyl ester	EtOH	205(3.44),236(3.82)	65-2359-84
$C_6H_{10}N_4O$ 1,2,4-Triazin-5(2H)-one, 6-(1-amino-ethyl)-3-methyl-, dihydrochloride	EtOH	235(3.95),260s(3.73)	39-0229-84C
$C_6H_{10}N_6O_2$ Urea, (4-amino-6-ethoxy-1,3,5-triazin-2-yl)-	H_2O	212.4(4.55)	18-2009-84
$C_6H_{10}OS$ 2-Butenal, 3-(ethylthio)- 3-Buten-2-one, 4-(ethylthio)-	heptane heptane	288(4.45) 283(4.20)	104-1643-84 104-1643-84
$C_6H_{10}O_2$ 2,5-Hexanedione	15% H_2SO_4	236(2.16)	97-0147-84
$C_6H_{10}O_3S$ 1,3-Butadiene-1-sulfonic acid, 2,3-di-methyl-, lithium salt, (Z)-	H_2O	240(3.00)	70-2519-84
$C_6H_{10}S_2$ 2-Butene(dithioic) acid, 3-methyl-, methyl ester	EtOH	300(4.22),322s(3.85), 495(1.78)	39-0859-84C
C_6H_{11} Cyclohexyl radical	C_6H_{12} C_6H_{12} C_6H_{12} C_6H_{12}	255(2.53) 240(3.32) 240(3.18) 240(3.00)	70-0936-84 70-0936-84 70-0936-84 70-0936-84
$C_6H_{11}NO_2$ 2(3H)-Furanone, 3(S)-aminodihydro-4(S),5(R)-dimethyl-, hydrochloride	MeOH	208(2.53)	44-2714-84
$C_6H_{11}NS$ Thiazole, 4,5-dihydro-2-(1-methyl-ethyl)-	EtOH	202(3.58),230(3.34)	78-2141-84
$C_6H_{12}N_2O$ 2-Propenamide, N,N-dimethyl-3-(methyl-amino)-, (Z)-	n.s.g.	282(4.19)	131-0019-84K
$C_6H_{12}N_2O_2$ 1-Butanamine, N-ethenyl-N-nitro- 2-Butanone, 3-[(2-hydroxyethyl)imino]-, oxime Propanal, 2-methyl-, O-[(methylamino)-carbonyl]oxime	MeCN CHCl$_3$ MeCN	282(3.98) 210(2.20),240(2.86), 325(3.20) <u>223(2.3)</u>	5-1494-84 34-0237-84 98-0873-84
$C_6H_{12}N_2O_4$ Ethanol, 1-[(ethyl-NNO-azoxy)oxy]-, acetate Ethanol, 1-(ethylnitroamino)-, acetate Methanol, [(1-methylethyl)nitroamino]-, acetate	MeCN MeCN MeCN	207(3.87) 236(3.78) 237(3.81)	5-1494-84 5-1494-84 5-1494-84

Compound	Solvent	$\lambda_{max}(\log \epsilon)$	Ref.
$C_6H_{12}N_2O_4S$			
2-Propenoic acid, 2-methyl-3-[[(methyl-amino)sulfonyl]amino]-, methyl ester, (E)-	50% EtOH + NaOH	256(3.98) 290(4.24)	73-0840-84 73-0840-84
(Z)-	50% EtOH + NaOH	264(4.04) 292(4.32)	73-0840-84 73-0840-84
$C_6H_{12}N_3O_3PS$			
Phosphonic acid, (5-amino-1,2,3-thia-diazol-4-yl)-, diethyl ester	MeCN	209(3.41),231(3.68), 284(3.80)	65-2359-84
$C_6H_{12}O$			
Butane, 1-(ethenyloxy)-	hexane	192(3.98)	104-0861-84
$C_6H_{12}OS_2$			
Butane(dithioic) acid, 3-hydroxy-3-methyl-, methyl ester	EtOH	308(4.21),454(1.26)	39-0859-84C
$C_6H_{12}S_4Si$			
5H-Silolo[3,4-e]-1,2,3,4-tetrathiin, tetrahydro-6,6-dimethyl-, trans	hexane	234.3(3.75),279(3.32)	70-0585-84
$C_6H_{12}Te_2$			
1,2-Ditellurocane	benzene	398(2.8)	48-0467-84
C_6H_{13}			
Hexyl radical	hexane	235(3.15)	70-0936-84
$C_6H_{13}NO_2$			
2-Butanol, 3-(ethylidenamino)-, N-oxide	pH 8.8	226(4.00)	94-2609-84
$C_6H_{13}NO_3$			
Isoleucine, γ-hydroxy-, (2S,3R,4R)-	H_2O	197(2.40)	44-2714-84
$C_6H_{14}NO_2S$			
Sulfonium, (3-amino-3-carboxypropyl)-dimethyl-, iodide, (S)-	MeOH	217.5(4.79)	36-1241-84
$C_6H_{14}O_4SSi$			
Ethanethioic acid, S-[(trimethoxysil-yl)methyl] ester	heptane Bu_2O MeCN CH_2Cl_2	234(3.58) 234(3.60) 234(3.64) 235(3.64)	70-0726-84 70-0726-84 70-0726-84 70-0726-84

Compound	Solvent	$\lambda_{max}(\log \epsilon)$	Ref.
$C_7ClF_5S_2$			
Carbonochloridodithioic acid, penta-fluorophenyl ester	hexane	212(4.11),281(3.83), 295(3.81),440(1.04)	78-4963-84
C_7F_6O			
2,4,6-Cycloheptatrien-1-one, 2,3,4,5,6,7-hexafluoro-	C_6H_{12}	225(4.38),309(3.88), 315(3.76)	155-0041-84A
	H_2O	230(3.68),262(3.66), 320(3.81)	155-0041-84A
	4M HCl	231(4.46),323(3.95)	155-0041-84A
C_7F_8			
1,3,5-Cycloheptatriene, 1,2,3,4,5,6,7,7-octafluoro-	C_6H_{12}	231(3.83),246(3.84)	155-0041-84A
$C_7HF_5O_2$			
2,4,6-Cycloheptatrien-1-one, 2,3,4,5,6-pentafluoro-7-hydroxy-	EtOH	240(4.45),328(4.02), 398(3.87)	155-0309-84B
2,4,6-Cycloheptatrien-1-one, 2,3,4,5,7-pentafluoro-6-hydroxy-	EtOH	259(4.50),310(3.94), 319(3.89)	155-0309-84B
2,4,6-Cycloheptatrien-1-one, penta-fluorohydroxy- (isomer mixture)	EtOH	240(4.14),258(4.19), 312(3.65),322(3.70), 366(3.74)	155-0309-84B
C_7HF_9			
1H-Cyclohepta-1,3-diene, nonafluoro-	n.s.g.	243(3.45)	155-0041-84A
2H-Cyclohepta-1,3-diene, nonafluoro-	n.s.g.	246(3.79)	155-0041-84A
5H-Cyclohepta-1,3-diene, nonafluoro-	n.s.g.	244(3.70)	155-0041-84A
$C_7H_2F_8$			
1,3-Cycloheptadiene, 2,3,5,5,6,6,7,7-octafluoro-	n.s.g.	238(3.60)	155-0041-84A
$C_7H_2N_2O_3$			
Propanedinitrile, (2-hydroxy-3,4-dioxo-1-cyclobuten-1-yl)-, potassium salt	H_2O	220(4.10),287s(3.91), 342(4.65)	24-2714-84
$C_7H_3BrCl_2O_2$			
Benzoic acid, 2-bromo-4,5-dichloro-	MeOH	237s(3.97),284s(2.86), 291.5(2.95),299s(2.89)	73-0992-84
$C_7H_3Br_2NO$			
Benzonitrile, 3,5-dibromo-4-hydroxy-(bromoxynil)	aq MeOH-H_2SO_4	217(4.63)	74-0053-84A
$C_7H_3Br_2NO_4$			
2,5-Cyclohexadiene-1,4-dione, 2,3-di-bromo-5-methyl-6-nitro-	$CHCl_3$	284(4.12),368(3.15)	12-0055-84
$C_7H_3Br_3O$			
2,5-Cyclohexadien-1-one, 4-(bromo-methylene)-2,6-dibromo-	dioxan	<u>240(3.7),337(4.4)</u>	39-0231-84B
$C_7H_3ClN_4$			
Propanedinitrile, (6-chloro-2(1H)-pyrazinylidene)-	H_2O	224(4.11),301(4.30), 390(3.85)	49-0179-84
$C_7H_3Cl_3O_2$			
3,5-Cyclohexadiene-1,2-dione, 3,4,6-trichloro-5-methyl-	$CHCl_3$	457(3.20)	12-2027-84

Compound	Solvent	λ_{max}(log ϵ)	Ref.
C$_7$H$_3$I$_2$NO Benzonitrile, 4-hydroxy-3,5-diiodo-	aq MeOH- H$_2$SO$_4$	237(4.52),255s(--)	74-0053-84A
C$_7$H$_4$BrNO$_4$ 2,5-Cyclohexadiene-1,4-dione, 5-bromo- 2-methyl-3-nitro-	CHCl$_3$	277(4.03)	12-0055-84
C$_7$H$_4$Br$_2$N$_2$O$_6$ 2,5-Cyclohexadien-1-one, 2,3-dibromo- 4-methyl-6-nitro-4-(nitrooxy)-	CHCl$_3$	299(3.74)	12-0055-84
C$_7$H$_4$Br$_2$O$_2$ Benzaldehyde, 3,5-dibromo-4-hydroxy-	95% dioxan	278(4.1)	39-0231-84B
C$_7$H$_4$Br$_4$O Phenol, 2,6-dibromo-4-(dibromomethyl)-	dioxan	255s(3.8),290s(3.5), 297s(3.4)	39-0231-84B
C$_7$H$_4$ClN$_3$O$_3$ 2,1-Benzisoxazol-3-amine, 4-chloro- 5-nitro-	EtOH	241(4.05),338(3.87), 360s(3.89)	97-0254-84
C$_7$H$_4$Cl$_4$ Bicyclo[4.1.0]hepta-2,4-diene, 2,3,4,5-tetrachloro-	C$_6$H$_{12}$	300(c.3.54)	88-4697-84
C$_7$H$_4$Cl$_4$O$_2$ 2,5-Cyclohexadien-1-one, 2,3,5,6- tetrachloro-4-hydroxy-4-methyl-	CHCl$_3$	256(4.15),298(3.38)	12-2027-84
C$_7$H$_4$D$_2$O Bicyclo[3.2.0]hepta-3,6-dien-2-one- 6,7-d$_2$, (1R)-	MeOH	330(1.95)	35-7261-84
C$_7$H$_4$F$_3$N$_3$O$_4$ Benzenamine, 2,6-dinitro-4-(trifluoro- methyl)-	MeOH anion	412(4.04) 412(4.36)	65-0289-84 65-0289-84
C$_7$H$_4$N$_4$ Propanedinitrile, 2(1H)-pyrazinyli- dene-	H$_2$O	225(4.32),293(4.30), 395(4.17)	49-0179-84
C$_7$H$_4$N$_4$O$_5$ 2,1-Benzisoxazol-3-amine, 5,7-dinitro-	EtOH	260(4.10),329(3.22), 346(3.26),406(3.82)	97-0254-84
C$_7$H$_4$O$_2$S$_4$ 1,3-Dithiole-4-carboxylic acid, 2-(1,3- dithiol-2-ylidene)-	MeCN	288(4.08),302(4.14), 313(4.15),431(3.30)	103-1342-84
C$_7$H$_4$O$_4$ 1,4-Cyclohexadiene-1-carboxylic acid, 3,6-dioxo-	HOAc	255(3.84),338(3.38)	44-4736-84
C$_7$H$_4$O$_5$ 1,4-Cyclohexadiene-1-carboxylic acid, 2-hydroxy-3,6-dioxo-	EtOH	207(4.12),255(4.17), 417(3.22)	44-4736-84
C$_7$H$_5$BrN$_2$OS 7H-Thiazolo[3,2-a]pyrimidin-7-one,	EtOH	224(4.92),274(4.26)	39-2221-84C

Compound	Solvent	$\lambda_{max}(\log \epsilon)$	Ref.
2-(bromomethyl)-, monohydrobromide			39-2221-84C
$C_7H_5Br_3O_2$ 2,5-Cyclohexadien-1-one, 2,3,6-tri-bromo-4-hydroxy-4-methyl-	CHCl$_3$	265(4.03),300(3.32)	12-0055-84
$C_7H_5Cl_2N_3$ 7H-Pyrrolo[2,3-d]pyrimidine, 2,4-di-chloro-7-methyl-	MeOH	231(4.46),277(3.62), 297s(3.53)	5-0722-84
$C_7H_5F_3O_2$ 2,4-Cyclohexadien-1-one, 4,6,6-tri-fluoro-3-methoxy-	hexane	224(4.04),294(3.61)	44-0806-84
$C_7H_5IN_2OS$ 7H-Thiazolo[3,2-a]pyrimidin-7-one, 2,3-dihydro-6-iodo-2-methylene-	EtOH	244(4.23),290s(3.93)	39-2221-84C
C_7H_5NOS 2(3H)-Benzoxazolethione	CH$_2$Cl$_2$	258(4.31),264(4.77), 296s(4.41),301(4.44), 306(4.42)	39-1383-84B
1:1 iodine adduct	CH$_2$Cl$_2$	271(4.55),318(4.84)	39-1383-84B
Thieno[2,3-b]pyridine, 7-oxide, hydro-chloride	EtOH	233(4.66),252s(3.73), 284(3.83),314(3.77)	4-1135-84
C_7H_5NOSe 2(3H)-Benzoxazoleselone	CH$_2$Cl$_2$	266(3.50),319(4.55)	39-1383-84B
1:1 iodine adduct	CH$_2$Cl$_2$	278(4.69),335(4.64)	39-1383-84B
$C_7H_5NO_3$ Benzaldehyde, 4-nitro-	H$_2$O	<u>265(4.0)</u>	35-4511-84
	M NaOH	<u>275(3.8)</u>	35-4511-84
1,2-Benzisoxazol-3(2H)-one, 6-hydroxy-	MeOH	219(4.27),252(3.84), 260s(--),280s(--), 283(3.81),290(3.78)	158-84-44
	MeOH-HCl	206(4.32),252(3.92), 260s(--),280s(--), 283(3.80),290(3.80)	158-84-44
	MeOH-NaOH	220s(--),237(4.26), 264(3.64),273s(--), 296(3.75)	158-84-44
$C_7H_5NS_2$ 2(3H)-Benzothiazolethione	CH$_2$Cl$_2$	281(3.29),328(4.51)	39-1383-84B
1:1 iodine adduct	CH$_2$Cl$_2$	284(4.49),343(4.66), 400s(3.97)	39-1383-84B
$C_7H_5N_3O_3$ 2,1-Benzisoxazol-3-amine, 5-nitro-	EtOH	264(4.05),289s(3.57), 321s(3.49)	97-0254-84
Imidazo[1,2-a]pyridin-2-ol, 3-nitro-	MeCN	270s(4.11),274(4.18), 330(3.65),368(4.13)	39-0069-84C
$C_7H_5N_3O_4$ Benzofurazan, 4-methoxy-7-nitro- (in 15% DMF)	pH 2.7	383(4.06)	48-0385-84
$C_7H_5N_3O_7$ 1(4H)-Pyridineacetic acid, 3,5-di-nitro-4-oxo-	EtOH	342(3.84)	18-1961-84

Compound	Solvent	$\lambda_{max}(\log \epsilon)$	Ref.
$C_7H_5N_4O_6$ Benzofurazan, 4,?-dihydro-4-methoxy- 5,7-dinitro-, ion(1-)	MeOH	465(4.44)	44-4176-84
$C_7H_5N_4O_7$ Benzofurazan, 4,?-dihydro-4-methoxy- 5,7-dinitro-, 3-oxide, ion(1-)	MeOH	460(4.47)	44-4176-84
$C_7H_5N_5S$ 1H-Pyrazolo[3,4-d]pyrimidine-3-carbo- nitrile, 4-(methylthio)-	EtOH	298(4.13),306(4.09)	103-0215-84
$C_7H_5S_2$ 1,3-Benzodithiol-1-ium, perchlorate	acid pH 5	238(3.65),253.2(3.85), 257.5(3.81),308.5(3.62) 226.5(4.49)	18-1128-84 18-1128-84
$C_7H_6BrIN_2OS$ 7H-Thiazolo[3,2-a]pyrimidin-7-one, 2-(bromomethyl)-2,3-dihydro-6-iodo-	EtOH	235(4.25),291s(3.84)	39-2221-84C
$C_7H_6Br_2O_2$ 2,5-Cyclohexadien-1-one, 2,6-dibromo- 4-hydroxy-4-methyl-	CHCl$_3$	258(3.97)	12-0055-84
$C_7H_6ClN_3O$ 1H-Pyrrolo[2,3-d]pyrimidine, 4-chloro- 2-methoxy-	MeOH	224(4.34),257(3.36), 297(3.70)	5-0273-84
$C_7H_6Cl_2N_2$ 1H-Pyrrolo[2,3-b]pyridine, 4,6-di- chloro-2,3-dihydro-	heptane PrOH	308(3.82) 315(3.83)	151-0343-84D 151-0343-84D
$C_7H_6N_2O$ Phenol, 3-(3H-diazirin-3-yl)-	C_6H_{12}	358(2.56)	23-1767-84
$C_7H_6N_2OS$ 7H-Thiazolo[3,2-a]pyrimidin-7-one, 2,3-dihydro-2-methylene- 7H-Thiazolo[3,2-a]pyrimidin-7-one, 2-methyl-	EtOH EtOH	247(3.82),283(3.52) 223(3.55),230(3.51), 275(3.47)	39-2221-84C 39-2221-84C
$C_7H_6N_2O_2S$ 7H-Thiazolo[3,2-a]pyrimidin-7-one, 2-(hydroxymethyl)-	EtOH	217(4.13),231s(3.91), 274(3.77)	39-2221-84C
$C_7H_6N_2O_5S$ 4-Benzofurazansulfonic acid, 7-meth- oxy-, sodium salt	H_2O	348(3.69)	96-1003-84
$C_7H_6N_2S$ 2H-Benzimidazole-2-thione, 1,3-di- hydro- 1:1 iodine complex	CH$_2$Cl$_2$ CH$_2$Cl$_2$	254(4.03),303s(4.44), 310(4.55) 284(4.39),316(4.66)	39-1383-84B 39-1383-84B
$C_7H_6N_4O_3$ 4-Benzofurazanamine, N-methyl- 7-nitro- (in 15% DMF)	pH 2.7	478(4.43)	48-0385-84
$C_7H_6N_4S$ 1H-Imidazo[1,2-b]pyrazole-7-carboni-	pH 1	257(3.87)	44-3534-84

Compound	Solvent	$\lambda_{max}(\log \epsilon)$	Ref.
nitrile, 6-(methylthio)- (cont.)	pH 7	257(3.90)	44-3534-84
	pH 11	260s(3.79)	44-3534-84
$C_7H_6O_2$			
Benzaldehyde, 2-hydroxy-	M HCl	255(4.08),324(3.53)	18-2508-84
	M NaOH	264(3.87),337(3.83)	18-2508-84
$C_7H_6O_3$			
Benzaldehyde, 2,3-dihydroxy-	pH 1	264(4.07)	3-1935-84
	borate	272(4.01)	3-1935-84
$C_7H_6O_4$			
Benzoic acid, 3,4-dihydroxy-	pH 1	259(4.00),293(3.71)	3-1935-84
	pH 10.68	277(3.98),300(4.04)	12-0885-84
	pH 14.8	288(3.82),325(4.03)	12-0885-84
	borate	262(3.85),295(3.84)	3-1935-84
$C_7H_7BrN_2OS$			
4(1H)-Pyrimidinone, 5-bromo-2,3-di-hydro-1-(2-propenyl)-2-thioxo-	MeOH	222(3.94),246s(3.86),298(3.97),324s(3.57)	39-2221-84C
7H-Thiazolo[3,2-a]pyrimidin-7-one, 2-(bromomethyl)-2,3-dihydro-, monohydrobromide	EtOH	233(4.32),265s(3.75)	39-2221-84C
$C_7H_7ClN_4$			
1H-Pyrrolo[2,3-d]pyrimidin-2-amine, 4-chloro-5-methyl-	MeOH	237(4.38),264(3.51),322(3.65)	5-0708-84
7H-Pyrrolo[2,3-d]pyrimidin-2-amine, 4-chloro-7-methyl-	MeOH	236(4.48),263s(3.57),317(3.76)	87-0981-84
$C_7H_7F_3N_2O_4$			
Alanine, 2,3-didehydro-N-[N-(trifluoro-acetyl)glycyl]-	EtOH	207(3.99),242(3.71)	5-0920-84
$C_7H_7IN_2OS$			
7H-Thiazolo[3,2-a]pyrimidin-7-one, 2,3-dihydro-2-(iodomethyl)-, hydriodide	EtOH	226(4.52),260(3.99)	39-2221-84C
C_7H_7Li			
Lithium, (phenylmethyl)- solution contg. sparteine)	heptane at -78°	329(4.11),380(3.08)	138-0757-84
	toluene	333(4.15),400s(3.23)	138-0757-84
C_7H_7NO			
Benzamide	EtOH	227(4.22)	105-0599-84
	EtOH	<u>226(4.2),268(3.6)</u>	70-2056-84
$C_7H_7NO_2$			
Benzamide, N-hydroxy-, color with 4-ni-troaniline and N-(1-naphthalenyl)-ethenediamine	NaOAc	545(4.40)	86-1013-84
$C_7H_7NO_3$			
3H-Pyrano[4,3-c]isoxazol-3-one, 6,7-dihydro-4-methyl-	EtOH	290(3.94)	142-0365-84
1(2H)-Pyridineacetic acid, 2-oxo-	pH 2 and 7	300(3.74)	73-2541-84
	pH 12	300(3.74)	73-2541-84
3-Pyridinecarboxylic acid, 1,2-dihydro-1-methyl-2-oxo-	EtOH	232(3.70),329(3.99)	4-0041-84

Compound	Solvent	$\lambda_{max}(\log \epsilon)$	Ref.
$C_7H_7NO_4S$			
Ethanone, 1-(5-methoxy-4-nitro-2-thienyl)-	MeOH	248(4.15),300(4.09)	39-0317-84B
$C_7H_7NO_5S$			
Benzenesulfonic acid, 2-methyl-5-nitro-	H_2O	278(3.92)	73-2751-84
2-Thiophenecarboxylic acid, 5-methoxy-4-nitro-, methyl ester	MeOH	246(4.15),284(4.05), 315(3.74)	39-0317-84B
$C_7H_7N_2$			
Benzenediazonium, 4-methyl-, tetrafluoroborate	DMSO	273(4.2)	123-0075-84
$C_7H_7N_2O$			
Benzenediazonium, 4-methoxy-, tetrafluoroborate	DMSO	312(4.34)	123-0075-84
$C_7H_7N_2O_4S$			
2-Thiophenecarbonitrile, 5,?-dihydro-5,5-dimethoxy-4-nitro-, ion(1-), sodium	MeOH	290(3.69),399(4.23)	39-0317-84B
$C_7H_7N_3O$			
Acetonitrile, [(2-amino-3-pyridinyl)-oxy]-	MeOH	234(4.00),297(3.78)	4-1081-84
Benzene, 1-azido-4-methoxy-	2% MeOH	254(4.11),290s(--)	149-0545-84B
Imidazo[1,2-a]pyrimidin-7(8H)-one, 8-methyl-	H_2O	221(4.09)	44-4021-84
2H-Pyrido[3,2-b]-1,4-oxazin-3-amine	MeOH	265(3.68),308(4.16)	4-1081-84
$C_7H_7N_3OS$			
6H-Thiopyrano[3,4-d]pyrimidin-5(8H)-one, 2-amino-	EtOH	285(4.205)	4-1437-84
$C_7H_7N_3O_2$			
4H-Furo[3,2-b]pyrrole-5-carboxylic acid hydrazide	n.s.g.	293(3.37)	73-0065-84
4H-Pyrrolo[2,3-d]pyrimidin-4-one, 1,7-dihydro-2-methoxy-	MeOH	214(4.22),252(3.97), 266s(3.88)	5-0273-84
$C_7H_7N_3O_3S$			
2-Propenoic acid, 2-[[(2,5-dihydro-5-oxo-1,2,4-triazin-3-yl)thio]-methyl]-	EtOH	231(4.21),299s(3.09)	142-1049-84
1,2,4-Triazine-4(3H)-propanoic acid, 2,5-dihydro-α-methylene-5-oxo-3-thioxo-	EtOH	215(4.04),269(4.07), 315s(3.36)	142-1049-84
$C_7H_7N_3O_5$			
Benzenamine, 4-methoxy-2,6-dinitro-	MeOH	473(3.88)	65-0289-84
	anion	440(4.08)	65-0289-84
$C_7H_7N_3S$			
1H-Pyrrolo[2,3-d]pyrimidine, 2-(methylthio)-	MeOH	247(4.37),270s(3.72), 308(3.76)	5-1719-84
$C_7H_7N_3S_2$			
4H-Pyrrolo[2,3-d]pyrimidine-4-thione, 3,7-dihydro-2-(methylthio)-	MeOH	241(4.31),274(3.95), 336(4.27)	5-1719-84

Compound	Solvent	$\lambda_{max}(\log \epsilon)$	Ref.
$C_7H_7N_5$ Cyanamide, (7,8-dihydroimidazo[1,2-c]- pyrimidin-5(6H)-ylidene)-	MeOH	244(4.21)	24-2597-84
$C_7H_7N_5O_2$ Benzenamine, 4-azido-N-methyl-2-nitro-	2% MeOH	260(4.30),480(3.66)	149-0545-84B
$C_7H_7N_5S_2$ 1H-Pyrazolo[3,4-d]pyrimidine-3-carbo- thioamide, 4-(methylthio)-	EtOH	330(3.82)	103-0210-84
C_7H_8 Spiro[2.4]hepta-4,6-diene Toluene	EtOH C_6H_{12}	223(3.80),257(3.43) 210(3.92),262(2.42)	70-1526-84 60-0383-84
$C_7H_8Br_2N_2O_2$ 2,4(1H,3H)-Pyrimidinedione, 5-bromo- 1-(3-bromopropyl)-	pH 2 H_2O pH 12	284(3.96) 281(3.92) 280(3.82)	142-2739-84 142-2739-84 142-2739-84
$C_7H_8F_2$ 1,3,6-Heptatriene, 5,5-difluoro-, cis	EtOH	224(4.45)	44-1033-84
C_7H_8N Pyridinium, 1-ethenyl-, bromide	H_2O	257(3.95)	142-0505-84
$C_7H_8N_2OS$ 4(3H)-Pyrimidinethione, 3-ethenyl- 5-hydroxy-2-methyl- 4(1H)-Pyrimidinone, 2,3-dihydro-1-(2- propenyl)-2-thioxo- Thiopyrano[3,4-c]pyrazol-4(5H)-one, 1,7-dihydro-1-methyl-	EtOH EtOH EtOH	277(3.62),354(4.01) 224(4.08),240s(3.96), 287(3.95) 249(3.92)	4-1149-84 39-2221-84C 4-1437-84
$C_7H_8N_2O_3$ 2,4(1H,3H)-Pyrimidinedione, 5-(2-oxo- propyl)- 1H-Pyrrole-2,5-dione, 3-acetyl- 4-amino-1-methyl-	MeOH MeOH	264(3.86) 247s(4.23),252(4.24), 360(3.51)	44-2738-84 39-2097-84C
$C_7H_8N_2O_4$ Propanedioic acid, (aminocyanomethyl- ene)-, dimethyl ester	$C_2H_4Cl_2$	304(4.00)	39-0965-84B
$C_7H_8N_2S$ 4,7-Methano-1,2,3-benzothiadiazole, 4,5,6,7-tetrahydro-, (±)-	EtOH	219(3.60),267(3.48)	44-4773-84
$C_7H_8N_3O_2S_2$ 5H-Pyrazolo[3,4-d]-1,2,3-dithiazol-2- ium, 6-(ethoxycarbonyl)-5-methyl-, chloride	CF_3COOH	328(3.91),442(3.17)	44-1224-84
$C_7H_8N_4$ 1H-Purine, 1,6-dimethyl-, hydriodide	pH 1 pH 13	268(3.89) 273(3.89)	39-0879-84C 39-0879-84C
$C_7H_8N_4O$ 3-Pyridinecarbonitrile, 2,4-diamino- 6-methoxy- 3-Pyridinecarbonitrile, 4,6-diamino-	n.s.g. n.s.g.	227(4.49),295(3.88) 224(4.42),270(4.14),	49-1421-84 49-1421-84

Compound	Solvent	λ_{max}(log ϵ)	Ref.
2-methoxy- (cont.)		280(4.13)	49-1421-84
4H-Pyrrolo[2,3-d]pyrimidin-4-one, 2-amino-1,7-dihydro-1-methyl-, mono-hydrochloride	MeOH	266(4.00)	87-0981-84
4H-Pyrrolo[2,3-d]pyrimidin-4-one, 2-amino-1,7-dihydro-3-methyl-	MeOH	215(4.26),260(4.03), 276(3.91)	87-0981-84
4H-Pyrrolo[2,3-d]pyrimidin-4-one, 2-amino-3,7-dihydro-5-methyl-	MeOH	222(4.26),264(3.98)	5-0708-84
4H-Pyrrolo[2,3-d]pyrimidin-4-one, 2-amino-3,7-dihydro-7-methyl-	MeOH	224(4.30),263(4.06), 281s(3.91)	87-0981-84
C$_7$H$_8$N$_4$O$_2$			
Theobromine	neutral	273(3.92)	41-0021-84
	anion	275(3.95)	41-0021-84
Theophylline	neutral	271(3.91)	41-0021-84
	anion	275(3.90)	41-0021-84
C$_7$H$_8$N$_4$O$_2$S$_2$			
2H-Thiazolo[3,2-b][1,2,4]thiadiazol-2-imine, 6-(1-methylethyl)-N-nitro-	isoPrOH	280(2.85),343(4.09)	94-3483-84
2H-Thiazolo[3,2-b][1,2,4]thiadiazol-2-imine, N-nitro-6-propyl-	isoPrOH	280(3.15),343(4.22)	94-3483-84
C$_7$H$_8$N$_4$O$_3$S			
1H-Pyrazino[2,3-c][1,2,6]thiadiazin-4(3H)-one, 6,7-dimethyl-, 2,2-dioxide	H$_2$O	258(4.16),395(3.95)	4-0861-84
	anion	244(4.04),335(3.79)	4-0861-84
	dianion	260(4.18),375(3.83)	4-0861-84
C$_7$H$_8$N$_6$O$_3$			
Carbamic acid, (4,6-diaminooxazolo-[5,4-d]pyrimidin-2-yl)-, methyl	pH 1	264(4.34)	4-1245-84
	pH 7	269(4.26)	4-1245-84
ester	pH 11	282(4.00),296(4.32)	4-1245-84
C$_7$H$_8$N$_6$S			
2,4-Pteridinediamine, 7-(methylthio)-	MeOH	241(4.43),263(4.19), 312(3.51),371(4.18)	138-1025-84
C$_7$H$_8$O$_2$			
1,2-Benzenediol, 4-methyl-	pH 1	280(3.39)	3-1935-84
	pH 7.2	280(4.41)	12-0885-84
	pH 11.6	237(3.80),293(3.61)	12-0885-84
	pH 14.7	250(3.75),307(3.73)	12-0885-84
	borate	287(3.61)	3-1935-84
C$_7$H$_8$O$_2$S			
Benzene, (methylsulfonyl)-	C$_6$H$_{12}$	217(4.00),252(2.58), 258(2.82),264(2.98), 270(2.78)	104-0473-84
C$_7$H$_8$O$_3$			
1,2-Benzenediol, 3-methoxy-	pH 1	266(2.94)	3-1935-84
	borate	273(3.27),277s(--)	3-1935-84
C$_7$H$_8$O$_3$S			
2H-Thiopyran-3-carboxylic acid, 3,4-dihydro-4-oxo-, methyl ester	n.s.g.	308(3.84)	39-2549-84C
2H-Thiopyran-5-carboxylic acid, 3,4-dihydro-4-oxo-, methyl ester	n.s.g.	312.5(4.04)	39-2549-84C
C$_7$H$_8$SSe			
4H-Selenin-4-thione, 2,6-dimethyl-	EtOH	238(4.4),338(4.2),	22-0241-84

Compound	Solvent	λ_{max}(log ϵ)	Ref.
(cont.)		426(4.39)	22-0241-84
C$_7$H$_9$BrO$_2$ 2,4-Hexadienoic acid, 5-bromo-, methyl ester, (E,E)-	MeCN	271(4.42)	22-0362-84
C$_7$H$_9$N 2,4-Hexadienenitrile, 5-methyl-	MeCN	266(4.36)	22-0362-84
C$_7$H$_9$NO Cyclopentanecarbonitrile, 1-methyl-2-oxo-	C$_6$H$_{12}$	301(1.64)	78-0613-84
C$_7$H$_9$NOS Pyridine, 2-(ethylthio)-, 1-oxide	MeOH-EtOH	242(4.5),265(3.9), 310(3.6)	39-2031-84B
1H-Pyrrole-2-carboxaldehyde, 1-methyl-5-(methylthio)-	MeOH	213(3.88),246(3.53), 320(4.31)	4-1041-84
C$_7$H$_9$NO$_2$ 1,2-Benzenediol, 4-(aminomethyl)-,	pH 1	278(3.41)	3-1935-84
	borate	285(3.64)	3-1935-84
2(1H)-Pyridinone, 1-ethoxy-	C$_6$H$_{12}$	234(2.72),305(2.62)	40-0001-84
	MeOH	228(2.79),300(2.68)	40-0001-84
	MeCN	230(2.78),302(2.66)	40-0001-84
4(1H)-Pyridinone, 1-(methoxymethyl)-	EtOH	264(4.27)	118-0485-84
C$_7$H$_9$NS$_3$ 6H-1,3-Thiazine-6-thione, 2-(ethylthio)-5-methyl-	MeCN	261(4.32),302(4.03), 420(4.13)	97-0435-84
C$_7$H$_9$N$_2$O$_5$S 2-Thiophenecarboxamide, 2,5-dihydro-5,5-dimethoxy-4-nitro-, ion(1-), sodium	MeOH	294(4.06),392(4.28)	39-0317-84B
C$_7$H$_9$N$_3$O$_2$ Hydrazine, (3-methyl-4-nitrophenyl)-	MeOH	237(3.82),382(4.17)	73-0086-84
C$_7$H$_9$N$_3$O$_2$S Pyrimidine, 2-[(1-methyl-1-nitroethyl)-thio]-	EtOH	239(4.43)	39-2327-84C
C$_7$H$_9$N$_3$O$_4$S Carbamic acid, (4-methyl-5-nitro-2-thi-azolyl)-, ethyl ester	isoPrOH	236(3.86),346(4.12), 420(3.98)	94-3483-84
C$_7$H$_9$N$_5$ Methanimidamide, N'-(4-cyano-1H-pyra-zol-3-yl)-N,N-dimethyl-	EtOH	278(4.11)	103-0215-84
1H-Purin-6-amine, 1,8-dimethyl-	pH 1	261(4.11)	39-0879-84C
	pH 13.0	273(4.19)	39-0879-84C
C$_7$H$_9$N$_5$O 1H-Purine-8-methanol, 6-amino-1-methyl-	pH 1.0	263(4.13)	39-0879-84C
	pH 13.0	273(4.20)	39-0879-84C
C$_7$H$_9$N$_5$O$_2$S 1H-Pyrazino[2,3-c][1,2,6]thiadiazin-4-amine, 6,7-dimethyl-, 2,2-dioxide	pH 1.0	252(4.02),339(3.74), 411(3.63)	4-0861-84
anion	pH 8	263(4.17),382(3.83)	4-0861-84

Compound	Solvent	$\lambda_{max}(\log \epsilon)$	Ref.
$C_7H_{10}ClNO$ Pyrrolidine, 1-(3-chloro-1-oxo-2-propenyl)-	EtOH	220(4.10),250s(3.72)	142-2369-84
$C_7H_{10}NO_6S_2$ Thiophene, 2,?-dihydro-2,2-dimethoxy-5-(methylsulfonyl)-3-nitro-, ion(1-), sodium	MeOH	285(3.68),385(4.17)	39-0317-84B
$C_7H_{10}N_2$ 1-Cyclohexene-1-carbonitrile, 2-amino-	MeOH	265(4.07)	107-0967-84
$C_7H_{10}N_2OS_2$ Carbamothioic acid, (4-methyl-2-thiazolyl)-, O-ethyl ester	isoPrOH	242s(3.73),298(4.02), 335s(3.57)	94-3483-84
$C_7H_{10}N_2O_3$ 2-Cyclohexen-1-one, 2-methyl-3-nitro-, oxime	EtOH	224(4.05),299(3.76)	12-1231-84
$C_7H_{10}N_2O_3S$ 4-Thiazoleacetic acid, 2,3-dihydro-3-hydroxy-2-imino-, ethyl ester, hydrochloride	EtOH EtOH-HCl EtOH-base	$\underline{278(3.7)}$ $\underline{228(4.3),286(4.3)}$ $\underline{230s(3.8),341(4.4)}$	4-1097-84 4-1097-84 4-1097-84
$C_7H_{10}N_4OS$ 5-Pyrimidinecarboxamide, 6-amino-1,2-dihydro-N,1-dimethyl-2-thioxo-	EtOH	224(4.05),291(4.11), 355(4.24)	142-2259-84
$C_7H_{10}N_4O_2$ Propanamide, N-(6-amino-1,4-dihydro-4-oxo-5-pyrimidinyl)-	pH 1 pH 13	260(3.93) 255(3.71)	39-0879-84C 39-0879-84C
$C_7H_{10}N_4O_2S$ Benzenesulfonamide, 4-amino-N-(aminoiminomethyl)-	MeOH	263.92(4.28)	83-0906-84
$C_7H_{10}N_4O_3S$ Urea, N,N-dimethyl-N'-(4-methyl-5-nitro-2-thiazolyl)-	isoPrOH	236(3.94),278(3.45), 356(4.13)	94-3483-84
$C_7H_{10}N_4O_5$ 2(1H)-Pyrimidinone, 4-amino-1-[(2-hydroxyethoxy)methyl]-5-nitro-	pH 1 pH 13	253(3.93),308(3.90) 335(4.24)	87-1486-84 87-1486-84
$C_7H_{10}N_5$ 7H-Purinium, 6-amino-7,9-dimethyl-, perchlorate iodide	MeOH MeOH	272(3.91) 270(4.08)	88-2989-84 150-3601-84M
$C_7H_{10}N_6O_3$ 1,2,4-Triazolo[3,4-f][1,2,4]triazin-8(7H)-one, 6-amino-3-[(2-hydroxyethoxy)methyl]-	EtOH	275(3.40)	4-0697-84
$C_7H_{10}O$ 1,3-Cyclopentadiene, 1(and 2)-ethoxy-	n.s.g.	259(3.17)	104-0061-84
$C_7H_{10}OS_2$ 1,4-Dithiaspiro[4.4]non-8-en-7-ol, (R)-	EtOH	219(3.34),243s(2.78)	39-2089-84C

Compound	Solvent	$\lambda_{max}(\log \epsilon)$	Ref.
$C_7H_{10}O_2$			
2(3H)-Furanone, dihydro-3-propyli-dene-, (E)-	EtOH	222(4.06)	95-0839-84
(Z)-	EtOH	224(4.01)	95-0839-84
Propanedial, cyclobutyl-	pH 13	276(4.42)	5-0649-84
$C_7H_{10}O_2Si$			
Silacyclohexa-2,4-dien-1-ol, acetate	C_6H_{12}	268(3.84)	24-2337-84
$C_7H_{10}O_3$			
Cyclopropanecarboxylic acid, 2-formyl-, ethyl ester (isomer mixture)	hexane	203(3.05),284(1.43)	44-1944-84
2(5H)-Furanone, 3-methoxy-4,5-dimethyl-	C_6H_{12}	201(3.43),227(3.84)	44-2714-84
2H-Pyran-3(6H)-one, 6-methoxy-2-methyl-	EtOH	211(3.90),278(1.30), 343(1.70)	94-4350-84
$C_7H_{10}O_3S_2$			
2-Propenoic acid, 3-[(ethoxothioxo-methyl)thio]-, methyl ester, (Z)-	isoPrOH	242(4.04),305(4.30), 370(2.22)	104-1082-84
$C_7H_{10}O_6$			
L-Ascorbic acid, 1-O-methyl-	pH 3-7	282(4.10)	98-0021-84
	pH 11	325(4.06)	98-0021-84
L-Ascorbic acid, 2-O-methyl-	pH 2	239(3.96)	98-0021-84
	pH 7	259(4.19)	98-0021-84
L-Ascorbic acid, 3-O-methyl-	pH 2	243(3.93)	98-0021-84
	pH 6	243(3.93)	98-0021-84
	pH 12	275(3.83)	98-0021-84
$C_7H_{11}ClO$			
Cyclopentanone, 2-(chloromethyl)-2-methyl-	C_6H_{12}	302(1.43)	78-0613-84
3-Hepten-2-one, 4-chloro-, (E)-	EtOH	244(3.70)	80-0193-84
$C_7H_{11}NO_2$			
Cyclohexene, 1-methyl-2-nitro-	hexane	257(3.62)	12-1231-84
3-Isoxazolol, 5-(1,1-dimethylethyl)-	EtOH	207(3.8)	23-1940-84
4(1H)-Pyridinone, 2,3-dihydro-1-(meth-oxymethyl)-	EtOH	312(4.25)	118-0485-84
$C_7H_{11}NO_2S$			
Formamide, N-(2,2,3-trimethyl-4-oxo-3-thietanyl)-, (R)-	EtOH	215(3.86),235s(3.40)	39-1127-84C
2,5-Pyrrolidinedione, 3-[(1-methyl-ethyl)thio]-	MeCN	201(3.53)	142-1179-84
$C_7H_{11}N_3O_2$			
1,2,4-Triazine-3,5(2H,4H)-dione, 6-(1,1-dimethylethyl)-	MeOH	208(3.7),261(3.7)	98-0749-84
$C_7H_{11}N_3O_2S$			
1H-Imidazole, 1-methyl-2-[(1-methyl-1-nitroethyl)thio]-	EtOH	234(4.67)	39-2327-84C
$C_7H_{11}N_3O$			
4(1H)-Pyrimidinone, 2-amino-1-[(2-hydroxyethoxy)methyl]-	pH 1	216(3.99),255(3.80)	87-1486-84
	pH 13	250s(3.79)	87-1486-84
$C_7H_{11}N_3O_5$			
1,2,4-Triazine-3,5(2H,4H)-dione, 2-[[2-hydroxy-1-(hydroxymethyl)eth-	pH 1	261(3.73)	23-0016-84
	H_2O	259(3.73)	23-0016-84

Compound	Solvent	λ_{max}(log ϵ)	Ref.
oxy]methyl]– (cont.)	pH 13	252(3.78)	23-0016-84
C$_7$H$_{11}$N$_3$S$_2$			
Carbamimidothioic acid, N-methyl-N'-(4-methyl-2-thiazolyl)-, methyl ester	isoPrOH	230(3.90),310(4.22)	94-2224-84
Thiourea, N,N-dimethyl-N'-(4-methyl-2-thiazolyl)-	isoPrOH	293(3.92),329(4.22)	94-2224-84
Thiourea, [4-(1-methylethyl)-2-thiazolyl]-	isoPrOH	252(3.83),293(4.32)	94-3483-84
Thiourea, (4-propyl-2-thiazolyl)-	isoPrOH	253(3.81),293(4.29)	94-3483-84
C$_7$H$_{11}$N$_4$S			
2H-Pyrimido[2,1-b][1,3]thiazin-5-ium, 6,8-diamino-3,4-dihydro-, bromide	pH 2	215(4.50),242(4.50), 280(4.17)	70-1285-84
	pH 7	215(4.54),242(4.53), 280(4.20)	70-1285-84
	pH 9	215(4.56),242(4.55), 280(4.22)	70-1285-84
C$_7$H$_{11}$N$_5$O$_2$S			
Guanidine, N,N-dimethyl-N'-(4-methyl-2-thiazolyl)-N''-nitro-	isoPrOH	240s(3.91),280s(3.73), 328(4.28)	94-3483-84
C$_7$H$_{12}$BrNO			
2-Propenamide, 3-bromo-N,N-diethyl-, (E)-	C$_6$H$_{12}$	221(3.94)	19-0085-84
	H$_2$O	221(3.92)	19-0085-84
	MeOH	220(3.96)	19-0085-84
	dioxan	220(3.97)	19-0085-84
	MeCN	218(3.95)	19-0085-84
2-Propenamide, 3-bromo-N-(1,1-dimethylethyl)-, (E)-	C$_6$H$_{12}$	224(4.13)	19-0085-84
	H$_2$O	227(4.16)	19-0085-84
	MeOH	225(4.15)	19-0085-84
	dioxan	224(4.10)	19-0085-84
	MeCN	222(4.13)	19-0085-84
(Z)-	C$_6$H$_{12}$	210(3.91)	19-0085-84
	H$_2$O	227(3.91)	19-0085-84
	MeOH	226(3.92)	19-0085-84
	dioxan	219(3.86)	19-0085-84
	MeCN	222(3.96)	19-0085-84
C$_7$H$_{12}$ClNO			
2-Propenamide, 3-chloro-N,N-diethyl-, (E)-	C$_6$H$_{12}$	217(4.00)	19-0085-84
	H$_2$O	220(4.02)	19-0085-84
	MeOH	220(4.06)	19-0085-84
	dioxan	216(4.06)	19-0085-84
	MeCN	215(4.05)	19-0085-84
2-Propenamide, 3-chloro-N-(1,1-dimethyl)-, (E)-	C$_6$H$_{12}$	216(4.22)	19-0085-84
	H$_2$O	218(4.18)	19-0085-84
	MeOH	216(4.24)	19-0085-84
	dioxan	215(4.19)	19-0085-84
	MeCN	214(4.21)	19-0085-84
(Z)-	C$_6$H$_{12}$	206(4.05)	19-0085-84
	H$_2$O	215(3.95)	19-0085-84
	MeOH	213(3.99)	19-0085-84
	dioxan	209(3.95)	19-0085-84
	MeCN	211(4.01)	19-0085-84
C$_7$H$_{12}$INO			
2-Propenamide, N,N-diethyl-3-iodo-, (E)-	C$_6$H$_{12}$	247(4.06)	19-0085-84
	H$_2$O	248(4.02)	19-0085-84

Compound	Solvent	$\lambda_{max}(\log \epsilon)$	Ref.
(cont.)	MeOH	248(4.04)	19-0085-84
	dioxan	245(4.03)	19-0085-84
	MeCN	247(4.06)	19-0085-84
2-Propenamide, N-(1,1-dimethylethyl)-	C_6H_{12}	245(4.01)	19-0085-84
3-iodo-, (E)-	H_2O	244(4.06)	19-0085-84
	MeOH	243(4.09)	19-0085-84
	dioxan	241(4.08)	19-0085-84
	MeCN	240(4.10)	19-0085-84
$C_7H_{12}N_2O$			
5-Isoxazolamine, 3-s-butyl-	EtOH	243(3.99)	39-1079-84C
2-Pyrrolidinone, 3-[(dimethylamino)- methylene]-	isoPrOH	212(3.64),290(4.38)	103-0047-84
$C_7H_{12}N_2O_2$			
5-Isoxazolamine, 3-(1-hydroxy-1-methyl- propyl)-	EtOH	244(3.92)	39-1079-84C
2H-Pyrrol-2-one, 4-(dimethylamino)- 1,5-dihydro-5-methoxy-	MeCN	213(3.88),280(4.08)	142-1179-84
$C_7H_{12}N_2O_3$			
2-Propenoic acid, 3-(methylamino)- 2-[(methylamino)carbonyl]-, methyl ester	MeCN	230(3.89),280(4.30)	88-1521-84
2(1H)-Pyrazinone, 3,6-dihydro-1-hydr- oxy-5,6,6-trimethyl-, 4-oxide	EtOH	237(3.99)	103-0097-84
2(1H)-Pyrazinone, 5,6-dihydro-1-hydr- oxy-5,6,6-trimethyl-, 4-oxide	EtOH	264(4.15)	103-0097-84
$C_7H_{12}N_2S$			
4H-Pyrazole-4-thione, 3,5-dihydro- 3,3,5,5-tetramethyl-	hexane	234.8(3.90),528(1.18), 554(1.18)	24-0277-84
$C_7H_{12}N_4$			
2,4-Pyrimidinediamine, 6-(1-methyl- ethyl)-	EtOH	232(3.94),281(3.99)	39-2677-84C
$C_7H_{12}N_4O$			
5H-Tetrazol-5-one, 1-(1,1-dimethyl- ethyl)-4-ethenyl-1,4-dihydro-	hexane	201(3.721),246(3.830)	24-2761-84
$C_7H_{12}N_4OS$			
1,2,4-Triazin-5(2H)-one, 4-amino- 6-(1,1-dimethylethyl)-3,4-dihydro- 3-thioxo-	MeOH	204(3.5),270(4.0)	98-0749-84
$C_7H_{12}N_4O_2$			
1,2,4-Triazine-3,5(2H,4H)-dione, 4-ami-	MeOH	212(3.7),262(3.7)	98-0749-84
no-6-(1,1-dimethylethyl)-	MeCN	260(3.62)	9-0556-84
in 10% MeOH	pH 3.5	211(3.76),258(3.59)	9-0556-84
in 10% MeOH	pH 11.5	220(3.63),291(3.66)	9-0556-84
$C_7H_{12}OS$			
5-Thiocanone	C_6H_{12}	227(3.46)	98-0873-83
$C_7H_{12}O_2$			
Cyclopentanone, 2-(hydroxymethyl)- 2-methyl-	C_6H_{12}	296(1.40)	78-0613-84
$C_7H_{12}O_2S_2$			
Butane(dithioic) acid, 3-acetoxy-,	EtOH	306(4.23),455(1.18)	39-0859-84C

Compound	Solvent	$\lambda_{max}(\log \epsilon)$	Ref.
methyl ester (cont.)			39-0859-84C
$C_7H_{12}S_2$			
3-Thietanethione, 2,2,4,4-tetramethyl-	hexane	206s(4.03),210.5(4.04), 235.5(3.57),292(2.59), 320s(2.51),488.5(0.78)	24-0277-84
$C_7H_{13}N$			
1-Azabicyclo[4.2.0]octane	MeCN	219(2.85)	126-0037-84
$C_7H_{13}NO$			
2-Propenamide, N,N-diethyl-	C_6H_{12}	244(3.48)	19-0085-84
	H_2O	237(3.62)	19-0085-84
	MeOH	240(3.71)	19-0085-84
	dioxan	243(3.49)	19-0085-84
	MeCN	243(3.52)	19-0085-84
$C_7H_{13}NS$			
Thiazole, 4,5-dihydro-2-(1-methyl-propyl)-	EtOH	202(3.62),232(3.36)	78-2141-84
$C_7H_{13}N_2O_4P$			
Phosphonic acid, (4,5-dihydro-5-oxo-1H-imidazol-4-yl)-, diethyl ester	H_2O	216(3.71),253(3.43)	65-2414-84
$C_7H_{14}FNSe$			
Carbamoselenoic fluoride, bis(1-methyl-ethyl)-	hexane	210(3.83),258(4.26), 358(2.40)	78-4963-84
$C_7H_{14}N_2O$			
2-Propenamide, 3-(dimethylamino)-N,N-dimethyl-, (E)-	n.s.g.	273(4.34)	131-0019-84K
$C_7H_{14}N_2O_2S$			
Propanal, 2-methyl-2-(methylthio)-, O-[(methylamino)carbonyl]oxime	MeCN	245(<u>4.3</u>)	98-0873-84
$C_7H_{14}N_2O_4$			
Ethanol, 1-[[(1-methylethyl)-NNO-azoxy]oxy]-, acetate	MeCN	209(3.84)	5-1494-84
Ethanol, 1-[(1-methylethyl)nitro-amino]-, acetate	MeCN	236(3.78)	5-1494-84
Methanol, (butylnitroamino)-, acetate	MeCN	237(3.81)	5-1494-84
$C_7H_{14}N_2O_4S$			
2-Propenoic acid, 3-[[(dimethylamino)-sulfonyl]amino]-2-methyl-, methyl ester, (E)-	50% EtOH + NaOH	256(4.20) 290(4.45)	73-0840-84 73-0840-84
(Z)-	50% EtOH + NaOH	265(4.11) 296(4.29)	73-0840-84 73-0840-84
$C_7H_{14}OS_3$			
Ethene, 1,1-bis(ethylthio)-2-(methyl-sulfinyl)-	EtOH	271(4.05)	39-0085-84C
Propane, 2-[[2-(methylsulfinyl)-1-(methylthio)ethenyl]thio]-, (Z)-	EtOH	269(4.06)	39-0085-84C
$C_7H_{14}O_3S_3$			
Ethane(dithioic) acid, [(2-hydroxy-2-methylpropyl)sulfonyl]-, methyl ester	EtOH	235(3.37),317(3.99)	39-0085-84C

Compound	Solvent	$\lambda_{max}(\log \epsilon)$	Ref.
$C_7H_{15}NO_2$			
2-Butanol, 3-(ethylidenamino)-3-methyl-, N-oxide	pH 8.8	225(3.88)	94-2609-84
$C_7H_{15}N_3O_3$			
Urea, N-butyl-N',N'-dimethyl-N-nitro-	hexane	242(3.67)	44-4860-84
$C_7H_{16}O_2Si$			
Silane, [(1-methoxy-1-propenyl)oxy]trimethyl-, (E)-	CH_2Cl_2	230(2.8)	12-2073-84
$C_7H_{18}ClNSi$			
Silanamine, N-chloro-N-(1,1-dimethylethyl)-1,1,1-trimethyl-	$C_2Cl_3F_3$	295(2.10)	39-1187-84B
Silanamine, N-chloro-1-(1,1-dimethylethyl)-N,1,1-trimethyl-	$C_2Cl_3F_3$	297(1.98)	39-1187-84B

Compound	Solvent	$\lambda_{max}(\log \epsilon)$	Ref.
$C_8Br_2F_8$ Bicyclo[4.2.0]octa-2,4-diene, 7,8-di-bromo-1,2,3,4,5,6,7,8-octafluoro-	C_6H_{12}	281(3.43)	35-8301-84
C_8F_8 Bicyclo[4.2.0]octa-2,4,7-triene, 1,2,3,4,5,6,7,8-octafluoro-	EtOH at $-60°$	241(3.3)	35-8301-84
$C_8F_8S_2Se$ Carbonoselenodithioic acid, S-(penta-fluorophenyl) Se-(trifluoromethyl) ester	hexane	210s(4.48),279(4.26), 308s(4.13),474(1.56)	78-4963-84
$C_8F_8S_3$ Carbonotrithioic acid, pentafluoro-phenyl trifluoromethyl ester	hexane	210(4.53),267(4.31), 306s(4.12),471(1.39)	78-4963-84
$C_8F_{18}S_2Se_4$ Ethane, 1,1,2,2-tetrakis[(trifluoro-methyl)seleno]1,2-bis[(trifluoro-methyl)thio]-	hexane	226(2.81)	78-4963-84
$C_8F_{18}S_4Se_2$ Ethane, 1,2-bis[(trifluoromethyl)sel-eno]-1,1,2,2-tetrakis[(trifluoro-methyl)thio]-	hexane	238(3.51)	78-4963-84
$C_8H_2Br_2N_2O$ Propanedinitrile, [(4,5-dibromo-2-furanyl)methylene]-	MeOH	212(3.03),267(2.72), 365(3.44)	73-0984-84
$C_8H_3ClO_3S_2$ Cyclohept[c][1,2]oxathiol-3(8H)-one, 5-chloro-4-hydroxy-8-thioxo-	MeCN	223(4.26),250(4.05), 289(4.07),304(4.17), 319(4.22),332(4.18), 354(3.87),368(3.77), 384(3.59),468(3.53)	88-0419-84
$C_8H_4BrN_3O$ Propanedinitrile, [(5-amino-4-bromo-2-furanyl)methylene]-	MeOH	225(3.12),275(2.42), 462(3.45)	73-0984-84
$C_8H_4N_2O_3$ Propanedinitrile, (2-methoxy-3,4-dioxo-1-cyclobuten-1-yl)-, compd. with N,N-diethylethanamine (1:1)	H_2O	204(4.19),334(4.71)	24-2714-84
$C_8H_4 N_4O_3$ 2,1-Benzisoxazole-5-carbonitrile, 3-amino-7-nitro-	EtOH	234(4.33),387(3.86)	97-0254-84
$C_8H_4O_3S_2$ Cyclohept[c][1,2]oxathiol-3(8H)-one, 4-hydroxy-8-thioxo- (thiotropocin)	MeOH	216.5(4.40),245(4.05), 306(4.21),356(3.79), 452(3.32)	88-0419-84
	MeOH	216.5(4.40),245(4.05), 307(4.21),356(3.79), 452(3.32)	158-1294-84
sodium salt	n.s.g.	214(4.40),243(4.10), 302(4.26),354(3.86), 430(3.42)	88-0419-84

Compound	Solvent	$\lambda_{max}(\log \epsilon)$	Ref.
$C_8H_4O_4S_4$			
1,3-Dithiole-4-carboxylic acid, 2-(4-carboxy-1,3-dithiol-2-ylidene)-	pH 12	287(4.20),302(4.26), 313(4.28),418(3.55)	103-1342-84
isomer B	pH 12	288(4.08),304(4.13), 315(4.15),404(3.41)	103-1342-84
$C_8H_4O_5S_2$			
Cyclohept[c][1,2]oxathiol-3(8H)-one, 4-hydroxy-8-thioxo-, S,1-dioxide	MeCN	216(4.17),279s(4.26), 288(4.29),345(3.64)	88-0419-84
C_8H_5Cl			
Benzene, 1-chloro-4-ethynyl-	MeOH	248(4.39),254(4.41)	118-0728-84
$C_8H_5Cl_2NO_2$			
1H,3H-Oxazolo[3,4-a]azepin-3-one, 5,9-dichloro-	MeOH	358(2.71)	119-0171-84
$C_8H_5NO_2$			
Benzene, 1-ethynyl-4-nitro-	MeOH	284(4.12)	118-0728-84
8H-Cycloheptoxazol-8-one	MeOH	237(4.56),310(3.94)	18-0609-84
$C_8H_5N_3O$			
2,1-Benzisoxazole-5-carbonitrile, 3-amino-	EtOH	223s(4.10),235s(4.10), 276s(3.54),297s(3.48)	97-0254-84
2,1-Benzisoxazole-7-carbonitrile, 3-amino-	EtOH	223(4.16),234s(3.95), 309(3.80)	97-0254-84
$C_8H_5N_3OS$			
Furo[2',3':4,5]pyrrolo[1,2-d][1,2,4]-triazine-8(7H)-thione	MeOH	281(3.13),364(3.21)	73-0065-84
$C_8H_5N_3O_2$			
Furo[2',3':4,5]pyrrolo[1,2-d][1,2,4]-triazin-8(7H)-one	n.s.g.	244(3.41),290(3.21)	73-0065-84
$C_8H_5N_3O_8$			
Benzeneacetic acid, 2,4,6-trinitro-, potassium salt	THF	372(--),520(4.09), 630(--)	44-0413-84
	DMSO	376(--),522(4.20), 630(--)	44-0413-84
$C_8H_5N_5O_3S_2$			
Thiazole, 4-azido-2-(methylthio)-5-(5-nitro-2-furanyl)-	dioxan	225(3.00),260(3.06), 301(2.90),416(3.23)	73-2285-84
C_8H_6BrClO			
Ethanone, 1-(4-bromophenyl)-2-chloro-	20% H_2SO_4	265(4.17)	48-0353-84
	96% H_2SO_4	352(4.26)	48-0353-84
$C_8H_6Br_2N_2$			
2H-Cyclopenta[d]pyridazine, 5,7-di-bromo-1-methyl-	ether	207(4.05),252(4.08), 267(4.06),318(3.48), 328s(3.42),400(3.76)	44-4769-84
C_8H_6ClFO			
Ethanone, 2-chloro-1-(4-fluorophenyl)-	20% H_2SO_4	253(4.11)	48-0353-84
	96% H_2SO_4	312(4.31)	48-0353-84
$C_8H_6ClNO_3$			
Ethanone, 2-chloro-1-(4-nitrophenyl)-	20% H_2SO_4	270(4.10)	48-0353-84
	96% H_2SO_4	279(4.23)	48-0353-84

Compound	Solvent	$\lambda_{max}(\log \epsilon)$	Ref.
$C_8H_6N_2O_2S$ 1,2,3-Benzothiadiazole-6-carboxylic acid, methyl ester	CH_2Cl_2	<u>263(4.0)</u>,330(3.5)	24-0107-84
$C_8H_6N_2O_6$ Benzeneacetic acid, 2,4-dinitro-, potassium salt	DMSO	408(4.15),660(--), 710(--)	44-0413-84
2,5-Cyclohexadiene-1,4-dione, 2,5-dimethyl-3,6-dinitro-	$CDCl_3$	270(4.12)	12-0055-84
$C_8H_6N_4O_2$ 1H-Pyrrolo[3,2-d]pyrimidine-7-carbonitrile, 2,3,4,5-tetrahydro-3-methyl-2,4-dioxo-	EtOH	221(4.43),269(3.91)	44-0908-84
$C_8H_6O_2S_4$ 1,3-Dithiole-4-carboxylic acid, 2-(1,3-dithiol-2-ylidene)-, methyl ester	MeCN	292(4.08),302(4.11), 313(4.14),430(3.30)	103-1342-84
$C_8H_6O_4$ Benzeneacetic acid, 2-hydroxy-α-oxo-	EtOH	213(3.88),257(3.86), 328(3.49)	39-2655-84C
$C_8H_6O_5$ Benzeneacetic acid, 2,3-dihydroxy-α-oxo-	EtOH	216(4.08),272(3.92), 346(3.28)	39-2655-84C
Benzeneacetic acid, 2,4-dihydroxy-α-oxo-	EtOH	209(4.13),228(3.83), 284(3.98),310(3.81)	39-2655-84C
$C_8H_6S_2Se_2$ 1,3-Dithiole, 2-[(1,3-diselenol-2-ylidene)ethylidene]-	$CHCl_3$	372(4.16),398(4.11)	88-4227-84
$C_8H_6S_4$ 1,3-Dithiole, 2,2'-(1,2-ethanediylidene)bis-	$CHCl_3$	387(4.31),406(4.33)	88-4227-84
$C_8H_6Se_4$ 1,3-Diselenone, 2,2'-(1,2-ethanediylidene)bis-	$CHCl_3$	365(4.06),390(4.06)	88-4227-84
$C_8H_7BrN_2$ 2H-Cyclopenta[d]pyridazine, 5-bromo-1-methyl-	ether	206.2(4.07),248(4.29), 264(4.15),312.5(3.60), 325s(3.54),390(2.93)	44-4769-84
C_8H_7ClO 2,4,6-Cycloheptatrien-1-one, 4-chloro-5-methyl-	n.s.g.	231(4.23),320(3.89)	70-2428-84
Ethanone, 2-chloro-1-phenyl-	20% H_2SO_4	251(4.05)	48-0353-84
	96% H_2SO_4	312(4.35)	48-0353-84
$C_8H_7Cl_2NO_2$ 2,4-Pentadienoic acid, 5,5-dichloro-2-cyano-, ethyl ester	EtOH	300(4.33)	70-2328-84
$C_8H_7F_3OSSi$ Benzenecarbothioic acid, S-[(trifluorosilyl)methyl] ester	heptane	242(3.98),273(3.92)	70-0726-84
	Bu_2O	239(3.96),273(3.89)	70-0726-84
	MeCN	246(3.95),281(4.02)	70-0726-84
	CH_2Cl_2	249(3.98),284(4.08)	70-0726-84

Compound	Solvent	$\lambda_{max}(\log \epsilon)$	Ref.
C_8H_7NOS			
4-Benzothiazolol, 7-methyl-	EtOH	258(3.30),316(3.25)	44-0997-84
2(3H)-Benzoxazolethione, 3-methyl-	CH_2Cl_2	259(3.68),264(3.63),	39-1383-84B
		303(4.14)	
compd. with iodine	CH_2Cl_2	270(4.28),315(4.49)	39-1383-84B
C_8H_7NOSe			
2(3H)-Benzoxazoleselone, 3-methyl-	CH_2Cl_2	267(3.56),318(4.53)	39-1383-84B
compd. with iodine	CH_2Cl_2	280(4.32),343(4.53),	39-1383-84B
		390s(4.14)	
$C_8H_7NO_4$			
2,5-Cyclohexadiene-1,4-dione, 2,5-di-methyl-3-nitro-	$CHCl_3$	260(4.16)	12-0055-84
C_8H_7NSSe			
2(3H)-Benzothiazoleselone, 3-methyl-	CH_2Cl_2	351(4.50),388(3.36)	39-1383-84B
compd. with iodine	CH_2Cl_2	267(4.21),295(4.58),	39-1383-84B
		357(4.63),410s(4.17)	
$C_8H_7NS_2$			
2(3H)-Benzothiazolethione, 3-methyl-	CH_2Cl_2	243(4.22),327(4.51)	39-1383-84B
compd. with iodine	CH_2Cl_2	258(4.27),284(4.41),	39-1383-84B
		338(4.67),396s(4.08)	
$C_8H_7N_3$			
1H-Isoindol-3-amine, 1-imino-	MeOH	251(4.10),256(4.10),	20-0459-84
		303(3.66)	
$C_8H_7N_3O$			
Ethanone, 1-(4-azidophenyl)-	2% MeOH	287(4.30)	149-0545-84B
1,3,4-Oxadiazol-2-amine, N-phenyl-,	H_2O	280(4.20)	39-0537-84B
anion	20% dioxan	286(4.23)	39-0537-84B
	40% dioxan	286(4.27)	39-0537-84B
	20% DMSO	285(4.26)	39-0537-84B
	40% DMSO	288(4.31)	39-0537-84B
	50% DMSO	292(4.33)	39-0537-84B
	60% DMSO	295(4.35)	39-0537-84B
	75% DMSO	300(4.40)	39-0537-84B
	90% DMSO	305(4.42)	39-0537-84B
2(1H)-Quinoxalinone oxime	EtOH	243(3.25),264(3.25),	33-1503-84
		372(2.98)	
$C_8H_7N_3O_2$			
Benzoic acid, 4-azido-, methyl ester	2% MeOH	273(4.34)	149-0545-84B
Imidazo[1,2-a]pyrimidine-3-carboxalde-hyde, 7,8-dihydro-8-methyl-7-oxo-	H_2O	264(4.18),280(4.16)	44-4021-84
Imidazo[1,2-c]pyrimidine-3-carboxalde-hyde, 5,6-dihydro-6-methyl-5-oxo-	H_2O	328(3.72)	44-4021-84
$C_8H_7N_3O_2S$			
Imidazo[1,2-a]pyridinium, 2-mercapto-1-methyl-3-nitro-, hydroxide, inner salt	n.s.g.	206(4.27),280s(4.08), 307(4.48),361(3.92), 409(3.95)	39-0069-84C
Pyrazolo[1,5-a]pyridine, 2-(methyl-thio)-3-nitro-	EtOH	244(4.54),365(4.24)	95-0440-84
$C_8H_7N_3O_3$			
Imidazo[1,2-a]pyridine, 2-methoxy-3-nitro-	MeCN	258(4.06),266(4.06), 274(4.04),320s(3.83), 358(4.23)	39-0069-84C

Compound	Solvent	$\lambda_{max}(\log \epsilon)$	Ref.
Imidazo[1,2-a]pyridinium, 2-hydroxy-1-methyl-3-nitro-, hydroxide, inner salt	MeCN	274s(4.12),278(4.21), 335s(3.89),370(4.23)	39-0069-84C
$C_8H_7N_3O_3S_2$ Carbonimidodithioic acid, cyano-, methyl (5-nitro-2-furanyl)methyl ester	dioxan	261(3.24),316(3.20)	73-2285-84
4-Thiazolamine, 2-(methylthio)-5-(5-nitro-2-furanyl)-	dioxan	243(3.00),272(2.95), 302(2.84),336(2.74), 472(3.36)	73-2285-84
$C_8H_7N_3O_5$ Acetamide, N-(2,4-dinitrophenyl)-	dioxan-base DMF-base	438(4.31) 436(4.37)	104-0565-84 104-0565-84
$C_8H_7N_3O_7$ 1(4H)-Pyridineacetic acid, α-methyl-3,5-dinitro-4-oxo-, (S)-	EtOH	341(3.63)	18-1961-84
$C_8H_7N_3O_8$ 1(4H)-Pyridineacetic acid, α-(hydroxymethyl)-3,5-dinitro-4-oxo-, (S)-	EtOH	341(3.68)	18-1961-84
$C_8H_7N_4O_7$ Benzofurazan, 4,?-dihydro-4,4-dimethoxy-5,7-dinitro-, ion(1-)	MeOH	450(4.44)	44-4176-84
$C_8H_7N_5O$ Furo[2',3':4,5]pyrrolo[1,2-d][1,2,4]-triazin-8(7H)-one, hydrazone	MeOH	300(3.92)	73-0065-84
$C_8H_8ClNO_3$ 3-Cyclobutene-1,2-dione, 3-chloro-4-morpholino-	pH 0.10	258(4.46)	88-0135-84
$C_8H_8Cl_3NO_2$ 2-Cyclobuten-1-one, 2,4,4-trichloro-3-morpholino-	pH 0.10	275(4.43)	88-0135-84
$C_8H_8F_6N_2OS$ 6H-1,3,4-Thiadiazine, 6-propoxy-2,5-bis(trifluoromethyl)-	CH_2Cl_2	278(3.4)	89-0890-84
$C_8H_8N_2$ 2H-Cyclopenta[d]pyridazine, 1-methyl-	$CHCl_3$	247.5(3.97),271.5s(3.64), 317s(3.19)	44-4769-84
2-Pyridinecarbonitrile, 4,5-dimethyl-	EtOH	230(4.03),263(3.52), 275(3.45),328(1.18)	44-0813-84
$C_8H_8N_2O$ 3H-Azepine-3(and 7)-carbonitrile, 2-methoxy-	EtOH	269(3.76)	39-0249-84C
3H-Azepine-5-carbonitrile, 2-methoxy-	EtOH	241(3.52),278(3.55)	39-0249-84C
$C_8H_8N_2OS$ 6H-Thiopyrano[3,4-d]pyrimidin-5(8H)-one, 2-methyl-	EtOH	239(3.87)	4-1437-84
$C_8H_8N_2O_4$ Methanamine, N-(benzoyloxy)-N-nitro-	CH_2Cl_2	237(4.12),270s(3.33), 280s(3.19)	44-4872-84

Compound	Solvent	$\lambda_{max}(\log \epsilon)$	Ref.
2-Propenoic acid, (1,2,3,4-tetrahydro-2,6-dioxo-4-pyrimidinyl)methyl ester	pH 1.5	262(4.00)	47-0813-84
	pH 8.4	263(3.96)	47-0813-84
	pH 12.4	262(3.83),281(3.91)	47-0813-84
polymer	pH 14	276(3.74)	47-0813-84
	DMSO	262(3.85)	47-0813-84
$C_8H_8N_2S$			
2H-Benzimidazole-2-thione, 1,3-dihydro-1-methyl-	CH_2Cl_2	250(4.19),304s(4.47), 312(4.59)	39-1383-84B
compd. with iodine	CH_2Cl_2	285(4.44),314(4.71), 370(4.23)	39-1383-84B
$C_8H_8N_2Se$			
2H-Benzimidazole-2-selone, 1,3-dihydro-1-methyl-	CH_2Cl_2	324(4.43)	39-1383-84B
compd. with iodine	CH_2Cl_2	390(4.18)	39-1383-84B
$C_8H_8N_4O$			
Acetamide, N-(4-azidophenyl)-	2% MeOH	265(4.26)	149-0545-84B
$C_8H_8N_4O_2$			
4(3H)-Pteridinone, 2-methoxy-3-methyl-	MeOH	237(4.28),261(4.20), 321(3.79)	136-0179-84F
$C_8H_8N_4O_3$			
4-Benzofurazanamine, N,N-dimethyl-7-nitro- (in 15% DMF)	pH 2.7	496(4.55)	48-0385-84
$C_8H_8N_4O_8$			
2-Butenedial, 1,4-dinitro-, bis(O-acetyloxime)	$CHCl_3$	260(4.53),270(4.60), 282(4.53)	104-1264-84
isomer 2b	$CHCl_3$	260(4.16),270(4.20), 280(4.08)	104-1264-84
$C_8H_8N_6$			
Methanimidamide, N'-(4,5-dicyano-1H-pyrazol-3-yl)-N,N-dimethyl-	EtOH	276(4.04)	103-0215-84
C_8H_8O			
Benzene, (ethenyloxy)-	C_6H_{12}	196(4.42),225(4.15), 263(3.03),269(3.14), 276(3.03)	104-0473-84
$C_8H_8O_2$			
Benzaldehyde, 2-methoxy-	pH 8.1	255(4.18),322(3.77)	18-2508-84
Ethanone, 1-(4-hydroxyphenyl)-	EtOH	280(4.13)	39-0825-84C
	EtOH-NaOH	238(3.84),332(4.33)	39-0825-84C
$C_8H_8O_3$			
2(6H)-Benzofuranone, 7,7a-dihydro-6-hydroxy-, cis-(-)-	MeOH	253(4.29)	102-2394-84
$C_8H_8O_4$			
Benzeneacetic acid, 3,4-dihydroxy-	pH 1	280(3.42)	3-1935-84
	borate	287(3.64)	3-1935-84
2,5-Cyclohexadiene-1,4-dione, 2,5-di-hydroxy-3,6-dimethyl-	aq base	327s(--),334(4.48), 550(2.35)	94-2406-84
2H-Pyran-2-one, 3-acetyl-4-hydroxy-6-methyl-	EtOH	230(3.90),295(4.23)	39-1317-84B
	dioxan	310(4.02)	39-1317-84B
	$CHCl_3$	310(4.03)	39-1317-84B
	80% DMSO	310(4.08)	39-1317-84B

Compound	Solvent	$\lambda_{max}(\log \epsilon)$	Ref.
4H-Pyran-4-one, 3-acetyl-6-hydroxy-2-methyl-	EtOH	260(3.97),295s(--)	39-1317-84B
	dioxan	265(3.95),295s(--)	39-1317-84B
	CHCl$_3$	268(4.00),295s(--)	39-1317-84B
	80% DMSO	258(3.87),290s(--)	39-1317-84B
$C_8H_9BrO_2$			
Ethanol, 2-(2-bromophenoxy)-	EtOH	205(3.61),275(3.36), 283(3.34)	12-2059-84
$C_8H_9ClN_2O$			
Benzoic acid, 4-chloro-, 1-methylhydrazide	EtOH	<u>225(3.9)</u>	70-2056-84
$C_8H_9ClN_2O_3$			
1(2H)-Pyrimidineacetic acid, 5-chloro-2-oxo-, ethyl ester	EtOH	208s(--),226(4.04), 332(3.60)	103-1185-84
$C_8H_9ClN_4$			
1H-Purine, 6-chloro-8-(1-methylethyl)-	pH 1	266(4.08)	39-0879-84C
	pH 13	278(4.05)	39-0879-84C
$C_8H_9ClN_4O$			
7H-Pyrrolo[2,3-d]pyrimidin-2-amine, 4-chloro-7-(methoxymethyl)-	MeOH	234(4.46),263s(3.76), 316(3.81)	87-0981-84
$C_8H_9ClO_4$			
Acetic acid, chloro(dihydro-4,4-dimethyl-5-oxo-2(3H)-furanylidene)-, (Z)-	MeOH	234(4.07)	35-3551-84
$C_8H_9ClO_6P_2$			
Phosphonic acid, [(4-chlorophenyl)ethenylidene]bis-, disodium salt	H$_2$O	271(4.09)	65-1987-84
$C_8H_9FN_2O$			
Benzoic acid, 4-fluoro-, 1-methylhydrazide	EtOH	<u>233s(4.0)</u>	70-2056-84
C_8H_9NO			
1-Cycloheptene-1-carbonitrile, 3-oxo-	EtOH	237(4.08),336(1.85)	44-2925-84
1-Cycloheptene-1-carbonitrile, 6-oxo-	EtOH	212(4.00),278(2.20)	44-2925-84
$C_8H_9NO_2$			
Acetamide, N-hydroxy-N-phenyl-	MeOH	268(3.11),272(3.40)	2-0060-84
$C_8H_9NO_3$			
1H-Azepine-3-carboxylic acid, 2,3-dihydro-2-oxo-, methyl ester	EtOH	256(3.74)	39-0249-84C
2,4-Furandicarboxaldehyde, 5-(dimethylamino)-	MeOH	302(3.22),351(3.46)	73-1600-84
3H-Pyrano[4,3-c]isoxazol-3-one, 6,7-dihydro-4,6-dimethyl-	EtOH	295(3.95)	142-0365-84
$C_8H_9NO_3S_2$			
Thiophene, 2-methoxy-5-(methylthio)-3-(2-nitroethenyl)-	EtOH	215(4.06),280(4.12), 385(4.08)	103-0962-84
Thiophene, 5-methoxy-2-(methylthio)-3-(2-nitroethenyl)-	EtOH	215(4.05),268(4.00), 308(4.11),380(3.85)	103-0962-84
$C_8H_9NO_4$			
Acetamide, N-(4-hydroxy-5-oxo-7-oxabicyclo[4.1.0]hept-3-en-2-yl)-	H$_2$O	212(4.04),275(3.54)	158-1149-84

Compound	Solvent	$\lambda_{max}(\log \epsilon)$	Ref.
3-Cyclobutene-1,2-dione, 3-hydroxy- 4-morpholino-	pH 7 20% H_2SO_4	291(4.49) 278(4.40)	88-0135-84 88-0135-84
$C_8H_9NO_8P_2$ Phosphonic acid, [(4-nitrophenyl)ethen- ylidene]bis-, disodium salt	H_2O	319(3.92)	65-1987-84
$C_8H_9N_3O$ 1H-Pyrrolo[3,4-d]pyridazin-1-one, 2,6- dihydro-5,7-dimethyl-	pH 7.00	202(4.48),232(4.34), 277(3.95),314(3.68)	78-3979-84
cation	acid	231(4.22),248(4.18), 283(3.79),349(3.78)	78-3979-84
dication	H_2SO_4	226(4.41),239(4.31), 304.5(4.17),373(3.37)	78-3979-84
$C_8H_9N_3O_2$ Imidazo[1,2-c]pyrimidin-5(6H)-one, 3-(hydroxymethyl)-6-methyl-	H_2O	277(4.16)	44-4021-84
1H-Pyrrolo[3,4-d]pyridazine-1,4(6H)- dione, 2,3-dihydro-5,7-dimethyl-	pH 7.0 cation	207(4.50),237.5(4.00) 204(4.48),218(4.43), 282(4.03),308s(3.63)	78-3979-84 78-3979-84
anion	pH 13	217(4.22),269(3.94), 305(3.77)	78-3979-84
1H-Pyrrolo[2,3-d]pyrimidine, 2,4-di- methoxy-	MeOH	216(4.33),258s(3.70), 271(3.79)	5-0273-84
$C_8H_9N_3O_3$ Benzoic acid, 4-nitro-, 1-methylhydra- zide	EtOH	<u>272(4.1)</u>	70-2056-84
$C_8H_9N_3O_3S$ 2-Propenoic acid, 2-[[(2,5-dihydro- 6-methyl-5-oxo-1,2,4-triazin-3-yl)- thio]methyl]-	EtOH	261(4.07),330(3.70)	142-1049-84
$C_8H_9N_5$ Cyanamide, (8,9-dihydro-7H-imidazo- [1,5-c][1,3]diazepin-3-yl)-	MeOH	250(4.19)	4-0753-84
Cyanamide, (7,8-dihydro-2-methylimid- azo[1,2-c]pyrimidin-5-yl)-	MeOH	247(4.23)	24-2597-84
7H-Imidazo[2,1-i]purine, 8,9-dihydro- 9-methyl-	pH 1 pH 7 pH 13	263(4.13) 265(4.14) 274(4.14)	4-0333-84 4-0333-84 4-0333-84
$C_8H_9N_5O$ Cyanamide, [7,8-dihydro-2-(hydroxy- methyl)imidazo[1,2-c]pyrimidin-5-yl]-	MeOH	247(4.22)	24-2597-84
$C_8H_9N_5O_3$ 9H-Purine-9-propanoic acid, 6-amino- α-hydroxy-, (R)-	pH 2	261(4.13)	73-2148-84
C_8H_{10} Spiro[2.4]hepta-4,6-diene, 1-methyl-	EtOH	227(3.63),258(3.19)	70-1526-84
$C_8H_{10}BrCl_3$ 3,5-Octadiene, 8-bromo-1,1,1-trichloro-	EtOH	236(4.44)	44-0495-84
$C_8H_{10}BrNO_2$ Cavernicolin, monobromo-	MeOH	250(3.78),315(1.54)	51-0425-84

Compound	Solvent	$\lambda_{max}(\log \epsilon)$	Ref.
$C_8H_{10}BrN_5O_3$			
6H-Purin-6-one, 2-amino-8-bromo-1,9-di- hydro-9-[(2-hydroxyethoxy)methyl]-	pH 1 pH 13	259(4.22) 270(4.14)	87-1486-84 87-1486-84
$C_8H_{10}ClN_5O_3$			
6H-Purin-6-one, 2-amino-8-chloro-1,9- dihydro-9-[(2-hydroxyethoxy)methyl]-	pH 1 pH 13	256(4.18) 268(4.12)	87-1486-84 87-1486-84
$C_8H_{10}IN_5O_3$			
6H-Purin-6-one, 2-amino-1,9-dihydro- 9-[(2-hydroxyethoxy)methyl]-8-iodo-	pH 1 pH 13	260(4.24) 270(4.18)	87-1486-84 87-1486-84
$C_8H_{10}N$			
Pyridinium, 1-ethenyl-3-methyl-, iodide	H_2O	265(3.89)	142-0505-84
Pyridinium, 1-ethenyl-4-methyl-, bromide	H_2O	256(3.89)	142-0505-84
$C_8H_{10}NO_5S$			
Ethanone, 1-(5,?-dihydro-5,5-dimeth- oxy-4-nitro-2-thienyl)-, ion(1-), sodium	MeOH	310(3.79),436(4.25)	39-0317-84B
$C_8H_{10}NO_6S$			
2-Thiophenecarboxylic acid, 5,?-dihy- dro-5,5-dimethyl-4-nitro-, methyl ester, ion(1-), sodium	MeOH	300(3.79),406(4.20)	39-0317-84B
$C_8H_{10}N_2$			
1H-Pyrrolo[2,3-b]pyridine, 2,3-dihydro- 4-methyl-	heptane PrOH	303(3.66) 308(3.66)	151-0343-84D 151-0343-84D
$C_8H_{10}N_2O$			
Benzoic acid 1-methylhydrazide	EtOH	<u>230s(4.0)</u>	70-2056-84
$C_8H_{10}N_2OS_2$			
1,2,3-Benzothiadiazole, 4,5,6,7-tetra- hydro-6,6-dimethyl-7-sulfinyl-, (Z)-	EtOH	232(3.87),282s(3.28), 339(4.03)	44-4773-84
$C_8H_{10}N_2O_2$			
Phenol, 2-[(methylnitrosoamino)methyl]-	hexane	235(3.56),274(3.29), 352(1.88)	150-0701-84M
Phenol, 4-[(methylnitrosoamino)methyl]-	EtOH	228(4.07),276(3.38), 350(1.92)	150-0701-84M
2,4(1H,3H)-Pyrimidinedione, 5-(2-meth- yl-2-propenyl)-	MeOH M NaOH	265(3.86) 286(3.81)	44-2738-84 44-2738-84
$C_8H_{10}N_2O_2S$			
Pyridine, 2-[(1-methyl-1-nitroethyl)- thio]-	EtOH	279(3.99)	39-2327-84C
Thiopyrano[3,4-c]pyrazol-4(5H)-one, 1,7-dihydro-1-(2-hydroxyethyl)-	EtOH	249(3.94)	4-1437-84
$C_8H_{10}N_2O_3$			
2-Furanamine, N,N-dimethyl-5-(2-nitro- ethenyl)-	MeOH	208(3.67),262(4.18), 428(4.42)	73-2496-84
1(2H)-Pyrimidineacetic acid, 2-oxo-, ethyl ester	EtOH	216(3.93),311(3.68)	103-1185-84
$C_8H_{10}N_2O_4$			
Isoxazolo[5,4-c]pyridine-6(2H)-carbox-	EtOH	260(3.9)	23-1940-84

Compound	Solvent	$\lambda_{max}(\log \epsilon)$	Ref.
ylic acid, 3,4,5,7-tetrahydro-3-oxo-, methyl ester (cont.)			23-1940-84
$C_8H_{10}N_2O_4S$ Pyridinium, 3-(aminocarbonyl)-1-(2-sulfoethyl)-, hydroxide, inner salt	H_2O anion	263(3.87) 255(3.53),261(3.61), 268(3.49)	56-1077-84 56-1077-84
$C_8H_{10}N_3$ Benzenediazonium, 4-(dimethylamino)-, tetrafluoroborate	DMSO	380(4.70)	123-0075-84
$C_8H_{10}N_4O$ 7H-Pyrrolo[2,3-d]pyrimidin-2-amine, 4-methoxy-5-methyl-	MeOH	228(4.43),264(3.79), 289(3.81)	5-0708-84
$C_8H_{10}N_4OS_2$ 6H-Purin-6-one, 1,2,3,9-tetrahydro-1,9-dimethyl-8-(methylthio)-2-thioxo-	pH 13	230s(4.04),300(4.13)	2-0316-84
$C_8H_{10}N_4O_2$ 4H-Pyrrolo[2,3-d]pyrimidin-4-one, 2-amino-1,7-dihydro-7-(methoxymethyl)-	MeOH	216(4.33),259(4.09), 284s(3.91)	87-0981-84
$C_8H_{10}N_4O_2S_2$ 2H-Thiazolo[3,2-b][1,2,4]thiadiazol-2-imine, 6-butyl-N-nitro-	isoPrOH	280(3.15),343(4.21)	94-3483-84
$C_8H_{10}N_4O_3$ 6H-Purin-6-one, 1,9-dihydro-9-[(2-hy-droxyethoxy)methyl]-	pH 1 pH 13	248(4.09) 252(4.12)	87-1486-84 87-1486-84
$C_8H_{10}N_4O_3S$ 1H-Pyrazino[2,3-a][1,2,6]thiadiazin-4(3H)-one, 1,6,7-trimethyl-, 2,2-dioxide	pH 4.0	248(4.09),338(3.81)	4-0861-84
$C_8H_{10}N_4O_{12}$ 2,3-Butanedial, 1,1,4,4-tetranitro-, diacetate	pH 13	365(4.18)(changing)	104-2277-84
$C_8H_{10}N_6O_3$ 9H-Purine-9-propanoic acid, 2,6-di-amino-α-hydroxy-, (±)-	pH 2 pH 12	246(3.95),290(3.88) 256(3.90),278(3.92)	73-2148-84 73-2148-84
$C_8H_{10}O_2$ 1,3-Cyclopentanedione, 2-(1-methyl-ethylidene)- 1,2-Ethanediol, 1-phenyl-, (+)-	CH_2Cl_2 MeOH	239(3.90),264(3.51), 390(1.08)(unstable) 230(2.70),252(2.33), 257(2.40),263(2.35), 280(2.28)	23-2612-84 94-1209-84
$C_8H_{10}O_3$ Benzeneethanol, 3,4-dihydroxy- 6(2H)-Benzofuranone, 3,3a,7,7b-tetra-hydro-3a-hydroxy- (halleridone) 2,4-Hexadienoic acid, 5-methyl-6-oxo-, methyl ester, (E,E)-	pH 1 borate MeOH EtOH	279(3.40) 286(3.64) 234(3.37) 282(4.58)	3-1935-84 3-1935-84 102-2617-84 104-1528-84

Compound	Solvent	$\lambda_{max}(\log \epsilon)$	Ref.
$C_8H_{10}O_4$ 1,5-Cyclohexadiene-1-carboxylic acid, 3,4-dihydroxy-, methyl ester, cis-(±)-	n.s.g.	272(3.40)	78-2461-84
$C_8H_{10}O_6P_2$ Phosphonic acid, (phenylethylidene)bis-	H_2O	265(3.63)	65-1987-84
$C_8H_{11}BrN_2O_5$ 2,4(1H,3H)-Pyrimidinedione, 5-bromo- 1-[[2-hydroxy-1-(hydroxymethyl)- ethoxy]methyl]-	pH 1 H_2O pH 13	276(3.90) 276(3.87) 274(3.74)	23-0016-84 23-0016-84 23-0016-84
$C_8H_{11}BrO_5$ Propanedioic acid, bromoformyl-, diethyl ester	EtOH	247.5(1.79)	118-0732-84
$C_8H_{11}ClN_4O_4S$ 1H-1,2,3-Triazole-5-carbothioamide, 4-chloro-1-β-D-ribofuranosyl-	EtOH	253(3.91),303(3.92)	103-1287-84
$C_8H_{11}ClN_4O_5$ 1H-1,2,3-Triazole-5-carboxamide, 4-chloro-1-β-D-ribofuranosyl-	EtOH	240(3.91)	103-1287-84
$C_8H_{11}Cl_3O_3$ 2-Hexenoic acid, 6,6,6-trichloro- 3-methoxy-5-methyl-, (E)-	EtOH	232(4.13)	44-3489-84
$C_8H_{11}FN_2O_5$ 2,4(1H,3H)-Pyrimidinedione, 5-fluoro- 1-[[2-hydroxy-1-(hydroxymethyl)eth- oxy]methyl]-	pH 1 H_2O pH 13	265(3.88) 266(3.88) 266(3.76)	23-0016-84 23-0016-84 23-0016-84
$C_8H_{11}IN_2O_5$ 2,4(1H,3H)-Pyrimidinedione, 1-[[2-hy- droxy-1-(hydroxymethyl)ethoxy]methyl]- 5-iodo-	pH 1 H_2O pH 13	283(3.83) 283(3.80) 275(3.68)	23-0016-84 23-0016-84 23-0016-84
$C_8H_{11}NO_2$ 1,2-Benzenediol, 4-(2-aminoethyl)-, hydrochloride 1H-Pyrrole-2-acetic acid, α-methyl-, methyl ester 1H-Pyrrole-3-propanoic acid, 5-methyl-, Ehrlich test product	pH 1 borate MeOH n.s.g.	279(3.42) 286(3.63) 211(3.78) 545(4.77)	3-1935-83 3-1935-84 107-0453-84 142-1747-84
$C_8H_{11}NO_2S$ 2H-Thiopyran-3,5(4H,6H)-dione, 4-[(di- methylamino)methylene]-	EtOH	277(4.19)	4-1437-84
$C_8H_{11}NO_3$ Acetic acid, (5-oxo-2-pyrrolidinyli- dene)-, ethyl ester 1H-Pyrrole-2-carboxylic acid, 2,3-di- hydro-2-methyl-3-oxo-, ethyl ester	EtOH EtOH-base MeCN	259(4.26) 297(4.40) 308(4.03)	118-0618-84 118-0618-84 33-1957-84
$C_8H_{11}N_3O_2$ 1(4H)-Pyridineacetamide, 3-(amino- carbonyl)- 2(1H)-Pyrimidinone, 4-amino-1-[2-(eth- enyloxy)ethyl]-	H_2O pH 1 H_2O	350(3.94) 280(--) 272(3.94)	47-2169-84 47-2841-84 47-2841-84

Compound	Solvent	λ_{max}(log ϵ)	Ref.
$C_8H_{11}N_3O_4$ 1H-Imidazole-1,4-dicarboxylic acid, 5-amino-, 4-ethyl 1-methyl ester	EtOH	275(3.60)	103-0204-84
$C_8H_{11}N_3O_4S$ 4-Thiazoleacetic acid, 2-amino-α-(meth- oxyimino)-, ethyl ester, 3-oxide, (Z)- 2-Thiophenamine, N-butyl-3,5-dinitro-	EtOH EtOH-base benzene	<u>220(4.2)</u>,259(4.10) 270(4.11) 410(4.20)	4-1097-84 4-1097-84 39-0781-84B
$C_8H_{11}N_3O_7$ 2,4(1H,3H)-Pyrimidinedione, 1-[[2-hy- droxy-1-(hydroxymethyl)ethoxy]- methyl]-5-nitro-	pH 1 H_2O pH 13	299(4.04) 307(3.99) 321(4.08)	23-0016-84 23-0016-84 23-0016-84
$C_8H_{11}N_5$ 1H-Purin-6-amine, 8-ethyl-1-methyl- 9H-Purin-6-amine, 9-ethyl-N-methyl-	pH 1 pH 13 MeOH	262(4.13) 273(4.20) 267(4.18)	39-0879-84C 39-0879-84C 150-3601-84M
$C_8H_{11}N_5O$ 8H-Purin-8-one, 6-amino-9-ethyl- 7,9-dihydro-7-methyl-	MeOH	274(4.07)	150-3601-84M
$C_8H_{11}N_5O_2S$ 1H-Pyrazino[2,3-c][1,2,6]thiadiazin- 4-amine, 1,6,7-trimethyl-, 2,2-di- oxide	pH 7 MeOH	253(4.15),348(3.89) 255(4.16),286s(3.38), 348(3.85)	4-0861-84 4-0861-84
$C_8H_{11}N_5O_3$ Acetamide, N,N'-(6-amino-1,4-dihydro- 4-oxo-2,5-pyrimidinediyl)bis- Imidazo[5,1-f][1,2,4]triazin-4(1H)-one, 2-amino-7-[(2-hydroxyethoxy)methyl]-	EtOH-HCl EtOH-NaOH EtOH	265(3.93) 220(4.32),238s(3.84), 262(3.91) 226(4.51),263(3.78)	78-0879-84 78-0879-84 4-0697-84
$C_8H_{11}N_5O_4$ 1H-Purine-6,8-dione, 2-amino-7,9-di- hydro-9-[(2-hydroxyethoxy)methyl]-	pH 1 pH 13	247(4.05),293(3.99) 256(4.03),281(4.01)	87-1486-84 87-1486-84
$C_8H_{11}P$ Phosphine, dimethylphenyl-	C_6H_{12}	252(3.75)	65-0489-84
$C_8H_{12}N_2$ 1-Cycloheptene-1-carbonitrile, 2-amino-	MeOH	265(4.00)	107-0967-84
$C_8H_{12}N_2O$ Pyrazine, (1,1-dimethylethoxy)- Pyrazine, (1-methylpropoxy)-	EtOH EtOH	214(4.08),279(3.71), 297(3.56) 213(4.08),281(3.68), 295(3.62)	39-0641-84B 39-0641-84B
$C_8H_{12}N_2O_2$ 2,5-Pyrazinediethanol 2,4(1H,3H)-Pyrimidinedione, 5-(1,1-di- methylethyl)- in 10% MeOH 1-Pyrrolidinecarboxaldehyde, 3-[(di- methylamino)methylene]-2-oxo-	H_2O MeCN pH 3.5 pH 11.5 isoPrOH	204(3.99),274(3.88), 294s(3.34) 260(3.60) 262(3.69) 254(3.69) 230(3.95),328(4.54)	39-1345-84C 9-0556-84 9-0556-84 9-0556-84 103-0047-84

Compound	Solvent	$\lambda_{max}(\log \epsilon)$	Ref.
$C_8H_{12}N_2O_2S$ 2-Thiophenamine, N-butyl-3-nitro-	benzene	390(3.99)	39-0781-84B
$C_8H_{12}N_2O_4$ 4-Isoxazolepropanoic acid, 2,3-dihydro- 5-methyl-α-(methylamino)-3-oxo-, monohydrobromide, (±)-	MeOH	219(3.71)	87-0585-84
$C_8H_{12}N_2O_5$ 2,4(1H,3H)-Pyrimidinedione, 1-[[2-hy- droxy-1-(hydroxymethyl)ethoxy]- methyl]-	pH 1 H_2O pH 13	258(4.03) 258(4.02) 259(3.86)	23-0016-84 23-0016-84 23-0016-84
$C_8H_{12}N_4O_2$ Propanamide, N-(4-amino-1,6-dihydro- 6-oxo-5-pyrimidinyl)-2-methyl-	pH 1 pH 13	260(3.92) 255(3.69)	39-0879-84C 39-0879-84C
$C_8H_{12}N_6O_3$ 6H-Purin-6-one, 2,8-diamino-1,9-di- hydro-9-[(2-hydroxyethoxy)methyl]-	pH 1 pH 13	248(4.33),285(4.13) 255(4.30),277s(4.21)	87-1486-84 87-1486-84
$C_8H_{12}O_2$ Cyclopropanecarboxylic acid, 2-ethen- yl-, ethyl ester, trans 2,4-Hexadienoic acid, 5-methyl-, methyl ester 2,4-Pentadienoic acid, 3-ethyl-, methyl ester, (E)- (Z)- Propanedial, cyclopentyl-, sodium salt	hexane MeCN MeOH MeOH pH 13	207(3.70) 272(4.24) 252(4.29) 251(4.25) 276(4.46)	44-1937-84 22-0362-84 78-4701-84 78-4701-84 5-0649-84
$C_8H_{12}O_3$ Cyclopentanecarboxylic acid, 1-methyl- 2-oxo-, methyl ester Cyclopropanecarboxylic acid, 2-formyl- 2-methyl-, ethyl ester	C_6H_{12} EtOH	300(1.60) 210(3.26),280(1.75)	78-0613-84 44-1944-84
$C_8H_{12}O_5$ 1-Cyclohexene-1-carboxylic acid, 3,4,5-trihydroxy-, methyl ester (±-5-epishikimic acid methyl ester)	n.s.g.	220(3.79)	78-2461-84
$C_8H_{12}Si$ Silacyclohexa-2,4-diene, 1-(2-propenyl)-	C_6H_{12}	272(3.68)	24-2337-84
$C_8H_{13}ClO$ 3-Octen-2-one, 4-chloro-, (E)-	EtOH	239(4.10)	80-0193-84
$C_8H_{13}N$ Pyrrolidine, 1-(1,3-butadienyl)-	MeCN	288(4.06)	149-0391-84A
$C_8H_{13}NO$ Methanamine, N-(3-methoxy-2-cyclohexen- 1-ylidene)-, hydriodide	MeOH	274(4.35)	44-0220-84
$C_8H_{13}NOS_2$ 3-Thiopheneethanamine, 2-methoxy- 5-(methylthio)-, hydrochloride 3-Thiopheneethanamine, 5-methoxy- 2-(methylthio)-, hydrochloride	EtOH EtOH	203(4.09),230(3.58), 280(3.86) 202(4.02),238(3.68), 275(3.86)	103-0962-84 103-0962-84

Compound	Solvent	$\lambda_{max}(\log \epsilon)$	Ref.
$C_8H_{13}NO_2$			
Cycloheptene, 1-methyl-2-nitro-	EtOH	268(3.34)	12-1231-84
4H-Pyran-4-one, 2,3,5,6-tetrahydro-3-[(dimethylamino)methylene]-, (E)-	EtOH	327.5(4.32)	4-1441-84
2,5-Pyrrolidinedione, 3,4-diethyl-, cis	MeOH	221(2.28),246(1.93)	35-2645-84
trans	MeOH	220(2.22),244(1.64)	35-2645-84
2H-Pyrrol-2-one, 1,5-dihydro-4-methoxy-5-(1-methylethyl)-, (±)-	EtOH	211(4.15)	44-3489-84
$C_8H_{13}NO_2S$			
2,5-Pyrrolidinedione, 3-[(1,1-dimethylethyl)thio]-	MeCN	208(2.84)	142-1179-84
2H-Pyrrol-2-one, 1,5-dihydro-5-methoxy-4-[(1-methylethyl)thio]-	MeCN	228(3.82),274(4.00)	142-1179-84
$C_8H_{13}NO_6$			
1,3-Dioxolane-2-butanoic acid, α-nitro-, methyl ester	EtOH	205(3.71),275(1.65)	78-4513-84
$C_8H_{13}N_3$			
4-Pyrimidinamine, 2-methyl-6-(1-methylethyl)-	EtOH	234(3.97),270(3.58)	39-2677-84C
$C_8H_{13}N_3OS$			
3(2H)-Thiazolecarboxamide, N,N,4-trimethyl-2-(methylimino)-	isoPrOH	253(3.90),288(3.20)	94-4292-84
5-Thiazolecarboxamide, N,N,4-trimethyl-2-(methylamino)-	isoPrOH	297(4.14)	94-4292-84
1,2,4-Triazin-5(2H)-one, 6-(1,1-dimethylethyl)-3-(methylthio)-	MeOH	236(3.1)	98-0749-84
1,2,4-Triazin-5(4H)-one, 6-(1,1-dimethylethyl)-3-(methylthio)-	MeCN	232(4.25),290(2.91)	9-0556-84
in 10% MeOH	pH 3.5	221(--),241(4.13)	9-0556-84
in 10% MeOH	pH 11.5	221(4.06),241(--), 287.5(3.66)	9-0556-84
Urea, (3,4-dimethyl-2(3H)-thiazolylidene)dimethyl-	isoPrOH	295(4.23)	94-4292-84
Urea, trimethyl(4-methyl-2-thiazolyl)-	isoPrOH	268(3.68)	94-4292-84
$C_8H_{13}N_3O_2$			
3H-Pyrazole, 3,3-dimethyl-5-(1-methylethyl)-4-nitro-	MeCN	280(3.84)	88-2909-84
$C_8H_{13}N_3O_2S$			
1H-Imidazole-4-carboxylic acid, 5-amino-1-methyl-2-(methylthio)-, ethyl ester	MeOH	256s(3.84),282(3.98)	2-0316-84
$C_8H_{13}N_3O_5$			
2,4(1H,3H)-Pyrimidinedione, 5-amino-1-[[2-hydroxy-1-(hydroxymethyl)-ethoxy]methyl]-	pH 1	261(3.86)	23-0016-84
	H_2O	291(3.76)	23-0016-84
	pH 13	286(3.67)	23-0016-84
$C_8H_{13}N_3S_2$			
Carbamimidothioic acid, N,N-dimethyl-N'-(4-methyl-2-thiazolyl)-, methyl ester	isoPrOH	245(3.91),316(3.86)	94-2224-84
Carbamimidothioic acid, N-(4,5-dimethyl-2-thiazolyl)-N'-methyl-, methyl ester	isoPrOH	225(3.85),310(4.28)	94-2224-84

Compound	Solvent	$\lambda_{max}(\log \epsilon)$	Ref.
3(2H)-Thiazolecarbothioamide, N,N,4-trimethyl-2-(methylimino)-	isoPrOH	277(4.28),325s(3.11)	94-4292-84
5-Thiazolecarbothioamide, N,N,4-trimethyl-2-(methylamino)-	isoPrOH	283(4.14),333s(3.88)	94-4292-84
Thiourea, (4-butyl-2-thiazolyl)-	isoPrOH	253(3.83),293(4.30)	94-3483-84
Thiourea, (3,4-dimethyl-2(3H)-thiazolylidene)dimethyl-	isoPrOH	296s(3.90),328(4.18)	94-4292-84
Thiourea, trimethyl(4-methyl-2-thiazolyl)-	isoPrOH	277(4.17)	94-4292-84
$C_8H_{13}N_5O_2$			
Pyrazolo[4,3-c]pyrazole, 3,3a,6,6a-tetrahydro-3,3,6,6-tetramethyl-3a-nitro-	MeCN	256(3.19),325(2.68)	88-2909-84
$C_8H_{13}N_5O_2S$			
Guanidine, N'-(4-ethyl-2-thiazolyl)-N,N-dimethyl-N"-nitro-	isoPrOH	240s(3.89),280s(3.74), 327(4.20)	94-3483-84
C_8H_{14}			
Hexene, 3,4-bis(methylene)-	hexane	228(4.11)	44-2981-84
$C_8H_{14}Ge$			
Germacyclopentane, 1,1-dimethyl-3,4-bis(methylene)-	hexane	227(3.81)	44-2981-84
$C_8H_{14}N_2O$			
5-Isoxazolamine, 3-(1-ethylpropyl)-	EtOH	242(4.08)	39-1079-84C
$C_8H_{14}N_2O_3$			
Pyrazinol, 2,3-dihydro-3,3,5,6-tetramethyl-, 1,4-dioxide	EtOH	242(3.97)	103-0097-84
$C_8H_{14}N_2O_4$			
Carbamic acid, nitro-2-propenyl-, butyl ester	EtOH	234(3.68)	104-0449-84
$C_8H_{14}N_4$			
2,4-Pyrimidinediamine, 6-(1-methylpropyl)-	EtOH	233(3.72),282(3.64)	39-2677-84C
$C_8H_{14}N_4O$			
5H-Tetrazol-5-one, 1-(1,1-dimethylethyl)-1,4-dihydro-4-(1-propenyl)-E/Z 85/15)	hexane	204(3.881),249(3.901)	24-2761-84
$C_8H_{14}N_4OS$			
1,2,4-Triazin-5(4H)-one, 4-amino-6-(1,1-dimethylethyl)-3-(methylthio)-	MeOH	227(3.9),293(3.9)	98-0749-84
	MeCN	227.5(3.90),292.5(3.95)	9-0556-84
in 10% MeOH	pH 0.3	260(3.97)	9-0556-84
in 10% MeOH	pH 3.5	210(4.12),232(4.08), 292(4.04)	9-0556-84
in 10% MeOH	pH 11.5	210s(--),231(4.03), 293(3.96)	9-0556-84
$C_8H_{14}OS$			
2-Heptenal, 2-(methylthio)-	EtOH	221(4.03)	78-2035-84
$C_8H_{14}S$			
Cyclobutanethione, 2,2,4,4-tetramethyl-	hexane	218(3.72),233(3.87), 500(1.08)	24-0277-84

Compound	Solvent	$\lambda_{max}(\log \epsilon)$	Ref.
$C_8H_{15}NOS$ Heptanenitrile, 7-(methylsulfinyl)- (diptocarpilidine)	EtOH	206(3.15)	105-0079-84
$C_8H_{15}NO_2$ 2-Propenoic acid, 3-(diethylamino)-, methyl ester	MeOH	280(4.48)	69-5833-84
$C_8H_{15}NO_6$ Pentanoic acid, 5,5-dimethoxy-2-nitro-, methyl ester	EtOH	203(3.76),306(2.72)	78-4513-84
$C_8H_{15}NS$ Thiazole, 4,5-dihydro-2-(1-methyl- butyl)-	EtOH	202(3.60),232(3.34)	78-2141-84
Thiazole, 2-(1-ethylpropyl)-4,5-di- hydro-	EtOH	202(3.61),232(3.38)	78-2141-84
$C_8H_{15}N_2O_4P$ Phosphonic acid, (4,5-dihydro-1-methyl- 5-oxo-1H-imidazol-4-yl)-, diethyl ester	MeCN	223(3.73),278(3.06)	65-2414-84
$C_8H_{15}N_2O_5P$ Acetic acid, diazo(diethoxyphosphin- yl)-, ethyl ester	EtOH	211(3.32),243(3.53)	65-2495-84
$C_8H_{15}N_3O$ 2H-Pyrrol-2-one, 4,5-bis(dimethyl- amino)-3,4-dihydro-	MeCN	242(4.30)	142-1179-84
$C_8H_{16}N_2O_3Si$ Butanenitrile, 2-methyl-3-nitro- 2-[(trimethylsilyl)oxy]-	CH_2Cl_2	257(3.86)	23-2936-84
$C_8H_{16}N_2O_4$ Ethanol, 1-(butylnitroamino)-, acetate	MeCN	238(3.77)	5-1494-84
$C_8H_{16}OS_3$ Butane, 1-[[2-(methylsulfinyl)-1-(meth- ylthio)ethenyl]thio]-	EtOH	225s(3.59),269(4.04)	39-0085-84C
1-Propene, 1,1-bis(ethylthio)-2-(meth- ylsulfinyl)-	EtOH	275(3.91)	39-0085-84C
$C_8H_{16}O_3S_3$ 2-Propanol, 1-[[2,2-bis(methylthio)- ethenyl]sulfonyl]-2-methyl-	EtOH	278(3.69)	39-0085-84C
$C_8H_{16}S_4$ Disulfide, 2,2-bis(ethylthio)ethenyl ethyl	EtOH	276(--)	39-0085-84C
$C_8H_{17}N$ 1-Buten-1-amine, N,N-diethyl-	gas	<u>167(4.0),180s(3.9), 213(3.7)</u>	149-0391-84A
$C_8H_{18}Te_2$ Ditelluride, dibutyl	benzene	395(2.8)	48-0467-84

Compound	Solvent	$\lambda_{max}(\log \epsilon)$	Ref.
$C_8H_{19}N$			
1-Butanamine, N-butyl- (phosphate)	H_2O	224(4.15)	94-2430-84
	MeOH	225(4.11)	94-2430-84
	25% MeOH	222(4.16)	94-2430-84
	MeCN	225(4.17)	94-2430-84
	25% MeCN	228(4.16)	94-2430-84
$C_8H_{19}N_3O$			
Guanidine, N-(6-hydroxyhexyl)-N-methyl-, monohydrochloride	MeOH	208(3.21)	158-1170-84
C_8O_8			
Benzo[1,2-d:4,5-d']bis[1,3]dioxole-2,4,6,8-tetrone	THF	223s(3.82),286(3.89)	78-4897-84

Compound	Solvent	$\lambda_{max}(\log \epsilon)$	Ref.
$C_9H_4Cl_4O_2$			
Tricyclo[3.2.2.0²,⁴]non-8-ene-6,7-di-one, 1,5,8,9-tetrachloro-	C_6H_{12}	431(2.09)	88-4697-84
Tricyclo[5.2.0.0²,⁴]non-5-ene-8,9-di-one, 1,5,6,7-tetrachloro-	C_6H_{12}	332(3.18),517(2.48)	88-4697-84
$C_9H_4F_5NO_2$			
Benzenamine, N-1,3-dioxolan-2-ylidene-2,3,4,5,6-pentafluoro-	MeOH	220(4.204)	104-1088-84
2-Oxazolidinone, 3-(pentafluorophenyl)-	MeOH	220(3.874),263s(3.061)	104-1088-84
$C_9H_5BrO_2$			
3-Benzofurancarboxaldehyde, 2-bromo-	EtOH	208(4.33),233(4.28), 250(3.87),285(3.93)	103-0957-84
$C_9H_5Br_2NO_3$			
2-Propenoic acid, 2-cyano-3-(4,5-di-bromo-2-furanyl)-, methyl ester	MeOH	207(3.01),248(2.58), 357(3.44)	73-0984-84
$C_9H_5ClN_2$			
Propanedinitrile, (4-cyanophenyl)-, potassium salt	MeCN	308(4.41)	35-6726-84
$C_9H_5ClN_4S$			
Pyrido[3',2':4,5]thieno[3,2-d]-1,2,3-triazine, 4-chloro-7-methyl-	EtOH	231(4.33),249(4.16), 279(4.41),288(4.06), 322(3.63)	87-1639-84
$C_9H_5N_3O$			
3-Cinnolinecarbonitrile, 4-hydroxy-	DMF	298(3.60),351(4.21)	5-1390-84
$C_9H_5N_3O_2$			
Propanedinitrile, (4-nitrophenyl)-, potassium salt	MeCN	475(4.48)	35-6726-84
tetrabutylammonium salt	MeCN	475(4.49)	35-6726-84
$C_9H_5N_5O$			
Furo[2',3':4,5]pyrrolo[1,2-d]-1,2,4-triazolo[3,4-f]-1,2,4-triazine	dioxan	250(3.46),303(3.13)	73-0065-84
C_9H_6BrNS			
Thiazole, 5-bromo-2-phenyl-	EtOH	213(4.00),298(4.21)	142-1077-84
$C_9H_6BrN_3O$			
Methanone, (5-bromo-1H-1,2,3-triazol-4-yl)phenyl-	EtOH	257(3.95)	103-1038-84
C_9H_6ClNOS			
2(1H)-Quinolinethione, 8-chloro-3-hy-droxy-	MeOH-KOH	263(4.51),376(4.05)	1-0109-84
$C_9H_6ClNO_2$			
2(1H)-Quinolinone, 8-chloro-3-hydroxy-	MeOH-KOH	260(4.16),329(3.84), 341(3.88)	1-0109-84
$C_9H_6ClN_3O$			
Methanone, (5-chloro-1H-1,2,3-triazol-4-yl)phenyl-	EtOH	203(3.97),258(4.07)	103-1038-84
$C_9H_6CrFNO_3$			
Chromium, tricarbonyl(η^6-2-fluoro-	EtOH	216(4.54)	112-0765-84

Compound	Solvent	$\lambda_{max}(\log \epsilon)$	Ref.
benzenamine)- (cont.)	CHCl$_3$	320(3.92)	112-0765-84
Chromium, tricarbonyl(η^6-3-fluoro-	EtOH	206(4.59)	112-0765-84
benzenamine)-	CHCl$_3$	313(3.96)	112-0765-84
Chromium, tricarbonyl(η^6-4-fluoro-	EtOH	216(4.41)	112-0765-84
benzenamine)-	CHCl$_3$	318(3.92)	112-0765-84
$C_9H_6CrO_3$			
Chromium, (η^6-benzene)tricarbonyl-	EtOH	217(4.34),255(3.71),	112-0765-84
	CHCl$_3$	315(4.02)	112-0765-84
$C_9H_6F_4O_3$			
2,4,6-Cycloheptatrien-1-one, 2,4,5,7-	C$_6$H$_{12}$	248(4.56),335(3.90)	155-0041-84A
tetrafluoro-3,6-dimethoxy-			
$C_9H_6IN_3O$			
Methanone, (5-iodo-1H-1,2,3-triazol-	H$_2$O	256(4.08)	103-1038-84
4-yl)phenyl-			
$C_9H_6N_2$			
Propanedinitrile, phenyl-, potassium	MeCN	298(4.34)	35-6726-84
salt			
$C_9H_6N_2OS$			
4H-[1]Benzopyrano[4,3-d][1,2,3]thiadia-	EtOH	255(4.13)	2-0203-84
zole			
$C_9H_6N_2O_3$			
5,8-Quinolinedione, 6-(hydroxyamino)-	EtOH	232(4.29),271(4.06),	95-0191-84
		328(3.71),519(3.32)	
5,8-Quinolinedione, 7-(hydroxyamino)-	EtOH	235(4.26),277(4.08),	95-0191-84
		334(3.80),525(3.36)	
$C_9H_6N_2S_2$			
4H-[1]Benzothiopyrano[4,3-d][1,2,3]-	EtOH	233(4.34),264(4.07)	2-0203-84
thiadiazole			
$C_9H_6N_4$			
3-Cinnolinecarbonitrile, 4-amino-	DMF	354(4.03)	5-1390-84
5H-Cyclopentapyrazine-2,3-dicarboni-	EtOH	248(3.90),291(3.90),	44-0813-84
trile, 6,7-dihydro-		310s(3.49)	
$C_9H_6O_3S_2$			
Cyclohept[c][1,2]oxathiole-3,4-dione,	MeCN	218(4.39),279s(4.18),	88-0419-84
8-(methylthio)-		291(4.24),364(3.64)	
$C_9H_6O_4$			
Esculetin	EtOH	230.0(4.13),258.0(3.72),	102-0699-84
		262s(3.71),300.7(3.76),	
		351.6(4.06)	
$C_9H_6O_4S_4$			
1,3-Dithiole-4-carboxylic acid, 2-(4-	MeCN	288(4.05),302(4.10),	103-1342-84
carboxy-1,3-dithiol-2-ylidene)-,		313(4.14),435(3.41)	
4-methyl ester			
$C_9H_6O_5$			
4H-Furo[3,2-c]pyran-2-carboxaldehyde,	MeOH	305(4.3)	83-0685-84
3-hydroxy-6-methyl-4-oxo-			
$C_9H_7BrN_2O_3$			
2-Propenoic acid, 3-(5-amino-4-bromo-	MeOH	215(3.03),343(2.59),	73-0984-84

Compound	Solvent	$\lambda_{max}(\log \epsilon)$	Ref.
2-furanyl)-2-cyano-, methyl ester		458(3.56)	73-0984-84
$C_9H_7BrO_2$			
3-Benzofuranmethanol, 2-bromo-	EtOH	214(4.17),251(4.04), 278(3.61),285(3.57)	103-0957-84
C_9H_7ClO			
2-Propyn-1-ol, 3-(4-chlorophenyl)-	MeOH	247(4.43),257(4.37)	118-0728-84
$C_9H_7Cl_2NO_2$			
1H-[1,3]Oxazino[3,4-a]azepin-1-one, 5,9-dichloro-3,4-dihydro-	MeOH	316(3.12)	119-0171-84
C_9H_7N			
2-Propenenitrile, 3-phenyl-, (E)-	CH_2Cl_2	275(4.41)	44-4352-84
with $AlEtCl_2$	CH_2Cl_2	315(4.47)	44-4352-84
	FSO_3H	317(4.48)	44-4352-84
2-Propenenitrile, 3-phenyl-, (Z)-	CH_2Cl_2	275(4.25)	44-4352-84
with $AlEtCl_2$	CH_2Cl_2	322(4.08)	44-4352-84
C_9H_7NO			
8-Quinolinol, EtHg$^+$ complex I	40% EtOH	340(3.43)	18-2266-84
complex II	40% EtOH	395(3.38)	18-2266-84
C_9H_7NOS			
2(3H)-Oxazolethione, 5-phenyl-	n.s.g.	222(4.09),306(4.39)	47-0739-84
$C_9H_7NO_2$			
8H-Cyclohept[d]oxazol-8-one, 2-methyl-	MeOH	243(4.44),312(3.68)	18-0609-84
1H-Isoindole-1,3(2H)-dione, 2-methyl-	n.s.g.	225(4.00),262(4.00), 366(3.30)	88-4917-84
$C_9H_7NO_3$			
1H-Indole-3-carboxylic acid, 1-hydroxy-	MeOH-acid	215(4.24),292(3.78)	150-1301-84M
1(2H)-Isoquinolinone, 6,8-dihydroxy-, hydrochloride (siaminine B)	MeOH	230(3.71),245(3.67), 252(3.74),297(3.12), 337(3.16),355(3.14)	100-0708-84
2-Propyn-1-ol, 3-(4-nitrophenyl)-	MeOH	294(4.20)	118-0728-84
$C_9H_7NO_4$			
2-Propenoic acid, 3-(4-nitrophenyl)-, trans	1.2M KOH	233(3.78),312(4.15)	65-1989-84
4(1H)-Quinolinone, 3,5,8-trihydroxy-	pH 1.0	240(4.18),321(3.42), 386(3.46)	88-2925-84
	pH 7.0	239(4.18),319(3.42), 365(3.46)	88-2925-84
	pH 8.0	239(4.12),304(3.53), 364(3.43),553(3.04)	88-2925-84
	pH 13	236(4.17),315(3.46), 604(3.48),641(3.49) (changing)	88-2925-84
$C_9H_7NO_4S$			
3(2H)-Isothiazolone, 4-hydroxy-5-phenyl-, S,S-dioxide	EtOH	224.5(4.06),306(4.01)	44-2212-84
$C_9H_7NO_5$			
4H-Furo[3,2-c]pyran-2-carboxaldehyde, 3-hydroxy-6-methyl-4-oxo-, 2-oxime	MeOH	305(4.23)	83-0685-84

Compound	Solvent	λ_{max}(log ϵ)	Ref.
C$_9$H$_7$NSe			
Isoselenazole, 5-phenyl-	isooctane	226s(3.65),278(4.17), 302s(3.76)	78-0931-84
C$_9$H$_7$N$_3$O			
6H-Imidazo[1,2-d]pyrido[3,2-b][1,4]- oxazine, monohydrochloride	MeOH	293(3.98)	4-1081-84
4H-[1,2,4]Triazolo[3,4-c][1,4]- benzoxazine	EtOH	242(2.83),250(2.68), 285(3.51)	2-0445-84
C$_9$H$_7$N$_3$OS			
Furo[2',3':4,5]pyrrolo[1,2-d][1,2,4]- triazine-8(7H)-thione, 5-methyl-	MeOH	280(3.10),362(3.17)	73-0065-84
1H-Pyrazolo[1,2-a]pyrazol-4-ium, 2- cyano-3-hydroxy-5,7-dimethyl-1- thioxo-, hydroxide, inner salt	MeCN	219(4.06),225(4.26), 306(4.12),428(3.71)	44-3672-84
C$_9$H$_7$N$_3$O$_2$			
Furo[2',3':4,5]pyrrolo[1,2-d][1,2,4]- triazin-8(7H)-one, 5-methyl-	n.s.g.	240(3.32),290(3.23)	73-0065-84
1H-Pyrazolo[1,2-a]pyrazol-4-ium, 2- cyano-3-hydroxy-5,7-dimethyl-1- oxo-, hydroxide, inner salt	MeCN	224(4.28),248s(4.25), 360(2.94)	44-3672-84
C$_9$H$_7$N$_3$O$_2$S			
1,3-Benzenediol, 4-(2-thiazolylazo)-, ferrous complex	pH 9	730(4.46)	74-0103-84C
C$_9$H$_7$N$_3$O$_9$			
Butanedioic acid, (3,5-dinitro-4-oxo- 1(4H)-pyridinyl)-, (S)-	EtOH	340(3.57)	18-1961-84
C$_9$H$_7$N$_3$S$_2$			
4H-1,3,5-Thiadiazine-4-thione, 2-amino- 6-phenyl-	EtOH	279(4.62),369(3.90)	150-0266-84S
C$_9$H$_8$Br$_2$N$_2$			
2H-Cyclopenta[d]pyridazine, 5,7-di- bromo-1,2-dimethyl-	ether	208(4.00),257(4.13), 272(4.11),325(3.54), 336(3.51),400(2.88)	44-4769-84
C$_9$H$_8$ClNO			
2,4-Pentadienal, 5-(3-chloro-1H-pyrrol- 2-yl)-, (E,E)-	EtOH	383(4.47)	39-1577-84C
2-Propenamide, 3-chloro-N-phenyl-	EtOH	215(4.01),275(3.85)	142-2369-84
C$_9$H$_8$ClNS			
Thiopyrano[2,3-c]pyrrole, 4-chloro- 5,7-dimethyl-, perchlorate	acetone	365(3.15),590(3.23)	12-1473-84
C$_9$H$_8$ClN$_5$O			
3,5-Isoxazolediamine, 4-[(4-chloro- phenyl)azo]-	isoPrOH	203s(4.23),236(4.15), 248s(4.05),370(4.29), 406s(4.12)	97-0256-84
3H-Pyrrol-3-one, 5-amino-4-[(4-chloro- phenyl)azo]-2,4-dihydro-	C$_6$H$_{12}$	230(3.81),245(3.76), 382(4.25)	22-0164-84
	EtOH	231(3.89),380(4.26)	22-0164-84
	ether	233(3.86),376(4.24)	22-0164-84
	CHCl$_3$	386(4.26)	22-0164-84
	DMF	395(4.23)	22-0164-84

Compound	Solvent	$\lambda_{max}(\log \epsilon)$	Ref.
$C_9H_8Cl_2O_2$			
2H-Oxocin-2-one, 6,7-dichloro-4,8-di-methyl-	MeOH	296(3.77)	138-1503-84
$C_9H_8N_2O$			
3-Isoxazolamine, 5-phenyl-	EtOH	206(4.20),260(4.16)	39-1079-84C
5-Isoxazolamine, 3-phenyl-	EtOH	206(4.23),233(4.08)	39-1079-84C
1,2,4-Oxadiazole, 5-methyl-3-phenyl-	EtOH	238(4.01),276(2.87), 282(2.60)	34-0231-84
2(1H)-Quinoxalinone, 1-methyl-	EtOH	206(4.21),230(4.37), 280(3.70),343.5(3.71)	44-0827-84
2(1H)-Quinoxalinone, 3-methyl-	EtOH	278(3.30),328(3.36), 338(3.24)	18-1653-84
$C_9H_8N_2OS$			
1,2,4-Thiadiazole, 3-methoxy-5-phenyl-	EtOH	248(4.08),293(3.98)	39-0075-84C
1,2,4-Thiadiazolium, 2,3-dihydro-4-methyl-3-oxo-5-phenyl-, hydroxide, inner salt	EtOH	225(3.91),238(3.88), 254(3.88),335(3.79)	39-0075-84C
1,2,4-Thiadiazol-3(2H)-one, 2-methyl-5-phenyl-	EtOH	225s(4.23),262(4.34), 300s(3.71),325(3.75)	39-0075-84C
4(5H)-Thiazolone, 2-(phenylamino)-, cesium salt	EtOH	265(4.20)	103-0734-84
potassium salt	EtOH	265(4.22)	103-0734-84
sodium salt	EtOH	265(4.22)	103-0734-84
$C_9H_8N_2O_2$			
2H-1,4-Benzoxazin-2-one, 7-amino-3-methyl-	EtOH	385(4.12)	4-0551-84
$C_9H_8N_2O_2S$			
1,2,3-Benzothiadiazole-6-carboxylic acid, ethyl ester	n.s.g.	230s(4.0),263(4.1), 330(3.6)	24-0107-84
$C_9H_8N_2O_2S_2$			
Carbonimidodithioic acid, cyano-, (5-formyl-2-furanyl)methyl methyl ester	dioxan	282(3.50)	73-2285-84
2-Furancarboxaldehyde, 5-[4-amino-2-(methylthio)-5-thiazolyl]-	dioxan	223(3.07),252(3.00), 277(2.78),410(3.38)	73-2285-84
$C_9H_8N_2O_3S$			
7H-Thiazolo[3,2-a]pyrimidin-7-one, 2-(acetoxymethyl)-	EtOH	218(4.27),230s(4.14), 273(3.61)	39-2221-84C
$C_9H_8N_2S$			
2(1H)-Quinoxalinethione, 3-methyl-	EtOH	216(4.23),278(3.97), 396(3.73)	18-1653-84
$C_9H_8N_4$			
1H-Cyclopentapyrazine-2,3-dicarbo-nitrile, 4,4a,5,7a-tetrahydro-	EtOH	325(4.13)	44-0813-84
1H-Tetrazole, 1-ethynyl-5-phenyl-	EtOH	239(4.07)	39-1933-84C
$C_9H_8N_4O_2S$			
1,3-Benzenediol, 4-[(5-methyl-1,3,4-thiadiazol-2-yl)azo]-	pH 1-3.7	399(4.32)	73-1360-84
	pH 6.70	480(4.49)	73-1360-84
	pH 9.25	490(4.57)	73-1360-84
$C_9H_8N_4O_3$			
2,4(1H,3H)-Pteridinedione, 7-(1-oxo-propyl)-	pH 5.0	240s(4.00),340(3.97)	5-1798-84
	pH 10.0	240(4.17),275(3.97),	5-1798-84

Compound	Solvent	$\lambda_{max}(\log \epsilon)$	Ref.
(cont.)	pH 14.0	372(3.72) 245(4.17),265s(4.11), 394(3.80)	5-1798-84 5-1798-84
C$_9$H$_8$N$_4$O$_3$S Pyrazolo[1,5-a]pyridine-6-carboxamide, 2-(methylthio)-3-nitro-	EtOH	260(4.47),290(4.25), 366(4.23)	95-0440-84
C$_9$H$_8$N$_4$O$_8$ 1(4H)-Pyridineacetic acid, α-(2-amino-2-oxoethyl)-3,5-dinitro-4-oxo-, (S)-	n.s.g.	340(3.65)	18-1961-84
C$_9$H$_8$N$_4$S 3-Cinnolinecarbothioamide, 4-amino-	EtOH	241(4.38),278(4.06), 307(4.08),359(4.11), 368s(4.09)	5-1390-84
C$_9$H$_8$N$_6$O$_2$S 2-Propenoic acid, 2-azido-3-(3-azido-2-thienyl)-, ethyl ester	EtOH	205(4.00),236(3.97), 250s(3.89),349(4.31)	39-0915-84C
C$_9$H$_8$N$_6$O$_3$ 3H-Pyrazol-3-one, 5-amino-4-[(4-nitro-phenyl)azo]-2,4-dihydro-	C$_6$H$_{12}$	228s(3.63),385(4.19), 460s(3.36)	22-0164-84
	EtOH	230s(3.81),388(4.36), 468(3.70)	22-0164-84
	ether	229(3.66),385(4.31), 476s(3.23)	22-0164-84
	DMF	460(4.21),520(4.19)	22-0164-84
	CHCl$_3$	390(4.41),453(3.33)	22-0164-84
C$_9$H$_8$O$_2$ Cinnamic acid, trans	MeOH	216(4.12),220(4.03), 272(4.15)	94-4620-84
C$_9$H$_8$O$_2$S$_4$ 1,3-Dithiole-4-carboxylic acid, 2-(1,3-dithiol-2-ylidene)-, ethyl ester	MeCN	292(4.07),303(4.11), 314(4.13),424(3.28)	103-1342-84
C$_9$H$_8$O$_3$ 2-Propenoic acid, 3-(2-hydroxyphenyl)-, cis	DMSO	268s(3.87),313s(3.78)	41-0243-84
trans	DMSO	276(4.23),328(4.02)	41-0243-84
	anion	260(4.13),313(3.90)	41-0243-84
	dianion	260(--),390(3.98)	41-0243-84
C$_9$H$_8$O$_4$ 1,4-Dioxaspiro[4.5]deca-6,7-diene-2,8-dione, 3-methyl-	MeOH	222(3.57),281(3.71), 366(1.93)	88-4471-84
C$_9$H$_8$O$_5$ Benzeneacetic acid, 2-hydroxy-5-meth-oxy-α-oxo-	EtOH	224(4.03),261(3.64), 289(3.15),360(3.36)	39-2655-84C
C$_9$H$_8$S 2,7-Methanothia[9]annulene	C$_6$H$_{12}$	236s(3.84),260s(3.57), 280s(3.34),350s(2.34)	35-5271-84
C$_9$H$_8$S$_3$ 1,3-Dithiolane-2-thione, 4-phenyl-	benzene	456(1.92)	18-1147-84
	C$_6$H$_{12}$	461(1.78)	18-1147-84
	1-C$_{10}$H$_7$Me	455(2.01)	18-1147-84

Compound	Solvent	$\lambda_{max}(\log \epsilon)$	Ref.
1,3-Dithiolane-2-thione, 4-phenyl- (cont.)	CCl_4	459(1.90)	18-1147-84
	CS_2	463(4.05)	18-1147-84
$C_9H_9BNO_5$			
Borate(1-), dihydroxy[N-[(2-hydroxy-phenyl)methylene]glycinato(2-)-N,O^N]-	pH 8.9	273(4.20),345(3.66)	18-2508-84
$C_9H_9BrN_2$			
2H-Cyclopenta[d]pyridazine, 5-bromo-1,2-dimethyl-	ether	208(4.27),249(4.51), 272(4.35),320(3.86), 330s(3.83),390(3.13)	44-4769-84
$C_9H_9BrN_2O_5$			
1H-Pyrazole-3,4-dicarboxylic acid, 5-(bromoacetyl)-, dimethyl ester	MeOH	232(4.44),285(4.24)	103-0416-84
$C_9H_9BrO_2S$			
2-Propenoic acid, 2-(bromomethyl)-3-(2-thienyl)-, methyl ester	EtOH	303(3.30)	73-1764-84
$C_9H_9ClN_2O_2$			
3H-Diazirine, 3-[2-chloro-3-(methoxy-methoxy)phenyl]-	C_6H_{12}	358(2.54)	23-1767-84
$C_9H_9ClN_6$			
1H-Pyrazole-3,5-diamine, 4-[(4-chloro-phenyl)azo]-	DMF	374(4.33),394s(4.28)	97-0256-84
C_9H_9ClO			
Benzeneethanol, 4-chloro-α-methylene-	EtOH	247(4.14)	44-2258-84
Ethanone, 2-chloro-1-(4-methylphenyl)-	20% H_2SO_4	263(4.16)	48-0353-84
	96% H_2SO_4	331(4.33)	48-0353-84
$C_9H_9ClO_2$			
Ethanone, 2-chloro-1-(4-methoxyphenyl)-	20% H_2SO_4	285(4.19)	48-0353-84
	96% H_2SO_4	357(4.46)	48-0353-84
$C_9H_9Cl_2N_3$			
6H-Pyrrolo[3,4-d]pyridazine, 1,4-di-chloro-5,6,7-trimethyl-	pH 7.0	206(4.30),235(4.70), 288(3.77),365(3.50)	78-3979-84
	H_0 -1.50	212(4.66),247.5(4.45), 299(3.87),405(3.42)	78-3979-84
$C_9H_9Cl_3N_2O$			
Benzoic acid, 4-(trichloromethyl)-, 1-methylhydrazide	EtOH	235(3.9)	70-2056-84
$C_9H_9FO_4$			
1,3-Benzodioxol-3a(7aH)-ol, 7a-fluoro-, acetate, cis	MeCN	249(3.57)	44-0806-84
Benzoic acid, 5-fluoro-2,3-dimethoxy-	EtOH	293(3.28)	104-0465-84
C_9H_9K			
Potassium, (1-phenyl-2-propenyl)-	THF	411(--)	101-0001-84C
C_9H_9N			
Isoquinoline, 3,4-dihydro-	hexane	208(4.49),212s(4.37), 246(3.91),250s(3.89), 256s(3.78),282(2.94), 289s(2.89)	32-0041-84

Compound	Solvent	$\lambda_{max}(\log \epsilon)$	Ref.
Isoquinoline, 3,4-dihydro- (cont.)	EtOH	211(4.44),215s(4.35), 255(3.99),260s(3.89), 293s(2.93)	32-0041-84
C$_9$H$_9$NOS			
Benzothiazole, 4-methoxy-7-methyl-	EtOH	257(3.61),266(3.51), 306(3.50)	44-0997-84
C$_9$H$_9$NO$_2$			
Acetamide, N-(2-oxobicyclo[3.2.0]hepta-3,6-dien-1-yl)-, (±)-	MeOH	345(2.07)	39-0891-84B
Acetamide, N-(2-oxobicyclo[3.2.0]hepta-3,6-dien-7-yl)-	MeOH	345(1.97)	39-0891-84B
2H-Cyclopenta[b]furan-4-carbonitrile, 3,3a,6,6a-tetrahydro-3a-methyl-2-oxo-, (3aR-cis)-	MeOH	212(3.91)	48-1016-84
2H-Indol-2-one, 1,3-dihydro-3-(hydroxymethyl)-	EtOH	250(3.94),280s(3.03)	69-4589-84
C$_9$H$_9$NO$_3$			
Benzofuran, 2,3-dihydro-2-methyl-5-nitro-	n.s.g.	209(4.00),237(3.82), 328(3.98)	103-0838-84
Glycine, N-[(2-hydroxyphenyl)methylene]-	pH 8.9	277(4.15),395(3.82)	18-2508-84
C$_9$H$_9$NO$_4$			
1H-Indole-2-carboxylic acid, 2,3-dihydro-5,6-dihydroxy-, (S)- (as glucoside)	acid	283(3.53)	33-1348-84
	pH 7.4	235s(3.71),302(3.60)	33-1348-84
	base	310(3.70)	33-1348-84
C$_9$H$_9$NS			
1H-Isoindole-1-thione, 2,3-dihydro-2-methyl-	EtOH	204(4.39),256(4.00), 306(4.00)	103-0978-84
Thiopyrano[2,3-c]pyrrole, 5,7-dimethyl-, perchlorate	EtOH	240(4.13),276(3.81), 310(3.56),596(3.11)	12-1473-84
C$_9$H$_9$N$_3$O			
1H-1,2,4-Triazole, 3-(4-methoxyphenyl)-	MeCN	257(4.28),293s(3.08)	48-0311-84
C$_9$H$_9$N$_3$O$_2$			
Acetic acid, cyano-2(1H)-pyrazinylidene)-, ethyl ester	pH 1,2,4	225(4.14),293(4.37), 403(4.00)	49-0179-84
	H$_2$O	227(4.09),294(4.35), 400(3.93)	49-0179-84
	pH 8	227(4.13),294(4.37), 400(4.00)	49-0179-84
	pH 10	240(4.31),296(4.36), 370(3.93)	49-0179-84
	pH 14	240(4.31),296(4.35), 370(3.92)	49-0179-84
Pyrido[2,3-d]pyrimidine-2,4(1H,3H)-dione, 1,3-dimethyl-	H$_2$O	245s(3.84),309(3.81)	4-1543-84
C$_9$H$_9$N$_3$O$_2$S			
Pyrazolo[1,5-a]pyridine, 4-methyl-2-(methylthio)-3-nitro-	EtOH	246(4.52),274(3.80), 374(4.08)	95-0440-84
Pyrazolo[1,5-a]pyridine, 6-methyl-2-(methylthio)-3-nitro-	EtOH	244(4.83),370(4.37)	95-0440-84
C$_9$H$_9$N$_3$O$_3$			
2-Propenoic acid, 2-(azidomethyl)-	EtOH	312(3.37)	73-1764-84

Compound	Solvent	$\lambda_{max}(\log \epsilon)$	Ref.
3-(2-furanyl)-, methyl ester (cont.)			73-1764-84
$C_9H_9N_3O_4$			
1H-1,2,3-Triazole-4,5-dicarboxylic acid, 1-(2-propynyl)-, dimethyl ester	MeCN	219(3.83)	1-0623-84
$C_9H_9N_3O_6S_2$			
L-Cysteine, S-(7-sulfo-4-benzofurazan-yl)-, disodium salt	H_2O	370(3.58)	96-1003-84
$C_9H_9N_3O_8$			
1(4H)-Pyridineacetic acid, α-(1-hydr-oxyethyl)-3,5-dinitro-4-oxo-, [R-(R* S*)]-	n.s.g.	340(3.67)	18-1961-84
$C_9H_9N_3S$			
1,2,3-Thiadiazolium, 3-methyl-5-(phen-ylamino)-, hydroxide, inner salt	isoPrOH	280(4.03),410(3.89)	4-1889-84
$C_9H_9N_5O$			
Furo[2',3':4,5]pyrrolo[1,2-d][1,2,4]-triazin-8(7H)-one, 5-methyl-, hydra-zone	MeOH	302(3.30)	73-0065-84
3,5-Isoxazolediamine, 4-(phenylazo)-	EtOH	238(4.18),363(4.28)	97-0256-84
3H-Pyrazol-3-one, 5-amino-2,4-dihydro-4-(phenylazo)-	C_6H_{12}	245(3.87),378(4.33)	22-0164-84
	EtOH	240s(3.90),379(4.34)	22-0164-84
	ether	240(3.90),373(4.30)	22-0164-84
	DMF	385(4.27)	22-0164-84
	$CHCl_3$	381(4.30)	22-0164-84
$C_9H_9N_5O_2$			
4(1H)-Pteridinone, 2-amino-6-(1-oxo-propyl)-	pH -1.0	230s(3.96),268(4.06), 320(4.04)	138-1025-84
	pH 4.0	239(3.96),303(4.17), 347(3.96)	138-1025-84
	pH 10.0	275(4.22),307s(3.87), 369(4.06)	138-1025-84
$C_9H_9N_5O_5$			
1H-Pyrazole-3,4-dicarboxylic acid, 5-(azidoacetyl)-, dimethyl ester	MeOH	235(4.22),282(4.11)	103-0416-84
1H-1,2,3-Triazole-4,5-dicarboxylic acid, 1-(3-diazo-2-oxopropyl)-, dimethyl ester	MeOH	250(4.38),265(4.36)	103-0416-84
C_9H_{10}			
Spiro[bicyclo[2.1.0]pentane-5,1'-[2,4]cyclopentadiene]	EtOH	201(3.70),229(3.82), 254(3.34)	35-3466-84
$C_9H_{10}Br_2N_2O_4$			
Uridine, 2',5'-dibromo-2',5'-dideoxy-	EtOH	258(3.98)	94-2544-84
$C_9H_{10}Br_2O_2$			
2,5-Cyclohexadien-1-one, 2,5-dibromo-4-hydroxy-2,4,6-trimethyl-	$CHCl_3$	254(4.26),291(3.66)	12-2027-84
2,5-Cyclohexadien-1-one, 3,5-dibromo-6-hydroxy-2,4,6-trimethyl-	$CHCl_3$	249(3.67),333(3.60)	12-2027-84
$C_9H_{10}ClFO_2$			
Benzene, 1-(chloromethyl)-2-fluoro-4,5-dimethoxy-	EtOH	209(4.23),237(3.89), 287(3.65)	104-0465-84

Compound	Solvent	λ_{max}(log ϵ)	Ref.
C$_9$H$_{10}$ClN$_5$O 7H,9H-[1,3,6]Oxadiazocino[3,4,5-gh]- purine, 2-chloro-10,11-dihydro-11- methyl-	MeOH	287(4.13)	4-0333-84
C$_9$H$_{10}$ClN$_5$O$_2$ 4(1H)-Pteridinone, 6-(1-chloro-2-hy- droxypropyl)-, [S-(R*,R*)]-	MeOH	235(4.02),280(4.25), 347(3.75)	5-1815-84
C$_9$H$_{10}$Cl$_2$N$_2$O$_4$ Uridine, 2',5'-dichloro-2',5'-dideoxy-	EtOH	258(3.99)	94-2591-84
C$_9$H$_{10}$Cl$_2$S Spiro[cyclopropane-1,7'-[4]thiatri- cyclo[3.2.1.0^{3,6}]octane], 2',8'- dichloro-	hexane	265(1.62)	104-0485-84
C$_9$H$_{10}$N$_2$ Benzeneacetonitrile, α-(methylamino)- 2H-Cyclopenta[d]pyridazine, 1,2-di- methyl-	EtOH hexane	245(4.28) 247(4.59),267s(4.35), 310(3.83),317(3.81), 323.5(3.77),388(3.24)	44-0282-84 44-4769-84
	MeOH	203(4.31),247(4.65), 267s(4.31),313(3.86), 378(3.17)	44-4769-84
5,8-Methanophthalazine, 5,6,7,8-tetra- hydro-	hexane	218(3.38),246(2.90), 332(2.67)	24-0534-84
C$_9$H$_{10}$N$_2$O 7H-Pyrrolo[3,4-b]pyridin-7-one, 5,6-di- hydro-2,3-dimethyl-	CH$_2$Cl$_2$	276.5(3.85)	83-0379-84
C$_9$H$_{10}$N$_2$OS 4(1H)-Quinazolinethione, 2,3-dihydro- 3-hydroxy-1-methyl- Urea, N-methyl-N'-thiobenzoyl-	n.s.g. EtOH	277(3.7),318(3.8), 375s(3.6) 260(4.12),412(2.48)	4-1615-84 39-0075-84C
C$_9$H$_{10}$N$_2$O$_2$ 3H-Diazirine, 3-[3-(methoxymethoxy)- phenyl]- 4H-Indazol-4-one, 1,5,6,7-tetrahydro- 5-(hydroxymethylene)-1-methyl-	C$_6$H$_{12}$ EtOH	358(2.56) 255(3.91),292(4.05)	23-1767-84 4-0361-84
C$_9$H$_{10}$N$_2$O$_3$ 2,4(1H,3H)-Pyrimidinedione, 5-(2-oxo- cyclopentyl)-	MeOH	264(3.89)	44-2738-84
C$_9$H$_{10}$N$_2$S 2H-Benzimidazole-2-thione, 1,3-dihydro- 1,3-dimethyl- compd. with iodine	CH$_2$Cl$_2$ CH$_2$Cl$_2$	247(4.29),304s(4.42), 313(4.55) 270(4.26),282s(4.21), 369(4.53)	39-1383-84B 39-1383-84B
C$_9$H$_{10}$N$_2$Se 2H-Benzimidazole-2-selone, 1,3-dihydro- 1,3-dimethyl- compd. with iodine	CH$_2$Cl$_2$ CH$_2$Cl$_2$	314s(4.40),325(4.51) 276(4.48),362(4.43)	39-1383-84B 39-1383-84B
C$_9$H$_{10}$N$_4$OS Guanidine, N-carbamoyl-N'-thiobenzoyl-	EtOH	219(4.20),281(4.22), 338(4.01),440(2.18)	39-0075-84C

Compound	Solvent	λ_{max}(log ϵ)	Ref.
C$_9$H$_{10}$N$_4$O$_2$			
Pyrido[2,3-d]pyrimidine-2,4(1H,3H)-dione, 5-amino-1,3-dimethyl-	EtOH	235(4.54),316(3.83)	94-0122-84
Pyrido[2,3-d]pyrimidine-2,4(1H,3H)-dione, 7-amino-1,3-dimethyl-	H$_2$O	220s(3.98),276(4.05), 317(4.34)	4-1543-84
	EtOH	223(4.43),277(3.93), 316(4.28),321(4.28), 328(4.24)	94-0122-84
Pyrido[3,4-d]-1,2,3-triazin-4(3H)-one, 3,6,8-trimethyl-, 7-oxide	EtOH	277(<u>4.5</u>),305s(<u>4.0</u>), 355(<u>3.2</u>)	95-1122-84
C$_9$H$_{10}$N$_4$O$_4$			
2,4(1H,3H)-Pteridinedione, 6-(1,2-dihydroxypropyl)-, [S-(R*,S*)]-	pH 5.0	231(4.11),243s(4.04), 329(3.91)	5-1815-84
	pH 10.0	221(4.09),239(4.13), 274(4.14),351(3.77), 370(3.72)	5-1815-84
	pH 14.0	255(4.39),276s(3.85), 373(3.87)	5-1815-84
C$_9$H$_{10}$N$_6$			
9H-Purine-9-propanenitrile, 1,6-dihydro-6-imino-1-methyl-, monohydriodide	EtOH	260(4.10)	47-3943-84
1H-Pyrazole-3,5-diamine, 4-(phenylazo)-	EtOH	226(4.17),361(4.13), 390s(4.13)	97-0256-84
C$_9$H$_{10}$OS			
Ethanone, 2-(methylthio)-1-phenyl-	EtOH	243(4.08),266(3.60)	78-2035-84
C$_9$H$_{10}$O$_2$			
Benzoic acid, ethyl ester	C$_6$H$_{12}$	228(4.14),273f(2.99)	65-0489-84
C$_9$H$_{10}$O$_2$S			
2-Propenoic acid, 2-methyl-3-(2-thienyl)-, methyl ester	EtOH	309(3.28)	73-1764-84
C$_9$H$_{10}$O$_3$			
2,4-Cyclohexadiene-1-carboxylic acid, 1-methyl-6-oxo-, methyl ester	MeOH	302(3.58)	44-4429-84
Ethanone, 1-(4-hydroxy-3-methoxyphenyl)-	EtOH	231(4.27),276(4.01), 305(3.93)	39-0825-84C
	EtOH-NaOH	250(--),347(--)	39-0825-84C
2,5-Furandione, 3-ethylidenedihydro-4-(1-methylethylidene)-, (E)-	C$_6$H$_{12}$	312(3.97)	48-0233-84
(Z)-	C$_6$H$_{12}$	315(4.03)	48-0233-84
C$_9$H$_{10}$O$_4$			
Benzenepropanoic acid, α,4-dihydroxy-	MeOH	225(3.55),278(3.03), 286s(2.94)	100-0179-84
Benzenepropanoic acid, 3,4-dihydroxy-	pH 1	279(3.41)	3-1935-84
	borate	286(3.63)	3-1935-84
Benzenepropanoic acid, 3,5-dihydroxy-	EtOH	202(4.63),274(3.34), 281(3.32)	100-0828-84
2-Propenoic acid, 3-(2-furanyl)-2-(methoxymethyl)-	EtOH	301(3.37)	73-1764-84
C$_9$H$_{11}$BrO$_2$			
Benzene, 1-(2-bromoethoxy)-2-methoxy-	MeOH	220.5(3.87),271.5(3.36)	36-1241-84
C$_9$H$_{11}$ClN$_2$O$_5$			
Pyridazinium, 4-chloro-2,3-dihydro-	MeOH	217(4.35),254(3.46),	87-1613-84

Compound	Solvent	λ_{max}(log ϵ)	Ref.
3-oxo-1-β-D-ribofuranosyl-, hydroxide, inner salt (cont.)		304(3.77)	87-1613-84
C$_9$H$_{11}$FN$_2$O$_3$ 2,4(1H,3H)-Pyrimidinedione, 5-fluoro- 1-[2-(2-propenyloxy)ethyl]-	pH 7 pH 12	274(4.00) 272(3.80)	161-0346-84 161-0346-84
C$_9$H$_{11}$FO$_2$ Benzene, 5-fluoro-1,2-dimethoxy- 3-methyl-	EtOH	272(3.35)	104-0465-84
C$_9$H$_{11}$NO 1-Cyclooctene-1-carbonitrile, 3-oxo-	EtOH	215(3.67),228(3.60), 290(2.26)	44-2925-84
1-Cyclooctene-1-carbonitrile, 7-oxo-	EtOH	211(4.13),285(1.89)	44-2925-84
1H-Inden-1-ol, 2-amino-2,3-dihydro-, (-)-	n.s.g.	210(3.96),215s(3.88), 252s(2.49),259(2.76), 265.5(2.94),272(2.98)	39-1655-84C
(S)-	n.s.g.	211s(3.96),214s(3.91), 259(3.74),265(2.92), 272(2.97)	39-1655-84C
2,4-Pentadien-1-ol, 5-(1H-pyrrol-2-yl)-, (E,E)-	EtOH	314(4.51)	39-1577-84C
4(1H)-Pyridinone, 1-(3-butenyl)-	EtOH	263(4.26)	118-0485-84
C$_9$H$_{11}$NO$_2$ Acetamide, N-(2-hydroxyphenyl)- N-methyl-	EtOH	277.5(3.48),282s(--)	94-2544-84
Acetamide, N-(4-hydroxyphenyl)- N-methyl-	EtOH	230(4.04),280(3.23), 285s(--)	94-2544-84
C$_9$H$_{11}$NO$_2$S 3-Pyridinecarboxylic acid, 1,2-dihydro- 1-(1-methylethyl)-2-thioxo-	EtOH	300(4.11),378(3.50)	4-0041-84
C$_9$H$_{11}$NO$_3$ 3H-Azepine-5-carboxylic acid, 2-meth- oxy-, methyl ester	EtOH	240s(3.65),275s(3.46), 279s(3.44)	39-0249-84C
3H-Azepine-6-carboxylic acid, 2-meth- oxy-, methyl ester	EtOH	279(3.94)	39-0249-84C
3H-Pyrano[4,3-c]isoxazol-3-one, 6,7-dihydro-4,6,6-trimethyl-	EtOH	295(3.98)	142-0365-84
3-Pyridinecarboxylic acid, 1,2-dihydro- 1-(1-methylethyl)-2-oxo-	EtOH	234(4.37),331(4.66)	4-0041-84
C$_9$H$_{11}$NO$_3$S$_2$ Thiophene, 2-methoxy-5-(methylthio)- 3-(2-nitro-1-propenyl)-	EtOH	215(4.13),282(4.09), 380(4.04)	103-0962-84
C$_9$H$_{11}$NO$_4$ L-Tyrosine, 3-hydroxy-	pH 1 borate	280(3.43) 287(3.66)	3-1935-84 3-1935-84
C$_9$H$_{11}$N$_2$O$_7$PS 4-Thiazolecarboxamide, 2-(3,5-O-phos- phinico-β-D-ribofuranosyl)-, monosodium salt	pH 1 and 11 pH 7	235(3.77) 235(3.79)	87-0266-84 87-0266-84
C$_9$H$_{11}$N$_3$O 1H-Pyrrolo[3,4-d]pyridazin-1-one, 2,6- dihydro-2,5,7-trimethyl-	pH 7	206.5(4.52),232(4.24), 279.5(4.01),319.5(3.72)	78-3979-84

Compound	Solvent	$\lambda_{max}(\log \epsilon)$	Ref.
(cont.)	cation	219(4.36),252(4.17), 285(3.83),355(3.85)	78-3979-84
	dication	205(4.58),226(4.35), 240.5(4.27),308(4.19), 380(3.38)	78-3979-84
1H-Pyrrolo[3,4-d]pyridazin-1-one, 2,6-dihydro-5,6,7-trimethyl-	pH 7	205.5(4.41),234(4.39), 282(3.84),315.5(3.62)	78-3979-84
	cation	218(4.24),235(4.27), 249(4.19),283(3.19), 348(3.74)	78-3979-84
	dication	232(3.43),244s(--), 306.5(4.05),374(3.26)	78-3979-84
$C_9H_{11}N_3OS$ 6H-Thiopyrano[3,4-d]pyrimidin-5(8H)-one, 2-(dimethylamino)-	EtOH	308.5(4.31)	4-1437-84
$C_9H_{11}N_3O_2$ 1H-Pyrrolo[3,4-d]pyridazine-1,4(6H)-dione, 2,3-dihydro-2,5,7-trimethyl-	pH 7	208(4.58),235(3.86)	78-3979-84
	cation	210(4.50),227(4.53), 285(4.02),310s(2.59)	78-3979-84
anion	pH 13	218(4.59),274(4.03), 302s(3.39)	78-3979-84
1H-Pyrrolo[3,4-d]pyridazine-1,4(6H)-dione, 2,3-dihydro-5,6,7-trimethyl-	pH 7.0	213(4.17),272(3.60)	78-3979-84
	cation	222(4.36),278(3.85), 290s(3.56)	78-3979-84
anion	pH 13	216.5(4.57),270(3.95), 307(3.82)	78-3979-84
$C_9H_{11}N_3O_2S$ 2-Thiophenecarbonitrile, 5-(butylamino)-4-nitro-	benzene	380(3.95)	39-0781-84B
$C_9H_{11}N_4S_2$ Thiazolium, 3-methyl-2-[[[(3-methyl-2(3H)-thiazolylidene)amino]methylene]amino]-, iodide	MeCN	390(4.59)	152-0475-84
perchlorate	MeCN	389(4.33)	152-0475-84
$C_9H_{11}N_5$ Cyanamide, (7,8-dihydro-2,3-dimethylimidazo[1,2-c]pyrimidin-5(6H)-ylidene)-	MeOH	247(4.18)	24-2597-84
Cyanamide, (7,8-dihydro-2,6-dimethylimidazo[1,2-c]pyrimidin-5(6H)-ylidene)-	MeOH	251(4.19)	24-2597-84
Cyanamide, (8,9-dihydro-2-methyl-7H-imidazo[1,2-c][1,3]diazepin-5-yl)-	MeOH	251(4.15)	4-0753-84
Cyanamide, (7,8,9,10-tetrahydroimidazo[1,2-c][1,3]diazocin-5-yl)-	MeOH	249(4.18)	4-0753-84
$C_9H_{11}N_5O$ 7H,9H-[1,3,6]Oxadiazocino[3,4,5-gh]purine, 10,11-dihydro-11-methyl-	MeOH	284.5(4.10)	4-0333-84
9H-Purin-6-amine, 9-[(2-ethenyloxy)ethyl]-	pH 1	258(--)	47-2841-84
	H_2O	260(4.18)	47-2841-84
	pH 13	260(--)	47-2841-84
$C_9H_{11}N_5O_2$ Carbamic acid, (9-ethyl-9H-purin-6-yl)-, methyl ester	MeOH	268(4.19)	150-3601-84M

Compound	Solvent	$\lambda_{max}(\log \epsilon)$	Ref.
4(1H)-Pteridinone, 2-amino-7,8-dihydro-6-(1-oxopropyl)-	pH -1.0	232(4.14),284(4.10), 330s(3.35),395(3.89)	138-1025-84
	pH 5.0	213(4.22),265(4.23), 286s(3.89),410(4.01)	138-1025-84
	pH 13.0	267(4.22),312(3.28), 430(4.10)	138-1025-84
4(1H)-Pteridinone, 2-amino-6-(1-hydroxypropyl)-	pH 0.0	247(4.04),322(3.92)	138-1025-84
	pH 5.0	236(4.03),273(4.15), 345(3.79)	138-1025-84
	pH 10.0	220s(3.92),254(4.35), 364(3.86)	138-1025-84
$C_9H_{11}N_5O_2S$ 4(1H)-Pteridinethione, 2-amino-6-(1,2-dihydroxypropyl)-, [S-(R*,S*)]-	pH -1.0	211(4.35),252(3.87), 276(3.82),375(3.96)	5-1815-84
	pH 4.0	230s(4.09),251s(3.90), 288(3.97),409(3.88)	5-1815-84
	pH 10.0	220(4.29),240s(4.10), 282(3.97),332(3.58), 406(3.82)	5-1815-84
Pyrimido[4,5-d]pyrimidine-2,4(1H,3H)-dione, 5-amino-1,3-dimethyl-7-(methylthio)-	EtOH	239(4.56),281(4.22), 292(4.21)	94-0122-84
$C_9H_{11}N_5O_3$ Biopterin	pH 0.0	228s(4.11),248(4.07), 320(3.92)	5-1815-84
	pH 5.0	218(4.06),235(4.08), 272(4.17),344s(3.79), 362s(3.70)	5-1815-84
	pH 10.0	254(4.37),277s(3.82), 363(3.86)	5-1815-84
Pterin, 6-(L-threo-1,2-dihydroxypropyl)-	pH 0.0	230s(4.11),248(4.08), 321(3.92)	5-1815-84
	pH 5.0	216(4.08),235(4.07), 274(4.17),345(3.78), 362s(3.69)	5-1815-84
	pH 10.0	254(4.36),277s(3.82), 363(3.85)	5-1815-84
9H-Purine-9-butanoic acid, 6-amino-β-hydroxy-, (S)-	pH 2	260(4.13)	73-2148-84
9H-Purine-9-propanoic acid, 6-amino-α-hydroxy-, methyl ester	MeOH	262(4.14)	73-2148-84
9H-Purine-9-propanoic acid, 6-amino-α-hydroxy-β-methyl-	pH 2	261(4.15)	73-2148-84
9H-Purine-9-propanoic acid, 6-amino-α-hydroxy-2-methyl-, (±)-	pH 2 and 7	263(4.10)	73-2148-84
	pH 12	264(4.12)	73-2148-84
9H-Purine-9-propanoic acid, 6-amino-α-methoxy-, (±)-	pH 7	261(4.15)	73-2148-84
$C_9H_{11}N_5O_4$ 3H-1,2,3-Triazolo[4,5-d]pyrimidine, 3-β-D-ribofuranosyl-	H_2O	262(3.77)	118-0401-84
C_9H_{12} Spiro[2.4]hepta-4,6-diene, 1,2-dimethyl-	EtOH	227(3.30),258(2.83)	70-1526-84
$C_9H_{12}BrN_3O$ 1-Triazene, 1-(4-bromophenyl)-3-(methoxymethyl)-3-methyl-	MeCN	219(4.05),281(4.34), 310s(4.18)	23-0741-84

Compound	Solvent	$\lambda_{max}(\log \epsilon)$	Ref.
$C_9H_{12}Br_2$			
Cyclopropane, [4-bromo-1-(1-bromoethenyl)-1-butenyl]-, (E)-	hexane	227(4.00)	44-0495-84
(Z)-	hexane	220(3.85)	44-0495-84
$C_9H_{12}ClN_3O_6$			
1H-1,2,3-Triazole-5-carboxylic acid, 4-chloro-1-ß-D-ribofuranosyl-, methyl ester	EtOH	240(3.91)	103-1287-84
$C_9H_{12}CuNOS$			
Copper, [4-amino-2-[(ethylthio)methyl]-phenolato-O,S]-	C_6H_{12}	255(4.52)	65-0142-84
$C_9H_{12}NO_5P$			
1-Aza-5-phosphabicyclo[4.2.0]oct-2-ene-2-carboxylic acid, 5-ethoxy-8-oxo-, 5-oxide, sodium salt	H_2O	250(3.89)	88-1733-84
$C_9H_{12}N_2$			
1,4-Ethano-1H-cyclopenta[d]pyridazine, 4,4a,5,6-tetrahydro-	EtOH	227(3.09),375(1.71)	35-3466-84
1,4-Ethano-1H-cyclopenta[d]pyridazine, 4,4a,5,7a-tetrahydro-	hexane	384(2.15)	24-0534-84
1H-Pyrrolo[2,3-b]pyridine, 2,3-dihydro-4,6-dimethyl-	heptane	308(3.87)	151-0343-84D
	PrOH	310(3.89)	151-0343-84D
$C_9H_{12}N_2O$			
Acetic acid, 1-methyl-2-phenylhydrazide	MeCN	234(3.88)	104-1609-84
Acetic acid, 2-methyl-2-phenylhydrazide	MeCN	242(3.80)	104-1609-84
Benzoic acid, 4-methyl-, 1-methyl-hydrazide	aq EtOH-HCl	230(4.0)	70-2056-84
$C_9H_{12}N_2OS$			
Thiourea, (2-methoxy-5-methylphenyl)-	EtOH	255(3.29),290(3.46)	44-0997-84
$C_9H_{12}N_2O_2$			
Benzoic acid, 4-methoxy-, 1-methyl-hydrazide	EtOH	250(4.1),310s(4.0)	70-2056-84
$C_9H_{12}N_2O_3$			
Acetic acid, (6-methyl-2-pyrimidinyl-oxy)-, ethyl ester	EtOH	210(3.76),263(3.78)	103-1185-84
Phenylalanine, 4-amino-3-hydroxy-	H_2O	275(3.28)	44-0997-84
2,4(1H,3H)-Pyrimidinedione, 1-[2-(ethenyloxy)ethyl]-5-methyl-	pH 1	271(--)	47-2841-84
	H_2O	270(4.00)	47-2841-84
	pH 13	269(--)	47-2841-84
2,4(1H,3H)-Pyrimidinedione, 1-[2-(2-propenyloxy)ethyl]-	pH 7	269(4.07)	161-0346-84
	pH 12	268(3.69)	161-0346-84
$C_9H_{12}N_2O_5$			
Uridine, 5'-deoxy-	MeOH	264(3.99)	18-2017-84
$C_9H_{12}N_2O_5S$			
6H-1,3,4-Thiadiazine-2,5-dicarboxylic acid, 6-ethoxy-, dimethyl ester	CH_2Cl_2	296(3.4)	89-0890-84
$C_9H_{12}N_2O_6$			
Pseudouridine	pH 1	263(3.88)	18-2515-84
	pH 13	285(3.87)	18-2515-84
	MeOH	264(3.68)	18-2515-84

Compound	Solvent	$\lambda_{max}(\log \epsilon)$	Ref.
$C_9H_{12}N_4$			
Pyrazolo[1,5-a]pyrimidin-7-amine, 5-(1-methylethyl)-	EtOH	226(4.49),286(3.87), 304(3.84)	4-1125-84
7,7'-Spirobi[2,3-diazabicyclo[2.2.1]-hept-2-ene]	EtOH	224(2.72),334(1.92), 342(1.93)	35-3466-84
$C_9H_{12}N_4OS$			
Ethanol, 2-[(2-amino-7-methyl-7H-pyrrolo[2,3-d]pyrimidin-4-yl)thio]-	MeOH	235(4.50),318(3.97)	87-0981-84
$C_9H_{12}N_4O_2$			
Benzamide, 4-(3-methoxy-3-methyl-1-triazenyl)-	EtOH	298(4.18)	87-0870-84
$C_9H_{12}N_4O_2S_2$			
2H-Thiazolo[3,2-b][1,2,4]thiadiazol-2-imine, N-nitro-6-pentyl-	isoPrOH	280(3.23),343(4.26)	94-3483-84
$C_9H_{12}N_4O_3$			
1-Triazene, 3-(methoxymethyl)-3-methyl-1-(4-nitrophenyl)-	MeCN	234(3.94),257(3.78), 350(4.26)	23-0741-84
$C_9H_{12}N_4O_6$			
2(1H)-Pyrimidinone, 1-[(2-acetoxyeth-oxy)methyl]-4-amino-5-nitro-	pH 1	253(3.93),307(3.96)	87-1486-84
	pH 13	334(4.23)	87-1486-84
$C_9H_{12}N_4O_7$			
1H-Pyrazole-3-carboxamide, 4-nitro-5-β-D-ribofuranosyl-	pH 1	275(3.92)	39-2367-84C
	pH 11	230s(3.57),310(4.07)	39-2367-84C
	MeOH	272(3.88)	39-2367-84C
$C_9H_{12}N_5O$			
7H-Purinium, 6-(acetylamino)-7,9-di-methyl-, iodide	H_2O	225(4.05),273(3.86)	150-3601-84M
$C_9H_{12}N_6O_2$			
4(1H)-Pteridinone, 2-amino-6-(1-amino-2-hydroxypropyl)-, [S-(R*,R*)]-	pH -1.0	232(4.12),246(4.09), 321(3.92)	5-1815-84
	pH 4.0	238(4.11),277(4.20), 345(3.79),362s(3.72)	5-1815-84
	pH 10.0	254(4.38),277s(3.86), 364(3.87)	5-1815-84
$C_9H_{12}O$			
2H-1-Benzopyran, 3,5,6,7-tetrahydro-	EtOH	230(3.98),249(3.97)	44-5000-83
3-Cyclohexen-1-one, 3-ethenyl-4-meth-yl- (98% pure)	EtOH	238(4.23)	12-2295-84
$C_9H_{12}O_2$			
1,2-Benzenediol, 3-(1-methylethyl)-	pH 1	271(3.28)	3-1935-84
	borate	280(3.56)	3-1935-84
2-Cyclohepten-1-one, 3-acetyl-	EtOH	244(3.92),315(2.09)	44-2925-84
3-Cyclohepten-1-one, 3-acetyl-	EtOH	223(3.86),233(3.88), 290(2.00)	44-2925-84
2,4,6-Octatrienoic acid, methyl ester	MeCN	295(4.63)	22-0362-84
Tricyclo[3.3.0.0^{2,8}]octan-3-one, 1-methoxy-	MeOH	264(2.23)	78-3749-84
Tricyclo[3.3.0.0^{2,8}]octan-3-one, 8-methoxy-	EtOH	270(2.26)	78-3749-84

Compound	Solvent	$\lambda_{max}(\log \epsilon)$	Ref.
$C_9H_{12}O_3$			
1,3-Benzenediol, 5-(3-hydroxypropyl)-	EtOH	202(4.50),275(3.16), 281(3.13)	100-0828-84
2-Butynal, 4-[(tetrahydro-2H-pyran-2-yl)oxy]-	hexane	213(3.73),220.6(3.87), 228.4(3.86),258s(2.21), 310s(1.34),319(1.38), 329.5(1.40),342(1.34), 357(1.66),375.5(0.78)	44-1898-84
1-Cycloheptene-1-carboxylic acid, 3-oxo-, methyl ester	EtOH	237(3.70),320(1.86)	44-2925-84
1,4-Dioxaspiro[4.5]decan-2-one, 3-methylene-	EtOH	238(3.98)	39-1531-84C
4H-Furo[3,2-c]pyran, 6,7-dihydro-4-methoxy-6-methyl-	MeCN	214(3.79),264(2.45)	24-2422-84
2,4-Hexadienoic acid, 5-methyl-6-oxo-, ethyl ester, (E,E)-	EtOH	282(4.58)	104-1528-84
$C_9H_{12}O_5$			
3-Furancarboxylic acid, 2,5-dihydro-2-methoxy-2-methyl-5-oxo-, ethyl ester	CCl_4	243(2.74)	136-0019-84I
6H-Furo[2,3-b]pyran-6-one, 2,3,3a,7a-tetrahydro-2-hydroxy-4-methoxy-3-methyl- (astepyrone)	EtOH	235(4.06)	158-84-32
$C_9H_{12}O_5S_2$			
2-Butenedioic acid, 2-[(ethoxythioxo-methyl)thio]-, dimethyl ester, (Z)-	isoPrOH	277(4.09),318s(--)	104-1082-84
$C_9H_{13}Br$			
Cyclopropane, [1-(1-bromoethenyl)-1-butenyl]-, (E)-	hexane	224(3.70)	44-0495-84
$C_9H_{13}ClO$			
1,3-Nonadien-5-one, 3-chloro-, (E)-	EtOH	267(4.32)	104-1884-84
(Z)-	EtOH	266(4.07)	104-1884-84
5,7-Octadien-4-one, 6-chloro-2-methyl-,	EtOH	267(4.19)	104-1884-84
(Z)-	EtOH	270(4.02)	104-1884-84
$C_9H_{13}NO$			
Benzenemethanol, 4-(dimethylamino)-	hexane-iso-PrOH	259(3.24)	39-1099-84B
1-Propanone, 2,2-dimethyl-1-(1H-2-pyrrolyl)-	EtOH	247(3.46),287(4.17)	131-0377-84J
4(1H)-Pyridinone, 1-(3-butenyl)-2,3-dihydro-	EtOH	323(4.19)	118-0485-84
$C_9H_{13}NOS$			
Phenol, 4-amino-2-[(ethylthio)methyl]-	C_6H_{12}	240(3.62),310(3.34)	65-0142-84
zinc chelate	C_6H_{12}	255(4.41)	65-0142-84
$C_9H_{13}NO_2$			
Acetamide, N-(2-oxobicyclo[3.2.0]hept-1-yl)-, (1S)-	EtOH	295(1.64)	39-0891-84B
1,2-Benzenediol, 4-[2-(methylamino)-ethyl]-, hydrochloride	pH 1	279(3.42)	3-1935-84
	borate	286(3.63)	3-1935-84
1H-Pyrrole-2-acetic acid, α,1-dimethyl-, methyl ester	MeOH	216(3.81)	107-0453-84
$C_9H_{13}NO_3$			
Acetic acid, (1-methyl-5-oxo-2-pyrroli-	EtOH	261(4.40)	118-0618-84

Compound	Solvent	$\lambda_{max}(\log \epsilon)$	Ref.
dinylidene)-, ethyl ester, (E)- (cont.)			118-0618-84
2-Cyclopenten-1-one, 2-(2-methyl-2-nitropropyl)-	EtOH	222(4.03)	12-0587-84
Epinephrine, (+)- (bitartrate)	pH 1	278(3.43)	3-1935-84
	borate	285(3.66)	3-1935-84
1H-Pyrrole-2-carboxylic acid, 2,3-di-hydro-1,2-dimethyl-3-oxo-, ethyl ester	MeCN	326(3.95)	33-1957-84
$C_9H_{13}NO_5$			
2(1H)-Pyridinone, 4-hydroxy-1-[[2-hydroxy-1-(hydroxymethyl)ethoxy]-methyl]-	pH 1	275(3.57)	23-0016-84
	H_2O	275(3.65)	23-0016-84
	pH 13	267(3.75)	23-0016-84
$C_9H_{13}N_2$			
Pyridinium, 4-(dimethylamino)-1-ethen-yl-, perchlorate	H_2O	310(4.40)	142-0505-84
$C_9H_{13}N_3O_3S$			
2-Thiophenecarboxamide, 5-(butylamino)-4-nitro-	benzene	388(4.00)	39-0781-84B
$C_9H_{13}N_3O_4$			
Cytidine, 2'-deoxy-, α-	pH 2	281(--)	94-1441-84
	EtOH	272(3.88)	94-1441-84
4(1H)-Pyrimidinone, 1-[(2-acetoxy-ethoxy)methyl]-2-amino-	pH 1	255(3.88)	87-1486-84
	pH 13	252s(3.71)	87-1486-84
$C_9H_{13}N_3O_5$			
Glycine, N-[2,5-dihydro-1-(1-methyl-ethyl)-4-nitro-2-oxo-1H-pyrrol-3-yl]-	EtOH	262(3.59),369(4.16)	44-1130-84
1H-Imidazole-4-carboxamide, 1-β-D-ribofuranosyl-	H_2O	239(3.56),278(4.02)	4-0529-84
	M HCl	243(3.80),278(3.64)	4-0529-84
	M NaOH	232(3.48),288(4.05)	4-0529-84
Pyridazinium, 4-amino-2,3-dihydro-3-oxo-1-β-D-ribofuranosyl-, hydroxide, inner salt	MeOH	228(4.14),311(4.14)	87-1613-84
4(1H)-Pyrimidinone, 2-amino-5-β-D-ribo-furanosyl-, monohydrochloride	pH 13	233(3.98),277(3.86)	18-2515-84
	MeOH	225(3.99),290(3.85)	18-2515-84
$C_9H_{13}N_3O_6$			
1H-Imidazole-4-carboxamide, 5-hydroxy-1-β-D-ribofuranosyl-	H_2O	244(3.81),278(4.16)	4-0525-84
	M HCl	243(3.81),279(4.09)	4-0525-84
	M NaOH	276(4.21)	4-0525-84
1H-Imidazole-4-carboxamide, 5-(β-D-ribofuranosyloxy)-	H_2O	251(4.07)	4-0849-84
	M HCl	242(3.91)	4-0849-84
	M NaOH	265(4.06)	4-0849-84
1H-Imidazole-5-carboxamide, 4-hydroxy-1-β-D-ribopyranosyl-	H_2O	236(3.60),283(3.92)	4-0529-84
	M HCl	242(3.74),279(3.63)	4-0529-84
	M NaOH	288(4.01)	4-0529-84
1H-Imidazole-5-carboxamide, 4-hydroxy-3-β-D-ribopyranosyl-	H_2O	245(3.80),278(4.15)	4-0529-84
	M HCl	244(3.83),280(4.08)	4-0529-84
	M NaOH	275(4.20)	4-0529-84
4-Pyrimidinecarboxamide, 1,2,3,6-tetra-hydro-N-[2-hydroxy-1,1-bis(hydroxy-methyl)ethyl]-2,6-dioxo-	pH 2 and 7	277(3.84)	73-2541-84
	pH 12	316(3.83)	73-2541-84
$C_9H_{13}N_5$			
Methanimidamide, N'-(4-cyano-1-ethyl-1H-pyrazol-5-yl)-N,N-dimethyl-	EtOH	286(4.12)	103-0215-84

Compound	Solvent	$\lambda_{max}(\log \epsilon)$	Ref.
Methanimidamide, N'-(4-cyano-1-ethyl-1H-pyrrol-3-yl)-N,N-dimethyl-	EtOH	263(4.11)	103-0215-84
$C_9H_{13}N_5O$ Acetamide, N-(8,9-dihydro-7,9-dimethyl-7H-purin-6-yl)-	MeOH	315(4.06)	150-3640-84M
$C_9H_{13}N_5O_2$ 1,2-Butanediol, 3-(6-amino-9H-purin-9-yl)-, (R*,S*)-	pH 2	262(4.12)	73-2148-84
1,3-Butanediol, 4-(6-amino-9H-purin-9-yl)-, (S)-	pH 2	261(4.15)	73-2148-84
$C_9H_{13}N_5O_3$ 6H-Purin-6-one, 2-amino-1,9-dihydro-9-[(2-hydroxyethoxy)methyl]-8-methyl-	pH 1 pH 13	255(4.15),275s(4.00) 260(4.13)	87-1486-84 87-1486-84
$C_9H_{13}N_5O_4$ 6H-Purin-6-one, 2-amino-9-[(2,3-di-hydroxypropoxy)methyl]-1,9-dihydro-	H_2O	253(4.10),272s(3.83)	88-0905-84
C_9H_{14} Cycloheptane, 1,2-bis(methylene)-	EtOH	235(3.76)	44-2981-84
$C_9H_{14}N_2$ 1,4-Ethano-1H-cyclopenta[d]pyridazine, 4,4a,5,6,7,7a-hexahydro-	hexane	387(2.03)	24-0534-84
Pyrazine, 2,5-diethyl-3-methyl-	EtOH	278(3.96),300s(3.42)	142-2317-84
$C_9H_{14}N_2O$ Pyrazine, (1,1-dimethylpropoxy)-	EtOH	213(4.12),280(3.73), 296(3.61)	39-0641-84B
Pyrazine, (1,2-dimethylpropoxy)-	EtOH	213(4.11),281(3.71), 294(3.65)	39-0641-84B
Pyrazine, (2,2-dimethylpropoxy)-	EtOH	212(4.10),280(3.73), 294(3.65)	39-0641-84B
Pyrazine, 2-methoxy-5-methyl-3-(1-methylethyl)-	MeOH	298(3.83)	95-0951-84
Pyrazine, 3-methoxy-5-methyl-2-(1-methylethyl)-	MeOH	294(3.82)	95-0951-84
$C_9H_{14}N_2O_3$ 1,3-Cyclohexanedicarboxamide, 1-methyl-2-oxo-	n.s.g.	250(4.12)	42-0975-84
$C_9H_{14}N_2O_4$ 2H-Pyrrol-2-one, 3-ethoxy-1,5-dihydro-1-(1-methylethyl)-4-nitro-	EtOH	293(3.99)	44-1130-84
$C_9H_{14}N_2O_4S_2$ 2-Thiophenamine, N-butyl-5-(methylsul-fonyl)-3-nitro-	benzene	374(3.96)	39-0781-84B
$C_9H_{14}N_2O_5$ 2,4(1H,3H)-Pyrimidinedione, 1-[[2-hy-droxy-1-(hydroxymethyl)ethoxy]meth-yl]-5-methyl-	pH 1 H_2O pH 13 EtOH	264(3.92) 264(3.91) 263(3.77) 264(--)	23-0016-84 23-0016-84 23-0016-84 23-0016-84
$C_9H_{14}N_3O_9P$ 1H-Imidazole-4-carboxylic acid, 5-ami-	pH 12	251(4.01)	65-2114-84

Compound	Solvent	$\lambda_{max}(\log \epsilon)$	Ref.
no-1-(5-O-phosphono-β-D-ribofurano- syl)- (cont.)			65-2114-84
$C_9H_{14}N_4O_5$ 1H-Pyrazole-3-carboxamide, 4-amino- 5-β-D-ribofuranosyl-	H_2O pH 11	230s(3.59),280(3.63) 230s(3.61),281(3.66)	39-2367-84C 39-2367-84C
$C_9H_{14}N_4O_6$ 1,2,4-Triazine-2(3H)-acetamide, 4,5-di- hydro-N-[2-hydroxy-1,1-bis(hydroxy- methyl)ethyl]-3,5-dioxo-	pH 2 and 7	268(3.79)	73-2541-84
$C_9H_{14}N_6O_3$ 6H-Purin-6-one, 2-amino-1,9-dihydro- 9-[(2-hydroxyethoxy)methyl]- 8-(methylamino)-	pH 1 pH 13	248(4.23),287(3.98) 259(4.14),280s(4.08)	87-1486-84 87-1486-84
$C_9H_{14}O_2$ Cyclopropanecarboxylic acid, 2-ethen- yl-2-methyl-, ethyl ester (cis/trans 3/7) Propanedial, cyclohexyl-, sodium salt	hexane pH 13	203(3.65) 275(4.43)	44-1937-84 5-0649-84
$C_9H_{14}O_3$ 2-Hexenoic acid, 5,5-dimethyl-4-oxo-, methyl ester, (E)-	MeOH	222(4.09)	5-0820-84
$C_9H_{14}O_4$ 1,3-Dioxane-4,6-dione, 2,2-dimethyl- 5-(1-methylethyl)-	MeOH	270(3.38)	12-1245-84
$C_9H_{14}O_5S_2$ Butanedioic acid, [(ethoxythioxometh- yl)thio]-, dimethyl ester	isoPrOH	222(4.06),278(3.90), 350(1.69)	104-1082-84
$C_9H_{14}Si$ Silane, trimethylphenyl-	C_6H_{12} C_6H_{12}	260(2.48) 211.5(4.02),260(2.48)	60-0341-84 60-0383-84
$C_9H_{15}NO$ 3H-Pyrrol-3-one, 1,2-dihydro-2,2-di- methyl-1-(1-methylethyl)-	benzene	322(3.97)	33-1402-84
$C_9H_{15}NOS_2$ 3-Thiopheneethanamine, 2-methoxy- α-methyl-5-(methylthio)-, hydro- chloride	EtOH	202(3.77),238(3.46), 280(3.99)	103-0962-84
$C_9H_{15}NO_2$ Bicyclo[6.1.0]nonane, 1-nitro-	MeOH	210(3.71)	12-1231-84
$C_9H_{15}NO_2S$ Pyrrolidine, 1-[(2-methyl-1,3-butadi- enyl)sulfonyl]-, (Z)-	EtOH	241(4.19)	70-2519-84
$C_9H_{15}NO_2S_2$ 2-Propenoic acid, 3-[[(diethylamino)- thioxomethyl]thio]-, (Z)-	isoPrOH	278(4.26),350(2.13)	104-1082-84
$C_9H_{15}NO_3S$ Morpholine, 4-(1,3-pentadienylsulfon-	EtOH	248(4.35)	70-2519-84

Compound	Solvent	$\lambda_{max}(\log \epsilon)$	Ref.
yl)-, (Z,E)- (cont.)			70-2519-84
$C_9H_{15}NS$ 3-Hexenenitrile, 3-(ethylthio)-4-methyl-	EtOH	206(3.95),253(3.54)	78-2141-84
$C_9H_{15}N_2O_2$ 1H-Imidazolidinyloxy, 3-formyl-2,2,5,5-tetramethyl-4-methylene-	EtOH	233(4.00)	70-2349-84
$C_9H_{15}N_3$ 4-Pyrimidinamine, 2-methyl-6-(1-methylpropyl)-	EtOH	234(4.05),270(3.65)	39-2677-84C
$C_9H_{15}N_3OS$ 1,2,4-Triazine, 6-(1,1-dimethylethyl)-5-methoxy-3-(methylthio)-	MeOH	208(3.9),248(4.1), 298(3.6)	98-0749-84
1,2,4-Triazin-5(2H)-one, 6-(1,1-dimethylethyl)-2-methyl-3-(methylthio)-	MeOH	236(4.1)	98-0749-84
1,2,4-Triazin-5(4H)-one, 6-(1,1-dimethylethyl)-4-methyl-3-(methylthio)-	MeOH	211(4.0),230(3.8), 295(3.9)	98-0749-84
$C_9H_{15}N_3O_5$ 2(1H)-Pyrimidinone, 4-amino-1-[[2-hydroxy-1,1-bis(hydroxymethyl)ethoxy]methyl]-	pH 1 pH 7 pH 14	276(4.00) 267(3.82) 267(3.86)	23-1622-84 23-1622-84 23-1622-84
$C_9H_{15}N_3O_5S$ Pyridazinium, 4-mercapto-3-oxido-1-β-D-ribofuranosyl-, ammonium salt	MeOH	376(4.38)	87-1613-84
$C_9H_{15}N_3S_2$ Carbamimidothioic acid, N'-(4,5-dimethyl-2-thiazolyl)-N,N-dimethyl-, methyl ester	isoPrOH	246(3.88),325(3.85)	94-2224-84
Thiourea, (4-pentyl-2-thiazolyl)-	isoPrOH	253(3.83),293(4.30)	94-3483-84
$C_9H_{15}N_4O_4$ Methanaminium, N-[5-(dimethylamino)-2,4-dinitro-2,4-pentadienylidene]-N-methyl-, perchlorate	CH_2Cl_2	349(3.90),421(4.20)	1-0117-84
$C_9H_{15}N_5$ 1H-Imidazo[4,5-b]pyrazin-2-amine, 5,6-dihydro-5,5,6,6-tetramethyl-	pH 1	274(4.37),283s(4.34), 296s(4.04)	35-7916-84
	pH 6.8	270(4.32),282s(4.22), 297s(3.84)	35-7916-84
$C_9H_{15}N_5O_2S$ Guanidine, N,N-dimethyl-N'-[4-(1-methylethyl)-2-thiazolyl]-N''-nitro-	isoPrOH	240s(3.97),280s(3.80), 328(4.32)	94-3483-84
Guanidine, N,N-dimethyl-N''-nitro-N'-(4-propyl-2-thiazolyl)-	isoPrOH	240s(3.94),280s(3.76), 329(4.30)	94-3483-84
$C_9H_{16}BrNO$ 2-Propenamide, 3-bromo-N,N-bis(1-methylethyl)-, (E)-	C_6H_{12} H_2O MeOH dioxan MeCN	226(4.09) 229(4.05) 227(4.05) 225(4.06) 224(4.05)	19-0085-84 19-0085-84 19-0085-84 19-0085-84 19-0085-84

Compound	Solvent	$\lambda_{max}(\log \epsilon)$	Ref.
$C_9H_{16}ClNO$			
2-Propenamide, 3-chloro-N,N-bis(1-methylethyl)-, (E)-	C_6H_{12}	215(4.15)	19-0085-84
	H_2O	218(3.97)	19-0085-84
	MeOH	217(4.12)	19-0085-84
	dioxan	213(4.11)	19-0085-84
	MeCN	213(4.11)	19-0085-84
$C_9H_{16}INO$			
2-Propenamide, 3-iodo-N,N-bis(1-methylethyl)-, (E)-	C_6H_{12}	245(4.03)	19-0085-84
	MeOH	245(4.00)	19-0085-84
	dioxan	244(3.99)	19-0085-84
	MeCN	241(3.98)	19-0085-84
$C_9H_{16}N_2$			
Cyclopentane, 2-diazo-1,1,3,3-tetramethyl-	hexane	505(0.70)	24-0277-84
$C_9H_{16}N_2O$			
5-Isoxazolamine, 3-(1,2,2-trimethylpropyl)-	EtOH	240(3.94)	39-1079-84C
2H-Pyran, 4-diazotetrahydro-3,3,5,5-tetramethyl-	hexane	492(0.85)	24-0277-84
$C_9H_{16}N_2O_2$			
Cyclopentanimine, 2,2,5,5-tetramethyl-N-nitro-	C_6H_{12}	272(2.41)	118-0479-84
1-Imidazolidinecarboxaldehyde, 3-hydroxy-2,2,4,4-tetramethyl-5-methylene-	EtOH	232(3.86)	70-2349-84
$C_9H_{16}N_2O_3$			
2(1H)-Pyrazinone, 3-ethyl-3,6-dihydro-1-hydroxy-5,6,6-trimethyl-, 4-oxide	EtOH	240(3.94)	103-0097-84
$C_9H_{16}N_3O_2$			
Methanaminium, N-[5-(dimethylamino)-4-nitro-2,4-pentadienylidene]-N-methyl-, perchlorate	EtOH	332(3.63),420(4.46)	1-0117-84
	CH_2Cl_2	404(4.68)	1-0117-84
$C_9H_{16}N_4$			
2,4-Pyrimidinediamine, 6-(1-ethylpropyl)-	EtOH	232(4.02),281(3.93)	39-2677-84C
$C_9H_{16}N_4O$			
5H-Tetrazol-5-one, 1-(1,1-dimethylethyl)-1,4-dihydro-4-(2-methyl-1-propenyl)-	hexane	205(3.853),247(3.861)	24-2761-84
$C_9H_{16}OS$			
2-Octenal, 2-(methylthio)-	EtOH	221(4.03)	78-2035-84
4H-Pyran-4-thione, tetrahydro-3,3,5,5-tetramethyl-	hexane	210(3.49),234(3.74), 265s(3.43),526(1.04)	24-0277-84
$C_9H_{16}O_2S$			
2-Butenethioic acid, 3-methoxy-, S-butyl ester, (E)-	EtOH	293(3.80)	18-3243-84
(Z)-	EtOH	273(3.84)	18-3243-84
3-Buten-2-one, 4-(butylthio)-4-methoxy-	EtOH	263(3.73)	18-3243-84
$C_9H_{16}S$			
Cyclopentanethione, 2,2,5,5-tetramethyl-	hexane	214(3.51),238(3.86), 499(0.90)	24-0277-84

Compound	Solvent	$\lambda_{max}(\log \epsilon)$	Ref.
$C_9H_{16}S_2$			
4H-Thiopyran-4-thione, tetrahydro- 3,3,5,5-tetramethyl-	hexane	213(3.68),238(3.77), 299(2.92),543(1.08)	24-0277-84
$C_9H_{16}Se$			
Cyclopentaneselone, 2,2,5,5-tetra- methyl-	C_6H_{12}	220(3.59),270(3.92), 670(1.36)	24-0277-84
$C_9H_{17}NO$			
2-Propenamide, N,N-bis(1-methylethyl)-	C_6H_{12}	242(3.51)	19-0085-84
	H_2O	239(3.63)	19-0085-84
	MeOH	240(3.59)	19-0085-84
	dioxan	244(3.49)	19-0085-84
	MeCN	241(3.54)	19-0085-84
$C_9H_{17}NO_2S$			
1,3-Butadiene-1-sulfonamide, N,N-dieth- yl-2-methyl-, (Z)-	EtOH	242(4.15)	70-2519-84
$C_9H_{17}N_2O_4P$			

Phosphonic acid, (4,5-dihydro-5-oxo-
  1H-imidazol-4-yl)-, bis(1-methyl
  ethyl) ester
	EtOH	219(3.73),264(3.16)	65-2414-84

Phosphonic acid, (1-ethyl-4,5-dihydro-
  5-oxo-1H-imidazol-4-yl)-, diethyl
  ester
	EtOH	218(3.67),268(3.47)	65-2414-84

$C_9H_{18}N_2OS$			
Morpholine, 4-[2-methyl-1-[(methyl- thio)imino]propyl]-	hexane	265(3.79)	39-2933-84C
$C_9H_{18}N_2O_2$			
2-Hexenimidic acid, 4-ethyl-3-(hydroxy- amino)-, methyl ester	EtOH	281(4.14)	39-1079-84C
5-Nonanimine, N-nitro-	C_6H_{12}	272(2.89)	118-0479-84
$C_9H_{18}OS_2$			
Carbonodithioic acid, S,S-dibutyl ester	EtOH	252(3.82)	12-1483-84
$C_9H_{18}O_3S_3$			
2-Pentanol, 1-[[2,2-bis(methylthio)eth- enyl]sulfonyl]-	EtOH	276(3.97)	39-0085-84C
$C_9H_{18}S$			
3-Pentanethione, 2,2,4,4-tetramethyl-	hexane	220s(3.64),236(3.81), 539(0.90)	24-0277-84
$C_9H_{18}S_3$			
Carbonotrithioic acid, dibutyl ester	EtOH	242(3.15),367(3.85), 432(2.23)	12-1483-84
$C_9H_{19}NSe_2$			
Carbamodiselenoic acid, dibutyl-, sodium salt (also metal chelates)	n.s.g.	288.5(4.37),311(4.23)	18-0591-84
$C_9H_{22}ClNSi$			
Silanamine, N-chloro-1-(1,1-dimethyl- ethyl)-1,1-dimethyl-N-(1-methylethyl)-	$CCl_2F.CClF_2$	297(1.81)	39-1187-84B
C_9O_9			
Benzo[1,2-d:3,4-d':5,6-d'']tris[1,3]- dioxole-2,5,8-trione	THF	259(3.85),317(3.74)	78-4897-84

Compound	Solvent	$\lambda_{max}(\log \epsilon)$	Ref.
$C_{10}Cl_8$ 1,3-Cyclopentadiene, 1,2,3,4-tetra-chloro-5-(2,3,4,5-tetrachloro-2,4-cyclopentadien-1-ylidene)-	isooctane	386(4.55),590(2.70)	88-2923-84
$C_{10}F_8O$ 1(2H)-Naphthalenone, 2,2,3,4,5,6,7,8-octafluoro-	heptane	229(4.40),291(3.42), 303(3.37),348(3.29)	104-0967-84
$C_{10}F_{12}$ Naphthalene, 1,1,2,3,4,4,5,5,6,7,8,8-dodecafluoro-	heptane	222(2.74)	104-0307-84
$C_{10}H_2N_4O_2$ Propanedinitrile, 2,2'-(3,4-dihydroxy-3-cyclobutene-1,2-diylidene)bis-, dipotassium salt	MeOH	305(3.68),429(5.03)	24-2714-84
$C_{10}H_3N_3O_2$ 1H-Isoindole-5,6-dicarbonitrile, 2,3-dihydro-1,3-dioxo-	MeCN	224(4.63),242(4.18), 250(4.08),303(3.32), 315(3.38)	40-0060-84
$C_{10}H_4Br_2O_4S$ 2-Furancarboxaldehyde, 5,5'-thiobis[4-bromo-	MeOH	226(3.48),248(3.52), 312(2.42)	73-0984-84
$C_{10}H_4Br_2O_4Se$ 2-Furancarboxaldehyde, 5,5'-seleno-bis[4-bromo-	MeOH	215(3.11),234(3.14), 318(2.42)	73-0984-84
$C_{10}H_4Cl_2Se_2$ Naphtho[1,8-cd]-1,2-diselenole, 5,6-dichloro-	C_6H_5Cl	409(3.88)	157-0732-84
$C_{10}H_4N_2O_2$ 5,6-Isobenzofurandicarbonitrile, 1,3-dihydro-1-oxo-	MeOH	239(4.43),244(4.43), 251(4.40)	40-0060-84
$C_{10}H_4N_4O$ 2,3-Pyrazinedicarbonitrile, 5-(2-fur-anyl)-	EtOH	222(4.07),253(3.60), 325s(4.25),350(4.31)	44-0813-84
$C_{10}H_4N_4OS$ Naphtho[2,3-c][1,2,5]thiadiazol-4(9H)-one, 9-diazo-	MeCN	252(3.91),295(3.68), 324(3.53),430(3.02)	33-0574-84
$C_{10}H_4N_4OSe$ Naphtho[2,3-c][1,2,5]selenadiazol-4(9H)-one, 9-diazo-	MeCN	256(3.66),318(3.70), 450(2.93)	33-0574-84
$C_{10}H_4S_4$ Naphtho[1,8-cd:4,5-c'd']bis[1,2]di-thiole	MeCN	251(4.16),420(4.26)	18-0022-84
$C_{10}H_4Se_4$ Naphtho[1,8-cd:4,5-c'd']bis[1,2]di-selenole	$C_2H_4Cl_2$	256(3.94),415(3.88), 436(3.96)	18-0022-84
	C_6H_5Br	417(4.20),441(4.34)	157-0732-84

Compound	Solvent	$\lambda_{max}(\log \epsilon)$	Ref.
$C_{10}H_5BrClNOS$ 6H-1,3-Thiazin-6-one, 2-(4-bromophen-yl)-4-chloro-	EtOH	228(4.23),303(3.85), 347(3.93)	103-1088-84
$C_{10}H_5ClN_2$ Benzo[3,4]cyclobuta[1,2-d]pyrimidine, 4-chloro-	EtOH	243(4.45),258(4.26), 314(3.54),323(3.71), 340(3.80)	44-0289-84
Cyclobuta[1,2-c:4,3-c']dipyridine, 3-chloro-	C_6H_{12}	237(4.79),245(4.95), 283.5(3.46),297.5(3.54), 311.5(3.55),327.5(3.37), 343s(2.73),355s(2.63), 368s(2.48),388s(2.18)	150-3001-84M
$C_{10}H_5ClN_2O_3S$ 6H-1,3-Thiazin-6-one, 4-chloro-2-(4-nitrophenyl)-	EtOH	230(3.94),260(3.77), 285(3.88),346(3.74)	103-1088-84
$C_{10}H_5ClO_3$ 1,4-Naphthalenedione, 6-chloro-2-hydroxy-	MeOH	237(4.29),242(4.27), 255s(4.16),264s(4.26), 272(4.33),281s(4.16), 325(3.81),460(3.40)	102-0313-84
$C_{10}H_5Cl_3O_2S$ Cyclopenta[c]thiopyran-3-carboxylic acid, 5,6,7-trichloro-, methyl ester	CH_2Cl_2	273(4.55),283(4.53), 313(4.21),326(4.27), 382(3.90),400(3.97), 474(3.24)	118-0262-84
$C_{10}H_5Cl_5O_2$ 2-Propenoic acid, 2-methyl-, penta-chlorophenyl ester	CCl_4	<u>265(0.3),299(0.4)</u>	93-1038-84
$C_{10}H_5N_3$ Propanedinitrile, (4-cyanophenyl)-, potassium salt	MeCN	357(4.66)	35-6726-84
$C_{10}H_5N_3O_3S$ 5H-Imidazo[2,1-b]pyrido[3,2-e][1,3]-thiazine-3-carboxylic acid, 5-oxo-	MeOH	209(4.29),229(4.32), 270(3.98),340(3.30)	142-0807-84
$C_{10}H_6BrNO$ 5H-Cyclohepta[b]pyridin-5-one, 9-bromo-	EtOH	206(4.44),220(4.5), 259s(--),263s(--), 342.5(3.84),350s(--)	39-2297-84C
$C_{10}H_6BrNO_2S$ 6H-1,3-Thiazin-6-one, 2-(4-bromo-phenyl)-4-hydroxy-	EtOH	250(4.30),268(4.26), 350(3.83)	103-1088-84
$C_{10}H_6BrNS_2$ 6H-1,3-Thiazine-6-thione, 2-(4-bromo-phenyl)-	$CHCl_3$	264(4.25),318(4.19), 430(3.74)	97-0435-83
$C_{10}H_6ClNOS$ 6H-1,3-Thiazin-6-one, 4-chloro-2-phenyl-	EtOH	228(4.15),290(3.83), 340(3.79)	103-1088-84
$C_{10}H_6ClNS_2$ 6H-1,3-Thiazine-6-thione, 2-(4-chloro-	$CHCl_3$	265(4.14),310(3.78),	97-0435-84

Compound	Solvent	$\lambda_{max}(\log \epsilon)$	Ref.
phenyl)- (cont.)		430(3.31)	97-0435-84
$C_{10}H_6Cl_4S$ Benzene, [(3,3,4,4-tetrachloro-1-butyn-yl)thio]-	heptane	208(4.50),267(4.36)	5-1873-84
$C_{10}H_6FNOS$ 6H-1,3-Thiazin-6-one, 4-fluoro-2-phen-yl-	EtOH	220(4.23),285(4.04), 340(3.96)	103-1088-84
$C_{10}H_6F_3NO_2$ 2H-1-Benzopyran-2-one, 7-amino-4-(trifluoromethyl)-	pentane	348(4.28)	135-0674-84
$C_{10}H_6F_3NO_3$ 3-Buten-2-one, 1,1,1-trifluoro-4-(4-nitrophenyl)-, (E)-	EtOH	246(3.54),298(4.22)	39-2863-84C
$C_{10}H_6N_2$ 1,6-Diazabiphenylene	MeOH	228.5(4.52),258(4.17), 289s(3.56),318.5(3.79), 335.5(3.80)	150-3001-84M
$C_{10}H_6N_2O$ 2,7-Diazabiphenylene, 2-oxide	MeOH	230(4.49),257(4.24), 284(4.31),360(3.65), 374(3.61)	150-3001-84M
$C_{10}H_6N_2O_2S_2$ 6H-1,3-Thiazine-6-thione, 2-(4-nitro-phenyl)-	CHCl$_3$	258(4.49),295(4.24), 435(3.62)	97-0435-84
$C_{10}H_6N_2O_4S$ 6H-1,3-Thiazin-6-one, 4-hydroxy-2-(4-nitrophenyl)-	EtOH	233(4.15),267(4.30), 370(3.62)	103-1088-84
$C_{10}H_6N_4$ 2,5,9,10-Tetraazaphenanthrene	MeOH	229(4.52),241(4.48), 288(3.87),325(3.46), 340(3.40),368(2.61), 395(2.42)	150-3001-84M
$C_{10}H_6N_4O_4$ 2,4'-Bipyridine, 3,3'-dinitro-	MeOH	240s(4.05),270s(3.85)	150-3001-84M
$C_{10}H_6N_4S$ Pyrazolo[1,5-a]pyridine-3,5-dicarbo-nitrile, 2-(methylthio)-	EtOH	221(4.33),260(4.58), 317(3.89),352(3.65)	95-0440-84
Pyrazolo[1,5-a]pyridine-3,6-dicarbo-nitrile, 2-(methylthio)-	EtOH	213(4.26),263(4.65), 313(3.98)	95-0440-84
$C_{10}H_6N_6O_3$ 1H-Imidazo[1,2-b][1,2,4]triazole, 5-(4-nitrophenyl)-6-nitroso-	EtOH	224(4.38),263s(3.31), 352(3.38)	4-1029-84
Methanone, (4-nitrophenyl)-1H-1,2,4-triazolo[4,3-b][1,2,4]triazol-3-yl-	EtOH	224(4.35)	4-1029-84
$C_{10}H_6O_2$ 1,5-Azulenedione	MeCN	216(3.89),254(4.36), 264(4.27),307s(3.54), 324(3.57),338(3.59),	35-4857-84

Compound	Solvent	λ_{max} (log ϵ)	Ref.
(cont.)		350(3.60),373(3.53), 389s(3.29)	35-4857-84
	EtOH	322(--),337(--), 350(--),373(--), 389s(--)(changing)	35-4857-84
1,7-Azulenedione	MeCN	236(4.38),319(3.84), 333s(3.75),345s(3.59), 384(3.54),401(3.51), 425s(3.21)	35-4857-84
	EtOH	319(--),323(--), 343s(--),384(--), 399(--),430s(--) (changing)	35-4857-84
$C_{10}H_6O_3$ 1,4-Naphthalenedione, 2-hydroxy-	MeOH	243(4.12),249(4.16), 274(4.10),333(3.38)	102-0313-84
1,4-Naphthalenedione, 5-hydroxy-	DMF	330(3.31),416(4.87)	90-0729-84
$C_{10}H_6O_4S$ 2-Furancarboxaldehyde, 5,5'-thiobis-	MeOH	205(3.05),225(2.98), 274(3.06),304(3.24)	73-0984-84
$C_{10}H_6O_4Se$ 2-Furancarboxaldehyde, 5,5'-selenobis-	MeOH	204(3.10),231(3.01), 271(3.14),308(3.23)	73-0984-84
$C_{10}H_6S_2$ Benzo[1,2-c:3,4-c']dithiophene, TCNE complex	CHCl$_3$	607(1.65)	44-1027-84
TCNQ complex	CHCl$_3$	627(1.72)	44-1027-84
$C_{10}H_7BF_2O_2S$ [1]Benzothieno[3,2-d][1,3,2]dioxaborine, 2,2-difluoro-4-methyl-	dioxan	258(4.09),266(4.13), 310(4.01),350(3.66)	83-0531-84
$C_{10}H_7BF_2O_3$ [1]Benzofuro[3,2-d][1,3,2]dioxaborine, 2,2-difluoro-4-methyl-	dioxan	234(4.08),305(4.38)	83-0526-84
$C_{10}H_7BrClNO$ Ethanone, 1-(3-bromo-5-chloro-1H-indol-2-yl)-	n.s.g.	214(4.37),242(4.27), 311(4.31)	103-0988-84
$C_{10}H_7Br_2NO$ Ethanone, 1-(3,5-dibromo-1H-indol-2-yl)-	n.s.g.	214(4.40),244(4.25), 311(4.30)	103-0988-84
$C_{10}H_7ClO$ Furan, 2-(4-chlorophenyl)-	C_6H_{12} MeOH	216(4.17),282(4.45) 216(4.53),282(4.50)	18-0844-84 18-0844-84
$C_{10}H_7Cl_2N_3$ 4H-Pyrazolo[1,5-a]benzimidazole, 6,7-dichloro-2-methyl-	EtOH	241(4.56),328(4.21)	48-0829-84
$C_{10}H_7Cl_3O_2$ 2-Propenoic acid, 2-methyl-, 2,4,5-tri-chlorophenyl ester	n.s.g.	265(0.4),283(0.6), 291(0.7)	93-1038-84

Compound	Solvent	$\lambda_{max}(\log \epsilon)$	Ref.
$C_{10}H_7CrFO_3$			
Chromium, tricarbonyl[(1,2,3,4,5,6-η)-	EtOH	216(4.38),255(3.92)	112-0765-84
1-fluoro-2-methylbenzene]-	CHCl$_3$	313(3.99)	112-0765-84
Chromium, tricarbonyl[(1,2,3,4,5,6-η)-	EtOH	216(4.32),255(3.87)	112-0765-84
1-fluoro-3-methylbenzene]-	CHCl$_3$	311(4.00)	112-0765-84
Chromium, tricarbonyl[(1,2,3,4,5,6-η)-	EtOH	217(4.25),255(3.77)	112-0765-84
1-fluoro-4-methylbenzene]-	CHCl$_3$	314(3.99)	112-0765-84
$C_{10}H_7CrFO_4$			
Chromium, tricarbonyl[(1,2,3,4,5,6-η)-	EtOH	211(4.44),255(3.84)	112-0765-84
1-fluoro-2-methoxybenzene]-	CHCl$_3$	315(3.95)	112-0765-84
Chromium, tricarbonyl[(1,2,3,4,5,6-η)-	EtOH	212(4.44),255(3.92)	112-0765-84
1-fluoro-3-methoxybenzene]-	CHCl$_3$	312(3.98)	112-0765-84
Chromium, tricarbonyl](1,2,3,4,5,6-η)-	EtOH	212(4.48),253(3.90)	112-0765-84
1-fluoro-4-methoxybenzene]-	CHCl$_3$	313(3.93)	112-0765-84
$C_{10}H_7F_3N_2O$			
Ethanone, 2,2,2-trifluoro-1-(1-methyl-	hexane	207.5(4.02),246.5(4.16),	44-4769-84
2H-cyclopenta[d]pyridazin-5-yl)-		296(3.80),321(3.75),	
		395(3.77)	
	MeOH	209(4.34),244(4.32),	44-4769-84
		256(4.32),294(4.16),	
		320(3.93),381(4.15)	
$C_{10}H_7F_3O$			
3-Buten-2-one, 1,1,1-trifluoro-	EtOH	233(3.78),248(3.59),	39-2863-84C
4-phenyl-		311(4.26)	
$C_{10}H_7F_6NO_2S$			
Isothiazolo[2,3-a]azepine, 2,3-dihydro-	hexane	222(2.634),354(0.422)	44-5124-84
5,7-bis(trifluoromethyl)-, 1,1-diox-			
ide			
$C_{10}H_7NO$			
Benzonitrile, 4-(3-hydroxy-1-propynyl)-	MeOH	258(4.40),268(4.39)	118-0728-84
5H-Cyclohepta[b]pyridin-5-one	EtOH	217.5(4.52),240(4.16),	39-2297-84C
		247.5(4.15),302.5(3.72),	
		340(3.78),351(3.73)	
5H-Cyclohepta[c]pyridin-5-one	EtOH	217(4.45),244s(--),	39-2297-84C
		252s(--),312(3.98),	
		325(3.98),348(3.84)	
$C_{10}H_7NO_2$			
Naphthalene, 1-nitro- (diprotonated)	CF$_3$SO$_3$H	524(4.05)	88-0325-84
Naphthalene, 2-nitro-	CF$_3$SO$_3$H at 0º	411(4.19),547(3.48)	88-0325-84
3-Quinolinecarboxylic acid	pH 2	312(2.72),320(2.73)	149-0057-84A
	pH 8	308(2.38),320(2.29)	149-0057-84A
	2M H$_2$SO$_4$	312(2.72),321(2.73)	149-0057-84A
$C_{10}H_7NO_6$			
Propanedioic acid, [(4-nitrophenyl)-	1.2M KOH	240(4.15),320(3.73)	65-1989-84
methylene]-, dipotassium salt			
$C_{10}H_7NS_2$			
6H-1,3-Thiazine-6-thione, 2-phenyl-	CHCl$_3$	257(4.32),320(4.20),	97-0435-84
		430(3.90)	
$C_{10}H_7N_3OS_2$			
2-Propanone, 1-(2-benzothiazolylthio)-	MeOH	225(4.07),247(3.91),	103-0505-84
3-diazo-		278(3.99)	

Compound	Solvent	$\lambda_{max}(\log \epsilon)$	Ref.
$C_{10}H_7N_3O_2S$ 2-Propanone, 1-(2-benzoxazolylthio)- 3-diazo-	MeOH	250(4.10),275(4.22)	103-0505-84
$C_{10}H_7N_3O_3S$ 1-Naphthalenesulfonic acid, 4-azido-	2% MeOH	231(4.60),308(4.04)	149-0545-84B
$C_{10}H_7N_3O_4S$ 3-Pyridinecarboxylic acid, 2-[(4-carb- oxy-1H-imidazol-2-yl)thio]-	MeOH	209(4.27),254(4.22), 295s(3.63)	142-0807-84
$C_{10}H_7N_3O_6S_2$ 1,5-Naphthalenedisulfonic acid, 2-azido-	2% MeOH	256(4.43),291(4.86), 325(3.30)	149-0545-84B
1,7-Naphthalenedisulfonic acid, 6-azido-	2% MeOH	222(4.30),252(4.45), 296(4.71)	149-0545-84B
$C_{10}H_7N_5O$ Furo[2',3':4,5]pyrrolo[1,2-d]-1,2,4- triazolo[3,4-f][1,2,4]triazine, 3-methyl-	dioxan	246(3.44),302(3.13)	73-0065-84
Furo[2',3':4,5]pyrrolo[1,2-d]-1,2,4- triazolo[3,4-f][1,2,4]triazine, 6-methyl-	dioxan	254(3.54),313(3.19)	73-0065-84
1H-Imidazo[1,2-b][1,2,4]triazole, 6-nitroso-5-phenyl-	EtOH	230(3.09),353(3.24), 407(2.39)	4-1029-84
Methanone, phenyl-1H-1,2,4-triazolo- [4,3-b][1,2,4]triazol-3-yl-	EtOH	243(3.10),331(0.87)	4-1029-84
$C_{10}H_7N_5O_2S_2$ Thiazole, 4-azido-2-(methylthio)- 5-(4-nitrophenyl)-	dioxan	263(3.17),387(3.23)	73-2285-84
$C_{10}H_8$ Azulene	CH_2Cl_2	233(3.96),280(3.65), 340(3.63),351(2.85), 575(2.49)	5-1905-84
Benzene, (1-methylene-2-propynyl)-	EtOH	256(2.61)	65-1400-84
Cyclobuta[a]pentalene, 2a,6a-dihydro-	C_6H_{12}	270(4.23),384(3.15)	35-7268-84
1,3-Cyclopentadiene, 5-(2,4-cyclo- pentadien-1-ylidene)-	pentane	278(3.93),289(4.30), 300(4.95),316(4.67), 416(2.01)	24-2006-84
Naphthalene (1-protonated)	$HF-BF_3$	390(4.04),410s(--)	88-0325-84
Tetracyclo[5.3.0.0^{2,4}.0^{3,5}]deca- 6,8,10-triene	C_6H_{12}	281(4.04),400(2.57)	35-7268-84
$C_{10}H_8BrNO_3$ 1H-Indole-2-carboxylic acid, 4-bromo- 5-methoxy-	EtOH	222(4.47),294(4.18)	44-4761-84
$C_{10}H_8BrN_3$ 4H-Pyrazolo[1,5-a]benzimidazole, 6-bromo-2-methyl-	EtOH	239(4.34),317(4.05)	48-0829-84
$C_{10}H_8BrN_3O_2$ 2-Propenal, 2-bromo-3-(5,6-dihydro- 6-methyl-5-oxoimidazo[1,2-c]pyri- midin-3-yl)-	EtOH	371(4.24)	44-4021-84
$C_{10}H_8ClNO$ Benzeneacetonitrile, 4-chloro-α-(meth-	MeOH	215(4.22),270(4.29),	56-0935-84

Compound	Solvent	$\lambda_{max}(\log \epsilon)$	Ref.
oxymethylene)-, (E)- (cont.)		276(4.28)	56-0935-84
(Z)-	MeOH	216(4.18),271(4.19)	56-0935-84
$C_{10}H_8ClN_3$			
4H-Pyrazolo[1,5-a]benzimidazole,	EtOH	236(4.17),316(3.94)	48-0829-84
6-chloro-2-methyl-			
4H-Pyrazolo[1,5-a]benzimidazole,	EtOH	237(4.50),319(4.00)	48-0829-84
7-chloro-2-methyl-			
$C_{10}H_8Cl_2Si$			
Silane, dichloroethenyl(phenylethynyl)-	hexane	242.5(4.73),251.3(4.89),	65-2317-84
		263.2(4.79)	
$C_{10}H_8CrFNO_3$			
Chromium, tricarbonyl[(1,2,3,4,5,6-η)-	EtOH	215(4.51)	112-0765-84
3-fluoro-2-methylbenzenamine]-	CHCl$_3$	314(3.94)	112-0765-84
Chromium, tricarbonyl[(1,2,3,4,5,6-η)-	EtOH	216(4.53)	112-0765-84
3-fluoro-4-methylbenzenamine]-	CHCl$_3$	314(3.92)	112-0765-84
Chromium, tricarbonyl[(1,2,3,4,5,6-η)-	EtOH	219(4.48)	112-0765-84
5-fluoro-2-methylbenzenamine]-	CHCl$_3$	313(3.89)	112-0765-84
$C_{10}H_8NOS_2$			
2H-[1,3]Thiazino[2,3-b]benzothiazolium,	10% MeCN	233(3.4),261(3.79),	103-0508-84
3,4-dihydro-3-oxo-, perchlorate		318(4.19)	
$C_{10}H_8NO_2S$			
2H-[1,3]Thiazino[2,3-b]benzoxazolium,	10% MeCN	230(3.70),282(3.89),	103-0508-84
3,4-dihydro-3-oxo-, perchlorate		288(3.87)	
$C_{10}H_8N_2$			
2,2'-Bipyridine	C_6H_{12}	232(4.2),246(4.1),	18-0341-84
		282(4.3)	
	pH 1.22	238(3.9),297(4.2)	18-0341-84
	3.87M H$_2$SO$_4$	289(--)	18-0341-84
4,4'-Bipyridine	C_6H_{12}	236(4.1)	18-0341-84
	pH 0.98	248(4.2),278s(4.0)	18-0341-84
Propanedinitrile, (4-methylphenyl)-,	MeCN	298(4.34)	35-6726-84
potassium salt			
$C_{10}H_8N_2O$			
[3,3'-Bipyridin]-2(1H)-one	MeOH	245(3.86),321(4.05)	150-3001-84M
[3,3'-Bipyridin]-4(1H)-one	H$_2$O	219(4.23),261(4.06)	150-3001-84M
1H-Indole-4-carbonitrile, 5-methoxy-	EtOH	204(4.31),228(4.34),	44-4761-84
		316(4.06)	
Propanedinitrile, (4-methoxyphenyl)-,	MeCN	293(4.32)	35-6726-84
potassium salt			
2-Pyridinamine, N-(2-furanylmethylene)-	C_6H_{12}	222(4.10),282(4.08),	18-0844-84
		315(4.21)	
	MeOH	231(4.21),294(3.97),	18-0844-84
		324(3.81)	
$C_{10}H_8N_2OS$			
[1]Benzoxepino[5,4-d][1,2,3]thiadia-	EtOH	250(4.16),288(3.71)	2-0203-84
zole, 4,5-dihydro-			
2H-[1,3]Thiazino[3,2-a]benzimidazol-	MeOH	212(4.2),252(3.9),	103-0508-84
3(4H)-one		274(4.01),280(4.01)	
$C_{10}H_8N_2O_2$			
1H-Indole, 4-(2-nitroethenyl)-	EtOH	218(4.44),256(3.98)	44-4761-84
1,4-Naphthalenedione, 2,3-diamino-	CH$_2$Cl$_2$	238(4.20),290(4.35),	83-0329-84
		518(3.36)	

Compound	Solvent	$\lambda_{max}(\log \epsilon)$	Ref.
$C_{10}H_8N_2O_2S$ 4H-[1]Benzopyrano[4,3-d][1,2,3]thiadiazole, 7-methoxy-	EtOH	260(4.12),299(3.89)	2-0203-84
$C_{10}H_8N_2O_4$ Pyrrolo[3,4-c]pyridine-1,3,4(2H,5H)-trione, 7-acetyl-6-methyl-	MeOH	219(4.20),266(3.90)	48-0594-84
$C_{10}H_8N_2S$ Pyridine, 3,3'-thiobis-	H_2O	195(4.21),211s(4.04), 249(3.94),278s(3.76)	4-0917-84
$C_{10}H_8N_2S_2$ [1]Benzothiepino[5,4-d][1,2,3]thiadiazole, 4,5-dihydro-	EtOH	239(4.31)	2-0203-84
$C_{10}H_8N_4O$ Benzaldehyde, 3-(1H-imidazol-2-ylazo)-	MeOH	194(4.20),235(3.98), 358(4.29)	56-0903-84
Benzaldehyde, 4-(1H-imidazol-2-ylazo)-	MeOH	194(4.27),234(3.77), 256(3.65),374(4.31)	56-0903-84
2,3-Pyrazinedicarbonitrile, 5-(2-furanyl)-1,4,5,6-tetrahydro-	EtOH	215(4.27),320(4.02)	44-0813-84
$C_{10}H_8N_4OS$ 2-Propanone, 1-(1H-benzimidazol-2-ylthio)-3-diazo-	MeOH	255(4.10),275(4.22)	103-0505-84
Pyrazolo[1,5-a]pyridine-5-carboxamide, 3-cyano-2-(methylthio)-	EtOH	215(4.33),257(4.58), 312(3.89),334(3.66)	95-0440-84
Pyrazolo[1,5-a]pyridine-6-carboxamide, 3-cyano-2-(methylthio)-	EtOH	265(4.65),311(3.97)	95-0440-84
Pyrido[3',2':4,5]thieno[3,2-d]-1,2,3-triazin-4(3H)-one, 3,7-dimethyl-	EtOH	236(4.52),276(4.13), 310(3.95)	87-1639-84
$C_{10}H_8N_4S$ Pyridinium, 1-[[2,2-dicyano-1-(methylthio)ethenyl]amino]-, hydroxide, inner salt	EtOH	247(4.17),279(4.17), 380(3.23)	95-0440-84
$C_{10}H_8N_4S_2$ Pyrido[3',2':4,5]thieno[3,2-d]-1,2,3-triazine, 7-methyl-4-(methylthio)-	EtOH	238(4.40),250(4.24), 295(4.40)	87-1639-84
$C_{10}H_8N_6O_4$ 1H-1,2,4-Triazol-5-amine, 1-[1-(hydroxyimino)-2-(4-nitrophenyl)-2-oxoethyl]-	EtOH	235(4.39),291(3.22), 410(2.69)	4-1029-84
$C_{10}H_8O$ Ethanone, 1-(4-ethynylphenyl)-	MeOH	269(4.17)	118-0728-84
Furan, 2-phenyl-	C_6H_{12}	216(4.20),276(4.39)	18-0844-84
	MeOH	216(4.02),276(4.26)	18-0844-84
$C_{10}H_8OS$ Spiro[benzo[b]thiophene-2(3H),1'-cyclopropan]-3-one	EtOH	208(5.46),237(4.53), 258(3.82),267(3.74), 350(3.42),367(3.43)	18-2325-84
$C_{10}H_8O_2$ 1,4-Azulenedione, 2,3-dihydro-	EtOH	233(4.35),325(3.93)	35-4852-84

Compound	Solvent	$\lambda_{max}(\log \epsilon)$	Ref.
1,6-Azulenedione, 2,3-dihydro-	EtOH	233(4.37),240s(--), 309s(--),316(3.95)	35-4852-84
Benzoic acid, 4-ethynyl-, methyl ester	MeOH	257(4.33)	118-0728-84
2H-1-Benzopyran-2-one, 4-bromo-	EtOH	270(4.04),311(3.79)	94-1178-84
$C_{10}H_8O_2S$			
Cyclopenta[c]thiopyran-3-carboxylic acid, methyl ester	CH_2Cl_2	263(4.39),272(4.32), 302(3.95),313(3.94), 370(3.79),386(3.86), 458(2.97)	118-0262-84
$C_{10}H_8O_2S_4$			
1,3-Dithiole-4-carboxylic acid, 2-(1,3-dithiol-2-ylidene)-, 2-propenyl ester	MeCN	288(4.11),300(4.12), 313(4.12),432(3.29)	103-1342-84
$C_{10}H_8O_2Se$			
2-Benzofurancarboxaldehyde, 3-(methyl-seleno)-	EtOH	235(3.88),305(4.23)	103-0957-84
3-Benzofurancarboxaldehyde, 2-(methyl-seleno)-	EtOH	207(4.39),236(4.19), 323(3.95)	103-0957-84
$C_{10}H_8O_3$			
2,4(3H,5H)-Furandione, 3-phenyl-	EtOH	260(4.30)	39-1539-84C
	EtOH-NaOH	260(4.33)	39-1539-84C
$C_{10}H_8O_3S$			
Spiro[benzo[b]thiophene-2(3H),1'-cyclo-propan]-3-one, 1,1-dioxide	MeOH	210(4.53),246(3.98), 283(2.89),292(2.82)	18-2325-84
$C_{10}H_8O_4$			
Cinnamic acid, 3,4-(methylenedioxy)-	EtOH	285(4.05),321(4.16)	102-1029-84
Isoscopoletin	EtOH	230.5(4.20),254.5(3.75), 259s(3.74),296.1(3.78), 347.6(4.01)	102-0699-84
Scopoletin	EtOH	229.2(4.18),253.7(3.71), 260s(3.68),298.8(3.74), 346.5(4.12)	102-0699-84
$C_{10}H_8O_4S_4$			
1,3-Dithiole-4-carboxylic acid, 2-[4-(methoxycarbonyl)-1,3-dithiol-2-ylidene]-, methyl ester	CH_2Cl_2	290(4.00),302(4.03), 315(4.04),444(3.39)	103-1342-84
$C_{10}H_9ClN_2O$			
2H-Pyrimido[2,1-b]benzoxazole, 7-chloro-3,4-dihydro-	$CHCl_3$	254(3.72),301(3.91)	94-3053-84
$C_{10}H_9ClN_2O_2S$			
2H-1,2,4-Benzothiadiazine, 7-chloro-3-(1-propenyl)-, 1,1-dioxide	EtOH	<u>300(4.1)</u>	103-0735-84
$C_{10}H_9ClO$			
3-Buten-2-one, 4-chloro-4-phenyl-, (E)-	EtOH	235(4.01)	80-0193-84
1(2H)-Naphthalenone, 7-chloro-3,4-di-hydro-	MeOH	245(4.01),305(3.30)	102-0313-84
$C_{10}H_9Cl_2NO_2$			
2,4,6-Heptatrienoic acid, 7,7-dichloro-2-cyano-, ethyl ester	EtOH	338(4.52)	70-2328-84

Compound	Solvent	$\lambda_{max}(\log \epsilon)$	Ref.
$C_{10}H_9Cl_2N_3O$ Diazene, (3,4-dichlorophenyl)(1-iso-cyanato-1-methylethyl)-	C_6H_{12}	276(4.08),389(2.28)	118-0315-84
$C_{10}H_9Cl_2N_3S$ Diazene, (3,4-dichlorophenyl)(1-iso-thiocyanato-1-methylethyl)-	C_6H_{12}	277(4.11),406(2.30)	118-0315-84
$C_{10}H_9CrNO_4$ Chromium, tricarbonyl[(1,2,3,4,5,6-η)-2-methoxybenzenamine]- Chromium, tricarbonyl[(1,2,3,4,5,6-η)-3-methoxybenzenamine]-	EtOH CHCl$_3$ EtOH CHCl$_3$	217(4.45) 320(3.88) 216(4.45) 312(3.91),355(3.39)	112-0765-84 112-0765-84 112-0765-84 112-0765-84
$C_{10}H_9FN_2O_3$ Ethanone, 2-diazo-1-(5-fluoro-2,3-di-methoxyphenyl)-	EtOH	218(3.53),262(3.18), 302(3.39)	104-0465-84
$C_{10}H_9FO_5$ 1,3-Benzodioxole-5-carboxaldehyde, 7a-acetoxy-3a-fluoro-3a,7a-dihydro-, cis	MeCN	267(3.22)	44-0806-84
$C_{10}H_9F_3O$ 3-Buten-2-ol, 1,1,1-trifluoro-3-phenyl-	EtOH	234(4.01)	44-2258-84
$C_{10}H_9F_5O$ Benzene, (1-ethoxyethyl)pentafluoro-	heptane	262(2.69)	104-0198-84
$C_{10}H_9FeNO$ Ferrocene, nitroso-	hexane	254(3.80),294(3.87), 379(3.01),523(3.12)	89-0521-84
$C_{10}H_9FeNO_2$ Ferrocene, nitro-	hexane	238(3.96),272(3.80), 363(3.11),468(2.87)	89-0521-84
$C_{10}H_9NO$ Benzeneacetonitrile, α-(methoxymethyl-ene)-, (E)- (Z)- 2H-Pyrrol-2-one, 1,5-dihydro-4-phenyl- 8-Quinolinol, 2-methyl-, complex with ethylenediamine and CoCl$_2$	MeOH MeOH EtOH MeOH	209(4.16),264(4.19) 213(4.13),267(4.09) 270(4.25) 326(3.30),342(3.35), 398(3.72)	56-0935-84 56-0935-84 83-0624-84 18-0047-84
$C_{10}H_9NO_2$ 2H-1-Benzopyran-2-one, 7-amino-4-meth-yl- 8H-Cycloheptoxazol-8-one, 2-ethyl- 1H-Indole-4-carboxaldehyde, 5-methoxy- 2H-Oxocin-6-carbonitrile, 4,8-dimethyl-2-oxo-	pentane MeOH EtOH MeOH	330(4.20) 245(4.49),312(3.83) 204(4.29),220(4.14), 240(4.07),353(4.06) 293(3.76)	135-0674-84 18-0609-84 44-4761-84 138-1503-84
$C_{10}H_9NO_2S$ 4-Benzothiazolol, 7-methyl-, acetate	EtOH	292(3.46)	44-0997-84
$C_{10}H_9NO_3$ 3-Butenoic acid, 4-amino-2-oxo-4-phenyl-, (Z)- 1H-Indole-3-carboxylic acid, 1-meth-oxy-	EtOH MeOH-acid	243(3.89),342(4.23) 213(4.30),287(3.88)	94-0497-84 150-1301-84M

Compound	Solvent	$\lambda_{max}(\log \epsilon)$	Ref.
5(4H)-Isoxazolone, 3-(2-methoxyphenyl)-	EtOH	252(4.25),300(4.18)	44-4419-84
$C_{10}H_9NO_3S$ 2-Propenoic acid, 3-(2-furanyl)-2-(thiocyanatomethyl)-, methyl ester	EtOH	310(3.36)	73-1764-84
$C_{10}H_9NO_4S$ 3(2H)-Isothiazolone, 4-hydroxy-2-methyl-5-phenyl-, S,S-dioxide	EtOH	231(4.12),316(3.97)	44-2212-84
$C_{10}H_9NO_5$ Cyclodopa, N-formyl (as glucoside)	H_2O + NaOH	253(4.05),307(4.02) 255s(3.94),324(4.01)	33-1348-84 33-1348-84
4H-Furo[3,2-c]pyran-2-carboxaldehyde, 3-hydroxy-6-methyl-4-oxo-, 2-(O-methyloxime)	MeOH	307(4.24)	83-0685-84
$C_{10}H_9NSe$ Isoselenazole, 3-methyl-5-phenyl-	isooctane	276(4.16)	78-0931-84
$C_{10}H_9N_3$ 4H-Pyrazolo[1,5-a]benzimidazole, 2-methyl-	EtOH	229(4.42),306(4.01)	48-0829-84
$C_{10}H_9N_3O_2S$ Propanedinitrile, [[1-methyl-5-(methylsulfonyl)-1H-pyrrol-2-yl]methylene]-	MeOH	223(3.94),360(4.58)	4-1041-84
$C_{10}H_9N_3O_2S_2$ Carbonimidodithioic acid, cyano-, methyl (4-nitrophenyl)methyl ester	dioxan	270(3.43)	73-2285-84
4-Thiazolamine, 2-(methylthio)-5-(4-nitrophenyl)-	dioxan	274(2.98),422(3.23)	73-2285-84
$C_{10}H_9N_3O_9$ Pentanedioic acid, 2-(3,5-dinitro-4-oxo-1(4H)-pyridinyl)-, (S)-	EtOH	342(3.63)	18-1961-84
$C_{10}H_9N_3S$ Propanedinitrile, [[1-methyl-5-(methylthio)-1H-pyrrol-2-yl]methylene]-	MeOH	210(3.90),432(4.54)	4-1041-84
4H-[1,2,4]Triazolo[3,4-c][1,4]benzothiazine, 1-methyl-	EtOH	225(3.30),264(2.63), 290(3.33)	2-0445-84
$C_{10}H_9N_3S_2$ 4H-1,3,5-Thiadiazine-4-thione, 2-amino-6-(3-methylphenyl)-	EtOH	282(4.65),368(3.85)	150-0266-84S
4H-1,3,5-Thiadiazine-4-thione, 2-(methylamino)-6-phenyl-	EtOH	282(4.56),354(3.85)	150-0266-84S
$C_{10}H_9N_5O$ 3H-Imidazo[2,1-i]purine-7-carboxaldehyde, 3-ethyl-	EtOH	230(4.31),328(4.18), 339(4.18)	44-4021-84
$C_{10}H_9N_5O_2$ Pyrido[2,3-d]pyrimidine-6-carbonitrile, 7-amino-1,2,3,4-tetrahydro-1,3-dimethyl-2,4-dioxo-	H_2O	242(4.10),289(4.09), 330(4.21)	4-1543-84

Compound	Solvent	λ_{max}(log ϵ)	Ref.
1H-1,2,4-Triazol-5-amine, 1-[1-(hy-droxyimino)-2-oxo-2-phenylethyl]-	EtOH	242(3.20),330(2.80), 371(2.53),407(2.67)	4-1029-84
C$_{10}$H$_9$N$_5$O$_2$S Benzenesulfonamide, 4-(5-amino-4-cyano-1H-imidazol-1-yl)-	EtOH	240(4.94),260s(4.11)	2-0870-84
C$_{10}$H$_9$N$_5$S Pyrido[3',2':4,5]thieno[3,2-d]-1,2,3-triazin-4-amine, N,7-dimethyl-	EtOH	232(3.28),242(4.33), 283(4.34),320(3.72)	87-1639-84
C$_{10}$H$_9$OS 2-Benzothiopyrylium, 4-hydroxy-3-meth-yl-, perchlorate	MeCN	275(2.84),294(3.04), 335(2.54),365(2.38)	1-0617-84
C$_{10}$H$_9$S 1-Benzothiopyrlium, 4-methyl-, perchlorate	MeCN-HClO$_4$	256(4.29),334(3.68), 376(3.60)	44-4192-84
C$_{10}$H$_{10}$Br$_2$N$_2$O$_5$ Uridine, 2',5'-dibromo-2',5'-dideoxy-, 3'-formate	EtOH	258(3.98)	94-2544-84
C$_{10}$H$_{10}$ClNO 2-Propenamide, 3-chloro-N-(phenyl-methyl)-	EtOH	212(4.12)	142-2369-84
C$_{10}$H$_{10}$ClN$_3$O 2-Butenamide, 2-chloro-3-(phenylazo)-	MeOH	205(4.0),225(3.9), 231(4.0),237(4.0), 320(4.3),322(4.3), 339s(--),442(2.5)	48-0367-84
C$_{10}$H$_{10}$ClN$_3$S Diazene, (4-chlorophenyl)(1-isothio-cyanato-1-methylethyl)-	C$_6$H$_{12}$	278(4.13),404(2.28)	118-0315-84
C$_{10}$H$_{10}$Cl$_2$N$_2$O$_5$ Uridine, 2',5'-dichloro-2',5'-dideoxy-, 3'-formate	EtOH	258(3.96)	94-2591-84
C$_{10}$H$_{10}$FNO$_2$ Benzeneacetonitrile, 2-fluoro-4,5-di-methoxy-	EtOH	207(3.87),229(3.82), 283(3.56)	104-0465-84
C$_{10}$H$_{10}$N$_2$ 1H-Cyclopenta[3,4]pyrazolo[1,5-a]pyri-dine, 2,3-dihydro-	H$_2$O	232(4.52),303(3.72)	56-1077-84
5,8-Ethanophthalazine, 5,8-dihydro-	hexane	208(3.47),248(3.03), 325(2.67)	24-0534-84
C$_{10}$H$_{10}$N$_2$O 1,2,4-Oxadiazole, 5-methyl-3-(2-methyl-phenyl)-	EtOH	238.5(4.02),279(2.98), 286(2.72)	34-0231-84
1,2,4-Oxadiazole, 5-methyl-3-(3-methyl-phenyl)-	EtOH	242(4.06),282(2.98), 288.5(2.71)	34-0231-84
1,2,4-Oxadiazole, 5-methyl-3-(4-methyl-phenyl)-	EtOH	245(4.20),273(3.13), 285(2.85)	34-0231-84
2(1H)-Quinoxalinone, 1,3-dimethyl-	EtOH	210(4.24),230(4.35), 279(3.75),329s(3.81), 338(3.81)	44-0827-84

Compound	Solvent	$\lambda_{max}(\log \epsilon)$	Ref.
$C_{10}H_{10}N_2O_2$			
Azobenzene, 2,2'-dihydroxy-, polymer with palladium	CH_2Cl_2	450(3.95),495(3.99)	77-0999-84
1H-Indole, 4-(2-nitroethyl)-	EtOH	218(4.48),274(3.78)	44-4761-84
3,6-Pyridazinedione, tetrahydro-1-phenyl-	MeOH	230(0.977),282(0.24)	2-0439-84
2,5-Pyrrolidinedione, 1-(phenylamino)-	MeOH	235(0.97),285(0.84)	2-0439-84
$C_{10}H_{10}N_2O_2S$			
2H-1,2,4-Benzothiadiazine, 3-(1-propenyl)-, 1,1-dioxide	EtOH	<u>240s(4.2)</u>,290(3.9)	103-0735-84
$C_{10}H_{10}N_2O_3$			
Imidazo[1,2-a]pyridine-3-acetic acid, 2-hydroxy-, methyl ester	EtOH	211(4.09),250(4.00), 307(3.51),348(3.79)	33-1897-84
5H-Pyrazolo[5,1-b][1,3]oxazin-5-one, 3-acetyl-2,7-dimethyl-	EtOH	238(4.27),286(4.10)	94-0930-84
1H,7H-Pyrazolo[1,2-a]pyrazole-1,7-dione, 2-acetyl-3,5-dimethyl-	EtOH	245(4.10),370(4.16)	94-0930-84
Pyrazolo[1,2-a]pyrazole-1,5-dione, 2-acetyl-1,5-dihydro-3,7-dimethyl-	EtOH	224(4.08),336(4.15)	94-0930-84
L-Tyrosine, 3-cyano-	pH 1	229(3.83),300(3.43)	56-0157-84
	pH 13	246(3.85),330(3.65)	56-0157-84
$C_{10}H_{10}N_2O_3S$			
7-Benzothiazolepropanoic acid, α-amino-4-hydroxy-, monohydrochloride	M HCl	268(3.25),280(3.29), 318(3.21)	44-0997-84
$C_{10}H_{10}N_2O_3S_2$			
2-Furancarboxylic acid, 5-[[[(cyanoimino)(methylthio)methyl]thio]-methyl]-, methyl ester	dioxan	268(3.44)	73-2285-84
$C_{10}H_{10}N_2O_4$			
Pyrrolo[3,2-b]pyrrole-1,4-dicarboxylic acid, dimethyl ester	C_6H_{12}	250s(3.91),256(4.23), 277(3.96),282s(3.95), 294s(3.72)	88-5669-84
$C_{10}H_{10}N_2O_6$			
Glycine, N-(methoxycarbonyl)-, 4-nitrophenyl ester	EtOH	211.5s(3.98),267(4.02)	94-1373-84
$C_{10}H_{10}N_2S$			
2H-2-Pyrindine-4-carbonitrile, 3,5,6,7-tetrahydro-1-methyl-3-thioxo-	dioxan	262(3.42),316(3.97), 409(3.30)	104-2216-84
Quinoxaline, 2-methyl-3-(methylthio)-	EtOH	235(4.03),262(4.03), 345(3.84),354(3.83)	18-1653-84
2(1H)-Quinoxalinethione, 1,3-dimethyl-	EtOH	217(4.35),274(4.26), 390(3.93)	18-1653-84
$C_{10}H_{10}N_4$			
[2,2'-Bipyridine]-6,6'-diamine	pH 10.0	244(<u>3.8</u>),325(<u>4.0</u>)	18-2121-84
2,4-Pyrimidinediamine, 6-phenyl-	EtOH	239(4.39),307(3.84)	39-2677-84C
1H-Tetrazole, 5-phenyl-1-(1-propenyl)-, (E)-	EtOH	240(4.08)	39-1933-84C
3H-1,2,4-Triazolo[4,3-b]indazole, 3,3-dimethyl-	CH_2Cl_2	247(3.69),270(3.74), 321s(3.53),361(3.83)	24-1726-84
$C_{10}H_{10}N_4O$			
3-Cinnolinecarboxamide, 4-amino-7-methyl-	DMF	342(3.93)	5-1390-84

Compound	Solvent	$\lambda_{max}(\log \epsilon)$	Ref.
$C_{10}H_{10}N_4O_2$			
3-Cinnolinecarboxamide, 4-amino-7-methoxy-	DMF	283(4.17),319(3.80), 344(3.80)	5-1390-84
$C_{10}H_{10}N_4O_2S$			
Diazene, (1-isothiocyanato-1-methyl-ethyl)(4-nitrophenyl)-	C_6H_{12}	275(4.23),403(2.40)	118-0315-84
$C_{10}H_{10}N_4O_3$			
Diazene, (1-isocyanato-1-methylethyl)-(4-nitrophenyl)-	C_6H_{12}	276(4.26),388(2.42)	118-0315-84
2,4(1H,3H)-Pteridinedione, 6-acetyl-1,3-dimethyl-	MeOH	248(4.11),281(4.04), 332(4.00)	5-1798-84
2,4(1H,3H)-Pteridinedione, 7-acetyl-1,3-dimethyl-	MeOH	248(4.08),348(3.89)	5-1798-84
2,4(1H,3H)-Pteridinedione, 1-methyl-7-(1-oxopropyl)-	pH 5.0	232(4.08),346(3.93)	5-1798-84
	pH 11.0	253(4.04),300s(3.55), 357(3.90)	5-1798-84
2,4(1H,3H)-Pteridinedione, 3-methyl-7-(1-oxopropyl)-	pH 5.0	237(4.07),339(3.94)	5-1798-84
	pH 11.0	245(4.23),265s(4.10), 384(3.75)	5-1798-84
1H-Pyrrole-2,5-dione, 1-[2-(4-amino-2-oxo-1(2H)-pyrimidinyl)ethyl]-	pH 1	281(--)	47-2841-84
	H_2O	271(3.95)	47-2841-84
	pH 13	273(--)	47-2841-84
$C_{10}H_{10}N_4O_4$			
1,2,4-Triazine-5,6-dicarboxylic acid, 3-cyano-, diethyl ester	MeOH	228(3.98),290(3.64), 384(2.60)	44-1040-84
$C_{10}H_{10}N_4O_8$			
1(4H)-Pyridineacetic acid, α-(3-amino-3-oxopropyl)-3,5-dinitro-4-oxo-, (S)-	DMF	354(3.62)	18-1961-84
$C_{10}H_{10}N_6$			
1H-Pyrazole, 1,1',1"-methylidynetris-	MeOH	218(4.23)	117-0299-84
$C_{10}H_{10}O$			
3-Buten-2-one, 4-phenyl-	MeCN	291(4.34)	23-1958-84
cis	CH_2Cl_2	281(4.18)	44-4352-84
	+ $AlEtCl_2$	372(4.57)	44-4352-84
trans	CH_2Cl_2	286(4.51)	44-4352-84
	+ $AlEtCl_2$	372(4.69)	44-4352-84
	FSO_3H	370(4.76)	44-4352-84
1H-Indene, 5-methoxy-	EtOH	252(3.74)	39-0687-84C
$C_{10}H_{10}O_2$			
1-Benzoxepin-5(2H)-one, 3,4-dihydro-	octane	298(3.52)	65-0148-84
	EtOH	249(3.86),304(3.34)	65-0148-84
$C_{10}H_{10}O_3$			
3(2H)-Benzofuranone, 5-hydroxy-2,2-di-methyl-	EtOH	253(3.89),370(3.65)	18-0791-84
2,5-Cyclohexadiene-1,4-dione, 2-(2-methyl-1-oxopropyl)-	MeCN	248(4.28),300s(2.66), 450(1.48)	18-0791-84
1H-3a,7-Epidioxyazulen-1-one, 2,3,4,7-tetrahydro- (decomposes)	EtOH	210(3.53),234(3.63), 257(3.60)	35-4857-84
1,4-Ethenocyclopenta[d][1,2]dioxepin-8(1H)-one, 4,5,6,7-tetrahydro-	EtOH	214(3.91),237(3.89)	35-4857-84
$C_{10}H_{10}O_4$			
Benzeneacetic acid, 2-hydroxy-3,6-di-	EtOH	218(4.10),269(3.91),	39-2655-84C

Compound	Solvent	$\lambda_{max}(\log \epsilon)$	Ref.
methyl-α-oxo- (cont.)		345(3.45)	39-2655-84C
1,4-Dioxaspiro[4.5]deca-6,9-diene-2,8-dione, 3,3-dimethyl-	MeOH	217(4.03),282(2.84)	88-4471-84
1(3H)-Isobenzofuranone, 6-hydroxy-4-methoxy-5-methyl- (silvaticol)	EtOH	215(4.56),256(3.81), 303(3.49)	94-2622-84
2-Propenoic acid, 3-(4-hydroxy-3-methoxyphenyl)-	EtOH	295s(--),320(4.25)	102-1029-84
$C_{10}H_{11}BrN_2$			
1H-Indole-3-ethanamine, 6-bromo-	EtOH	229(4.51),289(3.81), 294(3.76)	1-0709-84
	EtOH-HCl	228(4.51),287(3.81), 294(3.76)	1-0709-84
$C_{10}H_{11}BrN_2O_5$			
1H-Pyrrole-3,4-dicarboxylic acid, 5-(2-bromo-1-oxopropyl)-, dimethyl ester	MeOH	235(4.41),282(4.19)	103-0416-84
$C_{10}H_{11}BrN_4O_4Se$			
4H-Pyrazolo[3,4-d]pyrimidine-4-selone, 3-bromo-1,5-dihydro-1-β-D-ribofuranosyl-	pH 1	243s(3.43),352(4.01)	87-1026-84
	pH 7	234s(3.76),347(4.01)	87-1026-84
	pH 11	232s(3.95),345(4.13)	87-1026-84
$C_{10}H_{11}BrN_4O_5$			
4H-Pyrazolo[3,4-d]pyrimidin-4-one, 3-bromo-1,5-dihydro-1-β-D-ribofuranosyl-	pH 1 and 7	215(4.36),230s(3.97), 252(3.89)	87-1119-84
	pH 11	225(3.97),276(4.07)	87-1119-84
$C_{10}H_{11}BrO$			
1-Benzoxepin, 7-bromo-2,3,4,5-tetrahydro-	n.s.g.	225(4.06),266(3.44), 305s(--)	103-0838-84
$C_{10}H_{11}ClN_2O_3S$			
2-Butenamide, N-[2-(aminosulfonyl)-4-chlorophenyl]-	EtOH	$\underline{220(4.4),265(4.2)}$, $\underline{305s(3.9)}$	103-0735-84
$C_{10}H_{11}ClO$			
Ethanone, 2-chloro-1-(4-ethylphenyl)-	20% H_2SO_4	264(4.20)	48-0353-84
	96% H_2SO_4	333(4.34)	48-0353-84
$C_{10}H_{11}ClSi$			
Silane, chlorodimethyl)phenylethynyl-	hexane	238.5(4.19),248.5(4.39), 260.4(4.32)	65-2317-84
$C_{10}H_{11}Cl_3O_2$			
5-Hexen-3-ynoic acid, 5,6,6-trichloro-2,2-dimethyl-, ethyl ester	MeOH	245(4.13),251(4.14)	35-3551-84
$C_{10}H_{11}FO_4$			
Benzeneacetic acid, 2-fluoro-4,5-dimethoxy-	EtOH	207(3.88),226(3.73), 283(3.47)	104-0465-84
Benzeneacetic acid, 5-fluoro-2,3-dimethoxy-	EtOH	208(4.10),275(3.34)	104-0465-84
$C_{10}H_{11}N$			
Benzenamine, N-(1-methyl-2-propynyl)-	EtOH	200(3.60),243(4.12), 290(3.28)	65-0161-84
Isoquinoline, 3,4-dihydro-1-methyl-	hexane	203(4.47),207(4.46), 212s(4.29),239s(3.83), 246(3.88),281(3.04), 286s(3.00)	32-0041-84

Compound	Solvent	$\lambda_{max}(\log \epsilon)$	Ref.
Isoquinoline, 3,4-dihydro-1-methyl- (cont.)	EtOH	205(4.35),209(4.35), 214s(4.18),249(3.91), 259s(3.76),284s(3.12), 294s(2.98)	32-0041-84
Isoquinoline, 3,4-dihydro-6-methyl-	hexane	208(4.39),213(4.40), 218s(4.22),252(4.03), 258(4.02),264s(3.90), 283s(3.08),291s(2.93)	32-0041-84
	EtOH	211(4.42),215(4.43), 221s(4.32),262(4.06), 267s(3.88),286s(3.66), 298s(3.38)	32-0041-84
Isoquinoline, 3,4-dihydro-8-methyl-	hexane	210s(4.35),213(4.38), 218s(4.29),246s(3.80), 252(3.88),257(3.86), 264s(3.70),283(2.98), 294s(2.91)	32-0041-84
	EtOH	212s(4.30),215s(4.34), 221s(4.24),248s(3.78), 260(3.93),268s(3.83)	32-0041-84
$C_{10}H_{11}NO$ Cyclopentanone, 2-(2-pyridinyl)-	H_2O	262(3.50)	56-1077-84
$C_{10}H_{11}NOS$ Thiopyrano[2,3-c]pyrrole, 4-methoxy-5,7-dimethyl-, perchlorate	EtOH	380(3.00),510(3.28)	12-1473-84
$C_{10}H_{11}NO_2$ 2-Azabicyclo[4.2.0]octa-1,3,5-triene-4-carboxylic acid, 3-methyl-, methyl ester	EtOH	217s(3.67),222s(3.71), 228(3.72),283(3.81)	35-2672-84
Benzene, (1-methyl-2-nitro-1-propen-yl)-, (Z)-	MeOH	215(3.93),269(3.55)	12-1231-84
5H-Cyclohepta[c]pyridin-5-one, 6,7,8,9-tetrahydro-9-hydroxy-	EtOH	239s(--),251(3.46), 257(3.43),303s(--)	39-2297-84C
1H-[1,3]Oxazino[3,4-a]azepin-1-one, 3,4-dihydro-7-methyl-	MeOH	330(3.06)	119-0171-84
$C_{10}H_{11}NO_3$ Acetamide, N-acetoxy-N-phenyl-	MeOH	272(3.17)	2-0060-84
2,4-Furandicarboxaldehyde, 5-pyrroli-dino-	MeOH	301(3.30),358(3.48)	73-1600-84
Glycine, N-[(2-methoxyphenyl)methyl-ene]-	pH 9.8	252(4.26),307(3.87)	18-2508-84
$C_{10}H_{11}NO_4$ 2,4-Furandicarboxaldehyde, 5-morpho-lino-	MeOH	348(3.35)	73-1600-84
Pyrano[4,3-b]pyrrole-2,3,4(1H)-trione, tetrahydro-1-(2-propenyl)-	MeOH	248(4.34),308(3.67)	78-4513-84
$C_{10}H_{11}NO_4S$ Thieno[2,3-c]pyridin-3(2H)-one, 5-eth-yl-4-hydroxy-7-methyl-, S,S-dioxide	EtOH	225(4.15),348(3.68)	104-2032-84
$C_{10}H_{11}NO_5$ 5-Oxa-1-azabicyclo[4.2.0]oct-2-ene-2-carboxylic acid, 4-(2-methoxyethyl-idene)-8-oxo-, lithium salt, [R-(Z)]-	H_2O	295(4.21)	39-1599-84C

Compound	Solvent	$\lambda_{max}(\log \epsilon)$	Ref.
C$_{10}$H$_{11}$NS			
Benzothiazole, 2-(1-methylethyl)-	EtOH	217(4.34),250(3.91), 285(3.40)	78-2141-84
Thiopyrano[2,3-c]pyrrole, 4,5,7-tri- methyl-, perchlorate	EtOH	250(4.18),308(3.63), 572(3.28)	12-1473-84
C$_{10}$H$_{11}$N$_3$OS			
1,3,4-Thiadiazol-2(3H)-one, 3-(methyl- amino)methyl]-5-phenyl-	EtOH	280(4.15)	83-0547-84
C$_{10}$H$_{11}$N$_3$O$_2$			
Acetic acid, cyano(1-methyl-2(1H)- pyrimidinylidene)-, ethyl ester	EtOH	308(4.29),385(3.70)	103-1270-84
Acetic acid, cyano(1-methyl-4(1H)- pyrimidinylidene)-, ethyl ester	CF$_3$COOH	313(4.38),352(4.52)	103-1270-94
Pyrido[2,3-d]pyrimidine-2,4(1H,3H)- dione, 1,3,7-trimethyl-	H$_2$O	248s(3.81),308(3.94)	4-1543-84
Pyrido[3,4-d]pyrimidin-4(3H)-one, 3,6,8-trimethyl-, 7-oxide	EtOH	266(4.3),305s(3.8), 335(3.2),350(3.3)	95-1122-84
2-Pyrimidineacetic acid, α-cyano-α- methyl-, ethyl ester	EtOH	245(3.43),290(2.04)	103-1270-84
C$_{10}$H$_{11}$N$_3$O$_2$S			
2-Thiophenecarbonitrile, 4-nitro- 5-piperidino-	benzene	386(3.77)	39-0781-84B
C$_{10}$H$_{11}$N$_3$O$_3$S			
Pyrido[2,3-d]pyrimidine-2,4(1H,3H)- dione, 5-hydroxy-1,3-dimethyl- 7-(methylthio)-	EtOH	232(4.27),250(4.39), 283(4.04),318(4.25)	94-0122-84
2H-1,2,4,6-Thiatriazinium, 5,6-dihydro- 2,4-dimethyl-5-oxo-3-phenyl-, hydrox- ide, inner salt, 1,1-dioxide	MeCN	244(4.075)	142-0063-84
C$_{10}$H$_{11}$N$_3$O$_4$			
1H-Pyrrolo[3,2-d]pyrimidine-7-carbox- ylic acid, 2,3,4,5-tetrahydro-3- methyl-2,4-dioxo-, ethyl ester	EtOH	229.5(4.48),270(4.00)	44-0908-84
C$_{10}$H$_{11}$N$_3$O$_5$			
Glycine, N-[1-oxo-3-(1,2,3,4-tetra- hydro-6-methyl-2,4-dioxo-5-pyrim- idinyl)-2-propenyl]-, (E)-	H$_2$O	301(4.29)	94-4625-84
4-Pyridazinecarbonitrile, 2,3-dihydro- 3-oxo-2-β-D-ribofuranosyl-	MeOH	329(3.67)	87-1613-84
Pyridazinium, 4-cyano-3-hydroxy-1-β-D- ribofuranosyl-, hydroxide, inner salt	MeOH MeOH	223(4.28),363(3.65) 222(4.34),363(3.66)	77-0422-84 87-1613-84
4-Pyrimidinecarbonitrile, 1,2-dihydro- 2-oxo-1-β-D-ribofuranosyl-	MeOH	329(3.67)	77-0422-84
C$_{10}$H$_{11}$N$_3$O$_6$			
2,2'-Anhydrouridine-3'-carbamate	pH 2 H$_2$O pH 12	222(--),246.7(--) 221(3.82),246.6(3.72) 220(--),250(--)	44-2121-84 44-2121-84 44-2121-84
C$_{10}$H$_{11}$N$_3$O$_6$S$_2$			
L-Homocysteine, S-(7-sulfo-4-benzo- furazanyl)-, disodium salt	H$_2$O	380(3.84)	96-1003-84
C$_{10}$H$_{11}$N$_3$O$_7$			
1(4H)-Pyridineacetic acid, α-(1-methyl-	EtOH	342(3.67)	18-1961-84

Compound	Solvent	$\lambda_{max}(\log \epsilon)$	Ref.
ethyl)-3,5-dinitro-4-oxo-, (S)-			18-1961-84
$C_{10}H_{11}N_3O_7S$ 1(4H)-Pyridineacetic acid, α-[2-(methylthio)ethyl]-3,5-dinitro-4-oxo-, (S)-	EtOH	339(3.62)	18-1961-84
$C_{10}H_{11}N_3S_2$ 1,3,4-Thiadiazole-2(3H)-thione, 3-[(methylamino)methyl]-5-phenyl-	EtOH	340(4.20)	83-0547-84
$C_{10}H_{11}N_5O$ 3,5-Isoxazolediamine, 4-[(4-methylphenyl)azo]-	EtOH	241(4.13),287s(4.06), 363(4.26),394s(4.05)	97-0256-84
1H-Pyrazole-4,5-dione, 3-amino-, 4-[(4-methylphenyl)hydrazone]	C_6H_{12}	226(3.68),248(3.71), 385(4.23)	22-0164-84
	EtOH	227(3.66),249(3.68), 336(4.23)	22-0164-84
	ether	228s(3.65),248(3.71), 382(4.21)	22-0164-84
	DMF	394(4.21)	22-0164-84
	CHCl$_3$	391(4.22)	22-0164-84
$C_{10}H_{11}N_5O_2$ 3,5-Isoxazolediamine, 4-[(4-methoxyphenyl)azo]-	EtOH	243(4.24),366(4.36), 385s(4.27)	97-0256-84
1H-Pyrazole-4,5-dione, 3-amino-, 4-[(4-methoxyphenyl)hydrazone]	C_6H_{12}	232(3.74),254(3.63), 400(4.20)	22-0164-84
	EtOH	231(3.74),254(3.63), 400(4.22)	22-0164-84
	ether	231(3.74),250s(3.64), 392(4.19)	22-0164-84
	DMF	406(4.21)	22-0164-84
	CHCl$_3$	405(4.12)	22-0164-84
$C_{10}H_{11}N_5O_2S$ 4(1H)-Pteridinone, 2-amino-7-(methylthio)-6-(1-oxopropyl)-	pH 0.0	252(4.39),311(4.13), 365(4.18)	138-1025-84
	pH 4.0	273(4.46),310(4.18), 378(4.26)	138-1025-84
	pH 10.0	266(4.52),300s(4.11), 384(4.33)	138-1025-84
$C_{10}H_{11}N_5O_3$ Pyrido[2,3-d]pyrimidine-6-carboxamide, 7-amino-1,2,3,4-tetrahydro-1,3-dimethyl-2,4-dioxo-	H_2O	232(3.16),284(3.76), 332(3.85)	4-1543-84
$C_{10}H_{11}N_5O_3S$ 1H-Imidazole-4-carboxamide, 5-amino-1-[4-(aminosulfonyl)phenyl]-	H_2O	240(4.3),265s(4.18)	2-0870-84
$C_{10}H_{11}N_5O_5$ 1H-Pyrazole-3,4-dicarboxylic acid, 5-(2-azido-1-oxopropyl)-, dimethyl ester	MeOH	232(4.58)	103-0416-84
1H-1,2,3-Triazole-4,5-dicarboxylic acid, 1-(3-diazo-1-methyl-2-oxopropyl)-, dimethyl ester	MeOH	252(4.16),272(4.11)	103-0416-84
$C_{10}H_{11}N_5O_7$ 1H-Pyrazolo[3,4-d]pyrimidine-3,4-dione,	pH 1 and 7	250(3.91)	87-1119-84

Compound	Solvent	$\lambda_{max}(\log \epsilon)$	Ref.
2,5-dihydro-2-nitroso-1-β-D-ribo-furanosyl- (cont.)	pH 11	226s(3.91),264(3.86)	87-1119-84
$C_{10}H_{12}BrN_3O_2$			
Methanol, [3-(4-bromophenyl)-1-methyl-2-triazenyl]-, acetate	MeCN	219(4.06),280(4.38), 310s(4.16)	23-0741-84
$C_{10}H_{12}BrN_5O_4$			
1H-Pyrazolo[3,4-d]pyrimidin-4-amine, 3-bromo-1-β-D-ribofuranosyl-	pH 1	226(4.26),255(3.79)	87-1119-84
	pH 7 and 11	230(3.79),260(3.74), 280(3.87),290s(3.70)	87-1119-84
$C_{10}H_{12}Br_2$			
Cyclohexene, 1-cyclopropyl-6-(dibromo-methylene)-	hexane	251(4.04)	44-0495-84
Cyclopropane, 1,1-dibromo-2-(dicyclo-propylmethylene)-	hexane	202(4.06)	44-0495-84
$C_{10}H_{12}Br_2O_2$			
Benzene, 1-bromo-2-(2-bromoethyl)-4,5-dimethoxy-	n.s.g.	207(4.61),235(4.02), 285(3.58)	83-0488-84
$C_{10}H_{12}ClNO_2$			
3-Pyridinecarboxylic acid, 5-(chloro-methyl)-2,6-dimethyl-, methyl ester	EtOH	228(4.09),274(3.80)	35-2672-84
$C_{10}H_{12}ClNS$			
Ethanethioamide, 2-chloro-N-(2,6-di-methylphenyl)-	EtOH	276(4.31)	78-2047-84
$C_{10}H_{12}ClN_3O_5S$			
4-Thiazoleacetic acid, 2-[(chloroacet-yl)amino]-α-(methoxyimino)-, ethyl ester, 3-oxide, (Z)-	EtOH	263(4.2)	4-1097-84
	EtOH-HCl	237(4.1),280(4.2)	4-1097-84
	EtOH-KOH	280(4.1),316(4.1)	4-1097-84
$C_{10}H_{12}Cl_2O_4$			
Propanedioic acid, (3,3-dichloro-2-propenylidene)-, diethyl ester	EtOH	280(4.25)	70-2328-84
$C_{10}H_{12}FNO_4$			
Benzeneacetamide, 2-fluoro-N-hydroxy-4,5-dimethoxy-	EtOH	207(4.01),226(3.86), 284(3.60)	104-0465-84
Benzeneacetamide, 5-fluoro-N-hydroxy-2,3-dimethoxy-	EtOH	206(4.16),279(3.14)	104-0465-84
$C_{10}H_{12}N_2$			
5,8-Ethanophthalazine, 5,6,7,8-tetra-hydro-	hexane	214(3.35),249(3.07), 329(2.72)	24-0534-84
1H-Indol-3-ethanamine	EtOH	228(4.53),276(3.71), 284(3.73),292(3.65)	1-0709-84
	EtOH-HCl	228(4.53),276(3.71), 282(3.71),292(3.65)	1-0709-84
$C_{10}H_{12}N_2O_3$			
Butanedioic acid, mono(2-phenylhydra-zide)	MeOH	235(0.17),285(0.55)	2-0439-84
2,4-Furandicarboxaldehyde, 5-pipera-zino-	MeOH	343(3.46)	73-1600-84
2,4(1H,3H)-Pyrimidinedione, 5-(2-oxo-cyclohexyl)-	MeOH	264(3.89)	44-2738-84

Compound	Solvent	λ_{max} (log ϵ)	Ref.
$C_{10}H_{12}N_2O_3S$			
2-Butenamide, N-[2-(aminosulfonyl)-phenyl]-	EtOH	<u>215(4.4)</u>,260(4.1), <u>290(4.0)</u>	103-0735-84
$C_{10}H_{12}N_2O_4S$			
Hydroxylamine-O-sulfonic acid, N-[2-(2-pyridinyl)cyclopentylidene]-	H_2O anion	264(3.77) 256(3.43),262(3.47), 263(3.40)	56-1077-84 56-1077-84
$C_{10}H_{12}N_2O_5$			
2,4(1H,3H)-Pyrimidinedione, 1-[4,5-di-hydroxy-3-(hydroxymethyl)-2-cyclo-penten-1-yl]-, [1R-(1α,4β,5β)]-	H_2O	266(4.07)	87-1536-84
$C_{10}H_{12}N_2O_6$			
1',2-Anhydro-1-β-D-psicofuranosyl-uracil	pH 7.3	194(4.46),262(3.78)	1-0367-84
$C_{10}H_{12}N_4O$			
Benzonitrile, 4-[3-(methoxymethyl)-3-methyl-1-triazenyl]-	MeCN	224(3.99),300(4.28), 310s(4.27)	23-0741-84
2H-Tetrazole-2-ethanol, α-methyl-5-phenyl-	EtOH	230(3.98)	39-1933-84C
$C_{10}H_{12}N_4O_2S$			
Pyrido[2,3-d]pyrimidine-2,4(1H,3H)-dione, 5-amino-1,3-dimethyl-7-(methylthio)-	EtOH	250(4.47),295(4.27)	94-0122-84
$C_{10}H_{12}N_4O_3$			
2,4(1H,3H)-Pteridinedione, 6-(1-hy-droxypropyl)-1-methyl-	pH 5.0	234(4.14),248s(4.08), 335(3.88)	88-1031-84
	pH 13	245(4.29),285(3.44), 343(3.92)	88-1031-84
	MeOH	237(4.12),248s(4.08), 334(3.85)	88-1031-84
2,4(1H,3H)-Pteridinedione, 6-(1-hy-droxypropyl)-3-methyl-	pH 5.0	234(4.22),330(3.92)	88-1031-84
	pH 13	247(4.28),273(4.12), 366(3.82)	88-1031-84
	MeOH	235(4.21),331(3.86)	88-1031-84
9H-Purine, 9-(2-deoxy-β-D-erythro-pentofuranosyl)-	H_2O	264(3.87)	118-0401-84
$C_{10}H_{12}N_4O_4$			
Methanol, [1-methyl-3-(4-nitrophenyl)-2-triazenyl]-, acetate	MeCN	230(3.88),344(4.24)	23-0741-84
9H-Purine, 9-β-D-arabinofuranosyl-	H_2O	262(3.86)	118-0401-84
9H-Purine, 9-β-D-ribofuranosyl-	H_2O	262(3.83)	118-0401-84
$C_{10}H_{12}N_4O_4Se$			
4H-Pyrazolo[3,4-d]pyrimidine-4-selone, 1,5-dihydro-1-β-D-ribofuranosyl-	pH 1	347(4.33)	87-1026-84
	pH 7	233s(4.18),341(4.22)	87-1026-84
	pH 11	233s(4.18),335(4.20)	87-1026-84
$C_{10}H_{12}N_4O_5$			
Imidazo[4,5-d][1,3]oxazin-7(3H)-one, 5-amino-3-(2-deoxy-β-D-erythro-pentofuranosyl)-	MeOH	247(4.07),286(3.91)	158-0941-84
4H-Imidazo[4,5-d]pyridazin-4-one, 1,5-dihydro-1-β-D-ribofuranosyl-	pH 7	212(4.24),245(3.68), 253(3.68),269(3.67)	4-0481-84
	pH 12	253(3.69),282(3.72)	4-0481-84

Compound	Solvent	$\lambda_{max}(\log \epsilon)$	Ref.
$C_{10}H_{12}N_6$			
Methanimidamide, N'-(3,4-dicyano-1-eth- yl-1H-pyrazol-5-yl)-N,N-dimethyl-	EtOH	276(4.13)	103-0215-84
9H-Purine-9-propanenitrile, 1,6-di- hydro-1-methyl-6-(methylimino)-	EtOH	262(4.25)	47-3943-84
1H-Pyrazole-3,5-diamine, 4-[(4-methyl- phenyl)azo]-	EtOH	228(4.20),248s(3.88), 362(4.37),391s(4.20)	97-0256-84
$C_{10}H_{12}N_6OS$			
1-Propanone, 1-[2,4-diamino-7-(methyl- thio)-6-pteridinyl]-	MeOH	224(3.83),269(4.51), 306s(4.04),392(4.27)	138-1025-84
$C_{10}H_{12}O$			
Benzene, 1-methoxy-4-(1-propenyl)-, (E)-	n.s.g.	258(4.38)	154-1435-84
Benzeneethanol, 4-methyl-β-methylene-	EtOH	246(4.07)	44-2258-84
Spiro[3.6]deca-5,7-dien-1-one	MeCN	246s(3.88),255(3.91), 263s(3.81),271s(3.38), 310(2.74)	33-0774-84
Spiro[4.4]nona-2,6-dien-1-one, 6-meth- yl-	EtOH	217(4.02)	33-0073-84
Tricyclo[4.4.0.0^{1,5}]dec-7-en-4-one	MeCN	227(3.81),279(2.26)	33-0774-84
$C_{10}H_{12}OS$			
1-Propanone, 2-(methylthio)-1-phenyl-	EtOH	243(4.07),266(3.57)	78-2035-84
$C_{10}H_{12}O_2$			
Benzeneethanol, 4-methoxy-β-methylene-	EtOH	254(4.11)	44-2258-84
2,5-Cyclohexadiene-1,4-dione, 2,3,5,6- tetramethyl-	$CHCl_3$	262(4.40),268(4.39)	12-1489-84
1-Propanone, 1-(4-methoxyphenyl)-	$MeOCH_2CH_2OMe$	216(4.02),264(4.23)	151-0085-84D
$C_{10}H_{12}O_3$			
3-Cycloheptene-1-carboxaldehyde, 6-(hydroxymethylene)-4-methyl-5-oxo-, (Z)-	EtOH	260(3.57),312(3.76)	33-1506-84
2,4-Cyclohexadiene-1-carboxylic acid, 1,3-dimethyl-6-oxo-, methyl ester	MeOH	312(3.57)	44-4429-84
2,4-Cyclohexadiene-1-carboxylic acid, 1,4-dimethyl-6-oxo-, methyl ester	MeOH	308(3.29)	44-4429-84
2,4-Cyclohexadiene-1-carboxylic acid, 1,5-dimethyl-6-oxo-, methyl ester	MeOH	306(3.23)	44-4429-84
2,4-Cyclohexadiene-1-carboxylic acid, 1-ethyl-6-oxo-, methyl ester	MeOH	305(3.63)	44-4429-84
2-Cyclohexen-1-one, 2-(3-butenoyloxy)-	EtOH	230(3.87)	94-0538-84
2,5-Furandione, dihydro-3,4-bis(1-meth- ylethylidene)-	C_6H_{12}	276(4.02)	48-0233-84
1-Oxaspiro[2.5]octa-5,7-dien-4-one, 5-(hydroxymethyl)-6,8-dimethyl-	EtOH	333(3.80)	12-2525-84
$C_{10}H_{12}O_4$			
2,4-Cyclohexadiene-1-carboxylic acid, 2-methoxy-1-methyl-6-oxo-, methyl ester	MeOH	325(3.85)	44-4429-84
2,5-Cyclohexadien-1-one, 4-(2-acetoxy- ethyl)-4-hydroxy- (hallerone)	MeOH	231(3.92)	102-2617-84
2-Propenoic acid, 2-(ethoxymethyl)- 3-(2-furanyl)-	EtOH	307(3.36)	73-1764-84
4H,5H-Pyrano[4,3-b]pyran-4,5-dione, 2,3,7,8-tetrahydro-2,7-dimethyl-	EtOH	259(4.00)	4-0013-84
4H,5H-Pyrano[4,3-b]pyran-4,5-dione, 2,3,7,8-tetrahydro-2,8-dimethyl-	EtOH	260(4.05)	4-0013-84

Compound	Solvent	$\lambda_{max}(\log \epsilon)$	Ref.
4H,5H-Pyrano[4,3-b]pyran-4,5-dione, 2,3,7,8-tetrahydro-3,7-dimethyl-	EtOH	259(4.00)	4-0013-84
4H,5H-Pyrano[4,3-b]pyran-4,5-dione, 2,3,7,8-tetrahydro-3,8-dimethyl-	EtOH	258(4.00)	4-0013-84
$C_{10}H_{12}O_5$ Cyclopenta[c]pyran-4-carboxylic acid, 1,4a,5,7a-tetrahydro-1-hydroxy-7-(hydroxymethyl)- [1R-(1α,4aα,7aα)]-	EtOH	236(3.85)	102-0533-84
$C_{10}H_{12}S_2$ 1,3-Dithiepane, 2-(2,4-cyclopentadien-1-ylidene)-	EtOH	288(3.53),356(4.23)	33-1854-84
$C_{10}H_{12}S_4$ 1,3-Dithiolium, 2-(4,5-dimethyl-1,3-dithiol-2-ylidene)-4,5-dimethyl-, SOCl salt	MeCN	275(3.00),343(3.59), 462(4.10),528(3.52), 660(3.82)	70-0630-84
$C_{10}H_{13}BrN_2O_5$ 2,4(1H,3H)-Pyrimidinedione, 5-(2-bromoethenyl)-1-[[2-hydroxy-1-(hydroxymethyl)ethoxy]methyl]-	pH 1 H_2O pH 13	290(3.98) 289(3.95) 282(3.88)	23-0016-84 23-0016-84 23-0016-84
$C_{10}H_{13}Br_3$ Cyclopropane, 1,1-dibromo-2-(4-bromo-1-cyclopropyl-1-butenyl)-, (E)- (Z)-	hexane hexane	207(4.08) 205(4.04)	44-0495-84 44-0495-84
$C_{10}H_{13}Cl$ Benzene, 1-chloro-4-(1,1-dimethylethyl)-	C_6H_{12}	275(2.58)	77-0733-84
$C_{10}H_{13}ClN_2O_3S$ Benzenesulfonamide, 4-chloro-N-[(propylamino)carbonyl]-	MeOH	231.42(4.34)	83-0906-84
$C_{10}H_{13}ClN_2O_6$ Pyridazinium, 5-chloro-2,3-dihydro-4-methoxy-3-oxo-1-β-D-ribofuranosyl)-, hydroxide, inner salt	MeOH	228(4.41),320(3.72)	87-1613-84
$C_{10}H_{13}ClN_4O_4$ 1,3-Propanediol, 2-[(6-chloro-9H-purin-9-yl)methoxy]-2-(hydroxymethyl)-	pH 1 pH 7 pH 14	263(3.87) 264(3.92) 264(3.84)	23-1622-84 23-1622-84 23-1622-84
$C_{10}H_{13}ClO_2$ Benzene, 1-chloro-2-ethyl-4,5-dimethoxy-	n.s.g.	208(4.23),232(3.89), 284(3.48)	83-0488-84
$C_{10}H_{13}Cl_3O_3$ 5-Hexenoic acid, 5,6,6-trichloro-2,2-dimethyl-4-oxo-, ethyl ester	MeOH	260(3.74)	35-3551-84
$C_{10}H_{13}FN_2O_4$ 2,4(1H,3H)-Pyrimidinedione, 5-fluoro-1-[2-hydroxy-3-(2-propenyloxy)propyl]-	pH 7 pH 12	276(3.90) 274(4.08)	161-0346-84 161-0346-84

Compound	Solvent	$\lambda_{max}(\log \epsilon)$	Ref.
$C_{10}H_{13}F_3N_2O_4$			
Alanine, 2,3-didehydro-N-[N-(trifluoro-acetyl)-L-valyl]-	EtOH	207(4.01),242(3.70)	5-0920-84
$C_{10}H_{13}IO_2$			
Benzene, 1-(3-iodopropoxy)-2-methoxy-	MeOH	224.5(3.86),274s(3.43)	36-1241-84
$C_{10}H_{13}NO$			
1-Cyclononene-1-carbonitrile, 8-oxo- (end absorption)	EtOH	205(4.00),287(1.65)	44-2925-84
2-Cyclononene-1-carbonitrile, 8-oxo-	EtOH	205(4.15),276(1.64)	44-2925-84
5(1H)-Indolizinone, 2,3-dihydro-2,2-dimethyl-	MeOH	202(3.80),230(3.82), 302(3.80)	83-0143-84
1-Naphthalenol, 2-amino-1,2,3,4-tetra-hydro-, (+)-	n.s.g.	195(4.70),211(3.95), 216s(3.87),260s(3.43), 265(2.54),272.5(2.54)	39-1655-84C
4(1H)-Pyridinone, 1-(4-pentenyl)-	EtOH	265(4.25)	118-0485-84
$C_{10}H_{13}NO_2$			
4H-Cyclohept[d]isoxazole, 5,6-dihydro-3-methoxy-8-methyl-	MeOH	263(3.89)	87-0585-84
2,4-Hexadienoic acid, 2-cyano-5-meth-yl-, ethyl ester	EtOH	303(4.26)	70-2328-84
1H-Inden-1-ol, 2-amino-2,3-dihydro-5-methoxy-	n.s.g.	198(4.64),227(3.92), 277s(3.39),280(3.40), 286(3.36)	39-1655-84C
Methanamine, N-[[2-(methoxymethoxy)-phenyl]methylene]-	EtOH	248(4.10),300(3.61)	150-0701-84M
Phenol, 2-[[(2-methoxyethyl)imino]-methyl]-	M HCl	275(4.34),348(3.72)	18-2508-84
	pH 8.9	275(4.11),393(3.80)	18-2508-84
1H-Pyrrole-2-carboxaldehyde, 4-acetyl-1-ethyl-5-methyl-	EtOH	232(4.29),292(4.28)	136-0153-84I
$C_{10}H_{13}NO_3$			
Acetamide, N-[2-(3,4-dihydroxyphenyl)-ethyl]-	pH 1	279(3.39)	3-1935-84
	borate	287(3.61)	3-1935-84
2,4-Furandicarboxaldehyde, 5-(diethyl-amino)-	MeOH	309(3.21),351(3.46)	73-1600-84
3-Pyridinecarboxylic acid, 5-(hydroxy-methyl)-2,6-dimethyl-, methyl ester	EtOH	228(3.98),274(3.72)	35-2672-84
1H-Pyrrole-2-carboxylic acid, 2,3-di-hydro-3-oxo-2-(2-propenyl)-, ethyl ester	MeCN	309(3.95)	33-1535-84
1H-Pyrrole-2-carboxylic acid, 3-hy-droxy-1-(2-propenyl)-, ethyl ester	MeCN	263(4.18)	33-1535-84
1H-Pyrrole-2-carboxylic acid, 3-hy-droxy-4-(2-propenyl)-, ethyl ester	MeCN	265(4.13)	33-1535-84
$C_{10}H_{13}NO_4S$			
4-Thia-1-azabicyclo[3.2.0]heptane-2-carboxylic acid, 6-(methoxymethyl-ene)-3,3-dimethyl-7-oxo-, (E)-	n.s.g.	252(4.01)	69-5839-84
(Z)-	n.s.g.	245(4.09)	69-5839-84
$C_{10}H_{13}NO_6$			
2(1H)-Pyridinone, 1-β-D-arabino-furanosyl-3-hydroxy-, hydrochloride	MeOH	240(3.50),301(3.80)	87-0160-84
2(1H)-Pyridinone, 3-hydroxy-1-β-D-ribofuranosyl-	MeOH	238(3.48),302(3.77)	87-0160-84

Compound	Solvent	$\lambda_{max}(\log \epsilon)$	Ref.
$C_{10}H_{13}N_3O$			
6H-Pyrrolo[3,4-d]pyridazin-1-one, 1,2-dihydro-2,5,6,7-tetramethyl-	pH 7	211(4.51),234.5(4.37), 285.5(3.98),319(3.71)	78-3979-84
cation	acid	226(4.38),253(4.23), 283(3.78),356(3.36)	78-3979-84
dication	H_2SO_4	207.5(4.58),234(4.47), 246(4.35),311(4.17), 381(3.37)	78-3979-84
$C_{10}H_{13}N_3O_2$			
5,6-Diazatricyclo[5.2.0.0^{2,5}]non-8-ene-6-carboxamide, 3,9-dimethyl-4-oxo-, (1α,2α,3α,7α)-	MeOH	241(2.86)	23-2440-84
10-Oxa-2,3-diazabicyclo[4.3.1]deca-1,4,7-triene-3-carboxamide, 6,9-dimethyl-, endo	MeOH	243(3.94)	23-2440-84
1H-Pyrrolo[3,4-d]pyridazine-1,4(6H)-dione, 2,3-dihydro-2,5,6,7-tetramethyl-	pH 7	214(4.64),278(4.13)	78-3979-84
	cation	206(4.62),220(4.41), 283.5(4.01),310s(3.59)	78-3979-84
	anion	216(4.35),272(4.01), 300s(3.86)	78-3979-84
$C_{10}H_{13}N_3O_2S$			
4H-Pyrrolo[2,3-d]pyrimidin-4-one, 3,7-dihydro-7-(methoxymethyl)-3-methyl-2-(methylthio)-	MeOH	270(4.03),292(4.06)	87-0981-84
$C_{10}H_{13}N_3O_3S$			
2-Thiophenecarboxamide, 4-nitro-5-piperidino-	benzene	390(3.77)	39-0781-84B
$C_{10}H_{13}N_3O_4$			
2-Propenamide, N-(2-hydroxyethyl)-3-(1,2,3,4-tetrahydro-6-methyl-2,4-dioxo-5-pyrimidinyl)-, (E)-	H_2O	301(4.30)	94-4625-84
2(1H)-Pyrimidinone, 4-amino-1-[4,5-di-hydroxy-3-(hydroxymethyl)-2-cyclo-penten-1-yl]-	H_2O	274(3.91)	87-1536-84
$C_{10}H_{13}N_3O_6$			
Pyridazinium, 4-(aminocarbonyl)-2,3-dihydro-3-oxo-1-β-D-ribofuranosyl-, hydroxide, inner salt	MeOH	217(4.27),351(4.61)	87-1613-84
$C_{10}H_{13}N_5$			
Cyanamide, (8,9-dihydro-2,3-dimethyl-7H-imidazo[1,2-c][1,3]diazepin-5-yl)-	MeOH	249(4.12)	4-0753-84
7H-[1,4]Diazocino[1,2,3-gh]purine, 8,9,10,11-tetrahydro-11-methyl-	pH 1	290(4.25)	4-0333-84
	pH 7	285(4.18)	4-0333-84
	pH 13	285(4.17)	4-0333-84
$C_{10}H_{13}N_5O_2S$			
4(1H)-Pteridinone, 2-amino-6-(1-hy-droxypropyl)-7-(methylthio)-	pH 0.0	230(4.39),266s(3.76), 284(3.80),355(4.29)	138-1025-84
	pH 5.0	230s(4.23),243(4.33), 280(4.15),363(4.19)	138-1025-84
	pH 12.0	237(4.42),258(4.18), 369(4.18)	138-1025-84
$C_{10}H_{13}N_5O_3$			
Adenosine, 2'-deoxy-, α-	H_2O	262(4.15)	94-1441-84

Compound	Solvent	$\lambda_{max}(\log \epsilon)$	Ref.
$C_{10}H_{13}N_5O_3$			
9H-Purine-9-butanoic acid, 6-amino-β-hydroxy-γ-methyl-	pH 2	260(4.16)	73-2148-84
9H-Purine-9-propanoic acid, 6-amino-α-hydroxy-β-methyl-, methyl ester	pH 2	261(4.14)	73-2148-84
$C_{10}H_{13}N_5O_3S$			
7H-Pyrazolo[4,3-d]pyrimidine-7-thione, 5-amino-3-(2-deoxy-β-D-erythro-pentofuranosyl)-	pH 1	239(4.03),260(3.79), 340(4.21)	44-0528-84
	pH 13	232(4.27),281(3.86), 348(4.08)	44-0528-84
	MeOH	239(4.14),294(3.64), 363(4.19)	44-0528-84
$C_{10}H_{13}N_5O_4$			
Imidazo[4,5-d]pyridazin-4-amine, 3-β-D-ribofuranosyl-	pH 1	213(4.29)	4-0481-84
	pH 7	214(4.31),363(3.69)	4-0481-84
9H-Purin-6-amine, 9-β-D-arabino-furanosyl-	pH 7	260(4.16)	78-0125-84
Pyrazolo[4,3-d]pyrimidin-7-amine, 3-β-D-arabinofuranosyl-	EtOH	294(3.97)	78-0119-84
	EtOH-acid	236(3.83),296(3.96)	78-0119-84
	EtOH-base	237(4.21),304(3.81)	78-0119-84
$C_{10}H_{13}N_5O_4S$			
7H-Pyrazolo[4,3-d]pyrimidine-7-thione, 5-amino-3-β-D-ribofuranosyl-	pH 1	238(4.03),260(3.81), 339(4.23)	44-0528-84
	pH 7	232(4.16),296(3.77), 354(4.11)	44-0528-84
	pH 13	231(4.16),288(3.79), 347(4.16)	44-0528-84
	MeOH	237(4.13),295(3.67), 362(4.16)	44-0528-84
$C_{10}H_{13}N_5O_5$			
Formycin N^6-oxide	H_2O	231(4.37),244s(4.15), 298(3.78)	39-2421-84C
4H-Imidazo[4,5-d]-1,2,3-triazin-4-one, 1,5-dihydro-6-methyl-5-β-D-ribo-furanosyl-	H_2O	255(3.26),280(3.13)	4-1221-84
4H-Imidazo[4,5-d]-1,2,3-triazin-4-one, 1,5-dihydro-6-methyl-7-β-D-ribo-furanosyl-	H_2O	245(3.72),293(3.76)	4-1221-84
6H-Purin-6-one, 2-amino-9-β-D-ara-binofuranosyl-	pH 7	252(4.13)	78-0125-84
4H-Pyrazolo[3,4-d]pyrimidin-4-one, 3-amino-1,5-dihydro-1-β-D-ribo-furanosyl-	pH 1	223(4.35),279s(3.64)	87-1119-84
	pH 7	223(4.40),279s(3.70)	87-1119-84
	pH 11	217(4.41),279(3.85)	87-1119-84
5H-Pyrazolo[4,3-d]pyrimidin-5-one, 7-amino-1,4-dihydro-3-β-D-ribo-furanosyl-	pH 1	246s(3.70),305(3.70)	4-1865-84
	pH 7	244(3.94),294(3.76)	4-1865-84
	pH 11	264(3.95),293s(3.67)	4-1865-84
5H-Pyrazolo[4,3-d]pyrimidin-5-one, 7-amino-1,4-dihydro-3-β-D-ribo-furanosyl-	pH 1	253(3.72),310(3.71)	39-2421-84C
	pH 7	250(3.98),299(3.82)	39-2421-84C
	pH 13	266(3.96),304(3.70)	39-2421-84C
7H-Pyrazolo[4,3-d]pyrimidin-7-one, 5-amino-1,4-dihydro-3-β-D-ribo-furanosyl-	pH 7	242s(3.95),299(3.82)	39-2421-84C
Urea, (3-cyano-5-β-D-ribofuranosyl-1H-pyrazol-4-yl)-	pH 7	240s(3.49)	39-2421-84C
$C_{10}H_{13}OS$			
Sulfonium, dimethyl(2-oxo-2-phenyl-	EtOH	249(4.09),283(3.68)	78-2035-84

Compound	Solvent	$\lambda_{max}(\log \epsilon)$	Ref.
ethyl)-, bromide (cont.)			78-2035-84
$C_{10}H_{14}$			
Benzene, 1,2,4,5-tetramethyl-	heptane	196(4.75),217(3.98)	35-8024-84
Cyclohexene, 1-cyclopropyl-6-methylene-	hexane	234(4.20)	44-0495-84
Spiro[2.5]octa-4,6-diene, 8,8-di-methyl-	C_6H_{12}	273(3.7)	24-3134-84
$C_{10}H_{14}BNO_4$			
Boron, dihydroxy[2-[[(2-methoxyethyl)-imino]methyl]phenolato-N^2,O']-, (T-4)-	pH 8.9	271(4.18),345(3.61)	18-2508-84
$C_{10}H_{14}BrN_3O_6$			
1(2H)-Pyrimidineacetamide, 5-bromo-3,4-dihydro-N-[2-hydroxy-1,1-bis(hydroxymethyl)ethyl]-2,4-dioxo-	pH 2 and 7 pH 12	281(3.98) 278(3.83)	73-2541-84 73-2541-84
$C_{10}H_{14}Br_2$			
Cyclohexene, 2,4-dibromo-1-ethenyl-3,3-dimethyl-	MeOH	243(3.82)	102-1323-84
$C_{10}H_{14}ClN_3O_6$			
1(2H)-Pyrimidineacetamide, 5-chloro-3,4-dihydro-N-[2-hydroxy-1,1-bis(hydroxymethyl)ethyl]-2,4-dioxo-	pH 2 and 7 pH 12	279(3.99) 276(3.85)	73-2541-84 73-2541-84
1H-1,2,3-Triazole-5-carboxylic acid, 4-chloro-1-β-D-ribofuranosyl-, ethyl ester	EtOH	240(3.93)	103-1287-84
$C_{10}H_{14}F_3NO$			
3-Penten-2-one, 1,1,1-trifluoro-2-piperidino-	EtOH	324(4.40)	39-2863-84C
$C_{10}H_{14}IN_3O_6$			
1(2H)-Pyrimidineacetamide, 3,4-dihydro-N-[2-hydroxy-1,1-bis(hydroxymethyl)-ethyl]-5-iodo-2,4-dioxo-	pH 2 and 7 pH 12	291(3.89) 280(3.77)	73-2541-84 73-2541-84
$C_{10}H_{14}N_2$			
Dispiro[cyclopropane-1,5'-[2,3]diaza-bicyclo[2.2.2]oct-2-ene-6',1"-cyclo-propane]	C_6H_{12}	344(1.26),340(1.60), 382(1.72)	24-3134-84
1,4-Methano-1H-cyclopenta[d]pyridazine, 4,4a,5,7a-tetrahydro-8,8-dimethyl-, (1α,4α,4aα,7aα)-	hexane	355(2.44)	24-0517-84
$C_{10}H_{14}N_2O$			
Acetic acid, 1,2-dimethyl-2-phenyl-hydrazide	MeCN	242(4.06)	104-1609-84
$C_{10}H_{14}N_2OS$			
Thiopyrano[3,4-c]pyrazol-4(5H)-one, 1-(1,1-dimethylethyl)-1,7-dihydro-	EtOH	250(3.96)	4-1437-84
$C_{10}H_{14}N_2OS_2$			
1,2,3-Benzothiadiazole, 6-(1,1-dimeth-ylethyl)-4,5,6,7-tetrahydro-7-sul-finyl-, (Z)-	EtOH	234(3.86),285s(3.28), 343(4.02)	44-4773-84
$C_{10}H_{14}N_2O_2$			
Phenol, 2,6-dimethyl-4-(methylamino-	EtOH	222(4.12),272(3.43),	150-0701-84M

Compound	Solvent	$\lambda_{max}(\log \epsilon)$	Ref.
methyl)-N-nitroso- (cont.) 2,4(1H,3H)-Pyrimidinedione, 5-(1,1,2-trimethyl-2-propenyl)-	MeOH	346(1.92) 260(3.82)	150-0701-84M 44-2738-84
$C_{10}H_{14}N_2O_3$ Benzenemethanamine, 2-methoxymethoxy-N-methyl-N-nitroso-	hexane	232(3.78),273(3.29), 351(1.83),362(1.93), 374(1.79)	150-0701-84M
L-Phenylalanine, 4-amino-3-methoxy-, dihydrochloride	H_2O	272(3.34),278(3.30)	44-0997-84
1(2H)-Pyrimidineacetic acid, 4,6-dimethyl-2-oxo-, ethyl ester	EtOH	208s(--),217(3.86), 306(3.81)	103-1185-84
2,4(1H,3H)-Pyrimidinedione, 5-methyl-1-[2-(2-propenyloxy)ethyl]-	pH 7 pH 12	276(3.89) 272(3.81)	161-0346-84 161-0346-84
$C_{10}H_{14}N_2O_3S$ Ethanone, 1-[5-(butylamino)-4-nitro-2-thienyl]-	benzene	388(4.05)	39-0781-84B
$C_{10}H_{14}N_2O_4$ 2,4(1H,3H)-Pyrimidinedione, 1-[2-hydroxy-3-(2-propenyloxy)propyl]-	pH 7 pH 12	269(3.91) 266(4.11)	161-0346-84 161-0346-84
$C_{10}H_{14}N_2O_4S$ 2-Thiophenecarboxylic acid, 5-(butylamino)-4-nitro-, methyl ester	benzene	384(4.00)	39-0781-84B
$C_{10}H_{14}N_2O_4S_2$ Thiophene, 2-(methylsulfonyl)-4-nitro-5-piperidino-	benzene	382(3.74)	39-0781-84B
$C_{10}H_{14}N_2O_5$ Thymidine, α-	H_2O	268(3.99)	94-1441-84
$C_{10}H_{14}N_2O_6$ Pyridazinium, 2,3-dihydro-4-methoxy-3-oxo-1-β-D-ribofuranosyl-, hydroxide, inner salt	MeOH	218(4.36),325(3.58)	87-1613-84
$C_{10}H_{14}N_2O_7$ 2,4(1H,3H)-Pyrimidinedione, 1-α-D-fructofuranosyl-	pH 7.3	194(4.31),263(4.22)	1-0367-84
2,4(1H,3H)-Pyrimidinedione, 1-β-D-fructofuranosyl-	pH 7.3	194(4.27),222(4.09), 250(4.07)	1-0367-84
2,4(1H,3H)-Pyrimidinedione, 1-β-D-psicofuranosyl-	pH 7.3	194.5(4.43),263(3.91)	1-0367-84
$C_{10}H_{14}N_2S$ Benzenecarboximidamide, N,N-dimethyl-N'-(methylthio)-	hexane	314(3.34)	39-2933-84C
$C_{10}H_{14}N_3$ Benzenediazonium, 4-(diethylamino)-	H_2O EtOH	400(4.2) 400(4.2)	135-1280-84 135-1280-84
$C_{10}H_{14}N_4$ Pyrazolo[1,5-a]pyrimidin-7-amine, 5-(1-methylpropyl)-	EtOH	225(4.46),285(3.79), 304(3.76)	4-1125-84
$C_{10}H_{14}N_4O_2$ 4H-Pyrido[1,2-a]pyrimidine-3-carboxylic	EtOH	230(3.78),303(3.98)	118-0582-84

Compound	Solvent	$\lambda_{max}(\log \epsilon)$	Ref.
acid, 6,7,8,9-tetrahydro-6-methyl-4-oxo-, hydrazide (cont.)			118-0582-84
$C_{10}H_{14}N_4O_7$			
Glycine, N-(2-β-D-arabinofuranosyl-2,5-dihydro-5-oxo-1,2,4-triazin-3-yl)-	H_2O	216(4.36),250s(3.88)	73-2689-84
$C_{10}H_{14}N_6O$			
9H-Purin-6-amine, 9-(4-morpholinyl-methyl)-	MeOH	261(4.11)	78-3997-84
	MeOH-HCl	264(4.10)	78-3997-84
	MeOH-NaOH	271(4.08)	78-3997-84
$C_{10}H_{14}N_6O_3$			
7H-1,2,3-Triazolo[4,5-d]pyrimidin-7-one, 5-amino-3,4-dihydro-3-[3-hydroxy-4-(hydroxymethyl)cyclo-pentyl]-, (1α,3β,4α)-(±)-	pH 1	253(4.07),270s(3.91)	87-1416-84
	pH 7	253(4.08),270(3.94)	87-1416-84
	pH 13	255s(--),278(4.06)	87-1416-84
$C_{10}H_{14}N_6O_4$			
Adenosine, 8-amino-	pH 1	272(4.13)	2-0677-84
	pH 11	274(4.21)	2-0677-84
9H-Purine-2,6-diamine, 9-β-D-arabino-furanosyl-	pH 7	256(4.02),280(4.04)	78-0125-84
1H-Pyrazolo[3,4-d]pyrimidine-3,4-di-amine, 1-β-D-ribofuranosyl-	pH 1	237(4.33),282(3.45)	87-1119-84
	pH 7 and 11	224(4.19),285(3.60)	87-1119-84
$C_{10}H_{14}O$			
2H-1-Benzopyran, 3,5,6,7-tetrahydro-2-methyl-	EtOH	229(3.89),250(3.95)	44-5000-84
Bicyclo[5.1.0]oct-4-en-3-one, 8,8-di-methyl-, (+)-	EtOH	220.5(3.98)	94-3452-84
2,4-Cyclooctadien-1-one, 6,6-dimethyl-	pentane	268(3.83)	33-0774-84
2-Cyclopenten-1-one, 2,3-dimethyl-5-(1-methylethenyl)-	EtOH	237(3.87)	33-0073-84
2-Cyclopenten-1-one, 2,3-dimethyl-5-(1-methylethylidene)-	EtOH	260(4.24)	33-0073-84
2-Cyclopenten-1-one, 3,5-dimethyl-5-(1-methylethenyl)-	EtOH	225(4.12)	33-0073-84
$C_{10}H_{14}O_2$			
1,2-Benzenediol, 4-(1,1-dimethyl-ethyl)-	pH 1	278(3.42)	3-1935-84
	borate	285(3.62)	3-1935-84
2-Cycloocten-1-one, 2-acetyl-	EtOH	228(4.00),283(2.04)	44-2925-84
2-Cycloocten-1-one, 3-acetyl-(end absorption)	EtOH	207(3.78),234(3.70),298(2.35)	44-2925-84
2-Furancarboxaldehyde, 3-methyl-5-(2-methylpropyl)-	EtOH	237(3.1),290(4.4)	88-1061-84
2-Propenoic acid, 2-methyl-, 2,4-hexa-dienyl ester	isooctane	227(4.48)	116-1624-84
$C_{10}H_{14}O_2S$			
Benzene, 1-methoxy-2-[2-(methylthio)-ethoxy]-	MeOH	222.5(3.91),273s(3.42)	36-1241-84
$C_{10}H_{14}O_3$			
Bicyclo[4.2.1]non-3-en-2-one, 7,8-di-hydroxy-3-methyl-	EtOH	237(3.90)	33-1506-84
1-Cyclohexene-1-carboxylic acid, 4-acetyl-, methyl ester, (+)-	EtOH	220(3.79)	78-2961-84
2-Cyclohepten-1-one, 3-(acetoxymethyl)-	EtOH	232(4.06),314(1.90)	44-2925-84

Compound	Solvent	$\lambda_{max}(\log \epsilon)$	Ref.
1-Cyclooctene-1-carboxylic acid, 3-oxo-, methyl ester	EtOH	206(3.85),231(3.81), 300(2.16)	44-2925-84
4-Cyclopentene-1,3-dione, 4-hydroxy-	EtOH	276(4.17)	39-1555-84C
	EtOH-NaOH	232(4.00),327(4.05)	39-1555-84C
1,4-Dioxaspiro[4.5]decan-2-one, 3-ethylidene-	EtOH	247(4.04)	39-1531-84C
3(2H)-Furanone, 5-(3-hydroxy-1-methyl-1-propenyl)-2,2-dimethyl-	EtOH	243.5(4.10),292(4.42)	39-0535-84C
$C_{10}H_{14}O_4$			
2-Furanacetic acid, 2,3-dihydro-2,4,5-trimethyl-3-oxo-, methyl ester, (±)-	EtOH	270(3.95)	94-0457-84
1,2,3-Propanetriol, 2-(2-hydroxy-4-methylphenyl)-	MeOH	276(3.34),281s(3.32)	102-1947-84
	MeOH-NaOMe	293(3.62)	102-1947-84
$C_{10}H_{14}O_6$			
L-Ascorbic acid, 5,6-O-(1-methyleth-ylidene)-2-O-methyl-	pH 6 and 10	259(4.16)	98-0021-84
Propanedioic acid, (acetoxymethylene)-, diethyl ester	EtOH	227.5(2.82)	118-0732-84
$C_{10}H_{14}S$			
Tricyclo[3.3.1.1^{3,7}]decanethione	benzene	498(1.18),536(0.45)	39-1869-84C
$C_{10}H_{15}Br$			
Bicyclo[2.2.1]heptane, 3-(bromometh-ylene)-2,2-dimethyl-, (E)-	MeOH	210(3.60)	2-0331-84
$C_{10}H_{15}ClN_4O_2$			
Cyclopentanemethanol, 4-[(2-amino-6-chloro-4-pyrimidinyl)amino]-2-hy-droxy-, (1α,2β,4α)-(±)-	pH 1	214(4.27),237(4.10), 274(3.94),285s(--), 300(3.63)	87-1416-84
	pH 7	212(4.42),238(4.04), 287(4.00)	87-1416-84
	pH 13	238(4.03),286(4.00)	87-1416-84
$C_{10}H_{15}ClO$			
2-Cyclopenten-1-one, 5-(1-chloro-1-methylethyl)-3,5-dimethyl-	EtOH	227(4.09)	33-0073-84
$C_{10}H_{15}I$			
Bicyclo[2.2.1]heptane, 3-(iodomethyl-ene)-2,2-dimethyl-, (E)-	MeOH	220(3.60),260(2.98)	2-0331-84
$C_{10}H_{15}NO$			
Ethanone, 1-(9-azabicyclo[4.2.1]non-2-en-2-yl)- (anatoxin)	EtOH	226(4.03)	35-4539-84
4(1H)-Pyridinone, 2,3-dihydro-1-(4-pentenyl)-	EtOH	323(4.18)	118-0485-84
$C_{10}H_{15}NO_2$			
Benzenemethanamine, 2-(methoxymethoxy)-N-methyl-	EtOH	209(3.74),269(3.06), 274(3.01)	150-0701-84M
6,8(2H,7H)-Isoquinolinedione, hexa-hydro-2-methyl-	pH 1	255(4.14)	44-5109-84
	H_2O	280(4.34)	44-5109-84
4H-Pyran-4-one, 2,3,5,6-tetrahydro-3-(pyrrolidinomethylene)-, (E)-	EtOH	333(4.34)	4-1441-84
$C_{10}H_{15}NO_3$			
4H-Pyran-4-one, 2,3,5,6-tetrahydro-3-morpholinomethylene-, (E)-	EtOH	326(4.34)	4-1441-84

Compound	Solvent	λ_{max}(log ϵ)	Ref.
1H-Pyrrole-2-carboxylic acid, 3-hydroxy-1,4,5-trimethyl-, ethyl ester	MeCN	277(4.11)	33-1535-84
$C_{10}H_{15}NO_3S$ 3-Thia-1-azabicyclo[3.2.0]heptane-2-carboxylic acid, 7-oxo-, 1,1-dimethylethyl ester, endo	EtOH	215(3.31),253(2.40)	39-2785-84C
exo	EtOH	215(3.00),250s(2.30)	39-2785-84C
$C_{10}H_{15}NO_4$ 1H-Pyrrole-3-carboxylic acid, 4-ethoxy-2,5-dihydro-1-(1-methylethyl)-5-oxo-	EtOH	252(4.13)	44-1130-84
2H-Pyrrole-2-carboxylic acid, 3,4-dihydro-4-hydroxy-2,4,5-trimethyl-3-oxo-, ethyl ester	MeCN	328(1.98)	33-1957-84
$C_{10}H_{15}NO_5S$ 3-Thia-1-azabicyclo[3.2.0]heptane-2-carboxylic acid, 7-oxo-, 1,1-dimethylethyl ester, 3,3-dioxide, endo	EtOH	210(3.16),223s(2.85)	39-2785-84C
exo	EtOH	211(3.31),244s(3.04)	39-2785-84C
$C_{10}H_{15}NS$ 2-Hexenenitrile, 4-methyl-3-(2-propenylthio)-	EtOH	208(3.81),285(4.28)	78-2141-84
$C_{10}H_{15}N_2O_8P$ Uridine, 3'-deoxy-3'-(phosphonomethyl)-	H_2O	207(3.92),264(4.00)	78-0079-84
$C_{10}H_{15}N_3O_5$ 1(6H)-Pyridazineacetamide, N-[2-hydroxy-1,1-bis(hydroxymethyl)ethyl]-6-oxo-	pH 2 and 7	286(3.48)	73-2541-84
1(2H)-Pyrimidineacetamide, N-[2-hydroxy-1,1-bis(hydroxymethyl)ethyl]-2-oxo-	pH 2 and 7	300(3.75)	73-2541-84
$C_{10}H_{15}N_3O_6$ 1(2H)-Pyrimidineacetamide, 3,4-dihydro-N-[2-hydroxy-1,1-bis(hydroxymethyl)-ethyl]-2,4-dioxo-	pH 2 and 7 pH 12	263(4.16) 267(4.04)	73-2541-84 73-2541-84
$C_{10}H_{15}N_3O_7$ 1H-Imidazole-4-carboxamide, 5-(β-D-galactopyranosyloxy)-	M HCl H_2O M NaOH	242(3.91) 249(4.06) 264(4.08)	4-0849-84 4-0849-84 4-0849-84
1H-Imidazole-4-carboxamide, 5-(β-D-glucopyranosyloxy)-	M HCl H_2O M NaOH	241(3.92) 248(4.07) 265(4.10)	4-0849-84 4-0849-84 4-0849-84
1H-Imidazolium-4-olate, 5-carbamoyl-3-β-D-galactopyranosyl-	M HCl H_2O M NaOH	243(3.79),280(4.03) 243(3.79),278(4.08) 275(4.15)	4-0529-84 4-0529-84 4-0529-84
1H-Imidazolium-4-olate, 5-carbamoyl-3-β-D-glucopyranosyl-	M HCl H_2O M NaOH	243(3.80),280(4.05) 243(3.79),278(4.09) 276(4.16)	4-0529-84 4-0529-84 4-0529-84
3H-Imidazolium-4-olate, 5-carbamoyl-1-β-D-galactopyranosyl-	M HCl H_2O M NaOH	242(3.88),279(3.61) 235(3.71),283(4.05) 231(3.34),288(4.09)	4-0529-84 4-0529-84 4-0529-84
3H-Imidazolium-4-olate, 5-carbamoyl-1-β-D-glucopyranosyl-	M HCl H_2O M NaOH	242(3.80),281(3.66) 235(3.68),282(4.00) 232(3.32),288(4.07)	4-0529-84 4-0529-84 4-0529-84

Compound	Solvent	$\lambda_{max}(\log \epsilon)$	Ref.
$C_{10}H_{15}N_5O$			
4(1H)-Pteridinone, 2-amino-6,7-dihydro-6,6,7,7-tetramethyl-	M HClO$_4$	210(4.02),257(3.85), 341(3.56)	35-7916-84
$C_{10}H_{15}N_5O_2$			
Ethanol, 2-[[6-(dimethylamino)-9H-purin-9-yl]methoxy]-	pH 1	267(4.27)	87-1486-84
	pH 13	274(4.29)	87-1486-84
$C_{10}H_{15}N_5O_4$			
1,3-Propanediol, 2-[(2-amino-6-methoxy-7H-purin-7-yl)methoxy]-	pH 1	246(3.83),287(3.77)	23-2702-84
	pH 7	253(3.99),283(3.86)	23-2702-84
	pH 13	253(3.99),283(3.85)	23-2702-84
1,3-Propanediol, 2-[(2-amino-6-methoxy-9H-purin-9-yl)methoxy]-	pH 1	245(3.79),288(3.76)	23-2702-84
	pH 7	247(3.88),282(3.82)	23-2702-84
	pH 13	247(3.89),282(3.81)	23-2702-84
1,3-Propanediol, 2-[(6-amino-9H-purin-9-yl)methoxy]-2-(hydroxymethyl)-	pH 1	257(4.09)	23-1622-84
	pH 7	258(4.10)	23-1622-84
	pH 14	258(4.10)	23-1622-84
$C_{10}H_{15}N_5O_5$			
1,2-Ethanediol, 1-(1,3-dihydroxy-2-propyl)-1-(guanin-9-yl)-, (1R)-	pH 1	256(4.07)	136-0141-84C
	H$_2$O	253(4.11)	136-0141-84C
	pH 13	263(4.03)	136-0141-84C
6H-Purin-6-one, 2-amino-1,7-dihydro-7-[[2-hydroxy-1,1-bis(hydroxymethyl)-ethoxy]methyl]-	pH 1	254(4.07),274(3.91)	23-1622-84
	pH 7	250(4.09),270(3.95)	23-1622-84
	pH 14	254(4.05),266(4.04)	23-1622-84
6H-Purin-6-one, 2-amino-1,9-dihydro-9-[[2-hydroxy-1,1-bis(hydroxymethyl)-ethoxy]methyl]-	pH 1	250(3.98),268(3.88)	23-1622-84
	pH 7	240(3.83),284(3.89)	23-1622-84
	pH 14	282(3.89)	23-1622-84
$C_{10}H_{15}N_7O_2$			
Cyclopentanemethanol, 4-(5,7-diamino-3H-1,2,3-triazolo[4,5-d]pyrimidin-3-yl)-2-hydroxy-, (1α,2β,4α)-(±)-	pH 1	214(4.42),255(3.99), 285(3.89)	87-1416-84
	pH 7	223(4.41),258(3.76), 287(4.03)	87-1416-84
	pH 13	223(4.42),258(3.76), 287(4.03)	87-1416-84
$C_{10}H_{15}N_7O_3$			
1,2-Cyclopentanediol, 3-(5,7-diamino-3H-1,2,3-triazolo[4,5-d]pyrimidin-3-yl)-5-(hydroxymethyl)-, (1α,2α,3β,5β)-(±)-	pH 1	214(4.41),256(3.98), 286(3.89)	87-0670-84
	pH 7 and 13	223(4.42),258(3.76), 287(4.03)	87-0670-84
$C_{10}H_{15}P$			
Phosphine, diethylphenyl-	C_6H_{12}	254(3.56)	65-0489-84
$C_{10}H_{16}$			
1,3-Cycloheptadiene, 1,5,5-trimethyl-	EtOH	252(3.88)	104-0257-84
1,3-Cycloheptadiene, 2,6,6-trimethyl-	EtOH	254(3.72)	104-0257-84
Cyclohexene, 2-ethenyl-3,3-dimethyl-	EtOH	237(3.82)	88-4183-84
Cyclohexene, 2,6,6-trimethyl-3-methylene-	hexane	231(4.27),236(4.30), 243(4.15)	24-3473-84
Cyclooctane, 1,2-bis(methylene)-	EtOH	233(3.76)	44-2981-84
1,6,8-Decatriene	hexane	229(4.03)	105-0486-84
$C_{10}H_{16}ClN_5O_2$			
Cyclopentanemethanol, 4-[(2,5-diamino-6-chloro-4-pyrimidinyl)amino]-2-hydroxy-, (1α,2β,4α)-(±)-	pH 1	210(4.20),237(4.20), 298(3.89)	87-1416-84
	pH 7	205(4.29),225s(--), 240s(--),304(3.95)	87-1416-84

$C_{10}H_{16}ClN_5O_2-C_{10}H_{16}N_3O_7P$

Compound	Solvent	$\lambda_{max}(\log \epsilon)$	Ref.
(cont.)	pH 13	225s(--),240s(--), 304(3.95)	87-1416-84
$C_{10}H_{16}F_3NO_2$ 2-Pentene-2,4-diol, 5,5,5-trifluoro-4-piperidino-	C_6H_{12} EtOH	286(3.91) 294(4.19)	39-2863-84C 39-2863-84C
$C_{10}H_{16}NO_6$ Pyrimidinium, 1-β-D-arabinofuranosyl-3-hydroxy-2-methyl-4-hydroxy-, chloride	MeOH	285(3.90)	87-0160-84
$C_{10}H_{16}N_2$ Bicyclo[2.2.1]heptane, 2-diazo-1,3,3-trimethyl-, (1R)-	hexane	506(0.62)	24-0277-84
1,4-Methano-1H-cyclopenta[d]pyridazine, 4,4a,5,6,7,7a-hexahydro-8,8-dimethyl-, (1α,4α,4aα,7aα)-	hexane	357(2.33)	24-0517-84
Spiro[cyclopropane-1,5'-[2,3]diazabicyclo[2.2.2]oct-2-ene], 6',6'-dimethyl-	C_6H_{12}	335(0.78),345(1.15), 371(1.68),382(1.72)	24-3134-84
$C_{10}H_{16}N_2O$ Pyrazine, 2-methoxy-5-methyl-3-(1-methylpropyl)-	MeOH	300(3.81)	95-0951-84
Pyrazine, 2-methoxy-6-methyl-3-(1-methylpropyl)-	MeOH	296(3.82)	95-0951-84
Pyrazine, 2-(1,1,2-trimethylpropoxy)-	EtOH	216(4.05),282(3.70), 295(3.54)	39-0641-84B
Pyrazine, 2-(1,2,2-trimethylpropoxy)-	EtOH	214(4.06),280(3.68), 295(3.62)	39-0641-84B
$C_{10}H_{16}N_2O_2$ Bicyclo[2.2.1]heptan-2-imine, 1,3,3-trimethyl-N-nitro-, (1R)-	C_6H_{12}	272(2.82)	118-0479-84
Bicyclo[2.2.1]heptan-2-imine, 1,7,7-trimethyl-N-nitro-, (±)-	C_6H_{12}	270(2.67)	118-0479-84
2,5-Pyrazinediethanol, 3-ethyl-	H_2O	210(3.99),278(3.94), 296s(3.60)	39-1345-84B
$C_{10}H_{16}N_2O_2S$ 2H-Thiopyran-4-carboxaldehyde, tetrahydro-3,5-dioxo-, 4-[(1,1-dimethylethyl)hydrazone]	EtOH	256(3.97),306(4.17)	4-1437-84
$C_{10}H_{16}N_2O_4$ 2-Pentenedioic acid, 4-[(2,2-dimethylhydrazino)methylene]-, dimethyl ester, (Z,E)-	MeOH	205(3.72),285s(3.86), 322(4.02)	24-1620-84
$C_{10}H_{16}N_2O_6$ 2,4(1H,3H)-Pyrimidinedione, 1-[[2-hydroxy-1,1-bis(hydroxymethyl)ethoxy]-methyl]-5-methyl-	pH 1 pH 7 pH 14	265(4.06) 265(4.05) 264(3.96)	23-1622-84 23-1622-84 23-1622-84
$C_{10}H_{16}N_3S$ 1,2,3-Benzothiadiazole, 4,5,6,7-tetrahydro-5,5,7,7-tetramethyl-	EtOH	221(3.62),263(3.55)	44-4773-84
$C_{10}H_{16}N_3O_7P$ Cytidine, 3'-deoxy-3'-(phosphonomethyl)-	pH 1	213(3.99),280(4.13)	78-0079-84

Compound	Solvent	$\lambda_{max}(\log \epsilon)$	Ref.
Cytidine, 3'-deoxy-3'-(phosphono-meyhyl)- (cont.)	pH 12	220(3.92),271(3.95)	78-0079-84
$C_{10}H_{16}N_4OS$ 5-Pyrimidinecarboxamide, 6-amino-1-butyl-1,2-dihydro-N-methyl-2-thioxo-	EtOH	226(4.05),293(4.12), 358(4.23)	142-2259-84
$C_{10}H_{16}N_4O_2$ 1H-Pyrazole-4-carbonitrile, 5-amino-1-(2,2-diethoxyethyl)-	EtOH	228(3.97)	44-3534-84
$C_{10}H_{16}N_4O_2S_2$ 1H-Imidazole-4-carboxylic acid, 1-methyl-5-[[(methylamino)thioxomethyl]-amino]-2-(methylthio)-, ethyl ester	MeOH	247(4.17)	2-0316-84
$C_{10}H_{16}N_4O_5$ 1(2H)-Pyrimidineacetamide, 4-amino-N-[2-hydroxy-1,1-bis(hydroxymethyl)-ethyl]-2-oxo-	pH 2 and 7	283(4.19)	73-2541-84
$C_{10}H_{16}N_6O_3$ 6H-Purin-6-one, 2-amino-8-(dimethyl-amino)-1,9-dihydro-9-[(2-hydroxy-ethoxy)methyl]-	pH 1 pH 13	258(4.24),288(4.02) 268(4.18)	87-1486-84 87-1486-84
$C_{10}H_{16}O$ 5,7-Octadien-4-one, 2,6-dimethyl-, cis	EtOH	271(4.31)	104-1884-84
	EtOH	270(4.27)	104-1884-84
trans	EtOH	268(4.42)	104-1884-84
$C_{10}H_{16}O_2$ 2(3H)-Furanone, 3-hexylidenedihydro-, (E)-	EtOH	223(4.13)	95-0839-84
(Z)-	EtOH	226(4.03)	95-0839-84
$C_{10}H_{16}O_3$ Cyclohexanecarboxylic acid, 1,3-dimethyl-5-oxo- (cis:trans 4:1)	EtOH	224(1.26)	39-0261-84C
$C_{10}H_{16}O_3S$ 2-Hexenoic acid, 5,5-dimethyl-2-(methylthio)-4-oxo-, methyl ester	MeOH	300(3.94)	5-0820-84
$C_{10}H_{16}O_4S$ Bicyclo[2.2.1]heptane-1-methanesulfonic acid, 7,7-dimethyl-2-oxo-, ammonium salt, (1R)-	H_2O	285(1.61)	42-0985-84
	MeOH	287(1.49)	42-0985-84
	EtOH	288(1.54)	42-0985-84
	isoPrOH	290(1.40)	42-0985-84
	DMF	289(1.49)	42-0985-84
	pyridine	290(1.54)	42-0985-84
	DMSO	289(1.49)	42-0985-84
dimethylamine salt	H_2O	285(1.65)	42-0985-84
	MeOH	287(1.94)	42-0985-84
	EtOH	287.5(1.95)	42-0985-84
	isoPrOH	289(1.45)	42-0985-84
	DMF	287.5(1.51)	42-0985-84
	pyridine	297(3.10)	42-0985-84
	DMSO	287.5(1.21)	42-0985-84
ethylamine salt (also other salts)	H_2O	285(1.68)	42-0985-84
	MeOH	286(1.63)	42-0985-84

Compound	Solvent	$\lambda_{max}(\log \epsilon)$	Ref.
(cont.)	EtOH	287(1.49)	42-0985-84
	isoPrOH	288(1.33)	42-0985-84
	DMF	289(1.49)	42-0985-84
	pyridine	297(3.10)	42-0985-84
	DMSO	288(1.26)	42-0985-84
$C_{10}H_{16}Si$			
Silane, trimethyl(4-methylphenyl)-	C_6H_{12}	221.5(4.17),264(2.53)	60-0383-84
$C_{10}H_{17}NO$			
Piperidine, 1-[(dihydro-2(3H)-furan-ylidene)methyl]-	EtOH	220(3.78)	104-0861-84
Pyrrolidine, 1-[(tetrahydro-2H-pyran-2-ylidene)methyl]-	EtOH	235(3.90)	104-0861-84
$C_{10}H_{17}NO_2$			
3aH-Indol-3a-ol, 2,3,4,5,6,7-hexahydro-2,2-dimethyl-, 1-oxide	EtOH	232(4.08)	12-0587-84
4H-Pyran-4-one, 3-[(diethylamino)meth-ylene]-2,3,5,6-tetrahydro-, (E)-	EtOH	327.5(4.34)	4-1441-84
$C_{10}H_{17}NO_2S$			
1,3-Butadiene-1-sulfonamide, N-cyclo-hexyl-, (Z)-	H_2O or EtOH	232(3.70)	70-2519-84
$C_{10}H_{17}NO_3S$			
Morpholine, 4-[(2,3-dimethyl-1,3-buta-dienyl)sulfonyl]-, (Z)-	EtOH	240s(3.48)	70-2519-84
$C_{10}H_{17}NS$			
2-Hexenenitrile, 4-methyl-3-(propyl-thio)-	EtOH	210(3.78),285(4.26)	78-2141-84
3-Hexenenitrile, 4-ethyl-3-(ethylthio)-	EtOH	208(3.85),255(3.49)	78-2141-84
3-Hexenenitrile, 4-methyl-3-[(1-methyl-ethyl)thio]-	EtOH	208(3.81),? (3.48)	78-2141-84
$C_{10}H_{17}N_2O_2$			
1-Imidazolidinyloxy, 4-ethylidene-3-formyl-2,2,5,5-tetramethyl-	EtOH	242(3.90)	70-2348-84
$C_{10}H_{17}N_3$			
4-Pyrimidinamine, 6-(1-ethylpropyl)-2-methyl-	EtOH	233(3.96),268(3.54)	39-2677-84C
$C_{10}H_{17}N_3O_2$			
Urea, N-bicyclo[2.2.1]hept-1-yl-N',N'-dimethyl-N-nitroso-	EtOH	248(3.72),390(1.99)	44-4860-84
$C_{10}H_{17}N_5O$			
4(1H)-Pteridinone, 2-amino-5,6,7,8-tetrahydro-6,6,7,7-tetramethyl-	pH 1	264(4.15)	35-7916-84
$C_{10}H_{17}N_5O_2S$			
Guanidine, N'-(4-butyl-2-thiazolyl)-N,N-dimethyl-N''-nitro-	isoPrOH	240s(3.99),283s(3.82), 329(4.34)	94-3483-84
$C_{10}H_{18}$			
Octane, 4,5-bis(methylene)-	hexane	228(4.10)	44-2981-84
$C_{10}H_{18}N_2$			
Cyclohexane, 2-diazo-1,1,3,3-tetrameth-yl-	hexane	492(0.85)	24-0277-84

Compound	Solvent	$\lambda_{max}(\log \epsilon)$	Ref.
2,3-Diazabicyclo[2.2.2]oct-2-ene, 5,5,6,6-tetramethyl-	C_6H_{12}	334(1.08),344(1.26), 370(1.77),382(2.05)	24-3134-84
$C_{10}H_{18}N_2O_2$ Cyclohexanimine, 2,2,6,6-tetramethyl-N-nitro-	C_6H_{12}	273(2.77)	118-0479-84
$C_{10}H_{18}N_2O_6S$ 2(1H)-Pyrimidinethione, 1-β-D-galacto-pyranosyltetrahydro-6-hydroxy-	MeOH	252(4.01)	103-0447-84
2(1H)-Pyrimidinethione, 1-β-D-gluco-pyranosyltetrahydro-6-hydroxy-	MeOH	252(4.07)	103-0447-84
$C_{10}H_{18}N_4$ 2,4-Pyrimidinediamine, 6-(1,2,2-tri-methylpropyl)-	EtOH	232(3.99),282(3.87)	39-2677-84C
$C_{10}H_{18}O$ 6,9-Decadien-1-ol, (E,E)-	hexane	229(4.32)	105-0486-84
$C_{10}H_{18}S$ 1-Butene-1-thione, 2-(1,1-dimethyl-ethyl)-3,3-dimethyl-	C_6H_{12}	211(4.51),239(3.62), 570(0.90)	44-0393-84
Cyclohexanethione, 2,2,6,6-tetramethyl-	hexane	217s(3.61),238(3.92), 534(1.04)	24-0277-84
$C_{10}H_{19}NOSi$ 1H-Pyrrole, 1-[[2-(trimethylsilyl)eth-oxy]methyl]-	MeOH	214(3.80)	44-0203-84
$C_{10}H_{19}NO_4$ 2-Butenoic acid, 3-[(2,2-dimethoxyeth-yl)amino]-, ethyl ester	MeOH	286(4.36)	12-0389-84
$C_{10}H_{19}N_2$ Methanaminium, N-[2-cyclopropyl-3-(di-methylamino)-2-propenylidene]-N-methyl-, perchlorate	MeOH	319(4.64)	5-0649-84
$C_{10}H_{20}N_2O_2$ 2-Hexenimidic acid, 4-ethyl-3-(hydroxy-amino)-, ethyl ester	EtOH	281(4.10)	39-1079-84C
$C_{10}H_{21}N_3O$ 1H-1,2,3-Triazole, 4(and 5)-butyl-1-(2-ethoxyethyl)-4,5-dihydro-	hexane	240(3.61)	104-0580-84
1H-1,2,3-Triazole, 5-butyl-1-(2-ethoxyethyl)-4,5-dihydro-	hexane	240(3.66)	104-0580-84
$C_{10}H_{22}N_3O_4P$ Phosphonic acid, [2-[[(dimethylamino)-carbonyl]methylhydrazono]ethyl]-, diethyl ester	EtOH	205(3.96),241(4.15)	33-1547-84
$C_{10}H_{22}S_2Si$ Silane, (1,1-dimethylethyl)-1,3-di-thian-2-yldimethyl-	pentane	244(2.86)	33-1734-84
$C_{10}H_{22}Te_2$ Ditelluride, dipentyl	benzene	397(2.83)	48-0467-84

$C_{10}H_{24}ClNSi$

Compound	Solvent	$\lambda_{max}(\log \epsilon)$	Ref.

$C_{10}H_{24}ClNSi$
  Silanamine, N-chloro-N,1-bis(1,1-di- CCl_2FCF_2Cl 301(1.91)   39-1187-84B
    methylethyl)-1,1-dimethyl-
  Silanamine, N-chloro-N-(1,1-dimethyl- CCl_2FCF_2Cl 303(1.98)   39-1187-84B
    ethyl)-1,1,1-triethyl-

Compound	Solvent	$\lambda_{max}(\log \epsilon)$	Ref.
$C_{11}H_2Cl_6$			
7H-Cyclopenta[a]pentalene, 1,2,3,4,5,6-hexachloro-	C_6H_{12}	311(3.97),325(4.40), 341(4.75),358(4.85)	88-2923-84
$C_{11}H_2Cl_8$			
7H-Cyclopenta[a]pentalene, 1,2,3,3a,3b-4,5,6-octachloro-3a,3b-dihydro-	C_6H_{12}	271(3.32),323(3.19)	88-2923-84
$C_{11}H_2N_4O_3$			
Propanedinitrile, 2,2'-(4,5-dihydroxy-2-oxo-4-cyclopentene-1,3-diylidene)-bis-, dipotassium salt	H_2O EtOH + PVP	530(4.99) 538(5.00) 542(5.00)	151-0285-84A 151-0285-84A 151-0285-84A
$C_{11}H_4Cl_2N_2O$			
9H-Cyclopenta[1,2-b:4,3-b']dipyridin-9-one, 2,7-dichloro-	EtOH	208(4.23),250(4.10), 288(4.02),313(3.97), 378(3.30)	142-0073-84
$C_{11}H_5BrClNO_2S$			
6H-1,3-Thiazine-5-carboxaldehyde, 2-(4-bromophenyl)-4-chloro-6-oxo-	EtOH	230(4.15),290(3.79), 350(3.95)	103-1088-84
$C_{11}H_5ClN_2O_4S$			
6H-1,3-Thiazine-5-carboxaldehyde, 4-chloro-2-(4-nitrophenyl)-6-oxo-	EtOH	212(4.18),270(4.15), 285(4.18),345(3.60)	103-1088-84
$C_{11}H_5FN_4$			
Cyanamide, [2-(cyanoamino)-3-(4-fluorophenyl)-2-cyclopropen-1-ylidene]-, sodium salt	MeOH	241(4.32),307(4.27)	118-0686-84
$C_{11}H_5N_3O$			
Propanedinitrile, (1,2-dihydro-2-oxo-3H-indol-3-ylidene)-	CHCl₃	265(4.18),355(4.06), 487(3.18)	39-1331-84B
$C_{11}H_6ClNO_2S$			
6H-1,3-Thiazine-5-carboxaldehyde, 4-chloro-6-oxo-2-phenyl-	EtOH	228(3.98),290(3.72), 340(3.79)	103-1088-84
$C_{11}H_6ClNO_3$			
4H-[1]Benzopyrano[3,4-d]isoxazol-4-one, 8-chloro-3-methyl-	EtOH	266(4.08),278(4.08), 315(3.81)	44-4419-84
$C_{11}H_6Cl_2O_5$			
1,4-Naphthalenedione, 6,7-dichloro-5,8-dihydroxy-2-methoxy-	benzene	482s(3.79),495(3.81), 523s(3.72)	118-0953-84
$C_{11}H_6N_4O$			
1,2,3-Oxadiazolium, 5-(dicyanomethylene)-2,5-dihydro-3-phenyl-, hydroxide, inner salt	MeCN EtOH ether CH₂Cl₂	220(3.78),268(4.49), 417(3.99) 421(--) 442(--) 432(--)	138-1045-84 138-1045-84 138-1045-84 138-1045-84
$C_{11}H_6N_6OS$			
10H-Pyrido[1,2-a]thieno[3,2-d]pyrimidin-10-one, 7-(1H-tetrazol-5-yl)-	MeOH	257(4.57),355(3.92), 373(4.14)	87-0528-84
10H-Pyrido[1,2-a]thieno[3,4-d]pyrimidin-10-one, 7-(1H-tetrazol-5-yl)-	MeOH	226(4.25),268(4.56), 317(3.99),333(3.98), 349(3.81),400(3.50)	87-0528-84

Compound	Solvent	$\lambda_{max}(\log \epsilon)$	Ref.
$C_{11}H_6O_3$			
Psoralen	EtOH	213(4.245),241(4.395), 246.5(4.409),290(4.041), 330(3.798)	111-0365-84
$C_{11}H_7BrN_2O_2Se$			
4,6(1H,5H)-Pyrimidinedione, 5-[(3-bromophenyl)methylene]dihydro-2-selenoxo-	dioxan	281(3.87),340(4.07), 393(4.14)	104-0141-84
$C_{11}H_7ClN_2O_2Se$			
4,6(1H,5H)-Pyrimidinedione, 5-[(4-chlorophenyl)methylene]dihydro-2-selenoxo-	dioxan	280(3.90),394(4.32)	104-0141-84
$C_{11}H_7ClN_2S$			
Thiopyrano[4,3,2-de]quinazoline, 7-chloro-5-methyl-	MeOH	256(4.46),300(3.92), 392(3.83)	83-0027-84
	TFA	264(4.50),350(3.48), 446(3.82)	83-0027-84
$C_{11}H_7ClN_4O_2$			
1H-Pyrazolo[4,3-b]quinoline, 9-chloro-1-methyl-5-nitro-	EtOH	246(4.12),333(3.61), 392(3.54)	103-0918-84
1H-Pyrazolo[4,3-b]quinoline, 9-chloro-1-methyl-7-nitro-	EtOH	246(4.25),296(3.81), 328(3.41),341(3.39), 406(3.30)	103-0918-84
$C_{11}H_7ClO_3$			
1,4-Naphthalenedione, 2-chloro-5-hydroxy-6-methyl-	EtOH	278(4.04),438(3.66)	78-3455-84
1,4-Naphthalenedione, 2-chloro-5-hydroxy-7-methyl-	EtOH	254(3.95),262(3.96), 278(4.06),434(3.63)	78-3455-84
1,4-Naphthalenedione, 2-chloro-8-hydroxy-6-methyl-	EtOH	254(3.87),263(3.91), 277(4.01),369(2.78), 428(3.52)	78-3455-84
1,4-Naphthalenedione, 2-chloro-8-hydroxy-7-methyl-	EtOH	277(4.05),436(3.63)	78-3455-84
1,4-Naphthalenedione, 7-chloro-2-methoxy-	MeOH	248(4.37),253(4.38), 280.5(4.17),327(3.52)	102-0313-84
$C_{11}H_7ClO_4$			
1,4-Naphthalenedione, 2-chloro-5-hydroxy-6-methoxy-	EtOH	274(3.96),285s(3.86), 456(3.52)	78-3455-84
1,4-Naphthalenedione, 2-chloro-5-hydroxy-7-methoxy-	EtOH	270(4.04),286(3.86), 438(3.60)	78-3455-84
1,4-Naphthalenedione, 2-chloro-8-hydroxy-6-methoxy-	EtOH	271(3.54),287s(3.40), 434(3.09)	78-3455-84
1,4-Naphthalenedione, 2-chloro-8-hydroxy-7-methoxy-	EtOH	273(4.10),343(2.02), 462(3.61)	78-3455-84
$C_{11}H_7ClO_5$			
1,4-Naphthalenedione, 2-chloro-5,8-dihydroxy-3-methoxy-	benzene	489s(3.70),515(3.84), 553(3.64)	118-0953-84
$C_{11}H_7Cl_2NO_2$			
Methanone, (3,5-dichloro-2-hydroxyphenyl)(1H-pyrrol-2-yl)-	MeOH	227(4.22),260(3.70), 328s(4.04),342(4.06)	95-0238-84
Methanone, (3,5-dichloro-2-hydroxyphenyl)(1H-pyrrol-3-yl)-	MeOH	215(4.34),261(3.91), 272s(3.89),342(4.00)	95-0238-84

Compound	Solvent	λ_{max} (log ϵ)	Ref.
$C_{11}H_7Cl_2N_3$ 1H-Pyrazolo[4,3-b]quinoline, 7,9-dichloro-1-methyl-	EtOH	250(5.03),328(3.69), 341(3.93),375(3.77), 392(3.74)	103-0918-84
$C_{11}H_7NO$ 3H-Pyrrolo[2,1,5-de]quinolizin-3-one	EtOH	257(4.61),268s(4.43), 314(3.67),414(3.96), 433(3.99)	39-2553-84C
5H-Pyrrolo[2,1,5-de]quinolizin-5-one	EtOH	212(4.09),232(4.07), 279(3.76),297(3.76), 356(3.30),451s(3.72), 468s(3.82),475(3.83)	39-2553-84C
$C_{11}H_7NO_2S_2$ 2H-Pyran-3-carbonitrile, 4-(methylthio)-2-oxo-6-(2-thienyl)-	EtOH	218(3.99),267(4.11), 349(4.23),402(4.34)	94-3384-84
$C_{11}H_7NO_3$ 4H-[1]Benzopyrano[3,4-d]isoxazol-4-one, 3-methyl-	EtOH	265(4.08),272(4.12), 306(3.86)	44-4419-84
$C_{11}H_7NO_3S_2$ Acetonitrile, (methylthio)(3-oxobenzo-[b]thien-2(3H)-ylidene)-, S,S-dioxide	EtOH	221(4.14),253(4.02), 376(4.20)	95-0134-84
$C_{11}H_7N_3O$ Pyrimido[5,4-c]quinolin-4(1H)-one	MeOH	277(4.38),340s(3.78), 353(3.88),370s(3.72)	103-0439-84
$C_{11}H_7N_3O_2$ Imidazo[1,5-a]quinoline, 7-nitro-	n.s.g.	213(4.38),287(4.18)	11-0038-84B
$C_{11}H_7N_3O_2S$ 2H-Pyrimido[4,5-b][1,4]benzothiazine-2,4(3H)-dione, 3-methyl-	MeCN	226(4.30),258(4.00), 283(4.18),353(3.85), 450(3.70)	142-0013-84
$C_{11}H_7N_3O_4Se$ 4,6(1H,5H)-Pyrimidinedione, dihydro-5-[(3-nitrophenyl)methylene]-2-selenoxo-	dioxan	273(4.10),327(4.03), 393(3.99)	104-0141-84
$C_{11}H_7N_5O_2$ Imidazo[1,2-a]pyridinium, 2-(dicyanomethyl)-1-methyl-3-nitro-, hydroxide, inner salt	n.s.g.	206(4.40),241s(4.07), 283s(4.09),307(4.43), 369(3.95),428(4.00)	39-0069-84C
$C_{11}H_8BrNS_2$ 6H-1,3-Thiazine-6-thione, 2-(4-bromophenyl)-5-methyl-	CHCl$_3$	264(4.39),319(4.25), 435(3.95)	97-0435-84
$C_{11}H_8BrN_3O$ Pyrazolo[5,1-b]quinazolin-9(4H)-one, 7-bromo-2-methyl-	10% EtOH-NaOH	218(4.63),250(4.49), 324(4.20),402(3.59)	56-0411-84
3-Pyridinol, 2-[(3-bromophenyl)azo]-	MeCN	377(4.15)	103-1364-84
cation	MeCN	386(4.32)	103-1364-84
anion	MeCN	523(4.19)	103-1364-84
dication	MeNO$_2$	477(4.06)	103-1364-84
3-Pyridinol, 2-[(4-bromophenyl)azo]-	MeCN	379(4.32)	103-1364-84
cation	MeCN	396(4.51)	103-1364-84

Compound	Solvent	$\lambda_{max}(\log \epsilon)$	Ref.
(cont.)	anion	525(4.34)	103-1364-84
	dication	500(4.54),528(4.63)	103-1364-84
3-Pyridinol, 6-[(3-bromophenyl)azo]-	MeCN	343(4.20)	103-1364-84
cation	MeCN	360(4.26)	103-1364-84
anion	MeCN	483(4.45)	103-1364-84
dication	MeNO$_2$	470(4.34)	103-1364-84
3-Pyridinol, 6-[(4-bromophenyl)azo]-	MeCN	343(4.39)	103-1364-84
cation	MeCN	371(4.47)	103-1364-84
anion	MeCN	481(4.58)	103-1364-84
dication	MeNO$_2$	519(4.63)	103-1364-84
$C_{11}H_8BrN_3O_2$ Pyrazolo[5,1-b]quinazolin-9(4H)-one, 7-bromo-3-hydroxy-2-methyl-	10% EtOH-NaOH	223(4.45),244(4.41), 330(4.08),402(3.57)	56-0411-84
$C_{11}H_8ClNO$ Benzenamine, 2-chloro-N-(2-furanyl-methylene)-	C_6H_{12}	236(3.97),280(4.23), 324(3.90)	18-0844-84
	MeOH	234(3.95),282(4.27), 318(4.04)	18-0844-84
Benzenamine, 4-chloro-N-(2-furanyl-methylene)-	C_6H_{12}	225(4.02),285(4.27), 324(4.17)	18-0844-84
	MeOH	222(4.11),288(4.34), 318(4.37)	18-0844-84
$C_{11}H_8ClNOS$ 2(1H)-Quinolinethione, 8-chloro-1-ethenyl-3-hydroxy-	MeOH-KOH	260(4.47),385(3.85), 406(3.75)	1-0109-84
3-Quinolinol, 8-chloro-2-(ethenyl-thio)-	MeOH-KOH	268(4.44),351(3.88), 366(3.96)	1-0109-84
$C_{11}H_8ClNO_2S$ 6H-1,3-Thiazin-6-one, 4-chloro-2-(4-methoxyphenyl)-	EtOH	230(4.14),237(4.02), 370(4.15)	103-1088-84
$C_{11}H_8ClNO_4$ 5(4H)-Isoxazolone, 4-(5-chloro-2-hy-droxybenzoyl)-3-methyl-	EtOH	347(3.92)	44-4419-84
$C_{11}H_8ClNS_2$ 6H-1,3-Thiazine-6-thione, 2-(4-chloro-phenyl)-5-methyl-	CHCl$_3$	262(4.41),317(4.26), 435(3.92)	97-0435-84
$C_{11}H_8ClN_2$ Cyclobuta[1,2-c:4,3-c']dipyridinium, 6-chloro-2-methyl-, iodide	H$_2$O	230s(4.34),247s(4.72), 253(4.83),274s(3.86), 299.5(3.36),314.5(3.22), 328s(2.58),358(2.46)	150-3001-84M
$C_{11}H_8ClN_3$ 1H-Pyrazolo[4,3-b]quinoline, 9-chloro-1-methyl-	EtOH	246(5.08),318(3.71), 333(3.92),373(3.82), 390(3.81)	103-0918-84
2H-Pyrazolo[4,3-b]quinoline, 9-chloro-2-methyl-	EtOH	242(4.91),326(3.98), 340(4.07),389(3.94), 411(3.83)	103-0918-84
$C_{11}H_8ClN_3O$ Pyrazolo[5,1-b]quinazolin-9(4H)-one, 7-chloro-2-methyl-	10% EtOH-NaOH	220(4.43),255(4.51), 323(4.22),398(3.69)	56-0411-84

Compound	Solvent	$\lambda_{max}(\log \epsilon)$	Ref.
$C_{11}H_8ClN_3O_2$ Pyrazolo[5,1-b]quinazolin-9(4H)-one, 7-chloro-3-hydroxy-2-methyl-	10% EtOH- NaOH	221(4.47),328(4.03), 404(3.40)	56-0411-84
$C_{11}H_8Cl_4S$ 1-Butyne, 3,3,4,4-tetrachloro-1-[(4-methylphenyl)thio]-	hexane	198(4.45),225s(4.29), 262(4.16)	5-1873-84
$C_{11}H_8Cl_6$ Bicyclo[2.2.1]hept-2-ene, 5-(1,3-buta-dienyl)-1,2,3,4,7,7-hexachloro-	EtOH	220(4.18),230(4.23)	70-1743-84
$C_{11}H_8N$ Pyrrolo[2,1,5-de]quinolizinium per-chlorate	EtOH	229(4.81),269(3.89), 282s(3.74),295(3.58), 332(3.79)	39-2553-84C
$C_{11}H_8N_2$ 1H-Pyrrolo[2,3-f]isoquinoline	EtOH	210(4.48),267(4.78), 294(3.98)	103-0394-84
Tricyclo[3.3.1.0^{2,8}]nona-3,6-diene-2,6-dicarbonitrile	PrCN	201(4.087),239(3.926)	89-0631-84
$C_{11}H_8N_2O$ Benzeneacetonitrile, 4-cyano-α-(meth-oxymethylene)-, (E)-	MeOH	287(4.23)	56-0935-84
(Z)-	MeOH	220(4.19)	56-0935-84
6H-Pyrrolo[2,3-f]isoquinolin-6-one, 1,7-dihydro-	EtOH	206(3.85),259(4.68), 268(4.68),296(3.61), 312(3.55),327(3.71)	103-0399-84
$C_{11}H_8N_2OS$ 3-Pyridinecarbonitrile, 4-(2-furanyl)-1,2-dihydro-6-methyl-2-thioxo-	dioxan	236(3.68),268(3.62), 317(4.13),450(3.13)	104-1780-84
$C_{11}H_8N_2O_2$ Benz[cd]indole, 1,5-dihydro-4-nitro-	EtOH	204(4.22),224(4.31), 260(3.82),288(3.91)	44-4761-84
4H-[1]Benzoxepino[3,4-c]pyrazol-4-one, 1(or 2),10-dihydro-	EtOH	268(4.13)	4-0301-84
1H-Indole-3-carbonitrile, 1-acetoxy-	MeOH	213(4.29),231(4.42), 250s(4.12),280(3.58), 322(3.81)	150-1301-84M
$C_{11}H_8N_2O_2S$ 4,6(1H,5H)-Pyrimidinedione, dihydro-5-(phenylmethylene)-2-thioxo-	dioxan	<u>360(4.3)</u>	104-0141-84
$C_{11}H_8N_2O_2S_2$ 6H-1,3-Thiazine-6-thione, 5-methyl-2-(4-nitrophenyl)-	CHCl$_3$	262(4.37),301(4.24), 445(3.69)	97-0435-84
$C_{11}H_8N_2O_2Se$ 4,6(1H,5H)-Pyrimidinedione, dihydro-5-(phenylmethylene)-2-selenoxo-	dioxan	278(3.86),340(4.11), 389(4.20)	104-0141-84
$C_{11}H_8N_2O_3$ Acetamide, N-(5,8-dihydro-5,8-dioxo-6-quinolinyl)-	EtOH	229(4.21),250(4.26), 275(4.07),371(3.34)	95-0191-84
Benzenamine, N-(2-furanylmethylene)-4-nitro-	C$_6$H$_{12}$	218(4.00),323(4.17)	18-0844-84
	MeOH	218(4.15),341(4.35)	18-0844-84

Compound	Solvent	$\lambda_{max}(\log \epsilon)$	Ref.
Methanone, (3-nitrophenyl)-1H-pyrrol-2-yl-	MeOH	226(4.27),256s(4.02), 309(4.18)	95-0238-84
Methanone, (3-nitrophenyl)-1H-pyrrol-3-yl-	MeOH	222(4.31),259(4.19)	95-0238-84
2,4,6(1H,3H,5H)-Pyrimidinetrione, 5-(phenylmethylene)-	dioxan	<u>323(4.2)</u>	104-0141-84
$C_{11}H_8N_4$			
5,8-Methano-5H-imidazo[1,2-a]azepine-2,3-dicarbonitrile, 8,9-dihydro-	EtOH	255(4.02)	44-0813-84
$C_{11}H_8N_4OS$			
6H-Purin-6-one, 1,2,3,7-tetrahydro-1-phenyl-2-thioxo-	pH 13	234(4.40),285(4.19)	2-0316-84
$C_{11}H_8N_4O_2$			
2H-[1,2,4]Triazino[3,2-b]quinazoline-3,10(4H)-dione, 2-methyl-	EtOH	288(3.21),332(3.23)	24-1077-84
$C_{11}H_8N_4O_3$			
Pyrazolo[5,1-b]quinazolin-9(4H)-one, 2-methyl-7-nitro-	10% EtOH-NaOH	239(4.56),390(3.99), 460(4.13)	56-0411-84
3-Pyridinol, 2-[(3-nitrophenyl)azo]-	MeCN	375(4.14)	103-1364-84
cation	MeCN	382(4.35)	103-1364-84
anion	MeCN	527(4.25)	103-1364-84
dication	MeNO$_2$	466(4.35),490(4.38)	103-1364-84
3-Pyridinol, 2-[(4-nitrophenyl)azo]-	MeCN	391(4.22)	103-1364-84
cation	MeCN	391(4.41)	103-1364-84
anion	MeCN	587(4.39)	103-1364-84
dication	MeNO$_2$	469(4.42),492(4.70)	103-1364-84
3-Pyridinol, 6-[(3-nitrophenyl)azo]-	MeCN	343(4.09)	103-1364-84
cation	MeCN	357(4.19)	103-1364-84
anion	MeCN	492(4.36)	103-1364-84
dication	MeNO$_2$	448(3.86)	103-1364-84
3-Pyridinol, 6-[(4-nitrophenyl)azo]-	MeCN	360(4.27)	103-1364-84
cation	MeCN	367(4.37)	103-1364-84
anion	MeCN	551(4.52)	103-1364-84
dication	MeNO$_2$	477(4.29)	103-1364-84
$C_{11}H_8N_4O_4$			
Pyrazolo[5,1-b]quinazolin-9(4H)-one, 3-hydroxy-2-methyl-7-nitro-	10% EtOH-NaOH	240(4.50),390(4.02), 462(4.00)	56-0411-84
$C_{11}H_8N_6OS$			
10H-Pyrido[1,2-a]thieno[3,4-d]pyrimidin-10-one, 1,3-dihydro-7-(1H-tetrazol-5-yl)-	MeOH	239(4.37),346(4.08)	87-0528-84
$C_{11}H_8N_8$			
8H-Pyrazolo[3,4-e]tetrazolo[5,1-c]-[1,2,4]triazine, 6-methyl-8-phenyl-	EtOH	258(4.72),444(3.18)	4-1565-84
$C_{11}H_8O_2$			
1,4-Naphthalenedione, 6-methyl-	MeOH	208(3.93),249(4.34), 256(4.34),340(3.50)	44-3766-84
$C_{11}H_8O_3$			
1H-Indene-1,3(2H)-dione, 2-acetyl-	pH 1	275(4.41),285(4.54), 306(3.91)	65-1430-84
	pH 13	284(4.53),312(4.01), 325(3.98),380(3.20)	65-1430-84

Compound	Solvent	λ_{max}(log ϵ)	Ref.
(cont.) 1,4-Naphthalenedione, 5-hydroxy- 2-methyl- (plumbagin)	CHCl$_3$ MeOH	282(4.54) 208(4.46),240s(4.04), 264(4.06)	65-1430-84 102-2039-84
1,4-Naphthalenedione, 5-hydroxy- 7-methyl-	MeOH	213(4.26),249(4.16), 345(3.26),425(3.44)	44-3766-84
C$_{11}$H$_8$O$_3$S Cyclopenta[c]thiopyran-3-carboxylic acid, 6-formyl-, methyl ester	CH$_2$Cl$_2$	291(4.70),299s(4.61), 314(4.51),367(3.79), 382(3.77),485(2.97)	118-0262-84
3(2H)-Thiophenone, 2-(phenylmethyl- ene)-, 1,1-dioxide	EtOH	330(4.15)	104-0278-84
C$_{11}$H$_8$O$_4$ 1,4-Naphthalenedione, 2,7-dihydroxy- 5-methyl-	EtOH	265(4.31),296(4.05), 348(3.46)	44-1853-84
C$_{11}$H$_9$BF$_2$O$_2$ Boron, (2-acetyl-2,3-dihydro-1H-inden- 1-onato-0,0')difluoro-, (T-4)- (or cyclic isomer)	dioxan	238(3.65),261(3.75), 340(4.42)	83-0448-84
C$_{11}$H$_9$BrN$_2$O$_3$ 1,2,4-Oxadiazole-5-propanoic acid, 3- (4-bromophenyl)-	EtOH	250(4.32),280(3.43), 292.5(3.17)	4-1193-84
C$_{11}$H$_9$Cl Benzene, 1-chloro-3-(1-ethynyl-1-prop- enyl)-	EtOH	260(2.62)	65-1400-84
Benzene, 1-chloro-4-(1-ethynyl-1-prop- enyl)-	EtOH	258(2.60)	65-1400-84
C$_{11}$H$_9$ClN$_2$O$_2$S Thiopyrano[4,3,2-de]quinazoline, 7- chloro-4,5-dihydro-5-methyl-, 6,6- dioxide	MeOH	226(4.46),263s(3.76), 272(3.83),281s(3.75), 313(3.64),324s(3.63)	83-0027-84
	MeOH-H$_2$SO$_4$	225(4.09),348(3.67)	83-0027-84
C$_{11}$H$_9$ClN$_2$O$_3$ 1,2,4-Oxadiazole-5-propanoic acid, 3-(4-chlorophenyl)-	EtOH	246(4.34),276(3.31), 286(3.13)	4-1193-84
C$_{11}$H$_9$ClN$_2$S Thiopyrano[4,3,2-de]quinazoline, 7-chloro-4,5-dihydro-5-methyl-	MeOH	207(4.42),220s(4.24), 247(4.18),260s(4.09), 308s(3.23),360(3.64)	83-0027-84
	MeOH-H$_2$SO$_4$	207(4.14),240s(3.97), 275(4.01),320s(3.49), 350s(3.25)	83-0027-84
C$_{11}$H$_9$ClN$_4$ Pyridazino[3,4-b]quinoxaline, 3-chloro- 5,10-dihydro-10-methyl-	EtOH	246(4.68),368(3.90), 400s(3.78)	103-1281-84
C$_{11}$H$_9$CrFO$_3$ Chromium, tricarbonyl[(1,2,3,4,5,6-η)- 1-fluoro-2,3-dimethylbenzene]-	EtOH	217(4.29),255(3.78), 313(3.98)	112-0765-84
C$_{11}$H$_9$F$_3$N$_2$O Ethanone, 1-(1,2-dimethyl-2H-cyclo- penta[d]pyridazin-5-yl)-2,2,2-tri-	hexane	207(4.18),245(4.02), 257s(3.97),265(3.99),	44-4769-84

Compound	Solvent	$\lambda_{max}(\log \epsilon)$	Ref.
fluoro- (cont.)		298.5(4.21),360(3.86)	44-4769-84
	MeOH	210(4.29),245.5(4.07), 264(4.14),297(4.20), 374(4.10)	44-4769-84
Ethanone, 1-(1,2-dimethyl-2H-cyclo- penta[d]pyridazin-7-yl)-2,2,2-tri- fluoro-	hexane	209(4.25),245.5(4.42), 268s(4.16),284(3.95), 295(3.90),331(3.99), 394(4.07)	44-4769-84
	MeOH	221.5(4.16),248(4.17), 267s(3.99),294(3.71), 326(3.78),394(4.04)	44-4769-84
$C_{11}H_9F_3O$ 3-Buten-2-one, 1,1,1-trifluoro-4-(4- methylphenyl)-	EtOH	240(3.78),254(3.79), 324(4.24)	39-2863-84C
$C_{11}H_9F_3O_2$ 3-Buten-2-one, 1,1,1-trifluoro-4-(4- methoxyphenyl)-	EtOH	245(3.71),349(4.33)	39-2863-84C
$C_{11}H_9I$ Azulene, 4-(iodomethyl)-	CH_2Cl_2	235s(--),247(4.41), 287(4.37),341s(--), 590(2.79)	83-0984-84
$C_{11}H_9NO$ 4-Azulenecarboxaldehyde, oxime	CH_2Cl_2	280s(--),290(4.44), 295s(--),348(3.40), 600(2.58)	83-0984-84
Benzenamine, N-(2-furanylmethylene)-	C_6H_{12}	228(4.12),282(4.40), 318(4.24)	18-0844-84
	MeOH	228(4.14),285(4.44), 315(4.43)	18-0844-84
Benz[cd]indol-4(3H)-one, 1,5-dihydro-	EtOH	220(4.51),276(3.76)	44-4761-84
5H-Benzocyclohepten-5-one, 3-amino-	EtOH	243(4.28),268(4.41), 360(4.06),434(3.84)	39-2297-84C
Methanone, phenyl-1H-pyrrol-2-yl-	MeOH	246(3.93),305(4.21)	95-0238-84
Methanone, phenyl-1H-pyrrol-3-yl-	MeOH	248(4.07),280s(3.84)	95-0238-84
2-Oxa-3-azabicyclo[3.2.0]hepta-3,6-di- ene, 4-phenyl-	C_6H_{12}	268.5(4.01),275s(3.97), 287s(3.60),293s(3.08)	40-0158-84
	EtOH	269(4.04)	40-0158-84
1,3-Oxazepine, 2-phenyl-	EtOH	238(4.16),323(3.66)	40-0158-84
2-Propenal, 3-(3-phenyl-2H-azirin-2- yl)-	MeCN	256(4.49)	40-0158-84
1H-Pyrrole-3-carboxaldehyde, 2-phenyl-	MeOH	205(4.22),231(4.23), 304(4.00)	83-0143-84
8-Quinolinol, 2-ethenyl-	MeOH	262(4.66),285(3.76), 316(3.59)	138-1191-84
$C_{11}H_9NOS$ 2(1H)-Quinolinethione, 1-ethenyl- 3-hydroxy-	MeOH-KOH	248(4.37),282(4.11), 390s(4.38),400(4.43)	1-0109-84
$C_{11}H_9NOS_2$ 6H-1,3-Thiazine-6-thione, 2-(4-methoxy- phenyl)-	$CHCl_3$	280(4.20),326(4.09), 435(3.96)	97-0435-84
$C_{11}H_9NO_2$ Ethanone, 1-(4-hydroxy-3-quinolinyl)-	MeOH	245(4.00),314(4.06)	4-0759-84
1H-Pyrrole-2,3-dione, 1-methyl-5-phen- yl-	dioxan	255(3.59),285s(3.34), 405(3.81)	94-0497-84

Compound	Solvent	λ_{max}(log ϵ)	Ref.
1H-Pyrrole-2,3-dione, 1-methyl-5-phenyl-, hydrate	dioxan	228(4.00),285(3.43)	94-0497-84
3-Quinolinecarboxylic acid, methyl ester	2M H_2SO_4 pH 4	311(2.55),321(2.56) 286(2.49)	149-0057-84A 149-0057-84A
Spiro[cyclopropane-1,1'-[1H]indene], 2-nitro-	EtOH	242(4.24),298(3.30)	23-2506-84
isomer	EtOH	246(4.07),301(3.34)	23-2506-84
$C_{11}H_9NO_2S$			
1H-Indene-1,3(2H)-dione, 2-[amino(methylthio)methylene]-	EtOH	232(4.40),286(4.12), 296(4.46),373(4.37), 335(4.56)	95-0440-84
6H-1,3-Thiazin-6-one, 4-methoxy-2-phenyl-	EtOH	225(4.27),240(4.24), 270(4.20),350(3.79)	103-1088-84
$C_{11}H_9NO_3$			
2-Furancarboxamide, N-hydroxy-N-phenyl-	pH 0-6 base 12M HCl	270(4.15) 250(4.01),305(3.91) 285(4.12)	140-1598-84 140-1598-84 140-1598-84
1H-Indene-1,3(2H)-dione, 2-acetyl-4-amino-	M HCl pH 2.04 pH 13 CHCl₃	234(4.37),283(4.43) 221(4.34),283(4.45), 400(3.46) 250(4.14),297(4.46), 360(3.64) 283(4.40)	65-1430-84 65-1430-84 65-1430-84 65-1430-84
$C_{11}H_9NO_4$			
5(4H)-Isoxazolone, 4-(2-hydroxybenzoyl)-3-methyl-	EtOH	344(3.90)	44-4419-84
$C_{11}H_9NO_4S$			
Benzo[b]thiophen-3(2H)-one, 2-(2-oxazolidinylidene)-, 1,1-dioxide	EtOH	242(4.61),248(4.59), 256(4.59),264(4.61), 322(4.37)	95-0134-84
$C_{11}H_9NS_2$			
6H-1,3-Thiazine-6-thione, 5-methyl-2-phenyl-	CHCl₃	257(4.43),315(4.24), 430(4.00)	97-0435-84
$C_{11}H_9N_2O$			
Cyclobuta[1,2-c:4,3-c']dipyridinium, 2-methoxy-, iodide	H_2O	230(4.62),235s(4.59), 261(4.18),287(4.26), 360s(3.60),379(3.68)	150-3001-84M
$C_{11}H_9N_3$			
Imidazo[1,5-a]quinolin-7-amine	n.s.g.	214(4.23),258(4.20)	11-0038-84B
1H-Pyrrolo[2,3-f]isoquinolin-6-amine	EtOH	269(4.84),310(3.91)	103-0399-84
$C_{11}H_9N_3O$			
Pyrazolo[5,1-b]quinazolin-9(4H)-one, 2-methyl-	10% EtOH-NaOH	215(4.47),251(4.55), 325(4.14),400(3.61)	56-0411-84
Pyridine, 2-(phenyl-α-azoxy)-	EtOH	268(3.97),320(4.05)	23-1628-84
Pyridine, 3-(phenyl-α-azoxy)-	EtOH	225(4.03),318(4.22)	23-1628-84
Pyridine, 3-(phenyl-β-azoxy)-	EtOH	238(4.12),326(4.09)	23-1628-84
Pyridine, 4-(phenyl-α-azoxy)-	EtOH	225(4.11),302(3.60)	23-1628-84
Pyridine, 4-(phenyl-β-azoxy)-	EtOH	234(4.01),329(4.13)	23-1628-84
3-Pyridinol, 2-(phenylazo)-	MeCN	373(4.12)	103-1364-84
cation	MeCN	387(4.31)	103-1364-84
anion	MeCN	508(4.11)	103-1364-84
dication	MeNO₂	473(4.41),496(4.45)	103-1364-84

Compound	Solvent	$\lambda_{max}(\log \epsilon)$	Ref.
3-Pyridinol, 6-(phenylazo)-	MeCN	339(4.30)	103-1364-84
cation	MeCN	363(4.38)	103-1364-84
anion	MeCN	470(4.48)	103-1364-84
dication	MeNO$_2$	485(4.59)	103-1364-84
$C_{11}H_9N_3OS_2$			
2-Butanone, 3-(2-benzothiazolylthio)-1-diazo-	MeOH	230(4.32),249(4.16), 280(4.22)	103-0505-84
$C_{11}H_9N_3O_2$			
2-Azetidinone, 4-(diazoacetyl)-1-phenyl-	MeCN	249(3.42),344(1.74)	64-0095-84B
Pyrazolo[5,1-b]quinazolin-9(4H)-one, 3-hydroxy-2-methyl-	10% EtOH-NaOH	217(4.54),243(4.41), 317(3.96),390(3.41)	56-0411-84
Pyridine, 2-(phenyl-α-azoxy)-, N-oxide	EtOH	268(4.29),330(3.89)	23-1628-84
Pyridine, 3-(phenyl-α-azoxy)-, N-oxide	EtOH	278(4.34),308(4.20)	23-1628-84
Pyridine, 3-(phenyl-β-azoxy)-, N-oxide	EtOH	278(4.34),331(4.17)	23-1628-84
Pyridine, 4-(phenyl-α-azoxy)-, N-oxide	EtOH	289(3.91),367(4.38)	23-1628-84
Pyridine, 4-(phenyl-β-azoxy)-, N-oxide	EtOH	232(4.12),349(4.38)	23-1628-84
$C_{11}H_9N_3O_2S$			
2-Butanone, 3-(2-benzoxazolylthio)-1-diazo-	MeOH	212(4.11),255(3.95), 274(4.00)	103-0505-84
Pyrazolo[1,5-a]pyridine-6-carboxylic acid, 3-cyano-2-(methylthio)-, methyl ester	EtOH	267(4.60),315(4.00)	95-0440-84
$C_{11}H_9N_3O_3$			
Propanedinitrile, (2-hydroxy-3-oxo-1-morpholino-1-cyclobuten-4-ylidene)-, sodium salt	H$_2$O	217(4.30),315s(4.17), 358(4.58)	24-2714-84
$C_{11}H_9N_3O_4S$			
3-Pyridinecarboxylic acid, 2-[(4-carboxy-1H-imidazol-2-yl)thio]-, 3-methyl ester	MeOH	210(4.29),255(4.23), 295s(3.53)	142-0807-84
$C_{11}H_9N_3O_5$			
Butanoic acid, 2-diazo-3-oxo-, (4-nitrophenyl)methyl ester	CH$_2$Cl$_2$	260(4.24)	23-2936-84
$C_{11}H_9N_3S$			
3H-Naphtho[1,2-d][1,3]thiazol-2-imine, 3-amino-	MeOH	237(4.53),252(4.54), 313(3.76),327(3.75)	4-1571-84
6H-Pyrazolo[4,3-d]isothiazole, 4-methyl-6-phenyl-	EtOH	240(4.21),272(4.12)	65-0134-84
	CHCl$_3$	277(4.17),285s(--)	65-0134-84
$C_{11}H_9N_5O$			
Furo[2',3':4,5]pyrrolo[1,2-d]-1,2,4-triazolo[3,4-f][1,2,4]triazine, 3,6-dimethyl-	dioxan	248(3.37),303(3.16)	73-0065-84
$C_{11}H_9N_5O_2$			
[1,2,5]Oxadiazolo[3,4-d]pyrimidin-5-amine, 7-(phenylmethoxy)-	EtOH	218(4.15),256(3.74), 340(3.52)	78-0879-84
3H-1,2,4-Triazino[3,2-b]quinazoline-3,10(4H)-dione, 4-amino-2-methyl-	EtOH	313(3.34),350(3.24)	24-1083-84
$C_{11}H_9N_5O_4$			
1H-Imidazole-4-carboxamide, 5-[(4-nitrobenzoyl)amino]-	EtOH	270(4.42)	103-0204-84

Compound	Solvent	$\lambda_{max}(\log \epsilon)$	Ref.
$C_{11}H_{10}$ 3-Penten-1-yne, 3-phenyl-	EtOH	256(4.24)	65-1400-84
$C_{11}H_{10}BF_2NO_2$ Boron, (2-acetyl-1,2-dihydro-1-methyl-3H-indol-3-onato-0,0')difluoro-, (T-4)- (or cyclic isomer)	dioxan	245(4.04),266(3.70), 335(4.13),370(3.28)	83-0443-84
$C_{11}H_{10}BrNO$ Ethanone, 1-(3-bromo-5-methyl-1H-indol-2-yl)-	n.s.g.	210(4.33),242(4.13), 313(4.31)	103-0988-84
Ethanone, 1-(3-bromo-7-methyl-1H-indol-2-yl)-	n.s.g.	208(4.33),246(4.26), 311(4.30)	103-0988-84
$C_{11}H_{10}BrNO_3$ 1H-Indole-3-carboxylic acid, 5-bromo-2-(1-hydroxyethyl)-	n.s.g.	220(4.59),289(4.03), 296(4.00)	103-0988-84
$C_{11}H_{10}BrN_3O_4$ 2-Butenoic acid, 2-bromo-3-[(4-nitrophenyl)azo]-, methyl ester, (E,E)-	Tetra	330(4.4),342(4.3), 360s(--),454(2.6)	48-0367-84
$C_{11}H_{10}Br_2N_4O_4$ 2,4(1H,3H)-Pyrimidinedione, 1,1'-(1,3-propanediyl)bis[5-bromo-	pH 2 H_2O pH 12	284(4.22) 284(4.23) 281(4.10)	142-2739-84 142-2739-84 142-2739-84
$C_{11}H_{10}ClNO_2$ 2-Oxazolidinone, 3-(3-chlorophenyl)-5-methyl-4-methylene-	EtOH	242(3.46)	4-0985-84
$C_{11}H_{10}ClNO_2S$ 2(1H)-Quinolinone, 8-chloro-3-hydroxy-N-(2-mercaptoethyl)-	MeOH-KOH	257(4.21),316s(3.97), 328(4.02),342(4.05)	1-0109-84
$C_{11}H_{10}ClNO_3$ 1H-Indole-3-carboxylic acid, 5-chloro-2-(1-hydroxyethyl)-	n.s.g.	222(4.58),289(4.02), 296(3.98)	103-0988-84
$C_{11}H_{10}ClN_3O_4$ Pyrido[3,2-d]pyrimidine-6-carboxylic acid, 8-chloro-2,4-dimethoxy-, methyl ester	MeOH	288(3.93),319.8(3.85)	87-1710-84
$C_{11}H_{10}F_3N_3O_2$ L-Phenylalanine, 4-[3-(trifluoromethyl)-3H-diazirin-3-yl]-	H_2O	222(4.18),265(2.54), 271s(2.46),354(2.51)	35-7540-84
$C_{11}H_{10}N$ Pyridinium, 4-phenyl-	H_2O	386(4.24)	142-0505-84
$C_{11}H_{10}NOS_2$ 2H-[1,3]Thiazino[2,3-b]benzothiazolium, 3,4-dihydro-2-methyl-3-oxo-, perchlorate	10% MeCN	232(3.28),260(3.26), 318(3.75)	103-0508-84
$C_{11}H_{10}NO_2S$ 2H-[1,3]Thiazino[2,3-b]benzoxazolium, 3,4-dihydro-2-methyl-3-oxo-, perchlorate	10% MeCN	232(3.72),282(3.75), 289(3.73)	103-0508-84

Compound	Solvent	$\lambda_{max}(\log \epsilon)$	Ref.
$C_{11}H_{10}N_2$			
1H-Cyclopenta[b]quinoxaline, 2,3-di-hydro-	MeOH	206(4.36),241(4.39), 319(4.03)	56-0917-84
$C_{11}H_{10}N_2O$			
Benzonitrile, 2-[(3-oxo-1-butenyl)-amino]-, (Z)-	MeOH	228(4.03),338(4.37)	4-0759-84
1H-Cyclopenta[b]quinoxaline, 2,3-di-hydro-, 4-oxide	MeOH	213(4.29),247(4.76), 328(4.00)	56-0917-84
1H-Indole-3-acetonitrile, 4-methoxy-	MeOH	221(4.95),267(4.20), 281(4.04),291(4.00)	158-0931-84
2-Propenenitrile, 2-formyl-3-(methyl-amino)-3-phenyl-, (Z)-	EtOH	295(3.93),310(4.04)	44-0282-84
2(1H)-Pyrazinone, 1-methyl-5-phenyl-	EtOH	274(4.36),350(3.70)	39-0391-84C
1H-Pyrazole-4-carboxaldehyde, 3-methyl-1-phenyl-	EtOH	226(4.11),275(4.23)	65-0134-84
	CHCl$_3$	277(4.28),287s(--)	65-0134-84
Pyrrolo[2,1-b]quinazolin-9(1H)-one, 2,3-dihydro-	EtOH	224(4.42),265(3.89), 269s(3.87),293s(3.46), 302(3.59),314(3.52)	4-0219-84
$C_{11}H_{10}N_2OS$			
Naphtho[1,2-d][1,2,3]thiadiazole, 4,5-dihydro-7-methoxy-	EtOH	266(4.17)	2-0203-84
1H-Pyrazole-4-carboxaldehyde, 5-mer-capto-3-methyl-1-phenyl-	EtOH	230(4.18),266(4.19), 338(3.60)	65-0134-84
	CHCl$_3$	275(4.09)	65-0134-84
2H-[1,3]Thiazino[3,2-a]benzimidazol-3(4H)-one, 2-methyl-	MeOH	216(4.32),253(3.8), 272(3.89)	103-0508-84
$C_{11}H_{10}N_2O_2$			
Benz[cd]indole, 1,3,4,5-tetrahydro-4-nitro-	EtOH	224(4.50),280(3.73)	44-4761-84
Glycine, N-(8-quinolinyl)-, anion	pH 9.7	254(4.3),342(3.4)	18-1211-84
$C_{11}H_{10}N_2O_3$			
Acetamide, N-(3-methyl-2-oxo-2H-1,4-benzoxazin-7-yl)-	EtOH	343(4.15)	4-0551-84
2H-1-Benzopyran-4-carboxamide, 7-amino-N-methyl-2-oxo-	EtOH	373(4.10)	94-3926-84
1H-Indole, 5-methoxy-4-(2-nitroethen-yl)-	EtOH	206(4.40),225(4.36), 264(3.94)	44-4761-84
1,2,4-Oxadiazole-5-propanoic acid, 3-phenyl-	EtOH	235(4.16),275(3.24), 284(2.94)	4-1193-84
$C_{11}H_{10}N_2O_3S$			
Benzo[b]thiophen-3(2H)-one, 2-(2-imida-zolidinylidene)-, 1,1-dioxide	EtOH	247(4.14),257(4.26), 265(4.30),303(3.83), 328(3.91)	95-0134-84
$C_{11}H_{10}N_2O_5$			
1H-Pyrrolo[3,4-c]pyridine-7-carboxylic acid, 6-methyl-1,3,4-trioxo-, ethyl ester	MeOH	255(3.93),325(3.75)	48-0594-84
$C_{11}H_{10}N_2S$			
2H-[1]Benzothiepino[5,4-c]pyrazole, 4,5-dihydro-	EtOH	252(4.19)	2-0736-84
$C_{11}H_{10}N_2S_2$			
1H-Pyrazole, 4,5-dihydro-3,5-dithienyl-	EtOH	239(3.84),302(4.00)	34-0225-84

Compound	Solvent	$\lambda_{max}(\log \epsilon)$	Ref.
$C_{11}H_{10}N_4$			
2-Butenedinitrile, 2-amino-3-(3-aza-tricyclo[3.2.1.0^{2,4}]oct-6-en-3-yl)-	EtOH	303(4.15)	44-0813-84
Ethenetricarbonitrile, (4,5-dihydro-5,5-dimethyl-1H-pyrrol-3-yl)-	MeOH	203(3.77),208(3.77), 216(3.76),245(3.70), 273(3.50),439(4.61)	83-0143-84
$C_{11}H_{10}N_4O$			
Ethanone, 1-[3-(1H-imidazol-2-yl-azo)phenyl]-	MeOH	193(4.08),222(4.04), 359(4.27)	56-0903-84
Ethanone, 1-[4-(1H-imidazol-2-yl-azo)phenyl]-	MeOH	194(4.26),235(3.53), 252(3.54),371(4.27)	56-0903-84
$C_{11}H_{10}N_4OS$			
2-Butanone, 3-(1H-benzimidazol-2-yl-thio)-1-diazo-	MeOH	255(3.97),275(4.25)	103-0505-84
$C_{11}H_{10}N_4O_2$			
1H-Imidazole-4-carboxamide, 5-(benz-oylamino)-	EtOH	237(4.22),260(4.11)	103-0204-84
$C_{11}H_{10}N_4O_4$			
1H-Imidazole-4-carboxamide, 5-hydroxy-1-[(4-nitrophenyl)methyl]-	M HCl	240s(--),273(4.15)	4-0849-84
	H$_2$O	240s(--),277(4.29)	4-0849-84
	M NaOH	285(4.31)	4-0849-84
$C_{11}H_{10}N_4O_5$			
11H-Furo[2',3':4,5]oxazolo[2,3-b]pteri-din-11-one, 6a,7,8,9a-tetrahydro-7-hydroxy-8-(hydroxymethyl)-, [6aR-[(6aα,7β,8α,9aβ)]]-	MeOH	230(4.04),260(3.97), 319(3.78)	136-0179-84F
$C_{11}H_{10}N_4S$			
Pyridinium, 1-[[2,2-dicyano-1-(methyl-thio)ethenyl]amino]-2-methyl-, hydroxide, inner salt	EtOH	247(4.22),275(4.24)	95-0440-84
Pyridinium, 1-[[2,2-dicyano-1-(methyl-thio)ethenyl]amino]-3-methyl-, hydroxide, inner salt	EtOH	247(4.19),275(4.22), 378(3.28)	95-0440-84
$C_{11}H_{10}N_6O$			
7H-Pyrazolo[3,4-d]pyridazin-7-one, 3,4-diamino-2,6-dihydro-2-phenyl-	EtOH	224(4.39)	4-1049-84
$C_{11}H_{10}N_6O_2$			
1H-Pyrrole-2,5-dione, 1-[2-(6-amino-9H-purin-9-yl)ethyl]-	pH 1	258(--)	47-2841-84
	H$_2$O	260(4.21)	47-2841-84
	pH 13	260(--)	47-2841-84
polymer	pH 1	258(3.98)	47-3715-84
$C_{11}H_{10}O$			
2-Butenal, 4-(2,4,6-cycloheptatrien-1-ylidene)-, (E)-	MeCN	405(4.48)	24-2027-84
2-Butenal, 4-(tetracyclo[3.2.0.0^{2,7}-0^{4,6}]heptylidene)-, (E)-	MeCN	299(4.48)	24-2027-84
2-Butenal, 4-(tricyclo[4.1.0.0^{2,7}]-hept-4-en-3-ylidene)-, (E,E)-	MeCN	338(4.44)	24-2027-84
(E,Z)-	MeCN	344(4.42)	24-2027-84
Tricyclo[4.3.1.0^{3,10}]deca-1,4,7-triene-7-carboxaldehyde	MeCN	285(3.43)	24-2027-84

Compound	Solvent	$\lambda_{max}(\log \epsilon)$	Ref.
$C_{11}H_{10}O_2$			
1,2-Cyclopentanedione, 3-phenyl-	EtOH	309(4.46)	39-1531-84C
	EtOH-NaOH	356(4.37)	39-1531-84C
Ethanone, 1-[4-(3-hydroxy-1-propenyl)-phenyl]-	MeOH	272(4.38)	118-0728-84
2(3H)-Furanone, dihydro-3-(phenylmethylene)-, (E)-	EtOH	218(4.11),283(4.35)	95-0839-84
$C_{11}H_{10}O_2S_2$			
2-Propenoic acid, 3-[(phenylthioxomethyl)thio]-, methyl ester, (Z)-	isoPrOH	236(4.05),269(3.87), 315(4.20),355(4.10), 525(2.22)	104-1082-84
$C_{11}H_{10}O_3$			
Benzoic acid, 4-(3-hydroxy-1-propynyl)-, methyl ester	MeOH	264(4.43)	118-0728-84
4H-1-Benzopyran-4-one, 7-hydroxy-2,5-dimethyl-	MeOH	241(4.47),250(4.50), 289(4.28)	94-3493-84
	MeOH-AlCl$_3$	241(4.46),250(4.49), 289(4.26)	94-3493-84
$C_{11}H_{10}O_3S_3$			
Benzo[b]thiophen-3(2H)-one, 2-[bis-(methylthio)methylene]-, 1,1-dioxide	EtOH	221(4.21),245(3.97), 259(3.91),371(4.28)	95-0134-84
$C_{11}H_{10}O_4$			
2,4(3H,5H)-Furandione, 3-(4-methoxyphenyl)-	EtOH	260(4.17),276s(4.14)	39-1539-84C
Scoparone	EtOH	230.4(4.05),251s(3.57), 258s(3.53),289s(3.58), 294.6(3.60),343(3.78)	102-0699-84
$C_{11}H_{11}BrN_2O$			
2H-Pyrrol-2-one, 5-[(5-bromo-1H-pyrrol-2-yl)methylene]-1,5-dihydro-3,4-dimethyl-, (Z)-	CHCl$_3$	374(4.33),389(4.30)	49-0357-84
$C_{11}H_{11}BrN_2O_2$			
1H-Indole-2-carboxylic acid, 3-(2-aminoethyl)-6-bromo-	EtOH	229(4.39),299(4.05)	1-0709-84
	EtOH-HCl	229(4.32),306(4.15)	1-0709-84
$C_{11}H_{11}BrN_4O_4$			
2,4(1H,3H)-Pyrimidinedione, 5-bromo-1-[3-(3,4-dihydro-2,4-dioxo-1(2H)-pyrimidinyl)propyl]-	pH 2	273(4.20)	142-2739-84
	H$_2$O	273(4.19)	142-2739-84
	pH 12	271(4.08)	142-2739-84
$C_{11}H_{11}ClN_2$			
Pyrido[1,2-b]indazole, 9-chloro-1,2,3,4-tetrahydro-	H$_2$O	240(4.58),308(3.78)	56-1077-84
$C_{11}H_{11}ClN_2O_2$			
Benzoxazole, 5-chloro-2-morpholino-	CHCl$_3$	255(4.14),293(4.04)	94-3053-84
$C_{11}H_{11}ClN_4O_3$			
2,4(1H,3H)-Pteridinedione, 6-chloro-1,3-dimethyl-7-(1-oxopropyl)-	MeOH	244(4.15),348(3.85)	5-1798-84
$C_{11}H_{11}ClO$			
2-Oxabicyclo[3.1.0]hexane, 6-chloro-6-phenyl-	C$_6$H$_{12}$	216(3.93)	4-1237-84

Compound	Solvent	$\lambda_{max}(\log \epsilon)$	Ref.
Phenol, 4-chloro-2-(1,1-dimethyl-2-propynyl)-	50% MeCN-KOH	309(3.78)	39-1269-84B
$C_{11}H_{11}ClSi$			
Silane, chloroethenylmethyl(phenylethynyl)-	hexane	239.7(4.42),249.5(4.60), 261.2(4.52)	65-2317-84
$C_{11}H_{11}Cl_2N_3O_4$			
7H-Pyrrolo[2,3-d]pyrimidine, 7-β-D-arabinofuranosyl-	MeOH	231(4.46),278(3.63), 295s(3.61)	5-0722-84
$C_{11}H_{11}CrNO_3$			
Chromium, tricarbonyl[(1,2,3,4,5,6-η)-2,3-dimethylbenzenamine]-	EtOH CHCl$_3$	218(4.48) 318(3.92)	112-0765-84 112-0765-84
Chromium, tricarbonyl[(1,2,3,4,5,6-η)-2,4-dimethylbenzenamine]-	EtOH CHCl$_3$	218(4.58) 320(3.92)	112-0765-84 112-0765-84
Chromium, tricarbonyl[(1,2,3,4,5,6-η)-2,5-dimethylbenzenamine]-	EtOH CHCl$_3$	217(4.54) 318(3.92)	112-0765-84 112-0765-84
Chromium, tricarbonyl[(1,2,3,4,5,6-η)-2,6-dimethylbenzenamine]-	EtOH CHCl$_3$	217(4.60) 318(3.91)	112-0765-84 112-0765-84
Chromium, tricarbonyl[(1,2,3,4,5,6-η)-3,4-dimethylbenzenamine]-	EtOH CHCl$_3$	218(4.39) 318(3.89)	112-0765-84 112-0765-84
Chromium, tricarbonyl[(1,2,3,4,5,6-η)-3,5-dimethylbenzenamine]-	EtOH CHCl$_3$	217(4.59) 317(3.95)	112-0765-84 112-0765-84
$C_{11}H_{11}F_3O$			
Benzeneethanol, 4-methyl-β-methylene-α-(trifluoromethyl)-	EtOH	240(4.04)	44-2258-84
$C_{11}H_{11}F_3O_2$			
Benzeneethanol, 4-methoxy-β-methylene-α-(trifluoromethyl)-	EtOH	251(4.05)	44-2258-84
$C_{11}H_{11}NO$			
Benzeneacetonitrile, α-(methoxymethylene)-4-methyl-, (E)-	MeOH	212(4.19),267(4.24), 271(4.23)	56-0935-84
(Z)-	MeOH	217(4.16),270(4.07)	56-0935-84
$C_{11}H_{11}NO_2$			
Benzeneacetonitrile, 4-methoxy-α-(methoxymethylene)-, (E)-	MeOH	217(4.21),276(4.26)	56-0935-84
(Z)-	MeOH	223(4.18),277(4.13)	56-0935-84
Benzenepropanoic acid, α-cyano-, methyl ester	EtOH	305(3.34)	44-0282-84
Furo[3,4-d]isoxazole, 3a,4,6,6a-tetrahydro-3-phenyl-	MeOH	264(4.12)	73-1193-84
2H-Isoindole-4,7-dione, 2,5,6-trimethyl-	n.s.g.	223(4.00),265(4.02), 366(3.30)	88-4917-84
2H-1,3-Oxazine-5-carboxaldehyde, 3,6-dihydro-4-phenyl-	MeOH	238(4.08),310(4.12)	73-1193-84
2(1H)-Quinolinone, 4-methoxy-N-methyl-	EtOH	229(4.50),269(3.64), 279(3.67),318(3.56), 330(3.46)	105-0068-84
$C_{11}H_{11}NO_2S$			
2,5-Pyrrolidinedione, 3-[(phenylmethyl)thio]-	MeCN	202(4.08)	142-1179-84
2(1H)-Quinolinone, 3-hydroxy-1-(2-mercaptoethyl)-	MeOH-KOH	251(4.14),310s(3.82), 328(4.06),341(4.07)	1-0109-84

Compound	Solvent	$\lambda_{max}(\log \epsilon)$	Ref.
$C_{11}H_{11}NO_3$			
3-Butenoic acid, 4-amino-2-oxo-4-phenyl-, methyl ester	EtOH	253(3.85),354(4.28)	94-0497-84
1H-Indole-3-carboxylic acid, 1-methoxy-, methyl ester	MeOH	214(4.43),230s(4.20), 290(4.02)	150-1301-84M
2H-Isoindole-4,7-dione, 5-methoxy-2,6-dimethyl-	n.s.g.	232(4.08),273(4.04), 361(3.48)	88-4917-84
1(2H)-Isoquinolinone, 6,8-dihydroxy-3,4-dimethyl- (siaminine A)	MeOH-HCl	245(3.91),275(3.75), 290(3.81),325(3.19), 350(3.14)	100-0708-84
2(1H)-Quinolinone, 1,4-dimethoxy- (haplotusine)	EtOH	230(4.77),271(3.90), 280(3.90),320(3.84)	105-0599-84
$C_{11}H_{11}NO_4$			
4H-Furo[3,2-c]pyran-3,4(2H)-dione, 2-[(dimethylamino)methylene]-6-methyl-	MeOH	370(4.37)	83-0685-84
hydrochloride	MeOH	369(4.35)	83-0685-84
$C_{11}H_{11}NS$			
1H-Pyrrole, 1-methyl-2-(phenylthio)-	MeOH	207(4.32),248(4.25)	4-1041-84
$C_{11}H_{11}N_3$			
4H-Pyrazolo[1,5-a]benzimidazole, 2,6-dimethyl-	EtOH	233(4.30),309(3.98)	48-0829-84
4-Pyrimidinamine, 2-methyl-6-phenyl-	EtOH	239(4.41),287(3.77)	39-2677-84C
Pyrimido[2,1-a]isoindol-6(2H)-imine, 3,4-dihydro-	MeOH	258(4.15),293(3.93), 303s(3.88)	20-0459-84
$C_{11}H_{11}N_3O_2$			
1H-1,2,4-Triazole, 1-acetyl-3-(4-methoxyphenyl)-	MeCN	254s(4.01),276(4.18), 286s(4.03),296s(3.94)	48-0311-84
$C_{11}H_{11}N_3O_2S$			
Benzenesulfonamide, 4-amino-N-2-pyridinyl-	MeOH	268.92(4.32)	83-0906-84
$C_{11}H_{11}N_3O_3$			
5-Isoxazolidinecarbonitrile, N-methyl-4-nitro-3-phenyl-, (3α,4β,5α)-	EtOH	210(3.82),240(3.43)	44-0276-84
1H-Pyrazole, 3-(4-methoxy-3-nitrophenyl)-5-methyl-	MeOH	208(4.10),252(4.20)	18-1567-84
$C_{11}H_{11}N_3O_4$			
Pyrido[3,2-d]pyrimidine-6-carboxylic acid, 2,4-dimethoxy-, methyl ester	MeOH	275.6(3.94),314.8(3.89)	87-1710-84
2,4(1H,3H)-Pyrimidinedione, 1-[2-(2,5-dihydro-2,5-dioxo-1H-pyrrol-1-yl)-ethyl]-5-methyl-	pH 1	270(--)	47-2841-84
	H_2O	270(3.97)	47-2841-84
	pH 13	266(--)	47-2841-84
1H-Pyrrolo[3,4-c]pyridine-7-carboxylic acid, 4-amino-2,3-dihydro-6-methyl-1,3-dioxo-, ethyl ester	MeOH	219(4.30),262(4.19), 382(3.69)	48-0594-84
$C_{11}H_{11}N_3S$			
3H-Pyrazole-3-thione, 4-(aminomethylene)-2,4-dihydro-5-methyl-2-phenyl-	EtOH	222(4.28),278(4.22), 297s(--),381(3.66)	65-0134-84
	$CHCl_3$	280(4.18),302(4.11), 396(3.61)	65-0134-84
$C_{11}H_{11}N_3S_2$			
4H-1,3,5-Thiadiazine-4-thione,	EtOH	283(4.67),355(3.92)	150-0266-84S

Compound	Solvent	$\lambda_{max}(\log \epsilon)$	Ref.
2-(methylamino)-6-(3-methylphenyl)-			150-0266-84S
$C_{11}H_{12}N_5O_2S$			
Benzenesulfonamide, 4-[(5-cyano-1-methyl-1H-imidazol-4-yl)amino]-	EtOH	285(4.03)	2-0870-84
Pyrido[2,3-d]pyrimidine-6-carbonitrile, 5-amino-1,2,3,4-tetrahydro-1,3-dimethyl-7-(methylthio)-2,4-dioxo-	EtOH	264(4.71),304(4.19)	94-0122-84
Pyrido[2,3-d]pyrimidine-6-carbonitrile, 7-amino-1,2,3,4-tetrahydro-1,3-dimethyl-5-(methylthio)-2,4-dioxo-	EtOH	226(4.42),268(4.38), 277(4.40),337(4.16)	94-0122-84
$C_{11}H_{11}N_5O_4S$			
1H-Pyrazolo[3,4-d]pyrimidine-3-carbonitrile, 4,5-dihydro-1-β-D-ribofuranosyl-4-thioxo-	pH 1	248(3.85),330(4.09)	87-1119-84
	pH 7	248(3.85),333(4.03)	87-1119-84
	pH 11	248(3.85),336(4.02)	87-1119-84
$C_{11}H_{11}N_5O_5$			
1H-Pyrazolo[3,4-d]pyrimidine-3-carbonitrile, 4,5-dihydro-4-oxo-1-β-D-ribofuranosyl-	pH 1 and 7	227(4.02),261(3.92)	87-1119-84
	pH 11	230(4.00),281(3.97)	87-1119-84
$C_{11}H_{12}$			
1,3-Cyclopentadiene, methylenebis-	n.s.g.	248(3.85)	70-2437-84
$C_{11}H_{12}BrCl$			
Bicyclo[5.3.1]undeca-1(11),7,9-triene, 11-bromo-9-chloro-	C_6H_{12}	270(3.60),277(3.50), 330(2.30)	78-4401-84
$C_{11}H_{12}Br_2$			
Bicyclo[5.3.1]undeca-1(11),7,9-triene, 9,11-dibromo-	C_6H_{12}	243(3.90),286(3.50), 336(2.30)	78-4401-84
$C_{11}H_{12}ClN_5O_3$			
4(1H)-Pteridinone, 6-(2-acetoxy-1-chloropropyl)-2-amino-, [S-(R*,R*)]-	MeOH	235(4.07),281(4.25), 346(3.75)	5-1815-84
$C_{11}H_{12}Cl_2$			
Bicyclo[5.3.1]undeca-1(11),7,9-triene, 9,11-dichloro-	C_6H_{12}	238(4.20),272(3.70), 332(2.30)	78-4401-84
$C_{11}H_{12}FN_3O_4$			
2(1H)-Pyrimidinone, 4-amino-1-(2-deoxy-2-fluoro-β-D-arabinofuranosyl)-5-ethynyl-	MeOH	233(4.28),292(4.00)	87-0410-84
$C_{11}H_{12}N_2$			
Benz[cd]indol-4-amine, 1,3,4,5-tetrahydro-	EtOH	222(4.45),280(3.72)	44-4761-84
1H-Cyclopenta[3,4]pyrazolo[1,5-a]pyridine, 2,3-dihydro-8-methyl-	H_2O	232(4.58),305(3.76)	56-1077-84
Pyrido[1,2-b]indazole, 7,8,9,10-tetrahydro-	H_2O	232(4.52),303(3.65)	56-1077-84
$C_{11}H_{12}N_2O$			
Cyclopropaneacetaldehyde, α-(phenylhydrazono)-	MeCN	231(4.01),294.5(3.89), 332.5(4.30)	5-0649-84
Ethanone, 1-(2-methyl-1H-indol-3-yl)-, oxime	n.s.g.	228(4.38),273(4.00)	103-0055-84
1H-Pyrazole, 3-(4-methoxyphenyl)-5-methyl-	MeOH	205(4.09),258(4.16)	18-1567-84

Compound	Solvent	$\lambda_{max}(\log \epsilon)$	Ref.
3H-Pyrazol-3-one, 1,2-dihydro-1,5-di-methyl-2-phenyl-	EtOH CHCl$_3$	270(3.98) 330(--)	70-2482-84 70-2482-84
Pyrrolidinium, 1-imino-2-methyl-5-oxo-2-phenyl-, hydroxide, inner salt	ether	520(1.98)	35-1508-84
2(1H)-Quinoxalinone, 1-ethyl-3-methyl-	EtOH	209(3.91),230(3.96), 279(3.36),327s(3.40), 338(3.43)	44-0827-84
$C_{11}H_{12}N_2OS$			
Benzo[b]thiophen-3(2H)-one, 2-[(2,2-di-methylhydrazino)methylene]-	heptane PrOH DMSO	350(4.34) 350(3.96),440(3.93) 350(3.38),440(3.11)	104-1353-84 104-1353-84 104-1353-84
$C_{11}H_{12}N_2O_2$			
2-Butenoic acid, 3-(phenylazo)-, (E)-, methyl ester	MeOH	311(4.32),433(2.59), 540(1.97)	23-2456-84
(Z)-	MeOH	308(4.20),440(2.38), 549(1.76)	23-2456-84
Sydnone, 4-methyl-3-(1-phenylethyl)-, (S)-	EtOH	299(3.89)	35-3701-84
$C_{11}H_{12}N_2O_2S$			
2-Propenoic acid, 2-cyano-3-[1-methyl-5-(methylthio)-1H-pyrrol-2-yl]-, methyl ester	MeOH	213(3.95),422(4.50)	4-1041-84
$C_{11}H_{12}N_2O_3$			
L-Alanine, 3-(3-cyano-4-methoxy-phenyl)- (3-cyano-O-methyltyrosine)	pH 1	233(4.01),301(3.62)	56-0157-84
1H-Indole, 5-methoxy-4-(2-nitroethyl)-	EtOH	210(4.47),274(3.86), 296(3.73)	44-4761-84
$C_{11}H_{12}N_2O_3S$			
2H-1,2,6-Thiadiazin-3(6H)-one, 4-meth-yl-2-(phenylmethyl)-, 1,1-dioxide	50% EtOH-HCl + NaOH	266(3.84) 303(3.95)	73-0840-84 73-0840-84
2H-1,2,6-Thiadiazin-3(6H)-one, 4-meth-yl-6-(phenylmethyl)-, 1,1-dioxide	50% EtOH-HCl + NaOH	302(3.81) 284(3.91)	73-0840-84 73-0840-84
$C_{11}H_{12}N_2O_4$			
5H-Pyrazolo[5,1-b][1,3]oxazine-3-carb-oxylic acid, 2,7-dimethyl-5-oxo-, ethyl ester	EtOH	224(1.73),283(4.04)	94-0930-84
1H,5H-Pyrazolo[1,2-a]pyrazole-2-carb-oxylic acid, 3,7-dimethyl-1,5-di-oxo-, ethyl ester	EtOH	331(4.20)	94-0930-84
1H,7H-Pyrazolo[1,2-a]pyrazole-2-carb-oxylic acid, 3,5-dimethyl-1,7-di-oxo-, ethyl ester	EtOH	362(4.11)	94-0930-84
$C_{11}H_{12}N_2O_4S$			
2-Propenoic acid, 2-cyano-3-[1-methyl-5-(methylsulfonyl)-1H-pyrrol-2-yl]-, methyl ester	MeOH	224(3.86),355(4.51)	4-1041-84
$C_{11}H_{12}N_2O_5$			
Propanoic acid, 2-[(2-hydroxy-4-nitro-phenyl)imino]-, ethyl ester	EtOH	262(3.84),390(4.18)	4-0551-84
Pyridinium, 4-cyano-3-hydroxy-1-β-D-ribofuranosyl-, hydroxide, inner salt	MeOH	231(4.31),254s(--), 373(3.85)	77-0422-84

Compound	Solvent	$\lambda_{max}(\log \epsilon)$	Ref.
$C_{11}H_{12}N_2O_6$			
2,4(1H,3H)-Pyrimidinedione, 1-β-D-arabinofuranosyl-5-ethynyl-	MeOH	224(3.92),285(4.01)	87-0410-84
$C_{11}H_{12}N_2S$			
4-Isoquinolinecarbonitrile, 2,3,5,6,7,8-hexahydro-1-methyl-3-thioxo-	dioxan	260(3.90),320(3.70), 416(2.84)	104-2216-84
3H-Pyrazole-3-thione, 1,2-dihydro-1,5-dimethyl-2-phenyl-	EtOH	300(4.06)	70-2482-84
$C_{11}H_{12}N_4O$			
1,2,4-Triazin-5(2H)-one, 6-methyl-3-(methylphenylamino)-	EtOH	296(2.49)	24-1077-84
$C_{11}H_{12}N_4O_2$			
Benzonitrile, 4-(3-acetoxymethyl)-3-methyl-1-triazenyl]-	MeCN	223(4.05),295(4.36), 310s(4.30)	23-0741-84
Hydrazinecarboxamide, 2-[(5-methoxy-1H-indol-4-yl)methylene]-	EtOH	208(4.33),249(4.21), 344(4.24)	44-4761-84
$C_{11}H_{12}N_4O_2S$			
1H-1,2,4-Triazole, 3-[(1-methylethyl)-thio]-1-(4-nitrophenyl)-	MeCN	240(3.96),322(4.28)	48-0311-84
$C_{11}H_{12}N_4O_3$			
2,4(1H,3H)-Pteridinedione, 6-acetyl-1,3,7-trimethyl-	MeOH	247(4.12),278(4.04), 327(4.11)	5-1798-84
2,4(1H,3H)-Pteridinedione, 1,3-dimethyl-6-(1-oxopropyl)-	MeOH	252(4.12),280(4.08), 333(3.99)	5-1798-84
2,4(1H,3H)-Pteridinedione, 1,3-dimethyl-7-(1-oxopropyl)-	MeOH	245(4.11),346(3.89)	5-1798-84
$C_{11}H_{12}N_4O_4$			
1H-Imidazo[1,2-b]pyrazole-7-carbonitrile, 1-β-D-ribofuranosyl-	pH 1	242(4.25)	44-3534-84
	pH 7	242(4.27)	44-3534-84
	pH 11	242(4.26)	44-3534-84
5H-Imidazo[1,2-b]pyrazole-7-carbonitrile, 5-β-D-ribofuranosyl-	pH 1	255(4.08)	44-3534-84
	pH 7	276(4.01)	44-3534-84
	pH 11	276(4.08)	44-3534-84
2,4,6(3H)-Pteridinetrione, 1,5-dihydro-1,3-dimethyl-7-(1-oxopropyl)-	pH 1.0	211(4.33),251(4.15), 394(3.89)	5-1798-84
	pH 8.0	221(4.32),264s(4.06), 424(3.76)	5-1798-84
2,4,7(1H,3H,8H)-Pteridinetrione, 1,3-dimethyl-6-(1-oxopropyl)-	pH 1.0	213(4.36),255(3.98), 280(3.87),344(4.28)	5-1798-84
	pH 7.0	222(4.47),267(3.96), 287(4.00),357(4.19)	5-1798-84
$C_{11}H_{12}N_4O_6$			
2,4(1H,3H)-Pteridinedione, 3-β-D-ribofuranosyl-	pH 4.0	234(4.07),327(3.90)	136-0179-84F
	pH 11.0	212s(4.05),245(4.08), 272(4.10),366(3.78)	136-0179-84F
	MeOH	235(4.07),326(3.84)	136-0179-84F
6-Pteridinepropanoic acid, 1,2,3,4,7,8-hexahydro-8-(2-hydroxyethyl)-2,4,7-trioxo-	pH 1	281(4.08),328(4.10)	39-0953-84C
	pH 13	259(3.87),288(3.96), 352(4.12)	39-0953-84C
$C_{11}H_{12}N_6$			
1H-Cyclohepta[1,2-d:3,4-d']diimidazole-2,8-diamine, 1,4-dimethyl-	MeOH	252(4.19),298(4.77), 362(4.17),401(4.28)	94-3873-84
	MeOH-acid	241(4.25),287(4.70),	94-3873-84

Compound	Solvent	$\lambda_{max}(\log \epsilon)$	Ref.
(cont.)		292s(4.65),348(4.10), 390s(4.23),401(4.35)	94-3873-84
$C_{11}H_{12}O$			
Benzene, 1-ethynyl-4-propoxy-	MeOH	250(4.33)	118-0728-84
2-Oxabicyclo[3.1.0]hexane, 6-phenyl-,	C_6H_{12}	211.5(4.00),218s(3.93)	4-1237-84
endo			
exo	C_6H_{12}	225.5(4.09)	4-1237-84
Pentacyclo[5.4.0.0^{2,6}.0^{3,11}.0^{4,10}]un-	C_6H_{12}	301(1.28)	138-1741-84
decan-8-one			
Phenol, 2-(1,1-dimethyl-2-propynyl)-	50% MeCN-KOH	295(4.18)	39-1269-84B
$C_{11}H_{12}O_2$			
2H-1-Benzopyran, 6-hydroxy-2,2-dimethyl-	MeOH	260(3.45),329(3.45)	2-0885-84
1H-Cyclopenta[a]pentalene-1,5(2H)-	MeOH	241(4.06),298(2.99)	44-1728-84
dione, 3,3b,4,6,6a,7-hexahydro-, cis			
Ethanone, 1,1'-(4-methyl-1,3-phenyl-	MeOH	205(3.81),228(4.05),	24-1620-84
ene)bis-		247s(3.93)	
2(1H)-Pentalenone, 3,3a,4,6a-tetra-	MeOH	232(4.09)	44-1728-84
hydro-5-(3-hydroxy-1-propynyl)-, cis			
1,4-Pentanedione, 1-phenyl-	15% H_2SO_4	248(4.19)	97-0147-84
cation	95% H_2SO_4	298(4.32)	97-0147-84
2-Propenoic acid, 3-phenyl-, ethyl	CH_2Cl_2	269(4.03)	44-4352-84
ester, cis	+ $AlEtCl_2$	328(4.05)	44-4352-84
trans	CH_2Cl_2	277(4.35)	44-4352-84
	+ $AlEtCl_2$	328(4.46)	44-4352-84
$C_{11}H_{12}O_2S_4$			
1,3-Dithiole-4-carboxylic acid, 2-(1,3-	MeCN	289(4.04),304(4.11),	103-1342-84
dithiol-2-ylidene)-, butyl ester		315(4.13),435(3.31)	
$C_{11}H_{12}O_3$			
2-Propenoic acid, 3-(4-methoxyphenyl)-,	EtOH	213(4.04),246(4.00)	77-0405-84
methyl ester, (E)-			
$C_{11}H_{12}O_5$			
2-Propenoic-3^{14}C acid, 3-(4-hydroxy-	EtOH	320(4.25)	102-1029-84
3,5-dimethoxyphenyl)-			
$C_{11}H_{12}O_7$			
2-Propenoic acid, 3-[3,5-bis(hydroxy-	MeOH	233(4.45),335(4.11)	158-84-47
methyl)-4-methoxy-2-oxo-2H-pyran-6-			
yl]-, (E)- (islandic acid)			
$C_{11}H_{13}ClN_2O_2$			
Acetamide, 2-chloro-N-[3-(hydroxy-	EtOH	247(3.9)	5-1696-84
imino)-3-phenylpropyl]-, (E)-			
$C_{11}H_{13}ClN_2O_4S$			
Hydroxylamine-O-sulfonic acid, N-[2-(4-	H_2O	263(3.75)	56-1077-84
chloro-2-pyridinyl)cyclohexylidene]-			
anion	H_2O	257(3.33),263(3.45),	56-1077-84
		270(3.42)	
$C_{11}H_{13}ClN_4O_3$			
1H-Imidazo[4,5-c]pyridin-4-amine, 6-	pH 1	266(4.29),285s(4.13)	35-6379-84
chloro-1-(2-deoxy-β-D-erythro-	pH 7 and 12	271(4.31)	35-6379-84
pentofuranosyl)-			
7H-Pyrrolo[2,3-d]pyrimidin-4-amine,	pH 1	229(4.27),276(4.00)	35-6379-84
2-chloro-7-(2-deoxy-β-D-erythro-	pH 7 and 12	274(4.03)	35-6379-84
pentofuranosyl)-			

Compound	Solvent	$\lambda_{max}(\log \epsilon)$	Ref.
$C_{11}H_{13}ClN_4O_4$			
7H-Pyrrolo[2,3-d]pyrimidin-4-amine, 7-β-D-arabinofuranosyl-2-chloro-	MeOH	208(4.47),274(4.07)	5-0722-84
$C_{11}H_{13}ClO_3$			
2,4-Cyclohexadiene-1-carboxylic acid, 1-(3-chloropropyl)-6-oxo-, methyl ester	MeOH	305(3.59)	44-4429-84
$C_{11}H_{13}FO_4$			
Benzeneacetic acid, 2-fluoro-4,5-di-methoxy-, methyl ester	EtOH	206(4.00),226(3.87), 284(3.62)	104-0465-84
Benzeneacetic acid, 5-fluoro-2,3-di-methoxy-, methyl ester	EtOH	205(4.26),278(3.34)	104-0465-84
$C_{11}H_{13}N$			
Benzenamine, 2-methyl-N-(1-methyl-2-propynyl)-	EtOH	202(3.65),240(4.08), 287(3.42)	65-0161-84
Benzenamine, 4-methyl-N-(1-methyl-2-propynyl)-	EtOH	200(3.62),245(4.13), 295(3.24)	65-0161-84
1H-Indole, 3-ethyl-1-methyl-	EtOH	223(4.50),289(3.79)	78-4351-84
Isoquinoline, 3,4-dihydro-1,6-dimethyl-	hexane	208(4.34),211s(4.33), 217s(4.09),243s(3.91), 249(3.98),281(3.16), 289s(3.07)	32-0041-84
	EtOH	209(4.34),213(4.32), 219s(4.14),249s(3.99), 256(4.03),282s(3.39), 291s(3.24)	32-0041-84
Isoquinoline, 3,4-dihydro-6,8-dimethyl-	hexane	209s(4.27),213(4.36), 218(4.37),222s(4.23), 258(4.03),263(4.04), 269(3.90),295s(2.97)	32-0041-84
	EtOH	209s(4.17),215(4.24), 220(4.26),227s(4.13), 265(4.03)	32-0041-84
$C_{11}H_{13}NO$			
7-Azabicyclo[4.2.0]octa-1,3,5-trien-8-one, 7-(1,1-dimethylethyl)-	C_6H_{12}	222(4.47),273(3.32), 370(2.62)	44-3367-84
Benzamide, N-(1-isobutenyl)-	EtOH	225(3.99),268(3.92)	44-0714-84
Benzamide, N-methyl-N-(1-propenyl)-	MeOH	225(4.23),260(4.23), 300(3.90)	78-1835-84
2H-1-Benzazepin-2-one, 1,3,4,5-tetra-hydro-4-methyl-, (R)-(-)-	EtOH	239(4.1),275s(3.0)	103-0081-84
2H-1-Benzazepin-2-one, 1,3,4,5-tetra-hydro-5-methyl-, (S)-(-)-	EtOH	219(3.83),238(4.13), 275s(2.96)	103-0081-84
Benzenamine, 2-methoxy-N-(1-methyl-2-propynyl)-	EtOH	205(3.64),243(4.07), 287(3.49)	65-0161-84
2-Butenamide, N-(phenylmethyl)-	EtOH	208.5(3.87)	142-2369-84
Cyclopentanone, 2-(4-methyl-2-pyri-dinyl)-	H_2O	263(3.56)	56-1077-84
$C_{11}H_{13}NOS_2$			
Ethane(dithioic) acid, (benzoylamino)-	pH 7	<u>307(4.0)</u>	23-0763-84
	pH 12	<u>340(4.0)</u>	23-0763-84
$C_{11}H_{13}NO_2$			
Isoquinoline, 3,4-dihydro-7,8-dimeth-oxy-	MeOH and MeOH-base	227(4.42),262(3.92), 320(3.34)	36-1639-84
	MeOH-HCl	218s(4.17),237s(4.10)	36-1639-84

Compound	Solvent	$\lambda_{max}(\log \epsilon)$	Ref.
(cont.)		297(4.04),375(3.41)	36-1639-84
$C_{11}H_{13}NO_3$			
Carbamic acid, (phenylacetyl)-, ethyl ester	EtOH	217(3.08)	39-1127-84C
2,4-Furandicarboxaldehyde, 5-piperidino-	MeOH	350(3.42)	73-1600-84
$C_{11}H_{13}NO_4$			
1H-Pyrrole-3-carboxylic acid, 4-(3-methoxy-3-oxo-1-propenyl)-5-methyl-, methyl ester, (E)-	MeOH	217(4.12),313(4.01)	24-1620-84
Spiro[3-azabicyclo[4.2.0]oct-4-en-7,2'-oxetane]-2,4'-dione, 6-methoxy-3-methyl-	MeOH	258(3.70)	94-4707-84
$C_{11}H_{13}NO_4S_2$			
Thieno[2,3-c]pyridin-3(2H)-one, 4-hydroxy-7-methyl-5-[2-(methylthio)-ethyl]-, S,S-dioxide	EtOH	225(4.19),345(3.74)	104-2032-84
	EtOH-KOH	225(4.20),270(3.89), 385(4.01)	104-2032-84
$C_{11}H_{13}NO_5$			
1-Azabicyclo[3.2.0]hept-2-ene-3-acetic acid, 2-carboxy-6-ethyl-7-oxo-, disodium salt	n.s.g.	270(3.43)	44-5271-84
$C_{11}H_{13}NO_7$			
2,5-Cyclohexadiene-1,4-dione, 2-[2-[(aminocarbonyl)oxy]-1-methoxy-ethyl]-3,6-dihydroxy-5-methyl-	aq base	322(--),329(4.40), 521(2.37)	94-2406-84
$C_{11}H_{13}NS$			
Benzenecarbothioamide, N-(1-isobutenyl)-	EtOH	225(4.60),305(3.34), 385(2.76)	44-0714-84
Benzenecarbothioamide, N-methyl-N-(1-propenyl)-	hexane	295(3.86),408(2.49)	78-1835-84
Benzothiazole, 2-(1-methylpropyl)-	EtOH	218(4.36),249(3.96), 285(3.43)	78-2141-84
Thiopyrano[2,3-c]pyrrole, 2,4,5,7-tetramethyl-, perchlorate	EtOH	218(4.05),254(4.15), 320(3.88),560(3.26)	12-1473-84
$C_{11}H_{13}N_3O_2$			
Acetic acid, cyano(1,2-dimethyl-4(1H)-pyrimidinylidene)-, ethyl ester	CF_3COOH	313(4.37),352(4.48)	103-1270-84
2-Butenal, methyl(4-nitrophenyl)hydrazone	CH_2Cl_2	402(4.6)	88-0057-84
Pyrido[2,3-d]pyrimidine-2,4(1H,3H)-dione, 7-ethyl-1,3-dimethyl-	H_2O	249s(3.85),309(4.00)	5-1543-84
Pyrido[2,3-d]pyrimidine-2,4(1H,3H)-dione, 1,3,6,7-tetramethyl-	H_2O	248(3.88),316(3.93)	4-1543-84
4-Pyrimidineacetic acid, α-cyano-α,2-dimethyl-, ethyl ester	EtOH	253(3.62)	103-1270-84
$C_{11}H_{13}N_3O_3$			
Acetic acid, [[(4-methoxybenzoyl)amino]methylene]hydrazide	MeCN	283(4.39)	48-0311-84
2H-Pyrrole-2-carboxamide, 3,4-dihydro-4-hydroxy-5-(3-hydroxy-2-pyridinyl)-4-methyl-, (2R-cis)- (siderochelin A)	MeOH	316(3.92)	158-1260-84
(2S-trans)-	MeOH	315(3.93)	158-1260-84
7H-Pyrrolo[2,3-d]pyrimidine, 7-(2-deoxy-α-D-erythro-pentofuranosyl)-	MeOH	270(3.58)	5-1719-84

Compound	Solvent	$\lambda_{max}(\log \epsilon)$	Ref.
7H-Pyrrolo[2,3-d]pyrimidine, 7-(2-deoxy-β-D-erythro-pentofuranosyl)-	MeOH	270(3.59)	5-1719-84
$C_{11}H_{13}N_3O_4$			
7-Deazanebularine	H_2O	270(3.54)	118-0401-84
Imidazo[1,5-a]pyrazine, 3-β-D-ribofuranosyl-	EtOH	223(4.42),266(3.56), 275(3.61),286(3.54), 335(3.26)	39-0229-84C
$C_{11}H_{13}N_3O_5$			
L-Alanine, N-[1-oxo-3-(1,2,3,4-tetrahydro-6-methyl-2,4-dioxo-5-pyrimidinyl)-2-propenyl]-, (E)-	H_2O	302(4.31)	94-4625-84
Glycine, N-[1-oxo-3-(1,2,3,4-tetrahydro-6-methyl-2,4-dioxo-5-pyrimidinyl)-2-propenyl]-, methyl ester, (E)-	H_2O	302(4.29)	94-4625-84
$C_{11}H_{13}N_3O_6$			
L-Alanine, 3-(hydroxynitrosoamino)-N-[(phenylmethoxy)carbonyl]-	pH 1	227(3.84)	87-1295-84
	pH 13	251(3.97)	87-1295-84
	MeOH	230(3.83)	87-1295-84
1H-Pyrrolo[2,3-d]pyrimidine-2,4(3H,7H)-dione, 7-β-D-arabinofuranosyl-	pH 7.0	218(4.35),251(3.98), 283(3.81)	5-0273-84
D-Serine, N-[1-oxo-3-(1,2,3,4-tetrahydro-6-methyl-2,4-dioxo-5-pyrimidinyl)-2-propenyl]-, (E)-	H_2O	302(4.31)	94-4625-84
L-	H_2O	302(4.32)	94-4625-84
$C_{11}H_{13}N_3O_7$			
1(4H)-Pyridineacetic acid, α-(1-methylpropyl)-3,5-dinitro-4-oxo-, [S-(R*,R*)]-	n.s.g.	342(3.71)	18-1961-84
1(4H)-Pyridineacetic acid, α-(2-methylpropyl)-3,5-dinitro-4-oxo-, (S)-	n.s.g.	340(3.62)	18-1951-84
$C_{11}H_{13}N_4O_7P$			
1H-Imidazole-4-carboxamide, 5-(cyanomethyl)-1-(3,5-O-phosphinico-β-D-ribofuranosyl)-, monosodium salt	pH 1	218(3.86)	87-1389-84
	pH 7	233s(3.88)	87-1389-84
	pH 11	235s(3.96)	87-1389-84
4H-Imidazo[4,5-c]pyridin-4-one, 6-amino-1,5-dihydro-1-(3,5-O-phosphinico-β-D-ribofuranosyl)-	pH 1	285(4.13),309s(3.92)	87-1389-84
	pH 7	270(4.05),297s(3.93)	87-1389-84
	pH 11	270(4.01),296s(3.92)	87-1389-84
$C_{11}H_{13}N_5O_3$			
2,4(1H,3H)-Pteridinedione, 6-amino-1,3-dimethyl-7-(1-oxopropyl)-	pH -3.0	211(4.23),270(4.25), 306s(3.20),446(3.85)	5-1798-84
	pH 7.0	220(4.36),258s(4.08), 268(4.10),446(3.88)	5-1798-84
2,4(1H,3H)-Pteridinedione, 7-amino-1,3-dimethyl-6-(1-oxopropyl)-	pH -5.5	230s(3.99),270s(3.90), 285(3.99),383(4.03)	5-1798-84
	pH 3.0	231(4.58),273(4.06), 290(4.06),370(4.27)	5-1798-84
$C_{11}H_{13}N_5O_3S$			
1H-Imidazole-5-carboxamide, 4-[[4-(aminosulfonyl)phenyl]amino]-1-methyl-	EtOH	258(4.08),305(4.08)	2-0870-84
$C_{11}H_{13}N_5O_3S_2$			
4-Thia-1-azabicyclo[3.2.0]hept-2-ene-2-carboxylic acid, 6-ethyl-3-[[(1-methyl-1H-tetrazol-5-yl)thio]methyl]-7-oxo-, sodium salt	H_2O	258(3.72),314(3.67)	32-0319-84

Compound	Solvent	$\lambda_{max}(\log \epsilon)$	Ref.
$C_{11}H_{13}N_5O_4S_2$ 4-Thia-1-azabicyclo[3.2.0]hept-2-ene-2-carboxylic acid, 6-(1-hydroxyethyl)-3-[[(1-methyl-1H-tetrazol-5-yl)-methyl]thio]-7-oxo-, monosodium salt	H_2O	315(3.72)	32-0319-84
$C_{11}H_{13}N_5O_5S$ 1H-Pyrazolo[3,4-d]pyrimidine-3-carbothioamide, 4,5-dihydro-4-oxo-1-β-D-ribofuranosyl-	pH 1 and 7 pH 11	245(4.11),259s(4.11), 308(3.68) 257s(4.08),269(4.08), 317s(4.08)	87-1119-84 87-1119-84
$C_{11}H_{13}N_5O_6$ 1H-Pyrazolo[3,4-d]pyrimidine-3-carboxamide, 4,5-dihydro-4-oxo-1-β-D-ribofuranosyl-	pH 1 and 7 pH 11	257(3.86) 275(3.91)	87-1119-84 87-1119-84
$C_{11}H_{14}BrN$ Piperidine, 1-(4-bromophenyl)-	MeOH	261(3.76)	139-0253-84C
$C_{11}H_{14}BrNO_2$ Benzene, 1-(1-bromo-2,2-dimethylpropyl)-4-nitro- anion radical (decays to benzyl)	EtOH pH 7	278(4.00) 325(4.08)to 365(4.18)	35-3140-84 35-3140-84
$C_{11}H_{14}ClNO_2$ Benzene, 1-(1-chloro-2,2-dimethylpropyl)-2-nitro-, anion radical (decays to benzyl) Benzene, 1-(1-chloro-2,2-dimethylpropyl)-3-nitro-, anion radical	pH 7 pH 7	295(3.81)to 370(3.40) 310(4.15)	35-3140-84 35-3140-84
$C_{11}H_{14}ClN_3O$ 6H-Pyrrolo[3,4-d]pyridazine, 1-chloro-4-ethoxy-5,6,7-trimethyl-	pH 7 cation	226(4.64),275(3.71), 335(3.60) 207(4.40),237(4.60), 284(3.86),343.5(3.39)	78-3979-84 78-3979-84
$C_{11}H_{14}ClN_5O_2$ Cyclopentanemethanol, 4-(2-amino-6-chloro-9H-purin-9-yl)-2-hydroxy-, (1α,2β,4α)-(±)-	pH 1 pH 7 pH 13	219(4.44),242(3.75), 314(3.85) 223(4.45),246(3.66), 307(3.88) 223(4.43),246(3.65), 307(3.87)	87-1416-84 87-1416-84 87-1416-84
$C_{11}H_{14}ClN_5O_3$ 1,2-Cyclopentanediol, 3-(2-amino-6-chloro-9H-purin-9-yl)-5-(hydroxymethyl)-, (1α,2α,3β,5β)-(±)-	pH 1 pH 7	221(4.40),242(3.74), 313(3.85) 223(4.43),245(3.71), 307(3.88)	87-0670-84 87-0670-84
$C_{11}H_{14}NO_5S_2$ Morpholinium, 4-[4,5-bis(methoxycarbonyl)-1,3-dithiol-2-ylidene]-, bromide	MeOH	345(3.97)	104-2285-84
$C_{11}H_{14}NS$ 1H-Isoindolium, 3-(ethylthio)-2-methyl-, tetrafluoroborate	CHCl$_3$	244(4.36),300(3.70), 312(3.79),322(3.83), 330(3.66)	103-0978-84

Compound	Solvent	$\lambda_{max}(\log \epsilon)$	Ref.
$C_{11}H_{14}N_2$			
1,4-Ethano-5,8-methanophthalazine, 1,4,4a,5,8,8a-hexahydro-, (1α,4α,4aα,5β,8β,8aα)-	hexane	388(2.14)	24-0534-84
$C_{11}H_{14}N_2OS_3$			
Butanenitrile, 2-(dihydro-4-oxo-4H,8aH-thiazolo[2,3-d]-1,3,5-dithiazin-3-yl-idene)-3,3-dimethyl-	EtOH	281(2.7)	24-2205-84
$C_{11}H_{14}N_2O_2$			
Piperidine, 1-(4-nitrophenyl)-	MeOH	391(4.59)	139-0253-84C
2,4(1H,3H)-Pyrimidinedione, 5-(1-cyclo-hexen-1-ylmethyl)-	MeOH	266(3.78)	44-2738-84
	12M NaOH	287(3.80)	44-2738-84
$C_{11}H_{14}N_2O_3S$			
Benzenesulfonamide, 2-[N-(1-oxo-2-but-enyl)-N-methylamino]-	EtOH	220s(4.3),265s(3.4)	103-0735-84
Ethanone, 1-(4-nitro-5-piperidino-2-thienyl)-	benzene	390(3.81)	39-0781-84B
$C_{11}H_{14}N_2O_4$			
2H-Cyclohept[d]isoxazole-4-propanoic acid, α-amino-3,6,7,8-tetrahydro-3-oxo-, hydrobromide, (±)-	MeOH	250(3.45)	87-0585-84
2H-Cyclohept[d]isoxazole-8-propanoic acid, α-amino-3,4,5,6-tetrahydro-3-oxo-, hydrobromide, (±)-	MeOH	256(3.46)	87-0585-84
$C_{11}H_{14}N_2O_4S$			
Hydroxylamine-O-sulfonic acid, N-[2-(4-methyl-2-pyridinyl)cyclopentylidene]-	H_2O	262(3.79)	56-1077-84
	anion	255(3.45),261(3.49), 268(3.43)	56-1077-84
Hydroxylamine-O-sulfonic acid, N-[2-(2-pyridinyl)cyclohexylidene]-	H_2O	263(3.79)	56-1077-84
	anion	255(3.43),261(3.50), 268(3.41)	56-1077-84
2-Propenoic acid, 2-methyl-3-[[[(phen-ylmethyl)amino]sulfonyl]amino]-, (E)-	50% EtOH + NaOH	254(4.13) 274(4.17)	73-0840-84 73-0840-84
2(1H)-Pyrimidinone, 1-[4,5-dihydroxy-3-(hydroxymethyl)-2-cyclopenten-1-yl]-4-(methylthio)-, [1R-(1α,4β,5β)]-	MeOH	277(4.51),303(4.54)	87-1536-84
2-Thiophenecarboxylic acid, 4-nitro-5-piperidino-, methyl ester	benzene	388(3.77)	39-0781-84B
$C_{11}H_{14}N_4O_3$			
Acetamide, N-[4,7-dihydro-7-(methoxy-methyl)-3-methyl-4-oxo-3H-pyrrolo-[2,3-d]pyrimidin-2-yl]-	MeOH	265(3.94),287(3.96)	87-0981-84
2,4(1H,3H)-Pteridinedione, 6-(1-hydr-oxypropyl)-1,3-dimethyl-	MeOH	239(4.24),335(3.86)	5-1798-84
7H-Pyrrolo[2,3-d]pyrimidin-2-amine, 7-(2-deoxy-α-D-erythro-pento-furanosyl)-	MeOH	234(4.46),256(3.59), 314(3.69)	5-1719-84
β-	MeOH	234(4.49),256(3.61), 314(3.72)	5-1719-84
$C_{11}H_{14}N_4O_4$			
1H-Imidazole-4-carboxamide, 5-(cyano-methyl)-1-(2-deoxy-α-D-erythro-pentofuranosyl)-	pH 1 pH 7 and 11	217(4.02) 235(3.99)	87-1389-84 87-1389-84

Compound	Solvent	$\lambda_{max}(\log \epsilon)$	Ref.
1H-Imidazole-4-carboxamide, 5-(cyano-methyl)-1-(2-deoxy-β-D-erythro-pentofuranosyl)-	pH 1 pH 7 pH 11	214(4.12) 233(4.39) 230(4.17)	87-1389-84 87-1389-84 87-1389-84
Imidazo[1,5-b]pyridazin-2-amine, 7-β-D-ribofuranosyl-	EtOH	244(4.37),302(3.50), 335s(--)	39-0229-84C
4H-Imidazo[4,5-c]pyridin-4-one, 6-ami-no-1-(2-deoxy-α-D-erythro-pento-furanosyl)-1,5-dihydro-	pH 1 pH 7 pH 11	283(4.11),310s(3.84) 270(4.12),298(3.99) 271(4.10),297s(3.97)	87-1389-84 87-1389-84 87-1389-84
β-	pH 1 pH 7 pH 11	283(4.11),310s(3.83) 269(4.07),295s(3.93) 275(4.10)	87-1389-84 87-1389-84 87-1389-84
4H-Imidazo[4,5-c]pyridin-4-one, 6-ami-no-3-(2-deoxy-α-D-erythro-pento-furanosyl)-3,5-dihydro-	pH 1 pH 7 pH 11	276(3.94),317(3.72) 260(3.76),316(3.74) 258(3.76),316(3.72)	87-1389-84 87-1389-84 87-1389-84
β-	pH 1 pH 7 pH 11	277(4.12),317(3.81) 258(3.91),317(3.92) 258(3.90),316(3.89)	87-1389-84 87-1389-84 87-1389-84
7H-Pyrrolo[2,3-d]pyrimidin-4-amine, 7-α-D-ribofuranosyl- (α-tubercidin)	MeOH	271(4.06)	5-1972-84
Pyrrolo[2,3-d]pyrimidin-4-amine, 7-β-D-arabinofuranosyl-	pH 7	271(4.08)	78-0125-84
$C_{11}H_{14}N_4O_4S$ 4H-Pyrazolo[3,4-d]pyrimidine-4-thione, 1,5-dihydro-6-methyl-1-β-D-ribo-furanosyl-	pH 1 pH 7 pH 11	232s(3.88),318(4.36) 222s(3.91),318(4.36) 235(4.16),314(4.28)	87-1026-84 87-1026-84 87-1026-84
$C_{11}H_{14}N_4O_4Se$ 4H-Pyrazolo[3,4-d]pyrimidine-4-selone, 1,5-dihydro-6-methyl-1-β-D-ribo-furanosyl-	pH 1 pH 7 pH 11	240s(3.68),347(4.28) 240s(3.79),347(4.24) 237(4.15),332(4.19)	87-1026-84 87-1026-84 87-1026-84
$C_{11}H_{14}N_4O_5$ 1H-Imidazo[1,2-b]pyrazole-7-carbox-amide, 1-β-D-ribofuranosyl-	pH 1 pH 7 pH 11	248(4.02) 248(4.04) 248(4.05)	44-3534-84 44-3534-84 44-3534-84
5H-Imidazo[1,2-b]pyrazole-7-carbox-amide, 5-β-D-ribofuranosyl-	pH 1 pH 7 pH 11	213(4.09),258(3.73) 217(4.04),272(3.60) 220(4.06),279(3.67)	44-3534-84 44-3534-84 44-3534-84
4H-Imidazo[4,5-c]pyridin-4-one, 6-ami-no-1,5-dihydro-1-β-D-ribofuranosyl-	pH 1 pH 7 pH 11	284(4.13),309s(3.81) 270(4.00),298(3.90) 272(4.00),295s(3.90)	87-1389-84 87-1389-84 87-1389-84
Imidazo[5,1-f]-1,2,4-triazin-4(1H)-one, 2-methyl-7-β-D-ribofuranosyl-	EtOH	221(4.40),249s(3.92)	39-0229-84C
4H-Pyrazolo[3,4-d]pyrimidin-4-one, 1,5-dihydro-6-methyl-1-β-D-ribo-furanosyl-	pH 1 and 7 pH 11	248(3.97) 264(3.95)	87-1026-84 87-1026-84
$C_{11}H_{14}N_4O_5S$ 4H-Pyrazolo[3,4-d]pyrimidin-4-one, 1,5-dihydro-3-(methylthio)-1-β-D-ribofuranosyl-	pH 1 and 7 pH 11	228(4.56),243s(4.37) 250(4.34),273(4.21)	87-1119-84 87-1119-84
$C_{11}H_{14}N_4O_6$ 1H-Pyrazolo[4,3-d]pyrimidine-5,7(4H-6H)-dione, 1-methyl-3-β-D-ribo-furanosyl-	pH 1 and 7 pH 11	218(3.68),240s(3.26), 293(3.38) 212(3.99),240s(3.43), 312(3.40)	4-1865-84 4-1865-84
1H-Pyrazolo[4,3-d]pyrimidine-5,7(4H-6H)-dione, 6-methyl-3-β-D-ribo-furanosyl-	pH 1 pH 7 pH 11	284(3.83) 284(3.81) 295(3.66)	4-1865-84 4-1865-84 4-1865-84

Compound	Solvent	$\lambda_{max}(\log \epsilon)$	Ref.
2H-Pyrazolo[4,3-d]pyrimidine-5,7(4H-6H)-dione, 2-methyl-3-β-D-ribofuranosyl-	pH 1 pH 7 pH 11	239(3.32),290(3.92) 243(3.04),291(3.76) 244(3.51),307s(3.71)	4-1865-84 4-1865-84 4-1865-84
$C_{11}H_{14}N_6O_3$ Acetamide, N-[1-(2-amino-1,4-dihydro-4-oxo-6-pteridinyl)-2-hydroxypropyl]-, [S-(R*,R*)]-	MeOH	236(3.96),274(4.12), 345(3.67)	5-1815-84
$C_{11}H_{14}N_6O_4S$ 1H-Pyrazolo[3,4-d]pyrimidine-3-carbothioamide, 4-amino-1-β-D-ribofuranosyl-	pH 1 pH 7 pH 11	268(3.89) 233s(4.08),286(3.92) 233(4.08),281(3.93)	103-0210-84 103-0210-84 103-0210-84
$C_{11}H_{14}N_6O_5$ 6-Pteridinepropanoic acid, 1,2,3,4,7,8-hexahydro-8-(2-hydroxyethyl)-2,4,7-trioxo-, hydrazide	pH 1 pH 13	280(4.13),328(4.14) 261(4.01),288(3.97), 356(4.16)	39-0953-84C 39-0953-84C
1H-Pyrazolo[3,4-d]pyrimidine-3-carboximidamide, 4,5-dihydro-4-oxo-1-β-D-ribofuranosyl-, hydrochloride	pH 1 and 7 pH 11	233(4.13),261(3.95) 236(4.07),274(3.90)	87-1119-84 87-1119-84
$C_{11}H_{14}N_6O_6$ 1H-Pyrazolo[3,4-d]pyrimidin-3-carboximidamide, 4,5-dihydro-N-hydroxy-4-oxo-1-β-D-ribofuranosyl-	pH 1 pH 7 pH 11	233(4.15),253s(4.07) 253(4.03) 273(4.09)	87-1119-84 87-1119-84 87-1119-84
$C_{11}H_{14}N_6O_7$ 1(4H)-Pyridineacetic acid, α-[3-[(aminoiminomethyl)amino]propyl]-3,5-dinitro-4-oxo-, (S)-	6M HCl	346(3.60)	18-1961-84
$C_{11}H_{14}O$ 4,7-Methano-1H-indene, 3a,4,7,7a-tetrahydro-1-methoxy-	n.s.g.	205(3.65)	104-0061-84
Spiro[4.5]deca-2,6-dien-1-one, 6-methyl-	EtOH	217(4.02)	33-0073-84
Spiro[4.4]nona-2,6-dien-1-one, 3,6-dimethyl-	EtOH	225(4.14)	33-0073-84
Spiro[3.7]undeca-5,7-dien-1-one	pentane	237(3.79),304(2.28)	33-0774-84
Tricyclo[5.4.0.0^{1,8}]undec-5-en-9-one	pentane	214(4.00),284(1.69), 292s(1.68),302s(1.53), 313s(1.11)	33-0774-84
$C_{11}H_{14}OS_3$ Benzene, [[[2-(methylsulfinyl)-1-(methylthio)ethenyl]thio]methyl]-, (Z)-	EtOH	270(4.13)	39-0085-84C
$C_{11}H_{14}O_2$ 2-Butanone, 3-hydroxy-3-methyl-1-phenyl-	EtOH	215(3.54)	39-1531-84C
3-Butene-1,2-diol, 3-methyl-4-phenyl-, (E)-	EtOH	247(4.18)	44-2039-84
Cyclopentanone, 2-(2-cyclopenten-1-ylcarbonyl)-, (±)-	hexane CHCl$_3$	286(3.90) 291(3.87)	33-1154-84 33-1154-84
2-Propenoic acid, 2-methyl-, 1,3-cyclohexadien-1-ylmethyl ester	isooctane	261(3.85)	116-1624-84
$C_{11}H_{14}O_3$ 4-Pentenoic acid, 6-oxo-1-cyclohexen-1-yl ester	EtOH	230(3.91)	94-0538-84

Compound	Solvent	$\lambda_{max}(\log \epsilon)$	Ref.
$C_{11}H_{14}O_3S$ 2-Propenoic acid, 2-[(ethylthio)meth-yl]-3-(2-furanyl)-, methyl ester	EtOH	301(3.30)	73-1764-84
$C_{11}H_{14}O_3S_3$ Ethane(dithioic) acid, [(2-hydroxy-2-phenylethyl)sulfonyl]-, methyl ester	EtOH	212(4.06),317(4.04)	39-0085-84C
$C_{11}H_{14}O_4$ Benzoic acid, 2,4-dihydroxy-3,5,6-tri-methyl-, methyl ester	MeOH	226(4.71),270(4.66), 314(4.24)	102-0431-84
	MeOH-NaOH	221(4.66),246(4.58), 317(4.89)	102-0431-84
2-Propenoic acid, 3-(2-furanyl)-2-[(1-methylethoxy)methyl]-	EtOH	312(3.37)	73-1764-84
2H-Pyran-2-one, 5-(1-hydroxy-2-butenyl)-4-methoxy-6-methyl-, (E)-(+)-	MeOH	284(3.9)	102-2693-84
4H,5H-Pyrano[4,3-b]pyran-4,5-dione, 2,3,7,8-tetrahydro-2,2,7-trimethyl-	EtOH	260(3.99)	4-0013-84
4H,5H-Pyrano[4,3-b]pyran-4,5-dione, 2,3,7,8-tetrahydro-2,2,8-trimethyl-	EtOH	260(4.09)	4-0013-84
4H,5H-Pyrano[4,3-b]pyran-4,5-dione, 2,3,7,8-tetrahydro-2,3,7-trimethyl-	EtOH	262(4.00)	4-0013-84
4H,5H-Pyrano[4,3-b]pyran-4,5-dione, 2,3,7,8-tetrahydro-2,7,7-trimethyl-	EtOH	259(4.04)	4-0013-84
$C_{11}H_{14}O_5$ Acetic acid, (3-hydroxy-5-oxo-4-pentyl-2(5H)-furanylidene)-, (E)-	EtOH	259(4.18),315(3.74)	39-1555-84C
	EtOH-NaOH	258(4.17),325(3.88)	39-1555-84C
(Z)-	EtOH	261(4.24),302s(3.90)	39-1555-84C
	EtOH-NaOH	255(4.22),335(3.89)	39-1555-84C
Cyclopenta[c]pyran-4-carboxylic acid, 1,4a,5,7a-tetrahydro-1-hydroxy-7-(hydroxymethyl)-, methyl ester, [1R-(1α,4aα,7aα)]- (genipin)	EtOH	240(3.96)	102-0533-84
$C_{11}H_{14}Si$ Silane, trimethyl(phenylethynyl)-	hexane	238.1(4.27),248.5(4.45), 260.4(4.41)	65-2317-84
$C_{11}H_{15}ClN_2$ 10,11-Diazabicyclo[7.3.1]trideca-1(13),9,11-triene, 12-chloro-	EtOH	219(3.65),271(3.18), 305(2.54)	32-0289-84
$C_{11}H_{15}ClN_2S$ Benzenecarboximidamide, 4-chloro-N'-(ethylthio)-N,N-dimethyl-	hexane	324(3.32)	39-2933-84C
$C_{11}H_{15}ClO$ Bicyclo[2.2.1]heptan-2-one, 3-(chloro-methylene)-1,7,7-trimethyl-, [1R-(E)]-	EtOH	241(4.13)	41-0629-84
$C_{11}H_{15}N$ Piperidine, 1-phenyl-	MeOH	251(3.95)	139-0253-84C
$C_{11}H_{15}NO$ 1-Cyclodecene-1-carbonitrile, 9-oxo-, (E)-	EtOH	209(4.00),287(1.65)	44-2925-84

Compound	Solvent	$\lambda_{max}(\log \epsilon)$	Ref.
$C_{11}H_{15}NO_2$			
1,3-Cyclopentadiene-1-carboxylic acid, 5-[(dimethylamino)methylene]-, ethyl ester	dioxan	235s(4.14),244(3.22), 352(4.39)	157-0653-84
1,4-Cyclopentadiene-1-carboxylic acid, 3-[(dimethylamino)methylene]-, ethyl ester	dioxan	245(4.14),337(4.58)	157-0653-84
Ethanamine, 2-methoxy-N-[(2-methoxy-phenyl)methylene]-	M HCl	276(4.43),347(3.90)	18-2508-84
	pH 8.9	249(4.26),305(3.85)	18-2508-84
1H-Inden-2-amine, 2,3-dihydro-1,5-di-methoxy-, hydrochloride	n.s.g.	199(4.68),229(4.00), 275s(3.36),278(3.39), 284(3.36)	39-1655-84C
1H-Pyrrole-2-carboxaldehyde, 4-acetyl-5-methyl-1-propyl-	EtOH	296(4.19)	136-0153-84I
1H-Pyrrole-2,5-dione, 3-cyclohexyl-1-methyl-	MeOH	227(4.19)	39-2097-84C
2,5(1H,6H)-Quinolinedione, 3,4,7,8-tetrahydro-7,7-dimethyl-	pH 1	295(4.13)	39-0287-84C
	H_2O	295(4.13)	39-0287-84C
changing	pH 12	296(4.28),347(3.45)	39-0287-84C
$C_{11}H_{15}NO_3$			
Benzeneacetamide, 3,4-dihydroxy-α-propyl-	pH 1	279(3.48)	3-1935-84
	borate	287(3.68)	3-1935-84
2,5-Cyclohexadien-1-one, 2,3,4,5,6-pentamethyl-4-nitro-	$CHCl_3$	247(4.00)	12-1489-84
1H-Inden-1-ol, 2-amino-2,3-dihydro-4,5-dimethoxy-, (-)-	n.s.g.	203(4.63),230(3.91), 276(3.08),280(3.08)	39-1655-84C
1H-Pyrrole-2-carboxylic acid, 2,3-di-hydro-1-methyl-3-oxo-2-(2-propenyl)-, ethyl ester	MeCN	327(3.99)	33-1535-84
1H-Pyrrole-2-carboxylic acid, 3-hy-droxy-1-methyl-4-(2-propenyl)-, ethyl ester	MeCN	266(4.11)	33-1535-84
$C_{11}H_{15}NO_5$			
2,4-Hexadienoic acid, 6-methoxy-4-[(4-oxo-2-azetidinyl)oxy]-, methyl ester, [R-(Z,E)]-	n.s.g.	263(4.28)	39-1599-84C
$C_{11}H_{15}NO_5S$			
4,7-Epoxythieno[2,3-c]pyridin-3-ol, 5-ethyl-2,4,7,7a-tetrahydro-4-meth-oxy-7-methyl-, S,S-dioxide, (4α,7α-7aβ)-	EtOH	225(3.90),340(2.43), 390(2.48)	104-2032-84
	EtOH-HCl	225(3.93),315(3.32), 385(3.53)	104-2032-84
$C_{11}H_{15}NO_5S_2$			
2-Butenedioic acid, 2-[(4-morpholinyl-thioxomethyl)thio]-, dimethyl ester	MeOH	280(5.37)	104-2285-84
$C_{11}H_{15}NO_6$			
4(1H)-Pyridinone, 3-hydroxy-2-methyl-1-β-D-ribofuranosyl-	MeOH	286(4.15)	87-0160-84
$C_{11}H_{15}NO_6S$			
Propanedioic acid, (4-carboxy-2-thia-zolidinylidene)-, 1,3-diethyl ester, (R)-	EtOH	280(2.85)	118-0732-84
$C_{11}H_{15}N_3$			
Butanedinitrile, [(3,4-dihydro-3,3-di-methyl-2H-pyrrol-5-yl)methyl]-	MeOH	203(3.09)	83-0143-84

Compound	Solvent	$\lambda_{max}(\log \epsilon)$	Ref.
$C_{11}H_{15}N_3O$			
Benzaldehyde, 2,4,4-trimethylsemi-carbazone	EtOH	206(4.18),218(4.23), 283(4.33),291(4.33), 302s(4.19)	33-1547-84
4H-Indazol-4-one, 1,5,6,7-tetrahydro-5-[(dimethylamino)methylene]-1-methyl-, (E)-	EtOH	234(3.96),253(3.97), 352(4.38)	4-0361-84
6H-Pyrrolo[3,4-d]pyridazine, 1-ethoxy-5,6,7-trimethyl-	pH 7	230(4.32),260(3.80), 278.5(3.57)	78-3979-84
cation	acid	230(4.11),283(3.52), 350.5(3.61)	78-3979-84
dication	H_2SO_4	233(4.39),309(4.01), 370(3.16)	78-3979-84
$C_{11}H_{15}N_3O_2$			
Ethanone, 1-[4-[3-(methoxymethyl)-3-methyl-1-triazenyl]phenyl]-	MeCN	226(3.98),318(4.30)	23-0741-84
6H-Pyrrolo[3,4-d]pyridazine-1,4-dione, 1,2,3,4-tetrahydro-2,3,5,6,7-penta-methyl-	pH 7	215(4.57),281.5(4.10)	78-3979-84
cation	acid	213(4.46),228(4.44), 289(4.01),316s(3.72)	78-3979-84
6H-Pyrrolo[3,4-d]pyridazin-1-one, 1,2-dihydro-4-methoxy-2,5,6,7-tetramethyl-	pH 7	214(4.73),276(4.09), 300(3.85)	78-3979-84
cation	acid	206(4.39),225(4.48), 275(3.85),315(3.57)	78-3979-84
$C_{11}H_{15}N_3O_3$			
Benzoic acid, 4-[3-(methoxymethyl)-3-methyl-1-triazenyl]-, methyl ester	MeCN	224(4.03),300(4.36), 310s(4.34)	23-0741-84
$C_{11}H_{15}N_3O_4$			
2-Propenamide, N-(2-hydroxy-1-methyl-ethyl)-3-(1,2,3,4-tetrahydro-6-meth-yl-2,4-dioxo-5-pyrimidinyl)-, (E)-	H_2O	301(4.30)	94-4625-84
$C_{11}H_{15}N_3O_5$			
2-Propenamide, N-[2-hydroxy-1-(hydroxy-methyl)ethyl]-3-(1,2,3,4-tetrahydro-6-methyl-2,4-dioxo-5-pyrimidinyl)-, (E)-	H_2O	301(4.30)	94-4625-84
$C_{11}H_{15}N_3O_7$			
2-Cyclohexen-1-one, 2,3,4,5,6-penta-methyl-4,5,6-trinitro-, (4α,5α,6α)-	$CHCl_3$	246(4.10)	12-1489-84
(4α,5β,6β)-	$CHCl_3$	247(4.00)	12-1489-84
isomer	$CDCl_3$	245(4.04)	12-1489-84
$C_{11}H_{15}N_3O_8$			
β-D-Glucopyranosiduronic acid, 5-(ami-nocarbonyl)-1H-imidazol-4-yl methyl ester	M HCl	228(3.96)	4-0849-84
	H_2O	248(4.11)	4-0849-84
	M KOH	264(4.12)	4-0849-84
$C_{11}H_{15}N_5O_2S$			
6H-Purine-6-thione, 2-amino-1,9-dihy-dro-9-[3-hydroxy-4-(hydroxymethyl)-cyclopentyl]-, (1α,3β,4α)-(±)-	pH 1	208(4.41),227s(--), 263(3.87),350(4.33)	87-1416-84
	pH 7	208(4.34),231(4.24), 264(3.88),342(4.40)	87-1416-84
	pH 13	222(4.21),252(4.06), 271(3.85),318(4.31)	87-1416-84

Compound	Solvent	$\lambda_{max}(\log \epsilon)$	Ref.
$C_{11}H_{15}N_5O_3$			
4(1H)-Pteridinone, 6-(1,2-dihydroxy-propyl)-2-(dimethylamino)-,	pH -1.0	214(4.18),247(4.24), 327(3.90),418(2.82)	5-1815-84
[S-(R*,S*)]-	pH 5.0	223(4.14),245s(3.99), 287(4.29),352(3.75), 375s(3.64)	5-1815-84
	pH 10.0	228(4.08),273(4.34), 383(3.89)	5-1815-84
6H-Purin-6-one, 2-amino-1,9-dihydro-	pH 1	254(4.10),279(3.92)	87-1416-84
9-[3-hydroxy-4-(hydroxymethyl)cyclo-	pH 7	253(4.14),270s(4.01)	87-1416-84
pentyl]-, (1α,3β,4α)-(±)-	pH 13	257s(4.04),268(4.07)	87-1416-84
$C_{11}H_{15}N_5O_3S$			
6H-Purine-6-thione, 2-amino-1,9-dihy-	pH 1	207(4.39),225s(--), 264(3.88),348(4.34)	87-0670-84
dro-9-[2,3-dihydroxy-4-(hydroxymeth-			
yl)cyclopentyl]-, (1α,2β,3β,4α)-(±)-,	pH 7	207(4.34),231(4.25), 264(3.90),341(4.41)	87-0670-84
sulfate			
	pH 13	222(4.22),251(4.08), 269(3.88),318(4.32)	87-0670-84
$C_{11}H_{15}N_5O_4$			
1H-Pyrazolo[3,4-d]pyrimidin-4-amine,	pH 1	217(4.34),256(3.94)	87-1026-84
6-methyl-1-β-D-ribofuranosyl-	pH 7 and 11	265(3.97)	87-1026-84
$C_{11}H_{15}N_5O_5$			
1H-Imidazo[1,2-b]pyrazole-7-carboximid-	pH 1	251(3.99)	44-3534-84
amide, N-hydroxy-1-β-D-ribofuranosyl-	pH 7	251(3.95)	44-3534-84
	pH 11	247(3.85)	44-3534-84
5H-Imidazo[1,2-b]pyrazole-7-carboximid-	pH 1	261(3.90)	44-3534-84
amide, N-hydroxy-5-β-D-ribofuranosyl-	pH 7	267(3.92)	44-3534-84
	pH 11	274(3.89)	44-3534-84
4H-Imidazo[4,5-c]pyridin-4-one, 5,6-di-	pH 1	287(4.06),312s(3.75)	87-1389-84
amino-1,5-dihydro-1-β-D-ribofuranosyl-	pH 7	276(4.05),297s(3.95)	87-1389-84
	pH 11	276(4.04),297s(3.93)	87-1389-84
Imidazo[5,1-f][1,2,4]triazin-4(1H)-one,	EtOH	229(4.47),267(3.79)	39-0229-84C
2-amino-5-methyl-7-β-D-ribofuranosyl-			
9H-Purin-6-amine, 9-α-D-frucofuranosyl-	pH 7.3	203(4.41),260(4.22)	1-0367-84
9H-Purin-6-amine, 9-β-D-psicofuranosyl-	pH 7.3	200(4.25),260(4.03)	1-0367-84
4H-Pyrazolo[3,4-d]pyrimidin-4-one,	pH 1 and 7	227(4.09),243(3.77)	87-1119-84
1,5-dihydro-3-(methylamino)-1-β-	pH 11	223(4.14),242(3.83)	87-1119-84
D-ribofuranosyl-			
5H-Pyrazolo[4,3-d]pyrimidin-5-one,	pH 1	258(4.01),315(3.90)	4-1865-84
7-amino-1,4-dihydro-1-methyl-	pH 7	253(4.11),301(3.95)	4-1865-84
3-β-D-ribofuranosyl-	pH 11	250s(4.06),305(3.90)	4-1865-84
5H-Pyrazolo[4,3-d]pyrimidin-5-one,	pH 1	257(3.82),310(3.72)	4-1865-84
7-amino-2,4-dihydro-2-methyl-	pH 7	254(4.01),300(3.83)	4-1865-84
3-β-D-ribofuranosyl-	pH 11	254(3.99),300(3.81)	4-1865-84
5H-Pyrazolo[4,3-d]pyrimidin-5-one,	pH 1	250(3.76),308(3.72)	4-1865-84
7-amino-4,6-dihydro-6-methyl-	pH 7	278(3.77)	4-1865-84
3-β-D-ribofuranosyl-	pH 11	228(4.23),265(3.80), 327s(3.26)	4-1865-84
$C_{11}H_{15}N_5O_6$			
4H-Imidazo[4,5-d]-1,2,3-triazin-4-one,	H_2O	253(2.50),294(2.48)	4-1221-84
5-β-D-glucopyranosyl-1,5-dihydro- 6-methyl-			
4H-Imidazo[4,5-d]-1,2,3-triazin-4-one,	H_2O	245(2.60),293(2.70)	4-1221-84
7-β-D-glucopyranosyl-1,7-dihydro- 6-methyl-			

Compound	Solvent	$\lambda_{max}(\log \epsilon)$	Ref.
$C_{11}H_{15}OS$ Sulfonium, dimethyl(1-methyl-2-oxo-2-phenylethyl)-, bromide	EtOH	253(4.10),276(3.45)	78-2035-84
$C_{11}H_{15}O_2P$ Benzoic acid, 3-(dimethylphosphino)-, ethyl ester	C_6H_{12}	229(4.12),261s(3.51), 294s(3.04)	65-0489-84
Benzoic acid, 4-(dimethylphosphino)-, ethyl ester	C_6H_{12}	228(4.10),282(3.96)	65-0489-84
$C_{11}H_{16}$ 1,4-Cyclohexadiene, 1,5,6,6-tetrameth-yl-3-methylene-	EtOH	246(4.29),249(4.29), 254(4.28)	5-0340-84
$C_{11}H_{16}N_2$ 10,11-Diazabicyclo[7.3.1]trideca-1(13),9,11-triene	EtOH	218(3.46),260(3.10), 305(2.40)	32-0289-84
1,4-Ethano-5,8-methanophthalazine, 1,4,4a,5,6,7,8,8a-octahydro-	hexane	385(2.17)	24-0534-84
Piperazine, 1-methyl-4-phenyl-	C_6H_{12}	252(4.13),287(3.22)	35-1335-84
$C_{11}H_{16}N_2O$ Cyclohexaneacetamide, α-(cyanomethyl-ene)-N-methyl-, (E)-	MeOH	220(4.08)	39-2097-84C
10,11-Diazabicyclo[7.3.1]trideca-1(13),9-dien-12-one	EtOH	218(3.66),232s(3.52), 300(3.43)	32-0289-84
2H-Pyrrol-2-one, 3-cyclohexyl-1,5-di-hydro-5-imino-1-methyl-	MeOH	240(4.28),290(3.21)	39-2097-84C
$C_{11}H_{16}N_2OS$ Benzenecarboximidamide, 4-methoxy-N,N-dimethyl-N'-(methylthio)-	hexane	294s(3.56)	39-2933-84C
$C_{11}H_{16}N_2O_2$ Phenol, 2-[[(1-methylpropyl)nitroso-amino]methyl]-	hexane	236(3.56),276(3.42), 356(1.97)	150-0701-84M
Phenol, 4-[[(1-methylpropyl)nitroso-amino]methyl]-	hexane	365(1.98)	150-0701-84M
	EtOH	227(4.13),376(3.37), 354(1.88)	150-0701-94M
$C_{11}H_{16}N_2O_3S$ 1-Azabicyclo[3.2.0]hept-2-ene-2-carb-oxylic acid, 3-[(2-aminoethyl)thio]-6-ethyl-7-oxo-	H_2O	297(3.83)	44-5271-84
isomer	H_2O	294(3.83)	44-5271-84
$C_{11}H_{16}N_2O_4$ Benzoic acid, 4-nitro-, salt with sec-butylamine	EtOH	266(3.92)	150-0701-84M
2,4(1H,3H)-Pyrimidinedione, 1-[2-hy-droxy-3-(2-propenyloxy)propyl]-5-methyl-	pH 7 pH 12	273(4.08) 273(4.09)	161-0346-84 161-0346-84
4(1H)-Pyrimidinone, 1-(3,5-anhydro-β-D-threo-pentofuranosyl)-5,6-di-hydro-2-methoxy-5-methyl-	EtOH	243(4.04)	128-0415-84
$C_{11}H_{16}N_2O_5$ 1(2H)-Pyridineacetamide, N-[2-hydroxy-1,1-bis(hydroxymethyl)ethyl]-2-oxo-	pH 2 and 7 pH 12	300(3.75) 298(3.72)	73-2541-84 73-2541-84
2,4(1H,3H)-Pyrimidinedione, 1-[[2-hy-droxy-1-(hydroxymethyl)ethoxy]meth-	pH 1 H_2O	265(3.88) 265(3.86)	23-0016-84 23-0016-84

Compound	Solvent	$\lambda_{max}(\log \epsilon)$	Ref.
yl]-5-(2-propenyl)- (cont.)	pH 13	264(3.73)	23-0016-84
$C_{11}H_{16}N_2O_6S$ 2,5-Methano-5H,9H-pyrimido[2,1-b]-1,5,3- dioxazepin-9-one, 2,3,7,8-tetrahydro- 8-methyl-3-[[(methylsulfonyl)oxy]- methyl]-	EtOH	244(4.09)	128-0415-84
$C_{11}H_{16}N_4$ Pyrazolo[1,5-a]pyrimidin-7-amine, 5-(1- ethylpropyl)-	EtOH	226(4.55),286(3.89), 304(3.84)	4-1125-84
Pyrazolo[1,5-a]pyrimidin-7-amine, 5-(1- methylbutyl)-	EtOH	226(4.49),286(3.85), 304(3.85)	4-1125-84
$C_{11}H_{16}N_4O$ Benzamide, 4-(3-methyl-3-propyl-1-tria- zenyl)-	EtOH	322(4.31)	87-0870-84
$C_{11}H_{16}N_4OS$ 5-Pyrimidinecarboxamide, 4-amino-N- cyclohexyl-1,2-dihydro-2-thioxo-	EtOH	241(4.20),299(4.42)	142-2259-84
$C_{11}H_{16}N_5O_6P$ Adenine, 9-[(3-deoxy-3-dihydroxyphos- phinylmethyl)-β-D-ribofuranosyl]-	pH 1 pH 12	258(4.20) 259(4.18)	78-0079-84 78-0079-84
$C_{11}H_{16}N_6O_2$ Cyclopentanemethanol, 4-(2,6-diamino- 9H-purin-9-yl)-2-hydroxy-,	pH 1	218(4.35),253(3.98), 292(3.99)	87-1416-84
(1α,2β,4α)-(±)-	pH 7	216(4.46),250s(--), 255(3.91),280(4.02)	87-1416-84
	pH 13	250s(--),255(3.91), 280(4.02)	87-1416-84
$C_{11}H_{16}N_6O_3$ 1,2-Cyclopentanediol, 3-(2,6-diamino- 9H-purin-9-yl)-5-(hydroxymethyl)-,	pH 1	218(4.35),253(4.01), 291(4.02)	87-0670-84
(1α,2α,3β,5β)-(±)-, sulfate	pH 7	216(4.47),250s(--), 256(3.94),280(4.03)	87-0670-84
	pH 13	250s(--),256(3.94), 281(4.04)	87-0670-84
$C_{11}H_{16}N_6O_3S$ 1,2-Cyclopentanediol, 3-[5-amino- 7-(methylthio)-3H-1,2,3-triazolo- [4,5-d]pyrimidin-3-yl]-5-(hydroxy- methyl)-, (1α,2α,3β,5β)-(±)-	pH 1	223(4.23),245s(--), 278(3.97),313(4.03)	87-0670-84
	pH 7	225(4.24),247s(--), 277(3.93),318(4.02)	87-0670-84
	pH 13	223(4.23),245s(--), 275(3.93),317(4.03)	87-0670-84
$C_{11}H_{16}O$ 2H-1-Benzopyran, 3,5,6,7-tetrahydro- 2,2-dimethyl-	EtOH	230(3.92),250(3.97)	44-5000-84
2H-1-Benzopyran, 3,5,6,7-tetrahydro- 2,3-dimethyl-, trans	EtOH	229(3.96),250(3.91)	44-5000-84
Bicyclo[2.2.1]heptan-2-one, 1,7,7-tri- methyl-3-methylene-	EtOH	227(3.96)	41-0629-84
Tricyclo[3.3.1.1³,⁷]decanone, 4-methyl- isomer	EPA EPA	287(1.32) 290(1.34)	35-0934-84 35-0934-84

Compound	Solvent	$\lambda_{max}(\log \epsilon)$	Ref.
$C_{11}H_{16}O_2$			
Benzene, (diethoxymethyl)-	MeCN-H_2O	230s(--),236s(--), 241s(--),246(2.00), 251(2.17),257(2.28), 261(2.12),263(2.18), 267(1.91)	78-2651-84
1,3-Benzenediol, 5-pentyl- (olivetol)	EtOH	204(4.60),274(3.26), 280(3.26)	100-0828-84
3-Cyclononen-1-one, 3-acetyl-	EtOH	226(4.02),292(2.06)	44-2925-84
4-Cyclononen-1-one, 3-acetyl- (end absorption)	EtOH	205(4.00),275(2.00)	44-2925-84
1,4-Dioxaspiro[4.5]dec-6-ene, 7-ethen-yl-8-methyl-	EtOH	234(4.07)	12-2295-84
1,4-Dioxaspiro[4.5]dec-7-ene, 7-ethen-yl-8-methyl-	n.s.g.	237(4.18)	12-2295-84
2,5-Heptadienoic acid, 5-cyclopropyl-, methyl ester, (E,Z)-	EtOH	210(4.0471)	70-2160-84
(Z,Z)-	EtOH	208(4.0645)	70-2160-84
$C_{11}H_{16}O_2S$			
Benzene, 1-methoxy-2-[3-(methylthio)-propoxy]-	MeOH	222.5(3.88),273s(3.38)	36-1241-84
$C_{11}H_{16}O_3$			
1,3-Benzenediol, 5-(4-hydroxypentyl)-, (R)-	EtOH	202(4.56),274(3.17), 281(3.14)	100-0828-84
1-Cyclohexene-1-carboxaldehyde, 3-(ace-toxymethyl)-3-methyl-, (R)-	MeOH	229(4.15)	94-3417-84
1-Cyclohexene-1-carboxaldehyde, 5-(ace-toxymethyl)-5-methyl-, (R)-	MeOH	231(4.15)	94-3417-84
3-Cycloocten-1-one, 3-(acetoxymethyl)-	EtOH	275(2.02)	44-2925-84
$C_{11}H_{16}O_4$			
2-Cyclohexene-1-carboxylic acid, 1-(hy-droxymethyl)-2-methyl-4-oxo-, ethyl ester	EtOH	206s(2.90),241(3.96)	12-2037-84
2-Cyclopentene-1-carboxylic acid, 2-methoxy-1,3,5-trimethyl-4-oxo-, methyl ester	MeOH	248(4.06)	44-3762-84
$C_{11}H_{16}O_4SSi$			
Benzenecarbothioic acid, S-[(trimeth-oxysilyl)methyl] ester	heptane	236(4.01),273(3.81)	70-0726-84
	Bu_2O	237(4.02),273(3.83)	70-0726-84
	MeCN	238(4.06),273(3.93)	70-0726-84
	CH_2Cl_2	239(4.10),274(3.96)	70-0726-84
$C_{11}H_{16}O_5$			
1,3-Benzodioxole-5-carboxylic acid, 3a,6,7,7a-tetrahydro-7-hydroxy-2,2-dimethyl-, methyl ester, (±)-	n.s.g.	210(4.04)	78-2461-84
$C_{11}H_{16}S$			
Tricyclo[3.3.1.1³,⁷]decanethione, 4-methyl-, [1S-(1α,3β,4α,5α,7β)]-	heptane	242(4.10)	35-0934-84
	dioxan	484(1.08),505(1.05)	35-0934-84
[1S-(1α,3β,4β,5α,7β)]-	heptane	241(1.08)	35-0934-84
	dioxan	484(1.10),498(1.10)	35-0934-84
$C_{11}H_{17}Br$			
Bicyclo[2.2.1]heptane, 3-(1-bromoeth-ylidene)-2,2-dimethyl-	MeOH	209(3.29)	2-0331-84

Compound	Solvent	$\lambda_{max}(\log \epsilon)$	Ref.
$C_{11}H_{17}I$			
Bicyclo[2.2.1]heptane, 3-(1-iodoethylidene)-2,2-dimethyl-, (E)-	MeOH	219(3.97),260(2.59)	2-0331-84
$C_{11}H_{17}NO$			
2-Cyclohexen-1-one, 3-[methyl(2-methyl-1-propenyl)amino]-	MeOH	302(4.24)	44-0220-84
2-Cyclohexen-1-one, 3-[methyl(2-methyl-2-propenyl)amino]-	MeOH	298(4.48)	44-0220-84
2-Cyclopenten-1-one, 3-[methyl(3-methyl-2-butenyl)amino]-	MeOH	278(4.57)	44-0220-84
$C_{11}H_{17}NO_2$			
4H-Pyran-4-one, 2,3,5,6-tetrahydro-3-(piperidinomethylene)-, (E)-	EtOH	328.5(4.42)	4-1441-84
1H-Pyrrole-2-acetic acid, α-butyl-, methyl ester	MeOH	211(3.78)	107-0453-84
$C_{11}H_{17}NO_3$			
Acetic acid, (5-oxo-1-propyl-2-pyrrolidinylidene)-, ethyl ester, (E)-	EtOH	263(4.34)	118-0618-84
1H-Pyrrole-2-carboxylic acid, 2,3-dihydro-1,2,4,5-tetramethyl-3-oxo-, ethyl ester	MeCN	334(4.05)	33-1957-84
$C_{11}H_{17}NO_4S_2$			
2-Butenedioic acid, 2-[[(diethylamino)-thioxomethyl]thio]-, dimethyl ester, (Z)-	isoPrOH	252(4.2),269(4.24)	104-1082-84
$C_{11}H_{17}N_3O_6$			
1H-Imidazole-4-carboxylic acid, 5-amino-1-β-D-ribofuranosyl-, ethyl ester	pH 2	250s(3.96),268(4.08)	65-2114-84
	pH 6-7	240(3.76),269(4.12)	65-2114-84
	pH 12	240s(3.76),269(4.12)	65-2114-84
1(2H)-Pyrimidineacetamide, 3,4-dihydro-N-[2-hydroxy-1,1-bis(hydroxymethyl)-ethyl]-5-methyl-2,4-dioxo-	pH 2 and 7	269(4.00)	73-2541-84
	pH 12	267(3.88)	73-2541-84
$C_{11}H_{17}N_5O$			
1,3,5-Triazin-2(1H)-one, 4,6-dipyrrolidino-	MeOH	230(4.77)	24-1523-84
$C_{11}H_{17}N_5O_7$			
Yellow powder C, m. 275-6°	pH 1	268(4.30)	39-0959-84C
	pH 13	270(4.30)	39-0959-84C
$C_{11}H_{17}O_2S$			
Sulfonium, [2-(2-methoxyphenoxy)ethyl]-dimethyl-, iodide	MeOH	219(4.33),271.5s(3.40)	36-1241-84
$C_{11}H_{18}$			
Cyclononane, 1,2-bis(methylene)-	EtOH	225(3.87)	44-2981-84
$C_{11}H_{18}N_2$			
Pyrazine, 3-methyl-2,5-bis(1-methylethyl)-	EtOH	278(3.72),307s(2.95)	142-2317-84
$C_{11}H_{18}N_2O$			
10,11-Diazabicyclo[7.3.1]tridec-9-en-12-one	EtOH	245(3.81)	32-0289-84

Compound	Solvent	λ_{max}(log ϵ)	Ref.
Pyrazine, (1,1,2,2-tetramethylpropoxy)-	EtOH	215(4.04),283(3.68), 296(3.56)	39-0641-84B
C$_{11}$H$_{18}$N$_2$O$_2$ 4H-Pyran-4-one, 2,3,5,6-tetrahydro-3-(4-methylpiperazinomethylene)-, (E)-	EtOH	325.5(4.36)	4-1441-84
C$_{11}$H$_{18}$N$_2$O$_5$ 2,4(1H,3H)-Pyrimidinedione, 1-[[2-hydroxy-1-(hydroxymethyl)ethoxy]methyl]-5-propyl-	pH 1 H$_2$O pH 13	265(3.89) 265(3.86) 266(3.72)	23-0016-84 23-0016-84 23-0016-84
C$_{11}$H$_{18}$N$_3$O$_5$P 1H-Pyrazole-5-carboxylic acid, 3-cyano-5-(diethoxyphosphinyl)-4,5-dihydro-	MeCN	210(3.30),278(3.82)	65-2495-84
C$_{11}$H$_{18}$N$_3$O$_9$P 1H-Imidazole-4-carboxylic acid, 5-amino-1-(5-O-phosphono-β-D-ribofuranosyl)- ammonium salt	pH 2 pH 6-7 pH 12 pH 2 pH 6-7 pH 12	250s(3.98),267(4.09) 240s(3.76),271(4.14) 240s(3.72),271(4.14) 250s(3.96),267(4.08) 240s(3.76),269(4.11) 240(3.74),269(4.11)	65-2114-84 65-2114-84 65-2114-84 65-2114-84 65-2114-84 65-2114-84
C$_{11}$H$_{18}$N$_4$O$_2$S 1H-Pyrazole-4-carbonitrile, 5-amino-1-(2,2-diethoxyethyl)-3-(methylthio)-	pH 1,7 and 11	240s(3.76)	44-3534-84
C$_{11}$H$_{18}$N$_4$O$_7$ Pyridazinium, 4-(hydroxyamino)-3-oxido-1-β-D-ribofuranosyl-, acetamide complex	MeOH	233(3.98),318(4.25)	87-1613-84
C$_{11}$H$_{18}$N$_6$O$_2$ 9H-Purine-2,6-diamine, 9-(2,2-diethoxyethyl)-	MeOH	256(3.88),282(3.93)	73-2148-84
C$_{11}$H$_{18}$OS$_2$ Cyclohexanepropane(dithioic) acid, β-methyl-2-oxo-, methyl ester	EtOH	306(4.27),455(1.11)	39-0859-84C
C$_{11}$H$_{18}$O$_2$ 2-Cyclohexen-1-one, 3-(2-ethoxyethyl)-4-methyl-	n.s.g.	237(4.11)	12-2305-84
C$_{11}$H$_{18}$O$_3$ Cyclopentadieneethanol, α-(dimethoxymethyl)-β-methyl- Cyclopentadieneethanol, β-(dimethoxymethyl)-α-methyl-	EtOH EtOH	248(3.48) 248(3.49)	78-0695-84 78-0695-84
C$_{11}$H$_{18}$O$_3$S$_3$ 2-Cyclohexen-1-ol, 1-[[[2,2-bis(methylthio)ethenyl]sulfonyl]methyl]-	EtOH	253(3.68),276(3.84)	39-0085-84C
C$_{11}$H$_{18}$O$_4$Si 2,4-Hexadienedioic acid, mono[2-(trimethylsilyl)ethyl] ester, (E,E)- (E,Z)-	EtOH EtOH	261(4.39) 261(4.33)	44-1772-84 44-1772-84

Compound	Solvent	$\lambda_{max}(\log \epsilon)$	Ref.
$C_{11}H_{18}O_6$ Butanedioic acid, 2-acetyl-3-methoxy-, diethyl ester	MeOH	247(3.59)	136-0019-84I
$C_{11}H_{19}NO$ Piperidine, 1-[(tetrahydro-2H-pyran-2-ylidene)methyl]-	EtOH	230(3.81)	104-0861-84
$C_{11}H_{19}NO_2$ Cyclohepta[b]pyrrol-3a(3H)-ol, 2,4,5,6-7,8-hexahydro-2,2-dimethyl-, 1-oxide	EtOH	232(4.06)	12-0587-84
$C_{11}H_{19}NO_2S$ 1,3-Butadiene-1-sulfonamide, N-cyclohexyl-2-methyl-, (Z)-	H_2O or EtOH	242(4.23)	70-2519-84
$C_{11}H_{19}NO_2Si$ 1H-Pyrrole-2-carboxaldehyde, 1-[[2-(trimethylsilyl)ethoxy]methyl]-	MeOH	203(4.18),286(4.17)	44-0203-84
$C_{11}H_{19}NO_4S_2$ Butanedioic acid, [[(diethylamino)-thioxomethyl]thio]-, dimethyl ester	isoPrOH	215(3.98),248(3.87), 280(4.14),340(1.86)	104-1082-84
$C_{11}H_{19}NS$ 2-Hexenenitrile, 4-ethyl-3-(propylthio)-	EtOH	210(3.83),285(4.28)	78-2141-84
2-Octenenitrile, 3-[(1-methylethyl)-thio]-	EtOH	208(3.83),272(4.30)	78-2141-84
$C_{11}H_{19}NS_3$ Thiophene-1-carbo(dithioic) acid, triethylamine salt	MeOH	285.5(5.17)	104-2285-84
$C_{11}H_{19}N_3$ 4-Pyrimidinamine, 2-methyl-6-(1,2,2-trimethylpropyl)-	EtOH	233(4.28),271(3.83)	39-2677-84C
$C_{11}H_{19}N_3S_2$ Carbamimidothioic acid, N'-(4,5-dimethyl-2-thiazolyl)-N,N-diethyl-, methyl ester	isoPrOH	248(3.89),328(3.86)	94-2224-84
$C_{11}H_{19}N_5O_2S$ Guanidine, N,N-dimethyl-N'-nitro-N''-(4-pentyl-2-thiazolyl)-	isoPrOH	240s(3.94),283s(3.78), 329(4.31)	94-3483-84
$C_{11}H_{20}GeSi$ Silane, (dimethylphenylgermyl)trimethyl-	hexane	226(4.01)	138-1383-84
Silane, dimethylphenyl(trimethylgermyl)-	hexane	231(4.07)	138-1383-84
$C_{11}H_{20}GeSn$ Germane, dimethylphenyl(trimethylstannyl)-	hexane	235(4.11)	138-1383-84
$C_{11}H_{20}Ge_2$ Digermane, pentamethylphenyl-	hexane	227(4.03)	138-1383-84

Compound	Solvent	$\lambda_{max}(\log \epsilon)$	Ref.
$C_{11}H_{20}N_2$ Cycloheptane, 2-diazo-1,1,3,3-tetra-methyl-	hexane	490(0.48)	24-0277-84
$C_{11}H_{20}N_2O_2$ 4H-Pyran-4-one, 3-[[[2-(dimethylamino)-ethyl]methylamino]methylene]tetra-hydro-, (E)-	EtOH	326(4.10)	4-1441-84
$C_{11}H_{20}N_3O_8P$ Thymine, 1-(3-deoxy-3-dihydroxyphos-phinylmethyl-β-D-ribofuranosyl)-, ammonium salt	pH 1	260(4.01)	78-0079-84
$C_{11}H_{20}S$ Cycloheptanethione, 2,2,7,7-tetra-methyl-	hexane	217(3.60),239(3.91), 529(0.95)	24-0277-84
$C_{11}H_{20}SiSn$ Silane, dimethylphenyl(trimethyl-stannyl)-	hexane	242(4.07)	138-1383-84
$C_{11}H_{20}Si_2$ Disilane, pentamethylphenyl-	C_6H_{12} C_6H_{12} hexane	231(4.04),345(--) 231(4.04) 231(4.04)	60-0341-84 60-0383-84 138-1383-84
$C_{11}H_{21}N_2$ Methanaminium, N-[2-cyclobutyl-3-(di-methylamino)-2-propenylidene]-N-methyl-, perchlorate	MeOH	326(4.49)	5-0649-84
$C_{11}H_{21}N_2O_5P$ Acetic acid, [bis(1-methylethoxy)phos-phinyl]diazo-, 1-methylethyl ester	MeCN	211(2.93),243(3.02)	65-2495-84
$C_{11}H_{23}N_3O$ 1H-1,2,3-Triazole, 1-(2-ethoxyethyl)-4,5-dihydro-5-pentyl- containing 4-pentyl isomer	hexane hexane	240(3.66) 240(3.60)	104-0580-84 104-0580-84
$C_{11}H_{25}GeNO$ Germanecarboxamide, pentaethyl-	heptane or EtOH	253(2.69)	70-0310-84
$C_{11}H_{26}ClNSi$ Silanamine, N-chloro-1-(1,1-dimethyl-ethyl)-N-(2,2-dimethylpropyl)-1,1-dimethyl-	CCl_2FCF_2Cl	299(1.79)	39-1187-84B

Compound	Solvent	$\lambda_{max}(\log \epsilon)$	Ref.
$C_{12}F_{16}$			
1,4-Cyclohexadiene, 1,2,3,3,4,6,6-hep-tafluoro-5-(2,3,3,4,4,5,5,6,6-nona-fluoro-1-cyclohexen-1-yl)-	heptane	207(3.22),217(3.34)	104-0307-84
$C_{12}H_2F_5IN_2O_5$			
Iodonium, (2-hydroxy-3,5-dinitrophenyl)-(pentafluorophenyl)-, hydroxide, inner salt	DMSO	472(4.26)	39-0135-84C
$C_{12}H_5BrN_4OS$			
Propanedinitrile, [[4-bromo-5-(2-pyri-midinylthio)-2-furanyl]methylene]-	MeOH	234(3.25),278(2.90), 363(3.37)	73-0984-84
$C_{12}H_5NO_2S$			
4H-[1]Benzothieno[3,2-b]pyran-3-carbo-nitrile, 4-oxo-	MeOH	222(4.31),246(4.25), 290(4.17),320s(3.71)	83-0531-84
$C_{12}H_5NO_3$			
4H-Pyrano[3,2-b]benzofuran-3-carboni-trile, 4-oxo-	MeOH	252(3.37),305(3.19)	83-0526-84
$C_{12}H_6Cl_2N_2O_4S_2$			
3(2H)-Isothiazolone, 5-[2-(3,4-dichlo-rophenyl)-4-thiazolyl]-4-hydroxy-, 1,1-dioxide	EtOH	245(4.28),320.5(4.29)	44-2212-84
3(2H)-Isothiazolone, 5-[4-(3,4-dichlo-rophenyl)-2-thiazolyl]-4-hydroxy-, 1,1-dioxide	EtOH	246(4.31),357(4.02)	44-2212-84
$C_{12}H_6Cl_2O_2S$			
4H-[1]Benzothieno[3,2-b]pyran-4-one, 3-(dichloromethyl)-	MeOH	217(3.88),247(3.73), 290(3.71)	83-0531-84
$C_{12}H_6N_2O_2S$			
4H-[1]Benzothieno[3,2-b]pyran-3-carbo-nitrile, 2-amino-4-oxo-	MeOH	234(4.48),248s(4.44), 302(4.17)	83-0531-84
$C_{12}H_6OS_2$			
1H,3H-Naphtho[1,8-cd]pyran-1,3-dithione	$C_2H_4Cl_2$	258(4.33),330s(3.93), 414(4.26),556s(2.63)	35-6084-84
1H,3H-Naphtho[1,8-cd]thiopyran-1-one, 3-thioxo-	$C_2H_4Cl_2$	252(4.34),273s(4.06), 322s(4.00),335(4.06), 375s(4.08),410(4.27)	35-6084-84
	CS_2	532s(1.72),566(1.81), 600(1.81),640(1.66)	35-6084-84
$C_{12}H_6O_3S$			
4H-[1]Benzothieno[3,2-b]pyran-3-carbox-aldehyde, 4-oxo-	dioxan	250(4.34),290(4.12), 315(3.80)	83-0531-84
$C_{12}H_6O_4$			
4H-Pyrano[3,2-b]benzofuran-3-carbox-aldehyde, 4-oxo-	dioxan	228(4.24),248(4.25), 286(4.33)	83-0526-84
$C_{12}H_6S_3$			
1H,3H-Naphtho[1,8-cd]thiopyran-1,3-di-thione	$C_2H_4Cl_2$	250s(4.11),265(4.29), 360s(4.04),375(4.04), 379s(4.01),443(4.36), 523(2.79),553(2.81)	35-6084-84
	CS_2	533(2.55),562(2.66),	35-6084-84

Compound	Solvent	$\lambda_{max}(\log \epsilon)$	Ref.
(cont.)		633(1.94),681(1.92)	35-6084-84
$C_{12}H_7ClN_2O_2$ 1,6-Phenazinediol, 2-chloro-	MeOH	275(4.85),377(3.69), 440(3.53)	158-0943-84
	MeOH-NaOH	265(4.36),296(4.81)	158-0943-84
$C_{12}H_7Cl_2NO_2S$ 3-Pyridinecarboxylic acid, 1-(2,6-di-chlorophenyl)-1,2-dihydro-2-thioxo-	EtOH	243(4.29),267(4.05), 300(3.76),347(3.73)	4-0041-84
$C_{12}H_7IN_2O_5$ Iodonium, (2-hydroxy-3,5-dinitrophen-yl)phenyl-, hydroxide, inner salt	DMSO	372(4.15)	39-0135-84C
$C_{12}H_7NO_3S$ 4H-[1]Benzothieno[3,2-b]pyran-3-carbox-aldehyde, 2-amino-4-oxo-	dioxan	250(4.36),300(4.02)	83-0531-84
4H-[1]Benzothieno[3,2-b]pyran-3-carbox-aldehyde, 4-oxo-, oxime	MeOH	248(3.96),280(3.93), 320s(3.42)	83-0531-84
$C_{12}H_7NO_4$ 4H-Pyrano[3,2-b]benzofuran-3-carbox-aldehyde, 4-oxo-, 3-oxime, (E)-	MeOH	220(4.13),260(4.24), 290(4.33)	83-0526-84
$C_{12}H_7N_3O$ Propanedinitrile, (1,2-dihydro-1-meth-yl-2-oxo-3H-indol-3-ylidene)-	$CHCl_3$	270(4.22),354(4.02), 502(3.04)	39-1331-84B
$C_{12}H_7N_3OS$ Benzofuro[2',3':4,5]pyrrolo[1,2-d]-[1,2,4]triazine-1(2H)-thione	MeOH	278(3.28),377(3.39)	73-0065-84
$C_{12}H_7N_3S$ Naphtho[1',2':4,5]thiazolo[3,2-b]-[1,2,4]triazole	MeOH	243(4.64),293(3.97), 329(3.13)	4-1571-84
$C_{12}H_7N_5O$ Propanedinitrile, [3-(dicyanomethyl)-2-(dimethylamino)-4-oxo-2-cyclo-buten-1-ylidene]-, dimethylamine salt	H_2O	215(4.98),336(3.98), 445(4.46)	24-2714-84
$C_{12}H_7N_5O_2S$ 4H-[1]Benzothieno[3,2-b]pyran-4-one, 2-amino-3-(1H-tetrazol-5-yl)-	MeOH	252(4.49),315(4.15)	83-0531-84
$C_{12}H_7N_5O_3$ Imidazo[1,2-a]pyrazine, 2-(4-nitro-phenyl)-3-nitroso-	EtOH	227(4.59),258(4.97), 345(4.55),714(2.05)	4-1029-84
Imidazo[1,2-a]pyrimidine, 2-(4-nitro-phenyl)-3-nitroso-	EtOH	213(4.93),263(4.52), 350(4.26),684(2.19)	4-1029-84
$C_{12}H_7N_7O_2S$ 10H-Pyrido[1,2-a]thieno[3,2-d]pyrimi-dine-7-carboxamide, 10-oxo-N-1H-tetrazol-5-yl-	MeOH	261(4.48),350(4.09)	87-0528-84
$C_{12}H_8$ Acenaphthylene	n.s.g.	322(4.04),434(3.23)	78-0665-84

Compound	Solvent	λ_{max}(log ϵ)	Ref.
$C_{12}H_8Br_2N_4O$			
Diazene, 1,1'-oxybis[2-(4-bromophenyl)-	DMSO	308(3.60)	123-0075-84
$C_{12}H_8ClNO_2S$			
3-Pyridinecarboxylic acid, 4-[(4-chlorophenyl)thio]-	MeOH	225(4.44),267(4.00), 301(3.96)	73-1722-84
$C_{12}H_8ClNO_3S$			
6H-1,3-Thiazine-5-carboxaldehyde, 4-chloro-2-(4-methoxyphenyl)-6-oxo-	EtOH	231(4.04),241(3.94), 380(4.22)	103-1088-84
$C_{12}H_8ClN_3O_2$			
Pyrano[4,3-c]pyrazol-4(1H)-one, 3-amino-6-(4-chlorophenyl)-	EtOH	240(4.26),303(4.21)	94-3384-84
$C_{12}H_8ClN_5O_6$			
Benzenamine, 2-chloro-4-(1,4-dihydro-5,7-dinitro-4-benzofurazanyl)-, N-oxide	MeOH	239(4.13),270(4.02), 304(4.06),353(3.56), 461(4.47)	12-0985-84
	dioxan	283(3.86),426(3.57)	12-0985-84
potassium salt	MeOH	249(4.23),295(4.00), 363(3.81),474(4.40)	12-0985-84
$C_{12}H_8Cl_2F_3N_3$			
2-Pyrimidinamine, 4,5-dichloro-N-methyl-N-phenyl-6-(trifluoromethyl)-	MeOH	222(3.55),265(3.92), 344(2.99)	4-1161-84
4-Pyrimidinamine, 2,5-dichloro-N-methyl-N-phenyl-6-(trifluoromethyl)-	MeOH	220(3.81),272(3.58), 315(3.49)	4-1161-84
$C_{12}H_8Cl_2O_5$			
1,4-Naphthalenedione, 6,7-dichloro-2-ethoxy-5,8-dihydroxy-	benzene	482s(3.80),495(3.83), 523s(3.73)	118-0953-84
$C_{12}H_8F_3NO_4$			
2,4-Pentanedione, 1,1,1-trifluoro-3-[(4-nitrophenyl)methylene]-	EtOH	270(4.07),290(4.00)	39-2863-84C
$C_{12}H_8FeO_4$			
Iron, [η^4-6,7-bis(methylene)-2,4-cycloheptadien-1-one]tricarbonyl-	EtOH	212s(4.43),250s(4.04), 395(3.39)	88-5419-84
$C_{12}H_8N_2$			
Pyrido[3,2-g]quinoline	MeCN	207(4.143),246(5.179), 339(3.832),355(3.934), 375(3.416)	5-0133-84
$C_{12}H_8N_2O$			
Benzonitrile, 4-(2-oxa-3-azabicyclo-[3.2.0]hepta-3,6-dien-4-yl)-	C_6H_{12}	220.5(4.03),280s(4.09), 290(4.14),303(3.96)	40-0158-84
	EtOH	219(4.05),287(4.13), 304s(3.85)	40-0158-84
Benzonitrile, 4-(1,3-oxazepin-2-yl)-	MeCN	247(4.33),343(3.79)	40-0158-84
9H-Pyrido[3,4-b]indole-1-carboxaldehyde	EtOH	268(4.93),304(4.35), 366(3.89)	94-3579-84
1H-Pyrrolo[2,3-f]isoquinoline-3-carboxaldehyde	EtOH	204(4.18),259.7(4.66), 272(4.64)	103-0394-84
$C_{12}H_8N_2O_2$			
1,6-Phenazinediol	MeOH	272(4.73),373(3.50), 442(3.33)	158-0943-84
	MeOH-NaOH	262(4.12),291(4.48)	158-0943-84

Compound	Solvent	$\lambda_{max}(\log \epsilon)$	Ref.
3H-Phenazin-3-one, 2-amino-	CHCl$_3$	429(4.43)	103-0263-84
$C_{12}H_8N_2O_2S$			
4H-[1]Benzothieno[3,2-b]pyran-3-carbox-imidamide, 4-oxo-, monohydrochloride	MeOH	249(4.40),270(4.41), 315(3.80),325(3.82), 340(3.77)	83-0531-84
2H-Pyran-3-carbonitrile, 4-(methyl-thio)-2-oxo-6-(3-pyridinyl)-	EtOH	247(4.16),327(4.24), 365(4.10)	94-3384-84
$C_{12}H_8N_2O_3S$			
4H-[1]Benzothieno[3,2-b]pyran-3-carbox-aldehyde, 2-amino-4-oxo-, 3-oxime	MeOH	246(4.25),270(4.14), 325(3.93)	83-0531-84
4H-[1]Benzothieno[3,2-b]pyran-3-carbox-amide, 2-amino-4-oxo-	MeOH	235(3.81),248s(3.77), 310(3.44)	83-0531-84
$C_{12}H_8N_2O_4S$			
3-Pyridinecarboxylic acid, 1,2-dihydro-1-(4-nitrophenyl)-2-thioxo-	EtOH	238(4.17),300(4.16)	4-0041-84
$C_{12}H_8N_2O_6$			
Furan, 2-(2-nitroethenyl)-5-(3-nitro-phenoxy)-	MeOH	207(4.26),248(4.17), 370(4.35)	73-2496-84
$C_{12}H_8N_4$			
Pyrazolo[3,4-b]quinoline-4-carbo-nitrile, 3-methyl-	EtOH	418(3.69)	18-2984-84
$C_{12}H_8N_4O$			
Imidazo[1,2-a]pyrazine, 3-nitroso-2-phenyl-	EtOH	229(4.25),274(4.69), 353(4.57),704(2.14)	4-1029-84
Imidazo[1,2-a]pyrimidine, 3-nitroso-2-phenyl-	EtOH	229(4.24),278(4.62), 298(4.58),315(4.68), 674(2.06)	4-1029-84
$C_{12}H_8N_4O_3$			
4-Benzofurazanamine, 7-nitro-N-phenyl-	15% DMF-pH 2.7	490(4.48)	48-0385-84
$C_{12}H_8N_4O_6S_2$			
Carbonimidothioic acid, cyano-, bis(5-nitro-2-furanyl)methyl] ester	dioxan	263(3.23),313(3.44)	73-2285-84
4-Thiazolamine, 5-(5-nitro-2-furanyl)-2-[[(5-nitro-2-furanyl)methyl]thio]-	dioxan	315(3.23),474(3.29)	73-2285-84
$C_{12}H_8N_6O_2$			
1,2,3,5,6,7-Hexaazaacenaphthylene-4,8-dione, 2,3,5,7-tetrahydro-2-phenyl-	EtOH	238(5.13)	4-1049-84
Pyridazino[4',5':3,4]pyrazolo[1,5-a]-quinazoline-5,10-dione, 7-amino-6,9-dihydro-	EtOH	231(5.08)	4-1049-84
$C_{12}H_8N_6O_5$			
Diazene, 1,1'-oxybis[2-(4-nitrophenyl)-	DMSO	346(4.05),355s(--)	123-0075-84
$C_{12}H_8O_2$			
1-Dibenzofuranol	H$_2$O	223(4.51),256(4.13), 268(4.04),275(4.15), 295(3.69),306(3.85)	39-1213-84C
	pH 13	236(3.91),267(3.99), 275(4.04),325(4.00)	39-1213-84C

Compound	Solvent	λ_{max}(log ϵ)	Ref.
C$_{12}$H$_8$O$_3$			
2H,8H-Benzo[1,2-b:5,4-b']dipyran-2-one	EtOH	223(4.304),267(4.209), 347(3.957)	111-0365-84
7H-Furo[3,2-g][1]benzopyran-7-one, 9-methyl-	EtOH	213.5(4.349),245(4.390), 248(4.394),297.5(4.05)	111-0365-84
C$_{12}$H$_8$O$_3$S			
4H-[1]Benzothieno[3,2-b]pyran-4-one, 3-(hydroxymethyl)-	MeOH	250(4.21),290(4.15), 325s(3.69)	83-0531-84
C$_{12}$H$_8$O$_4$			
Furo[3,2-g][1]benzopyran-7-one, 9-methoxy-	EtOH	219(4.475),249(4.421), 300(4.122)	111-0365-84
	n.s.g.	219(4.32),249(4.35), 300(4.06)	156-0051-84B
2-Naphthaleneacetic acid, 1-hydroxy-α-oxo-	EtOH	216(4.37),260(4.37), 295(3.89),371(3.70)	39-2655-84C
4H-Pyrano[3,2-b]benzofuran-4-one, 3-(hydroxymethyl)-	MeOH	208(4.26),288(4.33)	83-0526-84
C$_{12}$H$_8$O$_5$			
Acetic acid, (3-hydroxy-5-oxo-4-phenyl-2(5H)-furanylidene)-, (E)-	EtOH	243s(4.26),263(4.43), 363(3.97)	39-1555-84C
	EtOH-NaOH	267(4.61),346(3.89)	39-1555-84C
(Z)-	EtOH	269(4.61),338-382(3.87)	39-1555-84C
	EtOH-NaOH	266(4.53),362(3.94)	39-1555-84C
C$_{12}$H$_8$S$_2$			
2,2'-Bithiophene, 5-(3-buten-1-ynyl)-	hexane	252(3.93),345.5(4.42)	78-2773-84
C$_{12}$H$_9$BrN$_2$OS			
Thiopyrano[3,4-c]pyrazol-4(5H)-one, 1-(4-bromophenyl)-1,7-dihydro-	EtOH	232(4.11),260(4.24)	4-1437-84
C$_{12}$H$_9$BrOS			
Methanone, (3-bromo-4-methylphenyl)-2-thienyl-	EtOH	337(3.18)	73-1764-84
C$_{12}$H$_9$ClF$_3$N$_3$			
2-Pyrimidinamine, 4-chloro-N-methyl-N-phenyl-6-(trifluoromethyl)-	MeOH	224(3.56),255(3.81), 322(2.87)	4-1161-84
4-Pyrimidinamine, 2-chloro-N-methyl-N-phenyl-6-(trifluoromethyl)-	MeOH	255(3.72),294(3.43)	4-1161-84
C$_{12}$H$_9$ClF$_3$N$_3$O			
4(1H)-Pyrimidinone, 5-chloro-2-(methylphenylamino)-6-(trifluoromethyl)-	MeOH	216(3.86),254(3.57), 315(3.59)	4-1161-84
C$_{12}$H$_9$ClF$_6$O			
Benzene, 1-chloro-4-[3,3,3-trifluoro-1-[(2,2,2-trifluoroethoxy)methyl]-1-propenyl]-, (E)-	EtOH	228(3.92)	44-2258-84
(Z)-	EtOH	254(4.11)	44-2258-84
C$_{12}$H$_9$ClN$_2$OS			
Thiopyrano[3,4-c]pyrazol-4(5H)-one, 1-(4-chlorophenyl)-1,7-dihydro-	EtOH	228(4.09),260(4.215)	4-1437-84
C$_{12}$H$_9$ClN$_4$O			
4H-Pyrazolo[3,4-d]pyrimidin-4-one, 5-(2-chlorophenyl)-1,5-dihydro-6-methyl-	EtOH	207(4.47),251(3.81)	11-0023-84A

Compound	Solvent	$\lambda_{max}(\log \epsilon)$	Ref.
$C_{12}H_9ClN_4S$			
1H-Pyridazino[4,5-e][1,3,4]thiadiazine, 5-chloro-1-methyl-3-phenyl-	EtOH	269(4.14),323(3.29)	94-4437-84
$C_{12}H_9ClO_3$			
2-Cyclopenten-1-one, 2-(1,3-benzodioxol-5-yl)-3-chloro-	MeCN	235(4.21),292(3.91)	44-0228-84
1,4-Naphthalenedione, 2-chloro-5-methoxy-6-methyl-	EtOH	252(3.89),274(3.77), 290s(3.61),348(3.31)	78-3455-84
1,4-Naphthalenedione, 2-chloro-5-methoxy-7-methyl-	EtOH	233(4.15),277(3.93), 348(3.33),402(3.34)	78-3455-84
1,4-Naphthalenedione, 2-chloro-8-methoxy-6-methyl-	EtOH	254(3.97),265(3.99), 273(4.00),340(2.91), 410(3.44)	78-3455-84
1,4-Naphthalenedione, 2-chloro-8-methoxy-7-methyl-	EtOH	254(4.11),272(4.03), 346(3.49),371(3.42)	78-3455-84
1,4-Naphthalenedione, 3-chloro-6-methoxy-8-methyl-	MeOH	254(4.20),275s(3.92), 330(3.48)	44-1853-84
2-Propen-1-one, 3-chloro-1-(3-methoxy-2-benzofuranyl)-, (Z)-	MeOH	235(3.79),251(3.68), 360(4.22)	83-0526-84
$C_{12}H_9ClO_4$			
1,4-Naphthalenedione, 2-chloro-5,7-dimethoxy-	EtOH	270(3.99),285(3.81), 424(3.44)	78-3455-84
1,4-Naphthalenedione, 2-chloro-6,8-dimethoxy-	EtOH	267(4.04),293(4.03), 360(3.52)	78-3455-84
$C_{12}H_9ClO_5$			
2H-Cyclohepta[b]furan-3-carboxylic acid, 6-chloro-5-methoxy-2-oxo-, methyl ester	MeOH	294(4.40),306(4.40), 402(4.28)	18-0621-84
1,4-Naphthalenedione, 2-chloro-3-ethoxy-5,8-dihydroxy-	benzene	489s(3.72),515(3.79), 553(3.59)	118-0953-84
$C_{12}H_9Cl_2NO_2$			
Methanone, (3,5-dichloro-2-methoxyphenyl)-1H-pyrrol-2-yl-	MeOH	220s(4.19),301(4.24)	95-0238-84
$C_{12}H_9FN_4$			
2-Butenedinitrile, 2-amino-3-[2-(4-fluorophenyl)-1-aziridinyl]-	EtOH	299(4.16)	44-0813-84
$C_{12}H_9F_3O_2$			
2,4-Pentanedione, 1,1,1-trifluoro-3-(phenylmethylene)-	EtOH	251s(3.57),295(4.00)	39-2863-84C
$C_{12}H_9FeO_3$			
Iron, tricarbonyl[[η^4-(7-methylene-1,3,5-cycloheptatrien-1-yl)methyl]-, tetrafluoroborate	CF_3COOH	522(3.68),713(2.20)	88-5419-84
$C_{12}H_9NO$			
9H-Carbazol-2-ol	EtOH	235(4.7),258(4.31), 303(4.19)	102-0471-84
$C_{12}H_9NO_2$			
Bullvalenedicarboximide	MeOH	218s(4.15),268(3.52)	24-0633-84
7bH-Cyclopent[cd]indene, 7b-methyl-1(and 2)-nitro-	EtOH	246(4.18),307(4.11), 332s(3.88),402(3.75), 485(3.19)	39-0175-84C

Compound	Solvent	$\lambda_{max}(\log \epsilon)$	Ref.
7bH-Cyclopent[cd]indene, 7b-methyl-5(and 6)-nitro-	EtOH	248s(3.95),275(4.15), 315(4.22),376(3.82), 444s(3.19),482(3.07)	39-0175-84C
1H-Indeno[1,2-b]pyridine-2,4(3H,5H)-dione, hydrochloride	MeOH	230(4.29),246s(4.06), 270s(3.78),340(4.22)	83-0448-84
$C_{12}H_9NO_2S$			
3-Pyridinecarboxylic acid, 1,2-dihydro-1-phenyl-2-thioxo-	EtOH	304(4.22),390(3.64)	4-0041-84
$C_{12}H_9NO_3$			
4H-[1]Benzopyrano[3,4-d]isoxazol-4-one, 3,7-dimethyl-	EtOH	270(4.07),280(4.15), 306(4.01),318(3.94)	44-4419-84
4H-[1]Benzopyrano[3,4-d]isoxazol-4-one, 3-ethyl-	EtOH	266(4.07),272(4.11), 306(3.86)	44-4419-84
2,5-Cyclohexadien-1-one, 2,5-dihydroxy-4-(phenylimino)-	EtOH	500(2.85)	12-0611-84
Furo[3,2-c]quinoline-2,4-dione, 3,5-dihydro-5-methyl-	EtOH	276(3.88),284(3.89), 305s(3.68),319(3.83), 329s(3.75)	4-1881-84
$C_{12}H_9NO_3S$			
Furan, 2-(2-nitroethenyl)-5-(phenylthio)-	MeOH	207(4.19),242(4.02), 370(4.14)	73-2496-84
3-Pyridinecarboxylic acid, 4-(phenylthio)-, 1-oxide	MeOH	340(4.34)	73-1722-84
$C_{12}H_9NO_4S$			
1H-Indene-1,3(2H)-dione, 2-[1-(methylthio)-2-nitroethylidene]-	EtOH	245(4.33),364(4.24), 463(4.01)	95-0127-84
$C_{12}H_9NO_5S$			
Furan, 2-(2-nitroethenyl)-5-(phenylsulfonyl)-	MeOH	208(3.13),241(3.12), 335(3.33)	73-2496-84
$C_{12}H_9NO_5S_2$			
4-Thiazolidinecarboxylic acid, 2-(3-oxobenzo[b]thien-2(3H)-ylidene)-, S,S-dioxide, (S)-	EtOH	214(4.51),250(4.23), 266(4.19),274(4.33), 341(4.36)	95-0134-84
$C_{12}H_9N_3O$			
9H-Pyrido[3,4-b]indole-1-carboxamide	EtOH	243(4.31),253(4.21), 271(4.29),291(4.04), 361(3.80)	94-0170-84
Pyrimido[5,4-c]quinolin-4(1H)-one, 8-methyl-	MeOH	278(4.41),340s(3.92), 352(3.96),376s(3.75)	103-0439-84
$C_{12}H_9N_3OS$			
2-Propanone, 1-diazo-3-(2-quinolinylthio)-	MeOH	213(3.9),252(3.80), 274(4.07)	103-0505-84
2-Propanone, 1-diazo-3-(8-quinolinylthio)-	MeOH	207(4.10),250(3.90), 280(4.20)	103-0505-84
$C_{12}H_9N_3O_2$			
2-Propenoic acid, 2-cyano-3-(phenylamino)-, cyanomethyl ester	DMF	287s(3.95),321(4.42)	5-1702-84
Pyrano[4,3-c]pyrazol-4(1H)-one, 3-amino-6-phenyl-	EtOH	236(4.26),300(4.17)	94-3384-84
1H-Pyrazolo[3,4-b]quinoline-4-carboxylic acid, 3-methyl-	EtOH	382(3.74)	18-2984-84

Compound	Solvent	$\lambda_{max}(\log \epsilon)$	Ref.
$C_{12}H_9N_3O_3$ 1,3-Benzenediol, 4-(5-hydroxy-2H-benzo-triazol-2-yl)-	MeCN	345(4.43)	126-2497-84
$C_{12}H_9N_3O_3S$ 5H-Imidazo[2,1-b]pyrido[3,2-e][1,3]-thiazine-3-carboxylic acid, 5-oxo-, ethyl ester	CHCl$_3$	274(4.15),350(3.52)	142-0807-84
$C_{12}H_9N_3O_4$ 1,3,5-Benzenetriol, 2-(5-hydroxy-2H-benzotriazol-2-yl)-	MeCN	351(4.38)	126-2497-84
2H-1-Benzopyran-4-carboxylic acid, 7-azido-2-oxo-, ethyl ester	EtOH	341(4.14)	94-3926-84
$C_{12}H_9N_5O$ Benzofuro[2',3':4,5]pyrrolo[1,2-d]-1,2,4-triazin-1(2H)-one, hydrazone	dioxan	332(3.36),435(3.36)	73-0065-84
$C_{12}H_9N_5O_5$ Benzenamine, 4-(1,4-dihydro-5,7-di-nitro-4-benzofurazanyl)- (inner salt)	MeOH	239(4.06),290(3.88), 350(3.80),477(4.40)	12-0985-84
$C_{12}H_9N_5O_6$ Benzenamine, 4-(1,4-dihydro-5,7-di-nitro-4-benzofurazanyl)-, N-oxide	MeOH	248(4.16),291s(3.90), 365(3.76),474(4.32)	12-0985-84
monopotassium salt	MeOH	249(4.21),290s(3.95), 365(3.80),475(4.37)	12-0985-84
$C_{12}H_9N_7$ 1H-Pyrazolo[3,4-e][1,2,4]triazolo-[3,4-c][1,2,4]triazine, 3-methyl-1-phenyl-	dioxan	266(4.88),482(3.41)	4-1565-84
$C_{12}H_9N_7O$ 8H-Pyrazolo[3,4-e][1,2,4]triazolo-[3,4-c][1,2,4]triazin-8-one, 1,7-dihydro-3-methyl-1-phenyl-	EtOH	226(3.94),264(4.24), 325s(2.65),532(2.80)	4-1565-84
$C_{12}H_9N_7S$ 8H-Pyrazolo[3,4-e][1,2,4]triazolo-[3,4-c][1,2,4]triazine-8-thione, 1,7-dihydro-3-methyl-1-phenyl-	EtOH	248(4.07),286(4.31), 404(2.78),560(2.08)	4-1565-84
$C_{12}H_{10}$ 1,3-Cyclopentadiene, 5,5'-(1,2-ethane-diylidene)bis-	pentane	310(4.20),324(4.54), 339(4.78),357(4.87), 419(2.69)	24-2006-84
2aH-Cyclopent[cd]indene, 2a-methyl-	EtOH	258(4.04),318s(2.50), 331s(2.41)	39-0165-84C
7bH-Cyclopent[cd]indene, 7b-methyl-	EtOH	249s(3.74),282(4.54), 335s(3.52),398s(2.11), 439s(2.57),450(2.64)	39-0165-84C
$C_{12}H_{10}BrNO_2S$ 2-Thiophenecarboxylic acid, 3-amino-5-(4-bromophenyl)-, methyl ester	MeOH	314(4.23),354(3.99)	118-0275-84
$C_{12}H_{10}Br_2$ 2aH-Cyclopent[cd]indene, 2a,4a-dibromo-	EtOH	326(3.46)	39-0175-84C

Compound	Solvent	$\lambda_{max}(\log \epsilon)$	Ref.
4a,7b-dihydro-7b-methyl- (cont.)			39-0175-84C
$C_{12}H_{10}ClNO_2S$ 2-Thiophenecarboxylic acid, 3-amino-5-(4-chlorophenyl)-, methyl ester	MeOH	299(4.30),352(3.98)	118-0275-84
$C_{12}H_{10}ClNO_3S$ 6H-1,3-Thiazin-6-one, 4-chloro-2-(3,4-dimethoxyphenyl)-	EtOH	219(4.24),247(3.89), 257(3.74),390(4.08)	103-1088-84
$C_{12}H_{10}ClN_3$ 1H-Pyrazolo[4,3-b]quinoline, 9-chloro-1,7-dimethyl-	EtOH	248(4.90),326(3.58), 339(3.81),371(3.64), 389(3.60)	103-0918-84
2H-Pyrazolo[4,3-b]quinoline, 9-chloro-2,7-dimethyl-	EtOH	246(4.91),330(3.98), 347(4.07),389(3.85), 411(3.83)	103-0918-84
$C_{12}H_{10}ClN_3O$ 1H-Pyrazolo[4,3-b]quinoline, 9-chloro-5-methoxy-1-methyl-	EtOH	251(4.84),349(4.54), 369(3.74),389(3.74), 406(3.30)	103-0918-84
1H-Pyrazolo[4,3-b]quinoline, 9-chloro-7-methoxy-1-methyl-	EtOH	251(4.84),349(4.10), 366(3.80),386(3.75)	103-0918-84
2H-Pyrazolo[4,3-b]quinoline, 9-chloro-7-methoxy-2-methyl-	EtOH	246(4.81),336(4.07), 352(4.31),390(3.88), 413(4.07)	103-0918-84
$C_{12}H_{10}Cl_2O_4$ 1-Oxaspiro[5.5]undeca-3,7,10-triene-2,9-dione, 7,10-dichloro-3-methoxy-5-methyl-	MeOH	240(4.16),290s(3.30)	78-5039-84
1-Oxaspiro[5.5]undeca-3,7,10-triene-2,9-dione, 8,10-dichloro-3-methoxy-5-methyl-	MeOH	243(4.02),297(3.29)	78-5039-84
$C_{12}H_{10}Cl_6$ Bicyclo[2.2.1]hept-2-ene, 1,2,3,4,7,7-hexachloro-5-(3-methyl-1,3-butadien-yl)-	EtOH	220(4.18),230(4.23)	70-1743-84
$C_{12}H_{10}FN_3$ Cyanamide, [2-(dimethylamino)-3-(4-fluorophenyl)-2-cyclopropen-1-yli-dene]-	CH_2Cl_2	306(4.31)	118-0686-84
$C_{12}H_{10}F_3N_3O$ 2(1H)-Pyrimidinone, 4-(methylphenyl-amino)-6-(trifluoromethyl)-	MeOH	207(4.09),265(3.52), 288(3.64)	4-1161-84
4(1H)-Pyrimidinone, 2-(methylphenyl-amino)-6-(trifluoromethyl)-	MeOH	216(3.75),233(3.56), 305(3.60)	4-1161-84
$C_{12}H_{10}F_6O$ Benzene, [3,3,3-trifluoro-1-[(2,2,2-trifluoroethoxy)methyl]-1-propen-yl]-, (E)-	EtOH	226(3.67)	44-2258-84
(Z)-	EtOH	248(4.01)	44-2258-84
$C_{12}H_{10}N$ Benzenamine, N-phenyl-, radical	toluene	770(3.59)	70-1603-84

Compound	Solvent	$\lambda_{max}(\log \epsilon)$	Ref.
$C_{12}H_{10}NOS$			
2H-Pyrido[1,2,3-ef]-1,5-benzothiazepin-ium, 3,4-dihydro-3-oxo-, perchlorate	10% MeCN	242(4.1),270(3.96), 318(3.69)	103-0343-84
1H-[1,3]Thiazino[3,2-a]quinolinium, 2,3-dihydro-2-oxo-, perchlorate	10% MeCN	242(3.82),265(4.01), 272(4.01),332(4.1)	103-0508-84
$C_{12}H_{10}NS$			
Pyrrolo[2,1,5-de]quinolizinium, 1-(methylthio)-, perchlorate	EtOH-HClO₄	226(4.66),300(3.88), 319s(3.69),347s(3.50), 428(3.79)	39-2553-84C
$C_{12}H_{10}N_2$			
Cyclooctatetraene-1,5-dicarbonitrile, 2,6-dimethyl-	MeCN	215s(4.375),224(4.401), 290s(2.749)	89-0631-84
Cyclopropa[cd]pentalene-1,2b(2aH)-di-carbonitrile, 4a,4b-dihydro-4a,4b-dimethyl- (semibullvalene)	C_6H_{12}	212(3.927),245(3.809), 360s(1.792)	89-0631-84
	EtOH	211(3.920),245(3.828), 360s(1.837)	89-0631-84
	MeCN	211(3.913),246(3.824), 360s(1.797)	89-0631-84
	PrCN	211(3.922),246(3.831), 360s(1.807)	89-0631-84
at 178°K	PrCN	247(3.848)	89-0631-84
at 366°K	PrCN	243(3.825),360s(2.127)	89-0631-84
Diazene, diphenyl-	pH 8 +C₅H₅N	422(2.62)	46-6185-84
1H-Pyrrole-3-acetonitrile, 2-phenyl-	MeOH	205(4.17),276(4.13)	83-0143-84
$C_{12}H_{10}N_2O$			
Pimprinine	EtOH	224(4.39),266(4.19), 282s(4.08),298s(4.04)	158-1153-84
9H-Pyrido[3,4-b]indole-1-methanol	EtOH	215(4.62),240(4.70), 285(4.50),340(3.80), 352(3.80)	94-3579-84
$C_{12}H_{10}N_2OS$			
3-Pyridinecarbonitrile, 4-(2-furanyl)-1,2-dihydro-5,6-dimethyl-2-thioxo-	dioxan	238(4.01),270(3.98), 316(4.20),455(3.18)	104-1780-84
4(3H)-Pyrimidinethione, 3-ethenyl-5-hydroxy-2-phenyl-	EtOH	296(4.75),376(4.68)	4-1149-84
Thiopyrano[3,4-c]pyrazol-4(5H)-one, 1,7-dihydro-1-phenyl-	EtOH	258(4.15)	4-1437-84
$C_{12}H_{10}N_2O_2$			
2H-Benzoxepino[3,4-c]pyrazol-10-one, 4,10-dihydro-1-methyl-	EtOH	270(4.08)	4-0301-84
2H-Benzoxepino[3,4-c]pyrazol-4-one, 4,10-dihydro-2-methyl-	EtOH	272(4.11)	4-0301-84
4H-[1]Benzoxepino[3,4-c]pyrazol-4-one, 1,10-dihydro-1-methyl-	EtOH	272(4.02)	4-0301-84
Diazene, diphenyl-, 1,2-dioxide, (Z)-	MeOH	218(3.95),235(3.91), 240(3.86),246(3.81), 258(3.82),264(3.87), 320(3.93),365(3.90)	2-0060-84
	MeOH-NaOH	218(4.16),235(4.09), 240(4.00),246(3.85), 258(3.88),264(4.00), 320(3.92)	2-0060-84
3H-Pyrido[3,4-b]indole-3-carboxylic acid, 4,9-dihydro- (anion)	10% MeOH	235s(4.18),280(4.06), 328(3.97),380(3.30)	78-0221-84
Pyrrolo[2,1-b]quinazoline-3-carboxalde-hyde, 1,2,3,9-tetrahydro-9-oxo-	EtOH	225(4.41),266(4.29), 293(3.58),303(3.70),	4-0219-84

Compound	Solvent	$\lambda_{max}(\log \epsilon)$	Ref.
Pyrrolo[2,1-b]quinazoline-3-carboxalde-hyde, 1,2,3,9-tetrahydro-9-oxo-(cont.)		314(3.67),336s(3.21), 355s(2.91)	4-0219-84
	dioxan	265(4.07),275(4.03), 294s(4.02),304(4.10), 317(4.07),330s(3.91), 350s(3.52)	4-0219-84
	DMSO	266(4.06),284s(3.99), 311(4.11),320s(4.10), 333s(4.00),350s(3.50)	4-0219-84
	$CHCl_3$	266(4.01),275(4.03), 289(3.94),306(4.14), 319(4.07),335(4.02), 350s(3.89)	4-0219-84
$C_{12}H_{10}N_2O_2S$ Thieno[2,3-b]pyridine-4-acetic acid, α-cyano-, ethyl ester	EtOH	228(4.42),362(4.23), 381s(4.18)	4-1135-84
$C_{12}H_{10}N_2O_3$ Benz[cd]indole, 1,5-dihydro-6-methoxy-4-nitro-	EtOH	206(4.25),230(4.26), 264(3.82),300(3.91)	44-4761-84
$C_{12}H_{10}N_2O_3Se$ 4,6(1H,5H)-Pyrimidinedione, dihydro-5-[(4-methoxyphenyl)methylene]-2-selenoxo-	dioxan	273(3.84),416(4.63)	104-0141-84
$C_{12}H_{10}N_2S$ 1H-Perimidine, 2-(methylthio)-, mono-hydriodide	MeOH	227(4.68),329(4.12)	4-0911-84
$C_{12}H_{10}N_4$ 2-Butenedinitrile, 2-amino-3-(2-phenyl-1-aziridinyl)-, (Z)-	EtOH	300(4.15)	44-0813-84
$C_{12}H_{10}N_4O$ 1H-Pyrazole-5-carbonitrile, 3-acetyl-4-amino-1-phenyl-	EtOH	220(4.04),290s(3.44)	4-1049-84
4H-Pyrazolo[3,4-d]pyridazin-4-one, 1,5-dihydro-5-methyl-3-phenyl-	EtOH	260(4.14),280(4.04)	94-4437-84
Pyrazolo[3,4-d]pyrimidin-4-one, 1,5-dihydro-6-methyl-5-phenyl-	EtOH	205.5(4.45),252.0(3.82)	11-0023-84A
1H-Pyrazolo[3,4-b]quinoline-4-carbox-amide, 3-methyl-	EtOH	388(3.74)	18-2984-84
1H-Tetrazole, 5-[5-(3-methylphenyl)-2-furanyl]-	dioxan	308(4.44),325s(4.20)	73-1699-84
$C_{12}H_{10}N_4OS$ 1H-Pyridazino[4,5-e][1,3,4]thiadiazin-5(6H)-one, 6-methyl-3-phenyl-	EtOH	269(4.11),297(3.98)	94-4437-84
$C_{12}H_{10}N_4O_2$ Benzo[g]pteridine-2,4(3H,10H)-dione, 3,10-dimethyl-	pH 7.27	341(3.94),434(4.02)	39-1227-84B
Imidazo[2,1-c][1,2,4]benzotriazine-1-carboxylic acid, ethyl ester	MeOH	238(4.49),342(4.08)	4-1081-84
2(1H)-Pyrazinimine, 1-(1,2-dioxoeth-yl)-, 1-oxime	EtOH	274(3.23),345(2.49), 421(2.80),440(2.74)	4-1029-84
1H-Tetrazole, 5-[5-(2-methoxyphenyl)-2-furanyl]-	dioxan	322(4.40),336s(4.28)	73-1699-84

Compound	Solvent	$\lambda_{max}(\log \epsilon)$	Ref.
3H-[1,2,4]Triazino[3,2-b]quinazoline-3,10(4H)-dione, 2,4-dimethyl-	EtOH	294(4.42),332(4.33)	24-1077-84
$C_{12}H_{10}N_8$			
9H-Purin-6-amine, 8-azido-1-(phenyl-methyl)-	EtOH	237(--),294(--)	39-0879-84C
	EtOH-acid	285(--)	39-0879-84C
	EtOH-base	291(--)	39-0879-84C
$C_{12}H_{10}O$			
Furobullvalene	MeOH	218(3.93)	24-0633-84
$C_{12}H_{10}OS_2$			
3-Butyn-1-ol, 4-[2,2'-bithiophene]-5-yl-	EtOH	241(3.79),327(4.33), 335.5(4.33)	78-2773-84
2-Propen-1-one, 2-methyl-1,3-di-2-thi-enyl-	EtOH	335(3.13)	73-1764-84
Spiro[1,3-dithiolane-2,1'(2'H)-naph-thalen]-2'-one	EtOH	220(4.07),239(4.05), 290s(3.57),318(3.67)	39-2069-84C
$C_{12}H_{10}O_2$			
1-Azulenol, acetate	EtOH	237(4.15),277(4.62), 344(3.53),605(2.46), 650s(2.38),725s(1.95)	35-4857-84
Bullvalenelactone	MeOH	218s(3.89),258(3.53)	24-0633-84
2(5H)-Oxepinone, 5-phenyl-	EtOH	201.5(4.28)	150-3501-84M
$C_{12}H_{10}O_3$			
Spiro[1,3-dioxolane-2,1'(2'H)-naphtha-len]-2'-one	EtOH	235(4.27),323(3.85)	39-2069-84C
$C_{12}H_{10}O_4$			
1,3-Cyclopentanedione, 2-(1,3-benzo-dioxol-5-yl)-	H_2O	266(4.06)	44-0228-84
sodium salt	H_2O	266(3.80)	44-0228-84
1,4-Naphthalenedione, 2,5-dihydroxy-3,8-dimethyl-	EtOH	208(4.62),236(4.59), 284(4.52),433(3.91)	100-0331-84
	EtOH-KOH	220(4.60),238(4.58), 270(4.59),394(3.91), 495(3.89)	100-0331-84
1,4-Naphthalenedione, 2-hydroxy-7-methoxy-5-methyl-	EtOH-HCOOH	262(4.33),295(4.09), 344(3.50),408(3.10)	44-1853-84
1,4-Naphthalenedione, 5-hydroxy-6-methoxy-2-methyl-	EtOH	266(4.02),444(3.54)	78-3455-84
1,4-Naphthalenedione, 7-hydroxy-2-methoxy-5-methyl-	MeOH	264(4.35),291(4.12), 344(3.36),402(3.25)	44-1853-84
1,4-Naphthalenedione, 8-hydroxy-2-methoxy-6-methyl-	EtOH	246(4.68),289(4.30), 369(3.26),412(3.75)	78-3455-84
$C_{12}H_{10}O_5$			
4H-1-Benzopyran-5-acetic acid, 7-hy-droxy-2-methyl-4-oxo-	MeOH	242(4.31),249(4.32), 290(5.05)	94-3493-84
	MeOH-AlCl$_3$	242(4.31),249(4.34), 290(4.05)	94-3493-84
1,4-Naphthalenedione, 5-hydroxy-2,7-dimethoxy-	EtOH	262(3.93),304(3.75), 434(3.33)	78-3455-84
$C_{12}H_{11}BrN_2O$			
2H-Pyrrol-2-one, 5-(bromo-3-pyridinyl-methylene)-1,5-dihydro-3,4-dimethyl-, (Z)-	MeOH	203(3.97),281(4.15), 308(4.12)	49-0357-84

Compound	Solvent	$\lambda_{max}(\log \epsilon)$	Ref.
2H-Pyrrol-2-one, 5-(bromo-4-pyridinyl-methylene)-1,5-dihydro-3,4-dimethyl-, (Z)-	MeOH	203(4.09),275(4.18), 308(4.07)	49-0357-84
$C_{12}H_{11}BrN_4$ 1,4,7,9b-Tetraazaphenalene, 3-bromo-2,5,8-trimethyl-	EtOH	234s(4.18),247(4.24), 350(4.24),366(4.19), 425s(1.98),453(2.19), 488(2.46),527(2.33), 572(2.28),608s(1.76)	11-0170-84A
$C_{12}H_{11}BrO$ 1-Pentyn-3-one, 4-bromo-4-methyl-1-phenyl-	EtOH	202(4.11),220(3.92), 280s(--),293(4.03)	39-0535-84C
$C_{12}H_{11}Br_2NO_3$ 1H-Indole-2-carboxylic acid, 3,4-di-bromo-5-methoxy-, ethyl ester	EtOH	234(4.38),304(4.31)	44-4761-84
$C_{12}H_{11}ClN_2OS$ 6H-1,3-Thiazin-6-one, 4-chloro-2-[4-(dimethylamino)phenyl]-	EtOH	230(4.23),262(3.65), 295(3.70),460(4.53)	103-1088-84
$C_{12}H_{11}ClO_3S$ Ethanone, 1-(5-chloro-2-hydroxy-benzo[b]thien-3-yl)-2-ethoxy-	MeOH	233(4.36),263(4.06), 277.5(4.08),310(4.06)	73-0603-84
Ethanone, 1-(5-chloro-3-hydroxy-benzo[b]thien-2-yl)-2-ethoxy-	MeOH	257.5(4.15),288(4.12), 312s(4.06),322.5(4.11), 404(3.85)	73-0603-84
$C_{12}H_{11}Cl_2NO_2$ 2,4,6,8-Nonatetraenoic acid, 9,9-di-chloro-2-cyano-, ethyl ester	EtOH	370(4.42)	70-2328-84
$C_{12}H_{11}Cl_2N_3$ 2-Pyrimidinamine, 4,5-dichloro-N,6-di-methyl-N-phenyl-	MeOH	236(3.65),264(3.92), 318(3.14)	4-1161-84
4-Pyrimidinamine, 2,5-dichloro-N,6-di-methyl-N-phenyl-	MeOH	228(3.72),268(3.47), 296(3.68)	4-1161-84
$C_{12}H_{11}Cl_3O$ 2-Cyclohexen-1-one, 2,6-dimethyl-6-(3,4,4-trichlorobut-3-en-1-ynyl)-	MeOH	245(4.28),323(2.70)	35-3551-84
$C_{12}H_{11}F_3O$ 3-Buten-2-one, 4-(2,4-dimethylphenyl)-1,1,1-trifluoro-, (E)-	EtOH	240(3.68),333(4.09)	39-2863-84C
$C_{12}H_{11}N$ Benzenamine, N-phenyl-	C_6H_{12}	282(4.32)	139-0259-84C
	MeOH	283(4.34)	139-0259-84C
	n.s.g.	241(--),286(41.028)[sic]	100-0775-84
$C_{12}H_{11}NO$ 4-Azulenecarboxaldehyde, O-methyloxime	CH_2Cl_2	231(4.14),261(4.41), 296(4.41),357s(--), 594(2.52)	83-0984-84
Benzenamine, N-(2-furanylmethylene)-2-methyl-	C_6H_{12}	234(3.97),279(4.34), 324(4.07)	18-0844-84
	MeOH	231(3.91),285(4.35), 312(4.09)	18-0844-84

Compound	Solvent	$\lambda_{max}(\log \epsilon)$	Ref.
Benzenamine, N-(2-furanylmethylene)-4-methyl-	C_6H_{12}	222(4.11),280(4.41), 323(4.30)	18-0844-84
	MeOH	222(4.07),282(4.34), 321(4.37)	18-0844-84
Ethanone, 1-(4-methyl-3-quinolinyl)-	MeOH	252(4.26),324(3.81)	4-0759-84
4(1H)-Pyridinone, 1-(phenylmethyl)-	EtOH	266(4.30)	118-0485-84
$C_{12}H_{11}NOS$			
1H-Pyrrole-2-carboxaldehyde, 1-methyl-5-(phenylthio)-	MeOH	206(4.62),244(4.34), 304(4.47)	4-1041-84
$C_{12}H_{11}NOS_2$			
6H-1,3-Thiazine-6-thione, 2-(4-methoxyphenyl)-5-methyl-	$CHCl_3$	279(4.35),325(4.17), 435(4.11)	97-0435-84
$C_{12}H_{11}NOS_3$			
1H-Thiopyran-4-carbonitrile, 1-methyl-3-(methylthio)-5-(2-thienyl)-, 1-oxide	EtOH	261.0(4.36),344.0(3.88)	118-0852-84
$C_{12}H_{11}NO_2$			
Benzenamine, N-(2-furanylmethylene)-2-methoxy-	C_6H_{12}	234(4.02),282(4.15), 333(3.84)	18-0844-84
	MeOH	228(4.05),280(4.18), 330(3.95)	18-0844-84
Benzenamine, N-(2-furanylmethylene)-4-methoxy-	C_6H_{12}	225(4.25),282(4.46), 335(4.43)	18-0844-84
	MeOH	220(4.21),282(4.36), 333(4.43)	18-0844-84
Benz[cd]indol-4(3H)-one, 1,5-dihydro-6-methoxy-	EtOH	220(4.41),268(3.76)	44-4761-84
5H-[1]Benzopyrano[3,4-c]pyridin-5-one, 1,2,3,4-tetrahydro-	MeOH	272(4.00),308(3.87)	4-1557-84
Methanone, (4-methoxyphenyl)-1H-pyrrol-2-yl-	MeOH	223(3.96),268s(3.90), 310(4.28)	95-0238-84
Methanone, (4-methoxyphenyl)-1H-pyrrol-3-yl-	MeOH	215s(4.15),279(4.20)	95-0238-84
Naphthalene, 4,8-dimethyl-1-nitro-(diprotonated)	CF_3SO_3H	554(3.76)	88-0325-84
1,2-Naphthalenedione, 4-(dimethylamino)-	MeOH	218(4.10),243(4.24), 278(4.12),470(3.69)	94-3093-84
2-Oxa-3-azabicyclo[3.2.0]hepta-3,6-diene, 4-(4-methoxyphenyl)-	C_6H_{12}	275(4.22),305s(3.38)	40-0158-84
	MeCN	275(4.21),303s(3.55)	40-0158-84
1,3-Oxazepine, 2-(4-methoxyphenyl)-	MeCN	251(4.08),284(3.92), 320(3.91)	40-0158-84
2-Propenal, 3-[3-(4-methoxyphenyl)-2H-azirin-2-yl]-, cis	C_6H_{12}	279(4.35),295s(4.29)	40-0158-84
trans	C_6H_{12}	277(4.25),293s(4.18)	40-0158-84
2(1H)-Pyridinone, 1-(phenylmethoxy)-	C_6H_{12}	233(2.77),305(2.72)	40-0001-84
	MeOH	227s(2.86),300(2.77)	40-0001-84
	MeCN	228(2.80),303(2.73)	40-0001-84
1H-Pyrrole-3-carboxaldehyde, 2-(4-methoxyphenyl)-	EtOH	239(4.29),313(3.96)	40-0158-84
1H-Pyrrole-2,5-dione, 3-methyl-1-(phenylmethyl)-	MeOH	215(4.16),222s(4.14)	39-2097-84C
1H-Pyrrole-2,5-dione, 1-(1-phenylethyl)-	THF	215(3.93)	47-2789-84
$C_{12}H_{11}NO_2S$			
1H-Thiopyran-4-carbonitrile, 3-(2-furanyl)-1-methyl-5-(methylthio)-, 1-oxide	EtOH	270.5(4.52),306s(3.90), 352(3.95)	118-0852-84

Compound	Solvent	$\lambda_{max}(\log \epsilon)$	Ref.
$C_{12}H_{11}NO_3$			
1H-Pyrrole-2,5-dione, 3-(3-methoxy-phenyl)-1-methyl-	MeOH	222(3.08),255(4.04), 275(3.75),345(3.52)	87-0628-84
$C_{12}H_{11}NO_4$			
2H-1-Benzopyran-4-carboxylic acid, 7-amino-2-oxo-, ethyl ester	EtOH	391.5(4.09)	94-3926-84
5(4H)-Isoxazolone, 4-acetyl-3-(2-meth-oxyphenyl)-	EtOH	282(4.13)	44-4419-84
5(4H)-Isoxazolone, 3-ethyl-4-(2-hydr-oxybenzoyl)-	EtOH	342(3.94)	44-4419-84
5(4H)-Isoxazolone, 4-(2-hydroxy-5-meth-ylbenzoyl)-3-methyl-	EtOH	344(3.90)	44-4419-84
$C_{12}H_{11}NO_4S$			
6H-1,3-Thiazin-6-one, 2-(3,4-dimeth-oxyphenyl)-4-hydroxy-	EtOH	240(4.28),250(3.95), 310(3.60)	103-1088-84
$C_{12}H_{11}NO_5$			
4H-3,1-Benzoxazin-4-one, 7-acetoxy-6-methoxy-2-methyl-	MeOH	231(4.31),261(3.74), 277s(3.48),318(3.46), 331s(3.40)	158-0200-84
$C_{12}H_{11}N_3O$			
Benzo[f]quinoxalin-6(2H)-one, 5-amino-3,4-dihydro-	CH_2Cl_2	246(4.20),293(4.21), 478(3.49)	83-0329-84
Pyrazolo[5,1-b]quinazolin-9(4H)-one, 2,4-dimethyl-	10% EtOH-NaOH	219(4.18),244(4.58), 282(3.90),365(3.63)	56-0411-84
Pyrazolo[5,1-b]quinazolin-9(4H)-one, 2,7-dimethyl-	10% EtOH-NaOH	221(4.36),250(4.50), 319(4.10),392(3.60)	56-0411-84
3-Pyridinol, 2-[(4-methylphenyl)azo]-	MeCN	379(4.24)	103-1364-84
	cation	403(4.41)	103-1364-84
	anion	510(4.18)	103-1364-84
dication	$MeNO_2$	496(4.59),523(4.76)	103-1364-84
3-Pyridinol, 6-[(4-methylphenyl)azo]-	MeCN	342(4.35)	103-1364-84
	cation	375(4.42)	103-1364-84
	anion	461(4.50)	103-1364-84
5-Pyrimidinecarbonitrile, 1,2,3,4-tetrahydro-1-methyl-2-oxo-4-phenyl-	EtOH	205(3.4),220s(3.28), 286(3.73)	103-0431-84
$C_{12}H_{11}N_3O_2$			
Bullvalenedicarboximide dioxime	MeOH	216s(4.21),245(4.29)	24-0633-84
Pyrazolo[5,1-b]quinazolin-9(4H)-one, 3-hydroxy-2,7-dimethyl-	10% EtOH-NaOH	250(4.37),320(3.51), 394(3.26)	56-0411-84
Pyrazolo[5,1-b]quinazolin-9(4H)-one, 7-methoxy-2-methyl-	10% EtOH-NaOH	221(4.39),243(4.46), 267(4.45),317(4.06), 402(3.72)	56-0411-84
$C_{12}H_{11}N_3O_2S$			
Pyrazolo[1,5-a]pyridine-5-carboxylic acid, 3-cyano-2-(methylthio)-, ethyl ester	EtOH	217(4.31),259(4.56), 315(3.92),350(3.65)	95-0440-84
$C_{12}H_{11}N_3O_3$			
2-Azetidinone, 4-(diazoacetyl)-1-(2-methoxyphenyl)-	MeCN	247(3.40),281(3.11), 351(1.54)	64-0095-84B
2-Azetidinone, 4-(diazoacetyl)-1-(3-methoxyphenyl)-	MeCN	249(3.34),282(3.03), 349(1.65)	64-0095-84B
2-Azetidinone, 4-(diazoacetyl)-1-(4-methoxyphenyl)-	MeCN	249(3.33),281(3.00), 350(1.60)	64-0095-84B

Compound	Solvent	$\lambda_{max}(\log \epsilon)$	Ref.
Pyrazolo[5,1-b]quinazolin-9(4H)-one, 3-hydroxy-7-methoxy-2-methyl-	10% EtOH-NaOH	222(4.37),262(4.14), 319(3.80),400(3.20)	56-0411-84
11H-Pyrido[2,1-b]quinazolin-11-one, 6,7,8,9-tetrahydro-2-nitro-	EtOH	212(4.35),224s(4.32), 324(4.08)	4-0219-84
$C_{12}H_{11}N_3O_3S_2$ Acetamide, N-[2-(methylthio)-5-(4-nitrophenyl)-4-thiazolyl]-	dioxan	257(3.01),357(3.13)	73-2285-84
$C_{12}H_{11}N_3O_4S$ 3-Pyridinecarboxylic acid, 2-[[4-(ethoxycarbonyl)-1H-imidazol-2-yl]thio]-	MeOH	208(4.31),254(4.28), 295s(3.61)	142-0807-84
$C_{12}H_{11}N_3O_5S_2$ Acetamide, N-acetyl-N-[2-(methylthio)-5-(5-nitro-2-furanyl)-4-thiazolyl]-	dioxan	298(2.90),397(3.21)	73-2285-84
$C_{12}H_{11}N_5O$ 1H-Pyrazolo[4,3-c]pyridin-4-one, 3,5-diamino-1,5-dihydro-6-phenyl-	EtOH	260(4.12),302(3.89)	94-3384-84
$C_{12}H_{11}P$ Phosphorin, 3-methyl-5-phenyl-	EtOH	210(4.31),241(4.46), 272s(4.02),312(2.67)	24-0763-84
$C_{12}H_{12}$ Naphthalene, 1,5-dimethyl-	hexane	227(5.06),274(3.86), 285(3.96),297(3.81)	35-8024-84
Naphthalene, 2,6-dimethyl-	hexane	226(5.11),274(3.69)	35-8024-84
Tetracyclo[8.2.0.0^{2,5}.0^{6,9}]dodeca-1,5,9-triene	C_6H_{12}	208(--),222(3.78), 252(2.35),265s(2.30), 274s(2.23),290s(2.00)	44-1412-84
$C_{12}H_{12}Br$ 1,4-Methano-1H-benzocycloheptenylium, 2-bromo-2,3,4,?-tetrahydro-, hexafluoroantimonate	MeCN	223(4.41),281(3.68), 300s(3.64)	44-2297-84
$C_{12}H_{12}BrNO_3$ 1H-Indole-2-carboxylic acid, 3-bromo-5-methoxy-, ethyl ester	EtOH	210(4.42),300(4.31)	44-4761-84
1H-Indole-2-carboxylic acid, 4-bromo-5-methoxy-, ethyl ester	EtOH	224(4.39),298(4.34)	44-4761-84
$C_{12}H_{12}BrN_3O_4$ 2-Butenoic acid, 2-bromo-3-[(4-nitrophenyl)azo]-, ethyl ester	Tetra	330(4.5),342(4.4), 360s(--),453(2.6)	48-0367-84
$C_{12}H_{12}ClN_3$ 2-Pyrimidinamine, 4-chloro-N,6-dimethyl-N-phenyl-	MeOH	227(3.55),262(3.78), 310(3.18)	4-1161-84
$C_{12}H_{12}ClN_3O$ 2(1H)-Pyrimidinone, 5-chloro-4-methyl-6-(methylphenylamino)-	MeOH	275(3.57),608(4.83)	4-1161-84
4(1H)-Pyrimidinone, 5-chloro-6-methyl-2-(methylphenylamino)-	MeOH	218(3.74),246(3.64), 307(3.59)	4-1161-84
$C_{12}H_{12}Cl_2N_2O_3$ 1H-Pyrrolo[3,2-c]pyridine, 4,6-dichloro-1-(2-deoxy-β-D-erythro-pentofuranosyl)-	pH 1,7 and 11	224(4.47),274(3.75), 289s(3.56)	35-6379-84 +35-6383-84

Compound	Solvent	λ_{max}(log ϵ)	Ref.
C₁₂H₁₂CrO₄			
Chromium, tricarbonyl[(1,2,3,4,5,6-η)-	EtOH	213(4.10),255(3.47)	112-0765-84
1-methoxy-2,3-dimethylbenzene]-	CHCl₃	317(3.95)	112-0765-84
Chromium, tricarbonyl[(1,2,3,4,5,6-η)-	EtOH	210(4.36),256(3.67)	112-0765-84
1-methoxy-2,4-dimethylbenzene]-	CHCl₃	317(3.94)	112-0765-84
Chromium, tricarbonyl[(1,2,3,4,5,6-η)-	EtOH	213(4.28),255(3.62)	112-0765-84
1-methoxy-2,5-dimethylbenzene]-	CHCl₃	316(3.96)	112-0765-84
Chromium, tricarbonyl[(1,2,3,4,5,6-η)-	EtOH	217(4.29),255(3.74)	112-0765-84
1-methoxy-2,6-dimethylbenzene]-	CHCl₃	318(3.96)	112-0765-84
Chromium, tricarbonyl[(1,2,3,4,5,6-η)-	EtOH	213(4.27),255(3.61)	112-0765-84
1-methoxy-3,4-dimethylbenzene]-	CHCl₃	318(3.95)	112-0765-84
Chromium, tricarbonyl[(1,2,3,4,5,6-η)-	EtOH	213(4.32),252(3.72)	112-0765-84
1-methoxy-3,5-dimethylbenzene]-	CHCl₃	313(3.96)	112-0765-84
C₁₂H₁₂FeO₄			
Iron, tricarbonyl[(1,3,4,5-η)-6,6-di-	hexane	218(4.23),266(3.75),	24-3473-84
methyl-2-oxo-4-cycloheptene-1,3-diyl]-		315(3.24)	
Iron, tricarbonyl[(1,3,4,5-η)-7,7-di-	hexane	217(4.23),270(3.71),	24-3473-84
methyl-2-oxo-4-cycloheptene-1,3-diyl]-		317(3.33)	
C₁₂H₁₂N₂			
1,4-Benzenediamine, N-phenyl-	n.s.g.	288(4.26)	30-0009-84
4,4'-Bipyridine, 2-ethyl-	EtOH	239(4.03)	39-0367-84C
Phenazine, 1,2,3,4-tetrahydro-	MeOH	206(4.45),238(4.44),	56-0917-84
		321(3.99)	
C₁₂H₁₂N₂O			
Phenazine, 1,2,3,4-tetrahydro-, 5-oxide	MeOH	210(4.28),244(4.75),	56-0917-84
		326(4.00)	
1H-Pyrido[3,4-b]indol-1-one, 2,3,4,9-	isoPrOH	230(4.38),302(4.16),	103-0047-84
tetrahydro-9-methyl-		328s(3.70)	
11H-Pyrido[2,1-b]quinazolin-11-one,	EtOH	225(4.43),268(3.87),	4-0219-84
6,7,8,9-tetrahydro-		275s(3.86),296s(3.46),	
		306(3.55),317(3.47)	
2H-Pyrrol-2-one, 1,5-dihydro-5-imino-	MeOH	212(4.04),235(4.19),	39-2097-84C
4-methyl-1-(phenylmethyl)-		294(3.19)	
C₁₂H₁₂N₂OS			
4H-Benzo[6,7]cyclohepta[1,2-d][1,2,3]-	EtOH	256(4.20)	2-0203-84
thiadiazole, 5,6-dihydro-8-methoxy-			
3H-[1]Benzothiepino[5,4-c]pyrazol-	EtOH	260(4.17)	2-0736-84
3-one, 2,3a,4,5-tetrahydro-3a-			
methyl-			
4H-1,3-Thiazin-6-one, 4-(dimethyl-	EtOH	218(3.81),265(4.41)	103-1088-84
amino)-2-phenyl-			
C₁₂H₁₂N₂O₂			
2H-[1]Benzoxepino[5,4-c]pyrazole,	EtOH	266(4.18)	2-0736-84
4,5-dihydro-8-methoxy-			
3H-[1]Benzoxepino[5,4-c]pyrazol-3-one,	EtOH	280(4.07)	2-0736-84
2,3a,4,5-tetrahydro-3a-methyl-			
2-Naphthalenol, 1-[(methylnitroso-	EtOH	227(4.82),267(3.72),	150-0701-84M
amino)methyl]-		278(3.80),290(3.73),	
		322(3.50),335(3.56)	
1H-Pyrano[3,4-b]quinoxaline, 3,4-di-	EtOH	214(4.04),239(4.38),	39-0733-84C
hydro-1-methoxy-, (S)-		324(3.85)	
C₁₂H₁₂N₂O₂S			
Benzenesulfonamide, N-(6-methyl-	EtOH	252(3.94),324(3.88)	152-0223-84
2-pyridinyl)-			
6H-1,3-Thiazin-6-one, 2-[4-(dimethyl-	EtOH	232(4.27),280(3.65)	103-1088-84

Compound	Solvent	$\lambda_{max}(\log \epsilon)$	Ref.
amino)phenyl]-4-hydroxy- (cont.)		328(3.83),410(4.35)	103-1088-84
$C_{12}H_{12}N_2O_3$			
Benz[cd]indole, 1,3,4,5-tetrahydro-6-methoxy-4-nitro-	EtOH	208(4.39),224(4.42), 268(3.77),298(3.70)	44-4761-84
1,8-Naphthyridine-3-carboxylic acid, 1-ethyl-1,4-dihydro-7-methyl-4-oxo-(nalidixic acid)	2M H_2SO_4 pH 4 pH 8	300(3.08),309(3.04) 315(3.06),323(3.05) 332(3.06)	149-0057-84A 149-0057-84A 149-0057-84A
1,2,4-Oxadiazole-5-propanoic acid, 3-(2-methylphenyl)-	EtOH	235(4.01),275(3.37), 285(3.17)	4-1193-84
1,2,4-Oxadiazole-5-propanoic acid, 3-(3-methylphenyl)-	EtOH	242(4.12),280(3.04), 288(2.94)	4-1193-84
1,2,4-Oxadiazole-5-propanoic acid, 3-(4-methylphenyl)-	EtOH	244(4.18),273(2.82), 285(2.65)	4-1193-84
5-Pyrimidinecarboxylic acid, 1,2,3,4-tetrahydro-1-methyl-2-oxo-4-phenyl-	EtOH	208(3.72),225s(3.65), 298(3.41)	103-0431-84
$C_{12}H_{12}N_2O_3S$			
Benzo[b]thiophen-3(2H)-one, 2-(tetra-hydro-2(1H)-pyrimidinylidene)-, 1,1-dioxide	EtOH	240(4.54),248(4.54), 258(4.69),265(4.76), 306(4.20),335(4.31)	95-0134-84
$C_{12}H_{12}N_2O_4$			
1,2,4-Oxadiazole-5-propanoic acid, 3-(4-methoxyphenyl)-	EtOH	256(4.26),284s(3.56), 290(3.38)	4-1193-84
$C_{12}H_{12}N_2S_2$			
1H-Pyrazole, 4,5-dihydro-1-methyl-3,5-dithienyl-	EtOH	235(4.02),320(4.07)	34-0225-84
$C_{12}H_{12}N_3O$			
Pyridinium, 1-methyl-2-(phenyl-α-azoxy)-, iodide	EtOH	384(4.01)	23-1628-84
Pyridinium, 1-methyl-3-(phenyl-α-azoxy)-, iodide	EtOH	318(4.21)	23-1628-84
Pyridinium, 1-methyl-4-(phenyl-α-azoxy)-, iodide	EtOH	333(4.15)	23-1628-84
$C_{12}H_{12}N_4$			
6,7-Indolizinedicarbonitrile, 1,2,3,5-tetrahydro-5-imino-2,2-dimethyl-	MeOH	210(4.58),238(4.29), 434(3.75)	83-0143-84
3H-Pyrazolo[5,1-c]-1,2,4-triazole, 3,3-dimethyl-6-phenyl-	CH_2Cl_2	253(4.21),291s(3.79), 340s(3.11)	24-1726-84
3H-Pyrazolo[5,1-c]-1,2,4-triazole, 3,3-dimethyl-7-phenyl-	CH_2Cl_2	242(3.86),304(3.43), 352(3.54)	24-1726-84
1,4,7,9b-Tetraazaphenalene, 2,5,8-tri-methyl-	MeOH and MeOH-NaOH	245(4.33),344.5(4.27), 361(4.23),448s(2.16), 477(2.34),514(2.41), 558(2.27)	11-0170-84A
	MeOH-HCl	218(4.25),240.5(4.35), 318(4.11),346(4.09), 360s(3.96),443(2.62), 468.5(2.67),497s(2.55)	11-0170-84A
$C_{12}H_{12}N_4OS$			
2-Butanone, 3-(1H-benzimidazol-2-yl-thio)-1-diazo-3-methyl-	MeOH	252(4.05),272(4.11)	103-0505-84
$C_{12}H_{12}N_4O_2$			
1H-Pyrazole-5-carboxamide, 3-acetyl-4-amino-1-phenyl-	EtOH	232(5.09),274(3.83)	4-1049-84

Compound	Solvent	$\lambda_{max}(\log \epsilon)$	Ref.
1H-Pyrazole-5-carboxamide, 3-acetyl-4-amino-1-phenyl-	EtOH	232(5.09),274(3.83)	4-1049-84
2,4(3H,5H)-Pyrimidinedione, 3-methyl-6-(methylamino)-5-(phenylimino)-	MeCN	228(4.33),277(3.95), 460(3.37)	77-1691-84
$C_{12}H_{12}N_4O_3$ 6H-Imidazo[1,2-d]pyrido[3,2-b][1,4]-oxazine-8-carboxylic acid, 9-amino-, ethyl ester	MeOH	270(3.94),289(4.20)	4-1081-84
$C_{12}H_{12}N_4O_3S$ 1H-Imidazole-4-carboxylic acid, 2-[[3-(aminocarbonyl)-2-pyridinyl]-thio]-, ethyl ester	MeOH	204(4.26),254(4.27), 290s(3.66)	142-0807-84
$C_{12}H_{12}N_4O_4S$ 1H-Imidazole-4-carboxylic acid, 2-[[3-[(hydroxyamino)carbonyl]-2-pyridin-yl]thio]-, ethyl ester	MeOH	203(4.21),254(4.26), 290s(3.67)	142-0807-84
$C_{12}H_{12}N_6$ 1H-Pyrazole-5-carbonitrile, 4-amino-3-(1-hydrazonomethyl)-1-phenyl-	EtOH	232(5.09),274(3.83)	4-1049-84
$C_{12}H_{12}O$ Azulene, 4-(methoxymethyl)-	CH_2Cl_2	238(3.94),278(4.19), 282s(--),330(3.14), 340(3.21),353s(--), 582(2.52)	83-0984-84
3H-Cyclopenta[a]pentalen-3-one, 3a,3b,4,6a,7,7a-hexahydro-4-methylene-	MeOH	224(4.22)	44-3848-84
2H-Cyclopent[cd]inden-2-ol, 2a,7b-di-hydro-7b-methyl-	EtOH	232(4.13),348(3.40)	39-0165-84C
2aH-Cyclopent[cd]inden-2a-ol, 4a,7b-di-hydro-7b-methyl-	EtOH	310(3.65)	39-0165-84C
1H-Inden-1-one, 4,5,7-trimethyl-	MeOH	400(3.42)	138-0631-84
1H-Inden-1-one, 4,6,7-trimethyl-	MeOH	400(3.33)	138-0631-84
Tricyclo[6.2.1.1^{2,7}]dodeca-2,4,6-trien-12-one	C_6H_{12}	223(3.93),245s(3.61), 271s(3.26),324(2.47)	88-3611-84
	MeCN	319(--)	88-3611-84
$C_{12}H_{12}OS$ Azulene, 4-[(methylsulfinyl)methyl]-	CH_2Cl_2	244(4.22),279(4.56), 289s(--),332(3.51), 345(3.60),357s(--), 580(2.42)	83-0984-84
$C_{12}H_{12}O_2$ 1,4-Benzocyclooctenedione, 5,6,7,8-tetrahydro-	C_6H_{12}	249(4.20),368(3.25), 439s(2.34),460s(2.09), 492s(1.64)	18-0615-84
2H-1-Benzopyran-6-carboxaldehyde, 2,2-dimethyl-	EtOH	252(4.4),300(3.8)	18-0442-84
Bullvalenelactone, dihydro-	MeOH	225s(3.74),254(3.48)	24-0633-84
2-Butenoic acid, 4-tetracyclo-[3.2.0.0^{2,7}.0^{4,6}]heptylidene, methyl ester, (E)-	MeCN	287(4.48)	24-2027-84
2-Cyclohexen-1-one, 2-hydroxy-3-phenyl-	ether	225(3.81),305(4.23)	44-2355-84
Cyclopent[a]inden-8(1H)-one, 2,3,3a,8a-tetrahydro-8a-hydroxy-	MeOH	248(4.14),294(3.39)	44-2050-84

Compound	Solvent	$\lambda_{max}(\log \epsilon)$	Ref.
4-Dibenzofuranol, 6,7,8,9-tetrahydro-	MeOH	250(4.12),255(4.13)	44-4399-84
2(5H)-Furanone, 5,5-dimethyl-4-phenyl-	C_6H_{12}	263(3.75)	44-4344-84
3(2H)-Furanone, 2,2-dimethyl-5-phenyl-	EtOH	202(4.13),219.5(3.97), 242.5(3.90),303.5(4.20)	39-0535-84C
1-Naphthalenol, 8-methoxy-3-methyl-	MeOH	228(4.49),304(3.56), 370(3.55),332(3.62)	102-2039-84
$C_{12}H_{12}O_3$			
2H-1-Benzopyran-6-carboxaldehyde, 7-hydroxy-2,2-dimethyl-	EtOH	228s(4.16),235(4.24), 259(4.48),282s(3.87), 293s(3.82),349(3.74)	18-0442-84
2H-1-Benzopyran-7-carboxaldehyde, 6-hydroxy-2,2-dimethyl-	EtOH	239(4.23),294(4.16), 391(3.76)	18-0442-84
2H-1-Benzopyran-6-carboxylic acid, 2,2-dimethyl- (anofinic acid)	EtOH	239.5(4.68),280.5(3.88), 306(3.78),318(3.67)	18-0442-84
1H-Indene-1,2(3H)-dione, 4-methoxy-6,7-dimethyl-	n.s.g.	265(3.0),285(2.5), 335(3.6)	2-0863-84
1H-Indene-1,2(3H)-dione, 6-methoxy-4,5-dimethyl-	n.s.g.	265(3.05),280(2.46), 335(3.53)	2-0863-84
1(3H)-Isobenzofuranone, 3-butylidene-7-hydroxy-, (Z)-	MeOH	221(3.06),260(3.79), 326(3.56)	102-2033-84
	MeOH-AlCl$_3$	230(--),340(--)	102-2033-84
1(3H)-Isobenzofuranone, 3-(2-hydroxy-butylidene)-, (Z)-	n.s.g.	270(3.93),306(3.59)	94-3770-84
2(1H)-Naphthalenone, 1,1-dimethoxy-	EtOH	234(4.25),322(3.89)	39-2069-84C
$C_{12}H_{12}O_4$			
2(5H)-Furanone, 4-methoxy-3-(4-methoxy-phenyl)-	CHCl$_3$	275(4.30)	39-1539-84C
$C_{12}H_{12}O_4S$			
1,2-Cyclohexanedione, 3-(phenylsulfon-yl)-	ether	222(4.09),278(3.85)	44-2355-84
2-Cyclohexen-1-one, 2-[(phenylsulfon-yl)oxy]-	ether	225(4.00),277(3.30), 313(1.60)	44-2355-84
$C_{12}H_{12}O_4S_4$			
1,3-Dithiole-4-carboxylic acid, 2-[4-(butoxycarbonyl)-1,3-dithiol-2-ylidene]-	MeCN	289(4.22),302(4.26), 313(4.30),439(3.61)	103-1342-84
isomer B	MeCN	287(4.18),301(4.22), 313(4.25),432(3.59)	103-1342-84
$C_{12}H_{12}O_5$			
Benzaldehyde, 2-(diacetoxymethyl)-	EtOH	245.5(3.94),283(3.21), 290s(3.15),321s(2.22)	44-1898-84
1,3-Benzenedicarboxylic acid, 5-acet-yl-, dimethyl ester	MeOH	216(4.03)	24-1620-84
$C_{12}H_{12}O_6S$			
Benzenepropanoic acid, 2-carboxy-3,4-dimethoxy-α-thioxo-	MeOH	241(4.48),326(4.08), 370(3.96)	103-0024-84
$C_{12}H_{12}S$			
Azulene, 4-[(methylthio)methyl]-	CH$_2$Cl$_2$	240(4.33),260(4.42), 331s(--),342(3.60), 356s(--),575(2.58)	83-0984-84
$C_{12}H_{13}BrN_2O_2$			
2-Butenoic acid, 2-bromo-3-(phenyl-azo)-, ethyl ester, (E,E)-	heptane	200(4.2),225(3.9), 231(4.0),237(3.9),	48-0367-84

Compound	Solvent	$\lambda_{max}(\log \epsilon)$	Ref.
(cont.)		314(4.4),324(4.4), 340s(--),445(2.5)	48-0367-84
L-Tryptophan, 5-bromo-, methyl ester	MeOH	226(4.54),282(3.70), 290(3.73),299(3.60)	94-2126-84
$C_{12}H_{13}BrN_2O_4S$ Cyclopentanecarboxamide, N-[2-(amino-sulfonyl)-5-bromophenyl]-2-oxo-	EtOH	228(4.41),251(4.07), 290(3.42)	104-0534-84
$C_{12}H_{13}BrN_4O_4$ 2,4(1H,3H)-Pyrimidinedione, 1-[3-(5-bromo-3,4-dihydro-2,4-dioxo-1(2H)-pyrimidinyl)propyl]-5-methyl-	pH 2 H_2O pH 12	276(4.20) 276(4.22) 274(4.07)	142-2739-84 142-2739-84 142-2739-84
$C_{12}H_{13}BrO_7$ 2-Cyclopenten-1-one, 4,5-diacetoxy-5-(acetoxymethyl)-2-bromo-, (4S-cis)-	EtOH	240(3.71)	39-2089-84C
$C_{12}H_{13}ClN_2O_2$ 2-Butenoic acid, 2-chloro-3-(phenyl-azo)-, ethyl ester, (E,E)-	heptane	200(4.1),225(4.0), 231(4.0),237(4.0), 314(4.4),320(4.4), 338s(--),446(2.5)	48-0367-84
L-Tryptophan, 5-chloro-, methyl ester	MeOH	227(4.54),282(3.72), 290(3.74),300(3.62)	94-2126-84
$C_{12}H_{13}ClN_2O_4S$ Cyclopentanecarboxamide, N-[2-(amino-sulfonyl)-5-chlorophenyl]-2-oxo-	EtOH	222(4.44),252(4.05), 290(3.41)	104-0534-84
$C_{12}H_{13}Cl_2N_3O_4$ Butanoic acid, 3-[(4,5-dichloro-2-ni-trophenyl)hydrazono]-, ethyl ester	EtOH	261(4.24),275(4.16), 290(4.10),440(3.77)	48-0829-84
$C_{12}H_{13}Cl_5O_4$ Propanedioic acid, methyl(1,2,3,4,4-pentachloro-1,3-butadienyl)-, diethyl ester	MeOH	220(4.11)	35-3551-84
$C_{12}H_{13}CrNO_3$ Chromium, tricarbonyl[(1,2,3,4,5,6-η)-2-ethyl-6-methylbenzenamine]- Chromium, tricarbonyl[(1,2,3,4,5,6-η)-2,4,6-trimethylbenzenamine]-	EtOH $CHCl_3$ EtOH $CHCl_3$	217(4.52) 320(3.91) 217(4.53) 323(3.89)	112-0765-84 112-0765-84 112-0765-84 112-0765-84
$C_{12}H_{13}F_3N_2O_5$ Uridine, 2'-deoxy-5-(3,3,3-trifluoro-1-propenyl)-	acid? pH 6.6 base	243(3.84),285(3.81) 243(3.85),285(3.81) 247(3.92),283(3.76)	87-0279-84 87-0279-84 87-0279-84
$C_{12}H_{13}N$ Benzonitrile, 4-(4-pentenyl)- 1-Naphthalenamine, N,N-dimethyl-	n.s.g. n.s.g.	267(2.91) 214(4.76),239(4.13), 303(3.73)	39-1833-84B 35-8024-84
1H-Pyrrole, 2-(2-phenylethyl)-	MeOH	208(4.16),268(2.54)	107-0453-84
$C_{12}H_{13}NO$ Benzamide, N-(1-ethyl-2-propynyl)-, (R)-	hexane MeOH	225(4.09),270s(--), 280s(--) 227(4.1),280s(--)	1-0141-84 1-0141-84

Compound	Solvent	$\lambda_{max}(\log \epsilon)$	Ref.
4(1H)-Pyridinone, 2,3-dihydro-1-(phen- ylmethyl)-	EtOH	324(4.28)	118-0485-84
$C_{12}H_{13}NO_2$			
Benzene, (2-nitro-1-cyclohexen-1-yl)-	MeOH	214(4.00),267(3.35)	12-1231-84
3-Buten-2-one, 4-[(2-acetylphenyl)- amino]-, (Z)-	MeOH	240(4.06),324(4.22), 364(4.30)	4-0759-84
$C_{12}H_{13}NO_2S$			
1,4-Oxathian-2-one, 3-[1-(phenylamino)- ethylidene]-, (E)-	EtOH	348(4.20)	137-0097-84
2H-Pyrrol-2-one, 1,5-dihydro-5-methoxy- 4-[(phenylmethyl)thio]-	EtOH	220(3.72),279(3.74)	142-1179-84
$C_{12}H_{13}NO_3$			
3-Azetidinecarboxylic acid, 1-methyl- 2-oxo-4-phenyl-, methyl ester, trans	EtOH	270(3.08)	44-0282-84
Benzoic acid, 2-[(3-oxo-1-butenyl)- amino]-, methyl ester	MeOH	232(3.97),333(4.18), 350(4.36)	4-0759-84
3-Butenoic acid, 4-amino-2-oxo-4-phen- yl-, ethyl ester, (Z)-	EtOH	254(3.80),353(4.21)	94-0497-84
4H-1,3,5-Dioxazocine-7-carboxaldehyde, 5,8-dihydro-6-phenyl-	MeOH	296(4.16)	88-2731-84
[1,3]Dioxepino[5,6-d]isoxazole, 3a,4,8,8a-tetrahydro-3-phenyl-, cis	MeOH	260(3.97)	88-2731-84
Hippuric acid, 3-butenyl ester	C_6H_{12}	224(3.97)	44-0399-84
1H-Indene-1,2(3H)-dione, 4-methoxy- 5,7-dimethyl-, 2-oxime	n.s.g.	285(4.21)	2-0863-84
1H-Indene-1,2(3H)-dione, 4-methoxy- 6,7-dimethyl-, 2-oxime	n.s.g.	285(4.2)	2-0863-84
1H-Indene-1,2(3H)-dione, 6-methoxy- 4,5-dimethyl-, 2-oxime	n.s.g.	280(4.2)	2-0863-84
1H-Indene-1,2(3H)-dione, 6-methoxy- 4,7-dimethyl-, 2-oxime	n.s.g.	280(4.15)	2-0863-84
1H-Indole-3-carboxylic acid, 2-(1-hy- droxyethyl)-5-methyl-	n.s.g.	217(4.55),233(4.30), 288(4.05),294(4.02)	103-0988-84
1H-Indole-3-carboxylic acid, 2-(1-hy- droxyethyl)-7-methyl-	n.s.g.	214(4.56),231(4.36), 285(4.05)	103-0988-84
$C_{12}H_{13}NO_4S$			
1-Pentanone, 4-methyl-4-nitro-1-phenyl- 2-sulfinyl-	EtOH	286(4.02)	12-0777-84
4H-Thieno[3,2-b]pyrrole-5,6-dicarbox- ylic acid, diethyl ester	EtOH	235(4.16),247s(4.06), 307(4.17)	39-0915-84C
$C_{12}H_{13}NO_6$			
Benzoic acid, 2-(acetylamino)-4-acet- oxy-5-methoxy-	MeOH	226(4.30),259(4.08), 316(3.61)	158-0200-84
$C_{12}H_{13}NS$			
2-Butenenitrile, 3-(ethylthio)-4-phen- yl-	EtOH	207(4.27),260(4.15)	78-2141-84
$C_{12}H_{13}N_2P$			
Propanenitrile, 3,3'-(phenylphosphin- idene)bis-	MeOH	244(3.61),264s(--), 271s(--)	139-0259-84C
$C_{12}H_{13}N_3O$			
2(1H)-Pyrimidinone, 4-methyl-6-(meth- ylphenylamino)-	MeOH	208(4.15),260(3.71), 278(3.76)	4-1161-84

Compound	Solvent	λ_{max}(log ϵ)	Ref.
4(1H)-Pyrimidinone, 6-methyl-2-(methyl-phenylamino)-	MeOH	220(3.77),240(3.63), 296(3.54)	4-1161-84
4H-[1,2,4]Triazolo[3,4-c][1,4]benzoxa-zine, 1-(1-methylethyl)-	EtOH	240(2.84),285(2.43)	2-0445-84
$C_{12}H_{13}N_3O_2$			
Pyridazine, 1,6-dihydro-3,6-dimethyl-1-(4-nitrophenyl)-	CH_2Cl_2	421(4.4)	88-0057-84
5-Pyrimidinecarboxamide, 1,2,3,4-tetra-hydro-1-methyl-2-oxo-4-phenyl-	EtOH	207(4.32),221s(3.95), 290(3.98)	103-0431-84
$C_{12}H_{13}N_3O_3$			
1H-1,2,4-Triazole-1-carboxylic acid, 3-(4-methoxyphenyl)-, ethyl ester	MeCN	271(4.38),286s(4.12), 294s(3.90)	48-0311-84
$C_{12}H_{13}N_3O_3S$			
Pyrazolo[1,5-a]pyridine-3-carboxylic acid, 6-(aminocarbonyl)-2-(methyl-thio)-, ethyl ester	EtOH	270(4.68),316(4.03)	95-0440-84
$C_{12}H_{13}N_3O_5$			
1H-Pyrrolo[3,4-c]pyridine-7-carboxylic acid, 4-amino-6-ethoxy-2,3-dihydro-1,3-dioxo-, ethyl ester	MeOH	218(4.20),267(4.13), 397(3.75)	48-0594-84
$C_{12}H_{13}N_3O_5S$			
Pyrido[2,3-d]pyrimidine-6-carboxylic acid, 1,2,3,4-tetrahydro-5-hydroxy-1,3-dimethyl-7-(methylthio)-2,4-di-oxo-, methyl ester	EtOH	262(4.50),319(4.10)	94-0122-84
$C_{12}H_{13}N_3O_6$			
Imidazo[1,2-c]pyrimidine-3-carboxalde-hyde, 5,6-dihydro-5-oxo-6-β-D-ribo-furanosyl-	H_2O	325(4.02)	44-4021-84
Pyrido[2.3-d]pyrimidine-6-propanoic acid, 1,2,3,4,7,8-hexahydro-8-(2-hydroxyethyl)-2,4,7-trioxo-	pH 1 pH 13	289(4.12),312(4.17) 250(3.85),286(3.91), 333(4.12)	39-0959-84C 39-0959-84C
$C_{12}H_{13}N_5O_4$			
7H-Pyrrolo[2,3-d]pyrimidin-4-amine, 7-β-D-arabinofuranosyl-5-cyano-	pH 1	272(4.09)	78-0125-84
$C_{12}H_{14}BrN_3O_4$			
Butanoic acid, 3-[(4-bromo-2-nitro-phenyl)hydrazono]-, ethyl ester	EtOH	256(4.28),293(4.15), 448(3.70)	48-0829-84
$C_{12}H_{14}ClN_3O_4$			
Butanoic acid, 3-[(4-chloro-2-nitro-phenyl)hydrazono]-, ethyl ester	EtOH	254(4.34),292(4.20), 446(3.79)	48-0829-84
Butanoic acid, 3-[(5-chloro-2-nitro-phenyl)hydrazono]-, ethyl ester	EtOH	273(4.24),287(4.11), 425(3.74)	48-0829-84
D-Ribitol, 1,4-anhydro-1-C-(2-chloro-5-methylimidazo[1,5-b]pyridazin-7-yl)-, (S)-	EtOH	237(4.43),266(3.57), 274s(3.56),284s(3.36), 388(3.30)	39-0229-84C
$C_{12}H_{14}ClN_3O_5$			
7H-Pyrrolo[2,3-d]pyrimidine, 7-β-D-arabinofuranosyl-2-chloro-4-methoxy-	MeOH	270(3.91)	5-0722-84

Compound	Solvent	$\lambda_{max}(\log \epsilon)$	Ref.
$C_{12}H_{14}F_3N_2O_8P$			
Uridine, 2'-deoxy-5'-O-(5,5-difluoro-1,3,2-dioxaphosphorinan-2-yl)-5-fluoro-, P-oxide	EtOH	268(3.88)	87-0440-84
$C_{12}H_{14}N$			
Pyridinium, 2,3-dihydro-1-methyl-4-phenyl-, bromide	H_2O	343(4.28)	33-2037-84
	CH_2Cl_2	356(5.52)	33-2037-84
$C_{12}H_{14}NO_2$			
Isoquinolinium, 6,7-dimethoxy-2-methyl-, iodide	EtOH-HCl	252(4.87),310(4.12)	24-1436-84
	$MeOCH_2CH_2OH$	312(3.78)	24-1436-84
$C_{12}H_{14}N_2$			
Benzonitrile, 4-piperidino-	MeOH	297.5(4.41)	139-0253-84C
Pyrido[1,2-b]indazole, 1,2,3,4-tetrahydro-9-methyl-	H_2O	232(4.6),306(3.78)	56-1077-84
$C_{12}H_{14}N_2O$			
Benz[cd]indol-4-amine, 1,3,4,5-tetrahydro-6-methoxy- (hemioxalate)	EtOH	221(4.40),275(3.76)	44-4761-84
Ethanone, 1-(1,2-dimethyl-1H-indol-3-yl)-, oxime	n.s.g.	231(4.24),284(3.93)	103-0055-84
1-Propanone, 1-(2-methyl-1H-indol-3-yl)-, oxime	n.s.g.	227(4.38),275(4.00)	103-0055-84
Pyrido[1,2-b]indazole, 1,2,3,4-tetrahydro-9-methoxy-	H_2O	235(4.60),300(3.78)	56-1077-84
$C_{12}H_{14}N_2OS$			
Ethanethione, 1-(2-hydroxyphenyl)(1-methyl-2-imidazolidinylidene)-	EtOH	240s(3.99),279(3.90), 362(4.14)	39-1631-84C
$C_{12}H_{14}N_2O_2S$			
4(3H)-Quinazolinethione, 6,7-dimethoxy-2,3-dimethyl-	EtOH	226s(4.47),233(4.53), 254(4.20),275(4.19), 343s(3.92),361(4.18), 375(4.19)	39-1143-84C
$C_{12}H_{14}N_2O_3$			
Carbamic acid, [2-(5-hydroxy-1H-indol-3-yl)ethyl]-, methyl ester	MeOH	223.5(4.33),278.5(3.79), 300.5(3.66)	94-2544-84
2(1H)-Pyrazinone, 3,6-dihydro-1-hydroxy-6,6-dimethyl-5-phenyl-, 4-oxide	EtOH	247(3.91)	103-0097-84
2(1H)-Pyrazinone, 5,6-dihydro-1-hydroxy-6,6-dimethyl-5-phenyl-, 4-oxide	EtOH	263(4.09)	103-0097-84
$C_{12}H_{14}N_2O_4$			
D-Ribitol, 1,4-anhydro-1-C-imidazo[1,5-a]pyridin-3-yl-, (S)-	EtOH	270s(3.76),279(3.90), 289(3.82),334(3.33)	39-0229-84C
$C_{12}H_{14}N_2O_4S$			
2H-1,2,4-Benzothiadiazine-3-pentanoic acid, 1,1-dioxide	EtOH	213(4.22),264(3.88)	104-0534-84
Cyclopentanecarboxamide, N-[2-(aminosulfonyl)phenyl]-2-oxo-	EtOH	211(4.45),247(4.06), 285(3.64)	104-0534-84
$C_{12}H_{14}N_2O_6$			
Alanine, N-(methoxycarbonyl)-2-methyl-, 4-nitrophenyl ester	EtOH	210s(3.98),267(4.03)	94-1373-84

Compound	Solvent	$\lambda_{max}(\log \epsilon)$	Ref.
$C_{12}H_{14}N_2S$			
4-Isoquinolinecarbonitrile, 1-ethyl-2,3,5,6,7,8-hexahydro-3-thioxo-	dioxan	253(3.81),310(3.61), 416(2.72)	104-2216-84
2-Pentenenitrile, 3-[(2-aminophenyl)-thio]-4-methyl-	EtOH	206(4.38),249(4.05), 270(3.99)	78-2141-84
3-Pentenenitrile, 3-[(2-aminophenyl)-thio]-4-methyl-	EtOH	206(4.41),240(4.08), 265(4.37)	78-2141-84
Pyridinium, 3,3'-thiobis[1-methyl-, diiodide	H_2O	195(4.65),226(4.53), 257s(3.88),298s(3.67)	4-0917-84
$C_{12}H_{14}N_4O_3$			
2,4(1H,3H)-Pteridinedione, 7-(1-oxobutyl)-1,3-dimethyl-	MeOH	248(4.09),349(3.90)	5-1798-84
2,4(1H,3H)-Pteridinedione, 1,3,7-tri-methyl-6-(1-oxopropyl)-	MeOH	248(4.13),278(4.04), 329(4.09)	5-1798-84
$C_{12}H_{14}N_4O_3S$			
2,4(1H,3H)-Pteridinedione, 1,3-dimethyl-7-(methylthio)-6-(1-oxopropyl)-	MeOH	255(4.31),308(4.09), 365(4.22)	5-1798-84
$C_{12}H_{14}N_4O_4$			
2,4(1H,3H)-Pteridinedione, 6-methoxy-1,3-dimethyl-7-(1-oxopropyl)-	MeOH	252(4.34),377(4.04)	5-1798-84
2,4(1H,3H)-Pteridinedione, 7-methoxy-1,3-dimethyl-6-(1-oxopropyl)-	MeOH	249(4.05),282(4.07), 334(4.36)	5-1798-84
Pyrido[2,3-d]pyrimidine-6-carboxylic acid, 7-amino-1,2,3,4-tetrahydro-1,3-dimethyl-2,4-dioxo-, ethyl ester	H_2O	243(4.11),286(4.20), 336(4.26)	4-1543-84
2,4(1H,3H)-Pyrimidinedione, 1-[3-(3,4-dihydro-2,4-dioxo-1(2H)-pyrimidinyl)propyl]-5-methyl-	H_2O	268(4.27)	142-2739-84
$C_{12}H_{14}N_4O_4S$			
1H-Imidazo[1,2-b]pyrazole-7-carbo-nitrile, 6-(methylthio)-1-β-D-ribofuranosyl-	pH 1	221(4.11),252s(3.94)	44-3534-84
	pH 7	221(4.19),252s(4.00)	44-3534-84
	pH 11	220s(4.28),252(4.06)	44-3534-84
2-Propenoic acid, 2-cyano-3-(methyl-thio)-3-[(1,2,3,6-tetrahydro-1,3-dimethyl-2,6-dioxo-4-pyrimidinyl)-amino]-, methyl ester	EtOH	268(4.18),302(4.13)	94-0122-84
Pyrido[2,3-d]pyrimidine-6-carboxylic acid, 5-amino-1,2,3,4-tetrahydro-1,3-dimethyl-7-(methylthio)-2,4-dioxo-	EtOH	267(4.69),308(4.24)	94-0122-84
$C_{12}H_{14}N_4O_6$			
6-Pteridinepropanoic acid, 1,2,3,4,7,8-hexahydro-8-(2-hydroxyethyl)-2,4,7-trioxo-, methyl ester	pH 1	283(4.12),328(4.11)	39-0953-84C
	pH 13	260(3.97),288(4.03), 353(4.17)	39-0953-84C
$C_{12}H_{14}N_4O_6S$			
1H-Imidazo[1,2-b]pyrazole-7-carboni-trile, 6-(methylsulfonyl)-1-β-D-ribofuranosyl-	pH 1 and 11	220(4.30),273(3.80)	44-3534-84
	pH 7	220(4.31),273(3.83)	44-3534-84
$C_{12}H_{14}N_6$			
1H-Cyclohepta[1,2-d:3,4-d']diimidazol-8-amine, 2,3-dihydro-2-imino-1,3,6-trimethyl-	MeOH	248(4.34),274s(4.45), 282(4.56),308(4.77), 368(4.30),406(4.09), 422(4.12)	94-3873-84
	MeOH-acid	248(4.38),289(4.92),	94-3873-84

Compound	Solvent	$\lambda_{max}(\log \epsilon)$	Ref.
(cont.)		350(4.10),388s(4.29), 400(4.44)	94-3873-84
1H-Cyclohepta[1,2-d:3,4-d']diimdazole-2,8-diamine, N,N',4-trimethyl-	MeOH	250(4.06),305s(4.71), 309(4.72),368(4.07), 404(4.04),416s(3.81)	94-3873-84
	MeOH-acid	241(4.12),297(4.76), 353(3.84),404(4.31), 425s(3.63)	94-3873-84
1H-Cyclohepta[1,2-d:3,4-d']diimidazole-2,8-diamine, N^8,N^8,4-trimethyl-	MeOH	250(4.09),303(4.65), 369(4.01),407(4.16), 420s(3.93)	94-3873-84
	MeOH-acid	244(4.23),294(4.70), 352(3.72),410(4.35)	94-3873-84
$C_{12}H_{14}N_6O_3$ 1H-Imidazole-4-carboximidamide, 5-amino-1-methyl-N-[(4-nitrophenyl)methoxy]-	pH 1	275(4.26)	94-4842-84
	pH 7	270(4.26)	94-4842-84
	pH 13	270(4.26)	94-4842-84
	EtOH	269(4.29)	94-4842-84
$C_{12}H_{14}N_6O_7$ D-Ribitol, 1-deoxy-1-(hexahydro-2,6,8-trioxo-4H-imidazo[4,5-g]pteridin-4-yl)-	pH 1	212(4.35),263(3.84), 296(3.75),360(4.22)	33-0550-84
	pH 1	213(4.39),264(3.87), 293(3.78),360(4.28)	33-0570-84
	H_2O	219(4.52),277(4.24), 296s(3.95),389(4.32)	33-0550-84
	H_2O	219(4.54),276(4.26), 295s(3.97),389(4.29)	33-0570-84
	pH 13	220(4.28),276(3.92), 310s(3.62),389(4.10)	33-0550-84
	pH 13	222(4.34),275(3.89), 308s(3.60),390(4.15)	33-0570-84
$C_{12}H_{14}O$ 5H-Benzocycloheptene, 6,7-dihydro-3-methoxy-	EtOH	262(4.24)	88-3007-84
1-Biphenylenol, 4b,5,6,7,8,8a-hexahydro-, cis	MeOH	271(2.93),277(2.92)	87-0792-84
2-Biphenylenol, 4b,5,6,7,8,8a-hexahydro-, cis	MeOH	284(3.59)	87-0792-84
1-Cyclopropenemethanol, 3,3-dimethyl-2-phenyl-	EtOH	268(4.11)	44-4344-84
4-Hexen-3-one, 5-phenyl-, (E)-	EtOH	210(3.69),200[sic](3.79), 280(4.20)	80-0857-84
4,8-Methano-1,3,6-methenoazulen-2(1H)-one, octahydro-	C_6H_{12}	298(1.42)	138-1741-84
2-Pentenal, 4-methyl-3-phenyl-, trans	C_6H_{12}	235(3.83)	44-4344-84
Phenol, 2-(1,1-dimethyl-2-butynyl)-	50% MeCN-KOH	294(4.15)	39-1269-84B
Phenol, 2-(1,1-dimethyl-2-propynyl)-4-methyl-	50% MeCN-KOH	303(4.08)	39-1269-84B
$C_{12}H_{14}OSi$ Silane, (ethenyloxy)dimethyl(phenylethynyl)-	hexane	238.1(4.30),248.3(4.49), 259.9(4.43)	65-2317-84
$C_{12}H_{14}O_2$ 1,8a(4bH)-Biphenylenediol, 5,6,7,8-tetrahydro-	MeOH	270(2.97),277(2.95)	87-0792-84

Compound	Solvent	λ_{max}(log ϵ)	Ref.
1,4-Ethano-2,3-benzodioxin, 1,4-di-hydro-1,4-dimethyl-	ether	246(2.33),252(2.35), 258(2.36),264s(2.19), 267(2.18)	22-0187-84
1H-Inden-1-one, 2,3-dihydro-4-methoxy-6,7-dimethyl-	n.s.g.	260(3.94),325(3.57)	2-0863-84
1H-Inden-1-one, 2,3-dihydro-6-methoxy-4,5-dimethyl-	n.s.g.	255(4.03),320(3.69)	2-0863-84
1H-Inden-1-one, 2,3-dihydro-6-methoxy-4,7-dimethyl-	n.s.g.	255(3.97),330(3.55)	2-0863-84
3-Penten-2-one, 1-hydroxy-4-methyl-3-phenyl-	C_6H_{12}	244(3.96)	44-4344-84
2-Propyn-1-ol, 3-(4-propoxyphenyl)-	MeOH	253(4.38)	118-0728-84
Tricyclo[6.2.0.0^{2,5}]deca-1,5,7-triene-6,7-dimethanol	EtOH	222s(4.06),232(3.70), 273(3.38),278(3.42), 282(3.40)	44-1412-84
$C_{12}H_{14}O_3$			
1,2-Dioxin-3-ol, 3,6-dihydro-6,6-di-methyl-5-phenyl-	EtOH	226(3.85)	44-4344-84
1(3H)-Isobenzofuranone, 4,5-dihydro-3-(2-hydroxybutylidene)-	n.s.g.	283(3.80),296(3.79), 323(3.86)	94-3770-84
1-Propanone, 2-acetoxy-2-methyl-1-phenyl-	C_6H_{12}	240(3.72),278(2.90)	44-4344-84
$C_{12}H_{14}O_4$			
Benzeneacetic acid, 3-(1,1-dimethyl-ethyl)-2-hydroxy-α-oxo-	EtOH	213(4.22),265(3.99), 340(3.54)	39-2655-84C
Benzoic acid, 4-hydroxy-2-methyl-5-(1-oxopropyl)-, methyl ester	MeOH	235(4.60),258(4.15), 318(3.56)	44-3791-84
1(3H)-Isobenzofuranone, 4,5-dihydro-3-hydroxy-3-(1-oxobutyl)- (senkyu-nolide D)	n.s.g.	287(3.42)	94-3770-84
2H,5H-Pyrano[4,3-b]pyran-5-one, 3,4-dihydro-4-hydroxy-2-methyl-7-(1-methylpropenyl)-	EtOH	225(4.48),262(3.47), 272(3.50),318(3.97)	102-0767-84
$C_{12}H_{14}O_5$			
2-Propenoic-3-^{14}C acid, 3-(3,4,5-tri-methoxyphenyl)-	EtOH	299(4.25)	102-1029-84
$C_{12}H_{14}O_7$			
2-Cyclopenten-1-one, 4,5-diacetoxy-5-(acetoxymethyl)-	EtOH	221(3.60)	39-2089-84C
$C_{12}H_{15}BrN_2O_5$			
Uridine, 5'-bromo-5'-deoxy-2',3'-O-(1-methylethylidene)-	EtOH	259(3.99)	94-2591-84
$C_{12}H_{15}BrO$			
1-Propanone, 1-[4-(bromomethyl)phenyl]-2,2-dimethyl-	MeOH	251(3.75)	35-3140-84
$C_{12}H_{15}ClN_2O$			
1-Butanamine, N-(5-chloro-3-methyl-2(3H)-benzoxazolylidene)-	CHCl$_3$	259(3.94),306(4.15)	94-3053-84
$C_{12}H_{15}ClN_2O_2$			
Acetamide, 2-chloro-N-[3-(hydroxy-imino)-3-(4-methylphenyl)propyl]-, (E)-	EtOH	252(4.1)	5-1696-84

Compound	Solvent	$\lambda_{max}(\log \epsilon)$	Ref.
$C_{12}H_{15}ClN_2O_3$ Acetamide, 2-chloro-N-[3-(hydroxy-imino)-3-(4-methoxyphenyl)prop-yl]-, (E)-	EtOH	262(4.2)	5-1696-84
$C_{12}H_{15}ClN_2O_5$ Uridine, 5'-chloro-5'-deoxy-2',3'-O-(1-methylethylidene)-	EtOH	259(4.00)	94-2591-84
$C_{12}H_{15}ClSi$ Silane, chlorodiethyl(phenylethynyl)-	hexane	238.8(4.33),249.3(4.51), 261.5(4.44)	65-2317-84
$C_{12}H_{15}IN_2O$ Piperazine, 1-(2-iodobenzoyl)-4-methyl-	MeOH	225s(4.13)	73-1009-84
$C_{12}H_{15}N$ 1H-1-Benzazepine, 2,3-dihydro-1,3-di-methyl-	MeOH	231(4.514),268(3.716), 320(3.518)	150-1441-84M
1H-1-Benzazepine, 2,3-dihydro-2,3-di-methyl-	MeOH	230(4.556),264(3.698), 328(3.505)	150-1441-84M
Isoquinoline, 3,4-dihydro-1,6,8-tri-methyl-	hexane	207s(4.32),212(4.39), 216(4.39),221s(4.23), 252(3.97),283s(3.18), 291s(4.07)	32-0041-84
	EtOH	208s(4.30),213(4.36), 217(4.34),223s(4.16), 259(4.01)	32-0041-84
$C_{12}H_{15}NO$ Benzaldehyde, 4-piperidino-	MeOH	342(4.51)	139-0253-84C
Benzamide, N-ethenyl-N-propyl-	MeOH	200(4.36),230(4.26), 263(4.27),304(3.86)	78-1835-84
Benzenamine, 3-ethoxy-N-(1-methyl-2-propynyl)-	EtOH	208(3.70),243(4.07), 289(3.46)	65-0161-84
Benzenamine, 4-ethoxy-N-(1-methyl-2-propynyl)-	EtOH	197(3.53),242(4.11), 304(3.31)	65-0161-84
Benzonitrile, 4-(2-methoxy-1,1-di-methylethyl)-	hexane	228(4.26),238(4.20), 258(2.63),260s(2.61), 266(2.73),269s(2.65), 272(2.66),277(2.60)	23-1785-84
Pyridine, 1,2,3,6-tetrahydro-1-methyl-4-phenyl-, 1-oxide	H_2O	242(4.21)	33-2037-84
$C_{12}H_{15}NO_2$ Cyclohexanone, 2-(4-methoxy-2-pyridin-yl)-	H_2O	240(3.98)	56-1077-84
$C_{12}H_{15}NO_2S$ 1,3-Butadiene-1-sulfonamide, 2-methyl-N-(phenylmethyl)-, (Z)-	H_2O or EtOH	242.5(4.32)	70-2519-84
1,3-Pentadiene-1-sulfonamide, N-(phen-ylmethyl)-, (Z,E)-	H_2O	243(4.30)	70-2519-84
$C_{12}H_{15}NO_3$ 3-Pyridinecarboxylic acid, 1-cyclo-hexyl-1,2-dihydro-2-oxo-	EtOH	231(4.74),332(5.08)	4-0041-84
$C_{12}H_{15}NO_4$ 1-Oxa-5-azaspiro[2.5]oct-6-ene-4,8-di-one, 6-methyl-2-(2-methyl-1-oxo-	EtOH	327(3.90)	158-84-98

Compound	Solvent	λ_{max}(log ϵ)	Ref.
butyl)-, [2S-[2α(R*),3α]]- (cont.)			158-84-98
$C_{12}H_{15}NO_4S$			
Thieno[2,3-c]pyridin-3(2H)-one, 4-hydroxy-7-methyl-5-(2-methylpropyl)-, 1,1-dioxide	EtOH	235(4.09),350(3.67)	104-2032-84
$C_{12}H_{15}NS$			
Benzenecarbothioamide, N-ethenyl-N-propyl-	hexane	313(4.10),402(2.99)	78-1835-84
Benzothiazole, 2-(1-ethylpropyl)-	EtOH	220(4.37),250(3.97), 285(3.46)	78-2141-84
Benzothiazole, 2-(1-methylbutyl)-	EtOH	220(4.36),250(3.98), 285(3.48)	78-2141-84
$C_{12}H_{15}N_3O_3$			
Ethanone, 1-[4-[3-(acetoxymethyl)-3-methyl-1-triazenyl]phenyl]-	MeCN	224(4.01),301(4.35), 320s(4.28)	23-0741-84
2H-Pyrrole-2-carboxamide, 4-ethyl-3,4-dihydro-4-hydroxy-5-(3-hydroxy-2-pyridinyl)- (siderochelin C)	MeOH	316(3.87)	158-1260-84
$C_{12}H_{15}N_3O_3S$			
7H-Pyrrolo[2,3-d]pyrimidine, 7-(2-deoxy-β-D-erythro-pentofuranosyl)-2-(methylthio)-	MeOH	249(4.36),270s(3.76), 307(3.77)	5-1719-84
$C_{12}H_{15}N_3O_4$			
Benzoic acid, 4-[3-(acetoxymethyl)-3-methyl-1-triazenyl]-, methyl ester	MeCN	222(4.01),293(4.36), 310s(4.29)	23-0741-84
Butanoic acid, 3-[(2-nitrophenyl)hydrazono]-, ethyl ester	EtOH	252(4.18),263(4.19), 282(4.10),433(3.75)	48-0829-84
$C_{12}H_{15}N_3O_5$			
L-Alanine, N-[1-oxo-3-(1,2,3,4-tetrahydro-6-methyl-2,4-dioxo-5-pyrimidinyl)-2-propenyl]-, methyl ester, (E)- (same spectrum for D-isomer)	H_2O	302(4.33)	94-4625-84
Glutamic acid, 5-[2-(4-carboxyphenyl)-hydrazide]	n.s.g.	258(4.05)	98-0676-84
$C_{12}H_{15}N_3O_5S$			
D-Cysteine, S-methyl-N-[1-oxo-3-(1,2-3,4-tetrahydro-6-methyl-2,4-dioxo-5-pyrimidinyl)-2-propenyl]-, (E)-	H_2O	303(4.31)	94-4625-84
L-	H_2O	303(4.30)	94-4625-84
$C_{12}H_{15}N_3O_6$			
D-Serine, N-[1-oxo-3-(1,2,3,4-tetrahydro-6-methyl-2,4-dioxo-5-pyrimidinyl)-2-propenyl]-, methyl ester, (E)-	H_2O	302.5(4.33)	94-4625-84
L-	H_2O	302.5(4.33)	94-4625-84
$C_{12}H_{15}N_3S_2$			
3-Thiazolidinecarbothioamide, N,N-dimethyl-2-(phenylimino)-	isoPrOH	273(4.06)	94-4292-84
$C_{12}H_{15}N_5$			
Cyanamide, (4,5,7,8,9,10-hexahydro-3H-[1,3]diazepino[1,7-a]benzimidazol-1-yl)-	MeOH	249(4.10)	4-0753-84

Compound	Solvent	λ_{max} (log ϵ)	Ref.
$C_{12}H_{15}N_5O_3$			
2,4(1H,3H)-Pteridinedione, 1,3-dimeth-yl-6-(methylamino)-7-(1-oxopropyl)-	pH –4.0 pH 3.0	274(4.31),456(3.80) 225(4.37),267s(4.18), 277(4.22),474(3.87)	5-1798-84 5-1798-84
2,4(1H,3H)-Pteridinedione, 1,3-dimeth-yl-7-(methylamino)-6-(1-oxopropyl)-	MeOH	234(4.55),285s(4.14), 298(4.22),377(4.19)	5-1798-84
$C_{12}H_{15}N_5O_3S$			
Propanamide, 3-[[(4-methylphenyl)sulfon-yl]amino]-N-1H-1,2,4-triazol-3-yl-, (±)-	EtOH	210(3.98),249(4.26)	120-0066-84
$C_{12}H_{15}N_5O_4S_2$			
1H-Pyrazolo[3,4-d]pyrimidine-3-carbo-thioamide, 4-(methylthio)-1-β-D-ribofuranosyl-	EtOH	292(4.05)	103-0210-84
$C_{12}H_{15}N_5O_5$			
4(1H)-Pyrimidinone, 2-(5-amino-1-β-D-ribofuranosyl-1H-imidazol-4-yl)-	pH 7	214(4.24),240(4.02), 318(4.22)	88-3471-84
9H-Purin-6-amine, 9-(2-O-acetyl-β-D-arabinofuranosyl)-	MeOH MeOH	258(4.15) 258(4.17)	31-0339-84 87-0270-84
Pyrrolo[2,3-d]pyrimidine-5-carboxamide, 4-amino-7-arabinofuranosyl-	pH 1	273(4.11)	78-0125-84
$C_{12}H_{15}N_5O_6$			
Butanedioic acid, mono[2-[(2-amino-1,6-dihydro-6-oxo-9H-purin-9-yl)-methoxy]ethyl] ester	MeOH	252(4.14),274s(--)	126-0687-84
$C_{12}H_{16}$			
Benzene, 1-methyl-2-(4-pentenyl)-	n.s.g.	263(2.71)	39-1833-84B
Benzene, 1-methyl-3-(4-pentenyl)-	n.s.g.	253(3.73)	39-1833-84B
Benzene, 1-methyl-4-(4-pentenyl)-	n.s.g.	266(2.89)	39-1833-84B
Benzene, (1-methyl-4-pentenyl)-	n.s.g.	248(3.83)	39-1833-84B
Benzene, (2-methyl-4-pentenyl)-	n.s.g.	253(2.80)	39-1833-84B
Benzene, (3-methyl-4-pentenyl)-	n.s.g.	259(2.49)	39-1833-84B
Benzene, (4-methyl-4-pentenyl)-	n.s.g.	261(2.52)	39-1833-84B
Tricyclo[3.3.0.0^{2,4}]oct-6-ene, 4-meth-yl-8-(1-methylethylidene)-, (1α,2β,4β,5α)-	EtOH	252(4.13)	33-1379-84
$C_{12}H_{16}Bi_2$			
1,1'-Bi-1H-bismole, 2,2',5,5'-tetra-methyl-	C_6H_{12}	230(4.16),320s(3.75)	157-0495-84
$C_{12}H_{16}FN_2O_8P$			
Uridine, 2'-deoxy-5-fluoro-5'-O-1,3,2-dioxaphosphorinan-2-yl-, P-oxide	EtOH	267(3.88)	87-0440-84
$C_{12}H_{16}FN_2O_9P$			
Uridine, 2'-deoxy-5-fluoro-5'-O-(5-hydroxy-1,3,2-dioxaphosphorinan-2-yl)-, P-oxide	EtOH	268(3.78)	87-0440-84
$C_{12}H_{16}NO_2$			
Isoquinolinium, 3,4-dihydro-7,8-di-methoxy-N-methyl-, iodide	MeOH	218(4.06),297(3.77), 375(3.03)	36-1639-84
$C_{12}H_{16}N_2$			
Benzeneacetonitrile, α-[(1,1-dimethyl-	EtOH	250(4.26)	44-0282-84

Compound	Solvent	$\lambda_{max}(\log \epsilon)$	Ref.
ethyl)amino]- (cont.)			44-0282-84
1,4:5,8-Dimethanophthalazine, 1,4,4a,5-8,8a-hexahydro-10,10-dimethyl-, endo	hexane	261(3.37),362(2.59)	24-0517-84
exo	hexane	360(2.35)	24-0517-84
$C_{12}H_{16}N_2O$			
6(2H)-Quinoxalinone, 3,4-dihydro-4,5,7,8-tetramethyl-	CH_2Cl_2	246(4.12),290(4.07), 462(3.32)	83-0743-84
$C_{12}H_{16}N_2OS$			
Morpholine, 4-[[(methylthio)imino]phenylmethyl]-	hexane	310(3.46)	39-2933-84C
$C_{12}H_{16}N_2O_2$			
2,4(1H,3H)-Pyrimidinedione, 5-(1-cyclohepten-1-ylmethyl)-	MeOH	266(3.88)	44-2738-84
$C_{12}H_{16}N_2O_3$			
5,6-Diazatricyclo[5.2.0.0^{2,5}]non-8-ene-6-carboxylic acid, 3,9-dimethyl-4-oxo-, ethyl ester	MeOH	238(2.89)	23-2440-84
10-Oxa-2,3-diazabicyclo[4.3.1]deca-1,4,7-triene-3-carboxylic acid, 6,9-dimethyl-, ethyl ester, endo	MeOH	240(3.92)	23-2440-84
4H-Pyran-4-one, tetrahydro-3-hydroxy-2-methoxy-, phenylhydrazone, (2S-cis)-	EtOH	278(4.28)	39-0733-84C
Pyrazolo[1,5-a]pyridine-1(3aH)-carboxylic acid, 4,7-dihydro-3a,6-dimethyl-7-oxo-, ethyl ester	MeOH	212(4.13),273s(3.23)	23-2440-84
Pyrazolo[1,5-a]pyridine-1(3aH)-carboxylic acid, 6,7-dihydro-3a,6-dimethyl-7-oxo-, ethyl ester	MeOH	230(3.96)	23-2440-84
$C_{12}H_{16}N_2O_4$			
2-Butenedioic acid, 2-(3,4,5-trimethyl-1H-pyrazol-1-yl)-, dimethyl ester, (Z)-	MeCN	295(4.27)	44-4647-84
$C_{12}H_{16}N_2O_4S$			
Hydroxylamine-O-sulfonic acid, N-[2-(4-methyl-2-pyridinyl)cyclohexylidene]-	H_2O	262(3.80)	56-1077-84
	anion	255(3.45),260(3.49), 267(3.43)	56-1077-84
2-Propenamide, 3-methoxy-2-methyl-N-[[(phenylmethyl)amino]sulfonyl]-, (E)-	50% EtOH	255(4.11)	73-0840-84
	+ NaOH	255(4.10)	73-0840-84
2-Propenoic acid, 2-methyl-3-[[[(phenylmethyl)amino]sulfonyl]amino]-, methyl ester, (E)-	50% EtOH	257(4.21)	73-0840-84
	+ NaOH	292(4.41)	73-0840-84
(Z)-	50% EtOH	259(4.06)	73-0840-84
	+ NaOH	292(4.40)	73-0840-84
$C_{12}H_{16}N_2O_5S$			
Hydroxylamine-O-sulfonic acid, N-[2-(4-methoxy-2-pyridinyl)cyclohexylidene]-	H_2O	240(4.01)	56-1077-84
	anion	250(3.00)	56-1077-84
4(1H)-Pyrimidinone, 2,3-dihydro-5-[2,3-O-(1-methylethylidene)-β-D-ribofuranosyl]-2-thioxo-	pH 13	264(4.11),290(4.02)	18-2515-84
	MeOH	276(4.14),292(4.11)	18-2515-84
$C_{12}H_{16}N_2O_6$			
2,4(1H,3H)-Pyrimidinedione, 5-[2,3-O-(1-methylethylidene)-β-D-ribofuranosyl]-	pH 13	284(3.85)	18-2515-84
	MeOH	263(3.78)	18-2515-84

Compound	Solvent	λ_{max}(log ϵ)	Ref.
$C_{12}H_{16}N_2O_6S$			
Dopa, 2-S-cysteinyl-	pH 1	293(3.53)	3-1935-84
	borate	301(3.65)	3-1935-84
Dopa, 5-S-cysteinyl-	pH 1	292(3.47)	3-1935-84
	borate	299(3.64)	3-1935-84
$C_{12}H_{16}N_4O_3$			
1,2-Cyclopentanediol, 3-(4-amino-7H-	pH 1	275(4.00)	87-0534-84
pyrrolo[2,3-d]pyrimidin-7-yl)-5-	pH 7	273(4.00)	87-0534-84
(hydroxymethyl)-, (1α,2α,3β,5β)-(±)-	pH 13	272(4.00)	87-0534-84
$C_{12}H_{16}N_4O_3S$			
7H-Pyrrolo[2,3-d]pyrimidin-4-amine,	pH 1	226(4.33),280(4.15)	35-6379-84
7-(2-deoxy-β-D-erythro-pentofurano-	pH 7 and 12	236(4.36),282(4.15)	35-6379-84
syl)-2-(methylthio)-			
$C_{12}H_{16}N_4O_4$			
Imidazo[1,5-b]pyridazin-2-amine, 5-	EtOH	250(4.36),304(3.44),	39-0229-84C
methyl-7-β-D-ribofuranosyl-		313s(--),343(3.37)	
4H-Pyrrolo[2,3-d]pyrimidin-4-one,	MeOH	226(4.34),264(4.01),	5-0708-84
2-amino-7-(2-deoxy-α-D-erythro-		285s(3.80)	
pentofuranosyl)-1,7-dihydro-5-methyl-			
β-	MeOH	224(4.35),264(4.02),	5-0708-84
		285s(3.81)	
4H-Pyrrolo[2,3-d]pyrimidin-4-one,	MeOH	225(4.34),264(4.03),	5-0708-84
2-amino-7-(2-deoxy-α-D-erythro-		285s(3.81)	
pentopyranosyl)-1,7-dihydro-5-methyl-			
β-	MeOH	225(4.35),264(4.02),	5-0708-84
		285s(3.80)	
$C_{12}H_{16}N_4O_4S$			
Imidazo[5,1-f][1,2,4]triazine-4(1H)-	EtOH	261(4.05),325(4.20)	39-0229-84C
thione, 2,5-dimethyl-7- -D-ribo-			
furanosyl-			
7H-Pyrrolo[2,3-d]pyrimidin-4-amine,	MeOH	234(4.36),282(4.13)	5-1972-84
2-(methylthio)-7-α-D-ribofuranosyl-			
β-	MeOH	234(4.35),281(4.12)	5-1972-84
$C_{12}H_{16}N_4O_5$			
Imidazo[5,1-f][1,2,4]triazin-4(1H)-	EtOH	223(4.38),255(3.95)	39-0229-84C
one, 2,5-dimethyl-7-β-D-ribo-			
furanosyl-			
7H-Pyrazolo[4,3-d]pyrimidin-7-one,	pH 1 and 7	273(3.87)	4-1865-84
1,6-dihydro-1,6-dimethyl-3-β-D-	pH 11	273(3.87)	4-1865-84
ribofuranosyl-			
2,6-Pyridinedicarboxylic acid, 4-	EtOH	265(3.99),376(4.50)	33-1547-84
[[(aminocarbonyl)hydrazono]ethyli-			
dene]-1,2,3,4-tetrahydro-, dimethyl			
ester			
$C_{12}H_{16}N_4O_6$			
1H-Pyrazolo[4,3-d]pyrimidine-5,7(4H-	pH 1 and 7	240s(3.77),287(3.79)	4-1865-84
6H)-dione, 1,6-dimethyl-3-β-D-ribo-	pH 11	315(3.84)	4-1865-84
furanosyl-			
$C_{12}H_{16}N_4O_7S$			
1H-Imidazo[1,2-b]pyrazole-7-carbox-	pH 1,7,11	240s(3.76),274(3.79)	44-3534-84
amide, 6-(methylsulfonyl)-1-β-D-			
ribofuranosyl-			

Compound	Solvent	$\lambda_{max}(\log \epsilon)$	Ref.
$C_{12}H_{16}N_6O_7$			
D-Ribitol, 1-deoxy-1-(1,2,5,6,7,8-hexa-hydro-2-hydroxy-6,8-dioxo-4H-imidazo-	pH 1	261(4.12),303(3.55), 383s(3.85),420(3.99)	33-0550-84
	H_2O	261(4.12),306(3.50), 412.5(4.04)	33-0550-84
	pH 13	219(4.23),263(4.14), 343(3.69),412.5(4.11)	33-0550-84
$C_{12}H_{16}O$			
Acetaldehyde, (3,4,4,5-tetramethyl-2,5-cyclohexadien-1-ylidene)-	hexane	309(4.48)	5-0340-84
Benzene, 1-methoxy-2-(4-pentenyl)-	n.s.g.	275(3.34)	39-1833-84B
Benzene, 1-methoxy-4-(4-pentenyl)-	n.s.g.	272(3.34)	39-1833-84B
Bicyclo[3.2.0]hept-2-en-6-one, 7,7-di-methyl-3-(1-methylethenyl)-	EtOH	234(4.26)	33-1379-84
Bicyclo[3.2.0]hept-2-en-6-one, 7,7-di-methyl-4-(1-methylethylidene)-	EtOH	253(4.14)	33-1379-84
4,7-Methano-1H-indene, 1-ethoxy-3a,4,7,7a-tetrahydro-	n.s.g.	206(3.47)	104-0061-84
Spiro[4.5]deca-2,6-dien-1-one, 3,6-di-methyl-	EtOH	225(4.14)	33-0073-84
Spiro[3.5]nona-5,7-dien-1-one, 5,9,9-trimethyl-	pentane	250s(3.54),260(3.58), 271(3.60),281s(3.54), 329(2.17),341(2.19), 359s(1.73)	33-0774-84
Tricyclo[4.3.0.0^{1,3}]non-4-en-7-one, 2,2,6-trimethyl-	pentane	275(2.27),285(2.34), 295(2.37),305(2.41), 316(2.41),328(2.27), 334(1.84)	33-0774-84
Tricyclo[4.3.0.0^{1,5}]non-7-en-4-one, 5,9,9-trimethyl-	pentane	226(3.85),276(1.94), 286(1.96),300s(1.80)	33-0774-84
$C_{12}H_{16}OS$			
Dispiro[4.1.4.1]dodecan-6-one, 12-thioxo-	$CHCl_3$	260(4.16),320(3.34), 520(1.18)	2-0498-84
$C_{12}H_{16}OS_2$			
1,3-Benzodithiole, 2-(3-methylbutoxy)-	H_2O	226(4.42)	18-1128-84
$C_{12}H_{16}O_2$			
Bicyclo[3.2.0]hept-2-en-6-one, 4-(eth-oxymethylene)-7,7-dimethyl-, (E)-(±)-	EtOH	252(3.98)	33-1854-84
(Z)-(±)-	EtOH	261(3.92)	33-1854-84
Cyclopentadiene, methoxy-, dimers	n.s.g.	213(3.46)	104-0061-84
1(3H)-Isobenzofuranone, 3-butyl-4,5-di-hydro- (sedanenolide)	MeOH	275(3.55)	64-0872-84C
10-Oxatricyclo[5.3.0.0^{2,4}]dec-7-en-9-one, 1,6,6-trimethyl-(1R*,2S*,4S*)-	pentane	210(4.09)	33-0136-84
	MeCN	211(4.15)	33-0136-84
$C_{12}H_{16}O_2S$			
3-Thiopheneacetaldehyde, α-cyclohexyl-α-hydroxy-, (S)-	MeOH	216s(3.79),234(3.85), 295(2.15)	22-0077-84
$C_{12}H_{16}O_2S_3$			
Ethane(dithioic) acid, [(2-hydroxy-2-phenylethyl)sulfinyl]-, ethyl ester	EtOH	316(4.09)	39-0085-84C
$C_{12}H_{16}O_3$			
1,2-Benzenediol, 3-methoxy-5-(3-methyl-2-butenyl)-	MeOH	207(4.11),265(3.76)	102-2396-84

Compound	Solvent	$\lambda_{max}(\log \epsilon)$	Ref.
Benzoic acid, 4-ethoxy-2-methyl-, ethyl ester	EtOH	220(4.15),259(4.11)	12-2037-84
Cyclopenta[c]pyran, 3-(dimethoxymethyl)-3,4-dihydro-4-methyl-	EtOH	290(3.92)	78-0695-84
5,5-Epoxymethano-2,2,6-trimethyl-7-oxabicyclo[4.3.2]non-9-en-8-one	pentane	213(4.06)	33-0815-84
2-Furancarboxylic acid, 4-(4-methyl-3-pentenyl)-, methyl ester	MeOH	220(3.67),260(3.96)	102-0301-84
1(3H)-Isobenzofuranone, 3-butyl-4,5-dihydro-3-hydroxy- (senkyunolide G)	n.s.g.	281(3.51)	94-3770-84
$C_{12}H_{16}O_3S$			
3-Thiopheneacetic acid, α-cyclohexyl-α-hydroxy-	MeOH	215s(3.74),238s(3.84), 232(3.85),244s(3.66)	22-0077-84
$C_{12}H_{16}O_4$			
2,5-Cyclohexadiene-1,4-dione, 2,5-dihydroxy-3,6-bis(1-methylethyl)-	aq base	327s(--),333(4.41), 550(2.36)	94-2406-84
1(3H)-Isobenzofuranone, 3-butylidene-4,5,6,7-tetrahydro-6,7-dihydroxy-(ligustilidiol)	MeOH	272(3.93)	64-0872-84C
cis	MeOH	272(4.02)	102-2033-84
2H-Pyran-5-carboxylic acid, 3,4-dihydro-2-methyl-6-(1-methylethenyl)-4-oxo-, ethyl ester	EtOH	272(4.05)	4-0013-84
2H-Pyran-5-carboxylic acid, 3,4-dihydro-3-methyl-4-oxo-6-(1-propenyl)-, ethyl ester	EtOH	240(3.90),296(4.21)	4-0013-84
4H-Pyran-3-carboxylic acid, 5,6-dihydro-5-methyl-2-(1-methylethenyl)-4-oxo-, ethyl ester	EtOH	276(3.91)	4-0013-84
4H,5H-Pyrano[4,3-b]pyran-4,5-dione, 2,3,7,8-tetrahydro-2,2,7,7-tetramethyl-	EtOH	261(3.99)	4-0013-84
$C_{12}H_{16}O_6$			
2H-Pyran-4-acetic acid, 3-ethyl-5-(methoxycarbonyl)-2-oxo-, methyl ester	MeOH	234(3.97)	95-1232-84
$C_{12}H_{16}O_7$			
β-D-Alloside, 4-hydroxyphenyl	MeOH	225(3.70),286(3.34)	102-0468-84
Cyclopentanone, 2,3-diacetoxy-2-(acetoxymethyl)-, (2S,3S)-	EtOH	213(2.86)	39-2089-84C
$C_{12}H_{17}ClN_2$			
11,12-Diazabicyclo[8.3.1]tetradeca-1(14),10,12-triene, 13-chloro-	EtOH	219(3.61),268(3.16), 302(2.48)	32-0289-84
$C_{12}H_{17}ClO_3$			
3-Cyclohexene-1-carboxylic acid, 1-chloro-4-ethoxy-2-methylene-, ethyl ester	CCl$_4$	316(3.43)	12-2037-84
$C_{12}H_{17}FN_3O_7P$			
Uridine, 2'-deoxy-5-fluoro-5'-O-(3-methyl-1,3,2-oxazaphospholidin-2-yl)-, P-oxide	EtOH	270(3.93)	39-1471-84C
$C_{12}H_{17}N$			
Piperidine, 1-(4-methylphenyl)-	MeOH	246(3.95)	139-0253-84C

Compound	Solvent	$\lambda_{max}(\log \epsilon)$	Ref.
Quinoline, 1,2,3,4-tetrahydro-2,2,4-trimethyl-	heptane	<u>250(4.0)</u>,300(3.3)	70-2005-84
$C_{12}H_{17}NO$			
Piperidine, 1-(4-methoxyphenyl)-	MeOH	237(4.13)	139-0253-84C
1-Propanone, 1-[5-[(dimethylamino)-methylene]-1,3-cyclopentadien-1-yl]-2-methyl-	EtOH	248(4.21),349(4.11), 387(4.13)	33-1854-84
6-Quinolinol, 1,2,3,4-tetrahydro-2,2,4-trimethyl-	heptane	<u>248(3.9)</u>,300(3.3)	70-2005-84
$C_{12}H_{17}NO_2$			
2-Butenamide, N-(5,5-dimethyl-3-oxo-1-cyclohexen-1-yl)-	pH 1	299(4.30)	39-0287-84C
	H_2O	299(4.31)	39-0287-84C
	pH 13	297(4.15),348(4.16)	39-0287-84C
after 90 minutes	pH 13	288(4.43)	39-0287-84C
Ethanone, 1-[3-[(dimethylamino)methyl]-2-hydroxy-5-methylphenyl]-	EtOH	250(3.81)	2-0904-84
2-Propenamide, N-(5,5-dimethyl-3-oxo-cyclohexen-1-yl)-2-methyl-	pH 1	293(4.16)	39-0287-84C
	H_2O	293(4.15)	39-0287-84C
	pH 13	290(3.99),341(3.94)	39-0287-84C
after 20 minutes	pH 13	288(4.36)	39-0287-84C
2,5(1H,3H)-Quinolinedione, 4,6,7,8-tetrahydro-3,7,7-trimethyl-	pH 1	299(4.16)	39-0287-84C
	H_2O	299(4.16)	39-0287-84C
	pH 13	297(4.30),348(3.38)	39-0287-84C
$C_{12}H_{17}NO_3$			
1H-Inden-2-amine, 2,3-dihydro-1,5,6-trimethoxy-	n.s.g.	202.5(4.66),233(3.86), 285(3.66),288(3.66), 293s(3.57)	39-1655-84C
1H-Indole-2-carboxylic acid, 2,3,4,5-6,7-hexahydro-2-methyl-3-oxo-, ethyl ester	MeCN	315(4.09)	33-1957-84
1H-Pyrrole-2-carboxylic acid, 2,3-di-hydro-2,5-dimethyl-3-oxo-4-(2-prop-enyl)-, ethyl ester	MeCN	316(4.07)	33-1957-84
1H-Pyrrole-2-carboxylic acid, 2,3-di-hydro-4,5-dimethyl-3-oxo-2-(2-prop-enyl)-, ethyl ester	MeCN	320(4.13)	33-1535-84
$C_{12}H_{17}NO_4$			
5-Isoquinolinecarboxylic acid, deca-hydro-2-methyl-6,8-dioxo-, methyl ester, $(4a\alpha,8a\beta)-(\pm)-$	H_2O	282(4.42)	44-5109-84
$C_{12}H_{17}NO_4S$			
DL-Homocysteine, S-[2-(2-hydroxyphen-oxy)ethyl]-	MeOH	215(3.86),274s(3.45)	36-1241-84
$C_{12}H_{17}NO_4S_2$			
2-Butenedioic acid, 2-[(1-piperidinyl-thioxomethyl)thio]-, dimethyl ester	MeOH	247(3.76)	104-2285-84
$C_{12}H_{17}NO_6$			
4(1H)-Pyridinone, 3-methoxy-2-methyl-1-β-D-ribofuranosyl-	MeOH	275(4.27)	87-0160-84
$C_{12}H_{17}N_3O$			
Piperazine, 1-(2-aminobenzoyl)-4-methyl-	MeOH	232s(4.06),252s(3.69), 305(3.41)	73-1009-84

Compound	Solvent	$\lambda_{max}(\log \epsilon)$	Ref.
$C_{12}H_{17}N_3O_3S$ 3H-1,2-Oxathiino[6,5-e]indazol-4-amine, 4,5,6,7-tetrahydro-N,N,7-trimethyl-, 2,2-dioxide	EtOH	250(3.95),255(3.95), 267s(3.85)	4-0361-84
$C_{12}H_{17}N_3O_4S$ 2-Propenamide, N-[1-(hydroxymethyl)- 2-(methylthio)ethyl]-3-(1,2,3,4- tetrahydro-6-methyl-2,4-dioxo-5- pyrimidinyl)-, [R-(E)]-	H_2O	301(4.31)	94-4625-84
[S-(E)]-	H_2O	301(4.30)	94-4625-84
$C_{12}H_{17}N_3O_5$ Isocytosine, 5-[2,3-O-(1-methylethyli- dene)-β-D-ribofuranosyl]-	pH 13 MeOH	233(4.10),277(3.99) 226(3.74),290(3.73)	18-2515-84 18-2515-84
$C_{12}H_{17}N_3O_7$ Glycine, N-(1-β-D-arabinofuranosyl- 1,2-dihydro-2-oxo-4-pyrimidinyl)-, methyl ester	H_2O	235(3.96),274(4.06)	73-2689-84
Glycine, N-(1,2-dihydro-2-oxo-1-β-D- ribofuranosyl-4-pyrimidinyl)-, methyl ester	H_2O	236(3.95),272(4.04)	73-2689-84
$C_{12}H_{17}N_3O_7S$ Heptofuranuronic acid, 6-amino-1,6-di- deoxy-1-(3,4-dihydro-3-methyl-2,4- dioxo-1(2H)-pyrimidinyl)-4-thio-	M HCl	210(3.92),266(3.85)	5-1399-84
$C_{12}H_{17}N_4O_9P$ Phosphorimidic acid, (2,4,6-trinitro- phenyl)-, triethyl ester	MeOH	270(4.46),335(4.18)	65-0289-84
$C_{12}H_{17}N_5O_3S$ 1,2-Cyclopentanediol, 3-[2-amino- 6-(methylthio)-9H-purin-9-yl]- 5-(hydroxymethyl)-, (1α,2α,3β,5β)-(±)-	pH 1	227(4.22),248(4.00), 261s(--),324(4.07)	87-0670-84
	pH 7	224(4.33),245(4.08), 253s(--),261s(--), 310(4.11)	87-0670-84
	pH 13	224(4.32),245(4.08), 253s(--),261s(--), 310(4.11)	87-0670-84
$C_{12}H_{17}N_5O_4$ D-Ribitol, 1-C-(4-amino-2,5-dimethyl- imidazo[5,1-f][1,2,4]triazin-7-yl)- 1,4-anhydro-, (S)-	EtOH	240(4.44),265s(3.76), 275s(3.60),310(3.45)	39-0229-84C
$C_{12}H_{17}N_5O_5$ 4H-Pyrazolo[3,4-d]pyrimidin-4-one, 3-(dimethylamino)-1,5-dihydro- 1-β-D-ribofuranosyl-	pH 1 pH 7 pH 11	220s(4.22),242s(4.07) 227(4.36),244s(4.23) 217(4.34),240s(4.12), 270(3.88)	87-1119-84 87-1119-84 87-1119-84
$C_{12}H_{18}$ 8,10-Dodecadien-1-yne, (E,E)- Naphthalene, 2,3,4,4a,5,6-hexahydro- 1,4a-dimethyl-, (S)-	hexane EtOH	229(4.26) 242(4.24),236s(4.03), 251s(4.09)	105-0486-84 39-1323-84C
$C_{12}H_{18}N_2$ 11,12-Diazabicyclo[8.3.1]tetradeca-	EtOH	215(3.46),255(3.11),	32-0289-84

Compound	Solvent	$\lambda_{max}(\log \epsilon)$	Ref.
1(14),10,12-triene (cont.)		304(2.38)	32-0289-84
1,4:5,8-Dimethanophthalazine, 1,4,4a,5,6,7,8,8a-octahydro-10,10-dimethyl-, endo	hexane	350(2.50),353(2.79)	24-0517-84
exo	hexane	457(2.43)	24-0517-84
Methanediamine, 1-(2,4,6-cyclohepta-trien-1-ylidene)-N,N,N',N'-tetra-methyl-	ether	335(3.4),430(2.45)	64-1586-84B
1,4-Methano-1H-cyclopenta[d]pyridazine, 4,4a,5,7a-tetrahydro-1,4,8,8-tetra-methyl-	hexane	361(2.47)	24-0517-84
$C_{12}H_{18}N_2O$			
11,12-Diazabicyclo[8.3.1]tetradeca-1(14),10-dien-13-one	EtOH	214(3.63),228(3.52), 295(3.48)	32-0289-84
Piperazine, 1-(4-methoxyphenyl)-4-methyl-	C_6H_{12}	248(4.09),302(3.30)	35-1335-84
$C_{12}H_{18}N_2O_4S$			
1-Azabicyclo[3.2.0]hept-2-ene-2-carbox-ylic acid, 3-[(2-aminoethyl)thio]-6-(1-hydroxy-1-methylethyl)-7-oxo-	H_2O	298(3.83)	44-5271-84
$C_{12}H_{18}N_2O_5S$			
1-Azabicyclo[3.2.0]hept-2-ene-2-carbox-ylic acid, 3-[(2-aminoethyl)thio]-6-(1-hydroxy-2-methoxyethyl)-7-oxo-, (R)-	H_2O	299(3.66)	44-5271-84
(S)-	H_2O	299(3.79)	44-5271-84
$C_{12}H_{18}N_4OS$			
5-Pyrimidinecarboxamide, 6-amino-N-cyclohexyl-1,2-dihydro-1-methyl-2-thioxo-	EtOH	227(3.98),291(4.13), 355(4.26)	142-2259-84
$C_{12}H_{18}N_6O_3$			
1,2-Cyclopentanediol, 3-[2-amino-6-(methylamino)-9H-purin-9-yl]-5-(hydroxymethyl)-, (1α,2α,3β,5β)-(±)-	pH 1	208(4.28),210(?), 255(4.02),291(4.06)	87-0670-84
	pH 7	215(4.38),262s(--), 281(4.13)	87-0670-84
	pH 13	262s(--),281(4.13)	87-0670-84
$C_{12}H_{18}N_6O_9$			
L-Alanine, N-[(5-amino-1-β-D-ribo-furanosyl-1H-imidazol-4-yl)carbo-nyl]-3-(hydroxynitrosoamino)-, dipotassium salt	H_2O	258(4.19)	87-1295-84
$C_{12}H_{18}O_2$			
3-Cyclodecen-1-one, 3-acetyl-, (E)-	EtOH	235(4.00),291(2.00)	44-2925-84
Hydroperoxide, 1,2,3,4,5,6-hexahydro-1,4-dimethyl-1-azulenyl, (1R-trans)-	MeOH	247s(3.92),255(3.93)	
2,5,7-Nonatrienoic acid, 4,4-dimethyl-, methyl ester, (E,E,E)-	MeCN	225(4.40)	22-0362-84
2-Propenoic acid, 2-methyl-, 2,4-octa-dienyl ester, (E,E)-	isooctane	230(4.52)	116-1624-84
$C_{12}H_{18}O_3$			
1,3-Cyclohexanedione, 2-(3-oxohexyl)-	EtOH	263(4.18)	18-3351-84
3-Cyclohexene-1-acetic acid, 2,4,6-trimethyl-5-oxo-, methyl ester	EtOH	234(4.25)	33-1208-84

Compound	Solvent	λ_{max}(log ϵ)	Ref.
3-Cyclononen-1-one, 3-(acetoxymethyl)-	EtOH	293(1.90)	44-2925-84
4H-Pyran-4-one, 2-ethyl-6-(2-hydroxy-1-methylethyl)-3,5-dimethyl-	MeOH	258(3.78)	44-0559-84
C$_{12}$H$_{18}$O$_6$			
4H-Pyran-4-acetic acid, 3-ethyl-2,3-dihydro-2-hydroxy-5-(methoxycarbonyl)-, methyl ester	MeOH	236(4.03)	95-1232-84
C$_{12}$H$_{18}$O$_6$S$_4$			
Butanedioic acid, 2,3-bis[(ethoxythioxomethyl)thio]-, dimethyl ester	isoPrOH	221(4.30),277(4.27), 358(1.96)	104-1082-84
C$_{20}$H$_{19}$Cl$_3$O$_3$S			
2-Hexenoic acid, 6,6,6-trichloro-3-methoxy-5-methyl-, 2-(ethylthio)-ethyl ester, (E)-(±)-	EtOH	235(4.15)	44-3489-84
C$_{12}$H$_{19}$N			
2-Cyclohexen-1-amine, N-cyclohexylidene-	C$_6$H$_{12}$	235s(2.76)	33-0748-84
C$_{12}$H$_{19}$NO			
2-Cyclohexen-1-one, 3-[methyl(3-methyl-2-butenyl)amino]-	MeOH	300(4.49)	44-0220-84
4-Piperidinone, 1,2,2,6,6-pentamethyl-3,5-bis(methylene)-	EtOH	242(4.04),252(3.88)	103-0761-84
C$_{12}$H$_{19}$NO$_7$			
α-D-Glucopyranose, 2-deoxy-2-[(2,2-diacetylethenyl)amino]-	EtOH	262(4.17),294(4.28)	136-0101-84L
C$_{12}$H$_{19}$NO$_9$			
α-D-Glucopyranose, 2-deoxy-2-[[2,2-bis(methoxycarbonyl)ethenyl]amino]-	EtOH	278(4.26)	136-0101-84L
C$_{12}$H$_{19}$N$_5$O$_2$			
9H-Purin-6-amine, 9-(2,2-diethoxyethyl)-2-methyl-	MeOH	264(4.12)	73-2148-84
9H-Purin-6-amine, 9-(2,2-diethoxy-1-methylethyl)-, (±)-	MeOH	261(4.13)	73-2148-84
C$_{12}$H$_{19}$O$_2$S			
Sulfonium, [3-(2-methoxyphenoxy)propyl]dimethyl-, iodide	MeOH	219.5(4.38),271.5s(3.62)	36-1241-84
C$_{12}$H$_{19}$P			
Phosphine, bis(1-methylethyl)phenyl-	C$_6$H$_{12}$	258(3.30)	65-0489-84
C$_{12}$H$_{20}$			
Cyclodecane, 1,2-bis(methylene)-	EtOH	233(3.84)	44-2981-84
1,8,10-Dodecatriene, (E,E)-	hexane	229(4.24)	105-0486-84
C$_{12}$H$_{20}$Ge			
Germane, (1,1-dimethylethyl)dimethylphenyl-	hexane	215(3.92)	138-1383-84
C$_{12}$H$_{20}$NO			
2-Buten-1-aminium, N-(3-methoxy-2-cyclopenten-1-ylidene)-N,3-dimethyl-, (E)-, perchlorate	MeOH	270(4.42)	44-0220-84

Compound	Solvent	$\lambda_{max}(\log \epsilon)$	Ref.
$C_{12}H_{20}N_2$ 1,4-Methano-1H-cyclopenta[d]pyridazine, 4,4a,5,6,7,7a-hexahydro-1,4,8,8-tetramethyl-, (1α,4α,4aα,7aα)-	hexane	364(2.23)	24-0517-84
$C_{12}H_{20}N_2O$ Pyrazine, 2-(octyloxy)-	EtOH	212(4.00),280(3.60), 295s(--)	39-0641-84B
$C_{12}H_{20}N_2O_7S$ 4(1H)-Pyrimidinone, 1-[2-deoxy-5-O-(methylsulfonyl)-β-D-threo-pento-furanosyl]-5,6-dihydro-2-methoxy-5-methyl-	EtOH	245(4.08)	128-0415-84
$C_{12}H_{20}N_3O_{10}P$ 1H-Imidazole-4-carboxylic acid, 5-amino-1-[5-O-[hydroxy(3-hydroxypropoxy)-hydroxyphosphinyl]-β-D-arabino-furanosyl]-	pH 2 pH 6-7 pH 12	250s(3.95),267(4.07) 240(3.76),269(4.12) 240s(3.75),269(4.12)	65-2114-84 65-2114-84 65-2114-84
$C_{12}H_{20}O$ Ethanone, 1-(4-butyl-1-cyclohexen-1-yl)-	EtOH	232(2.72),307(1.67)	103-0515-84
$C_{12}H_{20}O_3$ Citronellol, 10-oxo-, acetate	EtOH	228.5(4.29)	95-1232-84
$C_{12}H_{20}O_6$ Butanedioic acid, 2-acetyl-3-ethoxy-, diethyl ester	MeOH	248(3.59)	136-0019-84I
$C_{12}H_{20}Si$ Silane, (1,1-dimethylethyl)dimethyl-phenyl-	hexane	216(3.97)	138-1383-84
$C_{12}H_{20}Si_2$ Silane, 3-hexene-1,5-diyne-1,6-diyl-bis[trimethyl-, cis trans	THF THF	272.4(4.39),288.4(4.45) 271.3(4.57),287.5(4.61)	44-4733-84 44-4733-84
$C_{12}H_{21}I$ Cyclohexane, 2-(3-iodopropyl)-1,1-di-methyl-3-methylene-	pentane	257(2.49)	33-1734-84
$C_{12}H_{21}NO_2$ 3aH-Cycloocta[b]pyrrol-3a-ol, 2,3,4,5-6,7,8,9-octahydro-2,2-dimethyl-, 1-oxide	EtOH	233(4.08)	12-0587-84
$C_{12}H_{21}NO_2Si$ Ethanone, 1-[1-[[2-(trimethylsilyl)eth-oxy]methyl]-1H-pyrrol-2-yl]-	MeOH	284(4.14)	44-0203-84
$C_{12}H_{21}NS$ 2-Hexenenitrile, 3-(butylthio)-4-ethyl- 3-Hexenenitrile, 3-(butylthio)-4-ethyl-	EtOH EtOH	209(3.83),286(4.31) 208(3.92),255(3.56)	78-2141-84 78-2141-84
$C_{12}H_{21}N_2O_7P$ 1H-Pyrazole-3,5-dicarboxylic acid, 5-(diethoxyphosphinyl)-4,5-dihydro-,	EtOH	202(3.31),288(3.80)	65-2495-84

Compound	Solvent	$\lambda_{max}(\log \epsilon)$	Ref.
5-ethyl 3-methyl ester (cont.)			65-2495-84
$C_{12}H_{21}N_5O$			
1H-Imidazole-4-carboximidamide, 5-amino-N-(cyclohexylmethoxy)-1-methyl-	pH 1	225(3.88),280(3.98)	94-4842-84
	pH 7	221(4.03),263(3.99)	94-4842-84
	pH 13	263(3.99)	94-4842-84
	EtOH	220(4.03),263(4.06)	94-4842-84
$C_{12}H_{22}$			
Decane, 5,6-bis(methylene)-	hexane	229(4.08)	44-2981-84
	hexane	228(4.08)	44-2981-84
Octane, 2,7-dimethyl-4,5-bis(methylene)-	hexane	231(4.07)	44-2981-84
$C_{12}H_{22}N_2O_3Si$			
3-Butenoic acid, 2-diazo-3-[[(1,1-dimethylethyl)dimethylsilyl]oxy]-, ethyl ester	CH_2Cl_2	282(3.90)	23-2936-84
$C_{12}H_{22}N_4O_2$			
2,4(1H,3H)-Pyrimidinedione, 6-(propylamino)-1-[2-(propylamino)ethyl]-	H_2O	275(5.6)	103-1033-84
$C_{12}H_{22}O$			
8,10-Dodecadien-1-ol, (E,E)-	hexane	229(4.36)	105-0486-84
$C_{12}H_{22}OS_5$			
Ethene, 1,1'-sulfinylbis[2,2-bis(ethylthio)-	EtOH	270s(4.14),308(4.35)	39-0085-84C
$C_{12}H_{22}S_5$			
Ethene, 1,1'-thiobis[2,2-bis(ethylthio)-	EtOH	306(4.31)	39-0085-84C
$C_{12}H_{22}Si_2$			
Disilane, pentamethyl(4-methylphenyl)-	C_6H_{12}	233(4.18)	60-0383-84
$C_{12}H_{22}Te_2$			
Ditelluride, dicyclohexyl	benzene	389(--)	48-0467-84
$C_{12}H_{23}$			
Cyclododecyl	$C_{12}H_{24}$	260(3.20)	70-0936-84
$C_{12}H_{23}N_3O$			
2H-Pyrrol-2-one, 4,5-bis(diethylamino)-3,4-dihydro-	MeCN	243(4.38)	142-1179-84
$C_{12}H_{24}F_2$			
Diphosphirane, bis(1,1-dimethylethyl)-(1-methylethylidene)-, trans	hexane	228(4.27),255(3.85)	24-1542-84
$C_{12}H_{24}S_4Si_2$			
1H,6H-Bissilolo[3,4-c:3',4'-g][1,2,5,6]-tetrathiocin, octahydro-2,2,7,7-tetramethyl-	hexane	240(3.06)	70-0585-84
$C_{12}H_{24}S_5Si_2$			
1H,7H-Bissilolo[3,4-d:3',4'-h]-[1,2,3,6,7]pentathionin, octahydro-2,2,8,8-tetramethyl-	hexane	240(2.84)	70-0585-84

Compound	Solvent	$\lambda_{max}(\log \epsilon)$	Ref.
$C_{12}H_{25}$ Dodecyl	dodecane	255(3.20)	70-0936-84
$C_{12}H_{25}NO$ Pentanamide, N,N-diethyl-3,4,4-tri-methyl-	heptane or EtOH	238(1.84)	70-0310-84
$C_{12}H_{25}N_3O$ 1H-1,2,3-Triazole, 1-(2-ethoxyethyl)-4(and 5)-hexyl-4,5-dihydro-	hexane	240(4.60)	104-0580-84
$C_{12}H_{26}Ge_2$ Germane, [2,3-bis(methylene)-1,4-but-anediyl]bis[trimethyl-	hexane	241(4.14)	44-2981-84
$C_{12}H_{26}Si_2$ Silane, [2,3-bis(methylene)-1,4-butane-diyl]bis[trimethyl-	hexane	240(4.07)	44-2981-84
$C_{12}H_{26}Sn_2$ Stannane, 2,3-bis(methylene)-1,4-butane-diyl]bis[trimethyl-	hexane	247(4.10)	44-2981-84
$C_{12}H_{27}GeNO_2$ Butanamide, 3-hydroxy-N,N-dimethyl-3-(triethylgermyl)-	heptane or EtOH	263(2.24)	70-0310-84
$C_{12}H_{30}Si_2Te_2$ Silane, [ditellurobis(methylene)]-bis[dimethylpropyl-	benzene	420(2.72)	48-0467-84
$C_{12}H_{32}Si_4$ Cyclotetrasilane, 1,2,3,4-tetraethyl-1,2,3,4-tetramethyl-	isooctane	211s(4.23),229s(4.04), 250s(3.20),298(2.54)	101-0353-84L

Compound	Solvent	$\lambda_{max}(\log \epsilon)$	Ref.
$C_{13}F_{10}S_3$			
Carbonotrithioic acid, bis(pentafluoro- phenyl) ester	hexane	217(4.53),276(4.40), 304s(4.15),452(1.44)	78-4963-84
$C_{13}H_2N_6O$			
1-Cyclobutene-1,2-diacetonitrile, α,α'- dicyano-3-(dicyanomethylene)-4-oxo-, dipotassium salt	H_2O	226(4.22),250s(4.02), 274(4.05),307(4.06), 380(4.25),454(4.73)	24-2714-84
compd. with morpholine (1:2)	MeOH	227(4.14),253(4.00), 275(4.00),312(4.06), 380(4.22),461(4.64)	24-2714-84
$C_{13}H_4FN_4$			
Cyanamide, [2-(dicyanomethyl)-3-(4- fluorophenyl)-2-cyclopropen-1-yl- idene]-, ion(1-)-, sodium	MeOH	233(4.32),259(4.39), 334(4.30)	118-0686-84
$C_{13}H_5Cl_3OS$			
9H-Thioxanthen-9-one, 1,2,4-trichloro-	MeOH	262(4.56),309(3.80), 374(3.73)	73-2295-84
$C_{13}H_5D_5O$			
Methanone, (2,4-dihydroxyphenyl)- phenyl-d_5	MeOH	243(4.02),289(4.14), 324(4.01)	35-6155-84
$C_{13}H_6Cl_2N_2O_3S$			
1H-Pyrrole-2,5-dione, 3-[2-(3,4-di- chlorophenyl)-4-thiazolyl]-4-hydroxy-	EtOH	262(4.39),365(3.84)	44-2212-84
$C_{13}H_6Cl_3N_3S$			
Naphtho[1',2':4,5]thiazolo[3,2-b]- [1,2,4]triazole, 9-(trichloromethyl)-	MeOH	243(4.62),301(4.09), 330(3.53)	4-1571-84
$C_{13}H_6Cl_4N_6$			
7H-Pyrrolo[2,3-d]pyrimidine, 7,7'-meth- ylenebis[2,4-dichloro-	MeOH	230(4.79),276s(4.00), 286(4.02)	5-0722-84
$C_{13}H_6Cl_4O_2S$			
Benzoic acid, 3,5-dichloro-2-[(2,3-di- chlorophenyl)thio]-	MeOH	245s(4.15),290(3.72)	73-2295-84
$C_{13}H_6F_3N_3S$			
Naphtho[1',2':4,5]thiazolo[3,2-b]- [1,2,4]triazole, 9-(trifluoromethyl)-	MeOH	241(4.67),293(4.00), 329(3.31)	4-1571-84
$C_{13}H_6F_5NO$			
Benzenamine, N-[(pentafluorophenyl)- methylene]-, N-oxide	EtOH	298(4.11)	70-1268-84
$C_{13}H_6N_4$			
Pyrazolo[1,5-a]quinoline-2,3-dicarbo- nitrile	EtOH	231(4.30),257(4.49), 299(4.10),303(4.10), 310(4.13),324(3.99), 339(4.02)	95-0440-84
$C_{13}H_7BrClNO$			
Benzo[b]cyclohept[e][1,4]oxazine, 6-bromo-2-chloro-	MeOH	229(4.26),264(4.38), 274(4.35),310(3.81), 415(3.99)	138-1145-84
	MeOH-HCl	233(4.32),274(4.36), 325(3.90),440(3.91)	138-1145-84

Compound	Solvent	$\lambda_{max}(\log \epsilon)$	Ref.
$C_{13}H_7ClO$ 1-Acenaphthylenecarboxaldehyde, 2-chloro-	EtOH	246(4.18),335(4.05)	78-2959-84
$C_{13}H_7Cl_3O_2S$ Benzoic acid, 4,5-dichloro-2-[(4-chlorophenyl)thio]-	MeOH	229(4.50),262.5(4.13), 281s(3.92),327(3.61)	73-0992-84
Benzoic acid, 2-[(2,4,5-trichlorophenyl)thio]-	MeOH	252s(4.09),311.5(3.83)	73-2295-84
$C_{13}H_7NO_2$ Indeno[1,2-b]pyran-3-carbonitrile, 4,5-dihydro-4-oxo-	MeOH	221(4.06),240(3.94), 292(4.09)	83-0448-84
$C_{13}H_7N_3O_2$ 2H-Indol-2-one, 1-acetyl-3-(dicyanomethylene)-1,3-dihydro-	CHCl$_3$	266(4.03),362(4.07), 446(3.30)	39-1331-84B
$C_{13}H_7N_3O_3$ 9H-Pyrido[2,1-b]quinazoline-8-carbonitrile, 5,11-dihydro-7-hydroxy-9,11-dioxo-	EtOH	223(4.33),250s(3.98), 294(3.79),342(3.84)	83-0824-84
	EtOH-HCl	220(4.02),247s(3.70), 296(3.55),340(3.60)	83-0824-84
	EtOH-NaOH	242(4.55),260s(4.40), 296(4.17),350(4.05)	83-0824-84
$C_{13}H_7N_4$ 1,1,3,3-Propanetetracarbonitrile, 2-phenyl-, ion(1-)	H$_2$O	<u>227(4.2)</u>	44-0482-84
$C_{13}H_7N_5O$ Benzofuro[2',3':4,5]pyrrolo[1,2-d]-1,2,4-triazolo[3,4-f][1,2,4]triazine	dioxan	256(3.24),320(3.43)	73-0065-84
$C_{13}H_8BrNO_2S$ 2H-Pyran-3-carbonitrile, 6-(4-bromophenyl)-4-(methylthio)-2-oxo-	EtOH	243(4.03),262(4.18), 337(4.33),372(4.26)	94-3384-84
$C_{13}H_8BrNO_3S$ 2-Propenenitrile, 3-(5-bromo-2-furanyl)-2-(phenylsulfonyl)-	MeOH	210(3.11),235(2.86), 355(3.50)	73-2141-84
$C_{13}H_8BrN_3O_3$ 1H-Indazole-3,4,7(2H)-trione, 5-[(4-bromophenyl)amino]-	dioxan	261(4.29),356(3.77), 466(3.81)	4-0825-84
$C_{13}H_8BrN_3O_3S$ 2-Propenoic acid, 3-[4-bromo-5-(2-pyrimidinylthio)-2-furanyl]-2-cyano-, methyl ester	MeOH	235(3.25),277(2.90), 357(3.39)	73-0984-84
$C_{13}H_8ClFOS$ Benzaldehyde, 2-[(4-chloro-2-fluorophenyl)thio]-	MeOH	226(4.36),233(4.33), 252s(4.08),261s(4.03), 283s(3.77),328(--)	73-2531-84
$C_{13}H_8ClNO_2S$ 2H-Pyran-3-carbonitrile, 6-(4-chlorophenyl)-4-(methylthio)-2-oxo-	EtOH	243(4.07),261(4.23), 336(4.34),371(4.27)	94-3384-84

Compound	Solvent	λ_{max}(log ϵ)	Ref.
$C_{13}H_8ClNS$			
9(10H)-Acridinethione, 2-chloro-	50% EtOH	486(4.49)	103-1254-84
9(10H)-Acridinethione, 4-chloro-	acetone	480(4.45)	103-1254-84
$C_{13}H_8ClN_3S_2$			
4H-[1,2,4]Triazolo[3,4-c][1,4]benzo-thiazine, 7-chloro-1-(2-thienyl)-	EtOH	240(3.44),280(3.97), 290(2.80)	2-0445-84
$C_{13}H_8Cl_2N_2O_2$			
Spiro[9H-cyclopenta[1,2-b:4,3-b']di-pyridine-9,2'-[1,3]dioxolane], 2,7-dichloro-	EtOH	212(4.39),248(4.15), 323(3.97)	142-0073-84
$C_{13}H_8Cl_2O$			
2,5-Cyclohexadien-1-one, 2,6-dichloro-4-(2,4,6-cycloheptatrien-1-ylidene)-	MeCN	278(3.86),310(3.62), 541(4.33)	88-0077-84
$C_{13}H_8Cl_2O_2S$			
Benzoic acid, 4,5-dichloro-2-(phenyl-thio)-	MeOH	234.5(4.37),264(3.91), 270(3.91),329(3.51)	73-0992-84
$C_{13}H_8N_2$			
9H-Fluorene, 9-diazo-	MeOH	218(4.27),237(4.73), 256(4.01),267(3.91), 281(4.12),286(4.25), 291(4.25),302(4.09), 332(3.99),346(4.08), 462(1.61)	2-0802-84
$C_{13}H_8N_2OS$			
Imidazo[2,1-b]naphtho[1,2-d]thiazol-9(10H)-one	MeOH	242(4.77),305(4.19), 317(4.10),324(3.97)	83-0472-84
11H-Naphtho[2',1':5,6]pyrano[4,3-d]-1,2,3-thiadiazole	EtOH	262(4.63)	2-0203-84
$C_{13}H_8N_2O_2$			
Indeno[1,2-b]pyran-3-carbonitrile, 2-amino-4,5-dihydro-4-oxo-	MeOH	230(4.22),246(4.13), 310(4.18)	83-0448-84
Pyrano[3,2-b]indole-3-carbonitrile, 4,5-dihydro-5-methyl-4-oxo-	MeOH	224(4.44),250s(3.89), 300(4.22),340s(3.65)	83-0443-84
$C_{13}H_8N_2O_2S$			
9(10H)-Acridinethione, 2-nitro-	acetone	489(4.45)	103-1254-84
$C_{13}H_8N_2O_5S$			
2-Propenenitrile, 3-(5-nitro-2-furan-yl)-2-(phenylsulfonyl)-	MeOH	209(3.05),248(3.07), 354(3.36)	73-2141-84
$C_{13}H_8N_3O$			
Pyridinium, 1-[5-(2,2-dicyanoethenyl)-2-furanyl]-, bromide	H_2O	370(3.30)	73-2485-84
$C_{13}H_8N_4$			
Benzo[f]cinnoline-2-carbonitrile, 1-amino-	DMF	297(4.40),360(3.78), 380(3.86)	5-1390-84
2,3-Pyrazinedicarbonitrile, 5-(4-meth-ylphenyl)-	EtOH	222(4.12),253(3.64), 311(4.31),335s(4.26)	44-0813-84
$C_{13}H_8N_4O$			
2,3-Pyrazinedicarbonitrile, 5-(4-meth-oxyphenyl)-	EtOH	229(4.14),260s(3.74), 351(4.37)	44-0813-84

Compound	Solvent	$\lambda_{max}(\log \epsilon)$	Ref.
$C_{13}H_8N_4O_2$ Pyrazolo[1,5-a]quinazoline-3-carbo- nitrile, 2-acetyl-4,5-dihydro-5-oxo-	EtOH	238(4.30),245(3.84)	4-1049-84
$C_{13}H_8N_4O_3$ Imidazo[1,2-a]pyridine, 2-(4-nitro- phenyl)-3-nitroso-	EtOH	223(3.93),252(3.80), 342(3.44),380(3.39), 664(1.81)	4-1029-84
$C_{13}H_8N_4O_5$ 1H-Indazole-3,4,7(2H)-trione, 5-[(4-ni- trophenyl)amino]-	DMSO	259(4.13),361(4.04), 439(3.92)	4-0825-84
$C_{13}H_8N_4O_7$ Benzamide, N-(2,4-dinitrophenyl)- 4-nitro-	dioxan-base DMF-base	440(4.46) 444(4.46)	104-0565-84 104-0565-84
$C_{13}H_8O$ 3H-Benz[cd]azulen-3-one	CH_2Cl_2	247(4.13),252s(--), 289(4.44),262s(--), 377(3.90),398(3.96), 594(3.07)	83-0984-84
$C_{13}H_8OS$ 9H-Thioxanthen-9-one	MeOH	254.5(4.65),284s(3.72), 296s(3.57),374(3.89)	73-0603-84
$C_{13}H_8O_2$ Indeno[1,2-b]pyran-3-carboxaldehyde, 4,5-dihydro-4-oxo-	dioxan	225(4.24),260(4.33), 286(4.23)	83-0448-84
$C_{13}H_8O_4$ 2H,8H-Benzo[1,2-b:3,4-b']dipyran-2,8- dione, 4-methyl-	EtOH	291(4.39)	94-1178-84
2H,8H-Benzo[1,2-b:5,4-b']dipyran-2,8- dione, 4-methyl-	EtOH	260(4.51),325.5(4.23), 341(4.21)	94-1178-84
Indeno[1,2-b]pyran-3-carboxylic acid, 4,5-dihydro-4-oxo-	MeOH	250(3.78),310(3.82), 360(4.05)	83-0448-84
2H,5H-Pyrano[3,2-c][1]benzopyran-2,5- dione, 4-methyl-	EtOH	238(3.89),256.5(4.05), 266.5(4.03),334.5(4.07), 346.5(4.10)	94-1178-84
$C_{13}H_8O_6$ 9H-Xanthen-9-one, 1,2,6,8-tetrahydroxy-	MeOH	203(4.57),236(4.36), 264(4.38),327(4.09)	94-2290-84
9H-Xanthen-9-one, 1,3,6,7-tetrahydroxy- (norathyriol)	EtOH	255(4.46),312(4.23), 365(4.16)	94-4455-84
9H-Xanthen-9-one, 1,3,5,8-tetrahydroxy-	MeOH	202(4.41),252(4.37), 276(4.16),333(4.14)	94-2290-84
$C_{13}H_9BrClNO_2$ 2,4,6-Cycloheptatrien-1-one, 2-bromo- 7-[(5-chloro-2-hydroxyphenyl)amino]-	MeOH	257(4.26),346(3.97), 418(4.19)	138-1145-84
	MeOH-NaOH	243(4.33),265(4.21), 345(3.84),425(4.11), 460(4.02)	138-1145-84
$C_{13}H_9BrN_2S$ 3-Pyridinecarbonitrile, 4-(4-bromo- phenyl)-1,2-dihydro-6-methyl-2- thioxo-	dioxan	311(4.08),428(3.28), 480(4.17)	104-1837-84

Compound	Solvent	$\lambda_{max}(\log \epsilon)$	Ref.
$C_{13}H_9ClN_2S$ 3-Pyridinecarbonitrile, 4-(4-chloro-phenyl)-1,2-dihydro-6-methyl-2-thioxo-	dioxan	324(4.00),349(4.01), 458(3.15)	104-1837-84
$C_{13}H_9ClO$ 2,5-Cyclohexadien-1-one, 2-chloro-4-(2,4,6-cycloheptatrien-1-ylidene)-	MeCN	269(3.83),506(4.33), 532s(4.30)	88-0077-84
$C_{13}H_9FN_2S$ 3-Pyridinecarbonitrile, 4-(3-fluoro-phenyl)-1,2-dihydro-6-methyl-2-thioxo-	dioxan	259(4.24),314(3.26), 426(3.59)	104-1837-84
3-Pyridinecarbonitrile, 4-(4-fluoro-phenyl)-1,2-dihydro-6-methyl-2-thioxo-	dioxan	265(3.96),317(3.92), 426(3.22)	104-1837-84
$C_{13}H_9F_3N_2O_2$ 2H-1-Benzopyran-3-carbonitrile, 7-(di-methylamino)-2-oxo-4-(trifluoro-methyl)-	pentane	443(4.61)	135-0674-84
$C_{13}H_9FeN_3$ Ferrocene, [(dicyanomethylene)amino]-	hexane	274(3.82),349(4.12), 614(3.50)	89-0521-84
$C_{13}H_9Li$ Lithium, 9H-fluoren-9-yl-(-)-sparteine in solution	toluene at -78°	347(4.00),418s(3.04), 438(3.15),465(3.00)	138-0757-84
$C_{13}H_9N$ 7bH-Cyclopent[cd]indene-2-carbonitrile, 7b-methyl-	EtOH	251(3.87),297(4.66), 329(3.84),470(3.15)	39-0165-84C
$C_{13}H_9NO$ 3H-Benz[e]indole-1-carboxaldehyde	MeOH	190(4.61),216(4.67), 249(4.23),273(4.48), 310(3.85),320(3.78)	103-0275-84
3H-Benz[e]indole-2-carboxaldehyde	MeOH	210(4.45),231s(4.57), 236(4.64),260(4.00), 286(4.20),323(4.26), 343(4.26),362(4.23)	103-0275-84
Benzo[a]cyclopropa[c]cycloheptene-1a(1H)-carbonitrile, 2,8b-dihydro-2-oxo-, cis	EtOH	209(4.00),233s(--), 292(3.69)	77-0334-84
6(5H)-Phenanthridinone (8λ,9ϵ)	n.s.g.	224(5.05),230(5.22), 236.5(5.03),258(4.12), 269(4.25),308(3.99), 321(3.81),335.5(3.86), ?(3.81)	42-0975-84
$C_{13}H_9NO_2$ 1,2-Acenaphthylenedione, mono(O-methyl-oxime)	hexane	226(3.80),276(3.16), 304(2.90),352(2.18)	54-0023-84
9H-Carbazole-3-carboxaldehyde, 2-hy-droxy- (mukonal)	EtOH	234(4.42),247(4.21), 278(4.54),297(4.58), 342(4.06)	102-0471-84
$C_{13}H_9NO_2S$ 1H-Indeno[1,2-b]pyridine-2,5-dione, 4-(methylthio)-	EtOH	234(4.41),283(4.42), 304(4.23),336(3.83)	95-0127-84

Compound	Solvent	λ$_{max}$(log ε)	Ref.
C$_{13}$H$_9$NO$_2$S			
2H-Pyran-3-carbonitrile, 4-(methyl-thio)-2-oxo-6-phenyl-	EtOH	239(4.05),255(4.22), 332(4.29),368(4.23)	94-3384-84
C$_{13}$H$_9$NO$_3$			
5H-[1]Benzopyrano[3,4-c]pyridin-5-one, 7-methoxy-	EtOH	277(4.14)	44-0056-84
5H-[1]Benzopyrano[4,3-b]pyridin-5-one, 7-methoxy-	EtOH	273(4.24)	44-0056-84
Indeno[1,2-b]pyran-3-carboxaldehyde, 2-amino-4,5-dihydro-4-oxo-	dioxan	236(4.27),250(4.29), 305(4.30)	83-0448-84
Indeno[1,2-b]pyran-3-carboxaldehyde, 4,5-dihydro-4-oxo-, 3-oxime, (E)-	MeOH	222(3.59),263s(3.79), 286s(3.77)	83-0448-84
2H-Pyran-3-carbonitrile, 4-methoxy-2-oxo-6-phenyl-	EtOH	226(4.17),248(4.27), 320(4.18)	94-3384-84
Pyrano[3,2-b]indole-3-carboxaldehyde, 4,5-dihydro-5-methyl-4-oxo-	dioxan	232(4.53),325s(3.99)	83-0443-84
C$_{13}$H$_9$NO$_4$			
Pyrano[3,2-b]indole-3-carboxylic acid, 4,5-dihydro-5-methyl-4-oxo-	MeOH	225(4.39),250s(3.89), 305(4.21),340s(3.59)	83-0443-84
C$_{13}$H$_9$NS			
9(10H)-Acridinethione	toluene	480(4.36)	103-1254-84
	H$_2$O	474(4.53)	103-1254-84
	50% EtOH	478(4.49)	103-1254-84
	isoPrOH	479(4.48)	103-1254-84
	acetone	478(4.46)	103-1254-84
	dioxan	480(4.41)	103-1254-84
	MeCN	477(4.41)	103-1254-84
	CHCl$_3$	480(4.33)	103-1254-84
C$_{13}$H$_9$N$_3$			
Propanedinitrile, (2-phenyl-1H-pyrrol-3-yl)-	MeOH	208(4.15),270(4.12)	83-0143-84
C$_{13}$H$_9$N$_3$O			
Benzo[f]quinoxaline-5-carbonitrile, 2,3,4,6-tetrahydro-6-oxo-	CH$_2$Cl$_2$	233(4.25),266(4.21), 275s(--),398(3.61)	83-0855-84
Imidazo[1,2-a]pyridine, 3-nitroso-2-phenyl-	EtOH	227(3.74),260(3.84), 282(3.74),356(3.73), 650(2.22)	4-1029-84
C$_{13}$H$_9$N$_3$OS			
Benzo[b]furo[2',3':4,5]pyrrolo[1,2-d]-[1,2,4]triazine-1(2H)-thione	dioxan	380(3.38)	73-0065-84
4H-[1,2,4]Triazolo[3,4-c][1,4]benzo-thiazine, 1-(2-furanyl)-	EtOH	226(3.34),268(2.98), 290(2.76)	2-0445-8-
C$_{13}$H$_9$N$_3$O$_3$			
1H-Indazole-3,4,7(2H)-trione, 5-(phen-ylamino)-	EtOH	251(4.21),297(3.92), 444.5(4.08)	4-0825-84
C$_{13}$H$_9$N$_3$O$_5$			
Benzamide, N-(2,4-dinitrophenyl)-	dioxan-base	442(4.42)	104-0565-84
	DMF-base	445(4.45)	104-0565-84
C$_{13}$H$_9$N$_3$S			
Naphtho[1',2':4,5]thiazolo[3,2-b]-[1,2,4]triazole, 9-methyl-	MeOH	244(4.62),295(3.96), 330(3.08)	4-1571-84
Pyrazolo[1,5-a]quinoline-3-carboni-	EtOH	229(4.24),264(4.66)	95-0440-84

Compound	Solvent	$\lambda_{max}(\log \epsilon)$	Ref.
nitrile, 2-(methylthio)- (cont.)		292(3.96),313(4.03), 343(3.87)	95-0440-84
$C_{13}H_9N_4O$			
Pyridinium, 3-amino-1-[5-(2,2-dicyano-ethenyl)-2-furanyl]-, bromide	H_2O	231(3.30),257(3.13), 372(3.36)	73-2485-84
Pyridinium, 4-amino-1-[5-(2,2-dicyano-ethenyl)-2-furanyl]-, bromide	H_2O	230(2.79),400(3.23)	73-2485-84
$C_{13}H_9N_5O_2$			
Pyrano[3,2-b]indol-4(5H)-one, 5-methyl-2-(1H-tetrazol-5-yl)-	MeOH	214(4.24),266(4.04), 325(4.08),370s(3.54)	83-0443-84
$C_{13}H_{10}$			
1H-Cyclopent[e]azulene (30% 3H)	CH_2Cl_2	266(4.28),294(4.36), 340(3.43),356(3.49), 370(3.34),384(3.10), 580(2.63),624(2.59), 690(2.18)	35-6383-84
Fluorene	n.s.g.	260(4.28)	105-0467-84
$C_{13}H_{10}BrNO_2$			
Benzamide, N-(3-bromophenyl)-N-hydroxy-	MeOH	270(4.03)	118-0938-84
Benzamide, N-(4-bromophenyl)-N-hydroxy-	MeOH	272(4.12)	118-0938-84
$C_{13}H_{10}BrNO_2$			
1H-Pyrrole-3-carboxylic acid, 2-(4-bromophenyl)-4,5-dihydro-4,5-dioxo-, ethyl ester	dioxan	295(4.09),400(3.68)	94-0497-84
$C_{13}H_{10}ClNO_2S$			
2H-Thieno[2,3-h]-1-benzopyran-2-one, 3-chloro-4-(dimethylamino)-	EtOH	234(4.28),258s(4.26), 260(4.26),300s(3.92), 337(4.10)	161-0081-84
$C_{13}H_{10}ClNO_3S$			
1H-Indeno[1,7-cd]azepine-2-sulfonyl chloride, 2,9a-dihydro-9a-methyl-1-oxo-	EtOH	254(4.05),275s(3.95), 294s(3.71),326s(3.03)	39-0175-84C
1H-Indeno[1,7-cd]azepine-2-sulfonyl chloride, 2,9b-dihydro-9b-methyl-1-oxo-	EtOH	232(4.05),299(4.42), 484(3.02)	39-0175-84C
$C_{13}H_{10}ClNO_4S$			
6H-1,3-Thiazine-5-carboxaldehyde, 4-chloro-2-(3,4-dimethoxyphenyl)-6-oxo-	EtOH	218(4.16),246(3.98), 320(3.51),395(4.01)	103-1088-84
$C_{13}H_{10}Cl_2N_2O_3$			
Benzo[f]quinoxalin-6(2H)-one, 8,9-di-chloro-3,4-dihydro-7,10-dihydroxy-2(or 3)-methyl-	benzene	471s(3.20),508s(3.53), 550s(3.92),588(4.17), 635(4.14)	118-0953-84
$C_{13}H_{10}INO_2$			
Benzamide, N-hydroxy-N-(4-iodophenyl)-	MeOH	276(4.14)	118-0938-84
$C_{13}H_{10}N$			
Acridizinium	MeOH	260s(3.04),275(3.02)	48-0757-84
$C_{13}H_{10}N_2O$			
7H-Pyrrolo[3,4-b]pyridin-7-one, 5,6-	CH_2Cl_2	249(4.23),286(3.98)	83-0379-84

Compound	Solvent	$\lambda_{max}(\log \epsilon)$	Ref.
dihydro-2-phenyl- (cont.)			83-0379-84
$C_{13}H_{10}N_2OS$			
6H-Thiopyrano[3,4-d]pyrimidin-5(8H)-one, 2-phenyl-	EtOH	291.5(4.36)	4-1437-84
$C_{13}H_{10}N_2O_2$			
3H-Benz[e]indazole-1-carboxylic acid, methyl ester	EtOH	230(4.10),237(4.68), 255(4.25),270(4.20), 281(4.10),292(4.09), 317(3.84),333(3.80)	23-2506-84
3-Furancarbonitrile, 5-acetyl-4-amino-2-phenyl-	EtOH	253(4.18),281(4.17), 368(4.25)	5-1702-84
Propanoic acid, 2-diazo-3-(1H-inden-1-ylidene)-, methyl ester	EtOH	210(4.46),245(4.15), 252(4.17),287(4.10), 298(4.10),363(4.20)	23-2506-84
Pyrazolo[1,5-a]quinoline-3-carboxylic acid, methyl ester	EtOH	208(4.40),258(4.45), 265(4.47),320(3.84), 335(3.93)	23-2506-84
9H-Pyrido[3,4-b]indole-1-carboxylic acid, methyl ester	EtOH	240(4.26),258(4.26), 275(4.31),301(4.07), 370(3.83)	94-0170-84 +94-3579-84
$C_{13}H_{10}N_2O_3$			
Indeno[1,2-b]pyran-3-carboxaldehyde, 2-amino-4,5-dihydro-4-oxo-, 3-oxime, (E)-	MeOH	225(4.10),268(4.26), 330(4.14)	83-0448-84
Pyrano[3,2-b]indole-3-carboxaldehyde, 4,5-dihydro-5-methyl-4-oxo-, oxime, (E)-	MeOH	229(4.18),265(3.98), 300(4.11),335s(3.56)	83-0443-84
$C_{13}H_{10}N_2O_3S$			
2-Furancarboxylic acid, 5-(3-cyano-1,2-dihydro-6-methyl-2-thioxo-4-pyridinyl)-	dioxan	237(4.11),264(4.14), 316(4.35),462(3.23)	104-1780-84
$C_{13}H_{10}N_2O_4$			
Benzamide, N-hydroxy-N-(3-nitrophenyl)-	MeOH	266(4.36)	118-0938-84
$C_{13}H_{10}N_2O_6$			
1H-Pyrrole-3-carboxylic acid, 4,5-dihydro-2-(4-nitrophenyl)-4,5-dioxo-, ethyl ester	dioxan	233(4.10),271(4.15), 398(3.62)	94-0497-84
2,5-Pyrrolidinedione, 1-[[3-(2-nitroethenyl)benzoyl]oxy]-	CHCl₃	300(4.22)	94-5036-84
2,5-Pyrrolidinedione, 1-[[4-(2-nitroethenyl)benzoyl]oxy]-	CHCl₃	303(4.41)	94-5036-84
$C_{13}H_{10}N_2S$			
3-Pyridinecarbonitrile, 1,2-dihydro-6-methyl-4-phenyl-2-thioxo-	dioxan	262(3.60),314(3.64), 421(2.93)	104-1837-84
$C_{13}H_{10}N_2Se$			
3-Pyridinecarbonitrile, 1,2-dihydro-4-methyl-6-phenyl-2-selenoxo-	EtOH	217(4.14),280(4.07), 342(3.99),440(3.22)	70-2528-84
$C_{13}H_{10}N_3O_2$			
Pyridinium, 1-[5-(3-amino-2-cyano-3-oxo-1-propenyl)-2-furanyl]-, bromide	H₂O	242(3.93),259(3.83), 335s(4.08),370(4.25)	73-2485-84

Compound	Solvent	$\lambda_{max}(\log \epsilon)$	Ref.
$C_{13}H_{10}N_4$ 1H-Pyrazolo[3,4-b]quinoline-4-carbo- nitrile, 1,3-dimethyl-	EtOH	396(3.59)	18-2984-84
$C_{13}H_{10}N_4OS_2$ Ethanone, 1-(4,5,6,7-tetrahydro-1- phenyl-4,6-dithioxo-1H-pyrazolo- [3,4-d]pyrimidin-3-yl)-	EtOH	275(3.69),302(3.60)	4-1049-84
$C_{13}H_{10}N_4O_2$ Pyrrolo[2',3':4,5]furo[3,2-b]indole- 2-carboxylic acid, 1,9-dihydro-, hydrazide	dioxan	339(3.58)	73-1529-84
$C_{13}H_{10}N_4O_3$ Acetic acid, cyano[(2-methylphenyl)hy- drazono]-, 2-cyano-2-oxoethyl ester	EtOH	239(3.96),289s(3.48), 300(3.52),361(4.25)	5-1702-84
2,1,3-Benzoxadiazole, 4-(methylphenyl- amino)-7-nitro- (in 15% DMF)	pH 2.7	495(4.49)	48-0385-84
1H-Isoindole-1,3(2H)-dione, 2-[(2,5-di- hydro-3-methyl-5-oxo-1,2,4-triazin- 6-yl)methyl]-	EtOH	263s(--),291s(3.45), 301s(--)	39-0229-84C
Isoalloxazine, 6-acetyl-10-methyl-	pH 7.5	263(4.51),340(3.88), 433(3.03)	35-6778-84
Isoalloxazine, 8-acetyl-10-methyl-	pH 7.0	244(4.42),341(3.98), 442(4.01)	35-6778-84
1,3,7,8-Tetraazaspiro[4.5]deca-1,6,9- triene-10-carboxylic acid, 4-oxo-2- phenyl-	MeOH	233(4.20),258(3.91), 270(3.91)	39-2491-84C
$C_{13}H_{10}N_4O_3S$ 1,3,7,8-Tetraazaspiro[4.5]deca-6,9-di- ene-10-carboxylic acid, 4-oxo-3- phenyl-2-thioxo-	MeOH	275(4.25),302(3.80), 325s(3.52)	39-2491-84C
$C_{13}H_{10}N_6$ 1H-Benzotriazole, 1-(2H-benzotriazol- 2-ylmethyl)-	MeOH	279(4.23),286(4.19)	117-0299-84
1H-Benzotriazole, 1,1'-methylenebis-	MeOH	254(4.54),282(3.86)	117-0299-84
2H-Benzotriazole, 2,2'-methylenebis-	MeOH	282(4.37),289(4.37)	117-0299-84
$C_{13}H_{10}N_6O$ Pyridazino[4',5':3,4]pyrazolo[1,5-a]- quinazolin-5(6H)-one, 7-amino-10- methyl-	EtOH	233(5.05)	4-1049-84
$C_{13}H_{10}N_6O_3$ Pyrrolo[3,4-c]pyrazole-4,6(1H,5H)-di- one, 3-(azidoacetyl)-3a,6a-dihydro- 5-phenyl-	MeOH	242(4.37)	103-0416-84
Pyrrolo[3,4-d]-1,2,3-triazole-4,6(1H- 5H)-dione, 1-(3-diazo-2-oxopropyl)- 3a,6a-dihydro-5-phenyl-	MeOH	250(4.13),275(4.10)	103-0416-84
$C_{13}H_{10}O$ 3H-Benz[cd]azulen-3-one, 4,5-dihydro-, cis	CH_2Cl_2	239(4.26),263(4.08), 300s(--),308(4.43), 342s(--),356s(--), 371(4.04),496(2.63)	83-0984-84
2,5-Cyclohexadien-1-one, 4-(2,4,6- cycloheptatrien-1-ylidene)-	MeCN	250(3.83),270(3.85), 282s(3.77),320s(3.37),	88-0077-84

Compound	Solvent	λ_{max}(log ϵ)	Ref.
(cont.)		488(4.40)	88-0077-84
7bH-Cyclopent[cd]indene-2-carboxalde-hyde, 7b-methyl-	EtOH	257(4.07),312(4.51), 350(3.85),488(3.20)	39-0165-84C
Methanone, diphenyl-	C_6H_{12}	248(4.29),346(2.08)	88-4525-84
Propenal, 3-(4-azulenyl)-, (E)-	hexane	258(4.50),285(4.50), 310s(--),363(3.40), 383(3.27),405s(2.79), 610s(--),630(2.64), 663(2.59),700(2.51), 740s(--),780(2.11)	5-1905-84
Propenal, 3-(6-azulenyl)-, (E)-	CH_2Cl_2	250(4.09),316(4.89), 367(4.10),374(4.10), 385(4.28),500s(--), 651(2.57),712(2.47)	5-1905-84
$C_{13}H_{10}O_2$ 7bH-Cyclopent[cd]indene-2-carboxylic acid, 7b-methyl-	EtOH	250(3.89),299(4.62), 332(3.82),418s(2.15), 474(3.17)	39-0165-84C
$C_{13}H_{10}O_2S$ 1H-2-Benzopyran-1-one, 3,4-dihydro-3-(3-thienyl)-	EtOH	238(4.05),284(3.15)	44-0742-84
$C_{13}H_{10}O_3$ 2H,8H-Benzo[1,2-b:5,4-b']dipyran-2-one, 10-methyl-	EtOH	225(4.344),267(4.339), 348(4.036)	111-0365-84
1H-2-Benzopyran-1-one, 3-(2-furanyl)-3,4-dihydro-	EtOH	237(4.87),290(4.26)	44-0742-84
2H-1-Benzopyran-2-one, 3-(1-oxo-2-but-enyl)-	EtOH	305(3.54),340(3.43)	118-0859-84
Indeno[1,2-b]pyran-4(5H)-one, 3-(hy-droxymethyl)-	MeOH	224(4.11),290(4.31)	83-0448-84
4H,5H-Pyrano[3,2-c][1]benzopyran-4-one, 2-methyl-	EtOH	277(4.12),288(4.08), 337(3.90)	39-0503-84C
$C_{13}H_{10}O_4$ 2H,8H-Benzo[1,2-b:5,4-b']dipyran-2,8-dione, 3,4-dihydro-6-methyl-	EtOH	276.5(4.00),284(4.01), 318.5(4.06)	94-1178-84
2H,8H-Benzo[1,2-b:5,4-b']dipyran-2-one, 10-methoxy-	EtOH	227(4.491),268(4.543), 349(4.081)	111-0365-84
2,4(3H,5H)-Furandione, 3-acetyl-5-(phenylmethylene)-	EtOH	307(4.55)	39-1539-84C
	EtOH-NaOH	329(4.42)	39-1539-84C
2H-Pyran-2-one, 3-acetyl-4-hydroxy-6-phenyl- (pogopyrone B)	EtOH	212(4.17),242s(3.87), 276(3.62),340(4.11)	2-0611-84
2H-Pyran-2-one, 5-acetyl-4-hydroxy-6-phenyl-	EtOH	255(3.96),308(--)	39-1317-84B
	dioxan	260(4.03),315(--)	39-1317-84B
	$CHCl_3$	266(4.13),320(--)	39-1317-84B
	80% DMSO	255(4.05),310(--)	39-1317-84B
2H-Pyran-2-one, 5-benzoyl-4-hydroxy-6-methyl-	EtOH	248(4.49),270s(--)	39-1317-84B
	dioxan	250(4.50),280(--)	39-1317-84B
	$CHCl_3$	260(4.49),275s(--)	39-1317-84B
	80% DMSO	258(4.49),280s(--)	39-1317-84B
$C_{13}H_{11}BrN_2O$ Benzo[f]quinoxalin-6(2H)-one, 5-bromo-3,4-dihydro-4-methyl-	CH_2Cl_2	243(4.30),280(4.25), 435(4.50)	83-0743-84
$C_{13}H_{11}ClF_3N_3O$ 4-Pyrimidinamine, 5-chloro-2-methoxy-N-methyl-N-phenyl-6-(trifluoromethyl)-	MeOH	259(3.45),313(3.62)	4-1161-84

Compound	Solvent	$\lambda_{max}(\log \epsilon)$	Ref.
$C_{13}H_{11}ClF_3N_3S$			
4-Pyrimidinamine, 5-chloro-N-methyl-2-(methylthio)-N-phenyl-6-(trifluoromethyl)-	MeOH	255(3.99),326(3.53)	4-1161-84
$C_{13}H_{11}ClN_2$			
Benzenemethanamine, 4-chloro-α-(2-pyridinylmethylene)-, (Z)-	EtOH	236(4.11),267(3.93), 355(4.37)	142-1319-84
$C_{13}H_{11}ClN_2O_2S$			
6H-1,3-Thiazine-5-carboxaldehyde, 4-chloro-2-[4-(dimethylamino)phenyl]-6-oxo-	EtOH	230(3.94),317(3.32), 506(4.28)	103-1088-84
$C_{13}H_{11}ClN_2O_3$			
Benzo[f]quinoxalin-6(2H)-one, 5-chloro-3,4-dihydro-7,10-dihydroxy-2(or 3)-methyl-	benzene	444s(3.49),470s(3.59), 505(3.67),544(3.80), 585(3.92),632(3.72)	118-0953-84
$C_{13}H_{11}ClN_2S$			
1H-Pyrazole, 3-(4-chlorophenyl)-4,5-dihydro-5-(thienyl)-	EtOH	238(3.88),274(3.89)	34-0225-84
1H-Pyrazole, 5-(4-chlorophenyl)-4,5-dihydro-3-(thienyl)-	EtOH	220(4.11),257(3.79), 305(4.09)	34-0225-84
$C_{13}H_{11}F_3N_2O_4$			
Phenylalanine, α,β-didehydro-N-[N-(trifluoroacetyl)glycyl]-, (E)-	EtOH	206(4.12),282(4.11)	5-0920-84
(Z)-	EtOH	206(4.12),282(4.11)	5-0920-84
$C_{13}H_{11}F_3O_2$			
1,4-Hexadien-3-one, 6,6,6-trifluoro-5-hydroxy-1-(4-methylphenyl)-, (?,E)-	EtOH	243(3.88),311s(4.02), 365(4.31)	39-2863-84C
2,4-Pentanedione, 1,1,1-trifluoro-3-[(4-methylphenyl)methylene]-	EtOH	260(4.10),304(4.31)	39-2863-84C
$C_{13}H_{11}F_3O_3$			
1,4-Hexadien-3-one, 6,6,6-trifluoro-5-hydroxy-1-(4-methoxyphenyl)-, (?,E)-	EtOH	243(3.81),379(4.19)	39-2863-84C
$C_{13}H_{11}Li$			
Lithium, (diphenylmethyl)-(-)-sparteine also present	heptane	407(4.11)	138-0757-84
$C_{13}H_{11}NO$			
9H-Carbazol-3-ol, 6-methyl- (glycozolinol)	EtOH	225(4.30),255(4.02), 270(3.9),302(4.2)	2-0049-84
7bH-Cyclopent[cd]indene-2-carboxamide, 7b-methyl-	EtOH	245s(3.89),299(4.61), 333(3.81),471(3.06)	39-0165-84C
1H-Indeno[1,7-cd]azepin-1-one, 2,9a-dihydro-9a-methyl-	EtOH	270s(3.97),274(3.98), 286s(3.86),345s(2.11)	39-0175-84C
1H-Indeno[1,7-cd]azepin-1-one, 2,9b-dihydro-9b-methyl-	EtOH	229(3.72),275(4.11), 343(3.41),550(2.16)	39-0175-84C
$C_{13}H_{11}NO_2S$			
3H-Pyrrolizine, 2-(phenylsulfonyl)-	MeCN	201(4.08),221(3.96), 269(3.42),345(4.18)	24-1424-84
3H-Pyrrolizine, 6-(phenylsulfonyl)-	MeCN	210(4.09),230(4.13), 269(4.03)	24-1424-84

Compound	Solvent	$\lambda_{max}(\log \epsilon)$	Ref.
$C_{13}H_{11}NO_3$			
Benzofuro[2,3-c]pyridin-1(2H)-one, 7-methoxy-2-methyl-	MeOH	235(4.2),300(4.0)	2-0478-84
Pyrano[3,2-b]indol-4(5H)-one, 3-(hydroxymethyl)-5-methyl-	MeOH	250(3.56),300(3.62), 330s(3.19)	83-0443-84
3H-Pyrano[4,3-c]isoxazol-3-one, 6,7-dihydro-6-methyl-4-phenyl-	EtOH	264(3.99),342(4.03)	142-0365-84
$C_{13}H_{11}NO_4$			
Furo[2,3-b]quinolin-6-ol, 4,7-dimethoxy- (delbine)	n.s.g.	241(4.17),250(4.19), 310(3.51),326(3.88)	100-0379-84
1H-Pyrrole-2-carboxylic acid, 4-(2-hydroxybenzoyl)-, methyl ester	EtOH	215(4.27),242(4.40), 295(4.09)	2-1048-84
1H-Pyrrole-3-carboxylic acid, 4,5-dihydro-4,5-dioxo-2-phenyl-, ethyl ester	dioxan	274(3.97),395(3.63)	94-0497-84
$C_{13}H_{11}NO_4S$			
2H-Pyran-3-carboxylic acid, 4-(methylthio)-2-oxo-6-(3-pyridinyl)-, methyl ester	EtOH	246(4.20),325(4.23)	94-1665-84
$C_{13}H_{11}NO_5$			
1,3-Dioxolo[4,5-g]quinoline-7-carboxylic acid, 5-ethyl-5,8-dihydro-8-oxo- (oxolinic acid)	2M H_2SO_4 pH 4 pH 8	342(2.85) 320(2.86),335(2.85) 325(3.19),338(3.20)	149-0057-84A 149-0057-84A 149-0057-84A
2H-Isoindole-2-propanoic acid, 1,3-dihydro- ,1,3-trioxo-, ethyl ester	EtOH	241s(3.92),293(3.36)	39-0229-84C
2H-Pyran-3-carboxylic acid, 4-methoxy-2-oxo-6-(3-pyridinyl)-, methyl ester	EtOH	233(4.35),266(3.80), 338(4.23)	94-1665-84
5H-Pyrano[3,2-f]-1,2-benzisoxazol-5-one, 4,9-dimethoxy-7-methyl-	EtOH	230(4.44),252(4.40), 260(4.37),272s(3.92), 315(3.79)	88-2953-84
$C_{13}H_{11}NO_5S$			
Furan, 2-[(4-methylphenyl)sulfonyl]-5-(2-nitroethenyl)-, (E)-	MeOH	210(3.18),255(3.20), 335(2.93)	73-2496-84
$C_{13}H_{11}N_3O$			
Acetamide, N-imidazo[1,5-a]quinolin-7-yl-	n.s.g.	211(4.30),263(4.45)	11-0038-84B
Phenol, 2-(2H-benzotriazol-2-yl)-4-methyl-	heptane	<u>295(4.1)</u>,340(4.3)	46-5544-84
$C_{13}H_{11}N_3OS$			
2-Butanone, 1-diazo-3-(2-quinolinylthio)-	MeOH	237(3.60),255(3.50), 280(4.11)	103-0505-84
2-Butanone, 1-diazo-3-(8-quinolinylthio)-	MeOH	210(4.45),255(4.88), 275(4.30)	103-0505-84
$C_{13}H_{11}N_3O_3$			
1,3-Benzenediol, 4-(5-methoxy-2H-benzotriazol-2-yl)-	CHCl$_3$	348(4.48)	126-2497-84
$C_{13}H_{11}N_3O_4$			
1,3,5-Benzenetriol, 2-(5-methoxy-2H-benzotriazol-2-yl)-	CHCl$_3$	352(4.54)	126-2497-84
2H-1-Benzopyran-4-acetic acid, 7-azido-2-oxo-, ethyl ester	EtOH	328.5(4.26)	94-3926-84
7H-Pyrido[2,1-b]quinazoline-6-carboxaldehyde, 5,8,9,11-tetrahydro-2-	CHCl$_3$	300(4.00),388(4.47)	4-0219-84

Compound	Solvent	λ_{max} (log ϵ)	Ref.
nitro-11-oxo- (cont.)	dioxan	224(4.05),232s(3.99), 297(4.02),384(4.45)	4-0219-84
	DMSO	306(4.02),396(4.39)	4-0219-84
7H-Pyrido[2,1-b]quinazoline-6-carbox-aldehyde, 6,8,9,11-tetrahydro-2-nitro-11-oxo-	EtOH	224(4.29),234s(4.24), 299(3.92),384(4.39)	4-0219-84
$C_{13}H_{11}N_3S$ 4-Pyridazinecarbonitrile, 5,6-dimethyl-3-(phenylthio)-	MeOH	211.5(4.27),233(4.15), 258(4.04),329(3.30)	73-1722-84
$C_{13}H_{11}N_5$ Cyanamide, (7,8-dihydro-2-phenylimida-zo[1,2-c]pyrimidin-5-yl)-	MeOH	218(4.35)	24-2597-84
$C_{13}H_{11}N_5O$ Benzofuro[2',3':4,5]pyrrolo[1,2-d]-[1,2,4]triazin-1(2H)-one, hydrazone	dioxan	325(3.27),429(2.77)	73-0065-84
7H-Purin-6-amine, N-benzoyl-7-methyl-	MeOH	235(4.14),283(4.00), 329(4.00)	150-3601-84M
$C_{13}H_{11}N_5O_2S$ Pyrimido[4,5-d]pyrimidine-2,4(1H,3H)-dione, 5-amino-7-(methylthio)-1-phenyl-	EtOH	240(--),282(--)	94-0122-84
$C_{13}H_{11}N_5O_6$ Benzenamine, 4-(1,4-dihydro-5,7-dini-tro-4-benzofurazanyl)-N-methyl-, N-oxide	MeOH	259(4.36),295(4.02), 365(3.84),475(4.40)	12-0985-84
monopotassium salt	MeOH	257(4.34),293(4.00), 363(3.82),473(4.38)	12-0985-84
Benzenamine, 4-(1,4-dihydro-5,7-dini-tro-4-benzofurazanyl)-2-methyl-, N-oxide	MeOH	247(4.17),291s(3.95), 365(3.79),475(4.33)	12-0985-84
monopotassium salt	MeOH	249(4.14),290s(3.91), 365(3.76),475(4.36)	12-0985-84
$C_{13}H_{11}N_7$ 1H-Pyrazolo[3,4-e][1,2,4]triazolo-[3,4-c][1,2,4]triazine, 3,8-di-methyl-1-phenyl-	dioxan	268(4.63),498(3.20)	4-1565-84
$C_{13}H_{12}BF_2NO_2S$ Boron, [2-[3-(dimethylamino)-1-oxo-2-propenyl]benzo[b]thiophen-3(2H)-onato-O,O']difluoro-, [T-4-(E)]-	CHCl$_3$	231(4.21),270(3.94), 280s(3.88),370(4.29), 415(4.61)	83-0531-84
$C_{13}H_{12}BF_2NO_3$ Boron, [2-[3-(dimethylamino)-1-oxo-2-propenyl]-3(2H)-benzofuranonato-O^2,O^3]difluoro-, [T-4-(E)]-	CHCl$_3$	227(3.90),250(3.91), 275(3.80),290(3.92), 335(3.55),420(4.83)	83-0526-84
$C_{13}H_{12}BrNO_2S$ 2-Thiophenecarboxylic acid, 3-amino-5-(4-bromophenyl)-	MeOH	316(4.18),355(3.97)	118-0275-84
$C_{13}H_{12}BrNO_4$ 2-Cyclopenten-1-one, 2-[1-(3-bromo-phenyl)-2-nitroethyl]-3-hydroxy-	MeOH	245(4.0)	73-1421-84

Compound	Solvent	$\lambda_{max}(\log \epsilon)$	Ref.
2-Cyclopenten-1-one, 2-[1-(4-bromo-phenyl)-2-nitroethyl]-3-hydroxy-	MeOH	246(4.18)	73-1421-84
$C_{13}H_{12}BrN_5O_3$ Hydrazinium 5-(4-bromoanilino)-4,7-di-hydro-4,7-dioxoindazol-3-olate	EtOH	258(4.21),297(3.88), 418(4.02)	4-0825-84
$C_{13}H_{12}ClNOS$ Phenol, 4-amino-2-[[(4-chlorophenyl)-thio]methyl]- (also metal chelates)	C_6H_{12}	230(4.13),310(3.30)	65-0142-84
$C_{13}H_{12}ClNO_2$ 5H-[1]Benzopyrano[3,4-c]pyridin-5-one, 9-chloro-1,2,3,4-tetrahydro-8-methyl-	MeOH	274(4.03),318(3.94)	4-1557-84
2-Propen-1-one, 1-(3-chloro-2-benzo-furanyl)-3-(dimethylamino)-, (Z)-	MeOH	230(3.81),249(3.72), 350(4.31)	83-0526-84
$C_{13}H_{12}ClNO_2S$ 2H-Thieno[2,3-h]-1-benzopyran-2-one, 3-chloro-4-(dimethylamino)-5,6-di-hydro-	EtOH	218s(4.01),223(4.02), 243(4.07),272s(3.60), 350(4.12)	161-0081-84
2-Thiophenecarboxylic acid, 3-amino-5-(4-chlorophenyl)-, ethyl ester	MeOH	302(4.35),350(3.96)	118-0275-84
$C_{13}H_{12}ClNO_4$ 2-Cyclopenten-1-one, 2-[1-(4-chloro-phenyl)-2-nitroethyl]-3-hydroxy-	MeOH	246(4.10)	73-1421-84
$C_{13}H_{12}ClN_3O_2$ Pyrano[2,3-e]indazol-2(7H)-one, 3-chloro-4-(dimethylamino)-7-methyl-	EtOH	219(4.48),258(4.32), 299(4.10),329(4.19)	4-0361-84
$C_{13}H_{12}ClN_5$ 1H-Purin-6-amine, N-(4-chlorophenyl)-2,8-dimethyl-	EtOH	301(4.54)	11-0208-84B
$C_{13}H_{12}Cl_2N_2OS$ 2,4-Cyclohexadien-1-one, 2,4-dichloro-6-(4,5,6,7-tetrahydro-4-methyl-2H-isothiazolo[2,3-a]pyrimidin-2-yli-dene)-	EtOH	237(4.35),284(3.95), 295(3.98),402(4.18)	39-1631-84C
$C_{13}H_{12}FNO_4$ 2-Cyclopenten-1-one, 2-[1-(4-fluoro-phenyl)-2-nitroethyl]-3-hydroxy-	MeOH	249(4.12)	73-1421-84
$C_{13}H_{12}FN_5$ 1H-Purin-6-amine, N-(2-fluorophenyl)-2,8-dimethyl-	EtOH	297(4.18)	11-0208-84B
$C_{13}H_{12}F_3N_3S$ 4-Pyrimidinamine, N-methyl-2-(methyl-thio)-N-phenyl-6-(trifluoromethyl)-	MeOH	245(4.01),308(3.52)	4-1161-84
$C_{13}H_{12}F_6O$ Benzene, 1-methyl-4-[3,3,3-trifluoro-1-[(2,2,2-trifluoroethoxy)methyl]-1-propenyl]-, (E)-	EtOH	236(3.82)	44-2258-84
(Z)-	EtOH	257(3.98)	44-2258-84

Compound	Solvent	$\lambda_{max}(\log \epsilon)$	Ref.
$C_{13}H_{12}FeO_2$ Ferrocene, (2-carboxyethenyl)-	CH_2Cl_2	464(3.09)	30-0192-84

$C_{13}H_{12}N$
  Pyridinium, 1-ethenyl-4-phenyl-,
    perchlorate
	H_2O	386(4.24)	142-0505-84

$C_{13}H_{12}NOS$
  2H-Pyrido[1,2,3-ef]-1,5-benzothiaze-
    pinium, 3,4-dihydro-2-methyl-3-oxo-,
    perchlorate
	10% MeCN	245(4.05),273(3.95),
		  321(3.51)
			103-0343-84

1H-[1,3]Thiazino[3,2-a]quinolinium,
    2,3-dihydro-3-methyl-2-oxo-,
    perchlorate
	10% MeCN	242(3.80),265(4.47),
		  271(4.47),352(4.5),
		  367(4.47)
			103-0508-84

$C_{13}H_{12}N_2$
  1H-Indole, 1-methyl-3-(1H-pyrrol-2-yl)-
	EtOH	212(4.19),245(4.30),
		  320(3.72)
			103-0058-84

Pyridine, 2-(3-phenyl-2-aziridinyl)-,
    trans
	EtOH	210(4.13),232(4.20),
		  267(3.88)
			142-1319-84

$C_{13}H_{12}N_2O$
  Benzo[f]quinoxalin-6(2H)-one, 3,4-di-
    hydro-5-methyl-
	CH_2Cl_2
	  + CF_3COOH
		<u>290(4.3),421(3.5)</u>
		  <u>290(4.3),320(4.0),</u>
		  528(3.2)
			83-0855-84
			83-0855-84

9H-Pyrido[3,4-b]indole, 7-methoxy-
    1-methyl- (harmine)
	neutral	241(4.44),301(4.03),
		  325(3.55),338(3.50)
			151-0355-84A

	acid	250(4.30),326(4.08), 360(3.71)	151-0355-84A
	base	241(4.43),301(4.11), 325(3.89),337(3.86)	151-0355-84A

8-Quinolinecarbonitrile, 5-ethoxy-
    2-methyl-
	EtOH	220(4.38),250(4.59),
		  320(3.94)
			39-1795-84C

$C_{13}H_{12}N_2OS$
  1H-[1]Benzothiepino[5,4-c]pyrazole,
    1-acetyl-4,5-dihydro-
	EtOH	268(4.20)	2-0736-84

$C_{13}H_{12}N_2O_2$
  Benzo[f]quinoxalin-6(2H)-one, 3,4-di-
    hydro-5-methoxy-
	CH_2Cl_2	236(4.23),282(4.33),
		  440(3.50)
			83-0855-84

4H-[1]Benzoxepino[3,4-c]pyrazol-4-one, 1,10-dihydro-1,3-dimethyl-	EtOH	276(3.89)	4-0301-84
4H-[1]Benzoxepino[3,4-c]pyrazol-4-one, 2,10-dihydro-2,3-dimethyl-	EtOH	274(4.11)	4-0301-84
1H-1,2-Diazepine-1-carboxylic acid, phenylmethyl ester	MeOH	216(4.18),357(2.45)	23-2440-84
Pyridinium, 1-[[(phenylmethoxy)carbo- nyl]amino]-, hydroxide, inner salt	MeOH	312(3.81)	23-2440-84

7H-Pyrido[2,1-b]quinazoline-6-carbox-
    aldehyde, 5,8,9,11-tetrahydro-11-oxo-
	dioxan	242s(3.91),253s(3.77),
		  268s(4.20),275(4.24),
		  283s(3.85),343(4.32),
		  354s(4.28)
			4-0219-84

	$CHCl_3$	276(3.89),283s(3.77), 347(4.39),357s(4.36)	4-0219-84

	DMSO	270(3.80),278(3.84),
		  286s(3.72),345(4.31),
		  360s(4.25)
			4-0219-84

7H-Pyrido[2,1-b]quinazoline-6-carbox-
    aldehyde, 6,8,9,11-tetrahydro-11-oxo-
	EtOH	223(4.29),275(3.81),
		  282s(3.71),345(4.26),
		  355s(4.23)
			4-0219-84

Compound	Solvent	$\lambda_{max}(\log \epsilon)$	Ref.
$C_{13}H_{12}N_2O_3$			
Acetic acid, (2,3-dihydro-2-oxo-1-phenyl-4(1H)-pyrimidinylidene)-, methyl ester	n.s.g.	268(3.99),323(4.83)	4-0053-84
1H-Pyrrole-2,5-dione, 3-acetyl-4-amino-1-(phenylmethyl)-	MeOH	210(4.04),242(4.26), 360(3.52)	39-2097-84C
$C_{13}H_{12}N_2O_4$			
5H-[1]Benzopyrano[3,4-c]pyridin-5-one, 1,2,3,4-tetrahydro-8-methyl-9-nitro-	MeOH	262(4.35),314(3.85)	4-1557-84
$C_{13}H_{12}N_2O_4S$			
2-Thiophenecarboxylic acid, 3-amino-5-(4-nitrophenyl)-, ethyl ester	MeOH	320(4.15),394(3.82)	118-0275-84
$C_{13}H_{12}N_2O_4Se$			
4,6(1H,5H)-Pyrimidinedione, 5-[(3,4-dimethoxyphenyl)methylene]dihydro-2-selenoxo-	dioxan	273(3.99),438(4.58)	104-0141-84
$C_{13}H_{12}N_2O_5S$			
3-Thia-1-azabicyclo[3.2.0]heptane-2-carboxylic acid, 7-oxo-, (4-nitrophenyl)methyl ester, endo	EtOH	210(3.64),265(3.60)	39-2785-84C
exo	EtOH	212(3.64),265(3.77)	39-2785-84C
$C_{13}H_{12}N_2O_6$			
2-Cyclopenten-1-one, 3-hydroxy-2-[2-nitro-1-(3-nitrophenyl)ethyl]-	MeOH	250(4.17)	73-1421-84
2-Cyclopenten-1-one, 3-hydroxy-2-[2-nitro-1-(4-nitrophenyl)ethyl]-	MeOH	257(4.18)	73-1421-84
$C_{13}H_{12}N_2O_7S$			
3-Thia-1-azabicyclo[3.2.0]heptane-2-carboxylic acid, 7-oxo-, (4-nitrophenyl)methyl ester, 3,3-dioxide, cis-(±)-	EtOH	210(3.30),265(3.32)	39-2785-84C
$C_{13}H_{12}N_2S$			
1H-Pyrazole, 4,5-dihydro-3-phenyl-5-(thienyl)-	EtOH	229(3.78),284(3.85)	34-0225-84
$C_{13}H_{12}N_2S_2$			
Benzenamine, N-(4,6-dimethyl-5H-1,3-dithiolo[4,5-c]pyrrol-2-ylidene)-	EtOH	248(3.72),310(4.10)	12-2479-84
monohydrobromide	EtOH-HBr	245(3.72),343(4.24)	12-2479-84
$C_{13}H_{12}N_4$			
2-Butenedinitrile, 2-amino-3-[2-(4-methylphenyl)-1-aziridinyl]-, (Z)-	EtOH	300(4.17)	44-0813-84
$C_{13}H_{12}N_4O$			
2,3-Pyrazinedicarbonitrile, 1,4,5,6-tetrahydro-5-(4-methoxyphenyl)-	EtOH	224(4.34),324(4.01)	44-0813-84
4H-Pyrazolo[3,4-d]pyrimidin-4-one, 1,5-dihydro-1,6-dimethyl-5-phenyl-	EtOH	210.5(4.60),256(3.90)	11-0023-84A
4H-Pyrazolo[3,4-d]pyrimidin-4-one, 2,5-dihydro-2,6-dimethyl-5-phenyl-	EtOH	260.0(3.84)	11-0023-84A
1H-Pyrazolo[3,4-b]quinoline-4-carboxamide, 1,3-dimethyl-	EtOH	394(3.93)	18-2984-84

Compound	Solvent	$\lambda_{max}(\log \epsilon)$	Ref.
$C_{13}H_{12}N_4OS$			
6H-Purin-6-one, 1,2,3,7-tetrahydro-1-methyl-8-(phenylmethyl)-2-thioxo-	pH 13	234(4.51),285(4.35)	2-0316-84
$C_{13}H_{12}N_4O_2$			
Acetamide, N-(4,9-dihydro-2-methyl-9-oxopyrazolo[5,1-b]quinazolin-7-yl)-	10% EtOH-NaOH	219(4.31),249(4.38), 327(4.13),398(3.51)	56-0411-84
Benzo[g]pteridine-2,4(1H,3H)-dione, 3,7,8-trimethyl-	MeOH	220(4.60),250(4.59), c.259(4.55),343(3.93), 387(3.91)	39-0497-84C
1H-Pyrazole-3-carboxylic acid, 5-amino-4-cyano-1-phenyl-, ethyl ester	EtOH	229(3.74),280(3.20)	4-1049-84
4H-Pyrazolo[3,4-d]pyrimidin-4-one, 1,5-dihydro-5-(2-methoxyphenyl)-6-methyl-	EtOH	252.0(3.66)	11-0023-84A
$C_{13}H_{12}N_4O_3$			
Acetamide, N-(4,9-dihydro-3-hydroxy-2-methyl-9-oxopyrazolo[5,1-b]quin-azolin-7-yl)-	10% EtOH-NaOH	220(4.41),250(4.48), 331(4.33),404(3.64)	56-0411-84
$C_{13}H_{12}N_6O$			
3-Pyridinecarbonitrile, 2,4-diamino-6-methoxy-5-(phenylazo)-	n.s.g.	375(4.52)	49-1421-84
3-Pyridinecarbonitrile, 4,6-diamino-2-methoxy-5-(phenylazo)-	n.s.g.	385(4.34)	49-1421-84
$C_{13}H_{12}N_6O_3$			
9H-Purin-6-amine, 9-methyl-N-[(4-nitro-phenyl)methoxy]-	pH 1	276(4.30)	94-4842-84
	pH 7	272(4.35)	94-4842-84
	pH 13	286(4.30)	94-4842-84
	EtOH	272(4.29)	94-4842-84
6H-Purin-6-imine, 1,9-dihydro-9-methyl-1-[(4-nitrophenyl)methoxy]-, mono-perchlorate	pH 1	262(4.34)	94-4842-84
	pH 7	262(4.33)	94-4842-84
	pH 13	260(4.30),266(4.30) (unstable)	94-4842-84
	EtOH	261(4.34)	94-4842-84
$C_{13}H_{12}N_8$			
1H-Pyrazole, 1,1',1'',1'''-methanetetra-yltetrakis-	MeOH	218(4.36)	117-0299-84
$C_{13}H_{12}O$			
Benzenemethanol, α-phenyl-	n.s.g.	222(4.00),259(2.70)	110-0939-84
1,2,4-Methenodicyclobut[cd,f]inden-3(1H)-one, 1a,2,3a,4,4a,6a,6b,6c-octahydro-	C_6H_{12}	300(1.48)	138-1741-84
Tricyclo[6.2.2.1^{2,7}]trideca-2,4,6,9-tetraen-13-one	C_6H_{12}	220.5(4.06),239s(3.76), 262s(3.42),311s(2.38)	88-3611-84
$C_{13}H_{12}O_2$			
4-Azulenepropanoic acid	CH_2Cl_2	240(4.31),278(4.53), 282s(--),300s(--), 330(3.46),340(3.52), 353s(--),571(2.57)	83-0984-84
1,5-Azulenedione, 4,6,8-trimethyl-	MeOH	400(3.65)	138-0631-84
1,7-Azulenedione, 4,6,8-trimethyl-	MeOH	386(3.95)	138-0631-84
1,3-Benzodioxole, 3a,7a-dihydro-2-phenyl-	MeOH	202(4.23),252(3.64), 257(3.68),260(3.67), 266(3.62)	35-7310-84

Compound	Solvent	$\lambda_{max}(\log \epsilon)$	Ref.
4,6-Heptadienal, 2-oxo-7-phenyl-	EtOH	361(4.63)	39-1531-84C
	EtOH-NaOH	408(4.58)	39-1531-84C
2-Oxabicyclo[3.2.0]hept-6-en-3-one, 4-methyl-1-phenyl-, endo	EtOH	211(3.98)	150-3501-84M
exo	EtOH	225(3.91),264(3.72)	150-3501-84M
2(3H)-Oxepinone, 3-methyl-3-phenyl-	EtOH	202(4.15),240(3.71)	150-3501-84M
2(5H)-Oxepinone, 3-methyl-5-phenyl-	EtOH	202(c.4.28)	150-3501-84M
Spiro[cyclopropan-1,1'-[1H]indene]-2-carboxylic acid, methyl ester, cis	EtOH	227(4.42),245(3.76), 272(3.12),290(2.98)	23-2506-84
trans	EtOH	230(4.33),263(3.75), 290(3.27),300(3.18)	23-2506-84
Spiro[cyclopropane-1,9'-[1,4]methano-naphthalen]-2'(1'H)-one, 3',4'-di-hydro-3'-hydroxy-, (1'α,3'β,4'α)-	n.s.g.	255s(2.46),262s(2.70), 268(2.86),275(2.87), 316(2.86)	12-1035-84

$C_{13}H_{12}O_3$

Compound	Solvent	$\lambda_{max}(\log \epsilon)$	Ref.
3,5-Hexadienoic acid, 3-methyl-2-oxo-6-phenyl-, (E,E)-	MeOH	242s(3.76),247(3.80), 251s(3.76),346(4.56)	5-1616-84

$C_{13}H_{12}O_4$

Compound	Solvent	$\lambda_{max}(\log \epsilon)$	Ref.
4H-1-Benzopyran-4-one, 7-hydroxy-2-methyl-5-(2-oxopropyl)-	MeOH	243(4.32),250(4.34), 291(4.21)	94-3493-84
	MeOH-AlCl$_3$	243(4.34),250(4.37), 291(4.21)	94-3493-84
1,4-Epoxynaphthalen-1(4H)-ol, 5-meth-oxy-, acetate	MeOH	223(3.57),285(3.20), 291(3.16)	12-1699-84
1,4-Epoxynaphthalen-1(4H)-ol, 8-meth-oxy-, acetate	MeOH	223(3.62),240s(2.93), 285(3.14),291(3.13)	12-1699-84
2,4-Hexadien-1-one, 1-(6-hydroxy-1,3-benzodioxol-5-yl)-, (E,E)-	EtOH	203(4.28),239(3.97), 306(4.25),385(3.98)	78-4081-84
1,4-Naphthalenedione, 2,6-dimethoxy-8-methyl-	MeOH	266(4.25),290(4.17), 337(3.50),400(3.12)	44-1853-84
1,4-Naphthalenedione, 2,7-dimethoxy-5-methyl-	EtOH	263(4.36),291(4.13), 343(3.42),400(3.25)	44-1853-84
1,4-Naphthalenedione, 5,6-dimethoxy-2-methyl-	EtOH	260(4.32),392(3.59)	78-3455-84

$C_{13}H_{12}O_5$

Compound	Solvent	$\lambda_{max}(\log \epsilon)$	Ref.
2H-Cyclohepta[b]furan-2-one, 3-acetyl-5,6-dimethoxy-	MeOH	285(4.28),301(4.30), 414(4.28)	18-0621-84
1,4-Naphthalenedione, 2-hydroxy-7,8-dimethoxy-5-methyl-	MeOH	263(4.30),290(4.01), 361(3.52),408(3.42)	44-1853-84
1,4-Naphthalenedione, 8-hydroxy-2,7-dimethoxy-5-methyl-	MeOH	223(4.40),268(4.05), 295(3.98),450(3.67)	78-5039-84

$C_{13}H_{12}O_6$

Compound	Solvent	$\lambda_{max}(\log \epsilon)$	Ref.
2H-1-Benzopyran-5-carboxaldehyde, 3,4,8-trimethoxy-2-oxo-	EtOH	225(4.45),260(4.15), 301(4.21)	102-2094-84
4H-1-Benzopyran-6-carboxaldehyde, 7-hydroxy-5,8-dimethoxy-2-methyl-4-oxo-	EtOH	217(4.10),219s(4.09), 261(4.48),272(4.47), 300s(3.44),375s(3.08)	88-2953-84
2H-Cyclohepta[b]furan-3-carboxylic acid, 5,6-dimethoxy-2-oxo-, methyl ester	MeOH	278(4.36),294(4.32), 402(4.26)	18-0621-84
1,4-Naphthalenedione, 5-hydroxy-2,3,7-trimethoxy-	EtOH	266(3.66),312(3.48), 430(3.04)	78-3455-84

$C_{13}H_{13}ClO_3$

Compound	Solvent	$\lambda_{max}(\log \epsilon)$	Ref.
Naphthalene, 7-chloro-1,2,4-trimethoxy-	MeOH	225(4.49),248.5(4.78), 313(3.80),355(3.62)	102-0313-84

Compound	Solvent	$\lambda_{max}(\log \epsilon)$	Ref.
$C_{13}H_{13}Cl_2N_3OS$			
1,2,4-Triazin-5(2H)-one, 3-[(3,4-dichlorophenyl)thio]-6-(1,1-dimethylethyl)-	MeOH	234(4.2)	98-0749-84
$C_{13}H_{13}Cl_3N_2O_6$			
Alanine, 2-methyl-N-[(2,2,2-trichloroethoxy)carbonyl]-, 4-nitrophenyl ester	ether	209s(3.96),266(4.00)	94-1373-84
$C_{13}H_{13}N$			
Benzonitrile, 2-(2-cyclohexen-1-yl)-	hexane	223(4.09),268(3.06), 270s(3.08),274(3.26), 283(3.37)	23-1785-84
Benzonitrile, 4-(2-cyclohexen-1-yl)-	hexane	234(4.29),241(4.24), 258(2.70),266(2.79), 271(2.60),273(2.62), 277(2.47)	23-1785-84
$C_{13}H_{13}NOS$			
Benzo[b]thiophen-3(2H)-one, 2-[3-(dimethylamino)-2-propenylidene]-	hexane	278(4.46),343(4.61), 480(4.21)	103-0951-84
	DMSO	359(4.20),513(4.73)	103-0951-84
Cyclohexanone, 2-thieno[2,3-b]pyridin-4-yl-	EtOH	231(4.42),273(3.80), 282s(3.78),293s(3.68)	4-1135-84
Phenol, 4-amino-2-[(phenylthio)methyl]- (also metal chelates)	C_6H_{12}	245(3.48),310(3.20)	65-0142-84
$C_{13}H_{13}NO_2$			
3-Azabicyclo[4.4.1]undeca-1,3,5,7,9-pentaene-2-carboxylic acid, ethyl ester	MeOH	242(4.25),267(4.15), 335(3.57),377(3.36)	44-1040-84
2H-Azirine-2-carboxylic acid, 3-phenyl-, 3-butenyl ester	C_6H_{12}	238(3.95)	44-0399-84
5H-[1]Benzopyrano[3,4-c]pyridin-5-one, 1,2,3,4-tetrahydro-8-methyl-	MeOH	284(4.00),313(4.02)	4-1557-84
1H-Carbazol-1-one, 2,3,4,9-tetrahydro-6-hydroxy-3-methyl-	EtOH	235(4.29),305(4.03)	2-0049-84
6H-Furo[2',3':2,3]cyclobut[1,2-d]oxazole, 3a,4,4a,5-tetrahydro-2-phenyl-	C_6H_{12}	242(4.05)	44-0399-84
Oxazole, 5-(3-butenyloxy)-2-phenyl-	C_6H_{12}	275(4.12),284(4.12), 290(4.11),307(3.83)	44-0399-84
1H-Pyrrole-3-carboxylic acid, N-methyl-2-phenyl-, methyl ester	MeOH	272(3.99)	44-3314-84
1H-Pyrrole-3-carboxylic acid, N-methyl-5-phenyl-, methyl ester	MeOH	264(4.16)	44-3314-84
$C_{13}H_{13}NO_2S$			
3H-Azepine, ?-methoxysulfinyl-phenyl-	EtOH	263(3.91)	39-0249-84
2-Thiophenecarboxylic acid, 3-amino-5-phenyl-, ethyl ester	MeOH	293(4.25),350(3.95)	118-0275-84
$C_{13}H_{13}NO_3$			
Acetic acid, [(2-cyano-2-phenylethenyl)oxy]-, ethyl ester	EtOH	264(4.26)	5-1702-84
Cyclohepta[b]pyrrole-3-carboxylic acid, 1,2-dihydro-1-methyl-2-oxo-, ethyl ester	EtOH	223(4.19),232s(4.16), 280(4.51),410s(4.03), 430(4.19)	150-0264-84S
1H-Indole-2-carboxylic acid, 5-acetyl-, ethyl ester	EtOH	266(4.83),311(4.04), 325(3.92)	103-0280-84

Compound	Solvent	$\lambda_{max}(\log \epsilon)$	Ref.
Isoquinolinium, 3,4-dihydro-, 1-formyl- 2-methoxy-2-oxoethylide	benzene MeCN	456(3.98) 440(3.89)	142-0021-84 142-0021-84
2-Propen-1-one, 3-(dimethylamino)-1-(3- hydroxy-2-benzofuranyl)-, (E)-	MeOH	228(3.71),242(3.70), 300(3.72),380(4.32)	83-0526-84
$C_{13}H_{13}NO_3S$ 2-Thiophenecarboxylic acid, 3-amino- 5-(4-methoxyphenyl)-, methyl ester	MeOH	307(4.30),341(4.08)	118-0275-84
$C_{13}H_{13}NO_3S_2$ Sulfoximine, N-[(1,3-dihydro-1,3-dioxo- 2H-inden-2-ylidene)(methylthio)meth- yl]-S,S-dimethyl-	EtOH	233(4.40),242(4.38), 294(3.94),303(4.03), 364(4.39)	94-2910-84
$C_{13}H_{13}NO_4$ 2-Cyclopenten-1-one, 3-hydroxy-2-(2-ni- tro-1-phenylethyl)-	MeOH	247(4.13)	73-1421-84
5(2H)-Isoxazolone, 4-(2-methoxybenzoyl)- 2,3-dimethyl-	EtOH	244(3.94),294(4.17)	44-4419-84
$C_{13}H_{13}N_2OS_2$ 5H-Imidazo[1,2-c]thiazol-4-ium, 6,7- dihydro-5-methyl-3-(methylthio)-6- oxo-1-phenyl-, perchlorate	HOAc	277(4.10),349(4.07)	103-0029-84
$C_{13}H_{13}N_3$ 1H-Pyrazole, 4,5-dihydro-3-phenyl-5- pyrrolyl-	EtOH	224(3.65),287(3.78)	34-0225-84
$C_{13}H_{13}N_3O_2Se$ 4,6(1H,5H)-Pyrimidinedione, 5-[[4-(di- methylamino)phenyl]methylene]dihydro- 2-selenoxo-	dioxan	281(4.07),333(3.79), 495(4.75)	104-0141-84
$C_{13}H_{13}N_3O_3$ Acetic acid, cyano[(2-methylphenyl)- hydrazono]-, 2-oxopropyl ester	EtOH	247(3.97),252s(3.92), 288(3.44),370(4.40)	5-1702-84
Azepino[2,1-b]quinazolin-12(6H)-one, 7,8,9,10-tetrahydro-2-nitro-	EtOH	213(4.43),220s(4.38), 323(4.12)	4-0219-84
2,3-Furandione, 5-acetyl-4-amino-, 3-[(2-methylphenyl)hydrazone]	EtOH	247(4.01),255(4.01), 263(4.00),294(4.05), 394(4.21),475(4.08)	5-1702-84
$C_{13}H_{13}N_3O_4S$ 4a,10a-(Epoxyethanoxy)-5H-pyrimido- [4,5-b][1,4]benzothiazine-2,4(1H- 3H)-dione, 3-methyl-	MeCN	220(4.40),304(3.60)	142-0013-84
3-Pyridinecarboxylic acid, 2-[[4-(eth- oxycarbonyl)-1H-imidazol-2-yl]- thio]-, methyl ester	MeOH	209(4.27),255(4.24), 295s(3.58)	142-0807-84
$C_{13}H_{13}N_3O_4S_2$ Thiourea, (1,3-dihydro-4-mercapto- 8,9-dimethoxy-1,3-dioxo-2H-2- benzazepin-2-yl)-	MeOH	247(4.51),227(4.16), 370(3.99)	103-0024-84
$C_{13}H_{13}N_5$ 7H-Purin-6-amine, 7,8-dimethyl-N- phenyl-	EtOH	225s(--),300(4.26)	11-0208-84B
7H-Purin-6-amine, 7,8-dimethyl-2- phenyl-	EtOH	241(4.36),283(4.10)	11-0208-84B

Compound	Solvent	$\lambda_{max}(\log \epsilon)$	Ref.
7H-Purin-6-amine, 7-methyl-N-(phenyl-methyl)-	MeOH	273s(4.18),279(4.20), 288(4.03)	150-3601-84M
9H-Purin-6-amine, 2,8-dimethyl-N-phenyl-	EtOH	232s(--),238(3.96), 299(4.44)	11-0208-84B
9H-Purin-6-amine, 8,9-dimethyl-2-phenyl-	EtOH	239(4.42),280(4.14)	11-0208-84B
6H-Purin-6-imine, 1,7-dihydro-7,8-di-methyl-1-phenyl-	EtOH	257s(--),265(4.13), 274s(--),296(3.70)	11-0208-84B
$C_{13}H_{13}N_5O$ 8H-Purin-8-one, 6-amino-7,9-dihydro-7-methyl-9-(phenylmethyl)-	MeOH	273(4.09)	150-3601-84M
$C_{13}H_{13}N_5O_3$ Hydrazinium 4,7-dihydro-4,7-dioxo-5-(phenylamino)-2H-indazol-3-olate	EtOH	251(4.27),294(3.94), 417(4.07)	4-0825-84
$C_{13}H_{13}N_5O_5$ 3H-Imidazo[2,1-i]purine-7-carboxalde-hyde, 3-β-D-ribofuranosyl-	H_2O	228(4.38),325(4.24), 335(4.23)	44-4021-84
$C_{13}H_{13}N_7O$ Acetic acid, 2-(7-methyl-5-phenyl-5H-pyrazolo[3,4-e]-1,2,4-triazin-3-yl)-hydrazide	EtOH	260(4.22),334(3.70)	4-1565-84
$C_{13}H_{13}P$ Phosphine, methyldiphenyl-	C_6H_{12}	251(3.95)	151-0249-84A
Phosphorin, 3-ethyl-5-phenyl-	EtOH	211(4.29),241(4.38), 272s(4.00),306(2.86)	24-0763-84
$C_{13}H_{14}$ Azulene, trimethyl- (4,6,8?)	CH_2Cl_2	245(4.29),287(4.63), 334(3.60),347(3.85), 539(2.71),572(2.64)	5-1905-84
1H-Cycloprop[e]-as-indacene, 2,3,4,5,6,7-hexahydro-	pentane	270(2.96),279(2.98)	44-3436-84
1,4-Methanonaphthalene, 1,4-dihydro-9,9-dimethyl-	n.s.g.	263(2.76),271(2.79), 273s(2.73),279(2.68)	12-1035-84
$C_{13}H_{14}Br_2O_2$ 1,3-Benzodioxole, 5,6-dibromohexahydro-2-phenyl-	hexane	204(4.20),245(2.29), 251(2.32),257(2.38), 263(2.27),267(1.94)	35-7310-84
$C_{13}H_{14}ClNO_2$ 2,4,6-Heptatrienoic acid, 7-(3-chloro-1H-pyrrol-2-yl)-2-methyl-, methyl ester, (E,E,E)-	EtOH	382(4.56)	39-1577-84C
$C_{13}H_{14}ClN_3O$ 4-Pyrimidinamine, 5-chloro-2-methoxy-N,6-dimethyl-N-phenyl-	MeOH	228(3.68),256(3.33), 299(3.75)	4-1161-84
$C_{13}H_{14}ClN_3O_2$ Pyrano[2,3-e]indazol-2(5H)-one, 3-chloro-4-(dimethylamino)-6,7-di-hydro-7-methyl-	EtOH	230(3.94),256(3.99), 355s(4.17),358(4.19)	4-0361-84
$C_{13}H_{14}ClN_3O_6$ 2H-Pyrimido[1,6-a]pyrimidine-2,6(7H)-	pH 7	249(4.31),307(3.98)	88-3471-84

Compound	Solvent	$\lambda_{max}(\log \epsilon)$	Ref.
dione, 4-(chloromethyl)-7-β-D-ribo-furanosyl- (cont.)			88-3471-84
$C_{13}H_{14}ClN_3S$			
4-Pyrimidinamine, 5-chloro-N,6-dimethyl-2-(methylthio)-N-phenyl-	MeOH	252(4.00),308(3.68)	4-1161-84
$C_{13}H_{14}Cl_2$			
Benzene, (2,6-dichloro-3-methyl-1,3-hexadienyl)-	hexane	240(3.0),323(3.0)	104-1930-84
$C_{13}H_{14}Cl_2N_4OS$			
1,2,4-Triazin-5(4H)-one, 4-amino-3-[(3,4-dichlorophenyl)thio]-6-(1,1-dimethylethyl)-	MeOH	213(4.4),294(4.0)	98-0749-84
$C_{13}H_{14}FeO_4$			
Iron, [η⁴-5,6-bis(methylene)bicyclo-[2.2.2]octan-2-ol]tricarbonyl-, endo	isooctane	220(4.37),290(3.33)	33-0986-84
exo	isooctane	210(4.33),249(3.32)	33-0986-84
Iron, tricarbonyl[(1,3,4,5-η)-1,6,6-trimethyl-2-oxo-4-cycloheptene-1,3-diyl]-	hexane	219(4.27),270(3.79),320(3.33)	24-3473-84
Iron, tricarbonyl[(1,3,4,5-η)-3,6,6-trimethyl-2-oxo-4-cycloheptene-1,3-diyl]-	hexane	270(3.76),320(3.25)	24-3473-84
Iron, tricarbonyl[(1,3,4,5-η)-3,7,7-trimethyl-2-oxo-4-cycloheptene-1,3-diyl]-	hexane	276(3.76),320(3.36)	24-3473-84
Iron, tricarbonyl[(1,3,4,5-η)-4,6,6-trimethyl-2-oxo-4-cycloheptene-1,3-diyl]-	hexane	272(3.76),316(3.27)	24-3473-84
$C_{13}H_{14}N_2$			
6H-Cyclohepta[b]quinoxaline, 7,8,9,10-tetrahydro-	MeOH	206(4.43),238(4.42),317(3.91)	56-0917-84
2-Pyridinamine, N-(2-phenylethyl)-	MeOH	239(4.05),261(2.84),312(3.69)	24-1523-84
$C_{13}H_{14}N_2O$			
Azepino[2,1-b]quinazolin-12(6H)-one, 7,8,9,10-tetrahydro-	EtOH	225(4.03),267(3.95),274s(3.92),296s(3.51),305(3.62),317(3.51)	4-0219-84
Benzo[6,7]cyclohepta[1,2-c]pyrazole, 2,4,5,6-tetrahydro-8-methoxy-	EtOH	265(4.18)	2-0736-84
6H-Cyclohepta[b]quinoxaline, 7,8,9,10-tetrahydro-, 5-oxide	MeOH	210s(4.26),245(4.78),326(4.00)	56-0917-84
3-Pyridinecarboxamide, 1,4-dihydro-1-(phenylmethyl)-	EtOH	354(3.82)	35-6778-84
	5% isoPrOH	358(3.86)	47-2169-84
3H-Pyrido[3,4-b]indole, 4,9-dihydro-7-methoxy-1-methyl- (harmaline)	EtOH	261(4.15),337(4.35),365(3.71)	151-0355-84A
	EtOH-acid	259(4.08),383(4.46)	151-0355-84A
	EtOH-base	261(4.38),337(4.46)	151-0355-84A
hydrochloride	EtOH	255(3.84),380(4.15)	151-0355-84A
Strychnocarpene	MeOH	227(4.25),242(4.07),305(4.14)	142-2277-84
$C_{13}H_{14}N_2OS$			
2,4-Cyclohexadien-1-one, 6-(4,5,6,7-tetrahydro-4-methyl-2H-isothiazolo-	EtOH	230(4.35),281s(3.97),288(4.02).387(4.18)	39-1631-84C

Compound	Solvent	$\lambda_{max}(\log \epsilon)$	Ref.
[2,3-a]pyrimidin-2-ylidene)- (cont.)	EtOH-HCl	277(4.19),286(4.19), 337(4.25)	39-1631-84C
$C_{13}H_{14}N_2O_2$			
5H-[1]Benzopyrano[3,4-c]pyridin-5-one, 9-amino-1,2,3,4-tetrahydro-8-methyl-	MeOH	240(4.34),285(4.02), 355(3.70)	4-1557-84
2-Pentene-1,4-dione, 2-amino-3-(1-iminoethyl)-1-phenyl-	MeOH	260(4.17),323(3.80)	32-0261-84
$C_{13}H_{14}N_2O_3$			
3H-[1]Benzoxepino[5,4-c]pyrazol-3-one, 2,3a,4,5-tetrahydro-8-methoxy-3a-methyl-	EtOH	231(4.02),289(4.12), 308(4.13)	2-0736-84
1,8-Naphthyridine-3-carboxylic acid, 1-ethyl-1,4-dihydro-7-methyl-4-oxo-, methyl ester	2M H_2SO_4 pH 4	300(3.06),309(2.99) 319(3.15),328(3.16)	149-0057-84A 149-0057-84A
$C_{13}H_{14}N_2O_4$			
5H-Pyrrolo[2,1-c][1,4]benzodiazepin-5-one, 1,2,3,11a-tetrahydro-2,8-dihydroxy-7-methoxy- (chicamycin B)	MeCN	232(4.31),260s(3.84), 318(3.48)	158-0191-84
$C_{13}H_{14}N_2O_5$			
Alanine, 2,3-didehydro-N-[N-(phenylmethoxy)carbonyl]glycyl]-	EtOH	207(4.18),243(3.81)	5-0920-84
$C_{13}H_{14}N_3O_2$			
1H-Pyrazolium, 1,2-dimethyl-3-[2-(4-nitrophenyl)ethenyl]-, iodide	EtOH	327(4.39)	103-1266-84
$C_{13}H_{14}N_4$			
3H-Pyrazolo[5,1-c]-1,2,4-triazole, 3,3,6-trimethyl-7-phenyl-	CH_2Cl_2	242(4.06),309(3.72), 339(3.68)	24-1726-84
1H-Tetrazole, 1-(1-cyclohexen-1-yl)-5-phenyl-	EtOH	236(4.04)	39-1933-84C
$C_{13}H_{14}N_4O_2S$			
1H-Imidazole-4-carboxylic acid, 5-[[(phenylamino)thioxomethyl]-amino]-, ethyl ester	MeOH	260s(4.23),297(4.51)	2-0316-84
$C_{13}H_{14}N_4O_4$			
11H-Furo[2',3':4,5]oxazolo[2,3-b]-pteridin-11-one, 6a,7,8,9a-tetra-hydro-7-hydroxy-8-(hydroxymethyl)-2,3-dimethyl-, [6aR-(6aα,7β,8α,9aβ)-	MeOH	232(4.03),262(3.96), 322(3.84)	136-0179-84F
$C_{13}H_{14}N_5$			
7H-Purinium, 6-amino-7-methyl-9-(phenylmethyl)-, iodide	MeOH	271(4.19)	150-3601-84M
$C_{13}H_{14}N_6O_2$			
Pyrazolo[4,3-c][1,2,5]benzotriazepine, 1,5-dihydro-3-methyl-7-nitro-1-propyl-	C_6D_6	309s(3.842),345s(3.933), 378(4.075),447s(3.316), 693(2.544)	103-0930-84
	$CDCl_3$	333(4.208),377(3.924), 461s(2.161),696s(2.00)	103-0930-84
	DMSO-d_6	341(4.278),401(3.876), 522s(3.320),542(3.342), 572s(3.230)	103-0930-84

Compound	Solvent	$\lambda_{max}(\log \epsilon)$	Ref.
$C_{13}H_{14}N_6O_4$			
1H-Imidazole-4-carboximidamide, 5-(formylamino)-1-methyl-N-[(4-nitrophenyl)methoxy]-	pH 1	267(4.16)	94-4842-84
	pH 7	216(4.25),268(4.08)	94-4842-84
	pH 13	271(4.28)	94-4842-84
	EtOH	266(4.17)	94-4842-84
$C_{13}H_{14}O$			
2aH-Cyclopent[cd]indene, 4a,7b-dihydro-2a-methoxy-7b-methyl-, (2aα,4aα,7bα)-	EtOH	309(3.66)	39-0165-84C
1,4-Methanonaphthalen-2(1H)-one, 3,4-dihydro-9,9-dimethyl-	n.s.g.	267(2.75),274(2.76), 290s(2.79),299(2.89), 304s(2.88),308s(2.85), 315s(2.82),320s(2.58)	12-1035-84
Tricyclo[6.2.2.1^{2,7}]trideca-2,4,6-trien-13-one	C_6H_{12}	219.5(3.93),238s(3.66), 264s(3.23),316s(2.18)	88-3611-84
	MeCN	212(--)	88-3611-84
$C_{13}H_{14}O_2$			
Benzeneacetic acid, α-(2-methyl-1-propenylidene)-, methyl ester	EtOH	250(3.93)	44-4344-84
3-Butenoic acid, 3-methyl-2-(phenylmethylene)-, methyl ester, (E)-	EtOH	279(4.21)	44-4344-84
(Z)-	EtOH	279(4.21)	44-4344-84
Cyclopent[a]inden-8(1H)-one, 2,3,3a,8a-tetrahydro-8a-methoxy-	MeOH	249(4.14),295(3.42)	44-2050-84
2,4-Pentadienoic acid, 4-methyl-3-phenyl-, methyl ester, (E)-	EtOH	256(4.02)	44-4344-84
(Z)-	EtOH	256(4.02)	44-4344-84
Spiro[cyclopropane-1,9'-[1,4]methanonaphthalene]-2',3'-diol, 1',2',3',4'-tetrahydro-, (1'α,2'α,3'α,4'α)-	n.s.g.	248s(2.28),254s(2.56), 250(2.83),255(3.01), 273(3.04)[sic]	12-1035-84
$C_{13}H_{14}O_3$			
2H-1-Benzopyran-5-carboxaldehyde, 8-methoxy-2,2-dimethyl-	EtOH	252(4.19),293(4.03), 347(3.51)	18-0442-84
2H-1-Benzopyran-8-carboxaldehyde, 5-methoxy-2,2-dimethyl-	EtOH	264(4.24),277(4.16), 291(4.14)	18-0442-84
Bullvalenecarboxylic acid, (hydroxymethyl)-, methyl ester	MeOH	225(3.78)	24-0633-84
1H-Cyclohepta[c]furan-8-carboxaldehyde, 4,7-dihydro-1,1,5-trimethyl-4-oxo-	EtOH	245(3.92),333(4.05)	33-1506-84
2-Cyclohexen-1-one, 2-hydroxy-3-(4-methoxyphenyl)-	ether	225(3.60),320(4.30)	44-2355-84
Ethanone, 1-(6-hydroxy-2,2-dimethyl-2H-1-benzopyran-7-yl)-	EtOH	238(4.26),294(4.14), 388(3.68)	18-0442-84
Ethanone, 1-(7-hydroxy-2,2-dimethyl-2H-1-benzopyran-6-yl)-	EtOH	227s(4.16),234(4.27), 257(4.54),291s(3.81), 347(3.79)	18-0442-84
2,4-Hexadien-1-one, 1-(2-hydroxy-4-methoxyphenyl)-, (E,E)-	EtOH	207(4.30),311(4,31), 334(4.23)	78-4081-84
Naphtho[2,3-d]-1,3-dioxol-5(6H)-one, 7,8-dihydro-8,8-dimethyl-	MeCN	272(3.90)	87-0306-84
2H-Naphtho[1,8-bc]furan-2,6(2aH)-dione, 3-ethyl-5,5a,8a,8b-tetrahydro-	MeOH	210(3.98)	78-4701-84
$C_{13}H_{14}O_4$			
4H-1-Benzopyran-4-one, 7-hydroxy-2-(2-hydroxypropyl)-5-methyl-	MeOH	243(4.35),250(4.37), 289(4.19)	94-3493-84
	MeOH-AlCl$_3$	243(4.35),250(4.37), 289(4.18)	94-3493-84

Compound	Solvent	$\lambda_{max}(\log \epsilon)$	Ref.
2-Cyclopenten-1-one, 4,5-dihydroxy-5-[(phenylmethoxy)methyl]-	EtOH	221(3.58)	39-2089-84C
7,14-Dioxadispiro[5.1.5.2]pentadeca-9,12-diene-11,15-dione	MeOH	214(4.15),357(1.11)	88-4471-84
1(4H)-Naphthalenone, 3,4,4-trimethoxy-	EtOH	240(4.19),283(3.98)	39-2069-84C
2(1H)-Naphthalenone, 1,1,4-trimethoxy-	EtOH	231(4.32),234s(4.30), 317(3.90)	39-2069-84C
$C_{13}H_{14}O_4S$			
2-Cyclohexen-1-one, 2-[[(4-methoxyphenyl)sulfonyl]oxy]-	ether	225(4.36),274(3.48), 313(2.07)	44-2355-84
$C_{13}H_{14}O_5$			
Allamycin	MeOH	214(4.36)	94-2947-84
1,3-Benzenedicarboxylic acid, 5-acetyl-2-methyl-, dimethyl ester	MeOH	224(4.23)	24-1620-84
Propanedioic acid, [(4-methoxyphenyl)methylene]-, dimethyl ester	EtOH	225(3.85),318(4.29)	39-2863-84C
$C_{13}H_{14}O_5S$			
Benzenesulfonic acid, 4-methoxy-, 6-oxo-1-cyclohexen-1-yl ester	ether	235(4.48),320(1.70)	44-2355-84
$C_{13}H_{14}S_2$			
Azulene, 4-[bis(methylthio)methyl]-	CH_2Cl_2	239(4.36),281(4.61), 283s(--),346(3.62), 358s(--),583(2.68)	83-0984-84
$C_{13}H_{15}BrN_2O_2$			
1H-Indole-2-carboxylic acid, 3-(2-aminoethyl)-6-bromo-, ethyl ester, monohydrochloride	EtOH EtOH-HCl	231(4.32),309(4.18) 228(4.32),306(4.18)	1-0709-84 1-0709-84
2H-Pyrrol-2-one, 5-(bromo-2-pyridinylmethyl)-1,5-dihydro-5-methoxy-3,4-dimethyl-	MeOH	205(4.35),264(3.76)	49-0357-84
2H-Pyrrol-2-one, 5-(bromo-3-pyridinylmethyl)-1,5-dihydro-5-methoxy-3,4-dimethyl-	MeOH	201(4.22),262(3.43)	49-0357-84
2H-Pyrrol-2-one, 5-(bromo-4-pyridinylmethyl)-1,5-dihydro-5-methoxy-3,4-dimethyl-	MeOH	204(4.34),262(3.58)	49-0357-84
$C_{13}H_{15}BrN_4O_4$			
1H-Indazole-4,7-dione, 5-amino-3-[[(aminocarbonyl)oxy]methyl]-1-(3-bromopropyl)-6-methyl-	MeOH	210(4.20),258(4.12), 290(4.13),481(3.42)	5-1711-84
2,4(1H,3H)-Pyrimidinedione, 1-[3-(5-bromo-3,4-dihydro-2,4-dioxo-1(2H)-pyrimidinyl)propyl]-5-ethyl-	pH 2 H_2O pH 12	276(4.18) 276(4.17) 275(4.07)	142-2739-84 142-2739-84 142-2739-84
$C_{13}H_{15}ClN_2O_5$			
1,2-Propanediol, 3-[(2-chloro-6,7-dimethoxy-4-quinazolinyl)oxy]-	EtOH	238(4.70)	4-1189-84
$C_{13}H_{15}Cl_2N_3O_2$			
Pyrano[2,3-e]indazol-2(3H)-one, 3,3-dichloro-4-(dimethylamino)-4,5,6,7-tetrahydro-7-methyl-	EtOH	255s(3.89),279(4.09)	4-0361-84
$C_{13}H_{15}FN_3O_4$			
1H-Imidazol-1-yloxy, 4-(4-fluoro-2-	n.s.g.	245(4.20)	88-5809-84

Compound	Solvent	$\lambda_{max}(\log \epsilon)$	Ref.
nitrophenyl)-2,5-dihydro-2,2,5,5-tetramethyl-, 3-oxide (cont.)			88-5809-84
$C_{13}H_{15}N$			
Benzonitrile, 2-(2,3-dimethyl-2-but-enyl)-	hexane	223(3.96),268s(2.93), 275(3.10),283(3.16)	23-1785-84
Benzonitrile, 3-(2,3-dimethyl-2-but-enyl)-	hexane	212(4.11),266s(2.88), 274(3.05),282(3.09)	23-1785-84
Benzonitrile, 2-(1,1,2-trimethyl-2-propenyl)-	hexane	226(3.95),270s(2.82), 282(3.07)	23-1785-84
Benzonitrile, 3-(1,1,2-trimethyl-2-propenyl)-	hexane	220(4.06),226(4.06), 230(4.03),266(2.82), 273(2.99),281(3.02)	23-1785-84
1H-Pyrrole, 1-methyl-2-(2-phenylethyl)-	MeOH	206(4.13),268(2.41)	107-0453-84
$C_{13}H_{15}NO$			
3aH-[1]Benzoxepino[5,4-b]pyrrole, 2,3,4,5-tetrahydro-3a-methyl-	EtOH	245(3.84)	39-0005-84C
1H-Carbazol-6-ol, 2,3,4,9-tetrahydro-3-methyl-	EtOH	225(4.32),280(3.84)	2-0049-84
Cyclopentanone, 2-[1-(phenylimino)-ethyl]-	EtOH	207(--),240(--), 350(3.38)	34-0358-84
	EtOH-NaOEt	350(3.42)	34-0358-84
Indeno[1,2-b]pyrrol-3a(3H)-ol, 2,4-di-hydro-2,2-dimethyl-	EtOH	244(4.22),285(3.54), 293(3.52)	12-0577-84
2H-Indol-2-one, 1,3-dihydro-1-(2-meth-yl-3-butenyl)-	MeOH	251.5(3.58)	83-0639-84
2H-Indol-2-one, 1,3-dihydro-1-(3-meth-yl-2-butenyl)-	MeOH	250.0(3.99),284.0(3.21)	83-0639-84
2H-Indol-2-one, 1,3-dihydro-3-(3-meth-yl-2-butenyl)-	MeOH	249.0(4.21),278.0(3.46)	83-0639-84
2H-Indol-2-one, 1,3-dihydro-5-(3-meth-yl-2-butenyl)-	MeOH	252.0(3.82),284.0(2.94)	83-0639-84
$C_{13}H_{15}NOS$			
Morpholine, 4-(3-phenyl-1-thioxo-2-propenyl)-, cis	benzene	385(1.89)	151-0101-84D
	EtOH	272(4.18),365(2.87)	151-0101-84D
trans	benzene	297(4.30),345(3.70)	151-0101-84D
	EtOH	230.5(4.10),297.5(4.34), 345.5s(3.70)	151-0101-84D
$C_{13}H_{15}NO_2$			
2,4,6-Heptatrienoic acid, 2-methyl-7-(1H-pyrrol-2-yl)-, methyl ester, (E,E,E)-	EtOH	392(4.61)	39-1577-84C
Indeno[1,2-b]pyrrol-3a(3H)-ol, 2,4-di-hydro-2,2-dimethyl-, 1-oxide	EtOH	229(3.86),292(4.16), 303(4.16),314(4.12)	12-0587-84
2(1H)-Isoquinolinecarboxylic acid, 3-methyl-, ethyl ester	EtOH	278(3.82)	78-0311-84
4H-Pyran-4-one, 2,3,5,6-tetrahydro-3-[(methylphenylamino)methylene]-	EtOH	249(4.32),336(4.00)	4-1441-84
$C_{13}H_{15}NO_2S_4$			
6H-1,3-Thiazine-6-thione, 2,3-dihydro-4-(3-methylphenyl)-5-(methylsulfon-yl)-2-(methylthio)-	EtOH	258(3.78),348(4.20)	4-0953-84
$C_{13}H_{15}NO_3$			
5H-Benz[3,4]cyclobuta[1,2]cyclohepten-1-ol, 4b,6,7,8,9,9a-hexahydro-2-ni-tro-, cis	MeOH	233(3.65),315(4.14)	87-0792-84

Compound	Solvent	$\lambda_{max}(\log \epsilon)$	Ref.
Benzoic acid, 2-[(3-oxo-1-butenyl)-amino]-, ethyl ester	EtOH	234(3.99),332(4.26), 351(4.40)	4-0759-84
1,3-Butanedione, 2-[(dimethylamino)-methylene]-1-(2-hydroxyphenyl)-	CHCl$_3$	247(4.13),295(3.84), 357(3.95)	2-0668-84
4H-1,3,5-Dioxazocine-7-carboxaldehyde, 5,8-dihydro-2-methyl-6-phenyl-	MeOH	297(4.15)	88-2731-84
[1,3]Dioxepino[5,6-d]isoxazole, 3a,4,8,8a-tetrahydro-6-methyl-3-phenyl-, endo	MeOH	244s(3.88),265(3.97)	88-2731-84
exo	MeOH	261(4.00)	88-2731-84
1H-Inden-1-one, 2,3-dihydro-2-(2-methyl-2-nitropropyl)-	EtOH	244(4.15),293(3.48)	12-0117-84
1H-Indole-3-carboxylic acid, 2-(1-hydroxyethyl)-5-methyl-, methyl ester	n.s.g.	215(4.57),227(4.34), 282(4.05),288(4.04)	103-0988-84
$C_{13}H_{15}NO_4$			
1H-Inden-1-one, 2,3-dihydro-2-hydroxy-2-(2-methyl-2-nitropropyl)-	EtOH	248(4.11),293(3.40)	12-0587-84
2-Pyridinecarboxylic acid, 5-(3-methoxy-3-oxo-1-propenyl)-4,6-dimethyl-, methyl ester, (E)-	MeOH	227(4.46),269(4.43), 301(4.14)	24-1620-84
1H-Pyrrolo[1,2-a]indole-9-carboxylic acid, 2,3,5,6,7,8-hexahydro-2-hydroxy-6-methyl-8-oxo-	CHCl$_3$	303.5(3.327)	44-5164-84
$C_{13}H_{15}NS$			
2H-[1]Benzothiepino[5,4-b]pyrrole, 3,3a,4,5-tetrahydro-3a-methyl-	EtOH	228(4.01),267(3.62)	39-0005-84C
[1]Benzothiopyrano[2,3-c]pyrrole, 5,6,7,8-tetrahydro-1,3-dimethyl-, perchlorate (4λ,5ε)-	EtOH	214(4.05),245(4.09), 319(3.90),593(3.83), ?(3.00)	12-1473-84
[2]Benzothiopyrano[3,4-c]pyrrole, 6,7,8,9-tetrahydro-1,3-dimethyl-, perchlorate	EtOH	245(4.22),283(3.76), 327(3.99),592(3.23)	12-1473-84
2-Butenenitrile, 3-[(1-methylethyl)-thio]-4-phenyl-	EtOH	208(4.27),264(4.13)	78-2141-84
2-Butenenitrile, 4-phenyl-3-(propyl-thio)-	EtOH	207(4.26),260(4.12)	78-2141-84
2-Cyclohexene-1-thione, 3-[(phenyl-methyl)amino]-	EtOH	288(4.28),380(4.29)	44-3314-84
3-Hexenenitrile, 4-methyl-3-(phenyl-thio)-	EtOH	206(4.00),246(3.76)	78-2141-84
$C_{13}H_{15}NS_2$			
Benzothiazole, 2-[(1,3-dimethyl-2-butenyl)thio]-	EtOH	226(4.32),284(4.05), 292(4.04),302(3.99)	39-0101-84C
Benzothiazole, 2-[(1-ethyl-2-methyl-2-propenyl)thio]-	EtOH	225(4.27),245s(3.89), 283(4.05),290(4.04), 300.5(3.97)	39-0101-84C
Benzothiazole, 2-[(2-methyl-2-pentenyl)thio]-	EtOH	225(4.35),245s(4.00), 275s(4.04),282(4.08), 292(4.06),300(3.98)	39-0101-84C
$C_{13}H_{15}NS_3$			
Benzothiazole, 2-[(1,3-dimethyl-2-butenyl)dithio]-	EtOH	224(4.33),272(4.05), 289s(3.92),299s(3.81)	39-0101-84C
Benzothiazole, 2-[(2-methyl-2-pentenyl)dithio]-	EtOH	224(4.33),272(4.05), 289s(3.92),299s(3.87)	39-0101-84C

Compound	Solvent	$\lambda_{max}(\log \epsilon)$	Ref.
$C_{13}H_{15}N_2$ 1H-Pyrazolium, 1,2-dimethyl-3-(2-phen- ylethenyl)-, iodide	EtOH	306(4.43)	103-1266-84
$C_{13}H_{15}N_3$ Quinoline, 2-(1-piperazinyl)- (hydrate)	MeOH	248(4.53),345(3.73)	24-1523-84
$C_{13}H_{15}N_3O$ 2-Pyrimidinamine, 4-methoxy-N,6-di- methyl-N-phenyl-	MeOH	224(3.21),266(3.35)	4-1161-84
$C_{13}H_{15}N_3OS$ 3(2H)-Thiazolecarboxamide, N,N,4-tri- methyl-2-(phenylimino)-	isoPrOH	254(3.80),297(3.89)	94-4292-84
5-Thiazolecarboxamide, N,N,4-tri- methyl-2-(phenylamino)-	isoPrOH	314(4.28)	94-4292-84
Urea, dimethyl(4-methyl-3-phenyl- 2(3H)-thiazolylidene)-	isoPrOH	302(4.20)	94-4292-84
$C_{13}H_{15}N_3O_3$ 5-Pyrimidinecarboxamide, tetrahydro- 4-(4-methoxyphenyl)-1-methyl-2-oxo-	EtOH	206(3.72),225s(3.68), 298(3.41)	103-0431-84
$C_{13}H_{15}N_3O_4$ 1H-Pyrrolo[1,2-a]indole-5,8-dione, 2,7-diamino-2,3-dihydro-1-hydroxy- 9-(hydroxymethyl)-6-methyl-	MeOH	245(4.16),309(4.00), 350s(3.51),530(2.85)	35-7367-84
$C_{13}H_{15}N_3O_4S$ Benzenesulfonamide, N-[5-(2-methoxy- ethoxy)-2-pyrimidinyl]-	MeOH	240.02(4.19)	83-0906-84
2H-Pyrimido[4,5-b][1,4]benzothiazine- 2,4(3H)-dione, 1,4a,5,10a-tetra- hydro-4a,10a-dimethoxy-3-methyl-	MeCN	220(4.40),304(3.60)	142-0013-84
$C_{13}H_{15}N_3S$ 4-Pyrimidinamine, N,6-dimethyl-2-(meth- ylthio)-N-phenyl-, monohydrochloride	MeOH	237(3.97),246(3.99), 291(3.70)	4-1161-84
$C_{13}H_{15}N_3S_2$ 3(2H)-Thiazolecarbothioamide, N,N,4- trimethyl-2-(phenylimino)-	isoPrOH	280(4.26),328s(3.49)	94-4292-84
5-Thiazolecarbothioamide, N,N,4-tri- methyl-2-(phenylamino)-	isoPrOH	292(4.14),340s(3.85)	94-4292-84
Thiourea, dimethyl(4-methyl-3-phenyl- 2(3H)-thiazolylidene)-	isoPrOH	295s(3.90),333(4.24)	94-4292-84
Thiourea, N,N-dimethyl-N'-(4-methyl- 2-thiazolyl)-N'-phenyl-	isoPrOH	279(4.16)	94-4292-84
$C_{13}H_{15}N_4$ Pyridinium, 1-methyl-2-[[[(1-methyl- 2(1H)-pyridinylidene)amino]methyl- ene]amino]-, iodide	MeCN	390(4.77)	152-0475-84
perchlorate	MeCN	389(4.70)	152-0475-84
Pyridinium, 1-methyl-2-[[(1-methyl- 2(1H)-pyridinylidene)methyl]azo]-, iodide	MeCN	380(4.55)	152-0475-84
$C_{13}H_{15}N_4O_4$ 1H-Pyrazolo[1,2-a]indazol-4-ium, 7-ami- no-9-[[(aminocarbonyl)oxy]methyl]-	MeOH	212(4.26),271(4.15), 278(4.15),480(3.55)	5-1711-84

Compound	Solvent	$\lambda_{max}(\log \epsilon)$	Ref.
2,3,5,8-tetrahydro-6-methyl-5,8-di-oxo-, bromide (cont.)			5-1711-84
$C_{13}H_{15}N_5$			
7H-Purin-6-amine, 8,9-dihydro-7-methyl-9-(phenylmethyl)-	MeOH	298(3.99)	150-3601-84M
$C_{13}H_{15}N_5O$			
2H-Cyclohepta[1,2-d:3,4-d']diimidazol-2-one, 8-(dimethylamino)-1,3-dihydro-1,4-dimethyl-	MeOH	251(4.25),283s(4.39), 294(4.62),302s(4.48), 315s(4.24),380(4.17), 420(4.24)	94-3873-84
$C_{13}H_{15}N_5O_3$			
Hydrazinecarboxamide, N,N,1-trimethyl-2-(3-quinoxalinylmethylene)-, N^2,N^2-dioxide	EtOH	238(4.15),314(4.37), 327s(4.42),338(4.45), 386(4.24)	33-1547-84
$C_{13}H_{15}N_5O_5$			
4(1H)-Pteridinone, 2-amino-1-(1,2-di-acetoxypropyl)-, [S-(R*,S*)]-	pH 0.0	230(4.11),248(4.09), 320(3.94)	5-1815-84
	pH 5.0	216(4.08),236(4.10), 277(4.21),345(3.80), 360s(3.74)	5-1815-84
	pH 10.0	257(4.40),280s(3.83), 364(3.88)	5-1815-84
β-D-Ribofuranuronic acid, 1-(6-amino-9H-purin-9-yl)-1-deoxy-2,3-O-(1-methylethylidene)-	H_2O	222(4.00),241(3.93), 326(4.66)	35-3370-84
$C_{13}H_{15}N_5O_6$			
9H-Imidazo[1,2-a]purin-9-one, 3,4-di-hydro-6-(hydroxymethyl)-3-β-D-ribo-furanosyl-	pH 1	222(4.4),278(4.0), 288s(3.9)	88-0247-84
	pH 7	220(4.4),282(4.0)	88-0247-84
	pH 13	240(4.5),282(3.7), 297(3.8)	88-0247-84
$C_{13}H_{16}$			
Benzene, [1-(1,1-dimethylethyl)-2-cy-clopropen-1-yl]-	heptane	221(3.37),254(2.73)	104-1515-84
$C_{13}H_{16}FN_3O_4$			
1H-Imidazole, 4-(4-fluoro-2-nitrophen-yl)-2,5-dihydro-1-hydroxy-2,2,5,5-tetramethyl-, 3-oxide	n.s.g.	273(4.10),295(4.08)	88-5809-84
$C_{13}H_{16}FeO_3$			
Iron, tricarbonyl[(1,2,3,4-η)-1,3,5,5-tetramethyl-1,3-cyclohexadiene]-	hexane	222(4.30),284(3.42)	24-3473-84
$C_{13}H_{16}N$			
Pyridinium, 2,3-dihydro-1,5-dimethyl-4-phenyl-, bromide	H_2O	331(4.14)	33-2037-84
	CH_2Cl_2	338(5.45)	33-2037-84
$C_{13}H_{16}N_2$			
Benz[cd]indol-4-amine, 1,3,4,5-tetra-hydro-N,N-dimethyl-	EtOH	222(4.49),280(3.76)	44-4761-84
Cyclohepta[c]pyrrol-6-amine, N,N,1,3-tetramethyl- (changing)	CH_2Cl_2	333(--),379(--), 390(--),557(--)	118-0119-84
perchlorate	CH_2Cl_2	315(4.30),376(4.11), 473(2.57)	118-0119-84

Compound	Solvent	$\lambda_{max}(\log \epsilon)$	Ref.
1H-Indene, 2-diazo-2,3-dihydro-1,1,3,3-tetramethyl-	hexane	205(4.21),224s(3.88), 262(3.43),269(3.40), 277(3.20)	24-0277-84
$C_{13}H_{16}N_2O$			
1-Butanone, 1-(2-methyl-1H-indol-3-yl)-, oxime	n.s.g.	226(4.40),275(3.98)	103-0055-84
1-Propanone, 1-(1,2-dimethyl-1H-indol-3-yl)-, oxime	n.s.g.	229(4.27),283(3.95)	103-0055-84
1-Propen-2-ol, 1-[4,5-dihydro-1-(phenylmethyl)-1H-imidazol-2-yl]-, (Z)-	EtOH	289(4.43)	39-2599-84C
2-Pyrrolidinone, 3-[(dimethylamino)-methylene]-1-phenyl-	isoPrOH	236s(3.59),254(3.53), 318(4.52)	103-0047-84
$C_{13}H_{16}N_2O_2$			
1,2-Cyclohexanedione, 4-methyl-, 1-[(4-hydroxyphenyl)hydrazone]	EtOH	248(4.35),320(2.88)	2-0049-84
2H-Inden-2-imine, 1,3-dihydro-1,1,3,3-tetramethyl-N-nitro-	C_6H_{12}	273(3.13)	118-0479-84
L-Tryptophan, 5-methyl-, methyl ester	MeOH	222(4.51),275(3.77), 286(3.77),296(3.63)	94-2126-84
$C_{13}H_{16}N_2O_2S$			
L-Tryptophan, 5-(methylthio)-, methyl ester	MeOH	229(4.43),251(4.11), 286(3.65)	94-2126-84
$C_{13}H_{16}N_2O_3$			
2H-Pyran-3,4-dione, tetrahydro-2-methoxy-6-methyl-, 4-(phenylhydrazone)	EtOH	380(4.10)	39-0733-84C
4-Quinazolinemethanol, 6,7-dimethoxy-,2-dimethyl-	EtOH	222s(4.40),242(4.57), 322s(3.82),332(3.85)	39-1143-84C
L-Tryptophan, 5-methoxy-, methyl ester	MeOH	220(4.42),276(3.80), 297(3.69),308(3.57)	94-2126-84
$C_{13}H_{16}N_2O_5S$			
Cyclopentanecarboxamide, N-[2-(aminosulfonyl)-5-methoxyphenyl]-2-oxo-	EtOH	228(4.45),253(4.09)	104-0534-84
$C_{13}H_{16}N_2S$			
2-Hexenenitrile, 3-[(2-aminophenyl)-thio]-4-methyl-	EtOH	207(4.39),250(4.10), 270(4.03)	78-2141-84
3-Hexenenitrile, 3-[(2-aminophenyl)-thio]-4-methyl-	EtOH	206(4.42),240(4.02), 265(4.37)	78-2141-84
4-Isoquinolinecarbonitrile, 2,3,5,6,7,8-hexahydro-1-propyl-3-thioxo-	dioxan	263(4.25),324(3.67), 418(2.68)	104-2216-84
$C_{13}H_{16}N_3O_4$			
1H-Imidazol-1-yloxy, 2,5-dihydro-2,2,5,5-tetramethyl-4-(2-nitrophenyl)-, 3-oxide	n.s.g.	250(4.20)	88-5809-84
1H-Imidazol-1-yloxy, 2,5-dihydro-2,2,5,5-tetramethyl-4-(4-nitrophenyl)-, 3-oxide	n.s.g.	245(4.16),345(3.96)	88-5809-84
$C_{13}H_{16}N_4O_2S$			
Ethanol, 2,2'-[[4-(2-thiazolylazo)-phenyl]imino]bis-	acetone 50% acetone	483(4.58) 498(4.59)	7-0819-84 7-0819-84
$C_{13}H_{16}N_4O_3$			
2,4(1H,3H)-Pteridinedione, 1,3-diethyl-7-(1-oxopropyl)-	MeOH	248(4.09),350(3.89)	5-1798-84

Compound	Solvent	$\lambda_{max}(\log \epsilon)$	Ref.
2,4(1H,3H)-Pteridinedione, 1,3,7-tri-methyl-6-(1-oxobutyl)-	MeOH	247(4.10),278(4.02), 329(4.07)	5-1798-84
$C_{13}H_{16}N_4O_3S$ Pyrazolo[1,5-a]pyridine-6-carboxamide, N,N-diethyl-2-(methylthio)-3-nitro-	EtOH	253(4.51),286(4.09), 368(4.23)	95-0440-84
$C_{13}H_{16}N_4O_4$ 2,4(1H,3H)-Pyrimidinedione, 1-[3-(3,4-dihydro-2,4-dioxo-1(2H)-pyrimidin-yl)propyl]-5-ethyl-	H_2O	268(4.27)	142-2739-84
$C_{13}H_{16}N_4O_5$ 1H-Indazole-4,7-dione, 5-amino-3-[[(am-inocarbonyl)oxy]methyl]-1-(3-hydroxy-propyl)-6-methyl-	MeOH	211(4.15),257(4.10), 291(4.10),482(3.36)	5-1711-84
$C_{13}H_{16}N_4O_6$ 2,4(1H,3H)-Pteridinedione, 6,7-dimeth-yl-3-β-D-ribofuranosyl-	pH 4.0	235(3.92),330(3.91)	136-0179-84F
	pH 11.0	214s(4.08),244(4.01), 275(4.05),365(3.75)	136-0179-84F
	MeOH	237(4.07),330(3.99)	136-0179-84F
$C_{13}H_{16}N_4O_7$ 2,5-Pyrrolidinedione, 1-[[[[2,5-dihy-dro-1-(1-methylethyl)-4-nitro-2-oxo-1H-pyrrol-3-yl]amino]acetyl]oxy]-	EtOH	367(4.19)	44-1130-84
$C_{13}H_{16}N_4S$ Benzenamine, N,N-diethyl-4-(2-thiazo-lylazo)-	acetone 50% acetone	486(4.57) 506(4.60)	7-0819-84 7-0819-84
$C_{13}H_{16}N_6$ 1H-Cyclohepta[1,2-d:3,4-d']diimidazole-2,8-diamine, $N^2,N^8,N^8,4$-tetramethyl-(paragracine)	MeOH	252(4.08),309s(4.69), 314(4.72),372(4.10), 409(4.11),422s(3.82)	94-3873-84
	MeOH-acid	245(4.13),301(4.74), 363(3.84),410(4.33), 423s(4.12)	94-3873-84
1H-Cyclohepta[1,2-d:3,4-d']diimidazole-2,8-diamine, $N^8,N^8,1,4$-tetramethyl-	MeOH	260(4.31),306(4.69), 315s(4.60),376(4.26), 414(4.38)	94-3873-84
	MeOH-acid	248(4.37),298(4.68), 303s(4.64),360(4.18), 400s(4.35),414(4.45)	94-3873-84
$C_{13}H_{16}O$ 5H-Benzo[3,4]cyclobuta[1,2]cyclohepten-1-ol, 4b,6,7,8,9,9a-hexahydro-, cis	MeOH	263s(2.83),270(2.87), 275(2.81)	87-0792-84
5H-Benzo[3,4]cyclobuta[1,2]cyclohepten-2-ol, 4b,6,7,8,9,9a-hexahydro-, cis	MeOH	286(3.48)	87-0792-84
7H-Benzocycloheptene, 8,9-dihydro-2-methoxy-6-methyl-	EtOH	264(4.24)	88-3007-84
Ethanone, 1-[4-(4-pentenyl)phenyl]-	n.s.g.	252(3.46)	39-1833-84B
4-Hexen-3-one, 5-(2-methylphenyl)-	EtOH	215(3.83),225(3.83), 286(4.03)	80-0857-84
4-Hexen-3-one, 5-(4-methylphenyl)-	EtOH	225(3.40)	80-0857-84
1,4-Methanonaphthalen-2-ol, 1,2,3,4-tetrahydro-9,9-dimethyl-, (1α,2α,4α)-	n.s.g.	240s(2.20),255s(2.52), 261(2.79),267(2.98), 273(3.03)	12-1035-84

Compound	Solvent	$\lambda_{max}(\log \epsilon)$	Ref.
1,4-Methanonaphthalen-2-ol, 1,2,3,4-tetrahydro-9,9-dimethyl-, (1α,2β,4α)-	n.s.g.	242s(2.28),247s(2.43), 253s(2.60),259(2.81), 266(2.98),272(3.01)	12-1035-84
$C_{13}H_{16}OS$			
Ethanone, 1-(3,4-dihydro-4,4-dimethyl-2H-1-benzothiopyran-6-yl)-	MeCN	237(3.80),242(3.79), 310(4.26)	87-1516-84
$C_{13}H_{16}O_2$			
5H-Benzo[3,4]cyclobuta[1,2]cyclohept-ene-1,4b-diol, 4b,6,7,8,9,9a-hexa-hydro-	MeOH	270(2.99),277(2.95)	87-0792-84
5H-Benzo[3,4]cyclobuta[1,2]cyclohept-ene-4,4b-diol, 4b,6,7,8,9,9a-hexa-hydro-	MeOH	270(2.97),279(2.95)	87-0792-84
1H-Cyclopent[cd]inden-2-ol, 2,2a,7a,7b-tetrahydro-2a-methoxy-7b-methyl-	EtOH	304(3.89)	39-0165-84C
Ethanone, 1-(3,4-dihydro-4,4-dimethyl-2H-1-benzopyran-6-yl)-	MeCN	222(4.11),272(4.18)	87-1516-84
Ethanone, 1-(1,2,3,4-tetrahydro-8-meth-oxy-2-naphthalenyl)-	EtOH	271.5(3.17),278.5(3.20)	44-5116-84
1,4-Methanonaphthalene-2,3-diol, 1,2,3,4-tetrahydro-9,9-dimethyl-, (1α,2α,3α,4α)-	n.s.g.	249s(2.23),255s(2.54), 261(2.83),267(3.01), 274(3.04)	12-1035-84
Spiro[bicyclo[4.2.1]nona-3,7-dien-9,2'-oxiran]-2-one, 3,3',3'-tri-methyl-, (1α,6α,9R*)-	EtOH	236(3.62)	33-1506-84
$C_{13}H_{16}O_3$			
3-Cycloheptene-1-carboxaldehyde, 6-(hydroxymethylene)-4-methyl-7-(1-methylethylidene)-5-oxo-, (Z)-	EtOH	259(3.62),316(3.61)	33-1506-84
Methanone, (2-acetoxy-1-cyclopenten-1-yl)-2-cyclopenten-1-yl-	hexane	243(4.05)	33-1154-84
Spiro[4.5]deca-6,9-diene-1-carboxylic acid, 4-methyl-8-oxo-, methyl ester, cis	EtOH	242(4.14)	94-0447-84
trans	EtOH	243(4.15)	94-0447-84
Spiro[2.4]hepta-4,6-diene-1-propanoic acid, α-ethenyl-β-hydroxy-, methyl ester	MeOH	198(3.81),224(3.88), 258s(3.36)	23-1709-84
$C_{13}H_{16}O_4$			
2-Buten-1-one, 1-(2,4,6-trimethoxy-phenyl)-, (E)-	MeOH	224(4.20),302(3.43)	94-0325-84
Cyclopenta[c]pyran-7-carboxaldehyde, 4-(dimethoxymethyl)-3,4-dihydro-3-methyl-	EtOH	236(3.97),257(3.90), 328(4.01),388(3.66)	78-0695-84
$C_{13}H_{16}S$			
2H-Indene-2-thione, 1,3-dihydro-1,1,3,3-tetramethyl-	hexane	504(0.90)	24-0277-84
$C_{13}H_{16}Si$			
Silane, trimethyl-1-naphthalenyl-	C_6H_{12}	282.5(3.86)	60-0341-84
Silane, trimethyl-2-naphthalenyl-	C_6H_{12}	279(3.71)	60-0341-84
Silane, trimethyl(3-phenyl-3-buten-1-ynyl)-	EtOH	264(2.57)	65-1400-84
$C_{13}H_{17}BrN_2O_5$			
2,4(1H,3H)-Pyrimidinedione, 1-[5-(4-	pH 1 and 7	267(3.98)	87-0680-84

Compound	Solvent	$\lambda_{max}(\log \epsilon)$	Ref.
bromo-3-oxobutyl)tetrahydro-4-hydr-oxy-2-furanyl]-5-methyl-, [2R-(2α,4β,5α)]- (cont.)	pH 13	267(3.86)	87-0680-84
$C_{13}H_{17}BrO$			
Phenol, 4-bromo-2-(1,1-dimethyl-2-prop-enyl)-3,6-dimethyl-	MeCN-KOH	312(4.18)	39-1259-84B
$C_{13}H_{17}ClN_2O_5$			
2,4(1H,3H)-Pyrimidinedione, 1-[5-(4-chloro-3-oxobutyl)tetrahydro-4-hy-droxy-2-furanyl]-5-methyl-, [2R-(2α,4β,5α)]-	pH 1	267(3.99)	87-0680-84
	pH 7	268(3.99)	87-0680-84
	pH 13	267(3.87)	87-0680-84
$C_{13}H_{17}F_2N_2O_8P$			
Thymidine, 5'-O-(5,5-difluoro-1,3,2-dioxaphosphorinan-2-yl)-, P-oxide	EtOH	266(3.98)	87-0440-84
$C_{13}H_{17}F_3N_2O_6$			
Uridine, 2'-deoxy-5-(3,3,3-trifluoro-1-methoxypropyl)-	acid	264(3.96)	87-0279-84
	pH 6.6	264(3.98)	87-0279-84
	base	263(3.85)	87-0279-84
$C_{13}H_{17}IN_2O_5$			
2,4(1H,3H)-Pyrimidinedione, 5-methyl-1-[tetrahydro-4-hydroxy-5-(4-iodo-3-oxobutyl)-2-furanyl]-, [2R-(2α,4β,5α)]-	pH 1	267(3.99)	87-0680-84
	pH 7	267(3.98)	87-0680-84
	pH 13	267(3.86)	87-0680-84
$C_{13}H_{17}N$			
1H-1-Benzazepine, 2,3-dihydro-1,2,3-trimethyl-, (R*,R*)-	MeOH	234(4.389),269(3.645), 328(3.440)	150-1441-84M
(R*,S*)-	MeOH	236(4.329),274(3.478), 335(3.446)	150-1441-84M
1H-1-Benzazepine, 2,3-dihydro-1,3,4-trimethyl-	MeOH	234(4.531),272(3.875), 328(3.732)	150-1441-84M
1H-1-Benzazepine, 2,3-dihydro-1,3,5-trimethyl-	MeOH	229(4.437),260(3.760), 308(3.462)	150-1441-84M
1H-1-Benzazepine, 2,3-dihydro-2,3,5-trimethyl-, (R*,R*)-	MeOH	226(4.434),258(3.748), 308(3.079)	150-1441-84M
(R*,S*)-	MeOH	225(4.431),260(3.778), 306(3.204)	150-1441-84M
1H-1-Benzazepine, 2-ethyl-2,3-dihydro-3-methyl-	MeOH	232(4.445),266(3.651), 326(3.481)	150-1441-84M
1H-1-Benzazepine, 3-ethyl-2,3-dihydro-5-methyl-	MeOH	226(4.521),260(3.919), 310(3.556)	150-1441-84M
Isoquinoline, 1-(1,1-dimethylethyl)-3,4-dihydro-	n.s.g.	208(4.30),247(3.87)	83-0488-84
$C_{13}H_{17}NO$			
Benzamide, N-(1-methylethenyl)-N-propyl-	MeOH	230(4.33),264(4.11), 312(3.90)	78-1835-84
Benzamide, N-1-propenyl-N-propyl-	MeOH	225(4.26),265(4.27), 304(3.88)	78-1835-84
5H-Benzo[3,4]cyclobuta[1,2]cyclohepten-1-ol, 4-amino-4b,6,7,8,9,9a-hexahy-dro-, cis	MeOH	292(3.75)	87-0792-84
2,4,6-Cycloheptatrien-1-one, 2-(1-methyl-4-piperidinyl)-, 2-butene-dioate, (Z)-	MeOH	243(4.18)	87-0875-84
Ethanone, 1-[4-(1-piperidinyl)phenyl]-	MeOH	335(4.43)	139-0253-84C

Compound	Solvent	$\lambda_{max}(\log \epsilon)$	Ref.
Pyridine, 1,2,3,6-tetrahydro-1,5-di-methyl-4-phenyl-, N-oxide	H_2O	231(3.97)	33-2037-84
$C_{13}H_{17}NOS_2$ Ethane(dithioic) acid, [(1-oxo-3-phen-ylpropyl)amino]-, ethyl ester	pH 7.0 pH 12	310(3.9) 300(4.1)	23-0763-84 23-0763-84
$C_{13}H_{17}NO_2$ 1H-[1,3]Oxazino[3,4-a]azepin-1-one, 7-(1,1-dimethylethyl)-3,4-dihydro-	MeOH	320(3.07)	119-0171-84
$C_{13}H_{17}NO_2S$ 1,3-Butadiene-1-sulfonamide, N,2-di-methyl-N-(phenylmethyl)-, (Z)-	EtOH	243(4.18)	70-2519-84
1,3-Butadiene-1-sulfonamide, 2,3-di-methyl-N-(phenylmethyl)-, (Z)-	H_2O or EtOH	240(3.54)	70-2519-84
$C_{13}H_{17}NO_3$ Ethanone, 1-(4-hydroxy-3-methoxyphen-yl)-2-(2-pyrrolidinyl)-, (±)-	EtOH	228(4.05),274(3.91), 301(2.87),346(3.21)	39-0825-84C
	EtOH-NaOH	209(4.39),246(3.87), 302s(3.59),345(4.32)	39-0825-84C
1H-Pyrrole-2-carboxylic acid, 2,3-di-hydro-3-oxo-1,2-di-2-propenyl-, ethyl ester	MeCN	328(4.02)	33-1535-84
1H-Pyrrole-2-carboxylic acid, 3-hydr-oxy-1,4-di-2-propenyl-, ethyl ester	MeCN	265(4.08)	33-1535-84
$C_{13}H_{17}NO_3S$ 4-Morpholinepropanoic acid, α-(2-thi-enylmethylene)-, methyl ester	EtOH	310(3.20)	73-1764-84
$C_{13}H_{17}NO_4$ α-D-erythro-Pentopyranoside, methyl 3,4-dideoxy-3-[[(2-hydroxyphenyl)-methylene]amino]-	EtOH	255(3.94),282(3.58), 318(3.38),408(3.18)	39-0733-84C
$C_{13}H_{17}NO_5$ 1-Azabicyclo[3.2.0]hept-2-ene-3-acetic acid, 2-carboxy-6-ethyl-7-oxo-, α-ethyl ester, sodium salt	H_2O	273(3.51)	44-5271-84
2,3-Butanediol, 2,3-dimethyl-, mono(4-nitrobenzoate)	pH 11	265(4.0)(changing)	35-4511-84
$C_{13}H_{17}NS$ Benzenecarbothioamide, N-(1-methyleth-enyl)-N-propyl-	hexane	302(3.95),398(2.63)	78-1835-84
Benzenecarbothioamide, N-1-propenyl-N-propyl-	hexane	298(3.80),409(2.42)	78-1835-84
$C_{13}H_{17}NS_2$ Benzothiazole, 2-[(1,1-dimethylbutyl)-thio]-	EtOH	221.5(4.28),284(3.98), 292s(3.97),302.5s(3.93)	39-0101-84C
Benzothiazole, 2-[(1-ethyl-2-methyl-propyl)thio]-	EtOH	227(4.31),245(3.99), 283(4.13),291(4.11), 301.5(4.05)	39-0101-84C
$C_{13}H_{17}N_3O_2$ Pyrido[3,4-d]pyrimidin-4(3H)-one, 6,8-diethyl-2,3-dimethyl-, 7-oxide	EtOH	266(4.4),335(3.4), 350(3.4)	95-1122-84

Compound	Solvent	$\lambda_{max}(\log \epsilon)$	Ref.
$C_{13}H_{17}N_3O_4$			
Butanoic acid, 3-[(4-methyl-2-nitro-phenyl)hydrazono]-, ethyl ester	EtOH	253(4.27),286(4.07), 448(3.76)	48-0829-84
1H-Imidazole, 2,5-dihydro-1-hydroxy-2,2,5,5-tetramethyl-4-(2-nitro-phenyl)-, 3-oxide	n.s.g.	250(4.10)	88-5809-84
1H-Imidazole, 2,5-dihydro-1-hydroxy-2,2,5,5-tetramethyl-4-(4-nitro-phenyl)-, 3-oxide	n.s.g.	247(4.14),340(3.94)	88-5809-84
$C_{13}H_{17}N_3O_5S$			
D-Cysteine, S-methyl-N-[1-oxo-3-(1,2,3-4-tetrahydro-6-methyl-2,4-dioxo-5-pyrimidinyl)-2-propenyl]-, methyl ester, (E)-	H_2O	303(4.38)	94-4625-84
L-	H_2O	303(4.36)	94-4625-84
D-Methionine, N-[1-oxo-3-(1,2,3,4-tetrahydro-6-methyl-2,4-dioxo-5-pyrimidinyl)-2-propenyl]-, (E)-	H_2O	303(4.32)	94-4625-84
$C_{13}H_{17}N_3O_6$			
7H-Pyrrolo[2,3-d]pyrimidine, 7-β-D-arabinofuranosyl-2,4-dimethoxy-	MeOH	218(4.45),258(3.84), 271(3.85)	5-0273-84
$C_{13}H_{17}N_5O_2$			
1H-Imidazole-4-carboximidamide, 5-ami-no-N-[(4-methoxyphenyl)methoxy]-1-methyl-	pH 1	228(4.22),278(4.01)	94-4842-84
	pH 7	225(4.30),266(4.05)	94-4842-84
	pH 13	266(4.05)	94-4842-84
	EtOH	268(4.13)	94-4842-84
$C_{13}H_{17}N_5O_3$			
2,4(1H,3H)-Pteridinedione, 6-(dimeth-ylamino)-1,3-dimethyl-7-(1-oxo-propyl)-	pH -2.0	257(4.18),356(3.80)	5-1798-84
	pH 5.0	230(4.30),296(4.16), 466(3.72)	5-1798-84
2,4(1H,3H)-Pteridinedione, 7-(dimeth-ylamino)-1,3-dimethyl-6-(1-oxo-propyl)-	pH -4.9	258(4.13),288(4.07), 339(4.03)	5-1798-84
	pH 6.0	241(4.44),308(4.19), 378(4.12)	5-1798-84
$C_{13}H_{17}N_5O_5$			
Acetamide, N-[9-[2-deoxy-β-D-erythro-pentofuranosyl]-6-methoxy-9H-purin-2-yl]-	MeOH	265(4.21)	69-5686-84
9H-Purin-6-amine, 9-[2-O-(1-oxopropyl)-β-D-arabinofuranosyl]-	MeOH	257(4.20)	87-0270-84
$C_{13}H_{17}N_5O_6$			
Kanagawamicin	pH 1	231(4.15),272(3.80), 297(3.84)	158-84-3
	H_2O	233(4.20),272(3.80), 298(3.84)	158-84-3
	pH 13	232(4.15),295(3.74)	158-84-3
6H-Purin-6-one, 2-amino-9-[(2,3-di-acetoxypropoxy)methyl]-1,9-dihydro-	H_2O	253(4.10),270s(3.94)	88-0905-84
$C_{13}H_{18}ClN_7O_7S$			
Ascamycin	H_2O	263(4.09)	158-0670-84
$C_{13}H_{18}Cl_2O_5$			
3-Octenedioic acid, 3,4-dichloro-2,2,7,7-tetramethyl-5-oxo-,	MeOH	211(3.49),261(3.57)	35-3551-84

Compound	Solvent	$\lambda_{max}(\log \epsilon)$	Ref.
1-methyl ester, (Z)- (cont.)			35-3551-84
$C_{13}H_{18}F_3NO_5S$			
4-Thiazolidineacetic acid, 2-[(1,1-di-methylethoxy)carbonyl]-3-(trifluoro-acetyl)-, methyl ester, cis-(±)-	EtOH	220(3.46)	39-2785-84C
trans	EtOH	214(3.69)	39-2785-84C
$C_{13}H_{18}NO$			
5H-Oxazolo[2,3-a]isoquinolinium, 2,3,6,10b-tetrahydro-4,10b-di-methyl-, iodide	MeOH	214(4.23),263(2.50), 271(2.45)	12-1659-84
$C_{13}H_{18}N_2O$			
Ethanone, 1-(3-acetyl-4-methylphenyl)-, 1-(dimethylhydrazone)	MeOH	230(4.31),247(4.08), 312(3.40)	24-1620-84
2-Pyrrolidinone, 5-[(3,4-dimethyl-1H-pyrrol-2-yl)methylene]-3,3-dimethyl-, (E)-	EtOH	204(4.04),299(4.24)	49-1443-84
2-Pyrrolidinone, 5-[(3,4-dimethyl-1H-pyrrol-2-yl)methylene]-4,4-dimethyl-, (Z)-	EtOH	205(4.01),296(4.15)	49-1443-84
$C_{13}H_{18}N_2OS$			
Morpholine, 4-[1-[(methylthio)imino]-2-phenylethyl]-	hexane	265(3.88)	39-2933-84C
$C_{13}H_{18}N_2O_4S$			
1-Azabicyclo[3.2.0]hept-2-ene-2-carbox-ylic acid, 3-[[2-(acetylamino)ethyl]-thio]-6,6-dimethyl-7-oxo-, mono-sodium salt	H_2O	300(3.93)	44-5271-84
$C_{13}H_{18}N_2O_5$			
Thymine, 1-(2,5,6,8-tetradeoxy-β-D-erythro-oct-7-ulofuranosyl)-	pH 1	267(3.99)	87-0680-84
	pH 7	266(4.00)	87-0680-84
	pH 13	267(3.87)	87-0680-84
$C_{13}H_{18}N_2S_2$			
Cyclohexanamine, N-(4,6-dimethyl-5H-1,3-dithiolo[4,5-c]pyrrol-2-ylidene)-	EtOH	248s(3.52),278(3.92)	12-2479-84
monohydrobromide	EtOH-HBr	248(3.36),319(4.19), 365s(3.36)	12-2479-84
$C_{13}H_{18}N_4O_3$			
2H-Purin-2-one, 3,7-dihydro-3,7-di-methyl-6-[(5-oxohexyl)oxy]-	EtOH	212(4.22),240(3.63), 293(3.99)	106-0429-84
$C_{13}H_{18}N_4O_4$			
7H-Pyrrolo[2,3-d]pyrimidin-2-amine, 7-(2-deoxy-α-D-erythro-pentofurano-syl)-4-methoxy-5-methyl-	MeOH	228(4.43),265(3.89), 288(3.81)	5-0708-84
β-	MeOH	228(4.42),265(3.89), 287(3.81)	5-0708-84
7H-Pyrrolo[2,3-d]pyrimidin-2-amine, 7-(2-deoxy-α-D-erythro-pentopyrano-syl)-4-methoxy-5-methyl-	MeOH	229(4.42),265(3.88), 287(3.81)	5-0708-84
β-	MeOH	229(4.42),265(3.88), 287(3.81)	5-0708-84

Compound	Solvent	$\lambda_{max}(\log \epsilon)$	Ref.
$C_{13}H_{18}N_4O_6$			
1H-Pyrazolo[4,3-d]pyrimidine-5,7(4H-6H)-dione, 1,4,6-trimethyl-3-β-D-ribofuranosyl-	pH 1 pH 7 pH 11	241s(3.65),288(3.73) 240s(3.71),288(3.77) 241s(3.78),288(3.81)	4-1865-84 4-1865-84 4-1865-84
$C_{13}H_{18}O$			
5(1H)-Azulenone, 2,3,6,7-tetrahydro-2,2,8-trimethyl-	EtOH	315(3.80)	23-1954-84
Bicyclo[7.3.1]trideca-1(13),9,11-trien-10-ol	0.4% NaOH	241(4.01),302(3.52)	78-3117-84
3-Buten-2-one, 4-(2,6,6-trimethyl-1,3-cyclohexadien-1-yl)-, trans	hexane	220s(3.9),345(4.1)	44-3424-84
cis	hexane	225(--),310(--)	44-3424-84
1H-Inden-1-one, 4-ethenyl-2,3,3a,6,7,7a-hexahydro-3a,7a-dimethyl-, trans-(±)-	EtOH	235(4.08)	88-4183-84
Phenol, 2-(1,1-dimethyl-2-propenyl)-3,6-dimethyl-	MeCN-KOH	310(4.18)	39-1259-84B
Spiro[3.7]undeca-5,7-dien-1-one, 9,9-dimethyl-	MeCN	237(2.74),300(2.43)	33-0774-84
Tricyclo[5.4.0.0^{1,8}]undec-5-en-9-one, 4,4-dimethyl-	pentane	214(4.06),286(1.98)	33-0774-84
Tricyclo[6.3.0.0^{1,5}]undec-6-en-4-one, 9,9-dimethyl-	pentane	304(2.52)	33-0774-84
$C_{13}H_{18}OS_2$			
Bicyclo[2.2.1]heptan-2-one, 3-(1,3-dithiolan-2-ylidene)-1,7,7-trimethyl-, (+)-	EtOH	322(4.29),350(2.88)	78-2951-84
$C_{13}H_{18}OS_3$			
Benzene, [3-(ethylthio)-2-(methylsulfinyl)-3-(methylthio)-2-propenyl]-, (Z)-	EtOH	278(4.03)	39-0085-84C
$C_{13}H_{18}O_2$			
Tricyclo[6.3.1.0^{1,6}]dodecanone, hydroxy-	n.s.g.	234(3.94)	77-0930-84
$C_{13}H_{18}O_2S$			
2(5H)-Thiophenone, 3,5-diethyl-4-hydroxy-5-(2-methyl-1,3-butadienyl)-, (E)-(+)- (thiotetromycin)	EtOH	238(4.48),300(3.67)	158-84-26
$C_{13}H_{18}O_2S_3$			
Benzenemethanol, α-[[[2-(ethylthio)-2-(methylthio)ethenyl]sulfinyl]methyl]-, (E)-	EtOH	273(4.07)	39-0085-84C
Ethane(dithioic) acid, [[2-hydroxy-2-(4-methylphenyl)ethyl]sulfinyl]-, ethyl ester	EtOH	317(3.95)	39-0085-84C
Ethane(dithioic) acid, [(2-hydroxy-2-phenylpropyl)sulfinyl]-, ethyl ester	EtOH	317(4.04)	39-0085-84C
$C_{13}H_{18}O_3$			
Benzeneacetic acid, 2-methoxy-, 1,1-dimethylethyl ester	EtOH	271(3.23),277(3.20)	39-1547-84C
Bicyclo[4.2.1]non-3-en-2-one, 7,8-dihydroxy-3-methyl-9-(1-methylethylidene)-, (exo,exo)-	EtOH	233(3.90)	33-1506-84
2-Pentanone, 5-(3,5-dimethoxyphenyl)-	EtOH	203(4.72),272(3.48), 279(3.45)	100-0828-84

Compound	Solvent	$\lambda_{max}(\log \epsilon)$	Ref.
$C_{13}H_{18}O_3S$			
3-Thiopheneacetic acid, α-cyclohexyl-α-hydroxy-, methyl ester, (S)-(-)-	MeOH	218s(3.73),233(3.81), 238s(3.80),244s(3.60)	22-0077-84
$C_{13}H_{18}O_3S_3$			
Benzenemethanol, α-[[[2,2-bis(methylthio)ethenyl]sulfonyl]methyl]-α-methyl-	EtOH	211(4.19),280(4.15)	39-0085-84C
Ethane(dithioic) acid, [[2-hydroxy-2-(4-methoxyphenyl)ethyl]sulfinyl]-, ethyl ester	EtOH	226(4.36),276(3.68), 283(3.68),316(4.11)	39-0085-84C
$C_{13}H_{18}O_4$			
2-Cyclohexene-1-carboxylic acid, 1-(1-hydroxy-2-propenyl)-2-methyl-4-oxo-, ethyl ester	EtOH	203s(3.23),244(4.00)	12-2037-84
1,3-Dioxolane-4-methanol, 4-(2-hydroxy-4-methylphenyl)-2,2-dimethyl-	MeOH MeOH-MeONa	275(3.44),281s(3.42) 291(3.85)	102-1947-84 102-1947-84
2H-Pyran-5-carboxylic acid, 3,4-dihydro-2,2-dimethyl-6-(1-methylethenyl)-4-oxo-, ethyl ester	EtOH	286(3.95)	4-0013-84
2H-Pyran-5-carboxylic acid, 3,4-dihydro-2,2-dimethyl-6-(1-propenyl)-4-oxo-, ethyl ester	EtOH	240(3.88),300(4.15)	4-0013-84
2H-Pyran-5-carboxylic acid, 3,4-dihydro-2,3-dimethyl-6-(1-propenyl)-4-oxo-, ethyl ester	EtOH	240(3.92),296(4.19)	4-0013-84
2H-Pyran-5-carboxylic acid, 3,4-dihydro-2-methyl-6-(2-methyl-1-propenyl)-4-oxo-, ethyl ester	EtOH	252(3.78),303(4.15)	4-0013-84
Spiro[bicyclo[4.2.1]non-3-ene-9,2'-oxiran]-2-one, 7,8-dihydroxy-3,3',3'-trimethyl-, (1α,6α,7α,8α,9R*)-	EtOH	238(3.88)	33-1506-84
$C_{13}H_{18}O_5$			
Acetic acid, (3-methoxy-5-oxo-4-pentyl-2(5H)-furanylidene)-, methyl ester, (Z)-	EtOH	266(4.27)	39-1555-84C
1,3-Cyclopentadiene-1-carboxaldehyde, 2-(2-hydroxy-3,3-dimethoxy-1-methylpropyl)-5-(hydroxymethylene)-	EtOH	246(3.82),332(3.65), 380(3.65)	78-0695-84
D-arabino-Oct-3-en-2-ulose, 3-acetyl-5,8-anhydro-1,3,4-trideoxy-6,7-O-(1-methylethylidene)-, (4R,5S,8R)-	EtOH	265(4.42)	136-0217-84C
7-Oxabicyclo[4.1.0]hept-2-ene-1,3-dicarboxylic acid, 6-ethyl-2-methyl-, dimethyl ester	MeOH	232(3.85)	44-3791-84
$C_{13}H_{18}O_6$			
Spiro[furan-2(3H),4'(3'aH)-furo[3,4-d][1,3]dioxole]-4-carboxylic acid, 6',6'a-dihydro-2',2',5-trimethyl, methyl ester, [3'aR-(3'aα,4'α,6'aα)]-	EtOH	245(4.20)	136-0217-84C
[3'aR-(3'aα,4'β,6'aα)]-	EtOH	245(4.42)	136-0217-84C
$C_{13}H_{18}O_6S_2$			
Dispiro[bicyclo[2.2.1]heptane-2,2'-oxirane-3',2"-[1,3]dithiolan]-3-one, 4,7,7-trimethyl-, 1",1",3",3"-tetraoxide, [1S-(1α,2β,4α)]-	EtOH	250(3.49),310(1.81), 390(1.19)	78-2951-84

Compound	Solvent	$\lambda_{max}(\log \epsilon)$	Ref.
$C_{13}H_{18}O_7$ 2,4,6-Cycloheptatrien-1-one, 2,3,4,5,6,7-hexamethoxy-	EtOH	264(4.48),345(3.83)	155-0309-84B
$C_{13}H_{19}ClN_2$ 12,13-Diazabicyclo[9.3.1]pentadeca- 1(15),11,13-triene, 14-chloro-	EtOH	216(3.65),265(3.19), 306(2.54)	32-0289-84
$C_{13}H_{19}N$ Pyridine, 2-cyclohexyl-5-ethyl-	EtOH	214(3.88),269(3.56)	103-0515-84
$C_{13}H_{19}NO$ 1H,8H-Benzo[ij]quinolizin-8-one, 2,3,5,6,7,7a,10a,10b-octahydro- 10b-(hydroxymethyl)-, (7aα,10aα,10bα)- (±)-	EtOH	229(3.84)	77-0714-84
2-Butanamine, N-[[2-(methoxymethoxy)- phenyl]methylene]-	EtOH	247(4.15),298(3.64)	150-0701-84M
$C_{13}H_{19}NO_2S$ 2-Propenoic acid, 2-[(diethylamino)- methyl]-3-(2-thienyl)-, methyl ester	EtOH	312(3.23)	73-1764-84
2-Propenoic acid, 2-[[(1,1-dimethyl- ethyl)amino]methyl]-3-(2-thienyl)-, methyl ester	EtOH	311(3.20)	73-1764-84
$C_{13}H_{19}NO_3$ 3aH-Indene-3a-carboxylic acid, 1,2,3,4,5,6-hexahydro-6-(methyl- imino)-, ethyl ester, N-oxide	n.s.g.	286(4.23)	12-0381-84
1H-Indole-2-carboxylic acid, 2,3,4,5,6,7-hexahydro-1,2- dimethyl-3-oxo-, ethyl ester	MeCN	332(4.03)	33-1957-84
1-Isoquinolinemethanol, 1,2,3,4-tetra- hydro-6,7-dimethoxy-2-methyl-, hydrochloride, (±)-	MeOH	230(3.89),283(3.53), 287s(3.49)	12-1659-84
1H-Pyrrole-2-carboxylic acid, 4-(3- butenyl)-2,3-dihydro-2,5-dimethyl- 3-oxo-, ethyl ester	MeCN	317(4.06)	33-1957-84
1H-Pyrrole-2-carboxylic acid, 2,3-di- hydro-1,2,5-trimethyl-3-oxo-4-(2- propenyl)-, ethyl ester	MeCN	331(4.03)	33-1957-84
1H-Pyrrole-2-carboxylic acid, 2,3-di- hydro-1,4,5-trimethyl-3-oxo-2-(2- propenyl)-, ethyl ester	MeCN	336(4.15)	33-1535-84
$C_{13}H_{19}NO_4S$ DL-Homocysteine, S-[3-(2-hydroxyphen- oxy)propyl]-	MeOH	215(3.83),279(3.40), 274.5(sic)(3.46)	36-1241-84
DL-Homocysteine, S-[2-(2-methoxyphen- oxy)ethyl]-	MeOH	222.5(3.90),272s(3.42)	36-1241-84
2-Thietanone, 3-amino-3,4,4-trimethyl-, p-toluenesulfonate, (3R)-	EtOH	221(3.81),240s(3.19)	39-1127-84C
$C_{13}H_{19}NO_5$ D-allo-Heptonic acid, 3,6-anhydro-2- deoxy-2-[(dimethylamino)methylene]- 4,5-O-(1-methylethylidene)-, ζ-lac- tone	MeOH	296(4.13)	18-2515-84

Compound	Solvent	$\lambda_{max}(\log \epsilon)$	Ref.
$C_{13}H_{19}NO_5S_2$ 4-Thiazolidine-4-carboxylic acid, 5,5-dimethyl-, p-toluenesulfonate, (4S)-	EtOH	223(3.94),227s(3.86), 255(2.57),261(2.57), 268(2.40)	39-1127-84C
$C_{13}H_{19}NS$ Pyrrolidine, 1-[(3-methylbicyclo-[2.2.1]hept-5-en-2-yl)thioxo-methyl]-, (2-endo,3-exo)	EtOH	276(4.18),350(1.62)	39-0865-84C
$C_{13}H_{19}N_2$ Pyridinium, 1-ethenyl-4-(4-methyl-1-piperidinyl)-, bromide	H_2O	315(4.42)	142-0505-84
$C_{13}H_{19}N_2O_9P$ Thymidine, 5'-O-(5-hydroxy-1,3,2-dioxa-phosphorinan-2-yl)-, P-oxide	EtOH	266(3.92)	87-0440-84
$C_{13}H_{19}N_3O_4S$ 2-Propenamide, N-[1-(hydroxymethyl)-3-(methylthio)propyl]-3-(1,2,3,4-tetrahydro-6-methyl-2,4-dioxo-5-pyrimidinyl)-, [R-(E)]-	H_2O	301(4.29)	94-4625-84
[S-(E)]-	H_2O	301(4.28)	94-4625-84
$C_{13}H_{19}N_3O_6S$ Thymidine, 5'-deoxy-5'-(2-isothiazo-lidinyl)-, S,S-dioxide	EtOH	266(3.99)	78-0427-84
$C_{13}H_{19}N_5O$ 9H-Purin-6-imine, 1-(cyclohexylmeth-oxy)-1,9-dihydro-9-methyl-, mono-perchlorate	pH 1 pH 7 pH 13 EtOH	261(4.08) 261(4.08) 258(4.10),265s(4.06) 261(4.08)	94-4842-84 94-4842-84 94-4842-84 94-4842-84
6H-Purin-6-one, 1,9-dihydro-9-methyl-, O-(cyclohexylmethyl)oxime	pH 1 pH 7 pH 13 EtOH	271(4.15) 270(4.17) 285(4.05) 270(4.15)	94-4842-84 94-4842-84 94-4842-84 94-4842-84
$C_{13}H_{19}N_5O_4$ D-Ribitol, 1,4-anhydro-1-C-[2,5-dimeth-yl-4-(methylamino)imidazo[5,1-f]-1,2,4-triazin-7-yl]-, (S)-	EtOH	246(4.48),305(3.43)	39-0229-84C
$C_{13}H_{19}N_5O_7S$ Heptofuranuronic acid, 6-amino-1-[4-[(aminocarbonyl)imino]-3,4-dihydro-3-methyl-2-oxo-1(2H)-pyrimidinyl]-1,6-dideoxy-4-thio-	M HCl	214(4.02),235(3.91), 305(4.19)	5-1399-84
$C_{13}H_{19}O_2P$ Benzoic acid, 3-(diethylphosphino)-, ethyl ester	C_6H_{12}	230(4.02),262s(3.45), 290s(3.08)	65-0489-84
Benzoic acid, 4-(diethylphosphino)-, ethyl ester	C_6H_{12}	234(4.24),287(3.85)	65-0489-84
$C_{13}H_{20}$ Bicyclo[3.1.0]hex-2-ene, 6-ethyl-1,2,3,5-tetramethyl-4-methylene-, (1α,5α,6α)-	hexane	250(3.95)	64-1781-84B
(1α,5α,6β)-	hexane	253(4.06)	64-1781-84B

Compound	Solvent	$\lambda_{max}(\log \epsilon)$	Ref.
$C_{13}H_{20}ClN_3O_6S$			
Thymidine, 5'-[[(3-chloropropyl)sulfonyl]amino]-5'-deoxy-	EtOH	236(4.00)	78-0427-84
$C_{13}H_{20}FN_4O_6P$			
Uridine, 2'-deoxy-5-fluoro-5'-O-(1,3-dimethyl-1,3,2-diazaphospholidin-2-yl)-, P-oxide	EtOH	269(3.86)	39-1471-84C
$C_{13}H_{20}N_2$			
12,13-Diazabicyclo[9.3.1]pentadeca-1(15),11,13-triene	EtOH	213(3.51),255(3.20), 300(2.53)	32-0289-84
1H-Pyrazole, 3-methyl-5-(2,6,6-trimethyl-1-cyclohexen-1-yl)-	MeOH	205(3.85),213s(3.83)	18-1567-84
$C_{13}H_{20}N_2O$			
12,13-Diazabicyclo[9.3.1]pentadeca-1(15),11-dien-14-one	EtOH	213(3.66),227(3.54), 292(3.51)	32-0289-84
$C_{13}H_{20}N_2O_2$			
Phenol, 2,6-dimethyl-4-[[(1-methylpropyl)nitrosoamino]methyl]-	hexane	272(3.41),366(1.97)	150-0701-84M
$C_{13}H_{20}N_2O_3$			
Benzenemethanamine, 2-(methoxymethoxy)-N-(1-methylpropyl)-N-nitroso-	hexane	236(3.74),268(3.34), 356(1.83),366(1.90), 376(1.85)	150-0701-84M
$C_{13}H_{20}N_2O_4$			
1-Piperazinecarboxylic acid, 4-[(dihydro-4-oxo-2H-pyran-3(4H)-ylidene)-methyl]-, ethyl ester, (E)-	EtOH	325(4.33)	4-1441-84
$C_{13}H_{20}N_2O_5$			
2-Butenedioic acid, 2-[1-acetyl-2-(2,2-dimethylhydrazino)-1-propenyl]-, dimethyl ester, (Z,E)-	MeOH	210(3.67),319(3.63)	24-1620-84
$C_{13}H_{20}N_2O_6$			
1,3-Pentadiene-1,2,3-tricarboxylic acid, 4-(2,2-dimethylhydrazino)-, trimethyl ester, (Z,E)-	MeOH	205(3.86),248(3.85), 364(3.25)	24-1620-84
$C_{13}H_{20}N_3O$			
6H-Pyrrolo[3,4-d]pyridazinium, 4-ethoxy-2-ethyl-5,6,7-trimethyl-	cation	231(4.47),268(3.70), 360(3.68)	78-3979-84
dication	H_2SO_4	237(4.50),310(4.07), 377(3.10)	78-3979-84
$C_{13}H_{20}N_3O_7P$			
Thymidine, 5'-O-(3-methyl-1,3,2-oxazaphospholidin-2-yl)-, P-oxide	EtOH	267(3.95)	39-1471-84C
$C_{13}H_{20}N_4O_6S$			
5-Thiazolidineacetamide, 3-α-D-arabinofuranosyl-2-[(1-methylethylidene)hydrazono]-4-oxo-	MeOH	255(4.01)	128-0295-84
5-Thiazolidineacetamide, 2-[(1-methylethylidene)hydrazono]-4-oxo-3-β-D-ribofuranosyl-	MeOH	255(4.06)	128-0295-84

Compound	Solvent	$\lambda_{max}(\log \epsilon)$	Ref.
$C_{13}H_{20}N_5O_6P$ Adenosine, N,N-dimethyl-3'-(phosphono-methyl)-	pH 2	208(4.19),267(4.21)	78-0079-84
	pH 12	275(4.21)	78-0079-84
$C_{13}H_{20}N_6O_3$ 6H-Purin-6-one, 2-amino-1,9-dihydro-9-[(2-hydroxyethoxy)methyl]-8-(1-piperidinyl)-	pH 1	265(4.19),285s(4.03)	87-1486-84
	pH 13	272(4.12)	87-1486-84
$C_{13}H_{20}OS_2$ Bicyclo[2.2.1]heptan-2-one, 3-[bis-(methylthio)methylene]-1,7,7-tri-methyl-, (+)-	EtOH	323(4.11)	78-2951-84
$C_{13}H_{20}OSi$ Spiro[4.4]nona-2,6-dien-1-one, 6-methyl-2-(trimethylsilyl)-	EtOH	223(3.95)	33-0073-84
Spiro[4.4]nona-2,6-dien-1-one, 6-methyl-3-(trimethylsilyl)-	EtOH	232(4.16)	33-0073-84
$C_{13}H_{20}O_2$ Phenol, 4-(3-hydroxypropyl)-2-(1,1-di-methylethyl)-	EtOH	228(3.89),282(3.42)	104-2375-84
$C_{13}H_{20}O_3$ 3-Cyclodecen-1-one, 3-(acetoxymethyl)-, (E)-	EtOH	290(1.63)	44-2925-84
1,3-Cyclohexadiene-1-carboxylic acid, 2-methyl-4-(1-methylethoxy)-, ethyl ester	EtOH	215(3.78),323(3.97)	12-2037-84
2-Pentanol, 5-(3,5-dimethoxyphenyl)-	EtOH	203(4.72),272(3.30), 280(3.29)	100-0828-84
Spiro[5.5]undec-1-en-3-one, 8-hydroxy-11-(hydroxymethyl)-7-methyl-, (6α,7β,8β,11β)-	n.s.g.	236(4.11)	77-0930-84
$C_{13}H_{20}O_4$ 3-Cyclohexene-1-carboxylic acid, 4-eth-oxy-1-(hydroxymethyl)-2-methylene-, ethyl ester	EtOH	248(4.11)	12-2037-84
1,5-Dioxaspiro[5.5]undec-8-ene-9-carb-oxylic acid, 8-methyl-, ethyl ester	EtOH	228(3.91)	12-2037-84
$C_{13}H_{20}O_6S_2$ Spiro[bicyclo[2.2.1]heptane-2,2'-oxir-an]-3-one, 4,7,7-trimethyl-3',3'-bis(methylsulfonyl)-, [1S-(1α,2β,4α)]-	EtOH	312(1.65)	78-2951-84
$C_{13}H_{21}N$ Pyridine, 2,4-bis(1,1-dimethylethyl)-	MeOH	259(3.59),266(3.35)	44-1338-84
Pyridine, 3,4-bis(1,1-dimethylethyl)-	MeOH	225(3.11),266(3.42)	44-1338-84
$C_{13}H_{21}NO$ 1-Nonanone, 1-(1H-pyrrol-2-yl)-	EtOH	250(2.57),288(4.20)	12-0227-84
$C_{13}H_{21}NO_2$ Benzenemethanamine, 2-(methoxymethoxy)-N-(1-methylpropyl)-	EtOH	268(3.15),274(3.10)	150-0701-84M
$C_{13}H_{21}NO_3$ β-Alanine, N-(5,5-dimethyl-3-oxo-1-	pH 1	284(4.40)	39-0287-84C

Compound	Solvent	$\lambda_{max}(\log \epsilon)$	Ref.
cyclohexen-1-yl)-2-methyl-, methyl ester (cont.)	H_2O pH 13	293(4.47) 294(4.48)	39-0287-84C 39-0287-84C
$C_{13}H_{21}NO_5$ Ethanone, 1-[1-ethyl-2-methyl- 5-(1,2,3,4-tetrahydroxybutyl)-1H- pyrrol-3-yl]-, [1S-(1R*,2S*,3R*)]-	EtOH	248(3.94),281(3.79)	136-0153-84I
$C_{13}H_{21}NO_5S_2$ Butanoic acid, 2-amino-2,3-dimethyl- 3-mercapto-, p-toluenesulfonate, (2R)-	EtOH	223(3.85)	39-1127-84C
(2S)-	EtOH	224(4.00)	39-1127-84C
$C_{13}H_{21}NO_9$ α-D-Glucofuranoside, methyl 2-deoxy- 2-[[3-methoxy-2-(methoxycarbonyl)- 3-oxo-1-propenyl]amino]-	EtOH	284(4.32)	136-0101-84L
α-D-Glucopyranoside, methyl 2-deoxy- 2-[[3-methoxy-2-(methoxycarbonyl)- 3-oxo-1-propenyl]amino]-	EtOH	278(4.24)	136-0101-84L
β-	EtOH	288(3.98)	136-0101-84L
$C_{13}H_{21}N_3O$ 1,4-Cyclopentadiene-1-carboxamide, 3-[bis(dimethylamino)methylene]- N,N-dimethyl-	MeOH	257(4.27),349(4.52)	157-0653-84
$C_{13}H_{21}N_5O_2$ 1H-Imidazole-4-carboximidamide, N-(cy- clohexylmethoxy)-5-(formylamino)-1- methyl-	pH 1 pH 7 pH 13 EtOH	255(3.89) 220(4.08),249s(3.84) 257(4.04) 224(4.13),254s(3.84)	94-4842-84 94-4842-84 94-4842-84 94-4842-84
$C_{13}H_{22}N_2$ Pyrazine, 3-methyl-2,5-bis(2-methyl- propyl)-	EtOH	278(3.79),300s(3.25)	142-2317-84
$C_{13}H_{22}N_2O$ 12,13-Diazabicyclo[9.3.1]pentadec- 11-en-14-one	EtOH	245(3.92)	32-0289-84
$C_{13}H_{22}N_2O_3Si$ 3-Butenoic acid, 2-diazo-3-[[(1,1-di- methylethyl)dimethylsilyl]oxy]-, 2- propenyl ester	CH_2Cl_2	280(3.78)	23-2936-84
$C_{13}H_{22}N_4O_3S$ 1,1-Ethenediamine, N-[2-[[[5-[(dimeth- ylamino)methyl]-2-furanyl]methyl]- thio]ethyl]-N'-methyl-2-nitro-	pH 6.5 M HCl M NaOH	229(4.21),315(4.19) 226(--) 228(--),285(--)	39-1761-84B 39-1761-84B 39-1761-84B
$C_{13}H_{22}N_4O_5S$ Clithioneine	pH 2 pH 7 pH 12	254(4.24) 246(4.20) 248(4.18)	102-1003-84 102-1003-84 102-1003-84
$C_{13}H_{22}O$ Cyclohexanebutanal, 2,2-dimethyl- 6-methylene-	pentane	290(1.65)	33-1734-84

Compound	Solvent	$\lambda_{max}(\log \epsilon)$	Ref.
Ethanone, 1-(4-pentyl-1-cyclohexen-1-yl)-	EtOH	236(3.07),307(1.69)	103-0515-84
2-Propen-1-one, 1-(4-butylcyclohexyl)-, cis	EtOH	213(3.17),261(2.04), 333(1.66)	103-0515-84
trans	EtOH	214(3.75),266(3.27), 337(2.49)	103-0515-84
$C_{13}H_{23}N$			
Cyclohexanamine, N-(3-methylcyclohex-ylidene)-	C_6H_{12}	242s(2.35)	33-0748-84
1H-Pyrrole, 2-nonyl-	EtOH	216(3.87)	12-0227-84
$C_{13}H_{23}NOSi$			
2-Cyclopenten-1-one, 3-[methyl[2-(tri-methylsilyl)methyl]-2-propenyl]-amino]-	MeCN	273(4.50)	44-0220-84
$C_{13}H_{23}N_3O_5S$			
2,4(1H,3H)-Pyrimidinedione, 1-cyclohex-yl-3-methyl-5-sulfonyl-, dimethylam-ine salt	H_2O	273(4.07)	47-2455-84
$C_{13}H_{24}N_4O_2$			
2,4(1H,3H)-Pyrimidinedione, 6-(butyl-amino)-1-[2-(propylamino)ethyl]-	pH 7	275(5.4)	103-1033-84
$C_{13}H_{24}Si_2$			
Disilane, (2,5-dimethylphenyl)penta-methyl-	C_6H_{12}	234(4.04)	60-0341-84
$C_{13}H_{26}Si_3$			
Trisilane, 1,1,1,2,2,3,3-heptamethyl-2-phenyl-	C_6H_{12}	243(4.04)	60-0383-84
$C_{13}H_{27}N_3O$			
1H-1,2,3-Triazole, 1-(2-ethoxyethyl)-5(and 4)-heptyl-4,5-dihydro-	hexane	240(3.61)	104-0580-84
$C_{13}H_{30}ClNSi$			
Silanamine, N-chloro-N-(1,1-dimethyl-ethyl)-1,1,1-tris(1-methylethyl)-	CF_2ClCCl_2F	309(1.82)	39-1187-84B

Compound	Solvent	λ_{max}(log ϵ)	Ref.
$C_{14}H_2N_6O_2$ Propanedinitrile, 2,2',2"-(4,5-dihy-droxy-4-cyclopentene-1,2,3-triyli-dene)tris-, dipotassium salt	EtOH PV	605(4.95) 620(4.95)	151-0285-84A 151-0285-84A
$C_{14}H_4F_6O$ 1,4-Epoxyphenanthrene, 5,6,7,8,9,10-hexafluoro-1,4-dihydro-	EtOH	222.5(4.59),312(3.55), 335(3.35)	155-0263-84A
$C_{14}H_4O_2S_3$ Bis thieno[2',3':3,4]cyclopenta[1,2-b:1',2'-d]thiophene-4,9-dione	CH_2Cl_2	257(4.42),320s(4.27), 335(4.37),345s(4.35), 405(3.52),527(3.61)	139-0121-84B
$C_{14}H_6ClNO_5$ 1,4-Anthracenedione, 10-chloro-9-hy-droxy-5-nitro- 1,4-Anthracenedione, 10-chloro-9-hy-droxy-8-nitro-	benzene benzene	467(3.84) 467(3.84)	104-0604-84 104-0604-84
$C_{14}H_6Cl_2O_3$ 1,4-Anthracenedione, 2,9-dichloro-10-hydroxy- 1,4-Anthracenedione, 2,10-dichloro-9-hydroxy-	benzene benzene	464(3.88) 464(3.88)	104-1766-84 104-1766-84
$C_{14}H_6Cl_2O_5$ 9,10-Anthracenedione, 5,8-dichloro-1,2,4-trihydroxy-	DMF	283(4.18),387(3.88), 577(3.84)	117-0309-84
$C_{14}H_6FeN_2O_4S_4$ Ferrate(2-), bis[2,6-pyridinedicarbo-thioato(2-)-N^1,S^2,S^6]-, (OC-6-11')	MeOH	275s(4.23),355(4.18), 475s(3.68),565s(3.37), 755(3.01)	64-1607-84B
$C_{14}H_6O_8$ [1]Benzopyrano[5,4,3-cde][1]benzopyran-5,10-dione, 2,3,7,8-tetrahydroxy-(ellagic acid)	EtOH EtOH-NaOH	254(4.68),365(3.97) 253(4.55),277(4.65), 354(4.11)	94-4478-84 94-4478-84
$C_{14}H_6S_4$ Anthra[1,9-cd:4,10-c'd']bis[1,2]di-thiole	$CHCl_3$	288(4.36),437(3.71), 520(3.93),548(3.98)	18-0022-84
$C_{14}H_6Se_4$ Anthra[1,9-cd:4,10-c'd']bis[1,2]di-selenole	$CHCl_3$	297(4.20),430(3.57), 520(3.93),552(4.05)	18-0022-84
$C_{14}H_7BrN_2OS$ Propanedinitrile, [[4-bromo-5-(phenyl-thio)-2-furanyl]methylene]-	MeOH	210(3.03),248(2.86), 356(3.44)	73-0984-84
$C_{14}H_7BrN_2O_3S$ Propanedinitrile, [[4-bromo-5-(phenyl-sulfonyl)-2-furanyl]methylene]-	MeOH	205(3.22),247(3.29), 347(3.43)	73-0984-84
$C_{14}H_7ClO_2$ Cyclohepta[de]naphthalene-7,8-dione, 9-chloro-	$CHCl_3$	300(3.49),315(3.46), 360s(3.71),405(4.00)	39-1465-84C

Compound	Solvent	$\lambda_{max}(\log \epsilon)$	Ref.
$C_{14}H_7ClO_3$			
1,4-Anthracenedione, 9-chloro-10-hydroxy-	$CHCl_3$	270(3.7),304(3.5), 346(3.1),467(3.6)	150-0147-84M
$C_{14}H_7ClO_5$			
9,10-Anthracenedione, 2-chloro-1,4,5-trihydroxy-	EtOH	229(4.46),278(4.39), 302s(4.12),467(3.99), 482(3.96),516(3.72)	78-4561-84
9,10-Anthracenedione, 2-chloro-1,4,6-trihydroxy-	EtOH	278(4.36),303s(4.15), 467(3.96)	78-4561-84
$C_{14}H_7F_3N_2O_2S$			
Thiazolo[3,2-a]pyrimidin-4-ium, 3-hydroxy-7-phenyl-2-(trifluoroacetyl)-, hydroxide, inner salt	MeCN CCl_4	318(4.47),460(4.00) 500(--)	103-0791-84 103-0791-84
$C_{14}H_7NO_3$			
5H-Furo[3',2':6,7][1]benzopyrano-[3,4-c]pyridin-5-one	EtOH 2% EtOH	310s(--),327(3.81) 310s(--),330(3.83)	149-0145-84A 149-0145-84A
$C_{14}H_7N_5O_4$			
5H-Indolo[2,3-b]quinoxaline, 2,3-dinitro-	EtOH	213(4.40),275(4.41), 313(4.29),405(3.57)	103-1276-84
$C_{14}H_8BrNO_2$			
Benzenepropanenitrile, α-[(5-bromo-2-furanyl)methylene]-β-oxo-	MeOH	210(3.30),260(3.02), 368(3.50)	73-2141-84
$C_{14}H_8BrN_3$			
5H-Indolo[2,3-b]quinoxaline, 9-bromo-	EtOH	223(4.58),272(4.61), 358(4.38),408(3.58)	103-0537-84
$C_{14}H_8ClFOS$			
Dibenzo[b,e]thiepin-11(6H)-one, 2-chloro-8-fluoro-	MeOH	250(4.39),270s(4.10)	73-1800-84
$C_{14}H_8Cl_2O$			
Cyclobut[a]acenaphthylen-7(6aH)-one, 8,8-dichloro-8,8a-dihydro-	MeOH	287(4.18)	39-1465-84C
$C_{14}H_8Cl_2OS$			
Dibenzo[b,f]thiepin-10(11H)-one, 2,3-dichloro-	MeOH	234(4.33),265s(3.96), 328(3.55)	73-0992-84
$C_{14}H_8Cl_4O_2$			
Bicyclo[2.2.2]oct-5-ene-2,3-dione, 1,4,5,6-tetrachloro-7-phenyl-	C_6H_{12}	445(2.22)	62-0167-84C
$C_{14}H_8Co_2O_4$			
Cobalt, [μ-[(1,2,3,4,5-η):1',2',3',4'-5'-η)[bi-2,4-cyclopentadien-1-yl]-1,1'-diyl]tetracarbonyldi-	n.s.g.	278s(4.30),289(4.48), 301(4.71),315(4.77), 400s(3.70)	157-0082-84
$C_{14}H_8N_2$			
Pyrazino[2,1,6-cd:3,4,5-c'd']dipyrrolizine	EtOH	252s(4.30),273(4.84), 294(3.59),347(4.28), 511(3.49),522(3.58), 537(3.87),550(4.11), 565s(2.91)	77-0821-84
Pyridine, 4,4'-(1,3-butadiyne-1,4-diyl)bis-	MeOH	244(4.51),255(4.49), 268(4.18),285(4.35),	4-0607-84

Compound	Solvent	$\lambda_{max}(\log \epsilon)$	Ref.
(cont.)		302(4.45),323(4.39)	4-0607-84
$C_{14}H_8N_2O_4$ Benzenepropanenitrile, α-[(5-nitro-2-furanyl)methylene]-β-oxo-	MeOH	210(3.16),234(3.13), 356(3.30)	73-2141-84
2H-1,4-Benzoxazin-2-one, 7-nitro-3-phenyl-	EtOH	372(4.10)	4-0551-84
$C_{14}H_8N_2O_5$ 1H-Pyrrole-3-carboxamide, 4-(2,3-di-hydro-1,3-dioxo-1H-inden-2-yl)-2,5-dihydro-2,5-dioxo-	MeOH	245(3.30),480(3.75)	48-0594-84
$C_{14}H_8N_4OS$ 2H-1,2,4-Triazino[4",5":1',5']pyrrolo-[2',3':4,5]furo[3,2-b]indole-1(6H)-thione	dioxan	396(3.13)	73-1529-84
$C_{14}H_8N_4O_2$ 5H-Indolo[2,3-b]quinoxaline, 2-nitro-	EtOH	212(4.43),223(4.42), 257(4.61),330(4.50), 381(3.94)	103-0537-84
5H-Indolo[2,3-b]quinoxaline, 3-nitro-	EtOH	230(4.39),275(4.42), 317(4.31),405(3.56)	103-1276-84
2H-1,2,4-Triazino[4",5":1',5']pyrrolo-[2',3':4,5]furo[3,2-b]indol-1(6H)-one	dioxan	344(3.06)	73-1529-84
$C_{14}H_8N_4O_2S$ Furo[2',3':4,5]pyrrolo[1,2-d][1,2,4]-triazine-8(7H)-thione, 2-(2-nitro-phenyl)-	dioxan	390(3.02)	73-1529-84
$C_{14}H_8N_4O_4$ Furo[2',3':4,5]pyrrolo[1,2-d][1,2,4]-triazin-8(7H)-one, 2-(2-nitrophenyl)-	dioxan	341(3.43)	73-1529-84
$C_{14}H_8O_4$ 1,4-Anthracenedione, 5,6-dihydroxy-	EtOH	256(4.21),330(3.56), 452(4.01)	40-1164-84
9,10-Anthracenedione, 1,2-dihydroxy- (also metal chelates)	DMF	370s(3.58),430(3.70)	90-0729-84
$C_{14}H_8O_4Ru_2$ Ruthenium, [μ-[(1,2,3,4,5-η:1',2',3',4'-5'-η)-[bis-2,4-cyclopentadien-1-yl]-1,1'-diyl]]tetracarbonyldi-, (Ru-Ru)-	THF	243(3.99),273(4.04), 329(3.85),388s(3.18)	157-0082-84
$C_{14}H_8S$ Benzo[b]benzo[3,4]cyclobuta[1,2-d]-thiophene	MeOH	221(4.50),246(4.47), 264(4.33),269(4.32), 274(4.54),294(3.70), 307(3.77),332(3.71), 347(3.69),373(2.79)	88-4283-84
$C_{14}H_9BrO_2$ 2H-Naphtho[2,3-b]pyran-2-one, 4-meth-yl-, monobromo deriv.	EtOH	228(4.58),265.5(4.41), 276.5(4.49),326.5(4.22)	94-1178-84
$C_{14}H_9ClN_4O_2$ 1H-1,2,4-Triazole, 1-(4-chlorophenyl)-5-(4-methoxyphenyl)-	MeCN	231(4.29),246s(4.18), 353(2.06)	48-0311-84

Compound	Solvent	$\lambda_{max}(\log \epsilon)$	Ref.
$C_{14}H_9ClO_2$ Cyclohepta[de]naphthalen-7(10H)-one, 9-chloro-8-hydroxy-	CHCl$_3$	300(3.83),323(3.84), 340(3.85),360(3.83)	39-1465-84C
$C_{14}H_9NO$ Indolo[4,5-d]benzo[b]furan	EtOH	208(5.36),218(5.44), 241(5.14),267(4.52), 276(4.68),299(5.04), 310(5.22),325(5.24)	103-1123-84
Indolo[5,4-d]benzo[b]furan	EtOH	250(5.03),297(4.43), 301(4.36),315(4.19)	103-1123-84
Indolo[5,6-d]benzo[b]furan	EtOH	208(4.50),243(4.70), 252(4.80),268(4.15), 282(4.08),308(4.41), 320(4.43)	103-1123-84
Indolo[6,5-d]benzo[b]furan	EtOH	233(4.57),247(4.57), 315(4.56),334(4.38)	103-1123-84
$C_{14}H_9NO_2$ Benzo[h]quinoline-5,6-dione, 4-methyl-	EtOH	242(4.36),266s(4.03), 283(3.92),301(3.71), 325(3.45),352(3.08), 406(2.48)	12-1271-84
8H-Cycloheptoxazol-8-one, 2-phenyl-	MeOH	275(4.42),349s(3.84), 365s(3.63)	18-0609-84
1H-Indolizinium, 3-hydroxy-1-oxo-2-phenyl-, hydroxide, inner salt	MeCN	221(4.15),273(4.27), 346(3.93),515(3.00)	44-3672-84
$C_{14}H_9NO_3$ Benzo[h]quinoline-2,5,6(1H)-trione, 4-methyl-	EtOH	235(4.20),244(4.20), 257(4.19),272(4.13), 287(4.13),311(4.03), 335(3.60),410(3.11), 467s(2.78)	12-1271-84
5H-Furo[3',2':6,7][1]benzopyrano-[3,4-c]pyridin-5-one, 9,10-dihydro-	EtOH	328(4.06)	149-0145-84A
$C_{14}H_9NO_4S$ 2H-Pyran-3-carbonitrile, 6-(1,3-benzo-dioxol-5-yl)-4-(methylthio)-2-oxo-	EtOH	228(4.13),251(4.12), 360(4.14),402(4.21)	94-3384-84
$C_{14}H_9N_3$ 5H-Indolo[2,3-b]quinoxaline	EtOH	222(4.48),271(4.69), 358(4.32),397(3.62)	103-0537-84
6H-Pyrido[3',2':4,5]cyclopenta[1,2-b]-[1,8]naphthyridine	EtOH	208(4.56),240(4.34), 331(4.40),337(4.36), 347(4.56)	44-2208-84
$C_{14}H_9N_3O$ Spiro[2H-benzimidazole-2,3'-[3H]indol]-2'(1'H)-one	EtOH	215(4.42),231(4.54), 246(4.49),324(4.24), 338(4.19)	103-0537-84
$C_{14}H_9N_3O_2$ 1H-Pyrrole-3-carbonitrile, 2,5-dihydro-4-(2-methyl-1H-indol-3-yl)-2,5-dioxo-	MeOH	260(4.18),480(4.06)	48-0594-84
$C_{14}H_9N_3O_6$ Benzene, 1,1'-(1-nitro-1,2-ethenedi-yl)bis[4-nitro-, (E)-	dioxan	267(4.31)	12-1217-84

Compound	Solvent	$\lambda_{max}(\log \epsilon)$	Ref.
$C_{14}H_9N_4O_2$			
Pyridinium, 3-(aminocarbonyl)-1-[5-(2,2-dicyanoethenyl)-2-furanyl]-, bromide	H_2O	249(3.03),324s(2.98), 341s(3.05),375(3.22)	73-2485-84
Pyridinium, 4-(aminocarbonyl)-1-[5-(2,2-dicyanoethenyl)-2-furanyl]-, bromide	H_2O	242(2.94),272(2.95), 333s(3.04),347s(3.07), 388(3.26)	73-2485-84
$C_{14}H_9N_5$			
Ethenetricarbonitrile, (1,2-dimethyl-2H-cyclopenta[d]pyridazin-5-yl)-	ether	256(3.78),285s(3.58), 312s(3.17),348s(3.35), 390s(3.44),451s(4.29), 477(4.38)	44-4769-84
$C_{14}H_9N_5O$			
Benzofuro[2',3':4,5]pyrrolo[1,2-d]-1,2,4-triazolo[3,4-f][1,2,4]triazine, 3-methyl-	dioxan	256(3.39),324(3.36)	73-0065-84
Benzofuro[2',3':4,5]pyrrolo[1,2-d]-1,2,4-triazolo[3,4-f][1,2,4]triazine, 6-methyl-	dioxan	261(3.37),322(3.18)	73-0065-84
$C_{14}H_{10}$			
Anthracene	hexane	217(4.049),252(5.049), 339(3.74),356(3.929), 374(3.929)	5-0133-84
Phenanthrene	DMF	250(4.81)	40-0145-84
$(C_{14}H_{10})n$			
Poly(1,4-phenylenemethylidyne-2,5-cyclohexadiene-1,4-diylidenemethylidyne)	CS_2	378.4(3.896)	116-2934-84
$C_{14}H_{10}BrNO$			
Benzo[b]cyclohept[e][1,4]oxazine, 6-bromo-2-methyl-	MeOH	228(4.30),264(4.35), 272(4.31),306(3.84), 415(4.02)	138-1145-84
	MeOH-HCl	230(4.30),272(4.31), 276(4.30),332(3.92), 447(3.91)	138-1145-84
2-Propenal, 3-(4-bromophenyl)-3-(3-pyridinyl)-, (E)-	pH 1	226(4.26),248s(4.19)	44-4209-84
(Z)-	pH 1	236(4.11),312(4.21)	44-4209-84
$C_{14}H_{10}BrNO_3$			
9(10H)-Acridinone, 2-bromo-1,3-dihydroxy-10-methyl-	EtOH	222s(4.16),246s(4.39), 265s(4.52),272(4.64), 292s(4.12),325s(3.73), 392(3.71)	5-0031-84
$C_{14}H_{10}BrN_3O$			
2(1H)-Quinoxalinone, 3-(2-amino-5-bromophenyl)-	EtOH	222(4.57),305(3.94), 355(3.91),425(3.80)	103-0537-84
$C_{14}H_{10}BrN_3O_3$			
1H-Indazole-3,4,7(2H)-trione, 5-[(4-bromophenyl)amino]-2-methyl-	EtOH	257(4.22),308(3.87), 432(3.95)	4-0825-84
$C_{14}H_{10}Br_2N_2$			
2H-Cyclopenta[d]pyridazine, 5,7-dibromo-1-methyl-2-phenyl-	ether	207(4.41),255(4.42), 280(4.46),326(3.91),	44-4769-84

Compound	Solvent	λ_{max}(log ϵ)	Ref.
(cont.)		336s(3.89),412(3.32)	44-4769-84
C$_{14}$H$_{10}$ClN$_3$ 1H-1,2,4-Triazole, 5-(4-chlorophenyl)- 1-phenyl-	MeCN	221(3.66),255(3.60)	48-0311-84
C$_{14}$H$_{10}$Cl$_2$N$_2$ 5,10-Ethanopyrido[3,2-g]quinoline, 11,12-dichloro-5,10-dihydro-	MeCN	270(4.001),275(3.983)	5-0877-84
C$_{14}$H$_{10}$FN$_3$ Propanedinitrile, [2-(dimethylamino)- 3-(4-fluorophenyl)-2-cyclopropen-1- ylidene]-	CH$_2$Cl$_2$	252(4.17),336(4.34)	118-0686-84
C$_{14}$H$_{10}$INO$_3$ 9(10H)-Acridinone, 1,3-dihydroxy- 2-iodo-10-methyl-	EtOH	234(4.08),246s(4.27), 275(4.56),306s(3.93), 392(3.61)	5-0031-84
C$_{14}$H$_{10}$INO$_5$ Iodonium, [2-hydroxy-5-(methoxycarbo- nyl)-3-nitrophenyl]phenyl-, hydroxide, inner salt	DMSO	412(3.87)	39-0135-84C
C$_{14}$H$_{10}$N$_2$ 9cH-Cyclopenta[3,4]indeno[1,2-d]pyrida- zine, 9c-methyl-	EtOH	248(3.91),321(4.25), 406(3.34),478s(2.92)	39-0909-84C
Indolo[4,5-e]indole, 3,8-dihydro-	EtOH	201(4.40),232(4.38), 253s(4.38),262(4.70), 271(4.85),301(3.99), 309(3.88),323(3.34)	103-0283-84
Indolo[5,4-e]indole, 1,6-dihydro-	EtOH	204(4.53),215(4.61), 244(4.62),257s(4.33), 302(4.02),317(4.23), 326s(4.10),333(4.10), 341(3.96),348(3.98)	103-0283-84
C$_{14}$H$_{10}$N$_2$OS Imidazo[2,1-b]naphtho[1,2-d]thiazol- 9(10H)-one, 10-methyl-	MeOH	242(4.79),305(4.22), 317(4.12),324(4.00), 332(4.18)	83-0472-84
Naphth[1',2':2,3]oxepino[4,5-d]-1,2,3- thiadiazole, 11,12-dihydro-	EtOH	252(4.73)	2-0203-84
C$_{14}$H$_{10}$N$_2$O$_2$ 2H-1,4-Benzoxazin-2-one, 7-amino- 3-phenyl-	EtOH	266(4.03),425(4.29)	4-0551-84
Pyrrolo[2',3':4,5]furo[3,2-b]indole, 1-acetyl-1,9-dihydro-	MeOH	365(4.35)	73-1529-84
C$_{14}$H$_{10}$N$_2$O$_3$ Benzo[h]quinolin-6-ol, 4-methyl-5-ni- tro-	EtOH	247(4.43),268s(4.17), 300(3.80),341(3.36), 354(3.34),396(2.95)	12-1271-84
C$_{14}$H$_{10}$N$_2$O$_4$ Benzene, 1-nitro-4-(1-nitro-2-phenyl- ethenyl)-, (E)-	dioxan	265s(4.08),315(4.22)	12-1217-84
(Z)-	dioxan	252s(4.09),263s(4.07), 303(3.71)	12-1217-84

Compound	Solvent	$\lambda_{max}(\log \epsilon)$	Ref.
1,4,2-Dioxazole, 3-(4-nitrophenyl)-5-phenyl-	dioxan	305(4.02)	12-1217-84
4H-Furo[3,2-b]pyrrole, 4-acetyl-2-(2-nitrophenyl)-	MeOH	335(4.28)	73-1529-84
$C_{14}H_{10}N_2O_5$ 2H-Benz[g]indole-3,5-dione, 2,2-dimethyl-4-nitro-, 1-oxide	EtOH	276(3.78),335(4.06), 412(4.04)	12-0577-84
$C_{14}H_{10}N_2O_{10}S_2$ Benzenesulfonic acid, 2,2'-(1,2-ethenediyl)bis[5-nitro-	H_2O	357(4.42)	73-2751-84
$C_{14}H_{10}N_3O$ Pyridinium, 1-[5-(2,2-dicyanoethenyl)-2-furanyl]-3-methyl-, bromide	H_2O	369(3.31)	73-2485-84
$C_{14}H_{10}N_4O$ 2(1H)-Quinoxalinone, 3-(phenylazo)-	EtOH	263(4.49),315s(3.98), 334s(3.60)	4-0521-84
$C_{14}H_{10}N_4O_2$ 1H-1,2,4-Triazole, 1-(4-nitrophenyl)-3-phenyl-	MeCN	230s(4.10),247(4.23), 317(4.23)	48-0311-84
1H-1,2,4-Triazole, 1-(4-nitrophenyl)-5-phenyl-	MeCN	221s(4.28),241s(4.18), 292(4.00)	48-0311-84
$C_{14}H_{10}N_4O_3$ 2H-Indol-2-one, 3-[(2-amino-5-nitrophenyl)imino]-1,3-dihydro-	EtOH	210(4.20),249(4.22), 312(3.76),388(4.00), 490(3.68)	103-1276-84
Imidazo[1,2-a]pyridine, 6-methyl-2-(4-nitrophenyl)-3-nitroso-	EtOH	221(4.34),257(4.39), 345(3.76),680(2.42)	4-1029-84
Imidazo[1,2-a]pyridine, 7-methyl-2-(4-nitrophenyl)-3-nitroso-	EtOH	227(3.86),253(3.83), 340(3.47),384(3.50), 645(2.06)	4-1029-84
Pyrazolo[1,5-a]quinazoline-2-carboxylic acid, 3-cyano-4,5-dihydro-5-oxo-, ethyl ester	EtOH	267(4.44)	4-1049-84
2(1H)-Quinoxalinone, 3-(2-amino-5-nitrophenyl)-	EtOH	213(4.43),309(4.17), 360(4.17),393(4.17)	103-0537-84
2(1H)-Quinoxalinone, 3-(2-aminophenyl)-7-nitro-	EtOH	209(4.53),310(4.03), 370(4.00),465(3.67)	103-1276-84
$C_{14}H_{10}N_4O_7$ 2-Benzoxazolecarboxamide, 2,3-dihydro-N-(2-hydroxy-4-nitrophenyl)-6-nitro-	EtOH	352(4.22)	4-0551-84
$C_{14}H_{10}N_4O_7S_2$ Acetamide, N-[5-(5-nitro-2-furanyl)-2-[[(5-nitro-2-furanyl)methyl]-thio]-4-thiazolyl]-	dioxan	314(3.27),394(3.32)	73-2285-84
$C_{14}H_{10}N_4S$ Quinolinium, 1-[[2,2-dicyano-1-(methylthio)ethenyl]amino]-, hydroxide, inner salt	EtOH	238(4.42),265(4.10), 282(4.18),324(3.95)	95-0440-84
$C_{14}H_{10}O$ Benzo[3,4]cyclobuta[1,2-b]benzofuran, 4b,9b-dihydro-, cis	EtOH	209(4.16),261s(3.14), 267(3.31),273.5(3.43)	12-1283-84

Compound	Solvent	$\lambda_{max}(\log \epsilon)$	Ref.
9-Phenanthrenol	hexane	251(4.55),272(4.11), 294(3.80),304(3.75), 336(3.25),354(3.25)	39-0947-84B
	H_2O	252(4.23),295(3.59), 337(3.59),355(3.00)	39-0947-84B
	MeOH	253(4.50),274(4.01), 298(3.74),306(3.52), 340(3.08),367(2.78)	39-0947-84B
	MeCN	252(4.51),272(4.08), 297(3.67),308(3.54), 337(3.24),355(2.89)	39-0947-84B
	anion	252(4.21),264(4.10), 323(3.43),381(2.95)	39-0947-84B
$C_{14}H_{10}O_2$			
Acenaphtho[1,2-b][1,4]dioxin, 8,9-di-hydro-	n.s.g.	318(3.81),450(2.78)	78-0665-84
Anthra[4a,10-b:9,9a-b']bisoxirene, 6a,10b-dihydro-	ether	242s(3.63),273(3.48), 278.5(3.51),284.5(3.49), 292s(3.41),305s(3.14)	22-0145-84
7bH-Cyclopent[cd]indene-1,2-dicarbox-aldehyde, 7b-methyl-	EtOH	217(4.09),242(4.02), 318(4.30),375s(3.66), 498(3.30)	39-0165-84C
6,11-Epoxydibenz[b,e]oxepin, 6,11-di-hydro-	n.s.g.	270s(3.33),278(3.48), 285.5(3.52)	22-0145-84
1H-Naphtho[2,1-b]pyran-1-one, 3-methyl-	EtOH	229(4.34),259(4.26), 302(4.00),334(3.67)	94-1178-84
2H-Naphtho[2,3-b]pyran-2-one, 4-methyl-	EtOH	225.5(4.59),263(4.41), 273(4.43),320.5(4.19)	94-1178-84
3H-Naphtho[2,1-b]pyran-3-one, 1-methyl-	EtOH	230.5(4.68),248(4.25), 316.5(4.02),347(4.06), 362.5(3.92)	94-1178-84
Naphthopyranone, methyl-	EtOH	264.5(4.47),274.5(4.59), 292(3.81),304(3.90), 317(3.90),351.5(3.74)	94-1178-84
$C_{14}H_{10}O_2S$			
3H-Benz[cd]azulen-3-one, 4-hydroxy-5-(methylthio)-	CH_2Cl_2	238s(--),256(4.53), 280s[sic],252(4.22), 304(4.20),316(4.14), 380s(--),390(4.16), 426(3.46),585(2.86)	83-0984-84
$C_{14}H_{10}O_2S_4$			
1,3-Dithiole-4-carboxylic acid, 2-(1,3-dithiol-2-ylidene)-, phenylmethyl ester	MeCN	290(4.11),299(4.13), 313(4.15),430(3.45)	103-1342-84
$C_{14}H_{10}O_3$			
2(3H)-Benzofuranone, 3-hydroxy-3-phenyl-	MeOH	266(3.09),271(3.13)	35-7352-84
$C_{14}H_{10}O_3S$			
2-Anthracenesulfonic acid, potassium salt	H_2O	341(3.51),359(3.62), 379(3.54)	138-0053-84
compd. with cyclodextrin	H_2O	349(--),362(--), 381(--)	138-0053-84
$C_{14}H_{10}O_4$			
Anthra[1,2-b:3,4-b':4a,10-b":9,9a-b"']-tetrakisoxirene, 1a,1b,2a,3a,7a,8b-	THF	250s(2.22),256s(2.44), 264s(2.66),265(2.66),	22-0145-84

Compound	Solvent	$\lambda_{max}(\log \epsilon)$	Ref.
hexahydro-, anti (cont.)		267s(2.68),269s(2.73), 271(2.76),277(2.75)	22-0145-84
syn	THF	255(2.46),263s(2.61), 266(2.61),270(2.68), 276(2.64)	22-0145-84
2,5-Etheno-2H,5H-bisoxireno[1,2:3,4]-naphtho[2,3-d][1,2]dioxin, 6a,10b-dihydro-, anti	THF	250.5(2.32),256s(2.46), 263s(2.63),264.5(2.65), 265.5(2.66),268s(2.68), 270s(2.71),271(2.73), 277(2.73)	22-0145-84
syn	THF	255s(2.49),263s(2.68), 264.5(2.70),266(2.70), 268s(2.73),269.5(2.76), 271(2.78),277(2.79)	22-0145-84
9H-Fluoren-9-one, 1,5-dihydroxy-7-methoxy- (dengibsin)	EtOH	267(5.36),275(5.40), 338(4.34),479.5(4.08)	42-1010-84
	EtOH	291.5(5.45),549.5(4.08)	42-1010-84
$C_{14}H_{10}O_5$			
9H-Fluoren-9-one, 1,2,5-trihydroxy-7-methoxy- (dendroflorin)	ELOH	259(4.60),279(4.31), 383.5(4.31),485.5(3.36)	42-1010-84
	EtOH	272(4.65),520(3.54)	42-1010-84
7H-Furo[3,2-g][1]benzopyran-6-carbox-ylic acid, 7-oxo-, ethyl ester	EtOH	248(4.420),314(4.223)	111-0365-84
9H-Xanthen-9-one, 1,7-dihydroxy-6-methoxy-	MeOH	253(4.38),272s(4.10), 293(4.00),375(3.91)	102-2390-84
	MeOH-NaOH	262(4.35),280s(4.05)	102-2390-84
	MeOH-NaOAc	252(4.35),272s(4.15), 293(4.00),377(3.72)	102-2390-84
	MeOH-AlCl$_3$	231(4.28),260(4.24), 282(4.26),317(4.08)	102-2390-84
	MeOH-NaOAc-H$_3$BO$_3$	253(4.36),272s(4.06), 293(3.95),375(3.80)	102-2390-84
9H-Xanthen-9-one, 5,6-dihydroxy-1-methoxy-	EtOH	238(4.34),243(4.34), 265s(4.08),290s(3.76), 375(4.10)	102-2390-84
	EtOH-NaOH	238(4.35),248(4.36), 265(4.14),285s(3.76), 350s(3.94),385(4.07)	102-2390-84
	EtOH-NaOAc	235(4.34),243(4.33), 265(4.12),285s(3.76), 373(4.18)	102-2390-84
	EtOH-NaOAc-H$_3$BO$_3$	235(4.33),246(4.34), 290(3.72),325(3.79), 368(3.98)	102-2390-84
$C_{14}H_{10}O_6$			
1,4-Anthracenedione, 2,3-dihydro-5,6,9,10-tetrahydroxy-	EtOH	225(4.16),280(4.00), 345(3.73)	40-1164-84
5H-Furo[3,2-g][1]benzopyran-7-carbox-aldehyde, 4,9-dimethoxy-5-oxo-	EtOH	213s(4.28),216(4.29), 246(4.52),262s(4.35), 281s(3.64),334(3.64)	44-5035-84
9H-Xanthen-9-one, 1,2,8-trihydroxy-6-methoxy-	MeOH	203(4.44),237(4.28), 261(4.43),322(4.10)	94-2290-84
9H-Xanthen-9-one, 1,3,8-trihydroxy-2(or 4)-methoxy-	MeOH	222s(--),247(3.87), 274s(--),316(3.57), 368(3.15)	100-0868-84
	MeOH-NaOMe	233(--),254(--), 2 91(--),350(--)	100-0868-84
	MeOH-AlCl$_3$	221(--),245(--), 261(--),280(--), 340(--),420(--)	100-0868-84

Compound	Solvent	$\lambda_{max}(\log \epsilon)$	Ref.
9H-Xanthen-9-one, 1,4,7-trihydroxy-3-methoxy-	MeOH	233(4.39),270(4.50), 321(3.85),402(3.71)	39-1507-84C
9H-Xanthen-9-one, 1,5,8-trihydroxy-3-methoxy-	MeOH	202(4.54),253(4.50), 277(4.32),332(4.14)	94-2290-84
$C_{14}H_{10}O_8$ Quinhydrone from 3,6-dioxo-1,4-cyclo-hexadiene-1-carboxylic acid	ether	220(4.33),331(3.82)	44-4736-84
$C_{14}H_{10}O_{10}$ Quinhydrone from 2-hydroxy-3,6-dioxo-1,4-cyclohexadiene-1-carboxylic acid	EtOH	221(4.34),324(3.97), 559(3.24)	44-4736-84
$C_{14}H_{11}BrN_2$ 2H-Cyclopenta[d]pyridazine, 5-bromo-1-methyl-2-phenyl-	ether	207(4.38),249(4.48), 280(4.41),320(3.92), 331s(3.89),396(3.30)	44-4769-84
$C_{14}H_{11}BrN_2S$ 3-Pyridinecarbonitrile, 4-(4-bromo-phenyl)-1,2-dihydro-5,6-dimethyl-2-thioxo-	dioxan	260(4.16),319(3.84), 428(2.93)	104-1837-84
$C_{14}H_{11}BrOS$ 2-Propen-1-one, 2-(bromomethyl)-1-phenyl-3-(2-thienyl)-	EtOH	332(3.09)	73-1764-84
$C_{14}H_{11}BrO_2$ 2-Propen-1-one, 2-(bromomethyl)-3-(2-furanyl)-1-phenyl-	EtOH	329(3.42)	73-1764-84
$C_{14}H_{11}BrO_3$ 2,5-Furandione, 3-[(4-bromophenyl)-methylene]dihydro-4-(1-methyl-ethylidene)-, (E)-	C_6H_{12}	331(4.16)	48-0233-84
(Z)-	C_6H_{12}	343(4.29)	48-0233-84
$C_{14}H_{11}BrO_4$ 1,6-Azulenediol, 3-bromo-, diacetate	C_6H_{12}	236(4.11),291(4.56), 298(4.57),343s(--), 358(3.69),368(3.42), 377(3.53),624(2.42), 672s(--)	35-4851-84
	MeCN	612(2.57)	35-4851-84
3-Butenoic acid, 4-(5-bromo-2-furanyl)-2-(2-furanylmethylene)-, methyl ester, (E,E)-	n.s.g.	204(4.14),280(4.23), 370(4.43)	73-2502-84
(E,Z)-	n.s.g.	200(4.42),280(4.32), 337(4.38)	73-2502-84
$C_{14}H_{11}ClN_2S$ 3-Pyridinecarbonitrile, 4-(4-chloro-phenyl)-1,2-dihydro-5,6-dimethyl-2-thioxo-	dioxan	259(3.97),318(3.96), 430(3.24)	104-1837-84
$C_{14}H_{11}ClO$ Ethanone, 1-[1,1'-biphenyl]-4-yl-2-chloro-	20% H_2SO_4 96% H_2SO_4	299(4.21) 393(4.19)	48-0353-84 48-0353-84
$C_{14}H_{11}ClO_2S$ Benzene, 1-chloro-4-[[2-(phenylsulfin-	dioxan	241(4.42),275(3.56),	104-0473-84

Compound	Solvent	$\lambda_{max}(\log \epsilon)$	Ref.
yl)ethenyl]oxy]- (cont.)		281(3.32)	104-0473-84
Benzoic acid, 5-chloro-2-[(2-methyl-phenyl)thio]-	MeOH	260.5(4.08),327(3.64)	73-1009-84
$C_{14}H_{11}Cl_2NOS_2$			
1H-Thiopyran-4-carbonitrile, 3-(3,4-dichlorophenyl)-1-methyl-5-(methyl-thio)-, 1-oxide	EtOH	247.5(4.51),285s(3.95), 337.0(3.91)	118-0852-84
$C_{14}H_{11}FN_2S$			
3-Pyridinecarbonitrile, 4-(3-fluoro-phenyl)-1,2-dihydro-5,6-dimethyl-2-thioxo-	dioxan	316(3.77)	104-1837-84
3-Pyridinecarbonitrile, 4-(4-fluoro-phenyl)-1,2-dihydro-5,6-dimethyl-2-thioxo-	dioxan	264(4.34)	104-1837-84
$C_{14}H_{11}IO_4$			
3-Butenoic acid, 2-(2-furanylmethyl-ene)-4-(5-iodo-2-furanyl)-, methyl ester, (E,E)-	n.s.g.	207(4.05),297(4.21), 381(4.35)	73-2502-84
(E,Z)-	n.s.g.	207(4.07),294(4.13), 348(4.23)	73-2502-84
$C_{14}H_{11}IO_5$			
5H-Furo[3,2-g][1]benzopyran-5-one, 7-(iodomethyl)-4,9-dimethoxy-	EtOH	216(4.48),253(4.55), 340(3.63)	44-5035-84
$C_{14}H_{11}NOS$			
9(10H)-Acridinethione, 2-methoxy-	acetone	492(4.48)	103-1254-84
9(10H)-Acridinethione, 4-methoxy-	acetone	477(4.48)	103-1254-84
$C_{14}H_{11}NO_2$			
9(10H)-Anthracenone, 4-amino-1-hy-droxy-	EtOH	218(4.24),262(4.81), 363(4.02)	40-1164-84
Benzamide, N-(2-oxobicyclo[3.2.0]hepta-3,6-dien-1-yl)-, (±)-	MeOH	342(2.19)	39-0891-84B
Benzamide, N-(4-oxobicyclo[3.2.0]hepta-2,6-dien-6-yl)-, (-)-	MeOH	230(4.00),274(3.69)	39-0891-84B
Benzene, 1,1'-(1-nitroethenediyl)bis-, (Z)-	EtOH	281(4.28)	12-1217-84
$C_{14}H_{11}NO_3$			
2,4,6,8-Nonatetraynoic acid, 9-morpho-lino-, methyl ester	MeCN	231f(4.89),286f(4.83), 390f(3.89)	27-0157-84
3H-Pyrrolo[2,1,5-de]quinolizine-1-carb-oxylic acid, 3-oxo-, ethyl ester	EtOH	252(4.52),269(4.59), 302(3.51),314(3.67), 318(3.68),329(3.65), 392s(3.79),412(4.09), 434(4.14)	39-2553-84C
5H-Pyrrolo[2,1,5-de]quinolizine-1-carb-oxylic acid, 5-oxo-, ethyl ester	EtOH	225(4.26),240(4.28), 280(4.37),296s(4.19), 341s(3.48),353(3.55), 442(4.14),470(4.45)	39-2553-84C
$C_{14}H_{11}NO_3S$			
4H-[1]Benzothieno[3,2-b]pyran-3-carb-oximidic acid, 4-oxo-, ethyl ester, hydrochloride	MeOH	227(4.36),248(4.28), 295(4.19),330(3.85), 360s(3.42)	83-0531-84
2H-Pyran-3-carbonitrile, 6-(4-methoxy-phenyl)-4-(methylthio)-2-oxo-	EtOH	250(4.21),342(4.28), 395(4.45)	94-3384-84

Compound	Solvent	$\lambda_{max}(\log \epsilon)$	Ref.
$C_{14}H_{11}NO_4$			
4H-Furo[3,2-c][1]benzopyran-3,4(2H)-dione, 2-[(dimethylamino)methylene]-	MeOH	385(4.29)	83-0685-84
2H-Pyran-3-carbonitrile, 4-methoxy-6-(4-methoxyphenyl)-2-oxo-	EtOH	230(4.20),288(3.75), 300(3.74),376(4.45)	94-3384-84
$C_{14}H_{11}NO_5$			
Benzofuro[2,3-c]pyridine-4-carboxylic acid, 1,2-dihydro-7-methoxy-2-methyl-1-oxo-	MeOH	255(4.14),310(3.8)	2-0478-84
1,4-Cyclohexadiene-1-carboxylic acid, 2-[(4-methoxyphenyl)amino]-3,6-dioxo-	EtOH	231(4.16),264(3.99), 439(3.35),522(3.33)	44-4736-84
$C_{14}H_{11}NO_5S_2$			
[1]Benzothieno[3,2-b]pyridine-3-carboxylic acid, 1,2-dihydro-4-(methyl-thio)-2-oxo-, methyl ester, 5,5-dioxide	EtOH	221(4.51),249(4.27), 290(3.89),336(3.91)	95-0134-84
$C_{14}H_{11}NO_6$			
3-Butenoic acid, 2-(2-furanylmethylene)-4-(5-nitro-2-furanyl)-, methyl ester, (E,E)-	n.s.g.	200(4.61),326(4.41), 422(4.31)	73-2502-84
(E,Z)-	n.s.g.	207(4.10),324(4.35), 374(4.21)	73-2502-84
1H-Pyrrole-3-carboxylic acid, 2-(1,3-benzodioxol-5-yl)-4,5-dihydro-4,5-dioxo-, ethyl ester	dioxan	255(4.06),305s(3.65), 350(3.92),403(3.89)	94-0497-84
$C_{14}H_{11}NS$			
9(10H)-Acridinethione,10-methyl-	toluene	493(4.44)	103-1254-84
	heptane	485(4.55)	103-1254-84
	EtOH	490(4.48)	103-1254-84
	isoPrOH	489(4.49)	103-1254-84
	dioxan	492(4.46)	103-1254-84
	60% dioxan	490(4.50)	103-1254-84
	MeCN	491(4.52)	103-1254-84
	$CHCl_3$	492(4.53)	103-1254-84
1H-Isoindole-1-thione, 2,3-dihydro-2-phenyl-	EtOH	203(4.46),266(4.11), 294(3.87),329(3.87)	103-0978-84
$C_{14}H_{11}N_2O_3$			
Pyridinium, 1-[5-(2-cyano-3-methoxy-3-oxo-1-propenyl)-2-furanyl]-, bromide	H_2O	262(2.85),320s(3.00), 338s(3.11),371(3.28)	73-2485-84
$C_{14}H_{11}N_3$			
1H-1,2,4-Triazole, 1,3-diphenyl-	MeCN	266(4.47),278s(4.19), 285(4.03)	48-0311-84
1H-1,2,4-Triazole, 1,5-diphenyl-	MeCN	217(4.25),249(4.06)	48-0311-84
$C_{14}H_{11}N_3O$			
Imidazo[1,2-a]pyridine, 6-methyl-3-nitroso-2-phenyl-	EtOH	229(4.81),256(4.91), 356(4.85),641(2.42)	4-1029-84
Imidazo[1,2-a]pyridine, 7-methyl-3-nitroso-2-phenyl-	EtOH	229(4.76),258(4.62), 291(4.54),364(3.67), 635(2.00)	4-1029-84
2(1H)-Quinoxalinone, 3-(2-aminophenyl)-	EtOH	217(4.54),308(3.87), 350(3.92),405(3.79)	103-0537-84

Compound	Solvent	$\lambda_{max}(\log \epsilon)$	Ref.
$C_{14}H_{11}N_3OS$ 2H-1,4-Benzothiazine-2,3(4H)-dione, 3-(phenylhydrazone)	EtOH	278(4.04),356(4.41)	4-0521-84
$C_{14}H_{11}N_3O_2$ 2H-1,4-Benzoxazine-2,3(4H)-dione, 3-(phenylhydrazone)	EtOH	270(4.20),360(4.42)	4-0521-84
1H-Indazole-4,7-dione, 3-methyl- 5-(phenylamino)-	n.s.g.	261(4.29),314(3.86), 458(3.67)	4-0825-84
Pyrano[4,3-c]pyrazol-4(1H)-one, 3-amino-6-(2-phenylethenyl)-	EtOH	260(4.12),302(3.89)	94-3384-84
$C_{14}H_{11}N_3O_3$ 1H-Indazole-3,4,7(2H)-trione, 2-methyl- 5-(phenylamino)-	EtOH	246(4.26),309(3.99), 429(4.09)	4-0825-84
$C_{14}H_{11}N_3O_3S$ 2H-1,2,4,6-Thiatriazinium, 5,6-dihydro- 5-oxo-2,4-diphenyl-, hydroxide, inner salt, 1,1-dioxide	MeCN	235s(4.098),267s(3.950)	142-0063-84
$C_{14}H_{11}N_3O_4$ 1H-[1,2,4]Triazolo[1,2-a]pyridazine- 5-carboxylic acid, 2,3-dihydro-1,3- dioxo-2-phenyl-, methyl ester	EtOH	406(3.23)	44-0587-84
$C_{14}H_{11}N_3O_5$ Benzeneacetamide, α-(hydroxyimino)- N-(2-hydroxy-4-nitrophenyl)-	EtOH	381(4.13)	4-0551-84
$C_{14}H_{11}N_3O_7$ 1(4H)-Pyridineacetic acid, 3,5-dinitro- 4-oxo-α-(phenylmethyl)-, (S)-	EtOH	342(3.44)	18-1961-84
$C_{14}H_{11}N_3O_8$ 1(4H)-Pyridineacetic acid, α-[(4-hy- droxyphenyl)methyl]-3,5-dinitro-4- oxo-, (S)-	EtOH	340(3.60)	18-1961-84
$C_{14}H_{11}N_3S$ Naphtho[1',2':4,5]thiazolo[3,2-b]- [1,2,4]triazole, 9-ethyl-	MeOH	242(4.42),295(3.96), 330(3.10)	4-1571-84
Pyrazolo[1,5-a]quinoline-3-carboni- trile, 5-methyl-2-(methylthio)-	EtOH	229(4.32),242(4.32), 266(4.68),315(4.01), 342(3.90),354(3.82)	95-0440-84
$C_{14}H_{12}$ 7H-Benzocyclononene, 7-methylene-, (Z,Z,Z)-	hexane	255(3.89)	88-5599-84
1,3-Cyclopentadiene, 5,5'-(2-butene- 1,4-diylidene)bis-, (E)-	C_6H_{12}	255(3.34),332(4.12), 350(4.56),368(4.82), 390(4.81)	24-2006-84
	MeCN	257(3.35),332(4.12), 351(4.54),367(4.78), 386(4.81)	24-2006-84
(Z)-	C_6H_{12}	265(3.27),330(4.18), 346(4.49),366(4.71), 387(4.70)	24-2006-84
	MeCN	265(3.22),328(4.00), 348(4.34),364(4.55), 384(4.56)	24-2006-84

Compound	Solvent	λ_{max}(log ϵ)	Ref.
Phenanthrene, 9,10-dihydro-	n.s.g.	264(4.23)	105-0467-84
$C_{14}H_{12}BrNO$ 2-Propen-1-ol, 3-(4-bromophenyl)-3-(3-pyridinyl)-, (E)-	pH 1	225(4.19),240s(4.16)	44-4209-84
$C_{14}H_{12}BrNOS_2$ 1H-Thiopyran-4-carbonitrile, 3-(4-bromophenyl)-1-methyl-5-(methylthio)-, 1-oxide	EtOH	251.5(4.57),285s(3.92), 337.0(3.95)	118-0852-84
$C_{14}H_{12}BrNO_2$ 2,4,6-Cycloheptatrien-1-one, 2-bromo-7-[(2-hydroxy-5-methylphenyl)amino]-	MeOH	258(4.30),346(4.03), 418(4.18)	138-1145-84
	MeOH-NaOH	238(--),346(--), 423(--)	138-1145-84
$C_{14}H_{12}BrNO_3$ 2,4-Benzofurandione, 3-(4-bromophenyl)-3,5,6,7-tetrahydro-, 2-oxime	MeOH	267(4.111)	73-1421-84
$C_{14}H_{12}BrN_2$ Phenazinium, N-(2-bromoethyl)-, bromide (mixt. with radical cation)	EtOH	210(4.35),262(4.70), 390(4.09),451(3.64), 604(2.82),657(2.98), 723(2.91)	103-1162-84
$C_{14}H_{12}BrN_3O_3$ Pyrrolo[3,4-c]pyrazole-4,6(1H,5H)-di-one, 3-(2-bromo-1-oxopropyl)-3a,6a-dihydro-5-phenyl-, (3aα,6aα)-	MeOH	245(4.30),282(4.22)	103-0416-84
$C_{14}H_{12}ClNOS_2$ 1H-Thiopyran-4-carbonitrile, 3-(4-chlorophenyl)-1-methyl-5-(methylthio)-, 1-oxide	EtOH	251.0(4.54),285s(3.98), 336.0(3.93)	118-0852-84
$C_{14}H_{12}ClNO_2$ Benzamide, N-(2-chloro-5-methylphenyl)-N-hydroxy-	MeOH	262(3.90)	118-0938-84
1,1'-Biphenyl, 4-(1-chloroethyl)-4'-nitro-	EtOH	212(4.19),231(4.09), 314(4.24)	35-3140-84
$C_{14}H_{12}ClNO_3$ 2,4-Benzofurandione, 3-(4-chlorophenyl)-3,5,6,7-tetrahydro-, 2-oxime	MeOH	269(4.071)	73-1421-84
$C_{14}H_{12}ClN_2$ Phenazinium, N-(2-chloroethyl)-, chloride, (mixt. with radical cation)	EtOH	210(4.29),262(4.74), 390(4.13),451(3.66), 604(2.84),657(3.00), 723(2.93)	103-1162-84
$C_{14}H_{12}F_3N_5$ 1H-Purin-6-amine, 2,8-dimethyl-N-[3-(trifluoromethyl)phenyl]-	EtOH	297(4.52)	11-0208-84B
$C_{14}H_{12}FeO_4$ Ferrocene, 1,1'-[1,2-ethanediylbis(oxycarbonyl)]-	MeOH	455(2.46)	18-2435-84

Compound	Solvent	$\lambda_{max}(\log \epsilon)$	Ref.
$C_{14}H_{12}N$			
Benzo[b]quinolizinium, 9-methyl-	MeOH	263(2.85)	48-0757-84
$C_{14}H_{12}N_2$			
2H-Cyclopenta[d]pyridazine, 1-methyl-2-phenyl-	hexane	247.5(4.33),276.5(4.19), 311s(3.66),319s(3.64), 324.5(3.60),393.5(3.16)	44-4769-84
	MeOH	203.5(4.35),248(4.49), 272s(4.26),314(3.78), 385(3.16)	44-4769-84
5,10-Ethanophenazine	ether	261(3.18),274(3.18)	103-1162-84
$C_{14}H_{12}N_2OS$			
Ethanone, 1-[2-(methylthio)pyrazolo-[1,5-a]quinolin-2-yl]-	EtOH	230(4.17),263(4.64), 284(4.21),325(4.15), 349(4.12)	95-0440-84
1H-Pyrazolo[1,2-a]pyrazol-4-ium, 3-hydroxy-5,7-dimethyl-2-phenyl-1-thioxo-, hydroxide, inner salt	MeCN	245(4.16),287(4.30), 489(3.60)	44-3672-84
	MeCN	277(4.36),469(3.34)	44-3672-84
3-Quinolinecarbonitrile, 4-(2-furanyl)-1,2,5,6,7,8-hexahydro-2-thioxo-	dioxan	236(4.07),266(4.13), 316(4.32),460(3.25)	104-1780-84
$C_{14}H_{12}N_2O_2$			
3-Azabicyclo[4.4.1]undeca-1,3,5,7,9-pentaene-5-carboxylic acid, 2-cyano-, ethyl ester	MeOH	245(4.32),282(4.31), 354(3.64),397(3.53)	44-1040-84
4H-Indazol-4-one, 1,5,6,7-tetrahydro-5-(hydroxymethylene)-1-phenyl-	EtOH	228.5(4.17),257(4.16), 294(4.02)	4-0361-84
9H-Pyrido[3,4-b]indole-1-carboxylic acid, ethyl ester	EtOH	218(4.90),247(4.52), 258(4.52),288(4.56), 302(4.31),368(4.06)	94-3579-84
9H-Pyrido[3,4-b]indole-1-propanoic acid	EtOH	238(4.98),292(4.51), 336(4.38),350(4.41)	94-0170-84 +94-3579-84
$C_{14}H_{12}N_2O_2S$			
1H-Indazole, 1-[(4-methylphenyl)sulfonyl]-	MeOH	237(4.28),241(4.27), 269(3.92),277(4.01), 287(4.00)	18-1567-84
2H-Indazole, 2-[(4-methylphenyl)sulfonyl]-	MeOH	244(4.40),276(3.60), 288(3.73),296(3.68)	18-1567-84
$C_{14}H_{12}N_2O_2S_2$			
2(1H)-Pyridinethione, 4-(3-hydroxy-6-methyl-2-pyridinylthio)-1-ethenyl-3-hydroxy-6-methyl-	EtOH	270(4.02),382(4.14)	1-0293-84
$C_{14}H_{12}N_2O_2S_3$			
2(1H)-Pyridinethione, 6,6'-thiobis[1-ethenyl-3-hydroxy-	EtOH	267(4.03),294(3.72), 393(3.84)	1-0293-84
$C_{14}H_{12}N_2O_3$			
2H-1,4-Benzoxazine, 3,4-dihydro-7-nitro-4-phenyl-	EtOH	258(3.84),401(4.24)	78-1755-84
6-Quinolinecarboxylic acid, 8-cyano-5-hydroxy-2-methyl-, ethyl ester	EtOH	230(4.36),265(4.57), 310(3.78),345(3.65)	39-1795-84C
4H-Quinolizine-3-carboxylic acid, 1-cyano-6-methyl-4-oxo-, ethyl ester	EtOH	270(4.14),350(3.90), 430(4.19)	39-1795-84C
$C_{14}H_{12}N_2O_3S$			
Acetic acid, (7-acetylthieno[2,3-b]pyridin-4(7H)-ylidene)cyano-, ethyl ester	MeOH	229(4.63),263s(3.98) 354(4.46)	4-1135-84

Compound	Solvent	λ_{max}(log ϵ)	Ref.
C$_{14}$H$_{12}$N$_2$O$_5$S Benzene, 1-[[4-(methoxymethoxy)phenyl]-thio]-2,4-dinitro-	MeOH	220(4.43),245s(4.05), 266s(3.96),328(4.10)	73-1722-84
C$_{14}$H$_{12}$N$_2$O$_6$ 1H-Pyrrole-3-carboxylic acid, 4,5-di-hydro-1-methyl-2-(4-nitrophenyl)-4,5-dioxo-, ethyl ester	dioxan	228(4.19),264(4.09), 408(3.46)	94-0497-84
C$_{14}$H$_{12}$N$_2$S 3-Pyridinecarbonitrile, 1,2-dihydro-5,6-dimethyl-4-phenyl-2-thioxo-	dioxan	280(3.88),317(3.76), 428(3.00)	104-1837-84
C$_{14}$H$_{12}$N$_4$ 9-Phenanthrenamine, 10-azido-9,10-di-hydro-, trans	CH$_2$Cl$_2$	229.7(4.03),271.6(4.19)	4-1597-84
C$_{14}$H$_{12}$N$_4$O 2,3-Quinoxalinedione, mono(phenylhydra-zone)	EtOH	263(4.90),332(4.05), 336(4.02)	4-0521-84
C$_{14}$H$_{12}$N$_4$O$_2$ Imidazo[1,2-a]pyridinium, 1-methyl-3-nitro-2-(phenylamino)-, hydroxide, inner salt	n.s.g.	224(4.26),249(4.25), 281(4.23),310(4.22), 340s(4.01),360s(3.94), 432(3.79)	39-0069-84C
C$_{14}$H$_{12}$N$_4$O$_2$S 1H-Pyridazino[4,5-e][1,3,4]thiadiazin-5(6H)-one, 1-acetyl-6-methyl-3-phen-yl-	EtOH	263(3.92),306(3.80)	94-4437-84
C$_{14}$H$_{12}$N$_4$O$_2$S$_2$ 1H-Pyrazolo[3,4-d]pyrimidine-3-carbox-ylic acid, 4,5,6,7-tetrahydro-1-phenyl-4,6-dithioxo-, ethyl ester	EtOH	274(3.98),308(3.92)	4-1049-84
C$_{14}$H$_{12}$N$_4$O$_3$ Methanimidic acid, N-benzoyl-, 2-(4-ni-trophenyl)hydrazide	MeCN	231(4.25),293(3.87), 402(4.43)	48-0311-84
1,3,7,8-Tetraazaspiro[4.5]deca-1,6,9-triene-10-carboxylic acid, 4-oxo-2-phenyl-, methyl ester	MeOH	234(4.32),261(4.08), 325(3.69)	39-2491-84C
C$_{14}$H$_{12}$N$_6$O$_3$ Pyrrolo[3,4-c]pyrazole-4,6(1H,5H)-dione, 3-(2-azido-1-oxopropyl)-	MeOH	245(4.30),282(4.22)	103-0416-84
Pyrrolo[3,4-d]-1,2,3-triazole-4,6(1H-5H)-dione, 1-(3-diazo-1-methyl-2-oxopropyl)-3a,6a-dihydro-5-phenyl-	MeOH	255(4.2),270(4.0)	103-0416-84
C$_{14}$H$_{12}$O 9H-Fluorene, 9-methoxy-	MeOH	228(4.28),228(4.32), 235(4.26),274(4.13)	2-0802-84
C$_{14}$H$_{12}$OS 3H-Benz[cd]azulen-3-one, 4,5-dihydro-5-(methylthio)-	CH$_2$Cl$_2$	242(4.22),256s(--), 264(4.08),309(4.32), 346s(--),368s(--), 377(3.91),534(2.70)	83-0984-84
2-Propen-1-one, 2-methyl-1-phenyl-	EtOH	327(3.05)	73-1764-84

Compound	Solvent	$\lambda_{max}(\log \epsilon)$	Ref.
3-(2-thienyl)- (cont.)			73-1764-84
$C_{14}H_{12}O_2$			
Acenaphthylene, 1,2-dimethoxy-	n.s.g.	318(3.89),430(2.86)	78-0665-84
9,10-Anthracenedione, 1,2,3,4-tetra- hydro-	EtOH	228(4.36),245(4.20)	40-1164-84
2,5-Cyclohexadien-1-one, 4-(2,4,6-cy- clohexatrien-1-ylidene)-2-methoxy-	MeCN	271(3.79),285s(3.62), 336(3.37),488(4.32)	88-0077-84
7bH-Cyclopent[cd]indene-2-carboxylic acid, 7b-methyl-, methyl ester	EtOH	251(3.89),302(4.60), 336(3.81),475(3.16)	39-0165-84C
Ethanone, 1,1'-(2,3-naphthalenediyl)- bis-	ether	272(3.79),282(3.83), 292s(3.63),332(3.13)	22-0187-84
Phenol, 2,2'-(1,2-ethenediyl)bis-	MeOH	211(4.37),234(4.12), 280(4.14),331(4.30)	44-1627-84
2-Propen-1-one, 3-(2-furanyl)-2-methyl- 1-phenyl-	EtOH	327(3.48)	73-1764-84
$C_{14}H_{12}O_2S$			
Benzene, [(2-phenoxyethenyl)sulfinyl]-, (Z)-	dioxan	241(4.36),272s(3.68)	104-0473-84
$C_{14}H_{12}O_2S_2$			
2H,5H-Indeno[1,2-c]thiopyran-5-one, 2-methyl-4-(methylthio)-, 2-oxide	EtOH	242(4.56),271(4.32), 327(4.11),347(3.82), 364(3.60),420(3.30)	94-2910-84
$C_{14}H_{12}O_3$			
1(2H)-Anthracenone, 3,4-dihydro-9,10- dihydroxy-	EtOH	223(4.48),258(4.31), 275(4.46),330(3.52), 402(3.69)	40-1164-84
1,3-Benzenediol, 5-[2-(4-hydroxyphen- yl)ethenyl]-, (E)-	EtOH	221(4.11),310(4.25), 325(4.24)	78-4245-84
Benzoic acid, 4-(hydroxyphenylmethyl)-	n.s.g.	240(4.24),269(3.25), 281(2.93)	110-0939-84
2H-1-Benzopyran-2-one, 3-(3-methyl- 1-oxo-2-butenyl)-	EtOH	300(3.87),335(3.65)	118-0859-84
Dibenzofuran, 1,3-dimethoxy-	EtOH	214(4.56),223(4.53), 225s(4.53),227s(4.52), 261(4.18),283(4.27), 291s(4.17),301(4.11)	39-1605-84C
2,5-Furandione, dihydro-3-(1-methyl- ethylidene)-4-(phenylmethylene)-, (E)-	C_6H_{12}	321(4.08)	48-0233-84
(Z)-	C_6H_{12}	332(4.20)	48-0233-84
Furo[3,2-c][1]benzopyran-4-one, 2,3- dihydro-3,3-dimethyl-2-methylene-	MeOH	208(4.00),250(3.46), 258(3.42),294(3.44), 314(3.48),330(3.33)	107-0737-84
Naphtho[2,3-c]furan-1,3-dione, 4,4a- dihydro-4,4-dimethyl-	C_6H_{12}	455(3.90)	48-0233-84
$C_{14}H_{12}O_3S$			
Benzene, [(2-phenoxyethenyl)sulfonyl]-	dioxan	221s(4.12),247(4.16), 264(3.85),272s(3.65)	104-0473-84
1H-Indene-1,3(2H)-dione, 2-[1-(methyl- thio)-3-oxobutylidene]-	EtOH	231(4.20),248(4.33), 297(3.69),308(3.81), 361(4.42)	95-0127-84
$C_{14}H_{12}O_4$			
1,4-Azulenediol, diacetate	C_6H_{12}	240(4.42),281(4.66), 335(3.56),350(3.68), 359(3.46),368(3.52),	35-4852-84

Compound	Solvent	$\lambda_{max}(\log \epsilon)$	Ref.
(cont.)		598(2.49),648s(2.44), 716s(2.05)	35-4852-84
	MeCN	592(2.31)	35-4852-84
1,5-Azulenediol, diacetate	EtOH	217(2.34),240(4.23), 278(4.74),344(3.66), 623(2.58),670s(2.51), 747s(2.09)	35-4857-84
1,6-Azulenediol, diacetate	C_6H_{12}	236(4.19),282(4.78), 336(3.61),351(3.72), 369(3.39),599(2.58), 652s(2.51),722s(2.10)	35-4852-84
	MeCN	595(2.45)	35-4852-84
1,7-Azulenediol, diacetate	EtOH	240(4.18),278(4.65), 344(3.59),623(2.55), 675s(2.49),750s(2.06)	35-4857-84
1,8-Azulenediol, diacetate	EtOH	240(4.37),278(4.60), 342(3.60),590(2.61), 635s(2.53),700s(2.08)	35-4857-84
1,2-Benzenediol, 4-[2-(3,5-dihydroxy-phenyl)ethenyl]- (piceatannol)	MeOH	220(4.41),303(4.36), 323(4.48)	94-3501-84
3-Butenoic acid, 4-(2-furanyl)-2-(2-furanylmethylene)-, methyl ester, (E,E)-	n.s.g.	200(4.43),278(4.39), 363(4.42)	73-2502-84
(E,Z)-	n.s.g.	200(4.41),273(4.11), 330(4.30)	73-2502-84
Seselin-5-ol	MeOH	279(4.44),288(4.42), 318(4.24),350(3.98)	78-3129-84
Xanthyletin-5-ol	MeOH	250(4.40),256(4.43), 262(4.40),270(4.43), 322(4.09)	78-3129-84
$C_{14}H_{12}O_4S_4$ 1,3-Dithiole-4-carboxylic acid, 2-[4-[(2-propenyloxy)carbonyl]-1,3-dithiol-2-ylidene]-, 2-propenyl ester	MeCN	286(4.12),300(4.13), 313(4.12),430(3.47)	103-1342-84
$C_{14}H_{12}O_5$ Acetic acid, (3-methoxy-5-oxo-4-phenyl-2(5H)-furanylidene)-, methyl ester, (Z)-	EtOH EtOH-NaOH	266(4.30),310s(4.04) 259(4.76)	39-1555-84C 39-1555-84C
3-Furancarboxylic acid, 2,5-dihydro-4-hydroxy-2-oxo-5-(phenylmethylene)-, ethyl ester	EtOH	305(4.50)	39-1539-84C
1,4-Naphthalenedione, 2-acetoxy-7-meth-oxy-5-methyl-	MeOH	243(4.23),264s(4.04), 299s(3.73),333(3.55)	44-1853-84
1,4-Naphthalenedione, 7-acetoxy-2-meth-oxy-5-methyl-	MeOH	250(4.23),281(4.09), 341(3.45)	44-1853-84
$C_{14}H_{12}O_6$ 5H-Furo[3,2-g][1]benzopyran-5-one, 7-(hydroxymethyl)-4,9-dimethoxy-	EtOH	216(4.29),247(4.57), 279(3.68),332(3.67)	44-5035-84
Furo[3,4-b]furan-2,4-dione, tetrahydro-6-(hydroxymethyl)-3-[(4-hydroxyphen-yl)methylene]-, [3aR-(3E,3aα,6α,6aα)]-	MeOH	233(4.0),323(4.4)	94-1808-84
$C_{14}H_{12}O_8$ 1H,3H-Pyrano[4,3-b][1]benzopyran-9-car-boxylic acid, 4,10-dihydro-3,7,8-tri-hydroxy-3-methyl-10-oxo- (fulvic acid)	EtOH	224(4.50),317(4.07), 343(4.06)	77-1565-84

Compound	Solvent	$\lambda_{max}(\log \epsilon)$	Ref.
$C_{14}H_{13}BrOS$			
Ethanone, 1-[3-bromo-4-(phenylthio)-1,3-cyclohexadien-1-yl]-	EtOH	361(4.11)	44-2954-84
$C_{14}H_{13}ClN_2S$			
1H-Pyrazole, 3-(4-chlorophenyl)-4,5-dihydro-1-methyl-5-(thienyl)-	EtOH	231(4.05),307(4.04)	34-0225-84
1H-Pyrazole, 5-(4-chlorophenyl)-4,5-dihydro-1-methyl-3-(thienyl)-	EtOH	223(4.03),264(3.86), 290(3.86)	34-0225-84
$C_{14}H_{13}ClOS$			
Ethanone, 2-chloro-1-[4-[(methylthio)-methyl]-1-azulenyl]-	CH_2Cl_2	237(4.07),310(4.41), 380(3.88),545(2.58)	83-0984-84
$C_{14}H_{13}ClOSe$			
Ethanone, 1-[3-chloro-4-(phenylseleno)-1,3-cyclohexadien-1-yl]-	EtOH	361(4.11)	44-2954-84
$C_{14}H_{13}Cl_2NO_4$			
1,4-Naphthalenedione, 2-(butylamino)-6,7-dichloro-3,8-dihydroxy-	benzene	485s(3.91),505(3.93), 530s(3.84)	118-0953-84
$C_{14}H_{13}F_3N_2O_4$			
Alanine, 2,3-didehydro-N-[N-(trifluoro-acetyl)-L-phenylalanyl]-	EtOH	207(4.26),242(3.82)	5-0920-84
$C_{14}H_{13}F_3O_2$			
1,4-Hexadien-3-one, 1-(2,4-dimethyl-phenyl)-6,6,6-trifluoro-5-hydroxy-, (?,E)-	EtOH	244(3.51),370(4.12)	39-2863-84C
2,4-Pentanedione, 3-[(2,4-dimethyl-phenyl)methylene]-1,1,1-trifluoro-	EtOH	260(3.80),300(3.89)	39-2863-84C
$C_{14}H_{13}NOS_2$			
1H-Thiopyran-4-carbonitrile, 1-methyl-3-(methylthio)-5-phenyl-, 1-oxide	EtOH	249.5(4.49),282s(3.93), 332.0(3.93)	118-0852-84
$C_{14}H_{13}NO_2$			
Flindersine	MeOH	235(4.45),333(3.92), 348(4.02),365(3.83)	100-0391-84
1,2-Naphthalenedione, 4-(2-pyrroli-dinyl)-	pH 7	252(4.32),282(3.71), 390(3.83),532(3.91)	94-3093-84
$C_{14}H_{13}NO_3$			
2,4-Benzofurandione, 3,5,6,7-tetra-hydro-3-phenyl-, 2-oxime	MeOH	268(4.062)	73-1421-84
1,2-Naphthalenedione, 4-morpholino-	H_2O	244(4.30),275(4.05), 325(3.89),487(3.73)	94-3093-84
3H-Pyrano[4,3-c]isoxazol-3-one, 6,7-dihydro-6,6-dimethyl-4-phenyl-	EtOH	262(3.98),344(4.01)	142-0365-84
3-Pyridinecarboxylic acid, 1,2-dihydro-2-oxo-1-(1-phenylethyl)-	EtOH	232(3.66),331(3.94)	4-0041-84
3-Pyridinecarboxylic acid, 1-(2,6-di-methylphenyl)-1,2-dihydro-2-oxo-	EtOH	232(3.44),331(3.63)	4-0041-84
$C_{14}H_{13}NO_4$			
2H-Pyrano[3,2-g]quinoline-5,10-dione, 3,4-dihydro-2-methoxy-2-methyl-	EtOH	255(3.90),278(3.88), 368(2.85)	4-0865-84
1H-Pyrrole-3-carboxylic acid, 4,5-di-hydro-1-methyl-4,5-dioxo-2-phenyl-, ethyl ester	dioxan	285s(3.79),408(3.60)	94-0497-84

Compound	Solvent	$\lambda_{max}(\log \epsilon)$	Ref.
$C_{14}H_{13}NO_5$			
1,3-Dioxolo[4,5-g]quinoline-7-carbox-	$2M\ H_2SO_4$	342(2.79)	149-0057-84A
ylic acid, 5-ethyl-5,8-dihydro-8-	pH 4	322(2.98),336(2.97)	149-0057-84A
oxo-, methyl ester			
1H-Pyrrole-3-carboxylic acid, 4,5-di-	dioxan	255(3.63),325(3.87),	94-0497-84
hydro-2-(4-methoxyphenyl)-4,5-dioxo-,		402(3.59)	
ethyl ester			
$C_{14}H_{13}N_2O$			
Phenazinium, N-(2-hydroxyethyl)-,	EtOH	262(4.89),387(4.39),	103-1162-84
chloride		440(3.47)	
$C_{14}H_{13}N_3O$			
Methanimidic acid, N-benzoyl-,	MeCN	229(4.20),256s(4.02),	48-0311-84
2-phenylhydrazide		324(3.97)	
2H-Pyrido[2,3-b][1,4]diazepin-2-one,	EtOH	265(3.71),324(3.91)	161-0162-84
1,3,4,5-tetrahydro-3-phenyl-			
2H-Pyrido[2,3-b][1,4]diazepin-2-one,	EtOH	259(3.70),318(3.84)	161-0162-84
1,3,4,5-tetrahydro-4-phenyl-			
$C_{14}H_{13}N_3OS$			
2-Butanone, 1-diazo-3-methyl-3-(2-quin-	MeOH	235(3.70),250(3.90),	103-0505-84
olinylthio)-		280(4.20)	
2-Butanone, 1-diazo-3-methyl-3-(8-quin-	MeOH	208(3.95),252(4.15),	103-0505-84
olinylthio)-		278(4.28)	
$C_{14}H_{13}N_3O_2$			
Acetamide, N-(2,3,4,6-tetrahydro-6-oxo-	CH_2Cl_2	287(4.34),455(3.46)	83-0329-84
benzo[f]quinoxalin-5-yl)-			
$C_{14}H_{13}N_3O_2S_2$			
Sulfoximine, S,S-dimethyl-N-[2-(methyl-	EtOH	246(4.57),297(4.69)	94-2910-84
thio)-5-oxo-5H-indeno[1,2-d]pyrimi-			
din-4-yl]-			
$C_{14}H_{13}N_3O_4$			
Azepino[2,1-b]quinazoline-6-carboxalde-	EtOH	213(4.48),326(4.14),	4-0219-84
hyde, 6,7,8,9,10,12-hexahydro-2-ni-		390s(3.68)	
tro-12-oxo-	dioxan	225(4.36),300(3.96),	4-0219-84
		389(4.10)	
	$CHCl_3$	300(3.63),393(4.14)	4-0219-84
	DMSO	380(4.13)	4-0219-84
$C_{14}H_{13}N_5O$			
Acetamide, N-[9-(phenylmethyl)-9H-	MeOH	273(4.26)	150-3601-84M
purin-6-yl]-			
Benzamide, N-(9-ethyl-9H-purin-6-yl)-	MeOH	280(4.24)	150-3601-84M
$C_{14}H_{13}N_5O_2$			
1H-Imidazole-4-carboxamide, 5-amino-	EtOH	228(4.57),266(4.17),	2-0870-84
1-(6-methoxy-8-quinolinyl)-		335(3.74)	
$C_{14}H_{13}N_5O_2S$			
Pyrimido[4,5-d]pyrimidine-2,4(1H,3H)-	EtOH	240(4.54),283(4.23)	94-0122-84
dione, 5-amino-3-methyl-7-(methyl-			
thio)-1-phenyl-			
$C_{14}H_{13}N_5O_5$			
1H-Purine-2,6-dione, 3,7-dihydro-	MeCN	275(4.23),353(3.52)	44-3869-84
7-[(4-hydroxy-3-nitrophenyl)-			
methyl]-1,3-dimethyl- (phidolopin)			

Compound	Solvent	$\lambda_{max}(\log \epsilon)$	Ref.
$C_{14}H_{13}N_5O_6$			
Benzenamine, 4-(1,4-dihydro-5,7-di-nitro-4-benzofurazanyl)-N,N-dimethyl-, N-oxide	MeOH	263(4.31),292s(3.95), 364(3.75),475(4.30)	12-0985-84
monopotassium salt	MeOH	263(4.39),293(4.02), 365(3.82),474(4.39)	12-0985-84
$C_{14}H_{13}N_7$			
5,8-Methanoquinoxaline-2,3-dicarbo-nitrile, 1-(cyanodiiminoethyl)-1,4,4a,5,6,7,8,8a-octahydro-	EtOH	317(3.93),398(4.07)	44-0813-84
$C_{14}H_{13}N_7O_2$			
Pteroic acid, 4-amino-4-deoxy-	pH 1	243(4.27),297(4.47), 337s(4.14)	87-0600-84
	pH 13	261(4.48),371(3.96)	87-0600-84
$C_{14}H_{14}$			
4a,8a-[2]Butenonaphthalene	C_6H_{12}	250(4.00)	35-1518-84
$C_{14}H_{14}As_2Cl_2$			
Arsinous chloride, 1,2-ethanediyl-bis[phenyl-	EtOH	230s(4.30),264(3.64)	65-1597-84
$C_{14}H_{14}BrNO$			
2H-Pyrrol-2-one, 5-[bromo(4-methyl-phenyl)methylene]-1,5-dihydro-3,4-dimethyl-, (Z)-	MeOH	302(4.25)	49-0357-84
$C_{14}H_{14}BrNO_2$			
2H-Pyrrol-2-one, 5-[bromo(4-methoxy-phenyl)methylene]-1,5-dihydro-3,4-dimethyl-, (Z)-	MeOH	232(4.01),270(4.09), 313(4.20)	49-0357-84
$C_{14}H_{14}ClNO_4$			
2-Cyclohexen-1-one, 2-[1-(4-chloro-phenyl)-2-nitroethyl]-3-hydroxy-	MeOH	259(4.108)	73-1421-84
1,4-Naphthalenedione, 2-(butylamino)-3-chloro-5,8-dihydroxy-	benzene	480s(3.92),495(3.94), 530s(3.84)	118-0953-84
1,4-Naphthalenedione, 2-chloro-3-[(2,2-dimethoxyethyl)amino]-	MeOH	374(4.43)	12-0389-84
$C_{14}H_{14}ClN_5$			
1H-Purin-6-amine, N-(2-chloro-4-methyl-phenyl)-2,8-dimethyl-	EtOH	297(4.23)	11-0208-84B
7H-Purin-6-amine, N-(4-chlorophenyl)-2,7,8-trimethyl-	EtOH	304(4.18)	11-0208-84B
$C_{14}H_{14}FN_5$			
7H-Purin-6-amine, N-(2-fluorophenyl)-2,7,8-trimethyl-	EtOH	298(4.09)	11-0208-84B
$C_{14}H_{14}F_3NO_2$			
2H-1-Benzopyran-2-one, 7-(diethyl-amino)-4-(trifluoromethyl)-	n.s.g.	372(4.40)	135-0674-84
$C_{14}H_{14}FeO$			
Ferrocene, (3-oxo-1-butenyl)-, (E)-	CH_2Cl_2	478(3.30)	30-0192-84
$C_{14}H_{14}NOS$			
2H-Pyrido[1,2,3-ef]-1,5-benzothiaze-	10% MeCN	246(4.15),270(4.08),	103-0343-84

Compound	Solvent	λ_{max}(log ϵ)	Ref.
pinium, 3,4-dihydro-2,2-dimethyl-3-oxo-, perchlorate (cont.)		330(3.75)	103-0343-84
1H-[1,3]Thiazino[3,2-a]quinolinium, 2,3-dihydro-3,3-dimethyl-2-oxo-, perchlorate	10% MeCN	240(4.12),262(4.12), 268(4.09),352(4.20), 362(4.17)	103-0508-84
C$_{14}$H$_{14}$N$_2$			
1,3-Benzenedicarbonitrile, 4-(2,3-dimethyl-2-butenyl)-	EtOH	234(4.13),275(3.05), 283(3.10),293(3.13)	23-1785-84
4,5-Benzotryptamine, hydrochloride	EtOH	209(4.36),227(4.56), 250(4.49),305(3.90), 310s(3.88),318(3.87), 332(3.54)	103-0277-84
5,6-Benzotryptamine, hydrochloride	EtOH	221(4.28),247(4.90), 332(3.65),349(3.74), 363(3.72)	103-0277-84
6,7-Benzotryptamine, hydrochloride	EtOH	212(4.54),261(4.78), 280s(3.95),320(3.30), 336(2.70)	103-0277-84
1H-Imidazole, 4,5-dihydro-2-(1-naphthalenylmethyl)-, compd. with iodine	CHCl$_3$	285(4.46),360(3.97)	83-0246-84
1H-Indole, 1,2-dimethyl-3-(1H-pyrrol-2-yl)-	EtOH	211(4.24),240(4.31), 298(3.80)	103-0058-84
C$_{14}$H$_{14}$N$_2$O			
Acetamide, N-phenyl-N-(phenylamino)-	MeOH	275(3.19)	2-0060-84
Benzo[f]quinoxalin-6(2H)-one, 3,4-dihydro-4,5-dimethyl-	CH$_2$Cl$_2$	246(4.20),283(4.33), 436(3.65)	83-0743-84
5(10H)-Phenazineethanol, radical cation, chloride	EtOH	210(4.30),263(4.62), 390(4.33),452(3.59), 600(3.05),650(3.20), 720(3.13)	103-1162-84
3(2H)-Pyridazinone, 4-methyl-6-[2-(4-methylphenyl)ethenyl]-	EtOH	229(4.20),298(4.23), 312(4.22)	48-0799-84
WS-30581-A	EtOH	224(4.38),266(4.19), 282s(4.11),298s(4.05)	158-1153-84
C$_{14}$H$_{14}$N$_2$OS			
1H-Pyrazole, 4,5-dihydro-3-(4-methoxyphenyl)-5-(thienyl)-	EtOH	224(3.98),287(4.14)	34-0225-84
1H-Pyrazole, 4,5-dihydro-5-(4-methoxyphenyl)-3-(thienyl)-	EtOH	222(4.00),257(3.71), 304(3.96)	34-0225-84
C$_{14}$H$_{14}$N$_2$O$_2$			
5H-[1]Benzopyrano[3,4-c]pyridazine, 8-methoxy-5,5-dimethyl-	EtOH	209(4.53),221(4.59), 246s(4.04),257s(4.00), 294(4.36),343(4.51)	39-2815-84C
3(2H)-Pyridazinone, 6-[2-(4-methoxyphenyl)ethenyl]-4-methyl-	EtOH	228(4.11),302(4.45), 315(4.44)	48-0799-84
Pyrrolo[2,3-b]indole-2-carboxylic acid, 1,8-dihydro-8-methyl-, ethyl ester	EtOH	279(4.35),336(4.55)	39-2903-84C
C$_{14}$H$_{14}$N$_2$O$_2$S			
3-Pyridinecarboxylic acid, 1-[4-(dimethylamino)phenyl]-1,2-dihydro-2-thioxo-	EtOH	260(4.29),302(4.17), 384(3.58)	4-0041-84
2H-Pyrrol-2-one, 1,5-dihydro-3-morpholino-1-phenyl-5-thioxo-	CHCl$_3$	342(4.20),442(4.04)	78-3499-84
C$_{14}$H$_{14}$N$_2$O$_3$			
Acetic acid, (2,3-dihydro-2-oxo-1-	n.s.g.	268(4.00),323(4.39)	4-0053-84

Compound	Solvent	$\lambda_{max}(\log \epsilon)$	Ref.
phenyl-4(1H)-pyrimidinylidene)-, ethyl ester			4-0053-84
Acetic acid, 2-[(6-phenyl-2-pyrimidinyl)oxy]-, ethyl ester	EtOH	204(4.61),244s(--), 252(4.34),313(4.33)	103-1185-84
Benzenamine, N-[2-(3-nitrophenoxy)-ethyl]-	MeCN	275s(3.83),330s(3.32)	78-1755-84
1H-[1]Benzoxepino[5,4-c]pyrazole, 1-acetyl-4,5-dihydro-8-methoxy-	EtOH	237(4.06),275(4.08), 285(4.14),308(4.22)	2-0736-84
2-Butenoic acid, 3-(benzoylamino)-2-cyano-, ethyl ester	MeOH	242(3.95),304(4.38)	32-0431-84
4,9-Epoxy-1H-benz[f]indazole-3-carboxylic acid, 3a,4,9,9a-tetrahydro-, ethyl ester, (3aα,4β,9β,9aα)-	MeOH	298(3.02)	73-1990-84
5-Pyrimidinecarboxylic acid, 1,2-dihydro-1-methyl-2-oxo-4-phenyl-, ethyl ester	EtOH	205(3.7),218s(3.58), 290(3.17)	103-0431-84
Pyrrolo[2,1-b]quinazoline-3-carboxylic acid, 6,7,8,10-tetrahydro-10-oxo-, ethyl ester	EtOH	225(4.43),264s(3.84), 276(3.75),312(3.44), 322(3.89),336(3.49)	4-0219-84
C₁₄H₁₄N₂O₄			
2-Propenoic acid, 2-cyano-3-(phenylamino)-, 2-ethoxy-2-oxoethyl ester	EtOH	288s(4.00),321(4.39)	5-1702-84
3-Pyridinecarboxylic acid, 5-cyano-4-(2-furanyl)-1,4,5,6-tetrahydro-2-methyl-6-oxo-, ethyl ester	EtOH	216(4.23),283(4.27)	103-1241-84
C₁₄H₁₄N₂S			
1H-Pyrazole, 4,5-dihydro-1-methyl-3-phenyl-5-(thienyl)-	EtOH	228(4.18),300(4.07)	34-0225-84
C₁₄H₁₄N₄			
2,3-Pyrazinedicarbonitrile, 1,4,5,6-tetrahydro-5-methyl-5-(4-methyl-phenyl)-	EtOH	323(3.99)	44-0813-84
C₁₄H₁₄N₄O			
Diazene, 1,1'-oxybis[2-(4-methyl-phenyl)-	DMSO	310(3.37)	123-0075-84
C₁₄H₁₄N₄OS			
6H-Purin-6-one, 1,2,3,7-tetrahydro-1,7-dimethyl-8-(phenylmethyl)-2-thioxo-	pH 13	234(4.40),280(4.29)	2-0316-84
C₁₄H₁₄N₄O₃			
Diazene, 1,1'-oxybis[2-(4-methoxy-phenyl)-	DMSO	345(3.52)	123-0075-84
C₁₄H₁₄N₆O₃			
Carbamic acid, [2-amino-6,9-dihydro-6-oxo-9-(phenylmethyl)-1H-purin-8-yl]-, methyl ester	pH 1	259(4.56)	4-1245-84
	pH 1	259(4.26)	142-2439-84
	pH 7	266(4.53)	4-1245-84
	pH 7	266(4.23)	142-2439-84
	pH 11	264(4.45),273(4.42), 289(4.43)	4-1245-84
	pH 11	264(4.15),273(4.11), 289(4.15)	142-2439-84
Carbamic acid, [5-amino-7-[(phenylmethyl)amino]oxazolo[5,4-d]pyrimidin-2-yl]-, methyl ester	pH 1	268(4.40),306(4.15)	4-1245-84 +142-2439-84
	pH 7	284(4.34)	4-1245-84

Compound	Solvent	$\lambda_{max}(\log \epsilon)$	Ref.
(cont.)	pH 11	297(4.40)	4-1245-84
	pH 11	297(4.40)	142-2439-84
$C_{14}H_{14}O$			
Benzenemethanol, 4-methyl-α-phenyl-	n.s.g.	222(4.08),265(2.78)	110-0939-84
2-Cyclopenten-1-one, 3-phenyl-5-(2-propenyl)-	n.s.g.	224(3.82),287(4.21)	150-2060-84M
Ethanone, 1-(3,8-dimethyl-5-azulenyl)-	MeOH	575(2.72)	138-0627-84
$C_{14}H_{14}ORu$			
Ruthenocene, 1-formyl-1',3-(1,3-propanediyl)-	EtOH	253(4.02),345(3.47)	18-0719-84
$C_{14}H_{14}OS$			
1H-Cyclopent[e]azulene, 2,3-dihydro-3-(methylsulfinyl)- (and isomer)	CH$_2$Cl$_2$	244(4.09),277(4.41), 326(3.41),342(3.48), 354(2.82),384(2.41), 408(2.41),600(2.65)	35-6383-84
$C_{14}H_{14}O_2$			
[1,1'-Biphenyl]-2,2'-diol, 4,4'-dimethyl-	EtOH	219(4.29),292(3.65)	42-0142-84
1(2H)-Dibenzofuranone, 3,4-dihydro-3,3-dimethyl-	EtOH	229(4.35),248(3.92), 266(3.88),281(3.76)	39-1213-84C
3(4H)-Dibenzofuranone, 4a,9b-dihydro-8,9b-dimethyl-	EtOH	223(4.30),260(3.27), 301(3.50)	42-0142-84
Dicyclopenta[a,e]pentalene-1,5-dione, 3a,3b,4,4a,7a,7b,8,8a-octahydro-	MeOH	226(4.16),320(1.92)	44-1728-84
2H-Naphtho[2,3-b]pyran-2-one, 6,7,8,9-tetrahydro-4-methyl-	EtOH	280(3.85),324(3.72)	94-1178-84
Phenol, 4-methyl-2-(4-methylphenoxy)-	EtOH	212(--),278(--)	42-0142-84
$C_{14}H_{14}O_2S$			
4-Azulenepropanoic acid, β-(methylthio)-	CH$_2$Cl$_2$	238(4.36),280(4.62), 342(3.58),356s(--), 583(2.65)	83-0984-84
$C_{14}H_{14}O_2S_2$			
1,4-Dithiaspiro[4.4]non-8-en-7-ol, benzoate, (R)-	EtOH	211s(3.87),229(4.24)	39-2089-84C
$C_{14}H_{14}O_3$			
4-Azulenepropanoic acid, β-methoxy-	CH$_2$Cl$_2$	236(4.23),276(4.51), 280s(--),330s(--), 342(3.50),352s(--), 580(2.54)	83-0984-84
2H-1-Benzopyran-2-one, 3-(1,1-dimethyl-2-propenyl)-7-hydroxy- (angustifolin)	EtOH	258(3.66),269(3.64), 297s(3.83),325(4.01)	102-2095-84
1(2H)-Biphenylenone, 8b-acetoxy-3,4,4a,8b-tetrahydro-	MeOH	264s(3.21),268.5(3.36), 275(3.35)	44-2050-84
3-Buten-2-one, 4-(3,5-diacetylphenyl)-	MeOH	207(4.43),240(4.67), 276(4.45)	24-1620-84
2(5H)-Furanone, 3-methoxy-4-methyl-5-(2-phenylethenyl)-, (E)-	MeOH	205(4.44),209(4.41), 254(4.37),281(3.43), 292(3.26)	5-1616-84
3(2H)-Furanone, 4-acetyl-2,2-dimethyl-5-phenyl-	EtOH	204(4.05),253(4.15), 266(4.14)	39-0535-84C
4H-Furo[3,2-c][1]benzopyran-4-one, 2,3-dihydro-2,3,3-trimethyl-	MeOH	207(4.09),214(4.00), 217(3.99),276(3.64), 288(3.77),311(3.78), 326(3.64)	107-0737-84

Compound	Solvent	$\lambda_{max}(\log \epsilon)$	Ref.
3,5-Hexadienoic acid, 3-methyl-2-oxo-6-phenyl-, methyl ester, (E,E)-	MeOH	236(3.74),242(3.73), 339(4.56)	5-1616-84
$C_{14}H_{14}O_4$			
2H-1-Benzopyran-2-one, 7-hydroxy-8-[(3-methyl-2-butenyl)oxy]-	EtOH	260(4.0),327(4.33)	102-0859-84
	EtOH-NaOEt	280(3.97),388(4.47)	102-0859-84
2H-1-Benzopyran-2-one, 8-hydroxy-7-[(3-methyl-2-butenyl)oxy]-	EtOH	263(4.29),322(4.47)	102-0859-84
	EtOH-NaOEt	283(4.59),332(4.33)	102-0859-84
Naphthalene, 1,2:1,2-bis(ethylenedioxy)-1,2-dihydro-	EtOH	250(3.83)	39-2069-84C
Naphthalene, 1,1:2,2-bis(ethylenedioxy)-1,2-dihydro-	EtOH	268(3.80)	39-2069-84C
Tricyclo[6.2.0.0^{2,5}]deca-1,5,7-triene-6,7-dicarboxylic acid, dimethyl ester	EtOH	2.4(4.15),248(4.05), 297(3.71)[sic]	44-1412-84
$C_{14}H_{14}O_5$			
1,4-Naphthalenedione, 2,3,7-trimethoxy-5-methyl-	MeOH	266(4.39),298(4.12), 344(3.62),400(3.11)	44-1853-84
1,4-Naphthalenedione, 2,7,8-trimethoxy-5-methyl-	MeOH	263(4.32),287(4.05), 396(3.56)	44-1853-84
$C_{14}H_{14}O_7S$			
2H,8H-Benzo[1,2-b:3,4-b']dipyran-2-one, 9,10-dihydro-8,8-dimethyl-9-(sulfo-oxy)-, potassium salt, (R)-	MeOH	220s(4.1),246s(3.5), 257(3.5),325(4.0)	102-0863-84
$C_{14}H_{14}O_8S$			
2H,8H-Benzo[1,2-b:3,4-b']dipyran-2-one, 9,10-dihydro-10-hydroxy-8,8-dimethyl-9-(sulfooxy)-, cis	MeOH	219s(4.10),247(3.47), 257(3.43),299s(3.86), 325(4.09)	102-0863-84
Rutaretin 1"-(hydrogen sulfate), 2'(S)-, potassium salt	MeOH	230s(4.07),257s(3.59), 266(3.66),335(4.14)	102-0863-84
	MeOH-NaOAc	226(4.3),284(4.0), 340(4.1)	102-0863-84
$C_{14}H_{14}S$			
1H-Cyclopent[e]azulene, 2,3-dihydro-3-(methylthio)-	CH$_2$Cl$_2$	246(4.40),280(4.58), 314(3.67),326(3.62), 341(3.65),354(3.21), 382(2.67),404(2.65), 600(2.65)	35-6383-84
$C_{14}H_{15}BrN_2O_4$			
L-Tryptophan, 5-bromo-N-(methoxycarbonyl)-, methyl ester	MeOH	227(4.56),283(3.74), 290(3.76),299(3.64)	94-2126-84
$C_{14}H_{15}BrN_4O$			
Pyrazolo[5,1-b]quinazolin-9(4H)-one, 7-bromo-3-[(dimethylamino)methyl]-2-methyl-	10% EtOH-NaOH	217(4.65),255(4.56), 332(4.38),402(3.63)	56-0411-84
$C_{14}H_{15}Br_2NO$			
2,6-Methano-1-benzazocin-11-one, 8,10-dibromo-1,2,3,4,5,6-hexahydro-1,6-dimethyl-	EtOH	270(4.0),304(3.6)	142-1771-84
$C_{14}H_{15}ClN_2O$			
3-Azetidinecarbonitrile, 3-chloro-1-(1,1-dimethylethyl)-2-oxo-4-phenyl-, trans	EtOH	210(4.11)	44-0282-84

Compound	Solvent	$\lambda_{max}(\log \epsilon)$	Ref.
$C_{14}H_{15}ClN_2O_2$ 1,4-Naphthalenedione, 5-amino-3-(butyl-amino)-2-chloro-	benzene	458(3.90)	39-1297-84C
$C_{14}H_{15}ClN_2O_3$ 1,4-Naphthalenedione, 5-amino-3-(butyl-amino)-2-chloro-8-hydroxy-	benzene	497(3.84),527(3.85), 558(3.70)	39-1297-84C
$C_{14}H_{15}ClN_2O_4$ L-Tryptophan, 5-chloro-N-(methoxycarbo-nyl)-, methyl ester	MeOH	228(4.58),282(3.78), 290(3.81),300(3.71)	94-2126-84
$C_{14}H_{15}ClN_4$ Pyridazino[3,4-b]quinoxaline, 10-butyl-3-chloro-5,10-dihydro-	MeOH-HCl	213(4.21),252(4.60), 258s(4.54),282(4.26), 291s(4.18),315s(3.84), 370(3.84),474(3.64)	103-1281-84
$C_{14}H_{15}ClN_4O$ Pyrazolo[5,1-b]quinazolin-9(4H)-one, 7-chloro-3-[(dimethylamino)methyl]-2-methyl-	10% EtOH-NaOH	221(4.51),255(4.55), 332(4.36),404(3.66)	56-0411-84
$C_{14}H_{15}ClO_4$ Propanedioic acid, [(4-chlorophenyl)-methylene]-, diethyl ester	dioxan	287(4.30)	65-1987-84
$C_{14}H_{15}NO$ 1H-Inden-1-one, 2-[3-(dimethylamino)-2-propenylidene]-2,3-dihydro-, (E,E)-	EtOH	273(4.28),438(4.84)	118-0424-84
$C_{14}H_{15}NOS$ Phenol, 4-amino-2-[[(4-methylphenyl)-thio]methyl]- (also metal chelates)	C_6H_{12}	245(3.50),310(3.28)	65-0142-84
Phenol, 4-amino-2-[[(phenylmethyl)-thio]methyl]- (also metal chelates)	C_6H_{12}	230(4.20),310(3.48)	65-0142-84
$C_{14}H_{15}NO_2$ 1H-Indole-3-carboxylic acid, 2-(1,1-dimethyl-2-propenyl)-	EtOH	278(3.06),286(3.07)	39-0567-84C
Methanamine, N-[[2-(methoxymethoxy)-1-naphthalenyl]methylene]-	EtOH	225.5(4.57),296(3.72), 333(3.54)	150-0701-84M
1H-Pyrrole-2-acetic acid, α-(phenyl-methyl)-, methyl ester	MeOH	205(4.20)	107-0453-84
$C_{14}H_{15}NO_2S$ 1-Oxa-4-azaspiro[4.5]dec-3-en-2-one, 3-(phenylthio)-	EtOH	270.5(3.79)	88-5127-84
$C_{14}H_{15}NO_3S$ 2-Thiophenecarboxylic acid, 3-amino-5-(4-methoxyphenyl)-, ethyl ester	MeOH	308(4.32),348(4.12)	118-0275-84
$C_{14}H_{15}NO_4$ 2-Cyclohexen-1-one, 3-hydroxy-2-(2-ni-tro-1-phenylethyl)-	MeOH	258(4.102)	73-1421-84
2-Cyclopenten-1-one, 3-hydroxy-2-[1-(4-methylphenyl)-2-nitroethyl]-	MeOH	250(4.12)	73-1421-84
$C_{14}H_{15}NO_5$ Cyclohepta[b]pyrrol-2(1H)-one, 1-β-D-	EtOH	264(4.46),289s(4.13),	150-0264-84S

Compound	Solvent	$\lambda_{max}(\log \epsilon)$	Ref.
ribofuranosyl- (cont.)		389(3.86),410(3.89), 430s(3.53),452s(3.14), 482s(2.45)	150-0264-84S
$C_{14}H_{15}NO_6$			
Propanedioic acid, [(3-nitrophenyl)-methylene]-, diethyl ester	dioxan	262(4.41)	65-1987-84
Propanedioic acid, [(4-nitrophenyl)-methylene]-, diethyl ester	dioxan	299(4.26)	65-1987-84
$C_{14}H_{15}NO_8$			
Pancratistatin	MeOH	209s(--),219s(--), 233(4.32),278(3.91), 308s(--)	77-1693-84
$C_{14}H_{15}N_3$			
Benzenamine, N,N-dimethyl-4-(phenyl-azo)-	pH 8	447.1(3.54)	46-6185-84
1H-Pyrazole, 4,5-dihydro-1-methyl-3-phenyl-5-(1H-pyrrolyl)-	EtOH	221(3.06),300(3.98)	34-0225-84
$C_{14}H_{15}N_3O_3S$			
Methyl Orange	pH 8	463.5(4.30)	46-6185-84
sodium salt	pH 9	462(4.5)	136-0181-84C
+ γ-cyclodextrin	pH 9	430(4.5)	136-0181-84C
$C_{14}H_{15}N_3O_4$			
Acetic acid, cyano[(2-methylphenyl)hy-drazono]-, 2-ethoxy-2-oxoethyl ester	EtOH	248(3.88),255(3.82), 287s(3.18),370(4.33)	5-1702-84
$C_{14}H_{15}N_5$			
7H-Purin-6-amine, 2,7,8-trimethyl-N-phenyl-	EtOH	225s(--),302(4.28)	11-0208-84B
9H-Purin-6-amine, 2,8-dimethyl-N-(3-methylphenyl)-	EtOH	300(3.96)	11-0208-84B
$C_{14}H_{15}N_5O$			
8H-Purin-8-one, 6-amino-7-ethyl-7,9-dihydro-9-(phenylmethyl)-	MeOH	274(4.09)	150-3601-84M
$C_{14}H_{15}N_5O_2$			
9H-Purin-6-amine, N-[(4-methoxyphenyl)-methoxy]-9-methyl-	pH 1	273(4.21)	94-4842-84
	pH 7	272(4.25)	94-4842-84
	pH 13	282(4.13)	94-4842-84
	EtOH	272(4.24)	94-4842-84
6H-Purin-6-imine, 1,9-dihydro-1-[(4-methoxyphenyl)methoxy]-9-methyl-, monoperchlorate	pH 1	260(4.12)	94-4842-84
	pH 7	260(4.11)	94-4842-84
	pH 13	259(4.10),265s(4.07)	94-4842-84
	EtOH	233s(4.21),260(4.14)	94-4842-84
Pyrrolidine, 1-[(9-amino-6H-imidazo-[1,2-d]pyrido[3,2-b][1,4]oxazin-8-yl)carbonyl]-	MeOH	237(4.14),281(4.29)	4-1081-84
$C_{14}H_{15}N_5O_3$			
Pyrazolo[5,1-b]quinazolin-9(4H)-one, 3-[(dimethylamino)methyl]-2-methyl-7-nitro-	10% EtOH-NaOH	245(4.47),288(4.07), 317(4.03),495(4.08)	56-0411-84
$C_{14}H_{15}N_5O_4$			
1H-Pyrrolo[1,2-a]indole-5,8-dione, 7-amino-2-azido-2,3-dihydro-9-(hy-	MeOH	214(3.99),247(4.13), 306(3.95),350s(3.44),	44-5164-84

Compound	Solvent	$\lambda_{max}(\log \epsilon)$	Ref.
droxymethyl)-1-methoxy-6-methyl-, (1S-trans)- (cont.)		520(2.70)	44-5164-84
$C_{14}H_{15}N_5O_6S$ L-Histidine, N-acetyl-2-[(4-sulfophenyl)azo]-, barium salt	pH 1.5 pH 6 pH 13.1	364(4.44) 386(4.38) 443(4.46)	69-0589-84 69-0589-84 69-0589-84
$C_{14}H_{15}N_7O_2$ Hydrazinecarboxylic acid, 2-(7-methyl-5-phenyl-5H-pyrazolo[3,4-e]-1,2,4-triazin-3-yl)-, ethyl ester	EtOH	260(4.24),334(3.73)	4-1565-84
$C_{14}H_{16}$ Benzo[a]cyclopropa[cd]pentalene, 1,2,2a,2b,6b,6c-hexahydro-1,2-dimethyl-, (1α,2α,2aβ,2bβ,6bβ,6cβ)-	C_6H_{12}	230(3.70),262s(2.78), 267(2.99),273s(3.11), 275(3.15),282.5(3.22)	23-2769-84
Bicyclo[8.2.2]tetradeca-3,5,10,12,13-pentaene	EtOH	219(4.07),247s(3.55), 284s(2.45)	24-0474-84
Cyclopent[fg]acenaphthylene, 1,2,2a,3,4,4a,5,6-octahydro-	EtOH	255(3.10),263s(3.02), 270s(2.89)	24-0455-84
$C_{14}H_{16}AsN_5O_6$ L-Histidine, N-acetyl-2-[(4-arsonophenyl)azo]-, barium salt L-Histidine, N-acetyl-5-[(4-arsonophenyl)azo]-, barium salt	pH 6 pH 13.7 pH 6 pH 13.0	389(4.41) 440(4.45) 352(4.44) 386(4.49)	69-0589-84 69-0589-84 69-0589-84 69-0589-84
$C_{14}H_{16}As_2O_4$ Arsinic acid, 1,2-ethanediylbis[phenyl-	EtOH	211(3.93),217(3.96), 257(2.31),263(2.44), 269(2.36)	65-1597-84
$C_{14}H_{16}BrNO$ 2,6-Methano-1-benzazocin-11-one, 8-bromo-1,2,3,4,5,6-hexahydro-1,6-dimethyl-	EtOH	267(4.3),310(3.4), 335(3.3)	142-1771-84
$C_{14}H_{16}BrN_3O_2S$ Eudistomin C	MeOH	226(4.37),287(3.90)	35-1524-84
$C_{14}H_{16}ClNO_2$ 2,4,6-Heptatrienoic acid, 7-(3-chloro-1H-pyrrol-2-yl)-2-methyl-, ethyl ester, (E,E,E)-	EtOH	382(4.57)	39-1577-84C
$C_{14}H_{16}ClN_2O_3PS$ Phenyl phosphite, S-(4-chlorobenzyl)-thiouronium salt	MeOH	223(4.15)	118-0410-84
$C_{14}H_{16}FeO_4$ Iron, tricarbonyl[(1,3,4,5-η)-1,3,6,6-tetramethyl-2-oxo-4-cycloheptene-1,3-diyl]-	hexane	218(4.27),269(3.81), 318(3.37)	24-3473-84
Iron, tricarbonyl[(1,3,4,5-η)-1,4,6,6-tetramethyl-2-oxo-4-cycloheptene-1,3-diyl]-	hexane	267(3.79),318(3.36)	24-3473-84
Iron, tricarbonyl[(1,3,4,5-η)-1,4,7,7-tetramethyl-2-oxo-4-cycloheptene-1,3-diyl]-	hexane	272(3.75),323(3.41)	24-3473-84

Compound	Solvent	$\lambda_{max}(\log \epsilon)$	Ref.
Iron, tricarbonyl[(1,3,4,5-η)-3,4,6,6-tetramethyl-2-oxo-4-cycloheptene-1,3-diyl]-	hexane	270(3.77),314(3.33)	24-3473-84
Iron, tricarbonyl[(1,3,4,5-η)-3,4,7,7-tetramethyl-2-oxo-4-cycloheptene-1,3-diyl]-	hexane	276(3.74),315(3.41)	24-3473-84
Iron, tricarbonyl[(1,3,4,5-η)-6,6,7,7-tetramethyl-2-oxo-4-cycloheptene-1,3-diyl]-	hexane	218(4.25),269(3.77), 318(3.29)	24-3473-84
$C_{14}H_{16}N_2$			
4,9-Azo-5,8-methano-1H-benz[f]indene, 3a,4,4a,5,8,8a,9,9a-octahydro-, (3aα,4β,4aα,5α,8α,8aα,9β,9aα)-	hexane	396(2.10)	24-1455-84
Benzenepropanenitrile, 2-cyano-α,α,β,β-tetramethyl-	hexane	226(4.05),234s(3.93), 266(2.92),270s(2.96), 274(3.14),282(3.20)	23-1785-84
Benzenepropanenitrile, 4-cyano-α,α,β,β-tetramethyl-	EtOH	231(4.30),238(4.24), 260s(2.38),267s(2.89), 271(2.96),279(2.91)	23-1785-84
[1,1'-Biphenyl]-2,2'-diamine, 6,6'-di-methyl-, (S)-(-)-	heptane	206(4.75),236.5(4.18), 293(3.74)	39-2013-84C
	MeOH	205(4.77),235(4.16), 292.5(3.68)	39-2013-84C
dihydrochloride	MeOH	212s(4.35),263(3.02), 266s(3.99),271(2.98)	39-2013-84C
Cycloocta[b]quinoxaline, 6,7,8,9,10,11-hexahydro-	MeOH	206(4.43),238(4.46), 319(3.96)	56-0917-84
15,16-Diazapentacyclo[9.2.1.2^{3,9}.0^{1,10}.0^{4,8}]hexadeca-5,12,15-triene, (1S*,2R*,3S*,4S*)-	hexane	355(1.30),379(1.64), 385(1.63),395(1.48)	44-0001-84
2-Pyridinamine, N-(1-phenylethyl)-, (S)-(-)-	isooctane	250(4.46),313(3.72)	103-0077-84
	EtOH	204(4.23),250(4.37), 315(3.64)	103-0077-84
$C_{14}H_{16}N_2O$			
3-Azetidinecarbonitrile, 1-(1,1-dimeth-ylethyl)-2-oxo-4-phenyl-, cis	EtOH	210(3.45)	44-0282-84
trans	EtOH	210(4.04)	44-0282-84
Cycloocta[b]quinoxaline, 6,7,8,9,10,11-hexahydro-, 5-oxide	MeOH	211s(4.32),246(4.79), 328(4.00)	56-0917-84
Ethanone, 1-(1,2-dimethyl-1H-indol-3-yl)-, O-ethenyloxime, (E)-	EtOH	223(4.41),280(4.01)	103-0058-84
$C_{14}H_{16}N_2OS$			
Benzo[b]thiophene-3-ol, 2-[(1-piperi-dinylimino)methyl]-, (E)-	heptane	350(4.26),440(2.85)	104-1353-84
	PrOH	440(4.26)	104-1353-84
	DMSO	350(4.21),440(3.80)	104-1353-84
$C_{14}H_{16}N_2O_2$			
Benzo[6,7]cyclohepta[1,2-c]pyrazol-3(2H)-one, 3a,4,5,6-tetrahydro-8-methoxy-3a-methyl-	EtOH	286(4.17)	2-0736-84
$C_{14}H_{16}N_2O_2S$			
2,4-Cyclohexadien-1-one, 2-methoxy-6-(4,5,6,7-tetrahydro-4-phenyl-2H-isothiazolo[2,3-a]pyrimidin-2-yli-dene)-	EtOH	243(4.23),300(4.12), 390(4.12)	39-1631-84C

Compound	Solvent	λ_{max}(log ϵ)	Ref.
$C_{14}H_{16}N_2O_3$			
2-Butenoic acid, 3-amino-2-(1-imino-ethyl)-4-oxo-4-phenyl-, ethyl ester	MeOH	241(4.22),324(3.91)	32-0261-84
1-Naphthalenemethanamine, 2-(methoxy-methoxy)-N-methyl-N-nitroso-	EtOH	227(4.76),268(3.67), 279(3.77),291(3.71), 316(3.28),330(3.35)	150-0701-84M
5-Pyrimidinecarboxylic acid, 1,2,3,4-tetrahydro-1-methyl-2-oxo-4-phenyl-, ethyl ester	EtOH	205(4.15),228s(3.84), 296(3.97)	103-0431-84
$C_{14}H_{16}N_2O_4$			
Pyrrolo[2,3-b]indole-1(2H)-carboxylic acid, 8-acetyl-3,3a,8,8a-tetrahydro-5-hydroxy-, methyl ester, cis-(±)-	MeOH	251.5(4.14),294.5(3.52)	94-2544-84
Pyrrolo[2,3-b]indole-1(2H)-carboxylic acid, 8-acetyl-3,3a,8,8a-tetrahydro-6-hydroxy-, methyl ester, cis-(±)-	MeOH	215(4.40),246.5(3.99), 290(3.70)	94-2544-84
Pyrrolo[2,3-b]indole-1(2H)-carboxylic acid, 8-acetyl-3,3a,8,8a-tetrahydro-7-hydroxy-, methyl ester, cis-(±)-	MeOH	215(4.41),248(3.82), 285.5(3.42)	94-2544-84
$C_{14}H_{16}N_2O_5$			
Tryptophan, 5-hydroxy-N-(methoxycarbo-nyl)-, methyl ester	MeOH	220s(4.37),277(3.79), 300(3.66)	94-2544-84
$C_{14}H_{16}N_2O_6$			
[2,2'-Bi-2H-pyrrole]-2,2'-dicarboxylic acid, 1,1',3,3'-tetrahydro-3,3'-di-oxo-, diethyl ester	MeCN	305(4.02)	33-1957-84
2,6-Pyridinedicarboxylic acid, 4-[2-(2-carboxy-1-pyrrolidinyl)ethenyl]-2,3-dihydro- (indicaxanthin plus isomer)	H_2O	260(3.95),295s(3.65), 481(4.73)	33-1547-84
$C_{14}H_{16}N_4O$			
Pyrazolo[5,1-b]quinazolin-9(4H)-one, 3-[(dimethylamino)methyl]-2-methyl-	10% EtOH-NaOH	217(4.63),255(4.67), 326(4.37),400(3.82)	56-0411-84
$C_{14}H_{16}N_5$			
7H-Purinium, 6-amino-7-ethyl-9-(phenyl-methyl)-, iodide	MeOH	271(4.11)	150-3601-84M
$C_{14}H_{16}N_2O_3$			
1,2,4-Triazolo[3,4-f][1,2,4]triazin-8(7H)-one, 6-amino-3-[[2-(phenyl-methoxy)ethoxy]methyl]-	EtOH	242s(--),278(3.45)	4-0697-84
$C_{14}H_{16}O$			
2,4-Cyclohexadien-1-one, 4-methyl-6-(2-methyl-2-propenyl)-6-(2-propynyl)-	EtOH	224s(3.45),315(3.42)	33-1298-84
2,5-Cyclohexadien-1-one, 4-methyl-2-(2-methyl-1-propenyl)-4-(2-propynyl)-	EtOH	231(4.13),290(3.20)	33-1298-84
2,5-Cyclohexadien-1-one, 4-methyl-2-(2-methyl-2-propenyl)-4-(2-propynyl)-	EtOH	232(4.05)	33-1298-84
1H-Inden-1-one, 2,3-dihydro-7-methyl-3-methylene-5-(1-methylethyl)-	MeOH	330(3.79)	138-0627-84
$C_{14}H_{16}O_2$			
2H-1-Benzopyran, 4-ethenyl-7-methoxy-	EtOH	212(4.20),228s(3.04),	39-2815-84C

Compound	Solvent	$\lambda_{max}(\log \epsilon)$	Ref.
2,2-dimethyl- (cont.)		270(3.70),310(3.57)	39-2815-84C
Naphtho[1,2-b]furan-6(7H)-one, 2,3,8,9-tetrahydro-9,9-dimethyl-	MeCN	265(3.91)	87-0306-84
Naphtho[2,3-b]furan-8(5H)-one, 2,3,6,7-tetrahydro-5,5-dimethyl-	MeCN	259(3.96)	87-0306-84
$C_{14}H_{16}O_2S$			
Cyclopenta[c]thiopyran-3-carboxylic acid, 6-(1,1-dimethylethyl)-, methyl ester	CH_2Cl_2	270(4.60),277(4.57), 308(4.06),320(4.02), 375(3.90),389(3.93), 445(3.02)	118-0262-84
$C_{14}H_{16}O_3$			
1,8a(4bH)-Biphenylenediol, 5,6,7,8-tetrahydro-, 1-acetate, cis	MeOH	255(2.86),263(2.94), 271(2.90)	87-0792-84
Ethanone, 1-(8-methoxy-2,2-dimethyl-2H-1-benzopyran-5-yl)-	EtOH	253(4.36),285(4.06), 340(3.60)	18-0442-84
2,4-Hexadien-1-one, 1-(2,4-dihydroxy-3,5-dimethylphenyl)-, (E,E)-	EtOH	204(4.39),323(4.30)	78-4081-84
2(1H)-Naphthalenone, 1-hydroperoxy-1-(1-methylpropyl)-	EtOH	233(4.15),238(4.14), 314(3.91)	150-0701-84
Spiro[benzocyclooctene-6(5H),2'-[1,5]-dioxolan]-5-one, 7,8,9,10-tetrahydro-	MeOH	245s(3.25)	44-2050-84
Spiro[biphenylene-1(8bH),2'-[1,3]dioxolan]-8b-ol, 2,3,4,4a-tetrahydro-, cis	MeOH	260(3.07),266(3.25), 273(3.23)	44-2050-84
$C_{14}H_{16}O_4$			
1-Naphthalenecarboxylic acid, 2-ethyl-1,4,4a,5,8,8a-hexahydro-5,8-dioxo-, methyl ester	MeOH	224(4.01),293(2.61)	78-4701-84
4-Pentenoic acid, 5-(1,3-benzodioxol-5-yl)-, ethyl ester	MeCN	260(4.45),298(4.12)	44-0228-84
Propanedioic acid, (phenylmethylene)-, diethyl ester	dioxan	280(4.28)	65-1987-84
$C_{14}H_{16}O_5$			
Allamycin, 3-O-methyl-	MeOH	215(4.20)	94-2947-84
1,3-Benzodioxole-5-pentanoic acid, γ-oxo-, ethyl ester	MeCN	234(3.90),286(3.85)	44-0228-84
1,3-Dioxane-4,6-dione, 5-(4-methoxyphenyl)-2,2,5-trimethyl-	MeOH	234(4.00),275(3.15), 282(3.08)	12-1245-84
$C_{14}H_{16}O_5S$			
1,4-Dioxaspiro[4.5]dec-6-en-6-ol, benzenesulfonate	ether	218(4.03),265(3.00)	44-2355-84
$C_{14}H_{16}Se_2$			
4H-Selenin, 4-(2,6-dimethyl-4H-selenin-4-ylidene)-2,6-dimethyl-	$CHCl_3$	245(3.9),395(4.54), 415(4.62)	22-0241-84
$C_{14}H_{17}BrO_4$			
Naphtho[1,2-b]furan-2,8(4H,9H)-dione, 7-bromo-5,5a,6,7,9a,9b-hexahydro-3-hydroxy-5a,9-dimethyl-, [5aS-(5aα,7β,9β,9aβ,9bα)]-	EtOH	236(4.03)	32-0107-84
$C_{14}H_{17}Br_3$			
Bicyclo[8.3.1]tetradeca-1(14),10,12-triene, 11,12,13-tribromo-	n.s.g.	212(4.68),227s(4.19), 283(2.82)	24-0455-84

Compound	Solvent	$\lambda_{max}(\log \epsilon)$	Ref.
$C_{14}H_{17}ClN_2O$			
Cyclohexanamine, N-(5-chloro-3-methyl-2(3H)-benzoxazolylidene)-	$CHCl_3$	261(3.82),307(4.10)	94-3053-84
$C_{14}H_{17}ClN_2O_4S$			
1-Cyclopentene-1-carboxylic acid, 2-[[2-(aminosulfonyl)-5-chlorophenyl]-amino]-, ethyl ester	EtOH	214(4.21),227s(4.18),326(4.25)	104-0534-84
$C_{14}H_{17}ClO_6$			
2,4-Cyclohexadiene-1,1-dicarboxylic acid, 3-chloro-4-hydroxy-2-methyl-5-(1-oxopropyl)-, dimethyl ester	MeOH	230(3.74),256(3.89),337(3.90)	44-3791-84
$C_{14}H_{17}ClSi$			
Silane, [3-(3-chlorophenyl)-3-penten-1-ynyl]trimethyl-	EtOH	260(2.64)	65-1400-84
Silane, [3-(4-chlorophenyl)-3-penten-1-ynyl]trimethyl-	EtOH	258(2.65)	65-1400-84
$C_{14}H_{17}Cl_2N_7O_2$			
Byproduct, m. 215-8°	pH 1	237s(--),244(4.39),260s(--),323(4.31),330s(--)	87-1416-84
	pH 7	229(4.50),245s(4.42),260s(--),304(4.28)	87-1416-84
$C_{14}H_{17}IN_2O_3$			
1-Piperazinecarboxylic acid, 4-(2-iodo-benzoyl)-, ethyl ester	MeOH	226s(4.18)	73-1009-84
$C_{14}H_{17}N$			
2H-Benz[g]indole, 3,3a,4,5-tetrahydro-2,2-dimethyl-	EtOH	248(4.16),286(3.34),297(3.26)	12-0577-84
$C_{14}H_{17}NO$			
Azeto[1,2-b][2]benzazepin-4(9H)-one, 1,2,10,10a-tetrahydro-2,2-dimethyl-	EtOH	227(3.95)	12-0577-84
Benzamide, N-(cyclohexylidenemethyl)-	EtOH	225(4.03),270(4.02)	44-0714-84
2H-Benz[g]indole, 3,3a,4,5-tetrahydro-2,2-dimethyl-, 1-oxide	EtOH	229(3.90),300(4.20)	12-0117-84
2H-Benz[g]indole, 3,3a,4,5-tetrahydro-7-methoxy-3a-methyl-	EtOH	268(4.26)	39-0005-84C
2H-Benz[g]indol-3a-ol, 3,3a,4,5-tetra-hydro-2,2-dimethyl-	EtOH	248(4.22),287(3.34)	12-0577-84
1H-Benz[g]oxazirino[3,2-i]indole, 2,2a,3,4-tetrahydro-1,1-dimethyl-	EtOH	260(2.95),268(3.00),275(3.00)	12-0577-84
2,6-Methano-1-benzazocin-11-one, 1,2,3,4,5,6-hexahydro-1,6-dimethyl-	EtOH	258(4.3),304(3.6),330(3.2)	142-1771-84
2,4-Pentadien-1-one, 5-(dimethylamino)-2-methyl-1-phenyl-, (E,E)-	EtOH	249(3.96),387(4.62)	118-0424-84
$C_{14}H_{17}NO_2$			
2H-Benz[g]indol-3a-ol, 3,3a,4,5-tetra-hydro-2,2-dimethyl-, 1-oxide	EtOH	229(3.98),294(4.25)	12-0587-84
2H-1-Benzopyran-2-one, 7-(diethyl-amino)-4-methyl-	pentane	351(4.34)	135-0674-84
2aH-Benz[g]oxazirino[3,2-i]indol-2a-ol, 1,2,3,4-tetrahydro-1,1-dimethyl-	EtOH	255(3.08),268(3.00),275(3.00)	12-0599-84
3aH-[1]Benzoxepino[5,4-b]pyrrole, 2,3,4,5-tetrahydro-8-methoxy-3a-methyl-	EtOH	260(4.06)	39-0005-84C

Compound	Solvent	$\lambda_{max}(\log \epsilon)$	Ref.
2,4,6-Heptatrienoic acid, 2-methyl-7-(1H-pyrrol-2-yl)-, ethyl ester, (E,E,E)-	EtOH	393(4.61)	39-1577-84C
2H-Indol-3a-ol, 3,3a,4,5,6,7-hexahydro-3-phenyl-, 1-oxide	EtOH	232(4.06)	12-0587-84
1-Naphthalenemethanamine, 2-(methoxymethoxy)-N-methyl-	EtOH	228(4.83),269(3.63), 278(3.73),290(3.67), 316(3.25),330(3.31)	150-0701-84M
$C_{14}H_{17}NO_2S_4$			
6H-1,3-Thiazine-6-thione, 2-(ethylthio)-2,3-dihydro-4-(3-methylphenyl)-5-(methylsulfonyl)-	EtOH	259(3.93),347(4.19)	4-0953-84
$C_{14}H_{17}NO_3$			
Hippuric acid, 5-hexenyl ester	C_6H_{12}	224(4.04)	44-0399-84
1(2H)-Naphthalenone, 3,4-dihydro-2-(2-methyl-2-nitropropyl)-	EtOH	245(4.06),291(3.18)	12-0117-84
$C_{14}H_{17}NO_4$			
2-Butenedioic acid, 2-[1-methyl-5-(2-propenyl)-1H-pyrrol-2-yl]-, dimethyl ester	EtOH	238(3.96),355(4.21)	39-2541-84C
1(2H)-Naphthalenone, 3,4-dihydro-2-hydroxy-2-(2-methyl-2-nitropropyl)-	EtOH	250(4.06),292(3.28)	12-0587-84
$C_{14}H_{17}NO_5$			
L-Tyrosine, N-acetyl-3-formyl-, ethyl ester	EtOH	220(4.26),255(4.00)	56-0157-84
$C_{14}H_{17}NO_6S$			
1H-Pyrrolo[1,2-a]indole-9-carboxylic acid, 2,3,5,6,7,8-hexahydro-6-methyl-2-[(methylsulfonyl)oxy]-8-oxo-	CHCl$_3$	303.0(3.454)	44-5164-84
$C_{14}H_{17}NS$			
Benzenecarbothioamide, N-(cyclohexylidenemethyl)-	EtOH	223(4.54),312(4.30), 388(2.59)	44-0714-84
2-Butenenitrile, 3-(butylthio)-4-phenyl-	EtOH	207(4.27),259(4.10)	78-2141-84
$C_{14}H_{17}N_3O$			
1H-Isoindol-1-imine, 3-(2,6-dimethyl-4-morpholinyl)-, cis	MeOH	266(4.06),305(3.70), 332(3.66)	20-0459-84
$C_{14}H_{17}N_3O_3$			
4-Isoxazolidinecarbonitrile, 2-(1,1-dimethylethyl)-5-nitro-3-phenyl-, (3α,4β,5α)-	C_6H_{12}	220(3.53)	44-0276-84
5-Isoxazolidinecarbonitrile, 2-(1,1-dimethylethyl)-4-nitro-3-phenyl-, (3α,4β,5α)-	EtOH	210(3.79)	44-0276-84
L-Tryptophan, 5-(acetylamino)-, methyl ester	MeOH	238(4.43),311s(3.30)	94-2126-84
$C_{14}H_{17}N_3O_4$			
β-D-erythro-Pentofuranoside, methyl 3-azido-2,3-dideoxy-, 5-(4-methylbenzoate)	MeOH	237(4.11)	118-0961-84

Compound	Solvent	$\lambda_{max}(\log \epsilon)$	Ref.
$C_{14}H_{17}N_3S_2$			
1,3,4-Thiadiazole-2(3H)-thione, 5-phenyl-3-(1-piperidinylmethyl)-	EtOH	339(4.17)	83-0547-84
$C_{14}H_{17}N_4OS$			
Thiazolium, 5-(2-hydroxyethyl)-4-methyl-3-[(5-methylimidazo[1,2-c]pyrimidin-8-yl)methyl]-, perchlorate	pH 7	224(4.29),263(3.94), 296(3.56)	149-0111-84A
	MeOH	224(4.28),263(3.94), 298(3.57)	149-0111-84A
	MeOH-acid	214(4.31),264(4.12)	149-0111-84A
	10% MeOH- N H_2SO_4	214(4.29),267(4.09)	149-0111-84A
$C_{14}H_{17}N_5O_4S_2$			
4-Thia-1-azabicyclo[3.2.0]hept-2-ene-2-carboxylic acid, 6-ethyl-3-[[(1-methyl-1H-tetrazol-5-yl)thio]methyl]-7-oxo-, 2-oxopropyl ester, cis-(±)-	$CHCl_3$	332(3.77)	32-0319-84
$C_{14}H_{17}N_5O_5S_2$			
4-Thia-1-azabicyclo[3.2.0]heptane-2-carboxylic acid, 6-ethyl-3-[[(1-methyl-1H-tetrazol-5-yl)thio]-methylene]-7-oxo-, (acetyloxy)-methyl ester, (2α,5α,6α)-(±)-	MeOH	260(3.91)	32-0319-84
4-Thia-1-azabicyclo[3.2.0]hept-2-ene-2-carboxylic acid, 6-ethyl-3-[[(1-methyl-1H-tetrazol-5-yl)thio]methyl]-7-oxo-, (acetyloxy)methyl ester, cis-(±)-	MeOH	252(3.68),334(3.74)	32-0319-84
$C_{14}H_{18}Br_2$			
Bicyclo[8.2.2]tetradeca-10,12,13-triene, 5,6-dibromo-, cis	EtOH	225(3.83),265s(2.69), 272(2.71),280s(2.66)	24-0474-84
Bicyclo[8.3.1]tetradeca-1(14),10,12-triene, 11,13-dibromo-	n.s.g.	207(4.73),225(4.16), 279(2.88),287s(2.86)	24-0455-84
$C_{14}H_{18}Br_4$			
Benzene, 1,2,4-tribromo-5-(8-bromo-octyl)-	n.s.g.	207(4.77),225s(4.32), 278(3.14),287s(3.12)	24-0455-84
$C_{14}H_{18}N_2$			
4,9-Azo-5,8-methano-1H-benz[f]indene, 3a,4,4a,5,6,7,8,8a,9,9a-decahydro-, (3aα,4β,4aα,5α,8α,8aα,9β,9aα)-	hexane	381(2.19)	24-0534-84
$C_{14}H_{18}N_2O$			
Benz[cd]indol-4-amine, 1,3,4,5-tetra-hydro-6-methoxy-N,N-dimethyl-, oxalate salt	EtOH	221(4.37),276(3.74)	44-4761-84
1-Butanone, 3-methyl-1-(2-methyl-1H-indol-3-yl)-, oxime	n.s.g.	226(4.27),275(3.98)	103-0055-84
Ethanone, 1-(5,5a,6,7,8,9,9a,10-octa-hydro-1-phenazinyl)-, cis	MeOH	264(3.90),422(3.58)	18-0623-84
trans	MeOH	262(4.06),417(3.74)	18-0623-84
1-Pentanone, 1-(2-methyl-1H-indol-3-yl)-, oxime	n.s.g.	226(4.42),275(4.00)	103-0055-84
$C_{14}H_{18}N_2O_2$			
Acetic acid, [1-(phenylmethyl)-2-imid-azolidinylidene]-, ethyl ester	EtOH	272(4.51)	39-2599-84C

Compound	Solvent	$\lambda_{max}(\log \epsilon)$	Ref.
Ethanone, 1,1'-[5-[1-(dimethylhydra-zono)ethyl]-1,3-phenylene]bis-	MeOH	230(4.61),325(3.38)	24-1620-84
$C_{14}H_{18}N_2O_2S$ 4(3H)-Quinazolinethione, 2-(1,1-dimeth-ylethyl)-6,7-dimethoxy-	EtOH	224s(4.42),230(4.45), 253(4.11),274(4.11), 351s(3.92),366(4.15), 380(4.14)	39-1143-84C
$C_{14}H_{18}N_2O_3$ 4(1H)-Quinazolinone, 2-(1,1-dimethyl-ethyl)-6,7-dimethoxy-	EtOH	240(4.70),282s(3.85), 310(3.75),323(3.69)	39-1143-84C
$C_{14}H_{18}N_2O_4$ 1,3-Benzenedicarboxylic acid, 5-[1-(di-methylhydrazono)ethyl]-, dimethyl ester	MeOH	221(4.55),325(3.43)	24-1620-84
$C_{14}H_{18}N_2O_4S$ Cyclopentanecarboxylic acid, 2-[[2-(am-inosulfonyl)phenyl]imino]-, ethyl ester	pH 1 EtOH	208(4.28),248(3.86), 312(3.42) 210(4.45),250(4.05), 323(3.93)	104-0534-84 104-0534-84
1-Cyclopentene-1-carboxylic acid, 2-[[2-(aminosulfonyl)phenyl]amino]-, ethyl ester	pH 1 EtOH	208(4.37),246(3.79), 312(3.65) 205(4.22),245s(3.56), 317(4.27)	104-0534-84 104-0534-84
$C_{14}H_{18}N_2O_5$ 5H-Pyrrolo[2,1-c][1,4]benzodiazepin-5-one, 1,2,3,10,11,11a-hexahydro-2,8-dihydroxy-7,11-dimethoxy-(chicamycin A)	MeCN	232(4.40),260s(3.89), 320(3.59)	158-0191-84
$C_{14}H_{18}N_2O_7$ L-Threonine, N-[N-(2,3-dihydroxybenz-oyl)-L-alanyl]- (Bu-2743E)	EtOH and EtOH-HCl EtOH-NaOH	213(4.31),249.5(3.94), 315(3.51) 224(4.16),238s(4.10), 262s(3.90),328(3.59)	158-84-171 158-84-171
$C_{14}H_{18}N_2O_8$ Uridine, 2'-deoxy-5-ethyl-, 5'-(methyl ethanedioate)	H_2O	267(3.97)	83-0867-84
$C_{14}H_{18}N_2S$ 2-Heptenenitrile, 3-[(2-aminophenyl)-thio]-4-methyl-	EtOH	207(4.40),250(4.11), 270(4.04)	78-2141-84
$C_{14}H_{18}N_3O_4$ 1H-Imidazol-1-yloxy, 2,5-dihydro-2,2,5,5-tetramethyl-4-(4-methyl-3-nitrophenyl)-, 3-oxide	n.s.g.	270(4.11),296(4.10)	88-5809-84
$C_{14}H_{18}N_4O_2$ 5,2,4-(Iminometheno)dicyclopent[cd,g]-indole, tetradecahydro-1,9-dinitroso-	n.s.g.	237(4.09)	44-0001-84
4(1H)-Quinazolinone, 2-[4-(2-hydroxy-ethyl)-1-piperazinyl]-	MeOH	236(4.5),271(4.08), 307(3.2),316(3.2), 336(2.87)	24-1523-84

Compound	Solvent	$\lambda_{max}(\log \epsilon)$	Ref.
$C_{14}H_{18}N_4O_2S$			
Ethanol, 2,2'-[[3-methyl-4-(2-thiazo-lylazo)phenyl]imino]bis-	acetone	493(4.56)	7-0819-84
	50% acetone	506(4.59)	7-0819-84
$C_{14}H_{18}N_4O_3$			
2,4(1H,3H)-Pteridinedione, 1,3-dimeth-yl-6-(1-oxopropyl)-7-propyl-	MeOH	248(4.12),278(4.03), 329(4.08)	5-1798-84
$C_{14}H_{18}N_6$			
1H-Cyclohepta[1,2-d:3,4-d']diimidazole-2,8-diamine, N^2,N^8,N^8,1,4-pentamethyl-(9-methylparagracine)	MeOH	259(3.90),310(4.77), 317(4.66),374(3.94), 416(4.27)	94-3873-84
	MeOH-acid	243s(3.84),311(4.80), 368(3.90),415(4.35)	94-3873-84
$C_{14}H_{18}N_6O_4$			
Acetic acid, [2-(phenylmethoxy)ethoxy]-, 2-(3-amino-2,5-dihydro-5-oxo-1,2,4-triazin-6-yl)hydrazide	EtOH	240s(--),272s(3.60)	4-0697-84
$C_{14}H_{18}O$			
2-Butenal, 4-(3,4,4,5-tetramethyl-2,5-cyclohexadien-1-ylidene)-	hexane	347(4.74),356(4.73)	5-0340-84
3-Buten-2-one, 4-[2-methyl-5-(1-methyl-ethyl)phenyl]-, (Z)-	MeOH	279(4.36)	138-0627-84
1-Cyclopropenemethanol, α,α,3,3-tetra-methyl-2-phenyl-	EtOH	268(4.16)	44-4344-84
4,5-Epoxy[8]paracyclophane	EtOH	222(3.86),264s(2.55), 269(2.64),277s(2.61)	24-0474-84
4-Hexen-3-one, 2,5-dimethyl-4-phenyl-	C_6H_{12}	237(3.95),243(3.94), 248(3.91),254(3.79), 260(3.58)	44-4344-84
$C_{14}H_{18}OS_2$			
Bicyclo[3.2.0]hept-2-en-6-one, 4-(1,3-dithiepan-2-ylidene)-7,7-dimethyl-, (±)-	EtOH	218(3.85),304(4.25)	33-1854-84
$C_{14}H_{18}O_2$			
Benzaldehyde, 3-(1,1-dimethyl-2-prop-enyl)-4-hydroxy-2,5-dimethyl-	MeCN-KOH	360(4.24)	39-1259-84B
Benzeneacetaldehyde, α-cyclohexyl-α-hydroxy-, (S)-(+)-	MeOH	242s(3.11),297(2.36)	22-0077-84
Bicyclo[7.3.2]tetradec-9(14),10,12-tri-en-13-one, 14-hydroxy-	MeOH	253s(4.13),264s(4.17), 271(4.19),287s(3.68), 335s(3.57),370(3.78), 400s(3.23)	88-4761-84
	MeOH-H_2SO_4	244(4.04),274(4.48), 348(3.72)	88-4761-84
	MeOH-NaOH	257(4.01),288(4.04), 374s(3.80),403(3.91)	88-4761-84
2-Butenoic acid, 3-[2-methyl-5-(1-meth-ylethyl)phenyl]-, (Z)-	MeOH	278(4.23)	138-0627-84
$C_{14}H_{18}O_3$			
Benzeneacetic acid, α-cyclohexyl-α-hy-droxy-, (S)-	MeOH	227(2.70),251(2.42), 257(2.42),263(2.30)	22-0077-84
Benzenepentanoic acid, δ-oxo-α-propyl-	EtOH	242(4.09)	2-0821-84
5(6H)-Benzocyclooctenone, 7,8,9,10-tetrahydro-6,6-dimethoxy-	MeOH	244s(3.19)	44-2050-84

$C_{14}H_{18}O_3-C_{14}H_{19}NOS_2$

Compound	Solvent	$\lambda_{max}(\log \epsilon)$	Ref.
Bicyclo[4.2.0]octa-1,3,5-trien-7-ol, 8,8-dimethyl-7-(2-methyl-1,3-diox-olan-2-yl)-	MeOH	260(3.09),266(3.24), 272.5(3.23)	44-2050-84
Bicyclo[4.2.0]octa-1,3,5-trien-7-ol, 8-ethyl-7-(2-methyl-1,3-dioxolan-2-yl)-	MeOH	260.5(3.17),266.5(3.32), 272(3.30)	44-2050-84
4a(2H)-Biphenylenol, 1,3,4,8b-tetra-hydro-4,4-dimethoxy-, cis	MeOH	260(3.07),267(3.22), 274(3.20)	44-2050-84
Pterosin B	MeOH	216(4.55),260(4.18), 303(3.40)	94-4620-84
$C_{14}H_{18}O_4$			
Pterosin S	MeOH	216.5(4.58),259(4.12), 301(3.47)	94-4620-84
$C_{14}H_{18}O_5$			
Benzoic acid, 3-formyl-2,4-dihydroxy-5,6-dimethyl-, 1,1-dimethylethyl ester	MeOH	243(4.63),269(4.56), 290s(4.40)	102-0431-84
	MeOH-NaOH	220(4.54),292(4.47), 410(4.04)	102-0431-84
2-Buten-1-one, 1-(2,3,4,6-tetrameth-oxyphenyl)-, (E)-	MeOH	225(4.20),284(3.14)	94-1355-84
D-Xylofuranose, 1-O-methyl-5-(O-4-meth-ylbenzoyl)-2-deoxy-	MeOH	236(4.11)	118-0961-84
$C_{14}H_{18}Si$			
Silane, trimethyl(3-phenyl-3-penten-1-ynyl)-	EtOH	256(4.30)	65-1400-84
$C_{14}H_{19}Br$			
Bicyclo[8.3.1]tetradeca-1(14),10,12-triene, 11-bromo-	EtOH	220(3.99),273(2.62), 280s(2.56)	24-0455-84
$C_{14}H_{19}BrClN_3O_6S$			
Uridine, 5-(2-bromoethenyl)-5'-[[(3-chloropropyl)sulfonyl]amino]-2',5'-dideoxy-, (E)-	EtOH	250(4.12),287(4.03)	78-0427-84
$C_{14}H_{19}BrO$			
Phenol, 4-bromo-2-(1,1-dimethyl-2-but-enyl)-3,6-dimethyl-, (E)-	MeCN-KOH	312(4.18)	39-1259-84B
	MeCN-KOH	312(4.23)	39-1259-84B
$C_{14}H_{19}DO$			
Bicyclo[8.2.2]tetradeca-10,12,13-tri-en-5-d-5-ol	EtOH	226(3.58),266s(2.17), 272(2.25),279s(2.18)	24-0474-84
$C_{14}H_{19}NO$			
Benzonitrile, 2-(2-methoxy-1,1,2-tri-methylpropyl)-	hexane	226(4.01),233s(3.90), 266s(2.87),273(3.08), 282(3.09)	23-1785-84
Benzonitrile, 3-(2-methoxy-1,1,2-tri-methylpropyl)-	hexane	226(4.04),230(4.00), 265(2.75),272(2.91), 280(2.14)	23-1785-84
Benzonitrile, 4-(2-methoxy-1,1,2-tri-methylpropyl)-	hexane	232(4.26),239(4.21), 258(2.69),266(2.80), 269(2.26),272(2.75), 278(2.72)	23-1785-84
$C_{14}H_{19}NOS_2$			
1H-Thiopyran-4-carbonitrile, 3-cyclo-hexyl-1-methyl-5-(methylthio)-, 1-oxide	EtOH	248(4.46),276(3.99), 325(3.89)	118-0852-84

Compound	Solvent	$\lambda_{max}(\log \epsilon)$	Ref.
$C_{14}H_{19}NO_2$			
1-Propanone, 1-(3,4-dihydro-2H-1-benzo-pyran-6-yl)-3-(dimethylamino)-, hydrochloride	n.s.g.	208(4.09),228(3.98), 289(4.10)	103-0838-84
1-Propanone, 1-(2,3-dihydro-2-methyl-5-benzofuranyl)-3-(dimethylamino)-, hydrochloride	n.s.g.	208(4.13),229(4.04), 293(4.11)	103-0838-84
2H-Pyrrol-4-ol, 4-ethyl-3,4-dihydro-2,2-dimethyl-5-phenyl-, 1-oxide	EtOH	281(4.06)	12-0109-84
$C_{14}H_{19}NO_2S$			
1-Piperidinepropanoic acid, α-(2-thien-ylmethylene)-, methyl ester	EtOH	312(3.78)	73-1764-84
$C_{14}H_{19}NO_3$			
1H-2-Benzazepin-1-one, 2,3,4,5-tetra-hydro-4,9-dihydroxy-3-(2-methylpropyl)-	MeOH	230s(3.76),305(3.59)	78-2519-84
Ethanone, 1-[2-hydroxy-5-methyl-3-(4-morpholinylmethyl)phenyl]-	EtOH	253(3.39)	2-0904-84
Glyoxylamide, 2-hydroxy-N,N-dimethyl-3-tert-butylphenyl-	EtOH	216(4.27),265(4.09), 338(3.56)	39-2655-84C
1H-Indole-2-carboxylic acid, 2,3,4,5-6,7-hexahydro-3-oxo-2-(2-propenyl)-, ethyl ester	MeCN	320(3.96)	33-1535-84
$C_{14}H_{19}NO_4$			
Cyclohexanecarboxylic acid, 3-(1-cyano-2-ethoxy-2-oxoethylidene)-1-methyl-, methyl ester	EtOH	235(4.23)	39-0261-84C
α-L-ribo-Hexopyranoside, methyl 3,4,6-trideoxy-3-[[(2-hydroxyphenyl)meth-ylene]amino]-	EtOH	253(3.89),280(3.60), 315(3.34)	39-0733-84C
$C_{14}H_{19}N_3$			
Benzenamine, N-methyl-2-[1-methyl-1-(5-methyl-1H-pyrazol-3-yl)ethyl]-	EtOH	248(4.33),296(3.40)	150-3201-84M
$C_{14}H_{19}N_3O$			
4H-Indazol-4-one, 1,5,6,7-tetrahydro-5-(1-piperidinomethylene)-1-methyl-, (E)-	EtOH	235(3.94),253(3.95), 355(4.39)	4-0361-84
$C_{14}H_{19}N_3O_4$			
1H-Imidazole, 2,5-dihydro-1-hydroxy-2,2,5,5-tetramethyl-4-(4-methyl-2-nitrophenyl)-, 3-oxide	n.s.g.	244(4.10)	88-5809-84
$C_{14}H_{19}N_3O_5S$			
D-Methionine, N-[1-oxo-3-(1,2,3,4-tetrahydro-6-methyl-2,4-dioxo-5-pyrimidinyl)-2-propenyl]-, methyl ester	H_2O	303(4.31)	94-4625-84
L-	H_2O	303(4.30)	94-4625-84
$C_{14}H_{19}N_5O_2$			
1-Propanone, 1-[2-amino-4-(pentyloxy)-6-pteridinyl]-	MeOH	252(4.21),301(4.18), 363(4.12)	138-1025-84
$C_{14}H_{19}N_5O_5$			
Adenosine, 7,8-dihydro-7-methyl-2',3'-O-(1-methylethylidene)-8-oxo-	MeOH	272(4.08)	150-3601-84M

Compound	Solvent	$\lambda_{max}(\log \epsilon)$	Ref.
9H-Purin-6-amine, 9-[2-O-(1-oxobutyl)-β-D-arabinofuranosyl]-	MeOH	259(4.16)	87-0270-84
$C_{14}H_{20}$ Benzene, (4,5-dimethyl-4-hexenyl)-	n.s.g.	248(2.83)	39-1833-84B
$C_{14}H_{20}Br_2$ Benzene, 1-bromo-4-(8-bromooctyl)-	EtOH	217(4.07),225s(3.93), 259s(2.78),266(2.81), 274s(2.72)	24-0455-84
$C_{14}H_{20}ClN_3O_5$ 1H-Imidazole-4-carboxylic acid, 5-amino-1-[5-chloro-5-deoxy-2,3-O-(1-methylethylidene)-β-D-ribofuranosyl]-	pH 2 pH 6-7 pH 12	245s(3.94),266(4.06) 240s(3.77),268(4.10) 240s(3.78),268(4.10)	65-2114-84 65-2114-84 65-2114-84
$C_{14}H_{20}FeN_2$ Ferrocene, 1,1'-bis(dimethylamino)-	MeOH	214(4.26),285(3.80), 322(3.84),433(4.41)	101-0113-84R
$C_{14}H_{20}NO_3$ 3H-Oxazolo[4,3-a]isoquinolinium, 1,5,6,10b-tetrahydro-8,9-dimethoxy-4-methyl-, iodide	MeOH	220s(4.27),283(3.55), 286s(3.52)	12-1659-84
5H-Oxazolo[2,3-a]isoquinolinium, 2,3,6,10b-tetrahydro-8,9-dimethoxy-4-methyl-, iodide	MeOH	221s(4.33),282(3.60), 287s(3.54)	12-1659-84
$C_{14}H_{20}N_2$ 4,9-Azo-5,8-methano-1H-benz[f]indene, 2,3,3a,4,4a,5,6,7,8,8a,9,9a-dodecahydro-	hexane	384(2.21)	24-0534-84
15,16-Diazapentacyclo[9.2.1.2³,⁹.0²,¹⁰-0⁴,⁸]hexadec-15-ene, (1S*,2R*,3S*,4S*)-	hexane	354(1.59),382(1.83), 392(1.98)	44-0001-84
1,4:5,8-Dimethanophthalazine, 1,4,4a,5,8,8a-hexahydro-1,4,10,10-tetramethyl-, endo	hexane	260(2.30),369(2.53)	24-0517-84
exo	hexane	367(2.29)	24-0517-84
$C_{14}H_{20}N_2OS$ Benzo[b]thiophen-4(5H)-one, 5-[[[2-(dimethylamino)ethyl]methylamino]methylene]-6,7-dihydro-, (E)-	EtOH	220(4.04),253(4.18), 364(4.22)	161-0081-84
$C_{14}H_{20}N_2O_4$ 2,4-Pentanedione, 3,3'-[1,2-ethanediylbis(iminomethylidyne)]bis-	MeCN	252(4.46),286(4.40), 303(4.37)	62-0947-84A
$C_{14}H_{20}N_2O_5S$ 1-Azabicyclo[3.2.0]hept-2-ene-2-carboxylic acid, 3-[[2-(acetylamino)ethyl]thio]-6-(1-hydroxy-1-methylethyl)-7-oxo-, sodium salt	H_2O	302(3.94)	44-5271-84
$C_{14}H_{20}N_2O_6S$ 1-Azabicyclo[3.2.0]hept-2-ene-2-carboxylic acid, 3-[[2-(acetylamino)ethyl]thio]-6-(1-hydroxy-2-methoxyethyl)-7-oxo-, sodium salt, (R)-	H_2O	301(3.59)	44-5271-84
(S)-	H_2O	301(3.68)	44-5271-84
Carpetimycin C, sodium salt	H_2O	285(4.00)	158-84-129

Compound	Solvent	$\lambda_{max}(\log \epsilon)$	Ref.
$C_{14}H_{20}N_2O_9S_2$ Carpetimycin D, disodium salt	H_2O	285(4.10)	158-84-129
$C_{14}H_{20}N_6O_8$ D-Ribitol, 1-deoxy-1-[6-(3-hydrazino- 3-oxopropyl)-1,3,4,7-tetrahydro- 2,4,7-trioxo-8(2H)-pteridinyl]-	pH 1 pH 13	283(4.03),329(4.03) 261(3.96),286(3.84), 357(4.10)	39-0953-84C 39-0953-84C
$C_{14}H_{20}O$ 5(1H)-Azulenone, 2,3,6,7-tetrahydro- 2,2,6,8-tetramethyl-	EtOH	315(3.79)	23-1954-84
1,3-Undecadien-9-yn-5-one, 2,6,6-tri- methyl-	pentane	262(4.27)	33-0815-84
$C_{14}H_{20}OS$ Dispiro[5.1.5.1]tetradecan-7-one, 14-thioxo-	$CHCl_3$	260(3.35),319(3.26), 517(1.18)	2-0498-84
$C_{14}H_{20}OS_2$ Benzene, [[[2,2-dimethyl-1-[(methylsul- finyl)methylene]propyl]thio]methyl]-, (Z)-	EtOH	280s(3.57)	39-0085-84C
$C_{14}H_{20}O_2$ Bicyclo[4.1.0]heptan-2-one, 3,4,4-tri- methyl-3-(3-oxo-1-butenyl)-, (E)-	pentane	214(4.08),293(2.16), 303(2.25),311(2.27), 322(2.18),334(1.94)	33-0136-84
(Z)-	pentane	218(3.69),293(2.11)	33-0136-84
Bicyclo[8.2.2]tetradeca-10,12,13-tri- ene-4,7-diol	EtOH	206(3.89),232(3.44), 270s(2.00),278(2.12), 285s(2.03)	24-0455-84
Bicyclo[8.2.2]tetradeca-10,12,13-tri- ene-5,6-diol, cis	EtOH	225(3.83),265s(2.69), 272(2.71),280s(2.66)	24-0474-84
2-Butanone, 4-(5,5-dimethyl-2-methyl- ene-3-oxabicyclo[5.1.0]oct-4-yli- dene)-	pentane	213(3.93),281(2.00)	33-0136-84
3-Buten-2-one, 4-(2,5,5-trimethyl-3- oxatricyclo[5.1.0.0²,⁴]oct-4-yl)-	pentane	242(5.06)	33-0136-84
3-Buten-2-one, 4-(2,5,5-trimethyl-3- oxatricyclo[5.1.0.0²,⁴]oct-4-yl)-	pentane	231(3.95),329(1.51)	33-0136-84
2,5-Cyclohexadiene-1,4-dione, 2,6- bis(1,1-dimethylethyl)-	MeCN	320(2.66)	44-0491-84
1,3-Cyclohexanedione, 2-(2-cyclohex- 1-enyl)-5,5-dimethyl-	pH 1 H_2O pH 13	268(4.02) 270(3.98) 294(4.35)	39-1213-84C 39-1213-84C 39-1213-84C
Cyclopentadiene, ethoxy-, dimers	n.s.g.	214(3.45)	104-0061-84
1(2H)-Dibenzofuranone, 3,4,5a,6,7,8,9- 9a-octahydro-3,3-dimethyl-	EtOH	274(4.13)	39-1213-84C
Ethanone, 1-[2-[2-methyl-2-(5-methyl- 2-furanyl)propyl]cyclopropyl]-	pentane	218(4.01)	33-0136-84
Ethanone, 1-(1,6,6-trimethyl-10-oxa- tricyclo[5.3.0.0²,⁴]dec-7-en-9-yl)-, (1R*,2R*,4R*)-	pentane	299(2.15)	33-0136-84
(1R*,2S*,4S*)-	pentane	298(2.11)	33-0136-84
Furan, 2-ethoxy-5-(2-heptynylidene)- 2,5-dihydro-2-methyl-, cis	n.s.g.	226(3.97),294(4.03)	104-2302-84
trans	n.s.g.	222(3.75),300(3.61)	104-2302-84
6-Oxa-3,7,10-undecatrien-2-one, 7-meth- yl-5-(1-methylethylidene)-, (E)-	pentane	288(4.23)	33-0136-84

Compound	Solvent	$\lambda_{max}(\log \epsilon)$	Ref.
$C_{14}H_{20}O_3$			
Benzenepropanoic acid, 3-(1,1-dimethyl-ethyl)-4-hydroxy-, methyl ester	EtOH	230(3.84),280(3.40)	104-2375-84
Bicyclo[5.1.0]octan-2-one, 8-acetyl-6,6-(epoxymethano)-3,3,7-trimethyl-	pentane	281(2.25)	33-0815-84
2-Butanone, 4-(7,7-dimethyl-4-methyl-ene-1,5-dioxaspiro[2.6]non-6-yli-dene)-	pentane	280(2.02)	33-0815-84
3-Buten-2-one, 4-[1,2-epoxy-3,3-(epoxy-methano)-2,6,6-trimethyl-1-cyclo-hexyl]-, (E)-	pentane	229(4.09),331(1.61)	33-0815-84
isomer	pentane	227(4.10),330(1.65)	33-0815-84
3-Buten-2-one, 4-[(5,5-epoxymethano)-1,2,2-trimethyl-6-oxo-1-cyclohexyl]-, (E)-	MeCN	228(4.03),301(2.74)	33-0815-84
(Z)-	pentane	222(3.49),300(2.16)	33-0815-84
2,11-Dioxabicyclo[4.4.1]undeca-3,5-di-ene, 10,10-(epoxymethano)-1,3,7,7-tetramethyl-	pentane	259(3.92)	33-0815-84
Ethanone, 1-[2-[3-(3-acetyl-1-cyclo-propen-1-yl)-3-methylbutyl]oxir-anyl]-	MeCN	284(2.19)	33-0815-84
2-Heptanone, 3,3-(epoxymethano)-6-meth-yl-6-(5-methyl-2-furanyl)-	pentane	216s(4.14),279(1.70)	33-0815-84
1-Oxaspiro[2.6]nonan-4-one, 7,7-dimeth-yl-6-methylene-5-(2-oxo-1-propyl)-	MeCN	280(1.94)	33-0815-84
2,4-Pentadienal, 2,4-dimethyl-5-(2,4,5-trimethyl-3,6-dioxabicyclo[3.1.0]-hex-2-yl)-, [1α,2α(2E,4E),4α,5α]-	MeOH	277(4.07)	44-3762-84
10,12-Tetradecadiynoic acid, 14-hydroxy-	EtOH	233(2.43),245.6(2.61), 258.9(2.62)	118-0230-84
3,4-Undecadiene-2,10-dione, 9,9-(epoxy-methano)-6,6-dimethyl-	pentane	219(3.85),280s(3.08)	33-0815-84
$C_{14}H_{20}O_4$			
2H-1-Benzopyran-6-ol, 3,4-dihydro-7,8-dimethoxy-2,2,5-trimethyl-	EtOH	293(3.54)	88-1929-84
2H-Pyran-5-carboxylic acid, 3,4-dihy-dro-2,2-dimethyl-6-(2-methyl-1-propenyl)-4-oxo-, ethyl ester	EtOH	250(3.85),306(4.25)	4-0013-84
$C_{14}H_{20}O_5$			
Acetic acid, (3-oxo-1,4-dioxaspiro-[4.5]dec-2-ylidene)-, 1,1-dimethyl-ethyl ester, (Z)-	EtOH	258(4.05)	39-1555-84C
2-Cyclohexene-1-propanoic acid, 1-(eth-oxycarbonyl)-2-methyl-4-oxo-, methyl ester	EtOH	204s(3.42),240(4.01)	12-2037-84
$C_{14}H_{21}N$			
Pyridine, 2-cyclohexyl-5-propyl-	EtOH	214(3.81),270(3.53)	103-0515-84
$C_{14}H_{21}NO$			
2,4-Cyclohexadien-1-one, 3,5-bis(1,1-dimethylethyl)-6-imino-	DMSO	390(3.30)	44-3579-84
plus two electrons	DMSO	270(3.78),325(3.72)	44-3579-84
$C_{14}H_{21}NOSi$			
1H-Indole, 1-[[2-(trimethylsilyl)-ethoxy]methyl]-	MeOH	218(4.57),266(4.89), 277(4.82),289(4.65)	44-0203-84

Compound	Solvent	$\lambda_{max}(\log \epsilon)$	Ref.
$C_{14}H_{21}NO_2$			
Acetic acid, (dimethylamino)[3-(1,1-di-methylethyl)-2,4-cyclopentadien-1-ylidene]-, methyl ester	dioxan	260(3.29),335(4.42)	157-0653-84
Ethanone, 1-[3-[(diethylamino)methyl]-2-hydroxy-5-methylphenyl]-	EtOH	252(3.71),332(2.92)	2-0904-84
1-Propanone, 3-(dimethylamino)-1-(3-ethyl-4-methoxyphenyl)-, hydrochloride	n.s.g.	208(4.10),226(4.14), 280(4.17)	103-0838-84
2-Pyridinecarboxylic acid, 4,5-bis(1,1-dimethylethyl)-	MeOH	228(4.14),274(3.90)	44-1338-84
2-Pyridinecarboxylic acid, 4,6-bis(1,1-dimethylethyl)-	MeOH	227(3.30),269(3.37)	44-1338-84
$C_{14}H_{21}NO_3$			
3a(7aH)-Benzofurancarboxamide, N,N-di-ethyl-2,3-dihydro-7a-methoxy-	MeOH	266(3.84)	44-4429-84
Propanamide, 3-hydroxy-N-(2-hydroxy-1-methyl-2-phenylethyl)-N,2-dimethyl-, (R,S,S)-	EtOH	259(2.00),264(1.93)	44-5202-84
(S,R,S)-	EtOH	258(2.32),264(2.26)	44-5202-84
(1'S,2'R,2R)-	EtOH	259(2.05),264(1.97)	44-5202-84
1H-Pyrrole-2-carboxylic acid, 4-(3-but-enyl)-2,3-dihydro-1,2,5-trimethyl-3-oxo-, ethyl ester	MeCN	333(4.04)	33-1957-84
$C_{14}H_{21}NO_4$			
2,4-Cyclohexadien-1-one, 5-amino-6-hy-droxy-3-methoxy-2,6-dimethyl-4-(2-methyl-1-oxobutyl)- (aspersitin)	EtOH	214s(4.05),239(3.93), 316(4.07),399(3.67)	158-84-79
$C_{14}H_{21}NO_4S$			
DL-Homocysteine, S-[3-(2-methoxyphen-oxy)propyl]-	MeOH	222.5(3.88),273s(3.40)	36-1241-84
$C_{14}H_{21}NO_5S_2$			
4-Thiazolidinecarboxylic acid, 4,5,5-trimethyl-, p-toluenesulfonate	EtOH	226(3.64),256(2.58), 261(2.58),268(2.43)	39-1127-84C
$C_{14}H_{21}N_3O_5$			
1H-Imidazole-4-carboxylic acid, 5-ami-no-1-[5-deoxy-2,3-O-(1-methylethyli-dene)-β-D-ribofuranosyl]-, ethyl ester	pH 2	250s(3.99),267(4.09)	65-2114-84
	pH 6-7	240s(3.76),269(4.11)	65-2114-84
	pH 12	240s(3.76),269(4.12)	65-2114-84
$C_{14}H_{21}N_3O_7$			
Pyridazinium, 2,3-dihydro-3-oxo-1-β-D-ribofuranosyl-4-[[(tetrahydro-2H-pyran-2-yl)oxy]amino]-, hydroxide, inner salt	MeOH	232(3.99),316(4.25)	87-1613-84
$C_{14}H_{21}N_5OS$			
2-Pteridinamine, 4-(pentyloxy)-7-(prop-ylthio)-	MeOH	240(4.53),273(4.01), 370s(4.26),377(4.27)	138-1025-84
$C_{14}H_{21}N_5O_3$			
9H-Purine-9-propanoic acid, 6-amino-β-hexyl-α-hydroxy-	pH 2,7 and 12	262(4.14)	73-2148-84
$C_{14}H_{21}N_5O_7S$			
5-Thiazolidineacetamide, 3-[2(or 3)-O-(aminocarbonyl)-3-β-D-ribofuranosyl-	MeOH	257(4.04)	128-0295-84

Compound	Solvent	$\lambda_{max}(\log \epsilon)$	Ref.
2-[(1-methylethylidene)hydrazono]- 4-oxo- (cont.)			128-0295-85
$C_{14}H_{21}N_5O_{11}S_3$			
muco-Inositol, 3-(6-amino-9H-purin-9-	pH 1	257(2.95)	136-0033-84J
yl)-3-deoxy-, 1,5,6-trimethanesul-	pH 7	260(2.83)	136-0033-84J
fonate (in ethanol)	pH 11	258(3.24)	136-0033-84J
$C_{14}H_{22}$			
Benzene, 1,4-bis(1,1-dimethylethyl)-	EtOH	226(3.92),258s(2.64), 265(2.68),274(2.64)	24-0455-84
Bicyclo[3.1.0]hex-2-ene, 1,2,3,5-tetra- methyl-4-methylene-6-(1-methylethyl)-, endo	hexane	250(4.02)	64-1781-84B
exo	hexane	250(4.08)	64-1781-84B
Bicyclo[8.2.2]tetradeca-1(13),10-diene	EtOH	208(3.35),228(2.85)	24-0455-84
1,3-Cyclohexadiene, 1,2,3,4-tetramethyl- 5-methylene-6-(1-methylethyl)-	hexane	212(4.10),310(3.78)	64-1781-84B
$C_{14}H_{22}N$			
Methanaminium, N-methyl-N-[(3,4,4,5- tetramethyl-2,5-cyclohexadien-1- ylidene)ethylidene]-, perchlorate	MeCN	376(4.56)	5-0340-84
$C_{14}H_{22}NO_4S$			
Sulfonium, (3-amino-3-carboxypropyl)- [2-(2-methoxyphenoxy)ethyl]methyl-, iodide	MeOH	219(4.60),272s(3.69)	36-1241-84
$C_{14}H_{22}N_2$			
1,4:5,8-Dimethanophthalazine, 1,4,4a,5- 6,7,8,8a-octahydro-1,4,10,10-tetra- methyl-, endo	hexane	359(2.66)	24-0517-84
exo	hexane	363(2.36)	24-0517-84
Pyrrolo[3,2-b]pyrrole, 3,6-bis(1,1-di-	C_6H_{12}	248(4.2)	138-2033-84
methylethyl)-1,4-dihydro-	EtOH	248(4.2)	138-2033-84
	1.7M HClO₄	262(4.0),380(3.6)	138-2033-84
$C_{14}H_{22}N_2O_4$			
Cyclohexanone, 2,2'-azobis[2-methyl-, N,N'-dioxide	dioxan	295(3.65)	12-1231-84
$C_{14}H_{22}N_4$			
5,2,4-(Iminometheno)dicyclopent[cd,g]- indole-1,9(2H)-diamine, dodecahydro-, $(2\alpha,2a\beta,4\alpha,4a\beta,5\alpha,5a\beta,8a\beta,8b\beta,8c\beta,10S*)$-	EtOH	275(2.46)	44-0001-84
$C_{14}H_{22}N_4O$			
Benzamide, 4-(3-hexyl-3-methyl-1-tria- zenyl)-	EtOH	322(4.33)	87-0870-84
$C_{14}H_{22}N_4S_2$			
1,4-Benzenedicarboximidamide, N,N,N'',N''- tetramethyl-N',N'''-bis(methylthio)-	hexane	326(3.53)	39-2933-84C
$C_{14}H_{22}O$			
4H-1-Benzopyran, 4a,5,6,7,8,8a-hexa- hydro-2,5,5,8a-tetramethyl-4-meth- ylene-	pentane	259(4.11)	33-1175-84
2-Butanone, 4-(2,2-dimethyl-6-methyl- enecyclohexylidene)-3-methyl-	pentane	221(3.78),289(2.48), 296(2.48),306s(2.34), 317s(2.02)	33-1175-84

Compound	Solvent	$\lambda_{max}(\log \epsilon)$	Ref.
3-Buten-2-one, 4-[2-methyl-2-(4-methyl-3-pentenyl)cyclopropyl]-, (E)-	EtOH	257(4.24)	19-0207-84
2-Oxatricyclo[4.4.0.0^{1,5}]dec-3-ene, 3,5,6,10,10-pentamethyl-, (1R*,5S*-6S*)-	pentane	224(3.61)	33-1175-84
3-Penten-2-one, 4-(2,6,6-trimethyl-1-cyclohexen-1-yl)-, (E)-	pentane	233(4.14),331(1.63)	33-1175-84
(Z)-	pentane	231(4.03),332(1.82)	33-1175-84
3-Penten-2-one, 4-(2,6,6-trimethyl-2-cyclohexen-1-yl)-, (E)-	pentane	238(4.12),332(1.72)	33-1175-84
(Z)-	pentane	242(4.08),338(1.83)	33-1175-84
$C_{14}H_{22}OSi$			
Spiro[4.5]deca-2,6-dien-1-one, 6-methyl-2-(trimethylsilyl)-	EtOH	222(3.93)	33-0073-84
Spiro[4.5]deca-2,6-dien-1-one, 6-methyl-3-(trimethylsilyl)-	EtOH	231(4.16)	33-0073-84
Spiro[4.4]nona-2,6-dien-1-one, 3,6-di-methyl-2-(trimethylsilyl)-	EtOH	232(4.06)	33-0073-84
$C_{14}H_{22}O_2$			
1,2-Benzenediol, 3,5-bis(1,1-dimethyl-ethyl)-	pH 1	277(3.28)	3-1935-84
	borate	285(3.59)	3-1935-84
Benzeneoctanol, 4-hydroxy-	MeOH	224.5(3.84),278.5(3.23)	33-2111-84
Cyclohexanone, 4-(1,5-dimethyl-3-oxo-4-hexenyl)-	EtOH	238(4.04)	107-0925-84
2-Heptanone, 6-(3,5-dimethyl-2-furanyl)-6-methyl-	pentane	223(3.93),284(1.57)	33-0120-84
2-Oxabicyclo[4.4.0]dec-3-en-6-ol, 1,3,7,7-tetramethyl-5-methylene-	pentane	258(4.03)	33-0120-84
3-Penten-2-one, 4-(2,2,6-trimethyl-7-oxabicyclo[4.1.0]hept-1-yl)-, (E)-	pentane	241(4.08),335(1.63)	33-0120-84
(Z)-A	pentane	224(3.66),306(1.79)	33-0120-84
(Z)-B	pentane	237(3.88),335(1.60)	33-0120-84
2-Propenoic acid, 2-methyl-, 2,4-deca-dienyl ester, (E,E)-	isooctane	230(4.46)	116-1624-84
3,4-Undecadiene-2,10-dione, 5,6,6-tri-methyl- (90% pure)	pentane	220(4.00)	33-0120-84
$C_{14}H_{22}O_3$			
Bicyclo[4.1.0]heptane-7-carboxaldehyde, 1-acetoxy-2,2,6,7-tetramethyl-, (1α,6α,7α)- (90% pure)	pentane	275s(1.48)	33-1175-84
1-Cyclopentene-1-carboxylic acid, 2-hexyl-5-oxo-, ethyl ester	EtOH	232(4.02)	39-2049-84C
Ethanone, 1-(7-hydroxy-1,6,6-trimethyl-10-oxatricyclo[5.2.1.0^{2,4}]dec-9-yl)-	MeCN	276(1.57)	33-0136-84
2,5-Heptanedione, 7-(2-acetylcyclo-propyl)-6,6-dimethyl-	MeCN	281(2.04)	33-0136-84
1-Propen-1-ol, 2-(2,2,6-trimethyl-7-oxabicyclo[4.1.0]hept-1-yl)-, acetate	pentane	207(4.03),215s(3.97)	33-0120-84
3-Undecene-2,5,10-trione, 4,6,6-tri-methyl-	pentane	231(3.84)	33-0120-84
$C_{14}H_{23}NO$			
Cyclohexanone, 2-[1-(cyclohexylimino)-ethyl]-	EtOH	205(--),248s(--), 343(3.00)	34-0358-84
	EtOH-NaOEt	343(3.20)	34-0358-84
Phenol, 2-amino-3,5-bis(1,1-dimethyl-ethyl)-	DMSO	290(3.51)	44-3579-84

Compound	Solvent	$\lambda_{max}(\log \epsilon)$	Ref.
less two electrons (cont.)	DMSO	390(3.23)	44-3579-84
anion	DMSO	265(3.76),315(3.71)	44-3579-84
less one electron	DMSO	290(3.40),385(2.95)	44-3579-84
1H-Pyrrole-2-carboxaldehyde, 5-nonyl-	EtOH	205(3.45),250(3.48), 300(4.29)	12-0227-84
$C_{14}H_{23}NO_6$			
Pyrrole, 3-acetyl-1-ethyl-2-methyl-5-(D-galacto-pentitol-1-yl)-	EtOH	249(3.93),284(3.75)	136-0153-84I
Pyrrole, 3-acetyl-1-ethyl-2-methyl-5-(D-gluco-pentitol-1-yl)-	EtOH	249(3.92),282(3.76)	136-0153-84I
Pyrrole, 3-acetyl-1-ethyl-2-methyl-5-(D-manno-pentitol-1-yl)-	EtOH	249(3.93),284(3.74)	136-0153-84I
1H-Pyrrole-3-carboxylic acid, 1-ethyl-2-methyl-5-(1,2,3,4-tetrahydroxy-butyl)-, ethyl ester, [1S-(1R*,2S*-3R*)]-	EtOH	228(3.94),256(3.79)	136-0153-84I
$C_{14}H_{23}NO_9$			
α-D-Glucopyranose, 2-deoxy-2-[[3-eth-oxy-2-(ethoxycarbonyl)-3-oxo-1-propenyl]amino]-	EtOH	280(4.36)	136-0101-84L
$C_{14}H_{23}N_3O$			
2H-Pyrrol-2-one, 3,4-dihydro-4,5-di-piperidino-	MeCN	244(4.40)	142-1179-84
$C_{14}H_{23}N_4O_6P$			
Thymidine, 5'-O-(1,3-dimethyl-1,3-diaza-2-phospholidin-2-yl)-, P-oxide	EtOH	267(3.97)	39-1471-84C
$C_{14}H_{23}N_5O_2$			
1,2-Nonanediol, 3-(6-amino-9H-purin-9-yl)-	pH 2	261(4.13)	73-2148-84
$C_{14}H_{23}N_5O_3$			
2(1H)-Pyrimidinone, 1-(4-morpholinyl-methyl)-4-[(4-morpholinylmethyl)-amino]-	MeOH	270(3.89)	78-3997-84
	MeOH-HCl	282(4.03)	78-3997-84
	MeOH-NaOH	276(3.86)	78-3997-84
$C_{14}H_{23}N_5O_7$			
L-Lysine, N^6-(2-β-D-arabinofuranosyl)-2,5-dihydro-5-oxo-1,2,4-triazin-3-yl)-	H_2O	215(4.36),248s(3.90)	73-2689-84
copper complex	H_2O	641(1.78)	73-2689-84
$C_{14}H_{23}N_7O_7$			
L-Arginine, N^2-(2-β-D-arabinofurano-syl)-2,5-dihydro-5-oxo-1,2,3-tria-zin-3-yl)-	H_2O	216(4.40),250s(3.95)	73-2689-84
$C_{14}H_{23}P$			
Phosphine, bis(1,1-dimethylethyl)-phenyl-	C_6H_{12}	260(3.23)	65-0489-84
$C_{14}H_{24}O$			
2-Pentanone, 5-(2,2-dimethyl-6-meth-ylenecyclohexyl)-	pentane	280(1.30)	33-1734-84
$C_{14}H_{24}O_2$			
1-Dodecen-3-yne, 12,12-dimethoxy-	hexane	215(3.87),224(4.05),	105-0354-84

Compound	Solvent	$\lambda_{max}(\log \epsilon)$	Ref.
(cont.)		234(4.00)	105-0354-84
2(3H)-Furanone, 3-decylidenedihydro-, (E)-	EtOH	223(4.12)	95-0839-84
(Z)-	EtOH	224(4.05)	95-0839-84
3(2H)-Furanone, 4,5-bis(1,1-dimethyl-ethyl)-2,2-dimethyl-	n.s.g.	275(3.92)	23-2429-84
$C_{14}H_{24}O_4$			
Propanoic acid, 2,2-dimethyl-, 1,1,4,4-tetramethyl-2,3-dioxopentyl ester	n.s.g.	402(1.37)	23-2429-84
$C_{14}H_{24}S_2$			
Thietane, 2,2,4,4-tetramethyl-3-(2,2,4,4-tetramethyl-3-thietan-ylidene)-	C_6H_{12}	193.0(4.35),203.5(4.22), 223s(3.18)	24-0277-84 +24-0310-84
$C_{14}H_{25}N$			
2-Butanamine, N-[(2,6,6-trimethyl-1-cyclohexen-1-yl)methylene]-, (S)-	isopentane at 10°	241(4.14)	35-4621-84
	at -100°	241(3.85)	35-4621-84
	at -150°	241(3.77)	35-4621-84
	MeOH at 10°	241(3.71)	35-4621-84
	at -20°	241(3.73)	35-4621-84
	at -80°	241(3.64)	35-4621-84
$C_{14}H_{25}NO$			
2-Pyrrolemethanol, 5-nonyl-	EtOH	204(3.02),270(3.18)	12-0227-84
$C_{14}H_{25}NOSi$			
2-Cyclohexen-1-one, 3-[methyl[2-(tri-methylsilyl)methyl]-2-propenyl]-amino]-	MeCN	290(4.48)	44-0220-84
$C_{14}H_{25}NO_2Si_2$			
1H-Azepine-1-carboxylic acid, 2,5-bis(trimethylsilyl)-, methyl ester	C_6H_{12}	231(4.28),333(3.11)	88-5669-84
1H-Azepine-1-carboxylic acid, 3,6-bis(trimethylsilyl)-, methyl ester	C_6H_{12}	222(4.28),257(3.56), 315(2.76)	88-5669-84
$C_{14}H_{26}$			
Octane, 2,2,7,7-tetramethyl-4,5-bis(methylene)-	hexane	228(3.81)	44-2981-84
$C_{14}H_{26}NOSi$			
2-Propen-1-aminium, N-(3-methoxy-2-cyclopenten-1-ylidene)-N-methyl-2-[(trimethylsilyl)methyl]-, perchlorate	MeCN	270(4.32)	44-0220-84
$C_{14}H_{26}N_4O_2$			
2,4(1H,3H)-Pyrimidinedione, 6-(butyl-amino)-1-[2-(butylamino)ethyl]-	pH 7	275(5.3)	103-1033-84
$C_{14}H_{26}OS_5$			
1-Propene, 2,2'-sulfinylbis[1,1-bis(ethylthio)-	EtOH	286(4.13)	39-0085-84C
$C_{14}H_{26}Si_2$			
Disilane, pentamethyl(2,4,6-trimethyl-phenyl)-	C_6H_{12}	243(4.12)	60-0341-84

Compound	Solvent	$\lambda_{max}(\log \epsilon)$	Ref.
$C_{14}H_{27}NOSi$ 1H-Pyrrole, 2-butyl-1-[[2-(trimethyl-silyl)ethoxy]methyl]-	MeOH	218(3.84)	44-0203-84
$C_{14}H_{27}N_5O$ 2(1H)-Pyrimidinone, 1-[(diethylamino)-methyl]-4-[[(diethylamino)methyl]-amino]-	MeOH MeOH-acid MeOH-NaOH	270(3.85) 281(4.12) 279(3.86)	78-3997-84 78-3997-84 78-3997-84
$C_{14}H_{28}N_2O_7S$ Thiourea, N-(3,3-diethoxypropyl)-N'-β-D-galactopyranosyl-	MeOH	247(4.08)	103-0447-84
Thiourea, N-(3,3-diethoxypropyl)-N'-β-D-glucopyranosyl-	MeOH	247(4.08)	103-0447-84
$C_{14}H_{29}N_3O$ 1H-1,2,3-Triazole, 1-(2-ethoxyethyl)-4,5-dihydro-4-octyl-	hexane	240(3.59)	104-0580-84
$C_{14}H_{30}Te_2$ Ditelluride, diheptyl	benzene	397(2.85)	48-0467-84
$C_{14}H_{31}GeNO$ Butanamide, N,N-diethyl-3-(triethyl-germyl)-	heptane or EtOH	240(1.60)	70-0310-84

Compound	Solvent	$\lambda_{max}(\log \epsilon)$	Ref.
$C_{15}H_4FN_4Na$ Propanedinitrile, [2-(dicyanomethyl)-3-(4-fluorophenyl)-2-cyclopropen-1-ylidene]-, ion(1-)-, sodium	MeOH	240(4.19),282(4.53), 352(4.32)	118-0686-84
$C_{15}H_4F_{10}N_2O$ 2-Imidazolidinone, 1,3-bis(pentafluoro-phenyl)-	EtOH	229(4.146),241s(4.086)	104-1088-84
$C_{14}H_5F_{14}N_3$ 1,3,5-Triazine, 2,4-bis(heptafluoro-propyl)-6-phenyl-	$CHCl_3$	290(4.32)	70-0842-84
$C_{15}H_6BrN_3OS_2$ Propanedinitrile, [[5-(2-benzothiazo-lylthio)-4-bromo-2-furanyl]methylene]-	MeOH	225(3.49),275(3.22), 359(3.22)	73-0984-84
$C_{15}H_7BrN_4OS$ Propanedinitrile, [[5-(1H-benzimidazol-2-ylthio)-4-bromo-2-furanyl]methyl-ene]-	MeOH	205(3.52),251(3.16), 376(3.22)	73-0984-84
$C_{15}H_7Cl_3O_2$ 1,10-Anthracenedione, 2,4,9-trichloro-3-methyl-	benzene	482(4.07)	104-1766-84
$C_{15}H_7NO_4$ Anthra[1,2-c]isoxazole-3,6,11(1H)-tri-one	dioxan	435(3.60)	104-2012-84
$C_{15}H_7N_3$ [1,1'-Biphenyl]-2,4,5-tricarbonitrile	MeOH	242(4.32),283(4.00), 320(2.81)	40-0060-84
$C_{15}H_7N_3O_4$ 2-Anthracenecarboxylic acid, 1-azido-9,10-dihydro-9,10-dioxo-, sodium salt	DMF	384(3.63)	104-2012-84
$C_{15}H_8Br_2O_6$ Naphtho[1,8-bc]pyran-3-carboxylic acid, 8,9-dibromo-2,6-dihydro-7-hydroxy-2,6-dioxo-, ethyl ester	$CHCl_3$	251(4.09),298(3.96), 320(4.04),479(3.86)	39-1957-84C
$C_{15}H_8ClF_6NOS$ 2-Cyclohexen-1-one, 2-chloro-4,4,5,5,6-6-hexafluoro-3-[(1-methyl-2(3H)-benzothiazolylidene)methyl]-	EtOH	500(5.21)	104-0390-84
$C_{15}H_8ClNO_4$ 1(3H)-Isobenzofuranone, 3-[(4-chloro-phenyl)methylene]-6-nitro-	EtOH	363(4.09)	131-0239-84H
$C_{15}H_8FNO_4$ 1(3H)-Isobenzofuranone, 3-[(4-fluoro-phenyl)methylene]-6-nitro-	EtOH	360(4.06)	131-0239-84H
$C_{15}H_8F_4N_2OS$ 1-Cyclopentene-1-carbonitrile, 3,3,4,4-tetrafluoro-2-[(3-methyl-2(3H)-benzo-thiazolylidene)methyl]-5-oxo-	EtOH	457(4.69)	104-0390-84

Compound	Solvent	$\lambda_{max}(\log \epsilon)$	Ref.
$C_{15}H_8N_2O_3$			
6H-Pyrido[4,3,2-kl]acridin-6-one, 5,10-dihydroxy- (necatorone)	MeOH	212s(4.38),233(4.60), 265s(4.13),293(3.88), 310s(3.85),431(4.13)	88-3575-84
	+ NH₃	213s(--),234(--), 253s(--),307(--), 335s(--),420(--), 557(--)	88-3575-84
	xs NH₃	248(--),270s(--), 327(--),400(--), 520(--),610s(--)	88-3575-84
$C_{15}H_8N_2O_5S_2$			
7H-Azeto[2,1-c]furo[3,4-e][1,2,4]dithi-azine-1,7(3H)-dione, 6-(1,3-dihydro-1,3-dioxo-2H-isoindol-2-yl)-5a,6-di-hydro-, (5aR-trans)-	EtOH	223(4.57),277(3.89), 329(3.33)	88-4167-84
$C_{15}H_8N_2O_6$			
1(3H)-Isobenzofuranone, 6-nitro-3-[(4-nitrophenyl)methylene]-	EtOH	357(4.09)	131-0239-84H
$C_{15}H_8N_6O$			
8H-1,2,4-Triazolo[4"',3",:1",6"][1,2,4]-triazino[4",5":1',5']pyrrolo[2',3'-4,5]furo[3,2-b]indole	dioxan	345(3.01)	73-1529-84
$C_{15}H_8OS$			
6H-Anthra[9,1-bc]thiophen-6-one	EtOH	490(4.02)	104-1415-84
$C_{15}H_8O_2S_3$			
4H-Cyclopenta[2,1-b:3,4-c']dithiophene-7-carboxylic acid, 4-oxo-5-(2-thien-yl)-, methyl ester	CH₂Cl₂	245s(4.14),251s(4.19), 256s(4.22),263(4.22), 270(4.12),318s(4.24), 330(4.33)	139-0121-84B
7H-Cyclopenta[1,2-b:3,4-b']dithiophene-3-carboxylic acid, 7-oxo-2-(2-thien-yl)-, methyl ester	CH₂Cl₂	240s(4.27),245s(4.31), 250s(4.38),257s(4.47), 262(4.52),270(4.45), 283s(3.42),347(4.04), 485(2.97)	139-0121-84B
$C_{15}H_8O_4$			
2-Anthracenecarboxaldehyde, 9,10-di-hydro-4-hydroxy-9,10-dioxo-	MeOH	206(4.32),219(4.32), 252(4.47),276s(--), 325(3.51),401(3.80)	78-3677-84
$C_{15}H_9BrN_4S$			
2-Benzothiazoleacetonitrile, α-[(4-bromophenyl)hydrazono]-	EtOH	258(4.18),403(4.47)	104-0523-84
$C_{15}H_9BrO_3$			
1,4-Phenanthrenedione, 2-bromo-3-meth-oxy-	CHCl₃	285(4.35),291(4.41), 330(3.88),383(3.50), 446(3.09)	40-0090-84
$C_{15}H_9BrO_3S_2$			
Cyclohept[c][1,2]oxathiole-3,4-dione, 8-[[(4-bromophenyl)methyl]thio]-	MeCN	217(4.55),234s(4.46), 280s(4.24),292(4.29), 350(3.66)	88-0419-84

Compound	Solvent	$\lambda_{max}(\log \epsilon)$	Ref.
$C_{15}H_9BrO_6$ Naphtho[1,8-bc]pyran-3-carboxylic acid, 8-bromo-2,6-dihydro-7-hydroxy-2,6-dioxo-, ethyl ester	$CHCl_3$	250(3.82),300(4.00), 322(4.07),478(3.86)	39-1957-84C
$C_{15}H_9ClN_4O_2S$ 4H-[1,2,4]Triazolo[3,4-c][1,4]benzothiazine, 7-chloro-1-(4-nitrophenyl)-	EtOH	238(3.37),260(3.11), 285(3.12)	2-0445-84
$C_{15}H_9ClN_4S$ 2-Benzothiazoleacetonitrile, α-[(4-chlorophenyl)hydrazono]-	EtOH	255(4.23),404(4.39)	104-0523-84
$C_{15}H_9ClO_3$ 1,4-Anthracenedione, 9-chloro-10-hydroxy-2-methyl-	benzene	463(3.90)	104-1766-84
1,4-Anthracenedione, 10-chloro-9-hydroxy-2-methyl-	benzene	463(3.90)	104-1766-84
1,4-Anthracenedione, 9-chloro-10-methoxy-	$CHCl_3$	293(3.7),302(3.8), 329(3.3),416(3.5)	150-0147-84M
$C_{15}H_9ClS$ 1H-Indene-1-thione, 3-chloro-2-phenyl-	MeOH	263(4.41),366(3.72)	104-1574-84
$C_{15}H_9Cl_2N_3$ 4H-Pyrazolo[1,5-a]benzimidazole, 6,7-dichloro-2-phenyl-	EtOH	253(4.37),335(3.87)	48-0829-84
$C_{15}H_9FO_2$ 1H-2-Benzopyran-1-one, 4-(4-fluorophenyl)-	MeOH	224(4.47),228s(4.46), 266(3.85),275s(3.79), 320(3.67)	88-3025-84
$C_{15}H_9NO_3$ 5H-Furo[3',2':6,7][1]benzopyrano[3,4-c]pyridin-5-one, 7-methyl-	EtOH 2% EtOH	308(3.90),330s(--) 308(3.92),330s(--)	149-0145-84A 149-0145-84A
Indolo[4,5-d]benzo[b]furan-2-carboxylic acid	EtOH	208(4.55),220(4.54), 273(4.10),281(4.28), 324(4.59)	103-1123-84
isomer 16	EtOH	206(4.59),241(4.42), 253(4.53),263(4.67), 280(4.44),290(4.62), 328(4.38),342(4.38)	103-1123-84
Indolo[5,4-d]benzo[b]furan-2-carboxylic acid	EtOH	205(4.57),269(4.85), 288(4.61)	103-1123-84
Indolo[6,5-d]benzo[b]furan-2-carboxylic acid	EtOH	204(4.49),216(4.56), 242(4.54),250(4.59), 327(3.71)	103-1123-84
$C_{15}H_9NO_4$ 1(3H)-Isobenzofuranone, 6-nitro-3-(phenylmethylene)-	EtOH	358(4.20)	131-0239-84H
$C_{15}H_9NS$ [1]Benzothiopyrano[4,3,2-de]isoquinoline	MeOH	226(4.49),239(4.44), 273(3.89),281(3.89), 302.5(3.80),324(3.75), 370(3.78),397(3.83)	73-1021-84
$C_{15}H_9N_3O_2S_2$ 1,2-Benzodithiol-1-ium, 3-(3,5-dioxo-	dioxan	234(4.25),284(3.91),	104-1423-84

Compound	Solvent	$\lambda_{max}(\log \epsilon)$	Ref.
4-phenyl-1,2,4-triazolidin-1-yl)-, hydroxide, inner salt (cont.)		335s(3.34),480s(3.95), 505(4.01)	104-1423-84
$C_{15}H_9N_5O_4$			
6H-Indolo[2,3-b]quinoxaline, 6-methyl-2,3-dinitro-	EtOH	215(4.51),266(4.34), 325(4.53),415(3.58)	103-1276-84
6H-Indolo[2,3-b]quinoxaline, 6-methyl-2,9-dinitro-	EtOH	208(4.13),260(4.31), 326(4.65),405(3.59)	103-1276-84
$C_{15}H_{10}$			
1H-Cycloprop[b]anthracene	hexane	252(5.07),320(3.18), 334(3.54),351(3.72), 371(3.67)	35-0440-84
$C_{15}H_{10}BrNO$			
Benzo[h]quinoline-4-carboxaldehyde, 5-bromo-6-methoxy-	EtOH	245(4.55),279(4.11), 315(3.70),360(3.48)	12-1271-84
$C_{15}H_{10}BrNO_3S$			
2-Propenoic acid, 3-[4-bromo-5-(phenyl-thio)-2-furanyl]-2-cyano-, methyl ester	MeOH	205(3.14),247(3.17), 340(3.31)	73-0984-84
$C_{15}H_{10}BrNO_5S$			
2-Propenoic acid, 3-[4-bromo-5-(phenyl-sulfonyl)-2-furanyl]-2-cyano-, methyl ester	MeOH	208(3.21),247(3.26), 342(3.41)	73-0984-84
$C_{15}H_{10}BrN_2O_2$			
Benzenediazonium, 2-[2-(4-bromophenyl)-2-carboxyethenyl]-, (E)-, tetra-fluoroborate	MeCN	268(4.08),330(3.96), 370(3.60)	39-1093-84B
$C_{15}H_{10}BrN_3$			
6H-Indolo[2,3-b]quinoxaline, 9-bromo-6-methyl-	EtOH	228(4.50),286(4.57), 357(4.35),420(3.51)	103-0537-84
4H-Pyrazolo[1,5-a]benzimidazole, 6-bromo-2-phenyl-	EtOH	258(4.37),330(4.24)	48-0829-84
$C_{15}H_{10}Br_3NO$			
Benzo[h]quinoline, 5-bromo-4-(dibromo-methyl)-6-methoxy-	EtOH	225(4.32),246(4.48), 276(4.19),314(3.80), 343(3.34),356(3.40)	12-1271-84
$C_{15}H_{10}ClNO$			
1,3-Benzoxazepine, 2-(4-chlorophenyl)-	EtOH	246(4.24),280s(3.86), 333(3.84)	78-3567-84
$C_{15}H_{10}ClNO_2$			
1(3H)-Isobenzofuranone, 6-amino-3-[(4-chlorophenyl)methylene]-	EtOH	316(4.21),379(4.12)	131-0239-84H
$C_{15}H_{10}ClN_3$			
4H-Pyrazolo[1,5-a]benzimidazole, 6-chloro-2-phenyl-	EtOH	256(4.28),330(4.11)	48-0829-84
4H-Pyrazolo[1,5-a]benzimidazole, 7-chloro-2-phenyl-	EtOH	256(4.40),282(4.03), 333(4.14)	48-0829-84
$C_{15}H_{10}ClN_3O$			
4H-[1,2,4]Triazolo[3,4-c][1,4]benzoxa-zine, 1-(4-chlorophenyl)-	EtOH	245(3.11),280(2.71)	2-0445-84

Compound	Solvent	$\lambda_{max}(\log \epsilon)$	Ref.
$C_{15}H_{10}ClN_3S$			
4H-[1,2,4]Triazolo[3,4-c][1,4]benzo-thiazine, 1-(4-chlorophenyl)-	EtOH	220(3.39),266(3.04), 290(2.44)	2-0445-84
4H-[1,2,4]Triazolo[3,4-c][1,4]benzo-thiazine, 7-chloro-1-phenyl-	EtOH	242(3.45),270(3.08), 295(2.70)	2-0445-84
$C_{15}H_{10}FNO_2$			
1(3H)-Isobenzofuranone, 6-amino-3-[(4-fluorophenyl)methylene]-	EtOH	321(4.29),376(4.12)	131-0239-84H
$C_{15}H_{10}N_2O_2$			
Benzo[h]pyrrolo[4,3,2-de]quinolin-4(5H)-one, 6-methoxy-	EtOH	218(4.36),237(4.55), 282s(4.33),293(4.33), 302(4.32),406(3.79)	12-1271-84
6H-Indolo[3,2,1-de][1,5]naphthyridin-6-one, 5-(hydroxymethyl)-	EtOH	235(4.96),262(4.36), 270(4.30),298(4.23), 360(5.51),376(4.51)	94-0170-84
$C_{15}H_{10}N_2O_3$			
1,3-Benzoxazepine, 2-(4-nitrophenyl)-	EtOH	222(4.15),270(4.13), 370(3.82)	78-3567-84
6H-Indolo[3,2,1-de]naphthyridin-6-one, 5-hydroxy-4-methoxy-	EtOH	246(4.60),262s(4.50), 286(4.25),340(4.01), 356(4.05),374(4.00)	94-3579-84
$C_{15}H_{10}N_2O_3S$			
Benzo[b]thiophen-3(2H)-one, 2-(1,3-di-hydro-2H-benzimidazol-2-ylidene)-, 1,1-dioxide	EtOH	242(4.18),254(4.09), 263(4.04),284(4.21), 322(4.11),366(4.30)	95-0134-84
3H-Pyrrolo[2,3-c]phenothiazine-1-carbox-aldehyde, 11,11-dioxide	EtOH	212(4.80),266(4.78), 299(4.44),399(4.41)	103-1095-84
$C_{15}H_{10}N_2O_4$			
1(3H)-Isobenzofuranone, 3-[(4-nitro-phenyl)methylene]-6-nitro-	EtOH	452(4.37)	131-0239-84H
$C_{15}H_{10}N_2O_5S$			
2,4,6(1H,3H,5H)-Pyrimidinetrione, 5-[(1,3-dihydro-1,3-dioxo-2H-inden-2-ylidene)(methylthio)methyl]-	EtOH	241(4.48),258(4.42), 363(4.27),452(3.98)	95-0127-84
$C_{15}H_{10}N_2S$			
Phenanthro[9,10-d]thiazol-2-amine	CH_2Cl_2	261.3(4.56),318.1(3.98), 326(3.98),340(3.49), 364(2.97)	4-1597-84
$C_{15}H_{10}N_2S_2$			
Benzenamine, N-4H-1,3-dithiolo[4,5-b]-indol-2-ylidene-	EtOH	249(4.03),300(4.12), 347s(3.95)	12-2479-84
	EtOH-NaOH	254(4.09),275(4.22), 300s(3.96),385(3.88)	12-2479-84
monohydrobromide	EtOH-HBr	246(4.00),280(4.14), 320s(3.91),398(3.85)	12-2479-84
$C_{15}H_{10}N_3O_3$			
Pyridinium, 1-[5-(2,2-dicyanoethenyl)-2-furanyl-3-(methoxycarbonyl)-, brom-ide	H_2O	251(3.00),345s(3.04), 377(3.16),468(2.76)	73-2485-84
$C_{15}H_{10}N_4$			
Ethenetricarbonitrile, 4,5-dihydro-	MeOH	204(4.15),286(3.72),	83-0143-84

Compound	Solvent	λ_{max} (log ϵ)	Ref.
2-phenyl-1H-pyrrol-3-yl)- (cont.) 1,2,4-Triazolo[1,5-c]quinazoline, 5-phenyl-	EtOH	456(4.56) 249(4.22),292(3.89)	83-0143-84 18-1138-84
$C_{15}H_{10}N_4OS$ 2H-[1,2,4]Triazino[4",5":1',5']pyrrolo- [2',3':4,5]furo[3,2-b]indole-1-thione, 4-methyl-	dioxan	389(3.04)	73-1529-84
$C_{15}H_{10}N_4O_2$ 6H-Indolo[2,3-b]quinoxaline, 6-methyl- 2-nitro-	EtOH	215(4.46),280(4.50), 320(4.37),422(3.54)	103-1276-84
6H-Indolo[2,3-b]quinoxaline, 6-methyl- 9-nitro-	EtOH	209(4.33),229(4.37), 260(4.63),330(4.51), 392(3.88)	103-0537-84
2H-[1,2,4]Triazino[4",5":1',5']pyrrolo- [2',3':4,5]furo[3,2-b]indol-1(6H)- one, 4-methyl-	dioxan	340(3.62)	73-1529-84
$C_{15}H_{10}N_4O_2S_2$ Pyrimido[5,4-e]thiazolo[3,4-a]pyrimi- din-10-ium, 1,2,3,4-tetrahydro- 9-(methylthio)-2,4-dioxo-7-phenyl-, hydroxide, inner salt	DMF	385(4.10),396(4.10), 446(4.09)	103-0921-84
$C_{15}H_{10}N_4O_3S$ Furo[2',3':4,5]pyrrolo[1,2-d][1,2,4]- triazine-8(7H)-thione, 5-methyl- 2-(2-nitrophenyl)-	dioxan	387(3.06)	73-1529-84
$C_{15}H_{10}N_4O_4$ Furo[2',3':4,5]pyrrolo[1,2-d]-1,2,4- triazin-4(3H)-one, 1-methyl-7-(2- nitrophenyl)-	dioxan	343(3.46)	73-1529-84
$C_{15}H_{10}N_4O_5S$ 3H-Pyrazol-3-one, 2-(2,4-dinitrophen- yl)-2,4-dihydro-5-methyl-4-(thienyl- methylene)-	C_6H_{12} EtOH ether DMF DMSO	365(3.78) 224(3.18),320(2.88), 374(3.44) 365(3.85) 370(3.79) 374(3.79)	23-2841-84 23-2841-84 23-2841-84 23-2841-84 23-2841-84
$C_{15}H_{10}N_4S$ 2-Benzothiazoleacetonitrile, α-(phenyl- hydrazono)-	EtOH	258(4.20),402(4.08)	104-0523-84
1,2,4-Triazolo[4,3-c]quinazoline- 3(2H)-thione, 5-phenyl-	EtOH	290(3.97)	18-1138-84
$C_{15}H_{10}O$ 6,11-Methanobenzocyclodecen-13-one	CH_2Cl_2	273(4.88),344(3.71)	89-0719-84
$C_{15}H_{10}O_2$ 9,10-Anthracenedione, 2-methyl-	MeOH	255(4.65),265(4.32), 274(4.24),324(3.66)	102-0313-84
$C_{15}H_{10}O_3$ 9,10-Anthracenedione, 1-hydroxy- 2-methyl-	EtOH	253(4.17),408(3.52)	78-3455-84
9,10-Anthracenedione, 1-hydroxy- 3-methyl-	MeOH	211s(--),224(4.32), 245(4.41),253(4.43),	78-3677-84

Compound	Solvent	$\lambda_{max}(\log \epsilon)$	Ref.
9,10-Anthracenedione, 1-hydroxy-3-methyl- (Pachybasin) (cont.)		258(4.42),277s(--), 326(3.47),383s(--), 402(3.77),422s(--)	78-3677-84
	EtOH	246(3.95),254(3.97), 259(3.97),282s(3.66), 400(3.38)	78-3455-84
9,10-Anthracenedione, 1-hydroxy-7-methyl- (barleriaquinone)	EtOH	217(4.35),259(4.59), 285s(4.02),342(3.22), 410(3.63)	94-4137-84
4H-1-Benzopyran-4-one, 2-(2-hydroxyphenyl)- (in 15% MeOH)	pH 7	249(4.24),309(4.21), 330s(--)	61-0759-84
	MeOH	255(4.23),306(4.20), 325s(--)	61-0759-84
	95% H_2SO_4	248(4.34),275(4.11), 367(4.36)	61-0759-84
	pH 13	304(4.08),396(3.95)	61-0759-84
4H-1-Benzopyran-4-one, 2-(3-hydroxyphenyl)- (in 15% MeOH)	pH 7	250(4.23),305(4.29)	61-0759-84
	pH 13	245(4.36),309(4.32), 358(3.64)	61-0759-84
	MeOH	241(4.28),296(4.31)	61-0759-84
	95% H_2SO_4	238(4.11),258(4.21), 347(4.42)	61-0759-84
4H-1-Benzopyran-4-one, 2-(4-hydroxyphenyl)-	pH 7.0	326(4.46)	61-0759-84
	pH 13	308(3.99),383(4.52)	61-0759-84
	MeOH	325(4.45)	61-0759-84
	95% H_2SO_4	272(4.42),374(4.54)	61-0759-84
4H-1-Benzopyran-4-one, 3-hydroxy-2-phenyl-	pH 7	342(4.23)	61-0759-84
	pH 13	402(4.19)	61-0759-84
	MeOH	344(4.21),352s(--)	61-0759-84
	70% H_2SO_4	378(4.40)	61-0759-84
4H-1-Benzopyran-4-one, 5-hydroxy-2-phenyl-	pH 7	266(4.32),298(4.04), 332(3.79)	61-0759-84
	pH 13	274(4.44),372(3.64)	61-0759-84
	MeOH	267(4.45),328(3.81)	61-0759-84
	95% H_2SO_4	283(4.19),345(4.41)	61-0759-84
4H-1-Benzopyran-4-one, 6-hydroxy-2-phenyl-	pH 7	266(4.39),308(4.31)	61-0759-84
	pH 13	287(4.53),390(3.66)	61-0759-84
	MeOH	268(4.44),302(4.25)	61-0759-84
	95% H_2SO_4	277(4.21),355(4.42)	61-0759-84
4H-1-Benzopyran-4-one, 7-hydroxy-3-phenyl-	MeOH	296(4.19),305s(4.16)	64-0238-84B
4H-1-Benzopyran-4-one, 8-hydroxy-2-phenyl- (in 15% MeOH)	pH 7	264(4.44),302(4.23)	61-0759-84
	pH 13	284(4.52),362(3.35)	61-0759-84
	MeOH	266(4.47),295s(--)	61-0759-84
	95% H_2SO_4	263(4.21),289(4.14), 357(4.48)	61-0759-84
$C_{15}H_{10}O_4$ 9,10-Anthracenedione, 1,5-dihydroxy-2-methyl-	EtOH	255(4.32),279(3.93), 289(3.96),426s(3.98), 434(3.99)	78-3455-84
9,10-Anthracenedione, 1,5-dihydroxy-3-methyl- (ziganein)	EtOH	254(4.31),280(3.99), 290(4.01),430(4.01)	78-3455-84
9,10-Anthracenedione, 1,8-dihydroxy-2-methyl-	EtOH	256(4.35),287(3.99), 432(4.05)	78-3455-84
9,10-Anthracenedione, 1,8-dihydroxy-3-methyl- (chrysophanol)	EtOH	256(4.02),277(3.72), 287(3.74),430(3.76)	78-3455-84
9,10-Anthracenedione, 1-hydroxy-2-(hydroxymethyl)- (digiferruginol)	MeOH	223(4.24),253(4.47), 330(3.35),405(2.61)	102-1733-84
9,10-Anthracenedione, 1-hydroxy-6(or 7)-(hydroxymethyl)-	EtOH	210(4.67),256(4.78), 280s(4.36),335(4.32), 398(4.58)	102-1733-84

Compound	Solvent	$\lambda_{max}(\log \epsilon)$	Ref.
9,10-Anthracenedione, 1-hydroxy-2-meth-oxy-	EtOH	247(4.37),277(4.02), 376(3.36),424(3.69)	78-3455-84
9,10-Anthracenedione, 1-hydroxy-3-meth-oxy-	EtOH	240(4.14),243s(4.12), 266s(4.00),280(4.08), 408(3.52)	78-3455-84
9,10-Anthracenedione, 1-hydroxy-8-meth-oxy-	n.s.g.	223(4.52),255(4.32), 280(3.97),395(3.89), 415(3.93),440(3.79)	5-0306-84
4H-1-Benzopyran-4-one, 6,7-dihydroxy-2-phenyl-	MeOH	327(4.00)	64-0238-84B
4H-1-Benzopyran-4-one, 7-hydroxy-2-(4-hydroxyphenyl)-	MeOH	300s(4.05)	64-0238-84B
5H-Phenanthro[4,5-bcd]pyran-5-one, 9,10-dihydro-2,7-dihydroxy-	EtOH	220(4.32),246(4.15), 288(4.04),371(3.66)	102-0671-84
(oxoflavidin)	EtOH-NaOH	214(4.29),230s(4.19), 258(4.19),309(4.16)	102-0671-84

$C_{15}H_{10}O_4S$

Compound	Solvent	$\lambda_{max}(\log \epsilon)$	Ref.
Dibenz[b,c][1,4]oxathiepin-7-acetic acid, 11-oxo-	MeOH	251a(3.83)	73-2531-84

$C_{15}H_{10}O_4S_4$

Compound	Solvent	$\lambda_{max}(\log \epsilon)$	Ref.
1,3-Dithiole-4-carboxylic acid, 2-(4-carboxy-1,3-dithiol-2-ylidene)-, 4-(phenylmethyl) ester	MeCN	286(4.21),299(4.21), 312(4.20),442(3.52)	103-1342-84

$C_{15}H_{10}O_5$

Compound	Solvent	$\lambda_{max}(\log \epsilon)$	Ref.
9,10-Anthracenedione-9-^{13}C, 1,3-di-hydroxy-2-(hydroxymethyl)-(lucidin)	MeOH	242(4.34),246(4.35), 280(4.35),332(3.40), 410(3.75)	102-0307-84
9,10-Anthracenedione, 1,5-dihydroxy-2-(hydroxymethyl)-	EtOH	227(4.32),255(4.13), 277s(3.64),287.5(3.65), 420(3.64),430(3.65)	102-1733-84
9,10-Anthracenedione, 1,8-dihydroxy-3-methoxy-	EtOH	245(4.16),265(4.19), 284(4.18),432(4.04)	78-3455-84
9,10-Anthracenedione, 1,5,8-trihydroxy-3-methyl-	EtOH	254(3.84),288(3.50), 294s(3.49),478(3.66), 488(3.69),508(3.57), 522(3.49)	78-3455-84
4H-1-Benzopyran-4-one, 5,7-dihydroxy-3-(4-hydroxyphenyl)-	MeOH	334s(3.63)	64-0238-84B
4H-1-Benzopyran-4-one, 6,7-dihydroxy-3-(4-hydroxyphenyl)-	MeOH	329(3.90)	64-0238-84B

$C_{15}H_{10}O_6$

Compound	Solvent	$\lambda_{max}(\log \epsilon)$	Ref.
9,10-Anthracenedione, 1,2,5,7-tetra-hydroxy-4-methyl-	EtOH	229(4.42),264(4.26), 293(4.25),464(3.94)	78-5039-84
4H-1-Benzopyran-4-one, 2-(3,4-dihy-droxyphenyl)-5,7-dihydroxy-	MeOH	250(4.33),262(4.29), 288(4.05),345(4.41)	94-0295-84
Naphtho[1,8-bc]pyran-3-carboxylic acid, 2,6-dihydro-7-hydroxy-2,6-dioxo-, ethyl ester	$CHCl_3$	250(3.54),289(3.90), 323(4.00),484(3.71)	39-1957-84C
1H-Xanthene-1,4,9-trione, 3,7-dimeth-oxy-	MeOH	247s(4.29),277(4.11), 328(3.61),367(3.57)	39-1507-84C

$C_{15}H_{10}O_7$

Compound	Solvent	$\lambda_{max}(\log \epsilon)$	Ref.
9,10-Anthracenedione, 1,2,3,5,6-penta-hydroxy-8-methyl-	MeOH-HCOOH	285(4.44),322s(3.82), 454(3.96)	78-5039-84

$C_{15}H_{10}O_8$

Compound	Solvent	$\lambda_{max}(\log \epsilon)$	Ref.
4H-1-Benzopyran-4-one, 5,6,7-trihydr-	MeOH	266(4.3),294(4.1),	94-4935-84

Compound	Solvent	$\lambda_{max}(\log \epsilon)$	Ref.
oxy-2-(2,4,5-trihydroxyphenyl)-		376(4.2)	94-4935-84
(cont.)	MeOH-NaOMe	262(--),315(--),	94-4935-84
		420(--)	
	MeOH-NaOAc	265(--),414(--)	94-4935-84
	MeOH-AlCl$_3$	271(--),444(--)	94-4935-84
Lateropyrone	CH$_2$Cl$_2$	244(4.23),272(4.15),	23-2101-84
		282(4.19),351(4.16),	
		370(4.16)	
$C_{15}H_{11}Br$			
6,11-Methanobenzocyclodecene, 13-bromo-	CH$_2$Cl$_2$	280(4.77),349(3.68)	89-0719-84
$C_{15}H_{11}BrN_2S$			
1H-1-Pyrindine-3-carbonitrile, 4-(3-bromophenyl)-2,5,6,7-tetrahydro-2-thioxo-	dioxan	258(4.19),320(3.98), 442(3.28)	104-1402-84
$C_{15}H_{11}ClF_3N_3O_3S$			
Benzenepropanamide, 2-[(chloroacetyl)-amino]-β-oxo-N-(2-thiazolyl)-3-(trifluoromethyl)-	EtOH-HCl	272(3.98),312(2.30)	4-1345-84
2H-3,1-Benzoxazine-2-acetamide, 2-(chloromethyl)-1,4-dihydro-4-oxo-N-2-thiazolyl-8-(trifluoromethyl)-	EtOH-HCl	227(4.36),265(4.16), 339(3.62)	4-1345-84
$C_{15}H_{11}ClN_2S$			
1H-1-Pyrindine-3-carbonitrile, 4-(4-chlorophenyl)-2,5,6,7-tetrahydro-2-thioxo-	dioxan	263(4.42),333(4.46), 435(3.71)	104-1402-84
$C_{15}H_{11}ClO$			
4H-1-Benzopyran, 2-(4-chlorophenyl)-	ether	244(4.05),280(3.76)	78-3567-84
$C_{15}H_{11}ClO_3S$			
Benzoic acid, 2-[(2-acetyl-4-chloro-phenyl)thio]-	MeOH	235s(4.30),255(4.03), 317(3.70)	73-1009-84
$C_{15}H_{11}Cl_2NO$			
Benzo[h]quinoline, 2,6-dichloro-6-methoxy-4-methyl-	EtOH	216(4.45),222(4.45), 245(4.66),251(4.66), 268(4.44),276(4.51), 286s(4.15),298(4.08), 310(4.11),328(3.26), 343(3.42),353(3.42)	12-1271-84
$C_{15}H_{11}Cl_2NO_3$			
9(10H)-Acridinone, 2,4-dichloro-1-hydroxy-3-methoxy-10-methyl-	EtOH	213s(3.99),248s(4.01), 270(4.23),316(3.61), 392s(3.31),409(3.36)	5-0031-84
$C_{15}H_{11}F$			
Anthracene, 1-fluoro-4-methyl-	EtOH	252(4.93),314(3.24), 328(4.35),345(3.50), 356(3.62),365(3.54)	12-1769-84
Anthracene, 9-fluoro-10-methyl-	C$_6$H$_{12}$	248(5.0),256(5.49), 356(3.84),375(4.5), 396(4.01)	44-2803-84
$C_{15}H_{11}FOS$			
Dibenzo[b,f]thiepin-10(11H)-one, 3-fluoro-8-methyl-	MeOH	239(4.30),256s(4.08), 330(3.60)	73-2638-84

Compound	Solvent	$\lambda_{max}(\log \epsilon)$	Ref.
C₁₅H₁₁FS			
Dibenzo[b,f]thiepin, 7-fluoro-2-methyl-	MeOH	263(4.41),296(3.69), 339s(3.05)	73-2638-84
C₁₅H₁₁F₆NOS			
3-Penten-2-one, 4-[(3-ethyl-2(3H)-ben- zothiazolylidene)methyl]-1,1,1,5,5,5- hexafluoro-	benzene EtOH	482(4.47),518(4.50) 476(4.43),512(4.31)	104-0390-84 104-0390-84
C₁₅H₁₁N			
9-Phenanthrenecarbonitrile, 9,10-di- hydro-	MeCN	266(3.81),293s(3.00)	40-0060-84
C₁₅H₁₁NO			
1,3-Benzoxazepine, 2-phenyl-	EtOH	334(3.83)	78-3567-84
1H-Isoindole-1-carboxaldehyde, 3-phenyl-	EtOH	242(4.19),264(4.37), 384(4.47)	39-0833-84B
C₁₅H₁₁NOS			
2(3H)-Thiazolone, 4,5-diphenyl-	n.s.g.	226(4.21),280(4.24), 304(4.21)	47-0739-84
C₁₅H₁₁NO₂			
Benzo[h]quinoline-4-carboxaldehyde, 6-methoxy-	EtOH	220(4.35),232s(4.44), 244(4.54),268s(4.15), 273(4.17),281s(4.11), 293s(3.94),305(3.83), 325(3.42),341(3.54), 355(3.62),396(3.42)	12-1271-84
2,4,6-Heptatriynoic acid, 7-(methyl- phenylamino)-	MeCN	214f(4.70),262(4.56), 352f(4.09)	27-0157-84
1(3H)-Isobenzofuranone, 6-amino- 3-(phenylmethylene)-	EtOH	314(4.19),371(4.06)	131-0239-84H
2H-Isoindole-4,7-dione, 2-(phenyl- methyl)-	n.s.g.	226(4.30),242(4.18), 373(3.60)	88-4917-84
9,10-Phenanthrenedione, mono(O-methyl- oxime)	hexane	262(3.43),318(2.69), 385(2.30)	54-0023-84
C₁₅H₁₁NO₃			
4H-1-Benzopyran, 2-(4-nitrophenyl)-	ether	275(4.07),335(3.86)	78-3567-84
Benzo[h]quinoline-4-carboxylic acid, 6-methoxy-	EtOH	220(4.29),236s(4.42), 244(4.47),275(4.15), 284(4.17),308s(3.77), 349s(3.48),362(3.72)	12-1271-84
9H-Carbazole-3-carboxaldehyde, 2-acet- oxy-	EtOH	238(4.45),247(4.29), 274(4.50),290(4.56), 340(4.16)	102-0471-84
5H-Furo[3',2':6,7][1]benzopyrano[3,4- c]pyridin-5-one, 9,10-dihydro-7- methyl-	EtOH 2% EtOH	332(4.14) 330(--)	149-0145-84A 149-0145-84A
2H-Pyran-3-carbonitrile, 4-methoxy- 2-oxo-6-(2-phenylethenyl)-	EtOH	237(4.13),340(4.37)	94-3384-84
C₁₅H₁₁NS			
Benzothiazole, 2-(2-phenylethenyl)-	C₆H₁₂	326.4(4.45),337.8(4.45), 360.0(4.24)	131-0417-84H
C₁₅H₁₁N₂O₂			
Benzenediazonium, 2-(2-carboxy-2-phen- ylethenyl)-, (E)-, tetrafluoroborate	MeCN	235(3.89),265(4.00), 370(3.55)	39-1093-84B

Compound	Solvent	$\lambda_{max}(\log \epsilon)$	Ref.
$C_{15}H_{11}N_3$			
6H-Indolo[2,3-b]quinoxaline, 6-methyl-	EtOH	224(4.49),274(4.69), 358(4.33),410(3.57)	103-0537-84
4H-Pyrazolo[1,5-a]benzimidazole, 2-phenyl-	EtOH	249(4.40),279(4.14), 318(4.18)	48-0829-84
Pyrido[2,3-b][1,10]phenanthroline, 5,6-dihydro-	EtOH	218(4.50),335(4.30), 349(4.33)	44-2208-84
$C_{15}H_{11}N_3O$			
Spiro[2H-benzimidazole-2,3'-[3H]indol]-2'(1'H)-one, 1'-methyl-	EtOH	209(4.48),232(4.58), 248(4.57),327(4.29), 341(4.21)	103-0537-84
4H-[1,2,4]Triazolo[3,4-c][1,4]benzoxazine, 1-phenyl-	EtOH	242(2.99),255(2.90), 285(2.57)	2-0445-84
$C_{15}H_{11}N_3O_2S$			
Propanedinitrile, [[2-(1-methyl-5-(phenylsulfonyl)-1H-pyrrol-2-yl]methylene]-	MeOH	210(4.12),252(3.90), 367(4.59)	4-1041-84
$C_{15}H_{11}N_3O_3$			
2-Propenoic acid, 4-(2H-benzotriazol-2-yl)-3-hydroxyphenyl ester	DMAc	270(4.11),290(4.27)	49-0853-84
	CHCl$_3$	295s(4.13),334(4.39)	49-0153-84
polymer	DMAc	270(4.05),290(4.16)	49-0153-84
$C_{15}H_{11}N_3O_5$			
Benzaldehyde, 2-[2-[((2,4-dinitrophenyl)amino]ethenyl]-	MeCN	208(4.33),236(4.30), 260(4.20),392(4.43)	24-0702-84
	CH$_2$Cl$_2$	391(2.43)	24-0702-84
Benzo[h]quinoline, 6-methoxy-4-methyl-5,7-dinitro-	EtOH	251(4.51),267s(4.19), 293(3.91),304(3.87), 321s(3.51),337(3.45), 354(3.40)	12-1271-84
1(2H)-Isoquinolinone, 2-(2,4-dinitrophenyl)-	MeCN	279(4.18),395(4.09)	24-0702-84
	CH$_2$Cl$_2$	284(4.08),399(4.11)	24-0702-84
$C_{15}H_{11}N_3S$			
Propanedinitrile, [[1-methyl-5-(phenylthio)-1H-pyrrol-2-yl]methylene]-	MeOH	206(4.33),246(4.04), 399(4.40)	4-1041-84
3,5-Pyridinedicarbonitrile, 1,4-dihydro-2-methyl-6-(3-methylphenyl)-4-thioxo-	EtOH	267(4.31),323(4.28), 370(3.51)	4-1445-84
3,5-Pyridinedicarbonitrile, 1,4-dihydro-2-methyl-6-(4-methylphenyl)-4-thioxo-	EtOH	272(4.28),282(4.29), 323(4.25),370(3.43)	4-1445-84
$C_{15}H_{11}N_5$			
1H-Pyrazolo[3,4-b]quinoline-1-propanenitrile, 4-cyano-3-methyl-	EtOH	400(3.50)	18-2984-84
$C_{15}H_{11}N_5O$			
Benzofuro[2',3':4,5]pyrrolo[1,2-d]-[1,2,4]triazolo[3,4-f][1,2,4]triazine	dioxan	258(3.47),326(3.43)	73-0065-84
$C_{15}H_{11}N_5O_2$			
6H-Purin-6-one, 1,9-dihydro-9-(6-methoxy-8-quinolinyl)-	H$_2$O	221(4.07),332(3.36)	2-0870-84
3-Quinolinecarbonitrile, 2-amino-4-[(4,5-dihydro-3-methyl-5-oxo-1H-pyrazol-4-yl)carbonyl]-	BuOH	278(4.30),302(4.26), 318(4.38)	4-1233-84
3H-1,2,4-Triazolo[4,3-b]indazole, 3-methyl-3-(4-nitrophenyl)-	CH$_2$Cl$_2$	270(4.71),384(4.29)	24-1726-84

Compound	Solvent	$\lambda_{max}(\log \epsilon)$	Ref.
$C_{15}H_{11}N_5O_2S$			
Pyrido[2,3-d]pyrimidine-6-carbonitrile, 5-amino-1,2,3,4-tetrahydro-7-(methylthio)-2,4-dioxo-1-phenyl-	EtOH	264(4.68),304(4.20)	94-0122-84
$C_{15}H_{11}N_5O_5$			
2H-Indol-2-one, 3-[(2-amino-5-nitrophenyl)imino]-1,3-dihydro-1-methyl-5-nitro-	EtOH	215(4.23),250(4.20), 318(4.08),520(3.59)	103-1276-84
2(1H)-Quinoxalinone, 3-[2-(methylamino)-5-nitrophenyl]-6-nitro-	EtOH	213(4.42),310(4.26), 369(4.29),458(4.08)	103-1276-84
Spiro[2H-benzimidazole-2,3'-[3H]indol]-2'(1'H)-one, 1,3-dihydro-1'-methyl-5,5'-dinitro-	EtOH	213(4.42),282(4.26), 333(4.19),437(3.98)	103-1276-84
Spiro[2H-benzimidazole-2,3'-[3H]indol]-2'(1'H)-one, 1,3-dihydro-1'-methyl-5,6-dinitro-	EtOH	213(4.62),284(4.34), 465(3.71)	103-1276-84
$C_{15}H_{11}OP$			
Phosphorin, 3-(2-furanyl)-5-phenyl-	EtOH	216(4.16),235(4.16), 266(4.24),297(4.04), 306s(4.02)	24-0763-84
$C_{15}H_{11}PS$			
Phosphorin, 3-phenyl-5-(2-thienyl)-	EtOH	220(4.25),236(4.29), 264(4.43),306s(4.17)	24-0763-84
$C_{15}H_{11}S$			
1-Benzothiopyrylium, 4-phenyl-, perchlorate	MeCN	262(4.44),347(3.88), 408(3.95)	44-4192-84
$C_{15}H_{12}$			
Bicyclo[4.1.0]hepta-1,3,5-triene, 7-(1-phenylethylidene)-	C_6H_{12}	242(4.04),249(4.09), 256s(3.95),357s(4.15), 371(4.26),394(4.12)	35-6108-84
$C_{15}H_{12}$			
5H-Dibenzo[a,c]cycloheptene	EtOH	239(3.60)	44-4029-84
$C_{15}H_{12}BrN_3O$			
2(1H)-Quinoxalinone, 3-[5-bromo-2-(methylamino)phenyl]-	EtOH	226(4.55),310(3.92), 358(3.90),455(3.85)	103-0537-84
Spiro[2H-benzimidazole-2,3'-[3H]indol]-2'(1'H)-one, 5'-bromo-1,3-dihydro-1'-methyl-	EtOH	219(4.60),263(4.37), 310(4.00)	103-0537-84
$C_{15}H_{12}BrN_3O_3$			
1H-Indazole-4,7-dione, 5-[(4-bromophenyl)amino]-3-methoxy-1-methyl-	dioxan	259(4.29),358(3.94), 470(3.76)	4-0825-84
2H-Indazole-4,7-dione, 5-[(4-bromophenyl)amino]-3-methoxy-2-methyl-	EtOH	265(4.50),327s(3.92), 445(3.91)	4-0825-84
1H-Indazole-3,4,7(2H)-trione, 5-[(4-bromophenyl)amino]-1,2-dimethyl-	EtOH	265(4.50),327s(3.92), 520s(3.61)	4-0825-84
$C_{15}H_{12}ClNO_2$			
Benzo[h]quinolin-2(1H)-one, 5-chloro-6-methoxy-4-methyl-	EtOH	234(4.19),273(3.93), 286(3.94),298(3.84), 309(3.83),345(3.65), 380(2.78)	12-1271-84

Compound	Solvent	$\lambda_{max}(\log \epsilon)$	Ref.
$C_{15}H_{12}ClNO_3$ 9(10H)-Acridinone, 4-chloro-1-hydroxy- 3-methoxy-10-methyl-	EtOH	212s(4.18),249s(4.37), 271(4.58),303(3.99), 328s(3.81),392(3.69)	5-0031-84
$C_{15}H_{12}ClNO_3S$ 2H-Thieno[2,3-h]-1-benzopyran-2-one, 3-chloro-4-morpholino-	EtOH	234.5(4.30),257s(4.27), 261(4.28),305s(3.94), 337(4.10)	161-0081-84
$C_{15}H_{12}ClN_3O$ Diazene, (4-chlorophenyl)(1-isocyanato- 1-phenylethyl)-	C_6H_{12}	286(4.16),385(2.32)	118-0315-84
$C_{15}H_{12}ClN_3S$ Diazene, (4-chlorophenyl)(1-isothio- cyanato-1-phenylethyl)-	C_6H_{12}	287(4.16),402(2.34)	118-0315-84
$C_{15}H_{12}INO_3$ 9(10H)-Acridinone, 1-hydroxy-2-iodo- 3-methoxy-10-methyl-	EtOH	213s(4.09),237s(4.12), 246s(4.24),275(4.59), 292s(4.19),305s(3.98), 392(3.66)	5-0031-84
$C_{15}H_{12}N_2O$ Methanone, 2-benzofuranylphenyl-, hydrazone, (E)-	EtOH	224(2.37),310(2.21)	4-0937-84
(Z)-	EtOH	235(2.45),264(2.47), 310(2.48)	4-0937-84
2H-Naphth[1',2':2,3]oxepino[4,5-c]pyra- zole, 11,12-dihydro-	EtOH	259(4.63),266(4.59), 296(4.04)	2-0736-84
2(1H)-Quinoxalinone, 1-methyl-3-phenyl-	EtOH	213(4.42),224(4.40), 304(4.06),359(4.07)	44-0827-84
2(1H)-Quinoxalinone, 3-methyl-1-phenyl-	EtOH	210(4.36),230(4.30), 279(3.73),335(3.74)	44-0827-84
$C_{15}H_{12}N_2O_2$ 2H-1,4-Benzoxazpin-2-one, 7-amino- 3-(phenylmethyl)-	EtOH	250(3.87),394(4.08)	4-0551-84
1(3H)-Isobenzofuranone, 6-amino-3-[(4- aminophenyl)methylene]-	EtOH	336(4.42),400(4.31)	131-0239-84H
$C_{15}H_{12}N_2O_3$ Benzo[h]quinoline, 6-methoxy-4-methyl- 5-nitro-	EtOH	249(4.57),266s(4.29), 290(3.90),302(3.87), 320(3.18),345(3.38), 351(3.45)	12-1271-84
5H-Pyrazolo[5,1-b][1,3]oxazin-5-one, 3-acetyl-2-methyl-7-phenyl-	EtOH	247(4.35),298(4.02)	94-0930-84
1H,5H-Pyrazolo[1,2-a]pyrazole-1,5-di- one, 2-acetyl-3-methyl-7-phenyl-	EtOH	251(4.10),338(4.13)	94-0930-84
1H,7H-Pyrazolo[1,2-a]oyrazole-1,7-di- one, 2-acetyl-3-methyl-5-phenyl-	EtOH	255(4.09),370(3.90)	94-0930-84
$C_{15}H_{12}N_2O_4$ 3-Azabicyclo[4.4.1]undeca-1,3,5,7,9- pentaene-4,5-dicarboxylic acid, 2-cyano-, 5-ethyl ester	MeOH	257(4.32),273(4.31), 358(3.71)	44-1040-84
$C_{15}H_{12}N_2S$ 10H-Naphtho[1',2':6,7]cyclohepta[1,2-d]-	EtOH	246(4.63),278(3.99)	2-0203-84

Compound	Solvent	$\lambda_{max}(\log \epsilon)$	Ref.
1,2,3-thiadiazole, 11,12-dihydro-Phenanthro[9,10-d]thiazol-2-amine, 3a,11b-dihydro-, trans	CH_2Cl_2	229.2(4.04),269.2(4.12), 293.5s(3.58)	2-0203-84 4-1597-84
1H-1-Pyrindine-3-carbonitrile, 2,5,6,7-tetrahydro-4-phenyl-2-thioxo-	dioxan	258(4.22),320(3.81), 438(2.94)	104-1402-84
$C_{15}H_{12}N_3O$			
Pyridinium, 1-[5-(2,2-dicyanoethenyl)-2-furanyl]-3,5-dimethyl-, bromide	H_2O	371(3.30)	73-2485-84
Pyridinium, 1-[5-(2,2-dicyanoethenyl)-2-furanyl]-3-ethyl-, bromide	H_2O	370(3.37)	73-2485-84
$C_{15}H_{12}N_3O_4$			
Pyridinium, 3-(aminocarbonyl)-1-[5-(2-cyano-3-methoxy-3-oxo-1-propenyl)-2-furanyl]-, bromide	H_2O	248(3.19),326s(3.13), 339s(3.15),377(3.33)	73-2485-84
$C_{15}H_{12}N_4$			
1H-Tetrazole, 5-phenyl-1-(1-phenylethenyl)-	EtOH	243(4.31)	39-1933-84C
1H-Tetrazole, 5-phenyl-1-(2-phenylethenyl)-, (E)-	EtOH	243(4.12),284(4.25)	39-1933-84C
$C_{15}H_{12}N_4O$			
Hydrazinecarboxaldehyde, 2-(2-phenyl-4-quinazolinyl)-	EtOH	275(3.99),288s(3.86), 350(3.25)	18-1138-84
$C_{15}H_{12}N_4O_2$			
1H-1,2,4-Triazole, 1-(4-nitrophenyl)-5-(phenylmethyl)-	MeCN	222s(4.08),302(4.23)	48-0311-84
$C_{15}H_{12}N_4O_2S$			
2H-1,4-Benzothiazin-2-one, 3-methyl-, (4-nitrophenylhydrazone)	EtOH	236(4.36),265(4.50), 415(4.66)	4-0521-84
$C_{15}H_{12}N_4O_3$			
2H-1,4-Benzoxazin-2-one, 3-methyl-, (4-nitrophenylhydrazone)	EtOH	267(4.69),412(4.69)	4-0521-84
2,4(1H,3H)-Pteridinedione, 7-(1-oxo-propyl)-1-phenyl-	pH 5.0 pH 10.0	243s(4.07),343(3.89) 253(4.07),353(3.89)	5-1798-84 5-1798-84
1H-Pyrazole-5-carboxylic acid, 4-[(2-hydroxy-1-naphthalenyl)azo]-1-methyl-	EtOH	485(4.19)	104-0976-84
2(1H)-Quinoxalinone, 3-[2-(methylamino)-5-nitrophenyl]-	EtOH	215(4.49),308(4.10), 396(4.31)	103-0537-84
Spiro[2H-benzimidazole-2,3'-[3H]indol]-2'(1'H)-one, 1,3-dihydro-1'-methyl-5-nitro-	EtOH	212(4.52),283(4.20), 445(3.98)	103-1276-84
Spiro[2H-benzimidazole-2,3'-[3H]indol]-2'(1'H)-one, 1,3-dihydro-1'-methyl-5'-nitro-	EtOH	218(4.49),297(4.22), 330(4.20)	103-0537-84
1H-1,2,4-Triazole, 3-(4-methoxyphenyl)-1-(4-nitrophenyl)-	MeCN	230(4.05),260(4.33), 282s(3.98),295(3.98), 330(4.28)	48-0311-84
1H-1,2,4-Triazole, 5-(4-methoxyphenyl)-1-(4-nitrophenyl)-	MeCN	260(4.26),321s(3.86)	48-0311-84
$C_{15}H_{12}N_6O_2S$			
Formazan, 1-(2-benzothiazolyl)-3-methyl-5-(4-nitrophenyl)-	benzene	555(4.38),585s(4.36)	135-0577-84
photoinduced product	benzene	420(4.27)	135-0577-84

Compound	Solvent	$\lambda_{max}(\log \epsilon)$	Ref.
$C_{15}H_{12}O$			
6,11-Methanobenzocyclodecen-13-ol	CH_2Cl_2	278(4.88),351(3.70)	89-0719-84
$C_{15}H_{12}OS_2$			
9H-Thioxanthen-9-one, 2-[(methylthio)-methyl]-	MeOH	260(4.66),291s(3.83), 302s(3.69),386(3.75)	73-1722-84
$C_{15}H_{12}O_2$			
8H-Acenaphtho[1,2-b][1,4]dioxepin, 9,10-dihydro-	n.s.g.	318(3.85),436(2.82)	78-0665-84
1H-2-Benzopyran-1-one, 3,4-dihydro-3-phenyl-	EtOH	237(4.38),284(3.15)	44-0742-84
$C_{15}H_{12}O_2S$			
9H-Thioxanthen-9-one, 2-(methoxymeth-yl)-	MeOH	220(4.26),258(4.74), 280(3.81),301(3.58), 382(3.86)	73-1722-84
$C_{15}H_{12}O_3$			
2(3H)-Benzofuranone, 3-hydroxy-5-meth-yl-3-phenyl-	MeOH	268(3.09),274(3.15)	35-7352-84
2H-1-Benzopyran-2-one, 7-[(3-methyl-2-penten-4-ynyl)oxy]-, (E)-	EtOH	222.5(4.05),323(4.00)	39-0535-84C
4H-1-Benzopyran-4-one, 2,3-dihydro-3-hydroxy-2-phenyl-, (2R,3R)-(-)-	MeOH	212.5(4.65),252(4.00), 317.5(3.66)	94-4852-84
$C_{15}H_{12}O_4$			
6-Benzofuranol, 2-(2-hydroxy-4-methoxy-phenyl)-	MeOH	273s(4.11),281(4.15), 306s(4.34),320(4.59), 334.5(4.57)	94-3267-84
1H-2-Benzopyran-1-one, 3,4-dihydro-8-hydroxy-3-(4-hydroxyphenyl)-	EtOH	248s(3.89),286(3.40), 316(3.70)	44-0742-84
4H-1-Benzopyran-4-one, 2,3-dihydro-7-hydroxy-2-(4-hydroxyphenyl)-	MeOH	240s(--),280(4.30), 314(1.72)	42-0728-84
2aH-Cyclopent[cd]indene-1,2-dicarbox-ylic acid, dimethyl ester	EtOH	215(4.02),238(3.92), 245(3.91),275(3.79), 320(3.61)	23-2506-84
2-Propen-1-one, 1-(2,4-dihydroxyphen-yl)-3-(4-hydroxyphenyl)-	MeOH	245(3.96),325s(--), 370(4.45)	42-0728-84
Spiro[2-cyclopropen-1,1'-[1H]indene]-2,3-dicarboxylic acid, dimethyl ester	n.s.g.	227(4.22),266(3.86), 292(3.17)	23-2506-84
$C_{15}H_{12}O_5$			
Benzo[1,2-b:5,4-b']dipyran-3-carboxylic acid, 2-oxo-, ethyl ester	EtOH	229(4.436),282(4.286), 314(--),379(4.181)	111-0365-84
Butanedioic acid, 1H-inden-1-ylidene-oxo-, dimethyl ester	EtOH	245(3.77),247(4.22)	23-2506-84
7H-Furo[3,2-g][1]benzopyran-6-carbox-ylic acid, 9-methyl-7-oxo-, ethyl ester	EtOH	250(4.346),320(4.176)	111-0365-84
2-Propen-1-one, 1-(2,4-dihydroxyphen-yl)-3-(3,4-dihydroxyphenyl)- (butein)	MeOH	262(4.02),380(4.55)	95-0935-84
2H-Pyran-2-one, 3-(1,3-dioxo-3-phenyl-propyl)-4-hydroxy-6-methyl- (pogo-pyrone A)	EtOH	212(4.36),275(3.97), 312(4.06),368(4.40)	2-0611-84
$C_{15}H_{12}O_5S$			
Benzeneacetic acid, 3-[(2-carboxyphen-yl)thio]-4-hydroxy-	MeOH	254s(3.89),297(3.75), 320s(3.67)	73-2531-84

Compound	Solvent	$\lambda_{max}(\log \epsilon)$	Ref.
$C_{15}H_{12}O_6$			
1,4-Naphthalenedione, 2,7-diacetoxy-5-methyl-	MeOH	238(4.25),290s(3.66), 340s(3.45)	44-1853-84
9H-Xanthen-9-one, 1,4-dihydroxy-3,7-dimethoxy-	MeOH	233(4.38),272(4.50), 313s(3.85),323(3.86), 399(3.69)	39-1507-84C
9H-Xanthen-9-one, 1,8-dihydroxy-2,6-dimethoxy-	MeOH	202(4.48),236(4.38), 266(4.40),327(4.11)	94-2290-84
9H-Xanthen-9-one, 1,8-dihydroxy-3,5-dimethoxy-	MeOH	203(4.61),252(4.50), 276(4.30),330(4.15)	94-2290-84
	EtOH	240(4.28),254(4.43), 315(4.23),337(4.05)	102-1637-84
9H-Xanthen-9-one, 5,6-dihydroxy-1,3-dimethoxy-	EtOH	244(4.71),282(4.41), 312(4.50)	102-1816-84
	EtOH-NaOAc	244(4.71),284(4.42), 337(4.60)	102-1816-84
	EtOH-AlCl$_3$	244(4.75),315(4.20), 360(3.66)	102-1816-84
	EtOH-AlCl$_3$-HCl	244(4.83),282(4.40), 312(4.49)	102-1816-84
	EtOH-NaOAc-H$_3$BO$_3$	254(4.75),282(4.55), 324(4.57)	102-1816-84
$C_{15}H_{12}O_8$			
2H-Pyran carboxlic acid, 3-(2-carboxy-3-hydroxy-5-methoxyphenyl)-4-methyl-2-oxo-	n.s.g.	220(4.53),260(4.11), 305(4.16)	78-2451-84
$C_{15}H_{13}Br$			
6,11-Methanobenzocyclodecene, 13-bromo-5,12-dihydro-	CH$_2$Cl$_2$	245s(3.64),280s(3.01), 362s(1.46),385s(1.26)	89-0719-84
$C_{15}H_{13}BrN_4O_3$			
1H-Tetrazole-1-acetic acid, 5-[5-(4-bromophenyl)-2-furanyl]-, ethyl ester	dioxan	318(4.52),333s(4.33)	73-1699-84
2H-Tetrazole-2-acetic acid, 5-[5-(4-bromophenyl)-2-furanyl]-, ethyl ester	dioxan	314(4.51),329s(4.32)	73-1699-84
$C_{15}H_{13}ClN_2$			
1H-Pyrazole, 3-(4-chlorophenyl)-4,5-dihydro-5-phenyl-	EtOH	227(3.78),278s(3.89), 296(3.93)	34-0225-84
$C_{15}H_{13}ClN_2O$			
Benzenemethanamine, N-(5-chloro-3-methyl-2(3H)-benzoxazolylidene)-	CHCl$_3$	255(3.91),304(4.10)	94-3053-84
2-Benzoxazolamine, 5-chloro-N-methyl-N-(phenylmethyl)-	CHCl$_3$	257(4.25),294(4.14)	94-3053-84
$C_{15}H_{13}ClN_4O_3$			
1H-Tetrazole-1-acetic acid, 5-[5-(2-chlorophenyl)-2-furanyl]-, ethyl ester	dioxan	311(4.36),327s(4.11)	73-1699-84
1H-Tetrazole-1-acetic acid, 5-[5-(3-chlorophenyl)-2-furanyl]-, ethyl ester	dioxan	313(4.50),328s(4.30)	73-1699-84
1H-Tetrazole-1-acetic acid, 5-[5-(4-chlorophenyl)-2-furanyl]-, ethyl ester	dioxan	316(4.55),332s(4.34)	73-1699-84
2H-Tetrazole-2-acetic acid, 5-[5-(2-chlorophenyl)-2-furanyl]-, ethyl ester	dioxan	308(4.34),323s(4.11)	73-1699-84

Compound	Solvent	$\lambda_{max}(\log \epsilon)$	Ref.
2H-Tetrazole-2-acetic acid, 5-[5-(3-chlorophenyl)-2-furanyl]-, ethyl ester	dioxan	310(4.23),324s(4.23)	73-1699-84
2H-Tetrazole-2-acetic acid, 5-[5-(4-chlorophenyl)-2-furanyl]-, ethyl ester	dioxan	312(4.54),328s(4.34)	73-1699-84
$C_{15}H_{13}ClO_3$			
1,4-Naphthalenedione, 6-chloro-2-hydroxy-3-(3-methyl-2-butenyl)-	EtOH	259(4.356),285(4.222), 324(3.462),396(3.164)	87-0990-84
2H-Naphtho[1,2-b]pyran-5,6-dione, 8(or 9)-chloro-3,4-dihydro-2,2-dimethyl-	MeOH	259(4.124),265s(--), 285s(3.664),335(3.133)	87-0990-84
$C_{15}H_{13}DO_3$			
1,4-Naphthalenedione-7-d, 2-hydroxy-3-(1,1-dimethylallyl)-	MeOH	252(4.42),276(4.41), 325(3.63),491(3.12)	102-0313-84
$C_{15}H_{13}F_3O_3S_2$			
2-Thieten-1-ium, 1-methyl-3-(2-naphthalenyl)-, trifluoromethanesulfonate	MeCN	253(4.25),262(4.25), 277(4.03),291(4.02), 302(4.05)	44-4192-84
$C_{15}H_{13}N$			
2H-Azirine, 2-methyl-2,3-diphenyl-	hexane	245(4.30)	44-3174-84
2H-Isoindole, 1-methyl-3-phenyl- (absorbances approximate)	hexane	242(4.11),275(4.01), 395(3.56)	44-3174-84
$C_{15}H_{13}NO$			
Acetamide, N-(diphenylmethylene)-	C_6H_{12}	198.0(4.70),246.35(4.32)	24-1597-84
	CH_2Cl_2	253.05(4.29)	24-1597-84
Benzo[h]quinoline, 6-methoxy-4-methyl-	MeOH	216s(4.38),231(4.44), 241(4.53),247(4.61), 264s(4.31),277s(4.22), 295(3.88),306(3.89), 323(3.36),338(3.63), 351(3.67)	12-1271-84
$C_{15}H_{13}NOS$			
9(10H)-Acridinethione, 2-ethoxy-	50% EtOH	488(4.49)	103-1254-84
1H-Isoindole-1-thione, 2,3-dihydro-2-(4-methoxyphenyl)-	EtOH	203(4.69),246(4.20), 280(4.18),329(3.92)	103-0978-84
$C_{15}H_{13}NOS_2$			
Carbamothioic acid, phenyl(phenylthioxomethyl)-, O-methyl ester	MeOH	504(2.21)	97-0438-84
$C_{15}H_{13}NO_2$			
Benzo[h]quinolin-2(1H)-one, 6-methoxy-4-methyl-	EtOH	232(4.20),264s(3.89), 273(4.01),284(4.13), 347s(3.32),362(3.42), 370(3.30),480(2.48)	12-1271-84
9H-Carbazol-3-ol, 6-methyl-, acetate	EtOH	230(4.56),235(4.58), 262(4.20),300(4.22), 332(3.60)	2-0049-84
$C_{15}H_{13}NO_4S$			
2H-Pyran-3-carbonitrile, 6-(3,4-dimethoxyphenyl)-4-(methylthio)-2-oxo-	EtOH	225(4.19),249(4.14), 355(4.16),402(4.39)	94-3384-84

Compound	Solvent	$\lambda_{max}(\log \epsilon)$	Ref.
$C_{15}H_{13}NO_5$ 1H-Pyrano[3,4-b]benzofuran-1,3(4H)-di-one, 4-[(dimethylamino)methylene]-7-methoxy-	MeOH	230(4.0),355(4.1)	2-0478-84
$C_{15}H_{13}NS$ 1H-Isoindole-1-thione, 2,3-dihydro-2-(4-methylphenyl)-	EtOH	205(4.53),246(4.02), 270(4.10),329(3.86)	103-0978-84
Thiopyrano[2,3-c]pyrrole, 5,7-dimethyl-2-phenyl-, perchlorate	EtOH	220(3.79),245(3.92), 300(3.83),366(3.91), 616(2.78)	12-1473-84
$C_{15}H_{13}N_2O_3$ Benzo[h]quinolinium, 6-methoxy-4-methyl-1-nitro-, acetate	EtOH	237s(4.63),243(4.68), 259s(4.41),266(4.44), 291(3.99),302(3.98), 319(3.45),336(3.70), 349(3.73),378(2.48)	12-1271-84
Pyridinium, 1-[5-(2-cyano-3-methoxy-3-oxo-1-propenyl)-2-furanyl]-3-methyl-, bromide	H_2O	237(2.96),263(2.87), 321s(3.05),338s(3.18), 371(3.35)	73-2485-84
$C_{15}H_{13}N_3$ 1H-1,2,4-Triazole, 5-(4-methylphenyl)-	MeCN	219s(4.24),254(4.12)	48-0311-84
$C_{15}H_{13}N_3O$ Diazene, (1-isocyanato-1-phenylethyl)-phenyl-	C_6H_{12}	270(4.09),384(2.27)	118-0315-84
2H-Indol-2-one, 3-[(2-aminophenyl)-imino]-1,3-dihydro-1-methyl-	EtOH	210(4.50),253(4.25), 500(3.42)	103-0537-84
2(1H)-Quinoxalinone, 3-[2-(methyl-amino)phenyl]-	EtOH	222(4.54),300(3.89), 354(3.96),442(3.83)	103-0537-84
Spiro[2H-benzimidazole-2,3'-[3H]indol]-2'(1'H)-one, 1,3-dihydro-1'-methyl-	EtOH	218(4.54),244(4.14), 300(4.00)	103-0537-84
1H-1,2,4-Triazole, 1-(4-methoxyphenyl)-5-phenyl-	MeCN	229(4.26),247s(4.07)	48-0311-84
1H-1,2,4-Triazole, 3-(4-methoxyphenyl)-1-phenyl-	MeCN	273(4.43),298s(3.94)	48-0311-84
1H-1,2,4-Triazole, 5-(4-methoxyphenyl)-1-phenyl-	MeCN	226s(4.09),265(4.12)	48-0311-84
$C_{15}H_{13}N_3O_2$ 2,4-Imidazolidinedione, 3-amino-5,5-di-phenyl-	MeOH	202(4.08)	56-0585-84
2H-Indazole-4,7-dione, 2,3-dimethyl-5-(phenylamino)-	EtOH	264(4.35),310(4.86), 461(4.67)	4-0825-84
$C_{15}H_{13}N_3O_3$ 1H-Indazole-4,7-dione, 3-methoxy-1-methyl-5-(phenylamino)-	EtOH	246(4.26),359(3.89), 481.5(3.72)	4-0825-84
2H-Indazole-4,7-dione, 3-methoxy-2-methyl-5-(phenylamino)-	dioxan	263(4.40),332s(3.79), 444(3.71)	4-0825-84
1H-Indazole-3,4,7(2H)-trione, 1,2-di-methyl-5-(phenylamino)-	EtOH	247(4.25),400(4.11), 519s(3.47)	4-0825-84
3-Pyridazinecarboxylic acid, 5-cyano-1,6-dihydro-4-methyl-6-oxo-1-phenyl-, ethyl ester	EtOH	342(3.74)	118-0062-84
$C_{15}H_{13}N_3O_3S$ 2H-1,2,4,6-Thiatriazinium, 5,6-dihydro-4-methyl-5-oxo-2,3-diphenyl-,	MeCN	245(3.979)	142-0063-84

Compound	Solvent	$\lambda_{max}(\log \epsilon)$	Ref.
hydroxide, inner salt, 1,1-dioxide			142-0063-84
Thieno[2,3-d]pyrimidine-6-carboxylic acid, 5-amino-1,2-dihydro-2-oxo-4-(phenylmethyl)-, methyl ester	EtOH	288(4.35),320(4.04)	39-2447-84C
$C_{15}H_{13}N_3S$			
Diazene, (1-isothiocyanato-1-phenyl-ethyl)phenyl-	C_6H_{12}	280(4.11),400(2.22)	118-0315-84
$C_{15}H_{13}N_3S_2$			
2H-1,3-Thiazine-2-acetonitrile, 5-cyano-3,6-dihydro-2-methyl-4-(3-methylphenyl)-6-thioxo-	EtOH	267(3.96),324(3.57), 389(4.23)	4-1445-84
2H-1,3-Thiazine-2-acetonitrile, 5-cyano-3,6-dihydro-2-methyl-4-(4-methylphenyl)-6-thioxo-	EtOH	273(4.11),320(3.71), 389(4.26)	4-1445-84
$C_{15}H_{13}N_5O$			
1H-Pyrazolo[3,4-b]quinoline-4-carboxamide, 1-(2-cyanoethyl)-3-methyl-	EtOH	392(3.86)	18-2984-84
$C_{15}H_{13}N_5O_5$			
1H-Tetrazole-1-acetic acid, 5-[5-(2-nitrophenyl)-2-furanyl]-, ethyl ester	dioxan	292(4.26)	73-1699-84
2H-Tetrazole-2-acetic acid, 5-[5-(2-nitrophenyl)-2-furanyl]-, ethyl ester	dioxan	292(4.27)	73-1699-84
$C_{15}H_{13}N_5S$			
Formazan, 1-(2-benzothiazolyl)-3-methyl-5-phenyl-	benzene	412(4.38)	135-0577-84
photoinduced hydrohalide	benzene	385(4.58)	135-0577-84
$C_{15}H_{13}N_7O_3$			
Benzoic acid, 4-[[(2,4-dinitro-6-pteridinyl)methyl]formylamino]-	pH 1	247(4.36),337(3.99)	87-0600-84
	pH 7	258(4.45),370(3.86)	87-0600-84
$C_{15}H_{14}BrClN_2O$			
Benzenecarboximidamide, 4-bromo-N-(4-chlorophenyl)-N'-hydroxy-2,6-dimethyl-	MeOH	232(4.08),260(4.25)	32-0015-84
$C_{15}H_{14}BrNO_2$			
1-Cyclohexene-1-carbonitrile, 2-[2-(4-bromophenyl)-2-oxoethoxy]-	EtOH	254(4.48)	5-1702-84
Methanone, (3-amino-4,5,6,7-tetrahydro-2-benzofuranyl)(4-bromophenyl)-	EtOH	257(4.06),263(4.04), 359(4.15)	5-1702-84
$C_{15}H_{14}BrNO_3$			
2H-Cyclohepta[b]furan-2,4(3H)-dione, 3-(4-bromophenyl)-5,6,7,8-tetrahydro-, 2-oxime	MeOH	272(4.146)	73-1421-84
$C_{15}H_{14}BrNO_8S$			
1H-Pyrrolo[1,2-a]indole-9-carboxylic acid, 7-bromo-2,3,5,8-tetrahydro-1-methoxy-6-methyl-2-[(methylsulfonyl)-oxy]-5,8-dioxo-, cis	MeOH	231(4.16),287(4.06), 350(3.46),425(3.36)	44-5164-84
$C_{15}H_{14}BrN_5O_3$			
Benzoic acid, [1-[2-(4-bromo-2-nitrophenyl)hydrazino]ethylidene]hydrazide	20% acetone	440(3.28)	140-1028-84

Compound	Solvent	$\lambda_{max}(\log \epsilon)$	Ref.
$C_{15}H_{14}ClNO_3$ 2H-Cyclohepta[b]furan-2,4(3H)-dione, 3-(4-chlorophenyl)-5,6,7,8-tetra- hydro-, 2-oxime	MeOH	274(4.121)	73-1421-84
$C_{15}H_{14}ClNO_3S$ 2H-Thieno[2,3-h]-1-benzopyran-2-one, 3-chloro-5,6-dihydro-4-morpholino-	EtOH	217s(4.11),219(4.11), 242.5(4.11),273s(3.56), 357(4.14)	161-0081-84
$C_{15}H_{14}F_3N_5$ 7H-Purin-6-amine, 2,7,8-trimethyl- N-[3-(trifluoromethyl)phenyl]-	n.s.g.	301(4.37)	11-0208-84B
$C_{15}H_{14}NO$ Acridinium, 3-methoxy-N-methyl-, iodide	MeCN	366(4.45)	23-1780-84
$C_{16}H_{14}NP$ Phosphine, (2-isocyanoethyl)diphenyl-	MeOH	245(4.03),264s(--), 270s(--)	139-0259-84C
$C_{15}H_{14}N_2$ 5H-[1,4]Diazepino[1,2,3,4-1mn][1,10]- phenanthrolinediium, 6,7-dihydro-, bis(hexafluoroarsenate)	MeCN	312(3.94)	47-0069-84
1H-Pyrazole, 4,5-dihydro-3,5-diphenyl-	EtOH	222(3.91),287(3.91)	34-0225-84
$C_{15}H_{14}N_2O$ Benzo[h]quinolin-5-amine, 6-methoxy- 4-methyl-	EtOH	221(4.22),252(4.42), 284(4.26),346(3.71)	12-1271-84
1,2,4-Oxadiazole, 4,5-dihydro-4-methyl- 3,5-diphenyl-	EtOH	216(4.23),274(3.36)	44-0276-84
$C_{15}H_{14}N_2OS$ 4H-3,1-Benzothiazin-4-one, 1,2-dihydro- 1-(2-methylphenyl)-, oxime	n.s.g.	<u>285(3.9)</u>,340(3.7)	4-1615-84
3-Pyridinecarbonitrile, 1,2-dihydro- 4-(4-methoxyphenyl)-5,6-dimethyl- 2-thioxo-	dioxan	261(4.35),313(4.36), 422(3.60)	104-1837-84
4(1H)-Quinazolinethione, 2,3-dihydro- 3-hydroxy-1-(2-methylphenyl)-	n.s.g.	<u>318(3.8)</u>,375(3.6)	4-1615-84
$C_{15}H_{14}N_2O_2$ 9H-Pyrido[3,4-b]indole-1-propanoic acid, methyl ester (infractin)	MeOH	212(4.30),232.5(4.55), 238s(4.52),247s(4.36), 280s(4.01),287(4.21), 335(3.66),347(3.65)	88-2341-84
1H-Pyrrole-3-acetic acid, α-cyano-2- phenyl-, ethyl ester	MeOH	204(4.22),273(4.10)	83-0143-84
$C_{15}H_{14}N_2O_2S_2$ 4H-1,3-Dithiolo[4,5-b]pyrrole-6-carbox- ylic acid, 5-methyl-2-(phenylimino)-, ethyl ester	EtOH	248(3.75),307(3.71), 340(4.05)	12-2479-84
	EtOH-NaOH	263(4.24),305(3.93), 380(3.90)	12-2479-84
monohydrobromide	EtOH-HBr	270(3.85),385(3.95)	12-2479-84
$C_{15}H_{14}N_2O_2S_3$ 2(1H)-Pyridinethione, 1-ethenyl-4-[(1- ethenyl-1,2-dihydro-3-hydroxy-2-thi-	EtOH	294(3.83),388(3.94)	1-0293-84

Compound	Solvent	$\lambda_{max}(\log \epsilon)$	Ref.
oxo-6-pyridinyl)thio]-3-hydroxy- 6-methyl- (cont.)			1-0293-84
$C_{15}H_{14}N_2O_3$			
Benzoic acid, 4-(N-nitroso-methyl- aminomethyl)phenyl ester	hexane	340(1.64),352(1.83), 364(1.95),376(1.87)	150-0701-84M
	EtOH	233(4.40),347(1.88)	150-0701-84M
1,5-Benzoxazepine, 2,3,4,5-tetrahydro- 8-nitro-5-phenyl-	EtOH	262(4.02),405(4.13)	78-1755-84
9H-Pyrido[3,4-b]indole-1-propanoic acid, 6-hydroxy-, methyl ester	MeOH	214(4.36),231(4.53), 246(4.37),258(4.24), 290s(4.14),296(4.33), 360(3.72)	88-2341-84
$C_{15}H_{14}N_2O_3S$			
2-Propenamide, N-[2-(aminosulfonyl)- phenyl]-3-phenyl-	EtOH	<u>220s(4.3)</u>,300(4.4)	103-0735-84
$C_{15}H_{14}N_2O_4$			
Pyrrolo[2,1-b]quinazoline-6-carboxylic acid, 3-formyl-1,2,3,9-tetrahydro-9- oxo-, ethyl ester	EtOH	235(4.58),268(3.96), 278(3.87),312s(3.74), 322(3.81),336(3.73), 350s(3.24)	4-0219-84
	dioxan	277s(4.09),296s(4.06), 305(4.14),316(4.11), 331s(3.96),350s(3.75)	4-0219-84
	CHCl$_3$	277(4.15),294s(4.10), 305(4.19),317(4.17), 334(4.07),350s(3.95)	4-0219-84
	DMSO	266(4.04),298s(4.00), 309(4.09),319(4.07), 334(3.93),356(3.42)	4-0219-84
$C_{15}H_{14}N_2O_5$			
2H-Cyclohepta[b]furan-2,4(3H)-dione, 5,6,7,8-tetrahydro-3-(4-nitro- phenyl)-, 2-oxime	MeOH	271(4.076)	73-1421-84
$C_{15}H_{14}N_4$			
Ethenetricarbonitrile, [4-(diethyl- amino)phenyl]-	CH$_2$Cl$_2$	530(4.60)	88-0117-84
2(1H)-Quinoxalinone, 3-methyl-, phenylhydrazone	EtOH	245(4.44),283(4.25), 348(3.84),410s(3.82)	4-0521-84
$C_{15}H_{14}N_4O$			
1H-Pyrazolo[3,4-d]pyrimidin-4-one, 6-methyl-5-[2-(2-propenyl)phenyl]-	EtOH	205.5(4.38),251.5(3.88)	11-0023-84A
$C_{15}H_{14}N_4O_3$			
Methanimidic acid, N-(phenylacetyl)-, 2-(4-nitrophenyl)hydrazide	MeCN	279(3.98),398(4.38)	48-0311-84
1,3,7,8-Tetraazaspiro[4.5]deca-1,3,6,9- tetraene-10-carboxylic acid, 4-meth- oxy-2-phenyl-, methyl ester	MeOH	235(4.26),274(4.04), 327(3.68)	39-2491-84C
1,3,7,8-Tetraazaspiro[4.5]deca-1,6,9- triene-10-carboxylic acid, 3-methyl- 4-oxo-2-phenyl-, methyl ester	MeOH	230(4.28),268(3.75), 330(3.71)	39-2491-84C
1H-Tetrazole-1-acetic acid, 5-(5-phen- yl-2-furanyl)-, ethyl ester	dioxan	314(4.43),327s(4.22)	73-1699-84
2H-Tetrazole-2-acetic acid, 5-(5-phen- yl-2-furanyl)-, ethyl ester	dioxan	308(4.13),324s(3.90)	73-1699-84

Compound	Solvent	$\lambda_{max}(\log \epsilon)$	Ref.
$C_{15}H_{14}N_4S$			
Benzenamine, 4-(2-benzothiazolylazo)-N,N-dimethyl-	acetone	498(4.63)	7-0819-84
	50% acetone	520(4.68)	7-0819-84
$C_{15}H_{14}N_6O_5$			
Benzoic acid, [1-[2-(2,4-dinitrophenyl)hydrazino]ethylidene]hydrazide	20% acetone	350(4.60)	140-1028-84
Benzoic acid, 4-nitro-, [1-[2-(4-nitrophenyl)hydrazino]ethylidene]hydrazide	20% acetone	400(3.64)	140-1028-84
$C_{15}H_{14}N_6S$			
Propanenitrile, 3,3'-[[4-(2-thiazolylazo)phenyl]imino]bis-	acetone	438(4.51)	7-0819-84
	50% acetone	466(4.56)	7-0819-84
$C_{15}H_{14}N_8$			
[1,3'-Bi-1H-pyrazole]-5-carbonitrile, 4-amino-3-(1-hydrazonoethyl)-5'-phenyl-	EtOH	232(3.65)	4-1049-84
$C_{15}H_{14}O_2$			
Furan, 2-ethoxy-2,5-dihydro-5-(3-phenyl-2-propynylidene)-, (Z)-	n.s.g.	226(4.00),325(4.32)	104-2302-84
$C_{15}H_{14}O_2S$			
Benzene, 1-methyl-4-[[2-(phenylsulfinyl)ethenyl]oxy]-, (Z)-	dioxan	241(4.35),272s(3.62)	104-0473-84
$C_{15}H_{14}O_3$			
3H-Benz[e]inden-3-one, 1,2-dihydro-5,7-dimethoxy-	MeOH	245(4.46),280(4.69), 320(4.10)	2-0424-84
2H-1-Benzopyran-3,4-diol, 3,4-dihydro-2-phenyl-, (2R,3S,4R)-(+)-	MeOH	200.5(4.43),275(3.32), 283(3.31)	94-4852-84
(2R,3S,4S)-(+)-	MeOH	213.5(4.39),274.5(3.40), 282(3.34)	94-4852-84
2H-1-Benzopyran-7-ol, 3,4-dihydro-2-(4-hydroxyphenyl)-	MeOH	224(4.37),281(3.78), 290s(3.57)	138-0689-84
2,4-Cyclohexadiene-1-carboxylic acid, 6-oxo-1-(phenylmethyl)-, methyl ester	MeOH	308(3.40)	44-4429-84
2,5-Cyclohexadien-1-one, 4-(2,4,6-cycloheptatrien-1-ylidene)-2,6-dimethoxy-	MeCN	353(3.31),478(4.29)	88-0077-84
Dibenzofuran, 1,3-dimethoxy-2-methyl-	EtOH	223(4.54),258(4.09), 291(4.24),298(4.25)	39-1605-84C
Dibenzofuran, 1,3-dimethoxy-4-methyl-	EtOH	215(4.58),231(4.50), 264(4.10),284(4.24), 296(4.04),308(4.17)	39-1605-84C
Dibenzofuran, 2,4-dimethoxy-3-methyl-	EtOH	212(4.50),218(4.51), 255(4.14),290(4.23), 318s(3.67)	39-1605-84C
Ethanone, 1-(2,4-dihydroxy-3-(phenylmethyl)phenyl]-	MeOH	210(4.30),280(4.15)	2-1030-84
Ethanone, 1-[2,4-dihydroxy-5-(phenylmethyl)phenyl]-	MeOH	212(4.38),276(4.20)	2-1030-84
2,5-Furandione, dihydro-3-(1-methylethylidene)-4-(1-phenylethylidene)-, (E)-	C_6H_{12}	312(3.91)	48-0233-84
(Z)-	C_6H_{12}	296(4.06)	48-0233-84
4H-Furo[3,2-c][1]benzopyran-4-one, 2,3-dihydro-3,3,7-trimethyl-2-methylene-	MeOH	208(4.47),240(3.93), 254(3.97),260(3.96), 294(3.94),303(3.96), 316(4.05),324(3.88), 332(3.92)	107-0737-84

Compound	Solvent	$\lambda_{max}(\log \epsilon)$	Ref.
4H-Furo[3,2-c][1]benzopyran-4-one, 2,3-dihydro-3,3,9-trimethyl-2-methylene-	MeOH	219(4.22),222(4.20), 230(4.10),238(4.04), 255(3.98),303(4.02), 316(3.98),332(3.83)	107-0737-84
1,4-Naphthalenedione, 2-hydroxy-3-(3-methyl-2-butenyl)-	EtOH	267s(--),284(3.516), 332(2.931),430(2.970)	87-0990-84
1,4-Naphthalenedione, 2-[(3-methyl-2-butenyl)oxy]-	MeOH	242.5(4.31),248.5(4.33), 277(4.19),330(3.54)	102-0313-84
Naphtho[2,3-b]furan-3-carboxaldehyde, 5,6,7,8-tetrahydro-4,5-dimethyl-8-oxo- ((-)-viteralone)	EtOH	215(4.27),275(4.24)	142-2203-84
Naphtho[2,3-c]furan-1,3-dione, 4,4a-dihydro-4,4,9-trimethyl-	C_6H_{12}	470(3.87)	48-0233-84
2H-Naphtho[1,8-bc]furan-6,7-dione, 2,2,5,8-tetramethyl-	MeOH	440(3.73)	138-0627-84
2H-Naphtho[1,2-b]pyran-5,6-dione, 3,4-dihydro-2,2-dimethyl-	MeOH	255(3.721),262s(--), 287(3.292),340(2.642), 450(2.517)	87-0990-84
$C_{15}H_{14}O_3S$			
Benzoic acid, 2-[4-(methoxymethyl)-phenylthio]-	MeOH	220(4.43),263(3.97), 272(3.66),317(3.67)	73-1722-84
$C_{15}H_{14}O_4$			
1,3-Benzenediol, 5-[2-(3-hydroxy-4-methoxyphenyl)ethenyl]- (rhapontigenin)	MeOH	220(4.41),302(4.49), 322(4.53)	94-3501-84
3-Butenoic acid, 2-(2-furanylmethylene)-4-(5-methyl-2-furanyl)-, methyl ester, (E,E)-	n.s.g.	202(4.00),283(4.17), 372(4.32)	73-2502-84
(E,Z)-	n.s.g.	200(4.42),280(4.20), 346(4.21)	
2,5-Furandione, dihydro-3-[(4-methoxyphenyl)methylene]-4-(1-methylethylidene)-, (E)-	C_6H_{12}	347(4.11)	48-0233-84
(Z)-	C_6H_{12}	365(4.35)	48-0233-84
2H-Naphtho[2,3-b]pyran-5,10-dione, 3,4-dihydro-2-methoxy-2-methyl-	EtOH	245s(4.30),250(4.38), 280(4.15),332(3.45)	4-0865-84
1-Propanone, 3-phenyl-1-(2,4,6-trihydroxyphenyl)-	EtOH	212(4.14),222(4.12), 286(4.20),326s(3.57)	102-1198-84
$C_{15}H_{14}O_5$			
5H-Furo[3,2-g][1]benzopyran-5-one, 4,9-dimethoxy-6,7-dimethyl-	EtOH	220(4.18),247(4.65), 282(3.67),332(3.74)	44-5035-84
$C_{15}H_{14}O_5S_2$			
2,4-Pentanedione, 3-[(methylthio)(3-oxobenzo[b]thien-2(3H)-ylidene)-methyl]-, S,S-dioxide	EtOH	220(4.29),260(4.22), 363(4.22)	95-0134-84
$C_{15}H_{14}O_6$			
2H-1-Benzopyran-3,5,7-triol, 2-(3,4-dihydroxyphenyl)-3,4-dihydro- (catechin)	pH 1 borate	277(3.59) 285(3.77)	3-1935-84 3-1935-84
(2R,3S)-	pH 3.7 pH 12.0 pH 14.8	278(3.59) 290(3.86)(corrected) 305(3.85)	12-0885-84 12-0885-84 12-0885-84
epicatechin (2R,3R)-	pH 12.4 pH 14.9	291(3.84)(corrected) 305(3.85)	12-0885-84 12-0885-84
2,7-Dibenzofurandiol, 1,3,4-trimethoxy-	EtOH	210(4.41),227(4.43), 262(4.19),297(4.20), 308s(4.10)	39-1441-84C

Compound	Solvent	$\lambda_{max}(\log \epsilon)$	Ref.
2,7-Dibenzofurandiol, 3,4,6-trimethoxy-	EtOH	226(4.52),246s(4.38), 266(4.23),306(4.18), 318s(4.10)	39-1445-84C
1,4-Naphthalenedione, 2-acetoxy-7,8-di-methoxy-5-methyl-	MeOH	243(4.22),322(3.68)	44-1853-84
1-Propanone, 2,3-dihydroxy-3-phenyl-1-(2,4,6-trihydroxyphenyl)-	EtOH	228(4.38),335s(--)	106-0711-84
	EtOH-NaOH	331(--)	106-0711-84
	EtOH-AlCl$_3$	308(--),365(--)	106-0711-84
$C_{15}H_{14}O_6S$			
Benzoic acid, 2-hydroxy-5-[[(4-methyl-phenyl)sulfonyl]oxy]-, methyl ester	EtOH	228(4.40),311(3.64)	39-1507-84C
5H-Furo[3,2-g][1]benzopyran-5-one, 4,9-dimethoxy-7-[(methylsulfinyl)-methyl]-	EtOH	211(4.36),254(4.57), 286s(3.67),337(3.63)	44-5035-84
$C_{15}H_{14}O_6S_2$			
Butanoic acid, 2-[(methylthio)(3-oxo-benzo[b]thien-2(3H)-ylidene)methyl]-3-oxo-, methyl ester, S,S-dioxide	EtOH	219(4.29),260(4.22), 363(4.22)	95-0134-84
$C_{15}H_{14}O_7$			
1H-Cyclopenta[1,3]cyclopropa[1,2]benz-ene-2,3-dicarboxylic acid, 3a,3b,4,7-tetrahydro-1-methoxy-4,7-dioxo-, di-methyl ester, (1α,3aα,3bβ,7aS*)-(±)-	MeCN	217(4.2),330(2.8)	5-0773-84
4,8-Epoxyfuro[2,3-d]oxepin-2(3H)-one, hexahydro-4,5-dihydroxy-3-[(4-hydr-oxyphenyl)methylene]- (plagiogyrin A)	MeOH	232(4.01),321(4.41)	94-1808-84 +94-1815-84
$C_{15}H_{14}O_7S_2$			
Propanedioic acid, [(methylthio)(3-oxo-benzo[b]thien-2(3H)-ylidene)methyl]-, dimethyl ester, S,S-dioxide	EtOH	252(4.06),356(3.93), 422(3.78)	95-0134-84
$C_{15}H_{15}BrN_2$			
2-Propen-1-amine, 3-(4-bromophenyl)-N-methyl-3-(3-pyridinyl)-, (E)-	pH 1	220(4.32),236s(4.27)	44-4209-84
(Z)-	pH 1	248(4.28)	44-4209-84
$C_{15}H_{15}BrN_2O$			
Benzenecarboximidamide, 4-bromo-N-hydroxy-2,6-dimethyl-N'-phenyl-	MeOH	234(4.24),252(4.32)	32-0015-84
$C_{15}H_{15}BrO_4S$			
Ethanol, 2-(2-bromophenoxy)-, 4-methyl-benzenesulfonate	EtOH	264(3.34),271(3.43), 279(3.32)	12-2059-84
$C_{15}H_{15}ClFeN$			
Ferrocene, (1-chloro-3-(dimethylimmon-io)-prop-1-enyl)-, perchlorate	EtOH	265(4.18),385(3.60), 595(4.06)	65-1439-84
$C_{15}H_{15}ClN_2O$			
Benzenecarboximidamide, N-(4-chloro-phenyl)-N'-hydroxy-2,6-dimethyl-	MeOH	262(4.17)	32-0015-84
$C_{15}H_{15}ClN_2O_2S$			
Acetamide, N-[3-(2-chlorobenzoyl)-5-ethyl-2-thienyl]-	MeOH	237(4.21),269(4.04), 273(4.04),349(3.94)	73-0621-84
hydrochloride	MeOH	236(4.19),248s?(3.93?), 265(4.03),272(4.02)	73-0621-84

Compound	Solvent	$\lambda_{max}(\log \epsilon)$	Ref.
2H-Pyrrol-2-one, 4-chloro-1,5-dihydro-1-(4-methylphenyl)-3-(4-morpholinyl)-5-thioxo-	$CHCl_3$	340(4.01),467(3.90)	78-3499-84
$C_{15}H_{15}Cl_2NS_2$			
Benzenesulfenamide, 4-chloro-N-[(4-chlorophenyl)thio]-N-(1-methylethyl)-	hexane	252(4.34)	44-2724-84
$C_{15}H_{15}Cl_3$			
Benzene, 1-chloro-4-(2,6-dichloro-3-cyclopropyl-1,3-hexadienyl)-	hexane	248(3.1),313(3.3)	104-1930-84
$C_{15}H_{15}F_3O_2$			
1,4-Hexadien-3-one, 6,6,6-trifluoro-5-hydroxy-1-(2,4,6-trimethylphenyl)-, (?,E)-	EtOH	233s(3.28),360(3.81)	39-2863-84C
$C_{15}H_{15}NO$			
1-Naphthalenecarboxamide, N-(2-methyl-1-propenyl)-	EtOH	225(4.27),280(3.86)	44-0714-84
$C_{15}H_{15}NO_2$			
Benzeneethanamine, β-hydroxy-N-(phenylmethylene)-, N-oxide	dioxan	291(4.26)	94-2609-84
$C_{15}H_{15}NO_3$			
2,4-Benzofurandione, 3,5,6,7-tetrahydro-3-(4-methylphenyl)-, 2-oxime	MeOH	270(4.026)	73-1421-84
2H-Cyclohepta[b]furan-2,4(3H)-dione, 5,6,7,8-tetrahydro-3-phenyl-, 2-oxime	MeOH	273(4.044)	73-1421-84
2,4,6-Octatrienal, 7-methyl-5-nitro-6-phenyl-	MeCN	238(4.23),280s(4.06), 365(3.38)	88-2909-84
1-Propanone, 2,2-dimethyl-1-(5-nitro-1-naphthalenyl)-	MeOH	330(3.62)	35-3140-84
1H-Pyrrole-3-carboxylic acid, 2-(1-oxopropyl)-5-phenyl-, methyl ester	MeOH	248(3.88),340(4.08)	2-0289-84
$C_{15}H_{15}NO_3Se$			
Phenol, 4-[3-[(2-nitrophenyl)seleno]propyl]-	MeOH	226(4.14),255(4.13), 275s(3.88),389(3.56)	18-2893-84
$C_{15}H_{15}NO_4$			
3H-3-Benzazepine-3,7-dicarboxylic acid, 3-ethyl 7-methyl ester	C_6H_{12}	210(4.4),282(4.2), 330(3.5),400s(2.7)	18-3483-84
2,4-Benzofurandione, 3,5,6,7-tetrahydro-3-(4-methoxyphenyl)-, 2-oxime	MeOH	271(4.082)	73-1421-84
1H-Pyrrole-3-carboxylic acid, 1-ethyl-4,5-dihydro-4,5-dioxo-2-phenyl-, ethyl ester	dioxan	285s(3.65),408(3.49)	94-0497-84
1H-Pyrrole-3,4-dicarboxylic acid, N-methyl-2-phenyl-, dimethyl ester	MeOH	262(4.01)	44-3314-84
$C_{15}H_{15}NO_5$			
1H-Pyrrole-3-carboxylic acid, 4,5-dihydro-2-(4-methoxyphenyl)-1-methyl-4,5-dioxo-, ethyl ester	dioxan	333(3.93),415(3.62)	94-0497-84
$C_{15}H_{15}NS$			
1-Naphthalenecarbothioamide, N-(2-methyl-1-propenyl)-	EtOH	210(4.55),305(4.25), 385(2.76)	44-0714-84

Compound	Solvent	$\lambda_{max}(\log \epsilon)$	Ref.
$C_{15}H_{15}N_3$			
Pyridine, (4,5-dihydro-1-methyl-3-phenyl-1H-pyrazol-5-yl)-	EtOH	220(4.18),300(4.04)	34-0225-84
$C_{15}H_{15}N_3O_2$			
Acetamide, N-methyl-N-(2,3,4,6-tetrahydro-6-oxobenzo[f]quinoxalin-5-yl)-	CH_2Cl_2	234(4.25),266(4.36), 417(3.48)	83-0329-84
Methanimidic acid, N-(4-methoxybenzoyl)-, 2-phenylhydrazide	MeCN	260(4.30),316(4.10)	48-0311-84
2-Naphthalenecarbonitrile, 5-amino-3-(butylamino)-1,4-dihydro-1,4-dioxo-	benzene	418s(3.77),447(3.85), 484s(3.79)	39-1297-84C
1H-Pyrrole-2-carboxylic acid, 5-amino-4-cyano-3-methyl-1-phenyl-, ethyl ester	EtOH	303(4.58)	118-0062-84
1H-Pyrrole-2-carboxylic acid, 4-cyano-3-methyl-5-(phenylamino)-, ethyl ester	EtOH	304(4.33)	118-0062-84
$C_{15}H_{15}N_3O_3$			
2-Naphthalenecarbonitrile, 5-amino-3-(butylamino)-1,4-dihydro-8-hydroxy-1,4-dioxo-	benzene	510(3.79),540(3.92), 580(3.78)	39-1297-84C
1H-Pyrimido[5,4-a]carbazole-1,3(2H)-dione, 4,4a,5,6,11,11b-hexahydro-5-hydroxy-11b-methyl-	MeOH	283(3.94),292(3.85)	88-3243-84
$C_{15}H_{15}N_3S$			
1,4-Benzothiazine, 2-phenylhydrazono-3-phenyl-	EtOH	266(4.55),392(4.45)	4-0521-84
$C_{15}H_{15}N_4S_2$			
Thiazolium, 3-methyl-2-[[[(3-methyl-2(3H)-thiazolylidene)amino]phenyl-methylene]amino]-, iodide	MeCN	394(4.44)	152-0475-84
$C_{15}H_{15}N_5OS$			
1H-Pyridazino[4,5-e][1,3,4]thiadiazine, 5-morpholino-3-phenyl-	EtOH	266(4.03),320(3.44)	94-4437-84
$C_{15}H_{15}N_5O_2$			
Benzamide, N-(9-ethyl-8,9-dihydro-7-methyl-8-oxo-7H-purin-6-yl)-	MeOH	230(4.30),291(4.12)	150-3601-84M
1H-Imidazole-5-carboxamide, 4-[(6-methoxy-8-quinolinyl)amino]-1-methyl-	EtOH	255(4.7),303(3.59), 362(3.84)	2-0870-84
$C_{15}H_{15}N_5O_3$			
Benzoic acid, [1-[2-(2-nitrophenyl)-hydrazino]ethylidene]hydrazide	20% acetone acid neutral	435(3.56) 372(--) 435(--)	140-1028-84 140-1028-84 140-1028-84
Benzoic acid, [1-[2-(4-nitrophenyl)-hydrazino]ethylidene]hydrazide	20% acetone acid base	350(4.31) 320(--) 430(--)	140-1028-84 140-1028-84 140-1028-84
$C_{15}H_{15}N_5O_4$			
Imidazo[5,1-f][1,2,4]triazin-4(1H)-one, 2-amino-7-[[2-(benzoyloxy)ethoxy]-methyl]-	n.s.g.	226(4.61),264(3.78), 280s(3.67)	4-0697-84
$C_{15}H_{15}N_5O_5$			
Inosine, 2-(3-pyridinyl)-	pH 1 pH 7	249(4.07),306(3.94) 258(4.01),293(4.05)	87-0429-84 87-0429-84

Compound	Solvent	$\lambda_{max}(\log \epsilon)$	Ref.
Inosine, 2-(3-pyridinyl)- (cont.)	pH 13	259(4.06),289(3.98)	87-0429-84
$C_{15}H_{15}N_5O_6$ Benzofurazan, 4,6-dinitro-, 1-oxide, adduct with 2,4,6-trimethylbenzen- amine	MeOH dioxan	235(4.11),274(4.02), 355(3.41),458(4.23) 241(4.22),282(4.07), 330(3.40),422(3.81)	12-0985-84 12-0985-84
$C_{15}H_{15}N_5O_6S$ L-Histidine, N^α-acetyl-4-mono(sulfanil- azo)-	pH 1.5 pH 6 pH 13	337(4.37) 351(4.31) 388(4.33)	69-0589-84 69-0589-84 69-0589-84
$C_{15}H_{15}N_5O_9$ 1H-1,2,3-Triazole-4,5-dicarboxylic acid, 1-[2-[4,5-bis(methoxycarbonyl)- 1H-pyrazol-3-yl]-2-oxoethyl]-, dimethyl ester	MeOH	255(4.12),278(4.28)	103-0416-84
$C_{15}H_{16}As_2Cl_2$ Arsinous chloride, 1,3-propanediyl- bis[phenyl-	EtOH	218s(4.06),263(3.46), 269(3.40)	65-1597-84
$C_{15}H_{16}BrNO_7S$ 1H-Pyrrolo[1,2-a]indole-5,8-dione, 7-bromo-2,3-dihydro-9-(hydroxymethyl)- 1-methoxy-6-methyl-2-[(methylsulfon- yl)oxy]-, (1R-cis)-	MeOH	229(4.16),285.5(4.09), 348(3.59),425(3.35)	44-5164-84
$C_{15}H_{16}BrN_5O_3$ Methylhdrazinium 5-(4-bromoanilino)-2- methyl-4,7-dioxo-4,7-dihydro-2H-inda- zol-3-olate	EtOH	254(4.27),309(4.00), 429.5(4.09)	4-0825-84
$C_{15}H_{16}Br_2$ Benzene, (2,6-dibromo-3-cyclopropyl- 1,3-hexadienyl)-	hexane	246(3.1),320(3.3)	104-1930-84
$C_{15}H_{16}ClNO_2$ Naphthalene, 1-(1-chloro-2,2-dimethyl- propyl)-5-nitro-	EtOH	245(3.93),335(3.57)	35-3140-84
$C_{15}H_{16}ClN_5$ 7H-Purin-6-amine, N-(2-chloro-4-meth- ylphenyl)-2,7,8-trimethyl-	EtOH	301(4.30)	11-0208-84B
$C_{15}H_{16}Cl_2$ Benzene, (2,6-dichloro-3-cyclopropyl)- 1,3-hexadienyl)-	hexane	248(3.1),324(3.2)	104-1930-84
$C_{15}H_{16}Cl_2N_2O_8$ Pyridazinium, 4,5-dichloro-3-hydroxy- 1-(2,3,5-tri-O-acetyl-β-D-ribo- furanosyl)-, hydroxide, inner salt	MeOH	223(4.41),331(3.61)	87-1613-84
$C_{15}H_{16}Cl_2N_4O_4$ 1H-Pyrazole-3,4-dimethanol, 1-(3,4-di- chlorophenyl)-, bis(methylcarbamate)	MeCN	218(4.29)	87-1559-84
$C_{15}H_{16}FeO_5$ Iron, [[μ⁴-5,6-bis(methylene)bicyclo-	isooctane	220(4.27),293(3.34)	33-0986-84

Compound	Solvent	$\lambda_{max}(\log \epsilon)$	Ref.
[2.2.2]oct-2-yl]acetato]tricarbonyl-			33-0986-84
$C_{15}H_{16}N_2$			
1H-Indole, 1,2-dimethyl-3-(3-methyl-1H-pyrrol-2-yl)-	EtOH	212(4.31),237(4.39), 296(3.82)	103-0058-84
$C_{15}H_{16}N_2O$			
Benzenecarboximidamide, N-hydroxy-2,6-dimethyl-N'-phenyl-	MeOH	253(4.10)	32-0015-84
Benzo[f]quinoxalin-6(2H)-one, 4-ethyl-3,4-dihydro-5-methyl-	CH_2Cl_2	247(4.22),283(4.30), 439(3.68)	83-0743-84
Methanone, 1H-indol-2-yl(1,2,3,6-tetrahydro-1-methyl-4-pyridinyl)-	EtOH	248(4.00),269s(3.96), 310(4.05)	78-3339-84
Pyrido[3',4':3,4]cyclopent[1,2-b]indol-5(1H)-one, 2,3,4,4a,6,10c-hexahydro-2-methyl-	EtOH	206(4.20),234(4.18), 304(4.28)	78-3339-84
WS-30581B	EtOH	224(4.35),266(4.16), 282s(4.09),298s(4.02)	158-1153-84
$C_{15}H_{16}N_2OS$			
1H-Pyrazole, 4,5-dihydro-3-(4-methoxyphenyl)-1-methyl-5-(thienyl)-	EtOH	228(4.17),299(4.24)	34-0225-84
1H-Pyrazole, 4,5-dihydro-5-(4-methoxyphenyl)-1-methyl-3-(thienyl)-	EtOH	224(4.10),260(4.09), 302(4.13)	34-0225-84
2H-Pyrrol-2-one, 1,5-dihydro-1-(4-methylphenyl)-3-(1-pyrrolidinyl)-5-thioxo-	$CHCl_3$	344(4.14),450(4.23)	78-3499-84
$C_{15}H_{16}N_2O_2$			
Benzo[6,7]cyclohepta[1,2-c]pyrazole, 1-acetyl-1,4,5,6-tetrahydro-8-methoxy-	EtOH	285(4.22)	2-0736-84
2,4-Pentadienoic acid, 2-cyano-5-(methylphenylamino)-, ethyl ester	EtOH	393(4.74)	70-2328-84
$C_{15}H_{16}N_2O_2S$			
2H-Pyrrol-2-one, 1,5-dihydro-1-(4-methylphenyl)-3-(4-morpholinyl)-5-thioxo-	$CHCl_3$	340(4.20),442(4.06)	78-3499-84
$C_{15}H_{16}N_2O_3$			
Acetic acid, [2,3-dihydro-1-(4-methylphenyl)-2-oxo-4(1H)-pyrimidinylidene]-, ethyl ester	n.s.g.	266(3.99),323(4.26)	4-0053-84
Benzenamine, N-[3-(3-nitrophenoxy)propyl]-	MeCN	275s(3.81),330s(3.31)	78-1755-84
9H-Pyrido[3,4-b]indole-1-ethanol, β,4-dimethoxy-	EtOH	248(4.85),287(4.45), 336(3.99),350(3.90)	94-0170-84
Spiro[furan-2(5H),9'-[9H]imidazo[1,2-a]indole]-3',5(2'H)-dione, 1',3,4,9'a-tetrahydro-2',2'-dimethyl-	EtOH	246.5(4.07),284s(3.18)	94-1373-84
$C_{15}H_{16}N_2O_4$			
Benzenamine, N-[2-(2-methoxy-5-nitrophenoxy)ethyl]-	MeCN	300(3.88),341(3.85)	78-1755-84
$C_{15}H_{16}N_2O_5$			
Acetamide, N-[4-[1-(2-hydroxy-5-oxo-1-cyclopenten-1-yl)-2-nitroethyl]phenyl]-	MeOH	249(4.137)	73-1421-84
1H-Indole-3-propanoic acid, 1-[[(methoxycarbonyl)amino]acetyl]-	EtOH	240(4.29),262s(3.94), 271s(3.86),291(3.83), 299.5(3.85)	94-1373-84

Compound	Solvent	$\lambda_{max}(\log \epsilon)$	Ref.
Uridine, 2'-deoxy-5-phenyl-	pH 1	279(4.03)	88-2431-84
	pH 13	268(3.91)	88-2431-84
$C_{15}H_{16}N_2S_2$			
Cyclohexanamine, N-(4-H-1,3-dithiolo-[4,5-b]indol-2-ylidene)-	EtOH	248(3.85),305(4.16)	12-2479-84
monohydrobromide	EtOH-HBr	250s(3.78),277(4.06), 377(4.05)	12-2479-84
$C_{15}H_{16}N_4$			
Propanedinitrile, [3-amino-2-cyano-6,6-dimethyl-4-(1-methylethenyl)-2,4-cyclohexadien-1-yl]-	MeOH	228(4.02),327(3.65)	78-2829-84
1H-Pyrrolo[2,3-d]pyrimidin-4-amine, 2,5,6-trimethyl-N-phenyl-	EtOH	227(4.18),284(3.85)	11-0073-84A
$C_{15}H_{16}N_4O$			
4H-Pyrazolo[3,4-d]pyrimidin-4-one, 1,5-dihydro-6-methyl-5-(2-propylphenyl)-	EtOH	210(4.42),252(3.88)	11-0023-84A
$C_{15}H_{16}N_4O_2$			
3-Pyridazinecarboxylic acid, 6-amino-5-cyano-1,2-dihydro-4-methyl-1-phenyl-, ethyl ester	EtOH	267(4.22),356(3.49)	118-0062-84
$C_{15}H_{16}N_4O_4S_2$			
Acetic acid, [[(3,4,5,10-tetrahydro-3,10-dimethyl-2,4-dioxobenzo[g]-	MeCN	222(4.43),267s(--), 352(3.81)	35-3309-84
pteridin-4a(2H)-yl)methyl]dithio]-	6M HCl	213(4.36),262(4.06), 290s(--),381(3.40)	35-3309-84
$C_{15}H_{16}N_5O$			
7H-Purinium, 6-(acetylamino)-7-methyl-9-(phenylmethyl)-, iodide	H_2O	224(4.24),274(4.12)	150-3601-84M
7H-Purinium, 6-(benzoylamino)-9-ethyl-7-methyl-, iodide	H_2O	279(4.11),323(3.80)	150-3601-84M
$C_{15}H_{16}N_6O_5$			
1H-Pyrrolo[1,2-a]indole-5,8-dione, 7-amino-9-[[(aminocarbonyl)oxy]methyl]-2-azido-2,3-dihydro-1-methoxy-6-methyl-, (1S-trans)-	MeOH	208(4.28),245(4.39), 305(4.08),347(3.61), 525(2.95)	44-5164-84
$C_{15}H_{16}O$			
Benzocyclooctene, 5-methoxy-8,9-di-methyl-	C_6H_{12}	230s(4.12),253s(3.74), 270s(3.22)	23-2769-84
2-Cyclopenten-1-one, 5-(2-butenyl)-3-phenyl-, (E)-	n.s.g.	224.5(3.97),285.5(4.32)	150-2060-84M
2-Cyclopenten-1-one, 5-(2-methyl-2-propenyl)-3-phenyl-	n.s.g.	224(4.12),285(4.08)	150-2060-84M
1,4-Ethenonaphthalene, 1,4-dihydro-1-methoxy-2,3-dimethyl-	C_6H_{12}	248s(2.93),256.5(2.85), 263(2.91),271(3.02), 278(3.00)	23-2769-84
$C_{15}H_{16}ORu$			
Ruthenocene, 1,1'-(1,4-butanediyl)-2-formyl-	EtOH	248(4.02),351(3.04)	18-0719-84
Ruthenocene, 1,1'-(1,4-butanediyl)-3-formyl-	EtOH	257(4.12),359(3.31)	18-0719-84

Compound	Solvent	$\lambda_{max}(\log \epsilon)$	Ref.
$C_{15}H_{16}O_2$			
1,7-Azulenedione, 3,8-dimethyl-5-(1-methylethyl)-	MeOH	398(3.95)	138-0627-84
2-Cyclopenten-1-one, 3-(4-methoxyphenyl)-5-(2-propenyl)-	n.s.g.	233(4.13),316(4.43)	150-2060-84M
1(2H)-Dibenzofuranone, 3,4-dihydro-3,3,8-trimethyl-	EtOH	237(4.32),270(3.89)	39-1213-84C
1H-Indene-6-carboxaldehyde, 3,7-dimethyl-5-(1-methylethyl)-1-oxo-	MeOH	340(3.89),380s(3.42)	138-0627-84
1,4-Naphthalenedione, 2,5-dimethyl-8-(1-methylethyl)-	MeOH	360(3.71)	138-0627-84
$C_{15}H_{16}O_2S$			
Benzene, [[(1-phenylethyl)sulfonyl]methyl]-, (S)-	MeOH	252(2.67),256(2.80), 262(3.67),263(2.60), 268(2.55)	35-1779-84
$C_{15}H_{16}O_3$			
2H,6H-Benzo[1,2-b:5,4-b']dipyran-2-one, 7,8-dihydro-4,8,8-trimethyl-	EtOH	220(4.20),246(3.51), 256.5(3.40),329.5(4.39)	94-1178-84
2-Cyclohexene-1-carboxylic acid, 6-oxo-1-(phenylmethyl)-, methyl ester	MeOH	220(3.31)	44-4429-84
1,4-Dioxaspiro[4.5]decan-2-one, 3-(phenylmethylene)-, (Z)-	EtOH	295(4.57),308(4.50)	39-1531-84C
4H-Furo[3,2-c][1]benzopyran-4-one, 2,3-dihydro-2,3,3,7-tetramethyl-	MeOH	206(4.54),277(3.89), 288(4.05),313(4.14), 328(4.00)	107-0737-84
4H-Furo[3,2-c][1]benzopyran-4-one, 2,3-dihydro-2,3,3,9-tetramethyl-	MeOH	215(4.70),287(4.35), 298(4.12),313(4.01), 328(4.21)	107-0737-84
2-Propanol, 1,3-bis(4-hydroxyphenyl)-	MeOH	279(3.80)	102-0897-84
Spiro[1,3-dioxane-2,1'(2'H)-naphthalen]-2'-one, 5,5-dimethyl-	EtOH	233(4.28),322(3.86)	39-2069-84C
$C_{15}H_{16}O_3S$			
Ethanol, 2-[2-[(4-methylphenyl)sulfinyl]phenoxy]-, (R)-	EtOH	238(4.08),283(3.45)	12-2059-84
$C_{15}H_{16}O_4$			
2H-1-Benzopyran-2-one, 5-(1,1-dimethyl-2-propenyl)-8-hydroxy-7-methoxy-	MeOH	228(3.70),264(4.06), 325(4.10)	78-5229-84
(celerin)	MeOH-base	232(3.79),284(4.23), 332(3.99)	78-5229-84
2H-1-Benzopyran-2-one, 4-hydroxy-7-methoxy-6-(3-methyl-2-butenyl)-	MeOH	290(4.16),310(4.23), 325(4.15)	2-1028-84
2H-1-Benzopyran-2-one, 4-hydroxy-7-methoxy-8-(3-methyl-2-butenyl)-	MeOH	318(3.89)	2-1028-84
2H-1-Benzopyran-2-one, 7-methoxy-8-[(3-methyl-2-butenyl)oxy]-	EtOH	260(4.17),320(4.41)	102-0859-84
2H-1-Benzopyran-2-one, 8-methoxy-7-[(3-methyl-2-butenyl)oxy]-	EtOH	260(3.51),320(4.12)	102-0859-84
1β,10β-Epoxyfuranoeremophilane-6,9-dione, (±)-	EtOH	241(3.76),305(3.94)	94-3396-84
$C_{15}H_{16}O_5$			
9,10-Anthracenedione, 1,2,3,4,4a,9a-hexahydro-1,2,8-trihydroxy-6-methyl-, (1α,2β,4aα,9aβ)-(+)-	MeOH	234.9(4.46),277.9(3.81), 346.7(3.80)	158-84-90
1-Cyclohexene-1-carboxylic acid, 4-hydroxy-2-[2-(4-hydroxyphenyl)ethyl]-6-oxo-, (S)-	pH 2	222(4.08),240(3.98), 280s(--)	102-1607-84
	pH 7	220(4.00),257(4.02)	102-1607-84

Compound	Solvent	$\lambda_{max}(\log \epsilon)$	Ref.
after three hours (cont.) 2,4-Pentadienoic acid, 5-(6-methoxy-1,3-benzodioxol-5-yl)-, ethyl ester, (E,E)-	pH 12 MeOH	238(4.25),293(3.72) 225(3.80),250(3.97), 300(3.91),366(4.23)	102-1607-84 78-2541-84
$C_{15}H_{16}O_6S$ 2,2(3H)-Furandicarboxylic acid, 4-formyl-3-(2-thienyl)-, diethyl ester	EtOH	202(4.35),241(4.06), 285(4.04)	118-0974-84
$C_{15}H_{16}O_7$ 4H-1-Benzopyran-6-acetic acid, 7-hydroxy-5,8-dimethoxy-2-methyl-4-oxo-, methyl ester	EtOH	208(4.28),228(4.32), 248(4.31),254(4.33), 268s(3.89),293(3.91), 342s(3.32)	88-2953-84
1,4-Etheno-1H-cyclopenta[1,3]cyclopropa[1,2-d][1,2]dioxin-5,6-dicarboxylic acid, 4,4a,4b,7-tetrahydro-7-methoxy-, (1α,4α,4aβ,4bα,7α,7aR*)-	MeCN	215(3.8),245(3.7)	5-0773-84
$C_{15}H_{16}O_8$ Plagiogyrin B	MeOH	235.5(4.09),325.5(4.47)	94-1815-84
$C_{15}H_{16}O_9$ Esculin	EtOH	224(4.12),250(3.67), 298(3.79),336(4.07)	102-2839-84
$C_{15}H_{17}Br_2NO$ 2,6-Methano-1-benzazocin-11-one, 8,10-dibromo-1-ethyl-1,2,3,4,5,6-hexahydro-6-methyl-	EtOH	268(3.9),305(3.5)	142-1771-84
$C_{15}H_{17}ClFeN$ Methanaminium, N-(3-chloro-3-ferrocenyl)-2-propenylidene)-N-methyl-, iodide	EtOH	262(4.23),384(3.62), 596(4.02)	65-1439-84
$C_{15}H_{17}ClN_2O_8$ Pyridazinium, 4-chloro-2,3-dihydro-3-oxo-1-(2,3,5-tri-O-acetyl-β-D-ribofuranosyl)-	MeOH	221(4.35),249s(--), 328(3.60)	87-1613-84
$C_{15}H_{17}ClN_4$ Pyridazino[3,4-b]quinoxaline, 2-butyl-3-chloro-2,10-dihydro-10-methyl-	EtOH	260(4.46),278(4.25), 290s(4.10),316(3.60), 332(3.58),382(3.99), 400s(3.91),435s(3.78), 455(3.85),484s(3.73), 520s(3.32)	103-1281-84
	EtOH-HCl	212(4.22),252(4.54), 258s(4.50),280(4.23), 292s(4.15),315s(3.90), 372(3.78),470(3.73)	103-1281-84
Pyridazino[3,4-b]quinoxaline, 5-butyl-3-chloro-5,10-dihydro-10-methyl-	EtOH	247(4.63),360(3.90), 400s(3.74)	103-1281-84
	EtOH-HCl	212(4.18),252(4.52), 260s(4.45),280(4.18), 290s(4.15),315s(3.90), 370(3.78),468(3.56)	103-1281-84
$C_{15}H_{17}ClOS$ Cyclohexanone, 6-[[(4-chlorophenyl)-	C_6H_{12}	302(4.10)	5-0576-84

Compound	Solvent	$\lambda_{max}(\log \epsilon)$	Ref.
thio]methylene]-2,2-dimethyl-, (E)-			5-0576-84
(Z)-	C_6H_{12}	316(4.23)	5-0576-84
$C_{15}H_{17}ClO_4$			
2-Propenoic acid, 2-methyl-, 2-[4-(3-chloro-1-oxopropyl)phenoxy]ethyl ester	n.s.g.	270(4.49),323(2.36)	73-2635-84
$C_{15}H_{17}NO$			
3-Buten-2-one, 1-(1-methyl-2-pyrrolidinylidene)-4-phenyl-, (E,E)-	EtOH	288(4.17),296(4.16), 364(4.49)	78-2879-84
	EtOH-HCl	366(4.58)	78-2879-84
1(2H)-Naphthalenone, 2-[3-(dimethylamino)-2-propenylidene]-3,4-dihydro-, (E,E)-	EtOH	274(4.37),438(4.69)	118-0424-84
4(1H)-Phenanthridinone, 2,3,5,6-tetrahydro-6,6-dimethyl-	ether	236(3.68),340(3.56)	142-0097-84
4(1H)-Phenanthridinone, 2,3,5,6-tetrahydro-6,8-dimethyl-	ether	216(3.89),240(3.98), 384(3.68)	142-0097-84
$C_{15}H_{17}NO_2$			
Oxazole, 5-[(5-hexenyl)oxy]-2-phenyl-	C_6H_{12}	283(4.12),288(4.12), 296(4.12),310(3.83)	44-0399-84
1H-Pyrrole-2-acetic acid, 1-methyl-α-(phenylmethyl)-, methyl ester	MeOH	205(4.16)	107-0453-84
$C_{15}H_{17}NO_3$			
3H-Indol-3-one, 1-acetyl-1,2-dihydro-2-(1-methyl-3-oxobutyl)-	EtOH	205(3.96),233(4.43), 239(4.51),260(4.11), 266(4.06),335(3.59)	103-1374-84
1-Naphthalenemethanol, α-(1,1-dimethylethyl)-5-nitro-	EtOH	245(3.94),350(3.57)	35-3140-84
2H-Pyrano[3,2-c]quinolin-5-one, 3,4,5,6-tetrahydro-7-hydroxy-2,2,6-trimethyl-(ravesilone)	EtOH	216(4.41),230(4.4), 250(4.45),256(4.49), 280(3.90),292(3.92), 325(3.48)	102-1825-84
$C_{15}H_{17}NO_4$			
1,3-Cycloheptanedione, 2-(2-nitro-1-phenylethyl)-	MeOH	264(3.221)	73-1421-84
1H-Pyrrole-3,4-dicarboxylic acid, 4,5-dihydro-N-methyl-2-phenyl-, methyl ester	MeOH	307(4.13)	44-3314-84
$C_{15}H_{17}NO_5S$			
4-Thia-1-azabicyclo[3.2.0]heptane-2-carboxylic acid, 3,3-dimethyl-7-oxo-, phenylmethyl ester, 4,4-dioxide, (2S-cis)-	EtOH	215(3.69)	39-2785-84C
$C_{15}H_{17}NO_6$			
2-Pyridineacetic acid, 3-(methoxycarbonyl)-5-(3-methoxy-3-oxo-1-propenyl)-6-methyl-, methyl ester, (E)-	MeOH	209(4.24),226(4.17), 268(3.92)	24-1620-84
$C_{15}H_{17}N_2$			
16-Aza-15-azoniatricyclo[9.3.1.1^{4,8}]-hexadeca-1(15),4,6,8(16),11,13-hexaene, 15-methyl-, trifluoromethanesulfonate	EtOH	268(3.61),304(3.52)	35-2672-84

Compound	Solvent	$\lambda_{max}(\log \epsilon)$	Ref.
$C_{15}H_{17}N_3$			
Benzenamine, N,N,3-trimethyl-4-(phenyl-azo)- (with pyridine)	pH 8	453.5(3.11)	46-6185-84
$C_{15}H_{17}N_3O_6$			
Imidazo[1,2-a]pyridinium, 2-[2-ethoxy-1-(ethoxycarbonyl)-2-oxoethyl]-1-methyl-3-nitro-, hydroxide, inner salt	n.s.g.	244(4.38),278s(3.97), 318(4.21),374(3.70), 454(3.73)	39-0069-84C
1H-Indazole-4,7-dione, 1-(3-acetoxy-propyl)-3-[[(aminocarbonyl)oxy]-methyl]-6-methyl-	MeOH	221(4.03),260(4.15), 268s(4.12),323(3.45)	5-1711-84
$C_{15}H_{17}N_5$			
1H-Purin-6-amine, N-(3,5-dimethylphen-yl)-2,8-dimethyl-	EtOH	302(4.45)	11-0208-84B
7H-Purin-6-amine, 2,7,8-trimethyl-N-(3-methylphenyl)-	EtOH	304(4.12)	11-0208-84B
$C_{15}H_{17}N_5O$			
Acetamide, N-[8,9-dihydro-7-methyl-9-(phenylmethyl)-7H-purin-6-yl]-	MeOH	316(4.17)	150-3640-84M
$C_{15}H_{17}N_5OS$			
Pyridinium, 1-[[2,2-dicyano-1-(methyl-thio)ethenyl]amino]-3-[(diethylami-no)carbonyl]-, hydroxide, inner salt	EtOH	249(4.23),280(4.19)	95-0440-84
$C_{15}H_{17}N_5O_2S$			
Pyrimido[4,5-b]quinoxaline-2,4(3H,5H)-dione, 3,7,8,10-tetramethyl-4a-(thio-cyanomethyl)-	MeCN	221(4.53),272(4.20), 300s(--),357(3.87)	35-3309-84
	6M HCl	301(3.83),392(3.64)	35-3309-84
$C_{15}H_{17}N_5O_3$			
3,7-Ethano-4H,6H,11H-furo[3,4-d]dipyrr-olo[3,4-c:3',4'-e][1,2,4]triazolo-[1,2-a]pyridazine-11,13(2H)-dione, 3,7,9a,14-tetrahydro-12-methyl-	MeOH	269(3.58),388(2.10)	88-2459-84
Methylhydrazinium 4,7-dihydro-2-methyl-4,7-dioxo-5-(phenylamino)-2H-indazol-3-olate	EtOH	244(4.27),309.5(4.01), 426(4.09)	4-0825-84
$C_{15}H_{17}N_5O_4$			
5H,11H-1,4-Ethano-5,11-etheno-4a,11a-(methanoxymethano)-7H-pyridazino-[4,5-d][1,2,4]triazolo[1,2-a]pyrid-azine-7,9(8H)-dione, 1,4-dihydro-8-methyl-, 2-oxide	MeOH	233(4.02)	88-2459-84
3,7-Ethano-4H,6H,11H-furo[3,4-d]dipyrr-olo[3,4-c:3',4'-e][1,2,4]triazolo-[1,2-a]pyridazine-11,13(12H)-dione, 3,7,9a,14-tetrahydro-12-methyl-, 2-oxide	MeOH at 0°	225(3.83),265s(3.48)	88-2459-84
$C_{15}H_{17}N_5O_5S$			
Acetamide, N-[6-(1,2-diacetoxypropyl)-1,4-dihydro-4-thioxo-2-pteridinyl]-, [S-(R*,S*)]-	MeOH	231(4.42),258(4.08), 285(4.01),315s(3.88), 403(3.99)	5-1815-84
$C_{15}H_{17}N_5O_6$			
Acetamide, N-[6-(1,2-diacetoxypropyl)-	pH -4.0	230s(4.08),257(4.07),	5-1815-84

Compound	Solvent	$\lambda_{max}(\log \epsilon)$	Ref.
1,4-dihydro-4-oxo-2-pteridinyl]-, [S-(R*,S*)]- (cont.)		295(3.95),312(3.88), 376(3.60)	5-1815-84
	pH 3.0	233(4.15),281(4.21), 335(3.92)	5-1815-84
	pH 10.0	256(4.48),287s(3.76), 347(3.88)	5-1815-84
$C_{15}H_{17}P$			
Phosphorin, 3-butyl-5-phenyl-	EtOH	211(4.29),241(4.46), 270s(3.90),304(2.99)	24-0763-84
Phosphorin, 3-(1,1-dimethylethyl)- 5-phenyl-	EtOH	210(4.25),240(4.38), 270s(3.93),304(2.77)	24-0763-84
$C_{15}H_{18}$			
Bicyclo[9.2.2]pentadeca-5,6,11,13,14- pentaene	EtOH	218(3.73),261s(2.52), 268(2.57),277(2.54)	24-0474-84
4H-Cyclopenta[def]phenanthrene, 1,2,3,3a,8,9,9a,9b-octahydro-	C_6H_{12}	220(3.91),268(2.86), 276(2.86)	18-0725-84
lower melting isomer	C_6H_{12}	212(4.14),266(2.69)	18-0725-84
1H-Cyclopropa[l]phenanthrene, 2,3,4,5,6,7,8,9-octahydro-	pentane	273(2.96),283(2.96)	44-3436-84
$C_{15}H_{18}As_2O_4$			
Arsinic acid, 1,3-propanediylbis[phen- yl-	EtOH	211(4.37),217(4.36), 257(3.29),263(3.38), 270(3.32)	65-1597-84
$C_{15}H_{18}BrNO$			
2,6-Methano-1-benzazocin-11-one, 8-bromo-1-ethyl-1,2,3,4,5,6- hexahydro-6-methyl-	EtOH	270(4.2),312(3.3), 347(3.1)	142-1771-84
$C_{15}H_{18}Br_2$			
Tricyclo[9.2.2.0^{4,6}]pentadeca-11,13,14- triene, 5,5-dibromo-	EtOH	222(3.83),270(2.68), 277(2.61)	24-0474-84
$C_{15}H_{18}ClN_2O_3PS$			
Benzyl phsophite, S-(4-chlorobenzyl)- thiouronium salt	MeOH	205(3.93),223(4.16)	118-0410-84
$C_{15}H_{18}Cl_2$			
Tricyclo[9.2.2.0^{4,6}]pentadeca-11,13,14- triene, 5,5-dichloro-	EtOH	223(3.80),263s(2.67), 269(2.74),276s(2.70)	24-0474-84
$C_{15}H_{18}Cl_2O$			
2H-Oxocin, 5-chloro-2-(1-chloro-3-hex- en-5-ynyl)-8-ethyl-3,8-dihydro-	EtOH	214s(3.87),223(3.98), 233s(3.87)	12-1545-84
$C_{15}H_{18}N_2$			
4,9-Azo-5,8-ethano-1H-benz[f]indene, 3a,4,4a,5,8,8a,9,9a-octahydro-	hexane	394(2.03)	24-0534-84
1H-Indole, 3-[(1,2,3,6-tetrahydro- 1-methyl-4-pyridinyl)methyl]-	EtOH	276s(4.03),282(4.06), 291(4.00)	78-3339-84
2-Pyridinemethanamine, α-methyl-N-(1- phenylethyl)-	isooctane	260(3.58)	103-0077-84
	EtOH	260(3.65)	103-0077-84
6H-Pyrrolo[3,2,1-de]phenazine, 6a,7,8,9,10,10a-hexahydro-1- methyl-, cis	MeOH	226(4.47),283(3.94), 295s(3.91)	18-0623-84
trans	MeOH	227(4.41),280(3.91)	18-0623-84

Compound	Solvent	$\lambda_{max}(\log \epsilon)$	Ref.
$C_{15}H_{18}N_2O$			
Ethanone, 1-(2,3,4,4a,5,11a-hexahydro-1H-cyclohepta[b]quinoxalin-6-yl)-, cis	MeOH	255(4.30),390(3.83), 454(4.03)	18-0623-84
trans	MeOH	255(4.11),390(3.65), 448(3.81)	18-0623-84
2H-Indol-2-one, 3-[3-(dimethylamino)-2-methyl-2-propenylidene]-1,3-di-hydro-1-methyl-	hexane	368(4.45),420(4.48)	103-0951-84
	DMSO	431(4.62)	103-0951-84
1-Propanone, 1-(1,2-dimethyl-1H-indol-3-yl)-, O-ethenyloxime, (E)-	EtOH	223(4.42),284(4.04)	103-0058-84
Pyrano[3,2-b]indol-2-amine, 2,5-dihy-dro-N,N,3,5-tetramethyl-	hexane	326(4.40),368(4.40)	103-0951-84
	DMSO	324(4.19),365(4.23), 520(3.73)	103-0951-84
$C_{15}H_{18}N_2OS$			
1-Piperidinamine, N-[(3-methoxybenzo-[b]thien-2-yl)methylene]-	PrOH	335(4.44)	104-1353-84
2H-Pyrrol-2-one, 4-(butylamino)-1,5-dihydro-1-(4-methylphenyl)-5-thioxo-	CHCl$_3$	310(4.25),430(4.10)	78-3499-84
$C_{15}H_{18}N_2O_2$			
2-Naphthalenol, 1-[[(1-methylpropyl)-nitrosoamino]methyl]-	EtOH	229(4.80),267(3.69), 278(3.76),289(3.70), 323(3.46),335(3.54)	150-0701-84
1-Phenazinecarboxaldehyde, 4-acetyl-5,5a,6,7,8,9,9a,10-octahydro-, cis	MeOH	241(4.24),272s(3.96), 339(4.05),492(3.78)	18-0623-84
trans	MeOH	240(4.21),274(3.96), 327(3.99),488(3.72)	18-0623-84
$C_{15}H_{18}N_2O_4$			
L-Alanine, N-acetyl-3-(3-cyano-4-meth-oxyphenyl)-, ethyl ester	EtOH	306(3.71)	56-0157-84
2-Cyclopenten-1-one, 2-[1-[4-(dimethyl-amino)phenyl]-2-nitroethyl]-3-hydroxy-	MeOH	253(4.142)	73-1421-84
5-Pyrimidinecarboxylic acid, 1,4,5,6-tetrahydro-4-(4-methoxyphenyl)-1-methyl-2-oxo-, ethyl ester	EtOH	205(3.6),222s(3.52), 300(3.36)	103-0431-84
L-Tryptophan, N-(methoxycarbonyl)-5-methyl-, methyl ester	MeOH	224(4.48),279(3.74), 286(3.74),297(3.61)	94-2126-84
$C_{15}H_{18}N_2O_4S$			
L-Tryptophan, N-(methoxycarbonyl)-5-(methylthio)-, methyl ester	MeOH	231(4.44),252s(4.10), 287(3.64)	94-2126-84
$C_{15}H_{18}N_2O_5$			
DL-Tryptophan, 5-methoxy-N-(methoxy-carbonyl)-, methyl ester	MeOH	220.5(4.42),277(3.80), 297(3.70),309(3.57)	94-2544-84
DL-Tryptophan, 6-methoxy-N-(methoxy-carbonyl)-, methyl ester	MeOH	223(4.53),273(3.65), 293(3.72)	94-2544-84
$C_{15}H_{18}N_2O_9$			
Uridine, 2',3',5'-triacetate	MeOH	262(3.96)	31-0339-84
$C_{15}H_{18}N_2S_2$			
1H-Pyrrole-2,5-dithione, 3-(butyl-amino)-1-(4-methylphenyl)-	CHCl$_3$	255(4.01),390(4.36), 515(3.52)	78-3499-84
$C_{15}H_{18}N_4$			
1H-Imidazo[1,2-b]-1,2,4-triazepine, 7,8-dihydro-6,8,8-trimethyl-2-phenyl-	MeOH	274(4.58)	103-1152-84

Compound	Solvent	$\lambda_{max}(\log \epsilon)$	Ref.
$C_{15}H_{18}N_4O$			
Pyrazolo[5,1-b]quinazolin-9(4H)-one, 3-[(dimethylamino)methyl]-2,7-di-methyl-	10% EtOH-NaOH	221(4.47),256(4.61), 328(4.29),404(3.69)	56-0411-84
$C_{15}H_{18}N_4O_2$			
Pyrazolo[5,1-b]quinazolin-9(4H)-one, 3-[(dimethylamino)methyl]-7-methoxy-2-methyl-	10% EtOH-NaOH	223(4.45),249(4.49), 270(4.49),326(4.20), 408(3.69)	56-0411-84
$C_{15}H_{18}N_4O_2S$			
1H-Imidazole-4-carboxylic acid, 5-[[(methylamino)thioxomethyl]amino]-2-(phenylmethyl)-, ethyl ester	MeOH	254(4.49),295(4.38)	2-0316-84
$C_{15}H_{18}O$			
1(8aH)-Azulenone, 6,8a-dimethyl-3-(1-methylethyl)-, (R)-	ether	237(4.30),350(3.60)	78-5197-84
Marmelerin	MeOH	222(4.32),255(4.18), 278(3.51),289(3.42)	44-5154-84
$C_{15}H_{18}O_2$			
Azuleno[6,5-b]furan-2(3H)-one, deca-hydro-3,5,8-tris(methylene)-	MeOH	206(3.91)	102-0188-84
5H-Benzo[3,4]cyclobuta[1,2]cyclohepten-2-ol, 4b,6,7,8,9,9a-hexahydro-, acetate, cis	MeOH	270(3.45),276(3.40)	87-0792-84
2,5-Cyclohexadiene-1,4-dione, 2-(1,2-dimethylbicyclo[3.1.0]hex-2-yl)-5-methyl- (laurequinone)	MeOH	254(4.12),299(2.93)	102-2672-84
Cyclohexenecarboxylic acid, 1,3-dimeth-yl-5-phenyl-	EtOH	259(4.07)	39-0261-84C
2,4,6,8,10-Dodecapentaenedial, 2,6,11-trimethyl-, (all-E)-	benzene	284(3.81),291(3.77), 376(4.72),394(4.93), 416(4.92)	137-0102-84
1,4-Methanonaphthalen-2-ol, 1,2,3,4-tetrahydro-9,9-dimethyl-, acetate, ($1\alpha,2\alpha,4\alpha$)-	n.s.g.	248s(2.18),254s(2.52), 260(2.77),273(3.00)	12-1035-84
6-Oxabicyclo[3.2.1]octan-7-one, 1,3-dimethyl-5-phenyl-	EtOH	215(2.35)	39-0261-84C
2,4-Pentanedione, 3-[(2,4,6-trimethyl-phenyl)methylene]-	EtOH	286(3.40)	39-2863-84C
$C_{15}H_{18}O_3$			
3-Acenaphthylenecarboxylic acid, 1,2,6,7,8,8a-hexahydro-5-hydroxy-1,6-dimethyl-, ($1\alpha,6\alpha,8a\alpha$)-(-)-(ryomenin)	MeOH	211(4.51),252(3.98), 310(3.58)	94-1355-84
Benzeneacetic acid, α-1-cyclohexen-1-yl)-2-hydroxy-, methyl ester	ether	276(3.34),282(3.30)	35-3590-84
5H-Benzo[3,4]cyclobuta[1,2]cyclohep-tene-1,9a-diol, 4b,6,7,8,9,9a-hexa-hydro-, 1-acetate	MeOH	256s(2.76),263(2.83), 270(2.79)	87-0792-84
1,3-Benzodioxol-5-ol, 4-(3-methyl-2-butenyl)-6-(2-propenyl)-	EtOH	239(3.60),298(3.59)	94-0023-84
1,3-Benzodioxol-5(7aH)-one, 7a-(3-meth-yl-2-butenyl)-6-(2-propenyl)-, (S)-	EtOH	237(4.10),283(2.74)	94-0011-84
2-Butenal, 3-methyl-4-[tetrahydro-3-(3-methyl-1,3-butadienyl)-4-methylene-5-oxo-2-furanyl]-	MeOH	224(4.19)	102-2573-84
Furanoeremophil-9-en-6-one, 1β-hydroxy-	EtOH	240(3.99),329(3.85)	94-3396-84

Compound	Solvent	$\lambda_{max}(\log \epsilon)$	Ref.
Naphtho[2,3-b]furan-2,6(3H,4H)-dione, 3a,8a,9,9a-tetrahydro-3,5,8a-trimethyl-	EtOH	238(4.00),262s(3.72)	138-1021-84
Naphtho[2,3-b]furan-8(4H)-one, 4a,7,8a,9-tetrahydro-8a-hydroxy-3,4a,5-trimethyl-	MeOH	218(4.50)	102-1793-84
4,7-Nonadiene-1,6-dione, 1-(3-furanyl)-4,8-dimethyl-, (E)-	hexane	204(4.31),259(4.60)	102-0759-84
(Z)-	hexane	201(4.15),261(4.36)	102-0759-84
11-Oxabicyclo[8.2.1]trideca-1(13),4,7-triene-6,12-dione, 5,9,9-trimethyl- (asteriscunolide A)	EtOH	223(3.98),237(3.88), 243(3.85)	78-0873-84
(asteriscunolide B)	EtOH	222(3.29),232(2.32), 245(3.26)	78-0873-84
(asteriscunolide C)	EtOH	217(3.95)	78-0873-84
(asteriscunolide D)	EtOH	221(3.90),247(3.67)	78-0873-83
α-Santonin	MeOH	240(4.11),260(3.94)	23-2813-84
Spiro[benzofuran-2(3H),1'-cyclohexane]-3-carboxylic acid, methyl ester	ether	217(3.76),281(3.46), 287(3.38)	35-3590-84
$C_{15}H_{18}O_4$			
13-Aldophomenone	MeOH	226(4.23),244(4.28)	102-2781-84
1,3-Benzodioxol-5(7aH)-one, 7a-[(3,3-dimethyloxiranyl)methyl]-6-(2-propenyl)- (+)-(4S,11S)-illicinone C	EtOH	241(4.15),288(3.64)	94-0011-84
(-)-(4R,11S)-	EtOH	239(3.88),289(3.45)	94-0011-84
Elemanschkuhriolide	MeOH	218(3.91)	44-2994-84
1β,10β-Epoxyfuranoeremophilan-6-one, 9β-hydroxy-	EtOH	263(3.54)	94-3396-84
Ethanone, 1-[3-(4-acetoxy-3-methyl-2-butenyl)-4-hydroxyphenyl-, (Z)-	EtOH	226(4.19),280(4.10)	102-1819-84
isomer	EtOH	228(4.27),282(4.16)	102-1819-84
Naphtho[1,2-b]furan-2,8(3H,4H)-dione, 3a,5,5a,6,7,9b-hexahydro-6-hydroxy-5a,9-dimethyl-3-methylene-	MeOH	210(4.00),241(4.08)	102-1665-84
3H-Oxireno[8,8a]naphtho[2,3-b]furan-5(9H)-one, 1a,2,4,4a-tetrahydro-9-hydroxy-4,4a,6-trimethyl-	EtOH	268(3.52)	94-3396-84
Subcordatolide	EtOH	216(3.94),224(3.90)	100-0626-84
Urospermal A, 8-deoxy-	MeOH	224.0(4.01)	94-1724-84
$C_{15}H_{18}O_5$			
1,3-Dioxane-4,6-dione, 5-ethyl-5-(4-methoxyphenyl)-2,2-dimethyl-	MeOH	235(3.98),274(3.18), 281(3.08)	12-1245-84
Propanedioic acid, [(4-methoxyphenyl)-methylene]-, diethyl ester	dioxan	312(4.31)	65-1987-84
$C_{15}H_{18}O_6$			
1H-2-Benzopyran-5,6-dicarboxylic acid, 3,4-dihydro-1-methoxy-3-methyl-, methyl ester	MeCN	203(4.51),236(3.88), 280(3.17),288s(3.12)	24-2422-84
1H-2-Benzopyran-7,8-dicarboxylic acid, 3,4-dihydro-1-methoxy-3-methyl-, methyl ester	MeCN	206(4.59),237(3.97), 275(3.01),285s(2.91)	24-2422-84
$C_{15}H_{18}O_7$			
2-Propenoic acid, 3-acetoxy-2-[(3-oxo-1,4-dioxaspiro[4.5]dec-2-ylidene)-methyl]-, methyl ester	EtOH	222(4.14),288(4.21)	77-1008-84
6H-Pyrano[4,3-b]oxepin-4,5-dicarboxylic acid, 8,9-dihydro-6-methoxy-8-methyl-, dimethyl ester	MeCN	206(4.31),266s(3.34), 310(3.51)	24-2422-84

Compound	Solvent	$\lambda_{max}(\log \epsilon)$	Ref.
$C_{15}H_{18}O_8$			
D-Glucose, 4-[3-(4-hydroxyphenyl)-2-propenoate]	MeOH	235(3.98),325(4.36)	94-1808-84 +94-1998-84
$C_{15}H_{19}BrN_2O_4S$			
2H-Furo[2,3-d]imidazole-2-thione, 3-(4-bromophenyl)-5-(1,2-dihydroxyethyl)-1-ethylhexahydro-6-hydroxy-	EtOH	244(4.32)	136-0131-84E
2H-Furo[2,3-d]imidazole-2-thione, 3-(4-bromophenyl)-5-(1,2-dihydroxyethyl)-1-ethylhexahydro-6-hydroxy-, (S)-	EtOH	246(4.30)	136-0131-84E
$C_{15}H_{19}BrO$			
Phenol, 2-[2-(bromomethylene)-1,3-dimethylcyclopentyl]-5-methyl-, cis-(-)- (isolaurenisol)	EtOH	212(4.35),278(3.53), 284(3.51)	102-1951-84
$C_{15}H_{19}BrO_3$			
Naphtho[1,2-h]furan-2,8(3H,4H)-dione, 7-bromooctahydro-5a,9-dimethyl-3-methylene-, [3aS-(3aα,5aβ,7α,9α,9aα-9bβ)]-	EtOH	206(4.02)	32-0107-84
$C_{15}H_{19}ClN_2O_6$			
2,4(1H,3H)-Pyrimidinedione, 1-[5-[2-[2-(chloromethyl)-1,3-dioxolan-2-yl]ethenyl]tetrahydro-4-hydroxy-2-furanyl]-5-methyl-, [2R-[2α,4β,5α(E)]]-	pH 1	266(3.99)	87-0680-84
	pH 7	266(3.98)	87-0680-84
	pH 13	266(3.86)	87-0680-84
$C_{15}H_{19}ClO_2$			
2H-Oxocin-5(6H)-one, 8-(1-chloro-3-hexen-5-ynyl)-2-ethyl-7,8-dihydro-	n.s.g.	215s(4.13),223(4.19), 233s(4.04)	12-1545-84
$C_{15}H_{19}N$			
1H-Indole, 2-(2,2-dimethyl-1-methylenepropyl)-1-methyl-	EtOH	224(4.47),283(3.88)	78-4837-84
$C_{15}H_{19}NO$			
Benzeneacetamide, N-(cyclohexylidenemethyl)-	EtOH	237(4.03)	44-0714-84
Benzo[6,7]cyclohepta[1,2-b]pyrrole, 2,3,3a,4,5,6-hexahydro-8-methoxy-3a-methyl-	EtOH	255(4.06)	39-0005-84C
2,6-Methano-1-benzazocin-11-one, 1-ethyl-1,2,3,4,5,6-hexahydro-6-methyl-	EtOH	261(4.0),305(3.0), 335(2.9)	142-1771-84
$C_{15}H_{19}NO_2$			
Spiro[1H-2-benzopyran-1,1'-cyclohexane]-4-carbonitrile, 3,5,6,7,8,8a-hexahydro-3-oxo-	EtOH	236(3.88)	118-1075-84
$C_{15}H_{19}NO_2S_4$			
6H-1,3-Thiazine-6-thione, 2,3-dihydro-4-(3-methylphenyl)-5-(methylsulfonyl)-2-(propylthio)-	EtOH	258(3.77),349(4.20)	4-0953-84
$C_{15}H_{19}NO_3$			
Ethanone, 1-(3,4-dimethoxyphenyl)-2-(1-methyl-2-pyrrolidinylidene)-	EtOH	229(4.14),271(3.79), 342(4.45)	78-2879-84
	EtOH-HCl	236(3.97),278(3.83),	78-2879-84

Compound	Solvent	λ_{max} (log ϵ)	Ref.
(cont.) 4-Isoxazolecarboxylic acid, 2-(1,1-di-methylethyl)-2,3-dihydro-3-phenyl-, methyl ester	EtOH	354(4.29) 272(3.76)	78-2879-84 44-0282-84
5-Isoxazolecarboxylic acid, 2-(1,1-di-methylethyl)-2,3-dihydro-3-phenyl-, methyl ester	EtOH	282(3.42)	44-0282-84
Spiro[cyclohexan-1,4'(4'aH)-oxireno[d]-[2]benzopyran]-1'a(2'H)-carbonitrile, tetrahydro-2'-oxo-	MeOH	238(3.11)	118-1075-84
$C_{15}H_{19}NS$ Benzeneethanethioamide, N-(cyclohexyli-denemethyl)-	EtOH	225(4.60),305(4.34), 380(2.76)	44-0714-84
$C_{15}H_{19}N_3$ Pyrrolidinium, 1-(1,3-dimethylcyclo-hepta[c]pyrrol-6(2H)-ylidene)-, perchlorate	CH_2Cl_2	316(4.57),362(4.24), 376(4.37),447(2.66)	118-0119-84
$C_{15}H_{19}N_2O$ Morpholinium, 4-(1,3-dimethylcyclo-hepta[c]pyrrol-6(2H)-ylidene)-, perchlorate	CH_2Cl_2	321(4.57),381(4.38), 479(2.74)	118-0119-84
$C_{15}H_{19}N_3O_5$ 1H-Indazole-1-propanol, 3-[[(aminocarbo-nyl)oxy]methyl]-4-hydroxy-6-methyl-	MeOH	222(4.46),300(3.79)	5-1711-84
$C_{15}H_{19}N_3O_7S$ L-Cysteine, S-[1-[4-[[(1-carboxyethyl)-amino]carbonyl]phenyl]-2-nitroethyl]-	MeOH	230(4.21)	94-5036-84
$C_{15}H_{19}N_3O_8S$ 5-Thiazolidineacetic acid, 3-(3,5-O-carbonyl-β-D-ribofuranosyl)-2-[(1-methylethylidene)hydrazono]-4-oxo-, methyl ester	$CHCl_3$	259(4.08)	128-0295-84
$C_{15}H_{19}N_3O_9$ 1H-Imidazole-5-carboxamide, 4-hydroxy-3-(2,3,5-tri-O-acetyl-β-D-ribo-furanosyl)-	H_2O M HCl M NaOH	245(3.80),278(4.11) 244(3.77),280(4.06) 276(4.17)	4-0529-84 4-0529-84 4-0529-84
3H-Imidazole-5-carboxamide, 4-hydroxy-1-(2,3,5-tri-O-acetyl-β-D-ribo-furanosyl)-	H_2O M HCl M NaOH	235(3.61),279(3.96) 242(3.81),277(3.53) 234(3.42),288(4.00)	4-0529-84 4-0529-84 4-0529-84
1H-Imidazole-4-carboxamide, 5-[(2,3,5-tri-O-acetyl-β-D-ribofuranosyl)oxy]-	H_2O M HCl M NaOH	250(4.13) 241(3.99) 265(4.12)	4-0849-84 4-0849-84 4-0849-84
1H-Imidazole-5-carboxamide, 4-hydroxy-3-(2,3,4-tri-O-acetyl-β-D-ribopyrano-syl)-	H_2O M HCl M NaOH	245(3.82),278(4.11) 244(3.81),280(4.08) 275(4.18)	4-0529-84 4-0529-84 4-0529-84
3H-Imidazole-5-carboxamide, 4-hydroxy-1-(2,3,4-tri-O-acetyl-β-D-ribopyrano-syl)-	H_2O M HCl M NaOH	236(3.70),285(3.99) 240(3.76),282(3.79) 288(4.07)	4-0529-84 4-0529-84 4-0529-84
$C_{15}H_{19}N_3S_2$ 3(2H)-Thiazolecarbothioamide, N,N-di-ethyl-4-methyl-2-(phenylimino)-	isoPrOH	284(4.32),332s(3.58)	94-4292-84
5-Thiazolecarbothioamide, N,N-diethyl-4-methyl-2-(phenylamino)-	isoPrOH	294(4.12)	94-4292-84

Compound	Solvent	$\lambda_{max}(\log \epsilon)$	Ref.
Thiourea, diethyl(4-methyl-3-phenyl-2(3H)-thiazolylidene)-	isoPrOH	298s(3.97),333(4.28)	94-4292-84
$C_{15}H_{19}N_4$			
Pyridinium, 1-ethyl-2-[[[(1-ethyl-2(1H)-pyridinylidene)amino]methylene]amino]-, iodide	MeCN	390(4.70)	152-0475-84
perchlorate	MeCN	390(4.68)	152-0475-84
$C_{15}H_{20}$			
Tricyclo[9.2.2.0^{4,6}]pentadeca-11,13,14-triene	EtOH	224(3.84),265s(2.43), 271(2.63),278s(2.58)	24-0474-84
$C_{15}H_{20}BrClO$			
2H-Oxocin, 3-bromo-8-(1-chloro-3-hexen-5-ynyl)-2-ethyl-3,4,7,8-tetrahydro-(+)-intricenyne)	EtOH	222(4.18),231s(4.06)	12-1545-84
$C_{15}H_{20}BrN_3O_6$			
Uridine, 5'-bromo-5'-deoxy-5-[(dimethyl-amino)methylene]-5,6-dihydro-2',3'-O-(1-methylethylidene)-6-oxo-	EtOH	240(3.86),285(3.88), 313(4.09)	94-2591-84
$C_{15}H_{20}BrO_5P$			
2-Propenoic acid, 3-(3-bromophenyl)-2-(diethoxyphosphinyl)-, ethyl ester, (E)-	dioxan	262(4.18)	65-1987-84
$C_{15}H_{20}ClN_3O_6$			
Uridine, 5'-chloro-5'-deoxy-5-[(dimeth-ylamino)methylene]-5,6-dihydro-2',3'-O-(1-methylethylidene)-6-oxo-	EtOH	239(3.85),285s(3.85), 313(4.11)	94-2591-84
$C_{15}H_{20}NO_3S$			
1(2H)-Pyridinyloxy, 4-[5-(carboxymeth-yl)-2-thienyl]-3,6-dihydro-2,2,6,6-tetramethyl-	EtOH	204(3.98),288(4.06), 333(3.20)	70-1901-84
$C_{15}H_{20}NO_7P$			
2-Propenoic acid, 2-(diethoxyphosphin-yl)-3-(3-nitrophenyl)-, ethyl ester, (E)-	dioxan	255(4.38)	65-1987-84
2-Propenoic acid, 2-(diethoxyphosphin-yl)-3-(4-nitrophenyl)-, ethyl ester, (E)-	dioxan	293(4.25)	65-1987-84
$C_{15}H_{20}N_2$			
4,9-Azo-5,8-ethano-1H-benz[f]indene, 2,3,3a,4,4a,5,8,8a,9,9a-decahydro-, (3aα,4β,4aα,5α,8α,8aα,9β,9aα)-	hexane	395(2.04)	24-0534-84
4,9-Azo-5,8-ethano-1H-benz[f]indene, 3a,4,4a,5,6,7,8,8a,9,9a-decahydro-, (3aα,4β,4aα,8aα,9α,9aα)-	hexane	383(2.14)	24-0534-84
$C_{15}H_{20}N_2O$			
1-Butanone, 1-(1,2-dimethyl-1H-indol-3-yl)-3-methyl-, oxime, (E)-	n.s.g.	226(4.45),280(3.99)	103-0055-84
$C_{15}H_{20}N_2O_2S$			
4(3H)-Quinazolinethione, 3-butyl-6,7-dimethoxy-2-methyl-	EtOH	227s(4.48),234(4.52), 256(4.20),276(4.16),	39-1143-84C

Compound	Solvent	$\lambda_{max}(\log \epsilon)$	Ref.
(cont.)		349s(3.95),364(4.16), 381(4.16)	39-1143-84C
$C_{15}H_{20}N_2O_2S_2$			
4H-1,3-Dithiolo[4,5-b]pyrrole-6-carbox-ylic acid, 2-(cyclohexylimino)-5-methyl-, ethyl ester	EtOH EtOH-NaOH	248(3.62),296(3.98) 267(3.95),317(3.93)	12-2479-84 12-2479-84
monohydrobromide	EtOH-HBr	250s(3.54),265(3.78), 360(4.04)	12-2479-84
$C_{15}H_{20}N_2O_3$			
Pyrrolo[1,2-c]quinazoline-5,8,10(1H,6H-9H)-trione, 2,3,7,10b-tetrahydro-7,7,9,9-tetramethyl-, (±)- (syn-carpurea)	MeOH MeOH-NaOH	236(3.63),306(3.74) 362(--)	88-0371-84 88-0371-84
$C_{15}H_{20}N_2O_4$			
1,3-Benzenedicarboxylic acid, 5-[1-(di-methylhydrazono)ethyl]-2-methyl-, dimethyl ester	MeOH	221(4.54),330(3.32)	24-1620-84
β-L-erythro-Pentopyranosid-2-ulose, methyl 3,4-O-(1-methylethylidene)-, phenylhydrazone, syn	EtOH	280(4.30),302s(4.11)	39-0733-84C
$C_{15}H_{20}N_2O_4S$			
2H-Furo[2,3-d]imidazole-2-thione, 5-(1,2-dihydroxyethyl)-1-ethyl-hexahydro-6-hydroxy-3-phenyl-, [3aR-[3aα,5α(S*),6α,6aα]]-	EtOH	244(4.24)	136-0131-84E
[3aS-[3aα,5β(R*),6β,6aα]]-	EtOH	246(4.22)	136-0131-84E
$C_{15}H_{20}N_2O_5S$			
1-Cyclopentene-1-carboxylic acid, 2-[[2-(aminosulfonyl)-5-methoxy-phenyl]amino]-, ethyl ester	EtOH	218(4.07),240(4.07), 322(4.22)	104-0534-84
$C_{15}H_{20}N_2O_{10}$			
2,5,8,11-Tetraoxatridecan-13-ol, 1-(3,5-dinitrophenyl)-1-oxo-	EtOH	211(4.3)	24-1994-84
$C_{15}H_{20}N_3$			
1H-Pyrazolium, 3-[2-[4-(dimethylamino)-phenyl]ethenyl]-1,2-dimethyl-, iodide	EtOH	391(4.52)	103-1266-84
$C_{15}H_{20}N_4O_5$			
Imidazo[5,1-f][1,2,4]triazin-4(1H)-one, 2,5-dimethyl-7-[2,3-O-(1-methylethyl-idene)-β-D-ribofuranosyl]-	EtOH	222(4.37),254(3.94)	39-0229-84C
$C_{15}H_{20}N_4O_8S$			
Chitinovorin C	H_2O	260(3.92)	158-1486-84
$C_{15}H_{20}N_4O_9$			
D-Ribitol, 1-deoxy-1-[1,3,4,7-tetra-hydro-6-(3-methoxy-3-oxopropyl)-2,4,7-trioxo-8(2H)-pteridinyl]-	pH 1 pH 13	282(4.07),328(4.07) 260(3.91),288(3.89), 355(4.11)	39-0953-84C 39-0953-84C
$C_{15}H_{20}N_4O_{11}$			
D-Ribitol, 1-deoxy-1-[(1,2,3,6-tetra-hydro-5-nitro-2,6-dioxo-4-pyrimidin-yl)amino]-, 2,3,4-triacetate	$CHCl_3$	331(3.99)	64-0252-84B

Compound	Solvent	$\lambda_{max}(\log \epsilon)$	Ref.
$C_{15}H_{20}O$			
Furan, 4-methyl-2-(2-methyl-6-methyl-ene-2,7-octadienyl)-, (E)-	MeOH	222(4.07)	100-0877-84
4-Hexen-3-one, 5-(2,4,6-trimethyl-phenyl)-, (E)-	EtOH	223(4.38),273(3.51)	80-0857-84
(Z)-	EtOH	222(4.36),278(3.17)	80-0857-84
2H-Indeno[4,5-b]furan, 1,6,7,8-tetra-hydro-1,2,5,8-tetramethyl- (dihydro-marmelerin)	MeOH	211(3.89),221s(3.89), 235s(3.52),286(3.51)	44-5154-84
Naphtho[2,3-b]furan, 4,4a,7,8,8a,9-hexahydro-6,9,9-trimethyl- (furo-dysinin)	EtOH	223(3.92)	100-0076-84
2,4-Pentadienal, 3-methyl-5-(5-methyl-spiro[2.5]oct-4-en-4-yl)-, (E,E)-	isoPrOH	378(4.62)	89-0251-84
$C_{15}H_{20}O_2$			
1(7H)-Azulenone, 8,8a-dihydro-8-hy-droxy-6,8a-dimethyl-3-(1-methyl-ethyl)-, (8R-trans)-	ether	226(4.20),324(3.97)	78-5197-84
Bicyclo[7.3.2]tetradeca-9(14),10,12-trien-13-one, 14-methoxy-	MeOH	243(4.06),250s(4.02), 258s(4.02),310(3.39), 346s(3.36)	88-4761-84
	MeOH-H_2SO_4	246(4.02),275(4.21), 352(3.48)	88-4761-84
Ethanone, 1-[3-(1,1-dimethyl-2-propen-yl)-4-hydroxy-2,5-dimethylphenyl]-	MeCN-KOH	350(4.23)	39-1259-84C
1(2H)-Naphthalenone, 3,4-dihydro-7-methoxy-6-methyl-4-(1-methyl-ethyl)-	EtOH	227(4.22),265(4.00), 322(3.53)	44-0034-84
1(4H)-Naphthalenone, 4a,7,8,8a-tetra-hydro-8a-hydroxy-4a-methyl-8-meth-ylene-2-(1-methylethyl)-, (4aS-trans)- (stemonolone)	EtOH	245(3.78)	39-0937-84C
Naphtho[2,3-b]furan-2(3H)-one, hexa-hydro-3,5,8a-trimethyl-	EtOH	216(4.00),272(3.76)	138-1021-84
$C_{15}H_{20}O_3$			
Benzeneacetic acid, α-cyclohexyl-α-hy-droxy-, methyl ester, (S)-(+)-	MeOH	217s(3.78),251(2.36), 257(2.40),263(2.26)	22-0077-84
Benzenepentanoic acid, α-butyl-δ-oxo-	EtOH	242(4.15)	2-0821-84
Benzenepentanoic acid, γ-butyl-δ-oxo-	EtOH	242(4.15)	2-0821-84
1H-Benz[e]indene-3,9(2H,8H)-dione, 9b-hydroxy-3a,6-dimethyl-3a,4,5,6-7,9b-hexahydro-	EtOH	244(3.93)	12-2305-84
1,3-Benzodioxol-5(6H)-one, 7,7a-dihydro-7a-(3-methyl-2-butenyl)-6-(2-propen-yl)- (illicinone B)	EtOH	249(4.02),278s(3.23)	94-0011-84
Dendrolasinolide, 6-oxo-	MeCN	244(3.83)	102-0759-84
2(3H)-Furanone, dihydro-5-(4-methoxy-3-methylphenyl)-5-(1-methylethyl)-	EtOH	227(3.96),274(3.23)	44-0034-84
Furodysinin lactone	EtOH	221(3.94)	100-0076-84
Istanbulin B	MeOH	223(4.11),283(1.53)	73-1311-84
Laurenobiolide, 6-epi-deacetyl-	MeOH	205(4.12)	102-1971-84
2-Naphthalenemethanol, 1,2,3,4-tetra-hydro-8-methoxy-α-methyl-, acetate	EtOH	271(3.25),278.5(3.27)	44-5116-84
Naphtho[2,3-b]furan-2,6(3H,4H)-dione, 3a,7,8,8a,9,9a-hexahydro-3,5,8a-tri-methyl-	EtOH	244(4.18)	138-1021-84
Naphth[1,2-b]oxiren-2(1aH)-one, 4,5,6-7,7a,7b-hexahydro-6-hydroxy-7,7a-di-methyl-1a-(1-methylethenyl)-	MeOH	240(4.02)	88-5907-84

Compound	Solvent	$\lambda_{max}(\log \epsilon)$	Ref.
Onitin	MeOH	232(4.51),270.5(4.15), 324(3.71)	94-4620-84
Phomenone, 13-deoxy-	EtOH	245(4.34)	158-84-146
Pterosin D	MeOH	216(4.58),259(4.28), 303(3.28)	94-4620-84
6α-Santonin, 1,2-dihydro-	MeOH	244(4.18)	23-2813-84
$C_{15}H_{20}O_4$			
Armefolin	MeOH	205(4.25)	102-1665-84
Benzeneacetic acid, 2-hydroxy-3-(1,1-dimethylethyl)-α-oxo-, 1-methylethyl ester	EtOH	213(4.22),265(3.95), 340(3.48)	39-2655-84C
1,3-Benzodioxol-5(6H)-one, 7a-[(3,3-dimethyloxiranyl)methyl]-7,7a-dihydro-6-(2-propenyl)- (illicinone D)	EtOH	248(4.30)	94-0011-84
isomer	EtOH	248(4.15)	94-0011-84
isomer	EtOH	250(4.15)	94-0011-84
Benzoic acid, 4-(2,2-dimethyl-1-oxo-propoxy)-2-methyl-, ethyl ester	EtOH	213(3.91),241(4.08), 278s(2.68),284s(2.78)	12-2037-84
Furanoeremophilan-6-one, 9β,10α-di-hydroxy-	EtOH	265(3.49)	94-3396-84
Istanbulin A	MeOH	211(4.16),285(1.20)	73-1311-84
Ixerin A	MeOH	229.5(4.10)	94-1724-84
Speciformin, 6-epi-	MeOH	203(3.72)	102-1971-84
$C_{15}H_{20}O_5$			
1H-Inden-1-one, 2,3-dihydro-3-hydroxy-6-(2-hydroxyethyl)-2,7-bis(hydroxy-methyl)-2,5-dimethyl-, (2R-cis)- (jamesonin)	MeOH	218(4.31),261.5(3.91), 304(3.23)	94-4620-84
Koningic acid	MeOH	213(3.98)	158-84-150
2-Penten-1-one, 1-(2,3,4,6-tetrameth-oxy)-, (E)-	MeOH	225(4.19),284(3.14)	94-1355-84
Propanedioic acid, (4-methoxyphenyl)-methyl-, diethyl ester	MeOH	228(4.00),272(3.15), 279(3.08)	12-1245-84
$C_{15}H_{20}O_6$			
β-D-Glucopyranoside, 4-(2-propenyl)-phenyl	MeOH	222(3.83),273(2.95), 279(2.85)	18-2893-84
$C_{15}H_{20}O_7$			
7-Oxabicyclo[4.1.0]heptane-1,3,3-tri-carboxylic acid, 6-ethyl-2-methyl-ene-, trimethyl ester	MeOH	240(2.86)	44-3791-84
$C_{15}H_{20}O_8$			
2-Butenedioic acid, 2-acetyl-, 1-ethyl 4-(tetrahydro-2,2-dimethylfuro[3,4-d]-1,3-dioxol-4-yl) ester, [3aR-[3aα,4α(E),6aα]]-	CHCl₃	230(3.85)	136-0019-84I
$C_{15}H_{21}BrO_3$			
5αH,4,6βH-Eudesma-6,13-olide, 2α-bromo-3α-hydroxy-11-methylene-3β-	EtOH	207(4.09)	32-0107-84
	EtOH	207(4.23)	32-0107-84
$C_{15}H_{21}Cl$			
Bicyclo[5.3.1]undeca-1(11),7,9-triene, 9-chloro-11-(1,1-dimethylethyl)-	C_6H_{12}	270(3.45),280(3.40), 320(3.15)	77-0733-84

Compound	Solvent	$\lambda_{max}(\log \epsilon)$	Ref.
$C_{15}H_{21}ClO_4$			
3-Cyclohexene-1-carboxylic acid, 1-chloro-4-(2,2-dimethyl-1-oxopropoxy)-2-methylene-, ethyl ester	EtOH	239(4.07)	12-2037-84
$C_{15}H_{21}FeN_2$			
Methanaminium, N-[(dimethylamino)ferrocenylmethylene]-N-methyl-, hexafluorophosphate	MeCN	236(4.11),279(4.13), 348(3.3),482(3.00) (anomalous)	157-0653-84
$C_{15}H_{21}NO$			
1-Cyclohexene-1-carbonitrile, 4-(1,5-dimethyl-3-oxo-4-hexenyl)-	EtOH	240(4.00)	107-0925-84
1H-Indole-2-methanol, α-(1,1-dimethylethyl)-α,1-dimethyl-	EtOH	255(4.18),285(3.56)	78-4837-84
$C_{15}H_{21}NO_2$			
Ethanone, 1-[2-hydroxy-5-methyl-3-(1-piperidinylmethyl)phenyl]-	EtOH	225(3.99),255(3.69)	2-0904-84
Ethanone, 1-[1-methoxy-3-(1-piperidinylmethyl)phenyl]-	EtOH	265(3.99)	2-0904-84
1-Propanone, 3-(dimethylamino)-1-(2,3-4,5-tetrahydro-1-benzoxepin-7-yl)-, hydrochloride	n.s.g.	218s(--),266(4.09)	103-0838-84
Spiro[1H-2-benzopyran-1,1'-cyclohexane]-4-carbonitrile, octahydro-3-oxo-	EtOH	262(2.60)	118-1075-84
$C_{15}H_{21}NO_3$			
1H-Indole-2-carboxylic acid, 2,3,4,5-6,7-hexahydro-1-methyl-3-oxo-2-(2-propenyl)-, ethyl ester	MeCN	336(4.02)	33-1535-84
$C_{15}H_{21}NO_3S$			
Benzo[b]thiophen-4(5H)-one, 5-[[bis(2-methoxyethyl)amino]methylene]-6,7-dihydro-, (E)-	EtOH	220(4.04),252.5(4.22), 362(4.27)	161-0081-84
$C_{15}H_{21}NO_7$			
Sesbanimide	MeOH	end absorption	39-1311-84C
$C_{15}H_{21}N_2Ru$			
Methanaminium, N-[(dimethylamino)ruthenocenylmethylene]-N-methyl-, hexafluorophosphate	MeCN	222(3.92),269(4.25), 360(3.27),507(1.58)	157-0653-84
$C_{15}H_{21}N_3O_2$			
1H-Isoindole-1-acetamide, 3-(aminocarbonyl)-4,5,6,7-tetrahydro-1-methyl-6-(1-methylethenyl)- (manicoline B)	EtOH	228(3.32),272(3.23)	88-2359-84
4-Quinazolinamine, N-butyl-6,7-dimethoxy-2-methyl-	EtOH	207(4.39),245(4.49), 251s(4.43),280(3.77), 292(3.78),322(3.99), 333(3.95)	39-1143-84C
hydrochloride	EtOH	218(4.39),250(4.39), 278s(3.69),326(4.19), 340(4.17)	39-1143-84C
$C_{15}H_{21}N_3O_3S$			
3H-1,2-Oxathiino[6,5-e]imidazole, 4,5,6,7-tetrahydro-7-methyl-4-(1-	EtOH	250(3.96),255(3.97), 269s(3.82)	4-0361-84

Compound	Solvent	$\lambda_{max}(\log \epsilon)$	Ref.
piperidinyl)-, 2,2-dioxide (cont.)			4-0361-84
$C_{15}H_{21}N_5O_4$ Adenosine, N,N-dimethyl-2',3'-O-(1-methylethylidene)-	MeOH	275(4.28)	150-3601-84M
$C_{15}H_{21}N_5O_5$ 9H-Purin-6-amine, 9-[O-(1-oxopentyl)-β-D-arabinofuranosyl]-	MeOH	259(4.17)	87-0270-84
$C_{15}H_{21}N_5O_{10}$ D-Ribitol, 1-[(2-amino-1,6-dihydro-5-nitro-6-oxo-4-pyrimidinyl)amino]-1-deoxy-, 2,3,4-triacetate	CHCl$_3$	328(4.10)	64-0252-84B
$C_{15}H_{21}O_5P$ 2-Propenoic acid, 2-(diethoxyphosphin-yl)-3-phenyl-, ethyl ester	dioxan	265(4.21)	65-1987-84
$C_{15}H_{22}$ Naphthalene, 1,2,3,7,8,8a-hexahydro-5,8a-dimethyl-3-(1-methylethyli-dene)-, (S)-	EtOH	273s(4.05),283(4.18), 295(4.09)	39-1323-84C
$C_{15}H_{22}BrNO_3$ Propanamide, N-[2-(2-bromo-4,5-dimeth-oxyphenyl)ethyl]-2,2-dimethyl-	n.s.g.	205(4.62),231(3.97), 285(3.50)	83-0488-84
$C_{15}H_{22}NO_2$ Pyrrolo[2,1-a]isoquinolinium, 1,2,3,5-6,10b-hexahydro-8,9-dimethoxy-N-methyl-, iodide	MeOH	220s(4.30),283(3.56), 288s(3.50)	12-1203-84
$C_{15}H_{22}NO_3$ 5H-Oxazolo[2,3-a]isoquinolinium, 2,3,6,10b-tetrahydro-8,9-dimethoxy-4,10b-dimethyl-, iodide	MeOH	219(4.30),281(3.51), 286s(3.47)	12-1659-84
$C_{15}H_{22}N_2$ 4,9-Azo-5,8-ethano-1H-benz[f]indene, 2,3,3a,4,4a,5,6,7,8,8a,9,9a-dodeca-hydro-, (3aα,4β,4aα,8aα,9β,9aα)-	hexane	386(2.09)	24-0534-84
$C_{15}H_{22}N_2O$ 4H-Pyrrolo[2,3-g]isoquinolin-4-one, 3-ethyl-1,4a,5,6,7,8,8a,9-octa-hydro-2,6-dimethyl-	n.s.g.	252(4.12),290(3.64)	44-5109-84
4H-Pyrrolo[3,2-h]isoquinolin-4-one, 3-ethyl-1,5,5a,6,7,8,9,9a-octahydro-2,8-dimethyl-, trans-(±)-	n.s.g.	251(4.02),293(3.67)	44-5109-84
$C_{15}H_{22}N_2O_3$ 2-Propanone, 1-[3,5-diacetyl-1-(dimeth-ylamino)-1,2-dihydro-6-methyl-2-pyri-dinyl]-	MeOH	228(4.18),311(4.34), 377(4.06)	24-1620-84
2(1H)-Pyridinone, 6-[octahydro-1-(hy-droxymethyl)-2H-quinolizin-3-yl]-, N-oxide, (1α,3α,5β,9aα)-(-)-(mamanine N-oxide)	EtOH	228(3.95),304(4.01)	102-0887-84

Compound	Solvent	$\lambda_{max}(\log \epsilon)$	Ref.
$C_{15}H_{22}N_3O_8P$			
Uridine, 2',3'-O-(1-methylethylidene)-5'-O-(3-methyl-1,3,2-oxazaphospholidin-2-yl)-, P-oxide	EtOH	259(4.05)	39-1471-84C
$C_{15}H_{22}N_4O_5$			
Inosine, 2-pentyl-	pH 1	252(4.09)	87-0429-84
	pH 7	250(4.10)	87-0429-84
	pH 13	256(4.11)	87-0429-84
2,6-Pyridinedicarboxylic acid, 4-[[[(dimethylamino)carbonyl]methylhydrazono]ethylidene]-1,2,3,4-tetrahydro-, dimethyl ester, (E,E)-(±)-	EtOH	265(3.92),376(4.57)	33-1547-84
$C_{15}H_{22}N_4O_9S$			
L-glycero-α-L-ido-Heptofuranuronic acid, 6-[(2-amino-3-hydroxy-1-oxopropyl)amino]-1,6-dideoxy-1-(3,4-dihydro-3-methyl-2,4-dioxo-1(2H)-pyrimidinyl)-4-thio-	M HCl	211(4.00),266(3.95)	5-1399-84
$C_{15}H_{22}O$			
Africanone	EtOH	257(4.00),345(1.76)	102-0688-84
1H-Cyclopropa[a]naphthalene-1-carboxaldehyde, 1a,2,3,5,6,7,7a,7b-octahydro-1,7,7a-trimethyl-, [1S-(1α,1aα,7α,7aα,7bα)]-	ether	275(2.18)	102-1647-84
Isobicyclogermacrenal, (-)-	EtOH	261(4.06)	39-0203-84C
Lepidozenal	EtOH	265(4.15)	39-0203-84C
Vitrenal, (+)-	EtOH	242(4.12)	39-0215-84C
$C_{15}H_{22}OS_2$			
Bicyclo[2.2.1]heptan-2-one, 3-(1,3-dithiepan-2-ylidene)-1,7,7-trimethyl-, (+)-	EtOH	328(4.32),370(2.85)	78-2951-84
$C_{15}H_{22}O_2$			
2-Butenal, 3-(2,4,5,6,7,7a-hexahydro-4,4,7a-trimethyl-2-benzofuranyl)-, [2α(E),7aα]-(±)-	hexane	229.5(4.29)	33-0175-84
isomer 10B	hexane	229(4.23),270(3.38)	33-0175-84
3-Cyclohexen-1-ol, 4-(5-hydroxy-3-methyl-3-penten-1-ynyl)-3,5,5-trimethyl-	EtOH	270(4.18),284(4.08)	39-2147-84C
2H-Cyclopropa[a]naphthalen-2-one, 1,1a,4,5,6,7,7a,7b-octahydro-4-hydroxy-1,1,7,7a-tetramethyl-, [1aR-(1aα,4α,7α,7aα,7bα)]-	ether	228(4.05)	102-1647-84
1(2H)-Dibenzofuranone, 3,4,5a,6,7,8,9-9a-octahydro-3,3,5a-trimethyl-	EtOH	274(4.15)	39-1213-84C
2,4-Pentadienal, 3-methyl-5-(2,2,6-trimethyl-7-oxabicyclo[4.1.0]hept-1-yl)-	MeOH	284(4.43)	33-0184-84
	MeCN	282(4.42)	33-0184-84
Spiro[7-oxabicyclo[4.1.0]heptane-2,2'-oxiran], 1,5,5-trimethyl-6-(3-methyl-1,3-butadienyl)-	pentane	236(4.39)	33-0815-84
$C_{15}H_{22}O_3$			
Cyclohexanone, 4-hydroxy-4-(5-hydroxy-3-methyl-3-penten-1-ynyl)-3,3,5-trimethyl-, cis	EtOH	228(4.10)	39-2147-84C
trans	EtOH	230(4.10)	39-2147-84C

Compound	Solvent	$\lambda_{max}(\log \epsilon)$	Ref.
1-Cyclohexene-1-carboxylic acid, 4-(1,5-dimethyl-3-oxo-4-hexenyl)-, [S-(R*,S*)]-	EtOH	227(4.19)	78-2961-84
5H-Cyclopropa[a]naphthalen-5-one, 1,1a,2,3,6,7,7a,7b-octahydro-3-hydroperoxy-1,1,7,7a-tetramethyl-, [1aR-(1aα,3α,7α,7aα,7bα)]-	MeOH	231(4.09),320(2.85)	102-1647-84
2(5H)-Furanone, 5-(2-acetyl-5-methylcyclopentyl)-4-(1-methylethyl)- (alpinolide)	EtOH	217(3.85)	138-1687-84
1,4-Hexadien-3-one, 1-(1,4-dioxaspiro[4.5]dec-8-yl)-5-methyl-	EtOH	260(4.11)	107-0925-84
Laurenobiolide, 6-epideacetyl-11,13-dihydro-	MeOH	201(3.76)	102-1971-84
$C_{15}H_{22}O_4$			
[2,3'-Bifuran]-2'(5'H)-one, 2,3,4,5-tetrahydro-5-methyl-5-(4-methyl-2-oxopentyl)-, (2R-cis)- (ipomeamaronolide)	MeCN	210(3.73)	102-0759-84
3-Buten-2-one, 4-(4-acetoxy-1,2-epoxy-2,6,6-trimethylcyclohexyl)-, (1R,2S,4S)-	EtOH	232(4.09)	33-2043-84
2H-Cyclohepta[b]furan-2-one, 3,3a,4,5-6,8a-hexahydro-5-hydroxy-7-(3-hydroxybutyl)-6-methyl-3-methylene- (dicorin)	MeOH	228(3.82)	102-2553-84
1,3-Cyclohexadiene-1-carboxylic acid, 4-(2,2-dimethyl-1-oxopropoxy)-2-methyl-, ethyl ester	EtOH	208(3.76),234s(3.43), 298(3.92)	12-2037-84
3-Cyclohexene-1-carboxylic acid, 4-(2,2-dimethyl-1-oxopropoxy)-2-methylene-, ethyl ester	EtOH	243(3.89)	12-2037-84
7H-6a,2-(Epoxymethano)naphth[2,3-b]oxiran-7-one, octahydro-6-hydroxy-2a,3,9,9-tetramethyl-	EtOH	340(10)[sic]	83-0680-84
β-Ionone, 3-acetoxy-5,6-epoxy-5,6-dihydro-, (3S,5R,6S)-	EtOH	232(4.10)	33-2043-84
$C_{15}H_{22}O_5$			
Acetic acid, (3-hydroxy-5-oxo-4-pentyl-2(5H)-furanylidene)-, 1,1-dimethylethyl ester, (E)-	EtOH	261(4.19),313(3.80)	39-1555-84C
	EtOH-NaOH	258(4.21),348(3.69)	39-1555-84C
(Z)-	EtOH	262(4.24),305s(3.87)	39-1555-84C
	EtOH-NaOH	261(4.28),353(3.80)	39-1555-84C
Butanoic acid, 2-methyl-, 2,6-dimethoxy-1,3-dimethyl-4-oxocyclohexa-2,5-dien-1-yl ester	EtOH	213(4.41),315(3.73)	138-0339-84
Propanoic acid, 2-(3-oxo-1,4-dioxaspiro[4.5]dec-2-ylidene)-, 1,1-dimethylethyl ester	EtOH	256(4.08)	39-1531-84C
Vestolide	MeOH	212(4.09)	102-2379-84
$C_{15}H_{22}O_6$			
Pseudoanisatin	EtOH	304(1.32)	39-2511-84C
$C_{15}H_{22}O_6S_2$			
Dispiro[bicyclo[2.2.1]heptan-2,2'-oxirane-3',2"-[1,3]dithiepan]-3-one, 4,7,7-trimethyl-, 1",1",3",3"-tetraoxide, [1S-(1α,2β,4α)]-	EtOH	250(3.54),310(2.40), 390(1.62)	78-2951-84

Compound	Solvent	$\lambda_{max}(\log \epsilon)$	Ref.
$C_{15}H_{22}Si_2$			
Disilane, pentamethyl-1-naphthalenyl-	C_6H_{12}	287(4.00)	60-0341-84
Disilane, pentamethyl-2-naphthalenyl-	C_6H_{12}	272(3.82)	60-0341-84
$C_{15}H_{23}Br$			
2H-2,4a-Methanonaphthalene, 1-bromo-3,4,5,6,7,8-hexahydro-8,8,9,9-tetramethyl-, (2R)-	n.s.g.	216(3.66)	2-0339-84
$C_{15}H_{23}FeN_2$			
Methanaminium, N-[1'-(dimethylamino)-ferrocenyl]-N,N-dimethyl-, iodide	MeOH	221(4.48),304(3.67)	101-0113-84R
$C_{15}H_{23}I$			
2H-2,4a-Methanonaphthalene, 3,4,5,6,7,8-hexahydro-1-iodo-8,8,9,9-tetra-methyl-, (2R)-	n.s.g.	226(3.65),254(2.89)	2-0339-84
$C_{15}H_{23}NO_2$			
3,5-Cyclohexadiene-1,2-dione, 3,5-bis(1,1-dimethylethyl)-, 4-(O-methyloxime)	hexane	207(3.21),244(2.62),298(2.64),375(2.69)	54-0023-84
$C_{15}H_{23}N_3O$			
1-Deazapurine, 9-(2-hydroxy-3-nonyl)-, erythro-	pH 7.6	284(3.99)	87-0274-84
$C_{15}H_{23}N_3O_5$			
1,3,4-Triazabicyclo[4.2.0]oct-2-ene-2,4-dicarboxylic acid, 8-oxo-, bis(1,1-dimethylethyl) ester, (±)-	EtOH	287(4.16)	77-1289-84
$C_{15}H_{23}N_3O_6Si$			
Pyrrolo[3,2-b]pyrrole-1,4-dicarboxylic acid, 3a,6a-dihydro-3-[(methoxycarbonyl)amino]-6-(trimethylsilyl)-, dimethyl ester, cis	C_6H_{12}	233(4.35),237(4.35),265s(3.60)	88-5669-84
$C_{15}H_{23}N_5O_3$			
9H-Purine-9-propanoic acid, 6-amino-β-hexyl-α-hydroxy-, methyl ester, (R*,S*)-(±)-	pH 2	261(4.12)	73-2148-84
$C_{15}H_{23}N_5O_8$			
L-Lysine, N^6-[2-(β-D-arabinofuranosyl)-2,5-dihydro-5-oxo-1,2,4-triazin-3-yl]-N^2-formyl-	H_2O	212(4.39),245s(3.88)	73-2689-84
$C_{15}H_{23}N_7O_2$			
9H-Purin-6-amine, N,9-bis(4-morpholin-ylmethyl)-	MeOH	269(4.20)	78-3997-84
	MeOH-HCl	270(4.11)	78-3997-84
	MeOH-NaOH	274(4.18)	78-3997-84
$C_{15}H_{23}O_2P$			
Benzoic acid, 3-[bis(1-methylethyl)-phosphino]-, ethyl ester	C_6H_{12}	230(4.00),263s(3.20),281s(3.00)	65-0489-84
Benzoic acid, 4-[bis(1-methylethyl)-phosphino]-, ethyl ester	C_6H_{12}	235(4.29),288(3.62)	65-0489-84
$C_{15}H_{24}$			
Cyclohexene, 4-(1,5-dimethyl-1,3-hexa-	MeOH	238(4.28)	88-5401-84

Compound	Solvent	$\lambda_{max}(\log \epsilon)$	Ref.
dienyl)-1-methyl- (theonelline)(cont.)			88-5401-84
1,3,6,10-Dodecatetraene, 3,7,11-tri- methyl-, (Z,E)-	EtOH	237(4.28)	154-0641-84
Isobicyclogermacrene, (+)-	EtOH	212(3.93)	39-0203-84C
Lepidozene, (-)-	EtOH	214(3.89)	39-0203-84C
$C_{15}H_{24}NO_4$			
Benzeneethanaminium, 2-formyl-N-(2-hy- droxyethyl)-4,5-dimethoxy-N,N-dimeth- yl-, iodide	MeOH	232(4.53),281(4.00), 308(3.83)	12-1659-84
$C_{15}H_{24}NO_4S$			
Sulfonium, (3-amino-3-carboxypropyl)- [3-(2-methoxyphenoxy)propyl]- methyl-, iodide	MeOH	219.5(4.30),272.5(3.43), 298s(2.28)	36-1241-84
$C_{15}H_{24}N_2O_2$			
Pyrrolo[1,2-c]quinazoline-6,8(1H,6H)- dione, octahydro-7,7,9,9-tetramethyl-	MeOH	225(2.47),300(1.72)	88-0371-84
$C_{15}H_{24}N_2O_4S$			
2(1H)-Pyrimidinethione, 3,4-dihydro- 4,4,6-trimethyl-1-[2,3-O-(1-methyl- ethylidene)-β-D-ribofuranosyl]-	MeOH	264(4.19)	103-0201-84
$C_{15}H_{24}N_4O$			
Benzamide, 4-(3-heptyl-3-methyl- 1-triazenyl)-	EtOH	322(4.34)	87-0870-84
1-Deazaadenine, 9-(2-hydroxy-3-nonyl)-, erythro-	pH 7.6	262(4.04)	87-0274-84
3-Deazaadenine, 9-(2-hydroxy-3-nonyl)-, erythro-	pH 7.6	262(4.02)	87-0274-84
$C_{15}H_{24}N_4O_5$			
2,6-Piperidinedicarboxylic acid, 4-[[[(dimethylamino)carbonyl]- methylhydrazono]ethylidene]-, dimethyl ester, [2α,4(E),6α]-	EtOH	278(4.52)	33-1547-84
$C_{15}H_{24}O$			
Ethanone, 1-[2-methyl-3-(1,1,5-trimeth- yl-5-hexenyl)-2-cyclopropen-1-yl]-	MeCN	279(2.16)	33-0129-84
2-Heptanone, 6-methyl-6-[2-methyl-3-(1- methylethenyl)-1-cyclopropen-1-yl]-	MeCN	272(1.78),278(1.78)	33-0129-84
2-Hepten-1-ol, 2-methyl-6-(1-methyl- 1-cyclohexen-4-ylidene)-	MeOH	250(3.69)	31-0931-84
Isobicyclogermacrenol, (+)-	EtOH	215(3.90)	39-0203-84C
Lepidozenol, (-)-	EtOH	211(3.83)	39-0203-84C
1,3-Pentadiene, 4-(1,2-epoxy-2,6,6-tri- methyl-1-cyclohexyl)-2-methyl-, (E)-	pentane	237(4.04)	33-0129-84
(Z)-	pentane	234(3.92)	33-0129-84
isomer B	pentane	241(4.01)	33-0129-84
$C_{15}H_{24}OS_2$			
Bicyclo[2.2.1]heptan-2-one, 3-[bis(eth- ylthio)methylene]-1,7,7-trimethyl-	EtOH	325(4.08)	78-2951-84
$C_{15}H_{24}OSi$			
Spiro[4.5]deca-2,6-dien-1-one, 3,6-di- methyl-2-(trimethylsilyl)-	EtOH	232(4.05)	33-0073-84

Compound	Solvent	$\lambda_{max}(\log \epsilon)$	Ref.
$C_{15}H_{24}O_2$			
1(4H)-Naphthalenone, 4a,5,6,7,8,8a-hexahydro-8a-hydroxy-4a,8-dimethyl-2-(1-methylethyl)-	EtOH	245(3.78)	39-0937-84C
2,4-Pentadien-1-ol, 3-methyl-5-(1,2-epoxy-2,6,6-trimethyl-1-cyclohexyl)-	MeOH	238(4.45)	33-0184-84
$C_{15}H_{24}O_3$			
1,3-Cyclohexadiene-1-carboxylic acid, 2-methyl-4-(1-methylethoxy)-, 1,1-dimethylethyl ester	EtOH	320(3.29)	12-2037-84
1,4-Cyclohexanediol, 1-(5-hydroxy-3-methyl-3-penten-1-ynyl)-2,2,6-tri-methyl-, cis	EtOH	228(4.11)	39-2147-84C
trans	EtOH	230(4.13)	39-2147-84C
1-Cyclohexene-1-carboxylic acid, 4-(1,5-dimethyl-4-hexenyl)-3-hydroxy-	MeOH	220(3.71)	102-0186-84
7,10-Epoxydodeca-3,11-dien-5-one, 2-hydroxy-2,6,10-trimethyl-, (6S,7S-10R)- (hydroxydavanone)	EtOH	230(3.0)	102-2545-84
Geranial, 10-tetrahydropyranyloxy-	EtOH	238(4.23)	95-1232-84
1-Hexen-3-one, 1-(1,4-dioxaspiro[4.5]-dec-8-yl)-5-methyl-	EtOH	228(4.00)	107-0925-84
β-Ionone, 10,10-dimethoxy-	EtOH	236(3.83),307(3.91)	94-1709-84
2(1H)-Naphthalenone, 4a,5,6,7,8,8a-hexahydro-7-hydroxy-6-(1-hydroxy-1-methylethyl)-4,8a-dimethyl-, (4aR,6S,7S,8aS)-	EtOH	243(4.19)	12-0629-84
2(3H)-Naphthalenone, 4,4a,5,6,7,8-hexahydro-8-hydroxy-7-(2-hydroxy-1-methylethyl)-1,4a-dimethyl-	MeOH	252(4.10)	23-2813-84
2,4-Pentadienal, 5-(1,4-dihydroxy-2,2,6-trimethylcyclohexyl)-3-methyl-, cis	EtOH	283(4.11)	39-2147-84C
Punctatin A	EtOH	228(3.79)	77-0405-84
Punctatin B	EtOH	212(3.80)	77-0917-84
$C_{15}H_{24}O_4$			
Bicyclo[7.2.0]undec-4-ene-2,3,6,7-tetrol, 4,11,11-trimethyl-8-meth-ylene-	MeOH	270s(3.03),295s(3.14),332(3.24)	5-1332-84
Cyclohexanecarboxylic acid, 4-ethoxy-1-(1-hydroxy-2-propyl)-2-methylene-, ethyl ester	EtOH	207s(3.20),251(4.11)	12-2037-84
7,10-Epoxydodeca-3,11-dien-5-one, 2-hydroperoxy-2,6,10-trimethyl-, (6S,7S,10R)-	EtOH	232(3.2)	102-2545-84
$C_{15}H_{24}O_5$			
3-Buten-2-one, 4-(4-acetoxy-1,2-di-hydroxy-2,6,6-trimethylcyclo-hexyl)-, (1R,2R,4S)-	EtOH	231(4.02)	33-2043-84
β-Ionone, 3-acetoxy-5,6-dihydro-5,6-dihydroxy-, (3S,5S,6S)-	EtOH	230(4.00)	33-2043-84
$C_{15}H_{24}O_6S_2$			
Spiro[bicyclo[2.2.1]heptane-2,2'-oxir-an]-3-one, 3',3'-bis(ethylsulfonyl)-4,7,7-trimethyl-, [1S-(1α,2β,4α)]-	EtOH	312(1.68)	78-2951-84

Compound	Solvent	$\lambda_{max}(\log \epsilon)$	Ref.
$C_{15}H_{24}O_7$			
2,7-Octadienoic acid, 6-(α-L-arabino-pyranosyloxy)-2,6-dimethyl-, [S-(E)]-	EtOH	216(4.13)	94-2617-84
$C_{15}H_{25}NO$			
3-Buten-2-one, 4-[2-methyl-2-(4-methyl-3-pentenyl)cyclopropyl]-, O-methyl-oxime	EtOH	264(4.39)	19-0207-84
4-Piperidinone, 1-butyl-3,5-bis(1-methylethylidene)-	heptane	268(4.12)	103-0761-84
4-Piperidinone, 1-(1,1-dimethylethyl)-3,5-bis(1-methylethylidene)-	heptane	266(4.10)	103-0761-84
$C_{15}H_{25}NO_3$			
2,4-Cyclohexadiene-1-carboxamide, N,N-diethyl-6,6-dimethoxy-1,2-dimethyl-	MeOH	265(3.76)	44-4429-84
$C_{15}H_{25}NO_6$			
1H-Pyrrole, 3-acetyl-2-methyl-5-(D-galacto-pentitol-1-yl)-1-propyl-	EtOH	252(4.01),279(3.91)	136-0153-84I
1H-Pyrrole, 3-acetyl-2-methyl-5-(D-gluco-pentitol-1-yl)-1-propyl-	EtOH	252(3.95),287(3.76)	136-0153-84I
1H-Pyrrole, 3-acetyl-2-methyl-5-(D-manno-pentitol-1-yl)-1-propyl-	EtOH	252(3.95),287(3.76)	136-0153-84I
$C_{15}H_{25}NO_7$			
1H-Pyrrole-3-carboxylic acid, 1-ethyl-2-methyl-5-(D-galacto-pentitol-1-yl)-, ethyl ester	EtOH	228(3.95),256(3.80)	136-0153-84I
1H-Pyrrole-3-carboxylic acid, 1-ethyl-2-methyl-5-(D-gluco-pentitol-1-yl)-, ethyl ester	EtOH	228(3.91),265(3.77)	136-0153-84I
1H-Pyrrole-3-carboxylic acid, 1-ethyl-2-methyl-5-(D-manno-pentitol-1-yl)-, ethyl ester	EtOH	228(3.93),256(3.78)	136-0153-84I
$C_{15}H_{26}N$			
Ethanaminium, N,N,N-trimethyl-2-(3,4,4-5-tetramethyl-2,5-cyclohexadien-1-ylidene)-, iodide	EtOH	223s(--),264(4.41)	5-0340-84
$C_{15}H_{26}N_2$			
3H-Pyrazole, 4,5-dihydro-3,3,5,5-tetramethyl-4-(2,2,4,4-tetramethylcyclobutylidene)-	C_6H_{12}	196.0(4.20),325.0(2.26)	24-0277-84 24-0310-84
$C_{15}H_{26}O_2$			
3-Buten-2-one, 4-[2-(4-methoxy-4-methylpentyl)-2-methylcyclopropyl]-, [1α(E),2β]-($\pm$)-	EtOH	255(4.05)	19-0207-84
1-Cyclohexene-1-methanol, 4-(1,5-dimethyl-4-hexenyl)-3-hydroxy-	EtOH	207(4.07)	102-0186-84
$C_{15}H_{26}O_3$			
1,4-Cyclohexanediol, 1-(5-hydroxy-3-methyl-1,3-pentadienyl)-2,2,6-trimethyl-, cis	EtOH	238(4.08)	39-2147-84C
$C_{15}H_{26}O_9$			
β-D-Glucopyranoside, 2-(2-hydroxyethyl)-3,4-bis(hydroxymethyl)-3-cyclopenten-	EtOH	208(3.6)	102-1431-84

Compound	Solvent	$\lambda_{max}(\log \epsilon)$	Ref.
1-yl, (1R-trans)- (cont.)			102-1431-84
$C_{15}H_{28}NOS$			
2-Propen-1-aminium, N-(3-methoxy-2-cyclohexen-1-ylidene)-N-methyl-2-[(trimethylsilyl)methyl]-, perchlorate	MeCN	288(4.46)	44-0220-84
$C_{15}H_{30}N_2O_3Si_2$			
3-Butenoic acid, 2-diazo-3-[[(1,1-dimethylethyl)dimethylsilyl]oxy]-, 2-(trimethylsilyl)ethyl ester	CH_2Cl_2	280(3.88)	23-2936-84
$C_{15}H_{32}Si_4$			
Trisilane, 1,1,1,3,3,3-hexamethyl-2-phenyl-2-(trimethylsilyl)-	C_6H_{12} C_6H_{12}	241(4.12) 241(4.11)	60-0341-84 60-0383-84
$C_{15}H_{35}GeNO_2Si$			
Butanamide, N,N-dimethyl-3-(triethylgermyl)-3-[(trimethylsilyl)oxy]-	heptane or EtOH	227(1.88)	70-0310-84
$C_{15}H_{40}Si_5$			
Cyclopentasilane, 1,2,3,4,5-pentaethyl-1,2,3,4,5-pentamethyl-	isooctane	212s(4.36),228s(3.91), 268f(2.90)	101-0353-84L

Compound	Solvent	$\lambda_{max}(\log \epsilon)$	Ref.
$C_{16}H_2N_8$ 1-Cyclobutene-1,2-diacetonitrile, α,α'-dicyano-3,4-bis(dicyanometh- ylene)-, disodium salt	H_2O	227(4.32),308(4.43), 460s(4.63),482(4.65)	24-2714-84
bis(dimethylamine) salt	H_2O	228(4.28),308(4.46), 440s(4.46),484(4.69)	24-2714-84
$C_{16}H_8$ Dicyclopenta[a,e]dicyclopropa[c,g]cy- clooctene	CH_2Cl_2	241(4.63),328(4.51), 390(4.27),470(2.81)	89-0063-84
$C_{16}H_8Cr_2O_6$ Chromium, [μ-[(1,2,3,4,5-η;1',2',3',4'- 5')[bi-2,4-cyclopentadien-1-yl]-1,1'- diyl]]hexacarbonyldi-(Cr-Cr)-	ether	250s(4.38),370(3.73), 448(4.30)	157-0082-84
$C_{16}H_8F_6N_2OS$ 1-Cyclohexene-1-carbonitrile, 3,3,4,4- 5,5-hexafluoro-2-[(3-methyl-2(3H)- thiazolylidene)methyl]-6-oxo-	EtOH benzene	459(4.27) 499(--)	104-0390-84 104-0390-84
$C_{16}H_8N_2O_4$ Benzonitrile, 4-[(5-nitro-3-oxo-1(3H)- isobenzofuranylidene)methyl]-, (E)-	EtOH	357(4.12)	131-0239-84H
$C_{16}H_8O_3$ 9,10-Anthracenedicarboxylic acid anhydride	n.s.g.	263.5(2.73),271.5(2.68)	62-0167-84C
$C_{16}H_9BrN_2O_2$ 6H-Anthra[1,9-cd]isoxazol-6-one, 5-(1- aziridinyl)-3-bromo-	THF	256(4.48),260(4.38), 460(4.19),488(4.20)	103-0717-84
$C_{16}H_9BrN_2O_3S_2$ 2-Propenoic acid, 3-[5-(2-benzothiazo- lylthio)-4-bromo-2-furanyl]-2-cyano-, methyl ester	MeOH	224(3.46),273(3.18), 350(3.31)	73-0984-84
$C_{16}H_9ClF_2O_2S$ Ethanone, 2-chloro-1-(2,8-difluoro-11- hydroxydibenzo[b,f]thiepin-10-yl)-	MeOH	233s(4.32),267s(3.80), 300s(3.85),340(4.00)	73-0603-84
$C_{16}H_9ClF_6O_6S_2$ Methanesulfonic acid, trifluoro-, 1-(4- chlorophenyl)-2-phenyl-1,2-ethenediyl ester, (E)-	C_6H_{12}	266(4.32)	44-2273-84
(Z)-	C_6H_{12}	227.5(4.32),274(4.12)	44-2273-84
$C_{16}H_9ClN_2O_2$ 6H-Anthra[1,9-cd]isoxazol-6-one, 5-(1- aziridinyl)-3-chloro-	THF	253(4.39),260(4.42), 460(4.19),485(4.20)	103-0717-84
$C_{16}H_9ClN_4O_2$ 4H-3,1-Benzoxazine-2-acetonitrile, α-[(2-chlorophenyl)azo]-4-oxo-	EtOH	212(3.83),240(3.20), 370(4.12)	80-0345-84
	ether	214(3.72),280(3.64), 350(3.60),360(4.05)	80-0345-84
$C_{16}H_9D_{10}NS_2$ Benzene-d_5-sulfenamide, N-(1,1-dimeth- ylethyl)-N-(phenyl-d_5-thio)-	hexane	248(4.26)	44-2724-84

Compound	Solvent	$\lambda_{max}(\log \epsilon)$	Ref.
$C_{16}H_9F$ Pyrene, 1-fluoro-	C_6H_{12}	232(4.33),242(4.78), 264(4.30),272(4.53), 322(4.20),338(4.30)	44-2803-84
$C_{16}H_9NO$ 7H-Dibenzo[de,h]quinolin-7-one	benzene	<u>315(4.3)</u>,390(4.5), <u>400s(4.4)</u>	39-2177-84C

$C_{16}H_9NO_3$
  4H-[1]Benzopyrano[3,4-d]isoxazol-4-one,
  3-phenyl-
	EtOH	264(4.20),306(3.95)	44-4419-84

$C_{16}H_9NO_4$
  6H-Anthra[1,9-cd]isoxazole-3-carboxylic
  acid, 6-oxo-, methyl ester
	EtOH	460(4.05)	104-2012-84

$C_{16}H_9N_3$ Pyrene, 2-azido-	benzene	340(4.41)	35-5234-84

$C_{16}H_9N_3O$
  [1,1'-Biphenyl]-2,4,5-tricarbonitrile,
  4'-methoxy-
	MeCN	246(4.26),305(3.81),
		345(3.81)
			40-0060-84

$C_{16}H_9N_5O_4$
  4H-3,1-Benzoxazine-2-acetonitrile,
  α-[(3-nitrophenyl)azo]-4-oxo-
	EtOH
	ether
		246(3.95),367(4.06)
		242(4.26),350(4.05)
			80-0345-84
			80-0345-84

$C_{16}H_{10}$ Aceanthrylene	heptane	235(4.53),250(4.60), 343(3.34),360(3.58), 378(3.37),399(3.29), 422(3.16),455(2.77), 560(1.00)	44-2069-84
Anthracene, 9-ethynyl-	$CHCl_3$	255(5.22),263(5.38), 313(3.04),330(3.35), 348(3.65),364(3.94), 382(4.08),403(4.04)	80-0447-84
Fluoranthene	hexane	287(4.70),308(3.56), 323(3.80),341(3.90), 358(3.94)	35-8024-84
Pyrene	hexane	239(4.94),251(4.06), 261(4.41),272(4.73), 293(3.63),304(4.07), 318(4.49),334(4.75), 351(2.71)	35-8024-84

$C_{16}H_{10}BrCl_3$
  Cyclobuta[1]phenanthrene, 2a-bromo-
  1,1,2-trichloro-1,2,2a,10b-tetra-
  hydro-, syn
	CH_2Cl_2	335(4.02),350(4.08),
		357s(4.01),368(3.91),
		377(4.11),412s(3.38)
			78-5249-84

Cyclobuta[1]phenanthrene, 10b-bromo-
  1,1,2-trichloro-1,2,2a,10b-tetra-
  hydro-, anti
	CH_2Cl_2	236(4.11),244(3.91),
		271s(4.18),281(4.32),
		289s(4.18),308s(3.18)
			78-5249-84

$C_{16}H_{10}BrNS_2$
  6H-1,3-Thiazine-6-thione, 2-(4-bromo-
  phenyl)-5-phenyl-
	$CHCl_3$	275(4.51),324(4.62),
		455(4.10)
			97-0435-84

$C_{16}H_{10}BrN_3O_3S$ 2-Propenoic acid, 3-[5-(1H-benzimida-	MeOH	205(3.53),249(3.10),	73-0984-84

Compound	Solvent	$\lambda_{max}(\log \epsilon)$	Ref.
zol-2-ylthio)-4-bromo-2-furanyl]- 2-cyano-, methyl ester (cont.)		282(3.17),361(3.23)	73-0984-84
$C_{16}H_{10}Br_2N_2$ Pyrazine, 2,3-bis(4-bromophenyl)-	EtOH	224.5(4.24),278s(4.06), 286.5(4.07)	4-0103-84
$C_{16}H_{10}Br_2N_2O$ Pyrazine, 2,3-bis(4-bromophenyl)-, 1-oxide	EtOH	211s(4.35),226(4.40), 266.5(4.47),321.5(3.65)	4-0103-84
$C_{16}H_{10}Br_2N_2O_2$ Pyrazine, 2,3-bis(4-bromophenyl)-, 1,4-dioxide	EtOH	221.5(4.53),269(4.26), 284(4.28),318(4.39)	4-0103-84
$C_{16}H_{10}ClNS_2$ 6H-1,3-Thiazine-6-thione, 2-(4-chloro- phenyl)-5-phenyl-	$CHCl_3$	275(4.51),324(4.61), 455(4.11)	97-0435-84
$C_{16}H_{10}ClN_3$ Benzo[f]pyrido[2,3-h]quinoxaline, 7-chloro-5-methyl-	EtOH	260(4.68),285(4.12), 307(4.11),320s(3.90), 337(3.78),352(3.75)	12-1271-84
$C_{16}H_{10}Cl_2$ Cyclobuta[1]phenanthrene, 1,2-dichloro- 1,2-dihydro-, cis	CH_2Cl_2	230(4.41),249(4.71), 256(4.83),264s(4.41), 280(3.97),291(3.93), 303(4.03)	78-5249-84
Cyclobuta[1]phenanthrene, 1,2-dichloro- 2a,10b-dihydro-, cis	EtOH	216(4.68),234(4.26), 244(3.99),274(4.13), 281(4.19),292(4.06), 311(3.53)	78-5249-84
$C_{16}H_{10}Cl_2N_2$ Pyrazine, 2,3-bis(2-chlorophenyl)-	EtOH	211.5(4.27),245(3.87), 270(3.88)	4-0103-84
Pyrazine, 2,3-bis(4-chlorophenyl)-	EtOH	224(4.35),278s(4.12), 286(4.13)	4-0103-84
$C_{16}H_{10}Cl_2N_2O$ Pyrazine, 2,3-bis(2-chlorophenyl)-, 1-oxide	EtOH	220(4.24),268(4.18)	4-0103-84
Pyrazine, 2,3-bis(4-chlorophenyl)-, 1-oxide	EtOH	223.5(4.34),265.5(4.41), 322(3.58)	4-0103-84
$C_{16}H_{10}Cl_2N_2O_2$ 6H-Anthra[1,9-cd]isoxazol-6-one, 3- chloro-5-[(2-chloroethyl)amino]-	THF	255(4.43),261(4.53), 498(4.19),533(4.23)	103-0717-84
Pyrazine, 2,3-bis(2-chlorophenyl)-, 1,4-dioxide	EtOH	218(4.57),257(4.07), 318(4.48)	4-0103-84
Pyrazine, 2,3-bis(4-chlorophenyl)-, 1,4-dioxide	EtOH	221(4.28),267(3.97), 282(3.98),318(4.13)	4-0103-84
$C_{16}H_{10}Cl_2N_2O_4$ 9,10-Ethanoanthracene, 11,12-dichloro- 9,10-dihydro-2,6-dinitro-	MeCN	272.5(4.26)	44-2368-84
9,10-Ethanoanthracene, 11,12-dichloro- 9,10-dihydro-2,7-dinitro-, trans	MeCN	275(4.25)	44-2368-84

Compound	Solvent	$\lambda_{max}(\log \epsilon)$	Ref.
$C_{16}H_{10}F_3N_3$ 1,3,5-Triazine, 2,4-diphenyl-6-(trifluoromethyl)-	CHCl$_3$	284(4.54)	70-0842-84
$C_{16}H_{10}F_3N_3O_2S$ Furo[3,4-b]quinolin-9-ol, 1,3-dihydro-3-methyl-1-(2-thiazolylimino)-5-(trifluoromethyl)-	EtOH-HCl	319(4.32),352(4.51), 363(4.39)	4-1345-84
	EtOH-NaOH	270(3.90),304(4.41), 317(4.44),365(4.38), 381(4.30)	4-1345-84
$C_{16}H_{10}F_6O_6S_2$ Methanesulfonic acid, trifluoro-, 1,2-diphenyl-1,2-ethenediyl ester, (E)-	EtOH	259(4.27)	44-2273-84
(Z)-	EtOH	220(4.30),267(4.11)	44-2273-84
$C_{16}H_{10}N_2O$ 3H-Pyrazolo[4,5,1-ij]quinolin-3-one, 5-phenyl-	EtOH	236(4.42),244(4.39), 304(3.61),368s(3.77), 385(4.11),400s(3.66)	142-2467-84
5H-Pyrazolo[4,5,1-ij]quinolin-5-one, 3-phenyl-	EtOH	251(4.29),330(4.11), 452s(3.37),478(3.76), 514s(3.29)	142-2467-84
$C_{16}H_{10}N_2O_2$ 6H-Anthra[1,9-cd]isoxazol-6-one, 5-(1-aziridinyl)-	THF	249(4.38),257(4.41), 454(4.16),480(4.18)	103-0717-84
$C_{16}H_{10}N_2O_2S$ 2H-Pyran-3-carbonitrile, 4-(methylthio)-2-oxo-6-(2-quinolinyl)-	EtOH	258(4.16),304(3.68), 355(4.07),374(4.06)	94-3384-84
$C_{16}H_{10}N_2O_2S_2$ 6H-1,3-Thiazine-6-thione, 2-(4-nitrophenyl)-5-phenyl-	CHCl$_3$	295(4.54),338(4.45), 455(3.91)	97-0435-84
$C_{16}H_{10}N_2O_3$ 5,10[3',4']-Furanopyrido[3,2-g]quinoline-12,14-dione, 5,10,11,15-tetrahydro-	MeCN	269(3.953),274(3.945)	5-0877-84
$C_{16}H_{10}N_2O_3S$ Acetonitrile, (3-oxobenzo[b]thien-2(3H)-ylidene)(phenylamino)-, S,S-dioxide	EtOH	217(4.35),280(3.68), 371(4.23)	95-0134-84
$C_{16}H_{10}N_2O_4$ Indolo[4,5-e]indole-2,9-dicarboxylic acid, 3,8-dihydro-	EtOH	236(4.56),278s(4.36), 289s(4.60),298(4.75), 315s(4.27),333s(4.04), 350(4.05)	103-0283-84
Indolo[5,4-e]indole-2,7-dicarboxylic acid, 1,6-dihydro-	EtOH	238(3.34),253(3.14), 269(3.00),321(2.71), 333(2.88),351(2.87), 368(2.91)	103-0283-84
$C_{16}H_{10}N_2S$ 5-Thiazolecarbonitrile, 2,4-diphenyl-	EtOH	266(4.49),318(4.08)	18-0605-84
$C_{16}H_{10}N_4O_2$ Acridine, 9-(5-nitro-1H-imidazol-1-yl)-	EtOH	253(4.89),380(4.17),	42-0611-84

Compound	Solvent	$\lambda_{max}(\log \epsilon)$	Ref.
(cont.)		398(3.98)	42-0611-84
4H-3,1-Benzoxazine-2-acetonitrile,	EtOH	210(3.96),280(3.38),	80-0345-84
4-oxo-α-(phenylazo)-		245s(--),370(3.38)	
	ether	240(4.24),365(4.03)	80-0345-84
3H-1,2,4-Triazino[3,2-b]quinazoline-	EtOH	290(3.36),350(2.60)	24-1077-84
3,10(4H)-dione, 2-phenyl-			
$C_{16}H_{10}N_4O_8$			
9H-Cyclohepta[c]pyridin-9-one, picrate	EtOH	214(4.36),248(4.02),	39-2297-84C
		315(3.83),345s(--)	
$C_{16}H_{10}N_6$			
[1,2,4,5]Tetrazino[1,6-a:4,3-a']diquin-	CHCl₃	270(4.32),360(4.50),	33-1503-84
oxaline		460(2.62)	
$C_{16}H_{10}N_6O$			
8H-1,2,4-Triazolo[4"',3"':1",6"]-	dioxan	344(3.03)	73-1529-84
[1,2,4]triazino[4",5":1',3']pyrrolo-			
[2',3':4,5]furo[3,2-b]indole,			
3-methyl-			
8H-1,2,4-Triazolo[4"':3"':1",6"]-	dioxan	342(3.14)	73-1529-84
[1,2,4]triazino[4",5":1',3']pyrrolo-			
[2',3':4,5]furo[3,2-b]indole,			
6-methyl-			
$C_{16}H_{10}O$			
Cyclobuta[1]phenanthren-1(2H)-one	EtOH	209(3.96),243s(4.32),	78-5249-84
		248(4.41),263(4.24),	
		283(3.74),315(3.71),	
		343(3.31),360(3.23)	
$C_{16}H_{10}O_3$			
2,5-Furandione, 3,4-diphenyl-	CHCl₃	288(4.13)	77-1234-84
1H-Indene-1,3(2H)-dione, 2-benzoyl-	pH 1	300(4.04),337(4.04)	65-1430-84
	pH 13	290(4.18),335(4.11)	65-1430-84
	CHCl₃	302(4.33),338(4.42)	65-1430-84
$C_{16}H_{10}O_4$			
2-Propyn-1-one, 1-(6-hydroxy-1,3-benzo-	EtOH	218(4.25),267(4.11),	78-4081-84
dioxol-5-yl)-3-phenyl-		299(4.14),309(4.15),	
		391(4.09)	
$C_{16}H_{10}O_5$			
9,10-Anthracenedione, 1-acetoxy-8-hy-	n.s.g.	222(4.35),247(4.36),	5-0306-84
droxy-		272(4.04),340(3.52),	
		307(3.71),406(3.76),	
		430(3.64)	
$C_{16}H_{11}BrN_2O_2$			
6H-Anthra[1,9-cd]isoxazol-6-one,	THF	253(4.43),260(4.51),	103-0717-84
5-[(2-bromoethyl)amino]-		488(4.20),520(4.25)	
$C_{16}H_{11}BrN_4S$			
2-Benzothiazoleacetonitrile, α-[(4-	EtOH	263(4.18),405(4.47)	104-0523-84
bromophenyl)hydrazono]-6-methyl-			
$C_{16}H_{11}ClN_2O_2$			
6H-Anthra[1,9-cd]isoxazol-6-one,	THF	252(4.44),258(4.52),	103-0717-84
5-[(2-chloroethyl)amino]-		486(4.22),520(4.26)	
1H-1,5-Benzodiazepine-3-carboxaldehyde,	EtOH	217(4.55),243(4.29),	103-1035-84
8-chloro-2,3-dihydro-2-oxo-4-phenyl-		258(4.24),325(3.95)	

Compound	Solvent	$\lambda_{max}(\log \epsilon)$	Ref.
$C_{16}H_{11}ClOS$ 7,5-Metheno-5H-dicyclohepta[b,e]thio- pyran, 5a-chloro-5a,6a-dihydro-, 6-oxide	dioxan	237(3.72),303(5.05), 356(3.73),437s(2.04), 503(2.52)	89-0717-84
$C_{16}H_{11}ClO_3$ 9,10-Anthracenedione, 2-(chloromethyl)- 1-hydroxy-3-methyl-	MeOH	205(4.36),227(4.34), 230s(--),246(4.49), 256(4.46),260(4.46), 283s(--),330(3.53), 395s(--),410(3.86), 425s(--)	78-3677-84
$C_{16}H_{11}ClO_4$ 9,10-Anthracenedione, 2-(chloromethyl)- 1-hydroxy-8-methoxy-	n.s.g.	226(4.55),255(4.35), 277(4.02),357(3.96), 415(4.01),430(3.54)	5-0306-84
4H-1-Benzopyran-4-one, 2-(2-chlorophen- yl)-7-hydroxy-3-methoxy-	MeOH	203(4.44),240(4.23), 300(4.21)	2-1002-84
4H-1-Benzopyran-4-one, 2-(4-chlorophen- yl)-3-hydroxy-6-methoxy-	CHCl$_3$	257(4.47),333(4.56), 351s(4.47)	142-1943-84
$C_{16}H_{11}ClO_5$ 4H-1-Benzopyran-4-one, 2-(2-chlorophen- yl)-5,7-dihydroxy-3-methoxy-	MeOH	203(4.73),258(4.48), 300(3.99),330(3.89)	2-1002-84
$C_{16}H_{11}ClO_6$ 4H-1-Benzopyran-4-one, 2-(2-chlorophen- yl)-5,7,8-trihydroxy-3-methoxy-	MeOH	208(4.10),268(4.00), 360(3.18)	2-1002-84
$C_{16}H_{11}Cl_2NO_2$ 9,10-Ethanoanthracene, 11,12-dichloro- 9,10-dihydro-2-nitro-, (9α,10α- 11S*,12S*)-	MeCN	275(3.88)	44-2368-84
$C_{16}H_{11}Cl_3$ Cyclobuta[1]phenanthrene, 1,1,2-tri- chloro-1,2,2a,10b-tetrahydro-, anti	EtOH	233(3.85),252(3.95), 257s(3.92),268(3.89), 277(4.05),312s(3.36)	78-5249-84
syn	EtOH	235(3.98),243(3.76), 262s(3.68),270s(4.05), 280(4.22),289s(4.05), 306s(3.07)	78-5249-84
$C_{16}H_{11}F_3N_2O$ Ethanone, 2,2,2-trifluoro-1-(1-methyl- 2-phenyl-2H-cyclopenta[d]pyridazin- 5-yl)-	hexane	203(4.09),245(3.85), 257(3.85),265s(3.84), 300(4.11),365(3.73)	44-4769-84
	MeOH	208(4.58),234s(4.24), 247s(4.15),263(4.20), 301(4.23),376(4.05)	44-4769-84
Ethanone, 2,2,2-trifluoro-1-(1-methyl- 2-phenyl-2H-cyclopenta[d]pyridazin- 7-yl)-	hexane	204(4.43),251.5(4.46), 286s(4.02),297(4.03), 331(4.00),398(4.09)	44-4769-84
	MeOH	209(4.46),251(4.48), 297s(4.00),398(4.23)	44-4769-84
$C_{16}H_{11}F_3N_2O_3S$ Thiazolo[3,2-a]pyrimidin-4-ium, 3-hy- droxy-7-(4-methoxyphenyl)-5-methyl- 2-(trifluoroacetyl)-	MeCN CCl$_4$	358(4.53),455(4.14) 484(--)	103-0791-84 103-0791-84

Compound	Solvent	$\lambda_{max}(\log \epsilon)$	Ref.
$C_{16}H_{11}F_6NO$			
3-Penten-2-one, 1,1,1,5,5,5-hexafluoro-4-[(1-methyl-2(1H)-quinolinylidene)-methyl]-	benzene	543(4.30)	104-0390-84
	EtOH	523(4.14)	104-0390-84
$C_{16}H_{11}N$			
3-Fluoranthenamine	C_6H_{12}	225(4.74),245(4.72), 292.5(4.29),327.5(3.97), 368(3.87),398.5(3.93)	39-0943-84B
(also a band at 305(4.48) listed separately)	MeOH	225(4.54),245(4.50), 325(3.84),368(3.50), 414(3.81)	39-0943-84B
	EtOH	225(4.60),245(4.59), 325(3.91),368(3.55), 410(3.88)	39-0943-84B
	ether	227(4.53),246(4.52), 293(4.09),327(3.85), 368(3.40),420(3.78)	39-0943-84B
	MeCN	225(4.56),245.5(4.52), 293(4.09),326(3.79), 368(3.44),414(3.78)	39-0943-84B
	H_2O	220(--),244(--), 300(--),325(--), 366(--),406(--)	39-0943-84B
$C_{16}H_{11}NO$			
Dibenz[cd,g]indol-6(1H)-one, 1-methyl-	EtOH	552(4.27)	104-1415-84
$C_{16}H_{11}NO_2$			
Benzoic acid, 2-(3-isoquinolinyl)-	MeOH	219(4.64),242(4.48), 282(3.98),325(3.59)	83-0002-84
hydrochloride	2M HCl	230(4.69),276(3.78), 340(3.76)	83-0002-84
sulfate (1:1)	2N H_2SO_4	230(4.73),275(3.78), 339(3.83)	83-0002-84
Cyclopenta[4,5]azepino[2,1,7-cd]pyrro-lizine-1-carboxylic acid, methyl ester	EtOH	245(4.34),264(4.28), 282(4.47),302(4.45), 340(4.58),354(4.74), 388(3.92),425(3.83), 453(3.68),580(2.81), 625(2.82),688(2.48)	24-1649-84
Isoindolo[2,1-b]isoquinolin-5(7H)-one, 7-hydroxy-	MeOH	230(4.5),277(3.68), 297s(3.65),345(3.89), 375s(3.66)	83-0002-84
	H_2SO_4	270(4.67),272(4.67), 280s(4.58),454(4.08)	83-0002-84
Isoindolo[2,1-b]isoquinolin-7(5H)-one, 5-hydroxy-	MeOH	208(4.48),229(4.42), 302(4.12),315(4.23), 338(3.96),355(3.76)	83-0381-84
1(4H)-Naphthalenone, 4-(phenylimino)-	$CHCl_3$	264(4.19),279(4.14), 445(3.83)	104-0733-84
5(4H)-Oxazolone, 2-phenyl-4-(phenyl-methylene)-	toluene	360(4.59)	135-1330-84
$C_{16}H_{11}NO_3$			
Acetamide, N-(9,10-dihydro-9,10-dioxo-1-anthracenyl)-	benzene	417(3.78)	39-0529-84C
1H-Indene-1,3(2H)-dione, 4-amino-2-benzoyl-	6M HCl	240(4.31),306(4.20), 335(4.30)	65-1430-84
	M HCl	230(4.30),320(4.34), 408(3.78)	65-1430-84

Compound	Solvent	$\lambda_{max}(\log \epsilon)$	Ref.
(cont.)	pH 13	242(4.29),318(4.28)	65-1430-84
	CHCl$_3$	322(4.31)	65-1430-84
1H-Pyrrole-2,5-dione, 3-[1,1'-biphen-yl]-4-yl-4-hydroxy-	EtOH	272(4.35),370(3.83)	44-2212-84
$C_{16}H_{11}NO_4$			
9,10-Anthracenedione, 1-acetoxy-4-amino-	EtOH at 77°K	490(3.83)	104-1780-84
1(3H)-Isobenzofuranone, 3-[(4-methyl-phenyl)methylene]-6-nitro-	EtOH	374(4.39)	131-0239-84H
5(4H)-Isoxazolone, 4-(2-hydroxybenz-oyl)-3-phenyl-	EtOH	252(4.01),346(3.89)	44-4419-84
$C_{16}H_{11}NO_5$			
1(3H)-Isobenzofuranone, 3-[(4-methoxy-phenyl)methylene]-6-nitro-	EtOH	395(4.35)	131-0239-84H
$C_{16}H_{11}NS$			
Thieno[2',3':6,7]cyclohept[1,2-b]ind-ole, 2-methyl-	C$_6$H$_{12}$	255(4.23),265(4.24), 305(4.44),325(4.49), 354(4.20),400(4.08), 410(4.17)	4-1585-84
	EtOH	500(3.19)	4-1585-84
$C_{16}H_{11}NS_2$			
6H-1,3-Thiazine-6-thione, 2,5-diphenyl-	CHCl$_3$	270(4.23),322(4.35), 455(3.84)	97-0435-84
$C_{16}H_{11}N_3$			
Acridine, 8-(1H-imidazol-1-yl)-	EtOH	253(4.80),379(2.92), 397(3.90)	42-0611-84
Benzo[f]pyrido[2,3-h]quinoxaline, 5-methyl-	EtOH	257(4.72),280(4.23), 294(4.04),306(4.01), 334(3.77),349(3.72), 460(2.30)	12-1271-84
$C_{16}H_{11}N_3O$			
Benzo[f]pyrido[2,3-h]quinoxalin-7(8H)-one, 5-methyl-	EtOH	234(4.48),255(4.49), 271s(4.34),280(4.42), 291s(4.24),315(4.02), 328(3.96),354(3.81), 362(3.81)	12-1271-84
$C_{16}H_{11}N_3O_2S$			
2H-Naphtho[1,8-cd]isothiazole, 5-(phen-ylazo)-, 1,1-dioxide	dioxan	240(4.28),385(4.34)	4-0337-84
$C_{16}H_{11}N_3S$			
10H-Phenothiazine, 10-pyrazinyl-	hexane	247(4.75),276(4.28), 334(4.01)	4-0661-84
	20% HCl	259(4.37),305(--), 373(3.67)	4-0661-84
radical cation	96% H$_2$SO$_4$	259(5.14),305(3.59), 511(3.95)	4-0661-84
$C_{16}H_{11}N_5O_2$			
3H-1,2,4-Triazino[3,2-b]quinazoline-3,10(4H)-dione, 4-amino-2-phenyl-	EtOH	310(3.75),356(3.03)	24-1083-84
$C_{16}H_{12}$			
9cH-Cyclopenta[jk]fluorene, 9c-methyl-	EtOH	311(4.43),400(3.34), 464s(2.82)	39-0909-84C

Compound	Solvent	$\lambda_{max}(\log \epsilon)$	Ref.
Tetracyclo[3.2.0.0^{2,7}.0^{4,6}]heptane, [4-(2,4-cyclopentadien-1-ylidene)-2-butynylidene]-	MeCN	230(4.12),285s(4.07), 302(4.10),364(4.37)	24-2027-84
$C_{16}H_{12}BrNO_4S$ 2,5-Pyrrolidinedione, 3-[(4-bromophenyl)sulfonyl]-1-phenyl-	CH$_2$Cl$_2$	267s(--),270s(--), 278(2.97)	18-0219-84
$C_{16}H_{12}Br_2N_2$ Pyrazine, 2,3-bis(4-bromophenyl)-5,6-dihydro-	EtOH	231.5(4.09),285(3.67)	4-0103-84
$C_{16}H_{12}Br_2O_4S_2$ Benzene, 1,1'-[1,3-butadiene-1,4-diyl-bis(sulfonyl)]bis[4-bromo-, (E,E)-	CHCl$_3$	248(4.22),278(4.55)	139-0005-84C
(E,Z)-	CHCl$_3$	247(4.20),277(4.51)	139-0005-84C
$C_{16}H_{12}ClNO$ 1H-Indole-2-carbonyl chloride, 5-(phenylmethyl)-	THF	241(4.21),321(4.42)	109-0338-84
$C_{16}H_{12}ClNO_3S_2$ Benzo[b]thiophen-3(2H)-one, 2-[[(3-chlorophenyl)amino](methylthio)-methylene]-, 1,1-dioxide	EtOH	258(4.13),282(4.03), 376(4.27)	95-0134-84
Benzo[b]thiophen-3(2H)-one, 2-[[(4-chlorophenyl)amino](methylthio)-methylene]-, 1,1-dioxide	EtOH	218(4.38),257(4.13), 283(4.03),376(4.34)	95-0134-84
$C_{16}H_{12}ClNO_4S$ 2,5-Pyrrolidinedione, 3-[(4-chloro-phenyl)sulfonyl]-1-phenyl-	CH$_2$Cl$_2$	265s(--),269s(--), 278(2.92)	18-0219-84
$C_{16}H_{12}Cl_2$ Cyclobuta[1]phenanthrene, 1,1-dichloro-1,2,2a,10b-tetrahydro-, cis	EtOH	235(3.96),243(3.73), 262s(3.61),270s(3.84), 280(3.94),289s(3.81), 310(3.33),320(3.15), 357(2.95)	78-5249-84
$C_{16}H_{12}Cl_2N_2$ Pyrazine, 2,3-bis(2-chlorophenyl)-5,6-dihydro-	EtOH	208.5(4.22),272.5(3.36)	4-0103-84
Pyrazine, 2,3-bis(4-chlorophenyl)-5,6-dihydro-	EtOH	227.5(4.35),286(3.92)	4-0103-84
$C_{16}H_{12}F_3N$ 5H-Dibenz[b,f]azepine, 5-(2,2,2-tri-fluoroethyl)-	EtOH	253.5(4.33),286(3.83)	142-0379-84
$C_{16}H_{12}NO_2$ [1,3]Dioxolo[4,5-j]pyrrolo[3,2,1-de]-phenanthridinium, 4,5-dihydro-, chloride	MeOH	258(4.51),267(4.47), 279(4.46),341(4.08)	100-0796-84
$C_{16}H_{12}N_2$ 2,5'-Bi-1H-indole	EtOH	209(4.28),242(4.66), 284(4.13),320(4.32)	103-0280-84
1H,5H-Pyrrolo[2,3-f]indole, 2-phenyl-	EtOH	206(4.35),228(4.37), 259.7(4.23),325(4.43)	103-0376-84

Compound	Solvent	λ_{max}(log ϵ)	Ref.
$C_{16}H_{12}N_2O_2$			
[1]Benzazepino[6,5,4-def][1]benzaze- pine-5,11(4H,6H)-dione, 10,12-di- hydro-, (±)-	MeOH	239.0(4.21),296.5(3.27), 305.5(3.40)	39-2013-84C
1H-1,5-Benzodiazepine-3-carboxaldehyde, 2,3-dihydro-2-oxo-4-phenyl-	EtOH	206(4.46),270(4.29)	103-0183-84
2-Furanmethanol, 5-(5H-pyrido[4,3-b]- indol-1-yl)- (isoperlolyrine)	EtOH	239(4.29),254(4.22), 265s(4.17),274(4.19), 289s(4.20),293(4.23), 301s(4.12),309s(4.08), 372(4.03),385(4.02)	73-1536-84
	EtOH-HCl	220(4.30),234s(4.11), 278(4.20),316s(3.84), 349(4.17),413(4.01)	73-1536-84
$C_{16}H_{12}N_2O_3$			
Acetamide, N-(2-oxo-3-phenyl-2H-1,4- benzoxazin-7-yl)-	EtOH	379(4.40)	4-0551-84
6H-Indolo[3,2,1-de][1,5]naphthyridin- 6-one, 4,5-dimethoxy-	EtOH	240(4.54),248(4.59), 290(4.02),300(3.99), 349(3.95),357(4.01), 373(3.92)	94-3579-84
$C_{16}H_{12}N_2O_3S$			
Ethanone, 1-(3,6-dihydropyrrolo[2,3-c]- phenothiazin-2-yl)-, S,S-dioxide	EtOH	213(4.29),270(4.27), 339(3.83),353(3.72)	103-1095-84
Pyrrolo[2,3-c]phenothiazine, 3-acetyl- 3,6-dihydro-, S,S-dioxide	EtOH	208(4.23),222(4.21), 261(4.22),285(4.23), 319(3.75)	103-1095-84
$C_{16}H_{12}N_2O_4$			
Pyrido[3,2-g]quinoline-2,8-dicarboxylic acid, dimethyl ester	MeCN	258(5.079),282(4.429), 289(4.388),357(3.811), 373(3.831),395(3.412)	5-0133-84
$C_{16}H_{12}N_2O_4Te_2$			
2(3H)-Benzoxazolone, 3,3'-[ditelluro- bis(methylene)]bis-	benzene	386(2.8)	48-0467-84
$C_{16}H_{12}N_2S_2$			
1,3-Dithiolo[4,5-b]indole-1-S^{IV}-2- amine, N-methyl-N-phenyl-	$CHCl_3$	265s(3.97),277(4.07), 400(3.63),486(3.81)	12-2479-84
$C_{16}H_{12}N_3O_3$			
Pyridinium, 1-[5-(2,2-dicyanoethenyl)- 2-furanyl]-3-(ethoxycarbonyl)-, bromide	H_2O	251(3.00),345s(3.04), 377(3.16),468(2.76)	73-2485-84
$C_{16}H_{12}N_4$			
Dibenzo[b,h][1,4,7,10]tetraazacyclo- dodecine	DMF	295(4.39),318(4.54), 332(4.70),349(4.69), 438(3.87),481(3.78)	150-0168-84S
[1,2,4]Triazolo[1,5-c]quinazoline, 2-methyl-5-phenyl-	EtOH	259(4.38),264(4.25), 294(4.3)	18-1138-84
$C_{16}H_{12}N_4O_2$			
2-Quinoxalinecarboxamide, N-(2-amino- phenyl)-3-formyl-	EtOH	240(3.95),307(3.37)	136-0324-84J
$C_{16}H_{12}N_4O_2S_2$			
Pyrimido[5,4-e]thiazolo[3,4-a]pyrimi-	DMF	387(4.04),396(4.04),	103-0921-84

Compound	Solvent	$\lambda_{max}(\log \epsilon)$	Ref.
din-10-ium, 1,2,3,4-tetrahydro-3-methyl-9-(methylthio)-2,4-dioxo-7-phenyl-, hydroxide, inner salt (cont.)		440(4.01)	103-0921-84
$C_{16}H_{12}N_4O_4$			
Acetamide, N-[2-(3,4-dihydro-6-nitro-3-oxo-2-quinoxalinyl)phenyl]-	EtOH	226(4.23),308(3.95), 375(3.84)	103-1276-84
$C_{16}H_{12}N_4O_4S_2$			
Carbonimidodithioic acid, cyano-, bis[(4-nitrophenyl)methyl] ester	dioxan	275(3.39)	73-2285-84
4-Thiazolamine, 5-(4-nitrophenyl)-2-[[(4-nitrophenyl)methyl]thio]-	dioxan	273(3.27),418(3.24)	73-2285-84
$C_{16}H_{12}N_4O_7$			
1H-Indole-3-propanoic acid, α-(3,5-dinitro-4-oxo-1(4H)-pyridinyl)-, (S)-	EtOH	340(3.67)	18-1961-84
$C_{16}H_{12}N_4S$			
2-Benzothiazoleacetonitrile, α-[(4-methylphenyl)hydrazono]-	EtOH	258(4.32),405(4.60)	104-0523-84
2-Benzothiazoleacetonitrile, 6-methyl-α-(phenylhydrazono)-	EtOH	255(4.26),402(4.60)	104-0523-84
$C_{16}H_{12}N_6S_2$			
2(3H)-Benzothiazolone, [1-(2-benzothiazolylazo)ethylidene]hydrazone	benzene	460(4.56)	135-0577-84
photoinduced hydrobromide	benzene	420(4.68),620s(3.40)	135-0577-84
$C_{16}H_{12}O$			
5,10-Epoxybenzo[b]biphenylene, 4b,5,10,10a-tetrahydro-, (4bα,5β,10β,10aα)-	EtOH	209(4.17),229(4.30), 252s(4.01),271(3.84), 279.5(3.87),290s(3.71), 324s(3.34)	12-1283-84
9,10-Ethanoanthracen-11-one, 9,10-dihydro-	MeCN	253s(2.715),261s(2.906), 266(3.0113),273(3.005), 286s(2.650),295(2.787), 306(2.824),316(2.619)	5-0381-84
Furan, 2,4-diphenyl-	C_6H_{12}	228(4.48),300(2.62)	18-0844-84
	MeOH	216(4.37),300(2.68)	18-0844-84
Furan, 2,5-diphenyl-	C_6H_{12}	222(4.42),324(4.52)	18-0844-84
	EtOH	225(4.38),322(4.61)	18-0844-84
$C_{16}H_{12}O_2$			
2-Butenedial, 2,3-diphenyl-, (E)-	$CHCl_3$	317(3.99),408(3.23)	77-1234-84
6H-Cyclopenta[jk]fluorene-6,9-dione, 9,9c-dihydro-9c-methyl-	EtOH	275(4.25),295(4.17), 374(4.08),560(3.67)	39-0909-84C
2(5H)-Furanone, 4,5-diphenyl-	MeOH	217(3.7),272(3.2)	2-0514-84
$C_{16}H_{12}O_3$			
9,10-Anthracenedione, 1-methoxy-3-methyl-	EtOH	256(4.42),326(3.44), 388(3.70)	78-3455-84
4H-1-Benzopyran-4-one, 2-(2-methoxyphenyl)-	pH 7.0	250(4.30),308(4.21), 325s(--)	61-0759-84
	MeOH	255(4.28),306(4.26), 325s(--)	61-0759-84
	95% H_2SO_4	276(4.09),320(4.07), 372(4.34)	61-0759-84
4H-1-Benzopyran-4-one, 2-(3-methoxyphenyl)- (in 15% MeOH)	pH 7.0	252(4.33),301(4.32)	61-0759-84

Compound	Solvent	$\lambda_{max}(\log \epsilon)$	Ref.
(cont.)	MeOH	240(4.32),293(4.32), 305s(--)	61-0759-84
	95% H_2SO_4	281(4.54),349(4.40)	61-0759-84
4H-1-Benzopyran-4-one, 5-methoxy-2-phenyl-	pH 7	263(4.43),293(4.16), 322(4.05)	61-0759-84
	MeOH	262(4.45),289(4.15), 319(4.05)	61-0759-84
	95% H_2SO_4	283(4.24),344(4.42), 370s(--)	61-0759-84
4H-1-Benzopyran-4-one, 6-methoxy-2-phenyl-	pH 7	267(--),308(--)	61-0759-84
	MeOH	268(4.42),308(4.27)	61-0759-84
	95% H_2SO_4	280(4.28),351(4.38), 370s(--)	61-0759-84
4H-1-Benzopyran-4-one, 8-methoxy-2-phenyl-	MeOH	264(4.45),296s(--), 318s(--)	61-0759-84
	95% H_2SO_4	267(4.16),291(4.23), 354(4.47)	61-0759-84
2-Biphenylenecarboxylic acid, 6-acetyl-, methyl ester	EtOH	216(3.76),266(4.32), 279(4.42),326s(3.04), 339s(3.34),342(3.34), 354s(3.61),357(3.62), 374(3.87)	18-1914-84
$C_{16}H_{12}O_4$			
9,10-Anthracenedione, 1,2-dimethoxy-	EtOH	249(4.20),263(4.14), 385(3.47)	78-3455-84
9,10-Anthracenedione, 1-hydroxy-2-methoxy-4-methyl-	MeOH	249(4.51),277(4.12), 424(3.82)	78-5039-84
9,10-Anthracenedione, 1-hydroxy-8-methoxy-2-methyl-	n.s.g.	225(4.35),255(4.34), 277(3.98),398(3.91), 418(3.96),430(3.89)	5-0306-84
9,10-Anthracenedione, 1-hydroxy-8-methoxy-3-methyl-	EtOH	262(4.06),400(3.63)	78-3455-84
9,10-Anthracenedione, 5-hydroxy-1-methoxy-2-methyl-	EtOH	255(4.25),280(3.83), 402(3.68)	78-3455-84
9,10-Anthracenedione, 8-hydroxy-1-methoxy-2-methyl-	EtOH	257(4.31),282(3.76), 400(3.88)	78-3455-84
9,10-Anthracenedione, 8-hydroxy-1-methoxy-3-methyl-	EtOH	258(3.96),275s(3.75), 283(3.70),414(3.64)	78-3455-84
4H-1-Benzopyran-4-one, 5,7-dihydroxy-6-(phenylmethyl)-	MeOH	205(4.35),256(4.18), 297(3.43)	2-1036-84
4H-1-Benzopyran-4-one, 5,7-dihydroxy-8-(phenylmethyl)-	MeOH	202(4.49),258(4.09), 296(3.62)	2-1036-84
4H-1-Benzopyran-4-one, 3-hydroxy-6-methoxy-2-phenyl-	$CHCl_3$	257(4.17),330(4.23), 350s(4.13)	142-1943-84
4H-1-Benzopyran-4-one, 3-hydroxy-7-methoxy-2-phenyl-	$CHCl_3$	255(4.34),323(4.29), 340(4.28)	142-1943-84
4H-1-Benzopyran-4-one, 6-hydroxy-7-methoxy-3-phenyl-	MeOH	325(4.04)	64-0238-84B
4H-1-Benzopyran-4-one, 7-hydroxy-3-(4-methoxyphenyl)-	MeOH	302(3.99)	64-0238-84B
4H-1-Benzopyran-4-one, 7-hydroxy-6-methoxy-3-phenyl-	MeOH	321(4.04),360s(--)	64-0238-84B
2,6-Biphenylenedicarboxylic acid, dimethyl ester	EtOH	216(4.16),262s(4.67), 269(4.75),277s(4.62), 322s(3.25),330s(3.38), 334(3.51),338(3.52), 347(3.78),352(3.79), 369(3.98)	18-1914-84
4a,9a-Ethanoanthracene-1,4,5,8-tetrone, 9,10-dihydro-	EtOH	225(4.22),240(4.19), 248(4.19),340(2.88)	18-0615-84

Compound	Solvent	λ_{max}(log ϵ)	Ref.
2-Propen-1-one, 1-(6-hydroxy-1,3-benzo-dioxol-5-yl)-3-phenyl-	EtOH	204(4.38),228(4.05), 278(3.93),331(4.20), 390(3.96)	78-4081-84
Spiro[3-cyclohexene-1,2'(1'H)-naphtha-lene]-2,5,5',8'-tetrone, 3',4'-di-hydro-6-methylene-	EtOH	240(4.32),338(2.97)	18-0615-84

$C_{16}H_{12}O_5$

Compound	Solvent	λ_{max}(log ϵ)	Ref.
9,10-Anthracenedione, 1,3-dihydroxy-2-(methoxymethyl)-	EtOH	246(4.51),282(4.40), 315s(4.07),417(3.72)	102-1733-84
9,10-Anthracenedione, 1,4-dihydroxy-8-methoxy-6-methyl-	EtOH	255(4.02),476(3.84), 488(3.81),524(3.49)	78-3455-84
9,10-Anthracenedione, 1,5-dihydroxy-2-methoxy-4-methyl-	MeOH	231(4.46),253(4.39), 293(4.09),454(4.06)	78-5039-84
9,10-Anthracenedione, 1,5-dihydroxy-2-methoxy-6-methyl-	EtOH	260(4.38),290(3.93), 299(3.97),444(3.99)	78-3455-84
9,10-Anthracenedione, 1,8-dihydroxy-2-methoxy-4-methyl-	MeOH	232(4.46),251(4.34), 289(4.09),446(4.03)	78-5039-84
9,10-Anthracenedione, 1,8-dihydroxy-3-methoxy-6-methyl-	EtOH	253(3.93),265(3.94), 434(3.77)	78-3455-84
	EtOH	223.5(4.41),254(4.14), 264(4.16),286(4.14), 433(3.99)	102-1485-84
	EtOH-NaOH	212(--),229(--), 255(--),304(--), 503(--)	102-1485-84
9,10-Anthracenedione, 1-hydroxy-2-(hydroxymethyl)-8-methoxy-	n.s.g.	225(4.75),255(4.35), 279(4.00),396(3.94), 414(3.99),430(3.91)	5-0306-84
4H-1-Benzopyran-4-one, 5,7-dihydroxy-3-(4-methoxyphenyl)-	MeOH	332s(3.65)	64-0238-84B
4H-1-Benzopyran-4-one, 6,7-dihydroxy-3-(4-methoxyphenyl)-	MeOH	326(3.93)	64-0238-84B
4H-1-Benzopyran-4-one, 6-hydroxy-3-(4-hydroxyphenyl)-7-methoxy-	MeOH	325(4.09)	64-0238-84B
4H-1-Benzopyran-4-one, 7-hydroxy-3-(4-hydroxyphenyl)-6-methoxy-	MeOH	319(4.02)	64-0238-84B
9H-Fluoren-9-one, 5-acetoxy-1-hydroxy-7-methoxy- (dengibsin monoacetate)	EtOH	243(4.29),260(4.44), 271(4.42),334(3.64), 446.5(3.31)	42-1010-84

$C_{16}H_{12}O_6$

Compound	Solvent	λ_{max}(log ϵ)	Ref.
9,10-Anthracenedione, 1,3,5-trihydroxy-4-methoxy-2-methyl-	MeOH	225(4.58),255(4.39), 280(4.12),443(3.67)	102-2104-84
	MeOH-base	226(--),263(--), 325(--),480(--)	102-2104-84
4H-1-Benzopyran-4-one, 3-(2,4-dihy-droxyphenyl)-5-hydroxy-7-methoxy-	MeOH	260(4.63),285s(4.26), 330s(3.80)	102-0871-84
Naphtho[1,8-bc]pyran-3-carboxylic acid, 2,6-dihydro-7-hydroxy-4-methyl-2,6-dioxo-, ethyl ester	CHCl$_3$	247(3.83),293(4.10), 320(4.23),452s(3.91), 470(3.92)	39-1957-84C

$C_{16}H_{12}O_7$

Compound	Solvent	λ_{max}(log ϵ)	Ref.
9,10-Anthracenedione, 1,2,6,8-tetra-hydroxy-7-methoxy-3-methyl-	dioxan	282(4.56),310s(3.87), 320(3.90),410(3.96)	94-0860-84
1H-Xanthene-1,4,9-trione, 2,3,7-tri-methoxy-	MeOH	246(4.24),266(4.31), 332s(3.54),380(3.77)	39-1507-84C

$C_{16}H_{13}BrN_2O_2$

Compound	Solvent	λ_{max}(log ϵ)	Ref.
9,10-Anthracenedione, 1-amino-4-[(2-bromoethyl)amino]-	dioxan	566(4.10),609(4.10)	103-0717-84

Compound	Solvent	$\lambda_{max}(\log \epsilon)$	Ref.
Benzo[h]quinoline-4-carboxamide, 5-bromo-6-methoxy-N-methyl-	EtOH	218(4.47),245(4.61), 275(4.30),312(3.90), 347(3.56),357(3.58)	12-1271-84
$C_{16}H_{13}BrN_2S$ 3-Quinolinecarbonitrile, 4-(4-bromo-phenyl)-1,2,5,6,7,8-hexahydro-2-thioxo-	dioxan	254(4.06),314(4.20), 430(3.51)	104-1402-84
$C_{16}H_{13}Br_2NO_3$ 9(10H)-Acridinone, 2,4-dibromo-1,3-di-methoxy-10-methyl-	EtOH	218s(4.53),266(4.91), 318s(3.94),392(4.13)	5-0031-84
$C_{16}H_{13}ClN_2O$ 2H-1,4-Benzodiazepin-2-one, 7-chloro-1,3-dihydro-3-methyl-5-phenyl-	MeOH	204(4.51),224(4.53), 253s(4.15),320(3.30)	33-0916-84
$C_{16}H_{13}ClN_2OSe$ Methanone, (3-amino-4,6-dimethylselen-olo[2,3-b]pyridin-2-yl)(4-chloro-phenyl)-	EtOH	212(4.40),235(4.11), 241(4.02),296(4.62), 375(3.95)	103-0577-84
3-Pyridinecarbonitrile, 2-[[2-(4-chlo-rophenyl)-2-oxoethyl]seleno]-4,6-di-methyl-	EtOH	237(4.43),254(4.37), 268(4.19),314(3.79)	103-0577-84
$C_{16}H_{13}ClN_2O_2$ 9,10-Anthracenedione, 1-amino-4-[(2-chloroethyl)amino]-	dioxan	566(4.14),609(4.15)	103-0717-84
$C_{16}H_{13}ClN_2S$ Benzenamine, 4-chloro-N-(3-methyl-4-phenyl-2(3H)-thiazolylidene)-	EtOH	305(4.27)	56-0447-84
3-Quinolinecarbonitrile, 4-(4-chloro-phenyl)-1,2,5,6,7,8-hexahydro-2-thioxo-	dioxan	257(4.11),315(4.12), 428(3.39)	104-1402-84
$C_{16}H_{13}FN_2OSe$ Methanone, (3-amino-4,6-dimethyl-selenolo[2,3-b]pyridin-2-yl)(4-fluorophenyl)-	EtOH	235(4.05),248(4.00), 278(4.50),378(3.95)	103-0577-84
$C_{16}H_{13}FN_2S$ 3-Quinolinecarbonitrile, 4-(3-fluoro-phenyl)-1,2,5,6,7,8-hexahydro-2-thioxo-	dioxan	259(4.18),316(4.10), 427(3.25)	104-1402-84
3-Quinolinecarbonitrile, 4-(4-fluoro-phenyl)-1,2,5,6,7,8-hexahydro-2-thioxo-	dioxan	264(4.??),314(4.14), 428(3.47)	104-1402-84
$C_{16}H_{13}FOS$ Dibenzo[b,f]thiepin-10(11H)-one, 8-ethyl-3-fluoro-	MeOH	254s(4.10),331(3.61)	73-2638-84
$C_{16}H_{13}FS$ Dibenzo[b,f]thiepin, 2-ethyl-7-fluoro-	MeOH	262.5(4.42),296(3.70), 339s(2.94)	73-2638-84
$C_{16}H_{13}F_3$ Benzene, 1-methyl-2-(3,3,3-trifluoro-2-phenyl-1-propenyl)-, cis	MeOH	203(4.94),258(3.86)	151-0327-84D
trans	MeOH	248(4.01)	151-0327-84D

Compound	Solvent	λ_{max}(log ϵ)	Ref.
C$_{16}$H$_{13}$N			
1H-Pyrrole, 2,4-diphenyl-	EtOH	246(4.3),302(4.3)	44-0448-84
C$_{16}$H$_{13}$NO			
1,3-Benzoxazepine, 4-methyl-2-phenyl-	EtOH	242(4.32),337(3.83)	78-3567-84
2H-Isoindole-1-carboxaldehyde, 2-methyl-3-phenyl-	EtOH	243(4.13),265(4.31), 376(4.43)	39-0833-84B
Phenol, 2-(4-isoquinolinylmethyl)-	MeOH	203s(--),218(4.75), 274(3.78),284s(--), 309(3.35),322(3.68)	78-0215-84
Phenol, 4-(4-isoquinolinylmethyl)-	MeOH	217.5(4.70),272.5(3.73), 285s(--),309(3.51), 322.5(3.62)	78-0215-84
C$_{16}$H$_{13}$NO$_2$			
1(3H)-Isobenzofuranone, 6-amino-3-[(4-methylphenyl)methylene]-	EtOH	317(4.29),377(4.13)	131-0239-84H
C$_{16}$H$_{13}$NO$_2$S			
2-Thiophenecarboxylic acid, 3-amino-5-(2-naphthalenyl)-, methyl ester	MeOH	312(4.35),350(4.04)	118-0275-84
C$_{16}$H$_{13}$NO$_3$			
2,3-Azetidinedione, 1-(4-methoxyphenyl)-4-phenyl-	n.s.g.	354.0(3.95)	88-4733-84
Benzo[h]quinoline-4-carboxylic acid, 6-methoxy-, methyl ester	EtOH	244(4.48),277s(4.15), 285(4.18),291s(4.16), 306s(3.82),364(3.80)	12-1271-84
1(3H)-Isobenzofuranone, 6-amino-3-[(4-methoxyphenyl)methylene]-	EtOH	324(4.35),382(4.18)	131-0239-84H
C$_{16}$H$_{13}$NO$_3$S			
2H-Pyran-3-carbonitrile, 6-[2-(4-methoxyphenyl)ethenyl]-4-(methylthio)-2-oxo-	EtOH	242(4.13),415(4.30)	94-3384-84
C$_{16}$H$_{13}$NO$_3$S$_2$			
Benzo[b]thiophen-3(2H)-one, 2-[(methylthio)(phenylamino)methylene]-, 1,1-dioxide	EtOH	217(4.36),254(4.08), 282(4.01),372(4.31)	95-0134-84
C$_{16}$H$_{13}$NO$_4$			
2H-Pyran-3-carbonitrile, 4-methoxy-6-[2-(4-methoxyphenyl)ethenyl]-2-oxo-	EtOH	230(4.20),415(4.37)	94-3384-84
C$_{16}$H$_{13}$NO$_4$S			
2,5-Pyrrolidinedione, 1-phenyl-3-(phenylsulfonyl)-	CH$_2$Cl$_2$	260(3.28),266(3.26), 273(3.09)	18-0219-84
C$_{16}$H$_{13}$NO$_5$			
1,3-Dioxolo[4,5-c]acridin-6(11H)-one, 5-hydroxy-4-methoxy-11-methyl-(normelicopine)	EtOH	251(3.27),279(3.32), 315(3.11),431(2.62)	100-0285-84
C$_{16}$H$_{13}$N$_2$O$_3$			
Benzenediazonium, 2-[2-carboxy-2-(4-methoxyphenyl)ethenyl]-, (E)-, tetrafluoroborate	MeCN	235(4.10),270(4.11)	39-1093-84B

Compound	Solvent	$\lambda_{max}(\log \epsilon)$	Ref.
$C_{16}H_{13}N_3$			
4H-Pyrazolo[1,5-a]benzimidazole, 6-methyl-2-phenyl-	EtOH	251(4.42),279(4.16), 323(4.25)	48-0829-84
5H-Pyrido[3',2':6,7]cyclohepta[1,2-b]- [1,8]naphthyridine, 6,7-dihydro-	EtOH	218(4.59),315(4.12), 325(4.12)	44-2208-84
$C_{16}H_{13}N_3O$			
2H-1-Benzopyran, 2-azido-3-methyl- 2-phenyl-	ether	260(3.95),300(3.52)	78-3567-84
Methanone, (4-amino-3-cinnolinyl)(4- methylphenyl)-	DMF	346(3.87)	5-1390-84
4H-[1,2,4]Triazolo[3,4-c][1,4]benzoxa- zine, 1-(4-methylphenyl)-	EtOH	245(3.07),282(2.84)	2-0445-84
4H-[1,2,4]Triazolo[3,4-c][1,4]benzoxa- zine, 1-(phenylmethyl)-	EtOH	242(2.87),284(2.43)	2-0445-84
$C_{16}H_{13}N_3O_2$			
Acetamide, N-[2-(3,4-dihydro-3-oxo- 2-quinoxalinyl)phenyl]-	EtOH EtOH	303(4.34),358(4.58) 213(4.56),303(3.96), 362(4.05)	2-0114-84 103-0537-84
4H-1,2,4-Triazolo[3,4-c][1,4]benzoxa- zine, 1-(4-methoxyphenyl)-	EtOH	250(3.02),275(2.95)	2-0445-84
$C_{16}H_{13}N_3O_2S$			
Quinoxaline, 2-methyl-3-[[(4-nitro- phenyl)methyl]thio]-	EtOH	240(3.94),266(4.04), 343(3.63),355(3.61)	18-1653-84
$C_{16}H_{13}N_3O_3$			
2-Propenoic acid, 2-methyl-, 4-(2H- benzotriazol-2-yl)-3-hydroxyphenyl ester	CHCl$_3$ AcNMe$_2$	295s(4.13),334(4.38) 270(4.13),290(4.27)	49-0853-84 49-0853-84
polymer	AcNMe$_2$	270(4.05),290(4.16)	49-0853-84
$C_{16}H_{13}N_3O_5S$			
Pyrido[2,3-d]pyrimidine-6-carboxylic acid, 1,2,3,4-tetrahydro-5-hydroxy- 7-(methylthio)-2,4-dioxo-1-phenyl-, methyl ester	EtOH	261(4.50),317(4.10)	94-0122-84
$C_{16}H_{13}N_3S_2$			
4H-1,3,5-Thiadiazine-4-thione, 2-(3- methylphenyl)-6-(phenylamino)-	EtOH	279(4.60),368(3.87)	150-0266-84S
$C_{16}H_{13}N_5O_2$			
6H-Purin-6-one, 3,7-dihydro-3-(6-meth- oxy-8-quinolinyl)-7-methyl-	H$_2$O	225(4.56),307(3.78), 322s(3.72)	2-0870-84
$C_{16}H_{13}S$			
1-Benzothiopyrylium, 2-methyl-4-phenyl-, perchlorate	MeCN	264(4.40),343(3.94), 396(4.26)	44-4192-84
1-Benzothiopyrylium, 4-methyl-2-phenyl-, perchlorate	MeCN	264(4.36),292(4.00), 389(4.22)	44-4192-84
$C_{16}H_{14}$			
Benzene, 1,1'-(1,2-ethynediyl)bis[4- methyl-	90% EtOH	304(4.48)	101-0199-84A
1,3,5-Cycloheptatriene, 7-[4-(2,4- cyclopentadien-1-ylidene)-2-buten- ylidene]-, (E)-	C$_6$H$_{12}$	237(4.13),255s(4.00), 288(3.81),340s(3.89), 370s(4.13),398s(4.41), 428(4.65),450s(4.70), 462(4.72)	24-2027-84

Compound	Solvent	$\lambda_{max}(\log \epsilon)$	Ref.
(cont.)	MeOH	425s(--),454(--)	24-2027-84
	MeCN	244(--),292(--), 368s(--),394s(--), 426s(--),457(--)	24-2027-84
1,3-Cyclopentadiene, 5,5'-(2,4-hexadi- ene-1,6-diylidene)bis-, (E,E)-	C_6H_{12}	359(4.34),375(4.71), 395(5.00),419(5.06)	24-2006-84
	MeCN	359(--),377(--), 396(--),419(--)	24-2006-84
(Z,Z)-	C_6H_{12}	353(4.27),369(4.56), 389(4.78),412(4.75)	24-2006-84
	MeCN	351(--),372(--), 388(--),410(--)	24-2006-84
Tetracyclo[3.2.0.0^{2,7}.0^{4,6}]heptane, [4-(2,4-cyclopentadien-1-ylidene)- 2-butenylidene]-, (E)-	MeCN	371(4.73)	24-2027-84
Tricyclobuta[a,c,g]naphthalene, 1,2,3,4,6,7-hexahydro-	n.s.g.	240(4.68),244(4.68), 289(3.64),301(3.66), 326(2.85)	44-1412-84
Tricyclo[4.1.0.0^{2,7}]hept-3-ene, 5-[4-(2,4-cyclopentadien-1- ylidene)-2-butenylidene]-, (E,Z)-	C_6H_{12}	367s(4.56),383(4.76), 406(4.77)	24-2027-84
	MeCN	402(4.71)	24-2027-84
$C_{16}H_{14}BrNO_3$ 9(10H)-Acridinone, 2-bromo-1,3-dimeth- oxy-10-methyl-	EtOH	215s(4.41),222s(4.41), 245s(4.56),269s(4.93), 276(4.96),300s(3.98), 356s(3.72),370s(3.99), 388(4.10)	5-0031-84
9(10H)-Acridinone, 4-bromo-1,3-dimeth- oxy-10-methyl-	EtOH	216s(4.45),223s(4.44), 258(4.79),268s(4.74), 300(4.32),316s(4.21), 384(4.11)	5-0031-84
$C_{16}H_{14}ClNO_2S$ 2H-Thieno[2,3-h]-1-benzopyran-2-one, 3-chloro-4-piperidino-	EtOH	235.5(4.29),258s(4.26), 261(4.26),305s(3.93), 339(4.14)	161-0081-84
$C_{16}H_{14}Cl_2N_2O_3$ Benzo[a]phenazin-5(7H)-one, 2,3-di- chloro-7a,8,9,10,11,11a-hexahydro- 1,4-dihydroxy-	benzene	472s(3.20),509s(3.51), 550s(3.90),591(4.15), 637(4.11)	118-0953-84
$C_{16}H_{14}Cl_2N_2O_3$ Butanediamide, N,N'-bis(4-chlorophen- yl)-N,N'-dihydroxy-	MeOH	202(4.19),234(4.03), 330(4.24)	34-0345-84
$C_{16}H_{14}F_3N$ 5H-Dibenz[b,f]azepine, 10,11-dihydro- 5-(2,2,2-trifluoroethyl)-	EtOH	248(3.80)	142-0379-84
$C_{16}H_{14}INO_3$ 9(10H)-Acridinone, 2-iodo-1,3-dimeth- oxy-10-methyl-	EtOH	213(4.32),222s(4.24), 246s(4.39),269s(4.71), 276(4.73),300s(3.93), 372s(3.80),392(3.88)	5-0031-84
$C_{16}H_{14}NO_2P$ Anthra[1,9-de]-1,3,2-dioxaphosphorin- 2-amine, N,N-dimethyl-	EtOH	264(5.12),303(4.05), 255(3.81),374(4.11), 393(4.05)	40-1158-84

Compound	Solvent	$\lambda_{max}(\log \epsilon)$	Ref.
$C_{16}H_{14}N_2$			
Pyrimidine, 1,2-dihydro-4,6-diphenyl-	EtOH	207(4.24),253(4.18), 377(3.81)	138-1773-84
$C_{16}H_{14}N_2O$			
2(1H)-Quinoxalinone, 1-ethyl-3-phenyl-	EtOH	210(4.45),225(4.41), 303(4.06),359(4.08)	44-0827-84
2 (1H)-Quinoxalinone, 3-methyl- 1-(phenylmethyl)-	EtOH	282(3.43),330(3.46), 338(3.40)	18-1653-84
$C_{16}H_{14}N_2OS$			
Benzo[b]thiophen-3(2H)-one, 2-[(2-methyl-2-phenylhydrazino)methylene]-	heptane	370(4.44)	104-1353-84
	PrOH	370(4.44),440(3.58)	104-1353-84
	DMSO	380(4.51)	104-1353-84
$C_{16}H_{14}N_2OSe$			
Methanone, (3-amino-4,6-dimethylselenolo[2,3-b]pyridin-2-yl)phenyl-	EtOH	210(4.38),230(4.09), 245(4.00),290(4.54), 375(3.89)	103-0577-84
3-Pyridinecarbonitrile, 4,6-dimethyl-2-[(2-oxo-2-phenylethyl)seleno]-	EtOH	232(4.42),250(4.30), 275(4.18),312(3.78)	103-0577-84
$C_{16}H_{14}N_2O_2$			
1H-Naphth[1',2':2,3]oxepino[4,5-c]pyrazol-1-one, 2,11,12,12a-tetrahydro-12a-methyl-	EtOH	273(4.39),306(4.17)	2-0736-84
$C_{16}H_{14}N_2O_2S$			
2-Propenoic acid, 2-cyano-3-[1-methyl-5-(phenylthio)-1H-pyrrol-2-yl]-, methyl ester	MeOH	204(4.44),245(4.10), 394(4.39)	4-1041-84
$C_{16}H_{14}N_2O_3$			
3-Pyridinecarboxylic acid, 5-cyano-1,6-dihydro-2-methyl-6-oxo-4-phenyl-, ethyl ester	EtOH	254(4.30),340(4.06)	103-1241-84
$C_{16}H_{14}N_2O_3S$			
Benzo[b]thiophene-2-propanenitrile, 3-hydroxy-α-(4-morpholinomethylene)-β-oxo-, (E)-	MeOH	254(3.54),293(3.37), 345(3.66),370(3.67)	83-0531-84
$C_{16}H_{14}N_2O_3S_2$			
1H-Imidazo[1,2-c]thiazol-4-ium, 1,3-diacetyl-2-hydroxy-5-(methylthio)-7-phenyl-, hydroxide, inner salt	HOAc	320(3.96)	103-0029-84
$C_{16}H_{14}N_2O_4$			
Benzene, 1,1'-(3-methyl-1-propene-1,3-diyl)bis[4-nitro-, cis	heptane	280(4.29),310(4.03)	73-2912-84
	MeOH	295(4.26),320(4.14)	73-2912-84
	EtOH	295(4.26),320(4.14)	73-2912-84
	isoPrOH	295(4.25),320(4.10)	73-2912-84
	tert-BuOH	290(4.32),320(4.12)	73-2912-84
	MeCN	295(4.24),320(4.09)	73-2912-84
trans	heptane	295(4.35),310(4.28)	73-2912-84
	MeOH	313(4.29),320(4.26)	73-2912-84
	EtOH	313(4.29),320(4.26)	73-2912-84
	isoPrOH	310(4.31),320(4.26)	73-2912-84
	tert-BuOH	307(4.35),320(4.30)	73-2912-84
	MeCN	310(4.33),320(4.31)	73-2912-84
Benzoic acid, 4,4'-azobis-, dimethyl ester	base	359(4.52)	108-0302-84

Compound	Solvent	λ_{max}(log ϵ)	Ref.
1H-Indene-1-acetic acid, 1-[diazo(meth-oxycarbonyl)methyl]-α-methylene-, methyl ester	EtOH	210(4.42),250(4.04)	23-2506-84
5H-Pyrazolo[5,1-b][1,3]oxazine-3-carb-oxylic acid, 2-methyl-5-oxo-7-phen-yl-, ethyl ester	EtOH	226(4.29),256(4.15), 295(4.04)	94-0930-84
5H-Pyrazolo[5,1-b][1,3]oxazine-3-carb-oxylic acid, 7-methyl-5-oxo-2-phen-yl-, ethyl ester	EtOH	230s(--),291(4.19)	94-0930-84
1H,5H-Pyrazolo[1,2-a]pyrazole-2-carbox-ylic acid, 3-methyl-1,5-dioxo-7-phen-yl-, ethyl ester	EtOH	231(4.07),249(4.08), 341(4.18)	94-0930-84
1H,7H-Pyrazolo[1,2-a]pyrazole-2-carbox-ylic acid, 3-methyl-1,7-dioxo-5-phen-yl-, ethyl ester	EtOH	244(4.04),285(3.89), 362(3.90)	94-0930-84
1H,7H-Pyrazolo[1,2-a]pyrazole-2-carbox-ylic acid, 5-methyl-1,7-dioxo-3-phen-yl-, ethyl ester	EtOH	265(3.88),364(4.09)	94-0930-84
Spiro[2,3-diazabicyclo[3.1.0]hex-2-ene-6,1'-[1H]indene]-1,5-dicarboxylic acid, dimethyl ester, (1α,5α,6α)-	EtOH	215(4.34),242(4.16)	23-2506-84
(1α,5α,6β)-	EtOH	220(4.21),236(4.16)	23-2506-84
$C_{16}H_{14}N_2O_4S$ 2-Propenoic acid, 2-cyano-3-[1-methyl-5-(phenylsulfonyl)-1H-pyrrol-2-yl]-, methyl ester	MeOH	210(4.19),252(3.88), 361(4.57)	4-1041-84
$C_{16}H_{14}N_2O_6$ Furo[3,4-b]quinoxalin-1(3H)-one, 3-(1,2-diacetoxyethyl)-, [S-(R*,S*)]-	EtOH	246(4.56),318(3.12)	136-0324-84J
$C_{16}H_{14}N_2O_6S_2$ 2-Furancarboxylic acid, 5,5'-[(cyano-carbonimidoyl)bis(thiomethylene)]-bis-, dimethyl ester	dioxan	268(3.61)	73-2285-84
$C_{16}H_{14}N_2S$ Benzenamine, N-(3-methyl-4-phenyl-2(3H)-thiazolylidene)-	EtOH	298(4.23)	56-0447-84
3-Quinolinecarbonitrile, 1,2,5,6,7,8-hexahydro-4-phenyl-2-thioxo-	dioxan	254(4.12),345(4.14), 426(3.45)	104-1402-84
Quinoxaline, 2-methyl-3-[(phenylmeth-yl)thio]-	EtOH	238(3.89),266(3.85), 347(3.63),356(3.62)	18-1653-84
$C_{16}H_{14}N_4$ 1H-Perimidine, 2-(3,5-dimethyl-1H-pyrazol-1-yl)-	MeOH	234(4.54),257(4.42), 333(4.15)	4-0911-84
$C_{16}H_{14}N_4O$ Pyridinium, 2-methyl-, N-[(2,2-dicyano-2-oxido-1-phenyl)ethylamino]-	EtOH	236(4.02),300(3.80), 485(4.80),502(4.29)	88-4429-84
$C_{16}H_{14}N_4O_2$ Benzonitrile, 4-[3-[(benzoyloxy)meth-yl]-3-methyl-1-triazenyl]-	MeCN	225(4.34),288(4.42), 300s(4.31)	23-0741-84
1,2-Naphthalenedione, 4-(4-amino-2-methyl-5-pyrimidinylmethylamino)-	pH 7	236(4.50),264(4.46), 274(4.42),300(4.20), 340(3.82),460(4.03)	94-3093-84

Compound	Solvent	$\lambda_{max}(\log \epsilon)$	Ref.
$C_{16}H_{14}N_4O_3S_2$ 2,4,6(1H,3H,5H)-Pyrimidinetrione, 1-methyl-5-[[[2-(methylthio)-5-phenyl-4-thiazolyl]imino]methyl]-	DMF	280(4.07),364(4.40)	103-0921-84
$C_{16}H_{14}N_4O_4$ 4-Oxa-1-azabicyclo[3.2.0]heptane-2-carboxylic acid, 7-oxo-3-[2-(4-phenyl-1H-1,2,3-triazol-1-yl)ethylidene]-, potassium salt, [2R-(2α,3Z,5α)]-	pH 6	241.5(4.14)	158-0885-84
3-Pyridinecarboxylic acid, 6-amino-5-cyano-2-methyl-4-(4-nitrophenyl)-, ethyl ester	EtOH	269(4.20),338(3.75)	103-1241-84
$C_{16}H_{14}N_4O_4S$ 2-Propenoic acid, 2-cyano-3-(methylthio)-3-(1,2,3,6-tetrahydro-2,6-dioxo-3-phenyl-4-pyrimidinyl)amino]-, methyl ester	EtOH	273(4.22),307(4.17)	94-0122-84
$C_{16}H_{14}N_6O_5S_2$ 4-Thia-1-azabicyclo[3.2.0]hept-2-ene-2-carboxylic acid, 3-[[(1-methyl-1H-tetrazol-5-yl)thio]methyl]-7-oxo-, (4-nitrophenyl)methyl ester, (±)-	$CHCl_3$	265(4.14),334(3.89)	32-0319-84
$C_{16}H_{14}N_6O_{12}$ 1(4H)-Pyridinehexanoic acid, α-(3,5-dinitro-4-oxo-1(4H)-pyridinyl)-3,5-dinitro-4-oxo-, (S)-	DMF	358(3.91)	18-1961-84
$C_{16}H_{14}O$ Ethanone, 2-(2-ethenylphenyl)-1-phenyl-	MeOH	241(4.15),277s(3.25), 287(3.14),310(2.59)	151-0327-84D
$C_{16}H_{14}O_2$ Acenaphtho[1,2-b][1,4]dioxocin, 8,9,10,11-tetrahydro-	EtOH	222(5.33),260(5.51)	2-1289-84
Anthracene, 1,4-dimethoxy-	EtOH	239(4.81),260(4.56), 345(3.93),363(3.10), 385(4.04),410(3.92)	40-1164-84
1,4-Butanedione, 1,4-diphenyl-	15% H_2SO_4 95% H_2SO_4	252(4.28) 297(4.41)	97-0147-84 97-0147-84
1,4-Epoxyanthracen-5-ol, 1,2,3,4,4a,10-hexahydro-2,3-bis(methylene)-, (1α,4α,4aα)-(±)-	MeCN	215(4.45),238s(4.00), 256(3.90),288(4.05)	78-4549-84
1,4-Epoxyanthracen-5-ol, 1,2,3,4,9,9a-hexahydro-2,3-bis(methylene)-, (1α,4α,9aα)-(±)-	MeCN	226(4.43),248s(4.09), 267s(3.90),291s(3.90), 302(3.95),317(3.90), 326s(3.83)	78-4549-84
1,4-Epoxyanthracen-5-ol, 1,2,3,4,9,10-hexahydro-2,3-bis(methylene)-	MeCN	225(4.16),232s(4.08), 244s(3.81),274s(3.36), 280(3.29)	78-4549-84
2,4-Hexadien-1-one, 1-(2-hydroxy-1-naphthalenyl)-, (E,E)-	EtOH	225(4.56),282(4.36), 370(3.65)	78-4081-84
2,6-Phenanthrenediol, 1,7-dimethyl-	MeOH	229(4.49),258(4.65), 273(4.47),289(4.19), 344(3.37),361(3.50)	39-1913-84C
Phenol, 2-(dihydro-4-methylene-1-benzopyran-3-yl)-	MeOH	254.6(3.60),285(3.37), 312(3.28)	78-4473-84

Compound	Solvent	λ_{max}(log ϵ)	Ref.
$C_{16}H_{14}O_2S$			
Benzo[b]thiophene, 4-methoxy-2-(4-methoxyphenyl)-	EtOH	215(4.49),236(4.23), 258(4.10),304s(4.37), 312(4.38),327(4.30), 340(4.08)	87-1057-84
Benzo[b]thiophene, 6-methoxy-2-(4-methoxyphenyl)-	EtOH	236(4.34),262(4.07), 273(4.06),310s(4.46), 317(4.48)	87-1057-84
$C_{16}H_{14}O_3$			
9(10H)-Anthracenone, 1,4-dimethoxy-	EtOH	208(4.27),240(4.78), 258(4.59),363(3.21), 386(3.21)	40-1164-84
9(10H)-Anthracenone-10-^{13}C, 2,4-dimethoxy-	MeOH	255(4.16),275s(4.08), 343(3.42)	102-0307-84
1H-2-Benzopyran-1-one, 3,4-dihydro-3-(2-methoxyphenyl)-	EtOH	236(4.00),280(3.61)	44-0742-84
1H-2-Benzopyran-1-one, 3,4-dihydro-3-(3-methoxyphenyl)-	EtOH	226s(4.19),283(3.62)	44-0742-84
1H-2-Benzopyran-1-one, 3,4-dihydro-3-(4-methoxyphenyl)-	EtOH	230(4.37),276(3.45), 282(3.45)	44-0742-84
$C_{16}H_{14}O_4$			
1H-2-Benzopyran-1-one, 3,4-dihydro-8-hydroxy-3-(4-methylphenyl)-	EtOH	230s(--),252s(--), 284(--),318(--)	44-0742-84
2(5H)-Furanone, 5-[3-[3-(1,2-heptadiene-4,6-diynyl)oxiranyl]-3-hydroxy-1-propenyl]dihydro- (cepacin A)	EtOAc	248(4.01),261.5(4.18), 277(4.09)	158-0431-84
Pacharin	MeOH	220(4.10),310(4.42)	78-4245-84
	MeOH-NaOMe	295(4.12),340(4.38)	78-4245-84
2-Propen-1-one, 3-(4-hydroxy-2-methoxyphenyl)-1-(4-hydroxyphenyl)-	EtOH	235(4.11),350(4.41)	138-0689-84
$C_{16}H_{14}O_4S$			
4-Benzofuranol, 2-methyl-, 4-methylbenzenesulfonate	MeOH	248(4.12),272(3.54), 281(3.38)	44-4399-84
5-Benzofuranol, 2-methyl-, 4-methylbenzenesulfonate	MeOH	247(4.15),273(3.60), 279(3.61),286(3.59)	44-4399-84
6-Benzofuranol, 2-methyl-, 4-methylbenzenesulfonate	MeOH	249(4.09),274(3.60), 286(3.50)	44-4399-84
7-Benzofuranol, 2-methyl-, 4-methylbenzenesulfonate	MeOH	248(4.07),274(3.41), 283(3.24)	44-4399-84
1-Naphthalenesulfonic acid, 6-oxo-1-cyclohexen-1-yl ester	ether	225(4.15),285(3.76)	44-2355-84
$C_{16}H_{14}O_5$			
2H,8H-Benzo[1,2-b:5,4-b']dipyran-3-carboxylic acid, 8-methyl-2-oxo-, ethyl ester	EtOH	230(4.495),235(4.499), 260(4.200),267(4.238), 282(4.150),335(4.050), 371(4.156)	111-0365-84
1H-2-Benzopyran-1-one, 3,4-dihydro-8-hydroxy-3-(3-hydroxy-4-methoxyphenyl)-	EtOH	288(3.63),318(3.71)	44-0742-84
2(5H)-Furanone, 5-[[3'-(1,2-heptadiene-4,6-diynyl)[2,2'-bioxiran]-3-yl]hydroxymethyl]dihydro- (cepacin B)	EtOAc	248(3.99),262(4.18), 277(4.08)	158-0431-84
2-Propen-1-one, 3-(3,4-dihydroxyphenyl)-1-(4-hydroxy-2-methoxyphenyl)-(sappanchalcone)	MeOH	250(3.73),363(4.40)	95-0935-84
	MeOH-AlCl$_3$	250(--),278(--), 330s(--),402(--)	95-0935-84

Compound	Solvent	$\lambda_{max}(\log \epsilon)$	Ref.
$C_{16}H_{14}O_5S$			
Benzeneacetic acid, 3-[(2-carboxyphen-yl)thio]-4-methoxy-	MeOH	224(4.42),252(3.99), 295(3.79),311(3.75), 320s(3.74)	73-2531-84
$C_{16}H_{14}O_6$			
Benz[b]indeno[1,2-b]pyran-3,4,6a,9,10-(6H)-pentol, 7,11b-dihydro- (hematox-ylin)-	pH 1 borate	288(3.62) 297(3.75)	3-1935-84 3-1935-84
2H,8H-Benzo[1,2-b:5,4-b']dipyran-3-carboxylic acid, 10-methoxy-2-oxo-, ethyl ester	EtOH	235(4.382),268(4.202)	111-0365-84
2-Furancarboxylic acid, 5-[4-(2-furan-yl)-3-(methoxycarbonyl)-1,3-butadi-enyl]-, methyl ester, (E,E)-	n.s.g.	207(4.10),297(4.18), 379(4.28)	73-2502-84
(E,Z)-	n.s.g.	207(4.12),295(4.00), 357(4.00)	73-2502-84
Furo[3,2-g][1]benzopyran-5-one, 2-ace-tyl-4,9-dimethoxy-7-methyl-	EtOH	228s(4.21),245(4.33), 282(4.48),337(4.02)	88-2953-84
Usnic acid, 2-deacetyl-, (+)-	MeOH	218(4.28),287(4.03), 330(3.49)	23-0320-84
isomer	MeOH	216(4.32),289(4.04), 328(3.52)	23-0320-84
9H-Xanthen-9-one, 1-hydroxy-2,6,8-tri-methoxy-	MeOH	203(4.44),239(4.38), 257(4.48),315(4.13)	94-2290-84
9H-Xanthen-9-one, 1-hydroxy-3,4,7-tri-methoxy-	MeOH	233(4.30),264(4.45), 311(3.88),382(3.68)	39-1507-84C
9H-Xanthen-9-one, 1-hydroxy-3,5,8-tri-methoxy-	MeOH	203(4.58),250(4.48), 273(4.15),327(4.08)	94-2290-84
9H-Xanthen-9-one, 1-hydroxy-3,7,8-tri-methoxy-	EtOH	238(3.32),258(4.46), 310(4.10),374(4.00)	102-1637-84
9H-Xanthen-9-one, 8-hydroxy-1,2,6-tri-methoxy-	MeOH	202(4.36),238(4.42), 261(4.33),308(4.07)	94-2290-84
9H-Xanthen-9-one, 8-hydroxy-1,3,5-tri-methoxy-	MeOH	203(4.45),247(4.41), 275(4.21),315(4.10)	94-2290-84
$C_{16}H_{14}O_6S$			
Propanedioic acid, [(1,3-dihydro-1,3-dioxo-2H-inden-2-ylidene)(methyl-thio)methyl]-, dimethyl ester	EtOH	232(4.30),252(4.35), 310(3.86),367(4.32)	95-0127-84
$C_{16}H_{14}O_7$			
Benzoic acid, 5-methoxy-2-[(5-methoxy-3,6-dioxo-1,4-cyclohexadien-1-yl)-oxy]-, methyl ester	MeOH	231(4.10),283(4.23)	39-1507-84C
5H-Furo[3,2-g][1]benzopyran-5-one, 7-(acetoxymethyl)-4,9-dimethoxy-	EtOH	214(4.32),247(4.54), 282(3.66),334(3.64)	44-5035-84
9H-Xanthen-9-one, 1,4-dihydroxy-2,3,7-trimethoxy-	EtOH	234(4.45),277(4.53), 306s(3.95),400(3.68)	39-1507-84C
$C_{16}H_{14}S$			
Anthracene, 9-(ethylthio)-	EtOH	217(3.85),256(4.88), 336(2.91),354(3.20), 373(3.55),394(3.50)	40-1158-84
$C_{16}H_{15}BrO_2$			
Benzoic acid, 4-bromo-, 5,6-bis(methyl-ene)bicyclo[2.2.1]hept-2-yl-, (1S-endo)-	isooctane EtOH	200(4.37),244(4.45) 204(4.37),246(4.45)	33-0600-84 33-0600-84
exo	isooctane EtOH	202(4.44),245(4.49) 204(4.39),245(4.47)	33-0600-84 33-0600-84

Compound	Solvent	$\lambda_{max}(\log \epsilon)$	Ref.
$C_{16}H_{15}BrO_2S$ Benzoic acid, 5-bromo-2-[[4-(1-methyl-ethyl)phenyl]thio]-	MeOH	220(4.47),263(4.09), 330(3.68)	73-0086-84
$C_{16}H_{15}ClN_2$ 2H-Indazole, 5-chloro-3-(1-methyl-ethyl)-2-phenyl-	EtOH	212(4.5),227(4.5), 284(3.9),311(3.9)	35-6015-84
2H-Indazole, 2-(4-chlorophenyl)-3-(1-methylethyl)-	EtOH	208(4.4),230(4.3), 283(3.9),304(3.8)	35-6015-84
$C_{16}H_{15}ClN_2O$ Benzeneethanamine, N-(5-chloro-3-meth-yl-2(3H)-benzoxazolylidene)-	CHCl$_3$	259(3.90),305(4.12)	94-3053-84
$C_{16}H_{15}ClN_2O_2$ Pyrano[4,3-c]pyrazol-4(1H)-one, 1-(4-chlorophenyl)-6,7-dihydro-6-methyl-3-(1-propenyl)-	EtOH	256(4.31)	4-0017-84
$C_{16}H_{15}ClN_2O_3$ 3-Pyridinecarboxylic acid, 4-(4-chloro-phenyl)-5-cyano-1,4,5,6-tetrahydro-2-methyl-6-oxo-, ethyl ester	EtOH	219s(4.15),282(4.11)	103-1241-84
$C_{16}H_{15}ClN_4O_3$ 1H-Tetrazole-1-propanoic acid, 5-[5-(4-chlorophenyl)-2-furanyl]-, ethyl ester	dioxan	314(4.27),329s(4.23)	73-1699-84
2H-Tetrazole-2-propanoic acid, 5-[5-(4-chlorophenyl)-2-furanyl]-, ethyl ester	dioxan	309(4.17),323s(4.10)	73-1699-84
$C_{16}H_{15}ClOS$ 4bH,5aH-4a,5b-Methano-10H-bisbenzo-[2,3]cyclopropa[1,2-b:2',1'-e]-thiopyran, 4b-chloro-1,4-dihydro-,5-oxide	dioxan	225s(3.45),258(3.47), 280s(3.25)	89-0717-84
$C_{16}H_{15}ClO_2$ Benzoic acid, 5,6-bis(methylene)bi-cyclo[2.2.1]hept-2-yl ester, (1S-endo)-	isooctane EtOH	202(4.43),240(4.42) 204(4.36),242(4.40)	33-0600-84 33-0600-84
exo-	isooctane EtOH	202(4.34),240(4.46) 206(4.35),242(4.41)	33-0600-84 33-0600-84
$C_{16}H_{15}Cl_2N_3O_3$ 4H-Pyrido[1,2-a]pyrimidine-3-carbox-amide, N-(3,5-dichloro-4-hydroxy-phenyl)-6,7,8,9-tetrahydro-6-methyl-4-oxo-	EtOH	310(4.19)	118-0582-84
$C_{16}H_{15}Cl_3N_2O$ Benzenecarboximidamide, 3,5-dichloro-N-(4-chlorophenyl)-N'-hydroxy-2,4,6-trimethyl-	MeOH	260(4.22)	32-0015-84
$C_{16}H_{15}Cl_4NS_2$ Benzenesulfenamide, 3,5-dichloro-N-[(3,5-dichlorophenyl)thio]-N-(1,1-dimethylethyl)-	hexane	256(4.27)	44-2724-84

Compound	Solvent	λ_{max}(log ϵ)	Ref.
$C_{16}H_{15}FN_2$			
2H-Indazole, 5-fluoro-3-(1-methyl-ethyl)-2-phenyl-	EtOH	205(4.5),217(4.4), 274(4.0),308(4.0)	35-6015-84
2H-Indazole, 2-(4-fluorophenyl)-3-(1-methylethyl)-	EtOH	209(4.1),223(4.1), 280(3.8)	35-6015-84
$C_{16}H_{15}FO_2S$			
Benzoic acid, 5-fluoro-2-[[4-(1-methyl-ethyl)phenyl]thio]-	MeOH	223(4.37),253(3.94), 275s(3.67),329(3.56)	73-0086-84
$C_{16}H_{15}N$			
5H-Indeno[2,1-c]isoquinoline, 6,6a,7,11b-tetrahydro-, (6aR-cis)-	MeOH	194.5(4.85),217s(4.49), 260(3.02),266(3.18), 273(3.20)	39-1655-84C
hydrochloride	MeOH	195(4.83),211s(4.29), 215(4.23),260(2.86), 265(2.98),272(2.98)	39-1655-84C
1H-Indole, 5-(2-phenylethyl)-	EtOH	210(4.4),224(4.5), 273(3.7),286(3.6), 297(3.4)	103-0372-84
Methanamine, 1-(1H-cyclopent[e]azulen-1-ylidene)-N,N-dimethyl-	CH_2Cl_2	277(4.50),305(4.18), 387(4.30),413(4.27), 451s(3.90),560(2.82)	35-6383-84
Methanamine, 1-(3H-cyclopent[e]azulen-5-ylidene)-N,N-dimethyl-	CH_2Cl_2	255(4.26),296(4.22), 394(4.38),412(4.30), 450s(3.87),570(2.98)	35-6383-84
$C_{16}H_{15}NO$			
1H-Inden-1-ol, 2,3-dihydro-2-[(phenyl-methylene)amino]-, (1S-trans)-	MeOH	195.5(4.78),205(4.62), 210s(4.50),249(4.36), 272(3.79),279s(3.46), 287s(3.21)	39-1655-84C
$C_{16}H_{15}NOS_2$			
Carbamothioic acid, (3-methylphenyl)-(phenylthioxomethyl)-, O-methyl ester	MeOH	505(2.21)	97-0438-84
Carbamothioic acid, phenyl(phenylthi-oxomethyl)-, O-ethyl ester	MeOH	506(2.23)	97-0438-84
1H-Thiopyran-4-carbonitrile, 1-methyl-3-(methylthio)-5-(2-phenylethenyl)-, 1-oxide	EtOH	235.5(4.40),286.5(4.49), 320s(4.24),362(3.83)	118-0852-84
$C_{16}H_{15}NO_2$			
3H-Benz[e]indene-5-carbonitrile, 3a,4,5,9b-tetrahydro-7-methoxy-3a-methyl-3-oxo-	EtOH	227(4.12),279(3.44)	2-0395-84
2,4,6-Heptatrienoic acid, 2-cyano-7-phenyl-, ethyl ester	EtOH	382(4.52)	70-2328-84
$C_{16}H_{15}NO_2S_2$			
2H-Pyran-3-carbonitrile, 4-(methyl-thio)-6-[2-(methylthio)-2-phenyl-ethyl]-2-oxo-	EtOH	235(4.22),303(4.12), 350(4.06)	94-3384-84
$C_{16}H_{15}NO_4$			
3-Pyridinecarboxylic acid, 5-acetyl-1,4-dihydro-2,6-dimethyl-4-oxo-1-phenyl-	EtOH	262(4.17),300(--)	39-1317-84B
	dioxan	262(4.22),300(--)	39-1317-84B
	$CHCl_3$	262(4.22),300(--)	39-1317-84B
	80% DMSO	260(4.17),295s(--)	39-1317-84B

Compound	Solvent	$\lambda_{max}(\log \epsilon)$	Ref.
3,4-Pyridinedicarboxylic acid, 2-methyl-6-phenyl-, dimethyl ester	EtOH	262(4.28),297(4.33)	88-0959-84
1H-Pyrrole-3-carboxylic acid, 4,5-dihydro-4,5-dioxo-2-phenyl-1-(2-propenyl)-, ethyl ester	dioxan	288s(3.64),405(3.51)	94-0497-84
$C_{16}H_{15}NO_6$			
4,9-Epoxy-1H-benz[f]isoindole-1,3(2H)-dione, 4-acetoxy-3a,4,9,9a-tetrahydro-5-methoxy-2-methyl-, (3aα,4β,9α,9aα)-	MeOH	220(3.77),273(3.21), 280(3.22)	12-1699-84
$C_{16}H_{15}NS$			
2H-Isoindole, 1-(ethylthio)-2-phenyl-	EtOH	203(4.42),214(4.31), 236(4.32),287(3.54), 300(3.37),340(3.35)	103-0978-84
$C_{16}H_{15}N_3O$			
Diazene, (1-isocyanato-1-phenylethyl)-(4-methylphenyl)-	C_6H_{12}	280(4.11),400(2.35)	118-0315-84
1H-1,2,4-Triazole, 1-(4-methoxyphenyl)-5-(4-methylphenyl)-	MeCN	230(4.33),247s(4.22)	48-0311-84
1H-1,2,4-Triazole, 3-(4-methoxyphenyl)-1-(4-methylphenyl)-	MeOH	255(4.27),299s(4.06), 282(4.13)	48-0311-84
$C_{16}H_{15}N_3OS$			
1,3,4-Thiadiazol-2(3H)-one, 5-(4-methylphenyl)-3-[(phenylamino)methyl]-	EtOH	273(4.07)	83-0547-84
$C_{16}H_{15}N_3O_2$			
Acetic acid, cyano(1-methyl-2-phenyl-4(1H)-pyrimidinylidene)-, ethyl ester	EtOH	362(4.46)	103-1270-84
1H-1,5-Benzodiazepine, 2,3-dihydro-4-methyl-2-(4-nitrophenyl)-	EtOH	258(4.16),321s(--)	103-1370-84
3-Pyridinecarboxylic acid, 6-amino-5-cyano-2-methyl-4-phenyl-, ethyl ester	EtOH	256(4.34),333(3.92)	103-1241-84
4-Pyrimidineacetic acid, α-cyano-α-methyl-2-phenyl-, ethyl ester	EtOH	262(4.35)	103-1270-84
$C_{16}H_{15}N_3O_2S$			
1H-Pyrazole-3-propanoic acid, 5-amino-1-phenyl-4-(2-thienyl)-	EtOH	250(3.77)	4-1585-84
$C_{16}H_{15}N_3O_4$			
Benzoic acid, 2-[(5-cyano-4,6-dimethoxy-2-pyridinyl)amino]-, methyl ester	EtOH	221(4.14),248(3.61), 275(3.42),306(4.10), 340(4.08)	83-0824-84
Pyrano[4,3-c]pyrazol-4(1H)-one, 6,7-dihydro-6-methyl-1-(4-nitrophenyl)-3-(1-propenyl)-	EtOH	250(4.15)	4-0017-84
$C_{16}H_{15}N_3O_5$			
3-Pyridinecarboxylic acid, 5-cyano-1,4,5,6-tetrahydro-2-methyl-4-(3-nitrophenyl)-6-oxo-, ethyl ester	EtOH	217s(4.20),273(4.27)	103-1241-84
3-Pyridinecarboxylic acid, 5-cyano-1,4,5,6-tetrahydro-2-methyl-4-(4-nitrophenyl)-6-oxo-, ethyl ester	EtOH	219s(4.02),273(4.28)	103-1241-84
Uridine, 5-(4-cyanophenyl)-2'-deoxy-	pH 1	287(4.27)	88-2431-84
	pH 13	289(4.21)	88-2431-84

Compound	Solvent	$\lambda_{max}(\log \epsilon)$	Ref.
$C_{16}H_{15}N_3S$ Diazene, (1-isothiocyanato-1-phenyl-ethyl)(4-methylphenyl)-	C_6H_{12}	285(4.11),403(2.36)	118-0315-84
$C_{16}H_{15}N_3S_2$ 1,3,4-Thiadiazole-2(3H)-thione, 5-(4-methylphenyl)-3-[(phenylamino)-methyl]-	EtOH	337(4.01)	83-0547-84
$C_{16}H_{15}N_5$ 2,3-Pyrazinedicarbonitrile, 5-(diethyl-amino)-6-phenyl-	MeCN	223(4.13),247(4.02), 325(4.22),380(3.76)	44-0813-84
$C_{16}H_{15}N_5O_2$ Benzoic acid, 4-[2-(2,4-diaminopyrido-[3,2-d]pyrimidin-6-yl)ethyl]-	pH 13	238(4.53),343(3.72)	87-0376-84
$C_{16}H_{15}N_5O_6S$ Benzenepropanamide, α-[(methylsulfonyl)-amino]-N-(7-nitro-4-benzofurazanyl)-, (S)-	buffer	284(3.73),386(4.07)	94-0336-84
$C_{16}H_{16}$ Biphenylene, 2,6-diethyl-	C_6H_{12}	<u>247(4.8),258(5.0), 330s(3.6),347(3.9), 370(4.0)</u>	88-3603-84
$C_{16}H_{16}BrN_3$ Pyrrolidine, 1-[4-[(4-bromophenyl)azo]-phenyl]-	C_6H_{12} EtOH EtOH-HCl	418(4.59) 425(4.52) 535(4.79)	39-0149-84B 39-0149-84B 39-0149-84B
$C_{16}H_{16}Br_2O_2S$ Benzene, 1,1'-[thiobis(2,1-ethanediyl-oxy)]bis[2-bromo-	EtOH	211(3.79),275(3.30), 282(3.26)	12-2059-84
$C_{16}H_{16}Br_2O_3$ Benzene, 1,1'-[oxybis(2,1-ethanediyl-oxy)]bis[2-bromo-	EtOH	277(3.67)	12-2059-84
$C_{16}H_{16}ClIN_4O_7$ 9H-Purine, 6-chloro-2-iodo-9-(2,3,5-tri-O-acetyl-β-D-ribofuranosyl)-	MeOH	222.5(4.32),258(3.82), 281(3.97)	44-4340-84
$C_{16}H_{16}ClNO_2S$ 2H-Thieno[2,3-h]-1-benzopyran-2-one, 3-chloro-5,6-dihydro-4-piperidino-	EtOH	218(4.24),219.5(4.23), 243(4.24),273s(3.77), 351(4.33)	151-0081-84
$C_{16}H_{16}ClN_3$ Pyrrolidine, 1-[4-[(4-chlorophenyl)-azo]phenyl]-	C_6H_{12} EtOH EtOH-HCl	416(4.53) 423(4.49) 534(4.77)	39-0149-84B 39-0149-84B 39-0149-84B
$C_{16}H_{16}Cl_2N_2O$ Benzenecarboximidamide, 3,5-dichloro-N-hydroxy-2,4,6-trimethyl-N'-phenyl-	MeOH	252(4.14)	32-0015-84
$C_{16}H_{16}FN_3O_2$ 4H-Pyrido[1,2-a]pyrimidine-3-carbox-amide, N-(2-fluorophenyl)-6,7,8,9-	EtOH	225(4.12),310(4.26)	118-0582-84

Compound	Solvent	$\lambda_{max}(\log \epsilon)$	Ref.
tetrahydro-6-methyl-4-oxo- (cont.)			118-0582-84
$C_{16}H_{16}FeN$			
Pyridinium, 1-(ferrocenylmethyl)-, perchlorate	EtOH	260(3.84),330(3.34), 425(2.45)	65-1067-84
tetrafluoroborate	EtOH	260(3.87),330(3.49), 425(2.54)	65-1067-84
$C_{16}H_{16}FeO_2$			
Ferrocene, (2-acetyl-3-oxo-1-butenyl)-	CH_2Cl_2	492(3.30)	30-0192-84
$C_{16}H_{16}FeO_4S$			
Ferrocene, 1,1'-[thiobis(2,1-ethanediyloxycarbonyl)]-	dioxan	448(2.36)	18-2435-84
$C_{16}H_{16}FeO_5$			
Ferrocene, 1,1'-[oxybis(2,1-ethanediyloxycarbonyl)]-	MeOH	450(2.45)	18-2435-84
$C_{16}H_{16}NS$			
1H-Isoindolium, 3-(ethylthio)-2-phenyl-, tetrafluoroborate	$CHCl_3$	248(4.37),292(3.74), 304(3.65),344(3.77)	103-0978-84
$C_{16}H_{16}N_2$			
2H-Indazole, 3-(1-methylethyl)-2-phenyl-	EtOH	278(3.8),303(3.8)	35-6015-84
$C_{16}H_{16}N_2O$			
1H-Pyrazole, 4,5-dihydro-3-(4-methoxyphenyl)-5-phenyl-	EtOH	225(3.72),285(4.17)	34-0225-84
1H-Pyrazole, 4,5-dihydro-5-(4-methoxyphenyl)-3-phenyl-	EtOH	224(4.13),284(4.03), 290(4.02)	34-0225-84
Pyridine, 5-(4,5-dihydro-4,4-dimethyl-2-oxazolyl)-2-phenyl-	EtOH	245(4.05)	44-0056-84
$C_{16}H_{16}N_2OS$			
3-Pyridinecarbonitrile, 4-(4-ethoxyphenyl)-1,2-dihydro-5,6-dimethyl-2-thioxo-	dioxan	316(4.40),422(3.63)	104-1837-84
Thieno[2,3-g][1,4]oxazonine-8-carbonitrile, 6,7,9,10-tetrahydro-4-phenyl-	MeOH	246(3.74)	12-1043-84
$C_{16}H_{16}N_2O_2$			
5,6-Diazatricyclo[5.2.0.0^{2,5}]non-8-en-4-one, 6-benzoyl-3,9-dimethyl-, (1α,2α,3α,7α)-	MeOH	228(4.15)	23-2440-84
10-Oxa-2,3-diazabicyclo[4.3.1]deca-1,4,7-triene, 3-benzoyl-6,9-dimethyl-, endo	MeOH	204(4.17),227(3.87), 280(3.90)	23-2440-84
Phenazine, 1,6-dimethoxy-2,3-dimethyl-	EtOH	268.6(4.97),374.5(3.97)	44-5116-84
Phenazine, 1,9-dimethoxy-2,3-dimethyl-	EtOH	269(4.94),373(3.95)	44-5116-84
Pyrano[4,3-c]pyrazol-4(1H)-one, 6,7-dihydro-6-methyl-3-(1-methylethenyl)-1-phenyl-	EtOH	248(4.19)	4-0017-84
Pyrano[4,3-c]pyrazol-4(1H)-one, 6,7-dihydro-7-methyl-3-(1-methylethenyl)-1-phenyl-	EtOH	252(4.16)	4-0017-84
Pyrano[4,3-c]pyrazol-4(1H)-one, 6,7-dihydro-6-methyl-1-phenyl-3-(1-propenyl)-	EtOH	259(4.24)	4-0017-84

Compound	Solvent	$\lambda_{max}(\log \epsilon)$	Ref.
Pyrano[4,3-c]pyrazol-4(1H)-one, 6,7-dihydro-7-methyl-1-phenyl-3-(1-propenyl)-	EtOH	250(4.21)	4-0017-84
$C_{16}H_{16}N_2O_2S$			
Benzenamine, N-[2-(5,6-dimethyl-2H-thiopyran-2-ylidene)propylidene]-4-nitro-, hydrochloride	EtOH	513(4.30)	97-0183-84
Pyrazolo[1,5-a]quinoline-3-carboxylic acid, 5-methyl-2-(methylthio)-, ethyl ester	EtOH	233(4.21),265(4.70), 320(4.08),343(4.01), 358(3.97)	95-0440-84
$C_{16}H_{16}N_2O_2S_3$			
2(1H)-Pyridinethione, 4,4'-thiobis[1-ethenyl-3-hydroxy-6-methyl-	EtOH	268(3.65),415(4.20)	1-0293-84
$C_{16}H_{16}N_2O_3$			
1,2-Diazabicyclo[5.2.0]nona-3,5-diene-2-carboxylic acid, 8-methyl-9-oxo-, phenylmethyl ester, cis	MeOH	274(3.98)	23-2440-84
5,6-Diazatricyclo[5.2.0.0^{2,5}]non-8-ene-6-carboxylic acid, 3-methyl-4-oxo-, phenylmethyl ester, (1α,2α,3α,7α)-	MeOH	242(2.89)	23-2440-84
10-Oxa-2,3-diazabicyclo[4.3.1]deca-1,4,7-triene-3-carboxylic acid, 9-methyl-, phenylmethyl ester	MeOH	207(4.13),229(4.03)	23-2440-84
Phenazine, 1,9-dimethoxy-2,3-dimethyl-, 5-oxide	EtOH	279(5.13),368(3.67), 389(3.91),451(3.80)	44-5116-84
3-Pyridinecarboxylic acid, 5-cyano-1,4,5,6-tetrahydro-2-methyl-6-oxo-4-phenyl-, ethyl ester	EtOH	283(4.08)	103-1241-84
1H-Pyrrole-2-carboxylic acid, 4-cyano-5-methoxy-3-methyl-1-phenyl-, ethyl ester	EtOH	219(4.29),274(4.10)	118-0062-84
$C_{16}H_{16}N_2O_4$			
Butanediamide, N,N'-dihydroxy-N,N'-di-phenyl-	MeOH	210(4.26),240(4.29)	34-0345-84
1H-Perimidine-2-acetic acid, 2,3-di-hydro-2-(methoxycarbonyl)-, methyl ester	EtOH	214(4.45),250(3.66), 328(3.83),336(3.84)	150-0016-84S
Phenazine, 1,6-dimethoxy-2,3-dimethyl-, 5,10-dioxide	EtOH	261(4.30),281.8(4.39), 291.4(4.26)	44-5116-84
3-Pyridinecarboxylic acid, 5-(amino-carbonyl)-1,6-dihydro-2-methyl-6-oxo-4-phenyl-, ethyl ester	EtOH	252(4.26),320(3.95)	103-1241-84
$C_{16}H_{16}N_2O_7$			
4-Oxa-1-azabicyclo[3.2.0]heptane-2-carboxylic acid, 3-(2-methoxyethyli-dene)-7-oxo-, (4-nitrophenyl)methyl ester, [2R-(2α,3Z,5α)]-	EtOH	308(4.16)	39-1599-84C
$C_{16}H_{16}N_2O_8$			
3-Pentenoic acid, 5-methoxy-2-oxo-3-[(4-oxo-2-azetidinyl)oxy]-, (4-nitrophenyl)methyl ester, [R-(Z)]-	EtOH	267(4.09)	39-1599-84C
$C_{16}H_{16}N_2O_8S$			
Propanoic acid, 3-[[1-[4-[[(2,5-dioxo-1-pyrrolidinyl)oxy]carbonyl]phenyl]-	MeOH	238(4.32)	94-5036-84

Compound	Solvent	$\lambda_{max}(\log \epsilon)$	Ref.
2-nitroethyl]thio]- (cont.)			94-5036-84
$C_{16}H_{16}N_4O$ 3,6-Pyridazinedione, tetrahydro-1-phenyl-, 6-(phenylhydrazone)	dioxan	285(0.696)	2-0439-84
$C_{16}H_{16}N_4O_2$ Pyrrolidine, 1-[4-[(4-nitrophenyl)azo]-phenyl]-	C_6H_{12} EtOH EtOH-HCl	459(4.54) 488(4.54) 518(4.83)	39-0149-84B 39-0149-84B 39-0149-84B
$C_{16}H_{16}N_4O_2S_2$ Thiazolo[3,2-a]pyrimidin-4-ium, 2-[[(4,6-dimethyl-2-pyrimidinyl)-thio]acetyl]-3-hydroxy-5,7-dimethyl-, hydroxide, inner salt	MeCN CCl$_4$	285(4.12),430(3.93) 470(--)	103-0791-84 103-0791-84
$C_{16}H_{16}N_4O_3$ 1H-Tetrazole-1-acetic acid, 5-[5-(3-methylphenyl)-2-furanyl]-, ethyl ester	dioxan	313(4.48),328s(4.20)	73-1699-84
2H-Tetrazole-2-acetic acid, 5-[5-(3-methylphenyl)-2-furanyl]-, ethyl ester	dioxan	309(4.42),324s(4.20)	73-1699-84
$C_{16}H_{16}N_4O_3S$ Thiazolo[2,3-b]purine-4,8(1H,7H)-dione, 5a-ethoxy-5,5a-dihydro-5-methyl-7-phenyl-	MeOH	263(4.10),280(4.06)	2-0316-84
$C_{16}H_{16}N_4O_3S_2$ Benzeneacetic acid, α-[[6,9-dihydro-1,9-dimethyl-8-(methylthio)-6-oxo-1H-purin-2-yl]thio]-	MeOH	276(4.31),300s(4.19)	2-0316-84
$C_{16}H_{16}N_4O_4$ 1H-Tetrazole-1-acetic acid, 5-[5-(2-methoxyphenyl)-2-furanyl]-, ethyl ester	dioxan	324(4.38),340s(4.25)	73-1699-84
1H-Tetrazole-1-acetic acid, 5-[5-(4-methoxyphenyl)-2-furanyl]-, ethyl ester	dioxan	323(4.42),338s(4.28)	73-1699-84
2H-Tetrazole-2-acetic acid, 5-[5-(2-methoxyphenyl)-2-furanyl]-, ethyl ester	dioxan	319(4.43),335s(4.31)	73-1699-84
2H-Tetrazole-2-acetic acid, 5-[5-(4-methoxyphenyl)-2-furanyl]-, ethyl ester	dioxan	317(4.44),333s(4.24)	73-1699-84
$C_{16}H_{16}N_4O_4S_2$ Acetic acid, [[(2,3,4,10-tetrahydro-3,10-dimethyl-2,4-dioxobenzo[g]-pteridin-6-yl)methyl]dithio]-, methyl ester	6M HCl MeCN	217(4.41),266(4.32), 379(4.20) 225(4.24),268(4.42), 350(3.99),442(4.05)	35-3309-84 35-3309-84
2-Propenenitrile, 3-(methylthio)-2-(phenylsulfonyl)-3-[(1,2,3,4-tetrahydro-1,3-dimethyl-2,6-dioxo-4-pyrimidinyl)amino]-	EtOH	264(4.21),306(4.13), 357(4.26)	94-0122-84
Pyrido[2,3-d]pyrimidine-2,4(1H,3H)-dione, 5-amino-1,3-dimethyl-7-(methylthio)-6-(phenylsulfonyl)-	EtOH	224(4.25),268(4.61), 307(4.22)	94-0122-84

Compound	Solvent	$\lambda_{max}(\log \epsilon)$	Ref.
$C_{16}H_{16}N_4O_5$			
Inosine, 2-phenyl-	EtOH	260(3.69),290(3.76)	44-4340-84
$C_{16}H_{16}N_5O_2S$			
1,2,4,5-Tetrazin-1(2H)-yl, 3,4-dihydro-2,4-dimethyl-6-phenyl-3-[(phenylsulfonyl)imino]-	CHCl$_3$	460(2.60)	104-1224-84
$C_{16}H_{16}N_6O_8$			
2,3-Butanediamine, 1,1,4,4-tetranitro-N,N'-diphenyl-	CHCl$_3$	230-280(3.67),370(3.73) (changing)	104-2277-84
1H-1,2,3-Triazole-4,5-dicarboxylic acid, 1,1'-[1,2-bis(methylene)-1,2-ethanediyl]bis-, tetramethyl ester	MeCN	220(4.135)	89-0736-84
1H-1,2,3-Triazole-4,5-dicarboxylic acid, 1,1'-(2-butyne-1,4-diyl)bis-, tetramethyl ester	MeCN	221(4.01)	1-0623-84
$C_{16}H_{16}N_6S$			
Propanenitrile, 3,3'-[[3-methyl-4-(2-thiazolylazo)phenyl]imino]-	acetone	447(4.49)	7-0819-84
	50% acetone	477(4.49)	7-0819-84
$C_{16}H_{16}O$			
Bicyclo[2.2.1]hept-2-ene-2-carboxaldehyde, 3-(tetracyclo[3.2.0.0^{2,7}.0^{4,6}]heptylidenemethyl)-	MeCN	322(4.37)	24-2027-84
Dibenzofuran, 1,2,3,4-tetramethyl-	EtOH	287.5(4.28),312(3.51)	39-0799-84B
	diisopropyl ether	287.5(4.29),312.5(3.55)	39-0799-84B
	THF	287.5(4.30),312(3.51)	39-0799-84B
	CHCl$_3$	289.0(4.30),310s(3.30)	39-0799-84B
	CH$_2$Cl$_2$	289.0(4.30),310s(3.30)	39-0799-84B
	C$_2$H$_4$Cl$_2$	290.0(4.24),310s(3.53)	39-0799-84B
$C_{16}H_{16}ORu$			
Ruthenocene, 1,1'-(1-oxo-1,3-propanediyl)-3,3'-(1,3-propanediyl)-	EtOH	254(3.77),340s(3.06), 372(3.12)	18-0719-84
$C_{16}H_{16}O_2$			
Bicyclo[2.2.1]heptan-2-ol, 5,6-bis(methylene)-, (1S-endo)-	isooctane	230(4.30),248s(4.04), 256s(3.85)	33-0600-84
	EtOH	202(4.14),233(4.24), 242(4.12)	33-0600-84
(1S-exo)-	isooctane	198(4.52),230(4.34), 242s(4.06),252s(3.89)	33-0600-84
	EtOH	202(4.23),236(4.32), 254s(3.86)	33-0600-84
2,5-Epoxynaphtho[2,3-b]oxepin, 2,3,4,5-tetrahydro-2,5-dimethyl-	ether	257(3.51),267(3.67), 278(3.73),290(3.59), 307s(3.02),320(3.34), 334(3.47)	22-0187-84
2,6-Phenanthrenediol, 9,10-dihydro-1,7-dimethyl-	MeOH	218(4.45),277(4.14), 297(3.94),314s(3.88)	39-1913-84C
2-Propanone, 1-(2,3-dihydro-3-methylnaphtho[2,3-b]furan-2-yl)-, trans	ether	255(3.55),265(3.68), 275(3.72),287(3.55), 318(3.42),332(3.57)	22-0187-84
$C_{16}H_{16}O_2Ru$			
Ruthenocene, 1-(2-carboxyethenyl)-1',3-(1,3-propanediyl)-	EtOH	275(4.25),365(2.85)	18-0719-84

Compound	Solvent	λ_{max}(log ϵ)	Ref.
C$_{16}$H$_{16}$O$_3$			
Benzeneacetic acid, α-(2-hydroxy-5-methylphenyl)-α-phenyl-, methyl ester	MeOH	285(3.23)	35-7352-84
Benzeneacetic acid, α-(2-hydroxy-6-methylphenyl)-α-phenyl-, methyl ester	MeOH	285(3.22)	35-7352-84
1H-Benz[e]indene-5-carboxylic acid, 2,3,4,5-tetrahydro-3-oxo-, ethyl ester	EtOH	231(3.98),299(4.00)	2-0395-84
2,5-Cyclohexadiene-1,4-dione, 2-[4-(4-methyl-3-pentenyl)-2-furanyl]-, (echinofuran B)	EtOH	259(4.20),453(3.70)	102-0301-84
Dibenzofuran, 3,7-dimethoxy-1,9-dimethyl-	MeOH	218s(4.37),226(4.39), 264(4.01),298s(4.12), 308(4.17)	39-1613-84C
Ethanone, 1-[3-methoxy-4-(phenylmethoxy)phenyl]-	EtOH	231(4.01),275(3.93), 306(3.79)	39-0825-84C
1,4-Naphthalenedione, 6-methyl-3-(3-methyl-2-butenyl)-	MeOH	258(4.381),288(4.232), 332(3.472),390(3.108)	87-0990-84
2H-Naphtho[1,2-b]pyran-5,6-dione, 3,4-dihydro-2,2,8-trimethyl-	MeOH	258(4.437),266(4.422), 288(3.040),343(3.474), 434(3.176)	87-0990-84
4(1H)-Phenanthrenone, 2,3-dihydro-7,9-dimethoxy-	EtOH	206(3.94),224(4.17), 254(4.41),316(3.92), 350(3.78)	56-1071-84
C$_{16}$H$_{16}$O$_3$S			
1-Azuleneacetic acid, 4(and 8)-[(methylthio)methyl]-α-oxo-, ethyl ester	CH$_2$Cl$_2$	235(4.36),313(4.43), 386s(--),396(4.03), 512(2.85)	83-0984-84
Benzoic acid, 2-[[[4-(methoxymethyl)-phenyl]thio]methyl]-	MeOH	223s(4.24),259(3.94), 280s(3.67)	73-1722-84
2-Propenoic acid, 3-(2-furanyl)-2-[[(phenylmethyl)thio]methyl]-, methyl ester	EtOH	305(3.34)	73-1764-84
C$_{16}$H$_{16}$O$_4$			
9,10-Anthracenediol, 9,10-dihydro-1,4-dimethoxy-	EtOH	206(5.04),235(4.22), 293(4.17)	40-1164-84
Benzeneacetic acid, 4-methoxy-3-(phenylmethoxy)-	MeOH	227s(--),279(3.46)	44-5243-84
2,5-Furandione, dihydro-3-[1-(4-methoxyphenyl)ethylidene]-4-(1-methylethylidene)-, (E)-	C$_6$H$_{12}$	332(3.96)	48-0233-84
(Z)-	C$_6$H$_{12}$	333(4.09)	48-0233-84
Naphtho[1,2-b]furan-4,5-dione, 2,3-dihydro-7-hydroxy-2,2,3,9-tetramethyl-, (±)-	MeOH	275(4.36),280(4.37), 314(3.76),516(3.24)	44-1853-84
Naphtho[1,2-b]furan-4,5-dione, 2,3-dihydro-7-hydroxy-2,3,3,9-tetramethyl-, (±)-	MeOH	275(4.47),284(4.49), 314(3.80),522(3.38)	44-1853-84
1-Propanone, 1-(2,6-dihydroxy-4-methoxyphenyl)-3-phenyl-	EtOH	211s(4.14),225(4.15), 284(4.23),323s(3.50)	102-1198-84
Spiro[2,5-cyclohexadiene-1,1'(2'H)-naphthalen]-4-one, 3',4'-dihydro-6',7'-dihydroxy-2-methoxy- (spirobroussonin B)	EtOH	210(4.57),230(4.25), 260s(3.97)	138-0693-84
Spiro[3,5-cyclohexadiene-1,1'(2'H)-naphthalen]-2-one, 3',4'-dihydro-6',7'-dihydroxy-4-methoxy-	EtOH	211(4.61),240s(3.85), 288(3.83),294s(3.81), 310s(3.60)	138-0693-84

Compound	Solvent	$\lambda_{max}(\log \epsilon)$	Ref.
$C_{16}H_{16}O_5$			
Acetic acid, (3-hydroxy-5-oxo-4-phenyl-2(5H)-furanylidene)-, 1,1-dimethyl-ethyl ester, (E)-	EtOH	236(4.15),272(4.28), 363(4.12)	39-1555-84C
	EtOH-NaOH	268(4.49),370(3.85)	39-1555-84C
(Z)-	EtOH	270(4.66),337(4.02)	39-1555-84C
	EtOH-NaOH	269(4.78),384(3.89)	39-1555-84C
9(4H)-Anthracenone, 4a,10-dihydro-1,10-dihydroxy-4,4-dimethoxy-, cis	MeOH	238(3.97),262s(3.94), 310(3.51),377(4.06)	12-1699-84
2,5-Furandione, 3-[(3,5-dimethoxyphenyl)methylene]dihydro-4-(1-methyl-ethylidene)-, (E)-	C_6H_{12}	328(4.04)	48-0233-84
(Z)-	C_6H_{12}	346(4.26)	48-0233-84
1H-Naphtho[2,3-c]pyran-5,10-dione, 3,4-dihydro-9-hydroxy-3-(2-hydroxy-ethyl)-1-methyl-, (1S-trans)-	MeOH	248(3.70),273(3.80), 422(3.48)	158-84-166
2H-Naphtho[2,3-b]pyran-5,10-dione, 3,4-dihydro-2,7-dimethoxy-2-methyl-	EtOH	260s(4.29),266(4.39), 295(4.19),333(3.65), 390(3.13)	4-0865-84
$C_{16}H_{16}O_5S$			
5H-Furo[3,2-g][1]benzopyran-5-one, 4,9-dimethoxy-7-methyl-6-[(methylthio)methyl]-	EtOH	214(4.17),249(4.66), 281(3.68),284(3.69), 298s(3.55),328(3.71)	44-5035-84
1-Propanone, 2-hydroxy-3-[(methylsulfonyl)oxy]-1,2-diphenyl-	MeOH	251(4.19),332(2.37)	126-1795-84
2-Propanone, 1-[2-hydroxy-5-[[(4-methylphenyl)sulfonyl]oxy]phenyl]-	MeOH	273(3.44)	44-4399-84
$C_{16}H_{16}O_7$			
2(3H)-Benzofuranone, 7-acetyl-4,6-di-acetoxy-3,5-dimethyl-, (+)-	MeOH	213(4.14),245(3.65), 297(3.19)	23-0320-84
1,4-Naphthalenedione, 2,5,7,8-tetra-hydroxy-6-(1-oxohexyl)-	n.s.g.	225(3.70),267(3.67), 313(3.65),542(3.28)	158-0325-84
$C_{16}H_{16}O_7S$			
2-Dibenzofuranol, 1,3,4-trimethoxy-, methanesulfonate	EtOH	216(4.44),228(4.46), 259(4.04),282(4.11)	39-1441-84C
$C_{16}H_{16}S_2$			
Spiro[3H-benz[cd]azulene-3,2'-[1,3]-dithiane], 4,5-dihydro-	CH_2Cl_2	240(4.37),282s(--), 286(4.64),300s(--), 342(3.73),356s(--), 600(2.48)	83-0984-84
$C_{16}H_{17}BrN_2$			
1,4-Benzenediamine, N'-[1-(4-bromophenyl)ethylidene]-N,N-dimethyl-	MeOH	263(4.20),357(4.57)	118-0128-84
2-Propen-1-one, 3-(4-bromophenyl)-N,N-dimethyl-3-(3-pyridinyl)-, (E)-	pH 1	219(4.34),237s(4.26)	44-4209-84
(Z)-	pH 1	250(4.29)	44-4209-84
$C_{16}H_{17}BrN_2O$			
Benzenecarboximidamide, 4-bromo-N-hydroxy-2,6-dimethyl-N'-(4-methyl-phenyl)-	MeOH	234(4.13),251(4.18)	32-0015-84
$C_{16}H_{17}BrN_2O_2$			
Benzenecarboximidamide, 4-bromo-N-hydroxy-N'-(4-methoxyphenyl)-2,6-di-methyl-	MeOH	235(4.24),249(4.24)	32-0015-84

Compound	Solvent	$\lambda_{max}(\log \epsilon)$	Ref.
$C_{16}H_{17}BrO_2$ Benzene, 1-(3-bromopropoxy)-2-(phenyl-methoxy)-	MeOH	275(3.44)	36-1241-84
$C_{16}H_{17}Br_2NS_2$ Benzenesulfenamide, 4-bromo-N-[(4-bro-mophenyl)thio]-N-(1,1-dimethylethyl)-	hexane	256(4.40)	44-2724-84
$C_{16}H_{17}ClN_2$ 1,4-Benzenediamine, N'-[1-(4-chloro-phenyl)ethylidene]-N,N-dimethyl-	MeOH	260(4.52),333(3.78), 362(3.79)	118-0128-84
$C_{16}H_{17}ClN_2O$ Benzenecarboximidamide, N-(4-chloro-phenyl)-N'-hydroxy-2,4,6-trimethyl-	MeOH	264(4.21)	32-0015-84
$C_{16}H_{17}Cl_2NS_2$ Benzenesulfenamide, 4-chloro-N-[(4-chlorophenyl)thio]-N-(1,1-dimethyl-ethyl)-	hexane	231s(4.20),255(4.34)	44-2724-84
$C_{16}H_{17}F_3N_2O_4$ Phenylalanine, α,β-didehydro-N-[N-(tri-fluoroacetyl)-L-valyl]-, (Z)-	EtOH	206(4.25),281(4.19)	5-0920-84
$C_{16}H_{17}N$ Quinoline, 1,2,3,4-tetrahydro-6-methyl-4-phenyl-?	n.s.g.	246(3.8),318(3.5)	70-2005-84
$C_{16}H_{17}NO$ Benzeneethanamine, N-hydroxy-N-methyl-α-(2-phenylethenyl)-, (E)-	EtOH	253(4.27),284(3.68), 292(3.53),320s(3.48)	150-1531-84M
1H-Inden-1-ol, 2,3-dihydro-2-[(phenyl-methyl)amino]-, (1S-trans)-	MeOH	206(4.31),214s(4.17), 253s(2.74),259(2.93), 265.5(3.07)	39-1655-84C
4(1H)-Phenanthridinone, 6-ethenyl-2,3,5,6-tetrahydro-8-methyl-	ether	242(3.69),296(3.77), 376(3.13)	142-0097-84
$C_{16}H_{17}NO_2$ 1H-Benzo[a]furo[3,4-f]quinolizin-1-one, 3,5,6,10b,11,12-hexahydro-3-methyl-, (3R-cis)-	MeOH	214(4.00),281(4.39)	104-2335-84
$C_{16}H_{17}NO_2S_2$ 2-Propen-1-one, 2-(4-morpholinylmeth-yl)-1,3-di-2-thienyl-	EtOH	331(3.13)	73-1764-84
$C_{16}H_{17}NO_3$ 1H-1,3-Benzodioxolo[5,6-d][1]benzaze-pin-1-one, 2,3,4,5,6,7-hexahydro-5-methyl-	MeCN	250(3.98),320(4.30)	44-0228-84
$C_{16}H_{17}NO_3S$ Methanesulfonamide, N-(2-oxo-2-phenyl-ethyl)-N-(phenylmethyl)-	ether	243.2(4.23),318.0(1.78)	48-0177-84
$C_{16}H_{17}NO_4$ 3-Azabicyclo[4.4.1]undeca-1,3,5,7,9-pentaene-4,5-dicarboxylic acid, diethyl ester	MeOH	246(4.19),275(4.10), 333(3.57)	44-1040-84

Compound	Solvent	λ_{max}(log ϵ)	Ref.
2H-Cyclohepta[b]furan-2,4(3H)-dione, 5,6,7,8-tetrahydro-3-(4-methoxy-phenyl)-, 2-oxime	MeOH	274(4.181)	73-1421-84
1H-Pyrrole-3-carboxylic acid, 1-(1-methylethyl)-4,5-dioxo-2-phenyl-, ethyl ester	dioxan	282s(3.59),408(3.52)	94-0497-84
$C_{16}H_{17}NO_4S_2$			
1,3-Dithiane-2-carboxylic acid, 2-[(1,3-dihydro-1,3-dioxo-2H-isoindol-2-yl)-methyl]-, ethyl ester	EtOH	220(4.60),239s(4.00), 294(3.28)	39-0229-84C
$C_{16}H_{17}NO_5$			
5H-Furo[3,2-g][1]benzopyran-5-one, 7-[(dimethylamino)methyl]-4,9-dimethoxy-	EtOH	213(4.31),248(4.55), 280s(3.67),333(3.65)	44-5035-84
Glycine, N-[2-(1,3-benzodioxol-5-yl)-3-oxo-1-cyclopenten-1-yl]-, ethyl ester	MeCN	278(4.28)	44-0228-84
$C_{16}H_{17}NS$			
Benzenamine, N-[2-(5,6-dimethyl-2H-thiopyran-2-ylidene)propylidene]-, hydrochloride	EtOH	500(4.40)	97-0183-84
$C_{16}H_{17}N_3$			
Pyrrolidine, 1-[4-(phenylazo)phenyl]-	C_6H_{12}	407(4.54)	39-0149-84B
	EtOH	413(4.46)	39-0149-84B
	EtOH-HCl	530(4.76)	39-0149-84B
$C_{16}H_{17}N_3O$			
4H-Indazol-4-one, 5-[(dimethylamino)-methylene]-1,5,6,7-tetrahydro-1-phenyl-, (E)-	EtOH	228(4.21),258(4.21), 355(4.39)	4-0361-84
4H-Indazol-4-one, 1,5,6,7-tetrahydro-1-methyl-5-[(methylphenylamino)-methylene]-, (E)-	EtOH	237s(3.99),255(4.05), 364(4.40)	4-0361-84
1H-Pyrrolo[2,3-f]isoquinoline, 3-mor-pholinomethyl-	EtOH	212(4.39),269(4.58), 296(4.11)	103-0394-84
$C_{16}H_{17}N_3O_2$			
Dicyclopenta[c,e]pyridazine, 1,2,3,3a-4,6,7,8-octahydro-4-(4-nitrophenyl)-	CH_2Cl_2	431(4.4)	88-0057-84
4H-Pyrido[1,2-a]pyrimidine-3-carbox-amide, 6,7,8,9-tetrahydro-6-methyl-4-oxo-N-phenyl-	EtOH	228(4.12),312(4.23)	118-0582-84
$C_{16}H_{17}N_3O_2S_2$			
Guanidine, N-methyl-N'-(thiobenzoyl)-N"-(p-tolylsulfonyl)-	EtOH	227(4.33),275(4.25), 424(2.56)	39-0075-84C
$C_{16}H_{17}N_3O_5$			
Cytidine, N-benzoyl-5'-deoxy-	MeOH	261(4.35),306(--)	18-2017-84
$C_{16}H_{17}N_3O_5S$			
Acetic acid, α-cyano(tetrahydro-4-pyrimidinylidene)-, ethyl ester, p-toluenesulfonate	CF_3COOH	313(4.30)	103-1270-84
$C_{16}H_{17}N_3O_8$			
4-Pyridazinecarbonitrile, 2,3-dihydro-	MeOH	324(3.65)	77-0422-84

Compound	Solvent	$\lambda_{max}(\log \epsilon)$	Ref.
3-oxo-2-(2,3,5-tri-O-acetyl- -D-ribofuranosyl)- (cont.)	MeOH	324(3.65)	87-1613-84
1H-1,2,3-Triazole-4,5-dicarboxylic acid, 1-[4,5-bis(methoxycarbonyl)-1,4-cyclohexadien-1-yl]-, dimethyl ester	MeCN	218(3.98)	89-0736-84
$C_{16}H_{17}N_5O_2S$ Benzenesulfonamide, N-(4,5-dihydro-2,4-dimethyl-6-phenyl-1,2,4,5-tetrazin-3(2H)-ylidene)-	CHCl$_3$	325(4.05)	104-1224-84
$C_{16}H_{17}N_5O_4$ Benzoic acid, 4-methoxy-, [1-[2-(2-nitrophenyl)hydrazino]ethylidene]hydrazide	20% acetone	440(3.57)	140-1028-84
$C_{16}H_{17}N_5O_5$ Inosine, 2-(3-pyridinylmethyl)-	pH 1	256(4.24),260s(4.23)	87-0429-84
	pH 7	251(4.17),260s(4.10)	87-0429-84
	pH 13	256(4.22),260s(4.21)	87-0429-84
$C_{16}H_{17}N_5O_9$ 1H-1,2,3-Triazole-4,5-dicarboxylic acid, 1-[2-[4,5-bis(methoxycarbonyl)-1H-pyrazol-3-yl]-1-methyl-2-oxoethyl]-, dimethyl ester	MeOH	258(4.01),280(4.15)	103-0416-84
$C_{16}H_{17}N_5S$ 4H-Pyridazino[4,5-e][1,3,4]thiadiazine, 3-phenyl-5-piperidino-	EtOH	267(4.11),321(3.39)	94-4437-84
$C_{16}H_{18}As_2Cl_2$ Arsinous chloride, 1,4-butanediyl-bis[phenyl-	EtOH	218s(4.10),263(3.39)	65-1597-84
$C_{16}H_{18}As_2Cl_2O$ Arsinic chloride, (oxydi-2,1-ethanediyl)bis[phenyl-	EtOH	217s(4.31),263(3.70)	65-1597-84
$C_{16}H_{18}BrNO$ 1H-3a,7-Methanopyrrolo[1,2-a][1]benzazocin-13-one, 9-bromo-2,3,4,5,6,7-hexahydro-7-methyl-	EtOH	271(3.7),310(3.8)	142-1771-84
$C_{16}H_{18}Br_3NO$ 2,6-Methano-1-benzazocin-11-one, 8,10-dibromo-1-(3-bromopropyl)-1,2,3,4,5,6-hexahydro-6-methyl-	EtOH	265(3.8),300(3.0)	142-1771-84
$C_{16}H_{18}ClFN_2O_9$ 2,4(1H,3H)-Pyrimidinedione, 5-fluoro-1-(2,4,6-tri-O-acetyl-3-chloro-3-deoxy-α-D-allopyranosyl)-	EtOH	208(3.88),263(3.86)	65-2363-84
$C_{16}H_{18}ClN_3O_2$ Pyrano[2,3-e]indazol-2(5H)-one, 3-chloro-6,7-dihydro-7-methyl-4-piperidino-	EtOH	230(3.92),257(3.95), 339(4.17),360s(4.17)	4-0361-84
$C_{16}H_{18}Cl_2N_4O_4$ 1H-Pyrazole-3,4-dimethanol, 1-(3,4-di-	MeCN	251(4.27)	87-1559-84

Compound	Solvent	$\lambda_{max}(\log \epsilon)$	Ref.
chlorophenyl)-5-methyl-, bis(methyl-carbamate) (cont.)			87-1559-84
$C_{16}H_{18}Cl_2N_6O_2$ Cyclopentanemethanol, 4-[[2-amino-6-chloro-5-[(4-chlorophenyl)azo]-4-pyrimidinyl]amino]-2-hydroxy-, (1α,2β,4α)-(±)-	pH 1	239(4.26),280(3.90), 372(4.42)	87-1416-84
$C_{16}H_{18}Cl_2O$ Benzene, 1-(2,6-dichloro-3-cyclopropyl-1,3-hexadienyl)-4-methoxy-	hexane	240(3.1),318(3.2)	104-1930-84
$C_{16}H_{18}F_3N_3O_4$ L-Phenylalanine, N-[(1,1-dimethyleth-oxy)carbonyl]-4-[3-(trifluoromethyl)-3H-diazirin-3-yl]-	CHCl$_3$	265(2.61),272s(2.48), 358(2.52)	35-7540-84
$C_{16}H_{18}N_2$ Benzo[lmn][3,8]phenanthrolinium, 4,5,9,10-tetrahydro-2,7-dimethyl-, diiodide	MeOH	<u>208(4.4),290(4.1), 400s(2.0)</u>	88-1585-84
1,4:5,10-Dimethanobenzo[g]phthalazine, 1,4,4a,5,10,10a-hexahydro-12,12-di-methyl-, (1α,4α,4aα,5β,10β,10aα)-	hexane	252(2.60),258(2.78), 265(2.97),272(3.02), 358(2.42)	24-0517-85
$C_{16}H_{18}N_2O$ Benzenecarboximidamide, N-hydroxy-2,6-dimethyl-N'-(4-methylphenyl)-	MeOH	254(4.14)	32-0015-84
Benzenecarboximidamide, N-hydroxy-2,4,6-trimethyl-N'-phenyl-	MeOH	256(4.26)	32-0015-84
Methanone, (4,5-dihydro-1,2-dimethyl-1H-pyrrol-3-yl)(1-methyl-1H-indol-3-yl)-	EtOH	217(4.40),297s(3.9), 363(4.43)	150-2738-84
Pyrrolo[2,3-b]indol-2(1H)-one, 3a-(1,1-dimethyl-2-propenyl)-3,3a-dihydro-1-methyl-	EtOH	206(4.25),254(4.02), 312(3.83)	103-0376-84
$C_{16}H_{18}N_2OS$ 2H-Pyrrol-2-one, 1,5-dihydro-1-(4-meth-ylphenyl)-3-piperidino-5-thioxo-	CHCl$_3$	340(4.10),457(4.00)	78-3499-84
$C_{16}H_{18}N_2O_2$ Benzenecarboximidamide, N-hydroxy-4-methoxy-2,6-dimethyl-N'-phenyl-	MeOH	254(4.11)	32-0015-84
Benzenecarboximidamide, N-hydroxy-N'-(4-methoxyphenyl)-2,6-dimethyl-	MeOH	262(4.10)	32-0015-84
2,5-Piperazinedione, 1-methyl-6-(2-methylpropylidene)-3-(phenylmeth-ylene)-, (E,Z)-	MeOH	230(3.94),317(4.48)	102-0200-84
1H-Pyrano[4,3-c]pyrazol-4(1H)-one, 6,7-dihydro-6-methyl-1-phenyl-3-propyl-	EtOH	250(4.15)	4-0017-84
$C_{16}H_{18}N_2O_3$ Benzenamine, N-[4-(3-nitrophenoxy)-butyl]-	MeCN	275s(3.81),330(3.30)	78-1755-84
$C_{16}H_{18}N_2O_4$ Benzenamine, N-[3-(2-methoxy-5-nitro-phenoxy)propyl]-	MeCN	300(3.89),342(3.86)	78-1755-84

Compound	Solvent	$\lambda_{max}(\log \epsilon)$	Ref.
3-Pyridinecarboxylic acid, 5-(amino-carbonyl)-1,4,5,6-tetrahydro-2-methyl-6-oxo-4-phenyl-, ethyl ester	EtOH	287(4.20)	103-1241-84
$C_{16}H_{18}N_2O_5$			
2(1H)-Pyrimidinone, 1-(2-deoxy-β-D-erythro-pentofuranosyl)-5-methyl-4-phenoxy-	EtOH	291(3.87)	39-1263-84C
Uridine, 2'-deoxy-5-(4-methylphenyl)-	pH 1	283(4.01)	88-2431-84
	pH 13	274(3.99)	88-2431-84
$C_{16}H_{18}N_2O_6$			
Pyrrolo[2,3-b]indole-1(2H)-carboxylic acid. 8-acetyl-3a-acetoxy-3,3a,8,8a-tetrahydro-5-hydroxy-, methyl ester	EtOH	247(4.16),282(3.25)	142-0059-84
Pyrrolo[2,3-b]indole-1,2(2H)-dicarbox-ylic acid, 8-acetyl-3,3a,8,8a-tetra-hydro-7-hydroxy-, dimethyl ester	MeOH	215(4.42),248(3.81), 286.5(3.43)	94-2544-84
Uridine, 2'-deoxy-5-(3-methoxyphenyl)-	pH 1	281(4.03)	88-2431-84
	pH 13	277(4.00)	88-2431-84
$C_{16}H_{18}N_3S$			
Methylene blue	MeOH	657(4.89)	149-0031-84A
	EtOH	665(4.95)	35-5879-84
	EtOH	657(4.93)	149-0031-84A
$C_{16}H_{18}N_4$			
7H-Pyrrolo[2,3-d]pyrimidin-4-amine, N,2,5,6-tetramethyl-7-phenyl-	EtOH	206(4.52),232s(4.23), 284(4.14)	11-0073-84B
$C_{16}H_{18}N_4O$			
Pyrazolo[3,4-d]pyrimidin-4(1H)-one, 6-methyl-5-(2-butylphenyl)-	EtOH	204.0(4.23),252.0(3.67)	11-0023-84A
$C_{16}H_{18}N_4O_2$			
Butanedioic acid, bis(2-phenylhydra-zide)	MeOH	235(0.336),285(0.612)	2-0439-84
$C_{16}H_{18}N_4O_2S$			
1H,3H-Benzo[g]thiazolo[4,3-e]pteridine-4,6(5H,8H)-dione, 5,8,10,11-tetra-methyl-	MeCN	210(4.47),275(4.11), 307(3.70),363(3.71)	35-3309-84
$C_{16}H_{18}N_4O_4S_2$			
Acetic acid, [[(3,4,5,10-tetrahydro-3,10-dimethyl-2,4-dioxobenzo[g]-pteridin-4a(2H)-yl)methyl]dithio]-, methyl ester	MeCN	270(4.10),358(3.80)	35-3309-84
$C_{16}H_{18}N_4O_8$			
1H-Imidazolium, 5-(aminocarbonyl)-4-hydroxy-1-[(4-nitrophenyl)methyl]-3-β-D-ribofuranosyl-, hydroxide, inner salt	M HCl	245s(--),279(4.30)	4-0849-84
	H_2O	245s(--),279(4.31)	4-0849-84
	M NaOH	273(4.36)	4-0849-84
1H-1,2,3-Triazole-4,5-dicarboxylic acid, 1-[2-(2-carboxy-7-oxo-4-oxa-1-azabicyclo[3.2.0]hept-3-ylidene)-ethyl]-, 4,5-diethyl ester, sodium salt	H_2O	223(4.13)	158-0885-84

Compound	Solvent	$\lambda_{max}(\log \epsilon)$	Ref.
$C_{16}H_{18}N_4O_8S$ Thymidine, 5'-deoxy-5'-[[(4-nitrophenyl)sulfonyl]amino]-	EtOH	267(4.46)	78-0427-84
$C_{16}H_{18}N_6O_2$ Dibenzo[b,h][1,4,7,10]tetraazacyclododecine-6,7-dione, 5,8,13,14,15,16-hexahydro-, dioxime	EtOH	208(4.66),215(4.62), 228(4.51),319(4.20), 331(4.43),349(4.39)	150-0168-84S
$C_{16}H_{18}O$ Bicyclo[2.2.1]hept-2-ene-2-methanol, 3-(tetracyclo[3.2.0.0^{2,7}.0^{4,6}]heptylidenemethyl)-	MeCN	264(4.21)	24-2027-84
2,4-Cycloheptadien-1-one, 2,7,7-trimethyl-6-phenyl-	EtOH	205.5(4.16),244s(3.54), 302(3.57)	39-0769-84C
3,5-Cycloheptadien-1-one, 2,2,7-trimethyl-7-phenyl-	EtOH	212(4.08),256s(3.56), 296s(2.70)	39-0769-84C
2-Cyclopenten-1-one, 5-(3-methylbut-2-enyl)-3-phenyl-	n.s.g.	226(4.03),287(4.19)	150-2060-84M
2-Cyclopenten-1-one, 2,5,5-trimethyl-4-(2-phenylethenyl)-, (E)-	EtOH	212s(4.27),218s(4.21), 251(4.18),285s(3.63), 293s(3.43)	39-0769-84C
$C_{16}H_{18}O_2$ 1(2H)-Dibenzofuranone, 3,4-dihydro-3,3,6,8-tetramethyl-	EtOH	236(4.38),278(3.93)	39-1213-84C
Dicyclopenta[a,e]pentalene-3,7-dione, 3a,3b,4,4a,7a,7b,8,8a-octahydro-, (3aα,3bα,4aα,7aα,7bα,8aα)-	MeOH	229(4.41),310(2.11)	44-1728-84
8,11a-Methano-11aH-cyclohepta[a]naphthalene-5,10(7H,11H)-dione, 1,2,3,4-8,9-hexahydro-	EtOH	248(4.17)	78-0757-84
Tetracyclo[10.2.2.0^{2,7}.0^{1,10}]hexadecadiene-8,13-dione	EtOH	248(4.18)	78-0757-84
$C_{16}H_{18}O_2Ru$ Ruthenocene, 1-(2-carboxyethyl)-1',3-(1,3-propanediyl)-	EtOH	270s(2.81),324(2.95)	18-0719-84
$C_{16}H_{18}O_2S_3$ Ethane(dithioic) acid, [[2-hydroxy-2-(1-naphthalenyl)ethyl]sulfinyl]-, ethyl ester	EtOH	273(3.92),283(4.02), 294(3.98),314(4.00)	39-0085-84C
$C_{16}H_{18}O_3$ 3,5-Hexadienoic acid, 2-(methoxymethylene)-3-methyl-6-phenyl-, methyl ester (9E-strobilurin A)	MeOH	207(4.03),224(4.00), 229(4.02),237(3.98), 305(4.34),314s(4.32)	5-1616-84
Naphtho[2,3-b]pyran-4-one, 6,7,8,9-tetrahydro-5-hydroxy-10-propyl-	EtOH	227(4.20),261(4.24)	39-0487-84C
3-Phenanthrenecarboxylic acid, 1,2,3,4,9,10-hexahydro-7-methoxy-	EtOH	272(4.26)	2-1168-84
3-Phenanthrenecarboxylic acid, 1,2,3,9,10,10a-hexahydro-7-methoxy-	EtOH	263(4.41)	2-1168-84
3-Phenanthrenecarboxylic acid, 1,4,4a,9,10,10a-hexahydro-7-methoxy-, cis-(±)-	EtOH	278(3.6)	2-1168-84
$C_{16}H_{18}O_4$ 4H-Cyclopenta-1,3-dioxol-4-one, 3a,6a-dihydro-2,2-dimethyl-3a-[(phenylmeth-	EtOH	216(3.80)	39-2089-84C

Compound	Solvent	$\lambda_{max}(\log \epsilon)$	Ref.
oxy)methyl]-, (3aS-cis)- (cont.)			39-2089-84C
Phenol, 2-[3-(4-hydroxyphenyl)-2-hy-droxypropyl]-5-methoxy- (brousso-nin D)	EtOH	225(4.26),278(3.71), 284s(3.66)	138-0689-84
1-Propanone, 1-(8-hydroxy-2,4-dimeth-oxy-3-methyl-1-naphthalenyl)-	EtOH-HCl	224(4.59),240(4.58), 305(3.85)	88-1373-84
$C_{16}H_{18}O_4S_4$			
Carbonodithioic acid, S,S'-(3-oxo-1(3H)-isobenzofuranylidene) 0,0'-bis(1-methylethyl) ester	CH_2Cl_2	290(4.75),365(2.10)	2-0509-84
$C_{16}H_{18}O_5$			
1,3-Naphthalenedicarboxylic acid, 1,2-dihydro-7-methoxy-4-methyl-, dimethyl ester	EtOH	230(4.02),311(4.21)	2-0395-84
1-Naphthalenepropanoic acid, 4-carboxy-3,4-dihydro-6-methoxy-2-methyl-	EtOH	219(4.13),271(4.12)	2-0395-84
$C_{16}H_{18}O_9$			
Chlorogenic acid	pH 1	324(4.25)	3-1935-84
	borate	342(4.26)	3-1935-84
$C_{16}H_{18}Ru$			
Ruthenocene, 1,1':3,3'-bis(1,3-propane-diyl)-	EtOH	272s(2.68),322(2.89)	18-0719-84
$C_{16}H_{19}Br$			
Bicyclo[2.2.1]heptane, 3-(bromophenyl-methylene)-2,2-dimethyl-	MeOH	224(3.97),265(3.42)	2-0331-84
$C_{16}H_{19}BrN_2O_3$			
1H-Pyrrole-2-carboxylic acid, 5-[bromo-(1,5-dihydro-3,4-dimethyl-5-oxo-2H-pyrrol-2-ylidene)methyl]-3,4-di-methyl-, ethyl ester, (Z)-	MeOH	272(4.16),325(4.27)	49-0357-84
$C_{16}H_{19}BrN_2O_4S_2$			
Thiouronium, 2-[2-(2-bromophenoxy)eth-yl]-, p-toluenesulfonate	EtOH	272(3.44),278(3.30)	12-2059-84
$C_{16}H_{19}Br_2NO$			
2,6-Methano-1-benzazocin-11-one, 8-bromo-1-(3-bromopropyl)-1,2,3,4,5-6-hexahydro-6-methyl-	EtOH	268(3.9),313(3.0), 350(2.6)	142-1771-84
$C_{16}H_{19}ClN_2O_2S$			
2H-Thieno[2,3-h]-1-benzopyran-2-one, 3-chloro-4-[2-(dimethylamino)meth-ylamino]-5,6-dihydro-	EtOH	220(4.11),242.5(4.12), 275s(3.61),356(4.20)	161-0081-84
$C_{16}H_{19}ClN_2O_5$			
Quinazoline, 2-chloro-4-[(2,2-dimethyl-1,3-dioxolan-4-yl)methoxy]-6,7-di-methoxy-	EtOH	239(4.70)	4-1189-84
$C_{16}H_{19}ClOS$			
Cyclohexanone, 6-[1-[(4-chlorophenyl)-thio]ethylidene]-2,2-dimethyl-, (E)-	C_6H_{12}	263(3.89),294(3.99)	5-0576-84
(Z)-	C_6H_{12}	308(3.85)	5-0576-84

Compound	Solvent	$\lambda_{max}(\log \epsilon)$	Ref.
$C_{16}H_{19}Cl_2N_3O_2$			
Pyrano[2,3-e]indazol-2(3H)-one, 3,3-dichloro-4,5,6,7-tetrahydro-7-methyl-4-piperidino-	EtOH	255s(3.93),279(4.10)	4-0361-84
$C_{16}H_{19}NO$			
5H-Benzocyclohepten-5-one, 6-[3-(dimethylamino)-2-propenylidene]-6,7,8,9-tetrahydro-	EtOH	267(4.05),418(4.59)	118-0424-84
Bicyclo[2.2.1]heptan-2-one, 1,7,7-trimethyl-3-(4-pyridinylmethylene)-, [1R-(1α,3E,4α)]-	EtOH	280(4.28)	41-0629-84
Formamide, N-methyl-N-[(4,6,8-trimethyl-1-azulenyl)methyl]-	MeOH	600(--)	138-0631-84
Formamide, N-methyl-N-[(4,6,8-trimethyl-2-azulenyl)methyl]-	MeOH	550(2.67)	138-0631-84
1H-3a,7-Methanopyrrolo[1,2-a][1]benzazocin-13-one, 2,3,4,5,6,7-hexahydro-7-methyl-	EtOH	258(3.8),302(3.0)	142-1771-84
4(1H)-Phenanthridinone, 2,3,5,6-tetrahydro-6,6,8-trimethyl-	ether	214(4.13),242(4.17), 382(4.07)	142-0097-84
$C_{16}H_{19}NOS_2$			
2-Propen-1-one, 2-[(diethylamino)methyl]-1,3-di-2-thienyl-	EtOH	335(3.20)	73-1764-84
2-Propen-1-one, 2-[[(1,1-dimethylethyl)amino]methyl]-1,3-di-2-thienyl-	EtOH	331(3.28)	73-1764-84
$C_{16}H_{19}NO_2$			
4(1H)-Phenanthridinone, 2,3,5,6-tetrahydro-8-methoxy-6,6-dimethyl-	ether	224(3.98),244(4.04), 380(3.96)	142-0097-84
$C_{16}H_{19}NO_3$			
5H-Pyrano[3,2-c]quinolin-5-one, 2,3,4,6-tetrahydro-7-methoxy-2,2,6-trimethyl-	EtOH	230(4.43),245(4.37), 253(4.37),285(3.82), 290(3.81),325(3.39)	25-0352-84
2(1H)-Quinolinone, 4-hydroxy-8-methoxy-1-methyl-3-(3-methyl-2-butenyl)- (glycophylone)	EtOH	218(4.3),240(4.48), 252(4.40),287(3.90), 292(3.91),325(3.50)	25-0352-84
$C_{16}H_{19}NO_4$			
Galanthan-1,2,10-triol, 3,12-didehydro-9-methoxy-	EtOH	221(3.85),286(3.55)	100-1003-84
$C_{16}H_{19}NO_5$			
3-Azabicyclo[3.2.0]heptane-2,4-dione, 7,7-dimethoxy-1-(3-methoxyphenyl)-3-methyl-	MeOH	275(3.44),282(4.41)	87-0628-84
Naphth[1,2-b]oxiren-7b(1aH)-ol, 2,3-dihydro-1a-(2-methyl-2-nitropropyl)-, acetate	EtOH	265(2.70)	12-0587-84
$C_{16}H_{19}NO_6$			
1H-Pyrrolo[1,2-a]indole-7,9-dicarboxylic acid, 2,3,5,6,7,8-hexahydro-2-hydroxy-6-methyl-8-oxo-, dimethyl ester	CHCl₃	298.0(3.404)	44-5164-84
$C_{16}H_{19}NS_2$			
Benzenesulfenamide, N-(1,1-dimethylethyl)-N-(phenylthio)-	hexane	248(4.25)	44-2724-84

Compound	Solvent	$\lambda_{max}(\log \epsilon)$	Ref.
$C_{16}H_{19}N_3$ Benzenamine, 2-[(3,4-dimethylphenyl)-azo]-4,5-dimethyl-	EtOH	<u>330(4.2)</u>,440(4.0)	104-0381-84
$C_{16}H_{19}N_3O_3$ Piperazine, 1-[1-(2-furanyl)-2-nitro-ethyl]-4-phenyl-	MeOH	209(4.38),250(4.12)	73-2496-84
$C_{16}H_{19}N_5$ 7H-Purin-6-amine, N-(3,5-dimethyl-phenyl)-2,7,8-trimethyl-	EtOH	301(4.30)	11-0208-84B
$C_{16}H_{19}N_5O_2$ Acetamide, N-[3-[(dimethylamino)meth-yl]-4,9-dihydro-2-methyl-9-oxopyra-zolo[5,1-b]quinazolin-7-yl]-	10% EtOH-NaOH	220(4.45),255(4.55), 338(4.38),402(3.70)	56-0411-84
$C_{16}H_{19}N_5O_7$ Guanosine, N-acetyl-2'-deoxy-, 3',5'-diacetate	MeOH	255(4.13),278s(3.98)	69-5686-84
$C_{16}H_{19}N_5O_8$ 4H-Imidazo[4,5-d][1,2,3]triazin-4-one, 1,5-dihydro-6-methyl-5-(2,3,5-tri-O-acetyl-β-D-ribofuranosyl)-	H_2O	255(2.88),280s(2.74)	4-1221-84
$C_{16}H_{20}As_2O_4$ Arsinic acid, 1,4-butanediylbis[phenyl-	EtOH	211(4.16),217(4.15), 257(2.87),263(3.54), 270(2.90)	65-1597-84
$C_{16}H_{20}As_2O_5$ Arsinic acid, (oxydi-2,1-ethanediyl)-bis[phenyl-	EtOH	217s(2.70),257(1.41), 263(1.54),270(1.47)	65-1597-84
$C_{16}H_{20}N_2$ 9,10-Azo-1,4:5,8-dimethanoanthracene, 1,2,3,4,4a,5,8,8a,9,9a,10,10a-dodeca-hydro-, (1α,4α,4aα,5α,8α,8aα,9β,9aα-10β,10aα)-	hexane	387(2.36)	24-0534-84
(1α,4α,4aα,5β,8β,8aα,9β,9aα,10β,10aα)-	hexane	386(2.29)	24-0534-84
1,4-Benzenediamine, N-(2-methylpropyl)-N'-phenyl-	n.s.g.	290(4.30)	30-0009-84
4,4'-Bipyridine, 2,2',6-triethyl-	EtOH	236(4.06),273(3.67)	39-0367-84B
15,16-Diazoniatricyclo[9.3.1.1[4,8]]hexa-deca-1(15),4,6,8(16),11,13-hexaene, 15,16-dimethyl-, salt with 4-methyl-benzenesulfonic acid (1:2)	EtOH	255(3.23),261(3.25), 268(3.20)	35-2672-84
2-Pyridineethanamine, N-methyl-N-(1-phenylethyl)-, (S)-	isooctane	204s(4.26),260(3.59), 310s(1.40)	103-0077-84
	EtOH	204s(4.27),262(3.67), 312s(1.48)	103-0077-84
$C_{16}H_{20}N_2O_3S$ 1-Piperazinecarboxylic acid, 4-[(6,7-dihydro-4-oxobenzo[b]thien-5(4H)-ylidene)methyl]-, ethyl ester, (E)-	EtOH	220(4.03),253(4.22), 361(4.30)	161-0081-84
$C_{16}H_{20}N_2O_4$ 4H-[1]Benzopyrano[3,4-c]pyridazine-4-carboxylic acid, 2,3,4,4a-tetra-	EtOH	220(4.26),229s(4.11), 235s(4.00),262(4.15),	39-2815-84C

Compound	Solvent	λ_{max}(log ϵ)	Ref.
hydro-8-methoxy-5,5-dimethyl-, methyl ester (cont.)		267s(4.11),306(3.81), 311s(3.78)	39-2815-84C
C$_{16}$H$_{20}$N$_2$O$_5$			
Phenylalanine, α,β-didehydro-N-[N-[(1,1-dimethylethoxy)carbonyl]-glycyl]-, (Z)-	EtOH	207(4.13),243(3.81)	5-0920-84
2-Propenoic acid, 3,3'-[1-(dimethyl-amino)-1,2-dihydro-6-methyl-2-oxo-3,5-pyridinediyl]bis-, dimethyl ester, (E,E)-	MeOH	203(4.08),268(4.16), 319(4.20),369(4.03)	24-1620-84
C$_{16}$H$_{20}$N$_4$O$_2$S			
1H-Imidazole-5-carboxylic acid, 1-meth-yl-4-[[(methylamino)thioxomethyl]-amino]-2-(phenylmethyl)-, ethyl ester	MeOH	254(4.47),297(4.48)	2-0316-84
C$_{16}$H$_{20}$N$_4$O$_3$			
1H-Pyrazole-4-carboxylic acid, 5-[[eth-oxy(phenylamino)methylene]amino]-1-methyl-, ethyl ester	CHCl$_3$	265(4.00)	24-0585-84
C$_{16}$H$_{20}$N$_4$O$_6$			
5-Pyrimidinecarboxylic acid, 4-(amino-carbonyl)-2-[1-amino-2-(ethoxycarbo-nyl)-3-oxo-1-butenyl]-6-methyl-, ethyl ester	C$_2$H$_4$Cl$_2$	242(4.03),357(3.82)	39-0965-84B
C$_{16}$H$_{20}$N$_4$O$_6$S			
Thymidine, 5'-[[(4-aminophenyl)sulfon-yl]amino]-5'-deoxy-	EtOH	267(4.46)	78-0427-84
C$_{16}$H$_{20}$N$_4$O$_{11}$			
1H-1,2,4-Triazole, 5-nitro-1-(2,3,4,6-tetra-O-acetyl-β-D-glucopyranosyl)-	EtOH	230(3.63),253(3.60)	111-0089-84
C$_{16}$H$_{20}$O$_2$			
2-Oxabicyclo[9.3.1]pentadeca-1(15),4-8,11,13-pentaen-12-ol, 4,8-dimethyl-, (E,E)- (arnebinol)	n.s.g.	203(4.36),221s(--), 291.5(3.48)	88-1085-84
isoarnebinol	n.s.g.	203.5(4.49),231s(--), 290(3.54)	88-1085-84
C$_{16}$H$_{20}$O$_2$RuS$_2$			
Ruthenocene, 1,1'-[1,2-ethanediyl-bis(oxy-2,1-ethanediylthio)]-	MeCN	320(2.67)	88-1991-84
C$_{16}$H$_{20}$O$_4$			
4a(2H)-Biphenylenol, 1,3,4,8b-tetra-hydro-4,4-dimethoxy-, acetate	MeOH	261.5(3.13),267.5(3.29), 274(3.27)	44-2050-84
Cyclopenta[b]pyran-5(2H)-one, 3,4,6,7-tetrahydro-2-(2-hydroxy-5-oxo-1-cyclopenten-1-yl)-2,4,4-trimethyl-	EtOH	250(4.39)	23-2612-84
	EtOH-K$_2$CO$_3$	260(4.42)	23-2612-84
2-Naphthaleneacetic acid, 1,2,3,4,4a,7-hexahydro-α,4a,8-trimethyl-3,7-dioxo-, methyl ester	EtOH	241(3.73)	138-1021-84
8-Nonenoic acid, 9-(1,3-benzodioxol-5-yl)-, (E)-	MeOH	222(4.32),262(4.26), 308(3.93)	78-2541-84
(Z)-	MeOH	219(4.18),262(4.18), 268(3.86),302(3.26)	78-2541-84

Compound	Solvent	$\lambda_{max}(\log \epsilon)$	Ref.
$C_{16}H_{20}O_4S_4$			
1,3-Dithiole-4-carboxylic acid, 2-[4-(butoxycarbonyl)-1,3-dithiol-2-ylidene]-, butyl ester	MeCN	288(4.21),301(4.23), 312(4.26),434(3.60)	103-1342-84
isomer B	MeCN	288(4.12),300(4.15), 312(4.18),415(3.48)	103-1342-84
$C_{16}H_{20}O_5$			
4H-1-Benzopyran-4-one, 2-butyl-8-hydroxy-5,7-dimethoxy-3-methyl-	MeOH	228s(4.20),244(4.34), 255(4.34),261(4.34), 286(3.43),335(3.76)	5-1883-84
2H-3-Benzoxacyclododecine-2,10(1H)-dione, 4,5,6,7,8,9-hexahydro-11,13-dihydroxy-4-methyl-, (S)- (curvularin)	EtOH	273(3.90),303(3.85)	36-1846-84
1,3-Dioxane-4,6-dione, 5-(4-methoxyphenyl)-2,2-dimethyl-5-(1-methylethyl)-	MeOH	236(4.00),274(3.18), 280(3.11)	12-1245-84
Propanoic acid, 2-methyl-, 2-[2-(acetyloxy)methyl]oxiranyl]-5-methylphenyl ester	MeOH	276(3.42),281s(3.40)	102-1947-84
$C_{16}H_{20}O_6$			
2,2(3H)-Furandicarboxylic acid, 4-formyl-3-(1,3-pentadienyl)-, diethyl ester, (E,E)-	EtOH	223(4.32),255(4.14)	118-0974-84
Ligustilide, 6,7-diacetoxy-, cis	MeOH	275(3.80)	102-2033-84
$C_{16}H_{20}S_2$			
Naphthalene, 1,8-dimethyl-2,7-bis-[(methylthio)methyl]-	CHCl$_3$	241(4.96),304(3.93), 337s(2.89)	44-4128-84
$C_{16}H_{21}BrN_2O_5S$			
2H-Furo[2,3-d]imidazole-2-thione, 3-(4-bromophenyl)-1-ethylhexahydro-6-hydroxy-5-(1,2,3-trihydroxypropyl)-	EtOH	246(4.31)	136-0071-84G
$C_{16}H_{21}ClO_4$			
Benzaldehyde, 3-chloro-6-(3,4-dihydroxy-1-nonenyl)-2-hydroxy- (monilidiol)	MeOH	232(4.26),273(3.88), 357(3.60)	158-84-21
	MeOH-NaOH	217(--),287(--), 398(--)	158-84-21
$C_{16}H_{21}NO$			
Benzamide, N-cyclohexyl-N-1-propenyl-	MeOH	215(4.26),265(4.27), 305(3.88)	78-1835-84
$C_{16}H_{21}NO_3$			
2,4,6-Cycloheptatrien-1-one, 2-[4-(2-acetoxyethyl)-1-piperidinyl]-	MeOH	222(4.02),258(4.10), 356(3.91)	87-0875-84
$C_{16}H_{21}NO_4$			
Propanedioic acid, [[4-(dimethylamino)-phenyl]methylene]-, diethyl ester	dioxan	370(4.48)	65-1987-84
$C_{16}H_{21}NO_5$			
Spiro[4.5]deca-6,9-diene-6,7-dicarboxylic acid, 8-(dimethylamino)-1-oxo-, dimethyl ester	EtOH	283(4.53),412(3.46)	88-3547-84
$C_{16}H_{21}NO_6$			
α-D-Glucopyranoside, methyl 2-deoxy-	EtOH	243(3.95),343(4.27)	136-0101-84L

Compound	Solvent	$\lambda_{max}(\log \epsilon)$	Ref.
2-[(3-oxo-3-phenyl-1-propenyl)-amino]-, (Z)- (cont.)			136-0101-84L
$C_{16}H_{21}NS$ Benzenecarbothioamide, N-cyclohexyl-N-1-propenyl-	MeOH	305(4.09),395(3.00)	78-1835-84
$C_{16}H_{21}N_2$ Piperidinium, 1-(1,3-dimethylcyclo-hepta[c]pyrrol-6(2H)-ylidene)-, perchlorate	CH_2Cl_2	317(4.56),378(4.34), 468(2.66)	118-0119-84
$C_{16}H_{22}BrNO$ 1-Penten-3-one, 4-bromo-1-(diethyl-amino)-4-methyl-1-phenyl-	EtOH	203.5(4.22),328(4.35)	39-0535-84C
$C_{16}H_{22}ClNO_2$ Benzene, 1-[1-chloro-4-(1,1-dimethyl-ethyl)cyclohexyl]-4-nitro-, cis	EtOH	203(4.02),273(3.96)	35-3140-84
$C_{16}H_{22}Cl_2O_4$ 3-Octen-5-ynedioic acid, 3,4-dichloro-2,2,7,7-tetramethyl-, diethyl ester	MeOH	244(3.98),248(3.98)	35-3551-84
$C_{16}H_{22}Cl_3NO_4$ 2H-Pyrrol-2-one, 1,5-dihydro-4-methoxy-5-(1-methylethyl)-1-(6,6,6-trichloro-3-oxo-2-hexenyl)- (dysidin)	MeOH	228(3.91),264(4.64)	44-3489-84
$C_{16}H_{22}F_3NO_3$ 5-Quinolineacetic acid, 1,4,4a,5,6,7-8,8a-octahydro-1,7-dimethyl-3-(tri-fluoroacetyl)-, (4aα,5α,7β,8aα)-(±)-	ether	312(4.19)	5-1519-84
$C_{16}H_{22}NO_3S$ 1(2H)-Pyridinyloxy, 3,6-dihydro-4-[5-(2-methoxy-2-oxoethyl)-2-thienyl]-2,2,6,6-tetramethyl-	EtOH	204(3.97),290(4.17), 345(2.85)	70-1901-84
$C_{16}H_{22}N_2$ 9,10-Azo-1,4:5,8-dimethanoanthracene, tetradecahydro-, (1α,4α,4aα,5α,8α-8aα,9β,9aα,10β,10aα)-	hexane	363(2.25),373(2.43)	24-0534-84
(1α,4α,4aα,5β,8β,8aα,9β,9aα,10β,10aα)-	hexane	382(2.51)	24-0534-84
$C_{16}H_{22}N_2O$ 1H-Imidazole, 4,5-dihydro-2-(2-hydroxy-3-methyl-1-pentenyl)-1-(phenylmethyl)-	EtOH	291(4.45)	39-2599-84C
Piperazine, 1,2,5-trimethyl-4-(1-oxo-3-phenyl-2-propenyl)- (nigerazine A)	MeOH	280(4.34)	158-84-52
nigerazine B	MeOH	280(4.35)	158-84-52
$C_{16}H_{22}N_2OS$ Benzo[b]thiophen-3(2H)-one, 2-[[2,2-di-methyl-1-(3-methylbutyl)hydrazino]-methylene]-	CCl₄ DMSO	318(4.19),437(4.15) 315(4.11),440(4.17)	104-1353-84 104-1353-84
$C_{16}H_{22}N_2O_4$ 4H-[1]Benzopyrano[3,4-a]pyridazine-4-carboxylic acid, 1,2,3,4,4a,10b-hexahydro-8-methoxy-5,5-dimethyl-,	EtOH	210(4.34),224s(3.85), 283(3.30),289(3.28)	39-2815-84C

Compound	Solvent	$\lambda_{max}(\log \epsilon)$	Ref.
methyl ester, trans (cont.)			39-2815-84C
2-Butenedioic acid, 2-cyano-3-[(4,4-di-methyl-2-pyrrolidinylidene)methyl]-, diethyl ester	MeOH	219(3.98),246(4.13), 321(3.81)	83-0143-84
α-L-lyxo-Hexopyranosid-2-ulose, methyl 6-deoxy-3,4-O-(1-methylethylidene)-, phenylhydrazone, (Z)-	EtOH	283(3.28)	39-0733-84C
$C_{16}H_{22}N_2O_4S$			
2H-Furo[2,3-d]imidazole-2-thione, 5-(1,2-dihydroxyethyl)hexahydro-6-hydroxy-3-phenyl-1-propyl-, [3aR-[3aα,5α(S*),6α,6aα]]-	EtOH	251(3.91)	136-0131-84E
$C_{16}H_{22}N_2O_5S$			
2H-Furo[2,3-d]imidazole-2-thione, 1-ethylhexahydro-6-hydroxy-3-phenyl-5-(1,2,3-trihydroxypropyl)-, [3aR-[3aα,5α(1R*,2R*),6β,6aα]]-	EtOH	245(4.25)	136-0071-84G
[3aS-[3aα,5β(1S*,2S*),6α,6aα]]-	EtOH	245(4.25)	136-0071-84G
$C_{16}H_{22}N_2O_8$			
Uridine, 5'-(ethyl propanedioate)	H_2O	267(3.89)	83-0867-84
$C_{16}H_{22}N_4$			
Benzo[1,2-b:5,4-b']dipyrrole-3,5-dimeth-anamine, 1,7-dihydro-N,N,N',N'-tetra-methyl-	EtOH	206s(4.37),229.8(4.64), 304(4.13),322s(4.05), 333s(3.94)	103-0996-84
$C_{16}H_{22}N_4O_4$			
7H-Pyrrolo[2,3-d]pyrimidin-4-amine, N-(3-methyl-2-butenyl)-7-α-D-ribofuranosyl-	MeOH	275(4.21)	5-1972-84
β-	MeOH	275(4.22)	5-1972-84
$C_{16}H_{22}N_4O_9$			
1H-1,2,4-Triazol-3-amine, 1-(2,3,4,6-tetra-O-acetyl-β-D-glucopyranosyl)-	EtOH	231(3.57)	111-0089-84
$C_{16}H_{22}N_6$			
1H-Pyrazole, 1,1',1''-methylidyne-tris[3,5-dimethyl-	MeOH	223(4.31)	117-0299-84
$C_{16}H_{22}O$			
Naphthalene, 1,2-dihydro-6-methoxy-4,7-dimethyl-1-(1-methylethyl)-	EtOH	222(3.46),267(3.07)	44-0034-84
$C_{16}H_{22}O_2$			
1(2H)-Dibenzofuranone, 7-(1,1-dimethyl-ethyl)-3,4,6,7,8,9-hexahydro-	H_2O	284(3.50)	39-1213-84C
1,4-Ethanonaphthalene-2,5(1H,3H)-dione, 4,6,7,8-tetrahydro-8,8,9,9-tetrameth-yl-	C_6H_{12}	254(4.0),298f(3.0), 340s(1.9)	33-1493-84
6-Hepten-3-one, 5-hydroxy-2,2,5-tri-methyl-7-phenyl-, (E)-	EtOH	251(4.23)	78-4127-84
Tetracyclo[6.4.0.0^{1,5}.0^{2,8}]dodecane-7,12-dione, 4,4,9,9-tetramethyl-	C_6H_{12}	208(3.90),298(2.00)	33-1493-84
$C_{16}H_{22}O_2S$			
1,4-Dioxaspiro[4.5]decane, 7-methyl-7-[(phenylthio)methyl]-	EtOH	254(3.99)	94-3417-84

Compound	Solvent	$\lambda_{max}(\log \epsilon)$	Ref.
17-Thiabicyclo[12.2.1]heptadeca-1(16),14-diene-2,13-dione	EtOH	208(3.46),265(3.98), 290(4.03)	70-0195-84
$C_{16}H_{22}O_2S_2$ 3,10-Dithiabicyclo[10.3.2]heptadeca-12(17),13,15-trien-16-one, 17-methoxy-	MeOH	252(4.06),273s(3.87), 305(3.80),356s(3.58)	88-4761-84
$C_{16}H_{22}O_3$ Bicyclo[4.2.1]non-3-en-2-one, 7,8-O-(1-methylethylidene)-9-(1-methylethylidene)-6-methyl-	EtOH	234(3.87)	33-1506-84
2(3H)-Furanone, dihydro-5-methoxy-4-[2-(2,6,6-trimethyl-1-cyclohexen-1-yl)ethenyl]-, (E)-	EtOH	258(4.10),319(4.10)	94-1709-84
$C_{16}H_{22}O_4$ Benzaldehyde, 2-(3,4-dihydroxy-1-nonenyl)-6-hydroxy-, (dechloromonilidiol)	MeOH	228(4.38),274(4.04), 347(3.72)	158-84-21
	MeOH-NaOH	220(--),285(--), 389(--)	158-84-21
2(5H)-Furanone, 5-hydroxy-4-[2-(4-methoxy-2,6,6-trimethyl-1-cyclohexen-1-yl)ethenyl]-	EtOH	253(3.93),315(4.10)	94-1709-84
1-Naphthaleneacetic acid, 1,2,3,4,4a-5,6,8a-octahydro-4-hydroxy-4,7-dimethyl-α-methylene-6-oxo-, methyl ester	n.s.g.	239(4.17)	78-2189-84
2-Naphthaleneacetic acid, 1,2,3,4,4a,7-hexahydro-3-hydroxy-α,4a,8-trimethyl-7-oxo-, methyl ester, [2R-[2α(S*),3α-4aα]]-	EtOH	241(3.97)	138-1021-84
[2R-[2α(S*),3β,4aα]]-	EtOH	240(4.07)	138-1021-84
$C_{16}H_{22}O_4S$ 1,4-Dioxaspiro[4.5]decane, 7-methyl-7-[(phenylsulfonyl)methyl]-	EtOH	215(3.96)	94-3417-84
$C_{16}H_{22}O_5$ 2H-3-Benzoxacyclododecin-2-one, 1,4,5,6,7,8,9,10-octahydro-10,11,13-trihydroxy-4-methyl-(dihydrocurvularin)	EtOH	284(3.45)	36-1846-84
Naematolone	MeOH	217s(3.36),240(3.39), 318(2.53)	5-1332-84
$C_{16}H_{22}O_6$ 2-Cyclohexene-1-carboxylic acid, 2-methyl-4-oxo-6-(tetrahydro-2,2-dimethylfuro[3,4-d]-1,3-dioxol-4-yl)-, methyl ester	EtOH	235(4.25)	136-0217-84C
$C_{16}H_{22}O_6Pb$ Plumbane, triacetoxy[4-(2-methylpropyl)phenyl]-	MeOH	225(4.20)	12-1245-84
$C_{16}H_{22}O_{10}$ Loganic acid	MeOH	232(3.94)	33-0160-84
$C_{16}H_{22}O_{11}$ Secologanoside	MeOH	230(3.92)	102-2539-84

Compound	Solvent	$\lambda_{max}(\log \epsilon)$	Ref.
$C_{16}H_{22}Se$ Benzene, [[4-(1,1-dimethylethyl)-1-cyclohexen-1-yl]seleno]-	EtOH	255(3.97)	88-1287-84
$C_{16}H_{23}NO_2S$ 2-Thiopheneacetic acid, 5-(1,2,3,6-tetrahydro-2,2,6,6-tetramethyl-4-pyridinyl)-, methyl ester	EtOH	202(4.15),290(4.16)	70-1901-84
$C_{16}H_{23}N_3O$ Urea, N,N-diethyl-N'-[1-methyl-3-(phenylimino)-1-butenyl]-	EtOH	225(4.15),308(4.15)	39-0239-84C
$C_{16}H_{23}N_3OS$ Diazene, [4-(hexyloxy)phenyl](1-iso-thiocyanato-1-methylethyl)-	C_6H_{12}	309(4.21),392(2.53)	118-0315-84
$C_{16}H_{23}N_3O_2$ Diazene, [4-(hexyloxy)phenyl](1-iso-cyanato-1-methylethyl)-	C_6H_{12}	306(4.11),376(2.49)	118-0315-84
$C_{16}H_{23}O_6P$ 2-Propenoic acid, 2-(diethoxyphosphin-yl)-3-(4-methoxyphenyl)-, ethyl ester, (E)-	dioxan	294(4.28)	65-1987-84
$C_{16}H_{24}Cl_3N_3O_3$ Morpholine, 4,4',4"-[2-chloro-1-(di-chloromethylene)-2-propen-1-yl-3-ylidene]tris-	MeCN	255(3.88),355(--)	88-0135-84
$C_{16}H_{24}FeNP$ Ferrocene, [(triethylphosphoranyli-dene)amino]-	hexane	207(4.45),273(3.66), 293s(3.60),453(2.38)	89-0521-84
$C_{16}H_{24}NO_2$ Pyrrolo[2,1-a]isoquinolinium, 1,2,3,5-6,10b-hexahydro-8,9-dimethoxy-4,10b-dimethyl-, iodide	MeOH	219(4.29),282(3.54), 287s(3.45)	12-1203-84
$C_{16}H_{24}N_2$ 1H-Imidazole, 2-[[4-(1,1-dimethyleth-yl)-2,6-dimethylphenyl]methyl]-4,5-dihydro-, compd. with iodine (1:1)	$CHCl_3$	290(4.35),365(4.09)	83-0246-84
$C_{16}H_{24}N_2O$ Phenol, 3-[(4,5-dihydro-1H-imidazol-2-yl)methyl]-6-(1,1-dimethylethyl)-2,4-dimethyl-, compd. with iodine	$CHCl_3$	290(4.54),365(4.29)	83-0246-84
$C_{16}H_{24}N_2O_2$ 2H-[1]Benzopyrano[3,4-c]pyridazine, 1,3,4,4a,5,10b-hexahydro-8-methoxy-3,4,5,5-tetramethyl-, trans	EtOH	211(4.26),224s(3.89), 285(3.48),291(3.45)	39-2815-84C
[9,9'-Bi-9-azabicyclo[3.3.1]nonane]-3,3'-dione, radical ion (1+)	MeCN	236(3.40),331(3.70)	35-0791-84
Pyrrolo[3,2-b]pyrrole-1(4H)-carboxylic acid, 3,6-bis(1,1-dimethylethyl)-, methyl ester	C_6H_{12}	<u>255(4.1),280s(3.9)</u>	138-2033-84

Compound	Solvent	$\lambda_{max}(\log \epsilon)$	Ref.
$C_{16}H_{24}N_2O_7S$			
Thymidine, 5'-[4-[(methylthio)methoxy]-butanoate]	EtOH	265(3.96)	39-1785-84C
$C_{16}H_{24}N_4O_5$			
Inosine, 2-hexyl-	pH 1	250(4.09)	87-0429-84
	pH 7	251(4.10)	87-0429-84
	pH 13	255(4.13)	87-0429-84
$C_{16}H_{24}N_6O_6$			
L-Lysine, N-(9-β-D-ribofuranosyl-9H-purin-6-yl)-	H_2O	269(4.22)	73-2689-84
$C_{16}H_{24}N_6O_9S$			
L-glycero-α-L-ido-Heptofuranuronic acid, 1-[4-[(aminocarbonyl)imino]-3,4-dihydro-3-methyl-2-oxo-1(2H)-pyrimidinyl]-6-[(2-amino-3-hydroxy-1-oxopropyl)amino]-1,6-dideoxy-4-thio-, (S)-	M HCl	213(4.06),233(3.90), 306(4.17)	5-1399-84
$C_{16}H_{24}O_2$			
1(2H)-Dibenzofuranone, 3,4,5a,6,7,8,9-9a-octahydro-3,3,6,8-tetramethyl-	EtOH	274(4.11)	39-1213-84C
$C_{16}H_{24}O_3$			
1-Cyclohexene-1-carboxylic acid, 4-(1,5-dimethyl-4-hexenyl)-3-oxo-, methyl ester	MeOH	235(4.03)	102-0186-84
1-Cyclohexene-1-carboxylic acid, 4-(1,5-dimethyl-3-oxo-4-hexenyl)-, methyl ester	EtOH	230(4.12)	107-0925-84
Pallescensolide	pentane	205(4.34)	44-5160-84
$C_{16}H_{24}O_4$			
3-Cyclohexene-1-carboxylic acid, 4-(2,2-dimethyl-1-oxopropoxy)-1-methyl-2-methylene-, ethyl ester	EtOH	233(4.07)	12-2037-84
3-Cyclohexene-1-carboxylic acid, 1-(1-hydroxy-2-propenyl)-2-methyl-ene-4-(1-methylethoxy)-, ethyl ester	EtOH	255(4.04)	12-2037-84
Juvabione, 7-hydroxy-	EtOH	221(3.99)	78-2961-84
$C_{16}H_{24}O_5$			
3-Cyclohexene-1-propanoic acid, 4-eth-oxy-1-(ethoxycarbonyl)-2-methylene-, methyl ester	EtOH	213s(3.30),241(3.68)	12-2037-84
$C_{16}H_{24}O_5Si$			
1,2-Benzenedicarboxylic acid, 3-[[(1,1-dimethylethyl)dimethylsilyl]oxy]-, dimethyl ester	hexane	242s(4.68),292.3(4.35)	44-1898-84
$C_{16}H_{24}O_9$			
4-Pentulosonic acid, 3,5-dideoxy-3-(ethoxycarbonyl)-2-O-methyl-, tetrahydro-2,2-dimethylfuro[3,4-d]-1,3-dioxol-4-yl ester, [3aR-(3aα,4α,6aα)]-	MeOH	243(2.73)	136-0019-84I

Compound	Solvent	$\lambda_{max}(\log \epsilon)$	Ref.
$C_{16}H_{24}S$ 1-Butanethione, 2-(1,1-dimethylethyl)-3,3-dimethyl-1-phenyl-	C_6H_{12}	205(3.82),227(3.78), 250(3.52),315(3.85), 595(1.73)	44-0393-84
$C_{16}H_{25}ClO_6P_2$ Phosphonic acid, [(4-chlorophenyl)ethenylidene]bis-, tetraethyl ester	dioxan	279(4.28)	65-1987-84
$C_{16}H_{25}NO_2$ Ethanone, 1-[3-[[bis(1-methylethyl)-amino]methyl]-2-hydroxy-5-methyl-phenyl]-	EtOH	255(3.76),335(3.04)	2-0904-84
2-Pyridinecarboxylic acid, 4,5-bis(1,1-dimethylethyl)-, ethyl ester	MeOH	233(4.17),265(3.84)	44-1338-84
2-Pyridinecarboxylic acid, 4,6-bis(1,1-dimethylethyl)-, ethyl ester	MeOH	228(3.93),271(3.53)	44-1338-84
$C_{16}H_{25}NO_3$ 1H-3-Benzazonine, 2,3,4,5,6,7-hexahydro-7,9,10-trimethoxy-3-methyl-, hydrochloride	MeOH	233(4.01),282(3.52), 286s(3.49)	12-1203-84
$C_{16}H_{25}NO_4$ 2,5-Benzoxazonine, 1,3,4,5,6,7-hexahydro-1,9,10-trimethoxy-1,5-dimethyl-	MeOH	233(3.98),280(3.49), 285s(3.45)	12-1659-84
$C_{16}H_{25}NO_6$ Hordenine-4-O-β-D-glucoside	MeOH and MeOH-NaOH	225(3.92),266s(3.61), 275(3.52),305s(2.98)	102-1167-84
$C_{16}H_{25}NO_8P_2$ Phosphonic acid, [(4-nitrophenyl)ethenylidene]bis-, tetraethyl ester	dioxan	292(4.20)	65-1987-84
$C_{16}H_{25}NS$ Cyclohexane, 4-(1,5-dimethyl-1,3-hexadienyl)-1-isothiocyanato-1-methyl-, (theonellin isothiocyanate)	MeOH	238(4.48)	88-5401-84
$C_{16}H_{26}FeN_2$ Methanaminium, N,N'-1,1'-ferrocenediyl-bis[N,N-dimethyl-, diiodide	MeCN	208(4.39),233s(3.72), 268s(3.40)	101-0113-84R
$C_{16}H_{26}NO_2$ 2-Buten-1-aminium, N-[3-(2,2-dimethyl-1-oxopropoxy)-2-cyclopenten-1-ylidene]-N,3-dimethyl-, perchlorate	MeOH	275(4.40)	44-0220-84
$C_{16}H_{26}NO_3$ 1H-3-Benzazoninium, 2,3,4,5,6,7-hexahydro-7-hydroxy-9,10-dimethoxy-3,3-dimethyl-, iodide	MeOH	220(4.27),282(3.43), 287s(3.39)	12-1203-84
$C_{16}H_{26}NO_4$ 2,5-Benzoxazoninium, 1,3,4,5,6,7-hexahydro-1,9,10-trimethoxy-5,5-dimethyl-, iodide	MeOH	233(4.09),282(3.59), 286s(3.57)	12-1659-84
$C_{16}H_{26}N_2O$ [9,9'-Bi-9-azabicyclo[3.3.1]nonan]-	MeCN	250s(3.11),338(3.51)	35-0791-84

Compound	Solvent	$\lambda_{max}(\log \epsilon)$	Ref.
3-one, radical cation (cont.)			35-0791-84
$C_{16}H_{26}O$			
2-Cyclohexen-1-one, 3-cyclohexyl-4,4,6,6-tetramethyl-	C_6H_{12}	230(4.20),320(2.05)	33-0748-84
Naphthalene, 1,2,3,7,8,8a-hexahydro-3-(1-methoxy-1-methylethyl)-5,8a-dimethyl-, (3R-trans)-	EtOH	235s(4.19),240(4.34), 247s(4.15)	39-1323-84C
$C_{16}H_{26}O_3$			
1-Cyclohexene-1-carboxylic acid, 4-(1,5-dimethyl-4-hexenyl)-3-hydroxy-, methyl ester	MeOH	218(3.92)	102-0186-84
3-Cyclohexene-1-carboxylic acid, 4-(1-methylethoxy)-1-methyl-2-methylene-, 1,1-dimethylethyl ester	EtOH	249(4.21)	12-2037-84
2,6,11-Dodecatrienoic acid, 10-hydroxy-3,7,11-trimethyl-, methyl ester	MeCN	231(3.34)	102-0759-84
2-Hepten-4-one, 6-(1,4-dioxaspiro[4.5]dec-8-yl)-2-methyl-	EtOH	237(4.06)	107-0925-84
$C_{16}H_{26}O_4$			
3-Buten-2-one, 1,1-dimethoxy-4-(4-methoxy-2,6,6-trimethyl-1-cyclohexen-1-yl)-, (E)-	EtOH	232(3.86),304(4.03)	94-1709-84
$C_{16}H_{26}O_6P_2$			
Phosphonic acid, (phenylethenylidene)-bis-, tetraethyl ester	dioxan	273(4.22)	65-1987-84
$C_{16}H_{27}NO$			
Formamide, N-[4-(1,5-dimethyl-1,3-hexadienyl)-1-methylcyclohexyl]-, [1α,4β(1E,3E)]- (theonellin formamide)	MeOH	238(4.38)	88-5401-84
$C_{16}H_{27}NO_9$			
α-D-Glucopyranoside, ethyl 2-deoxy-2-[(2,2-diethoxycarbonylethenyl)-amino]-	EtOH	282(4.36)	136-0101-84L
$C_{16}H_{27}N_2O$			
Pyridinium, 3-(aminocarbonyl)-1-decyl-, iodide	H_2O	262(3.60)	47-2169-84
$C_{16}H_{27}N_3O_5$			
Acetamide, N-[acetyl[6-(acetoxyhexyl)-amino](acetylimino)methyl]-N-methyl-	MeOH	223(4.09),254(4.22)	158-1170-84
$C_{16}H_{27}N_5O$			
2(1H)-Pyrimidinone, 1-(1-piperidinyl-methyl)-4-[(1-piperidinylmethyl)-amino]-	MeOH	272(3.80)	78-3997-84
$C_{16}H_{28}$			
Cyclobutane, 1,1,3,3-tetramethyl-2-(2,2,4,4-tetramethylcyclo-butylidene)-	C_6H_{12}	199.0(4.16),204.5(4.16)	24-0277-84 +24-0310-84
$C_{16}H_{28}NO_3$			
Benzeneethanaminium, N-(4-hydroxybutyl)-3,4-dimethoxy-N,N-dimethyl-, iodide	MeOH	222(4.29),279(3.46), 284s(3.40)	12-1203-84

Compound	Solvent	$\lambda_{max}(\log \epsilon)$	Ref.
$C_{16}H_{28}N_2$			
9,9'-Bi-9-azabicyclo[3.3.1]nonane, radical cation	MeCN	260s(3.11),340(3.60)	35-0791-84
hexafluorophosphate	MeCN	260(3.11),340(3.60)	35-3366-84
$C_{16}H_{28}N_2O$			
3-Pyridinecarboxamide, 1-decyl-1,4-dihydro-	isoPrOH	354(3.85)	47-2169-84
$C_{16}H_{28}N_2O_4S_4$			
Butanedioic acid, 2,3-bis[[(diethylamino)thioxomethyl]thio]-, dimethyl ester	isoPrOH	252(4.03),281(4.02), 345(2.20)	104-1082-84
$C_{16}H_{28}N_2O_4Si_2$			
Pyrrolo[3,2-b]pyrrole-1,4-dicarboxylic acid, 3a,6a-dihydro-3,6-bis(trimethylsilyl)-, dimethyl ester, cis	C_6H_{12}	224(4.54)	88-5669-84
$C_{16}H_{28}O_2$			
3-Buten-2-one, 4-[2-(4-ethoxy-4-methylpentyl)-2-methylcyclopropyl]-, [1 (E),2]-(±)-	EtOH	255(4.04)	19-0207-84
$C_{16}H_{28}N_2S_2$			
2,8-Dithia-5,6-diazaspiro[3.4]oct-5-ene, 7,7-bis(1,1-dimethylethyl)-1,1,3,3-tetramethyl-	C_6H_{12}	210.5(3.41),294.0(2.93), 339s(2.30)	24-0277-84
$C_{16}H_{30}S$			
Thietane, 3-[1-(1,1-dimethylethyl)-2,2-dimethylpropylidene]-	C_6H_{12}	209.0(4.08),240s(3.20), 277.5(2.60)	24-0277-84 +24-0310-84
$C_{16}H_{34}O_2Si_2$			
4,9-Dioxa-3,10-disiladodeca-5,7-diene, 2,2,3,3,10,10,11,11-octamethyl-, (E,E)-	hexane	210(3.92)	44-1898-84
$C_{16}H_{36}As_2N_2S$			
Sulfur diimide, bis[bis(1,1-dimethylethyl)arsino]-	hexane	240(4.03),300(3.76), 390s(3.17)	24-1999-84
$C_{16}H_{36}N_2P_2S$			
Sulfur diimide, bis[bis(1,1-dimethylethyl)phosphino]-, (Z,Z)-	hexane	228(3.89),303(3.59), 398(3.46)	24-1999-84

Compound	Solvent	$\lambda_{max}(\log \epsilon)$	Ref.
$C_{17}H_9ClO_4$			
6H-Anthra[9,1-bc]furan-2-carboxylic acid, 5-chloro-6-oxo-, methyl ester	EtOH	497(4.10)	104-0745-84
$C_{17}H_9N_3O_3$			
Pyrrolo[2,1-a]isoquinoline-1-carbonitrile, 3-(5-nitro-2-furanyl)-	dioxan	432(4.12)	73-0533-84
$C_{17}H_9N_5O_4$			
12H-Benz[de]pyrazolo[3',4':5,6]pyrimido[2,1-a]isoquinoline-4,12(3H)-dione, 3-methyl-8-nitro-	EtOH	373(4.27)	104-0976-84
1H-Pyrazole-5-carbonitrile, 1-methyl-4-(6-nitro-1,3-dioxo-1H-benz[de]isoquinolin-2(3H)-yl)-	EtOH	350(4.06)	104-0976-84
$C_{17}H_{10}Br_2N_2$			
Cyclohepta[4,5]pyrrolo[1,2-a]imidazole, 3,10-dibromo-2-phenyl-	$CHCl_3$	248(4.26),285(4.28), 326(4.48),392(3.64), 413(3.76),438(3.77), 513(3.83),556(3.81), 604s(3.68),655s(3.78)	150-3465-84
$C_{17}H_{10}Br_2N_4O_2$			
1H-[1,2,4]Triazino[4,3-a]quinazoline-1,6(5H)-dione, 8-bromo-5-(4-bromophenyl)-2-methyl-	MeOH	227(4.62),315(3.91)	106-0717-84
1H-[1,2,4]Triazino[4,3-a]quinazoline-1,6(5H)-dione, 8,10-dibromo-2-methyl-5-phenyl-	MeOH	232(4.38),289(4.18)	106-0717-84
$C_{17}H_{10}ClNS$			
Thieno[2,3-b]quinoline, 4-chloro-3-phenyl-	EtOH	226(4.27),261(4.79), 344(3.65),358s(3.61)	48-0917-84
$C_{17}H_{10}ClN_3S$			
5-Pyrimidinecarbonitrile, 2-(4-chlorophenyl)-1,4-dihydro-6-phenyl-4-thioxo-	EtOH	270(4.46),316(4.23)	39-2447-84C
$C_{17}H_{10}CrN_2O_5$			
Chromium, pentacarbonyl[N-(4-imino-2,5-cyclohexadien-1-ylidene)benzenamine-N']-, (OC-6-22)-	acetone	390(3.50),530(4.00), 610(4.00)	101-0379-84R
$C_{17}H_{10}N_2$			
7H-Benz[de]anthracene, 7-diazo-	ether	269(4.53),285(4.37), 327(3.93),407(4.14), 425(4.10),600(2.42)	18-3526-84
$C_{17}H_{10}N_4O_2$			
12H-Benz[de]pyrazolo[3,4:5',6']pyrimido[2,1-a]isoquinoline-4,12(3H)-dione, 3-methyl-	DMF	367(4.34)	104-0976-84
$C_{17}H_{10}O$			
6H-Benz[de]anthracen-6-one	benzene	<u>305(4.1)</u>,385(4.2)	39-2177-84C
$C_{17}H_{10}O_4$			
6H-Anthra[9,1-bc]furan-2-carboxylic acid, 6-oxo-, methyl ester	EtOH	494(4.19)	104-0745-84

Compound	Solvent	$\lambda_{max}(\log \epsilon)$	Ref.
Anthra[1,2-b]furan-6,11-dione, 5-hydroxy-2-methyl-	EtOH	253(4.28),282(4.11), 424(3.82)	12-1511-84
$C_{17}H_{10}O_6$ 6H-[1,3]Dioxolo[5,6]benzofuro[3,2-c]-[1]benzopyran-6-one, 3-methoxy-	MeOH	244(4.20),309(3.95), 347(4.34)	102-0167-84
$C_{17}H_{11}BrN_4O_2$ 1H-[1,2,4]Triazino[4,3-a]quinazoline-1,6(5H)-dione, 5-(4-bromophenyl)-2-methyl-	MeOH	221(4.73),293(3.63)	106-0717-84
$C_{17}H_{11}ClN_4O_5$ 3H-Pyrazol-3-one, 4-[(4-chlorophenyl)-methylene]-2-(2,4-dinitrophenyl)-2,4-dihydro-5-methyl-	C_6H_{12} EtOH ether DMF DMSO	334(4.27) 218(4.38),344(4.04) 335(4.33) 356(4.14) 352(4.14)	23-2841-84 23-2841-84 23-2841-84 23-2841-84 23-2841-84
$C_{17}H_{11}Cl_2N$ 9,10-Ethanoanthracene-2-carbonitrile, 11,12-dichloro-9,10-dihydro-, $(9\alpha,10\alpha,11R*,12R*)$-	MeCN	242(3.81),254(3.52), 271(2.92),284(2.85)	44-2368-84
$C_{17}H_{11}Cl_2NO_4$ 1,4-Naphthalenedione, 6,7-dichloro-5,8-dihydroxy-2-[(phenylmethyl)-amino]-	benzene	482s(3.88),505(3.90), 540s(3.75)	118-0953-84
$C_{17}H_{11}NO$ 5H-Benzo[6,7]cyclohept[1,2-b]indolizin-5-one	EtOH	262(4.49),271(4.60), 283(4.41),313(3.93), 336(3.93),353(3.97), 447(4.13)	142-0791-84
$C_{17}H_{11}NOS$ Thieno[2,3-b]quinolin-4(9H)-one, 3-phenyl-	EtOH	224(--),255(--), 260s(--),301s(--), 312(--),336(--), 358s(--)	48-0917-84
$C_{17}H_{11}NO_4$ 6H-Anthra[1,9-cd]isoxazole-3-carboxylic acid, 6-oxo-, ethyl ester	EtOH	460(4.07)	104-2012-84
1H-[1,3]Dioxolo[4,5-g]indolo[7a,1-a]-isoquinoline-2,6-dione, (±)-	EtOH	246(4.28),276(4.31), 418(3.22)	142-2255-84
$C_{17}H_{11}N_3OS$ Isothiazolo[5,4-d]pyrimidin-3(2H)-one, 4,6-diphenyl-	EtOH	276(4.56)	18-0605-84
$C_{17}H_{11}N_3O_2$ Cyclohepta[4,5]pyrrolo[1,2-a]imidazole, 3-nitro-2-phenyl-	EtOH	228(4.39),240(4.40), 310s(4.39),325(4.36), 400s(3.72),428(3.72), 535s(3.52),552(3.54), 590s(3.41)	150-3465-84
$C_{17}H_{11}N_3O_3$ 1H-Isoindole-1,3(2H)-dione, 2-[2-(1H-indazol-3-yl)-2-oxoethyl]-	MeOH	207(4.84),222(4.38), 242s(4.26),303(3.79)	103-0066-84

Compound	Solvent	λ_{max} (log ϵ)	Ref.
Propanedinitrile, [[(4-nitrophenyl)-methoxy]phenylmethylene]-	EtOH	272(4.33)	5-1702-84
$C_{17}H_{11}N_5O_2$ 12H-Benzo[de]pyrazolo[3',4':5,6]pyrimido[2,1-a]isoquinoline-4,12(3H)-dione, 8-amino-3-methyl-	EtOH dioxan	495(3.85) 461(3.91)	104-0976-84 104-0976-84
$C_{17}H_{11}N_5O_3$ 1H-Imidazo[1,2-b]pyrazole, 2-(4-nitrophenyl)-3-nitroso-6-phenyl-	EtOH	217(4.30),256(3.70), 353(3.24),367(3.20)	4-1029-84
$C_{17}H_{11}N_5O_5$ 1H-Pyrazole-5-carboxamide, 1-methyl-4-(6-nitro-1,3-dioxo-1H-benz[de]-isoquinolin-2(3H)-yl)-	dioxan	347(4.05)	104-0976-84
$C_{17}H_{11}N_5O_7$ 3H-Pyrazol-3-one, 2-(2,4-dinitrophenyl)-2,4-dihydro-5-methyl-4-[(4-nitrophenyl)methylene]-	C_6H_{12} EtOH ether DMF DMSO	335(4.28) 218(4.38),275(4.15), 335s(4.00) 336(4.32) 355(4.43) 350(4.43)	23-2841-84 23-2841-84 23-2841-84 23-2841-84 23-2841-84
$C_{17}H_{12}$ 1,10-Etheno-11H-cyclohepta[a]naphthalene	EPA at 77°K	212(4.3),237(4.2), 272(4.6),340(4.1), 352(4.1)	33-0305-84
4,6-Etheno-5H-dibenzo[a,d]cycloheptene	EPA at 77°K	225(4.5),237(4.4), 247(4.4),310(4.1)	33-0305-84
$C_{17}H_{12}BrNO_6$ Propanedioic acid, [6-(4-bromophenyl)-3-cyano-2-oxo-2H-pyran-4-yl]-, dimethyl ester	EtOH	244(3.84),288(4.04), 330(3.99),400(3.73)	94-3384-84
$C_{17}H_{12}Cl_2O_5$ 4H-1-Benzopyran-4-one, 2-(3,5-dichloro-2-methoxyphenyl)-7-hydroxy-3-methoxy-	MeOH	203(4.52),235(4.43), 300(4.28)	2-1002-84
$C_{17}H_{12}Cl_2O_6$ 11H-Dibenzo[b,e][1,4]dioxepin-4-carbox-aldehyde, 2,7-dichloro-3-hydroxy-8-methoxy-1,6-dimethyl-11-oxo- (eriodermin)	MeOH MeOH-NaOH	234(4.18),254s(4.06) 224(4.15),262(4.15), 306(3.98),370(3.56)	102-0857-84 102-0857-84
$C_{17}H_{12}CrN_2O_5$ Chromium, pentacarbonyl(N-phenyl-1,4-benzenediamine-N')-, (OC-6-22)-	acetone	415(3.47)	101-0379-84R
$C_{17}H_{12}F_6O_7S_2$ Methanesulfonic acid, trifluoro-, 1-(4-methoxyphenyl)-2-phenyl-1,2-ethenediyl ester, (E)- (Z)-	EtOH EtOH	280(4.29) 232(4.29),289.5(4.11)	44-2273-84 44-2273-84
$C_{17}H_{12}N_2O_2$ 6H-Anthra[1,9-cd]isoxazol-6-one, 5-(1-aziridinyl)-3-methyl-	THF	250(4.41),257(4.40), 452(4.10),480(4.11)	103-0717-84

Compound	Solvent	$\lambda_{max}(\log \epsilon)$	Ref.
4H-[1]Benzoxepino[3,4-c]pyrazol-4-one, 1,10-dihydro-1-phenyl-	EtOH	280(4.20)	4-0301-84
[2,5'-Bi-1H-indole]-2'-carboxylic acid	EtOH	209(4.38),251(4.52), 298(4.44),320s(3.33), 337s(4.18)	103-0280-84
1,4-Naphthalenedione, 5-(1,3-dimethyl-1H-pyrazol-4-yl)ethynyl]-	EtOH	251s(4.32),314(4.23), 431(3.67)	70-2345-84
$C_{17}H_{12}N_2O_3$			
2-Propen-1-one, 1-(1H-indol-3-yl)-3-(4-nitrophenyl)-	EtOH	298(4.33),366(4.29)	65-0685-84
$C_{17}H_{12}N_2O_3S$			
Benzo[b]thiophen-3(2H)-one, 2-[3-[(4-nitrophenyl)amino]-2-propenylidene]-	toluene	384(4.01),512(4.68)	103-0951-84
$C_{17}H_{12}N_2S$			
10H-Pyrido[3,2-b][1,4]benzothiazine, 10-phenyl-	hexane	251(4.82),325(3.90)	4-0661-84
$C_{17}H_{12}N_4O$			
1H-Imidazo[1,2-b]pyrazole, 3-nitroso-2,6-diphenyl-	EtOH	222(4.50),264(3.67), 349(3.32),421(4.23)	4-1029-84
1H-Pyrazole-4-carbonitrile, 5-amino-3-benzoyl-1-phenyl-	EtOH	219(3.96),268(3.78)	4-1049-84
1H-Pyrazole-5-carbonitrile, 4-amino-3-benzoyl-1-phenyl-	EtOH	252(3.20)	4-1049-84
1H-Pyrazolo[1,2-a]pyrazol-4-ium, 3-(dicyanomethyl)-5,7-dimethyl-1-oxo-2-phenyl-, hydroxide, inner salt	MeCN	252(4.19),324(3.83), 497(3.83)	44-3672-84
4H-Pyrazolo[3,4-d]pyridazin-4-one, 1,5-dihydro-3,5-diphenyl-	EtOH	258(4.10),289(4.04)	94-4437-84
$C_{17}H_{12}N_4OS$			
1H-Pyridazino[4,5-e][1,3,4]thiadiazin-5(6H)-one, 3,6-diphenyl-	EtOH	273(3.99),302(3.79)	94-4437-84
$C_{17}H_{12}N_4O_2$			
Benzo[g]pteridine-2,4(3H,10H)-dione, 3-methyl-10-phenyl-	pH 7.27	348(4.00),438(4.01)	39-1227-84B
4H-3,1-Benzoxazine-2-acetonitrile, α-[(4-methylphenyl)azo]-4-oxo-	C_6H_{12}	230(4.03),255(3.78), 360(3.87)	80-0345-84
	EtOH	212(4.58),230(4.42), 261(4.25),305(3.89), 373(4.48)	80-0345-84
	ether	250(4.27),295(4.05), 363(4.45)	80-0345-84
	dioxan	362(4.54)	80-0345-84
	$CHCl_3$	370(4.18)	80-0345-84
	CCl_4	365(4.47)	80-0345-84
3H-[1,2,4]Triazino[3,2-b]quinazoline-3,10(4H)-dione, 4-methyl-2-phenyl-	EtOH	295(3.93),352(3.16)	24-1077-84
3H-[1,2,4]Triazino[3,2-b]quinazoline-3,10(4H)-dione, 2-(phenylmethyl)-	EtOH	290(3.65)	24-1077-84
3H-[1,2,4]Triazino[3,2-b]quinazoline-3,10(5H)-dione, 5-methyl-2-phenyl-	EtOH	296(3.86)	24-1077-84
$C_{17}H_{12}N_4O_3$			
4H-3,1-Benzoxazine-2-acetonitrile, α-[(4-methoxyphenyl)azo]-4-oxo-	C_6H_{12}	255(3.78),355(3.87)	80-0345-84
	EtOH	208(4.40),230(4.17), 250(4.09),305(--),	80-0345-84

Compound	Solvent	$\lambda_{max}(\log \epsilon)$	Ref.
(cont.)		368(3.88)	80-0345-84
	ether	240(--),300(3.60), 362(4.31)	80-0345-84
	acetone	364(4.32)	80-0345-84
	$CHCl_3$	368(4.32)	80-0345-84
	CCl_4	365(4.28)	80-0345-84
$C_{17}H_{12}N_4O_3S$			
1H-Pyrazino[2,3-c][1,2,6]thiadiazin-4(3H)-one, 6,7-diphenyl-, 2,2-dioxide	acid	222s(4.27),285(4.28), 428(4.06)	4-0861-84
	pH 4.0	276(4.23),363(3.99)	4-0861-84
dianion	pH 10.0	238(4.19),285(4.31), 396(3.98)	4-0861-84
$C_{17}H_{12}N_4O_5$			
3H-Pyrazol-3-one, 2-(2,4-dinitrophenyl)-2,4-dihydro-5-methyl-4-(phenylmethylene)-	C_6H_{12}	320(4.04)	23-2841-84
	EtOH	212(4.41),338(4.04)	23-2841-84
	ether	335(4.24)	23-2841-84
	DMF	360(4.28)	23-2841-84
	DMSO	355(4.29)	23-2841-84
$C_{17}H_{12}N_4O_6$			
3H-Pyrazol-3-one, 2-(2,4-dinitrophenyl)-2,4-dihydro-4-[(2-hydroxyphenyl)-methylene]-5-methyl-	C_6H_{12}	325(4.10)	23-2841-84
	EtOH	218(4.38),334(4.08), 355s(4.05)	23-2841-84
	ether	335(4.13)	23-2841-84
	DMF	368(4.19)	23-2841-84
	DMSO	365(4.19)	23-2841-84
3H-Pyrazol-3-one, 2-(2,4-dinitrophenyl)-2,4-dihydro-4-[(4-hydroxyphenyl)-methylene]-5-methyl-	C_6H_{12}	426(--)	23-2841-84
	ether	378(4.32)	23-2841-84
	DMF	380(3.93),492(4.26)	23-2841-84
	DMSO	384(4.17),492(4.00)	23-2841-84
	EtOH	215(3.97),255s(3.74), 315s(3.54),385(4.15), 492(3.95)	23-2841-84
$C_{17}H_{12}N_4S$			
4-Pyrimidinesulfenamide, 5-cyano-2,6-diphenyl-	EtOH	279(4.63)	18-0605-84
$C_{17}H_{12}N_6O$			
8H-1,2,4-Triazolo[4"',3"':1",6"][1,2,4]-triazino[4",5":1',5']pyrrolo[2',3'-4,5]furo[3.2-b]indole, 3,6-dimethyl-	dioxan	339(3.59)	73-1529-84
$C_{17}H_{12}O_3$			
2,5-Furandione, 3-phenyl-4-(phenylmethyl)-	EtOH	313(3.68)	158-84-42
2(5H)-Furanone, 4-hydroxy-3-phenyl-5-(phenylmethylene)-, (Z)-	EtOH	358(3.97)	39-1539-84C
	EtOH-NaOH	368(3.93)	39-1539-84C
$C_{17}H_{12}O_4$			
9,10-Anthracenedione, 1-hydroxy-3-(2-methoxyethenyl)-, (E)-	MeOH	209(4.43),250(4.45), 297(4.43),429(3.98)	78-3677-84
(Z)-	MeOH	211(4.44),252(4.46), 296(4.28),424(3.87)	78-3677-84
2(5H)-Furanone, 4-hydroxy-3-(4-hydroxyphenyl)-5-(phenylmethylene)-, (Z)-	EtOH	300(4.58),380(4.05)	39-1539-84C

Compound	Solvent	$\lambda_{max}(\log \epsilon)$	Ref.
$C_{17}H_{12}O_5$			
2-Anthraceneacetic acid, 9,10-dihydro-4-hydroxy-9,10-dioxo-, methyl ester	MeOH	206(4.23),223(4.23), 253(4.38),277s(--), 328(3.47),403(3.73)	78-3677-84
9,10-Anthracenedione, 1-acetoxy-8-methoxy-	n.s.g.	219(4.46),252(4.41), 267(4.17),350(3.57), 383(3.80)	5-0306-84
9,10-Anthracenedione, 1,4-dihydroxy-2-(2-oxopropyl)-	EtOH	251(4.57),257(4.53), 286(4.04),326(3.53)	12-1518-84
Spiro[anthracene-9(10H),2'(5'H)-furan]-5',10-dione, 3',4'-dihydro-4,5-dihydroxy-	n.s.g.	212(4.56),220(4.47), 258(4.02),266(4.14), 296(4.13),373(4.25)	5-0306-84
$C_{17}H_{12}O_5S$			
1,4-Naphthalenedione, 2-[[(4-methylphenyl)sulfonyl]oxy]-	EtOH	226(4.34),330(3.48)	44-2355-84
$C_{17}H_{12}O_6$			
2,5-Cyclohexadiene-1,4-dione, 2,2'-(1,5-dioxo-1,5-pentanediyl)bis-	MeCN	246(4.56),300s(2.99), 450(1.86)	18-0791-84
Cyclopenta[k]phenanthrene-1,4,5,8(12bH)-tetrone, 5,6,7,7a-tetrahydro-9,12-dihydroxy-, (4aS*,7aα,12bα)-(±)-	MeCN	260s(3.85),373(3.70)	18-0791-84
2(5H)-Furanone, 5-[(3,4-dihydroxyphenyl)methylene]-4-hydroxy-3-(4-hydroxyphenyl)-, (Z)-	EtOH EtOH-NH₃	242(4.30),381(4.48) 259(--),386(--)	39-1539-84C 39-1539-84C
2-Propen-1-one, 3-(1,3-benzodioxol-5-yl)-1-(6-hydroxy-1,3-benzodioxol-5-yl)-, (E)-	EtOH	203(4.49),237(4.03), 294(4.03),340(4.01), 395(3.83)	78-4081-84
$C_{17}H_{12}O_8$			
8H-1,3-Dioxolo[4,5-g][1]benzopyran-8-one, 6-(3,4-dihydroxyphenyl)-9-hydroxy-7-methoxy-	MeOH	252(4.16),281(4.16), 348(4.37)	102-2043-84
	MeOH-NaOAc	243(4.15),262(4.16), 270(4.17),353(4.27), 405s(4.01)	102-2043-84
	+ H₃BO₃	245(4.17),264(4.30), 373(4.38)	102-2043-84
	MeOH-AlCl₃ + HCl	277(4.24),433(4.42) 243(4.13),269(4.16), 283(4.15),380(4.36)	102-2043-84 102-2043-84
	MeOH-NaOMe	244(4.13),274(4.20), 402(4.36)	102-2043-84
Ellagic acid, 3,3',4-tri-O-methyl-	EtOH	248(4.48),360s(3.85), 373(3.92)	94-4478-84
	EtOH-NaOAc	256(4.76),305(3.90), 412(3.85)	94-4478-84
$C_{17}H_{12}O_9$			
4H,6H-Benzo[1,2-b:4,5-c']dipyran-8-carboxylic acid, 5-acetoxy-9-methoxy-2-methyl-4,6-dioxo-	CH₂Cl₂	274(4.43),313(4.04), 328(4.16),352(4.02)	23-2101-84
$C_{17}H_{12}SSe$			
4H-Selenin-4-thione, 2,6-diphenyl-	CHCl₃	238(4.4),338(4.2), 426(4.39)	22-0241-84
$C_{17}H_{13}BrN_2O_2$			
6H-Anthra[1,9-cd]isoxazol-6-one, 5-[(2-bromoethyl)amino]-3-methyl-	THF	252(4.44),260(4.48), 483(4.23),515(4.28)	103-0717-84

Compound	Solvent	$\lambda_{max}(\log \epsilon)$	Ref.
$C_{17}H_{13}BrN_4O_3$ Propanoic acid, 2-[[3-(4-bromophenyl)- 3,4-dihydro-4-oxo-2-quinazolinyl)- hydrazono]-	MeOH	222(4.54),343(4.07)	106-0717-84
$C_{17}H_{13}BrO_2$ 1-Benzoxepin-5(2H)-one, 4-[(4-bromo- phenyl)methylene]-3,4-dihydro-, (E)-	octane EtOH	292(4.27) 306(4.29)	65-0148-84 65-0148-84
$C_{17}H_{13}BrO_5$ 9-Anthracenecarboxylic acid, 2-bromo- 9,10-dihydro-4-hydroxy-1-methoxy- 10-oxo-, methyl ester, (±)-	MeOH	273(4.22),294(4.20), 364(3.81)	44-0318-84
$C_{17}H_{13}ClN_2$ Pyrimidine, 4-chloro-2-phenyl-5-(phen- ylmethyl)-	EtOH	205(4.16),268(4.18)	103-0431-84
$C_{17}H_{13}ClN_2O_2$ 6H-Anthra[1,9-cd]isoxazol-6-one, 5-[(2-chloroethyl)amino]-3-methyl-	THF	253(4.47),259(4.98), 485(4.23),515(4.29)	103-0717-84
$C_{17}H_{13}ClO_2$ 1-Benzoxepin-5(2H)-one, 4-[(4-chloro- phenyl)methylene]-3,4-dihydro-, (E)-	octane EtOH	294(4.26),296(4.29) 306(4.25)	65-0148-84 65-0148-84
$C_{17}H_{13}ClO_4$ 4H-1-Benzopyran-4-one, 2-(2-chloro- phenyl)-3,7-dimethoxy-	MeOH	200(4.51),210s(4.37), 240s(4.25),298(4.19)	2-1002-84
$C_{17}H_{13}ClO_5$ 4H-1-Benzopyran-4-one, 2-(5-chloro- 2-methoxyphenyl)-7-hydroxy-3-methoxy-	MeOH	204(4.55),222(4.54), 303(4.30)	2-1002-84
4H-1-Benzopyran-4-one, 2-(2-chloro- phenyl)-5-hydroxy-3,7-dimethoxy-	MeOH	210(4.42),257(4.27), 296(3.86)	2-1002-84
$C_{17}H_{13}ClO_6$ 4H-1-Benzopyran-4-one, 2-(5-chloro- 2-methoxyphenyl)-5,7-dihydroxy-3- methoxy-	MeOH	210(4.57),256(4.37), 302(4.02),330s(3.98)	2-1002-84
$C_{17}H_{13}ClO_7$ 4H-1-Benzopyran-4-one, 2-(5-chloro- 2-methoxyphenyl)-5,7,8-trihydroxy- 3-methoxy-	MeOH	208(4.41),230s(4.56), 268(4.37),304(4.20), 360(3.50)	2-1002-84
$C_{17}H_{13}F_3$ Benzene, 1-ethenyl-2-(3,3,3-trifluoro- 2-phenyl-1-propenyl)-, cis	MeOH	214(4.75),234(4.05), 260s(3.97),278(3.80), 315(2.71)	151-0327-84D
trans	MeOH	200(4.24),247(4.13), 270s(3.80),315s(2.37)	151-0327-84D
$C_{17}H_{13}N$ 9-Phenanthrenecarbonitrile, 1-ethenyl- 9,10-dihydro-	MeOH	258(4.53),279(4.28), 304(3.84)	151-0327-84D
$C_{17}H_{13}NO$ Ethanone, 1-(4-phenyl-3-quinolinyl)-	MeOH	241(4.38),290(3.84)	4-0759-84

Compound	Solvent	$\lambda_{max}(\log \epsilon)$	Ref.
$C_{17}H_{13}NOS$			
Benzo[b]thiophen-3(2H)-one, 2-[3-(phenylamino)-2-propenylidene]-	toluene	365(4.03),488(4.48)	103-0951-84
	DMSO	380(4.05),520(4.66)	103-0951-84
2H-Pyrrol-2-one, 1,5-dihydro-3-(4-methylphenyl)-1-phenyl-5-thioxo-	CHCl₃	355(4.36)	78-3499-84
$C_{17}H_{13}NOS_2$			
6H-1,3-Thiazine-6-thione, 2-(4-methoxyphenyl)-	CHCl₃	284(4.25),324(4.39), 459(3.89)	97-0435-84
$C_{17}H_{13}NO_2$			
Benzeneacetonitrile, α-[(2-oxo-2-phenylethoxy)methylene]-	EtOH	260(4.43)	5-1702-84
5H-Benzo[g]-1,3-benzodioxolo[6,5,4-de]quinoline, 6,7-dihydro- (dehydroanonaine)	MeOH	253(4.69),258(4.70), 332(4.13),380(3.85)	39-1273-84C
3(2H)-Benzofuranone, 2-[3-(phenylamino)-2-propenylidene]-	DMSO	353(3.83),482(4.67)	103-0951-84
Benzoic acid, 2-(3-isoquinolinyl)-, methyl ester	MeOH	220(4.65),283(3.96), 324(3.56)	83-0002-84
Cyclopenta[4,5]azepino[2,1,7-cd]pyrrolizine-2-carboxylic acid, ethyl ester	EtOH	273(4.18),307(4.56), 331(4.63),345(4.82), 402(3.80),425(3.85), 455(3.90),526(2.55), 544(2.58),564(2.53), 586(2.48),643(2.15)	24-1649-84
4,9-Epoxynaphth[2,3-d]isoxazole, 3a,4,9,9a-tetrahydro-3-phenyl-, (3aα,4β,9β,9aα)-	MeOH	267(3.39)	73-1990-84
Methanone, (3-amino-4-phenyl-2-furanyl)phenyl-	EtOH	243(4.20),349(4.11)	5-1702-84
1(4H)-Naphthalenone, 5-hydroxy-4-[(4-methylphenyl)imino]-	CHCl₃	258(4.30),292(4.13), 457(3.94)	104-0733-84
$C_{17}H_{13}NO_3$			
Indolo[4,5-d]benzo[b]furan-2-carboxylic acid, ethyl ester	EtOH	211(4.67),219(4.66), 240(4.46),253(4.23), 273(4.28),281(4.46), 331(4.60),323(4.77), 346(4.40)	103-1123-84
Indolo[5,4-d]benzo[b]furan-2-carboxylic acid, ethyl ester	EtOH	253(4.50),273(4.91), 286(4.66),295(4.65)	103-1123-84
Indolo[5,6-d]benzo[b]furan-2-carboxylic acid, ethyl ester	EtOH	205(4.65),235(4.42), 241(4.47),256(4.51), 274(4.72),290(4.78), 330(4.47),345(4.54)	103-1123-84
Indolo[6,5-d]benzo[b]furan-2-carboxylic acid, ethyl ester	EtOH	219(4.65),242(4.59), 252(4.65),335(4.83), 352(4.57)	103-1123-84
Propanamide, N-(9,10-dihydro-9,10-dioxo-1-anthracenyl)-	benzene	418(3.79)	39-0529-84C
$C_{17}H_{13}NO_4$			
1-Benzoxepin-5(2H)-one, 3,4-dihydro-4-[(4-nitrophenyl)methylene]-, (E)-	octane	301(4.39)	65-0148-84
	EtOH	309(4.28)	65-0148-84
Carbamic acid, (9,10-dihydro-9,10-dioxo-1-anthracenyl)-, ethyl ester	benzene	417(3.77)	39-0529-84C
Erythrinan-8-one, 1,2,6,7,10,11-hexadehydro-3-hydroxy-15,16-bis[methylenebis(oxy)]-	EtOH	228(4.35),265(4.38), 360(3.44)	142-2255-84

Compound	Solvent	$\lambda_{max}(\log \epsilon)$	Ref.
$C_{17}H_{13}NO_6$			
Propanedioic acid, (3-cyano-2-oxo-6-phenyl-2H-pyran-4-yl)-, dimethyl ester	EtOH	254(4.07),274(3.99), 368(4.31)	94-3384-84
$C_{17}H_{13}N_2O$			
Indolo[3,2,1-de]pyrido[3,2,1-ij][1,5]-naphthyridin-4-ium, 1,2,3,12-tetra-hydro-12-oxo-, chloride (infractopicrin)	MeOH	213(4.34),236s(4.06), 254s(3.89),261.5(4.00), 270(3.96),313.5(3.88), 326.5(3.90),349s(3.69), 368(3.84),387(3.83)	88-2341-84
$C_{17}H_{13}N_3$			
Acridine, 9-(2-methyl-1H-imidazolyl)-	EtOH	251(5.02),359(3.95), 383(3.66)	42-0611-84
$C_{17}H_{13}N_3O$			
3H-Pyrazolo[4,3-b]quinolin-3-one, 1,2-dihydro-1-methyl-2-phenyl-	EtOH	322(4.15),329(4.15), 336(4.27)	94-1604-84
$C_{17}H_{13}N_3O_2$			
3H-[2]Benzoxepino[3,4-d]-1,2,3-triazol-10(5H)-one, 3-(phenylmethyl)-	EtOH	261(3.89),297(3.66), 347(4.00)	150-3755-84M
4H-Pyrrolo[1,2-a]benzimidazole, 4-methyl-2-(3-nitrophenyl)-	DMSO	274(4.4),306(4.5), 340(3.8)	103-0150-84
4H-Pyrrolo[1,2-a]benzimidazole, 4-methyl-2-(4-nitrophenyl)-	DMSO	260(4.22),380(4.3)	103-0150-84
4H-Pyrrolo[1,2-a]benzimidazole, 4-methyl-7-nitro-2-phenyl-	DMSO	260(4.16),290(4.48), 333(3.86),455(3.95)	103-0150-84
$C_{17}H_{13}N_3O_2S$			
[1]Benzothiopyrano[2,3-d]-1,2,3-tria-zol-9(1H)-one, 3-[(4-methoxyphenyl)-methyl]-	EtOH	237(4.54),242(4.54), 332(3.86)	87-0223-84
$C_{17}H_{13}N_3O_3$			
Ethanone, 1-(1H-indol-3-yl)-2-[[(4-ni-trophenyl)methylene]amino]-	EtOH	263(4.08),290(4.03)	65-0685-84
Ethanone, 1-[3-[[(2-nitrophenyl)methyl-ene]amino]-1H-indol-2-yl]-, (E)-	MeOH	203(4.39),234(4.37), 297(4.22),404(3.98)	83-1029-84
2(1H)-Quinoxalinone, 1-methyl-3-[2-(2-nitrophenyl)ethenyl]-	EtOH	394(3.78)	18-1653-84
2(1H)-Quinoxalinone, 1-methyl-3-[2-(4-nitrophenyl)ethenyl]-	EtOH	402(3.97)	18-1653-84
$C_{17}H_{13}N_3O_4S$			
3-Pyridinecarboxylic acid, 2-[[5-carb-oxy-1-(phenylmethyl)-1H-imidazol-2-yl]thio]-	MeOH	207(4.52),253(4.41), 295s(3.69)	142-0807-84
1H-Thieno[3,2-c][1,2,4]triazolo[1,2-a]-pyridazine-5-carboxylic acid, 2,3-di-hydro-1,3-dioxo-2-phenyl-, ethyl ester	EtOH	220(3.95),242(3.98), 283(3.63),405(3.46)	39-0915-84C
$C_{17}H_{13}N_3O_7$			
1H-Pyrazole-3,4-dicarboxylic acid, 5-[(1,3-dihydro-1,3-dioxo-2H-iso-indol-2-yl)acetyl]-, dimethyl ester	MeOH	219(4.84),239s(4.29), 276(3.68),301(3.24)	103-0066-84
$C_{17}H_{13}N_5O$			
1H-Pyrazole-3-carboxamide, 5-amino-	EtOH	219(3.94),266(3.69)	4-1049-84

Compound	Solvent	$\lambda_{max}(\log \epsilon)$	Ref.
4-cyano-N,1-diphenyl- (cont.)			4-1049-84
$C_{17}H_{13}N_5O_2S$			
1H-Pyrazino[2,3-c][1,2,6]thiadiazin-4-amine, 6,7-diphenyl-, 2,2-dioxide anion	pH 1.0	276(4.26),369(3.97), 441(3.45)	4-0861-84
	pH 6.0	285(4.33),405(3.98)	4-0861-84
$C_{17}H_{13}O$			
Pyrylium, 2,6-diphenyl-, TCNQ salt	MeCN	478(3.05)	80-0817-84
perchlorate	MeCN	415(--)	80-0817-84
$C_{17}H_{13}P$			
Phosphorin, 3,5-diphenyl-	EtOH	220s(4.29),252(4.48), 276s(4.23),320s(2.79)	24-0763-84
$C_{17}H_{14}N_2$			
12H-Indolo[2,3-a]quinolizin-5-ium, 3-ethyl-, hydroxide, inner salt (flavopereirine)	MeOH	233(4.58),244(4.50), 288(4.22),318(4.21), 348(4.33),388(4.23)	142-0233-84
	MeOH-KOH	238(4.45),252(4.24), 285(4.50),313(4.13), 360(4.35)	142-0233-84
Pyrazine, 2-methyl-3,5-diphenyl-	EtOH	235(4.26),254s(4.20), 317(4.05)	142-2317-84
$C_{17}H_{14}N_2O$			
Ethanone, 1-[2-[(1H-indol-3-ylmethylene)amino]phenyl]-, (E)-	MeOH	238(4.31),264(4.25), 304(4.29),380(4.05)	83-1029-84
2(1H)-Pyrazinone, 1-methyl-5,6-diphenyl-	EtOH	266(4.08),347(3.86)	39-0391-84C
1H-Pyrido[3,4-b]indol-1-one, 2,3,4,9-tetrahydro-2-phenyl-	isoPrOH	226(4.43),240s(4.30), 310(4.44)	103-0047-84
1H-Pyrido[3,4-b]indol-1-one, 2,3,4,9-tetrahydro-9-phenyl-	isoPrOH	226(4.46),302(4.22), 328s(3.87)	103-0047-84
4(1H)-Pyrimidinone, 2-phenyl-5-(phenylmethyl)-	EtOH	207(4.43),244(4.12), 300(4.18)	103-0431-84
2(1H)-Quinoxalinone, 1-methyl-3-(2-phenylethenyl)-, (E)-	EtOH	384(3.87)	18-1653-84
$C_{17}H_{14}N_2OS$			
Methanone, [5-(methylthio)-2-phenyl-1H-imidazol-4-yl]phenyl-	MeOH	270(4.54),350(4.23)	77-0430-84
1H-Pyrazole, 3-(furanyl)-4,5-dihydro-1-phenyl-5-(thienyl)-	EtOH	252(4.19),364(4.13)	34-0225-84
3-Thiophenecarboxamide, 4-phenyl-2-(phenylamino)-	EtOH	231(4.31),334(4.20)	48-0917-84
$C_{17}H_{14}N_2O_2$			
Indolo[2,3-a]quinolizin-4(6H)-one, 3-acetyl-7,12-dihydro-	MeOH	253(3.53),261(3.55), 274(3.27),281(3.04), 318(3.17),426(4.16)	142-0233-84
4-Quinolinamine, N-(1,3-benzodioxol-5-ylmethyl)-	MeOH	214(4.61),234(4.33), 327(4.15)	24-1523-84
2(1H)-Quinoxalinone, 3-[2-(4-hydroxyphenyl)ethenyl]-1-methyl-	EtOH	402(3.84)	18-1653-84
$C_{17}H_{14}N_2O_4$			
1(3H)-Isobenzofuranone, 3-[[4-(dimethylamino)phenyl]methylene]-6-nitro-	EtOH	476(4.31)	131-0239-84H

Compound	Solvent	$\lambda_{max}(\log \epsilon)$	Ref.
$C_{17}H_{14}N_2O_5$ Heterodimer of thymine and 8-methoxy-psoralen, cis-syn	MeCN	222(4.47),257.5(4.18), 294s(3.44)	156-0051-84B
$C_{17}H_{14}N_2O_5S$ 2,4,6(1H,3H,5H)-Pyrimidinetrione, 5-[(1,3-dihydro-1,3-dioxo-2H-inden-2-ylidene)(methylthio)methyl]-1,3-dimethyl-	EtOH	245(4.49),364(4.31), 460(3.97)	95-0127-84
$C_{17}H_{14}N_2S_2$ 1H-Pyrazole, 4,5-dihydro-1-phenyl-3,5-dithienyl-	EtOH	252(4.21),364(4.11)	34-0225-84
$C_{17}H_{14}N_4O$ Acetamide, N-(2-phenyl-4H-pyrazolo-[1,5-a]benzimidazol-6-yl)-	EtOH	251(4.43),288(4.28), 331(4.39)	48-0829-84
$C_{17}H_{14}N_4O_3$ Benzoic acid, 2-[(2,5-dihydro-2-methyl-5-oxo-6-phenyl-1,2,4-triazin-3-yl)-amino]-	EtOH	292(4.14)	24-1077-84
$C_{17}H_{14}N_4S$ 2-Benzothiazoleacetonitrile, 6-methyl-α-[(4-methylphenyl)hydrazono]-	EtOH	258(3.92),405(9.19)[sic]	104-0523-84
$C_{17}H_{14}N_4S_2$ 4H,6H-1,5,3-Dithiazepino[6,7-b]indol-2-amine, 4-imino-N-methyl-N-phenyl-	EtOH	289(4.45),380s(3.49)	12-2479-84
$C_{17}H_{14}N_{10}$ 5,8-Methanoquinoxaline-1(4H)-ethanimid-amide, N-(2-amino-1,2-dicyanoethen-yl)-2,3-dicyano-4a,5,8,8a-tetrahy-dro-α-imino-	EtOH	240s(3.99),320(4.04), 395(4.20)	44-0813-84
$C_{17}H_{14}OS_3$ Ethane(dithioic) acid, (methylthio)-9H-xanthen-9-ylidene-, methyl ester	CCl₄	278(2.78),332(2.92), 428(2.22),525(1.20)	54-0152-84
$C_{17}H_{14}O_2$ Benzeneacetic acid, α-[(2-ethenylphen-yl)methylene]-, cis	MeOH	221(4.29),243s(4.21), 287(4.08)	151-0327-84D
1-Benzoxepin-5(2H)-one, 3,4-dihydro-4-(phenylmethylene)-, (E)-	octane	292(4.36)	65-0148-84
	EtOH	226(4.17),304(4.18)	65-0148-84
2-Naphthalenol, 1-[(4-hydroxyphenyl)-methyl]-	EtOH	229.5(5.02),273(3.93), 279(4.06),290(3.95), 324(3.67),335(3.70)	150-0701-84
$C_{17}H_{14}O_3$ 1-Benzoxepin-5(2H)-one, 8-methoxy-3-phenyl-	MeOH	202(4.27),307(3.83)	2-1211-84
2(5H)-Furanone, 4-(4-methoxyphenyl)-5-phenyl-	MeOH	217(3.9),303(4.1)	2-0514-84
$C_{17}H_{14}O_4$ 9,10-Anthracenedione, 1,2-dimethoxy-4-methyl-	MeOH	225(4.31),244(4.34), 270(4.38),376(3.70)	78-5039-84
9,10-Anthracenedione, 1,8-dimethoxy-3-methyl-	EtOH	257(4.24),394(3.75)	78-3455-84

Compound	Solvent	$\lambda_{max}(\log \epsilon)$	Ref.
4H-1-Benzopyran-4-one, 3-acetoxy-2,3-dihydro-2-phenyl-	MeOH	211.5(4.52),252(3.96), 318(3.47)	94-4852-84
4H-1-Benzopyran-4-one, 6,7-dimethoxy-3-phenyl-	MeOH	319(3.45)	64-0238-84B
4H-1-Benzopyran-4-one, 7,8-dimethoxy-3-phenyl-	MeOH	301(3.92)	64-0238-84B
4H-1-Benzopyran-4-one, 3-hydroxy-6-methoxy-2-(4-methylphenyl)-	CHCl₃	256(4.24),335(4.32), 351s(4.28)	142-1943-84
4H-1-Benzopyran-4-one, 3-hydroxy-7-methoxy-2-(4-methylphenyl)-	CHCl₃	256(4.25),322(4.26), 345(4.32)	142-1943-84
4H-1-Benzopyran-4-one, 6-methoxy-3-(4-methoxyphenyl)-	MeOH	330(3.18)	64-0238-84B
Oxoflavidin dimethyl ether	EtOH	222(4.43),234s(4.39), 246(4.27),287(4.19), 363(3.86)	102-0671-84

$C_{17}H_{14}O_4S_2$

Compound	Solvent	$\lambda_{max}(\log \epsilon)$	Ref.
1,3-Dithiole-4,5-dicarboxylic acid, 2-(1-naphthalenyl)-, dimethyl ester	MeOH	290(5.56)	104-2285-84

$C_{17}H_{14}O_5$

Compound	Solvent	$\lambda_{max}(\log \epsilon)$	Ref.
9-Anthracenecarboxylic acid, 9,10-dihydro-4-hydroxy-1-methoxy-10-oxo-, methyl ester	MeOH	270(4.19),286s(4.02), 390(3.59)	44-0318-84
9,10-Anthracenedione, 2-(ethoxymethyl)-1,3-dihydroxy-	EtOH	245(4.21),277(4.11), 313s(3.71),410(3.33)	102-1733-84
9,10-Anthracenedione, 1-hydroxy-2,5-dimethoxy-4-methyl-	MeOH	230(4.41),251(4.42), 283(4.11),430(3.92)	78-5039-84
9,10-Anthracenedione, 1-hydroxy-3,8-dimethoxy-6-methyl-	EtOH	252(4.22),268(4.24), 283(4.29),424(3.97)	78-3455-84
9,10-Anthracenedione, 1-hydroxy-5,6-dimethoxy-2-methyl-	EtOH	267(4.33),292(4.07), 410(3.94)	78-3455-84
9,10-Anthracenedione, 1-hydroxy-5,7-dimethoxy-3-methyl-	EtOH	251(4.20),277(4.39), 298s(4.16),404(4.06)	78-3455-84
9,10-Anthracenedione, 1-hydroxy-6,8-dimethoxy-3-methyl-	EtOH	271(4.20),280(4.21), 422(3.87)	78-3455-84
9,10-Anthracenedione, 1,4,5-trihydroxy-2-propyl-	EtOH	233(4.60),253(4.39), 292(3.89),466(3.99), 481(4.06),493(4.11), 514(3.98),527(3.95)	23-1922-84
4H-1-Benzopyran-4-one, 3-hydroxy-6-methoxy-2-(4-methoxyphenyl)-	CHCl₃	255(4.26),340s(4.35), 352(4.38)	142-1943-84
4H-1-Benzopyran-4-one, 3-hydroxy-7-methoxy-2-(4-methoxyphenyl)-	CHCl₃	259(4.49),320(4.40), 352(4.55)	142-1943-84
4H-1-Benzopyran-4-one, 6-hydroxy-7-methoxy-3-(4-methoxyphenyl)-	MeOH	325(3.96)	64-0238-84B
4H-1-Benzopyran-4-one, 7-hydroxy-6-methoxy-3-(4-methoxyphenyl)-	MeOH	319(4.00)	64-0238-84B

$C_{17}H_{14}O_6$

Compound	Solvent	$\lambda_{max}(\log \epsilon)$	Ref.
9,10-Anthracenedione, 1,5-dihydroxy-2,7-dimethoxy-4-methyl-	MeOH	229(4.55),262(4.40), 292(4.30),456(4.09)	78-5039-84
9,10-Anthracenedione, 1,8-dihydroxy-2,6-dimethoxy-4-methyl-	MeOH	230(4.53),249(4.19), 278(4.41),307s(4.08), 440(4.15)	78-5039-84
5H-Phenanthro[4,5-bcd]pyran-5-one, 9,10-dihydro-2,6-dihydroxy-7,8-dimethoxy- (coeloginin)	EtOH	202s(4.37),208(4.41), 250(4.44),287s(4.00), 364(4.00)	39-1919-84C

$C_{17}H_{14}O_6S$

Compound	Solvent	$\lambda_{max}(\log \epsilon)$	Ref.
Benzeneacetic acid, 4-acetoxy-	MeOH	254(3.95),278(3.69),	73-2531-84

Compound	Solvent	$\lambda_{max}(\log \epsilon)$	Ref.
3-[(2-carboxyphenyl)thio]-, benzene solvate (cont.)		314(3.67)	73-2531-84
$C_{17}H_{14}O_7$			
9,10-Anthracenedione, 1,2,8-trihydroxy-6,7-dimethoxy-3-methyl-	dioxan	282(4.56),310s(3.87), 320(3.90),410(3.96)	94-0860-84
9,10-Anthracenedione, 1,3,5-trihydroxy-4,7-dimethoxy-2-methyl-	MeOH	227(4.56),270s(4.13), 285(4.38),422(4.05), 440(3.78)	102-2104-84
	MeOH-base	228(--),267(--), 302(--),490(--)	102-2104-84
$C_{17}H_{14}O_8$			
4H,6H-Benzo[1,2-b:4,5-c']dipyran-8-carboxylic acid, 5,9-dimethoxy-2-methyl-4,6-dioxo-, methyl ester	CH_2Cl_2	234(4.04),276(4.29), 292(4.16),320(4.06)	23-2101-84
4H-1-Benzopyran-4-one, 2-(2,4-dihydroxy-5-methoxyphenyl)-5,7-dihydroxy-6-methoxy-	MeOH	266(4.2),292(3.9), 372(4.3)	94-4935-84
	MeOH-NaOMe	274(--),316(--), 425(--)	94-4935-84
	MeOH-NaOAc	270(--),295(--), 382(--)	94-4935-84
	MeOH-AlCl$_3$	274(--),297(--), 408(--)	94-4935-84
	+ HCl	268(--),296(--), 402(--)	94-4935-84
$C_{17}H_{14}S_4$			
Ethane(dithioic) acid, (methylthio)-9H-thioxanthen-9-ylidene-, methyl ester	CCl$_4$	278(3.15),323(3.11), 412(2.43),510(1.38)	54-0152-84
$C_{17}H_{15}BrN_4O$			
2H-Imidazol-2-one, 4-[(4-bromophenyl)-methylamino]-1,5-dihydro-1-methyl-5-(phenylimino)-	MeCN	237(4.23),250(4.23), 323(3.79)	142-2509-84
$C_{17}H_{15}BrOS$			
Dibenzo[b,f]thiepin-10(11H)-one, 2-bromo-8-(1-methylethyl)-	MeOH	240(4.49),265s(4.04), 335(3.59)	73-0086-84
$C_{17}H_{15}BrO_4$			
1,3-Benzodioxole, 5-[2-(2-bromo-4,5-dimethoxyphenyl)ethenyl]-	CHCl$_3$	244(4.12),299(4.11), 342(4.22)	142-1217-84
$C_{17}H_{15}BrO_5$			
Ethanone, 1-(1,3-benzodioxol-5-yl)-2-(2-bromo-4,5-dimethoxyphenyl)-	CHCl$_3$	242(4.06),278(4.08), 296(3.99),308s(3.96)	142-1217-84
$C_{17}H_{15}ClN_2O$			
1H-Pyrazole, 1-acetyl-3-(4-chlorophenyl)-4,5-dihydro-5-phenyl-	EtOH	222(4.21),295(4.45)	34-0225-84
1H-Pyrazole, 1-acetyl-5-(4-chlorophenyl)-4,5-dihydro-3-phenyl-	EtOH	221(4.33),286(4.29), 294(4.29)	34-0225-84
$C_{17}H_{15}ClN_2S$			
Benzenamine, 4-chloro-N-(3-ethyl-4-phenyl-2(3H)-thiazolylidene)-	EtOH	308(4.31)	56-0447-84
$C_{17}H_{15}ClOS$			
Dibenzo[b,f]thiepin-10(11H)-one,	MeOH	234(4.40),260s(4.07),	73-0086-84

Compound	Solvent	$\lambda_{max}(\log \epsilon)$	Ref.
2-chloro-8-(1-methylethyl)- (cont.)		337(3.62)	73-0086-84
$C_{17}H_{15}ClO_6$ 2-Butenoic acid, 4-(3-chloro-5-hydroxy-7-methoxy-1,4-dioxo-2-naphthalenyl)-3-methyl-, methyl ester	EtOH	268(4.08),297(3.98), 430(3.59)	78-3455-84
$C_{17}H_{15}ClO_7$ Spiro[anthracene-2(1H),2'-[1,3]dioxolane]-9,10-dione, 6-chloro-3,4-dihydro-5,8-dihydroxy-4-methoxy-, (±)-	EtOH	281(3.86),498(3.70)	78-4561-84
$C_{17}H_{15}Cl_2N_3O_4$ Benzenepropanoic acid, β-[(4,5-dichloro-2-nitrophenyl)hydrazono]-, ethyl ester	EtOH	239(4.28),335(4.25), 450(3.84)	48-0829-84
$C_{17}H_{15}FOS$ Dibenzo[b,f]thiepin-10(11H)-one, 2-fluoro-8-(1-methylethyl)-	MeOH	241(4.29),260s(4.07), 335(3.57)	73-0086-84
Dibenzo[b,f]thiepin-10(11H)-one, 3-fluoro-8-(1-methylethyl)-	MeOH	240(4.64),255s(4.25), 333(4.27)	73-2638-84
$C_{17}H_{15}F_3O_5S$ Methanesulfonic acid, trifluoro-, 2-methoxy-2-(4-methoxyphenyl)-1-phenylethenyl ester, (Z)-	C_6H_{12}	234(4.21),289(4.07)	44-2273-84
$C_{17}H_{15}F_6NO$ 3-Penten-2-one, 4-[(1,3-dihydro-1,3,3-trimethyl-2H-indol-2-ylidene)methyl]-1,1,1,5,5,5-hexafluoro-	EtOH benzene	466(3.88) 470(--)	104-0390-84 104-0390-84
$C_{17}H_{15}IOS$ Dibenzo[b,f]thiepin-10(11H)-one, 2-iodo-8-(1-methylethyl)-	MeOH	240(4.55),265s(4.10), 335(3.79)	73-0086-84
$C_{17}H_{15}N$ 1H-Indole, 1-methyl-2-(1-phenylethenyl)-	EtOH	216(4.68),289(4.11)	78-4837-84
1H-Indole, 1-methyl-5-(2-phenylethenyl)-, trans	EtOH	206(4.5),234(4.2), 240(4.1),268(4.4), 275(4.5),286(4.5), 324(4.5)	103-0372-84
1H-Phenanthro[9,10-b]azirine, 1a,9b-dihydro-1-(2-propenyl)-	CH_2Cl_2	232.1(4.02),241s(3.91), 271s(4.06),277.1(4.07), 282s(4.06),296s(3.78), 305.7(3.52)	4-0001-84
$C_{17}H_{15}NO$ Isoquinoline, 4-[(2-methoxyphenyl)-methyl]-	MeOH	217(4.60),273(3.62), 310(3.39),323(3.51)	78-0215-84
4(1H)-Pyridinone, 2,3-dihydro-2,6-diphenyl-	EtOH	244(4.14),338(4.10)	33-1547-84
$C_{17}H_{15}NO_2$ 9,10-Anthracenedione, 2-[(dimethylamino)methyl]-	EtOH	256(4.57),327(3.57)	42-0611-84
1,3-Benzoxazepine, 2-(4-methoxyphenyl)-4-methyl-	EtOH	254(4.24),280(4.00), 333(3.95)	78-3567-84

Compound	Solvent	$\lambda_{max}(\log \epsilon)$	Ref.
3-Buten-2-one, 4-[(2-benzoylphenyl)-amino]-, (Z)-	MeOH	248(4.30),327(4.32), 364(4.25)	4-0759-84
2H-Isoindole-4,7-dione, 5,6-dimethyl-2-(phenylmethyl)-	n.s.g.	226(4.28),264(4.04), 366(3.46)	88-4917-84
5a,10b-Propano-6H-[1,4]benzodioxino-[2,3-b]indole	MeOH	246(3.82),279(3.46), 307(3.36)	4-1841-84
$C_{17}H_{15}NO_2S$ 2-Thiophenecarboxylic acid, 3-amino-5-(2-naphthalenyl)-, ethyl ester	MeOH	311(4.31),351(4.10)	118-0275-84
$C_{17}H_{15}NO_2S_4$ 6H-1,3-Thiazine-6-thione, 2,3-dihydro-2-(methylthio)-4-phenyl-5-(phenyl-sulfonyl)-	EtOH	258(3.95),352(4.16)	4-0953-84
$C_{17}H_{15}NO_4$ 1H-[1,3]Dioxolo[7,8][1]benzoxepino-[2,3,4-ij]isoquinolin-6-ol, 2,3,13,13a-tetrahydro-, (S)-(norcularicine)	MeOH	206(4.57),224s(4.23), 287(3.79)	100-0753-84
$C_{17}H_{15}NO_4S$ 2,5-Pyrrolidinedione, 3-[(4-methyl-phenyl)sulfonyl]-1-phenyl-	CH_2Cl_2	262s(--),275(2.96)	18-0219-84
$C_{17}H_{15}NO_5$ 2-Propenoic acid, 2-(benzoylamino)-3-(4-hydroxy-3-methoxyphenyl)-	EtOH	214(4.77),246(4.82), 252(4.96),258(5.01), 264(4.88),324(4.23)	118-0418-84
Spiro[1H-inden-1,6'-[2]oxa[3]azabicy-clo[3.1.0]hex[3]ene]-1',5'-dicarb-oxylic acid, 4'-methyl-, dimethyl ester, (1'α,5'α,6'α)-	EtOH	210(4.17),247(4.14)	23-2506-84
(1'α,5'α,6'β)-	EtOH	221(4.03),242(4.04)	23-2506-84
$C_{17}H_{15}NO_5S$ L-Cysteine, S-(9,10-dihydro-4,5-di-hydroxy-10-oxo-9-anthracenyl)-, hydrochloride	EtOH-acid	261(3.99),286(3.80), 375(4.04)	88-4837-84
	EtOH-base	366(3.91),385(4.29), 438(3.85),462(--), 491(3.63)	88-4837-84
$C_{17}H_{15}N_2$ 4,4'-Bipyridinium, 1-(phenylmethyl)-, bromide	H_2O	260(4.33)	39-0367-84B
$C_{17}H_{15}N_3O$ Benzamide, N-[2-(1-methyl-1H-imidazol-2-yl)phenyl]-	EtOH	229(4.17),255(4.26), 280(4.04)	78-1919-84
Isoxazole, 5-ethyl-3-phenyl-4-(phenyl-azo)-	EtOH	320(4.23)	104-1844-84
Methanone, (4-amino-7-methyl-3-cinno-linyl)(4-methylphenyl)-	DMF	283s(4.01),333s(3.80), 359(3.85)	5-1390-84
2-Quinolinecarboxamide, 3-(methyl-amino)-N-phenyl-	EtOH	263(5.25),319(3.86)	94-1604-84
$C_{17}H_{15}N_3O_2$ 2H-1-Benzopyran, 2-azido-2-(4-methoxy-phenyl)-3-methyl-	ether	262(3.98),300(3.50)	78-3567-84

Compound	Solvent	$\lambda_{max}(\log \epsilon)$	Ref.
$C_{17}H_{15}N_3O_2S_3$ 1H,5H-Imidazo[1,2-c]thiazol-2(3H)-one, 5-(3-ethyl-4-oxo-2-thioxo-5-thiazolidinylidene)-3-methyl-7-phenyl-	DMF	488(4.44)	103-0029-84
$C_{17}H_{15}N_3O_3$ 3-Oxepincarboxylic acid, 5-cyano-6,7-dihydro-7-methyl-2-[(phenylcarbonimidoyl)amino]-, methyl ester	CHCl$_3$	260(3.86),267(3.84), 332(4.15)	24-0585-84
$C_{17}H_{15}N_3O_6$ 1-Azabicyclo[3.2.0]hept-2-ene-2-carboxylic acid, 3-cyano-6-ethyl-7-oxo-, (4-nitrophenyl)methyl ester, 3-oxide, (5R-cis)-	CH$_2$Cl$_2$	267(4.08),324(4.03)	77-1513-84
$C_{17}H_{15}N_3O_7S_2$ Acetic acid, [[2-[(4,5-dihydro-4-oxo-2-thiazolyl)amino]-2,3-dihydro-8,9-dimethoxy-1,3-dioxo-1H-2-benzazepin-4-yl]thio]-	MeOH	250(4.32),339(3.96), 389(3.92)	103-0024-84
$C_{17}H_{15}N_5O$ 7H-Pyrido[2,1-b]quinazoline-6,11-dione, 8,9-dihydro-, 6-(2-pyridinylhydrazone)	EtOH	229(4.34),244(4.24), 297(4.02),369(4.34)	4-1301-84
7H-Pyrido[2,1-b]quinazoline-6,11-dione, 8,9-dihydro-, 6-(3-pyridinylhydrazone)	EtOH	228s(4.23),246s(4.07), 310(3.93),365(4.39)	4-1301-84
$C_{17}H_{15}N_5O_2$ Benzonitrile, 5-nitro-2-[[4-(1-pyrrolidinyl)phenyl]azo]-	C$_6$H$_{12}$ EtOH EtOH-HCl	503(4.61) 536(4.65) 498(4.81)	39-0149-84B 39-0149-84B 39-0149-84B
1H-Pyrrolo[2,3-d:4,5-d']dipyridazine-1,6(2H)-dione, 5,7-dihydro-2,7-dimethyl-5-(phenylmethyl)-	EtOH	252(4.25),261(4.24), 281(4.19),312(4.09)	94-1423-84
$C_{17}H_{15}N_5O_2S$ 1H-Dipyridazino[4,5-b:4',5'-e][1,4]-thiazine-1,9(7H)-dione, 8,10-dihydro-2,8-dimethyl-10-(phenylmethyl)-	EtOH	236(4.03),272(4.15), 315(4.10)	94-1423-84 +142-0675-84
3H-Dipyridazino[4,5-b:4',5'-e][1,4]-thiazine-4,6(7H,10H)-dione, 3,7-dimethyl-10-(phenylmethyl)-	EtOH	233(4.48),273(4.49), 315(4.21)	142-0675-84
$C_{17}H_{15}N_5O_3$ 2H-1-Benzopyran-2-one, 7-[3-(6-amino-9H-purin-9-yl)propoxy]-	H$_2$O	263(4.14),325(4.14)	23-2006-84
$C_{17}H_{15}N_5O_8$ Benzenepropanoic acid, β-[(2,4-dinitrophenyl)hydrazono]-4-nitro-, ethyl ester	EtOH	231(4.18),387(4.30)	48-0829-84
$C_{17}H_{16}$ Benzene, 1-ethenyl-2-(2-phenyl-1-propenyl)-, cis	MeOH	247(3.91)	151-0327-84D
trans	MeOH	246(4.24),278(3.99)	151-0327-84D
1H-Dicyclobuta[a,c]cyclopenta[g]naphthalene, 2,3,4,6,7,8-hexahydro-	n.s.g.	219(4.43),242(4.69), 245(4.71),266(3.34), 279s(3.46),288(3.58), 299(3.60),307s(3.43),	44-1412-84

Compound	Solvent	$\lambda_{max}(\log \epsilon)$	Ref.
(cont.) 1H-Indene, 2,3-dihydro-1-(1-phenyl- ethenyl)-	MeOH	326s(2.70) 243s(3.83)	44-1412-84 151-0327-84D
$C_{17}H_{16}BrN_3O_4$ Benzenepropanoic acid, β-[(4-bromo- 2-nitrophenyl)hydrazono]-, ethyl ester	EtOH	236(4.32),323(4.16), 452(3.89)	48-0829-84
$C_{17}H_{16}ClNO$ 4-Piperidinone, 1-chloro-2,6-diphenyl-, cis	MeCN	227(3.51),253(2.92), 258(2.93),265s(2.86), 269s(2.76),280s(2.48)	33-1547-84
$C_{17}H_{16}ClN_3O_4$ Benzenepropanoic acid, β-[(4-chloro- 2-nitrophenyl)hydrazono]-, ethyl ester	EtOH	236(4.36),323(4.23), 452(3.84)	48-0829-84
Benzenepropanoic acid, β-[(5-chloro- 2-nitrophenyl)hydrazono]-, ethyl ester	EtOH	233(4.27),288(4.12), 318(4.21),430(3.92)	48-0829-84
$C_{17}H_{16}ClN_5O_7S$ Inosine, 2-[[(2-chloro-4-nitrophenyl)- methyl]thio]-	pH 1 pH 7 pH 13	270(4.29) 265(4.26) 271(4.30)	87-0429-84 87-0429-84 87-0429-84
$C_{17}H_{16}Cl_2N_2O_4$ Pentanediamide, N,N'-bis(4-chlorophen- yl)-N,N'-dihydroxy-	MeOH	204(4.50),222(4.45), 244(4.33)	34-0345-84
$C_{17}H_{16}F_3N_3$ Pyrrolidine, 1-[4-[[4-(trifluoromethyl- yl)phenyl]azo]phenyl]-	C_6H_{12} EtOH EtOH-HCl	424(4.54) 434(4.50) 514(4.77)	39-0149-84B 39-0149-84B 39-0149-84B
$C_{17}H_{16}N_2$ Flavopereirine, 6,7-dihydro-	MeOH	251(3.84),313(4.13), 391(4.11)	142-0233-84
	MeOH-KOH	250(3.85),263(3.71), 316(4.04),360(4.03), 405(4.10)	142-0233-84
$C_{17}H_{16}N_2O$ Ethanone, 1-(2-methyl-1H-indol-3-yl)- 2-phenyl-, oxime	n.s.g.	226(4.31),277(3.98)	103-0055-84
Indolo[2,3-a]quinolizin-4(6H)-one, 3-ethyl-7,12-dihydro-	MeOH	249(4.24),258(4.25), 276(4.20),284(4.19), 288(4.21),350(4.52), 365(4.60),383(4.54)	142-0233-84
1H-Pyrazole, 1-acetyl-4,5-dihydro- 3,5-diphenyl-	EtOH	286(4.27),294(4.27)	34-0225-84
2(1H)-Quinoxalinone, 3-methyl-1-[(4- methylphenyl)methyl]-	EtOH	282(3.46),330(3.49), 338(3.40)	18-1653-84
$C_{17}H_{16}N_2OS$ Benzenamine, 4-methoxy-N-(3-methyl- 4-phenyl-2(3H)-thiazolylidene)-	EtOH	297(4.30)	56-0447-84
$C_{17}H_{16}N_2O_2$ Acetamide, N-(6-methoxy-4-methylbenzo-	EtOH	244s(4.59),251(4.62),	12-1271-84

Compound	Solvent	$\lambda_{max}(\log \epsilon)$	Ref.
[h]quinolin-5-yl)- (cont.)		271(4.39),294(3.99), 303(3.99),323s(3.23), 337(3.34),347(3.34)	12-1271-84
1,2-Benzenediol, 4-[2-(4-quinolinyl-amino)ethyl]- (hydrate)	MeOH	216(4.53),232(4.27), 289(3.67),329(2.51)	24-1523-84
Indolo[2,3-a]quinolizin-4(6H)-one, 7,12-dihydro-3-(1-hydroxyethyl)-	MeOH	247(4.13),256(4.16), 272(4.07),284(4.06), 348(4.25),366(4.54), 386(4.87)	142-0233-84
1(3H)-Isobenzofuranone, 6-amino-3-[[4-(dimethylamino)phenyl]methy;ene]-	EtOH	349(4.30),412(4.25)	131-0239-84H
Pyrano[4,3-b]pyrrol-6(1H)-one, 2,3-di-hydro-1-methyl-4-(1-methyl-1H-indol-3-yl)-	EtOH	230(4.55),279s(4.05), 298s(4.03),343(4.11)	150-2738-84M

$C_{17}H_{16}N_2O_2S$

Compound	Solvent	$\lambda_{max}(\log \epsilon)$	Ref.
Benzenepropanimidothioic acid, α-(hy-droxyimino)-β-oxo-N-(phenylmethyl)-, methyl ester	MeOH	242(4.64)	77-0430-84
Benzo[b]thiophene-2-propanenitrile, 3-hydroxy-β-oxo-α-(1-piperidinyl-methylene)-, (E)-	MeOH	250(3.20),295(3.05), 345(3.34),370(3.37)	83-0531-84

$C_{17}H_{16}N_2O_3$

Compound	Solvent	$\lambda_{max}(\log \epsilon)$	Ref.
Ethanone, 1,1'-[2,6-dimethyl-4-(2-ni-trosophenyl)-3,5-pyridinediyl]bis-	MeOH	204(4.35),222s(4.28), 279(4.03),310(3.83)	83-1029-84
Propanoic acid, 2-(2-dibenzofuranyl-hydrazono)-, ethyl ester, anti	EtOH	208(4.55),224(4.47), 235(4.48),250(4.42), 258(4.35),282(4.48), 322(4.51)	103-1123-84
syn	EtOH	206(4.59),221(4.51), 240(4.46),250(4.43), 256(4.35),291(4.23), 330(4.35)	103-1123-84
Propanoic acid, 2-(3-dibenzofuranyl-hydrazono)-, ethyl ester, anti	EtOH	217(4.84),250(4.27), 266(4.03),300(3.88), 257[sic](4.44)	103-1123-84
syn	EtOH	202(4.28),217(4.60), 251(4.10),260(4.10), 299(3.92)	103-1123-84
4H-Pyrido[1,2-a]pyrimidine-3-carboxylic acid, 6,7,8,9-tetrahydro-6-methyl-4-oxo-9-(phenylmethylene)-	EtOH	358(4.35)	118-0582-84

$C_{17}H_{16}N_2O_3S$

Compound	Solvent	$\lambda_{max}(\log \epsilon)$	Ref.
2-Propenenitrile, 2-(phenylsulfonyl)-3-[5-(1-pyrrolidinyl)-2-furanyl]-	MeOH	227(3.02),285(2.71), 357(2.77),470(3.25)	73-2141-84

$C_{17}H_{16}N_2O_4$

Compound	Solvent	$\lambda_{max}(\log \epsilon)$	Ref.
3-Azabicyclo[4.4.1]undeca-1,3,5,7,9-pentaene-4,5-dicarboxylic acid, 2-cyano-, diethyl ester	MeOH	251(4.29),283(4.27), 360(3.69)	44-1040-84
Ethanone, 1,1'-[2,6-dimethyl-4-(2-ni-trophenyl)-3,5-pyridinediyl]bis-	MeOH	208(4.49),265(3.91)	83-1029-84
5H-Pyrazolo[5,1-b][1,3]oxazine-3-carb-oxylic acid, 7-methyl-5-oxo-2-(phen-ylmethyl)-, ethyl ester	EtOH	224(4.25),282(4.09)	94-0930-84
Pyrazolo[1,2-a]pyrazole-2-carboxylic acid, 5-methyl-1,7-dioxo-3-(phenyl-methyl)-, ethyl ester	EtOH	235(4.10),364(4.11)	94-0930-84
Pyrazolo[1,2-a]pyrazole-2-carboxylic	EtOH	331(4.16)	94-0930-84

Compound	Solvent	$\lambda_{max}(\log \epsilon)$	Ref.
acid, 7-methyl-1,5-dioxo-3-(phenyl-methyl)-, ethyl ester (cont.)			94-0930-84
$C_{17}H_{16}N_2O_4S$			
2-Propenenitrile, 3-[5-(4-morpholinyl)-2-furanyl]-2-(phenylsulfonyl)-	MeOH	230(3.10),273(2.67), 464(3.56)	73-2141-84
$C_{17}H_{16}N_2O_5$			
Ethanone, 1,1'-[2,6-dimethyl-4-(2-nitrophenyl)-3,5-pyridinediyl]bis-, N-oxide	MeOH	220(4.41),270(4.20)	83-1029-84
2,4(1H,3H)-Pyrimidinedione, 5-methyl-1-[3-[(2-oxo-2H-1-benzopyran-7-yl)oxy]propyl]-	H_2O	278(4.09),324(4.18)	23-2006-84
$C_{17}H_{16}N_2S$			
Benzenamine, N-(3-ethyl-4-phenyl-2(3H)-thiazolylidene)-	EtOH	296(4.22)	56-0447-84
Benzenamine, 4-methyl-N-(3-methyl-4-phenyl-2(3H)-thiazolylidene)-	EtOH	298(4.23)	56-0447-84
4-Isoquinolinecarbonitrile, 2,3,5,6,7-8-hexahydro-1-(phenylmethyl)-3-thioxo-	dioxan	230(4.11),320(4.10), 405(3.76)	104-2225-84
3-Quinolinecarbonitrile, 1,2,5,6,7,8-hexahydro-4-(4-methylphenyl)-2-thioxo-	dioxan	260(3.84),312(3.79), 424(3.06)	104-1402-84
$C_{17}H_{16}N_2S_2$			
2H-3,1-Benzothiazine-2-thione, 1,4-dihydro-4-[(1-methylethyl)imino]-1-phenyl-	EtOH-1% dioxan	258(4.32),309s(4.17), 321(4.20)	97-0328-84
$C_{17}H_{16}N_4$			
Benzonitrile, 2-[[4-(1-pyrrolidinyl)-phenyl]azo]-	C_6H_{12}	438(4.52)	39-0149-84B
	EtOH	458(4.49)	39-0149-84B
	EtOH-HCl	505(4.53)	39-0149-84B
Benzonitrile, 4-[[4-(1-pyrrolidinyl)-phenyl]azo]-	C_6H_{12}	438(4.57)	39-0149-84B
	EtOH	460(4.55)	39-0149-84B
	EtOH-HCl	518(4.79)	39-0149-84B
$C_{17}H_{16}N_4O_3$			
1-Imidazolidineacetic acid, 2,5-dioxo-4,4-diphenyl-, hydrazide	MeOH	201(4.38)	56-0585-84
2,4(1H,3H)-Pteridinedione, 1,3-dimethyl-7-(1-oxo-3-phenylpropyl)-	MeOH	248(4.08),349(3.92)	5-1798-84
$C_{17}H_{16}N_4O_4$			
3-Quinolinecarboxylic acid, 2-amino-4-[(4,5-dihydro-2-methyl-5-oxo-1H-pyrazol-4-yl)carbonyl]-, ethyl ester	BuOH	275(4.21),307(4.19), 323(4.25)	4-1233-84
$C_{17}H_{16}N_4O_6$			
Benzenepropanoic acid, β-[(2,4-dinitro-phenyl)hydrazono]-, ethyl ester	EtOH	221(4.24),374(4.32)	48-0829-84
$C_{17}H_{16}N_6O_9S$			
Inosine, 2-[[(3,5-dinitrophenyl)methyl]thio]-	pH 1	250(4.46)	87-0429-84
	pH 7	250(4.43)	87-0429-84
	pH 13	250(4.44)	87-0429-84

Compound	Solvent	$\lambda_{max}(\log \epsilon)$	Ref.
$C_{17}H_{16}O$			
Benzene, 1-ethenyl-2-(1-methoxy-2-phen-ylethenyl)-, cis	MeOH	247(--),289(--)	151-0327-84D
trans	MeOH	250(4.57),291(4.50)	151-0327-84D
$C_{17}H_{16}O_2$			
1,4-Epoxyanthracene, 1,2,3,4,4a,10-hexahydro-5-methoxy-2,3-bis(methylene)-, (1α,4α,4aα)-(±)-	MeCN	217(4.48),237s(4.10), 256(3.89),286(4.03)	78-4549-84
1,4-Epoxyanthracene, 1,2,3,4,4a,10-hexahydro-8-methoxy-2,3-bis(methylene)-, (1α,4α,4aα)-(±)-	MeCN	212(4.43),236(4.16), 254s(3.93),280s(3.91), 292(3.96),304(3.96), 312(3.94)	78-4549-84
1,4-Epoxyanthracene, 1,2,3,4,9,10-hexahydro-5-methoxy-2,3-bis(methylene)-, (±)-	MeCN	224(4.19),244s(3.82), 270(3.42),278(3.30)	78-4549-84
2,7-Phenanthrenediol, 5-ethenyl-9,10-dihydro-1-methyl- (effusol)	EtOH	215(4.39),248(4.14), 318s(3.67),326s(3.66)	12-2111-84
3-Phenanthrenol, 2-methoxy-7,8-dimethyl-	MeOH	191(4.30),220(4.32), 258(4.74),281(4.32), 307s(3.86),339(3.45), 355(3.54)	39-1913-84C
$C_{17}H_{16}O_3$			
2H-1-Benzopyran-2-one, 3-(1-cyclohexen-1-ylacetyl)-	EtOH	283(3.81),314(3.70)	118-0859-84
2H-1-Benzopyran-2-one, 3-(cyclohexyli-deneacetyl)-	EtOH	305(3.95),334(3.79)	118-0859-84
$C_{17}H_{16}O_4$			
1,3-Benzenediol, 5-[2-(2,2-dimethyl-1,3-benzodioxol-5-yl)ethenyl]-, (E)-	EtOH	302(4.61),328(4.76)	94-0801-84
1H-2-Benzopyran-1-one, 3,4-dihydro-8-methoxy-3-(4-methoxyphenyl)-	EtOH	228s(4.72),284(3.46), 310(3.75)	44-0742-84
Ethanone, 1-[2,4-dihydroxy-3-[(3-phen-yloxiranyl)methyl]phenyl]-	n.s.g.	234(4.09),282(4.37)	2-1001-84
Methanone, [2-(methoxymethoxy)phenyl]-(3-phenyloxiranyl)-, (2R-trans)-	MeOH	206(4.44),254(4.12), 306.5(3.74)	94-4852-84
2-Naphthalenecarboxylic acid, 1,4-di-hydro-3-(3-methyl-2-butenyl)-1,4-dioxo-, methyl ester	MeOH	247(4.24),252(4.24), 265s(4.10),334(3.53)	102-0307-84
Phenol, 5-methoxy-2-(7-methoxy-2H-1-benzopyran-3-yl)-	MeOH	203(4.49),280(3.98), 322(3.93)	39-2767-84C
$C_{17}H_{16}O_5$			
Benzoic acid, 2-(benzoyl-CO-^{13}C)-3,5-dimethoxy-, methyl ester	MeOH	249(4.08),310(3.47)	102-0307-84
1H-Naphtho[2,3-c]pyran-1-one, 7,9,10-trimethoxy-3-methyl-	EtOH	262s(--),271(4.75), 281(4.78),294(4.55), 380(3.72)	39-1053-84C
$C_{17}H_{16}O_6$			
4H-1-Benzopyran-4-one, 2,3-dihydro-5-hydroxy-3-(4-hydroxy-2-methoxy-phenyl)-7-methoxy-	MeOH	229(4.5),287(4.49), 335(3.68)	102-0871-84
1(9bH)-Dibenzofuranone, 6-acetyl-7,9-dihydroxy-3-methoxy-8,9b-dimethyl-, (R)-	MeOH	218(4.23),282(4.00), 330(3.25),370(2.60)	23-0320-84
1H-Naphtho[2,3-c]pyran-3-acetic acid, 3,4,5,10-tetrahydro-9-hydroxy-1-methyl-5,10-dioxo-, methyl ester	MeOH	248(4.00),273(4.04), 422(3.56)	158-84-166

Compound	Solvent	$\lambda_{max}(\log \epsilon)$	Ref.
9H-Xanthen-9-one, 1,2,6,8-tetramethoxy-	MeOH	202(4.42),239(4.54), 251(4.55),302(4.16)	94-2290-84
9H-Xanthen-9-one, 1,3,4,7-tetramethoxy-	EtOH	234(4.48),258(4.67), 300s(4.03),310(4.08), 371(3.95)	39-1507-84C
$C_{17}H_{16}O_7$			
4a,10a-Epoxy-1H-naphtho[2,3-c]pyran-3-acetic acid, 3,4,5,10-tetrahydro-9-hydroxy-1-methyl-5,10-dioxo-, methyl ester, [1S-(1α,3β,4aβ,10aβ)]-	MeOH	236(4.23),280s(3.47), 364(3.71)	158-84-167
9H-Xanthen-9-one, 1-hydroxy-2,3,4,7-tetramethoxy-	EtOH	235(4.46),270(4.55), 303(4.05),390(3.76)	39-1507-84C
$C_{17}H_{16}O_7S$			
5H-Furo[3,2-g][1]benzopyran-5-one, 7-[(acetyloxy)(methylthio)methyl]-4,9-dimethoxy-	EtOH	213(4.37),251(4.52), 285(3.70),337(3.61)	44-5035-84
$C_{17}H_{16}O_8$			
Benzoic acid, 2-[(4,5-dimethoxy-3,6-di-oxo-1,4-cyclohexadien-1-yl)oxy]-5-methoxy-, methyl ester	EtOH	228s(4.25),287(4.31)	39-1507-84C
2H-Pyran-6-carboxylic acid, 3-[3-hy-droxy-5-methoxy-2-(methoxycarbonyl)-phenyl]-4-methyl-2-oxo-, methyl ester	n.s.g.	215(4.48),260(4.18), 306(4.16)	78-2451-84
9H-Xanthen-9-one, 1,4-dihydroxy-2,3,6,8-tetramethoxy-	MeOH	242s(4.23),248(4.24), 272(4.26),324(4.11), 378(3.36)	39-1507-84C
9H-Xanthen-9-one, 1,6-dihydroxy-3,5,7,8-tetramethoxy-	MeOH	253(4.09),321(4.19), 360(4.24)	100-0123-84
	MeOH-NaOMe	247(--),265s(--), 368(--)	100-0123-84
9H-Xanthen-9-one, 1,8-dihydroxy-2,3,4,6-tetramethoxy-	MeOH	235(4.33),260(4.53), 333(4.35),381s(3.63)	39-1507-84C
	MeOH and MeOH-NaOH	234(4.01),259(4.05), 333(4.16)	100-0123-84
	MeOH-AlCl$_3$	277(--),330s(--), 373(--)	100-0123-84
$C_{17}H_{16}O_9$			
Propanedioic acid, [5-methoxy-3-(meth-oxycarbonyl)-2-oxo-2H-cyclohepta[b]-furan-6-yl]-, dimethyl ester	MeOH	283(4.40),294(4.40), 393(4.28)	18-0621-84
$C_{17}H_{16}S$			
Anthracene, 9-[(1-methylethyl)thio]-	EtOH	215(3.95),256(5.31), 336(3.21),354(3.66), 373(3.78),392(3.70)	40-1158-84
Anthracene, 9-(propylthio)-	EtOH	215(3.98),256(4.90), 335(2.38),354(3.38), 372(3.62),393(3.55)	40-1158-84
$C_{17}H_{17}BrN_4O$			
2H-Imidazol-2-one, 4-[(4-bromophenyl)-methylamino]-1,5-dihydro-1-methyl-5-(phenylamino)-	MeCN	257(4.23),232(4.26)	142-2509-84
$C_{17}H_{17}ClN_2O$			
2H-Indazole, 5-chloro-2-(4-methoxy-phenyl)-3-(1-methylethyl)-	EtOH	213(4.5),283(3.8), 310(3.8)	35-6015-84

Compound	Solvent	$\lambda_{max}(\log \epsilon)$	Ref.
2H-Indazole, 2-(4-chlorophenyl)-5-methoxy-3-(1-methylethyl)-	EtOH	215(4.6),287(3.9), 316(3.8)	35-6015-84
$C_{17}H_{17}ClN_2O_5S$			
Inosine, 2-[[(2-chlorophenyl)methyl]thio]-	pH 1	269(4.46)	87-0429-84
	pH 7	263(4.40),280s(4.33)	87-0429-84
	pH 13	265(4.47)	87-0429-84
$C_{17}H_{17}ClOS$			
Ethanone, 1-[5-chloro-2-[[4-(1-methylethyl)phenyl]thio]phenyl]-	MeOH	272(4.01),345(3.63)	73-0086-84
$C_{17}H_{17}ClO_6$			
2,2(3H)-Furandicarboxylic acid, 3-(4-chlorophenyl)-4-formyl-, diethyl ester	EtOH	201(4.43),223(4.13), 255(4.12)	118-0974-84
$C_{17}H_{17}N$			
Benzo[a]phenanthridine, 5,6,6a,7,8,12b-hexahydro-, (6aS-cis)-	MeOH	195(4.81),215s(3.32), 259s(2.79),266(2.88), 273(2.81)	39-1655-84C
1H-Indole, 2,3-dihydro-1-methyl-5-(2-phenylethenyl)-, (E)-	EtOH	206(4.5),240(4.2), 263(3.9),345(4.4)	103-0372-84
1H-Phenanthro[9,10-b]azirine, 1a,9b-dihydro-1-(1-methylethyl)-	CH_2Cl_2	232.2(4.04),243s(3.94), 272.5(4.11),276.6(4.11), 281.5(4.11),294s(3.80), 306.5(3.63)	4-0001-84
$C_{17}H_{17}NO$			
[1,1'-Biphenyl]-2-carboxamide, N-(2-methyl-1-propenyl)-	EtOH	260(4.14)	44-0714-84
5H-Indeno[2,1-c]isoquinoline, 6,6a,7,11b-tetrahydro-3-methoxy-, (6aS-cis)-	MeOH	217s(4.65),260s(3.15), 267(3.34),273(3.44), 280(3.33),287(3.06)	39-1655-84C
hydrochloride	MeOH	195(4.79),202s(4.71), 226(4.06),252s(2.82), 266(3.25),272(3.37), 280(3.30),286(3.22)	39-1655-84C
1H-Indole-2-methanol, α,1-dimethyl-α-phenyl-	EtOH	224(3.77),263(3.10)	78-4837-84
$C_{17}H_{17}NOS$			
Dibenzo[b,f]thiepin-10(11H)-one, 2-amino-8-(1-methylethyl)-	MeOH	264(4.31),351(3.50)	73-0086-84
2H-Isoindole, 1-(ethylthio)-2-(4-methoxyphenyl)-	EtOH	203(4.33),223(4.43), 246(4.29),287(3.59), 300(3.46),340(3.61)	103-0978-84
2-Propen-1-one, 3-(methylthio)-1-phenyl-3-[(phenylmethyl)amino]-	MeOH	245(4.35),342(4.58)	77-0430-84
$C_{17}H_{17}NOS_2$			
Carbamothioic acid, (3-methylphenyl)-(phenylthioxomethyl)-, O-ethyl ester	MeOH	507(2.22)	97-0438-84
Carbamothioic acid, phenyl(phenylthioxomethyl)-, O-propyl ester	MeOH	509(2.23)	97-0438-84
$C_{17}H_{17}NO_2$			
1H-Inden-1-ol, 2,3-dihydro-2-[[(3-methoxyphenyl)methylene]amino]-, (1S-trans)-	MeOH	212(4.52),253(4.23), 257s(4.21),271(3.77), 303(3.57)	39-1655-84C

Compound	Solvent	$\lambda_{max}(\log \epsilon)$	Ref.
1H-Inden-1-ol, 2,3-dihydro-5-methoxy-2-[(phenylmethylene)amino]-, (1R-trans)-	MeOH	201(4.82),212s(4.40), 234s(4.23),249(4.34), 280s(3.80),286(3.69)	39-1655-84C
9-Phenanthrenecarbonitrile, 1,2,3,9-10,10a-hexahydro-7-methoxy-10a-methyl-1-oxo-	EtOH	265(4.15)	2-0395-84
$C_{17}H_{17}NO_3$			
1H-Benzo[a]furo[3,4-f]quinolizine-1,12(3H)-dione, 5,6,10b,11-tetra-hydro-3,3-dimethyl-	MeOH	239(4.18),295(4.14)	104-2335-84
1-Propanone, 2,2-dimethyl-1-(4'-nitro-[1,1'-biphenyl]-4-yl)-	EtOH	304(4.33)	35-3140-84
$C_{17}H_{17}NO_4$			
9(10H)-Acridinone, 1,2,3-trimethoxy-10-methyl-	EtOH	272(3.29),393(2.60)	100-0285-84
$C_{17}H_{17}NO_4S$			
Benzeneacetic acid, 2-[[4-(1-methyl-ethyl)phenyl]thio]-5-nitro-	MeOH	223(4.31),253(3.94), 340(4.09)	73-0086-84
$C_{17}H_{17}NO_5$			
5-Oxa-1-azabicyclo[4.2.0]oct-2-ene-2-carboxylic acid, 4-(2-methoxyethyl-idene)-8-oxo-, phenylmethyl ester, [R-(Z)]-	EtOH	304(4.25)	39-1599-84C
$C_{17}H_{17}NO_6$			
4,7-Epoxy-1,2-benzisoxazole-5,6-dicarb-oxylic acid, 3a,4,5,6,7,7a-hexahydro-3-phenyl-, dimethyl ester	MeOH	264(4.12)	73-1193-84
8-Oxa-2-azabicyclo[3.2.1]oct-3-ene-6,7-dicarboxylic acid, 4-formyl-3-phenyl-, dimethyl ester	MeOH	238(4.02),309(4.15)	73-1193-84
1H-Pyrrole-1-acetic acid, 4-(ethoxy-carbonyl)-2,3-dihydro-2,3-dioxo-5-phenyl-, ethyl ester	dioxan	280s(3.64),395(3.48)	94-0497-84
$C_{17}H_{17}NS$			
Benzenecarbothioamide, N-(phenylmeth-yl)-N-1-propenyl-, (E)-	MeOH	314(4.11),418(3.37)	78-1835-84
[1,1'-Biphenyl]-2-carbothioamide, N-(2-methyl-1-propenyl)-	EtOH	233(4.24),264s(4.01), 319(3.96),390(2.63)	44-0714-84
$C_{17}H_{17}N_2O_4$			
Indobetalain (chloride)	H_2O	268(3.83),518(4.77)	33-1547-84
1H-Pyrazolium, 3-(2-acetyl-3-oxo-1-butenyl)-2-ethyl-4,5-dihydro-4,5-dioxo-1-phenyl-, iodide	EtOH	400(4.24)	48-0695-84
$C_{17}H_{17}N_3$			
Propanedinitrile, [2-(4,4-dimethyl-2-pyrrolidinylidene)-1-phenyleth-ylidene]-	MeOH	205(4.28),250(4.30), 285(3.95),337(3.73), 391(3.48)	83-0143-84
$C_{17}H_{17}N_3O$			
5-Pyrazolone, 1-phenyl-4-(2-phenyl-aminoethyl)-	isoPrOH	247(4.48),284s(3.00)	103-0047-84

Compound	Solvent	λ$_{max}$(log ϵ)	Ref.
C$_{17}$H$_{17}$N$_3$O$_2$S			
Pyrido[2,3-d]pyrimidine-2,4(1H,3H)-dione, 1,3-dimethyl-7-(4-methylphenyl)-5-(methylthio)-	EtOH	212(4.28),254(4.39), 323(4.24),340(4.17), 353(4.12)	94-0122-84
Pyrrolo[2,3-c]phenothiazine-1-methanamine, 3,6-dihydro-N,N-dimethyl-, 11,11-dioxide	EtOH	211(4.78),267.4(4.66), 342(4.25),357(4.09)	103-1095-84
C$_{17}$H$_{17}$N$_3$O$_3$			
Ethanone, 1-[4-[3-[(benzoyloxy)methyl]-3-methyl-1-triazenyl]phenyl]-	MeCN	227(4.43),293(4.50), 320s(4.39)	23-0741-84
C$_{17}$H$_{17}$N$_3$O$_4$			
Benzenepropanoic acid, β-[(2-nitrophenyl)hydrazono]-, ethyl ester	EtOH	232(4.39),317(4.19), 438(3.89)	48-0829-84
1H-Benz[de]isoquinoline-1,3(2H)-dione, 2-[3-(dimethylamino)propyl]-5-nitro-	buffer	336(3.96)	87-1677-84
DNA bound	buffer	340(3.67)	87-1677-84
1H-Benz[de]isoquinoline-1,3(2H)-dione, 2-[3-(dimethylamino)propyl]-6-nitro-	buffer	355(4.06)	87-1677-84
DNA bound	buffer	361(3.74)	87-1677-84
Benzoic acid, 4-[3-[(benzoyloxy)methyl]-3-methyl-1-triazenyl]-, methyl ester	MeCN	224(3.94),300(4.18)	23-0741-84
C$_{17}$H$_{17}$N$_3$O$_5$S			
2H-1,2,4,6-Thiatriazinium, 5,6-dihydro-2,4-bis(4-methoxyphenyl)-3-methyl-5-oxo-, hydroxide, inner salt, 1,1-dioxide	MeCN	230(4.506),263(3.805), 278s(3.726)	142-0063-84
C$_{17}$H$_{17}$N$_3$O$_7$			
1-Azabicyclo[3.2.0]hept-2-ene-2-carboxylic acid, 6-ethyl-3-(nitromethyl)-7-oxo-, (4-nitrophenyl)methyl ester, (5R-cis)-	CH$_2$Cl$_2$	268(4.08)	77-1513-84
C$_{17}$H$_{17}$N$_3$O$_7$S			
L-Tyrosine, N-acetyl-3-[(4-sulfophenyl)azo]-, barium salt (1:1)	pH 7	326(4.33),380s(3.92)	69-0589-84
	pH 11	328(4.12)	69-0589-84
C$_{17}$H$_{17}$N$_3$O$_8$			
1-Azabicyclo[3.2.0]hept-2-ene-2-carboxylic acid, 6-(1-hydroxyethyl)-3-(nitromethyl)-7-oxo-, (4-nitrophenyl)methyl ester, [5R-[5α,6α(S*)]]-	CH$_2$Cl$_2$	268(4.12)	77-1513-84
C$_{17}$H$_{17}$N$_3$S$_2$			
1,3,4-Thiadiazole-2(3H)-thione, 5-(4-methylphenyl)-3-[[(phenylmethyl)amino]methyl]-	EtOH	339(4.15)	83-0547-84
C$_{17}$H$_{17}$N$_5$O$_2$			
Benzoic acid, 4-[2-(2,4-diaminopyrido[3,2-d]pyrimidin-6-yl)-1-methylethyl]-	pH 13	238(4.56),343(3.87)	87-0376-84
Oxepino[2,3-d]pyrimidine-6-carbonitrile, 2-hydrazino-3,4,7,8-tetrahydro-7,8-dimethyl-4-oxo-3-phenyl-	CHCl$_3$	270(4.00),340(4.35)	24-0585-84
Oxepino[2,3-d]pyrimidine-6-carbonitrile, 2-hydrazino-3,4,7,8-tetra-	CHCl$_3$	273(3.96),338(4.37)	24-0585-84

Compound	Solvent	$\lambda_{max}(\log \epsilon)$	Ref.
hydro-8,8-dimethyl-4-oxo-3-phenyl-			24-0585-84
C₁₇H₁₇N₅O₇S			
Inosine, 2-[[(3-nitrophenyl)methyl]-	pH 1	267(4.36)	87-0429-84
thio]-	pH 7	262(4.37)	87-0429-84
	pH 13	270(4.35)	87-0429-84
Inosine, 2-[[(4-nitrophenyl)methyl]-	pH 1	270(4.14)	87-0429-84
thio]-	pH 7	263(4.21),275s(4.21)	87-0429-84
	pH 13	271(4.46)	87-0429-84
C₁₇H₁₇N₅O₈			
L-Alanine, 3-[[(4-nitrophenyl)methoxy]- NNO-azoxy]-, (4-nitrophenyl)methyl ester, monohydrochloride	MeOH	261(4.35)	87-1295-84
C₁₇H₁₇N₅O₉			
D-Ribitol, 1-deoxy-1-[1,3,4,7-tetra-	pH 1	204(4.32),275(3.90), 371(4.24)	33-0570-84
hydro-6-(4-nitrophenyl)-2,4,7-tri- oxo-8(2H)-pteridinyl]-	H₂O	213(4.40),266(4.00), 394(4.33)	33-0570-84
	pH 13	218(4.33),266(4.15), 413(4.37)	33-0570-84
C₁₇H₁₇N₇O₃			
Acetic acid, 1,2-diacetyl-2-(7-methyl- 5-phenyl-5H-pyrazolo[3,4-e]-1,2,4- triazin-3-yl)hydrazide	EtOH	268(4.43),316(3.85), 360s(2.78)	4-1565-84
C₁₇H₁₈AsN₃O₇			
L-Tyrosine, N-acetyl-3-[(4-amino-	pH 7	326(4.41)	69-0589-84
phenyl)azo]-	pH 11	327(4.23)	69-0589-84
C₁₇H₁₈BrN₃			
Piperidine, 1-[4-[(4-bromophenyl)azo]-	C₆H₁₂	404(4.48)	39-0149-84B
phenyl]-	EtOH	413(4.42)	39-0149-84B
	EtOH-HCl	534(3.46)	39-0149-84B
C₁₇H₁₈ClNO₂			
1,1'-Biphenyl, 4-(1-chloro-2,2-dimeth- ylpropyl)-4'-nitro-	EtOH	311(4.11)	35-3140-84
C₁₇H₁₈ClNO₄S			
2H-Thieno[2,3-h]-1-benzopyran-2-one, 4-[bis(2-methoxyethyl)amino]-3- chloro-	EtOH	236(4.38),261(4.33), 274s(4.04),310s(3.99), 342(4.15)	161-0081-84
C₁₇H₁₈ClN₃			
Piperidine, 1-[4-[(4-chlorophenyl)azo]-	C₆H₁₂	403(4.45)	39-0149-84B
phenyl]-	EtOH	412(4.46)	39-0149-84B
	EtOH-HCl	534(3.42)	39-0149-84B
C₁₇H₁₈ClN₅O₅			
Guanosine, N-[(4-chlorophenyl)methyl]-	pH 1	261(4.20),281s(3.97)	44-0363-84
	pH 6.9	255(4.18),274s(4.03)	44-0363-84
	pH 13	260(4.12),270s(4.11)	44-0363-84
9H-Purin-2-amine, 6-[(4-chlorophenyl)-	pH 1	243(3.88),288(4.01) (changing)	44-0363-84
methoxy]-9-β-D-ribofuranosyl-	pH 6.9	247(3.99),282(4.02)	44-0363-84
	pH 13	247(3.96),282(4.02)	44-0363-84

Compound	Solvent	$\lambda_{max}(\log \epsilon)$	Ref.
$C_{17}H_{18}Cl_2N_2O$			
Benzenecarboximidamide, 3,5-dichloro-N-hydroxy-2,4,6-trimethyl-N'-(4-methylphenyl)-	MeOH	252(4.22)	32-0015-84
$C_{17}H_{18}Cl_2N_2O_2$			
Benzenecarboximidamide, 3,5-dichloro-N-hydroxy-N'-(4-methoxyphenyl)-2,4,6-trimethyl-	MeOH	249(4.16)	32-0015-84
$C_{17}H_{18}FeO_3$			
Ferrocene, [2-(ethoxycarbonyl)-3-oxo-1-butenyl]-	CH_2Cl_2	482(3.19)	30-0192-84
$C_{17}H_{18}NOS$			
1H-Isoindolium, 3-(ethylthio)-2-(4-methoxyphenyl)-, tetrafluoroborate	$CHCl_3$	243(4.03),277(3.80), 355(3.10),380(3.00)	103-0978-84
$C_{17}H_{18}NS$			
1H-Isoindolium, 3-(ethylthio)-2-(4-methylphenyl)-, tetrafluoroborate	$CHCl_3$	244(4.05),277(3.86), 360(3.18),380(3.12)	103-0978-84
$C_{17}H_{18}N_2$			
1-Naphthalenecarbonitrile, 4-(1-piperidinylmethyl)-	C_6H_{12}	298(3.95)	35-1335-84
$C_{17}H_{18}N_2O$			
2H-Indazole, 5-methoxy-3-(1-methylethyl)-2-phenyl-	EtOH	199(4.6),225(4.5), 276(4.0),317(4.0)	35-6015-84
2H-Indazole, 2-(4-methoxyphenyl)-3-(1-methylethyl)-	EtOH	210(4.6),223(4.4), 278(4.2),301(4.1)	35-6015-84
1H-Pyrazole, 4,5-dihydro-3-(4-methoxyphenyl)-1-methyl-5-phenyl-	EtOH	221s(3.58),299(4.02)	34-0225-84
1H-Pyrazole, 4,5-dihydro-5-(4-methoxyphenyl)-1-methyl-3-phenyl-	EtOH	226(4.08),286s(3.92), 304(4.07)	34-0225-84
$C_{17}H_{18}N_2OS$			
4H-Thieno[2,3-h][1,5]oxazecine-9(6H)-carbonitrile, 7,8,10,11-tetrahydro-4-phenyl-	MeOH	247(3.68)	12-1043-84
$C_{17}H_{18}N_2O_2$			
1H-Benz[de]isoquinoline-1,3(2H)-dione, 2-[3-(dimethylamino)propyl]-	buffer	345(4.13)	87-1677-84
DNA bound	buffer	347(3.78)	87-1677-84
Ethanone, 1-[1,4,5,6-tetrahydro-4-(1H-indol-3-ylcarbonyl)-1-methyl-3-pyridinyl]-	EtOH	241(3.56),256s(4.06), 302(4.56)	78-3339-84
1H-Indole, 1-acetyl-3-[(1,2,3,6-tetrahydro-1-methyl-4-pyridinyl)carbonyl]-	EtOH	248s(4.08),253s(4.06), 304(3.99)	78-3339-84
1H-Pyrano[4,3-c]pyrazol-4-one, 6,7-dihydro-6-methyl-1-(4-methylphenyl)-3-(1-propenyl)-	EtOH	248(4.32)	4-0017-84
1H-Pyrano[4,3-c]pyrazol-4-one, 6,7-dihydro-6-methyl-3-(1-methyl-1-propenyl)-1-phenyl-	EtOH	250(4.21)	4-0017-84
1H-Pyrano[4,3-c]pyrazol-4-one, 6,7-dihydro-6-methyl-3-(2-methyl-1-propenyl)-1-phenyl-	EtOH	260(4.31)	4-0017-84
1H-Pyrano[4,3-c]pyrazol-4-one, 6,7-dihydro-7-methyl-3-(2-methyl-	EtOH	259(4.26)	4-0017-84

Compound	Solvent	λ_{max}(log ϵ)	Ref.
1-propenyl)-1-phenyl- (cont.)			4-0017-84
1H-Pyrano[4,3-c]pyrazol-4-one, 6,7-di-hydro-6-methyl-1-(phenylmethyl)-3-(1-propenyl)-	EtOH	250(4.06)	4-0017-84
$C_{17}H_{18}N_2O_3$			
Phenol, 2,6-dimethyl-4-[(methylnitroso-amino)methyl]-, benzoate	hexane	228(4.46),343(1.69),352(1.86),364(1.99),376(1.91)	150-0701-84M
Pyrrolo[2,3-b]indole-2-carboxylic acid, 2,8-dihydro-8-(methoxymethyl)-2-(2-propenyl)-, methyl ester	MeOH	248s(4.34),253(4.38),278(3.44),290(3.39)	39-2903-84C
$C_{17}H_{18}N_2O_3S$			
4-Oxazolecarboxamide, 2-(phenylmethyl)-N-(2,2,3-trimethyl-4-oxo-3-thietan-yl)-, (R)-	EtOH	227(3.81)	39-1127-84C
5(4H)-Oxazolone, 2-(phenylmethyl)-4-[[(2,3,3-trimethyl-4-oxo-3-thie-tanyl)amino]methylene]-, (R)-	EtOH	210(4.30),322(4.55)	39-1127-84C
$C_{17}H_{18}N_2O_4$			
Ethanone, 1,1'-[1,4-dihydro-2,6-di-methyl-4-(2-nitrophenyl)-3,5-pyridinediyl]bis-	MeOH	252(4.06),302(3.99),376(3.36)	83-1029-84
Pentanediamide, N,N'-dihydroxy-N,N'-di-phenyl-	MeOH	212(4.31),230(4.24)	34-0345-84
3-Pyridinecarboxylic acid, 5-cyano-1,4,5,6-tetrahydro-4-(4-methoxy-phenyl)-2-methyl-6-oxo-, ethyl ester	EtOH	226(4.00),280(4.02)	103-1241-84
$C_{17}H_{18}N_2O_4S$			
2-Propenoic acid, 3-[[2-(aminosulfon-yl)phenyl]amino]-3-phenyl-, ethyl ester	EtOH	205(4.34),253(4.02),290s(3.90),323(4.19)	104-0534-84
$C_{17}H_{18}N_2O_5$			
Spiro[furan-2(3H),9'-[9H]imidazo[1,2-a]indole]-1'(9'aH)-carboxylic acid, 2',3',4,5-tetrahydro-2',2'-dimethyl-3',5-dioxo-, methyl ester, trans	EtOH	244.5(4.02),276(3.15),285s(3.06)	94-1373-84
$C_{17}H_{18}N_2O_6$			
Diacetyldemethanolchicamycin A	MeOH	220(4.43),243s(4.21),320(3.58)	158-0200-84
Uridine, 5-(2-acetylphenyl)-2'-deoxy-	pH 1	275(4.03)	88-2431-84
	pH 13	273(3.94)	88-2431-84
$C_{17}H_{18}N_2O_7$			
1H-Pyrrolo[2,1-c][1,4]benzodiazepine-5,11(10H,11aH)-dione, 2,8-diacetoxy-2,3-dihydro-7-methoxy-	MeOH	228(4.37),257s(4.00),307(3.56)	158-0200-84
$C_{17}H_{18}N_4$			
Pyrido[3,4-d]pyridazine, 5,6,7,8-tetrahydro-1-(1H-indol-3-yl)-4,6-dimethyl-	EtOH	217(4.41),277s(3.94),280(3.95),288(3.94),296s(3.87),325s(3.57),366s(3.17)	78-3339-84
5H-Pyrimido[4,5-b]indol-4-amine, 6,7,8,9-tetrahydro-2-methyl-9-phenyl-	EtOH	227(4.20),284(3.87)	11-0073-84A

Compound	Solvent	λ_{max} (log ϵ)	Ref.
$C_{17}H_{18}N_4O_2$			
Piperidine, 1-[4-[(4-nitrophenyl)azo]-	C_6H_{12}	444(4.47)	39-0149-84B
phenyl]-	EtOH	470(4.44)	39-0149-84B
	EtOH-HCl	522(4.12)	39-0149-84B
$C_{17}H_{18}N_4O_2S$			
Ethanol, 2,2'-[[4-(2-benzothiazolyl-	acetone	503(4.69)	7-0819-84
azo)phenyl]imino]bis-	50% acetone	518(4.71)	7-0819-84
Ethanol, 2-[[2-[[4-(dimethylamino)-	pH 1.2	547(4.58),617s(4.42)	48-0151-84
phenyl]azo]-6-benzothiazolyl]oxy]-	pH 6.2	527(4.69)	48-0151-84
	EtOH	518(4.68)	48-0151-84
$C_{17}H_{18}N_4O_3S$			
Thiepino[2,3-d]pyrimidine-6-carboxylic	CHCl$_3$	273(4.38),360(4.19)	24-0585-84
acid, 2-hydrazino-3,4,7,8-tetrahydro-			
4-oxo-3-phenyl-, ethyl ester			
$C_{17}H_{18}N_4O_5$			
Inosine, 2-(phenylmethyl)-	pH 1	251(4.10)	87-0429-84
	pH 7	251(4.09)	87-0429-84
	pH 13	256(4.10)	87-0429-84
$C_{17}H_{18}N_4O_6$			
Inosine, 2-(4-methoxyphenyl)-	pH 1	265s(4.09),302(4.32)	87-0429-84
	pH 7	260(4.13),300(4.28)	87-0429-84
	pH 13	249(4.31),270(4.19),	87-0429-84
		290(4.23)	
$C_{17}H_{18}N_4S$			
Benzenamine, 4-(2-benzothiazolylazo)-	acetone	507(4.75)	7-0819-84
N,N-diethyl-	50% acetone	526(4.76)	7-0819-84
$C_{17}H_{18}N_5O_2S$			
1,2,4,5-Tetrazin-1(2H)-yl, 3,4-dihydro-	CHCl$_3$	460(2.62)	104-1224-84
2,4-dimethyl-3-[[(4-methylphenyl)sul-			
fonyl]imino]-6-phenyl-			
$C_{17}H_{18}N_6O_6S$			
L-Cysteic acid, N-(4-nitro-4-deoxy-	pH 7.4	260(4.43),282(4.41),	87-0600-84
pteroyl)-		370(3.94)	
$C_{17}H_{18}N_6O_7$			
Guanosine, N-[(4-nitrophenyl)methyl]-	pH 1	263(4.30),275s(4.28)	44-0363-84
	pH 6.9	258(4.29),279(4.29)	44-0363-84
	pH 13	267(4.29),274(4.31)	44-0363-84
9H-Purin-2-amine, 6-[(4-nitrophenyl)-	pH 1	248(4.05),285(4.23)	44-0363-84
methoxy]-9-β-D-ribofuranosyl-		(changing)	
	pH 6.9	252(4.14),282(4.26)	44-0363-84
	pH 13	252(4.13),282(4.27)	44-0363-84
$C_{17}H_{18}ORu$			
Ruthenocene, 1,1'-(1,4-butanediyl)-	EtOH	259(3.68),333s(2.90)	18-0719-84
3,3'-(1-oxo-1,3-propanediyl)-			
$C_{17}H_{18}O_2$			
Benzoic acid, 4-methyl-, 3,6-bis(meth-	isooctane	202(4.43),238(4.36),	33-0600-84
ylene)bicyclohept[2.2.1]-2-yl ester		254s(3.89)	
	EtOH	204(4.35),241(4.35)	33-0600-84
Benzoic acid, 4-methyl-, 5,6-bis(meth-	isooctane	202(4.44),238(4.45)	33-0600-84
ylene)bicyclohept[2.2.1]-2-yl ester	EtOH	204(4.40),240(4.42)	33-0600-84

Compound	Solvent	$\lambda_{max}(\log \epsilon)$	Ref.
Bicyclo[2.2.1]hept-2-ene-2-carboxylic acid, 3-(tetracyclo[3.2.0.0^{2,7}.0^{4,6}]-heptylidenemethyl)-, methyl ester	MeCN	302(4.33)	24-2027-84
1,4a-Cyclopropa-3,10a-ethanophenan-thren-2(1H)-one, 3,4,4a,9,10,10a-hexahydro-7-methoxy-	EtOH	250(4.17)	2-1168-84
1H-3,10a-Ethanophenanthren-12-one, 2,3,9,10-tetrahydro-7-methoxy-, (±)-	EtOH	274(4.32)	2-1168-84
3-Phenanthrenol, 9,10-dihydro-2-meth-oxy-7,8-dimethyl-	MeOH	219(4.17),277(4.14), 293s(3.94),314(3.96)	39-1913-84C
3-Phenanthrenol, 9,10-dihydro-7-meth-oxy-2,8-dimethyl-	EtOH	218(4.57),277(4.27), 296(4.03),316(4.02)	39-1913-84C
$C_{17}H_{18}O_2Ru$			
Ruthenocene, 1,1'-(1,4-butanediyl)-3-(2-carboxyethenyl)-	EtOH	287(4.27),354(3.68)	18-0719-84
$C_{17}H_{18}O_3$			
2,5-Furandione, dihydro-3-(1-methyleth-ylidene)-4-[(2,4,6-trimethylphenyl)-methylene]-, (E)-	C_6H_{12}	323(3.95)	48-0233-84
(Z)-	C_6H_{12}	344(3.94)	48-0233-84
$C_{17}H_{18}O_3S_3$			
Ethane(dithioic) acid, [(2-hydroxy-2,2-diphenylethyl)sulfonyl]-, methyl ester	EtOH	211(4.18),317(3.97)	39-0085-84C
$C_{17}H_{18}O_4$			
3H-Benz[e]indene-5-carboxylic acid, 3a,4,5,9b-tetrahydro-7-methoxy-3a-methyl-3-oxo-, methyl ester, cis	EtOH	227(4.11),276(3.50)	2-0395-84
1,3-Cyclohexadiene-1-carboxylic acid, 4-(benzoyloxy)-2-methyl-, ethyl ester	EtOH	205(4.18),235(4.14), 285s(2.70),301(3.90)	12-2037-84
Naphtho[1,2-b]furan-4,5-dione, 2,3-di-hydro-7-methoxy-2,2,3,9-tetramethyl-, (±)-	MeOH	278(4.43),313(3.81), 504(3.30)	44-1853-84
Naphtho[1,2-b]furan-4,5-dione, 2,3-di-hydro-7-methoxy-2,3,3,9-tetramethyl-, (±)-	MeOH	273(4.43),283(4.45), 314(3.76),500(3.35)	44-1853-84
2H-Naphtho[1,2-b]pyran-5-carboxylic acid, 3,4-dihydro-6-hydroxy-2,2-dimethyl-, methyl ester (dihydro-mollugin)	MeOH	262s(4.40),270(4.47), 324(3.53),377(3.52)	102-0307-84
Oxiranemethanol, α-[2-(methoxymethoxy)-phenyl]-3-phenyl-	MeOH	214.5(4.13),267(3.45), 300(2.89)	94-4852-84
4-Phenanthrenemethanol, 9,10-dihydro-5-hydroxy-2,7-dimethoxy- (flavidin)	EtOH	217(4.55),274(4.29), 301(4.10)	102-0671-84
	EtOH-NaOH	222(4.44),243s(4.24), 272(3.97),278(3.99), 323(4.02)	102-0671-84
Spiro[cyclopropane-1,1'-[1H]indene]-2,3-dicarboxylic acid, diethyl ester, (1α,2α,3β)-	EtOH	232(4.41),260(3.47), 295(3.15)	23-2506-84
$C_{17}H_{18}O_5$			
1(2H)-Anthracenone, 3,4-dihydro-3,8-di-hydroxy-6,9-dimethoxy-3-methyl-	dioxan	223(4.29),272(4.77), 307s(3.53),319(3.67), 333(3.66),368(3.84)	94-3436-84

Compound	Solvent	$\lambda_{max}(\log \epsilon)$	Ref.
1(2H)-Anthracenone, 3,4-dihydro-3,9-di-hydroxy-6,8-dimethoxy-3-methyl-, (S)-(8-methyltorosachrysone)	dioxan	227(4.25),269(4.66), 302s(3.61),312(3.74), 324(3.56),381(4.04), 395s(4.01)	94-3436-84
Benzeneacetic acid, α-(3-oxo-1,4-dioxa-spiro[4.5]dec-2-ylidene)-, methyl ester, (E)-	n.s.g.	297(4.21),309s(4.10)	39-1547-84C
(Z)-	n.s.g.	269(4.04)	39-1547-84C
2,4,6-Cycloheptatrien-1-one, 2-methoxy-5-(3,4,5-trimethoxyphenyl)-	buffer	238(4.53),343(4.25)	69-1742-84
1,4-Naphthalenedione, 2-hydroxy-6,7-di-methoxy-3-(3-methyl-2-butenyl)-	MeOH	225s(--),268s(4.151), 275(4.180),315(3.836)	87-0990-84
2H-Naphtho[1,2-b]pyran-5,6-dione, 3,4-dihydro-8,9-dimethoxy-2,2-dimethyl-	MeOH	228s(--),272s(--), 279(4.228),324(3.667)	87-0990-84
$C_{17}H_{18}O_5S$			
1,4-Epoxyanthracen-5(1H)-one, 2,3,4,8-8a,9,10,10a-octahydro-2,3-bis(meth-ylene)-8-[(methylsulfonyl)oxy]-, (1α,4α,8α,8aβ,10aβ)-(±)-	MeCN	212(4.39),234s(4.06), 244(3.89),258s(3.46)	78-4549-84
$C_{17}H_{18}O_6$			
2,5-Cyclohexadiene-1-carboxylic acid, 4-oxo-1-[(3-oxo-1,4-dioxaspiro[4.5]dec-2-ylidene)methyl]-, methyl ester	EtOH	242(4.35)	77-1008-84
Dibenzofuran, 1,2,3,4,7-pentamethoxy-	EtOH	229(4.61),262(4.32), 292(4.31),296s(4.30), 308(4.16)	39-1441-84C
Dibenzofuran, 2,3,4,6,7-pentamethoxy-	EtOH	226(4.57),229(4.48), 264(4.26),302(4.23), 314s(4.13)	39-1445-84C
1β,10β-Epoxyfuranoeremophilane-6,9-di-one, 3,3-(ethylenedioxy)-	EtOH	245.5(3.83),305(4.08)	94-3396-84
2,2(3H)-Furandicarboxylic acid, 4-form-yl-3-phenyl-, diethyl ester	EtOH	201(4.36),220(4.14), 255(4.13)	118-0974-84
2,5-Furandione, dihydro-3-(1-methyleth-ylidene)-4-[(3,4,5-trimethoxyphenyl)-methylene]-, (E)-	C_6H_{12}	355(4.11)	48-0233-84
(Z)-	C_6H_{12}	375(4.26)	48-0233-84
$C_{17}H_{18}O_7$			
1H-Naphtho[2,3-c]pyran-3-acetic acid, 3,4,4a,5,10,10a-hexahydro-9,10a-di-hydroxy-1-methyl-5,10-dioxo-	MeOH	230(4.25),245s(3.88), 265(3.60),350(3.71)	158-84-167
$C_{17}H_{18}O_8$			
Islandic acid I	MeOH	234(4.51),260(4.36), 335(4.16)	94-1583-84
$C_{17}H_{18}O_{10}S_2$			
2,7-Dibenzofurandiol, 1,3,4-trimeth-oxy-, dimethanesulfonate	EtOH	219(4.59),229(4.59), 260(4.25),286(4.29)	39-1441-84C
$C_{17}H_{19}BrO$			
Bicyclo[2.2.1]heptan-2-one, 3-[(4-bromophenyl)methylene]-1,7,7-tri-methyl-, [1R-(1α,3E,4α)]-	EtOH	298(4.48)	41-0629-84
$C_{17}H_{19}Br_2NO_2$			
Ethanamine, 2-(2-bromophenoxy)-N-[2-(2-bromophenoxy)ethyl]-N-methyl-	EtOH	233(3.79),276(3.63), 284(3.60)	12-2059-84

Compound	Solvent	$\lambda_{max}(\log \epsilon)$	Ref.
$C_{17}H_{19}ClN_4O_7$ 1H-Pyrazolo[3,4-d]pyrimidine, 4-chloro-6-methyl-1-(2,3,5-tri-O-acetyl-β-D-ribofuranosyl)-	MeOH	260(3.79)	87-1026-84
$C_{17}H_{19}ClN_5O_5$ 1H-Purinium, 2-amino-7-[(4-chlorophenyl)methyl]-6,9-dihydro-6-oxo-9-β-D-ribofuranosyl-, iodide	pH 1 pH 6.9 pH 13	257(4.07),280s(3.86) 258(3.94),285(3.84) 266(4.04)	44-0363-84 44-0363-84 44-0363-84
$C_{17}H_{19}ClO$ Bicyclo[2.2.1]heptan-2-one, 3-[(4-chlorophenyl)methylene]-1,7,7-trimethyl-, [1R-(1α,3E,4α)]-	EtOH	293(4.42)	41-0629-84
$C_{17}H_{19}FO$ Bicyclo[2.2.1]heptan-2-one, 3-[(4-fluorophenyl)methylene]-1,7,7-trimethyl-, [1R-(1α,3E,4α)]-	EtOH	292(4.31)	41-0629-84
$C_{17}H_{19}I_2N_3$ 1H-Imidazole-2-methanamine, 4,5-dihydro-N-phenyl-N-(phenylmethyl)-, compd. with iodine (1:1)	$CHCl_3$	290(4.40),365(4.16)	83-0246-84
$C_{17}H_{19}N$ Acridine, 9,10-dihydro-2,7,9,9-tetramethyl-	EtOH	286(4.29)	24-2703-84
$C_{17}H_{19}NO$ Benzenemethanamine, N-hydroxy-N,α-dimethyl-α-(2-phenylethenyl)-, (E)-	EtOH	252(4.20),284s(3.52), 297(3.53)	150-1531-84
3,5-Hexadien-2-one, 1-(1-methyl-2-pyrrolidinylidene)-6-phenyl-, (?,E,E)-	EtOH EtOH-HCl	274(3.83),326(4.32), 377(4.54) 266(3.72),394(4.65)	78-2879-84 78-2879-84
1-Naphthalenol, 1,2,3,4-tetrahydro-2-[(phenylmethyl)amino]-, (1S-trans)-	MeOH	195s(4.83),206s(4.29), 216s(4.10),254(2.56), 259(2.68),265(2.72), 272.5(2.61)	39-1655-84C
Phenol, 2-[(1,2,3,4-tetrahydro-2-methyl-4-isoquinolinyl)methyl]-	MeOH	204(4.48),214s(--), 273(3.53),281s(--)	78-0215-84
Phenol, 4-[(1,2,3,4-tetrahydro-2-methyl-4-isoquinolinyl)methyl]-	MeOH	204(4.38),225s(--), 272.5(3.03),280s(--)	78-0215-84
$C_{17}H_{19}NOS$ Benzenamine, N-[2-(5,6-dimethyl-2H-thiopyran-2-ylidene)propylidene]-4-methoxy-, perchlorate	EtOH	513(4.53)	97-0183-84
2H-Pyrrol-2-one, 1-cyclohexyl-1,5-dihydro-3-(4-methylphenyl)-5-thioxo-	$CHCl_3$	347(4.45)	78-3499-84
$C_{17}H_{19}NOS_2$ 2-Propen-1-one, 2-(1-piperidinylmethyl)-1,3-di-2-thienyl-	EtOH	330(3.20)	73-1764-84
$C_{17}H_{19}NOSi$ 3H-Benz[cd]azulene-3-carbonitrile, 4,5-dihydro-3-[(trimethylsilyl)oxy]-	CH_2Cl_2	277s(--),284(4.56), 288(4.57),291s(--), 255s[sic],338(3.67), 354(3.46),571(2.37)	83-0984-84

Compound	Solvent	λ_{max}(log ϵ)	Ref.
$C_{17}H_{19}NO_2$			
Benzenemethanamine, N-hydroxy-4-meth-oxy-N-methyl-α-(2-phenylethenyl)-, (E)-	EtOH	258(4.33),282s(3.96), 292(3.59)	150-1531-84M
Benzenemethanamine, N-hydroxy-α-[2-(4-methoxyphenyl)ethenyl]-N-methyl-, (E)-	EtOH	226(4.42),305s(3.68), 347s(3.47)	150-1531-84M
1H-Benzo[a]furo[3,4-f]quinolizin-1-one, 3,5,6,10b,11,12-hexahydro-3,3-di-methyl-, (R)-	MeOH	214(3.94),283(4.38)	104-2335-84
1H-Inden-1-ol, 2,3-dihydro-2-[[(3-meth-oxyphenyl)methyl]amino]-, (1S-trans)-	MeOH	211s(4.23),215s(4.19), 260s(3.14),266(3.35), 272.5(3.47),281(3.27)	39-1655-84C
1H-Inden-1-ol, 2,3-dihydro-5-methoxy-2-[(phenylmethyl)amino]-, (1S-trans)-	MeOH	198(4.75),227(4.19), 278s(3.41),280(3.42), 286(3.38)	39-1655-84C
9-Phenanthrenecarbonitrile, 1,2,3,9,10-10a-hexahydro-1-hydroxy-7-methoxy-10a-methyl-	EtOH	262(4.17)	2-0395-84
Phenol, 2,6-dimethyl-4-[(methylamino)-methyl]-, benzoate	EtOH	228(4.31)	150-0701-84M
$C_{17}H_{19}NO_3$			
Bicyclo[2.2.1]heptan-2-one, 1,7,7-tri-methyl-3-[(4-nitrophenyl)methylene]-, [1R-(1α,3E,4α)]-	EtOH	314(4.33)	41-0629-84
[1,1'-Biphenyl]-4-methanol, α-(1,1-di-methylethyl)-4'-nitro-	EtOH	315(4.25)	35-3140-84
$C_{17}H_{19}NO_4$			
Hippamine (2-O-methyllycorine)	EtOH	236(3.59),292(3.65)	100-1061-84
Laudanosoline, hydrobromide	pH 1	282(3.80)	3-1935-84
	borate	289(4.02)	3-1935-84
$C_{17}H_{19}NO_5$			
2H-Isoindole-2-butanoic acid, 1,3-di-hydro-γ-(1-methylethyl)-β,1,3-tri-oxo-, ethyl ester, (±)-	EtOH	221(4.61)	44-3489-84
$C_{17}H_{19}NO_7$			
1-Azetidineacetic acid, 2-[(1-formyl-3-methoxy-1-propenyl)oxy]-α-hydroxy-4-oxo-, phenylmethyl ester, [R-[R*,S*-(Z)]]-	EtOH	243(3.90)	39-1599-84C
Cyclohepta[b]pyrrole-3-carboxylic acid, 1,2-dihydro-2-oxo-1-β-D-ribofurano-syl-, ethyl ester	EtOH	225(4.18),236(4.17), 281(4.35),408s(3.98), 426(4.11)	150-0264-84S
$C_{17}H_{19}N_2$			
2H-Indazolium, 1-methyl-3-(1-methyl-ethyl)-2-phenyl-, salt with tri-fluoromethanesulfonic acid	EtOH	217(4.2),261(3.9), 268(4.0),307(3.9)	35-6015-84
$C_{17}H_{19}N_3$			
1H-Imidazole-2-methanamine, 4,5-di-hydro-N-phenyl-N-(phenylmethyl)-, compd. with iodine (1:1)	CHCl$_3$	290(4.40),365(4.16)	83-0246-84
Piperidine, 1-[4-(phenylazo)phenyl]-	C$_6$H$_{12}$	392(4.42)	39-0149-84B
	EtOH	400(4.40)	39-0149-84B
	EtOH-HCl	528(3.54)	39-0149-84B
1,3-Propanediamine, N-9-acridinyl-N'-methyl-	pH 7.0	409(4.04)	77-0509-84
	+DNA	416(3.80)	77-0509-84

Compound	Solvent	λ_{max}(log ϵ)	Ref.
Pyrrolidine, 1-[4-[(4-methylphenyl)-azo]phenyl]-	C_6H_{12}	407(4.54)	39-0149-84B
	EtOH	412(4.48)	39-0149-84B
	EtOH-HCl	540(4.78)	39-0149-84B
1H-Pyrrolo[2,3-f]isoquinoline, 3-(piperidinomethyl)-	EtOH	214(4.35),270(4.46), 300(3.82)	103-0394-84
$C_{17}H_{19}N_3O$			
Pyrrolidine, 1-[4-[(4-methoxyphenyl)-azo]phenyl]-	C_6H_{12}	407(4.54)	39-0149-84B
	EtOH	411(4.47)	39-0149-84B
	EtOH-HCl	559(4.77)	39-0149-84B
$C_{17}H_{19}N_3O_2$			
1H-Benz[de]isoquinoline-1,3(2H)-dione, 5-amino-2-[3-(dimethylamino)propyl]-	buffer	346(3.99)	87-1677-84
	+ DNA	353(3.61)	87-1677-84
1H-Benz[de]isoquinoline-1,3(2H)-dione, 6-amino-2-[3-(dimethylamino)propyl]-	buffer	432(4.08)	87-1677-84
	+ DNA	445(3.81)	87-1677-84
4H-Pyrido[1,2-a]pyrimidine-3-carboxamide, 6,7,8,9-tetrahydro-6-methyl-N-(2-methylphenyl)-4-oxo-	EtOH	228(4.09),310(4.20)	118-0582-84
$C_{17}H_{19}N_3O_3S$			
3H-1,2-Oxathiino[6,5-e]indazol-4-amine, 4,5,6,7-tetrahydro-N,N-dimethyl-7-phenyl-, 2,2-dioxide	EtOH	224(4.20),252(4.00), 261(3.98),276(3.94)	4-0361-84
$C_{17}H_{19}N_3O_4$			
L-Alanine, 3-[(phenylmethoxy)-NNO-azoxy]-, phenylmethyl ester, monohydrochloride	MeOH	238(4.05)	87-1295-84
$C_{17}H_{19}N_3O_5S$			
Propanoic acid, 2-(tetrahydro-2-methyl-4-pyrimidinylidene)-, ethyl ester, p-toluenesulfonate	CF_3COOH	315(4.36)	103-1270-84
$C_{17}H_{19}N_4OS$			
Benzothiazolium, 2-[[4-(dimethylamino)-phenyl]azo]-5-methoxy-3-methyl-, chloride	H_2O	615(4.7)	131-0363-84I
dimer	H_2O	555(4.7)	131-0363-84I
$C_{17}H_{19}N_5O_2S$			
Benzenesulfonamide, N-(1,4-dihydro-2,4-dimethyl-6-phenyl-1,2,4,5-tetrazin-3(2H)-ylidene)-4-methyl-	CHCl$_3$	325(4.03)	104-1224-84
$C_{17}H_{19}N_6O_7$			
1H-Purinium, 2-amino-6,9-dihydro-7-[(4-nitrophenyl)methyl]-6-oxo-9-β-D-ribofuranosyl-, bromide	pH 1	262(4.29),280s(4.18)	44-0363-84
	pH 6.9	263(4.23),285(4.17)	44-0363-84
	pH 13	266(4.27)	44-0363-84
$C_{17}H_{20}Br_3NO$			
2,6-Methano-1-benzazocin-11-one, 8,10-dibromo-1-(4-bromobutyl)-1,2,3,4,5,6-hexahydro-6-methyl-	EtOH	270(3.0),310(2.5)	142-1771-84
$C_{17}H_{20}ClNO_2$			
4,6,8,10-Undecatetraen-3-one, 11-(3-chloro-1H-pyrrol-2-yl)-5-hydroxy-2,6-dimethyl-, (all-E)-	EtOH	425(4.70)	39-1577-84C

Compound	Solvent	$\lambda_{max}(\log \epsilon)$	Ref.
$C_{17}H_{20}ClNO_4S$			
2H-Thieno[2,3-h]-1-benzopyran-2-one, 4-[bis(2-methoxyethyl)amino]-3-chloro-5,6-dihydro-	EtOH	218(4.14),242(4.07), 275s(3.50),360(4.22)	161-0081-84
$C_{17}H_{20}F_3NO_5$			
L-Phenylalanine, N-[(1,1-dimethylethoxy)carbonyl]-4-(trifluoroacetyl)-, methyl ester	n.s.g.	<u>225(4.1),260s(2.7), 350(2.4)</u>	35-7540-84
$C_{17}H_{20}NO_3$			
2H-[1,3]Benzodioxolo[5,6-d][1]benzazepinium, 3,4,6,7-tetrahydro-1-methoxy-5-methyl-, perchlorate	MeCN	253(3.91),300(3.96), 330(3.76)	44-0228-84
$C_{17}H_{20}N_2$			
9,10-Azo-1,4-ethano-5,8-methanoanthracene, 1,4,4a,5,8,8a,9,9a,10,10a-decahydro-, (1α,4α,4aα,5α,8α,8aα,9β,9aα-10β,10aα)-	hexane	396(1.79)	24-0534-84
(1α,4α,4aα,5β,8β,8aα,9β,9aα,10β,10aα)-	hexane	398(1.90)	24-0534-84
1,4-Benzenediamine, N,N-dimethyl-N'-[1-(4-methylphenyl)ethylidene]-	MeOH	262(4.39),332(3.67), 348(3.67)	118-0128-84
1H-Indole, 1,2-dimethyl-3-[3-(1-methylethyl)-1H-pyrrol-2-yl]-	EtOH	210(4.36),242(4.41), 295(3.84)	103-0058-84
$C_{17}H_{20}N_2O$			
Benzamide, 4-(1,1-dimethylethyl)-N-(6-methyl-2-pyridinyl)-	EtOH	250(4.16),288(4.25)	152-0223-84
Benzenecarboximidamide, N-hydroxy-2,4,6-trimethyl-N'-(4-methylphenyl)-	MeOH	253(4.21)	32-0015-84
1,4-Benzenediamine, N'-[1-(4-methoxyphenyl)ethylidene]-N,N-dimethyl-	MeOH	270(4.42),328(3.80), 346(3.80)	118-0128-84
1H-1,2-Diazepine, 1-[4-(1,1-dimethylethyl)benzoyl]-3-methyl-	EtOH	234(4.15)	152-0223-84
Pyridinium, 1-[[4-(1,1-dimethylethyl)-benzoyl]amino]-2-methyl-, hydroxide, inner salt	MeCN	335(3.59)	152-0223-84
$C_{17}H_{20}N_2O_2$			
Benzenecarboximidamide, N-hydroxy-N'-(4-methoxyphenyl)-2,4,6-trimethyl-	MeOH	252(4.13)	32-0015-84
$C_{17}H_{20}N_2O_2S_2$			
Butanenitrile, 2-[5,6-dihydro-4-(4-methoxyphenyl)-5-methyl-6-oxo-4H-1,3,5-dithiazin-2-ylidene]-3,3-dimethyl-	EtOH	233(3.8),285(3.9), 295(3.9)	24-2205-84
1H-Pyrrole-3-carbodithioic acid, 4-(butylamino)-2,5-dihydro-1-(4-methylphenyl)-2,5-dioxo-, methyl ester	CHCl_3	308(4.32),415(3.92)	48-0401-84
$C_{17}H_{20}N_2O_4$			
Benzenamine, N-[4-(2-methoxy-5-nitrophenoxy)butyl]-	MeCN	302(3.88),342(3.87)	78-1755-84
Propanoic acid, 3-[[4-(1H-indol-3-yl)-4-oxobutyl]methylamino]-3-oxo-, methyl ester	EtOH	241(--),258s(--), 300(--)	150-2738-84M

Compound	Solvent	$\lambda_{max}(\log \epsilon)$	Ref.
$C_{17}H_{20}N_2O_5$ 1H-Indole-3-propanoic acid, 1-[2-[(methoxycarbonyl)amino]-2-methyl-1-oxopropyl]-	EtOH	243(4.27),265s(3.93), 274s(3.85),293(3.81), 301(3.85)	94-1373-84
$C_{17}H_{20}N_2O_5S$ Benzo[b]thiophen-3(2H)-one, 2-(dimorpholinomethylene)-, 1,1-dioxide	EtOH	214(4.44),241(4.11), 277(4.18),349(4.13)	95-0134-84
$C_{17}H_{20}N_2O_6$ Pyrrolo[2,3-b]indole-1,2(2H)-dicarboxylic acid, 8-acetyl-3,3a,8,8a-tetrahydro-5-methoxy-, dimethyl ester isomer	MeOH MeOH	252.5(4.15),294.5(3.46) 250.5(4.16),292(3.51), 299s(3.45)	94-2544-84 94-2544-84
Pyrrolo[2,3-b]indole-1,2(2H)-dicarboxylic acid, 8-acetyl-3,3a,8,8a-tetrahydro-6-methoxy-, dimethyl ester isomer	MeOH MeOH	216(3.95),245(4.00), 287(3.66),293(3.64) 218(4.39),248(4.01), 288.5(3.65),294.5(3.65)	94-2544-84 94-2544-84
$C_{17}H_{20}N_2O_7$ 3,5-Pyridinedicarboxylic acid, 1,4-dihydro-2,6-dimethyl-4-(5-nitro-2-furanyl)-, diethyl ester	EtOH	218(4.12),317(4.03)	33-2270-84
$C_{17}H_{20}N_2O_9$ Benzenebutanoic acid, 5-acetoxy-,2-bis[(methoxycarbonyl)amino]-γ-oxo-, methyl ester, (S)-	EtOH	231(4.54),256(4.07), 262(4.03),338(3.68)	142-0059-84
$C_{17}H_{20}N_4$ 7H-Pyrrolo[2,3-d]pyrimidin-4-amine, N-ethyl-2,5,6-trimethyl-7-phenyl-	EtOH	206(4.35),216s(4.31), 232s(4.08),286(3.96)	11-0073-84B
$C_{17}H_{20}N_4O_4S$ Acetic acid, [[(3,4,5,10-tetrahydro-3,7,8,10-tetramethyl-2,4-dioxobenzo[g]pteridin-4a(2H)-yl)methyl]-thio]-	pH 7.0	224(4.41),268(4.17), 300s(--),366(3.80)	35-3309-84
$C_{17}H_{20}N_4O_4S_2$ Acetic acid, [[(3,4,5,10-tetrahydro-3,7,8,10-tetramethyl-2,4-dioxobenzo[g]pteridin-4a(2H)-yl)methyl]-dithio]-	6M HCl MeCN	265(4.06),297s(--), 395(3.42) 222(4.69),273(4.18), 300s(--),358(3.81)	35-3309-84 35-3309-84
$C_{17}H_{20}N_4O_6$ Acetamide, N-[7-amino-9-[[(aminocarbonyl)oxy]methyl]-2,3,5,8-tetrahydro-1-methoxy-6-methyl-5,8-dioxo-1H-pyrrolo[1,2-a]indol-2-yl]-	MeOH	209(4.18),248(4.32), 305(4.19),350s(3.71), 524(3.08)	44-5164-84
$C_{17}H_{20}N_4O_7S$ 4H-Pyrazolo[3,4-d]pyrimidine-4-thione, 1,5-dihydro-6-methyl-1-(2,3,5-tri-O-acetyl-β-D-ribofuranosyl)-	pH 1 pH 7 pH 11	230s(3.71),318(4.22) 230(3.83),318(4.29) 230(4.08),315(4.18)	87-1026-84 87-1026-84 87-1026-84
$C_{17}H_{20}O$ Bicyclo[2.2.1]heptan-2-one, 1,7,7-trimethyl-3-(phenylmethylene)-	EtOH	292(4.36)	41-0629-84

Compound	Solvent	$\lambda_{max}(\log \epsilon)$	Ref.
$C_{17}H_{20}O_2Ru$			
Ruthenocene, 1,1'-(1,4-butanediyl)-3-(2-carboxyethyl)-	EtOH	313(2.49)	18-0719-84
$C_{17}H_{20}O_3$			
2H-1-Benzopyran-6-carboxaldehyde, 5-hydroxy-2-methyl-2-(4-methyl-4-pentenyl)-	EtOH	256s(4.23),260s(4.24), 274(4.30),302(3.97), 314s(3.93)	142-1791-84
	EtOH-AlCl$_3$	260s(4.19),272(4.27), 286s(4.10),319(3.93), 330(3.92),386(2.85)	142-1791-84
3-Phenanthrenecarboxylic acid, 1,2,3,4-9,10-hexahydro-7-methoxy-, methyl ester	EtOH	272(4.27)	2-1168-84
3-Phenanthrenecarboxylic acid, 1,2,3,9-10,10aβ-hexahydro-7-methoxy-, methyl ester	EtOH	263(4.38)	2-1168-84
3-Phenanthrenecarboxylic acid, 1,4,4a-9,10,10a-hexahydro-7-methoxy-, methyl ester, cis-(±)-	EtOH	278(3.6)	2-1168-84
$C_{17}H_{20}O_4$			
Mucidin	MeOH	222(4.33),227(4.33), 299(4.47)	158-84-63
Phenol, 2-[3-(3-hydroxy-4-methoxyphenyl)propyl]-5-methoxy- (broussonin E)	EtOH	220s(4.06),280(3.65), 285s(3.60)	138-0689-84
$C_{17}H_{20}O_5$			
Epoxydecompositin, (±)-	EtOH	250(3.62),285.5(3.99)	94-3396-84
3-Furanmethanol, 5-[(2,3,4,5-tetrahydro-5-methyl[2,3'-bifuran]-5-yl)-methyl]-, acetate, (2R-cis)-	MeCN	237(3.18)	102-0759-84
Furanoeremophil-1(10)-en-6-one, 3,3-(ethylenedioxy)-9β-hydroxy-	EtOH	265.5(3.57)	94-3396-84
Perymeniolide oxidn. product, m. 158-160°	EtOH	223(4.22)	102-0813-84
$C_{17}H_{20}O_6$			
1H-Cyclobuta[7,8]cyclonona[1,2-d:5,6-d']bis[1,3]dioxole-4,9-dione, 2,2a-2b,5c,7a,10a,11,11a-octahydro-2,2,6-trimethyl-11-methylene-	CHCl$_3$	236(2.53),317(1.37)	5-1332-84
Furanoeremophilan-6-one, 3,3-(ethylenedioxy)-1β,10β-epoxy-9β-hydroxy-	EtOH	261(3.49)	94-3396-84
Naphtho[1,2-b]furan-2,8(3H,4H)-dione, 4-acetoxy-3a,5,5a,6,7,9b-hexahydro-6-hydroxy-5a,9-dimethyl-3-methylene-	MeOH	209(3.99),237(4.13)	102-1665-84
$C_{17}H_{20}O_9$			
D-Glucose, 2-acetate 4-[3-(4-hydroxyphenyl)-2-propenoate]-, (E)-	MeOH	230(4.09),316(4.42)	94-1998-84
$C_{17}H_{20}Ru$			
Ruthenocene, 1,1'-(1,4-butanediyl)-3,3'-(1,3-propanediyl)-	EtOH	266s(2.81),319(2.78)	18-0719-84
$C_{17}H_{21}BrO_2$			
Isolaurenisol acetate	EtOH	212(3.944),267(2.799), 275(2.799)	102-1951-84

Compound	Solvent	$\lambda_{max}(\log \epsilon)$	Ref.
$C_{17}H_{21}Br_2NO_2$ 2,6-Methano-1-benzazocin-11-one, 8,10-dibromo-1,2,3,4,5,6-hexahydro-1-(3-methoxypropyl)-6-methyl-	EtOH	262(4.6),300(4.0)	142-1771-84
$C_{17}H_{21}NO$ 2H-Indole, 3,3a,4,5,6,7-hexahydro-2,2-dimethyl-7-(phenylmethylene)-, 1-oxide	EtOH	234(4.11),317(4.08)	12-0117-84
3aH-Indol-3a-ol, 2,3,4,5,6,7-hexahydro-2,2-dimethyl-7-(phenylmethylene)-	EtOH	273(4.23)	12-0577-84
4a,8-Methano-4aH-pyrido[1,2-a][1]benzazocin-14-one, 1,2,3,4,5,6,7,8-octahydro-8-methyl-	EtOH	260(4.3),304(3.5), 330(3.3)	142-1771-84
$C_{17}H_{21}NO_2$ 1H-Azonine-2,7-dione, hexahydro-9,9-dimethyl-3-(phenylmethylene)-	EtOH	250(4.11)	12-0599-84
2-Butanamine, N-[[2-(methoxymethoxy)-1-naphthalenyl]methylene]-	EtOH	226(4.66),296(3.80), 333(3.62)	150-0701-84M
4aH-Oxazirino[3,2-i]indol-4a-ol, hexahydro-3,3-dimethyl-8-(phenylmethylene)-, (4aR*,8aR*)-	EtOH	246(4.36)	12-0599-84
1H-Pyrrole-2-carboxylic acid, 3-butyl-4-methyl-, phenylmethyl ester	EtOH	205(4.54),275(4.23)	23-2054-84
4,6,8,10-Undecatetraen-3-one, 5-hydroxy-2,6-dimethyl-11-(1H-pyrrol-2-yl)-, (Z,E,E,E)- (wallemia A)	EtOH	428(4.71)	39-1577-84C
$C_{17}H_{21}NO_2S$ Benzenesulfonamide, N,4-dimethyl-N-(2,3,5-trimethylphenyl)-	n.s.g.	222(4.13)	12-0381-84
$C_{17}H_{21}NO_3$ Cyclohexanone, 2-(2-methyl-2-nitropropyl)-6-(phenylmethylene)-	EtOH	285(4.20)	12-0117-84
Spiro[cyclohexane-1,1'(2'H)-cyclopenta-[ij][1,3]dioxolo[4,5-g]isoquinolin]-4-ol, 2'a,3',4',5'-tetrahydro-(lauformine)	EtOH	240s(3.32),290(3.36)	142-1031-84
$C_{17}H_{21}NO_4$ Pseudolycorine, methyl-	EtOH	283(3.51)	100-1003-84
$C_{17}H_{21}NO_7$ 2,6-Piperidinedione, 4-[6-[3-(acetoxymethyl)-2-methyl-1-oxo-3-butenyl]-4H-1,3-dioxin-4-yl]-	MeOH	257(3.46)	102-2789-84
$C_{17}H_{21}NS_3$ Benzothiazole, 2-[[4-methyl-1-(1-methyl-1-propenyl)-3-pentenyl]dithio]-	EtOH	223(4.24),245s(3.83), 254(3.93),289s(3.85), 299s(3.75)	39-0101-84C
$C_{17}H_{21}N_3O_{11}$ β-D-Glucopyranosiduronic acid, 5-(aminocarbonyl)-1H-imidazol-4-yl methyl ester 2,3,4-triacetate	H_2O M HCl M NaOH	247(4.12) 230(3.97) 264(4.14)	4-0849-84 4-0849-84 4-0849-84
$C_{17}H_{21}N_5$ 1H-Purin-6-amine, N-(4-butylphenyl)-	EtOH	302(4.30)	11-0208-84B

Compound	Solvent	$\lambda_{max}(\log \epsilon)$	Ref.
2,8-dimethyl- (cont.)			11-0208-84B
$C_{17}H_{22}BrNO_2$			
2,6-Methano-1-benzazocin-11-one, 8-bromo-1,2,3,4,5,6-hexahydro-1-(3-methoxypropyl)-6-methyl-	EtOH	268(4.6),314(3.7), 340(3.5)	142-1771-84
$C_{17}H_{22}N_2$			
9,10-Azo-1,4-ethano-5,8-methanoanthracene, 1,2,3,4,4a,5,8,8a,9,9a,10,10a-dodecahydro-, cis	hexane	386(2.21)	24-0534-84
trans	hexane	390(2.33)	24-0534-84
9,10-Azo-1,4-ethano-5,8-methanoanthracene, 1,4,4a,5,6,7,8,8a,9,9a,10,10a-dodecahydro-	hexane	394(2.08)	24-0534-84
Benzenamine, 4,4'-methylenebis[N,N-dimethyl-	hexane-isoPrOH	263(3.53)	39-1099-84B
$C_{17}H_{22}N_2O$			
1-Butanone, 1-(1,2-dimethyl-1H-indol-3-yl)-3-methyl-, O-ethenyloxime, (E)-	EtOH	225(4.45),287(4.05)	103-0058-84
$C_{17}H_{22}N_2O_2$			
5H-[1]Benzopyrano[3,4-c]pyridin-5-one, 3-[2-(ethylamino)propyl]-1,2,3,4-tetrahydro-	MeOH	271(4.00),308(3.89)	4-1561-84
$C_{17}H_{22}N_2O_3$			
1-Naphthalenemethanamine, 2-(methoxymethoxy)-N-(1-methylpropyl)-N-nitroso-	EtOH	228(4.82),269(3.71), 280(3.80),292(3.76), 316(3.35),330(3.40)	150-0701-84M
1H-Pyrrole-3-propanoic acid, 5-[(3-ethyl-1,5-dihydro-4-methyl-5-oxo-2H-pyrrol-2-ylidene)methyl]-4-methyl-, methyl ester	MeOH	231(4.77),261(4.83), 395(4.32)	35-2645-84
1H-Pyrrole-3-propanoic acid, 5-[(4-ethyl-1,5-dihydro-3-methyl-5-oxo-2H-pyrrol-2-ylidene)methyl]-4-methyl-, methyl ester	MeOH	230(4.74),259(4.80), 394(5.30)	35-2645-84
Xanthobilirubic acid	MeOH	416(4.53)	4-0139-84
	CHCl$_3$	408(4.45)	4-0139-84
$C_{17}H_{22}N_2O_4$			
1H-Pyrrole-3-propanoic acid, 5-[(3-ethyl-2,5-dihydro-4-methyl-5-oxo-1H-pyrrol-2-yl)methyl]-2-formyl-4-methyl-, (±)-	MeOH	311(4.21)	5-1441-84
1H-Pyrrole-3-propanoic acid, 5-[(4-ethyl-2,5-dihydro-3-methyl-5-oxo-1H-pyrrol-2-yl)methyl]-2-formyl-4-methyl-, (±)-	MeOH	311(4.15)	5-1441-84
$C_{17}H_{22}N_2O_4S$			
Tricyclo[3.3.1.1^{3,7}]decan-2-amine, N-[[(4-methylphenyl)sulfonyl]oxy]-N-nitroso-	EtOH	275(2.92)	39-1693-84B
$C_{17}H_{22}N_6O_5$			
Propanamide, N-[6-(2-cyanoethoxy)-9-(2-deoxy-β-D-erythro-pentofuranosyl)-9H-purin-2-yl]-2-methyl-	MeOH	220(4.35),268(4.24)	78-0003-84

Compound	Solvent	λ_{max}(log ϵ)	Ref.
$C_{17}H_{22}O$			
Benzaldehyde, 2-methyl-4-(2,6,6-tri-methyl-1-cyclohexen-1-yl)-	MeCN	258(4.18)	87-1516-84
4,7-Methano-1H-cyclopentacycloocten-8(2H)-one, 3,3a,4,7-tetrahydro-2,9-dimethyl-10-(1-methylethylidene)-	EtOH	263(3.74)	33-1506-84
$C_{17}H_{22}O_2$			
2H-1-Benzopyran-5-ol, 2,7-dimethyl-2-(4-methyl-3-pentenyl)-	MeOH	230(4.10),281(3.56), 291s(3.48)	102-1909-84
$C_{17}H_{22}O_3$			
Cyclodeca[b]furan-8-ol, 4,7,8,11-tetra-hydro-3,6,10-trimethyl-, acetate, [S-(E,E)]-	MeOH	218(4.45)	102-1793-84
2,4-Cyclohexadien-1-one, 5-(5-acetoxy-3-methyl-1,3-pentadienyl)-4,6,6-tri-methyl-, (E,E)-	MeOH	255(4.10),357(4.21)	1-0043-84
$C_{17}H_{22}O_4$			
1,3-Dioxane-4,6-dione, 2,2,5-trimethyl-5-[4-(2-methylpropyl)phenyl]-	EtOH	223(4.00)	12-1245-84
6-Epilaurenobiolide	MeOH	207(3.86)	102-1971-84
8-Nonenoic acid, 9-(1,3-benzodioxol-5-yl)-, methyl ester, (E)-	MeOH	214(4.39),260(4.18), 267(4.17),304(3.75)	78-2541-84
$C_{17}H_{22}O_5$			
1,3-Benzodioxole-5-nonenoic acid, θ-oxo-, methyl ester	MeOH	210(3.65),228(4.21), 273(3.32),308(3.52)	78-2541-84
1-Cyclohexene-1-carboxylic acid, 4,6-dihydroxy-2-[2-(4-methoxyphenyl)-ethyl]-, methyl ester, equatorial equatorial-axial	MeOH	225(4.17),269s(--), 278(3.24),284(3.17)	102-1607-84
	MeOH	224(4.19),269s(--), 278(3.28),284(3.21)	102-1607-84
2-Furancarboxylic acid, 4-[4-methyl-1-[(3-methyl-1-oxo-2-butenyl)oxy]-3-pentenyl]-, methyl ester	MeOH	218(4.16),252(4.01)	102-0301-84
Furanoeremophilan-6-one, 9β-acetoxy-10α-hydroxy-	EtOH	265.5(3.65)	94-3396-84
Ivalbatin monoacetate	MeOH	230(3.84)	102-2553-84
Perymeniolide	EtOH	212(4.07)	102-0813-84
$C_{17}H_{22}O_6$			
α-Epoxyludalbin	MeOH	205(4.04)	102-1665-84
Furanoeremophilan-6-one, 3,3-(ethylene-dioxy)-9β,10α-dihydroxy-	EtOH	266(3.51)	94-3396-84
$C_{17}H_{22}O_{11}$			
Ixoside 11-methyl ester	EtOH	216(4.12),233(3.99)	32-0049-84
$C_{17}H_{23}NO_2$			
2H-1,4-Benzoxazin-2-one, 6,8-bis(1,1-dimethylethyl)-3-methyl-	hexane	290(1.90),340(1.47), 413(0.42)	54-0023-84
Butanamide, N-(4b,6,7,8,9,9a-hexahydro-4-hydroxy-5H-benzo[3,4]cyclobuta[1,2]-cyclohepten-1-yl)-, cis	MeOH	249(4.17)	87-0792-84
1-Naphthalenemethanamine, 2-(methoxy-methoxy)-N-(1-methylpropyl)-	EtOH	227(4.87),269(3.65), 278(3.76),290(3.68), 316(3.24),360(3.31)	150-0701-84M

Compound	Solvent	λ_{max}(log ϵ)	Ref.
C$_{17}$H$_{23}$NO$_2$Si Methanone, phenyl[1-[[2-(trimethyl- silyl)ethoxy]methyl]-1H-pyrrol-2-yl]-	MeOH	203(4.26),248(4.99), 302(4.16)	44-0203-84
C$_{17}$H$_{23}$NO$_3$ C-Homo-D-nor-16-oxaerythrinan-15-metha- nol, 1,6-didehydro-3-methoxy-, (3β)-	EtOH	209(3.66),226(3.66), 285(2.66)	78-2685-84
C$_{17}$H$_{23}$NO$_4$ 2-Butenedioic acid, 2-[5-(2,2-dimethyl- 1-methylenepropyl)-1-methyl-1H-pyrrol- 2-yl]-, dimethyl ester, cis	EtOH	346(4.21)	39-2541-84C
trans	EtOH	380(3.03)	39-2541-84C
C$_{17}$H$_{23}$NO$_6$ α-D-Glucopyranoside, methyl 2-deoxy- 2-[(1-methyl-3-oxo-3-phenyl-1-prop- enyl)amino]-, (Z)-	EtOH	241(3.90),340(4.05)	136-0101-84L
C$_{17}$H$_{23}$NO$_7$ 1H-Pyrrolizine-6,7-dicarboxylic acid, 2,3-dihydro-2-hydroxy-5-(4-methoxy- 2-methyl-4-oxobutyl)-, dimethyl ester	CHCl$_3$	283.7(3.337)	44-5164-84
C$_{17}$H$_{23}$N$_3$O$_2$ Xanthobilirubic acid amide	MeOH CHCl$_3$	414(4.43) 407(4.45)	4-0139-84 4-0139-84
C$_{17}$H$_{23}$N$_3$O$_5$Si 3-Butenoic acid, 2-diazo-3-[[(1,1-di- methylethyl)dimethylsilyl]oxy]-, (4-nitrophenyl)methyl ester	CH$_2$Cl$_2$	270(4.19)	23-2936-84
C$_{17}$H$_{24}$ClN$_3$O$_4$ 2-Propanol, 1-[(2-chloro-6,7-dimethoxy- 4-quinazolinyl)oxy]-3-[(1,1-dimethyl- ethyl)amino]-, hydrochloride	EtOH	255(4.60)	4-1189-84
C$_{17}$H$_{24}$N$_2$ 9,10-Azo-1,4-ethano-5,8-methanoanthra- cene, 1,2,3,4,4a,5,6,7,8,8a,9,9a- 10,10a-tetradecahydro-	hexane	374(2.52)	24-0534-84
C$_{17}$H$_{24}$N$_2$O 1-Hepten-2-ol, 1-[4,5-dihydro-1-(phen- ylmethyl)-1H-imidazol-2-yl]-, (Z)-	EtOH	291(4.43)	39-2599-84C
C$_{17}$H$_{24}$N$_2$OSi$_2$ Benzo[3,4]cyclobuta[1,2-d]pyrimidine, 4-methoxy-6,7-bis(trimethylsilyl)-	EtOH	262(4.56),314(3.40), 328(3.61),344(3.66)	44-0289-84
C$_{17}$H$_{24}$N$_2$O$_3$ 3-Buten-2-one, 4-[5-acetyl-1-(dimethyl- amino)-1,6-dihydro-2-methyl-6-(2-oxo- propyl)-3-pyridinyl]-, (E)-	MeOH	214(4.30),273(4.06)	24-1620-84
C$_{17}$H$_{24}$N$_2$O$_4$ 1H-Pyrrole-3-carboxylic acid, 5-[(3- ethyl-4-methyl-5-oxo-2-pyrrolidinyl)- methyl]-2-formyl-4-methyl-, (2α,3α- 4β)-(±)-	MeOH	310(4.21)	5-1441-84

Compound	Solvent	$\lambda_{max}(\log \epsilon)$	Ref.
(2α,3β,4α)-(±)-	MeOH	312(4.25)	5-1441-84
1H-Pyrrole-3-carboxylic acid, 5-[(4-ethyl-3-methyl-5-oxo-2-pyrrolidinyl)-methyl]-2-formyl-4-methyl-, (2α,3α-4β)-(±)-	MeOH	311(4.23)	5-1441-84
(2α,3β,4α)-(±)-	MeOH	311(3.91)	5-1441-84
$C_{17}H_{24}N_2O_5S$			
2H-Furo[2,3-d]imidazole-2-thione, 6-hydroxy-3-phenyl-1-propyl-5-(1,2,3-tri-hydroxypropyl)-, [3aR-[3aα,5α-(1R*,2R*),6β,6aα]]-	EtOH	253(3.90)	136-0071-84G
$C_{17}H_{24}N_2O_6$			
2-Pyridineacetic acid, 1-(dimethylamino)-1,2-dihydro-3-(methoxycarbonyl)-5-(3-methoxy-3-oxo-1-propenyl)-6-methyl-, methyl ester, (E)-	MeOH	207(3.71),229(3.63), 349(3.95)	24-1620-84
$C_{17}H_{24}N_2O_8$			
Uridine, 2'-deoxy-5-ethyl-, 5'-(methyl pentanedioate)	H_2O	267(3.97)	83-0867-84
$C_{17}H_{24}N_6O_7$			
L-Lysine, N^2-formyl-N^6-(9-β-D-ribo-furanosyl-9H-purin-6-yl)-	H_2O	266(4.24)	73-2689-84
L-Lysine, N^6-formyl-N^2-(9-β-D-ribo-furanosyl-9H-purin-6-yl)-	H_2O	268(4.26)	73-2689-84
$C_{17}H_{24}O_2$			
2-Hepten-1-ol, 2-methyl-6-(4-methyl-phenyl)-, acetate, (E)-	MeOH	247(3.34)	31-0931-84
5-Hepten-2-one, 5-hydroxy-2,2,4,5-tetramethyl-7-phenyl-, erythro-	EtOH	251.5(4.24)	78-4127-84
threo-	EtOH	252(4.26)	78-4127-84
5H-Inden-5-one, 1,2,3,6,7,7a-hexahydro-7a-methyl-1-(5-methyl-3-oxo-1-hexen-yl)-, [1R-[1α(Z),7aα]]-	n.s.g.	237(4.30)	44-0948-84
2,6-Octanedione, 5,7,7-trimethyl-4-phenyl-, erythro-	EtOH	252(2.50)	78-4127-84
threo-	EtOH	252(2.48)	78-4127-84
7-Octen-4-one, 6-hydroxy-2,2,6-tri-methyl-8-phenyl-, (E)-	EtOH	251(4.13)	78-4127-84
2,4-Pentadien-1-ol, 3-methyl-5-(2,6,6-trimethyl-1,3-cyclohexadien-1-yl)-, acetate, (E,E)-	MeOH	240(3.99),314(4.02)	1-0043-84
8a(2H)-Phenanthrenecarboxylic acid, 1,3,4,6,7,8,9,10-octahydro-, ethyl ester	EtOH	245(4.28)	44-3988-84
8a(2H)-Phenanthrenecarboxylic acid, 3,4,5,6,7,8,9,10-octahydro-, ethyl ester	EtOH	246(4.25)	44-3988-84
Tricyclo[3.2.2.0^{2,4}]non-8-ene-6,7-di-one, 1,9-bis(1,1-dimethylethyl)-	C_6H_{12}	447(2.05)	88-4697-84
Tricyclo[5.2.0.0^{2,4}]non-5-ene-8,9-di-one, 5,7-bis(1,1-dimethylethyl)-, (1α,2β,4β,7α)-	C_6H_{12}	330(3.00),530(2.51)	88-4697-84
$C_{17}H_{24}O_2S_2$			
3,11-Dithiabicyclo[11.3.2]octadeca-13(18),14,16-trien-17-one, 18-methoxy-	MeOH	249.5(4.15),309(3.87), 352s(3.67)	88-4761-84

Compound	Solvent	$\lambda_{max}(\log \epsilon)$	Ref.
$C_{17}H_{24}O_3$			
2,4,6,8,10-Dodecapentaenal, 12,12-di-methoxy-2,7,11-trimethyl-, (all-E)-	pet ether	362s(--),375(4.77), 396(4.74)	137-0102-84
Spiro[bicyclo[4.1.0]heptane-2,1'-[4]-cyclopenten]-3'-one, 4'-(acetoxy-methyl)-3,7,7-trimethyl-, [1R-(1α,2β,3β,6α)]-	EtOH	227(3.95)	39-0215-84C
2,5,9-Undecatrienal, 2-(2-acetoxyethen-yl)-6,10-dimethyl-, (E,E,E)-	MeOH	229(3.45)	78-2913-84
$C_{17}H_{24}O_4$			
Cyclohexanone, 4-(5-acetoxy-3-methyl-3-penten-1-ynyl)-4-hydroxy-3,5,5-trimethyl-, [4α,4(E),5β]-(±)-	EtOH	230(4.16)	39-2147-84C
[4α,4(Z),5β]-(±)-	EtOH	228(4.15)	39-2147-84C
$C_{17}H_{24}O_5$			
Bicyclo[7.2.0]undec-5-en-3-one, 4-acet-oxy-7,8-dihydroxy-6,10,10-trimethyl-2-methylene- (naematolin)	MeOH	210(3.94),242s(3.75), 326(2.94)	5-1332-84
Propanedioic acid, (4-methoxyphenyl)(1-methylethyl)-, diethyl ester	n.s.g.	227(3.99),271(3.08), 277(3.00)	12-1245-84
$C_{17}H_{24}O_6$			
Ethanone, 1-[2-hydroxy-4,6-bis(methoxy-methoxy)-3-(3-methyl-2-butenyl)-phenyl]-	EtOH	285(4.09),318s(3.39)	142-0997-84
	EtOH-AlCl$_3$	285(4.05),318s(3.37)	142-0997-84
Ethanone, 1-[6-hydroxy-2,4-bis(methoxy-methoxy)-3-(3-methyl-2-butenyl)-phenyl]-	EtOH	280(4.05),328(3.56)	142-0997-84
	EtOH-AlCl$_3$	285(3.94),302(3.93), 332s(3.43),364s(3.19)	142-0997-84
1,6-Nonanedione, 9-acetoxy-1-(3-furan-yl)-4-hydroxy-4,8-dimethyl-	MeCN	248(3.56)	102-0759-84
$C_{17}H_{24}O_{11}$			
Gardenosine	MeOH	232(4.10)	94-2947-84
Secoxyloganin	MeOH	233(4.04)	102-2539-84
$C_{17}H_{25}NOSi$			
1H-Pyrrole, 2-(phenylmethyl)-1-[[2-(trimethylsilyl)ethoxy]methyl]-	MeOH	205(4.14),269(2.63)	44-0203-84
$C_{17}H_{25}N_5O_2S$			
1-Propanone, 1-[2-amino-4-(pentyloxy)-7-(propylthio)-6-pteridinyl]-	MeOH	247s(4.18),272(4.45), 304(4.11),388(4.28)	138-1025-84
$C_{17}H_{25}O_7P$			
2-Propenoic acid, 2-(diethoxyphosphin-yl)-3-(3,4-dimethoxyphenyl)-, ethyl ester, (E)-	dioxan	290(4.06),315(4.11)	65-1987-84
$C_{17}H_{26}NO_5P$			
2-Propenoic acid, 2-(diethoxyphosphin-yl)-3-[4-(dimethylamino)phenyl]-, ethyl ester, (E)-	dioxan	350(4.42)	65-1987-84
$C_{17}H_{26}N_2O_3$			
Pyrrolo[1,2-c]quinazoline-5,8(1H,6H)-dione, 6-acetyloctahydro-7,7,9,9-tetramethyl-	MeOH	237(2.70),300s(1.85)	88-0371-84

Compound	Solvent	$\lambda_{max}(\log \epsilon)$	Ref.
$C_{17}H_{26}N_4O_5$			
Inosine, 2-heptyl-	pH 1	250(4.10)	87-0429-84
	pH 7	250(4.07)	87-0429-84
	pH 13	255(4.10)	87-0429-84
$C_{17}H_{26}O_2$			
5H-Inden-5-one, 1,2,3,6,7,7a-hexahydro-1-(3-hydroxy-5-methyl-1-hexenyl)-7a-methyl-, [1R-[1α(1Z,3R*),7aα]]-	n.s.g.	239(4.15)	44-0948-84
$C_{17}H_{26}O_3$			
2,4-Pentadienoic acid, 3-methyl-5-(2,2,6-trimethyl-7-oxabicyclo[4.1.0]hept-1-yl)-, ethyl ester, [1S-[1α(2E,4E),6α]]-	EtOH	267(4.44)	33-0184-84
	EPA	264.5(4.45)	33-0184-84
[1S-[1α(2Z,4E),6α]]-	EtOH	268(4.25)	33-0184-84
	EPA	264(4.33)	33-0184-84
$C_{17}H_{26}O_4$			
1,3-Cyclohexadiene-1-carboxylic acid, 4-(2,2-dimethyl-1-oxopropoxy)-2-methyl-, 1,1-dimethylethyl ester	EtOH	205(3.63),295(3.82)	12-2037-84
1,4-Dioxaspiro[4.5]decan-8-ol, 8-(5-hydroxy-3-methyl-3-penten-1-ynyl)-7,7,9-trimethyl-, [8α,8(E),9β]-(±)-	EtOH	230(4.15)	39-2147-84C
[8α,8(Z),9β]-(±)-	EtOH	228(4.08)	39-2147-84C
$C_{17}H_{26}O_5$			
Acetaldehyde, 5,6-dihydro-3,5,6-tri-hydroxy-β-ionylidene-, (3S,5R,6R)-	EtOH	286(4.37)	33-2043-84
(3S,5S,6S)-	EtOH	286(4.35)	33-2043-84
$C_{17}H_{26}O_{10}$			
Loganin	MeOH	235(4.01)	33-0160-84
$C_{17}H_{26}O_{12}$			
Morroniside, 10-hydroxy-	MeOH	238(4.19)	102-2535-84
$C_{17}H_{27}N$			
Pyridine, 2-(4-butylcyclohexyl)-5-ethyl-	EtOH	214(3.72),269(3.39)	103-0515-84
Pyrrolidine, 1-[3-(2,6,6-trimethyl-1-cyclohexen-1-yl)-1-methylene-2-propenyl]-	n.s.g.	337(<u>3.8</u>)	149-0391-84A
$C_{17}H_{27}NO_2$			
Ethanone, 1-[3-[[bis(1-methylethyl)-amino]methyl]-2-methoxy-5-methyl-phenyl]-	EtOH	250(3.36),330(2.34)	2-0904-84
Pyridinium, 1-(1-carboxyundecyl)-, hydroxide, inner salt	n.s.g.	261.5(3.65)	46-6041-84
$C_{17}H_{27}N_5O_4S_2Si$			
4-Thia-1-azabicyclo[3.2.0]hept-2-ene-2-carboxylic acid, 6-(1-hydroxyethyl)-3-[3-[(1-methyl-1H-tetrazol-5-yl)-thio]methyl]-7-oxo-, 2-[dimethyl-(1,1-dimethylethyl)silyl] ester	hexane	334(3.78)	32-0319-84
$C_{17}H_{27}N_7$			
9H-Purin-6-amine, N,9-bis(1-piperidin-	MeOH	267(4.22)	78-3997-84

Compound	Solvent	$\lambda_{max}(\log \epsilon)$	Ref.
ylmethyl)- (cont.)	MeOH-HCl	270(4.16)	78-3997-84
	MeOH-NaOH	274(4.19)	78-3997-84
$C_{17}H_{27}O_2P$			
Benzoic acid, 3-[bis(1,1-dimethylethyl)-phosphino]-, ethyl ester	C_6H_{12}	228(3.97),263s(3.18), 276s(3.00)	65-0489-84
Benzoic acid, 4-[bis(1,1-dimethylethyl)-phosphino]-, ethyl ester	C_6H_{12}	236(4.32),288(3.48)	65-0489-84
$C_{17}H_{28}NO_2$			
2-Buten-1-aminium, N-[3-(2,2-dimethyl-1-oxopropoxy)-2-cyclohexen-1-yli-dene]-N,3-dimethyl-, (E)-, per-chlorate	MeOH	287(4.28)	44-0220-84
$C_{17}H_{28}NO_3$			
1H-3-Benzazoninium, 2,3,4,5,6,7-hexa-hydro-7,9,10-trimethoxy-3,3-dimeth-yl-, iodide	MeOH	218s(4.29),282(3.43), 287s(3.41)	12-1203-84
$C_{17}H_{28}OS_2$			
Bicyclo[2.2.1]heptan-2-one, 3-[bis(pro-pylthio)methylene]-1,7,7-trimethyl-	EtOH	325(4.06)	78-2951-84
$C_{17}H_{28}O_5$			
1,2,4-Cyclohexanetriol, 1-(5-hydroxy-3-methyl-1,3-pentadienyl)-2,6,6-trimethyl-, 4-acetate	EtOH	236(4.41)	33-2043-84
isomer	EtOH	236(4.44)	33-2043-84
$C_{17}H_{28}O_5S$			
1,3-Cyclohexanedione, 2-[4-(1,1-dimeth-ylethyl)cyclohex-1-enyl]-, methane-sulfonic acid salt (1:1)	pH 1	265(4.03)	39-1213-84C
	H_2O	264(4.05)	39-1213-84C
	pH 13	292(4.34)	39-1213-84C
$C_{17}H_{28}O_6S_2$			
Spiro[bicyclo[2.2.1]heptane-2,2'-oxir-an]-3-one, 4,7,7-trimethyl-3',3'-bis(propylsulfonyl)-, [1S-(1α,2β,4α)]-	EtOH	312(1.62)	78-2951-84
$C_{17}H_{28}O_7P_2$			
Phosphonic acid, [(4-methoxyphenyl)eth-enylidene]bis-, tetraethyl ester	dioxan	307(4.35)	65-1987-84
$C_{17}H_{29}GeNO$			
Benzenepropanamide, N,N-dimethyl-β-(triethylgermyl)-	heptane or EtOH	227(--)	70-0310-84
$C_{17}H_{29}N_3O_5SSi$			
1,2,4-Triazin-5(2H)-one, 6-[5-O-[(1,1-dimethylethyl)dimethylsilyl]-2,3-O-(1-methylethylidene)-β-D-ribo-furanosyl]-3,4-dihydro-3-thioxo-	pH 1	213(3.87),269(4.15)	18-2515-84
	pH 13	226(4.21),258(4.06), 313(3.51)	18-2515-84
	MeOH	213(3.57),272(4.08)	18-2515-84
$C_{17}H_{29}N_3O_6Si$			
1,2,4-Triazine-3,5(2H,4H)-dione, 6-[5-O-[(1,1-dimethylethyl)dimethylsil-yl]-2,3-O-(1-methylethylidene)-β-D-ribofuranosyl]-	pH 1	262(3.87)	18-2515-84
	pH 13	258(3.70),293(3.51)	18-2515-84
	MeOH	264(3.84)	18-2515-84

Compound	Solvent	$\lambda_{max}(\log \epsilon)$	Ref.
$C_{17}H_{29}N_5O_2$ 9H-Purin-6-amine, 9-[1-(diethoxymethyl)- heptyl]-, (±)-	MeOH	261(4.13)	73-2148-84
$C_{17}H_{29}O_8P$ -D-ribo-Hexofuranose, 3-deoxy-3-[(di- ethoxyphosphinyl)methylene]-1,2:5,6- bis-O-(1-methylethylidene)-, (E)-	MeOH	215(3.60)	78-0079-84
$C_{17}H_{30}$ Cyclopentane, 1,1,3,3-tetramethyl- 2-(2,2,4,4-tetramethylcyclobut- ylidene)-	C_6H_{12}	203.0(4.13)	24-0277-84 +24-0310-84
$C_{17}H_{30}N_2S$ 5-Thia-11,12-diazadispiro[3.1.4.2]do- dec-11-ene, 1,1,3,3,7,7,10,10-octa- methyl-	C_6H_{12}	203.0(3.15),222s(2.48), 289.0(3.09),328s(2.48)	24-0277-84
$C_{17}H_{30}N_2S_2$ 5,9-Dithia-12,13-diazadispiro[3.1.5.2]- tridec-12-ene, 1,1,3,3,7,7,11,11- octamethyl-	C_6H_{12}	202.0(3.43),225s(2.76), 295.5(2.96),338s(2.23)	24-0277-84
$C_{17}H_{30}O$ 2H-Pyran, tetrahydro-3,3,5,5-tetrameth- yl-4-(2,2,4,4-tetramethylcyclobutyli- dene)-	C_6H_{12}	201.5(4.18)	24-0277-84 +24-0310-84
$C_{17}H_{30}OS$ Cyclododecanone, 2-[(butylthio)methyl- ene]-	MeCN	294(4.29)	24-2293-84
$C_{17}H_{30}S$ 2H-Thiopyran, tetrahydro-3,3,5,5-tetra- methyl-4-(2,2,4,4-tetramethylcyclo- butylidene)-	C_6H_{12}	203.0(4.13)	24-0277-84
$C_{17}H_{30}S_2$ 1,3-Dithiane, 2-[3-(2,2-dimethyl- 6-methylenecyclohexyl)propyl]- 2-methyl-	pentane	226s(2.79),250(2.88)	33-1734-84
$C_{17}H_{32}$ Cyclobutane, 2-[1-(1,1-dimethylethyl)- 2,2-dimethylpropylidene]-1,1,3,3- tetramethyl-	C_6H_{12}	207.0(4.07)	24-0277-84 +24-0310-84
$C_{17}H_{32}N_2S$ 8-Thia-5,6-diazaspiro[3.4]oct-5-ene, 7,7-bis(1,1-dimethylethyl)-1,1,3,3- tetramethyl-	C_6H_{12}	206.0(3.15),230s(2.28), 298.0(2.95),329s(2.53)	24-0277-84

Compound	Solvent	$\lambda_{max}(\log \epsilon)$	Ref.
$C_{18}F_{22}S_6$			
Benzene, 1,1'-[[1,1,2,2-tetrakis[(tri-fluoromethyl)thio]-1,2-ethanediyl]-bis(thio)]bis[pentafluoro-	hexane	225(4.43),275(3.88)	78-4963-84
$C_{18}H_8Br_2O_3$			
Naphtho[1,8-bc]pyran-2,6-dione, 5,9-di-bromo-3-phenyl-	CHCl$_3$	254(3.96),279(3.81), 323(4.04),384(4.12)	39-1957-84C
$C_{18}H_8Cl_2O_8$			
2H,5H-1,4,7,9-Tetraoxabenzo[fg]cyclo-pent[b]anthracene-2,5-dione, 3,6-dichloro-8,12-dimethoxy-	n.s.g.	252(4.17),432(3.99), 459(3.87)	83-0970-84
$C_{18}H_8Cl_2Se_2$			
Naphthaceno[5,6-cd]-1,2-diselenole, 7,8-dichloro-	DMF	325(4.93),438(3.78), 625(3.90)	104-0810-84
$C_{18}H_8N_2O_4$			
Pyrazino[1,2-a:4,3-a']diindole-6,7,13,14-tetrone	xylene	439(3.93)	18-0470-84
$C_{18}H_8S_2Se_2$			
[1,2]Diselenolo[3',4',5':11,12]naph-thaceno[5,6-cd]-1,2-dithiole	DMF	465(3.89),658(3.83), 711(4.03)	104-0810-84
$C_{18}H_8S_4$			
Naphthaceno[5,6-cd:11,12-c'd']bis-[1,2]dithiole-1,2-diium	CHCl$_3$	316(4.62),462(4.0), 596(2.67),649(3.45)	18-0022-84
$C_{18}H_8Se_2Te_2$			
[1,2]Ditellurolo[3',4',5':11,12]naph-thaceno[5,6-cd]-1,2-diselenole	DMF	465(3.80),670(3.90), 730(4.10)	104-0810-84
$C_{18}H_8Se_4$			
Naphthaceno[5,6-cd:11,12-c'd']bis[1,2]-diselenole	1,2,4- C$_6$H$_3$Cl$_3$	321(4.13),465(3.29), 598(1.73),656(2.75)	18-0022-84
$C_{18}H_9BrO_3$			
Naphtho[1,8-bc]pyran-2,6-dione, 9-bromo-3-phenyl-	CHCl$_3$	249(3.97),266s(3.99), 277(4.04),294(4.05), 305s(4.04),376(4.12)	39-1957-84C
$C_{18}H_9Br_3O_4$			
2H-Pyran-2-one, 5-bromo-3-(4-bromo-benzoyl)-6-(4-bromophenyl)-4-hydroxy-	EtOH	209(3.39),246(3.34), 338(1.00)	104-1595-84
$C_{18}H_9ClO_4$			
5,12-Naphthacenedione, 6-chloro-1,11-dihydroxy-	Ac$_2$O + H$_3$BO$_3$	448(4.24) 468(4.17),495(4.43), 527(4.47)	150-0147-84M 150-0147-84M
$C_{18}H_9NO_4$			
1,4-Naphthalenedione, 5-[(4-nitrophen-yl)ethynyl]-	EtOH	251(4.00),327(4.00), 394(3.52)	70-2345-84
$C_{18}H_{10}$			
Benzene, 1,1'-(1,3,5-hexatriyne-1,6-diyl)bis-	MeOH	<u>254(4.9),270(4.8), 282(4.8),312(4.3), 339(4.4),356(4.3)</u>	18-2905-84

Compound	Solvent	$\lambda_{max}(\log \epsilon)$	Ref.
$C_{18}H_{10}BrNO_3$ 2H-Indol-2-one, 5-bromo-1,3-dihydro- 3-(2-oxo-5-phenyl-3(2H)-furanyli- dene)-	CHCl$_3$	272(4.34),450(4.39), 466s(4.37)	39-1331-84B
$C_{18}H_{10}Br_2N_4S_2$ 1,2,5-Thiadiazole, 3,3'-(1,2-dibromo- 1,2-ethenediyl)bis[4-phenyl-, cis trans	C$_6$H$_{12}$ C$_6$H$_{12}$	205(4.54),225(4.36), 250(4.28),285(4.27) 205(4.52),230(4.33), 250(4.27),290(4.26)	4-1157-84 4-1157-84
$C_{18}H_{10}Cl_2O_4$ 2H-1-Benzopyran-2-one, 4-chloro-, anti-head-tail dimer	CHCl$_3$	251(3.42),278(3.44), 282(3.43)	2-0502-84
$C_{18}H_{10}F_2$ Triphenylene, 1,4-difluoro-	C$_6$H$_{12}$	254(4.99),262(5.02), 278(4.27),289(4.29)	44-2803-84
$C_{18}H_{10}F_4$ Triphenylene, 1,1,4,4-tetrafluoro- 1,4-dihydro-	isooctane	224(4.35),248(4.57), 256(4.71),282(3.80), 292(3.79),304(3.8)	44-2803-84
$C_{18}H_{10}N_2O_5$ 2H-Indol-2-one, 1,3-dihydro-5-nitro- 3-(2-oxo-5-phenyl-3(2H)-furanyli- dene)-	CHCl$_3$	261(4.18),462(4.29), 500s(4.19)	39-1331-84B
$C_{18}H_{10}N_4O_2$ Pyrazolo[1,5-a]quinazoline-3-carbo- nitrile, 2-benzoyl-4,5-dihydro-5-oxo-	EtOH	276(4.97)	4-1049-84
$C_{18}H_{10}O_2$ 1,4-Naphthalenedione, 5-(phenylethyn- yl)-	EtOH	251(4.32),298(4.21), 314(4.17),410(3.49)	70-2345-84
$C_{18}H_{10}O_4Se_2$ 2-Benzofurancarboxaldehyde, 3,3'-di- selenobis- 3-Benzofurancarboxaldehyde, 2,2'-di- selenobis-	CHCl$_3$ CHCl$_3$	238(3.78),286(3.69), 305(3.71) 236(4.42),306(4.05)	103-0957-84 103-0957-84
$C_{18}H_{10}O_5$ Anthra[1,2-b]furan-4-carboxaldehyde, 6,11-dihydro-5-hydroxy-2-methyl- 6,11-dioxo-	EtOH	212(4.38),259(4.40), 266(4.38),286(4.20), 442(3.97)	12-1511-84
$C_{18}H_{10}O_6$ Anthra[1,2-b]furan-3-carboxylic acid, 6,11-dihydro-5-hydroxy-2-methyl- 6,11-dioxo-	EtOH	238(4.28),258(4.54), 282s(4.23),301s(3.91), 426(3.91)	12-1518-84
$C_{18}H_{11}ClN_4O_2$ 3H-[1,2,4]Triazino[3,2-b]quinazoline- 3,10(4H)-dione, 2-[2-(4-chlorophenyl)- ethenyl]-	EtOH	276(3.86),311(3.89), 320(3.83)	24-1077-84
$C_{18}H_{11}F$ Triphenylene, 1-fluoro-	C$_6$H$_{12}$	252(4.9),260(5.1), 276(4.27),287(4.2)	44-2803-84

Compound	Solvent	$\lambda_{max}(\log \epsilon)$	Ref.
$C_{18}H_{11}NO_3$			
2(1H)-Indolone, 2,3-dihydro-3-(dihydro-2-oxo-5-phenylfuran-3-ylidene)-	$CHCl_3$	270(4.29),445s(4.34), 461(4.34)	39-1331-84B
$C_{18}H_{11}NO_5$			
8H-Benzo[g]-1,3-benzodioxolo[6,5,4-de]-quinolin-8-one, 10-hydroxy-12-meth-oxy- (oxoisocalycinine)	EtOH	252(4.07),280(3.96), 320s(3.18)	100-0353-84
	EtOH-acid	264(4.32),294(4.26), 360(3.66),390(3.66)	100-0353-84
	EtOH-base	256(4.06),295(4.12), 328s(3.85),380s(3.30)	100-0353-84
2H-1-Benzopyran-2-one, 3-[3-(4-nitro-phenyl)-1-oxo-2-propenyl]-	EtOH	317(4.43),343(4.36)	118-0859-84
2,5-Furandione, dihydro-3-[(4-nitro-phenyl)methylene]-4-(phenylmethyl-ene)-, (E,E)-	toluene	382(3.81)	48-0233-84
Oxocompostelline	EtOH	208(4.67),254(4.41), 292s(--),397(3.61)	100-0753-84
	EtOH-HCl	208(4.67),261(4.34), 410(3.47),460(3.23)	100-0753-84
$C_{18}H_{11}N_3$			
5H-Benz[g]indolo[2,3-b]quinoxaline	EtOH	238(4.53),293(4.86), 395(4.34),461(3.47)	103-1155-84
$C_{18}H_{11}N_3O_2$			
1,1(8aH)-Azulenedicarbonitrile, 2-(4-nitrophenyl)- (photochromic)	MeCN	235(4.0),385(4.2)	89-0960-84
Propanedinitrile, [2-(2,4,6-cyclohepta-trien-1-ylidene)-1-(4-nitrophenyl)-ethylidene]-	MeCN	483(4.1)	89-0960-84
$C_{18}H_{11}N_3O_4$			
9,10-Anthracenedione, 2-[(5-nitro-1H-imidazol-1-yl)methyl]-	EtOH	254(4.55),320(3.78)	42-0611-84
$C_{18}H_{11}N_3S$			
Naphtho[1',2':4,5]thiazolo[3,2-b]-[1,2,4]triazole, 9-phenyl-	MeOH	234(4.67),305(4.34), 332(3.76)	4-1571-84
$C_{18}H_{11}N_5$			
Ethenetricarbonitrile, [4-[(phenylmeth-ylene)hydrazino]phenyl]-, (E)-	benzene	509(4.30)	32-0111-84
$C_{18}H_{11}N_5O_2$			
Pyrazolo[1,5-a]quinazoline-2-carbox-amide, 3-cyano-4,5-dihydro-5-oxo-N-phenyl-	EtOH	263(4.31)	4-1049-84
$C_{18}H_{11}N_5O_3S$			
Thiazolo[3,2-a]pyrimidin-4-ium, 3-hy-droxy-2-[(4-nitrophenyl)azo]-7-phenyl-, hydroxide, inner salt	MeCN	322(4.30),412(4.29), 580(4.55)	103-0791-84
	CCl_4	606(--)	103-0791-84
$C_{18}H_{12}BrNO_2$			
5H-[1]Benzopyrano[2,3-b]pyridin-5-ol, 7-bromo-5-phenyl-	96% H_2SO_4	381(4.54)	103-0607-84
$C_{18}H_{12}Br_2N_2$			
Cyclohepta[4,5]pyrrolo[1,2-a]imidazole, 3,10-dibromo-2-(4-methylphenyl)-	EtOH	246(4.37),267s(4.34), 287s(4.33),323(4.51),	150-3465-84

Compound	Solvent	λ_{max}(log ϵ)	Ref.
(cont.)		389(3.71),409(3.79), 434(3.76),505(3.04), 545(3.01),600s(2.90), 655s(2.65)	150-3465-84M
$C_{18}H_{12}ClF_6NO$ 2-Cyclohexen-1-one, 2-chloro-3-[(1-ethyl-2(1H)-quinolinylidene)methyl]-4,4,5,5,6,6-hexafluoro-	EtOH benzene	534(4.58) 547(--)	104-0390-84 104-0390-84
$C_{18}H_{12}ClNO_2$ 5H-[1]Benzopyrano[2,3-b]pyridin-5-ol, 7-chloro-5-phenyl-	96% H_2SO_4	381(4.54)	103-0607-84
$C_{18}H_{12}ClNO_2S$ 2H-Thieno[2,3-h]-1-benzopyran-2-one, 3-chloro-4-(methylphenylamino)-	EtOH	216(4.27),239(4.37), 260s(4.27),275s(4.04), 305s(3.87),318(3.92), 353(4.01)	161-0081-84
$C_{18}H_{12}Cl_2N_2$ Cyclohepta[4,5]pyrrolo[1,2-a]imidazole, 3,10-dichloro-2-(4-methylphenyl)-	EtOH	250s(4.15),268(4.17), 324(4.35),392(3.51), 408(3.62),434(3.60), 506(2.69),546(2.68), 580s(2.55),650s(2.21)	150-3465-84
$C_{18}H_{12}Cl_2N_2S_3$ 2H-2a,3-Dithia(2a-S^{IV})-1,2-diazacyclopenta[cd]pentalene, 2-(4-chlorophenyl)-4-[(4-chlorophenyl)thio]-5,6-dihydro-	MeCN	556(4.50)	97-0251-84
$C_{18}H_{12}FNO_2$ 5H-[1]Benzopyrano[2,3-b]pyridin-5-ol, 7-fluoro-5-phenyl-	96% H_2SO_4	372(4.51)	103-0607-84
$C_{18}H_{12}F_2NO_2P$ Phosphine, bis(4-fluorophenyl)(3-nitrophenyl)-	C_6H_{12} MeOH	255(4.17),323(3.99) 252(4.29),327(4.05)	139-0259-84C 139-0259-84C
$C_{18}H_{12}F_7NO$ 2-Cyclohexen-1-one, 3-[(1-ethyl-2(1H)-quinolinylidene)methyl]-2,4,4,5,5,6,6-heptafluoro-	EtOH benzene	535(4.49) 540(--)	104-0390-84 104-0390-84
$C_{18}H_{12}N_2$ Phenanthridine, 6-(4-pyridinyl)-	ether	294(3.77),333(3.32), 350(3.22)	4-1321-84
$C_{18}H_{12}N_2O$ Dibenz[b,f][1,4]oxazepine, 11-(4-pyridinyl)- 4-Pyridinamine, N-9H-xanthen-9-ylidene-	ether ether	240(3.40),260(3.30), 350(2.79) 240(4.50),340(3.80)	4-1321-84 4-1321-84
$C_{18}H_{12}N_2OS$ Thieno[2,3-d]pyrimidin-4(1H)-one, 1,5-diphenyl-	DMF	282(4.10)	48-0917-84
$C_{18}H_{12}N_2O_2$ 9,10-Anthracenedione, 2-(1H-imidazol-	EtOH	256(4.49),325(3.66)	42-0611-84

Compound	Solvent	$\lambda_{max}(\log \epsilon)$	Ref.
1-ylmethyl)- (cont.) 1H-Benz[f]isoquino[8,1,2-hi]quinazo- line-1,3(2H)-dione, 5,6-dihydro-	MeOH	230(4.34),248s(4.62), 255(4.73),264(4.54), 295(3.98),319(3.81), 334(3.88),351(3.88), 369(3.95)	42-0611-84 78-4003-84
3-Furancarbonitrile, 4-amino-5-benzoyl- 2-phenyl-	EtOH	283(4.39),318(3.97), 369(4.20)	5-1702-84
$C_{18}H_{12}N_2O_3$ 1H-Anthra[1',2':4,5]imidazo[2,1-c]- [1,4]oxazine-3,13-dione, 3,4-dihydro-	EtOH	370(3.92)	103-0571-84
$C_{18}H_{12}N_2S$ Dibenzo[b,f][1,4]thiazepine, 11-(4- pyridinyl)-	ether	245(4.65),360(3.90)	4-1321-84
4-Pyridinamine, N-9H-thioxanthen- 9-ylidene-	ether	245(4.55),265(4.46), 360(3.76)	4-1321-84
$C_{18}H_{12}N_2Se$ 4-Pyridinamine, N-9H-selenoxanthen- 9-ylidene-	ether	260(4.32),360(3.88)	4-1321-84
$C_{18}H_{12}N_3O_6P$ Phosphine, tris(4-nitrophenyl)-	MeOH	253(4.43),315(4.31)	139-0259-84C
$C_{18}H_{12}N_3O_7P$ Phosphine oxide, tris(4-nitrophenyl)-	MeOH	259(4.65)	139-0259-84C
$C_{18}H_{12}N_4$ Dipyrido[2,3-b:3',2'-j][1,10]phenan- throline, 6,7-dihydro-	EtOH	236(4.63),365(4.47)	44-2208-84
1H-Pyrazolo[3,4-b]quinoline-4-carbo- nitrile, 3-methyl-1-phenyl-	EtOH	428(3.36)	18-2984-84
$C_{18}H_{12}N_4O$ Pyridine, 4-(9-azido-9H-xanthen-9-yl)-	ether	254(4.10),295(3.36)	4-1321-84
$C_{18}H_{12}N_4OS_2$ Methanone, phenyl(4,5,6,7-tetrahydro- 1-phenyl-3,4-dithioxo-1H-pyrazolo- [3,4-d]pyrimidin-3-yl)-	EtOH	275(3.69),302(3.60)	4-1049-84
$C_{18}H_{12}N_4O_2$ 3H-1,2,4-Triazino[3,2-b]quinazoline- 3,10(4H)-dione, 2-(2-phenylethenyl)-	EtOH	296(3.84),332(3.86), 336(3.83)	24-1077-84
$C_{18}H_{12}N_4S$ Pyridine, 4-(9-azido-9H-thioxanthen- 9-yl)-	ether	265(4.33),295(3.86)	4-1321-84
$C_{18}H_{12}N_4Se$ Pyridine, 4-(9-azido-9H-selenoxanthen- 9-yl)-	ether	260(4.20),295(3.70)	4-1321-84
$C_{18}H_{12}N_6O$ Pyridazino[4',5':3,4]pyrazolo[1,5-a]- quinazolin-5(6H)-one, 7-amino-10- phenyl-	EtOH	233(5.05)	4-1049-84

Compound	Solvent	$\lambda_{max}(\log \epsilon)$	Ref.
$C_{18}H_{12}N_6O_4$			
1,3-Benzenediol, 2,4-bis(5-hydroxy-2H-benzotriazol-2-yl)-	MeCN	345(4.55)	126-2497-84
$C_{18}H_{12}N_6O_5$			
1,3,5-Benzenetriol, 2,4-bis(5-hydroxy-2H-benzotriazol-2-yl)-	MeCN	349(4.56)	126-2497-84
$C_{18}H_{12}O_2$			
11H-Benzo[a]fluoren-11-one, 7-methoxy-	MeCN	226(5.71),275(5.98)	2-0603-84
11H-Benzo[a]fluoren-11-one, 8-methoxy-	MeCN	229(5.51),275(5.88)	2-0603-84
11H-Benzo[a]fluoren-11-one, 9-methoxy-	MeCN	224(5.5),274(5.91)	2-0603-84
1-Dibenzofuranol, 7-phenyl-	H₂O	213(4.56),290(4.47),316(4.24)	39-1213-84C
	pH 13	223(4.62),247(4.36),292(4.35),340(4.26)	39-1213-84C
5,12-Epidioxynaphthacene, 5,12-dihydro-	CH₂Cl₂	275f(4.0),290s(3.7),315f(2.5)	151-0017-84D
$C_{18}H_{12}O_3$			
2H-1-Benzopyran-2-one, 3-(1-oxo-3-phenyl-2-propenyl)-	EtOH	313(4.05),340(4.03)	118-0859-84
2,5-Furandione, dihydro-3,4-bis(phenylmethylene)-, (E,E)-	toluene	355(3.97)	48-0233-84
(E,Z)-	toluene	378(4.07)	48-0233-84
(Z,Z)-	toluene	382(4.26)	48-0233-84
$C_{18}H_{12}O_5$			
Benzeneacetic acid, α-(3-hydroxy-5-oxo-4-phenyl-2(5H)-furanylidene)-, (E)-	EtOH	212(3.99),254(4.18),366(3.96)	39-1547-84C
2,5-Cyclohexadiene-1,4-dione, 2,5-dihydroxy-3-(4-hydroxyphenyl)-6-phenyl-(ascocorynin)	dioxan	262(4.59),353(3.71),470(2.69)	158-84-41
$C_{18}H_{12}O_8$			
Benzeneacetic acid, α-[4-(3,4-dihydroxyphenyl)-3-hydroxy-5-oxo-2(5H)-furanylidene]-4-hydroxy-, (E)-	EtOH	257(4.42),395(4.10)	39-1547-84C
	EtOH-NaOH	243(4.21),383(4.40)	39-1547-84C
$C_{18}H_{12}O_9$			
2-Furancarboxylic acid, 5-(hydroxymethyl)-, cyclic trimer	CHCl₃	254.5(4.59)	126-2347-84
$C_{18}H_{12}O_9S_2$			
Pyrylium, 4-carboxy-2,6-bis(4-sulfophenyl)-, hydroxide, inner salt	pH 4.8	250(4.44)	39-0867-84B
	pH 13.0	480(4.74)	39-0867-84B
	17.25M H₂SO₄	215(4.04),430(4.37)	39-0867-84B
$C_{18}H_{13}BrN_2$			
Cyclohepta[4,5]pyrrolo[1,2-a]imidazole, 3-bromo-2-(4-methylphenyl)-	EtOH	267(4.25),321(4.52),383(3.62),401(3.68),426(3.64),504(2.86),545(2.84),590s(2.70),640s(2.40)	150-3465-84M
Cyclohepta[4,5]pyrrolo[1,2-a]imidazole, 10-bromo-2-(4-methylphenyl)-	EtOH	266(4.40),328(4.68),397(3.72),416(3.81),442(3.75),512(2.83),546(2.80),595s(2.65),635s(2.44)	150-3465-84

Compound	Solvent	$\lambda_{max}(\log \epsilon)$	Ref.
$C_{18}H_{13}BrN_2O_4$			
6H-Anthra[1,9-cd]isoxazol-6-one, 5-[(2-acetoxyethyl)amino]-3-bromo-	THF	255(4.42),262(4.54), 496(4.19),532(4.29)	103-0717-84
$C_{18}H_{13}ClN_2$			
Cyclohepta[4,5]pyrrolo[1,2-a]imidazole, 3-chloro-2-(4-methylphenyl)-	EtOH	252s(4.23),266(4.24), 322(4.47),383(3.60), 402(3.66),427(3.63), 500(2.77),561(2.77), 589s(2.64),640s(3.35)	150-3465-84M
Cyclohepta[4,5]pyrrolo[1,2-a]imidazole, 10-chloro-2-(4-methylphenyl)-	EtOH	268(4.28),327(4.53), 395s(3.60),413(3.82), 439(3.58),490s(2.76), 537s(2.70),599(2.53), 653(2.21)	150-3465-84M
$C_{18}H_{13}ClN_2O_2$			
Methanone, (4-chlorophenyl)(5-hydroxy-6-methyl-2-phenyl-4-pyrimidinyl)-	MeOH	277(4.30),420(3.63)	32-0431-84
Methanone, [2-(4-chlorophenyl)-5-hydroxy-6-methyl-4-pyrimidinyl]phenyl-	MeOH	282(4.38),415(3.70)	32-0431-84
$C_{18}H_{13}ClN_2O_4$			
6H-Anthra[1,9-cd]isoxazol-6-one, 5-[(2-acetoxyethyl)amino]-3-chloro-	THF	255(4.43),261(4.53), 498(4.19),533(4.23)	103-0717-84
$C_{18}H_{13}ClN_4S$			
1H-Pyridazino[4,5-e][1,3,4]thiadiazine, 5-chloro-3-phenyl-1-(phenylmethyl)-	EtOH	271(4.04),319(3.29)	94-4437-84
$C_{18}H_{13}ClO_4$			
5,12-Naphthacenedione, 6-chloro-1,2,3,4-tetrahydro-1,11-dihydroxy-	EtOH	278(4.48),345(3.54), 441(4.18),460(4.22)	150-0147-84M
$C_{18}H_{13}FNO_2P$			
Phosphine, (4-fluorophenyl)(4-nitrophenyl)phenyl-	MeOH	253(4.23),328(3.96)	139-0259-84C
$C_{18}H_{13}NO$			
5H-Benzo[6,7]cyclohept[1,2-b]indolizin-5-one, 11-methyl-	EtOH	273(4.63),286(4.39), 300(4.24),314(4.19), 337(3.95),354(3.96), 457(4.21)	142-0791-84
$C_{18}H_{13}NOS$			
9H-Thioxanthen-9-ol, 9-(4-pyridinyl)-	ether	265(4.04),293(3.40)	4-1321-84
$C_{18}H_{13}NOSe$			
9H-Selenoxanthen-9-ol, 9-(4-pyridinyl)-	ether	255(3.81),270(3.75)	4-1321-84
$C_{18}H_{13}NO_2$			
5H-[1]Benzopyrano[2,3-b]pyridin-5-ol, 5-phenyl-	96% H_2SO_4	367(4.53)	103-0607-84
Oxazole, 4,5-diphenyl-2-(2-propynyloxy)-	C_6H_{12}	298(4.08)	44-0399-84
2(3H)-Oxazolone, 4,5-diphenyl-3-(1,2-propadienyl)-	C_6H_{12}	223(4.21),299(4.19)	44-0399-84
9H-Xanthen-9-ol, 9-(4-pyridinyl)-	ether	226(4.06),281(3.32), 290(3.44)	4-1321-84

Compound	Solvent	$\lambda_{max}(\log \epsilon)$	Ref.
$C_{18}H_{13}NO_3$			
6H-Anthra[9,1-bc]pyrrole-2-carboxylic acid, 1-methyl-6-oxo-, methyl ester	EtOH	495(4.08)	104-1415-84
$C_{18}H_{13}NO_3S_2$			
[1]Benzothieno[3,2-b]pyridin-2(1H)-one, 4-(methylthio)-3-phenyl-, 5,5-dioxide	EtOH	219(4.52),245(4.29), 340(3.99)	95-0134-84
$C_{18}H_{13}NO_4$			
6H-Anthra[9,1-bc]furan-2-carboxylic acid, 5-(methylamino)-6-oxo-, methyl ester	EtOH	514(4.26),552(4.38)	104-0745-84
6H-Anthra[1,9-cd]isoxazole-3-carboxylic acid, 6-oxo-, propyl ester	EtOH	460(4.04)	104-2012-84
$C_{18}H_{13}NO_5$			
Acetamide, N-(1-acetoxy-9,10-dihydro-9,10-dioxo-2-anthracenyl)-	EtOH at 77°K	337(3.75),373s(3.25)	104-1780-84
Acetamide, N-(4-acetoxy-9,10-dihydro-9,10-dioxo-1-anthracenyl)-	EtOH at 77°K	410(3.79)	104-1780-84
Acetamide, N-(9-acetoxy-1,10-dihydro-1,10-dioxo-2-anthracenyl)-	EtOH at 77°K	546(0.36)(relative absorbance)	104-1780-84
12H-[1]Benzoxepino[2,3,4-ij]isoquinolin-12-one, 6-hydroxy-8,9-dimethoxy-(oxosarcophylline)	EtOH	218(4.48),252(4.27), 330(3.54),396(3.61)	88-5933-84
	EtOH-HCl	218(4.48),260(4.24), 395(3.53),470(3.25)	88-5933-84
Oxosarcocapnidine	EtOH	252(4.26),432(3.34), 396(3.59)	100-0753-84
	EtOH-acid	217(4.28),265(4.05), 458(3.55)	100-0753-84
	EtOH-base	243(4.26),340(3.34), 400(3.57)	100-0753-84
$C_{18}H_{13}NO_7$			
Phenanthro[3,4-d]-1,3-dioxole-5-carboxylic acid, 8-methoxy-6-nitro-, methyl ester (methylaristocholate)	EtOH	222(4.56),255(4.30), 315(4.01)	100-0331-84
$C_{18}H_{13}NS$			
10H-Phenothiazine, 10-phenyl-	hexane	237s(--),256(5.05), 314(4.90)	4-0661-84
	96% H_2SO_4	239(4.68),274(5.23), 314(3.80),514(4.45)	4-0661-84
	38% HCl	238(3.85),272(3.78), 314(--),515(--)	4-0661-84
$C_{18}H_{13}N_2O_4P$			
Phosphine, bis(4-nitrophenyl)phenyl-	C_6H_{12}	251(4.57),317(4.33)	139-0259-84C
	MeOH	253(4.34),322(4.18)	139-0259-84C
$C_{18}H_{13}N_2O_4PS$			
Phosphine sulfide, bis(4-nitrophenyl)-phenyl-	MeOH	255(4.49)	139-0259-84C
$C_{18}H_{13}N_2O_5P$			
Phosphine oxide, bis(4-nitrophenyl)-phenyl-	MeOH	258(4.48)	139-0259-84C
$C_{18}H_{13}N_3O$			
Benzo[g]quinoxalin-2(1H)-one, 3-(2-aminophenyl)-	EtOH	241(4.66),298(4.29), 360(4.15),425(3.90)	103-1155-84

Compound	Solvent	$\lambda_{max}(\log \epsilon)$	Ref.
3H-Pyrazole-5-carbonitrile, 3-acetyl-1,4-diphenyl-	EtOH	236(3.73)	4-1049-84
$C_{18}H_{13}N_3O_2$			
Cyclohepta[4,5]pyrrolo[1,2-a]imidazole, 2-(4-methylphenyl)-3-nitro-	EtOH	228s(4.42),242(4.45), 310(4.36),323s(4.35), 428(3.69),534s(3.51), 553(3.53),590s(3.40)	150-3465-84M
1H-Pyrazolo[3,4-b]quinoline-4-carboxylic acid, 3-methyl-1-phenyl-	EtOH	386(3.81)	18-2984-84
$C_{18}H_{13}N_3O_3$			
1H-Isoindole-1,3(2H)-dione, 2-[3-(1H-indazol-3-yl)-3-oxopropyl]-	MeOH	207.5(4.40),234s(4.47), 241s(3.99),292(3.85)	103-0066-84
$C_{18}H_{13}N_5OS_2$			
1H-Pyrazolo[3,4-d]pyrimidine-3-carboxamide, 4,5,6,7-tetrahydro-N,1-diphenyl-4,6-dithioxo-	EtOH	274(3.65),303(3.60)	4-1049-84
$C_{18}H_{13}N_5O_2$			
3H-[1,2,4]Triazino[3,2-b]quinazoline-3,10(4H)-dione, 2-methyl-4-[(phenylmethylene)amino]-	EtOH	350(3.36)	24-1083-84
$C_{18}H_{13}N_7$			
1H-Pyrazolo[3,4-e][1,2,4]triazolo[3,4-c][1,2,4]triazine, 3-methyl-1,8-diphenyl-	dioxan	285(4.74),516(3.34)	4-1565-84
$C_{18}H_{14}$			
1,3,5-Cycloheptatriene, 7,7'-(2-butyne-1,4-diylidene)bis-	C_6H_{12}	310(3.99),415(4.59), 434s(4.55),450s(4.50)	24-2045-84
	MeCN	310(--),418(--), 432s(--),445s(--)	24-2045-84
Cycloocta[def]biphenylene, 7,8-dimethyl-	C_6H_{12}	226(4.49),233s(4.45), 257(4.36),271s(4.11), 282s(4.02),297(3.98), 310(3.70),324s(3.05), 336(2.89),343s(2.82), 351(2.73),362(2.74), 382(2.64),408(2.36), 418(2.41),447(2.50), 475(2.43),511(2.00), 553(1.59),617(0.77)	35-7195-84
5,8-Ethenocyclohept[fg]acenaphthylene, 1,2,5,8-tetrahydro-	MeCN	229(4.57),299(3.89), 311(4.03),325(3.89)	44-1146-84
$C_{18}H_{14}ClNOS$			
2H-Pyrrol-2-one, 4-chloro-1,5-dihydro-1,3-bis(4-methylphenyl)-5-thioxo-	CHCl$_3$	355(4.32)	78-3499-84
$C_{18}H_{14}ClNO_2S$			
2H-Thieno[2,3-h]-1-benzopyran-2-one, 3-chloro-5,6-dihydro-4-(methylphenylamino)-	EtOH	233s(4.28),240(4.28), 280s(3.67),375(4.32)	161-0081-84
$C_{18}H_{14}ClN_3O_2$			
Pyrano[2,3-e]indazol-2(7H)-one, 3-chloro-4-(dimethylamino)-7-phenyl-	EtOH	215s(4.37),242(4.33), 260(4.36),304s(4.10), 333(4.30)	4-0361-84

Compound	Solvent	λ_{max} (log ϵ)	Ref.
$C_{18}H_{14}Cl_2N_2O_4$			
5,10-Ethanopyrido[3,2-g]quinoline-2,8-dicarboxylic acid, 11,12-dichloro-5,10-dihydro-, dimethyl ester, trans	MeCN	277(4.090),281s(4.076)	5-0877-84
$C_{18}H_{14}Cl_2N_4O$			
7H-Pyrido[2,1-b]quinazoline-6,11-dione, 2-chloro-8,9-dihydro-, 6-[(4-chlorophenyl)hydrazone]-, (E)-	EtOH	230(4.35),250(4.18), 303(4.00),390(4.20)	4-1301-84
$C_{18}H_{14}Cl_2O$			
Ethanone, 1-(11,12-dichloro-9,10-dihydro-9,10-ethenoanthracen-2-yl)-, trans-anti-syn	MeCN	254(4.08),292(3.04)	44-2368-84
trans-syn-anti-	MeCN	254(4.08),292(3.0)	44-2368-84
$C_{18}H_{14}FeO_4S$			
Ferrocene, 1,1'-[2,5-thiophenediylbis(methyleneoxycarbonyl)]-	dioxan	450(2.31)	18-2435-84
$C_{18}H_{14}FeO_5$			
Ferrocene, 1,1'-[2,5-furandiylbis(methyleneoxycarbonyl)]-	dioxan	450(2.52)	18-2435-84
$C_{18}H_{14}NO_2P$			
Phosphine, (4-nitrophenyl)diphenyl-	C_6H_{12}	257(4.13),325(3.72)	139-0259-84C
	MeOH	255(4.20),332(3.92)	139-0259-84C
$C_{18}H_{14}NO_2PS$			
Phosphine sulfide, (4-nitrophenyl)diphenyl-	MeOH	255(4.19)	139-0259-84C
$C_{18}H_{14}NO_3P$			
Phosphine oxide, (4-nitrophenyl)diphenyl-	MeOH	258(4.14),265s(--), 272s(--)	139-0259-84C
$C_{18}H_{14}N_2$			
Benzenamine, N-(phenyl-2-pyridinylmethylene)-	ether	260(4.64),340(4.00)	4-1313-84
Benzenamine, N-(phenyl-3-pyridinylmethylene)-	ether	254(4.46),342(3.64)	4-1313-84
Benzenamine, N-(phenyl-4-pyridinylmethylene)-	ether	254(4.42),347(3.56)	4-1313-84
2-Pyridinamine, N-(diphenylmethylene)-	ether	254(4.60),330(3.67)	4-1313-84
3-Pyridinamine, N-(diphenylmethylene)-	ether	250(4.82),325(3.96)	4-1313-84
4-Pyridinamine, N-(diphenylmethylene)-	ether	254(4.54),325(3.54)	4-1313-84
$C_{18}H_{14}N_2O$			
3(2H)-Pyridazinone, 4-phenyl-6-(2-phenylethenyl)-	EtOH	222(4.12),228(4.35), 308(4.34)	48-0799-84
$C_{18}H_{14}N_2OS$			
7H-Thiazolo[3,2-a]pyrimidin-7-one, 5,6-dihydro-5,6-diphenyl-, cis	EtOH	307(4.21)	142-2529-84
trans	EtOH	305(4.21)	142-2529-84
$C_{18}H_{14}N_2O_2$			
Benzenamine, 4-nitro-N,N-diphenyl-	C_6H_{12}	288(3.92),380(4.18)	139-0259-84C
	MeOH	256(3.83),280(3.79), 392(4.14)	139-0259-84C
4H-[1]Benzoxepino[3,4-c]pyrazol-4-one,	EtOH	282(4.18)	4-0301-84

Compound	Solvent	$\lambda_{max}(\log \epsilon)$	Ref.
1,10-dihydro-3-methyl-1-phenyl- (cont.)			4-0301-84
4H-[1]Benzoxepino[3,4-c]pyrazol-4-one, 1,10-dihydro-1-(phenylmethyl)-	EtOH	273(4.10)	4-0301-84
4H-[1]Benzoxepino[3,4-c]pyrazol-4-one, 2,10-dihydro-2-(phenylmethyl)-	EtOH	273(4.19)	4-0301-84
Methanone, (5-hydroxy-6-methyl-2-phenyl-4-pyrimidinyl)phenyl-	MeOH	275(4.33),415(3.67)	32-0431-84
2H-Pyrimido[6,1-a]isoquinoline-2,4(3H)-dione, 6,7-dihydro-1-phenyl-	MeOH	238(4.29),273(3.57), 303s(4.08),320(4.16)	78-4003-84
$C_{18}H_{14}N_2O_3$			
9,10-Anthracenedione, 1-amino-2-(2,3-dihydro-4H-1,4-oxazin-4-yl)-	EtOH	513(3.99)	103-0571-84
2(3H)-Furanone, 4-amino-5-benzoyl-3-[(phenylamino)methylene]-	EtOH	230(4.15),247(4.07), 341(4.15),437(4.15)	5-1702-84
2-Propenoic acid, 2-cyano-3-(phenylamino)-, 2-oxo-2-phenylethyl ester	EtOH	231(4.15),240(4.15), 277s(3.82),288(4.02), 321(4.44)	5-1702-84
$C_{18}H_{14}N_2O_4$			
6H-Anthra[1,9-cd]isoxazol-6-one, 5-[(2-acetoxyethyl)amino]-	THF	253(4.48),260(4.55), 490(4.27),522(4.35)	103-0717-84
$C_{18}H_{14}N_2O_4S$			
Pyrrolo[2,3-c]phenothiazine, 2,3-diacetyl-3,6-dihydro-, 11,11-dioxide	EtOH	212(4.47),246(4.09), 286(4.63),318(3.95), 347(3.97)	103-1095-84
$C_{18}H_{14}N_2O_6S$			
Imidodicarbonic acid, (3-cyano-4-oxo-4H-[1]benzothieno[3,2-b]pyran-2-yl)-, diethyl ester	MeOH	228(4.23),245(4.26), 295(4.13),330s(3.67)	83-0531-84
$C_{18}H_{14}N_2S_3$			
2H-2a,3-Dithia(2a-S^{IV})-1,2-diazacyclopenta[cd]pentalene, 5,6-dihydro-2-phenyl-4-(phenylthio)-	MeOH	551(4.12)	97-0251-84
$C_{18}H_{14}N_4$			
Pyridine, 2-(azidodiphenylmethyl)-	ether	260(3.41)	4-1313-84
Pyridine, 3-(azidodiphenylmethyl)-	ether	260(3.55)	4-1313-84
Pyridine, 4-(azidodiphenylmethyl)-	ether	260(3.41)	4-1313-84
1H-Tetrazolium, 5-(1,3-cyclopentadien-1-yl)-1,3-diphenyl-, hydroxide, inner salt	MeCN	258(4.26),335(4.49), 475(3.08)	39-2545-84C
$C_{18}H_{14}N_4O$			
1H-Pyrazole-4-carbonitrile, 5-amino-3-(4-methylbenzoyl)-1-phenyl-	EtOH	220(4.04),254(3.84)	4-1049-84
1H-Pyrazole-5-carbonitrile, 4-amino-3-(4-methylbenzoyl)-1-phenyl-	EtOH	220(3.76),266(3.57)	4-1049-84
4H-Pyrazolo[3,4-d]pyridazin-4-one, 1,5-dihydro-3-phenyl-5-(phenylmethyl)-	EtOH	261(4.23),281(4.14)	94-4437-84
1H-Pyrazolo[3,4-b]quinoline-4-carboxamide, 3-methyl-1-phenyl-	EtOH	394(3.54)	18-2984-84
$C_{18}H_{14}N_4OS$			
1H-Pyridazino[4,5-e][1,3,4]thiadiazin-5(6H)-one, 3-phenyl-6-(phenylmethyl)-	EtOH	270(4.20),296(4.06)	94-4437-84

Compound	Solvent	$\lambda_{max}(\log \epsilon)$	Ref.
$C_{18}H_{14}N_4O_5S_2$			
Acetamide, N-[5-(4-nitrophenyl)-2-[[(4-nitrophenyl)methyl]thio]-4-thiazolyl]-	dioxan	265(2.97),355(2.85)	73-2285-84
$C_{18}H_{14}O_2$			
Benz[a]anthracene-1,2-diol, dihydro-, trans	75% MeOH	263(4.49),297(4.36), 364(3.32),383(3.52), 406(3.39)	44-3621-84
Benz[a]anthracene-8,9-diol, dihydro-, trans	75% MeOH	265(4.51),306(4.05), 320(4.10)	44-3621-84
Benz[a]anthracene-10,11-diol, 10,11-di-hydro-, trans	75% MeOH	275(4.57),310(3.90)	44-3621-84
$C_{18}H_{14}O_3$			
9,10-Anthracenedione, 1-[(2-methyl-2-propenyl)oxy]-	EtOH	252(3.95),275s(3.67), 378(3.09)	12-1518-84
5,12-Naphthacenedione, 1,2,3,4-tetra-hydro-6-hydroxy-	EtOH	239(4.59),278(4.12), 456(3.80)	150-0147-84M
1-Naphthalenecarboxylic acid, 2-(2-methoxyphenyl)-	MeCN	223(5.34),274(5.43)	2-0603-84
1-Naphthalenecarboxylic acid, 2-(3-methoxyphenyl)-	MeCN	223(5.72),275(5.61)	2-0603-84
1-Naphthalenecarboxylic acid, 2-(4-methoxyphenyl)-	MeCN	223(5.71),276(5.44)	2-0603-84
$C_{18}H_{14}O_4$			
6H-Benzofuro[3,2-c]furo[3,2-g][1]benzo-pyran, 6a,11a-dihydro-9-methoxy-, (6aR-cis)-	MeOH	248(4.34)	4-0845-84
Cyclohepta[de]naphthalene-7,8-diol, diacetate	MeOH	229.5(4.28),247.5(4.09), 336(3.74)	39-1465-84C
$C_{18}H_{14}O_4S_2$			
Benzo[b]thiophen-3(2H)-one, 2-[1-(meth-ylthio)-3-oxo-3-phenylpropylidene]-, 1,1-dioxide	EtOH	247(4.30),361(4.18), 428(4.00)	95-0134-84
$C_{18}H_{14}O_6$			
5H-Benz[e]anthracene-1,4,5,9(13bH)-tetrone, 6,7,8,8a-tetrahydro-10,13-dihydroxy-, (4aS*,8aα,13bα)-(±)-	MeCN	257s(3.74),371(3.69)	18-0791-84
2,5-Cyclohexadiene-1,4-dione, 2,2'-(1,6-dioxo-1,6-hexanediyl)bis-	MeCN	246(4.52),300s(2.98), 445(1.83)	18-0791-84
9H-Fluoren-9-one, 1,5-diacetoxy-7-methoxy-	EtOH	258(4.46),267(4.47), 426(2.96)	42-1010-84
2H-Pyrano[2',3':6,7]xanthone, 1,3,5-trihydroxy-6',6'-dimethyl-	MeOH	273(4.46),330(3.95), 366s(3.72)	100-0620-84
$C_{18}H_{14}O_7$			
9H-Fluoren-9-one, 2,5-diacetoxy-1-hydroxy-1-methoxy- (dendro-florin diacetate)	EtOH	244(4.16),257(4.15), 272(4.09),447(3.20)	42-1010-84
$C_{18}H_{14}O_8$			
8H-1,3-Dioxolo[4,5-g][1]benzopyran-8-one, 9-hydroxy-6-(4-hydroxy-3-methoxyphenyl)-7-methoxy-	MeOH	240(4.22),255(4.19), 276(4.19),347(4.44)	102-2043-84
	MeOH-NaOMe	249(4.20),269(4.21), 398(4.43)	102-2043-84
	MeOH-NaOAc	244(4.17),258(4.16), 274(4.17),350(4.30), 406(4.07)	102-2043-84

Compound	Solvent	$\lambda_{max}(\log \epsilon)$	Ref.
(cont.)	MeOH-AlCl$_3$	244(4.18),266(4.17), 292(4.21),381(4.48)	102-2043-84
$C_{18}H_{15}BrN_2O$ 3H-Pyrazol-3-one, 2-(4-bromophenyl)-2,4-dihydro-4-(1-methylethylidene)-5-phenyl-	EtOH	267(4.23)	94-2146-84
$C_{18}H_{15}BrN_2O_2$ 1,4-Naphthalenedione, 2-bromo-3-[[2-(phenylamino)ethyl]amino]-	CH$_2$Cl$_2$	221(4.43),243(4.44), 276(4.49),330(3.48), 468(3.64)	83-0743-84
$C_{18}H_{15}BrN_4O$ 7H-Pyrido[2,1-b]quinazoline-6,11-dione, 6-[(4-bromophenyl)hydrazone]	EtOH	230(4.38),256(4.30), 300(4.07),386(4.41)	4-1301-84
$C_{18}H_{15}BrO_4$ 9,10-Anthracenedione, 2-(bromomethyl)-1-ethoxy-8-methoxy-	n.s.g.	224(4.50),257(4.41), 378(3.80)	5-0306-84
$C_{18}H_{15}ClN_2O$ 3H-Pyrazol-3-one, 4-(1-methylethylidene)-5-phenyl-	EtOH	265(4.28)	94-2146-84
$C_{18}H_{15}ClN_4O$ 7H-Pyrido[2,1-b]quinazoline-6,11-dione, 2-chloro-8,9-dihydro-, 6-(phenylhydrazone)	EtOH	230(4.36),248(4.23), 298(3.98),391(4.39)	4-1301-84
7H-Pyrido[2,1-b]quinazoline-6,11-dione, 6-[(3-chlorophenyl)hydrazone]	EtOH	230(4.37),252(4.25), 300(3.94),380(4.35)	4-1301-84
7H-Pyrido[2,1-b]quinazoline-6,11-dione, 6-[(4-chlorophenyl)hydrazone]	EtOH	230(4.34),256(4.26), 300(3.96),387(4.36)	4-1301-84
$C_{18}H_{15}ClO_6$ 4H-1-Benzopyran-4-one, 2-(2-chlorophenyl)-5-hydroxy-3,7,8-trimethoxy-	MeOH	203(3.72),260(3.53), 340(2.83)	2-1002-84
$C_{18}H_{15}DO$ Pyrazine, (2-d-1,2-diphenylethoxy)-, erythro-(±)-	EtOH	281(3.93),296(3.84)	39-0641-84B
threo-(±)-	EtOH	281(3.93),296(3.84)	39-0641-84B
$C_{18}H_{15}FN_4O$ 7H-Pyrido[2,1-b]quinazoline-6,11-dione, 6-[(4-fluorophenyl)hydrazone]	EtOH	230(4.26),246(4.07), 298(3.85),388(4.17)	4-1301-84
$C_{18}H_{15}N$ Benzenamine, N,N-diphenyl-	C$_6$H$_{12}$	301(4.42)	139-0259-84C
	MeOH	295(4.40)	139-0259-84C
$C_{18}H_{15}NO$ 4(1H)-Pyridinone, 1-(diphenylmethyl)-	EtOH	270(4.37)	118-0485-84
$C_{18}H_{15}NOS$ Benzo[b]thiophen-3(2H)-one, 2-[2-methyl-3-(phenylamino)-2-propenylidene]-	DMSO	354(4.18),503(4.72)	103-0951-84
2H-Pyrrol-2-one, 1,5-dihydro-1,3-bis(4-methylphenyl)-5-thioxo-	CHCl$_3$	355(4.37)	78-3499-84

Compound	Solvent	$\lambda_{max}(\log \epsilon)$	Ref.
$C_{18}H_{15}NO_2$			
Benz[c]acridine-1,2-diol, 1,2-dihydro-7-methyl-, trans	MeOH	255(4.62),280(4.68), 288(4.63),344s(3.60), 363(3.91),381(4.05)	44-4446-84
Benz[c]acridine-3,4-diol, 3,4-dihydro-7-methyl-, trans	MeOH	261(5.01),348(3.75), 366(4.00),393(3.83)	44-4446-84
Benz[c]acridine-8,9-diol, 8,9-dihydro-7-methyl-, trans	MeOH	250(4.67),277(4.41), 313(4.15),327s(4.02), 343(3.81),360(3.82)	44-4446-84
Benz[c]acridine-10,11-diol, 10,11-di-hydro-7-methyl-, trans	MeOH	226(4.27),268(4.69), 277(3.70),300(4.30), 312(4.06),332(3.30), 347(3.52),366(3.54)	44-4446-84
Benzeneacetonitrile, α-[(2-oxopropoxy)-phenylmethylene]-	EtOH	282(4.15)	5-1702-84
Benzoic acid, 2-(3-isoquinolinyl)-, ethyl ester	MeOH	220(4.61),282(3.90), 325(3.52)	83-0002-84
Ethanone, 1-(3-amino-4,5-diphenyl-2-furanyl)-	EtOH	267(4.07),327(4.10)	5-1702-84
4H-Furo[3,2-c][1]benzopyran-2-carbo-nitrile, 2,3,3a,9b-tetrahydro-2-phenyl- (isomer mixture)	EtOH	217s(4.04),275(3.36), 283(3.30)	24-2157-84
Furo[2,3-b]quinoline, 2,3-dihydro-4-(phenylmethoxy)-	EtOH	272f(3.9),280(3.8), 310(3.5),320(3.6)	103-0304-84
Oxazole, 4,5-diphenyl-2-(2-propenyl-oxy)-	C_6H_{12}	302(4.34)	44-0399-84
2(3H)-Oxazolone, 4,5-diphenyl-3-(2-propenyl)-	C_6H_{12}	278(4.16)	44-0399-84
2(5H)-Oxazolone, 4,5-diphenyl-5-(2-propenyl)-	C_6H_{12}	215(4.28),266(4.24)	44-0399-84
Oxiranecarbonitrile, 2-phenyl-3-[2-(2-propenyloxy)phenyl]-, trans	EtOH	234s(4.09),274(3.59), 282(3.62)	24-2157-84
$C_{18}H_{15}NO_2S$			
Benzo[b]thiophen-3(2H)-one, 2-[3-[(4-methoxyphenyl)amino]-2-propenyli-dene]-	toluene DMSO	369(3.95),498(4.47) 375(3.97),524(4.62)	103-0951-84 103-0951-84
$C_{18}H_{15}NO_3$			
Butanamide, N-(9,10-dihydro-9,10-dioxo-1-anthracenyl)-	benzene	419(3.79)	39-0529-84C
2,4-Pentadienoic acid, 2-(benzoyl-amino)-5-phenyl-	EtOH	222(3.92),234(3.91), 327(4.30)	118-0418-84
geometric isomer	EtOH	222(3.92),234(4.03), 327(4.38)	118-0418-84
$C_{18}H_{15}NO_4$			
9,10-Anthracenedione, 1-acetoxy-2-(dimethylamino)-	EtOH at 77°K	463(3.70)	104-1780-84
9,10-Anthracenedione, 1,4-dihydroxy-2-pyrrolidino-	n.s.g.	223(4.28),235(4.21), 281(4.18),513(3.94), 545(4.20),577(4.06)	23-1922-84
Erythrinan-8-one, 1,2,6,7,10,11-hexa-dehydro-3-methoxy-15,16-[methylene-bis(oxy)]-, trans?	EtOH	225(4.37),255(4.24), 368(3.28)	142-2255-84
isomer	EtOH	228(4.28),264(4.29), 357(3.35)	142-2255-84
1H-Indolo[7a,1-a]isoquinoline-2,6-di-one, 11,12-dimethoxy-, (±)-	EtOH	246(4.13),277(4.15)	142-2255-84

Compound	Solvent	$\lambda_{max}(\log \epsilon)$	Ref.
$C_{18}H_{15}NO_4S_2$			
Benzeneacetamide, α-[(methylthio)(3-oxo-benzo[b]thien-2(3H)-ylidene)methyl]-, S,S-dioxide	EtOH	233(4.51)	95-0134-84
$C_{18}H_{15}NO_5$			
1H-Pyrano[3,4-c]pyridine-5-carboxylic acid, 7,8-dihydro-6-methyl-1,8-di-oxo-3-phenyl-, ethyl ester	EtOH	231(4.10),271(4.39), 350(4.40)	94-3384-84
$C_{18}H_{15}N_3$			
Benzo[a]phenazin-5-amine, N,N-dimethyl-	MeOH	228(3.64),275(3.56), 300(3.38),424(3.31)	94-3093-84
$C_{18}H_{15}N_3O_2$			
Compd. m. 140-142°	MeOH	285(4.46)	2-0060-84
$C_{18}H_{15}N_3O_3$			
Acetic acid, cyano[(2-methylphenyl)hy-drazono]-, 2-oxo-2-phenylethyl ester	EtOH	248(4.30),284(3.47), 290s(3.44),372(4.31)	5-1702-84
2,3-Furandione, 4-amino-5-benzoyl-, 3-[(2-methylphenyl)hydrazone]	EtOH	230(4.08),260s(4.01), 266(4.02),324(3.90), 405(4.20),493(4.17)	5-1702-84
2H-Indol-2-one, 1,3-dihydro-1-methyl-3-[3-[(4-nitrophenyl)amino]-2-prop-enylidene]-	DMSO	450(4.58)	103-0951-84
Pyrrolo[2,1-a]isoquinolin-2-ol, 2,3,5,6-tetrahydro-3-imino-2-(4-nitrophenyl)-	EtOH	247(4.42),270s(4.12)	94-1170-84
$C_{18}H_{15}N_3O_7$			
1H-Pyrazole-3,4-dicarboxylic acid, 5-[3-(1,3-dihydro-1,3-dioxo-2H-isoindol-2-yl)-1-oxopropyl]-, dimethyl ester	MeOH	210s(4.65),222(4.80), 241(4.28)	103-0066-84
$C_{18}H_{15}N_4O_4$			
1H-Pyrazolium, 2-ethyl-4,5-dihydro-3-[[(4-nitrophenyl)imino]methyl]-4,5-dioxo-1-phenyl-, iodide	EtOH	350(3.94)	48-0965-84
$C_{18}H_{15}N_5$			
1H-Purin-6-amine, 8-methyl-N,2-di-phenyl-	EtOH	266(4.38),279s(--), 300(4.42)	11-0208-84B
1H-Purin-6-amine, 8-methyl-1,2-di-phenyl-	EtOH	238(4.42),283(4.08)	11-0208-84B
1H-Pyrazole-5-carbonitrile, 3-(1-hy-drazonoethyl)-1,4-diphenyl-	EtOH	232(3.77)	4-1049-84
$C_{18}H_{15}N_5O_2$			
Oxepino[3,2-e][1,2,4]triazolo[1,5-a]-pyrimidine-7-carbonitrile, 4,5,8,9-tetrahydro-8,9-dimethyl-5-oxo-4-phenyl-	CHCl₃	260(3.93),307(4.19), 315(4.17),335s(3.90)	24-0585-84
$C_{18}H_{15}N_5O_3$			
7H-Pyrido[2,1-b]quinazoline-6,11-dione, 8,9-dihydro-, 6-[(4-nitrophenyl)hy-drazone]	EtOH	227(4.30),314(3.92), 408(4.48)	4-1301-84
$C_{18}H_{15}N_5S$			
1H-Pyridazino[4,5-e][1,3,4]thiadiazin-	EtOH	265(4.06),325(3.32)	94-4437-84

Compound	Solvent	$\lambda_{max}(\log \epsilon)$	Ref.
5-amine, 3-phenyl-N-(phenylmethyl)-			94-4437-84
$C_{18}H_{15}P$			
Phosphine, triphenyl-	C_6H_{12}	<u>260(4.1)</u>	151-0249-84A
Phosphorin, 3-(4-methylphenyl)-5-phenyl-	EtOH	239s(4.27),254(4.37), 277s(4.19),312(3.15)	24-0763-84
$C_{18}H_{16}$			
Bicyclo[5.4.0]undeca-1,3,5,9-tetraene, 8-(2,4,6-cycloheptatrien-1-ylidene)-	C_6H_{12}	224(4.09),239(4.03), 325(3.73),444s(2.26)	24-2045-84
	MeCN	224s(4.03),239(3.95), 325(3.71),444s(2.75)	24-2045-84
1,3,5-Cycloheptatriene, 7,7'-(2-butene-1,4-diylidene)bis-, (E)-	C_6H_{12}	247(4.06),377s(4.16), 406s(4.51),427(4.71), 452(4.73)	24-2045-84
	MeCN	247(3.97),378s(4.16), 400s(4.48),424(4.72), 449(4.76)	24-2045-84
(Z)-	C_6H_{12}	225s(4.17),251s(3.92), 379s(4.09),401s(4.36), 424(4.55),449(4.52)	24-2045-84
	MeCN	225s(4.12),250s(3.88), 378s(4.14),400s(4.42), 423(4.60),447(4.60)	24-2045-84
2-Hexen-4-yne, 2,3-diphenyl-, (E)-	C_6H_{12}	275(4.10)	44-0856-84
2-Hexen-4-yne, 2,3-diphenyl-, (Z)-	C_6H_{12}	229(4.23),283(4.10)	44-0856-84
1H-Indene, 1-(2,3-dihydro-1H-1-inden-ylidene)-2,3-dihydro-	EtOH-CH$_2$Cl$_2$	238(4.11),246(3.95), 322(4.49),338(4.47)	24-2300-84
Pyrene, 10b,10c-dihydro-10b,10c-di-methyl-	C_6H_{12}	337.5(4.94),377(4.57), 463(3.78),641(2.52)	35-7776-84
$C_{18}H_{16}BrNO$			
2H-Pyrrole, 3-bromo-2,2-dimethyl-4,5-diphenyl-, 1-oxide	EtOH	236(3.20),257(4.28), 337(3.60)	12-0095-84
$C_{18}H_{16}BrN_5OS$			
1H-1,2,4-Triazolium, 4-[acetyl[[(4-bromophenyl)amino]thioxomethyl]-amino]-1-(phenylmethyl)-, hydroxide, inner salt	MeOH	310(2.96)	73-1713-84
$C_{18}H_{16}ClN_3O$			
1H-1,5-Benzodiazepin-2-one, 8-chloro-3-[(dimethylamino)methylene]-2,3-dihydro-4-phenyl- (cation)	EtOH	221(4.56),260(4.38), 297(4.26),381(3.96)	103-1035-84
$C_{18}H_{16}ClN_3O_2$			
Pyrano[2,3-e]indazol-2(5H)-one, 3-chloro-6,7-dihydro-7-methyl-4-(methylphenylamino)-	EtOH	238(4.23),265s(3.88), 376(4.32)	4-0361-84
Pyrano[2,3-e]indazol-2(5H)-one, 3-chloro-4-(dimethylamino)-6,7-di-hydro-7-phenyl-	EtOH	228(4.23),256(4.13), 335s(4.23),355(4.25)	4-0361-84
$C_{18}H_{16}Cl_3NS$			
1-Propanamine, N,N-dimethyl-3-(1,2,4-trichloro-9H-thioxanthen-9-ylidene)-	MeOH	212(4.46),230(4.46), 262s(4.15),285s(3.96), 323(3.52)	73-2295-84
$C_{18}H_{16}N$			
Pyridinium, 1-methyl-2,6-diphenyl-,	EtOH	<u>308(4.0)</u>	103-1245-84

Compound	Solvent	$\lambda_{max}(\log \epsilon)$	Ref.
perchlorate (cont.)			103-1245-84
$C_{18}H_{16}NO$			
Quinolinium, 2-[2-(4-hydroxyphenyl)eth-enyl]-1-methyl-, iodide	H_2O	408(4.47)	62-0154-84A
	MeOH	430(4.48),544(4.01)	62-0154-84A
	EtOH	436(4.40),565(3.45), 606(4.17)	62-0154-84A
	isoPrOH	440(4.50),570(3.60), 603(3.77)	62-0154-84A
	acetone	427(4.49),530(3.11), 573(3.36),620(3.40)	62-0154-84A
	dioxan	388(4.11),530(4.47), 570(4.60),618(4.46)	62-0154-84A
	DMF	430(4.38),528s(--), 570(4.30),618(4.38)	62-0154-84A
	DMSO	432(4.48),530s(--), 576(3.66),616(3.68)	62-0154-84A
	CH_2Cl_2	439(4.50)	62-0154-84A
	$CHCl_3$	440(4.49)	62-0154-84A
	CCl_4	452(4.50)	62-0154-84A
Quinolinium, 4-[2-(4-hydroxyphenyl)eth-enyl]-1-methyl-, iodide	H_2O	416(4.47)	62-0154-84A
	MeOH	441(4.49),590(4.01)	62-0154-84A
	EtOH	450(4.40),604(3.45), 616(4.17)	62-0154-84A
	isoPrOH	455(4.50),604(3.61), 616(3.76)	62-0154-84A
	acetone	437(4.49),568(3.11), 612(3.34),660(3.40)	62-0154-84A
	dioxan	394(4.13),570(4.47), 615(4.61),668(4.46)	62-0154-84A
	DMF	440(4.34),570s(--), 618(4.31),664(4.41)	62-0154-84A
	DMSO	442(4.47),570s(--), 615(3.66),662(3.73)	62-0154-84A
	CH_2Cl_2	447(4.50)	62-0154-84A
	$CHCl_3$	450(4.48)	62-0154-84A
	CCl_4	470(4.52)	62-0154-84A
$C_{18}H_{16}N_2$			
Pyrazine, 2,3-bis(4-methylphenyl)-	EtOH	223(4.33),281(4.07), 315(3.95)	4-0103-84
Pyrazine, 2,6-dimethyl-3,5-diphenyl-	EtOH	213(4.01),231(4.06), 294(3.91),310(3.90)	142-2317-84
$C_{18}H_{16}N_2O$			
2H-Indol-2-one, 1,3-dihydro-1-methyl-3-[3-(phenylamino)-2-propenylidene]-	toluene DMSO	423(4.42) 443(4.60)	103-0951-84 103-0951-84
Pyrazine, 2,3-bis(4-methylphenyl)-, 1-oxide	EtOH	222(4.38),268(4.40), 332(3.67)	4-0103-84
2(1H)-Pyrazinone, 1-ethyl-5,6-diphenyl-	EtOH	264(4.04),345(3.81)	39-0391-84C
3H-Pyrazol-3-one, 2,4-dihydro-4-(1-methylethylidene)-2,5-diphenyl-	EtOH	259(4.20)	94-2146-84
Pyrido[3,4-b]indol-1-one, 2,3,4,9-tetrahydro-9-methyl-2-phenyl-	isoPrOH	228s(4.45),233(4.46), 309(4.39)	103-0047-84
Pyrido[3,4-b]indol-1-one, 2,3,4,9-tetrahydro-9-(phenylmethyl)-	isoPrOH	230(4.32),302(4.13), 328s(3.70)	103-0047-84
1H,5H-Pyrrolo[2,3-f]indole, 5-acetyl-6,7-dihydro-2-phenyl-	EtOH	205(4.53),235.5(4.63), 333(4.62)	103-0376-84
3H,6H-Pyrrolo[3,2-e]indole, 6-acetyl-7,8-dihydro-2-phenyl-	EtOH	206.9(4.19),270(4.63), 317.5(4.43)	103-0376-84

Compound	Solvent	$\lambda_{max}(\log \epsilon)$	Ref.
$C_{18}H_{16}N_2OS$			
Methanone, [4-methyl-2-(methylphenyl-amino)-5-thiazolyl]phenyl-	EtOH	242(4.03),356(4.22)	77-0837-84
2H-Pyrrol-2-one, 1,5-dihydro-1-(4-meth-ylphenyl)-4-[(4-methylphenyl)amino]-5-thioxo-	CHCl$_3$	250(4.46),303(4.20), 465(3.58)	78-3499-84
5-Thiazolamine, 2-acetyl-N-methyl-N,3-diphenyl-	EtOH	252(4.13),344(4.12)	77-0837-84
7H-Thiazolo[3,2-a]pyrimidin-7-one, 2,3,5,6-tetrahydro-5,6-diphenyl-, cis	EtOH	251(4.25)	142-2529-84
trans	EtOH	253(4.27)	142-2529-84
3-Thiophenecarboxamide, 2-[(4-methyl-phenyl)amino]-4-phenyl-	EtOH	230(4.25),337(4.15)	48-0917-84
$C_{18}H_{16}N_2O_2$			
[1]Benzazocino[7,6,5-efg][1]benzazo-cine-5,12-dione, 4,6,7,11,13,14-hexahydro-, (R)-	MeOH	213.5(4.68),268s(3.08)	39-2013-84C
(S)-	MeOH	213.5(4.68),244s(3.93), 268s(3.08)	39-2013-84C
Benzenepropanenitrile, β-oxo-α-[[5-(1-pyrrolidinyl)-2-furanyl]methylene]-	MeOH	208(3.10),238(2.93), 283(2.52),515(3.61)	73-2141-84
1H-1,5-Benzodiazepine, 2,4-di-2-furan-yl-2,3-dihydro-2-methyl-	EtOH	283(4.17),367(3.86)	103-0106-84
2-Butene-1,4-dione, 2-amino-3-(1-imino-ethyl)-1,4-diphenyl-	MeOH	274(4.09),330(3.96)	32-0261-84
1,4-Naphthalenedione, 2-[[2-(phenyl-amino)ethyl]amino]-	CH$_2$Cl$_2$	241(4.35),262s(--), 270(4.38),330(3.43), 445(3.52)	83-0743-84
1-Phenazinecarboxylic acid, 6-(3-meth-yl-2-butenyl)-	MeOH	255(4.81),355s(4.02), 367(4.15)	158-84-64
Pyrazine, 2,3-bis(4-methoxyphenyl)-	EtOH	230(4.25),290(4.14), 325.5(3.96)	4-0103-84
Pyrazine, 2,3-bis(4-methylphenyl)-, 1,4-dioxide	EtOH	220(4.34),293s(4.11), 317(4.22)	4-0103-84
2(1H)-Quinoxalinone, 3-[2-(4-methoxy-phenyl)ethenyl]-1-methyl-, (E)-	EtOH	404(3.94)	18-1653-84
$C_{18}H_{16}N_2O_3$			
Benzenepropanenitrile, α-[[5-(4-morpho-linyl)-2-furanyl]methylene]-β-oxo-	MeOH	210(3.21),238(3.04), 280(2.65),500(3.71)	73-2141-84
Pyrazine, 2,3-bis(4-methoxyphenyl)-, 1-oxide	EtOH	227(4.40),277(4.40), 343(3.79)	4-0103-84
2,4,6(1H,3H,5H)-Pyrimidinetrione, 5,5-bis(4-methylphenyl)-	MeOH	245s(4.46),260s(3.98)	12-1245-84
$C_{18}H_{16}N_2O_4$			
Pyrazine, 2,3-bis(4-methoxyphenyl)-, 1,4-dioxide	EtOH	227(4.65),306(4.67)	4-0103-84
$C_{18}H_{16}N_2O_5$			
2-Butenoic acid, 4-(4,8-dimethoxy-9H-pyrido[3,4-b]indol-1-yl)-4-oxo-, methyl ester, (E)- (picrasidine E)	EtOH	240(4.85),280s(4.25), 310(4.35),410(4.35)	94-3579-84
2,4,6(1H,3H,5H)-Pyrimidinetrione, 5,5-bis(4-methoxyphenyl)-	MeOH	225(4.59),254s(4.56), 275s(4.38)	12-1245-84
$C_{18}H_{16}N_2O_5S_2$			
Benzenesulfonic acid, 5-[[3-(aminocar-bonyl)-4-phenyl-2-thienyl]amino]-2-methoxy-, monosodium salt	DMF	332(4.09)	48-0917-84

Compound	Solvent	$\lambda_{max}(\log \epsilon)$	Ref.
$C_{18}H_{16}N_2O_8$			
2,6-Pyridinedicarboxylic acid, 4-[2-(2-carboxy-2,3-dihydro-5,6-dihydroxy-1H-indol-1-yl)ethenyl]-2,3-dihydro-	H_2O	272(3.83),299s(3.75), 540(4.65)	33-1547-84
$C_{18}H_{16}N_2S_2$			
1H-Pyrrole-2,5-dithione, 1-(4-methyl-phenyl)-3-[(4-methylphenyl)amino]-	$CHCl_3$	257(4.42),395(4.43), 550(3.97)	78-3499-84
$C_{18}H_{16}N_3O_2$			
1H-Pyrazolium, 2-ethyl-4,5-dihydro-4,5-dioxo-1-phenyl-3-[(phenylimino)-methyl]-, iodide	EtOH	370(4.14)	48-0695-84
Pyridinium, 2-[2-(4,5-dihydro-4,5-dioxo-1-phenyl-1H-pyrazol-3-yl)ethenyl]-1-ethyl-, iodide	EtOH	390(4.20)	48-0695-84
$C_{18}H_{16}N_4$			
3H-Pyrazolo[5,1-c]-1,2,4-triazole, 3,6-dimethyl-3,7-diphenyl-	CH_2Cl_2	245(4.51),313(4.19), 347(4.16)	24-1726-84
$C_{18}H_{16}N_4O$			
7H-Pyrido[2,1-b]quinazoline-6,11-dione, 8,9-dihydro-, 6-(phenylhydrazone)	EtOH	231(4.40),250(4.30), 296(3.94),388(4.36)	4-1301-84
$C_{18}H_{16}N_4O_2$			
Acetamide, N-[2-(4-methoxyphenyl)-4H-pyrazolo[1,5-a]benzimidazol-6-yl]-	EtOH	257(4.42),288(4.27), 331(4.42)	48-0829-84
7H-Pyrido[2,1-b]quinazoline, 6,11-dione, 8,9-dihydro-, 6-[(4-hydroxyphenyl)hy-drazone]	EtOH	231(4.32),255(4.14), 304(3.90),417(4.21)	4-1301-84
$C_{18}H_{16}N_4O_3S$			
1,2,4-Triazolo[3,2-b]thiepino[2,3-d]-pyrimidine-7-carboxylic acid, 4,5,8,9-tetrahydro-9-methyl-5-oxo-4-phenyl-, ethyl ester	$CHCl_3$	334(4.12)	24-0585-84
$C_{18}H_{16}N_6$			
2H-Pyrazolo[3,4-d]pyridazine-3,4-di-amine, 7-(4-methylphenyl)-2-phenyl-	EtOH	226(5.06)	4-1049-84
$C_{18}H_{16}N_6O_8$			
1H-1,2,3-Triazole-4,5-dicarboxylic acid, 1,1'-(2,4-hexadiyne-1,6-diyl)bis-, dimethyl ester	MeCN	219(4.26)	1-0623-84
$C_{18}H_{16}O_2$			
Benzeneacetic acid, α-[(2-ethenylphen-yl)methylene]-, methyl ester, cis	MeOH	222(4.10),241(4.00), 290(3.97)	151-0327-84D
Benzo[b]naphtho[2,1-d]furan-7(8H)-one, 9,10-dihydro-9,9-dimethyl-	EtOH	245(4.87),317(3.58), 326(3.57),331(3.56)	39-1213-84C
Benzo[b]naphtho[2,3-d]furan-1(2H)-one, 3,4-dihydro-3,3-dimethyl-	EtOH	241(4.85),293(4.23), 321(4.21),327(4.16)	39-1213-84C
1-Benzoxepin-5(2H)-one, 3,4-dihydro-4-[(4-methylphenyl)methylene]-	octane	303(4.30)	65-0148-84
	EtOH	227(4.16),314(4.23)	65-0148-84
$C_{18}H_{16}O_2S$			
Naphthalene, 1-[[(phenylmethyl)sulfon-yl]methyl]-	MeCN	265(3.68),274.5(3.88), 284(3.97),295(3.80), 313(2.78),317(2.40)	35-1779-84

Compound	Solvent	$\lambda_{max}(\log \epsilon)$	Ref.
Naphthalene, 2-[[(phenylmethyl)sulfonyl]methyl]-	MeCN	260(3.59),268.5(3.737), 277(3.73),287(3.54), 304.5(2.56),311(2.43), 318(2.32)	35-1779-84
$C_{18}H_{16}O_3$			
1-Benzoxepin-5(2H)-one, 3,4-dihydro-4-[(4-methoxyphenyl)methylene]-, (E)-	octane	322(4.32)	65-0148-84
	EtOH	253(4.17),333(4.19)	65-0148-84
$C_{18}H_{16}O_4$			
1-Benzoxepin-5(2H)-one, 8-methoxy-3-(4-methoxyphenyl)-	MeOH	202(4.54),334(4.16)	2-1211-84
1,3-Dioxane-4,6-dione, 2,2-dimethyl-5,5-diphenyl-	EtOH	257(2.81)	12-1245-84
1,4-Ethanoanthracene-9,10-dione, 1,4-dihydro-8-hydroxy-1-methoxy-6-methyl-	MeOH	216(4.44),250(3.95), 278(3.93),425(3.62)	44-3766-84
1-Naphthalenecarboxylic acid, 1,2,3,4-tetrahydro-2-(2-methoxyphenyl)-4-oxo-	MeCN	233(5.75),274(5.55)	2-0603-84
1-Naphthalenecarboxylic acid, 1,2,3,4-tetrahydro-2-(3-methoxyphenyl)-4-oxo-	MeCN	229(5.89),278(5.33)	2-0603-84
1-Naphthalenecarboxylic acid, 1,2,3,4-tetrahydro-2-(4-methoxyphenyl)-4-oxo-	MeCN	224(5.72),278(5.54)	2-0603-84
$C_{18}H_{16}O_5$			
9-Anthracenecarboxylic acid, 9,10-dihydro-4-hydroxy-1-methoxy-2-methyl-10-oxo-, methyl ester, (±)-	MeOH	258(4.15),284(4.12), 366(3.72)	44-0318-84
9,10-Anthracenedione, 1,2,8-trimethoxy-4-methyl-	MeOH	223(4.46),266(4.35), 370(3.94)	78-5039-84
4H-1-Benzopyran-4-one, 7,8-dimethoxy-3-(4-methoxyphenyl)-	MeOH	305s(3.95)	64-0238-84B
$C_{18}H_{16}O_6$			
9,10-Anthracenedione, 1-hydroxy-2-(hydroxymethyl)-5,8-dimethoxy-3-methyl-	MeOH	230(4.65),259(4.27), 280s(4.03),451(4.06)	78-3677-84
9,10-Anthracenedione, 1-hydroxy-2,5,7-trimethoxy-4-methyl-	MeOH	227(4.51),273(4.36), 436(3.92)	78-5039-84
9,10-Anthracenedione, 5-hydroxy-1,2,3-trimethoxy-6-methyl-	EtOH	276(4.42),408(3.80)	78-3455-84
9-Anthracenepropanoic acid, 9,10-dihydro-4,5,9-trihydroxy-10-oxo-, methyl ester	n.s.g.	212(4.69),257(4.06), 266(4.16),298(4.21), 374(4.26)	5-0306-84
4H-1-Benzopyran-4-one, 7-hydroxy-3-(2,4,5-trimethoxyphenyl)-	MeOH	244(3.88),297(3.76)	102-0167-84
$C_{18}H_{16}O_7$			
9,10-Anthracenedione, 1,2-dihydroxy-6,7,8-trimethoxy-3-methyl-	dioxan	278(4.76),316(4.12), 400(3.86)	94-0860-84
9,10-Anthracenedione, 1,5-dihydroxy-2,3,6-trimethoxy-8-methyl-	MeOH	229(4.45),275(4.53), 318(3.97),456(4.11)	78-5039-84
4H-1-Benzopyran-4-one, 2,3-dihydro-2-(1,3-benzodioxol-5-yl)-6-hydroxy-5,7-dimethoxy-, (S)-	MeOH	238(4.5),281(4.4), 342(3.8)	94-1472-84
4H-1-Benzopyran-4-one, 5,7-dihydroxy-3-(3,4,5-trimethoxyphenyl)-	MeOH	332s(3.63)	64-0238-84B
1(9bH)-Dibenzofuranone, 6-acetyl-3-acetoxy-7,9-dihydroxy-8,9b-dimethyl-, (R)-	MeOH	214(4.41),282(4.26), 338(3.58),386(3.23)	23-0320-84
2-Propen-1-one, 3-(1,3-benzodioxol-5-yl)-1-(3,6-dihydroxy-2,4-dimethoxyphenyl)-, (E)- (agestricin A)	MeOH	259(4.1),314s(4.1), 367(4.3)	94-1472-84

Compound	Solvent	λ_{max} (log ϵ)	Ref.
$C_{18}H_{16}O_8$			
4H-1-Benzopyran-4-one, 5,6-dihydroxy-2-(4-hydroxy-2,5-dimethoxyphenyl)-7-methoxy-	MeOH	256(4.2),284(4.3), 365(4.4)	94-4935-84
	MeOH-NaOMe	270(--),372(--)	94-4935-84
	MeOH-NaOAc	267(--),380(--)	94-4935-84
	MeOH-AlCl$_3$	277(--),296(--), 332(--),410(--)	94-4935-84
	+ HCl	276(--),295(--), 328(--),403(--)	94-4935-84
4H-1-Benzopyran-4-one, 5,7-dihydroxy-2-(2-hydroxy-4,5-dimethoxyphenyl)-6-methoxy-	MeOH	264(4.3),288(4.1), 370(4.4)	94-4935-84
	MeOH-AlCl$_3$	270(--),295(--), 404(--)	94-4935-84
4H-1-Benzopyran-4-one, 5,7-dihydroxy-2-(4-hydroxy-2,5-dimethoxyphenyl)-6-methoxy-	MeOH	275(4.2),368(4.3)	94-4935-84
	MeOH-NaOMe	275(--),416(--)	94-4935-84
	MeOH-NaOAc	265(--),295(--), 378(--)	94-4935-84
	MeOH-AlCl$_3$	270(--),296(--), 402(--)	94-4935-84
	+ HCl	265(--),295(--), 394(--)	94-4935-84
4H-1-Benzopyran-4-one, 2-(2,4-dihydroxy-5-methoxyphenyl)-5-hydroxy-6,7-dimethoxy-	MeOH	265(4.2),372(4.3)	94-4935-84
	MeOH-AlCl$_3$	276(--),295(--), 410(--)	94-4935-84
4H-1-Benzopyran-4-one, 2-(2,5-dihydroxy-4-methoxyphenyl)-5-hydroxy-6,7-dimethoxy-	MeOH	265(4.3),312(4.0), 372(4.3)	94-4935-84
	MeOH-AlCl$_3$	277(--),296(--), 332(--),410(--)	94-4935-84
1(9bH)-Dibenzofuranone, 6-acetyl-2-acetoxy-3,7,9-trihydroxy-8,9b-dimethyl-, (R)-	MeOH	223(4.17),289(3.93), 334(3.39)	23-0320-84
Mycoversilin	EtOH	269(3.11)	158-0733-84
	EtOH-base	289(3.09)	158-0733-84
$C_{18}H_{16}Si$			
Silane, dimethylbis(phenylethynyl)-	hexane	239.9(4.68),250(4.86), 263.7(4.88)	65-2317-84
$C_{18}H_{17}ClN_2O_4S$			
1-Piperazinecarboxylic acid, 4-(3-chloro-2-oxo-2H-thieno[2,3-h]-1-benzopyran-4-yl)-, ethyl ester	EtOH	234(4.42),261(4.37), 274s(4.10),305s(4.04), 336(4.17)	161-0081-84
$C_{18}H_{17}Cl_2N_3O_2$			
Pyrano[2,3-e]indazol-2(3H)-one, 3,3-dichloro-4-(dimethylamino)-4,5,6,7-tetrahydro-7-phenyl-	EtOH	238(4.18),284(4.19)	4-0361-84
Pyrano[2,3-e]indazol-2(3H)-one, 3,3-dichloro-4,5,6,7-tetrahydro-7-methyl-4-(methylphenylamino)-	EtOH	250.5(4.27),290s(3.82)	4-0361-84
$C_{18}H_{17}IN_4O$			
3H-Pyrazolo[3,4-c]isoxazolium, 3-(1-ethyl-2(1H)-pyridinylidene)-4-methyl-6-phenyl-, iodide	EtOH	320(3.76)	48-0811-84
$C_{18}H_{17}NO$			
2H-Benz[g]indole, 3,3a,4,5-tetrahydro-3-phenyl-, 1-oxide	EtOH	227(4.00),300(4.18)	12-0117-84
4(1H)-Pyridinone, 1-(diphenylmethyl)-2,3-dihydro-	EtOH	320(4.31)	118-0485-84

Compound	Solvent	$\lambda_{max}(\log \epsilon)$	Ref.
$C_{18}H_{17}NOS$			
2-Propenamide, 2-methyl-N-(phenylmethyl)-N-(phenylthioxomethyl)-	n.s.g.	298(3.96),322(4.00), 462(2.28)	44-0396-84
$C_{18}H_{17}NO_2$			
5a,10b-Butano-6H-[1,4]benzodioxino[2,3-b]indole	MeOH	281(3.65),288(3.65)	4-1841-84
1H-Indole-3-propanoic acid, phenylmethyl ester	EtOH	220.5(4.55),274s(3.75), 281(3.78),289.5(3.71)	94-1373-84
Noraporphine, 1,2-dimethoxy-6a,7-dehydro-	MeOH	252(4.65),261(4.62), 293(3.79),326(4.08), 380(3.40)	39-1273-84C
5a,10b-Propano-6H-[1,4]benzodioxino[2,3-b]indole, 2-methyl-	MeOH	246(3.83),283(3.54), 306(3.37)	4-1841-84
4H-Pyran-4-one, 3-[(diphenylamino)methylene]-2,3,5,6-tetrahydro-, (E)-	EtOH	232(4.00),281(3.96), 350(4.39)	4-1441-84
$C_{18}H_{17}NO_2S_4$			
6H-1,3-Thiazine-6-thione, 2-(ethylthio)-2,3-dihydro-4-phenyl-5-(phenylsulfonyl)-	EtOH	263(4.04),352(4.23)	4-0953-84
$C_{18}H_{17}NO_3$			
Benzamide, N-(1,1-dimethyl-2,3-dioxo-3-phenylpropyl)-	EtOH	227(4.09),260(4.05)	12-0109-84
Benzoic acid, 2-[(3-oxo-1-phenyl-1-butenyl)amino]-, methyl ester, (±)-	MeOH	228(4.03),302(3.99), 350(3.97)	4-0759-84
4H-1,3,5-Dioxazocine-7-carboxaldehyde, 5,8-dihydro-2,6-diphenyl-	MeOH	299(4.16)	88-2731-84
Heptazolicine	EtOH	242(4.65),275(4.62), 300(4.40)	102-2409-84
1(2H)-Naphthalenone, 3,4-dihydro-2-(2-nitro-1-phenylethyl)-	EtOH	248(3.85),292(3.00)	12-0117-84
Norstephanine, (±)-	EtOH	222(4.28),264s(3.90), 273(4.03),282s(4.02), 302(3.54),325s(3.31)	100-0504-84
6-Oxa-1-azabicyclo[3.1.0]hexan-3-one, 4-hydroxy-2,2-dimethyl-4,5-diphenyl-	EtOH	256(3.36)	12-0109-84
3H-Pyrrol-3-one, 2,4-dihydro-4-hydroxy-2,2-dimethyl-4,5-diphenyl-, 1-oxide	EtOH	292(4.26)	12-0095-84
$C_{18}H_{17}NO_4$			
Cularicine, (+)-	n.s.g.	288(3.79)	100-0753-84
Erythrinan-3,8-dione, 1,2,6,7,10,11-hexadehydro-15,16-dimethoxy-	EtOH	230(4.39),319(4.04)	142-2255-84
Erythrinan-8-one, 1,2,6,7,10,11-hexadehydro-3-hydroxy-15,16-dimethoxy-	EtOH	267(4.47),360(3.34)	142-2255-84
Isocalycinine, (-)-	EtOH	218(4.42),282(4.26), 302(4.14)	100-0353-84
	EtOH-base	240(4.23),320(4.33)	100-0353-84
1(2H)-Naphthalenone, 3,4-dihydro-2-hydroxy-2-(2-nitro-1-phenylethyl)-	EtOH	252(4.21)	12-0587-84
$C_{18}H_{17}NO_7$			
4,7-Epoxy-1,2-benzisoxazole-5,6-dicarboxylic acid, 3-benzoyl-3a,4,5,6,7,7a-hexahydro-, dimethyl ester	MeOH	274(3.94)	73-1193-84
5-Oxa-1-azabicyclo[4.2.0]oct-2-ene-2,3-dicarboxylic acid, 4-(2-methoxyethylidene)-8-oxo-, 3-(phenylmethyl) ester, lithium salt, [R-(Z)]-	H_2O	303(4.25)	39-1599-84C

Compound	Solvent	$\lambda_{max}(\log \epsilon)$	Ref.
7-Oxabicyclo[2.2.1]heptane-2,3-dicarb-oxylic acid, 5-(1-amino-2-oxo-2-phen-ylethylidene)-6-oxo-, dimethyl ester, (exo,exo)-	MeOH	264(4.11),315(3.61)	73-1193-84
$C_{18}H_{17}NO_8$ 1H-Pyrrole-1-acetic acid, 5-(1,3-benzo-dioxol-5-yl)-4-(ethoxycarbonyl)-2,3-dihydro-2,3-dioxo-, ethyl ester	dioxan	262s(3.88),360(3.76), 400(3.76)	94-0497-84
$C_{18}H_{17}N_3O$ 2H-1,5-Benzodiazepin-2-one, 3-[(dimeth-ylamino)methylene]-1,3-dihydro-4-phenyl-	EtOH	205(4.54),296(4.15), 252(4.36)[sic],370(3.85)	103-0183-84
$C_{18}H_{17}N_3O_2$ 1,3-Diazabicyclo[3.1.0]hex-3-ene, 6-(4-nitrophenyl)-2,2-dimethyl-4-phenyl-	n.s.g.	250(4.15),289(4.08)	116-1895-84
2,4-Imidazolidinedione, 3-[(1-methyl-ethylidene)amino]-5,5-diphenyl-	MeOH	201(4.39)	56-0585-84
3H-Indol-5-amine, 3,3-dimethyl-2-[2-(4-nitrophenyl)ethenyl]-	EtOH	438(4.33)	103-0976-84
3H-Indol-5-amine, 2,3,3-trimethyl-N-[(4-nitrosophenyl)methylene]-	EtOH	375(3.55)	103-0976-84
$C_{18}H_{17}N_4O$ 3H-Pyrazolo[3,4-c]isoxazolium, 3-(1-ethyl-2(1H)-pyridinylidene)-4-methyl-6-phenyl-, iodide	EtOH	320(3.76)	48-0811-84
$C_{18}H_{17}N_5O_2$ Benzonitrile, 5-nitro-2-[[4-(1-piperi-dinyl)phenyl]azo]-	C_6H_{12}	499(4.52)	39-0149-84B
	EtOH	529(4.48)	39-0149-84B
	EtOH-HCl	498(3.59)	39-0149-84B
1,2-Ethanediamine, N-[4-[(4-nitrophen-yl)azo]-1-naphthalenyl]-	NaOAc	545(4.82)	86-1013-84
$C_{18}H_{17}N_5O_4$ 1H-Benzimidazo[5,6-g]quinazolin-9-am-ine, 1-β-D-ribofuranosyl-	EtOH	256(4.67),277(4.45), 341(3.54),357(3.49), 396(3.52),418(3.66), 442(3.52)	44-2158-84
	EtOH-HCl	246(4.59),267(4.51), 338(3.81),354(3.98), 389(3.65),411(3.64), 436(3.48)	44-2158-84
	EtOH-NaOH	256(4.50),272s(--), 340(3.34),358(3.52), 396(3.61),419(3.75), 445(3.72)	44-2158-84
3H-Benzimidazo[5,6-g]quinazolin-9-am-ine, 3-β-D-ribofuranosyl-	EtOH	257(4.59),266(4.62), 344(3.52),362(3.58), 392(3.64),414(3.77), 438(3.61)	44-2158-84
	EtOH-HCl	241(4.46),266(4.49), 289s(--),297s(--), 335(3.68),352(3.91), 387(3.65),408(3.76), 433(3.63)	44-2158-84
	EtOH-NaOH	264(4.54),347(3.28),	44-2158-84

Compound	Solvent	$\lambda_{max}(\log \epsilon)$	Ref.
(cont.)		362(3.42),392(3.50), 414(3.70),439(3.58)	44-2158-84
Oxepino[2,3-d]pyrimidine-5-carboxylic acid, 6-cyano-2-hydrazino-3,4,7,8-tetrahydro-8-methyl-4-oxo-3-phenyl-, methyl ester	$CHCl_3$	273(4.00),335(4.40)	24-0585-84
$C_{18}H_{17}O_4$			
1-Benzopyrylium, 5,7-dihydroxy-2-(4-methoxyphenyl)-6,8-dimethyl-glucoside	MeOH-HCl	282(4.02),333(3.64), 442(4.15),480s(4.04)	94-0490-84
	MeOH-HCl	244(4.09),278(4.30), 331(3.79),472(4.65)	94-0490-84
$C_{18}H_{18}$			
Cycloheptatrienylium, 1,1'-(2-butene-1,4-diyl)bis-, bis(tetrafluoroborate)	CF_3COOH	276(4.11),295s(4.04), 350s(3.18),432s(2.70), 456(2.70),506s(2.48)	24-2045-84
Dicyclobuta[a,c]anthracene, octahydro-	n.s.g.	220(4.29),242(4.63), 245(4.62),264(3.77), 271s(3.53),282s(3.45), 292(3.53),304(3.53), 316s(3.34),328s(2.70)	44-1412-84
5,8-Ethanocyclohept[fg]acenaphthylene, 1,2,5,6,7,8-hexahydro-	MeCN	236(4.90),291(3.96), 303(4.04),314(3.91), 320(3.89),334(3.74)	44-1146-84
6H-1,11-Ethano-7,11-methenobenzocyclo-dodecene, 5,12-dihydro-13-methylene-	C_6H_{12}	225(3.92),272(3.46), 290(3.48)	39-2165-84C
$C_{18}H_{18}BrNO_5$			
Formamide, N-[1-(1,3-benzodioxol-5-yl)-2-(2-bromo-4,5-dimethoxyphenyl)-ethyl]-	$CHCl_3$	244(4.11),288(4.05)	142-1217-84
$C_{18}H_{18}BrN_3O_6$			
β-D-erythro-Pentofuranuronamide, N-[4-(bromoacetyl)phenyl]-1,2-dideoxy-1-(3,4-dihydro-5-methyl-2,4-dioxo-1-(2H)-pyrimidinyl)-	pH 1 pH 7 pH 13	276(4.31) 276(4.32) 273(4.22)	87-0680-84 87-0680-84 87-0680-84
$C_{18}H_{18}Br_2O_4$			
Benzene, 1,1'-(1,2-ethenediyl)bis[2-bromo-4,5-dimethoxy-, (E)-	EtOH	301(4.28),337(4.38)	20-1099-84
$C_{18}H_{18}Br_2O_5$			
Ethanone, 1,2-bis(2-bromo-4,5-dimeth-oxyphenyl)-	EtOH	280(4.32),310(3.97)	20-1099-84
$C_{18}H_{18}ClN_3O_6$			
β-D-erythro-Pentofuranuronamide, N-[4-(chloroacetyl)phenyl]-1,2-dideoxy-1-[3,4-dihydro-5-methyl-2,4-dioxo-1(2H)-pyrimidinyl]-	pH 1 and 7 pH 13	267(4.37) 275(4.26)	87-0680-84 87-0680-84
$C_{18}H_{18}Cl_2N_2O_4$			
Hexanediamide, N,N'-bis(4-chlorophenyl)-	MeOH	222(4.25),246(4.21)	34-0345-84
$C_{18}H_{18}F_3N_3$			
Piperidine, 1-[4-[[4-(trifluoromethyl)-phenyl]azo]phenyl]-	C_6H_{12} EtOH EtOH-HCl	411(4.43) 423(4.43) 515(3.81)	39-0149-84B 39-0149-84B 39-0149-84B

Compound	Solvent	λ_{max}(log ϵ)	Ref.
$C_{18}H_{18}F_6N_2O_6S_2$			
4,9-Ethanodipyrido[1,2-a:2',1'-h][1,4]-diazocinediium, 6,7,13,14-tetrahydro-, salt with trifluoromethanesulfonic acid, (1:2)	EtOH	225(3.87),276(4.16)	35-2672-84
$C_{18}H_{18}IN_3O_6$			
β-D-erythro-Pentofuranuronamide, 1,2-dideoxy-1-(3,4-dihydro-5-methyl-2,4-dioxo-1(2H)-pyrimidinyl)-N-[4-(iodoacetyl)phenyl]-	pH 1	275(4.28),297s(4.24)	87-0680-84
	pH 7	275(4.29),295s(4.25)	87-0680-84
	pH 13	276(4.28),296s(4.20)	87-0680-84
$C_{18}H_{18}N_2$			
5,12-Diaza[2₄](1,2,4,5)cyclophane	EtOH	300(3.79),307(3.79)	35-2672-84
5,15-Diaza[2₄](1,2,4,5)cyclophane	EtOH	298s(3.64),307(3.72)	35-2672-84
Dibenzo[e,g][1,4]diazocine, 1,6,7,12-tetramethyl-, (S)-	MeOH	213(4.55),276(3.50)	39-2013-84C
[6,7:13,14]Dicyclobuta-5,12-diaza-[2₂](1,2)cyclophane	EtOH	280(4.17),286s(4.16)	35-2672-84
Pyrazine, 2,3-dihydro-5,6-bis(4-methylphenyl)-	EtOH	227(4.23),295(3.84)	4-0103-84
1H-Quindoline, 2,10-dihydro-1,3,11-trimethyl-	EtOH	238(4.52),270(4.35), 291(4.49),348(4.11), 364(4.09)	103-1374-84
$C_{18}H_{18}N_2O$			
5,12-Diaza[2₄](1,2,4,5)cyclophane, N-oxide	EtOH	228(3.88),250(3.91), 308(3.79)	35-2672-84
5,15-Diaza[2₄](1,2,4,5)cyclophane, N-oxide	EtOH	232(4.02),252(3.98), 304(3.83)	35-2672-84
Ethanone, 1-(1,2-dimethyl-1H-indol-3-yl)-2-phenyl-, oxime	n.s.g.	226(4.52),283(4.02)	103-0058-84
$C_{18}H_{18}N_2OS$			
Benzenamine, N-(3-ethyl-4-phenyl-2(3H)-thiazolylidene)-4-methoxy-	EtOH	298(4.32)	56-0447-84
$C_{18}H_{18}N_2O_2$			
Cyclohexanecarboxylic acid, 3-(dicyanomethylene)-1-methyl-5-phenyl-, methyl ester	EtOH	240(4.3)	39-0261-84C
5,12-Diaza[2₄](1,2,4,5)cyclophane, N,N'-dioxide	EtOH	228(4.31),245s(4.09), 292(4.21),325s(3.24)	35-2672-84
5,15-Diaza[2₄](1,2,4,5)cyclophane, N,N'-dioxide	EtOH	230(--),262(--), 280s(--)	35-2672-84
Ethanol, 2,2'-(9,10-phenanthrenediylidenedinitrilo)bis-	CHCl₃	211(2.70),254(2.63), 301(3.70),334(3.26), 351(3.29),400(2.46)	34-0237-84
Naphtho[1',2':6,7]cyclohepta[1,2-c]-pyrazole, 2,10,11,12-tetrahydro-5,7-dimethoxy-	EtOH	261(4.69),309(4.07)	2-0736-84
4H-1,2,6-Oxadiazocin-7(8H)-one, 5,6-dihydro-3-phenyl-6-(phenylmethyl)-	EtOH	239(4.0)	5-1696-84
Pyrazine, 2,3-dihydro-5,6-bis(4-methoxyphenyl)-	EtOH	235(4.12),289s(3.67), 312.5(3.75)	4-0103-84
3-Pyridinecarbonitrile, 2-[3-(3,4-dimethoxyphenyl)-3-butenyl]-	MeOH	258(4.10)	35-7175-84
$C_{18}H_{18}N_2O_3S$			
2-Propenenitrile, 2-(phenylsulfonyl)-3-[5-(1-piperidinyl)-2-furanyl]-	MeOH	208(3.08),233(3.13), 338(2.87),467(3.72)	73-2141-84

Compound	Solvent	$\lambda_{max}(\log \epsilon)$	Ref.
$C_{18}H_{18}N_2O_4$			
Spiro[2,3-diazabicyclo[3.1.0]hex-2-en-6,1'-[1H]indene]-1,5-dicarboxylic acid, 4,4-dimethyl-, dimethyl ester, (1α,5α,6α)-	EtOH	232(4.28)	23-2506-84
$C_{18}H_{18}N_2O_4S$			
3,5-Thiepindicarboxylic acid, 6,7-dihydro-2-[(phenylcarbonimidoyl)amino]-, 5-ethyl 3-methyl ester	CHCl$_3$	264(4.25),267(4.28), 350(4.02)	24-0585-84
$C_{18}H_{18}N_2O_5$			
3-Quinolinecarboxylic acid, 4-(2-acetyl-1,3-dioxobutyl)-2-amino-, ethyl ester	BuOH	278(4.31),304(4.27), 316(4.39)	4-1233-84
$C_{18}H_{18}N_2O_{10}S$			
2-Butenoic acid, 2-isothiocyanato-, 4'-ester with 5-hexopyranosyl-2-hydroxy-3,6-dioxo-1,4-cyclohexadiene-1-carboxamide	MeOH	231(4.19),266(4.26), 438(3.23)	158-1273-84
$C_{18}H_{18}N_2S$			
Benzenamine, N-(3-ethyl-4-phenyl-2(3H)-thiazolylidene)-4-methyl-	EtOH	298(4.29)	56-0447-84
$C_{18}H_{18}N_3O$			
Pyridinium, 2-[(1,5-dihydro-3-methyl-5-oxo-1-phenyl-4H-pyrazol-4-ylidene)methyl]-1-ethyl-, iodide	EtOH	390(4.08),490(3.88)	48-0811-84
$C_{18}H_{18}N_4$			
Benzonitrile, 2-[[4-(1-piperidinyl)-phenyl]azo]-	C$_6$H$_{12}$	428(4.44)	39-0149-84B
	EtOH	442(4.45)	39-0149-84B
	EtOH-HCl	498(--)	39-0149-84B
Benzonitrile, 4-[[4-(1-piperidinyl)-phenyl]azo]-	C$_6$H$_{12}$	427(4.49)	39-0149-84B
	EtOH	442(4.46)	39-0149-84B
	EtOH-HCl	521(4.00)	39-0149-84B
1H-Pyrazole, 3,3'-(1,4-phenylenedi-2,1-ethenediyl)bis[1-methyl-, trans,trans	EtOH	344(4.80)	103-1266-84
2,4-Pyrimidinediamine, N,N'-bis(phenyl-methyl)-	MeOH	217(4.5),289(3.93)	24-1523-84
$C_{18}H_{18}N_4O_2$			
1-Triazenium, 1-(1,3-dihydro-1,3-dioxo-2H-isoindol-2-yl)-2-(1,1-dimethylethyl)-3-phenyl-, hydroxide, inner salt, (Z)-	EtOH	217.5(4.54),238s(4.19), 295(3.71)	104-0408-84
$C_{18}H_{18}N_4O_3$			
1-Imidazolidinepropanoic acid, 2,5-dioxo-4,4-diphenyl-, hydrazide	MeOH	203(4.46)	56-0583-84
2,4(1H,3H)-Pteridinedione, 1,3,7-trimethyl-6-(1-oxo-3-phenylpropyl)-	MeOH	248(4.10),279(4.01), 329(4.05)	5-1798-84
$C_{18}H_{18}N_4O_4$			
2-Quinoxalinecarboxamide, N-(2-amino-phenyl)-3-(1,2,3-trihydroxypropyl)-, erythro-	EtOH	240(3.77),282(3.13)	136-0324-84J

Compound	Solvent	$\lambda_{max}(\log \epsilon)$	Ref.
$C_{18}H_{18}N_4O_5$			
Inosine, 2-(2-phenylethenyl)-	pH 1	326(4.35)	87-0429-84
	pH 7	269(4.08),329(4.32)	87-0429-84
	pH 13	261(4.20),288(4.23), 318(4.32)	87-0429-84
$C_{18}H_{18}N_4O_7$			
Benzenepropanoic acid, β-[(2,4-dinitro-phenyl)hydrazono]-4-methoxy-, ethyl ester	EtOH	223(3.94),292(3.92), 390(4.34)	48-0829-84
$C_{18}H_{18}O_2$			
Ethanone, 1-(1,2,3,4-tetramethyl-7-dibenzofuranyl)-	EtOH	314(4.42)	39-0799-84B
Ethanone, 1-(1,2,3,4-tetramethyl-8-dibenzofuranyl)-	EtOH	248(4.39),270(4.51)	39-0799-84B
Phenanthrene, 2,6-dimethoxy-1,7-di-methyl-	EtOH	225(4.51),260(4.66), 271(4.60),283(4.31), 303s(4.04),327s(3.27), 343(3.42),360(3.50)	39-1913-84C
$C_{18}H_{18}O_3$			
[1,1'-Biphenyl]-2,2'-diol, 5-(3-hydr-oxy-1-propenyl)-5'-(2-propenyl)- (randainol)	MeOH	245(3.52),270s(3.29), 285s(3.17)	102-2329-84
Ethanone, 1-(9,10-dihydro-3,7-dihydr-oxy-2,8-dimethyl-4-phenanthrenyl)- (juncunone)	MeOH	211(4.54),252(4.20), 293s(3.79),375(3.81)	39-1913-84C
$C_{18}H_{18}O_4$			
1,4,5,8-Anthracenetetrone, 2,3,6,7-tetrahydro-2,2,6,6-tetramethyl-	MeOH	238(4.76),262(4.60), 324(3.48)	138-2031-84
[Bi-2,4,6-cycloheptatrien-1-yl]-3,3'-dicarboxylic acid, dimethyl ester	MeCN	222(4.3),260s(3.7)	24-0809-84
Methanone, phenyl[2,4,6-trihydroxy-3-(3-methyl-2-butenyl)phenyl]-	MeOH	252s(3.68),311(4.01)	39-1413-84C
	MeOH-base	253s(4.03),344(4.15)	39-1413-84C
$C_{18}H_{18}O_4S$			
Thiopyrano[4,3,2-cd]benzofuran-4-carb-oxylic acid, 2-acetyl-6-propyl-, ethyl ester	EtOH	246(4.35),322(3.98), 428(3.78)	39-0487-84C
$C_{18}H_{18}O_4S_2$			
Benzene, 1,1'-[1,3-butadiene-1,4-diyl-bis(sulfonyl)]bis[2-methyl-, (E,E)-	CHCl$_3$	269(4.53)	139-0005-84C
(E,Z)-	CHCl$_3$	267(4.45)	139-0005-84C
Benzene, 1,1'-[1,3-butadiene-1,4-diyl-bis(sulfonyl)]bis[4-methyl-, (E,E)-	CHCl$_3$	242(3.92),276(4.46)	139-0005-84C
(E,Z)-	CHCl$_3$	269(4.42)	139-0005-84C
$C_{18}H_{18}O_5$			
2-Dibenzofurancarboxylic acid, 3,7-di-methoxy-1,9-dimethyl-, methyl ester	MeOH	225s(4.16),226(4.52), 306(4.30)	39-1613-84C
$C_{18}H_{18}O_6$			
9,10-Epoxyanthracene-1(4H)-one, 9-acet-oxy-4a,9,9a,10-tetrahydro-4,4-di-methoxy-, (4aα,9β,9aα,10α)-	MeOH	223(3.60),260s(2.94)	12-1699-84
$C_{18}H_{18}O_6S_2$			
Benzene, 1,1'-[1,3-butadiene-1,4-diyl-	CHCl$_3$	246(4.23),294(4.44)	139-0005-84C

Compound	Solvent	$\lambda_{max}(\log \epsilon)$	Ref.
bis(sulfonyl)bis[4-methoxy-, (E,E)- (E,Z)-	 $CHCl_3$	 247(4.30),294(4.22)	139-0005-84C 139-0005-84C
$C_{18}H_{18}O_7$			
4H-1-Benzopyran-4-one, 2,3-dihydro- 5-hydroxy-2-(3-hydroxy-4-methoxy- phenyl)-6,7-dimethoxy-	EtOH	210(4.52),228(4.47), 285(4.45),340(3.74)	102-0703-84
	EtOH-NaOH	210(4.75),238s(4.51), 290(4.41),370(4.11)	102-0703-84
	EtOH-NaOAc	210(4.75),230(4.54), 285(4.45),335(3.88)	102-0703-84
	EtOH-AlCl$_3$	218(4.75),306(4.57), 358(4.06)	102-0703-84
	+ HCl	222(4.76),310(4.59), 358(4.18)	102-0703-84
4H-1-Benzopyran-4-one, 2,3-dihydro- 5-hydroxy-2-(3-hydroxy-4-methoxy- phenyl)-7,8-dimethoxy-	EtOH	210(4.07),230s(3.92), 288(3.86),340(3.26)	102-0703-84
	EtOH-NaOH	210(4.19),285(3.84), 360(3.58)	102-0703-84
	EtOH-NaOAc	210(4.98),285(4.22), 340(3.68)	102-0703-84
	EtOH-AlCl$_3$	215(4.37),310(4.13), 365(3.86)	102-0703-84
	+ HCl	215(4.35),308(4.12), 360(3.90)	102-0703-84
4H-1-Benzopyran-4-one, 2,3-dihydro- 6-hydroxy-2-(3-hydroxy-4-methoxy- phenyl)-5,7-dimethoxy-	MeOH	237(4.4),280(4.3), 344(3.7)	94-1472-84
4H-1-Benzopyran-4-one, 2,3-dihydro- 6-hydroxy-2-(4-hydroxy-3-methoxy- phenyl)-5,7-dimethoxy-	MeOH	237(4.3),280(4.1), 345(3.5)	94-1472-84
9H-Xanthen-9-one, 1,2,3,4,7-penta- methoxy-	EtOH	240(4.51),261(4.63), 287(4.00),309s(3.83), 368(3.85)	39-1507-84C
$C_{18}H_{18}O_8$			
Crotepoxide	EtOH	229(4.10),274(3.12), 282(3.04)	12-0221-84
9H-Xanthen-9-one, 1-hydroxy-2,3,4,6,8- pentamethoxy-	MeOH	250(4.42),260s(4.36), 323(4.28),369s(3.56)	39-1507-84C
9H-Xanthen-9-one, 1-hydroxy-3,5,6,7,8- pentamethoxy-	MeOH	256(4.45),314(4.54), 358(4.60)	100-0123-84
	MeOH-NaOMe	247(--),257(--), 274s(--),318(--)	100-0123-84
$C_{18}H_{18}O_9$			
Benzoic acid, 2-[(4,5-dimethoxy-3,6-di- oxo-1,4-cyclohexadien-1-yl)oxy]-4,6- dimethoxy-, methyl ester	MeOH	273(4.24),283(4.22)	39-1507-84C
$C_{18}H_{18}S$			
Anthracene, 9-(butylthio)-	EtOH	216(3.23),256(4.51), 335(2.61),354(2.96), 373(3.17),394(3.05)	40-1158-84
Anthracene, 9-[(2-methylpropyl)thio]-	EtOH	215(3.62),256(4.76), 334(3.02),354(3.36), 374(3.52),392(3.44)	40-1158-84
$C_{18}H_{19}BrO_2$			
Phenanthrene, 5-bromo-9,10-dihydro- 2,6-dimethoxy-1,7-dimethyl-	EtOH	216(4.56),282(4.29)	39-1913-84C

Compound	Solvent	$\lambda_{max}(\log \epsilon)$	Ref.
$C_{18}H_{19}BrO_4$ Benzene, 1-bromo-2-[2-(3,4-dimethoxy-phenyl)ethenyl]-4,5-dimethoxy-, (E)-	EtOH	298(4.25),324(4.26)	20-1099-84
$C_{18}H_{19}BrO_5$ Phenol, 6-[2-(2-bromo-4,5-dimethoxy-phenyl)ethenyl]-2,3-dimethoxy-	CHCl$_3$	246(4.21),298(4.20), 336(4.23)	142-1217-84
$C_{18}H_{19}BrO_6$ Ethanone, 2-(2-bromo-4,5-dimethoxyphen-yl)-1-(2-hydroxy-3,4-dimethoxyphenyl)-	CHCl$_3$	243(4.18),287(4.40)	142-1217-84
$C_{18}H_{19}ClN_2O_2$ Acetamide, 2-chloro-N-[3-(hydroxyimi-no)-3-phenylpropyl]-N-(phenylmethyl)-	EtOH	244(4.1)	5-1696-84
$C_{18}H_{19}ClN_2O_2S$ Acetamide, N-[3-(2-chlorobenzoyl)-5-ethyl-2-thienyl]-2-[(1-methylethyli-dene)amino]-	MeOH	239(3.92),265(3.71), 271s(3.70),350(3.66)	73-0621-84
$C_{18}H_{19}ClN_2O_4S$ 1-Piperazinecarboxylic acid, 4-(3-chloro-5,6-dihydro-2-oxo-2H-thieno-[2,3-h]-1-benzopyran-4-yl)-, ethyl ester	EtOH	224(4.14),243(4.19), 275s(3.65),362(4.18)	161-0081-84
$C_{18}H_{19}ClO_2$ Benzene, 1-chloro-5-methoxy-2-[2-(3-methoxy-2-methylphenyl)ethenyl]-4-methyl-, (E)-	EtOH	217(4.44),299(4.40)	39-1913-84C
(Z)-	EtOH	197(4.61),277(4.18)	39-1913-84C
$C_{18}H_{19}Cl_2NS_2$ Benzenesulfenamide, 4-chloro-N-[(4-chlorophenyl)thio]-N-cyclohexyl-	hexane	252(4.34)	44-2724-84
$C_{18}H_{19}Cl_3N_2O_5$ 1H-Indole-3-propanoic acid, 1-[2-meth-yl-1-oxo-2-[[(2,2,2-trichloroethoxy)-carbonyl]amino]propyl]-	EtOH	242(4.27),263(3.94), 273s(3.84),293(3.81), 301(3.84)	94-1373-84
$C_{18}H_{19}F_3N_4O_2$ Pyrimido[4,5-b]quinoline-2,4(3H,10H)-dione, 8-(dimethylamino)-10-ethyl-3,7-dimethyl-5-(trifluoromethyl)-	MeCN	254(4.43),489(4.13)	83-0042-84
$C_{18}H_{19}NO$ 1H-Indole, 1-acetyl-2,3-dihydro-5-(2-phenylethyl)-	EtOH	212(4.5),257(4.3), 262(4.2),286(3.7), 297(3.6)	103-0372-84
1H-Indole, 2-(1-methoxy-1-phenylethyl)-1-methyl-	EtOH	230(3.84),275(3.84), 285(3.81)	78-4837-84
1-Naphthalenecarboxamide, N-(cyclohex-ylidenemethyl)-	EtOH	225(4.27),285(3.80)	44-0714-84
6-Oxa-1-azabicyclo3.1.0]hexane, 2,2-di-methyl-4,5-diphenyl-	EtOH	248(3.28)	12-0095-84
Pyrano[2,3-a]carbazole, 2,3,4,11-tetra-hydro-2,2,5-trimethyl-	EtOH	225(4.45),252(4.70), 290(4.02),305(4.28), 325(4.12),355(3.75), 372(3.70)	25-0301-84

Compound	Solvent	$\lambda_{max}(\log \epsilon)$	Ref.
2H-Pyrrole, 3,4-dihydro-2,2-dimethyl-4,5-diphenyl-, 1-oxide	EtOH	293(4.00)	12-0095-84
2H-Pyrrol-4-ol, 3,4-dihydro-2,2-dimethyl-4,5-diphenyl-	EtOH	246(4.20)	12-0095-84
$C_{18}H_{19}NO_2$			
Benzamide, N-(1,1-dimethyl-3-oxo-3-phenylpropyl)-	EtOH	242(4.25)	12-0109-84
Bharatamine	pH 12	208(4.85),295(3.91), 310(2.71)	100-0397-84
	EtOH	206(5.08),290(2.77)	100-0397-84
Hydroperoxide, 3,4-dihydro-2,2-dimethyl-4,5-diphenyl-2H-pyrrol-4-yl	EtOH	245(4.11)	12-0095-84
Hydroxylamine, N-(1,3-diphenyl-2-propenyl)-O-acetyl-, trans	EtOH	254(4.26),285(3.75), 293(3.72)	150-1531-84
5H-Indeno[2,1-c]isoquinoline, 6,6a,7-11b-tetrahydro-2,3-dimethoxy-, (6aR-cis)-	MeOH	203s(4.53),231s(3.99), 268(3.40),273(3.56), 284(3.64),286(3.64), 292s(3.55)	39-1655-84C
hydrochloride	MeOH	194(4.77),204(4.75), 232(3.96),260s(3.15), 267(3.34),273(3.50), 283(3.60),286(3.60), 291s(3.53)	39-1655-84C
5H-Indeno[2,1-c]isoquinoline, 6,6a,7-11b-tetrahydro-3,4-dimethoxy-, (6aS-cis)-	MeOH	204s(4.69),230s(3.96), 261(3.07),267(3.25), 273(3.36),282(3.14)	39-1655-84C
hydrochloride	MeOH	194(4.73),203(4.69), 230s(3.95),260s(3.02), 266(3.21),272(3.34), 283(3.26)	39-1655-84C
5H-Indeno[2,1-c]isoquinoline, 6,6a,7-11b-tetrahydro-3,9-dimethoxy-, (6aR-cis)-	MeOH	198(4.82),231(4.26), 279(3.64),287(3.59)	39-1655-84C
hydrochloride	MeOH	199(4.87),232(4.31), 280(3.66),287(3.60)	39-1655-84C
6-Oxa-1-azabicyclo[3.1.0]hexan-4-ol, 2,2-dimethyl-4,5-diphenyl-	EtOH	252(3.08)	12-0109-84
Phenol, 4-[[(1,1-dimethylethyl)imino]-methyl]-, benzoate	EtOH	248(4.37)	150-0701-84M
$C_{18}H_{19}NO_2S$			
2-Propen-1-one, 2-(4-morpholinylmethyl)-1-phenyl-3-(2-thienyl)-	EtOH	328(3.28)	73-1764-84
$C_{18}H_{19}NO_3$			
2H-Benz[e]indene-5-carbonitrile, 3-acetoxy-3,3a,4,5-tetrahydro-7-methoxy-3a-methyl-	EtOH	268(4.22)	2-0395-84
1H-Inden-1-ol, 2,3-dihydro-5-methoxy-2-[[(3-methoxyphenyl)methylene]-amino]-, (1R-trans)-	MeOH	199(4.75),220(4.51), 253(4.24),280s(3.77), 287(3.74),302(3.58)	39-1655-84C
1H-Inden-1-ol, 2-[[(2,3-dimethoxyphenyl)methylene]amino]-2,3-dihydro-, (1S-trans)-	MeOH	218(4.55),258(4.24), 265s(4.19),271(4.02), 306(3.39)	39-1655-84C
1H-Inden-1-ol, 2-[[(3,4-dimethoxyphenyl)methylene]amino]-2,3-dihydro-, (1S-trans)-	MeOH	209(4.52),225s(4.30), 268(4.29),272(4.30), 303(4.12)	39-1655-84C
1-Pentanone, 4-methyl-4-nitro-1,2-diphenyl-	EtOH	247(3.98)	12-0095-84

Compound	Solvent	λ_{max}(log ϵ)	Ref.
2-Propen-1-one, 3-(2-furanyl)-2-(4-morpholinylmethyl)-1-phenyl-	EtOH	328(3.35)	73-1764-84
3-Pyridinecarboxaldehyde, 2-[3-(3,4-dimethoxyphenyl)-3-butenyl]-	n.s.g.	260(4.11),290(3.79)	35-7175-84
Quettamine, N-demethyl-, (±)-	MeOH	208(4.49),231(4.26), 275(4.04)	100-0753-84
$C_{18}H_{19}NO_4$			
Breoganine	EtOH	218(4.18),284(3.86)	88-1829-84
	EtOH-base	218(4.34),240(4.19), 300(4.01)	88-1829-84
2-Butenedioic acid, 2-[1-methyl-2-(2-propenyl)-1H-indol-3-yl]-, dimethyl ester, (E)-	EtOH	226(4.50),281(4.22), 363(3.87)	78-4837-84
(Z)-	EtOH	255(4.34),287(3.97)	78-4837-84
Caranine, acetyl-	MeOH	236(3.52),294(3.55)	100-0796-84
Celtisine	MeOH	225(4.10),283(3.85)	88-1829-84
	EtOH-base	225(4.50),294(3.91)	88-1829-84 +100-0753-84
Claviculine, (+)-	MeOH	227s(4.08),275s(3.48), 281(3.52)	142-0101-84
	+ base	244s(4.00),290(2.74)	142-0101-84
	EtOH	218(4.59),276(4.10)	100-0753-84
	+ base	240(5.66),292(4.55)	100-0753-84
Codeine, 10-oxo-	MeOH	220(3.99),250(3.98), 295(3.99),334(3.64)	106-0687-84
Culacorine, (+)-	MeOH	209(4.50),225s(4.19), 285(3.80),296s(3.64)	100-0753-84
Erythrinine, 8-oxo-	EtOH	238(4.10),272(3.83), 357(3.01)	102-0449-84
Norcularidine, (+)-	MeOH	228s(4.06),279s(3.70), 285(3.78),293s(3.60)	142-0101-84
	MeOH-base	246s(3.89),289(3.79)	142-0101-84
$C_{18}H_{19}NO_5$			
2-Butanone, 1-[(4,7-dimethoxyfuro-[2,3-b]quinolin-6-yl)oxy]-3-methyl-	MeOH	245(4.80),251(4.81), 279(4.11),308(4.07), 320(4.05),333(3.95)	142-2821-84
$C_{18}H_{19}NO_5S$			
5H-Furo[3,2-g][1]benzopyran-5-one, 7-[2-(dimethylamino)-1-(methylthio)-ethenyl]-4,9-dimethoxy-, (E)-	EtOH	214(4.42),239(4.38), 254(4.45),381(4.56)	44-5035-84
$C_{18}H_{19}NO_6$			
3-Azabicyclo[4.4.1]undeca-1,3,5,7,9-pentaene-2,4,5-tricarboxylic acid, 4,5-diethyl 2-methyl ester	MeOH	253(4.38),281(4.24), 364(3.69)	44-1040-84
1H-Pyrrole-1-propanoic acid, 4-(ethoxy-carbonyl)-2,3-dihydro-2,3-dioxo-5-phenyl-, ethyl ester	dioxan	283s(3.65),395(3.49)	94-0497-84
$C_{18}H_{19}NS$			
1-Naphthalenecarbothioamide, N-(cyclo-hexylidenemethyl)-	EtOH	210(4.55),305(4.25), 380s(2.74)	44-0714-84
2-Naphthalenecarbothioamide, N-(cyclo-hexylidenemethyl)-	EtOH	220(4.49),311(4.30), 392(2.31)	44-0714-84
$C_{18}H_{19}N_2O_5$			
1H-Pyrazolium, 3-(2-acetyl-3-ethoxy-3-oxo-1-propenyl)-2-ethyl-4,5-di-	EtOH	365(4.26)	48-0695-84

Compound	Solvent	$\lambda_{max}(\log \epsilon)$	Ref.
hydro-4,5-dioxo-1-phenyl-, iodide			48-0695-84
$C_{18}H_{19}N_3O$			
Ethanone, 1-[4-[[4-(1-pyrrolidinyl)- phenyl]azo]phenyl]-	C_6H_{12}	434(4.57)	39-0149-84B
	EtOH	454(4.51)	39-0149-84B
	EtOH-HCl	531(4.80)	39-0149-84B
1H-Indole, 1-acetyl-2,3-dihydro-5-[(1- phenylethylidene)hydrazino]-	EtOH	206.8(4.50),272.5(4.05), 320(4.24),347.5(4.31)	103-0376-84
1H-Indole, 1-acetyl-2,3-dihydro-6-[(1- phenylethylidene)hydrazino]-	EtOH	205.2(4.47),246(4.43), 318(4.37)	103-0376-84
$C_{18}H_{19}N_3O_3$			
4H-Pyrido[1,2-a]pyrimidine-3-carboxylic acid, 6,7,8,9-tetrahydro-6-methyl-4- oxo-9-(3-pyridinylmethylene)-, ethyl ester, (E)-	EtOH	294(4.03),348(4.26)	118-0582-84
$C_{18}H_{19}N_3O_4$			
Benzoic acid, 4-[3-[(benzoyloxy)methyl]- 3-methyl-1-triazenyl]-, ethyl ester	MeCN	227(4.33),295(4.35), 320s(4.22)	23-0741-84
Benzenepropanoic acid, β-[(4-methyl- 2-nitrophenyl)hydrazono]-, ethyl ester	EtOH	235(4.28),323(4.18), 453(3.84)	48-0829-84
$C_{18}H_{19}N_3O_4S$			
Benzenamine, 2-[5-(3,4-dimethoxyphenyl)- 1,3,4-thiadiazol-2-yl]-4,5-dimethoxy-	EtOH	214(4.48),320(4.13), 392(4.19)	39-1143-84C
$C_{18}H_{19}N_5O_2$			
Benzoic acid, 4-[1-[(2,4-diaminopyrido- [3,2-d]pyrimidin-6-yl)methyl]propyl]-	pH 13	238(4.55),343(3.71)	87-0376-84
$C_{18}H_{20}$			
1,3,5-Cycloheptatriene, 2-butene-1,4- diylbis- (isomer mixture)	isooctane	264(3.88)	24-2045-84
1,3,5-Cycloheptatriene, 7,7'-(2-butene- 1,4-diyl)bis-	isooctane	258(3.85)	24-2045-84
$C_{18}H_{20}ClN_3O_4$			
4H-Pyrido[1,2-a]pyrimidine-3-carbox- amide, N-(3-chloro-2,4-dimethoxy- phenyl)-6,7,8,9-tetrahydro-6-methyl- 4-oxo-	EtOH	232(4.15),330(4.14)	118-0582-84
$C_{18}H_{20}F_6$			
Tricyclo[10.4.0.0^{3,14}]hexadeca- 2,12,15-triene, 15,16-bis(tri- fluoromethyl)-	EtOH	225s(2.59),232s(2.54), 275(1.80)	24-0455-84
$C_{18}H_{20}FeO_4$			
Ferrocene, [3-ethoxy-2-(ethoxycarbon- yl)-3-oxo-1-propenyl]-	CH_2Cl_2	482(3.23)	30-0192-84
$C_{18}H_{20}FeO_6$			
Ferrocene, 1,1'-(1,12-dioxo-2,5,8,11- tetraoxadodecane-1,12-diyl)-	MeOH	452(2.53)	18-2435-84
$C_{18}H_{20}N_2$			
[1]Benzazocino[7,6,5-efg][1]benzazo- cine, 4,5,6,7,11,12,13,14-octa- hydro-, (S)-	MeOH	222.5(4.52),250s(3.74), 274(3.24),324(4.05)	39-2013-84C

Compound	Solvent	$\lambda_{max}(\log \epsilon)$	Ref.
dihydrochloride, (S)- (cont.)	MeOH	210(4.56),234(3.96), 262s(2.89),270(2.96), 278.5(3.02)	39-2013-84C
2,4,6-Cycloheptatriene-1-acetaldehyde, azine	isooctane	257(3.84)	24-2045-84
Methanediamine, 1-(3H-cyclopent[e]azulen-3-ylidene)-N,N,N',N'-tetramethyl-	CH_2Cl_2	254(4.39),300s(4.13), 336s(3.74),420(4.55), 482s(3.87),600(2.99)	35-6383-84
Tricyclo[10.4.0.0^{3,14}]hexadeca-2,12,15-triene-15,16-dicarbonitrile	EtOH	238(3.75),246s(3.70), 337(2.59)	24-0455-84
$C_{18}H_{20}N_2OS$			
Morpholine, 4-[phenyl[[(phenylmethyl)-thio]imino]methyl]-	hexane	308(3.53)	39-2933-84C
$C_{18}H_{20}N_2O_2$			
1,2-Ethanediimine, N,N'-bis(2-hydroxyethyl)-1,2-diphenyl-	$CHCl_3$	220(2.83),260(2.85), 305(3.23)	34-0237-84
Ethanone, 1-[1,4,5,6-tetrahydro-1-methyl-4-[(1-methyl-1H-indol-3-yl)carbonyl]-3-pyridinyl]-	EtOH	210(4.38),246(4.06), 303(4.41)	78-3339-84
Pyrano[4,3-c]pyrazol-4(1H)-one, 6,7-dihydro-6,6-dimethyl-3-(2-methyl-1-propenyl)-1-phenyl-	EtOH	259(4.28)	4-0017-84
$C_{18}H_{20}N_2O_3$			
Acetic acid, [1-methyl-3-[(1-methyl-1H-indol-3-yl)carbonyl]-2-pyrrolidinylidene]-, methyl ester	EtOH	246s(4.24),292(4.45)	150-2738-84M
Phenol, 4-[[(1,1-dimethylethyl)nitrosoamino]methyl]-, benzoate	hexane	355(1.65),366(1.94), 378(1.89)	150-0701-84M
	EtOH	231(4.38),352(1.83)	150-0701-84M
Pyridine, 3-(4,5-dihydro-4,4-dimethyl-2-oxazolyl)-4-(2,3-dimethoxyphenyl)-	EtOH	255(3.95)	44-0056-84
Pyridine, 5-(4,5-dihydro-4,4-dimethyl-2-oxazolyl)-2-(2,3-dimethoxyphenyl)-	EtOH	250(3.98)	44-0056-84
1H-Pyrrole-3-propanoic acid, 4,5-dihydro-1-methyl-2-(1-methyl-1H-indol-3-yl)-β-oxo-, methyl ester	EtOH	217(4.57),283s(3.98), 290s(3.99),295s(3.98), 347(4.38)	150-2738-84
$C_{18}H_{20}N_2O_4$			
1H-3-Benzazepin-7-ol, 8-methoxy-3-methyl-2-(4-nitrophenyl)-2,3,4,5-tetrahydro-	EtOH	211(4.17),233s(3.92), 279(4.03)	44-0581-84
Butanediamide, N,N'-dihydroxy-N,N'-bis(2-methylphenyl)-	MeOH	214(4.30),228(4.23)	34-0345-84
Butanediamide, N,N'-dihydroxy-N,N'-bis(4-methylphenyl)-	MeOH	205(4.47),216(4.40), 240(4.28)	34-0345-84
Hexanediamide, N,N'-dihydroxy-N,N'-diphenyl-	MeOH	218(4.15),248(4.14)	34-0345-84
Phenol, 2,2'-[1,2-ethanediylbis(nitrilomethylidyne)]bis[6-methoxy-	$CHCl_3$	223(2.76),263(2.80), 310(3.21),368(3.21), 400(2.85),420(2.64)	34-0237-84
$C_{18}H_{20}N_2O_6$			
Antibiotic SF-2140	MeOH	222(4.54),258s(3.88), 265(3.91),284(3.80), 294(3.84)	158-0931-84
Dityrosine	acid	283(3.73)	149-0731-84B
	base	317(--)	149-0731-84B

Compound	Solvent	$\lambda_{max}(\log \epsilon)$	Ref.
$C_{18}H_{20}N_2S$ 1,3,4-Thiadiazole, 2,5-bis(2,4,6-cyclo-heptatrien-1-ylmethyl)-2,5-dihydro-	MeCN	257(3.85),322s(2.68)	24-2045-84
$C_{18}H_{20}N_4$ 9H-Pyrimido[4,5-b]indol-4-amine, N,2-dimethyl-9-phenyl-5,6,7,8-tetrahydro-	EtOH	209(4.56),233s(4.29), 288(4.13)	11-0073-84B
$C_{18}H_{20}N_4O_2S$ Ethanol, 2,2'-[[4-(2-benzothiazolyl-azo)-3-methylphenyl]imino]bis-	acetone 50% acetone	514(4.72) 526(4.72)	7-0819-84 7-0819-84
$C_{18}H_{20}N_4O_3S$ Ethanol, 2-[[4-[[6-(2-hydroxyethoxy)-2-benzothiazolyl]azo]phenyl]methyl-amino]-	pH 1.2 pH 6.2 EtOH	538(4.63),618s(4.35) 536(4.71) 521(4.70)	48-0151-84 48-0151-84 48-0151-84
$C_{18}H_{20}N_4O_3S_2$ Thiazolo[2,3-b]purine-4,8(1H,7H)-dione, 5a-ethoxy-5,5a-dihydro-1,5-dimethyl-2-(methylthio)-7-phenyl-	MeOH	278(4.34),300(4.35)	2-0316-84
$C_{18}H_{20}N_4O_4$ Oxepino[2,3-d]pyrimidine-6-carboxylic acid, 2-hydrazino-3,4,7,8-tetrahydro-8-methyl-4-oxo-3-phenyl-, ethyl ester	CHCl$_3$	277(3.97),343(4.24)	24-0585-84
$C_{18}H_{20}N_4O_4S_2$ Acetic acid, [[(2,3,4,10-tetrahydro-3,7,8,10-tetramethyl-2,4-dioxobenzo-[g]pteridin-6-yl)methyl]dithio]-, methyl ester	MeCN 6M HCl	227(4.49),272(4.64), 363(4.02),448(4.10) 224(4.48),273(4.42), 394(4.32)	35-3309-84 35-3309-84
$C_{18}H_{20}N_4O_5S_2$ Acetic acid, [[(5-formyl-3,4,5,10-tetrahydro-3,7,8,10-tetramethyl-2,4-dioxobenzo[g]pteridin-4a(2H)-yl)methyl]dithio]-	MeCN	242(4.01),260s(--), 321(3.98)	35-3309-84
$C_{18}H_{20}N_4O_6$ Inosine, 2-[(4-methoxyphenyl)methyl]- 4H-Pyrazolo[3,4-d]pyrimidin-4-one, 1,5-dihydro-5-(2-methoxyphenyl)-6-methyl-1-β-D-ribofuranosyl- 4H-Pyrazolo[3,4-d]pyrimidin-4-one, 2,5-dihydro-5-(2-methoxyphenyl)-6-methyl-2-β-D-ribofuranosyl-	pH 1 pH 7 pH 13 EtOH EtOH	253(4.14) 251(4.12) 256(4.16) 210.5(4.66),254.0(3.97), 272(4.01) 268.0(3.83)	87-0429-84 87-0429-84 87-0429-84 11-0023-84A 11-0023-84A
$C_{18}H_{20}N_4O_8$ Carbamic acid, [1-(2-deoxy-β-D-erythro-pentofuranosyl)-1,2-dihydro-2-oxo-4-pyrimidinyl]-, 2-(4-nitrophenyl)ethyl ester	MeOH	242(4.24),281(4.16)	78-0059-84
$C_{18}H_{20}N_4O_9$ Carbamic acid, (1,2-dihydro-2-oxo-1-β-D-ribofuranosyl-4-pyrimidinyl)-, 2-(4-nitrophenyl)ethyl ester	MeOH	243(4.23),280(4.14)	78-0059-84

Compound	Solvent	$\lambda_{max}(\log \epsilon)$	Ref.
$C_{18}H_{20}N_6O_6$			
9H-Purin-2-amine, 9-(2-deoxy-β-D-erythro-pentofuranosyl)-6-[2-(4-nitrophenyl)ethoxy]-	MeOH	250(4.16),279(4.25)	78-0059-84
$C_{18}H_{20}N_6O_7$			
9H-Purin-2-amine, 6-[2-(4-nitrophenyl)-ethoxy]-9-β-D-ribofuranosyl-	MeOH	251(4.17),278(4.26)	78-0059-84
$C_{18}H_{20}N_8O_6S$			
L-Cysteic acid, N-(4-amino-4-deoxy-N^{10}-methylpteroyl)-	pH 1	242(4.25),306(4.35)	87-0600-84
	pH 7.4	258(4.37),300(4.38), 372(3.90)	87-0600-84
L-Homocysteic acid, N-(4-amino-4-deoxy-pteroyl)-	pH 1	243(4.26),290(4.30)	87-0600-84
	pH 7.4	260(4.44),282(4.42), 370(3.95)	87-0600-84
$C_{18}H_{20}O$			
15(14→8α)-abeo-16-Norestra-1,3,5(10),6-tetraen-17-one, 4-methyl-	MeOH	223(4.38),278(4.01)	39-2173-84C
$C_{18}H_{20}ORu$			
Ruthenocene, 1,1'-(1,4-butanediyl)-3,3'-(2-oxo-1,4-butanediyl)-	EtOH	280s(3.17)	18-0719-84
$C_{18}H_{20}O_2$			
[1,1'-Biphenyl]-3-ol, 5-methoxy-4-(3-methyl-2-butenyl)-	MeOH	205(6.9),262(6.06)[sic]	102-0765-84
1,3-Cyclohexanedione, 2-(4-phenyl-1-cyclohexen-1-yl)-, compd. with methanesulfonic acid	pH 1	264(4.07)	39-1213-84C
	H_2O	264(4.07)	39-1213-84C
	pH 13	291(4.34)	39-1213-84C
Phenanthrene, 9,10-dihydro-2,6-dimethoxy-1,7-dimethyl-	EtOH	218(4.56),278(4.23), 295(4.05),315(4.00)	39-1913-84C
$C_{18}H_{20}O_2S_3$			
Ethane(dithioic) acid, [(2-hydroxy-2,2-diphenylethyl)sulfinyl]-, ethyl ester	EtOH	317(4.09)	39-0085-84C
$C_{18}H_{20}O_3$			
9-Epiallogibberic acid	n.s.g.	266(2.54)	23-1996-84
$C_{18}H_{20}O_3S_3$			
Benzenemethanol, α-[[[2,2-bis(methylthio)ethenyl]sulfonyl]methyl]-α-phenyl-	EtOH	211(4.27),282(4.12)	39-0085-84C
$C_{18}H_{20}O_4$			
Methanone, phenyl[2,4,6-trihydroxy-3-(3-methylbutyl)phenyl]-	MeOH	251(3.88),311(4.16)	39-1413-84C
	MeOH-base	342(4.22)	39-1413-84C
4a(2H)-Phenanthreneacetic acid, 3,4,9,10-tetrahydro-5-methoxy-2-oxo-, methyl ester	EtOH	230(4.41),282(3.43)	39-1515-84C
4a(2H)-Phenanthreneacetic acid, 3,4,9,10-tetrahydro-6-methoxy-2-oxo-, methyl ester	EtOH	230(4.39),282(3.60)	39-1515-84C
$C_{18}H_{20}O_5$			
Dibenzofuran, 1,3,7,9-tetramethoxy-2,6-dimethyl-	EtOH	221s(4.56),232(4.57), 268(4.18),287(4.23), 297s(4.09),309(4.17)	39-1605-84C

Compound	Solvent	$\lambda_{max}(\log \epsilon)$	Ref.
Dibenzofuran, 1,3,7,9-tetramethoxy-2,8-dimethyl-	EtOH	227(4.60),263(4.13), 299(4.27)	39-2573-84C
1-Dibenzofuranol, 3,7,9-trimethoxy-2,4,6-trimethyl-	EtOH	218(4.56),234(4.57), 278(4.32),288s(4.30), 305s(3.45)	39-1605-84C
2,5-Furandione, 3-[(3,5-dimethoxyphenyl)methylene]-4-(1-ethylpropylidene)dihydro-, (E)-	C_6H_{12}	332(4.06)	48-0233-84
(Z)-	C_6H_{12}	346(4.23)	48-0233-84
Naphtho[1,2-b]furan-4,5-dione, 2,3-dihydro-6,7-dimethoxy-2,2,3,9-tetramethyl-, (±)-	MeOH	276(4.37),310s(3.69), 369(3.35),470(3.43)	44-1853-84
Naphtho[1,2-b]furan-4,5-dione, 2,3-dihydro-6,7-dimethoxy-2,3,3,9-tetramethyl-, (±)-	MeOH	276(4.45),310s(3.72), 369(3.42),474(3.56)	44-1853-84

$C_{18}H_{20}O_6$

Benzo[b]cyclopropa[e]pyran-1a(1H)-carboxylic acid, 1-acetyl-7,7a-dihydro-6-hydroxy-7-oxo-3-propyl-, (1α,1aα,7aα)-	EtOH	215(4.29),275(3.96)	39-0487-84C
(1α,1aβ,7aβ)-	EtOH	208(4.23),278(3.94)	39-0487-84C
4H-1-Benzopyran-2-carboxylic acid, 4-oxo-5-(2-oxopropoxy)-8-propyl-, ethyl ester	EtOH	220(4.24),238(4.18), 266(3.97),275(3.97)	39-0487-84C
2H-1-Benzopyran-3-ol, 3,4-dihydro-2-(4-hydroxy-3-methoxyphenyl)-5,7-dimethoxy-, (2R-trans)-	MeOH	204(3.79),228(4.36), 278(4.60)	102-0675-84
Nagilactone D	EtOH	303(3.85)	154-0547-84

$C_{18}H_{20}O_7$

| 2-Benzofuranbutanoic acid, 7-acetyl-4,6-dihydroxy-3,5-dimethyl-β-oxo-, ethyl ester | MeOH | 243(3.98),301(3.86), 346(3.40) | 23-0320-84 |
| 2,2(3H)-Furandicarboxylic acid, 4-formyl-3-(4-methoxyphenyl)-, ethyl ester | EtOH | 201(4.40),230(4.02), 255(4.02) | 118-0974-84 |

$C_{18}H_{20}S$

| Thiirane, 2,3-bis(2,4,6-cycloheptatrien-1-ylmethyl)- | isooctane | 258(3.73) | 24-2045-84 |

$C_{18}H_{20}S_2$

| Spiro[3H-benz[cd]azulene-3,2'-[1',3]-dithiane], 4,5-dihydro-7,9-dimethyl- | CH_2Cl_2 | 242(4.31),294(4.65), 338s(--),349(3.81), 363s(--),532(2.68) | 83-0984-84 |

$C_{18}H_{21}BrN_2O_6$

| Carbamic acid, [2-(4a-bromo-3,4,4a,9a-tetrahydro-9a-methoxy-2-oxopyrano-[2,3-b]indol-9(2H)-yl)-1,1-dimethyl-2-oxoethyl]-, methyl ester | EtOH | 238(4.10) | 94-1373-84 |

$C_{18}H_{21}ClO_2$

| Estra-1,4-diene-3,17-dione, 10β-chloro- | EtOH | 243(4.2) | 48-0941-84 |

$C_{18}H_{21}F_3N_4O_2$

Pyrimido[4,5-b]quinoline-2,4(1H,3H)-dione, 8-(dimethylamino)-10-ethyl-5,10-dihydro-3,7-dimethyl-5-(trifluoromethyl)-	pH 1	231(3.96),303(4.04)	83-0042-84
	pH 6	228(4.24),305(4.01)	83-0042-84
	pH 13	264s(4.04),300(4.00)	83-0042-84

Compound	Solvent	$\lambda_{max}(\log \epsilon)$	Ref.
$C_{18}H_{21}F_3N_4O_3$			
2,4(1H,3H)-Pyrimidinedione, 6-[[3-(di-methylamino)-4-methylphenyl]ethyl-amino]-3-methyl-5-(trifluoroacetyl)-	MeOH	263(4.24)	83-0042-84
Pyrimido[4,5-b]quinoline-2,4(1H,3H)-di-one, 8-(dimethylamino)-10-ethyl-5,10-dihydro-5-hydroxy-3,7-dimethyl-5-(tri-fluoromethyl)-	CHCl$_3$	299(4.01)	83-0042-84
$C_{18}H_{21}Li$			
Lithium, (1,1-diphenylhexyl)-	heptane at -78°	440(4.23)	138-0757-84
(all solutions contg. sparteine)	toluene	446(4.18)	138-0757-84
(at -78°)	toluene	450(4.28)	138-0757-84
$C_{18}H_{21}NO$			
Benzenemethanol, α-[1-[(dimethylamino)-methyl]ethenyl]-α-phenyl-	MeOH	207(4.20),275(3.43)	78-0215-84
Isoquinoline, 1,2,3,4-tetrahydro-4-[(2-methoxyphenyl)methyl]-2-methyl-	MeOH	209(4.31),247(3.31), 272(3.54)	78-0215-84
$C_{18}H_{21}NOS$			
2-Propen-1-one, 2-[(diethylamino)meth-yl]-1-phenyl-3-(2-thienyl)-	EtOH	326(3.28)	73-1764-84
2-Propen-1-one, 2-[[(1,1-dimethyleth-yl)amino]methyl]-1-phenyl-3-(2-thi-enyl)-	EtOH	326(3.28)	73-1764-84
$C_{18}H_{21}NO_2$			
Phenol, 4-[[(1-methylpropyl)amino]meth-yl]-, benzoate, hydrochloride	EtOH	232(4.24)	150-0701-84M
2-Propen-1-one, 2-[(diethylamino)meth-yl]-3-(2-furanyl)-1-phenyl-	EtOH	330(3.35)	73-1764-84
2-Propen-1-one, 2-[[(1,1-dimethyleth-yl)amino]methyl]-3-(2-furanyl)-1-phenyl-	EtOH	329(3.37)	73-1764-84
$C_{18}H_{21}NO_3$			
1H-Inden-1-ol, 2,3-dihydro-4,5-dimeth-oxy-2-[(phenylmethyl)amino]-, (1S-trans)-	MeOH	203(4.74),227s(3.95), 274(3.12),276(3.13), 281(3.12)	39-1655-84C
1H-Inden-1-ol, 2-[[(2,3-dimethoxyphen-yl)methyl]amino]-2,3-dihydro-, (1S-trans)-	MeOH	196(4.73),215s(4.26), 266(3.31),272(3.42), 278s(3.21)	39-1655-84C
1H-Inden-1-ol, 2-[[(3,4-dimethoxyphen-yl)methyl]amino]-2,3-dihydro-, (1S-trans)-	MeOH	201s(4.79),216s(4.16), 230(3.95),266(3.39), 273(3.54),278(3.48), 280(3.48),285s(3.42)	39-1655-84C
8-Isoquinolinol, 1,2,3,4-tetrahydro-1-(4-hydroxyphenyl)methyl]-7-methoxy-2-methyl-, (R)- (norjuziphine)	MeOH	234(3.52),280(3.26), 284s(3.22)	100-0459-84
$C_{18}H_{21}NO_4$			
2-Butenedioic acid, 2-(1-cyclohexen-1-ylphenylamino)-, dimethyl ester, cis	EtOH	258(4.12),320(4.03)	103-0295-84
trans	EtOH	260(4.09),370(4.03)	103-0295-84
$C_{18}H_{21}NO_5$			
Ambelline	MeOH	287(3.17)	100-0796-84

Compound	Solvent	$\lambda_{max}(\log \epsilon)$	Ref.
3-Azabicyclo[4.4.1]undeca-1,3,5,7,9-pentaene-4,5-dicarboxylic acid, 2-ethoxy-, diethyl ester	MeOH	252(4.27),273(4.20), 330(3.73)	44-1040-84
2-Hexenoic acid, 4-(1,3-dihydro-1,3-dioxo-2H-isoindol-2-yl)-3-methoxy-5-methyl-, ethyl ester, (E)-(±)-	EtOH	221(4.61)	44-3489-84
Spiro[benzofuran-3(2H),4'-piperidine]-2-carboxylic acid, 7-methoxy-1'-methyl-3'-methylene-2'-oxo-, ethyl ester	EtOH	275(3.46)	44-0056-84
Sterbergine	MeCN	222(3.90),286(3.51)	100-1003-84
Undulatine	MeOH	287(3.25)	100-0796-84

$C_{18}H_{21}NO_6$

Compound	Solvent	$\lambda_{max}(\log \epsilon)$	Ref.
2,3-Butanediol, 1-[(4,7-dimethoxyfuro-[2,3-b]quinolin-6-yl)oxy]-3-methyl-, (+)- (nkolbisine)	MeOH	245(4.83),251(4.84), 308(4.12),321(4.13), 334(3.98)	142-2821-84
	MeOH-HCl	217(4.22),251(4.72), 335(4.23)	142-2821-84
1,2-β-Epoxyambelline	MeOH	240s(3.04),288(3.18)	150-0232-84S
Montrifoline	n.s.g.	244(4.68),252(4.73), 309(3.99),321(3.99), 333(3.82)	100-0379-84
Nigdenine, (+)-	MeOH	249(4.78),320(3.74), 332(3.73),348s(3.54)	142-2821-84
	MeOH-HCl	253(4.68),321(3.71), 350(3.73)	142-2821-84

$C_{18}H_{21}NO_7$

Compound	Solvent	$\lambda_{max}(\log \epsilon)$	Ref.
1-Propanone, 1-(2,4,5,8-tetramethoxy-3-methyl-6-nitro-1-naphthalenyl)-	EtOH	232(4.57),266(4.07), 378(3.58)	88-1373-84

$C_{18}H_{21}NO_{10}$

Compound	Solvent	$\lambda_{max}(\log \epsilon)$	Ref.
2(1H)-Pyridinone, 3-acetoxy-1-(2,3,5-tri-O-acetyl-β-D-ribofuranosyl)-	MeOH	227(3.60),302(3.80)	87-0160-84

$C_{18}H_{21}N_2$

Compound	Solvent	$\lambda_{max}(\log \epsilon)$	Ref.
5,12-Diaza[2₄](1,2,4,5)cyclophanonium, N-methyl-, triflate	EtOH	286(3.44),318(3.70)	35-2672-84
5,15-Diaza[2₄](1,2,4,5)cyclophanonium, N-methyl-, triflate	EtOH	292(3.37),320(3.55)	35-2672-84

$C_{18}H_{21}N_3$

Compound	Solvent	$\lambda_{max}(\log \epsilon)$	Ref.
Piperidine, 1-[4-[(4-methylphenyl)azo]phenyl]-	C_6H_{12}	391(4.45)	39-0149-84B
	EtOH	398(4.39)	39-0149-84B
	EtOH-HCl	541(3.45)	39-0149-84B

$C_{18}H_{21}N_3O$

Compound	Solvent	$\lambda_{max}(\log \epsilon)$	Ref.
Piperidine, 1-[4-[(4-methoxyphenyl)-azo]phenyl]-	C_6H_{12}	392(4.43)	39-0149-84B
	EtOH	398(4.45)	39-0149-84B
	EtOH-HCl	558(3.08)	39-0149-84B

$C_{18}H_{21}N_3O_2$

Compound	Solvent	$\lambda_{max}(\log \epsilon)$	Ref.
Benzo[c]cinnoline, 1,2,3,4,4a,5,7,8-9,10-decahydro-5-(4-nitrophenyl)-	CH_2Cl_2	442(4.4)	88-0057-84
4H-Pyrido[1,2-a]pyrimidine-3-carboxamide, 6,7,8,9-tetrahydro-6-methyl-4-oxo-N-(2-phenylethyl)-	EtOH	230(3.85),300(3.99)	118-0582-84
2-Pyrrolidinecarboxamide, 1-[[1,4-dihydro-1-(phenylmethyl)-3-pyridinyl]-carbonyl]-, (S)-	EtOH	349(3.70)	35-1481-84
2-Pyrrolidinecarboxamide, 1-[[1,6-di-	$CHCl_3$	355(3.63)	35-1481-84

Compound	Solvent	$\lambda_{max}(\log \epsilon)$	Ref.
hydro-1-(phenylmethyl)-3-pyridinyl]- carbonyl]-, (S)- (cont.)			35-1481-84
$C_{18}H_{21}N_3O_3$ 2-Naphthalenecarbonitrile, 5-amino- 3-(butylamino)-1,4-dihydro-8-hy- droxy-1,4-dioxo-6(or 7)-propyl-	benzene	510s(--),540(--), 580(--)	39-1297-84C
$C_{18}H_{21}N_3O_5S$ Acetic acid, α-cyano-α-(dihydro-1,2- dimethyl-4-pyrimidinylidene)-, ethyl ester, p-toluenesulfonate	CF$_3$COOH	315(4.38)	103-1270-84
$C_{18}H_{21}N_5O$ 4H-Pyrazolo[3,4-d]pyrimidin-4-one, 1,5-dihydro-6-methyl-5-phenyl-1-(4- piperidinylmethyl)- (or isomer)	EtOH	205.5(4.57),252.0(3.91)	11-0023-84A
$C_{18}H_{21}N_5O_5$ Guanosine, N-[(4-methylphenyl)methyl]-	pH 1	259(4.16),282s(3.94)	44-0363-84
	pH 6.9	253(4.16),274s(4.00)	44-0363-84
	pH 13.0	257(4.11),270s(4.07)	44-0363-84
Guanosine, 1-[(4-methylphenyl)methyl]-	pH 1	260(4.07),277s(3.91)	44-0363-84
	pH 6.9	256(4.13),270s(3.99)	44-0363-84
	pH 13	257(4.13),270s(4.00)	44-0363-84
Guanosine, 8-[(4-methylphenyl)methyl]-	pH 1	262(4.25),275s(4.14)	44-0363-84
	pH 6.9	255(4.31),272s(4.14)	44-0363-84
	pH 13	260(4.24)	44-0363-84
9H-Purin-2-amine, 6-[(4-methylphenyl)- methoxy]-9-β-D-ribofuranosyl-	pH 6.9	247(3.98),281(4.00)	44-0363-84
	pH 13	248(3.98),281(4.03)	44-0363-84
$C_{18}H_{21}N_5O_7S_2$ 1H-Pyrazolo[3,4-d]pyrimidine-3-carbo- thioamide, 4-(methylthio)-1-(2,3,5- tri-O-acetyl-β-D-ribofuranosyl)-	EtOH	292(4.01)	103-0210-84
$C_{18}H_{21}N_5O_8$ Acetamide, N-[9-(2,3,5-tri-O-acetyl- β-D-arabinofuranosyl)-9H-purin-6- yl]-	MeOH	268(3.88)	31-0339-84
$C_{18}H_{22}Br_2N_2O_2$ 2H-Pyrrole-3-propanoic acid, 2-[(5- bromo-3,4-diethyl-1H-pyrrol-2-yl)- methylene]-5-(bromomethyl)-4- methyl-, monohydrobromide	CH$_2$Cl$_2$	374(3.83),505(4.89)	78-0579-84
$C_{18}H_{22}ClNO_2$ Phenol, 4-[[(1-methylpropyl)amino]- methyl]-, benzoate, hydrochloride	EtOH	232(4.24)	150-0701-84M
$C_{18}H_{22}N_2$ Dibenzo[e,g][1,4]diazocine, 5,6,7,8- tetrahydro-1,6,7,12-tetramethyl-, (S)-(-)-	MeOH	215(4.59),226s(4.46), 251s(3.99),290(3.28)	39-2013-84C
dihydrochloride	MeOH	199(4.59),212s(4.51), 233s(4.10),267s(3.30), 275s(3.20)	39-2013-84C
1,4:5,10-Dimethanobenzo[g]phthalazine, 1,4,4a,5,10,10a-hexahydro-1,4,11,11- tetramethyl-	hexane	254(2.57),260(2.77), 267(2.95),273(2.93), 365(2.27)	24-0517-84

Compound	Solvent	$\lambda_{max}(\log \epsilon)$	Ref.
$C_{18}H_{22}N_2O$			
Appacicine, 16,22-dihydro-16-hydroxy-, (S)-	EtOH	220(4.43),284(3.79), 292s(--)	100-0835-84
$C_{18}H_{22}N_2O_2$			
Cycloocta[1,2-b:5,6-b']dipyridine-3,9-dimethanol, 5,6,11,12-tetrahydro-2,8-dimethyl-	EtOH	214(4.12),271(3.96), 276(3.96),280s(3.91)	35-2672-84
Cycloocta[1,2-b:6,5-b']dipyridine-2,8-dimethanol, 5,6,11,12-tetrahydro-3,9-dimethyl-	EtOH	216(4.16),271s(3.97), 276(3.98),280(3.93)	35-2672-84
Pentanoic acid, 4-cyano-2-(1-cyano-1-methylethyl)-4-methyl-, 4-methyl-phenyl ester	dioxan	266(2.88),272(2.87)	126-0499-84
2-Pyrrolidinemethanol, 1-[[1,4-dihydro-1-(phenylmethyl)-3-pyridinyl]carbo-nyl]-, (S)-	MeCN	342(3.65)	35-1481-84
6H-Pyrrolo[3,2-f]indolizine-2-carbox-ylic acid, 3,4,6,7,8-pentamethyl-, ethyl ester, monohydrochloride	MeOH	260(4.70),314(4.20), 322(4.23)	65-2350-84
Spiro[3H-indole-3,1'(5'H)-indolizine]-8'-carboxaldehyde, 8'-ethyl-1,2,2'-3',6',7',8',8'a-octahydro-2-oxo-, [1'R-(1'α,8'α,8'aβ)]-	MeOH	227(4.15),254(3.94), 285(3.32)	44-0547-84
$C_{18}H_{22}N_2O_3$			
Spiro[3H-indole-3,1'(5'H)-indolizine]-8'-carboxaldehyde, 8'-ethyl-1,2,2'-3',6',7',8',8'a-octahydro-2-oxo-, 4'-oxide	MeOH	224(4.27),258(4.00), 290(3.31)	44-0547-84
$C_{18}H_{22}N_2O_5$			
1H-Indole-3-propanoic acid, 1-[2-[(methoxycarbonyl)amino]-2-methyl-1-oxopropyl]-, methyl ester	EtOH	242.5(4.28),265s(3.92), 273s(3.86),293(3.82), 301(3.85)	94-1373-84
$C_{18}H_{22}N_2O_6$			
2H-[1]Benzopyrano[3,4-c]pyridazine-3,4-dicarboxylic acid, 4a,5-dihydro-8-methoxy-5,5-dimethyl-, dimethyl ester	EtOH	219(4.36),262(4.17), 265s(4.16),303(3.83), 309s(3.81)	39-2815-84C
$C_{18}H_{22}N_2S$			
1,3,4-Thiadiazolidine, 2,5-bis(2,4,6-cycloheptatrien-1-ylmethyl)-	MeCN	258(3.81)	24-2045-84
$C_{18}H_{22}N_4$			
7H-Pyrrolo[2,3-d]pyrimidin-4-amine, 2,5,6-trimethyl-7-phenyl-N-propyl-	EtOH	206(4.69),216s(4.65), 230(4.43),285(4.28)	11-0073-84B
$C_{18}H_{22}N_4O_2$			
2-Naphthalenecarbonitrile, 5-amino-3-(butylamino)-1,4-dihydro-1,4-dioxo-8-(propylamino)-	benzene	440(3.52),612(4.05), 662(4.01)	39-1297-84C
$C_{18}H_{22}N_4O_4S_2$			
Acetic acid, [[(3,4,5,10-tetrahydro-3,7,8,10-tetramethyl-2,4-dioxo-benzo[g]pteridin-4a(2H)-yl)methyl]-dithio]-, methyl ester	MeCN	218(4.52),275(4.14), 300s(--),363(3.81)	35-3309-84

Compound	Solvent	$\lambda_{max}(\log \epsilon)$	Ref.
$C_{18}H_{22}N_4O_5S$			
Acetic acid, [[(3,4,5,10-tetrahydro-3,7,8,10-tetramethyl-2,4-dioxo-benzo[g]pteridin-4a(2H)-yl)methyl]-sulfinyl]-, methyl ester	MeOH	228(4.23),274(4.09), 300s(--),362(3.79)	35-3309-84
diastereoisomer B	MeOH	248(4.14),275(4.09), 302s(--),364(3.79)	35-3309-84
$C_{18}H_{22}N_4O_5S_2$			
2-Propenamide, 2-methyl-N-[[(phenyl-methyl)amino]sulfonyl]-3-[[[(phen-ylmethyl)amino]sulfonyl]amino]-	50% EtOH + NaOH	265(4.09) 298(4.27)	73-0840-84 73-0840-84
$C_{18}H_{22}N_4O_6S$			
5-Thiazolidineacetamide, 3-α-D-arabino-furanosyl-4-oxo-2-[(1-phenylethyli-dene)hydrazono]-	MeOH	725?(4.00)	128-0295-84
5-Thiazolidineacetamide, 4-oxo-2-[(1-phenylethylidene)hydrazono]-3-β-D-ribofuranosyl-	MeOH	290(4.24)	128-0295-84
$C_{18}H_{22}N_4O_7S$			
Thymidine, 5'-[[[4-(acetylamino)phenyl]-sulfonyl]amino]-5'-deoxy-	EtOH	262.5(4.45)	78-0427-84
$C_{18}H_{22}N_5O_5$			
1H-Purinium, 2-amino-6,9-dihydro-7-[(4-methylphenyl)methyl]-6-oxo-9-β-D-ribofuranosyl-, bromide	pH 1 pH 6.9 pH 13	256(4.07),280s(3.88) 258(3.84),285(3.86) 266(3.95)	44-0363-84 44-0363-84 44-0363-84
$C_{18}H_{22}O$			
Bicyclo[2.2.1]heptan-2-one, 1,7,7-tri-methyl-3-[(4-methylphenyl)methylene]-	EtOH	300(4.42)	41-0629-84
17,19,20-Trinorkaura-1,3,5(10)-trien-16-one, 13-methyl-, (8β,9α,13β)-	MeOH	230(2.78),267(2.21), 272(2.13)	39-2173-84C
$C_{18}H_{22}O_2$			
Bicyclo[2.2.1]heptan-2-one, 3-[(4-meth-oxyphenyl)methylene]-1,7,7-trimethyl-, methyl ester	EtOH	313(4.42)	41-0629-84
Cyclopentanone, 2-(2-hydroxy-3-phenyl-2-propenylidene)-3,3,5,5-tetramethyl-	EtOH	412(4.30)	1-0679-84
1(2H)-Dibenzofuranone, 7-(1,1-dimethyl-ethyl)-3,4-dihydro-3,3-dimethyl-	EtOH	231(4.46),270(3.87)	39-1213-84C
2(1H)-Pentalenone, 4,5,6,6a-tetrahydro-6a-hydroxy-4,4,6,6-tetramethyl-1-phenyl-	EtOH	312(1.82)	1-0679-84
$C_{18}H_{22}O_3$			
2H-1-Benzopyran-6-carboxaldehyde, 5-methoxy-2-methyl-2-(4-methyl-3-pen-tenyl)-	EtOH	234(4.29),264(4.56), 299(3.92)	142-1791-84
$C_{18}H_{22}O_4$			
2H-1-Benzopyran-6-carboxylic acid, 5-hydroxy-2,7-dimethyl-2-(4-methyl-3-pentenyl)-	MeOH	247(4.20),255(4.14), 290(3.55),327(3.28)	102-1909-84
9-Phenanthrenecarboxylic acid, 1,2,3,9-10,10a-hexahydro-1-hydroxy-7-methoxy-10a-methyl-, methyl ester	EtOH	261(4.26)	2-0395-84

Compound	Solvent	$\lambda_{max}(\log \epsilon)$	Ref.
$C_{18}H_{22}O_5$			
1,3-Naphthalenedicarboxylic acid, 1,2-dihydro-7-methoxy-4-methyl-, diethyl ester	EtOH	230(4.01),310(4.17)	2-0395-84
1-Naphthalenepropanoic acid, 3,4-dihydro-6-methoxy-4-(methoxycarbonyl)-2-methyl-, methyl ester	EtOH	224(4.12),274(4.09)	2-0395-84
$C_{18}H_{22}O_6$			
Naematolin 2,3-O-carbonate	CHCl$_3$	224(3.63),240s(3.53), 333(3.06)	5-1332-84
$C_{18}H_{22}Ru$			
Ruthenocene, 1,1':3,3'-bis(1,4-butanediyl)-	EtOH	280s(2.47),309(2.42)	18-0719-84
$C_{18}H_{23}Br_3N_2O_2$			
2H-Pyrrole-3-propanoic acid, 2-[(5-bromo-3,4-diethyl-1H-pyrrol-2-yl)-methylene]-5-(bromomethyl)-4-methyl-, monohydrobromide	CH$_2$Cl$_2$	374(3.83),505(4.89)	78-0579-84
$C_{18}H_{23}ClO_2$			
Estra-4,14-dien-3-one, 4-chloro-17β-hydroxy-	EtOH	256(4.12)	106-0092-84
$C_{18}H_{23}ClO_5$			
Oudemansin B	dioxan	203s(4.19),219(4.34), 248(4.25),261(4.22), 270s(4.11),299(3.67), 312(3.49)	158-84-85
$C_{18}H_{23}NO$			
Formamide, N-[[1,4-dimethyl-7-(1-methylethyl)-2-azulenyl]methyl]-N-methyl-	MeOH	608(2.64)	138-0627-84
2H-Indol-2-one, 1,3-dihydro-1,3-bis(3-methyl-2-butenyl)-	MeOH	252.0(3.64)	83-0639-84
2H-Indol-2-one, 1,3-dihydro-3,3-bis(3-methyl-2-butenyl)-	MeOH	251.0(3.84),268(3.26), 280.0(3.13),296(2.45)	83-0639-84
4(1H)-Quinolinone, 2,3,5,6,7,8-hexahydro-3-methyl-1-(1-phenylethyl)-	heptane	315(4.07)	103-0315-84
	MeOH	337(4.26)	103-0315-84
diastereomer	heptane	316(4.19)	103-0315-84
	MeOH	338(4.32)	103-0315-84
$C_{18}H_{23}NO_3$			
Lauformine, N-methyl-	EtOH	240s(3.18),292(3.23)	142-1031-84
$C_{18}H_{23}NO_6$			
2,3-Butanediol, 1-[(2,3-dihydro-4,7-dimethoxyfuro[2,3-b]quinolin-6-yl)oxy]-3-methyl-, (+)- (dihydronkolbisine)	MeOH	223(4.74),252s(3.99), 264(3.89),273(3.89), 288s(3.75),305s(3.70), 317(3.93),331(3.98)	142-2821-84
	MeOH-HCl	226(4.58),238s(4.49), 296(3.81),323s(4.11), 337(4.17)	142-2821-84
$C_{18}H_{23}NS_2$			
Benzenesulfenamide, N-(1,1-dimethylethyl)-4-methyl-N-[(4-methylphenyl)-thio]-	hexane	249(4.29)	44-2724-84

Compound	Solvent	λ_{max}(log ϵ)	Ref.
$C_{18}H_{23}N_3O_2$			
Phellinamide	EtOH	232(3.99),272(3.65)	142-2453-84
$C_{18}H_{23}N_3O_2S$			
3,5-Pyridinedicarboxamide, N,N'-dieth-yl-1,4-dihydro-1-methyl-4-(phenyl-thio)-	CH_2Cl_2	238(4.15),265(4.25), 347(4.02)	35-6029-84
$C_{18}H_{23}N_3O_{11}$			
1H-Imidazole-5-carboxamide, 4-hydroxy-	M HCl	244(3.79),281(4.04)	4-0529-84
3-(2,3,4,6-tetra-O-acetyl-β-D-galac-	H_2O	244(3.81),279(4.05)	4-0529-84
topyranosyl)-	M NaOH	275(4.15)	4-0529-84
1H-Imidazole-5-carboxamide, 4-hydroxy-	M HCl	244(3.86),280(4.11)	4-0529-84
3-(2,3,4,6-tetra-O-acetyl-β-D-gluco-	H_2O	247(3.85),277(4.14)	4-0529-84
pyranosyl)-	M NaOH	276(4.20)	4-0529-84
3H-Imidazole-5-carboxamide, 4-hydroxy-	M HCl	239(3.74),284(3.78)	4-0529-84
1-(2,3,4,6-tetra-O-acetyl-β-D-galac-	H_2O	236(3.74),285(3.97)	4-0529-84
topyranosyl)-	M NaOH	288(4.07)	4-0529-84
3H-Imidazole-5-carboxamide, 4-hydroxy-	M HCl	243(3.79),280(3.49)	4-0529-84
1-(2,3,4,6-tetra-O-acetyl-β-D-gluco-	H_2O	236(3.59),280(3.85)	4-0529-84
pyranosyl)-	M NaOH	228(3.45),289(3.97)	4-0529-84
3H-Imidazole-4-carboxamide, 5-(2,3,4,6-	M HCl	234(3.92)	4-0849-84
tetra-O-acetyl-β-D-galactopyranosyl-	H_2O	248(4.11)	4-0849-84
oxy)-	M NaOH	264(4.13)	4-0849-84
3H-Imidazole-4-carboxamide, 5-(2,3,4,6-	M HCl	231(3.95)	4-0849-84
tetra-O-acetyl-β-D-glucopyranosyloxy)-	H_2O	248(4.13)	4-0849-84
	M NaOH	264(4.16)	4-0849-84
$C_{18}H_{23}N_5$			
7H-Purin-6-amine, N-(4-butylphenyl)-2,7,8-trimethyl-	EtOH	305(4.22)	11-0208-84B
$C_{18}H_{23}N_5O_3$			
Acetamide, N-[4-[4-(2-acetoxyethyl)-1-piperazinyl]-2-quinazolinyl]-	MeOH	247(4.48),368(3.58)	24-1523-84
$C_{18}H_{24}$			
6H-1,11-Ethano-7,11-methanobenzocyclo-decene, 5,7,8,9,10,12-hexahydro-13-methyl-	C_6H_{12}	217(4.20),272(2.63)	39-2165-84C
$C_{18}H_{24}BrNO_2$			
2,6-Methano-1-benzazocin-11-one, 8-bromo-1,2,3,4,5,6-hexahydro-1-(4-methoxybutyl)-6-methyl-	EtOH	267(4.5),313(3.6), 347(3.0)	49-0333-84
$C_{18}H_{24}Br_2O$			
5,9-Methanobenzocycloocten-4-ol, 3,5-dibromo-5,6,7,8,9,10-hexahydro-1,2,7,7,9-pentamethyl-	n.s.g.	213(4.52),292(3.66)	49-0333-84
$C_{18}H_{24}ClN_3O_2$			
1,4-Naphthalenedione, 5-amino-3,8-bis(butylamino)-2-chloro-	benzene	550s(3.78),595(4.08), 642(4.11)	39-1297-84C
$C_{18}H_{24}N_2$			
Benzenamine, 2-[[4-(dimethylamino)phen-yl]methyl]-N,N,4-trimethyl-	hexane-iso-PrOH	256(3.07)	39-1099-84B
Cyclododeca[b]quinoxaline, 6,7,8,9,10-11,12,13,14,15-decahydro-	MeOH	206(4.48),239(4.57), 322(4.05)	56-0917-84

Compound	Solvent	$\lambda_{max}(\log \epsilon)$	Ref.
$C_{18}H_{24}N_2O$			
Benzenamine, 4,4'-[oxybis(methylene)]-bis[N,N-dimethyl-	hexane-iso-PrOH	262(3.51)	39-1099-84B
Cyclododeca[b]quinoxaline, 6,7,8,9,10-11,12,13,14,15-decahydro-, 5-oxide	MeOH	214(4.36),245(4.88),328(4.01)	56-0917-84
$C_{18}H_{24}N_2OS$			
6H-1,3-Thiazin-6-one, 4-[bis(2-methylpropyl)amino]-2-phenyl-	EtOH	222(4.02),227(4.00),237(3.85),285(3.79),350(3.60)	103-1088-84
$C_{18}H_{24}N_2O_2$			
1H-Pyrrole-2-carboxaldehyde, 5-[(3,4-diethyl-1,3-dihydro-5-oxo-2H-pyrrol-2-ylidene)methyl]-3,4-diethyl-, (Z)-	MeOH	161s(--),268(3.94),394(3.94),415s(--)	35-2645-84
1H-Pyrrole-3-propanoic acid, 2-(3,4-diethyl-2H-pyrrol-2-ylidene)methyl]-4,5-dimethyl-, monohydrobromide	CHCl₃	371(3.21),456(4.49),482(4.83)	78-0579-84
$C_{18}H_{24}N_2O_3$			
1H-Pyrrole-3-propanoic acid, 5-[(3-ethyl-1,5-dihydro-4-methyl-5-oxo-2H-pyrrol-2-ylidene)methyl]-2,4-dimethyl-, methyl ester, (Z)- (xanthobilirubic acid methyl ester)	MeOH CHCl₃	411(4.43) 403(4.45)	4-0139-84 4-0139-84
$C_{18}H_{24}N_2O_4$			
1H-Pyrrole-3-propanoic acid, 5-[(3-ethyl-4-methyl-5-oxo-2-pyrrolidinylidene)methyl]-2-formyl-4-methyl-, methyl ester, cis	MeOH	240(4.10),350(4.11)	35-2645-84
trans	MeOH	242(4.09),351(4.11)	35-2645-84
1H-Pyrrole-3-propanoic acid, 5-[(4-ethyl-3-methyl-5-oxo-2-pyrrolidinylidene)methyl]-2-formyl-4-methyl-, methyl ester, cis	MeOH	241(4.11),354(4.12)	35-2645-84
trans	MeOH	241(4.10),353(4.13)	35-2645-84
$C_{18}H_{24}N_2O_5$			
2(3H)-Furanone, dihydro-3-[2-(hydroxymethyl)-5-[[(tetrahydro-4,5-dimethyl-2-oxo-3-furanyl)imino]methyl]-1H-pyrrol-1-yl]-4,5-dimethyl- (funebrine)	DMSO	265s(4.30),293(4.48)	44-2714-84
L-Proline, 1-[N-(1-carboxy-3-phenylpropyl)-L-alanyl]-	pH 1	257(2.25)	44-2816-84
$C_{18}H_{24}N_2O_6$			
2H-[1]Benzopyrano[3,4-c]pyridazine-3,4-dicarboxylic acid, 1,4a,5,10b-tetrahydro-8-methoxy-5,6-dimethyl-, dimethyl ester, trans	EtOH	211(4.18),225s(3.70),284(3.34),290(3.30)	39-2815-84C
[2,2'-Bi-2H-pyrrole]-2,2'-dicarboxylic acid, 1,1',3,3'-tetrahydro-4,4',5,5'-tetramethyl-3,3'-dioxo-, diethyl ester	MeCN	309(4.08)	33-1957-84
$C_{18}H_{24}N_4O_{10}$			
Acetamide, N-[1-(2,3,4,6-tetra-O-acetyl-β-D-glucopyranosyl)-1H-1,2,4-triazol-3-yl]-	EtOH	224(4.00)	111-0089-84

Compound	Solvent	$\lambda_{max}(\log \epsilon)$	Ref.
$C_{18}H_{24}N_6$ 2a,4a,6a,8a,10a,12a-Hexaazacoronene, 1,2,3,4,5,6,7,8,9,10,11,12-dodeca- hydro-, cation radical	CH_2Cl_2	304(4.43),373(3.92), 716(4.02)	35-6453-84
dication	CH_2Cl_2	316(4.45),360s(--), 525(4.18)	35-6453-84
$C_{18}H_{24}N_6O_{12}$ L-Alanine, N-[[5-amino-1-(2,3,5-tri-O- acetyl-β-D-ribofuranosyl)-1H-imida- zol-4-yl]carbonyl]-3-(hydroxynitroso- amino)-	MeOH	235(4.07),265(4.15)	87-1295-84
$C_{18}H_{24}O_2$ Estra-4,14-dien-3-one, 17α-hydroxy-	EtOH	239(4.22)	106-0092-84
$C_{18}H_{24}O_2S$ Cyclopenta[c]thiopyran-3-carboxylic acid, 5,7-bis(1,1-dimethylethyl)-	CH_2Cl_2	266(4.53),275(4.46), 310(4.11),322(4.11), 405(4.12),475(2.89)	118-0262-84
$C_{18}H_{24}O_3$ Estr-4-ene-3,17-dione, 14β-hydroxy-	EtOH	240(4.20)	106-0092-84
Estr-4-en-3-one, 14β,15β-epoxy-17α- hydroxy-	EtOH	240(4.19)	106-0092-84
(14β,15β,17β)-	EtOH	238(4.22)	106-0092-84
(15α,17α)-	EtOH	238(4.23)	106-0092-84
(15α,17β)-	EtOH	238(4.23)	106-0092-84
2(1H)-Naphthalenone, 1,1-bis(1,1-di- methylethoxy)-	EtOH	238(4.20),321(3.86)	39-2069-84C
1H-4-Oxabenzo[f]cyclobut[cd]indene, 1a,2,3,3a,8b,8c-hexahydro-6,8-di- methoxy-1,1,3a-trimethyl-, (±)-	EtOH	218(4.56),232(4.56)	44-1793-84
$C_{18}H_{24}O_3RuS_2$ Ruthenocene, 1,1'-[oxybis(2,1-ethane- diyloxy-2,1-ethanediylthio)]-	MeCN	320(2.67)	88-1991-84
$C_{18}H_{24}O_6S$ Estra-1,3,5(10)-triene-3,4,17-triol, 17-(hydrogen sulfate), monopotassium salt, (17β)-	MeOH	255s(3.85),280(3.28)	94-2745-84
$C_{18}H_{24}O_{11}$ 3-Buten-2-one, 4-[(2,3,4,6-tetra-O- acetyl-β-D-glucopyranosyl)oxy]-, (E)-	EtOH	237(4.20)	78-4657-84
$C_{18}H_{25}BrO_2$ 5,9-Methanobenzocyclooctene-4,5(6H)- diol, 3-bromo-7,8,9,10-tetrahydro- 1,2,7,7,9-pentamethyl-	n.s.g.	213(4.39),292(3.56)	49-0333-84
$C_{18}H_{25}N$ 1-Azulenemethanamine, N,N,3,8-tetra- methyl-5-(1-methylethyl)-	$CHCl_3$	247(4.37),293(4.56), 307s(4.27),355(3.75), 372(3.69),612(2.66)	88-4707-84
hydriodide	$CHCl_3$	242(4.45),293(4.50), 305s(--),351(3.64), 368(3.64),620(2.62)	88-4707-84

Compound	Solvent	$\lambda_{max}(\log \epsilon)$	Ref.
$C_{18}H_{25}N_2O_4S$ 1(2H)-Pyridinyloxy, 3,6-dihydro-4-[5-[2-[(2-methoxy-2-oxoethyl)amino]-2-oxoethyl]-2-thienyl]-2,2,6,6-tetramethyl-	EtOH	204(4.07),290(4.10)	70-1901-84
$C_{18}H_{25}N_3O_2$ Xanthobilirubic acid amide, N-methyl-	MeOH CHCl$_3$	413(4.55) 405(4.52)	4-0139-84 4-0139-84
$C_{18}H_{26}Ge_3$ 5H-Dibenzo[d,f][1,2,3]trigermepin, 6,7-dihydro-5,5,6,6,7,7-hexamethyl-	hexane	225s(4.40),250(4.00)	138-1379-84
$C_{18}H_{26}N_2$ Cyclopenta[cd]pentalene-2a,4a-diamine, N,N,N',N'-tetraethyl-	hexane	215s(4.012),235(3.644), 285(3.584)	88-1693-84
$C_{18}H_{26}N_2O_2$ 2H-Imidazol-2-one, 4-[3,5-bis(1,1-dimethylethyl)-4-hydroxyphenyl]-1,3-dihydro-5-methyl-	MeOH	273(4.04)	95-0909-84
1H-Pyrrole-2-carboxaldehyde, 5-[(3,4-diethyl-5-oxo-2-pyrrolidinylidene)-methyl]-3,4-diethyl-, cis	MeOH	242(4.16),360(4.10)	35-2645-84
trans	MeOH	239(4.14),366(4.09)	35-2645-84
$C_{18}H_{26}N_2O_4$ Pyrrolo[3,2-b]pyrrole-1,4-dicarboxylic acid, 3,6-bis(1,1-dimethylethyl)-, dimethyl ester	C$_6$H$_{12}$	<u>256s(4.3),260(4.3),</u> <u>295(3.9)</u>	138-2033-84
$C_{18}H_{26}N_2O_4S$ Benzenesulfonamide, N-[[(3-hydroxy-4,7,7-trimethylbicyclo[2.2.1]hept-2-yl)amino]carbonyl]-4-methyl-	MeOH	227.25(4.13)	83-0906-84
$C_{18}H_{26}N_2O_8$ Uridine, 2'-deoxy-5-ethyl-, 5'-(ethyl ethylpropanedioate)	H$_2$O	268(3.98)	83-0867-84
Uridine, 2'-deoxy-5-ethyl-, 5'-(methyl hexanedioate)	H$_2$O	267(3.97)	83-0867-84
$C_{18}H_{26}N_4O_4$ Butanedioic acid, 2,3-bis(3,4,5-trimethyl-1H-pyrazol-1-yl)-, dimethyl ester, (R*,S*)-	MeOH	235(3.99)	44-4647-84
$C_{18}H_{26}N_6O_{17}P_2$ 1H-Imidazole-4-carboxylic acid, 5-amino-1-β-D-ribofuranosyl-, 5',5'''-P,P'-dihydrogen diphosphate)	pH 12	248(4.26)	65-2114-84
$C_{18}H_{26}O_2$ 1,3-Cyclohexanedione, 5,5-dimethyl-2-(3,4,4a,5,6,7,8,8a-octahydro-2-naphthalenyl)-	pH 1 H$_2$O pH 13	268(4.01) 270(3.98) 294(4.32)	39-1213-84C 39-1213-84C 39-1213-84C
5,9-Methanobenzocyclooctene-4,5(6H)-diol, 7,8,9,10-tetrahydro-1,2,7,7,9-pentamethyl-	n.s.g.	214(3.98),223s(3.86), 287(3.31)	49-0333-84

Compound	Solvent	$\lambda_{max}(\log \epsilon)$	Ref.
$C_{18}H_{26}O_2S$ 1,4-Dioxaspiro[4.5]decane, 2,3,7-tri-methyl-7-[(phenylthio)methyl]-	EtOH	256(3.95)	94-3417-84
$C_{18}H_{26}O_3$ 2,4-Cyclohexadiene-1-carboxylic acid, 1-methyl-6-oxo-, 5-methyl-2-(1-meth-ylethyl)cyclohexyl ester	MeOH	302(3.62)	44-4429-84
Estr-4-en-3-one, 14β,17β-dihydroxy-	EtOH	240(4.25)	106-0092-84
$C_{18}H_{26}O_4$ Propanedioic acid, methyl[4-(2-methyl-propyl)phenyl]-, diethyl ester	EtOH	221(3.94)	12-1245-84
$C_{18}H_{26}O_5$ 3-Cyclohexene-1-carboxylic acid, 4-(2,2-dimethyl-1-oxopropoxy)-1-(1-hydroxy-2-propenyl)-2-methylene-, ethyl ester	EtOH	239(3.79)	12-2037-84
$C_{18}H_{26}Si_2Te_2$ Silane, [ditellurobis(methylene)]-bis[dimethylphenyl-	benzene	410(2.70)	48-0467-84
$C_{18}H_{26}Si_3$ 5H-Dibenzo[d,f][1,2,3]trisilepin, 6,7-dihydro-5,5,6,6,7,7-hexamethyl-	hexane	220s(4.45),250(3.97)	138-1379-84
$C_{18}H_{27}BrO_2$ 5,9-Methanobenzocycloocten-4(1H)-one, 3-bromo-2,3,5,6,7,8,9,10-octahydro-5-hydroxy-2,2,7,7,9-pentamethyl-	n.s.g.	207(3.46),263(3.97)	49-0793-84
$C_{18}H_{27}ClN_7O_8PS$ 5'-Azido-5'deoxythymidine-3'-phosphite S-(p-chlorobenzyl)thiouronium salt	MeOH	223(4.30),263(--)	118-0410-84
$C_{18}H_{27}NO_3$ Protoemetinol, 10-demethyl-	EtOH	227(4.01),286(3.77)	100-0397-84
$C_{18}H_{27}N_3O$ Urea, N'-[1-methyl-3-(phenylimino)-1-butenyl]-N,N-dipropyl-	EtOH	225(4.15),308(4.15)	39-0239-84C
$C_{18}H_{28}$ Bicyclo[3.3.1]nonane, 9-bicyclo[3.3.1]-non-9-ylidene-, radical cation	CH_2Cl_2 at -78°	450(2.9+)	35-0791-84
Cyclopentene, 3,3,5,5-tetramethyl-4-(2,2,5,5-tetramethyl-3-cyclo-penten-1-ylidene)-	C_6H_{12}	219.6(3.52)	24-0310-84
$C_{18}H_{28}N_2O_8$ Propanedioic acid, 2,2'-[1,2-ethanedi-ylbis(iminomethylidyne)]bis-	MeCN	222(4.36),272(4.44), 292(4.43)	62-0947-84A
$C_{18}H_{28}N_4O_5$ Inosine, 2-octyl-	pH 1 pH 7 pH 13	251(4.08) 251(4.08) 253(4.16)	87-0429-84 87-0429-84 87-0429-84

Compound	Solvent	$\lambda_{max}(\log \epsilon)$	Ref.
$C_{18}H_{28}O$			
3-Buten-2-one, 4-[2,6,6-trimethyl-5-(3-methyl-2-butenyl)-1-cyclohexen-1-yl]-, (E)-	EtOH	222(3.87),294(4.09)	78-1545-84
3-Buten-2-one, 4-[2,6,6-trimethyl-5-(3-methyl-2-butenyl)-2-cyclohexen-1-yl]-, [1α(E),5α]-	EtOH	225(4.13)	78-1545-84
[1α(E),5β]-	EtOH	225(4.18)	78-1545-84
$C_{18}H_{28}O_2$			
2-Butanone, 4-[3,5-bis(1,1-dimethyl-ethyl)-4-hydroxyphenyl]-	EtOH	205(4.41),232.5(3.86), 281(3.32)	104-0294-84
1(2H)-Dibenzofuranone, 7-(1,1-dimethyl-ethyl)-3,4,5a,6,7,8,9,9a-octahydro-3,3-dimethyl-	EtOH	274(4.10)	39-1213-84C
$C_{18}H_{28}O_4$			
Albocycline	EtOH	212s(3.83)	158-1187-84
1-Cyclohexene-1-carboxylic acid, 3-acetoxy-4-(1,5-dimethyl-4-hexenyl)-, methyl ester	MeOH	217(4.02)	102-0186-84
Oxacyclohexadeca-4,6-dien-2-one, 7-acetoxy-3-methyl-	MeOH	242(4.34)	89-0440-84
$C_{18}H_{28}O_5$			
Albocycline, 8,9-epoxy-8,9-dihydro-	EtOH	220s(4.20)	158-1187-84
stereoisomer	EtOH	215s(4.66)	158-1187-84
Albocycline, 11(or 12)-hydroxy-	EtOH	215(3.77)	158-1187-84
Albocycline, 18-hydroxy-	EtOH	220s(3.78)	158-1187-84
$C_{18}H_{29}NO_2$			
4a-Azaandrostan-3-one, 17-hydroxy-, (5β17β)-	n.s.g.	204(3.69)	111-0465-84
$C_{18}H_{29}NO_4$			
2,5-Benzoxazonine, 1,3,4,5,6,7-hexa-hydro-9,10-dimethoxy-1,5-dimethyl-1-propoxy-	MeOH	233(3.98),280(3.49), 285s(3.45)	12-1659-84
$C_{18}H_{29}NO_6Si$			
1H-Pyrrole-2,5-dione, 3-[5-O-[(1,1-di-methylethyl)dimethylsilyl]-2,3-O-(1-methylethylidene)-β-D-ribofuranosyl]-	MeOH	221(4.14)	18-2515-84
$C_{18}H_{29}PS_2$			
Phosphine sulfide, thioxo[2,4,6-tris(1,1-dimethylethyl)phenyl]-	hexane	243(3.93),328(3.40)	138-0317-84
	MeCN	331(3.23)	139-0105-84C
$C_{18}H_{29}PSe_2$			
Phosphine selenide, selenoxo[2,4,6-tris(1,1-dimethylethyl)phenyl]-	C_6H_{12}	260(3.99),405(3.32) (impure)	138-0603-84
$C_{18}H_{30}$			
Bicyclo[2.2.1]heptane, 1,3,3-trimethyl-2-(2,2,4,4-tetramethylcyclobutyli-dene)-, (1R)-	C_6H_{12}	203.0(4.21)	24-0277-84
$C_{18}H_{30}NO_3$			
1H-3-Benzazoninium, 2,3,4,5,6,7-hexa-hydro-7,9,10-trimethoxy-3,3,7-tri-methyl-, iodide	MeOH	217s(4.29),282(3.43), 287s(3.41)	12-1203-84

Compound	Solvent	$\lambda_{max}(\log \epsilon)$	Ref.
$C_{18}H_{30}N_2O_2$ 10-Oxa-9,12-diazapentacyclo[9.3.3.0-$0^{2,6}.0^{3,9}$]heptadecan-12-ol, 8,8,13,13-tetramethyl-	EtOH	232(3.18)	12-0117-84
$C_{18}H_{30}O$ 3-Buten-2-ol, 4-[2,6,6-trimethyl-5-(3-methyl-2-butenyl)-1-cyclohexen-1-yl]-	EtOH	204(4.08),233(3.61)	78-1545-84
$C_{18}H_{30}O_5$ 4H-Pyran-4-one, 2-(2,4-dihydroxy-1,3-dimethylbutyl)-6-(2-hydroxy-1-methyl-butyl)-3,5-dimethyl-, [1R*(1S*,2S*)-2R*,3S*]-	MeOH	260(3.86)	44-0559-84
$C_{18}H_{31}NO$ Piperidine, 1-(6-cyclohexyl-3-ethyl-3,4-dihydro-2H-pyran-2-yl)-	EtOH	220(3.61)	103-0515-84
$C_{18}H_{31}NO_2Sn$ Stannane, tributyl(4-nitrophenyl)-	MeOH	278(3.99)	70-1044-84
$C_{18}H_{31}NO_6P_2$ Phosphonic acid, [[4-(dimethylamino)-phenyl]ethenylidene]bis-, tetraethyl ester	dioxan	370(4.48)	65-1987-84
$C_{18}H_{32}$ Cyclohexane, 1,1,3,3-tetramethyl-2-(2,2,4,4-tetramethylcyclo-butylidene)-	C_6H_{12}	202.0(4.16)	24-0277-84 +24-0310-84
Cyclopentane, 1,1,3,3-tetramethyl-2-(2,2,5,5-tetramethylcyclo-pentylidene)-	C_6H_{12}	203.0(4.18)	24-0277-84 +24-0310-84
$C_{18}H_{32}NO_2Si$ 2-Propen-1-aminium, N-[3-(2,2-dimethyl-1-oxopropoxy)-2-cyclopenten-1-yli-dene]-N-methyl-2-[(trimethylsilyl)-methyl]-, perchlorate	MeCN	273(4.45)	44-0220-84
$C_{18}H_{32}N_2S$ 5-Thia-12,13-diazadispiro[3.1.5.2]tri-dec-12-ene, 1,1,3,3,7,7,11,11-octa-methyl-	C_6H_{12}	202.5(3.25),294.0(3.02), 334s(2.34)	24-0277-84
$C_{18}H_{32}N_4O_2$ Piperidine, 1,1'-[1-(1,1-dimethyleth-yl)-2,3-dihydro-4-nitro-1H-pyrrole-2,3-diyl]bis-	MeOH	230(3.7),379(4.31)	150-0018-84S
$C_{18}H_{32}O_2$ 2(3H)-Furanone, dihydro-3-tetradecyli-dene-, (E)-	EtOH	223.5(4.13)	95-0839-84
(Z)-	EtOH	226(4.07)	95-0839-84
$C_{18}H_{32}S$ 11-Thiadispiro[4.0.4.1]undecane, 1,1,4,4,7,7,10,10-octamethyl-	C_6H_{12}	214s(2.72),270.0(1.48)	24-0277-84

Compound	Solvent	$\lambda_{max}(\log \epsilon)$	Ref.
$C_{18}H_{33}N$			
2-Pentanamine, 2,4,4-trimethyl-N-(4,4,6,6-tetramethyl-2-cyclo-hexen-1-ylidene)-, (Z)-	C_6H_{12}	223(4.24),309(2.22)	33-0748-84
$C_{18}H_{33}N_3O_2S$			
Guanidine, [2-[(1-ethenyl-1,5,9-tri-methyl-4,8-decadienyl)sulfonyl]-ethyl]-, [S-(E)]- (agelasidine A)	EtOH	227(3.73),265(3.26)	35-1819-84
	EtOH-HCl	226(3.60),270(3.30)	35-1819-84
$C_{18}H_{33}N_2O_6SSi$			
D-allo-2-Heptulosonic acid, 3,6-anhy-dro-7-O-[(1,1-dimethylethyl)dimeth-ylsilyl]-4,5-O-(1-methylethylidene)-, methyl ester, 2-[(aminothioxomethyl)-hydrazone]	MeOH	268(3.83),314(3.72)	18-2515-84
$C_{18}H_{33}N_3O_7Si$			
D-allo-2-Heptulosonic acid, 3,6-anhy-dro-7-O-[(1,1-dimethylethyl)dimeth-ylsilyl]-4,5-O-(1-methylethylidene)-, methyl ester, 2-[(aminocarbonyl)hy-drazone]	MeOH	265(3.94)	18-2515-84
$C_{18}H_{35}ClN_4O_9$			
Antibiotic U-64846	H_2O	295(3.76)	158-0096-84
$C_{18}H_{35}NOSi$			
1H-Pyrrole, 2,5-dibutyl-1-[[2-(tri-methylsilyl)ethoxy]methyl]-	MeOH	225(3.83)	44-0203-84
$C_{18}H_{36}O_3Si_2$			
2-Cyclopenten-1-one, 4-[[(1,1-dimethyl-ethyl)dimethylsilyl]oxy]-2-[[[(1,1-dimethylethyl)dimethylsilyl]oxy]-methyl]-	EtOH	220(3.86)	39-2089-84C
$C_{18}H_{38}O_5Si_2$			
Cyclopentanone, 4-[[(1,1-dimethyleth-yl)dimethylsilyl]oxy]-2-[[[(1,1-di-methylethyl)dimethylsilyl]oxy]-methyl]-3,3-dihydroxy-	EtOH	222(2.86),275s(2.26)	39-2089-84C
$C_{18}H_{38}Si_4$			
Trisilane, 1,1,1,3,3,3-hexamethyl-2-(2,4,6-trimethylphenyl)-2-(tri-methylsilyl)-	C_6H_{12}	247(4.15)	60-0341-84
$C_{18}H_{38}Te_2$			
Ditelluride, dinonyl	benzene	397(2.96)	48-0467-84
$C_{18}H_{48}Si_6$			
Cyclohexasilane, 1,2,3,4,5,6-hexaethyl-1,2,3,4,5,6-hexamethyl-	isooctane	239(3.72),260s(3.08)	101-0353-84L

Compound	Solvent	$\lambda_{max}(\log \epsilon)$	Ref.
$C_{19}Cl_{12}O$			
2,5-Cyclohexadien-1-one, 2,3,5,6-tetra-chloro-4-(1,2,3,4,5,6,7,8-octachloro-9H-fluoren-9-ylidene)-	C_6H_{12}	217(4.81),290(4.62), 303(4.64),325s(4.08), 404(3.49),427(3.57), 452(3.45),480s(2.60)	44-0770-84
3H-Fluoren-3-one, 1,2,4,5,6,7,8-hepta-chloro-9-(pentachlorophenyl)-	C_6H_{12}	217(4.93),305(4.42), 400s(3.54),428(3.79), 452(3.83),485s(3.38)	44-0770-84
$C_{19}Cl_{13}$			
9H-Fluoren-9-yl, 1,2,3,4,5,6,7,8-octa-chloro-9-(pentachlorophenyl)-	C_6H_{12}	217(4.87),230s(4.67), 293(4.66),320s(4.13), 345s(3.74),373(3.71), 388(3.79),408s(3.58), 460(3.12),498(3.07), 530(2.81),573(3.31), 622(3.71)	44-0770-84
$C_{19}Cl_{14}$			
9H-Fluorene, 1,2,3,4,5,6,7,8,9-nona-chloro-9-(pentachlorophenyl)-	C_6H_{12}	218(4.85),237(4.73), 272s(4.34),310(4.19)	44-0770-84
$C_{19}Cl_{14}O$			
2,5-Cyclohexadien-1-one, 2,3,6-tri-chloro-4-[chloro(pentachlorophenyl)-methylene]-5-(pentachlorophenyl)-	C_6H_{12}	215(4.97),323(4.05), 400s(2.73)	44-2884-84
$C_{19}Cl_{16}$			
1,1'-Biphenyl, 2,2',3,3',4,4',5,5',6-nonachloro-6'-[dichloro(pentachloro-phenyl)methyl]-	C_6H_{12}	219(5.03),300s(3.56), 308(3.57)	44-2884-84
$C_{19}F_{15}N$			
Benzenamine, N-[bis(pentafluorophenyl)-methylene]-2,3,4,5,6-pentafluoro-	EtOH	246(4.15),325(3.44)	70-1268-84
$C_{19}F_{15}NO$			
Benzenamine, N-[bis(pentafluorophenyl)-methylene]-2,3,4,5,6-pentafluoro-, N-oxide	EtOH	290(4.07)	70-1268-84
$C_{19}HCl_{13}$			
9H-Fluorene, 1,2,3,4,5,6,7,8-octachloro-9-(pentachlorophenyl)-	C_6H_{12}	218(4.81),230s(4.78), 297(4.40)	44-0770-84
$C_{19}HCl_{15}$			
1,1'-Biphenyl, 2,2',3,3',4,4',5,5',6-nonachloro-6'-[chloro(pentachloro-phenyl)methyl]-	C_6H_{12}	214(5.03),293(3.14), 303(3.19)	44-0770-84
$C_{19}H_2Cl_{12}O$			
9H-Fluoren-3-ol, 1,2,4,5,6,7,8-hepta-chloro-9-(pentachlorophenyl)-	C_6H_{12}	219(4.83),228(4.83), 297(4.40),305s(4.35)	44-0770-84
$C_{19}H_2Cl_{14}$			
1,1'-Biphenyl, 2,2',3,3',4,4',5,5',6-nonachloro-6'-[(pentachlorophenyl)-methyl]-	C_6H_{12}	213(5.05),250s(4.34), 293(3.11)	44-2884-84
$C_{19}H_4Cl_4I_4O_3$			
3H-Xanthen-3-one, 6-hydroxy-2,4,5,7-	pH 1	425(4.09),505(4.30)	88-4285-84

Compound	Solvent	$\lambda_{max}(\log \epsilon)$	Ref.
tetraiodo-9-(2,3,4,5-tetrachloro-phenyl)-, sodium salt (cont.) spectra in 50% dioxan	pH 8	523(4.52),561(5.02)	88-4285-84
$C_{19}H_4F_{11}NO$ Benzenamine, N-[bis(pentafluorophenyl)-methylene]-2-fluoro-, N-oxide	EtOH	286(4.09)	70-1268-84
$C_{19}H_5F_{10}N$ Benzenamine, 2,3,4,5,6-pentafluoro-N-[(pentafluorophenyl)phenylmeth-ylene]-	EtOH	268(4.26),320(3.50)	70-1268-84
$C_{19}H_5F_{10}NO$ Benzenamine, N-[bis(pentafluorophenyl)-methylene]-, N-oxide	EtOH	288(4.05)	70-1268-84
Benzenamine, 2,3,4,5,6-pentafluoro-N-[(pentafluorophenyl)phenylmeth-ylene]-, N-oxide	EtOH	312(4.20)	70-1268-84
$C_{19}H_9NO_4$ 12H,14H-Bis[1]benzopyrano[2,3-b:3',2'-e]pyridine-12,14-dione	MeOH	200(4.27),252(4.44), 284(4.34),345(3.92)	83-0377-84
$C_{19}H_{10}Br_2O_4$ Naphtho[1,8-bc]pyran-2,6-dione, 5,9-di-bromo-3-(4-methoxyphenyl)-	CHCl$_3$	256(4.10),324(4.16), 333(4.16),444(4.13)	39-1957-84C
$C_{19}H_{10}ClN$ Acenaphtho[1,2-b]quinoline, 10-chloro-	CHCl$_3$	255(4.38),285(4.37), 314(4.43),384(3.40)	78-2959-84
$C_{19}H_{10}Cl_2N_2O_2$ Cyclohepta[1,2-b:1,7-b']bis[1,4]benz-oxazine, 3,12-dichloro-	MeOH	237(4.33),279(4.29), 375(3.85)	138-1145-84
isomer	MeOH	280(4.38),380(3.94)	138-1149-84
$C_{19}H_{10}F_5NO$ Benzenamine, N-[(pentafluorophenyl)-phenylmethylene]-, N-oxide	EtOH	304(4.20)	70-1268-84
$C_{19}H_{10}N_2O_2$ Acenaphtho[1,2-b]quinoline, 10-nitro-	CHCl$_3$	253(4.11),300(4.11), 367(3.98)	78-2959-84
$C_{19}H_{10}O_6$ 4H,10H-Benzo[1,2-b:3,4-b']dipyran-8-carboxylic acid, 4,10-dioxo-2-phenyl-	MeOH	250(4.29),304(4.26)	2-0969-84
$C_{19}H_{11}BrO_3$ 1,4-Chrysenedione, 2-bromo-3-methoxy-	CHCl$_3$	258(4.76),299(4.34), 312(4.45),324(4.51), 443(3.79)	40-0090-84
$C_{19}H_{11}BrO_4$ Naphtho[1,8-bc]pyran-2,6-dione, 9-bromo-3-(4-methoxyphenyl)-	CHCl$_3$	254(3.98),280(4.07), 305(4.10),429(4.11)	39-1957-84C
$C_{19}H_{11}N$ Acenaphtho[1,2-b]quinoline	CHCl$_3$	254(4.04),300(4.04),	78-2959-84

Compound	Solvent	$\lambda_{max}(\log \epsilon)$	Ref.
(cont.)		380(3.43)	78-2959-84
$C_{19}H_{11}NO_4$			
Avicine, de-N-methyl-	n.s.g.	230(4.59),275(4.86), 327s(4.26),370(3.50)	100-0001-84
Norsanguinarine	EtOH	212(4.33),243(4.67), 251s(4.50),275s(4.50), 281(4.55),293s(4.41), 328(4.22),341s(4.00), 382(3.18),399(3.65)	100-0001-84
$C_{19}H_{12}BBrO_6$			
Boron, [5-(4-bromophenyl)-1-phenyl-4-pentene-1,3-dionato-O,O'][ethane-dioato(2-)-O,O']-	CH_2Cl_2	434(4.78)	97-0292-84
$C_{19}H_{12}BClO_6$			
Boron, [5-(4-chlorophenyl)-1-phenyl-4-pentene-1,3-dionato-O,O'][ethane-dioato(2-)-O,O']-	CH_2Cl_2	434(4.80)	97-0292-84
$C_{19}H_{12}BrNO_3$			
2H-Indol-2-one, 5-bromo-1,3-dihydro-1-methyl-3-(2-oxo-5-phenyl-3(2H)-furanylidene)-	$CHCl_3$	273(4.28),452(4.35), 476(4.29)	39-1331-84B
$C_{19}H_{12}BrN_3$			
5H-Benz[g]indolo[2,3-b]quinoxaline, 2-bromo-5-methyl-	EtOH	238(4.48),299(4.93), 394(4.25),476(3.36)	103-1155-84
$C_{19}H_{12}BrN_3O_3$			
1H-Indazole-3,4,7(2H)-trione, 5-[(4-bromophenyl)amino]-2-phenyl-	dioxan	270(4.51),322.5(4.05), 438(3.85)	4-0825-84
$C_{19}H_{12}Br_2O$			
2,5-Cyclohexadien-1-one, 2,6-dibromo-4-(diphenylmethylene)-	hexane?	212(4.48),265(4.09), 275(4.10),388(4.38)	73-1949-84
$C_{19}H_{12}F_6N_2O$			
1-Cyclohexene-1-carbonitrile, 2-[(1-ethyl-2(1H)-quinolinylidene)methyl]-3,3,4,4,5,5-hexafluoro-6-oxo-	EtOH benzene	467(4.53) 528(--)	104-0390-84 104-0390-84
$C_{19}H_{12}N_2O_2$			
Cyclohepta[1,2-b:1,7-b']bis[1,4]benz-oxazine	MeOH	285(4.37),378(3.90)	138-1149-84
Cyclohepta[4,5]pyrrolo[1,2-a]imidazole-3,10-dicarboxaldehyde	$CHCl_3$	267(4.04),328(4.29), 446(3.80),492s(3.29), 522s(3.24),566s(3.07), 630s(2.46)	150-3465-84M
$C_{19}H_{12}N_2O_5$			
3-Furancarbonitrile, 2-amino-4,5-bis(1,3-benzodioxol-5-yl)-	MeOH	207(4.69),256(4.21), 293(4.17),333(4.23)	73-1788-84
2H-Indol-2-one, 1,3-dihydro-1-methyl-5-nitro-3-(2-oxo-5-phenyl-3(2H)-furanylidene)-	$CHCl_3$	262(4.21),446s(4.30), 466(4.32)	39-1331-84B
$C_{19}H_{12}N_3O_2$			
Pyridinium, 1-[5-(2,2-dicyanoethenyl)-2-furanyl]-4-phenoxy-, bromide	H_2O	264(2.89),379(3.31)	73-2485-84

Compound	Solvent	$\lambda_{max}(\log \epsilon)$	Ref.
$C_{19}H_{12}N_4O_2$			
5H-Benzo[g]indolo[2,3-b]quinoxaline, 5-methyl-2-nitro-	EtOH	238(4.53),290(4.93), 355(4.65),450(3.50)	103-1155-84
$C_{19}H_{12}N_4S$			
[1,2,4]Triazolo[1,5-c]quinazoline, 5-phenyl-2-(2-thienyl)-	EtOH	310(3.69)	18-1138-84
$C_{19}H_{12}O_3$			
7H-Benzo[c]fluoren-7-one, 9,10-dimeth-oxy-	MeCN	223(5.5),273(5.8)	2-0603-84
1,4-Naphthalenedione, 5-[(4-methoxy-phenyl)ethynyl]-	EtOH	251(4.45),323(4.35), 431(3.58)	70-2345-84
$C_{19}H_{12}O_5$			
Anthra[1,2-b]furan-4-carboxaldehyde, 6,11-dihydro-5-methoxy-2-methyl-6,11-dioxo-	EtOH	230s(4.19),256(4.25), 284s(4.11),393(3.63)	12-1511-84
1,3-Benzodioxole-5,6-dione, 4-(4-hy-droxyphenyl)-7-phenyl-	MeOH	273(4.22),352(2.87), 500(3.13)	64-0695-84C
	MeOH-NaOH	585(--)	64-0695-84C
$C_{19}H_{12}O_6$			
Anthra[1,2-b]furan-4-acetic acid, 6,11-dihydro-10-hydroxy-6,11-dioxo-, methyl ester	MeOH	226(4.57),273(4.54), 295s(4.13),391(4.09)	78-4685-84
Anthra[1,2-b]furan-3-carboxylic acid, 6,11-dihydro-5-methoxy-2-methyl-6,11-dioxo-	EtOH	254(4.61),276s(4.37), 405(3.94)	12-1518-84
1H-Anthra[2,3-c]pyran-1,6,11-trione, 12-hydroxy-5-methoxy-3-methyl-	EtOH	263(3.60),312(3.03), 454(3.24)	12-1511-84
1,3-Benzodioxole-5,6-dione, 4,7-bis(4-hydroxyphenyl)-	MeOH	276(4.33),382(3.19), 530(3.38)	64-0695-84C
	MeOH-NaOH	628(--)	64-0695-84C
[6,8'-Bi-2H-1-benzopyran]-2,2'-dione, 7'-hydroxy-7-methoxy-	MeOH	211(4.08),326(3.95)	142-0333-84
	MeOH-base	213(4.15),356(3.80), 375(3.81)	142-0333-84
$C_{19}H_{12}O_7$			
1,3-Benzodioxole-5,6-dione, 4-(3,4-di-hydroxyphenyl)-7-(4-hydroxyphenyl)-	MeOH	277(4.51),543(3.58)	64-0695-84C
$C_{19}H_{13}BO_6$			
Boron, (1,5-diphenyl-4-pentene-1,3-di-onato-O,O')[ethanedioato(2-)-O,O']-	CH_2Cl_2	430(4.73)	97-0292-84
$C_{19}H_{13}BrO$			
2,5-Cyclohexadien-1-one, 2-bromo-4-(diphenylmethylene)-	hexane?	211(4.30),262(4.01), 270(4.02),373(4.30)	73-1949-84
$C_{19}H_{13}BrO_5S$			
Bromophenol Red (sodium salt)	acid	436(4.16)	35-0265-84
	base	575(4.56)	35-0265-84
in hexadecyltrimethylammonium bromide	acid	427(4.17)	35-0265-84
	base	586(4.59)	35-0265-84
$C_{19}H_{13}NO$			
Methanone, 9-anthracenyl-1H-pyrrol-2-yl-	EtOH	255(5.10),302(4.17), 329(3.67),347(3.81), 366(3.97),385(3.95)	64-1393-84B

Compound	Solvent	λ_{max}(log ϵ)	Ref.
Methanone, 9-anthracenyl-1H-pyrrol-3-yl-	EtOH	255(5.16),331(3.38), 347(3.79),365(3.98), 385(3.96)	64-1393-84B
$C_{19}H_{13}NOS_2$ 2(1H)-Pyridinone, 3-phenyl-4,6-di-2-thienyl-	EtOH	284(4.09),369(4.09)	4-1473-84
$C_{19}H_{13}NO_3$ 1H-Benz[f]isoindole-1,3(2H)-dione, 4-(2-hydroxyphenyl)-2-methyl-	ether	249s(4.70),257(4.77), 282s(3.99),292(3.93), 314s(3.81),340(3.51), 357(3.61)	22-0145-84
2,5-Cyclohexadien-1-one, 4-(diphenyl-methylene)-2-nitro-	hexane	262(3.95),272(3.94), 381(4.26),344(4.41)	73-1949-84
2H-Indol-2-one, 1,3-dihydro-1-methyl-3-(4,5-dihydro-5-oxo-2-phenyl-furan-3-ylidene)-	CHCl₃	272(4.33),447(4.39), 470s(4.34)	39-1331-84B
$C_{19}H_{13}NO_3S_2$ 2-Propenenitrile, 2-(phenylsulfonyl)-3-[5-(phenylthio)-2-furanyl]-	MeOH	212(3.22),240(3.05), 398(3.25)	73-2141-84
$C_{19}H_{13}NO_4$ Decarine	EtOH	249(4.54),257(4.55), 277(4.67),285s(4.52), 326(4.20),335s(4.11), 384(3.46)	100-0001-84
	EtOH-NaOH	253(4.31),297(4.37), 330(4.06),384(3.46)	100-0001-84
$C_{19}H_{13}NO_5$ Luguine	dioxan	255(4.44),334(4.36), 345(4.39),384(3.65)	100-0001-84
	dioxan-HCl	250(4.33),269(4.29), 352(4.27),362(4.30), 458(3.80)	100-0001-84
Oxysanguinarine, N-demethyl-5,6-di-hydro-	MeOH	225(4.08),258s(3.96), 347(4.39),378(4.13), 400s(3.98)	100-0001-84
2-Propenoic acid, 2-(benzoylamino)-3-(4-oxo-4H-1-benzopyran-3-yl)-	EtOH	232(4.38),275(4.32)	2-1048-84
$C_{19}H_{13}NO_5S_2$ 2-Propenenitrile, 2-(phenylsulfonyl)-3-[5-(phenylsulfonyl)-2-furanyl]-	MeOH	210(3.18),238(3.15), 340(3.40)	73-2141-84
$C_{19}H_{13}N_3$ 5H-Benz[g]indolo[2,3-b]quinoxaline, 5-methyl-	EtOH	237(4.51),296(4.92), 395(4.26),467(3.44)	103-1155-84
$C_{19}H_{13}N_3O$ Spiro[3H-indole-3,2'-[2H]naphtho[2,3-d]imidazol]-2(1H)-one, 1-methyl-	EtOH	232(4.62),275(4.73), 348(4.39),365(4.18)	103-1155-84
$C_{19}H_{13}N_3OS_2$ Thiazolo[3,2-a]pyrimidin-4-ium, 3-hy-droxy-7-phenyl-2-[(phenylamino)-thioxomethyl]-, hydroxide, inner salt	MeCN	296(4.41),328(4.45), 506(4.27)	103-0791-84
	CCl₄	545(--)	103-0791-84

Compound	Solvent	$\lambda_{max}(\log \epsilon)$	Ref.
$C_{19}H_{13}N_3O_3$			
1H-Indazole-3,4,7(2H)-trione, 2-phenyl-5-(phenylamino)-	EtOH	258(4.47),325(3.99), 440(4.00)	4-0825-84
$C_{19}H_{13}N_3O_4$			
9,10-Anthracenedione, 2-[(2-methyl-5-nitro-1H-imidazol-1-yl)methyl]-	EtOH	254(4.50),308(3.87)	42-0611-84
$C_{19}H_{13}N_5$			
Ethenetricarbonitrile, [4-[methyl(phenylmethylene)hydrazino]phenyl]-, (E)-	benzene	523(4.54)	32-0111-84
Ethenetricarbonitrile, [4-[(1-phenylethylidene)hydrazino]phenyl]-, (E)-	benzene	510(4.35)	32-0111-84
$C_{19}H_{13}N_9$			
1H-Benzotriazole, 1,1'-(2H-benzotriazol-2-ylmethylene)bis-	MeOH	258(4.26),282(4.30), 288(4.27)	117-0299-84
1H-Benzotriazole, 1-[bis(2H-benzotriazol-2-yl)methyl]-	MeOH	212(4.69),282(4.45), 290(4.44)	117-0299-84
1H-Benzotriazole, 1,1',1"-methylidynetris-	MeOH	253(4.33),284(4.02)	117-0299-84
$C_{19}H_{14}$			
1H-Cyclopropa[b]naphthalene, 1-(1-phenylethylidene)-	C_6H_{12}	230(4.70),242s(4.38), 253s(4.24),289(4.38), 377s(4.32),294(4.62), 412s(4.49),422(4.72)	35-6108-84
$C_{19}H_{14}BrNO_2$			
5H-[1]Benzopyrano[2,3-b]pyridin-5-ol, 3-bromo-2-methyl-5-phenyl-	96% H_2SO_4	391(4.41)	103-0607-84
5H-[1]Benzopyrano[2,3-b]pyridin-5-ol, 7-bromo-2-methyl-5-phenyl-	96% H_2SO_4	392(4.56)	103-0607-84
$C_{19}H_{14}BrN_3O$			
Benzo[g]quinoxalin-2(1H)-one, 3-[5-bromo-2-(methylamino)phenyl]-	EtOH	244(4.68),298(4.33), 360(4.17),470(3.94)	103-1155-84
Spiro[3H-indole-3,2'-[2H]naphth[2,3-d]-imidazol]-2(1H)-one, 5-bromo-1',3'-dihydro-1-methyl-	EtOH	218(4.61),254(4.81), 352(4.04)	103-1155-84
$C_{19}H_{14}ClNO_2$			
5H-[1]Benzopyrano[2,3-b]pyridin-5-ol, 3-chloro-2-methyl-5-phenyl-	96% H_2SO_4	387(4.43)	103-0607-84
5H-[1]Benzopyrano[2,3-b]pyridin-5-ol, 7-chloro-2-methyl-5-phenyl-	96% H_2SO_4	387(4.59)	103-0607-84
$C_{19}H_{14}ClNO_6$			
Propanedioic acid, [6-[2-(4-chlorophenyl)ethenyl]-3-cyano-2-oxo-2H-pyran-4-yl]-, dimethyl ester	EtOH	224(4.25),269(4.05), 304(3.94),368(4.24)	94-3384-84
$C_{19}H_{14}Cl_2N_2S_3$			
[1,2]Dithiolo[4,5,1-hi][1,2,3]benzothiadiazole-3-S^{IV}, 2-(4-chlorophenyl)-5-[(4-chlorophenyl)thio]-2,6,7,8-tetrahydro-	MeCN	524(4.51)	97-0251-84
$C_{19}H_{14}FNO_2$			
5H-[1]Benzopyrano[2,3-b]pyridin-5-ol, 7-fluoro-2-methyl-5-phenyl-	96% H_2SO_4	377(4.53)	103-0607-84

Compound	Solvent	$\lambda_{max}(\log \epsilon)$	Ref.
$C_{19}H_{14}NO_4$			
Coptisine	MeOH	209(4.39),229(4.46), 244(4.44),267(4.41), 353(4.34),460(3.61)	73-0704-84
$C_{19}H_{14}N_2O$			
1,1(8aH)-Azulenedicarbonitrile, 2-(4-methoxyphenyl)- (photochromic)	MeCN	220(4.2),275(4.2), 365(4.4)	89-0960-84
Propanedinitrile, [2-(2,4,6-cyclohepta-trien-1-ylidene)-1-(4-methoxyphenyl)-ethylidene]-	MeCN	237(4.2),467(4.5)	89-0960-84
7H-Pyrrolo[3,4-b]pyridin-7-one, 5,6-di-hydro-2,3-diphenyl-	CH_2Cl_2	286(3.97),300(3.99)	83-0379-84
$C_{19}H_{14}N_2OS$			
Thieno[2,3-d]pyrimidin-4(1H)-one, 1-(4-methylphenyl)-5-phenyl-	DMF	290(4.12)	48-0917-84
$C_{19}H_{14}N_2O_2$			
9,10-Anthracenedione, 2-[(2-methyl-1H-imidazol-1-yl)methyl]-	EtOH	254(4.49),308(3.86)	42-0611-84
1H-Benz[f]isoquino[8,1,2-hij]quinazo-line-1,3(2H)-dione, 5,6-dihydro-2-methyl-	MeOH	229s(4.33),246s(4.60), 254(4.72),263(4.56), 294(3.99),318(3.83), 332(3.88),350(3.90), 368(3.98)	78-4003-84
Pyrazolo[1,5-a]quinoline-3-carboxylic acid, 2-phenyl-, methyl ester	EtOH	265(4.61),321(3.76), 337(3.81)	23-2506-84
$C_{19}H_{14}N_2O_5$			
Pyrrolo[2,1-a]isoquinoline-1-carboxylic acid, 3-(5-nitro-2-furanyl)-, ethyl ester	dioxan	444(4.27)	73-0533-84
$C_{19}H_{14}N_4$			
6H-Cyclohepta[2,1-b:3,4-b']di[1,8]-naphthyridine, 7,8-dihydro-	EtOH	239(4.52),329(4.22)	44-2208-84
1-Naphthalenamine, 4-(8-quinolinyl-azo)-	aq EtOH	244(--),280s(--), 460(4.32)	160-1009-84
	pH 1-4	250(--),290(--), 540(4.48)	160-1009-84
$C_{19}H_{14}N_4O_2$			
4(1H)-Pteridinone, 2-methoxy-6,7-di-phenyl-	MeOH	222(4.33),245(4.21), 277(4.23),362(4.05)	136-0179-84F
$C_{19}H_{14}N_4O_2S$			
1H-Pyridazino[4,5-e][1,3,4]thiadiazin-5(6H)-one, 1-acetyl-3,6-diphenyl-	EtOH	264(4.15),312(4.00)	94-4437-84
$C_{19}H_{14}N_4O_3$			
Ethanone, 1-[4-methyl-2,6-bis(2-ni-trosophenyl)-5-pyrimidinyl]-	MeOH	208(3.91),228(3.97), 278(3.87),330(3.38)	83-1029-84
Spiro[3H-indole-3,2'-[2H]naphth[2,3-d]-imidazol]-2(1H)-one, 1',3'-dihydro-1-methyl-5-nitro-	EtOH	215(4.45),253(4.79), 350(4.25)	103-1145-84
3H-[1,2,4]Triazino[3,2-b]quinazoline-3,10(4H)-dione, 2-[2-(4-methoxy-phenyl)ethenyl]-	EtOH	286(3.61),291(3.66), 336(3.66)	24-1077-84

Compound	Solvent	$\lambda_{max}(\log \epsilon)$	Ref.
$C_{19}H_{14}N_4O_3S$			
Benzeneacetic acid, α-[[6,9-dihydro-6-oxo-1-phenyl-1H-purin-2-yl]thio]-	MeOH	264(4.03),284(4.01)	2-0316-84
1,3,7,8-Tetraazaspiro[4.5]deca-6,9-diene-10-carboxylic acid, 4-oxo-1,3-diphenyl-2-thioxo-	MeOH	235(4.22),280(4.17), 308(3.76)	39-2491-84C
$C_{19}H_{14}N_4O_4$			
Ethanone, 1-[4-methyl-2-(nitrophenyl)-6-(nitrosophenyl)-5-pyrimidinyl]-(or isomer)	MeOH	226(4.55),283(4.70), 343(3.96)	83-1029-84
$C_{19}H_{14}N_4O_5$			
Ethanone, 1-[4-methyl-2,6-bis(2-nitrophenyl)-5-pyrimidinyl]-	MeOH	204(4.25),228s(4.17), 260s(3.26)	83-1029-84
$C_{19}H_{14}N_4S$			
2-Thiophenecarboxaldehyde, (2-phenyl-4-quinazolinyl)hydrazone	EtOH	381(3.77)	18-1138-84
$C_{19}H_{14}N_4S_2$			
4H,6H-[1,4]Dithiepino[2,3-b]indole-3-carbonitrile, 4-imino-2-(methylphenylamino)-	EtOH	249(4.05),290(4.47), 330s(4.08)	12-2479-84
$C_{19}H_{14}N_6O$			
1,2,3,5,6,7-Hexaazaacenaphthylen-4(3H)-one, 2,5-dihydro-8-(4-methylphenyl)-2-phenyl-	EtOH	207(4.69)	4-1049-84
$C_{19}H_{14}O$			
2,5-Cyclohexadien-1-one, 4-(diphenylmethylene)-	hexane?	211(4.15),260(4.20), 267(4.19),358(4.50)	73-1949-84
$C_{19}H_{14}O_2$			
2,5-Cyclohexadien-1-one, 4-(diphenylmethylene)-2-hydroxy-	hexane?	211(4.34),264(4.08), 272(4.06),389f(4.19)	73-1949-84
$C_{19}H_{14}O_3S$			
1H-Indene-1,3(2H)-dione, 2-[1-(methylthio)-3-oxo-3-phenylpropylidene]-	EtOH	248(4.54),293(3.70), 309(3.77),361(4.40)	95-0127-84
$C_{19}H_{14}O_4$			
2,5-Furandione, dihydro-3-[(4-methoxyphenyl)methylene]-4-(phenylmethylene)-, (E,E)-	toluene	395(4.05)	48-0233-84
(Z,E)-	toluene	406(4.32)	48-0233-84
(Z,Z)-	toluene	411(4.40)	48-0233-84
$C_{19}H_{14}O_5$			
9,10-Anthracenedione, 1-hydroxy-8-methoxy-2-(3-oxo-1-butenyl)-	n.s.g.	212(4.51),217(4.50), 265(4.57),282(4.44), 417(4.11),434(4.14)	5-0306-84
Benzeneacetic acid, α-(3-hydroxy-5-oxo-4-phenyl-2(5H)-furanylidene)-, methyl ester, (E)-	EtOH	289(4.21),366(3.94)	39-1547-84C
$C_{19}H_{14}O_6$			
Benzeneacetic acid, α-(3-hydroxy-5-oxo-4-phenyl-2(5H)-furanylidene)-2-methoxy-, (E)-	EtOH	267(4.40),363(4.05)	39-1547-84C

Compound	Solvent	$\lambda_{max}(\log \epsilon)$	Ref.
Oxoflavidin diacetate	EtOH	220(4.49),226(4.51), 232s(4.46),240s(4.29), 279(4.45),332(3.81)	102-0671-84
$C_{19}H_{15}BrN_2O_2S$ 5H-Imidazo[1,2-a][3,1]benzothiazine-2-carboxylic acid, 1-bromo-5-phenyl-, ethyl ester	MeOH	233(4.31)	4-1081-84
$C_{19}H_{15}BrN_2O_4$ Pyrimidine, 4-(1,3-benzodioxol-5-yl)-5-(2-bromo-4,5-dimethoxyphenyl)-	CHCl$_3$	246(4.27),279(4.31), 290(4.29)	142-1217-84
$C_{19}H_{15}BrO$ Benzenemethanol, 2-bromo-α,α-diphenyl-	EtOH	249s(2.80),254(2.88), 259(2.94),266(2.87)	104-1921-84
	80% H$_2$SO$_4$	406s(4.39),449(4.62)	104-1921-84
Benzenemethanol, 4-bromo-α,α-diphenyl-	EtOH	254(3.08),260(3.10), 266(3.03),277(2.67)	104-1921-84
	80% H$_2$SO$_4$	415s(4.37),448(4.46)	104-1921-84
$C_{19}H_{15}ClN_2S$ Benzenecarbothioamide, N-(4-chlorophenyl)-2-(phenylamino)-	EtOH	277(4.34),291(4.32), 408s(3.51)	97-0328-84
1H-Pyrazole, 3-(4-chlorophenyl)-4,5-dihydro-1-phenyl-5-(thienyl)-	EtOH	242(4.16),312s(3.69), 358(4.12)	34-0225-84
1H-Pyrazole, 5-(4-chlorophenyl)-4,5-dihydro-1-phenyl-3-(thienyl)-	EtOH	225(4.31),253(4.33), 365(4.29)	34-0225-84
$C_{19}H_{15}ClO$ Benzenemethanol, 2-chloro-α,α-diphenyl-	EtOH	250s(2.76),254(2.94), 259(2.97),265s(2.91)	104-1921-84
	80% H$_2$SO$_4$	406s(4.17),449(4.39)	104-1921-84
Benzenemethanol, 4-chloro-α,α-diphenyl-	EtOH	254(2.90),260(2.95), 266(2.88),277(2.52)	104-1921-84
	80% H$_2$SO$_4$	417s(4.40),443(4.53)	104-1921-84
$C_{19}H_{15}ClO_4$ 5,12-Naphthacenedione, 6-chloro-1,2,3,4-tetrahydro-11-hydroxy-1-methoxy-	EtOH	245(4.71),277(4.10), 459(3.89)	150-0147-84M
$C_{19}H_{15}F_3N_4O$ 7H-Pyrido[2,1-b]quinazoline-6,11-dione, 8,9-dihydro-, 6-[[(3-trifluoromethyl)phenyl]hydrazone]	EtOH	227(4.35),253(4.27), 298(3.98),379(4.33)	4-1301-84
$C_{19}H_{15}FeNO_4$ Ferrocene, 1,1'-[pyridinediylbis(methyleneoxycarbonyl)]-	dioxan	467(2.29)	18-2435-84
$C_{19}H_{15}Li$ Lithium, (triphenylmethyl)- (contg. sparteine at -78°)	heptane	400s(3.76),437(3.88)	138-0757-84
$C_{19}H_{15}NOS_2$ 2(1H)-Pyridinone, 3,4-dihydro-3-phenyl-4,6-di-2-thienyl-	EtOH	242(4.38),290(4.14), 363(3.24)	4-1473-84
$C_{19}H_{15}NO_2$ 5H-[1]Benzopyrano[2,3-b]pyridin-5-ol,	96% H$_2$SO$_4$	374(4.51)	103-0607-84

Compound	Solvent	$\lambda_{max}(\log \epsilon)$	Ref.
2-methyl-5-phenyl- (cont.)			103-0607-84
5H-[1]Benzopyrano[2,3-b]pyridin-5-ol, 7-methyl-5-phenyl-	96% H_2SO_4	380(4.51)	103-0607-84
$C_{19}H_{15}NO_3$			
Acetamide, N-[3-(3-oxo-2(3H)-benzofuranylidene)-1-propenyl]-N-phenyl-	DMSO	487(4.55)	103-0951-84
$C_{19}H_{15}NO_4$			
6H-Anthra[9,1-bc]furan-2-carboxylic acid, 5-(dimethylamino)-6-oxo-, methyl ester	EtOH	510(4.14),541(4.16)	104-0745-84
6H-Anthra[1,9-cd]isoxazole-3-carboxylic acid, 6-oxo-, butyl ester	EtOH	460(4.08)	104-2012-84
4,9[1',2']-Benzeno-1H-benz[f]isoindole-1,3(2H)-dione, 3a,4,9,9a-tetrahydro-4,9-dihydroxy-2-methyl-	ether	243s(3.80),249.5(3.95), 261(3.56),270(3.49), 321(2.94),333s(2.80)	22-0145-84
1H-Benz[f]isoindole-1,3(2H)-dione, 9,9a-dihydro-9-hydroxy-4-(2-hydroxyphenyl)-2-methyl-, cis	ether	257.5(4.06),300(4.16)	22-0145-84
Chileninone	MeOH	238(4.06),277(4.03), 328(3.59),389(3.60)	78-3957-84
	MeOH-acid	232(4.02),272(4.03), 355(3.84),438(3.28)	78-3957-84
4,9-Epoxy-1H-benz[f]isoindole-1,3(2H)-dione, 3a,4,9,9a-tetrahydro-4-(2-hydroxyphenyl)-2-methyl-, endo	ether	249s(2.78),263s(3.24), 269s(3.41),276(3.50), 283(3.47)	22-0145-84
exo	ether	249.5(2.40),257(3.10), 263s(3.26),271(3.39), 272.5(3.40),281(3.35)	22-0145-84
Guattescine, dehydro-	EtOH	206(4.39),245(4.44), 263s(4.32),306(3.76), 320(3.80),362(3.65)	28-0311-84B
	EtOH-HCl	208(4.35),257(4.42), 280(4.36),400(3.50)	28-0311-84B
1H-Pyrrole-3-carboxylic acid, 4,5-dihydro-4,5-dioxo-1,2-diphenyl-, ethyl ester	dioxan	240(4.13),305(3.77), 405(3.49)	94-0497-84
$C_{19}H_{15}NO_5$			
Oxocularine	MeOH	214(4.53),254(4.40), 302s(3.62),402(3.71)	100-0753-84
	MeOH-acid	224(4.47),267(4.37), 331s(3.70),345s(3.60), 486(3.61)	100-0753-84
Oxosarcocapnine	EtOH	254(4.19),330(3.16), 400(3.53)	100-0753-84
	EtOH-HCl	266(--),398(--), 462(--)	100-0753-84
$C_{19}H_{15}NO_6$			
Linaresine, (±)-	MeOH	236(4.71),298(4.27), 334(4.09)	35-6099-84 +100-0753-84
	MeOH-acid	245(4.61),312(4.23), 346(4.11)	35-6099-84 +100-0753-84
Propanedioic acid, [3-cyano-2-oxo-6-(2-phenylethenyl)-2H-pyran-4-yl]-, dimethyl ester	ether	240(4.16),268(4.14), 283(4.13),428(4.43)	94-3384-84
$C_{19}H_{15}N_3O$			
Benzo[g]quinoxalin-2(1H)-one, 3-[2-	EtOH	241(4.66),298(4.30),	103-1155-84

Compound	Solvent	$\lambda_{max}(\log \epsilon)$	Ref.
(methylamino)phenyl]- (cont.) Spiro[3H-indole-3,2'-[2H]naphth[2,3-d]-imidazol]-2(1H)-one, 1',3'-dihydro-1-methyl-	EtOH	360(4.08),461(3.86) 215(4.62),253(4.80), 352(4.03)	103-1155-84 103-1155-84
$C_{19}H_{15}N_3O_3$ 1H-Isoindole-1,3(2H)-dione, 2-[4-(1H-indazol-3-yl)-4-oxobutyl]-	MeOH	220(4.56),234(4.27), 242(4.27),272(4.44)	103-0066-84
$C_{19}H_{15}N_3O_6P$ Phosphonium, methyltris(4-nitrophenyl)-, iodide	MeOH	253(4.58)	139-0259-84C
$C_{19}H_{15}N_5$ Cyanamide, (7,8-dihydro-2,3-diphenyl-imidazo[1,2-c]pyrimidin-5-yl)-	MeOH	229(4.42),251(4.35)	24-2597-84
4-Pteridinamine, 7-phenyl-N-(phenyl-methyl)-	MeOH	231(4.57),255s(4.38), 280(4.31),319(4.09), 362(4.36)	150-3601-84M
$C_{19}H_{15}N_5O$ Benzamide, N-[9-(phenylmethyl)-9H-purin-6-yl]-	MeOH	280(4.24)	150-3601-84M
Benzonitrile, 4-[(8,9-dihydro-11-oxo-7H-pyrido[2,1-b]quinazolin-6(11H)-ylidene)hydrazino]-	EtOH	227(4.32),275(4.08), 378(4.42)	4-1301-84
$C_{19}H_{15}N_5O_3$ 1,3,7,8-Tetraazaspiro[4.5]deca-1,6,9-triene-10-carboxylic acid, 4-oxo-3-phenyl-2-(phenylamino)-	MeOH	265s(3.84),304(3.75)	39-2491-84C
$C_{19}H_{16}$ Naphthalene, 2-methyl-3-(2-phenyleth-enyl)-, trans	$C_6H_{11}Me$	<u>280(4.6),330(4.3)</u>	35-7624-84
$C_{19}H_{16}ClF_3N_4$ 2,4-Pyrimidinediamine, 5-chloro-N,N'-dimethyl-N,N'-diphenyl-6-(trifluoro-methyl)-	MeOH	234(3.82),270(3.98), 314(3.64)	4-1161-84
$C_{19}H_{16}N_2OS$ 7H-Thiazolo[3,2-a]pyrimidin-7-one, 5,6-dihydro-3-methyl-5,6-diphenyl-, cis	EtOH	310(4.22)	142-2529-84
trans	EtOH	309(4.15)	142-2529-84
$C_{19}H_{16}N_2O_2$ 4H-[1]Benzoxepino[3,4-c]pyrazol-4-one, 1,10-dihydro-3-methyl-1-(phenyl-methyl)-	EtOH	276(4.11)	4-0301-84
4H-[1]Benzoxepino[3,4-c]pyrazol-4-one, 2,10-dihydro-3-methyl-2-(phenyl-methyl)-	EtOH	276(4.18)	4-0301-84
[2,5'-Bi-1H-indole]-2'-carboxylic acid, ethyl ester	EtOH	233(4.41),253(4.73), 307(4.58),322s(4.34), 347(4.04)	103-0280-84
Methanone, (5-hydroxy-6-methyl-2-phen-yl-4-pyrimidinyl)(4-methylphenyl)-	MeOH	285(4.37),418(3.76)	32-0431-84
Methanone, (5-methoxy-6-methyl-2-phen-yl-4-pyrimidinyl)phenyl-	MeOH	256(3.51)	32-0431-84

Compound	Solvent	$\lambda_{max}(\log \epsilon)$	Ref.
$C_{19}H_{16}N_2O_2S$			
5H-Imidazo[1,2-a][3,1]benzothiazine-2-carboxylic acid, 5-phenyl-, ethyl ester, hydrobromide	n.s.g.	251(4.25)	4-1081-84
2H,5H-Indeno[2,1-d][1,2]thiazin-5-one, 2-methyl-4-[(phenylmethyl)amino]-, 2-oxide	EtOH	239(4.55),250(4.56), 290(4.13),311(4.28), 326(4.14),340(4.02), 410(3.72)	94-2910-84
$C_{19}H_{16}N_2O_3$			
Alangimarine	EtOH	220(4.37),261(4.11), 290s(3.83),365(4.42)	100-0397-84
$C_{19}H_{16}N_2O_4$			
9H-Anthra[1,9-cd]isoxazol-6-one, 5-(2-acetoxyethylamino)-3-methyl-	THF	252(4.50),259(4.52), 485(4.25),538(4.30)	103-0717-84
$C_{19}H_{16}N_2O_4P$			
Phosphonium, methylbis(4-nitrophenyl)-phenyl-, iodide	MeOH	251(4.49)	139-0259-84C
$C_{19}H_{16}N_2O_6$			
Imidodicarbonic acid, (3-cyano-4,5-dihydro-4-oxoindeno[1,2-b]pyran-2-yl)-, diethyl ester	MeOH	222(4.21),246(4.19), 305(4.17)	83-0448-84
$C_{19}H_{16}N_2S$			
1H-1,5-Benzodiazepine, 2,3-dihydro-2-phenyl-4-(2-thienyl)-	EtOH	263(4.4),384(3.84)	103-1370-84
1H-1,5-Benzodiazepine, 2,3-dihydro-4-phenyl-2-(2-thienyl)-	EtOH	246(4.43),367(3.76)	103-1370-84
1H-Pyrazole, 4,5-dihydro-1,3-diphenyl-5-(thienyl)-	EtOH	241(4.30),300s(3.92), 350(4.18)	34-0225-84
$C_{19}H_{16}N_2S_3$			
[1,2]Dithiolo[4,5,1-hi][1,2,3]benzothiadiazole-3-S^{IV}, 2,6,7,8-tetrahydro-2-phenyl-5-(phenylthio)-	MeCN	520(4.39)	97-0251-84
$C_{19}H_{16}N_3O_4$			
1H-Pyrazolium, 2-ethyl-4,5-dihydro-3-[2-(4-nitrophenyl)ethenyl]-4,5-dioxo-1-phenyl-, iodide	EtOH	345(4.11)	48-0695-84
$C_{19}H_{16}N_4O_3$			
Benzoic acid, 4-[(8,9-dihydro-11-oxo-7H-pyrrolo[2,1-b]quinazolin-6(11H)-ylidene)hydrazino]-	EtOH	227(4.26),280(4.14), 385(4.40)	4-1301-84
7H-1,3-Dioxolo[4,5-g]pyrido[2,1-b]-quinazoline-6,11-dione, 8,9-dihydro-, 6-(phenylhydrazone)	EtOH	226s(4.32),243(4.50), 302(3.71),389(4.30)	4-1301-84
7H-Pyrido[2,1-b]quinazoline-2-carboxylic acid, 6,8,9,11-tetrahydro-11-oxo-6-(phenylhydrazono)-	EtOH	213(4.53),248s(4.23), 301(3.91),398(4.36)	4-1301-84
$C_{19}H_{16}N_4S_2$			
1,3,4-Thiadiazole-2(3H)-thione, 5-(4-methylphenyl)-3-[(5-quinolinyl-amino)methyl]-	EtOH	340(4.09)	83-0547-84

Compound	Solvent	$\lambda_{max}(\log \epsilon)$	Ref.
$C_{19}H_{16}N_6O_4$			
Oxepino[2,3-d]pyrimidine-5-carboxylic acid, 2-azido-6-cyano-3,4,7,8-tetra-hydro-8,8-dimethyl-4-oxo-3-phenyl-, methyl ester	CHCl$_3$	340(4.23)	24-0585-84
$C_{19}H_{16}N_6S$			
Propanenitrile, 3,3'-[[4-(2-benzothia-zolylazo)phenyl]imino]bis-	acetone	472(4.60)	7-0819-84
	50% acetone	488(4.61)	7-0819-84
$C_{19}H_{16}O$			
6,9-Ethano-1H-cyclohepta[cd]phenalen-1-one, 6,7,8,9-tetrahydro-	MeCN	267(4.34),324(3.47), 366(3.82),411(4.13)	44-1146-84
	60% H$_2$SO$_4$	210(4.47),223(4.49), 263(3.97),365(4.13), 415(4.10),459(4.17), 477(4.21)	44-1146-84
$C_{19}H_{16}O_2$			
Benzenepropanoic acid, α-(1H-inden-1-ylidene)-, methyl ester	EtOH	250(4.31),310(3.61)	23-2506-84
1,2-Chrysenediol, 1,2-dihydro-11-meth-yl-, trans	n.s.g.	222(4.73),253(4.57), 270(4.66),318(4.20)	44-0381-84
2,5-Cyclohexadien-1-one, 2-methoxy-4,4-diphenyl-	MeOH	204(4.75),217s(4.54), 240s(4.28)	94-4721-84
1,1'-Spirobi[1H-indene]-3-carboxylic acid, 3a,7a-dihydro-, methyl ester	EtOH	218(4.44),258(4.05)	23-2506-84
$C_{19}H_{16}O_4$			
9,10-Anthracenedione, 1-hydroxy-3-meth-yl-2-(3-oxobutyl)-	MeOH	208(4.20),232s(--), 246(4.46),263(4.48), 280s(--),326(3.52), 391s(--),410(3.83), 432s(--)	78-3677-84
5,12-Naphthacenedione, 1,2,3,4-tetra-hydro-11-hydroxy-1-methoxy-	EtOH	241(4.93),275(4.29), 462(4.04)	150-0147-84M
1-Naphthalenecarboxylic acid, 2-(3,4-dimethoxyphenyl)-	MeCN	224(5.66),278(5.29)	2-0603-84
$C_{19}H_{16}O_5$			
9,10-Anthracenedione, 1-hydroxy-8-meth-oxy-2-(3-oxobutyl)-	n.s.g.	214(4.56),224(4.56), 254(4.48),275(4.16), 312(3.60),324(3.64), 400(3.96),415(4.01), 437(3.90)	5-0306-84
5H-Indeno[5,6-d]-1,3-dioxol-5-one, 7-(3,4-dimethoxyphenyl)-6-methyl-	MeOH	240s(4.24),265(4.51), 340(3.88)	102-2021-84
$C_{19}H_{16}O_6$			
9,10-Anthracenedione, 1,4-dihydroxy-8-methoxy-2-(3-oxobutyl)-	MeOH	239(4.53),245(4.30), 284(3.92),476(4.00), 493(3.99),527(3.71)	78-4609-84
6-Benzofuranol, 2-(2-acetoxy-4-methoxy-phenyl)-, acetate	MeOH	242s(3.82),277s(4.24), 286s(4.32),307(4.54), 320s(4.41)	94-3267-84
2,5-Cyclohexadiene-1,4-dione, 2,2'-(1,5-dioxo-1,5-pentanediyl)bis-[5-methyl-	MeCN	250(4.54),314s(3.11), 450(2.00)	18-0791-84
Cyclopenta[k]phenanthrene-1,4,5,8(12bH)-tetrone, 5,6,7,7a-tetrahydro-9,12-di-hydroxy-2,11-dimethyl-, (4aS*,7aα,12bα)-	MeCN	234(4.32),268s(3.85), 361(3.70)	18-0791-84

Compound	Solvent	$\lambda_{max}(\log \epsilon)$	Ref.
5,12-Naphthacenedione, 7,8,9,10-tetra-hydro-1,6,8,11-tetrahydroxy-8-meth-yl-, (±)-	MeOH	219(4.30),235(4.47), 254(4.52),293(3.94), 411(4.53),465(4.07), 490(4.20),510(4.05), 525(4.08)	78-4609-84
5,12-Naphthacenedione, 7,8,9,10-tetra-hydro-6,7,9,11-tetrahydroxy-9-meth-yl-, (7S-cis)-	CHCl$_3$	258(4.63),290(4.02), 486(4.03)	78-4649-84
$C_{19}H_{16}O_7$			
4H-1-Benzopyran-4-one, 5,7-dimethoxy-2-(7-methoxy-1,3-benzodioxol-5-yl)-	MeOH	242(4.22),268s(--), 336(4.17)	94-0166-84
5,12-Naphthacenedione, 7,8,9,10-tetra-hydro-1,6,8,10,11-pentahydroxy-8-methyl-, (cis-(±)-	MeOH	210(4.19),234(4.55), 254(4.45),292(3.91), 472(4.08),480(4.12), 492(4.18),510(4.94), 526(4.07),560(2.90)	78-4609-84
$C_{19}H_{16}O_8$			
8H-1,3-Dioxolo[4,5-g][1]benzopyran-8-one, 7,9-dimethoxy-6-(4-hydroxy-3-methoxyphenyl)-	MeOH	245(4.28),270s(4.17), 336(4.46)	102-2043-84
	MeOH-NaOAc	245(4.28),270s(4.17), 336(4.39),400s(3.81)	102-2043-84
	MeOH-NaOMe	245(4.28),258(4.24), 335s(4.03),397(4.49)	102-2043-84
$C_{19}H_{17}BrN_2O_2S$			
5H-Imidazo[1,2-a][3,1]benzothiazine-2-carboxylic acid, 5-phenyl-, ethyl ester, hydrobromide	n.s.g.	251(4.25)	4-1081-84
$C_{19}H_{17}BrN_4O_2$			
2,4(3H,5H)-Pyrimidinedione, 6-[(4-bromophenyl)methylamino]-3-methyl-5-[(4-methylphenyl)imino]-	MeCN	231(4.32),243(4.30), 334(3.78),400-600s(--)	142-1021-84
$C_{19}H_{17}BrO_4$			
2H-1-Benzopyran-2-one, 7-[(7-bromo-3,7-dimethyl-6-oxo-2-octen-4-ynyl)oxy]-, (E)-	MeOH	215(4.39),292(4.22), 319(4.27)	39-0535-84C
$C_{19}H_{17}BrO_6$			
4H-1-Benzopyran-4-one, 3-(2-bromo-4,5-dimethoxyphenyl)-7,8-dimethoxy-	CHCl$_3$	256(4.42),294(4.17)	142-1217-84
$C_{19}H_{17}ClN_2O_3$			
6-Phthalazinecarboxylic acid, 3-(2-chlorophenyl)-3,4-dihydro-5,7-di-methyl-4-oxo-, ethyl ester	EtOH	217(4.64),309(4.05)	87-1300-84
$C_{19}H_{17}ClN_4O$			
7H-Pyrido[2,1-b]quinazoline-6,11-dione, 2-chloro-8,9-dihydro-, 6-[(4-methyl-phenyl)hydrazone], (E)-	EtOH	230(4.32),247(4.20), 303(3.92),400(4.32)	4-1301-84
$C_{19}H_{17}ClO_4$			
1,4-Ethanoanthracene-9,10-dione, 7-(chloromethyl)-1,4-dihydro-8-hydroxy-1-methoxy-6-methyl-	MeOH	212(4.16),248(4.22), 293(3.52),369(3.69)	44-3766-84

Compound	Solvent	$\lambda_{max}(\log \epsilon)$	Ref.
$C_{19}H_{17}F_3N_4$			
2,4-Pyrimidinediamine, N,N'-dimethyl-N,N'-diphenyl-6-(trifluoromethyl)-	MeOH	224(3.66),257(3.63), 311(3.29)	4-1161-84
$C_{19}H_{17}NOS$			
Benzo[b]thiophen-3(2H)-one, 2-[3-(dimethylamino)-2-phenyl-2-propenylidene]-	hexane	281(4.22),356(4.45), 476(4.49)	103-0951-84
	DMSO	364(4.17),509(4.71)	103-0951-84
$C_{19}H_{17}NO_2$			
Furo[3,2-d][1]benzoxepin-2-carbonitrile, 2,3,3a,4,5,10b-hexahydro-2-phenyl-, (2α,3aα,10bβ)-	EtOH	244s(2.53),250s(2.72), 258(2.88),264(2.95), 271(2.92)	24-2157-84
1,5-Methano-3-benzazocine-4,6(1H,5H)-dione, 2,3-dihydro-3-methyl-1-phenyl-	n.s.g.	247(3.05),295(3.40)	4-0611-84
Oxazole, 2-[(2-butenyl)oxy]-4,5-diphenyl-	C_6H_{12}	224(4.33),302(4.14)	44-0399-84
2(3H)-Oxazolone, 3-(1-methyl-2-propenyl)-4,5-diphenyl-	C_6H_{12}	217(4.20),292(4.13), 302(4.15)	44-0399-84
Oxiranecarbonitrile, 3-[2-(3-butenyloxy)phenyl]-2-phenyl-	EtOH	231s(4.08),274s(3.58), 282(3.63)	24-2157-84
1H-Pyrrole-3-carboxylic acid, 2-phenyl-N-(phenylmethyl)-, methyl ester	MeOH	265(4.04)	44-3314-84
1H-Pyrrole-3-carboxylic acid, 5-phenyl-N-(phenylmethyl)-, methyl ester	MeOH	265(4.21)	44-3314-84
$C_{19}H_{17}NO_2P$			
Phosphonium, methyl(4-nitrophenyl)diphenyl-, iodide	MeOH	253(4.21)	139-0259-84C
$C_{19}H_{17}NO_2S$			
2-Thiophenecarboxylic acid, 3-amino-5-[1,1'-biphenyl]-4-yl-, ethyl ester	MeOH	310(4.46),355(4.16)	118-0275-84
$C_{19}H_{17}NO_3$			
Acronycine, de-N-methyl-	EtOH	252s(3.60),266(3.76), 283(3.55),293(3.58), 307(3.18),335(2.73), 391(2.96)	100-0285-84
9,10-Anthracenedione, 2-(morpholinomethyl)-	EtOH	256(4.59),327(3.69)	42-0611-84
Belemine	EtOH	224(4.25),270(4.47), 290s(4.05),323(3.82), 375s(2.78)	28-0311-84B
	EtOH-NaOH	212(4.69),290(4.64), 324s(3.76),364(3.48), 384(3.39)	28-0311-84B
Guadiscidine	EtOH	234(4.12),270(4.38), 306(4.13),322s(3.92), 348(3.78)	100-0353-84
	EtOH-base	224(4.16),288(4.22), 312(4.17),278(3.88)	100-0353-84
	EtOH-acid	214(4.35),278(4.43), 368(3.76),420(3.68)	100-0353-84
Noracronycine	EtOH	227(4.22),256(4.41), 284(4.68),295s(4.63), 312(4.37),342s(3.68), 410(3.69)	100-0285-84
$C_{19}H_{17}NO_4$			
Cularimine, 1,2,3,4-tetradehydro-	EtOH	228(4.61),284(3.92),	100-0753-84

Compound	Solvent	$\lambda_{max}(\log \epsilon)$	Ref.
(cont.)		348(3.74)	100-0753-84
	EtOH-HCl	251(4.51),284(3.76), 396(3.73)	100-0753-84
1H-Indole-4,5-dicarboxylic acid, 7-methyl-1-phenyl-, dimethyl ester	EtOH	214(4.43),248(3.74)	39-2541-84C
Noraporphine, 6a,7-dehydro-1,2-dimethoxy-9,10-(methylenedioxy)-	MeOH	240s(4.43),260(4.66), 294(4.13),335(4.03), 382(3.54)	39-1273-84C
Stylopine, (-)-	MeOH	209(4.62),236s(4.02), 287(3.93)	73-0704-84
$C_{19}H_{17}NO_5$			
Guacolidine	EtOH	224(4.19),274(4.12), 328(3.54),361(3.57)	100-0353-84
	EtOH-acid	228(4.12),278(4.45), 364(3.74),428(3.44)	100-0353-84
	EtOH-base	215(4.16),234s(4.04), 290(4.19),323s(4.16), 386(3.66)	100-0353-84
Norchelidonine, (-)-	EtOH	239(3.86),288(3.91)	100-0001-84
$C_{19}H_{17}NO_6$			
Linaresine, dihydro-, (±)-	MeOH	230(4.37),299(4.23), 325(4.07)	35-6099-84
	MeOH-acid	238(4.22),250(4.22), 260(4.06),307(4.14), 381(3.86)	35-6099-84 +100-0753-84
Rugosinone, dihydro-	MeOH	228s(4.71),299(4.50)	100-1050-84
$C_{19}H_{17}N_2O_2$			
Benzoxazolium, 3-methyl-2-[3-(3-methyl-2(3H)-benzoxazolylidene)-1-propenyl]-, perchlorate	MeOH	482(5.13)	48-1034-84
$C_{19}H_{17}N_2O_3$			
1H-Pyrazolium, 2-ethyl-4,5-dihydro-3-[2-(4-hydroxyphenyl)ethenyl]-4,5-dioxo-1-phenyl-, iodide	EtOH	360(4.28),540(3.44)	48-0695-84
$C_{19}H_{17}N_2S_2$			
Benzothiazolium, 3-methyl-2-[3-(3-methyl-2(3H)-benzothiazolylidene)-1-propenyl]-, perchlorate	MeOH	557(5.11)	48-1034-84
$C_{19}H_{17}N_3OS$			
7H-1,3,4-Thiadiazolo[3,2-a]pyrimidin-7-one, 2-ethyl-5,6-dihydro-5,6-diphenyl-, cis	EtOH	292(4.21)	142-2529-84
trans	EtOH	292(4.21)	142-2529-84
$C_{19}H_{17}N_3O_3$			
Pyrrolo[2,1-a]isoquinolin-2-ol, 2,3,5,6-tetrahydro-3-imino-1-methyl-2-(4-nitrophenyl)-	EtOH	240(4.15),270s(3.90)	94-1170-84
$C_{19}H_{17}N_3O_4S$			
3-Pyridinecarboxylic acid, 2-[[5-(ethoxycarbonyl)-1-(phenylmethyl)-1H-imidazol-2-yl]thio]-	MeOH	206(4.43),254(4.32), 295s(3.61)	142-0807-84

Compound	Solvent	$\lambda_{max}(\log \epsilon)$	Ref.
$C_{19}H_{17}N_3O_5$			
3,4-Oxepindicarboxylic acid, 5-cyano-6,7-dihydro-7-methyl-2-[(phenyl-carbonimidoyl)amino]-, dimethyl ester	CHCl$_3$	260(3.84),267(3.83), 323(4.33)	24-0585-84
Oxepino[2,3-d]pyrimidine-5-carboxylic acid, 6-cyano-1,2,3,4,7,8-hexahydro-8,8-dimethyl-2,4-dioxo-3-phenyl-, methyl ester	CHCl$_3$	260(3.79),318(4.15)	24-0585-84
$C_{19}H_{17}N_3O_7$			
1H-Pyrazole-3,4-dicarboxylic acid, 5-[4-(1,3-dihydro-1,3-dioxo-2H-isoindol-2-yl)-1-oxobutyl]-, dimethyl ester	MeOH	217s(4.61),221(4.62), 225s(4.55),234(4.23), 243(4.12),318(3.82)	103-0066-84
$C_{19}H_{17}N_3O_8$			
4,9:5,8-Diepoxy-1H-naphtho[2,3-d]tria-zole-4a,8a-dicarboxylic acid, 1-(5-formyl-2-furanyl)-3a,4,5,8,9,9a-hexahydro-, dimethyl ester	MeOH	348(3.31)	73-1990-84
$C_{19}H_{17}N_3S$			
4-Isoquinolinecarbonitrile, 3-[(cyano-methyl)thio]-5,6,7,8-tetrahydro-1-(phenylmethyl)-	dioxan	266(4.31),305(3.99), 345(3.60)	104-2225-84
Thieno[2,3-c]isoquinoline-2-carboni-trile, 1-amino-6,7,8,9-tetrahydro-5-(phenylmethyl)-	dioxan	277(4.63),298(4.19), 346(4.04)	104-2225-84
$C_{19}H_{17}N_5$			
7H-Purin-6-amine, 7-methyl-8-phenyl-N-(phenylmethyl)-	MeOH	237(4.38),292(4.24)	150-3601-84M
9H-Purin-6-amine, 8,9-dimethyl-N,2-di-phenyl-	EtOH	235(4.31),263(4.33), 306(4.34)	11-0208-84B
6H-Purin-6-imine, 1,7-dihydro-7,8-di-methyl-1,2-diphenyl-	EtOH	255s(--),264(4.03), 272s(--),309(3.64)	11-0208-84B
6H-Purin-6-imine, 1,9-dihydro-8,9-di-methyl-1,2-diphenyl-	EtOH	255s(--),268(4.04), 305(3.63)	11-0208-84B
$C_{19}H_{17}N_5O$			
Benzamide, N-[8,9-dihydro-9-(phenyl-methyl)-7H-purin-6-yl]-	MeOH	284(3.98),336(3.95)	150-3601-84M
$C_{19}H_{17}N_5O_2$			
Acetamide, N-[2-[4-(acetylamino)phen-yl]-4H-pyrazolo[1,5-a]benzimidazol-6-yl]-	EtOH	248(4.43),298(4.40), 330(4.52)	48-0829-84
$C_{19}H_{17}N_5O_6$			
3-Penten-2-one, 4-[[(2-nitrophenyl)-methylene]amino]-3-[(2-nitrophen-yl)(nitrosoamino)methyl]-, (E,E)-	MeOH	204(4.41),285(4.14)	83-1029-84
$C_{19}H_{17}OS$			
Thiopyrylium, 2-(4-methoxyphenyl)-6-methyl-4-phenyl-, perchlorate	MeCN	260(3.89),300(4.07), 355(4.20),430(4.21)	103-1226-84
Thiopyrylium, 4-(4-methoxyphenyl)-2-methyl-6-phenyl-, perchlorate	MeCN	258(4.21),370s(3.89), 421(4.20)	103-1226-84
$C_{19}H_{17}O_2$			
Pyrylium, 2-(4-methoxyphenyl)-6-methyl-	MeCN	275(4.20),328(4.24),	103-1226-84

Compound	Solvent	$\lambda_{max}(\log \epsilon)$	Ref.
4-phenyl-, perchlorate (cont.)		418(4.36)	103-1226-84
Pyrylium, 4-(4-methoxyphenyl)-2-methyl-6-phenyl-, perchlorate	MeCN	250(4.25),350(4.22), 400(4.58)	103-1226-84
$C_{19}H_{18}$			
Azulene, 1-(2,4,6-trimethylphenyl)-	hexane	533(1.93),549(2.08), 573(2.19),593(2.30), 620(2.19),650(2.22), 683(1.74),719(1.70)	5-1605-84
Azulene, 2-(2,4,6-trimethylphenyl)-	hexane	537(2.10),567(2.28), 614(2.20),673(1.70)	5-1605-84
Azulene, 4-(2,4,6-trimethylphenyl)-	hexane	513(2.11),533(2.29), 553(2.43),575(2.59), 597(2.51),626(2.53), 652(2.10),690(2.02)	5-1605-84
Azulene, 5-(2,4,6-trimethylphenyl)-	hexane	521(1.95),543(2.15), 550(2.28),583(2.45), 606(2.39),637(2.45), 667(2.02),705(1.98)	5-1605-84
Azulene, 6-(2,4,6-trimethylphenyl)-	hexane	512(1.90),532(2.04), 550(2.19),572(2.27), 595(2.20),625(2.19), 651(1.78),690(1.70)	5-1605-84
9H-Fluorene, 9-(2-cyclohexen-1-yl)-	MeOH	225(4.30),230(4.24), 261(4.16),269(4.22), 293(3.86),304(3.94)	2-0802-84
9H-Fluorene, 9-(1,3-dimethyl-2-buten-ylidene)-	hexane	234(4.65),251(4.42), 260(4.44),319(4.20)	39-2871-84C
Spiro[bicyclo[4.1.0]heptane-2,9'-[9H]-fluorene]	MeOH	226(4.32),229(4.31), 260(4.03),271(4.08), 292(3.82),303(3.85)	2-0802-84
$C_{19}H_{18}ClN_3O$			
5-Pyrimidinecarbonitrile, 5-chlorohexa-hydro-1,3-dimethyl-4-oxo-2,6-di-phenyl-	EtOH	208(4.06)	44-0282-84
$C_{19}H_{18}FeNO$			
Benzoxazolium, 3-(ferrocenylmethyl)-2-methyl-, tetrafluoroborate	EtOH	260(4.02),341(3.72), 421(2.58)	65-1067-84
$C_{19}H_{18}FeNS$			
Benzothiazolium, 3-(ferrocenylmethyl)-2-methyl-, tetrafluoroborate	EtOH	262(4.11),345(3.83), 432(2.60)	65-1067-84
$C_{19}H_{18}N_2O$			
2(1H)-Pyrazinone, 1-methyl-5,6-bis(4-methylphenyl)-	EtOH	265(4.18),351(3.89)	39-0391-84C
3H-Pyrazol-3-one, 2,4-dihydro-4-(1-methylethylidene)-2-(4-methylphen-yl)-5-phenyl-	EtOH	260(4.40)	94-2146-84
$C_{19}H_{18}N_2O_2$			
Benzenepropanenitrile, β-oxo-α-[[5-(1-piperidinyl)-2-furanyl]methylene]-	MeOH	208(4.08),238(2.93), 285(2.52),500(3.62)	73-2141-84
1,4-Naphthalenedione, 2-methyl-3-[[2-(phenylamino)ethyl]amino]-	CH₂Cl₂	242(4.37),276(4.42), 330(3.36),465(3.50)	83-0743-84
$C_{19}H_{18}N_2O_2S$			
Acetic acid, [[4-cyano-5,6,7,8-tetra-	dioxan	226(4.46),266(4.25),	104-2225-84

Compound	Solvent	$\lambda_{max}(\log \epsilon)$	Ref.
hydro-1-(phenylmethyl)-3-isoquinolin-yl]thio]- (cont.)		317(3.86)	104-2225-84
Thieno[2,3-c]isoquinoline-2-carboxylic acid, 1-amino-6,7,8,9-tetrahydro-5-(phenylmethyl)-	dioxan	282(4.46),302(4.04), 361(3.84)	104-2225-84
$C_{19}H_{18}N_2O_3$			
Alangimaridine	EtOH	220(4.53),255(4.00), 284(3.84)	100-0397-84
6-Phthalazinecarboxylic acid, 3,4-di-hydro-5,7-dimethyl-4-oxo-3-phenyl-, ethyl ester	EtOH	228(4.67),322(4.33)	87-1300-84
Santiagonamine	MeOH	225(5.02),253(5.03), 273s(4.70),300(4.60), 310(4.61),366(4.45)	88-3163-84
	MeOH-acid	229(5.03),258(4.99), 267s(4.94),278s(4.73), 298(4.39),309(4.43), 329(4.32),368(4.26), 391(4.28),396(4.28), 400(4.27),434(4.22)	88-3163-84
$C_{19}H_{18}N_2O_3S$			
Acetic acid, [[2-(3,4-dihydro-4-methyl-3-oxo-2-quinoxalinyl)-1-phenylethyl]-thio]-	EtOH	343(3.30)	18-1653-84
Sulfoximine, N-[(1,3-dihydro-1,3-dioxo-2H-inden-2-ylidene)[(phenylmethyl)-amino]methyl]-S,S-dimethyl-	EtOH	226(4.54),293(4.53), 320(4.37),332(4.50)	94-2910-84
$C_{19}H_{18}N_2O_4$			
Alamarine	EtOH	220(4.37),253(4.20), 363(4.44)	100-0397-84
$C_{19}H_{18}N_2O_4S_2$			
4,6(1H,5H)-Pyrimidinedione, 5-[(1,3-di-hydro-1,3-dioxo-2H-inden-2-ylidene)-(methylthio)methyl]-1,3-diethyldi-hydro-2-thioxo-	EtOH	246(4.50),292(4.18), 364(4.34),466(3.96)	95-0127-84
$C_{19}H_{18}N_2O_5$			
Phenylalanine, α,β-didehydro-N-[N-[(phenylmethoxy)carbonyl]glycyl]-, (Z)-	EtOH	206(4.48),281(4.26)	5-0920-84
$C_{19}H_{18}N_3O_3$			
1H-Pyrazolium, 2-ethyl-4,5-dihydro-3-[[[(4-methoxyphenyl)imino]methyl]-4,5-dioxo-1-phenyl-, iodide	EtOH	390(4.28)	48-0695-84
$C_{19}H_{18}N_4O$			
3H-Pyrazolo[5,1-c]-1,2,4-triazole, 3-(4-methoxyphenyl)-3,6-dimethyl-7-phenyl-	CH_2Cl_2	248(4.11),284(3.81), 312(3.91),350(3.85)	24-1726-84
7H-Pyrido[2,1-b]quinazoline-6,11-dione, 8,9-dihydro-, 6-[(4-methylphenyl)hy-drazone]	EtOH	230(4.34),250(4.26), 297(3.91),395(4.29)	4-1301-84
7H-Pyrido[2,1-b]quinazoline-6,11-dione, 8,9-dihydro-9-methyl-, 6-(phenylhy-drazone)	EtOH	228(4.36),248(4.26), 305(3.96),380(4.33)	4-1301-84

Compound	Solvent	λ_{max}(log ϵ)	Ref.
C$_{19}$H$_{18}$N$_4$O$_2$			
7H-Pyrido[2,1-b]quinazoline-6,11-dione, 8,9-dihydro-, 6-[(4-methoxyphenyl)hydrazone]	EtOH	232(4.33),254(4.20), 304(4.23),413(4.29)	4-1301-84
C$_{19}$H$_{18}$N$_4$O$_3$S			
3-Pyridinecarboxamide, 2-[5-(ethoxycarbonyl)-1-(phenylmethyl)-2-imidazolylthio]-	MeOH	254(4.32),290s(3.70)	142-0807-84
C$_{19}$H$_{18}$N$_4$O$_5$			
Ethanone, 1-(1,2,3,4-tetrahydro-6-methyl-2,4-bis(2-nitrophenyl)-5-pyrimidinyl]-, cis-(±)-	MeOH	204(4.36),250(3.98), 302(4.30)	83-1029-84
C$_{19}$H$_{18}$O$_2$			
2-Cyclohexen-1-one, 2-methoxy-4,4-diphenyl-	MeOH	261(4.09),266s(4.08)	94-4721-84
2-Cyclohexen-1-one, 2-methoxy-5,5-diphenyl-	MeOH	262(3.89)	94-4721-84
2-Naphthalenol, 1-[(4-hydroxy-3,5-dimethylphenyl)methyl]-	EtOH	230(4.71),270(3.60), 290(3.60),326(3.34), 334(3.38)	150-0701-84
C$_{19}$H$_{18}$O$_2$S			
Naphthalene, 1-[1-[(phenylmethyl)sulfonyl]ethyl]-, (S)-	EtOH	218(4.63),264(3.85), 274(4.01),282(4.04), 296(3.88),310(3.68), 315(2.24)	35-1779-84
Naphthalene, 2-[1-[(phenylmethyl)sulfonyl]ethyl]-, (R)-	EtOH	222(4.62),260(3.73), 266(3.82),276(3.82), 285(3.65),310(2.63), 318(2.16)	35-1779-84
C$_{19}$H$_{18}$O$_3$			
Benzenepropanoic acid, β-(2-oxo-1-phenylpropylidene)-, methyl ester	EtOH	248(3.92)	23-2592-84
Sakyomicin D	EtOH	213(4.57),249(3.92), 273(3.94),425(3.56)	158-84-24
C$_{19}$H$_{18}$O$_4$S			
1-Dibenzofuranol, 6,7,8,9-tetrahydro-, 4-methylbenzenesulfonate	MeOH	256(4.10),284s(3.46)	44-4399-84
2-Dibenzofuranol, 6,7,8,9-tetrahydro-, 4-methylbenzenesulfonate	MeOH	256(4.02),281s(3.59), 288s(3.52)	44-4399-84
3-Dibenzofuranol, 6,7,8,9-tetrahydro-, 4-methylbenzenesulfonate	MeOH	257(4.08),272s(3.79), 280s(3.69),289s(3.55)	44-4399-84
4-Dibenzofuranol, 6,7,8,9-tetrahydro-, 4-methylbenzenesulfonate	MeOH	255(4.13),284s(3.33)	44-4399-84
1H-Indene-1,3(2H)-dione, 2-[(2-hydroxy-4,4-dimethyl-6-oxo-1-cyclohexen-1-yl)(methylthio)methylene]-	EtOH	233(4.40),248(4.46), 284(4.30),365(4.43)	95-0127-84
C$_{19}$H$_{18}$O$_5$			
2H-1-Benzopyran-2-one, 7-[[3-(4,5-dihydro-5,5-dimethyl-4-oxo-2-furanyl)-2-butenyl]oxy]-, (E)- (geiparvarin)	EtOH	216(4.27),236(4.15), 300s(--),312(4.43)	39-0535-84C
1-Benzoxepin-5(2H)-one, 3-(3,4-dimethoxyphenyl)-8-methoxy-	MeOH	202(4.37),347(3.75)	2-1211-84
1-Benzoxepin-5(2H)-one, 6,8-dimethoxy-3-(4-methoxyphenyl)-	MeOH	224(4.28),260(3.68), 325(4.32)	2-1211-84

Compound	Solvent	$\lambda_{max}(\log \epsilon)$	Ref.
1,3-Dioxane-4,6-dione, 5-(4-methoxy-phenyl)-2,2-dimethyl-5-phenyl-	MeOH	216(4.08),233(4.04), 274(3.23),282(3.11)	12-1245-84
1,4-Ethanoanthracene-9,10-dione, 1,4-dihydro-8-hydroxy-7-(hydroxymethyl)-1-methoxy-6-methyl-	MeOH	216(4.21),249(4.14), 369(3.60),438(3.13)	44-3766-84
Ethanone, 1-(3-acetoxy-3,4-dihydro-5-hydroxy-2-phenyl-2H-benzopyran-6-yl)-, trans	n.s.g.	232(4.24),280(4.40)	2-1001-84
1-Naphthalenecarboxylic acid, 2-(3,4-dimethoxyphenyl)-1,2,3,4-tetrahydro-4-oxo-	MeCN	228(5.74),276(5.44)	2-0603-84
Naphtho[2,3-d]-1,3-dioxol-6-ol, 5,6,7,8-tetrahydro-7-(4-methyl-1,3-dioxol-5-yl)-, (6S-cis)-	MeOH	205(5.13),230s(3.93), 289(3.92)	83-0223-84
C₁₉H₁₈O₆			
9,10-Anthracenedione, 1,2,3,5-tetra-methoxy-6-methyl-	EtOH	276(4.42),360(3.71)	78-3455-84
9-Anthracenepropanoic acid, 9,10-dihy-dro-4,9-dihydroxy-5-methoxy-10-oxo-, methyl ester	n.s.g.	213(4.27),264(3.78), 292(3.84),350(3.94)	5-0306-84
4H-1-Benzopyran-4-one, 3-(3,4-dimeth-oxyphenyl)-6,7-dimethoxy-	MeOH	318(4.03)	64-0238-84B
C₁₉H₁₈O₇			
9,10-Anthracenedione, 5-hydroxy-1,2,3,6-tetramethoxy-8-methyl-	MeOH	224(4.36),274(4.54), 426(3.86)	78-5039-84
C₁₉H₁₈O₈			
4H-1-Benzopyran-4-one, 5,7-dihydroxy-6-methoxy-2-(2,4,5-trimethoxyphenyl)-(tabularin)	MeOH	258(4.3),272(4.2), 365(4.7)	94-4935-84
	MeOH-AlCl₃	266(--),296(--), 402(--)	94-4935-84
4H-1-Benzopyran-4-one, 5-hydroxy-2-(2-hydroxy-3,5-dimethoxyphenyl)-6,7-di-methoxy-	MeOH	268s(4.7),277s(4.5), 310(4.3),360s(4.0)	94-2296-84
	MeOH-AlCl₃	268s(--),280(--), 294s(--),335(--), 390(--)	94-2296-84
4H-1-Benzopyran-4-one, 5-hydroxy-2-(5-hydroxy-2,3-dimethoxyphenyl)-6,7-di-methoxy-	MeOH	282(4.2),322(4.1)	94-2296-84
	MeOH-AlCl₃	288(--),352(--)	94-2296-84
4H-1-Benzopyran-4-one, 5-hydroxy-2-(3-hydroxy-4,5-dimethoxyphenyl)-6,7-di-methoxy-	MeOH	278(4.1),335(4.3)	94-2296-84
	MeOH-AlCl₃	290(--),300s(--), 358(--)	94-2296-84
4H-1-Benzopyran-4-one, 5-hydroxy-2-(4-hydroxy-3,5-dimethoxyphenyl)-6,7-di-methoxy-	MeOH	269s(4.6),278s(4.3), 349(4.5)	94-2296-84
	MeOH-AlCl₃	269s(--),280s(--), 368(--)	94-2296-84
2,7-Dibenzofurandiol, 1,3,4-trimethoxy-, diacetate	EtOH	230(4.54),263(4.23), 289(4.18),300s(4.12), 311(3.99)	39-1441-84C
2,7-Dibenzofurandiol, 3,4,6-trimethoxy-, diacetate	EtOH	234(4.63),266(4.28), 290(4.26),308s(3.61)	39-1445-84C
C₁₉H₁₈O₉			
4H-1-Benzopyran-4-one, 2-(2,4-dihydr-oxy-5-methoxyphenyl)-5-hydroxy-6,7,8-trimethoxy- (agecorynin D)	MeOH	275(4.2),378(4.3)	94-4935-84
	MeOH-NaOMe	274(--),454(--)	94-4935-84
	MeOH-NaOAc	275(--),388(--)	94-4935-84
	MeOH-AlCl₃	282(--),305(--), 424(--)	94-4935-84
	+ HCl	280(--),305(--), 410(--)	94-4935-84

Compound	Solvent	$\lambda_{max}(\log \epsilon)$	Ref.
$C_{19}H_{19}BrN_4O_2$			
2,4(1H,3H)-Pyrimidinedione, 6-[(4-bromophenyl)methylamino]-3-methyl-5-[(4-methylphenyl)amino]-	MeCN	210(4.40),247(4.30), 299(4.11)	142-1021-84
$C_{19}H_{19}BrN_4O_3$			
1H-Imidazole-5-carboxylic acid, 4-[(4-bromophenyl)methylamino]-2,5-dihydro-1-methyl-2-oxo-5-(phenylamino)-, methyl ester	MeCN	232(4.23),255s(4.11)	142-2509-84
$C_{19}H_{14}BrO_4$			
Benzoic acid, 2-bromo-5-methoxy-3-[2-(3-methoxy-2-methylphenyl)ethenyl]-, methyl ester, (E)-	EtOH	225(4.35),238s(4.26), 294(4.23)	12-2111-84
(Z)-	EtOH	201(4.61),218s(4.39), 248(4.25)	12-2111-84
$C_{19}H_{19}ClN_4$			
2,4-Pyrimidinediamine, 5-chloro-N,N',6-trimethyl-N,N'-diphenyl-	MeOH	234(3.82),270(3.98), 314(3.64)	4-1161-84
$C_{19}H_{19}NO$			
Benzo[j]phenanthridin-4(1H)-one, 2,3,5,6-tetrahydro-6,6-dimethyl-	ether	215(4.61),280(3.86), 400(3.49)	142-0097-84
Benzo[k]phenanthridin-4(1H)-one, 2,3,5,6-tetrahydro-6,6-dimethyl-	ether	212(4.70),290(3.79), 403(3.32)	142-0097-84
$C_{19}H_{19}NO_2$			
9,10-Anthracenedione, 2-[(diethylamino)methyl]-	EtOH	256(4.59),327(3.60)	42-0611-84
1-Benzoxepin-5(2H)-one, 4-[[4-(dimethylamino)phenyl]methylene]-3,4-dihydro-, (E)-	octane	257(4.26),384(4.36)	65-0148-84
	EtOH	260(4.18),405(4.21)	65-0148-84
5a,10b-Butano-6H-[1,4]benzodioxino-[2,3-b]indole, 2-methyl-	MeOH	285(3.78)	4-1841-84
1H-Pyrrole-3-carboxylic acid, 4,5-dihydro-2-phenyl-1-(phenylmethyl)-, methyl ester	MeOH	265(3.94),317(4.18)	44-3314-84
2H-Pyrrole-3-carboxylic acid, 3,4-dihydro-2-methyl-2,5-diphenyl-, methyl ester, cis	MeOH	245(4.12)	44-3174-84
trans	hexane	242(c.4.19)	44-3174-84
2H-Pyrrole-3-carboxylic acid, 3,4-dihydro-5-methyl-2,2-diphenyl-, methyl ester	hexane	258(c.3.67)	44-3174-84
$C_{19}H_{19}NO_2S_4$			
6H-1,3-Thiazine-6-thione, 2,3-dihydro-4-(3-methylphenyl)-5-(methylsulfonyl)-2-[(phenylmethyl)thio]-	EtOH	257(3.88),348(4.19)	4-0953-84
6H-1,3-Thiazine-6-thione, 2,3-dihydro-4-phenyl-5-(phenylsulfonyl)-2-(propylthio)-	EtOH	249(3.89),352(4.20)	4-0953-84
$C_{19}H_{19}NO_3$			
1-Propanone, 2,2-dimethyl-1-[4-[2-(4-nitrophenyl)ethenyl]phenyl]-, (E)-	MeOH	265(4.04),332(4.03)	35-3140-84
Thebinan, 6,7,8,9,10,14-hexahydro-4,5-epoxy-3,6-dimethoxy-17-methyl-	CH_2Cl_2	272(3.94),306(3.58), 376(3.62)	39-1701-84C
	+ CF_3COOH	326(3.89),496(3.18)	39-1701-84C

Compound	Solvent	$\lambda_{max}(\log \epsilon)$	Ref.
$C_{19}H_{19}NO_4$			
Boldine, dehydro-6a,7-dehydro-	EtOH	263(4.72),327(4.06)	100-0525-84
	EtOH-NaOH	285(4.83)	100-0525-84
2-Butenedioic acid, 2-[1-methyl-5-(1-phenylethenyl)-1H-pyrrol-2-yl]-, dimethyl ester, cis	EtOH	235(4.17),352(4.26)	39-2541-84C
trans	EtOH	240(3.95)	39-2541-84C
Cheilanthifoline, (±)-	MeOH	287.5(3.98)	2-0818-84
	MeOH-NaOH	291(3.96),302s(3.7)	2-0818-84
Cularimine, 1,2-dehydro-	EtOH	284(3.88),351(3.61)	100-0753-84
	EtOH-HCl	226(4.34),285(3.95), 368(3.53),394(3.49)	100-0753-84
6H-1,3-Dioxolo[4,5-h]isoindolo[1,2-b]-[3]benzazepin-9-ol, 5,8,12b,13-tetrahydro-10-methoxy-	MeOH	229s(4.10),286(3.74), 298s(3.59)	78-3957-84
Erythrinan-8-one, 1,2,6,7,10,11-hexadehydro-3,15,16-trimethoxy-, (3α)-(±)-	EtOH	225(4.32),266(4.32), 355(3.25)	142-2255-84
(3β)-(±)-	EtOH	230(4.27),255(4.11), 370(2.90)	142-2255-84
Groenlandicine, tetrahydro-, (±)-	MeOH	288(3.9)	2-0818-84
	MeOH-NaOH	290(3.9),302s(3.7)	2-0818-84
1H-Indole-4,5-dicarboxylic acid, 6,7-dihydro-1-methyl-7-phenyl-, dimethyl ester	EtOH	245(4.08)	39-2541-84C
1H-Indole-4,5-dicarboxylic acid, 6,7-dihydro-7-methyl-1-phenyl-, dimethyl ester	EtOH	235(3.60),290(2.83)	39-2541-84C
Isoquinoline, 3-(3,4-dimethoxyphenyl)-6,7-dimethoxy-	EtOH	240(4.53),270(4.57), 310(4.20)	4-0525-84
$C_{19}H_{19}NO_5$			
6H-Dibenzo[de,g]quinoline-6-carboxaldehyde, 4,5,6a,7-tetrahydro-1,9-dihydroxy-2,10-dimethoxy-	MeOH	280(3.85),302(3.95)	44-1439-84
Norisosalutaridine, N-formyl-	MeOH	240(4.25),288(3.86)	44-1439-84
$C_{19}H_{19}N_3O$			
2-Quinolinecarboxamide, 3-(dimethylamino)-N-methyl-N-phenyl-	EtOH	263(4.86),356(3.54)	94-1604-84
2(1H)-Quinoxalinone, 3-[2-[4-(dimethylamino)phenyl]ethenyl]-1-methyl-, (E)-	EtOH	452(3.97)	18-1653-84
$C_{19}H_{19}N_3OS$			
Acetamide, 2-[[[4-cyano-5,6,7,8-tetrahydro-1-(phenylmethyl)-3-isoquinolinyl]thio]-	dioxan	272(4.20),306(3.80), 356(4.48)	104-2225-84
Thieno[2,3-c]isoquinoline-2-carboxamide, 1-amino-6,7,8,9-tetrahydro-5-(phenylmethyl)-	dioxan	285(4.38),306(3.90), 358(3.78)	104-2225-84
$C_{19}H_{19}N_3O_2$			
Benzeneacetonitrile, α-[[4-(diethylamino)phenyl]methylene]-4-nitro-	EtOH	459(4.31)	48-0063-84
2H-1,5-Benzodiazepin-2-one, 3-[(dimethylamino)methylene]-1,3-dihydro-8-methoxy-4-phenyl-	EtOH	206(4.37),252(4.29), 341(3.99),269(3.86)	103-1035-84
1H-Indole-2-carboxylic acid, 5-[1-(phenylhydrazono)ethyl]-, ethyl ester	EtOH	207(4.38),252(4.45), 302(4.51),331s(4.41)	103-0280-84

Compound	Solvent	$\lambda_{max}(\log \epsilon)$	Ref.
$C_{19}H_{19}N_3O_3$			
3-Oxepincarboxylic acid, 5-cyano-6,7-dihydro-6,7-dimethyl-2-[(phenyl-carbonimidoyl)amino]-, ethyl ester	CHCl$_3$	263(3.89),333(4.24)	24-0585-84
3-Oxepincarboxylic acid, 5-cyano-6,7-dihydro-7,7-dimethyl-2-[(phenyl-carbonimidoyl)amino]-, ethyl ester	CHCl$_3$	262(3.75),267(3.71), 275(3.10)	24-0585-84
$C_{19}H_{19}N_3O_3S$			
Pyrrolo[2,3-c]phenothiazine, 3,6-dihydro-1-(morpholinomethyl)-, 11,11-dioxide	EtOH	211(4.49),271.5(4.37), 311.5(3.96),328(3.94), 345(3.97),343(3.89), 358(3.81)	103-1095-84
$C_{19}H_{19}N_3O_7$			
L-Alanine, N-[N-[(phenylmethoxy)carbonyl]glycyl]-, 4-nitrophenyl ester	MeOH	274(3.80)	65-0836-84
Glycine, N-[N-[(phenylmethoxy)carbonyl]-L-alanyl-, 4-nitrophenyl ester	MeOH	274(3.79)	65-0836-84
$C_{19}H_{19}N_3S$			
2(1H)-Quinoxalinethione, 3-[2-[4-(dimethylamino)phenyl]ethenyl]-1-methyl-	CHCl$_3$	442(4.13)	18-1653-84
$C_{19}H_{19}N_4$			
Benzenediazonium, 4-[1-cyano-2-[4-(diethylamino)phenyl]ethenyl]-, tetrafluoroborate	EtOH	573(4.52)	48-0063-84
	H$_2$O	573(--)	48-0063-84
	MeOH	573(--)	48-0063-84
	dioxan	550(--)	48-0063-84
	DMF	565(--)	48-0063-84
(also other solvents)	MeNO$_2$	595(--)	48-0063-84
$C_{19}H_{19}N_5O_2S$			
Propanenitrile, 3-[[4-[[6-(2-hydroxyethoxy)-2-benzothiazolyl]azo]phenyl]-methylamino]-	pH 1.2	517(4.56),617s(3.78)	48-0151-84
	pH 6.2	512(4.68)	48-0151-84
	EtOH	505(4.65)	48-0151-84
$C_{19}H_{19}N_5O_3$			
2H-1-Benzopyran-2-one, 7-[[5-(6-amino-9H-purin-9-yl)pentyl]oxy]-	H$_2$O	264(4.16),325(4.15)	23-2006-84
$C_{19}H_{19}N_5O_4$			
Oxepino[2,3-d]pyrimidine-5-carboxylic acid, 6-cyano-2-hydrazino-3,4,7,8-tetrahydro-7,8-dimethyl-4-oxo-3-phenyl-, methyl ester	CHCl$_3$	274(3.82),337(4.37)	24-0585-84
Oxepino[2,3-d]pyrimidine-5-carboxylic acid, 6-cyano-2-hydrazino-3,4,7,8-tetrahydro-8,8-dimethyl-4-oxo-3-phenyl-, methyl ester	CHCl$_3$	273(3.93),337(4.44)	24-0585-84
Oxepino[2,3-d]pyrimidine-6-carboxylic acid, 2-azido-3,4,7,8-tetrahydro-7,8-dimethyl-4-oxo-3-phenyl-, ethyl ester	CHCl$_3$	265(3.79),280(3.69), 325(3.81)	24-0585-84
$C_{19}H_{19}N_5O_7$			
Propanedial, [2-[(3,7-dihydro-3-β-D-ribofuranosylpyrimido[2,1-i]purin-7-yl)oxy]cyclopropylidene]-	H$_2$O	237(4.43),260s(4.14), 327(4.47)	35-3370-84

Compound	Solvent	$\lambda_{max}(\log \epsilon)$	Ref.
$C_{19}H_{19}O_2P$			
Phosphorin, 1,1-dihydro-1,1-dimethoxy-3,5-diphenyl-	EtOH	241(4.45),265s(4.20), 368(4.14)	24-0763-84
$C_{19}H_{20}$			
Anthracene, 9-(1,1-dimethylpropyl)-	heptane	260(5.06),342(3.40), 355(3.66),373(3.81), 393(3.75)	61-0042-84
9H-Fluorene, 9-cyclohexyl-	MeOH	218(4.28),228(3.85), 257(4.25),265(4.24), 291(3.77),302(3.86)	2-0802-84
$C_{19}H_{20}BrN_3O_6$			
Thymidine, 5'-[[4-(bromoacetyl)benz-oyl]amino]-5'-deoxy-	pH 1	259(4.37)	87-0680-84
	pH 7	259(4.38)	87-0680-84
$C_{19}H_{20}Br_2O_2$			
Androsta-4,9(11),15-triene-3,17-dione, 15,16-dibromo-, (14β)-	EtOH	244(4.36)	23-1103-84
$C_{19}H_{20}ClNO_2$			
Benzene, 1-(1-chloro-2,2-dimethylprop-yl)-4-[2-(4-nitrophenyl)ethenyl]-, (E)-	MeOH	248(4.16),340(4.10)	35-3140-84
$C_{19}H_{20}N_2$			
1H-Indole, 1-butyl-5-[2-(4-pyridinyl)-ethenyl]-	EtOH	341(4.43)	44-2546-84
$C_{19}H_{20}N_2O_2$			
4H-1,2,6-Oxadiazocin-7(8H)-one, 5,6-di-hydro-3-(4-methylphenyl)-6-(phenyl-methyl)-	EtOH	247(4.1)	5-1696-84
2-Propenoic acid, 3-(1,2,6,7,12,12b-hexahydroindolo[2,3-a]quinolizin-3-yl)-, methyl ester	EtOH	225(4.50),273(3.87), 290(3.78),358(4.68)	44-1832-84
2-Pyrrolidinecarboxamide, N-(2-formyl-phenyl)-1-(phenylmethyl)-, (S)-	CHCl₃	226(4.02),273(4.00), 338(3.68)	70-0738-84
$C_{19}H_{20}N_2O_3$			
1H-Indole, 1-acetyl-3-[(5-acetyl-1,2,3,4-tetrahydro-1-methyl-4-pyridinyl)carbonyl]-	EtOH	217(4.35),227s(--), 301(4.45)	78-3339-84
4H-1,2,6-Oxadiazocin-7(8H)-one, 5,6-di-hydro-3-(4-methoxyphenyl)-6-(phenyl-methyl)-	EtOH	265(4.1)	5-1696-84
4H-Pyrido[1,2-a]pyrimidine-3-carboxylic acid, 6,7,8,9-tetrahydro-6-methyl-4-oxo-9-(phenylmethylene)-, ethyl ester, (E)-	EtOH	354(4.33)	118-0582-84
Rosicine	n.s.g.	203(3.64),223(3.62), 295(3.60),325(3.64)	88-6051-84
$C_{19}H_{20}N_2O_4$			
2,6-Pyridinedicarboxylic acid, 4-[2-(2,3-dihydro-1H-indol-1-yl)-ethenyl]-2,3-dihydro-, dimethyl ester, (±)-, perchlorate	MeOH	267(3.94),278(3.93), 527(4.58)	33-1547-84
$C_{19}H_{20}N_2O_4S$			
2H-Thiocin-5,7-dicarboxylic acid, 3,4-	CHCl₃	263(4.32),335(4.02)	24-0585-84

Compound	Solvent	$\lambda_{max}(\log \epsilon)$	Ref.
dihydro-8-[(phenylcarbonimidoyl)-amino]-, 5-ethyl 7-methyl ester (cont.)			24-0585-84
$C_{19}H_{20}N_2O_5$			
3aH-Indazole-3a,4-dicarboxylic acid, 1,4,5,7a-tetrahydro-1,6-dimethyl-5-oxo-7-phenyl-, dimethyl ester, (1λ,2ε)	MeOH	?(4.13),286(3.92)	44-1261-84
3,5-Oxepindicarboxylic acid, 6,7-dihydro-7-methyl-2-[(phenylcarbonimidoyl)amino]-, 5-ethyl 3-methyl ester	CHCl₃	267(3.87),333(3.96)	24-0585-84
Oxepino[2,3-d]pyrimidine-6-carboxylic acid, 1,2,3,4,7,8-hexahydro-7,8-dimethyl-2,4-dioxo-3-phenyl-, ethyl ester	CHCl₃	263(4.12),316(4.19)	24-0585-84
Oxepino[2,3-d]pyrimidine-6-carboxylic acid, 1,2,3,4,7,8-hexahydro-8,8-dimethyl-2,4-dioxo-3-phenyl-, ethyl ester	CHCl₃	263(4.10),314(4.16)	24-0585-84
$C_{19}H_{20}N_4$			
2,4-Pyrimidinediamine, N,N',6-trimethyl-N,N'-diphenyl-	MeOH	229(3.66),265(3.62), 296(3.48)	4-1161-84
$C_{19}H_{20}N_4O_2S$			
Ethanol, 2-[[2-[[4-(1-pyrrolidinyl)-phenyl]azo]-6-benzothiazolyl]oxy]-	pH 1.2	559(4.65),611s(4.57)	48-0151-84
	pH 6.2	536(4.76)	48-0151-84
	EtOH	528(4.75)	48-0151-84
$C_{19}H_{20}N_4O_3$			
Oxepino[2,3-d]pyrimidine-6-carbonitrile, 3,4,7,8-tetrahydro-2-[(2-hydroxyethyl)amino]-7,8-dimethyl-4-oxo-3-phenyl-	CHCl₃	272(4.04),340(4.44)	24-0585-84
$C_{19}H_{20}N_6O_7$			
Carbamic acid, [9-(2-deoxy-β-D-erythro-pentofuranosyl)-9H-purin-6-yl]-, 2-(4-nitrophenyl)ethyl ester	MeOH	267(4.44)	78-0059-84
$C_{19}H_{20}N_6O_8$			
Carbamic acid, (9-β-D-ribofuranosyl-9H-purin-6-yl)-, 2-(4-nitrophenyl)-ethyl ester	MeOH	267(4.45)	78-0059-84
$C_{19}H_{20}O_2$			
6H-Dibenzo[h,l][1,7]dioxacyclotridecin, 7,8,9,10-tetrahydro-, (E)-	MeOH	209(4.35),233(4.17), 288(4.23),299(4.23), 321(4.34),335(4.20)	44-1627-84
(Z)-	MeOH	210s(4.30),282(3.68)	44-1627-84
Naphthalene, 3,4-dihydro-6-methoxy-1-(4-methoxyphenyl)-2-methyl-	EtOH	274.5(4.23)	12-2279-84
Naphthalene, 3,4-dihydro-6-methoxy-2-(4-methoxyphenyl)-1-methyl-	EtOH	201.5(4.51),225(4.14), 289(4.36)	12-2279-84
1-Propanone, 1-(6,7,8,9-tetramethyl-2-dibenzofuranyl)-	EtOH	248(4.41),269(4.52)	39-0799-84B
1-Propanone, 1-(6,7,8,9-tetramethyl-3-dibenzofuranyl)-	EtOH	314(4.50)	39-0799-84B
$C_{19}H_{20}O_3$			
2H,8H-Benzo[1,2-b:5,4-b']dipyran-2-one,	EtOH	226(4.3),267(4.3),	42-0650-84

Compound	Solvent	$\lambda_{max}(\log \epsilon)$	Ref.
3-(1,1-dimethyl-2-propenyl)-8,8-di-methyl- (cont.)		346(4.1)	42-0650-84
1,4-Dioxaspiro[4.5]decan-2-one, 3-(5-phenyl-2,4-pentadienylidene)-	EtOH	353(4.70)	39-1531-84C
3-Phenanthrenol, 9,10-dihydro-7-methoxy-2,5-dimethyl-, acetate	EtOH	215(4.57),278(4.32), 287s(4.25)	39-1913-84C
$C_{19}H_{20}O_3Si$			
6H-Cyclopenta[jk]fluorene-6,9(5aH)-di-one, 9b,9c-dihydro-9c-methyl-9b-[(trimethylsilyl)oxy]-	EtOH	247(4.19),302(3.67), 472(1.90)	39-0909-84C
$C_{19}H_{20}O_4$			
Benzaldehyde, 2,2'-[1,5-pentanediyl-bis(oxy)]bis-	MeOH	214(4.51),253(4.17), 319(3.80)	44-1627-84
[1,1'-Biphenyl]-2-carboxylic acid, 3-hydroxy-5-methoxy-4-(3-methyl-2-butenyl)-	MeOH	231(6.6),268(6.1), 305(5.8)[sic]	102-0765-84
Carbonic acid, 2-(2-cyclopenten-1-yl-carbonyl)-1-cyclopenten-1-yl) phenyl-methyl ester	hexane C_6H_{12} CHCl$_3$	242(3.84) 241(4.1),329(1.9) 252(3.95)	33-1154-84 33-1154-84 33-1154-84
$C_{19}H_{20}O_5$			
7(4H)-Benzofuranone, 2-(1,3-benzodiox-ol-5-yl)-2,3,3a,7a-tetrahydro-7a-hy-droxy-3-methyl-3a-(2-propenyl)-, cis	MeOH	235(3.56),287(3.31)	102-2101-84
trans	MeOH	235(3.92),287(3.62)	102-2101-84
12-Epiteucvin	EtOH	224(4.00)	44-1789-84
$C_{19}H_{20}O_5S$			
Cyclohexanone, 2-[2-hydroxy-3-[[(4-methylphenyl)sulfonyl]oxy]phenyl]-	MeOH	274(3.52)	44-4399-84
Cyclohexanone, 2-[2-hydroxy-5-[[(4-methylphenyl)sulfonyl]oxy]phenyl]-	MeOH	276(3.51)	44-4399-84
Cyclohexanone, 2-[3-[[(4-methylphenyl)-sulfonyl]oxy]phenoxy]-	MeOH	268(3.39),273(3.42)	44-4399-84
$C_{19}H_{20}O_6$			
Diosbulbin D	MeOH	207(3.84)	102-0623-84
2,2(3H)-Furandicarboxylic acid, 4-form-yl-3-(2-phenylethenyl)-, diethyl ester, (E)-	EtOH	204(4.26),256(4.34)	118-0974-84
Isoteuflidin	EtOH	223(3.93)	102-1465-84
$C_{19}H_{20}O_7$			
4H-1-Benzopyran-4-one, 2-(3,4-dimeth-oxyphenyl)-2,3-dihydro-6-hydroxy-5,7-dimethoxy-, (S)- (agestricin C)	MeOH	237(4.3),280(4.2), 345(3.6)	94-1472-84
$C_{19}H_{20}O_8$			
9H-Xanthen-9-one, 1,2,3,4,6,8-hexa-methoxy-	MeOH	250(4.60),304(4.30), 337s(3.79)	39-1507-84C
	EtOH	204(4.35),249(4.22), 303(3.85),336s(--)	100-0123-84
β-D-Xylopyranoside, 5-[2-(3,5-dihydr-oxyphenyl)ethenyl]-2-hydroxyphenyl, (E)-	MeOH	217(4.46),305(4.45), 322(4.49)	94-3501-84
$C_{19}H_{20}O_{10}$			
Methanone, (3-β-D-glucopyranosyl-2,4,6-trihydroxyphenyl)(4-hydroxyphenyl)-	MeOH	296(4.06),314(4.10)	94-2676-84

Compound	Solvent	$\lambda_{max}(\log \epsilon)$	Ref.
$C_{19}H_{20}O_{11}$			
Methanone, (3,4-dihydroxyphenyl)(3-β-D-glucopyranosyl-2,4,6-trihydroxy-phenyl)-	MeOH	296(3.93),326(4.03)	94-2676-84
$C_{19}H_{21}Br_2NO_5$			
Formamide, N-[1,2-bis(2-bromo-4,5-di-methoxyphenyl)ethyl]-	EtOH	283(3.98)	20-1099-84
$C_{19}H_{21}ClN_2O_2$			
Acetamide, 2-chloro-N-[3-(hydroxyimi-no)-3-(4-methylphenyl)propyl]-N-(phenylmethyl)-	EtOH	252(4.1)	5-1696-84
$C_{19}H_{21}ClN_2O_3$			
Acetamide, 2-chloro-N-[3-(hydroxyimi-no)-3-(4-methoxyphenyl)propyl]-N-(phenylmethyl)-	EtOH	260(4.6)	5-1696-84
$C_{19}H_{21}ClN_4O_2S$			
Ethanol, 2-[[2-[[4-[(2-chloroethyl)-ethylamino]phenyl]azo]-6-benzo-thiazolyl]oxy]-	pH 1.2	524(4.67),618s(3.98)	48-0151-84
	pH 6.2	521(4.71)	48-0151-84
	EtOH	513(4.70)	48-0151-84
$C_{19}H_{21}NO$			
[1,1'-Biphenyl]-2-carboxamide, N-1-propenyl-N-propyl-, (E)-	MeOH	230(4.49),247(4.15), 263(4.22),304(3.10), 316(3.95)	78-1835-84
9H-Carbazole, 1-methoxy-3-methyl-2-(3-methyl-2-butenyl)-	EtOH	224(4.55),240(4.59), 250(4.50),290(4.29), 320(5.39),335(3.50)	25-0301-84
$C_{19}H_{21}NOS$			
2-Propen-1-one, 1-phenyl-2-(1-piperi-dinylmethyl)-3-(2-thienyl)-	EtOH	324(3.36)	73-1764-84
$C_{19}H_{21}NO_2$			
2-Propen-1-one, 3-(2-furanyl)-1-phenyl-2-(1-piperidinylmethyl)-	EtOH	329(3.37)	73-1764-84
$C_{19}H_{21}NO_3$			
Aporphine, 1,2-dimethoxy-11-hydroxy-	EtOH	220(4.42),265(4.12), 270(4.15)	100-1040-84
Benzenemethanol, α-(1,1-dimethylethyl)-4-[2-(4-nitrophenyl)ethenyl]-, (E)-	MeOH	252(4.10),345(4.08)	35-3140-84
1H-Inden-2-amine, 2,3-dihydro-1,5,6-trimethoxy-N-(phenylmethylene)-,(1S-cis)-	MeOH	203(4.85),243(4.33), 284(3.88),288s(3.87)	39-1655-84C
5H-Indeno[2,1-c]isoquinoline, 6,6a,7,11b-tetrahydro-2,3,9-trimethoxy-, (6aS-cis)-	MeOH	199(4.81),229(4.21), 283(3.82),287(3.80)	39-1655-84C
hydrochloride	MeOH	200(4.80),203(4.80), 232(4.23),282(3.81), 287s(3.79)	39-1655-84C
5H-Indeno[2,1-c]isoquinoline, 6,6a,7,11b-tetrahydro-3,4,9-trimethoxy-, (6aS-cis)-	MeOH	199(4.82),230(4.27), 279(3.59),282(3.59), 288(3.45)	39-1655-84C
5H-Indeno[2,1-c]isoquinoline, 6,6a,7,11b-tetrahydro-3,8,9-trimethoxy-, (6aS-cis)-	MeOH	199(4.81),230s(4.27), 279(3.53),286s(3.44)	39-1655-84C

Compound	Solvent	$\lambda_{max}(\log \epsilon)$	Ref.
5H-Indeno[2,1-c]isoquinoline, 6,6a,7,11b-tetrahydro-3,9,10-trimethoxy-, (6aR-cis)-	MeOH	200(4.82),227(4.21), 282s(3.79),287(3.83), 297s(3.54)	39-1655-84C
hydrochloride	MeOH	200(4.80),203s(4.79), 231(4.21),281s(3.77), 287(3.81),295s(3.59)	39-1655-84C
Isothebaine	n.s.g.	276(4.18)	105-0467-84
9-Phenanthrenecarbonitrile, 1-acetoxy-1,2,3,9,10,10a-hexahydro-7-methoxy-10a-methyl-	EtOH	260(4.20)	2-0395-84
Secoquettamine	MeOH	205(4.19),250(3.75), 300(4.12),308(4.11), 323s(3.85)	100-0753-84
	MeOH-base	220(4.08),250(3.50), 324(4.17)	100-0753-84

$C_{19}H_{21}NO_4$

Compound	Solvent	$\lambda_{max}(\log \epsilon)$	Ref.
Amurinine, (+)-	MeOH	235(4.27),294(3.85)	100-0342-84
Celtine, (+)-	EtOH	216(4.11),228s(3.95), 282(3.58)	88-1829-84 +100-0753-84
	EtOH-base	216(4.59),298(3.60)	88-1829-84
Corytuberine	MeOH	224(4.72),272(4.06), 310(3.78)	73-0704-84
Cularidine	n.s.g.	284(3.85)	100-0753-84
Discretine, 10-demethyl-, (-)-	EtOH	208(4.49),226s(4.06), 289(3.71)	100-0353-84
	EtOH-base	219(4.41),249(3.98), 312(3.82)	100-0353-84
Epiamurinine, (-)-	MeOH	230(4.12),290(3.85)	100-0342-84
C-Homoerythrinan-2-one, 1,6-didehydro-3-methoxy-15,16-[methylenebis(oxy)]-, (3β)-	EtOH	230(3.97),288(3.04)	78-2677-84
1H-Inden-1-ol, 2-[[(2,3-dimethoxyphenyl)methylene]amino]-5-methoxy-, (1S-trans)-	MeOH	198(4.72),223(4.58), 258(4.23),286(3.75), 305(3.40)	39-1655-84C
1H-Inden-1-ol, 2-[[(3,4-dimethoxyphenyl)methylene]amino]-2,3-dihydro-5-methoxy-, (1S-trans)-	MeOH	199(4.72),227(4.42), 272(4.29),287s(4.15), 303(4.12)	39-1655-84C
Isocularine, 5'-O-demethyl-, (±)-	EtOH	276s(3.77),283(3.78)	100-0753-84
Sarcocapnidine, (+)-	EtOH	238(3.8),281(3.5)	100-0753-84
	EtOH-base	250(3.73),294(3.62)	100-0753-84
Stepholide	MeOH	228(4.11),288(3.64)	95-0946-84

$C_{19}H_{21}NO_5$

Compound	Solvent	$\lambda_{max}(\log \epsilon)$	Ref.
1H-Benzo[a]furo[3,4-f]quinolizine-1,12(3H)-dione, 5,6,10b,11-tetrahydro-8,9-dimethoxy-3,3-dimethyl-, (R)-	MeOH	206(4.52),237(4.28), 293(4.15)	104-2335-84
Hippamine, 1-acetoxy-	EtOH	250(3.40),290(3.28)	100-1061-84
Limousamine, (+)-	MeOH	210(4.54),230s(4.14), 283(3.82),292s(3.69)	100-0753-84
	MeOH-base	211(4.60),251s(4.06), 291(3.92)	100-0753-84

$C_{19}H_{21}NO_5S$

Compound	Solvent	$\lambda_{max}(\log \epsilon)$	Ref.
5H-Furo[3,2-g][1]benzopyran-5-one, 7-[2-(dimethylamino)ethenyl]-4,9-dimethoxy-6-[(methylthio)methyl]-, (E)-	EtOH	216(4.40),228(4.36), 241(4.41),256(4.40), 284s(4.11),390(4.65)	44-5035-84

Compound	Solvent	$\lambda_{max}(\log \epsilon)$	Ref.
$C_{19}H_{21}NO_6$ 3-Azabicyclo[4.4.1]undeca-1,3,5,7,9-pentaene-2,4,5-tricarboxylic acid, triethyl ester	MeOH	244(4.39),280(4.23), 364(3.69)	44-1040-84
$C_{19}H_{21}NS$ [1,1'-Biphenyl]-2-carbothioamide, N-(1-propenyl)-N-propyl-, (E)-	MeOH	235(4.65),265(4.26), 313(4.02),320(4.54), 415(2.88)	78-1835-84
$C_{19}H_{21}N_2$ Methanaminium, N-[9-anthracenyl(dimethylamino)methylene]-N-methyl-, iodide	MeCN	252(5.21),360(3.75), 375(3.74),392(3.68)	64-1586-84B
$C_{19}H_{21}N_2O$ Melinonine E (chloride)	MeOH	218s(4.18),252(4.48), 260s(4.43),286s(3.91), 300s(4.19),307(4.36), 307(4.33)[sic],366(3.65)	33-0455-84
	MeOH-NaOH	217(4.22),250s(4.13), 276s(4.61),283(4.70), 320s(4.00),328(4.05), 417(3.49)	33-0455-84
$C_{19}H_{21}N_3O$ Ethanone, 1-[4-[[4-(1-piperidinyl)-phenyl]azo]phenyl]-	C_6H_{12}	420(4.44)	39-0149-84B
	EtOH	436(4.49)	39-0149-84B
	EtOH-HCl	530(3.98)	39-0149-84B
4H-Indazol-4-one, 1,5,6,7-tetrahydro-5-(1-piperidinylmethylene)-1-phenyl-, (E)-	EtOH	229(4.18),258(4.19), 358(4.43)	4-0361-84
$C_{19}H_{21}N_5$ 1,1,2-Ethanetricarbonitrile, 2-(4,5-di-hydro-2-phenyl-1H-pyrrol-3-yl)-2-[(1,1-dimethylethyl)amino]-	MeOH	205(4.14),403(4.23), 456(4.15)	83-0143-84
$C_{19}H_{22}FN_2O_9P$ Uridine, 2'-deoxy-5-fluoro-5'-O-[5-(phenylmethoxy)-1,3,2-dioxa-phosphorinan-2-yl]-, P-oxide	EtOH	269(3.92)	87-0440-84
$C_{19}H_{22}NO_3$ 2H-Furo[2,3,4-ij]isoquinolinium, 2a,3,4-tetrahydro-2-(4-hydroxy-phenyl)-8-methoxy-3,3-dimethyl-, chloride, cis-(±)- (quettamine)	MeOH	223s(3.89),280(3.16)	100-0753-84
	MeOH-base	248(3.83),283(3.32)	100-0753-84
$C_{19}H_{22}N_2O_2$ Cyclohexanecarboxylic acid, 3-(dicyano-methyl)-1,3-dimethyl-5-phenyl-, methyl ester, (1α,3α,5β)-	EtOH	238(3.92)	39-0261-84C
Indolo[2,3-a]quinolizin-4(3H)-one, 2-ethyl-2,6,7,12-tetrahydro-3-(2-hydroxyethyl)-, (2S*,3R*)-	EtOH	227(4.50),308(4.30), 319(4.27)	39-1237-84C
20-Epipseudoaspidospermidine, 5,17-di-oxo-	EtOH	246(3.85),299(3.40)	44-3733-84
14-Iso-20-epipseudoaspidospermidine, 5,17-dioxo-	EtOH	242(3.87),297(3.45)	44-3733-84
14-Isopseudoaspidospermidine, 5,17-di-oxo-	EtOH	245(3.87),300(3.39)	44-3733-84

Compound	Solvent	$\lambda_{max}(\log \epsilon)$	Ref.
Pyrano[4,3-c]pyrazol-4(1H)-one, 6,7-di-hydro-6,6-dimethyl-3-(2-methyl-1-pro-penyl)-1-(phenylmethyl)-	EtOH	248(4.12)	4-0017-84
$C_{19}H_{22}N_2O_3$			
Leuconolam	EtOH	205(3.95),218(3.96), 292s(3.14)	88-3483-84
Pyrrolo[2,3-b]indole-2-carboxylic acid, 2,8-dihydro-8-(methoxymethyl)-2-(3-methyl-2-butenyl)-	MeOH	247s(4.41),252(4.44), 276(3.67),290(3.57)	39-2903-84C
$C_{19}H_{22}N_2O_4$			
1H-3-Benzazepin-7-ol, 2,3,4,5-tetra-hydro-8-methoxy-1,3-dimethyl-2-(4-nitrophenyl)-, cis	EtOH	279(3.95)	44-0581-84
Benzenemethanamine, 2-ethenyl-4,5-di-methoxy-N-methyl-N-[(4-nitrophenyl)-methyl]-	EtOH	215(3.99),266(4.09)	44-0581-84
Pentanediamide, N,N'-dihydroxy-N,N'-bis(2-methylphenyl)-	MeOH	210(4.52),228(4.37)	34-0345-84
Pentanediamide, N,N'-dihydroxy-N,N'-bis(4-methylphenyl)-	MeOH	218(4.53),250(4.35)	34-0345-84
Pyridine, 3-(4,5-dihydro-4,4-dimethyl-2-oxazolyl)-2-[3-methoxy-2-(methoxy-methoxy)phenyl]-	EtOH	255(3.97)	44-0056-84
Pyridine, 3-(4,5-dihydro-4,4-dimethyl-2-oxazolyl)-4-[3-methoxy-2-(methoxy-methoxy)phenyl]-	EtOH	249(3.99)	44-0056-84
$C_{19}H_{22}N_2O_5$			
Phenol, 2,2'-[(2-hydroxy-1,3-propane-diyl)bis(nitriloethylidyne)bis[6-methoxy-	$CHCl_3$	226(2.77),248(2.78), 262(2.79),295(3.18), 299(3.20),368(3.19), 418(2.72)	34-0237-84
$C_{19}H_{22}N_2O_7S$			
Propanedioic acid, (acetylamino)[(4-acetoxy-7-benzothiazolyl)methyl]-, diethyl ester	EtOH	254(3.56),290(3.20), 298(3.14)	44-0997-84
$C_{19}H_{22}N_2S_2$			
5H-Cyclopenta[e]dicyclopropa[c,g][1,2]-diazocine, 4,7-bis[(1,1-dimethyleth-yl)thio]-	CH_2Cl_2	242(4.31),280(4.04), 340s(4.08),374(4.28)	88-4223-84
$C_{19}H_{22}N_4$			
5H-Pyrimido[4,5-b]indol-4-amine, N-ethyl-6,7,8,9-tetrahydro-2-methyl-9-phenyl-	EtOH	208(4.53),233(4.28), 288(4.11)	11-0073-84B
$C_{19}H_{22}N_4O_2S$			
Ethanol, 2-[[2-[[4-(diethylamino)phen-yl]azo]-6-benzothiazolyl]oxy]-	pH 1.2	558(4.56),612s(4.46)	48-0151-84
	pH 6.2	535(4.73)	48-0151-84
	EtOH	528(4.71)	48-0151-84
$C_{19}H_{22}N_4O_3S$			
Ethanol, 2-[ethyl[4-[[6-(2-hydroxyeth-oxy)-2-benzothiazolyl]azo]phenyl]-amino]-	pH 1.2	548(4.62),617s(4.44)	48-0151-84
	pH 6.2	530(4.74)	48-0151-84
	EtOH	524(4.73)	48-0151-84

Compound	Solvent	$\lambda_{max}(\log \epsilon)$	Ref.
$C_{19}H_{22}N_4O_4$			
Oxepino[2,3-d]pyrimidine-6-carboxylic acid, 2-hydrazino-3,4,7,8-tetrahydro-7,8-dimethyl-4-oxo-3-phenyl-, ethyl ester	CHCl$_3$	270(4.01),343(4.42)	24-0585-84
$C_{19}H_{22}N_4O_5S_2$			
Acetic acid, [[(5-formyl-3,4,5,10-tetrahydro-3,7,8,10-tetramethyl-2,4-dioxobenzo[g]pteridin-4a(2H)-yl)methyl]dithio]-, methyl ester	MeCN	218(4.24),260s(--), 320(3.96)	35-3309-84
$C_{19}H_{22}N_4O_{10}$			
5-Pyrimidinecarbonitrile, 1,2,5,6-tetrahydro-6-imino-2-oxo-1-(2,3,4,6-tetra-O-acetyl-β-D-glucopyranosyl)-	H$_2$O	247(4.19),310(4.10)	103-0443-84
	pH 1	224(4.15),285(4.10)	103-0443-84
	pH 13	247(4.09),318(4.19)	103-0443-84
5-Pyrimidinecarbonitrile, 1,2,5,6-tetrahydro-6-imino-2-oxo-1-(2,3,4,6-tetra-O-acetyl-β-D-mannopyranosyl)-	EtOH	248(4.16),309(4.07)	103-0443-84
Urea, N-(2,2-dicyanoethenyl)-N'-(2,3,4-6-tetra-O-acetyl-β-D-galactopyrano-syl)-	EtOH	278(3.92)	103-0443-84
Urea, N-(2,2-dicyanoethenyl)-N'-(2,3,4-6-tetra-O-acetyl-β-D-glucopyranosyl)-	EtOH	280(4.29),314s(4.01)	103-0443-84
	EtOH-HCl	280(4.08)	103-0443-84
	EtOH-NaOH	315(4.10)	103-0443-84
Urea, N-(2,2-dicyanoethenyl)-N'-(2,3,4-6-tetra-O-acetyl-β-D-mannopyranosyl)-	EtOH	280(4.16),313(2.63)	103-0443-84
	EtOH-HCl	270(4.16)	103-0443-84
	EtOH-NaOH	310(4.17)	103-0443-84
$C_{19}H_{22}N_8O_6S$			
Butanoic acid, 2-[[4-[[(2,4-dinitro-6-pteridinyl)methyl]methylamino]-benzoyl]amino]-4-sulfo-, mono-ammonium salt	pH 1	242(4.27),305(4.36)	87-0600-84
	pH 7.4	259(4.39),302(4.40), 373(3.91)	87-0600-84
$C_{19}H_{22}O$			
9-Anthracenol, 9-(1,1-dimethylpropyl)-9,10-dihydro-	EtOH	255(2.94),263(2.98), 270.5(2.89)	61-0042-84
$C_{19}H_{22}OS_2$			
Bicyclo[2.2.1]heptan-2-one, 3-(1,5-di-hydro-2,4-benzodithiepin-3-ylidene)-1,7,7-trimethyl-	EtOH	322(4.27)	78-2951-84
$C_{19}H_{22}O_2$			
Androsta-4,9(11),15-triene-3,17-dione, (14β)-	n.s.g.	232.5(4.20)	23-1103-84
$C_{19}H_{22}O_3$			
Benzeneacetic acid, 4-(phenylmethoxy)-, 1,1-dimethylethyl ester	EtOH	224(4.16),275(3.23), 282(3.18)	39-1547-84C
3H-3,10a-Ethanophenanthren-3-ol, 4,4a,9,10-tetrahydro-7-methoxy-, acetate, (3α,4aβ,10aβ)-(±)-	EtOH	278(3.46)	2-1168-84
$C_{19}H_{22}O_4$			
Cyclopentaneacetaldehyde, 1-methyl-2-oxo-5-(1,2,3,4-tetrahydro-6-meth-oxy-1-oxo-2-naphthalenyl)-	EtOH	205(4.07),224(4.00), 274(4.12)	33-0332-84
Spiro[furan-3(2H),1'(2'H)-naphthalen]-2-one, 5-(3-furanyl)-4,5,6',7',8',8'a-	EtOH	225s(4.13),233s(4.18), 238(4.21)	102-0843-84

Compound	Solvent	λ_{max}(log ϵ)	Ref.
5'-(hydroxymethyl)-2'-methyl- (cont.)			102-0843-84
C$_{19}$H$_{22}$O$_5$			
Podolide	EtOH	218(4.11)	154-0547-84
C$_{19}$H$_{22}$O$_6$			
Furanoeremophil-1(10)-en-6-one, 3,3-(ethylenedioxy)-9β-acetoxy-	EtOH	267(3.64)	94-3396-84
3-Heptanone, 1,7-bis(3,4-dihydroxyphenyl)-5-hydroxy-, (±)-	EtOH	220(4.00),283(3.67)	39-1635-84C
(S)-	EtOH	221(4.07),283(3.75)	39-1635-84C
Zexbrevin B, 1,2-dehydro-	EtOH	205(4.28)	102-1967-84
C$_{19}$H$_{22}$O$_7$			
Nagilactone C	EtOH	300(3.48)	154-0547-84
Shinjulactone B	EtOH	279(3.91)	18-2484-84
	EtOH-base	330(--)	18-2484-84
C$_{19}$H$_{22}$O$_8$			
2-Cyclohexene-1-carboxylic acid, 6-acetoxy-4-oxo-1-[(3-oxo-1,4-dioxaspiro[4.5]dec-2-ylidene)methyl]-, methyl ester	EtOH	224(3.91),248(3.91)	77-1008-84
C$_{19}$H$_{22}$O$_9$			
Allamcidin A	MeOH	213(4.30)	94-2947-84
Allamcidin B	MeOH	213(4.26)	94-2947-84
C$_{19}$H$_{23}$			
7H-Cyclopenta[a]pentalene, 2,5-bis(1,1-dimethylethyl)-, anion	THF at -45°	473(4.0),493(3.9), 1010(2.9),1175(2.9), 1500s(2.5)	88-1445-84
C$_{19}$H$_{23}$NO			
Benzenemethanamine, 2-[1-[(2-methoxyphenyl)methyl]ethenyl]-N,N-dimethyl-	MeOH	205(4.46),272(3.44), 278(3.39)	78-0215-84
C$_{19}$H$_{23}$NO$_2$			
2,5-Cyclohexadien-1-one, 3-(methoxymethyl)-2,6-dimethyl-4-[(2,4,6-trimethylphenyl)imino]-	n.s.g.	209(4.35),277(4.35), 515(3.15)	44-1613-84
C$_{19}$H$_{23}$NO$_3$			
Secoquettamine, dihydro-, (±)-	MeOH	275(3.51),300s(2.72)	100-0753-84
C$_{19}$H$_{23}$NO$_4$			
Crassifoline, (+)-	MeOH	228s(4.67),282(4.26)	142-0101-84
	MeOH-base	237s(4.59),287(4.25), 294s(4.23)	142-0101-84
2-Butenedioic acid, 2-[1-cyclohexen-1-yl(4-methylphenyl)amino]-, dimethyl ester, cis	EtOH	262(4.20),314(4.30)	103-0295-84
trans	EtOH	260(4.40),364(3.96)	103-0295-84
Homoerythratine	EtOH	219(3.44),288(3.11)	78-2677-84
2-epi-	EtOH	240(3.74),290(3.70)	78-2677-84
C-Homoerythrinan-2-one, 1,6-didehydro-15-hydroxy-3,16-dimethoxy-, (3β)-	EtOH	230(4.10),283(3.62)	78-2677-84
C-Homoerythrinan-2-one, 1,6-didehydro-16-hydroxy-3,15-dimethoxy-, (3β)-	EtOH	230(4.10),283(3.62)	78-2677-84
1H-Inden-1-ol, 2-[[(3,4-dimethoxyphenyl)methyl]amino]-5-methoxy-, (1S-trans)-	MeOH	202(4.32),230s(4.19), 287(3.96)	39-1655-84C

Compound	Solvent	$\lambda_{max}(\log \epsilon)$	Ref.
Nudaurine, dihydro-, (+)-	MeOH	233(3.97),292(3.85)	100-0342-84
$C_{19}H_{23}NO_5$			
2-Butenedioic acid, 2-[1-cyclohexen-1-yl(4-methoxyphenyl)amino]-, dimethyl ester, cis	EtOH	250(4.10),306(4.20)	103-0295-84
trans	EtOH	242(4.47),257(4.37), 324(4.43)	103-0295-84
1H-2-Pyrindine-4-carboxylic acid, 2,4a,5,6,7,7a-hexahydro-6-hydroxy-2-[2-(4-hydroxyphenyl)ethyl]-7-methyl-1-oxo-, methyl ester (dinklageine)	n.s.g.	288(4.05)	88-2783-84
$C_{19}H_{23}N_3$			
1,2-Ethanediamine, N'-9-acridinyl-N,N-diethyl-	MeOH	267(4.77),337(3.03), 411(4.03),432(3.91)	24-1523-84
$C_{19}H_{23}N_5O_2$			
4H-Pyrazolo[3,4-d]pyrimidin-4-one, 1,5-dihydro-5-(2-methoxyphenyl)-6-methyl-1-(1-piperidinylmethyl)-(or isomer)	EtOH	204(4.53),251.5(3.86), 265(3.14),272(3.12)	11-0023-84A
$C_{19}H_{23}N_5O_{10}$			
4H-Imidazo[4,5-d]-1,2,3-triazin-4-one, 1,5-dihydro-6-methyl-5-(2,3,4,6-tetra-O-acetyl-β-D-glucopyranosyl)-	EtOH	253(3.15),288(3.11)	4-1221-84
4H-Imidazo[4,5-d]-1,2,3-triazin-4-one, 1,7-dihydro-6-methyl-7-(2,3,4,6-tetra-O-acetyl-β-D-glucopyranosyl)-	H_2O	245(3.00),292(3.00)	4-1221-84
$C_{19}H_{23}S_2$			
Thiopyrylium, 6-[3-(5,6-dimethyl-2H-thiopyran-2-ylidene)-1-methyl-1-butenyl]-2,3-dimethyl-, perchlorate	$CHCl_3$	761(4.56)	97-0146-84
$C_{19}H_{24}$			
7H-Cyclopenta[a]pentalene, 2,5-bis(1,1-dimethylethyl)-	hexane	291s(4.03),304s(4.40), 316(4.65),332(4.67), 436(2.56)	88-1445-84
anion at -45°	THF	473(4.0),493(3.9), 1010(2.9),1175(2.9), 1500s(2.5)	88-1445-84
$C_{19}H_{24}N_2$			
20-Epidehydropseudoaspidospermidine, (±)-	EtOH	222(4.20),227s(4.12), 265(3.71)	44-3733-84
14-Iso-20-epidehydropseudoaspido-spermidine	EtOH	222(4.19),227s(4.10), 264(3.73)	44-3733-84
$C_{19}H_{24}N_2O$			
14-Isopseudoaspidospermidine, 5-oxo-	EtOH	242(3.87),297(3.45)	44-3733-84
20-epi-	EtOH	245(3.86),300(3.41)	44-3733-84
Pseudoaspidospermidine, 17-oxo-	EtOH	246(3.80),299(3.46)	44-3733-84
20-epi-	EtOH	245(3.82),297(3.45)	44-3733-84
$C_{19}H_{24}N_2O_2S$			
Benzenesulfonic acid, 4-methyl-, [7,7-dimethyl-4-(1-methylethylidene)bi-cyclo[3.2.0]hept-2-en-6-ylidene)-	EtOH	233(4.28)	33-1379-84

Compound	Solvent	λ_{max}(log ϵ)	Ref.
hydrazide (cont.)			33-1379-84
isomer B	EtOH	235(4.27)	33-1379-84
C$_{19}$H$_{24}$N$_2$O$_3$			
Erythramide	EtOH	240(3.96),297(3.51)	95-0946-84
Holidine	EtOH	233(3.93),272(3.63)	142-2453-84
C$_{19}$H$_{24}$N$_2$O$_3$			
Spiro[3H-indole-3,1'(5'H)-indolizine]-	MeOH	224(4.37),260(4.07),	44-0547-84
8'-carboxaldehyde, 8'-ethyl-1,2,2'-		290(3.32)	
3',6',7',8',8'a-octahydro-1-methyl-			
2-oxo-, 4'-oxide, anti			
syn	MeOH	222(4.45),258(4.03),	44-0547-84
		285(3.38)	
C$_{19}$H$_{24}$N$_2$O$_4$			
Pyridine, 3-(4,5-dihydro-4,4-dimethyl-	EtOH	275(3.88)	44-0056-84
2-oxazolyl)-1,4-dihydro-4-[3-methoxy-			
2-(methoxymethyl)phenyl]-			
Pyrrolo[2,1-a]isoquinoline-3-carboxylic	EtOH	213(3.86),285(3.24)	78-0369-84
acid, 1-cyano-8,9-diethoxy-1,2,3,5,6-			
10b-hexahydro-, methyl ester, (1α,3β-			
10bβ)-			
C$_{19}$H$_{24}$N$_4$O$_2$			
2-Naphthalenecarbonitrile, 5-amino-	benzene	565s(3.78),612(4.04),	39-1297-84C
3,8-bis(butylamino)-1,4-dihydro-		660(4.01)	
1,4-dioxo-			
C$_{19}$H$_{24}$N$_4$O$_3$			
9H-Cyclopenta[1,2-b:4,3-b']dipyridin-	H$_2$O	254(4.60),306(4.17),	142-0073-84
9-one, 2,7-bis[2-(dimethylamino)-		440(2.66)	
ethoxy]-			
C$_{19}$H$_{24}$N$_4$O$_4$S$_2$			
Acetic acid, [[(3,4,5,10-tetrahydro-	MeCN	221(4.47),275(4.13),	35-3309-84
3,7,8,10-tetramethyl-2,4-dioxo-		300s(--),358(3.81)	
benzo[g]pteridin-4a(2H)-yl)methyl]-			
dithio]-, ethyl ester			
C$_{19}$H$_{24}$N$_4$O$_6$S$_2$			
Benzothiazolium, 2-[[4-(dimethylamino)-	pH 1.13	588(4.69)	48-0151-84
phenyl]azo]-6-(2-hydroxyethoxy)-	pH 6.88	589(4.70)	48-0151-84
3-methyl-, methyl sulfate	70% EtOH-	620(4.77)	48-0151-84
	pH 6.2		
C$_{19}$H$_{24}$O$_3$			
3H-3,10a-Ethanophenanthren-3-ol,	EtOH	278(3.66)	2-1168-84
1,2,4,4a,9,10-hexahydro-7-methoxy-,			
acetate, (±)-			
C$_{19}$H$_{24}$O$_4$			
Cannabichromeorcinic acid, methyl ester	MeOH	225(4.39),261(4.40),	102-1909-84
		283(3.43)	
6ζ,9α-Epoxy-9,10-secoandrosta-1,3,5(10)-	MeOH	221(3.64),282(3.27)	13-0175-84B
trien-17-one, 3,9β-dihydroxy-	MeOH-base	218(3.66),246(3.67),	13-0175-84
		300(3.34)	
1H-Indene-1,5(4H)-dione, hexahydro-	MeOH	220(3.75),281(3.31)	13-0175-84B
4-[1-hydroxy-2-(5-hydroxy-2-methyl-	MeOH-base	217(3.81),241(3.82),	13-0175-84B
phenyl)ethyl]-7a-methyl-		300(3.47)	

Compound	Solvent	$\lambda_{max}(\log \epsilon)$	Ref.
$C_{19}H_{24}O_5$			
1H-Indene-1,5(4H)-dione, hexahydro-7-hydroxy-4-[1-hydroxy-2-(5-hydroxy-2-methylphenyl)ethyl]-7a-methyl-	MeOH MeOH-base	217(4.41),280(4.04) 217(--),241(--), 300(--)	13-0175-84B 13-0175-84B
Podolide, 2,3-dihydro-	EtOH	218(4.02)	154-0547-84
Rupicolin A, 8-epi-isobutyryl-	MeOH	214(3.77)	102-1289-84
Rupicolin B, 8-epi-isobutyryl-	MeOH	215(3.81)	102-1289-84
$C_{19}H_{24}O_5S$			
1,3-Cyclohexanedione, 2-cyclohex-1-enyl-, 4-methylbenzenesulfonic acid	pH 1 H₂O pH 13	262(4.00) 262(4.00) 289(4.30)	39-1213-84C 39-1213-84C 39-1213-84C
$C_{19}H_{24}O_6$			
Nagilactone F	EtOH	260(4.18)	154-0547-84
Subcordatolide A	MeOH	206(3.94)	102-1289-84
$C_{19}H_{24}O_7$			
Furanoeremophilan-6-one, 9β-acetoxy-3,3-(ethylenedioxy)-10α-hydroxy-	EtOH	265.5(3.61)	94-3396-84
$C_{19}H_{24}Si_2$			
Disilane, 9-anthracenylpentamethyl-	C_6H_{12}	373(3.93)	60-0341-84
Disilane, pentamethyl-9-phenanthrenyl-	C_6H_{12}	301.5(4.21)	60-0341-84
$C_{19}H_{25}NO$			
Bicyclo[2.2.1]heptan-2-one, 3-[[4-(dimethylamino)phenyl]methylene]-1,7,7-trimethyl-, [1R-(1α,3E,4α)]-	EtOH	362(4.56)	41-0629-84
$C_{19}H_{25}NO_2S$			
1-Azaspiro[4.5]dec-6-ene-4-carboxylic acid, 7-(methylthio)-1-(phenylmethyl)-, methyl ester	EtOH	226(4.34)	44-3314-84
$C_{19}H_{25}NO_3$			
C-Homoerythrinan-15-ol, 1,6-didehydro-3,16-dimethoxy-, (3β)-	EtOH	219(4.06),230(4.04), 283(3.74)	78-2677-84
Lucidinine	EtOH EtOH-base	234(3.83),283(3.52) 252(3.91),299(3.67)	142-2453-84 142-2453-84
Taxodine	EtOH	209(3.67),230(3.28), 291(2.94)	78-2677-84
$C_{19}H_{25}NO_3Si$			
2-Cyclopenten-1-one, 2-(1,3-benzodioxol-5-yl)-3-[[2-[(trimethylsilyl)-methyl]-2-propenyl]amino]-	MeCN	278(4.28)	44-0228-84
$C_{19}H_{25}NO_4$			
2-Isotaxodinol	EtOH	240(3.28),283(2.93)	78-2677-84
Ryllistine	MeOH	234(4.30),280(3.77), 285s(3.68)	150-0412-84S
2-Taxodinol	EtOH	212(3.61),230(3.26), 283(2.88)	78-2677-84
$C_{19}H_{25}NO_8$			
D-Gluconic acid, 2-deoxy-2-[(3-methoxy-3-oxo-2-propenyl)amino]-4,6-O-(phenylmethylene)-, ethyl ester, (±)-	MeOH	268(4.38)	136-0081-84B
1H-Pyrrolizine-6,7-dicarboxylic acid, 2-acetoxy-2,3-dihydro-5-(4-methoxy-	CHCl₃	282.9(3.369)	44-5164-84

Compound	Solvent	λ_{max} (log ϵ)	Ref.
2-methyl-4-oxobutyl)-, dimethyl ester, [R-(R*,R*)]- (cont.)			44-5164-84
$C_{19}H_{25}N_3O_2S$ 3,5-Pyridinedicarboxamide, N,N'-diethyl-1,4-dihydro-1-methyl-4-[(phenylmethyl)thio]-	CH_2Cl_2	234(4.31),267(4.26), 348(4.10)	35-6029-84
$C_{19}H_{25}N_3O_4$ D-Ribitol, 1-deoxy-1-[[4,5-dimethyl-2-(phenylazo)phenyl]amino]-	EtOH	<u>330(4.2),482(4.0)</u>	104-0381-84
$C_{19}H_{25}Na$ Sodium, [1-(1,1-dimethylethyl)-3-[[3-(1,1-dimethylethyl)-2,4-cyclopentadien-1-ylidene]methyl]-2,4-cyclopentadien-1-yl]-	THF	458(4.7)	88-1445-84
$C_{19}H_{26}$ 7H-Cyclopenta[a]pentalene, 2,5-bis(1,1-dimethylethyl)-3a,3b-dihydro-	hexane	239(3.53),260(3.56)	88-1445-84
$C_{19}H_{26}N_2$ 20-Epipseudoaspidospermidine, (±)- 14-Iso-20-epipseudoaspidospermidine 14-Isopseudoaspidospermidine	EtOH EtOH EtOH	245(3.86),297(3.47) 246(3.82),297(3.45) 245(3.80),298(3.50)	44-3733-84 44-3733-84 44-3733-84
$C_{19}H_{26}N_2O$ Corynantheol, dihydro-, (±)-	EtOH	277(4.55),281(3.85), 292s(3.81)	39-1237-84C
$C_{19}H_{26}N_2O_4$ 5H-[1]Benzopyrano[3,4-c]pyridin-5-one, 3-[2-(dimethylamino)propyl]-1,2,3,4-tetrahydro-8,9-dimethoxy-, dihydrochloride	MeOH	229(4.26),280(3.74), 338(4.14)	4-1561-84
$C_{19}H_{26}N_2O_7$ Propanedioic acid, (acetylamino)(5,6-dihydro-3-methoxy-4H-cyclohept[d]-isoxazol-8-yl)methyl]-, diethyl ester	MeOH	259(4.10)	87-0585-84
$C_{19}H_{26}N_2O_9$ 1H-Cyclobuta[b]pyrrole-3a,4,5,5a-tetra-carboxylic acid, 3-acetyl-1-(dimethylamino)-3a,4,5,5a-tetrahydro-2-methyl-, tetramethyl ester, (3aα,4α,5β-5aα)- (3aα,4β,5α,5aα)-	MeOH MeOH	207(4.18),320(4.40) 205(4.07),323(4.43)	24-1620-84 24-1620-84
$C_{19}H_{26}N_4O$ Urea, N'-[(8α)-ergolin-8-yl]-N,N-diethyl-	MeOH	225(4.50),276s(3.81), 282(3.83),293(3.75)	73-2828-84
$C_{19}H_{26}N_4O_2S_2$ Benzo[g]pteridine-2,4(3H,4aH)-dione, 4a-[[(1,1-dimethylethyl)dithio]-methyl]-5,10-dihydro-3,7,8,10-tetramethyl-	MeCN 6M HCl	218(4.61),274(4.20), 300s(--),358(3.89) 215(4.47),230s(--), 270(4.16),300s(--), 401(3.51)	35-3309-84 35-3309-84

Compound	Solvent	$\lambda_{max}(\log \epsilon)$	Ref.
$C_{19}H_{26}N_6O_6$ Triethylammonium 4-(4-amino-3-methyl-phenyl)-5,7-dinitro-4,5-dihydro-benzofurazanide 3-oxide	MeOH	247(4.20),290s(3.97), 364(3.83),473(4.40)	12-0985-84
$C_{19}H_{26}O$ Androsta-2,4-dien-1-one (at -90°)	$C_6H_{11}Me-$ isopentane	<u>314(3.9)</u>	89-0440-84
$C_{19}H_{26}O_2$ Androsta-4,6-dien-17-one, 3β-hydroxy-	n.s.g.	234(4.42),240(4.45), 248(4.31)	39-2941-84C
$C_{19}H_{26}O_4$ 3-Cyclohexen-1-ol, 4-(5-acetoxy-3-meth-yl-3-penten-1-ynyl)-3,5,5-trimethyl-, acetate, (Z)-(±)-	EtOH	271(4.22),283(4.11)	39-2147-84C
5,9-Methanobenzocyclooctene-5(1H)-carb-oxylic acid, 2,3,4,6,7,8,9,10-octa-hydro-2,2,7,7,9-pentamethyl-3,4-di-oxo-	n.s.g.	206(3.52),257(3.66)	49-0793-84
5,9-Methanobenzocyclooctene-5(1H)-carb-oxylic acid, 2,3,4,6,7,8,9,10-octa-hydro-2,2,7,7,9-pentamethyl-4,10-di-oxo-	n.s.g.	220s(3.55),270(3.84)	49-0809-84
$C_{19}H_{26}O_5$ Subcordatolide C	MeOH	213(4.02)	100-0920-84
$C_{19}H_{26}O_5Si$ 2,4-Hexadienedioic acid, (4-methoxy-phenyl)methyl 2-(trimethylsilyl)ethyl ester, (Z,E)-	EtOH	227(4.24),264(4.48)	44-1772-84
$C_{19}H_{26}O_6$ 2H-Cyclohepta[b]furan-2-one, 5-acetoxy-7-(3-acetoxybutyl)-3,3a,4,5,6,8a-hexahydro-6-methyl-3-methylene-(dicorin diacetate)	MeOH	229(3.79)	102-2553-84
$C_{19}H_{26}O_{10}$ Ptelatoside A	MeOH	254(4.29),288s(3.23), 299s(3.00)	138-0397-84
$C_{19}H_{27}BrO_3$ 5,9-Methanobenzocyclooctene-5(1H)-carb-oxylic acid, 3-bromo-2,3,4,5,7,8,9,10-octahydro-2,2,7,7,9-pentamethyl-4-oxo-, (3α,5β,9β)-	n.s.g.	260(3.72)	49-0793-84
5,9-Methanobenzocyclooctene-5(1H)-carb-oxylic acid, 10-bromo-2,3,4,6,7,8,9-10-octahydro-2,2,7,7,9-pentamethyl-4-oxo-, (5α,9β,10β)-	n.s.g.	209(3.59),256(4.05)	49-0809-84
$C_{19}H_{27}NO_3$ Protoemetine (as perchlorate)	EtOH	232(3.92),283(3.61)	100-0397-84
$C_{19}H_{27}NO_5$ Alancine	MeOH	230s(3.56),273(2.86), 282(2.84)	142-1965-84
	MeOH-base	282(--),293s(--)	142-1965-84

Compound	Solvent	$\lambda_{max}(\log \epsilon)$	Ref.
$C_{19}H_{27}NO_{12}$			
α-D-Glucofuranoside, methyl 2-deoxy-2-[[3-methoxy-2-(methoxycarbonyl)-3-oxo-1-propenyl]amino]-, 3,5,6-triacetate	EtOH	280(4.16)	136-0101-84L
α-D-Glucopyranoside, methyl 2-deoxy-2-[[3-methoxy-2-(methoxycarbonyl)-3-oxo-1-propenyl]amino]-, 3,4,6-triacetate	EtOH	277(4.11)	136-0101-84L
$C_{19}H_{27}NO_{13}$			
D-ido-Heptonic acid, 2-acetyl-2,3-dideoxy-3-(nitromethyl)-, methyl ester, 4,5,6,7-tetraacetate, (2ξ)-	EtOH	256(3.64)	136-0063-84K
$C_{19}H_{27}N_3O_2$			
Xanthobilirubic acid amide, N,N-dimethyl-	MeOH	412(4.52)	4-0139-84
	$CHCl_3$	405(4.56)	4-0139-84
$C_{19}H_{28}$			
1,3-Cyclopentadiene, 1-(1,1-dimethylethyl)-3-[[4-(1,1-dimethylethyl)-1,3-cyclopentadien-1-yl]methyl]-	hexane	250(3.80),311(3.03), 322(3.05),333s(2.97)	88-1445-84
$C_{19}H_{28}N_2O_3$			
2-Imidazolidinone, 5-[3,5-bis(1,1-dimethylethyl)-4-oxo-2,5-cyclohexadien-1-ylidene]-4-methoxy-4-methyl-	MeOH	262(3.41),383(4.39)	95-0839-84
$C_{19}H_{28}N_2O_{12}$			
D-ido-Heptonic acid, 2-(1-aminoethylidene)-2,3-dideoxy-3-(nitromethyl)-, methyl ester, 4,5,6,7-tetraacetate, (Z)-	EtOH	279(4.20)	136-0063-84K
$C_{19}H_{28}O_3$			
5,9-Methanobenzocyclooctene-5(1H)-carboxylic acid, 2,3,4,6,7,8,9,10-octahydro-2,2,7,7,9-pentamethyl-4-oxo-	n.s.g.	206s(3.04),247(3.96)	49-0793-84
$C_{19}H_{28}O_3S$			
Cyclohexanepropanol, 2,2-dimethyl-6-methylene-, 4-methylbenzenesulfonate	pentane	256(2.53),261(2.63), 267(2.60),272(2.54)	33-1734-84
$C_{19}H_{28}O_4$			
3-Cyclohexene-1-carboxylic acid, 4-(2,2-dimethyl-1-oxopropoxy)-2-methylene-1-(2-methyl-2-propenyl)-, ethyl ester	EtOH	241(3.88)	12-2037-84
5,9-Methanobenzocyclooctene-5(1H)-carboxylic acid, 2,3,4,6,7,8,9,10-octahydro-3-hydroxy-2,2,7,7,9-pentamethyl-4-oxo-	n.s.g.	210s(3.28),247(3.85)	49-0793-84
$C_{19}H_{28}O_5$			
1,4-Cyclohexanediol, 1-(5-acetoxy-3-methyl-3-penten-1-ynyl)-2,2,6-trimethyl-, 4-acetate, [1α,1(Z),4α,6β]-(±)-	EtOH	228(4.11)	39-2147-84C

Compound	Solvent	$\lambda_{max}(\log \epsilon)$	Ref.
2,4-Pentadienoic acid, 5-(4-acetoxy-2,2,6-trimethyl-7-oxabicyclo[4.1.0]-hept-1-yl]-3-methyl-, ethyl ester, [1R-[1α(2E,4E),4β,6α]]-	EtOH	263(4.46)	33-2043-84
isomer	EtOH	264(4.45)	33-2043-84
$C_{19}H_{28}O_6$			
3-Cyclohexene-1-propanoic acid, 4-(2,2-dimethyl-1-oxopropoxy)-1-(ethoxycarbonyl)-2-methylene-, methyl ester	EtOH	238(3.09)	12-2037-84
Dicorin, dihydro-, diacetate	MeOH	224(3.80)	102-2553-84
	MeOH	229(3.80)	102-2553-84
$C_{19}H_{28}O_{12}$			
Shanzhiside, 8-acetyl-, methyl ester	EtOH	236(3.98)	32-0049-84
$C_{19}H_{29}NO_2$			
4-Azaandrost-5-en-3-one, 17β-hydroxy-N-methyl-	n.s.g.	222(4.33)	111-0465-84
$C_{19}H_{29}NO_4$			
Ankorine	EtOH	272(2.96)	100-0397-84
$C_{19}H_{29}N_2O_2S$			
1(2H)-Pyridinyloxy, 4-[5-[2-(diethyl-amino)-2-oxoethyl]-2-thienyl]-3,6-dihydro-2,2,6,6-tetramethyl-	EtOH	205(4.15),290(4.11),345(2.23)	70-1901-84
$C_{19}H_{29}N_7O_4$			
L-Histidine, N-[2-[[1-carboxy-4-[(4,6-dimethyl-2-pyrimidinyl)amino]butyl]-amino]ethyl]-, (S)-	H_2O	235(4.23),300(3.62)	158-0984-84
$C_{19}H_{30}NO_2$			
Phenoxy, 2,6-bis(1,1-dimethylethyl)-4-[[(1,1-dimethylethyl)imino]meth-yl]-, N-oxide	benzene	375(4.83),728(3.57)	48-0579-84
$C_{19}H_{30}N_2OS$			
2-Thiopheneacetamide, N,N-diethyl-5-(1,2,3,4-tetrahydro-2,2,6,6-tetramethyl-4-pyridinyl)-	EtOH	209(4.15),290(4.09)	70-1901-84
$C_{19}H_{30}N_4O_5$			
Inosine, 2-nonyl-	pH 1	251(4.09)	87-0429-84
	pH 7	251(4.10)	87-0429-84
	pH 13	256(4.11)	87-0429-84
$C_{19}H_{30}O_2$			
2,4,10-Dodecatrienoic acid, 3,7,11-tri-methyl-6,7-methylene-, 1-methylethyl ester	EtOH	287.5(4.43)	19-0207-84
$C_{19}H_{30}O_3$			
Bicyclo[3.1.1]heptane-2-acetic acid, 6,6-dimethyl-3-(3-oxo-1-octenyl)-, [1S-[1α,2β,3α(E),5α]]-	EtOH	229(4.14)	39-1069-84C
3,5,11-Tridecatrien-2-one, 9-[1-(form-yloxy)-1-methylethyl]-6,12-dimethyl-	EtOH	290(4.00)	78-1545-84

Compound	Solvent	$\lambda_{max}(\log \epsilon)$	Ref.
$C_{19}H_{30}O_5$			
Acetic acid, 3,10',10'-trimethoxy-β-ionylidene-, methyl ester	EtOH	252(3.96),294(4.02)	94-1709-84
1,4-Cyclohexanediol, 1-(5-acetoxy-3-methyl-1,3-pentadienyl)-2,2,6-trimethyl-, 4-acetate, [1α,1(E,Z)-4α,6β]-(±)-	EtOH	237(4.08)	39-2147-84C
$C_{19}H_{30}O_6$			
Acetic acid, (3-acetoxy-5,6-dihydro-5,6-dihydroxy-β-ionylidene)-, ethyl ester, (3S,5S,6S)-	EtOH	265(4.45)	33-2043-84
2,4-Pentadienoic acid, 5-(4-acetoxy-1,2-dihydroxy-2,6,6-trimethylhexyl)-3-methyl-, ethyl ester, (1R,2R,4S)-	EtOH	264(4.45)	33-2043-84
$C_{19}H_{31}NO_2$			
Pyridinium, 1-(1-carboxytridecyl)-, hydroxide, inner salt	n.s.g.	261.5(3.64)	46-6041-84
$C_{19}H_{32}O_2$			
1,3-Benzenediol, 5-tridecyl-	EtOH	276(3.34),282(3.33)	100-0530-84
1-Octen-3-one, 1-[2-(2-hydroxyethyl)-6,6-dimethylbicyclo[3.1.1]hept-3-yl]-, [1S-[1α,2β,3α(E),5α]]-	EtOH	230(4.12)	39-1069-84C
$C_{19}H_{32}O_3$			
3(2H)-Furanone, 2-hydroxy-2,4-dimethyl-5-(1,3,5,7-tetramethyl-1-nonenyl)-	EtOH	243(3.69),308(3.92)	44-2506-84
$C_{19}H_{32}O_6$			
10,12,14-Octadecatrienoic acid, 9,16-dihydroperoxy-, methyl ester, trans-cis-trans	n.s.g.	259(4.53)	77-0067-84
isomer	n.s.g.	268(4.64)	77-0067-84
all-trans	n.s.g.	278(4.53)	77-0067-84
$C_{19}H_{34}$			
Cycloheptane, 1,1,3,3-tetramethyl-2-(2,2,4,4-tetramethylcyclobut-ylidene)-	C_6H_{12}	205.0(4.13)	24-0277-84 +24-0310-84
Cyclohexane, 1,1,3,3-tetramethyl-2-(2,2,5,5-tetramethylcyclopen-tylidene)-	C_6H_{12}	210(3.68),228(3.69)	24-0277-84 +24-0310-84
$C_{19}H_{34}NO_2Si$			
2-Propen-1-aminium, N-[3-(2,2-dimethyl-1-oxopropoxy)-2-cyclohexen-1-ylidene]-N-methyl-2-[(trimethylsilyl)methyl]-, perchlorate	MeCN	300(4.37)	44-0220-84
$C_{19}H_{34}N_2S$			
5-Thia-13,14-diazadispiro[3.1.6.2]-tetradec-13-ene, 1,1,3,3,7,7,12,12-octamethyl-	C_6H_{12}	204.0(3.22),297.5(2.93),332s(2.42)	24-0277-84
6-Thia-13,14-diazadispiro[4.1.5.2]-tetradec-13-ene, 1,1,4,4,8,8,12,12-octamethyl-	C_6H_{12}	205.5(3.05),292.5(2.88),335s(2.24)	24-0277-84
$C_{19}H_{36}N_2O$			
2-Nonadecanone, 1-diazo-	hexane	245(4.11),377(1.26)	35-4531-84

Compound	Solvent	$\lambda_{max}(\log \epsilon)$	Ref.
$C_{19}H_{36}N_2O_3$ Diazenecarboxylic acid, acetyl-, hexadecyl ester	C_6H_{12}	430(1.47)	23-0574-84
$C_{19}H_{36}N_2O_5$ Neoenactin A, sulfate (2:1)	MeOH MeOH-HCl MeOH-NaOH	211(3.77) 211(3.77) 238(3.77)	158-84-168 158-84-168 158-84-168
$C_{19}H_{36}OSi$ 1-Butanone, 1-[(1,1-dimethylethyl)dimethylsilyl]-4-(2,2-dimethyl-6-methylenecyclohexyl)-	pentane	357(2.08),372(2.20), 387(2.13)	33-1734-84
$C_{19}H_{36}O_4Si_2$ 2-Cyclohexen-1-one, 2,5-bis[[(1,1-dimethylethyl)dimethylsilyl]oxy]-4-hydroxy-6-methylene-, trans-(±)-	hexane	244s(3.11),279(3.29)	44-1898-84

Compound	Solvent	$\lambda_{max}(\log \epsilon)$	Ref.
$C_{20}HF_{13}$			
1H-2,2'-Binaphthalene, tridecafluoro-	EtOH	216(4.62),230(3.60), 243(3.61),280(4.03), 330(3.77)	155-0263-84A
$C_{20}H_3Cl_{12}$			
9H-Fluoren-9-yl, 1,2,3,4,5,7,8-hepta-chloro-6-methyl-9-(pentachlorophenyl)-	CCl_4	287(4.62),317s(4.14), 345(3.63),367(3.80), 382(3.86),400s(3.66), 457(3.19),494(3.21), 525(2.93),566(4.41), 612(3.78)	44-0770-84
$C_{20}H_3Cl_{12}O$			
9H-Fluoren-9-yl, 1,2,3,4,5,7,8-hepta-chloro-6-methoxy-9-(pentachloro-phenyl)-	C_6H_{12}	219(4.74),226s(4.72), 259s(4.36),291(4.64), 320s(4.11),341(3.59), 369(3.75),383(3.83), 401s(3.65),457(3.23), 494(3.26),519(3.00), 563(3.43),609(3.79)	44-0770-84
$C_{20}H_4Cl_4I_4O_5$			
Rose Bengal	MeOH	518(4.51),558(5.02)	35-5879-84
	MeOH	210(4.76),320(4.05), 519(4.51),558(5.02)	64-0474-84B
	EtOH	558(5.04)	46-2297-84
	EtOH-pyri-dine	558(5.04)	46-2297-84
	CH_2Cl_2	246(4.76)	64-0474-84B
triethylamine salt	CH_2Cl_2	518(4.51),556(4.88)	35-5879-84
$C_{20}H_4Cl_{12}$			
9H-Fluorene, 1,2,3,4,5,7,8-heptachloro-6-methyl-9-(pentachlorophenyl)-	C_6H_{12}	216(4.92),228s(4.83), 297(4.40),308s(4.34)	44-0770-84
$C_{20}H_4Cl_{12}O$			
9H-Fluorene, 1,2,3,4,5,7,8-heptachloro-6-methoxy-9-(pentachlorophenyl)-	C_6H_{12}	218(4.83),228(4.82), 296(4.40),306s(4.34)	44-0770-84
$C_{20}H_6Cl_2N_4$			
Propanedinitrile, 2,2'-(1,5-dichloro-9,10-anthracenediylidene)bis-	MeCN	282(4.33),300s(4.25), 325(4.24)	5-0618-84
$C_{20}H_7F_{10}NO$			
Benzenamine, N-[bis(pentafluorophenyl)-methylene]-2-methyl-, N-oxide	EtOH	283(4.26)	70-1268-84
Benzenamine, N-[bis(pentafluorophenyl)-methylene]-4-methyl-, N-oxide	EtOH	286(4.08)	70-1268-84
$C_{20}H_7F_{10}NO_2$			
Benzenamine, N-[bis(pentafluorophenyl)-methylene]-4-methoxy-, N-oxide	EtOH	282(4.09),328s(3.89)	70-1268-84
$C_{20}H_8Br_4O_5$			
Eosin	pH 5.8	515(5.0)	35-4336-84
	aq NH_3	516(5.01)	86-1125-84
$C_{20}H_8I_4O_5$			
Erythrosine (disodium salt)	MeOH	530(4.97)	35-5879-84
Erythrosin B (disodium salt)	EtOH	534(4.48)	46-2297-84

Compound	Solvent	$\lambda_{max}(\log \epsilon)$	Ref.
Erythrosin B (cont.)	EtOH-pyridine	534(4.48)	46-2297-84
	EtOH-PVP	600(4.52)	46-2297-84
$C_{20}H_8N_4$ Propanedinitrile, 2,2'-(9,10-anthracenediylidene)bis-	MeCN	279(4.45),302(4.20), 342(4.39)	5-0618-84
	CH_2Cl_2	285(4.48),305(4.21), 350(4.42)	44-5002-84
$C_{20}H_8N_4O_2S_2$ Naphtho[2,3-c][1,2,5]thiadiazol-4(9H)-one, 9-(9-oxonaphtho[2,3-c][1,2,5]-thiadiazol-4(9H)-ylidene)-	DMF	286(4.09),394(3.49), 420s(3.46)	33-0574-84
$C_{20}H_9Br_3O_5$ Fluorescein, 2,4,5-tribromo-	aq NH$_3$	510(4.96)	86-1125-84
$C_{20}H_9F_3N_2O$ Benzenamine, N-(3,4,5-trifluoro-1H-furo[2,3,4-kl]acridin-1-ylidene)-	EtOH	207(4.54),270(4.77), 365(3.81),383(4.00), 443(4.16),461s(4.13)	104-0385-84
$C_{20}H_{10}BrNO_3S$ 8H-Naphtho[2,3-b]phenothiazine-8,12(3H)-dione, 3-bromo-7-hydroxy-	toluene	633(4.17),684(4.11)	104-2234-84
$C_{20}H_{10}BrNO_5S$ 6H-Anthra[1,9-cd]isoxazol-6-one, 3-[(4-bromophenyl)sulfonyl]-5-hydroxy-	toluene	300(4.05),480(3.90), 508(3.88)	104-2234-84
5H-Naphtho[2,3-a]phenothiazine-8,12(13H)-dione, 3-bromo-7-hydroxy-, 5,5-dioxide	toluene	340(4.11),575(3.97)	104-2234-84
$C_{20}H_{10}BrN_3O_5S$ 9,10-Anthracenedione, 1-azido-2-[(4-bromophenyl)sulfonyl]-4-hydroxy-	toluene	452(3.70)	104-2234-84
$C_{20}H_{10}Br_2O_5$ Fluorescein, 4,5-dibromo-	aq NH$_3$	504(4.93)	86-1125-84
$C_{20}H_{10}ClNO_4S$ 6H-Anthra[1,9-cd]isoxazol-6-one, 3-[(4-chlorophenyl)sulfonyl]-	toluene	305(4.08),468(4.01)	104-2234-84
8H-Naphtho[2,3-a]phenothiazine-8,12(13H)-dione, 3-chloro-, 5,5-dioxide	toluene	328(4.24),492(3.75)	104-2234-84
$C_{20}H_{10}ClNO_5S$ 6H-Anthra[1,9-cd]isoxazol-6-one, 3-[(4-chlorophenyl)sulfonyl]-5-hydroxy-	toluene	300(4.08),480(3.93), 508(3.90)	104-2234-84
8H-Naphtho[2,3-a]phenothiazine-8,12-(13H)-dione, 3-chloro-7-hydroxy-, 5,5-dioxide	toluene	340(4.06),575(3.94)	104-2234-84
$C_{20}H_{10}ClN_3O_4S$ 9,10-Anthracenedione, 1-azido-2-[(4-chlorophenyl)sulfonyl]-	toluene	338(3.69)	104-2234-84

Compound	Solvent	$\lambda_{max}(\log \epsilon)$	Ref.
$C_{20}H_{10}ClN_3O_5S$			
9,10-Anthracenedione, 1-azido-2-[(4-chlorophenyl)sulfonyl]-4-hydroxy-	toluene	452(3.69)	104-2234-84
$C_{20}H_{10}F_4N_2$			
Benzenamine, N-[(1,2,3,4-tetrafluoro-9-acridinyl)methylene]-	EtOH	217(4.52),255(4.98), 350s(3.79),363(3.92), 385s(3.86)	104-0385-84
$C_{20}H_{10}F_4N_2O$			
Benzenamine, N-[(1,2,3,4-tetrafluoro-9-acridinyl)methylene]-, N-oxide	EtOH	221(4.48),254(4.98), 353s(3.83),367(3.96), 387s(3.88)	104-0385-84
$C_{20}H_{10}N_2O_4$			
Perylene, 3,6-dinitro-	MeCN	263(3.77),473(3.71)	1-0701-84
Perylene, 3,7-dinitro-	MeCN	258(3.67),464.5(3.56)	1-0701-84
Perylene, 3,9-dinitro- (contains 3,10-isomer)	MeCN	259.5(4.03),330(--), 482(4.03)	1-0701-84
$C_{20}H_{10}O_6$			
6,6'-Bijuglone	EtOH	230s(4.35),252s(4.23), 445(3.80)	5-0319-84
	EtOH-NaOH	260(4.06),295s(3.96), 355(3.44),607(3.90)	5-0319-84
	$CHCl_3$	450(--)	5-0319-84
$C_{20}H_{11}BrO_5$			
Fluorescein, 4-bromo-	aq NH_3	496(4.97)	86-1125-84
$C_{20}H_{11}ClN_2$			
Naphtho[1',8':3,4,5]cyclohepta[1,2-b]-quinoxaline, 8-chloro-	MeOH	213(4.69),238(4.74), 250(4.69),262(4.70), 316(4.10),438(4.30)	39-1465-84C
$C_{20}H_{11}Cl_5$			
9H-Fluorene, 3-methyl-9-(pentachloro-phenyl)- ($5\lambda,6\epsilon$)	C_6H_{12}	230s(5.02),240s(4.59), 266(4.38),293(4.09), 304(3.81),?(3.85)	44-0770-84
$C_{20}H_{11}Cl_5O$			
9H-Fluoren-9-ol, 3-methyl-9-(penta-chlorophenyl)-	C_6H_{12}	216(4.91),232s(4.76), 240s(4.62),268(4.07), 277s(4.06),290s(3.97), 300s(3.86),312(3.59)	44-0770-84
$C_{20}H_{11}F_3N_2O_2S$			
Thiazolo[3,2-a]pyrimidin-4-ium, 5,7-diphenyl-2-(trifluoroacetyl)-, hydroxide, inner salt	MeCN CCl_4	327(4.46),480(4.05) 516(--)	103-0791-84 103-0791-84
$C_{20}H_{11}NO_2$			
1H-Benzo[h]naphtho[1,2,3-de]quinoline-2,8-dione	N-Mepyrro-lidine	430(3.92),457(3.80)	104-0980-84
Benzo[a]pyrene, 1-nitro-	MeOH	224(4.09),264(4.50), 307(4.11),438(4.04)	1-0309-84
Benzo[a]pyrene, 3-nitro-	MeOH	250(4.54),268(4.52), 299(4.14),379(4.04), 398(4.24),438(4.04)	1-0309-84
Benzo[a]pyrene, 6-nitro-	MeOH	217(4.42),253(4.60), 263(4.63),286(4.51),	1-0309-84

Compound	Solvent	$\lambda_{max}(\log \epsilon)$	Ref.
Benzo[a]pyrene, 6-nitro- (cont.)		297(4.57),354(4.01), 370(4.21),389(4.19), 404(3.98)	1-0309-84
Benzo[e]pyrene, 1-nitro-	MeOH	225(4.50),271(4.47), 306(4.15),376(3.66)	1-0309-84
Benzo[e]pyrene, 3-nitro-	MeOH	231(4.51),239(4.51), 256(4.34),273(4.28), 299(4.31),373(4.02)	1-0309-84
$C_{20}H_{11}NO_4S$ 6H-Anthra[1,9-cd]isoxazol-6-one, 3-(phenylsulfonyl)-	toluene	300(4.08),466(3.98)	104-2234-84
8H-Naphtho[2,3-a]phenothiazine-8,13(14H)-dione, 5,5-dioxide	toluene	328(4.19),493(3.75)	104-2234-84
$C_{20}H_{11}NO_5S$ 6H-Anthra[1,9-cd]isoxazol-6-one, 5-hydroxy-3-(phenylsulfonyl)-	toluene	300(3.97),480(3.90), 505(3.88)	104-2234-84
8H-Naphtho[2,3-a]phenothiazine-8,13(14H)-dione, 7-hydroxy-, 5,5-dioxide	toluene	340(3.96),577(3.94)	104-2234-84
$C_{20}H_{11}N_3O_4S$ 9,10-Anthracenedione, 1-azido-2-(phenylsulfonyl)-	toluene	338(3.62)	104-2234-84
$C_{20}H_{11}N_3O_5S$ 9,10-Anthracenedione, 1-azido-4-hydroxy-2-(phenylsulfonyl)-	toluene	455(3.72)	104-2234-84
$C_{20}H_{12}$ Dicycloocta[1,2,3,4-def:1',2',3',4'-jkl]biphenylene	C_6H_{12}	208?(5.16),249(4.29), 260(4.27),271(4.42), 282(4.47),294s(3.94), 309(3.94),322(3.84), 338s(3.55),358(3.23), 386s(2.86),425(2.80), 456(2.82),484(2.74), 557(2.74),596(2.77), 648s(2.64),702s(2.17)	35-7195-84
	CS_2	388(2.99),423(2.81), 455(2.84),484(2.80), 523s(2.64),567(2.70), 608(2.71),660s(2.33)	35-7195-84
Perylene	heptane	386(4.13),408(4.46), 435(4.60)	35-8024-84
$C_{20}H_{12}BrNO_4$ 2H-Indol-2-one, 1-acetyl-5-bromo-1,3-dihydro-3-(2-oxo-5-phenyl-3(2H)-furanylidene)-	$CHCl_3$	270(4.32),472(4.40), 505s(4.33)	39-1331-84B
$C_{20}H_{12}ClNO_3$ Acetamide, 2-chloro-N-(6,11-dihydro-6,11-dioxo-5-naphthacenyl)-	toluene	425(3.74)	104-0980-84
$C_{20}H_{12}ClN_3$ Benzo[a]pyrido[2,3-c]phenazine, 3-chloro-1-methyl- $(8\lambda,7\epsilon)$	EtOH	223(3.96),266(4.43), 302(3.86),333s(3.45), 350s(3.57),358s(3.81), 370(3.76),388(?)	12-1271-84

Compound	Solvent	$\lambda_{max}(\log \epsilon)$	Ref.
$C_{20}H_{12}Cl_2O_8$			
9,10-Anthracenedione, 1,2,4-triacetoxy-5,8-dichloro-	MeOH	252(4.49),349(3.76)	117-0309-84
$C_{20}H_{12}N_2$			
Naphtho[1',8':3,4,5]cyclohepta[1,2-b]-quinoxaline	MeOH	260(4.69),314(4.41), 428(4.61)	39-1465-84C
$C_{20}H_{12}N_2O_3$			
2H-Indol-2-one, 1,3-dihydro-3-[5-(1H-indol-3-yl)-2-oxo-3(2H)-furanylidene]-	CHCl$_3$	265(4.20),307s(3.83), 524(4.20)	39-1331-84B
$C_{20}H_{12}N_2O_6$			
2H-Indol-2-one, 1-acetyl-1,3-dihydro-5-nitro-3-(2-oxo-5-phenyl-3(2H)-furanylidene)-	CHCl$_3$	268(4.25),340(3.68), 490(4.41)	39-1331-84B
$C_{20}H_{12}N_4O_2$			
4-Quinolinepropanenitrile, 2-amino-α-benzoyl-3-cyano-β-oxo-	BuOH	278(4.81),314(4.90)	4-1233-84
$C_{20}H_{12}N_6O_{10}$			
1,4-Benzenedicarboxamide, N,N'-bis(2,4-dinitrophenyl)-	dioxan-base DMF-base	450(4.81) 450(4.82)	104-0565-84 104-0565-84
$C_{20}H_{12}O_4$			
Anthra[1,2-b:4,3-b']difuran-7,12-dione, 2,5-dimethyl-	EtOH	248(4.34),276s(4.23), 294(4.4),388(3.93)	12-1518-84
5,12-Naphthacenedione, 8-acetyl-6-hydroxy-	MeOH	206(4.25),255(4.65), 305(4.19),438(3.98)	78-3677-84
9H-Xanthen-9-one, 2-(2-hydroxybenzoyl)-	EtOH	220(4.38),260(4.49), 337(3.99)	77-1319-84
$C_{20}H_{12}O_5$			
Fluorescein	aq NH$_3$	490(5.00)	86-1125-84
$C_{20}H_{12}O_6$			
4H,10H-Benzo[1,2-b:3,4-b']dipyran-8-carboxylic acid, 3-methyl-4,10-dioxo-2-phenyl-	MeOH	240(4.30),302(4.27)	2-0969-84
$C_{20}H_{12}S$			
Benzo[b]phenanthro[2,1-d]thiophene	C$_6$H$_{12}$	220(4.43),253(4.73), 260(4.68),270(4.67), 277(4.69),287(4.7), 305(4.57),310(4.44), 320(3.72),330(3.59), 346(3.59),362(3.64)	4-0353-84
Benzo[b]phenanthro[2,3-d]thiophene	C$_6$H$_{12}$	220(4.35),253(4.46), 262(4.55),271(4.76), 285(4.87),292(4.91), 317(4.31),329(3.83), 359(2.82),378(3.20)	4-0353-84
Benzo[b]phenanthro[3,2-d]thiophene	C$_6$H$_{12}$	234(4.55),244(4.58), 267(4.60),278(4.75), 288(4.93),299(4.78), 320(4.07),338(4.11), 353(3.31),372(3.19)	4-0353-84
Benzo[b]phenanthro[3,4-d]thiophene	C$_6$H$_{12}$	231(4.73),248(4.82), 272(4.93),280(4.89), 295(4.65),308(4.70),	4-0353-84

Compound	Solvent	$\lambda_{max}(\log \epsilon)$	Ref.
Benzo[b]phenanthro[3,4-d]thiophene (cont.)		330(3.86),348(3.81), 365(3.98)	4-0353-84
$C_{20}H_{13}BrN_2O_2$ 9,10-Anthracenedione, 1-amino-2-bromo-4-(phenylamino)-	C_6H_5Cl	584(4.12),621(4.11)	34-0482-84
$C_{20}H_{13}ClN_2S_2$ 2,4(1H,3H)-Quinazolinedithione, 3-(4-chlorophenyl)-1-phenyl-	EtOH	258(4.23),287(4.47), 312s(4.23),339(4.11), 374(3.76)	97-0328-84
$C_{20}H_{13}F_3N_2O$ Ethanone, 2,2,2-trifluoro-1-[2-(4-methylphenyl)cyclohepta[4,5]pyrrolo[1,2-a]imidazol-3-yl]-	EtOH	239(4.43),311(4.34), 384s(3.67),413(3.64), 510(3.30),528s(3.23), 578s(3.13),638s(2.63)	150-3465-84M
$C_{20}H_{13}N$ Acenaphtho[1,2-b]quinoline, 8-methyl-	$CHCl_3$	256(4.44),305(4.41), 317(4.41),368(3.51), 388(3.34)	78-2959-84
Acenaphtho[1,2-b]quinoline, 10-methyl-	$CHCl_3$	258(4.10),303(4.12), 383(3.37)	78-2959-84
Cyclopent[4,5]azepino[2,1,7-cd]pyrrolizine, 1-phenyl-	EtOH	257(4.27),304(4.38), 312(4.31),349(4.54), 395(3.63),420(3.68), 446(3.49),512(2.31), 534(2.32),550(2.28), 578(2.23),642(1.80)	24-1649-84
$C_{20}H_{13}NO_2S$ Benzenepropanenitrile, β-oxo-α-[[5-(phenylthio)-2-furanyl]methylene]-	MeOH	212(3.25),400(3.19)	73-2141-84
Cyclopent[4,5]azepino[2,1,7-cd]pyrrolizine, 1-(phenylsulfonyl)-	EtOH	275(4.38),295(4.39), 301(4.40),337(4.51), 349(4.65),386(3.87), 424(3.72),432(3.65), 452(3.58),578(2.34), 625(2.35),687(2.05)	24-1649-84
$C_{20}H_{13}NO_4$ 2H-Indol-2-one, 1-acetyl-1,3-dihydro-3-(5-oxo-4,5-dihydro-2-phenylfuran-4-yl)-	$CHCl_3$	272(4.25),483(4.38)	39-1331-84B
$C_{20}H_{13}NO_4S$ Benzenepropanenitrile, β-oxo-α-[(5-(phenylsulfonyl)-2-furanyl]methylene]-	MeOH	211(3.31),231(3.22), 343(3.36)	73-2141-84
$C_{20}H_{13}NO_5$ Oxyavicine	EtOH	248(4.50),278(4.70), 289(4.76),322(4.21), 332(4.19)	100-0001-84
$C_{20}H_{13}N_3$ Benzo[c]pyrido[2,3-c]phenazine, 1-methyl-	EtOH	218(4.31),261(4.74), 297(4.14),308(4.08), 350(3.90),370(4.07), 388(4.08)	12-1271-84

Compound	Solvent	λ_{max}(log ϵ)	Ref.
C$_{20}$H$_{13}$N$_3$O			
5H-Benz[g]indolo[2,3-b]quinoxaline, 5-acetyl-	EtOH	238(4.75),302(4.93), 394(4.46),425(3.57)	103-1155-84
Benzo[a]pyrido[2,3-c]phenazin-3(4H)-one, 1-methyl-	EtOH	232(4.34),263s(4.49), 273(4.50),285s(4.47), 308(4.15),334(3.95), 349(3.97),364(3.91), 406(3.92)	12-1271-84
C$_{20}$H$_{13}$N$_3$O$_2$			
2H-Indol-2-one, 3-[1,2-dihydro-5-(1H-indol-3-yl)-2-oxo-3H-pyrrol-3-yli-dene]-1,3-dihydro- (deoxyviolacein)	CHCl$_3$	548(4.27)	39-1331-84B
C$_{20}$H$_{13}$N$_3$O$_3$			
1H-Isoindole-1,3(2H)-dione, 2-(2,3,4,6-tetrahydro-6-oxobenzo[f]quinoxalin-5-yl)-	CH$_2$Cl$_2$	231(4.47),265(4.33), 290s(--),408(3.53)	83-0329-84
C$_{20}$H$_{13}$N$_3$O$_5$			
Furo[2,3-d]pyrimidin-4-amine, 5,6-bis(1,3-benzodioxol-5-yl)-	MeOH	207(4.79),261(4.04), 298(4.36),328(4.54), 340(4.41)	73-1788-84
C$_{20}$H$_{13}$N$_3$S			
3,5-Pyridinedicarbonitrile, 1,4-dihy-dro-2-(3-methylphenyl)-6-phenyl-4-thioxo-	EtOH	268(4.49),328(4.26), 297(3.30)	4-1445-84
3,5-Pyridinedicarbonitrile, 1,4-dihy-dro-2-(4-methylphenyl)-6-phenyl-4-thioxo-	EtOH	271(4.51),286(4.43), 327(4.26),395(3.41)	4-1445-84
C$_{20}$H$_{14}$			
1,1'-Biazulene	CH$_2$Cl$_2$	260(4.69),297(4.49), 307(4.53),380(4.16), 618(2.70)	5-1905-84
2,2'-Biazulene	CH$_2$Cl$_2$	241(4.28),248s(4.26), 314(4.92),326(4.88), 407(4.63),433(4.87), 580(3.04),685(2.46)	5-1905-84
4,4'-Biazulene	CH$_2$Cl$_2$	246(4.63),273(4.74), 289s(4.53),299s(4.34), 333(3.78),344(3.83), 357s(3.53),545s(2.82), 587s(2.95),600(2.96), 633(2.97),690(2.64)	5-1905-84
4,6'-Biazulene	CH$_2$Cl$_2$	242(4.39),274(4.87), 282s(4.73),303s(4.60), 330(3.99),341s(3.98), 363s(3.62),580(3.00), 620s(2.96),690s(2.56)	5-1905-84
5,5'-Biazulene	CH$_2$Cl$_2$	264(4.64),305(4.82), 383(4.15),643(3.05), 710(2.73)	5-1905-84
6,6'-Biazulene	CH$_2$Cl$_2$	248(4.44),313(5.08), 335s(4.38),381(4.23), 592(2.88)	5-1905-84
Bicyclo[4.1.0]hepta-1,3,5-triene, 7-(diphenylmethylene)-	C$_6$H$_{12}$	249(4.34),265s(4.25), 385(4.41),405s(4.32)	35-6108-84
	MeCN	380(4.49)	35-6108-84
1,3,5-Cycloheptatriene, 7,7'-(2,4-hexa-	C$_6$H$_{12}$	340s(4.32),400s(4.56),	24-2045-84

Compound	Solvent	$\lambda_{max}(\log \epsilon)$	Ref.
diyne-1,6-diylidene)bis- (cont.)		432(4.64),474s(4.52)	24-2045-84
	MeCN	340s(--),400s(--), 430(--),470s(--)	24-2045-84
$C_{20}H_{14}BrN_3O_3$			
2H-Indazole-4,7-dione, 5-[(4-bromophen-yl)amino]-3-methoxy-2-phenyl-	EtOH	276(4.51),316s(4.02), 452(3.86)	4-0825-84
1H-Indazole-3,4,7(2H)-trione, 5-[(4-bromophenyl)amino]-1-methyl-2-phenyl-	dioxan	252(4.82),413(4.65), 519s(3.62)	4-0825-84
$C_{20}H_{14}Br_2N_2O_4$			
1,4-Cyclohexadiene-1-carboxylic acid, 2,5-bis[(4-bromophenyl)amino]-3,6-dioxo-, methyl ester	EtOH	256(4.32),268(4.24), 377(4.29)	4-0825-84
$C_{20}H_{14}ClN_3O_2S$			
Thieno[2,3-d]pyrimidine-6-carboxylic acid, 5-amino-2-(4-chlorophenyl)-4-phenyl-, methyl ester	EtOH	265(4.54),300(4.68), 395(3.92)	39-2447-84C
$C_{20}H_{14}Cl_2N_2OS$			
3-Quinolinecarbonitrile, 4-[5-(2,4-di-chlorophenyl)-2-furanyl]-1,2,5,6,7,8-hexahydro-2-thioxo-	dioxan	245(4.05),276(4.11), 324(4.00),349(4.01), 458(3.15)	104-1780-84
3-Quinolinecarbonitrile, 4-[5-(3,4-di-chlorophenyl)-2-furanyl]-1,2,5,6,7,8-hexahydro-2-thioxo-	dioxan	242(4.29),282(4.30), 348(4.37),454(3.20)	104-1780-84
$C_{20}H_{14}Cl_2N_3S_2$			
Benzothiazolium, 2-[1,3-dichloro-2-cyano-3-(3-methyl-2(3H)-benzo-thiazolylidene)-1-propenyl]-3-methyl-, tetrafluoroborate	n.s.g.	640(4.70)	104-2045-84
$C_{20}H_{14}Cl_2O_4$			
2H-Pyran-2,4(3H)-dione, 3,5-dichloro-3-(4-methylbenzoyl)-6-(4-methyl-phenyl)-	EtOH	209(c.3.23),264(c.3.21)	104-1595-84
$C_{20}H_{14}NO_4$			
Sanguinarine (chloride)	MeOH	234(4.45),284(4.48), 325(4.14)	73-0704-84
	EtOH	236(4.49),285(4.53), 328(4.25),352s(3.85), 400(3.07),476(3.12)	100-0001-84
$C_{20}H_{14}N_2$			
Quino[3,2-b]acridine, 5,14-dihydro-	MeCN	245(4.696),292(4.452), 430(3.987)	5-0133-84
$C_{20}H_{14}N_2O_2$			
9,10-Anthracenedione, 1-amino-4-(phenylamino)-	C_6H_5Cl	454s(3.93),573(4.11), 610(4.12)	34-0482-84
Cyclohepta[1,2-b:1,7-b']bis[1,4]benz-oxazine, 3-methyl-	MeOH	287(4.25),380(3.81)	138-1149-84
Cyclohepta[4,5]pyrrolo[1,2-a]imidazole-3,10-dicarboxaldehyde, 2-(4-methyl-phenyl)-	CHCl$_3$	254(4.45),264s(4.39), 330(4.61),355s(4.45), 448(3.83),495s(3.32), 526s(3.26),570s(3.07), 620s(2.58)	150-3465-84M

Compound	Solvent	λ_{max}(log ϵ)	Ref.
C$_{20}$H$_{14}$N$_2$O$_3$			
1H-Pyrrolo[3,4-c]pyridine-1,3,4(2H,5H)-trione, 7-phenyl-6-(phenylmethyl)-	MeOH	240(4.36),280(4.26), 330(4.13),440(4.07)	48-0594-84
C$_{20}$H$_{14}$N$_2$O$_4$			
3H-Indol-3-one, 1-acetyl-2-(1-acetyl-1,3-dihydro-3-oxo-2H-indol-2-ylidene)-1,2-dihydro-, cis	toluene benzene	432(3.61) 438(3.65)	23-2478-84 39-2305-84C
trans	toluene benzene xylene	560(3.81) 562(3.85) 562(3.82)	23-2478-84 39-2305-84C 18-0470-84
C$_{20}$H$_{14}$N$_2$O$_7$			
5,10[3',4']-Furanopyrido[3,2-g]quinoline-2,8-dicarboxylic acid, 5,10-11,12,14,15-hexahydro-12,14-dioxo-, dimethyl ester	MeCN	276(4.132),281s(4.121)	5-0877-84
Pyrrolo[2,1-a]isoquinoline-1,2-dicarboxylic acid, 3-(5-nitro-2-furanyl)-, dimethyl ester	dioxan	423(4.17)	73-0533-84
C$_{20}$H$_{14}$N$_2$S$_2$			
2,4(1H,3H)-Quinazolinedithione, 1,3-diphenyl-	EtOH-1% dioxan	258(4.24),287(4.43), 313s(4.20),333(4.16), 379(3.74)	97-0328-84
C$_{20}$H$_{14}$N$_4$			
3H-1,2,4-Triazolo[4,3-b]indazole, 3,3-diphenyl-	CH$_2$Cl$_2$	250s(3.98),294(4.03), 326(3.66),380(3.89)	24-1726-84
C$_{20}$H$_{14}$N$_4$O$_6$			
Benzoic acid, [1,4-phenylenebis(azo)]-bis[2-hydroxy-	DMF	390(--),430(--)	121-0021-84
C$_{20}$H$_{14}$N$_4$O$_8$			
1,4-Cyclohexadiene-1-carboxylic acid, 2,5-bis[(4-nitrophenyl)amino]-3,6-dioxo-, methyl ester	dioxan	224(4.41),265(4.18), 381(4.49)	4-0825-84
C$_{20}$H$_{14}$O			
9,13b-Epoxy-13bH-benzo[3,4]cyclobuta-[1,2-k]phenanthrene, (4aR*,8bα,9β-13bβ)-	EtOH	214(4.30),267s(3.81), 273(3.92),279(3.88)	12-1283-84
C$_{20}$H$_{14}$OS$_3$			
Spiro[2H-thiete-2,9'-[9H]xanthene], 3-(methylthio)-4-(2-thienyl)-	CCl$_4$	313(3.04),343(2.95), 489(1.51),620(0.59)	54-0152-84
2-Thiopheneethane)dithioic) acid, α-9H-xanthen-9-ylidene-, methyl ester	CCl$_4$	285(3.11),331(3.20), 390s(2.54),490(1.49)	54-0152-84
C$_{20}$H$_{14}$O$_2$			
Benzo[b]fluoranthene-1,2-diol, 1,2-dihydro-, trans	MeOH	244(4.86),252(4.95), 273(4.52),286(4.59), 297(4.62),340(3.77)	44-1091-84
Benzo[b]fluoranthene-11,12-diol, 11,12-dihydro-, trans	MeOH	236(4.45),260(4.65), 270(4.65),278(4.68), 288(4.40),300(4.02), 352(3.95),370(3.95)	44-1091-84
Benzophenone, 3-benzoyl-	C$_6$H$_{12}$	247(4.57),345(2.36)	88-4525-84
Benzophenone, 4-benzoyl-	C$_6$H$_{12}$	260(4.54),348(2.56)	88-4525-84

Compound	Solvent	$\lambda_{max}(\log \epsilon)$	Ref.
$C_{20}H_{14}O_3$ 1,3-Cyclopentadiene-1-carboxaldehyde, 4-benzoyl-3-(hydroxyphenylmethylene)-	EtOH	244(4.20),268(4.28), 298(4.29),360(3.81)	44-4165-84
$C_{20}H_{14}O_4$ 1-Naphthacenecarboxaldehyde, 3,4,6,11-tetrahydro-5-hydroxy-2-methyl-6,11-dioxo-	MeOH	210(4.38),250s(--), 260(4.47),275s(--), 325(3.61),395s(--), 417(3.93),437s(--)	78-3677-84
Phenolphthalein	pH 11.77	553(4.53)	140-1105-84
$C_{20}H_{14}O_5$ 4H-1-Benzopyran-4-one, 3,3'-(methoxymethylene)bis-	EtOH	228(4.41),275(4.20), 307(4.31)	44-2812-84
$C_{20}H_{14}O_6$ Anthra[1,2-b]furan-3-carboxylic acid, 6,11-dihydro-5-hydroxy-2-methyl-6,11-dioxo-, ethyl ester	EtOH	240s(4.35),258(4.61), 278s(4.33),305s(3.95), 426(3.98)	12-1518-84
[6,8'-Bi-2H-1-benzopyran]-2,2'-dione, 7,7'-dimethoxy-	MeOH	209(3.94),324(3.80)	142-0333-84
5,12-Naphthacenedione, 9-ethynyl-7,8,9,10-tetrahydro-6,7,9,11-tetrahydroxy-, (7S-cis)-	EtOH	206(4.45),251(4.67), 256s(4.62),285(3.98)	78-4657-84
3,10-Perylenedione, 1,2,12a,12b-tetrahydro-1,4,9,12a-tetrahydroxy-(alteichin)	n.s.g.	254(3.68),285(3.53), 260[sic](2.95)	31-1248-84
$C_{20}H_{14}O_8$ [3,8'-Bi-2H-1-benzopyran]-2,2'-dione, 7,7'-dihydroxy-6,6'-dimethoxy-(ipomopsin)	MeOH	203s(3.68),213(4.38), 257s(4.23),303s(4.24), 350(4.39)	100-0106-84
	MeOH-NaOH	397(4.91)	100-0106-84
$C_{20}H_{14}S_4$ Ethanethione, 2-(methylthio)-1-(2-thienyl)-2-(9H-thioxanthen-9-ylidene)-	CCl_4	328(3.04),382(3.00), 480(2.18),610(1.18)	54-0152-84
2-Thiopheneethane(dithioic) acid, α-9H-thioxanthen-9-ylidene-, methyl ester	CCl_4	275(3.28),324(3.15), 486(1.34)	54-0152-84
$C_{20}H_{15}BO_7$ Boron, [ethanedioato(2-)-O,O']-[5-(4-methoxyphenyl)-1-phenyl-4-pentene-1,3-dionato-O^1,O^3]-	CH_2Cl_2	466(4.79)	97-0292-84
$C_{20}H_{15}BrO_4$ 2H-Pyran-2-one, 5-bromo-4-hydroxy-3-(4-methylbenzoyl)-6-(4-methylphenyl)-	EtOH	211(3.51),242(3.43), 360(1.04)	104-1595-84
$C_{20}H_{15}ClN_2$ Benzenamine, N,N'-[1-(4-chlorophenyl)-1,2-ethanediylidene]bis-	MeOH	223(4.36),263(4.17)	118-0128-84
$C_{20}H_{15}ClN_2O$ Benzenamine, N-[1-(4-chlorophenyl)-2-(phenylimino)ethylidene]-, N-oxide	MeOH	267(4.34)	118-0128-84
Benzenamine, 4-chloro-N-[1-phenyl-2-(phenylimino)ethylidene]-, N-oxide	MeOH	229(4.27),262(4.31)	118-0128-84
2-Benzoxazolamine, 5-chloro-N-phenyl-N-(phenylmethyl)-	$CHCl_3$	263(4.13),299(4.29)	94-3053-84

Compound	Solvent	$\lambda_{max}(\log \epsilon)$	Ref.
$C_{20}H_{15}ClN_3S_2$			
Benzothiazolium, 2-[1-chloro-2-cyano-3-(3-methyl-2(3H)-benzothiazolyli-dene)-1-propenyl]-, tetrafluoroborate	n.s.g.	610(4.63)	104-2045-84
$C_{20}H_{15}ClO_3S_2$			
Spiro[1,3-dithiolane-2,2'(1'H)-naphtha-cene]-5',12'-dione, 11'-chloro-3',4'-dihydro-6'-hydroxy-	EtOH	244(4.69),280(4.19), 459(3.95)	150-0147-84M
$C_{20}H_{15}ClO_5$			
Spiro[1,3-dioxolane-2,2'(1'H)-naphtha-cene]-5',12'-dione, 11'-chlorodihy-dro-6'-hydroxy-	EtOH	244(4.69),280(4.14), 458(3.91)	150-0147-84M
$C_{20}H_{15}ClSi$			
Silane, chlorodiphenyl(phenylethynyl)-	hexane	242.0(4.36),252.8(4.53), 264.6(4.43)	65-2317-84
$C_{20}H_{15}Cl_2N_5O$			
3-Formazancarboxamide, 1,5-bis(2-chlorophenyl)-N-phenyl- (in 40% dioxan)	pH 1 pH 13	435(4.42) 521(4.39)	86-0755-84 86-0755-84
$C_{20}H_{15}Cl_2N_5O_7S_2$			
Benzenesulfonic acid, 3,3'-[3-[(phenyl-amino)carbonyl]-1,5-formazandiyl]-bis[4-chloro-	pH 1 pH 13	441(4.15) 513(4.27)	86-0755-84 86-0755-84
$C_{20}H_{15}FN_2$			
1H-Indol-1-amine, N-(4-fluorophenyl)-3-phenyl-	EtOH	224(4.66),265(4.28), 289(4.25)	103-1119-84
$C_{20}H_{15}N$			
Benzo[h]quinoline, 5-methyl-2-phenyl-	EtOH	220(4.6),242(5.01), 286(4.83),345(4.18), 362(4.14)	103-0519-84
$C_{20}H_{15}NO_2$			
1,4-Naphthalenedione, 5-[[4-(dimethyl-amino)phenyl]ethynyl]-	EtOH	248(4.23),368(4.20), 446(3.44),510(3.43)	70-2345-84
$C_{20}H_{15}NO_3$			
Benzo[6,7]cyclohept[1,2-b]indole-6-carboxylic acid, 5,12-dihydro-5-oxo-, ethyl ester	EtOH	208(4.63),224s(4.32), 253(4.18),281(4.22), 300s(3.99),452(4.39)	142-0791-84
5H-Benzo[6,7]cyclohept[1,2-b]indoliz-ine-6-carboxylic acid, 5-oxo-, ethyl ester	EtOH	231(4.30),269(4.46), 289(4.28),317(4.08), 449(4.21)	142-0791-84
Ethanone, 2-nitro-1,2,2-triphenyl-	C_6H_{12}	252(4.14)	12-1217-84
Oxazole, 2-(2-furanylmethoxy)-4,5-diphenyl-	C_6H_{12}	224(4.52),298(4.15)	44-0399-84
2(3H)-Oxazolone, 3-(2,3-dihydro-2-methylene-3-furanyl)-4,5-diphenyl-	C_6H_{12}	217(?),267(4.02), 271(4.01)	44-0399-84
2(3H)-Oxazolone, 3-(2-furanylmethyl)-4,5-diphenyl-	C_6H_{12}	216(4.40),299(4.12)	44-0399-84
2(3H)-Oxazolone, 3-(2-methyl-3-furan-yl)-4,5-diphenyl-	C_6H_{12}	218(4.40),295(4.12)	44-0399-84
2(5H)-Oxazolone, 5-(2-furanylmethyl)-4,5-diphenyl-	C_6H_{12}	212(4.36),265(4.11)	44-0399-84

Compound	Solvent	λ_{max} (log ϵ)	Ref.
7H-Pyrrolo[3,2,1-ij]quinoline-2-carboxylic acid, 7-oxo-9-phenyl-, ethyl ester	EtOH	220(3.48),297(4.65), 323(4.29),335s(4.23), 510(3.84),536s(3.77)	142-2467-84
9H-Pyrrolo[3,2,1-ij]quinoline-2-carboxylic acid, 9-oxo-7-phenyl-, ethyl ester	EtOH	223(4.42),254(4.44), 278(4.25),288(4.26), 304s(3.81),372s(3.85), 409(4.03),428s(3.92)	142-2467-84
$C_{20}H_{15}NO_4$			
Avicine, dihydro-	EtOH	232(4.60),278(4.50), 322(4.33)	100-0001-84
Benzo[c]phenanthridine, 7,8-dimethoxy-2,3-(methylenedioxy)-	MeOH	243(4.58),256(4.55), 276(4.67),324(4.10)	73-1412-84
2-Butenedioic acid, 2-cyano-3-(9H-fluoren-9-yl)-, dimethyl ester (plus 10% isomer)	DMSO-NaOEt	460(4.16)	94-2910-84
3-Indolizinecarboxylic acid, 2-(2,3-dihydro-1,3-dioxo-1H-inden-2-yl)-, ethyl ester	EtOH	221(4.66),254(4.50), 296(4.39),338(4.20)	94-2910-84
Norchelerythrine	EtOH	211(4.30),243(4.57), 256(4.54),276(4.67), 287s(4.51),324(4.18), 338s(4.00),363s(3.56), 385(4.49)	100-0001-84
N-Nornitidine	EtOH	229(4.36),274(4.73), 278s(4.71),311(4.15), 330s(3.89),348(3.60), 367(3.46)	100-0001-84
Sanguinarine, dihydro-	EtOH	237(4.55),284(4.57), 322(4.21),335s(4.12), 350s(3.72)	100-0001-84
	EtOH-HCl	238(4.53),250s(4.49), 265(4.40),274(4.40), 307(4.20),321(4.36), 338(4.20),355(4.30)	100-0001-84
	EtOH-KOH	235(4.53),282(4.56), 322(4.23),335s(4.12), 350s(3.72)	100-0001-84
$C_{20}H_{15}NO_5$			
2-Butenoic acid, 4-[2-(3-cyano-3-phenyl-2-oxiranyl)phenoxy]-4-oxo-, methyl ester	EtOH	272s(3.65),282s(3.54)	24-2157-84
$C_{20}H_{15}NO_6$			
Turkiyenine	MeOH	218s(4.42),258(4.62), 281s(4.11),352(3.55)	35-6101-84
$C_{20}H_{15}N_3O$			
2H-1,4-Benzoxazin-2-one, 3-phenyl-, phenylhydrazone	EtOH	262(4.18),384(4.32)	4-0521-84
$C_{20}H_{15}N_3O_2$			
Benzaldehyde, 4-nitro-, (diphenylmethylene)hydrazone	C_6H_{12}	332(4.45)	97-0021-84
$C_{20}H_{15}N_3O_3$			
2H-Indazole-4,7-dione, 3-methoxy-2-phenyl-5-(phenylamino)-	dioxan	260s(4.39),280(4.46), 446(3.72)	4-0825-84
1H-Indazole-3,4,7(2H)-trione, 1-methyl-2-phenyl-5-(phenylamino)-	EtOH	253(4.81),400(4.41), 510s(3.79)	4-0825-84

Compound	Solvent	$\lambda_{max}(\log \epsilon)$	Ref.
$C_{20}H_{15}N_3O_3S$			
2H-1,2,4,6-Thiatriazinium, 5,6-dihydro-5-oxo-2,3,4-triphenyl-, hydroxide, inner salt, 1,1-dioxide	MeCN	230(3.943),266s(3.493)	142-0063-84
$C_{20}H_{15}N_3O_4$			
Cyclohepta[4,5]pyrrolo[1,2-a]imidazole-10-carboxylic acid, 3-nitro-2-phenyl-, ethyl ester	$CHCl_3$	334(4.56),352(4.49), 433(3.72),533(3.52), 576s(3.40),625s(3.04)	150-3465-84M
1H-Isoindole-1,3(2H)-dione, 2-[2-[(3-amino-1,4-dihydro-1,4-dioxo-2-naphthalenyl)amino]ethyl]-	CH_2Cl_2	230(4.46),238(4.37), 293(4.37),535(3.10)	83-0329-84
$C_{20}H_{15}N_3S_2$			
2H-1,3-Thiazine-2-acetonitrile, 5-cyano-3,6-dihydro-4-(3-methylphenyl)-2-phenyl-6-thioxo-	dioxan	263(4.10),323(3.60), 386(4.09)	4-1445-84
2H-1,3-Thiazine-2-acetonitrile, 5-cyano-3,6-dihydro-4-(4-methylphenyl)-2-phenyl-6-thioxo-	EtOH	269(4.16),324(3.55), 386(4.11)	4-1445-84
$C_{20}H_{15}N_5$			
Ethenetricarbonitrile, [4-[methyl(1-phenylethylidene)hydrazino]phenyl]-, (E)-	benzene	489(4.51)	32-0111-84
$C_{20}H_{15}N_5S$			
Formazan, 1-(2-benzothiazolyl)-3,5-diphenyl-	benzene	480(4.09)	135-0577-84
$C_{20}H_{16}$			
9,10[1',2']-Benzenoanthracene, 4a,9,9a,10-tetrahydro-	MeOH	260(3.41),266(3.51), 272(3.56),279(3.36), 291(3.06)	35-7310-84
9,10[1',4']-Benzenoanthracene, 9,10,11,14-tetrahydro-	MeOH	219(4.38),256(3.06), 268(3.03),276(3.33), 284.5(3.59)	35-7310-84
Benz[e]indene, 2,3-dihydro-1-(phenylmethylene)-, trans	n.s.g.	280(4.4),290(4.4), 310(4.5),325(4.5)	35-7624-84
Benzo[j]fluoranthene, 1,2,3,12c-tetrahydro-	EtOH	274(4.84)	44-1030-84
Biphenylene, 1,4,5,8-tetraethenyl-	C_6H_{12}	224(4.62),232(4.61), 254(4.69),262(4.69), 283s(4.68),295(4.76), 304s(4.69),384s(3.86), 407(3.9),431(3.89)	35-7195-84
$C_{20}H_{16}BrNO_4$			
Naphtho[2,3-d]-1,3-dioxol-5-amine, N-[(6-bromo-2,3-dimethoxyphenyl)methylene]-	MeOH	216(4.65),243(4.69)	73-1412-84
$C_{20}H_{16}Br_2N_2O_4$			
2,3-Pyrazinediol, 5,6-bis(4-bromophenyl)-2,3-dihydro-, diacetate, trans	EtOH	234(4.37),298(4.06)	4-0103-84
$C_{20}H_{16}ClNO_3$			
2H,5H-Pyrano[4,3-b]pyran-2-one, 3-chloro-4-(diphenylamino)-7,8-dihydro-	EtOH	249.5(4.15),278(4.16), 316(3.92),367(3.91)	4-1441-84

Compound	Solvent	$\lambda_{max}(\log \epsilon)$	Ref.
$C_{20}H_{16}Cl_2N_2O_4$			
2,3-Pyrazinediol, 5,6-bis(2-chlorophen-yl)-2,3-dihydro-, diacetate	EtOH	209(4.27),280(3.51)	4-0103-84
2,3-Pyrazinediol, 5,6-bis(4-chlorophen-yl)-2,3-dihydro-, diacetate	EtOH	230.5(4.34),297.5(4.00)	4-0103-84
$C_{20}H_{16}FeO_4$			
Ferrocene, 1,1'-[1,3-phenylenebis(meth-yleneoxycarbonyl)]-	dioxan	450(2.54)	18-2435-84
$C_{20}H_{16}N$			
Benzenemethanaminium, α-phenyl-N-(phen-ylmethylene)-, (OC-6-11)-hexachloro-antimonate	CH_2Cl_2	260(4.21),294(4.40)	24-3222-84
$C_{20}H_{16}NO_4$			
Corysamine	MeOH	208(4.43),231(4.50), 244s(4.43),268(4.44), 346(4.32),452(3.67)	73-0704-84
Decarine, N-methyl-	EtOH	223(--),256(--), 276(--),289(--), 325(--)	100-0001-84
Fagaridine	EtOH	228(4.55),284(4.65), 322s(4.14)	100-0001-84
$C_{20}H_{16}N_2$			
Benzenamine, N,N'-(1-phenyl-1,2-ethane-diylidene)bis-	MeOH	230(4.29),267(4.15)	118-0128-84
1H-Indol-1-amine, N,3-diphenyl-	EtOH	225(4.56),265(4.19), 280(4.17)	103-1119-84
$C_{20}H_{16}N_2O$			
Benzenamine, N-[1-phenyl-2-(phenyl-imino)ethylidene]-, N-oxide, (Z,Z)-	MeOH	262(4.29)	118-0128-84
$C_{20}H_{16}N_2O_2S_2$			
1H-Imidazo[1,2-c]thiazol-4-ium, 1-acet-yl-2-hydroxy-5-(methylthio)-3,7-di-phenyl-, hydroxide, inner salt	HOAc	350(3.95)	103-0029-84
$C_{20}H_{16}N_2O_3$			
Methanone, (5-acetoxy-6-methyl-2-phen-yl-4-pyrimidinyl)phenyl-	MeOH	260(4.49)	32-0431-84
$C_{20}H_{16}N_4$			
Cycloocta[2,1-b:3,4-b']di[1,8]naphthyr-idine, 6,7,8,9-tetrahydro-	EtOH	215(4.92),323(4.34)	44-2208-84
2(1H)-Quinoxalinone, 3-phenyl-, phenyl-hydrazone	EtOH	256(4.55),364(4.40), 448(4.12)	4-0521-84
$C_{20}H_{16}N_4O_2S$			
1H-Pyridazino[4,5-e][1,3,4]thiadiazin-5(6H)-one, 1-acetyl-3-phenyl-6-(phenylmethyl)-	EtOH	263(4.35),308(4.23)	94-4437-84
$C_{20}H_{16}N_4O_3$			
2H-1,2,4-Triazino[3,2-b]quinazoline-2,6(11H)-dione, 3-[2-(4-methoxy-phenyl)ethenyl]-11-methyl-	EtOH	330(2.38),346(2.19)	24-1077-84

Compound	Solvent	$\lambda_{max}(\log \epsilon)$	Ref.
$C_{20}H_{16}N_4O_3S$ 1,3,7,8-Tetraazaspiro[4.5]deca-6,9- diene-10-carboxylic acid, 4-oxo- 1,3-diphenyl-2-thioxo-, methyl ester	MeOH	235(4.24),280(4.18), 308(3.85)	39-2491-84C
$C_{20}H_{16}N_6O_4$ 1,3-Benzenediol, 2,4-bis(5-methoxy- 2H-benzotriazol-2-yl)-	CHCl₃	344(4.62)	126-2497-84
$C_{20}H_{16}N_6O_5$ 1,3,5-Benzenetriol, 2,4-bis(5-methoxy- 2H-benzotriazol-2-yl)-	CHCl₃	352(4.65)	126-2497-84
$C_{20}H_{16}O$ 9,10[1',2']-Benzenoanthracen-2(1H)-one, 4a,9,9a,10-tetrahydro-	MeOH	219(4.00),265(3.16), 272(3.12)	35-7310-84
2,5-Cyclohexadien-1-one, 4-(diphenyl- methylene)-2-methyl-	hexane?	210(4.24),260(4.07), 266(4.07),360(4.35)	73-1949-84
Furan, 3-[1-(9H-fluoren-9-ylidene)- ethyl]-2-methyl-	hexane	234(4.58),259(4.44), 320(4.04)	39-2877-84C
$C_{20}H_{16}O_2$ 2,5-Cyclohexadien-1-one, 4-(diphenyl- methylene)-2-methoxy-	hexane?	210(4.30),261(4.09), 269(4.08),371(4.30)	73-1949-84
$C_{20}H_{16}O_2S$ Methanone, 2,5-thiophenediylbis[(4- methylphenyl)-	C₆H₁₂	305(4.42)	44-1177-84
$C_{20}H_{16}O_3$ 2,5-Furandione, dihydro-3,4-bis(1-phen- ylethylidene)-, (E,E)-	toluene	340(3.78)	48-0233-84
(E,Z)-	toluene	332(3.98)	48-0233-84
(Z,Z)-	toluene	325(4.12)	48-0233-84
2,5-Furandione, 3-(diphenylmethylene)- dihydro-4-(1-methylethylidene)-	C₆H₁₂	352(4.00)	48-0233-84
$C_{20}H_{16}O_4$ 4H,8H-Benzo[1,2-b:5,4-b']dipyran-4-one, 3-hydroxy-8,8-dimethyl-2-phenyl-	MeOH	263(4.26),368(3.90)	2-0168-84
4H,8H-Benzo[1,2-b:5,6-b']dipyran-4-one, 3-hydroxy-8,8-dimethyl-2-phenyl-	MeOH	228(4.25),350(4.33)	2-0168-84
2H-1-Benzopyran-2-one, 7-methyl-, dimer, syn-head-head	CHCl₃	246(3.54),275(3.27), 283(3.25)	2-0502-84
$C_{20}H_{16}O_6$ 2-Anthracenecarboxylic acid, 9,10-di- hydro-4-hydroxy-9,10-dioxo-3-(3-oxo- butyl)-, methyl ester	MeOH	211(4.35),229(4.34), 249(4.49),256(4.50), 329(3,52),408(3.87)	78-3677-84
1H-Anthra[2,3-c]pyran-6,11-dione, 12- hydroxy-1,5-dimethoxy-3-methyl-	EtOH	268(4.47),299s(4.20), 520(3.99)	12-1511-84
Benzeneacetic acid, α-(3-hydroxy-5-oxo- 4-phenyl-2(5H)-furanylidene)-2-meth- oxy-, methyl ester, (E)-	EtOH EtOH-NaOH	273(4.28),370(4.10) 270(4.33),371(3.95)	39-1547-84C 39-1547-84C
2H-1-Benzopyran-2-one, 4-methoxy-, dimer, anti-head-tail	CHCl₃	244(3.54),258(3.12), 273(3.39),278(3.40)	2-0502-84
2H-1-Benzopyran-2-one, 7-methoxy-, dimer, syn-head-tail	CHCl₃	247(3.84),279(3.59), 287(3.55)	2-0502-84
Cyclohepta[de]naphthalene-7,8,10-triol, triacetate	MeOH	250(4.36),256.5(4.33), 322.5(3.75),370(4.11),	39-1465-84C

Compound	Solvent	$\lambda_{max}(\log \epsilon)$	Ref.
(cont.)		395(4.10)	39-1465-84C
Gadain	EtOH	235(4.12),293(4.05), 337(4.20)	102-2323-84
Konyanin	MeOH	231s(3.99),289(3.82)	78-1145-84
	MeOH-NaOH	239s(4.01),292(3.88)	78-1145-84
Savinin	MeOH	215(4.55),237(4.45), 293(4.35),334(4.48)	100-0331-84
$C_{20}H_{16}O_7$			
9,10-Anthracenedione, 1,4,5-trihydroxy-2-[(tetrahydro-2-methyl-5-oxo-3-furanyl)methyl]-, trans	EtOH	233(3.86),254(3.64), 290(3.32),462(3.27), 482(3.32),498(3.35), 513(3.30),528(3.28)	78-4579-84
5,12-Naphthacenedione, 9-acetyl-7,8,9,10-tetrahydro-6,7,9,11-tetrahydroxy-, (7S-cis)-	EtOH	207(4.35),227s(4.32), 252(4.69),257s(4.64), 287(4.07)	78-4657-84
$C_{20}H_{16}O_8$			
9H-Fluoren-9-one, 1,2,5-triacetoxy-7-methoxy- (dendroflorin triacetate)	EtOH	258(4.32),267(4.30), 327.5(3.21),424.5(3.26)	42-1010-84
2-Naphthacenecarboxylic acid, 1,2,3,4-6,11-hexahydro-2,4,5,12-tetrahydroxy-6,11-dioxo-, methyl ester, (S-cis)-	n.s.g.	251(4.59),288(4.00), 485(4.01)	78-4649-84
$C_{20}H_{16}O_{11}S_2$			
Pyrylium, 4-carboxy-2,6-bis(4-methoxy-3-sulfophenyl)-, hydroxide, inner salt	pH 4.8	280(4.34)	39-0867-84B
	pH 13.0	480(4.60)	39-0867-84B
	17.5M H_2SO_4	370(3.90),510(4.40)	39-0867-84B
$C_{20}H_{17}BrN_2$			
1H-Indol-1-amine, N-(4-bromophenyl)-2,3-dihydro-3-phenyl-	EtOH	252(4.41),292(3.73)	103-1119-84
$C_{20}H_{17}BrO_4$			
Propanoic acid, acid, 2,2-dimethyl-, 3-(bromomethyl)-9,10-dihydro-9,10-dioxo-1-anthracenyl ester	MeOH	212(4.42),256(4.66), 273s(--),333(3.77)	78-3677-84
$C_{20}H_{17}ClN_2$			
1H-Indol-1-amine, N-(4-chlorophenyl)-2,3-dihydro-3-phenyl-	EtOH	252(4.38),294(3.74)	103-1119-84
$C_{20}H_{17}FN_2$			
1H-Indol-1-amine, N-(4-fluorophenyl)-2,3-dihydro-3-phenyl-	EtOH	243(4.16),296(4.20)	103-1119-84
$C_{20}H_{17}F_3N_2O_4$			
Phenylalanine, α,β-didehydro-N-[N-(trifluoroacetyl)-L-phenylalanyl]-, (Z)-	EtOH	207(4.40),281(4.23)	5-0920-84
$C_{20}H_{17}F_3O_2$			
1,3-Butanedione, 4,4,4-trifluoro-1-phenyl-2-[(2,4,6-trimethylphenyl)-methylene]-	EtOH	248(3.54),307(3.37)	39-2863-84C
$C_{20}H_{17}IN_4O_2$			
Benzoxazolium, 3-ethyl-2-(4-methyl-6-phenyl-6H-pyrazolo[3,4-c]isoxazol-3-yl)-, iodide	EtOH	330(3.75)	48-0811-84

Compound	Solvent	$\lambda_{max}(\log \epsilon)$	Ref.
$C_{20}H_{17}N$			
Acridine, 9,10-dihydro-9-methyl-9-phenyl-	EtOH	292(4.24)	24-2703-84
5H-Indeno[1,2-b]pyridine, 5,5-dimethyl-2-phenyl-	EtOH	206(4.8),240(4.6), 270(4.18),324(4.52)	103-0519-84
$C_{20}H_{17}NO_2$			
5H-[1]Benzopyrano[2,3-b]pyridin-5-ol, 2,3-dimethyl-5-phenyl-	96% H_2SO_4	382(4.51)	103-0607-84
5H-[1]Benzopyrano[2,3-b]pyridin-5-ol, 2,7-dimethyl-5-phenyl-	96% H_2SO_4	384(4.54)	103-0607-84
Phenanthro[9,10-b]oxazole, 2-(4-pentenyloxy)-	C_6H_{12}	237(4.63),253(4.19), 264(4.67),297(4.08), 309(4.17)	44-0399-84
$C_{20}H_{17}NO_3$			
Furo[2,3-a]acridin-11(6H)-one, 4-methoxy-6-methyl-2-(1-methylethenyl)-	EtOH	217(4.31),249(4.47), 272s(4.43),288(4.57), 303(4.49),320s(4.06), 333s(3.99),379(3.83)	5-0031-84
$C_{20}H_{17}NO_4$			
6H-Anthra[1,9-cd]isoxazole-3-carboxylic acid, 6-oxo-, 3-methylbutyl ester	EtOH	461(4.08)	104-2012-84
Benzo[c]phenanthridin-2-ol, 3,8,9-trimethoxy-	EtOH	231(4.76),253(4.76), 276(4.96),285(4.96), 315(4.63)	100-0453-84
[2]Benzopyrano[4,3-b]indole-3-carboxylic acid, 5,11-dihydro-2,4-dimethyl-5-oxo-, ethyl ester	EtOH	241.5(4.60),323(4.38), 378(4.11)	111-0223-84
2-Butenoic acid, 4-[2-(3-cyano-trans-3-phenyloxiranyl)phenoxy]-, methyl ester, (2E)-	EtOH	262s(3.46),274s(3.59), 280(3.61)	24-2157-84
Fagaronine, N-demethyl-	n.s.g.	227(4.30),272(4.67), 280(4.67),315s(4.04)	100-0001-84
4H-Furo[3,2-c][1]benzopyran-3-carboxylic acid, 2-cyano-2,3,3a,9b-tetrahydro-2-phenyl-, methyl ester, ($2\alpha,3\beta,3a\alpha,9b\beta$)-	EtOH	257s(2.93),268s(3.26), 276(3.37),283(3.32)	24-2157-84
Noraporphine-6-carboxylic acid, 1,2-(methylenedioxy)-6a,7-dehydro-, ethyl ester	MeOH	227(4.15),255(4.74), 287(3.96),318(4.03), 329(4.04),357(3.51), 376(3.51)	39-1273-84C
1H-Pyrrole-3-carboxylic acid, 4,5-dihydro-4,5-dioxo-2-phenyl-1-(phenylmethyl)-, ethyl ester	dioxan	285s(3.62),405(3.53)	94-0497-84
$C_{20}H_{17}NO_4S$			
Pyridinium, 1-(ethoxycarbonyl)-2-(1,3-dihydro-1,3-dioxo-2H-inden-2-ylidene)-2-(methylthio)ethylide	EtOH	251(4.46),310(4.05), 364(4.03)	94-2910-84
$C_{20}H_{17}NO_5$			
Benzo[g]-1,3-benzodioxolo[5,6-a]quinolizinium, 5,6-dihydro-13-hydroxy-9,10-dimethoxy-, hydroxide, inner salt (berberinophenolbetaine)	MeOH	236(4.48),258s(4.24), 312(4.05),365(4.02), 444(4.16)	94-2230-84
8H-Benzo[g]-1,3-benzodioxolo[5,6-a]quinolizin-8-one, 5,6-dihydro-9,10-dimethoxy-	MeOH	226(4.65),255s(4.23), 280(4.02),337(4.46), 367(4.19),389(4.00)	78-3957-84
Grandiflorine	EtOH	238s(3.79),297(3.78)	23-0258-84

Compound	Solvent	$\lambda_{max}(\log \epsilon)$	Ref.
$C_{20}H_{17}NO_6$			
9,10-Anthracenedione, 1,4-diacetoxy-2-(dimethylamino)-	EtOH at 77°K	463(3.81)	104-1780-84
Chilenine, 13-deoxy-	MeOH	213s(4.20),291(3.44), 308(3.92)	78-3957-84
Spiro[1,3-dioxolo[4,5-g]isoquinoline-5(6H),2'-[2H]indene]-1',3'-dione, 7,8-dihydro-4',5'-dimethoxy-, (±)-	MeOH	247.5(4.63),291.5(4.13), 334(3.73)	94-2230-84
Turkiyenine, dihydro-, (-)-	MeOH	223(4.47),288(3.92), 309s(3.83),319(3.76)	35-6101-84
(+)-epi-	MeOH	226(4.44),286(3.86), 313(3.79)	35-6101-84
$C_{20}H_{17}N_3$			
1H-1,5-Benzodiazepine, 2,3-dihydro-2-phenyl-4-(3-pyridinyl)-	EtOH	250(4.32),391(3.68)	103-1370-84
$C_{20}H_{17}N_3O_2S_2$			
Sulfoximine, S,S-dimethyl-N-[5-oxo-2-[(phenylmethyl)thio]-5H-indeno-[1,2-d]pyrimidin-4-yl]-	EtOH	247(4.57),297(4.70)	94-2910-84
$C_{20}H_{17}N_4O_2$			
Benzoxazolium, 3-ethyl-2-(4-methyl-6-phenyl-6H-pyrazolo[3,4-c]isoxazol-3-yl)-, iodide	EtOH	330(3.75)	48-0811-84
$C_{20}H_{17}N_5O_2$			
Benzamide, N-[8,9-dihydro-7-methyl-8-oxo-9-(phenylmethyl)-7H-purin-6-yl]-	MeOH	231(4.33),291(4.15)	150-3601-84M
$C_{20}H_{17}N_5O_3$			
1,3,7,8-Tetraazaspiro[4.5]deca-1,6,9-triene-10-carboxylic acid, 4-oxo-3-phenyl-2-(phenylamino)-, methyl ester	MeOH	310(3.84)	39-2491-84C
1,3,7,8-Tetraazaspiro[4.5]deca-2,6,9-triene-10-carboxylic acid, 4-oxo-1-phenyl-2-(phenylamino)-, methyl ester	MeOH	245(4.38),315(3.76)	39-2491-84C
$C_{20}H_{18}$			
Benzo[j]fluoranthene, 1,2,3,6b,7,8-hexahydro-	EtOH	305(4.37),314(4.45)	44-1030-84
1,3,5-Cycloheptatriene, 7,7'-(2,4-hexadiyne-1,6-diyl)bis-	MeCN	256(3.75)	24-2045-84
Spiro[3-cyclopentene-1,9'-[9H]fluorene], 2-cyclopropyl-	hexane	225(4.37),232.5(4.18), 260s(4.27),267(4.31), 270(4.30),278s(4.20), 295(3.86),306(4.02)	44-0495-84
Spiro[cyclopropane-1,9'-[9H]fluorene], 2-(2-cyclopropylethenyl)-, cis	hexane	207(5.11),230s(4.35), 239s(4.21),256s(4.24), 260(4.25),270(4.25), 293(3.96),304(3.98)	44-0495-84
trans	hexane	212(4.81),231s(4.33), 260(4.26),265s(4.25), 270(4.25),293(3.94), 305(3.97)	44-0495-84
$C_{20}H_{18}BrNO_4$			
Naphtho[2,3-d]-1,3-dioxol-5-amine, N-[6-bromo-2,3-dimethoxyphenyl)methyl]-	MeOH	241(4.43),264(4.48)	73-1412-84

Compound	Solvent	$\lambda_{max}(\log \epsilon)$	Ref.
$C_{20}H_{18}ClNO_3$ 3H-Indol-3-one, 1-acetyl-2-[1-(4-chlorophenyl)-3-oxobutyl]-1,2-dihydro-	EtOH	206(4.38),224(4.50), 241(4.55),260(4.18), 333(3.57),351(3.54)	103-1374-84
$C_{20}H_{18}Cl_2O_6$ 9,10-Anthracenedione, 2-(4,4-dichloro-3-hydroxy-3-methylbutyl)-1,4-dihydroxy-8-methoxy-, (±)-	MeOH	231(4.59),249(4.38), 284(4.06),477(4.06), 494(4.05),527(3.79)	78-4609-84
$C_{20}H_{18}NO_2P$ Benzamide, N-[(diphenylphosphinyl)-methyl]-	EtOH	265(3.35),272(3.19)	42-0430-84
$C_{20}H_{18}N_2$ 1H-Indol-1-amine, 2,3-dihydro-N,3-diphenyl-	EtOH	238(4.23),288(4.27)	103-1119-84
1H-Pyrazole, 1-methyl-3,5-bis(2-phenylethenyl)-	EtOH	310(4.68)	103-1266-84
9H-Pyrido[3,4-b]indole, 1-(3-propylphenyl)- (komaroine)	EtOH	215(4.44),235(4.42), 280(4.12),290(4.12), 350(3.81)	105-0378-84
	EtOH-acid	216(--),255(--), 265(--),310-318(--), 382(--)	105-0378-84
Quino[3,2-b]acridine, 1,2,3,4,7,12-hexahydro-	MeCN	227(4.741),275(4.403), 299s(4.267),372(3.992)	5-0133-84
$C_{20}H_{18}N_2O$ Phenol, 4,5-dimethyl-4-(9-methyl-2H-benz[g]indazol-2-yl)-	EtOH	242(4.51),249(4.50), 282(3.98),294(3.92), 303s(--),318(3.77), 332(3.79)	33-0113-84
Yohimban-19-one, 17,18,20,21-tetradehydro-14-methylene-, (3β)-	MeOH	220(4.50),275(3.78), 287(3.70),367(3.98)	78-4843-84
$C_{20}H_{18}N_2OS$ 1H-Pyrazole, 4,5-dihydro-3-(4-methoxyphenyl)-1-phenyl-5-(thienyl)-	EtOH	258(4.27),346(3.97)	34-0225-84
1H-Pyrazole, 4,5-dihydro-5-(4-methoxyphenyl)-1-phenyl-3-(thienyl)-	EtOH	226(4.23),253(4.33), 366(4.32)	34-0225-84
$C_{20}H_{18}N_2O_2S_2$ 1H-Pyrrole-3-carbodithioic acid, 2,5-dihydro-1-(4-methylphenyl)-4-[(4-methylphenyl)amino]-2,5-dioxo-, methyl ester	CHCl₃	304(4.37),435(3.97)	48-0401-84
$C_{20}H_{18}N_2O_2S_3$ 2H-2a,3-Dithia(2a-S^{IV})-1,2-diazacyclopenta[cd]pentalene, 5,6-dihydro-2-(4-methoxyphenyl)-4-[(4-methoxyphenyl)thio]-	MeCN	559(4.43)	97-0251-84
$C_{20}H_{18}N_2O_4$ Indolo[4,5-e]indole-2,9-dicarboxylic acid, 3,8-dihydro-, diethyl ester	EtOH	202(4.07),236(4.20), 280s(3.98),293s(4.22), 303(4.36),322(3.87), 335s(3.70),353(3.77), 370(3.82)	103-0283-84

Compound	Solvent	$\lambda_{max}(\log \epsilon)$	Ref.
Indolo[5,4-e]indole-2,7-dicarboxylic acid, 3,8-dihydro-, diethyl ester	EtOH	253(4.58),262s(4.42), 270(4.32),279(4.31), 288(4.26),338(4.36), 354(4.39),372(4.39)	103-0283-84
$C_{20}H_{18}N_2S_3$			
2H-2a,3-Dithia(2a-S^{IV})-1,2-diazacyclopenta[cd]pentalene, 5,6-dihydro-2-(4-methylphenyl)-4-[(4-methylphenyl)-thio]-	MeCN	557(4.48)	97-0251-84
$C_{20}H_{18}N_3O_2$			
Benzoxazolium, 2-[(1,5-dihydro-3-methyl-5-oxo-1-phenyl-4H-pyrazol-4-ylidene)methyl]-3-ethyl-, iodide	EtOH	410(3.94)	48-0811-84
$C_{20}H_{18}N_4O_2$			
3H-Pyrazolo[5,1-c]-1,2,4-triazole-7-carboxylic acid, 6-methyl-3,3-diphenyl-, ethyl ester	CH_2Cl_2	293(3.95)	24-1726-84
7H-Pyrido[2,1-b]quinazoline-6,11-dione, 8,9-dihydro-, 6-[(4-acetylphenyl)hydrazone]	EtOH	227(4.37),300(4.24), 392(4.53)	4-1301-84
$C_{20}H_{18}N_4O_3$			
7H-Pyrido[2,1-b]quinazoline-2-carboxylic acid, 6,8,9,11-tetrahydro-11-oxo-6-(phenylhydrazono)-, methyl ester	EtOH	215(4.58),248(4.25), 290(4.67),300(4.05), 408(4.46)	4-1301-84
$C_{20}H_{18}N_4O_8$			
4,9:5,8-Diepoxy-1H-naphtho[2,3-d]triazole-4a,8a-dicarboxylic acid, 3a,4-5,8,9,9a-hexahydro-1-(4-nitrophenyl)-, dimethyl ester	MeOH	357(3.35)	73-1990-84
$C_{20}H_{18}N_5O$			
7H-Purinium, 6-(benzoylamino)-7-methyl-9-(phenylmethyl)-, iodide	H_2O	226(4.42),280(4.16), 320s(3.48)	150-3601-84M
$C_{20}H_{18}N_6S$			
Propanenitrile, 3,3'-[[4-(2-benzothiazolylazo)-3-methylphenyl]imino]bis-	acetone 50% acetone	483(4.56) 498(4.58)	7-0819-84 7-0819-84
$C_{20}H_{18}O$			
Benzenemethanol, 2-methyl-α,α-diphenyl-	EtOH	249(2.85),254(2.92), 259(2.97),266(2.91)	104-1921-84
	60% H_2SO_4	432(4.53)	104-1921-84
$C_{20}H_{18}O_2$			
Benzenemethanol, 2-methoxy-α,α-diphenyl-	EtOH	254s(2.92),259s(3.13), 266s(3.26),271(3.34), 278(3.30)	104-1921-84
	60% H_2SO_4	405s(4.40),432(4.50), 554(3.87)	104-1921-84
9,10[1',2']-Benzenoanthracene-1,2-diol, 1,2,4a,9,9a,10-hexahydro-	MeOH	251(2.62),259(2.76), 265(2.96),272(3.08)	35-7310-84
1(2H)-Dibenzofuranone, 3,4-dihydro-3,3-dimethyl-7-phenyl-	EtOH	245(4.29),273(4.42)	39-1213-84C

Compound	Solvent	$\lambda_{max}(\log \epsilon)$	Ref.
$C_{20}H_{18}O_3$			
1,4-Phenanthrenedione, 3-hydroxy-5,7,8-trimethyl-2-(2-propenyl)- (plectranthone A)	EtOH	222.5(4.62),274(4.33), 299(4.35),346(3.80), 386s(3.59),391s(3.55), 396s(3.51),510s(2.89)	33-1003-84
	EtOH-NaOH	276(4.33),299(4.37), 350(3.79),385(3.66), 391(3.63),400s(3.56), 520s(2.87)	33-1003-84
	ether	228s(4.54),236.5(4.57), 282s(4.17),289s(3.29), 300(4.40),341(3.64), 381(3.46),386(3.45), 450s(3.11)	33-1003-84
$C_{20}H_{18}O_4$			
9,10-Anthracenedione, 1-hydroxy-3-methyl-2-(3-oxopentyl)-	MeOH	208(4.20),246(4.46), 263(4.48),326(3.52), 410(3.83)	78-3677-84
Hexacyclo[7.5.1.0^{3,13}.0^{5,12}.0^{7,11}0^{10,14}]- pentadecane-2,6-dione, 4-(2,5-dioxo-cyclopentylidene)-	CH$_2$Cl$_2$	286(4.09),310(3.94), 415(2.35)	23-2612-84
Propanoic acid, 2,2-dimethyl-, 9,10-di-hydro-3-methyl-9,10-dioxo-1-anthra-cenyl ester	MeOH	205(4.44),223(4.44), 236s(--),243(4.56), 253(4.57),258(4.56), 280s(--),325(3.66), 403(3.93)	78-3677-84
2-Propenoic acid, 3-phenyl-, 1,2-eth-anediyl ester	MeCN	274(4.65)	40-0022-84
$C_{20}H_{18}O_5$			
9,10-Anthracenedione, 1-hydroxy-3-(hy-droxymethyl)-2-(3-oxopentyl)-	MeOH	206(4.22),246(4.48), 257(4.47),410(3.85)	78-3677-84
Butanedioic acid, 9H-fluoren-9-ylidene-methoxy-, dimethyl ester	CHCl$_3$	253(4.51),262(4.68), 288(4.13),302(4.07), 312(4.08),366s(--)	24-2409-84
2-Butenedioic acid, 2-(9H-fluoren-9-yl)-3-methoxy-, dimethyl ester	CHCl$_3$	263(4.35),291(3.82), 302.5(3.68)	24-2409-84
5,12-Naphthacenedione, 8-ethyl-7,8,9,10-tetrahydro-1,6,8-tri-hydroxy-, (±)-	CHCl$_3$	259(4.40),266(4.40), 284(4.00),294(4.00), 432(4.00)	78-4633-84
5,12-Naphthacenedione, 7,8,9,10-tetra-hydro-6,11-dihydroxy-1-methoxy-8-methyl-, (±)-	MeOH	219(4.36),234(4.41), 252(4.50),287(3.94), 366(3.50),470(4.04), 492(4.10),528(3.92)	78-4609-84
$C_{20}H_{18}O_6$			
2-Anthracenepropanoic acid, 9,10-dihy-dro-1,8-dimethoxy-9,10-dioxo-, methyl ester	n.s.g.	223(4.57),252(4.46), 270(4.21),324(3.70), 396(3.71),414(3.70), 430(3.60)	5-0306-84
5H-Benz[e]anthracene-1,4,5,9(13bH)-tetrone, 6,7,8,8a-tetrahydro-10,13-dihydroxy-2,12-dimethyl-, (4aS*,8aα-13bα)-(±)-	MeCN	236(4.29),255s(3.85), 363(3.74)	18-0791-84
2,5-Cyclohexadiene-1,4-dione, 2,2'-(1,6-dioxo-1,6-hexanediyl)bis[5-methyl-	MeCN	250(4.55),315s(3.09), 435(1.88)	18-0791-84
5,12-Naphthacenedione, 7,8,9,10-tetra-hydro-6,8,11-trihydroxy-1-methoxy-8-methyl-, (±)-	MeOH	216(4.30),234(4.41), 252(4.43),289(3.86), 469(3.99),481(3.99),	78-4609-84

Compound	Solvent	$\lambda_{max}(\log \epsilon)$	Ref.
(cont.) 6,11-Naphthacenedione, 1,2,3,4,6,11-hexahydro-1,3,5,12-tetrahydroxy-3-ethyl-, (S-cis)-	CHCl$_3$	499(4.03),529(3.82) 253(4.60),289(4.01), 328s(3.43),460s(3.97), 483(4.03),517(3.81)	78-4609-84 78-4649-84
$(C_{20}H_{18}O_6S_2)_n$ Poly[1,2-bis[[[(4-methylphenyl)sulfonyl]oxy]methyl]-1-buten-3-yne-1,4-diyl]	CHCl$_3$	465(4.2)	126-1727-84
$C_{20}H_{18}O_7$ 5-Benzofuranol, 2-(2-acetoxy-4-methoxyphenyl)-6-methoxy-, acetate	MeOH	280(4.16),289s(4.19), 302s(4.35),315(4.54), 329(4.46)	94-3267-84
5,12-Naphthacenedione, 7,8,9,10-tetrahydro-6,8,10,11-tetrahydroxy-1-methoxy-8-methyl-, cis-(±)- trans isomer has same spectrum	MeOH	218(4.25),234(4.42), 250(4.35),288(3.81), 482(3.96),494(3.99), 529(3.79)	78-4609-84
$C_{20}H_{18}O_8$ 4H-1-Benzopyran-4-one, 5,6,7-trimethoxy-2-(7-methoxy-1,3-benzodioxol-5-yl)-	MeOH	239(4.39),270s(--), 332(4.37)	94-0166-84
1H,4H-Furo[3,4-c]furan-1,4-dione, tetrahydro-3,6-bis(4-hydroxy-3-methoxyphenyl)-, exo	EtOH	234(4.19),280(3.92)	39-1159-84C
5,12-Naphthacenedione, 7,8,9,10-tetrahydro-6,7,8,10,11-pentahydroxy-1-methoxy-8-methyl-, (7α,8α,10β)-(±)-	MeOH	207(4.21),232(4.58), 249(4.35),286(3.91), 477(4.05),493(4.06), 529(3.81)	78-4609-84
$C_{20}H_{19}As_2N_7O_9$ L-Histidine, N$^\alpha$-acetylbis(arsanilazo)-, calcium salt	pH 6 pH 11	419(4.50) 498(4.61)	69-0589-84 69-0589-84
$C_{20}H_{19}BrN_2O_5$ Phenol, 6-[5-(2-bromo-3,5-dimethoxyphenyl)-4-pyrimidinyl]-2,3-dimethoxy-	CHCl$_3$	246(4.25),307(4.18), 330s(4.15)	142-1217-84
$C_{20}H_{19}ClN_4O_3$ 1-Oxa-2-azaspiro[4.5]dec-2-ene, 4-[(2-chlorophenyl)azo]-4-nitro-3-phenyl-	EtOH	383(4.48)	104-1844-84
$C_{20}H_{19}ClN_4O_7S$ 9H-Purine, 6-chloro-2-(2-thienyl)-9-(2,3,5-tri-O-acetyl-β-D-ribofuranosyl)-	EtOH	245(3.96),270(3.81), 316(4.15)	44-4340-84
$C_{20}H_{19}ClO_4$ 2-Propenoic acid, 3-phenyl-, 2-[4-(3-chloro-1-oxopropyl)phenoxy]ethyl ester	n.s.g.	280(4.58),323(2.48)	73-2635-84
$C_{20}H_{19}NO$ 4(1H)-Phenanthridinone, 2,3,5,6-tetrahydro-8-methyl-6-phenyl-	ether	240(4.10),280(3.53)	142-0097-84
$C_{20}H_{19}NO_2$ 9,10-Anthracenedione, 2-(piperidinomethyl)-	EtOH	256(4.63),327(3.66)	42-0611-84
4H-Furo[3,2-e][1]benzoxocin-2-carbo-	EtOH	244s(2.59),250s(2.79),	24-2157-84

Compound	Solvent	$\lambda_{max}(\log \epsilon)$	Ref.
nitrile, 2,3,3a,5,6,11b-hexahydro-2-phenyl-, (2α,3aα,11bβ)- (cont.)		258(2.95),264(3.04), 271(2.95)	24-2157-84
Indeno[1,2-b]pyrrol-3a(3H)-ol, 2,4-di-hydro-2,2-dimethyl-, benzoate	EtOH	237(4.37),285(3.61), 295(3.54)	12-0577-84
Oxazole, 2-[(3-methyl-2-butenyl)oxy]-4,5-diphenyl-	C_6H_{12}	215(4.48),284(4.15)	44-0399-84
Oxazole, 2-[(4-pentenyl)oxy]-4,5-di-phenyl-	C_6H_{12}	224(4.33),301(4.13)	44-0399-84
2(3H)-Oxazolone, 3-(1,1-dimethyl-2-propenyl)-4,5-diphenyl-	C_6H_{12}	213(4.20),287(4.23), 297(4.15)	44-0399-84
2-Oxiranecarbonitrile, 3-[2-(4-penten-yloxy)phenyl]-, trans	EtOH	231s(4.09),274s(3.59), 282(3.64)	24-2157-84
$C_{20}H_{19}NO_3$			
1H,3H-Benz[f]isoquino[8,1,2-hij][3,1]-benzoxazine, 5,6-dihydro-8,9-dimeth-oxy-	MeOH	255(4.59),263(4.59), 327(4.06),380(3.40)	39-1273-84C
Guadiscine	EtOH	232s(3.95),265(4.33), 310(3.97),316s(3.84), 355s(3.66)	100-0353-84
	EtOH-acid	274(4.38),364(3.88), 408(3.62)	100-0353-84
$C_{20}H_{19}NO_4$			
1,3-Dioxolo[4,5-g]isoquinoline-6(5H)-carboxylic acid, 7,8-dihydro-5-(phen-ylmethylene)-, ethyl ester	MeOH	228(4.35),300(4.24), 328(4.29)	39-1273-84C
Heptazolicine acetate	EtOH	245(4.52),255(4.30), 280(4.48),300(4.58)	102-2409-84
$C_{20}H_{19}NO_5$			
Chelidonine, (±)- (diphylline)	MeOH	238(4.0),290(3.0)	100-0001-84
Chelidonine, (+)- (stylophorine)	EtOH	206(4.85),238(3.96), 289(3.90)	100-0001-84
	MeOH	210(4.66),238(4.00), 288(3.91)	73-0704-84
	MeOH	208(4.69),239(3.88), 289(3.89)	83-0223-84
	MeOH-HCl	208(4.73),242(3.91), 290(3.91)	83-0223-84
Gouregine	EtOH	229(4.28),247(4.25), 291(3.50),348(3.45)	100-0753-84
	EtOH-acid	231(4.22),274(4.27), 304(3.35),404(3.35)	100-0753-84
	EtOH-base	260(4.47),307(4.15), 379(3.98)	100-0753-84
Guacoline	EtOH	222(4.32),266(4.26), 324(3.85),353s(3.63)	100-0353-84
	EtOH-acid	225(4.25),273(4.16), 372(3.55)	100-0353-84
Protopine	MeOH	209(4.64),240(3.95), 287(3.90)	73-0704-84
	MeOH	209(4.71),240(3.96), 288(3.90)	73-1318-84
7H-Pyrano[2,3-c]acridin-7-one, 3,12-dihydro-6,11-dihydroxy-10-methoxy-3,3,12-trimethyl-	EtOH	205(4.26),268(4.45), 277(4.42),340(3.87), 390(3.84)	100-0325-84
$C_{20}H_{19}NO_6$			
Chelamine	n.s.g.	287(3.8)	100-0001-84
16,17-Dioxa-10-azatricyclo[11.2.1.1^{4,7}]-	EtOH	217(4.28),249(4.40),	77-0423-84

Compound	Solvent	$\lambda_{max}(\log \epsilon)$	Ref.
heptadeca-2,4,6,8,11,13,15-heptaene-9,11-dicarboxylic acid, diethyl ester (cont.)		292(4.88),307s(4.62), 370(3.90),566(2.53)	77-0423-84
Linaresinone, dihydro-N-methyl-	MeOH	235(4.09),295(4.11), 337(3.36)	35-6099-84 +100-0753-84
	MeOH-acid	238(4.09),253(4.10), 309(4.08),369(3.72)	35-6099-84
Spiro[1,3-dioxolo[4,5-g]isoquinoline-5(6H),2'-[2H]inden]-1'(3'H)-one, 7,8-dihydro-3'-hydroxy-6',7'-dimethoxy-, trans-(±)-	MeOH	224(4.47),259.5(4.05), 294(3.74),331.5(3.57)	94-2230-84

$C_{20}H_{19}N_2O_3$

Compound	Solvent	$\lambda_{max}(\log \epsilon)$	Ref.
1H-Pyrazolium, 2-ethyl-4,5-dihydro-3-[2-(4-methoxyphenyl)ethenyl]-4,5-dioxo-1-phenyl-, iodide	EtOH	380(4.30)	48-0695-84

$C_{20}H_{19}N_3O_3$

Compound	Solvent	$\lambda_{max}(\log \epsilon)$	Ref.
Acetamide, N-[3-[2-[4-(dimethylamino)-phenyl]ethenyl]-2-oxo-2H-1,4-benzoxazin-7-yl]-	EtOH	483(4.23)	4-0551-84

$C_{20}H_{19}N_3O_5$

Compound	Solvent	$\lambda_{max}(\log \epsilon)$	Ref.
3,4-Oxepindicarboxylic acid, 5-cyano-6,7-dihydro-6,7-dimethyl-2-[(phenylcarbonimidoyl)amino]-, dimethyl ester	CHCl$_3$	262(3.84),267(3.81), 330(4.30)	24-0585-84
3,4-Oxepindicarboxylic acid, 5-cyano-6,7-dihydro-7,7-dimethyl-2-[(phenylcarbonimidoyl)amino]-, dimethyl ester	CHCl$_3$	262(4.08),267(4.07), 323(4.22)	24-0585-84

$C_{20}H_{19}N_5O$

Compound	Solvent	$\lambda_{max}(\log \epsilon)$	Ref.
Benzamide, N-[8,9-dihydro-7-methyl-9-(phenylmethyl)-7H-purin-6-yl]-	MeOH	320(4.03)	150-3640-84M

$C_{20}H_{19}N_5O_2$

Compound	Solvent	$\lambda_{max}(\log \epsilon)$	Ref.
Benzamide, N-[5-(formylmethylamino)-6-[(phenylmethyl)amino]-4-pyrimidinyl]-	MeOH	247(4.48),285s(3.79)	150-3601-84M

$C_{20}H_{19}N_5O_5$

Compound	Solvent	$\lambda_{max}(\log \epsilon)$	Ref.
1-Oxa-2-azaspiro[4.5]dec-2-ene, 4-nitro-4-[(4-nitrophenyl)azo]-3-phenyl-	EtOH	370(4.41)	104-1844-84

$C_{20}H_{19}O_2Se$

Compound	Solvent	$\lambda_{max}(\log \epsilon)$	Ref.
Selenonium, bis(4-methoxyphenyl)phenyl]-, hexafluoroantimonate	MeOH	250(4.35)	11-0098-84A
hexafluorophosphate	MeOH	253(4.45)	11-0098-84A

$C_{20}H_{20}$

Compound	Solvent	$\lambda_{max}(\log \epsilon)$	Ref.
3,4-Benzotricyclo[4.3.0.0^{2,4}]nonane, 9-methyl-1-phenyl-	EtOH	228(4.24)	44-1353-84
Bicyclo[2.2.1]hept-2-ene, 2-methyl-1,3-diphenyl-	EtOH	260(3.91)	44-1353-84
Bicyclo[4.1.0]hept-2-ene, 3-methyl-1,2-diphenyl-	EtOH	290(--)(end abs.)	44-1353-84
Bicyclo[4.1.0]hept-3-ene, 1-methyl-6,7-diphenyl-	EtOH	226(4.22)	44-1353-84
1-Butene, 4-(1-methyl-2,3-diphenyl-2-cyclopropen-1-yl)-	EtOH	230(4.22),238(4.12), 320(4.47),338(4.35)	44-1353-84

Compound	Solvent	$\lambda_{max}(\log \epsilon)$	Ref.
1-Butene, 4-(2-methyl-1,3-diphenyl-2-cyclopropen-1-yl)-	EtOH	264(4.20)	44-1353-84
1H-Indene, 1-(3-butenyl)-2-methyl-3-phenyl-	EtOH	223(4.29),259(3.96)	44-1353-84
1H-Indene, 3-(3-butenyl)-2-methyl-1-phenyl-	EtOH	262(4.04)	44-1353-84
1H-Indene, 1-(2,3-dihydro-3-methyl-1H-inden-1-ylidene)-2,3-dihydro-3-methyl-	isopentane	<u>215(4.5),238f(4.1),</u> <u>315(4.6),340(4.6)</u>	77-1696-84
Naphthalene, 1-(3,4-dihydro-1(2H)-naphthalenylidene)-1,2,3,4-tetrahydro-, (E)-	EtOH-CH$_2$Cl$_2$	285(4.14)	24-2300-84
(Z)-	EtOH-CH$_2$Cl$_2$	231(3.45),295(3.31)	24-2300-84
Tricyclo[3.2.0.0^{2,7}]heptane, 2-methyl-1,7-diphenyl-	EtOH	235(4.15)	44-1353-84
Tricyclo[3.2.0.0^{2,7}]heptane, 7-methyl-1,2-diphenyl-	EtOH	234(4.13)	44-1353-84

$C_{20}H_{20}BrNO_3$

Compound	Solvent	$\lambda_{max}(\log \epsilon)$	Ref.
2H-Pyrrole, 3-bromo-4,5-bis(4-methoxyphenyl)-2,2-dimethyl-, 1-oxide	EtOH	232(4.26),268(4.40)	12-0095-84

$C_{20}H_{20}ClNOS$

Compound	Solvent	$\lambda_{max}(\log \epsilon)$	Ref.
1-Piperidineethanol, 4-(2-chloro-9H-thioxanthen-9-ylidene)-	MeOH	257s(3.81),273(4.10)	73-2295-84

$C_{20}H_{20}Cl_6Fe_4O$

Compound	Solvent	$\lambda_{max}(\log \epsilon)$	Ref.
Diferrice nium μ-oxo-bis(trichloroferrate)	MeCN	250(4.34),288(4.19), 320s(3.97),350s(3.69), 378s(3.51),470s(2.60), 520(2.54),562(2.51), 617(2.63)	119-0187-84

$C_{20}H_{20}CoN_2O_8S_2$

Compound	Solvent	$\lambda_{max}(\log \epsilon)$	Ref.
Cobalt, bis(5-ethyl-4-hydroxy-7-methyl-thieno[2,3-c]pyridin-3(2H)-one 1,1-dioxidato-O^3,O^4)-	EtOH	205(4.39),235(4.24), 270(3.82),420(3.92)	104-2032-84

$C_{20}H_{20}NO_4$

Compound	Solvent	$\lambda_{max}(\log \epsilon)$	Ref.
β-Stylopinium, N-methyl-, iodide, (-)-	MeOH	209(4.46),244s(3.81), 287(3.78)	73-0704-84

$C_{20}H_{20}N_2$

Compound	Solvent	$\lambda_{max}(\log \epsilon)$	Ref.
Quino[3,2-b]acridine, 1,2,3,4,8,9,10,11-octahydro-	MeCN	218(4.199),258(5.240), 350(4.023),361(4.005), 368(4.197)	5-0133-84

$C_{20}H_{20}N_2O$

Compound	Solvent	$\lambda_{max}(\log \epsilon)$	Ref.
Phenol, 4-(2,8-dimethyl-1-naphthalenyl-azo)-3,5-dimethyl-	EtOH	326(4.25)	33-0113-84

$C_{20}H_{20}N_2O_3$

Compound	Solvent	$\lambda_{max}(\log \epsilon)$	Ref.
Carbamic acid, [[3,4-dihydro-1-(phenyl-methylene)-2(1H)-isoquinolinyl]carbonyl]-, ethyl ester, (Z)-	MeOH	231(4.19),250(4.08), 302(4.24)	78-4003-84
Naphtho[1',2':6,7]cyclohepta[1,2-c]-pyrazole, 2-acetyl-2,10,11,12-tetra-hydro-5,7-dimethoxy-	EtOH	252(4.59),262(4.59), 315(4.12)	2-0736-84

$C_{20}H_{20}N_2O_4$

Compound	Solvent	$\lambda_{max}(\log \epsilon)$	Ref.
Benzo[b]phenazine-8-carboxylic acid,	EtOH	270.5(4.96),376(3.97)	44-5116-84

Compound	Solvent	$\lambda_{max}(\log \epsilon)$	Ref.
7,8,9,10-tetrahydro-1,6-dimethoxy-, methyl ester (cont.)			44-5116-84
Benzo[b]phenazine-8-carboxylic acid, 7,8,9,10-tetrahydro-4,6-dimethoxy-, methyl ester	EtOH	271.4(4.96),374.5(3.97)	44-5116-84
1H-2-Benzopyran-7-carboxylic acid, 3,4-dihydro-6,8-dimethyl-1-oxo-4-(phenylhydrazono)-, ethyl ester	EtOH	265(4.52),282(3.96), 300(3.76),377(4.35)	111-0223-84
$C_{20}H_{20}N_2O_5$			
Alanine, 2,3-didehydro-N-[N-(phenyl-methoxy)carbonyl]-L-phenylalanyl]-	EtOH	208(4.36),243(3.88)	5-0920-84
Benzo[b]phenazine-8-carboxylic acid, 7,8,9,10-tetrahydro-4,6-dimethoxy-, methyl ester, 12-oxide	EtOH	281(5.11),392(3.90), 460(3.76)	44-5116-84
4H-Pyrido[1,2-a]pyrimidine-3-carboxylic acid, 9-(1,3-benzodioxol-5-ylmethyl-ene)-6,7,8,9-tetrahydro-6-methyl-4-oxo-, ethyl ester, (E)-	EtOH	237(3.98),264(3.90), 380(4.26)	118-0582-84
Tryptophan, N-(methoxycarbonyl)-5-(phenylmethoxy)-	MeOH	223(4.48),276(3.83), 296(3.74),305(3.54)	94-2126-84
$C_{20}H_{20}N_2O_7$			
3,4,5-Oxepintricarboxylic acid, 6,7-di-hydro-2-[(phenylcarbonimidoyl)amino]-, 5-ethyl 3,4-dimethyl ester	CHCl₃	265(3.77),317(4.13)	24-0585-84
$C_{20}H_{20}N_4O$			
7H-Pyrido[2,1-b]quinazoline-6,11-dione, 8,9-dihydro-, 6-[(2,6-dimethylphen-yl)hydrazone]	EtOH	228s(4.28),254(4.18), 294(3.89),389(4.20)	4-1301-84
$C_{20}H_{20}N_4O_3$			
1-Imidazolidineacetic acid, 2,5-dioxo-4,4-diphenyl-, (1-methylethylidene)-hydrazide	MeOH	204(4.56)	56-0585-84
$C_{20}H_{20}N_4O_4$			
Benzoic acid, 2-hydroxy-, 1,2-cyclohex-anediylidenedihydrazide	benzene	350(4.25)	86-1075-84
	EtOH	350(4.22)	86-1075-84
	EtOAc	350(4.28)	86-1075-84
	acetone	350(4.21)	86-1075-84
	IBMK	350(4.22)	86-1075-84
	CHCl₃	350(4.25)	86-1075-84
(also other solvents)	DMF	400(4.20)	86-1075-84
Benzoic acid, 3-hydroxy-, 1,2-cyclohex-anediylidenedihydrazide	benzene	385(4.48)	86-1075-84
	EtOH	385(4.47)	86-1075-84
	EtOAc	385(4.48)	86-1075-84
	acetone	390(4.47)	86-1075-84
(also other solvents)	CHCl₃	385(4.48)	86-1075-84
Oxepino[3,2-e][1,2,4]triazolo[1,5-a]-pyrimidine-7-carboxylic acid, 4,5,8,9-tetrahydro-2,9-dimethyl-5-oxo-4-phenyl-, ethyl ester	CHCl₃	265(4.00),312(4.24), 317(4.23)	24-0585-84
Oxepino[3,2-e][1,2,4]triazolo[1,5-a]-pyrimidine-7-carboxylic acid, 4,5,8,9-tetrahydro-8,9-dimethyl-5-oxo-4-phenyl-, ethyl ester	CHCl₃	260(4.01),310(4.21), 319(4.20),338s(3.98)	24-0585-84
$C_{20}H_{20}N_5O$			
Pyrazolo[3,4-c]pyrazolium, 5-acetyl-	EtOH	360(3.54)	48-0811-84

Compound	Solvent	$\lambda_{max}(\log \epsilon)$	Ref.
4-(1-ethyl-2(1H)-pyridinylidene)-4,5-dihydro-3-methyl-1-phenyl-, iodide (cont.)			48-0811-84
$C_{20}H_{20}OS_2$			
Spiro[2H-thiete-2,9'-[9H]xanthene], 4-(1,1-dimethylethyl)-3-(methyl-thio)-	hexane or MeOH	241(3.18),270(2.88), 279s(2.83),302(2.60)	54-0152-84
$C_{20}H_{20}O_2$			
1H-Indene, 1-(2,3-dihydro-5-methoxy-1H-inden-1-ylidene)-2,3-dihydro-5-methoxy-, (E)-	EtOH-CH$_2$Cl$_2$	227(3.86),318(4.20), 334(4.41),351(4.41)	24-2300-84
Phenol, 4-[(2-methoxy-1-naphthalenyl)-methyl]-2,5-dimethyl-	EtOH	229(4.79),270(3.74), 281(3.85),291(3.74), 322(3.45),336(3.49)	150-0701-84M
$C_{20}H_{20}O_3$			
2-Pentenoic acid, 5-(2,3-diphenyloxir-anyl)-, methyl ester	MeCN	258(2.85),266s(2.72), 272s(2.46)	88-1137-84
$C_{20}H_{20}O_4$			
Crotmadine	MeOH	372(5.3)	100-0585-84
	MeOH-NaOMe	435(--)	100-0585-84
Crotmarine	MeOH	285(5.2)	100-0585-84
	MeOH-NaOMe	295(--)	100-0585-84
Isobavachin	EtOH	288(3.94),310(3.74)	105-0233-84
2-Propen-1-one, 3-(1,3-benzodioxol-5-yl)-1-[3-(1,1-dimethylethyl)-2-hydroxyphenyl]-, (E)-	EtOH	206(4.31),266(4.01), 375(4.33)	78-4081-84
$C_{20}H_{20}O_5$			
4H-1-Benzopyran-4-one, 2,3-dihydro-5,7-dihydroxy-2-[4-hydroxy-3-(3-methyl-2-butenyl)phenyl]-	MeOH	235(3.66)	2-0887-84
[2,6'-Bi-4H-1-benzopyran]-4-one, 2,2',3,3'-tetrahydro-5,7-dihydroxy-2',2'-dimethyl-	MeOH	240(3.06),305(3.48)	2-0887-84
Corynoline, de-N-methyl-, (+)-	MeOH	206(4.66),235(3.70), 290(3.88)	83-0223-84
Naphtho[2,3-d]-1,3-dioxol-5(6H)-one, 7,8-dihydro-8-(4-hydroxy-3-methoxy-phenyl)-6,7-dimethyl-, [6S-(6α,7α,8β)]-(enshicine)	MeOH	210(4.08),236(4.12), 279(3.73),320(3.55)	102-1143-84
$C_{20}H_{20}O_6$			
1-Benzoxepin-5(2H)-one, 3-(3,4-dimeth-oxyphenyl)-6,8-dimethoxy-	MeOH	272(3.67),342(2.77), 405(4.22)	2-1211-84
1,3-Dioxane-4,6-dione, 5,5-bis(4-meth-oxyphenyl)-2,2-dimethyl-	MeOH	236(4.27),274(3.53), 281(3.48)	12-1245-84
$C_{20}H_{20}O_7$			
9,10-Anthracenedione, 1,2,3,5,6-penta-methoxy-8-methyl-	MeOH	218(4.41),279(4.55), 354(3.77)	78-5039-84
9,10-Anthracenedione, 1,2,3,7,8-penta-methoxy-5-methyl-	MeOH	219(4.38),280(4.58), 358(3.86)	78-5039-84
4H-1-Benzopyran-4-one, 2-(3,4-dimeth-oxyphenyl)-5,6,7-trimethoxy-	MeOH	243(4.41),266(4.21), 330(4.45)	94-0166-84
4H-1-Benzopyran-4-one, 5,7-dimethoxy-2-(3,4,5-trimethoxyphenyl)-	MeOH	240s(--),268(4.30), 322(4.37)	94-0166-84

Compound	Solvent	$\lambda_{max}(\log \epsilon)$	Ref.
$C_{20}H_{20}O_8$			
4H-1-Benzopyran-4-one, 5-hydroxy-6,7-dimethoxy-2-(2,3,5-trimethoxyphenyl)-	MeOH	278(4.4),322(4.1)	94-2296-84
4H-1-Benzopyran-4-one, 5-hydroxy-6,7-dimethoxy-2-(2,4,5-trimethoxyphenyl)-	MeOH	275(4.2),364(4.3)	94-4935-84
	MeOH-NaOMe	288(--),350(--)	94-4935-84
	MeOH-NaOAc	275(--),365(--)	94-4935-84
	MeOH-AlCl$_3$	270(--),295(--), 398(--)	94-4935-84
	MeOH-AlCl$_3$-HCl	265(--),294(--), 398(--)	94-4935-84
1H,3H-Furo[3,4-c]furan-1-one, tetrahydro-4-hydroxy-3,6-bis(4-hydroxy-3-methoxyphenyl)-	EtOH	233(4.38),275(4.07)	39-1159-84C
$C_{20}H_{20}O_9$			
4H-1-Benzopyran-4-one, 5-hydroxy-2-(2-hydroxy-3,4,6-trimethoxyphenyl)-6,7-dimethoxy-	MeOH	264(4.2),321(4.1)	94-4217-84
	MeOH-AlCl$_3$	276(--),295s(--), 344(--)	94-4217-84
4H-1-Benzopyran-4-one, 5-hydroxy-2-(3-hydroxy-2,4,6-trimethoxyphenyl)-6,7-dimethoxy-	MeOH	262(4.3),300(4.0), 331s(4.0)	94-4217-84
	MeOH-AlCl$_3$	276(--),326(--), 379(--)	94-4217-84
4H-1-Benzopyran-4-one, 5-hydroxy-2-(6-hydroxy-2,3,4-trimethoxyphenyl)-6,7-dimethoxy-	MeOH	262(4.3),326(4.1)	94-4217-84
	MeOH-AlCl$_3$	275(--),293s(--), 368(--)	94-4217-84
$C_{20}H_{20}S$			
Thiophene, 2,5-bis[(4-methylphenyl)methyl]-	C_6H_{12}	245(4.08)	44-1177-84
$C_{20}H_{20}S_3$			
Spiro[2H-thiete-2,9'-[9H]thioxanthene], 4-(1,1-dimethylethyl)-3-(methylthio)-	hexane or MeOH	238s(3.28),268(3.15), 305s(2.60)	54-0152-84
$C_{20}H_{21}ClFNOS$			
Dibenzo[b,e]thiepin-11-ol, 2-chloro-8-fluoro-6,11-dihydro-11-(1-methyl-4-piperidinyl)-	MeOH	232s(4.01),265(3.99), 295s(3.24),305(3.01)	73-1800-84
$C_{20}H_{21}ClN_2O_2S$			
Piperazine, 1-[2-[[(2-acetyl-4-chlorophenyl)thio]benzoyl]-4-methyl-, monohydrochloride	MeOH	232.5(4.32),263(3.89), 286s(3.64),340(3.50)	73-1009-84
$C_{20}H_{21}NO$			
[1,1'-Biphenyl]-2-carboxamide, N-(cyclohexylidenemethyl)-	EtOH	278(4.10)	44-0714-84
$C_{20}H_{21}NO_2$			
Benzamide, N-[2-(2,2-dimethyl-2H-1-benzopyran-6-yl)ethyl]-	n.s.g.	224(4.56),264s(3.70), 273s(3.53),313(3.36)	142-1009-84
$C_{20}H_{21}NO_3$			
Guadiscine, dihydro-	EtOH	218(4.47),238s(4.20), 280(4.36),290s(4.31), 320s(3.74)	100-0353-84
Methanone, (2,3-dihydro-7,8-dimethoxy-3-methyl-1H-3-benzazepin-4-yl)phenyl-	EtOH	259(3.72),290(3.70), 388(3.87)	44-0581-84
Orientidine	EtOH	215(4.45),271(4.39), 340(3.38)	105-0076-84

Compound	Solvent	$\lambda_{max}(\log \epsilon)$	Ref.
$C_{20}H_{21}NO_4$			
Chilenamine	MeOH	228s(4.05),284(3.75), 286s(3.73)	78-3957-84
Dehydronorglaucine	MeOH	242s(4.45),260(4.68), 270(4.64),335(4.03), 381(3.52)	39-1273-84C
Dehydroprednicentrine	EtOH	215(4.14),243s(4.28), 262(4.49),270s(4.45), 294s(4.04),329(3.89), 380s(3.31)	100-0504-84
Isoquinoline, 3-(3,4-dimethoxyphenyl)- 6,7-dimethoxy-1-methyl-	EtOH	239(4.50),273(4.62), 312(4.15),350s(3.49)	4-0525-84
$C_{20}H_{21}NO_5$			
Codeine, 6-acetyl-10-oxo-	MeOH	248(4.00),294(3.95), 330(3.58)	106-0687-84
Marshaline	n.s.g.	290(2.75)	105-0644-84
$C_{20}H_{21}NS$			
Benzenecarbothioamide, N-1-cyclohexen- 1-yl-N-(phenylmethyl)-	MeOH	250(4.38),310(4.10), 425(3.00)	78-1835-84
[1,1'-Biphenyl]-2-carbothioamide, N-(cyclohexylidenemethyl)-	EtOH	235(4.26),265s(4.04), 316(3.96),385(2.51)	44-0714-84
$C_{20}H_{21}N_2$			
1H-1,4-Diazepin-4-ium, 6-cyclopropyl- 2,3-dihydro-1,4-diphenyl-, perchlor- ate	MeOH	400(4.41)	5-0649-84
$C_{20}H_{21}N_3O_2$			
Pyridinium, 1-ethyl-2-[2-(2-ethyl-4,5- dihydro-4,5-dioxo-1-phenyl-1H-pyra- zolium-3-yl)ethenyl]-, diiodide	EtOH	365(--),395(4.04)	48-0695-84
$C_{20}H_{21}N_3O_2S$			
Pyrrolo[2,3-c]phenothiazine, 3,6-di- hydro-1-(piperidinomethyl)-, 11,11- dioxide	EtOH	211(4.46),266.8(4.33), 298.5(3.78),326(3.98), 342(3.94),357(3.83)	103-1095-84
$C_{20}H_{21}N_3O_5S$			
Acetamide, N-[2-[5-(3,4-dimethoxyphen- yl)-1,3,4-thiadiazol-2-yl]-4,5-di- methoxyphenyl]-	EtOH	211(4.33),256(4.15), 364(4.21)	39-1143-84C
$C_{20}H_{21}N_3O_8S$			
5-Thiazolidineacetic acid, 3-(2,3-O- carbonyl-β-D-ribofuranosyl)-4-oxo- 2-[(1-phenylethylidene)hydrazono]-, methyl ester	CHCl₃	295(4.40)	128-0295-84
$C_{20}H_{21}N_5O$			
7H-Pyrido[2,1-b]quinazoline-6,11-dione, 8,9-dihydro-, 6-[[4-(dimethylamino)- phenyl]hydrazone]	EtOH	232(4.47),256(4.34), 308(4.05),421(4.38)	4-1301-84
$C_{20}H_{21}N_5O_3$			
1H-1,2,3-Triazole-4-carboxylic acid, 5-[[ethoxy(phenylamino)methylene]- amino]-1-phenyl-, ethyl ester	CHCl₃	293(4.19)	24-0585-84

Compound	Solvent	$\lambda_{max}(\log \epsilon)$	Ref.
$C_{20}H_{21}N_5O_4$ Adenosine, 2',3'-O-(1-methylethyli-dene)-N-(phenylmethylene)-	MeOH	228s(4.12),280(4.30)	150-3601-84M
$C_{20}H_{21}O_2S$ 1-Benzothiepinium, 3,5-dimethoxy-1,2-dimethyl-4-phenyl-, fluorosulfate	MeOH	236(4.13),260s(4.06), 290s(3.79)	64-0985-84B
$C_{20}H_{22}Cl_2N_2O_4$ Octanediamide, N,N'-bis(4-chlorophen-yl)-N,N'-dihydroxy-	MeOH	210(4.28),226(4.29), 246(4.41)	34-0345-84
$C_{20}H_{22}Cl_2O_4$ Spiro[2H-1-benzopyran-2,1'-[3,5]cyclo-hexadien]-2'-one, 3',8-bis(chloro-methoxy)-3,4-dihydro-4',5,6',7-tetramethyl-	EtOH	332(3.47)	12-2599-84
$C_{20}H_{22}N_2$ Methanediamine, 1-(5H-dibenzo[a,d]-cyclohepten-5-ylidene)-N,N,N',N'-tetramethyl-	hexane	221(4.5),276(4.3), 290(4.25)	64-1586-84B
radical cation	MeCN	545(3.2)	64-1586-84B
$C_{20}H_{22}N_2O$ Peduncularstine	n.s.g.	228(4.56),278s(3.89), 286(3.93),294(3.88)	33-0804-84
$C_{20}H_{22}N_2O_2$ Propanamide, N,N-diethyl-2-[2-(methoxy-imino)-1(2H)-acenaphthylenylidene]-	hexane	235(3.59),345(2.95), 367(2.60)	54-0023-84
$C_{20}H_{22}N_2O_3$ Strictamine N-oxide	MeOH	213(3.15),262(2.66)	102-0709-84
$C_{20}H_{22}N_2O_4$ Cycloocta[1,2-b:5,6-b']dipyridine-3,9-dicarboxylic acid, 5,6,11,12-tetrahydro-2,8-dimethyl-, dimethyl ester	EtOH	227(4.25),278(4.09)	35-2672-84
Cycloocta[1,2-b:6,5-b']dipyridine-3,8-dicarboxylic acid, 5,6,11,12-tetrahydro-2,9-dimethyl-, dimethyl ester	EtOH	227(4.29),277(4.05)	35-2672-84
$C_{20}H_{22}N_2O_5$ 3,5-Oxepindicarboxylic acid, 6,7-dihy-dro-6,7-dimethyl-2-[(phenylcarbon-imidoyl)amino]-, 5-ethyl 3-methyl ester	$CHCl_3$	270(4.01),350(4.29)	24-0585-84
3,5-Oxepindicarboxylic acid, 6,7-dihy-dro-7,7-dimethyl-2-[(phenylcarbon-imidoyl)amino]-, 5-ethyl 3-methyl ester	$CHCl_3$	263(3.86),330(4.27)	24-0585-84
2,3-Phthalazinedicarboxylic acid, 1,4-dihydro-1-hydroxy-1-phenyl-, diethyl ester	EtOH	228(4.15),253(4.03), 267s(3.97),281s(3.87), 286s(3.85),299s(3.79)	12-0893-84
$C_{20}H_{22}N_3O$ 1H-Pyrazolium, 4-[[4-(dimethylamino)-phenyl]methylene]-4,5-dihydro-2,3-	MeOH	351(2.77),484s(3.77), 514(4.06)	49-0197-84

Compound	Solvent	$\lambda_{max}(\log \epsilon)$	Ref.
dimethyl-5-oxo-1-phenyl-, tetra-fluoroborate (cont.)			49-0197-84
$C_{20}H_{22}N_4O_2$ [2,2'-Biquinoxaline]-3,3'(2H,2'H)-di-one, 1,1',4,4'-tetrahydro-2,2',4,4'-tetramethyl-	EtOH	214(4.63),227(4.67), 314(3.86)	77-1293-84
$C_{20}H_{22}N_4O_3$ 1H-Isoindole-1,3(2H)-dione, 5-[(2-amino-4-butoxy-5-methylphenyl)azo]-2-methyl-	toluene	477(1.75)	24-2275-84
Oxepino[2,3-d]pyrimidine-6-carboni-trile, 3,4,7,8-tetrahydro-2-[(3-hydroxypropyl)amino]-7,8-dimethyl-4-oxo-3-phenyl-	CHCl$_3$	260(3.77),338(4.26)	24-0585-84
$C_{20}H_{22}N_4O_5$ 1H-Indazole-4,7-dione, 3-[[(aminocarbo-nyl)oxy]methyl]-1-(3-hydroxypropyl)-6-methyl-5-[(phenylmethyl)amino]-	MeOH	212(4.27),264(4.15), 297(4.04),497(3.54)	5-1711-84
$C_{20}H_{22}N_6O_3$ 4-Morpholinecarboxamide, N-[5-cyano-2-(4-morpholinyl)-6-phenyl-4-pyri-midinyl]-	1,2-$C_6H_4Cl_2$-BuOH	323(4.03),460(2.49)	18-2144-84
$C_{20}H_{22}O_2$ 1(2H)-Dibenzofuranone, 3,4,6,7,8,9-hexahydro-3,3-dimethyl-7-phenyl-	H$_2$O	286(3.51)	39-1213-84C
1,4-Dioxocin, 5,6,7,8-tetrahydro-2,3-bis(4-methylphenyl)-	EtOH	250(4.08),315(3.68), 340(3.66)	2-1289-84
Phenanthrene, 2-methoxy-1,7-dimethyl-6-(1-methylethoxy)-	EtOH	228(4.55),259(4.67), 272(4.58),286(4.32), 327s(3.35),343(3.47), 360(3.54)	39-1913-84C
Phenanthrene, 7-methoxy-1,2-dimethyl-6-(1-methylethoxy)-	EtOH	214(4.36),230(4.23), 292(4.29),323(4.26)	39-1913-84C
1-Propanone, 2-methyl-1-(1,2,3,4-tetramethyl-7-dibenzofuranyl)-	EtOH	315(4.44)	39-0799-84B
1-Propanone, 2-methyl-1-(1,2,3,4-tetramethyl-8-dibenzofuranyl)-	EtOH	248(4.46),270(4.56)	39-0799-84B
$C_{20}H_{22}O_3$ Dibenzo[h,1][1,4,7]trioxacyclotridecin, 6,7,9,10-tetrahydro-16,17-dimethyl-, (E)-	MeOH	208(4.42),274(3.56), 278(3.53)	44-1627-84
(Z)-	MeOH	210s(4.39),288(3.70)	44-1627-84
Ethanone, 1-(9,10-dihydro-3,7-dimeth-oxy-2,8-dimethyl-4-phenanthrenyl)-	MeOH	213(4.57),278(4.24), 300(4.05)	39-1913-84C
Spiro[5.5]undeca-1,4-diene-3,8-dione, 7-methyl-11-[(phenylmethoxy)methyl]-, cis	n.s.g.	240(4.12)	77-0930-84
$C_{20}H_{22}O_4$ 2H-1-Benzopyran-3,7-diol, 3,4-dihydro-2-[4-hydroxy-3-(3-methyl-2-butenyl)-phenyl]-, (2R-trans)- (broussinol)	EtOH	224(4.27),282(3.77), 288s(3.65)	138-0689-84
$C_{20}H_{22}O_5$ 1H-2-Benzopyran, 3-(3,4-dimethoxyphen-	EtOH	215(4.55),270(4.23),	23-2435-84

Compound	Solvent	$\lambda_{max}(\log \epsilon)$	Ref.
yl)-6,7-dimethoxy-1-methyl- (cont.)		300(3.98)	23-2435-84
Cyclopentanone, 2,3-dihydroxy-4-(phen-ylmethoxy)-2-[(phenylmethoxy)meth-yl]-, (2S,3S,4R)-	EtOH	218(3.38)	39-2089-84C
Ethanone, 1,1'-[oxybis(2,1-ethanediyl-oxy-2,1-phenylene)]bis-	MeOH	212(4.58),246(4.16), 305(3.83)	44-1627-84
Propanedioic acid, (4-methoxyphenyl)-phenyl-, diethyl ester	MeOH	231(4.08),272(3.23), 279(3.04)	12-1245-84
$C_{20}H_{22}O_5S$			
Cyclohexanone, 2-[2-methoxy-4-[[(4-methylphenyl)sulfonyl]oxy]phenyl]-	MeOH	273(3.58),278s(3.49)	44-4399-84
Cyclohexanone, 2-[2-methoxy-5-[[(4-methylphenyl)sulfonyl]oxy]phenyl]-	MeOH	269(3.47),273(3.48)	44-4399-84
$C_{20}H_{22}O_6$			
$\Delta^{4,15}$-Isoatripliciolide tiglate	MeOH	215(4.00),268(3.52)	102-2557-84
Matairesinol	MeOH	230(4.21),280(3.90)	94-1612-84
Pinoresinol	MeOH	230(4.12),280(3.76)	94-1612-84
Pinoresinol, (+)-	EtOH	232.0(4.14),281.0(3.75)	94-4482-84
β-D-glucoside	EtOH	228.3(4.14),280.5(3.66)	94-4482-84
$C_{20}H_{22}O_7$			
Ethanone, 1,1',1",1"'-[oxybis[methyl-ene(5-methyl-2,3,4-furantriyl)]-tetrakis-	MeOH	268(4.05)	4-0569-84
16H-1,17-Metheno-2H-cyclohepta[p]-[1,4,7,10,13]pentaoxacyclonona-decin-2,16-dione, 4,5,7,8,10,11,13,14-octahydro-	MeCN	497(<u>2.8</u>)	24-2839-84
with calcium thiocyanate	MeOH	495(<u>2.8</u>)	24-2839-84
with barium thiocyanate	MeOH	493(<u>2.8</u>)	24-2839-84
Pinoresinol, 1-hydroxy-, (±)-	EtOH	232(4.27),281(3.88)	94-2730-84
	EtOH-NaOH	253(--),293.3(--)	94-2730-84 +94-4482-84
Shinjulactone F	EtOH	244(3.87),293(3.53)	18-2885-84
	EtOH	244(3.87),291(3.53)	138-0555-84
$C_{20}H_{22}O_8$			
2-Benzofuranbutanoic acid, 7-acetyl-4,6-dihydroxy-α-(1-hydroxyethyli-dene)-3,5-dimethyl-β-oxo-, ethyl ester, (Z)-	MeOH	241(4.01),296(3.91), 347(3.41)	23-0320-84
1H,3H-Furo[3,4-c]furan-1,4-diol, tetra-hydro-3,6-bis(4-hydroxy-3-methoxy-phenyl)-, (1α,3α,3aα,4α,6α,6aα)-	EtOH	230(4.42),280(4.02)	39-1159-84C
$C_{20}H_{22}O_8S$			
Podolactone C	MeOH	218(4.16)	44-0942-84
$C_{20}H_{22}O_9$			
β-D-Glucopyranoside, 3-[2-(3,4-dihy-droxyphenyl)ethenyl]-5-hydroxyphenyl, (E)-	MeOH	219(4.53),303(4.45), 326(4.54)	94-3501-84
β-D-Glucopyranoside, 5-[2-(3,5-dihy-droxyphenyl)ethenyl]-2-hydroxyphenyl, (E)-	MeOH	217(4.38),302(4.41), 320(4.48)	94-3501-84
$C_{20}H_{22}O_{11}$			
2H-1-Benzopyran-2-one, 7-[(2,6-di-O-acetyl-β-D-glucopyranosyl)oxy]-6-methoxy-	MeOH	227(4.03),286(3.75), 328(3.88)	102-0467-84

Compound	Solvent	$\lambda_{max}(\log \epsilon)$	Ref.
$C_{20}H_{23}ClN_2O$			
D-Homoeburnamenin-14(15H)-one, 15-chloro-, (3α,15α,16α)-	EtOH	202(4.46),221s(4.15), 254(4.15),258(4.15), 267(4.05)	39-1629-84B
monohydrochloride	EtOH	248(4.22),252(4.22), 265s(4.03),278s(3.83), 294(3.78)	39-1629-84C
$C_{20}H_{23}ClN_2O_2S$			
Piperazine, 1-[2-[[4-chloro-2-(1-hydroxyethyl)phenyl]thio]benzoyl]-4-methyl-	MeOH	251(4.08),275(3.78)	73-1009-84
$C_{20}H_{23}ClO_2$			
Benzene, 1-chloro-2-[2-(2,3-dimethylphenyl)ethenyl]-4-methoxy-5-(1-methylethoxy)-, (E)-	EtOH	214(4.36),230(4.23), 292(4.29),323(4.26)	39-1913-84C
(Z)-	EtOH	197(4.59),215(4.46), 276(4.11)	39-1913-84C
Benzene, 1-chloro-2-[2-(3-methoxy-2-methylphenyl)ethenyl]-4-methyl-5-(1-methylethoxy)-, (E)-	EtOH	217(4.42),300(4.39)	39-1913-84C
(Z)-	EtOH	278(4.20)	39-1913-84C
$C_{20}H_{23}NO_2$			
Benzamide, N,N-diethyl-2-[2-(4-methoxyphenyl)ethenyl]-	EtOH	228(3.97),324(3.84)	44-0742-84
Phenol, 2,6-dimethyl-4-[[(1-methylpropyl)imino]methyl]-, benzoate	EtOH	250(4.47)	150-0701-84M
2H-Pyrrole, 3,4-dihydro-4,5-bis(4-methoxyphenyl)-2,2-dimethyl-	EtOH	265(4.26)	12-0095-84
$C_{20}H_{23}NO_3$			
Atherospermine, N-oxide	EtOH	213(4.30),234(4.33), 252(4.60),258(4.63), 279s(4.01),304(4.04), 313(4.04),346(3.21), 364(3.21)	100-0353-84
Benzamide, 4-methoxy-N-[3-(4-methoxyphenyl)-1,1-dimethyl-2-propenyl]-	EtOH	260(4.59)	12-0095-84
Isothebaine, O-methyl-	EtOH	273(4.26),303(3.31)	105-0076-84
Orientine	EtOH	278(4.25),310s(3.21)	105-0076-84
6-Oxa-1-azabicyclo[3.1.0]hexane, 4,5-bis(4-methoxyphenyl)-2,2-dimethyl-	EtOH	228(4.34),275(3.48)	12-0095-84
2H-Pyrrole, 4,5-bis(4-methoxyphenyl)-2,2-dimethyl-, 1-oxide	EtOH	303(4.34)	12-0095-84
$C_{20}H_{23}NO_4$			
Corydine	MeOH	221(4.61),269(4.16), 305(3.80)	73-1318-84
	n.s.g.	270(4.00)	105-0467-84
Cularine, (+)-	EtOH	206s(5.21),229s(4.12), 274(3.63),283(3.79), 295s(3.48)	100-0753-84
Discretine, (±)-	MeOH	285.5(3.92)	2-0268-84
	MeOH-NaOH	290(3.9),303s(3.70)	2-0268-84
1H-Inden-2-amine, 2,3-dihydro-1,5,6-trimethoxy-N-[(3-methoxyphenyl)methylene]-, (1S-cis)-	MeOH	203(4.77),219s(4.50), 252(4.21),289(3.91), 310s(3.55)	39-1655-84C
5H-Indeno[2,1-c]isoquinoline, 6,6a,7-11b-tetrahydro-2,3,8,9-tetramethoxy-,	MeOH	202(4.84),230s(3.69), 277s(3.69),283(3.76),	39-1655-84C

Compound	Solvent	λ_{max}(log ϵ)	Ref.
(6aR-cis)- (cont.)		291s(3.61)	39-1655-84C
hydrochloride	MeOH	205(5.85),231s(4.21), 277s(3.71),283(3.77), 291s(3.62)	39-1655-84C
5H-Indeno[2,1-c]isoquinoline, 6,6a,7-11b-tetrahydro-2,3,9,10-tetrameth-oxy-, (6aR-cis)-	MeOH	203(4.79),231(4.13), 286.5(3.90)	39-1655-84C
hydrochloride	MeOH	203(4.84),233(4.18), 286(3.94)	39-1655-84C
5H-Indeno[2,1-c]isoquinoline, 6,6a,7-11b-tetrahydro-3,4,8,9-tetrameth-oxy-, (6aS-cis)-	MeOH	202(4.82),230s(4.25), 276(3.45),278(3.45), 282(3.45)	39-1655-84C
hydrochloride	MeOH	204(4.82),228s(4.24), 279(3.54),283(3.54)	39-1655-84C
5H-Indeno[2,1-c]isoquinoline, 6,6a,7-11b-tetrahydro-3,4,9,10-tetrameth-oxy-, (6aR-cis)-	MeOH	202(4.84),229s(4.20), 285(3.80),291s(3.72)	39-1655-84C
hydrochloride	MeOH	204(4.79),233(4.12), 285(3.78)	39-1655-84C
Isocorydine	n.s.g.	270(4.16)	105-0467-84
Sarcocapnine, (+)-	MeOH	232(4.12),283(3.14)	100-0753-84
Sebiferine, (9S)-	EtOH	208(4.62),240(4.29), 285(3.96)	22-0139-84
Secocularidine	EtOH	216(3.88),236s(3.83), 296s(3.41),320(3.51)	88-0889-84 +100-0753-84
	EtOH-base	216(4.13),330(3.57)	88-0889-84
Secocularine, 3'-O-demethyl-	EtOH	226(3.95),298s(3.45), 320(3.50)	100-0753-84
	EtOH-base	220(4.19),274(3.91), 302(3.65),350(3.42)	100-0753-84
Secocularine, 10-O-demethyl-	EtOH	226(3.95),298s(3.45), 320(3.50)	88-0889-84
	EtOH-base	220(4.19),274(3.91), 302(3.65),350(3.42)	88-0889-84
$C_{20}H_{23}NO_5$			
Codeine, 6-acetyl-10-hydroxy-	MeOH	209(4.50),231s(3.86), 285(3.17)	106-0687-84
Gouregine, tetrahydro-, (±)-	EtOH	208(4.44),279(3.97)	100-0753-84
	EtOH-NaOH	216(4.78),241s(4.31), 290(4.09)	100-0753-84
1-Pentanone, 1,2-bis(4-methoxyphenyl)-4-methyl-4-nitro-	EtOH	279(4.03)	12-0095-84
$C_{20}H_{23}NO_6$			
Benzenepropanamide, 3,4,5-trimethoxy-N-(4-methoxy-5-oxo-1,3,6-cyclo-heptatrien-1-yl)-	neutral	325(4.14)	69-1742-84
$C_{20}H_{23}NO_7$			
L-ribo-Hex-4-enar-1-amic acid, 4,5-di-deoxy-N-[1-(3,4-dihydro-8-hydroxy-1-oxo-1H-2-benzopyran-3-yl)-3-methyl-butyl]-, γ-lactone, [S-(R*,R*)]-	MeOH	245(3.79),312(3.62)	78-2519-84
Nigdenine, O-acetyl-	MeOH	249(4.71),271s(4.03), 319(3.78),332(3.77), 349s(3.56)	142-2821-84
	MeOH-HCl	253(4.67),276s(4.03), 322(3.74),351(3.75)	142-2821-84
Nkolbisine, O-acetyl-	MeOH	244(4.76),251(4.78), 274(4.07),294s(4.03),	142-2821-84

Compound	Solvent	$\lambda_{max}(\log \epsilon)$	Ref.
Nkolbisine, O-acetyl- (cont.)		308(4.03),320(3.99), 333(3.86)	142-2821-84
	MeOH-HCl	251(4.69),335(4.15)	142-2821-84
$C_{20}H_{23}NS_2$ 1-Propanamine, N,N-dimethyl-3-[2-[(methylthio)methyl]-9H-thioxanthen-9-ylidene]-	MeOH	233.5(4.42),272(4.06), 327(3.43)	73-1722-84
$C_{20}H_{23}N_3O_3S$ 3H-1,2-Oxathiino[6,5-e]indazole, 4,5,6,7-tetrahydro-7-phenyl-4-piperidino-, 2,2-dioxide	EtOH	224(4.22),251(4.03), 261(4.00),276(3.95)	4-0361-84
$C_{20}H_{23}N_4O_4$ 1H-Pyrazolium, 3-[(5,5-diethyltetrahydro-4,6-dioxo-2(1H)-pyrimidinylidene)methyl]-2-ethyl-4,5-dihydro-4,5-dioxo-1-phenyl-, iodide	EtOH	450(4.30)	48-0695-84
$C_{20}H_{23}N_7O_6$ Chryscandin	H_2O	260(4.51)	88-4689-84 158-1284-84
$C_{20}H_{24}ClN_3O$ Phenol, 2-(5-chloro-2H-benzotriazol-2-yl)-4,6-bis(1,1-dimethylethyl)-	MeOH	311(4.17),347(4.21)	12-2489-84
$C_{20}H_{24}FeO_7$ Ferrocene, 1,1'-(1,15-dioxo-2,5,8,11,14-pentaoxapentadecane-1,15-diyl)-	MeOH	452(2.55)	18-2435-84
$C_{20}H_{24}NO_3$ 5H-Oxazolo[2,3-a]isoquinolinium, 2,3,6,10b-tetrahydro-8,9-dimethoxy-4-methyl-10b-phenyl-, iodide	MeOH	237s(4.11),280(3.64), 284s(3.62)	12-1659-84
$C_{20}H_{24}NO_4$ Isoquinolinium, 3-(3,4-dimethoxyphenyl)-3,4-dihydro-6,7-dimethoxy-2-methyl-, iodide	EtOH	253(3.92),315(3.72), 372(3.68)	4-0525-84
Magnoflorine	MeOH	226(4.66),274(3.87), 320(3.77)	73-0704-84
$C_{20}H_{24}N_2$ Pyrido[3,2-g]quinoline, 2,8-bis(1,1-dimethylethyl)-	MeCN	212(4.204),252(5.209), 340(3.942),348(3.902), 358(4.064)	5-0133-84
$C_{20}H_{24}N_2O$ Aristoserratine	n.s.g.	228(4.42),275s(3.79), 282(3.82),290(3.75)	33-0804-84
D-Homoeburnamenin-14(15H)-one, (3α,16α)-	EtOH	242(4.23),269(4.05), 276s(4.00),292(3.68)	39-1629-84C
hydrochloride	EtOH	239(4.27),264(4.08), 270(4.04),291(3.73)	39-1629-84C
$C_{20}H_{24}N_2O_2$ [Bi-2,4,6-cycloheptatrien-1-yl]-3,3'-dicarboxamide, N,N,N',N'-tetramethyl-	MeCN	211(4.7),239(4.2), 260(4.0),271(3.9)	24-0809-84

Compound	Solvent	λ_{max}(log ϵ)	Ref.
$C_{20}H_{24}N_2O_2$			
[Bi-2,4,6-cycloheptatrien-1-yl]-3,4'-dicarboxamide, N,N,N',N'-tetramethyl-	MeCN	211(4.6),240(4.0), 260(4.0)	24-0809-84
16-Epiaffinine	EtOH	214(4.16),238(4.07), 318(4.28)	102-2359-84
Indolo[2,3-a]quinolizin-4(3H)-one, 2,6,7,12-tetrahydro-3-(2-hydroxyethyl-2-isopropyl-, trans-(±)-	EtOH	231(4.48),312(4.30), 320(4.29)	39-1237-84C
Triabunnine	n.s.g.	219s(4.31),223(4.35), 229s(4.26),266(3.74), 280s(3.68),290s(3.60), 299s(3.48),313s(3.11)	33-0804-84
$C_{20}H_{24}N_2O_3$			
Echitamidine	EtOH	237(4.35),300(4.26), 334(4.47)	102-2708-84
Phenol, 2,6-dimethyl-4-[[(1-methylpropyl)nitrosoamino]methyl]-, benzoate	hexane	228(4.51),356(1.89), 366(1.96),385(1.89)	150-0701-84M
Quinidine, 3-hydroxy-, (3S)-	MeOH	281(3.60),333(3.69)	100-0882-84
$C_{20}H_{24}N_2O_4$			
Diazaquinomycin B	MeOH	277(4.30),310s(4.17), 325s(4.07),356(3.79), 373(3.78)	158-84-130
Hexanediamide, N,N'-dihydroxy-N,N'-bis(2-methylphenyl)-	MeOH	222(4.19),262(3.90)	34-0345-84
Hexanediamide, N,N'-dihydroxy-N,N'-bis(4-methylphenyl)-	MeOH	220(4.26),250(4.31)	34-0345-84
Octanediamide, N,N'-dihydroxy-N,N'-diphenyl-	MeOH	218(4.42),258(4.10)	34-0345-84
$C_{20}H_{24}N_2O_6$			
3,5-Pyridinedicarboxylic acid, 1,2-dihydro-1,2,6-trimethyl-4-(2-nitrophenyl)-, diethyl ester	EtOH	206(4.37),222s(4.33), 292(4.25),378(3.90)	103-0522-84
3,5-Pyridinedicarboxylic acid, 1,2-dihydro-1,2,6-trimethyl-4-(4-nitrophenyl)-, diethyl ester	EtOH	204(4.48),281(4.40), 344(4.01),378s(3.93)	103-0522-84
$C_{20}H_{24}N_2O_7S$			
Thymidine, 5'-[2-[[(methylthio)methoxy]-methyl]benzoate]	EtOH	267(3.96)	39-1785-84C
$C_{20}H_{24}N_4$			
1H-Pyrazolium, 3,3'-(1,4-phenylenedi-2,1-ethenediyl)bis[1,2-dimethyl-, diiodide	EtOH	362(4.70)	103-1266-84
7H-Pyrrolo[2,3-d]pyrimidine, 2,5,6-trimethyl-7-phenyl-4-piperidino-	EtOH	227(4.28),300(3.98)	11-0073-84B
$C_{20}H_{24}N_4O_3$			
L-erythro-Pentos-2-ulose, 3,4-0-(1-methylethylidene)-, bis(phenylhydrazone)	EtOH	258(3.93),393(3.93)	39-0733-84C
$C_{20}H_{24}N_4O_3S$			
Ethanol, 2-[[2-[[4-(diethylamino)-2-methoxyphenyl]azo]-6-benzothiazolyl]oxy]-	pH 1.2	514(4.62),618s(3.74)	48-0151-84
	pH 6.2	515(4.64)	48-0151-84
	EtOH	511(4.64)	48-0151-84

Compound	Solvent	$\lambda_{max}(\log \epsilon)$	Ref.
$C_{20}H_{24}N_4O_4$			
Propanoic acid, 2,2'-(2,6-naphthalene-diyldi-2-hydrazinyl-1-ylidene)bis-, diethyl ester, anti-anti	EtOH	243(4.39),292(4.26), 375(4.55)	103-0283-84
syn-anti	EtOH	236(4.52),291(4.28), 395(4.57)	103-0283-84
syn-syn	EtOH	238(4.02),263s(3.89), 294(3.67),406(4.22)	103-0283-84
Propanoic acid, 2,2'-(2,7-naphthalene-diyldi-2-hydrazinyl-1-ylidene)bis-, diethyl ester, anti-anti	EtOH	210(4.42),248(4.40), 312(4.49),344(4.73)	103-0283-84
syn-anti	EtOH	212(4.37),252(4.37), 316s(4.41),357(4.62)	103-0283-84
syn-syn	EtOH	203(4.36),218*4.33), 256(4.34),321s(4.22), 370(4.57)	103-0283-84
$C_{20}H_{24}O_2$			
1,3-Cyclohexanedione, 5,5-dimethyl-2-(4-phenyl-1-cyclohexen-1-yl)-	pH 1	268(4.04)	39-1213-84C
	H_2O	290(4.07)	39-1213-84C
	pH 13	294(4.28)	39-1213-84C
1(2H)-Dibenzofuranone, 3,4,5a,6,7,8-9,9a-octahydro-3,3-dimethyl-7-phenyl-	EtOH	273(4.06)	39-1213-84C
Phenanthrene, 9,10-dihydro-2-methoxy-1,7-dimethyl-6-(1-methylethoxy)-	EtOH	219(4.56),278(4.26), 294(4.05),313(3.97)	39-1917-84C
Phenanthrene, 9,10-dihydro-7-methoxy-1,2-dimethyl-6-(1-methylethoxy)-	EtOH	219(4.66),278(4.31), 290s(4.16),311(4.04)	39-1913-84C
$C_{20}H_{24}O_3$			
Estra-1,3,5(10),9(11)-tetraen-17-one, 3-(2-hydroxyethoxy)-	MeOH	264(4.31)	48-0941-84
8H-2,5a-Methano-1-benzoxepin-8-one, 2,3,4,5,9,9a-hexahydro-10-methyl-5-[(phenylmethoxy)methyl]-, (2α,5β,5aα,9aβ,10S*)-	n.s.g.	234(3.92)	77-0930-84
1,4-Naphthalenedione, 3-hydroxy-6-meth-yl-2-(1-methylethyl)-5-(4-methyl-4-pentenyl)-	MeOH	250s(4.23),256(4.29), 282(4.09),288(4.08), 356(3.53)	102-1805-84
	MeOH-NaOMe	223(4.32),277(4.33), 289s(4.18),347(3.31), 472(3.13)	102-1805-84
	MeOH-AlCl$_3$	258(4.23),264s(4.19), 271s(4.13),314(3.89), 394(3.61)	102-1805-84
	+ HCl	258(4.21),265s(4.18), 273s(4.13),313(3.81), 390(3.53)	102-1805-84
4-Phenanthrenemethanol, 9,10-dihydro-3,7-dimethoxy-α,2,8-trimethyl-	MeOH	215(4.62),275(4.31), 290(4.10)	39-1913-84C
Spiro[5.5]undeca-1,5-dien-3-one, 8-hy-droxy-7-methyl-11-[(phenylmethoxy)-methyl]-, (7α,8α,11α)-	n.s.g.	248(3.97)	77-0930-84
$C_{20}H_{24}O_3S_2$			
Benzo[b]thiophen-3(2H)-one, 3a-(butyl-thio)-3a,4,7,7a-tetrahydro-5(and 6)-methyl-2-(phenylmethylene)-, 1,1-di-oxide	EtOH	206(4.09),260(3.00), 315(2.80)	104-0278-84
$C_{20}H_{24}O_4$			
2H-9,4a-(Epoxymethano)phenanthrene-	MeOH	240(3.97),275(3.90),	102-0919-84

Compound	Solvent	$\lambda_{max}(\log \epsilon)$	Ref.
5,8,12-trione, 1,3,4,9,10,10a-hexa-hydro-1,1-dimethyl-7-(1-methylethyl)-		418(3.85)	102-0919-84
$C_{20}H_{24}O_5$			
Heliangin, 2,3-dehydro-3-deoxy-	MeOH	216(3.87),235(3.59)	102-1439-84
1,4,10(4bH)-Phenanthrenetrione, 5,6,7,8-tetrahydro-3,9-dihydroxy-4b,8,8-trimethyl-2-(1-methyl-ethyl)-, (S)- (coleon Q quinone)	ether	240(4.27),272(4.03), 289s(3.79),431(3.38)	33-1523-84
D-Xylofuranoside, methyl 3,5-bis-O-(phenylmethyl)-	MeOH	266(2.60)	118-0961-84
$C_{20}H_{24}O_6$			
4a,10a-Epoxyphenanthrene-1,4,10(4bH)-trione, 5,6,7,8-tetrahydro-3,9-di-hydroxy-4b,8,8-trimethyl-2-(1-methyl-ethyl)-, [4aS-(4aα,4bβ,10aα)]-	ether	299(4.02)	33-1523-84
$C_{20}H_{24}O_6$ (cont.)			
3-Furanmethanol, tetrahydro-2-(4-hy-droxy-3-methoxyphenyl)-4-[(4-hy-droxy-3-methoxyphenyl)methyl]- (olivil)	EtOH	230.4(4.10),281.5(3.72)	94-2730-84 +94-4482-84
Phenol, 2-methoxy-5-[2-(3,4,5-trimeth-oxyphenyl)ethyl]-, acetate	EtOH	216(4.17),274(3.31)	2-1040-84
$C_{20}H_{24}O_7$			
2,3-Naphthalenedimethanol, 1,2,3,4-tetrahydro-7-hydroxy-1-(4-hydroxy-3-methoxyphenyl)-6-methoxy-	EtOH	230.5(4.03),284(3.69)	94-2730-84
Niveusin, 1,2-dehydro-, C-2',3'-epoxide	EtOH	220(4.32)	102-1281-84
$C_{20}H_{25}NO_2$			
Benzeneacetamide, α-ethyl-N-(2-hydroxy-1-methyl-2-phenylethyl)-N-methyl-, [1R-[1R*(S*),2S*]]-	EtOH	252(2.71),259(2.76), 264(2.61)	44-5202-84
Phenol, 2,6-dimethyl-4-[[(1-methylprop-yl)amino]methyl]-, benzoate	EtOH	228(4.34)	150-0701-84M
hydrochloride	EtOH	229(4.26)	150-0701-84M
$C_{20}H_{25}NO_3$			
Benzamide, N,N-diethyl-2-[2-hydroxy-2-(4-methoxyphenyl)ethyl]-	EtOH	228(4.29),278(3.35), 285(3.25)	44-0742-84
$C_{20}H_{25}NO_4$			
Erythroculine	EtOH	238(4.04),304(3.62)	95-0946-84
$C_{20}H_{25}NO_5$			
1H-Inden-1-ol, 2-[[(3,4-dimethoxyphen-yl)methyl]amino]-2,3-dihydro-4,5-di-methoxy-, (1S-trans)-	MeOH	202(4.96),231(4.25), 276s(3.60),280(3.61)	39-1655-84C
$C_{20}H_{25}NO_7$			
α-D-ribo-Hexopyranoside, methyl 2-(ace-tylamino)-2,3-dideoxy-3-(2-ethoxy-2-oxoethylidene)-4,6-O-(phenylmethyl-ene)-	EtOH	229(3.76)	159-0475-84
Oxostephasunoline	MeOH	286(3.32)	100-0465-84
$C_{20}H_{25}NO_8$			
Pyrrolo[2,1-a]isoquinoline-1,2,3-tri-	EtOH	212(4.05),285(3.89)	78-0369-84

Compound	Solvent	$\lambda_{max}(\log \epsilon)$	Ref.
carboxylic acid, 1,2,3,5,6,10b-hexa-hydro-8,9-dimethoxy-, trimethyl ester, endo (cont.)			78-0369-84
exo	EtOH	213(4.06),283(3.90)	78-0369-84
$C_{20}H_{25}NO_8S$			
Acetic acid, [[[(17β)-3-hydroxy-17-(sulfooxy)estra-1,3,5(10)-trien-6-ylidene]amino]oxy]-, mono-potassium salt	MeOH	261(4.06),311(3.62)	94-1885-84
$C_{20}H_{25}N_2O_9P$			
Thymidine, 5'-O-[5-(phenylmethoxy)-1,3,2-dioxaphosphorinan-2-yl]-, P-oxide	EtOH	266(3.99)	87-0440-84
$C_{20}H_{25}N_3O_4$			
4H-Pyrido[1,2-a]pyrimidine-3-carbox-amide, N-[2-(3,4-dimethoxyphenyl)-ethyl]-6,7,8,9-tetrahydro-6-methyl-4-oxo-	EtOH	229(4.19),287(4.04), 300(4.05)	118-0582-84
$C_{20}H_{25}N_5O$			
Urea, N'-[(8α)-6-cyanoergolin-8-yl]-N,N-diethyl-	MeOH	224(4.51),277s(3.81), 281(3.83),292(3.75)	73-2828-84
$C_{20}H_{25}N_5O_4$			
Guanosine, N-(4-butylphenyl)-2'-deoxy-	pH 2	272(4.31)	87-0175-84
	H_2O	273(4.30)	87-0175-84
	pH 12	231(4.22),278(4.37)	87-0175-84
7α-isomer	pH 2	265(4.26)	87-0175-84
	H_2O	264(4.16)	87-0175-84
	pH 12	239(4.14),267(4.31)	87-0175-84
$C_{20}H_{25}N_5O_{10}$			
Methanol, [[[9-(2,3,5-tri-O-acetyl-β-D-arabinofuranosyl)-9H-purin-6-yl]-amino]methoxy]-, acetate	EtOH	263(4.17)	44-1453-84
$C_{20}H_{26}$			
Benzene, 1,1'-(1,8-octanediyl)bis-	EtOH	204(4.25),247(2.53), 253s(2.63),258(2.68), 267(2.59)	24-0455-84
$C_{20}H_{26}N_2$			
Aristoserratenine	MeOH	226(4.38),259(3.57)	78-4359-84
Neohobartine	EtOH	244s(3.86),292(3.88)	33-1878-84
$C_{20}H_{26}N_2O$			
D-Homoeburnamenin-14-ol, 14,15-dihydro-	EtOH	230(4.52),276(3.87), 285(3.90),293(3.83)	39-1629-84B
hydrochloride	EtOH	223(4.55),271(3.92), 282(3.86),293(3.67)	39-1629-84B
1H-Indole-3-acetamide, N-[1-methyl-1-(4-methyl-3-cyclohexen-1-yl)-ethyl]-, (S)-	EtOH	222(4.51),276s(3.77), 283(3.81),292(3.74)	33-1040-84
$C_{20}H_{26}N_2O_2$			
Aristolarine	n.s.g.	233s(4.238),236(4.240), 256s(3.80),266s(3.63), 290s(3.07),325s(2.89),	33-0804-84

Compound	Solvent	$\lambda_{max}(\log \epsilon)$	Ref.
Aristolarine (cont.)		403(3.33)	33-0804-84
$C_{20}H_{26}N_2O_3S_2$			
Estr-2-eno[3,2-d][1,3,2]thiadiazol-17-ol, 1-sulfinyl-, acetate, (1Z,5α,17β)-	EtOH	230(3.83),285s(3.29), 346(3.94)	44-4773-84
$C_{20}H_{26}N_2O_6$			
2H-[1]Benzopyrano[3,4-c]pyridazine-3,4-dicarboxylic acid, 4a,5-dihydro-8-methoxy-2,2,5,5-tetramethyl-, dimethyl ester	EtOH	219(4.37),259(4.20), 265(4.20),300(3.87), 309(3.83)	39-2815-84C
$C_{20}H_{26}N_2S$			
1,4-Benzenediamine, N'-[2-(5,6-dimethyl-2H-thiopyran-2-ylidene)propylidene]-N,N-diethyl-, sulfate (1:1)	EtOH	550(4.57)	97-0183-84
$C_{20}H_{26}N_4O_{11}$			
Acetamide, N-acetyl-N-[1-(2,3,4,6-tetra-O-acetyl-β-D-glucopyranosyl)-1H-1,2,4-triazol-3-yl]-	EtOH	218(3.78)	111-0089-84
$C_{20}H_{26}O$			
Benzene, 1-(4-ethenyl-4,8-dimethyl-1,5,7-nonatrienyl)-4-methoxy-	EtOH	242(4.42),261(4.39)	154-1435-84
Benzene, 1-methoxy-4-[6-methyl-4-(2-methyl-1-propenyl)-1,5,7-octatrienyl]- (juvocimene I)	EtOH	237(4.36),260(4.34)	154-1435-84
$C_{20}H_{26}O_2$			
Oxirane, 3-[1-[3-(4-methoxyphenyl)-2-propenyl]-3-methyl-2,4-pentadienyl]-2,2-dimethyl- (juvocimene II)	EtOH	227(4.46),262(4.40)	154-1435-84
epimer II'	EtOH	227(4.40),262(4.44)	154-1435-84
4,10-Phenanthrenedione, 8-ethyl-4,4a,10,10a-tetrahydro-1,1,4a,7-tetramethyl-, (4aS-trans)-	MeOH	218(4.09),255(3.85), 304(3.08)	102-1293-84
$C_{20}H_{26}O_3$			
Benzaldehyde, 3-(2-acetoxy-4,8-dimethyl-3,7-nonadienyl)-, (E)-(+)-	MeOH	249(3.83),286(3.54)	78-3053-84
2H-9,4a-(Epoxymethano)phenanthrene-5,6-dione, 1,3,4,9,10,10a-hexahydro-1,1-dimethyl-7-(1-methylethyl)-, [4aR-(4aα,9α,10aβ)]-	C_6H_{12}	222(3.76),428(3.28)	102-1677-84
Halimedalactone	MeOH	236(3.86)	78-3053-84
Halimedatrial	MeOH	236(4.06)	78-3053-84
Taxodione	MeOH	320(4.40),332(4.42), 400(3.30)	2-0177-84
$C_{20}H_{26}O_4$			
2H-9,4a-(Epoxymethano)phenanthren-12-one, 1,3,4,9,10,10a-hexahydro-5,8-dihydroxy-1,1-dimethyl-7-(1-methylethyl)- (isocarnosol)	MeOH	235(4.34),295(4.01), 320(2.76)	102-0919-84
Estr-4-en-3-one, 17-acetoxy-14,15-epoxy-, (14β,15β,17β)-	EtOH	238(4.25)	106-0092-84
(15α,17β)-	EtOH	238(4.10)	106-0092-84
6H-Indeno[5,4-b]furan-6-one, decahydro-2-(5-hydroxy-2-methylphenyl)-3a-meth-	MeOH	206(3.85),219(3.77), 282(3.30)	13-0175-84B

Compound	Solvent	$\lambda_{max}(\log \epsilon)$	Ref.
oxy-5a-methyl- (cont.)	MeOH-base	217(4.00),243(3.88), 302(3.46)	13-0175-84B
Oxireno[13,14]cyclotetradeca[1,2-b]-furan-11,13-diene, 1a,2,3,5,6,10-11a,12,14a,14b-decahydro-1a,5,9-trimethyl-12-methylene-	n.s.g.	215(3.69),225(3.67)	12-0545-84
$C_{20}H_{26}O_6$			
3-Epinobilin, 9β-hydroxy-	MeOH	201(4.18)	102-0817-84
Royleanone	MeOH	273(3.79),423(2.69)	33-0201-84
Tirotundifolin A	MeOH	205(4.05)	102-0823-84
Tirotundifolin B	EtOH	220(3.69)	102-0823-84
Tirotundifolin C	$CHCl_3$	220(3.81)	102-0823-84
Trichomatolide A	MeOH	220(4.06)	102-0910-84
Trichomatolide B	MeOH	216(2.69),225(2.56), 240(2.34)	102-1439-84
Trichomatolide C	MeOH	220(3.98)	102-1439-84
Trichomatolide D	MeOH	215(3.86)	102-1439-84
$C_{20}H_{26}O_7$			
Leptocarpin, 15-hydroxy-	EtOH	212(4.27)	102-0823-84
Stizolicin	MeOH	208(4.33)	31-0930-84
$C_{20}H_{26}O_8$			
Niveusin C-2',3'-epoxide	EtOH	209(4.13)	102-1281-84
$C_{20}H_{26}O_9$			
Yadanziolide C	MeOH	244(4.04)	94-4698-84
$C_{20}H_{26}O_{10}$			
Yadanziolide A	MeOH	240(3.97)	94-4698-84
$C_{20}H_{26}O_{11}$			
Yadanziolide B	MeOH	244(4.00)	94-4698-84
$C_{20}H_{27}NO_3$			
Androstane-2-carbonitrile, 4,5-epoxy-17-hydroxy-3-oxo-, (2α,4α,5α,17β)-	EtOH	254.5(3.91)	87-0928-84
(2β,4β,5β,17β)-	EtOH	253(3.95)	87-0928-84
$C_{20}H_{27}NO_4$			
Athrocupressine	EtOH	227(3.62),280(2.97)	78-2677-84
Holidinine	EtOH	276(3.28),284(3.30)	142-2453-84
	EtOH-base	249(3.83),298(3.48)	142-2453-84
$C_{20}H_{27}NO_{11}$			
α-D-Glucopyranose, 2-[(2-acetyl-3-oxo-1-butenyl)amino]-2-deoxy-, 1,3,4,6-tetraacetate	EtOH	249(3.80),288(4.16)	136-0101-84L
$C_{20}H_{27}NO_{13}$			
α-D-Glucopyranose, 2-deoxy-2-[[3-meth-oxy-2-(methoxycarbonyl)-3-oxo-1-pro-penyl]amino]-, 1,3,4,6-tetraacetate	EtOH	277(4.17)	136-0101-84L
$C_{20}H_{27}O_2P$			
2-Pentenal, 3,3'-(phenylphosphinidene)-bis[4,4-dimethyl-	MeOH	241(4.14),317(3.76), 371(3.31),386(3.16)	89-0894-84
$C_{20}H_{27}P$			
1H-Phosphepin, 2,7-bis(1,1-dimethyl-	hexane	220(4.65),255(4.30),	89-0894-84

Compound	Solvent	$\lambda_{max}(\log \epsilon)$	Ref.
ethyl)-1-phenyl- (cont.)		294(3.86)	89-0894-84
$C_{20}H_{28}$ 1,1'-Bi-2,5-cyclohexadienylidene, 3,3',4,4,4',4',5,5'-octamethyl-	CH_2Cl_2	263(3.93),322s(--), 337(4.77),353(4.77)	5-0340-84
Phenanthrene, 8-ethyl-1,4,4a,9,10,10a- hexahydro-1,1,4a,7-tetramethyl-, (4aS-trans)-	MeOH	220(3.9),267(2.63)	102-1293-84
$C_{20}H_{28}F_2O_2$ Prosta-5,10,13-trien-1-oyl fluoride, 15-fluoro-9-oxo-, (5Z,13E,15R)-	EtOH	217(3.78)	30-0386-84
$C_{20}H_{28}HgO_4RuS_2$ Mercury(2+), [1,1'-[3,6,9,12-tetraoxa- tetradecane-1,14-diylbis(thio)ruthe- nocene-0,0',0'',0''',S,S']-, dichloride	MeCN	310(2.66)	88-1991-84
$C_{20}H_{28}N_2$ 1,4-Benzenediamine, N^1,2-bis(2-methyl- propyl)-N^4-phenyl-	n.s.g.	287(3.93)	30-0009-84
1H-Indole-3-ethanamine, N-[1-methyl- 1-(4-methyl-3-cyclohexen-1-yl)- ethyl]-, (S)-	EtOH	224(4.51),276s(3.73), 284(3.77),293(3.71)	33-1040-84
$C_{20}H_{28}N_2O_3$ 5H-[1]Benzopyrano[3,4-c]pyridin-5-one, 3-[2-(diethylamino)propyl]-1,2,3,4- tetrahydro-8-methoxy-	MeOH	250(3.41),320(4.22)	4-1561-84
$C_{20}H_{28}N_2O_5$ 1H-Pyrrole-2-carboxylic acid, 5-[[3-(2- methoxy-2-oxoethyl)-3-methyl-5-oxo-2- pyrrolidinylidene]methyl]-3,4-dimeth- yl-, 1,1-dimethylethyl ester, (Z)-(±)-	$CHCl_3$	289s(4.08),303(4.09)	49-0101-84
$C_{20}H_{28}N_2O_6$ 2H-[1]Benzopyrano[3,4-c]pyridazine- 3,4-dicarboxylic acid, 1,4a,5,10b- tetrahydro-8-methoxy-2,2,5,5-tetra- methyl-, dimethyl ester, trans	EtOH	211(4.28),283(3.67), 289(3.63)	39-2815-84C
$C_{20}H_{28}N_2O_8$ L-ribo-Hexar-1-amic acid, 4-amino-4,5- dideoxy-N-[1-(3,4-dihydro-8-hydroxy- 1-oxo-1H-2-benzopyran-3-yl)-3-methyl- butyl]-, [S-(R*,R*)]-	MeOH	246(3.80),314(3.65)	78-2519-84
$C_{20}H_{28}O$ Abieta-8,11,13-triene, 7-oxo-	EtOH	254(4.08),302(3.36)	23-2822-84
Benzene, 1-methoxy-4-[6-methyl-4-(2- methyl-1-propenyl)-5,7-octadienyl]-	EtOH	229(4.38)	154-1435-84
Cleistantha-8,11,13-trien-3-one	MeOH	219(3.85),267(2.62)	102-1293-84
Retinal	isoPrOH	380(4.60)	89-0081-84
7-cis	hexane	359(4.65)	78-0473-84
	EtOH	377(4.58)	78-0473-84
7-cis,9-cis	hexane	351(4.63)	78-0473-84
7-cis,11-cis	hexane	355(4.27)	78-0473-84
	EtOH	374(4.20)	78-0473-84
9-cis,11-cis	hexane	352(4.49)	78-0473-84
	EtOH	368(4.43)	78-0473-84

Compound	Solvent	$\lambda_{max}(\log \epsilon)$	Ref.
Retinal, 7-cis,9-cis,11-cis	hexane	346(4.34)	78-0473-84
Retinal, 7-cis,9-cis,13-cis	hexane	346(4.56)	78-0473-84
Retinal, all-trans	isoPrOH	388(4.63)	89-0251-84
$C_{20}H_{28}O_2$			
Androsta-4,6-dien-17-one, 3β-methoxy-	n.s.g.	234(4.30),240(4.34), 249(4.18)	39-2941-84C
2(1H)-Phenanthrenone, 8-ethyl-3,4,4a-9,10,10a-hexahydro-9-hydroxy-1,1,4a-7-tetramethyl-, [4aS-(4aα,9α,10a)]-	MeOH	220(3.85),267(2.68)	102-1293-84
$C_{20}H_{28}O_3$			
2H-9,4a-(Epoxymethano)phenanthrene-5,6-diol, 1,3,4,9,10,10a-hexahydro-1,1-dimethyl-7-(1-methylethyl)-, [4aR-(4aα,9α,10aβ)]- (20-deoxo-carnosol)	MeOH MeOH-base	224(3.78),278(3.15) 224(3.83),248(3.74), 291(3.43)	102-1677-84 102-1677-84
2(5H)-Furanone, 4-[2-(1,2,3,4,4a,7,8-8a-octahydro-1,2,4a,5-tetramethyl-4-oxo- (solidagolactone V)	EtOH	205.3(4.32)	88-2809-84
Retinal, 5,8-epidioxy-	hexane	316(4.32),332(4.28)	138-1673-84
$C_{20}H_{28}O_4$			
Bicyclo[10.2.2]hexadeca-12,14,15-tri-ene-13,14-dicarboxylic acid, dimethyl ester	EtOH	211(4.56),291(3.30)	24-2293-84
5,9-Methanobenzocyclooctene-5(1H)-carb-oxylic acid, 2,3,4,6,7,8,9,10-octa-hydro-2,2,7,7,9-pentamethyl-3,4-dioxo-, methyl ester	n.s.g.	210(3.48),257(3.87)	49-0793-84
5,9-Methanobenzocyclooctene-5(1H)-carb-oxylic acid, 2,3,4,6,7,8,9,10-octa-hydro-2,2,7,7,9-pentamethyl-4,10-dioxo-, methyl ester	n.s.g.	218(3.65),272(3.94)	49-0809-84
$C_{20}H_{28}O_4RuS_2$			
Ruthenocene, 1,1'-[3,6,9,12-tetraoxa-tetradecane-1,14-diylbis(thio)]-	MeCN	320(2.67)	88-1991-84
$C_{20}H_{28}O_5$			
Bicyclo[10.2.2]hexadeca-12,14,15-tri-ene-13,14-dicarboxylic acid, 15-hy-droxy-, dimethyl ester	EtOH	219(4.25),257(3.72), 296(3.36)	24-2293-84
1,4-Epoxybenzocyclododecene-2,3-dicarb-oxylic acid, 1,4,5,6,7,8,9,10,11,12-13,14-dodecahydro-, dimethyl ester	EtOH	204(3.70),238(3.09)	24-2293-84
4β,5α-Epoxy-trans-germacra-1(10)-en-12,6β-olide, 8α-[(2-methylbutanoyl)-oxy]-	MeOH	201(4.11)	102-1063-84
14-Oxatricyclo[10.3.2.0^{13,15}]heptadeca-1(16),12(17)-diene-16,17-dicarboxylic acid, dimethyl ester	EtOH	213(4.00),286(3.56)	24-2293-84
Oxireno[4,5]cyclotetradeca[1,2-b]furan-12(1aH)-one, 2,3,6,7,10,10a,13,13a-14,14a-decahydro-10-hydroxy-1a-(hy-droxymethyl)-5,9-dimethyl-13-methyl-ene-	n.s.g.	215(3.13),225(3.28)	12-0545-84
9(1H)-Phenanthrenone, 2,3,4,4a,10,10a-hexahydro-5,6,8,10-tetrahydroxy-1,1,4a-trimethyl-7-(1-methylethyl)-	ether	212(4.18),242s(3.87), 291(4.12),355(3.72)	33-1523-84

Compound	Solvent	λ_{max}(log ϵ)	Ref.
Spiro[cyclopropane-1,2'(1'H)-phenan-threne]-1',4'(3'H)-dione, 4'b,5'-6',7',8',8'a,9',10'-octahydro-3',9',10'-trihydroxy-2,4'b,8',8'-tetramethyl- (spirocoleon 12)	MeOH	232(3.98)	33-0201-84
C$_{20}$H$_{28}$O$_6$			
Tirotundifolin E	MeOH	202(4.74)	102-0823-84
C$_{20}$H$_{28}$O$_7$			
Shinjulactone G	EtOH	237(3.72)	18-2013-84
C$_{20}$H$_{28}$O$_{10}$			
Miyaginin	MeOH	221(4.00),272(3.08), 279(3.00)	18-2893-84
Ptelatoside B	MeOH	254(4.25),288s(3.28), 299s(3.04)	138-0397-84
C$_{20}$H$_{29}$BrO$_3$			
5,9-Methanobenzocyclooctene-5(1H)-carb-oxylic acid, 3-bromo-2,3,4,6,7,8,9,10-octahydro-2,2,7,7,9-pentamethyl-4-oxo-, methyl ester	n.s.g.	261(3.76)	49-0793-84
C$_{20}$H$_{29}$NO$_5$			
Benzoic acid, 2-acetoxy-6-[3-(acetyl-amino)-2,2,5-trimethylhexyl]-	MeOH	236(3.52),290(3.22)	78-2519-84
C$_{20}$H$_{29}$N$_3$O$_2$			
4-Quinolinecarboxamide, 2-butoxy-N-[2-(diethylamino)ethyl]-(dibucaine)	pH 0.6	245.5(4.42),317.2(3.96)	149-0755-84A
	pH 7.0	239.1(4.23),326(3.64)	149-0755-84A
	pH 12.6	238(4.28),326(3.64)	149-0755-84A
C$_{20}$H$_{29}$O$_8$P			
α-D-Ribofuranose, 3-deoxy-3-[(diethoxy-phosphinyl)methyl]-1,2-O-(1-methyl-ethylidene)-, benzoate	MeOH	226(4.07),273(2.90), 279(2.85)	78-0079-84
C$_{20}$H$_{30}$			
Phenanthrene, 8-ethyl-1,2,3,4,4a,9,10-10a-octahydro-1,1,4a,7-tetramethyl-, (4aS-trans)-	MeOH	226(3.30),270(2.16)	102-1293-84
C$_{20}$H$_{30}$N$_2$O$_2$			
Guettardine	EtOH	226(4.54),277(3.89), 284(3.91),292(3.85)	88-2767-84
C$_{20}$H$_{30}$N$_2$O$_3$			
Butanamide, N-(5,5-dimethyl-3-oxo-1-cyclohexen-1-yl)-3-[(5,5-dimethyl-3-oxo-1-cyclohexen-1-yl)amino]-	pH 1	285(4.59)	39-0287-84C
	H$_2$O	291(4.58)	39-0287-84C
	pH 13	299(4.65),342(3.98) (changing)	39-0287-84C
Propanamide, N-(5,5-dimethyl-3-oxo-1-cyclohexen-1-yl)amino]-3-[(5,5-di-methyl-3-oxo-1-cyclohexen-1-yl)-amino]-2-methyl-	pH 1	287(4.57)	39-0287-84C
	H$_2$O	296(4.59)	39-0287-84C
	pH 13	300(4.62),342(4.05) (changing)	39-0287-84C
C$_{20}$H$_{30}$N$_2$O$_5$			
1H-Pyrrole-2-carboxylic acid, 5-[[3-(2-methoxy-2-oxoethyl)-3-methyl-5-oxo-2-pyrrolidinyl]methyl]-3,4-dimethyl-,	CHCl$_3$	276(4.21)	49-0101-84

Compound	Solvent	$\lambda_{max}(\log \epsilon)$	Ref.
1,1-dimethylethyl ester (cont.)			49-0101-84
$C_{20}H_{30}N_2O_8$			
Propanedioic acid, 2,2'-(1,4-pipera-zinediyldimethylidyne)bis-, tetra-ethyl ester	MeCN	314(4.52)	62-0947-84A
$C_{20}H_{30}O$			
Dolabella-4(Z),8(E),18-trien-16-al, (1R,11S,12R)-	EtOH	242(3.88)	102-1681-84
Ferruginol	EtOH	205(4.00),281(3.38)	2-0177-84
	EtOH-NaOH	217(3.75),290(3.34)	2-0177-84
2-Phenanthrenol, 8-ethyl-1,2,3,4,4a,9-10,10a-octahydro-1,1,4a,7-tetramethyl-	MeOH	220(3.84),271(2.66)	102-1293-84
$C_{20}H_{30}O_2$			
1H-2-Benzopyran-8-methanol, 3,4-dihy-dro-4-methyl-5-(1,3,3-trimethylcyclo-hexyl)-, [R-(R*,S*)]-	EtOH	231(2.90),265(2.48), 299(2.00)	12-1081-84
4-Isobenzofuranethanol, 1,3-dihydro-β-methyl-5-(1,3,3-trimethylcyclo-hexyl)-, [R-(R*,S*)]-	EtOH	227(2.85),243(2.85), 278(2.48),288(2.40)	12-1081-84
$C_{20}H_{30}O_3$			
1H-Cyclonona[c]furan-1-one, 4-(1,5-di-methyl-4-hexenyl)-3,3a,4,5,6,9-hexa-hydro-5-hydroxy-7-methyl-	MeOH	206(4.00),225s(3.81)	138-0231-84
2-Cyclopentene-1-acetaldehyde, α-(4,8-dimethyl-3,7-nonadienyl)-2-formyl-3-(hydroxymethyl)-	MeOH	251(3.74)	78-2913-84
Dictyodial, hydroxy-	MeOH	228(3.70)	138-0231-84
2(5H)-Furanone, 4-[2-(1,2,3,4,4a,7,8-8a-octahydro-4-hydroxy-1,2,4a,5-tetramethyl-1-naphthalenyl)ethyl]-(solidagolactone IV)	EtOH	204.9(4.26)	88-2809-84
$C_{20}H_{30}O_4$			
Jolkinol, 17-hydroxy-	MeOH	278(4.11)	102-1461-84
5,9-Methanobenzocyclooctene-5(1H)-carb-oxylic acid, 2,3,4,6,7,8,9,10-octahy-dro-3-hydroxy-2,2,7,7,9-pentamethyl-4-oxo-, methyl ester	n.s.g.	206s(3.33),248(3.98)	49-0793-84
5,9-Methanobenzocyclooctene-5(1H)-carb-oxylic acid, 2,3,4,6,7,8,9,10-octahy-dro-10-hydroxy-2,2,7,7,9-pentamethyl-4-oxo-, methyl ester	n.s.g.	247(3.97)	49-0809-84
5,9-Methanobenzocyclooctene-5(1H)-carb-oxylic acid, 2,3,4,6,7,8,9,10-octahy-dro-3-methoxy-2,2,7,7,9-pentamethyl-4-oxo-, (3α,5α,9β)-	n.s.g.	205(3.40),250(3.96)	49-0809-84
$C_{20}H_{30}O_5$			
2(5H)-Furanone, 4-[2-[octahydro-8-hy-droxy-8a-(hydroxymethyl)-5,6-dimeth-ylspiro[naphthalen-1(2H),2'-oxiren]-4-yl]ethyl]- (deacetylajugarin II)	EtOH	218.5(3.95)	102-0849-84
Glutaric acid, 2-[[3,5-bis(1,1-dimeth-ylethyl)-4-hydroxyphenyl)methyl]-	EtOH	202.5(4.672),232.5(3.88), 277.8(2.699)	104-1732-84
Teumassilin	EtOH	212(3.61)	102-0849-84

Compound	Solvent	$\lambda_{max}(\log \epsilon)$	Ref.
$C_{20}H_{31}ClO_2Si_2$			
Cyclobutanone, 3-(4-chlorophenyl)-3-[[(1,1-dimethylethyl)dimethyl-silyl]oxy]-2,2-diethyl-	MeOH	268(2.36)	35-4566-84
3-Hexen-1-one, 1-(4-chlorophenyl)-3-[[(1,1-dimethylethyl)dimethyl-silyl]oxy]-4-ethyl-	MeOH	250(4.16)	35-4566-84
$C_{20}H_{31}NO_3$			
2-Cyclohexen-1-one, 3-methoxy-5,5-di-methyl-2-(7-morpholinobicyclo[4.1.0]-hept-7-yl)-, (1α,6α,7β)-	MeOH	235(4.0)	24-2910-84
$C_{20}H_{31}NO_5$			
Oxysanguinarine	EtOH	241(4.27),281s(4.61), 289(4.70),331(4.17), 348(4.18),370(4.06), 385(4.02)	100-0001-84
$C_{20}H_{32}N_4O_5$			
Inosine, 2-decyl-	pH 1	250(3.94)	87-0429-84
	pH 7	251(3.93)	87-0429-84
	pH 13	253(3.97)	87-0429-84
$C_{20}H_{32}O$			
5-Hexen-2-one, 1-[3-(1,5-dimethylhex-yl)-2-methyl-1,3-cyclopentadien-1-yl]-	EtOH	307(4.43)	23-0121-84
$C_{20}H_{32}O_2$			
Ambliol C	pentane	211(3.61)	44-5160-84
$C_{20}H_{32}O_3$			
Norpectinatone	EtOH	232(4.12),300(3.88)	44-2506-84
$C_{20}H_{32}O_4$			
Benzoic acid, 2,4-dihydroxy-6-tridecyl-	EtOH	260(3.90),300(3.55)	100-0530-84
2,4-Cyclohexadiene-1-carboxylic acid, 6,6-dimethoxy-1-methyl-, 5-methyl-2-(1-methylethyl)cyclohexyl ester, [1S-[1α(R*),2β,5α]]-	MeOH	260(3.51)	44-4429-84
[1S-[1α(S*),2β,5α]]-	MeOH	260(3.52)	44-4429-84
Oxacyclooctadeca-4,6-dien-2-one, 7-ace-toxy-3-methyl-, (E,Z)-(±)-	MeOH	241(4.37)	89-0440-84
$C_{20}H_{32}O_5$			
2(5H)-Oxepinone, 7-(1,6-dihydroxy-4-methyl-4-hexenyl)-6,7-dihydro-2-(5-hydroxy-4-methyl-3-pentenyl)-7-methyl-	MeOH	203(4.07),220s(3.70)	102-0829-84
$C_{20}H_{32}O_6$			
Melfusanolide, 1,10,17-trihydroxy-	MeOH	203(3.90)(end abs.)	102-0833-84
$C_{20}H_{33}N$			
Pyridine, 2-(4-hexylcyclohexyl)-5-propyl-	EtOH	215(3.94),271(3.62)	103-0515-84
$C_{20}H_{34}Cl_2N_2O_2Zn$			
Zinc, dichlorobis(3,3a,4,5,6,7-hexahy-dro-2,2-dimethyl-2H-indole 1-oxide-O)-, (T-4)-	EtOH	235(4.32)	12-0117-84

Compound	Solvent	$\lambda_{max}(\log \epsilon)$	Ref.
$C_{20}H_{34}N_4O$ Benzamide, 4-(3-dodecyl-3-methyl- 1-triazenyl)-	EtOH	322(4.33)	87-0870-84
$C_{20}H_{34}O_2Sn$ Benzoic acid, 4-(tributylstannyl)-, methyl ester	MeOH	242(4.14)	70-1044-84
$C_{20}H_{34}O_3Si$ 5H-4a,9-Methanocyclonona[1,6]benz[1,2- b]oxiren-2(1aH)-one, 12-[[(1,1-di- methylethyl)dimethylsilyl]oxy]octa- hydro-	EtOH	296(1.51)	35-1446-84
3a,8-Methano-3aH-cyclopentacyclodecen- 1(4H)-one, 12-[[(1,1-dimethylethyl)- dimethylsilyl]oxy]-2,3,5,6,7,8,9,10- octahydro-11-hydroxy-, (3aα,8α,12R*)-	EtOH	287(3.95)	35-1446-84
$C_{20}H_{35}NO_3Si$ Propanamide, 3-[[(1,1-dimethylethyl)di- methylsilyl]oxy]-N-(2-hydroxy-1-meth- yl-2-phenylethyl)-N,2-dimethyl-, [1R-[1R*(S*),2S*]]-	EtOH	256(2.36),262(2.28)	44-5202-84
isomer	EtOH	259(1.93),264(1.93)	44-5202-84
$C_{20}H_{36}N_2O_4$ Neoenactin B_1, sulfate (2:1)	MeOH and MeOH-HCl	211(3.70)	158-84-168
	MeOH-NaOH	238(3.76)	158-84-168
$C_{20}H_{38}N_2O_3$ Diazenecarboxylic acid, (1-oxoocta- decyl)-, methyl ester	C_6H_{12}	429(1.51)	23-0574-84
$C_{20}H_{38}N_2O_4$ Diazenedicarboxylic acid, ethyl hexa- decyl ester	C_6H_{12}	404(1.55)	23-0574-84
$C_{20}H_{38}N_2O_5$ Neoenactin B_2, sulfate (2:1)	MeOH and MeOH-HCl	211(3.73)	158-84-168
	MeOH-NaOH	238(3.78)	158-84-168
$C_{20}H_{40}O_6S_2Si_2$ Silane, (1,1-dimethylethyl)[[8-[[(1,1- dimethylethyl)dimethylsilyl]oxy]- 1,4-dithiaspiro[4.4]non-6-en-6-yl]- methoxy]dimethyl]-, S,S,S',S'-tetra- oxide, (R)-	EtOH	214(3.75)	39-2089-84C
$C_{20}H_{40}Si_5$ 5,6,11,16,21-Pentasilapentaspiro- [4.0.4.0.4.0.4.0.4.0]pentacosane	isooctane	212s(4.52),290(3.02)	35-5521-84
$C_{20}H_{48}Si_4$ Cyclotetrasilane, 1,2,3,4-tetrakis(1,1- dimethylethyl)-1,2,3,4-tetramethyl-	C_6H_{12}	247s(3.32),262s(3.26), 301(2.28)	157-0141-84
$C_{20}H_{50}Si_5$ Cyclopentasilane, decaethyl-	C_6H_{12}	265s(3.15)	157-0141-84

Compound	Solvent	$\lambda_{max}(\log \epsilon)$	Ref.
$C_{21}F_{39}N_3$ 1,3,5-Triazine, 2,4,6-tris(trideca-fluorohexyl)-	heptane	220(2.40)	70-0842-84
$C_{21}H_4Cl_{12}O_2$ 9H-Fluoren-3-ol, 1,2,4,5,6,7,8-hepta-chloro-9-(pentachlorophenyl)-, acetate	C_6H_{12}	219(4.85),226s(4.82), 293(4.40),304s(4.37)	44-0770-84
$C_{21}H_5F_{14}NO$ Benzenamine, N-[bis[2,3,5,6-tetraflu-oro-4-(trifluoromethyl)phenyl]meth-ylene]-	EtOH	292(4.21)	70-1268-84
$C_{21}H_6Cl_4I_4O_5$ Benzoic acid, 2,3,4,5-tetrachloro-6-(6-hydroxy-2,4,5,7-tetraiodo-3-oxo-3H-xanthen-9-yl)-, methyl ester, sodium salt (in 50% dioxan)	pH 1 pH 8	425(4.11),505(4.26) 529(4.53),568(5.00)	88-4285-84 88-4285-84
$C_{21}H_{10}N_4$ Propanedinitrile, 2,2'-(2-methyl-9,10-anthracenediylidene)bis-	MeCN	284(4.46),305(4.27), 345(4.39)	5-0618-84
$C_{21}H_{12}Br_2N_4O_3$ Isoxazole, 4-[(2,6-dibromo-4-nitrophen-yl)azo]-3,5-diphenyl-	EtOH	342(3.88)	104-1844-84
$C_{21}H_{12}N_2O_3$ 6H-Anthra[1,9-cd]isoxazole-3-carbox-amide, 6-oxo-N-phenyl- 1H-Anthra[1,2-d]pyrazole-3,6,11(2H)-trione, 2-phenyl-	toluene dioxan	459(4.20) 462(3.48)	104-0795-84 104-0795-84
$C_{21}H_{12}N_2O_5$ Benzamide, N-(9,10-dihydro-9,10-dioxo-1-anthracenyl)-4-nitro-	benzene	416(3.82)	39-0529-84C
$C_{21}H_{12}N_4$ Benzimidazo[1,2-b]phenanthro[9,10-e]-[1,2,4]triazine	DMF	330(3.82),420(3.31)	104-1223-84
$C_{21}H_{12}O_4$ Dinaphtho[1,2-d:2',1'-d]furan-5,6-di-one, 8-methoxy-	$CHCl_3$	262s(4.56),277(4.69), 310s(4.06),327(3.96), 345(3.93),425(3.19), 505s(3.46),530(3.50)	5-1367-84
$C_{21}H_{13}BrN_2O_2$ 6H-Anthra[1,9-cd]isoxazol-6-one, 3-bromo-5-[(3-methylphenyl)amino]-	toluene	536(4.25)	104-1592-84
$C_{21}H_{13}BrN_4$ [1,2,4]Triazolo[1,5-c]quinazoline, 2-(4-bromophenyl)-5-phenyl-	EtOH	300(3.51)	18-1138-84
$C_{21}H_{13}ClN_2O_4S$ 2H-Isoindole-2-acetamide, N-[3-(2-chlorobenzoyl)-2-thienyl]-1,3-dihydro-1,3-dioxo-	MeOH	231s(4.50),240(4.45), 274(4.00),340(4.00)	73-0621-84

Compound	Solvent	λ_{max}(log ϵ)	Ref.
$C_{21}H_{13}ClN_4$			
[1,2,4]Triazolo[1,5-c]quinazoline, 2-(3-chlorophenyl)-5-phenyl-	EtOH	305(3.35)	18-1138-84
[1,2,4]Triazolo[1,5-c]quinazoline, 2-(4-chlorophenyl)-5-phenyl-	EtOH	300(3.78)	18-1138-84
$C_{21}H_{13}Cl_2NS$			
Benzothiazole, 2-[2-(2,4-dichlorophenyl)-1-phenylethenyl]-, cis	C_6H_{12}	336.9(4.31)	131-0417-84H
trans	C_6H_{12}	337.8(4.42)	131-0417-84H
Benzothiazole, 2-[2-(2,6-dichlorophenyl)-1-phenylethenyl]-	C_6H_{12}	311.7(4.29)	131-0417-84H
$C_{21}H_{13}NO_2$			
8H-Benzo[h]naphtho[1,2,3-de]quinolin-8-one, 2-methoxy-	N-Mepyrrolidine	405(3.91),425(3.97)	104-0980-84
$C_{21}H_{13}NO_3$			
Benzamide, N-(9,10-dihydro-9,10-dioxo-1-anthracenyl)-	benzene	420(3.80)	39-0529-84C
$C_{21}H_{13}NO_4$			
Naphtho[2,3-b]phenoxazine-7,12-dione, 6-hydroxy-2-methyl-	toluene	600(4.29)	104-2234-84
$C_{21}H_{13}NO_4S$			
6H-Anthra[1,9-cd]isoxazol-6-one, 3-[(4-methylphenyl)sulfonyl]-	toluene	300(4.08),466(3.81)	104-2234-84
8H-Naphtho[2,3-a]phenothiazine-8,12(13H)-dione, 3-methyl-, 5,5-dioxide	toluene	330(4.24),500(3.75)	104-2234-84
$C_{21}H_{13}NO_5S$			
6H-Anthra[1,9-cd]isoxazol-6-one, 5-hydroxy-3-[(4-methylphenyl)sulfonyl]-	toluene	300(4.01),480(3.88), 505(3.84)	104-2234-84
8H-Naphtho[2,3-a]phenothiazine-8,12(13H)-dione, 7-hydroxy-3-methyl-, 5,5-dioxide	toluene	340(3.97),585(3.91)	104-2234-84
$C_{21}H_{13}N_3O_4S$			
9,10-Anthracenedione, 1-azido-2-[(4-methylphenyl)sulfonyl]-	toluene	355(3.67)	104-2234-84
$C_{21}H_{13}N_3O_5$			
Isoxazole, 3,5-bis(4-nitrophenyl)-4-phenyl-	dioxan	305(4.39)	12-1217-84
$C_{21}H_{13}N_3O_5S$			
9,10-Anthracenedione, 1-azido-4-hydroxy-2-[(4-methylphenyl)sulfonyl]-	toluene	455(3.72)	104-2234-84
$C_{21}H_{13}N_5O_2$			
[1,2,4]Triazolo[1,5-c]quinazoline, 2-(3-nitrophenyl)-5-phenyl-	EtOH	301(3.90)	18-1138-84
[1,2,4]Triazolo[1,5-c]quinazoline, 2-(4-nitrophenyl)-5-phenyl-	EtOH	299(4.17)	18-1138-84
$C_{21}H_{13}N_5O_5$			
Isoxazole, 4-[(2,4-dinitrophenyl)azo]-3,5-diphenyl-	EtOH	370(4.05)	104-1844-84

Compound	Solvent	$\lambda_{max}(\log \epsilon)$	Ref.
$C_{21}H_{14}BrNS$			
Benzothiazole, 2-[2-(2-bromophenyl)-1-phenylethenyl]-	C_6H_{12}	331.5(4.28)	131-0417-84H
Benzothiazole, 2-[2-(4-bromophenyl)-1-phenylethenyl]-	C_6H_{12}	341.5(4.52)	131-0417-84H
$C_{21}H_{14}Br_2O_5$			
4H-1-Benzopyran-4-one, 6-bromo-3-[(6-bromo-4-oxo-4H-1-benzopyran-3-yl)-methylene]-2-ethoxy-2,3-dihydro-	EtOH	225(4.50),319(4.10), 352(4.02)	44-2812-84
4H-1-Benzopyran-4-one, 3,3'-(ethoxy-methylene)bis[6-bromo-	CHCl$_3$	247(4.66),310(4.08)	44-2812-84
$C_{21}H_{14}Br_4O_5S$			
Bromcresol Green	acid	440(4.20)	35-0265-84
	base	615(4.59)	35-0265-84
with hexadecyltrimethylammonium	acid	420(4.20)	35-0265-84
bromide	base	627(4.59)	35-0265-84
$C_{21}H_{14}ClNOS$			
2(1H)-Pyridinone, 4-(4-chlorophenyl)-3-phenyl-6-(2-thienyl)-	EtOH	246(4.11),276(4.14), 363(4.20)	4-1473-84
2(1H)-Pyridinone, 6-(4-chlorophenyl)-3-phenyl-4-(2-thienyl)-	EtOH	274(4.19),322(4.11), 355(4.14)	4-1473-84
$C_{21}H_{14}ClNS$			
Benzothiazole, 2-[2-(2-chlorophenyl)-1-phenylethenyl]-	C_6H_{12}	330.7(4.33)	131-0417-84H
Benzothiazole, 2-[2-(3-chlorophenyl)-1-phenylethenyl]-	C_6H_{12}	338.7(4.45)	131-0417-84H
Benzothiazole, 2-[2-(4-chlorophenyl)-1-phenylethenyl]-	C_6H_{12}	340.6(4.49)	131-0417-84H
$C_{21}H_{14}Cl_2N_2O_3S$			
Benzo[b]thiophen-3(2H)-one, 2-[bis[(3-chlorophenyl)amino]methylene]-, 1,1-dioxide	EtOH	214(4.67),247(4.25), 278(4.24),310(4.02), 360(4.37)	95-0134-84
Benzo[b]thiophen-3(2H)-one, 2-[bis[(4-chlorophenyl)amino]methylene]-, 1,1-dioxide	EtOH	277(--),311(--), 368(--)	95-0134-84
$C_{21}H_{14}F_3N_3O_2$			
1H-Benzimidazole-1-carboxamide, 2,3-dihydro-2-oxo-N,3-diphenyl-6-(tri-fluoromethyl)-	MeOH	212(4.42),247(4.33)	39-2587-84C
$C_{21}H_{14}N_2O_2S$			
Benzothiazole, 2-[2-(3-nitrophenyl)-1-phenylethenyl]-	C_6H_{12}	336.0(4.16)	131-0417-84H
Benzothiazole, 2-[2-(4-nitrophenyl)-1-phenylethenyl]-	C_6H_{12}	357.1(4.51),372(4.47), 390.0(4.25)	131-0417-84H
$C_{21}H_{14}N_2O_3$			
2H-Indol-2-one, 1,3-dihydro-3-[5-(1H-indol-3-yl)-2-oxo-3(2H)-furanyli-dene]-1-methyl-	CHCl$_3$	265(4.17),315(3.51), 506(4.39)	39-1331-84B
$C_{21}H_{14}N_2O_4$			
2H-Indol-2-one, 1,3-dihydro-3-[5-(1H-indol-3-yl)-2-oxo-3(2H)-furanyli-dene]-5-methoxy-	CHCl$_3$	277(4.29),508(4.16)	39-1331-84B

Compound	Solvent	$\lambda_{max}(\log \epsilon)$	Ref.
$C_{21}H_{14}N_2O_4$ Phenol, 2,5-bis(2-benzoxazolyl)-4-methoxy-	heptane	<u>312(4.5)</u>,324(4.6), <u>409(4.3)</u>	131-0337-84H
$C_{21}H_{14}N_4$ [1,2,4]Triazolo[1,5-c]quinazoline, 2,5-diphenyl-	EtOH	300(3.80)	18-1138-84
$C_{21}H_{14}N_4O_2S_2$ Pyrimido[5,4-e]thiazolo[3,4-a]pyrimidin-10-ium, 1,2,3,4-tetrahydro-9-(methylthio)-2,4-dioxo-3,7-diphenyl-, hydroxide, inner salt	DMF	387(4.16),398(4.15), 445(4.11)	103-0921-84
$C_{21}H_{14}N_4O_5$ Pyrrolo[3,4-c]pyrazole-4,6(1H,5H)-dione, 3-[(1,3-dihydro-1,3-dioxo-2H-isoindol-2-yl)acetyl]-3a,6a-dihydro-5-phenyl-	MeOH	218(4.89),238(4.56), 274(4.17),300(4.12)	103-0066-84
$C_{21}H_{14}N_6S_2$ 2(3H)-Benzothiazolone, [(2-benzothiazolylazo)phenylmethylene]hydrazone monohydrobromide	benzene benzene	565(3.58) 520(3.56),610s(3.30)	135-0577-84 135-0577-84
$C_{21}H_{14}O$ 2,5-Cyclohexadien-1-one, 4-(2,2-diphenyl-2-cyclopropen-1-ylidene)-	MeCN dioxan	275(4.41),372s(4.58), 388(4.79),476s(3.50) <u>376(4.7)</u>,<u>400(4.7)</u>	88-0073-84 88-0073-84
$C_{21}H_{14}O_4$ [2,2'-Binaphthalene]-1,4-dione, 1'-hydroxy-4'-methoxy-	CHCl$_3$	251(4.51),322(4.12), 577(3.19)	5-1367-84
$C_{21}H_{14}O_6$ 4H,10H-Benzo[1,2-b:3,4-b']dipyran-8-carboxylic acid, 3-methyl-4,10-dioxo-2-phenyl-, methyl ester	MeOH	283(4.28)	2-0969-84
$C_{21}H_{15}BO_6$ Boron, (1,7-diphenyl-1,6-heptadiene-3,5-dionato-O,O')[ethanedioato(2-)-O,O']-, (T-4)-	CH$_2$Cl$_2$	467(4.88)	97-0292-84
$C_{21}H_{15}BrN_4$ Benzaldehyde, 4-bromo-, (2-phenyl-4-quinazolinyl)hydrazone	EtOH	370(4.10)	18-1138-84
$C_{21}H_{15}BrN_4O$ Benzamide, 3-bromo-N-(2,5-diphenyl-2H-1,2,3-triazol-4-yl)-	50% dioxan	290(4.35)	39-0785-84B
Benzamide, 4-bromo-N-(2,5-diphenyl-2H-1,2,3-triazol-4-yl)-	50% dioxan	295(4.38)	39-0785-84B
Methanone, [5-(3-bromophenyl)-1,2,4-oxadiazol-3-yl]phenyl-, phenylhydrazone	50% dioxan	366(4.20)	39-0785-84B
Methanone, [5-(4-bromophenyl)-1,2,4-oxadiazol-3-yl]phenyl-, phenylhydrazone	50% dioxan	366(4.20)	39-0785-84B
$C_{21}H_{15}BrO_2$ 9H-Xanthene, 9-[bromo(4-methoxyphenyl)-	EtOH	277(3.94),320.5(4.10)	44-0080-84

Compound	Solvent	$\lambda_{max}(\log \epsilon)$	Ref.
methylene]- (cont.)			44-0080-84
$C_{21}H_{15}ClN_2S$			
Benzenamine, 4-chloro-N-(3,4-diphenyl-2(3H)-thiazolylidene)-	EtOH	308(4.24)	56-0447-84
$C_{21}H_{15}ClN_4$			
Benzaldehyde, 3-chloro-, (2-phenyl-4-quinazolinyl)hydrazone	EtOH	364(4.14)	18-1138-84
Benzaldehyde, 4-chloro-, (2-phenyl-4-quinazolinyl)hydrazone	EtOH	368(4.06)	18-1138-84
$C_{21}H_{15}ClN_4O$			
Benzamide, 3-chloro-N-(2,5-diphenyl-2H-1,2,3-triazol-4-yl)-	50% dioxan	290(4.39)	39-0785-84B
Benzamide, 4-chloro-N-(2,5-diphenyl-2H-1,2,3-triazol-4-yl)-	50% dioxan	295(4.39)	39-0785-84B
Methanone, [5-(3-chlorophenyl)-1,2,4-oxadiazol-3-yl]phenyl-, phenylhydra-zone, (Z)-	50% dioxan	366(4.21)	39-0785-84B
Methanone, [5-(4-chlorophenyl)-1,2,4-oxadiazol-3-yl]phenyl-, phenylhydra-zone, (Z)-	50% dioxan	366(4.22)	39-0785-84B
$C_{21}H_{15}ClO_4$			
Anthra[1,2-b]furan-6,11-dione, 4-(2-chloro-2-propenyl)-5-methoxy-2-methyl-	EtOH	240.5(4.25),280(4.25), 359s(3.64),362(3.77)	12-1518-84
$C_{21}H_{15}D_3$			
5H-Tribenzo[a,d,g]cyclononene-2,7,12-d_3, 10,15-dihydro-	MeCN	253s(2.71),259s(3.83), 266.5(3.93),274.5(2.92)	35-5997-84
$C_{21}H_{15}NO$			
1,3-Benzoxazepine, 2,4-diphenyl-	EtOH	260(4.49),346(3.89)	78-3567-84
1,3-Benzoxazepine, 2,5-diphenyl-	EtOH	240(4.35),342(4.05)	78-3567-84
$C_{21}H_{15}NO_2$			
1H-Isoindol-1-one, 3-benzoyl-2,3-di-hydro-2-phenyl-	EtOH	236(4.28),251(4.25), 273s(4.07),281s(4.04)	4-1499-84
	EtOH-KOH	263(4.06),413(3.95)	4-1499-84
$C_{21}H_{15}NS$			
Benzothiazole, 2-(1,2-diphenylethenyl)-	C_6H_{12}	337.8(4.36)	131-0417-84H
Benzothiazole, 2-(2,2-diphenylethenyl)-	C_6H_{12}	328.1(4.47),339.7(4.47), 360.0(4.28)	131-0417-84H
$C_{21}H_{15}N_3$			
2-Pyridinecarboxaldehyde, (9-anthra-cenylmethyl)hydrazone	C_6H_{12}	404(4.14)	97-0021-84
1,3,5-Triazine, 2,4,6-triphenyl-	$CHCl_3$	270(4.81)	70-0842-84
$C_{21}H_{15}N_3O$			
2H-1-Benzopyran, 2-azido-2,3-diphenyl-	ether	280(4.09),312(3.9)	78-3567-84
Isoxazole, 3,5-diphenyl-4-(phenylazo)-	EtOH	343(4.19)	104-1844-84
$C_{21}H_{15}N_3OS$			
3,5-Pyridinedicarbonitrile, 1,4-dihy-dro-2-(4-methoxyphenyl)-6-(3-methyl-phenyl)-4-thioxo-	EtOH	281(4.43),318(4.44), 376(3.48)	4-1445-84

Compound	Solvent	$\lambda_{max}(\log \epsilon)$	Ref.
$C_{21}H_{15}N_3O_2$			
Phenol, 2-[3-phenyl-4-(phenylazo)-5-isoxazolyl]-	EtOH	343(4.22)	104-1844-84
$C_{21}H_{15}N_3O_3$			
Acetic acid, cyano(2-naphthalenylhydrazono)-, 2-oxo-2-phenylethyl ester	EtOH	240s(4.56),278(4.02), 289(4.06),300(4.04), 345s(4.21),381(4.47)	5-1702-84
2,3-Furandione, 4-amino-5-benzoyl-, 3-(2-naphthalenylhydrazone)	EtOH	287(--),297(--), 311(--),370s(--), 417(--),487(--)	5-1702-84
2H-Indol-2-one, 3-[1,2-dihydro-5-(1H-indol-3-yl)-2-oxo-3H-pyrrol-3-ylidene]-1,3-dihydro-5-methoxy-	DMSO-d_6	282(4.25),375(3.91), 550(4.33)	39-1331-84B
$C_{21}H_{15}N_3S$			
3,5-Pyridinedicarbonitrile, 1,4-dihydro-2,6-bis(3-methylphenyl)-4-thioxo-	EtOH	269(4.48),328(4.26), 395(3.39)	4-1445-84
3,5-Pyridinedicarbonitrile, 1,4-dihydro-2,6-bis(4-methylphenyl)-4-thioxo-	EtOH	278(4.50),323(4.25), 393(3.39)	4-1445-84
3,5-Pyridinedicarbonitrile, 1,4-dihydro-2-(3-methylphenyl)-6-(4-methylphenyl)-4-thioxo-	EtOH	272(4.51),328(4.26), 393(3.40)	4-1445-84
$C_{21}H_{15}N_5O_2$			
Benzaldehyde, 3-nitro-, (2-phenyl-4-quinazolinyl)hydrazone	EtOH	370(3.74)	18-1138-84
Benzaldehyde, 4-nitro-, (2-phenyl-4-quinazolinyl)hydrazone	EtOH	376(3.98)	18-1138-84
1H-Imidazole, 2-[(2-nitrophenyl)azo]-4,5-diphenyl-	C_6H_{12}	460(4.35)	62-0302-84A
	EtOH	225(4.37),288(4.32), 438(4.33)	62-0302-84A
	DMF	484(4.37),496s(4.36)	62-0302-84A
	DMSO	450(4.26)	62-0302-84A
1H-Imidazole, 2-[(3-nitrophenyl)azo]-4,5-diphenyl-	C_6H_{12}	450(4.32)	62-0302-84A
	EtOH	223(4.48),292(4.40), 432(4.35)	62-0302-84A
	DMF	466(4.40),510s(4.10)	62-0302-84A
	DMSO	458(4.30)	62-0302-84A
1H-Imidazole, 2-[(4-nitrophenyl)azo]-4,5-diphenyl-	C_6H_{12}	475(4.19)	62-0302-84A
	EtOH	223(4.34),290(4.30), 464(4.28)	62-0302-84A
	DMF	500(4.37),570(4.19)	62-0302-84A
	DMSO	480(4.19)	62-0302-84A
3-Quinolinecarbonitrile, 2-amino-4-[(4,5-dihydro-3-methyl-5-oxo-1-phenyl-1H-pyrazol-4-yl)carbonyl]-	BuOH	277(4.23),303(4.26), 318(4.67)	4-1233-84
$C_{21}H_{15}N_5O_3$			
Benzamide, N-(2,5-diphenyl-2H-1,2,3-triazol-4-yl)-3-nitro-	50% dioxan	290(4.42)	39-0785-84B
Benzamide, N-(2,5-diphenyl-2H-1,2,3-triazol-4-yl)-4-nitro-	50% dioxan	290(4.46)	39-0785-84B
Methanone, [5-(3-nitrophenyl)-1,2,4-oxadiazol-3-yl]phenyl-, phenylhydrazone	50% dioxan	350(4.25)	39-0785-84B
Methanone, [5-(4-nitrophenyl)-1,2,4-oxadiazol-3-yl]phenyl-, phenylhydrazone	50% dioxan	354(4.27)	39-0785-84B

Compound	Solvent	$\lambda_{max}(\log \epsilon)$	Ref.
$C_{21}H_{15}O$			
Cyclopropenylium, (4-hydroxyphenyl)diphenyl-, tetrafluoroborate	MeCN	272(4.27),342(4.78)	88-0073-84
$C_{21}H_{16}Cl_2N_2$			
1H-1,5-Benzodiazepine, 2,4-bis(4-chlorophenyl)-2,3-dihydro-	EtOH	261(4.40),375(3.88)	103-1370-84
$C_{21}H_{16}NO_5$			
Chelirubine (chloride)	EtOH	231.5(4.50),281(4.46), 305s(3.99),341(4.24), 353.5(4.25),413(2.91), 508(3.17)	100-0001-84
$C_{21}H_{16}N_2OS$			
Benzo[b]thiophen-3(2H)-one, 2-[(2,2-diphenylhydrazino)methylene]-	heptane	380(4.43)	104-1353-84
	PrOH	375(4.43),440(3.42)	104-1353-84
	acetone	370(4.49)	104-1353-84
$C_{21}H_{16}N_2O_2$			
Cyclohepta[1,2-b:1,7-b']bis[1,4]benzoxazine, 3,12-dimethyl-isomer?	MeOH	237(4.28),286(4.28), 375(3.86)	138-1145-84
	MeOH	287(4.34),380(3.90)	138-1149-84
14H-Cyclohepta[1,2-b:4,3-b']bis[1,4]-benzoxazine, 2,12-dimethyl-	MeOH	265(--),277(--), 320(--),365(--), 400(--),425(--), 505(--)	138-1145-84
	MeOH-HCl	282(--),328(--), 418(--),545(--)	138-1145-84
9H-Fluorene-4-carboxylic acid, 9-(dicyanomethylene)-, butyl ester	$C_2H_4Cl_2$	352(4.33),440s(--)	23-2546-84
	PVC	352(4.14)	23-2546-84
$C_{21}H_{16}N_2O_3$			
9,10-Anthracenedione, 1-amino-2-methoxy-4-(phenylamino)-	C_6H_5Cl	530s(4.00),552(4.18), 582(4.20)	34-0482-84
$C_{21}H_{16}N_2O_3S$			
Benzo[b]thiophen-3(2H)-one, 2-[bis(phenylamino)methylene]-, 1,1-dioxide	EtOH	244(4.18),277(4.24), 310(4.01),357(4.29)	95-0134-84
$C_{21}H_{16}N_2O_4$			
1H-Benz[f]isoquino[8,1,2-hi]quinazoline-2(3H)-carboxylic acid, 5,6-dihydro-1,3-dioxo-, ethyl ester	MeOH	231s(4.34),249s(4.63), 256(4.74),265(4.56), 271s(4.50),285s(4.00), 296(4.00),321(3.81), 336(3.89),354(3.89), 372(3.95)	78-4003-84
$C_{21}H_{16}N_4$			
Benzaldehyde, (2-phenyl-4-quinazolinyl)hydrazone	EtOH	362(3.96)	18-1138-84
Benzimidazo[1,2-b]phenanthro[9,10-e]-[1,2,4]triazine, 11,12,13,14-tetrahydro-	DMF	319(4.23),470(3.87)	104-1223-84
1H-Imidazole, 4,5-diphenyl-2-(phenylazo)-	C_6H_{12}	420(4.36),450s(4.37)	62-0302-84A
	EtOH	223(4.35),288(4.32), 414(4.35),440s(4.27)	62-0302-84A
	DMF	420(4.32),450s(4.28), 520(3.40)	62-0302-84A
	DMSO	424(4.30),450s(4.27)	62-0302-84A

Compound	Solvent	$\lambda_{max}(\log \epsilon)$	Ref.
$C_{21}H_{16}N_4O$			
Benzoic acid, 2-(2-phenyl-4-quinazolin-yl)hydrazide	EtOH	248(4.59),320(4.0), 356(3.84)	18-1138-84
Benzoic acid, (2-phenyl-4(1H)-quinazo-linylidene)hydrazide	EtOH	232(4.56),256(4.5), 308(4.06)	18-1138-84
Methanone, phenyl(5-phenyl-1,2,4-oxa-diazol-3-yl)-, phenylhydrazone	50% dioxan	366(4.22)	39-0785-84B
Phenol, 2-[(4,5-diphenyl-1H-imidazol-2-yl)azo]-	C_6H_{12}	430s(4.24),470(4.41)	62-0302-84A
	EtOH	223(4.51),295(4.38), 434s(4.34),454(4.40), 472s(4.37)	62-0302-84A
	DMF	440s(4.31),468(4.37), 570(3.93)	62-0302-84A
	DMSO	442s(4.30),464(4.36), 484s(4.33)	62-0302-84A
Phenol, 4-[(4,5-diphenyl-1H-imidazol-2-yl)azo]-	C_6H_{12}	420(4.36),444s(4.37)	62-0302-84A
	EtOH	225(4.23),295(4.08), 430(4.23),460s(4.21)	62-0302-84A
	DMF	420(4.33),444s(4.34), 520(3.74)	62-0302-84A
	DMSO	420(4.31),444s(4.31)	62-0302-84A
$C_{21}H_{16}N_4O_3S_2$			
2,4,6(1H,3H,5H)-Pyrimidinetrione, 5-[[[2-(methylthio)-5-phenyl-4-thia-zolyl]imino]methyl]-1-phenyl-	DMF	282(4.10),366(4.41)	103-0921-84
$C_{21}H_{16}OSe$			
1,3-Oxaselenole, 2,4,5-triphenyl-	MeCN	226(4.39),304(3.81), 358(3.53)	44-2057-84
$C_{21}H_{16}O_2$			
9H-Xanthene, 9-[(4-methoxyphenyl)meth-ylene]-	EtOH	247(4.19),287(3.83), 345(4.08)	44-0080-84
$C_{21}H_{16}O_2S$			
22-Thiatetracyclo[13.2.2.2^{8,11}.1^{3,6}]-docosa-3,5,8,10,15,17,18,20-octaene-2,7-dione	C_6H_{12}	301(4.10)	44-1177-84
$C_{21}H_{16}O_3$			
Dibenz[b,f]oxepin-10(11H)-one, 11-(4-methoxyphenyl)-	EtOH	285s(4.23),297(4.23)	44-0080-84
$C_{21}H_{16}O_4$			
Anthra[1,2-b]furan-6,11-dione, 5-hy-droxy-2-methyl-4-(2-methyl-2-prop-enyl)-	EtOH	256(4.01),288(3.91), 334(3.61),434(3.61), 460s(3.57),488s(3.39)	12-1511-84
Anthra[1,2-b]furan-6,11-dione, 2-meth-yl-5-[(2-methyl-2-propenyl)oxy]-	EtOH	250(4.59),280(4.47), 403(4.14)	12-1511-84
$C_{21}H_{16}O_5$			
Anthra[1,2-b]furan-6,11-dione, 5-meth-oxy-2-methyl-4-(2-oxopropyl)-	EtOH	242(4.47),282(4.49), 376(4.00)	12-1511-84
$C_{21}H_{16}O_6$			
Anthra[1,2-b]furan-3-carboxylic acid, 6,11-dihydro-5-methoxy-2-methyl-6,11-dioxo-, ethyl ester	EtOH	254(4.52),275s(4.31), 406(3.86)	12-1518-84
4H,8H-Benzo[1,2-b:5,4-b']dipyran-4-one, 3-hydroxy-8,8-dimethyl-2-(1,3-benzo-	MeOH	230(4.16),266(4.22), 370(4.19)	2-0168-84

Compound	Solvent	$\lambda_{max}(\log \epsilon)$	Ref.
dioxol-5-yl)- (cont.)			2-0168-84
4H,8H-Benzo[1,2-b:5,6-b']dipyran-4-one, 2-(1,3-benzodioxol-5-yl)-3-hydroxy-8,8-dimethyl-	MeOH	230(4.31),266(4.33), 370(4.46)	2-0168-84
2,5-Furandione, 3-[(4-acetoxy-3-methoxyphenyl)methylene]dihydro-4-(phenylmethylene)-, (E,E)-	toluene	373(3.94)	48-0233-84
(E,Z)-	toluene	392(4.15)	48-0233-84
Retrochinensin	EtOH	248(4.60),312(3.96), 348(3.56)	102-2323-84
$C_{21}H_{16}O_7$			
1H-Anthra[2,3-c]pyran-6,11-dione, 12-acetoxy-1-hydroxy-5-methoxy-	EtOH	257(3.35),306(3.08), 510(2.88)	12-1511-84
2-Butenedioic acid, 2-[(4-oxo-2-phenyl-4H-1-benzopyran-7-yl)oxy]-, dimethyl ester	MeOH	252(4.20),300(4.23)	2-0969-84
Diphyllin	MeOH	229(4.42),267(4.63), 308(3.96),321(3.94), 357(3.71)	78-1145-84
	MeOH-NaOH	229(4.47),278(4.50), 323(3.73),384(3.87)	78-1145-84
4-Pentenoic acid, 5-(9,10-dihydro-1-hydroxy-8-methoxy-9,10-dioxo-2-anthracenyl)-, methyl ester	n.s.g.	223(4.30),265(4.28), 280(4.23),291(4.17), 308(4.07),434(4.04)	5-0306-84
$C_{21}H_{17}BrN_2O_2$			
Cyclohepta[4,5]pyrrolo[1,2-a]imidazole-10-carboxylic acid, 3-bromo-2-(4-methylphenyl)-, ethyl ester	EtOH	253(4.55),280s(4.24), 345(4.45),440(3.79), 450s(3.78),530s(2.96), 570s(2.82)	150-3465-84M
$C_{21}H_{17}ClN_2$			
1H-1,5-Benzodiazepine, 2-(4-chlorophenyl)-2,3-dihydro-4-phenyl-	EtOH	256(4.28),370(3.66)	103-1370-84
$C_{21}H_{17}ClN_2O$			
Benzenemethanamine, N-[5-chloro-3-(phenylmethyl)-2(3H)-benzoxazolylidene]-	CHCl$_3$	256(3.83),304(4.02)	94-3053-84
2-Benzoxazolamine, 5-chloro-N.N-bis(phenylmethyl)-	CHCl$_3$	257(4.28),294(4.18)	94-3053-84
2-Benzoxazolamine, 5-chloro-N-phenyl-N-(2-phenylethyl)-	CHCl$_3$	261(4.15),298(4.32)	94-3053-84
$C_{21}H_{17}ClN_2O_2$			
Cyclohepta[4,5]pyrrolo[1,2-a]imidazole-10-carboxylic acid, 3-chloro-2-(4-methylphenyl)-, ethyl ester	EtOH	254(4.53),282s(4.24), 344(4.45),437(3.78), 449s(3.78),526s(2.98), 565s(2.83)	150-3465-84M
$C_{21}H_{17}NO_4$			
4H-1-Benzopyran-4-one, 2-(dimethylamino)-3-[(4-oxo-4H-1-benzopyran-3-yl)methyl]-	EtOH	228(4.31),307(3.87), 370(4.05)	44-2812-84
$C_{21}H_{17}NO_5$			
Chelirubine, dihydro-	MeOH	231(4.59),280(4.52), 338(4.30)	100-0001-84
Sanguinarine, dihydro-8-(hydroxymethyl)-	MeOH	212(4.24),235(4.39), 284(4.42),322(4.02), 350s(3.54)	100-0001-84

Compound	Solvent	$\lambda_{max}(\log \epsilon)$	Ref.
Sanguinarine, dihydro-8-methoxy-, (-)-	n.s.g.	210(4.16),235(4.23), 280(4.28),327(4.12)	100-0001-84
$C_{21}H_{17}NO_7$			
2H-[1,3]Dioxolo[4,5-g]indeno[2',1'-4,5]oxazolo[4,3-a]isoquinoline-2,11(15bH)-dione, 4,5-dihydro-14,15-dimethoxy-	MeOH	234(4.40),294(4.35)	94-2230-84
$C_{21}H_{17}N_3O$			
2H-1,4-Benzoxazin-2-one, 3-phenyl-, (4-methylphenyl)hydrazone	EtOH	264(4.21),388(4.36)	4-0521-84
$C_{21}H_{17}N_3OS_2$			
1H-Imidazo[1,2-c]thiazol-4-ium, 5-[(3-ethyl-2(3H)-benzothiazolylidene)-methyl]-2-hydroxy-7-phenyl-, hydroxide, inner salt	DMF	496(4.43)	103-0029-84
2H-1,3-Thiazine-2-acetonitrile, 5-cyano-3,6-dihydro-2-(4-methoxyphenyl)-4-(3-methylphenyl)-6-thioxo-	EtOH	266(4.09),326(3.63), 390(4.18)	4-1445-84
2H-1,3-Thiazine-2-acetonitrile, 5-cyano-3,6-dihydro-2-(4-methoxyphenyl)-4-(4-methylphenyl)-6-thioxo-	EtOH	271(4.18),324(3.66), 389(4.15)	4-1445-84
$C_{21}H_{17}N_3O_2$			
1H-Benzimidazole-1-carboxamide, 2,3-dihydro-5-methyl-2-oxo-N,3-diphenyl-	MeOH	214(4.42),245(4.44)	39-2587-84C
$C_{21}H_{17}N_3O_3$			
1H-Benzimidazole-1-carboxamide, 2,3-dihydro-5-methoxy-2-oxo-N,3-diphenyl-	MeOH	230(4.45),245.5(4.39)	39-2587-84C
$C_{21}H_{17}N_3O_4$			
Cyclohepta[4,5]pyrrolo[1,2-a]imidazole-10-carboxylic acid, 2-(4-methylphenyl)-3-nitro-, ethyl ester	CHCl$_3$	335s(4.43),365(4.53), 432(3.72),533(3.53), 579s(3.42),627s(2.93)	150-3465-84M
$C_{21}H_{17}N_3S$			
2H-1,4-Benzothiazin-2-one, 3-phenyl-, (4-methylphenyl)hydrazone	EtOH	266(4.55),396(4.44)	4-0521-84
$C_{21}H_{17}N_3S_2$			
2H-1,3-Thiazine-2-acetonitrile, 5-cyano-3,6-dihydro-2,4-bis(3-methylphenyl)-6-thioxo-	EtOH	263(4.15),324(3.63), 386(4.09)	4-1445-84
2H-1,3-Thiazine-2-acetonitrile, 5-cyano-3,6-dihydro-2,4-bis(4-methylphenyl)-6-thioxo-	EtOH	270(4.16),324(3.68), 389(4.13)	4-1445-84
2H-1,3-Thiazine-2-acetonitrile, 5-cyano-3,6-dihydro-2-(3-methylphenyl)-4-(4-methylphenyl)-6-thioxo-	EtOH	269(4.17),324(3.56), 386(4.11)	4-1445-84
2H-1,3-Thiazine-2-acetonitrile, 5-cyano-3,6-dihydro-2-(4-methylphenyl)-4-(3-methylphenyl)-6-thioxo-	EtOH	265(4.08),327(3.65), 390(4.21)	4-1445-84
$C_{21}H_{17}P$			
Phosphine, (2,6-dimethylphenyl)-9H-fluoren-9-ylidene-	THF	272(4.8),303(3.96), 360(4.43)	24-0915-84

Compound	Solvent	$\lambda_{max}(\log \epsilon)$	Ref.
$C_{21}H_{18}$			
Bicyclo[5.4.1]dodeca-2,5,7,9,11-pentaene, 4-(2,4,6-cycloheptatrien-1-ylidenethylidene)-	isooctane	429(4.41)	88-1785-84
Pentacyclo[14.4.1.0^{1,3}.0^{4,14}.0^{7,12}]heneicosa-5,7,9,11,13,15,17,19-octaene, (1 ,4)-	EtOH	351(4.05)	88-1785-84
Pentacyclo[14.4.1.0^{1,3}.0^{4,14}.0^{7,13}]heneicosa-4,6,8,11,13,15,17,19-octaene	EtOH	335(3.81)	88-1785-84
$C_{21}H_{18}BNO_6$			
Boron, [5-[4-(dimethylamino)phenyl]-1-phenyl-4-pentene-1,3-dionato-0,0'][ethanedioato(2-)-0,0']-	benzene	543(4.93)	97-0292-84
$C_{21}H_{18}ClNO_6$			
Propanedioic acid, [6-[2-(4-chlorophenyl)ethenyl]-3-cyano-2-oxo-2H-pyran-4-yl]-, diethyl ester	EtOH	270(4.16),288(4.12), 305(4.06),372(4.34), 308(4.36)	94-3384-84
$C_{21}H_{18}IN_5OS$			
4H-Pyrazolo[3,4-d]pyrimidinium, 4-(3-ethyl-2(3H)-benzoxazolylidene)-5,6-dihydro-3-methyl-1-phenyl-6-thioxo-, iodide	EtOH	360(4.04)	48-0811-84
$C_{21}H_{18}NO_4$			
Chelerythrine (chloride)	EtOH	228(4.26),272(4.54), 283s(4.47),302s(4.37), 343(4.17)	100-0001-84
Nitidine (chloride)	MeOH	231(4.42),272(4.49), 281s(4.48),303s(4.40), 329(4.38)	100-0001-84
$C_{21}H_{18}N_2$			
Benzenamine, N,N'-[1-(4-methylphenyl)-1,2-ethanediylidene]bis-	MeOH	238(4.34),265(4.12)	118-0128-84
Benzenamine, 4-methyl-N-[1-phenyl-2-(phenylimino)ethylidene]-	MeOH	232(4.32),270(4.14)	118-0128-84
1H-Indol-1-amine, N-(4-methylphenyl)-3-phenyl-	EtOH	225(4.27),265(4.19), 296(3.86)	103-1119-84
1H-Quindoline, 2,3,4,10-tetrahydro-11-phenyl-	EtOH	225(4.63),264(4.19), 313(4.09)	103-1374-84
$C_{21}H_{18}N_2O$			
Benzenamine, N,N'-[1-(4-methoxyphenyl)-1,2-ethanediylidene]bis-, (Z,Z)-	MeOH	234(4.39),282(4.11)	118-0128-84
Benzenamine, N-[1-(4-methylphenyl)-2-(phenylimino)ethylidene]-, N-oxide	MeOH	268(4.28)	118-0128-84
Benzenamine, 4-methyl-N-[1-phenyl-2-(phenylimino)ethylidene]-, N-oxide	MeOH	222(4.26),263(4.27)	118-0128-84
1H-Indol-1-amine, N-(4-methoxyphenyl)-3-phenyl-	EtOH	225.5(4.50),265(4.14), 292(4.13)	103-1119-84
$C_{21}H_{18}N_2O_2$			
Benzenamine, N-[1-(4-methoxyphenyl)-2-(phenylimino)ethylidene]-, N-oxide, (Z,Z)-	MeOH	282(4.38)	118-0128-84
Benzenamine, 4-methoxy-N-[1-phenyl-2-(phenylimino)ethylidene]-, N-oxide, (Z,Z)-	MeOH	237(4.27),263(4.27), 366(3.72)	118-0128-84

Compound	Solvent	$\lambda_{max}(\log \epsilon)$	Ref.
2-Butene-1,4-dione, 2-(3,5-dimethyl-1H-pyrazol-1-yl)-1,4-diphenyl-	MeOH	258(4.40),326(4.36)	44-4647-84
$C_{21}H_{18}N_2O_3S$ Benzenesulfonic acid, 4-(4,5-dihydro-3,5-diphenyl-1H-pyrazol-1-yl)-	MeOH	236(4.16),310(3.90), 359(4.30)	103-0787-84
$C_{21}H_{18}N_2O_5$ 2(3H)-Furanone, 5-(1,2-dihydroxyethyl)-dihydro-3-[(4-hydroxyphenyl)methylene]-4-(2-quinazolinyl)-	MeOH	212(4.55),238(4.67), 318(4.55)	94-1808-84
$C_{21}H_{18}N_2S$ 1H-1,5-Benzodiazepine, 2,3-dihydro-2-phenyl-4-[2-(2-thienyl)ethenyl]-, (E)-	EtOH	233(4.30),391(3.95)	103-1370-84
$C_{21}H_{18}N_3OS_2$ 1H-Imidazo[1,2-c]thiazol-4-ium, 5-[(3-ethyl-2(3H)-benzothiazolylidene)methyl]-2,3-dihydro-2-oxo-7-phenyl-, perchlorate	HOAc	450(4.73)	103-0029-84
$C_{21}H_{18}N_4$ 2(1H)-Quinoxalinone, 3-phenyl-, (4-methylphenyl)hydrazone	EtOH	255(4.63),360(4.36), 454(4.04)	4-0521-84
$C_{21}H_{18}N_4O_2$ 1H-Imidazo[1,2-b][1,2,4]triazepine, 6,8-di-2-furanyl-7,8-dihydro-8-methyl-2-phenyl-	MeOH	288(4.59),342(3.96)	103-1152-84
$C_{21}H_{18}N_4O_3S$ Benzeneacetic acid, α-[[6,7-dihydro-1-methyl-6-oxo-8-(phenylmethyl)-1H-purin-2-yl]thio]-	MeOH	266(4.21),286s(4.13)	2-0316-84
Thiazolo[2,3-b]purine-4,8(1H,7H)-dione, 5a-ethoxy-5,5a-dihydro-5,7-diphenyl-	MeOH	263(4.00),282(3.96)	2-0316-84
$C_{21}H_{18}N_4S_2$ 1H-Imidazo[1,2-b][1,2,4]triazepine, 7,8-dihydro-8-methyl-2-phenyl-6,8-di-2-thienyl-	MeOH	293(4.53),354(3.95)	103-1152-84
$C_{21}H_{18}N_5OS$ 4H-Pyrazolo[3,4-d]pyrimidinium, 4-(3-ethyl-2(3H)-benzoxazolylidene)-5,6-dihydro-3-methyl-1-phenyl-6-thioxo-, iodide	EtOH	360(4.04)	48-0811-84
$C_{21}H_{18}O$ 2,5-Cyclohexadien-1-one, 4-(diphenylmethylene)-2,6-dimethyl-	hexane?	211(4.26),260(4.19), 268(4.19),362(4.49)	73-1949-84
2,5-Cyclohexadien-1-one, 4-(diphenylmethylene)-2-ethyl-	hexane?	210(4.25),260(4.08), 267(4.07),361(4.35)	73-1949-84
4H-Cyclopenta[b]furan, 4,6-dimethyl-4,5-diphenyl-	EtOH	232(4.08),310(4.01)	39-2671-84C
Furan, 3-[1-(9H-fluoren-9-ylidene)ethyl]-2,5-dimethyl-	hexane	234(4.66),250(4.54), 259(4.48),280(4.13), 331(4.04)	39-2877-84C

Compound	Solvent	$\lambda_{max}(\log \epsilon)$	Ref.
Furan, 3-methyl-2-(1-methyl-2,3-diphenyl-2-cyclopropen-1-yl)-	EtOH	228(4.44),235s(4.40), 305s(4.38),313(4.47), 330(4.26)	39-2671-84C
Furan, 3-methyl-2-(2-methyl-1,3-diphenyl-2-cyclopropen-1-yl)-	EtOH	262(4.40)	39-2671-84C
Furan, 3-methyl-2-[3-phenyl-1-(phenylmethyl)-1,2-propadienyl]-	EtOH	250(3.88),282s(3.06)	39-2671-84C
2-Propanone, 1,3-dimethyl-2-fluoranthenyl)-	hexane	221(4.50),247(4.17), 268(4.61),283(4.24), 294(4.44),332(3.87), 350(3.92),368(3.89)	39-2877-84C
$C_{21}H_{18}O_3$			
2,5-Cyclohexadien-1-one, 4-(diphenylmethylene)-2,6-dimethoxy-	hexane?	211(4.44),264(4.12), 271(4.12),374(4.46)	73-1949-84
Ethanone, 1-[2,4-dihydroxy-3-(phenylmethyl)phenyl]-2-phenyl-	MeOH	202(4.46),284(4.15)	2-1030-84
Ethanone, 1-[2,4-dihydroxy-5-(phenylmethyl)phenyl]-2-phenyl-	MeOH	202(4.44),284(4.07)	2-1030-84
$C_{21}H_{18}O_4$			
5,12-Naphthacenedione, 2-acetyl-1,2,3,4-tetrahydro-7-methoxy-	MeCN	202(4.42),246(4.75), 272(4.26),284s(4.21), 296(4.04),320(3.46)	78-4549-84
$C_{21}H_{18}O_5$			
4H-1-Benzopyran-4-one, 3-acetyl-7-acetoxy-2-methyl-6-(phenylmethyl)-	MeOH	208(4.62),240(4.06)	2-1030-84
4H-1-Benzopyran-4-one, 3-acetyl-7-acetoxy-2-methyl-8-(phenylmethyl)-	MeOH	206(4.51),242(4.04)	2-1030-84
$C_{21}H_{18}O_6$			
2-Anthraceneacetic acid, 9,10-dihydro-4-hydroxy-9,10-dioxo-3-(3-oxobutyl)-, methyl ester	MeOH	207(4.27),227(4.30), 237s(--),246(4.46), 257(4.48),280s(--), 329(3.53),390s(--), 409(3.83),430(3.72)	78-3677-84
2-Anthracenecarboxaldehyde, 9,10-dihydro-4-hydroxy-3-[2-(2-methyl-1,3-dioxolan-2-yl)ethyl]-9,10-dioxo-	MeOH	203(4.38),229s(--), 245(4.54),262(4.57), 280s(--),327(3.61), 390s(--),410(3.92), 425s(--)	78-3677-84
2-Anthracenecarboxylic acid, 9,10-dihydro-4-hydroxy-9,10-dioxo-3-(3-oxopentyl)-, methyl ester	MeOH	211(4.35),229(4.34), 249(4.49),256(4.50), 329(3.52),408(3.87)	78-3677-84
2H-Anthra[1,2-b]pyran-2-carboxaldehyde, 2-ethyl-3,4,7,12-tetrahydro-6-hydroxy-11-methoxy-7,12-dioxo-, (±)-	MeOH	229(4.53),249(4.30), 270(4.05),465(3.95)	78-4609-84
Benzeneacetic acid, 2-methoxy-α-(3-methoxy-5-oxo-4-phenyl-2(5H)-furanylidene)-, methyl ester, (E)-	MeOH	230(4.10),260(3.97), 332(4.27)	39-1547-84C
6H-Benzofuro[3,2-c][1]benzopyran-6-one, 8,9-dihydroxy-3-methoxy-2-(3-methyl-2-butenyl)-	MeOH	285(3.78),305(3.86), 335(3.51)	2-1028-84
Daunomycinone, 7,9-dideoxy-	n.s.g.	255(3.56),292(3.23), 405(2.91),440(3.09), 465(3.13),493(3.04), 515(2.93),530(2.86)	78-4579-84
7H-Furo[3,2-g]benzopyran-7-one, 4-[[4-(2,5-dihydro-4-methyl-5-oxo-2-furanyl)-3-methyl-2-butenyl]oxy]-	MeOH	224(4.67),243(4.44), 250(4.57),256(4.30), 267(4.39),307(4.33)	102-2629-84

Compound	Solvent	$\lambda_{max}(\log \epsilon)$	Ref.
7H-Furo[3,2-g][1]benzopyran-7-one, 9-[[4-(2,5-dihydro-4-methyl-5-oxo-2-furanyl)-3-methyl-2-butenyl]oxy]- (indicolactone)	MeOH	212(3.71),252(3.48), 268(3.29),304(3.20)	102-2629-84
1-Naphthacenecarboxylic acid, 1,2,3,4-6,11-hexahydro-2,4-dihydroxy-2-methyl-6,11-dioxo-, methyl ester	MeOH	208(4.16),230s(--), 245(4.44),257s(--), 263(4.49),280s(--), 326(3.46),390s(--), 408(3.82)	78-3677-84
Spiro[1,3-dioxolane-2,2'(1'H)-naphthacene]-6',11'-dione, 3',4'-dihydro-12'-hydroxy-7'-methoxy-	CHCl$_3$	260(4.60),266(4.60), 287(4.20),399(4.10), 412(4.01),435(4.00)	78-4633-84
$C_{21}H_{18}O_7$			
7H-Furo[3,2-g][1]benzopyran-7-one, 4-[[3-[(2,5-dihydro-4-methyl-5-oxo-2-furanyl)methyl]-3-methyloxiranyl]-methoxy]- (anisolactone epoxide)	MeOH	224(4.55),242(4.42), 249(4.46),258(4.39), 266(4.35),305(4.30)	102-2629-84
1-Naphthacenecarboxylic acid, 1,2,3,4-6,11-hexahydro-2,4,5-trihydroxy-2-methyl-6,11-dioxo-, methyl ester, (1α,2α,4β)-	MeOH	208(4.22),227(4.28), 247(4.42),255(4.43), 260s(--),278s(--), 327(3.44),395s(--), 405(3.78),425s(--)	78-3677-84
(1α,2β,4α)-	MeOH	208(4.22),227(4.28), 247(4.42),255(4.43), 327(3.44),405(3.78)	78-3677-84
5,12-Naphthacenedione, 9,9-(ethylenedioxy)-7,8,9,10-tetrahydro-6,11-dihydroxy-7-methoxy-	EtOH	252(4.58),257s(4.55), 285(3.91),458s(3.95), 485(4.00),518(3.80)	78-4561-84
$C_{21}H_{18}O_8$			
Coeloginin diacetate	EtOH	225(4.42),248(4.49), 278(4.12),338(3.91)	39-1919-84C
6,11-Naphthacenedione, 1,2,3,4,6,11-hexahydro-3-(acetoxymethyl)-1,3,5,12-tetrahydroxy-, (S)-cis-	n.s.g.	253(4.63),259(4.59), 289(4.03),486(4.02)	78-4649-84
5,8,12-Naphthacenetrione, 7,8,9,10-tetrahydro-6,11-dihydroxy-1-(2-hydroxyethoxy)-10-methoxy-	EtOH	253(3.90),472(3.42), 488(3.44),497(3.45), 533(3.29)	78-4561-84
$C_{21}H_{18}S$			
4H-Cyclopenta[b]thiophene, 4,6-dimethyl-4,5-diphenyl-	EtOH	308(4.04)	39-2671-84C
6H-Cyclopenta[b]thiophene, 2,4-dimethyl-5,6-diphenyl-	EtOH	225(3.84),320(3.86)	39-2671-84C
6aH-Cyclopenta[b]thiophene, 4,6a-dimethyl-5,6-diphenyl-	C$_6$H$_{12}$	312(4.06)	39-2671-84C
Thiophene, 3-[1-(9H-fluoren-9-ylidene)-ethyl]-2,5-dimethyl-	hexane	234(4.67),252(4.55), 261(4.57)	39-2877-84C
Thiophene, 2-methyl-3-(2-methyl-1,3-diphenyl-3-cyclopropenyl)-	C$_6$H$_{12}$	258(4.31),300s(3.53)	39-2671-84C
Thiophene, 2-methyl-3-(3-methyl-1,2-diphenyl-3-cyclopropenyl)-	C$_6$H$_{12}$	227(4.24),235s(4.18), 250(3.93),298s(4.22), 328(4.13)	39-2671-84C
Thiophene, 2-methyl-4-(3-methyl-1,2-diphenyl-3-cyclopropenyl)-	C$_6$H$_{12}$	226(4.32),232s(4.28), 294s(4.28),310(4.38), 326(4.22)	39-2671-84C
Thiophene, 3-methyl-2-(1-methyl-2,3-diphenyl-2-cyclopropen-1-yl)-	EtOH	302(3.93),314(4.02), 333(3.88)	39-2671-84C
Thiophene, 3-methyl-2-(2-methyl-1,3-diphenyl-3-cyclopropenyl)-	EtOH	262(4.03)	39-2671-84C

Compound	Solvent	$\lambda_{max}(\log \epsilon)$	Ref.
$C_{21}H_{19}BrN_2O$ Pyrano[2,3-c]pyrazole, 1-(4-bromophen- yl)-1,6-dihydro-4,6,6-trimethyl-3- phenyl-	EtOH	267(4.25)	94-2146-84
$C_{21}H_{19}ClN_2O$ Pyrano[2,3-c]pyrazole, 1-(4-chlorophen- yl)-1,6-dihydro-4,6,6-trimethyl-3- phenyl-	EtOH	268(4.24)	94-2146-84
$C_{21}H_{19}ClN_4$ 7H-Pyrrolo[2,3-d]pyrimidin-4-amine, N-(2-chlorophenyl)-2,5,6-trimethyl- 7-phenyl-	EtOH	207(4.56),242s(4.24), 263s(4.10),309(4.36)	11-0073-84B
7H-Pyrrolo[2,3-d]pyrimidin-4-amine, N-(4-chlorophenyl)-2,5,6-trimethyl- 7-phenyl-	EtOH	205(4.57),224(4.31), 247s(4.18),266s(4.07), 310(4.38)	11-0073-84B
4H-Pyrrolo[2,3-d]pyrimidin-4-imine, 3-(2-chlorophenyl)-3,7-dihydro- 2,5,6-trimethyl-7-phenyl-	EtOH	245(4.24),272s(4.01), 283s(3.95),307(3.90)	11-0073-84A
4H-Pyrrolo[2,3-d]pyrimidin-4-imine, 3-(4-chlorophenyl)-3,7-dihydro- 2,5,6-trimethyl-7-phenyl-	EtOH	243(4.37),283s(3.89), 303(3.80)	11-0073-84A
$C_{21}H_{19}Cl_2NO_2S$ Ethanone, 1-(2,8-dichloro-11-hydroxy- dibenzo[b,f]thiepin-10-yl)-2-piper- idino-	MeOH	226(4.50),244s(4.43), 271(4.03),325(4.02)	73-0603-84
$C_{21}H_{19}FN_4$ 7H-Pyrrolo[2,3-d]pyrimidin-4-amine, N-(2-fluorophenyl)-2,5,6-trimethyl- 7-phenyl-	EtOH	203(4.64),225s(4.39), 239s(4.37),262(4.18), 307(4.44)	11-0073-84B
7H-Pyrrolo[2,3-d]pyrimidin-4-amine, N-(3-fluorophenyl)-2,5,6-trimethyl- 7-phenyl-	EtOH	203(4.47),228s(4.23), 242s(4.15),264s(3.96), 309s(4.26)	11-0073-84B
7H-Pyrrolo[2,3-d]pyrimidin-4-amine, N-(4-fluorophenyl)-2,5,6-trimethyl- 7-phenyl-	EtOH	203(4.58),221s(4.32), 241s(4.22),303(4.31)	11-0073-84B
4H-Pyrrolo[2,3-d]pyrimidin-4-imine, 3-(2-fluorophenyl)-3,7-dihydro- 2,5,6-trimethyl-7-phenyl-	EtOH	246(4.26),269s(4.05), 308(3.77)	11-0073-84A
4H-Pyrrolo[2,3-d]pyrimidin-4-imine, 3-(3-fluorophenyl)-3,7-dihydro- 2,5,6-trimethyl-7-phenyl-	EtOH	244(4.27),270s(4.02), 283(3.86),304(3.78)	11-0073-84A
4H-Pyrrolo[2,3-d]pyrimidin-4-imine, 3-(4-fluorophenyl)-3,7-dihydro- 2,5,6-trimethyl-7-phenyl-	EtOH	242(4.31),280s(3.89), 303(3.79)	11-0073-84A
$C_{21}H_{19}NO_4$ Anhydrocorynoline, (+)-	MeOH	207(4.32),225(4.38), 289(3.99)	83-0223-84
	MeOH-HCl	206(4.21),233(4.37), 292(4.00)	83-0223-84
14-epi-	MeOH	206(4.60),225(4.41), 288(3.92),308(3.83)	83-0223-84
	MeOH-HCl	206(4.64),228(4.40), 293(4.02)	83-0223-84
9,10-Anthracenedione, 1-acetoxy- 2-piperidino-	EtOH at 77°K	445(3.85)	104-1780-84
6H-Anthra[1,9-cd]isoxazole-3-carbox-	EtOH	461(4.09)	104-2012-84

Compound	Solvent	$\lambda_{max}(\log \epsilon)$	Ref.
ylic acid, 6-oxo-, hexyl ester (cont.)			104-2012-84
Chelerythrine, dihydro-	EtOH	227(4.53),282(4.65), 318(4.18),350s(3.52)	100-0001-84
	EtOH-HCl	231(4.48),255s(4.47), 265(4.58),273(4.62), 291s(4.16),304(4.31), 317(4.39),335(4.13), 352(4.25)	100-0001-84
	EtOH-KOH	281(4.65),318(4.21), 350s(3.54)	100-0001-84
2-Naphthalenol, 1-(1-methylpropyl)-, 4-nitrobenzoate	EtOH	222(4.87),257(4.31)	150-0701-84M
Nitidine, dihydro-	EtOH	229.5(4.65),280.5(4.59), 312(4.34)	100-0001-84
1H-Pyrrole-3,4-dicarboxylic acid, 2-phenyl-1-(phenylmethyl)-, dimethyl ester	MeOH	260(4.02)	44-3314-84
$C_{21}H_{19}NO_4S$			
Pyridinium, 3-methyl-, 1-(ethoxycarbo-nyl)-2-(1,3-dihydro-1,3-dioxo-2H-inden-2-ylidene)-2-(methylthio)-ethylide	EtOH	252(4.46),311(4.06), 366(4.00)	94-2910-84
$C_{21}H_{19}NO_5$			
Benzo[g]-1,3-benzodioxolo[5,6-a]quino-lizinium, 5,6-dihydro-13-hydroxy-9,10-dimethoxy-8-methyl-, hydroxide, inner salt	MeOH	237.5(4.53),259s(4.30), 315.5(4.08),371(4.05), 456(4.18)	94-2230-84
Corynolone	MeOH	205(4.57),235s(3.96), 288.5(3.94)	83-0223-84
Corynoloxine, (+)-	MeOH	205.5(4.69),225s(3.99), 290(3.93)	83-0223-84
	MeOH-HCl	205.5(--),236.5(4.13), 297(4.25),397(3.49)	83-0223-84
8H-Dibenzo[a,g]quinolizin-8-one, 13-acetoxy-5,6-dihydro-2,3-dimethoxy-	MeOH	227(4.51),257s(4.13), 332(4.40),350s(4.29), 366s(4.06)	44-2642-84
Fagaridine, 5,6-dihydro-6-(hydroxy-methyl)-	EtOH	228(--),284(--), 324s(--),353s(--)	100-0001-84
	EtOH-HCl	223(--),255(--), 268s(--),277(--), 307(--),321(--), 339(--),356(--)	100-0001-84
$C_{21}H_{19}NO_6$			
Arnottianamide	EtOH	236(4.73),280s(4.01), 321s(3.63),324(3.65), 332(3.81)	39-1769-84C +100-0001-84
8H-Benzo[g]-1,3-benzodioxolo[5,6-a]-quinolizin-8-one, 5,6-dihydro-9,10,13-trimethoxy-	MeOH	233(3.66),255s(3.31), 332(3.49),341(3.48), 344(3.48),346(3.47), 374s(3.20),394s(3.00)	78-3957-84
Isoarnottianamide	EtOH	237.5(4.73),290(4.00), 332(3.86)	39-1769-84C
Pictonamine	MeOH	235(3.35),274(2.92), 305(3.02),325(2.95), 379(3.51),399s(3.43)	78-3957-84
$C_{21}H_{19}NO_7$			
6H-Benzo[g]-1,3-benzodioxolo[5,6-a]-	MeOH	235(4.37),261s(4.01),	78-3957-84

Compound	Solvent	$\lambda_{max}(\log \epsilon)$	Ref.
quinolizine-8,13-dione, 5,13a-di-hydro-9,10,13a-trimethoxy- (cont.)		289(3.92),325(3.86)	78-3957-84
4,9:5,8-Diepoxynaphth[2,3-d]isoxazole-4a,8a-dicarboxylic acid, 3a,4,5,8-9,9a-hexahydro-3-phenyl-, dimethyl ester, (3aα,4β,4aβ,5α,8α,8aβ,9β,9aα)-	MeOH	268(3.15)	73-1990-84

$C_{21}H_{19}N_2O_2$
| Benzoxazolium, 3-methyl-2-[3-(3-methyl-2(3H)-benzoxazolylidene)-1-cyclopent-en-1-yl]-, perchlorate | MeOH | 535(5.14) | 48-1034-84 |

$C_{21}H_{19}N_2S_2$
| Benzothiazolium, 3-methyl-2-[3-(3-meth-yl-2(3H)-benzothiazolylidene)-1-cy-clopenten-1-yl]-, perchlorate | MeOH | 590(5.17) | 48-1034-84 |

$C_{21}H_{19}N_3O$
| 4H-Indazol-4-one, 5-[(diphenylamino)-methylene]-1,5,6,7-tetrahydro-1-methyl-, (E)- | EtOH | 243.5(4.12),277(4.10), 372(4.43) | 4-0361-84 |
| 4H-Indazol-4-one, 1,5,6,7-tetrahydro-5-(methylphenylamino)methylene]-1-phenyl-, (E)- | EtOH | 233(4.24),260(4.18), 366(4.41) | 4-0361-84 |

$C_{21}H_{19}N_3OS_2$
| 3H-Imidazo[1,2-c]thiazol-4-ium, 3-[[4-(dimethylamino)phenyl]methyl-ene]-2-hydroxy-5-(methylthio)-7-phenyl-, hydroxide, inner salt | DMF | 464(4.49) | 103-0029-84 |

$C_{21}H_{19}N_3O_3$
| 1H-Isoindole-1,3(2H)-dione, 2-[6-(1H-indazol-3-yl)-6-oxohexyl]-, hydrox-ide, inner salt | MeOH | 220(4.56),233s(4.27), 241s(4.18),299(3.94) | 103-0066-84 |

$C_{21}H_{19}N_3O_3S$
| 1H-Pyrazolo[1,2-a]pyrazol-4-ium, 3-hy-droxy-5,7-dimethyl-1-[[(4-methyl-phenyl)sulfonyl]imino]-2-phenyl-, hydroxide, inner salt | MeCN | 224(4.44),263(4.26), 428(3.61) | 44-3672-84 |

$C_{21}H_{19}N_3O_4$
| Benzamide, N-[1-(dimethylamino)-1,2-dihydro-5-(2-hydroxybenzoyl)-2-oxo-3-pyridinyl]- | EtOH | 235(4.39),322(4.40) | 2-1048-84 |

$C_{21}H_{19}N_5O_2$
| Benzamide, N-[7-ethyl-8,9-dihydro-8-oxo-9-(phenylmethyl)-7H-purin-6-yl]- | MeOH | 234(4.25),290(4.10) | 150-3601-84M |

$C_{21}H_{19}N_5O_3$
| 1,3,7,8-Tetraazaspiro[4.5]deca-6,9-di-ene-10-carboxylic acid, 3-methyl-4-oxo-1-phenyl-2-(phenylimino)-, methyl ester | MeOH | 235s(4.25),265(4.01), 295(3.92) | 39-2491-84C |

$C_{21}H_{19}N_5O_6$
| Carbonic acid, (7-amino-2-azido-2,3,5,8-tetrahydro-1-methoxy-6-methyl-5,8-dioxo-1H-pyrrolo[1,2-a]- | MeOH | 207(4.31),246(4.34), 304(4.12),350s(3.57), 522(2.93) | 44-5164-84 |

Compound	Solvent	$\lambda_{max}(\log \epsilon)$	Ref.
indol-9-yl)methyl phenyl ester, (1S-trans)- (cont.)			44-5164-84
$C_{21}H_{19}P$ Phosphine, 2,6-dimethylphenyl-(diphenylmethylene)-	THF	282s(2.42),322(2.60)	78-0765-84
$C_{21}H_{20}$ Tetracyclo[3.2.0.0^{2,7}.0^{4,6}]heptane, [[3-(2,4-cyclopentadien-1-ylidene-methyl)bicyclo[2.2.1]hept-2-en-2-yl]methylene]-	MeCN	396(c.4.60)	24-2027-84
$C_{21}H_{20}ClN_3O_2$ Pyrano[2,3-e]indazol-2(5H)-one, 3-chloro-6,7-dihydro-7-phenyl-4-piperidino-	EtOH	228(4.25),257(4.12), 345(4.26)	4-0361-84
$C_{21}H_{20}ClN_3O_2S$ 2-Thiophenecarboxylic acid, 5-(4-chloro-rophenyl)-3-[[4-(dimethylamino)phen-yl]azo]-, ethyl ester	MeOH	284(4.40),318(4.27), 444(4.50)	118-0275-84
$C_{21}H_{20}Cl_2N_2O_2S$ Ethanone, 1-(2,8-dichloro-11-hydroxy-dibenzo[b,f]thiepin-10-yl-2-(4-methyl-1-piperazinyl)-	MeOH	241s(4.44),272(4.04), 312s(4.03),326(4.04), 340(3.99)	73-0603-84
$C_{21}H_{20}Cl_2O_6$ 9,10-Anthracenedione, 2-[3-(dichloro-methyl)-3-hydroxypentyl]-1,4-dihy-droxy-8-methoxy-, (±)-	MeOH	231(4.59),249(4.38), 284(4.06),477(4.06), 494(4.05),527(3.79)	78-4609-84
$C_{21}H_{20}NO_2P$ Acetamide, 2-(diphenylphosphinyl)-N-(phenylmethyl)-	EtOH	260(2.90),273(2.84)	42-0430-84
Benzamide, N-[(diphenylphosphinyl)-methyl]-4-methyl-	EtOH	265(3.34),272(3.18)	42-0430-84
$C_{21}H_{20}NO_3P$ Benzamide, N-[(diphenylphosphinyl)-methyl]-4-methoxy-	EtOH	265(3.30),272(3.24)	42-0430-84
$C_{21}H_{20}NO_4$ Fagaronine (chloride)	n.s.g.	233(4.29),272(4.53), 305s(4.44)	100-0001-84
	base	346(4.31)	100-0001-84
Punctatine	EtOH-acid	270(--),280s(--), 305s(--),318(--), 325s(--),340s(--), 390(--)	100-0001-84
	EtOH-base	255(--),288(--), 302(--),333s(--)	100-0001-84
$C_{21}H_{20}N_2$ 1H-Indol-1-amine, 2,3-dihydro-N-(4-methylphenyl)-3-phenyl-	EtOH	234(4.37),288(3.99)	103-1119-84
$C_{21}H_{20}N_2O$ 6H-Cyclopentapyrazin-6-one, 1,2,3,4-tetrahydro-1,4-dimethyl-5,7-diphenyl-	MeCN	305(4.49),510(2.51)	89-0053B-84

Compound	Solvent	λ_{max}(log ϵ)	Ref.
1H-Indol-1-amine, N-(4-methoxyphenyl)-2,3-dihydro-3-phenyl-	EtOH	244(4.29),300(3.83)	103-1119-84
Phenol, 4-(3,9-dimethyl-2H-benz-[g]indazol-2-yl)-3,5-dimethyl-	EtOH	244(4.49),282(3.96), 297(3.96),317(3.84), 333(3.85),348(3.11)	33-0113-84
Pyrano[2,3-c]pyrazole, 1,6-dihydro-4,6,6-trimethyl-1,3-diphenyl-	EtOH	261(4.11)	94-2146-84
$C_{21}H_{20}N_2O_2S$			
2H-Pyrrol-2-one, 1,5-dihydro-3-(4-methylphenyl)-4-morpholino-1-phenyl-5-thioxo-	CHCl₃	340(4.00),500(4.00)	78-3499-84
2H-Pyrrol-2-one, 1,5-dihydro-4-(4-methylphenyl)-3-morpholino-1-phenyl-5-thioxo-	CHCl₃	343(4.16),465(3.96)	78-3499-84
$C_{21}H_{20}N_2O_2S_3$			
[1,2]Dithiolo[4,5,1-hi][1,2,3]benzo-thiadiazole-3-S^{IV}, 2,6,7,8-tetra-hydro-2-(4-methoxyphenyl)-5-[(4-methoxyphenyl)thio]-	MeCN	525(4.53)	97-0251-84
$C_{21}H_{20}N_2O_3$			
1H-Pyrazole-1-acetic acid, 3,5-dimeth-yl-α-(2-phenoxy-2-phenylethenyl)-, (Z)-	MeOH	252(4.27)	44-4647-84
$C_{21}H_{20}N_2S$			
6H-Cyclopentapyrazine-6-thione, 1,2,3,4-tetrahydro-1,4-dimethyl-5,7-diphenyl-	MeCN	305(4.13),388(4.49), 645(2.72)	89-0053B-84
$C_{21}H_{20}N_2S_3$			
[1,2]Dithiolo[4,5,1-hi][1,2,3]benzo-thiadiazole-3-S^{IV}, 2,6,7,8-tetrahy-dro-2-(4-methylphenyl)-5-[(4-methyl-phenyl)thio]-	MeCN	522(4.39)	97-0251-84
$C_{21}H_{20}N_3OS_2$			
1H-Imidazo[1,2-c]thiazol-4-ium, 3-[[4-(dimethylamino)phenyl]meth-ylene]-2,3-dihydro-5-(methylthio)-2-oxo-7-phenyl-, perchlorate	HOAc	516(4.61)	103-0029-84
$C_{21}H_{20}N_4$			
7H-Pyrrolo[2,3-d]pyrimidin-4-amine, 2,5,6-trimethyl-N,7-diphenyl-	EtOH	203(4.56),222s(4.32), 241s(4.21),307(4.34)	11-0073-84B
4H-Pyrrolo[2,3-d]pyrimidin-4-imine, 3,7-dihydro-2,5,6-trimethyl-3,7-diphenyl-	EtOH	228s(3.92),245(4.34), 303(3.84)	11-0073-84A
$C_{21}H_{20}N_4O_3$			
Benzoic acid, 4-[(8,9-dihydro-11-oxo-7H-pyrrolo[2,1-b]quinazolin-6(11H)-ylidene)hydrazino]-, ethyl ester	EtOH	225(4.24),283(4.19), 385(4.39)	4-1301-84
3,5-Pyridinedicarboxamide, 1,2-dihydro-N,N'-dimethyl-2-oxo-1-phenyl-6-(phen-ylamino)-	MeOH	209(4.42),291(4.04), 361(4.36)	94-1761-84
7H-Pyrido[2,1-b]quinazoline-2-carbox-ylic acid, 6,8,9,11-tetrahydro-11-oxo-6-(phenylhydrazono)-, ethyl ester	EtOH	215(4.59),248(4.26), 290(4.08),300(4.08), 406(4.46)	4-1301-84

Compound	Solvent	$\lambda_{max}(\log \epsilon)$	Ref.
7H-Pyrido[2,1-b]quinazoline-3-carboxylic acid, 6,8,9,11-tetrahydro-11-oxo-6-(phenylhydrazono)-, ethyl ester	EtOH	240(4.68),300(4.01), 392(4.47)	4-1301-84
$C_{21}H_{20}N_4O_7S_2$ 6-Thia-1,4-diazabicyclo[5.2.0]non-3-ene-2-carboxylic acid, 3-methoxy-9-oxo-8-[(2-thienylacetyl)amino]-, (4-nitrophenyl)methyl ester	EtOH	235(4.13),277(4.11)	88-2531-84
$C_{21}H_{20}N_5O$ 7H-Purinium, 6-(benzoylamino)-7-ethyl-9-(phenylmethyl)-, iodide	H_2O	250(4.12),280(4.30), 323(3.62)	150-3601-84M
$C_{21}H_{20}N_6O_2S$ Propanenitrile, 3,3'-[[4-[[6-(2-hydroxyethoxy)-2-benzothiazolyl]azo]-phenyl]imino]bis-	pH 1.2 pH 6.2 EtOH	524(4.62),618s(3.00) 521(4.63) 493(4.62)	48-0151-84 48-0151-84 48-0151-84
$C_{21}H_{20}O$ Furan, 2,5-dimethyl-3-(1-methyl-2,2-diphenylethenyl)-	hexane	209(4.51)	39-2877-84C
2-Propanone, 1-(1,3-dimethyl-4-phenyl-2-naphthalenyl)-	hexane	238(4.72),290(3.85)	39-2877-84C
$C_{21}H_{20}O_2$ Benzoic acid, 4-[2-(1,2,3,4-tetrahydro-1,4-methano-6-naphthalenyl)-1-propenyl]-, (E)-	EtOH	227(4.15),301(4.40)	87-1516-84
$C_{21}H_{20}O_3$ Ethanone, 1-(1,2,3,4,5,12-hexahydro-7-methoxy-5,12-epoxynaphthacen-2-yl)-, $(2\alpha,5\alpha,12\alpha)-(\pm)-$	MeCN	212s(4.42),226(4.51), 250(4.49),272s(3.89), 280(3.87),314(3.08), 328(2.90)	78-4549-84
[1,1':3',1"-Terphenyl]-2,2',2"-triol, 5,5',5"-trimethyl-	EtOH	246(4.42),292(4.01), 332(4.00)	42-0142-84
$C_{21}H_{20}O_4$ 7-Oxabicyclo[2.2.1]heptane-2-carboxylic acid, 3-(phenoxyphenylmethylene)-, methyl ester, endo	MeOH	260(4.26)	44-4165-84
2-Propenoic acid, 2-methyl-, 2-[4-(1-oxo-3-phenyl-2-propenyl)phenoxy]-ethyl ester	n.s.g.	317(4.8)	73-2635-84
2-Propenoic acid, 3-phenyl-, 1,3-propanediyl ester	MeCN	275(4.67)	40-0022-84
$C_{21}H_{20}O_4S$ Sulfonium, [2-(9H-fluoren-9-yl)-3-methoxy-1-(methoxycarbonyl)-3-oxo-1-propenyl]dimethyl-, hydroxide, inner salt	CHCl$_3$	435(4.16)	24-2409-84
$C_{21}H_{20}O_5$ 1,3-Benzodioxole, 5-[5-(3,4-dimethoxyphenyl)-3,4-dimethyl-2-furanyl]-	MeOH	230(4.03),280(3.85), 310(3.76)	102-2647-84
5-Naphthacenecarboxylic acid, 5,7,8,9-10,12-hexahydro-11-hydroxy-6-methoxy-12-oxo-, methyl ester, (±)-	MeOH	268(4.06),296(4.14), 384(3.71)	44-0318-84

Compound	Solvent	$\lambda_{max}(\log \epsilon)$	Ref.
$C_{21}H_{20}O_6$			
9,10-Anthracenedione, 1-hydroxy-3-(hydroxymethyl)-2-[2-(2-methyl-1,3-dioxolan-2-yl)ethyl]-	MeOH	207(4.20),230s(--), 246(4.48),257(4.47) 262(4.47),278s(--), 328(3.50),392s(--), 413(3.85),437s(--)	78-3677-84
1H-3b,11-Ethanophenanthro[1,2-c]furan-1,3(3aH)-dione, 11-acetoxy-4,5,11,11a-tetrahydro-7-methoxy-, (3aα,3bα,11β-11aα)-(±)-	EtOH	270(4.35)	2-1168-84
1(2H)-Naphthalenone, 4-(1,3-benzodioxol-5-yl)-2-hydroxy-6,7-dimethoxy-2,3-dimethyl-	MeOH	237(4.20),265(3.84), 284(3.88),315(3.67)	102-2021-84
1(4H)-Naphthalenone, 4-(1,3-benzodioxol-5-yl)-4-hydroxy-6,7-dimethoxy-2,3-dimethyl-	MeOH	245(4.53),289(4.06)	102-2021-84
$C_{21}H_{20}O_7$			
1H-Inden-1-one, 2-acetyl-2,3-dihydro-3-hydroxy-2-methyl-5,6-(methylenedioxy)-3-veratryl-	MeOH	236(4.38),280(4.05), 310s(3.90)	102-2021-84
5,12-Naphthacenedione, 8-ethyl-7,8,9,10-tetrahydro-6,7,8,11-tetrahydroxy-1-methoxy-, cis-(±)-	MeOH	219(4.33),234(4.51), 251(4.44),288(3.90), 468(4.06),481(4.07), 495(4.10),529(3.88)	78-4609-84
trans	MeOH	221(4.31),233(4.49), 250(4.39),288(3.88), 481(4.02),499(4.05), 529(3.83),570(3.80)	78-4609-84
$C_{21}H_{20}O_8$			
Podophyllotoxin, 4'-demethyl-	EtOH	282(3.64),291(3.61), 296s(--)	102-1147-84
$C_{21}H_{20}O_{10}$			
Afzelin	MeOH	266(4.33),344(4.17)	94-0490-84
Conyzorigun	MeOH	274(4.27),336(4.33)	39-2945-84C
Cosmosiin (apigenin 7-glucoside)	MeOH	268(4.31),335(4.38)	102-0468-84
$C_{21}H_{20}O_{11}$			
Astragalin	MeOH	267(4.41),353(4.31)	95-0142-84
Glucoluteolin	MeOH	253(4.43),265(4.40), 348(4.45)	94-0295-84
Isoorientin	MeOH	254(4.37),269(4.42), 348(4.47)	94-4003-84
Luteolin 7-glucoside	MeOH	254(4.27),265(4.23), 347(4.28)	94-1724-84
	MeOH-NaOMe	264(4.30),398(4.32)	94-1724-84
	MeOH-NaOAc	258(4.34),404(4.29)	94-1724-84
	+ H_3BO_3	258(4.41),371(4.35)	94-1724-84
	MeOH-AlCl_3	273(4.34),298(3.96), 330(3.74),427(4.41)	94-1724-84
	+ HCl	270(4.23),295(4.10), 356(4.17),386(4.19)	94-1724-84
Quercitrin	MeOH	257(4.36),264s(4.33), 350(4.21)	94-0490-84
$C_{21}H_{20}O_{11}$			
Nikkoshidin	MeOH	258(4.34),360(4.24)	95-0142-84
	MeOH-NaOMe	274(--),414(--)	95-0142-84

Compound	Solvent	$\lambda_{max}(\log \epsilon)$	Ref.
$C_{21}H_{20}S$			
22-Thiatetracyclo[13.2.2.2^{8,11}.1^{3,6}]do-cosa-3,5,8,10,15,17,18,20-octaene	n.s.g.	<u>258s(3.6)</u>,285s(3.32)	44-1177-84
$C_{21}H_{21}ClN_4O_8$			
9H-Purine, 6-chloro-2-(5-methyl-2-fur-anyl)-9-(2,3,5-tri-O-acetyl-β-D-ribofuranosyl)-	EtOH	245(3.89),302(4.11), 324(4.18)	44-4340-84
$C_{21}H_{21}Cl_2N_3O_2$			
Pyrano[2,3-e]indazol-2(3H)-one, 3,3-di-chloro-4,5,6,7-tetrahydro-7-phenyl-4-piperidino-	EtOH	237(4.19),283(4.16)	4-0361-84
$C_{21}H_{21}IN_8O_3$			
Piperazine, 1-(4-amino-6,7-dimethoxy-2-quinazolinyl)-4-(4-azido-3-iodo-benzoyl)-	MeOH	253(4.83),273(4.59), 340(3.84)	69-3765-84
$C_{21}H_{21}NO$			
Oxazole, 2-(5-hexenyl)-4,5-diphenyl-	C_6H_{12}	224(4.21),292(4.00)	44-0399-84
$C_{21}H_{21}NO_2$			
3aH-Benz[g]indol-3a-ol, 2,3,4,5-tetra-hydro-2,2-dimethyl-, benzoate	EtOH	236(4.33)	12-0577-84
Oxazole, 2-(5-hexenyloxy)-4,5-diphenyl-	C_6H_{12}	224(4.32),301(4.13)	44-0399-84
2-Oxiranecarbonitrile, 3-[2-(5-hexenyl-oxy)phenyl]-2-phenyl-, trans	EtOH	231s(4.09),274s(3.60), 282(3.65)	24-2157-84
$C_{21}H_{21}NO_3$			
2aH-Benz[g]oxazirino[3,2-i]indol-2a-ol, 1,2,3,4-tetrahydro-1,1-dimethyl-, benzoate	EtOH	225(4.26),275(3.32)	12-0599-84
3H-Indol-3-one, 1-acetyl-1,2-dihydro-2-(3-oxo-1-phenylpentyl)-	EtOH	206(4.49),241(4.70), 260(4.30),333(3.88)	103-1374-84
$C_{21}H_{21}NO_4$			
5,9[1',2']-Benzeno-5H-bisoxireno[1,6:-2,3]benz[1,2-f]isoindole-6,8(5aH,7H)-dione, 1a,2,3,3a,8a,9-hexahydro-1a,3a,7-trimethyl-, (1aα,3aα,4aS,5α-5aβ,8aβ,9α,9aR*)-	CH_2Cl_2	249(2.58),254(2.61), 260(2.63),268(2.49)	22-0187-84
Bulgaramine	MeOH	227(4.14),260(3.98), 341(4.23)	100-1048-84
	MeOH-acid	224s(3.98),238(4.07), 302s(4.01),331(4.18)	100-1048-84
Guadiscoline	EtOH	221(4.33),268(4.22), 320(3.89),356s(3.63)	100-0353-84
	EtOH-acid	224(4.27),259s(4.06), 273(4.15),368(3.82), 410s(3.50)	100-0353-84
$C_{21}H_{21}NO_4S_2$			
2-Butenedioic acid, 2-[[[bis(phenyl-methyl)amino]thioxomethyl]thio]-, dimethyl ester	MeOH	260(4.16)	104-2285-84
$C_{21}H_{21}NO_5$			
Berberine, dihydro-8-methoxy-	EtOH	276(4.09),286s(4.04), 352(4.30)	142-0101-84

Compound	Solvent	$\lambda_{max}(\log \epsilon)$	Ref.
Corynoline, (+)-	MeOH	207(4.62),238(3.90), 289(3.88)	83-0223-84
	MeOH-HCl	208(4.79),238(3.92), 290(3.94)	83-0223-84
	MeCN	206(4.85),238(3.92), 289(3.91)	83-0223-84
Corynoline, (±)-	MeOH	238(4.06),289(3.97)	100-0001-84
6-epi-	MeOH	238s(4.03),290(3.95)	100-0001-84
14-epi-, (+)-	MeOH	205(4.69),236(3.89), 290(3.88)	83-0223-84
	MeOH-HCl	207(4.75),240(3.89), 291(3.92)	83-0223-84
Deoxoarnottianamide	EtOH	229(4.76),263s(4.40), 315s(3.79),331(3.75)	39-1769-84C
Pseudopalmanine	MeOH	252(4.61),289s(3.99), 330s(3.74)	78-3957-84
$C_{21}H_{21}NO_6$			
Corynoline, 5-hydroxy-, (±)-	MeOH	240(3.84),289(3.74)	100-0001-84
Spiro[1,3-dioxolo[4,5-g]isoquinoline-5(6H),2'-[2H]indol]-1'(3'H)-one, 3'-hydroxy-4',5'-dimethoxy-6-methyl-, trans-(±)-	MeOH	225(4.50),261.5(4.05), 294(3.72),335(3.49)	94-2230-84
$C_{21}H_{21}NO_7$			
Benzaldehyde, 2-[(7,8-dihydro-5-oxo-1,3-dioxolo[4,5-g]isoquinolin-6(5H)-yl)methoxymethyl]-3,4-dimethoxy-	EtOH	234(4.32),269(4.17), 318(3.74)	142-0101-84
Palmanine	MeOH	216(4.44),235(4.32), 279(3.84),314(3.83)	78-3957-84
$C_{21}H_{21}NO_8$			
Propanedioic acid, [3-cyano-6-(3,4-dimethoxyphenyl)-2-oxo-2H-pyran-4-yl]-, diethyl ester	EtOH	220(4.32),262(3.99), 414(4.38)	94-3384-84
$C_{21}H_{21}NO_8S$			
Corynoline-6-O-sulfate, (+)-	MeOH	237(4.13),289(4.02)	100-0001-84
$C_{21}H_{21}N_2$			
1H-Pyrazolium, 1,2-dimethyl-3,5-bis(2-phenylethenyl)-, iodide	EtOH	310(4.65)	103-1266-84
$C_{21}H_{21}N_2S_2$			
Benzothiazolium, 3-ethyl-2-[3-(3-ethyl-2(3H)-benzothiazolylidene)-1-propenyl]-, iodide	$C_6H_{11}Me$ EtOH DMF	560(5.18) 558(5.14) 562(5.10)	99-0415-84 99-0415-84 99-0415-84
$C_{21}H_{21}N_3OS$			
5-Pyrimidinecarboxamide, N,N-diethyl-1,4-dihydro-2,6-diphenyl-4-thioxo-	EtOH	256(4.41),314(4.10), 381(3.61)	18-0605-84
$C_{21}H_{21}N_3O_7$			
1H-Pyrazole-3,4-dicarboxylic acid, 5-[6-(1,3-dihydro-1,3-dioxo-2H-isoindol-2-yl)-1-oxohexyl]-, dimethyl ester	MeOH	218s(4.86),221(4.87), 235s(4.53),242(4.44), 284(4.40)	103-0066-84
$C_{21}H_{21}N_3O_8$			
1-Azabicyclo[3.2.0]hept-2-ene-2-carboxylic acid, 3-[4,5-dihydro-5-(meth-	THF	251(4.09),257.5(4.13), 262.5(4.15),269(4.08),	77-1513-84

Compound	Solvent	$\lambda_{max}(\log \epsilon)$	Ref.
oxycarbonyl)-3-isoxazolyl]-6-ethyl-7-oxo-, (4-nitrophenyl)methyl ester (cont.)		325(4.09)	77-1513-84
$C_{21}H_{21}N_3O_{10}$ Hexanedioic acid, 2-[[[(4-nitrophenyl)-methoxy]carbonyl]amino]-, 1-[(4-nitrophenyl)methyl] ester, (S)-	MeCN	268(4.29)	78-1907-84
$C_{21}H_{21}N_5O_6$ 5,8-Dideazaisofolic acid	pH 7.0	235(4.53)	87-0232-84
$C_{21}H_{21}O_2Se$ Selenonium, bis(4-methoxyphenyl)(4-methylphenyl)-, hexafluorophosphate	MeOH	251(4.45)	11-0098-84A
$C_{21}H_{22}BrClN_6OS$ 6H-Thieno[3,2-f][1,2,4]triazolo[4,3-a]-[1,4]diazepine, 2-bromo-4-(2-chlorophenyl)-9-[4-(2-methoxyethyl)-1-piperazinyl]-	MeOH	241s(4.30),300(3.58)	73-0621-84
$C_{21}H_{22}BrClN_6S_2$ 6H-Thieno[3,2-f][1,2,4]triazolo[4,3-a]-[1,4]diazepine, 2-bromo-4-(2-chlorophenyl)-9-[4-(2-methylthio)ethyl]-1-piperazinyl]-	MeOH	240s(4.29),300(3.58)	73-0621-84
$C_{21}H_{22}Br_2O$ 3,7b-Etheno-7bH-cyclobut[e]inden-1(3H)-one, 7,8-dibromo-2,4-bis(1,1-dimethylethyl)-	EtOH	209(4.15),226s(3.95), 273(3.91),314(3.74)	33-1386-84
$C_{21}H_{22}ClN_5O_7$ 9H-Purine, 6-chloro-2-(1-methyl-1H-pyrrol-2-yl)-9-(2,3,5-tri-O-acetyl-β-D-ribofuranosyl)-	EtOH	245(3.95),298(4.00), 337(4.04)	44-4340-84
$C_{21}H_{22}N_2$ Piperazine, 1-(2-naphthalenylmethyl)-4-phenyl-	C_6H_{12}	252(4.26),275(3.84), 285s(3.73)	35-1335-84
$C_{21}H_{22}N_2O$ Phenol, 4-[(2-ethyl-8-methyl-1-naphthalenyl)azo]-3,5-dimethyl-	DMF-HCl	328(4.30),467(3.36)	33-0113-84
$C_{21}H_{22}N_2O_2S$ Acetic acid, [[4-cyano-5,6,7,8-tetrahydro-1-(phenylmethyl)-3-isoquinolinyl]thio]-, ethyl ester	dioxan	226(4.37),266(4.15), 317(3.76)	104-2225-84
Thieno[2,3-c]isoquinoline-2-carboxylic acid, 1-amino-6,7,8,9-tetrahydro-5-(phenylmethyl)-, ethyl ester	dioxan	285(4.55),303(4.16), 363(3.96)	104-2225-84
$C_{21}H_{22}N_2O_3$ Propanamide, N-[4-(butylamino)-9,10-dihydro-9,10-dioxo-1-anthracenyl]-	benzene	565(4.03),603(3.98)	39-0529-84C
4H-Pyrido[1,2-a]pyrimidine-3-carboxylic acid, 6,7,8,9-tetrahydro-6-methyl-4-oxo-9-(3-phenyl-2-propenylidene)-, ethyl ester, (E,?)-	EtOH	242(4.06),388(4.44)	118-0582-84

Compound	Solvent	λ_{max}(log ϵ)	Ref.
Rhazimine	MeOH	222(3.391),265(2.864), 290(2.773)	88-3913-84
$C_{21}H_{22}N_2O_4$			
Carbamic acid, [4-(butylamino)-9,10-di- hydro-9,10-dioxo-1-anthracenyl]-, ethyl ester	benzene	563(4.05),600(3.99)	39-0529-84C
$C_{21}H_{22}N_2O_6$			
Carbonic acid, [1-(3-acetoxypropyl)- 4-hydroxy-6-methyl-1H-indazol-3- yl]methyl phenyl ester	MeOH	222(4.54),300(3.88)	5-1711-84
$C_{21}H_{22}N_2O_6S$			
2H-Thiocin-5,6,7-tricarboxylic acid, 3,4-dihydro-8-[(phenylcarbonimid- oyl)amino]-, 5-ethyl 6,7-dimethyl ester	CHCl$_3$	257(4.37),313(4.01)	24-0585-84
$C_{21}H_{22}N_2O_7$			
Oxepino[2,3-d]pyrimidine-5,6-dicarbox- ylic acid, 1,2,3,4,7,8-hexahydro-7,8- dimethyl-2,4-dioxo-3-phenyl-, 6-ethyl 5-methyl ester	CHCl$_3$	263(3.89),312(4.07)	24-0585-84
Oxepino[2,3-d]pyrimidine-5,6-dicarbox- ylic acid, 1,2,3,4,7,8-hexahydro-8,8- dimethyl-2,4-dioxo-3-phenyl-, 6-ethyl 5-methyl ester	CHCl$_3$	267(4.08),303(4.15)	24-0585-84
3,4,5-Oxepintricarboxylic acid, 6,7-di- hydro-7-methyl-2-[(phenylcarbonimid- oyl)amino]-, 5-ethyl 3,4-dimethyl ester	CHCl$_3$	265(3.77),315(4.13)	24-0585-84
$C_{21}H_{22}N_3O_2$			
1H-Pyrazolium, 3-[2-[4-(dimethylamino)- phenyl]ethenyl]-2-ethyl-4,5-dihydro- 4,5-dioxo-1-phenyl-, iodide	EtOH	360(4.30),410(4.30)	48-0695-84
$C_{21}H_{22}N_4OS$			
5-Pyrimidinecarboxamide, 4-(aminothio)- N,N-diethyl-2,6-diphenyl-	EtOH	262(4.56)	18-0605-84
$C_{21}H_{22}N_4O_3$			
1-Imidazolidinepropanoic acid, 2,5-di- oxo-4,4-diphenyl-, (1-methylethyli- dene)hydrazide	MeOH	203(4.49)	56-0585-84
1-Oxa-2-azaspiro[4.5]dec-2-ene, 4-[(3- methylphenyl)azo]-4-nitro-3-phenyl-	EtOH	388(4.48)	104-1844-84
1H-Pyrazole-4-carboxylic acid, 5-[[eth- oxy(phenylamino)methylene]amino]-1- phenyl-, ethyl ester	CHCl$_3$	270(4.22)	24-0585-84
$C_{21}H_{22}N_4O_4$			
7H-Pyrido[2,1-b]quinazoline-6,11-dione, 8,9-dihydro-2,3,4-trimethoxy-, 6-(phenylhydrazone)	EtOH	246(4.60),321(3.83), 392(4.10)	4-1301-84
1,2,4-Triazolo[3,2-b]oxepino[2,3-d]- pyrimidine-7-carboxylic acid, 4,5,8,9- tetrahydro-2,8,9-trimethyl-5-oxo-4- phenyl-, ethyl ester	CHCl$_3$	263(4.03),312(4.19), 318(4.18),335s(3.96)	24-0585-84

Compound	Solvent	$\lambda_{max}(\log \epsilon)$	Ref.
$C_{21}H_{22}N_4O_8S_2$ 9H-Fluorene, 9,9-bis(butylthio)-2,4,5,7-tetranitro-	MeCN	300(4.2),340(4.2), 410(3.6)	77-0266-84
$C_{21}H_{22}O_2S$ Benzoic acid, 4-[2-(3,4-dihydro-4,4-di-methyl-2H-1-benzothiopyran-6-yl)-1-propenyl]-, (E)-	EtOH	233(4.04),319(4.40)	87-1516-84
$C_{21}H_{22}O_3$ Benzoic acid, 4-[2-(3,4-dihydro-4,4-di-methyl-2H-1-benzopyran-6-yl)-1-prop-enyl]-, (E)-	EtOH	231(4.11),307(4.40)	87-1516-84
$C_{21}H_{22}O_5$ 2-Butenoic acid, 3-methyl-, 1-[5-(3,6-dioxo-1,4-cyclohexadien-1-yl)-3-fur-anyl]-4-methyl-3-pentenyl ester, (-)-	EtOH	253(4.31),438(3.62)	102-0301-84
$C_{21}H_{22}O_6$ 1,4-Butanedione, 1-(1,3-benzodioxol-5-yl)-4-(3,4-dimethoxyphenyl)-2,3-dimethyl-	MeOH	232(4.57),273(4.29), 304(4.25)	102-2647-84
6aH-Cyclopropa[b]furo[4,3,2-de][1]-benzopyran-6a-carboxylic acid, 1,7-diacetyl-7,7a-dihydro-5-propyl-, ethyl ester, (6aα,7α,7aα)-	EtOH	243(4.13),300(4.19)	39-0487-84C
1H-3b,11-Ethanophenanthro[1,2-c]furan-1,3(3aH)-dione, 11-acetoxy-4,5,9b-10,11,11a-hexahydro-7-methoxy-, (3aα,3bα,9bα,11β,11aα)-(±)-	EtOH	227(3.86),278(3.23)	2-1168-84
2(3H)-Furanone, 3-(1,3-benzodioxol-5-ylmethyl)-4-[(3,4-dimethoxyphen-yl)methyl]dihydro- (kusunokinin)	MeOH	232(3.85),282s(3.64), 285(3.65),395s(3.35)	78-1145-84
4-Phenanthrenemethanol, 5-acetoxy-9,10-dihydro-2,7-dimethoxy-, acetate	EtOH	213(4.44),276(4.27)	102-0671-84
$C_{21}H_{22}O_7$ Bicyclo[2.2.2]oct-2-ene-2,3,5-tricarb-oxylic acid, 6-oxo-5-(phenylmethyl)-, trimethyl ester, (1α,4α,5α)-	MeOH	238(4.39)	44-4429-84
(1α,4α,5β)-	MeOH	228(3.68)	44-4429-84
1(2H)-Naphthalenone, 4-(1,3-benzodiox-ol-5-yl)-2,4-dihydroxy-6,7-dimethoxy-2,3-dimethyl-	MeOH	235(4.38),275(4.06), 315(3.86)	102-2021-84
Naphtho[2,3-d]-1,3-dioxol-5(6H)-one, 8-(3,4-dimethoxyphenyl)-7,8-dihydro-6,8-dihydroxy-6,7-dimethyl-	MeOH	235(4.38),248(4.06), 254(4.04),260(4.06), 277(4.11),310(3.77)	102-2021-84
$C_{21}H_{22}O_8$ 4H-1-Benzopyran-4-one, 5,6,7-trimeth-oxy-2-(2,3,5-trimethoxyphenyl)-	MeOH	266(4.2),310(4.3)	94-2296-84
4H-1-Benzopyran-4-one, 5,6,7-trimeth-oxy-2-(2,4,5-trimethoxyphenyl)-	MeOH	254(4.3),305(4.1), 354(4.3)	94-4935-84
4H-1-Benzopyran-4-one, 5,6,7-trimeth-oxy-2-(3,4,5-trimethoxyphenyl)-	MeOH	237s(--),258(4.19), 317(4.49)	94-0166-84
Pseudocyphellarin A	MeOH	218(4.15),250(4.36), 270s(4.21),290s(3.96), 342(3.30)	102-0431-84
	MeOH-NaOH	221(4.29),286(4.19), 302s(4.13)	102-0431-84

Compound	Solvent	λ_{max} (log ϵ)	Ref.
Pumilin, 8-acetyl-9-deacyl-, 9-methacrylate	MeOH	203(4.26)	102-0817-84
$C_{21}H_{22}O_{10}$ 4H-1-Benzopyran-4-one, 5,7-dihydroxy-3,6,8-trimethoxy-2-(3,4,5-trimethoxyphenyl)-	MeOH MeOH-NaOAc MeOH-AlCl$_3$	267(4.35),363(4.01) 278(--),377(--) 278(--),370(--)	150-3786-84M 150-3786-84M 150-3786-84M
$C_{21}H_{23}BrN_2O_5S$ D-Arabinitol, 5-C-[1-(4-bromophenyl)-2-[(phenylmethyl)thio]-1H-imidazol-4-yl]-	EtOH	271(3.81)	136-0091-84B
1,2,3-Propanetriol, 1-[3-(4-bromophenyl)-3a,5,6,6a-tetrahydro-6-hydroxy-2-[(phenylmethyl)thio]-3H-furo[2,3-d]imidazol-5-yl]-	EtOH	259(4.10)	136-0091-84B
$C_{21}H_{23}BrN_2O_6$ Carbonic acid, [1-(3-acetoxypropyl)-7-bromo-4,5,6,7-tetrahydro-6-methyl-4-oxo-1H-indazol-3-yl]methyl phenyl ester	MeOH	213(4.26),255(4.01)	5-1711-84
$C_{21}H_{23}ClN_2$ Hapalindole A	n.s.g.	222(4.58),280(3.85), 291(3.76)	35-6456-84
$C_{21}H_{23}NO_4$ 9-Azabicyclo[6.2.2]dodeca-8,10,11-triene-11,12-dicarboxylic acid, 10-phenyl-, dimethyl ester	EtOH	249(4.12),285(3.90), 342(3.80)	88-0959-84
Guadiscoline, dihydro-	EtOH	220(4.45),270s(4.17), 278(4.23),302(4.07)	100-0353-84
$C_{21}H_{23}NO_5$ Allocryptopine	MeOH	209(4.69),231(4.03), 286(3.79)	73-1318-84
Homochelidonine, (+)-	EtOH	286(3.74)	100-0001-84
$C_{21}H_{23}NO_6$ Chelamidine	n.s.g.	287(3.9)	100-0001-84
Norisoboldine, N-(carbethoxy)-	MeOH	280(3.81),303(3.88)	44-1439-84
Norisosalutaridine, N-(carbethoxy)-	MeOH	240(4.34),287(3.92)	44-1439-84
$C_{21}H_{23}NO_7$ Benzaldehyde, 3,4-dimethoxy-2-[ethoxy-(dihydro-5-oxo-1,3-dioxolo[4,5-g]-isoquinolin-6-yl)ethyl]-	EtOH	264(3.00),311(2.74)	142-0101-84
$C_{21}H_{23}N_3$ 6,4-Metheno-4H-benzo[b]pyrido[4,3-h]-[1,4]benzodiazepine, 2-(1,1-dimethylethyl)-1,2,3,4a,5,5a-hexahydro-	EtOH	249(4.74),257(4.80), 295(4.20),390(3.80)	39-0753-84C
$C_{21}H_{23}N_3O_2$ 3,12-Metheno-4,11-nitrilo-7H-dicyclo-octa[c,f][1,2]diazepine-7-carboxylic acid, 1,2,5,6,9,10,13,14-octahydro-, ethyl ester	EtOH	267(3.58),293s(3.43)	35-2672-84
4,11-Metheno-3,12-nitrilo-7H-dicyclo-octa[c,f][1,2]diazepine-7-carboxylic	MeOH	263(3.58),298(3.47)	35-2672-84

Compound	Solvent	$\lambda_{max}(\log \epsilon)$	Ref.
acid, 1,2,5,6,9,10,13,14-octahydro-, ethyl ester (cont.)			35-2672-84
3,11-Metheno-4,10-nitrilodicycloocta-[b,e]pyridinium, 7-[(ethoxycarbonyl)-amino]-1,2,5,6,8,9,12,13-octahydro-, hydroxide, inner salt	EtOH	253(3.67),311(3.78), 332s(3.48)	35-2672-84
4,10-Metheno-3,11-nitrilodicycloocta-[b,e]pyridinium, 7-[(ethoxycarbonyl)-amino]-1,2,5,6,8,9,12,13-octahydro-, hydroxide, inner salt	EtOH	255(3.58),315(3.69), 338s(3.40)	35-2672-84
$C_{21}H_{23}N_3O_6S_2$			
9H-Fluorene, 9,9-bis(butylthio)-2,4,7-trinitro-	MeCN	<u>324(4.3)</u>	77-0266-84
$C_{21}H_{23}N_5O_6$			
Adenosine, N-benzoyl-7,8-dihydro-7-methyl-2',3'-O-(1-methylethylidene)-8-oxo-	MeOH	232(4.31),291(4.14)	150-3601-84M
$C_{21}H_{23}N_5O_7$			
10,11-(Methanoxymethano)-3,6,10-meth-eno-7H,10H-3a,6a,8,9a,10b-pentaaza-dicyclopenta[b,def]phenanthrene-4,5-dicarboxylic acid, 1,2,3,5a,6,8,9-9b,10a,10c-octahydro-8-methyl-7,9-dioxo-, dimethyl ester	MeCN	350(3.08)	88-2459-84
$C_{21}H_{24}$			
2,9-Ethanodibenzo[a,e]cyclooctene, 5,6,11,12-tetrahydro-1,3,4-trimethyl-	$CHCl_3$	303(2.53)	78-4823-84
$C_{21}H_{24}Br_2O_3$			
Androsta-5,9(11),15-triene-3,17-dione, 15,16-dibromo-, cyclic 3-(1,2-ethane-diyl acetal)	EtOH	257(3.90)	23-1103-84
$C_{21}H_{24}NO_4$			
β-Canadinium, N-methyl-, iodide	MeOH	210(4.71),220s(4.54), 288(3.98)	73-1318-84
$C_{21}H_{24}N_2O$			
Ethanone, 1-[(20α)-17,18-didehydro-yohimban-18-yl]-	EtOH	283(3.73),290(3.66)	44-2708-84
$C_{21}H_{24}N_2O_3$			
Vincarine	n.s.g.	243(3.69),294(3.25)	105-0509-84
$C_{21}H_{24}N_2O_4$			
Echitamidine, N_a-formyl-	EtOH	210(4.24),252(3.95), 290s(3.4)	102-2708-84
$C_{21}H_{24}N_2O_4S_2$			
9H-Fluorene, 9,9-bis(butylthio)-2,7-dinitro-	MeCN	<u>325(4.4)</u>	77-0266-84
$C_{21}H_{24}N_2O_5S$			
D-Arabinitol, 5-C-[1-phenyl-2-[(phenyl-nethyl)thio]-1H-imidazol-4-yl]-, (S)-	EtOH	265(3.78)	136-0091-84B
1,2,3-Propanetriol, 1-[3a,5,6,6a-tetra-hydro-6-hydroxy-3-phenyl-2-[(phenyl-	EtOH	250(4.04)	136-0091-84B

Compound	Solvent	$\lambda_{max}(\log \epsilon)$	Ref.
methyl)thio]-3H-furo[2,3-d]imidazol-5-yl]-, [3aR-[3aα,5α(1S*,2R*),6α-6aα]]- (cont.)			136-0091-84B
$C_{21}H_{24}N_4O_6$ Oxepino[2,3-d]pyrimidine-5,6-dicarboxylic acid, 2-hydrazino-3,4,7,8-tetrahydro-7,8-dimethyl-4-oxo-3-phenyl-, 6-ethyl 5-methyl ester	CHCl$_3$	273(3.86),325(4.22)	24-0585-84
$C_{21}H_{24}N_5O_5$ 7H-Purinium, 6-(benzoylamino)-7-methyl-9-[2,3-O-(1-methylethylidene)-β-D-ribofuranosyl]-, iodide	H$_2$O	278(4.05),320(3.11)	150-3601-84M
$C_{21}H_{24}N_6O_5$ Adenosine, N-benzoyl-2'-deoxy-2-[(2-methyl-1-oxopropyl)amino]-	EtOH	245(4.55),297(4.18)	78-0003-84
$C_{21}H_{24}O$ Tetracyclo[6.3.2.0^{3,7}.0^{8,11}]trideca-2,4,6,10,12-pentaen-9-one, 2,10-bis(1,1-dimethylethyl)-	EtOH	213(3.94),224(3.93), 290(3.79),406(2.60)	33-1386-84
Tricyclo[8.2.2.2^{4,7}]hexadeca-4,6,10,12-13,15-hexaene-5-carboxaldehyde, 11,12,13,14-tetramethyl-	CHCl$_3$	254s(3.91),302(2.53)	78-4823-84
$C_{21}H_{24}O_2$ 6H-Dibenzo[h,l][1,7]dioxacyclotridecin, 7,8,9,10-tetrahydro-16,17-dimethyl-, (E)-	MeOH	208(4.41),276(3.56), 281(3.57)	44-1627-84
(Z)-	MeOH	208(4.46),274(3.60), 281(3.58)	44-1627-84
1,5-Heptanedione, 6,6-dimethyl-1,3-diphenyl-	EtOH	242.5(4.10)	78-4127-84
6-Hepten-3-one, 5-hydroxy-2,2-dimethyl-5,7-diphenyl-, (E)-	EtOH	252.5(4.17)	78-4127-84
1,5-Hexanedione, 3-phenyl-1-(2,4,6-trimethylphenyl)-	EtOH	247(3.42)	78-4127-84
4-Penten-1-one, 3-hydroxy-3-methyl-5-phenyl-1-(2,4,6-trimethylphenyl)-, (E)-	EtOH	251.5(4.32)	78-4127-84
$C_{21}H_{24}O_4$ Ethanone, 1,1'-[1,5-pentanediylbis(oxy-2,1-phenylene)]bis-	MeOH	212(4.64),246(4.19), 305(3.89)	44-1627-84
D-Homo-17a-oxaestra-1,3,5(10)-trien-17-one, 3-acetoxy-16-methylene-	EtOH	217(3.84)	13-0283-84A
$C_{21}H_{24}O_5$ Myricanone	EtOH	215(4.59),261(4.14), 297(3.95)	95-0037-84
$C_{21}H_{24}O_6$ Armenin C	MeOH	258(3.80)	102-0661-84
1,4-Butanediol, 2-(1,3-benzodioxol-5-ylmethylene)-3-[(3,4-dimethoxyphenyl)methyl]-, [S-(E)]-	EtOH	257(3.82)	102-2323-84
1-Butanone, 1-(1,3-dihydroxy-7,9-dimethoxy-2,8-dimethyldibenzofuran-4-yl)-3-methyl-	EtOH	220(4.48),262s(4.41), 269(4.50),290(4.40), 351(3.54)	39-2573-84C

Compound	Solvent	$\lambda_{max}(\log \epsilon)$	Ref.
Pseudocyphellarin B	MeOH	226(4.30),275(4.25), 318(3.93)	102-0431-84
	MeOH-NaOH	228(4.30),245(4.20), 322(4.41)	102-0431-84
$C_{21}H_{24}O_7$			
1H-3,10a-Ethanophenanthrene-1,2-dicarb-oxylic acid, 3-acetoxy-2,3,4,4a,9,10-hexahydro-7-methoxy-, ($1\alpha,2\alpha,3\alpha$-$4a\beta,10a\beta$)-($\pm$)-	EtOH	228(4.11),278(3.41)	2-1168-84
1H,3H-Furo[3,4-c]furan-3a(4H)-ol, 1-(3,4-dimethoxyphenyl)dihydro-4-(4-hydroxy-3-methoxyphenyl)-, [1S-($1\alpha,3a\alpha,4\alpha,6a\alpha$)]-	EtOH EtOH-NaOH	232.5(4.21),280(3.75) 235(--),253(--), 285(--),295(--)	94-2730-84 94-2730-84 +94-4482-84
Picrasa-3,9(11)-diene-1,2,12,16-tetr-one, 20-hydroxy-11-methoxy-, (5β)-	EtOH	256(3.86)	18-2885-84
Picrasa-3,9(11)-diene-2,12,16-trione, 1,11-epoxy-20-hydroxy-1-methoxy-, ($1\beta,5\beta$)-	EtOH	245(3.91),295(3.48)	18-2885-84
Pukalide, $11\beta,12\beta$-epoxy-	EtOH	238(3.73)	100-1009-84
Rishirilide A	EtOH	207(4.29),223(4.14), 319(4.23)	158-1091-84
$C_{21}H_{24}O_8$			
β-D-Glucopyranoside, 3-hydroxy-5-[2-(4-methoxyphenyl)ethenyl]phenyl, cis	MeOH	215(4.30),284(4.04)	94-3501-84
trans	MeOH	215(4.40),306(4.52), 319(4.52)	94-3501-84
$C_{21}H_{24}O_9$			
β-D-Glucopyranoside, 5-[2-(3,5-dihy-droxyphenyl)ethenyl]-2-methoxyphenyl, (E)-	MeOH	217(4.40),303(4.43), 330(4.51)	94-3501-84
β-D-Glucopyranoside, 5-[2-(3-hydroxy-4-methoxyphenyl)ethenyl]phenyl, (Z)-	MeOH	216(4.46),288(4.08)	94-3501-84
Rhaponticin	MeOH	219(4.40),302(4.38), 324(4.48)	94-3501-84
$C_{21}H_{24}O_{10}$			
Trifolirhizin	EtOH	280(3.56),286(3.61), 312(3.80)	105-0233-84
$C_{21}H_{24}Si_2$			
Disilane, pentamethyl-1-pyrenyl-	C_6H_{12}	350.5(4.71)	60-0341-84
$C_{21}H_{25}BrO_3$			
Androsta-5,9(11),15-triene-3,17-dione, 15-bromo-, cyclic 3-(1,2-ethanediyl acetal), (14β)-	EtOH	240(4.08)	23-1103-84
$C_{21}H_{25}Br_2NO_6$			
Carbamic acid, [1,2-bis(2-bromo-4,5-di-methoxyphenyl)ethyl]-, ethyl ester	EtOH	285(4.02)	20-1099-84
$C_{21}H_{25}NO_2S_2$			
9H-Fluorene, 9,9-bis(butylthio)-3-nitro-	MeCN	<u>258(4.4)</u>	77-0266-84
$C_{21}H_{25}NO_4$			
2-Butenedioic acid, 2-[2-(2,2-dimethyl-1-methylenepropyl)-1-methyl-1H-indol-	EtOH	224(4.66),282(3.97)	78-4837-84

Compound	Solvent	$\lambda_{max}(\log \epsilon)$	Ref.
3-yl]-, dimethyl ester, (E)- (cont.)			78-4837-84
1H-Carbazole-3,4-dicarboxylic acid, 2,9-dihydro-9-methyl-1-(1,1-dimethyl-ethyl)-, dimethyl ester	EtOH	231(4.37),275(4.04), 367(4.07)	78-4837-84
Secocularine	EtOH	220(4.32),235s(4.25), 296s(3.78),320(3.88)	88-0889-84 +100-0753-84
$C_{21}H_{25}NO_5$			
1H-Inden-2-amine, N-[(2,3-dimethoxy-phenyl)methylene]-1,5,6-trimethoxy-, (1S-cis)-	MeOH	202(4.73),222(4.57), 257(4.20),285(3.93), 310s(3.40)	39-1655-84C
1H-Inden-2-amine, N-[(3,4-dimethoxy-phenyl)methylene]-1,5,6-trimethoxy-, (1S-cis)-	MeOH	203(4.76),226(4.38), 272(4.27),295(4.18), 310s(4.04)	39-1655-84C
$C_{21}H_{25}NO_6$			
Ethanedione, [2-[2-(dimethylamino)eth-yl]-4,5-dimethoxyphenyl](3-hydroxy-4-methoxyphenyl)- (saxoguattine)	EtOH	230(4.33),282(4.08), 322(4.03)	100-0353-84
	EtOH-base	234(4.28),253(4.22), 280(4.05),376(3.52)	100-0353-84
$C_{21}H_{25}NO_7$			
Oxoepistephamiersine	EtOH	284.5(3.29)	100-0858-84
$C_{21}H_{25}N_3O$			
Urea, N,N-diethyl-N'-[1-phenyl-3-(phenylimino)-1-butenyl]-	EtOH	206(4.37),226(4.24), 322(4.22)	39-0239-84C
$C_{21}H_{25}N_3O_3S$			
4H-Thiocino[2,3-d]pyrimidine-6-carbox-ylic acid, 3,7,8,9-tetrahydro-2-[(1-methylethyl)amino]-4-oxo-3-phenyl-, ethyl ester	CHCl$_3$	272(4.36),350(4.17)	24-0585-84
$C_{21}H_{25}N_3O_4$			
Oxepino[2,3-d]pyrimidine-6-carboxylic acid, 3,4,7,8-tetrahydro-8-methyl-2-[(1-methylethyl)amino]-4-oxo-3-phenyl-, ethyl ester	CHCl$_3$	280(3.89),347(4.34)	24-0585-84
$C_{21}H_{25}N_5O_2$			
2H-Pyrido[2,3-b][1,4]diazepin-2-one, 1,3,4,5-tetrahydro-5-[(4-methyl-1-piperazinyl)acetyl]-3-phenyl-	EtOH	257(3.98),286(3.79)	161-0162-84
2H-Pyrido[2,3-b][1,4]diazepin-2-one, 1,3,4,5-tetrahydro-5-[(4-methyl-1-piperazinyl)acetyl]-4-phenyl-	EtOH	255(3.79),286(3.64)	161-0162-84
$C_{21}H_{25}N_5O_5$			
Adenosine, N-benzoyl-7,8-dihydro-7-methyl-, 2',3'-O-(1-methylethyli-dene)-	MeOH	324(3.99)	150-3640-84M
$C_{21}H_{25}N_5O_6S_2$			
Benzothiazolium, 2-[[4-[(2-cyanoethyl)-methylamino]phenyl]azo]-6-(2-hydroxy-ethoxy)-3-methyl-, methyl sulfate	pH 1.13 pH 6.88 70% EtOH-pH 6.2	609(4.75) 609(4.75) 626(4.79)	48-0151-84M 48-0151-84M 48-0151-84M
$C_{21}H_{26}$			
2,9-Ethanodibenzo[a,e]cyclooctene,	C$_6$H$_{12}$	221(3.98),294(2.15)	78-4823-84

Compound	Solvent	λ_{max}(log ϵ)	Ref.
5,6,7,10,11,12-hexahydro-1,3,4-tri-methyl- (cont.)			78-4823-84
$C_{21}H_{26}NO_4$			
Corypalminium, N-methyl-, iodide	MeOH	209(4.72),225s(4.34), 285(3.82)	73-1318-84
Isoquinolinium, 3-(3,4-dimethoxyphen-yl)-3,4-dihydro-6,7-dimethoxy-1,2-dimethyl-, iodide	EtOH	253(4.25),309(4.03), 363(3.98)	4-0525-84
Isoquinolinium, 3-[(3,4-dimethoxyphen-yl)methyl]-3,4-dihydro-6,7-dimeth-oxy-2-methyl-, perchlorate	MeOH	249(4.30),312(4.01), 368(3.94)	24-1436-84
	MeCH$_2$CH$_2$OH	249.5(4.29),311.5(3.98), 368(3.93)	24-1436-84
$C_{21}H_{26}N_2O_2$			
Coronaridine	EtOH	225(1.05),284(1.05), 292(1.01)	102-2359-84
20-Epipseudoaspidospermidine, 2,16-de-hydro-1-(methoxycarbonyl)-14-iso-	EtOH	250(3.96),282(3.08)	44-3733-84
	EtOH	248(3.96),281(3.05)	44-3733-84
20-Epipseudoaspidospermidine, dehydro-16α-(methoxycarbonyl)-14-iso-	EtOH	221(4.15),266(3.66)	44-3733-84
20-Epipseudovincadifformine, 14-iso-	EtOH	225(4.10),297(4.00), 328(4.20)	44-3733-84
$C_{21}H_{26}N_2O_3$			
Catharinensine	EtOH	250(3.63),282(3.29)	102-2359-84
Heyneanine	EtOH	220(4.24),284(3.85)	102-2359-84
Scandine, tetrahydro-	MeOH	213(4.40),254(3.94), 281(3.41),290(3.29)	44-3275-84
Vincamine	n.s.g.	225(4.50),280(3.95)	105-0509-84
$C_{21}H_{26}N_2O_8$			
1,2-Pyrrolidinedicarboxylic acid, 3-(2-methoxy-2-oxoethyl)-4-[1-methyl-2-(3-nitrophenyl)ethenyl]-, 1-ethyl 2-methyl ester, (E)-	MeOH	246(4.26)	87-0052-84
(Z)-	MeOH	232(4.28)	87-0052-84
$C_{21}H_{26}N_4O_2$			
2-Naphthalenecarbonitrile, 5-amino-3-(butylamino)-8-(cyclohexylamino)-1,4-dihydro-1,4-dioxo-	benzene	441(3.43),569(3.67), 613(3.93),663(3.93)	39-1297-84C
$C_{21}H_{26}O$			
2-Butenal, 3-[4-[2-(2,6,6-trimethyl-1-cyclohexen-1-yl)ethenyl]phenyl]-, (E,E)-	isoPrOH	330(4.30)	89-0081-84
(E,Z)-	isoPrOH	305(4.04)	89-0081-84
(Z,E)-	isoPrOH	323(4.00)	89-0081-84
Schiff's base with bacteriorhodopsin	H$_2$O	490(4.31)	89-0081-84
18-Norpregna-4,6,8(14),13(17)-tetraen-3-one, 20-methyl-	EtOH	412(4.32)	94-2486-84
$C_{21}H_{26}O_2$			
[Bi-1-cyclohexen-1-yl]-6-one, 2-meth-oxy-4,4-dimethyl-4'-phenyl-	H$_2$O	275(3.93)	39-1213-84C
$C_{21}H_{26}O_3$			
Androsta-5,9(11),15-triene-3,17-dione, cyclic 3-(1,2-ethanediyl acetal)-, (14β)-	EtOH	224(3.91)	23-1103-84

Compound	Solvent	$\lambda_{max}(\log \epsilon)$	Ref.
Pregna-1,4-diene-3,12,20-trione	MeOH	244(4.163)	73-1617-84
$C_{21}H_{26}O_4$			
Androsta-5,9(11)-diene-3,15,17-trione, cyclic 3-(1,2-ethanediyl acetal)	EtOH	251(4.14)(changing)	23-1103-84
	EtOH-acid	247(4.01)	23-1103-84
	EtOH-base	262(4.40)	23-1103-84
2-Butenoic acid, 3-methyl-4-[5-oxo-4-[2-(2,2,6-trimethyl-1-cyclohexen-1-yl)ethenyl]-2(5H)-furanylidene]-, methyl ester, (Z,E,E)-	EtOH	356(4.55)	94-1709-84
(Z,Z,E)-	EtOH	356(4.55)	94-1709-84
Strobilurin C	MeOH	226(4.92),231s(4.90), 240s(4.86),262s(4.65), 297(4.92)	158-84-87
2,4,12-Tridecatrienoic acid, 13-(1,3-benzodioxol-5-yl)-, methyl ester	MeOH	216(4.51),261(4.71), 305(4.00)	78-2541-84
$C_{21}H_{26}O_5$			
Androsta-1,4-diene-2-carboxylic acid, 17β-hydroxy-17-methyl-3,11-dioxo-	MeOH	248(4.02)	13-0271-84A
Androsta-2,5-diene-1,17-dione, 4β-acet-oxy-14-hydroxy-	EtOH	210(3.6)	105-0182-84
Tricyclo[12.3.1.1^{2,6}]nonadeca-1(18),2-4,6(19),14,16-hexaene-3,9,15-triol, 16,17-dimethoxy-, (R)- (myricanol)	EtOH	213(4.58),259(3.97), 296(3.92)	95-0037-84
$C_{21}H_{26}O_7$			
1,3-Naphthalenedicarboxylic acid, 1,2-dihydro-7-methoxy-4-(3-methoxy-3-oxo-propyl)-, diethyl ester	EtOH	233(4.01),309(4.20)	2-0395-84
$C_{21}H_{26}O_9$			
β-D-Glucopyranoside, 4-(2-propenyl)-phenyl, 2,3,4-triacetate	MeOH	221(4.02),272(3.11), 278(3.00)	18-2893-84
$C_{21}H_{26}O_{13}$			
Plumiepoxide	MeOH	222(4.43)	94-2947-84
$C_{21}H_{27}BrO$			
Benzene, 1-(8-bromooctyl)-4-(phenyl-methoxy)-	MeOH	210.0(4.18),225.0(4.11), 276.5(3.23),284.0(3.15)	33-2111-84
$C_{21}H_{27}BrO_3Se$			
Naphtho[1,2-b]furan-2(3H)-one, 7-bromo-decahydro-8-hydroxy-3,5a,9-trimethyl-3-(phenylseleno)-, [3R-(3α,3aβ,5aα,7β-8α,9β,9aβ,9bα)]-	EtOH	220(3.95)	32-0107-84
8β-	EtOH	220(3.98)	32-0107-84
$C_{21}H_{27}ClN_4O_6S_2$			
Benzothiazolium, 2-[[4-[(2-chloroeth-yl)ethylamino]phenyl]azo]-6-(2-hy-droxyethoxy)-3-methyl-, methyl sul-fate	pH 1.13	615(4.83)	48-0151-84
	pH 6.88	617(4.83)	48-0151-84
	70% EtOH-pH 6.2	629(4.92)	48-0151-84
$C_{21}H_{27}N$			
Piperidine, 1-(1,1-dimethylethyl)-4,4-diphenyl-, hydrochloride (budipin)	H_2O	259(2.69)	145-0233-84
$C_{21}H_{27}NO_2S_2$			
4-Thiazolidinone, 5-[3-(4-methoxyphen-	n.s.g.	430(4.59)	48-0457-84

Compound	Solvent	$\lambda_{max}(\log \epsilon)$	Ref.
yl)-2-propenylidene]-3-octyl-2-thioxo- (cont.)			48-0457-84
$C_{21}H_{27}NO_4S$			
1-Azaspiro[4.5]dec-6-ene-3,4-dicarboxylic acid, 7-(methylthio)-1-(phenylmethyl)-, dimethyl ester, cis	EtOH	227(4.36)	44-3314-84
trans	EtOH	230(4.28)	44-3314-84
diastereomer	EtOH	230(4.15)	44-3314-84
$C_{21}H_{27}NO_5$			
3,5-Pyridinedicarboxylic acid, 1,2-dihydro-4-(4-methoxyphenyl)-1,2,6-trimethyl-, diethyl ester	EtOH	203(4.47),216s(4.35), 280(4.36),386(3.97)	103-0522-84
$C_{21}H_{27}N_3O_6$			
Benzamide, N-[3-[[4-[(3,4-dihydroxybenzoyl)amino]butyl]amino]propyl]-3,4-dihydroxy-	H_2O	207(4.70),255(4.23), 288(3.95)	64-0010-84C
$C_{21}H_{28}N_2$			
Benzenamine, N,N-dibutyl-4-[2-(4-pyridinyl)ethenyl]-, (E)-	EtOH	387(4.59)	44-2546-84
Pyrido[3,2-g]quinoline, 2,8-bis(1,1-dimethylethyl)-1,2-dihydro-2-methyl-	MeCN	256(4.715),266(4.750), 288(4.152),306(3.900), 396(3.842)	5-0133-84
$C_{21}H_{28}N_2O_3S$			
Androst-2-eno[3,2-d][1,2,3]thiadiazol-1-one, 17-acetoxy-, (5α,17β)-	EtOH	236(3.72),277(3.62)	44-4773-84
$C_{21}H_{28}N_2O_3S_2$			
Androst-2-eno[3,2-d][1,2,3]thiadiazol-17-ol, 1-sulfinyl-, acetate, (1Z,5α-17β)-	EtOH	231(3.87),283s(3.22), 342(4.00)	44-4773-84
$C_{21}H_{28}N_4O_6S_2$			
Benzothiazolium, 2-[[4-(diethylamino)-phenyl]azo]-6-(2-hydroxyethoxy)-3-methyl-, methyl sulfate	pH 1.13	585(4.80)	48-0151-84
	pH 6.88	585(4.80)	48-0151-84
	70% EtOH-pH 6.2	621(4.90)	48-0151-84
$C_{21}H_{28}O$			
18-Norpregna-4,6,8(14)-trien-3-one, 20-methyl-, (17ξ)-	EtOH	349(4.41)	94-2486-84
Oxirane, 2-(4-pentenyl)-3-(1,3-tetradecadiene-5,8-diynyl)-, [2α,3β(1E-3E)]-(±)-	EtOH	263(4.53),276(4.60), 289(4.57)	104-2076-84
$C_{21}H_{28}O_2$			
Benzoic acid, 4-[2-methyl-4-(2,6,6-trimethyl-1-cyclohexen-1-yl)-1-butenyl]-, (E)-	EtOH	217(4.23),272(4.43)	87-1516-84
2,5-Cyclohexadiene-1,4-dione, 2-[(decahydro-1,2,4a-trimethyl-5-methylene-1-naphthalenyl)methyl]- (arenarone)	MeOH	245(3.68),330(2.39)	44-0241-84
Spiro[cyclobutane-1,6'(4'H)-[1,4]ethanonaphthalene]-2,10'-dione, 1',4'a,5',8'a-tetrahydro-1',5',5',7',9',9'-hexamethyl-	MeCN	232(3.83),232[sic](2.13), 344s(1.81)	33-0774-84

Compound	Solvent	$\lambda_{max}(\log \epsilon)$	Ref.
Spiro[cyclobutane-1,9'-[1,4]ethanonaphthalene]-2,7'(1'H)-dione, 4',4'a,8'-8'a-tetrahydro-4',6',8',8',10',10'-hexamethyl-	MeCN	304(2.11),312s(2.08), 324s(1.78)	33-0774-84
$C_{21}H_{28}O_3$			
Oxiranebutanoic acid, 3-(1,3-tetradeca-diene-5,8-diynyl)-, methyl ester, (+)-	EtOH	265(4.57),274(4.61), 282(4.53)	70-0442-84
(±)-	EtOH	265(4.57),274(4.61), 282(4.53)	104-2076-84
Spiro[cyclopentane-1,9'-[9H]xanthene]-1',8'(2'H,5'H)-dione, 3',4',6',7'-tetrahydro-3',3',6',6'-tetramethyl-	EtOH	232(4.07),306(3.64)	39-1213-84C
$C_{21}H_{28}O_4$			
Androsta-1,4-diene-2-carboxaldehyde, 11β,17β-dihydroxy-17-methyl-3-oxo-	H$_2$O	220(4.28),255s(--)	13-0271-84A
	MeOH	226(4.36),255s(--)	13-0271-84A
Androsta-1,4-diene-2-carboxylic acid, 17β-hydroxy-17-methyl-3-oxo-	MeOH	254(4.04)	13-0271-84A
Estr-4-en-3-one, 14α,15α-epoxy-17β-(1-oxopropoxy)-	EtOH	238(4.23)	106-0092-84
$C_{21}H_{28}O_5$			
Androsta-1,4-diene-2-carboxylic acid, 11,17-dihydroxy-17-methyl-3-oxo-, (11α,17β)-	MeOH	258(4.08)	13-0271-84A
(11β,17β)-	MeOH	254(4.02)	13-0271-84A
2-Cyclohexene-1-carboxylic acid, 2-(7-ethoxy-3-methyl-7-oxo-1,3,5-heptatri-enyl)-1,3-dimethyl-4-oxo-, ethyl ester, all-E-	EtOH	270(--),357(4.67)	104-0381-84 104-1528-84
$C_{21}H_{28}O_6$			
1,4-Phenanthrenedione, 9-(formyloxy)-4b,5,6,7,8,8a,9,10-octahydro-3,10-dihydroxy-4b,8,8-trimethyl-2-(1-methylethyl)-	ether	271(4.10),397(2.90)	33-1523-84
$C_{21}H_{28}O_7$			
Naematolin, 2,3-di-O-acetate	MeOH	239(3.52),298(3.36), 323(3.48)	5-1332-84
$C_{21}H_{28}O_9$			
Ixerin B	MeOH	226(4.15)	94-1724-84
$C_{21}H_{28}O_{10}$			
Coumestrin (rel. abs. given)	MeOH	244(1.33),265s(--), 303(0.46),341(1.36)	102-1204-84
$C_{21}H_{28}O_{12}$			
Allamcidin B, β-D-glucoside	MeOH	212(4.40)	94-2947-84
$C_{21}H_{29}ClN_2O_2S$			
Androst-2-eno[3,2-d][1,2,3]thiadiazol-17-ol, 1-chloro-, acetate, (5α,17β)-	EtOH	224(3.69),264(3.60), 320s(2.51)	44-4773-84
$C_{21}H_{29}NO_3$			
3,5-Nonadienamide, N-(2-hydroxy-1-meth-yl-2-phenylethyl)-N,2,6-trimethyl-8-oxo-, [1R-[1R*(S*),2S*]]-	EtOH	244(4.33)	44-5202-84

Compound	Solvent	$\lambda_{max}(\log \epsilon)$	Ref.
$C_{21}H_{29}NO_4$			
3,5-Heptadienamide, 6-acetoxy-N-(2-hy-droxy-1-methyl-2-phenylethyl)-N,2,4-trimethyl-, [1R-[1R*(S*)m2S*]]-	EtOH	229.5(3.72)	44-5202-84
$C_{21}H_{29}N_2O_2$			
Corynanium, 18,19-didehydro-17-hydroxy-10-methoxy-4-methyl-, chloride, (4α)-	EtOH	210(5.52),271(4.94), 296(4.69),307(4.61)	100-0687-84
$C_{21}H_{29}O_{10}P$			
β-D-Ribofuranose, 3-deoxy-3-[(diethoxy-phosphinyl)methyl]-, 1,2-diacetate 5-benzoate	MeOH	226(4.07),273(2.95), 280(2.88)	78-0079-84
$C_{21}H_{30}$			
1H-Indene, 2,3-dihydro-1,1,3,3-tetra-methylcyclobutylidene)-	C_6H_{12}	195.0(4.71),210s(4.18), 250s(2.54),258.0(2.85), 264.2(3.05),271.0(3.08)	24-0277-84
$C_{21}H_{30}N_2O_2S$			
Androst-2-eno[3,2-d][1,2,3]thiadiazol-17-ol, acetate, (5α,17β)-	EtOH	221(3.68),262(3.56)	44-4773-84
Androst-3-eno[3,4-d][1,2,3]thiadiazol-17-ol, acetate, (5α,17β)-	EtOH	219(3.71),264(3.60)	44-4773-84
$C_{21}H_{30}N_2O_5$			
2H-Pyrrole-4-acetic acid, 5-[[5-(1,1-dimethylethoxy)carbonyl]-3,4-dimeth-yl-2H-pyrrol-2-ylidene]methyl]-3,4-dihydro-2-methoxy-4-methyl-, methyl ester	CHCl₃	262(3.96),344(4.23)	49-0101-84
1H-Pyrrole-2-carboxylic acid, 5-[[3-(2-methoxy-2-oxoethyl)-1,3-dimethyl-5-oxo-2-pyrrolidinylidene]methyl]-3,4-dimethyl-, 1,1-dimethylethyl ester, (Z)-	CHCl₃	289(4.34)	49-0101-84
$C_{21}H_{30}N_2S$			
Dispiro[cyclobutane-1,2'(5'H)-[1,3,4]-thiadiazole-5',2"-[2H]indene], 1",3"-dihydro-1",1",2,2,3",3"4,4-octa-methyl-	C_6H_{12}	214(3.78),251s(2.83), 258s(3.68),264.5(3.27), 271(3.33),285(3.11), 325s(2.40)	24-0277-84
$C_{21}H_{30}N_4O$			
Urea, N,N-diethyl-N'-[(8α)-6-ethylergo-lin-8-yl]-	MeOH	234(3.75),275s(3.79), 282(3.81),293(3.74)	73-2828-84
$C_{21}H_{30}O_2$			
Arenarol	MeOH	220(3.61),293(3.44), 312(2.71)	44-0241-84
1,3-Benzenediol, 2-[3-methyl-6-(1-meth-ylethenyl)-3-cyclohexen-1-yl]-5-pentyl-	EtOH	274(2.91),281(2.89)	39-2881-84C
Benzo[b]cyclohepta[d]pyran-1-ol, 6,6a,7,8,9,11a-hexahydro-6,6-di-methyl-3-pentyl-, (6aR-trans)-	EtOH	273(3.17),282(3.14)	78-3839-84
2H-1-Benzopyran-5-ol, 2-methyl-2-(4-methyl-3-pentenyl)-7-pentyl- (canna-bichromene)	EtOH	225(4.44),280(3.99)	44-1793-84
6H-Dibenzo[b,d]pyran-1-ol, 6a,7,8,9-	EtOH	227(4.41),265(4.20),	39-2881-84C

$C_{21}H_{30}O_2-C_{21}H_{32}O_4$

Compound	Solvent	$\lambda_{max}(\log \epsilon)$	Ref.
tetrahydro-6,6,9-trimethyl-3-pentyl-, (6aR-cis)- (cont.)		271s(4.13)	39-2881-84C
(6aS-trans)- (Δ^2-THC)	EtOH	226(4.46),264(4.21), 272s(4.13)	39-2881-84C
Δ^3-THC	EtOH	272(4.15)	39-2881-84C
$C_{21}H_{30}O_3$			
Retinol, 12-carboxy-, (11-cis,13-cis)	MeOH	327(4.43)	44-0649-84
Xanthene-1,8(2H,7H)-dione, 3,4,5,6-tetrahydro-3,3,6,6,9-pentamethyl-9-propyl-	EtOH	307(3.64)	39-1213-84C
$C_{21}H_{30}O_4$			
Androst-4-en-3-one, 11,17-dihydroxy-2-(hydroxymethylene)-17-methyl-, (11β,17β)-	MeOH	256(3.99),312(3.78)	13-0271-84A
Cyclohexaneacetic acid, 2,2,6-trimethyl-6-[4-(1-methylethyl)-5,6-dioxo-1,3-cyclohexadien-1-yl]-, methyl ester	EtOH	274(2.85),418(3.20), 586(--)	23-2822-84
$C_{21}H_{30}O_6$			
1,3-Butadiene-1,4-diol, 2-(1-acetoxy-4,8-dimethyl-3,7-nonadienyl)-, di-acetate, (E,E,E)-(+)-	MeOH	248(4.11)	78-2913-84
$C_{21}H_{30}S$			
Dispiro[cyclobutane-1,2'-thiiran-3',2"-[2H]indene], 1",3"-dihydro-1",1"2,2-3",3",4,4-octamethyl-	C_6H_{12}	211.5(3.83),251s(2.82), 257.4(2.95),263.5(3.07), 270.3(3.05)	24-0277-84
$C_{21}H_{30}Si_3$			
1,2,3-Trisilacyclopent-4-ene, 2-ethyl-1,1,2,3,3-pentamethyl-4,5-diphenyl-	hexane	223s(4.58)	138-0393-84
$C_{21}H_{31}NO$			
4-Piperidinone, 3,5-bis(1-methylethyli-dene)-1-(tricyclo[3.3.1.1^{3,7}]dec-1-yl)-	heptane	268(4.16)	103-0761-84
$C_{21}H_{32}N_2O_8$			
Pentanedioic acid, 3-[2-[5-[(1,1-dimeth-ylethoxy)carbonyl]-3,4-dimethyl-1H-pyrrol-2-yl]-1-nitroethyl]-3-methyl-, dimethyl ester	CHCl₃	273(4.07),315s(3.08)	49-0101-84
$C_{21}H_{32}O_2$			
1,3-Benzenediol, 2-(3,7-dimethyl-1,6-octadienyl)-5-pentyl-	EtOH	224(4.47),266(4.28)	39-2881-84C
1,3-Benzenediol, 2-[3-methyl-6-(1-meth-ylethyl)-3-cyclohexen-1-yl]-5-pentyl-(1,6-didehydrotetrahydrocannabidiol)	EtOH	273(3.16)	39-2881-84C
Retinol, 12-(hydroxymethyl)-	MeOH	303(4.32)	44-0649-84
11-cis,13-cis	MeOH	300(4.66)	44-0649-84
13-cis	EtOH	305(4.24)	44-0649-84
$C_{21}H_{32}O_3$			
7,9,11,14-Eicosatetraenoic acid, 5,6-epoxy-, (E,E,Z,Z)-	EtOH	260(4.43),270(4.53), 282(4.45)	70-0442-84
$C_{21}H_{32}O_4$			
Cannabinol, dihydro-7α-oxo-, acetate	EtOH	276s(3.33),282(3.37)	87-1370-84

Compound	Solvent	$\lambda_{max}(\log \epsilon)$	Ref.
axial (cont.)			87-1370-84
equatorial	EtOH	278s(3.37),283(3.40)	87-1370-84
5,9-Methanobenzocyclooctene-5(1H)-carb-oxylic acid, 2,3,4,6,7,8,9,10-octahy-dro-3-methoxy-2,2,7,7,9-pentamethyl-4-oxo-, methyl ester, (3α,5α,9β)-	n.s.g.	251(3.85)	49-0809-84
$C_{21}H_{32}O_{11}Si$			
β-D-Glucopyranoside, 3-[(trimethylsil-yl)oxy]-1,3-butadienyl, tetraacetate, (E)-	EtOH	211s(3.78),237(4.16)	78-4657-84
$C_{21}H_{32}O_{12}$			
Deacylmartynoside	MeOH	219(3.94),281(3.58)	102-2313-84
$C_{21}H_{34}O_3SSi$			
Silane, (1,1-dimethylethyl)dimethyl[[3-methyl-3-[(phenylsulfonyl)methyl]-1-cyclohexen-1-yl]methoxy]-, (R)-	MeOH	218(4.08)	94-3417-84
$C_{21}H_{34}O_4$			
Spiro[4.5]dec-7-ene-8-carboxylic acid, 1-(1,5-dimethyl-4-oxohexyl)-9-hy-droxy-4-methyl-, methyl ester, (1R,4S,5S,9S,1'R)-	EtOH	204(3.94),214(3.96)	12-0635-84
$C_{21}H_{34}O_5$			
Betaenone A	EtOH	278(3.80)	158-84-59
$C_{21}H_{34}O_6$			
Stemphyloxin I (in ethanol)	pH 6.5	282(3.85)	158-84-120
	pH 8.8	303(4.08)	158-84-120
Stemphyloxin II	EtOH	276(3.66)	102-2193-84
$C_{21}H_{35}NO_2$			
Pyridinium, 1-(1-carboxypentadecyl)-, hydroxide, inner salt	n.s.g.	261.5(3.63)	46-6041-84
$C_{21}H_{36}N_2O_7Si_2$			
Uridine, 2'-deoxy-2'-oxo-3',5'-O-[1,1,3,3-tetrakis(1-methyl-ethyl)-1,3-disiloxanediyl]-	MeOH	261(3.99)	78-0125-84
$C_{21}H_{36}O_3Si$			
5H-4a,9-Methanocyclonona[1,6]benz[1,2-b]oxiren-2(1aH)-one, 12-[[(1,1-di-methylethyl)dimethylsilyl]oxy]octa-hydro-3-methyl-	EtOH	296.5(1.60)	35-1446-84
3a,8-Methano-3aH-cyclopentacyclodecen-1(4H)-one, 12-[[(1,1-dimethylethyl)-dimethylsilyl]oxy]-2,3,5,6,7,8,9,10-octahydro-11-hydroxy-2-methyl-, (2α,3aα,8α,12S*)-	EtOH	287(4.12)	35-1446-84
$C_{21}H_{36}O_5$			
1(2H)-Naphthalenone, octahydro-2,7-di-hydroxy-4-(3-hydroxy-1-oxopropyl)-2,4,5,7-tetramethyl-3-(1-methyl-propyl)- (betaenone B)	EtOH	259(2.12)	158-84-59

Compound	Solvent	$\lambda_{max}(\log \epsilon)$	Ref.
$C_{21}H_{36}O_7$			
Erythronolide, 5,12-dideoxy-5,8-epoxy-, (8R)-	MeOH	298(1.82)	78-2177-84
$C_{21}H_{36}O_8$			
Erythronolide, 5-deoxy-5,8-epoxy-, (8R)-	MeOH	300(1.70)	78-2177-84
$C_{21}H_{37}FO_7$			
Erythronolide A, 12-deoxy-8-fluoro-, (8S)- (8-fluoroerythronolide B)	MeOH	286(1.42)	78-2177-84
$C_{21}H_{37}FO_8$			
Erythronolide A, 8-fluoro-, (8S)-	MeOH	287(1.40)	78-2177-84
$C_{21}H_{37}N_3O_6Si_2$			
Cytidine, 2'-deoxy-2'-oxo-3',5'-O-[1,1,3,3-tetrakis(1-methylethyl)-1,3-disiloxanediyl]-	MeOH	271(3.95)	78-0125-84
$C_{21}H_{38}N_2O_6Si_2$			
Uridine, 2'-deoxy-3',5'-bis-O-[(1,1-dimethylethyl)dimethylsilyl]-2'-oxo-	MeOH	261(3.99)	78-0125-84
Uridine, 3'-deoxy-2',5'-bis-O-[(1,1-dimethylethyl)dimethylsilyl]-3'-oxo-	MeOH	261(4.00)	78-0125-84
$C_{21}H_{40}N_2O_6Si_2$			
2,4(1H,3H)-Pyrimidinedione, 1-[3,5-bis-O-[(1,1-dimethylethyl)dimethylsilyl]-β-D-arabinofuranosyl]-	MeOH	261(3.98)	78-0125-84
2,4(1H,3H)-Pyrimidinedione, 1-[3,5-bis-O-[(1,1-dimethylethyl)dimethylsilyl]-β-D-xylofuranosyl]-	MeOH	261(3.99)	78-0125-84
$C_{21}H_{44}N_2OS$			
Hydrazinecarbothioic acid, 2,2-dimethyl-, O-octadecyl ester	EtOH	246(3.99)	39-1005-84C
$C_{21}H_{56}Si_7$			
Cycloheptasilane, 1,2,3,4,5,6,7-heptaethyl-1,2,3,4,5,6,7-heptamethyl-	isooctane	223s(3.90),245(3.60)	101-0353-84L

Compound	Solvent	$\lambda_{max}(\log \epsilon)$	Ref.
$C_{22}H_6F_{16}$ Octafluoronaphthalene di-tert-butyl peroxide products	heptane	280(3.67)	104-0967-84
$C_{22}H_8Cl_4I_4O_5$ Rose Bengal, ethyl ester, salt with triethylamine	MeOH CH_2Cl_2	524(4.48),563(4.94) 525(4.53),563(4.87)	35-5879-84 35-5879-84
$C_{22}H_{11}NO_2$ Benzo[ghi]perylene, 5-nitro-	MeOH	218(4.51),278f(4.25), 421(3.96)	1-0309-84
Benzo[ghi]perylene, 7-nitro-	MeOH	272(4.42),298(4.31), 357(3.85),377(3.84), 407(3.66)	1-0309-84
$C_{22}H_{12}Cl_2$ Benzo[g]chrysene, 12,14-dichloro-, ($\pm$)-	CHCl$_3$	257(4.56),276(4.60), 287(4.69),297(4.77), 328(4.00),350(3.88), 383(2.71)	24-0336-84
$C_{22}H_{12}Cl_2O_6$ [2,2'-Binaphthalene]-5,5',8,8'-tetrone, 4,4'-dichloro-1,1'-dihydroxy-3,3'-di-methyl-	CHCl$_3$	260(4.45),347(3.53), 325s(3.73),420s(3.93), 444(4.03),465s(4.01)	5-0319-84
$C_{22}H_{12}N_4$ Propanedinitrile, 2,2'-(2,3-dimethyl-9,10-anthracenediylidene)bis-	MeCN	288(4.46),312(4.34), 350(4.37)	5-0618-84
$C_{22}H_{12}O_2$ 1,4-Naphthalenedione, 5-(2-naphthalen-ylethynyl)-	n.s.g.	233(4.70),281(4.34), 318(4.37),329(4.41), 420(3.67)	70-2345-84
$C_{22}H_{13}BrN_2O_4$ 1H-Indole, 1-acetyl-3-[4-(5-bromo-1,2-dihydro-2-oxo-3H-indol-3-ylidene)-4,5-dihydro-5-oxo-2-furanyl]-	CHCl$_3$	265(4.26),303(3.74), 484(4.41)	39-1331-84B
$C_{22}H_{13}N_3O_5$ Pyrrolo[2,1-a]isoquinoline, 1-nitro-3-(5-nitro-2-furanyl)-2-phenyl-	dioxan	465(4.00)	73-0533-84
$C_{22}H_{13}N_3O_6$ 1H-Indole, 1-acetyl-3-[4-(1,2-dihydro-5-nitro-2-oxo-3H-indol-3-ylidene)-4,5-dihydro-5-oxo-2-furanyl]-	CHCl$_3$	258(4.23),305(3.93), 482s(4.26),512(4.27), 550s(4.10)	39-1331-84B
$C_{22}H_{14}$ Anthracene, 9-(phenylethynyl)- dimer	C_6H_{12} C_6H_{12}	395(4.2),420(4.2) 338(3.3),352(3.8), 370(4.0),392(4.0)	151-0075-84C 151-0075-84C
$C_{22}H_{14}ClNO$ [1,1'-Biphenyl]-4-propanenitrile, α-[(4-chlorophenyl)methylene]-β-oxo-	EtOH	291(4.34)	73-0421-84
$C_{22}H_{14}Cl_2N_2$ Benzo[1,2-b:5,4-b']dipyrrole, 3,5-bis[(4-chlorophenyl)azo]-1,7-dihydro-	EtOH	204.5(3.37),208.5(3.34), 224(2.38),225(3.44),	103-0996-84

Compound	Solvent	$\lambda_{max}(\log \epsilon)$	Ref.
(cont.)		238s(3.18),256(3.15), 303(3.28),355(3.32)	103-0996-84
$C_{22}H_{14}FNO$			
[1,1'-Biphenyl]-4-propanenitrile, α-[(4-fluorophenyl)methylene]-β-oxo-	EtOH	289(4.30)	73-0421-84
$C_{22}H_{14}F_3N_3O$			
1,4-Benzenediamine, N,N-dimethyl-N'-(3,4,5-trifluoro-1H-furo[2,3,4-kl]acridin-1-ylidene)-	EtOH	209(4.00),256(4.09), 289s(3.75),365(3.08), 385(3.11),566(3.66)	104-0385-84
$C_{22}H_{14}INO$			
[1,1'-Biphenyl]-4-propanenitrile, α-[(4-iodophenyl)methylene]-β-oxo-	EtOH	284(4.48)	73-0421-84
$C_{22}H_{14}N_2O_3$			
[1,1'-Biphenyl]-4-propanenitrile, α-[(4-nitrophenyl)methylene]-β-oxo-	EtOH	282(4.44)	73-0421-84
Pyrrolo[2,1-a]isoquinoline, 3-(5-nitro-2-furanyl)-2-phenyl-	dioxan	465(4.00)	73-0533-84
$C_{22}H_{14}N_2O_4$			
1H-Indole, 1-acetyl-3-[4-(1,2-dihydro-2-oxo-3H-indol-3-ylidene)-4,5-dihydro-5-oxo-2-furanyl]-	CHCl$_3$	264(4.34),310(3.89), 480(4.45)	39-1331-84B
$C_{22}H_{14}N_4$			
Spiro[9H-fluorene-9,3'-[3H]pyrazolo-[5,1-c][1,2,4]triazole, 7'-phenyl-	CH$_2$Cl$_2$	240(4.17),276(3.92), 287(3.89),351(3.67)	24-1726-84
$C_{22}H_{14}N_8O_4$			
Benzo[1,2-b:5,4-b']dipyrrole, 1,7-dihydro-3,5-bis[(4-nitrophenyl)azo]-	EtOH	206(4.44),228(4.37), 234(4.34),295(4.24), 333(4.25)	103-0996-84
$C_{22}H_{14}O_4$			
5H-1,3-Dioxolo[4,5-g][2]benzopyran-5-one, 7,8-diphenyl-	EtOH	253(2.86),300(3.07)	2-0889-84
$C_{22}H_{14}O_6$			
[6,6'-Binaphthalene]-1,1',4,4'-tetrone, 5,5'-dihydroxy-7,7'-dimethyl-	EtOH	255(4.46),436(4.01)	5-0319-84
	EtOH-NaOH	296(4.21),470(3.84), 567(3.95)	5-0319-84
Elliptinone	EtOH	262(4.42),441(4.07)	5-0319-84
	EtOH-NaOH	266(4.28),365(3.46), 595(4.13)	5-0319-84
Isodiospyrin	EtOH	219(4.56),257(4.42), 435(3.94)	5-0319-84
	EtOH-NaOH	567(4.00)	5-0319-84
$C_{22}H_{15}ClO_2$			
2(3H)-Furanone, 5-(4-chlorophenyl)-3,3-diphenyl- (class spectrum)	EtOH	230-235(3.3),250-280(3.4)	104-0370-84
$C_{22}H_{15}ClO_4S_2$			
Ethanethioic acid, S-[2-[(1H-chloro-5,12-dihydro-6-hydroxy-5,12-dioxo-2-naphthacenyl)thio]ethyl] ester	EtOH	248(4.65),299(4.32), 450(4.12)	150-0147-84M

Compound	Solvent	$\lambda_{max}(\log \epsilon)$	Ref.
$C_{22}H_{15}F_4N_3O$			
1,4-Benzenediamine, N,N-dimethyl-N'-[(1,2,3,4-tetrafluoro-9-acridinyl)-methylene]-, N'-oxide	EtOH	203(4.41),254(4.88), 312(4.08),364(4.00), 410(3.95)	104-0385-84
$C_{22}H_{15}NO$			
[1,1'-Biphenyl]-4-propanenitrile, β-oxo-α-(phenylmethylene)-	EtOH	294(4.38)	73-0421-84
$C_{22}H_{15}NO_2$			
Acenaphtho[1,2-b]quinoline-10-carboxylic acid, ethyl ester	CHCl₃	256(4.28),275(4.32), 315(4.33),340(4.21)	78-2959-84
[1,1'-Biphenyl]-4-propanenitrile, α-[(4-hydroxyphenyl)methylene]-β-oxo-	EtOH	369(4.47)	73-0421-84
$C_{22}H_{15}NO_4$			
9,10-Anthracenedione, 1-acetoxy-4-(phenylamino)-	EtOH at 77°K	516(3.93)	104-1780-84
$C_{22}H_{15}N_5O$			
Benzamide, 4-cyano-N-(2,5-diphenyl-2H-1,2,3-triazol-4-yl)-	50% dioxan	285(4.45)	39-0785-84B
Benzonitrile, 4-[3-[phenyl(phenylhydrazono)methyl]-1,2,4-oxadiazol-5-yl]-, (Z)-	50% dioxan	352(4.22)	39-0785-84B
$C_{22}H_{15}N_5O_2S$			
2H-Naphth[1,8-cd]isothiazole, 3,5-bis(phenylazo)-, 1,1-dioxide	dioxan	345(4.24),410(4.20)	4-0337-84
$C_{22}H_{16}$			
Azulene, 1-[2-(4-azulenyl)ethenyl]-, (E)-	hexane	232s(4.44),260(4.58), 315(4.39),430s(4.51), 443(4.52),610(3.06), 670s(2.95),710s(2.66), 740s(2.52)	5-1905-84
protonated	CH₂Cl₂-TFA	640s(4.72),670(4.87)	5-1936-84
Azulene, 1-[2-(4-azulenyl)ethenyl]-, (Z)-	hexane	231(4.48),278(4.70), 395s(4.08),412(4.12), 565s(2.96),600(3.02), 630s(2.98),700s(2.61)	5-1905-84
Azulene, 1-[2-(6-azulenyl)ethenyl]-, (E)-	hexane	274(4.59),321(4.54), 350s(4.15),415s(4.35), 445(4.63),470(4.66), 570s(2.92),595(2.98), 623(2.98),650s(2.94), 685(2.79),770s(1.94)	5-1905-84
protonated	CH₂Cl₂-TFA	650s(4.80),682(4.94)	5-1936-84
Azulene, 1-[2-(6-azulenyl)ethenyl]-, (Z)-	hexane	235(4.45),278(4.68), 315s(4.47),345s(4.15), 438(4.27),608(2.86), 680s(2.66)	5-1905-84
Azulene, 1,1'-(1,2-ethenediyl)bis-, (E)-	CH₂Cl₂	228(4.45),250(4.62), 280(4.32),320(4.48), 425s(4.49),436(4.58), 464(4.56),635(2.90), 655(2.92),750s(2.69)	5-1905-84
radical cation	CH₂Cl₂	640(--),790(--), 935(--)	5-1936-84
dication	CH₂Cl₂	237(4.67),262(4.46), 304(4.32),326s(4.18),	5-1936-84

Compound	Solvent	$\lambda_{max}(\log \epsilon)$	Ref.
(cont.)		375s(4.25),396s(4.36), 428s(4.50),452(4.65), 493s(4.53)	5-1936-84
Azulene, 1,1'-(1,2-ethenediyl)bis-, (Z)-	CH_2Cl_2	228(4.53),272(4.56), 324(4.56),412(4.33), 625(2.83)	5-1905-84
Azulene, 4,4'-(1,2-ethenediyl)bis-	CH_2Cl_2	230s(4.37),270(4.76), 317(4.48),360s(4.22), 405s(3.92),610(3.11), 650s(3.05)	5-1905-84
diprotonated	CH_2Cl_2-TFA	432(3.96),476s(3.76)	5-1936-84
Azulene, 6,6'-(1,2-ethenediyl)bis-, (E)-	CH_2Cl_2	250(4.35),255(4.37), 315s(4.68),328(4.84), 344s(4.49),417(4.67), 438s(4.58),625(2.98), 690s(2.81)	5-1905-84
diprotonated	CH_2Cl_2-TFA	475s(4.53),503(4.61)	5-1936-84
$C_{22}H_{16}BrN_3O$ Isoxazole, 5-(4-bromophenyl)-4-[(3- methylphenyl)azo]-3-phenyl-	EtOH	350(4.28)	104-1844-84
$C_{22}H_{16}BrN_5O$ 1H-Pyrazole, 4-[(4-bromophenyl)azo]- 3-methyl-5-phenyl-1-(2-pyridinyl- carbonyl)-	n.s.g.	355(4.24)	48-1021-84
$C_{22}H_{16}ClN_5O$ 1H-Pyrazole, 4-[(2-chlorophenyl)azo]- 3-methyl-5-phenyl-1-(2-pyridinyl- carbonyl)-	n.s.g.	360(4.18)	48-1021-84
1H-Pyrazole, 4-[(3-chlorophenyl)azo]- 3-methyl-5-phenyl-1-(2-pyridinyl- carbonyl)-	n.s.g.	361(4.18)	48-1021-84
1H-Pyrazole, 4-[(4-chlorophenyl)azo]- 3-methyl-5-phenyl-1-(2-pyridinyl- carbonyl)-	n.s.g.	360(4.18)	48-1021-84
$C_{22}H_{16}Cl_2S_3$ 1,3-Butadiene, 1,2-dichloro-1,4,4- tris(phenylthio)-	hexane	206(4.70),262(4.27), 325(4.18)	5-1873-84
$C_{22}H_{16}N_2O_2$ 1(4H)-Naphthalenone, 5-hydroxy-2-(phen- ylamino)-	$CHCl_3$	266(4.37),307(4.00), 311(3.90),461(4.00)	104-0733-84
Quino[2,3-b]acridine-7,14-dione, 5,12- dihydro-5,12-dimethyl-	H_2SO_4	528(3.68),568(3.97), 616(4.09)	104-1771-84
$C_{22}H_{16}N_2O_3$ 2H-Indol-2-one, 1,3-dihydro-1-methyl- 3-[5-(1-methyl-1H-indol-3-yl)-2-oxo- 3(2H)-furanylidene]-	$CHCl_3$	263(4.28),320(3.56), 520(4.47)	39-1331-84B
$C_{22}H_{16}N_2O_4$ 2(3H)-Furanone, dihydro-4,5-bis(8-hy- droxy-4-quinolinyl)-, cis-(±)- (broussonetine)	MeOH MeOH-NaOH	252(4.19),333(3.70) 271(4.29),344(3.99), 386(4.07)	102-0929-84 102-0929-84
$C_{22}H_{16}N_4$ 1-Azacarbazole dimer	10% EtOH + KOH	<u>294(4.2),323(3.6)</u> 278(4.4),313(4.1)	46-1160-84 46-1160-84

Compound	Solvent	$\lambda_{max}(\log \epsilon)$	Ref.
1-Azacarbazole dimer (cont.)	$+ H_2SO_4$	270(4.2),303(4.1)	46-1160-84
3H-Pyrazolo[5,1-c]-1,2,4-triazole, 3,3,6-triphenyl-	CH_2Cl_2	254(4.35),340s(3.15)	24-1726-84
3H-Pyrazolo[5,1-c]-1,2,4-triazole, 3,3,7-triphenyl-	CH_2Cl_2	243(4.21),315(3.68), 352(3.85)	24-1726-84
Spiro[5H-dibenzo[a,d]cycloheptene-5,3'-[3H-1,2,4]triazolo[4,3-b]-indazole], 10,11-dihydro-	CH_2Cl_2	247s(4.00),260s(4.05), 270(4.08),318(3.64), 375(3.89)	24-1726-84
1H-Tetrazolium, 5-(1H-inden-1-yl)-1,3-diphenyl-, hydroxide, inner salt	MeCN	213s(4.44),242s(4.24), 277(4.47),363s(4.36), 376(4.49),530(3.08)	39-2545-84C
[1,2,4]Triazolo[1,5-c]quinazoline, 2-(4-methylphenyl)-5-phenyl-	EtOH	303(3.34)	18-1138-84
$C_{22}H_{16}N_4O$			
[1,2,4]Triazolo[4,3-a]quinazolin-5(4H)-one, 1-(4-methoxyphenyl)-4-phenyl-	EtOH	310(3.74)	18-1138-84
$C_{22}H_{16}N_4O_2$			
Benzoic acid, 4-[(4,5-diphenyl-1H-imidazol-2-yl)azo]-	C_6H_{12}	435(4.35)	62-0302-84A
	EtOH	223(4.39),292(4.34), 430(4.38)	62-0302-84A
	DMF	450(4.28),520(3.95)	62-0302-84A
	DMSO	446(4.35)	62-0302-84A
$C_{22}H_{16}N_4O_4$			
2,4-Imidazolidinedione, 3-[[(4-nitro-phenyl)methylene]amino]-5,5-diphenyl-	MeOH	204(4.64),308(4.28)	56-0585-84
$C_{22}H_{16}N_4O_5$			
Pyrrolo[3,4-c]pyrazole-4,6(1H,5H)-di-one, 3-[3-(1,3-dihydro-1,3-dioxo-2H-isoindol-2-yl)-1-oxopropyl]-3a,6a-dihydro-5-phenyl-	MeOH	209(4.53),221(4.63), 232(4.58),242(4.30), 315(4.17)	103-0066-84
$C_{22}H_{16}N_6O_3$			
1H-Pyrazole, 3-methyl-4-[(2-nitrophen-yl)azo]-5-phenyl-1-(2-pyridinyl-carbonyl)-	n.s.g.	354(4.07)	48-1021-84
1H-Pyrazole, 3-methyl-4-[(2-nitrophen-yl)azo]-5-phenyl-1-(3-pyridinyl-carbonyl)-	n.s.g.	354(4.07)	48-1021-84
1H-Pyrazole, 3-methyl-4-[(2-nitrophen-yl)azo]-5-phenyl-1-(4-pyridinyl-carbonyl)-	n.s.g.	353(4.07)	48-1021-84
$C_{22}H_{16}OS_2$			
Benzeneethane(dithioic) acid, α-9H-xanthen-9-ylidene-, methyl ester	CCl_4	280(3.08),326(3.20), 380s(2.51),492(1.46)	54-0152-84
Spiro[2H-thiete-2,9'-[9H]xanthene], 3-(methylthio)-4-phenyl-	CCl_4	312(3.04),342(2.88)	54-0152-84
	CH_2Cl_2	480(0.36),605(-0.54)	54-0152-84
$C_{22}H_{16}O_2$			
1H-2-Benzopyran-1-one, 7-methyl-3,4-diphenyl-	EtOH	255(2.87),292(3.06), 315(2.94)	2-0889-84
$C_{22}H_{16}O_3$			
1H-2-Benzopyran-1-one, 7-methoxy-3,4-diphenyl-	EtOH	253(2.94),300(3.1)	2-0889-84
1-Phenanthrenecarboxylic acid, 2-(2-methoxyphenyl)-	MeCN	228(5.77),277(5.51)	2-0603-84

Compound	Solvent	$\lambda_{max}(\log \epsilon)$	Ref.
1-Phenanthrenecarboxylic acid, 2-(3-methoxyphenyl)-	MeCN	230(5.71),276(5.44)	2-0603-84
1-Phenanthrenecarboxylic acid, 2-(4-methoxyphenyl)-	MeCN	231(5.78),275(5.71)	2-0603-84
$C_{22}H_{16}O_4$			
2H-1-Benzopyran-2-one, 4,7-dihydroxy-3-phenyl-6-(phenylmethyl)-	MeOH	206(4.07),318(3.90)	2-1030-84
[2,2'-Binaphthalene]-1,4-dione, 1',4'-dimethoxy-	CH₂Cl₂	228(4.56),246(4.64), 462(3.12)	5-1367-84
	acetone	447(3.10)	5-1367-84
	hexane	457(--)	5-1367-84
	CHCl₃	475(--)	5-1367-84
Dinaphtho[1,2-b:2',1'-d]furan-5(13aH)-one, 8,13a-dimethoxy-	CHCl₃	247(4.31),292(4.31), 336s(3.72),455(4.01)	5-1367-84
$C_{22}H_{16}O_4S_4$			
1,3-Dithiole-4-carboxylic acid, 2-[4-[(phenylmethoxy)carbonyl]-1,3-dithiol-2-ylidene]-, phenylmethyl ester	MeCN	286(4.19),299(4.18), 312(4.19),440(3.50)	103-1342-84
$C_{22}H_{16}O_5$			
[2,2'-Binaphthalene]-5,8-dione, 1,1',8'-trihydroxy-6,6'-dimethyl- (ebenone)	MeOH	206(4.42),235(4.76), 265s(4.35),310(3.93), 325(3.91),340(4.00), 435(3.70)	102-2039-84
$C_{22}H_{16}O_6$			
[2,2'-Binaphthalene]-1,1',4,4'-tetrone, 2,3-dihydro-5,5'-dihydroxy-7,7'-di-methyl-	MeOH	215(4.48),241(4.56), 355(3.88),415s(3.76), 430(3.77)	5-0319-84
$C_{22}H_{16}O_7$			
2,5-Furandione, 3,4-bis[2-(1,3-benzo-dioxol-5-yl)ethylidene]dihydro-, (E,E)-	toluene	380(4.04),405s(3.99)	48-0233-84
(E,Z)-	toluene	435(4.27)	48-0233-84
(Z,Z)-	toluene	445(4.46)	48-0233-84
5,12-Naphthacenedione, 1,6,11-trihy-droxy-8-(2-methyl-1,3-dioxolan-2-yl)-	EtOH	460(3.53),490(3.79), 525(3.81)	78-4579-84
$C_{22}H_{16}O_8$			
2H-1-Benzopyran-2-one, 7-acetoxy-, anti-head-head dimer	CHCl₃	248(3.65),258s(--), 273(3.56),282s(--)	2-0502-84
$C_{22}H_{16}O_8S_4$			
9H-Indeno[1,2-b]-1,4-dithiin-2,3-dicar-boxylic acid, 9-[4,5-bis(methoxycarb-onyl)-1,3-dithiol-2-ylidene]-, di-methyl ester	MeCN	235(4.221),258(4.253), 300s(3.855),400(4.309)	44-0726-84
$C_{22}H_{16}S_3$			
Benzeneethane(dithioic) acid, α-9H-thioxanthen-9-ylidene-, methyl ester	CCl₄	275(c.3.23),322(3.04), 485(1.34)	54-0152-84
Spiro[2H-thiete-2,9'-[9H]thioxanthene], 3-(methylthio)-4-phenyl-	CCl₄	333(3.15),475(1.83), 620(0.73)	54-0152-84
$C_{22}H_{17}BrN_2O_2$			
1,4-Naphthalenedione, 2-bromo-3-[[2-(1-naphthalenylamino)ethyl]-amino]-	CH₂Cl₂	226(4.38),246(4.40), 275(4.29),330(3.90), 466(3.47)	83-0743-84

Compound	Solvent	$\lambda_{max}(\log \epsilon)$	Ref.
$C_{22}H_{17}ClN_2OS$ Benzenamine, 4-chloro-N-[3-(4-methoxy-phenyl)-4-phenyl-2(3H)-thiazolyli-dene]-	EtOH	305(4.25)	56-0447-84
$C_{22}H_{17}ClN_2S$ Benzenamine, 4-chloro-N-[3-(4-methyl-phenyl)-4-phenyl-2(3H)-thiazolyli-dene]-	EtOH	305(4.23)	56-0447-84
$C_{22}H_{17}ClO_6$ 5,12-Naphthacenedione, 1,11-diacetoxy-6-chloro-1,2,3,4-tetrahydro-	EtOH	238(4.78),272(4.35), 296(4.23),403(3.76)	150-0147-84M
$C_{22}H_{17}N$ 2H-Cycloocta[c]pyrrole, 1,3-diphenyl-	EtOH	221(4.45),330(4.49)	44-0062-84
$C_{22}H_{17}NO$ Benzoxazole, 2-[2-(4-methylphenyl)-1-phenylethenyl]-, cis	C_6H_{12}	322(4.36)	131-0417-84H
trans	C_6H_{12}	327(2.442),336.9(4.39), 357.1(4.23)	131-0417-84H
$C_{22}H_{17}NOS$ Benzothiazole, 2-[2-(2-methoxyphenyl)-1-phenylethenyl]-	C_6H_{12}	347.2(4.33)	131-0417-84H
Benzothiazole, 2-[2-(3-methoxyphenyl)-1-phenylethenyl]-	C_6H_{12}	342.5(4.41)	131-0417-84H
Benzothiazole, 2-[2-(4-methoxyphenyl)-1-phenylethenyl]-	C_6H_{12}	350.1(4.48)	131-0417-84H
$C_{22}H_{17}NO_2$ Oxazole, 4,5-diphenyl-2-(phenylmeth-oxy)-	C_6H_{12}	224(4.38),302(4.15)	44-0399-84
2(3H)-Oxazolone, 4,5-diphenyl-3-(phen-ylmethyl)-	C_6H_{12}	215(4.41),265(4.20)	44-0399-84
$C_{22}H_{17}NO_2S$ 2(1H)-Pyridinone, 4-(4-methoxyphenyl)-3-phenyl-6-(2-thienyl)-	EtOH	262(4.12),356(4.00)	4-1473-84
$C_{22}H_{17}NS$ Benzothiazole, 2-[2-(2-methylphenyl)-1-phenylethenyl]-	C_6H_{12}	332.5(4.33)	131-0417-84H
Benzothiazole, 2-[2-(4-methylphenyl)-1-phenylethenyl]-	C_6H_{12}	340.6(4.44)	131-0417-84H
$C_{22}H_{17}N_3O$ Isoxazole, 4-[(3-methylphenyl)azo]-3,5-diphenyl-	EtOH	346(4.30)	104-1844-84
Isoxazole, 4-[(4-methylphenyl)azo]-3,5-diphenyl-	EtOH	350(4.29)	104-1844-84
$C_{22}H_{17}N_3O_2S_3$ 1H,5H-Imidazo[1,2-c]thiazol-2(3H)-one, 5-(3-ethyl-4-oxo-2-thioxo-5-thiazol-idinylidene)-3,7-diphenyl-	DMF	490(4.45)	103-0029-84
$C_{22}H_{17}N_3O_4$ 1H-Benzimidazole-5-carboxylic acid, 2,3-dihydro-2-oxo-3-phenyl-1-[(phen-	MeOH	211.5(--),239(4.37)	39-2587-84C

Compound	Solvent	$\lambda_{max}(\log \epsilon)$	Ref.
ylamino)carbonyl]-, methyl ester (cont.)			39-2587-84C
$C_{22}H_{17}N_3O_7S$			
Spiro[2H-1,3-benzoxazine-2,7'-[5]thia-[1]azabicyclo[4.2.0]oct[2]ene]-2'-carboxylic acid, 3,4-dihydro-3'-methyl-4,8'-dioxo-, (4-nitro-phenyl)methyl ester, (6'R-cis)-	EtOH	240(4.23)	39-2117-84C
$C_{22}H_{17}N_5O$			
1H-Pyrazole, 3-methyl-5-phenyl-4-(phen-ylazo)-1-(2-pyridinylcarbonyl)-	n.s.g.	358(4.34)	48-1021-84
$C_{22}H_{17}N_5O_2$			
1H-Pyrazole, 4-[(2-hydroxyphenyl)azo]-3-methyl-5-phenyl-1-(2-pyridinyl-carbonyl)-	n.s.g.	356(4.22)	48-1021-84
1H-Pyrazole, 4-[(3-hydroxyphenyl)azo]-3-methyl-5-phenyl-1-(2-pyridinyl-carbonyl)-	n.s.g.	355(4.22)	48-1021-84
1H-Pyrazole, 4-[(4-hydroxyphenyl)azo]-3-methyl-5-phenyl-1-(2-pyridinyl-carbonyl)-	n.s.g.	336(4.21)	48-1021-84
$C_{22}H_{18}Br_2N_2$			
1H-1,5-Benzodiazepine, 2,4-bis(4-bromo-phenyl)-2,3-dihydro-2-methyl-	EtOH	269(4.35),374(3.74)	103-0106-84
$C_{22}H_{18}Cl_2N_2$			
1H-1,5-Benzodiazepine, 2,4-bis(4-chlo-rophenyl)-2,3-dihydro-2-methyl-	EtOH	267(4.32),374(3.73)	103-0106-84
$C_{22}H_{18}NO_6$			
Macarpine	MeOH	222(4.53),240s(4.34), 287(4.49),318(3.97), 346(4.02)	73-0704-84
chloride	MeOH	285(4.4),342(4.2)	100-0001-84
$C_{22}H_{18}N_2$			
7H-Quino[3',2':4,5]pyrrolo[1,2-a][1]-benzazepine, 5,9-dimethyl- (10λ,9ϵ)	C_6H_{12}	214(4.54),221s(4.50), 230s(4.45),244(4.49), 266(4.50),284(4.38), 320.5(3.72),337(3.79), 351.5(3.72),368(?)	39-2529-84C
$C_{22}H_{18}N_2O$			
3-Pentenal, 5-(phenylimino)-2-(1-phen-yl-4(1H)-pyridinylidene)-	EtOH	204(4.57),260(4.17), 435(4.61)	73-0597-84
$C_{22}H_{18}N_2OS$			
Benzenamine, N-[3-(4-methoxyphenyl)-4-phenyl-2(3H)-thiazolylidene]-	EtOH	290(4.23)	56-0447-84
$C_{22}H_{18}N_2O_2$			
2H-Isoindole-4,7-dione, 2-(4-methyl-phenyl)-1-[(4-methylphenyl)amino]-	$C_2H_4Cl_2$	245s(4.36),296(4.15), 404(3.32),604(3.85)	5-1003-84
$C_{22}H_{18}N_2O_3$			
9,10-Anthracenedione, 1-amino-2-meth-oxy-4-[(4-methylphenyl)amino]-	C_6H_5Cl	530s(4.00),554(4.19), 593(4.21)	34-0482-84

Compound	Solvent	λ_{max}(log ϵ)	Ref.
Cyclohepta[4,5]pyrrolo[1,2-a]imidazole-10-carboxylic acid, 3-formyl-2-(4-methylphenyl)-, ethyl ester	CHCl$_3$	248(4.43),277(4.19),327(4.52),433(3.69),500(3.19),533s(3.17),570s(3.00)	150-3465-84M
C$_{22}$H$_{18}$N$_2$O$_4$			
9,10-Anthracenedione, 1-amino-2-methoxy-4-[(4-methoxyphenyl)amino]-	C$_6$H$_5$Cl	530s(3.90),556(4.09),592(4.11)	34-0482-84
3H-Indol-3-one, 2-[1,3-dihydro-3-oxo-1-(1-oxopropyl)-2H-indol-2-ylidene]-1,2-dihydro-1-(1-oxopropyl)-, cis	benzene	440(3.64)	39-2305-84C
trans	benzene	568(3.86)	39-2305-84C
C$_{22}$H$_{18}$N$_2$O$_7$			
Pyrrolo[2,1-a]isoquinoline-1,2-dicarboxylic acid, 3-(5-nitro-2-furanyl)-, diethyl ester	dioxan	418(4.10)	73-0533-84
C$_{22}$H$_{18}$N$_2$S			
Benzenamine, N-(3,4-diphenyl-2(3H)-thiazolylidene)-4-methyl-	EtOH	300(4.25)	56-0447-84
Benzenamine, N-[3-(4-methylphenyl)-4-phenyl-2(3H)-thiazolylidene]-	EtOH	300(4.20)	56-0447-84
C$_{22}$H$_{18}$N$_3$O$_2$			
Quinolinium, 2-[2-(4,5-dihydro-4,5-dioxo-1-phenyl-1H-pyrazol-3-yl)ethenyl]-1-ethyl-, iodide	EtOH	390(4.15),550(4.30)	48-0695-84
C$_{22}$H$_{18}$N$_4$			
1H-Imidazole, 2-[(4-methylphenyl)azo]-4,5-diphenyl-	C$_6$H$_{12}$	418(4.36),450s(4.32)	62-0302-84A
	EtOH	223(4.33),292(4.27),414(4.40),440s(4.32)	62-0302-84A
	DMF	422(4.35),450s(4.33),520(3.95)	62-0302-84A
	DMSO	422(4.39),450s(4.30)	62-0302-84A
4-Methylbenzaldehyde, (2-phenyl-4-quinazolinyl)hydrazone	EtOH	364(3.96)	18-1138-84
C$_{22}$H$_{18}$N$_4$O			
Benzaldehyde, 4-methoxy-, (2-phenyl-4-quinazolinyl)hydrazone	EtOH	382(4.00)	18-1138-84
Benzamide, N-(2,5-diphenyl-2H-1,2,3-triazol-4-yl)-3-methyl-	50% dioxan	290(4.38)	39-0785-84B
Benzamide, N-(2,5-diphenyl-2H-1,2,3-triazol-4-yl)-4-methyl-	50% dioxan	295(4.40)	39-0785-84B
1H-Imidazole, 2-[(2-methoxyphenyl)azo]-4,5-diphenyl-	C$_6$H$_{12}$	446(4.31),464s(4.29)	62-0302-84A
	EtOH	223(4.36),293(4.37),430(4.23),460s(4.21)	62-0302-84A
	DMF	430(4.06),470s(4.04),540(3.90)	62-0302-84A
	DMSO	432(4.28),470s(4.31)	62-0302-84A
1H-Imidazole, 2-[(4-methoxyphenyl)azo]-4,5-diphenyl-	C$_6$H$_{12}$	418(4.42),460s(4.37)	62-0302-84A
	EtOH	223(4.32),294(4.16),414(4.46),450s(4.39)	62-0302-84A
	DMF	420(4.37),460s(4.35),530(3.74)	62-0302-84A
	DMSO	430(4.35),460s(4.31)	62-0302-84A
Methanone, [5-(3-methylphenyl)-1,2,4-oxadiazol-3-yl]phenyl-, phenylhydrazone	50% dioxan	366(4.24)	39-0785-84B

Compound	Solvent	$\lambda_{max}(\log \epsilon)$	Ref.
Methanone, [5-(4-methylphenyl)-1,2,4-oxadiazol-3-yl]phenyl-, phenylhydrazone	50% dioxan	366(4.25)	39-0785-84B
7H-Pyrido[2,1-b]quinazoline-6,11-dione, 8,9-dihydro-, 6-(1-naphthalenyl-hydrazone)	EtOH	248(4.38),310(4.01), 410(4.19)	4-1301-84
7H-Pyrido[2,1-b]quinazoline-6,11-dione, 8,9-dihydro-, 6-(2-naphthalenyl-hydrazone)	EtOH	232(4.58),246(4.41), 294(4.09),398(4.38)	4-1301-84
$C_{22}H_{18}N_4O_2$			
Benzamide, N-(2,5-diphenyl-2H-1,2,3-triazol-4-yl)-4-methoxy-	50% dioxan	285(4.44)	39-0785-84B
Methanone, [5-(4-methoxyphenyl)-1,2,4-oxadiazol-3-yl]phenyl-, phenylhydrazone, (Z)-	50% dioxan	366(4.23)	39-0785-84B
$C_{22}H_{18}N_4O_2S$			
Thiazolo[2,3-b]purinium, 3,4-dihydro-8-hydroxy-3,5-dimethyl-4-oxo-7-phenyl-2-(phenylmethyl)-, hydroxide, inner salt	MeCN	224(4.34),273(4.29), 415(4.06)	2-0316-84
	benzene	442(--)	2-0316-84
	CHCl$_3$	434(--)	2-0316-84
	pyridine	433(--)	2-0316-84
$C_{22}H_{18}N_4O_2S_2$			
1H-Pyrazole-4-carboxaldehyde, 5,5-dithiobis[3-methyl-1-phenyl-	EtOH	225(4.45),250(4.33), 320s(--)	65-0134-84
	CHCl$_3$	250(4.34),290s(--)	65-0134-84
$C_{22}H_{18}N_4O_4$			
1H-1,5-Benzodiazepine, 2,3-dihydro-2-methyl-2,4-bis(3-nitrophenyl)-	EtOH	256(4.49),379(3.75)	103-0106-84
1H-1,5-Benzodiazepine, 2,3-dihydro-2-methyl-2,4-bis(4-nitrophenyl)-	EtOH	278(4.38),420(3.72)	103-0106-84
$C_{22}H_{18}N_6O_3S$			
1H-Pyrazole, 4-[[2-(aminosulfonyl)phenyl]azo]-3-methyl-5-phenyl-1-(2-pyridinylcarbonyl)-	n.s.g.	357(4.35)	48-1021-84
$C_{22}H_{18}O_2S$			
23-Thiatetracyclo[14.2.2.2^{8,11}.1^{3,6}]-tricosa-3,5,8,10,16,18,19,21-octa-ene-2,7-dione	C_6H_{12}	303(4.2)	44-1177-84
$C_{22}H_{18}O_4$			
Anthra[1,2-b]furan-6,11-dione, 5-methoxy-2-methyl-4-(2-methyl-2-propenyl)-	EtOH	240(4.59),282(4.61), 380(4.18)	12-1511-84
Benzaldehyde, 2,2'-[1,2-phenylene-bis(methyleneoxy)]bis-	MeOH	215(4.66),253(4.20), 317(3.86)	44-1627-84
1-Phenanthrenecarboxylic acid, 1,2,3,4-tetrahydro-2-(2-methoxyphenyl)-4-oxo-	MeCN	227(5.84),275(5.87)	2-0603-84
1-Phenanthrenecarboxylic acid, 1,2,3,4-tetrahydro-2-(3-methoxyphenyl)-4-oxo-	MeCN	230(5.82),274(5.73)	2-0603-84
1-Phenanthrenecarboxylic acid, 1,2,3,4-tetrahydro-2-(4-methoxyphenyl)-4-oxo-	MeCN	224(5.82),274(5.59)	2-0603-84
$C_{22}H_{18}O_5$			
9,10-Anthracenedione, 1-acetoxy-4-hydroxy-2,3-di-2-propenyl-	EtOH	251(4.57),261(4.48), 279(4.18),412(3.92), 430(3.90)	12-1518-84

Compound	Solvent	$\lambda_{max}(\log \epsilon)$	Ref.
$C_{22}H_{18}O_6$			
Benz[a]anthracene-1-carboxylic acid, 3,4,7,12-tetrahydro-2-hydroxy-11-methoxy-7,12-dioxo-, ethyl ester	n.s.g.	224(4.37),237(4.39), 254(4.42),301(3.75), 377(3.72)	5-0306-84
4H,8H-Benzo[1,2-b:5,4-b']dipyran-4-one, 2-(1,3-benzodioxol-5-yl)-3-methoxy-8,8-dimethyl-	MeOH	224(4.28),256(4.21), 344(3.93)	2-0168-84
4H-1-Benzopyran-4-one, 8-(4,5-dihydro-5,5-dimethyl-4-oxo-3-furanyl)-7-hydroxy-5-methoxy-2-phenyl-	EtOH	215(4.45),244(4.25), 270(4.44),328s(3.87)	5-1068-84
$C_{22}H_{18}O_7$			
1H-Anthra[2,3-c]pyran-6,11-dione, 12-acetoxy-1,5-dimethoxy-3-methyl-	EtOH	253(3.93),297(3.75), 407(3.42)	12-1511-84
2-Butenedioic acid, 2-[(3-methyl-4-oxo-2-phenyl-4H-1-benzopyran-7-yl)oxy]-, dimethyl ester	MeOH	248(4.18),294(4.22)	2-0969-84
Mycotoxin MT81	EtOH	206(3.78),225(4.34), 242(4.52),260(4.30), 334(3.75)	2-0393-84
$C_{22}H_{18}O_8$			
4-Pentenoic acid, 5-(9,10-dihydro-1-hydroxy-4,8-dimethoxy-9,10-dioxo-2-anthracenyl)-3-oxo-, methyl ester	n.s.g.	210(4.33),225(4.40), 264(4.57),288(4.20), 322(3.68),482(4.12), 510(3.97)	5-0306-84
$C_{22}H_{18}O_8S_4$			
1,3-Dithiole-4,5-dicarboxylic acid, 2,2'-(1,2-phenylenedimethylidyne)-bis-, tetramethyl ester	MeCN	235(4.083),310s(4.106), 365(4.208)	44-0726-84
$C_{22}H_{18}O_9S$			
Benzoic acid, 2-[(4-methoxy-3,6-dioxo-1,4-cyclohexadien-1-yl)oxy]-5-[[(4-methylphenyl)sulfonyl]oxy]-, methyl ester	EtOH	227(4.35),283(4.30), 312s(3.57)	39-1507-84C
Benzoic acid, 2-[(5-methoxy-3,6-dioxo-1,4-cyclohexadien-1-yl)oxy]-5-[[(4-methylphenyl)sulfonyl]oxy]-, methyl ester	EtOH	226(4.54),282(4.21), 378(2.95)	39-1507-84C
$C_{22}H_{19}BO_6$			
Boron, [ethanedioato(2-)-O,O'][1-phenyl-5-(2,4,6-trimethylphenyl)-4-pentene-1,3-dionato-O,O']-	CH_2Cl_2	444(4.59)	97-0292-84
$C_{22}H_{19}BrN_2O_5$			
1H-Pyrrolo[1,2-a]indole-9-carboxylic acid, 2-(acetylamino)-7-bromo-2,3,5,8-tetrahydro-6-methyl-5,8-dioxo-, phenylmethyl ester, (S)-	$CHCl_3$	294(4.187),334(3.594), 430(3.174)	44-5164-84
$C_{22}H_{19}ClN_2O$			
Benzenemethanamine, N-[5-chloro-3-(2-phenylethyl)-2(3H)-benzoxazolylidene]-	$CHCl_3$	259(3.87),305(4.06)	94-3053-84
$C_{22}H_{19}F_3N_4$			
4H-Pyrrolo[2,3-d]pyrimidin-4-imine, 3,7-dihydro-2,5,6-trimethyl-7-phenyl-	EtOH	245(4.28),305(3.71)	11-0073-84A

Compound	Solvent	$\lambda_{max}(\log \epsilon)$	Ref.
3-[3-(trifluoromethyl)phenyl]- (cont.)			11-0073-84A
$C_{22}H_{19}IN_4O$			
Quinolinium, 1-ethyl-2-(4-methyl-6-phenyl-6H-pyrazolo[3,4-c]isoxazol-3-yl)-, iodide	EtOH	330(3.85),350(3.70)	48-0811-84
$C_{22}H_{19}N$			
1H-Indole, 1,5-dimethyl-2,3-diphenyl-	EtOH	235(4.26),305(3.98)	78-4351-84
Indolizine, 3,7-dimethyl-1,2-diphenyl-	EtOH	254(4.70),280s(4.30), 314(4.02),324(4.04), 374(3.58)	103-1128-84
Spiro[acridine-9(10H),1'(2'H)-naphthalene], 3',4'-dihydro-	EtOH	294(4.21)	24-2703-84
$C_{22}H_{19}NO$			
1H-Indole, 1-(methoxymethyl)-2,3-diphenyl-	EtOH	237(4.51),293(3.66)	78-4351-84
1H-Indole, 7-methoxy-4-methyl-2,3-diphenyl-	EtOH	256(4.31),310(4.29)	2-1021-84
$C_{22}H_{19}NO_5$			
Benzamide, N-(5-ethoxy-9-methyl-2-oxo-2H,5H-Pyrano[3,2-c][1]benzopyran-3-yl)-	CHCl$_3$	243(4.08),385(4.46)	2-1048-84
2-Hexenoic acid, 6-[2-(3-cyano-trans-3-phenyl-2-oxiranyl)phenoxy]-4-oxo-, methyl ester	EtOH	275(3.48)	24-2157-84
Sanguinarine, ethoxydihydro-	n.s.g.	234(4.30),281(4.39), 322(3.91)	100-0001-84
$C_{22}H_{19}NO_6$			
2-Butenedioic acid, (E)-, 2-[2-(3-cyano-3-phenyloxiranyl)phenoxy]ethyl methyl ester, trans	EtOH	275s(3.45),281(3.61)	24-2157-84
$C_{22}H_{19}NO_7$			
8H-Benzo[g]-1,3-benzodioxolo[5,6-a]-quinolizin-8-one, 13-acetoxy-5,6-dihydro-9,10-dimethoxy-	MeOH	227(4.66),312s(4.11), 331s(4.38),341(4.42), 369s(4.29),386s(3.97)	44-2642-84
Spiro[1,3-dioxolo[4,5-g]isoquinoline-5(6H),2'-[2H]indene]-1',3'-dione, 6-acetyl-7,8-dihydro-4',5'-dimethoxy-, (±)-	MeOH	248(4.57),292.5(4.11), 332.5(3.69)	94-2230-84
$C_{22}H_{19}N_3OS_2$			
1H-Imidazo[1,2-c]thiazol-4-ium, 5-[(3-ethyl-2(3H)-benzothiazolylidene)methyl]-2-hydroxy-3-methyl-7-phenyl-, hydroxide, inner salt	DMF	500(4.45)	103-0029-84
$C_{22}H_{19}N_3O_2$			
Acetic acid, cyano(1-methyl-4,6-diphenyl-2(1H)-pyrimidinylidene)-, ethyl ester	EtOH	312(4.55),424(3.83)	103-1270-84
2-Pyrimidineacetic acid, α-cyano-α-methyl-4,6-diphenyl-, ethyl ester	EtOH	252(4.41),290(4.31)	103-1270-84
$C_{22}H_{19}N_3O_3$			
1H-1,5-Benzodiazepine, 2,3-dihydro-4-(4-methoxyphenyl)-2-(4-nitrophenyl)-	EtOH	270(4.45),361(3.86)	103-1370-84

Compound	Solvent	$\lambda_{max}(\log \epsilon)$	Ref.
$C_{22}H_{19}N_4O$			
Quinolinium, 1-ethyl-2-(4-methyl-6-phenyl-6H-pyrazolo[3,4-c]isoxazol-3-yl)-, iodide	EtOH	330(3.85),350(3.70)	48-0811-84
$C_{22}H_{19}N_5O_5$			
1H-Pyrrolo[1,2-a]indole-9-carboxylic acid, 2-(acetylamino)-7-azido-2,3,5,8-tetrahydro-6-methyl-5,8-dioxo-, phenylmethyl ester, (S)-	n.s.g.	314(3.89),340(3.52), 470(2.73)	44-5164-84
$C_{22}H_{20}$			
Spiro[cyclopropane-1,9-[9H]fluorene], 2-(dicyclopropylmethylene)-	EtOH	231(4.87),247(4.56), 257(4.73),315(4.37)	44-0495-84
$C_{22}H_{20}BrNO_3S$			
Benzenesulfonamide, N-[2-(4-bromophenyl)-2-oxoethyl]-4-methyl-N-(phenylmethyl)-	ether	256.5(4.39),324.4(2.05)	48-0177-84
$C_{22}H_{20}BrN_3O$			
1,4-Benzenediamine, N'-[1-(4-bromophenyl)-2-(phenylimino)ethylidene]-N,N-dimethyl-, N-oxide, (Z,Z)-	MeOH	261(4.33),434(3.73)	118-0128-84
$C_{22}H_{20}ClNO_3S$			
Benzenesulfonamide, N-[2-(4-chlorophenyl)-2-oxoethyl]-4-methyl-N-(phenylmethyl)-	ether	250.0(4.44),325.2(1.99)	48-0177-84
$C_{22}H_{20}ClN_3O$			
1,4-Benzenediamine, N'-[1-[(4-chlorophenyl)imino]-2-phenylethylidene]-N,N-dimethyl-, N-oxide, (Z,Z)-	MeOH	260(4.32),442(3.90)	118-0128-84
1,4-Benzenediamine, N'-[2-[(4-chlorophenyl)imino]-1-phenylethylidene]-N,N-dimethyl-, N-oxide, (Z,Z)-	MeOH	253(4.32),444(3.72)	118-0128-84
$C_{22}H_{20}FNO_3S$			
Benzenesulfonamide, N-[2-(4-fluorophenyl)-2-oxoethyl]-4-methyl-N-(phenylmethyl)-	ether	238.0(4.80),321.8(1.88)	48-0177-84
$C_{22}H_{20}NO_5$			
Benzo[c]phenanthridinium, 2,3-(methylenedioxy)-1,9,10-trimethoxy-N-methyl-, chloride	EtOH	229(4.42),254(4.34), 288(4.73),323(4.03), 355(3.65)	100-0001-84
Chelilutine	EtOH	230(4.45),241s(4.39), 280.5(4.56),340(4.29), 420s(3.41),470(3.55)	100-0001-84
Sanguirubine, chloride	EtOH	230.5(4.49),281(4.48), 343s(4.24),353(4.23), 422(2.98),510(3.23)	100-0001-84
$C_{22}H_{20}N_2$			
1H-1,5-Benzodiazepine, 2,3-dihydro-2-methyl-2,4-diphenyl-	EtOH	254(4.39),369(3.78)	103-0106-84
$C_{22}H_{20}N_2O_2$			
2-Butene-1,4-dione, 1,4-diphenyl-2-(3,4,5-trimethyl-1H-pyrazol-1-yl)-, (Z)-	MeOH	257(4.46),330(4.38)	44-4647-84

Compound	Solvent	$\lambda_{max}(\log \epsilon)$	Ref.
1H-Indazole, 1-[2-[4-(phenylmethoxy)-phenoxy]ethyl]-	EtOH	255(3.98),263(3.96), 291(4.19),303(3.97)	87-0503-84
2H-Indazole, 2-[2-[4-(phenylmethoxy)-phenoxy]ethyl]-	EtOH	277(4.22),291(4.23)	87-0503-84
1H-Indole, 5,5'-(1,2-ethanediyl)bis[1-acetyl-	EtOH	204(4.36),246.9(4.65), 274(4.16),294(4.08), 303(4.11)	103-0062-84

$C_{22}H_{20}N_2O_6$

Compound	Solvent	$\lambda_{max}(\log \epsilon)$	Ref.
1H-Benz[f]isoquino[8,1,2-hij]quinazo-line-1,3(2H)-dione, 5,6-dihydro-8,9,11,12-tetramethoxy-	CH$_2$Cl$_2$	255s(4.42),264(4.62), 279(4.61),306s(4.11), 335(3.76),351(3.81), 373(3.88),392(3.97)	78-4003-84

$C_{22}H_{20}N_2O_7$

Compound	Solvent	$\lambda_{max}(\log \epsilon)$	Ref.
4,5-Isoxazoledicarboxylic acid, 2,3-di-hydro-2-methyl-3-[2-(4-nitrophenyl)-ethenyl]-3-phenyl-, dimethyl ester	EtOH	310(4.21)	150-1531-84M
4,5-Isoxazoledicarboxylic acid, 2,3-di-hydro-2-methyl-3-(4-nitrophenyl)-3-(2-phenylethenyl)-, dimethyl ester	EtOH	258(4.18)	150-1531-84M
Methanaminium, N-[1-(4-nitrophenyl)-3-phenyl-2-propenylidene]-3-methoxy-1-(methoxycarbonyl)-2,3-dioxopropylide, (E,?)-	EtOH	278(4.24),318(4.03), 436(3.74)	150-1531-84M

$C_{22}H_{20}N_3O$

Compound	Solvent	$\lambda_{max}(\log \epsilon)$	Ref.
Quinolinium, 2-[(1,5-dihydro-3-methyl-5-oxo-1-phenyl-4H-pyrazol-4-ylidene)-methyl]-1-ethyl]-, iodide	EtOH	535(4.19)	48-0811-84

$C_{22}H_{20}N_3OS_2$

Compound	Solvent	$\lambda_{max}(\log \epsilon)$	Ref.
1H-Imidazo[1,2-c]thiazol-4-ium, 5-[(3-ethyl-2(3H)-benzothiazolylidene)-methyl]-2,3-dihydro-3-methyl-2-oxo-7-phenyl-, perchlorate	HOAc	456(4.70)	103-0029-84

$C_{22}H_{20}N_4$

Compound	Solvent	$\lambda_{max}(\log \epsilon)$	Ref.
4,4'-Bipyridinium, 1,1''-(1,2-ethane-diyl)bis-, dibromide	H$_2$O	268(4.56)	64-0074-84B
1,4-Phthalazinediamine, N,N'-bis(phen-ylmethyl)-	MeOH	343(3.81)	24-1523-84
1H-Pyrido[3,4-b]indole-1-propane-nitrile, 3-cyano-2,3,4,9-tetrahydro-2-(phenylmethyl)-, (1S-trans)-	MeOH	222.5(4.59),270s(3.89), 273(3.90),279(3.89), 282s(3.88),290(3.79)	94-1313-84

$C_{22}H_{20}N_4O_3S$

Compound	Solvent	$\lambda_{max}(\log \epsilon)$	Ref.
Benzeneacetic acid, α-[[6,7-dihydro-1,7-dimethyl-6-oxo-8-(phenylmethyl)-1H-purin-2-yl]thio]-	MeOH	267(4.15),286(4.03)	2-0316-84

$C_{22}H_{20}N_5O$

Compound	Solvent	$\lambda_{max}(\log \epsilon)$	Ref.
7H-Purinium, 6-(butylamino)-9-(phenyl-methyl)-7-(2-propenyl)-, bromide	H$_2$O	253(4.04),281(4.18), 323(3.73)	150-3601-84M

$C_{22}H_{20}N_5O_2$

Compound	Solvent	$\lambda_{max}(\log \epsilon)$	Ref.
Benzoxazolium, 2-(2-acetyl-2,6-dihydro-4-methyl-6-phenylpyrazolo[3,4-c]pyra-zol-3-yl)-3-ethyl-, iodide	EtOH	370(3.53)	48-0811-84

Compound	Solvent	$\lambda_{max}(\log \epsilon)$	Ref.
$C_{22}H_{20}O$			
2,5-Cyclohexadien-1-one, 4-(diphenyl-methylene)-2-(1-methylethyl)-	hexane?	210(4.27),261(4.08), 267(4.07),361(4.36)	73-1949-84
[1,1':4',1"-Terphenyl]-2'-carboxalde-hyde, 3',5',6'-trimethyl-	EtOH	222(4.56),267(4.11), 307(3.57)	23-2592-84
$C_{22}H_{20}O_3$			
Ethanone, 1-[2,4-dihydroxy-3,5-bis-(phenylmethyl)phenyl]-	MeOH	202(4.44),278(4.00)	2-1030-84
1-Phenanthrenecarboxylic acid, 1,2,3,4-tetrahydro-2-(2-methoxyphenyl)-	MeCN	228(5.71),276(5.48)	2-0603-84
1-Phenanthrenecarboxylic acid, 1,2,3,4-tetrahydro-2-(3-methoxyphenyl)-	MeCN	229(5.65),277(5.39)	2-0603-84
1-Phenanthrenecarboxylic acid, 1,2,3,4-tetrahydro-2-(4-methoxyphenyl)-	MeCN	228(5.71),275(5.28)	2-0603-84
$C_{22}H_{20}O_4$			
2-Propenoic acid, 3-[[1,2,3,4-tetra-hydro-2,3-bis(methylene)-1,4-epoxy-anthracen-5-yl]oxy]propyl ester	MeCN	218(4.66),252(4.68), 300(3.56)	89-0074-84
$C_{22}H_{20}O_5$			
Benzeneacetic acid, α-(3-hydroxy-5-oxo-4-phenyl-2(5H)-furanylidene)-, 1,1-dimethylethyl ester, (E)-	EtOH	287(4.20),365(4.07)	39-1547-84C
2,5-Furandione, 3,4-bis[1-(4-methoxy-phenyl)ethylidene]-, (E,E)-	toluene	364(4.06),400s(4.00)	48-0233-84
(E,Z)-	toluene	421(4.29)	48-0233-84
(Z,Z)-	toluene	430(4.60)	48-0233-84
$C_{22}H_{20}O_5S$			
1-Propanone, 2-hydroxy-3-[[(4-methyl-phenyl)sulfonyl]oxy]-1,2-diphenyl-	MeOH	225(4.32),251(4.22), 332(2.36)	126-1795-84
$C_{22}H_{20}O_6$			
2-Anthracenepropanoic acid, α-acetyl-9,10-dihydro-1-hydroxy-3-methyl-9,10-dioxo-, ethyl ester	MeOH	206(4.35),228(4.32), 246(4.48),263(4.50), 328(3.59),407(3.86)	78-3677-84
2H-1-Benzopyran-2-one, 4-ethoxy-, anti-head-tail dimer	CHCl₃	247(3.40),257(3.04), 273(3.39),278(3.38)	2-0502-84
1-Naphthacenecarboxylic acid, 2-ethyl-1,2,3,4,6,11-hexahydro-2,4-dihydroxy-6,11-dioxo-, methyl ester, trans-(±)-	MeOH	207(4.22),246(4.48), 260(4.52),409(3.85)	78-3677-84
$C_{22}H_{20}O_7$			
2-Anthracenepropanoic acid, α-acetyl-9,10-dihydro-1-hydroxy-8-methoxy-9,10-dioxo-, ethyl ester	n.s.g.	226(3.59),256(4.35), 279(4.01),403(3.96), 416(3.99),430(3.92)	5-0306-84
2H,6H-Benzo[1,2-b:5,4-b']dipyran-6-one, 5-hydroxy-7-(4-hydroxy-2,5-dimeth-oxyphenyl)-2,2-dimethyl-	EtOH	228(4.43),281(4.57)	142-0709-84
2,5-Furandione, 3,4-bis[(3,4-dimeth-oxyphenyl)methylene]dihydro-, (E,E)-	toluene	352(3.97)	48-0233-84
1-Naphthacenecarboxylic acid, 2-ethyl-1,2,3,4,6,11-hexahydro-2,4,5-trihy-droxy-6,11-dioxo-, methyl ester	MeOH	208(4.21),227(4.30), 247(4.46),255(4.46), 279s(--),330(3.47), 405(3.82)	78-3677-84
1-Naphthacenecarboxylic acid, 2-ethyl-1,2,3,4,6,11-hexahydro-2,5,7-trihy-droxy-6,11-dioxo-, methyl ester	MeOH	229(4.54),259(4.44), 278s(4.10),290(3.97), 432(4.09)	78-4701-84

Compound	Solvent	$\lambda_{max}(\log \epsilon)$	Ref.
$C_{22}H_{20}O_8$			
2-Anthraceneacetic acid, 3-(4,5-dimethyl-1,3-dioxolan-2-yl)-9,10-dihydro-4,5-dihydroxy-9,10-dioxo-, methyl ester, [4R-(2α,4α,5β)]-	MeOH	228(4.82),253(4.37), 278(4.07),287(4.07), 432(4.12)	78-4685-84
5,12-Naphthacenedione, 9-(1-acetoxyethyl)-7,8,9,10-tetrahydro-6,7,9,11-tetrahydroxy-, (R)-	CHCl$_3$	251(4.60),287(4.02), 330s(3.43),460s(3.98), 483(4.02),518(3.80)	78-4649-84
(S)-	dioxan	251(4.61),286(3.97), 330s(3.42),460s(3.98), 482(4.02),516s(3.82)	78-4649-84
5,12-Naphthacenedione, 9-(2-acetoxyethyl)-7,8,9,10-tetrahydro-6,7,9,11-tetrahydroxy-, cis-(S)-	CHCl$_3$	250(4.58),286(4.03), 330s(3.42),460s(3.96), 483(4.00),520(3.78)	78-4649-84
Zoapatanolide D	MeOH	205(4.20)	102-0125-84
$C_{22}H_{20}O_9$			
9,10-Anthracenedione, 1,2-diacetoxy-6,7,8-trimethoxy-3-methyl-	dioxan	243s(4.20),249(4.22), 278(4.62),343(3.74)	94-0860-84
$C_{22}H_{20}Si$			
Silane, 9-anthracenyldimethylphenyl-	heptane	257(5.18),335(3.49), 351(3.80),368(3.98), 388(3.94)	61-0042-84
$C_{22}H_{21}ClN_4O_7$			
9H-Purine, 6-chloro-2-phenyl-9-(2,3,5-tri-O-acetyl-β-D-ribofuranosyl)-	EtOH	236(4.23),286(4.20), 274(4.18)	44-4340-84
$C_{22}H_{21}N$			
Acridine, 9,10-dihydro-2,7,9-trimethyl-9-phenyl-	EtOH	291(4.21)	24-2703-84
$C_{22}H_{21}NO_2$			
Ethanone, 1-[3-[(diphenylamino)methyl]-2-hydroxy-5-methylphenyl]-	EtOH	293(4.32),340s(2.88)	2-0904-84
$C_{22}H_{21}NO_3S$			
Benzenesulfonamide, 4-methyl-N-(2-oxo-2-phenylethyl)-N-(phenylmethyl)-	ether	238.5(4.31),324.0(1.85)	48-0177-84
$C_{22}H_{21}NO_4$			
9,10-Anthracenedione, 1-acetoxy-4-(cyclohexylamino)-	EtOH at 77°K	527(3.90)	104-1780-84
4H-Furo[3,2-e][1]benzoxocin-3-carboxylic acid, 2-cyano-2,3,3a,5,6,11b-hexahydro-2-phenyl-, methyl ester, (2α,3β,3aα,11bβ)-	EtOH	251(2.72),258(2.91), 264(3.00),271(2.92)	24-2157-84
2-Hexenoic acid, 6-[2-(3-cyano-trans-3-phenyl-2-oxiranyl)phenoxy]-, methyl ester, (2E)-	EtOH	266s(3.38),277s(3.57), 281(3.57)	24-2157-84
1,5-Methano-3-benzazocine-3(2H)-acetic acid, 1,4,5,6-tetrahydro-4,6-dioxo-1-phenyl-, ethyl ester	n.s.g.	246(4.02),295(3.37)	4-0611-84
$C_{22}H_{21}NO_5$			
Angoline	EtOH	210(4.43),226(4.47), 282(4.55),320(4.10)	100-0001-84
Berberinophenol betaine, 8-ethyl-	MeOH	240(4.46),262s(4.24), 312.5(4.02),370(3.99), 458(4.14)	94-2230-84

Compound	Solvent	$\lambda_{max}(\log \epsilon)$	Ref.
Bocconoline	MeOH	230(4.52),283(4.61), 320(4.11)	100-0001-84
Chelilutine, dihydro-	MeOH	230(4.52),280(4.56), 325(4.21)	100-0001-84
4,5-Isoxazoledicarboxylic acid, 2,3-di- hydro-2-methyl-3-phenyl-3-(2-phenyl- ethenyl)-, dimethyl ester, (E)-	EtOH	252(4.14),278(3.83), 285(3.72),295(3.66)	150-1531-84M
Methanaminium, N-(1,3-diphenyl-2-prop- enylidene)-, 3-methoxy-1-(methoxy- carbonyl)-2,3-dioxopropylide, (E,?)-	EtOH	278(4.10),417(3.85)	150-1531-84M
Nitidine, 7,8-dihydro-8-methoxy-	CHCl$_3$	238(4.53),283(4.60), 310(4.40),325s(4.29)	100-0001-84

$C_{22}H_{21}NO_6$

6H-Benzo[de][1,3]benzodioxolo[5,6-g]- quinoline-6-carboxylic acid, 4,5- dihydro-1,2-dimethoxy-, ethyl ester	MeOH	237s(4.25),262(4.68), 267(4.72),291(4.26), 318(4.05),331(4.06), 355(3.42),374(3.34)	39-1273-84C

$C_{22}H_{21}NO_9$

Spiro[anthracene-2(1H),1'-cyclopent- ane]-1,4(3H)-dione, 8,9,10-trihydroxy- 3'-(2-methyl-1,3-dioxolan-2-yl)-2'- nitro-	EtOH	395(3.88),418(4.12), 440(4.20)	78-4579-84

$C_{22}H_{21}N_2O_2$

Benzoxazolium, 3-methyl-2-[3-(3-methyl- 2-[(3H)-benzoxazolylidene)-1-cyclo- hexen-1-yl]-, perchlorate	MeOH	507(5.10)	48-1034-84

$C_{22}H_{21}N_2S_2$

Benzothiazolium, 3-methyl-2-[3-(3-meth- yl-2(3H)-benzothiazolylidene)-1- cyclohexen-1-yl]-, perchlorate	MeOH	570(5.07)	48-1034-84

$C_{22}H_{21}N_3$

Benzaldehyde, 4-(dimethylamino)-, (diphenylmethylene)hydrazone	C$_6$H$_{12}$	366(4.52)	97-0021-84
1-Naphthalenecarbonitrile, 4-[(4-phen- yl-1-piperazinyl)methyl]-	C$_6$H$_{12}$	298(4.03)	35-1335-84

$C_{22}H_{21}N_3O$

1,4-Benzenediamine, N,N-dimethyl-N'-[1- phenyl-2-(phenylimino)ethylidene]-, N-oxide, (Z,Z)-	MeOH	260(4.39),435(3.93)	118-0128-84

$C_{22}H_{21}N_3O_3$

Benzoxazolium, 3-ethyl-2-[2-(2-ethyl- 4,5-dihydro-4,5-dioxo-1-phenyl-1H- pyrazolium-3-yl)ethenyl]-, diiodide	EtOH	385(3.77)	48-0695-84

$C_{22}H_{21}N_3O_5$

Ellipticinium, 10-[(carboxymethylene)- amino]-9-hydroxy-2-methyl-, acetate	H$_2$O	243(4.26),300(4.50), 360(3.48),437(3.41)	87-1161-84
1H-Pyrrolo[1,2-a]indole-9-carboxylic acid, 2-(acetylamino)-7-amino- 2,3,5,8-tetrahydro-6-methyl- 5,8-dioxo-, phenylmethyl ester, (S)-	MeOH	269(4.074),312(4.230), 530(3.165)	44-5164-84

$C_{22}H_{21}N_3O_8$

Anhydrotetracycline, 9-nitroso-,	MeOH	225(4.46),242s(4.53),	23-2583-84

Compound	Solvent	$\lambda_{max}(\log \epsilon)$	Ref.
hydrochloride (cont.)		258(4.64),275(4.63), 292s(4.28),298s(4.17), 322(4.00),341(3.91), 445(4.14)	23-2583-84
$C_{22}H_{21}N_4O_{12}P$			
5'-Thymidylic acid, bis(3-nitrophenyl) ester	H_2O	263(4.24)	87-1733-84
5'-Thymidylic acid, bis(4-nitrophenyl) ester	H_2O	269(4.36)	87-1733-84
$C_{22}H_{21}N_5O$			
Benzamide, N-[8,9-dihydro-9-(phenyl-methyl)-7-(2-propenyl)-7H-purin-6-yl]-	MeOH	320(4.02)	150-3640-84M
Triethylaminium dicyano[3-(dicyanometh-ylene)-4-oxo-2-phenyl-1-cyclobuten-1-yl]methanide	H_2O	255(4.03),395(4.67)	24-2714-84
$C_{22}H_{21}N_5O_7$			
5,8-Dideazaisofolic acid, 9-formyl-	pH 7.0	230(4.65)	87-0232-84
$C_{22}H_{21}O_4S$			
1-Benzothiepinium, 3,5-diacetoxy-1,2-dimethyl-4-phenyl-, fluorosulfate	MeOH	224(4.11),264(4.12), 285s(4.06)	64-0985-84B
$C_{22}H_{21}P$			
Phosphine, (diphenylmethylene)(2,4,6-trimethylphenyl)-	THF	254s(3.20),324(2.84)	78-0765-84
$C_{22}H_{22}$			
Bicyclo[3.2.0]hepta-2,6-diene, 1,2,5-trimethyl-6,7-diphenyl-	pentane	222(4.43),256s(4.08), 280(4.18)	23-2592-84
Bicyclo[3.2.0]hepta-2,6-diene, 1,2,7-trimethyl-5,6-diphenyl-	EtOH	217s(4.20),268(4.04)	23-2592-84
Bicyclo[3.2.0]hepta-2,6-diene, 2,5,6-trimethyl-1,7-diphenyl-	EtOH	216(4.40),266(4.23), 295s(3.11)	23-2592-84
Bicyclo[3.2.0]hepta-2,6-diene, 2,6,7-trimethyl-1,5-diphenyl-	EtOH	254(2.26),261(2.32), 267(2.19),272(2.05)	23-2592-84
1,3-Cyclopentadiene, 5-ethenyl-1,4,5-trimethyl-2,3-diphenyl-	EtOH	226s(4.20),234(4.24), 256s(3.71)	23-2592-84
1,3-Cyclopentadiene, 5-ethenyl-3,4,5-trimethyl-1,2-diphenyl-	EtOH	232(4.28),310(3.91)	23-2592-84
$C_{22}H_{22}ClNO_2S$			
1-Piperidinecarboxylic acid, 4-(2-chlorodibenzo[b,e]thiepin-11(6H)-ylidene)-, ethyl ester	MeOH	210(4.51),233(4.38), 271(4.07),313(3.59)	73-1800-84
$C_{22}H_{22}Cl_2N_6$			
1H-Pyrazolium, 4,4'-azobis[3-chloro-1,5-dimethyl-2-phenyl-, bis(methyl sulfate)	EtOH	226(4.29),312s(4.32), 320(4.38),332s(4.27), 413(3.28),422(3.28)	103-0196-84
$C_{22}H_{22}F_6N_2O_6S_2$			
5,12-Diaza[2_5][(1,2,4,5)cyclophane-bis(onium)] bis(triflate)	EtOH	304s(3.66),308(3.68)	35-2672-84
$C_{22}H_{22}N_2O$			
2H-Benz[g]indazole, 2-(4-methoxy-2,6-dimethylphenyl)-3,9-dimethyl-	EtOH	250(4.60),288(3.98), 302(4.01),305s(3.95),	33-0113-84

Compound	Solvent	λ_{max}(log ϵ)	Ref.
(cont.) Pyrano[2,3-c]pyrazole, 1,6-dihydro-4,6,6-trimethyl-1-(4-methylphenyl)-3-phenyl-	EtOH	321(3.87),339(3.84) 265(4.29)	33-0113-84 94-2146-84
C$_{22}$H$_{22}$N$_2$O$_2$S 1H-Pyrrole-2,5-dione, 3-(butylamino)-1-(4-methylphenyl)-4-(phenylthioxo-methyl)-	CHCl$_3$	318(4.00),430(3.79)	78-3499-84
C$_{22}$H$_{22}$N$_2$O$_3$ 1H-Pyrazole-1-acetic acid, 3,5-dimeth-yl-α-(2-phenoxy-2-phenylethenyl)-, methyl ester, (Z)-	MeOH	250(4.35)	44-4647-84
1H-Pyrazole-1-acetic acid, 3,4,5-tri-methyl-α-(2-phenoxy-2-phenylethen-yl)-, (Z)-	MeOH	255(4.24)	44-4647-84
C$_{22}$H$_{22}$N$_2$O$_3$S 4,9-Imino-1H-benz[f]isoindole-1,3(2H)-dione, 4-(ethylthio)-3a,4,9,9a-tetra-hydro-2-(4-methoxyphenyl)-10-methyl-, (3aα,4β,9α,9aα)-	EtOH	206(4.38),227(4.52), 253(3.78),336(3.68), 345(3.60)	103-0978-84
C$_{22}$H$_{22}$N$_2$O$_4$ 2,3-Pyrazinediol, 2,3-dihydro-5,6-bis(4-methylphenyl)-, diacetate	EtOH	228(4.21),297(3.87)	4-0103-84
C$_{22}$H$_{22}$N$_2$O$_5$ 1H-Indole-3-propanoic acid, 1-[[(meth-oxycarbonyl)amino]acetyl]-, phenyl-methyl ester	EtOH	241(4.28),263s(3.94), 273s(3.85),292(3.82), 301(3.85)	94-1373-84
18-Norcorynan-5-carboxylic acid, 16,17,20,21-tetradehydro-16-formyl-17-methoxy-19-oxo-, methyl ester	MeOH	218(3.95),292(3.81)	78-7853-84
C$_{22}$H$_{22}$N$_2$O$_6$ 2,3-Pyrazinediol, 2,3-dihydro-5,6-bis(4-methoxyphenyl)-, diacetate, trans	EtOH	237(4.28),290(3.91), 326(3.92)	4-0103-84
C$_{22}$H$_{22}$N$_2$O$_8$ 2-Butenedioic acid, [2,3-dihydro-2-(methoxycarbonyl)-2-(2-methoxy-2-oxoethyl)-1H-perimidin-1-yl]-, dimethyl ester, (E)-	EtOH	215(4.12),232(4.38), 328(3.76),341(3.77)	150-0016-84S
2-Butenedioic acid, 2,2'-(1,8-naphtha-lenediyldiimino)bis-, tetramethyl ester, (Z,Z)-	EtOH	206(4.37),226(4.34), 324(4.17)	150-0016-84S
C$_{22}$H$_{22}$N$_4$ Benzenamine, 4,4'-(2,3-dihydro-2-meth-yl-1H-1,5-benzodiazepine-2,4-diyl)-bis-	EtOH	230(4.32),313(4.17)	103-0106-84
7H-Pyrrolo[2,3-d]pyrimidin-4-amine, N,2,5,6-tetramethyl-N,7-diphenyl-	EtOH	229(4.44),321(4.12)	11-0073-84B
7H-Pyrrolo[2,3-d]pyrimidin-4-amine, 2,5,6-trimethyl-N-(2-methylphenyl)-7-phenyl-	EtOH	203(4.61),220s(4.42), 241s(4.30),303(4.28)	11-0073-84B
7H-Pyrrolo[2,3-d]pyrimidin-4-amine, 2,5,6-trimethyl-N-(3-methylphenyl)-	EtOH	207(4.52),242s(4.14), 307(4.29)	11-0073-84B

Compound	Solvent	$\lambda_{max}(\log \epsilon)$	Ref.
7-phenyl- (cont.)			11-0073-84B
7H-Pyrrolo[2,3-d]pyrimidin-4-amine, 2,5,6-trimethyl-N-(4-methylphenyl)-7-phenyl-	EtOH	222s(4.61),238(4.50), 308(4.60)	11-0073-84B
4H-Pyrrolo[2,3-d]pyrimidin-4-imine, 3,7-dihydro-2,5,6-trimethyl-3-(3-methylphenyl)-	EtOH	245(4.37),270s(3.96), 282s(3.91),302(3.86)	11-0073-84A
4H-Pyrrolo[2,3-d]pyrimidin-4-imine, 3,7-dihydro-2,5,6-trimethyl-3-(4-methylphenyl)-	EtOH	243(4.47),271s(3.99), 283s(3.97),305(3.91)	11-0073-84A
$C_{22}H_{22}N_4O_6$			
Oxepino[3,2-e][1,2,4]triazolo[1,5-a]-pyrimidine-6,7-dicarboxylic acid, 4,5,8,9-tetrahydro-8,9-dimethyl-5-oxo-4-phenyl-, 7-ethyl 6-methyl ester	CHCl$_3$	292(4.04)	24-0585-84
$C_{22}H_{22}N_4O_{15}$			
2,5,8,11-Tetraoxatridecan-13-ol, 1-(3,5-dinitrophenyl)-1-oxo-, 3,5-dinitrobenzoate	EtOH	213(4.56)	24-1994-84
$C_{22}H_{22}O_2$			
Benzoic acid, 4-[2-(1,2,3,4-tetrahydro-1,4-methano-6-naphthalenyl)ethenyl]-, ethyl ester, (E)-	EtOH	208(4.36),238(4.11), 332(4.60)	87-1516-84
Bicyclo[3.2.0]hept-6-en-3-one, 2-hydroxy-1,2,7-trimethyl-5,6-diphenyl-, (1α,2α,5α)-	EtOH	216s(4.08),224s(3.96), 256s(3.95),265(4.02), 277s(3.84),292s(2.74)	23-2592-84
$C_{22}H_{22}O_4$			
1,4-Dioxaspiro[4.5]decan-2-one, 3-[[4-(phenylmethoxy)phenyl]-methylene]-, (Z)-	EtOH	313(4.38)	39-1531-84C
2-Propenoic acid, 3-phenyl-, 1,4-butanediyl ester	MeCN	275(4.68)	40-0022-84
$C_{22}H_{22}O_5$			
1,4-Phenanthrenedione, 2-(2-acetoxypropyl)-3-hydroxy-5,7,8-trimethyl-	ether	225s(4.48),236.5(4.54), 292s(4.32),299(4.37), 341(3.63),380s(3.46), 395s(3.40),450s(3.12)	33-1003-84
$C_{22}H_{22}O_6$			
9,10-Anthracenedione, 1-hydroxy-5,8-dimethoxy-3-methyl-2-(3-oxopentyl)-	MeOH	210s(--),231(4.62), 261(4.35),280s(--), 451(4.08)	78-3677-84
4H-1-Benzopyran-4-one, 7-hydroxy-3-[2-hydroxy-4,6-dimethoxy-3-(3-methyl-2-butenyl)phenyl]-	EtOH	248.5(4.34),264(4.20), 297s(4.01),304.5s(3.97)	142-0261-84
	EtOH-NaOAc	261(4.40),289s(4.06), 341(3.96)	142-0261-84
4H-1-Benzopyran-4-one, 7-hydroxy-3-[2-hydroxy-4,6-dimethoxy-5-(3-methyl-2-butenyl)phenyl]- (licoricone)	EtOH	240(4.39),248(4.38), 285(4.11),305s(4.04)	142-0261-84
	EtOH-NaOAc	249.5(4.38),256s(4.37), 307s(3.95),337.5(3.93)	142-0261-84
2-Propenoic acid, 3-(3-methoxyphenyl)-, 1,2-ethanediyl ester	MeCN	276(4.59)	40-0022-84
2-Propenoic acid, 3-(4-methoxyphenyl)-, 1,2-ethanediyl ester	MeCN	308(4.70)	40-0022-84

Compound	Solvent	$\lambda_{max}(\log \epsilon)$	Ref.
$C_{22}H_{22}O_7$			
9,10-Anthracenedione, 1-hydroxy-3-(hydroxymethyl)-5,8-dimethoxy-2-(3-oxopentyl)-	MeOH	230(4.57),258(4.30), 450(3.99)	78-3677-84
$C_{22}H_{22}O_8$			
Podophyllotoxin	EtOH	287(3.63),291(3.64), 295s(--)	102-1147-84
Volubinol	MeOH	230(4.3),240(4.25), 294(4.4)	2-0680-84
$C_{22}H_{22}O_{10}$			
Javanin	MeOH	266(4.41),346(4.24)	2-0543-84
	MeOH-NaOAc	275(--),330(--)	2-0543-84
	MeOH-AlCl$_3$	277(--),335(--)	2-0543-84
$C_{22}H_{22}S$			
23-Thiatetracyclo[14.2.2.2^{8,11}.1^{3,6}]-tricosa-3,5,8,10,16,18,19,21-octaene	C$_6$H$_{12}$	<u>249(3.7),275s(2.9)</u>	44-1177-84
$C_{22}H_{22}Si$			
Silane, (9,10-dihydro-9-anthracenyl)-dimethylphenyl-	heptane	204(4.72),232(4.11), 264(3.26)	61-0042-84
$C_{22}H_{23}BrO_2$			
1,4-Epoxyanthracene, 5-[(6-bromohexyl)-oxy]-1,2,3,4-tetrahydro-2,3-bis(methylene)-	MeCN	218(4.65),252(4.68), 302(2.52)	89-0074-84
$C_{22}H_{23}Br_2NS_2$			
Benzenesulfenamide, 4-bromo-N-[(4-bromophenyl)thio]-N-tricyclo-[3.3.1.1^{3,7}]dec-1-yl-	hexane	257(4.41)	44-2724-84
$C_{22}H_{23}Cl_2NS_2$			
Benzenesulfenamide, 3-chloro-N-[(3-chlorophenyl)thio]-N-tricyclo-[3.3.1.1^{3,7}]dec-1-yl-	hexane	229s(4.17),253(4.36)	44-2724-84
Benzenesulfenamide, 4-chloro-N-[(4-chlorophenyl)thio]-N-tricyclo-[3.3.1.1^{3,7}]dec-1-yl-	hexane	233(4.22),256(4.38)	44-2724-84
$C_{22}H_{23}NO_5$			
1H,3H-Benz[f]isoquino[8,1,2-hij][3,1]-benzoxazine, 5,6-dihydro-8,9,11,12-tetramethoxy-	CH$_2$Cl$_2$	264(4.66),274(4.63), 339(4.08),384(3.54)	39-1273-84C
Gouregine, O,O-dimethyl-	EtOH	225(4.09),237s(4.03)	100-0753-84
$C_{22}H_{23}NO_6$			
Hydrastine, N-methyl-, (Z)-	CHCl$_3$	262s(3.88),310(4.06), 385(4.26)	78-1971-84
2(1H)-Isoquinolinecarboxylic acid, 1-(1,3-benzodioxol-5-ylmethylene)-3,4-dihydro-6,7-dimethoxy-, ethyl ester, (Z)-	MeOH	222(4.41),295s(4.15), 332(4.41)	39-1273-84C
Saülatine	EtOH	229(4.40),269(4.04), 294(3.94)	100-0539-84
$C_{22}H_{23}NO_7$			
Spiro[1,3-dioxolo[4,5-g]isoquinoline-5(6H),2'-[2H]inden]-1'(3'H)-one,	MeOH	230(4.33),289(4.29)	94-2230-84

Compound	Solvent	$\lambda_{max}(\log \epsilon)$	Ref.
7,8-dihydro-3',3',4',5'-tetrameth-oxy-, (±)- (cont.)			94-2230-84
$C_{22}H_{23}N_3O_2S$ 1H-Pyrrole-3-carbothioamide, 4-(butyl-amino)-2,5-dihydro-1-(4-methylphen-yl)-2,5-dioxo-N-phenyl-	CHCl$_3$	282(4.67),405(3.77)	48-0401-84
$C_{22}H_{23}N_3O_3S$ Piperazine, 1-[2-nitro-1-[5-(phenyl-thio)-2-furanyl]ethyl]-4-phenyl-	MeOH	207(4.25),244(4.11)	73-2496-84
$C_{22}H_{23}N_3O_7$ Anhydrotetracycline, 7-amino-, dihydro-chloride	MeOH	230s(4.44),268(4.88), 340s(3.53),408(3.89)	23-2583-84
Anhydrotetracycline, 9-amino-, dihydro-chloride	MeOH	226(4.37),244s(4.42), 274(4.62),305s(4.22), 315s(4.13),327s(4.00), 418(3.70),474s(3.40), 494s(3.31)	23-2583-84
$C_{22}H_{23}N_3O_9S_2$ DL-Alanine, N-[4-[(1-carboxyethyl)-thio]-1,3-dihydro-8,9-dimethoxy-1,3-dioxo-2H-2-benzazepin-2-yl]-N-(4,5-dihydro-5-methyl-4-oxo-2-thiazolyl)-	MeOH	251(4.59),342(4.11), 402(4.14)	103-0024-84
$C_{22}H_{23}N_5O_6$ 5,8-Dideazaisofolic acid, 5-methyl-	pH 7.0	240(4.60)	87-0232-84
5,8-Dideazaisofolic acid, 9-methyl-	pH 7.0	235(4.59)	87-0232-84
5,8-Dideazaisofolic acid, 10-methyl-	pH 7.0	230(4.65)	87-0232-84
Oxepino[2,3-d]pyrimidine-5-carboxylic acid, 6-cyano-2-[2-(ethoxycarbonyl)-hydrazino]-3,4,7,8-tetrahydro-7,8-dimethyl-4-oxo-3-phenyl-, methyl ester	CHCl$_3$	270(3.85),335(4.38)	24-0585-84
$C_{22}H_{23}N_7O_4$ 5'-Adenosineacetamide, N- 2-(1H-indol-3-yl)ethyl-5'-deoxy-	pH 6.83	<u>264(4.1)</u>,280s(3.6)	4-0259-84
$C_{22}H_{24}$ 1H-Indene, 1-(2,3-dihydro-2,2-dimethyl-1H-inden-1-ylidene)-2,3-dihydro-2,2-dimethyl-, (E)-	EtOH-CH$_2$Cl$_2$	240(3.90),247(3.85), 258(3.42),326(4.31)	24-2300-84
(Z)-	EtOH-CH$_2$Cl$_2$	228(4.05),260(4.31)	24-2300-84
1H-Indene, 1-(2,3-dihydro-4,7-dimethyl-1H-inden-1-ylidene)-2,3-dihydro-4,7-dimethyl-, (E)-	EtOH-CH$_2$Cl$_2$	236(4.37),307(4.58)	24-2300-84
$C_{22}H_{24}BrClN_6OS$ 6H-Thieno[3,2-f][1,2,4]triazolo[4,3-a][1,4]diazepine, 2-bromo-4-(2-chlorophenyl)-9-[4-(2-ethoxyethyl)-piperazino]-	MeOH	242(4.33),300(3.57)	73-0621-84
6H-Thieno[3,2-f][1,2,4]triazolo[4,3-a][1,4]diazepine, 2-bromo-4-(2-chlorophenyl)-9-[4-(3-methoxy-propyl)piperazino]-	MeOH	240s(4.30),300(3.56)	73-0621-84

Compound	Solvent	$\lambda_{max}(\log \epsilon)$	Ref.
$C_{22}H_{24}Cl_2N_2O_2$			
4-Quinolinemethanol, 2-(3,4-dichloro-phenyl)-α-[(diethylamino)methyl]-6-methoxy-	EtOH	<u>221(4.6),268(4.7),</u> <u>333(4.1),345s(4.1)</u>	149-0469-84A
$C_{22}H_{24}N_2O$			
Diazene, (2-ethyl-8-methyl-1-naphtha-lenyl)(4-methoxy-2,6-dimethylphenyl)-	DMF	324(4.26),470(3.27)	33-0113-84
Piperazine, 1-(4-methoxyphenyl)-4-(2-naphthalenylmethyl)-	C_6H_{12}	248(4.31),275s(3.90), 285s(3.80),300s(3.53)	35-1335-84
$C_{22}H_{24}N_2O_2$			
Propanamide, N,N-diethyl-2-[10-(meth-oxyimino)-9(10H)-phenanthrenylidene]-	hexane	243(3.31),300(2.72), 334(2.35)	54-0023-84
$C_{22}H_{24}N_2O_3$			
Butanamide, N-[4-(butylamino)-9,10-di-hydro-9,10-dioxo-1-anthracenyl]-	benzene	565(4.03),603(3.98)	39-0529-84C
$C_{22}H_{24}N_2O_4$			
Corynan-5-carboxylic acid, 20,21-dide-hydro-16-methyl-17,19-dioxo-, methyl ester	MeOH	255(4.31),290s(4.25), 298(4.27)	78-4853-84
$C_{22}H_{24}N_2O_5$			
Phenylalanine, α,β-didehydro-N-[N-[(phenylmethoxy)carbonyl]-L-valyl]-, (Z)-	EtOH	206(4.35),281(4.21)	5-0920-84
$C_{22}H_{24}N_2O_7$			
3,4,5-Oxepintricarboxylic acid, 6,7-di-hydro-6,7-dimethyl-2-[(phenylcarbon-imidoyl)amino]-, 5-ethyl 3,4-dimeth-yl ester	$CHCl_3$	264(3.81),315(4.10)	24-0585-84
3,4,5-Oxepintricarboxylic acid, 6,7-di-hydro-7,7-dimethyl-2-[(phenylcarbon-imidoyl)amino]-, 5-ethyl 3,4-dimeth-yl ester	$CHCl_3$	267(3.62),317(4.07)	24-0585-84
$C_{22}H_{24}N_2O_8$			
Antibiotic SF-2140, diacetate	MeOH	222(4.83),260s(4.15), 266(4.18),285(4.01), 294(4.10)	158-0931-84
$C_{22}H_{24}N_4O_2$			
1H-Pyrido[3,4-b]indole-1-propanamide, 3-(aminocarbonyl)-2,3,4,9-tetrahydro-2-(phenylmethyl)-, (1S-trans)-	MeOH	225.5(4.62),275s(3.90), 282(3.92),290(3.83)	94-1313-84
$C_{22}H_{24}N_4O_2S_2$			
Benzo[g]pteridine-2,4(3H,4aH)-dione, 5,10-dihydro-3,7,8,10-tetramethyl-4a-[[(phenylmethyl)dithio]methyl]-	MeCN	222(4.44),274(4.10), 295s(--),359(3.79)	35-3309-84
	6M HCl	219(4.37),272(4.03), 300s(--),406(3.42)	35-3309-84
$C_{22}H_{24}N_4O_4$			
1,9-Dioxa-2,6,10,14-tetraazacyclohexa-deca-2,10-diene-7,15-dione, 3,11-di-phenyl-	EtOH	247(4.3)	5-1696-84
Oxepino[2,3-d]pyrimidine-5-carboxylic acid, 6-cyano-3,4,7,8-tetrahydro-	$CHCl_3$	280(3.81),342(3.42)	24-0585-84

Compound	Solvent	λ_{max}(log ϵ)	Ref.
8,8-dimethyl-2-[(1-methylethyl)amino]-4-oxo-3-phenyl-, methyl ester (cont.)			24-0585-84
$C_{22}H_{24}N_4O_6$			
1H-Indazole-4,7-dione, 1-(3-acetoxy-propyl)-3-[[(aminocarbonyl)oxy]-methyl]-6-methyl-5-[(phenylmethyl)-amino]-	MeOH	213(4.34),265(4.22), 296(4.11),496(3.62)	5-1711-84
$C_{22}H_{24}N_4O_{11}$			
1H-Imidazole-5-carboxamide, 1-[(4-ni-trophenyl)methyl]-4-(2,3,5-tri-O-acetyl-β-D-ribofuranosyl)oxy]-	H_2O	256(4.22)	4-0849-84
	M HCl	262(4.12)	4-0849-84
1H-Imidazolium, 5-(aminocarbonyl)-4-hydroxy-1-[(4-nitrophenyl)methyl]-3-(2,3,5-tri-O-acetyl-β-D-ribofur-anosyl)-, hydroxide, inner salt	H_2O	245s(--),280(3.68)	4-0849-84
	M HCl	245s(--),281(3.68)	4-0849-84
	M NaOH	274(4.31)	4-0849-84
$C_{22}H_{24}N_4S_3$			
Benzenamine, 4-[[2-[4-(dimethylamino)-phenyl]-5,6-dihydro-2H-2a,3-dithia-(2a-S^{IV})-1,2-diazacyclopenta[cd]-pentalen-4-yl]thio]-N,N-dimethyl-	MeCN	569(4.08)	97-0251-84
$C_{22}H_{24}N_6O_5$			
L-Glutamic acid, N-[4-[2-(2,4-diamino-pyrido[3,2-d]pyrimidin-6-yl)-1-meth-ylethyl]benzoyl]-	pH 13	238(4.55),343(3.74)	87-0376-84
$C_{22}H_{24}O$			
3,5-Cycloheptadien-1-ol, 2,5,6-trimeth-yl-3,4-diphenyl-, trans	EtOH	234(4.15),269s(3.83)	23-2592-84
2,9-Ethanodibenzo[a,e]cyclooctene-3-carboxaldehyde, 5,6,11,12-tetra-hydro-7,8,10-trimethyl-	$CHCl_3$	256s(3.85),296(3.45)	78-4823-84
$C_{22}H_{24}O_2$			
[1,1'-Biphenyl]-4-carboxylic acid, 4'-(2,6,6-trimethyl-1-cyclohexen-1-yl)-	EtOH	278(4.34)	87-1516-84
$C_{22}H_{24}O_3$			
2-Heptenoic acid, 7-(2,3-diphenyl-oxiranyl)-, methyl ester, cis	MeCN	258(2.74),266s(2.58), 272s(2.30)	88-1137-84
Benzoic acid, 2-(bicyclo[8.3.1]tetra-decal(14),10,12-trien-11-ylcarbonyl)-	MeCN	206(4.60),260(4.18), 284s(3.81)	24-0455-84
$C_{22}H_{24}O_6$			
Dibenzo[4,5:6,7]cycloocta[1,2-c]furan-1(3H)-one, 3a,4,13,13a-tetrahydro-6,7,10,11-tetramethoxy-, (8bR*)- (8bS*)-	$CHCl_3$	240.2(4.28),258s(4.08), 282.5(4.08),295s(3.93)	12-1775-84
	$CHCl_3$	242.5(4.29),284.5(4.14)	12-1775-84
$C_{22}H_{24}O_7$			
Santhemoidin A	MeOH	204(4.38),230(3.82)	102-2911-84
$C_{22}H_{24}O_8$			
Pinoresinol, 1-acetoxy-, (+)-	EtOH	233(4.18),281(3.78)	94-2730-84
	EtOH-NaOH	255.5(--),294(--)	94-2730-84
Pumilin, 8-acetyl-	MeOH	203(4.35)	102-0817-84

Compound	Solvent	$\lambda_{max}(\log \epsilon)$	Ref.
$C_{22}H_{24}O_9$			
4H-1-Benzopyran-4-one, 5,6,7,8-tetra-methoxy-2-(2,4,5-trimethoxyphenyl)- (agecorynin C)	MeOH	256(4.3),270(4.4), 360(4.3)	94-4935-84
4H-1-Benzopyran-4-one, 5,6,7-trimeth-oxy-2-(2,3,4,5-tetramethoxyphenyl)-	MeOH	239s(4.4),267s(4.3), 312(4.4)	94-3354-84
4H-1-Benzopyran-4-one, 5,6,7-trimeth-oxy-2-(2,3,4,6-tetramethoxyphenyl)-	MeOH	259(4.5),308(4.5)	94-3354-84
3α-Epoxypumilin, 8-acetyl-	MeOH	203(4.23)	102-0817-84
$C_{22}H_{24}O_{10}$			
4H-1-Benzopyran-4-one, 5-hydroxy-3,6,7,8-tetramethoxy-2-(2,4,5-trimethoxyphenyl)-	MeOH	267(4.32),360(3.93)	150-3786-84M
	MeOH-AlCl₃	275(--),380(--)	150-3786-84M
	MeOH-NaOAc	266(--),362(--)	150-3786-84M
$C_{22}H_{25}NO_2$			
5a,10b-Butano-6H-[1,4]benzodioxino[2,3-b]indole, 2-(1,1-dimethylethyl)-	MeOH	284(3.84),289(3.83)	4-1841-84
$C_{22}H_{25}NO_4S$			
1H-2-Benzazepine-1,3(2H)-dione, 4-mer-capto-8,9-dimethoxy-2-tricyclo-[3.3.1.1³,⁷]dec-1-yl-	MeOH	246(4.47),320(4.02), 374(3.91)	103-0024-84
$C_{22}H_{25}NO_7$			
Spiro[1,3-dioxolo[4,5-g]isoquinoline-5(6H),2'-[2H]inden]-1'-ol, 1',3',7,8-tetrahydro-3',3',4',5'-tetramethoxy-, trans-(±)-	MeOH	238(4.10),289(3.82)	94-2230-84
Sternbergine, 2,9-O,O'-diacetyl-	MeCN	280(3.40)	100-1003-84
$C_{22}H_{25}NS_2$			
Benzenesulfenamide, N-(phenylthio)-N-tricyclo[3.3.1.1³,⁷]dec-1-yl-	hexane	249(4.28)	44-2724-84
$C_{22}H_{25}N_3O_4$			
Piperazine, 1-(3,4-dimethoxybenzoyl)-4-(1,2,3,4-tetrahydro-2-oxo-6-quin-olinyl)- (OPC-8212)	MeOH	271(4.40)	145-0334-84
	EtOH	271(4.40)	145-0334-84
	CHCl₃	271(4.36)	145-0334-84
$C_{22}H_{25}N_5O_3S$			
Propanenitrile, 3-[2-[ethyl[4-[[6-(2-hydroxyethoxy)-2-benzothiazolyl]-azo]phenyl]amino]ethoxy]-	pH 1.2	536(4.70),608s(4.24)	48-0151-84
	pH 6.2	530(4.76)	48-0151-84
	EtOH	524(4.74)	48-0151-84
$C_{22}H_{26}Cl_2N_2$			
5,10-Ethanopyrido[3,2-g]quinoline, 11,12-dichloro-2,8-bis(1,1-dimethyl-ethyl)-5,10-dihydro-, trans	MeCN	276(4.113),281(4.124)	5-0877-84
$C_{22}H_{26}Cl_2N_2O_4$			
Decanediamide, N,N'-bis(4-chlorophen-yl)-N,N'-dihydroxy-	MeOH	226(4.29),238(4.30), 330(4.21)	34-0345-84
$C_{22}H_{26}N_2O$			
5,10-Ethanopyrido[3,2-g]quinolin-12-one, 2,8-bis(1,1-dimethylethyl)-5,10-dihydro-	MeCN	276(4.031),306(3.334), 317(3.082)	5-0877-84
$C_{22}H_{26}N_2O_3$			
16-Epiaffinine acetate	EtOH	215(4.10),237(4.01),	102-2359-84

Compound	Solvent	λ_{max}(log ϵ)	Ref.
(cont.)		318(4.13)	102-2359-84
Vincamajine	n.s.g.	243(3.93),293(3.53)	105-0509-84
$C_{22}H_{26}N_2O_3S$			
Pyridinium, 4-[2-(1-butyl-1H-indol-5-yl)ethenyl]-1-(3-sulfopropyl)-, hydroxide, inner salt	EtOH	495(4.48)	44-2546-84
$C_{22}H_{26}N_2O_4$			
Akuammine	n.s.g.	244(3.88),312(3.60)	105-0509-84
5H-2,3-Benzodiazepine, 1-(3,4-dimethoxyphenyl)-5-ethyl-7,8-dimethoxy-4-methyl-	EtOH	239(4.42),272s(4.06), 311(4.21)	24-1476-84
$C_{22}H_{26}N_2O_4S$			
1,5-Benzothiazepin-4(5H)-one, 3-acetoxy-5-[2-(dimethylamino)ethyl]-2,3-dihydro-2-(4-methoxyphenyl)-, (2S,3S)-(hydrochloride)	MeOH	210(4.41),239(4.26), 280(3.26)	24-1476-84
$C_{22}H_{26}N_2O_5$			
Dehydroreserpic acid, hydrochloride	EtOH	264(4.24),295(3.84), 386(3.73)	151-0355-84A
Pseudoindoxyl of tetraphylline	EtOH	230(3.55),247(3.54), 282(3.19),377(2.73)	100-0687-84
$C_{22}H_{26}N_2O_5S$			
D-Arabinitol, 5-C-[1-(4-methylphenyl)-2-[(phenylmethyl)thio]-1H-imidazol-4-yl]-	EtOH	264(3.78)	136-0091-84B
1,2,3-Propanetriol, 1-[3a,5,6,6a-tetrahydro-6-hydroxy-3-(4-methylphenyl)-2-[(phenylmethyl)thio]-3H-furo[2,3-d]imidazol-5-yl]-	EtOH	248(4.10)	136-0091-84B
$C_{22}H_{26}N_4O_6$			
Oxepino[2,3-d]pyrimidine-6-carboxylic acid, 2-[2-(ethoxycarbonyl)hydrazino]-3,4,7,8-tetrahydro-7,8-dimethyl-4-oxo-3-phenyl-, ethyl ester	CHCl$_3$	215(3.80),332(4.32)	24-0585-84
$C_{22}H_{26}N_4O_6S$			
Piperazine, 1-[4-(2,3-dihydroxypropoxy)-6,7-dimethoxy-2-quinazolinyl]-4-(2-thienylcarbonyl)-, monohydrochloride	EtOH	216(4.50),249(4.70)	4-1189-84
$C_{22}H_{26}N_4O_7$			
Piperazine, 1-[4-(2,3-dihydroxypropoxy)-6,7-dimethoxy-2-quinazolinyl]-4-(2-furanylcarbonyl)-, monohydrochloride	EtOH	217(4.60),251(4.90)	4-1189-84
$C_{22}H_{26}N_6O_4S_2$			
Thiazolo[3,2-a]pyrimidin-4-ium, 3-hydroxy-5-methyl-2-[[[4-methyl-6-(4-morpholinyl)-2-pyrimidinyl]thio]acetyl]-7-(4-morpholinyl)-, hydroxide, inner salt	MeCN	240(4.42),292(4.44), 385(4.16)	103-0791-84
	CCl$_4$	410(--)	103-0791-84

Compound	Solvent	$\lambda_{max}(\log \epsilon)$	Ref.
$C_{22}H_{26}N_6O_7$			
Propanamide, N-[9-(2-deoxy-β-D-erythro-pentofuranosyl)-6-[2-(4-nitrophenyl)-ethoxy]-9H-purin-2-yl]-2-methyl-	MeOH	218(4.44),269(4.42)	78-0059-84
$C_{22}H_{26}N_6O_8$			
Propanamide, 2-methyl-N-[6-[2-(4-nitro-phenyl)ethoxy]-9-β-D-ribofuranosyl-9H-purin-2-yl]-	MeOH	222(4.43),268(4.40), 282s(4.25)	78-0059-84
$C_{22}H_{26}O_2$			
1,5-Heptanedione, 4,4,6-trimethyl-1,3-diphenyl-	EtOH	242(4.11)	78-4127-84
1,5-Heptanedione, 4,6,6-trimethyl-1,3-diphenyl-	EtOH	242.5(4.08)	78-4127-84
threo-	EtOH	242.5(4.11)	78-4127-84
6-Hepten-3-one, 5-hydroxy-2,2,4-tri-methyl-5,7-diphenyl-	EtOH	254(4.34)	78-4127-84
threo-	EtOH	254(4.37)	78-4127-84
$C_{22}H_{26}O_4$			
2,5,8,11-Tetraoxadodecane, 1-(9-anthra-cenyl)-	$C_6H_{11}Me$	248(4.88),256(5.16), 315(3.08),330(3.42), 345(3.73),365(3.92), 385(3.91)	150-1901-84M
$C_{22}H_{26}O_5$			
Furan, 2,5-dihydro-3,5-bis(3,4-dimeth-oxyphenyl)-2,2-dimethyl-	MeOH	264(4.13)	4-0881-84
α-D-Xylofuranose, 1,2-O-(1-methyleth-ylidene)-3,5-bis-O-(phenylmethyl)-	MeOH	266(2.60)	118-0961-84
$C_{22}H_{26}O_6$			
1-Butanone, 1-(1-hydroxy-3,7,9-trimeth-oxy-2,8-dimethyl-4-dibenzofuranyl)-3-methyl-	EtOH	216(4.63),265(4.50), 283(4.37),304s(4.22)	39-2573-84C
Cyclopropa[b]naphtho[2,3-e]pyran-1a-carboxylic acid, 1-acetyl-4,5,6,7,9-9a-hexahydro-8-hydroxy-9-oxo-3-prop-yl-, ethyl ester, (1α,1aα,9aα)-	EtOH	211(4.29),285(4.10)	39-0487-84C
(1α,1aβ,9aβ)-	EtOH	210(4.31),287(4.08)	39-0487-84C
2(3H)-Furanone, 3,4-bis[(3,4-dimeth-oxyphenyl)methyl]dihydro- (dimeth-ylmatairesinol)	CHCl₃	241(4.20),282(4.02), 287s(3.98)	12-1775-84
4H-Naphtho[2,3-b]pyran-2-carboxylic acid, 6,7,8,9-tetrahydro-4-oxo-5-(2-oxopropoxy)-10-propyl-, ethyl ester	EtOH	212(4.47),256(4.21), 275(4.04)	39-0487-84C
1,4,10(4bH)-Phenanthrenetrione, 3-acet-oxy-9-hydroxy-3,6,7,8-tetrahydro-4b,8,8-trimethyl-2-(1-methylethyl)-, (S)-	ether	262(4.17),290s(3.58), 300s(3.43),421(3.21)	33-1523-84
$C_{22}H_{26}O_7$			
Canellin E	MeOH	262(4.10)	102-0661-84
Plectrin	EtOH	237(4.26)	138-1513-84
$C_{22}H_{26}O_8$			
2-Butenoic acid, 2-methyl-, 3-acetoxy-4-formyl-1a,2,3,5a,7,8,8a,9,10,10a-decahydro-10a-methyl-8-methylene-7-oxooxireno[5,6]cyclodeca[1,2-b]-	EtOH	217(4.09)	102-0823-84

Compound	Solvent	$\lambda_{max}(\log \epsilon)$	Ref.
furan-9-yl ester (cont.)			102-0823-84
Ethanone, 1,1'-(2,2'-dihydroxy-4,4',6-6'-tetramethoxy-5,5'-dimethyl-[1,1'-biphenyl]-3,3'-diyl)bis-(contortin)	EtOH	229(4.35),277(4.31), 345(3.98)	12-1531-84
Lirioresinol B	MeOH	240(4.00),270(3.42)	94-1612-84
19H-1,20-Metheno-2H-cyclohepta[s]-[1,4,7,10,13,16]hexaoxacyclodocosin-2,19-dione, 4,5,7,8,10,11,13,14,16,17-decahydro-	MeCN	497(2.8)	24-2839-84
with barium thiocyanate	MeCN	493(2.8)	24-2839-84
with calcium thiocyanate	MeCN	488(2.8)	24-2839-84
$C_{22}H_{26}O_9$			
Syringaresinol, 1-hydroxy-, (+)-	EtOH	237.0(4.10),272.2(3.40), 281.0(3.32)	94-4482-84
Zoapatanolide C	MeOH	207(4.06)	102-0125-84
$C_{22}H_{27}BrN_2O_3$			
Eburnamenine-14-carboxylic acid, 15-bromo-14,15-dihydro-14-methoxy-, methyl ester, (3α,14β,15α,16α)-	EtOH	227(4.43),276(3.88), 284(3.83)	22-0435-84
$C_{22}H_{27}ClN_2O_8$			
Propanedioic acid, 2,2'-[(4-chloro-1,2-phenylene)bis(iminomethylidyne)]bis-, tetraethyl ester	MeCN	218(4.35),290(4.53), 330(4.44)	62-0947-84A
$C_{22}H_{27}NO_9$			
α-D-Glucopyranoside, methyl 2-deoxy-2-[(3-oxo-3-phenyl-1-propenyl)amino]-, 3,4,6-triacetate, (Z)-	EtOH	250(3.88),333(4.26)	136-0101-84L
Lycorine 1-O-β-D-glucoside	MeOH	235s(3.5),288(3.61)	102-1167-84
$C_{22}H_{27}N_3O_4$			
Oxepino[2,3-d]pyrimidine-6-carboxylic acid, 3,4,7,8-tetrahydro-7,8-dimethyl-2-[(1-methylethyl)amino]-4-oxo-3-phenyl-, ethyl ester	CHCl$_3$	277(3.99),345(4.44)	24-0585-84
Oxepino[2,3-d]pyrimidine-6-carboxylic acid, 3,4,7,8-tetrahydro-8,8-dimethyl-2-[(1-methylethyl)amino]-4-oxo-3-phenyl-, ethyl ester	CHCl$_3$	278(3.91),343(4.37)	24-0585-84
$C_{22}H_{27}N_3O_5$			
Oxepino[2,3-d]pyrimidine-6-carboxylic acid, 3,4,7,8-tetrahydro-2-[(3-hydroxypropyl)amino]-7,8-dimethyl-4-oxo-3-phenyl-, ethyl ester	CHCl$_3$	275(3.97),343(4.39)	24-0585-84
$C_{22}H_{27}N_5O_6S_2$			
Benzothiazolium, 2-[[4-[(2-cyanoethyl)-ethylamino]phenyl]azo]-6-(2-hydroxyethoxy)-3-methyl-, methyl sulfate	pH 1.13 pH 6.88 70% EtOH-pH 6.2	620(4.74) 620(4.76) 626(4.78)	48-0151-84 48-0151-84 48-0151-84
$C_{22}H_{28}NO_6$			
Ethanaminium, N-[2-(1,3-benzodioxol-5-yl)-3-(2,2-dimethyl-1-oxopropoxy)-2-cyclopenten-1-ylidene]-2-ethoxy-N-methyl-2-oxo-, perchlorate	MeCN	285(4.46)	44-0228-84

Compound	Solvent	$\lambda_{max}(\log \epsilon)$	Ref.
$C_{22}H_{28}N_2O_2$			
5H-[1]Benzopyrano[3,4-c]pyridin-5-one, 3-[2-(3-azabicyclo[3.2.2]non-3-yl)-ethyl]-1,2,3,4-tetrahydro-, mono-hydrochloride	MeOH	272(4.02),309(3.88)	4-1561-84
$C_{22}H_{28}N_2O_3$			
Isoretuline, 11-methoxy-	MeOH	223(4.32),250(4.05), 292(3.69)	102-2659-84
Vincamine, O-methyl-	EtOH	230(4.47),279(3.93), 283(3.93),294(3.78)	22-0435-84
Voacangine, iso-	EtOH	224(3.7),277(3.09), 298(3.13)	102-2359-84
$C_{22}H_{28}N_2O_4$			
Decanediamide, N,N'-dihydroxy-N,N'-di-phenyl-	MeOH	219(4.38),270(4.02)	34-0345-84
Octanediamide, N,N'-dihydroxy-N,N'-bis(2-methylphenyl)-	MeOH	210(4.33),224(4.05)	34-0345-84
Octanediamide, N,N'-dihydroxy-N,N'-bis(4-methylphenyl)-	MeOH	220(4.53),258(4.12)	34-0345-84
2-Propenoic acid, 3-(8'-ethyl-1,2,2'-3',6',7',8',8'a-octahydro-1-methyl-2-oxospiro[3H-indole-3,1'(5'H)-ind-olizin]-8'-yl)-, methyl ester, N-oxide	MeOH	224(4.47),258(4.00), 285(3.45)	44-0547-84
$C_{22}H_{28}N_2O_5$			
Tabernoxidine	EtOH	220(4.55),258(3.68), 286(3.48),293(3.37)	2-0101-84
$C_{22}H_{28}N_2O_6$			
[2,2'-Bi-2H-indole]-2,2'-dicarboxylic acid, 1,1',3,3',4,4',5,5',6,6',7,7'-dodecahydro-3,3'-dioxo-, diethyl ester	MeCN	309(4.00)	33-1957-84
[2,2'-Bi-2H-pyrrole]-2,2'-dicarboxylic acid, 1,1',3,3'-tetrahydro-5,5'-di-methyl-3,3'-dioxo-4,4'-di-2-propen-yl-, diethyl ester	MeCN	311(3.95)	33-1957-84
$C_{22}H_{28}N_2O_8$			
Propanedioic acid, 2,2'-[1,2-phenyl-enebis(iminomethylidyne)]bis-, tetraethyl ester	MeCN	219(4.36),288(4.55), 329(4.45)	62-0947-84A
12H-Pyrano[4,3-j]-1,2-benzodioxepin-10-ol, decahydro-3,6,9-trimethyl-3,12-epoxy-, (2-nitrophenyl)carbamate	CHCl$_3$	242(4.06)	33-1515-84
(3-nitrophenyl)carbamate	CHCl$_3$	242(4.09)	33-1515-84
(4-nitrophenyl)carbamate	CHCl$_3$	312(4.26)	33-1515-84
$C_{22}H_{28}N_4O_6$			
1H-Pyrrol-1-yloxy, 3,3'-[(2,4-dioxo-1,3(2H,4H)-pyrimidinediyl)dicarbonyl]-bis[2,5-dihydro-2,2,5,5-tetramethyl-	MeCN	212(4.30),234(4.32), 265s(4.00),392(1.95)	70-1668-84
$C_{22}H_{28}O$			
Methanone, phenyl[2,4,6-tris(1-methyl-ethyl)phenyl]-	C_6H_{12}	242(4.20),349(1.80)	88-4525-84

Compound	Solvent	$\lambda_{max}(\log \epsilon)$	Ref.
$C_{22}H_{28}O_2$			
2,4-Pentadienoic acid, 3-methyl-5-[2-methyl-4-(2,6,6-trimethyl-1-cyclohexen-1-yl)phenyl]-, (E,E)-	EtOH	236(4.00),313(4.48)	87-1516-84
$C_{22}H_{28}O_2S_2$			
1,3-Dithiane, 2-(4-butoxyphenyl)-5-(4-ethoxyphenyl)-, trans	EtOH-acid	248(4.36)	23-1103-84
	EtOH-base	263.5(4.48)	23-1103-84
$C_{22}H_{28}O_3$			
2,4-Pentadien-1-ol, 3-methyl-5-(2,2,6-trimethyl-7-oxabicyclo[4.1.0]hept-1-yl)-, benzoate, (E,E)-(±)-	EtOH	237.5(4.41)	33-0175-84
$C_{22}H_{28}O_4$			
2H-1-Benzopyran-8-propanol, 3,4-dihydro-5,7-dimethoxy-α,α-dimethyl-2-phenyl-, (S)- (nitenin)	EtOH	207(4.26),270(3.03)	25-0632-84
Naphthalene, 1-(3,4-dimethoxyphenyl)-1,2,3,4-tetrahydro-6,7-dimethoxy-2,3-dimethyl- (isogalbuline)	MeOH	220(3.54),285(3.36)	102-2647-84
$C_{22}H_{28}O_5$			
2-Butenoic acid, 4-[4-[2-(4-methoxy-2,6,6-trimethyl-1-cyclohexen-1-yl)-ethenyl]-5-oxo-2(5H)-furanylidene]-3-methyl-, methyl ester, (Z,E,E)-(Z,Z,E)-	EtOH	368(4.48)	94-1709-84
	EtOH	364(4.48)	94-1709-84
Furan, 3,5-bis(3,4-dimethoxyphenyl)-tetrahydro-2,2-dimethyl-	MeOH	229(3.99),276(3.52)	4-0881-84
Furo[4,3,2-ij][2]benzopyran-7(2H)-one, 2-acetoxy-6,8a-dihydro-6-methyl-5-(1,3,3-trimethylcyclohexyl)-(aplysulphurin)	EtOH	220(4.00),266(2.48), 274(2.27)	12-1081-84
$C_{22}H_{28}O_6$			
Canellin D	MeOH	260(4.26)	102-0661-84
4a,10a-Epoxyphenanthrene-1,4,10(4bH)-trione, 3-acetoxy-5,6,7,8,8a,9-hexahydro-4b,8,8-trimethyl-2-(1-methylethyl)-	ether	227s(3.81),258(3.69)	33-1116-84
$C_{22}H_{28}O_7$			
1-Butanone, 1,3-bis(2,4,6-trimethoxyphenyl)-	MeOH	225s(4.31),276(3.83)	94-0325-84
Germacra-1(10),4-dien-12,6β-olide, 3β-acetoxy-8α-angeloyloxy-9β-hydroxy-, trans,trans	MeOH	202(4.04)(end abs.)	102-1063-84
Leptocarpin, acetyl-	MeOH	205(4.09)	102-0675-84
Spiro[cyclohexane-1,4'(3'H)-[1H-7,9a]-methanocyclohepta[c]pyran]-2-carboxaldehyde, 6-acetoxyhexahydro-5'-hydroxy-3,3-dimethyl-8'-methylene-1',9'-dioxo-	MeOH	229(3.95)	142-1701-84
Spiro[isobenzofuran-4(1H),4'(3'H)-[1H-7,9a]methanocyclohepta[c]pyran]-1',9'(4'aH)-dione, 3-acetoxydecahydro-5'-hydroxy-7a-methyl-8'-methylene-	EtOH	228(3.97)	142-1701-84

Compound	Solvent	$\lambda_{max}(\log \epsilon)$	Ref.
$C_{22}H_{28}O_8$			
Coleon I	ether	278(4.16),323(4.21), 405(4.02)	33-0201-84
Leptocarpin, 3-acetyl-15-hydroxy-	EtOH	221(4.30)	102-0823-84
Santhemoidin C	MeOH	205(4.47)	102-2911-84
Teuflavin	EtOH	212(3.56)	102-0843-84
Teumarin	EtOH	211(3.54)	102-0611-84
$C_{22}H_{29}ClN_2O_8$			
2,3-Quinoxalinediacetic acid, 6-chloro-α,α'-bis(ethoxycarbonyl)-1,2,3,4-tetrahydro-, diethyl ester	MeCN	226(4.50),323(3.82)	62-0947-84A
$C_{22}H_{29}NO$			
1,4,14-Conatrienin-3-one	EtOH	232(4.51),240(4.49), 326(4.29)	22-0071-84
$C_{22}H_{29}NO_3$			
2,4,10-Undecatrienamide, 11-(1,3-benzodioxol-5-yl)-N-(2-methylpropyl)-, (E,E,E)-	EtOH	212(4.41),261(4.63), 268s(4.59),304(3.78)	88-4267-84
(E,E,Z)-	EtOH	209(4.27),261(4.44), 297s(3.46)	88-4267-84
$C_{22}H_{29}NO_3Si$			
1H-[1,3]Benzodioxolo[5,6-d][1]benzazepin-1-one, 2,3,4,5,6,7-hexahydro-5-[2-methyl-3-(trimethylsilyl)-2-propenyl]-	MeCN	255(3.72),324(4.01)	44-0228-84
1H-[1,3]Benzodioxolo[5,6-d][1]benzazepin-1-one, 2,3,4,5,6,7-hexahydro-5-[2-[(trimethylsilyl)methyl]-2-propenyl]-	MeCN	245(3.88),315(4.14)	44-0228-84
$C_{22}H_{29}NO_6$			
3,12-Epoxy-12H-pyrano[4,3-j][1,2]benzodioxepin-10-ol, decahydro-3,6,9-trimethyl-, phenylcarbamate, [3R-(3α,5aβ,6β,8aβ,9α,10β,12β,12aR*)]-	CHCl$_3$	250(4.40)	33-1515-84
$C_{22}H_{29}NO_9$			
Pseudolycorine 1-O-β-D-glucoside	MeOH	238s(3.38),288(3.55)	102-1167-84
	MeOH-NaOMe	252s(--),294(--)	102-1167-84
$C_{22}H_{29}N_3O_5$			
Oxepino[2,3-d]pyrimidine-6-carboxylic acid, 1,2,3,4,7,8-hexahydro-8-methyl-2,4-dioxo-3-phenyl-, ethyl ester, diethylamine salt	CHCl$_3$	263(4.15),317(4.25)	24-0585-84
$C_{22}H_{30}$			
1,4-Cyclohexadiene, 3,3'-(1,2-ethanediylidene)bis[1,5,6,6-tetramethyl-	CH$_2$Cl$_2$	258(3.92),353s(--), 369.5(4.88),391.5(4.93)	5-0340-84
$C_{22}H_{30}N_2O_4$			
Spiro[3H-indole-3,1'(5;H)-indolizine]-8'-propanoic acid, 8'-ethyl-1,2,2'-3',6',7',8',8'a-octahydro-1-methyl-2-oxo-, methyl ester, 4'-oxide	MeOH	222(4.54),256(4.04), 284(3.47)	44-0547-84

Compound	Solvent	$\lambda_{max}(\log \epsilon)$	Ref.
$C_{22}H_{30}N_2O_5$			
1H-Pyrrole-3-propanoic acid, 2-[(1,1-dimethylethoxy)carbonyl]-5-[(3-ethyl-1,5-dihydro-4-methyl-5-oxo-2H-pyrrol-2-ylidene)methyl]-4-methyl-, methyl ester, (Z)-	MeOH	250s(--),257(4.32), 380(4.38),400s(--)	35-2645-84
$C_{22}H_{30}N_2O_8$			
2,3-Quinoxalinediacetic acid, α,α'-bis(ethoxycarbonyl)-1,2,3,4-tetra-hydro-, diethyl ester	MeCN	222(4.45),313(3.72)	62-0947-84A
$C_{22}H_{30}N_4O$			
Urea, N,N-diethyl-N'-[(8α)-6-(2-prop-enyl)ergolin-8-yl]-	MeOH	234(3.77),277s(3.81), 282(3.83),293(3.75)	73-2828-84
$C_{22}H_{30}N_4O_7S_2$			
Benzothiazolium, 2-[[4-(dimethylamino)-2-methoxyphenyl]azo]-6-(2-hydroxyeth-oxy)-3-methyl-, methyl sulfate	pH 1.13	609(4.76)	48-0151-84
	pH 6.88	612(4.76)	48-0151-94
	70% EtOH-pH 6.2	595(4.85)	48-0151-84
$C_{22}H_{30}O$			
14'-Apo-β,ψ-carotenal, 11-cis	3-Mepentane	386(4.60),412(4.72)	149-0313-84A
$C_{22}H_{30}O_2$			
2,5-Cyclohexadien-1-one, 2,6-bis(1,1-dimethylethyl)-4-[(4-methyl-2-oxo-3-cyclohexen-1-yl)methylene]-	EtOH	204(4.407),250(4.23), 345(4.146)	104-0294-84
2-Cyclohexen-1-one, 6-[[3,5-bis(1,1-di-methylethyl)-4-hydroxyphenyl]methyl-ene]-	EtOH	202(4.431),250(4.255), 345(4.176)	104-0294-84
23,24-Dinorchola-1,4,6-trien-3-one, 22-hydroxy-	EtOH	224(4.08),252(3.96), 299(4.11)	94-3866-84
$C_{22}H_{30}O_3$			
B-Norpregna-5,16-dien-20-one, 3β-acet-oxy-	MeOH	208(3.84),242(3.89), 322(1.65)	80-0755-84
$C_{22}H_{30}O_5$			
Androsta-1,4-diene-2-carboxylic acid, 11β,17β-dihydroxy-17-methyl-3-oxo-, methyl ester	MeOH	245(4.09)	13-0271-84A
19,1-Metheno-1H-cyclohepta[q][1,4,7-10,13]pentaoxacycloheneicosin, 2,3,5,6,8,9,11,13,14,15,17,18-dodecahydro-	MeCN	618(<u>2.5</u>),680s(2.4)	24-2839-84
with barium thiocyanate	MeCN	612(<u>2.6</u>)	24-2839-84
with calcium thiocyanate	MeCN	610(<u>2.5</u>)	24-2839-84
Oxireno[13,14]cyclotetradeca[1,2-b]-furan-13(1aH)-one, 11-acetoxy-2,3,6,7,10,11,11a,12,14a,14b-deca-hydro-1a,5,9-trimethyl-12-methylene-, [1aR-(1aR*,4E,8E,11S*,11aR*,14aS*-14bR*)]-	n.s.g.	210(<u>3.88</u>)	12-0545-84
$C_{22}H_{30}O_6$			
Coleon Q, 7-O-acetyl-12-O-deacetyl-19-hydroxy-	MeOH	235(3.97)	33-0201-84

Compound	Solvent	$\lambda_{max}(\log \epsilon)$	Ref.
$C_{22}H_{31}BrO_4$ 3-Hexen-2-one, 1-(2-acetoxy-2,6,6-tri- methylcyclohexyl)-6-(5-bromo-3-furan- yl)-3-methyl-, [1S-[1α(E),2β]]-	MeOH	227(3.95)	44-3204-84
$C_{22}H_{31}NO_3$ 3,5-Nonadienamide, N-(2-hydroxy-1-meth- yl-2-phenylethyl)-N,2,4,6-tetrameth- yl-8-oxo-, (1'R,2'S,2S)-	EtOH	225(3.88)	44-5202-84
(1'S,2'R,2R)- B-Norpregna-5,16-dien-20-one, 3β-acet- oxy-, 20-oxime	EtOH MeOH	228.5(3.87) 208(3.95),238(4.14)	44-5202-84 80-0755-84
$C_{22}H_{31}NO_{15}$ D-glycero-D-ido-Octonic acid, 2-acetyl- 2,3-dideoxy-3-(nitromethyl)-, methyl ester, 4,5,6,7,8-pentaacetate	EtOH	254(3.73)	136-0063-84K
$C_{22}H_{31}NS_2$ Benzenesulfenamide, N-(1,1-dimethyleth- yl)-2,4,6-trimethyl-N-[(2,4,6-tri- methylphenyl)thio]-	hexane	211(--),249s(4.29)	44-2724-84
$C_{22}H_{31}N_3O_2S$ 3,5-Pyridinedicarboxamide, N,N,N',N'- tetraethyl-1,4-dihydro-1-methyl- 4-(phenylthio)-	CH_2Cl_2	244(3.89),270(3.67), 350(3.08)	35-6029-84
$C_{22}H_{31}N_5O_7$ Guanosine, 2'-deoxy-N-(2-methyl-1-oxo- propyl)-, 3',5'-bis(2-methylpropan- oate)	MeOH	227(3.60),259(4.27), 280s(4.14)	78-0003-84
$C_{22}H_{32}$ 1H-Indene, 2,3-dihydro-1,1,3,3-tetra- methyl-2-(2,2,5,5-tetramethylcyclo- pentylidene)-	C_6H_{12}	193.5(4.71),209s(4.32), 225s(3.57),251s(2.70), 258.1(2.92),263.9(3.11), 270.7(3.14)	24-0277-84
$C_{22}H_{32}F_3NO_7$ 2-Propene-1,1,3-tricarboxylic acid, 1-[(trifluoroacetyl)amino]-, 1,1- diethyl 3-[5-methyl-2-(1-methyl- ethyl)cyclohexyl] ester, (E)-	hexane	202(4.11)	78-1391-84
(Z)-	hexane	210(4.05)	78-1391-84
$C_{22}H_{32}N_2O_4$ Spiro[3,6a-methano-6aH-naphth[2,1-b]- oxocin-4(5H),3'-[3H]pyrazole-12a(7H)- carboxylic acid, 1,2,3,4',5',8,8a,9- 10,11,12,12b-dodecahydro-9,9-dimeth- yl-5-oxo-, methyl ester	MeOH	217(4.84),330(4.19)	100-0055-84
$C_{22}H_{32}N_2O_{14}$ D-glycero-D-gulo-Octonic acid, 2-(1- aminoethylidene)-2,3-dideoxy-3-(ni- tromethyl)-, methyl ester, 4,5,6,7,8- pentaacetate	EtOH	279(4.36)	136-0063-84K
D-glycero-D-ido-Octonic acid, 2-(1- aminoethylidene)-2,3-dideoxy-3-(ni- tromethyl)-, methyl ester, 4,5,6,7,8-	EtOH	277(4.29)	136-0063-84K

Compound	Solvent	$\lambda_{max}(\log \epsilon)$	Ref.
pentaacetate (cont.)			136-0063-84K
isomer	EtOH	279(4.28)	136-0063-84K
$C_{22}H_{32}N_4O$			
Urea, N,N-diethyl-N'-[(8α)-6-(1-methyl-ethyl)ergolin-8-yl]-	MeOH	223(4.35),275s(3.65), 281(3.67),291(3.53)	73-2828-84
Urea, N,N-diethyl-N'-[(8α)-6-propyl-ergolin-8-yl]-	MeOH	275(3.79),281(3.80), 292(3.72)	73-2828-84
$C_{22}H_{32}N_4O_3$			
1H-Pyrrole-3-propanamide, N-[3-(dimeth-ylamino)propyl]-5-[(3-ethyl-1,5-dihy-dro-4-methyl-5-oxo-2H-pyrrol-2-yli-dene)methyl]-2-formyl-4-methyl-, (Z)-	CHCl₃	273(4.30),405(4.40), 424s(4.34)	49-1071-84
$C_{22}H_{32}O_2$			
2-Cyclohexen-1-one, 4-[[3,5-bis(1,1-dimethylethyl)-4-hydroxyphenyl]-methyl]-3-methyl-	EtOH	203(4.64),231(4.29), 278(3.60)	104-0294-84
2-Cyclohexen-1-one, 6-[[3,5-bis(1,1-dimethylethyl)-4-hydroxyphenyl]-methyl]-3-methyl-	EtOH	204(4.524),233(4.28), 278(3.52),300(2.945)	104-0294-84
23,24-Dinorchola-1,4-dien-3-one, 22-hydroxy-	EtOH	247(4.19)	73-2713-84
Retinoic acid, 13,14-dihydro-13-de-methyl-13,14-methylene-, ethyl ester	hexane	299(4.47)	44-1937-84
cis	hexane	296(4.42)	44-1937-84
Δ^2-THC, methyl ether, (1R,4R)-	EtOH	269(4.19)	39-2881-84C
(1S,4R)-	EtOH	268(4.18)	39-2881-84C
$C_{22}H_{32}O_3$			
Abieta-8,11,13-triene, 11,12-dimethoxy-7-oxo-	EtOH	273(4.37)	23-2822-84
Abieta-8,11,13-triene, 11,14-dimethoxy-7-oxo-	EtOH	228(3.89),261(3.51), 331(3.89)	23-2822-84
Dolabella-4,8,18-trien-16-al, 3-acet-oxy-, 1R,11S,12R,4E,8E	EtOH	224(4.00)	100-0615-84
1S,11R,12S,4Z,8E	EtOH	230(3.91)	78-0799-84
1,4-Ethanonaphthalen-9-one, 1,4,4a,7-8,8a-hexahydro-7-hydroxy-7-(1-meth-oxycyclopropyl)-4,6,8,8,10,10-hexa-methyl-	MeCN	302(1.97),312s(1.85)	33-0774-84
Oxiranebutanoic acid, 3-(1,3,5,8,11-pentadecapentaenyl)-, methyl ester, [2α,3β(1E,3E,5Z,8Z,11Z)]-(±)-	EtOH	272(4.59),281(4.62), 293(4.48)	104-0806-84
$C_{22}H_{32}O_4$			
Androst-5-en-17-one, 3-(acetoxymeth-oxy)-, (3β)-	EtOH	200(3.46)	56-0297-84
2-Cyclopentene-1-acetaldehyde, 3-(acet-oxymethyl)-α-(4,8-dimethyl-3,7-nona-dienyl)-2-formyl-	MeOH	248(3.81)	78-2913-84
Macrophorin A	MeOH	237(3.76)	158-84-4
9(1H)-Phenanthrenone, 2,3,4,4a,10,10a-hexahydro-7-(2-hydroxy-1-methyleth-yl)-5,8-dimethoxy-1,1,4a-trimethyl-	EtOH	231(4.01),261(3.62), 330(3.40)	23-2822-84
Radbohakusin	EtOH	210(3.88),288(1.46)	138-1613-84
$C_{22}H_{32}O_5$			
Crassin acetate	MeOH	215(3.70)	44-1417-84

Compound	Solvent	$\lambda_{max}(\log \epsilon)$	Ref.
Pentanedioic acid, 2-[[3,5-bis(1,1-di-methylethyl)-4-oxo-2,5-cyclohexadien-1-ylidene]methyl]-, dimethyl ester	EtOH	227(3.716),304(4.358)	104-1732-84
$C_{22}H_{32}O_6$			
Euphorbia factor H_1	MeOH	194(4.27),211(4.22), 296(2.57)	64-0683-84B
$C_{22}H_{32}S$			
Dispiro[cyclopentane-1,2'-thiirane-3',2"-[2H]indene], 1",3"-dihydro-1",1",2,2,3",3",5,5-octamethyl-	C_6H_{12}	212.0(3.81),251s(2.72), 257.9(2.90),263.8(3.05), 270.6(3.04)	24-0277-84
$C_{22}H_{33}NO_{12}$			
α-D-Glucopyranoside, ethyl 2-deoxy-2-[[3-ethoxy-2-(ethoxycarbonyl)-3-oxo-1-propenyl]amino]-, 3,4,6-tri-acetate	EtOH	236(3.88),282(4.24)	136-0101-84L
$C_{22}H_{34}N_2O_4S$			
Benzenesulfonic acid, 4-methyl-, [7-(1,2-dihydroxy-1-methylethyl)octa-hydro-1,4a-dimethyl-2(1H)-naphtha-lenylidene]hydrazide	MeOH	225(4.12),274(2.99)	23-1407-84
$C_{22}H_{34}N_6O_{17}P_2$			
1H-Imidazole-4-carboxylic acid, 5-ami-no-1-β-D-ribofuranosyl-, ethyl ester, 5',5"'-(P,P'-dihydrogen diphosphate)	pH 2	250s(4.24),267(4.34)	65-2114-84
	pH 6 - 7	240s(4.05),270(4.37)	65-2114-84
	pH 12	240s(4.05),270(4.37)	65-2114-84
$C_{22}H_{34}O_2$			
23,24-Dinorchol-4-en-3-one, 22-hydroxy-	EtOH	242(4.18)	73-2713-84
Phenol, 2-(3,7-dimethyl-1,6-octadien-yl)-3-methoxy-5-pentyl-	EtOH	268(4.32)	39-2881-84C
Phenol, 2-(3,7-dimethyl-2,6-octadien-yl)-3-methoxy-5-pentyl-	EtOH	273(3.11),279s(3.10)	39-2881-84C
$C_{22}H_{34}O_4$			
Cyclohexaneacetic acid, 2-[2-hydroxy-5-methoxy-4-(1-methylethyl)phenyl]-2,6,6-trimethyl-, methyl ester, (1S-trans)-	EtOH	228(3.85),290(3.83)	23-2822-84
$C_{22}H_{34}O_5$			
Cyclohexaneacetic acid, 2-[4-(2-hydroxy-1-methylethyl)-2,5-dimethoxyphenyl]-2,6,6-trimethyl-	EtOH	232s(3.66),289(3.43)	23-2822-84
Pentanedioic acid, 2-[[3,5-bis(1,1-di-methylethyl)-4-hydroxyphenyl]methyl]-, dimethyl ester	EtOH	206(4.38),277(3.48)	104-1732-84
$C_{22}H_{34}O_6$			
2(5H)-Oxepinone, 3-(5-acetoxy-4-methyl-3-pentenyl)-7-(2,6-dihydroxy-4-meth-yl-4-hexenyl)-6,7-dihydro-7-methyl-	MeOH	202(4.11),220s(3.88)	102-0829-84
$C_{22}H_{34}O_7$			
Melfusanolide, 17-acetoxy-1,10-di-hydroxy-	MeOH	203(3.99)(end abs.)	102-0833-84

Compound	Solvent	$\lambda_{max}(\log \epsilon)$	Ref.
$C_{22}H_{34}O_{12}$			
Deacylacteoside dimethyl ether	MeOH	228(3.83),279(3.42), 284s(3.37)	94-1209-84
$C_{22}H_{34}O_{13}$			
Deacylacteoside, β-hydroxy-, dimethyl ether	MeOH	229(3.87),278(3.43), 282s(3.40)	94-1209-84
$C_{22}H_{35}NO_2$			
Cannabinol, 7α-(methylamino)hexahydro-7β-	EtOH EtOH	276(3.04),282(3.04) 278(2.76),285(2.77)	87-1370-84 87-1370-84
$C_{22}H_{35}NO_3S$			
6a,10-Propano-6aH-cyclopenta[4,5]cyclo-oct[1,2-d]oxazol-5(6H)-one, 4-ethyl-decahydro-9-methoxy-4,6,10,11-tetra-methyl-2-thioxo-, [3aS-(3aα,4β,6α-6aβ,9β,9aβ,10β,10aβ,11S*)]-	MeOH	245(4.40)	78-0919-84
$C_{22}H_{36}N_2O_2$			
2H-1,4-Benzoxazin-2-amine, 6,8-bis(1,1-dimethylethyl)-N,N-diethyl-2-methoxy-3-methyl-	hexane	272(2.80),311(2.38), 396(0.92)	54-0023-84
$C_{22}H_{36}N_2O_8$			
Propanedioic acid, 2,2'-[1,2-ethanediyl-bis(iminomethylidyne)]bis-, tetra-propyl ester	MeCN	222(4.35),273(4.42), 292(4.41)	62-0947-84A
$C_{22}H_{36}N_2O_{11}S$			
Thiourea, N-(3,3-diethoxypropyl)-N'-(2,3,4,6-tetra-O-acetyl-β-D-galactopyranosyl)-	MeOH	249(4.09)	103-0447-84
Thiourea, N-(3,3-diethoxypropyl)-N'-(2,3,4,6-tetra-O-acetyl-β-D-glucopyranosyl)-	MeOH	249(4.08)	103-0447-84
$C_{22}H_{36}N_4O_6Si_2$			
Inosine, 2'-deoxy-2'-oxo-3',5'-O-[1,1,3,3-tetrakis(1-methyl-ethyl)-1,3-disiloxanediyl]-	MeOH	249(4.11)	78-0125-84
$C_{22}H_{36}O_6$			
Tomexanthin (end absorptions given)	MeOH	213(3.72),220s(--)	102-0464-84
$C_{22}H_{38}O_6Si_2$			
1,2-Benzenedicarboxylic acid, 3,6-bis[[(1,1-dimethylethyl)dimethyl-silyl]oxy]-, dimethyl ester	hexane	306.5(3.54)	44-1898-84
$C_{22}H_{39}NO$			
1-Octadecanone, 1-(1H-pyrrol-2-yl)-	EtOH	248(3.58),288(4.17)	64-1393-84B
1-Octadecanone, 1-(1H-pyrrol-3-yl)-	EtOH	244(3.95),268(3.78)	64-1393-84B
$C_{22}H_{39}N_5O_4Si_2$			
9H-Purin-6-amine, 9-[(2,5-bis-O-tert-butyldimethylsilyl)-β-D-erythro-pentofuran-3-ulosyl]-	MeOH	259(4.19)	78-0125-84
9H-Purin-6-amine, 9-[(3,5-bis-O-tert-butyldimethylsilyl)-β-D-erythro-pentofuran-2-ulosyl]-	MeOH	259(4.17)	78-0125-84

Compound	Solvent	$\lambda_{max}(\log \epsilon)$	Ref.
$C_{22}H_{40}O_5Si_2$			
1,4-Cyclohexadiene-1-carboxaldehyde, 2-(acetoxymethyl)-3,6-bis[[(1,1-dimethylethyl)dimethylsilyl]oxy]-, cis-(±)-	hexane	226.5(4.04),332.5(2.69)	44-1898-84
$C_{22}H_{40}O_6Si_2$			
1,4-Cyclohexadiene-1,2-dicarboxylic acid, 3,6-bis[[(1,1-dimethylethyl)-dimethylsilyl]oxy]-, dimethyl ester, cis	hexane	273.5s(2.69),282.5(2.76), 306.5(2.77)	44-1898-84
7-Oxabicyclo[4.1.0]hept-3-ene-3-carbox-aldehyde, 4-(acetoxymethyl)-2,5-bis-[[(1,1-dimethylethyl)dimethylsilyl]-oxy]-, (1α,2α,5α,6α)-(±)-	hexane	229.5(3.91)	44-1898-84
$C_{22}H_{41}N$			
1H-Pyrrole, 2-octadecyl-	hexane	215(4.02)	64-1393-84B
$C_{22}H_{41}NO$			
1H-Pyrrole-2-methanol, α-heptadecyl-	hexane	215(4.24)	64-1393-84B
$C_{22}H_{41}N_5O_4Si_2$			
9H-Purin-6-amine, 9-[2,5-bis-O-[(1,1-dimethylethyl)dimethylsilyl]-β-D-arabinofuranosyl]-	MeOH	259(4.19)	87-0270-84
9H-Purin-6-amine, 9-[3,5-bis-O-[(1,1-dimethylethyl)dimethylsilyl]-β-D-arabinofuranosyl]-	MeOH	259(4.19)	87-0270-84
$C_{22}H_{42}S_2Si$			
Silane, (1,1-dimethylethyl)[2-[3-(2,2-dimethyl-6-methylenecyclohexyl]prop-yl]-1,3-dithian-2-yl]dimethyl-	pentane	232(2.95),245(2.99)	33-1734-84

Compound	Solvent	$\lambda_{max}(\log \epsilon)$	Ref.
$C_{23}H_{12}N_2O_3S_2$ 3-Furancarbonitrile, 4,5-bis(2-thien-yl)-2-[[(4-oxo-4H-1-benzopyran-3-yl)methylene]amino]-	MeOH	418(4.07)	73-1788-84
$C_{23}H_{12}N_2O_5$ 3-Furancarbonitrile, 4,5-bis(2-furan-yl)-2-[[(4-oxo-4H-1-benzopyran-3-yl)methylene]amino]-	MeOH	420(4.10)	73-1788-84
$C_{23}H_{12}N_4O_5$ Pyrrolo[2,1-a]isoquinoline-1-carboni-trile, 3-(5-nitro-2-furanyl)-2-(4-nitrophenyl)-	dioxan	425(4.12)	73-0533-84
$C_{23}H_{14}ClNO_2S$ 2H-Thieno[2,3-h]-1-benzopyran-2-one, 3-chloro-4-(diphenylamino)-	EtOH	212(4.29),233s(4.22), 239(4.24),256s(4.28), 261(4.29),277s(4.16), 310s(3.76),320(3.80), 353(3.90)	161-0081-84
$C_{23}H_{14}N_4O$ Spiro[anthracene-9(10H),3'-[3H]pyrazo-lo[5,1-c][1,2,4]triazol]-10-one, 7'-phenyl-	CH$_2$Cl$_2$	244(4.42),276(4.21), 315(3.88),335(3.88)	24-1726-84
$C_{23}H_{15}BO_6$ Boron, [ethanedioato(2-)-O,O'][5-(2-naphthalenyl)-1-phenyl-4-pentene-1,3-dionato-O,O']-	CH$_2$Cl$_2$	459(4.72)	97-0292-84
$C_{23}H_{15}Br_2O$ Pyrylium, 2,6-bis(4-bromophenyl)-4-phenyl-, perchlorate	MeCN	250(4.24),285.5(4.30), 362(4.61),415(4.50)	48-0287-84
$C_{23}H_{15}Cl_2N_6O_6P$ Phosphine, (3,5-dichloro-2,4,6-tri-nitrophenyltriazenylidene)triphenyl-	MeOH	346(4.15)	65-0289-84
$C_{23}H_{15}Cl_2O$ Pyrylium, 2,6-bis(4-chlorophenyl)-4-phenyl-, perchlorate	CH$_2$Cl$_2$	303(3.87),377(4.15), 432(3.97)	88-5143-84
$C_{23}H_{15}F_2O$ Pyrylium, 2,6-bis(4-fluorophenyl)-4-phenyl-, perchlorate	MeCN	237(4.15),277(4.26), 357(4.55),405(4.41)	48-0287-84
$C_{23}H_{15}I_2O$ Pyrylium, 2,6-bis(4-iodophenyl)-4-phenyl-, perchlorate	MeCN	276(4.26),293(4.26), 369(4.64),427(4.53)	48-0287-84
$C_{23}H_{15}NO_3$ 3H-Dibenz[f,ij]isoquinoline-2,7-dione, 4-methyl-6-phenoxy-	toluene	412(3.78),426(3.81)	104-0980-84
2-Propenamide, N-(9,10-dihydro-9,10-di-oxo-1-anthracenyl)-3-phenyl-	benzene	424(3.83)	39-0529-84C
4(1H)-Quinolinone, 2,3-dibenzoyl-	MeOH	247(4.48),322(4.00)	4-0759-84
$C_{23}H_{15}NO_4$ 6H-Anthra[9,1-bc]furan-2-carboxylic	EtOH	513(4.23),550(4.29)	104-0745-84

Compound	Solvent	$\lambda_{max}(\log \epsilon)$	Ref.
acid, 6-oxo-5-(phenylamino)-, methyl ester (cont.)			104-0745-84
$C_{23}H_{15}N_3O_8$ Pyrrolo[2,1-a]isoquinoline-1-carboxylic acid, 2,3-bis(5-nitro-2-furanyl)-, ethyl ester	dioxan	409(4.18)	73-0533-84
$C_{23}H_{15}N_5O_2$ 3H-[1,2,4]Triazino[3,2-b]quinazoline-3,10(4H)-dione, 2-phenyl-4-[(phenylmethylene)amino]-	EtOH	306(3.78),360(3.32)	24-1083-84
$C_{23}H_{16}BrN$ Pyridine, 2-(4-bromophenyl)-4,6-diphenyl-	dioxan	258(4.71),312.5(4.01)	48-0287-84
$C_{23}H_{16}BrNO$ 2(1H)-Pyridinone, 4-(4-bromophenyl)-3,6-diphenyl-	EtOH	260(4.63),344(4.36)	4-1473-84
$C_{23}H_{16}BrO$ Pyrylium, 2-(4-bromophenyl)-4,6-diphenyl)-, perchlorate	MeCN	242.5(4.18),280(4.27), 356(4.56),407(4.45)	48-0287-84
$C_{23}H_{16}Br_2O_2$ 2-Pentene-1,5-dione, 1,5-bis(4-bromophenyl)-3-phenyl-	dioxan	259.5(4.43),303(4.34)	48-0287-84
$C_{23}H_{16}ClN$ Pyridine, 2-(4-chlorophenyl)-4,6-diphenyl-	dioxan	257.5(4.70),312.5(4.03)	48-0287-84

$C_{23}H_{16}ClNO$
  2(1H)-Pyridinone, 3-(4-chlorophenyl)-4,6-diphenyl-
  2(1H)-Pyridinone, 4-(4-chlorophenyl)-3,6-diphenyl-
  2(1H)-Pyridinone, 6-(4-chlorophenyl)-3,4-diphenyl-
	EtOH
	EtOH
	EtOH
		250(4.37),350(4.34)
		254(4.19),344(4.12)
		259(4.10),350(4.00)
			4-1473-84
			4-1473-84
			4-1473-84

$C_{23}H_{16}ClNO_2S$
  2-Pyridinol, 6-(4-chlorophenyl)-3-phenyl-4-(2-thienyl)-, acetate
  2H-Thieno[2,3-h]-1-benzopyran-2-one, 3-chloro-4-(diphenylamino)-5,6-dihydro-
	EtOH
	
	EtOH
		266(4.51),286(4.39), 292(4.33)
		214(4.26),223s(4.20), 248(4.02),277(4.06), 382(4.19)
			4-1473-84
			
			161-0081-84

$C_{23}H_{16}ClO$ Pyrylium, 2-(4-chlorophenyl)-4,6-diphenyl-, perchlorate	MeCN	241(4.17),279.5(4.28), 355(4.55),406(4.44)	48-0287-84
$C_{23}H_{16}Cl_2O_2$ 2-Pentene-1,5-dione, 1,5-bis(4-chlorophenyl)-3-phenyl-	dioxan	255(4.39),303(4.32)	48-0287-84
$C_{23}H_{16}FN$ Pyridine, 2-(4-fluorophenyl)-4,6-diphenyl-	dioxan	255(4.66),311(3.90)	48-0287-84

Compound	Solvent	$\lambda_{max}(\log \epsilon)$	Ref.

$C_{23}H_{16}FO$
  Pyrylium, 2-(4-fluorophenyl)-4,6-di-
  phenyl-, perchlorate
	MeCN	237(4.13),276.5(4.26),
		  352(4.53),403(4.40)
			48-0287-84

$C_{23}H_{16}F_2O_2$
  2-Pentene-1,5-dione, 1,5-bis(4-fluoro-
  phenyl)-3-phenyl-
	dioxan	239(4.27),299(4.28)	48-0287-84

$C_{23}H_{16}IN$
  Pyridine, 2-(4-iodophenyl)-4,6-diphen-
  yl-
	dioxan	261.5(4.69),313(4.07)	48-0287-84

$C_{23}H_{16}IO$
  Pyrylium, 2-(4-iodophenyl)-4,6-di-
  phenyl-, perchlorate
	MeCN	252(4.16),284(4.20),
		  360(4.56),417(4.47)
			48-0287-84

$C_{23}H_{16}I_2O_2$
  2-Pentene-1,5-dione, 1,5-bis(4-iodo-
  phenyl)-3-phenyl-
	dioxan	278(4.47),303s(4.43)	48-0287-84

$C_{23}H_{16}NO_3$
  Pyrylium, 2-(4-nitrophenyl)-4,6-di-
  phenyl-, perchlorate
	MeCN	250(4.13),281(4.28),
		  360(4.56),376s(4.48)
			48-0287-84

Pyrylium, 4-(3-nitrophenyl)-2,6-di- phenyl-, TCNQ salt	MeCN	492(2.48)	80-0817-84

$C_{23}H_{16}N_2O_2$
  Benzonitrile, 2-[1-benzoyl-3-oxo-3-
  phenyl-1-propenyl)amino]-, (Z)-
	MeOH	258(3.97),308(3.66),
		  372(3.94)
			4-0759-84

Pyridine, 2-(4-nitrophenyl)-4,6-di- phenyl-	dioxan	268(4.51),324(4.21)	48-0287-84
4-Quinolinamine, 2,3-dibenzoyl-	MeOH	249(4.50),276(4.33)	4-0759-84

$C_{23}H_{16}N_2O_4$
  1H-Indole, 1-acetyl-3-[4-(1,2-dihydro-
  1-methyl-2-oxo-3H-indol-3-ylidene)-
  4,5-dihydro-5-oxo-2-furanyl]-
	CHCl$_3$	280(4.28),311(3.72),
		  480(4.42)
			39-1331-84B

2H-Indol-2-one, 1-acetyl-1,3-dihydro-
  3-[5-(1-methyl-1H-indol-3-yl)-2-oxo-
  3(2H)-furanylidene]-
	CHCl$_3$	262(4.25),565(4.50)	39-1331-84B

$C_{23}H_{16}N_2O_5$
  1H-Indole, 1-acetyl-3-[4-(1,2-dihydro-
  5-methoxy-2-oxo-3H-indol-3-ylidene)-
  4,5-dihydro-5-oxo-2-furanyl]-
	CHCl$_3$	255(4.22),275s(4.21),
		  465(4.37),487(4.37)
			39-1331-84B

$C_{23}H_{16}N_2O_6$
  2-Pentene-1,5-dione, 1,5-bis(4-nitro-
  phenyl)-3-phenyl-
	dioxan	267(4.49),306(4.26)	48-0287-84

$C_{23}H_{16}N_4$
  Spiro[9H-fluorene-9,3'-[3H]pyrazolo-
  [5,1-c][1,2,4]triazole], 6'-methyl-
  7'-phenyl-
	CH$_2$Cl$_2$	239(4.59),277(4.16),
		  288(4.13),344(3.87)
			24-1726-84

$C_{23}H_{16}N_4O$
  Spiro[3H-pyrazolo[5,1-c]-1,2,4-triazol-
  3,9'-[9H]xanthene], 6-methyl-7-phen-
  yl-
	CH$_2$Cl$_2$	241(4.36),296(3.79),
		  316(3.72),354(3.65)
			24-1726-84

Compound	Solvent	$\lambda_{max}(\log \epsilon)$	Ref.
$C_{23}H_{16}N_4O_2S$ Spiro[3H-pyrazolo[5,1-c]-1,2,4-triazol-3,9'-[9H]thioxanthene], 6-methyl-7-phenyl-, S,S-dioxide	CH_2Cl_2	238(4.47),276(3.83), 284(3.88),315(3.86), 354(3.85)	24-1726-84
$C_{23}H_{16}O$ Benzo[b]naphtho[2,1-b]furan, 6-methyl-5-phenyl-	hexane	256(4.63),265(4.79), 290(4.25)	39-2877-84C
$C_{23}H_{16}O_2$ 2,5-Cyclohexadien-1-one, 4-(2,6-diphenyl-4H-pyran-4-ylidene)-	0.4M NaOH	474(4.58)	4-1673-84
$C_{23}H_{16}O_5$ [2,2'-Binaphthalene]-1,4-dione, 1'-acetoxy-4'-methoxy-	CH_2Cl_2	242(4.67),302(4.13), 442(3.13)	5-1367-84
$C_{23}H_{16}O_6$ Benzo[a]naphthacene-8,13-dione, 5,6-dihydro-4,7,9,12-tetrahydroxy-2-methyl-	EtOH	265(3.91),311(4.18), 470(3.84)	23-2818-84
	EtOH-NaOH	255(--),300s(--), 318(--),348(--), 550(--)	23-2818-84
	EtOH-NH₃	292(--),328(--), 515(--)	23-2818-84
$C_{23}H_{16}O_8S_2$ Pyrylium, 4-(4-hydroxyphenyl)-2,6-bis(4-sulfophenyl)-, hydroxide, inner salt	MeOH-HClO₄	406(4.54),428(4.57)	4-1673-84
$C_{23}H_{16}O_9$ 1,3-Naphthacenedicarboxylic acid, 6,11-dihydro-2,5-dihydroxy-7-methoxy-6,11-dioxo-, dimethyl ester	n.s.g.	218(4.24),252(4.36), 271(4.36),338(3.95), 384(3.85),454(4.00), 475(3.89)	5-0306-84
$C_{23}H_{17}BrO_2$ 2-Pentene-1,5-dione, 5-(4-bromophenyl)-1,3-diphenyl-	dioxan	255(4.34),297(4.31)	48-0287-84
$C_{23}H_{17}ClN_2O_4S$ 2H-Isoindole-2-acetamide, N-[3-(2-chlorobenzoyl)-5-ethyl-2-thienyl]-1,3-dihydro-1,3-dioxo-	MeOH	230s(4.47),238s(4.42), 265(4.07),274(4.09), 348(4.01)	73-0621-84
$C_{23}H_{17}ClO_2$ 2-Pentene-1,5-dione, 5-(4-chlorophenyl)-1,3-diphenyl-	dioxan	248(4.34),296.5(4.30)	48-0287-84
$C_{23}H_{17}Cl_2NO$ 2(1H)-Pyridinone, 3,4-bis(4-chlorophenyl)-3,4-dihydro-6-phenyl-	EtOH	240(4.03),276(3.76)	4-1473-84
2(1H)-Pyridinone, 3,6-bis(4-chlorophenyl)-3,4-dihydro-4-phenyl-	EtOH	242(4.29),280(3.98)	4-1473-84
$C_{23}H_{17}FO_2$ 2-Pentene-1,5-dione, 5-(4-fluorophenyl)-1,3-diphenyl-	dioxan	240(4.31),296.5(4.29)	48-0287-84

Compound	Solvent	λ_{max}(log ϵ)	Ref.
$C_{23}H_{17}F_3N_2O_3$			
Cyclohepta[4,5]pyrrolo[1,2-a]imidazole-10-carboxylic acid, 2-(4-methylphenyl)-3-(trifluoroacetyl)-	EtOH	226(4.65),242s(4.61), 320(4.63),420(3.90), 508(3.56)	150-3465-84M
$C_{23}H_{17}IO_2$			
2-Pentene-1,5-dione, 5-(4-iodophenyl)-1,3-diphenyl-	dioxan	277(4.37),303s(4.33)	48-0287-84
$C_{23}H_{17}NO$			
[1,1'-Biphenyl]-4-propanenitrile, α-[(4-methylphenyl)methylene]-β-oxo-	EtOH	301(4.41)	73-0421-84
2(1H)-Pyridinone, 3,4,6-triphenyl-	EtOH	256(4.45),344(4.39)	4-1473-84
$C_{23}H_{17}NO_2$			
9-Anthracenol, 10-[(methylimino)methyl]-, benzoate	EtOH	256(5.09),340(3.46)	150-0701-84M
Benzeneacetonitrile, α-[(2-oxo-2-phenylethoxy)phenylmethylene]-	EtOH	248(3.27),282(3.24)	5-1702-84
[1,1'-Biphenyl]-4-propanenitrile, α-[(4-methoxyphenyl)methylene]-β-oxo-	EtOH	359(4.16)	73-0421-84
Methanone, (3-amino-4,5-diphenyl-2-furanyl)phenyl-	EtOH	253(4.18),281(4.17), 368(4.25)	5-1702-84
$C_{23}H_{17}NO_3$			
Furo[2,3-a]acridin-11(6H)-one, 4-methoxy-6-methyl-2-phenyl-	EtOH	220(4.13),234(4.13), 253(4.25),271(4.24), 296s(4.28),307(4.37), 343(4.08),382(3.61)	5-0031-84
$C_{23}H_{17}NO_4$			
2-Pentene-1,5-dione, 5-(4-nitrophenyl)-1,3-diphenyl-	dioxan	265(4.37),295(4.26)	48-0287-84
$C_{23}H_{17}N_3O_2$			
4H-Pyrrolo[1,2-a]benzimidazole, 2-(4-nitrophenyl)-4-(phenylmethyl)-	DMSO	263(4.18),380(4.3)	103-0150-84
$C_{23}H_{17}N_3O_3$			
Methanone, [3-methyl-1-(4-nitrophenyl)-4-phenyl-1H-pyrazol-5-yl]phenyl-	EtOH	254(4.25),328(4.20)	4-1013-84
Methanone, [3-methyl-1-(4-nitrophenyl)-5-phenyl-1H-pyrazol-4-yl]phenyl-	EtOH	253(4.26),298(4.30)	4-1013-84
$C_{23}H_{17}N_6O_6P$			
Phosphine, triphenyl(2,4,6-trinitrophenyltriazenylidene)-	MeOH	377(4.30)	65-0289-84
$C_{23}H_{17}O$			
Pyrylium, 2,4,6-triphenyl-, TCNQ salt	MeCN	490(2.30)	80-0817-84
perchlorate	MeCN	408(--)	80-0817-84
$C_{23}H_{17}O_2$			
Pyrylium, 4-(4-hydroxyphenyl)-2,6-diphenyl-, perchlorate	MeOH-HClO	420(4.76)	4-1673-84
$C_{23}H_{17}S$			
Thiopyrylium, 2,4,6-triphenyl-	DMSO	375(4.3)	35-7082-84
$C_{23}H_{18}$			
Dibenzo[2,3:4,5]pentaleno[1,6-ab]-	hexane	263.5(3.37),269(3.44),	89-0508-84

Compound	Solvent	$\lambda_{max}(\log \epsilon)$	Ref.
indene, 4b,8b,12b,12d-tetrahydro-12d-methyl- (cont.)		276.0(3.66)	89-0508-84
2-Penten-4-yne, 2,3,5-triphenyl-, (E)-	C_6H_{12}	226(4.23),247(4.31), 301(4.27)	44-0856-84
(Z)-	C_6H_{12}	230(4.31),249(4.19), 265(4.19),303(4.34)	44-0856-84
$C_{23}H_{18}BrNO$			
2(1H)-Pyridinone, 4-(4-bromophenyl)-3,4-dihydro-3,6-diphenyl-	EtOH	218(4.49),260(3.89), 342(3.09)	4-1473-84
$C_{23}H_{18}ClNO$			
2(1H)-Pyridinone, 4-(3-chlorophenyl)-3,4-dihydro-3,6-diphenyl-	EtOH	270(3.79),324(2.74)	4-1473-84
2(1H)-Pyridinone, 4-(4-chlorophenyl)-3,4-dihydro-3,6-diphenyl-	EtOH	219(4.11),266(3.64), 342(2.91)	4-1473-84
2(1H)-Pyridinone, 6-(4-chlorophenyl)-3,4-dihydro-3,4-diphenyl-	EtOH	244(3.86),284(3.63)	4-1473-84
$C_{23}H_{18}ClN_3O_2$			
Pyrano[2,3-e]indazol-2(5H)-one, 3-chloro-6,7-dihydro-4-(methylphenyl-amino)-7-phenyl-	EtOH	234(4.34),375(4.38)	4-0361-84
Pyrano[2,3-e]indazol-2(5H)-one, 3-chloro-4-(diphenylamino)-6,7-di-hydro-7-methyl-	EtOH	235(4.17),275.5(4.29), 380(4.35)	4-0361-84
$C_{23}H_{18}N_2O$			
1H-Pyrido[3,4-b]indol-1-one, 2,3,4,9-tetrahydro-2,9-diphenyl-	isoPrOH	220(4.54),240(4.42), 308(4.46)	103-0047-84
$C_{23}H_{18}N_2OS$			
2H-Pyrrol-2-one, 1,5-dihydro-3-(4-meth-ylphenyl)-1-phenyl-4-(phenylamino)-5-thioxo-	CHCl$_3$	256(4.50),330(4.34), 490(3.43)	78-3499-84
3-Thiophenecarboxamide, N,4-diphenyl-2-(phenylamino)-	MeOH	260(4.28),344(4.23)	48-0917-84
$C_{23}H_{18}N_2O_3$			
9-Anthracenol, 10-[(methylnitroso-amino)methyl]-, benzoate	EtOH	249(4.88),256(5.09), 324(3.09),338(3.44), 356(3.76),396(3.92)	150-0701-84M
$C_{23}H_{18}N_2O_4S$			
2H-Isoindole-2-acetamide, N-(3-benz-oyl-5-ethyl-2-thienyl)-1,3-dihydro-1,3-dioxo-	MeOH	230s(4.41),238(4.36), 250s(4.26),265s(4.20), 349(3.96)	73-0621-84
$C_{23}H_{18}N_4$			
3H-Pyrazolo[5,1-c]-1,2,4-triazole, 6-methyl-3,3,7-triphenyl-	CH$_2$Cl$_2$	235(4.56),317(3.79), 349(3.77),373s(3.74)	24-1726-84
$C_{23}H_{18}N_4O_5$			
11H-Furo[2',3':4,5]oxazolo[2,3-b]pteri-din-11-one, 6a,7,8,9a-tetrahydro-7-hydroxy-8-(hydroxymethyl)-2,3-di-phenyl-, [6aR-(6aα,7β,8α,9aβ)]-	MeOH	220s(4.50),245(4.32), 280(4.28),357(4.15)	136-0179-84F
Pyrrolo[3,4-c]pyrazole-4,6(1H,5H)-di-one, 3-[4-(1,3-dihydro-1,3-dioxo-2H-isoindol-2-yl)-1-oxobutyl]-3a,6a-di-hydro-5-phenyl-	MeOH	209(4.71),216s(4.78), 220(4.81),225(4.75), 243s(4.56),279(4.53)	103-0066-84

Compound	Solvent	$\lambda_{max}(\log \epsilon)$	Ref.
$C_{23}H_{18}N_8O$			
Carbonic dihydrazide, bis(di-2-pyridin-ylmethylene)-, copper complex in 60% ethanol	pH 1.2 pH 11.1	400(4.17) 465(4)	74-0407-84B 74-0407-84B
$C_{23}H_{18}O$			
Ethanone, 1-(1,2,3-triphenyl-2-cyclopropen-1-yl)-	EtOH	228(4.05),285(4.19), 297(4.29),327(4.30), 337(4.30)	35-1065-84
$C_{23}H_{18}O_2$			
1H-Indene-1,3(2H)-dione, 2-[1,1'-biphenyl]-4-yl-2-ethyl-	EtOH dioxan	350(2.74) 350(2.72)	135-0312-84 135-0312-84
2-Pentene-1,5-dione, 1,3,5-triphenyl-	dioxan	238(4.30),293(4.27)	48-0287-84
$C_{23}H_{18}O_3$			
4H-1-Benzopyran-4-one, 7-hydroxy-2-methyl-3-phenyl-6-(phenylmethyl)-	MeOH	234(4.44),300(4.08)	2-1030-84
$C_{23}H_{18}O_4$			
Benzenemethanol, 4-methoxy-α-9H-xanthen-9-ylidene-, acetate	EtOH	223(4.70),286.5(4.04), 333(4.16)	44-0080-84
1H-2-Benzopyran-1-one, 6,7-dimethoxy-3,4-diphenyl-	EtOH	255(2.9),295(3.07)	2-0889-84
1H-2-Benzopyran-1-one, 7,8-dimethoxy-3,4-diphenyl-	EtOH	254(2.91),300(3.07)	2-0889-84
4H-1-Benzopyran-4-one, 5,7-dihydroxy-6,8-bis(phenylmethyl)-	MeOH	204(4.59),261(4.30), 299(3.61)	2-1036-84
4H-1-Benzopyran-4-one, 5,7-dihydroxy-2-methyl-3-phenyl-8-(phenylmethyl)-	MeOH	204(4.47),258(4.25), 298(3.62)	2-1036-84
1-Phenanthrenecarboxylic acid, 2-(3,4-dimethoxyphenyl)-	MeCN	229(5.73),277(5.89)	2-0603-84
$C_{23}H_{18}O_{11}$			
4H-1-Benzopyran-4-one, 5-acetoxy-2-(3,4-diacetoxyphenyl)-3-methoxy-6,7-(methylenedioxy)-	MeOH	258(4.24),317(4.41)	102-2043-84
$C_{23}H_{19}BO_6$			
Boron, [1,7-bis(4-methylphenyl)-1,6-heptadiene-3,5-dionato-O,O'](ethanedioato(2-)-O,O']-	CH_2Cl_2	484(4.96)	97-0292-84
$C_{23}H_{19}BO_8$			
Boron, [1,7-bis(4-methoxyphenyl)-1,6-heptadiene-3,5-dionato-O^3,O^5]-(ethanedioato(2-)-),O']-	CH_2Cl_2	518(4.97)	97-0292-84
$C_{23}H_{19}ClO_2$			
Methanone, (4-chlorophenyl)(6,7,8,9-tetramethyl-2-dibenzofuranyl)-	EtOH	263(4.49),278(4.57)	39-0799-84B
Methanone, (4-chlorophenyl)(6,7,8,9-tetramethyl-3-dibenzofuranyl)-	EtOH	260(4.27),327(4.37)	39-0799-84B
$C_{23}H_{19}Cl_2N_3O_2$			
Pyrano[2,3-e]indazol-2(3H)-one, 3,3-dichloro-4-(diphenylamino)-4,5,6,7-tetrahydro-7-methyl-	EtOH	248(4.23),260s(4.20), 275s(4.11)	4-0361-84
Pyrano[2,3-e]indazol-2(3H)-one, 3,3-dichloro-4,5,6,7-tetrahydro-4-(methylphenylamino)-7-phenyl-	EtOH	245(4.40),282s(4.10)	4-0361-84

Compound	Solvent	$\lambda_{max}(\log \epsilon)$	Ref.
$C_{23}H_{19}Cl_2N_3O_3$			
4H-Pyrido[1,2-a]pyrimidine-3-carbox-amide, N-(3,5-dichloro-4-hydroxy-phenyl)-6,7,8,9-tetrahydro-6-methyl-4-oxo-9-(phenylmethylene)-	EtOH	300(4.10),310(4.10), 364(4.49)	118-0582-84
$C_{23}H_{19}FO_2$			
Methanone, (3-fluorophenyl)(6,7,8,9-tetramethyl-2-dibenzofuranyl)-	EtOH	256(4.37),278(4.48)	39-0799-84B
Methanone, (3-fluorophenyl)(6,7,8,9-tetramethyl-3-dibenzofuranyl)-	EtOH	328(4.36)	39-0799-84B
Methanone, (4-fluorophenyl)(6,7,8,9-tetramethyl-2-dibenzofuranyl)-	EtOH	254(4.35),275(4.45)	39-0799-84B
Methanone, (4-fluorophenyl)(6,7,8,9-tetramethyl-3-dibenzofuranyl)-	EtOH	324(4.49)	39-0799-84B
$C_{23}H_{19}N$			
1H-Indole, 5-(2-phenylethenyl)-1-(phenylmethyl)-, (E)-	EtOH	208(4.5),234(4.2), 240(4.1),266(4.4), 276(4.4),322(4.4)	103-0372-84
$C_{23}H_{19}NO$			
2(1H)-Pyridinone, 3,4-dihydro-3,4,6-triphenyl-	EtOH	216(4.20),275(3.65), 340(2.69)	4-1473-84
$C_{23}H_{19}NO_2$			
4,9-Epoxynaphth[2,3-d]isoxazole, 2,3,3a,4,9,9a-hexahydro-2,3-diphenyl-	MeOH	253(3.06)	73-1990-84
$C_{23}H_{19}NO_2S$			
Benzothiazole, 2-[2-(2,4-dimethoxyphenyl)-1-phenylethenyl]-	C_6H_{12}	360.2(4.45)	131-0417-84H
6H-1,3-Thiazine-6-thione, 2,3-dihydro-4-phenyl-2-[(phenylmethyl)thio]-5-(phenylsulfonyl)-	EtOH	260(3.94),353(4.21)	4-0953-84
$C_{23}H_{19}NO_3$			
Benzoic acid, 2-[(3-oxo-1,3-diphenyl-1-propenyl)amino]-, methyl ester	MeOH	258(4.21),332(4.00), 382(4.41)	4-0759-84
$C_{23}H_{19}NO_4$			
9H-Carbazole-3,4-dicarboxylic acid, 9-methyl-1-phenyl-, dimethyl ester	EtOH	234(3.98),283(4.25)	78-4837-84
$C_{23}H_{19}NO_5$			
2-Propanone, 1-(13,14-dihydro-13-methyl[1,3]benzodioxolo[5,6-c]-1,3-dioxolo[4,5-i]phenanthridin-14-yl)-	EtOH	240(4.43),288(4.47), 325(3.60)	100-0001-84
$C_{23}H_{19}N_3O_2$			
Deoxyviolacein, trimethyl-	$CHCl_3$	276(4.36),380(4.03), 562(4.32)	39-1331-84B
$C_{23}H_{19}N_3O_3$			
Ethanone, 1-[4-[4,5-dihydro-3-(4-nitrophenyl)-5-phenyl-1H-pyrazol-1-yl]-phenyl]-	MeOH	274(4.26),323(4.26), 460(3.97)	103-0787-84
Isoviolacein, N,N',O-trimethyl-	$CHCl_3$	283(4.27),390(3.86), 556(4.20)	39-1331-84B
Isoviolacein, N,N'',O-trimethyl-	$CHCl_3$	283(4.34),370(3.86), 553(4.30)	39-1331-84B

Compound	Solvent	$\lambda_{max}(\log \epsilon)$	Ref.
Methanone, [4,5-dihydro-3-methyl-1-(4-nitrophenyl)-4-phenyl-1H-pyrazol-5-yl]phenyl-, trans	EtOH	228(4.93),334(5.00)	4-1013-84
$C_{23}H_{19}N_5O$			
1H-Pyrazole, 3-methyl-4-[(2-methylphen-yl)azo]-5-phenyl-1-(2-pyridinylcarbo-nyl)-	n.s.g.	370(4.0)	48-1021-84
1H-Pyrazole, 3-methyl-4-[(3-methylphen-yl)azo]-5-phenyl-1-(2-pyridinylcarbo-nyl)-	n.s.g.	370(4.0)	48-1021-84
1H-Pyrazole, 3-methyl-4-[(4-methylphen-yl)azo]-5-phenyl-1-(2-pyridinylcarbo-nyl)-	n.s.g.	370(4.0)	48-1021-84
$C_{23}H_{19}N_5O_2$			
1H-Pyrazole, 4-[(2-methoxyphenyl)azo]-3-methyl-5-phenyl-1-(2-pyridinyl-carbonyl)-	n.s.g.	361(4.22)	48-1021-84
1H-Pyrazole, 4-[(3-methoxyphenyl)azo]-3-methyl-5-phenyl-1-(2-pyridinyl-carbonyl)-	n.s.g.	363(4.22)	48-1021-84
1H-Pyrazole, 4-[(4-methoxyphenyl)azo]-3-methyl-5-phenyl-1-(2-pyridinyl-carbonyl)-	n.s.g.	361(4.24)	48-1021-84
$C_{23}H_{20}BrNO_2$			
Benzamide, 4-bromo-N-(1-methyl-2-oxo-2-phenylethyl)-N-(phenylmethyl)-	ether	231(4.46),328(2.14)	48-0177-84
$C_{23}H_{20}ClFe_2$			
2-Propenylium, 3-chloro-1,3-diferro-cenyl-, perchlorate	CH_2Cl_2	364(4.98),395(5.01), 776(4.95)	65-1439-84
$C_{23}H_{20}ClNO_2$			
Benzamide, 4-chloro-N-(1-methyl-2-oxo-2-phenylethyl)-N-(phenylmethyl)-	ether	225(4.43),328(2.12)	48-0177-84
$C_{23}H_{20}Cl_2N_6O_2$			
3-Cyclohexene-1,1-dicarboxylic acid, 4-amino-3-cyano-, bis[[(2-chloro-phenyl)methylene]hydrazide]	n.s.g.	263(4.19)	42-0146-84
3-Cyclohexene-1,1-dicarboxylic acid, 4-amino-3-cyano-, bis[[(4-chloro-phenyl)methylene]hydrazide]	n.s.g.	260(4.18)	42-0146-84
$C_{23}H_{20}FN_3O_2$			
4H-Pyrido[1,2-a]pyrimidine-3-carbox-amide, N-(2-fluorophenyl)-6,7,8,9-tetrahydro-6-methyl-4-oxo-9-(phen-ylmethylene)-	EtOH	226(4.39),362(4.44)	118-0582-84
$C_{23}H_{20}F_3N_5O_2S$			
Pyridinium, 4-(dimethylamino)-1-[2-[4-hydroxy-3-[(2-thiazolylamino)carbo-nyl]-8-(trifluoromethyl)-2-quinolin-yl]ethyl]-, hydroxide, inner salt	EtOH and EtOH-NaOH	290(4.65),326(4.19), 332(4.19),349(4.05)	4-1345-84
	EtOH-HCl	295(4.61),327(4.19)	4-1345-84
$C_{23}H_{20}N_2$			
1H-Quindoline, 2,10-dihydro-3,11-di-methyl-1-phenyl-	EtOH	217(4.46),240(4.50), 271(4.37),293(4.47),	103-1374-84

Compound	Solvent	$\lambda_{max}(\log \epsilon)$	Ref.
(cont.)		349(4.08),364(4.08)	103-1374-84
$C_{23}H_{20}N_2O$			
Ethanone, 1-[4-[4,5-dihydro-3,5-diphenyl-1H-pyrazol-1-yl)phenyl]-	MeOH	245(4.15),309(3.83), 379(4.37)	103-0787-84
$C_{23}H_{20}N_2OS$			
Benzenamine, N-[3-(4-methoxyphenyl)-4-phenyl-2(3H)-thiazolylidene]-4-methyl-	EtOH	290(4.26)	56-0447-84
$C_{23}H_{20}N_2O_3S$			
Benzo[b]thiophen-3(2H)-one, 2-[bis-[(phenylmethyl)amino]methylene]-, 1,1-dioxide	EtOH	241(4.11),269(4.17), 235(3.92)[sic]	95-0134-84
$C_{23}H_{20}N_2O_4$			
1H-Pyrrole-2-carboxylic acid, 4-cyano-3-methyl-5-(2-oxo-2-phenylethoxy)-1-phenyl-	EtOH	220(4.39),241s(4.27), 276(4.20)	118-0062-84
$C_{23}H_{20}N_4O_3$			
2,4(1H,3H)-Pteridinedione, 7-(1-oxo-propyl)-1,3-bis(phenylmethyl)-	MeOH	250(4.11),348(3.90)	5-1798-84
$C_{23}H_{20}N_4O_4$			
3-Quinolinecarboxylic acid, 2-amino-4-[(4,5-dihydro-3-methyl-5-oxo-1-phenyl-1H-pyrazol-4-yl)carbonyl]-, ethyl ester	BuOH	278(4.72),302(4.49), 320(4.81)	4-1233-84
$C_{23}H_{20}N_4O_6$			
2,4(1H,3H)-Pteridinedione, 3-β-D-ara-binofuranosyl-6,7-diphenyl-	MeOH	220s(4.50),272(4.21), 363(4.20)	136-0179-84F
2,4(1H,3H)-Pteridinedione, 6,7-diphenyl-3-β-D-ribofuranosyl-	pH 1.0	220s(4.43),272(4.19), 363(4.19)	136-0179-84F
	pH 11.0	220s(4.4),240s(4.28), 292(4.39),388(4.07)	136-0179-84F
	MeOH	222(4.41),272(4.17), 364(4.18)	136-0179-84F
$C_{23}H_{20}N_5S$			
Quinolinium, 2-(5,6-dihydro-3-methyl-1-phenyl-6-thioxo-1H-pyrazolo[3,4-d]pyrimidin-4-yl)-1-ethyl-, iodide	EtOH	380(4.08)	48-0811-84
$C_{23}H_{20}N_8O_6$			
3-Cyclohexene-1,1-dicarboxylic acid, 4-amino-3-cyano-, bis[[(2-nitro-phenyl)methylene]hydrazide	n.s.g.	260(4.12)	42-0146-84
3-Cyclohexene-1,1-dicarboxylic acid, 4-amino-3-cyano-, bis[[(3-nitro-phenyl)methylene]hydrazide]	n.s.g.	264(4.11)	42-0146-84
3-Cyclohexene-1,1-dicarboxylic acid, 4-amino-3-cyano-, bis[[(4-nitro-phenyl)methylene]hydrazide]	n.s.g.	265(4.23)	42-0146-84
$C_{23}H_{20}O_2$			
Methanone, phenyl(6,7,8,9-tetramethyl-2-dibenzofuranyl)-	EtOH	254(4.14),275(4.54)	39-0799-84B

Compound	Solvent	$\lambda_{max}(\log \epsilon)$	Ref.
Methanone, phenyl(6,7,8,9-tetramethyl-3-dibenzofuranyl)-	EtOH	322(4.42)	39-0799-84B
$C_{23}H_{20}O_2S$			
24-Thiatetracyclo[15.2.2.2^{8,11}.1^{3,6}]-tetracosa-3,5,8,10,17,19,20,22-octaene-2,7-dione	C_6H_{12}	301(4.2)	44-1177-84
$C_{23}H_{20}O_3$			
9(10H)-Anthracenone, 10-hydroxy-10-(2-methoxyphenyl)-1,4-dimethyl-	ether	274(4.36),305s(3.67)	22-0195-84
9H-Xanthene, 9-[ethoxy(4-methoxyphenyl)methylene]-	EtOH	233(4.62),289(4.00), 337(4.13)	44-0080-84
$C_{23}H_{20}O_4$			
1H-2-Benzopyran-1-one, 3,4-dihydro-3-[4-methoxy-3-(phenylmethoxy)phenyl]-	EtOH	234(4.30),282(3.66)	44-0742-84
$C_{23}H_{20}O_5$			
Benzeneacetic acid, α-oxo-3,4-bis(phenylmethoxy)-, methyl ester	EtOH	233(4.16),284(4.03), 310(3.98)	39-1547-84C
1-Phenanthrenecarboxylic acid, 2-(3,4-dimethoxyphenyl)-1,2,3,4-tetrahydro-4-oxo-	MeCN	228(5.69),274(5.54)	2-0603-84
$C_{23}H_{20}O_6$			
2H,6H,10H-Dipyrano[3,2-b:2',3'-i]xanthen-6-one, 5,12-dihydroxy-2,2,N,N-tetramethyl-	MeOH	287(4.71),296(4.72), 346(4.10),395s(3.72)	102-1757-84
	MeOH-NaOMe	300(--)	102-1757-84
1-Naphthacenecarboxylic acid, 2-ethyl-1,4,6,11-tetrahydro-5-hydroxy-7-methoxy-6,11-dioxo-, methyl ester	MeOH	224(4.57),253s(4.34), 262(4.40),415(4.02)	78-4701-84
Pyrano[2',3':6,7]furano[2",3":3,4]-xanthone, 4",5"-dihydro-1,5-dihydroxy-4",4",5",6,6'-pentamethyl-	MeOH	286(4.50),326(4.35), 363s(3.92)	100-0620-84
$C_{23}H_{20}O_7$			
2H,8H-Benzo[1,2-b:3m4-b']dipyran-2-one, 9-acetoxy-10-(benzoyloxy)-9,10-dihydro-8,8-dimethyl-	MeOH	221(4.34),254s(3.66), 298s(3.95),332(4.16)	33-1729-84
2,5-Furandione, 3-[(4-acetoxy-3-methoxyphenyl)methylene]-4-[(3-methoxy-4-methylphenyl)methylene]-, (E,E)-	toluene	385(4.00)	48-0233-84
(E,Z)-	toluene	409(4.20)	48-0233-84
1-Naphthacenecarboxylic acid, 2-ethyl-3,4-epoxy-1,2,3,4,6,11-hexahydro-5-hydroxy-7-methoxy-6,11-dioxo-, methyl ester	MeOH	228(4.59),260(4.42), 285(3.98),416(4.04)	78-4701-84
$C_{23}H_{20}O_7S_2$			
4,6-Benzofurandiol, 2-methyl-, bis(4-methylbenzenesulfonate)	MeOH	254(4.14),272(3.75), 285s(3.52)	44-4399-84
$C_{23}H_{20}O_9$			
Pentanedioic acid, 2-[(9,10-dihydro-1-hydroxy-8-methoxy-9,10-dioxo-2-anthracenyl)methyl]-, dimethyl ester	n.s.g.	226(4.62),255(4.41), 277(4.08),396(3.98), 416(4.03),436(3.93)	5-0306-84
2-Propen-1-one, 3-(2,4-diacetoxyphenyl)-1-(3,4-diacetoxyphenyl)-	CH_2Cl_2	233(4.00),314(4.34)	5-1024-84

Compound	Solvent	$\lambda_{max}(\log \epsilon)$	Ref.
$C_{23}H_{20}O_{10}$			
9,10-Anthracenedione, 1,2,8-triacetoxy-6,7-dimethoxy-3-methyl-	dioxan	243(4.14),250(4.17), 276(4.63),340(3.71)	94-0860-84
$C_{23}H_{20}O_{14}$			
Flavone, 3',4',5-trihydroxy-3-methoxy-6,7-(methylenedioxy)-, 4'-β-D-glucuronoside	MeOH	250(4.21),277(4.29), 336(4.41)	102-2043-84
	MeOH-NaOAc	246(4.20),274(4.29), 337(4.35)	102-2043-84
	+ H_3BO_3	250(4.20),276(4.28), 338(4.40)	102-2043-84
	MeOH-NaOMe	270(4.35),327(4.19), 370s(4.09)	102-2043-84
	MeOH-AlCl$_3$	240s(4.23),263(4.17), 291(4.32),367(4.43)	102-2043-84
	+ HCl	240s(4.21),265s(4.17), 282(4.25),355(4.34)	102-2043-84
$C_{23}H_{21}FN_4$			
5H-Pyrimido[4,5-b]indol-4-amine, N-(4-fluorophenyl)-6,7,8,9-tetrahydro-2-methyl-9-phenyl-	EtOH	203(4.49),222s(4.32), 241s(4.26),303(4.30)	11-0073-84B
$C_{23}H_{21}F_3O_6$			
Benzeneacetic acid, α-methoxy-α-(trifluoromethyl)-, 5-hydroxy-4-oxo-5-[(phenylmethoxy)methyl]-2-cyclopenten-1-yl ester	EtOH	214(4.03)	39-2089-84C
$C_{23}H_{21}N$			
1H-Indole, 2,3-dihydro-5-(2-phenylethenyl)-1-(phenylmethyl)-, (E)-	EtOH	210(4.5),244(4.1), 260(3.9),357s(4.4)	103-0372-84
Spiro[acridine-9(10H),1'-[1H]indene], 2',3'-dihydro-3',3'-dimethyl-	EtOH	294(4.19)	24-2703-84
$C_{23}H_{21}NO$			
[1,1'-Biphenyl]-2-carboxamide, N-(phenylmethyl)-N-1-propenyl-, (E)-	MeOH	200(4.57),244(4.08), 264(4.20),306(4.10), 316(3.95)	78-1835-84
$C_{23}H_{21}NO_2$			
Benzamide, N-(3-oxo-3-phenylpropyl)-N-(phenylmethyl)-	ether	235(4.33),328(2.20)	48-0177-84
$C_{23}H_{21}NO_4$			
6H-Anthra[9,1-bc]furan-2-carboxylic acid, 5-(cyclohexylamino)-6-oxo-, methyl ester	EtOH	513(4.28),556(4.41)	104-0745-84
1H-Carbazole-3,4-dicarboxylic acid, 2,9-dihydro-9-methyl-1-phenyl-, dimethyl ester	EtOH	231(4.41),263(4.09), 275(4.06)	78-4837-84
$C_{23}H_{21}NO_5$			
Chelerythrinyl-8-acetaldehyde, dihydro-	EtOH	230(4.54),284(4.63), 320s(1.16)	100-0001-84
Chelerythrinyl-8-acetone, O-demethyl-dihydro-	EtOH	229(4.55),284(4.62), 320(4.17)	100-0001-84
	EtOH-NaOH	234(4.48),297(4.52), 338(4.33)	100-0001-84

Compound	Solvent	$\lambda_{max}(\log \epsilon)$	Ref.
$C_{23}H_{21}NO_6$			
9,10-Anthracenedione, 1,4-diacetoxy-2-piperidino-	EtOH at 77°K	444(3.83)	104-1780-84
2-Butenedioic acid, 3-[2-(3-cyano-3-phenyl-2-oxiranyl)phenoxy]propyl methyl ester, trans	EtOH	275s(3.43),281(3.62)	24-2157-84
$C_{23}H_{21}N_3O_2$			
4H-Pyrido[1,2-a]pyrimidine-3-carboxamide, 6,7,8,9-tetrahydro-6-methyl-4-oxo-N-phenyl-9-(phenylmethylene)-, (E)-	EtOH	226(4.22),362(4.38)	118-0582-84
$C_{23}H_{21}N_3O_3S$			
2-Propenenitrile, 3-[5-(4-phenyl-1-piperazinyl)-2-furanyl]-2-(phenyl-sulfonyl)-	MeOH	208(3.01),240(2.91), 343(2.82),455(2.90)	73-2141-84
$C_{23}H_{21}N_3O_5$			
Oxepino[2,3-d]pyrimidine-4-carboxylic acid, 8-hydroxy-2,3-dimethyl-6-oxo-7-phenyl-, diethylamine salt	CHCl₃	263(4.15),317(4.29)	24-0585-84
$C_{23}H_{21}N_3O_6S$			
4-Thia-2,6-diazabicyclo[3.2.0]hept-2-ene-6-acetic acid, α-(1-methyl-ethenyl)-7-oxo-3-(phenoxymethyl)-, (4-nitrophenyl)methyl ester, [1R-[1α,5α,6(R*)]]-	MeCN	267(4.06)	78-1907-84
$C_{23}H_{21}N_3O_{10}$			
4,5-Isoxazoledicarboxylic acid, 3-[6-ethyl-2-[[(4-nitrophenyl)methoxy]-carbonyl]-7-oxo-1-azabicyclo[3.2.0]-hept-2-en-3-yl]-, dimethyl ester, (5R-cis)-	THF	248(4.05),253(4.08), 260(4.11),268(4.13)	77-1513-84
$C_{23}H_{21}N_5O_{10}S_2$			
L-Tyrosine, N-acetyl-3,5-bis[(4-sulfo-phenyl)azo]-, barium salt (1:1)	pH 6	332(4.58),420s(4.09)	65-0589-84
	pH 11	318(4.40)	65-0589-84
$C_{23}H_{22}$			
Benzene, 2-(1,1-diphenylethenyl)-1,3,5-trimethyl-	THF	267(3.06),273s(3.02), 290s(2.70)	78-0765-84
$C_{23}H_{22}BrNO_6$			
14-Epicorynoline bromoacetate	MeOH	206.5(4.78),238(3.74), 290(3.83)	83-0223-84
	MeOH-HCl	206.5(4.82),242(3.75), 292(3.89)	83-0223-84
$C_{23}H_{22}ClNO_7$			
Spiro[1,3-dioxolo[4,5-g]isoquinoline-5(6H),2'-[2H]indene]-6-carboxylic acid, 3'-chloro-1',3',7,8-tetrahydro-4',5'-dimethoxy-1'-oxo-, ethyl ester	MeOH	234(4.42),289(4.34)	94-2230-84
$C_{23}H_{22}Cl_2O_2$			
Spiro[5.5]undecane-1,9-dione, 7,11-bis(4-chlorophenyl)-	EtOH	223(4.36),227(4.37)	117-0115-84

Compound	Solvent	$\lambda_{max}(\log \epsilon)$	Ref.
$C_{23}H_{22}N_2$			
Benzaldehyde, 2,4,6-trimethyl-, (diphenylmethylene)hydrazone	C_6H_{12}	318(4.37)	97-0021-84
$C_{23}H_{22}N_2O_3$			
1H-Indolo[3,2,1-de][1,5]naphthyridine-2-carboxylic acid, 2,3,3a,4,5,6-hexahydro-6-oxo-3-(phenylmethyl)-, methyl ester, cis	MeOH	240.5(4.31),265(4.04), 271s(4.01),293(3.65), 301(3.63)	94-1313-84
trans	MeOH	240.5(4.30),265(4.02), 271s(3.99),293(3.62), 301(3.61)	94-1313-84
$C_{23}H_{22}N_2O_4$			
Benzamide, N-[1-[(diethylamino)carbonyl]-2-(4-oxo-4H-1-benzopyran-3-yl)ethenyl]-	EtOH	227(3.76),283(3.69)	2-1048-84
Oxazolidine, 2-methoxy-5-(3-nitrophenyl)-2-phenyl-3-(phenylmethyl)-	MeOH	228(4.21),263(4.03)	44-3314-84
$C_{23}H_{22}N_2O_5$			
Acetamide, N-(4-acetoxy-9,10-dihydro-9,10-dioxo-3-piperidino-1-anthracenyl)-	EtOH at 77°K	463(3.88)	104-1780-84
$C_{23}H_{22}N_2O_6$			
1H-Benz[f]isoquino[8,1,2-hi]quinazoline-1,3(2H)-dione, 5,6-dihydro-8,9,11,12-tetramethoxy-2-methyl-	MeOH	235(4.24),262(4.59), 273(4.57),290s(4.39), 302s(4.10),330(3.72), 344(3.76),368(3.76), 382(3.80)	78-4003-84
$C_{23}H_{22}N_3O_4P$			
Piperidine, 1-[4-[bis(4-nitrophenyl)phosphino]phenyl]-	MeOH	281(4.50),430s(--)	139-0267-84C
$C_{23}H_{22}N_3O_4PS$			
Piperidine, 1-[4-[bis(4-nitrophenyl)phosphinothioyl]phenyl]-	MeOH	260s(4.49),287(4.61), 420s(--)	139-0267-84C
$C_{23}H_{22}N_3O_5P$			
Piperidine, 1-[4-[bis(4-nitrophenyl)phosphinyl]phenyl]-	MeOH	286.5(4.59),270s(4.54), 420s(--)	139-0267-84C
$C_{23}H_{22}N_4$			
4,4'-Bipyridinium, 1,1"-(1,3-propanediyl)bis-, dibromide	H_2O	265(4.53)	64-0074-84B
4H-Pyrimido[4,5-b]indol-4-imine, 3,5,6,7,8,9-hexahydro-2-methyl-3,9-diphenyl-	EtOH	245(4.44),283s(3.97), 306(3.91)	11-0073-84A
$C_{23}H_{22}N_4O_3S$			
Thiazolo[2,3-b]purine-4,8(1H,7H)-dione, 5a-ethoxy-5,5a-dihydro-5-methyl-7-phenyl-2-(phenylmethyl)-	MeOH	256(4.21),282s(4.12)	2-0316-84
$C_{23}H_{22}N_4O_5$			
Carbamic acid, [[1,2-dihydro-5-[(methylamino)carbonyl]-2-oxo-1-phenyl-6-(phenylamino)-3-pyridinyl]carbonyl]methyl-, methyl ester	MeOH	210(4.47),297(4.05), 367(4.31)	94-1761-84

Compound	Solvent	$\lambda_{max}(\log \epsilon)$	Ref.
$C_{23}H_{22}N_6O_2$			
3-Cyclohexene-1,1-dicarboxylic acid, 4-amino-3-cyano-, bis[(phenylmethylene)hydrazide]	n.s.g.	270(4.10)	42-0146-84
7H-Pyrido[2,1-b]quinazoline-6,11-dione, 8,9-dihydro-, 6-[(2,3-dihydro-1,5-dimethyl-3-oxo-2-phenyl-1H-pyrazol-4-yl)hydrazone]	EtOH	227(4.38),245s(4.18), 298(4.00),400(4.10)	4-1301-84
$C_{23}H_{22}N_6O_4$			
3-Cyclohexene-1,1-dicarboxylic acid, 4-amino-3-cyano-, bis[[(2-hydroxyphenyl)methylene]hydrazide]	n.s.g.	270(4.18)	42-0146-84
3-Cyclohexene-1,1-dicarboxylic acid, 4-amino-3-cyano-, bis[[(4-hydroxyphenyl)methylene]hydrazide]	n.s.g.	270(4.15)	42-0146-84
$C_{23}H_{22}O$			
2,5-Cyclohexadien-1-one, 2-(1,1-dimethylethyl)-4-(diphenylmethylene)-	hexane?	210(4.25),261(4.07), 267(4.07),362(4.36)	73-1949-84
2,5-Cyclohexadien-1-one, 4-(diphenylmethylene)-2,6-diethyl-	hexane?	211(4.23),261(4.20), 268(4.19),363(4.49)	73-1949-84
$C_{23}H_{22}O_2$			
6,11[1',2']-Benzenoanthra[1,2-d]-1,3-dioxole, 3a,5a,6,11,11a,11b-hexahydro-2,2-dimethyl-	hexane	208(4.65),251(2.74), 259(2.90),265(3.13), 272(3.23)	35-7310-84
5,10[1',2']-Benzeno-4,11-ethenobenzo-[5,6]cycloocta[1,2-d]-1,3-dioxole, 3a,4,5,10,11,11a-hexahydro-2,2-dimethyl-	hexane	211(4.49),255(2.45), 264(2.88),271(3.12), 276(3.06),279.5(3.28)	35-7310-84
$C_{23}H_{22}O_4$			
1-Phenanthrenecarboxylic acid, 2-(3,4-dimethoxyphenyl)-1,2,3,4-tetrahydro-	MeCN	229(4.47),277(5.19)	2-0603-84
$C_{23}H_{22}O_6$			
2-Anthracenepropanoic acid, 9,10-dihydro-1-hydroxy-3-methyl-9,10-dioxo-α-(1-oxopropyl)-, ethyl ester	MeOH	207(4.26),245(4.45), 263(4.49),408(3.83)	78-3677-84
Benzeneacetic acid, α-(3-hydroxy-5-oxo-4-phenyl-2(5H)-furanylidene)-2-methoxy-, 1,1-dimethylethyl ester, (E)-	EtOH EtOH-NaOH	274(4.30),370(4.19) 273(4.42),368(3.97)	39-1547-84C 39-1547-84C
(Z)-	EtOH EtOH-NaOH	271(4.30),344(4.10) 273(4.42),379(3.91)	39-1547-84C 39-1547-84C
6H-Benzofuro[3,2-c][1]benzopyran-6-one, 3,8,9-trimethoxy-4-(3-methyl-2-butenyl)-	MeOH	284(3.68),310(3.81), 332(3.75)	2-1028-84
1H,7H,11H-Dipyrano[3,2-b:3',2'-h]xanthen-7-one, 2,3-dihydro-6,13-dihydroxy-3,3,11,11-tetramethyl-	MeOH	274(4.70),330(4.18), 364s(3.74)	100-0620-84
	MeOH-NaOMe	280(--),332(--), 380s(--)	100-0620-84
2H,6H,10H-Dipyrano[3,2-b:2',3'-i]xanthen-6-one, 3,4-dihydro-5,12-dihydroxy-2,2,10,10-tetramethyl-	MeOH	276(4.70),332(4.18), 372s(3.78)	100-0620-84
	MeOH-NaOMe	282(--),329(--), 380s(--)	100-0620-84
2H,10H,14H-Dipyrano[3,2-b:2',3'-i]xanthen-14-one, 3,4-dihydro-5,8-dihydroxy-2,2,10,10-tetramethyl-	MeOH-AlCl₃	275(4.70),306(4.18), 356s(3.87)	100-0620-84
2H,6H-Pyrano[3,2-b]xanthen-6-one, 10-(1,1-dimethyl-2-propenyl)-	MeOH	277(4.62),327(4.11), 364s(3.88)	100-0620-84

Compound	Solvent	$\lambda_{max}(\log \epsilon)$	Ref.
7,9,12-trihydroxy-2,2-dimethyl- (cont.)			100-0620-84
$C_{23}H_{22}O_7$			
2-Anthracenepentanoic acid, 9,10-dihydro-1,8-dimethoxy-β,9,10-trioxo-, ethyl ester	n.s.g.	220(4.48),256(4.42), 371(3.80)	5-0306-84
1-Naphthacenecarboxylic acid, 2-ethyl-1,2,3,4,6,11-hexahydro-2,5-dihydroxy-7-methoxy-6,11-dioxo-, methyl ester	MeOH	226(4.49),260(4.35), 284(4.00),347s(3.68), 418(3.91)	78-4701-84
$C_{23}H_{22}O_8$			
2-Anthracenecarboxylic acid, 9,10-dihydro-4-hydroxy-5,8-dimethoxy-9,10-dioxo-3-(3-oxopentyl)-, methyl ester	MeOH	233(4.47),253(4.30), 440(3.81)	78-3677-84
Propanedioic acid, [(9,10-dihydro-1-hydroxy-8-methoxy-9,10-dioxo-2-anthracenyl)methyl]-, diethyl ester	n.s.g.	225(4.81),256(4.56), 282(4.28),330(3.25), 398(4.24),415(4.25), 429(4.21)	5-0306-84
$C_{23}H_{22}O_{11}$			
2H-Furo[2,3-c]pyran-2-one, 5,6-diacetoxy-7-(acetoxymethyl)-3-[(4-acetoxyphenyl)methylene]-3,5,7,7a-tetrahydro-	dioxan	229(4.23),329(4.26)	94-1815-84
$C_{23}H_{23}ClN_2O_2$			
3-Pyridinecarboxamide, 4-[1-(4-chlorophenyl)-3-oxobutyl]-1,4-dihydro-1-(phenylmethyl)-	MeCN	338(3.72)	44-0026-84
$C_{23}H_{23}NO_3$			
3H-Indol-3-one, 1-acetyl-1,2-dihydro-2-[(2-oxocyclohexyl)phenylmethyl]-	EtOH	206(4.34),241(4.54), 266(4.18),336(3.65)	103-1374-84
$C_{23}H_{23}NO_3S$			
Benzenesulfonamide, 4-methyl-N-(1-methyl-2-oxo-2-phenylethyl)-N-(phenylmethyl)-	ether	238(4.26),326(2.15)	48-0177-84
Benzenesulfonamide, 4-methyl-N-[2-(4-methylphenyl)-2-oxoethyl]-N-(phenylmethyl)-	ether	248.3(4.39),320.7(1.97)	48-0177-84
$C_{23}H_{23}NO_4S$			
Benzenesulfonamide, N-[2-(4-methoxyphenyl)-2-oxoethyl]-4-methyl-N-(phenylmethyl)-	ether	271.3(4.33),324.0(2.33)	48-0177-84
$C_{23}H_{23}NO_5$			
Chelerythrine, 8-ethoxydihydro-	EtOH	228(4.53),284(4.67), 320s(4.18)	100-0001-84
1H-Pyrrole-2,3-dicarboxylic acid, 4-[(4-methoxyphenyl)methyl]-1-methyl-5-phenyl-, dimethyl ester	EtOH	223(4.55),269(4.08)	150-1531-84M
Spiro[1,3-dioxolo[4,5-g]isoquinoline-5(6H),2'-[2H]inden]-1'(3'H)-one, 3'-ethylidene-7,8-dihydro-4',5'-dimethoxy-6-methyl-	MeOH	249s(4.52),257(4.60), 294(4.28)	94-2230-84
Tylohirsutinidine	MeOH	258(4.7),287(4.4), 301(3.7),338(1.8), 354(1.6)	102-1765-84
	MeOH-NaOH	238(4.3),254(4.6),	102-1765-84

Compound	Solvent	$\lambda_{max}(\log \epsilon)$	Ref.
Tylohirsutinidine (cont.)		284(4.1),298(3.5), 336(1.7),354(1.5)	102-1765-84
$C_{23}H_{23}NO_6$			
Corynolamine, (±)-	MeOH	237(3.25),288(3.24)	100-0001-84
Corynoline acetate, (±)-	MeOH	204(4.89),235s(3.95), 289(3.90)	83-0223-84
	MeOH	236s(3.95),290(3.86)	100-0001-84
6a,7-Dehydroboldine, diacetyl-	EtOH	246(4.79),260s(4.66), 290s(4.02),333(4.11)	100-0525-84
14-Epicorynoline, acetate, (+)-	MeOH	207(4.80),238(3.81), 290(3.92)	83-0223-84
	MeOH-HCl	208(4.86),244(3.83), 292(4.00)	83-0223-84
Isocorynoline, acetyl-, (+)-	MeOH	237(3.86),289(3.85)	100-0001-84
4,5-Isoxazoledicarboxylic acid, 2,3-di- hydro-3-[2-(4-methoxyphenyl)ethenyl]- 2-methyl-3-phenyl-, dimethyl ester, (E)-	EtOH	266(4.23),306(4.23)	150-1531-84M
4,5-Isoxazoledicarboxylic acid, 2,3-di- hydro-3-(4-methoxyphenyl)-2-methyl- 3-(2-phenylethenyl)-, dimethyl ester, (E)-	EtOH	250(4.14),275(3.94), 282(3.85),292(3.72)	150-1531-84
Methanaminium, N-[1-(4-methoxyphenyl)- 3-phenyl-2-propenylidene]-, 3-meth- oxy-1-(methoxycarbonyl)-2,3-dioxo- propylide, (E)-	EtOH	265(4.10),346(3.98), 412(3.91)	150-1531-84M
Methanaminium, N-[3-(4-methoxyphenyl)- 1-phenyl-2-propenylidene]-, 3-meth- oxy-1-(methoxycarbonyl)-2,3-dioxo- propylide, (E)-	EtOH	277(4.18),428(4.13)	150-1531-84M
$C_{23}H_{23}NO_7$			
8H-Dibenzo[a,g]quinolizin-8-one, 13- acetoxy-5,6-dihydro-2,3,10,11-tetra- methoxy-	MeOH	231.5(4.54),263(4.45), 333(4.39),345(4.38), 364s(4.14)	44-2642-84
$C_{23}H_{23}N_2O_2P$			
Piperidine, 1-[4-[(4-nitrophenyl)phen- ylphosphino]phenyl]-	MeOH	284.5(4.53),450s(--)	139-0267-84C
$C_{23}H_{23}N_2O_2PS$			
Piperidine, 1-[4-[(4-nitrophenyl)phen- ylphosphinothioyl]phenyl]-	MeOH	260s(4.24),290.5(4.45)	139-0267-84C
$C_{23}H_{23}N_2O_3P$			
Piperidine, 1-[4-[(4-nitrophenyl)phen- ylphosphinyl]phenyl]-	MeOH	288.5(4.50),420s(--)	139-0267-84C
$C_{23}H_{23}N_2S_2$			
Benzothiazolium, 3-ethyl-2-[5-(3-ethyl- 2(3H)-benzothiazolylidene)-1,3-penta- dienyl]-, iodide	$C_6H_{11}Me$ EtOH DMF	665(5.43) 658(5.40) 660(5.32)	99-0415-84 99-0415-84 99-0415-84
$C_{23}H_{23}N_3O$			
1,4-Benzenediamine, N,N-dimethyl-N'- [1-(4-methylphenyl)-2-(phenylimino)- ethylidene]-, N-oxide, (Z,Z)-	MeOH	253(4.29),430(3.79)	118-0128-84
$C_{23}H_{23}N_3O_2$			
1,4-Benzenediamine, N'-[1-(4-methoxy-	MeOH	253(4.40),294(4.34),	118-0128-84

Compound	Solvent	$\lambda_{max}(\log \epsilon)$	Ref.
phenyl)-2-(phenylimino)ethylidene]- N,N-dimethyl-, N-oxide, (Z,Z)- (cont.)		425(3.93)	118-0128-84
$C_{23}H_{23}N_3O_3S$ Benzenesulfonic acid, 4-[3-[4-(dimeth- ylamino)phenyl]-4,5-dihydro-5-phenyl- 1H-pyrazol-1-yl]-	MeOH	310(4.00),370(4.28)	103-0787-84
$C_{23}H_{23}N_3O_5$ Ellipticinium, 10-[(1-carboxyethyli- dene)amino]-9-hydroxy-2-methyl-, acetate	H_2O	243(4.26),300(4.49), 360(3.48),440(3.44)	87-1161-84
$C_{23}H_{23}N_3O_6S$ 4-Thia-2,6-diazabicyclo[3.2.0]hept- 2-ene-6-acetic acid, α-(1-methyl- ethyl)-7-oxo-3-(phenoxymethyl)-, (4-nitrophenyl)methyl ester, [1R-[1α,5α,6(R*)]]-	MeCN	268(4.05)	78-1907-84
$C_{23}H_{23}N_3O_7S$ 4-Thia-1-azabicyclo[3.2.0]heptane-2- carboxylic acid, 6-[(hydroxyphenyl- acetyl)amino]-3,3-dimethyl-7-oxo-, (4-nitrophenyl)methyl ester	EtOH	264(3.98)	39-2117-84C
$C_{23}H_{23}N_3O_8S$ 4-Thia-1-azabicyclo[3.2.0]heptane- 2-carboxylic acid, 6-[[[(2,3-dihy- droxybenzoyl)amino](4-hydroxyphenyl)- acetyl]amino]-3,3-dimethyl-7-oxo-, (4-nitrophenyl)methyl ester, 4- oxide, [2S-(2α,4β,5α,6β)]-	MeCN	268.5(4.08)	78-1907-84
$C_{23}H_{23}N_5O_4$ Meleagrin	EtOH	232(4.44),285s(3.92), 349(4.44)	94-0094-84
$C_{23}H_{23}N_5O_8$ Adenosine, N-benzoyl-, 2',3',5'-tri- acetate	MeOH	232(4.10),279(4.30)	150-3601-84M
$C_{23}H_{24}NOP$ Piperidine, 1-[4-(diphenylphosphinyl)- phenyl]-	MeOH	290(4.43)	139-0253-84C
$C_{23}H_{24}NO_5$ Sanguilutine	EtOH	229(4.47),279(4.58), 336(4.29),435s(3.31), 472(3.41)	100-0001-84
$C_{23}H_{24}NP$ Piperidine, 1-[4-(diphenylphosphino)- phenyl]-	MeOH	285(4.39)	139-0253-84C
$C_{23}H_{24}NPS$ Piperidine, 1-[4-(diphenylphosphino- thioyl)phenyl]-	MeOH	291(4.38)	139-0253-84C
$C_{23}H_{24}N_2O_2$ 3-Piperidinecarboxamide, 1,4-dihydro-	MeCN	345(3.70)	44-0026-84

Compound	Solvent	$\lambda_{max}(\log \epsilon)$	Ref.
4-(3-oxo-1-phenylbutyl)-1-(phenyl-methyl)- (cont.)			44-0026-84
$C_{23}H_{24}N_2O_3$ 1H-Pyrazole-1-acetic acid, 3,4,5-tri-methyl-α-(2-phenoxy-2-phenylethenyl)-, methyl ester, (Z)-	MeOH	248(4.20)	44-4647-84
$C_{23}H_{24}N_2O_4$ 9H-Pyrido[3,4-b]indole-1-propanoic acid, 1,2,3,4-tetrahydro-3-(meth-oxycarbonyl)-2-(phenylmethyl)-, (1S-trans)-	MeOH	225(4.60),275s(3.90), 282(3.91),290(3.82)	94-1313-84
$C_{23}H_{24}N_2O_6$ 2H-Pyrimido[6,1-a]isoquinoline-2,4(3H)-dione, 1-(3,4-dimethoxyphenyl)-6,7-dihydro-9,10-dimethoxy-3-methyl-	MeOH	225(4.46),244s(4.32), 278(4.02),332(4.21)	78-4003-84
$C_{23}H_{24}N_2O_7$ 2H-Pyrimido[6,1-a]isoquinoline-2,4(3H)-dione, 6,7-dihydro-9,10-dimethoxy-1-(3,4,5-trimethoxyphenyl)-	MeOH	220(4.53),243s(4.28), 275s(3.92),332(4.21)	78-4003-84
$C_{23}H_{24}N_3O_2$ Methylium, bis[4-(dimethylamino)phenyl]-(4-nitrophenyl)-, chloride	HOAc	576(4.86)	98-0596-84
$C_{23}H_{24}N_4$ 7H-Pyrrolo[2,3-d]pyrimidin-4-amine, N-(2,6-dimethylphenyl)-2,5,6-tri-methyl-7-phenyl-	EtOH	218s(4.42),236s(4.18), 289(4.41)	11-0073-84B
7H-Pyrrolo[2,3-d]pyrimidin-4-amine, N-(2-ethylphenyl)-2,5,6-trimethyl-7-phenyl-	EtOH	205(4.50),217s(4.36), 230s(4.23),301(4.17)	11-0073-84B
4H-Pyrrolo[2,3-d]pyrimidin-4-imine, 3-(2,6-dimethylphenyl)-3,7-dihydro-2,5,6-trimethyl-7-phenyl-	EtOH	244(4.41),272s(4.01), 283s(3.95),307(3.90)	11-0073-84A
4H-Pyrrolo[2,3-d]pyrimidin-4-imine, 3-(4-ethylphenyl)-3,7-dihydro-2,5,6-trimethyl-7-phenyl-	EtOH	245(4.44),284s(4.00), 308(3.86)	11-0073-84A
$C_{23}H_{24}N_4O_4$ 6H-Purin-6-one, 7,9-dihydro-9-[[2-(phenylmethoxy)-1-[(phenylmethoxy)-methyl]ethoxy]methyl]-	pH 1 pH 7 pH 13	247(3.96) 247(4.00) 253(4.03)	23-2702-84 23-2702-84 23-2702-84
$C_{23}H_{24}N_6O_9$ Carbamic acid, [9-(3,5-di-O-acetyl-2-deoxy-β-D-erythro-pentofuranosyl)-9H-purin-6-yl]-, 2-(4-nitrophenyl)-ethyl ester	MeOH	267(4.45)	78-0059-84
$C_{23}H_{24}O_2$ Benzoic acid, 4-[2-(1,2,3,4-tetrahydro-1,4-methanonaphthalen-6-yl)cyclo-propyl]-, ethyl ester, cis	EtOH	207(4.43),234(4.11), 251(4.08)	87-1516-84
trans	EtOH	207(4.45),263(4.26)	87-1516-84
Benzoic acid, 4-[2-(1,2,3,4-tetrahydro-1,4-methanonaphthalen-6-yl)-1-propen-yl]-, ethyl ester, (E)-	EtOH	233(4.17),308(4.43)	44-5265-84 +87-1516-84

Compound	Solvent	$\lambda_{max}(\log \epsilon)$	Ref.
Benzoic acid, 4-[2-(1,2,3,4-tetrahydro-1,4-methanonaphthalen-6-yl)-1-propenyl]-, ethyl ester, (Z)-	EtOH	239(4.22),300(4.23)	44-5265-84 +87-1516-84
Spiro[5.5]undecane-1,9-dione, 7,11-diphenyl-	EtOH	253(2.49),260(2.57), 265(2.53),300(2.04)	117-0115-84
14,20-Tricosadiene-3,5,10,12,22-pentayne-1,2-diol (siphonodiol)	n.s.g.	215(4.84),228(4.18), 241(4.05),254(4.14), 268(4.29),284(4.17)	138-0779-84
$C_{23}H_{24}O_4$			
2-Propenoic acid, 3-phenyl-, 1,5-pentanediyl ester	MeCN	274(4.66)	40-0022-84
$C_{23}H_{24}O_6$			
2-Propenoic acid, 3-(3-methoxyphenyl)-, 1,3-propanediyl ester	MeCN	276(4.60)	40-0022-84
2-Propenoic acid, 3-(4-methoxyphenyl)-, 1,3-propanediyl ester	MeCN	308(4.69)	40-0022-84
Vismione B, acetyl-	CHCl₃	255(4.15),301(4.28), 398(3.88)	102-1737-84
9H-Xanthen-9-one, 4-(1,1-dimethyl-2-propenyl)-1,3,5,6-tetrahydroxy-7-(3-methyl-2-butenyl)-	MeOH	253(4.55),288(4.02), 328(4.30)	100-0620-84
$C_{23}H_{24}O_{12}$			
2H-Furo[2,3-c]pyran-2-one, 4,5-diacetoxy-7-(acetoxymethyl)-3-[(4-acetoxyphenyl)methylene]hexahydro-3a-hydroxy- (plagiogyrin B tetraacetate)	dioxan	223(4.22),297(4.48)	94-1815-84
$C_{23}H_{24}S$			
24-Thiatetracyclo[15.2.2.2^{8,11}.1^{3,6}]-tetracosa-3,5,8,10,17,19,20,22-octaene	C₆H₁₂	<u>248(3.8),275s(2.9)</u>	44-1177-84
$C_{23}H_{25}As_2N_5O_{11}$			
L-Tyrosine, N-acetylbis(arsanilazo)-	pH 6	329(3.59),420s(4.10)	69-0589-84
	pH 11	317(4.44)	69-0589-84
$C_{23}H_{25}NO_5$			
Sanguilutine, dihydro-	MeOH	238(4.21),262(4.38), 275(4.42),325(4.22)	100-0001-84
$C_{23}H_{25}NO_7$			
6H-Dibenzo[a,g]quinolizine-8,13-dione, 13a-ethoxy-5,13a-dihydro-2,3,10,11-tetramethoxy-, (±)-	MeOH	254(4.65),311(3.86), 329(3.86)	44-2642-84
Narceine enol, lactone, (E)-	CHCl₃	272(4.11),348(4.05)	78-1971-84
(Z)-	CHCl₃	286(4.05),305s(3.97), 357(4.10)	78-1971-84
$C_{23}H_{25}N_2O$			
Quinolinium, 6-acetyl-4-[2-[4-(dimethylamino)phenyl]ethenyl]-1-ethyl-, iodide	MeOH	582(4.62)	103-0767-84
	C₆H₄Cl₂	635(4.70)	103-0767-84
	MeCN	574(--)	103-0767-84
$C_{23}H_{25}N_2O_2$			
Quinolinium, 4-[2-[4-(dimethylamino)phenyl]ethenyl]-1-ethyl-6-(methoxycarbonyl)-, iodide	MeOH	581(4.62)	103-0767-84
	C₆H₄Cl₂	632(4.73)	103-0767-84
	MeCN	574(--)	103-0767-84

Compound	Solvent	$\lambda_{max}(\log \epsilon)$	Ref.
$C_{23}H_{25}N_3O_3$ 9H-Pyrido[3,4-b]indole-3-carboxylic acid, 1-(3-amino-3-oxopropyl)-2,3,4,9-tetrahydro-2-(phenylmethyl)-, methyl ester, (1S-trans)-	MeOH	225(4.60),275s(3.91), 282(3.92),290(3.83)	94-1313-84
$C_{23}H_{25}N_3O_4$ 1,3,5-Triazabicyclo[3.2.0]hept-6-ene-2,4-dione, 6,7-bis(4-methoxy-2,6-di-methylphenyl)-3-methyl-	MeCN	270(4.06)	44-2917-84
$C_{23}H_{25}N_5O_4$ Meleagrin, 22,23-dihydro-	EtOH	232(4.06),348(4.08)	94-0094-84
$C_{23}H_{25}N_5O_6$ 5,8-Dideazaisofolic acid, 5,9-dimethyl-	pH 7.0	235(4.59)	87-0232-84
$C_{23}H_{25}N_5O_7$ Adenosine, N-benzoyl-7,8-dihydro-7-methyl-2',3'-O-(1-methylethylidene)-8-oxo-, 5'-acetate	MeOH	231(4.30),291(4.14)	150-3601-84M
$C_{23}H_{25}N_5O_8$ Adenosine, N-benzoyl-7,8-dihydro-, 2',3',5'-triacetate	MeOH	234(4.25),279(3.88), 330(3.96)	150-3601-84M
$C_{23}H_{26}N$ 1-Pyrenebutanaminium, N,N,N-trimethyl-, bromide	H_2O	342(4.53)	47-3001-84
$C_{23}H_{26}N_3$ 1H-Cyclopentapyrazinium, 6-(dimethyl-amino)-2,3-dihydro-1,4-dimethyl-5,7-diphenyl-, tetrafluoroborate	CH_2Cl_2	286(4.01),353(3.40), 438s(--),624(2.71)	89-0053B-84
$C_{23}H_{26}N_6O_5$ L-Glutamic acid, N-[4-[1-[(2,4-diamino-pyrido[3,2-d]pyrimidin-6-yl)methyl]-propyl]benzoyl]-	pH 13	238(4.58),343(3.72)	87-0376-84
$C_{23}H_{26}N_6O_6S_2$ Benzothiazolium, 2-[4-[bis(2-cyanoeth-ylamino)phenyl]azo]-6-(2-hydroxyeth-oxy)-3-methyl-, methyl sulfate	pH 1.13 pH 6.88 70% EtOH- pH 6.2	603(4.67) 603(4.67) 612(4.65)	48-0151-84 48-0151-84 48-0151-84
$C_{23}H_{26}O_2S$ Benzoic acid, 4-[2-(3,4-dihydro-4,4-dimethyl-2H-1-benzothiopyran-6-yl)-1-propenyl]-, ethyl ester, (E)-	EtOH	244(4.08),326(4.42)	87-1516-84
$C_{23}H_{26}O_3$ Benzoic acid, 4-[2-(3,4-dihydro-4,4-dimethyl-2H-1-benzopyran-6-yl)-1-propenyl]-, ethyl ester, (E)-	EtOH	236(4.15),316(4.38)	87-1516-84
$C_{23}H_{26}O_4$ 2,4-Cyclohexadien-1-one, 2-benzoyl-3,5-dihydroxy-6-(3-methyl-2-butenyl)-	MeOH MeOH-base	238s(4.04),287(3.80), 349(4.11) 243(4.44),347(4.23)	39-1413-84C 39-1413-84C

Compound	Solvent	$\lambda_{max}(\log \epsilon)$	Ref.
Methanone, [3,4-dihydro-5,7-dihydroxy-2,2-dimethyl-3-(3-methyl-2-butenyl)-2H-1-benzopyran-6-yl]phenyl-	EtOH EtOH-base	255(3.76),314(4.13) 319(4.08),405(3.43)	39-1413-84C 39-1413-84C
Methanone, [3,4-dihydro-5,7-dihydroxy-2,2-dimethyl-3-(3-methyl-2-butenyl)-2H-1-benzopyran-8-yl]phenyl-	EtOH EtOH-base	254(3.61),312(4.08) 247s(--),347(4.32)	39-1413-84C 39-1413-84C
$C_{23}H_{26}O_5$ 2,4-Cyclohexadien-1-one, 2-benzoyl-3,5,6-trihydroxy-4,6-bis(3-methyl-2-butenyl)-	MeOH-acid MeOH-base	251(3.87),307(3.65), 350s(3.44) 270(3.87),345(3.62)	39-1413-84C 39-1413-84C
$C_{23}H_{26}O_7$ 1H-3,10a-Ethanophenanthrene-1,2-dicarb-oxylic acid, 3-acetoxy-2,3,9,10-tetrahydro-7-methoxy-, dimethyl ester, (1α,2α,3α,10aβ)-(±)-	EtOH	266(4.35)	2-1168-84
$C_{23}H_{26}O_8$ 1H,3H-Furo[3,4-c]furan-3a(4H)-ol, 1-(3,4-dimethoxyphenyl)dihydro-4-(4-hydroxy-3-methoxyphenyl)-, 3a-acetate, [1S-(1α,3aα,4α,6aα)]-	EtOH EtOH-NaOH	233(4.16),280(3.67) 233(--),256(--), 285(--),292(--)	94-2730-84 94-2730-84
Picrasa-3,9(11)-diene-1,2,12,16-tetr-one, 20-acetoxy-11-methoxy-, (5β)-	EtOH	258(3.96)	18-2885-84
Picrasa-3,9(11)-diene-2,12,16-trione, 20-acetoxy-1,11-epoxy-1-methoxy-	EtOH	243(4.08),291(3.65)	18-2885-84 138-0555-84
$C_{23}H_{26}O_9$ Shinjulactone B, 11,20-di-O-acetyl-	EtOH	240(3.91)	18-2484-84
$C_{23}H_{26}O_{10}$ Sergeolide, 15-deacetyl-	EtOH	278(4.42)	100-0994-84
$C_{23}H_{27}ClN_2OS$ 1-Piperazineethanol, 4-[3-(2-chloro-dibenzo[b,e]thiepin-11(6H)-ylidene)-propyl]-, (E)-	MeOH	231(4.34),272(4.02), 311(3.41)	73-1816-84
$C_{23}H_{27}F_3O_5$ Benzeneacetic acid, α-methoxy-α-[(tri-fluoromethyl)-, 4-(3,5-dimethoxyphen-yl)-1-methylbutyl ester, (±)-	EtOH	204(4.61),274(3.25), 280(3.22)	100-0828-84
(R)-	EtOH	204(4.66),274(3.23), 280(3.21)	100-0828-84
$C_{23}H_{27}N_2O_5$ Dehydroreserpiline, perchlorate	EtOH	322(4.24),336(4.26), 400(3.51)	151-0355-84A
$C_{23}H_{27}N_3O_6$ Oxepino[2,3-d]pyrimidine-5,6-dicarbox-ylic acid, 3,4,7,8-tetrahydro-8-meth-yl-2-[(1-methylethyl)amino]-4-oxo-3-phenyl-, 6-ethyl 5-methyl ester	CHCl$_3$	283(3.86),333(4.30)	24-0585-84
$C_{23}H_{27}N_3O_7$ L-Alanine, N-Z-L-phenylalanylglycyl-, methyl ester	MeOH	260(2.90)	65-0836-84
Oxepino[2,3-d]pyrimidine-5,6-dicarbox-ylic acid, 3,4,7,8-tetrahydro-2-[(2-	CHCl$_3$	270(3.93),327(4.20)	24-0585-84

Compound	Solvent	$\lambda_{max}(\log \epsilon)$	Ref.
hydroxyethyl)amino]-7,8-dimethyl-4-oxo-3-phenyl-, 6-ethyl 5-methyl ester (cont.)			24-0585-84
$C_{23}H_{28}Br_2O_6$ 3-Heptanone, 1,7-bis(2-bromo-4,5-dimethoxyphenyl)-5-hydroxy-, (S)-	EtOH	230(4.32),286(3.89)	39-1635-84C
$C_{23}H_{28}ClN_3O_5S$ Benzamide, 5-chloro-N-[2-[4-[[[(cyclohexylamino)carbonyl]amino]sulfonyl]-phenyl]ethyl]-2-methoxy-	MeOH	227.0(4.47)	83-0906-84
$C_{23}H_{28}N_2O_3S$ Pyridinium, 4-[2-(1-butyl-1H-indol-5-yl)ethenyl]-1-(4-sulfobutyl)-, hydroxide, inner salt	EtOH	417(4.53)	44-2546-84
$C_{23}H_{28}N_2O_4$ 18,19-Secoyohimban-5-carboxylic acid, 16-(methoxymethylene)-17-oxo-, methyl ester, (3β,5β,15α,16E,20ξ)-	MeOH	225(4.35),258(4.08), 280(3.85),290(3.67)	78-4853-84
$C_{23}H_{28}N_2O_6$ Majdine	n.s.g.	225(4.55)	105-0509-84
$C_{23}H_{28}N_4O_6$ Oxepino[2,3-d]pyrimidine-5,6-dicarboxylic acid, 2-[(2-aminoethyl)amino]-3,4,7,8-tetrahydro-7,8-dimethyl-4-oxo-3-phenyl-, 6-ethyl 5-methyl ester	CHCl₃	275(3.92),328(4.30)	24-0585-84
Oxepino[2,3-d]pyrimidine-5,6-dicarboxylic acid, 2-[(2-aminoethyl)amino]-3,4,7,8-tetrahydro-8,8-dimethyl-4-oxo-3-phenyl-, 6-ethyl 5-methyl ester	CHCl₃	275(3.92),328(4.30)	24-0585-84
$C_{23}H_{28}O_2$ Benzoic acid, 4-[4-methyl-6-(2,6,6-trimethyl-1-cyclohexen-1-yl)-1,3,5-hexatrienyl]-, (E,E,E)-	EtOH	222(4.00),239(3.96), 349(4.72)	87-1516-84
Tetracyclo[6.3.2.0³,⁷.0⁸,¹¹]trideca-2,4,6,10,12-pentaene, 9-acetoxy-2,1-bis(1,1-dimethylethyl)-	EtOH	215(3.82),241(3.85), 305(3.71),407(2.70)	33-1386-84
$C_{23}H_{28}O_4$ 2,4-Cyclohexadien-1-one, 2-benzoyl-3,5-dihydroxy-6-(3-methyl-2-butenyl)-6-(3-methylbutyl)-	MeOH-acid	238(4.00),286(3.80), 351(3.98)	39-1413-84C
	MeOH-base	240(4.20),349(4.04)	39-1413-84C
2,4-Cyclohexadien-1-one, 6-benzoyl-3,5-dihydroxy-6-(3-methyl-2-butenyl)-2-(3-methylbutyl)-	MeOH-acid	254(3.75),356(4.07)	39-1413-84C
	MeOH-base	235s(3.80),350(4.01)	39-1413-84C
Methanone, phenyl[2,4,6-trihydroxy-3-(3-methyl-2-butenyl)-5-(3-methylbutyl)phenyl]-	MeOH	253(3.87),314(4.09)	39-1413-84C
	MeOH-base	255s(3.97),362(4.24)	39-1413-84C
$C_{23}H_{28}O_5$ 4-Hepten-3-one, 1,7-bis(3,4-dimethoxyphenyl)-, (E)-	EtOH	227(4.24),279(3.73)	39-1635-84C

Compound	Solvent	λ_{max}(log ϵ)	Ref.
$C_{23}H_{28}O_6$			
1-Butanone, 3-methyl-1-(1,3,7,9-tetra-methoxy-2,8-dimethyl-4-dibenzofuran-yl)-	EtOH	219(4.94),264(4.70), 295(3.30)	39-2573-84C
Rishirilide B	EtOH	218(4.35),264(4.40), 305(3.64),370(3.40)	158-1091-84
$C_{23}H_{28}O_7$			
1H-3,10a-Ethanophenanthrene-1,2-dicarb-oxylic acid, 3-acetoxy-2,3,4,4a,9,10-hexahydro-7-methoxy-, dimethyl ester, (1α,2α,3α,4aβ,10aβ)-(±)-	EtOH	228(4.03),278(3.33)	2-1168-84
Pinoresinol, 1-hydroxy-, 4'-O-ethyl 4"-O-methyl ether, (+)-	EtOH	232.5(4.31),320(3.81)	94-2730-84
$C_{23}H_{28}O_8$			
Divinorin A	MeOH	211(3.72)	44-4716-84
1H,3H-Furo[3,4-c]furan-3a(4H)-ol, 1-(3,4-dimethoxyphenyl)dihydro-4-(3,4,5-trimethoxyphenyl)- (fraxiresinol dimethyl ether)	EtOH	229.6(4.22),277.9(3.58)	94-4482-84
	EtOH-NaOH	229.6s(--),277.9(--)	94-4482-84
Santhemoidin B	MeOH	205(4.49)	102-2911-84
Sylvone	EtOH	207(4.48),219(4.42), 283(4.10)	78-5047-84
	EtOH-NaOH	223(4.34),282.5(4.05)	78-5047-84
$C_{23}H_{28}O_9$			
1,4-Phenanthrenedione, 8-(acetoxymeth-yl)-7-(formyloxy)-4b,5,6,7,8,8a,9,10-octahydro-3,9,10-trihydroxy-4b,8-di-methyl-2-(2-propenyl)- (allylroylea-none)	ether	270(3.87),400(2.80)	33-0201-84
Repandin E	MeOH	216(4.24)	102-0829-84
$C_{23}H_{29}NO_9$			
α-D-Glucopyranoside, methyl 2-deoxy-2-[(1-methyl-3-oxo-3-phenyl-1-prop-enyl)amino]-, 3,4,6-triacetate (Z)-	EtOH	240(3.90),337(4.30)	136-0101-84L
$C_{23}H_{29}NO_{10}$			
D-Gluconic acid, 2-deoxy-2-[(3-methoxy-3-oxo-2-propenyl)amino]-4,6-O-(phen-ylmethylene)-, ethyl ester, 3,5-di-acetate	MeOH	268(4.28)	136-0081-84B
$C_{23}H_{29}N_2O_{10}P$			
Uridine, 3'-deoxy-3'-[(diethoxyphos-phinyl)methyl]-, 2'-acetate 5'-benzoate	MeOH	230(4.17),260(3.99)	78-0079-84
$C_{23}H_{30}NO_5$			
5H-Oxazolo[2,3-a]isoquinolinium, 10b-[(3,4-dimethoxyphenyl)methyl]-2,3,6,10b-tetrahydro-8,9-dimethoxy-4-methyl-, iodide	MeOH	227s(4.35),280(3.72), 285s(3.69)	12-1659-84
$C_{23}H_{30}N_2O$			
1H-Imidazole, 1-[3,7-dimethyl-1-oxo-9-(2,6,6-trimethyl-1-cyclohexen-1-yl)-2,4,6,8-nonatetraenyl]-	MeOH	392(4.57)	107-0725-84
geometric isomer	MeOH	393(4.61)	107-0725-84

Compound	Solvent	$\lambda_{max}(\log \epsilon)$	Ref.
$C_{23}H_{30}N_2O_8$			
L-ribo-Hexar-1-amic acid, 4,5-dideoxy-N-[1-(3,4-dihydro-8-hydroxy-1-oxo-1H-2-benzopyran-3-yl)-3-methylbutyl]-4-[(1-oxopropyl)amino]-, γ-lactone, [S-(R*,R*)]-	MeOH	246(3.76),314(3.61)	78-2519-84
Propanedioic acid, 2,2'-[(4-methyl-1,2-phenylene)bis(iminomethylidyne)]bis-, tetraethyl ester	MeCN	220(4.37),290(4.58), 330(4.48)	62-0947-84A
$C_{23}H_{30}N_4O_6$			
1H-Pyrrol-1-yloxy, 3,3'-[(5-methyl-2,4-dioxo-1,3(2H,4H)-pyrimidinediyl)dicarbonyl]bis[2,5-dihydro-2,2,5,5-tetramethyl-	MeOH	209(4.29),232(4.32), 273s(3.90),392(1.97)	70-1668-84
$C_{23}H_{30}O_2$			
Benzoic acid, 3-[2-methyl-4-(2,6,6-trimethyl-1-cyclohexen-1-yl)-1,3-butadienyl]-, ethyl ester, (E,E)-	EtOH	229(4.15),250s(4.11), 287(4.32)	87-1516-84
$C_{23}H_{30}O_4$			
Benzeneacetic acid, α-ethyl-, 4-(3,5-dimethoxyphenyl)-1-methylbutyl ester	EtOH	204(4.80),272(3.34), 280(3.33)	100-0828-84
Trichodermadiene, 12,13-deoxy-	MeOH	265(3.42)	158-84-147
$C_{23}H_{30}O_5S$			
1-Octanesulfonic acid, 2-hydroxy-3-oxo-2,3-diphenylpropyl ester	MeOH	251(4.20),332(2.36)	126-1795-84
$C_{23}H_{30}O_6$			
3-Heptanone, 1,7-bis(3,4-dimethoxyphenyl)-5-hydroxy-	EtOH	229(4.13),280(3.70)	39-1635-84C
$C_{23}H_{31}NO$			
2H-Indol-2-one, 1,3-dihydro-1,3,3-tris(3-methyl-2-butenyl)-	MeOH	240(3.85),253(3.91), 280(3.32)	83-0639-84
$C_{23}H_{31}NO_5Si$			
Glycine, N-[2-(1,3-benzodioxol-5-yl)-3-oxo-1-cyclopenten-1-yl]-N-[2-[(trimethylsilyl)methyl]-2-propenyl]-, ethyl ester	MeCN	278(4.39)	44-0228-84
$C_{23}H_{31}NO_6$			
Myxopyronin A	MeOH	213(4.50),298(4.31)	158-84-165
$C_{23}H_{31}N_5O_{10}$			
Propanoic acid, 2,2-dimethyl-, [[[9-(2,3,5-tri-O-acetyl-β-D-arabinofuranosyl)-9H-purin-6-yl]amino]methoxy]methyl ester	EtOH	263(4.27)	44-1453-84
$C_{23}H_{32}NO_3Si$			
2H-[1,3]Benzodioxolo[5,6-d][1]benzazepinium, 3,4,6,7-tetrahydro-1-methoxy-3-[2-[(trimethylsilyl)methyl]-2-propenyl]-, perchlorate	MeCN	253(3.90),300(3.98), 340(3.76)	44-0228-84

Compound	Solvent	$\lambda_{max}(\log \epsilon)$	Ref.
$C_{23}H_{32}N_2O_8$ 2,3-Quinoxalinediacetic acid, α,α'- bis(ethoxycarbonyl)-1,2,3,4-tetra- hydro-6-methyl-, diethyl ester	MeCN	222(4.45),315(3.75)	62-0947-84A
$C_{23}H_{32}N_4O_3$ 9H-Cyclopenta[1,2-b:4,3-b']dipyridin- 9-one, 2,7-bis[2-(diethylamino)eth- oxy]-	hexane	252(4.63),323(4.30), 438(2.62)	142-0073-84
dihydrochloride	H_2O	256(4.44),308(4.02), 460(2.51)	142-0073-84
$C_{23}H_{32}N_4O_7$ 6,9,12-Trioxa-3,15,19-triazabicyclo- [15.3.1]heneicosa-17,20-diene-21- carbonitrile, 19-methyl-4,14-bis(1- methylethyl)-2,5,13,16-tetraoxo-, [4S-(4R*,14R*)]-	$CHCl_3$	246(4.38),365(4.20)	35-6029-85
$C_{23}H_{32}O$ 2-Propanone, 1-[2,2-dimethyl-3-(2,4- pentadecadiene-6,9-diynyl)cyclo- propyl]-, [1S-[1α,3α(2E,4E)]]-	EtOH	263(4.46),272(4.54), 289(4.44)	104-2076-84
$C_{23}H_{32}O_2$ 24-Norchola-1,4-diene-3,22-dione	n.s.g.	246(4.17)	73-2713-84
$C_{23}H_{32}O_3$ Benzo[b]cyclohepta[d]pyran-1-ol, 6,6a,7,8,9,11a-hexahydro-6,6-di- methyl-3-pentyl-, acetate, (6aR- trans)-	EtOH	274(3.13),280(3.15)	78-3839-84
Δ^2-THC, acetate, (1R,4R)-	EtOH	257(4.04),265s(3.98), 298(3.57),305(3.55)	39-2881-84C
(1S,4R)-	EtOH	256(4.09),265s(4.02), 300(3.61),307(3.58)	39-2881-84C
Δ^3-THC, acetate, (1R)-	EtOH	266(3.94),303(3.69)	39-2881-84C
$C_{23}H_{32}O_5$ 2,4-Pentadecadienoic acid, 15-(3-meth- oxy-2,5-dioxo-3,5-cyclohexadien-1- yl)-	MeOH	265(4.66),376(2.94)	73-1622-84
$C_{23}H_{32}O_6$ 1,3-Cyclohexadiene-1-carboxylic acid, 4,4'-[1,3-propanediylbis(oxy)]bis- [2-methyl-, diethyl ester	EtOH	210(4.07),318(4.25)	12-2037-84
$C_{23}H_{32}O_7$ 1,2,4-Butanetriol, 3-[(3,4-dimethoxy- phenyl)methyl]-2-[(4-ethoxy-3-meth- oxyphenyl)methyl]-, [S-(R*,R*)]-	EtOH	230(4.20),280.5(3.74)	94-2730-84
Trichodermadienediol A, 16-hydroxy-	MeOH	260(4.25)	158-84-46
Trichodermadienediol B, 16-hydroxy-	MeOH	260(4.27)	158-84-46
$C_{23}H_{32}O_8$ 1,2,4-Butanetriol, 3-[(3,4-dimethoxy- phenyl)methyl]-2-[(3,4,5-trimethoxy- phenyl)methyl]-	EtOH	224.5(4.16),278.5(3.52)	94-4482-84

Compound	Solvent	$\lambda_{max}(\log \epsilon)$	Ref.
$C_{23}H_{33}NO_2$			
Pregna-1,4-dien-18-al, 20-(dimethyl-amino)-3-oxo-, (20S)-	EtOH	244(4.11)	22-0071-84
$C_{23}H_{33}NO_{15}$			
D-glycero-L-gluco-Octonic acid, 2-acet-yl-2,3-dideoxy-3-(nitromethyl)-, ethyl ester, 4,5,6,7,8-penta-acetate, (2ξ)-	EtOH	252(3.35)	136-0063-84K +136-0075-84K
$C_{23}H_{34}N_2O_3$			
Acetamide, N-[(3β)-3-acetoxy-17-aza-D-homoandrosta-5,17-dien-17a-yl]-	n.s.g.	255-260(4.04)	70-0961-84
Acetamide, N-[(3β)-3-acetoxy-17a-aza-D-homoandrosta-5,17-dien-17-yl]-	n.s.g.	255-260(3.94)	70-0961-84
$C_{23}H_{34}N_2O_{14}$			
D-glycero-D-ido-Octonic acid, 2-(1-aminoethylidene)-2,3-dideoxy-3-(ni-tromethyl)-, ethyl ester, 4,5,6,7,8-pentaacetate	EtOH	280(4.43)	136-0063-84K
isomer	EtOH	280(4.25)	136-0063-84K
$C_{23}H_{34}N_4O$			
Urea, N'-[(8α)-6-butylergolin-8-yl]-N,N-diethyl-	MeOH	225(4.48),277s(3.80), 282(3.82),293(3.74)	73-2828-84
$C_{23}H_{34}N_4O_2$			
Ethanamine, 2,2'-[9H-cyclopenta[1,2-b:4,3-b']dipyridine-2,7-diylbis(oxy)]-bis[N,N-diethyl-	hexane	214(4.53),308(4.56), 338(4.38)	142-0073-84
dihydrochloride	H_2O	225(4.21),306(4.22)	142-0073-84
$C_{23}H_{34}O_2$			
24-Norchol-4-ene-3,22-dione	n.s.g.	243(4.21)	73-2713-84
Retinoic acid, 13,14-dihydro-13,14-methylene-, ethyl ester, cis	hexane	296(4.43)	44-1937-84
trans	hexane	298(4.43)	44-1937-84
$C_{23}H_{34}O_3$			
6H-Benzo[b]cyclohepta[d]furan-1-ol, 5a,7,8,9,10,10a-hexahydro-5a-(1-methylethyl)-3-pentyl-, acetate	EtOH	275s(3.25),281(3.24)	78-3839-84
Benzo[b]cyclohepta[d]pyran-1-ol, 6,6a,7,8,9,10,11,11a-octahydro-6,6-dimethyl-3-pentyl-, acetate, cis-(±)-	EtOH	275(3.30),281(3.32)	78-3839-84
trans-(±)-	EtOH	277s(3.32),285(3.34)	78-3839-84
$C_{23}H_{34}O_4$			
Pregna-5,17(20)-dien-21-oic acid, 3- hydroxy-20-methoxy-, methyl ester, (3β,17E)-	EtOH	237(4.02)	118-0132-84
$C_{23}H_{34}O_5$			
Androst-5-en-17-one, 3-[(acetoxymeth-oxy)methoxy]-, (3β)-	EtOH	200(3.57)	56-0297-84
$C_{23}H_{35}NO$			
Pregna-1,4-dien-3-one, 20-(dimethyl-amino)-, (20S)-	EtOH	242(4.29),322(3.88)	22-0071-84

Compound	Solvent	$\lambda_{max}(\log \epsilon)$	Ref.
$C_{23}H_{35}NO_2$ Pregna-1,4-dien-3-one, 20-(dimethyl- amino)-18-hydroxy-, (20S)-	EtOH	241(4.39),322(4.07)	22-0071-84
$C_{23}H_{35}NO_3$ Myxalamid D (shoulders at 340 and 370 nm)	EtOH	203(3.98),264(3.87), 355(4.31)	158-84-113
$C_{23}H_{36}O_2$ Cannabigerol, O,O-dimethyl-	EtOH	272(2.94),280s(2.86)	39-2881-84C
$C_{23}H_{36}O_5$ Cyclohexaneacetic acid, 2-[4-(2-hy- droxy-1-methylethyl)-2,5-dimethoxy- phenyl]-2,6,6-trimethyl-, methyl ester	EtOH	227(3.75),290(3.51)	23-2822-84
Pentadecanoic acid, 15-(3-methoxy-2,5- dioxo-3,5-cyclohexadien-1-yl)-, methyl ester	MeOH	269(3.44),366(2.86)	73-1622-84
$C_{23}H_{36}O_9$ β-D-Xypopyranoside, phenylmethyl 4-O- methyl-2,3-O-3,6,9,12-tetraoxa- tetradecane-1,14-diyl-	EtOH	258(2.62)	5-1036-84
$C_{23}H_{37}NO_2$ Cannabinol, dihydro-7β-(dimethylamino)-	EtOH	275(3.12),282(3.14)	87-1370-84
$C_{23}H_{45}N_3O_5Si_2$ Cytidine, 2'-deoxy-3',5'-bis-O-[(1,1- dimethylethyl)dimethylsilyl]-N-(2- hydroxyethyl)-	EtOH	273(3.98)	88-3195-84

Compound	Solvent	$\lambda_{max}(\log \epsilon)$	Ref.
$C_{24}H_{10}Cl_3N_3S_3$ 17H-Bis[1,4]benzothiazino[2,3-a:3',2'-c]phenothiazine, 3,8,14-trichloro-	DMF	594(4.15)	40-0522-84
$C_{24}H_{10}Cl_4I_4O_6$ Benzoic acid, 2-(6-acetoxy-2,4,5,7-tetraiodo-3-oxo-3H-xanthen-9-yl)-3,4,5,6-tetrachloro-, ethyl ester	CH_2Cl_2 MeOH	395(4.22),494(4.03) 400(--),494(--)	64-0474-84B 64-0474-84B
$C_{24}H_{13}N_3O_8$ 3-Furancarbonitrile, 4,5-bis(1,3-benzo-dioxol-5-yl)-2-[[(5-nitro-2-furanyl)-methylene]amino]-	MeOH CHCl$_3$	212(4.59),249(4.20), 311(4.45),450(4.14) 307(4.48),357(4.21), 500(4.28)	73-1788-84 73-1788-84
$C_{24}H_{13}N_3S_3$ 17H-Bis[1,4]benzothiazino[2,3-a:3',2'-c]phenothiazine	DMF	333(4.64),582(4.42)	40-0522-84
$C_{24}H_{14}N_2O_6$ 3-Furancarbonitrile, 4,5-bis(1,3-benzo-dioxol-5-yl)-2-[(2-furanylmethylene)-amino]-	MeOH CHCl$_3$	212(4.56),236(4.30), 294(4.31),318(4.38), 420(4.28) 302(4.28),324(4.35), 432(4.25)	73-1788-84 73-1788-84
$C_{24}H_{14}N_2O_9$ 1(2H)-Naphthalenone, 2,2-bis[(4-nitro-benzoyl)oxy]- 2(1H)-Naphthalenone, 1,1-bis[(4-nitro-benzoyl)oxy]-	MeCN MeCN	239(4.63),258(4.55), 332s(3.61) 241(4.56),260(4.58), 319(4.06)	39-2077-84C 39-2077-84C
$C_{24}H_{14}N_2S_2$ Benzothiazole, 2,2'-(1,4-naphthalene-diyl)bis-	DMF	354(4.35)	5-1129-84
$C_{24}H_{14}N_6$ 23,24,25,26-Tetraazapentacyclo[17.3.1-1^2,^{6}.1^8,^{12}.1^{13},^{17}]hexacosa-1(23),2,4-6,8,10,12(25),13,15,17,19,21-dodeca-ene-7,18-dicarbonitrile	1-$C_{10}H_7Cl$	357(4.56),375(4.57), 508(3.85),541(3.75), 592(3.48)	35-5760-84
$C_{24}H_{14}O_4$ 1H,3H-Benzo[1,2-c:4,5-c']difuran-1,7(5H)-dione, 3,5-bis(phenyl-methylene)- 1H,5H-Benzo[1,2-c:4,5-c']difuran-1,5-dione, 3,7-dihydro-3,7-bis(phen-ylmethylene)- s-Indacene-1,3,5,7(2H,6H)-tetrone, 2,6-diphenyl-	dioxan dioxan EtOH $C_2H_4Cl_2$	270(4.10),337(4.63), 384s(4.27) 250(4.41),351(4.54), 390s(4.35) 263(4.51),312(4.71), 337(4.42),395(4.43), 515s(3.20) 242(4.61),292(4.22), 356(3.76),455s(3.76)	131-0235-84H 131-0235-84H 131-0275-84H 131-0275-84H
$C_{24}H_{15}Br_2N_3O$ 2(10H)-Phenazinone, 10-(4-bromophenyl)-3-[(4-bromophenyl)amino]-	CHCl$_3$	281(4.68),470(4.52)	103-0263-84
$C_{24}H_{15}Cl_2N_3O$ 2(10H)-Phenazinone, 10-(4-chlorophen-	CHCl$_3$	280(4.62),471(4.44)	103-0263-84

Compound	Solvent	$\lambda_{max}(\log \epsilon)$	Ref.
yl)-3-[(4-chlorophenyl)amino]- (cont.)			103-0263-84
$C_{24}H_{15}NO_5$ 2,5-Furandione, 3-(diphenylmethylene)-dihydro-4-[(4-nitrophenyl)methylene]-, (Z)-	toluene	408(3.85)	48-0233-84
$C_{24}H_{15}N_3O$ 1,2,4-Benzenetricarbonitrile, 5-(9,10-dihydro-10-methoxy-9-phenanthrenyl)-	MeCN	253(4.40),257(4.40), 307(3.59)	40-0060-84
Benzoxazole, 2-[4-(1H-benzimidazol-2-yl)-1-naphthalenyl]-	DMF	363(4.50)	5-1129-84
$C_{24}H_{15}N_3S$ 3,5-Pyridinedicarbonitrile, 1,4-dihydro-2-(3-methylphenyl)-6-(2-naphthlenyl)-4-thioxo-	EtOH	269(4.65),310(4.40), 323(4.40),400(3.40)	4-1445-84
$C_{24}H_{15}N_5$ Ethenetricarbonitrile, [4-[(diphenylmethylene)hydrazino]phenyl]-	benzene	514(4.64)	32-0111-84
$C_{24}H_{16}$ 1H-Cyclopropa[b]naphthalene, 1-(diphenylmethylene)-	C_6H_{12}	230(4.61),251(4.34), 260s(4.32),270s(4.28), 291(4.14),412(4.48), 438(4.52)	35-6108-84
$C_{24}H_{16}BrN_3O$ [1,2,4]Triazolo[1,5-a]pyridinium, 1-(4-bromophenyl)-2,3-dihydro-2-oxo-5,7-diphenyl-, hydroxide, inner salt	EtOH	267(4.80)	39-1891-84C
$C_{24}H_{16}BrN_3S$ [1,2,4]Triazolo[1,5-a]pyridinium, 1-(4-bromophenyl)-2,3-dihydro-5,7-diphenyl-2-thioxo-, hydroxide, inner salt	EtOH	264(4.31),296(4.27)	39-1891-84C
$C_{24}H_{16}Br_2N_3O_4P$ Benzenamine, 3,5-dibromo-2,6-dinitro-N-(triphenylphosphoranylidene)-	MeOH	375(3.70)	65-0289-84
$C_{24}H_{16}Br_2N_5O_4P$ 1-Triazene, 1-(3,5-dibromo-2,6-dinitrophenyl)-3-(triphenylphosphoranylidene)-	MeOH	332(4.39)	65-0289-84
$C_{24}H_{15}ClN_3O$ [1,2,4]Triazolo[1,5-a]pyridinium, 1-(4-chlorophenyl)-2,3-dihydro-2-oxo-5,7-diphenyl-, hydroxide, inner salt	EtOH	265(4.84)	39-1891-84C
$C_{24}H_{16}ClN_3S$ [1,2,4]Triazolo[1,5-a]pyridinium, 1-(4-chlorophenyl)-2,3-dihydro-5,7-diphenyl-2-thioxo-, hydroxide, inner salt	EtOH	265(4.26),300(4.25)	39-1891-84C
$C_{24}H_{16}N_2$ 4,7-Phenanthroline, 1,3-diphenyl-	n.s.g.	225(4.62),255(4.70), 291(4.78),337(--), 354(3.40)	30-0094-84

Compound	Solvent	$\lambda_{max}(\log \epsilon)$	Ref.
$C_{24}H_{16}N_2O_5$			
2H-Indol-2-one, 1-acetyl-3-[5-(1-acet-yl-1H-indol-3-yl)-2-oxo-3(2H)-furan-ylidene]-1,3-dihydro-	CHCl$_3$	260(4.37),310(3.80), 510(4.51)	39-1331-84B
$C_{24}H_{16}N_4$			
1H-Benzimidazole, 2,2'-(1,4-naphtha-lenediyl)bis-	DMF	352(4.47)	5-1129-84
1H-Benzimidazole, 2,2'-(1,7-naphtha-lenediyl)bis-	DMF	303(4.55),337(4.35)	5-1129-84
1H-Benzimidazole, 2,2'-(2,6-naphtha-lenediyl)bis-	DMF	353(4.79),372(4.69)	5-1129-84
$C_{24}H_{16}N_4O$			
Spiro[anthracene-9(10H),3'-[3H]pyrazolo-[5,1-c][1,2,4]triazol]-10-one, 6'-methyl-7'-phenyl-	CH$_2$Cl$_2$	237(4.53),275(4.31), 316(4.09),351(3.93)	24-1726-84
$C_{24}H_{16}O_3$			
2,5-Furandione, 3-(diphenylmethylene)-dihydro-4-(phenylmethylene)-, (E)-	toluene	392(3.95)	48-0233-84
(Z)-	toluene	393(4.18)	48-0233-84
$C_{24}H_{16}O_4$			
Benzo[g]benzo[5,6]cyclohepta[1,2,3-cd]-benzofuran-3,8-dione, 9-hydroxy-2-(2-methyl-1-propenyl)- (radermachol)	CHCl$_3$	252(4.59),276(4.4), 305(4.34),330(4.21), 363(4.31),460(3.59)	88-5847-84
	MeOH	250(--),275(--), 300(--),330(--), 360(--),455(--)	88-5847-84
	MeOH-NaOH	240(--),265(--), 330(--),350(--), 470(--)	88-5847-84
$C_{24}H_{16}O_5$			
1(2H)-Naphthalenone, 2,2-bis(benzoyl-oxy)-	MeCN	241(4.84),277(3.82), 283s(3.80),333(3.43)	39-2077-84C
2(1H)-Naphthalenone, 1,1-bis(benzoyl-oxy)-	MeCN	238(4.61),324(3.90)	39-2077-84C
$C_{24}H_{16}O_6$			
Sphagnorubin B	MeOH-HCl	241(4.78),290(4.73), 330s(4.43),420s(4.12), 540(4.88)	5-1024-84
$C_{24}H_{16}O_{12}$			
Cyclic tetramer from 5-(hydroxymethyl)-2-furancarboxylic acid	CHCl$_3$	256.2(4.80)	126-2347-84
$C_{24}H_{17}Br_2N_3O$			
2,5-Cyclohexadien-1-one, 2,5-bis[(4-bromophenyl)amino]-4-(phenylimino)-	CHCl$_3$	291(4.48),397(4.29)	103-0263-84
$C_{24}H_{17}Br_3NP$			
Benzenamine, 2,4,6-tribromo-N-(tri-phenylphosphoranylidene)-	MeOH	254(4.11)	65-0289-84
$C_{24}H_{17}Br_3N_3P$			
1-Triazene, 1-(2,4,6-tribromophenyl)-3-(triphenylphosphoranylidene)-	MeOH	286(4.23)	65-0289-84

Compound	Solvent	$\lambda_{max}(\log \epsilon)$	Ref.
$C_{24}H_{17}ClN_3O_4P$ Benzenamine, 4-chloro-2,6-dinitro- N-(triphenylphosphoranylidene)-	MeOH	391(3.36)	65-0289-84
$C_{24}H_{17}ClN_5O_4P$ 1-Triazene, 1-(4-chloro-2,6-dinitro- phenyl)-3-(triphenylphosphoranyli- dene)-	MeOH	316(4.07)	65-0289-84
$C_{24}H_{17}ClO_7$ 5,12-Naphthacenedione, 6-acetoxy-2-(2- acetoxyethoxy)-11-chloro-	EtOH	239(4.55),288(4.48), 296(4.52),394(3.78)	150-0147-84M
$C_{24}H_{17}Cl_2N_3O$ 2,5-Cyclohexadien-1-one, 2,5-bis[(4- chlorophenyl)amino]-4-(phenylimino)-	CHCl$_3$	289(4.43),394(4.23)	103-0263-84
$C_{24}H_{17}Cl_2N_3O_7S_2$ Acetic acid, [[2-[[5-[(2,4-dichloro- phenyl)methylene]-4,5-dihydro-4-oxo- 2-thiazolyl]amino]-2,3-dihydro-8,9- dimethoxy-1,3-dioxo-1H-2-benzazepin- 4-yl]thio]-	MeOH	250(4.62),335(4.36), 392(4.34)	103-0024-84
$C_{24}H_{17}NO_2$ Methanone, (4-methyl-2,3-quinoline- diyl)bis[phenyl-	MeOH	252(3.63)	4-0759-84
$C_{24}H_{17}NO_3$ 7H-Dibenz[f,ij]isoquinolin-7-one, 2-methoxy-4-methyl-6-phenoxy-	toluene	396(4.02)	104-0980-84
$C_{24}H_{17}N_3O$ 2(10H)-Phenazinone, 10-phenyl-3-(phen- ylamino)-	CHCl$_3$	279(4.69),468(4.51)	103-0263-84
$C_{24}H_{17}N_3S_2$ 2H-1,3-Thiazine-2-acetonitrile, 5-cya- no-3,6-dihydro-4-(3-methylphenyl)- 2-(2-naphthalenyl)-6-thioxo-	EtOH	265(4.36),322(3.75), 388(4.09)	4-1445-84
2H-1,3-Thiazine-2-acetonitrile, 5-cya- no-3,6-dihydro-4-(4-methylphenyl)- 2-(2-naphthalenyl)-6-thioxo-	EtOH	269(4.39),322(3.65), 388(4.13)	4-1445-84
$C_{24}H_{17}N_4O_6P$ Benzenamine, 2,4,6-trinitro-N-(tri- phenylphosphoranylidene)-	MeOH	363(4.41)	65-0289-84
$C_{24}H_{17}N_5O_2S_2$ Pyrimido[5,4-e]thiazolo[3,4-a]pyrimi- din-10-ium, 9-[(3-ethyl-2(3H)-benzo- thiazolylidene)methyl]-2-hydroxy- 3,4-dihydro-4-oxo-7-phenyl-, hydroxide, inner salt	DMF	430(4.16),454(4.10), 555(4.51)	103-0921-84
$C_{24}H_{18}$ Azulene, 1-[4-(4-azulenyl)-1,3-butadi- enyl]-	hexane	242s(4.44),265(4.59), 280s(4.53),323(4.32), 420s(4.44),446(4.62), 465s(4.54),610(3.19), 625s(3.17),655s(3.12),	5-1905-84

Compound	Solvent	$\lambda_{max}(\log \epsilon)$	Ref.
Azulene, 1-[4-(4-azulenyl)-1,3-butadi-enyl]- (cont.)		730s(2.73)	5-1905-84
protonated	CH_2Cl_2-TFA	700s(4.87),755(5.03)	5-1936-84
Azulene, 1-[4-(6-azulenyl)-1,3-butadi-enyl]-	hexane	289(4.64),330(4.48), 410s(4.29),438s(4.59), 462(4.76),497(4.71), 580s(3.05),605(3.09), 630s(3.05),670s(2.99), 700s(2.85),750s(2.55)	5-1905-84
protonated	CH_2Cl_2-TFA	710s(4.83),765(5.03)	5-1936-84
Azulene, 1,1'-(1,3-butadiene-1,4-diyl)-bis-	CH_2Cl_2	232(4.42),260(4.63), 292(4.26),327(3.47), 430s(4.64),449(4.77), 476(4.71),635(3.02), 670s(3.01),750s(2.78)	5-1905-84
dication	CH_2Cl_2	249(4.69),267s(4.65), 323(4.60),425(4.59), 500s(4.75),527(4.79), 542s(4.78)	5-1936-84
radical cation	CH_2Cl_2	732(--),875(--), 1030(--),1070s(--)	5-1936-84
Azulene, 4,4'-(1,3-butadiene-1,4-diyl)-bis-	CH_2Cl_2	240(4.63),269(4.62), 288(4.58),326(4.43), 394(4.51),424s(4.32), 620(3.24),680s(3.15)	5-1905-84
dication	CH_2Cl_2-TFA	476(4.40),500s(4.35)	5-1936-84
Azulene, 6,6'-(1,3-butadiene-1,4-diyl)-bis-	CH_2Cl_2	250s(4.28),269(4.35), 295(4.23),320s(4.50), 332(4.56),351s(4.33), 410s(4.56),434(4.76), 452(4.72),603(3.08), 625s(3.07),695s(2.87)	5-1905-84
dication	CH_2Cl_2-TFA	515s(4.59),545(4.66)	5-1936-84
Benzo[g]chrysene, 12,14-dimethyl-, (±)-	$CHCl_3$	253(4.56),276s(4.70), 285(3.83),295(4.86), 330(4.06),346(3.98), 382(2.89)	24-0336-84
$C_{24}H_{18}BrN$ Pyridine, 2-(4-bromophenyl)-6-(4-meth-ylphenyl)-4-phenyl-	dioxan	263(4.67),317(4.02)	48-0287-84
$C_{24}H_{18}BrNO$ Pyridine, 2-(4-bromophenyl)-6-(4-meth-oxyphenyl)-4-phenyl-	dioxan	262(4.70),323(4.03)	48-0287-84
$C_{24}H_{18}BrN_3$ 1H-1,5-Benzodiazepine, 4-(4-bromophen-yl)-2,3-dihydro-2-(2-quinolinyl)-	EtOH	264(4.40),380(3.76)	103-1370-84
$C_{24}H_{18}BrO$ Pyrylium, 2-(4-bromophenyl)-6-(4-meth-ylphenyl)-4-phenyl-, perchlorate	MeCN	249(4.16),286.5(4.27), 358(4.57),421(4.47)	48-0287-84
$C_{24}H_{18}BrO_2$ Pyrylium, 2-(4-bromophenyl)-6-(4-meth-oxyphenyl)-4-phenyl-, perchlorate	MeCN	257(4.26),313(4.35), 361(4.57),447(4.49)	48-0287-84
$C_{24}H_{18}ClN$ Pyridine, 2-(4-chlorophenyl)-6-(4-meth-ylphenyl)-4-phenyl-	dioxan	259(4.72),317(3.99)	48-0287-84

Compound	Solvent	$\lambda_{max}(\log \epsilon)$	Ref.
$C_{24}H_{18}ClNO$ Pyridine, 2-(4-chlorophenyl)-6-(4-methoxyphenyl)-4-phenyl-	dioxan	260.5(4.69),321(4.01)	48-0287-84

$C_{24}H_{18}ClNO_2$
  2(1H)-Pyridinone, 3-(4-chlorophenyl)-6-(4-methoxyphenyl)-4-phenyl-
  2(1H)-Pyridinone, 6-(4-chlorophenyl)-3-(4-methoxyphenyl)-4-phenyl-
	EtOH
	
	EtOH
		266(3.97),350(4.06)
		
		262(4.12),356(4.05)
			4-1473-84
			
			4-1473-84

$C_{24}H_{18}ClN_3$ 1H-1,5-Benzodiazepine, 2-(4-chlorophenyl)-2,3-dihydro-4-(2-quinolinyl)-	EtOH	248(4.42),388(3.94)	103-1370-84
$C_{24}H_{18}ClN_3O_7S_2$ Acetic acid, [[2-[[(4-chlorophenyl)-methylene]-4,5-dihydro-4-oxo-2-thiazolyl]amino]-2,3-dihydro-8,9-dimethoxy-1,3-dioxo-1H-2-benzazepin-4-yl]thio]-	MeOH	250(4.61),334(4.32), 389(4.34)	103-0024-84
$C_{24}H_{18}ClO$ Pyrylium, 2-(4-chlorophenyl)-6-(4-methylphenyl)-4-phenyl-, perchlorate	MeCN	247(4.21),285(4.29), 357(4.58),418(4.48)	48-0287-84
$C_{24}H_{18}ClO_2$ Pyrylium, 2-(4-chlorophenyl)-6-(4-methoxyphenyl)-4-phenyl-, perchlorate	MeCN	255(4.25),313(4.34), 359(4.57),446(4.49)	48-0287-84
$C_{24}H_{18}Cl_2N_4O_2Pd_2$ Palladium, di-μ-chlorobis[2-(phenylazo)phenolato]di-	CH_2Cl_2	495(3.46)	77-0999-84
$C_{24}H_{18}Cl_2O_6$ 1(2H)-Naphthalenone, 4-chloro-2-(4-chloro-5,8-dimethoxy-1-oxo-2(1H)-naphthalenylidene)-3,8-dimethoxy-	$CHCl_3$	247(4.41),262(4.35), 303(4.17),598(4.33)	5-0319-84
$C_{24}H_{18}FN$ Pyridine, 2-(4-fluorophenyl)-6-(4-methylphenyl)-4-phenyl-	dioxan	255.5(4.67),314(3.94)	48-0287-84
$C_{24}H_{18}FNO$ Pyridine, 2-(4-fluorophenyl)-6-(4-methoxyphenyl)-4-phenyl-	dioxan	257(4.64),321(3.99)	48-0287-84
$C_{24}H_{18}FO$ Pyrylium, 2-(4-fluorophenyl)-6-(4-methylphenyl)-4-phenyl-, perchlorate	MeCN	243(4.15),281.5(4.27), 353(4.56),416(4.45)	48-0287-84
$C_{24}H_{18}FO_2$ Pyrylium, 2-(4-fluorophenyl)-6-(4-methoxyphenyl)-4-phenyl-, perchlorate	MeCN	251(4.18),310(4.30), 356(4.53),443(4.47)	48-0287-84
$C_{24}H_{18}IN$ Pyridine, 2-(4-iodophenyl)-6-(4-methylphenyl)-4-phenyl-	dioxan	260(4.77),317(4.08)	48-0287-84
$C_{24}H_{18}INO$ Pyridine, 2-(4-iodophenyl)-6-(4-methoxyphenyl)-4-phenyl-	dioxan	267(4.71),323(4.07)	48-0287-84

Compound	Solvent	$\lambda_{max}(\log \epsilon)$	Ref.
$C_{24}H_{18}IO$ Pyrylium, 2-(4-iodophenyl)-6-(4-methyl- phenyl)-4-phenyl-, perchlorate	MeCN	260s(4.13),279s(4.10), 292(4.33),363(4.55), 449(4.48)	48-0287-84
$C_{24}H_{18}IO_2$ Pyrylium, 2-(4-iodophenyl)-6-(4-meth- oxyphenyl)-4-phenyl-, perchlorate	MeCN	272(4.23),313(4.35), 366(4.58),449(4.50)	48-0287-84
$C_{24}H_{18}NO_3$ Pyrylium, 2-(4-methylphenyl)-6-(4-ni- trophenyl)-4-phenyl-, perchlorate	MeCN	252(4.16),289(4.28), 305s(4.21),356(4.52), 414(4.44)	48-0287-84
$C_{24}H_{18}NO_4$ Pyrylium, 2-(4-methoxyphenyl)-6-(4-ni- trophenyl)-4-phenyl-, perchlorate	MeCN	257(4.14),272(4.12), 328(4.43),353(4.47), 447(4.44)	48-0287-84
$C_{24}H_{18}N_2O_2$ Acetamide, N-[4-(3-[1,1'-biphenyl]- 4-yl-2-cyano-3-oxo-1-propenyl)- phenyl]-	EtOH	360(4.32)	73-0421-84
Pyridine, 2-(4-methylphenyl)-6-(4-ni- trophenyl)-4-phenyl-	dioxan	273(4.55),328(4.18)	48-0287-84
$C_{24}H_{18}N_2O_3$ Pyridine, 2-(4-methoxyphenyl)-6-(4-ni- trophenyl)-4-phenyl-	dioxan	278(4.55),334(4.07)	48-0287-84
$C_{24}H_{18}N_2O_4$ Dispiro[1H-indene-1,4'-[2,3]diazabi- cyclo[3.1.0]hex[2]ene-6',1"-[1H]- indene]-1',5'-dicarboxylic acid, dimethyl ester	EtOH	229(4.43)	23-2506-84
$C_{24}H_{18}N_3O_4P$ Benzenamine, 2,4-dinitro-N-(triphenyl- phosphoranylidene)-	MeOH	377(4.40)	65-0289-84
Benzenamine, 2,6-dinitro-N-(triphenyl- phosphoranylidene)-	MeOH	380(3.48)	65-0289-84
$C_{24}H_{18}N_3O_7PS$ Benzenesulfonic acid, 3,5-dinitro- 4-[(triphenylphosphoranylidene)- amino]-, potassium salt	MeOH	380(3.30)	65-0289-84
$C_{24}H_{18}N_4O_7$ 2,4(1H,3H)-Pteridinedione, 3-(2,3-O- carbonyl-β-D-ribofuranosyl)-6,7- diphenyl-	MeOH	222(4.41),276(4.18), 363(4.13)	136-0179-84F
$C_{24}H_{18}N_4O_9S_2$ Acetic acid, [[2-[[4,5-dihydro-5-[(2- nitrophenyl)methylene]-4-oxo-2-thia- zolyl]amino]-2,3-dihydro-8,9-dimeth- oxy-1,3-dioxo-1H-2-benzazepin-4-yl]- thio]-	MeOH	247(4.65),334(4.32), 390(4.31)	103-0024-84
Acetic acid, [[2-[[4,5-dihydro-5-[(3- nitrophenyl)methylene]-4-oxo-2-thia- zolyl]amino]-2,3-dihydro-8,9-dimeth-	MeOH	251(4.62),334(4.32), 391(4.25)	103-0024-84

Compound	Solvent	$\lambda_{max}(\log \epsilon)$	Ref.
oxy-1,3-dioxo-1H-2-benzazepin-4-yl]-thio]- (cont.)			103-0024-84
$C_{24}H_{18}N_5O_4P$ 1-Triazene, 1-(2,4-dinitrophenyl)-3-(triphenylphosphoranylidene)-	MeOH	305(4.13)	65-0289-84
$C_{24}H_{18}N_5O_7PS$ Benzenesulfonic acid, 3,5-dinitro-4-[3-(triphenylphosphoranylidene)-1-triazenyl]-, potassium salt	MeOH	323(c.4.11)	65-0289-84
$C_{24}H_{18}N_8O_4$ Benzo[g]pteridine-2,4(3H,10H)-dione, 10,10'-(1,2-ethanediyl)bis[3-methyl-	pH 7.27	345(4.10),440(4.20)	39-1227-84B
$C_{24}H_{18}O_2$ 1H-Inden-1-one, 2,3-dihydro-2-(3-oxo-1,3-diphenylpropylidene)-	dioxan	242(4.24),295(4.25)	48-0647-84
2-Propen-1-one, 1,1'-(1,4-phenylene)-bis[3-phenyl-	CH_2Cl_2	322(4.65)	88-0561-84
$C_{24}H_{18}O_4$ Butanedioic acid, di-1H-inden-1-yli-dene-, dimethyl ester	EtOH	250(4.16),355(3.64)	23-2506-84
2(5H)-Furanone, 4-hydroxy-3-[4-(phenyl-methoxy)phenyl]-5-(phenylmethylene)-, (Z)-	EtOH	300(4.26),380(3.78)	39-1539-84C
$C_{24}H_{18}O_7$ 4H-1-Benzopyran-4-one, 7-(benzoyloxy)-3-(2-hydroxy-4,6-dimethoxyphenyl)-	EtOH	260s(4.37),292s(3.91), 302(3.87)	142-0261-84
	EtOH-NaOAc	262s(4.34),292s(3.92), 302(3.87)	142-0261-84
$C_{24}H_{18}O_8$ 9,10-Anthracenedione, 1-(3-acetyl-2,6-dihydroxy-4-methoxyphenyl)-4,5-dihy-droxy-2-methyl- (knipholone)	EtOH	224(4.49),254(4.26), 288(4.26),432(3.84)	102-1729-84
	EtOH-NaOH	232(4.45),250(4.30), 330(4.28),500(2.78)	102-1729-84
$C_{24}H_{18}O_{10}$ 1,3-Naphthacenedicarboxylic acid, 6,11-dihydro-2,5-dihydroxy-7,12-dimethoxy-6,11-dioxo-, dimethyl ester	n.s.g.	229(4.66),251(4.32), 265(4.17),282(4.04), 430(3.91),467(4.01), 490(3.85)	5-0306-84
$C_{24}H_{19}BrN_2O_4$ 1H-Indazole-6,7-dicarboxylic acid, 1-(4-bromophenyl)-4-methyl-3-phen-yl-, dimethyl ester	EtOH	234(4.89),335(3.94)	94-2146-84
$C_{24}H_{19}BrO_2$ 2-Pentene-1,5-dione, 1-(4-bromophenyl)-5-(4-meth yl phenyl)-3-phenyl- (with isomer switching 1 and 5 substituents)	dioxan	257(4.39),296(4.33)	48-0287-84
$C_{24}H_{19}BrO_3$ 2-Pentene-1,5-dione, 1-(4-bromophenyl)-5-(4-methoxyphenyl)-3-phenyl- (with isomer)	dioxan	263(4.39),277.5(4.39), 304(4.37)	48-0287-84

Compound	Solvent	$\lambda_{max}(\log \epsilon)$	Ref.
$C_{24}H_{19}ClN_2O_4$ 1H-Indazole-6,7-dicarboxylic acid, 1-(4-chlorophenyl)-4-methyl-3-phenyl-, dimethyl ester	EtOH	250(4.56),332(4.08)	94-2146-84
$C_{24}H_{19}ClO_2$ 2-Pentene-1,5-dione, 1-(4-chlorophenyl)-3-(4-methylphenyl)-3-phenyl- (with isomer)	dioxan	255(4.39),297(4.32)	48-0287-84
$C_{24}H_{19}ClO_3$ 2-Pentene-1,5-dione, 1-(4-chlorophenyl)-5-(4-methoxyphenyl)-3-phenyl- (with isomer)	dioxan	264s(4.34),279(4.37), 303(4.36)	48-0287-84
$C_{24}H_{19}FO_2$ 2-Pentene-1,5-dione, 1-(4-fluorophenyl)-5-(4-methylphenyl)-3-phenyl- (with isomer)	dioxan	246(4.31),297(4.31)	48-0287-84
$C_{24}H_{19}FO_3$ 2-Pentene-1,5-dione, 1-(4-fluorophenyl)-5-(4-methoxyphenyl)-3-phenyl- (with isomer)	dioxan	233.5(4.25),279(4.34), 303(4.34)	48-0287-84
$C_{24}H_{19}IO_2$ 2-Pentene-1,5-dione, 1-(4-iodophenyl)-5-(4-methylphenyl)-3-phenyl- (with isomer)	dioxan	278(4.41),296s(4.40)	48-0287-84
$C_{24}H_{19}IO_3$ 2-Pentene-1,5-dione, 1-(4-iodophenyl)-5-(4-methoxyphenyl)-3-phenyl-	dioxan	278(4.46),304(4.38)	48-0287-84
$C_{24}H_{19}N$ Pyridine, 2-(4-methylphenyl)-4,6-diphenyl-	dioxan	255(4.67),314(3.93)	48-0287-84
$C_{24}H_{19}NO$ Dibenz[cd,g]indol-6(1H)-one, 1-(2,4,6-trimethylphenyl)-	EtOH	550(4.12)	104-1415-84
Pyridine, 2-(4-methoxyphenyl)-4,6-diphenyl-	dioxan	257(4.64),321(3.99)	48-0287-84
2(1H)-Pyridinone, 4-(4-methylphenyl)-3,6-diphenyl-	EtOH	258(4.54),344(4.42)	4-1473-84
$C_{24}H_{19}NO_2$ [1,1'-Biphenyl]-4-propanenitrile, α-[(4-ethoxyphenyl)methylene]-β-oxo-	EtOH	361(4.36)	73-0421-84
2(1H)-Pyridinone, 4-(3-methoxyphenyl)-3,6-diphenyl-	EtOH	254(4.26),344(4.21)	4-1473-84
$C_{24}H_{19}NO_3$ 2-Butene-1,4-dione, 2-[(2-acetylphenyl)amino]-1,4-diphenyl-, (Z)-	MeOH	265(3.26),327(3.18), 402(3.29)	4-0759-84
Methanone, (2,3,3a,4,9,9a-hexahydro-2-phenyl-4,9-epoxynaphth[2m3-d]isoxazol-3-yl)phenyl-	MeOH	247(3.34)	73-1990-84
2(1H)-Pyridinone, 4-(1,3-benzodioxol-5-yl)-3,4-dihydro-3,6-diphenyl-	EtOH	286(4.09),338(3.43)	4-1473-84

Compound	Solvent	$\lambda_{max}(\log \epsilon)$	Ref.
$C_{24}H_{19}NO_4$			
Benzoic acid, 2-[(1-benzoyl-3-oxo-3-phenyl-1-propenyl)amino]-, methyl ester	MeOH	258(4.21),332(4.00), 382(4.41)	4-0759-84
2-Pentene-1,5-dione, 1-(4-methylphenyl)-5-(4-nitrophenyl)-3-phenyl- (with isomer)	dioxan	262.5(4.41),295(4.31)	48-0287-84
$C_{24}H_{19}NO_5$			
2-Pentene-1,5-dione, 1-(4-methoxyphenyl)-5-(4-nitrophenyl)-3-phenyl- (with isomer)	dioxan	271(4.46),307(4.35)	48-0287-84
$C_{24}H_{19}N_3O$			
2,5-Cyclohexadien-1-one, 2,5-bis(phenylamino)-4-(phenylimino)-	$CHCl_3$	284(4.40),392(4.27)	103-0263-84
$C_{24}H_{19}N_3O_2$			
1H-Pyrazole-3-carboxamide, 5-benzoyl-N-methyl-1,4-diphenyl-	EtOH	254(4.51)	4-1013-84
$C_{24}H_{19}N_3O_4$			
1,3,5-Triazabicyclo[3.2.0]hept-6-ene-2,4-dione, 6,7-bis(4-methoxyphenyl)-3-phenyl-	MeCN	252(4.22),330(4.29)	44-2917-84
$C_{24}H_{19}N_5O_5$			
1-Imidazolidineacetic acid, 2,5-dioxo-4,4-diphenyl-, [(4-nitrophenyl)methylene]hydrazide	MeOH	202(4.48),320(4.20)	56-0585-84
$C_{24}H_{19}O$			
Pyrylium, 2-(4-methylphenyl)-4,6-diphenyl-, perchlorate	MeCN	240(4.16),280.5(4.28), 353(4.56),413(4.45)	48-0287-84
Pyrylium, 3-methyl-2,4,6-triphenyl-, iodide	MeCN	245(4.39),273(4.25), 342(4.27),384(4.37)	48-0657-84
$C_{24}H_{19}O_2$			
Pyrylium, 2-(4-methoxyphenyl)-4,6-diphenyl-, perchlorate	MeCN	250(4.19),310(4.33), 354(4.53),443(4.48)	48-0287-84
$C_{24}H_{19}O_{11}S_3$			
Pyrylium, 4-(4-methoxy-3-sulfophenyl)-2,6-bis(4-sulfophenyl)-, perchlorate	pH 1	238(4.46),274(4.44), 407(4.81)	39-0841-84B
$C_{24}H_{19}O_{12}S_3$			
Benzenesulfonic acid, 2-methoxy-5-[3-oxo-1-[2-oxo-2-(4-sulfophenyl)ethyl]-3-(4-sulfophenyl)-1-propenyl]-, ion(1-)	pH 13	260(4.36),487(4.60)	39-0841-84B
$C_{24}H_{19}S_4$			
Naphtho[1,2-d]-1,3-dithiol-1-ium, 2-[1-(4,5-dihydronaphtho[1,2-d]-1,3-dithiol-2-ylidene)ethyl]-4,5-dihydro-, perchlorate	acetone	557(4.89)	48-0479-84
$C_{24}H_{20}$			
Pyrene, 10b,10c-dihydro-10b,10c-dimethyl-2-phenyl-	C_6H_{12}	348(5.15),390(4.52), 493(4.18),648(2.51)	35-7776-84

Compound	Solvent	$\lambda_{max}(\log \epsilon)$	Ref.

$C_{24}H_{20}F_6O_6$
  3,4-Methanocyclopenta[cd]pentaleno-
  [2,1,6-gha]pentalene-1,5-dione,
  6-(2,5-dioxocyclopentylidene)do-
  decahydro-2,7-bis(2,2,2-trifluoro-
  ethoxy)-, (1aα,2α,3β,3aα,3bα,4β,5aα-
  6aα,6bα,6cα,7aα)-
	CH_2Cl_2	281(3.94),315(3.65),
		  420(c.2.18)
			23-2612-84

$C_{24}H_{20}N$
  Pyridinium, 1-methyl-2,4,6-triphenyl-,
  perchlorate
	EtOH	<u>311(4.6)</u>	103-1245-84

$C_{24}H_{20}N_2O$
  1H-Pyrido[3,4-b]indol-1-one, 2,3,4,9-
  tetrahydro-2-phenyl-9-(phenylmethyl)-
	isoPrOH	234(4.47),310(4.43)	103-0047-84

$C_{24}H_{20}N_2O_2$
  1,5-Naphthalenedione, 4,8-bis[(4-meth-
  ylphenyl)amino]-
  1(4H)-Naphthalenone, 5-hydroxy-2-[(4-
  methylphenyl)amino]-4-[(4-methyl-
  phenyl)imino]-
	CHCl₃
	
	CHCl₃
		288(4.22),346(3.90),
		  621(4.12),683(4.33)
		277(4.38),286(4.32),
		  335(3.78),472(4.02)
			104-0733-84
			
			104-0733-84

$C_{24}H_{20}N_2O_4$
  1H-Indazole-6,7-dicarboxylic acid,
  4-methyl-1,3-diphenyl-, dimethyl
  ester
	EtOH	246(4.62),333(3.96)	94-2146-84

$C_{24}H_{20}N_4$
  Propanedinitrile, (1,2,3,4-tetrahydro-
  1,4-dimethyl-5,7-diphenyl-6H-cyclo-
  pentapyrazin-6-ylidene)-
	CHCl₃	391(4.58),405s(--),
		  510s(--),602(2.86)
			89-0053B-84

$C_{24}H_{20}N_4O$
  7H-Pyrido[2,1-b]quinazoline-6,11-dione,
  8,9-dihydro-,6-([1,1'-biphenyl]-4-yl-
  hydrazone)
	EtOH	222(4.35),272(4.25),
		  401(4.57)
			4-1301-84

$C_{24}H_{20}N_4O_2$
  7H-Pyrido[2,1-b]quinazoline-6,11-dione,
  8,9-dihydro-, 6-[(4-phenoxyphenyl)-
  hydrazone]
	EtOH	232(4.36),257(4.28),
		  300(4.12),394(4.34)
			4-1301-84

$C_{24}H_{20}N_4O_4$
  4-Pyrimidinecarboxamide, 2-(1-amino-
  2-benzoyl-3-oxo-1-butenyl)-5-benz-
  oyl-6-methyl-
	$C_2H_4Cl_2$	248(4.62),371(4.00)	39-0965-84B

$C_{24}H_{20}N_6O_4S$
  Acetamide, N-[[2-[(3-methyl-5-phenyl-
  1-(2-pyridinylcarbonyl)-1H-pyrazol-
  4-yl]azo]phenyl]sulfonyl]-
	n.s.g.	366(4.46)	48-1021-84

$C_{24}H_{20}O_2$
  1H-Indene-1,3(2H)-dione, 2-[1,1'-bi-
  phenyl]-4-yl-2-propyl-
  2-Pentene-1,5-dione, 1-(4-methylphen-
  yl)-3,5-diphenyl- (with isomer)
	EtOH
	dioxan
	dioxan
		349(2.76)
		350(2.73)
		245(4.30),295(4.31)
			135-0312-84
			135-0312-84
			48-0287-84

$C_{24}H_{20}O_3$ 2-Pentene-1,5-dione, 1-(4-methoxyphen-	dioxan	233.3(4.26),279(4.35),	48-0287-84

Compound	Solvent	$\lambda_{max}(\log \epsilon)$	Ref.
yl)-3,5-diphenyl- (with isomer)- (cont.)		302(4.33)	48-0287-84
$C_{24}H_{20}O_5$ 1H-2-Benzopyran-1-one, 5,6,7-trimeth- oxy-3,4-diphenyl-	EtOH	255(2.87),300(3.08)	2-0889-84
$C_{24}H_{20}O_8$ 2H-1-Benzopyran-2-one, 7-acetoxy- 4-methyl-, anti-head-tail dimer	CHCl$_3$	247(3.63),257s(--), 273(3.56),280s(--)	2-0502-84
$C_{24}H_{20}O_{11}$ 9,10-Anthracenedione, 1,2,6,8-tetra- acetoxy-7-methoxy-3-methyl-	dioxan	285s(4.76),335(3.84)	94-0860-84
$C_{24}H_{20}O_{12}S_3$ 2-Pentene-1,5-dione, 1,5-bis(4-sulfo- phenyl)-3-(3-methoxy-4-sulfophenyl)-	pH 8	252(4.55),339(4.29)	39-0841-84B
$C_{24}H_{21}ClN_2O_4$ Benzamide, N-[2-(6-chloro-4-oxo-4H-1- benzopyran-3-yl)-1-(1-piperidinyl- carbonyl)ethenyl]-	EtOH	223(4.34),245(4.34), 297(4.41)	2-1048-84
$C_{24}H_{21}Cl_2N_5O_5$ Benzeneacetamide, N-[9-(2-deoxy-β-D- erythro-pentofuranosyl)-6-(3,5-di- chlorophenoxy)-9H-purin-2-yl]-	EtOH	223s(4.49),276(4.24)	39-1263-84C
$C_{24}H_{21}F_3N_4$ 4H-Pyrimido[4,5-b]indol-4-imine, 3,5,6,7,8,9-hexahydro-2-methyl- 9-phenyl-3-[3-(trifluoromethyl)- phenyl]-	EtOH	245(4.44),284s(4.00), 308(3.86)	11-0073-84A
$C_{24}H_{21}NO$ 2(1H)-Pyridinone, 3,4-dihydro-4-(4- methylphenyl)-3,6-diphenyl-	EtOH	217(4.15),270(3.70)	4-1473-84
$C_{24}H_{21}NO_2$ 2(1H)-Pyridinone, 3,4-dihydro-4-(3- methoxyphenyl)-3,6-diphenyl-	EtOH	211(4.31),275(3.82)	4-1473-84
2(1H)-Pyridinone, 3,4-dihydro-4-(4- methoxyphenyl)-3,6-diphenyl-	EtOH	220(4.64),278(3.94), 282(3.92)	4-1473-84
$C_{24}H_{21}NO_3$ Benzoic acid, 2-[(3-oxo-1,3-diphenyl- 1-propenyl)amino]-, ethyl ester	EtOH	258(4.21),332(4.00), 382(4.42)	4-0759-84
$C_{24}H_{21}N_2O$ 1H-Pyrazolium, 4-(diphenylmethylene)- 4,5-dihydro-2,3-dimethyl-5-oxo-1- phenyl-, tetrafluoroborate	CH$_2$Cl$_2$	497s(4.43),527(4.70)	49-0197-84
$C_{24}H_{21}N_3O_2$ Benzenepropanenitrile, β-oxo-α-[[5-(4- phenyl-1-piperazinyl)-2-furanyl]meth- ylene]-	MeOH	210(3.39),243(3.27), 280(2.69),500(3.84)	73-2141-84
$C_{24}H_{21}N_3O_3$ Isoviolacein, tetramethyl-	CHCl$_3$	287(4.31),385(3.86), 562(4.21)	39-1331-84B

Compound	Solvent	λ_{max}(log ϵ)	Ref.
Isoviolacein, tetramethyl- (cont.)	hexane	281(--),363(--), 528(--)	39-1331-84B
Violacein, tetramethyl-	hexane	230s(4.20),241(4.20), 274(4.20),297s(3.83), 315(3.76),363(3.82), 547(4.18)	39-1331-84B
	CHCl$_3$	274(4.37),378(4.09), 568(4.33)	39-1331-84B
$C_{24}H_{21}N_3O_6$ Hyperectine	EtOH	230s(4.29),292(3.81), 363(3.35)	105-0592-84
$C_{24}H_{22}BrN_3O_3$ Quinoxaline, 3-(4-bromophenyl)-1,2-di-hydro-1-[(4-nitrophenyl)methyl]-2-propoxy-	MeOH	237(4.34),266(4.44), 291s(--),376(3.89)	103-0542-84
$C_{24}H_{22}N_2$ 7H-Quino[3',2':4,5]pyrrolo[1,2-a][1]-benzazepine, 2,5,9,12-tetramethyl-	C$_6$H$_{12}$	236(4.49),242s(4.49), 247.5(4.50),269(4.46), 286s(4.35),328(3.79), 341.5(3.86),351(3.78), 365(3.70)	39-2529-84C
$C_{24}H_{22}N_2O$ Ethanone, 1-[4-[4,5-dihydro-3-(4-meth-ylphenyl)-5-phenyl-1H-pyrazol-1-yl]-phenyl]-	MeOH	250(4.30),305(4.22), 379(4.43)	103-0787-84
$C_{24}H_{22}N_2O_2$ Cyclopent[b]indole-2-carboxylic acid, 1,4-dihydro-4-methyl-3-(N-methyl-1H-indol-3-yl)-, ethyl ester	EtOH	220(4.71),260(4.22), 367(4.20)	94-4410-84
Ethanone, 1-[4-[4,5-dihydro-3-(4-meth-oxyphenyl)-5-phenyl-1H-pyrazol-1-yl]phenyl]-	MeOH	251(4.17),305(3.94), 382(4.47)	103-0787-84
7H-Quino[3',2':4,5]pyrrolo[1,2-a][1]-benzazepine, 2,12-dimethoxy-5,9-di-methyl-	C$_6$H$_{12}$	218(4.65),236(4.73), 249s(4.64),271(4.55), 290s(4.44),328s(3.96), 336s(4.02),342(4.09), 351(3.99),368(3.89)	39-2529-84C
$C_{24}H_{22}N_2O_2S$ 2(1H)-Quinoxalinone, 3-[2-(4-methoxy-phenyl)-2-(phenylthio)ethyl]-1-methyl-	EtOH	340(3.26)	18-1653-84
$C_{24}H_{22}N_2O_3$ 1H-Indole, 1,3-diacetyl-5-[2-(1-acetyl-1H-indol-5-yl)ethyl]-	EtOH	224.7(4.64),244(4.72), 294(4.27),303(4.34)	103-0062-84
$C_{24}H_{22}N_2O_4$ Benzamide, N-[2-(4-oxo-4H-1-benzopyran-3-yl)-1-(1-piperidinylcarbonyl)ethen-yl]-	EtOH	235(3.05),283(3.06)	2-1048-84
3H-Indol-3-one, 2-[1,3-dihydro-1-(2-methyl-1-oxopropyl)-3-oxo-2H-indol-2-ylidene]-1,2-dihydro-1-(2-methyl-1-oxopropyl)-, cis	benzene	477(3.58)	39-2305-84C
trans	benzene	576(3.84)	39-2305-84C

Compound	Solvent	$\lambda_{max}(\log \epsilon)$	Ref.

$C_{24}H_{22}N_2O_5$
  Benzamide, N-[2-(6-methyl-4-oxo-1-(4-
    morpholinylcarbonyl)ethenyl]-
	EtOH	230(4.55),295(4.66)	2-1048-84

$C_{24}H_{22}N_2O_6$
  Cyclobuta[1,2-c:3,4-c']dipyrrole-
    1,3,4,6(2H,5H)-tetrone, tetrahydro-
    3a,6a-bis(3-methoxyphenyl)-2,5-di-
    methyl-
	MeOH	270(3.58)	87-0628-84

$C_{24}H_{22}N_4O_5$
  Quinoxaline, 1,2-dihydro-3-(4-nitro-
    phenyl)-1-[(4-nitrophenyl)methyl]-
    2-propoxy-
	MeOH	232(4.45),272(4.38),
		331s(--),413(3.94)
			103-0542-84

$C_{24}H_{22}N_5$
  Pyrazolo[3,4-c]pyrazolium, 4-(1-ethyl-
    2(1H)-pyridinylidene)-4,5-dihydro-3-
    methyl-1,5-diphenyl-, iodide
	EtOH	320(4.03)	48-0811-84

$C_{24}H_{22}N_5O$
  Quinolinium, 2-(2-acetyl-2,6-dihydro-
    4-methyl-6-phenylpyrazolo[3,4-c]-
    pyrazol-3-yl)-1-ethyl-, iodide
	EtOH	480(3.20),540(3.13)	48-0811-84

$C_{24}H_{22}O_2$
  Methanone, (4-methylphenyl)(6,7,8,9-
    tetramethyl-2-dibenzofuranyl)-
	EtOH	258s(4.43),275(4.57)	39-0799-84B

Methanone, (4-methylphenyl)(6,7,8,9- tetramethyl-3-dibenzofuranyl)-	EtOH	260(4.17),323(4.36)	39-0799-84B
Tribenzo[c,g,k][1,6]dioxacyclododecin, 6,11-dihydro-17,18-dimethyl-, (E)-	MeOH	208(4.53),266(3.20), 271(3.23)	44-1627-84
(Z)-	MeOH	208(4.55),275(3.48), 281(3.49)	44-1627-84

$C_{24}H_{22}O_2S$
  25-Thiatetracyclo[16.2.2.2^{8,11}.1^{3,6}]-
    pentacosa-3,5,8,10,18,20,21,23-octa-
    ene-2,7-dione
	C_6H_{12}	<u>300(4.2)</u>	44-1177-84

$C_{24}H_{22}O_3$
  Methanone, (4-methoxyphenyl)(6,7,8,9-
    tetramethyl-2-dibenzofuranyl)-
	EtOH	254(4.29),275(4.50)	39-0799-84B

Methanone, (4-methoxyphenyl)(6,7,8,9- tetramethyl-3-dibenzofuranyl)-	EtOH	256s(4.03),322(4.44)	39-0799-84B

$C_{24}H_{22}O_4$
  Ethanone, 1,1'-[1,2-phenylenebis(meth-
    yleneoxy-2,1-phenylene)]bis-
	MeOH	212(4.56),248(4.07),
		307(3.85)
			44-1627-84

$C_{24}H_{22}O_5$
  1H-2-Benzopyran-1-one, 3,4-dihydro-
    8-methoxy-3-[4-methoxy-3-(phenyl-
    methoxy)phenyl]-
	EtOH	232s(4.20),288(3.70),
		310(3.78)
			44-0742-84

$C_{24}H_{22}O_9$ Zoapatanolide D acetate	MeOH	205(4.15)	102-0125-84

$C_{24}H_{22}O_{10}$
  Pentanedioic acid, 2-[(9,10-dihydro-
    1-hydroxy-4,8-dimethoxy-9,10-dioxo-
	n.s.g.	205(4.28),229(4.60),
		251(4.30),280(4.04),
			5-0306-84

Compound	Solvent	$\lambda_{max}(\log \epsilon)$	Ref.
2-anthracenyl)methyl]-3-oxo-, dimethyl ester (cont.)		422(3.82),447(3.97), 465(3.98),500(3.75)	5-0306-84
$C_{24}H_{22}O_{14}$ Flavone, 4',5-dihydroxy-3,3'-dimethoxy-6,7-(methylenedioxy)-, 4'-β-D-glucuronide	MeOH	246(4.14),278(4.22), 338(4.39)	102-2043-84
	MeOH-NaOMe	283(4.22),337(4.33)	102-2043-84
	MeOH-AlCl$_3$	262(4.06),292(4.23), 367(4.38)	102-2043-84
$C_{24}H_{23}BrO_5$ Propanoic acid, 2,2-dimethyl-, 3-(bromomethyl)-9,10-dihydro-9,10-dioxo-2-(3-oxobutyl)-1-anthracenyl ester	MeOH	213(4.45),260(4.68), 337(3.80)	78-3677-84
$C_{24}H_{23}ClN_2$ Pyrrolo[1,2-a:5,4-b']diquinoline, 5-(chloromethyl)-5,6-dihydro-2,5,8,11-tetramethyl-	C_6H_{12}	232(4.34),243s(4.31), 252s(4.26),277(4.48), 288(4.60),303s(3.96), 316(3.86),335s(3.78), 341(3.78),362(3.72), 414s(2.97),432s(2.89)	39-2529-84C
$C_{24}H_{23}ClN_2O_2$ Pyrrolo[1,2-a:5,4-b']diquinoline, 5-(chloromethyl)-5,6-dihydro-2,11-dimethoxy-5,8-dimethyl-	C_6H_{12}	235(4.46),247s(4.33), 270s(4.35),279(4.59), 291(4.68),297s(4.45), 330.5(3.93),345.5(4.04), 363(3.81)	39-2529-84C
$C_{24}H_{23}NO_3$ Benzamide, 4-methoxy-N-(1-methyl-2-oxo-2-phenylethyl)-N-(phenylmethyl)-	ether	235(4.42),328(2.17)	48-0177-84
$C_{24}H_{23}NO_4S$ 1H-Pyrrole-3,4-dicarboxylic acid, 2-[2-(methylthio)-2-phenylethenyl]-1-(phenylmethyl)-, dimethyl ester	MeOH	303(3.98)	44-3314-84
$C_{24}H_{23}NO_5$ Chelerythrine, 8-acetonyldihydro-	EtOH	231(4.53),284(4.62), 319s(4.16)	100-0001-84
$C_{24}H_{23}N_3O_2$ 4H-Pyrido[1,2-a]pyrimidine-3-carboxamide, 6,7,8,9-tetrahydro-6-methyl-N-(2-methylphenyl)-4-oxo-9-(phenylmethylene)-	EtOH	225(4.24),362(4.39)	118-0582-84
Quinolinium, 1-ethyl-2-[2-(2-ethyl-4,5-dihydro-4,5-dioxo-1-phenyl-1H-pyrazolium-3-yl)ethenyl]-, diiodide	EtOH	500(3.70)	48-0695-84
$C_{24}H_{23}N_3O_3$ Quinoxaline, 1,2-dihydro-1-[(4-nitrophenyl)methyl]-3-phenyl-2-propoxy-	MeOH	236(4.33),260(4.40), 291s(--),369(3.67)	103-0542-84
Quinoxaline, 1,2-dihydro-3-(4-nitrophenyl)-1-(phenylmethyl)-2-propoxy-	MeOH	236(4.37),276(4.31), 333(4.22),412(3.59)	103-0542-84
$C_{24}H_{23}N_3O_7S_2$ 5-Thia-1-azabicyclo[4.2.0]oct-2-ene-2-	EtOH	264(4.12)	39-2117-84C

Compound	Solvent	$\lambda_{max}(\log \epsilon)$	Ref.
carboxylic acid, 7-[(hydroxyphenyl-acetyl)amino]-3-methyl-7-(methylthio)-8-oxo-, (4-nitrophenyl)methyl ester (cont.)			39-2117-84C
$C_{24}H_{23}N_3O_9$			
Galanthan-1,2-diol, 3,12-didehydro-9-(2,4-dinitrophenoxy)-10-methoxy-, 1-acetate, (1α,2β)-	EtOH	250(3.70),285(3.79)	100-1003-84
$C_{24}H_{24}Br_3NO_3$			
Ethanamine, 2-(2-bromophenoxy)-N,N-bis[2-(2-bromophenoxy)ethyl]-	EtOH	235(3.80),278(3.77),284(3.76)	12-2059-84
$C_{24}H_{24}N_2$			
1H-1,5-Benzodiazepine, 2,3-dihydro-2-methyl-2,4-bis(4-methylphenyl)-	EtOH	267(4.33),364(3.75)	103-0106-84
$C_{24}H_{24}N_2O$			
Quinoxaline, 1,2-dihydro-3-phenyl-1-(phenylmethyl)-2-propoxy-	MeOH	240(4.34),262s(--),296(4.09),372(3.84)	103-0542-84
$C_{24}H_{24}N_2O_2$			
1H-1,5-Benzodiazepine, 2,3-dihydro-2,4-bis(4-methoxyphenyl)-2-methyl-	EtOH	277(4.26),365(3.81)	103-0106-84
$C_{24}H_{24}N_2O_5S$			
4-Thia-2,6-diazabicyclo[3.2.0]hept-2-ene-6-acetic acid, α-(1-methyl-ethenyl)-7-oxo-3-(phenoxymethyl)-, (4-methoxyphenyl)methyl ester [1R-[1α,5α,6(R*)]]-	MeCN	223(4.26),269(3.43),274(3.41)	78-1907-84
$C_{24}H_{24}N_4$			
4,4'-Bipyridinium, 1,1''-(1,4-butane-diyl)bis-, dibromide	H_2O	262(4.59)	64-0074-84B
$C_{24}H_{24}N_4Ni$			
Nickel, [2,3,7,8-tetrahydro-5,10,15,20-tetramethyl-21H,23H-porphinato(2-)-$N^{21},N^{22},N^{23},N^{24}$]-, (SP-4-2)-	benzene	386s(4.36),404(4.57),418(4.62),481(3.63),511(3.48),555(3.72),594(4.20)	35-5164-84
$C_{24}H_{24}N_4O_3S$			
Thiazolo[2,3-b]purine-4,8(1H,7H)-dione, 5a-ethoxy-5,5a-dihydro-3,5-dimethyl-7-phenyl-2-(phenylmethyl)-	MeOH	268(4.19),286s(4.07)	2-0316-84
$C_{24}H_{24}N_6$			
1,3,5-Triazine-2,4,6-triamine, N,N',N''-tris(phenylmethyl)-	MeOH	261(3.01),268(2.8)	24-1523-84
$C_{24}H_{24}N_6O_9$			
Adenosine, 8-(aminocarbonyl)-N-benz-oyl-, 2',3',5'-triacetate	MeOH	235(4.29),293(4.25)	150-3601-84M
$C_{24}H_{24}O$			
2,5-Cyclohexadien-1-one, 2-(1,1-dimeth-ylethyl)-4-(diphenylmethylene)-6-methyl-	hexane?	210(4.27),261(4.07),268(4.06),363(4.36)	73-1949-84

Compound	Solvent	$\lambda_{max}(\log \epsilon)$	Ref.
$C_{24}H_{24}O_2$			
Benzoic acid, 4-[2-(7,8-dimethyl-2-naphthalenyl)-1-propenyl]-, ethyl ester, (E)-	EtOH	220(4.56),240(4.43), 288(4.46),318(4.43)	44-5265-84
(Z)-	EtOH	231(4.74),287(4.32)	44-5265-84
5,12:6,11-Diethenodibenzo[a,e]cyclooct-ene, 13,16-diethoxy-5,6,11,12-tetra-hydro-, (5α,6α,11α,12α)-	CHCl$_3$	274(4.06),283(4.05)	44-0536-84
$C_{24}H_{24}O_2S_2$			
1,3-Dithiane, 2-[3,4-bis(phenylmeth-oxy)phenyl]-	EtOH	230(4.28),279(3.69)	39-1547-84C
$C_{24}H_{24}O_5$			
Propanoic acid, 2,2-dimethyl-, 9,10-di-hydro-3-methyl-9,10-dioxo-2-(3-oxo-butyl)-1-anthracenyl ester	MeOH	209(4.33),258(4.68), 333(3.76)	78-3677-84
$C_{24}H_{24}O_7$			
6H-Benzo[d]naphtho[1,2-b]pyran-11-ol, 2,3,8,9-tetramethoxy-6-methyl-, acetate	EtOH	230(4.46),280(4.42), 320(3.96)	23-2435-84
$C_{24}H_{24}O_9$			
2,10-Anthracenedicarboxylic acid, 10-[(ethoxycarbonyl)oxy]-9,10-di-hydro-1-methoxy-3-methyl-9-oxo-, 10-ethyl 2-methyl ester	MeOH	224(4.16),259(5.01), 336(3.49),359(3.77), 377(3.86),397(3.81)	78-3677-84
2,5-Furandione, dihydro-3,4-bis[(3,4,5-trimethoxyphenyl)methylene]-, (E,E)-	toluene	410(4.00)	48-0233-84
(E,Z)-	toluene	435(4.20)	48-0233-84
(Z,Z)-	toluene	445(4.34)	48-0233-84
Podophyllotoxin acetate	EtOH	288(3.60),291(3.61), 296s(--)	102-1029-84
$C_{24}H_{24}O_{12}$			
Flavone, 4',5,7-trihydroxy-3,3',6-tri-methoxy-, 4'-β-D-glucuronide	MeOH	250(4.18),271(4.23), 337(4.29)	102-2043-84
	MeOH-NaOMe	273(4.39),305(4.05), 374(4.21)	102-2043-84
	MeOH-NaOAc	273(4.38),308(4.19), 370(4.19)	102-2043-84
	MeOH-AlCl$_3$	263(4.30),280(4.36), 358(4.31)	102-2043-84
	+ HCl	263(4.29),274(4.30), 354(4.34)	102-2043-84
$C_{24}H_{24}S_2$			
Dicyclopenta[a,e]dicyclopropa[c,g]-cyclooctene, 1,5-bis[(1,1-dimethyl-ethyl)thio]-	CH$_2$Cl$_2$	239(4.51),266(4.49), 337(4.74),372(4.43), 413(4.39),525(2.74)	89-0063-84
$C_{24}H_{25}NO_2$			
3aH-Indol-3a-ol, 2,3,4,5,6,7-hexahydro-2,2-dimethyl-7-(phenylmethyl)-, benzoate	EtOH	229(4.29),273(4.22)	12-0577-84
$C_{24}H_{25}NO_3$			
3aH-Indol-3a-ol, 1-benzoyl-1,2,3,4,5,6-hexahydro-2,2-dimethyl-, benzoate	EtOH	224(4.37),270(3.78)	12-0577-84

Compound	Solvent	λ_{max}(log ϵ)	Ref.
C$_{24}$H$_{25}$NO$_3$S			
Benzenesulfonamide, 2,4,6-trimethyl-N-(2-oxo-2-phenylethyl)-N-(phenylmethyl)-	ether	240.6(4.43),324.2(1.85)	48-0177-84
C$_{24}$H$_{25}$NO$_4$			
4H-Furo[3,2-g][1]benzoxecin-3-carboxylic acid, 2-cyano-2,3,3a,5,6,7,8-13b-octahydro-2-phenyl-, methyl ester, (2α,2β,3aα,13bβ)-	EtOH	258s(3.10),273(3.42), 278(3.38)	24-2157-84
2-Octenoic acid, 8-[2-(3-cyano-3-phenyloxiranyl)phenoxy]-, methyl ester, [2α(E),3α]-	EtOH	267s(3.36),276s(3.56), 281(3.58)	24-2157-84
Tylohirsutinine	MeOH	254(4.6),262(4.4), 278(3.4),284(3.1), 304(2.6),315(2.5), 343(1.9),360(1.8)	102-1765-84
C$_{24}$H$_{25}$NO$_4$S			
1H-Pyrrole-3,4-dicarboxylic acid, 2,5-dihydro-2-[2-(methylthio)-2-phenylethenyl]-1-(phenylmethyl)-, dimethyl ester	MeOH	277(3.81)	44-3314-84
C$_{24}$H$_{25}$NO$_6$			
Corynoline, 8-ax'-acetonyl-, (±)-	MeOH	237(3.30),288(3.20)	100-0001-84
Ethanaminium, N-[3-(4-methoxyphenyl)-1-phenyl-2-propenylidene]-, 3-methoxy-1-(methoxycarbonyl)-2,3-dioxopropylide, (E,?)-	EtOH	276(4.19),428(4.13)	150-1531-84M
4,5-Isoxazoledicarboxylic acid, 2-ethyl-2,3-dihydro-3-[2-(4-methoxyphenyl)ethenyl]-3-phenyl-, dimethyl ester, (E)-	EtOH	266(4.23),306(3.85)	150-1531-84M
C$_{24}$H$_{25}$NO$_8$			
Spiro[1,3-dioxolo[4,5-g]isoquinoline-5(6H),2'-[2H]inden]-1'(3'H)-one, 6-acetyl-7,8-dihydro-3',3',4',5'-tetramethoxy-, (±)-	MeOH	226.5(4.36),286(4.28)	94-2230-84
C$_{24}$H$_{25}$N$_3$O$_4$P			
Phosphonium, methylbis(4-nitrophenyl)-[4-(1-piperidinyl)phenyl]-, iodide	MeOH	251(4.48),303(4.57)	139-0267-84C
C$_{24}$H$_{25}$N$_5$O$_2$			
Benzaldehyde, 4-nitro-, [bis[4-(dimethylamino)phenyl]methylene]hydrazone	C$_6$H$_{12}$	404(4.30)	97-0021-84
C$_{24}$H$_{25}$N$_5$O$_3$S			
7H-Pyrazolo[4,3-d]pyrimidine-7-thione, 5-amino-3-[2-deoxy-3,5-bis-O-(phenylmethyl)-β-D-erythro-pentofuranosyl]-1,4-dihydro-	MeOH	235(4.11),294(3.54), 366(4.14)	44-0528-84
C$_{24}$H$_{25}$N$_5$O$_4$			
Meleagrin, N(14)-methyl-	MeOH	229(4.02),285s(3.63), 347(4.06)	94-0094-84
C$_{24}$H$_{25}$N$_5$O$_9$			
Adenosine, N-benzoyl-8-(hydroxymethyl)-,	MeOH	233(4.18),282(4.31)	150-3601-84M

Compound	Solvent	λ_{max} (log ϵ)	Ref.
2',3',5'-triacetate (cont.)			150-3601-84M
$C_{24}H_{26}N_2O_2P$ Phosphonium, methyl(4-nitrophenyl)-phenyl[4-(1-piperidinyl)phenyl]-, iodide	MeOH	252(4.24),304(4.53)	139-0267-84C
$C_{24}H_{26}N_2O_3$ 5,10[3',4']-Furanopyrido[3,2-g]quino-line-12,14-dione, 2,8-bis(1,1-di-methylethyl)-5,10,11,15-tetrahydro-	MeCN	276(4.073),280(4.080)	5-0877-84
3-Pyridinecarboxamide, 1,4-dihydro-4-[1-(4-methoxyphenyl)-3-oxobutyl]-1-(phenylmethyl)-	MeCN	338(3.81)	44-0026-84
$C_{24}H_{26}N_2O_4$ 2,4,6-Cycloheptatrien-1-one, 2,2'-(1,2-cyclohexanediyldiimino)bis[7-acetyl-, cis	MeOH	248(4.43),355(4.05), 421(4.21)	18-0623-84
trans	MeOH	249(4.46),358(4.06), 421(4.20)	18-0623-84
1H-Pyrido[3,4-b]indole-1-propanoic acid, 2,3,4,9-tetrahydro-3-(meth-oxycarbonyl)-9-methyl-2-(phenyl-methyl)-, methyl ester, (1S-cis)-	MeOH	225(4.60),275s(3.90), 282(3.92),290(3.83)	94-1313-84
$C_{24}H_{26}N_2O_5$ 1H-Indole-3-propanoic acid, 1-[2-(meth-oxycarbonyl)amino]-2-methyl-1-oxo-propyl]-, phenylmethyl ester	EtOH	243(4.28),265s(3.94), 273s(3.86),293(3.82), 301(3.86)	94-1373-84
Macrostomine N-oxide	EtOH	243(4.52),290(3.43), 320(3.69),333(3.48)	105-0071-84
$C_{24}H_{26}N_2O_5S$ 4-Thia-2,6-diazabicyclo[3.2.0]hept-2-ene-6-acetic acid, α-(1-methyl-ethyl)-7-oxo-3-(phenoxymethyl)-, (4-methoxyphenyl)methyl ester, [1R-[1α,5α,6(R*)]]-	MeCN	224(4.20),268(3.38), 274(3.36)	78-1907-84
$C_{24}H_{26}N_2O_7S$ 4-Thia-1-azabicyclo[3.2.0]heptane-2-carboxylic acid, 3,3-dimethyl-7-oxo-6-[(phenoxyacetyl)amino]-, (4-methoxyphenyl)methyl ester, 4-oxide, [2S-(2α,4β,5α,6β)]-	MeCN	222(4.34),268(3.39), 274(3.38)	78-1907-84
$C_{24}H_{26}N_4$ 4,4'-Bipyridinium, 1,1"-(1,2-ethanedi-yl)bis[1-methyl-, tetraperchlorate	H_2O	265(4.64)	64-0074-84B
$C_{24}H_{26}N_4O_4$ 9H-Purine, 6-methyl-9-[[2-(phenylmeth-oxy)-1-[(phenylmethoxy)methyl]eth-oxy]methyl]-	pH 1 pH 7 pH 13	247(3.91) 247(3.94) 247(3.93)	23-2702-84 23-2702-84 23-2702-84
$C_{24}H_{26}N_4O_{13}$ 1H-Imidazole-5-carboxamide, 1-[(4-ni-trophenyl)methyl]-4-(2,3,4,6-tetra-O-acetyl-β-D-galactopyranosyloxy)-	M HCl H_2O M NaOH	253(4.02) 252(4.16) 256(4.07)	4-0849-84 4-0849-84 4-0849-84

Compound	Solvent	λ_{max}(log ϵ)	Ref.
1H-Imidazole-5-carboxamide, 1-[(4-ni-trophenyl)methyl]-4-(2,3,4-tri-O-acetyl-β-D-glucopyranosyloxyuronic acid methyl ester)	M HCl H$_2$O M NaOH	253(4.12) 253(4.21) 250(4.22)	4-0849-84 4-0849-84 4-0849-84
C$_{24}$H$_{26}$O$_4$ Mulberrofuran L	MeOH	244s(3.85),254s(3.79), 280s(3.81),315(4.30), 330s(4.19)	142-2805-84
C$_{24}$H$_{26}$O$_6$ 6H-Benzo[d]naphtho[1,2-b]pyran, 11-eth-oxy-2,3,8,9-tetramethoxy-6-methyl-	EtOH	215(4.88),260(4.51), 300(4.30)	23-2435-84
2-Propenoic acid, 3-(3-methoxyphenyl)-, 1,4-butanediyl ester	MeCN	276(4.60)	40-0022-84
2-Propenoic acid, 3-(4-methoxyphenyl)-, 1,4-butanediyl ester	MeCN	308(4.72)	40-0022-84
C$_{24}$H$_{26}$O$_8$ 2-Propenoic acid, 3-(3,4-dimethoxyphen-yl)-, 1,2-ethanediyl ester	MeCN	323(4.62)	40-0022-84
C$_{24}$H$_{26}$O$_9$ 1H,3H-Furo[3,4-c]furan-3a(4H)-ol, 1,4-bis(4-acetoxy-3-methoxyphenyl)dihy-dro-, [1S-(1α,3aα,4α,6aα)]-	EtOH	218.2(4.26),273.8(3.74), 279.2(3.72)	94-2730-84
Shinjulactone F, 1,20-di-O-acetyl-	EtOH	245(3.90),285(3.46)	18-2885-84
C$_{24}$H$_{26}$S 25-Thiatetracyclo[16.2.2.2^{8,11}.1^{3,6}]-pentacosa-3,5,8,10,18,20,21,23-octa-ene	C$_6$H$_{12}$	<u>250(3.9),275s(2.9)</u>	44-1177-84
C$_{24}$H$_{27}$NO$_3$ Hypoestestatin 1	MeOH	258(4.45),286(4.19), 342(2.94),360(2.71)	100-0913-84
C$_{24}$H$_{27}$NO$_4$ Hypoestestatin 2	MeOH	259(4.62),285(4.41), 340(3.2036)	100-0913-84
C$_{24}$H$_{27}$NO$_5$ Tylohirsutinedine, 13a-methyl-	MeOH	258(4.9),278(3.5), 286(3.3),314(2.1), 342(1.8),360(1.6)	102-1765-84
	MeOH-NaOH	238(4.0),256(3.9), 280(2.8),296(2.5), 336(1.8)	102-1765-84
Tylophorinicine	MeOH	258(4.4),287(4.2), 302(3.6),339(3.1), 335(2.9)	102-1206-84
C$_{24}$H$_{27}$NO$_8$ 1-Azaspiro[5.5]undeca-2,4-diene-2,3,4,5-tetracarboxylic acid, 1-phenyl-, tetramethyl ester	EtOH	240s(4.56),280(4.34), 353(4.32)	103-0295-84
C$_{24}$H$_{27}$NP Phosphonium, methyldiphenyl[4-(1-pip-eridinyl)phenyl]-, iodide	MeOH	305(4.37)	139-0253-84C

Compound	Solvent	$\lambda_{max}(\log \epsilon)$	Ref.
$C_{24}H_{27}N_3O_7$			
2-Naphthacenecarboxamide, 4,7-bis(di-methylamino)-1,4,4a,5,12,12a-hexahy-dro-3,10,11,12a-tetrahydroxy-6-methyl-7,12-dioxo- (7-dimethylaminoanhydro-tetracycline)	MeOH	230s(4.38),270(4.81), 329(3.55),410(3.43)	23-2583-84
2-Naphthacenecarboxamide, 4,9-bis(di-methylamino)-1,4,4a,5,12,12a-hexahy-dro-3,10,11,12a-tetrahydroxy-6-methyl-7,12-dioxo-	MeOH	244(4.38),278(4.68), 320s(3.74),408(3.77)	23-2583-84
$C_{24}H_{27}N_5O_4$			
7H-Purin-2-amine, 6-methoxy-7-[[2-(phenylmethoxy)-1-[(phenylmethoxy)-methyl]ethoxy]methyl]-	pH 1	247(3.86),288(3.82)	23-2702-84
	pH 7	253(4.00),283(3.86)	23-2702-84
	pH 13	253(3.99),283(3.83)	23-2702-84
9H-Purin-2-amine, 6-methoxy-9-[[2-(phenylmethoxy)-1-[(phenylmethoxy)-methyl]ethoxy]methyl]-	pH 1	244(3.93),289(3.93)	23-2702-84
	pH 7	247(4.05),282(4.00)	23-2702-84
	pH 13	247(4.05),282(3.99)	23-2702-84
$C_{24}H_{27}N_7O_7$			
8-Deazafolic acid, N^{10}-nitroso-, diethyl ester	pH 1	250(4.39),285(4.18)	87-1710-84
	pH 7	272(4.35),294s(4.11)	87-1710-84
	pH 13	274(4.32),334s(3.89)	87-1710-84
$C_{24}H_{28}$			
1H-Indene, 1-(2,3-dihydro-2,2,3-tri-methyl-1H-inden-1-ylidene)-2,3-di-hydro-2,2,3-trimethyl-	isopentane	238(4.1),330(4.4)	77-1696-84
Naphthalene, 1-(3,4-dihydro-2,2-dimeth-yl-1(2H)-naphthalenylidene)-1,2,3,4-tetrahydro-2,2-dimethyl-, (E)-	EtOH-CH₂Cl₂	222(4.29),310(3.50)	24-2300-84
$C_{24}H_{28}Br_2O_2$			
2,12-Ethano-2,6-methano-2H-1-benzoxe-cin-15-one, 3,5-dibromo-4,10-bis(1,1-dimethylethyl)-7,8-dihydro-	CHCl₃	260(4.24),310(3.72), 358(2.64)	94-1641-84
$C_{24}H_{28}ClN_3O_4$			
Quinazoline, 2-chloro-4-[[3-(1,1-di-methylethyl)-2-phenyl-5-oxazolidin-yl]methoxy]-6,7-dimethoxy-	EtOH	238(4.70)	4-1189-84
isomer	EtOH	236(4.70)	4-1189-84
$C_{24}H_{28}N_2O_2$			
1H-Imidazole, 1-[9-(4-methoxy-2,3,6-trimethylphenyl)-3,7-dimethyl-1-oxo-2,4,6,8-nonatetraenyl]-	MeOH	212(4.33),262(4.33), 394(4.60)	107-0725-84
$C_{24}H_{28}N_2O_5$			
3H-2,3-Benzodiazepine, 3-acetyl-1-(3,4-dimethoxyphenyl)-5-ethyl-4,5-dihydro-7,8-dimethoxy-4-methylene-	EtOH	208(4.56),238(4.45), 319(4.18)	24-1476-84
3H-2,3-Benzodiazepine, 3-acetyl-1-(3,4-dimethoxyphenyl)-5-ethyl-7,8-dimeth-oxy-4-methyl-	EtOH	208(4.52),246(4.52), 316(4.06)	24-1476-84
$C_{24}H_{28}N_2O_6$			
Pyrrolo[2,1-a]isoquinoline-3-carboxylic acid, 8,9-diethoxy-1,2,3,5,6,10b-hexa-hydro-1-nitro-2-phenyl-, methyl ester	EtOH	200(4.06),285(2.60)	78-0369-84

Compound	Solvent	$\lambda_{max}(\log \epsilon)$	Ref.
$C_{24}H_{28}N_2O_7$			
Carbamic acid, [[1-[(3,4-dimethoxyphenyl)methylene]-3,4-dihydro-6,7-dimethoxy-2(1H)-isoquinolinyl]carbonyl]-, ethyl ester, (Z)-	MeOH	296s(4.13),334(4.42)	78-4003-84
$C_{24}H_{28}N_3$			
Methanaminium, N-[4-[[4-(dimethylamino)phenyl][4-(methylamino)phenyl]methylene]-2,5-cyclohexadien-1-ylidene]-N-methyl-, chloride (pentamethylpararosaniline)	HOAc	578(5.00)	98-0596-84
$C_{24}H_{28}N_4O_2$			
Oxazolium, 3,3'-(1,2-ethanediyl)-bis[5-amino-4-ethyl-2-phenyl-, diperchlorate	MeOH	214(4.11),336(3.81)	2-0289-84
$C_{24}H_{28}N_4O_4$			
1,9-Dioxa-2,6,10,14-tetraazacyclohexadeca-2,10-diene-7,15-dione, 3,11-bis(4-methylphenyl)-	EtOH	255(4.1)	5-1696-84
$C_{24}H_{28}N_4O_6$			
1,9-Dioxa-2,6,10,14-tetraazacyclohexadeca-2,10-diene-7,15-dione, 3,11-bis(4-methoxyphenyl)-	EtOH	267(4.5)	5-1696-84
$C_{24}H_{28}N_6O_5$			
L-Glutamic acid, N-[4-[1-[(2,4-diaminopyrido[3,2-d]pyrimidin-6-yl)methyl]-butyl]benzoyl]-	pH 13	238(4.57),343(3.70)	87-0376-84
$C_{24}H_{28}O_2$			
Benzoic acid, 4-[2-(5,6,7,8-tetrahydro-8,8-dimethyl-2-naphthalenyl)-1-propenyl]-, ethyl ester, (E)-	EtOH	233(4.15),305(4.42)	44-5265-84
$C_{24}H_{28}O_3$			
Spiro[bicyclo[4.1.0]heptane-2,1'-[4]-cyclopenten]-3'-one, 4'-(acetoxymethyl)-3,7,7-trimethyl-2'-(phenylmethylene)-, [1R-(1α,2α,3β,6α)]-	EtOH	229(3.87),309(4.06)	39-0215-84C
$C_{24}H_{28}O_4$			
6,6';7,3'a-Diligustilide	MeOH	272(4.10)	64-0872-84C
Levistolide A	MeOH	275(4.17)	5-0397-84
$C_{24}H_{28}O_5$			
2H-1-Benzopyran-2-one, 7-[(1,4,4a,5,6-7,8,8a-octahydro-6-hydroxy-2,5,5,8a-tetramethyl-4-oxo-1-naphthalenyl)-methoxy]-	EtOH	204(4.43),215(4.18), 240(3.59),251s(3.46), 290s(3.96),321(4.18)	49-1207-84
$C_{24}H_{28}O_7$			
Benzeneacetic acid, 2-(6,7-dimethoxy-1-methyl-1H-2-benzopyran-3-yl)-4,5-dimethoxy-, ethyl ester	EtOH	215(4.82),270(4.42), 300(4.18)	23-2435-84
1-Butanone, 2-methyl-1-[1,3,7,9-tetrahydroxy-2,8-dimethyl-6-(3-methyl-1-oxobutyl)-4-dibenzofuranyl]-	EtOH	223s(4.23),266(4.59), 293(4.37),356(3.86)	39-2573-84C

Compound	Solvent	$\lambda_{max}(\log \epsilon)$	Ref.

$C_{24}H_{28}O_8$
1H,3H-Furo[3,4-c]furan-3a(4H)-ol,
  1,4-bis(3,4-dimethoxyphenyl)di-
  hydro-, acetate, [1S-(1α,3aα,4α,6aα)]-
	EtOH	233(4.31),279(3.81)	94-2730-84

$C_{24}H_{28}O_9$
  Eruberin A
	MeOH	225(4.29),275(3.24),
		  282(3.21)
			94-0490-84

$C_{24}H_{28}O_{10}$ Zoapatanolide C acetate	MeOH	206(4.71)	102-0125-84

$C_{24}H_{29}ClO_2S$
  Dibenz[b,f]oxepin-1(2H)-one, 11-[(4-
  chlorophenyl)thio]-3,4,6,7,8,9,10,11-
  octahydro-2,2,6,6-tetramethyl-
	C_6H_{12}	250(4.11)	5-0576-84

$C_{24}H_{29}NO_5$
  Septicine, 13a-hydroxy-
	MeOH	236s(4.1),248s(4.3),
		  256(4.9),287(4.0),
		  302s(3.7),338(1.8),
		  355(1.6)
			102-1765-84

$C_{24}H_{29}NO_8$
  Benzoic acid, 6-[[6-[2-(dimethylamino)-
  ethyl]-4-methoxy-1,3-benzodioxol-5-
  yl]acetyl]-2,3-dimethoxy-, methyl
  ester
	MeOH	216(4.40),231s(4.17),
		  274(4.08),298s(3.97)
			78-1971-84

$C_{24}H_{29}N_3O_6$
  Oxepino[2,3-d]pyrimidine-5,6-dicarbox-
  ylic acid, 3,4,7,8-tetrahydro-7,8-
  dimethyl-2-[(1-methylethyl)amino]-
  4-oxo-3-phenyl-, 6-ethyl 5-methyl
  ester
	CHCl₃	277(3.92),330(4.31)	24-0585-84

Oxepino[2,3-d]pyrimidine-5,6-dicarbox-
  ylic acid, 3,4,7,8-tetrahydro-8,8-
  dimethyl-2-[(1-methylethyl)amino]-
  4-oxo-3-phenyl-, 6-ethyl 5-methyl
  ester
	CHCl₃	287(3.84),334(4.27)	24-0585-84

$C_{24}H_{30}$
  Phenanthrene, 2,4-bis(1,1-dimethyleth-
  yl)-5,7-dimethyl-, (±)-
	CHCl₃	278(4.66),313(3.96),
		  325(3.86),360(2.50),
		  378s(2.31)
			24-0336-84

$C_{24}H_{30}N_2$
  4,4'-Bipyridinium, 1-heptyl-1'-(phen-
  ylmethyl)-, dibromide
	H_2O	260(4.38)	39-0367-84B

$C_{24}H_{30}N_2O_4$
  Indolo[2,3-a]quinolizin-4(3H)-one,
  2,6,7,12-tetrahydro-3-(2-hydroxy-
  ethyl)-2-[2-(tetrahydro-2H-pyran-
  2-yloxy)ethyl]-, (2S*,3R*)-
	EtOH	231(4.46),309(4.28),
		  321(4.30)
			39-1237-84C

$C_{24}H_{30}N_2O_6$
  Acetic acid, N-[(3,4-dimethoxyphenyl)-
  [2-(1-ethyl-2-oxopropyl)-4,5-dimeth-
  oxyphenyl]methylene]-
	EtOH	211(4.58),231s(4.43),
		  292(4.26),312(4.26)
			24-1476-84

Compound	Solvent	$\lambda_{max}(\log \epsilon)$	Ref.
$C_{24}H_{30}N_2O_9$ L-ribo-Hexar-1-amic acid, 4-(acetyl-amino)-4,5-dideoxy-N-[1-(3,4-dihy-dro-8-hydroxy-1-oxo-1H-2-benzopyran-3-yl)-3-methylbutyl]-, γ-lactone, 2-acetate	MeOH	246(3.73),314(3.59)	78-2519-84
$C_{24}H_{30}N_6O_2S_4$ 19,22,27,30-Tetrathia-1,8,9,16,33,34-hexaazatetracyclo[14.8.8.1^{3,7}.1^{10,14}]-tetratriaconta-3,5,7(34),8,10,12-14(33)-heptaene-2,15-dione, trans	MeOH $C_6H_4Cl_2$-BuOH	323(4.12),460(2.15) 328(4.10),450(2.34)	18-2144-84 18-2144-84
$C_{24}H_{30}O_2$ 2,12-Ethano-2,6-methano-2H-1-benzoxe-cin-15-one, 4,10-bis(1,1-dimethyl-ethyl)-7,8-dihydro-	C_6H_{12}	280(3.31),355(2.39), 370(2.32),380(2.19)	94-1641-84
$C_{24}H_{30}O_5$ 2H-1-Benzopyran-2-one, 7-[(decahydro-4-hydroxy-1,2,4a,5-tetramethyl-6-oxo-1-naphthalenyl)methoxy]- (kama-lonol)	EtOH	205(4.42),215s(4.15), 240s(3.56),251s(3.42), 296s(3.94),322(4.18)	49-1207-84
$C_{24}H_{30}O_5S$ 3-Cyclohexene-1-carboxylic acid, 4-(2,2-dimethyl-1-oxopropoxy)-2-methylene-1-[3-oxo-3-(phenyl-thio)propyl]-, ethyl ester	EtOH	236(4.24),250(4.23)	12-2037-84
$C_{24}H_{30}O_8$ Trichomatolide E diacetate	MeOH	218(2.81),225(2.74)	102-1439-84
$C_{24}H_{30}O_9$ 1H,3H-Furo[3,4-c]furan-3a(4H)-ol, di-hydro-1,4-bis(3,4,5-trimethoxyphen-yl)-, [1S-(1α,3aα,4α,6aα)]-	EtOH	226s(4.19),275.7(3.53)	94-4482-84
$C_{24}H_{30}O_{10}$ Dauroside L	EtOH	229(4.11),251s(3.65), 259s(3.57),288(3.67), 343(3.74)	105-0226-84
$C_{24}H_{31}NO_6$ 3-Pyridinecarboxylic acid, 4-[2-(2-eth-oxy-2-oxoethoxy)-3-methoxyphenyl]-, 1,1-diethylpropyl ester	EtOH	261(3.77)	44-0056-84
$C_{24}H_{31}N_7O_7$ 8-Deazafolic acid, 5,6,7,8-tetrahydro-N^{10}-nitroso-, diethyl ester	pH 1 pH 7 pH 13	273(4.28) 282(4.20) 285(4.22)	87-1710-84 87-1710-84 87-1710-84
$C_{24}H_{32}$ [8,8'-Bispiro[4.5]deca-7,9-diene]-6,6'-diylium, 6,6',10,10'-tetra-methyl-, (deloc-6,7,8,9,10:6',7'-8',9',10')-	CH_2Cl_2	263(3.84),316s(--), 330(4.53),345(4.76), 363(4.78)	5-0340-84
1,4-Cyclohexadiene, 3,3'-(2-butene-1,4-diylidene)bis[1,5,6,6-tetra-methyl-	CH_2Cl_2	262(3.90),355s(--), 374.5(4.75),396.5(5.04), 421.5(5.09)	5-0340-84

Compound	Solvent	$\lambda_{max}(\log \epsilon)$	Ref.
$C_{24}H_{32}N_2$			
Bicyclo[3.2.0]hept-2-en-6-one, 7,7-dimethyl-4-(1-methylethylidene)-, azine	EtOH	220(4.46),247(4.49)	33-1379-84
1b isomer	EtOH	221(4.39),247(4.42)	33-1379-84
$C_{24}H_{32}N_2O_2$			
Pyridinium, 1-(2-carboxyethyl)-4-[2-[4-(dibutylamino)phenyl]ethenyl]-, hydroxide, inner salt	EtOH	486(4.59)	44-2546-84
$C_{24}H_{32}N_2O_6$			
[2,2'-Bi-2H-pyrrole]-5,5'-dicarboxylic acid, 3,3'-di-3-butenyl-4,4'-dihydroxy-2,2'-dimethyl-, diethyl ester	MeCN	309(4.04)	33-1957-84
$C_{24}H_{32}N_2O_8$			
Propanedioic acid, 2,2'-[(4,5-dimethyl-1,2-phenylene)bis(iminomethylidyne)]-bis-, tetraethyl ester	MeCN	222(4.31),292(4.53), 333(4.44)	62-0947-84A
$C_{24}H_{32}N_6O_6$			
8-Deazafolic acid, 5,6,7,8-tetrahydro-, diethyl ester	pH 1	294(4.12)	87-1710-84
	pH 7	300(4.43)	87-1710-84
	pH 13	197(4.43)	87-1710-84
$C_{24}H_{32}O_2$			
Benzoic acid, 4-[2-[2-(2,6,6-trimethyl-1-cyclohexen-1-yl)cyclopropyl]-1-propenyl]-, ethyl ester, (E)-	EtOH	294(4.36)	87-1516-84
(Z)-	EtOH	290(4.26)	87-1516-84
Dispiro[cyclobutane-1,6'(4'H)-[1,4]-ethanonaphthalene-10',1"-cyclobutane]-2,2"-dione, 1',4'a,5',8'a-tetrahydro-1',5',5',7',9',9'-hexamethyl-	MeCN	301(2.36),311(2.35), 322s(2.20),350s(1.43)	33-0774-84
2,4-Pentadienoic acid, 3-methyl-5-[2-methyl-4-(2,6,6-trimethylcyclohexen-yl)phenyl]-, diethyl ester, (E,E)-	EtOH	237(4.18),322(4.59)	87-1516-84
(Z,E)-	EtOH	235(4.28),326(4.51)	87-1516-84
$C_{24}H_{32}O_3$			
[1,1'-Biphenyl]-2-ol, 6-methoxy-5'-methyl-2'-(1-methylethyl)-4-pentyl-, acetate	EtOH	208(4.51),275(3.58)	39-2881-84C
$C_{24}H_{32}O_5$			
21,24-Dinorchol-4-ene-6,20-dione, 17α-acetoxy-16β,23-oxido-	EtOH	239(3.53)	70-0663-84
2H-9,4a-(Epoxymethano)phenanthrene-5,6-diol, 1,3,4,9,10,10a-hexahydro-1,1-dimethyl-7-(1-methylethyl)-, di-acetate, [4aR-(4aα,9α,10aß)]-(20-deoxocarnosol diacetate)	C_6H_{12}	222(3.59),260(2.69)	102-1677-84
$C_{24}H_{32}O_6$			
Benzo[4,5]cyclodeca[1,2-b]furan-2(3aH)-one, 8,9-diacetoxy-6,7,8,8a,9,10-12a,13-octahydro-1,5,8a,12-tetramethyl-, [3aS-(3aR*,4Z,8R,8aR*,9R*-12aS*)]- (brianthein W)	EtOH	228(3.88)	44-3398-84

Compound	Solvent	$\lambda_{max}(\log \epsilon)$	Ref.
Verrucosidin	MeOH	241(4.32),294(4.11)	44-3762-84
$C_{24}H_{32}O_7$ Ethanone, 2-[8-acetoxy-8a-(acetoxymeth-yl)octahydro-5,6-dimethylspiro[naph-thalen-1(2H),2'-oxiran]-5-yl]-1-(3-furanyl)-, [1R-(1α,4aα,5β,6α,8α,8aα)]-	EtOH	214(3.79),255(3.59)	102-0849-84
$C_{24}H_{32}O_8$ 1,12-Dioxacyclodocosa-4,6,15,17-tetra-ene-2,13-dione, 7,18-diacetoxy-,(E,E,Z,Z)-	MeOH	235(4.62)	89-0440-84
$C_{24}H_{33}NO_6$ Myxopyronin B	MeOH	213(4.48),297(4.31)	158-84-165
$C_{24}H_{34}N_2$ 3,8-Phenanthrolinium, 3,8-dihexyl-,dibromide	H_2O	250(4.73)	39-0367-84B
$C_{24}H_{34}N_2O_3$ Corynan-18-ol, 17-[(tetrahydro-2H-pyran-2-yl)oxy]-, (±)-	EtOH	228(4.58),281(3.88), 290s(3.78)	39-1237-84C
$C_{24}H_{32}N_2O_4$ Decanediamide, N,N'-dihydroxy-N,N'-bis(2-methylphenyl)-	MeOH	219(3.99),262(3.26)	34-0345-84
Decanediamide, N,N'-dihydroxy-N,N'-bis(4-methylphenyl)-	MeOH	220(4.41),256(3.97)	34-0345-84
1,4,10,13-Tetraoxa-7,16-diazacycloocta-decane, 7,16-diphenyl-	MeOH	257(4.53),301(3.64)	78-0793-84
$C_{24}H_{34}N_2O_4S$ 2'H-Androst-1-eno[3,2-d][1,2,3]thiadia-zole-2'-carboxylic acid, ethyl ester,(5α,17β)-	EtOH	337(4.18)	44-4773-84
$C_{24}H_{34}N_2O_6$ 2H-[1]Benzopyrano[3,4-c]pyridazine-3,4-dicarboxylic acid, 4a,5-dihydro-8-methoxy-5,5-dimethyl-, bis(1,1-dimethylethyl) ester	EtOH	219(4.56),262(4.18), 266s(4.15),303(3.84), 311s(3.84)	39-2815-84C
$C_{24}H_{34}N_2O_8$ 2,3-Quinoxalinediacetic acid, α,α'-bis(ethoxycarbonyl)-1,2,3,4-tetra-hydro-6,7-dimethyl-, diethyl ester	MeCN	222(4.42),320(3.81)	62-0947-84A
$C_{24}H_{34}N_4O$ Urea, N'-[(8α)-6-cyclopentyl-8-ergoli-nyl)-N,N-diethyl-	MeOH	223(4.53),275s(3.80), 281(3.82),291(3.74)	73-2828-84
$C_{24}H_{34}O$ Acetaldehyde, [2-[3-methyl-5-(2,6,6-trimethyl-1-cyclohexen-1-yl)-2,4-pentadienylidene]cycloheptylidene]-,("all-trans except 12-cis locked")	hexane EtOH	365(4.3) 380(--)	88-1007-84 88-1007-84
("9-cis")	hexane	350(4.3)	88-1007-84
("13-cis")	hexane	340(4.3)	88-1007-84
("9-cis,13-cis")	hexane	330(4.2)	88-1007-84

Compound	Solvent	$\lambda_{max}(\log \epsilon)$	Ref.
$C_{24}H_{34}O_5$			
2,6,11-Dodecatrienal, 12-acetoxy-10-(acetoxymethylene)-6-methyl-2-(4-methyl-3-pentenyl)-, (all-E)-	MeOH	243(3.86)	78-2913-84
(E,E,Z,E)-	MeOH	246(3.99)	78-2913-84
$C_{24}H_{34}O_7$			
Picras-2-ene-1,16-dione, 11-acetoxy-2,12-dimethoxy-, (11α,12β)- (picrasinol B)	EtOH	262(3.66)	138-0221-84
Teumassilin, 6,19-diacetyl-	EtOH	213(3.50)	102-0849-84
$C_{24}H_{35}NO_2$			
β-Alanine, N-[3,7-dimethyl-9-(2,6,6-trimethyl-1-cyclohexen-1-yl)-2,4,6,8-nonatetraenylidene]-, methyl ester, hydrochloride, (E,E,E,?,E)-	MeOH	365(4.60),445(4.62)	104-1669-84
$C_{24}H_{36}N_2$			
1,4-Benzenediamine, 2,6-bis(2-methylpropyl)-N⁴-[4-(2-methylpropyl)phenyl]-	n.s.g.	290(4.09)	30-0009-84
$C_{24}H_{36}N_2O_5S$			
Androst-2-eno[3,2-d][1,2,3]thiadiazole-2'(5'H)-carboxylic acid, 17-acetoxy-, ethyl ester, 1'-oxide, (2β,5α,17β)-	EtOH	219(3.68),262(3.50), 335(3.45)	44-4773-84
$C_{24}H_{36}N_2O_6$			
2H-[1]Benzopyrano[3,4-c]pyridazine-3,4-dicarboxylic acid, 8-methoxy-5,5-dimethyl-1,2,3,4,4a,10b-hexahydro-, bis(1,1-dimethylethyl) ester, trans	EtOH	217(4.38),261(4.08), 266(4.08),303(3.81), 310(3.79)	39-2815-84C
$C_{24}H_{36}N_4O_4$			
Benzenamine, 2,2'-(1,4,10,13-tetraoxa-7,16-diazacyclooctadecane-7,16-diyl)-bis-	MeOH	221(4.33),293(3.68)	78-0793-84
tetraprotonated	MeOH	236(3.71)	78-0793-84
complex with CuCl₂	MeOH	232(3.92),615(2.15)	78-0793-84
$C_{24}H_{36}O_5$			
Hexanorcucurbitacin F	MeOH	206(3.92)	100-0988-84
Udoteal, dihydro-	MeOH	250(3.76)	78-2913-84
$C_{24}H_{36}O_7$			
2(5H)-Oxepinone, 7-(6-acetoxy-1-hydroxy-4-methyl-4-hexenyl)-3-(5-acetoxy-4-methyl-3-pentenyl)-6,7-dihydro-7-methyl-	MeOH	202(3.95),220s(3.67)	102-0829-84
Picras-2-en-1-one, 11-acetoxy-16-hydroxy-2,12-dimethoxy-, (11α,12β)-(picrasinol A)	EtOH	261(3.58)	138-0221-84
$C_{24}H_{37}NO_3$			
Myxalamid C	EtOH	204(4.12),265(4.06), 340s(--),356(4.54), 370s(--)	158-84-113
$C_{24}H_{38}N_2$			
4,4'-Bipyridinium, 2-ethyl-1,1'-di-	H_2O	260(4.22)	39-0367-84B

Compound	Solvent	$\lambda_{max}(\log \epsilon)$	Ref.
hexyl-, dibromide (cont.)			39-0367-84B
$C_{24}H_{38}N_2O_2$ Spiro[piperidine-3,2'-[2H]pyrano[3,2-c]pyridin]-4-one, 3',4',5',6',7',8'-hexahydro-1,2,2,5',5',6,6,6',7',7'-decamethyl-5,8'-bis(methylene)-	heptane	250(4.00)	103-0761-84
$C_{24}H_{40}O$ 1-Octadecanone, 1-phenyl-	heptane	320(1.70)	35-7033-84
$C_{24}H_{40}O_3$ Bicyclo[3.1.1]heptane-2-acetaldehyde, 6,6-dimethyl-3-[3-[(tetrahydro-2H-pyran-2-yl)oxy]-1-octenyl]-, [1S-(1α,2β,3α(1E,3R*),5α]]-	EtOH	230(4.17)	39-1069-84C
$C_{24}H_{42}N_4O_5$ Tetrabutylammonium N-acetyl-2,4-dinitro-phenylaminide	dioxan MeCN	438(4.28) 240(--),390(--), 439(--)	104-0565-84 104-0565-84
$C_{24}H_{43}N$ 2-Pentanamine, N-(3-cyclohexyl-4,4,6,6-tetramethyl-2-cyclohexen-1-ylidene)-2,4,4-trimethyl-, (Z)-	C_6H_{12}	234(4.25),307(2.33)	33-0748-84
$C_{24}H_{46}N_2O_2$ Diazene, bis(1-oxododecyl)-	THF	452(1.39)	23-0574-84
$C_{24}H_{48}Si_6$ 5,6,11,16,21,26-Hexasilahexaspiro-[4.0.4.0.4.0.4.0.4.0]triacontane	isooctane	208s(4.50),225s(3.93), 252(3.71),279s(3.68)	35-5521-84
$C_{24}H_{50}Te_2$ Ditelluride, didodecyl	benzene	397(2.96)	48-0467-84
$C_{24}H_{55}N_3Si_2$ 1-Triazene, 1,3-bis[tris(1,1-dimethyl-ethyl)silyl]-	hexane	259(4.20),384(1.78)	89-0059-84
$C_{24}H_{60}Si_6$ Cyclohexasilane, 1,2,3,4,5,6-hexameth-yl-1,2,3,4,5,6-hexapropyl-	C_6H_{12} C_6H_{12}	232(3.89),257s(3.45) 235(3.89),255s(3.38)	157-0141-84 157-0141-84
$C_{24}H_{64}Si_8$ Cyclooctasilane, 1,2,3,4,5,6,7,8-octa-ethyl-1,2,3,4,5,6,7,8-octamethyl-	isooctane	217s(4.18),238(4.15), 248s(3.91)	101-0353-84L

Compound	Solvent	$\lambda_{max}(\log \epsilon)$	Ref.
$C_{25}Cl_{16}$			
9,9'-Spirobi[9H-fluorene], hexadeca-chloro-	C_6H_{12}	240(4.87),290s(4.41), 303(4.57),320s(4.41)	44-0770-84
$C_{25}Cl_{17}$			
9H-Fluoren-9-yl, 1,2,3,4,5,6,7,8-octachloro-9-(2,2',3,3',4',5,5',6,6'-nonachloro[1,1'-biphenyl]-4-yl)-	C_6H_{12}	214(5.00),230s(4.85), 292(4.61),346(3.71), 371(3.79),387(3.85), 407(3.67),460(4.30), 496(3.26),525(3.01), 569(3.39),618(3.75)	44-0770-84
9H-Fluoren-9-yl, 1,2,3,4,5,6,7,8-octachloro-9-(2',3,3',4,4',5,5',6,6'-nonachloro[1,1'-biphenyl]-2-yl)-	C_6H_{12}	215(5.14),299(4.66), 382s(3.63),391(3.72), 455s(3.26),495(3.05), 572(3.09),620(3.41)	44-0770-84
$C_{25}HCl_{17}$			
9H-Fluorene, 1,2,3,4,5,6,7,8-octachloro-9-(2',3,3',4,4',5,5',6,6'-nonachloro[1,1'-biphenyl]-2-yl)-	C_6H_{12}	216(5.02),302(4.32), 310s(4.27)	44-0770-84
$C_{25}HCl_{19}$			
1,1'-Biphenyl, 2,2',3,3',4,4',5,5',6-nonachloro-6'-[chloro(2,2',3,3',4'-5,5',6,6'-nonachloro[1,1'-biphenyl]-4-yl)methyl]-	C_6H_{12}	215(5.21),295s(3.40), 301(3.47)	44-0770-84
$C_{25}H_{14}O_6$			
4H,10H-Benzo[1,2-b:3,4-b']dipyran-8-carboxylic acid, 4,10-dioxo-2,3-diphenyl-	MeOH	232(4.25),305(4.28)	2-0969-84
$C_{25}H_{15}N_3O_5$			
5(4H)-Oxazolone, 4-[[4-[5-(4-nitrophenyl)-2-oxazolyl]phenyl]methylene]-2-phenyl-	toluene	415(4.73)	135-1330-84
5(4H)-Oxazolone, 2-(4-nitrophenyl)-4-[[4-(5-phenyl-2-oxazolyl)phenyl]-methylene]-	toluene	345(4.45)	135-1330-84
$C_{25}H_{16}N_2O_3$			
5(4H)-Oxazolone, 4-[[3-(5-phenyl-2-oxazolyl)phenyl]methylene]-	toluene	320(4.51),365(4.58)	135-1330-84
5(4H)-Oxazolone, 2-phenyl-4-[[4-(5-phenyl-2-oxazolyl)phenyl]methylene]-	toluene	305(4.15),415(4.70)	135-1330-84
$C_{25}H_{16}N_2O_5S$			
Benzo[h]quinoline-10-sulfonic acid, 2-(4-nitrophenyl)-4-phenyl-	MeOH	<u>270(4.6),380(3.9)</u>	104-0149-84
$C_{25}H_{16}S_4$			
1,3-Benzodithiole, 2,2'-bicyclo[4.4.1]-undeca-3,5,8,10-tetraene-2,7-diylidenebis-	CH_2Cl_2	239(4.14),262s(3.71), 310s(3.30),490(3.98)	33-0574-84
compd. with 4,5-dichloro-3,6-dioxo-1,4-cyclohexadiene-1,2-dicarbonitrile	DMF	340(3.41),480(3.52), 594s(2.97)	33-0574-84
1,3-Benzodithiol-1-ium, 2,2'-bicyclo-[4.4.1]undeca-1,3,5,7,9-pentaene-2,7-diylbis-, (+), bis(tetrafluoro-borate)	MeCN	266(3.76),320(3.48), 517(3.86)	33-0574-84

Compound	Solvent	$\lambda_{max}(\log \epsilon)$	Ref.
$C_{25}H_{17}F_3N_3O_4P$ Benzenamine, 2,6-dinitro-4-(trifluoro-methyl)-N-(triphenylphosphoranyli-dene)-	MeOH	386(3.51)	65-0289-84
$C_{25}H_{17}F_3N_5O_4P$ 1-Triazene, 1-[2,6-dinitro-4-(trifluo-romethyl)phenyl]-3-(triphenylphos-phoranylidene)-	MeOH	328(4.13)	65-0289-84
$C_{25}H_{17}N$ 5,10[1',2']-Benzenobenzo[g]quinoline, 5,10-dihydro-2-phenyl-	MeCN	262s(4.077),270s(4.107), 277s(4.120),293s(4.266), 297(4.274)	5-0381-84
Spiro[acridine-9(10H),9'-fluorene]	EtOH	273(4.37),297(4.16), 308(4.19),320s(3.9)	24-2703-84
$C_{25}H_{17}NS$ Benzothiazole, 2-[2-(1-naphthalenyl)-1-phenylethenyl]-	C_6H_{12}	351.1(4.27)	131-0417-84H
Benzothiazole, 2-[2-(2-naphthalenyl)-1-phenylethenyl]-	C_6H_{12}	354.1(4.46)	131-0417-84H
$C_{25}H_{17}N_2S_2$ Benzothiazolium, 2-[4-(2-benzothiazo-lyl)-1-naphthalenyl]-3-methyl-, methyl sulfate	DMF	346(4.26)	5-1129-84
$C_{25}H_{17}N_3$ 3-Quinolinecarboxaldehyde, (9-anthra-cenylmethylene)hydrazone	C_6H_{12}	412(4.14)	97-0021-84
$C_{25}H_{17}N_3O$ Benzoxazole, 2-[4-(1-methyl-1H-benz-imidazol-2-yl)-1-naphthalenyl]-	DMF	343(4.43)	5-1129-84
$C_{25}H_{17}N_3O_4S_2$ 1-Naphthalenesulfonic acid, 3-[[4-[2-(2-benzothiazolyl)ethenyl]-phenyl]azo]-4-hydroxy-	10% pyri-dine	530(4.43)	7-0027-84
$C_{25}H_{17}N_3O_7S_3$ 2,7-Naphthalenedisulfonic acid, 4-[[4-[2-(2-benzothiazolyl)ethenyl]phen-yl]azo]-3-hydroxy-	10% pyri-dine	531(4.50)	7-0027-84
$C_{25}H_{17}N_3O_8S_3$ 2,7-Naphthalenedisulfonic acid, 3-[[4-[2-(2-benzothiazolyl)ethenyl]phenyl]-azo]-4,5-dihydroxy-	10% pyri-dine	550(4.59)	7-0027-84
$C_{25}H_{17}N_5$ Ethenetricarbonitrile, [4-[(diphenyl-methylene)methylhydrazino]phenyl]-	benzene	537(4.60)	32-0111-84
$C_{25}H_{17}N_5O_8S_2$ 2,7-Naphthalenedisulfonic acid, 4,5-dihydroxy-3-(phenylazo)-6-(8-quin-olinylazo)-	n.s.g.	570(4.6)	86-1041-84
calcium chelate (also other metal chelates)	pH 4.5	726(5.39)	86-1041-84

Compound	Solvent	$\lambda_{max}(\log \epsilon)$	Ref.
$C_{25}H_{17}S_4$ 1,3-Benzodithiol-1-ium, 2-[7-(1,3-ben- zodithiol-2-yl)bicyclo[4.4.1]undeca- 1,3,5,7,9-pentaen-2-yl]-, tetra- fluoroborate	MeCN	230(3.97),252(3.97), 383(3.27),386(3.09), 515(3.67)	33-0574-84
$C_{25}H_{18}ClNO_2$ 2-Pyridinol, 6-(4-chlorophenyl)-3,4-di- phenyl-, acetate	EtOH	256(4.28),286(4.09), 350(3.66)	4-1473-84
$C_{25}H_{18}NO_8S_3$ Pyrylium, 2-(2-benzothiazolyl)-4-(4- methoxy-3-sulfophenyl)-6-(4-sulfo- phenyl)-, perchlorate	H_2O	242(4.3),277(4.21), 408(4.47)	39-0879-84B
$C_{25}H_{18}N_2$ 2-Pyridinamine, 3-(9-anthracenyl)- 6-phenyl-	MeCN	254(5.154),322s(4.052), 328(4.058),347(4.029), 366(4.125),385(4.079)	5-0381-84
9,9'(10H,10'H)-Spirobiacridine	EtOH	293(4.34),321(4.17)	24-2703-84
$C_{25}H_{18}N_4O_3S_2$ 1-Naphthalenesulfonic acid, 4-amino- 3-[[4-[2-(2-benzothiazolyl)ethen- yl]phenyl]azo]-	10% pyri- dine	498(4.60)	7-0027-84
$C_{25}H_{18}N_4O_4S_2$ 2-Naphthalenesulfonic acid, 6-amino- 3-[[4-[2-(2-benzothiazolyl)ethen- yl]phenyl]azo]-4-hydroxy-	10% pyri- dine	545(4.55)	7-0027-84
2-Naphthalenesulfonic acid, 7-amino- 3-[[4-[2-(2-benzothiazolyl)ethen- yl]phenyl]azo]-4-hydroxy-	10% pyri- dine	523(4.42)	7-0027-84
$C_{25}H_{18}N_4O_7S_3$ 2,7-Naphthalenedisulfonic acid, 5-ami- no-3-[[4-[2-(2-benzothiazolyl)ethen- yl]phenyl]azo]-4-hydroxy-	10% pyri- dine	566(4.53)	7-0027-84
$C_{25}H_{18}N_4O_8S_2$ Acetic acid, [[2-[[5-(1,2-dihydro-2- oxo-3H-indol-3-ylidene)-4,5-dihydro- 4-oxo-2-thiazolyl]amino]-2,3-dihydro- 8,9-dimethoxy-1,3-dioxo-1H-2-benzaze- pin-4-yl]thio]-	MeOH	250(4.63),335s(4.29), 386(4.40)	103-0024-84
$C_{25}H_{18}O$ Pyrene, 1-[2-(4-methoxyphenyl)ethen- yl]-, cis	MeCN	<u>280(4.5),350(4.3)</u>	46-2714-84
trans	MeCN	<u>295(4.3),375(4.7)</u>	46-2714-84
$C_{25}H_{18}O_3$ 2,5-Furandione, 3-(diphenylmethylene)- dihydro-4-(1-phenylethylidene)-, (E)-	toluene	375(3.90)	48-0233-84
(Z)-	toluene	367(3.90)	48-0233-84
$C_{25}H_{18}O_4$ 2,5-Furandione, 3-(diphenylmethylene)- dihydro-4-[(4-methoxyphenyl)methyl- ene]-, (E)-	toluene	406(4.03)	48-0233-84
(Z)-	toluene	407(4.30)	48-0233-84

Compound	Solvent	$\lambda_{max}(\log \epsilon)$	Ref.
$C_{25}H_{18}O_6$ Sphagnorubin C	MeOH-HCl	240(4.80),291(4.76), 330s(4.50),420s(4.16), 537(4.88)	5-1024-84

$C_{25}H_{18}O_9S_2$
  Pyrylium, 4-(4-acetoxyphenyl)-2,6-
  bis(4-sulfophenyl)-, hydroxide,
  inner salt
	MeOH-HClO$_4$	398(4.59)	4-1673-84

$C_{25}H_{18}S_4$
  1,3-Benzodithiole, 2,2'-bicyclo[4.4.1]-
  undeca-1,3,5,7,9-pentaene-2,7-diyl-
  bis-, (+)-
	CH$_2$Cl$_2$	233(4.13),268(4.6),
		324(3.45)
			33-0574-84

$C_{25}H_{19}BO_4$
  1,5-Methano-1H-anthra[2,3-e][1,3,2]-
  dioxaborocin-7,14-dione, 5,6-dihy-
  dro-5-methyl-3-phenyl-, (1S)-
	CHCl$_3$	242(4.48),288(4.28),
		300(4.30),419(3.71)
			78-4649-84

$C_{25}H_{19}BO_6$
  Boron, (1,11-diphenyl-1,3,8,10-undeca-
  tetraene-5,7-dionato-O,O')[ethanedi-
  oato(2-)-O,O']-
	CH$_2$Cl$_2$	530(4.85)	97-0292-84

$C_{25}H_{19}ClO_3$
  Benzoic acid, 4-chloro-, 1,2,3,4-tetra-
  hydro-1-hydroxybenz[a]anthracen-2-yl
  ester, (1S-cis)-
	MeOH	247(4.86),256(5.04),
		358(3.76),377(3.74)
			44-3621-84

$C_{25}H_{19}Cl_2N_3$
  Benzenecarboximidamide, N,N'-bis(4-
  chlorophenyl)-2-(phenylamino)-
	EtOH	242(4.39),290(4.48)	97-0328-84

$C_{25}H_{19}N_3O$
  [1,2,4]Triazolo[1,5-a]pyridinium, 2,3-
  dihydro-1-(4-methylphenyl)-2-oxo-
  5,7-diphenyl-, hydroxide, inner salt
	EtOH	266(4.84)	39-1891-84C

$C_{25}H_{19}N_3O_2$
  [1,2,4]Triazolo[1,5-a]pyridinium, 2,3-
  dihydro-1-(4-methoxyphenyl)-2-oxo-
  5,7-diphenyl-, hydroxide, inner salt
	EtOH	266(4.86)	39-1891-84C

$C_{25}H_{19}N_3O_4$
  Deoxyviolacein, N,N''-diacetyl-N'-meth-
  yl-
	CHCl$_3$	261(4.26),546(4.16)	39-1331-84B

$C_{25}H_{19}N_3O_8S$
  1-Naphthalenesulfonic acid, 8-[[1,3-
  bis(4-nitrophenyl)-3-oxopropyl]-
  amino]-
	n.s.g.	265(4.6),380s(3.5)	104-0149-84

$C_{25}H_{19}N_3S$
  [1,2,4]Triazolo[1,5-a]pyridinium, 2,3-
  dihydro-1-(2-methylphenyl)-5,7-di-
  phenyl-2-thioxo-, hydroxide,
  inner salt
	EtOH	260(4.24)	39-1891-84C

[1,2,4]Triazolo[1,5-a]pyridinium, 2,3-
  dihydro-1-(4-methylphenyl)-5,7-di-
  phenyl-2-thioxo-, hydroxide, inner salt
	EtOH	246(4.27)	39-1891-84C

Compound	Solvent	$\lambda_{max}(\log \epsilon)$	Ref.
$C_{25}H_{19}N_5O_2S_2$ Pyrimido[5,4-e]thiazolo[3,4-a]pyrimidin-10-ium, 9-[(3-ethyl-2(3H)-benzothiazolylidene)methyl]-1,2,3,4-tetrahydro-3-methyl-2,4-dioxo-7-phenyl-, hydroxide, inner salt	DMF	430(4.05),454(3.97), 554(4.40)	103-0921-84
$C_{25}H_{19}N_5O_3$ 1,3,7,8-Tetraazaspiro[4.5]deca-6,9-diene-10-carboxylic acid, 4-oxo-1,3-diphenyl-2-(phenylimino)-	MeOH	269(4.01),297s(3.91)	39-2491-84C
$C_{25}H_{19}N_7O_3S_2$ 1H-Pyrazole, 3-methyl-5-phenyl-1-(2-pyridinylcarbonyl)-4-[[2-[(2-thiazolylamino)sulfonyl]phenyl]azo]-	n.s.g.	370(4.42)	48-1021-84
$C_{25}H_{19}O$ 5H-Indeno[1,2-b]pyrylium, 2-(4-methylphenyl)-4-phenyl-, perchlorate	MeCN	283(4.21),352(4.41), 429(4.56)	48-0647-84
$C_{25}H_{19}O_3$ Pyrylium, 4-(4-acetoxyphenyl)-2,6-diphenyl-, perchlorate	MeOH-HClO$_4$	405(4.76)	4-1673-84
$C_{25}H_{20}$ [5]Helicene, 1,3,6-trimethyl-, (±)-	CHCl$_3$	238(4.65),275(4.36), 315(4.52),388(2.63), 408(2.45)	24-0336-84
$C_{25}H_{20}Br_2N_4$ 1H-Imidazo[1,2-b][1,2,4]triazepine, 7,8-dihydro-6,8-bis(4-bromophenyl)-8-methyl-2-phenyl-	MeOH	277(4.65),345(3.95)	103-1152-84
$C_{25}H_{20}Cl_2N_4$ 1H-Imidazo[1,2-b][1,2,4]triazepine, 6,8-bis(4-chlorophenyl)-7,8-dihydro-8-methyl-2-phenyl-	MeOH	279(4.65),347(3.91)	103-1152-84
$C_{25}H_{20}NO_2P$ Benzamide, N-[4-(diphenylphosphinyl)-phenyl]-	EtOH	277(4.50)	42-0430-84
$C_{25}H_{20}N_2O_3$ 1H-Pyrazole-3-carboxylic acid, 4-benzoyl-1,5-diphenyl-, ethyl ester	EtOH	250(4.36)	4-1013-84
1H-Pyrazole-3-carboxylic acid, 5-benzoyl-1,4-diphenyl-, ethyl ester	EtOH	249(4.53),344(4.19)	4-1013-84
1H-Pyrrole-2,5-dione, 3-benzoyl-1-(4-methylphenyl)-4-[(4-methylphenyl)-amino]-	CHCl$_3$	280(4.61),415(3.83)	48-0401-84
$C_{25}H_{20}N_4$ Spiro[5H-dibenzo[a,d]cycloheptene-5,3'-[3H]pyrazolo[5,1-c][1,2,4]-triazole], 10,11-dihydro-6'-methyl-7'-phenyl-	CH$_2$Cl$_2$	241(4.29),321(3.92), 352(3.91)	24-1726-84
$C_{25}H_{20}N_4O_6$ 11H-Furo[2',3':4,5]oxazolo[2,3-b]-	MeOH	220(4.48),245(4.29),	136-0179-84F

Compound	Solvent	$\lambda_{max}(\log \epsilon)$	Ref.
pteridin-11-one, 7-acetoxy-6a,7,8,9a-tetrahydro-8-(hydroxymethyl)-2,3-diphenyl-, [6aR-(6aα,7β,8α,9aβ)]- (cont.)		280(4.27),357(4.13)	136-0179-84F
$C_{25}H_{20}N_6O_4$			
1H-Imidazo[1,2-b][1,2,4]triazepine, 7,8-dihydro-8-methyl-6,8-bis(4-nitrophenyl)-2-phenyl-	MeOH	284(4.62),383(3.87)	103-1152-84
$C_{25}H_{20}N_8O_4$			
Benzo[g]pteridine-2,4(3H,10H)-dione, 10,10'-(1,3-propanediyl)bis[3-methyl-	pH 7.27	342(4.09),436(4.15)	39-1227-84B
	glycol	265(4.83),338(4.20), 434(4.27)	88-1035-84
$C_{25}H_{20}O$			
6a,10b[1',2']-Benzenocyclopent[fg]aceanthrylene, 1,2,5,6-tetrahydro-3-methoxy-	C_6H_{12}	262s(3.40),275(3.51), 282(3.60)	88-1505-84
$C_{25}H_{20}O_2$			
9,10-Ethanoanthracen-11-one, 9,10-dihydro-12-(3-oxo-3-phenylpropyl)-	MeCN	234(4.195),266(3.274), 273(3.292),285s(3.026), 306(2.843),317(2.692)	5-0381-84
1H-Inden-1-one, 2,3-dihydro-2-[3-(4-methylphenyl)-3-oxo-1-phenylpropylidene]-	dioxan	256(4.34),294(4.25)	48-0647-84
$C_{25}H_{20}O_4$			
Spiro[dinaphtho[1,2-b:1',2'-d]furan-11(10H),2'-[1,3]dioxan]-10-one, 5',5'-dimethyl-	EtOH	223(4.51),252(4.56), 287(3.89),324(4.12), 355(4.15)	39-2069-84C
$C_{25}H_{21}ClO_9$			
Benzoic acid, 3-[2-chloro-4-(methoxycarbonyl)phenoxy]-4-methoxy-5-[4-(methoxycarbonyl)phenoxy]-, methyl ester	MeOH	206(4.78),253(4.66)	44-0635-84
$C_{25}H_{21}N$			
Pyridine, 2,3-dimethyl-4,5,6-triphenyl-	C_6H_{12}	235(4.34),286(3.96)	35-1065-84
Pyridine, 2,4-dimethyl-3,5,6-triphenyl-	C_6H_{12}	282(3.96)	35-1065-84
Pyridine, 2,5-dimethyl-3,4,6-triphenyl-	C_6H_{12}	238(3.83),287(3.95), 290(4.32)	35-1065-84
$C_{25}H_{21}NO$			
1H-Benz[g]indole, 1-benzoyl-2,3,4,5-tetrahydro-3-phenyl-	EtOH	223(4.40),272(4.11)	12-0117-84
[1,1'-Biphenyl]-4-propanenitrile, α-[[4-(1-methylethyl)phenyl]-methylene]-β-oxo-	n.s.g.	325(4.39)	73-0421-84
Pyridine, 2-(4-methoxyphenyl)-6-(4-methylphenyl)-4-phenyl-	dioxan	261(4.68),323(4.01)	48-0287-84
$C_{25}H_{21}NO_4$			
Benzoic acid, 2-[(1-benzoyl-3-oxo-3-phenyl-1-propenyl)amino]-, ethyl ester	EtOH	259(4.15),322(3.91), 391(4.15)	4-0759-84
$C_{25}H_{21}N_3$			
Benzenecarboximidamide, N,N'-diphenyl-	EtOH	288(4.31),392s(3.27)	97-0328-84

Compound	Solvent	$\lambda_{max}(\log \epsilon)$	Ref.
2-(phenylamino)- (cont.)			97-0328-84
$C_{25}H_{21}N_3O_2S$ 1H-Pyrrole-3-carbothioamide, 2,5-dihy- dro-1-(4-methylphenyl)-4-[(4-methyl- phenyl)amino]-2,5-dioxo-N-phenyl-	CHCl$_3$	282(4.53),430(3.88)	48-0401-84
$C_{25}H_{21}N_3O_5$ Benzamide, N-[4-(butylamino)-9,10-di- hydro-9,10-dioxo-1-anthracenyl]-4- nitro-	benzene	572(4.04),608(3.98)	39-0529-84C
$C_{25}H_{21}N_3O_9S_2$ Acetic acid, [[2-[[4,5-dihydro-5-[(3- hydroxy-4-methoxyphenyl)methylene]- 4-oxo-2-thiazolyl]amino]-2,3-dihy- dro-8,9-dimethoxy-1,3-dioxo-1H-2- benzazepin-4-yl]thio]-	MeOH	224(4.65),251(4.70), 335s(4.45),388(4.38)	103-0024-84
$C_{25}H_{21}N_3O_{11}$ 5-Oxa-1-azabicyclo[4.2.0]oct-2-ene- 2,3-dicarboxylic acid, 4-(2-methoxy- ethylidene)-8-oxo-, bis[(4-nitro- phenyl)methyl] ester, [R-(Z)]-	EtOH	265(4.30),314(4.22)	39-1599-84C
$C_{25}H_{21}N_5O_5$ 1-Imidazolidinepropanoic acid, 2,5-di- oxo-4,4-diphenyl-, [(4-nitrophenyl)- methylene]hydrazide	MeOH	203(4.63),325(4.30)	56-0585-84
$C_{25}H_{21}O$ Pyrylium, 2,6-bis(4-methylphenyl)- 4-phenyl-, perchlorate	CH$_2$Cl$_2$	373(4.61),445(4.48), 532(3.55)	88-5143-84
$C_{25}H_{21}O_2$ Pyrylium, 2-(4-methoxyphenyl)-6-(4- methylphenyl)-4-phenyl-, perchlorate	MeCN	255(4.13),265s(4.10), 312(4.33),361(4.55), 449(4.48)	48-0287-84
$C_{25}H_{22}$ Benzene, 1-(1-methyl-2,3-diphenyl- 2-cyclopropen-1-yl)-2-(2-propenyl)-	EtOH	227s(4.36),317(4.30)	44-1353-84
Benzene, 1-(2-methyl-1,3-diphenyl- 2-cyclopropen-1-yl)-2-(2-propenyl)-	EtOH	263(4.27)	44-1353-84
Benzo[a]cyclopropa[c]cycloheptene, 1,1a,2,8b-tetrahydro-8b-methyl- 1,1a-diphenyl-, (1α,1aα,8bα)-	EtOH	229(4.49),260s(3.90)	44-1353-84
1H-Cycloprop[a]indene, 6-ethenyl- 1,1a,6,6a-tetrahydro-1a-methyl- 1,6a-diphenyl-, (1α,1aα,6β,6aα)-	EtOH	241(4.27)	44-1353-84
Naphthalene, 1-ethenyl-1,2-dihydro- 4-methyl-2,3-diphenyl-, trans	EtOH	226(4.34),283(4.14)	44-1353-84
1,1'-Spirobi[1H-indene], 2,3-dihydro- 2,2'-dimethyl-3'-phenyl-	EtOH	224(4.47),266(3.95), 273(3.96)	44-1353-84
$C_{25}H_{22}Cl_2S_3$ 1,3-Butadiene, 1,2-dichloro-1,4,4- tris[(4-methylphenyl)thio]-	heptane	200(4.90),225s(4.69), 259(4.41),317(4.30)	5-1873-84
$C_{25}H_{22}N_2OS$ 4-Isoquinolinecarbonitrile, 5,6,7,8-	dioxan	277(4.39),312(4.25),	104-2225-84

Compound	Solvent	$\lambda_{max}(\log \epsilon)$	Ref.
3-[(2-oxo-2-phenylethyl)thio]-1-(phenylmethyl)- (cont.)		360(3.91)	104-2225-84
Methanone, [1-amino-6,7,8,9-tetrahydro-5-(phenylmethyl)thieno[2,3-c]isoquinolin-2-yl]phenyl-	dioxan	282(4.52),315(4.40), 327(4.43),402(4.20)	104-2225-84
$C_{25}H_{22}N_2O_3$			
Benzamide, N-[9,10-dihydro-4-[(1-methylpropyl)amino]-9,10-dioxo-1-anthracenyl]-	benzene	572(4.06),610(4.00)	39-0529-84C
Benzamide, N-[9,10-dihydro-4-[(1,1-dimethylethyl)amino]-9,10-dihydro-9,10-dioxo-1-anthracenyl]-	benzene	573(4.03),612(3.98)	39-0529-84C
1H-Pyrazole-3-carboxylic acid, 5-benzoyl-4,5-dihydro-1,4-diphenyl-, ethyl ester, trans	EtOH	228(4.95),278(4.74)	4-1013-84
$C_{25}H_{22}N_2O_4$			
Cyclopent[b]indole-1,2-dicarboxylic acid, 1,4-dihydro-3-(1H-indol-3-yl)-, diethyl ester	EtOH	210(4.72),255(4.20), 360(4.27)	94-4410-84
6H-[1,4]Diazacyclotridecino[1,2-a:4,3-a']diindole-6,14,20,21(7H)-tetrone, 8,9,10,11,12,13-hexahydro-, (E)-	xylene	569(3.81)	18-0470-84
1H-Indazole-6,7-dicarboxylic acid, 4-methyl-1-(4-methylphenyl)-3-phenyl-, dimethyl ester	EtOH	253(4.70),339(4.09)	94-2146-84
$C_{25}H_{22}N_2O_7$			
2(3H)-Furanone, 5-(2-acetoxy-1-hydroxyethyl)-3-[(4-acetoxyphenyl)methylene]-dihydro-4-(2-quinoxalinyl)-	MeOH	239(4.46),294(4.33)	94-1808-84
$C_{25}H_{22}N_2O_9$			
5-Oxa-1-azabicyclo[4.2.0]oct-2-ene-2,3-dicarboxylic acid, 4-(2-methoxyethylidene)-8-oxo-1-[(4-nitrophenyl)methyl] 2-(phenylmethyl) ester, [R-(Z)]-	EtOH	265(4.15),313(4.19)	39-1599-84C
$C_{25}H_{22}N_4$			
1H-Imidazo[1,2-b][1,2,4]triazepine, 7,8-dihydro-8-methyl-2,6,8-triphenyl-	MeOH	276(4.53),341(3.85)	103-1152-84
$C_{25}H_{22}N_4O_5$			
Pyrrolo[3,4-c]pyrazole-4,6(1H,5H)-dione, 3-[6-(1,3-dihydro-1,3-dioxo-2H-isoindol-2-yl)-1-oxohexyl]-3a,6a-dihydro-5-phenyl-	MeOH	208(4.50),217s(4.58), 221(4.60),224s(4.54), 234(4.32),242(4.26), 302(4.09)	103-0066-84
$C_{25}H_{22}N_4O_9S_2$			
11H-Furo[2',3':4,5]oxazolo[2,3-b]pteridin-11-one, 6a,7,8,9a-tetrahydro-7-[(methylsulfonyl)oxy]-8-[[(methylsulfonyl)oxy]methyl]-2,3-diphenyl-, [6aR-(6aα,7β,8α,9aβ)]-	MeOH	220(4.44),246(4.26), 280(4.22),357(4.10)	136-0179-84F
$C_{25}H_{22}N_6$			
1H-Benzimidazole, 1,1',1"-methylidynetris[2-methyl-	MeOH	243(4.43),274(4.02), 281(3.99)	117-0299-84

Compound	Solvent	$\lambda_{max}(\log \epsilon)$	Ref.
$C_{25}H_{22}OS_2$			
1,4-Epoxynaphthalene-2-carbodithioic acid, 1,2,3,4-tetrahydro-3-methyl-1,4-diphenyl-, methyl ester	$CHCl_3$	315(4.16),455(1.43)	39-0859-84C
$C_{25}H_{22}O_2$			
Benz[a]anthracen-10-ol, 5,6,8,9,10,11-hexahydro-, benzoate, (R)-	MeOH-CHCl$_3$	226(4.41),268(4.14), 305(3.71)	44-3621-84
6a,10b[1',2']-Benzenocyclopent[fg]ace-anthrylen-3(1H)-one, 2,2a,5,6,10c,10d-hexahydro-10d-methoxy-	MeOH	243(3.90),249(3.80), 254(3.80),260(3.70)	88-1505-84
1H-Indene-1,3(2H)-dione, 2-[1,1'-bi-phenyl]-4-yl-2-butyl-	EtOH	348(2.74)	135-0312-84
	dioxan	348(2.71)	135-0312-84
2-Pentene-1,5-dione, 1,5-bis(4-methyl-phenyl)-3-phenyl-	dioxan	255(4.35),294(4.35)	48-0287-84
2-Pentene-1,5-dione, 2-methyl-1-(4-methylphenyl)-3,5-diphenyl-	dioxan	243(4.40),258s(4.26), 288s(3.72)	48-0647-84
$C_{25}H_{22}O_3$			
2-Pentene-1,5-dione, 1-(4-methoxyphen-yl)-5-(4-methylphenyl)-3-phenyl- (with isomer)	dioxan	279.5(4.39),300.5(4.37)	48-0287-84
$C_{25}H_{22}O_4$			
2-Pentene-1,5-dione, 1,5-bis(4-methoxy-phenyl)-3-phenyl-	dioxan	277.5(4.46),306.5(4.43)	48-0287-84
$C_{25}H_{22}O_6$			
9-Phenanthrenecarboxylic acid, 3,4,6-trimethoxy-7-(phenylmethoxy)-	MeOH	265(3.83),281s(--), 317(4.03),345(3.45), 364(3.26)	44-5243-84
$C_{25}H_{22}O_8$			
5,12-Naphthacenedione, 8-acetyl-6,11-diacetoxy-7,8,9,10-tetrahydro-1-methoxy-, (±)-	MeCN	218(4.53),260(4.57), 370(3.84)	78-4549-84
$C_{25}H_{22}P_2$			
Phosphine, methylenebis[diphenyl-	C_6H_{12}	251(4.26)	151-0249-84A
$C_{25}H_{23}ClN_4O_3$			
4-Imidazolidinecarboxamide, 1-(4-chlo-rophenyl)-3-[4-(dimethylamino)phen-yl]-4-methyl-2,5-dioxo-N-phenyl-	EtOH	207(4.39),251(4.46), 261(4.50)	49-0187-84
$C_{25}H_{23}NO$			
2,4-Pentadien-1-one, 5-(dimethylamino)-1,3,5-triphenyl-	dioxan	438(4.06)	48-0657-84
$C_{25}H_{23}NO_2$			
2H-Pyrrol-4-ol, 3,4-dihydro-2,2-di-methyl-4,5-diphenyl-, benzoate	EtOH	233(4.39)	12-0095-84
$C_{25}H_{23}NO_4$			
Isoquinoline, 3-(3,4-dimethoxyphenyl)-6,7-dimethoxy-1-phenyl-	EtOH	245(4.61),273(4.68), 312(4.28),360s(3.67)	4-0525-84
$C_{25}H_{23}NO_8$			
Benzeneacetic acid, α-[(3,4-dimethoxy-2-nitrophenyl)methylene]-4-methoxy-3-(phenylmethoxy)-, (E)-	MeOH	229s(--),284(4.08)	44-5243-84

Compound	Solvent	$\lambda_{max}(\log \epsilon)$	Ref.
3,4,5,6-Pyridinetetracarboxylic acid, 1,2-dihydro-1,2-diphenyl-, tetramethyl ester	EtOH	264(4.14),390(3.50)	103-0295-84
$C_{25}H_{23}N_5O_5$ 4-Imidazolidinecarboxamide, 3-[4-(dimethylamino)phenyl]-4-methyl-1-(4-nitrophenyl)-2,5-dioxo-N-phenyl-	EtOH	206(4.52),263(4.37), 299(4.17)	49-0187-84
$C_{25}H_{23}N_5O_{10}$ L-Alanine, 3-[[(4-nitrophenyl)methoxy]-NNO-azoxy]-N-(phenylmethoxy)carbonyl]-, (4-nitrophenyl)methyl ester	MeOH	260(4.46)	87-1295-84
$C_{25}H_{24}ClN_3O_4$ 4H-Pyrido[1,2-a]pyrimidine-3-carboxamide, N-(3-chloro-2,4-dimethoxyphenyl)-6,7,8,9-tetrahydro-6-methyl-4-oxo-9-(phenylmethylene)-, (E)-	EtOH	300(4.10),366(4.50)	118-0582-84
$C_{25}H_{24}N_2$ 4,4'-Bipyridinium, 2-methyl-1,1'-bis(phenylmethyl)-, dibromide	H_2O	257(4.34)	39-0367-84B
$C_{25}H_{24}N_2O_5$ 1H-Indole-2-carboxylic acid, 5-[2-[2-(ethoxycarbonyl)-1H-indol-5-yl]ethyl]-3-formyl-, ethyl ester	EtOH	235(3.68),259.7(3.52), 303(3.61),322(3.50)	103-0062-84
$C_{25}H_{24}N_2O_8$ 1H-Benz[f]isoquino[8,1,2-hij]quinazoline-2(3H)-carboxylic acid, 5,6-dihydro-8,9,11,12-tetramethoxy-1,3-dioxo-, ethyl ester	MeOH	265(4.73),280(4.72), 309s(4.27),339(3.88), 354(3.93),387(4.01), 402(4.01)	78-4003-84
$C_{25}H_{24}N_4O_6$ 1-Azabicyclo[3.2.0]hept-2-ene-2-carboxylic acid, 3-(4,5-dihydro-4-methyl-5-phenyl-1,2,4-oxadiazol-3-yl)-6-ethyl-7-oxo-, (4-nitrophenyl)methyl ester	THF	255(3.94),261(4.00), 264(4.03),270(4.04)	77-1513-84
$C_{25}H_{24}N_6O_8$ Benzamide, N-[6-[2-(4-nitrophenyl)ethoxy]-9-β-D-ribofuranosyl-9H-purin-2-yl]-	MeOH	230(4.31),271(4.49)	78-0059-84
$C_{25}H_{24}O_2S$ 26-Thiatetracyclo[17.2.2.2^{8,11}.1^{3,6}]-hexacosa-3,5,8,10,19,21,22,24-octaene-2,7-dione	C_6H_{12}	<u>301(4.2)</u>	44-1177-84
$C_{25}H_{24}O_4$ 2-Propen-1-one, 1-(5-hydroxy-2,2,8,8-tetramethyl-2H,8H-benzo[1,2-b:3,4-b']dipyran-6-yl)-3-phenyl-, (E)-(flemiculosin)	n.s.g.	210(3.08),239s(3.07), 255s(3.14),266s(3.19), 278(3.24),308(3.24), 360(3.15)	142-0249-84
$C_{25}H_{24}O_6$ 2H,6H-Benzo[1,2-b:5,4-b']dipyran-6-one, 7,8-dihydro-5-hydroxy-8-(7-hydroxy-2,2-dimethyl-2H-1-benzopyran-6-yl)-	EtOH	267s(4.61),273(4.66), 295(4.25),305s(4.19), 332s(3.47)	142-0997-84

Compound	Solvent	$\lambda_{max}(\log \epsilon)$	Ref.
2,2-dimethyl- (cont.)	EtOH-AlCl$_3$	267s(4.53),273(4.61), 282s(4.40),307(4.18), 343s(3.32)	142-0997-84
Ethanone, 1-(5,12-diacetoxy-1,2,3,4-tetrahydro-7-methoxy-2-naphthacen-yl)-, (±)-	MeCN	222(4.25),244s(4.55), 262(4.12),363(3.26), 380(3.54),400(3.32)	78-4549-84
Spiro[5.5]undecane-1,9-dione, 7,11-bis(1,3-benzodioxol-5-yl)-	EtOH	242(3.52),246(3.54), 289(3.93)	117-0115-84
$C_{25}H_{24}O_7$			
Sanggenon M	EtOH	232(4.19),269s(4.40), 277(4.43),319(3.99), 375(3.40)	142-1791-84
	EtOH-AlCl$_3$	232(4.19),277(4.44), 333(4.05),435(3.34)	142-1791-84
$C_{25}H_{25}BN_2O_6$			
Boron, [1,7-bis[4-(dimethylamino)phen-yl]-1,6-heptadiene-3,5-dionato-0,0']-[ethanedioato(2-)-0,0']-	CH$_2$Cl$_2$	646(5.13)	97-0292-84
$C_{25}H_{25}BrO_5$			
Propanoic acid, 2,2-dimethyl-, 3-(bro-momethyl)-9,10-dihydro-9,10-dioxo-2-(3-oxopentyl)-1-anthracenyl ester	MeOH	213(4.45),260(4.68), 337(3.80)	78-3677-84
$C_{25}H_{25}ClFeN$			
3H-Indolium, 2-(4-chloro-4-ferrocenyl-1,3-butadienyl)-1,3,3-trimethyl-, perchlorate	EtOH	269(4.18),391(3.55), 685(4.20)	65-1439-84
$C_{25}H_{25}Cl_3N_2O_5$			
1H-Indole-3-propanoic acid, 1-[2-meth-yl-1-oxo-2-[[(2,2,2-trichloroethoxy)-carbonyl]amino]propyl]-, phenylmethyl ester	EtOH	242(4.27),263s(3.94), 273s(3.84),292(3.81), 301(3.85)	94-1373-84
$C_{25}H_{25}N$			
Spiro[acridine-9(10H),1'-[1H]indene], 2',3'-dihydro-2,3',3',7-tetramethyl-	EtOH	291(4.13)	24-2703-84
$C_{25}H_{25}NO_5$			
Valachine, (±)-	MeOH	282(4.23),291s(4.23)	77-1371-84
$C_{25}H_{25}NO_6$			
Benzeneacetic acid, α-[(2-amino-3,4-di-methoxyphenyl)methylene]-4-methoxy-3-(phenylmethoxy)-, (E)-	MeOH	224s(--),254s(--), 296(4.00),330(4.01)	44-5243-84
$C_{25}H_{25}N_2S_2$			
Benzothiazolium, 3-ethyl-2-[7-(3-ethyl-2(3H)-benzothiazolylidene)-1,3,5-hep-tatrienyl]-, iodide	C$_6$H$_{11}$Me	770(5.49)	99-0415-84
	EtOH	765(5.34)	99-0415-84
	DMF	765(5.26)	99-0415-84
$C_{25}H_{25}N_3O$			
Ethanone, 1-[4-[3-[4-(dimethylamino)-phenyl]-4,5-dihydro-5-phenyl-1H-pyrazol-1-yl]phenyl]-	MeOH	250(3.97),305(4.07), 394(4.34)	103-0787-84
Urea, N-[1,3-diphenyl-3-(phenylimino)-1-propenyl]-N'-propyl-	EtOH	206(4.49),221s(4.28), 250s(4.19),331(4.11), 358s(4.09)	39-0239-84C

Compound	Solvent	$\lambda_{max}(\log \epsilon)$	Ref.

$C_{25}H_{25}N_3O_2$
   4H-Pyrido[1,2-a]pyrimidine-3-carbox-
    amide, 6,7,8,9-tetrahydro-6-methyl-
    4-oxo-N-(2-phenylethyl)-9-(phenyl-
    methylene)-, (E)-
	EtOH	352(4.41)	118-0582-84

$C_{25}H_{25}N_3O_6$
   L-Alanine, 3-[(phenylmethoxy)-NNO-
    azoxy]-N-[(phenylmethoxy)carbon-
    yl]-, phenylmethyl ester
	MeOH	237(4.10)	87-1295-84

$C_{25}H_{25}N_5O_5$ Meleagrin, 9-O-acetyl-	EtOH	204(4.68),223s(4.45), 295s(4.17),347(4.50)	94-0094-84

$C_{25}H_{26}N_2O_4$
   Phenol, 2-methoxy-5-[2-[methyl[(4-ni-
    trophenyl)methyl]amino]ethyl]-4-(2-
    phenylethenyl)-, (E)-
	EtOH	244(4.05),293(4.16),
		329(4.17)
			44-0581-84

$C_{25}H_{26}N_2O_8$
   2H-Pyrimido[6,1-a]isoquinoline-3(4H)-
    carboxylic acid, 1-(3,4-dimethoxy-
    phenyl)-6,7-dihydro-9,10-dimethoxy-
    2,4-dioxo-, ethyl ester
	MeOH	244(4.34),281(4.04),
		338(4.20)
			78-4003-84

$C_{25}H_{26}N_6O_2$
   3-Cyclohexene-1,1-dicarboxylic acid,
    4-amino-3-cyano-, bis[[(2-methyl-
    phenyl)methylene]hydrazide]
	n.s.g.	268(4.21)	42-0146-84

$C_{25}H_{26}N_6O_4$
   3-Cyclohexene-1,1-dicarboxylic acid,
    4-amino-3-cyano-, bis[[(2-methoxy-
    phenyl)methylene]hydrazide]
	n.s.g.	263(4.17)	42-0146-84

$C_{25}H_{26}N_6O_{11}$
   Carbamic acid, [9-(2,3,5-tri-O-acetyl-
    β-D-ribofuranosyl)-9H-purin-6-yl]-,
    2-(4-nitrophenyl)ethyl ester
	MeOH	267(4.46)	78-0059-84

$C_{25}H_{26}O$
   2,5-Cyclohexadien-1-one, 4-(diphenyl-
    methylene)-2,6-bis(1-methylethyl)-
	hexane?	211(4.12),261(4.17),
		268(4.17),364(4.46)
			73-1949-84

1H-Inden-1-one, 2,3-dihydro-4,5,7-tri-
    methyl-3-(4,6,8-trimethyl-1-azulen-
    yl)-,
	MeOH	563(3.53)	138-0631-84

1H-Inden-1-one, 2,3-dihydro-4,6,7-tri-
    methyl-3-(4,6,8-trimethyl-1-azulen-
    yl)-
	MeOH	563(3.53)	138-0631-84

$C_{25}H_{26}O_4$
   6H-Benzofuro[3,2-c][1]benzopyran-3,9-
    diol, 2,10-bis(3-methyl-2-butenyl)-
    (erycristagallin)
	MeOH	214(4.47),244(4.19),
		251(4.18),291s(3.82),
		330s(4.34),339(4.39),
		356(4.34)
			142-1673-84

	MeOH-NaOH	306(3.82),354(4.47), 370(4.41)	142-1673-84

6,11-Epoxy-1,8-(epoxyhexanoxymethano)-
    naphthacen-14-one, 6,7,8,9,10,11-
    hexahydro-, (6α,8β,11α)-
	EtOH	230(4.31),252(4.39),
		274s(3.76),280s(3.75),
		292s(4.61),317(2.85),
			89-0074-84

Compound	Solvent	$\lambda_{max}(\log \epsilon)$	Ref.
(cont.)		334(2.62)	89-0074-84
2-Propenoic acid, 6-[[1,2,3,4-tetra-hydro-2,3-bis(methylene)-1,4-epoxy-anthracen-5-yl]oxy]hexyl ester	MeCN	218(4.63),252(4.65), 302(3.53)	89-0074-84
Rubranine	EtOH	220(4.05),317(3.86), 347(3.99)	12-0449-84
$C_{25}H_{26}O_5$			
Propanoic acid, 2,2-dimethyl-, 9,10-di-hydro-3-methyl-9,10-dioxo-2-(3-oxo-pentyl)-1-anthracenyl ester	MeOH	208(4.20),246(4.46), 263(4.48),326(3.52), 410(3.83)	78-3677-84
$C_{25}H_{26}O_6$			
Cudraflavanone A	EtOH	294(4.45),314s(4.15), 332s(3.69)	142-0997-84
	EtOH-AlCl$_3$	295(4.41),314s(4.23), 334s(3.61)	142-0997-84
Sanggenon C	EtOH	213s(4.55),229(4.70), 288(4.40),324s(3.82)	142-1791-84
	EtOH-AlCl$_3$	228(4.72),307(4.38), 373(3.50)	142-1791-84
$C_{25}H_{26}O_8$			
2-Anthracenepropanoic acid, 9,10-dihy-dro-1-hydroxy-5,8-dimethoxy-3-methyl-9,10-dioxo-α-(1-oxopropyl)-, ethyl ester, ($\pm$)-	MeOH	231(4.55),261(4.35), 280s(--),450(4.08)	78-3677-84
$C_{25}H_{26}O_9$			
Sakyomicin C	EtOH	216(4.30),240(4.06), 310(3.54),415(3.64)	158-84-24
$C_{25}H_{26}O_{10}$			
Aquayamycin	90% MeOH	220(4.42),320(3.76), 430(3.72)	94-4350-84
	+ NaOH	230(4.42),280(4.17), 320(3.98),395(3.55), 540(3.81)	94-4350-84
Sakyomicin A	EtOH	216(4.45),238(4.20), 310(3.72),415(3.65)	158-84-24
$C_{25}H_{26}O_{13}$			
Plagiogyrin B pentaacetate	dioxan	223(4.12),296(4.29)	94-1815-84
$C_{25}H_{27}NO_5$			
Valachine, dihydro-	MeOH	219s(4.05),292(3.50)	77-1371-84
$C_{25}H_{27}N_2PS$			
3-Thietanone, 2,2,4,4-tetramethyl-, (triphenylphosphoranylidene)hydra-zone	C_6H_{12}	220(4.62),288(3.76)	24-0277-84
$C_{25}H_{27}N_3O_5$			
Ellipticinium, 10-[(1-carboxy-2-methyl-propylidene)amino]-9-hydroxy-2-meth-yl-, acetate	H_2O	243(4.30),300(4.52), 360(3.53),440(3.50)	87-1161-84
$C_{25}H_{28}N_2O_4$			
Pyrrolo[2,1-a]isoquinoline-3-carboxylic acid, 1-cyano-8,9-diethoxy-1,2,3,5-6,10b-hexahydro-2-phenyl-, methyl	EtOH	200(4.27),232(3.56), 285(2.85)	78-0369-84

Compound	Solvent	λ_{max}(log ϵ)	Ref.

ester, (1α,2β,3β,10bβ)- (cont.)   78-0369-84

C$_{25}$H$_{28}$N$_4$

4,4'-Bipyridinium, 1,1"-(1,3-propanedi- H$_2$O   262(4.64)   64-0074-84B
yl)bis[1'-methyl-, tetraperchlorate

7H-Pyrrolo[2,3-d]pyrimidin-4-amine, EtOH   221s(4.30),241s(4.17),   11-0073-84B
N-(4-butylphenyl)-2,5,6-trimethyl-   308(4.31)
7-phenyl-

4H-Pyrrolo[2,3-d]pyrimidin-4-imine, EtOH   243(4.40),280s(3.95),   11-0073-84A
3-(4-butylphenyl)-3,7-dihydro-   302(3.84)
2,5,6-trimethyl-7-phenyl-

C$_{25}$H$_{28}$N$_4$O$_{13}$

1H-Imidazole-5-carboxamide, 1-[(4-ni- M HCl   253(4.15)   4-0849-84
trophenyl)methyl]-4-[(2,3,4,6-tetra- H$_2$O   252(4.24)   4-0849-84
O-acetyl-β-D-glucopyranosyl)oxy]- M NaOH 255(4.24)   4-0849-84

C$_{25}$H$_{28}$O$_3$

Estra-1,3,5(10)-trien-6-one, 17β-hy- MeOH   256(3.96),322(3.53)   94-1885-84
droxy-3-(phenylmethoxy)-

C$_{25}$H$_{28}$O$_4$

Glabrol   EtOH   288(4.05),311(3.77)   105-0233-84
Spiro[5.5]undecane-1,9-dione, 7,11- EtOH   213(3.57),230(4.27),   117-0115-84
bis(4-methoxyphenyl)-   278(3.44),284(3.41)

C$_{25}$H$_{28}$O$_5$

2H,6H-Benzo[1,2-b:5,4-b']dipyran-6-one, MeOH   240(3.21),310(3.46)   2-0887-84
8-(3,4-dihydro-2,2-dimethyl-2H-1-
benzopyran-6-yl)-3,4,7,8-tetrahy-
dro-5-hydroxy-2,2-dimethyl-

4H-1-Benzopyran-4-one, 2,3-dihydro- MeOH   245(3.71),310(3.69)   2-0887-84
5,7-dihydroxy-2-[4-hydroxy-3-(3-
methyl-2-butenyl)phenyl]-6-(3-
methyl-2-butenyl)-

2(5H)-Furanone, 5-[9-[4-(3-furanyl- EtOH   234(4.56),265s(4.34)   88-4941-84
methyl)-3-furanyl]-2,6-dimethyl-
4,6-nonadienylidene]-4-hydroxy-
3-methyl-

C$_{25}$H$_{28}$O$_5$S

1,3-Cyclohexanedione, 2-(4-phenylcyclo- pH 1   264(4.05)   39-1213-84C
hex-1-enyl)-, p-toluenesulfonate H$_2$O   264(4.06)   39-1213-84C
               pH 13   290(4.30)   39-1213-84C

C$_{25}$H$_{28}$O$_5$S$_2$

Bicyclo[2.2.1]heptan-2-one, 3-[bis- EtOH   240(3.88),390(2.00)   78-2951-84
[(phenylmethyl)sulfonyl]methylene]-
1,7,7-trimethyl-, (1R)-

C$_{25}$H$_{28}$O$_6$

3aH-Benzo[g]cyclopropa[b]furo[4,3,2- EtOH   252(4.16),312(4.16)   39-0487-84C
de][1]benzopyran-3a-carboxylic acid,
ethyl ester, (2bα,3α,3aα)-

2-Propenoic acid, 3-(3-methoxyphenyl)- MeCN   276(4.60)   40-0022-84
1,5-pentanediyl ester

2-Propenoic acid, 3-(4-methoxyphenyl)- MeCN   307(4.71)   40-0022-84
1,5-pentanediyl ester

C$_{25}$H$_{28}$O$_8$

2-Propenoic acid, 3-(3,4-dimethoxy- MeCN   322(4.58)   40-0022-84

Compound	Solvent	$\lambda_{max}(\log \epsilon)$	Ref.
phenyl)-, 1,3-propanediyl ester (cont.)			40-0022-84
$C_{25}H_{28}O_9$ 1H,3H-Furo[3,4-c]furan-3a(4H)-ol, 4-(4-acetoxy-3-methoxyphenyl)-1-(3,4-dimethoxyphenyl)dihydro-, acetate, [1S-(1α,3aα,4α,6aα)]-	EtOH	224.8(4.18),279.5(3.74)	94-2730-84
Gibb-3-ene-1-carboxylic acid, 2,7-diacetoxy-10-(acetoxymethyl)-4a-hydroxy-1-methyl-8-methylene-9-oxo-, γ-lactone, (1α,2β,4aα,4bβ,10β)-	MeOH	345(1.80)	97-0331-84
$C_{25}H_{28}O_{10}$ Aquayamycin, dihydro-	MeOH	217.5(4.60),248(4.00), 276(3.93),430(3.73)	94-4350-84
$C_{25}H_{28}S$ 26-Thiatetracyclo[17.2.2.2^{8,11}.1^{3,6}]-hexacosa-3,5,8,10,19,21,22,24-octaene	C_6H_{12}	<u>251(3.9),275s(2.9)</u>	44-1177-84
$C_{25}H_{29}D_5O_3$ Methanone, [4-(dodecyloxy)-2-hydroxyphenyl]phenyl-d$_5$	MeOH	243(3.97),289(4.14), 324(3.97)	35-6155-84
$C_{25}H_{29}NO_4$ Tylohirsutine, 13a-methyl-	MeOH	254(4.6),262(4.4), 278(3.4),284(3.1), 306(2.6),315(2.5), 343(1.9),360(1.8)	102-1765-84
$C_{25}H_{29}NO_8$ 1-Azaspiro[5.5]undeca-2,4-diene-2,3,4,5-tetracarboxylic acid, 1-(4-methylphenyl)-, tetramethyl ester	EtOH	242s(4.13),280(4.03), 350(3.98)	103-0295-84
$C_{25}H_{29}N_2$ 3H-Indolium, 2-[3-(1,3-dihydro-1,3,3-trimethyl-2H-indol-2-ylidene)-1-propenyl]-1,3,3-trimethyl-, iodide	$C_6H_{11}Me$ EtOH DMF	550(5.13) 545(5.13) 550(5.11)	99-0415-84 99-0415-84 99-0415-84
$C_{25}H_{29}N_3O_2$ 5,10-Ethanopyrido[3,2-g]quinoline-11-carbonitrile, 11-acetoxy-2,8-bis(1,1-dimethylethyl)-5,10-dihydro-	MeCN	276(4.087),280(4.096)	5-0877-84
$C_{25}H_{30}Br_2O_7$ 3-Heptanone, 5-acetoxy-1,7-bis(2-bromo-4,5-dimethoxyphenyl)-	EtOH	231(4.25),287(3.79)	39-1635-84C
$C_{25}H_{30}Cl_2O$ 1H-Benz[e]inden-1-one, 2,3-dichloro-4,6,8-tris(1,1-dimethylethyl)-	n.s.g.	294(4.00),423(3.55)	88-3133-84
$C_{25}H_{30}N_2$ 2-Naphthalenamine, N,N-dibutyl-6-[2-(4-pyridinyl)ethenyl]-	EtOH	396(4.54)	44-2546-84
$C_{25}H_{30}N_2O_5$ 3H-2,3-Benzodiazepine, 1-(3,4-dimeth-	EtOH	211(4.49),234(4.43),	24-1476-84

Compound	Solvent	λ_{max}(log ϵ)	Ref.
oxyphenyl)-5-ethyl-4,5-dihydro-7,8-dimethoxy-4-methylene-3-(1-oxo-propyl)-		320(4.19)	24-1476-84
3H-2,3-Benzodiazepine, 1-(3,4-dimeth-oxyphenyl)-5-ethyl-7,8-dimethoxy-4-methyl-3-(1-oxopropyl)-	EtOH	213(4.45),244(4.48), 315(4.00)	24-1476-84
C$_{25}$H$_{30}$N$_2$O$_7$			
Pyrrolo[2,1-a]isoquinoline-3-carboxylic acid, 8,9-dimethoxy-1,2,3,5,6,10b-hexahydro-2-(4-methoxyphenyl)-1-nitro-, methyl ester, (1α,2β,3β,10bβ)-	EtOH	200(4.04),285(3.15)	78-0369-84
C$_{25}$H$_{30}$N$_2$O$_8$			
Carbamic acid, [[3,4-dihydro-6,7-di-methoxy-1-[(3,4,5-trimethoxyphenyl)-methylene]-2(1H)-isoquinolinyl]-carbonyl]-, ethyl ester, (Z)-	MeOH	330(4.36)	78-4003-84
C$_{25}$H$_{30}$N$_3$			
Crystal Violet F (salt of Crystal Violet and TCNE)	CHCl$_3$	305(4.48),398(4.34), 416(4.34),586(5.15)	18-0001-84
C$_{25}$H$_{30}$N$_4$O$_6$S			
Piperazine, 1-[4-[(2,2-dimethyl-1,3-dioxolan-4-yl)methoxy]-6,7-dimeth-oxy-2-quinazolinyl]-4-(2-thienyl-carbonyl)-	EtOH	217(4.60),251(5.00)	4-1189-84
C$_{25}$H$_{30}$N$_4$O$_7$			
Piperazine, 1-[4-[(2,2-dimethyl-1,3-dioxolan-4-yl)methoxy]-6,7-dimethoxy-2-quinazolinyl]-4-(2-furanylcarbonyl]-	EtOH	218(4.60),250(4.80)	4-1189-84
C$_{25}$H$_{30}$O$_2$			
4,8'-Diapocarotenene-4,8'-dial	benzene	298(4.38),457(4.97), 482(5.13),518(5.10)	137-0102-84
	acetone	474(--),505(--)	137-0102-84
	CHCl$_3$	299(--),489(--), 521(--)	137-0102-84
	CS$_2$	473s(--),501(--), 536(--)	137-0102-84
C$_{25}$H$_{30}$O$_4$			
Linderatin	EtOH	225s(4.11),290(4.15)	94-3747-84
C$_{25}$H$_{30}$O$_7$			
Talassin A	n.s.g.	224(4.56),250(4.49)	105-0113-84
C$_{25}$H$_{30}$O$_{13}$			
Liquiritigenin 4'-apiosyl(1→2)-glucoside	MeOH	274(4.25),312(4.04)	102-2108-84
C$_{25}$H$_{31}$NO$_6$			
Acetamide, N-[1,4-dihydro-6,8-dimeth-oxy-7-methyl-5-(2-methyl-1-oxo-2-nonenyl)-1,4-dioxo-2-naphthalenyl]-	EtOH	231(4.42),256(4.38), 301(4.10),367(3.54)	88-1373-84
1-Azaspiro[4.4]non-6-ene-1-acetic acid, 6-(1,3-benzodioxol-5-yl)-7-(2,2-di-methyl-1-oxopropoxy)-3-methylene-, ethyl ester	MeCN	288(3.61)	44-0228-84

Compound	Solvent	$\lambda_{max}(\log \epsilon)$	Ref.
$C_{25}H_{31}NO_{10}$ Alangiside	MeOH	237(4.27),285(3.68)	100-0397-84
$C_{25}H_{31}NO_{12}$ D-Gluconic acid, 2-deoxy-2-[[3-methoxy-1-(methoxycarbonyl)-3-oxo-1-propen-yl]amino]-4,6-O-(phenylmethylene)-, methyl ester, 3,5-diacetate, (Z)-	MeOH	307(4.04)	136-0081-84B
$C_{25}H_{31}N_3O_4$ 1H-Pyrrole-2-carboxylic acid, 5-[[5-[(5-carboxy-4-ethyl-3-methyl-2H-pyrrol-2-ylidene)methyl]-3-ethyl-4-methyl-1H-pyrrol-2-yl]methyl]-3-ethyl-4-methyl-, monohydrobromide, (Z)-	CH_2Cl_2	490(4.46)	44-4602-84
$C_{25}H_{32}Br_2O_8$ Pregn-4-ene-3,20-dione, 6,21-bis(bromo-acetoxy)-11,17-dihydroxy-, (6β,11β)-	MeOH	238(4.13)	44-2634-84
$C_{25}H_{32}N_2$ 4,4'-Bipyridinium, 1-heptyl-1'-[(2-methylphenyl)methyl]-, dibromide	H_2O	260(4.35)	39-0367-84B
$C_{25}H_{32}N_2O_6$ Propanoic acid, [(3,4-dimethoxyphenyl)-[2-(1-ethyl-2-oxopropyl)-4,5-dimeth-oxyphenyl]methylene]hydrazide	EtOH	213(4.60),233s(4.43), 292(4.30),314(4.29)	24-1476-84
$C_{25}H_{32}N_4O_5$ 9H-Cyclopenta[1,2-b:4,3-b']dipyridin-9-one, 2,7-bis[3-(4-morpholinyl)-propoxy]-	H_2O	257(4.48),308(4.09), 460(2.51)	142-0073-84
$C_{25}H_{32}OSe$ Androsta-5,16-dien-3-ol, 17-(phenyl-seleno)-, (3β)-	EtOH	257(3.80)	88-1287-84
$C_{25}H_{32}O_2$ Benzoic acid, [4-methyl-6-(2,6,6-tri-methyl-1-cyclohexen-1-yl)-1,3,5-heptatrienyl]-, ethyl ester, (E,E,E)-	EtOH	223(3.92),246(3.97), 361(4.75)	87-1516-84
$C_{25}H_{32}O_4$ 2H-1-Benzopyran-2-one, 6-methoxy-7-[(3,7,11-trimethyldodeca-2,6,10-trienyl)oxy]- (scopfarnol)	ether	204(4.74),228(4.30), 251(3.87),258s(3.76), 287s(3.74),295(3.79), 342(4.06)	102-0181-84
Cannabinodiol, 8,9-dihydro-, diacetate	EtOH	212(4.51),267(3.08), 273s(3.02)	39-2881-84C
$C_{25}H_{32}O_5$ Scopodrimol A	ether	205(4.72),227(4.30), 250(3.86),258s(3.74), 286s(3.78),295(3.83), 343(4.05)	102-0181-84
$C_{25}H_{32}O_7$ 3-Heptanone, 5-acetoxy-1,7-bis(3,4-di-methoxyphenyl)-, (S)-	EtOH	229(4.25),280(3.80)	39-1635-84C

Compound	Solvent	$\lambda_{max}(\log \epsilon)$	Ref.
$C_{25}H_{32}O_8$ Phyllanthocindiol, methyl ester	MeOH	216(4.24),221(4.15), 279(4.32)	44-4258-84
$C_{25}H_{32}O_{13}$ Oleuropein	MeOH	228(4.25),280(3.74)	102-2839-84
$C_{25}H_{32}O_{15}$ Loganin, 2',3',4',7-tetra-O-acetyl- 6'-carboxy-	MeOH	233(4.04)	102-1917-84
$C_{25}H_{33}NO_6$ Spiro[benzofuran-3(2H),4'(1'H)-pyri- dine]-2,3'-dicarboxylic acid, 7- methoxy-1'-methyl-3'-(1,1-diethyl- propyl) 2-ethyl ester	EtOH	284(3.20),347(3.76)	44-0056-84
$C_{25}H_{34}NO_6$ Pyridinium, 3-[(1,1-diethylpropoxy)- carbonyl]-4-[2-(2-ethoxy-2-oxoeth- oxy)-3-methoxyphenyl]-1-phenyl-, iodide	EtOH	305(3.79)	44-0056-84
$C_{25}H_{34}NP$ Benzenamine, N-[[[2,4,6-tris(1,1-di- methylethyl)phenyl]phosphinidene]- methylene]-	hexane	258(4.51),288(3.94), 415(2.69)	88-1809-84
$C_{25}H_{34}N_3S$ Pyridinium, 1-(3-isothiocyanatopropyl)- 4-[p-(dibutylamino)styryl]-, bromide	EtOH	500(4.72)	44-2546-84
$C_{25}H_{34}O_2$ 12'-Apo-β,ψ-carotenal, 5,8-epoxy-5,8- dihydro-(5R,8R)-($\pm$)-	EPA	272.5(4.03),385(4.80)	33-0471-84
(5R,8S)-($\pm$)-	EPA	272.5(4.08),385(4.76)	33-0471-84
Ophiobolin G	EtOH	235(4.56)	98-0778-84
$C_{25}H_{34}O_4$ Cannabidiol, 1,2-dihydro-1,6-didehydro-, diacetate	EtOH	266(3.14),274(3.08)	39-2881-84C
$C_{25}H_{34}O_5$ Spiro[2H-1-benzopyran-2,3'-bicyclo- [3.2.0]heptane]-6,8-dicarboxalde- hyde, 3,4-dihydro-5,7-dimethoxy- 2',2'-dimethyl-4-(2-methylpropyl)-	MeOH	261(4.06),280(3.96), 320(3.46)	35-0734-84
$C_{25}H_{34}O_6$ Hellebrigenin, 3-O-acetyl-	MeOH	299(3.78)	39-1573-84C
$C_{25}H_{34}O_7$ 11(12)-Kempen-20-oic acid, 5β,9β-di- acetoxy-2-oxo-, methyl ester	MeOH	223.5(3.96),300(0.90)	73-2024-84
$C_{25}H_{34}O_8$ Pregn-4-ene-3,20-dione, 6,21-diacetoxy- 11,17-dihydroxy-, (6β,11β)-	MeOH	239(4.21)	44-2634-84
$C_{25}H_{35}N$ Butanamine, N-[2-[4-(2,6,6-trimethyl-	isoPrOH	380(4.42)	89-0081-84

Compound	Solvent	$\lambda_{max}(\log \epsilon)$	Ref.
1-cyclohexen-1-yl)ethenyl]phenyl]-2-butenylidene]-, protonated (cont.)			89-0081-84
$C_{25}H_{35}NO_2$ 1H-Isoindole, 2-[2-(1,1-dimethyleth-oxy)-1-methyl-1-phenylethoxy]-2,3-dihydro-1,1,3,3-tetramethyl-	EtOH	258(2.94),264(3.05), 270(3.03)	126-1809-84
$C_{25}H_{35}NO_6$ Spiro[benzofuran-3(2H),4'(1'H)-pyri-dine]-2,3'-dicarboxylic acid, 5',6'-dihydro-7-methoxy-1'-methyl-, 3'-(1,1-dimethylpropyl) 2-ethyl ester	EtOH	290(4.22)	44-0056-84
$C_{25}H_{35}N_2O_2$ Pyridinium, 4-[2-[4-(dibutylamino)phen-yl]ethenyl]-1-(2-ethoxy-2-oxoethyl)-, bromide	EtOH	507(4.73)	44-2546-84
$C_{25}H_{36}N_4O_4$ Ethanamine, 2,2'-[spiro[9H-cyclopenta-[1,2-b:4,3-b']dipyridine-9,2'-[1,3]-dioxolane]-2,7-diylbis(oxy)]bis-[N,N-diethyl-, dihydrochloride	H_2O	214(4.51),292(4.32), 337(3.64)	142-0073-84
$C_{25}H_{36}O_2$ Cyclopropanebutanoic acid, 2,2-dimeth-yl-3-(1,3-tetradecadiene-5,8-diyn-yl)-, ethyl ester, [1S-[1α,3α(1E-3E)]]-	EtOH	265(4.51),274(4.59), 286(4.45)	104-2076-84
$C_{25}H_{36}O_3$ Chola-4,6-dien-24-oic acid, 3-oxo-, methyl ester	n.s.g.	286(4.39)	13-0095-84B
$C_{25}H_{36}O_5$ Solidagolactone VIII cis isomer	EtOH	209.1(4.41)	88-2809-84
trans (solidagolactone VII)	EtOH	211.8(4.42)	88-2809-84
$C_{25}H_{36}O_6$ 1H-Cyclopentacyclododecene-4,9-dione, 3a-acetoxy-2,3,3a,7,8,10,11,13a-octa-hydro-2,5,8,8,12-pentamethyl-1-(1-oxopropoxy)-	MeOH	193(4.11),234(3.89)	102-1689-84
Euphorbia factor H, acetonide	MeOH	194(4.28),209(4.23), 298(2.60)	64-0683-84B
$C_{25}H_{37}NO_2$ Butanoic acid, 4-[[3,7-dimethyl-9-(2,6,6-trimethyl-1-cyclohexen-1-yl)-2,4,6,8-nonatetraenylidene)-amino]-, methyl ester, hydrochlor-ide, (E,E,E,?,E)-	MeOH	360(4.60),440(4.61)	104-1669-84
$C_{25}H_{37}NO_6$ Spiro[benzofuran-3(2H),4'-piperidine]-2,3'-dicarboxylic acid, 7-methoxy-1'-methyl-, 3'-(1,1-diethylpropyl) 2-ethyl ester	EtOH	278(3.32)	44-0056-84

Compound	Solvent	$\lambda_{max}(\log \epsilon)$	Ref.
$C_{25}H_{37}NO_{15}$			
D-glycero-L-gluco-Octonic acid, 2-acetyl-2,3-dideoxy-3-(nitromethyl)-, 1,1-dimethylethyl ester, 4,5,6,7,8-pentaacetate, (2ξ)-	EtOH	255(3.24)	136-0063-84K +136-0075-84K
$C_{25}H_{37}N_3O_{13}$			
D-Streptamine, N,N'-bis(ethoxycarbonyl)-4-O-[3-[(ethoxycarbonyl)amino]-6-formyl-3,4-dihydro-2H-pyran-2-yl]-, 5,6-diacetate, (2S-cis)-	n.s.g.	252(3.76)	158-0143-84
$C_{25}H_{38}NO_7PSi$			
1-Aza-5-phosphabicyclo[4.2.0]oct-2-ene-2-carboxylic acid, 7-[1-[[(1,1-dimethylethyl)dimethylsilyl]oxy]ethyl]-5-ethoxy-8-oxo-, (4-methoxyphenyl)-methyl ester, 5-oxide, (R)-	EtOH	227(4.17),265(3.92)	158-0143-84
$C_{25}H_{38}N_2O_{14}$			
D-glycero-L-gluco-Octonic acid, 2-(1-aminoethylidene)-2,3-dideoxy-3-(nitromethyl)-, 1,1-dimethylethyl ester, 4,5,6,7,8-pentaacetate	EtOH	280(4.52)	136-0063-84K
$C_{25}H_{38}O_3$			
Ophiobolin H	EtOH	243(4.29)	98-0778-84
Pregna-7,9(11)-diene-20-carboxylic acid, 3-hydroxy-4,4,14-trimethyl-, (3β,5α,20S)-	MeOH	236(4.22),243(4.29), 252(4.12)	39-0497-84C
$C_{25}H_{38}O_4$			
Chol-4-ene-24-carboxylic acid, 7α-hydroxy-3-oxo-	n.s.g.	243(4.17)	13-0095-84B
Compd., m. 229-233° (hydroxyacid)	MeOH	239(4.20),246(4.27), 255(4.10)	39-0497-84C
Compd., double m. 230-242 and 266°	MeOH	255(3.81)	39-0497-84C
Pregna-5,17(20)-dien-21-oic acid, 20-ethoxy-3-hydroxy-, ethyl ester, (3β,17E)-	EtOH	237(3.86)	118-0132-84
$C_{25}H_{38}O_5$			
2-Butenoic acid, 2-methyl-, 1,2,3,3a-4,7,8,8a-octahydro-1-hydroxy-3a,6-dimethyl-1-(1-methylethyl)-4,8-azulenediyl ester, [1R-[1α,3aα,4β(Z)-8β(Z),8aβ]]-	ether	229(4.15)	78-5197-84
$C_{25}H_{39}NO_3$			
Myxalamid B	EtOH	207(4.38),263(4.04), 340s(--),357(4.67), 370s(--)	158-84-113
$C_{25}H_{39}NO_8$			
Debenzoylpyroaconitine	EtOH	270(2.31)	33-2017-84
16-epi-	EtOH	270(2.2)	33-2017-84
$C_{25}H_{40}N_2$			
4,4'-Bipyridinium, 1,1'-diheptyl-2-methyl-, dibromide	H_2O	261(4.29)	39-0367-84B

Compound	Solvent	$\lambda_{max}(\log \epsilon)$	Ref.

$C_{25}H_{40}O_4$
    Compd. double m. 140 and 233°
      (a ketoalcohol)
	MeOH	255(3.70)	39-0497-84C

$C_{25}H_{40}O_5$ Compd. m. 226-30°	MeOH	257(4.10)	39-0497-84C
$C_{25}H_{41}N_3O_2$ Benzoyl azide, 4-(octadecyloxy)-	C_6H_{12}	277(4.39),284(4.41), 289(4.35)	35-4531-84

$C_{25}H_{50}Si_5$
    6,7,13,19,25-Pentasilapentaspiro-
    [5.0.5.0.5.0.5.0.5.0]triacontane
	isooctane	209s(4.48),269(2.99),
		 282(2.99)
			35-5521-84

Compound	Solvent	$\lambda_{max}(\log \epsilon)$	Ref.
$C_{26}H_{12}Br_2O_4$			
Naphtho[1,8-bc:5,4-b'c']dipyran-2,7-di-one, 5,10-dibromo-3,8-diphenyl-	CHCl$_3$	272.5(4.48),598(4.36), 639(4.45)	39-1957-84C
$C_{26}H_{14}O_4$			
Naphtho[1,2-b:5,6-b']difuran-2,7-dione, 3,8-diphenyl-	CHCl$_3$	267.5(5.00),552(4.81)	39-1957-84C
Naphtho[2,1-b:3,4-b']difuran-2,9-dione, 3,8-diphenyl-	CHCl$_3$	276(4.23),422(4.53)	39-1957-84C
Naphtho[1,8-bc:5,4-b'c']dipyran-2,7-di-one, 3,8-diphenyl-	CHCl$_3$	270(4.50),592(4.37), 634(4.46)	39-1957-84C
$C_{26}H_{15}NO_3$			
1H-Benzo[h]naphtho[1,2,3-de]quinoline-2,8-dione, 9-phenoxy-	toluene	425(3.90),452(3.81)	104-0980-84
$C_{26}H_{15}N_3O_7$			
3-Furancarbonitrile, 4,5-bis(1,3-benzo-dioxol-5-yl)-2-[[(4-nitrophenyl)meth-ylene]amino]-	CHCl$_3$	267(4.52),305(4.26), 365(4.06)	73-1788-84
	MeOH	213(--),222(--), 265(--),269(--), 322(--),333(--), 455(--)	73-1788-84
	dioxan	462(--)	73-1788-84
$C_{26}H_{16}$			
9H-Fluorene, 9-(9H-fluoren-9-ylidene)-	MeOH	213(4.36),245(4.54), 257(4.55),271(4.27), 455(3.89)	2-0802-84
$C_{26}H_{16}ClNO_4$			
Acetamide, 2-chloro-N-(6,11-dihydro-6,11-dioxo-12-phenoxy-5-naphtha-cenyl)-	toluene	425(3.10)	104-0980-84
$C_{26}H_{16}N$			
9H-Fluoren-9-iminium, N-9H-fluoren-9-ylidene-, (OC-6-11)-, hexachloro-antimonate	CH$_2$Cl$_2$	257(4.99),264(5.10), 283(4.51),342(4.48)	24-3222-84
$C_{26}H_{16}N_2O$			
Acridine, 9,9'-oxybis-	MeCN	340(3.93),357(4.13), 372(4.13),390(4.11)	64-1399-84B
$C_{26}H_{16}O_{14}$			
Antibiotic DK-7814-C	CHCl$_3$	311(4.29),354(4.04), 370(4.01),475(3.85), 508(3.87),547(3.65)	158-84-28
$C_{26}H_{17}NO$			
Spiro[acridin-9(10H),9'(10'H)-anthra-cen]-10'-one	EtOH	253(4.26),276(4.19), 322(3.90)	24-2703-84
$C_{26}H_{17}N_3O_3$			
1H-Benzimidazole-5-carboxylic acid, 2-[4-(2-benzoxazolyl)-1-naphthal-enyl]-1-methyl-	DMF	343(4.47)	5-1129-84
$C_{26}H_{18}$			
9,9'-Bi-9H-fluorene	MeOH	218(4.68),260(4.44), 267(4.49),293(3.97),	2-0802-84

Compound	Solvent	$\lambda_{max}(\log \epsilon)$	Ref.
9,9'-Bi-9H-fluorene (cont.)		304(4.00)	2-0802-84
$C_{26}H_{18}N_2O_3$ 9,10-Anthracenedione, 1-amino-2-phen- oxy-4-(phenylamino)-	C_6H_5Cl	528s(3.96),536(4.18), 600(4.19)	34-0482-84
$C_{26}H_{18}N_2O_4$ 9,10-Anthracenedione, 1-amino-2-(4-hy- droxyphenoxy)-4-(phenylamino)-	C_6H_5Cl	534s(3.94),562(4.14), 600(4.14)	34-0482-84
4H-Imidazol-4-one, 3,5-dihydro-3-(4- methoxyphenyl)-5-[(4-oxo-4H-1-benzo- pyran-3-yl)methylene]-2-phenyl-	$CHCl_3$	245(4.52),393(4.36)	2-1048-84
5(4H)-Oxazolone, 4-[[4-[5-(4-methoxy- phenyl)-2-oxazolyl]phenyl]methylene]- 2-phenyl-	toluene	320(4.51),426(4.64)	135-1330-84
$C_{26}H_{18}N_2S_3$ 2H-2a,3-Dithia(2a-S^{IV})-1,2-diazacyclo- penta[cd]pentalene, 5,6-dihydro-2-(2- naphthalenyl)-4-(2-naphthalenylthio)-	MeCN	563(4.45)	97-0251-84
$C_{26}H_{18}N_4$ 1H-Tetrazolium, 5(9H)-fluoren-9-yl)- 1,3-diphenyl-, hydroxide, inner salt	MeCN	246(4.57),257(4.55), 276(4.60),300s(4.33), 372s(4.14),393s(4.28), 412(4.35),589(3.17)	39-2545-84C
$C_{26}H_{18}N_4O$ 5H-Pyrano[2,3-d]pyrimidine-6-carbo- nitrile, 4-amino-7-[1,1'-biphenyl]- 4-yl]-5-phenyl-	EtOH	259(4.32),290(4.40)	73-2309-84
$C_{26}H_{18}O_{14}$ Antibiotic DK-7814-B	$CHCl_3$	314(4.23),348(3.96), 365(3.91),480(3.72), 510(3.74),550(3.49)	158-84-28
$C_{26}H_{18}O_{15}$ Antibiotic DK-7814-A	$CHCl_3$	312(4.32),352(4.06), 366(4.01),480(3.86), 505(3.87),545(3.64)	158-84-28
$C_{26}H_{19}BO_6$ 1,5-Methano-5H-anthra[2,3-e][1,3,2]di- oxaborocin-5-carboxylic acid, 1,6,7,14-tetrahydro-7,14-dioxo- 3-phenyl-, methyl ester, (1S-trans)-	dioxan	285(4.29),296(4.29), 402(3.73)	78-4649-84
$C_{26}H_{19}BO_8$ 1,5-Methano-5H-anthra[2,3-e][1,3,2]di- oxaborocin-5-carboxylic acid, 1,6,8,13-tetrahydro-7,14-dihydroxy- 8,13-dioxo-3-phenyl-, methyl ester, (1S)-	n.s.g.	250(4.60),286(4.00), 481(4.01)	78-4649-84
$C_{26}H_{19}N$ Spiro[acridine-9(10H),9'(10'H)-anthra- cene]	EtOH	296(4.05),326(3.99)	24-2703-84
$C_{26}H_{19}NO_2S_3$ 3-Thiophenecarboxylic acid, 5-(3H-1,2-	MeCN	593(4.63)	48-0917-84

Compound	Solvent	$\lambda_{max}(\log \epsilon)$	Ref.
benzodithiol-3-ylidene)-2,5-dihydro-4-phenyl-2-(phenylimino)-, ethyl ester, perchlorate (cont.)			48-0917-84
$C_{26}H_{19}N_3O_3$			
9,10-Anthracenedione, 1-amino-2-(4-aminophenoxy)-4-(phenylamino)-	C_6H_5Cl	534s(3.94),561(4.15), 600(4.16)	34-0482-84
$C_{26}H_{19}N_5O$			
5H-Pyrano[2,3-d]pyrimidine-6-carbonitrile, 7-[1,1'-biphenyl]-4-yl-4-hydrazino-5-phenyl-	EtOH	246(4.26),271(4.35), 298(4.33)	73-2309-84
$C_{26}H_{19}O_9S$			
Pyrylium, 2,6-bis(4-carboxyphenyl)-4-(4-methoxy-3-sulfophenyl)-, perchlorate	pH 2.0	275(4.35),409(4.59)	39-0867-84B
pseudobase	pH 8.0	250(4.45),328(4.15)	39-0867-84B
anion	pH 13.0	485(4.43)	39-0867-84B
$C_{26}H_{20}$			
Azulene, 1-[6-(6-azulenyl)-1,3,5-hexatrienyl]-	hexane	296(4.62),330(4.46), 455s(4.77),480(4.87), 509s(4.74),608(3.19), 635s(3.16),675s(3.07), 705s(2.95),750s(2.67)	5-1905-84
	CH_2Cl_2-TFA	780s(5.02),837(5.11)	5-1936-84
Azulene, 1,1'-(1,3,5-hexatriene-1,6-diyl)bis-	CH_2Cl_2	229(4.41),268(4.58), 335(4.37),445s(4.78), 462(4.90),491(4.83), 630(3.11),680s(3.06), 770s(2.71)	5-1905-84
radical cation	CH_2Cl_2	823(--),930s(--), 1135(--),1210s(--)	5-1936-84
dication	CH_2Cl_2	261(4.57),273s(4.46), 335(4.23),440s(4.10), 510(4.84),550s(4.93), 582(4.96),610s(4.90)	5-1936-84
[2](2,6)Biphenyleno[2](2,6)naphthalenophane	C_6H_{12}	231(5.1),275(4.3), 335(3.3),360(3.7), 385(3.9)	88-3603-84
$C_{26}H_{20}N$			
Benzenemethanaminium, N-(diphenylmethylene)-2-phenyl-, (OC-6-11)-, hexachloroantimonate	CH_2Cl_2	293(4.62),303s(--)	24-3222-84
$C_{26}H_{20}N_2O_2$			
2-Butene-1,4-dione, 2-(5-methyl-3-phenyl-1H-pyrazol-1-yl)-1,4-diphenyl-, (Z)-	MeOH	252(4.56),340(4.44)	44-4647-84
$C_{26}H_{20}N_2O_5$			
Benzamide, N-[1-[[(4-methoxyphenyl)-amino]carbonyl]-2-(4-oxo-4H-1-benzopyran-3-yl)ethenyl]-	EtOH	223(4.39),280(4.39)	2-1048-84
$C_{26}H_{20}N_3O$			
1H-Benzimidazolium, 2-[4-(2-benzoxazolyl)-1-naphthalenyl]-1,3-dimethyl-, iodide	DMF	342(4.29)	5-1129-84

Compound	Solvent	$\lambda_{max}(\log \epsilon)$	Ref.
methyl sulfate (cont.)	DMF	344(4.41)	5-1129-84

$C_{26}H_{20}N_3O_5P$
  Phosphine, (4-methoxy-2,6-dinitrophen-
    ylimino)triphenyl-
	MeOH	394(3.18)	65-0289-84

$C_{26}H_{20}N_4$
  1H-Benzimidazole, 2,2'-(1,4-naphtha-
    lenediyl)bis[1-methyl-
	DMF	314(4.33)	5-1129-84

1H-Benzimidazole, 2,2'-(1,7-naphtha- lenediyl)bis[1-methyl-	DMF	290(4.48)	5-1129-84
1H-Benzimidazole, 2,2'-(1,7-naphtha- lenediyl)bis[5-methyl-	DMF	310(4.56),342(4.39)	5-1129-84
1H-Benzimidazole, 2,2'-(2,6-naphtha- lenediyl)bis[1-methyl-	DMF	330(4.61)	5-1129-84
9H-Carbazole, 3,3'-azobis[9-methyl-, (E)-	dioxan	410(4.5)	97-0286-84
(Z)-	dioxan	466(3.63)	97-0286-84

$C_{26}H_{20}N_4O_2$
  1H-Benzimidazole, 2,2'-(1,4-naphtha-
    lenediyl)bis[5-methoxy-
	DMF	370(4.49)	5-1129-84

1H-Benzimidazole, 2,2'-(1,7-naphtha- lenediyl)bis[5-methoxy-	DMF	320(4.48)	5-1129-84
1H-Benzimidazole, 2,2'-(2,6-naphtha- lenediyl)bis[5-methoxy-	DMF	370(4.77)	5-1129-84

$C_{26}H_{20}N_4O_2S_2$
  Thiazolo[3,2-a]pyrimidin-4-ium, 3-hy-
    droxy-5-methyl-2-[[(4-methyl-6-phen-
    yl-2-pyrimidinyl)thio]acetyl]-7-
    phenyl-, hydroxide, inner salt
	MeCN	257(4.40),316(4.56),
		482(4.04)
			103-0791-84

$C_{26}H_{20}N_4O_4$
  Imidazo[2,1-c][1,2,4]benzotriazine-
    2-carboxylic acid, 4,5-dibenzoyl-
    4,5-dihydro-, ethyl ester
	MeOH	226(4.55)	4-1081-84

$C_{26}H_{20}N_4O_8S_2$
  Acetic acid, [[2-[[5-(1,2-dihydro-1-
    methyl-2-oxo-3H-indol-3-ylidene)-
    4,5-dihydro-4-oxo-2-thiazolyl]-
    amino]-2,3-dihydro-8,9-dimethoxy-
    1,3-dioxo-1H-2-benzazepin-4-yl]thio]-
	MeOH	250(4.54),335s(4.17),
		386(4.31)
			103-0024-84

$C_{26}H_{20}N_4O_9S_2$
  Acetic acid, [[2-[[4,5-dihydro-5-[3-(4-
    nitrophenyl)-2-propenylidene]-4-oxo-
    2-thiazolyl]amino]-2,3-dihydro-8,9-
    dimethoxy-1,3-dioxo-1H-2-benzazepin-
    4-yl]thio]-
	MeOH	250(4.55),340s(4.34),
		382(4.41)
			103-0024-84

$C_{26}H_{20}N_8O_3S$
  1H-Pyrazole, 3-methyl-5-phenyl-1-(2-
    pyridinylcarbonyl)-4-[[2-[(2-pyrimi-
    dinylamino)sulfinyl]phenyl]azo]-
	n.s.g.	368(4.28)	48-1021-84

$C_{26}H_{20}N_8O_4$
  10H,14H-6,9:13,16-Dimethano-8H,23H-
    [1,3,5,9,11,13]hexaazacyclohexa-
    decino[1,2-a:13,12-a']diquinoxaline-
	glycol	269(4.79),341(4.25),
		430(4.26)
			88-1035-84

Compound	Solvent	$\lambda_{max}(\log \epsilon)$	Ref.
8,14,27,28-tetrone, 11,12,24,25-tetrahydro- (cont.)			88-1035-84
radical ion	glycol	562(2.33),583(2.33)	88-1035-84
$C_{26}H_{20}O_5$			
2,5-Furandione, 3-[(3,5-dimethoxyphenyl)methylene]-4-(diphenylmethylene)-dihydro-, (E)-	toluene	369(3.96)	48-0233-84
(Z)-	toluene	400(4.18)	48-0233-84
$C_{26}H_{20}O_{11}$			
β-D-Glucopyranosiduronic acid, 6'-hydroxy-3-oxospiro[isobenzofuran-1(3H)-9'-[9H]xanthen]-3'-yl-	MeOH	223(4.81),274(3.85)	94-2832-84
$C_{26}H_{21}BO_4$			
1,5-Methano-1H-anthra[2,3-e][1,3,2]dioxaborocin-7,14-dione, 5-ethyl-5,6-dihydro-3-phenyl-, (1S)-	$CHCl_3$	242(4.49),277(4.26), 288(4.29),300(4.30), 415(3.73)	78-4649-84
$C_{26}H_{21}BO_6$			
1,5-Methano-1H-anthra[2,3-e][1,3,2]-dioxaborocin-8,13-dione, 5-ethyl-5,6-dihydro-7,14-dihydroxy-3-phenyl-, (1S)-	$CHCl_3$	252(4.63),257(4.61), 288(4.03),326s(3.45), 465s(4.00),487(4.05), 506s(3.92),521(3.86)	78-4649-84
$C_{26}H_{21}N$			
5H-Benzo[6,7]cyclohepta[1,2-b]pyridine, 6,7-dihydro-2,4-diphenyl-	dioxan	245.5(4.60),285s(4.12), 301(4.03)	48-0647-84
Benzo[h]quinoline, 5,6-dihydro-2-(4-methylphenyl)-4-phenyl-	dioxan	251(4.62),273s(4.32), 323(4.13)	48-0647-84
$C_{26}H_{21}NO_2$			
5(2H)-Oxazolone, 2-methyl-2-(1-methyl-2,3-diphenyl-2-cyclopropen-1-yl)-4-phenyl- (2λ,3ε)	EtOH	228(4.25),338(4.44), ?(4.33)	35-1065-84
5(2H)-Oxazolone, 2-methyl-2-(2-methyl-1,3-diphenyl-2-cyclopropen-1-yl)-4-phenyl-	C_6H_{12}	213(4.34),238(4.02), 265(4.24)	35-1065-84
5(4H)-Oxazolone, 4-methyl-4-(1,2-diphenyl-1,3-butadien-1-yl)-2-phenyl-, cis	C_6H_{12}	247(4.38),278(4.26)	35-1065-84
trans	C_6H_{12}	234(4.45),275(4.16)	35-1065-84
$C_{26}H_{21}NS$			
Benzenamine, N-[2-(5,6-diphenyl-2H-thiopyran-2-ylidene)propylidene]-, hydrochloride	EtOH	515(4.31)	97-0183-84
Benzenamine, N-[2-(6-methyl-5-phenyl-2H-thiopyran-2-ylidene)-2-phenylethylidene]-, hydrochloride	$CHCl_3$	523(4.47)	97-0183-84
$C_{26}H_{21}N_3O$			
4H-Indazol-4-one, 5-[(diphenylamino)-methylene]-1,5,6,7-tetrahydro-1-phenyl-, (E)-	EtOH	235(4.32),272(4.26), 374.5(4.43)	4-0361-84
2(10H)-Phenazinone, 10-(4-methylphenyl)-3-[(4-methylphenyl)amino]-	$CHCl_3$	281(4.67),470(4.41)	103-0263-84
$C_{26}H_{21}N_3O_3$			
2(10H)-Phenazinone, 10-(4-methoxyphen-	$CHCl_3$	282(4.67),471(4.42)	103-0263-84

Compound	Solvent	$\lambda_{max}(\log \epsilon)$	Ref.
yl)-3-[(4-methoxyphenyl)amino]- (cont.)			103-0263-84
C$_{26}$H$_{21}$N$_3$O$_3$S			
Benzamide, N-[[2,5-dihydro-1-(4-methyl-phenyl)-4-[(4-methylphenyl)amino]-2,5-dioxo-1H-pyrrol-3-yl]thioxo-methyl]-	CHCl$_3$	242(4.38),256(4.39), 306(4.41),440(3.85)	48-0401-84
C$_{26}$H$_{21}$N$_3$O$_7$S$_2$			
Acetic acid, [[2-[[4,5-dihydro-4-oxo-5-(3-phenyl-2-propenylidene)-2-thia-zolyl]amino]-2,3-dihydro-8,9-dimeth-oxy-1,3-dioxo-1H-2-benzazepin-4-yl]-thio]-	MeOH	250(4.61),345(4.33), 387(4.34)	103-0024-84
C$_{26}$H$_{21}$N$_5$O			
Ethanone, 1-phenyl-2-[8-phenyl-6-[(phenylmethyl)amino]-7H-purin-7-yl]-	MeOH	239(4.47),288(4.20)	150-3601-84M
C$_{26}$H$_{21}$N$_5$O$_3$			
5H-[1]Benzopyrano[3,4-c]pyridin-5-one, 1,2,3,4-tetrahydro-3-(phenylmethyl)-8-[(1-phenyl-1H-tetrazol-5-yl)oxy]-	MeOH	271(4.03),310(4.08)	4-1557-84
1,3,7,8-Tetraazaspiro[4.5]deca-6,9-di-ene-10-carboxylic acid, 4-oxo-1,3-diphenyl-2-(phenylimino)-, methyl ester	MeOH	295(3.93)	39-2491-84C
C$_{26}$H$_{21}$O			
5H-Benzo[6,7]cyclohepta[1,2-b]pyrylium, 6,7-dihydro-2,4-diphenyl-, perchlor-ate	MeCN	234(4.20),277.5(4.26), 342(4.26),394(4.42)	48-0647-84
Naphtho[1,2-b]pyrylium, 5,6-dihydro-2-(4-methylphenyl)-4-phenyl-, perchlorate	MeCN	240s(4.20),291.5(4.26), 346(4.24),432(4.50)	48-0647-84
C$_{26}$H$_{22}$ClF$_3$N$_2$O$_3$			
DL-Valine, N-[2-chloro-4-(trifluoro-methyl)phenyl]-, cyano(3-phenoxy-phenyl)methyl ester	MeOH	204(4.75),254(4.26)	98-1134-84
C$_{26}$H$_{22}$Cl$_2$O$_6$			
1(2H)-Naphthalenone, 4-chloro-2-(4-chloro-5,8-dimethoxy-6-methyl-1-oxo-2(1H)-naphthalenylidene)-5,8-dimethoxy-6-methyl-	CHCl$_3$	260s(4.41),290(4.22), 562(4.28)	5-0319-84
C$_{26}$H$_{22}$N			
Pyridinium, 2,4,6-triphenyl-1-(2-prop-enyl)-	C$_6$H$_5$Cl	306(4.59)	120-0062-84
C$_{26}$H$_{22}$NO$_2$P			
Benzamide, N-[4-(diphenylphosphinyl)-phenyl]-4-methyl-	EtOH	277(4.48)	42-0430-84
Benzamide, 4-(diphenylphosphinyl)phen-yl]-N-(phenylmethyl)-	EtOH	265(3.75),273(3.54)	42-0430-84
C$_{26}$H$_{22}$NO$_3$P			
Benzamide, N-[4-(diphenylphosphinyl)-phenyl]-4-methoxy-	EtOH	287(4.47)	42-0430-84

Compound	Solvent	$\lambda_{max}(\log \epsilon)$	Ref.
$C_{26}H_{22}N_2O_2S_2$ 1H-Pyrrole-3-carbodithioic acid, 2,5-dihydro-1-(4-methylphenyl)-4-[(4-methylphenyl)amino]-2,5-dioxo-, phenylmethyl ester	CHCl$_3$	306(4.34),435(4.00)	48-0401-84
$C_{26}H_{22}N_2O_3$ 1H-Pyrazole-1-acetic acid, 5-methyl-α-(2-phenoxy-2-phenylethenyl)-3-phenyl-, (Z)-	MeOH	255(4.47)	44-4647-84
$C_{26}H_{22}N_2O_8$ 1H-Indole-1-propanoic acid, 2-[1-(3-ethoxy-1,3-dioxopropyl)-1,3-dihydro-3-oxo-2H-indol-2-ylidene]-2,3-dihydro-β,3-dioxo-, ethyl ester, cis	benzene	424(3.60)	39-2305-84C
trans	benzene	552(3.81)	39-2305-84C
$C_{26}H_{22}N_2S$ 1,4-Benzenediamine, N-[2-(5,6-diphenyl-2H-thiopyran-2-ylidene)propylidene]-, monohydrochloride	EtOH	544(4.31)	97-0183-84
$C_{26}H_{22}N_4$ 4H-Pyrrolo[2,3-d]pyrimidin-4-imine, 3,7-dihydro-5,6-dimethyl-2,3,7-triphenyl-	EtOH	242(4.39),325(3.89)	11-0073-84A
$C_{26}H_{22}N_5O$ Benzoxazolium, 2-(2,6-dihydro-4-methyl-2,6-diphenylpyrazolo[3,4-c]pyrazol-3-yl)-3-ethyl-, iodide	EtOH	330(4.03)	48-0811-84
7H-Purinium, 6-(benzoylamino)-7,9-bis(phenylmethyl)-, bromide	H$_2$O	284(3.99),318(4.02), 328s(3.99)	150-3601-84M
$C_{26}H_{22}N_6O_7$ Pyrazolo[1,5-a]quinazoline-4(5H)-acetic acid, 3-cyano-2-(ethoxycarbonyl)-α-[[2-(methoxycarbonyl)phenyl]hydrazono]-5-oxo-, ethyl ester	EtOH	202(4.25)	4-1049-84
$C_{26}H_{22}N_8O_4$ Benzo[g]pteridine-2,4(3H,10H)-dione, 10,10'-(1,4-butanediyl)bis[3-methyl-	pH 7.27	336(4.10),438(4.06)	39-1227-84B
$C_{26}H_{22}O_2$ 5H-Benzocyclohepten-5-one, 6,7,8,9-tetrahydro-6-(3-oxo-1,3-diphenylpropylidene)-	dioxan	245(4.30)	48-0647-84
1(2H)-Naphthalenone, 3,4-dihydro-2-[3-(4-methylphenyl)-3-oxo-1-phenylpropylidene]-	dioxan	254.5(4.36)	48-0647-84
$C_{26}H_{22}O_{15}$ Benzoic acid, 3,4,5-trihydroxy-, 6'-ester with 2-β-D-glucopyranosyl-1,3,6,7-tetrahydroxy-9H-xanthen-9-one	MeOH	218(4.33),240(4.19), 257(4.27),271(4.06), 315(3.91)	94-2676-84
$C_{26}H_{23}BrN_2O_4$ 1H-Indazole-6,7-dicarboxylic acid,	EtOH	240(4.52),338(4.11)	94-2146-84

Compound	Solvent	λ_{max} (log ϵ)	Ref.
1-(4-bromophenyl)-4-methyl-3-phenyl-, diethyl ester (cont.)			94-2146-84
$C_{26}H_{23}ClN_2O_4$ 1H-Indazole-6,7-dicarboxylic acid, 1-(4-chlorophenyl)-4-methyl-3-phenyl-, diethyl ester	EtOH	249(4.78),336(4.34)	94-2146-84
$C_{26}H_{23}ClN_4O_5$ 1H-Pyrazolo[3,4-d]pyrimidine, 4-chloro-1-[2-deoxy-3,5-bis-O-(4-methylbenzoyl)-β-D-erythro-pentofuranosyl]-	MeOH	240(4.25)	35-6379-84
$C_{26}H_{23}NO$ Cyclopenta[b]pyrrol-4-ol, 3,3a,4,6a-tetrahydro-3-methyl-4,5,6a-triphenyl-	EtOH	251(4.2)	44-0448-84
2,6-Methano-2H-cyclopent[d]oxazole, 3,3a,6,6a-tetrahydro-7-methyl-3a,5,6a-triphenyl-	EtOH	262(4.2)	44-0448-84
1,3-Oxazepine, 4-(1-methylethyl)-2,5,7-triphenyl-	ether	263(4.40),350(3.86)	78-3559-84
$C_{26}H_{23}NO_2$ 9-Anthracenol, 10-[[(1-methylpropyl)-imino]methyl]-, benzoate	EtOH	256(5.08),338(3.44), 356(3.75),372(3.93), 391(3.92)	150-0701-84M
$C_{26}H_{23}N_3O$ 2,5-Cyclohexadien-1-one, 2,5-bis[(4-methylphenyl)amino]-4-(phenylimino)-	CHCl$_3$	284(4.40),400(4.27)	103-0263-84
$C_{26}H_{23}N_3O_3$ 2,5-Cyclohexadien-1-one, 2,5-bis[(4-methoxyphenyl)amino]-4-(phenylimino)-	CHCl$_3$	280(4.34),410(4.23)	103-0263-84
1H-Pyrrole-3-carboxamide, 2,5-dihydro-N,1-bis(4-methylphenyl)-4-[(4-methylphenyl)amino]-2,5-dioxo-	CHCl$_3$	260(4.58),430(3.88)	48-0401-84
$C_{26}H_{23}O$ Pyrylium, 2,4,6-tris(4-methylphenyl)-, perchlorate	HOAc	273(4.21),364(4.44)	78-3539-84
$C_{26}H_{24}$ Azulene, 4,4'-(1,2-ethenediyl)bis[6,8-dimethyl-, (E)-	CH$_2$Cl$_2$	235(4.38),280(4.84), 320(4.57),395s(3.87), 586(3.17)	5-1905-84
(Z)-	CH$_2$Cl$_2$	245(4.53),284(4.78), 352s(4.08),377s(3.70), 573(3.12)	5-1905-84
$C_{26}H_{24}BrClN_6S_2$ 6H-Thieno[3,2-f][1,2,4]triazolo[4,3-a]-[1,4]diazepine, 2-bromo-4-(2-chlorophenyl)-9-[4-(2-phenylthioethyl)-1-piperazinyl]-	MeOH	250(4.42),297.5(3.61)	73-0621-84
$C_{26}H_{24}BrNO$ 2H-Pyran-2-amine, 2-(4-bromophenyl)-N,N,3-trimethyl-4,6-diphenyl-	dioxan	240(4.37),248s(4.34), 342(4.06)	48-0657-84
2H-Pyran-2-amine, 4-(4-bromophenyl)-N,N,3-trimethyl-2,6-diphenyl-	dioxan	243s(4.31),248(4.32), 343(4.06)	48-0657-84

Compound	Solvent	$\lambda_{max}(\log \epsilon)$	Ref.
2H-Pyran-2-amine, 6-(4-bromophenyl)-N,N,3-trimethyl-2,4-diphenyl-	dioxan	248(4.30),251(4.31), 346(4.15)	48-0657-84
$C_{26}H_{24}ClNO$ 2H-Pyran-2-amine, 2-(4-chlorophenyl)-N,N,3-trimethyl-4,6-diphenyl-	dioxan	240(4.53),248s(4.31), 342(4.06)	48-0657-84
2H-Pyran-2-amine, 4-(4-chlorophenyl)-N,N,3-trimethyl-2,6-diphenyl-	dioxan	243s(4.29),248(4.30), 343(4.05)	48-0657-84
2H-Pyran-2-amine, 6-(4-chlorophenyl)-N,N,3-trimethyl-2,4-diphenyl-	dioxan	247s(4.30),251(4.31), 346(4.11)	48-0657-84
$C_{26}H_{24}FNO$ 2H-Pyran-2-amine, 2-(4-fluorophenyl)-N,N,3-trimethyl-4,6-diphenyl-	dioxan	243(4.29),248(4.29), 340(4.07)	48-0657-84
2H-Pyran-2-amine, 6-(4-fluorophenyl)-N,N,3-trimethyl-2,4-diphenyl-	dioxan	245(4.25),340(4.02)	48-0657-84
$C_{26}H_{24}N$ Pyridinium, 1-(1-methylethyl)-2,4,6-triphenyl-	C_6H_5Cl	305(4.54)	120-0062-84
Pyridinium, 2,4,6-triphenyl-1-propyl-	C_6H_5Cl	307(4.48)	120-0062-84
$C_{26}H_{24}NO$ 2-Azoniabicyclo[3.3.0]octa-2,7-diene, 6-hydroxy-4-methyl-1,6,7-triphenyl-, tetrafluoroborate	EtOH	254(4.2)	44-0448-84
Pyrylium, 2-[4-(dimethylamino)phenyl]-3-methyl-4,6-diphenyl-, perchlorate	$CHCl_3$	236(4.13),261(4.24), 342(4.51),537(4.39)	97-0287-84
Pyrylium, 4-[4-(dimethylamino)phenyl]-3-methyl-2,6-diphenyl-, perchlorate	$CHCl_3$	229(4.13),265s(4.26), 275(4.28),368(4.28), 531(4.63)	97-0287-84
Pyrylium, 6-[4-(dimethylamino)phenyl]-3-methyl-2,4-diphenyl-, perchlorate	$CHCl_3$	236(4.24),255s(4.14), 335(4.41),529(4.57)	97-0287-84
$C_{26}H_{24}N_2O$ [1,1'-Biphenyl]-4-propanenitrile, α-[[4-(diethylamino)phenyl]methylene]- -oxo-	EtOH	458(4.71)	73-0421-84
$C_{26}H_{24}N_2O_3$ 2-Anthracenol, 10-[[(1-methylpropyl)-nitrosoamino]methyl]-, benzoate	EtOH	246(6.00),256(5.07), 324(2.95),338(3.36), 356(3.70),374(3.91), 396(3.91)	150-0701-84M
2H-Pyran-2-amine, N,N,3-trimethyl-4-(4-nitrophenyl)-2,6-diphenyl-	dioxan	235(4.22),250(4.21), 269(4.24),334(4.10)	48-0657-84
$C_{26}H_{24}N_2O_4$ Cyclopent[b]indole-1,2-dicarboxylic acid, 1,4-dihydro-3-(1H-indol-3-yl)-4-methyl-, diethyl ester	EtOH	218(4.68),260(4.23), 355(4.15)	94-4410-84
Cyclopent[b]indole-1,2-dicarboxylic acid, 1,4-dihydro-3-(1-methyl-1H-indol-3-yl)-, diethyl ester	EtOH	218(4.70),255(4.15), 365(4.28)	94-4410-84
[1,4]Diazacyclotetradecino[1,2-a:4,3-a']diindole-6,15,21,22-tetrone, 7,8,9,10,11,12,13,14-octahydro-, (E)-	xylene	572(3.89)	18-0470-84
1H-Indazole-6,7-dicarboxylic acid, 4-methyl-1,3-diphenyl-, diethyl ester	EtOH	244(4.69),335(4.05)	94-2146-84

Compound	Solvent	$\lambda_{max}(\log \epsilon)$	Ref.
$C_{26}H_{24}N_2O_5$ Phenylalanine, α,β–didehydro–N– [N–[(phenylmethoxy)carbonyl]– L–phenylalanyl]–, (Z)–	EtOH	207(4.45),282(4.15)	5-0920-84
$C_{26}H_{24}O_2$ 1H–Indene–1,3(2H)–dione, 2–[1,1'–bi– phenyl]–4–yl–2–pentyl–	EtOH dioxan	350(2.79) 350(2.76)	135-0312-84 135-0312-84
$C_{26}H_{24}O_6$ 1(3H)–Isobenzofuranone, 6,7–bis(benz– oyloxy)–3–butylidene–4,5,6,7–tetra– hydro–, cis	MeOH	230(4.39),273(3.28)	102-2033-84
$C_{26}H_{24}O_7S_2$ 1,3–Dibenzofurandiol, 6,7,8,9–tetrahy– dro–, bis(4–methylbenzenesulfonate)	MeOH	262(4.30),280s(3.94), 290s(3.75)	44-4399-84
$C_{26}H_{24}O_8$ Bicyclo[2.2.1]hepta–2,5–diene–2,3–di– carboxylic acid, 7,7'–(2–butene–1,4– diylidene)bis–, tetramethyl ester, (E)– (Z)–	MeCN MeCN	256s(3.36),268(3.62), 278(3.76),291(3.66) 257s(3.36),269(3.58), 279(3.68),292(3.54)	24-2006-84 24-2006-84
$C_{26}H_{24}O_9$ 2H,6H–Benzo[1,2–b:5,4–b']dipyran–6–one, 5–acetoxy–7–(4–acetoxy–2,5–dimethoxy– phenyl)–2,2–dimethyl–	EtOH	227(4.45),263(4.57), 239(4.22),334s(3.94)	142-0709-84
$C_{26}H_{24}O_{11}$ Tuberculatin	CHCl$_3$	263(4.66),294(3.99), 310(3.99),352(3.66)	102-0151-84
$C_{26}H_{24}P_2$ Phosphine, 1,2–ethanediylbis[diphenyl–	C$_6$H$_{12}$	251(4.27)	151-0249-84A
$C_{26}H_{25}NO$ 2H–Pyran–2–amine, N,N,3–trimethyl– 2,4,6–triphenyl–	dioxan	242(4.28),249(4.28), 343(4.02)	48-0657-84
$C_{26}H_{25}NO_2$ 9–Anthracenol, 10–[[(1–methylpropyl)– amino]methyl]–, benzoate	EtOH	255(5.03),321(3.05), 336(3.43),353(3.77), 372(3.98),392(3.97)	150-0701-84M
$C_{26}H_{26}N_2$ 4,4'–Bipyridinium, 2–ethyl–1,1'– bis(phenylmethyl)–, dibromide	H$_2$O	260(4.22)	39-0367-84B
$C_{26}H_{26}N_2O_2$ 3,7–Epoxy–1,7–methano–7H–oxonino[4,3– b]indole, 1,3,4,5,6,8–hexahydro– 5–(1H–indol–3–yl)–1,3,5–trimethyl–, (1α,3β,5β,7β)–	EtOH	224(3.78),283(4.12), 290(4.06)	2-1223-84
$C_{26}H_{26}N_2O_9$ 1H–Benz[f]isoquino[8,1,2–hij]quinazo– line–2(3H)–carboxylic acid, 5,6–di– hydro–8,9,10,11,12–pentamethoxy–	MeOH	236s(4.28),243s(4.32), 269(4.68),290(4.46), 317s(4.14),368(4.00),	78-4003-84

Compound	Solvent	$\lambda_{max}(\log \epsilon)$	Ref.
1,3-dioxo-, ethyl ester (cont.)		388(3.95)	78-4003-84
$C_{26}H_{26}N_4O_2$			
[4,4'-Bipyridine]-3,3'-dicarboxamide, 1,1',4,4'-tetrahydro-1,1'-bis(phenylmethyl)-	MeOH	356(3.84)	44-0026-84
1,4-Butanedione, 2,3-bis(3,5-dimethyl-1H-pyrazol-1-yl)-1,4-diphenyl-, (R*,S*)-	MeOH	248(4.42)	44-4647-84
$C_{26}H_{26}N_4O_3$			
4-Imidazolidinecarboxamide, 3-[4-(dimethylamino)phenyl]-4-methyl-1-(4-methylphenyl)-2,5-dioxo-N-phenyl-	EtOH	208(4.45),262(4.40)	49-0187-84
$C_{26}H_{26}N_4O_4$			
4-Imidazolidinecarboxamide, 3-[4-(dimethylamino)phenyl]-1-(4-methoxyphenyl)-4-methyl-2,5-dioxo-N-phenyl-	EtOH	207(4.38),236(4.12), 263(4.31)	49-0187-84
$C_{26}H_{26}N_4O_{12}S_3$			
2,4(1H,3H)-Pteridinedione, 6,7-diphenyl-3-[2,3,5-tris-O-(methylsulfonyl)-β-D-ribofuranosyl]-	MeOH	222(4.52),276(4.27), 363(4.21)	136-0179-84F
$C_{26}H_{26}N_6$			
7,18,23,24,25,26-Hexaazapentacyclo-[17.3.1.1^{2,6}.1^{8,12}.1^{13,17}]hexacosa-1(23),2,4,6(26),8,10,12(25),13,15-17(24),19,21-dodecaene, 7-hexyl-	MeOH	<u>275(4.9),300s(4.8), 325(4.1),410(4.0)</u>	39-2023-84C
$C_{26}H_{26}O_2$			
1,5-Pentanedione, 1,3-diphenyl-5-(2,4,6-trimethylphenyl)-	EtOH	242.5(4.20)	78-4127-84
4-Penten-1-one, 3-hydroxy-3,5-diphenyl-1-(2,4,6-trimethylphenyl)-, (E)-	EtOH	254(4.37)	78-4127-84
$C_{26}H_{26}O_2S$			
27-Thiatetracyclo[18.2.2.2^{8,11}.1^{3,6}]-heptacosa-3,5,8,10,20,22,23,25-octaene-2,7-dione	C_6H_{12}	<u>299(4.3)</u>	44-1177-84
$C_{26}H_{26}O_3P$			
Phosphonium, (3-oxo-1,4-dioxaspiro-[4.5]dec-2-yl)triphenyl-, bromide	EtOH	225(4.48),268(3.67)	39-1531-84C
$C_{26}H_{26}O_4S_2$			
1,3-Dithiane-2-carboxylic acid, 2-[3,4-bis(phenylmethoxy)phenyl]-, methyl ester	EtOH	245(3.87),281(3.49)	39-1547-84C
$C_{26}H_{26}O_7$			
Propanoic acid, 2,2-dimethyl-, 3-(acetoxymethyl)-9,10-dihydro-9,10-dioxo-2-(3-oxobutyl)-1-anthracenyl ester	MeOH	209(4.39),257(4.68), 332(3.79)	78-3677-84
$C_{26}H_{26}O_8$			
Licoricone diacetate	EtOH	246s(4.41),294(3.86), 302(3.86)	142-0261-84

Compound	Solvent	λ_{max}(log ϵ)	Ref.

C$_{26}$H$_{26}$O$_{14}$

  Flavanone, 4'-hydroxy-3,3',5-trimeth-
oxy-6,7-(methylenedioxy)-, 4'-β-D-
glucuronide methyl ester — MeOH — 245(4.20),270(4.17), 328(4.40) — 102-2043-84

  α-Sorinin — EtOH — 219(3.55),250s(4.09), 257(4.24),348(3.36) — 102-1485-84

   — EtOH-NaOH — 241(--),264(--), 365(--) — 102-1485-84

C$_{26}$H$_{27}$BrO$_6$

  Propanoic acid, 2,2-dimethyl-, 3-(bro-
momethyl)-9,10-dihydro-2-[2-(2-meth-
yl-1,3-dioxolan-2-yl)ethyl]-9,10-di-
oxo-1-anthracenyl ester — MeOH — 213(4.44),260(4.23), 278(4.21),336(3.80) — 78-3677-84

C$_{26}$H$_{27}$NO$_5$

  Coyhaiquine — MeOH or EtOH — 210(4.44),231s(4.24), 283(3.63) — 100-0565-84

  4H-Dibenzo[de,g]quinolin-10-ol,
5,6,6a,7-tetrahydro-9-[4-(hydroxy-
methyl)phenoxy]-1,2-dimethoxy-6-
methyl- — MeOH or EtOH — 217(4.36),227(3.99), 304(3.84) — 100-0565-84

C$_{26}$H$_{27}$NO$_8$

  5,12-Naphthacenedione, 8-acetyl-10-[(3-
amino-2,3,6-trideoxy-α-L-lyxo-hexo-
pyranosyl)oxy]-7,8,9,10-tetrahydro-
6,8-dihydroxy-, (8S-cis)-, hydro-
chloride — MeOH-HCl — 208(4.70),248(4.43), 262(4.47),328(3.48), 408(3.81) — 78-4677-84

C$_{26}$H$_{27}$N$_3$O

  Urea, N-(1,1-dimethylethyl)-N'-[1,3-
diphenyl-3-(phenylimino)-1-propen-
yl]- — EtOH — 206(4.54),223s(4.33), 250s(4.27),350(4.17) — 39-0239-84C

  Urea, N'-[1,3-diphenyl-3-(phenylimino)-
1-propenyl]-N,N-diethyl- — EtOH — 206(4.52),225s(4.25), 250(4.21),340s(4.14), 370(4.24) — 39-0239-84C

C$_{26}$H$_{27}$N$_5$O$_8$

  9H-Purine, 2-(5-methyl-2-furanyl)-6-(1-
methyl-1H-pyrrol-2-yl)-9-(2,3,5-tri-
O-acetyl-β-D-ribofuranosyl)- — EtOH — 250(3.92),352(4.38), 336(4.43),326(4.45), 310(4.48)[sic] — 44-4340-84

C$_{26}$H$_{28}$

  Benzo[c]phenanthrene, 1,3-bis(1,1-di-
methylethyl)-, (±)- — CHCl$_3$ — 290s(4.52),300(4.62), 335(3.76),373(2.57), 395(2.20) — 24-0336-84

C$_{26}$H$_{28}$NO$_4$

  Isoquinolinium, 3-(3,4-dimethoxyphen-
yl)-3,4-dihydro-6,7-dimethoxy-2-
methyl-1-phenyl-, iodide — EtOH — 259(4.06),319(3.99), 380(3.94) — 4-0525-84

C$_{26}$H$_{28}$N$_2$

  5,10[1',2']-Benzenopyrido[3,2-g]quino-
line, 2,8-bis(1,1-dimethylethyl)-
5,10-dihydro- — MeCN — 280(4.230),287.5(4.228) — 5-0381-84

C$_{26}$H$_{28}$N$_2$O$_9$

  2H-Pyrimido[6,1-a]isoquinoline-3(4H)-
carboxylic acid, 6,7-dihydro-9,10- — MeOH — 227s(4.44),245(4.29), 278(3.88),340(4.20) — 78-4003-84

Compound	Solvent	$\lambda_{max}(\log \epsilon)$	Ref.
dimethoxy-2,4-dioxo-1-(3,4,5-trimeth-oxyphenyl)-, ethyl ester (cont.)			78-4003-84
$C_{26}H_{28}N_6O_7$			
9H-Purine, 2,6-bis(1-methyl-1H-pyrrol-2-yl)-9-(2,3,5-tri-O-acetyl-β-D-ribofuranosyl)-	EtOH	331.5(4.49)	44-4340-84
$C_{26}H_{28}O_2$			
Benzoic acid, 4-[2-(5,6,7,8-tetrameth-yl-2-naphthalenyl)-1-propenyl]-, ethyl ester, (E)-	EtOH	224(4.52),245(4.45), 296(4.53),325(4.43)	44-5265-84
$C_{26}H_{28}O_4$			
Boesenbergin B, (±)-	EtOH	234(4.17),293(4.20), 304(4.23),352(4.28)	12-1739-84
2-Butenedioic acid, 2,3-bis(2-methyl-1-phenyl-1-propenyl)-, dimethyl ester, (E)-	C_6H_{12}	238(4.24),283(3.65)	44-4344-84
(Z)-	C_6H_{12}	235(4.23),287(3.53)	44-4344-84
1,3-Cyclohexadiene-2,3-dicarboxylic acid, 5,5,6,6-tetramethyl-1,4-di-phenyl-, dimethyl ester	C_6H_{12}	227(3.61),288(3.63)	44-4344-84
8,12-Methano-4H,8H-pyrano[2,3-h][1]-benzoxocin-4-one, 2,3,9,10,11,12-hexahydro-5-methoxy-8-methyl-11-(1-methylethylidene)-2-phenyl-	EtOH	220(4.41),232(4.34), 240(4.24),293(4.36), 350(3.56)	12-1739-84
2-Propen-1-one, 3-phenyl-1-[3,4,5,6-tetrahydro-7-hydroxy-9-methoxy-2-methyl-5-(1-methylethylidene)-2,6-methano-2H-1-benzoxocin-8-yl]-, (E)-(±)-	EtOH	220(4.50),235(4.22), 300(4.22),353(4.51)	12-1739-84
$C_{26}H_{28}O_6$			
Chapelieric acid, methyl ester	EtOH	266(4.40),274(4.60), 298(4.00),312(4.10), 370(3.65)	102-0323-84
Isochapelieric acid, methyl ester	EtOH	265(4.20),270(4.50), 292(4.05),310(4.10), 370(3.60)	102-0323-84
$C_{26}H_{28}O_9$			
2-Anthraceneacetic acid, 9,10-dihydro-4,5-dihydroxy-3-[1-(2-hydroxy-1-meth-ylpropoxy)-3-oxopentyl]-9,10-dioxo-, methyl ester	MeOH	228(4.74),257(4.40), 277s(4.05),288(4.03), 432(4.08)	78-4685-84
2,5-Furandione, dihydrobis[1-[3-meth-oxy-4-(methoxymethoxy)phenyl]ethyl-idene]-, (E,E)-	toluene	417(4.02)	48-0233-84
(E,Z)-	toluene	437(4.28)	48-0233-84
1-Naphthacenecarboxylic acid, 2-ethyl-1,2,3,4,6,11-hexahydro-2,5,7-trihy-droxy-4-(2-hydroxy-1-methylpropoxy)-6,11-dioxo-, methyl ester, [1R-[1α,2α,4β(1R*,2R*)]]-	MeOH	228(4.66),256(4.35), 278(4.00),287s(3.97), 430(4.02)	78-4685-84
[1R-[1α,2β,4β(1R*,2R*)]]-	MeOH	228(4.63),258(4.44), 278s(3.98),287s(3.92), 431(4.08)	78-4685-84
$C_{26}H_{28}O_{10}$			
1H,3H-Furo[3,4-c]furan-1-ol, 3,6-bis(4-	EtOH	218(4.20),275(3.69),	94-2730-84

Compound	Solvent	$\lambda_{max}(\log \epsilon)$	Ref.
acetoxy-3-methoxyphenyl)tetrahydro-, acetate (cont.)		279.8(3.68)	94-2730-84
$C_{26}H_{28}O_{14}$ Kaempferol, 3-0-α-L-arabinofuranoside 7-0-α-L-rhamnopyranoside	MeOH	268(4.45),348(4.36)	95-0142-84
$C_{26}H_{29}NO_2$ Carbazole, 4a,9b-(3,5-di-tert-butyl-phenylenedioxy)-1,2,3,4,4a,9b-hexahydro-	MeOH	286(3.74)	4-1841-84
$C_{26}H_{29}NO_4S_2$ Benzoic acid, 4-hexyloxy-, 4-[(4-oxo-3-propyl-2-thioxo-5-thiazolidinylidene)methyl]phenyl ester	n.s.g.	391(4.35)	48-0457-84
$C_{26}H_{29}N_3O_5$ Ellipticinium, 10-[(1-carboxy-3-methyl-butylidene)amino]-9-hydroxy-2-methyl-, acetate	H_2O	243(4.27),300(4.51), 340(3.48),440(3.48)	87-1161-84
$C_{26}H_{30}Fe_2Hg_2N_2$ Mercury, bis[μ-[2-[(dimethylamino)methyl]-1,1'-ferrocenediyl]]di-, bis(tetrafluoroborate)	MeCN	390(2.66),629(2.79)	101-0283-84E
$C_{26}H_{30}N_4$ 4,4'-Bipyridinium, 1,1"-(1,4-butane-diyl)bis[1'-methyl-, tetraperchlorate	H_2O	262(4.65)	64-0074-84B
4,4'-Bipyridinium, 1,1"-(1,2-ethanedi-yl)bis[1'-ethyl-, tetraperchlorate	H_2O	265(4.64)	64-0074-84B
$C_{26}H_{30}N_4O_6S_2$ Benzothiazolium, 6-(2-hydroxyethoxy)-3-methyl-2-[3-(2-hydroxyethoxy)-4-ethylphenyl]azo]-, benzenesulfonate	pH 1.13 pH 6.88 70% EtOH-pH 6.2	615(4.93) 615(4.92) 626(4.91)	48-0151-84 48-0151-84 48-0151-84
$C_{26}H_{30}N_6O_7$ Adenosine, N-benzoyl-2'-deoxy-2-[(2-methyl-1-oxopropyl)amino]-, 3'-(4-oxopentanoate)	EtOH	245(4.53),297(4.16)	78-0003-84
$C_{26}H_{30}O_4$ Panduratin A	EtOH	228(4.18),290(4.28)	12-0449-84
$C_{26}H_{30}O_5$ 4H-1-Benzopyran-4-one, 2,3-dihydro-5-hydroxy-2-(4-hydroxyphenyl)-7-methoxy-6,8-bis(3-methyl-2-buten-yl)-, (S)- (amoradin)	MeOH	285(4.40),362(3.81)	102-1818-84
$C_{26}H_{30}O_6$ 4H-1-Benzopyran-4-one, 2-(3,4-dihy-droxyphenyl)-2,3-dihydro-5-hydroxy-7-methoxy-6,8-bis(3-methyl-2-but-enyl)-, (S)- (amoradicin)	MeOH	286(4.25),363(3.68)	102-1818-84
2-Propenoic acid, 2-(3-methoxyphenyl)-, 1,6-hexanediyl ester	MeCN	276(4.63)	40-0022-84

Compound	Solvent	$\lambda_{max}(\log \epsilon)$	Ref.
$C_{26}H_{30}O_8$ 2-Propenoic acid, 3-(3,4-dimethoxyphenyl)-, 1,4-butanediyl ester	MeCN	321(4.59)	40-0022-84
$C_{26}H_{30}S$ 27-Thiatetracyclo[18.2.2.2^{8,11}.1^{3,6}]-heptacosa-3,5,8,10,20,22,23,25-octaene	C_6H_{12}	<u>250(3.9),275s(2.9)</u>	44-1177-84
$C_{26}H_{31}NO_2$ 2-Oxiranecarbonitrile, 2-phenyl-3-[2-(10-undecenyloxy)phenyl]-, trans	EtOH	231s(4.09),274s(3.59), 282(3.63)	24-2157-84
$C_{26}H_{31}NO_8$ D-Gluconic acid, 2-deoxy-2-[(3-ethoxy-3-oxo-1-phenyl-1-propenyl)amino]-4,6-O-(phenylmethylene)-, ethyl ester, (Z)-	MeOH	230(3.75),293(4.04)	136-0081-84B
$C_{26}H_{32}$ 2,2'-Bi-2H-indene, 1,1',3,3'-tetrahydro-1,1,1',1',3,3,3',3'-octamethyl-	C_6H_{12}	199(4.51),209s(4.41), 227s(3.66),233s(3.57), 257(3.17),263.9(3.33), 270.7(3.32)	24-0277-84 +24-0310-84
$C_{26}H_{32}N_2O_{10}$ L-ribo-Hexaramic acid, 4-(acetylamino)-N-[1-[8-(acetyloxy)-3,4-dihydro-1-oxo-1H-2-benzopyran-3-yl]-3-methylbutyl]-4,5-dideoxy-, γ-lactone, 2-acetate	MeOH	236(3.85),288(3.26)	78-2519-84
$C_{26}H_{32}N_2S$ Dispiro[2H-indene-2,2'(3'H)-[1,3,4]-thiadiazole-5',2"-[2H]indene],1,1",3,3"-tetrahydro-1,1,1",1"-3,3,3",3"-octamethyl-	C_6H_{12}	219(3.53),258.2(3.30), 264.4(3.46),271(3.46), 282s(3.07),333(2.32)	24-0277-84
$C_{26}H_{32}N_3O$ Quinolinium, 6-[(diethylamino)carbonyl]-4-[2-[4-(dimethylamino)phenyl]ethenyl]-1-ethyl-, iodide	MeOH $C_6H_4Cl_2$ MeCN	572(4.60) 626(4.67) 558(--)	103-0767-84 103-0767-84 103-0767-84
$C_{26}H_{32}N_4O_2S_4$ 19,22,27,30-Tetrathia-1,8,9,16-tetra-azatetracyclo[14.8.8.1^{3,7}.1^{10,14}]-tetratriaconta-3,5,7(34),8,10,12-14(33)-heptaene-2,15-dione, trans	$C_6H_4Cl_2$-BuOH	325(4.24),440(2.57)	18-2144-84
$C_{26}H_{32}O_2$ Benzoic acid, 4-[2-(5,6,7,8-tetrahydro-5,5,8,8-tetramethyl-2-naphthalenyl)-1-propenyl]-, ethyl ester, (E)-	EtOH	304(4.32)	44-5265-84
$C_{26}H_{32}O_6$ 2H-1-Benzopyran-2-one, 1-[[5-(2,2-dimethyl-6-methylene-3-oxocyclohexyl)-3-methyl-2-pentenyl]oxy]-6,8-dimethoxy-	EtOH	206s(4.69),227s(4.36), 297(4.06),340(3.90)	49-0477-84
$C_{26}H_{32}O_7$ 4H-1-Benzopyran-4-one, 2,3-dihydro-	MeOH	235(3.71),285(3.76)	2-0887-84

Compound	Solvent	$\lambda_{max}(\log \epsilon)$	Ref.
5,7-bis(methoxymethoxy)-2-[4-(meth-oxymethoxy)-3-(3-methyl-2-butenyl)-phenyl]- (cont.)			2-0887-84
2-Propen-1-one, 1-[2,4-bis(methoxymeth-oxy)phenyl]-3-[4-(methoxymethoxy)-3-(3-methyl-2-butenyl)phenyl]-	MeOH	240(3.78),335(3.79)	2-0887-84
$C_{26}H_{32}O_8$			
2-Buten-1-one, 1-[2,4,6-trimethoxy-3-[1-methyl-3-oxo-3-(2,4,6-trimeth-oxyphenyl)propyl]phenyl]-, (E)-	MeOH	224(4.45),279(3.90)	94-0325-84
$C_{26}H_{32}O_{12}$			
β-D-Glucopyranoside, dihydro-1,4-bis(4-hydroxy-3-methoxyphenyl)-1H,3H-furo-[3,4-c]furan-3a(4H)-yl, (+)-	EtOH	228(4.11),279.8(3.66)	94-4482-84 +102-2839-84
	EtOH-NaOH	252(--),280(--),292s(--)	94-4482-84
$C_{26}H_{32}O_{14}$			
Tarennine	EtOH	233(4.30),298s(4.11),323(4.24)	32-0049-84
$C_{26}H_{32}S$			
1H-Indene, 2,3-dihydro-1,1,3,3-tetra-methyl-2-[[[1,1,3(or 2)-trimethyl-1H-inden-2(or 3)-yl]methyl]thio]-	C_6H_{12}	194.5(4.74),204s(4.51),224s(4.08),263.3(4.03),269.5(3.95)	24-0277-84
$C_{26}H_{33}N$			
2-Cyclohexen-1-yl, 1-(p-dimethylamino-styryl)-4-isopropyl-3-p-tolyl-, di-	H_2SO_4	539(5.00)	83-0724-84
	MeOH-HClO	600(4.63+)(decomp.)	83-0724-84
$C_{26}H_{33}N_3$			
Benzenamine, 2,4-bis[[4-(dimethylami-no)phenyl]methyl]-N,N-dimethyl-	hexane-iso-PrOH	260(3.60)	39-1099-84B
$C_{26}H_{33}N_3O_{16}$			
1H-Imidazolium, 4-(aminocarbonyl)-5-hy-droxy-1,3-bis(2,3,5-tri-O-acetyl-β-D-ribofuranosyl)-, hydroxide, inner salt	M HCl	244(3.66),282(3.93)	4-0529-84
	H_2O	241(3.60),284(3.90)	4-0529-84
	M NaOH	268(4.25)	4-0529-84
1H-Imidazolium, 4-(aminocarbonyl)-5-hy-droxy-1,3-bis(2,3,4-tri-O-acetyl-β-D-ribopyranosyl)-, hydroxide, inner salt	M HCl	241(3.43),288(3.57)	4-0529-84
	H_2O	240(3.47),288(3.58)	4-0529-84
	M NaOH	262(4.35)	4-0529-84
$C_{26}H_{34}$			
1,4-Cyclohexadiene, 3,3'-(2,4-hexadi-ene-1,6-diylidene)bis[1,5,6,6-tetra-methyl-	CH_2Cl_2	265(3.98),356s(--),376(4.48),397.5(4.86),421(5.13),449(5.19)	5-0340-84
6H-1,11-Ethano-7,11-methanobenzocyclo-decene, 3,9-bis(1,1-dimethylethyl)-5,12-dihydro-13-methylene-	C_6H_{12}	225(4.08),268(2.43),289(3.39)	39-2165-84C
$C_{26}H_{34}N_{10}O_6$			
L-Lysine, N^6-[N-[4-[[2,4-diamino-6-pteridinyl]methyl]methylamino]ben-zoyl]-L-γ-glutamyl]-	pH 1	243(4.30),306(4.37)	87-0888-84
	pH 7.4	259(4.44),302(4.41),372(4.05)	87-0888-84
$C_{26}H_{34}O_4S$			
A-Norandrostan-3-one, 2-methylene-17-[[(4-methylphenyl)sulfonyl]oxy]-,	EtOH	228(3.79)	13-0179-84A

Compound	Solvent	$\lambda_{max}(\log \epsilon)$	Ref.
(5β,17β)- (cont.) A-Norandrost-1-en-3-one, 2-methyl-17- [[(4-methylphenyl)sulfonyl]oxy]-, (5β,17β)-	EtOH	228(4.58)	13-0179-84A 13-0179-84A
$C_{26}H_{34}O_6$ Albartol	EtOH	206s(4.68),228s(4.37), 296(4.05),339(3.89)	49-0477-84
2H-1-Benzopyran-2-one, 7-[[5-(3-hy- droxy-2,2-dimethyl-6-methylene- cyclohexyl)-3-methyl-2-pentenyl]- oxy]-6,8-dimethoxy- (deparnol)	EtOH	206s(4.68),227s(4.36), 297(4.04),338(3.89)	49-0477-84
$C_{26}H_{35}BrO_7$ Pregn-4-ene-3,20-dione, 17,21-[(1-meth- ylethylidene)bis(oxy)]-, (6β,11β)-	MeOH	235(4.19)	44-2634-84
$C_{26}H_{35}N_5O_5S$ Piperazine, 1-[4-[3-[(1,1-dimethyleth- yl)amino]-2-hydroxypropoxy]-6,7-di- methoxy-2-quinazolinyl]-4-(2-thien- ylcarbonyl)-, dihydrochloride	EtOH	220(4.50),252(4.70)	4-1189-84
$C_{26}H_{35}N_5O_6$ Piperazine, 1-[4-[3-[(1,1-dimethyleth- yl)amino]-2-hydroxypropoxy]-6,7-di- methoxy-2-quinazolinyl]-4-(2-furan- ylcarbonyl)-, dihydrochloride	EtOH	218(4.60),250(4.80)	4-1189-84
$C_{26}H_{36}$ Spiro[5.5]undeca-1,4-diene, 3-(1,5-di- methylspiro[5.5]undeca-1,4-dien-3- ylidene)-1,5-dimethyl-	CH_2Cl_2	263(3.91),331s(--), 347(4.69),364(4.71)	5-0340-84
$C_{26}H_{36}O_2Si$ Cyclobutanone, 3-(1,1-dimethylethyl)- 3-[[(1,1-dimethylethyl)dimethylsil- yl]oxy]-2,3-diphenyl-	MeOH	269(4.22)	35-4566-84
5-Hexen-3-one, 5-[[(1,1-dimethylethyl)- dimethylsilyl]oxy]-2,2-dimethyl-6,6- diphenyl-	MeOH	256(4.06)	35-4566-84
$C_{26}H_{36}O_4$ Pregna-7,9(11)-diene-20-carboxylic acid, 4,4,14-trimethyl-3,12-dioxo-, methyl ester, (5α,20S)-	n.s.g.	291(4.08)	39-1219-84C
$C_{26}H_{36}O_7$ Waixenicin A	EtOH	224(3.06)	78-3823-84
$C_{26}H_{36}O_8$ Waixenicin B	EtOH	221(3.41),284(2.57)	78-3823-84
$C_{26}H_{36}O_9SSi$ D-threo-Pentonic acid, 2,3-anhydro-4- deoxy-3-C-methyl-, 2-[(4-methylphen- yl)sulfonyl]ethyl ester, 2-(trimeth- ylsilyl)ethyl 2,4-hexadienoate, (Z,E)-	EtOH	226(4.25),263(4.42)	44-4332-84
$C_{26}H_{36}O_{15}$ Secologanoside, methyl ester, tetraacet- ate	MeOH	232(4.04)	95-1232-84

Compound	Solvent	$\lambda_{max}(\log \epsilon)$	Ref.
$C_{26}H_{37}N_5O_7$			
L-Leucine, N-[N-[N-[2,5-dihydro-1-(1-methylethyl)-4-nitro-2-oxo-1H-pyrrol-3-yl]glycyl]-L-phenylalanyl]-, ethyl ester	EtOH	369(4.14)	44-1130-84
$C_{26}H_{38}N_2O_6$			
5H-[1]Benzopyrano[3,4-c]pyridazine-3,4-dicarboxylic acid, 4,4a-dihydro-8-methoxy-2,2,5,5-tetramethyl-, bis(1,1-dimethylethyl) ester	EtOH	218(4.42),255(4.22), 264(4.22),300(3.87), 310(3.82)	39-2815-84C
$C_{26}H_{38}O_4$			
Atis-16-en-18-oic acid, 15-[(3-methyl-1-oxo-2-butenyl)oxy]-, methyl ester	EtOH	221(4.04)	102-0195-84
$C_{26}H_{38}O_5$			
Androsta-5,7-diene-3,17,19-triol, 3-acetate 17-(2,2-dimethylprop-anoate)	EtOH	273(4.01),282.5(4.01)	56-0711-84
Androsta-5,7-diene-3,17,19-triol, 19-acetate 17-(2,2-dimethylprop-anoate), (3β,17β)-	EtOH	271(4.04),281(4.05)	56-0711-84
Atis-16-en-18-oic acid, 15-hydroxy-7-[(3-methyl-1-oxo-2-butenyl)oxy]-, methyl ester	EtOH	220(4.27)	102-2075-84
$C_{26}H_{38}O_6$			
Androst-5-en-7-one, 3-acetoxy-17-(2,2-dimethyl-1-oxopropoxy)-19-hydroxy-, (3β,17β)-	EtOH	237(4.10)	56-0711-84
$C_{26}H_{38}O_7$			
1,3,5-Hexanetricarboxylic acid, 6-[3,5-bis(1,1-dimethylethyl)-4-oxo-2,5-cy-clohexadien-1-ylidene)-, trimethyl ester	EtOH	207(4.08),305(2.34), 422(2.77)	104-1732-84
Isocharaciol, 15-O-acetyl-5β-hydroxy-, 3-isobutyrate	MeOH	193(3.99),215s(3.81), 274(2.93)	102-1689-84
$C_{26}H_{38}O_8$			
2(5H)-Oxepinone, 3-(5-acetoxy-4-methyl-3-pentenyl)-7-(1,6-diacetoxy-4-meth-yl-4-hexenyl)-6,7-dihydro-7-methyl-	MeOH	203(4.00),220(3.77)	102-0829-84
$C_{26}H_{39}NO_2$			
Pentanoic acid, 5-[[3,7-dimethyl-9-(2,6,6-trimethyl-1-cyclohexen-1-yl)-2,4,6,8-nonatetraenylidene]-amino]-, methyl ester, hydrochloride	MeOH	360(4.60),443(4.61)	104-1669-84
$C_{26}H_{39}N_5$			
7H-Purin-6-amine, 7-[3,7-dimethyl-9-(1,2,6-trimethyl-2-cyclohexen-1-yl)-2,6-nonadienyl]-9-methyl-(ageline Λ)(cation)	EtOH EtOH-HCl	212(4.26),272(4.00) 212(4.26),272(4.00)	35-1819-84 35-1819-84
7H-Purin-6-amine, 7-[3,7-dimethyl-9-(1,2,6-trimethyl-2-cyclohexen-1-yl)-2,6-nonadienyl]-N-methyl-	EtOH EtOH-HCl	218(4.31),277(4.18) 210(4.24),283(4.27)	35-1819-84 35-1819-84

Compound	Solvent	$\lambda_{max}(\log \epsilon)$	Ref.
$C_{26}H_{40}$			
6H-1,11-Ethano-7,11-methanobenzocyclo-decene, 3,9-bis(1,1-dimethylethyl)-5,7,8,9,10,12-hexahydro-13-methyl-	C_6H_{12}	215(4.17),270(2.60)	39-2165-84C
$C_{26}H_{40}N_2O_6$			
2H-[1]Benzopyrano[3,4-c]pyridazine-3,4-dicarboxylic acid, 1,4a,5,10b-tetrahydro-8-methoxy-2,2,5,5-tetra-methyl-, bis(1,1-dimethylethyl) ester, trans	EtOH	219(4.29),228(4.14), 280(3.72),283(3.76), 295(3.71)	39-2815-84C
$C_{26}H_{40}N_5$			
Agelasine A (chloride)	MeOH	272(3.95)	88-2989-84
Agelasine B (chloride)	MeOH	272(3.92)	88-2989-84
Agelasine C (chloride)	MeOH	272(3.92)	88-2989-84
Agelasine D (chloride)	MeOH	272(3.96)	88-2989-84
7H-Purinium, 6-amino-7-[9-(2,2-dimeth-yl-6-methylenecyclohexyl)-3,7-dimeth-yl-2,6-nonadienyl]-9-methyl-, chlor-ide (agelasine E)	MeOH	272(3.99)	88-3719-84
7H-Purinium, 6-amino-7-[3,7-dimethyl-9-(1,2,6-trimethyl-2-cyclohexen-1-yl)-2,6-nonadienyl]-9-methyl-, chloride (agelasine F)	MeOH	272(3.89)	88-3719-84
$C_{26}H_{40}OSi_2$			
Oxadisilacyclopropane, 2,3-bis(1,1-di-methylethyl)-2,3-bis(2,4,6-tri-methylphenyl)-	3-Mepentane	203(4.80),247(4.30)	77-1525-84
$C_{26}H_{40}O_2Si_2$			
1,2-Dioxa-3,4-disilacyclobutane, 3,4-bis(1,1-dimethylethyl)-3,4-bis(2,4,6-trimethylphenyl)-	3-Mepentane	198(4.85),247(4.17)	77-1525-84
$C_{26}H_{40}O_4$			
Chol-4-ene-24-carboxylic acid, 7-hy-droxy-3-oxo-, methyl ester, (7α)-	n.s.g.	244(4.20)	13-0095-84B
$C_{26}H_{40}O_7$			
1,3,5-Hexanetricarboxylic acid, 6-[3,5-bis(1,1-dimethylethyl)-4-hydroxy-phenyl]-, trimethyl ester	EtOH	206(4.43),233(3.81), 280(3.26),323(2.64)	104-1732-84
$C_{26}H_{40}Si_2$			
Disilene, 1,2-bis(1,1-dimethylethyl)-1,2-bis(2,4,6-trimethylphenyl)-, (E)-	benzene	338(<u>4.0</u>),400s(<u>3.3</u>)	35-0821-84
$C_{26}H_{41}NO_3$			
Myxalamid A	EtOH	204(4.21),265(4.06), 340s(--),357(4.59), 370s(--)	158-84-113
$C_{26}H_{41}NO_5S$			
7,9,11,14,17-Heneicosapentaenoic acid, 6-[(2-amino-3-methoxy-3-oxopropyl)-thio]-4-hydroxy-, methyl ester	EtOH	278(4.49),282(4.55), 290(4.45)	104-0806-84
$C_{26}H_{41}N_5O$			
Formamide, N-[4-amino-6-(methylamino)-	EtOH	206(4.46),224(4.58),	35-1819-84

Compound	Solvent	λ_{max}(log ϵ)	Ref.
5-pyrimidinyl]-N-(3,7-dimethyl- 9-(1,2,6-trimethyl-2-cyclohexen- 1-yl)-2,6-nonadienyl]- (cont.)	EtOH-HCl	260(3.73) 204(4.43),225(4.39), 270(4.05)	35-1819-84 35-1819-84
C$_{26}$H$_{41}$N$_9$O$_{11}$S Chitinovorin A	H$_2$O	261(3.90)	158-1486-84

C$_{26}$H$_{42}$N$_2$O$_2$
  Ethanone, 2-diazo-1-[4-(octadecyloxy)-
    phenyl]-
	C$_6$H$_{12}$	286(4.35),375(1.56)	35-4531-84

C$_{26}$H$_{42}$O$_9$ Pseudomonic acid D	EtOH	220(4.19)	158-84-153

C$_{26}$H$_{42}$O$_{13}$
  2,7-Octadienoic acid, 6-[[3-O-(β-D-
    glucopyranosyl)-4-O-(2-methyl-4-oxo-
    butyl)-α-L-arabinopyranosyl]oxy]-2,6-
    dimethyl-, trans
	EtOH	216(4.04)	94-2617-84

C$_{26}$H$_{44}$O 1-Eicosanone, 1-phenyl-	heptane	320(1.70)	35-7033-84

C$_{26}$H$_{50}$O$_4$Si$_2$
  7-Oxabicyclo[4.1.0]hept-3-ene-3-metha-
    nol, 2,5-bis[[(1,1-dimethylethyl)di-
    methylsilyl]oxy]-4-(1-heptenyl)-,
    [1α,2α,4(E),5α,6α]-($\pm$)-
	hexane	242.5(4.28)	44-1898-84

Compound	Solvent	$\lambda_{max}(\log \epsilon)$	Ref.
$C_{27}H_{10}Cl_4I_4O_5$			
Benzoic acid, 2,3,4,5-tetrachloro-6-(6-	MeOH	524(4.50),564(5.01)	35-5879-84
hydroxy-2,4,5,7-tetraiodo-3-oxo-3H-	MeOH	210(4.83),524(4.50),	64-0474-84B
xanthen-9-yl)-, phenylmethyl ester,		564(5.01)	
sodium salt	CH_2Cl_2	407(4.19),496(4.19)	64-0474-84B
triethylamine salt	MeOH	524(4.49),563(4.96)	35-5879-84
	CH_2Cl_2	528(4.44),569(4.79)	35-5879-84
$C_{27}H_{15}N_3O_7$			
Methanone, [2,3-bis(5-nitro-2-furanyl)-	dioxan	410(4.34)	73-0533-84
pyrrolo[2,1-a]isoquinolin-1-yl]phenyl-			
$C_{27}H_{16}N_2O_3$			
3-Furancarbonitrile, 4,5-diphenyl-	MeOH	411(4.34)	73-1788-84
2-[(4-oxo-4H-1-benzopyran-3-yl)			
methyleneamino]-			
$C_{27}H_{16}O_3S$			
7H-Phenaleno[2,1-b]thiophene-10-carbox-	$CHCl_3$	258(5.123),297(3.813),	139-0315-84C
ylic acid, 9-(1-naphthalenyl)-7-oxo-,		377(4.94)	
methyl ester			
$C_{27}H_{17}BO_6$			
Boron, [5-(9-anthracenyl)-1-phenyl-	benzene	560(4.30)	97-0292-84
4-pentene-1,3-dionato-O,O'][ethane-			
dionato(2-)-O,O']-			
$C_{27}H_{17}NO_3$			
8H-Benzo[h]naphtho[1,2,3-de]quinolin-	toluene	400(3.98),422(4.29)	104-0980-84
8-one, 2-methoxy-9-phenoxy-			
$C_{27}H_{18}BrN_5O$			
1H-Pyrazole, 4-[(4-bromophenyl)azo]-	n.s.g.	369(4.52)	48-1021-84
3,5-diphenyl-1-(2-pyridinylcarbonyl)-			
$C_{27}H_{18}ClN_5O$			
1H-Pyrazole, 4-[(2-chlorophenyl)azo]-	n.s.g.	370(4.42)	48-1021-84
3,5-diphenyl-1-(2-pyridinylcarbonyl)-			
1H-Pyrazole, 4-[(3-chlorophenyl)azo]-	n.s.g.	370(4.46)	48-1021-84
3,5-diphenyl-1-(2-pyridinylcarbonyl)-			
1H-Pyrazole, 4-[(4-chlorophenyl)azo]-	n.s.g.	370(4.42)	48-1021-84
3,5-diphenyl-1-(2-pyridinylcarbonyl)-			
$C_{27}H_{18}N_6O_3$			
1H-Pyrazole, 4-[(2-nitrophenyl)azo]-	n.s.g.	357(4.81)	48-1021-84
3,5-diphenyl-1-(2-pyridinylcarbonyl)-			
1H-Pyrazole, 4-[(3-nitrophenyl)azo]-	n.s.g.	357(4.81)	48-1021-84
3,5-diphenyl-1-(2-pyridinylcarbonyl)-			
1H-Pyrazole, 4-[(4-nitrophenyl)azo]-	n.s.g.	358(4.81)	48-1021-84
3,5-diphenyl-1-(2-pyridinylcarbonyl)-			
$C_{27}H_{19}N_3$			
9-Acridinecarboxaldehyde, (diphenyl-	C_6H_{12}	412(4.05)	97-0021-84
methylene)hydrazone			
$C_{27}H_{19}N_3O_3$			
1H-Benzimidazole-5-carboxylic acid,	DMF	343(4.47)	5-1129-84
2-[4-(2-benzoxazolyl)-1-naphtha-			
lenyl]-1-methyl-, methyl ester			

Compound	Solvent	$\lambda_{max}(\log \epsilon)$	Ref.
$C_{27}H_{19}N_3O_3S_3$ 17H-Bis[1,4]benzothiazino[2,3-a:3',2'-c]phenothiazine, 3,8,14-trimethoxy-	DMF	594(4.48)	40-0522-84
$C_{27}H_{19}N_3S_3$ 17H-Bis[1,4]benzothiazino[2,3-a:3',2'-c]phenothiazine, 3,8,14-trimethyl-	DMF	587(4.38)	40-0522-84
$C_{27}H_{19}N_5O$ 1H-Pyrazole, 3,5-diphenyl-4-(phenyl-azo)-1-(2-pyridinylcarbonyl)-	n.s.g.	365(3.99)	48-1021-84
$C_{27}H_{19}N_5O_2$ 1H-Pyrazole, 4-[(2-hydroxyphenyl)azo]-3,5-diphenyl-1-(2-pyridinylcarbonyl)-	n.s.g.	370(4.42)	48-1021-84
1H-Pyrazole, 4-[(3-hydroxyphenyl)azo]-3,5-diphenyl-1-(2-pyridinylcarbonyl)-	n.s.g.	370(4.42)	48-1021-84
1H-Pyrazole, 4-[(4-hydroxyphenyl)azo]-3,5-diphenyl-1-(2-pyridinylcarbonyl)-	n.s.g.	370(4.42)	48-1021-84
1H-Pyrrolo[2,3-d:4,5-d']dipyridazine-1,6(2H)-dione, 5,7-dihydro-2,7-di-phenyl-5-(phenylmethyl)-	EtOH	250(4.32),259(4.26), 275(4.37),310(4.02)	94-1423-84
$C_{27}H_{19}N_5O_2S$ 1H-Dipyridazino[4,5-b:4',5'-e][1,4]-thiazine-1,6(7H)-dione, 2,10-dihy-dro-2,7-diphenyl-10-(phenylmethyl)-	EtOH	232(4.33),269(4.39), 337(4.36)	94-1423-84
1H-Dipyridazino[4,5-b:4',5'-e][1,4]-thiazine-1,9(2H)-dione, 8,10-dihy-dro-2,9-diphenyl-10-(phenylmethyl)-	EtOH	228(4.13),267(4.08), 312(4.03)	142-0675-84
3H-Dipyridazino[4,5-b:4',5'-e][1,4]-thiazine-4,6(7H,10H)-dione, 3,7-diphenyl-10-(phenylmethyl)-	EtOH	232(4.43),271(4.49), 312(4.20)	142-0675-84
$C_{27}H_{20}N$ Pyridinium, 1-(1-naphthalenyl)-2,4-di-phenyl-, perchlorate	EtOH	<u>313(4.8)</u>	103-1262-84
Pyridinium, 1-(2-naphthalenyl)-2,4-di-phenyl-, perchlorate	EtOH	<u>314(4.6)</u>	103-1262-84
$C_{27}H_{20}NO$ 9H-Fluoren-9-iminium, N-[(4-methoxy-phenyl)phenylmethylene]-, (OC-6-11)-, hexachloroantimonate	CH_2Cl_2	256(4.75),262(4.82), 280(4.43),346(4.46)	24-3222-84
$C_{27}H_{20}N_2O_3$ 9,10-Anthracenedione, 1-amino-2-(4-methylphenoxy)-4-(phenylamino)-	C_6H_5Cl	530s(3.93),562(4.15), 600(4.16)	34-0482-84
9,10-Anthracenedione, 1-amino-4-[(4-methylphenyl)amino]-2-phenoxy-	C_6H_5Cl	568(3.15),604(4.15)	34-0482-84
$C_{27}H_{20}N_2O_4$ 9,10-Anthracenedione, 1-amino-2-(3-methoxyphenoxy)-4-(phenylamino)-	C_6H_5Cl	534s(3.92),564(4.12), 600(4.13)	34-0482-84
9,10-Anthracenedione, 1-amino-2-(4-methoxyphenoxy)-4-(phenylamino)-	C_6H_5Cl	534s(4.02),562(4.20), 600(4.21)	34-0482-84
9,10-Anthracenedione, 1-amino-4-[(4-methoxyphenyl)amino]-2-phenoxy-	C_6H_5Cl	536s(3.98),568(4.18), 600(4.18)	34-0482-84
$C_{27}H_{20}N_2S_3$ [1,2]Dithiolo[4,5,1-hi][1,2,3]benzo-	MeCN	501(4.26)	97-0251-84

Compound	Solvent	λ_{max}(log ϵ)	Ref.
thiadiazole-3-S^{IV}, 2,6,7,8-tetra-hydro-2-(1-naphthalenyl)-5-(1-naphthalenylthio)- (cont.)			97-0251-84
$C_{27}H_{20}N_4O$ 5H-Pyrano[2,3-d]pyrimidine-6-carbo-nitrile, 7-[1,1'-biphenyl]-4-yl-4-(methylamino)-5-phenyl-	EtOH	246(4.28),264(4.31), 299(4.35)	73-2309-84
$C_{27}H_{20}N_6O_3S$ 1H-Pyrazole, 4-[[2-(aminosulfonyl)phen-yl]azo]-3,5-diphenyl-1-(2-pyridinyl-carbonyl)-	n.s.g.	365(4.27)	48-1021-84
$C_{27}H_{20}O_3$ 1-Naphthalenecarboxylic acid, 1,2,3,4-9,9a-hexahydro-2,3-bis(methylene)-1,4-epoxyanthracen-5-yl ester, ($1\alpha,4\alpha,9a\alpha$)-(±)-	MeCN	219(4.76),228s(4.66), 241s(4.43),244s(4.34), 287(4.23),300s(4.14), 320s(3.77)	78-4549-84
1-Naphthalenecarboxylic acid, 1,2,3,4-9,10-hexahydro-2,3-bis(methylene)-1,4-epoxyanthracen-5-yl ester, (±)-	MeCN	213(4.69),217(4.68), 296(3.93),310s(3.82)	78-4549-84
$C_{27}H_{20}O_7$ 2-Butenedioic acid, 2-[(4-oxo-2,3-di-phenyl-4H-1-benzopyran-2-yl)oxy]-, dimethyl ester	MeOH	247(4.17),302(4.20)	2-0969-84
$C_{27}H_{20}O_7S_2$ Dibenzo[c,h]xanthylium, 7-(2,4-disulfo-phenyl)-5,6,8,9-tetrahydro-, hydrox-ide, inner salt	pH 2.0 pH 8.0	413(3.42),455(3.18) 282(4.29),290(4.25)	39-0867-84B 39-0867-84B
$C_{27}H_{21}BO_6$ 1,5-Methano-1H-anthra[2,3-e][1,3,2]di-oxaborocin-7,14-dione, 5-(acetoxy-methyl)-5,6-dihydro-3-phenyl-, (1S)-	dioxan	240(4.31),285(4.29), 296(4.28),403(3.70)	78-4649-84
$C_{27}H_{21}Br_2Se_2$ Seleninium, 2-(4-bromophenyl)-6-[3-[6-(4-bromophenyl)-2H-selenin-2-ylidene]-1-methyl-1-butenyl]-, perchlorate	CHCl$_3$	805(4.16)	97-0146-84
$C_{27}H_{21}Cl_2Se_2$ Seleninium, 2-(4-chlorophenyl)-6-[3-[6-(4-chlorophenyl)-2H-selenin-2-ylidene]-1-methyl-1-butenyl]-, bromide	CHCl$_3$	802(4.67)	97-0146-84
$C_{27}H_{21}N$ Spiro[acridine-9(10H),9'-[9H]fluorene], 2,7-dimethyl-	EtOH	274(4.35),297(4.18), 309(4.19),320s(3.9)	24-2703-84
$C_{27}H_{21}N_3O$ Benzoxazole, 2-[4-(1,5,6-trimethylbenz-imidazol-2-yl)-1-naphthalenyl]-	DMF	346(4.40)	5-1129-84
$C_{27}H_{21}N_3OS_2$ 1H-Imidazo[1,2-c]thiazol-4-ium, 5-[(3-ethyl-2(3H)-benzothiazolylidene)-	DMF	500(4.41)	103-0029-84

Compound	Solvent	$\lambda_{max}(\log \epsilon)$	Ref.
methyl]-2-hydroxy-3,7-diphenyl-, hydroxide, inner salt (cont.)			103-0029-84
$C_{27}H_{21}N_3O_2S$			
Benzenesulfonamide, N-(2,4-diphenyl-3H-pyrido[1,2-b]pyridazin-3-ylidene)-4-methyl-	EtOH	269(4.29),369(4.30)	142-1709-84
$C_{27}H_{21}N_3O_3$			
5(4H)-Oxazolone, 2-[4-(dimethylamino)-phenyl]-4-[[4-(5-phenyl-2-oxazolyl)-phenyl]methylene]-	toluene	316(4.62),335(4.56)	135-1330-84
$C_{27}H_{21}N_7O_3S$			
1H-Pyrazole, 3-methyl-5-phenyl-4-[[2-[(2-pyridinylamino)sulfonyl]phenyl]-azo]-1-(2-pyridinylcarbonyl)-	n.s.g.	365(4.26)	48-1021-84
$C_{27}H_{21}O_7S$			
Naphtho[1,2-b]pyrylium, 2-(4-carboxy-phenyl)-5,6-dihydro-4-(4-methoxy-3-sulfophenyl)-, perchlorate	pH 2.0	284(4.32),418(4.51)	39-0867-84B
	pH 8.0	285(3.79),330(4.40)	39-0867-84B
	pH 13.0	515(4.43)	39-0867-84B
$C_{27}H_{22}N_2$			
1H-1,5-Benzodiazepine, 4-[1,1'-biphen-yl]-4-yl-2,3-dihydro-2-phenyl-	EtOH	287(4.46),374(3.96)	103-1370-84
$C_{27}H_{22}N_2O_3$			
1H-Pyrrole-2,5-dione, 1-(4-methylphen-yl)-3-[(4-methylphenyl)amino]-4-(3-oxo-3-phenyl-1-propenyl)-	CHCl$_3$	298(4.31),440(3.95)	48-0401-84
$C_{27}H_{22}N_3OS_2$			
1H-Imidazo[1,2-e]thiazol-4-ium, 5-[(3-ethyl-2(3H)-benzothiazolylidene)meth-yl]02,3-dihydro-2-oxo-3,7-diphenyl-, perchlorate	HOAc	458(4.73)	103-0029-84
$C_{27}H_{22}N_4O_7$			
11H-Furo[2',3':4,5]oxazolo[2,3-b]pteri-din-11-one, 7-acetoxy-8-(acetoxymeth-yl)-6a,7,8,9a-tetrahydro-2,3-diphenyl-	MeOH	220s(4.44),246(4.27), 280(4.25),357(4.12)	136-0179-84F
$C_{27}H_{22}N_5O_2$			
7H-Purinium, 6-(benzoylamino)-7-(2-oxo-2-phenylethyl)-9-(phenylmethyl)-, bromide	MeOH	245(4.31),284(4.09), 319(4.09),330(4.06)	150-3601-84M
$C_{27}H_{22}O_2$			
6,11[1',2']-Benzenoanthra[1,2-d]-1,3-dioxole, 3a,5a,6,11,11a,11b-hexahy-dro-2-phenyl-	MeOH	256.5(2.90),263(2.94), 268(2.94),271(3.06), 275(3.02),279(3.18)	35-7310-84
isomer	MeOH	256.5(2.88),261(2.90), 262.5(2.92),267(2.95), 271(3.05),275.5(2.99), 279(3.17)	35-7310-84
$C_{27}H_{22}O_6$			
Mulberrofuran H	EtOH	220s(4.48),290s(4.15), 321(4.52),333s(4.45)	94-0808-84

Compound	Solvent	$\lambda_{max}(\log \epsilon)$	Ref.
$C_{27}H_{23}N$			
Acridine, 9,10-dihydro-2,7-dimethyl-9,9-diphenyl-	EtOH	293(4.09)	24-2703-84
5H-Benzo[6,7]cyclohepta[1,2-b]pyridine, 6,7-dihydro-2-(4-methylphenyl)-4-phenyl-	dioxan	247(4.58),268s(4.42), 303(4.03)	48-0647-84
$C_{27}H_{23}NO_5$			
Benzo[g]-1,3-benzodioxolo[5,6-a]quino-lizinium, 5,6-dihydro-13-hydroxy-9,10-dimethoxy-8-(phenylmethyl)-, hydroxide, inner salt	MeOH	240.5(4.53),262s(4.34), 323(4.09),373(4.08), 461(4.26)	94-2230-84
$C_{27}H_{23}N_2$			
1H-Indolizinium, 1-cyano-2,3,8,8a-tetrahydro-5,7,8a-triphenyl-, tetra-fluoroborate	MeOH	242(4.18),361(3.95)	39-0941-84C
$C_{27}H_{23}N_5O_3$			
Benzamide, N-[5-[formyl(2-oxo-2-phenyl-ethyl)amino]-6-[(phenylmethyl)amino]-4-pyrimidinyl]-	MeOH	249(4.60),290s(3.90)	150-3601-84M
$C_{27}H_{23}O$			
5H-Benzo[6,7]cyclohepta[1,2-b]pyrylium, 6,7-dihydro-2-(4-methylphenyl)-4-phenyl-, perchlorate	MeCN	282.5(4.30),341(4.30), 406(4.46)	48-0647-84
$C_{27}H_{23}S_2$			
Thiopyrylium, 2-[1-methyl-3-(6-phenyl-2H-thiopyran-2-ylidene)-1-butenyl]-6-phenyl-, bromide	CHCl$_3$	800(4.71)	97-0146-84
$C_{27}H_{23}Se_2$			
Seleninium, 2-[1-methyl-3-(6-phenyl-2H-selenin-2-ylidene)-1-butenyl]-6-phenyl-, iodide	CHCl$_3$	808(4.04)	97-0146-84
$C_{27}H_{24}F_3NO_{10}$			
1,6,11(2H)-Naphthacenetrione, 3,4-di-hydro-2,5,7-trihydroxy-2-methyl-4-[[2,3,6-trideoxy-3-[(trifluoroacet-yl)amino]-α-L-lyxo-hexopyranosyl]-oxy]-	MeOH	210(4.38),239(4.45), 261(4.30),438(4.04)	44-3766-84
$C_{27}H_{24}N_2O_3$			
Benzamide, N-[4-(cyclohexylamino)-9,10-dihydro-9,10-dioxo-1-anthra-cenyl]-	benzene	573(4.05),612(4.00)	39-0529-84C
2-Propenamide, N-[4-(butylamino)-9,10-dihydro-9,10-dioxo-1-anthracenyl]-3-phenyl-	benzene	572(4.07),610(4.01)	39-0529-84C
1H-Pyrazole-1-acetic acid, 5-methyl-α-(2-phenoxy-2-phenylethenyl)-3-phenyl-, methyl ester, (Z)-	MeOH	245(4.22)	44-4647-84
$C_{27}H_{24}N_2O_3S$			
4,9-Imino-1H-benz[f]isoindole-1,3(2H)-dione, 4-(ethylthio)-3a,4,9,9a-tetra-hydro-2-(4-methoxyphenyl)-10-phenyl-, (3aα,4β,9α,9aβ)-	EtOH	206(4.57),234(4.31), 289(3.39),300(3.31), 336(3.27)	103-0978-84

Compound	Solvent	$\lambda_{max}(\log \epsilon)$	Ref.
4,9-Imino-1H-benz[f]isoindole-1,3(2H)-dione, 4-(ethylthio)-3a,4,9,9a-tetra-hydro-10-(4-methoxyphenyl)-2-phenyl-, (3aα,4β,9α,9aβ)-	EtOH	206(4.54),222(4.51), 244(4.35),289(3.88), 300(3.78),336(3.68)	103-0978-84
$C_{27}H_{24}N_2O_8$ 2(3H)-Furanone, 3-[(4-acetoxyphenyl)-methylene]-5-(1,2-diacetoxyethyl)-dihydro-4-(2-quinoxalinyl)-	MeOH	219(4.28),239(4.48), 296(4.32)	94-1808-84
$C_{27}H_{24}N_4$ 4H-Pyrrolo[2,3-d]pyrimidin-4-imine, 3-[1,1'-biphenyl]-2-yl-3,7-dihydro-2,5,6-trimethyl-7-phenyl-	EtOH	243(4.53),303(3.85)	11-0073-84A
$C_{27}H_{24}O_2$ 5H-Benzocyclohepten-5-one, 6,7,8,9-tetrahydro-6-[3-(4-methylphenyl)-3-oxo-1-phenylpropylidene]-	dioxan	255.5(4.43)	48-0647-84
$C_{27}H_{24}DN_2O_4$ Cyclopent[b]indole-1,2-dicarboxylic acid, 1,4-dihydro-1-d-4-methyl-3-(1-methyl-1H-indol-3-yl)-, diethyl ester	EtOH	220(4.70),255(4.25), 360(4.20)	94-4410-84
$C_{27}H_{25}NO$ 1,3-Oxazepine, 4-(1-methylethyl)-2-(4-methylphenyl)-5,7-diphenyl-	ether	264(4.06),352(3.41)	78-3559-84
$C_{27}H_{25}NOS$ Morpholine, 4-(2,4,6-triphenyl-2H-thiopyran-2-yl)-	DMSO	<u>350(3.7)</u>	35-7082-84
$C_{27}H_{25}NO_4S$ Methanone, [6-hydroxy-2-(4-hydroxyphen-yl)benzo[b]thien-3-yl][4-[2-(1-pyrro-lidinyl)ethoxy]phenyl]-	EtOH	290(4.51)	87-1057-84
$C_{27}H_{25}N_3O_5$ 7H-Pyrrolo[2,3-d]pyrimidine, 7-[2-de-oxy-3,5-bis-O-(4-methylbenzoyl)-α-D-erythro-pentofuranosyl]-	MeOH	224(4.61),240(4.55), 268(3.83)	5-1719-84
β-	MeOH	225(4.61),240(4.53), 268(3.78)	5-1719-84
$C_{27}H_{26}N$ Pyridinium, 1-(1-methylpropyl)-2,4,6-triphenyl-	C_6H_5Cl	306(4.59)	120-0062-84
$C_{27}H_{26}N_2O_4$ Cyclopent[b]indole-1,2-dicarboxylic acid, 1,4-dihydro-4-methyl-3-(1-methyl-1H-indol-3-yl)-, diethyl ester	EtOH	218(4.73),260(4.25), 369(4.25)	94-4410-84
1H-Indazole-6,7-dicarboxylic acid, 4-methyl-1-(4-methylphenyl)-3-phenyl-, diethyl ester	EtOH	254(4.83),330(4.15)	94-2146-84
$C_{27}H_{26}N_2O_5$ Cyclopent[b]indole-1,2-dicarboxylic	EtOH	218(4.72),260(4.24),	94-4410-84

Compound	Solvent	$\lambda_{max}(\log \epsilon)$	Ref.
acid, 1,4-dihydro-1-hydroxy-4-methyl-3-(1-methyl-1H-indol-3-yl)-, diethyl ester (cont.)		367(4.25)	94-4410-84
$C_{27}H_{26}N_2O_6$			
Cyclopent[b]indole-1,2-dicarboxylic acid, 3,4-dihydro-3-hydroxy-3-(2-hydroxy-1-methyl-1H-indol-3-yl)-4-methyl-, diethyl ester	EtOH	220(4.64),260s(--), 345(4.16)	94-4410-84
$C_{27}H_{26}N_2O_9$			
Spiro[9H-fluorene-9,3'-[3H]pyrazole]-4',5'-dicarboxylic acid, 1'-[2,3-dimethoxy-1-(methoxycarbonyl)-3-oxo-1-propenyl]-1',2'-dihydro-2'-methyl-, dimethyl ester	CHCl$_3$	271(4.21),285(4.10), 308(3.78),333(3.79)	24-2409-84
$C_{27}H_{26}N_4O_2$			
1H-Imidazo[1,2-b][1,2,4]triazepine, 7,8-dihydro-6,8-bis(4-methoxyphenyl)-8-methyl-2-phenyl-	MeOH	287(4.63),334(4.02)	103-1152-84
$C_{27}H_{26}N_4O_5$			
7H-Pyrrolo[2,3-d]pyrimidin-2-amine, 7-[2-deoxy-3,6-bis-O-(4-methyl-benzoyl)-α-D-erythro-pentofuranosyl]-	MeOH	236(4.76),282(3.45), 312(3.71)	5-1719-84
β-	MeOH	236(4.76),282(3.49), 312(3.73)	5-1719-84
$C_{27}H_{26}N_6O_6$			
Spiro[1,3-dioxolane-2,10'(7'H)-[1H]-[1,2,4]triazolo[1,2-a]cinnoline]-1',3'(2'H)-dione, 7'-[3,5-dioxo-4-phenyl-1,2,4-triazolidin-1-yl)-5',8',9',10'a-tetrahydro-7'-methyl-2'-phenyl-	EtOH	220(3.45)	12-2295-84
$C_{27}H_{26}N_6O_8$			
1H-Pyrazole-3-carboxylic acid, 4-cyano-5-[[2-ethoxy-1-[[2-(methoxycarbonyl)-phenyl]hydrazono]-2-oxoethyl]amino]-1-[2-(methoxycarbonyl)phenyl]-, ethyl ester	EtOH	225(4.14),262(4.34), 333(3.59)	4-1049-84
$C_{27}H_{26}O_2$			
1H-Indene-1,3(2H)-dione, 2-[1,1'-bi-phenyl]-4-yl-2-hexyl-	EtOH	348(2.75)	135-0312-84
	dioxan	348(2.73)	135-0312-84
$C_{27}H_{26}O_8$			
Sanggenon M acetate	EtOH	228s(4.02),237s(3.98), 268s(4.29),277(4.32), 319(3.92),375(3.30)	142-1791-84
	EtOH-AlCl$_3$	231s(4.00),275(4.34), 333(3.93),417(3.17)	142-1791-84
$C_{27}H_{27}NO$			
2H-Pyran-2-amine, N,N,3,5-tetramethyl-2,4,6-triphenyl-	dioxan	328(3.90)	48-0657-84
2H-Pyran-2-amine, N,N,3-trimethyl-6-(4-methylphenyl)-2,4-diphenyl-	dioxan	243(4.27),249(4.28), 342(4.07)	48-0657-84

Compound	Solvent	$\lambda_{max}(\log \epsilon)$	Ref.
$C_{27}H_{27}NO_2$			
2H-Pyran-2-amine, 2-(4-methoxyphenyl)-N,N,3-trimethyl-4,6-diphenyl-	dioxan	245s(4.28),249(4.29), 341(4.07)	48-0657-84
2H-Pyran-2-amine, 4-(4-methoxyphenyl)-N,N,3-trimethyl-2,6-diphenyl-	dioxan	250(4.23),264(4.22), 341(4.04)	48-0657-84
2H-Pyran-2-amine, 6-(4-methoxyphenyl)-N,N,3-trimethyl-2,4-diphenyl-	dioxan	250(4.21),260(4.22), 344(4.14)	48-0657-84
$C_{27}H_{27}NO_6$			
Isoquinoline, 1,3-bis(3,4-dimethoxyphenyl)-6,7-dimethoxy-	EtOH	238(4.18),273(4.25), 315(4.14),360s(3.41)	4-0525-84
$C_{27}H_{27}N_2O$			
Quinolinium, 6-acetyl-1-ethyl-4-[3-(1-ethyl-4(1H)-quinolinylidene)-1-propenyl]-, iodide	MeOH	718(5.20)	103-0767-84
	$C_6H_4Cl_2$	737(5.23)	103-0767-84
	MeCN	717(--)	103-0767-84
$C_{29}H_{27}N_2O_2$			
Quinolinium, 1-ethyl-4-[3-(1-ethyl-4(1H)-quinolinylidene)-1-propenyl]-6-(methoxycarbonyl)-, iodide	MeOH	715(5.19)	103-0767-84
	$C_6H_4Cl_2$	734(5.23)	103-0767-84
	MeCN	714(--)	103-0767-84
$C_{27}H_{27}N_2S_2$			
Benzothiazolium, 3-ethyl-2-[2-[9-(3-ethyl-2(3H)-benzothiazolylidene)-1,3,5,7-nonatetraenyl]-, iodide	$C_6H_{11}Me$	770(5.49)	99-0415-84
	EtOH	765(5.34)	99-0415-84
	DMF	765(5.26)	99-0415-84
$C_{27}H_{28}Br_2O_5S$			
Bromthymol Blue	acid	430(3.98)	35-0265-84
	base	615(4.56)	35-0265-84
with hexadecyltrimethylaminium ion	acid	418(4.05)	35-0265-84
	base	625(4.55)	35-0265-84
$C_{27}H_{28}F_3N_4O_4S_2$			
1H-Benzimidazolium, 2-[[5-[(acetylphenyl)amino]methylene]-3-ethyl-4-oxothiazolidin-2-ylidene]methyl]-1,3-diethyl-5-[(trifluoromethyl)-sulfonyl]-, perchlorate	EtOH	310(4.66),348(4.69)	103-0665-84
$C_{27}H_{28}N_2$			
4,4'-Bipyridinium, 2-methyl-1,1'-bis[(2-methylphenyl)methyl]-, dibromide	H_2O	258(4.43)	39-0367-84B
$C_{27}H_{28}O_2$			
Methanone, [4-(1,1-dimethylethyl)phenyl](6,7,8,9-tetramethyl-2-dibenzofuranyl)-	EtOH	256s(4.45),275(4.59)	39-0799-84B
Methanone, [4-(1,1-dimethylethyl)phenyl](6,7,8,9-tetramethyl-3-dibenzofuranyl)-	EtOH	261(4.26),324(4.43)	39-0799-84B
$C_{27}H_{28}O_7$			
Propanoic acid, 2,2-dimethyl-, 3-(acetoxymethyl)-9,10-dihydro-9,10-dioxo-2-(3-oxopentyl)-1-anthracenyl ester	MeOH	209(4.39),257(4.68), 332(3.79)	78-3677-84
$C_{27}H_{29}BrO_7$			
Propanoic acid, 2,2-dimethyl-, 3-(bromomethyl)-9,10-dihydro-5,8-dimeth-	MeOH	224(4.54),260(4.53), 332(3.49)	78-3677-84

Compound	Solvent	$\lambda_{max}(\log \epsilon)$	Ref.
oxy-9,10-dioxo-2-(3-oxopentyl)-1-anthracenyl ester (cont.)		425(3.81)	78-3677-84
$C_{27}H_{29}N_3O$ Urea, N,N-diethyl-N'-[3-[(4-methylphenyl)imino]-1,3-diphenyl-1-propenyl]-	EtOH	205(4.23),225s(3.95), 250(3.91),335s(3.80), 376(3.93)	39-0239-84C
$C_{27}H_{29}N_3O_2$ Urea, N,N-diethyl-N'-[3-[(4-methoxyphenyl)imino]1,3-diphenyl-1-propenyl]-	EtOH	206(4.52),230(4.25), 255s(4.18),332(4.06), 381(4.20)	39-0239-84C
$C_{27}H_{29}N_3O_4$ 4H-Pyrido[1,2-a]pyrimidine-3-carboxamide, N-[2-(3,4-dimethoxyphenyl)-ethyl]-6,7,8,9-tetrahydro-6-methyl-4-oxo-9-(phenylmethylene)-	EtOH	228(4.36),286(4.07), 354(4.41)	118-0582-84
$C_{27}H_{29}N_3O_9$ Isotrityrosine	acid pH 7.5 base	283(3.90) 283(3.90) 303(3.98)	149-0731-84B 149-0731-84B 149-0731-84B
Trityrosine	acid base	286(3.90) 322(--)	149-0731-84B 149-0731-84B
$C_{27}H_{29}N_5O_{11}$ 2H,8H-1,8,17-Metheno-7,2-(iminoetheno)-7a,13b-(methanoxymethano)-10H-[1,2,4]triazolo[1',2':1,2]pyridazino-[3,4,5-ef][1]benzazepine-2,3,18,19-tetracarboxylic acid, 1,4a,5,6,7,11-12,13a-octahydro-11-methyl-10,12-dioxo-, tetramethyl ester	MeCN	360(3.58)	88-2459-84
$C_{27}H_{30}N_4$ 4H-Pyrimido[4,5-b]indol-4-imine, 3-(4-butylphenyl)-3,5,6,7,8,9-hexahydro-2-methyl-9-phenyl-	EtOH	245(4.30),283s(3.78), 305(3.75)	11-0073-84A
$C_{27}H_{30}N_6O_6$ 3-Cyclohexene-1,1-dicarboxylic acid, 4-amino-3-cyano-, bis[[(2,4-dimethoxyphenyl)methylene]hydrazide]	n.s.g.	268(4.11)	42-0146-84
$C_{27}H_{30}O$ 2,5-Cyclohexadien-1-one, 2,6-bis(1,1-dimethylethyl)-4-(diphenylmethylene)-	hexane?	211(4.14),261(4.13), 268(4.12),366(4.41)	73-1949-84
$C_{27}H_{30}O_4$ 2-Propen-1-one, 1-[5,7-dimethoxy-2-methyl-2-(4-methyl-3-pentenyl)-2H-1-benzopyran-6-yl]-3-phenyl-, (E)-(±)-	EtOH	232(4.53),240(4.43), 293(4.42)	12-1739-84
2-Propen-1-one, 3-phenyl-1-[3,4,5,6-tetrahydro-7,9-dimethoxy-2-methyl-5-(1-methylethylidene)-2,6-methano-2H-1-benzoxocin-8-yl]-, (E)-(±)-	EtOH	218(4.37),240(3.97), 270s(4.01),300(4.14), 340(3.76)	12-1739-84
$C_{27}H_{30}O_5S$ Thymol Blue	acid base	435(4.12) 595(4.53)	35-0265-84 35-0265-84

Compound	Solvent	$\lambda_{max}(\log \epsilon)$	Ref.
Thymol Blue, with hexadecyltrimethyl- aminium bromide	acid base	437(4.15) 605(4.55)	35-0265-84 35-0265-84

$C_{27}H_{30}O_7$
  Propanoic acid, 2,2-dimethyl-, 9,10-
  dihydro-5,8-dimethoxy-3-methyl-9,10-
  dioxo-2-(3-oxopentyl)-1-anthracenyl
  ester
	MeOH	221(4.50),260(4.52),
		332(3.48),421(3.81)
			78-3677-84

$C_{27}H_{30}O_8$ Abyssinin	EtOH	228s(3.89),256(4.13), 296s(3.85)	88-4601-84
$C_{27}H_{30}O_{14}$ Physcion, 8-O-β-primeveroside	MeOH MeOH-NaOH	222.5(4.52),269(4.39), 277(4.38),416(3.94) 266(--),442(--)	102-1485-84 102-1485-84

$C_{27}H_{30}O_{15}$
  Apigenin 4',7-bis-O-β-D-allopyranoside
 	
  Luteolin 7-O-rutinoside
	MeOH
	MeOH-NaOAc
	MeOH-AlCl₃
	
	MeOH
		270.5(4.57),318.5(4.54)
		270(4.55),319(4.53)
		279.5(4.51),299.5(4.50),
		338.5(4.56),382(4.32)
		257(4.26),265(4.28),
		342(4.29)
			94-5023-84
			94-5023-84
			94-5023-84
			
			94-0295-84

$C_{27}H_{31}NO_4$ Ryosenamine	EtOH	230(4.10),273.5(2.97), 281(2.88)	95-0222-84

$C_{27}H_{31}NO_5$
  Acetic acid, [[[(17β)-17-hydroxy-
  3-(phenylmethoxy)estra-1,3,5(10)-
  trien-6-ylidene]amino]oxy]-
	MeOH	263(4.07),308(3.64)	94-1885-84

$C_{27}H_{31}NO_8S$
  Acetic acid, [[[(17β)-3-(phenylmeth-
  oxy)-17-(sulfooxy)estra-1,3,5(10)-
  trien-6-ylidene]amino]oxy]-, mono-
  potassium salt
	MeOH	261(4.04),308(3.59)	94-1885-84

$C_{27}H_{31}N_2$
  3H-Indolium, 2-[5-(1,3-dihydro-1,3,3-
  trimethyl-2H-indol-2-ylidene)-1,3-
  pentadienyl]-1,3,3-trimethyl-,
  iodide
	$C_6H_{11}Me$
	EtOH
	DMF
		652(5.38)
		640(5.33)
		642(5.31)
			99-0415-84
			99-0415-84
			99-0415-84

$C_{27}H_{31}S_2$
  Thioxanthylium, 4-[(2,3,5,6,7,8-hexa-
  hydro-1H-thioxanthen-4-yl)methyl-
  ene]-1,2,3,4,5,6,7,8-octahydro-,
  perchlorate
	CHCl₃	806(4.34)	97-0146-84

$C_{27}H_{32}N_2O_7$ Antibiotic X-14885-A	EtOH	204(4.49),257(4.18), 306(4.34)	158-84-162
$C_{27}H_{32}N_2O_8$ Raucaffricine	MeOH	220(4.34),258(3.72)	33-2078-84

Compound	Solvent	$\lambda_{max}(\log \epsilon)$	Ref.
$C_{27}H_{32}N_4$ 4,4'-Bipyridinium, 1,1''-(1,5-pentane-diyl)bis[1'-methyl-, tetraperchlorate	H_2O	262(4.64)	64-0074-84B
$C_{27}H_{32}N_4O_6$ Piperazine, 1-benzoyl-4-[4-[(2,2-dimethyl-1,3-dioxolan-4-yl)methoxy]-6,7-dimethoxy-2-quinazolinyl]-	EtOH	217(4.60),251(4.80)	4-1189-84
$C_{27}H_{32}N_5Ni$ Nickel(1+), (3,7,8,9,12,13,17,18-octa-hydro-2,2,12,12,17,17-hexamethyl-2H,21H-porphine-5-carbonitrilato-$N^{21},N^{22},N^{23},N^{24}$)-, (SP-4-3)-, perchlorate	EtOH	267(4.19),305s(4.25), 320s(4.35),332(4.40), 394(3.77),422(3.82), 466(3.90)	77-1365-84
$C_{27}H_{32}N_8O_2$ 3-Cyclohexene-1,1-dicarboxylic acid, 4-amino-3-cyano-, bis[[[4-(dimethylamino)phenyl]methylene]hydrazide]	n.s.g.	261(4.19)	42-0146-84
$C_{27}H_{32}O_4$ Methanone, (2-hydroxy-4,6-dimethoxy-phenyl)[3-methyl-2-(3-methyl-2-butenyl)-6-phenyl-3-cyclohexen-1-yl]-, (1α,2α,6α)-(±)-	EtOH	228(4.14),292(4.28)	12-0449-84
$C_{27}H_{32}O_6$ 4H-1-Benzopyran-4-one, 2,3-dihydro-5-hydroxy-2-[3(or 4)-hydroxy-4(or 3)-methoxyphenyl]-7-methoxy-	MeOH	286(4.40),362(3.70)	102-1818-84
$C_{27}H_{32}O_8$ 2-Propenoic acid, 3-(3,4-dimethoxy-phenyl)-, 1,5-pentanediyl ester	MeCN	322(4.61)	40-0022-84
Verrucarin J	EtOH	218(4.44),262(4.31)	44-1772-84
isomer 40	EtOH	217(4.30),262(4.17)	44-1772-84
$C_{27}H_{32}O_9$ Verrucarin J, 12'-hydroxy-, 2'(E)-	MeOH	217(4.37),262(4.28)	158-0823-84
$C_{27}H_{32}O_{14}$ Liquiritigenin 4',7-diglucoside	MeOH	270(4.36),313(3.97)	102-2108-84
$C_{27}H_{33}BrO_5$ Benzoic acid, 4-bromo-, 4-[2-(2,5-dihydro-5-oxo-3-furanyl)ethyl]decahydro-4,5,7a,7b-tetramethylnaphtho-[1,2-b]oxiren-7-yl ester	EtOH	200.6(4.53),242.9(4.24)	88-2809-84
$C_{27}H_{33}NO_6$ Pyrrolo[2,1-a]isoquinoline-3-carboxylic acid, 1-acetyl-8,9-diethoxy-1,2,3,5,6,10b-hexahydro-2-(4-methoxyphenyl)-, methyl ester	EtOH	213(4.07),230(4.03), 285(3.37)	78-0369-84
$C_{27}H_{33}N_2O_7$ Aspidospermidinium, 3,4-diacetoxy-6,7,8,9-tetradehydro-16-methoxy-3-(methoxycarbonyl)-1-methyl-	MeOH	251(4.07),301(3.77)	87-0749-84

Compound	Solvent	$\lambda_{max}(\log \epsilon)$	Ref.
$C_{27}H_{34}N_2O_4$			
Demethylpsychotrine	pH 13	243(4.41),307(4.28), 326(4.32)	100-0397-84
	EtOH	223(3.95),277(3.83), 310(3.34),410(3.96)	100-0397-84
$C_{27}H_{34}N_2O_7$			
Vindoline, 16-O-acetyl-	EtOH	217(4.60),252(3.86), 304(3.71)	87-0749-84
$C_{27}H_{34}O_8$			
Abyssinin, tetrahydro-	EtOH	248(3.99)	88-4601-84
$C_{27}H_{34}O_9$			
Verrucarin J seco acid	EtOH	221(4.46),263(4.53)	44-1772-84
(E,E,E)-	EtOH	219(4.20),262(4.34)	44-1772-84
(Z,E,Z)-	EtOH	218(4.23),263(4.36)	44-1772-84
$C_{27}H_{34}O_{12}$			
Pinoresinol, 1-hydroxy-, 4"-O-methyl ether 4'-β-D-glucoside, (+)-	EtOH	230(4.26),277(3.76)	94-4482-84
$C_{27}H_{34}O_{16}$			
Gardenosine pentaacetate	MeOH	233(4.00)	94-2947-84
$C_{27}H_{35}BrO_2$			
Benzoic acid, 4-bromo-, 3,9,13-trimethyl-6-(1-methylethyl)-2,7,9,12-cyclotetradecatetraen-1-yl ester, [1R-(1R*,2E,6S*,7E,9Z,12E)]-	hexane	244(4.36)	102-1681-84
$C_{27}H_{35}NO_{12}$			
Ipecoside	EtOH	227(4.14),287(3.58)	100-0397-84
$C_{27}H_{36}N_2O_4$			
Demethylcephaeline	EtOH	211(4.28),225s(3.91), 286(3.69)	100-0397-84
	EtOH-NaOH	213(3.48),227(3.96), 247(4.03),301(3.81)	100-0397-84
$C_{27}H_{36}O_3$			
2,4,6,8,10,12,14,16,18-Eicosanonaenal, 20,20-dimethoxy-2,6,11,15,19-pentamethyl-	pet ether	461.5(5.09),491.5(4.99)	137-0102-84
$C_{27}H_{36}O_4$			
5-Heptenoic acid, 7-[6,6-dimethyl-3-(3-oxo-4-phenoxy-1-butenyl)bicyclo-[3.1.1]hept-2-yl]-, methyl ester, [1S-[1α,2β(Z),3α(E),5α]-	EtOH	219(4.18),235(4.06), 268(3.46),278(3.26)	39-1069-84C
$C_{27}H_{36}O_5$			
2,5-Cyclohexadiene-1,4-dione, 2-[(1,2-3,6,7,8,8a,9,10,11,12,12a-dodecahydro-6,6,9,10,12a-pentamethyl-3-oxo-naphtho[2,1-d][1,2]dioxocin-9-yl)-methyl]-6-methyl-, [8aR-(8aα,9α,10α-12aβ)]-	EtOH	257(3.40)	138-1649-84
$C_{27}H_{36}O_9$			
3-Heptanone, 7-(3,4-dimethoxyphenyl)-	EtOH	225(4.24),287(3.70),	39-1635-84C

Compound	Solvent	$\lambda_{max}(\log \epsilon)$	Ref.
1-(4-methoxyphenyl)-5-(β-D-xylo-pyranosyloxy)-, (S)- (cont.)		284(3.65)	39-1635-84C
$C_{27}H_{36}O_{10}$ Myricanol glucoside	EtOH	213(4.49),253(4.04), 296(3.77)	95-0037-84
$C_{27}H_{37}N_5O_7$ L-Leucine, N-[N-[N-(1-cyclohexyl-2,5-dihydro-4-nitro-2-oxo-1H-pyrrol-3-yl)glycyl]-L-phenylalanyl]-	EtOH	369(4.17)	44-1130-84
$C_{27}H_{38}NO_4Si$ 2H-[1,3]Benzodioxolo[5,6-d][1]benzaze-pinium, 1-(2,2-dimethyl-1-oxopropoxy)-3,4,6,7-tetrahydro-5-[2-[(trimethyl-silyl)methyl]-2-propenyl]-, per-chlorate	EtOH	290(3.91),315s(3.39)	44-0228-84
$C_{27}H_{38}N_4O_{10}S$ 1-Azabicyclo[3.2.0]heptane-2-carboxylic acid, 3-[[2-[[3-[(2,4-dihydroxy-3,3-dimethyl-1-oxobutyl)amino]-1-oxopropyl]amino]ethyl]thio]-6-(1-hydroxyeth-yl)-7-oxo-, (4-nitrophenyl)methyl ester	CH_2Cl_2	268(3.74)	158-0211-84
$C_{27}H_{38}O_3$ Furosta-3,5,20(22)-trien-7-one, 26-hy-droxy-	MeOH	274(4.2)	2-0435-84
$C_{27}H_{38}O_5S$ 1-Dodecanesulfonic acid, 2-hydroxy-3-oxo-2,3-diphenylpropyl ester	MeOH	251(4.21),332(2.36)	126-1795-84
$C_{27}H_{38}O_6$ Chola-7,14-dien-24-oic acid, 3-[(meth-oxycarbonyl)oxy]-12-oxo-, methyl ester, (3α,5β)-	EtOH	250(3.78)	70-2377-84
Chola-8,14-dien-24-oic acid, 3-[(meth-oxycarbonyl)oxy]-12-oxo-, methyl ester, (3α,5β)-	EtOH	245(4.07),294(2.85)	70-2377-84
$C_{27}H_{38}O_7$ Characiol, 15-O-acetyl-3-O-tigloyl-, 5β,6β-epoxide	MeOH	193(3.97),217s(4.21), 223(4.24),229s(4.20), 307(2.20)	102-1689-84
Isocharaciol, 15-O-acetyl-5β-hydroxy-, 3-tiglate	MeOH	194(4.09),215(4.13), 302(2.54)	102-1689-84
$C_{27}H_{38}O_{10}$ Pregna-1,4-diene-3,20-dione, 21-(β-D-glucopyranosyloxy)-11,17-dihydroxy-, (11β)-	n.s.g.	242(4.12)	87-0261-84
$C_{27}H_{39}N_3O_{15}$ D-Streptamine, 4-O-(3-O-acetyl-2,4-di-deoxy-2-[(ethoxycarbonyl)amino]-α-D-threo-hex-4-enodialdo-1,5-pyranosyl]-2-deoxy-N,N'-bis(ethoxycarbonyl)-, 5,6-diacetate	n.s.g.	254(3.36)	158-0143-84

Compound	Solvent	$\lambda_{max}(\log \epsilon)$	Ref.
$C_{27}H_{39}S_4$ Cyclopropenylium, [5-[2,3-bis[(1,1-di-methylethyl)thio]-2-cyclopropen-1-ylidene]-1,3-cyclopentadien-1-yl]-bis[(1,1-dimethylethyl)thio]-, perchlorate	CH_2Cl_2	273(4.22),350(3.85), 436(4.01)	88-0345-84
$C_{27}H_{40}O_4$ Cholesta-1,4,22-trien-3-one, 16,18,20-trihydroxy-, (16β,20ξ,22E)-	MeOH	244(4.07)	31-0246-84
1-Pentanone, 1-(1-acetoxy-6a,7,8,9,10-10a-hexahydro-6,6-dimethyl-3-pentyl-6H-dibenzo[b,d]pyran-9-yl)-, [6aR-(6aα,9β,10aβ)]-	EtOH	275(3.26),286(3.26)	78-3839-84
Pregna-7,9(11)-diene-20-carboxylic acid, 3-acetoxy-4,4,14-trimethyl-, (3β,5α,20S)-	MeOH	235(4.09),243(4.16), 252(4.01)	39-0497-84C
$C_{27}H_{40}O_5$ Androsta-5,7-diene-3,17-diol, 19-meth-oxy-, 3-acetate 17-(2,2-dimethyl-propanoate), (3β,17β)-	EtOH	271(4.03),280.5(4.03)	56-0711-84
$C_{27}H_{40}O_6$ Androst-5-en-7-one, 3-acetoxy-17-(2,2-dimethyl-1-oxopropoxy)-19-methoxy-, (3β,17β)-	EtOH	233.5(4.14)	56-0711-84
$C_{27}H_{40}P_2S$ Diphosphene, (2,4,6-trimethylphenyl)-[2,4,6-tris(1,1-dimethylethyl)-phenyl]-, 1-sulfide	CH_2Cl_2	244(4.14),267s(3.94), 368(3.48)	142-0681-84
$C_{27}H_{41}NO_2$ Hexanoic acid, 6-[[3,7-dimethyl-9-(2,6,6-trimethyl-1-cyclohexen-1-yl)-2,4,6,8-nonatetraenylidene]-amino]-, methyl ester, hydrochloride	MeOH	363(4.61),440(4.60)	104-1669-84
$C_{27}H_{42}N_4O_4$ Discarine F	MeOH	233(3.66)	64-1825-84B
$C_{27}H_{42}O_3$ 6H-Dibenzo[b,d]pyran-1-ol, 6a,7,8,9,10-10a-hexahydro-6,6-dimethyl-3,9-di-pentyl-, acetate, [6aR-(6aα,9β,10aβ)]-	EtOH	277s(3.26),283(3.30)	78-3839-84
[6aR-(6aα,9α,10aβ)]-	EtOH	277s(3.34),282(3.35)	78-3839-84
$C_{27}H_{43}FO$ 9,10-Secocholesta-5,7,10(19)-trien-3-ol, 1-fluoro-, (1α,3β,5Z,7E)-	EtOH	243(<u>4.2</u>),271(<u>4.2</u>)	94-3518-84
(1β,3β,5Z,7E)-	EtOH	265(<u>4.2</u>)	94-3518-84
$C_{27}H_{44}N_2O$ Harappamine	MeOH	238(4.22),246(4.25)	64-0524-84B
$C_{27}H_{44}O_3$ Cholest-4-en-3-one, 7α,25-dihydroxy-	n.s.g.	243(4.22)	13-0095-84B
$C_{27}H_{44}O_4$ Cholest-4-en-3-one, 7α,12α,25-trihy-	n.s.g.	243(4.17)	13-0095-84B

Compound	Solvent	$\lambda_{max}(\log \epsilon)$	Ref.
droxy- (cont.)			13-0095-84B
$C_{27}H_{44}O_6$			
Ponasterone A	EtOH	244(4.08),327(2.18)	154-0547-84
$C_{27}H_{44}O_7$			
Cucurbitacin F	MeOH	232(4.08)	100-0988-84
Ecdysterone	EtOH	244(4.05)	105-0300-84
$C_{27}H_{44}O_8$			
Polypodine B	EtOH	244(4.09)	105-0300-84
$C_{27}H_{46}O_5$			
4,6-Decadienal, 8-ethyl-10-[4-hydroxy-8-(2-hydroxypropyl)-3,9-dimethyl-1,7-dioxaspiro[5.5]undec-2-yl]-2-methyl-	EtOH	226(4.22),232(4.23), 240s(4.08)	33-1208-84
$C_{27}H_{46}O_7$			
Cucurbitacin F, 23,24-dihydro-	MeOH	208(4.0)	100-0988-84

Compound	Solvent	$\lambda_{max}(\log \epsilon)$	Ref.
$C_{28}H_{14}Cl_2O_2$ Dibenzo[a,j]anthracene-7,14-dione, 2-chloro-5-(4-chlorophenyl)-	CHCl$_3$	249(4.57),287(4.45), 307(4.74),334(4.02), 368(3.67),440(3.52)	40-0090-84
$C_{28}H_{14}F_2O_2$ Dibenzo[a,j]anthracene-7,14-dione, 2-fluoro-5-(4-fluorophenyl)-	CHCl$_3$	248(4.52),287(4.47), 304(4.77),331(4.01), 367(3.74),441(3.54)	40-0090-84
$C_{28}H_{16}N_4O_4$ 3H-Indol-3-one, 2-[1,3-dihydro-3-oxo-1-(3-pyridinylcarbonyl)-2H-indol-2-ylidene]-1,2-dihydro-1-(3-pyridinyl-carbonyl)-, trans	benzene	569(3.60)	39-2305-84C
$C_{28}H_{16}O_2$ Dibenz[a,j]anthracene-7,14-dione, 5-phenyl-	CHCl$_3$	285(4.24),305(4.54), 332(3.80),363(3.54), 407(3.29),434(3.26)	40-0090-84
$C_{28}H_{17}ClN_2O_2$ 1,6-Pyrenedione, 3-chloro-5,8-bis(phen-ylamino)-	C$_6$H$_5$Cl	608(4.14),645(4.15)	104-0738-84
1,6-Pyrenedione, 8-chloro-3,5-bis(phen-ylamino)-	C$_6$H$_5$Cl	404(4.17),510(3.79), 540(3.81),584(3.76)	104-0738-84
$C_{28}H_{17}NO_4$ 1H-Benzo[h]naphtho[1,2,3-de]quinoline-2,8-dione, 3-acetyl-9-phenoxy-	toluene	433(3.92),454(3.92)	104-0980-84
$C_{28}H_{18}Cl_2N_2$ 10H-Quindoline, 1,1-bis(4-chlorophen-yl)-3-methyl-	EtOH	208(4.63),230(4.67), 356(4.90),356(4.21) [sic]	103-1374-84
$C_{28}H_{18}N_2O_2$ 1,6-Pyrenedione, 3,5-bis(phenylamino)-	C$_6$H$_5$Cl	540(3.89),580(3.90)	104-0738-84
1,6-Pyrenedione, 3,10-bis(phenyl-amino)-	C$_6$H$_5$Cl	600(4.11),640(4.14)	104-0738-84
$C_{28}H_{18}N_2O_5$ 3-Furancarbonitrile, 4,5-bis(1,3-benzo-dioxol-5-yl)-2-[(3-phenyl-2-propen-ylidene)amino]-	MeOH	212(4.60),247(4.28), 300(4.46),323(4.49), 435(4.39)	73-1788-84
	CHCl$_3$	306(4.43),327(4.47), 444(4.37)	73-1788-84
$C_{28}H_{18}N_4$ 9-Acridinecarboxaldehyde, (9-acridin-ylmethylene)hydrazone	C$_6$H$_{12}$	426(4.20)	97-0021-84
Spiro[9H-fluorene-9,3'-[3H]pyrazolo-[5,1-c][1,2,4]triazole, 6',7'-di-phenyl-	CH$_2$Cl$_2$	236(4.66),260s(4.42), 278s(4.32),290s(4.19), 318(3.94),354s(3.80)	24-1726-84
1H-Tetrazolium, 5-(4H-cyclopenta[def]-phenanthren-4-yl)-1,3-diphenyl-, hydroxide, inner salt	MeCN	227(4.74),245(4.67), 266s(4.39),313(4.20), 335(4.20),386s(4.32), 406(4.57),568(3.18)	39-2545-84C
$C_{28}H_{18}N_4O$ Spiro[3H-pyrazolo[5,1-c]-1,2,4-triazole-	CH$_2$Cl$_2$	237(4.59),298(3.99),	24-1726-84

Compound	Solvent	$\lambda_{max}(\log \epsilon)$	Ref.
3,9'-[9H]xanthene], 6,7-diphenyl-		316(3.93),380s(3.58)	24-1726-84
$C_{28}H_{18}N_8O_4$ Benzo[g]pteridine-2,4(3H,10H)-dione, 10,10'-(1,4-phenylene)bis[3-methyl-	pH 7.27	347(4.04),437(4.18)	39-1227-84B
$C_{28}H_{18}O$ Ethanone, 1-[9-(9H-fluoren-9-ylidene)- 9H-fluoren-2-yl]-	DMF	462(4.37)	40-0145-84
$C_{28}H_{19}NO$ Methanone, (2,4-diphenyl-3-quinolinyl)- phenyl-	MeOH	252(4.58),290(4.10)	4-0759-84
$C_{28}H_{19}N_3O$ [1,2,4]Triazolo[1,5-a]pyridinium, 2,3- dihydro-1-(1-naphthalenyl)-2-oxo- 5,7-diphenyl-, hydroxide, inner salt	EtOH	267(4.91)	39-1891-84C
$C_{28}H_{20}$ [2.2](2,6)Biphenylenophane	C_6H_{12}	230(5.0),247(4.7), 270(4.2),360(3.8), 385(3.9)	88-3603-84
$C_{28}H_{20}ClNO_2S_3$ 3-Thiophenecarboxylic acid, 5-(5-chlo- ro-4-phenyl-3H-1,2-dithiol-3-ylidene)- 2,5-dihydro-4-phenyl-2-(phenylimino)-, ethyl ester, perchlorate	MeCN	561(4.59)	48-0917-84
$C_{28}H_{20}ClN_3O_2$ Pyrano[2,3-e]indazol-2(5H)-one, 3-chlo- ro-6,7-dihydro-4-(diphenylamino)-7- phenyl-	EtOH	223s(4.34),257s(4.25), 271(4.28),379(4.37)	4-0361-84
$C_{28}H_{20}Cl_2N_2$ 1H-Quindoline, 1,11-bis(4-chlorophen- yl)-2,10-dihydro-3-methyl-	EtOH	208(4.63),233(4.67), 241(4.67),276(4.51), 295(4.54),357(4.07), 371(4.11)	103-1374-84
$C_{28}H_{20}Cl_4I_4O_5$ Benzoic acid, 2,3,4,5-tetrachloro-6-(6- hydroxy-2,4,5,7-tetraiodo-3-oxo-3H- xanthen-9-yl)-, octyl ester, tri- butylamine salt	MeOH	525(4.52),564(5.01)	35-5879-84
$C_{28}H_{20}N_2$ 9-Anthracenecarboxaldehyde, (diphenyl- methylene)hydrazone	C_6H_{12}	412(4.15)	97-0021-84
Benzo[1,2,3-kl:6,5,4-k'l']diacridine, 7,16-dimethyl-	benzene	321(4.08),391(3.96), 443(3.93),520(4.30)	39-0441-84B
10H-Quindoline, 3-methyl-1,11-diphenyl-	EtOH	207(4.65),230(4.68), 283(4.89),354(4.24)	103-1374-84
$C_{28}H_{20}N_2O$ 1H-Pyrazole, 4-benzoyl-1,3,5-triphenyl-	EtOH	245(4.85)	4-1013-84
1H-Pyrazole, 5-benzoyl-1,3,4-triphenyl-	EtOH	260(4.55)	4-1013-84
$C_{28}H_{20}N_2O_3$ Benzamide, N-[9,10-dihydro-9,10-dioxo-	benzene	563(4.05),596(3.98)	39-0529-84C

Compound	Solvent	$\lambda_{max}(\log \epsilon)$	Ref.
4-[(phenylmethyl)amino]-1-anthra-cenyl]- (cont.)			39-0529-84C
$C_{28}H_{20}N_4$			
3H-Pyrazolo[5,1-c]-1,2,4-triazole, 3,3,6,7-tetraphenyl-	CH_2Cl_2	234(4.40),256s(4.24), 319(3.89),377s(3.61)	24-1726-84
$C_{28}H_{20}N_4O_2S$			
Benzenesulfonamide, N-(5-cyano-2,4-di-phenyl-3H-pyrido[1,2-b]pyridazin-3-ylidene)-4-methyl-	MeCN	286(4.07),376(4.22)	142-1709-84
Benzenesulfonamide, N-(7-cyano-2,4-di-phenyl-3H-pyrido[1,2-b]pyridazin-3-ylidene)-4-methyl-	MeCN	258(4.17),390(4.29)	142-1709-84
$C_{28}H_{20}N_6O_4$			
9cH-2a,5b:6,9b-Dietheno-3H,6H,7H-2b,4,5a,6a,8,9a-hexaazadicyclo-pent[a,cd]-s-indacene-1,5,7,9-(4H,8H)-tetrone, 9c-methyl-4,8-diphenyl-	EtOH	242s(3.96),290s(3.49)	39-0175-84C
$C_{28}H_{20}O_2S_2$			
Methanone, (2,4-diphenyl-1,3-dithie-tane-2,4-diyl)bis[phenyl-	CH_2Cl_2	245(4.40),350(3.00)	44-4752-84
$C_{28}H_{20}O_2Se$			
Methanone, phenyl(2,4,5-triphenyl-1,3-oxaselenol-2-yl)-	MeCN	230(4.33),300(3.75), 336(3.60)	44-2057-84
$C_{28}H_{21}ClN_2$			
1H-Quindoline, 1-(4-chlorophenyl)-2,10-dihydro-3-methyl-11-phenyl-	EtOH	210(4.48),244(4.28), 275(3.99),295(4.04), 357(3.62),361(3.63)	103-1374-84
1H-Quindoline, 11-(4-chlorophenyl)-2,10-dihydro-3-methyl-1-phenyl-	EtOH	200(4.99),239(4.97), 275(4.84),295(4.89), 357(4.47),371(4.51)	103-1374-84
3-Quinolinecarbonitrile, 2-[1,1'-bi-phenyl]-4-yl-4-(4-chlorophenyl)-5,6,7,8-tetrahydro-	EtOH	207(4.73),218s(4.60), 238s(4.29),287(4.49)	73-1395-84
$C_{28}H_{21}Cl_2N_3O_2$			
Pyrano[2,3-e]indazol-2(3H)-one, 3,3-dichloro-4-(diphenylamino)-7-phenyl-4,5,6,7-tetrahydro-	EtOH	225(4.38),242(4.35), 265s(4.25),280s(4.19)	4-0361-84
$C_{28}H_{21}FN_2$			
3-Quinolinecarbonitrile, 2-[1,1'-bi-phenyl]-4-yl-4-(4-fluorophenyl)-5,6,7,8-tetrahydro-	EtOH	207(4.73),288(4.44)	73-1395-84
$C_{28}H_{21}NO$			
Spiro[acridine-9(10H),9'(10'H)-anthra-cen]-10'-one, 2,7-dimethyl-	EtOH	259(4.36),273(4.34), 331(3.91)	24-2703-84
$C_{28}H_{21}NO_2S_3$			
3-Thiophenecarboxylic acid, 2,5-dihy-dro-4-phenyl-5-(5-phenyl-3H-1,2-di-thiol-3-ylidene)-2-(phenylimino)-, ethyl ester, perchlorate	MeCN	554(4.67)	48-0917-84

Compound	Solvent	$\lambda_{max}(\log \epsilon)$	Ref.
$C_{28}H_{21}N_3O_2$			
1,5-Naphthalenediol, 2,6-dihydro-4-(phenylamino)-2,6-bis(phenylimino)-	CHCl$_3$	274(4.66),332(4.00), 424(4.03),628(4.29), 683(4.30)	104-0733-84
1,5-Naphthalenedione, 2,4,8-tris(phenylamino)-	CHCl$_3$	285(4.50),360(3.78?), 514(3.78),628(4.23), 675(4.45)	104-0733-84
4H-Pyran-3,5-dicarbonitrile, 2-[1,1'-biphenyl]-4-yl-6-[(ethoxymethylene)amino]-4-phenyl-	EtOH	255(4.37),267s(4.38), 288(4.44)	73-2309-84
$C_{28}H_{21}N_3O_7$			
Violacein, tetraacetyl-	CHCl$_3$	266(4.32),302(4.00), 560(4.28)	39-1331-84B
$C_{28}H_{21}N_3O_{10}$			
4-Benzoxazolecarboxylic acid, 2-[2-[acetyl[(3-acetoxy-2-pyridinyl)-carbonyl]amino]-3-(acetoxyphenyl)-, anhydride with acetic acid	dioxan dioxan-base	<u>255s(4.3),320(4.4)</u> <u>250(4.6),360(4.0)</u>, <u>400(4.0)</u>	158-0441-84 158-0441-84
$C_{28}H_{21}N_5O$			
1H-Pyrazole, 4-[(2-methylphenyl)azo]-3,5-diphenyl-1-(2-pyridinylcarbonyl)-	n.s.g.	356(4.16)	48-1021-84
1H-Pyrazole, 4-[(3-methylphenyl)azo]-3,5-diphenyl-1-(2-pyridinylcarbonyl)-	n.s.g.	356(4.16)	48-1021-84
1H-Pyrazole, 4-[(4-methylphenyl)azo]-3,5-diphenyl-1-(2-pyridinylcarbonyl)-	n.s.g.	356(4.16)	48-1021-84
$C_{28}H_{21}N_5O_2$			
1H-Pyrazole, 4-[(2-methoxyphenyl)azo]-3,5-diphenyl-1-(2-pyridinylcarbonyl)-	n.s.g.	373(4.22)	48-1021-84
1H-Pyrazole, 4-[(3-methoxyphenyl)azo]-3,5-diphenyl-1-(2-pyridinylcarbonyl)-	n.s.g.	373(4.22)	48-1021-84
1H-Pyrazole, 4-[(4-methoxyphenyl)azo]-3,5-diphenyl-1-(2-pyridinylcarbonyl)-	n.s.g.	373(4.22)	48-1021-84
$C_{28}H_{21}N_5O_5$			
1H-Indole-4-carboxaldehyde, 7-methoxy-2,3-diphenyl-, 2,4-dinitrophenylhydrazone	THF	243(4.41),260(4.25), 293(3.99),425(4.50)	2-1021-84
$C_{28}H_{22}$			
Azulene, 1,1'-(1,3,5,7-octatetraene-1,8-diyl)bis-	CH$_2$Cl$_2$	225(4.50),271(4.55), 342(4.29),454s(4.85), 476(4.99),505(4.90), 615(3.19),675s(3.12), 770s(2.71)	5-1905-84
radical cation	CH$_2$Cl$_2$	898(--),1020s(--), 1260(--),1350(--)	5-1936-84
dication	CH$_2$Cl$_2$	240(4.86),268(4.86), 315(4.81),360s(4.80), 540(4.76),584s(4.96), 623(5.03),652s(4.98)	5-1936-84
Benz[a]indeno[1,2,3-fg]aceanthrylene, 8b,9,10,10a-tetrahydro-8b,10a-dimethyl-	EtOH	252(4.54),261(4.51), 363(4.36),382(4.36)	22-0195-84
Coronene, 12c,12d-dihydro-1,6,12c,12d-tetramethyl-, (12cS-trans)-	hexane	256(4.78),265(4.84), 274(4.95),294(4.87), 320(4.49),336(4.55), 353(4.53),372(4.57),	138-1977-84

Compound	Solvent	$\lambda_{max}(\log \epsilon)$	Ref.
(cont.)		445(3.89),471(4.00), 498(3.89)	138-1977-84
$C_{28}H_{22}Cl_2Te$ Tellurium, dichlorobis(2,2-diphenyl-ethenyl)-, (T-4)-	$CDCl_3$	302(4.26)	157-1308-84
$C_{28}H_{22}NOP$ Ethanone, 1-(1H-indol-3-yl)-2-(triphen-ylphosphoranylidene)-	EtOH	221(4.64),261(4.16), 312(4.18)	65-0685-84
$C_{28}H_{22}NO_2$ 9H-Fluoren-9-iminium, N-[bis(4-methoxy-phenyl)methylene]-, hexachloroanti-monate	CH_2Cl_2	256(4.77),262(4.83), 349(4.53),372(4.49)	24-3222-84
$C_{28}H_{22}N_2$ Benzo[b]quinolizinium, 9,9'-(1,2-eth-anediyl)bis-	n.s.g.	365(3.3),382(3.3), 405(3.2)	62-0081-84A
cyclomer	MeOH	265(4.04)	48-0757-84
1H-Quindoline, 2,10-dihydro-3-methyl-1,11-diphenyl-	EtOH	209(4.59),241(4.63), 274(4.49),293(4.66), 354(4.12),368(4.16)	103-1374-84
3-Quinolinecarbonitrile, 2-[1,1'-bi-phenyl]-4-yl-5,6,7,8-tetrahydro-4-phenyl-	EtOH	206(4.77),238s(4.41), 289(4.48)	73-1395-84
$C_{28}H_{22}N_2O$ Methanone, [4-(4,5-dihydro-3,5-diphen-yl-1H-pyrazol-1-yl)phenyl]phenyl-	MeOH	246(4.46),311(4.18), 382(4.44)	103-0787-84
Methanone, (4,5-dihydro-1,3,4-triphen-yl-1H-pyrazol-5-yl)phenyl-	EtOH	230(4.85),282(4.60)	4-1013-84
$C_{28}H_{22}N_2O_3$ 9,10-Anthracenedione, 1-amino-2-(4-methylphenoxy)-4-[(4-methylphenyl)-amino]-	C_6H_5Cl	566(4.16),602(4.18)	34-0482-84
$C_{28}H_{22}N_2O_4$ 9,10-Anthracenedione, 1-amino-2-(4-methoxyphenoxy)-4-[(4-methyl-phenyl)amino]-	C_6H_5Cl	534s(3.91),565(4.16), 602(4.18)	34-0482-84
9,10-Anthracenedione, 1-amino-4-[(4-methoxyphenyl)amino]-2-(4-methyl-phenoxy)-	C_6H_5Cl	536s(3.98),602(4.18)	34-0482-84
1H-Furo[3,4-c]pyrrole-1-carbonitrile, 3,3a,4,5,6,6a-hexahydro-4,6-dioxo-1,5-diphenyl-3-[2-(2-propenyloxy)-phenyl]-	MeCN	220s(4.32),274(3.43), 280s(3.40)	24-2157-84
$C_{28}H_{22}N_2O_5$ 9,10-Anthracenedione, 1-amino-2-(4-methoxyphenoxy)-4-[(4-methoxyphen-yl)amino]-	C_6H_5Cl	536s(4.00),568(4.18), 600(4.18)	34-0482-84
$C_{28}H_{22}N_4O$ 5H-Pyrano[2,3-d]pyrimidine-6-carbo-nitrile, 7-[1,1'-biphenyl]-4-yl-4-(ethylamino)-5-phenyl-	EtOH	247(4.28),265(4.32), 299(4.34)	73-2309-84

Compound	Solvent	$\lambda_{max}(\log \epsilon)$	Ref.
$C_{28}H_{22}N_4O_3$ 1H-Indole-4-carboxaldehyde, 7-methoxy-2,3-diphenyl-, (4-nitrophenyl)hydrazone	THF	242(4.40),267(4.26), 292(4.00),432(4.52)	2-1021-84
$C_{28}H_{22}N_8O_3S$ 1H-Pyrazole, 4-[[2-[[(aminoiminomethyl)amino]sulfonyl]phenyl]azo]-3,5-diphenyl-1-(2-pyridinylcarbonyl)-	n.s.g.	358(4.06)	48-1021-84
$C_{28}H_{22}O$ 1,7a,2,3a-[2]Butene[1,4]diylidene-3,8-epoxycyclobuta[b]naphthalene, 1,2,2a-3,8,8a-hexahydro-3,8-diphenyl-	EtOH	275(3.51)	88-2573-84
5,10-Epoxybenzo[b]biphenylene, 4a,4b,5,10,10a,10b-hexahydro-5,10-diphenyl-	EtOH	266(3.66),273(3.43)	88-2573-84
isomer	EtOH	266(3.70),273(3.72)	88-2573-84
Naphtho[3,2,1-kl]xanthene, 5a,6-dihydro-5a,8-dimethyl-9-phenyl-	ether	252(4.35),277(4.41), 331(4.21),344s(4.17)	22-0195-84
Phenol, 2-(1,4-dimethyl-10-phenyl-9-anthracenyl)-	ether	267(4.88),345s(3.53), 364(3.85),383(4.05), 404(4.02)	22-0195-84
$C_{28}H_{22}O_4$ Anthra[1,2-b]furan-6,11-dione, 2-methyl-4-(2-methyl-2-propenyl)-5-(phenylmethoxy)-	EtOH	246(4.45),288(4.51), 381(3.96)	12-1511-84
$C_{28}H_{22}O_5S$ Benzenemethanol, 4-methoxy-α-9H-xanthen-9-ylidene-, 4-methylbenzenesulfonate	EtOH	223(4.72),285(4.00), 335(4.11)	44-0080-84
$C_{28}H_{23}BO_6$ 1,5-Methano-1H-anthra[2,3-e][1,3,2]dioxaborocin-7,14-dione, 5-(1-acetoxyethyl)-5,6-dihydro-3-phenyl-, [1S-[1α,5β(R*)]]-	CHCl₃	240(4.56),277(4.26), 287(4.28),299(4.30), 416(3.75)	78-4649-84
[1S-[1α,5β(S*)]]-	CHCl₃	239(4.51),276(4.14), 286(4.15),300(4.18), 417(3.62)	78-4649-84
1,5-Methano-1H-anthra[2,3-e][1,3,2]dioxaborocin-7,14-dione, 5-(2-acetoxyethyl)-5,6-dihydro-3-phenyl-, (1S)-	MeOH	242(4.58),276(4.27), 287(4.29),300(4.31), 416(3.74)	78-4649-84
$C_{28}H_{23}FN_2$ 3-Quinolinecarbonitrile, 2-[1,1'-biphenyl]-4-yl-4-(4-fluorophenyl)-1,4,5,6,7,8-hexahydro-	EtOH	207(4.56),245s(4.07), 277(4.38),340(3.45)	73-1395-84
$C_{28}H_{23}N$ Spiro[acridine-9(10H),9'(10'H)-anthracene], 2,7-dimethyl-	EtOH	293(4.06),325(3.84)	24-2703-84
$C_{28}H_{23}NOP$ Phosphonium, [2-(1H-indol-3-yl)-2-oxoethyl]triphenyl-, chloride	EtOH	218(4.59),262(4.10), 312(4.18)	65-0685-84
$C_{28}H_{23}NS$ Benzenamine, N-[(6,7-dihydro-2,3-di-	CHCl₃	588(4.20)	97-0183-84

Compound	Solvent	$\lambda_{max}(\log \epsilon)$	Ref.
phenyl-5H-1-benzothiopyran-8-yl)meth-ylene]-, hydrochloride (cont.)			97-0183-84
$C_{28}H_{23}N_2OP$ Ethanone, 1-(1H-indol-3-yl)-2-[(tri-phenylphosphoranylidene)amino]-	EtOH	260(3.94),297(4.02)	65-0685-84
$C_{28}H_{23}N_3O_2$ 3-Quinolinecarbonitrile, 2-[1,1'-bi-phenyl]-4-yl-1,4,5,6,7,8-hexahydro-4-(4-nitrophenyl)-	EtOH	206(4.66),245(4.21), 282(4.49)	73-1395-84
$C_{28}H_{23}N_3O_2S$ Benzenesulfonamide, 4-methyl-N-(5-meth-yl-2,4-diphenyl-3H-pyrido[1,2-b]pyri-dazin-3-ylidene)-	EtOH	269(4.37),371(4.36)	142-1709-84
Benzenesulfonamide, 4-methyl-N-(7-meth-yl-2,4-diphenyl-3H-pyrido[1,2-b]pyri-dazin-3-ylidene)-	EtOH	273(4.36),370(4.33)	142-1709-84
$C_{28}H_{24}$ [2](2,7)Naphthalenophane, 4,5,14,15-tetramethyl-, anti	$CHCl_3$	246(4.96),280s(4.28), 375(3.35)	44-4128-84
$C_{28}H_{24}Cl_2N_2O_5$ 1H-Pyrrolo[3,2-c]pyridine, 4,6-dichlo-ro-1-[2-deoxy-3,5-bis-O-(4-methyl-benzoyl)-β-D-erythro-pentofuranosyl]-	EtOH	227(4.90),240(4.73), 274(4.15)	35-6379-84
$C_{28}H_{24}Fe_2N_2O_4$ Iron, [[1-acetyl-4,5-dihydro-3-[(1,2-3,4-η)-4-phenyl-1,3-butadienyl]-1H-pyrazol-3-yl]ferrocene]tricarbonyl-	n.s.g.	228(5.10),264(4.92), 299(4.87),440(2.80)	101-0201-84C
$C_{28}H_{24}Fe_2O_8$ Ferrocene, 1,1":1',1"'-bis[1,2-ethane-diylbis(oxycarbonyl)]bis-	dioxan	448(2.71)	18-2435-84
$C_{28}H_{24}N_2$ 3-Quinolinecarbonitrile, 2-[1,1'-biphen-yl]-4-yl-1,4,5,6,7,8-hexahydro-4-phenyl-	EtOH	206(4.75),276(4.44), 357(3.80)	73-1395-84
$C_{28}H_{24}N_3O$ 1H-Benzimidazolium, 1,3,5-trimethyl-2-[4-(5-methyl-2-benzoxazolyl)-1-naphthalenyl]-, iodide	DMF	343(4.37)	5-1129-84
$C_{28}H_{24}N_4$ 1H-Benzimidazole, 2,2'-(1,7-naphtha-lenediyl)bis[5,6-dimethyl-	DMF	316(4.55),345(4.38)	5-1129-84
1H-Benzimidazole, 2,2'-(2,6-naphtha-lenediyl)bis[5,6-dimethyl-	DMF	366(4.74),388(4.69)	5-1129-84
1H-Benzimidazole, 2,2'-(1,7-naphtha-lenediyl)bis[1-ethyl-	DMF	286(4.46)	5-1129-84
1H-Benzimidazole, 2,2'-(2,6-naphtha-lenediyl)bis[1-ethyl-	DMF	326(4.56)	5-1129-84
9H-Carbazole, 3,3'-azobis[9-ethyl-, (E)- (Z)-	dioxan dioxan	412(4.52) 462(3.68)	97-0286-84 97-0286-84

Compound	Solvent	$\lambda_{max}(\log \epsilon)$	Ref.
$C_{28}H_{24}N_4O_2$			
1H-Benzimidazole, 2,2'-(1,4-naphtha-lenediyl)bis[5-methoxy-1-methyl-	DMF	328(4.33)	5-1129-84
1H-Benzimidazole, 2,2'-(1,7-naphtha-lenediyl)bis[5-methoxy-1-methyl-	DMF	290(4.39)	5-1129-84
1H-Benzimidazole, 2,2'-(2,6-naphtha-lenediyl)bis[5-methoxy-1-methyl-	DMF	345(4.60)	5-1129-84
$C_{28}H_{24}N_4O_4S_2$			
Thiazolo[3,2-a]pyrimidin-4-ium, 3-hy-droxy-7-(4-methoxyphenyl)-2-[[[4-(4-methoxyphenyl)-6-methyl-2-pyrimidin-yl]thio]acetyl]-5-methyl-, hydroxide, inner salt	MeCN CCl₄	262(4.31),285(4.38), 346(4.53),480(4.07) 526(--)	103-0791-84 103-0791-84
$C_{28}H_{24}N_5$			
Quinolinium, 2-(2,6-dihydro-4-methyl-2,6-diphenylpyrazolo[3,4-c]pyrazol-3-yl)-1-ethyl-, iodide	EtOH	330(3.93),340(3.94)	48-0811-84
$C_{28}H_{24}N_8O_3S$			
1H-Pyrazole, 4-[[2-[[(4,6-dimethyl-2-pyrimidinyl)amino]sulfonyl]phenyl]-azo]-3-methyl-5-phenyl-1-(2-pyridin-ylcarbonyl)-	n.s.g.	372(3.98)	48-1021-84
$C_{28}H_{24}O$			
1,4-Epoxyanthracene, 1,2,3,4-tetrahy-dro-1,4-dimethyl-9,10-diphenyl-	ether	233(4.71),274(3.91), 283(4.00),293(3.92)	22-0195-84
$C_{28}H_{24}O_2$			
2-Butanone, 4-(6-hydroxy-7-methyl-5-phenylbenzo[c]fluoren-7-yl)-	ether	235(4.26),250s(4.38), 297s(3.72)3,14s(4.04), 324(4.23),349(4.56)	22-0195-84
2,5-Epoxynaphtho[2,3-b]oxepin, 2,3,4,5-tetrahydro-2,5-dimethyl-6,11-diphenyl-	ether	240(4.76),276(4.84), 287(3.92),299(3.98), 331(3.57),344(3.69)	22-0195-84
Naphtho[3,2,1-kl]xanthen-8-ol, 5a,6,7,8-tetrahydro-5a,8-dimethyl-9-phenyl-	ether	243(4.62),280(3.55), 294s(3.68),306s(3.89), 332(4.19),347(4.14)	22-0195-84
2-Propanone, 1-(2,3-dihydro-3-methyl-4,9-diphenylnaphtho[2,3-b]furan-2-yl)-, trans	ether	240(4.64),278s(3.83), 297(3.99),327s(3.69), 341(3.80)	22-0195-84
$C_{28}H_{24}O_3$			
Ethanone, 1-[2,4-dihydroxy-3,5-bis-(phenylmethyl)phenyl]-2-phenyl-	MeOH	202(4.55),280(4.03)	2-1030-84
$C_{28}H_{24}O_5$			
4H-1-Benzopyran-4-one, 3-acetyl-7-ace-toxy-2-methyl-6,8-bis(phenylmethyl)-	MeOH	206(4.72),234(4.56)	2-1030-84
$C_{28}H_{24}O_8$			
1,2-Azulenedicarboxylic acid, 3-[3-[2,3-bis(methoxycarbonyl)bicyclo-[2.2.1]hepta-2,5-dien-7-ylidene]-1-propenyl]-, dimethyl ester, (E,Z)-	MeCN	233(4.35),280(4.38), 290s(4.42),310(4.50), 365s(3.88),400s(3.76), 590(2.79)	24-2027-84
$C_{28}H_{25}NO_3S$			
Benzenesulfonamide, N-(2-[1,1'-biphen-yl]-4-yl-2-oxoethyl)-4-methyl-	ether	283.5(4.44)	48-0177-84

Compound	Solvent	λ_{max}(log ϵ)	Ref.
N-(phenylmethyl)- (cont.)			48-0177-84
C$_{28}$H$_{26}$			
Azulene, 4,4'-(1,3-butadiene-1,4-diyl)-bis[6,8-dimethyl-, (E,E)-	CH$_2$Cl$_2$	245(4.40),283(4.66), 336(4.50),378(4.48), 395s(4.45),559(3.25)	5-1905-84
C$_{28}$H$_{26}$F$_3$NO$_9$			
5,12-Naphthacenedione, 8-acetyl-7,8,9,10-tetrahydro-6,8-dihydroxy-10-[[2,3,6-trideoxy-3-[(trifluoro-acetyl)amino]-α-L-lyxo-hexopyrano-syl]oxy]-, cis	MeOH	204(4.46),246(4.43), 262(4.47),328(3.48), 408(3.82)	78-4677-84
C$_{28}$H$_{26}$F$_3$NO$_{10}$			
5,12-Naphthacenedione, 8-acetyl-7,8,9,10-tetrahydro-1,6,8-trihydroxy-10-[[2,3,6-trideoxy-3-[(trifluoro-acetyl)amino]-α-L-lyxo-hexopyrano-syl]oxy]-, cis	MeOH	208(4.73),230(4.54), 258(4.44),290(3.93), 432(4.07)	78-4677-84
C$_{28}$H$_{26}$NO$_2$			
1H-Indolizinium, 2,3,8,8a-tetrahydro-1-(methoxycarbonyl)-5,7,8a-tri-phenyl-, tetrafluoroborate	MeOH	238(4.07),320s(--), 365(4.12)	39-0941-84C
C$_{28}$H$_{26}$N$_2$O$_2$			
3-Pyridinecarboxamide, 1,4-dihydro-4-(3-oxo-1,3-diphenylpropyl)-1-(phenylmethyl)-	MeCN	344(3.68)	44-0026-84
C$_{28}$H$_{26}$N$_2$O$_2$S			
3-Thiophenecarboxylic acid, 5-[[4-(di-methylamino)phenyl]methylene]-2,5-dihydro-4-phenyl-2-(phenylimino)-, ethyl ester	EtOH	235(4.41),262s(4.20), 346(4.25)	48-0917-84
C$_{28}$H$_{26}$N$_2$O$_3$S			
4,9-Imino-1H-benz[f]isoindole-1,3(2H)-dione, 4-(ethylthio)-3a,4,9,9a-tetra-hydro-2-(4-methoxyphenyl)-10-(4-meth-ylphenyl)-, (3aα,4β,9α,9aβ)-	EtOH	206(4.58),234(4.35), 289(3.45),300(3.32), 336(3.32)	103-0978-84
C$_{28}$H$_{26}$N$_2$O$_4$			
Spiro[2H-1-benzopyran-2,2'-[2H]indole], 1',3'-dihydro-8-methoxy-1',3',3'-tri-methyl-6-nitro-5'-(2-phenylethenyl)-	n.s.g.	252(4.37),281(3.88), 333(4.39),357(3.90), 435(3.18)	103-1222-84
C$_{28}$H$_{26}$N$_4$O$_7$			
9H-Purine, 2,6-diphenyl-9-(2,3,5-tri-O-acetyl-β-D-ribofuranosyl)-	EtOH	266(4.43),310(4.00)	44-4340-84
C$_{28}$H$_{26}$O$_4$			
[1,1':3',1":3",1"'-Quaterphenyl]-4,4",4"',6'-tetrol, 3,3"',5',5"-tetramethyl-	EtOH	250(4.59)	42-0142-84
C$_{28}$H$_{27}$NO$_2$			
Morpholine, 4-(3-methyl-2,4,6-triphen-yl-2H-pyran-2-yl)-	dioxan	243s(4.27),247(4.28), 342(4.07)	48-0657-84

Compound	Solvent	$\lambda_{max}(\log \epsilon)$	Ref.
$C_{28}H_{27}NO_4S$			
Methanone, [6-hydroxy-2-(4-hydroxyphen-yl)benzo[b]thien-3-yl][4-[2-(1-piper-idinyl)ethoxy]phenyl]-	EtOH	290(4.53)	87-1057-84
hydrochloride	EtOH	286(4.52)	87-1057-84
$C_{28}H_{27}NS$			
Piperidine, 1-(2,4,6-triphenyl-2H-thio-pyran-2-yl)-	DMSO	357(3.7)	35-7082-84
$C_{28}H_{27}N_3O_4$			
1,3,5-Triazabicyclo[3,2,0]hept-6-ene-2,4-dione, 6,7-bis(4-methoxy-2,6-dimethylphenyl)-3-phenyl-	MeCN	267(4.16)	44-2917-84
$C_{28}H_{27}N_3O_7$			
1H-Imidazole-4-carboxylic acid, 5-(cya-nomethyl)-1-[2-deoxy-3,5-bis-O-(4-methylbenzoyl)-α-D-erythro-pento-furanosyl]-, methyl ester	pH 1	243(4.53)	87-1389-84
	pH 7	243(4.56)	87-1389-84
	pH 11	243(4.33)	87-1389-84
β-	pH 1	244(4.49)	87-1389-84
	pH 7 and 11	243(4.34)	87-1389-84
1H-Imidazole-5-carboxylic acid, 4-(cya-nomethyl)-1-[2-deoxy-3,5-bis-O-(4-methylbenzoyl)-α-D-erythro-pento-furanosyl]-, methyl ester	pH 1	247(4.31)	87-1389-84
	pH 11	254(4.25)	87-1389-84
	MeOH	247(4.29)	87-1389-84
β-	pH 1	247(4.35)	87-1389-84
	pH 11	255(4.27)	87-1389-84
	MeOH	247(4.29)	87-1389-84
$C_{28}H_{28}$			
[2_2](2,7)Naphthalenophane, 4,5,14,15-tetramethyl-, anti	C_6H_{12}	225(5.10),307(3.77)	44-4128-84
$C_{28}H_{28}NO_{10}S_3$			
Pyridinium, 1-butyl-4-(4-methoxy-3-sul-fophenyl)-2,6-bis(4-sulfophenyl)-, perchlorate	H_2O	222(4.55),337(4.40)	39-0849-84B
$C_{28}H_{28}N_4O_8$			
L-Alanine, N-[N-[N-[(phenylmethoxy)-carbonyl]-L-phenylalanyl]glycyl]-, 4-nitrophenyl ester	$CHCl_3$	270(3.90)	65-0836-84
Glycine, N-[N-[N-[(phenylmethoxy)carbo-nyl]-L-phenylalanyl]-L-alanyl]-, 4-nitrophenyl ester	MeOH	270(3.99)	65-0836-84
$C_{28}H_{28}O_{13}$			
β-D-Glucopyranoside, 3-hydroxy-5-[2-(3-hydroxy-4-methoxyphenyl)ethenyl]phen-yl, 2-(3,4,5-trihydroxybenzoate), cis	MeOH	217(4.46),280(4.04)	94-3501-84
$C_{28}H_{28}S_2$			
2,13-Dithia[3.3](2,7)naphthalenophane, 5,6,16,17-tetramethyl-, anti	$CHCl_3$	298s(4.39)	44-4128-84
$C_{28}H_{29}NO_{11}S_3$			
Benzenesulfonic acid, 2-methoxy-5-[2-[(1-methylpropyl)amino]-2,6-bis(4-sulfophenyl)-2H-pyran-4-yl]-	H_2O	221(4.22),259(3.95)	39-0849-84B

Compound	Solvent	$\lambda_{max}(\log \epsilon)$	Ref.
$C_{28}H_{29}N_2O$ Pyrylium, 4,6-bis[4-(dimethylamino)-phenyl]-3-methyl-2-phenyl-, per-chlorate	CHCl$_3$	330(4.12),362(4.05), 478(4.25),568(4.57)	97-0287-84
$C_{28}H_{30}N_4O_2$ 1,4-Butanedione, 1,4-diphenyl-2,3-bis(3,4,5-trimethyl-1H-pyrazol-1-yl)-	MeOH	245(4.60)	44-4647-84
$C_{28}H_{30}O$ Phenol, 4-(9-anthracenyl)-2,6-bis(1,1-dimethylethyl)-	CH$_2$Cl$_2$	255(4.93),284s(3.64), 320s(3.26),355s(3.58), 350[sic](3.08), 368(4.08),388(4.04)	44-0965-84
$C_{28}H_{30}O_2S$ 29-Thiatetracyclo[20.2.2.2^{8,11}.1^{3,6}]-nonacosa-3,5,8,10,22,24,25,27-octa-ene-2,7-dione	C$_6$H$_{12}$	297(4.36)	44-1177-84
$C_{28}H_{30}O_7$ 3(2H)-Benzofuranone, 2-[(5,7-dimethoxy-2,2-dimethyl-2H-1-benzopyran-6-yl)-carbonyl]-6-methoxy-2-(3-methyl-2-butenyl)-, (+)- (sanggenon A trimethyl deriv.)	EtOH	236(4.38),276(4.42), 324(3.78),391s(3.36)	142-1791-84
2H,6H-Benzofuro[3,2-b]pyrano[3,2-g][1]-benzopyran-6-one, 6a,11b-dihydro-5,9,11b-trimethoxy-2,2-dimethyl-6a-(3-methyl-2-butenyl)-	EtOH	232(4.27),255s(4.26), 271(4.36),290s(4.20), 300s(4.04),338(3.48)	142-1791-84
$C_{28}H_{30}O_8$ Propanoic acid, 2,2-dimethyl-, 3-(acet-oxymethyl)-9,10-dihydro-2-[2-(2-meth-yl-1,3-dioxolan-2-yl)ethyl]-2,10-di-oxo-1-anthracenyl ester	MeOH	210(4.40),257(4.69), 277(4.17),333(3.79)	78-3677-84
$C_{28}H_{30}O_{12}$ 1H,3H-Furo[3,4-c]furan-1,4-diol, 3,6-bis(4-acetoxy-3-methoxyphenyl)tetra-hydro-, diacetate, (1α,3α,3aα,4aα-6α,6aα)-	EtOH	220(4.59),270(4.40)	39-1159-84C
$C_{28}H_{30}O_{15}$ 4H-1-Benzopyran-4-one, 3-[(3-O-acetyl-α-L-arabinofuranosyl)oxy]-7-[(6-deoxy-α-L-mannopyranosyl)oxy]-5-hydroxy-2-(4-hydroxyphenyl)-	MeOH MeOH-NaOMe	267(4.47),349(4.38) 277(--),390(--)	95-0142-84
$C_{28}H_{31}N_3O$ Urea, N'-[1,3-diphenyl-3-(phenylimino)-1-propenyl]-N,N-dipropyl-	EtOH	206(4.53),250(4.19), 340s(4.13),373(4.22)	39-0239-84C
$C_{28}H_{32}NO_6$ Isoquinolinium, 1,3-bis(3,4-dimethoxy-phenyl)-3,4-dihydro-6,7-dimethoxy-2-methyl-, iodide	EtOH	258(4.27),316(4.03), 372(4.01)	4-0525-84
$C_{28}H_{32}N_3O_2P$ Piperidine, 1,1'-[[(4-nitrophenyl)phos-phinidene]di-4,1-phenylene]bis-	MeOH	288(4.65),440s(--)	139-0267-84C

Compound	Solvent	$\lambda_{max}(\log \epsilon)$	Ref.
$C_{28}H_{32}N_3O_2PS$ Piperidine, 1,1'-[[(4-nitrophenyl)phosphinothioylidene]di-4,1-phenylene]-bis-	MeOH	290.5(4.71),420s(--)	139-0267-84C
$C_{28}H_{32}N_3O_3P$ Piperidine, 1,1'-[[(4-nitrophenyl)phosphinylidene]di-4,1-phenylene]bis-	MeOH	288.5(4.70),410s(--)	139-0267-84C
$C_{28}H_{32}O$ Benzenemethanol, α-[bis(2,4,6-trimethylphenyl)methylene]-2,6-dimethyl-	MeOH	240s(4.36),274(4.12)	35-0477-84
$C_{28}H_{32}O_2S$ 3-Buten-1-one, 3-[[(1,1-dimethylethyl)-dimethylsilyl]oxy]-1,4,4-triphenyl-	MeOH	245(4.37)	35-4566-84
$C_{28}H_{32}O_6$ 2-Propen-1-one, 1-(2-hydroxy-4,6-dimethoxyphenyl)-3-[5-methoxy-2-methyl-2-(4-methyl-3-pentenyl)-2H-1-benzopyran-6-yl]-	EtOH	233s(3.99),300(3.87), 377(4.21)	142-1791-84
	EtOH-AlCl₃	223s(4.15),246s(3.89), 315s(3.70),337s(3.71), 419(4.25)	142-1791-84
$C_{28}H_{32}O_{12}$ 1H,3H-Furo[3,4-c]furan-3a(4H)-ol, 1,4-bis(4-acetoxy-3,5-dimethoxyphenyl)-dihydro-, acetate, [1S-(1α,3aα,4α-6aα)]- (1-hydroxysyringaresinol triacetate)	EtOH	223.5s(4.23),274.4(3.40), 277.8(3.39)	94-4482-84
Shinjulactone E, 1,6,12,20-tetra-O-acetyl-	EtOH	243(4.08)	18-2484-84
$C_{28}H_{32}O_{16}$ Flavone, 3,4',5,7-tetrahydroxy-8-methoxy-, 3-glucoside 7-rhamnoside	MeOH	225(4.22),245s(4.14), 272(4.35),330(4.16), 357(4.16)	102-1199-84
	MeOH-NaOMe	258(--),272(--), 404(--)	102-1199-84
	MeOH-AlCl₃	239(--),281(--), 309(--),353(--), 412(--)	102-1199-84
$C_{28}H_{33}NO_5S$ Carbazole, 1,2,3,4,4a,9b-hexahydro-4a,9b-(phenylenedioxy)-, D-camphor-10-sulfonate	MeOH	281(3.87)	4-1841-84
$C_{28}H_{33}NO_8$ Pyrrolo[2,1-a]isoquinoline-1,1,3(5H)-tricarboxylic acid, 2,3,6,10b-tetrahydro-8,9-dimethoxy-2-phenyl-, 1,1-diethyl 3-methyl ester	EtOH	212(4.08),285(3.00)	78-0369-84
$C_{28}H_{33}N_2OP$ Piperidine, 1,1'-[(phenylphosphinylidene)di-4,1-phenylene]bis-	MeOH	291(4.69)	139-0253-84C
$C_{28}H_{33}N_2P$ Piperidine, 1,1'-[(phenylphosphinidene)di-4,1-phenylene]bis-	MeOH	287(4.68)	139-0253-84C

Compound	Solvent	$\lambda_{max}(\log \epsilon)$	Ref.
$C_{28}H_{33}N_2PS$			
Piperidine, 1,1'-[(phenylphosphinothi-oylidene)di-4,1-phenylene]bis-	MeOH	292(4.72)	139-0253-84C
$C_{28}H_{34}N_4$			
4,4'-Bipyridinium, 1,1"-(1,4-butanedi-yl)bis[1'-ethyl-, tetraperchlorate	H_2O	262(4.66)	64-0074-84B
$C_{28}H_{34}O_4$			
2,4-Cyclohexadien-1-one, 4-benzoyl-3,5-dihydroxy-2,6,6-tris(3-methyl-2-buten-yl)-	MeOH	245s(3.99),284(3.80),358(4.15)	39-1413-84C
	MeOH-base	365(4.28)	39-1413-84C
Methanone, [3-methyl-2-(3-methyl-2-but-enyl)-6-phenyl-3-cyclohexen-1-yl]-(2,4,6-trimethoxyphenyl)-, (1α,2α,6α)-(±)-	EtOH	228(4.00),273(3.77)	12-0449-84
2-Propenoic acid, 3-phenyl-, 1,10-dec-anediyl ester	MeCN	275(4.69)	40-0022-84
$C_{28}H_{34}O_5$			
Wightianone (as palmitic acid clathrate)	MeOH	232(4.14),254s(4.13),258(4.14),298(3.92),324s(3.50)	39-1755-84C
	MeOH-NaOH	231(4.39),265(4.14),270(4.16),342(3.99)	39-1755-84C
$C_{28}H_{34}O_6$			
Withaphysalin A	EtOH	224(4.26)	100-0527-84
Withaphysalin D	EtOH	228(3.89)	100-0527-84
$C_{28}H_{34}O_{13}$			
Pinoresinol, 1-acetoxy-, 4'-β-D-gluco-side, (+)-	EtOH	231(4.31),279.5(3.83)	102-2839-84
	EtOH-NaOH	254(--),280(--),292(--)	102-2839-84
$C_{28}H_{34}O_{17}$			
Physcion 8-β-gentiobioside	MeOH	221(4.62),245s(4.15),268(4.48),280s(4.42),296s(3.69),395s(3.89),416(3.98),435s(3.92)	94-3436-84
$C_{28}H_{34}S$			
29-Thiatetracyclo[20.2.2.$2^{8,11}.1^{3,6}$]-nonacosa-3,5,8,10,22,24,25,27-octa-ene	C_6H_{12}	<u>250(4.0)</u>,274(2.98)	44-1177-84
$C_{28}H_{35}ClO_7$			
Ergosta-2,14,24-trien-26-oic acid, 6-chloro-5,17,20,22-tetrahydroxy-1,4-dioxo-, δ-lactone, (5β,6α,17α-22R)-	MeOH	231(4.06)	100-0648-84
$C_{28}H_{35}N_3O_3$			
Tubulosine, 10-demethyl-	MeOH	278(4.10)	100-0397-84
$C_{28}H_{36}$			
1,4-Cyclohexadiene, 3,3'-(2,4,6-octa-triene-18-diylidene)bis[1,5,6,6-tetramethyl-	CH_2Cl_2	255(4.19),315(3.90),372s(--),395(4.59),417(4.94),442.5(5.18),472.5(5.25)	5-0340-84

Compound	Solvent	$\lambda_{max}(\log \epsilon)$	Ref.
$C_{28}H_{36}N_2O_4$			
4-Isoquinolinecarboxylic acid, 1,2,4a-7,8,8a-hexahydro-2-[2-(1H-indol-3-yl)ethyl]-7-(2-methyl-1,3-dioxolan-2-yl)-, 1,1-dimethylethyl ester, [4aR-(4aα,7α,8aα)]-	EtOH	290(4.25)	44-2708-84
Psychotrine	pH 1	240(4.14),288(3.76), 306(3.80),356(3.83)	100-0397-84
$C_{28}H_{36}N_2O_5$			
Alangicine	pH 13	238(4.17),292(3.82), 328(4.07)	100-0397-84
	EtOH	275(3.84),312(3.42), 408(4.09)	100-0397-84
$C_{28}H_{36}N_4O_6$			
Safracin A, dihydrochloride	MeOH	271(3.93)	158-84-8
$C_{28}H_{36}N_4O_7$			
Safracin B, dihydrochloride	MeOH	270(3.89)	158-84-8
$C_{28}H_{36}O_2$			
Retinoic acid, 1-phenylethyl ester	MeOH	357(4.64)	107-0725-84
$C_{28}H_{36}O_6$			
Withanicandrin	n.s.g.	222(4.22)	102-1717-84
$C_{28}H_{36}O_7$			
2H-1-Benzopyran-2-one, 7-[[5-(3-acet-oxy-2,2-dimethyl-6-methylenecyclo-hexyl]-3-methyl-2-pentenyl]oxy]-6,8-dimethoxy- (acetyldeparnol)	EtOH	206s(4.70),226s(4.37), 297(4.05),340(3.90)	49-0477-84
1-Butanone, 2-methyl-1-[1,3,7,9-tetra-methoxy-2,8-dimethyl-6-(3-methyl-1-oxobutyl)-4-dibenzofuranyl]-	EtOH	217(4.59),251(4.42), 262s(4.35),296s(4.13)	39-2573-84C
$C_{28}H_{36}O_8$			
Withaperuvin E	MeOH	225.5(4.03)	102-0853-84
$C_{28}H_{36}O_{11}$			
Phorbol 12,13,16,20-tetraacetate	MeOH	194(4.16),254(3.48)	102-0129-84
$C_{28}H_{36}O_{15}$			
Torasachrysone 8-β-gentiobioside	MeOH	222(4.47),268(4.76), 300s(3.77),312(3.90), 325(3.85),385(4.04)	94-3436-84
$C_{28}H_{37}ClO_7$			
Physalolactone C	MeOH	238(3.81)	100-0648-84
$C_{28}H_{37}N_3O_7S$			
6,9,12-Trioxa-3,15,19-triazabicyclo-[5.3.1]heneicosa-17,20-diene-2,5-1,319-tetrone, 19-methyl-4,14-bis(1-methylethyl)-21-(phenylthio)-, [4S-(4R*,14R*)]-	CHCl₃	247(4.17),350s(--)	35-6029-84
$C_{28}H_{37}N_3O_{14}$			
Furan-2-methanamine, 4-[N-(4,5,7-tri-carboxyheptanoyl-γ-L-glutamyl-γ-L-glutamyl)-p-(2-aminoethyl)phenoxy-	pH 2 and 7	218(4.04),275(3.11), 282s(3.04)	35-3636-84
	pH 12	237(4.04),291(3.38)	35-3636-84

Compound	Solvent	$\lambda_{max}(\log \epsilon)$	Ref.
methyl]- (cont.)			35-3636-84
$C_{28}H_{37}N_5O_5$			
Piperazine, 1-benzoyl-4-[4-[3-[(1,1-di- methylethyl)amino]-2-hydroxypropoxy]- 6,7-dimethoxy-2-quinazolinyl]-, di- hydrochloride	EtOH	216(4.50),252(4.70)	4-1189-84
$C_{28}H_{38}O_4$			
Phenol, 4-methoxy-2-methyl-6-[3-methyl- 4-(5,6,7,7a-tetrahydro-4a,5',5',7a- tetramethylspiro[cyclopenta[c]pyran- 1(4aH),2'(5'H)-furan]-3-yl)-2-buten- yl]- (cystoketal)	EtOH	220(4.02),287(3.50)	100-0947-84
Spiro[cyclopenta[c]pyran-1(4aH),2'(5'H)- furan], 3-[(3,4-dihydro-6-methoxy-2,8- dimethyl-2H-1-benzopyran-2-yl)meth- yl]-5,6,7,7a-tetrahydro-4a,5',5',7a- tetramethyl-	EtOH	220(3.98),295(3.51)	100-0947-84
$C_{28}H_{38}O_5$			
Ergosta-2,24-dien-26-oic acid, 6,7- epoxy-5,22-dihydroxy-1-oxo-, δ- lactone, (5α,6α,7α,22R)-	n.s.g.	224(4.19)	102-1717-84
$C_{28}H_{38}O_6$			
Ergosta-2,24-dien-26-oic acid, 6,7- epoxy-5,12,22-trihydroxy-1-oxo-, δ-lactone, (5α,6α,7α,12α,22R)-	n.s.g.	222(4.21)	102-1717-84
(5α,6α,7α,12β,22R)-	n.s.g.	222(4.22)	102-1717-84
Ferucrin isobutyrate	EtOH	216.5(4.17),242(3.47), 253(3.27),324(4.21)	105-0617-84
Nicandrin B	MeOH	226(4.12)	102-0853-84
$C_{28}H_{38}O_7$			
Withastramonolide	n.s.g.	219(4.28)	102-1717-84
$C_{28}H_{38}O_8$			
Withanolide E, 4β-hydroxy-	EtOH	220(4.10)	105-0182-84
Withaphysanolide, 28-hydroxy-	EtOH	212(4.10)	105-0182-84
$C_{28}H_{38}O_9$			
Opuntial (4,9-diacetoxyudoteal)	EtOH	236s(4.36),254(4.43)	102-1331-84
5,9-Undecadienal, 4-acetoxy-2-[3,6- diacetoxy-4-(acetoxymethylene)-5- hexenylidene]-6,10-dimethyl-	MeOH	235(4.19)	78-3053-84
$C_{28}H_{38}O_{10}$			
3-Heptanone, 1,7-bis(3,4-dimethoxy- phenyl)-5-(β-D-xylopyranosyloxy)-, (S)-	EtOH	228(4.07),280(3.69)	39-1635-84C
Phorbol, 3-deoxo-12-deoxy-3ξ,16-di- hydroxy-, tetraacetate	MeOH	193.5(4.30),316(2.00)	102-0129-84
$C_{28}H_{39}FO_{10}$			
Pregna-1,4-diene-3,20-dione, 9-fluoro- 21-(β-D-glucopyranosyloxy)-11,17-di- hydroxy-16-methyl-, (11β,16α)-	n.s.g.	239(4.16)	87-0261-84
$C_{28}H_{40}NO_6Si$			
2-Propen-1-aminium, N-[2-(1,3-benzodi-	MeCN	294(4.38)	44-0228-84

Compound	Solvent	$\lambda_{max}(\log \epsilon)$	Ref.
oxol-5-yl)-3-(2,2-dimethyl-1-oxoprop-oxy)-2-cyclopenten-1-ylidene]-N-(2-ethoxy-2-oxoethyl)-2-[(trimethylsil-yl)methyl]-, perchlorate	MeCN	294(4.38)	44-0228-84
$C_{28}H_{40}N_4O_9S$ Antibiotic OA-6129E, (4-nitrophenyl)-methyl ester (as acetonide)	n.s.g.	268(4.06)	158-0211-84
$C_{28}H_{40}O_4$ 4H-Inden-4-one, 1,2,3,3a,7,7a-hexahy-dro-6-[4-(2-hydroxy-5-methoxy-3-methylphenyl)-2-methyl-2-butenyl]-5-(2-hydroxy-2-methylpropyl)-3a,7a-dimethyl-, [3aα,6(E),7aα]-(+)-	EtOH	222(3.98),246(3.86), 287(3.45)	102-2017-84
$C_{28}H_{40}O_5$ Balearone	EtOH	218(3.98),289(3.42)	78-1721-84
$C_{28}H_{40}O_6$ Androsta-5,7-diene-3,17,19-triol, 3,19-diacetate 17-(2,2-dimethylpropan-oate), (3β,17β)-	EtOH	271(4.04),281(4.04)	56-0711-84
Atis-16-en-18-oic acid, 15-acetoxy-7-[(3-methyl-1-oxo-2-butenyl)oxy]-, methyl ester, (4α,5β,7β,8α,9β,10α-12α,15β)-	EtOH	218(4.04)	102-2075-84
Ergost-24-en-26-oic acid, 6,7-epoxy-5,12,22-trihydroxy-1-oxo-, δ-lac-tone, (5α,6α,7α,12α,22R)-	n.s.g.	213(4.09)	102-1717-84
$C_{28}H_{40}O_7$ Androst-5-en-7-one, 3,19-diacetoxy-17-(2,2-dimethyl-1-oxopropoxy)-, (3β,17β)-	EtOH	232.5(4.16)	56-0711-84
Ergosta-5,24-dien-26-oic acid, 3,14,20,22,27-pentahydroxy-1-oxo-, δ-lactone, (3β,22R)-	EtOH	218(3.99)	102-0143-84
$C_{28}H_{40}O_{11}$ Picrasa-2,12-diene-1,11-dione, 16-(β-D-glucopyranosyloxy)-2,12-dimethoxy-, (16α)- (picrasinoside B)	EtOH	255(4.05)	138-0221-84
$C_{28}H_{41}N_3O_2$ Olivoretin D	MeOH	234(4.56),287(4.03), 298s(3.98)	94-0354-84
Olivoretin C, de-O-methyl-	MeOH	233(4.52),289(3.96), 298s(3.94)	94-4233-84
Teleocidin B-1	MeOH	233(4.54),286(3.99), 298s(3.91)	94-4233-84
Teleocidin B-2	MeOH	233(4.51),287(3.98), 298s(3.93)	94-4233-84
Teleocidin B-3	MeOH	232(4.54),286(4.00), 298s(3.92)	94-4233-84
Teleocidin B-4 (olivoretin D)	MeOH	232(4.54),287(4.01), 298s(3.94)	94-4233-84
$C_{28}H_{42}O_3$ Ergosta-2,4-diene-1,6-dione, 11α-hy-droxy-	n.s.g.	310(3.75)	88-5925-84

Compound	Solvent	$\lambda_{max}(\log \epsilon)$	Ref.
Ergosta-2,22-dien-1-one, 5,6-epoxy-11-hydroxy-, (5β,6β,11α,22E)-(stoloniferone-b)	n.s.g.	225(3.90)	88-5925-84
Ergosta-2,24(28)-dien-1-one, 5,6-epoxy-11-hydroxy-, (5β,6β,11α)- (stoloniferone-a)	n.s.g.	225(3.91)	88-5925-84
$C_{28}H_{42}O_4$			
Cholesta-1,4,22E-trien-3-one, 16β,18,20ξ-trihydroxy-24-methyl-	MeOH	244(4.08)	31-0246-84
$C_{28}H_{42}O_5$			
Chol-1-en-3-one, 6,24-diacetoxy-, (5α,6β)-	EtOH	233(3.92)	94-3525-84
4H-Inden-4-one, pctahydro-6-hydroxy-6-[4-(2-hydroxy-5-methoxy-3-methyl-phenyl)-2-methyl-2-butenyl]-5-(2-hydroxy-2-methylpropyl)-3a,7a-di-methyl-	EtOH	215(3.97),289(3.46)	102-2017-84
Pregna-5,17(20)-dien-21-oic acid, 20-methoxy-3-[(tetrahydro-2H-pyran-2-yl)oxy]-, methyl ester, (3β,17E)-	EtOH	237(3.89)	118-0132-84
$C_{28}H_{42}O_8$			
Siphonarin A	MeOH	260(3.60)	35-6748-84
$C_{28}H_{42}O_{11}$			
Picrasinoside C	EtOH	261(3.64)	138-0221-84
$C_{28}H_{44}O_2$			
Cholest-4-ene-3,6-dione, 24(S)-methyl-	MeOH	250(3.99),325(2.68)	106-0117-84
Ergocalciferol, 25-hydroxy-	EtOH	265(4.25)	44-2148-84
24-epi	EtOH	265(4.24)	44-2148-84
$C_{28}H_{44}O_3$			
Cholestane-3,6-dione, 5α-hydroxy-24-methylene-	MeOH	209(3.03),250(2.51)	106-0117-84
Stoloniferone-c	n.s.g.	225(3.89)	88-5925-84
$C_{28}H_{44}O_4$			
Cholesta-4,22E-dien-3-one, 16β,18,20ξ-trihydroxy-24-methyl-	MeOH	241(4.18)	31-0246-84
$C_{28}H_{44}O_{11}$			
Picrasinoside F	EtOH	271(3.64)	138-0221-84
$C_{28}H_{44}O_{12}$			
Picras-2-en-1-one, 16-(β-D-glucopyrano-syloxy)-11,13-dihydroxy-2,12-dimeth-oxy-, (11α,12β,16α)- (picrasinoside G)	EtOH	271(3.79)	138-0221-84
$C_{28}H_{45}BrO_3$			
Ergost-2-en-1-one, 5-bromo-6,11-dihy-droxy-, (5α,6β,11α)-	n.s.g.	225(3.90)	88-5925-84
$C_{28}H_{46}N_2O$			
Moenjodaramine	MeOH	237(4.23),245(4.26)	64-0524-84B
$C_{28}H_{46}O_3$			
Cholestane-3,6-dione, 5α-hydroxy-24(S)-	dioxan	218(2.43),253(2.11)	106-0117-84

Compound	Solvent	$\lambda_{max}(\log \epsilon)$	Ref.
methyl- (cont.)			106-0117-84
$C_{28}H_{48}O$ 1-Docosanone, 1-phenyl-	heptane	320(1.70)	35-7033-84
$C_{28}H_{52}O_5Si_2$ 7-Oxabicyclo[4.1.0]hept-3-ene-3-metha-nol, 2,5-bis[[(1,1-dimethylethyl)di-methylsilyl]oxy]-4-(1-heptenyl)-, acetate, $(1\alpha,2\alpha,4(E),5\alpha,6\alpha)-(\pm)-$	hexane	243(4.32)	44-1898-84
$C_{28}H_{54}N_2O_2$ Diazene, bis(1-oxotetradecyl)-	THF	452(1.39)	23-0574-84
$C_{28}H_{56}Si_7$ 5,6,7,12,17,22,27-Heptasilaheptaspiro-[4.0.4.0.4.0.4.0.4.0.4.0.4.0]penta-triacontane	isooctane	236s(4.00),260s(3.72)	35-5521-84

Compound	Solvent	$\lambda_{max}(\log \epsilon)$	Ref.
$C_{29}H_{16}N_2O_7$			
3-Furancarbonitrile, 4,5-bis(1,3-benzo-dioxol-5-yl)-2-[[(4-oxo-4H-1-benzo-pyran-3-yl)methylene]amino]-	MeOH	215(4.69),257(4.32), 324(4.36),474(4.49)	73-1788-84
	CHCl$_3$	303(4.43),320(4.42), 433(4.26)	73-1788-84
$C_{29}H_{17}N_3O_3$			
9,10-Anthracenedione, 2-[(3,5-diphenyl-4-isoxazolyl)azo]-	EtOH	325(4.16)	104-1844-84
$C_{29}H_{18}N_2O_3$			
5(4H)-Oxazolone, 4-[[4-[5-(1-naphtha-lenyl)-2-oxazolyl]phenyl]methylene]-2-phenyl-	toluene	325(4.57),420(4.74)	135-1330-84
5(4H)-Oxazolone, 2-(1-naphthalenyl)-4-[[4-(5-phenyl-2-oxazolyl)phenyl]-methylene]-	toluene	345(4.51),430(--)	135-1330-84
$C_{29}H_{18}N_2O_4$			
Methanone, [3-(5-nitro-2-furanyl)-2-phenylpyrrolo[2,1-a]isoquinolin-1-yl]phenyl-	dioxan	452(4.23)	73-0533-84
$C_{29}H_{18}N_4O$			
Spiro[anthracene-9(10H),3'-[3H]pyra-zolo[5,1-c][1,2,4]triazol]-10-one, 6',7'-diphenyl-	CH$_2$Cl$_2$	235(4.56),259(4.47), 318(4.04),385s(3.58)	24-1726-84
$C_{29}H_{19}Br_2ClN_6O_2$			
Pyrazolo[5,1-b]quinazolin-9(4H)-one, 3,3'-[(4-chlorophenyl)methylene]-bis[7-bromo-2-methyl-	10% EtOH-NaOH	218(5.01),255(4.83), 332(4.57),400(3.91)	56-0411-84
$C_{29}H_{19}Br_2N_7O_4$			
Pyrazolo[5,1-b]quinazolin-9(4H)-one, 3,3'-[(4-nitrophenyl)methylene]-bis[7-bromo-2-methyl-	10% EtOH-NaOH	218(5.04),255(4.87), 330(4.60),400(4.01)	56-0411-84
$C_{29}H_{19}NO_4$			
Benzo[h]naphtho[1,2,3-de]quinolin-8-one, 3-acetyl-2-methoxy-9-phenoxy-	toluene	425(3.16)	104-0980-84
$C_{29}H_{19}N_3$			
9-Acridinecarboxaldehyde, (9-anthracen-ylmethylene)hydrazone	C$_6$H$_{12}$	435(4.02)	97-0021-84
$C_{29}H_{20}BrNO$			
1,3-Oxazepine, 7-(4-bromophenyl)-2,4,5-triphenyl-	ether	273(4.58),375(4.18)	78-3559-84
$C_{29}H_{20}Br_2N_6O_2$			
Pyrazolo[5,1-b]quinazolin-9(4H)-one, 3,3'-(phenylmethylene)bis[7-bromo-2-methyl-	10% EtOH-NaOH	217(5.00),249(4.88), 332(4.57),400(3.79)	56-0411-84
$C_{29}H_{20}N_4$			
Spiro[9H-fluorene-9,3'-[3H]pyrazolo-[5,1-c][1,2,4]triazole, 7'-phenyl-6'-(phenylmethyl)-	CH$_2$Cl$_2$	238(4.59),278(4.17), 289(4.14),348(3.86)	24-1726-84

Compound	Solvent	$\lambda_{max}(\log \epsilon)$	Ref.
$C_{29}H_{20}O_2$			
Methanone, (3-hydroxy-1,4-diphenyl-2-naphthalenyl)phenyl-	EtOH	230(4.63),287(4.51), 340(4.26)	44-4165-84
Phenanthro[9,10-b]furan-2(3H)-one, 3-phenyl3-(phenylmethyl)-	MeOH	214(3.29),254(3.32), 273s(2.90),298(2.70), 310(2.70),338(1.90), 356(2.00)	35-7352-84
$C_{29}H_{21}ClN_6O_2$			
Pyrazolo[5,1-b]quinazolin-9(4H)-one, 3,3'-[(4-chlorophenyl)methylene]-bis[2-methyl-	10% EtOH-NaOH	217(4.94),249(4.91), 325(4.56),398(4.04)	56-0411-84
$C_{29}H_{21}N_3O_6S$			
12H-Indolo[2,3-a]pyrrolo[3,4-c]pyra-zole-12-carboxylic acid, 5,6,7,13-tetrahydro-13-[(4-methoxyphenyl)-sulfonyl]-7-oxo-, methyl ester	MeOH	236(4.06),272(3.75), 284(3.71),310(3.84), 316(3.83),342(3.39)	78-2795-84
$C_{29}H_{21}N_3S$			
2H-Thiopyran, 2-azido-2,3,5,6-tetra-phenyl-	ether	225(4.50),260(4.34), 365(3.60)	78-3539-84
$C_{29}H_{21}N_7O_4$			
Pyrazolo[5,1-b]quinazolin-9(4H)-one, 3,3'-[(4-nitrophenyl)methylene]-bis[2-methyl-	10% EtOH-NaOH	217(4.48),243(4.92), 330(4.68),364(4.49)	56-0411-84
$C_{29}H_{21}O$			
Pyrylium, 2,3,4,6-tetraphenyl-, TCNQ salt	n.s.g.	492(1.92)	80-0817-84
Pyrylium, 2,3,5,6-tetraphenyl-, TCNQ salt	n.s.g.	480(3.86)	80-0817-84
$C_{29}H_{21}S$			
Thiopyrylium, 2,3,4,6-tetraphenyl-, perchlorate	MeCN	215(4.11),243(4.20), 293(3.99),385(4.18)	78-3539-84
Thiopyrylium, 2,3,5,6-tetraphenyl-, perchlorate	MeCN	225(4.29),245(4.31), 290(4.19),412(4.10)	78-3539-84
$C_{29}H_{22}$			
Dibenzo[2,3:4,5]pentaleno[1,6-ab]ind-ene, 4b,8b,12b,12d-tetrahydro-12d-(phenylmethyl)-	hexane	263.0(3.45),269.2(3.62), 276.3(3.67)	89-0508-84
$C_{29}H_{22}Cl_2O_4$			
Anthra[1,2-b]furan-6,11-dione, 4-[(2,2-dichloro-1-methylcyclopropyl)methyl]-2-methyl-5-(phenylmethoxy)-	EtOH	244(4.45),288(4.51), 382(3.97)	12-1511-84
$C_{29}H_{22}N_4$			
3H-Pyrazolo[5,1-c]-1,2,4-triazole, 3,3,7-triphenyl-6-(phenylmethyl)-	CH_2Cl_2	237(4.24),316(3.82), 343(3.78)	24-1726-84
$C_{29}H_{22}N_6O_2$			
Pyrazolo[5,1-b]quinazolin-9(4H)-one, 3,3'-(phenylmethylene)bis[2-methyl-	10% EtOH-NaOH	215(4.47),251(4.55), 325(4.14),400(3.61)	56-0411-84
$C_{29}H_{22}N_6O_4S$			
Acetamide, N-[[2-[[3,5-diphenyl-1-(2-pyridinylcarbonyl)-1H-pyrazol-4-yl]-	n.s.g.	372(3.75)	48-1021-84

Compound	Solvent	$\lambda_{max}(\log \epsilon)$	Ref.
azo]phenyl]sulfonyl]- (cont.)			48-1021-84
$C_{29}H_{22}O_2$			
2(3H)-Furanone, 3,4,5-triphenyl- 3-(phenylmethyl)-	MeOH	215(3.55),278(3.04)	35-7352-84
$C_{29}H_{22}O_5$			
2H-1-Benzopyran-3,4-diol, 3,4-dihydro- 2-phenyl-, dibenzoate, (+)-cis-	MeOH	217.5(4.42),274.5(3.51), 280(3.45)	94-4852-84
(-)-trans-	MeOH	227.0(4.39),274.0(3.48), 280(3.43)	94-4852-84
$C_{29}H_{22}S$			
4H-Thiapyran, 2,3,5,6-tetraphenyl-	ether	225(4.70),265(4.40)	78-3539-84
$C_{29}H_{23}NO$			
Spiro[acridine-9(10H),9'(10'H)-anthra- cen]-10'-one, 2,7,10-trimethyl-	EtOH	257(4.35),275(4.34), 296s(4.2),327(3.96)	24-2703-84
$C_{29}H_{23}N_2$			
Pyridinium, 1-(4-aminophenyl)-2,4,6- triphenyl-, perchlorate	EtOH	250s(4.3),312(4.6), 425(3.3)	103-1245-84
Pyridinium, 2,4,6-triphenyl-1-(2- pyridinylmethyl)-	C_6H_5Cl	307(4.51)	120-0062-84
Pyridinium, 2,4,6-triphenyl-1-(3- pyridinylmethyl)-	C_6H_5Cl	315(4.36)	120-0062-84
$C_{29}H_{23}N_2O_{10}S_3$			
1,2'-Bipyridinium, 4-(4-methoxy-3-sul- fophenyl)-2,6-bis(4-sulfophenyl)-, perchlorate	H_2O	222s(4.52),354(4.49)	39-0849-84B
$C_{29}H_{23}N_3O_2$			
5H-Cyclohepta[b]pyridine-3-carboni- trile, 2-[1,1'-biphenyl]-4-yl- 6,7,8,9-tetrahydro-4-(4-nitrophenyl)-	EtOH	209(4.61),219s(4.48), 244(4.25),289(4.54)	73-1395-84
1H-Pyrazole-3-carboxamide, 5-benzoyl- 4,5-dihydro-N,1,4-triphenyl-, trans	EtOH	228(4.99),284(4.69)	4-1013-84
$C_{29}H_{23}N_5O_2$			
1H-Pyrrolo[2,3-d:4,5-d']dipyridazine- 1,6(2H)-dione, 5,7-dihydro-2,5,7- tris(phenylmethyl)-	EtOH	249(4.30),256(4.22), 282(4.05),316(3.92)	94-1423-84
$C_{29}H_{23}N_5O_2S$			
1H-Dipyridazino[4,5-b:4',5'-e][1,4]- thiazine-1,6(7H)-dione, 2,10-dihy- dro-2,7,10-tris(phenylmethyl)-	EtOH	255(4.33),270(4.24), 321(4.20)	94-1423-84
1H-Dipyridazino[4,5-b:4',5'-e][1,4]- thiazine-1,9(2H)-dione, 8,10-dihy- dro-2,9,10-tris(phenylmethyl)-	EtOH	226(4.18),268(4.17), 310(4.08)	142-0675-84
3H-Dipyridazino[4,5-b:4',5'-e][1,4]- thiazine-4,6(7H,10H)-dione, 3,7,10-tris(phenylmethyl)-	EtOH	234(4.42),276(4.47), 316(4.16)	142-0675-84
$C_{29}H_{24}ClNO_2$			
Benzenebutanamide, 4-chloro-γ-oxo-α,α- diphenyl-N-(phenylmethyl)-	EtOH	254(3.18)	104-0370-84
$C_{29}H_{24}N_2$			
Benzo[b]quinolizinium, 9,9'-(1,3-prop-	MeOH	265(4.05)	48-0757-84

Compound	Solvent	$\lambda_{max}(\log \epsilon)$	Ref.
anediyl)bis- (cont.)			48-0757-84
5H-Cyclohepta[b]pyridine-3-carbonitrile, 2-[1,1'-biphenyl]-4-yl-6,7,8,9-tetrahydro-4-phenyl-	EtOH	207(4.66),218s(4.47), 242(4.21),290(4.46)	73-1395-84
3-Quinolinecarbonitrile, 2-[1,1'-biphenyl]-4-yl-5,6,7,8-tetrahydro-4-(4-methylphenyl)-	EtOH	207(4.86),218s(4.70), 234s(4.36),285(4.55)	73-1395-84
$C_{29}H_{24}N_2O$			
3-Quinolinecarbonitrile, 2-[1,1'-biphenyl]-4-yl-5,6,7,8-tetrahydro-4-(4-methoxyphenyl)-	EtOH	207(4.72),217s(4.61), 235(4.22),287(4.54)	73-1395-84
$C_{29}H_{24}N_4O$			
5H-Pyrano[2,3-d]pyrimidine-6-carbonitrile, 7-[1,1'-biphenyl]-4-yl-5-phenyl-4-(propylamino)-	EtOH	247(4.29),265(4.33), 299(4.36)	73-2309-84
$C_{29}H_{24}O$			
Anthracene, 9-(2-methoxyphenyl)-1,4-dimethyl-10-phenyl-	ether	266(4.87),350s(3.55), 362(3.82),382(4.02), 403(3.98)	22-0195-84
$C_{29}H_{25}Br_2N_3$			
2-Propen-1-amine, 3-(4-bromophenyl)-3-(3-pyridinyl)-2-propenyl]-N-methyl-3-(3-pyridinyl)-, (E,E)-	pH 1	220(4.52),237s(4.47)	44-4209-84
(Z,Z)-	pH 1	251(4.58)	44-4209-84
$C_{29}H_{25}N_5O_3$			
Adenosine, 2',3'-dideoxy-2'-oxo-5'-O-(triphenylmethyl)-	MeOH	259(4.18)	78-0125-84
$C_{29}H_{26}N_2$			
9,9'(10H,10'H)-Spirobiacridine, 2,2',7,7'-tetramethyl-	EtOH	294(4.37),320(4.1)	24-2703-84
$C_{29}H_{26}N_2O$			
3-Quinolinecarbonitrile, 2-[1,1'-biphenyl]-4-yl-1,4,5,6,7,8-hexahydro-4-(4-methoxyphenyl)-	EtOH	206(4.67),243(4.18), 278(4.43),345(4.26)	73-1395-84
$C_{29}H_{26}N_2O_5$			
Thymidine, 3'-deoxy-3'-oxo-5'-O-(triphenylmethyl)-	MeOH	266(3.95)	78-0125-84
$C_{29}H_{26}N_4O_9$			
2,4(1H,3H)-Pteridinedione, 6,7-diphenyl-3-(2,3,5-tri-O-acetyl-β-D-ribofuranosyl)-	MeOH	222(4.46),272(4.26), 363(4.17)	136-0179-84
$C_{29}H_{26}O_3$			
9,10-Anthracenediol, 9,10-dihydro-9-(2-methoxyphenyl)-1,4-dimethyl-10-phenyl-	ether	273(3.47),281(3.42)	22-0195-84
$C_{29}H_{26}O_9$			
α-D-Xylofuranose, 5-C-(9,10-dihydro-1,4-dihydroxy-9,10-dioxo-2-anthracenyl)-1,2-O-(1-methylethylidene)-3-O-(phenylmethyl)-, (S)-	EtOH	208(4.43),230(4.28), 251(4.55),280(3.99), 480(3.90)	39-1279-84C

Compound	Solvent	$\lambda_{max}(\log \epsilon)$	Ref.
$C_{29}H_{27}N_5O_3$			
Adenosine, 3'-deoxy-5'-O-(triphenyl-methyl)-	MeOH	259(4.18)	78-0125-84
9H-Purin-6-amine, 9-[3-deoxy-5-O-(tri-phenylmethyl)-β-D-threo-pentofurano-syl]-	MeOH	259(4.19)	78-0125-84
$C_{29}H_{27}O_2S_2$			
Thiopyrylium, 2-(4-methoxyphenyl)-6-[3-[6-(4-methoxyphenyl)-2H-thio-pyran-2-ylidene]-1-methyl-1-buten-yl]-, bromide	CHCl$_3$	782(4.80)	97-0146-84
$C_{29}H_{28}N_2O_2$			
3-Pyridinecarboxamide, 1,4-dihydro-4-(3-oxo-1,2-diphenylbutyl)-1-(phenylmethyl)-	MeCN	339(3.65)	44-0026-84
$C_{29}H_{28}N_2O_4$			
5,9-Methano-9H-8-oxa-3,9a-diazacyclo-nona[jk]fluorene-2-carboxylic acid, 6-acetyl-1,2,3,3a,4,5-hexahydro-14-methylene-3-(phenylmethyl)-, methyl ester, [2S-(2α,3aα,5β,9β)]-	MeOH	225(4.20),258(3.98), 282(3.78),292(3.49)	78-4853-84
$C_{29}H_{28}O_5$			
1,4-Dioxaspiro[4.5]decan-2-one, 3-[(3,4-bis(phenylmethoxy)phenyl]methyl-ene]-, (Z)-	EtOH	306s(4.24),324(4.37)	39-1539-84C
$C_{29}H_{28}O_9$			
Sanggenon M diacetate	EtOH	257(3.99),272(3.98), 305s(3.45),318(3.50), 362s(2.80)	142-1791-84
$C_{29}H_{29}NO$			
Piperidine, 1-(3-methyl-2,4,6-triphen-yl-2H-pyran-2-yl)-	dioxan	244(4.26),249(4.27), 344(4.06)	48-0657-84
$C_{29}H_{29}NO_4S$			
Methanone, [4-[2-(hexahydro-1H-azepin-1-yl)ethoxy]phenyl][6-hydroxy-2-(4-hydroxyphenyl)benzo[b]thien-3-yl]-	EtOH	290(4.51)	87-1057-84
$C_{29}H_{29}N_2O_2$			
Quinolinium, 6-acetyl-4-[3-(6-acetyl-1-ethyl-4(1H)-quinolinylidene)-1-propenyl]-1-ethyl-, iodide	MeOH	727(5.33)	103-0767-84
	MeCN	726(5.26)	103-0767-84
	C$_6$H$_4$Cl$_2$	745(5.25)	103-0767-84
$C_{29}H_{29}N_2O_4$			
Quinolinium, 1-ethyl-4-[3-[1-ethyl-6-(methoxycarbonyl)-4(1H)-quinol-inylidene]-1-propenyl]-6-(methoxy-carbonyl)-, iodide	MeOH	725(5.33)	103-0767-84
	MeCN	724(5.26)	103-0767-84
	C$_6$H$_4$Cl$_2$	743(5.25)	103-C767-84
$C_{29}H_{29}N_2S_2$			
Benzothiazolium, 3-ethyl-2-[11-(3-eth-yl-2(3H)-benzothiazolylidene)-1,3,5,7,9-undecapentaenyl]-, iodide	C$_6$H$_{11}$Me	1000(5.29)	99-0415-84
	EtOH	806(4.27),980(4.24)	99-0415-84
	DMF	850(4.30),990(4.11)	99-0415-84

Compound	Solvent	$\lambda_{max}(\log \epsilon)$	Ref.
$C_{29}H_{30}F_3N_4O_3S_3$ Benzothiazolium, 2-[[5-[[1,3-diethyl-1,3-dihydro-5-[(trifluoromethyl)-sulfonyl]-2H-benzimidazol-2-ylid-ene]ethylidene]-3-ethyl-4-oxo-2-thiazolidinylidene]methyl]-3-ethyl-, iodide	EtOH	592(5.01)	103-0665-84
$C_{29}H_{30}F_3N_4O_4S_2$ Benzoxazolium, 2-[[5-[[1,3-diethyl-1,3-dihydro-5-[(trifluoromethyl)-sulfonyl]-2H-benzimidazol-2-ylid-ene]ethylidene]-3-ethyl-4-oxo-2-thiazolidinylidene]methyl]-3-ethyl-, iodide	EtOH	578(5.06)	103-0665-84
$C_{29}H_{30}N_2O_3$ Benzamide, N-[9,10-dihydro-4-(octyl-amino)-9,10-dioxo-1-anthracenyl]-	benzene	571(4.02), 608(3.96)	39-0529-84C
$C_{29}H_{30}N_2O_5$ 1H-Pyrido[3,4-b]indole-3-carboxylic acid, 1-[(5-acetyl-3,4-dihydro-2-hydroxy-3-methylene-2H-pyran-4-yl)-methyl]-2,3,4,9-tetrahydro-2-(phen-ylmethyl)-, methyl ester	MeOH MeOH-NaOH	225(4.64),258(4.24), 278s(4.08),290(3.93) 285(5.00)	78-4853-84 78-4853-84
$C_{29}H_{30}O_8$ Sanggenon C diacetate	EtOH EtOH-AlCl$_3$	209s(4.66),227(4.89), 272(4.32),320(3.83), 342s(3.70) 208s(4.64),226(4.92), 300(4.34),391(3.81)	142-1791-84 142-1791-84
$C_{29}H_{31}NO_{12}$ Arugorol	MeOH	236(4.63),258(4.30), 292(3.88),475(4.11)	88-1937-84
$C_{29}H_{32}Cl_4$ Spiro[cyclopropane-1,17'-tricyclo-[10.4.1.1^{4,9}]octadeca[2,4,6,8,12-14,16]heptaene], 2,2,8',16'-tetra-chloro-6',14'-bis(1,1-dimethyl-ethyl)-18'-methylene-	C_6H_{12}	217s(4.26),252s(4.19), 294s(3.93),415(4.03)	94-4220-84
$C_{29}H_{32}N_2O$ 2,5-Heptadien-4-one, 1,7-bis(1,3-dihy-dro-1,3,3-trimethyl-2H-indol-2-yli-dene)- (ethanol complex)	hexane	470(3.69)	110-0556-84
$C_{29}H_{32}N_2O_4$ 6H,16H-Dibenzo[b,e][1,7]dioxacyclo-dodecin-7-propanenitrile, 9-(1-cyano-1-methylethyl)-7,8,9,10-tetrahydro-α,α,2,14-tetramethyl-6,10-dioxo-	dioxan	269(3.11),276(3.10)	126-0499-84
$C_{29}H_{32}N_2O_7$ Antibiotic U-56407	MeOH MeOH-HCl MeOH-NaOH	265s(4.55),315(4.78) 265s(4.55),315(4.78) 265s(4.64),307(4.78)	158-84-142 158-84-142 158-84-142

Compound	Solvent	$\lambda_{max}(\log \epsilon)$	Ref.
$C_{29}H_{32}N_2O_9S$ 1,2,3-Propanetriol, 1-[6-acetoxy- 3a,5,6,6a-tetrahydro-3-phenyl-2- [(phenylmethyl)thio]-3H-furo[2,3-d]- imidazol-5-yl]-, triacetate	EtOH	248(4.06)	136-0091-84B
$C_{29}H_{32}O$ 1H-Inden-1-one, 2-[3,8-dimethyl-5-(1- methylethyl)-1-azulenyl]-3,7-dimeth- yl-5-(1-methylethyl)-	MeOH	520(3.51)	138-0627-84
$C_{29}H_{32}O_2$ Methanone, (3-benzoylphenyl)[2,4,6- tris(1-methylethyl)phenyl]-	C_6H_{12}	233(4.58),351(2.18)	88-4525-84
Methanone, (4-benzoylphenyl)[2,4,6- tris(1-methylethyl)phenyl]-	C_6H_{12}	260(4.48),349(2.23)	88-4525-84
1,5-Pentanedione, 3-phenyl-1,5- bis(2,4,6-trimethylphenyl)-	EtOH	245(3.45)	78-4127-84
$C_{29}H_{32}O_9$ Malaphyll	n.s.g.	221(4.61),260(4.46), 292(3.95)	105-0113-84
Propanoic acid, 2,2-dimethyl-, 3-(acet- oxymethyl)-9,10-dihydro-5,8-dimeth- oxy-9,10-dioxo-2-(3-oxopentyl)-1- anthracenyl ester	MeOH	222(4.48),255(4.47), 325(3.45),420(3.79)	78-3677-84
$C_{29}H_{32}O_{16}$ Apigenin 4',7-bis-O-β-D-allopyranoside acetate	MeOH MeOH-NaOAc MeOH-AlCl$_3$	264.0(4.03),311(4.01) 264.5(4.02),313.0(4.00) 274.0(3.99),290.5(3.96), 333.0(4.05),374.5(3.83)	94-5023-84 94-5023-84 94-5023-84
$C_{29}H_{33}Cl$ Benzene, 1,1',1''-(1-chloro-1-ethenyl- 2-ylylidene)tris[2,4,6-trimethyl-	MeOH	228(4.33),240s(4.30), 278(4.12)	35-0477-84
$C_{29}H_{33}N_2$ 3H-Indolium, 2-[7-(1,3-dihydro-1,3,3- trimethyl-2H-indol-2-ylidene)-1,3,5- heptatrienyl]-1,3,3-trimethyl-	$C_6H_{11}Me$ EtOH DMF	758(5.48) 742(5.37) 746(5.30)	99-0415-84 99-0415-84 99-0415-84
$C_{29}H_{34}N_2$ Benzenamine, N,N-dibutyl-4-[2-[4-[2-(4- pyridinyl)ethenyl]phenyl]ethenyl]-	CH_2Cl_2	412(4.61)	44-2546-84
$C_{29}H_{34}N_3O_2S$ Quinolinium, 6-[(diethylamino)sulfon- yl]-1-ethyl-4-[3-(1-ethyl-4(1H)- quinolinylidene)-1-propenyl]-, iodide	MeOH $C_6H_4Cl_2$ MeCN	714(5.12) 732(5.19) 714(--)	103-0767-84 103-0767-84 103-0767-84
$C_{29}H_{34}O_3Si$ 3-Buten-1-one, 3-[[(1,1-dimethylethyl)- dimethylsilyl]oxy]-1-(4-methoxy- phenyl)-4,4-diphenyl-	MeOH	268(4.40)	35-4566-84
$C_{29}H_{34}O_{11}$ Ixerin C	MeOH	224.5(4.32),277.5(3.75)	94-1724-84

Compound	Solvent	$\lambda_{max}(\log \epsilon)$	Ref.
$C_{29}H_{35}NO_4$			
12,15-Methano-15H-1,4-benzodioxacyclo-heptadecin-20-carboxylic acid, 13-cyano-2,3,4,5,6,7,8,9,10,11,12,13-dodecahydro-13-phenyl-, methyl ester	EtOH	215s(4.15),272(3.54), 278(3.52)	24-2157-84
2-Tridecenoic acid, 13-[2-(3-cyano-3-phenyloxiranyl)phenoxy]-, methyl ester, [2α(E),3β]-	EtOH	266s(3.38),277s(3.58), 281(3.58)	24-2157-84
$C_{29}H_{35}N_3O_2P$			
Phosphonium, methyl(4-nitrophenyl)-bis[4-(1-piperidinyl)phenyl]-, iodide	MeOH	297s(4.73),309(4.76)	139-0267-84C
$C_{29}H_{36}N_2O_4$			
Emetamine	EtOH	236(4.85),283(3.86)	100-0397-84
$C_{29}H_{36}N_2P$			
Phosphonium, methylphenylbis[4-(1-piperidinyl)phenyl]-, iodide	MeOH	310(4.76)	139-0253-84C
$C_{29}H_{36}N_4O_2$			
21H-Biline-1(2H)-one, 17-ethyl-3,23-di-hydro-19-methoxy-3,3,7,8,12,13,18-heptamethyl-, (4Z,9Z,15Z)-	$CHCl_3$	269(4.32),350(4.58), 576(4.19),610(4.23)	49-1443-84
21H-Biline-19(3H)-one, 17-ethyldihydro-1-methoxy-3,3,7,8,12,13,18-hepta-methyl-, (4Z,10Z,15Z)-	$CHCl_3$	277(4.15),346(4.38), 626s(3.99),671(4.12)	49-1443-84
$C_{29}H_{36}O_5$			
Tovophenone A	MeOH and MeOH-NaOAc	235(4.07),266(3.89), 304(4.08)	32-0055-84
	MeOH-NaOMe	235(--),304(--), 428(--)	32-0055-84
$C_{29}H_{36}O_6$			
Tovophenone B	MeOH	242(4.0),262s(3.51), 305(4.03)	32-0055-84
$C_{29}H_{36}O_7$			
Characiol, 15-O-acetyl-3-O-benzoyl-5β,6β-epoxy-	MeOH	195(4.69),230(4.45), 273(3.09),280(3.01)	102-1689-84
Isocharaciol, 15-O-acetyl-3-O-benzoyl-5β-hydroxy-	MeOH	194(4.69),227(4.23) 266s(3.00),273(3.04), 280(3.00),297(2.66)	102-1689-84
$C_{24}H_{36}O_9$			
Satratoxin H (stereoisomer)	MeOH	227(4.29),255s(4.10)	158-0823-84
$C_{29}H_{36}O_{13}$			
β-D-Glucopyranoside, 4-[6a-acetoxy-4-(3,4-dimethoxyphenyl)tetrahydro-1H,3H-furo[3,4-c]furan-1-yl]-2-methoxyphenyl-, (+)-	EtOH	231(4.22),279(3.72)	102-2839-84
Osmanthuside B	MeOH	225(4.28),313(4.29)	94-3880-84
$C_{29}H_{36}O_{15}$			
Acteoside	MeOH	216s(4.28),248s(3.98), 290(4.00),332(4.10)	94-1209-84

Compound	Solvent	$\lambda_{max}(\log \epsilon)$	Ref.
$C_{29}H_{36}O_{16}$			
Acteoside, β-hydroxy-	MeOH	218s(4.33),231s(4.21), 289(3.98),331(4.00)	94-1209-84
	MeOH-NaOH	297(--),379(--)	94-1209-84
$C_{29}H_{37}BrO_4$			
Benzoic acid, 4-bromo-, 6-(acetoxymethyl)-1,2,3,3a,4,5,8,9,12,12a-decahydro-3a,10-dimethyl-1-(1-methylethenyl)-1-cyclopentacycloundecenyl ester, [1R-(1R*,3aR*,5*,6E,10E-12aS*)]-	hexane	244(4.26)	100-0615-84
$C_{29}H_{37}N_3O_3$			
Alangimarckine	EtOH	226(4.28),283(4.18)	100-0397-84
Isotubulosine	MeOH	279(4.08)	100-0397-84
Tubulosine	MeOH	223(4.60),279(4.16), 310s(3.67)	100-0397-84
$C_{29}H_{38}N_2O_3S$			
Pyridinium, 4-[2-[6-(dibutylamino)-2-naphthalenyl]ethenyl]-1-(4-sulfobutyl)-, hydroxide, inner salt	EtOH	495(4.61)	44-2546-84
$C_{29}H_{38}N_2O_4$			
Psychotrine, O-methyl-	pH 1	241.5(4.26),288.5(3.86), 305(3.92),354(3.91)	100-0397-84
$C_{29}H_{38}N_2O_5$			
Thymine, 1-(2-deoxy-5-O-trityl-β-D-threo-pentofuranosyl)-	MeOH	266(3.96)	78-0125-84
$C_{29}H_{38}N_2O_{15}S$			
Pyrocatechol sulfonphthalein complexan	pH 2.41	445(--)	86-1121-84
	pH 6.10	615(--)	86-1121-84
magnesium chelate	pH 10.0	550(4.10)	86-1121-84
$C_{29}H_{38}O_7$			
4H-1-Benzopyran-4,5,8-trione, 2-(4,6-dimethoxy-3,5,11-trimethyl-7,9,11-tridecatrienyl)-7-methoxy-3-methyl-, (all-E)-	MeOH	239s(4.34),248(4.38), 258(4.41),268(4.55), 279(4.39),325(3.34)	5-1883-84
$C_{29}H_{38}O_9$			
Miotoxin B	EtOH	252(3.9),266(4.0), 270(3.9)	5-1746-84
Roridin L-2	MeOH	259(4.39)	158-84-7
$C_{29}H_{38}O_{10}$			
Roridin L-2, 16-hydroxy-	MeOH	261(3.90)	158-84-134
$C_{29}H_{38}O_{11}$			
Fumigatonin	EtOH	219(4.00)	88-3233-84
$C_{29}H_{38}O_{12}$			
Hydrangenoside C	MeOH	227.0(4.21),280.0(3.24)	33-2111-84
Hydrangenoside D	MeOH	227.0(4.27),278.0(3.22)	33-2111-84
Hydrangenoside G	MeOH	227.0(4.34),278.5(3.33)	33-2111-84
$C_{29}H_{38}O_{14}$			
2-Naphthalenecarboxylic acid, 1,4-	MeOH	231(4.57),285(3.79),	102-0307-84

Compound	Solvent	$\lambda_{max}(\log \epsilon)$	Ref.
bis(β-D-glucopyranosyloxy)-3-(3-methyl-2-butenyl)-, methyl ester (cont.)		295(3.72),330(3.29)	102-0307-84
$C_{29}H_{38}O_{17}$ Loganin, 10-hydroxy-, hexaacetate	EtOH	232(4.02)	102-2535-84
$C_{29}H_{38}O_{18}$ Morroniside, 10-hydroxy-, hexaacetate, (7R)-	EtOH	235(4.09)	102-2535-84
(7S)-	EtOH	235(4.02)	102-2535-84
$C_{29}H_{39}N_{3}O_{7}S$ 6,9,12-Trioxa-3,15,19-triazabicyclo-[15.3.1]heneicosa-17,20-diene-2,5,13,19-tetrone, 19-methyl-4,14-bis(1-methylethyl)-21-[(phenyl-methyl)thio]-, [4S-(4R*,14R*)]-	CHCl$_3$	249(4.08),365(3.78)	35-6029-84
$C_{29}H_{39}N_{3}O_{8}S$ 6,9,12-Trioxa-3,15,19-triazabicyclo-[15.3.1]heneicosa-17,20-diene-2,5,13,19-tetrone, 19-methyl-21-[(4-methoxyphenyl)thio]-4,14-bis-(1-methylethyl)-, [4S-(4R*,14R*)]-	CH$_2$Cl$_2$	240(4.30),275(4.03), 351(3.88)	35-6029-84
$C_{29}H_{40}O_{4}$ Furosta-3,5,20(22)-trien-7-one, 26-acetoxy-	MeOH	275(4.32)	2-0435-84
$C_{29}H_{40}O_{11}$ 3-Heptanone, 1,7-bis(3,4-dimethoxy-phenyl)-5-(β-D-glucopyranosyloxy)-, (S)-	EtOH	228(4.14),280(3.73)	39-1635-84C
$C_{29}H_{40}O_{12}$ Hydrangenoside E	MeOH	225(4.12),277.5(3.20), 285(3.11)	33-2111-84
Hydrangenoside F	MeOH	226.5(4.18),277.5(3.24)	33-2111-84
$C_{29}H_{41}NO_{4}$ 7a-Aza-B-homofurosta-3,5,20(22)-trien-7-one, 26-acetoxy-	MeOH	272(4.2)	2-0435-84
$C_{29}H_{41}N_{5}O_{7}$ L-Leucine, N-[N-[N-(1-cyclohexyl-2,5-dihydro-4-nitro-2-oxo-1H-pyrrol-3-yl)glycyl]-L-phenylalanyl]-, ethyl ester	EtOH	369(4.20)	44-1130-84
$C_{29}H_{42}O_{4}$ Cystalgerone	EtOH	220(4.06),250(4.03), 280(3.52)	32-0169-84
$C_{29}H_{42}O_{5}$ Pregna-7,9(11)-diene-20-carboxylic acid, 3-acetoxy-4,4,14-trimethyl-, anhydride with acetic acid, (3β,5α-20S)-	MeOH	236(4.14),243(4.20), 252(4.03)	39-0497-84C
$C_{29}H_{42}O_{9}$ Physalactone	EtOH	228(3.94)	105-0182-84

Compound	Solvent	$\lambda_{max}(\log \epsilon)$	Ref.
$C_{29}H_{42}O_{10}$			
17α-Digitoxigenin, 16β,17β-epoxy-, 3-O-β-galactoside	MeOH	219(4.39)	88-2241-84
Strophalloside	EtOH	217(4.22)	33-0054-84
Strophanthidin 3-β-O-(4-O-β-D-allo-pyranosyl-B-6-deoxy-D-allo-pyranoside)	EtOH	217(4.14)	33-0054-84
$C_{29}H_{42}O_{11}$			
Antialloside	EtOH	217(4.14)	33-0054-84
Strophanthidin 3-O-β-galactoside	MeOH	219(4.48)	88-2241-84
$C_{29}H_{42}O_{12}$			
Hydrangenoside F, dihydro-	MeOH	228(4.22),278(3.28), 285(3.28)	33-2111-84
$C_{29}H_{43}BrCl_2N_3O_3P$			
Phosphoramidohydrazidic acid, N,N-bis(2-chloroethyl)-2-[(5α,17β)-17-hydroxyandrostan-3-ylidene]-, 4-bromophenyl ester	EtOH	271(2.69)	65-0389-84
cation	EtOH	273(--)	65-0389-84
$C_{29}H_{43}Cl_2FN_3O_3P$			
Phosphoroamidohydrazidic acid, N,N-bis(2-chloroethyl)-2-[(5α,17β)-17-hydroxyandrostan-3-ylidene]-, 4-fluorophenyl ester	EtOH	267(3.05)	65-0389-84
cation	EtOH	268(--)	65-0389-84
$C_{29}H_{43}Cl_3N_3O_3P$			
Phosphoramidohydrazidic acid, N,N-bis(2-chloroethyl)-2-[(5α,17β)-17-hydroxyandrostan-3-ylidene]-, 4-chlorophenyl ester	EtOH	271(2.78)	65-0389-84
cation	EtOH	273(--)	65-0389-84
$C_{29}H_{43}NO_{16}$			
Paulomenol A	EtOH	241(3.93),318(3.95)	158-1273-84
$C_{29}H_{43}N_3O_2$			
Olivoretin A	MeOH	232(4.55),286(4.01), 297s(3.95)	94-0354-84
Olivoretin B	MeOH	232(4.65),286.5(4.11), 298s(4.03),310s(3.84)	94-3774-84
Olivoretin C	MeOH	233(4.56),289(4.00), 298s(3.98)	94-3774-84
$C_{29}H_{44}Cl_2N_3O_3P$			
Phosphoramidohydrazidic acid, N,N-bis(2-chloroethyl)-2-[(5α,17β)-17-hydroxyandrostan-3-ylidene]-, phenyl ester	EtOH	268(2.53)	65-0389-84
cation	EtOH	269(--)	65-0389-84
$C_{29}H_{44}N_2O_3S$			
Pyridinium, 4-[2-[4-(dihexylamino)-phenyl]ethenyl]-1-(4-sulfobutyl)-, hydroxide, inner salt	EtOH	495(4.53)	44-2546-84

Compound	Solvent	$\lambda_{max}(\log \epsilon)$	Ref.
$C_{29}H_{44}N_4O_5$			
Tetrabutylaminium 2,4'-dinitrobenz- anilide	dioxan	448(4.40)	104-0565-84
$C_{29}H_{44}O_3$			

27-Norergosta-5,7,22-trien-25-one,
  3-hydroxy-, cyclic 1,2-ethane-
  diyl acetal, (3β,22E,24ξ)-
	EtOH	282(4.04)	44-2148-84

33-Norgorgost-2-en-1-one, 5,6-epoxy-
  11-hydroxy-, (5β,6β,11α)- [name
  as given by Chemical Abstracts]
	n.s.g.	225(3.90)	88-5925-84

27-Nor-9,10-seco-ergosta-5,7,10(19),22-
  tetraen-25-one, 3-hydroxy-, cyclic
  1,2-ethanediyl acetal, (3β,5Z,7E-
  22E,24ξ)-
	EtOH	263(3.95)	44-2148-84

$C_{29}H_{44}O_5$			

4-Hexen-2-one, 6-(2,5-dimethoxy-3-meth-
  ylphenyl)-1-[2-(4-hydroxy-4-methyl-1-
  oxopentyl)-1,2-dimethylcyclopentyl]-
  4-methyl-, [1α(E),2α]-(+)-
	EtOH	220(4.14),285(3.46)	102-2017-84

[1α(Z),2α]-(+)-	EtOH	225(4.15),283(3.49)	102-2017-84

4H-Inden-4-one, 6-[4-(2,5-dimethoxy-
  3-methylphenyl)-2-methyl-2-butenyl]-
  octahydro-6-hydroxy-5-(2-hydroxy-2-
  methylpropyl)-3a,7a-dimethyl-
	EtOH	220(3.98),283(3.34)	102-2017-84

$C_{29}H_{44}O_7$			

4H-1-Benzopyran-4,5,8-trione, 2-(4,6-
  dimethoxy-3,5,11-trimethyltridecyl)-
  7-methoxy-3-methyl-
	MeOH	231(4.35),238(4.36),
		247s(4.29),266(3.78),
		281(3.79),325(3.43)
			5-1883-84

$C_{29}H_{44}O_8$			
Siphonarin B	MeOH	260(3.42)	35-6748-84
$C_{29}H_{44}O_9$			
Digitoxigenin 3-O-β-galactoside	MeOH	219(4.23)	88-2241-84
Digitoxigenin 3-O-β-glucoside	MeOH	219(4.05)	88-2241-84
$C_{29}H_{44}O_{10}$			
Gitoxigenin 3-O-β-galactoside	MeOH	219(4.37)	88-2241-84
Gitoxigenin 3-O-β-glucoside	MeOH	219(4.37)	88-2241-84
$C_{29}H_{46}N_2$			
Plakinamine A	MeOH	246(3.85)	44-5157-84
$C_{29}H_{46}N_{10}O_{12}S$			
Chitinovorin B	H_2O	261(3.90)	158-1486-84
$C_{29}H_{46}O$			
Cyclonervilasterol	EtOH	210(3.72)	94-1256-84
24-Epicyclonervilasterol	EtOH	210(3.72)	94-1256-84
$C_{29}H_{46}O_2$			
30-Norlupa-12,18-diene-3β,28-diol	C_6H_{12}	236(3.88)	73-0141-84
$C_{29}H_{46}O_3$			
Furosta-4,20(22)-diene, 3,26-dimeth- oxy-, (3β)-	EtOH	205(4.01),220s(3.82)	94-2111-84

Compound	Solvent	$\lambda_{max}(\log \epsilon)$	Ref.
$C_{29}H_{46}O_9Si$			
Verrucarin J seco acid, 2-(trimethyl-silyl)ethyl ester	EtOH	218(4.27),263(4.39)	44-1772-84
(E,E,E)-	EtOH	219(4.29),264(4.52)	44-1772-84
$C_{29}H_{47}NO_7$			
Mycinamicin VII	MeOH	215(4.29),281(4.32)	158-84-22
$C_{29}H_{48}O$			
Cyclonervilasterol, dihydro-	EtOH	211(3.75)	94-1256-84
24-Epicyclonervilasterol, dihydro-	EtOH	211(3.75)	94-1256-84
$C_{29}H_{50}O_2$			
α-Tocopherol, (±)-	hexane	298(3.56)	44-0491-84

Compound	Solvent	$\lambda_{max}(\log \epsilon)$	Ref.
$C_{30}H_{14}N_2O_2$ 5,14-Diazaviolanthrene-9,18-dione	H_2SO_4	663(4.43)	39-2177-84C
$C_{30}H_{16}Cl_2N_2O_4$ Indigo, N,N'-bis(4-chlorobenzoyl)-, cis	toluene	460(3.42)	23-2478-84
trans	toluene	575(3.92)	23-2478-84
$C_{30}H_{16}O_3$ 2,5-Furandione, 3,4-di-9H-fluoren-9-ylidenedihydro-	toluene	502(4.09)	48-0233-84
$C_{30}H_{18}Br_2$ Anthracene, 9,9'-(1,2-ethenediyl)-bis[10-bromo-, (E)-	C_6H_{12}	255(5.2),400(4.3)	12-1329-84
$C_{30}H_{18}Cl_2N_2$ 9-Anthracenecarboxaldehyde, 10-chloro-, [(10-chloro-9-anthracenyl)methylene]-hydrazone	C_6H_{12}	433(4.26)	97-0021-84
$C_{30}H_{18}N_2O_4$ 3H-Indol-3-one, 1-benzoyl-2-(1-benzoyl-1,3-dihydro-3-oxo-2H-indol-2-ylidene)-1,2-dihydro-, cis	toluene	460(3.68)	23-2478-84
trans	toluene	577(3.89)	23-2478-84
$C_{30}H_{19}N_5$ Ethenetricarbonitrile, [4-[(diphenyl-methylene)phenylhydrazino]phenyl]-	benzene	529(4.49)	32-0111-84
$C_{30}H_{20}$ Anthracene, 9,9'-(1,2-ethenediyl)bis-	C_6H_{12}	252s(5.1),260(5.1), 410(4.3)	12-1329-84
$C_{30}H_{20}Cl_2N_4$ 1,2,4-Triazolo[1,5-a]pyridinium, 1-(4-chlorophenyl)-2-[(4-chlorophenyl)-amino]-5,7-diphenyl-, hydroxide, inner salt	EtOH	262(4.86)	39-1891-84C
$C_{30}H_{20}N_2$ 5,10[1',2']-Benzenopyrido[3,2-g]quino-line, 5,10-dihydro-2,8-diphenyl-	MeCN	231(4.328),266(4.317), 286(4.446),306(4.480)	5-0381-84
9-Phenanthrenecarboxaldehyde, (9-phen-anthrenylmethylene)hydrazone	C_6H_{12}	361(4.61)	97-0021-84
$C_{30}H_{20}O_2$ Dibenz[a,j]anthracene-7,14-dione, 2-methyl-5-(4-methylphenyl)-	$CHCl_3$	287(4.75),309(4.98), 335(4.12),363(4.01), 442(3.81)	40-0090-84
Ethanone, 1-[9-(2-acetyl-9H-fluoren-9-ylidene)-9H-fluoren-2-yl]-	DMF	465(4.38)	40-0145-84
Methanone, (4b,8d-dihydrodibenzo[a,f]-cyclopropa[cd]pentalene-8b,8c-diyl)-bis[phenyl-	EtOH	230(4.40),250(4.47), 279(3.88),324(2.85)	44-4923-84
$C_{30}H_{20}O_3$ 3-Butene-1,2-dione, 3-(9-anthracenyl)-4-hydroxy-1,4-diphenyl-	C_6H_{12}	236(3.15),260(4.00), 316(4.15),332(4.15), 352(3.93),372(3.95),	44-4165-84

Compound	Solvent	$\lambda_{max}(\log \epsilon)$	Ref.
(cont.) 2,5-Furandione, 3,4-bis(diphenylmeth- ylene)dihydro-	toluene	392(3.90),404(3.38) 420(4.01)	44-4165-84 48-0233-84
Methanone, (9,10-dihydro-9,10-endo- oxiranoanthracene-11,13-diyl)bis- [phenyl-	EtOH	252(4.42),330(2.93)	44-4165-84
$C_{30}H_{20}O_4$ Dibenz[a,j]anthracene-7,14-dione, 2-methoxy-5-(4-methoxyphenyl)-	$CHCl_3$	288(4.57),313(4.59), 369(3.84),473(3.68)	40-0090-84
Dibenz[a,j]anthracene-7,14-dione, 3-methoxu-5-(3-methoxyphenyl)-	$CHCl_3$	288(4.38),316(4.63), 346(3.92),387(3.91)	40-0090-84
Dibenz[a,j]anthracene-7,14-dione, 4-methoxy-5-(2-methoxyphenyl)-	$CHCl_3$	287(4.95),312(4.97), 480(3.73)	40-0090-84
$C_{30}H_{20}O_{15}$ 3,10,17,24,31,36,37,38,39,40-Decaoxa- hexacyclo[31.2.1.1^{5,8}.1^{12,15}1^{19,22}- 1^{26,29}]tetraconta-5,7,12,14,19,21- 26,28,33,35-decaene-2,9,16,23,30- pentone	$CHCl_3$	258.0(4.73)	126-2347-84
$C_{30}H_{21}N_5O_2S_2$ Pyrimido[5,4-e]thiazolo[3,4-a]pyrimi- din-10-ium, 9-[(3-ethyl-2(3H)-benzo- thiazolylidene)methyl]-1,2,3,4- tetrahydro-2,4-dioxo-3,7-diphenyl-, hydroxide, inner salt	DMF	430(4.10),456(4.05), 556(4.50)	103-0921-84
$C_{30}H_{22}$ 5,5'-Bi-5H-dibenzo[a,c]cycloheptene	EtOH	239(3.89)	44-4029-84
m,m,o,p,o-Pentaphenylene	C_6H_{12}	190(4.82),211(4.89), 237(4.73),256s(4.56)	18-3494-84
$C_{30}H_{22}Cl_2N_6$ 1H-Indole, 5,5'-(1,2-ethanediyl)- bis[3-[(4-chlorophenyl)azo]-	EtOH	222(3.42),285(3.99), 465(3.45)	103-0062-84
$C_{30}H_{22}N_2O$ Methanone, [1,4-diphenyl-3-(2-phenyl- ethenyl)-1H-pyrazol-5-yl]phenyl-	EtOH	258(4.25),304(4.20)	4-1013-84
Methanone, [1,5-diphenyl-3-(2-phenyl- ethenyl)-1H-pyrazol-4-yl]phenyl-	EtOH	257(4.49),350(4.47)	4-1013-84
$C_{30}H_{22}N_2O_4S$ 6-Azabicyclo[3.2.0]hept-2-ene-6-sulfon- amide, 4,7-dioxo-1,2,3,5-tetraphenyl-	EtOH	320(4.36)	2-1046-84
$C_{30}H_{22}N_4$ Spiro[5H-dibenzo[a,d]cycloheptene- 5,3'-[3H]pyrazolo[5,1-c][1,2,4]- triazole], 10,11-dihydro-6',7'- diphenyl-	CH_2Cl_2	234(4.60),253(4.42), 321(4.06),375s(3.76)	24-1726-84
[1,2,4]Triazolo[1,5-a]pyridinium, 1,5,7-triphenyl-2-(phenylamino)-, hydroxide, inner salt	EtOH	258(4.84)	39-1891-84C
$C_{30}H_{22}N_4O_4Zn$ Zincate(2-), [[3,3'-(3,7,12,17-tetra- methyl-21H,23H-porphine-2,18-diyl)- bis[2-propenoato]](4-)-N^{21},N^{22},N^{23}- N^{24}]-, dihydrogen	MeOH	433(5.06),557(4.06), 598(4.09)	5-1386-84

Compound	Solvent	$\lambda_{max}(\log \epsilon)$	Ref.
$C_{30}H_{22}N_8O_3S$ 1H-Pyrazole, 3-methyl-5-phenyl-1-(2-pyridinylcarbonyl)-4-[[2-[(2-quinoxalinylamino)sulfonyl]phenyl]azo]-	n.s.g.	370(4.42)	48-1021-84
$C_{30}H_{22}N_8O_4$ 1H-Indole, 5,5'-(1,3-ethanediyl)-bis[3-[(4-nitrophenyl)azo]-	EtOH	229.8(--),289.8(--), 436.8(--)	103-0062-84
$C_{30}H_{22}O_2$ Methanone, (4b,10-dihydroindeno[1,2-a]-indene-9,9a(9H)-diyl)bis[phenyl-, (4bα,9aα)-	MeOH	250(4.44),290s(3.46)	44-4923-84
$C_{30}H_{22}O_{10}$ [3,3'-Bi-4H-1-benzopyran]-4,4'-dione, 2,2',3,3'-tetrahydro-5,5',7,7'-tetrahydroxy-2,2'-bis(4-hydroxyphenyl)- (chamajasmine)	EtOH	293(4.59)	138-1587-84
Chamaechromone	MeOH	225(4.61),260(4.30), 296(4.34)	88-3735-84
Isochamaejasmine	MeOH	296(4.48)	138-1587-84
Neochamaejasmine A	MeOH	297(4.48)	138-0539-84
Neochamaejasmine B	MeOH	298(4.49)	138-0539-84
$C_{30}H_{23}BO_{10}$ 1,5-Methano-5H-anthra[2,3-e][1,3,2]di-oxaborocin-5-carboxylic acid, 1,14-diacetoxy-1,6,8,13-tetrahydro-8,13-dioxo-3-phenyl-, methyl ester, (1S)-	dioxan	257(4.64),339(3.77)	78-4649-84
$C_{30}H_{23}ClN$ Pyridinium, 1-[(4-chlorophenyl)methyl]-2,4,6-triphenyl-	C_6H_5Cl	313(4.48)	120-0062-84
$C_{30}H_{23}N$ 2H-Azirine, 3-methyl-2-phenyl-2-(1,2,3-triphenyl-2-cyclopropen-1-yl)-	EtOH	227(4.50),315(4.32), 334(4.23)	35-1065-84
Pyridine, 2-methyl-3,4,5,6-tetraphenyl-	C_6H_{12}	243(4.44),295(4.04)	35-1065-84
$C_{30}H_{23}NO$ 2,5-Cyclohexadien-1-one, 4-[2,6-diphenyl-1-(phenylmethyl)-4(1H)-pyridinylidene]-	0.4M NaOH	417(4.86)	4-1673-84
$C_{30}H_{23}NO_2$ Spiro[acridine-9(10H),9'(10'H)-anthracen]-10'-one, 10-acetyl-2,7-dimethyl-	EtOH	259(4.35)	24-2703-84
$C_{30}H_{23}N_2OPPd$ Palladium, [2-(phenylazo)phenolato-(2-)](triphenylphosphine)-, (SP-4-3)-	CH_2Cl_2	660(3.78)	77-0999-84
$C_{30}H_{23}N_5O_5S$ Benzamide, N-[6-[1,2-bis(benzoyloxy)-propyl]-1,4-dihydro-4-thioxo-2-pteridinyl]-	MeOH	233(4.64),250s(4.50), 295(4.26),408(4.00)	5-1815-84
$C_{30}H_{23}N_5O_6$ Benzamide, N-[6-[1,2-bis(benzoyloxy)-propyl]-1,4-dihydro-4-oxo-2-pteri-	pH -4.3	232(4.62),268(4.39), 300(4.15),312s(4.07),	5-1815-84

Compound	Solvent	$\lambda_{max}(\log \epsilon)$	Ref.
dinyl]- (cont.)		376(3.40)	5-1815-84
	pH 3.0	232(4.63),286(4.32), 330(3.96)	5-1815-84
	pH 10.0	231(4.58),267(4.44), 332(3.92),350s(3.90)	5-1815-84
$C_{30}H_{24}$ Benzene, 1,3,5-tris(2-phenylethenyl)-	CH_2Cl_2	317(4.96)	24-2452-84
$C_{30}H_{24}ClN_2PPd$ Palladium, chloro[2-(phenylazo)phenyl]- (triphenylphosphine)-, (SP-4-4)-	EtOH	305(4.03),355(4.03), 430(3.57)	101-0091-84I
$C_{30}H_{24}N$ Pyridinium, 2,4,6-triphenyl-1-(phenyl-methyl)-	C_6H_5Cl	312(4.53)	120-0062-84
$C_{30}H_{24}NO$ Pyridinium, 1-(4-methoxyphenyl)-2,4,6-triphenyl-, perchlorate	EtOH	250s(4.3),313(4.6)	103-1245-84
$C_{30}H_{24}NO_2P$ Phosphoramidous acid, dimethyl-, di-9-anthracenyl ester	EtOH	256(5.08),335(3.88), 354(3.98),372(4.06), 393(4.05)	40-1158-84
$C_{30}H_{24}NO_{10}S_3$ Pyridinium, 4-(4-methoxy-3-sulfophen-yl)-1-phenyl-2,6-bis(4-sulfophenyl)-, perchlorate	H_2O	220s(4.61),246s(4.35), 345(4.50)	39-0849-84B
$C_{30}H_{24}N_2O$ 1H-Indole, 1,1'-[oxybis(methylene)]-bis[3-phenyl-	EtOH	235(4.43),267(4.33), 293(4.16)	78-4351-84
Methanone, [4,5-dihydro-1,4-diphenyl-3-(2-phenylethenyl)-1H-pyrazol-5-yl]phenyl-	EtOH	254(4.37),372(4.38)	4-1013-84
$C_{30}H_{24}N_4O_4$ 1H-Benzimidazole-5-carboxylic acid, 2,2'-(1,4-naphthalenediyl)bis[1-methyl-	DMF	315(4.40)	5-1129-84
1H-Benzimidazole-5-carboxylic acid, 2,2'-(2,6-naphthalenediyl)bis[1-methyl-	DMF	332(4.67)	5-1129-84
$C_{30}H_{24}N_4O_5$ 11H-Furo[2',3':4,5]oxazolo[2,3-b]pteri-din-11-one, 6a,7,8,9a-tetrahydro-7-hydroxy-8-(triphenylmethoxy)methyl]-	MeOH	225s(4.26),260(3.95), 319(3.76)	136-0179-84F
$C_{30}H_{24}N_6$ 1H-Indole, 5,5'-(1,2-ethanediyl)-bis[3-(phenylazo)-	EtOH	227(4.65),232(4.60), 281(4.47),473(4.67)	103-0062-84
$C_{30}H_{24}O$ Benzofuran, 3-(1-methyl-2,2-diphenyl-ethenyl)-2-(phenylmethyl)-	hexane	210(4.65)	39-2877-84C
$C_{30}H_{24}O_4$ 4H-1-Benzopyran-4-one, 5,7-dihydroxy-	MeOH	206(4.63),260(4.34),	2-1036-84

Compound	Solvent	$\lambda_{max}(\log \epsilon)$	Ref.
2-methyl-3-phenyl-6,8-bis(phenyl-methyl)- (cont.)		301(3.61)	2-1036-84
$C_{30}H_{25}BO_8$ 1,5-Methano-5H-anthra[2,3-e][1,3,2]di-oxaborocin-5-carboxylic acid, 7,16-diacetoxy-1,6-dihydro-3-phenyl-, methyl ester, (1S)-	dioxan	261(5.25),335(3.56), 351(3.81),370(3.92), 391(3.72)	78-4649-84
1,5-Methano-1H-anthra[2,3-e][1,3,2]di-oxaborocin-8,13-dione, 7,14-diacet-oxy-5-ethyl-5,6-dihydro-3-phenyl-, (1S)-	CHCl$_3$	259(4.65),340(3.79)	78-4649-84
$C_{30}H_{25}ClN_4O_3$ 4-Imidazolidinecarboxamide, 1-(4-chlo-rophenyl)-3-[4-(dimethylamino)phen-yl]-2,5-dioxo-N,4-diphenyl-	EtOH	207(4.59),266(4.36)	49-0187-84
$C_{30}H_{25}N_3O$ Acetamide, N-[4-(2-[1,1'-biphenyl]-4-yl-3-cyano-5,6,7,8-tetrahydro-4-quinolinyl)phenyl]-	EtOH	206(4.75),218s(4.54), 240(4.41),289(4.57)	73-1395-84
$C_{30}H_{25}N_3O_2$ Cycloocta[b]pyridine-3-carbonitrile, 2-[1,1'-biphenyl]-4-yl-5,6,7,8,9,10-hexahydro-4-(4-nitrophenyl)-	EtOH	207(4.80),218s(4.64), 244(4.33),288(4.54)	73-1395-84
$C_{30}H_{25}N_5O_5$ 4-Imidazolidinecarboxamide, 3-[4-(di-methylamino)phenyl]-1-(4-nitrophen-yl)-2,5-dioxo-N,4-diphenyl-	EtOH	208(4.45),219(4.16), 267(4.30),305(3.99)	49-0187-84
$C_{30}H_{26}N_2$ Benzo[b]quinolizinium, 9,9'-(1,4-but-anediyl)bis-	MeOH	260(3.12)	48-0757-84
Cycloocta[b]pyridine-3-carbonitrile, 2-[1,1'-biphenyl]-4-yl-5,6,7,8,9,10-hexahydro-4-phenyl-	EtOH	208(4.64),218s(4.48), 244(3.96),291(4.47), 369(3.34)	73-1395-84
1,3-Diazabicyclo[3.1.0]hex-3-ene, 2,6-dimethyl-2,4,5,6-tetraphenyl-	C$_6$H$_{12}$	245(4.01),310(3.34)	44-3174-84
1H-Indole, 2,3-dihydro-1-methyl-3-phenyl-3-[(3-phenyl-1H-indol-1-yl)methyl]-	EtOH	236(4.38),262(4.26), 283(4.18)	78-4351-84
1H-Quindoline, 3-ethyl-2,10-dihydro-4-methyl-1,11-diphenyl-	EtOH	208(4.49),242(4.58), 272(4.38),289(4.48), 353(4.07),367(4.11)	103-1374-84
$C_{30}H_{26}N_2O$ 3-Quinolinecarbonitrile, 2-[1,1'-bi-phenyl]-4-yl-4-(4-ethoxyphenyl)-5,6,7,8-tetrahydro-	EtOH	207(4.69),218s(4.58), 239(4.19),288(4.53)	73-1395-84
$C_{30}H_{26}N_2O_2$ 1H-Quindoline, 2,10-dihydro-1,11-bis(4-methoxyphenyl)-3-methyl-	EtOH	207(4.59),222(4.66), 240(4.59),284(4.62), 355(4.16),371(4.19)	103-1374-84
$C_{30}H_{26}N_4O$ 5H-Pyrano[2,3-d]pyrimidine-6-carbo-nitrile, 7-[1,1'-biphenyl]-4-yl-	EtOH	247(4.29),265(4.33), 298(4.35)	73-2309-84

Compound	Solvent	$\lambda_{max}(\log \epsilon)$	Ref.
4-(butylamino)-5-phenyl- (cont.)			73-2309-84
$C_{30}H_{26}N_4O_3$ 1H-Benzimidazole-5-carboxylic acid, 2-[4-(2-benzoxazolyl)-1-naphthalenyl]-1-methyl-, 2-(dimethylamino)-ethyl ester	DMF	343(4.48)	5-1129-84
$C_{30}H_{26}N_4O_6$ 2,4(1H,3H)-Pteridinedione, 3-[5-O-(triphenylmethyl)-β-D-ribofuranosyl]-	MeOH	228s(4.28),326(3.87)	136-0179-84F
$C_{30}H_{26}O_5$ 4H-1-Benzopyran-4-one, 2-[3,4-bis(phenylmethoxy)phenyl]-2,3-dihydro-7-methoxy-	MeOH	234(4.4),276(4.2), 313(3.8)	94-1472-84
2-Propen-1-one, 3-[3,4-bis(phenylmethoxy)phenyl]-1-(2-hydroxy-4-methoxyphenyl)-	MeOH	248(4.0),260(4.0), 315s(4.1),377(4.5)	94-1472-84
$C_{30}H_{27}BO_6$ 1,5-Methano-1H-anthra[2,3-e][1,3,2]-dioxaborocin-7,14-diol, 5-ethyl-5,6-dihydro-3-phenyl-, diacetate, (1S)-	CHCl$_3$	263(5.18),320(3.19), 336(3.46),351(3.69), 370(3.84),391(3.78)	78-4649-84
$C_{30}H_{28}$ Azulene, 4,4'-(1,3,5-hexatriene-1,6-diyl)bis[6,8-dimethyl-, (E,E,E)-	CH$_2$Cl$_2$	244(4.44),284(4.70), 338(4.49),414(4.70), 594(3.38)	5-1905-84
$C_{30}H_{28}N_2O$ 3-Quinolinecarbonitrile, 2-[1,1'-biphenyl]-4-yl-4-(4-ethoxyphenyl)-1,4,5,6,7,8-hexahydro-	EtOH	206(4.67),218s(4.48), 279(4.45),354(3.76)	73-1395-84
$C_{30}H_{28}N_3$ 9H-Fluoren-9-iminium, N-[bis[4-(dimethylamino)phenyl]methylene]-, hexachloroantimonate	CH$_2$Cl$_2$	254(4.80),261(4.87), 297(4.32),304(4.30), 425(4.54),556(4.87)	24-3222-84
$C_{30}H_{28}N_3O_6S_2$ Benzoxazolium, 3-ethyl-5-phenyl-2-[(5-anilinomethylene-4-oxo-3-ethylthiazolidin-2-ylidene)methyl]-, ethyl sulfate	EtOH	476(4.81)	103-0665-84
$C_{30}H_{29}Cl_2NO_5$ 4H-Cyclohepta[b]furan-4-one, 2,2'-(hydroxyimino)bis[3-(4-chlorophenyl)-2,3,4,5,6,7,8-hexahydro-	MeOH	272(4.370)	73-1421-84
$C_{30}H_{30}$ Benzo[g]chrysene, 12,14-bis(1,1-dimethylethyl)-, (±)-	CHCl$_3$	255(4.52),273s(4.45), 282s(4.51),292(4.67), 304(4.80),338(3.94), 386(2.98)	24-0336-84
$C_{30}H_{30}Cl_2N_4O_2Pd_2$ Palladium, di-μ-chlorobis[2-[(2,6-dimethylphenyl)azo]-4-methylphenolato]di-	CH$_2$Cl$_2$	495(3.53)	77-0999-84

Compound	Solvent	$\lambda_{max}(\log \epsilon)$	Ref.
$C_{30}H_{30}N_2O_4$ 3H-Indol-3-one, 1-(cyclohexylcarbonyl)-2-[1-(cyclohexylcarbonyl)-1,3-dihydro-3-oxo-2H-indol-2-ylidene]-1,2-dihydro-, cis	benzene	476(3.61)	39-2305-84C
trans	benzene	579(3.88)	39-2305-84C
$C_{30}H_{30}N_2O_6$ Thymidine, 5'-O-[(4-methoxyphenyl)diphenylmethyl]-	MeOH	231s(4.15),266(4.00)	118-0965-84
$C_{30}H_{30}N_2O_7$ Uridine, 5'-O-[bis(4-methoxyphenyl)phenylmethyl]-2'-deoxy-	MeOH	234(4.36),264(4.03)	78-0003-84
$C_{30}H_{30}N_3$ Benzenemethanaminium, 4-(dimethylamino)-α-[4-(dimethylamino)phenyl]-N-(diphenylmethylene)-, hexachloroantimonate	CH_2Cl_2	263(4.47),284(4.43), 423(4.52),542(4.69)	24-3222-84
$C_{30}H_{30}O_4S_3$ Benzene, 1,1'-[thiobis(2,1-ethanediyloxy)]bis[2-[(4-methylphenyl)sulfinyl]-	EtOH	233(4.56),290(3.93)	12-2059-84
$C_{30}H_{30}O_5S_2$ Benzene, 1,1'-[oxybis(2,1-ethanediyloxy)]bis[2-[(4-methylphenyl)sulfinyl]-, [R-(R*,R*)]-	EtOH	229(4.50),273(3.92)	12-2059-84
$C_{30}H_{30}O_8$ [2,2'-Binaphthalene]-1,1',4,4'-tetrone, 6,6',7,7'-tetraethoxy-3,3'-dimethyl-	$CHCl_3$	277(4.79),284(4.80), 350(3.84),425(3.52)	64-0244-84B
$C_{30}H_{30}O_{11}$ β-D-Glucopyranoside, 3-hydroxy-5-[2-(3-hydroxy-4-methoxyphenyl)ethenyl]phenyl 2-[3-(4-hydroxyphenyl)-2-propenoate], (E,E)-	MeOH	220(4.32),302s(4.43), 315(4.49)	94-3501-84
$C_{30}H_{31}NO_4S$ Methanone, [6-methoxy-2-(4-methoxyphenyl)benzo[b]thien-3-yl][4-[2-(1-piperidinyl)ethoxy]phenyl]-, hydrochloride	MeOH	286(4.53)	87-1057-84
$C_{30}H_{31}NO_5$ 4H-Cyclohepta[b]furan-4-one, 2,2'-(hydroxyimino)bis[2,3,5,6,7,8-hexahydro-3-phenyl-	MeOH	272(4.367)	73-1421-84
$C_{30}H_{31}N_2O_{12}S_3$ Pyridinium, 1-(5-amino-5-carboxypentyl)-4-(4-methoxy-3-sulfophenyl)-2,6-bis(4-sulfophenyl)-, hydroxide, (S)-	H_2O	223(4.56),340(4.43), 408(3.66)	39-0849-84B
$C_{30}H_{31}N_2S_2$ Benzothiazolium, 3-ethyl-2-[11-(3-ethyl-2(3H)-benzothiazolylidene)-3-methyl-1,3,5,7,9-undecapentaenyl]-, iodide	EtOH	1010(3.9912)	104-0395-84

$C_{30}H_{31}N_2S_2-C_{30}H_{33}NO_{12}$

Compound	Solvent	$\lambda_{max}(\log \epsilon)$	Ref.
Benzothiazolium, 3-ethyl-2-[[11-(3-ethyl-2(3H)-benzothiazolylidene)-4-methyl-1,3,5,7,9-undecapenta-enyl]-, iodide	EtOH	1025(4.0492)	104-0395-84
Benzothiazolium, 3-ethyl-2-[[11-(3-ethyl-2(3H)-benzothiazolylidene)-5-methyl-1,3,5,7,9-undecapenta-enyl]-, iodide	EtOH	1010(4.0128)	104-0395-84
$C_{30}H_{32}Cl_6$ Dispiro[cyclopropane-1,17'-tricyclo-[10.4.1.1^{4,9}]octadeca[2,4,6,8,12-14,16]heptaene-18',1''-cycloprop-ane], 2,2,2'',2'',5',13'-hexachloro-7',15'-bis(1,1-dimethylethyl)-	C_6H_{12}	218(4.39),272(4.31), 350s(3.88)	94-4220-84
$C_{30}H_{32}N_4O_8S_2$ 1H-Benzimidazolium, 2,2'-(1,4-naphtha-lenediyl)bis[1,3-dimethyl-, bis(methyl sulfate)	DMF	279(4.31),305s(4.27)	5-1129-84
$C_{30}H_{32}O_2$ 1H-Indene-6-carboxaldehyde, 2-[3,8-di-methyl-5-(1-methylethyl)-1-azulenyl]-3,7-dimethyl-5-(1-methylethyl)-1-oxo-	MeOH	537(3.50)	138-0627-84
$C_{30}H_{32}O_7$ Sanggenon L	EtOH	229(4.10),273s(4.32), 281(4.35),323(3.96), 370(3.23)	142-1791-84
	EtOH-AlCl$_3$	228(4.11),272s(4.29), 281(4.33),325(3.93), 383(3.17)	142-1791-84
$C_{30}H_{32}O_8$ 1(2H)-Naphthalenone, 2-(6,7-diethoxy-4-methoxy-1-oxo-2(1H)-naphthalenyl-idene)-6,7-diethoxy-4-methoxy-	CHCl$_3$	303(4.47),426(3.98), 670(4.21)	64-0244-84B
$C_{30}H_{32}P$ Phosphonium, diphenyl[2-(3,4,4,5-tetra-methyl-2,5-cyclohexadien-1-ylidene)-ethyl]-, iodide	EtOH	240(4.19),266(4.26), 274s(--)	5-0340-84
$C_{30}H_{32}S_2$ Pentacyclo[11.5.3.3^{4,10}.0^{7,23}.0^{16,20}]-tetracosa-1(19),4,6,8,10(22),13,15-17,20,23-decaene, 19,21,22,24-tetra-methyl-2,11(or 12)-bis(methylthio)-	CHCl	281(4.39)	44-4128-84
$C_{30}H_{33}NO_4$ Benzeneacetamide, α-(1,2-dihydroxy-3-phenyl-2-propenylidene)-N,N-bis(1-methylethyl)-4-(phenyl-methoxy)-, (E,Z)-	EtOH	300(4.41),382(3.92)	39-1539-84C
$C_{30}H_{33}NO_{12}$ Spiro[2-azabicyclo[2.2.2]octa-5,7-di-ene-3,1'-cyclohexane]-1,4,5,6,7,8-hexacarboxylic acid, 2-phenyl-, hexamethyl ester	EtOH	240s(4.50),280(4.36), 334(4.06)	103-0295-84

Compound	Solvent	$\lambda_{max}(\log \epsilon)$	Ref.
$C_{30}H_{33}N_5O_9$ D-Ribitol, 1-[(2-amino-1,6-dihydro-5-nitro-6-oxo-4-pyrimidinyl)amino]-5-O-[bis(4-methoxyphenyl)phenylmethyl]-1-deoxy-	$CHCl_3$	333(4.05)	64-0252-84B
$C_{30}H_{34}$ 1,2'-Biazulene, 1',3,4',8-tetramethyl-5,7'-bis(1-methylethyl)-	MeOH	612(2.79)	138-0627-84
$C_{30}H_{34}Cl_6$ Dispiro[cyclopropane-1,17'-tricyclo-[10.4.1.1^{4,9}]octadeca[4,6,8,12,14,16]-hexaene-18',1"-cyclopropane], 2,2,2"-2",5',13'-hexahydro-7',15'-bis(1,1-dimethylethyl)-	C_6H_{12}	213(4.32),237(4.33), 283(3.70)	94-4220-84
$C_{30}H_{34}N_2O_9S$ 1,2,3-Propanetriol, 1-[6-acetoxy-3a,5-6,6a-tetrahydro-3-(4-methylphenyl)-2-[(phenylmethyl)thio]-3H-furo[2,3-d]imidazol-5-yl]-, triacetate, [3aR-[3aα,5α(1S*,2R*),6α,6aα]]-	EtOH	246(4.06)	136-0091-84B
$C_{30}H_{34}N_3O$ Quinolinium, 6-[(diethylamino)carbo-nyl]-1-ethyl-4-[3-(1-ethyl-4(1H)-quinolinylidene)-1-propenyl]-, iodide	MeOH MeCN $C_6H_4Cl_2$	718(5.22) 714(--) 735(5.26)	103-0767-84 103-0767-84 103-0767-84
$C_{30}H_{34}N_4$ Porphine, 2,4,7-triethyl-1,3,5,8-tetra-methyl-	CH_2Cl_2	385(5.17),496(4.28), 530(4.10),566(3.96), 620(3.78)	44-4602-84
$C_{30}H_{34}N_8O_4S_2$ Benzo[g]pteridine-2,4(3H,4aH)-dione, 4a,4'a-[dithiobis(methylene)]bis-[5,10-dihydro-3,7,8,10-tetramethyl-	6M HCl MeCN	234(4.53),265(3.48), 300s(--),401(4.58) 225(4.68),270(4.38), 300s(--),360(3.98)	35-3309-84 35-3309-84
$C_{30}H_{34}O_{11}$ Ainsliaside A	MeOH	245s(4.02),302s(4.10), 332(4.24)	94-3043-84
$C_{30}H_{35}NO_{10}$ D-Gluconic acid, 2-deoxy-2-[(3-ethoxy-3-oxo-1-phenyl-1-propenyl)amino]-4,6-O-(phenylmethylene)-, ethyl ester, 3,5-diacetate, (Z)-	MeOH	228(3.89),291(4.18)	136-0081-84B
$C_{30}H_{36}N_4O_4$ 21H-Biline-3-acetic acid, 17-ethyl-1,2,3,19,23,24-hexahydro-3,7,8,12-13,18-hexamethyl-1,19-dioxo-, methyl ester, (4Z,9Z,15Z)-(±)-	$CHCl_3$	275(4.28),348(4.57), 589(4.20)	49-0101-84
Furo[2,3-b]-21H-biline-1,5',19-trione, 17-ethyl-1,2,3,4,4',5,5',19,22,24-decahydro-3,7,8,12,13,18,21-hepta-methyl-, (10Z,15Z)-(±)-	$CHCl_3$	324(4.26),553(4.11)	49-0101-84

Compound	Solvent	$\lambda_{max}(\log \epsilon)$	Ref.
Furo[2,3-b]-21H-biline-1,5',19-trione, 17-ethyl-1,2,3,4,4',5,5',19,22,24-decahydro-2,2,7,8,12,13,18-hepta-methyl-, (10Z,15Z)-(±)-	CHCl$_3$	323(4.39),364s(4.05), 514s(4.08),553(4.20)	49-0101-84
$C_{30}H_{36}N_8O_{11}$			
L-Glutamic acid, N-[4-[[[2-amino-3,4,5-6-tetrahydro-4-oxo-5-[(2-deoxyuridin-5-yl)methyl]pyrido[3,2-d]pyrimidin-6-yl]methyl]amino]benzoyl]-	pH 1 pH 7 pH 13	269(4.23) 278(4.36),299(4.33) 291(4.42)	87-1710-84 87-1710-84 87-1710-84
$C_{30}H_{37}NO_3S$			
Benzenesulfonamide, 2,4,6-tris(1-meth-ylethyl)-N-(2-oxo-2-phenylethyl)-N-(phenylmethyl)-	ether	239.7(4.44),325.0(1.89)	48-0177-84
$C_{30}H_{38}N_6O_9$			
Guanosine, N^2,3',5-triisobutyryl-O^6-p-nitrophenylethyl-2'-deoxy-	MeOH	218(4.42),269(4.40)	78-0059-84
$C_{30}H_{38}O_6$			
2-Propenoic acid, 3-(3-methoxyphenyl)-, 1,10-decanediyl ester	MeCN	276(4.63)	40-0022-84
2-Propenoic acid, 3-(4-methoxyphenyl)-, 1,10-decanediyl ester	MeOH	308(4.70)	40-0022-84
$C_{30}H_{38}O_{15}$			
Cistanoside C	MeOH	222(4.18),245s(3.90), 290s(4.00),332(4.15)	94-3880-84
$C_{30}H_{40}Cl_2N_2O_3$			
4-Aza-5-androsten-3-one, 17-[[[4-[bis-(2-chloroethyl)amino]phenyl]acetyl]-oxy]-, (17β)-	n.s.g.	206(4.37),256(4.29), 300(3.04)	111-0465-84
$C_{30}H_{40}Cl_2N_2O_4$			
4-Aza-5-androsten-3-one, 17-[[[4-[bis(2-chloroethyl)amino]phenoxy]-acetyl]oxy]-, (17β)-	n.s.g.	205(4.33),256(4.29), 315(3.43)	111-0465-84
$C_{30}H_{40}N_2O_5S$			
Benzenesulfonic acid, 4-methyl-, [(1-acetoxy-6a,7,8,9,10,10a-hexahydro-6,6-dimethyl-3-pentyl-6H-dibenzo-[b,d]pyran-9-yl)methylene]hydra-zide, [6aR(6aα,9α,10aβ)]-	EtOH	274(3.48),282(3.41)	78-3839-84
9β-	EtOH	274(3.39),283(3.33)	78-3839-84
$C_{30}H_{40}N_4O_4S_4$			
1-Oxa-4,10-dithia-7-azacyclododecane, 7,7'-[azobis(4,1-phenylenecarbo-nyl)]bis-	CH$_2$Cl$_2$	330(4.46),445(2.89)	18-2879-84
in propylene carbonate		330(4.46),435(2.94)	18-2879-84
$C_{30}H_{40}O_6$			
Methanone, [3,4-dihydro-7-hydroxy-5-methoxy-3-(3-methoxy-3-methylbutyl)-2,2-dimethyl-6-(3-methyl-2-butenyl)-2H-1-benzopyran-8-yl](3-hydroxy-phenyl)-	MeOH	326(3.70)	32-0055-84

Compound	Solvent	$\lambda_{max}(\log \epsilon)$	Ref.
$C_{30}H_{40}O_7$ Methanone, (3-hydroxyphenyl)[3,4,7,8-tetrahydro-3-hydroxy-5-methoxy-7-(3-methoxy-3-methylbutyl)-2,2,8,8-tetramethyl-2H,6H-benzo[1,2-b:5,4-b']dipyran-10-yl]-	MeOH	326(3.58)	32-0055-84
$C_{30}H_{40}O_8$ Excoecaria toxin	MeOH	232(4.40),280(2.71)	100-0270-84
$C_{30}H_{40}O_{15}$ Eruberin B	MeOH	225.5(4.50),276.5(3.69), 282(3.69)	94-0490-84
$C_{30}H_{41}N_3O_5$ Alangamide	EtOH and EtOH-HCl EtOH-NaOH	211(4.66),227s(4.13), 286(3.85) 222(4.74),230s(4.19), 289(3.93),302s(3.82)	100-0397-84 100-0397-84
$C_{30}H_{42}Cl_2N_2O_3$ 4-Azaandrostan-3-one, 17-[[[4-[bis(2-chloroethyl)amino]phenyl]acetyl]-oxy]-, (5α,17β)-	n.s.g.	203(4.42),255(4.16), 315(3.32)	111-0465-84
$C_{30}H_{42}Cl_2N_2O_4$ 4-Azanadrostan-3-one, 17-[[[4-[bis(2-chloroethyl)amino]phenoxy]acetyl]-oxy]-, (5α,17β)-	n.s.g.	204(4.42),256(4.16), 315(3.32)	111-0465-84
$C_{30}H_{42}N_2O_{10}$ 8,9-Epoxyherbimycin A Herbimycin A, 7,9-cyclic carbamate	MeOH MeOH	272(4.38) 274(4.55)	158-1264-84 158-1264-84
$C_{30}H_{42}O_7$ 4H-1-Benzopyran-4-one, 2-(4,6-dimethoxy-3,5,11-trimethyl-7,9,11-tridecatrienyl)-5-hydroxy-7,8-dimethoxy-3-methyl-	MeOH	246(4.56),258(4.65), 265(4.65),279(4.47), 340(3.69)	5-1883-84
4H-1-Benzopyran-4-one, 2-(4,6-dimethoxy-3,5,11-trimethyl-7,9,11-tridecatrienyl)-8-hydroxy-5,7-dimethoxy-3-methyl- (stigmatellin A)	MeOH	248s(4.61),258(4.77), 267(4.82),279(4.62), 335(3.72)	5-1883-84
	MeOH-HCl	243s(4.47),264(4.80), 274(4.81),283s(4.66), 329(3.68),420(4.32)	5-1883-84
	MeOH-KOH	249s(4.66),258(4.78), 268(4.86),277(4.73), 320(3.40),370(3.48)	5-1883-84
4H-1-Benzopyran-4-one, 2-(4,12-dimethoxy-3,5,11-trimethyl-6,8,10-tridecatrienyl)-8-hydroxy-5,7-dimethoxy-3-methyl-, (6E,8E,10E)-	MeOH	227s(4.28),246(4.50), 264(4.74),275(4.71), 287(4.60),330(3.80)	5-1883-84
Stigmatellin	MeOH	266f(<u>4.8</u>),330(<u>3.7</u>)	158-0454-84
$C_{30}H_{44}N_4O_2S_4$ 1-Oxa-4,10-dithia-7-azacyclododecane, 7,7'-[azobis(4,1-phenylenemethylene)]bis]- in propylene carbonate	CH_2Cl_2 	325(4.30),436(2.72) 330(4.36),440(2.89)	18-2879-84 18-2879-84

Compound	Solvent	$\lambda_{max}(\log \epsilon)$	Ref.
$C_{30}H_{44}O_3$			
Lupa-1,20(29)-dien-26-oic acid, 3-oxo-	MeOH	230(3.81)	78-2069-84
$C_{30}H_{44}O_4$			
Lup-20(29)-en-26-oic acid, 1,3-dioxo-	MeOH	257(4.00)	78-2069-84
	MeOH-NaOH	288(4.42)	78-2069-84
Uvariastrol	MeOH	241(4.29)	102-2077-84
$C_{30}H_{44}O_6$			
Androsta-5,7-diene-3,17-diol, 19-[(tetrahydro-2-furanyl)oxy]-, 3-acetate, 17-(2,2-dimethylpropanoate)-, (3β,17β)-	EtOH	271(4.01),281(4.01)	56-0711-84
5β-Chola-8,14-diene-3,12,24-triol, triacetate, (3α,5β,12α)-	EtOH	246(3.81)	70-2580-84
$C_{30}H_{44}O_7$			
Androst-5-en-7-one, 3-acetoxy-17-(2,2-dimethyl-1-oxopropoxy)-19-[(tetrahydro-2-furanyl)oxy]-, (3β,17β)-	EtOH	234(4.12)	56-0711-84
$C_{30}H_{46}Cl_2N_3O_3P$			
Phosphoramidohydrazidic acid, N,N-bis(2-chloroethyl)-2-[(5α,17β)-17-hydroxyandrostan-3-ylidene]-, 4-methylphenyl ester	EtOH	268(2.98)	65-0389-84
	cation	267(--)	65-0389-84
$C_{30}H_{46}Cl_2N_3O_4P$			
Phosphoramidohydrazidic acid, N,N-bis(2-chloroethyl)-2-[(5α,17β)-17-hydroxyandrostan-3-ylidene]-, 4-methoxyphenyl ester	EtOH	267(2.88)	65-0389-84
	cation	267(--)	65-0389-84
$C_{30}H_{46}O$			
Oleana-9(11),12-dien-3-one	CH_2Cl_2	283(3.93)	94-3674-84
Oleana-9(11),12,15-trien-3-ol, (3β)-	MeOH	282(3.81)	2-0712-84
$C_{30}H_{46}O_2$			
Lup-20(29)-en-30-al, 3-oxo-	MeOH	227(3.30)	102-0631-84
$C_{30}H_{46}O_5$			
Pregna-5,17(20)-dien-21-oic acid, 20-ethoxy-3β-[(tetrahydro-2H-pyran-2-yl)oxy]-, ethyl ester, (3β,17E)-	EtOH	237(3.88)	118-0132-84
$C_{30}H_{46}O_7$			
Androsta-5,7-diene-3,17-diol, 19-[(2-methoxyethoxy)methoxy]-, 3-acetate 17-(2,2-dimethylpropanoate)-, (3β,17β)-	EtOH	272(4.02),282(4.01)	56-0711-84
$C_{30}H_{46}O_8$			
Androst-5-en-7-one, 3-acetoxy-17-(2,2-dimethyl-1-oxopropoxy)-19-[(2-methoxyethoxy)methoxy]-, (3β,17β)-	EtOH	234(4.15)	56-0711-84
$C_{30}H_{46}O_{12}$			
Picrasinoside D	EtOH	262(3.58)	138-0221-84
$C_{30}H_{46}O_{13}$			
Picrasinoside E	EtOH	262(3.54)	138-0221-84

Compound	Solvent	λ_{max}(log ϵ)	Ref.
C$_{30}$H$_{47}$NO$_2$S$_2$ 4-Thiazolidinone, 3-decyl-5-[[4-(decyl-oxy)phenyl]methylene]-2-thioxo-	n.s.g.	403(4.54)	48-0457-84
C$_{30}$H$_{48}$O Benzo[4,5,6]cholest-4-en-5'(6'H)-one, 4α,5α-dihydro-	EtOH	248(4.07),254(4.03)	39-0397-84C
4α,5β-	EtOH	243(4.02)	39-0397-84C
C$_{30}$H$_{48}$O$_2$ Lanost-8-ene-3,7-dione, (5α)-	EtOH	254(3.99)	56-0705-84
Lup-20(29)-en-30-al, 3β-hydroxy-	MeOH	228(3.89)	102-0631-84
C$_{30}$H$_{48}$O$_4$ Cholestan-3-one, 5α,6β-dihydroxy-24-methylene-, 6-acetate	dioxan	225(2.73)	106-0117-84
C$_{30}$H$_{48}$O$_7$ 4H-1-Benzopyran-4-one, 2-(4,6-dimeth-oxy-3,5,11-trimethyltridecyl)-8-hy-droxy-5,7-dimethoxy-3-methyl-	MeOH	228(4.19),244(4.34), 255(4.32),261(4.32), 286(3.42),335(3.74)	5-1883-84
C$_{30}$H$_{50}$O$_3$ 5α-Cholest-9(11)-en-12-one, 3β-acetoxy-7a,8-seco-B-homo-	EtOH	237(4.23)	73-0301-84
C$_{30}$H$_{50}$O$_4$ Cholestan-3-one, 6β-acetoxy-5α-hydroxy-24-methyl-, (S)-	dioxan	226(2.74)	106-0117-84
Cholestan-6-one, 3β-acetoxy-5α-hydroxy-24-methyl-, (S)-	dioxan	218(2.38),250(2.18)	106-0117-84
5α-Lanost-8-en-7-one, 11ξ-hydroperoxy-3β-hydroxy-	EtOH	251(3.92)	56-0705-84
C$_{30}$H$_{50}$O$_7$ 4H-1-Benzopyran-4-one, 2-(4,6-dimeth-oxy-3,5,11-trimethyltridecyl)-2,3-dihydro-8-hydroxy-5,7-dimethoxy-3-methyl-	MeOH	212(4.29),243(4.07), 286(4.12),345(3.66)	5-1883-84
C$_{30}$H$_{53}$NOS Cholestan-3-ol, dimethylcarbamothioate, (3β,5α)-	EtOH	251(4.30)	39-1005-84C
C$_{30}$H$_{54}$N$_2$OS Cholestan-3-ol, 2,2-dimethylhydrazine-carbothioate, (3β,5α)-	EtOH	248(4.07)	39-1005-84C
C$_{30}$H$_{58}$N$_2$Si$_6$ 1,2-Disilenediamine, 1,2-bis(2,4,6-trimethylphenyl)-N,N,N',N'-tetra-kis(trimethylsilyl)-, (E)-	benzene	351(<u>3.7</u>),483(<u>3.7</u>)	35-0821-84
(Z)-	benzene	362(<u>3.9</u>),468(<u>3.5</u>)	35-0821-84
C$_{30}$H$_{60}$Si$_6$ 6,7,13,19,25,31-Hexasilahexaspiro[5.0-5.0.5.0.5.0.5.0.5.0.]hexatriacontane	isooctane	204s(4.64),244(3.81), 267s(3.20)	35-5521-84
C$_{30}$H$_{70}$Si$_5$ Cyclopentasilane, decapropyl-	C$_6$H$_{12}$	260(3.15)	157-0141-84

Compound	Solvent	$\lambda_{max}(\log \epsilon)$	Ref.
$C_{31}H_{16}O_3$			
10H-Phenanthro[9,10-b]phenanthro-[9',10':4,5]furo[2,3-d]pyran-10-one	$C_6H_4Cl_2$	379(4.61),400(4.56)	24-0666-84
$C_{31}H_{17}Br_3O_3$			
Dispiro[naphthalene-1(2H),2'(3'H)-[1H]-naphtho[2,1-b]pyran-3',1"(2"H)-naph-thalene]-2,2'-dione, 6,6",8'-tri-bromo-, cis	CHCl$_3$	247(4.96),284(4.17), 294(4.14),310(4.17), 344(3.88),360(3.68)	39-2375-84C
trans	CHCl$_3$	246(4.84),270(4.18), 280(4.15),296(4.10), 328(3.94),346(3.77), 360(3.52)	39-2375-84C
$C_{31}H_{20}N_2O$			
Furo[3,4-c]pyrazolo[5,1-a]isoquinoline, 2,8,10-triphenyl-	MeOH	255(4.62),260(4.42), 300(4.44),340s(--)	44-4647-84
$C_{31}H_{20}N_2O_3$			
5(4H)-Oxazolone, 4-[[4-(5-[1,1'-bi-phenyl]-4-yl-2-oxazolyl)phenyl]-methylene]-2-phenyl-	toluene	320(4.36),420(4.70)	135-1330-84
5(4H)-Oxazolone, 2-[1,1'-biphenyl]-4-yl-4-[[4-(5-phenyl-2-oxazolyl)phen-yl]methylene]-	toluene	350(4.54),430(--)	135-1330-84
$C_{31}H_{20}O_3$			
Dispiro[naphthalene-1(2H),2'(3'H)-[1H]-naphtho[2,1-b]pyran-3',1"-naphtha-lene]-2,2"(2"H)-dione (same for cis and trans isomers)	CHCl$_3$	245(4.37),268(3.90), 280(3.88),290(3.80), 316(3.78),334(3.67), 360(3.04)	39-2375-84C
$C_{31}H_{20}O_{10}$			
Podocarpusflavone A	MeOH	215(4.60),270(4.54), 335(4.43)	154-0547-84
$C_{31}H_{22}Cl_2N_4$			
[1,2,4]Triazolo[1,5-a]pyridinium, 2-[(4-chlorophenyl)methyl]-1-[(4-chlorophenyl)methyl]-5,7-diphenyl-, hydroxide, inner salt	EtOH	261(4.93)	39-1891-84C
$C_{31}H_{22}N_2O_2$			
2-Butene-1,4-dione, 2-(3,5-diphenyl-1H-pyrazol-1-yl)-1,4-diphenyl-, (Z)-	MeCN	248(4.58),345(4.12)	44-4647-84
Methanone, (5,6-dihydro-2-phenylpyra-zolo[5,1-a]isoquinoline-5,6-diyl)-bis[phenyl-, cis	MeOH	225s(4.34),250(4.50)	44-4647-84
$C_{31}H_{22}N_8O_3S$			
1H-Pyrazole, 3,5-diphenyl-1-(2-pyri-dinylcarbonyl)-4-[[2-[(2-pyrimidin-ylamino)sulfonyl]phenyl]azo]-	n.s.g.	378(3.78)	48-1021-84
$C_{31}H_{22}O_2$			
Methanone, (4b,8d-dihydro-4b-methyl-dibenzo[a,f]cyclopropa[cd]penta-lene-8b,8c-diyl)bis[phenyl-	MeOH	251(4.45),275s(3.89), 300s(3.15)	44-4923-84
$C_{31}H_{22}O_{11}$			
Dechlorogilmaniellin	MeCN	236(4.64),272s(4.10),	33-0690-84

Compound	Solvent	$\lambda_{max}(\log \epsilon)$	Ref.
(cont.)		304(3.75)	33-0690-84
$C_{31}H_{23}BrN_4O_8$ 4H-Pyrazolo[3,4-d]pyrimidin-4-one, 3-bromo-1,5-dihydro-1-(2,3,5-tri-O-benzoyl-β-D-ribofuranosyl)-	EtOH	224(4.31),263s(3.56)	87-1119-84
$C_{31}H_{23}NO_2$ 5(4H)-Oxazolone, 4-(1,2-diphenyl-1H-inden-3-yl)-4-methyl-2-phenyl-	C_6H_{12}	228(4.49),302(4.06), 316(4.12),332(3.89)	35-1065-84
5(4H)-Oxazolone, 4-(1-methyl-2,3-di-phenyl-2-cyclopropen-1-yl)-2,4-di-phenyl-	C_6H_{12}	237(4.45),316(4.43), 334(4.34)	35-1065-84
5(4H)-Oxazolone, 4-methyl-2-phenyl-4-(1,2,3-triphenyl-2-cyclopropen-1-yl)-	C_6H_{12}	239(4.55),318(4.41), 335(4.29)	35-1065-84
$C_{31}H_{23}NS$ Benzenamine, N-[2-(5,6-diphenyl-2H-thiopyran-2-ylidene)-2-phenyleth-ylidene)-2-phenylethylidene]-, hydrochloride	EtOH	520(4.40)	97-0183-84
$C_{31}H_{24}$ Bicyclo[2.2.1]hepta-2,5-diene, 2,3,5,6-tetraphenyl-	C_6H_{12}	233(4.43),249(4.27), 289(4.43)	44-0856-84
Methane, bis(2,3-diphenylcyclopropen-2-yl)-	C_6H_{12}	225(4.50),230(4.51), 240(4.44),313(4.59), 323(4.66),330(3.61), 342(4.55)	44-0856-84
$C_{31}H_{24}BrN_5O_7$ 1H-Pyrazolo[3,4-d]pyrimidin-4-amine, 3-bromo-1-(2,3,5-tri-O-benzoyl-β-D-ribofuranosyl)-	EtOH	228(3.86),271(4.19), 288s(3.93)	87-1119-84
$C_{31}H_{24}N_2$ Cyclopentapyrazole, 6-(diphenylmethyl-ene)-1,3a,6,6a-tetrahydro-1,3-di-phenyl-	MeOH	235(4.88),289(4.86), 335(4.65)	78-1585-84
$C_{31}H_{24}N_2O_2$ 1,4-Butanedione, 2-(3,5-diphenyl-1H-pyrazol-1-yl)-	$CHCl_3$	248(4.41)	44-4647-84
$C_{31}H_{24}N_2O_3$ 1H-Pyrazole-1-acetic acid, α-[2-phen-oxy-2-(phenylethenyl)-3,5-diphenyl-, (Z)-	MeOH	252(4.68)	44-4647-84
$C_{31}H_{24}N_4$ 1,2,4-Triazolo[1,5-a]pyridinium, 5,7-diphenyl-2-(phenylamino)-1-(phenyl-methyl)-, hydroxide, inner salt	EtOH	257(4.75)	39-1891-84C
$C_{31}H_{24}N_4O_7$ 2,4(1H,3H)-Pteridinedione, 3-[2,3-O-carbonyl-5-O-(triphenylmethyl)-β-D-ribofuranosyl]-	MeOH	228s(4.28),326(3.85)	136-0179-84F

Compound	Solvent	$\lambda_{max}(\log \epsilon)$	Ref.
$C_{31}H_{24}O_2$ Methanone, 4b,10-dihydro-4b-methyl- indeno[1,2-a]indene-9,9a(9H)-diyl)- bis[phenyl-, (4bα,9aα)-	MeOH	245(4.37),285(3.49)	44-4923-84
$C_{31}H_{24}O_{10}$ Benzo[a]naphthacene-8,13-dione, 4,7,9,12-tetraacetoxy-5,6-di- hydro-2-methyl-	EtOH	300(4.34),382(3.18)	23-2818-84
$C_{31}H_{25}BO_{10}$ 1,5-Methano-1H-anthra[2,3-e][1,3,2]di- oxaborocin-8,13-dione, 7,14-diacet- oxy-5-(acetoxymethyl)-5,6-dihydro- 3-phenyl-, (1S)-	CHCl$_3$	260(4.66),343(3.79)	78-4649-84
$C_{31}H_{25}Br_2N_7O_2$ Pyrazolo[5,1-b]quinazolin-9(4H)-one, 3,3'-[[4-(dimethylamino)phenyl]- methylene]bis[7-bromo-2-methyl-	10% EtOH- NaOH	218(5.04),255(4.91), 332(4.60),402(3.98)	56-0411-84
$C_{31}H_{25}ClN$ Pyridinium, 1-[2-(4-chlorophenyl)eth- yl]-2,4,6-triphenyl-	C$_6$H$_5$Cl	310(4.48)	120-0062-84
$C_{31}H_{25}ClN_6O_4$ Pyrazolo[5,1-b]quinazolin-9(4H)-one, 3,3'-[(4-chlorophenyl)methylene]- bis[7-methoxy-2-methyl-	10% EtOH- NaOH	223(4.81),248(4.81), 326(4.44),402(4.02)	56-0411-84
$C_{31}H_{25}N_7O_6$ Pyrazolo[5,1-b]quinazolin-9(4H)-one, 3,3'-[(4-nitrophenyl)methylene]- bis[7-methoxy-2-methyl-	10% EtOH- NaOH	220(4.74),244(4.68), 325(4.34),398(4.03)	56-0411-84
$C_{31}H_{26}NO$ Pyridinium, 1-[(4-methoxyphenyl)meth- yl]-2,4,6-triphenyl-	C$_6$H$_5$Cl	312(4.49)	120-0062-84
$C_{31}H_{26}NO_{10}S_3$ Pyridinium, 4-(4-methoxy-3-sulfophen- yl)-1-(phenylmethyl)-2,6-bis(4-sul- fophenyl)-, hydroxide	H$_2$O	222(4.60),343(4.46)	39-0849-84B
$C_{31}H_{26}N_6$ Methanone, (5,6-dihydro-2-phenylpyrazo- lo[5,1-a]isoquinoline-5,6-diyl)bis- [phenyl-, dihydrazone, trans	MeOH	260(4.33)	44-4647-84
$C_{31}H_{26}N_6O_4$ Pyrazolo[5,1-b]quinazolin-9(4H)-one, 3,3'-(phenylmethylene)bis[7-methoxy- 2-methyl-	10% EtOH- NaOH	221(4.80),249(4.80), 268(4.73),325(4.45), 405(4.06)	56-0411-84
$C_{31}H_{26}O_2$ 2,4-Pentadien-1-one, 5-ethoxy-1,2,4,5- tetraphenyl-	MeOH	204(4.70),233(4.30), 353(4.11)	80-0817-84
$C_{31}H_{26}O_5$ 1-Naphthalenecarboxylic acid, 2-acetyl- 1,2,3,4,5,11,11a,12-octahydro-4a,12a-	MeCN	224s(4.61),238s(4.34), 270(4.09),284s(3.98),	78-4549-84

Compound	Solvent	$\lambda_{max}(\log \epsilon)$	Ref.
5,12-diepoxynaphthacen-7-yl ester, (2α,4aα,5α,11aα,12α,12aα)-(±)- (cont.)		300s(3.87),324s(3.62)	78-4549-84
$C_{31}H_{27}BO_8$ 1,5-Methano-1H-anthra[2,3-e][1,3,2]di-oxaborocin-7,14-diol, 5-(acetoxy-methyl)-5,8-dihydro-3-phenyl-, diacetate,)1S)-	CHCl₃	264(5.28),352(3.70), 371(3.86),392(3.78)	78-4649-84
$C_{31}H_{27}N_3O_2$ 1,5-Naphthalenediol, 2,6-dihydro-4-[(4-methylphenyl)amino]-2,6-bis-[(4-methylphenyl)imino]-	CHCl₃	275(4.61),332(4.05), 438(4.02),634(4.31), 683(4.32)	104-0733-84
1,5-Naphthalenedione, 2,4,8-tris[(4-methylphenyl)amino]-	CHCl₃	286(4.50),360(3.83), 532(3.86),628(4.27), 675(4.45)	104-0733-84
$C_{31}H_{27}N_7O_2$ Pyrazolo[5,1-b]quinazolin-9(4H)-one, 3,3'-[[4-(dimethylamino)phenyl]-methylene]bis[2-methyl-	10% EtOH-NaOH	217(4.94),254(4.95), 317(4.54),388(4.13)	56-0411-84
$C_{31}H_{28}N_2$ Benzo[b]quinolizinium, 9,9'-(1,5-pent-anediyl)bis-	MeOH	260(3.06)	48-0757-84
3-Quinolinecarbonitrile, 2-[1,1'-bi-phenyl]-4-yl-5,6,7,8-tetrahydro-4-[4-(1-methylethyl)phenyl]-	EtOH	208(4.70),217s(4.56), 236(4.25),286(4.50)	73-1395-84
$C_{31}H_{28}N_2O$ 5H-Cyclohepta[b]pyridine-3-carboni-trile, 2-[1,1'-biphenyl]-4-yl-4-(4-ethoxyphenyl)-6,7,8,9-tetrahydro-	EtOH	209(4.61),219s(4.48), 244(4.24),289(4.54)	73-1395-84
$C_{31}H_{28}N_4O_3$ 4-Imidazolidinecarboxamide, 3-[4-(di-methylamino)phenyl]-1-(4-methyl-phenyl)-2,5-dioxo-N,4-diphenyl-	EtOH	207(4.46),267(4.51)	49-0187-84
$C_{31}H_{28}N_4O_4$ 4-Imidazolidinecarboxamide, 3-[4-(di-methylamino)phenyl]-1-(4-methoxy-phenyl)-2,5-dioxo-N,4-diphenyl-	EtOH	208(4.54),253(4.47), 267(4.35)	49-0187-84
$C_{31}H_{29}NO_6$ Pentacyclo[19.3.1.1³,⁷.1⁹,¹³.1¹⁵,¹⁹]-octacosa-1(25),3,5,7(28),9,11,13(27)-15,17,19(26),21,23-dodecaene-25,26,27,28-tetrol, 5,11,17-trimethyl-23-nitro-	MeOH-HCl pH 7.9 + LiOH + NaOH + KOH	322(3.99) 400(3.93) 430(3.95) 451(4.05) 453(4.06)	126-0221-84B 126-0221-84B 126-0221-84B 126-0221-84B 126-0221-84B
$C_{31}H_{30}N_2O_8S_2$ 3-Cephem-4-carboxylic acid, 7β-amino-2-[(Z)-(methoxycarbonyl)methylene]-3-methyl-, diphenylmethyl ester, salt with p-toluenesulfonic acid	EtOH	350(4.05)	87-1225-84
$C_{31}H_{30}O_7$ Benzeneacetic acid, α-(3-oxo-1,4-di-oxaspiro[4.5]dec-2-ylidene)-3,4-bis(phenylmethoxy)-, methyl ester, (E)-	EtOH	238(4.04),328(4.26)	39-1547-84C

Compound	Solvent	$\lambda_{max}(\log \epsilon)$	Ref.
$C_{31}H_{31}FeN_2$ 3H-Indolium, 2-[4-ferrocenyl-4-(phenyl-amino)-1,3-butadienyl]-1,3,3-tri-methyl-, perchlorate	EtOH	264(4.20),395(3.55), 482(3.42)	65-1439-84
$C_{31}H_{31}NO_3$ 2H-Naphtho[1',2':4,5]pyrano[3,2-h]iso-quinoline, 1,3,4,6-tetrahydro-8,9-di-methoxy-2,4-dimethyl-3-(phenylmeth-yl)-, cis-(±)-	CHCl$_3$	331(4.2),345(4.3), 361(4.2),367(3.0)	88-2537-84
$C_{31}H_{31}N_5O_4S$ 7H-Pyrazolo[4,3-d]pyrimidine-7-thione, 5-amino-1,4-dihydro-3-[2,3,5-tris-0-(phenylmethyl)-β-D-arabinofurano-syl]-	MeOH	235(4.14),258(3.78), 305(3.70),359(4.11)	44-0528-84
7H-Pyrazolo[4,3-d]pyrimidine-7-thione, 5-amino-1,4-dihydro-3-[2,3,5-tris-0-(phenylmethyl)-β-D-ribofuranosyl]-	MeOH	233(4.15),258(3.80), 292(3.63),366(4.15)	44-0528-84
$C_{31}H_{32}N_3O$ Pyrylium, 2,6-bis[4-(dimethylamino)-phenyl]-4-[[4-(dimethylamino)phenyl]-ethynyl]-, perchlorate	CH$_2$Cl$_2$	590(4.89),640(4.80)	88-5143-84
$C_{31}H_{32}N_4O_4$ 21H,23H-Porphine-2,18-dipropanoic acid, 3,7,12,17-tetramethyl-, 2-methyl ester	CHCl$_3$	398(5.47),495(4.32), 528(4.01),566(3.90), 618(3.26)	65-1610-84
$C_{31}H_{32}O_7$ Naematolin 2,3-di-O-benzoate	CHCl$_3$	231(4.41),274(3.68), 283s(3.65)	5-1332-84
$C_{31}H_{32}O_9$ Sanggenon C triacetate	EtOH	229(5.04),260(4.30), 311(3.93)	142-1791-84
$C_{31}H_{33}NO_3$ 8-Isoquinolinol, 7-(4,5-dimethoxy-2-methyl-1-naphthalenyl)-1,2,3,4-tetrahydro-1,3-dimethyl-2-(phen-ylmethyl)-	CHCl$_3$	305(4.0),319(4.0), 334(3.9)	88-2537-84
$C_{31}H_{33}NO_4S_2$ Ethanamine, N-methyl-2-[2-[(4-methyl-phenyl)sulfinyl]phenoxy]-N-[2-[2-[(4-methylphenyl)sulfinyl]phenoxy]-ethyl]-, [R-(R*,R*)]-	EtOH	220(4.47),239(4.44), 285(3.89)	12-2059-84
$C_{31}H_{33}NO_5S_2$ Ethanamine, N-methyl-2-[2-[(4-methyl-phenyl)sulfinyl]phenoxy]-N-[2-[2-[(4-methylphenyl)sulfinyl]phenoxy]-ethyl]-, N-oxide, [R-(R*,R*)]-	EtOH	241(4.17),283(3.65)	12-2059-84
$C_{31}H_{34}N_2O_{10}S$ D-Arabinitol, 3-C-(1-phenyl-2-[(phenyl-methyl)thio]-1H-imidazol-4-yl]-, 1,2,3,4,5-pentaacetate	EtOH	262(3.74)	136-0091-84B

Compound	Solvent	$\lambda_{max}(\log \epsilon)$	Ref.
$C_{31}H_{34}N_4$ 4,7:14,17-Diimino-2,21-metheno-9,12-nitrilo-18H-1-benzazacyclononadecine, 11-ethyl-19,20-dihydro-3,6,10-15,18,22-hexamethyl- (absorbances approximate)	CH_2Cl_2	400(5.12),498(4.02), 533(3.51),568(3.60), 623(2.85)	78-4033-84
$C_{31}H_{34}N_5O_9P$ Adenosine, N-benzoyl-3'-deoxy-3'-[(diethoxyphosphinyl)methyl]-, 2'-acetate 5'-benzoate	MeOH	229(4.39),279(4.30)	78-0079-84
$C_{31}H_{34}O_{17}$ 4H-1-Benzopyran-4-one, 2-[4-(β-D-allopyranosyloxy)phenyl]-7-[(4,6-di-O-acetyl-β-D-allopyranosyl)oxy]-5-hydroxy-	MeOH MeOH-NaOAc MeOH-AlCl$_3$	269.5(4.47),317.5(4.45) 268(4.47),316.5(4.45) 279(4.42),299(4.00), 337.5(4.48),382(4.24)	94-5023-84 94-5023-84 94-5023-84
$C_{31}H_{35}N_2$ 3H-Indolium, 2-[9-(1,3-dihydro-1,3,3-trimethyl-2H-indol-2-ylidene)-1,3,5,7-nonatetraenyl]-1,3,3-trimethyl-, iodide	$C_6H_{11}Me$ EtOH DMF	865(5.63) 850(5.32) 855(5.18)	99-0415-84 99-0415-84 99-0415-84
$C_{31}H_{35}N_4O_3S_2$ Benzothiazolium, 2-[[5-[[5-(ethoxycarbonyl)-1,3-diethyl-1,3-dihydro-2H-benzimidazol-2-ylidene]ethylidene]-3-ethyl-4-oxo-2-thiazolidinylidene]methyl]-3-ethyl-, iodide	EtOH	591(4.93)	103-0665-84
$C_{31}H_{35}N_4O_4S$ Benzoxazolium, 2-[[5-[[5-(ethoxycarbonyl)-1,3-diethyl-1,3-dihydro-2H-benzimidazol-2-ylidene]ethylidene]-3-ethyl-4-oxo-2-thiazolidinylidene]methyl]-3-ethyl-, iodide	EtOH	577(4.98)	103-0665-84
$C_{31}H_{37}NO_4S_2$ 2-Propenoic acid, 3-[4-(octyloxy)phenyl]-, 4-[(3-butyl-4-oxo-2-thioxo-5-thiazolidinylidene)methyl]phenyl ester	n.s.g.	394(4.35)	48-0457-84
$C_{31}H_{37}NO_6$ 2-Butenedioic acid, 11-[2-(3-cyano-3-3-phenyloxiranyl)phenoxy]undecyl methyl ester	EtOH	275s(3.62),281(3.65)	24-2157-84
$C_{31}H_{37}N_3O_5$ 1,2,4-Triazolidine-3,5-dione, 1-(1-acetoxy-6a,7,8,9-tetrahydro-6,8,9-trimethyl-3-pentyl-6H-dibenzo[b,d]-pyran-9-yl)-4-phenyl-, (6aR-trans)-	EtOH	222(4.30),255(4.16), 300(3.65)	39-2881-84C
$C_{31}H_{38}N_4O_4$ 21H-Biline-3-acetic acid, 17-ethyl-1,2,3,19,23,24-hexahydro-3,7,8,12-13,18,21-heptamethyl-1,19-dioxo-, methyl ester (4E/Z,10Z,15Z)-(±)-	CHCl$_3$	329(4.41),576(4.19)	49-0101-84

Compound	Solvent	$\lambda_{max}(\log \epsilon)$	Ref.
Furo[2,3-b]-21H-biline-1,5',19-trione, 17-ethyl-1,2,3,4,4',5,5',19,22,24-decahydro-2,2,7,8,12,13,18,21-octa-methyl-, (10Z,15Z)-(±)-	CHCl$_3$	320(4.42),546(4.12)	49-0101-84
$C_{31}H_{38}O_9$ [2,6'-Bi-2H-1-benzopyran-4(3H)-one, 5,7,7'-tris(methoxymethoxy)-2',2'-dimethyl-6-(3-methyl-2-butenyl)-	EtOH	234s(4.51),293(4.28), 330s(4.25),335(4.26), 394(4.53)	142-0997-84
	EtOH-AlCl$_3$	234s(4.51),293(4.26), 330s(4.23),335(4.25), 394(4.52)	142-0997-84
2-Propen-1-one, 1-[2-hydroxy-4,6-bis(methoxymethoxy)-3-(3-methyl-2-butenyl)phenyl]-3-[7-(methoxy-methoxy)-2,2-dimethyl-2H-1-benzo-pyran-6-yl]-	EtOH	291(4.18),316s(4.01), 325s(4.00),396(4.31)	142-0997-84
	EtOH-AlCl$_3$	298(4.07),324s(3.95), 440(4.24)	142-0997-84
$C_{31}H_{38}O_{16}$ Acteoside, 2'-acetyl-	MeOH	222(4.05),245s(3.89), 295s(3.99),334(4.05)	94-3880-84
$C_{31}H_{40}N_4$ a,c-Biladiene, 3,5,8-triethyl-1',2,4-6,7,8'-hexamethyl-, dihydrobromide	CH$_2$Cl$_2$	446(5.02),522(5.09)	44-4602-84
$C_{31}H_{40}N_4O_2$ a,c-Biladiene-6'-carboxylic acid, 3,5,8-triethyl-1',2,4,6,7-penta-methyl-, dihydrobromide	CH$_2$Cl$_2$	444(5.09),520(5.02)	44-4602-84
$C_{31}H_{40}O_{13}$ Hydrangenoside A	MeOH	226(4.10),238s(4.04), 276s(2.82)	33-2111-84
Hydrangenoside B	MeOH	227.5(4.20),279(3.29)	33-2111-84
$C_{31}H_{40}O_{15}$ Cistanoside D	MeOH	237(4.03),285(4.02), 328(4.02)	94-3880-84
Isomartynoside	MeOH	218(4.37),288(4.18), 327(4.31)	102-2313-84
$C_{31}H_{41}N_5O_6$ Nummularine N	MeOH	267(4.00),320(3.8)	102-2118-84
$C_{31}H_{41}S_4$ Cyclopropenylium, [1-[2,3-bis[(1,1-di-methylethyl)thio]-2-cyclopropen-1-ylidene]-1H-inden-3-yl]bis[(1,1-di-methylethyl)thio]-, perchlorate	CH$_2$Cl$_2$	241(4.53),260s(4.32), 278(4.23),304s(4.12), 340s(3.70),436s(4.47), 463(4.96)	88-0345-84
$C_{31}H_{42}Cl_2N_2O_3$ 4-Azaandrost-5-en-3-one, 17-[[[4-[bis(2-chloroethyl)amino]phenyl]-acetyl]oxy]-4-methyl-, (17β)-	n.s.g.	206(4.38),258(4.36), 300(3.30)	111-0465-84
$C_{31}H_{42}Cl_2N_2O_4$ 4-Azaandrost-5-en-3-one, 17-[[[4-[bis(2-chloroethyl)amino]phenoxy]-acetyl]oxy]-4-methyl-, (17β)-	n.s.g.	205(4.38),250(4.30), 310(3.18)	111-0465-84

Compound	Solvent	$\lambda_{max}(\log \epsilon)$	Ref.
$C_{31}H_{42}D_9O_2Ta$ Tantalum, bis[2,6-bis(1,1-dimethyleth-yl)phenolato]tri(methyl-d_3)-, (TB-5-11)-	C_6H_{12}	307(3.86)	35-1847-84
$C_{31}H_{42}N_2O_2$ 1H-Pyrrole-2-carboxylic acid, [3,4,4a-5,6,7,8,8a-octahydro-5,6,8a-trimeth-yl-5-[3-methyl-5-[6-(methylamino)-7H-purin-7-yl]-3-pentenyl]-1-naphthalen-yl]methyl ester, [4aS-[4aα,5α(E),6α-8aα]]-	EtOH EtOH-HCl	268(4.24) 217(4.20)	35-1819-84 35-1819-84
$C_{31}H_{42}O_5$ Butanedioic acid, 3-decyl-2-(3,3-di-phenyl-2-propenyl)-2-hydroxy-, dimethyl ester	EtOH	251(4.2)	1-0837-84
$C_{31}H_{42}O_7$ Milbemycin J	EtOH	240(4.45)	158-84-49
$C_{31}H_{42}O_{11}$ Miotoxin C	EtOH	251(4.1),266(4.2), 271(4.2)	5-1746-84
$C_{31}H_{42}O_{15}$ Eruberin C	MeOH	226(4.42),276.5(3.58), 282(3.58)	94-0490-84
$C_{31}H_{44}N_4O_5$ Discarine G	n.s.g.	230(3.78),274(3.14)	63-1351-84
$C_{31}H_{44}N_6O_3$ 1H-Pyrrole-2-carboxylic acid, [5-[5-[[4-amino-6-(methylamino)-5-pyrim-idinyl]formylamino]-3-methyl-3-pentenyl]-3,4,4a,5,6,7,8,8a-octa-hydro-5,6,8-trimethyl-1-naphthal-enyl]methyl ester, [4aS-[4aα,5α(E)-6β,8aα]]-	EtOH EtOH-HCl	224(4.62),263(4.03) 227(4.50),266(4.46)	35-1819-84 35-1819-84
$C_{31}H_{44}O_7$ Anhydride diacetate, m. 156-160°	MeOH	237(4.08),245(4.17), 254(4.00)	39-0497-84C
$C_{31}H_{46}O_4$ D-Friedoolean-14-en-28-oic acid, 2,3-dioxo-, methyl ester	EtOH EtOH-NaOH	272(3.68) 324(3.56)	102-2593-84 102-2593-84
$C_{31}H_{46}O_5S$ 1-Hexadecanesulfonic acid, 2-hydroxy-3-oxo-2,3-diphenylpropyl ester	MeOH	251(4.20),332(2.36)	126-1795-84
$C_{31}H_{46}O_6$ Androsta-5,7-diene-3,17-diol, 19-[(tet-rahydro-2H-pyran-2-yl)oxy]-, 3-acet-ate 17-(2,2-dimethylpropanoate)	EtOH	271(4.04),281(4.04)	56-0711-84
$C_{31}H_{46}O_7$ Androst-5-en-7-one, 3-acetoxy-17-(2,2-dimethyl-1-oxopropoxy)-19-[(tetra-	EtOH	233(4.18)	56-0711-84

Compound	Solvent	$\lambda_{max}(\log \epsilon)$	Ref.
hydro-2H-pyran-2-yl)oxy]-, (3β,17α)- (cont.)			56-0711-84
$C_{31}H_{48}O_5$			
Androsta-1,4-diene-2-carboxylic acid, 11β,17β-dihydroxy-17-methyl-3-oxo-, decyl ester	MeOH	246(4.02)	13-0271-84A
9,10-Cyclolanostan-28-oic acid, 1,3-di-hydroxy-24-methylene-23-oxo-, (1α,3β,4α)- (jessic acid)	EtOH	219(3.82)	102-0635-84
$C_{31}H_{50}N_2$			
Plakinamine B	MeOH	241(3.43)	44-5157-84
$C_{31}H_{52}N_2O_3$			
Pregn-5-ene-15,18-diol, 20-(dimethyl-amino)-3-(methylamino)-, 18-(3,4-dimethyl-2-pentenoate), (3β,13β,20S)-	EtOH	223(3.93)	22-0071-84
$C_{31}H_{53}NO_8$			
5-O-Mycaminosyltylactone	EtOH	282(4.32)	158-84-50

Compound	Solvent	$\lambda_{max}(\log \epsilon)$	Ref.
$C_{32}H_{16}N_2O$			
5,17-Diazaviolanthrone B	xylene	653(4.38)	39-2177-84C
5,10-Diazaviolanthrone C	H_2SO_4	522(4.30),628(4.07)	39-2177-84C
$C_{32}H_{16}O_4$			
Phenanthro[9,10-b]phenanthro[9',10'-5,6]pyrano[3,4-d]pyran-10,20-dione	$C_6H_4Cl_2$	458(4.14),483(4.13)	24-0666-84
$C_{32}H_{18}$			
5,14[1',2']-Benzenopentacene, 1,17-di-ethynyl-5,14-dihydro-, (5S)-	EtOH	227(4.74),265.3(5.23), 293.2(4.24),321.4(3.51), 337.2(3.80),354.8(3.98), 374.8(3.85)	44-4266-84
1,1'-Bipyrene	C_6H_{12}	244(5.09),280(4.85), 350(4.67),c.410(--)	35-7776-84
$C_{32}H_{18}O_2$			
Naphtho[2,1-b]chrysene-7,14-dione, 9-phenyl-	$CHCl_3$	262(4.57),298(4.53), 315(4.44),327(4.52), 373(3.71),424(3.64)	40-0090-84
$C_{32}H_{20}N_2O_2$			
[2,2'-Bidibenz[cd,g]indole[-6,6'-(1H,1'H)-dione, 1,1'-dimethyl-	C_6H_5Cl	612(4.24)	104-1415-84
$C_{32}H_{21}N_3O_3$			
1H-Benzimidazole-5-carboxylic acid, 2-[4-(2-benzoxazolyl)-1-naphthal-enyl]-1-(phenylmethyl)-	DMF	343(4.46)	5-1129-84
$C_{32}H_{22}$			
5,14[1',2']-Benzenopentacene, 1,17-di-ethynyl-5,5a,6,13,13a,14-hexahydro-, [5S-(5α,5aα,13aα,14α)]-	EtOH	232.5(5.14),249.5(4.60), 277.0(3.98)	44-4266-84
$C_{32}H_{22}N_2O_4$			
3H-Indol-3-one, 2-[1,3-dihydro-3-oxo-1-(1-phenylacetyl)-2H-indol-2-yli-dene]-1,2-dihydro-1-(phenylacetyl)-, cis	benzene	439(3.62)	39-2305-84C
trans	benzene	568(3.86)	39-2305-84C
$C_{32}H_{22}N_2O_6$			
Indigo, N,N'-bis(4-methoxybenzoyl)-, trans	toluene	580(3.95)	23-2478-84
$C_{32}H_{22}O_4$			
5,14[1',2']-Benzenopentacene-1,17-di-carboxylic acid, 5,14-dihydro-, dimethyl ester, (5S)-	EtOH	228(4.59),262.6(5.19), 322(3.57),338.1(3.74), 355.7(3.91),375.7(3.78)	44-4266-84
3-Butene-1,2-dione, 4-acetoxy-3-(9-anthracenyl)-1,4-diphenyl-	MeOH	256(4.91),332(4.63), 380(4.51)	44-4165-84
$C_{32}H_{22}O_{10}$			
Etheramentoflavone, 7",4'''-dimethyl-	MeOH	225(4.72),270(4.57), 340(4.47)	154-0547-84
$C_{32}H_{22}O_{14}$			
Strychnobiflavone	MeOH	256(4.27),265s(3.92), 302(3.49),352(3.98)	100-0953-84
	MeOH-AlCl$_3$	263(4.09),308s(3.98),	100-0953-84

Compound	Solvent	λ_{max}(log ϵ)	Ref.
Strychnobiflavone (cont.)		356s(3.94),420(4.43)	100-0953-84
	MeOH–AlCl$_3$–HCl	265(3.54),302s(3.63),360(3.47),400(3.95)	100-0953-84
$C_{32}H_{23}N_5O_8$ 1H-Pyrazolo[3,4-d]pyrimidine-3-carbonitrile, 4,5-dihydro-4-oxo-1-(2,3,5-tri-O-benzoyl-β-D-ribofuranosyl)-	MeOH	262(4.09)	87-1119-84
$C_{32}H_{23}N_7O_3S$ 1H-Pyrazole, 3,5-diphenyl-4-[[2-[(2-pyridinylamino)sulfonyl]phenyl]azo]-1-(2-pyridinylcarbonyl)-	n.s.g.	360(3.88)	48-1021-84
$C_{32}H_{24}ClN_3O_7$ D-Ribitol, 1,4-anhydro-1-C-(2-chloroimidazo[1,5-b]pyridazin-7-yl)-, 2,3,5-tribenzoate, (S)-	EtOH	233(4.98),264(3.97),272(3.97),281(3.81),366(3.42)	39-0229-84C
$C_{32}H_{24}F_6O_9S_2$ Methanesulfonic acid, trifluoro-, oxybis[2-(4-methoxyphenyl)-1-phenyl-2,1-ethenediyl] ester, (Z,Z)-	C_6H_{12}	232(4.48),294(4.33)	44-2273-84
$C_{32}H_{24}N_2$ 9-Anthracenecarboxaldehyde, 10-methyl-, [(10-methyl-9-anthracenyl)methylene]hydrazone	C_6H_{12}	435(4.32)	97-0021-84
$C_{32}H_{24}N_2O_2$ 9-Anthracenecarboxaldehyde, 10-methoxy-, [(10-methoxy-9-anthracenyl)methylene]hydrazone	C_6H_{12}	437(4.34)	97-0021-84
2-Butene-1,4-dione, 2-(4-methyl-3,5-diphenyl-1H-pyrazol-1-yl)-1,4-diphenyl-	MeCN	244(4.59),345(4.37)	44-4647-84
Methanone, (5,6-dihydro-1-methyl-2-phenylpyrazolo[5,1-a]isoquinoline-5,6-diyl)bis[phenyl-	MeOH	246(4.49)	44-4647-84
$C_{32}H_{24}N_4O_9$ 2,4(1H,3H)-Pteridinedione, 1-(2,3,5-tri-O-benzoyl-β-D-ribofuranosyl)-	MeOH	231(4.72),315(3.84)	136-0179-84F
2,4(1H,3H)-Pteridinedione, 3-(2,3,5-tri-O-benzoyl-β-D-ribofuranosyl)-	MeOH	230(4.71),274(3.60),281(3.61),325(3.90)	136-0179-84F
$C_{32}H_{24}N_6O_6$ 1,2,4-Triazolidinium, 1,1'-[1,2-bis-(4-methoxyphenyl)-1,2-ethanediylidene]bis[3,5-dioxo-4-phenyl-, dihydroxide, bis(inner salt)	CH_2Cl_2	256(4.14),332(4.24),460(4.48)	44-2917-84
$C_{32}H_{24}O_2$ 1H-Indeno[2',1':2,3]indeno[1,2-c]furan-1-ol, 7b,12-dihydro-7b-methyl-12-methylene-1,3-diphenyl-	MeOH	255(4.39)	44-4923-84
Methanone, (5,10-dimethylbenzo[b]biphenylene-4b,10a-diyl)bis[phenyl-	MeOH	243(4.46),313(4.26),335s(4.06)	44-4923-84
Methanone, (1,11-dimethyldibenzo[a,e]cyclooctene-5,12-diyl)bis[phenyl-	MeOH	250(3.56),282s(3.91)	44-4923-84

Compound	Solvent	λ_{max}(log ϵ)	Ref.
C$_{32}$H$_{25}$BrN$_4$O			
[1,2,4]Triazolo[1,5-a]pyridinium, 2-[(4-bromophenyl)amino]-1-[(4-methoxyphenyl)methyl]-5,7-diphenyl-, hydroxide, inner salt	EtOH	261(4.92)	39-1891-84C
C$_{32}$H$_{26}$			
Bicyclo[2.2.2]octa-2,5-diene, 2,3,5,6-tetraphenyl-	C$_6$H$_{12}$	224(4.48),270(4.38)	44-0856-84
Cyclopropene, 3-methyl-1-(1-methyl-2,3-diphenyl-2-cyclopropen-1-yl)-2,3-diphenyl-	EtOH	226(4.48),296(4.47), 313(4.46),331(4.33)	44-0856-84
Ethane, 1,2-bis(2,3-diphenyl-2-cyclopropen-1-yl)-	EtOH	224(4.44),229(4.46), 238(4.40),300(4.56), 320(4.62),339(4.47)	44-0856-84
Naphthalene, 1-methyl-3-(1-methyl-2-phenylethenyl)-2,4-diphenyl-, (E)-	C$_6$H$_{12}$	242(4.71),323(4.19), 340(4.05)	44-0856-84
(Z)-	C$_6$H$_{12}$	235(4.78),267(4.27), 286(4.12),308(3.75)	44-0856-84
C$_{32}$H$_{26}$NO$_2$			
1-Naphthalenemethaniminium, N-[bis(4-methoxyphenyl)methylene]- -phenyl-, (OC-6-11)-, hexachloroantimonate	CH$_2$Cl$_2$	271(4.39),355(4.53)	24-3222-84
Pyridinium, 4-(4-acetoxyphenyl)-2,6-diphenyl-1-(phenylmethyl)-, perchlorate	MeOH-HClO$_4$	320(4.56),358(4.59)	4-1673-84
C$_{32}$H$_{26}$N$_2$O$_3$			
1H-Pyrazole-1-acetic acid, 4-methyl-α-(2-phenoxy-2-phenylethenyl)-3,5-diphenyl-, (Z)-	MeOH	260(4.55)	44-4647-84
1H-Pyrazole-1-acetic acid, α-(2-phenoxy-2-phenylethenyl)-3,5-diphenyl-, methyl ester, (Z)-	MeOH	248(4.71)	44-4647-84
C$_{32}$H$_{26}$N$_2$O$_4$S			
1H,8H-Azeto[2,1-b][1,3]benzothiazine-8-carboxylic acid, 2,2a-dihydro-1-oxo-2-[(phenylacetyl)amino]-, diphenylmethyl ester, [2R-(2α,2aα,8α)]-	EtOH	287(2.98)	77-1705-84
C$_{32}$H$_{26}$N$_{10}$			
Dibenzo[c,m]dipyrazolo[3,4-f:4',3'-j]-[1,2,5,8,9,12]hexaazacyclotetradecine, 1,8,9,20-tetrahydro-3,6-dimethyl-1,8-diphenyl-, (E,E)-	hexane	227(4.43),255(4.32), 304(4.36),328(4.32), 384(4.38),409s(4.29), 503s(3.34)	103-0196-84
C$_{32}$H$_{26}$O$_4$			
5,14[1',2']-Benzenopentacene-1,17-dicarboxylic acid, 5,5a,6,13,13a,14-hexahydro-, [5S-(5α,5aα,13aα,14α)]-	EtOH	231.5(5.00),267(3.96), 277.8(3.98),287.5(3.96)	44-4266-84
C$_{32}$H$_{26}$O$_{10}$			
[7,10'-Bi-4H-naphtho[2,3-b]pyran]-4,4'-dione, 5,5'-dihydroxy-6,6',8,8'-tetramethoxy-2,2'-dimethyl-	EtOH	223(4.62),255s(4.66), 278(4.93),325(3.98), 401(4.01)	78-3617-84
Chamaejasmenin A	MeOH	296(4.46)	94-0362-84
Chamaejasmenin B	MeOH	298(4.54)	94-0362-84
4H-Naphtho[2,3-b]pyran-4-one, 5-hydroxy-10-(5-hydroxy-8,10-dimethoxy-	EtOH	228(4.66),256(4.65), 278(4.83),325s(4.06),	78-3617-84

Compound	Solvent	$\lambda_{max}(\log \epsilon)$	Ref.
2-methyl-4-oxo-4H-naphtho[1,2-b]-pyran-9-yl)-6,8-dimethoxy-2-methyl-(cont.)		398(3.85)	78-3617-84
$C_{32}H_{27}ClN_{10}$ 1H-Pyrazol-5-amine, N-[2-[(2-aminophen-yl)azo]phenyl]-4-[(5-chloro-3-methyl-1-phenyl-1H-pyrazol-4-yl)azo]-3-meth-yl-1-phenyl-	hexane	229(4.56),252s(4.43), 333(4.45),382(4.39), 403s(4.33),463s(4.10)	103-0196-84
$C_{32}H_{27}N_3O_8$ D-Allonamide, 2,5-anhydro-N-(pyrazin-ylmethyl)-, 3,4,6-tribenzoate	EtOH	230.5(4.56),267(3.93), 272s(3.90),282s(3.52)	39-0229-84C
$C_{32}H_{27}N_3O_9$ D-Allonamide, 2,5-anhydro-N-[(1,6-di-hydro-6-oxo-3-pyridazinyl)methyl]-, 3,4,6-tribenzoate	EtOH	229(4.62),276(3.63), 283(3.61)	39-0229-84C
$C_{32}H_{28}NO_2P$ Phosphoramidous acid, dimethyl-, bis(2-methyldi-9-anthracenyl) ester	EtOH	220(3.69),250(4.05), 257(4.13),338(2.94), 355(3.01),374(3.06), 395(2.95)	40-1158-84
$C_{32}H_{28}N_2O_5S$ 1H-Azeto[2,1-b][1,3]benzothiazine-8-carboxylic acid, 2,2a,3a,4-tetrahy-dro-1-oxo-2-[(phenylacetyl)amino]-, diphenylmethyl ester, 3-oxide, [2R-(2α,2aβ,3aβ)]-	EtOH	317(4.20)	77-1705-84
$C_{32}H_{28}N_4$ Schiff's base from benzil and ethylene-diamine	CHCl₃	215(2.75),255(2.78), 310(3.23),402(2.90)	34-0237-84
$C_{32}H_{28}N_4O_5$ 11H-Furo[2',3':4,5]oxazolo[2,3-b]pter-idin-11-one, 6a,7,8,9a-tetrahydro-7-hydroxy-2,3-dimethyl-8-[(triphenyl-methoxy)methyl]-, [6aR-(6aα,7β,8α 9aβ)]-	MeOH	229s(4.25),262(3.98), 322(3.84)	136-0179-84F
$C_{32}H_{28}N_4O_8$ D-Allonamide, N-[(3-aminopyrazinyl)-methyl]-2,5-anhydro-, 3,4,6-tri-benzoate	EtOH	231(4.64),276(3.61), 283(3.61),321(3.75)	39-0229-84C
$C_{32}H_{28}N_4O_{12}S_2$ Pyridinium, 3-[[2-[1,3-dihydro-1-[(1-methylpyridinium-3-yl)carbonyl]-3-oxo-2H-indol-2-ylidene]-2,3-dihydro-3-oxo-1H-indol-1-yl]carbonyl]-1-met hyl-, (E)-, bis(methyl sulfate)	MeCN	548(3.70)	39-2305-84C
$C_{32}H_{28}O_6$ 2-Butene-1,4-dione, 1,2,3,4-tetrakis(4-methoxyphenyl)-	EtOH	220(4.42),291(4.62)	44-4752-84
Diacenaphtho[1,2-b:1',2'-k][1,4,7,10-13,16]hexaoxacyclooctadecin, 8,9,11-12,21,22,24,25-octahydro-	n.s.g.	318(4.25),434(3.23)	78-0665-84

Compound	Solvent	$\lambda_{max}(\log \epsilon)$	Ref.
$C_{32}H_{28}O_{10}$ [2,2'-Binaphthalene]-1,1',5,8,8'-pent-ol, 6,6'-dimethyl-, pentaacetate	MeOH	222(4.68),239(4.80), 250s(4.75),280(4.18), 327(3.53)	102-2039-84
$C_{32}H_{28}O_{11}$ [7,10'-Bi-4H-naphtho[2,3-b]pyran]-4,4'-dione, 2,3-dihydro-2,5,5'-trihydroxy-6,6',8,8'-tetramethoxy-2,2'-dimethyl-	EtOH	229(4.59),255s(4.62), 280(4.85),320(4.15), 328s(4.15),403(3.96)	78-3617-84
[7,10'-Bi-4H-naphtho[2,3-b]pyran]-4,4'-dione, 2',3'-dihydro-2',5,5'-trihy-droxy-6,6',8,8'-tetramethoxy-2,2'-dimethyl-	EtOH	227(4.43),279(4.69), 316(3.91),328s(3.84), 403(3.80)	78-3617-84
4H-Naphtho[2,3-b]pyran-4-one, 2,3-dihy-dro-2,5-dihydroxy-10-(5-hydroxy-8,10-dimethoxy-2-methyl-4-oxo-4H-naphtho-[1,2-b]pyran-9-yl)-6,8-dimethoxy-2-methyl-	EtOH	234(4.71),254(4.58), 279(4.78),315s(4.29), 327s(4.17),398(3.85)	78-3617-84
$C_{32}H_{29}BO_8$ 1,5-Methano-1H-anthra[2,3-e][1,3,2]di-oxaborocin-7,14-diol, 5-(2-acetoxy-ethyl)-5,6-dihydro-3-phenyl-, di-acetate, (1S)-	$CHCl_3$	262(5.16),320(3.16), 336(3.38),352(3.64), 371(3.79),391(3.73)	78-4649-84
$C_{32}H_{29}Cl_2N_3O_4$ 7H-Pyrrolo[2,3-d]pyrimidine, 2,4-di-chloro-7-[2,3,5-tris-O-(phenylmeth-yl)-α-D-arabinofuranosyl]-	MeOH	232(4.46),282(3.72), 293s(3.63)	5-0722-84
β-	MeOH	232(4.47),276(3.72), 293s(3.64)	5-0722-84
$C_{32}H_{29}N_3O_4$ [3,3'-Bi-1H-pyrrole]-2,2',5,5'-tetrone, 4-(butylamino)-1,4'-bis(4-methyl-phenyl)-1'-phenyl-	$CHCl_3$	280(4.20),440(3.60)	48-0401-84
$C_{32}H_{30}$ Azulene, 4,4'-(1,3,5,7-octatetraene-1,8-diyl)bis[6,8-dimethyl-	CH_2Cl_2	245(4.44),286(4.71), 348(4.45),433(4.82), 450s(4.75),591(3.49)	5-1905-84
$C_{32}H_{30}Cl_2N_5O_{13}P$ 3'-Cytidylic acid, 2'-deoxy-N-[[2-(4-nitrophenyl)oxy]carbonyl]-, 3,5-di-chlorophenyl 2-(4-nitrophenyl)ethyl ester	MeOH	245(4.38),273(4.40)	78-0059-84
$C_{32}H_{30}N_2$ Benzo[b]quinolizinium, 9,9'-(1,6-hex-anediyl)bis-	MeOH	260(3.05)	48-0757-84
$C_{32}H_{30}N_2O$ Cycloocta[b]pyridine-3-carbonitrile, 2-[1,1'-biphenyl]-4-yl-4-(4-ethoxy-phenyl)-5,6,7,8,9,10-hexahydro-	EtOH	207(4.82),218s(4.72), 243(4.13),289(4.59)	73-1395-84
$C_{32}H_{30}N_2O_4$ 2,5-Cyclohexadiene-1,4-dione, 2,5-bis[2-(1,1-dimethyl-2-propenyl)-1H-indol-3-yl]-3,6-dihydroxy-	EtOH	280(4.44),288(4.24), 506(2.98)	39-0567-84C
	EtOH-NaOH	280(4.27),288(4.29),	39-0567-84C

Compound	Solvent	$\lambda_{max}(\log \epsilon)$	Ref.
(hinnuliquinone) (cont.)		322(4.31),516(3.17)	39-0567-84C
$C_{32}H_{30}N_3O_7S_2$ Benzoxazolium, 3-ethyl-5-phenyl-2-[(5-acetanilinomethylene-4-oxo-3-ethyl-thiazolidin-2-ylidene)methyl]-, ethyl sulfate	EtOH	440(4.84)	103-0665-84
$C_{32}H_{30}N_4$ 9-Anthracenecarboxaldehyde, [bis[4-(di-methylamino)phenyl]methylene]hydra-zone	C_6H_{12}	429(4.40)	97-0021-84
$C_{32}H_{30}N_4O_4$ Porphyrin, 13,17-bis[2-(methoxycarbo-nyl)ethenyl]-2,7,12,18-tetramethyl-	CHCl$_3$	425(5.19),514(4.10), 553(4.08),584(3.87), 639(3.76)	5-1386-84
$C_{32}H_{30}N_4O_6$ Deuteroporphyrin, 3,8-diformyl-	DMSO	428(4.83),518(3.93), 554(3.82),587(3.76), 645(3.42)	5-1057-84
2,4(1H,3H)-Pteridinedione, 6,7-dimeth-yl-3-[5-O-(triphenylmethyl)-β-D-ribofuranosyl]-	MeOH	228s(4.28),247s(4.06), 331(3.99)	136-0179-84F
$C_{32}H_{30}O_{12}$ [7,10'-Bi-4H-naphtho[2,3-b]pyran]-4,4'-dione, 2,2',3,3'-tetrahydro-2,2',5,5'-tetrahydroxy-6,6',8,8'-tetramethoxy-2,2'-dimethyl-	EtOH	233(4.63),270s(4.73), 280(4.84),318(4.22), 331(4.22),404(4.01)	78-3617-84
$C_{32}H_{30}O_{14}$ Tuberculatin, triacetyl-	CHCl$_3$	263(4.88),295(3.98), 313(3.77),352(3.68)	102-0151-84
$C_{32}H_{30}O_{16}$ 5a,11a-Epoxynaphthacene-5,6,8,11,12-(7H)-pentone, 6a,9,10,10a-tetrahy-dro-10-[(2,3,4,6-tetra-O-acetyl-β-D-glucopyranosyl)oxy]-, [5aR-(5aα,6aα,10β,10aα,11aα)]-	EtOH	209(4.33),233(4.42), 265s(4.32),415(4.08)	78-4657-84
$C_{32}H_{31}ClN_4O_4$ 7H-Pyrrolo[2,3-d]pyrimidin-2-amine, 4-chloro-7-[2,3,5-tris-O-(phenyl-methyl)-α-D-arabinofuranosyl]-	MeOH	236(4.41),258(3.73), 314(3.69)	5-0722-84
β-	MeOH	236(4.43),259(3.72), 316(3.70)	5-0722-84
7H-Pyrrolo[2,3-d]pyrimidin-4-amine, 2-chloro-7-[2,3,5-tris-O-(phenyl-methyl)-β-D-arabinofuranosyl]-	MeOH	275(4.06)	5-0722-84
$C_{32}H_{31}Cl_2N_4O_{12}P$ 2(1H)-Pyrimidinone, 1-[2-deoxy-3-O-[(2,5-dichlorophenoxy)[2-(4-nitro-phenyl)ethoxy]phosphinyl]-β-D-erythro-pentofuranosyl]-5-methyl-4-[2-(4-nitrophenyl)ethoxy]-	MeOH	275(4.38),280(4.36)	78-0059-84

Compound	Solvent	$\lambda_{max}(\log \epsilon)$	Ref.
$C_{32}H_{31}N_3$ 3-Quinolinecarbonitrile, 2-[1,1'-bi-phenyl]-4-yl-4-[4-(diethylamino)-phenyl]-5,6,7,8-tetrahydro-	EtOH	207(4.71),242(4.25), 269(4.39),289(4.47), 362(4.04)	73-1395-84
$C_{32}H_{31}N_5$ 1,3-Propanediamine, N-9-acridinyl-N'-[3-(9-acridinylamino)propyl]-	pH 7.0	406(4.26)	77-0509-84
DNA bound	pH 7.0	408(4.06)	77-0509-84
$C_{32}H_{32}Fe_2O_8S_2$ Ferrocene, 1,1":1',1'''-bis[thiobis(2,1-ethanediyloxycarbonyl)]bis-	dioxan	448(2.77)	18-2435-84
$C_{32}H_{32}Fe_2O_{10}$ Ferrocene, 1,1":1',1'''-bis[oxybis(2,1-ethanediyloxycarbonyl)]bis-	dioxan	448(2.76)	18-2435-84
$C_{32}H_{32}O_{15}$ 2,5,12(1H)-Naphthacenetrione, 3,4,4a-12a-tetrahydro-6,11-dihydroxy-4-[(2,3,4,6-tetra-O-acetyl-β-D-glucopyranosyl)oxy]-, [4S-(4α,4aβ-12aβ)]-	EtOH	239(4.42),250(4.42), 278(4.39),286(4.37), 378s(3.94),396(4.11), 416(4.10)	78-4657-84
$C_{32}H_{33}N_3O_7$ 1H-Pyrazole-4-carboxylic acid, 5-(aminocarbonyl)-3-[2,3,5-tris-O-(phenylmethyl)-β-D-arabinofuranosyl]-, methyl ester	MeOH-base	257(4.03)	44-0528-84
$C_{32}H_{33}N_5O_4$ 7H-Pyrrolo[2,3-d]pyrimidine-2,4-di-amine, 7-[2,3,5-tris-O-(phenylmethyl)-β-D-arabinofuranosyl]-	MeOH	224(4.45),264(4.07), 284(3.95)	5-0722-84
$C_{32}H_{35}N_2O_2S_2$ Benzothiazolium, 3-ethyl-2-[11-(3-ethyl-5-methoxy-2(3H)-benzothiazolyli-dene]-5-methyl-1,3,5,7,9-undecapenta-enyl]-5-methoxy-, iodide	EtOH	1020(4.0170)	104-0395-84
$C_{32}H_{36}$ Azulene, 1,1'-(1,2-ethenediyl)bis[3,8-dimethyl-5-(1-methylethyl)-	CH_2Cl_2	232(4.55),266(4.58), 329(4.53),415s(--), 456(4.62),485s(--), 665(3.09)	5-1905-84
$C_{32}H_{36}N_2O_{10}S$ D-Arabinitol, 5-C-[1-(4-methylphenyl)-2-[(phenylmethyl)thio]-1H-imidazol-4-yl]-, 1,2,3,4,5-pentaacetate	EtOH	262(3.74)	136-0091-84B
$C_{32}H_{36}N_4$ Deoxophylloerythroetioporphyrin	CH_2Cl_2	400(5.32),498(4.20), 534(3.59),564(3.81), 618(3.82)	44-4602-84
4,7:14,17-Diimino-2,21-metheno-9,12-nitrilo-18H-1-benzazacyclononadec-ine, 6,11-diethyl-19,20-dihydro-5,10,15,18,22-pentamethyl-	CH_2Cl_2	400(5.13),498(4.02), 533(3.51),568(3.62), 622(2.99)	78-4033-84

Compound	Solvent	$\lambda_{max}(\log \epsilon)$	Ref.
$C_{32}H_{36}N_4O_2$			
Amauromine	EtOH	245(4.04),300(3.62)	88-4673-84 +158-1320-84
21H,23H-Porphine-2-propanoic acid, 8,13-diethyl-3,7,12,17-tetra-methyl-, methyl ester	CH_2Cl_2	398(4.26),496(4.16), 530(3.95),566(3.83), 620(3.65)	44-4602-84
$C_{32}H_{36}N_4O_9$			
Parabactin	EtOH	250(4.44),309(4.05)	39-0183-84C
$C_{32}H_{36}N_4O_{10}S_2$			
1H-Benzimidazolium, 2,2'-(2,6-naphtha-lenediyl)bis[5-methoxy-1,3-dimethyl-, bis(methyl sulfate)	$DMF-H_2O$	320(4.48)	5-1129-84
$C_{32}H_{36}N_6O_{12}$			
L-Alanine, N-[[5-amino-1-(2,3,5-tri-O-acetyl-β-D-ribofuranosyl)-1H-imid-azol-4-yl]carbonyl]-3-[(phenylmeth-oxy)-N,N,O-azoxy]-, phenylmethyl ester	MeOH	240(4.27),266(4.15)	87-1295-84
$C_{32}H_{36}O_{11}$			
1-Butanone, 2-methyl-1-[1,3,7,9-tetra-acetoxy-2,8-dimethyl-6-(3-methyl-2-oxobutyl)-4-dibenzofuranyl]-	EtOH	239(4.46),246s(4.43), 257s(4.36),300(4.21)	39-2573-84C
$C_{32}H_{37}NO_{10}$			
Maymyrsine	EtOH	230(4.13),258s(3.51), 265(3.55),271s(3.47), 282s(3.06)	142-2221-84
$C_{32}H_{37}N_4O_4S_2$			
Benzothiazolium, 2-[[5-[[5-(ethoxycarb-onyl)-1,3-diethyl-1,3-dihydro-2H-benzimidazol-2-ylidene]ethylidene]-3-ethyl-4-oxo-2-thiazolidinylidene]-methyl]-3-ethyl-5-methoxy-, iodide	EtOH	594(4.97)	103-0665-84
$C_{32}H_{39}N_3O_5$			
Gentiacraline	MeOH	222(4.37),255(3.85), 295(4.3)	102-2407-84
$C_{32}H_{39}N_5O_5$			
Glycine, N-[(17-ethyl-1,2,3,19,23,24-hexahydro-2,2,7,8,12,13,18-heptameth-yl-1,19-dioxo-21H-bilin-3-yl)acetyl]-, (±)-	$CHCl_3$	272(4.16),330(4.45), 362s(4.33),615(4.30)	49-1071-84
$C_{32}H_{39}P$			
Phosphine, (diphenylethenylidene)-[2,4,6-tris(1,1-dimethylethyl)-phenyl]-	hexane	242(4.48),263(4.44), 337(3.61)	88-1809-84
$C_{32}H_{40}N_2O_3S$			
Pyridinium, 4-[2-[4-[2-[4-(dibutyl-amino)phenyl]ethenyl]phenyl]eth-enyl]-1-(3-sulfopropyl)-, hydroxide, inner salt, (E,E)-	EtOH	482(4.58)	44-2546-84

Compound	Solvent	$\lambda_{max}(\log \epsilon)$	Ref.
$C_{32}H_{40}N_4O_4$			
21H-Biline-3-acetic acid, 17-ethyl-1,2,3,19,22,24-hexahydro-2,2,7,8-12,13,18,22-octamethyl-1,19-dioxo-, methyl ester, (4E,10Z,15Z)-(±)-	$CHCl_3$	273(4.17),342(4.53), 584(4.28),656s(4.09)	49-0101-84
$C_{32}H_{40}O_{13}$			
Flindercarpin-2	EtOH	212(3.78)(end abs.)	12-1461-84
$C_{32}H_{40}O_{16}$			
Miyaginin hexaacetate	MeOH	220(4.03),272(3.00), 278(2.90)	18-2893-84
$C_{32}H_{41}N_3O_{20}$			
1H-Imidazolium, 4-(aminocarbonyl)-5-hy-droxy-1,3-bis(2,3,4,6-tetra-O-acetyl-β-D-galactopyranosyl)-, hydroxide, inner salt	M HCl H_2O M NaOH	246(3.57),290(3.76) 241(3.76),290(3.96) 266(4.34)	4-0529-84 4-0529-84 4-0529-84
1H-Imidazolium, 4-(aminocarbonyl)-5-hy-droxy-1,3-bis(2,3,4,6-tetra-O-acetyl-β-D-glucopyranosyl)-, hydroxide, inner salt	M HCl H_2O M NaOH	241(3.43),286(3.60) 241(3.78),290(3.94) 266(4.35)	4-0529-84 4-0529-84 4-0529-84
$C_{32}H_{42}N_2O_7$			
Corynan-5-carboxylic acid, 20,21-dide-hydro-16-(5,5-dimethyl-1,3-dioxan-2-yl)-17-[(2,2-dimethyl-1,3-propane-diyl)bis(oxy)-19-oxo-, methyl ester	MeOH	300(4.64)	78-4853-84
$C_{32}H_{42}O_8$			
2-Propenoic acid, 3-(3,4-dimethoxyphen-yl)-, 1,10-decanediyl ester	MeCN	322(4.62)	40-0022-84
$C_{32}H_{42}O_{13}$			
α-D-Galactopyranoside, 3,3'-O-(oxydi-2,1-ethanediyl)bis[methyl 4,6-O-(phenylmethylene)-, (S,S)-	EtOH	256(2.64)	5-1046-84
α-D-Mannopyranoside, 3,3'-O-(oxydi-2,1-ethanediyl)bis[methyl 4,6-O-(phenylmethylene)-, (R,R)-	MeCN	206(4.23),230(2.94), 256(2.72)	5-1046-84
$C_{32}H_{43}NO_{10}$			
Aconine, benzoyl-	EtOH	234(4.08),270(2.92)	33-2017-84
$C_{32}H_{44}N_4O_4$			
9,10-Anthracenedione, 1,4-dihydroxy-5,8-bis[(2,2,6,6-tetramethylpiper-idinyl)amino]-	MeOH	240(4.67),272(4.16), 620(4.33),673(4.42)	88-0269-84
$C_{32}H_{44}N_4O_6$			
9,10-Anthracenedione, 1,4-dihydroxy-5,8-bis[(2,2,6,6-tetramethylpiper-idinyl)amino]-, di-N-oxide	MeOH	240(4.54),278(4.16), 570s(--),616(4.23), 670(4.35)	88-0269-84
$C_{32}H_{44}N_4O_{10}$			
Butanoic acid, 4,4'-[1,4,10,13-tetra-oxa-7,16-diazacyclooctadecane-7,16-diylbis(2,1-phenyleneimino)]bis[4-oxo-	H_2O	232(4.43),252s(4.31)	78-0793-84
$CuCl_2$ complex	H_2O	235s(4.34),250(4.38), 280s(3.86),395(2.03), 680(1.85)	78-0793-84

Compound	Solvent	$\lambda_{max}(\log \epsilon)$	Ref.
$C_{32}H_{44}O_7$ Milbemycin K	EtOH	240(4.47)	158-84-49
$C_{32}H_{44}O_9$ Ergosta-5,24-dien-26-oic acid, 3,27-di- acetoxy-14,20,22-trihydroxy-1-oxo-, δ-lactone, (3β,22R)-	EtOH	216(4.04)	102-0143-84
$C_{32}H_{44}O_{11}$ Lanceotoxin B	MeOH	299(3.79)	39-1573-84C
$C_{32}H_{44}O_{12}$ Lanceotoxin A	MeOH	299(3.68)	39-1573-84C
$C_{32}H_{46}N_{12}O_7$ L-Lysine, N^2-[N^6-[N-[4-[[2,4-diamino- 6-pteridinyl)methyl]methylamino]- benzoyl]-L-γ-glutamyl]-L-lysyl]-	pH 1 pH 7.4	242(4.21),306(4.34) 258(4.33),304(4.35), 366(3.93)	87-0888-84 87-0888-84
$C_{32}H_{46}O_6$ Leptomycin A	EtOH	225(4.28),243s(4.23)	158-84-82
$C_{32}H_{46}O_7$ 4H-Pyran-4-one, 2-[7-(6-ethyl-3,5-di- methyl-4-oxo-4H-pyran-2-yl)-2-hydr- oxy-1,3,5-trimethyl-4-oxo-5-octen- yl]-6-(2-hydroxy-1-methylbutyl)-3,5- dimethyl-, [1R*(1S*,2S*),2R*,3R*- 5E,7R*]-	MeOH	242(4.14),275(4.12)	44-0559-84
$C_{32}H_{46}O_8$ Cucurbitacin B Isocucurbitacin B 3-epi-	MeOH MeOH MeOH	229(4.47) 232(4.32) 229(4.47)	36-0411-84 36-0411-84 36-0411-84
$C_{32}H_{46}O_{10}Si$ Trichothec-9-ene-4,15-diol, 12,13-ep- oxy-15-[3-[2-[(1,6-dioxo-6-[2-(tri- methylsilyl)ethoxy]ethoxy]-2,4-hexadienyl]- oxy]ethyl]-3-methyloxiranecarboxyl- ate], [4β,15[2S,3R(2E,4Z)]]-	EtOH	263(4.34)	44-4332-84
$C_{32}H_{47}N_3O_9$ Geldanamycin, 17-demethoxy-19-(dimeth- ylamino)-15-methoxy-11-O-methyl-, (15R)-	MeOH	259(4.31)	158-1264-84
$C_{32}H_{47}N_3O_{10}$ Geldanamycin, 17-demethoxy-19-(dimeth- ylamino)-8,9-epoxy-8,9-dihydro-15- methoxy-11-O-methyl-	MeOH	268(4.28),325(3.00)	158-1264-84
11-Oxa-9,19-diazatricyclo[18.3.1.1^{8,11}]- pentacosa-1(23),14,16,20-tetraene- 10,18,22,24-tetrone, 21-(dimethyl- amino)-25-hydroxy-2,5,6,13-tetra- methoxy-3,7,17,25-tetramethyl-	MeOH	258(4.36)	158-1264-84
$C_{32}H_{48}O_2$ Oleana-9(11),12,15-trien-3-ol, acetate, (3β)-	MeOH	280(3.79)	2-0712-84

Compound	Solvent	$\lambda_{max}(\log \epsilon)$	Ref.
$C_{32}H_{48}O_5$ Lanosta-7,9(11),24-triene-3,15,26-triol, 22,26-epoxy-, 15-acetate, (3β,15α-22S,26S)- (perenniporiol)	EtOH	232(4.11),242(4.17), 251(4.00)	102-1129-84
$C_{32}H_{48}O_6$ Oleana-12,18-diene-23,28-dioic acid, 2,3-dihydroxy-, dimethyl ester, (2α,3β,4β)-	n.s.g.	244(4.30),252(4.40), 261(4.22)	102-2962-84
$C_{32}H_{48}O_7$ Ilikonapyrone	MeOH	260(4.10)	44-0559-84
$C_{32}H_{49}BrO_2$ Oleana-9(11),12-dien-3-ol, 15-bromo-, acetate, (3β)-	MeOH	277(3.78)	2-0712-84
$C_{32}H_{49}BrO_3$ Olean-12-en-11-one, 3β-acetoxy-15-bromo-	MeOH	245(4.05)	2-0712-84
$C_{32}H_{50}O_2$ Alnusa-1(10),5-dien-3β-ol, acetate	EtOH	232(4.15),238(4.20), 245(4.00)	18-2490-84
$C_{32}H_{50}O_3$ 16-Oxotaraxeryl acetate	MeOH	248(4.04)	2-0712-84
$C_{32}H_{50}O_5$ Jessic acid, methyl ester	EtOH	221(3.85)	102-0635-84
$C_{32}H_{52}O_2$ Bauerenyl acetate	EtOH	211(2.70)	100-0363-84
$C_{32}H_{54}O_5$ 4,8,10-Tetradecatrien-3-one, 12-ethyl-14-[4-hydroxy-8-(2-hydroxypropyl)-3,9-dimethyl-1,7-dioxaspiro[5.5]-undec-2-yl]-4,6-dimethyl-	EtOH	226(4.21),231(4.20), 238s(4.11)	33-1208-84
$C_{32}H_{55}NO_8$ Myxovirescin E	MeOH	232(4.32),238(4.30), 280(2.48)	158-84-83
$C_{32}H_{56}$ Cyclooctacosane, 1,2,15,16-tetrakis-(methylene)- (absorbance per diene)	hexane	228(4.08)	44-2981-84
$C_{32}H_{62}N_2O_2$ Diazene, bis(1-oxohexadecyl)-	hexane-CH_2Cl_2	454(1.36)	23-0574-84
$C_{32}H_{64}Si_8$ 5,6,11,16,21,26,31,36-Octasilaocta-spiro[4.0.4.0.4.0.4.0.4.0.4.0.4.0-4.0]tetracontane	isooctane	228s(4.23),254(4.04), 265s(3.85)	35-5521-84
$C_{32}H_{72}Si_4$ Cyclotetrasilane, octakis(1-methyl-propyl)-	C_6H_{12}	290s(2.30)	157-0141-84

Compound	Solvent	$\lambda_{max}(\log \epsilon)$	Ref.
$C_{33}H_{23}ClN_2$			
Spiro[9H-fluorene-9,4'-[4H]pyrazole], 5'-(4-chlorophenyl)-1',5'-dihydro-1',3'-diphenyl-	MeOH	230(4.70),270(4.87), 357(4.77)	78-1585-84
$C_{33}H_{23}N_3O_3$			
1H-Benzimidazole-5-carboxylic acid, 2-[4-(2-benzoxazolyl)-1-naphthalenyl]-1-(phenylmethyl)-, methyl ester	DMF	343(4.46)	5-1129-84
$C_{33}H_{24}N$			
Pyridinium, 1-(1-naphthalenyl)-2,4,6-triphenyl-, perchlorate	EtOH	<u>312(4.6)</u>	103-1262-84
Pyridinium, 1-(2-naphthalenyl)-2,4,6-triphenyl-, perchlorate	EtOH	<u>312(4.5)</u>	103-1262-84
$C_{33}H_{24}N_2$			
1H-Quindoline, 2,10-dihydro-1,3,11-triphenyl-	EtOH	208(4.62),230(4.52), 287(4.40),320(4.55), 354(4.38),368(4.41)	103-1374-84
Spiro[9H-fluorene-9,4'-[4H]pyrazole], 1',5'-dihydro-1',3',5'-triphenyl-	MeOH	222(4.2),270(4.04), 352(3.91)	78-1585-84
$C_{33}H_{24}N_4O$			
5H-Pyrano[2,3-d]pyrimidine-6-carbonitrile, 7-[1,1'-biphenyl]-4-yl-5-phenyl-4-[(phenylmethyl)amino]-	EtOH	247(4.39),258(4.36), 297(4.37)	73-2309-84
$C_{33}H_{24}O_{10}$			
Abiesin	EtOH	272(4.52),332(4.46)	102-0704-84
	EtOH-NaOMe	226(4.59),286(4.62), 391(4.20)	102-0704-84
	EtOH-NaOAc	272(4.52),283(4.51), 315s(4.43),350(4.42)	102-0704-84
	EtOH-AlCl$_3$	283(4.81),339(4.78), 395(4.56)	102-0704-84
$C_{33}H_{26}ClN_3O_7$			
D-Ribitol, 1,4-anhydro-1-C-(2-chloro-5-methylimidazo[1,5-b]pyridazin-7-yl)-, 2,3,5-tribenzoate, (S)-	EtOH	233(4.63),276(3.76), 382(3.79)	39-0229-84C
$C_{33}H_{26}N_2O_7$			
D-Ribitol, 1,4-anhydro-1-C-imidazo[1,5-a]pyridin-3-yl-, 2,3,5-tribenzoate, (S)-	EtOH	225(4.67),281(4.03), 332(3.32)	39-0229-84C
$C_{33}H_{26}N_8O_3S$			
1H-Pyrazole, 4-[[2-[[(4,6-dimethyl-2-pyrimidinyl)amino]sulfonyl]phenyl]azo]-3,5-diphenyl-1-(2-pyridinylcarbonyl)-	n.s.g.	363(4.07)	48-1021-84
$C_{33}H_{26}O_2Sn$			
Tin, (1,3-diphenyl-1,3-propanedionato-0,0')triphenyl-	benzene	346(4.29)	101-0047-84N
	C$_6$H$_{12}$	337(4.34)	101-0047-84N
	EtOH	343(4.42)	101-0047-84N
	CHCl$_3$	341(4.35)	101-0047-84N

Compound	Solvent	$\lambda_{max}(\log \epsilon)$	Ref.
$C_{33}H_{26}O_{17}$ Guavin B	MeOH	208(4.66),221(4.56), 283(4.27)	94-3787-84
$C_{33}H_{27}ClN_8O_4$ Acetamide, N,N'-[[(4-chlorophenyl)meth- ylene]bis(4,9-dihydro-2-methyl-9-oxo- pyrazolo[5,1-b]quinazoline-3,7- diyl)]bis-	10% EtOH- NaOH	220(4.78),249(4.78), 335(4.56),400(3.90)	56-0411-84
$C_{33}H_{27}N_9O_6$ Acetamide, N,N'-[[(4-nitrophenyl)meth- ylene]bis(4,9-dihydro-2-methyl-9-oxo- pyrazolo[5,1-b]quinazoline-3,7- diyl)]bis-	10% EtOH- NaOH	220(4.79),238(4.80), 318(4.44),392(4.13)	56-0411-84
$C_{33}H_{28}N_2O_3$ 1H-Pyrazole-1-acetic acid, 4-methyl- α-(2-phenoxy-2-phenylethenyl)-3,5- diphenyl-, methyl ester, (Z)-	MeOH	250(4.77)	44-4647-84
$C_{33}H_{28}N_2O_5$ Kurramine, (+)-	MeOH	224s(4.71),261s(4.25), 290(3.86),338(3.77)	78-2513-84
	MeOH-acid	222s(4.67),263s(4.26), 329(3.80),389(3.85)	78-2513-84
$C_{33}H_{28}N_4O_7$ 2,4(1H,3H)-Pteridinedione, 3-[2,3-O- carbonyl-5-O-(triphenylmethyl)-β-D- ribofuranosyl]-6,7-dimethyl-	MeOH	229s(4.24),245s(3.99), 330(3.97)	136-0179-84F
$C_{33}H_{28}N_8O_4$ Acetamide, N,N'-[(phenylmethylene)- bis(4,9-dihydro-2-methyl-9-oxo- pyrazolo[5,1-b]quinazoline-3,7- diyl)]bis-	10% EtOH- NaOH	219(4.78),254(4.83), 335(4.68),400(4.03)	56-0411-84
$C_{33}H_{28}O_{10}$ Chamaejasmenin C	MeOH	295(4.52)	94-0362-84
$C_{33}H_{28}O_{11}$ 2-Propen-1-one, 1-(4-acetoxy-3-methoxy- phenyl)-3-(2,6,7-triacetoxy-4-meth- oxy-1-phenanthrenyl)-	CH_2Cl_2	256(4.20),273s(4.15), 306s(3.75),330(3.50), 344(3.49),362(3.47), 500(2.76)	5-1024-84
$C_{33}H_{28}O_{14}$ β-D-Glucopyranosiduronic acid, 6'-hy- droxy-3-oxospiro[isobenzofuran- 1(3H),9'-[9H]xanthen]-3'-yl, methyl ester, 2,3,4-triacetate	MeOH	223(4.78),274(3.87)	94-2832-84
$C_{33}H_{28}O_{17}$ Benzoic acid, 6'-ester with (3,4-dihy- droxyphenyl)(3-β-D-glucopyranosyl- 2,4,6-trihydroxyphenyl)methanone, 2'-(4-hydroxybenzoate)	MeOH	258(4.47),270(4.48), 295(4.30),325(4.17)	94-2676-84
Benzoic acid, 3,4,5-trihydroxy-, 2'- ester with (3,4-dihydroxyphenyl)(3- β-D-glucopyranosyl-2,4,6-trihydroxy-	MeOH	258(4.38),270(4.38), 295(4.21),325(4.08)	94-2676-84

Compound	Solvent	$\lambda_{max}(\log \epsilon)$	Ref.
phenyl)methanone, 6'-(4-hydroxy- benzoate) (cont.)			94-2676-84

$C_{33}H_{28}O_{18}$
  Benzoic acid, 3,4,5-trihydroxy-, 2',6'-
  diester with (3-β-D-glucopyranosyl-
  2,4,6-trihydroxyphenyl)(4-hydroxy-
  phenyl)methanone
	MeOH	278(4.53),324(4.17)	94-2676-84

$C_{33}H_{29}N_3O_9$
  D-Allonamide, 2,5-anhydro-N-[1-(1,6-di-
  hydro-6-oxo-3-pyridazinyl)ethyl]-,
  3,4,6-tribenzoate
	EtOH	229(4.59),276(3.61),
		  283(3.60),300s(3.30)
			39-0229-84C

$C_{33}H_{30}Cl_2N_7O_{12}P$
  3'-Adenylic acid, 2'-deoxy-N-[[2-(4-ni-
  trophenyl)ethoxy]carbonyl]-, 2,5-di-
  chlorophenyl 2-(4-nitrophenyl)ethyl-
  ester
	MeOH	267(4.58)	78-0059-84

$C_{33}H_{30}N_4O_9$
  D-Allonamide, 2,5-anhydro-N-[1-(2,5-di-
  hydro-3-methyl-5-oxo-1,2,4-triazin-6-
  yl)ethyl]-, 3,4,6-tribenzoate (isomer
  mixture)
	EtOH	230(4.65),266(3.84),
		  274s(3.81)
			39-0229-84C

$C_{33}H_{30}O_7$
  Asticolorin A
	MeOH	225(4.91),263(4.45),
		  290(4.37),304(4.10),
		  316(4.31)
			77-0764-84

$C_{33}H_{31}N_7O_4$
  Pyrazolo[5,1-b]quinazolin-9(4H)-one,
  3,3'-[[4-(dimethylamino)phenyl]-
  methylene]bis[7-methoxy-2-methyl-
	10% EtOH-
	NaOH
		221(4.77),244(4.75),
		  326(4.45),404(4.03)
			56-0411-84

$C_{33}H_{32}N_2O_8$
  Cyclohept[b]indole-7,8,9,10-tetracarb-
  oxylic acid, 5,6,7,8,9,10-hexahydro-
  5-methyl-6-(1-methyl-1H-indol-3-yl)-,
  9,10-diethyl 7,8-dimethyl ester
	EtOH	250s(--),270(3.97),
		  290(3.84),450(3.88)
			94-2456-84

$C_{33}H_{32}N_2O_{10}S_2$
  3-Cephem-4-carboxylic acid, 7β-amino-
  2-[(Z)-(methoxycarbonyl)methylene]-
  3-(acetoxymethyl)-, diphenylmethyl
  ester, p-toluenesulfonic acid salt
	EtOH	345(4.00)	87-1225-84

$C_{33}H_{33}N_3$
  5H-Cyclohepta[b]pyridine-3-carboni-
  trile, 2-[1,1'-biphenyl]-4-yl-4-
  [4-(diethylamino)phenyl]-6,7,8,9-
  tetrahydro-
	EtOH	208(4.69),244(4.27),
		  269(4.46),291(4.54),
		  349(3.93)
			73-1395-84

$C_{33}H_{34}N_4O_5Zn$
  Zinc, [2-(2-hydroxyethyl) 18-methyl
  3,7,12,17-tetramethyl-21H,23H-por-
  phine-2,18-dipropanoato(2-)-
  $N^{21},N^{22},N^{23},N^{24}$]-, (SP-4-2)-
	CHCl$_3$	404(5.39),532(4.18),
		  566(4.09)
			65-1610-84

Compound	Solvent	$\lambda_{max}(\log \epsilon)$	Ref.
$C_{33}H_{36}N_4O_2$ 1H-Benzimidazolium, 2,2'-(5,10-dihy- droxybicyclo[4.4.1]undeca-1,3,5,7,9- pentaene-2,7-diyl)bis[1,3,5,6-tetra- methyl-, diiodide	CH_2Cl_2	292(4.55),377(4.15), 446(3.85)	33-2192-84
$C_{33}H_{36}N_4O_5$ 21H,23H-Porphine-2,18-dipropanoic acid, 3,7,12,17-tetramethyl-, 2-hydroxyeth- yl methyl ester	$CHCl_3$	400(5.32),495(4.38), 528(4.07),567(3.93), 622(3.43),650(3.28)	65-1610-84
$C_{33}H_{37}N_2$ 3H-Indolium, 2-[[11-(1,3-dihydro-1,3,3- trimethyl-2H-indol-2-ylidene)-1,3,5- 7,9-undecapentaenyl]-1,3,3-tri- methyl-, iodide	$C_6H_{11}Me$ EtOH DMF	980(5.45) 958(5.05) 960(4.82)	99-0415-84 99-0415-84 99-0415-84
$C_{33}H_{38}N_4O_{18}$ α-D-Ribofuranose, 1,2-O-[1-[[[1-[(4-ni- trophenyl)methyl]-4-[(2,3,5-tri-O- acetyl-β-D-ribofuranosyl)oxy]-1H- imidazol-5-yl]carbonyl]amino]ethyli- dene]-, 3,5-diacetate	M HCl H_2O	257(4.16) 258(4.29)	4-0849-84 4-0849-84
$C_{33}H_{38}OP$ Phosphonium, [3-methyl-5-(2,2,6-tri- methyl-7-oxabicyclo[4.1.0]hept-1- yl)-2,4-pentadienyl]triphenyl-, bromide	EtOH	206(4.73),227(4.61), 336(3.81)	33-0471-84
$C_{33}H_{39}N_2O_9$ Dehydroreserpine trifluoroacetate	EtOH EtOH-acid EtOH-base	265(4.21),319(3.85), 332(3.91),390(4.17) 265(4.19),390(4.27) 267(4.33),290(4.24), 319(4.32),332(3.30)	151-0355-84A 151-0355-84A 151-0355-84A
$C_{33}H_{42}N_3OP$ Piperidine, 1,1',1"-(phosphinylidyne- tri-4,1-phenylene)tris-	MeOH	291(4.90)	139-0253-84C
$C_{33}H_{42}N_3P$ Piperidine, 1,1',1"-(phosphinidynetri- 4,1-phenylene)tris-	MeOH	291(4.81)	139-0253-84C
$C_{33}H_{42}N_3PS$ Piperidine, 1,1',1"-(phosphinothioyl- idynetris-4,1-phenylene)tris-	MeOH	292(4.91)	139-0253-84C
$C_{33}H_{42}N_4O_6$ 21H-Biline-8,12-dipropanoic acid, 2,18-diethyl-1,4,5,15,16,19,22,24- octahydro-3,7,13,17-tetramethyl- 1,19-dioxo-, monohydrochloride	CH_2Cl_2	498(4.74)	5-1454-84
21H-Biline-8,12-dipropanoic acid, 3,17-diethyl-1,4,5,15,16,19,22,24- octahydro-2,7,13,18-tetramethyl- 1,19-dioxo-, monohydrochloride	CH_2Cl_2	498(4.71)	5-1454-84
21H-Biline-8,12-dipropanoic acid, 3,18-diethyl-1,4,5,15,16,19,22,24- octahydro-2,7,13,17-tetramethyl-	CH_2Cl_2	398(4.82)	5-1454-84

Compound	Solvent	$\lambda_{max}(\log \epsilon)$	Ref.
1,19-dioxo- (cont.)			5-1454-84
$C_{33}H_{43}N_4O_4S_2$			
Quinolinium, 6-[(diethylamino)sulfon-	MeOH	730(5.25)	103-0767-84
yl]-4-[3-[6-[(diethylamino)sulfonyl]-	MeCN	730(5.27)	103-0767-84
1-ethyl-4(1H)-quinolinylidene]-1-pro-	$C_6H_4Cl_2$	747(5.33)	103-0767-84
penyl]-1-ethyl-, iodide			
$C_{33}H_{44}N_4O_6$			
21H-Biline-8,12-dipropanoic acid,	CH_2Cl_2	498(4.86)	5-1454-84
2,17-diethyl-1,2,3,4,5,15,16,19-			
22,24-decahydro-3,7,13,18-tetrameth-			
yl-1,19-dioxo-, hydrochloride			
21H-Biline-8,12-dipropanoic acid,	CH_2Cl_2	498(4.98)	5-1454-84
2,18-diethyl-1,2,3,4,5,15,16,19-			
22,24-decahydro-3,7,13,17-tetrameth-			
yl-1,19-dioxo-, hydrochloride			
isomer	CH_2Cl_2	498(4.76)	5-1454-84
21H-Biline-8,12-dipropanoic acid,	CH_2Cl_2	498(4.91)	5-1454-84
3,17-diethyl-1,2,3,4,5,15,16,19-			
22,24-decahydro-2,7,13,18-tetrameth-			
yl-1,19-dioxo-, hydrochloride			
21H-Biline-8,12-dipropanoic acid,	CH_2Cl_2	498(4.82)	5-1454-84
3,18-diethyl-1,2,3,4,5,15,16,19-			
22,24-decahydro-2,7,13,17-tetrameth-			
yl-1,19-dioxo-, hydrochloride			
$C_{33}H_{44}O_{18}$			
Periclymenoside	MeOH	220(4.15),234(4.22),	33-0160-84
		293(4.08),318(4.08)	
$C_{33}H_{46}N_4O_6$			
Methylstercobiline IXα, hydrochloride	CH_2Cl_2	496(4.65)	5-1454-84
isomer 13	CH_2Cl_2	496(4.66)	5-1454-84
isomer 15	CH_2Cl_2	496(4.48)	5-1454-84
isomer 16	CH_2Cl_2	496(4.79)	5-1454-84
isomer 18	CH_2Cl_2	496(4.70)	5-1454-84
$C_{33}H_{47}N_3O_9$			
Herbimycin A, 17-(cyclopropylamino)-	MeOH	245(4.08)	158-1264-84
Herbimycin A, 17-(2-propenylamino)-	MeOH	247(4.10),338(4.07)	158-1264-84
Herbimycin A, 19-(2-propenylamino)-	MeOH	247(4.12),335(3.95)	158-1264-84
$C_{33}H_{47}N_3O_{10}$			
Geldanamycin, 19-(cyclopropylamino)-	MeOH	265(4.33)	158-1264-84
17-demethoxy-8,9-epoxy-8,9-dihydro-			
15-methoxy-11-O-methyl- (19-cyclo-			
propylamino-8,9-epoxyherbimycin A)			
11-Oxa-9,19-diazatricyclo[18.3.1.1^{8,11}]-	MeOH	246(4.40)	158-1264-84
pentacosa-1(23),14,16,20-tetraene-			
10,18,22,24-tetrone, 21-(cycloprop-			
ylamino)-25-hydroxy-2,5,6,13-tetra-			
methoxy-4,7,17,25-tetramethyl-			
$C_{33}H_{48}O_5$			
Lanosta-7,9(11),24-trien-3-one, 15-	EtOH	235(4.01),241(4.06),	102-1129-84
acetoxy-22,26-epoxy-26-methoxy-,		251(3.91)	
(15α,22S,26S)-			
$C_{33}H_{48}O_6$			
Antibiotic PD-114,720	MeOH	237(4.50)	77-1450-84

Compound	Solvent	$\lambda_{max}(\log \epsilon)$	Ref.
Leptomycin B	EtOH	225(4.30),240s(4.18)	158-84-82
Milbemycin H	EtOH	237(4.40)	158-84-48
$C_{33}H_{48}O_7$			
Antibiotic PD-114,721	MeOH	237(4.50)	77-1450-84
Kazusamycin	MeOH	232(4.52),245s(4.48)	158-0706-84
$C_{33}H_{49}BrO_4$			
30-Norlupa-18,21-diene-3,28-diol, 22-bromo-, diacetate, (3β)-	C_6H_{12}	272(3.76)	73-0141-84
$C_{33}H_{50}OSe$			
Cholest-5-en-3-ol, 6-(phenylseleno)-	EtOH	259(3.91)	88-1287-84
$C_{33}H_{50}O_4$			
30-Norlupa-12,18-diene-3β,28-diol, diacetate	C_6H_{12}	236(3.88)	73-0141-84
30-Norlupa-18,21-diene-3β,28-diol, diacetate	C_6H_{12}	256(3.59)	73-0141-84
$C_{33}H_{50}O_5$			
Lanosta-7,9(11),24-triene-3,15-diol, 22,26-epoxy-26-methoxy-, 15-acetate, (3β,15α,22S,26S)-	EtOH	235(4.28),243(4.34), 251(4.17)	102-1129-84
$C_{33}H_{50}O_7$			
Androst-5-en-7-one, 3-acetoxy-17-(2,2-dimethyl-1-oxopropoxy)-18-[(1-methoxycyclohexyl)oxy]-, (3β,17β)-	EtOH	235(4.12)	56-0711-84
$C_{33}H_{52}O_6$			
Cholesta-5,7-diene-1,3-diol, bis(ethyl carbonate), (1α,3β)-	EtOH	262s(3.88),271(4.03), 281(4.06),293(3.83)	94-3244-84
	ether	262s(3.95),271(4.09), 281(4.11),293(3.89)	94-3244-84
$C_{33}H_{52}P_2S$			
Diphosphene, [2,4-bis(1,1-dimethylethyl)-6-methylphenyl][2,4,6-tris(1,1-dimethylethyl)phenyl]-, 1-sulfide	CH_2Cl_2	253(4.28),373(3.89)	142-0681-84
$C_{33}H_{54}O_{12}$			
Sileneoside D	EtOH	247(4.15)	105-0700-84
$C_{33}H_{60}O_8$			
2,10,18,26-Tritriacontanetetrone, 4,12,20,28-tetrahydroxy-	MeOH	276(2.29)	158-84-139
$C_{33}H_{62}O_8$			
8,16,24-Tritriacontanetrione, 6,14,22,30,32-pentahydroxy-	MeOH	275(2.14)	158-84-139
$C_{33}H_{62}O_9$			
2,10,18-Tritriacontanetrione, 4,12,20,24,26,28-hexahydroxy- (PM-Toxin D)	MeOH	276(2.07)	158-84-139

Compound	Solvent	$\lambda_{max}(\log \epsilon)$	Ref.
$C_{34}H_{22}$			
5,14[1',2']-Benzenopentacene, 5,14-di-hydro-1,17-di-1-propynyl-, (5S)-(+)-	EtOH-MeCN	228(4.64),266.7(5.07), 293.2(4.13),337.1(3.71), 354.8(3.87),375(3.71)	44-4266-84
$C_{34}H_{22}O_2$			
Naphtho[2,1-b]chrysene-7,14-dione, 11-methyl-9-(4-methylphenyl)-	CHCl$_3$	260(4.75),302(4.67), 316(4.65),327(4.14), 376(3.88),421(3.85)	40-0090-84
Naphtho[2,1-b]chrysene-7,14-dione, 12-methyl-9-(4-methylphenyl)-	CHCl$_3$	263(4.35),301(4.17), 316(4.15),327(4.26), 386(3.34),425(3.35), 441(3.32)	40-0090-84
$C_{34}H_{22}O_4$			
Naphtho[2,1-b]chrysene-7,14-dione, 10-methoxy-9-(2-methoxyphenyl)-	CHCl$_3$	264(4.58),312(4.40), 327(4.47),379(3.65), 420(3.66?)	40-0090-84
Naphtho[2,1-b]chrysene-7,14-dione, 11-methoxy-9-(3-methoxyphenyl)-	CHCl$_3$	261(4.69),328(4.72), 420(3.83)	40-0090-84
Naphtho[2,1-b]chrysene-7,14-dione, 12-methoxy-9-(4-methoxyphenyl)-	CHCl$_3$	265(4.65),303(4.44), 317(4.44),329(4.51), 404(3.71),427(3.71), 468(3.65)	40-0090-84
$C_{34}H_{22}O_8$			
Albanol B	EtOH	222(4.71),275s(4.12), 285(4.18),318s(4.34), 335s(4.54),350(4.68), 365(4.65)	142-0473-84
$C_{34}H_{22}N_6O_4$			
1,1'-Bi-4H-pyrrolo[1,2-a]benzimidazole, 4,4'-dimethyl-2,2'-bis(3-nitrophenyl)-	DMSO	260(4.28),368(4.43)	103-0150-84
1,1'-Bi-4H-pyrrolo[1,2-a]benzimidazole, 4,4'-dimethyl-2,2'-bis(4-nitrophen-yl)-	DMSO	260(4.40),368(4.58)	103-0150-84
1,1'-Bi-4H-pyrrolo[1,2-a]benzimidazole, 4,4'-dimethyl-7,7'-dinitro-2,2'-di-phenyl-	DMSO	260(4.35),290(4.50), 368(4.09),480(3.92)	103-0150-84
$C_{34}H_{24}O_8$			
Mulberrofuran I	EtOH	250s(4.10),290s(4.10), 330s(4.38),338(4.39)	94-1260-84
	EtOH-acid	296s(4.09),330s(4.23), 340(4.26),456(3.79), 548(4.08)	94-1260-84
$C_{34}H_{24}O_9$			
Plumbazeylanone	MeOH	218(4.64),232s(4.41), 249s(4.26),276(4.18), 354(3.76),418(3.71)	88-4801-84
$C_{34}H_{24}Se_2$			
4H-Selenin, 4-(2,6-diphenyl-4H-selenin-4-ylidene)-2,6-diphenyl-	CHCl$_3$	246(5.01),300(4.28), 488(4.64)	22-0241-84
$C_{34}H_{25}ClN_3OP$			
Ethanone, 2-[(4-chlorophenyl)azo]-1-(1H-indol-3-yl)-2-(triphenyl-phosphoranylidene)-	EtOH	416(4.26)	65-0685-84

Compound	Solvent	$\lambda_{max}(\log \epsilon)$	Ref.
$C_{34}H_{25}N_4O_3P$ Ethanone, 1-(1H-indol-3-yl)-2-[(4-ni-trophenyl)azo]-2-(triphenylphosphor-anylidene)-	EtOH	482(4.31)	65-0685-84
$C_{34}H_{26}N_2$ Spiro[9H-fluorene-9,4'-[4H]pyrazole], 1',5'-dihydro-5'-(4-methylphenyl)- 1',3'-diphenyl-	MeOH	226(4.56),268(4.10), 352(4.15)	78-1585-84
$C_{34}H_{26}N_2O_2$ 7H-Quino[3',2':4,5]pyrrolo[1,2-a][1]-benzazepine, 2,12-dimethoxy-5,9-di-phenyl-	C_6H_{12}	240(4.93),280(4.49), 296s(4.29),329s(3.90), 344(3.99),366s(3.85), 440s(2.56)	39-2529-84C
$C_{34}H_{26}N_2O_4$ 3H-Indol-3-one, 2-[1,3-dihydro-3-oxo-1-(1-oxo-3-phenylpropyl)-2H-indol-2-ylidene]-1,2-dihydro-1-(1-oxo-3-phenylpropyl)-, cis	benzene	435(3.61)	39-2305-84C
trans	benzene	565(3.84)	39-2305-84C
$C_{34}H_{26}N_2O_6$ 3H-Indol-3-one, 2-[1,3-dihydro-1-[(4-methoxyphenyl)acetyl]-3-oxo-2H-indol-2-ylidene]-1,2-dihydro-1-[(4-methoxyphenyl)acetyl]-, cis	benzene	430(3.62)	39-2305-84C
trans	benzene	570(3.86)	39-2305-84C
$C_{34}H_{26}N_3OP$ Ethanone, 1-(1H-indol-3-yl)-2-(phenyl-azo)-2-(triphenylphosphoranylidene)-	EtOH	408(4.20)	65-0685-84
$C_{34}H_{26}O_8$ Mulberrofuran G	EtOH	223(4.63),285(4.29), 295s(4.24),306s(4.37), 321(4.57),335(4.50)	142-0473-84
$C_{34}H_{27}NO_2$ 2H-Pyrrole-3-carboxylic acid, 2-methyl-5-phenyl-2-(1,2,3-triphenyl-2-cyclo-propen-1-yl)-, methyl ester	C_6H_{12}	227(4.43),322(4.30), 335(4.18)	35-1065-84
$C_{34}H_{28}N_4O_9$ 2,4(1H,3H)-Pteridinedione, 6,7-dimeth-yl-1-(2,3,5-tri-O-benzoyl-β-D-ribo-furanosyl)-	MeOH	229(4.75),326(4.01)	136-0179-84F
2,4(1H,3H)-Pteridinedione, 6,7-dimeth-yl-3-(2,3,5-tri-O-benzoyl-β-D-ribo-furanoyl)-	MeOH	229(4.68),272s(3.54), 281s(3.51),330(3.98)	136-0179-84F
$C_{34}H_{28}O_9$ Mulberrofuran J	EtOH	209(4.97),283(4.51), 293s(4.46),310s(4.57), 321(4.67),337(4.56)	142-1007-84
	EtOH-AlCl$_3$	210(4.78),222s(4.68), 310s(4.63),320(4.65), 335s(4.52),370s(3.52)	142-1007-84

Compound	Solvent	$\lambda_{max}(\log \epsilon)$	Ref.
$C_{34}H_{28}O_{12}$ 2-Propen-1-one, 1-(3,4-diacetoxyphen- yl)-3-(2,6,7-triacetoxy-4-methoxy-1- phenanthrenyl)-	CH_2Cl_2	254(4.37),274s(4.30), 306s(3.79),331(3.52), 345(3.55),362(3.56), 500(2.35)	5-1024-84
$C_{34}H_{30}O_3$ Anthracene, 9,9'-[oxybis(2,1-ethanediyl- oxymethylene)]bis-	$C_6H_{11}Me$	249(5.28),256(5.43), 315(3.42),330(3.80), 346(4.10),364(4.29), 384(4.27)	150-1901-84M
$C_{34}H_{30}S_4$ Benzene, 1,2,4,5-tetrakis[(phenyl- methyl)thio]-	CH_2Cl_2	268(4.48)	35-7131-84
$C_{34}H_{32}N_2O$ 1H-Indole, 1,1'-[oxybis(methylene)]- bis[3-ethyl-2-phenyl-	EtOH	238(4.55),294(4.42)	78-4351-84
$C_{34}H_{32}N_2O_6$ Kohatine, (+)-	MeOH	233(4.59),288(3.77)	78-2513-84
$C_{34}H_{32}N_3$ 1-Naphthalenemethanaminium, N-[bis[4- (dimethylamino)phenyl]methylene]-α- phenyl-, (OC-6-11)-, hexachloro- antimonate	CH_2Cl_2	262(4.44),282(4.43), 424(4.51),546(4.68)	24-3222-84
$C_{34}H_{32}N_6O_4$ 21H,23H-Porphine-2,18-dipropanoic acid, 7,12-dicyano-3,8,13,17-tetramethyl-, dimethyl ester	n.s.g.	422(5.11),517(4.05), 552(3.88),586(3.72), 639(3.43)	5-1057-84
$C_{34}H_{32}O$ 9-Anthracenemethanol, α-[bis(2,4,6-tri- methylphenyl)methylene]-	MeOH	250(5.00),397(3.98)	35-0477-84
$C_{34}H_{32}O_{15}$ 5,12-Naphthacenedione, 9-ethynyl- 7,8,9,10-tetrahydro-6,9,11-tri- hydroxy-7-[(2,3,4,6-tetra-O-acetyl- β-D-glucopyranosyl)oxy]-, (7S-cis)-	EtOH	208(4.21),251(4.57), 257s(4.52),285(3.90)	78-4657-84
$C_{34}H_{34}BNOS$ Pyridinium, ω-dimethylsulfuranylidene- acetonyl-, tetraphenylborate	20% acetone	385(2.20)	104-1911-84
$C_{34}H_{34}BNOSe$ Pyridinium, ω-dimethylselenuranyli- deneacetonyl-, tetraphenylborate	20% acetone	402(2.92)	104-1911-84
$C_{34}H_{34}N_2$ Benzo[h]quinolizinium, 9,9'-(1,8-oct- anediyl)bis-	MeOH	252(3.30)	48-0757-84
$C_{34}H_{34}N_2O_4$ 2,5-Cyclohexadiene-1,4-dione, 2,5- bis[2-(1,1-dimethyl-2-propenyl)- 1H-indol-3-yl]-3,6-dimethoxy-	EtOH	280(4.51),287(4.49), 488(3.01)	39-0567-84C

Compound	Solvent	$\lambda_{max}(\log \epsilon)$	Ref.
$C_{34}H_{34}O_{12}S_2$ 1H,3H-Furo[3,4-c]furan-1,4-diol, tetrahydro-3,6-bis[3-methoxy-4-[[(4-methylphenyl)sulfonyl]oxy]phenyl]-, (1α,3α,3aα,4α,6α,6aα)-	EtOH	223(3.96),275(4.60)	39-1159-84C
$C_{34}H_{34}O_{15}$ 5,12-Naphthacenedione, 3-ethynyl-1,2,3,4,4a,12a-hexahydro-3,6,11-trihydroxy-1-[(2,3,4,6-tetra-O-acetyl-β-D-glucopyranosyl)oxy]-, [1S-(1α,3α,4aα,12aβ)]-	EtOH	237(4.50),253(4.47), 278(4.41),286(4.35), 380s(3.94),399(4.17), 417(4.16),421(4.13)	78-4657-84
$C_{34}H_{34}O_{16}$ 5,12-Naphthacenedione, 9-acetyl-7,8,9,10-tetrahydro-6,9,11-trihydroxy-7-[(2,3,4,6-tetra-O-acetyl-β-D-glucopyranosyl)oxy]-, (7S-cis)-	EtOH	207(4.13),250(4.52), 256(4.47),284(3.86)	78-4657-84
$C_{34}H_{35}NO_6$ Pentacyclo[19.3.1.1^{3,7}.1^{9,13}.1^{15,19}]-octacosa-1(25),3,5,7(28),9,11,13(27)-15,17,19(26),21,23-dodecaene-25,26,27,28-tetrol, 5-(1,1-dimethylethyl)-11,23-dimethyl-17-nitro-	aq MeOH-HCl	320(3.94)	126-0221-84B
	pH 6.3	400(3.93)	126-0221-84B
with LiOH	pH 6.3	432(3.95)	126-0221-84B
with NaOH	pH 6.3	451(4.09)	126-0221-84B
with KOH	pH 6.3	453(4.10)	126-0221-84B
$C_{34}H_{35}N_3O_6$ 7H-Pyrrolo[2,3-d]pyrimidine, 2,4-dimethoxy-7-[2,3,5-tris-O-(phenylmethyl)-α-D-arabinofuranosyl]-	MeOH	217(4.52),258s(3.85), 263(3.83)	5-0273-84
β-	MeOH	217(4.52),258s(3.85), 263(3.83)	5-0273-84
$C_{34}H_{35}N_5O_6$ Guanosine, 2'-deoxy-5'-O-[(4-methoxyphenyl)diphenylmethyl]-N-(2-methyl-1-oxopropyl)-	MeOH	235(4.26),254(4.24), 259(4.24),275(4.11), 281(4.11)	118-0965-84
$C_{34}H_{36}N_6O_6$ 21H,23H-Porphine-2,18-dipropanoic acid, 7,12-bis[(hydroxyimino)methyl]-3,8,13,17-tetramethyl-, dimethyl ester	CHCl₃	418(5.20),512(4.13), 548(4.03),582(3.80), 637(3.63)	5-1057-84
$C_{34}H_{38}N_4O_2$ Deoxophylloerythroin methyl ester	CH₂Cl₂	400(5.37),498(4.23), 532(3.60),564(3.82), 616(3.84)	44-4602-84
21-Phorbinecarboxylic acid, 3,4-didehydro-3,9,14-triethyl-4,8,13,18-tetramethyl-, methyl ester	CH₂Cl₂	399(5.39),498(4.24), 536(3.72),564(3.90), 616(3.85)	44-4602-84
2-Propenoic acid, 3-(7,12,18-triethyl-3,8,13,17-tetramethyl-21H,23H-porphin-2-yl)-, methyl ester	CH₂Cl₂	412(5.21),508(4.04), 550(4.27),574(4.05), 636(3.43)	44-4602-84
$C_{34}H_{38}N_4O_6$ Hematoporphyrin	EtOH	529(2.50)	35-5879-84

Compound	Solvent	$\lambda_{max}(\log \epsilon)$	Ref.
$C_{34}H_{38}N_4O_6S$ 5-Thiazolidineacetamide, 2-[(1-methyl-ethylidene)hydrazono]-4-oxo-3-[2,3,5-tris-O-(phenylmethyl)-α-D-arabino-furanosyl]-	MeOH	257(4.11)	128-0295-84
$C_{34}H_{38}F_3N_5O_5S_2$ 1H-Benzimidazolium, 2-[[5-[[5-(ethoxy-carbonyl)-1,3-diethyl-1,3-dihydro-2H-benzimidazol-2-ylidene)ethylidene]-3-ethyl-4-oxo-2-thiazolidinylidene]-methyl]-1,3-diethyl-5-[(trifluoro-methyl)sulfonyl]-, perchlorate	EtOH	547(4.68)	103-0665-84
$C_{34}H_{39}NO_{11}$ Acetylmaymyrsine	EtOH	230(4.12),258s(3.51), 265(3.54),271s(3.47), 282s(3.06)	142-2221-84
$C_{34}H_{40}Br_2O_4$ Verrucosane, 2,8-bis(4-bromobenzoyl)-	EtOH	244(4.73)	44-4644-84
$C_{34}H_{40}N_2O_7$ 1H-Pyrido[3,4-b]indole-3-carboxylic acid, 1-[[5-acetyl-3-(5,5-dimethyl-1,3-dioxan-2-yl)-3,4-dihydro-2-hy-droxy-2H-pyran-4-yl]methyl]-2,3,4,9-tetrahydro-2-(phenylmethyl)-, methyl ester	MeOH	225(4.02),262(4.14), 285(3.97),290(3.88)	78-4853-84
$C_{34}H_{40}N_4O_8S_2$ 2,2'-(1,4-Naphthalenylene)dibenzimida-zolium, 1,1',3,3',5,5',6,6'-octa-methyl-, bis(methyl sulfate)	DMF	310(4.36)	5-1129-84
$C_{34}H_{44}N_2O_{16}S$ Paulomycinone A	MeOH	232(4.18),264(4.27), 440(3.32)	158-1273-84
$C_{34}H_{44}N_6O_{11}$ Propanoic acid, 2-methyl-, 2',3',4'-triester with 2-methyl-N-[6-[2-(4-nitrophenyl)ethoxy]-9-β-D-ribo-furanosyl-9H-purin-2-yl]propanoate	MeOH	217(4.46),268(4.46)	78-0059-84
$C_{34}H_{45}N_3P$ Phosphonium, methyltris[4-(1-piperidin-yl)phenyl]-, iodide	MeOH	307(5.01)	139-0253-84C
$C_{34}H_{46}N_2O_8$ Ansatrienin A_2	MeOH	230(4.39),264s(--), 271(4.66),279(4.57), 387(3.26)	158-84-19
	MeOH-NaOH	261(4.64),269(4.69), 278(4.64),481(3.28)	158-84-19
$C_{34}H_{47}BrO_3$ 9,10-Secocholesta-5,10(19)-dien-3-ol, 7,8-epoxy-, 4-bromobenzoate, (3β,5Z,7R,8α)-	hexane-EtOH	243(4.64)	44-1537-84

Compound	Solvent	$\lambda_{max}(\log \epsilon)$	Ref.
$C_{34}H_{47}N_3O_{18}$ 2H-1-Benzopyran-2-one, 7-[[O-2-(acetyl-amino)-2-deoxy-β-D-glucopyranosyl-(1-4)-O-2-(acetylamino)-2-deoxy-β-D-glucopyranosyl-(1→4)-2-(acetylamino)-2-deoxy-β-D-glucopyranosyl]oxy]-4-methyl-	DMSO	288(3.92),318(4.10)	94-1597-84
$C_{34}H_{50}O_4S$ Cholest-4-en-3-one, 4-[[(4-methylphen-yl)sulfonyl]oxy]-	ether	231(4.31),274(2.70), 295(1.74)	44-2355-84
$C_{34}H_{50}O_6$ Perenniporiol, 3-acetyl-	EtOH	236(4.31),243(4.36), 252(4.20)	102-1129-84
$C_{34}H_{50}O_7$ Milbemycin D, 5-methoxy- Milbemycin G	n.s.g. EtOH	237s(4.47),244(4.48) 238s(--),244(4.48), 253s(--)	158-84-112 158-84-48
$C_{34}H_{51}CoN_6$ Cobalt, (3,4,7,8,12,13,17,18-octahydro-2,2,3,3,7,7,8,8,12,12,13,13,17,17,18-18-hexadecamethyl-2H,21H-5,15-diaza-porphinato-N^{21},N^{22},N^{23},N^{24})-, (SP-4-3)-	MeOH	370(4.01),468(3.57), 508s(3.44),616(3.29), 722(3.32)	33-1801-84
$C_{34}H_{52}O_7$ Milbemycin E Perenniporiol, 12β-acetoxy-	EtOH EtOH	241(4.42) 237(4.48),244(4.55), 253(4.37)	158-84-48 102-2885-84
$C_{34}H_{54}O_7$ Irumanolide I	MeOH	232(4.07)	158-84-111
$C_{34}H_{58}O_8$ Phomenoic acid PLM I PLM II	MeOH MeOH MeOH	236(4.06) 232(3.96) 233.2(3.34)	39-2133-84C 158-84-125 158-84-125
$C_{34}H_{66}N_2O_4$ Diazenedicarboxylic acid, dihexadecyl ester	hexane	401(1.55)	23-0574-84

Compound	Solvent	$\lambda_{max}(\log \epsilon)$	Ref.
$C_{35}H_{20}N_2O_3$ 1H-Isoindole-1,3(2H)-dione, 2-(1a,2-di- hydro-2-oxo-1a,3-diphenyl-1H-naphtho- [1',8':4,5,6]pentaleno[1,6a-b]azirin- 1-yl)-	EtOH	266(4.36),315(4.43), 420(4.26)	142-1369-84
$C_{35}H_{24}N_8O_3S$ 1H-Pyrazole, 3,5-diphenyl-1-(2-pyridin- carbonyl)-4-[[2-[(2-quinoxalinylami- no)sulfonyl]phenyl]azo]-	n.s.g.	356(3.49)	48-1021-84
$C_{35}H_{25}N_3S$ 2H-Thiopyran, 2-azido-2,3,4,5,6-penta- phenyl-	ether	230(4.24),320(3.68)	78-3539-84
$C_{35}H_{25}O$ Pyrylium, pentaphenyl-, TCNQ salt	MeCN	480(1.95)	80-0817-84
$C_{35}H_{25}P$ Spiro[9H-fluorene-9,2'(1'H)-phosphor- in], 1',3',6'-triphenyl-	MeCN	261(4.38),353(4.20)	77-1217-84
$C_{35}H_{25}S$ Thiopyrylium, pentaphenyl-, perchlorate	MeCN	213(4.29),252(4.28), 290(3.91),392(3.83)	80-0317-84
$C_{35}H_{26}$ Dibenzo[2,3:4,5]pentaleno[1,6-ab]ind- ene, 12d-(diphenylmethyl)-4b,8b,12b- 12d-tetrahydro-	hexane	263.5(3.48),269.7(3.61), 276.8(3.65)	89-0508-84
$C_{35}H_{26}S$ Thiopyran, 2,3,4,5,6-pentaphenyl-	ether	230(4.39),260s(4.13), 325s(3.80)	78-3539-84
$C_{35}H_{27}N_3O_4$ [3,3'-Bi-1H-pyrrole]-2,2',5,5'-tetrone, 1,4-bis(4-methylphenyl)-4-[(4-methyl- phenyl)amino]-1'-phenyl-	CHCl$_3$	290(4.27),450(3.62)	48-0401-84
$C_{35}H_{28}N_2O_4$ 9,10-Anthracenedione, 1-amino-2-[4- [1-(4-hydroxyphenyl)-1-methyleth- yl]phenoxy]-4-(phenylamino)-	C_6H_5Cl	536s(3.90),564(4.15), 600(4.16)	34-0482-84
$C_{35}H_{28}N_3OP$ Ethanone, 1-(1H-indol-3-yl)-2-[(4-meth- ylphenyl)azo]-2-(triphenylphosphoran- ylidene)-	EtOH	412(4.17)	65-0685-84
$C_{35}H_{28}N_3O_2P$ Ethanone, 1-(1H-indol-3-yl)-2-[(4-meth- oxyphenyl)azo]-2-(triphenylphosphor- anylidene)-	EtOH	414(4.22)	65-0685-84
$C_{35}H_{29}N_3O_4S_2$ Isoquinoline, 1,1'-(1-methyl-1H-pyr- role-2,5-diyl)bis[1,2-dihydro- 2-(phenylsulfonyl)-	n.s.g.	263(4.18)	103-1136-84

Compound	Solvent	$\lambda_{max}(\log \epsilon)$	Ref.
$C_{35}H_{32}O_4$			
Anthracene, 9-[2-[2-[2-(9-anthracenyl-methoxy)ethoxy]ethoxy]ethoxy]-	$C_6H_{11}Me$	248(5.27),256(5.49), 317(3.56),332(3.83), 347(4.04),365(4.21), 385(4.16)	150-1901-84M
11a,16[1',2']:17,21b[1'',2'']-Dibenzeno-11H-dibenzo[3,4:7,8]cycloocta[1,2-k]-1,4,7,10-tetraoxacyclotridecin, 2,3,5,6,8,9,16,17-octahydro-	$C_6H_{11}Me$	224(4.2),247(3.6), 256(3.6),271(3.1), 285(2.8)	150-1901-84M
$C_{35}H_{32}O_{15}$			
β-Naphthocyclinone epoxide	MeOH	206(4.54),212s(--), 252(4.36),264s(--), 352(3.76),423(3.80)	158-84-36
	MeOH-NaOH	248(4.44),282s(--), 342(3.70),500(3.80)	158-84-36
$C_{35}H_{33}ClO_{15}$			
β-Naphthocyclinone chlorohydrin	$CHCl_3$	256(4.46),278(4.19), 350(3.73),413(4.01), 432s(--)	158-84-36
$C_{35}H_{34}F_3N_3O_9S$			
Glycine, N-[S-[10-(benzoyloxy)-9,10-di-hydro-9-phenanthrenyl]-N-[N-(tri-fluoroacetyl)-L-γ-glutamyl]-L-cys-teinyl]-, dimethyl ester, (9R-trans)-(9S)-	MeOH	270(4.20)	77-1491-84
	MeOH	270(4.18)	77-1491-84
$C_{35}H_{34}F_3N_4O_4S_2$			
Benzoxazolium, 2-[[5-[[1,3-diethyl-1,3-dihydro-5-[(trifluoromethyl)sulfon-yl]-2H-benzimidazol-2-ylidene]ethyl-idene]-3-ethyl-4-oxo-2-thiazolidin-ylidene]methyl]-2-ethyl-5-phenyl-, iodide	EtOH	582(5.14)	103-0665-84
$C_{35}H_{34}N_2O_6$			
Cocsuline N-2-oxide	MeOH	272(3.36),282s(3.76), 303s(3.24)	142-0993-84
	MeOH-NaOH	285(3.48),300(3.45)	142-0993-84
$C_{35}H_{36}N_2O_3$			
9,10-Anthracenedione, 1-amino-2-[3,5-bis(1,1-dimethylethyl)-4-hydroxy-phenyl]-4-[(4-methylphenyl)amino]-	toluene	372(3.87),578(4.21), 615(4.22)	104-1592-84
$C_{35}H_{37}N_5O_6$			
Adenosine, N-[(4-methoxyphenyl)diphen-ylmethyl]-2'-O-(tetrahydro-2H-pyran-2-yl)-	EtOH	275(4.35)	78-0153-84
$C_{35}H_{37}N_5O_7$			
Guanosine, 5'-O-[bis(4-methoxyphenyl)-phenylmethyl]-2'-deoxy-N-(2-methyl-1-oxopropyl)-	MeOH	236(4.36),252(4.26), 260(4.24),273(4.12), 280(4.11)	118-0965-84
Guanosine, N-[(4-methoxyphenyl)diphen-ylmethyl]-2'-O-(tetrahydro-2H-pyran-2-yl)-	EtOH	260(4.15)	78-0153-84

Compound	Solvent	$\lambda_{max}(\log \epsilon)$	Ref.
$C_{35}H_{38}O_{16}Si$			
5a,11a-Epoxynaphthacene-5,6,11,12-tetrone, 1,4,4a,12a-tetrahydro-1-[(2,3,4,6-tetra-O-acetyl-β-D-glucopyranosyl)oxy]-3-[(trimethylsilyl)-oxy]-, [1S-(1α,4aβ,5aβ,11aβ,12aβ)-	EtOH	211(3.98),233(4.18), 260s(3.69),307(3.18)	78-4657-84
$C_{35}H_{39}NO_5$			
2-Propenoic acid, 3-[2-[[10-[2-(3-cyano-3-phenyloxiranyl)phenoxy]-decyl]oxy]phenyl]-, methyl ester, (2R*,4S*,5R*,29S*)-	MeCN	267(3.38),273s(3.42), 286s(3.48),298(3.49), 309(3.51)	24-2157-84
$C_{35}H_{40}Cl_2N_4O_6$			
21H-Biline-8,12-dipropanoic acid, 3,18-bis(2-chloroethyl)-1,19,22,24-tetra-hydro-2,7,13,17-tetramethyl-1,19-di-oxo-, dimethyl ester	CH_2Cl_2	372(4.62),654(4.06)	78-1749-84
$C_{35}H_{40}N_4O_6$			
21H-Biline-8,12-dipropanoic acid, 3-ethenyl-18-ethyl-1,19,22,24-tetra-hydro-2,7,13,17-tetramethyl-1,19-dioxo-, dimethyl ester	CH_2Cl_2	373(4.62),654(4.13)	78-1749-84
21H-Biline-8,12-dipropanoic acid, 18-ethenyl-3-ethyl-1,19,22,24-tetra-hydro-2,7,13,17-tetramethyl-1,19-dioxo-, dimethyl ester	CH_2Cl_2	370(4.61),650(4.12)	78-1749-84
$C_{35}H_{41}ClN_4O_6$			
21H-Biline-8,12-dipropanoic acid, 3-(2-chloroethyl)-18-ethyl-1,19,22,24-tet-rahydro-2,7,13,17-tetramethyl-1,19-dioxo-, dimethyl ester	CH_2Cl_2	373(4.69),636(4.16)	78-1749-84
21H-Biline-8,12-dipropanoic acid, 18-(2-chloroethyl)-3-ethyl-1,19,22-24-tetrahydro-2,7,13,17-tetramethyl-1,19-dioxo-, dimethyl ester	CH_2Cl_2	370(4.63),636(4.16)	78-1749-84
$C_{35}H_{41}N_5O_5$			
Bilirubin IXα 8-dimethylamide	CHCl$_3$	444(4.71)	78-4253-84
	DMSO	460(4.66)	78-4253-84
Bilirubin IXα 12-dimethylamide	CHCl$_3$	443(4.70)	78-4253-84
	DMSO	456(4.71)	78-4253-84
$C_{35}H_{42}N_4O_2Ru$			
Ruthenium, carbonyl(ethanol)[5,10,15-20-tetrapropyl-21H,23H-porphinato-(2-)-$N^{21},N^{22},N^{23},N^{24}$]-, (OC-6-42)-	CH_2Cl_2	411(5.17),533(4.07), 567(3.54)	35-5151-84
$C_{35}H_{42}O$			
4'-Apo-χ,ψ-carotenal	benzene	318(--),508(5.06)	137-0102-84
$C_{35}H_{42}O_{18}$			
Acteoside acetate	EtOH	217s(4.31),284(4.22)	94-1209-84
$C_{35}H_{42}O_{19}$			
Acteoside, β-hydroxy-, acetate	EtOH	216s(4.16),283(4.07)	94-1209-84
$C_{35}H_{43}NO_{14}$			
Shaunadimycin	MeOH	234(4.63),257(4.31),	100-0698-84

Compound	Solvent	$\lambda_{max}(\log \epsilon)$	Ref.
Shaunadimycin (cont.)		291(3.86),481s(4.08), 492(4.12),510s(3.99), 524(3.90)	100-0698-84
$C_{35}H_{43}N_4O_2$ Quinolinium, 6-[(diethylamino)carbonyl]-4-[3-[6-[(diethylamino)carbonyl]-1-ethyl-4(1H)-quinolinylidene]-1-propenyl]-1-ethyl-, iodide	MeOH	723(5.28)	103-0767-84
$C_{35}H_{44}N_4O_6$ 21H-Biline-8,12-dipropanoic acid, 3,17-diethyl-1,10,19,22,23,24-hexahydro-2,7,13,18-tetramethyl-1,10-dioxo-, methyl ester	MeOH	272(4.25),343(4.45), 583(4.06)	35-2645-84
$C_{35}H_{44}O_9$ β-D-Ribofuranoside, methyl 2,3-O-(1,2-ethanediylbis(oxy-2,1-ethanediyloxy-2,1-ethanediyl)]-5-O-(triphenylmethyl)-	EtOH	255(2.81),260(2.81)	5-1036-84
$C_{35}H_{46}O_{11}$ Trichoverritone	MeOH	259(4.57)	158-84-134
$C_{35}H_{46}O_{14}$ Pregna-1,4-diene-3,20-dione, 11,17-dihydroxy-21-[(2,3,4,6-tetra-O-acetyl-β-D-glucopyranosyl)oxy]-, (11β)-	n.s.g.	242(4.13)	87-0261-84
$C_{35}H_{46}O_{20}$ Echinacoside	MeOH	220(4.23),244s(3.98), 292(4.07),334(4.21)	94-3009-84
$C_{35}H_{47}NO_9$ Rhizoxin	MeOH	295(4.57),308(4.68), 325(4.54)	158-0354-84
$C_{35}H_{48}N_4O_2$ 21H-Biline-1,19-dione, 2,3,23,24-tetrahydro-2,3,7,8,12,13,17,18-octaethyl-, cis	MeOH	270(4.26),358(4.50), 580(4.01)	35-2645-84
trans	MeOH	274(4.30),354(4.54), 584(4.00)	35-2645-84
$C_{35}H_{48}N_6O_3$ 21H-Biline-8-propanamide, N-[3-(dimethylamino)propyl]-3-ethyl-1,17,18,19-22,24-hexahydro-2,7,12,13,17,17-hexamethyl-1,19-dioxo-, bis(trifluoroacetate)	CHCl3	280(4.20),352(4.36), 610(4.19),668s(3.88)	49-1071-84
$C_{35}H_{52}N_4O_9$ Herbimycin A, 19-(4-methyl-1-piperazinyl)-	MeOH	257(4.33)	158-1264-84
$C_{35}H_{52}N_4O_{10}$ Herbimycin A, 8,9-epoxy-19-(4-methyl-1-piperazinyl)-	MeOH	261(4.36)	158-1264-84

Compound	Solvent	$\lambda_{max}(\log \epsilon)$	Ref.
$C_{35}H_{52}O_6$ Perenniporiol, 26-O-methyl-, acetate, (26S)-	MeOH	236(4.27),243(4.34), 252(4.16)	102-1129-84
$C_{35}H_{52}O_{10}$ Antiarigenin-3-β-O-[4-O-β-D-allopyrano-syl-β-6-deoxy-D-allopyranoside]	EtOH	217(4.14)	33-0054-84
Antiarigenin-3-β-O-[3-O-β-D-gulopyrano-syl-β-6-deoxy-D-talopyranoside]	EtOH	217(4.17)	33-0054-84
$C_{35}H_{52}O_{15}$ Strophanthidin-3-β-O-[4-O-β-D-allo-pyranosyl-β-6-deoxy-D-allopyranoside]	EtOH	217(4.21)	33-0054-84
Strophanthidin-3-β-O-[3-O-β-D-gluco-pyranosyl-β-6-deoxy-D-talopyranoside]	EtOH	217(4.19)	33-0054-84
$C_{35}H_{56}ClN_2O_3PS$ Cholesteryl phosphite 3-(4-chlorobenz-yl)thiouronium salt	H_2O	223(4.15)	118-0410-84
$C_{35}H_{57}NO_{11}$ Mycinamicin VI	MeOH	215(4.31),281.5(4.33)	158-84-22
$C_{35}H_{58}O_9$ Bafilomycin A_1	MeOH	245(4.40),280(4.08)	158-0110-84
$C_{35}H_{59}NO_8$ M-230B	MeOH	230(4.40),238(4.37)	158-0013-84
$C_{35}H_{61}NO_8$ Myxovirescin A	MeOH	238(4.38)	158-0013-84
$C_{35}H_{63}NO_7$ Myxovirescin C	MeOH	238(4.32)	158-84-83
Myxovirescin D	MeOH	238(4.32)	158-84-83
$C_{35}H_{66}O_9$ 8,16,24-Pentatricontanetrione, 6,14,22,30,32,34-hexahydroxy-	MeOH	276(2.11)	158-84-139

Compound	Solvent	$\lambda_{max}(\log \epsilon)$	Ref.
$C_{36}H_{18}O_4$			
[11,11'-Binaphth[2,3-a]azulene]-5,5',12,12'-tetrone, (±)-	n.s.g.	$\underline{300(4.8),440(4.2),}$ $\underline{660(3.6)}$	89-0314-84
$C_{36}H_{19}N_3S_3$			
14H-Benzo[h]bisnaphtho[2',1':5,6][1,4]-thiazino[2,3-a:3',2'-c]phenothiazine	DMF	653(4.18)	40-0522-84
$C_{36}H_{22}N_2O_6$			
Polybeccarine	MeOH	216(4.66),234(4.70), 248s(4.68),273(4.55), 308(4.17),334(4.05), 406(4.24),425(4.23)	100-0504-84
	MeOH-HCl	216(--),234s(--), 266(--),280s(--), 340(--),396(--), 470(--)	100-0565-84
$C_{36}H_{24}$			
m,m,m,m,m,m-Hexaphenylene	C_6H_{12}	191(5.00),198s(4.94), 247(5.07)	18-3494-84
m,m,m,o,m,o-Hexaphenylene	C_6H_{12}	190(4.94),240(5.03), 252s(4.82)	18-3494-84
m,m,o,m,m,o-Hexaphenylene	C_6H_{12}	188(4.88),207(4.84), 236(4.97),248s(4.77)	18-3494-84
m,o,m,o,m,o-Hexaphenylene	C_6H_{12}	189(4.93),207(4.86), 233(4.91),254s(4.56)	18-3494-84
m,o,o,m,o,o-Hexaphenylene	C_6H_{12}	235(4.98),242(4.94)	18-3494-84
m,o,p,m,o,p-Hexaphenylene	C_6H_{12}	195(4.89),206(4.89), 246(4.91),264s(4.64)	18-3494-84
m,o,o,o,p,o-Hexaphenylene	C_6H_{12}	198(4.94),205s(4.93), 237(4.67)	18-3494-84
o,o,p,o,o,p-Hexaphenylene	C_6H_{12}	198(4.84),203s(4.82), 236(4.63),284(4.53), 288s(4.53)	18-3494-84
o,p,o,p,o,p-Hexaphenylene	C_6H_{12}	188s(4.96),200(4.98), 245(4.64),276s(4.13)	18-3494-84
$C_{36}H_{24}N_4$			
2H-Imidazole, 2,2'-(2,5-cyclohexadiene-1,4-diylidene)bis[4,5-diphenyl-(solvatochromic)	benzene	620(5)	70-0948-84
	hexane	595(5)	70-0948-84
Spiro[2,4-cyclopentadiene-1,3'-[3H-1,2,4]triazolo[4,3-b]indazole], 2,3,4,5-tetraphenyl-	CH_2Cl_2	249(4.51),279s(4.41), 363(4.03)	24-1726-84
$C_{36}H_{24}O_3$			
4,9-Epoxynaphtho[2,3-c]furan-1(4H)-one, 9,9a-dihydro-3,4,9,9a-tetraphenyl-, (4α,9α,9aα)-	EtOH	256(4.66),300(3.48), 347(3.54)	44-4165-84
Methanone, [3-(benzoyloxy)-1,4-diphenyl-2-naphthalenyl]phenyl-	$CHCl_3$	248(3.26),284(3.99)	44-4165-84
$C_{36}H_{25}N_3O_7S$			
Indolo[2,3-a]carbazole-11(12H)-carboxylic acid, 5-[(1,3-dihydro-1,3-dioxo-2H-isoindol-2-yl)methyl]-12-[(4-methoxyphenyl)sulfonyl]-, methyl ester	EtOH	222(4.64),242(4.63), 277(4.36),289(4.33), 313(4.32),344(3.67)	78-2795-84

Compound	Solvent	$\lambda_{max}(\log \epsilon)$	Ref.
$C_{36}H_{26}N_8O_4$ Benzo[g]pteridine-2,4(3H,10H)-dione, 10,10'-(1,2-ethanediyldi-4,1-phenylene)bis[3-methyl-	pH 7.27	347(4.16),438(4.18)	39-1227-84B
$C_{36}H_{26}O_5$ 9aH-Phenanthro[9,10-b]phenanthro[9',10'-4,5]furo[3,2-e]pyran-19-carboxylic acid, 9a-ethoxy-, ethyl ester	EtOH	391(4.18),414(4.28)	24-0666-84
$C_{36}H_{27}NO_2$ Spiro[acridine-9(10H),9'(10'H)-anthracen]-10'-one, 2,7-dimethyl-10-(phenylacetyl)-	EtOH	260(4.41)	24-2703-84
$C_{36}H_{28}Fe_2O_8S_2$ Ferrocene, 1,1":1',1"'-bis[2,5-thiophenediylbis(methyleneoxycarbonyl)]-bis-	dioxan	450(2.84)	18-2435-84
$C_{36}H_{28}Fe_2O_{10}$ Ferrocene, 1,1":1',1"'-bis[2,5-furandiylbis(methyleneoxycarbonyl)]bis-	dioxan	450(2.80)	18-2435-84
$C_{36}H_{28}N_4$ [1,2,4]Triazolo[1,5-a]pyridinium, 1-[(4-methylphenyl)methyl]-2-(1-naphthalenylamino)-5,7-diphenyl-, hydroxide, inner salt	EtOH	262(4.95)	39-1891-84C
$C_{36}H_{28}Ti$ Titanium, [1,1',1",1"'-(η^4-1,3-cyclobutadiene-1,2,3,4-tetrayl)tetrakis[benzene]](η^8-1,3,5,7-cyclooctatetraene)-	hexane	451(3.83),780(3.08)	101-0199-84A
$C_{36}H_{30}$ 2,2'-Bipyrene, 10b,10b',10c,10c'-tetrahydro-10b,10b',10c,10c'-tetramethyl-, [trans(trans)]-	C_6H_{12}	332s(4.35),345(4.42), 368(4.72),400(4.12), 436(3.83),480s(3.80), 544s(4.13),577(4.20)	35-7776-84
$C_{36}H_{30}N_2O_4$ 3H-Indol-3-one, 2-[1,3-dihydro-3-oxo-1-(1-oxo-2-phenylbutyl)-2H-indol-2-ylidene]-1,2-dihydro-1-(1-oxo-2-phenylbutyl)-, cis	benzene	464(3.55)	39-2305-84C
trans	benzene	578(3.85)	39-2305-84C
$C_{36}H_{30}O_{12}$ 2-Anthracenecarboxylic acid, 10,10'-dioxybis[9,10-dihydro-3,8-dihydroxy-1,7,10-trimethyl-9-oxo-, (R*,R*)-(-)-(oxanthromicin)	MeOH	315(4.36),354(4.19)	77-0473-84
$C_{36}H_{32}N_2O_7$ Dehydrocancentrine A	EtOH	216(4.77),269(4.36), 296s(4.29),445(3.87)	100-0753-84
Dehydrocancentrine B	EtOH	216s(4.86),242(4.78), 270s(4.23),310s(4.16), 370(3.90),446(4.00),	100-0753-84

Compound	Solvent	$\lambda_{max}(\log \epsilon)$	Ref.
(cont.)		492(3.95),525s(3.85)	100-0753-84
$C_{36}H_{33}N_3O_6$ Cytidine, N-benzoyl-2'-deoxy-5'-O- [(4-methoxyphenyl)diphenylmethyl]-	MeOH	231(4.41),258(4.36), 304(4.04)	118-0965-84
$C_{36}H_{34}N_2O_6$ 2,5-Cyclohexadiene-1,4-dione, 2,5-di- acetoxy-3,6-bis[2-(1,1-dimethyl-2- propenyl)-1H-indol-3-yl]-	EtOH	268(4.4),275(4.39), 286(4.33),536(3.26)	39-0567-84C
$C_{36}H_{34}N_2O_7$ Cancentrine	EtOH	213(4.80),230s(4.63), 268(4.32),291s(4.22), 330s(3.62),435(3.82)	100-0753-84
Cheratamine, (+)-	MeOH MeOH-acid MeOH-base	227(4.28),289(3.99) 287(3.87),337(3.95) 288(3.95),349(3.91)	78-2513-84 78-2513-84 78-2513-84
$C_{36}H_{34}O_4$ 2,5,8,11-Tetraoxadodecane, 1,12-di-9- anthracenyl-	$C_6H_{11}Me$	247(5.16),256(5.33), 330(3.82),346(4.05), 364(4.21),384(4.18)	150-1901-84M
$C_{36}H_{34}O_8$ [1,1':3',1":3",1'''-Quaterphenyl]- 2,2',2",2'''-tetrol, 5,5',5",5'''- tetramethyl-, tetraacetate	n.s.g.	220(4.71)	42-0142-84
$C_{36}H_{35}O_7PS_2$ Phosphonic acid, phenyl-, bis[2-[2- [(4-methylphenyl)sulfinyl]phenoxy]- ethyl] ester, [R-(R*,R*)]-	EtOH	235s(4.31),242(4.37), 282(3.79)	12-2059-84
$C_{36}H_{36}Cl_2N_7O_{13}P$ Phosphoric acid, 2,5-dichlorophenyl 2-(4-nitrophenyl)ethyl ester, 3'- ester with N-[9-(2-deoxy-β-D-erythro- pentofuranosyl)-6-[2-(4-nitrophenyl)- ethoxy]-9H-purin-2-yl]-2-methyl- propanamide	MeOH	217(4.62),270(4.54), 282s(4.40)	78-0059-84
$C_{36}H_{36}N_2O_5$ Isotrilobine, (+)-	MeOH	234(4.56),287(3.65)	78-2513-84
$C_{36}H_{36}O_3$ Methanone, (3-benzoylphenyl)[3-[2,4,6- tris(1-methylethyl)benzoyl]phenyl]-	C_6H_{12}	234(4.62),348(2.46)	88-4525-84
$C_{36}H_{36}O_8$ 2H-1-Benzopyran, 4-[2,4-dimethoxy-5-(7- methoxy-2H-1-benzopyran-3-yl)phenyl]- 3-(2,4-dimethoxyphenyl)-3,4-dihydro- 7-methoxy-, (3S-trans)-	MeOH	203(4.55),283(3.86), 330(3.84)	39-2767-84C
4,6'-Bi-2H-1-benzopyran, 3,3'-bis(2,4- dimethoxyphenyl)-3,4-dihydro-7,7'- dimethoxy-, (3S-trans)-	MeOH	205(4.51),285(3.47), 330(3.78)	39-2767-84C
$C_{36}H_{36}O_{10}$ Resiniferonol-9,13,14-ortho-benzoate, 12-O-cinnamoyl-5-hydroxy-6,7-epoxy-	MeOH	204(4.74),215s(--), 280(4.63),303s(--)	31-0808-84

Compound	Solvent	$\lambda_{max}(\log \epsilon)$	Ref.
$C_{36}H_{37}N_5O_6$			
21H,23H-Porphine-2,8,12-tripropanoic acid, 17-cyano-3,7,13-trimethyl-, trimethyl ester	$CHCl_3$	407(5.33),512(4.01), 551(4.31),572(4.05), 626(3.17)	44-3327-84
$C_{36}H_{38}N_2O_6$			
Berbivaldine	MeOH or EtOH	214(4.72),231s(4.63), 284(4.03)	100-0565-84
Chitraline	MeOH or EtOH	220s(4.51),268s(4.03), 278(4.10),292s(3.94), 304(3.96)	100-0565-84
4H-Dibenzo[de,g]quinoline-1,10-diol, 5,6,6a,7-tetrahydro-2-methoxy-6-methyl-9-[4-[(1,2,3,4-tetrahydro-6-hydroxy-7-methoxy-2-methyl-1-isoquinolinyl)methyl]phenoxy]-	MeOH	222s(4.55),273(4.13), 295s(3.98),309s(3.93)	142-2231-84
Epiberbivaldine, (+)-	MeOH	233(4.47),286(3.98)	142-2231-84
Khyberine	MeOH or EtOH	220s(4.53),264s(3.97), 272(4.02),292s(3.80), 304(3.70)	100-0565-84
Norpenduline, (+)-	MeOH	239s(4.36),283(3.95), 293s(3.83),311(3.44)	78-2513-84
Porveniramine	MeOH or EtOH	225(4.66),267s(4.19), 277(4.26),307(3.90)	100-0565-84
Valdiberine	MeOH or EtOH	212(4.71),231s(4.57), 284(4.02)	100-0565-84
epi-	MeOH or EtOH	210(4.65),232s(4.47), 284(3.86)	100-0565-84
$C_{36}H_{38}N_4O_7$			
21H,23H-Porphine-2,8,12-tripropanoic acid, 17-formyl-3,7,13-trimethyl-, trimethyl ester	$CHCl_3$	414(5.22),516(3.91), 557(4.20),579(3.99), 638(3.32)	44-3327-84
$C_{36}H_{38}N_4O_{13}$			
D-Ribitol, 5-O-[bis(4-methoxyphenyl)-phenylmethyl]-1-deoxy-1-[(1,2,3,6-tetrahydro-5-nitro-2,6-dioxo-4-pyrimidinyl)amino]-, 2,3,4-triacetate	$CHCl_3$	324(4.06)	64-0252-84B
$C_{36}H_{38}O_8$			
4,6'-Bi-2H-1-benzopyran, 3,3'-bis(2,4-dimethoxyphenyl)-3,3',4,4'-tetrahydro-7,7-dimethoxy-	MeOH	209(4.71),285(3.98)	39-2767-84C
$C_{36}H_{39}N_5O_{12}$			
D-Ribitol, 1-[(2-amino-1,6-dihydro-5-nitro-6-oxo-4-pyrimidinyl)amino]-5-O-[bis(4-methoxyphenyl)phenylmethyl]-1-deoxy-, 2,3,4-triacetate	$CHCl_3$	330(4.18)	64-0252-84B
$C_{36}H_{40}Fe_2O_{12}$			
Ferrocene, 1,1":1',1"'-bis(1,12-dioxo-2,5,8,11-tetraoxa-1,12-dodecanediyl)-bis-	dioxan	449(2.73)	18-2435-84
$C_{36}H_{40}N_4NiO_2$			
Nickel, [3,7,12,18-tetramethyl-13,17-dipentyl-21H,23H-porphine-2,8-dicarboxaldehydato(2-)-N^{21},N^{22},N^{23}-N^{24}]-, (SP-4-2)-	n.s.g.	412(5.08),509(3.74), 607(4.54)	35-3943-84

Compound	Solvent	$\lambda_{max}(\log \epsilon)$	Ref.
Nickel, [3,8,13,18-tetramethyl-7,17-dipentyl-21H,23H-porphine-2,12-dicarboxaldehydato(2-)-N^{21},N^{22},N^{23}-N^{24}]-, (SP-4-1)-	n.s.g.	423(5.10),535(4.06), 578(4.32)	35-3943-84
$C_{36}H_{40}N_4O_4$			
3-Phorbinepropanoic acid, 3,4-didehydro-9,14-diethyl-21-(methoxycarbonyl)-4,8,13,18-tetramethyl-, methyl ester	CH_2Cl_2	400(5.26),498(4.14), 532(3.58),568(3.74), 616(3.78)	44-4602-84
21H,23H-Porphine-2-propanoic acid, 8,13-diethyl-18-(3-methoxy-3-oxo-1-propenyl)-3,7,12,17-tetramethyl-, methyl ester	CH_2Cl_2	414(5.22),510(4.14), 552(4.29),576(4.11), 638(3.69)	44-4602-84
$C_{36}H_{42}O_{20}$			
Protoplumericin B	MeOH	206(4.53),230(4.34), 288(4.13),320(4.07)	94-2947-84
$C_{36}H_{44}CrN_7$			
Chromium, azido[2,3,7,8,12,13,17,18-octaethyl-21H,23H-porphinato(2-)-$N^{21},N^{22},N^{23},N^{24}$]-, (SP-5-12)-	CH_2Cl_2	348(4.58),370(4.54), 386(4.47),432(4.90), 535(3.93),564(3.77), 651(3.19)	64-0222-84B
$C_{36}H_{44}MnN_7$			
Manganese, azido[2,3,7,8,12,13,17,18-octaethyl-21H,23H-porphinato(2-)-$N^{21},N^{22},N^{23},N^{24}$]-	CH_2Cl_2	365(4.99),422(4.26), 476(4.59),560(4.00), 591s(3.59),686(2.85)	64-0222-84B
$C_{36}H_{44}Si_2$			
Disilene, tetrakis(2,4,6-trimethylphenyl)-	benzene	<u>360(3.7),423(4.0)</u>	35-0821-84
$C_{36}H_{46}N_4O_2$			
21H,23H-Porphine-2,12-dipropanol, 7,8,17,18-tetraethyl-3,13-dimethyl-	pyridine	403(4.89),499(4.13), 533(3.97),569(3.79), 625(3.45)	78-0579-84
$C_{36}H_{47}FO_{14}$			
Pregna-1,4-diene-3,20-dione, 9-fluoro-11,17-dihydroxy-16-methyl-21-[(2,3-4,6-tetra-O-acetyl-β-D-glucopyranosyl)oxy]-, (11β,16α)-	n.s.g.	239(4.16)	87-0261-84
$C_{36}H_{47}MnN_4O_2$			
Manganese, aquahydroxy[2,3,7,8,12,13-17,18-octaethyl-21H,23H-porphinato(2-)-$N^{21},N^{22},N^{23},N^{24}$]-, (OC-6-23)-	CH_2Cl_2	356(4.87),391s(4.18), 423(4.09),473(4.70), 557(3.95),587(3.60), 684(2.82)	64-0222-84B
$C_{36}H_{47}N_5O_5$			
Glycine, N-[(17-ethyl-1,2,3,19,23,24-hexahydro-2,2,7,8,12,13,18-heptamethyl-1,19-dioxo-21H-bilin-3-yl)acetyl]-, 1,1-dimethylethyl ester, (±)-	$CHCl_3$	272(4.27),348(4.51), 588(4.15)	49-1071-84
$C_{36}H_{48}Li_2N_2O_2$			
Lithium, bis[1,1'-oxybis[ethane]]bis[μ-[N-(phenylmethyl)benzeneethanaminato]]di-	benzene	520(--)	77-0287-84

Compound	Solvent	$\lambda_{max}(\log \epsilon)$	Ref.
$C_{36}H_{48}N_6O_5$ L-Lysine, N^2-[(17-ethyl-1,2,3,19,23,24-hexahydro-2,2,7,8,12,13,18-heptamethyl-1,19-dioxo-21H-bilin-3-yl)acetyl]-, bis(trifluoroacetate), (4Z,9Z,15Z)-	CHCl₃	328(4.46),615(4.31), 680s(4.25)	49-1071-84
$C_{36}H_{48}O_{14}$ α-D-Galactopyranoside, 2,2':3,3'-bis-O-(oxydi-2,1-ethanediyl)bis[methyl 4,6-O-(phenylmethylene)-	MeCN	206(4.23),256(2.65)	5-1046-84
α-D-Mannopyranoside, 2,2':3,3'-bis-O-(oxydi-2,1-ethanediyl)bis[methyl 4,6-O-(phenylmethylene)-	DMF	206(4.28),255(2.69)	5-1046-84
$C_{36}H_{48}O_{20}$ Cistanoside A	MeOH	223(4.11),246s(3.93), 290(4.00),333(4.11)	94-3009-84
$C_{36}H_{52}O_2$ 6,8,10,12,14,16,18,20-Heneicosaoctaen-5-one, 2-methoxy-2,6,10,15,19-penta-methyl-21-(2,6,6-trimethyl-1-cyclo-hexen-1-yl)-, (R)-	benzene hexane CHCl₃	460(4.99) 425s(--),448(--), 473(--) 462(--)	137-0102-84 137-0102-84 137-0102-84
$C_{36}H_{54}O_2$ 6,8,10,12,14,16,18,20-Heneicosaoctaen-5-ol, 2-methoxy-2,6,10,15,19-penta-methyl-21-(2,6,6-trimethyl-1-cyclo-hexen-1-yl)-, (all-E)-(±)-	benzene	414s(--),436(5.03), 463(4.99)	137-0102-84
$C_{36}H_{54}O_7$ 9,19-Cyclolanostan-28-oic acid, 1,3-diacetoxy-24-methylene-23-oxo-, methyl ester, (1α,3β,4α)- (methyl jessate diacetate)	EtOH	221(3.61)	102-0635-84
$C_{36}H_{54}O_8$ Medicagenic acid, dimethyl ester, diacetate	EtOH	210(3.8)	20-0323-84
$C_{36}H_{54}O_{11}$ Ergosta-5,24-dien-26-oic acid, 1-acet-oxy-3-(β-D-glucopyranosyloxy)-20,22-dihydroxy-, δ-lactone, (1α,3β,22R)-	MeOH	230(3.87)	102-2293-84
$C_{36}H_{56}O_9$ 9,19-Cyclolanostan-28-oic acid, 3-(α-L-arabinopyranosyloxy)-1-hydroxy-24-methylene-23-oxo-, (1α,3β,4α)-	EtOH	220(3.78)	102-0635-84
$C_{36}H_{58}P_2S$ Diphosphene, bis[2,4,6-tris(1,1-di-methylethyl)phenyl]-, 1-sulfide	CH₂Cl₂	267(4.26),384(3.82)	142-0681-84
$C_{36}H_{58}P_2Se$ Selenadiphosphirane, bis[2,4,6-tris(1,1-dimethylethyl)phenyl]-	CH₂Cl₂	248(4.25),294(3.99)	138-0603-84
$C_{36}H_{60}N_2O_9$ 3,4'-Dideoxy-2-eno-neospiramycin I 12-(Z)-	MeOH MeOH	226(4.09) 229(4.31)	158-0738-84 158-0738-84

Compound	Solvent	$\lambda_{max}(\log \epsilon)$	Ref.
$C_{36}H_{60}O_9$			
Bafilomycin A_2	MeOH	245(4.40),280(4.08)	158-0110-84
L-681,110 B_1	MeOH	246(4.58),284.5(4.23)	158-84-72
$C_{36}H_{61}ClN_2O_{10}$			
4'-Epichloro-4'-deoxyneospiromycin I	MeOH	232(4.29)	158-0738-84
$C_{36}H_{62}N_2O_{10}$			
4'-Deoxyneospiramycin I	MeOH	231(4.08)	158-0738-84
12-Z	MeOH	236(4.32)	158-0738-84
$C_{36}H_{63}NO_{13}$			
8,19-Epoxyerythronolide B, 3-O-olean-drosyl-5-O-desosaminyl-, (8R)-	MeOH	294(1.51)	158-84-53
$C_{36}H_{64}FNO_{12}$			
Erythromycin D, 8-fluoro-, (8S)-	MeOH	285(1.49)	158-84-108
Erythronolide B, 3-O-oleandrosyl-5-O-desosaminyl-8-fluoro-, (8S)-	MeOH	285(1.46)	158-84-116
$C_{36}H_{64}FNO_{13}$			
Erythromycin C, 8-fluoro-, (8S)-	MeOH	284(1.37)	158-84-108
Erythronolide A, 3-O-oleandrosyl-5-O-desosaminyl-8-fluoro-, (8S)-	MeOH	283(1.31)	158-84-116
$C_{36}H_{65}NO_{12}$			
Erythronolide B, 3-O-oleandrosyl-5-O-desosaminyl-	MeOH	290(1.78)	158-84-53
$C_{36}H_{65}NO_{13}$			
Erythronolide B, 3-O-oleandrosyl-5-O-desosaminyl-8-hydroxy-, (8S)-	MeOH	280(1.59)	158-84-53
Erythronolide B, 3-O-oleandrosyl-5-O-desosaminyl-15-hydroxy-	MeOH	288(1.62)	158-84-53
$C_{36}H_{70}N_2O_2$			
Diazene, bis(1-oxooctadecyl)-	C_6H_{12}	450(1.36)	23-0574-84
$C_{36}H_{72}Si_9$			
5,6,11,16,21,26,31,36,41-Nonasilanona-spiro[4.0.4.0.4.0.4.0.4.0.4.0.4.0.4-0.4.0]pentatetracontane	isooctane	246(4.30),264s(3.08)	35-5521-84

Compound	Solvent	$\lambda_{max}(\log \epsilon)$	Ref.
$C_{37}H_{22}N_2O_9$			
Beccapolydione	MeOH	218s(4.53),224(4.70), 250(4.67),276(4.47), 314(4.19),325s(4.19), 370s(4.02),438(4.29)	100-0504-84 +100-0565-84
$C_{37}H_{24}N_2O_7$			
Beccapoline	EtOH	218s(4.52),231(4.63), 250s(4.53),279(4.40), 310(4.01),366s(4.07), 428(4.22),440(4.21)	100-0565-84
$C_{37}H_{27}O_2$			
Pyrylium, 2-[3-(4,6-diphenyl-2H-pyran- 2-ylidene)-1-propenyl]-4,6-diphenyl-, perchlorate	MeCN CH_2Cl_2	725(4.76),790(4.80) 730(4.77),800(4.93)	103-1226-84 103-1226-84
Pyrylium, 4-[3-(2,6-diphenyl-4H-pyran- 4-ylidene)-1-propenyl]-2,6-diphenyl-, perchlorate	CH_2Cl_2	686(5.40)	103-0359-84
$C_{37}H_{27}S_2$			
Thiopyrylium, 2-[3-(4,6-diphenyl-2H- thiopyran-2-ylidene)-1-propenyl]- 4,6-diphenyl-, perchlorate	MeCN CH_2Cl_2	838(4.72) 865(4.85)	103-1226-84 103-1226-84
Thiopyrylium, 4-[3-(2,6-diphenyl-4H- thiopyran-4-ylidene)-1-propenyl]- 2,6-diphenyl-, perchlorate	CH_2Cl_2	762(5.39)	103-0359-84
$C_{37}H_{28}$			
Benzo[c]phenanthrene, 5,6-dihydro-8- methyl-5,6,7-triphenyl-, trans	C_6H_{12}	226(4.59),241(4.68), 251(4.68),311(4.06), 339(4.09)	44-0856-84
Cyclopropane, 3-methyl-1-(1,2,3-tri- phenyl-2-cyclopropen-1-yl)-2,3-di- phenyl-	C_6H_{12}	227(4.62),287(4.49), 297(4.52),311(4.46), 328(4.30)	44-0856-84
Naphthalene, 2-(1,2-diphenylethenyl)- 4-methyl-1,3-diphenyl-	C_6H_{12}	234(4.86),287(4.34), 304(4.26)	44-0856-84
$C_{37}H_{29}AsO_4$			
Arsonium, [2-(9H-fluoren-9-yl)-3-meth- oxy-1-(methoxycarbonyl)-3-oxo-1-pro- penyl]triphenyl-, hydroxide, inner salt	$CHCl_3$	460(4.14)	24-2409-84
$C_{37}H_{30}O_2$			
2,4-Pentadien-1-one, 5-ethoxy- 1,2,3,4,5-pentaphenyl-	MeOH	204(4.79),298(4.32), 409(4.19)	80-0817-84
$C_{37}H_{33}BN_2O$			
Pyridinium, ω-pyridinioacetonyl-, tetraphenylborate	20% acetone	388(3.03)	104-1911-84
$C_{37}H_{34}$			
Benzene, 1,1',1'',1''',1''''-(1-ethyl- 2-pentene-1,2,3,4,5-pentayl)- pentakis-	MeCN	253s(3.34),258(3.29), 262s(3.21),269(3.02)	44-1937-84
$C_{37}H_{34}N_2O_6$			
Deoxydecahydrobeccapoline	EtOH	225(4.59),244s(4.35), 276(4.35),304s(3.98), 328s(3.76)	100-0504-84

Compound	Solvent	$\lambda_{max}(\log \epsilon)$	Ref.
Glycobismine A	MeOH	219s(4.64),235s(4.72), 246(4.75),282(4.73), 300s(4.62),336s(4.13), 372(4.20),423(4.02)	94-1647-84
$C_{37}H_{34}N_2O_{12}$ Indeno[1,2-b]indole-1,2,3,4,10,10a(4aH)- hexacarboxylic acid, 5,10-dihydro- 4a-(1H-indol-3-yl)-, 10,10a-diethyl 1,2,3,4-tetramethyl ester	EtOH	240(4.40),293(4.07), 405(3.68)	94-2456-84
$C_{37}H_{35}N_3O_7$ Cytidine, N-benzoyl-5'-O-[bis(4-meth- oxyphenyl)phenylmethyl]-2'-deoxy-	MeOH	235(4.53),258(4.40), 304(4.05)	118-0965-84
$C_{37}H_{35}N_3O_9$ 2(1H)-Pyrimidinone, 1-[5-O-[(4-methoxy- phenyl)diphenylmethyl]-β-D-ribofuran- osyl]-4-[2-(4-nitrophenyl)ethoxy]-	MeOH	274(4.25),280(4.23)	78-0059-84
$C_{37}H_{36}N_2O_7$ Cancentrine O-methyl ether	EtOH	210(4.83),230s(4.66), 269(4.22),330s(3.56), 433(3.71)	100-0753-84
$C_{37}H_{36}O_{10}$ 2-Propen-1-one, 1,1'-[methylenebis- [oxy(4,6-dimethoxy-2,1-phenylene)]]- bis[3-(4-methoxyphenyl)-	MeOH	253(3.96),288(4.51)	2-1211-84
$C_{37}H_{38}N_2O_6S_2$ 5,12-Diaza[2₄](1,2,4,5)cyclophane- bis(onium), N,N'-dimethyl-, bis- (p-toluenesulfonate)	EtOH	260s(3.46),268s(3.38), 298s(3.32),331(3.79)	35-2672-84
5,15-Diaza[2₄](1,2,4,5)cyclophane- bis(onium), N,N'-dimethyl-, bis- (p-toluenesulfonate)	EtOH	256(4.00),319(3.86)	35-2672-84
$C_{37}H_{38}N_2O_8$ Secantioquine	EtOH	206(4.77),225(4.84), 272(4.32),288(4.36)	28-0591-84A
	EtOH-base	226(5.28),299(4.50), 346(4.56)	28-0591-84A
$C_{37}H_{40}N_2O_6$ Chitraline, 1-O-methyl-	MeOH or EtOH	210(4.60),224(4.54), 267s(4.04),278(4.17), 304(3.97)	100-0565-84
4H-Dibenzo[de,g]quinolin-10-ol, 5,6,6a,7-tetrahydro-1,2-dimethoxy- 6-methyl-9-[4-[(1,2,3,4-tetrahydro- 6-hydroxy-7-methoxy-2-methyl-1-iso- quinolinyl)methyl]phenoxy]-, (+)-	MeOH	229s(4.87),277(4.49), 302(4.33)	142-2231-84
Kalashine	MeOH or EtOH	220(4.54),272(4.04), 290s(3.74),304(3.70)	100-0565-84
Patagonine	MeOH or EtOH	204(4.70),231s(4.50), 283(4.01)	100-0565-84
Rupancamine, (+)-	MeOH	231(4.27),285(3.49)	142-2231-84
Thaliphylline, (+)-	MeOH	211(4.71),279(3.89), 290(3.75)	78-1975-84

Compound	Solvent	$\lambda_{max}(\log \epsilon)$	Ref.
Valdivianine	MeOH or EtOH	214(4.74),232s(4.72), 282(4.12)	100-0565-84
$C_{37}H_{40}N_4O_6$ 21H,23H-Porphine-2,8,12-tripropanoic acid, 17-ethenyl-3,7,13-trimethyl-, trimethyl ester	$CHCl_3$	406(5.23),504(4.14), 542(4.08),573(3.86), 631(3.65)	44-3327-84
$C_{37}H_{40}N_4O_7$ 21H,23H-Porphine-2,8,12-tripropanoic acid, 17-acetyl-3,7,13-trimethyl-, trimethyl ester	$CHCl_3$	412(5.32),515(4.03), 555(4.26),579(4.05), 638(3.36)	44-3327-84
$C_{37}H_{42}N_2O_6$ Thaligrisine, (+)-	MeOH	226s(4.48),284(3.98)	78-1975-84
$C_{37}H_{42}N_4NiO$ Nickel, [12-ethenyl-3,8,13,18-tetra-methyl-7,17-dipentyl-21H,23H-porph-ine-2-carboxaldehydato(2-)-$N^{21},N^{22},N^{23},N^{24}$]-, (SP-4-2)-	CH_2Cl_2	409(5.06),516(3.77), 542(3.84),589(4.32)	35-3943-84
$C_{37}H_{42}N_4O_2$ 1H-Benzimidazole, 2,2'-[5,10-bis(1,1-dimethylethyl)bicyclo[4.4.1]undeca-1,3,5,7,9-pentaene-2,7-diyl]bis[5,6-dimethyl-	CH_2Cl_2	286(4.41),308(4.38), 427(4.46),484(4.29)	33-2192-84
$C_{37}H_{44}Cl_2N_4O_8$ 21H-Biline-8,12-dipropanoic acid, 1,19-dicarboxy-2,17-bis(2-chloroethyl)-5,15,22,24-tetrahydro-3,7,13,18-tetramethyl-, α,α'-dimethyl ester, hydrobromide	CH_2Cl_2	504(4.60)	78-1749-84
$C_{37}H_{44}O_{10}$ Gnidilatidin	MeOH	232(4.76),266(3.16), 280(2.94)	100-0270-84
$C_{37}H_{44}O_{20}$ Protoplumericin B monomethylate	MeOH	206(4.54),230(4.34), 283(3.97),317(3.89)	94-2947-84
$C_{37}H_{45}ClN_4O_8$ 21H-Biline-8,12-dipropanoic acid, 1,19-dicarboxy-2-(2-chloroethyl)-17-ethyl-5,15,22,24-tetrahydro-3,7,13,18-tetramethyl-, α,α'-dimethyl ester, monohydrobromide	CH_2Cl_2	504(4.66)	78-1749-84
21H-Biline-8,12-dipropanoic acid, 1,19-dicarboxy-17-(2-chloroethyl)-2-ethyl-5,15,22,24-tetrahydro-3,7,13,18-tetramethyl-, α,α'-dimethyl ester, monohydrobromide	CH_2Cl_2	504(4.66)	78-1749-84
$C_{37}H_{45}NO_{12}$ 3-Pyridinecarboxylic acid, ester with 2,3,3a,5,8,10,11,12,13,13a-decahydro-2,3a,10,13-tetrahydroxy-2,5,8,8-tetramethyl-12-methylene-1-[(2-meth-yl-1-oxo-2-butenyl)oxy]-1H-cyclo-	MeOH	196(4.37),218(4.26), 258(3.42),263(3.43), 270s(3.33),302(2.57)	102-1689-84

Compound	Solvent	$\lambda_{max}(\log \epsilon)$	Ref.
pentacyclododecene-4,9-dione tri-acetate (cont.)			102-1689-84
$C_{37}H_{46}N_6O_4$			
21H-Biline-8,12-dipropanamide, 2,18-di-ethenyl-1,10,19,22,23,24-hexahydro-N,N,N',N',3,7,13,17-octamethyl-1,19-dioxo-	CHCl$_3$ DMSO	446(4.84) 464(4.85)	78-4253-84 78-4253-84
21H-Biline-8,12-dipropanamide, 3,17-di-ethenyl-1,10,19,22,23,24-hexahydro-N,N,N',N',2,7,13,18-octamethyl-1,19-dioxo-	CHCl$_3$ DMSO	441(4.79) 460(4.79)	78-4253-84 78-4253-84
21H-Biline-8,12-dipropanamide, 3,18-di-ethenyl-1,10,19,22,23,24-hexahydro-N,N,N',N',2,7,13,17-octamethyl-1,19-dioxo-	CHCl$_3$ DMSO	435(4.72) 456(4.75)	78-4253-84 78-4253-84
$C_{37}H_{47}N_{11}O_8$			
L-Leucinamide, L-tyrosyl-D-alanylgly-cyl-L-phenylalanyl-N-[2-[(4-azido-2-nitrophenyl)amino]ethyl]-	EtOH	260(4.32),455(3.66)	87-0836-84
$C_{37}H_{48}O_{10}$			
Gnidilatin	MeOH	231(4.23),266(3.01), 280(2.78)	100-0270-84
$C_{37}H_{49}NO_6$			
Janithrem E	MeOH	228(4.25),258s(4.44), 265(4.48),330(4.23)	39-0697-84C
$C_{37}H_{49}NO_{10}$			
Rhizoxin acetate	MeOH	296(4.58),308(4.69), 323(4.56)	158-0354-84
$C_{37}H_{50}O_9$			
Pimelea factor P$_2$	MeOH	230(4.05),266(2.84), 280(2.66)	100-0270-84
$C_{37}H_{50}O_{10}$			
Kraussianin	MeOH	230(4.08),266(2.89), 280(2.79)	100-0270-84
$C_{37}H_{50}O_{20}$			
Cistanoside B	MeOH	220s(4.02),232s(3.95), 289(3.78),330(3.93)	94-3009-84
$C_{37}H_{52}N_2O_6$			
Kopsirachine	n.s.g.	233s(4.35),281(3.88)	33-0237-84
$C_{37}H_{56}O_9$			
Myriantic acid, methyl ester, triacet-ate	MeOH	212(4.54)	102-1125-84
$C_{37}H_{58}P_2$			
Phosphine, methanetetraylbis[[2,4,6-tris(1,1-dimethylethyl)phenyl]-	hexane	262(4.69),358(3.11)	77-0689-84
$C_{37}H_{66}FNO_{12}$			
Erythromycin, 12-deoxy-8-fluoro-	MeOH	285(1.47)	158-84-108

Compound	Solvent	$\lambda_{max}(\log \epsilon)$	Ref.
$C_{37}H_{66}FNO_{13}$			
Erythromycin A, 8-fluoro-, (8S)-	MeOH	283(1.25)	158-84-108
$C_{37}H_{66}O_7$			
Isorollinicin	CH_2Cl_2	231(4.30)	102-2013-84
Rollinicin	CH_2Cl_2	231(4.31)	102-2013-84
	CH_2Cl_2	231(--),280(4.18)	100-0652-84

Compound	Solvent	$\lambda_{max}(\log \epsilon)$	Ref.
$C_{38}H_{26}N_4$			
Spiro[2,4-cyclopentadiene-1,3'-[3H]-pyrazolo[5,1-c][1,2,4]triazole], 2,3,4,5,7'-pentaphenyl-	CH_2Cl_2	246(4.30),357(3.86)	24-1726-84
$C_{38}H_{27}N_2O_7$			
Beccapolinium	MeOH or EtOH	218(4.71),235s(4.67), 262(4.57),280s(4.50), 338s(3.97),384(4.18), 440s(4.05),480s(3.90)	100-0565-84
$C_{38}H_{30}Fe_2N_2O_8$			
Ferrocene, 1,1":1',1'''-bis[2,6-pyridinediylbis(methyleneoxycarbonyl)]-bis-	dioxan	450(2.74)	18-2435-84
$C_{38}H_{30}N_2O_{11}$			
2,4(1H,3H)-Pyrimidinedione, 1-(1,3,4,6-tetra-O-benzoyl-α-D-fructofuranosyl)-	EtOH	202(4.63),231(4.60), 265(4.01)	1-0367-84
2,4(1H,3H)-Pyrimidinedione, 1-(1,3,4,6-tetra-O-benzoyl-β-D-psicofuranosyl)-	EtOH	202(4.73),230(4.77), 265(4.01)	1-0367-84
$C_{38}H_{30}O_6$			
2(5H)-Furanone, 5-[[3,4-bis(phenylmethoxy)phenyl]methylene]-4-hydroxy-3-[4-(phenylmethoxy)phenyl]-, (Z)-	EtOH	257(4.09),304(4.13), 324(4.16),372(3.97)	39-1539-84C
$C_{38}H_{32}N_4O_2$			
Benzo[1,2-c:4,5-c']dipyrrole-4,8(2H,6H)-dione, 2,6-bis(4-methylphenyl)-1,5-bis[(4-methylphenyl)amino]-	$C_2H_4Cl_2$	293(4.54),602(4.30)	5-1003-84
Benzo[1,2-c:4,5-c']dipyrrole-4,8(2H,6H)-dione, 2,6-bis(4-methylphenyl)-1,7-bis[(4-methylphenyl)amino]-	$C_2H_4Cl_2$	250(4.60),297(4.44), 376s(3.16),608(4.25)	5-1003-84
$C_{38}H_{34}N_2O_6$			
[1,5'-Bi-7H-pyrano[2,3-c]acridine-7,7'-dione, 1,2,3,3',12,12'-hexahydro-6,6'-dihydroxy-3,3,3',3',12,12'-hexamethyl- (noracronycine dimer)	$CHCl_3$	255(3.76),283(3.92), 303s(3.79),412(3.11)	100-0143-84
isomer	$CHCl_3$	256(3.68),281(3.81), 302s(3.67),347s(3.20), 412(3.06)	100-0143-84
$C_{38}H_{34}N_4O_{12}$			
21H,23H-Porphine-2,18-dipropanoic acid, 7,12-bis(2,2-dicarboxyethenyl)-3,8,13,17-tetramethyl-	DMSO	422(5.16),513(4.13), 550(4.10),581(3.91), 636(3.64)	5-1057-84
zinc chelate	pH 12	420(5.29),547(4.15), 587(4.16)	5-1057-84
$C_{38}H_{34}O_{14}$			
Strychnobiflavone hexamethyl ether	MeOH	261(4.52),272s(4.27), 299s(3.98),342(4.33)	100-0953-84
$C_{38}H_{35}N_5O_5S$			
Benzamide, N-[4,7-dihydro-7-thioxo-3-[2,3,5-tris-O-(phenylmethyl)-β-D-ribofuranosyl]-1H-pyrazolo[4,3-d]pyrimidin-5-yl]-	MeOH	244(4.22),309(3.65)	44-0528-84

Compound	Solvent	$\lambda_{max}(\log \epsilon)$	Ref.
$C_{38}H_{36}N_2O_6$ [1,5'-Bi-7H-pyrano[2,3-c]acridine-7,7'-dione, 1,1',2,2',3',3',12,12'-octahydro-6,6'-dihydroxy-3,3,3',3',12-12'-hexamethyl-	CHCl$_3$	253(4.08),283(4.22), 338(3.66),408(3.45)	100-0143-84
Bistephanine	EtOH	225(4.61),244s(4.34), 278(4.32),306s(3.94), 330s(3.67)	100-0504-84
$C_{38}H_{36}N_4O_9$ Carbamic acid, [1-[2-deoxy-5-O-[(4-methoxyphenyl)diphenylmethyl]-β-D-erythro-pentofuranosyl]-1,2-dihydro-2-oxo-4-pyrimidinyl]-, 2-(4-nitrophenyl)ethyl ester	MeOH	235(4.41),277s(4.16), 281(4.17)	78-0059-84
$C_{38}H_{37}N_3O_8$ 2(1H)-Pyrimidinone, 1-[2-deoxy-5-O-[(4-methoxyphenyl)diphenylmethyl]-β-D-erythro-pentofuranosyl]-5-methyl-4-[2-(4-nitrophenyl)ethoxy]-	MeOH	275(4.18),280s(4.17)	78-0059-84
$C_{38}H_{40}N_2O_6$ 1,4-Benzenediol, 2,5-bis[2-(1,1-dimethyl-2-propenyl)-1H-indol-3-yl]-3,6-dimethoxy-, diacetate	EtOH	280(3.32),288(3.29)	39-0567-84C
$C_{38}H_{40}N_6O_{15}S$ 1-Azetidineacetic acid, 2-(acetylthio)-α-(1-methylethyl)-3-[6-[(4-nitrophenyl)methoxy]-5-[[[(4-nitrophenyl)methoxy]carbonyl]amino]-1,6-dioxohexyl]-amino]-4-oxo-, (4-nitrophenyl)methyl ester, [2R-[1(R*),2α,3α(S*)]]-	MeOH	268(4.52)	78-1907-84
$C_{38}H_{42}N_2O_6$ Kalashine, 1-O-methyl-	MeOH or EtOH	222(4.51),272(4.00), 302(3.68)	100-0565-84
Lumipakistanine	MeOH or EtOH	220s(4.87),282(4.55)	100-0565-84
$C_{38}H_{42}N_4O$ Matopensine	EtOH	217(4.46),292(3.92), 315(3.95)	102-2659-84
$C_{38}H_{42}N_4O_8$ 21H,23H-Porphine-2,8,12-tripropanoic acid, 17-(ethoxycarbonyl)-3,7,13-trimethyl-, trimethyl ester	CHCl$_3$	409(5.33),510(4.06), 551(4.21),574(3.97), 633(3.43)	44-3327-84
$C_{38}H_{44}N_2O_6$ Cuspidaline, 7'-O-methyl-, (R,R)-(-)-	MeOH MeOH-NaOH	285(3.98) 285(--),305(--)	102-2706-84 102-2706-84
$C_{38}H_{44}O_{18}$ β-D-Glucopyranoside, 4-[6a-acetoxy-4-(4-acetoxy-3-methoxyphenyl)tetrahydro-1H,3H-furo[3,4-c]furan-1-yl]-2-methoxyphenyl tetraacetate, [1R-(1α,3α,4α,6aα)]-	EtOH	220(4.26),275(3.75), 279(3.74)	94-4482-84

Compound	Solvent	$\lambda_{max}(\log \epsilon)$	Ref.
$C_{38}H_{46}N_4$ 1,17-Metheno-5,8:10,13-dinitrilo-8H-di-pyrrolo[1,2-d:2',1'-q][1,4]diazacy-cloheptadecine, 2,3,6,7,11,12,15,16-octaethyl-	$CHCl_3$	393(5.07),535(3.89), 570(4.00),615(3.19)	157-0440-84
$C_{38}H_{46}N_4O_8$ 21H,23H-Porphine-2,18-dipropanoic acid, 7,12-bis(dimethoxymethyl)-3,8,13,17-tetramethyl-, dimethyl ester	$CHCl_3-CF_3-$ COOH + MeOH	566(3.18),613(2.78) 554(3.15),595(2.73)	5-1057-84 5-1057-84
$C_{38}H_{47}N_9O_{14}$ L-Glutamic acid, N-nitroso-N-[4-[[[2-amino-3,4,5,6,7,8-hexahydro-4-oxo-5-[(3,5-diacetyl-2-deoxyuridin-5-yl)methyl]pyrido[3,2-d]pyrimidin-6-yl]methyl]amino]benzoyl]-, diethyl ester	pH 1 pH 7	269(4.34) 269(4.33),282s(4.26)	87-1710-84 87-1710-84
$C_{38}H_{48}N_4O$ 21H,23H-Porphine-21-acetaldehyde, 2,3,7,8,12,13,17,18-octaethyl-	$CHCl_3$	379s(4.84),407(5.11), 501(4.12),531(3.86), 580(3.71),638(3.54)	157-0440-84
$C_{38}H_{48}N_4O_2Ru$ Ruthenium, carbonyl(methanol)[2,3,7,8-12,13,17,18-octaethyl-21H,23H-porph-inato(2-)-$N^{21},N^{22},N^{23},N^{24}$]-, (OC-6-42)-	benzene	393(5.16),517(4.06), 549(4.38)	35-3500-84 +35-5151-84
$C_{38}H_{48}N_4O_8S_2$ Benzenesulfonamide, N,N'-(1,4,10,13-tetraoxa-7,16-diazacyclooctadecane-7,16-diyldi-2,1-phenylene)bis[4-methyl-	MeOH	223(4.55)	78-0793-84
disodium salt	MeOH	223(4.55),248(4.36), 285s(3.96)	78-0793-84
$CuCl_2$ yellow complex	MeOH	223(4.54),245s(4.36), 345(3.55),444s(2.00)	78-0793-84
$CuCl_2$ blue complex	MeOH	223(4.49),243(4.43)	78-0793-84
$C_{38}H_{48}N_8O_{13}$ L-Glutamic acid, N-[4-[[[2-amino-3,4,5,6,7,8-hexahydro-4-oxo-5-[(3,5-diacetyl-2-deoxyuridin-5-yl)methyl]-pyrido[3,2-d]pyrimidin-6-yl]methyl]-amino]benzoyl]-, diethyl ester	pH 1 pH 7 pH 13	269(4.32),294s(4.15) 270(4.40),299(4.41) 292(4.40)	87-1710-84 87-1710-84 87-1710-84
$C_{38}H_{50}N_4$ 28H-19,22-Imino-2,15-(methano[2,5]-endo-pyrrolometheno-5H-pyrrolo-[2,3-i]azacyclodocosene, 21,31-diethyl-4,6,7,8,9,10,11,12,13,18-decahydro-3,20,25,30-tetramethyl-	CH_2Cl_2	384(5.06),407(5.21), 465(3.56),487(4.20), 520(3.98),670(3.72), 700(4.00),738(5.09)	35-6457-84
$C_{38}H_{50}N_4O$ 21H,23H-Porphine-21-ethanol, 2,3,7,8,12,13,17,18-octaethyl-	$CHCl_3$	380s(4.62),409(4.95), 503(3.98),534(3.81), 582(3.73),639(3.41)	157-0440-84

Compound	Solvent	$\lambda_{max}(\log \epsilon)$	Ref.
$C_{38}H_{50}O_2$			
Methanone, 1,3-phenylenebis[[2,4,6-tris(1-methylethyl)phenyl]-	C_6H_{12}	232(4.57),347(2.27)	88-4525-84
Methanone, 1,4-phenylenebis[[2,4,6-tris(1-methylethyl)phenyl]-	C_6H_{12}	262(4.51),350(2.48)	88-4525-84
$C_{38}H_{51}NO_9$			
Milbemycin F	EtOH	238s(--),245(4.38), 253(4.33),266s(--)	158-84-48
$C_{38}H_{52}N_4$			
21H,23H-Porphine, 2,3,7,8,12,13,17,18-octaethyl-5,15-dihydro-5,15-dimethyl-	CH_2Cl_2	420(4.91)	5-1259-84
$C_{38}H_{55}NO_3$			
Benzenamine, 4-[[(3β,23Z)-3,26-dimethoxyfurosta-5,20(22)-dien-23-yli-dene]methyl]-N,N-dimethyl-	$CHCl_3$	207(4.23),320(4.42)	94-2111-84
$C_{38}H_{56}O_{10}$			
Olean-12-ene-28,29-dioic acid, 2,3,23-triacetoxy-, dimethyl ester, (2β,3β,4α,20β)-	EtOH	209(3.6)	20-0323-84
$C_{38}H_{58}N_{14}O_8$			
L-Lysine, N^2-[N^2-[N^6-[N-[4-[[(2,4-di-amino-6-pteridinyl)methyl]methyl-amino]benzoyl]L-γ-glutamyl]-L-lysyl]-L-lysyl]-	pH 1 pH 7.4	242(4.36),306(4.25) 258(4.36),304(4.38), 365(3.95)	87-0888-84 87-0888-84
$C_{38}H_{59}NO_{10}$			
Hygrolidin, 37-(aminocarbonyl)-37-de-carboxy-	MeOH	245(4.15),276(3.76)	158-0610-84
$C_{38}H_{62}O_{10}$			
Octanoic acid, 3,6-dihydroxy-1,2,4,5-benzenetetrayl ester	pentane	207(4.47),217s(4.22), 263(2.60)	65-1657-84
$C_{38}H_{69}NO_{13}$			
Erythromycin A, 6-O-methyl-	$CHCl_3$	288(1.45)	158-0187-84

Compound	Solvent	$\lambda_{max}(\log \epsilon)$	Ref.
$C_{39}H_{28}N_4$ Spiro[2,4-cyclopentadiene-1,3'-[3H]-pyrazolo[5,1-c][1,2,4]triazole], 6'-methyl-2,3,4,5,7'-pentaphenyl-	CH_2Cl_2	247(4.63),358(4.13)	24-1726-84
$C_{39}H_{29}ClN_4O_9$ 9H-Purine, 6-chloro-9-(1,3,4,6-tetra-O-benzoyl-α-D-fructofuranosyl)-	EtOH	202(4.75),231(4.73), 266(4.14)	1-0367-84
9H-Purine, 6-chloro-9-(1,3,4,6-tetra-O-benzoyl-β-D-psicofuranosyl)-	EtOH	202(4.75),230(4.70), 266(4.10)	1-0367-84
$C_{39}H_{29}O_2$ Pyrylium, 4-[5-(2,6-diphenyl-4H-pyran-4-ylidene)-1,3-pentadienyl]-2,6-diphenyl-, perchlorate	CH_2Cl_2	806(5.50)	103-0359-84
$C_{39}H_{29}S_2$ Thiopyrylium, 4-[5-(2,6-diphenyl-4H-thiopyran-4-ylidene)-1,3-pentadienyl]-2,6-diphenyl-, perchlorate	CH_2Cl_2	889(5.45)	103-0359-84
$C_{39}H_{31}O_2S_2$ Thiopyrylium, 2-(4-methoxyphenyl)-6-[3-[6-(4-methoxyphenyl)-4-phenyl-2H-thiopyran-2-ylidene]-1-propenyl]-4-phenyl-, perchlorate	CH_2Cl_2 MeCN	890(4.86) 855(4.81)	103-1226-84 103-1226-84
Thiopyrylium, 4-(4-methoxyphenyl)-2-[3-[4-(4-methoxyphenyl)-6-phenyl-2H-thiopyran-2-ylidene]-1-propenyl]-6-phenyl-, perchlorate	CH_2Cl_2 MeCN	890(4.86) 858(4.80)	103-1226-84 103-1226-84
$C_{39}H_{31}O_4$ Pyrylium, 2-(4-methoxyphenyl)-6-[3-[6-(4-methoxyphenyl)-4-phenyl-2H-pyran-2-ylidene]-1-propenyl]-4-phenyl-, perchlorate	CH_2Cl_2 MeCN	746(4.73),822(4.92) 738(4.76),810(4.82)	103-1226-84 103-1226-84
Pyrylium, 4-(4-methoxyphenyl)-2-[3-[4-(4-methoxyphenyl)-6-phenyl-2H-pyran-2-ylidene]-1-propenyl]-6-phenyl-, perchlorate	CH_2Cl_2 MeCN	734(4.85),810(5.05) 728(4.73),800(4.82)	103-1226-84 103-1226-84
$C_{39}H_{31}S_2$ Thiopyrylium, 2-methyl-6-[3-(6-methyl-5-phenyl-2H-thiopyran-2-ylidene)-1,3-diphenyl-1-propenyl]-3-phenyl-, perchlorate	$CHCl_3$	781(4.60)	97-0146-84
$C_{39}H_{32}N_4O_9$ D-Allonamide, 2,5-anhydro-N-[[3-(benzoylamino)pyrazinyl]methyl]-, 3,4,6-tribenzoate	EtOH	231(4.62),276(4.06), 304(3.56)	39-0229-84C
$C_{39}H_{32}O_8$ Mulberrofuran K	EtOH	225(4.72),286(4.29), 306s(4.45),320(4.56), 334(4.47)	142-2729-84
	EtOH-AlCl_3	225(4.72),286(4.29), 306s(4.45),320(4.56), 334(4.47)	142-2729-84

Compound	Solvent	$\lambda_{max}(\log \epsilon)$	Ref.
$C_{39}H_{33}N_2$ Pyridinium, 1-methyl-2-[3-(1-methyl-4,6-diphenyl-2(1H)-pyridinylidene)-1-propenyl]-4,6-diphenyl-, perchlorate	CH_2Cl_2 MeCN	630(4.97) 626(4.88)	103-1226-84 103-1226-84
$C_{39}H_{34}N_5OS_3$ Benzothiazolium, 2-[[5-[[5-(2-benzothiazolyl)-3-ethyl-1,3-dihydro-1-phenyl-2H-benzimidazol-2-ylidene]ethylidene]-3-ethyl-4-oxo-2-thiazolidinylidene]methyl]-3-ethyl-, iodide	EtOH	602(4.97)	103-0665-84
$C_{39}H_{34}N_5O_2S_2$ Benzoxazolium, 2-[[5-[[5-(2-benzothiazolyl)-3-ethyl-1,3-dihydro-1-phenyl-2H-benzimidazol-2-ylidene]ethylidene]-3-ethyl-4-oxo-2-thiazolidinylidene]methyl]-3-ethyl-, iodide	EtOH	588(4.97)	103-0665-84
$C_{39}H_{34}O_8$ Mulberrofuran F	EtOH	230s(4.51),285(4,15), 296s(4.12),306s(4.26), 321(4.46),335(4.39)	142-0473-84
$C_{39}H_{36}N_6O_8$ Carbamic acid, [9-[2-deoxy-5-O-[(4-methoxyphenyl)diphenylmethyl]-β-D-erythro-pentofuranosyl]-9H-purin-6-yl]-, 2-(4-nitrophenyl)ethyl ester	MeOH	268(4.45)	78-0059-84
$C_{39}H_{36}O_8$ Mulberrofuran E	EtOH	220(4.64),243s(4.21), 287s(4.37),297s(4.42), 310s(4.51),321(3.59), 334(4.50)	94-0350-84
$C_{39}H_{38}CuN_4O_7$ Copper, [trimethyl 5,11,15,22-tetramethyl-18-oxo-4,7:14,17-diimino-2,21-metheno-9,12-nitrilo-18H-1-benzazacyclononadecine-6,10,16-tripropanoate(2-)-N^1,N^{23},N^{24},N^{25}]-, (SP-4-2)-	$CHCl_3$	332(4.54),388(4.60), 445(5.04),554(3.45), 704(3.96)	12-0143-84
$C_{39}H_{40}CuN_4O_7$ Copper, [trimethyl 19,20-dihydro-5,11-15,22-tetramethyl-18-oxo-4,7:14,17-diimino-2,21-metheno-9,12-nitrilo-18H-1-benzazacyclononadecine-6,10,16-tripropanoato(2-)-N^1,N^{23},N^{24},N^{25}]-, (SP-4-2)-	$CHCl_3$	409(5.26),542.5(3.88), 586.5(4.14)	k2-0143-84
$C_{39}H_{40}N_4O_7$ 4,7:14,17-Diimino-2,21-metheno-9,12-nitrilo-18H-1-benzazacyclononadecine-6,10,16-tripropanoic acid, 5,11,15,22-tetramethyl-18-oxo-, trimethyl ester	$CHCl_3$	337.5(4.64),381(4.65), 432(4.87),635(3.74), 701(4.04)	12-0143-84
$C_{39}H_{42}N_4O_7$ 4,7:14,17-Diimino-2,21-metheno-9,12-nitrilo-18H-1-benzazacyclononadecine-	$CHCl_3$	410(5.18),512.5(4.04), 550(3.91),583.5(3.77),	12-0143-84

Compound	Solvent	λ_{max}(log ϵ)	Ref.
6,10,16-tripropanoic acid, 19,20-di-hydro-5,11,15,22-tetramethyl-18-oxo-, trimethyl ester (cont.)		637(3.96)	12-0143-84
C$_{39}$H$_{43}$NO$_{12}$ 3-Pyridinecarboxylic acid, ester with 1-(benzoyloxy)-2,3,3a,5,8,10,11,12-13,13a-decahydro-2,3a,10,13-tetrahy-droxy-2,5,8,8-tetramethyl-12-methyl-ene-1H-cyclopentacyclododecene-4,9-dione triacetate	MeOH	194(4.76),226(4.33), 257s(3.56),264(3.60), 270s(3.54),282s(3.18), 300(2.81)	102-1689-84
C$_{39}$H$_{44}$ClNO$_9$ Naphthomycin B	EtOH	235(4.63),307(4.56), 360s(4.05)	158-84-65
	EtOH-NaOH	235(4.59),297(4.54), 345s(4.14),434(4.17), 570(3.08)	158-84-65
C$_{39}$H$_{44}$N$_2$O$_8$ Calafatine 2α-N-oxide	MeOH	210(4.95),231s(4.67), 281(3.91)	39-0651-84C
Calafatine 2β-N-oxide	MeOH	206(4.91),229s(4.63), 280(3.85)	39-0651-84C
Istanbulamine	MeOH or EtOH	205(4.87),225s(4.75), 270s(4.20),282(4.34), 304s(4.15),313(4.12)	100-0565-84
Uskudaramine	MeOH or EtOH	209(4.78),221s(4.73), 286(4.35),300s(4.18), 312s(4.06)	100-0565-84
C$_{39}$H$_{44}$N$_4$O$_7$ 4,7:14,17-Diimino-2,21-metheno-9,12-nitrilo-18H-1-benzazacyclononadecine-6,10,16-tripropanoic acid, 19,20-di-hydro-18-hydroxy-5,11,15,22-tetra-methyl-, trimethyl ester	CHCl$_3$	405(4.23),503(3.86), 537.5(3.84),572(3.79), 626.5(3.65)	12-0143-84
C$_{39}$H$_{46}$N$_4$O$_2$ 1H-Benzimidazole, 2,2'-[5,10-bis(1,1-dimethylethoxy)bicyclo[4.4.1]undeca-1,3,5,7,9-pentaene-2,7-diyl)bis[1,5,6-trimethyl-, ($\pm$)-	CH$_2$Cl$_2$	270s(4.45),295(4.51), 396(4.43),456(4.12)	33-2192-84
C$_{39}$H$_{46}$O$_{15}$ 2H-3,9a-Methano-1-benzoxepin-10-one, 5,6-bis(benzoyloxy)-7-[(2,6-di-O-acetyl-β-D-glucopyranosyl)oxy]octa-hydro-9-hydroxy-2,2,5a,9-tetramethyl-, [2S-(2α,5β,5aα,6α,7β,9β,9aα)]-	EtOH	228(4.62)	102-1651-84
C$_{39}$H$_{48}$O$_{17}$ Hydrangenoside C pentaacetate	MeOH	220.0(4.12)	33-2111-84
Hydrangenoside D pentaacetate	MeOH	219.5(4.42)	33-2111-84
2H-Pyran-5-carboxylic acid, 4α-[[2β-[2-(4-acetoxyphenyl)ethyl]-3,4-di-hydro-4-oxo-2H-pyran-6-yl]methyl]-3α-ethyl-3,4-dihydro-2β-(2,3,4,6-tetra-O-acetyl-β-D-glucopyranosyl-oxy)-, methyl ester	MeOH	217.5(4.11),227.0(4.13), 265.0(4.18)	33-2111-84

Compound	Solvent	$\lambda_{max}(\log \epsilon)$	Ref.
2H-Pyran-5-carboxylic acid, 4-[[6-[2-(4-acetoxyphenyl)ethyl]-3,4-dihydro-4-oxo-2H-pyran-2-yl]methyl]-3-ethyl-3,4-dihydro-2-[(2,3,4,6-tetra-O-acetyl-β-D-glucopyranosyl)oxy]-, methyl ester	MeOH	233.0(4.09),263.0(4.10)	33-2111-84
$C_{39}H_{50}O_{16}$			
2H-Pyran-5-carboxylic acid, 4α-[8-(4-acetoxyphenyl)-4-oxo-5-octenyl]-3α-ethyl-3,4-dihydro-2-[(2,3,4,6-tetra-O-acetyl-β-D-glucopyranosyl)oxy]-, methyl ester, [2S-(2α,3β,4β)]-	MeOH	227.5(4.42),270.0(2.71)	33-2111-84
$C_{39}H_{50}O_{17}$			
Hydrangenoside C, dihydro-, pentaacetate	MeOH	222.5(4.16),270.0(2.85)	33-2111-84
Hydrangenoside E pentaacetate	MeOH	219.5(4.08),271.0(2.72)	33-2111-84
13-epi-	MeOH	225.0(4.14),277.5(3.02)	33-2111-84
2H-Pyran-5-carboxylic acid, 4α-[[6-[2-(4-acetoxyphenyl)ethyl]-3,4-dihydro-4α-hydroxy-2H-pyran-2a-yl]methyl]-3α-ethyl-3,4-dihydro-2β-[(2,3,4,6-tetra-O-acetyl-β-D-glucopyranosyl)oxy]-, methyl ester	MeOH	217.5(4.14),265.0(3.30)	33-2111-84
$C_{39}H_{51}NO_6$			
Janithrem G	MeOH	228(4.36),259s(4.18), 263(4.20),331(4.00)	39-0697-84C
$C_{39}H_{51}NO_7$			
Janithrem F	MeOH	228(4.15),258s(4.31), 265(4.35),330(4.09)	39-0697-84C
$C_{39}H_{52}O_{17}$			
13-Epidihydrohydrangenoside E pentaacetate	MeOH	226.0(4.16),279.0(3.11)	33-2111-84
Hydrangenoside F, dihydro-, pentaacetate	MeOH	219.0(4.09),229.5(4.12), 270.5(2.82)	33-2111-84
13-epi-	MeOH	220.0(4.09),224.0(4.09), 271.0(2.82)	33-2111-84
15-epi-	MeOH	215.0(4.04),226.0(4.11), 268.0(2.94)	33-2111-84
$C_{39}H_{56}N_2O_6$			
Kopsirachine, O,O'-dimethyl-	n.s.g.	230s(4.31),277(3.66), 282s(3.63)	33-0237-84
$C_{39}H_{58}CrO_3P_2S$			
Chromium, tricarbonyl[2-[(1,2,3,4,5,6-η)-2,4,6-tris(1,1-dimethylethyl)phenyl]-1-[2,4,6-tris(1,1-dimethylethyl)-phenyl]diphosphene 1-sulfide]	CH_2Cl_2	262(4.21),325(3.90), 378(3.99)	142-0681-84
Chromium, tricarbonyl[[(1,2,3,4,5,6-η)-2,4,6-tris(1,1-dimethylethyl)phenyl]-[2,4,6-tris(1,1-dimethylethyl)phenyl]thiadiphosphirane]	CH_2Cl_2	325(3.89)	142-0681-84
$C_{39}H_{60}O_{12}$			
Bafilomycin C_1	MeOH	245(4.34),280(4.08), 335(3.68),350(3.40)	158-0110-84

Compound	Solvent	$\lambda_{max}(\log \epsilon)$	Ref.
Antibioltic L-681,110 A$_2$	MeOH	210(4.42),246(4.55), 284.5(4.20)	158-84-72
$C_{39}H_{71}NO_{13}$ Erythromycin A, 6,11-di-O-methyl-	CHCl$_3$	292(1.37)	158-0187-84
$C_{39}H_{74}N_2O_4$ Propanedioic acid, diazo-, dioctadecyl ester	hexane	250(3.91),351(1.32)	35-4531-84

Compound	Solvent	$\lambda_{max}(\log \epsilon)$	Ref.
$C_{40}H_{24}O_4$ Dibenzo[a,e]cyclooctene-5,6,11,12- tetrone, 1,4,7,10-tetraphenyl-	CH_2Cl_2	229(4.6),303(4.1)	89-0622-84
$C_{40}H_{26}N_2O_4$ 3H-Indol-3-one, 2-[1,3-dihydro-1-(1- naphthalenylacetyl)-3-oxo-2H-indol- 2-ylidene]-1,2-dihydro-1-(1-naph- thalenylacetyl)-, cis	benzene	431(3.67)	39-2305-84C
trans	benzene	566(3.82)	39-2305-84C
$C_{40}H_{26}O_5$ 5,11-Epoxydibenzo[a,e]cyclooctene- 6,12(5H,11H)-dione, 5,11-dihydroxy- 1,4,7,10-tetraphenyl-	CH_2Cl_2	231(4.4),310(3.5)	89-0622-84
$C_{40}H_{28}N_2S_2$ Indolizine, 3,3'-dithiobis[1,2-diphen- yl-	EtOH	252(4.40),310s(3.97), 380(3.52),450(3.50)	103-1128-84
Indolizine, 3,3'-dithiobis[2,8-diphen- yl-	EtOH	262(4.42),320s(3.84), 380(3.46),450(3.50)	103-1128-84
$C_{40}H_{28}N_6$ 1H-Benzimidazole, 1,1',1"-methylidyne- tris[2-phenyl-	MeOH	243(4.42),281(4.07)	117-0299-84
$C_{40}H_{30}N_4$ 29H,31H-Tetrabenzo[b,g,l,q]porphine, 6,13,20,27-tetramethyl-, tribenzyl- amine adduct	$CHCl_3$	390(4.56),418(5.25), 432(5.35),562(4.12), 600(4.60),606(4.69), 614(4.63),662(4.35)	103-0052-84
zinc chelate	$CHCl_3$	404s(4.44),430(5.33), 576(3.82),628(4.80)	103-0052-84
in polymethylmethacrylate		338s(--),405s(--), 425(--),580(--), 626(--)	103-0052-84
$C_{40}H_{32}BF_{24}N$ Tetraethylammonium tetrakis[3,5-bis- (trifluoromethyl)phenylborate	n.s.g.	269(3.61),278(3.51)	18-2600-84
$C_{40}H_{32}Fe_2O_8$ Ferrocene, 1,1":1',1"'-bis[1,3-phenyl- enebis(methyleneoxycarbonyl)]bis-	dioxan	450(2.76)	18-2435-84
$C_{40}H_{37}N_6O_2P$ Phosphorin, 1,1-dihydro-1,1-dimethoxy- 2,4,6-tris[(4-methyl phenyl)azo]-3,5- diphenyl-	EtOH	224s(4.38),238s(4.29), 270s(4.19),364(4.29), 570(4.23)	24-0763-84
$C_{40}H_{37}N_6O_5P$ Phosphorin, 1,1-dihydro-1,1-dimethoxy- 2,4,6-tris[(4-methoxyphenyl)azo]-3,5- diphenyl-	EtOH	238s(4.40),272(4.22), 298s(4.27),318s(4.27), 334s(4.33),380(4.52), 574(4.46)	24-0763-84
$C_{40}H_{38}O_8$ Kuwanon V	EtOH	226s(3.85),296(3.46), 370(3.59)	94-0350-84

Compound	Solvent	$\lambda_{max}(\log \epsilon)$	Ref.
$C_{40}H_{38}O_9$			
Kuwanon Q	EtOH	222s(4.37),266s(3.94), 300(4.06),389(4.26)	94-0350-84
Kuwanon R	EtOH	224s(4.47),299(4.22), 370(4.35)	94-0350-84
$C_{40}H_{41}N_4O_3S_2$			
Thiazolium, 5-[[5-(ethoxycarbonyl)-1,3-diethyl-1,3-dihydro-2H-benzimidazol-2-ylidene]ethylidene]-2-[(3-ethyl-4,5-diphenyl-2(3H)-thiazolylidene)-methyl]-4,5-dihydro-4-oxo-3-(2-propenyl)-, iodide	EtOH	588(4.89)	103-0665-84
$C_{40}H_{42}N_2O_6$			
24H-9,19,23-(Methanoxymethyno)-6H,16H-dibenzo[j,s][1,9]dioxacycloeicosin-7-propanenitrile, 11-(1-cyano-1-methylethyl)-7,8,9,10,11,12-hexahydro-α,α,2,16,21-pentamethyl-6,12,27-trioxo-	dioxan	269(3.29),276(3.28)	126-0499-84
$C_{40}H_{42}N_4O_8$			
4,8-Phorbinedipropanoic acid, 3,4-didehydro-21-(methoxycarbonyl)-14-(3-methoxy-3-oxo-1-propenyl)-3,9,13,18-tetramethyl-, dimethyl ester	CH_2Cl_2	416(5.21),508(4.19), 546(3.90),580(3.83), 632(4.10)	44-4602-84
21H,23H-Porphine-2,18-dipropanoic acid, 7,12-bis(3-methoxy-3-oxo-1-propenyl)-3,8,13,17-tetramethyl-, dimethyl ester	CH_2Cl_2	416(5.28),508(4.24), 544(3.92),580(3.86), 636(4.18)	44-4602-84
$C_{40}H_{44}N_4O_2$			
16,17-Dehydroisostrychnobiline	MeOH	218(4.38),250(4.19), 280(3.75),290(3.75)	102-2659-84
$C_{40}H_{44}N_4O_8$			
4,9,13-Phorbinetripropanoic acid, 3,4-didehydro-21-(methoxycarbonyl)-3,8,14,18-tetramethyl-, trimethyl ester	CH_2Cl_2	399(5.48),500(4.38), 534(3.85),564(3.96), 619(3.94)	44-4602-84
21H,23H-Porphine-2,8,12-tripropanoic acid, 17-(3-methoxy-3-oxo-1-propenyl)-3,7,13,18-tetramethyl-, trimethyl ester	CH_2Cl_2	400(5.37),500(4.27), 534(3.70),564(3.89), 618(3.88)	44-4602-84
$C_{40}H_{44}N_8$			
1H-Pyrazol-5-amine, 4,4'-azobis[N-(4-butylphenyl)-3-methyl-1-phenyl]-	hexane	230s(4.67),254s(4.52), 277s(4.32),331(4.17), 398s(4.19),450(4.56)	103-0196-84
$C_{40}H_{46}N_2O_2$			
Propanedinitrile, [4-[2,3-bis[3,5-bis(1,1-dimethylethyl)-4-hydroxyphenyl]-2-cyclopropen-1-ylidene]-3,5-cyclohexadien-1-ylidene]-	THF	236(4.28),280(3.86), 305(4.04),330(4.31), 344(4.38),510(4.80)	35-0355-84
$C_{40}H_{46}N_2O_8$			
Faberidine	MeOH	281?(4.30),302(4.17), 313s(4.06)	78-2133-84
	MeOH-base	281(4.30),302(4.17),	78-2133-84

Compound	Solvent	$\lambda_{max}(\log \epsilon)$	Ref.
(cont.)		313s(4.07)	78-2133-84
Northalicarpine	MeOH or EtOH	282(4.21),303s(4.08), 314s(3.97)	100-0565-84
Thalifarapine	MeOH	283(4.33),310s(4.10)	78-2133-84
	MeOH-base	293(4.19),320(4.26), 328(4.26)	78-2133-84
$C_{40}H_{46}N_2O_9$			
Bursanine	MeOH or EtOH	209(4.81),221s(4.75), 283(4.34),304s(4.24), 314(4.19)	100-0565-84
Iznikine	MeOH or EtOH	208(4.68),222s(4.54), 281(4.17),301s(4.03), 312(3.98)	100-0565-84
Thalifasine	MeOH	282(4.29),310s(4.05)	78-2133-84
	MeOH-base	283(4.12),312(4.17), 330(4.15)	78-2133-84
$C_{40}H_{46}N_4O_8$			
21H,23H-Porphine-2,7,12,17-tetraprop-anoic acid, 3,8,13,18-tetramethyl-, tetramethyl ester	$CHCl_3$	400(5.24),498(4.16), 533(3.99),568(3.84), 621(3.71)	44-3327-84
$C_{40}H_{46}N_{10}O_8$			
1,2-Cyclobutanedicarboxylic acid, 3,4-bis[4-[[5-(6-amino-9H-purin-9-yl)pentyl]oxy]-2-hydroxyphenyl]-, dimethyl ester, (1α,2β,3β,4α)-	MeOH	237(4.32),263(4.45), 285s(3.85)	23-2006-84
$C_{40}H_{48}Fe_2O_{14}$			
Ferrocene, 1,1":1',1"'-bis(1,15-dioxo-2,5,8,11,14-pentaoxapentadecane-1,15-diyl)bis-	dioxan	447(2.71)	18-2435-84
$C_{40}H_{48}N_4O_2$			
Isotabernamine	EtOH	232(4.31),290(3.81), 294(3.79)	100-0478-84
Kosoffine	EtOH	234(4.44),259(3.96), 287(3.84),292(3.84)	100-0117-84
Tabernamine	EtOH	235(4.48),286(4), 295(3.96)	100-0478-84
$C_{40}H_{50}N_4O_4$			
21H,23H-Porphine-2,12-dipropanoic acid, 7,8,17,18-tetraethyl-3,13-dimethyl-,	$CHCl_3$	400(5.15),499(3.98), 567(3.68),620(3.55)	78-0579-84
diethyl ester	pyridine	401(5.22),497(4.15), 530(3.99),566(3.83), 620(3.66)	78-0579-84
$C_{40}H_{52}Ge_2$			
Digermane, tetrakis(2,6-diethylphenyl)-	hexane	263(4.11),412(3.93)	88-4191-84
$C_{40}H_{52}N_2O_{11}$			
25-O-Deacetyl deriv. of a pyrimido[4,5-b]rifamycin	pH 7.0	228(4.52),271(4.31), 310s(4.18),355(4.17), 565(3.99)	94-4388-84
	MeOH	227(4.51),275(4.33), 313s(4.16),360(4.20), 592(3.98)	94-4388-84

Compound	Solvent	$\lambda_{max}(\log \epsilon)$	Ref.
$C_{40}H_{52}O$			
Mutatochrome, (5S,8R)-	EPA	250(4.35),312(3.97), 402(4.98),425(5.15), 451(5.10)	33-0170-84
Mutatochrome, (5S,8S)-	EPA	250(4.27),312(3.72), 402(4.97),425(5.14), 451(5.09)	33-0170-84
$C_{40}H_{52}O_2$			
Alloxanthin, (3RS,3'RS)-	benzene	462(5.12),492(--)	39-2147-84C
9-cis	benzene	352(--),456(4.82), 486(--)	39-2147-84C
	EtOH	347(--),446(--), 474(--)	39-2147-84
β,β-Carotene-3,3'-diol, 7,7',8,8'- tetradehydro-, (9-cis,9'-cis)-	benzene	427(4.85),452(5.01), 480(4.96)	39-2147-84C
	EtOH	346(--),439(--), 466(--)	39-2147-84C
$C_{40}H_{52}O_{16}$			
1H,3H-Pyrano[3,4-c]pyran-1-one, 3α- [8-(4-acetoxyphenyl)-4-oxooctyl]- 5α-ethyl-4,4aβ,5,6-tetrahydro-6β- [(2,3,4,6-tetra-O-acetyl-β-D-gluco- pyranosyl)oxy]-	MeOH	210(4.22),219(4.26), 227(4.29),238(4.24)	33-2111-84
$C_{40}H_{52}O_{17}$			
Phyllanthoside	MeOH	216(4.25),222(4.19), 277(4.34)	44-4258-84
Phyllanthostatin 1	MeOH	216(4.19),222(4.12), 277(4.29)	44-4258-84
$C_{40}H_{52}O_{18}$			
Phyllanthostatin 2	MeOH	216(4.28),222(4.21), 278(4.38)	44-4258-84
$C_{40}H_{54}N_4O_2$			
21H,23H-Porphine, 21-(2,2-dimethoxy- ethyl)-2,3,7,8,12,13,17,18-octa- ethyl-	$CHCl_3$	379s(4.67),410(5.03), 503(3.99),535(3.76), 582(3.65),639(3.40)	157-0440-84
$C_{40}H_{54}O_{15}$			
1H,3H-Pyrano[3,4-c]pyran-1-one, 3β- [8-(4-acetoxyphenyl)octyl]-5α-ethyl- 4,4aβ,5,6-tetrahydro-6β-[(2,3,4,6- tetrao-O-acetyl-β-D-glucopyranosyl)- oxy]-	MeOH	210(3.98),219(3.99), 240(3.95)	33-2111-84
(7S)-	MeOH	210(3.99),218(4.01), 241(4.00)	33-2111-84
$C_{40}H_{54}O_{18}$			
Phyllanthostatin 3	MeOH	205(4.19),216(4.25), 222(4.18),278(4.35)	44-4258-84
$C_{40}H_{56}N_4$			
21H,23H-Porphine, 2,3,5,7,8,12,13,15- 17,18-decaethyl-5,15-dihydro-, a	CH_2Cl_2	421(4.90)	5-1259-84
b	CH_2Cl_2	414(4.79)	5-1259-84
$C_{40}H_{56}O_2$			
Aurochrome C	EPA	235(4.29),380(4.87),	33-0471-84

Compound	Solvent	$\lambda_{max}(\log \epsilon)$	Ref.
Aurochrome C (cont.)		401(5.07),426(5.07)	33-0471-84
Aurochrome D	EPA	235(4.17),380(4.85), 401(5.05),426(5.04)	33-0471-84
Aurochrome E	EPA	235(4.33),380(4.92), 401(5.11),426(5.11)	33-0471-84
Aurochrome F	EPA	235(4.26),380(4.90), 401(5.10),426(5.09)	33-0471-84
Aurochrome G	EPA	235(4.34),380(4.81), 401(5.10),427(5.07)	33-0471-84
β,β-Carotene, 5,6;5',6'-diepoxy-, (5R,5'R,6S,6'S)-	hexane	267(4.84),415.5(5.00), 439(5.19),469.2(5.19)	33-0184-84
Luteochrome, (5R,5'R,6S,8'R)-	EPA	251.5(4.48),398.5(5.03), 421.5(5.22),448.5(5.21)	33-2226-84
(5R,5'R,6S,8'S)-	EPA	251.5(4.38),398.5(4.97), 421.5(5.16),448.5(5.15)	33-2226-84
$C_{40}H_{56}O_4$			
Neochrome, (8'R)-(all-E)-	EPA	250(4.45),312(3.89), 379(4.71),399(5.01), 422(5.21),449(5.20)	33-0461-84
Neochrome, (8'S)-(all-E)-	EPA	250(4.34),312(3.86), 379s(4.70),399(5.00), 422(5.20),449(5.20)	33-0461-84
Sidnyaxanthin	hexane	468(4.47)	88-3087-84
$C_{40}H_{60}O_6$			
β,β-Carotene, 5,5',6,6'-tetrahydro- 3,3',5,5',6,6'-hexahydroxy-, (3S,3'S,5R,5'R,6R,6'R)-	EtOH	265(4.40),317(3.72), 327(3.80),395s(4.63), 416(4.91),440(5.08), 469(5.07)	33-2043-84
(3S,3'S,5S,5'S,6S,6'S)-	EtOH	265(4.33),313(4.04), 327(4.10),395s(4.61), 415(4.82),438(4.96), 467(4.92)	33-2043-84
Olean-12-en-28-oic acid, 21-[[3-(4-eth- ylidenetetrahydro-2-furanyl)-2-meth- yl-1-oxo-2-propenyl]oxy]-3-hydroxy-, (3β,21β)-	MeOH	215(4.02)	100-0547-84
$C_{40}H_{62}O_8$			
Olean-12-ene-3,15,16,21,22,28-hexol, 21,22-bis(2-methyl-2-butenoato)-, [3β,15α,16α,21β(Z),22α(Z)]-	MeOH	215(4.39)	94-3378-84
$C_{40}H_{68}N_2O_{11}$			
Leucomycin V, 4^A-de[(2,6-dideoxy-3-C- methyl-α-L-ribo-hexopyranosyl)oxy]- 9-O-[5-(dimethylamino)tetrahydro-6- methyl-2H-pyran-2-yl]-3-O-(tetrahy- dro-2-furanyl)-	MeOH	232(4.32)	158-0738-84
12Z-	MeOH	233(4.04)	158-0738-84
$C_{40}H_{68}N_2O_{12}$			
Neospiramycin I, 3-O-tetrahydrofuran- yl-, 4'epi-	MeOH	232(4.34)	158-0738-84
$C_{40}H_{80}Si_{10}$			
5,6,11,16,21,26,31,36,41,46-Decasila- decaspiro[4.0.4.0.4.0.4.0.4.0.- 4.0.4.0.4.0.4.0]pentacontane	isooctane	225s(4.49),237(4.45), 268(4.64),282s(3.93)	35-5521-84

Compound	Solvent	$\lambda_{max}(\log \epsilon)$	Ref.
$C_{40}H_{90}Si_5$			
Cyclopentasilane, decabutyl-	C_6H_{12}	260(3.11)	157-0141-84
Cyclopentasilane, decakis(2-methyl-propyl)-	C_6H_{12}	260s(3.32)	157-0141-84

Compound	Solvent	$\lambda_{max}(\log \epsilon)$	Ref.

$C_{41}H_{28}$
  9,10[1',2']-Benzenoanthracene, 2,2'-
  methylenebis[9,10-dihydro-
	$CHCl_3$	273(3.91),280(3.98),
		282s(3.96),310(1.85),
		324(1.91)
			44-5084-84

$C_{41}H_{28}O_{26}$ Potentillin	MeOH	222(4.81),258(4.55)	94-2165-84

$C_{41}H_{31}N_4$
  1H-Benzimidazolium, 2-[3-(1,3-dihydro-
  1,3-diphenyl-2H-benzimidazol-2-yli-
  dene)-1-propenyl]-1,3-diphenyl-
	$C_6H_{11}Me$
	EtOH
	DMF
		526(5.08)
		520(5.11)
		520(5.05)
			99-0415-84
			99-0415-84
			99-0415-84

$C_{41}H_{31}O_2$
  5H-Cyclopenta[c]pyrylium, 5-[(6,7-di-
  hydro-1,3-diphenylcyclopenta[c]pyr-
  an-5-yl)methylene]-6,7-dihydro-1,3-
  diphenyl-, perchlorate
	CH_2Cl_2	750(5.42)	103-0359-84

$C_{41}H_{31}S_2$
  5H-Cyclopenta[c]thiopyrylium, 5-[(6,7-
  dihydro-1,3-diphenylcyclopenta[c]-
  thiopyran-5-yl)methylene]-6,7-dihy-
  dro-1,3-diphenyl-, perchlorate
	CH_2Cl_2	818(5.32)	103-0359-84

Thiopyrylium, 4-[7-(2,6-diphenyl-4H-1-
  thiopyran-4-ylidene)-1,3,5-heptatri-
  enyl]-2,6-diphenyl-, perchlorate
	CH_2Cl_2	1020(5.47)	103-0359-84

$C_{41}H_{35}O_6$
  Pyrylium, 2-[3-[4,6-bis(4-methoxyphen-
  yl)-2H-pyran-2-ylidene]-1-propenyl]-
  4,6-bis(4-methoxyphenyl)-, perchlor-
  ate
	CH_2Cl_2	746(4.79),824(5.02)	103-1226-84

$C_{41}H_{37}N_2O_2$
  Pyridinium, 2-(4-methoxyphenyl)-6-[3-
  [6-(4-methoxyphenyl)-1-methyl-4-phen-
  yl-2(1H)-pyridinylidene]-1-propenyl]-
  1-methyl-4-phenyl-, perchlorate
	CH_2Cl_2
	MeCN
		632(4.97)
		626(4.93)
			103-1226-84
			103-1226-84

Pyridinium, 4-(4-methoxyphenyl)-2-[3-
  [4-(4-methoxyphenyl)-1-methyl-6-phen-
  yl-2(1H)-pyridinylidene]-1-propenyl]-
  1-methyl-6-phenyl-, perchlorate
	CH_2Cl_2
	MeCN
		628(4.97)
		625(4.95)
			103-1226-84
			103-1226-84

$C_{41}H_{44}N_4O_8$
  1H-Pyrrole-2-carboxylic acid, 5,5'-
  [(2,6-dinitrophenyl)methylene]bis[3-
  butyl-4-methyl-, bis(phenylmethyl)
  ester
	EtOH	209(4.87),240(4.50),
		280(4.50)
			23-2054-84

$C_{41}H_{46}N_2O_8$ Dehydrothalifaberine	MeOH	256(4.56),272(4.56), 332(4.06)	78-2133-84
$C_{41}H_{46}N_2O_9$ Thalifabine	MeOH	282(4.30),310s(4.03)	78-2133-84 +100-0565-84
$C_{41}H_{48}N_2O_8$ Thalifaberine	MeOH	282(4.36),310s(3.98)	78-2133-84 +100-0565-84

Compound	Solvent	$\lambda_{max}(\log \epsilon)$	Ref.
$C_{41}H_{48}N_2O_9$			
Faberanine	MeOH	281(4.33),301(4.18), 312s(4.12)	78-2133-84
2'-Noradiantifoline	MeOH or EtOH	208(4.77),220s(4.69), 280(4.31),296s(4.24), 302s(4.17),314(4.10)	100-0565-84
Thalifabatine	MeOH	281(4.26),311s(3.93)	78-2133-84
$C_{41}H_{50}N_4O_3$			
Voacamine, decarbomethoxy-	EtOH	233(4.35),288(3.84), 295(3.85)	102-2359-84
$C_{41}H_{50}O_{16}$			
2H-Pyran-5-carboxylic acid, 4α-[[6β- [4-(4-acetoxyphenyl)butenyl]dihydro- 2H-pyran-2α-yl)methyl]-3α-ethenyl- 3,4-dihydro-2β-[(2,3,4,6-tetra-0- acetyl-β-D-glucopyranosyl)oxy]-, methyl ester	MeOH	210.0(4.13),220.0(4.17)	33-2111-84
7-epi-	MeOH	213.0(4.07),220.5(4.15), 270.5(2.95)	33-2111-84
$C_{41}H_{50}O_{18}$			
Bussein H	MeOH	207(3.84),268(4.00)	33-0885-84
	MeOH-NaOH	288(4.24)	33-0885-84
Hydrangenoside A pentaacetate	MeOH	222.5(4.01)	33-2111-84
Hydrangenoside B pentaacetate	MeOH	220.0(4.12)	33-2111-84
Hydrangenoside G hexaacetate	MeOH	227.5(4.29)	33-2111-84
$C_{41}H_{51}N_5Ni$			
Nickel, [N-[(12-ethenyl-3,8,13,18-tet- ramethyl-7,17-dipentyl-21H,23H-por- phin-2-yl)methylene]-1-butanamin- ato(2-)-$N^{21},N^{22},N^{23},N^{24}$]-, (SP-4-2)- monohydriodide	n.s.g.	404(5.17),515(3.66), 538(3.74),577(4.20)	35-3943-84
	CH_2Cl_2	382(4.83),438(4.65), 635(4.40)	35-3943-84
monohydrobromide	CH_2Cl_2	378(4.74),435(4.66), 626(4.34)	35-3943-84
monohydrochloride	CH_2Cl_2	377(4.72),432(4.68), 620(4.32)	35-3943-84
	MeCN	374(4.73),427(4.75), 617(4.38)	35-3943-84
	THF	374(4.63),417(4.75), 611(4.26)	35-3943-84
monoperchlorate	CH_2Cl_2	378(4.81),440(4.62), 635(4.36)	35-3943-84
	MeCN	373(4.73),427(4.75), 617(4.38)	35-3943-84
	THF	376(4.71),428(4.77), 618(4.38)	35-3943-84
$C_{41}H_{52}N_4O_2$			
1H-Benzimidazolium, 2,2'-[5,10-bis(1,1- dimethylethoxy)bicyclo[4.4.1]undeca- 1,3,5,7,9-pentaene-2,7-diyl]bis- [1,3,5,6-tetramethyl-, diiodide	CH_2Cl_2	294(4.59),385(4.33), 449(3.99)	33-2192-84
$C_{41}H_{52}O_{17}$			
Bussein K	MeOH	207(3.84),268(4.00)	33-0885-84
	MeOH-NaOH	288(4.24)	33-0885-84

Compound	Solvent	λ_{max} (log ϵ)	Ref.
$C_{41}H_{52}O_{18}$			
Hydrangenoside E hexaacetate	MeOH	220.0(4.11),270.0(2.74)	33-2111-84
13-epi-	MeOH	225.0(4.14),277.5(2.97)	33-2111-84
$C_{41}H_{54}O_2$			
Okenone	acetone	464s(--),485(5.18), 515(5.13)	137-0102-84
	EtOH	466s(--),487(--), 508(--)	137-0102-84
$C_{41}H_{54}O_{18}$			
Hydrangenoside F, dihydro-, hexaacetate	MeOH	219.0(4.07),230.0(4.10), 270.0(2.69)	33-2111-84
13-epi-	MeOH	220.0(4.10),228(4.11), 270.0(2.81)	33-2111-84
Hydrangenoside G, tetrahydro-, hexaacetate	MeOH	227.5(4.13),270.0(2.81)	33-2111-84
2H-Pyran-5-carboxylic acid, 4α-[[6β-[4-(4-acetoxyphenyl)-2-hydroxybutyl]-3,4,5,6-tetrahydro-4-hydroxy-2H-pyran-2α-yl]methyl]-3α-ethenyl-3,4-dihydro-2β-[(2,3,4,6-tetra-O-acetyl-β-D-glucopyranosyl)oxy]-, methyl ester	MeOH	221.0(4.13),271.0(2.91)	33-2111-84
$C_{41}H_{55}N_3O_7Si_2$			
Cytidine, N-[(4-methoxyphenyl)diphenylmethyl]-3',5'-O-[1,1,3,3-tetrakis(1-methylethyl)-1,3-disiloxanediyl]-	EtOH	285(4.23)	78-0153-84
$C_{41}H_{56}O_2$			
χ,ψ-Caroten-4'-ol, 1',2'-dihydro-1'-methoxy- (okenol)	acetone	449s(5.05),472(5.20), 503(5.14)	137-0102-84
$C_{41}H_{56}O_6$			
Pregn-4-ene-3,20-dione, 11,17,21-trihydroxy-, 21-retinoate	MeOH	240(4.31),359(4.67)	107-0725-84
$C_{41}H_{56}O_{16}$			
2H-Pyran-5-carboxylic acid, 4α-[[6 -[4-(4-acetoxyphenyl)butyl]-3,4,5,6-tetrahydro-2H-pyran-2α-yl]methyl]-3α-ethyl-3,4-dihydro-2β-[(2,3,4,6-tetra-O-acetyl-β-D-glucopyranosyl)-oxy]-, methyl ester, cis	MeOH	220.0(4.09),229.0(4.10)	33-2111-84
trans	MeOH	220.0(4.03),228.5(4.03)	33-2111-84
7-epi-cis-	MeOH	220.0(4.08),227.5(4.10), 270.0(2.78)	33-2111-84
$C_{41}H_{58}O_{15}$			
2H-Pyran-5-carboxylic acid, 4α-[10-(4-acetoxyphenyl)decyl]-3α-ethyl-3,4-dihydro-2β-[(2,3,4,6-tetra-O-acetyl-β-D-glucopyranosyl)oxy]-, methyl ester	MeOH	220.0(4.10),229.5(4.11)	33-2111-84
$C_{41}H_{58}O_{16}$			
2H-Pyran-5-carboxylic acid, 4α-[(2R)-10-(4-acetoxyphenyl)-2-hydroxydecyl]-3α-ethyl-3,4-dihydro-2β-(2,3,4,6-tetra-O-acetyl-β-D-glucopyranosyl)oxy]-, methyl ester	MeOH	220.0(4.08),233.0(4.11), 270.0(2.82)	33-2111-84

Compound	Solvent	$\lambda_{max}(\log \epsilon)$	Ref.
$C_{42}H_{28}$			
m,m,m,m,p,o,p-Heptaphenylene	C_6H_{12}	202(4.59),256(5.03), 284s(4.38)	18-3494-84
m,o,p,p,o,m,p-Heptaphenylene	C_6H_{12}	198(4.94),205s(4.93), 267(4.85),312s(4.23)	18-3494-84
$C_{42}H_{28}O_2$			
5,12-Epidioxynaphthacene, 5,12-dihydro-5,6,11,12-tetraphenyl-	CH_2Cl_2	<u>300s(4.0)</u>	151-0017-84D
$C_{42}H_{29}ClN_2O$			
4-Oxa-1,2-diazaspiro[4.4]nona-2,6,8-triene, 3-(2-chlorophenyl)-1,6,7,8,9-pentaphenyl-	MeOH	235(4.69),320(4.58)	2-0766-84
4-Oxa-1,2-diazaspiro[4.4]nona-2,6,8-triene, 3-(4-chlorophenyl)-1,6,7,8,9-pentaphenyl-	MeOH	245(4.69),360(4.31)	2-0766-84
$C_{42}H_{30}Br_2N_6O_5$			
2-Quinoxalinol, 3-(4-bromophenyl)-1-[[3-(4-bromophenyl)-1,2-dihydro-1-[(4-nitrophenyl)methyl]-2-quinoxalinyl](4-nitrophenyl)methyl]-1,2-dihydro-	MeOH	270s(--),306s(--), 364(4.35),434s(--)	103-0542-84
$C_{42}H_{30}N_2O$			
4-Oxa-1,2-diazaspiro[4.4]nona-2,6,8-triene, 1,3,6,7,8,9-hexaphenyl-	MeOH	235(4.50),347(4.17)	2-0766-84
$C_{42}H_{30}N_6O_4$			
Pyrazino[1,2-a:4,5-a']diquinoxaline, 6a,7,14a,15-tetrahydro-7,15-bis(4-nitrophenyl)-6,14-diphenyl-	MeOH	262(4.67),305s(--), 392(3.43)	103-0542-84
$C_{42}H_{32}N_2O$			
1H-Indole, 1,1'-[oxybis(methylene)]-bis[2,3-diphenyl-	EtOH	239(4.66),297(4.44)	78-4351-84
$C_{42}H_{32}N_2S_2$			
Indolizine, 3,3'-dithiobis[6-methyl-2,7-diphenyl-	EtOH	260(4.80),380-440(4.10)	103-1128-84
$C_{42}H_{32}N_4O_4$			
1H-Benzimidazole-5-carboxylic acid, 2,2'-(1,4-naphthalenediyl)bis[1-(phenylmethyl)-, dimethyl ester	DMF	312(4.40)	5-1129-84
$C_{42}H_{32}N_4O_5$			
1H-Furo[2',3':4,5]oxazolo[2,3-b]pteridin-11-one, 6a,7,8,9a-tetrahydro-7-hydroxy-2,3-diphenyl-8-[(triphenylmethoxy)methyl]-, [6aR(6aα,7β,8α,9aβ)-	MeOH	220s(4.58),246(4.25), 282(4.22),357(4.09)	136-0179-84F
$C_{42}H_{32}N_6O_5$			
2-Quinoxalinol, 1-[[1,2-dihydro-3-(4-nitrophenyl)-1-(phenylmethyl)-2-quinoxalinyl]phenylmethyl]-1,2-dihydro-3-(4-nitrophenyl)-	MeOH	242(4.57),279(4.55), 333(4.42),413(3.32)	103-0542-84
$C_{42}H_{34}N_4O$			
2-Quinoxalinol, 1-[[1,2-dihydro-3-phen-	MeOH	255(4.52),314(3.94),	103-0542-84

Compound	Solvent	$\lambda_{max}(\log \epsilon)$	Ref.
yl-1-(phenylmethyl)-2-quinoxalinyl]-phenylmethyl]-1,2-dihydro-3-phenyl-(cont.)		397(3.00)	103-0542-84
$C_{42}H_{34}N_4O_6$ 2,4(1H,3H)-Pteridinedione, 6,7-diphen-yl-3-[5-0-(triphenylmethyl)-β-D-arabinofuranosyl]-	MeOH	222s(4.56),271(4.20), 365(4.16)	136-0179-84F
2,4(1H,3H)-Pteridinedione, 6,7-diphen-yl-3-[5-0-(triphenylmethyl)-β-D-ribofuranosyl]-	MeOH	220s(4.64),272(4.26), 364(4.20)	136-0179-84F
$C_{42}H_{36}O_{10}$ Dimeric neolignan	MeOH	223(4.68),255(4.89), 290(4.41)	102-2647-84
$C_{42}H_{42}Li_3N_3$ Lithium, tris[μ-[N-(phenylmethyl)benz-enemethanaminato]tri-, cyclo	benzene	525(2.0)(anom.)	77-0287-84
$C_{42}H_{42}N_6O_8$ Propanamide, N-[9-[2-deoxy-5-0-[(4-methoxyphenyl)diphenylmethyl]-β-D-erythro-pentofuranosyl]-6-[2-(4-nitrophenyl)ethoxy]-9H-purin-2-yl]-2-methyl-	MeOH	236s(4.32),269(4.45)	78-0059-84
$C_{42}H_{48}N_2O_9$ Dehydrohuangshanine	MeOH	257(4.54),267s(4.51), 275s(4.49),332(3.85)	78-2133-84
$C_{42}H_{48}N_4O_6$ 14-Dehydrotetrastachyne	neutral	282s(3.72),295(3.80), 335(3.79)	28-0627-84A
	base	281(3.76),295(3.76), 324(3.89),341s(3.81)	28-0627-84A
$C_{42}H_{48}N_4O_7$ 4,7:14,17-Diimino-2,21-methano-9,12-nitrilo-12H-1-benzazacyclononadecine-6,10,16-tripropanol, 22-hydroxy-5,11,15,22-tetramethyl-, 6,10,16-triacetate	CHCl$_3$	396(4.87),411(4.93), 429(4.80),530(3.99), 564(4.06),610(4.09), 665(4.47)	12-0143-84
$C_{42}H_{49}FeN_4$ Iron, [2,3,7,8,12,13,17,18-octaethyl-21H,23H-porphinato(2-)-N^{21},N^{22},N^{23}-N^{24}]phenyl-	PhCN	393(5.11),527s(--), 554(4.34)	35-4472-84
cation	PhCN	392(5.08),530(4.30), 555s(--)	35-4472-84
anion	PhCN	350(4.62),408(4.99), 502(4.28),542(4.75), 758(3.85)	35-4472-84
$C_{42}H_{50}N_2O_9$ Huangshanine	MeOH	281(4.41),302(4.20), 312s(4.13)	78-2133-84
$C_{42}H_{52}ClFeN_6O_2$ Iron, chloro[24,25,34,35-tetraethyl-4,5,7,8,9,10,11,12,13,14,16,17-	CHCl$_3$	380(4.77),506(3.87), 535(3.88),638(3.51)	78-0579-84

Compound	Solvent	$\lambda_{max}(\log \epsilon)$	Ref.
dodecahydro-3,29-dimethyl-32H-23,26-imino-21,18-metheno-2,19-(metheno-[2,5]-endo-pyrrolometheno)-2H-pyrrolo[3,2-r][1,7,14]triazacyclohexacosine-6,15-dionato(2-)-N^1,N^{20},N^{28}-N^{32}]-, (SP-5-12)- (cont.)	pyridine	381(4.77),505(3.87), 535(3.88),638(3.51)	78-0579-84
$C_{42}H_{52}N_2O_{19}$ AG1	MeOH	237(4.80),255(4.54), 290(4.06),478(4.18)	88-1937-84
$C_{42}H_{52}N_6O_2Zn$ Zinc, [24,25,34,35-tetraethyl-4,5,7,8-9,10,11,12,13,14,16,17-dodecahydro-3,29-dimethyl-32H-23,26-imino-21,18-metheno-2,19-(metheno[2,5]-endo-pyrrolometheno)-2H-Pyrrolo[3,2-r][1,7-14]triazacyclohexacosine-6,15-dionato(2-)-N^1,N^{20},N^{28},N^{32}]-	CHCl$_3$	406(5.24),537(4.15), 573(4.22)	78-0579-84
$C_{42}H_{52}O_{18}$ Bussein F	MeOH MeOH-NaOH	207(3.84),268(4.00) 288(4.24)	33-0885-84 33-0885-84
$C_{42}H_{54}N_2O_{12}$ Rifamycin, 1,4-dideoxy-1,4-dihydro-1,4-dioxo-3-(1-piperidinyl)-	MeOH	226(4.52),275(4.32), 310s(4.17),358(4.18), 595(3.94)	94-4388-84
$C_{42}H_{54}N_6O_2Zn$ Zinc, [7,8,17,18-tetraethyl-3,13-dimethyl-N,N-dipropyl-21H,23H-porphine-2,12-dipropanamidato(2-)-N^{21},N^{22},N^{23}-N^{24}]-	CHCl$_3$	401(5.24),535(4.14), 572(4.23)	78-0579-84
$C_{42}H_{54}O_{11}$ Pallidine	EtOH	282(3.27)	102-2931-84
$C_{42}H_{55}N_5O_6Si_2$ Adenosine, N-[(4-methoxyphenyl)diphenylmethyl]-3',5'-O-[1,1,3,3-tetrakis-(1-methylethyl)-1,3-disiloxanediyl]-	EtOH	273(4.42)	78-0153-84
$C_{42}H_{55}N_5O_7Si_2$ Guanosine, N-[(4-methoxyphenyl)diphenylmethyl]-3',5'-O-[1,1,3,3-tetrakis-(1-methylethyl)-1,3-disiloxanediyl]-	EtOH	260(4.29)	78-0153-84
$C_{42}H_{60}N_4$ 21H,23H-Porphine, 2,3,7,8,12,13,17,18-octaethyl-5,15-dihydro-5,15-bis(1-methylethyl)-	CH$_2$Cl$_2$	422(4.80)	5-1259-84
stereoisomer b 21H,23H-Porphine, 2,3,7,8,12,13,17,18-octaethyl-5,15-dihydro-5,15-dipropyl-	CH$_2$Cl$_2$ CH$_2$Cl$_2$	416(4.85) 422(4.78)	5-1259-84 5-1259-84
$C_{42}H_{60}N_4O_{23}$ 4-Methylcoumarin-7-yloxy tetra-N-acetyl-β-chitotetraoside	DMSO	289(3.94),314(4.12)	94-1597-84

Compound	Solvent	$\lambda_{max}(\log \epsilon)$	Ref.
$C_{42}H_{62}O_2$ 5,9,13,17-Nonadecatetraenal, 2-[4,8-di-methyl-10-(phenylmethoxy)-4,8-decadi-enylidene]-6,10,14,18-tetramethyl-, (E,Z,E,Z,E,E)-	EtOH	203(4.56),230(3.86), 294(2.20)	70-0134-84
$C_{42}H_{65}N_7O_9$ Rhizonin A	MeOH	206(4.52)	158-84-6
$C_{42}H_{67}NO_{16}$ Carbomycin A	MeOH	240(4.20)	158-0118-84
$C_{42}H_{69}NO_{16}$ Carbomycin AP1	MeOH	end absorption	158-0118-84
Carbomycin AP2	MeOH	232.5(3.94)	158-0118-84
Carbomycin AP3	MeOH	234(3.92)	158-0118-84
$C_{42}H_{98}Si_7$ Cycloheptasilane, tetradecapropyl-	C_6H_{12}	242(3.80)	157-0141-84

Compound	Solvent	$\lambda_{max}(\log \epsilon)$	Ref.
$C_{43}H_{28}N_4O_2$ 21H-2-Oxaporphin-3(23H)-one, 5,10,15,20-tetraphenyl-	CHCl$_3$	423(5.43),525(4.07), 562(4.09),592(3.87), 643(3.49)	77-0920-84
$C_{43}H_{32}N_2$ 2,3-Diazaspiro[4.4]nona-2,6,8-triene, 1,3,6,7,8,9-hexaphenyl-	MeOH	242(4.67),340(4.34)	78-1585-84
$C_{43}H_{32}N_2O$ 4-Oxa-1,2-diazaspiro[4.4]nona-2,6,8-triene, 3-(4-methylphenyl)-1,6,7,8,9-pentaphenyl-	MeOH	240(4.70),345(4.42)	2-0766-84
$C_{43}H_{32}N_2O_2$ 4-Oxa-1,2-diazaspiro[4.4]nona-2,6,8-triene, 3-(4-methoxyphenyl)-1,6,7,8,9-pentaphenyl-	MeOH	230(4.52),335(4.23)	2-0766-84
$C_{43}H_{32}N_2O_9$ Chelidimerine, (±)-	MeOH	232(4.59),285(4.43), 324(3.89)	100-0001-84
Sanguidimerine	EtOH	235(4.65),284(4.63), 323(4.23)	100-0001-84
$C_{43}H_{34}N_4O_7$ 2,4(1H,3H)-Pteridinedione, 3-[2,3-O-carbonyl-5-O-(triphenylmethyl)-β-D-ribofuranosyl]-	MeOH	220s(4.61),277(4.22), 366(4.15)	136-0179-84F
$C_{43}H_{33}N_4$ 1H-Benzimidazolium, 2-[5-[1,3-dihydro-1,3-diphenyl-2H-benzimidazol-2-ylidene]-1,3-pentadienyl]-1,3-diphenyl-, iodide	C$_6$H$_{11}$Me EtOH DMF	625(5.08) 616(5.14) 620(5.05)	99-0415-84 99-0415-84 99-0415-84
$C_{43}H_{33}O_2$ 5H-Cyclopenta[c]pyrylium, 5-[3-(6,7-dihydro-1,3-diphenylcyclopenta[c]pyran-5-yl)-2-propenylidene]-6,7-dihydro-1,3-diphenyl-, perchlorate	CH$_2$Cl$_2$	855(5.59)	103-0359-84
$C_{43}H_{33}S_2$ 5H-Cyclopenta[c]thiopyrylium, 5-[3-(6,7-dihydro-1,3-diphenylcyclopenta[c]thiopyran-5-yl)-2-propenylidene]-6,7-dihydro-1,3-diphenyl-, perchlorate	CH$_2$Cl$_2$	925(5.42)	103-0359-84
$C_{43}H_{35}O_2$ 2-Benzopyrylium, 5-[(7,8-dihydro-1,3-diphenyl-6H-2-benzopyran-5-yl)methylene]-5,6,7,8-tetrahydro-1,3-diphenyl-, perchlorate	CH$_2$Cl$_2$	728(5.15)	103-0359-84
$C_{43}H_{35}S_2$ 1-Benzothiopyrylium, 7-[(6,7-dihydro-2,3-diphenyl-5H-1-benzothiopyran-7-yl)methylene]-5,6,7,8-tetrahydro-2,3-diphenyl-, perchlorate	CHCl$_3$	837(4.48)	97-0146-84

Compound	Solvent	$\lambda_{max}(\log \epsilon)$	Ref.
$C_{43}H_{36}N_2O_5S$ 5-Thia-1-azabicyclo[4.2.0]oct-2-ene- 2-carboxylic acid, 4-(2-methoxy-2- oxoethylidene)-3-methyl-8-oxo-7- [(triphenylmethyl)amino]-, diphen- ylmethyl ester, cis	EtOH	320(3.86)	87-1225-84
trans	EtOH	348(4.04)	87-1225-84
$C_{43}H_{38}O_8$ Benzeneacetic acid, α-[4-[3,4-bis(phen- ylmethoxy)phenyl]-3-hydroxy-5-oxo- 2(5H)-furanylidene]-4-(phenylmeth- oxy)-, 1,1-dimethylethyl ester, (E)-	EtOH	307(4.27),382(3.91)	39-1547-84C
$C_{43}H_{41}N_2O_4$ Pyridinium, 6-[3-[4,6-bis(4-methoxy- phenyl)-1-methyl-2(1H)-pyridinyli- dene]-1-propenyl]-2,4-bis(4-meth- oxyphenyl)-1-methyl-, perchlorate	CH_2Cl_2 MeCN	632(4.79) 628(4.73)	103-1226-84 103-1226-84
$C_{43}H_{44}N_6O_9$ Propanamide, N-[9-[5-O-[bis(4-methoxy- phenyl)phenylmethyl]-2-deoxy-β-D- erythro-pentofuranosyl]-6-[2-(4- nitrophenyl)ethoxy]-9H-purin-2- yl]-2-methyl-	MeOH	270(4.50),335(4.45)	78-0003-84
$C_{43}H_{46}N_6O_9$ Pentanoic acid, 4-oxo-, ester with N- [9-[5-O-[bis(4-methoxyphenyl)phenyl- methyl]-2-deoxy-β-D-erythro-pento- furanosyl]-6-(2-cyanoethoxy)-9H- purin-2-yl]-2-methylpropanamide	MeOH	235(4.42),270(4.32)	78-0003-84
$C_{43}H_{50}O_6$ Retinoic acid, (16α)-21-acetoxy-17-hy- droxy-16-methyl-20-oxo-19-norpregna- 1,3,5,7,9,14-hexaen-3-yl ester	MeOH	245(4.26),255(4.31), 264(4.22),295(4.09), 307(4.12),367(4.45)	107-0725-84
$C_{43}H_{50}O_{17}$ Deacetylmortonol B, 2-O-β-(2,6-diacet- ylglucopyranosyl)-, tetraacetate	EtOH	228(4.60)	102-1651-84
$C_{43}H_{50}O_{20}$ Osmanthuside B, heptaacetate	MeOH	218(4.39),282(4.43)	94-3880-84
$C_{43}H_{52}N_4O_5$ Conodurine	EtOH	220(3.73),284(3.35), 290(3.37)	102-2359-84
$C_{43}H_{53}FO_7$ Pregna-1,4-diene-3,20-dione, 9-fluoro- 11,17-dihydroxy-21-[[9-(4-methoxy- 2,3,6-trimethylphenyl)-3,7-dimethyl- 1-oxo-2,4,6,8-nonatetraenyl]oxy]-16- methyl-	MeOH	237(4.36),360(4.63)	107-0725-84
$C_{43}H_{54}O_{17}$ Bussein J	MeOH MeOH-NaOH	207(3.84),268(4.00) 288(4.24)	33-0885-84 33-0885-84

Compound	Solvent	$\lambda_{max}(\log \epsilon)$	Ref.
$C_{43}H_{54}O_{18}$			
Bussein B	MeOH	207(3.84),268(4.00)	33-0885-84
	MeOH–NaOH	288(4.24)	33-0885-84
$C_{43}H_{54}O_{19}$			
Bussein L	MeOH	207(3.84),268(4.00)	33-0885-84
	MeOH–NaOH	288(4.24)	33-0885-84
$C_{43}H_{60}O_{17}$			
2H-Pyran-5-carboxylic acid, 4α-[(2R)-2-acetoxy-10-(4-acetoxyphenyl)decyl]-3α-ethyl-3,4-dihydro-2β-(2,3,4,6-tetra-O-acetyl-D-glucopyranosyloxy)-, methyl ester	MeOH	220.0(4.10),227.5(4.10), 270(2.80)	33-2111-84
2H-Pyran-5-carboxylic acid, 4α-[(2R)-6-acetoxy-10-(4-acetoxyphenyl)decyl]-3α-ethyl-3,4-dihydro-2β-(2,3,4,6-tetra-O-acetyl-D-glucopyranosyloxy)-, methyl ester	MeOH	220.0(4.06),227.5(4.07), 270.0(2.76)	33-2111-84
$C_{43}H_{74}OP_2Si$			
Phosphine, [[[(1,1-dimethylethyl)di-methylsilyl]oxy][2,4,6-tris(1,1-dimethylethyl)phenyl]phosphinidene]-methyl][2,4,6-tris(1,1-dimethyleth-yl)phenyl]-, (E)-	hexane	245(4.22),292s(2.91)	77-0689-84

Compound	Solvent	$\lambda_{max}(\log \epsilon)$	Ref.
$C_{44}H_{28}CrN_5$			
Chromium, nitrido[5,10,15,20-tetraphenyl-21H,23H-porphinato(2-)-N^{21},$N^{22}N^{23}$-N^{24}]-, (SP-5-31)-	toluene	401s(4.56),422(5.57), 480(3.29),504(3.38), 544(4.29),578(3.23)	64-0222-84B
$C_{44}H_{28}MnN_5$			
Manganese(V), nitrido[5,10,15,20-tetraphenyl-21H,23H-porphinato(2-)-N^{21},N^{22}-N^{23},N^{24}]-, hydrate	toluene	396s(4.42),422(5.54), 474(3.40),499(3.44), 535(4.17),566s(3.44)	64-0222-84B
$C_{44}H_{30}CrN_7O$			
Chromium, aquaazido[5,10,15,20-tetraphenyl-21H,23H-porphinato(2-)-N^{21},N^{22},N^{23},N^{24}]-, (OC-6-23)-	CH_2Cl_2	340(4.42),355(4.43), 398(4.64)	64-0222-84B
$C_{44}H_{32}N_4O_9$			
2,4(1H,3H)-Pteridinedione, 6,7-diphenyl-1-(2,3,5-tri-O-benzoyl-β-D-ribofuranosyl)-	MeOH	228(4.84),272(4.35), 358(4.16)	136-0179-84F
2,4(1H,3H)-Pteridinedione, 6,7-diphenyl-3-(2,3,5-tri-O-benzoyl-β-D-ribofuranosyl)-	MeOH	229(4.83),273(4.30), 364(4.21)	136-0179-84F
$C_{44}H_{34}N_2$			
2,3-Diazaspiro[4.4]nona-1,6,8-triene, 4-methyl-1,3,6,7,8,9-hexaphenyl-	MeOH	240(4.51),340(4.39)	78-1585-84
$C_{44}H_{36}N_2O$			
1H-Indole, 1,1'-[oxybis(methylene)]-bis[5-methyl-2,3-diphenyl-	EtOH	237(4.65),301(4.45)	78-4351-84
$C_{44}H_{36}N_2O_2$			
1H-Indole, 7-methoxy-5-[(7-methoxy-2,3-diphenyl-1H-indol-4-yl)methyl]-4-methyl-2,3-diphenyl-	EtOH	255(4.69),311(4.59)	2-1021-84
1H-Indole, 7-methoxy-6-[(7-methoxy-2,3-diphenyl-1H-indol-4-yl)methyl]-4-methyl-2,3-diphenyl-	EtOH	252(4.70),311(4.59)	2-1021-84
$C_{44}H_{36}N_4Zn$			
Tetrabenzoporphine, tetraethyl-, zinc chelate, tribenzylamine salt	$CHCl_3$	404s(4.49),430(5.43), 576(3.83),630(4.95)	103-0052-84
$C_{44}H_{38}N_2O_9$			
Toddalidimerine	EtOH	232(5.68),286(5.71), 325(5.16)	100-0001-84
$C_{44}H_{38}N_4O$			
2-Quinoxalinol, 1-[[1,2-dihydro-3-(4-methylphenyl)-1-(phenylmethyl)-2-quinoxalinyl]phenylmethyl]-1,2-dihydro-3-(4-methylphenyl)-	MeOH	256s(--),314(4.03), 397(3.48)	103-0542-84
$C_{44}H_{38}N_8$			
Pyridinium, 1,1',1'',1'''-(21H,23H-porphine-5,10,15,20-tetrayl)tetrakis-[1-methyl-	pH 0.5	444(5.51),591(4.22), 640(4.25)	138-1871-84
	pH 9.3	423(5.42),518(4.26), 556(3.89),586(4.01), 636(3.63)	138-1871-84

Compound	Solvent	$\lambda_{max}(\log \epsilon)$	Ref.
$C_{44}H_{40}Cl_2N_3O_{11}P$ 3'-Thymidylic acid, 5'-O-[(4-methoxy-phenyl)diphenylmethyl]-, 2,5-di-chlorophenyl 2-(4-nitrophenyl)-ethyl ester	MeOH	262(4.29)	118-0965-84
$C_{44}H_{44}N_2$ 5H-Benzo[b]carbazole, 11,11'-(1,10-dec-anediyl)bis[6-methyl-	EtOH	233(4.28),269(4.52), 297.5(4.34),321(3.74), 335(3.61),382(3.51), 401(3.59)	77-0168-84
$C_{44}H_{44}N_4O_{12}Zn$ Zinc, [dimethyl 7,12-bis(3-methoxy-2-(methoxycarbonyl)-3-oxo-1-prop-enyl]-3,8,13,17-tetramethyl-21H,23H-porphine-2,18-dipropanoato(2-)-$N^{21},N^{22},N^{23},N^{24}$]-, (SP-4-2)-	$CHCl_3$	439(5.12),557(4.16), 596(4.24)	5-1057-84
$C_{44}H_{46}N_4O_{12}$ 21H,23H-Porphine-2,18-dipropanoic acid, 7,12-bis[3-methoxy-2-(methoxycarbo-nyl)-3-oxo-1-propenyl]-3,8,13,17-tetramethyl-, dimethyl ester	$CHCl_3$	426(5.12),514(4.15), 550(4.08),583(3.87), 638(3.73)	5-1057-84
$C_{44}H_{54}O_{18}$ Bussein E	MeOH MeOH-NaOH	216(4.16),268(4.00) 288(4.24)	33-0885-84 33-0885-84
$C_{44}H_{54}O_{19}$ Bussein D	MeOH MeOH-NaOH	207(3.84),268(4.00) 288(4.24)	33-0885-84 33-0885-84
$C_{44}H_{54}O_{22}$ β-D-Glucopyranoside, 1,1'-dihydroxy-6,6'-dimethyl[2,2'-binaphthalene]-8,8'-diyl bis[6-O-D-apio-β-D-furan-osyl-	EtOH	226(4.69),240(4.68), 260(4.60),320(4.17)	88-0523-84
$C_{44}H_{56}ClFeN_6O_2$ Iron, chloro[26,27,36,37-tetraethyl-4,5,7,8,9,10,11,12,13,14,15,16,18,19-tetradecahydro-3,31-dimethyl-34H-25,28-imino-23,20-metheno-2,21-(meth-eno[2,5]-endo-pyrrolometheno)-2H-pyrrolo[3,2-t][1,7,16]triazacyclo-octacosine-6,17-dionato(2-)-N^1,N^{22}-N^{30},N^{34}]-, (SP-5-12)-	$CHCl_3$ pyridine	380(4.89),508(3.81), 537(3.82),638(3.51) 400(4.76),506(3.65), 536(3.64),636(3.31)	78-0579-84 78-0579-84
$C_{44}H_{56}Cl_2N_{10}O_6$ Deuteroporphyrin, 3,8-bis[[(trimethyl-ammoniomethyl)carbonyl](1-hydrazin-yl-2-ylidene)methyl]-, dimethyl ester, dichloride	H_2O DMSO	375(4.42),410(4.47), 530(3.42),568(3.36), 588(3.30),646(3.00) 298(4.47),428(5.13), 519(4.16),557(4.15), 587(3.93),643(3.70)	5-1057-84 5-1057-84
$C_{44}H_{56}N_6O_2Zn$ Zinc, [26,27,36,37-tetraethyl-4,5,7,8-9,10,11,12,13,14,15,16,18,19-tetra-decahydro-3,31-dimethyl-34H-25,28-	$CHCl_3$	403(5.22),537(4.15), 573(4.24)	78-0579-84

Compound	Solvent	$\lambda_{max}(\log \epsilon)$	Ref.
imino-23,20-metheno-2,21-(metheno-[2,5]-endo-pyrrolometheno)-2H-pyrrolo[3,2-t][1,7,16]triazacyclooctacosine-6,17-dionato(2-)-N^1,N^{22},N^{30}-N^{34}]- (cont.)			78-0579-84
$C_{44}H_{56}O_{18}$			
Bussein A	MeOH	207(3.84),268(4.00)	33-0885-84
	MeOH-NaOH	288(4.24)	33-0885-84
$C_{44}H_{56}O_{19}$			
Bussein G	MeOH	207(3.84),268(4.00)	33-0885-84
	MeOH-NaOH	288(4.24)	33-0885-84
$C_{44}H_{56}O_{20}$			
Bussein M	MeOH	207(3.84),268(4.00)	33-0885-84
	MeOH-NaOH	288(4.24)	33-0885-84
$C_{44}H_{59}NO_{27}$			
D-Arabinitol, 1,1'-C-[(hydroxyimino)-bis[2,3-dihydro-4-(methoxycarbonyl)-5-methyl-2,3-furandiyl]bis-, 1,1'-2,2',3,3',4,4',5,5'-decaacetate	EtOH	248(4.02)	136-0075-84K
$C_{44}H_{62}I_2N_4$			
21H,23H-Porphine, 2,3,7,8,12,13,17,18-octaethyl-5,15-dihydro-5,15-bis(4-iodobutyl)-	benzene	424(4.94)	5-1259-84
$C_{44}H_{62}N_2O_4$			
3H-Indol-3-one, 2-[1,3-dihydro-3-oxo-1-(1-oxotetradecyl)-2H-indol-2-ylidene]-1,2-dihydro-1-(1-oxotetradecyl)-, (E)-	benzene	569(3.87)	151-0383-84A
	hexane	565(3.88)	151-0383-84A
	C_6H_{12}	565(3.86)	151-0383-84A
	$C_{14}H_{30}$	567(3.85)	151-0383-84A
	PrOH	564(3.81)	151-0383-84A
	2-BuOH	564(3.84)	151-0383-84A
	$C_6H_{11}OH$	567(3.85)	151-0383-84A
	m-cresol	563(3.81)	151-0383-84A
	ether	566(3.86)	151-0383-84A
	dioxan	561(3.82)	151-0383-84A
	THF	560(3.87)	151-0383-84A
	acetone	556(3.87)	151-0383-84A
	butanone	560(3.88)	151-0383-84A
	MeCN	557(3.88)	151-0383-84A
	DMF	555(3.85)	151-0383-84A
	pyridine	565(3.84)	151-0383-84A
	$MeCONMe_2$	556(3.84)	151-0383-84A
	PhCN	565(3.89)	151-0383-84A
	DMSO	555(3.85)	151-0383-84A
	$CHCl_3$	567(3.85)	151-0383-84A
	$C_2H_2Br_4$	574(3.86)	151-0383-84A
(also other solvents)	$C_{10}H_7Cl$	571(3.90)	151-0383-84A
$C_{44}H_{62}N_4$			
21H,23H-Porphine, 2,3,7,8,12,13,17,18-octapropyl-	$CHCl_3$	401(5.35),499(4.16),503s(4.14),536(4.04),568(3.86),595s(3.09),621(3.73)	103-0748-84
$C_{44}H_{64}I_2N_{10}O_{13}S$			
1-Carbaoxytocin-1-butanoic acid,	H_2O	287(3.54),295(3.53)	73-1921-84

Compound	Solvent	$\lambda_{max}(\log \epsilon)$	Ref.
2-(3,5-diiodo-L-tyrosine)-4-L-gluta-mic acid (cont.)	base	314(3.83)	73-1921-84
$C_{44}H_{64}N_4$ 21H,23H-Porphine, 5,15-bis(1,1-dimeth-ylethyl)-2,3,7,8,12,13,17,18-octa-ethyl-5,15-dihydro-	CH_2Cl_2	423(4.87)	5-1259-84
$C_{44}H_{64}O_{24}$ Crocin	H_2O	410s(4.74),443(4.95), 462(4.92)	32-0189-84
cis	H_2O	328(4.42),410s(4.68), 437(4.80),462s(4.75)	32-0189-84
$C_{44}H_{65}NO_{13}$ Bafilomycin B_1	MeOH	248(4.54),285(4.27), 355(3.40)	158-0110-84
$C_{44}H_{72}O_{16}$ Rutamycin B	EtOH	220s(4.49),225(4.52), 232(4.49),240s(4.28), 295s(2.53)	33-1208-84
$C_{44}H_{76}N_2O_{16}S$ Spiramycin I, 4''-O-(methanesulfonyl)-	MeOH	239(4.49)	158-0750-84
$C_{44}H_{88}Si_{11}$ 5,6,11,16,21,26,31,36,41,46,51-Undeca-silaundecaspiro[4.0.4.0.4.0.4.0.4.0-4.0.4.0.4.0.4.0.4.0.4.0]pentapenta-contane	isooctane	235s(4.43),254(4.46), 268(4.38),283s(4.00)	35-5521-84

Compound	Solvent	$\lambda_{max}(\log \epsilon)$	Ref.
$C_{45}H_{29}Br_3N_4O$			
21H,23H-Porphine, 5,10,15-tris(4-bromo-phenyl)-20-(4-methoxyphenyl)- (or isomer)	benzene	423(5.70),518(4.32), 552(4.00),595(3.78), 652(3.82)	39-1483-84C
	+ TFA	447(5.56),677(4.69)	39-1483-84C
$C_{45}H_{32}N_4$			
Spiro[2,4-cyclopentadiene-1,3'-[3H]-pyrazolo[5,1-c][1,2,4]triazole], 2,3,4,5,7'-pentaphenyl-6'-(phenyl-methyl)-	CH_2Cl_2	246(4.64),355(4.13)	24-1726-84
$C_{45}H_{35}N_4$			
1H-Benzimidazolium, 2-[7-(1,3-dihydro-1,3-diphenyl-2H-benzimidazol-2-yli-dene)-1,3,5-heptatrienyl]-1,3-di-phenyl-, iodide	$C_6H_{11}Me$ EtOH DMF	730(5.01) 720(5.02) 715(5.20)	99-0415-84 99-0415-84 99-0415-84
$C_{45}H_{35}O_2$			
5H-Cyclopenta[c]pyrylium, 5-[5-(6,7-di-hydro-1,3-diphenylcyclopenta[c]pyran-5-yl)-2,4-pentadienylidene]-6,7-dihy-dro-1,3-diphenyl-, perchlorate	CH_2Cl_2	980(5.49)	103-0359-84
$C_{45}H_{35}S_2$			
5H-Cyclopenta[c]thiopyrylium, 5-[5-(6,7-dihydro-1,3-diphenylcyclopenta[c]-thiopyran-5-yl)-2,4-pentadien-1-ylidene]-6,7-dihydro-1,3-diphenyl-, perchlorate	CH_2Cl_2	1052(5.44)	103-0359-84
$C_{45}H_{37}O_2$			
2-Benzopyrylium, 5-[3-(7,8-dihydro-1,3-diphenyl-6H-2-benzopyran-5-yl)-2-pro-penylidene]-5,6,7,8-tetrahydro-1,3-diphenyl-, perchlorate	CH_2Cl_2	822(5.45)	103-0359-84
$C_{45}H_{38}N_2O_7S$			
5-Thia-1-azabicyclo[4.2.0]oct-2-ene-2-carboxylic acid, 3-(acetoxymethyl)-4-(2-methoxy-2-oxoethylidene)-8-oxo-7-[(triphenylmethyl)amino]-, diphen-ylmethyl ester, [6R-(4Z,6α,7β)]-	EtOH	355(3.98)	87-1225-84
$C_{45}H_{40}N_2O_8S$			
5-Thia-1-azabicyclo[4.2.0]oct-2-ene-4-acetic acid, 3-(acetoxymethyl)-2-[(diphenylmethoxy)carbonyl]-8-oxo-7-[(triphenylmethyl)amino]-, methyl ester, 5-oxide, [4R-)4α,6α,7β)]-	EtOH	354(3.98)	87-1225-84
$C_{45}H_{41}N_3O_9$			
2(1H)-Pyrimidinone, 1-[3-O-benzoyl-2-deoxy-5-O-[(4-methoxyphenyl)diphen-ylmethyl]-β-D-erythro-pentofurano-syl]-5-methyl-4-[2-(4-nitrophenyl)-ethoxy]-	MeOH	227s(4.54),275(4.25), 280s(4.25)	78-0059-84
$C_{45}H_{42}Cl_2N_3O_{12}P$			
3'-Thymidylic acid, 5'-O-[bis(4-meth-oxyphenyl)phenylmethyl]-, 2,5-di-	MeOH	229s(4.46),266(4.31)	118-0965-84

Compound	Solvent	$\lambda_{max}(\log \epsilon)$	Ref.
chlorophenyl 2-(4-nitrophenyl)ethyl ester (cont.)			118-0965-84
$C_{45}H_{45}NO_6S_3$ Ethanamine, 2-[2-[(4-methylphenyl)sulfinyl]phenoxy]-N,N-bis[2-[2-[(4-methylphenyl)sulfinyl]phenoxy]ethyl]-	EtOH	226s(4.52),242(4.56), 287(4.04)	12-2059-84
$C_{45}H_{45}NO_7S_3$ Ethanamine, 2-[2-[(4-methylphenyl)sulfinyl]phenoxy]-N,N-bis[2-[2-[(4-methylphenyl)sulfinyl]phenoxy]ethyl]-, N-oxide	EtOH	219(4.58),240(4.58), 283(4.06)	12-2059-84
$C_{45}H_{52}O_{23}$ Acteoside	MeOH	218s(4.27),247s(4.03), 292(4.13),334(4.25)	94-3009-84
$C_{45}H_{54}O_3$ Methanone, bis[3-[2,4,6-tris(1-methylethyl)benzoyl]phenyl]-	C_6H_{12}	233(4.66),350(2.42)	88-4525-84
Methanone, bis[4-[2,4,6-tris(1-methylethyl)benzoyl]phenyl]-	C_6H_{12}	265(4.83),346(2.65)	88-4525-84
Methanone, [3-[2,4,6-tris(1-methylethyl)benzoyl]phenyl][4-[2,4,6-tris(1-methylethyl)benzoyl]phenyl]-	C_6H_{12}	238(4.21),260(4.46), 352(2.53)	88-4525-84
$C_{45}H_{54}O_{22}$ Cistanoside D heptaacetate	MeOH	228(3.88),270(3.88)	94-3880-84
$C_{45}H_{59}FO_7$ Retinoic acid, 9-fluoro-11-hydroxy-3,20-dioxo-16-(1-oxobutoxy)pregna-1,4-dien-21-yl ester	MeOH	238(4.33),359(4.58)	107-0725-84
Retinoic acid, 9-fluoro-11-hydroxy-16-methyl-3,20-dioxo-17-(1-oxopropoxy)pregna-1,4-dien-21-yl ester	MeOH	238(4.32),360(4.69)	107-0725-84
$C_{45}H_{64}N_6O_7$ L-Lysine, N^6-[(1,1-dimethylethoxy)carbonyl]-N^2-[(17-ethyl-1,2,3,19,23,24-hexahydro-2,2,7,8,12,13,18-heptamethyl-1,19-dioxo-21H-bilin-3-yl)acetyl]-, 1,1-dimethylethyl ester, (3R)-	$CHCl_3$	272(4.27),348(4.51), 582(4.16)	49-1071-84
(3S)-	$CHCl_3$	272(4.27),347(4.53), 580(4.18)	49-1071-84
(15E)-	$CHCl_3$	270(4.32),348(4.38), 540(4.25)	49-1081-84
$C_{45}H_{65}NO_{16}$ Deltamycin, 4"-phenylacetyl-	MeOH	240(4.19)	158-0127-84
$C_{45}H_{67}NO_{16}$ Leucomycin V, 9-deoxydihydro-13-hydroxy-9-oxo-, 3-acetate, 4^B-benzeneacetate	MeOH	230(4.10)	158-0127-84
$C_{45}H_{73}NO_{14}$ Concanamycin B	isoPrOH	245(4.58),285(4.27)	158-1333-84

Compound	Solvent	$\lambda_{max}(\log \epsilon)$	Ref.
$C_{45}H_{74}O_{13}$ Concanamycin C	isoPrOH	245(4.61),285(4.29)	158-1333-84
$C_{45}H_{78}N_2O_{19}$ Spiramycin I, 4"-O-ethyl-	MeOH	231(4.23)	158-0750-84
$C_{45}H_{78}N_2O_{16}S$ Spiramycin I, 4"-O-(ethanesulfonyl)-	MeOH	236(4.47)	158-0750-84
$C_{46}H_{29}Cl_3N_4O_2$ Benzoic acid, 4-[10,15,20-tris(4-chlo- rophenyl)-21H,23H-porphin-5-yl]-, methyl ester	benzene	420(5.55),514(4.22), 549(3.86),591(3.66), 650(3.57)	39-1483-84C
	+ TFA	442(5.54),656(4.60)	39-1483-84C
$C_{46}H_{32}Br_2N_4O_2$ 21H,23H-Porphine, 5,10-bis(4-bromophen- yl)-15,20-bis(4-methoxyphenyl)- (or isomer)	benzene	421(5.66),517(4.26), 552(4.00),594(3.70), 652(3.82)	39-1483-84C
	+ TFA	449(5.51),675(4.78)	39-1483-84C
$C_{46}H_{32}N_6O_4$ 1,1'-Bi-4H-pyrrolo[1,2-a]benzimidazole, 2,2'-bis(4-nitrophenyl)-4,4'- bis(phenylmethyl)-	DMSO	270(4.92),305(4.89), 368(4.18)	103-0150-84
$C_{46}H_{36}N_6O_{11}$ Benzamide, N-[6-[2-(4-nitrophenyl)eth- oxy]-9-(2,3,5-tri-O-benzoyl-β-D-ribo- furanosyl)-9H-purin-2-yl]-	MeOH	230(4.77),270(4.55)	78-0059-84
$C_{46}H_{38}N_4O_8$ 3H-Indol-3-one, 1,1'-(1,10-dioxo-1,10- decanediyl)bis[2-(1-acetyl-1,3-dihy- dro-3-oxo-2H-indol-2-ylidene)-1,2- dihydro-, (E,E)-	xylene	566(4.10)	18-0470-84
$C_{46}H_{42}N_2O_6$ 1,2-Ethenediamine, N,N'-bis[bis(4-meth- oxyphenyl)methylene]-1,2-bis(4-meth- oxyphenyl)-	EtOH	278(4.55),413(3.68)	44-4755-84
$C_{46}H_{50}N_8O_4$ 21H,23H-Porphine-2,18-dipropanoic acid, 7,12-bis[[(4-aminophenyl)amino]meth- yl]-3,8,13,17-tetramethyl-, dimethyl ester	$CHCl_3$	403(5.16),501(4.17), 535(4.00),569(3.89), 624(3.65)	5-1057-84
$C_{46}H_{50}O_{15}$ α-D-Galactopyranoside, 3,3'-O-(oxydi- 2,1-ethanediyl)bis[methyl 4,6- O-(phenylmethylene)-, dibenzoate, (S,S)-	MeCN	230(4.43),274(3.27)	5-1046-84
$C_{46}H_{54}O_{23}$ Cistanoside C octaacetate	MeOH	225(4.17),272(4.18)	94-3880-84
$C_{46}H_{60}ClFeN_6O_2$ Iron, chloro[28,29,38,39-tetraethyl- 4,5,7,8,9,10,11,12,13,14,15,16,17- 18,20,21-hexadecahydro-3,33-dimethyl-	$CHCl_3$	380(4.76),508(3.79), 537(3.81),639(3.51)	78-0579-84

Compound	Solvent	$\lambda_{max}(\log \epsilon)$	Ref.
36H-27,30-imino-25,22-metheno-2,23-(metheno[2,5]-endo-pyrrolometheno)-2H-pyrrolo[3,2-v][1,7,18]triazacyclo-triacontine-6,19-dionato(2-)-N^1,N^{24},N^{32},N^{36}]- (cont.)			78-0579-84
$C_{46}H_{60}N_2O_{12}$			
2H-2,7-(Epoxypentadeca[1,11,13]trieno)-azonino[2,1-b]benzofuro[4,5-g]quina-zoline-1,17,18-trione, 28-acetoxy-7a,8,9,10,11,12,13,14-octahydro-5,6,24,26-tetrahydroxy-30-methoxy-2,4,19,23,25,27,29-pentamethyl-, (2'R)-	MeOH	227(4.51),252s(4.36), 278s(4.29),326s(4.13), 364(4.19),614(3.94)	94-4388-84
(2'S)-	MeOH	226(4.55),281(4.29), 312s(4.16),366(4.24), 620(4.01)	94-4388-84
$C_{46}H_{60}N_6O_2Zn$			
Zinc, [28,29,38,39-tetraethyl-4,5,7,8-9,10,11,12,13,14,15,16,17,18,20,21-hexadecahydro-3,33-dimethyl-36H-27,30-imino-25,22-metheno-2,23-(meth-eno[2,5]-endo-pyrrolometheno)-2H-pyrrolo[3,2-v][1,7,18]triazacyclo-triacontine-6,19-dionato(2-)-N^1,N^{24},N^{32},N^{36}]-	CHCl$_3$	401(5.24),535(4.15), 573(4.23)	78-0579-84
$C_{46}H_{61}NO_{28}$			
D-Arabinitol, 1,1'-C-[[(acetyloxy)imi-no]bis[2,3-dihydro-4-(methoxycarbo-nyl)-5-methyl-2,5-furandiyl]]bis-, 1,1',2,2',3,3',4,4',5,5'-decaacetate	EtOH	245(4.02)	136-0075-84K
$C_{46}H_{63}N_3O_8Si_2$			
Cytidine, N-[(4-methoxyphenyl)diphenyl-methyl]-2'-O-(tetrahydro-2H-pyran-2-yl)-3',5'-O-[1,1,3,3-tetrakis(1-meth-ylethyl)-1,3-disiloxanediyl]-	EtOH	278(4.32)	78-0153-84
$C_{46}H_{64}O_{17}$			
Dunawithanine B	MeOH	228(3.86)	102-2293-84
$C_{46}H_{68}N_6$			
6,9,33,34,35,36-Hexaazahexacyclo-[12.9.9.1^{15,18}.1^{20,23}.1^{24,27}1^{29,32}]-hexatriaconta-15,17,19,21,23(35),24-26,28,30,32-decaene, 16,17,21,22,25-26,30,31-octaethyl-	benzene	426(4.62)	5-1259-84
$C_{46}H_{70}O_{17}$			
Bryostatin 4	MeOH	228(4.56)	35-6768-84
$C_{46}H_{72}O_{12}$			
Napoleogenin B	MeOH	215(4.04)	94-3378-84
$C_{46}H_{72}O_{19}$			
Deacetyldunawithanine A	MeOH	228(3.93)	102-2293-84
$C_{46}H_{75}NO_{14}$			
Concanamycin A	isoPrOH	245(4.61),285(4.30)	158-1333-84

Compound	Solvent	$\lambda_{max}(\log \epsilon)$	Ref.
$C_{46}H_{76}O_{18}$ Vitexin	MeOH	266(4.39),330(4.39)	94-4003-84
$C_{46}H_{80}N_2O_{14}$ Spiramycin I, 4"-O-propyl-	MeOH	237(4.41)	158-0750-84
$C_{46}H_{80}N_2O_{16}S$ Spiramycin I, 4"-O-(propylsulfonyl)-	MeOH	233(4.51)	158-0750-84
$C_{47}H_{35}BrN_4O_3$ 21H,23H-Porphine, 5-(4-bromophenyl)- 10,15,20-tris(4-methoxyphenyl)-	benzene	422(4.43),518(4.07), 553(3.85),595(3.49), 650(3.58)	39-1483-84C
	+ TFA	450(5.2),680(4.66)	39-1483-84C
$C_{47}H_{39}O_2$ 2-Benzopyrylium, 5-[5-(7,8-dihydro-1,3- diphenyl-6H-2-benzopyran-5-yl)-2,4- pentadienylidene]-5,6,7,8-tetrahydro- 1,3-diphenyl-, perchlorate	CH_2Cl_2	942(5.41)	103-0359-84
$C_{47}H_{54}N_6Zn$ Zinc, [1-methyl-1'-(2,3,7,8,12,13,17,18- octaethyl-21H,23H-porphin-5-yl)-4,4'- pyridiniumato(2-)-$N^{21},N^{22},N^{23},N^{24}$]-, dichloride	MeOH	260(4.34),341(4.42), 413(5.33),542(4.20), 578(4.20)	5-0426-84
dication	MeOH	404(5.07),505(3.94), 536(3.88),570(3.61), 622(3.60)	5-0426-84
trication	MeOH-HOAc	340(4.53),404(5.20), 535(4.00),568(4.09)	5-0426-84
tetracation	MeOH- CF_3COOH	403(5.17),562(4.06), 608(3.88)	5-0426-84
$C_{47}H_{54}O_{24}$ Acteoside, 2'-acetyl-, octaacetate	MeOH	281(4.51)	94-3009-84 +94-3880-84
$C_{47}H_{63}FO_7$ Retinoic acid, 9-fluoro-11-hydroxy-16- methyl-3,20-dioxo-17-[(1-oxopentyl)- oxy]pregna-1,4-dien-21-yl ester, cis? trans?	MeOH MeOH	238(4.30),360(4.66) 239(4.38),353(4.28)	107-0725-84 107-0725-84
$C_{47}H_{63}N_5O_8Si_2$ Guanosine, N-[(4-methoxyphenyl)diphen- ylmethyl]-2'-O-(tetrahydro-2H-pyran- 2-yl]-3',5'-O-[1,1,3,3-tetrakis(1- methylethyl)-1,3-disiloxanediyl]-	EtOH	264(4.29)	78-0153-84
$C_{47}H_{72}N_2O_5$ Cholest-5-en-3-ol, (3β)-, [2-[[5-[(1,1- dimethylethoxy)carbonyl]-3,4-dimeth- yl-1H-pyrrol-2-yl)methyl]-4,4-dimeth- yl-5-oxo-3-pyrrolidinyl] acetate, (Z)-	$CHCl_3$	305(4.16)	49-0837-84
$C_{47}H_{78}N_2O_{16}$ Spiramycin I, 2',4"-di-O-acetyl-	MeOH	232(4.41)	158-0760-84
$C_{48}H_{32}CdN_8O_8$ Cadmium, [[4,4',4",4"'-(21H,23H-porph-	n.s.g.	443(5.25)	138-1871-84

Compound	Solvent	$\lambda_{max}(\log \epsilon)$	Ref.
ine-5,10,15,20-tetrayl)tetrakis-[1-(carboxymethyl)pyridiniumato]]-(6-)-N^{21},N^{22},N^{23},N^{24}]-, (SP-4-1)-(cont.)			138-1871-84
$C_{48}H_{32}Cl_2N_4O_4$			
Benzoic acid, 4,4'-[10,20-bis(4-chloro-phenyl)-21H,23H-porphine-5,15-diyl]-bis-, dimethyl ester	benzene	420(5.63),515(4.30), 549(3.94),592(3.74), 651(3.74)	39-1483-84C
	+ TFA	442(5.60),655(4.64)	39-1483-84C
Benzoic acid, 4,4'-[15,20-bis(4-chloro-phenyl)-21H,23H-porphine-5,10-diyl]-bis-, dimethyl ester	benzene	420(5.63),515(4.31), 549(3.95),592(3.75), 651(3.70)	39-1483-84C
	+ TFA	442(5.61),655(4.67)	39-1483-84C
$C_{48}H_{32}CoN_8O_8$			
Cobalt, [[4,4',4'',4'''-(21H,23H-porph-ine-5,10,15,20-tetrayl)tetrakis-[1-(carboxymethyl)pyridiniumato]]-(6-)-N^{21},N^{22},$N^{23}N^{24}$]-, (SP-4-1)-	n.s.g.	436(5.02)	138-1871-84
$C_{48}H_{32}CuN_8O_8$			
Copper, [[4,4',4'',4'''-(21H,23H-porph-ine-5,10,15,20-tetrayl)tetrakis-[1-(carboxymethyl)pyridiniumato]]-(6-)-N^{21},N^{22},N^{23},N^{24}]-, (SP-4-1)-	n.s.g.	427(5.29)	138-1871-84
$C_{48}H_{32}O_{31}$			
Agrimonic acid B	MeOH	202(5.29),275(4.41)	94-2165-84
$C_{48}H_{34}N_8O_8$			
Pyridinium, 4,4',4'',4'''-(21H,23H-porph-ine-5,10,15,20-tetrayl)tetrakis[1-(carboxymethyl)-, tetrahydroxide, tetrakis(inner salt)	pH 0.5	442(4.39),590(4.11), 636(4.18)	138-1871-84
	pH 9.3	422(5.31),518(4.14), 555(3.81),585(3.83), 640(3.42)	138-1871-84
$C_{48}H_{36}ClCoN_4O_4$			
Cobalt, chloro[5,10,15,20-tetrakis(4-methoxyphenyl)-21H,23H-porphinato-(2-)-N^{21},N^{22},N^{23},N^{24}]-, (SP-5-12)-	toluene	413(4.92),527(3.83)	94-4252-84
$C_{48}H_{36}FeN_7$			
Iron, azido[5,10,15,20-tetrakis(4-meth-ylphenyl)porphinato(2-)-N^{21},N^{22},N^{23}-N^{24}]-, (SP-5-12)-	CH_2Cl_2	360(4.77),412(5.05), 505(4.12),571(3.70), 694(3.53)	64-0222-84B
$C_{48}H_{36}N_2O_4$			
Ethanone, 1,1',1'',1'''-([7',7'''-bispiro-[9H-fluorene-9,1'(8'aH)-indolizine]]-2',2''',3',3'''-tetrayl)tetrakis-	CH_2Cl_2	420(4.13)	89-0241-84
$C_{48}H_{36}N_2O_6$			
[7',7'''-Bispiro[9H-fluorene-9,1'(8'aH)-indolizine]]-2',3'-dicarboxylic acid, 2''',3'''-diacetyl-, dimethyl ester	CH_2Cl_2	407(4.22)	89-0241-84
photochromic form	CH_2Cl_2	585(--)	89-0241-84
$C_{48}H_{36}N_2O_8$			
[7',7'''-Bispiro[9H-fluorene-9,1'(8'aH)-indolizine]]-2',2''',3',3'''-tetracarb-	CH_2Cl_2	396(4.31)	89-0241-84

Compound	Solvent	$\lambda_{max}(\log \epsilon)$	Ref.
oxylic acid, tetramethyl ester (cont.)			89-0241-84
$C_{48}H_{36}O_2$ 5,14:8,13-Diepoxynaphtho[2',3':3,4]-cyclobut[1,2-a]anthracene, 5,5a,7a-7b,8,13,13a,13b,13c,14-decahydro-5,8,13,14-tetraphenyl-, (5α,5aα-7aβ,7bα,8α,13α,13aα,13bβ,13cα,14α)-	EtOH	265(3.73),272(3.50)	88-2573-84
$C_{48}H_{36}O_4$ Pentacyclo[16.2.2.2^{8,11}.0^{3,6}.0^{13,16}]-tetracosa-8,10,18,20,21,23-hexaene-2,7,12,17-tetrone, 4,5,14,15-tetra-phenyl-, (±)-	CH_2Cl_2	232(4.28),267(4.41)	88-0561-84
$C_{48}H_{39}MnN_4O_2$ Manganese, aquahydroxy[5,10,15,20-tetrakis(4-methylphenyl)-21H,23H-porphinato(2-)-N^{21},N^{22},N^{23},N^{24}]-, ()C-6-23)-	CH_2Cl_2	379(4.63),421(4.73), 475(4.89),535(3.72), 586(3.89),624(4.00)	64-0222-84B
$C_{48}H_{40}N_5OS_3$ Thiazolium, 5-[[5-(2-benzothiazolyl)-3-ethyl-1,3-dihydro-1-phenyl-2H-benzimidazol-2-ylidene]ethylidene]-2-[(3-ethyl-4,5-diphenyl-2(3H)-thi-azolylidene)methyl]-4,5-dihydro-4-oxo-3-(2-propenyl)-, iodide	EtOH	599(5.09)	103-0665-84
$C_{48}H_{42}N_2O_7$ 2,4(1H,3H)-Pyrimidinedione, 1-[1,6-bis-O-(triphenylmethyl)-β-D-fructo-furanosyl]-	EtOH	206(4.93),253(4.07), 258(4.06)	1-0367-84
2,4(1H,3H)-Pyrimidinedione, 1-[1,6-bis-O-(triphenylmethyl)-β-D-psico-furanosyl]-	EtOH	205(4.92),263(3.92)	1-0367-84
$C_{48}H_{42}N_8O_4$ Pyridinium, 4,4',4",4"'-(21H,23H-porph-ine-5,10,15,20-tetrayl)tetrakis-[1-(2-hydroxyethyl)-, tetrahydroxide, tetrakis(inner salt)	pH 0.5 pH 9.3	444(4.35),591(3.91), 637(4.00) 423(5.29),517(4.13), 554(3.74),583(3.78), 640(3.23)	138-1871-84 138-1871-84
$C_{48}H_{42}N_8O_{12}S_4$ Pyridinium, 4,4',4",4"'-(21H,23H-porph-ine-5,10,15,20-tetrayl)tetrakis-[1-(2-sulfoethyl)-, tetrahydroxide, tetrakis(inner salt)	pH 0.5 pH 9.3	446(5.40),592(4.03), 638(4.00) 423(5.30),518(4.18), 557(3.83),583(3.83), 641(3.32)	138-1871-84 138-1871-84
$C_{48}H_{44}N_4Zn$ Tetrabenzoporphine, meso-tetrapropyl-, zinc chelate, tribenzylamine adduct	CHCl$_3$	404s(4.65),430(5.51), 576(3.97),630(4.97)	103-0052-84
$C_{48}H_{45}Cl_2N_6O_{11}P$ 3'-Guanylic acid, 2'-deoxy-5'-O-[(4-methoxyphenyl)diphenylmethyl]-N-(2-methyl-1-oxopropyl)-, 2,5-dichloro-phenyl 2-(4-nitrophenyl)ethyl ester	MeOH	228s(4.48),253s(4.42), 260(4.46),273s(4.39), 280s(4.35)	118-0965-84

Compound	Solvent	$\lambda_{max}(\log \epsilon)$	Ref.
$C_{48}H_{48}N_2O_2$			
Propanedinitrile, [10-[bis[3,5-bis(1,1-dimethylethyl)-4-oxo-2,5-cyclohexadien-1-ylidene)cyclopropylidene]-9(10H)-anthracenylidene]-	THF	658(4.43)	35-0355-84
	CHCl$_3$	276(4.22),307(4.35), 380(4.25),420(4.24), 610s(4.48),660(4.54)	35-0355-84
$C_{48}H_{50}N_2O_2$			
Propanedinitrile, [10-[2,3-bis[3,5-bis(1,1-dimethylethyl)-4-hydroxyphenyl]-2-cyclopropen-1-ylidene]-9(10H)-anthracenylidene]-	THF	256(4.04),267(4.04), 278(4.06),288(4.09), 307(4.14),355(3.85), 370(3.87),410s(3.44), 575(3.92)	35-0355-84
	CHCl$_3$	583(3.87)	35-0355-84
$C_{48}H_{50}N_8$			
29H,31H-Phthalocyanine, tetrakis(1,1-dimethylethyl)-	pentane	340(4.86),654(5.04), 695(5.21)	59-0584-84
	hexane	338(4.90),654(5.11), 695(5.33)	59-0584-84
	heptane	338(4.90),653(5.11), 696(5.34)	59-0584-84
	octane	340(4.92),656(5.17), 695(5.37)	59-0584-84
	toluene	345(4.84),662(5.18), 699(5.26)	59-0584-84
	2,2-Me$_2$but-ane	336(4.85),653(4.94), 695(5.07)	59-0584-84
	2,2-Me$_2$pent-ane	338(4.82),654(5.02), 695(5.16)	59-0584-84
	2,2-Me$_2$hept-ane	339(4.87),654(5.09), 695(5.26)	59-0584-84
	PrOH	336(4.82),658(4.92), 697(4.93)	59-0584-84
	BuOH	338(4.85),660(4.97), 698(5.01)	59-0584-84
	tert-BuOH	335(4.74),655(4.78), 694(4.79)	59-0584-84
	pentanol	340(4.86),660(5.01), 698(5.10)	59-0584-84
	HOAc	338(4.73),660(4.86), 697(4.89)	59-0584-84
	pinacolone	340(4.87),657(5.10), 696(5.15)	59-0584-84
	tert-BuOAc	340(4.88),657(5.10), 696(5.15)	59-0584-84
	pyridine	345(4.86),665(5.15), 700(5.19)	59-0584-84
	3,5-lutidine	345(4.78),665(5.10), 700(5.16)	59-0584-84
	2,6-lutidine	663(5.15),700(5.22)	59-0584-84
	4-tert-butyl-pyridine	342(4.87),663(5.17), 700(5.23)	59-0584-84
	Me$_2$SO$_2$	342(4.76),666(5.02), 698(5.02)	59-0584-84
	C$_{10}$H$_7$Cl	345(4.84),668(5.16), 703(5.22)	59-0584-84
$C_{48}H_{54}O_{27}$			
Safflor Yellow B	MeOH	240(4.04),335(3.79), 408(4.35)	88-2471-84

Compound	Solvent	$\lambda_{max}(\log \epsilon)$	Ref.
$C_{48}H_{61}N_3O_{25}$ 2H-1-Benzopyran-2-one, 4-methyl-, 7-[[O-(3,4,6-tri-O-acetyl-2-(acetyl-amino)-2-deoxy-β-D-glucopyranosyl)-(1→4)-O-[3,6-di-O-acetyl-2-(acetyl-amino)-2-deoxy-β-D-glucopyranosyl}-(1→4)-3,6-di-O-acetyl-2-(acetylami-no)-2-deoxy-β-D-glucopyranosyl]oxy]-	DMSO	288(4.01),316(4.19)	94-1597-84
$C_{48}H_{66}ClFeN_6O_2$ Iron, chloro[7,8,17,18-tetraethyl-N,N'-dihexyl-3,13-dimethyl-21H,23H-porph-ine-2,12-dipropanamidato(2-)-$N^{21},N^{22},N^{23},N^{24}$]-	CHCl	381(4.76),505(3.87), 638(3.52)	78-0579-84
$C_{48}H_{74}N_6$ 21H,23H-Porphine-5,15-dibutanamine, N,N',2,3,7,8,12,13,17,18-decaeth-ane-5,15-dihydro-	benzene	427(5.00)	5-1259-84
$C_{48}H_{74}O_{20}$ Dunawithanine A	MeOH	228(3.67)	102-2293-84
$C_{48}H_{80}N_2O_{16}$ Spiramycin I, 3"-O-acetyl-4"-O-propan-oyl-	MeOH	232(4.25)	158-0760-84
Spiramycin I, 4"-O-acetyl-3"-O-propan-oyl-	MeOH	232(4.36)	158-0760-84
$C_{48}H_{96}Si_{12}$ 5,6,11,16,21,26,31,36,41,46,51,56-Do-decasiladodecaspiro[4.0.4.0.4.0.4.0-4.0.4.0.4.0.4.0.4.0.4.0.4.0]hexa-contane	isooctane	230s(4.36),247(4.28), 265(4.28),283s(3.89)	35-5521-84
$C_{49}H_{35}S_2$ Thiopyrylium, 6-[3-(5,6-diphenyl-2H-thiopyran-2-ylidene)-1,3-diphenyl-1-propenyl]-2,3-diphenyl-, per-chlorate	CHCl$_3$	766(4.78)	97-0146-84
$C_{49}H_{36}N_2$ 2,3-Diazaspiro[4.4]nona-1,6,8-triene, 1,3,4,6,7,8,9-heptaphenyl-	MeOH	245(4.56),354(4.39)	78-1585-84
$C_{49}H_{47}Cl_2N_6O_{12}P$ 3'-Guanylic acid, 5'-O-[bis(4-methoxy-phenyl)phenylmethyl]-2'-deoxy-N-(2-methyl-1-oxopropyl)-, 2,5-dichloro-phenyl 2-(4-nitrophenyl)ethyl ester	MeOH	225(4.58),230(4.58), 237s(4.47),255s(4.41), 261(4.42),273(4.38), 280s(4.35)	118-0965-84
$C_{49}H_{58}O_{18}$ P-1894B	90% MeOH	218(4.61),318(3.74), 438(3.82)	94-4350-84
	+ HCl	218(4.60),318(3.76), 435(3.82)	94-4350-84
	+ NaOH	228(4.44),282(4.13), 530(3.86)	94-4350-84
$C_{49}H_{64}N_2O_{21}$ Arugorol, 4'-(L-diginosyl-L-decilo-	MeOH	235(4.74),248(4.45),	88-1941-84

Compound	Solvent	$\lambda_{max}(\log \epsilon)$	Ref.
nitrosyl-2-deoxyfucosyl)-7-deoxy- (cont.)		292(4.08),478(4.20)	88-1941-84
$C_{49}H_{64}N_2O_{22}$ 　AG3	MeOH	235(4.68),257(4.35), 290(3.92),478(4.17)	88-1937-84
$C_{49}H_{64}O_{18}$ 　P-1894B, hexahydro-	MeOH	219(4.59),250(4.01), 275(3.94),430(3.69)	94-4350-84
$C_{49}H_{66}O_{14}$ 　Pallidinine	EtOH	282(3.24)	102-2931-84
$C_{49}H_{80}N_2O_{17}$ 　Leucomycin V, 9-O-[5-(dimethylamino)- 　　tetrahydro-6-methyl-2H-pyran-2-yl]-, 　　$2^A,3,4^B$-triacetate (2',3,4"-tri-O- 　　acetylspiramycin I)	MeOH	239(4.26)	158-0760-84
$C_{49}H_{82}N_2O_{16}$ 　Leucomycin V, 9-O-[5-(dimethylamino)- 　　tetrahydro-6-methyl-2H-pyran-2-yl]-, 　　3^B-acetate 4^B-butanoate	MeOH	231(4.20)	158-0760-84
4^B-acetate 3^B-butanoate	MeOH	232(4.34)	158-0760-84
$2^A,4^B$-dipropanoate	MeOH	230(4.29)	158-0760-84
$3^B,4^B$-dipropanoate	MeOH	233(4.52)	158-0760-84
$C_{49}H_{88}N_2O_{14}Si$ 　Spiramycin I 3,18-(O-tert-butylsilyl)- 　　acetal	MeOH	234(4.39)	158-0750-84
$C_{50}H_{35}ClN_4O_6$ 　Benzoic acid, 4,4',4"-[20-(4-chloro- 　　phenyl)-21H,23H-porphine-5,10,15- 　　triyl]tris-, trimethyl ester	benzene	421(5.61),515(4.29), 550(3.94),592(3.74), 652(3.76)	39-1483-84C
	+ TFA	442(5.58),653(4.63)	39-1483-84C
$C_{50}H_{36}N_4$ 　21H,23H-Porphine, 5,15-dihydro- 　　5,10,15,20-tetraphenyl-5,15- 　　di-2-propynyl- (with isomers)	benzene	426(4.86)	35-5217-84
$C_{50}H_{38}FeN_4$ 　Iron, [5,15-dihydro-5,10,15,20-tetra- 　　phenyl-5,15-di-2-propenyl-21H,23H- 　　porphinato(2-)-$N^{21},N^{22},N^{23},N^{24}$]-, 　　(SP-4-1)-	benzene	471(4.90)	35-5217-84
$C_{50}H_{38}N_2O_8$ 　Spiro[9H-fluorene-9,1'(8'aH)-indoli- 　　zine]-2',3'-dicarboxylic acid, 　　7',7"'-(1,2-ethenediyl)bis-, 　　tetramethyl ester	CH_2Cl_2	394(4.18)	89-0241-84
photochromic form	CH_2Cl_2	702(--)	89-0241-84
$C_{50}H_{40}N_2O_6$ 　[7',7"'-Bispiro[9H-fluorene-9,1'(8'aH)- 　　indolizine]]-2',3'-dicarboxylic acid, 　　2"'.3"'-diacetyl-5',5"'-dimethyl-, 　　dimethyl ester	CH_2Cl_2	404(4.35)	89-0241-84

Compound	Solvent	$\lambda_{max}(\log \epsilon)$	Ref.
$C_{50}H_{40}N_2O_8$			
[7',7"'-Bispiro[9H-fluorene-9,1'(8'aH)-indolizine]]-2',2"',3',3"'-tetracarboxylic acid, 5',5"'-dimethyl-, tetramethyl ester	CH_2Cl_2	396(4.32)	89-0241-84
[7',7"'-Bispiro[9H-fluorene-9,1'(8'aH)-indolizine]]-2',2"',3',3"'-tetracarboxylic acid, 8',8"'-dimethyl-, tetramethyl ester	CH_2Cl_2	399(4.16)	89-0241-84
Spiro[9H-fluorene-9,1'(8'aH)-indolizine]-2',3'-dicarboxylic acid, 7',7"'-(1,2-ethanediyl)bis-, tetramethyl ester	CH_2Cl_2	381(4.39)	89-0241-84
photochromic form	CH_2Cl_2	596(--)	89-0241-84
$C_{50}H_{40}N_4$			
21H,23H-Porphine, 5,10,15,20-tetraphenyl-5,15-di-2-propenyl-	benzene	424(4.85)	35-5217-84
$C_{50}H_{43}AsBNO$			
Pyridinium, ω-triphenylarsenylidene-acetonyl-, tetraphenylborate	20% acetone	400(2.68)	104-1911-84
$C_{50}H_{43}Cl_2N_4O_{11}P$			
3'-Cytidylic acid, N-benzoyl-2'-deoxy-5'-O-[(4-methoxyphenyl)diphenylmethyl]-, 2,5-dichlorophenyl 2-(4-nitrophenyl)ethyl ester	MeOH	260(4.54),305(4.12)	118-0965-84
$C_{50}H_{80}N_2O_{14}$			
Spiramycin I, 4"-O-(phenylmethyl)-	MeOH	238(4.47)	158-0750-84
$C_{50}H_{80}N_2O_{16}S$			
Spiramycin I, 4"-O-[(phenylmethyl)sulfonyl]-	MeOH	242(4.54)	158-0750-84
Spiramycin I, 4"-O-(p-toluenesulfonyl)-	MeOH	238(4.51)	158-0750-84
$C_{50}H_{82}N_2O_{17}$			
Spiramycin I, 3,3"-di-O-acetyl-4"-O-propanoyl-	MeOH	231(4.39)	158-0760-84
Spiramycin I, 3,3"-di-O-acetyl-3"-O-propanoyl-	MeOH	234(4.55)	158-0760-84
$C_{50}H_{84}N_2O_{16}$			
Spiramycin I, 3"-O-acetyl-4"-O-(3-methylbutanoyl)-	MeOH	231(4.08)	158-0760-84
Spiramycin I, 4"-O-acetyl-3"-O-(3-methylbutanoyl)-	MeOH	232(4.54)	158-0760-84
Spiramycin I, 3"-O-butanoyl-4"-O-propanoyl-	MeOH	232(4.46)	158-0760-84
Spiramycin I, 4"-O-butanoyl-3"-O-propanoyl-	MeOH	232(4.40)	158-0760-84
$C_{51}H_{42}CrN_7O$			
Chromium, azido(2-propanone)[5,10,15,20-tetrakis(4-methylphenyl)-21H,23H-porphinato(2-)-$N^{21},N^{22},N^{23},N^{24}$]-, (as hydrate)	CH_2Cl_2	344(4.40),359(4.40), 399(4.65),448(5.35), 525(3.73),565(4.06), 602(4.05)	64-0222-84B
$C_{51}H_{42}OP_2$			
Phosphonium, [carbonylbis(4,1-phenyl-	n.s.g.	366(2.19)	116-2468-84

Compound	Solvent	$\lambda_{max}(\log \epsilon)$	Ref.
enemethylene)]bis[triphenyl- (cont.)			116-2468-84
$C_{51}H_{43}Cl_2N_6O_{10}P$ 3'-Adenylic acid, N-benzoyl-2'-deoxy-5'-O-[(4-methoxyphenyl)diphenyl-methyl]-, 2,5-dichlorophenyl 2-(4-nitrophenyl)ethyl ester	MeOH	229(4.54),277(4.44)	118-0965-84
$C_{51}H_{43}N_3O_{11}$ 2(1H)-Pyrimidinone, 1-[2,3-di-O-benzoyl-5-O-[(4-methoxyphenyl)diphenylmeth-yl]-β-D-ribofuranosyl]-4-[2-(4-nitro-phenyl)ethoxy]-	MeOH	273(4.31),280s(4.28)	78-0059-84
$C_{51}H_{45}Cl_2N_4O_{12}P$ 3'-Cytidylic acid, N-benzoyl-5'-O-[bis(4-methoxyphenyl)phenylmethyl]-2'-deoxy-, 2,5-dichlorophenyl 2-(4-nitrophenyl)ethyl ester	MeOH	229s(4.60),238s(4.55), 261(4.54),300(4.12)	118-0965-84
$C_{51}H_{80}N_2O_{16}S$ Spiramycin I, 4"-O-(2-phenylethane-sulfonyl)-	MeOH	206(4.38),234(4.88)	158-0750-84
$C_{51}H_{84}N_2O_{17}$ Spiramycin I, 3"-O-acetyl-3,4"-di-O-propanoyl-	MeOH	239(4.59)	158-0760-84
Spiramycin I, 4"-O-acetyl-3,3"-di-O-propanoyl-	MeOH	232(4.47)	158-0760-84
Spiramycin I, 3,3"-di-O-acetyl-4"-O-butanoyl-	MeOH	231(4.34)	158-0760-84
Spiramycin I, 3,4"-di-O-acetyl-3"-O-butanoyl-	MeOH	231(4.39)	158-0760-84
$C_{51}H_{86}N_2O_{16}$ Spiramycin I, 2',4"-di-O-butanoyl-	MeOH	233(4.56)	158-0760-84
Spiramycin I, 3",4"-di-O-butanoyl-	MeOH	232(4.30)	158-0760-84
Spiramycin I, 3"-(3-methylbutanoyl)-4"-O-propanoyl-	MeOH	231(4.58)	158-0760-84
Spiramycin I, 4"-O-(3-methylbutanoyl)-3"-O-propanoyl-	MeOH	231(4.49)	158-0760-84
$C_{51}H_{90}N_2O_{15}Si$ Leucomycin V, 18-deoxo-3-deoxy-9-O-[5-(dimethylamino)tetrahydro-6-methyl-2H-pyran-2-yl]-N-[[(1,1-di-methylethyl)dimethylsilyl]oxy]-3,18-epoxy-, 2^A-acetate	MeOH	235(4.54)	158-0750-84
$C_{51}H_{90}O_7$ Sitoindoside	MeOH	none above 210nm	150-0965-84M
$C_{52}H_{28}N_{12}O_4Zn$ Zinc, [2,9,16,23-tetrakis(3-pyridinyl-oxy)-29H,31H-phthalocyaninato(2-)-$N^{29},N^{30},N^{31},N^{32}$]-, (SP-4-1)-	pyridine	678(5.27)	142-2047-84
$C_{52}H_{30}N_{12}O_4$ 29H,31H-Phthalocyanine, 2,9,16,23-tetrakis(3-pyridinyloxy)-	pyridine	666(5.03)	142-2047-84

Compound	Solvent	$\lambda_{max}(\log \epsilon)$	Ref.
$C_{52}H_{45}Cl_2N_6O_{11}P$ 3'-Adenylic acid, N-benzoyl-5'-O-[bis(4-methoxyphenyl)phenylmethyl]-2'-deoxy-, 2,5-dichlorophenyl 2-(4-nitrophenyl)ethyl ester	MeOH	225(4.64),229(4.64), 277(4.50)	118-0965-84
$C_{52}H_{46}Cl_2N_5O_{14}P$ Carbamic acid, [1-[2-deoxy-3-O-[(2,5-dichlorophenoxy)[2-(4-nitrophenyl)-ethoxy]phosphinyl]-5-O-[(4-methoxy-phenyl)diphenylmethyl]-β-D-erythro-pentofuranosyl]-1,2-dihydro-2-oxo-4-pyrimidinyl]-, 2-(4-nitrophenyl)-ethyl ester	MeOH	229(4.56),273(4.51)	78-0059-84
$C_{52}H_{47}Cl_2N_4O_{13}P$ 2(1H)-Pyrimidinone, 1-[2-deoxy-3-O-[(2,5-dichlorophenoxy)[2-(4-nitrophenyl)ethoxy]phosphinyl]-5-O-[(4-methoxyphenyl)diphenylmethyl]-β-D-erythro-pentofuranosyl]-5-meth-yl-4-[2-(4-nitrophenyl)ethoxy]-	MeOH	273(4.35),280(4.32)	78-0059-84
$C_{52}H_{58}O_4$ Methanone, 1,3-phenylenebis[3-(2,4,6-tris(1-methylethyl)benzoyl]phenyl]-	C_6H_{12}	235(4.80),350(3.59)	88-4525-84
$C_{52}H_{59}N_6O_{14}P$ 2'-Deoxycytidine, 5'-O-(methoxytriphen-yl)methyl-4-N-[2-(4-nitrophenyl)eth-oxycarbonyl]-3'-[2-(4-nitrophenyl)-ethyl] phosphate, triethylamine salt	MeOH	274(4.41)	78-0059-84
$C_{52}H_{60}N_5O_{13}P$ Thymidine, 5'-O-(methoxytriphenyl)meth-yl-4-[2-(4-nitrophenyl)ethoxy]-3'-[2-(4-nitrophenyl)ethyl]phosphate, triethylamine salt	MeOH	275(4.38),280s(4.37)	78-0059-84
$C_{52}H_{60}N_8O_{16}$ Takaomycin	MeOH MeOH-HCl MeOH-NaOH	279(4.23) 273(4.27) 281(4.26)	158-0700-84 158-0700-84 158-0700-84
$C_{52}H_{60}O_6Si_4$ Silane, [(1,4,7,10-tetraphenyl-5,12-6,11-diepoxydibenzo[a,e]cyclooct-ene-5,6,11,12-tetrayl)tetrakis(oxy)-tetrakis[trimethyl-, (5α,6β,11β,12α)-	CH_2Cl_2	245(4.7),265(4.8)	89-0622-84
$C_{52}H_{70}N_{20}O_{16}$ Adenosine, N,N'-1,4-butanediylbis[N-[4-[(9-β-D-ribofuranosyl-9H-purin-6-yl)amino]butyl]-	pH 7	271(4.71)	88-5619-84
$C_{52}H_{86}N_2O_{17}$ Spiramycin I, 2',3,4"-tri-O-propanoyl- Spiramycin I, 3,3",4"-tri-O-propanoyl-	MeOH MeOH	231(4.39) 231(4.35)	158-0760-84 158-0760-84
$C_{52}H_{88}N_2O_{16}$ Spiramycin I, 4"-O-butanoyl-3"-O-(3-	MeOH	234(4.60)	158-0760-84

Compound	Solvent	$\lambda_{max}(\log \epsilon)$	Ref.
methylbutanoyl)- (cont.)			158-0760-84
$C_{53}H_{44}N_4O_8$ 1-Isoquinolinecarbonitrile, 2-benzoyl-1-[[4-[5-[(2-benzoyl-1-cyano-1,2-di-hydro-6,7-dimethoxy-1-isoquinolinyl)-methyl]-2-methoxyphenoxy]phenyl]meth-yl]-1,2-dihydro-6,7-dimethoxy-	MeOH	207(4.86),227(4.82), 285(4.30),313(4.32)	83-0092-84
$C_{53}H_{46}Cl_2N_7O_{13}P$ 3'-Adenylic acid, 2'-deoxy-5'-O-[(4-methoxyphenyl)diphenylmethyl]-N-[[2-(4-nitrophenyl)ethoxy]carbonyl]-,2,5-dichlorophenyl 2-(4-nitrophenyl)-ethyl ester	MeOH	267(4.59)	78-0059-84
$C_{53}H_{59}N_8O_{13}P$ 2'-Deoxyadenosine, 5'-O-(methoxytri-phenyl)methyl-6-N-[2-(4-nitrophenyl)-ethoxycarbonyl]-3'-[2-(4-nitrophen-yl)ethyl]phosphate, triethylamine salt	MeOH	267(4.58)	78-0059-84
$C_{53}H_{62}O_{20}$ P-1894B, diacetate	MeOH	213(4.65),255(4.37), 310(3.73),352(3.72)	94-4350-84
$C_{53}H_{80}N_2O_{16}S$ Spiramycin I, 4"-O-(2-naphthalene-sulfonyl)-	MeOH	239(4.65),276(4.10), 313(3.42),324(3.53)	158-0750-84
$C_{53}H_{88}N_2O_{17}$ Spiramycin I, 3"-O-butanoyl-3,4"-di-O-propanoyl)-	MeOH	235(4.35)	158-0760-84
Spiramycin I, 4"-O-butanoyl-3,3"-di-O-propanoyl)-	MeOH	233(4.58)	158-0760-84
$C_{54}H_{36}$ o,p,p,p,p,p,p,p-Nonaphenylene	C_6H_{12}	203(5.13),279(4.94), 320s(4.34)	18-3494-84
$C_{54}H_{57}N_7$ 1,6-Hexanediamine, N'-9-acridinyl-N-[6-(9-acridinyl)amino]hexyl]-N-[3-(9-acridinylamino)propyl]-	pH 7.0	410(4.44)	77-0509-84
DNA bound	pH 7.0	415(4.30)	77-0509-84
$C_{54}H_{59}BrN_4PRu$ Ruthenium, bromo[2,3,7,8,12,13,17,18-octaethyl-21H,23H-porphinato(2-)-$N^{21},N^{22},N^{23},N^{24}$](triphenylphos-phine)-	CH_2Cl_2	360s(4.50),405(4.80), 510(4.05)	23-1238-84
$C_{54}H_{59}ClN_4PRu$ Ruthenium, chloro[2,3,7,8,12,13,17,18-octaethyl-21H,23H-porphinato(2-)-$N^{21},N^{22},N^{23},N^{24}$](triphenylphos-phine)-	CH_2Cl_2	403(4.90),510(3.97)	23-1238-84
$C_{54}H_{80}O_{20}$ Avenacin B-2	EtOH	228(4.16),274(3.11),	77-0244-84

Compound	Solvent	$\lambda_{max}(\log \epsilon)$	Ref.
(cont.)		281(3.10)	77-0244-84
$C_{54}H_{80}O_{21}$ Avenacin A-2	EtOH	230(4.17),274(3.08), 281(3.04)	77-0244-84
$C_{55}H_{55}N_3O_9$ Cytidine, 5'-O-[bis(4-methoxyphenyl)- phenylmethyl]-N-[(4-methoxyphenyl)- diphenylmethyl]-2-O-(tetrahydro-2H- pyran-2-yl)-	EtOH	278(4.22)	78-0153-84
$C_{55}H_{83}NO_{20}$ Avenacin B-1	EtOH	223(4.37),255(3.90), 356(3.72)	77-0244-84
$C_{55}H_{83}NO_{21}$ Avenacin A-1	EtOH	223(4.40),255(3.90), 357(3.74)	77-0244-84
$C_{55}H_{102}N_2O_{14}Si_2$ Spiramycin I, 2'-O-tert-butyldimethyl- silyl-, 3,18-(O-tert-butyldimethyl- silyl)acetal	MeOH	234(4.35)	158-0750-84
$C_{56}H_{52}Cl_2N_7O_{13}P$ Phosphoric acid, 2,5-dichlorophenyl 2-(4-nitrophenyl)ethyl ester, ester with N-[9-[2-deoxy-5-O-[(4-methoxy- phenyl)diphenylmethyl]-β-D-erythro- pentofuranosyl]-6-[2-(4-nitrophen- yl)ethyl]-9H-purin-2-yl]-2-methyl- propanamide	MeOH	222s(4.71),270(4.56), 282s(4.43)	78-0059-84
$C_{56}H_{55}N_5O_8$ Adenosine, 5'-O-[bis(4-methoxyphenyl)- phenylmethyl]-N-[(4-methoxyphenyl)- diphenylmethyl]-2'-O-(tetrahydro- 2H-pyran-2-yl)-	EtOH	275(4.54)	78-0153-84
$C_{56}H_{55}N_5O_9$ Guanosine, 5'-O-[bis(4-methoxyphenyl)- phenylmethyl]-N-[(4-methoxyphenyl)- diphenylmethyl]-2'-O-(tetrahydro-2H- pyran-2-yl]-	EtOH	262(4.40)	78-0153-84
$C_{56}H_{58}O_{26}$ Potentillin, pentadeca-O-methyl-	MeOH	215(5.13),255s(4.70), 290s(4.26)	94-2165-84
$C_{56}H_{60}N_4Zn$ Zinc, [6,13,20,27-tetrapentyl-29H,31H- tetrabenzo[b,g,l,q]porphinato(2-)- $N^{29},N^{30},N^{31},N^{32}$]-, (SP-4-1)-	CHCl$_3$	404s(4.65),429(4.92), 576(3.55),629(4.40)	103-0052-84
$C_{56}H_{65}N_8O_{13}P$ 2'-Deoxyguanosine, 2-N-isobutyryl-5'- O-(methoxytriphenyl)methyl-6-[2-(4- nitrophenyl)ethoxy]-3'-[[2-(4-nitro- phenyl)ethyl]phosphate, triethylamine salt	MeOH	236s(4.35),270(4.58), 282s(4.47)	78-0059-84

Compound	Solvent	$\lambda_{max}(\log \epsilon)$	Ref.
$C_{56}H_{76}N_2O_{25}$ AG2	MeOH	235(4.68),259(4.35), 293(3.87),478(4.14), 546(4.12)	88-1937-84
$C_{56}H_{82}O_{17}$ Napoleogenin B pentaacetate	MeOH	215(4.31)	94-3378-84
$C_{57}H_{70}O_{30}$ Cistanoside B decaacetate	MeOH	281(4.01)	94-3009-84

$C_{57}H_{80}N_4O_4$
  21H-Biline-3-acetic acid, 17-ethyl-
  1,2,3,19,23,24-hexahydro-2,7,8,12-
  13,18-heptamethyl-1,19-dioxo-, (3β)-
  cholest-5-en-3-yl ester (4Z,9Z,15Z)-
  (15E)-
	MeOH
	
	CHCl$_3$
	
	CHCl$_3$
		268(4.41),339(4.56),
		 582(4.16)
		271(4.31),344(4.54),
		 580(4.17)
		348(4.39),550(4.26)
			49-0837-84
			
			49-0837-84
			
			49-0837-84

$C_{57}H_{103}N_{19}O_{23}P_4$
  Guanosine-5'-phosphate, P^3-[5'-(2',3'-
  O-isopropylideneadenylyl(5'→3')aden-
  osyl]-
	pH 7	257(4.53)	5-0867-84

$C_{58}H_{44}N_4O_{16}$
  2,4(1H,3H)-Pteridinedione, 1,3-
  bis(2,3,5-tri-O-acetyl-β-D-
  ribofuranosyl)-
	MeOH	229(4.95),315(3.96)	136-0179-84F

$C_{58}H_{46}FeN_6$
  Iron, bis(pyridine)[5,10,15,20-tetra-
  kis(4-methylphenyl)-21H,23H-porph-
  inato(2-)-$N^{21},N^{22},N^{23},N^{24}$]-
	pyridine	423(5.25),470s(4.06),
		 531(4.24),561(3.71),
		 609(3.12)
			64-0222-84B

$C_{58}H_{58}O_2$
  Phenol, 4,4'-(1,2-ethynediyldi-10,9-
  anthracenediyl)bis[2,6-bis(1,1-di-
  methylethyl)-
	CHCl$_3$	252(4.96),265(4.90),
		 352(3.64),372(3.77),
		 410s(4.07),453(4.37),
		 476s(4.26)
			44-0965-84

$C_{58}H_{70}O_{31}$ Cistanoside A undecaacetate	MeOH	280(4.31)	94-3009-84
$C_{58}H_{94}O_{20}$ Dunawithanine A, deca-O-methyl-	MeOH	228(3.69)	102-2293-84

$C_{59}H_{58}O_3$
  2-Cyclopropen-1-one, 2,3-bis[10-[3,5-
  bis(1,1-dimethylethyl)-4-hydroxy-
  phenyl)-9-anthracenyl]-
	CHCl$_3$	268(4.79),360s(3.96),
		 390s(4.08),408(4.21),
		 482(4.41)
			44-0965-84

$C_{59}H_{68}O_3$
  2,5-Cyclohexadien-1-one, 4,4'-[3-[10-
  [3,5-bis(1,1-dimethylethyl)-4-oxo-
  2,5-cyclohexadien-1-ylidene]-9(10H)-
  anthracenylidene]-1,2-cyclopropane-
  diylidene]bis[2,6-bis(1,1-dimethyl-
  ethyl)-
	CHCl$_3$	245(4.20),320(4.07),
		 365(4.26),396(4.27),
		 432s(3.98),672(4.35)
			44-0965-84

$C_{59}H_{70}O_3$
  2,5-Cyclohexadien-1-one, 4-[2-[3,5-
  bis(1,1-dimethylethyl)-4-hydroxy-
	THF	268(4.65),272(4.61),
		 350(4.23),367(4.44),
			44-0965-84

Compound	Solvent	$\lambda_{max}(\log \epsilon)$	Ref.
phenyl]-3-[10-[3,5-bis(1,1-dimethyl-ethyl)-4-hydroxyphenyl]-9-anthracen-yl]-2-cyclopropen-1-ylidene]-2,6-bis(1,1-dimethylethyl)-		405s(4.49),422(4.55), 445(4.50),522(3.72)	44-0965-84
$C_{59}H_{70}O_{32}$ Echinacoside dodecaacetate	MeOH	283(4.25)	94-3009-84
$C_{59}H_{103}N_3O_{18}$ Niphimycin I	EtOH	228s(4.48),233(4.50), 240s(4.35)	158-94-5
$C_{60}H_{28}CoN_{12}O_6$ Cobalt, [[5,5'-(29H,31H-phthalocyanine-diyldicarbonyl)bis[2-(2-pyridinyl)-1H-isoindole-1,3(2H)-dionato](2-)-$N^{29},N^{30},N^{31},N^{32}$]-	NMP	345(4.65),620(4.43), 640(4.47),673(4.82)	126-1395-84
$C_{60}H_{28}CuN_{12}O_6$ Copper, [[5,5'-(29H,31H-phthalocyanine-diyldicarbonyl)bis[2-(2-pyridinyl)-1H-isoindole-1,3(2H)-dionato(2-)-$N^{29},N^{30},N^{31},N^{32}$]-	NMP	345(4.67),616(4.45), 640(4.49),673(4.85)	126-1395-84
$C_{60}H_{28}N_{12}NiO_6$ Nickel, [[5,5'-(29H,31H-phthalocyanine-diyldicarbonyl)bis[2-(2-pyridinyl)-1H-isoindole-1,3(2H)-dionato(2-)-$N^{29},N^{30},N^{31},N^{32}$]-	NMP	330(4.52),614(4.39), 635(4.41),665(4.68)	126-1395-84
$C_{60}H_{46}$ Dicyclobuta[e,l]pyrene, 3b,4,5,5a,8b-9,10,10a-octahydro-4,5,9,10-tetra-phenyl-2,7-bis(2-phenylethenyl)-	CH_2Cl_2	383(4.60)	24-2452-84
$C_{60}H_{48}$ Heptacyclo[8.8.4.1^{3,17}.1^{8,12}.0^{4,7}-0^{13,16}.0^{19,22}]tetracosa-1,3(23)-8,10,12(24),17-hexaene (plus shoulders)	CH_2Cl_2	270(3.70)	24-2452-84
$C_{60}H_{48}N_4O_{16}$ 2,4(1H,3H)-Pteridinedione, 6,7-dimeth-yl-1,3-bis(2,3,5-tri-O-benzoyl-β-D-ribofuranosyl)-	MeOH	229(5.02),323(4.09)	136-0179-84F
$C_{60}H_{52}N_6O_4$ Benzeneacetamide, 2,5-dimethoxy-N-[2-[[4-[10,15,20-tris(4-methylphenyl)-21H,23H-porphin-5-yl]benzoyl]amino]-ethyl]-	CH_2Cl_2	419(5.65),515(4.26), 552(3.97),592(3.73), 646(3.67)	23-0967-84
dication	CH_2Cl_2-4% MeOH	443(5.58),665(4.72)	23-0967-84
$C_{60}H_{78}Ge_3$ Cyclotrigermane, hexakis(2,6-diethyl-phenyl)-	C_6H_{12}	210(5.30),272(4.67), 309s(--)	88-4191-84
$C_{60}H_{78}N_4O_{33}$ 2H-1-Benzopyran-2-one, 4-methyl-7-[[O-3,4,6-tri-O-acetyl-2-(acetylamino)-2-deoxy-β-D-glucopyranosyl]-(1→)-O-	DMSO	289(4.03),316(4.20)	94-1597-84

Compound	Solvent	$\lambda_{max}(\log \epsilon)$	Ref.
3,6-di-O-acetyl-2-(acetylamino)-2-deoxy-β-D-glucopyranosyl-(1→4)-O-3,6-di-O-acetyl-2-(acetylamino)-2-deoxy-β-D-glucopyranosyl-(1→4)-O-3,6-di-O-acetyl-2-(acetyamino)-2-deoxy· β-D-glucopyranosyl]oxy]- (cont.)			94-1597-84
$C_{60}H_{82}N_4O_{26}$ Decilorubicin	MeOH	220(4.45),235(4.54), 254(4.44),290(3.83), 380(3.51),476(3.97), 496(4.01),535(3.88), 586(3.63)	158-84-43
	+ HCl	220(4.45),235(4.57), 255(4.51),292(3.92), 383(3.53),475(4.07), 498(4.10),535(3.85)	158-84-43
	+ NaOH	253(4.55),295s(3.81), 360(3.70),560(4.13), 597(4.13)	158-84-43
$C_{60}H_{100}O_{20}$ Dunawithanine A, dodeca-O-methyl-	MeOH	228(3.58)	102-2293-84
$C_{61}H_{54}N_6O_4$ Benzeneacetamide, 2,5-dimethoxy-N-[3-[[4-[10,15,20-tris(4-methylphenyl)-21H,23H-porphin-5-yl]benzoyl]amino]-propyl]-	CH_2Cl_2	419(5.64),515(4.23), 552(3.93),591(3.72), 646(3.63)	23-0967-84
dication	CH_2Cl_2-MeOH	443(5.53),665(4.62)	23-0967-84
$C_{61}H_{59}Cl_3N_4O_2$ Benzoic acid, 4-[10,15,20-tris(4-chlorophenyl)-21H,23H-porphin-5-yl]-, hexadecyl ester	benzene	420(5.61),514(4.17), 548(3.80),590(3.60), 646(3.43)	39-1483-84C
	+ TFA	442(5.60),606s(3.76), 655(4.54)	39-1483-84C
$C_{61}H_{60}Cl_3N_5O$ Benzamide, N-hexadecyl-4-[10,15,20-tris(4-chlorophenyl)-21H,23H-porphin-5-yl]-	benzene	419(5.52),514(4.00), 548(3.59),592(3.36), 650(3.36)	39-1483-84C
	+ TFA	441(5.49),608s(3.64), 655(4.38)	39-1483-84C
$C_{61}H_{93}N_3O_{21}$ Carbamic acid, [(5S)-2,3-di-O-acetyl-5-C-[[(3β,20β)-20-[[(1-methylethoxy)-carbonyl]amino]-11-oxo-13-norolean-12-en-3-yl]oxy]-4-O-[(5S)-2,3,4-tri-O-acetyl-5-C-[[(1-methylethoxy)carbonyl]amino]-β-D-xylopyranosyl]-β-L-xylopyranosyl]-, 1-methylethyl ester	EtOH	205(4.14),246(4.12), 305(3.73)	65-2299-84
isomer	EtOH	205(4.15),245(4.11), 310(3.74)	65-2299-84
$C_{61}H_{109}N_{11}O_{13}$ Leucinostatin B, hydrochloride	EtOH	204(4.38),213(4.27)	158-84-135
$C_{61}H_{116}N_2O_{14}Si_3$ Spiramycin I, 2',4"-di-O-tert-butyl-	MeOH	236(4.34)	158-0750-84

Compound	Solvent	$\lambda_{max}(\log \epsilon)$	Ref.
dimethylsilyl-, 3,18-(O-tert-butyldimethylsilyl)acetal (cont.)			158-0750-84
Spiramycin I enol, 2',4",18-tri-O-tert-butyldimethylsilyl-	MeOH	232(4.23)	158-0750-84
$C_{62}H_{43}BrN_4PRu$			
Ruthenium, bromo[5,10,15,20-tetraphenyl-21H,23H-porphinato(2-)-N^{21},N^{22}-N^{23},N^{24}](triphenylphosphine)-, (OC-6-23)-	CH_2Cl_2	345(4.47),413(5.10), 525(4.03)	23-1238-84
$C_{62}H_{56}N_6O_4$			
Benzeneacetamide, 2,5-dimethoxy-N-[4-[[4-[10,15,20-tris(4-methylphenyl)-21H,23H-porphin-5-yl]benzoyl]amino]-butyl]-	CH_2Cl_2	419(5.70),515(4.24), 552(3.95),591(3.71),	23-0967-84
dication	CH_2Cl_2-MeOH	443(5.67),665(4.79)	23-0967-84
$C_{62}H_{86}N_2O_{28}$			
AG4 (from arugomycin)	MeOH	235(4.72),257(4.39), 290(3.98),478(4.21)	88-1941-84
$C_{63}H_{43}N_4OPRu$			
Ruthenium, carbonyl[5,10,15,20-tetraphenyl-21H,23H-porphinato(2-)-N^{21},N^{22},N^{23},N^{24}](triphenylphosphine)-, (OC-6-42)-	CH_2Cl_2	418(5.34),535(4.38), 573(3.92)	23-0755-84
$C_{64}H_{44}N_4Zn$			
Zinc, [6,13,20,27-tetrakis(phenylmethyl)-29H,31H-tetrabenzo[b,g,l,q]porphinato(2-)-N^{29},N^{30},N^{31},N^{32}]-, (SP-4-1)-	$CHCl_3$	402s(4.64),429(5.35), 576(3.92),628(4.76)	103-0052-84
$C_{64}H_{66}N_8O_8$			
Deuteroporphyrin IX dimer VII	$CHCl_3$	440(5.65),498(4.56), 530(4.32),570(4.20), 624(3.74)	65-1610-84
$C_{64}H_{92}N_{12}O_{12}$			
23,26,51,54,59,62,71,74-Octaoxa-1,4,10-11,17,20,29,32,38,39,43,48-dodecaaza-heptacyclo[46.8.8.8^{20,29}.2^{6,9}.2^{12,15}-2^{34,37}.2^{40,43}]octaconta-6,8,10,12-14,34,36,38,40,42,65,67,77,79-tetra-decaene-5,16,33,34-tetrone, 4,17,32-45-tetramethyl-, (E,E)-	$C_6H_4Cl_2$-BuOH	330(4.72)	108-0302-84
$C_{64}H_{96}N_6O_{24}$			
L-Alanine, N-[[[(5S)-2,3-di-O-acetyl-5-C-[[(3β,20β(S)]-20-[[[(2-methoxy-1-methyl-2-oxoethyl)amino]carbonyl]-amino]-11-oxo-2-oxoethyl)amino]carb-onyl]amino]-11-oxo-30-norolean-12-en-3-yl]oxy]-4-O-[[5S-[5R*(R*)]]-2,3,4-tri-O-acetyl-5-C-[[[(2-methoxy-1-methyl-2-oxoethyl)amino]carbonyl]-amino]-β-D-xylopyranosyl]-β-L-xylo-pyranosyl]amino]carbonyl]-, methyl ester	EtOH	204(4.19),244(4.15), 307(3.84)	65-2299-84

Compound	Solvent	$\lambda_{max}(\log \epsilon)$	Ref.
$C_{64}H_{99}N_3O_{21}$ Carbamic acid, [(5S)-2,3-di-O-acetyl-5-C-[[(3β,20β)-20-[(butoxycarbonyl)-amino]-11-oxo-30-norolean-12-en-3-yl]oxy]-4-O-[(5S)-2,3,4-tri-O-acetyl-5-C-[(butoxycarbonyl)amino]-β-D-xylo-pyranosyl]-β-L-xylopyranosyl]-, butyl ester	EtOH	207(4.17),244(4.14), 305(3.76)	65-2299-84
$C_{64}H_{106}O_{31}$ Methyl protorhapissaponin (El$_3$)	MeOH	268(4.18),330(4.18)	94-4003-84
$C_{66}H_{51}MnN_4P$ Manganese(1+), [5,10,15,20-tetrakis(4-methylphenyl)-21H,23H-porphinato(2-)-$N^{21},N^{22},N^{23},N^{24}$](triphenylphosphine)-, hexafluorophosphate	CH$_2$Cl$_2$	388(4.80),407s(4.72), 434s(4.49),479(4.76), 520(3.78),575(3.98), 610(3.99)	24-2261-84
$C_{66}H_{53}MnN_5OP$ Manganese, aqua[5,10,15,20-tetrakis(4-methylphenyl)-21H,23H-porphinato(2-)-$N^{21},N^{22},N^{23},N^{24}$](P,P,P-triphenyl-phosphine imidato-N)-, (OC-6-23)-	toluene	380(4.51),423(4.80), 446(4.88),474(4.73), 536(3.73),579(3.92), 620(3.99)	24-2261-84
$C_{66}H_{68}O_{31}$ Agrimoniin, octadecamethyl deriv.	MeOH	219(5.36),249s(5.05), 287s(4.57)	94-2165-84
isomer	MeOH	217(5.35),249s(5.03), 284s(4.55)	94-2165-84
$C_{66}H_{78}N_8O_3Ru_2$ Ruthenium, dimethoxy-μ-oxobis[5,10,15-20-tetrapropyl-21H,23H-porphinato-(2-)-$N^{21},N^{22},N^{23},N^{24}$]di-	CH$_2$Cl$_2$	394(4.94),538(4.05), 576(4.00)	35-5151-84
$C_{67}H_{64}N_3O_{10}PS_2$ Cytidine, 5'-O-[bis(4-methoxyphenyl)-phenylmethyl]-N-[(4-methoxyphenyl)-diphenylmethyl]-2'-O-(tetrahydro-2H-pyran-2-yl)-, 3'-(S,S-diphenyl phos-phorodithioate)	EtOH	274(4.18)	78-0153-84
$C_{67}H_{105}N_3O_{21}$ Carbamic acid, [(5S)-2,3-di-O-acetyl-5-C-[[(3β,20β)-11-oxo-20-[[(pentyl-oxy)carbonyl]amino]-13-norolean-12-en-3-yl]oxy]-4-O-[(5S)-2,3,4-tri-O-acetyl-5-C-[[(pentyloxy)carbonyl]-amino]-β-D-xylopyranosyl-β-L-xylo-pyranosyl]-, pentyl ester	EtOH	205(4.17),245(4.12), 310(3.91)	65-2299-84
$C_{68}H_{50}O_{44}$ Cornusiin A	MeOH	218(5.14),271(4.81)	94-4662-84
$C_{68}H_{51}N_4P_2Ru$ Ruthenium(II), (meso-tetrapropyl-porphyrinato)bis(triphenylphos-phine)-	CH$_2$Cl$_2$	414(5.01),433(5.40), 506(3.98)	35-5151-84
$C_{68}H_{64}N_5O_9PS_2$ Adenosine, 5'-O-[bis(4-methoxyphenyl)-	EtOH	271(4.84)	78-0153-84

Compound	Solvent	$\lambda_{max}(\log \epsilon)$	Ref.
phenylmethyl]-N-[(4-methoxyphenyl)di-phenylmethyl]-2'-O-(tetrahydro-2H-pyran-2-yl]-, 3'-(S,S-diphenyl phosphorodithioate) (cont.)			78-0153-84
$C_{68}H_{64}N_5O_{10}PS_2$ Guanosine, 5'-O-[bis(4-methoxyphenyl)-phenylmethyl]-N-[(4-methoxyphenyl)-diphenylmethyl]-2'-O-(tetrahydro-2H-pyran-2-yl]-, 3'-(S,S-diphenyl phosphorodithioate)	EtOH	261(4.52)	78-0153-84
$C_{68}H_{73}N_{10}O_{15}PSSi_2$ Adenosine, N,N'-dibenzoyl-2',3'-di-O-benzoyl-P(O)-methyl-P-thioadenylyl-(5'→2')-N-benzoyl-3',5'-O-[1,1,3,3-tetrakis(1-methylethyl)-1,3-disil-oxanediyl]-	MeOH	277(4.60)	44-2314-84
$C_{68}H_{74}F_6N_8O_6Ru_2$ Ruthenium(IV), μ-oxobis(5,10,15,20-tetrapropylporphinato)(trifluoro-acetato)-	CH_2Cl_2	388(4.84),536(3.42), 580(3.55),615(3.53)	35-5151-84
$C_{68}H_{82}N_4P_2Ru$ Ruthenium, [5,10,15,20-tetraphenyl-21H,23H-porphinato(2-)-N^{21},N^{22},N^{23}-N^{24}]bis(tributylphosphine)-, (OC-6-12)-	CH_2Cl_2	354(4.38),399(4.38), 437(5.12),478(3.40), 525(3.82),559(3.56)	23-0755-84
$C_{68}H_{104}N_4O_4$ Tetraspiro[7,13,23,29-tetraoxapenta-cyclo[28.2.2.$2^{3,6}.2^{14,17}.2^{19,22}$]-tetraconta-3,5,14,16,19,21,30,32-33,35,37,39-dodecaene-2,4':10,4"-18,4''':26,4""-tetrapiperidinium], hexadecamethyl-, tetrachloride	H_2O	196(5.05),232s(4.24), 268(3.30),274s(3.22), 302s(1.66)	35-8024-84
$C_{70}H_{52}N_4O_{16}$ 2,4(1H,3H)-Pteridinedione, 6,7-diphen-yl-1,3-bis(2,3,5-tri-O-benzoyl-β-D-ribofuranosyl)-	MeOH	229(4.98),273(4.32), 360(4.13)	136-0179-84F
$C_{70}H_{90}N_6O_{18}$ Urea, N-[(5S)-2,3-di-O-acetyl-5-C-[[(3β,20β)-11-oxo-20-[[(phenylam-ino)carbonyl]amino]-30-norolean-12-en-3-yl]oxy]-4-O-[(5S)-2,3,4-tri-O-acetyl-5-C-[[(phenylamino)carbonyl]-amino]-β-D-xylopyranosyl]-β-L-xylo-pyranosyl]-N'-phenyl-	EtOH	203(4.61),241(4.44), 282(3.94),315(3.85)	65-2299-84
$C_{70}H_{108}N_6O_{18}$ Urea, N-cyclohexyl-N'-[[(5S)-2,3-di-O-acetyl-5-C-[[(3β,20β)-20-[[(cyclohex-ylamino)carbonyl]amino]-11-oxo-30-norolean-12-en-3-yl]oxy]-4-O-[(5S)-2,3,4-tri-O-acetyl-5-C-[[(cyclohexyl-amino)carbonyl]amino]-β-D-xylopyrano-syl-β-L-xylopyranosyl]-	EtOH	204(4.27),250(4.13), 319(3.84)	65-2299-84

Compound	Solvent	$\lambda_{max}(\log \epsilon)$	Ref.
$C_{70}H_{108}N_6O_{24}S_3$ DL-Methionine, N-[[[(5S)-2,3-di-O-acetyl-5-C-[[[(3β,20β)-20-[[[1-(methoxycarbonyl)-3-(methylthio)propyl]amino]-carbonyl]amino]-11-oxo-30-norolean-12-en-3-yl]oxy]-4-O-[(5S)-2,3,4-tri-O-acetyl-5-C-[[[[1-(methoxycarbonyl)-3-(methylthio)propyl]amino]carbonyl]-amino]-β-D-xylopyranosyl-β-L-xylo-pyranosyl]amino]carbonyl]-, methyl ester	EtOH	204(4.37),245(4.13), 294(4.01),313(4.00)	65-2299-84
$C_{71}H_{106}N_4O_2$ 2,5-Cyclohexadiene-1,4-dione, 2-(10,15,20-tripentadecyl-21H,23H-porphin-5-yl)-	benzene	418(5.77),516(4.24), 548(3.80),600(3.59), 660(3.49)	39-1483-84C
	+ TFA	425(5.65),587(3.95), 645(4.36)	39-1483-84C
$C_{71}H_{108}N_4O_2$ 1,4-Benzenediol, 2-(10,15,20-tripenta-decyl-21H,23H-porphin-5-yl)-	benzene	417(5.69),518(4.23), 551(3.95),596(3.63), 625(3.73)	39-1483-84C
	+ TFA	425(5.59),554(3.49), 586(3.93),638(4.38)	39-1483-84C
$C_{72}H_{48}$ m,o,o,m,o,o,m,o,o,m,o,o-Dodecaphenyl-ene	C_6H_{12}	195(5.14),230(5.09), 252s(4.90)	18-3494-84
$C_{72}H_{69}N_5O_{13}$ D-Ribitol, 1-[(2-amino-1,6-dihydro-5-nitro-6-oxo-4-pyrimidinyl)amino]-tris-O-[bis(4-methoxyphenyl)phenyl-methyl]-1-deoxy-	CHCl	332(4.07)	64-0252-84B
$C_{72}H_{74}N_4P_2Ru$ Ruthenium, [2,3,7,8,12,13,17,18-octa-ethyl-21H,23H-porphinato(2-)-N^{21}-N^{22},N^{23},N^{24}]bis(triphenylphos-phine)-, (OC-6-12)-	CH_2Cl_2	355(4.55),420(5.30), 512(4.02),530(4.01)	23-0755-84
$C_{72}H_{74}N_{12}O_{28}S_2$ 1-Azetidineacetic acid, 2,2'-dithio-bis[α-(1-methylethyl)-3-[[6-[(4-nitrophenyl)methoxy]-5-[[[(4-nitro-phenyl)methoxy]carbonyl]amino]-1,6-dioxohexyl]amino]-4-oxo-, bis(4-ni-trophenyl)methyl] ester, [2R-[1(R*),2α[1'(R*),2'R*,3'R*(S*)],3α(S*)]]-	MeCN	267(4.82)	78-1907-84
$C_{72}H_{88}N_8Ru_2$ Ruthenium, bis[2,3,7,8,12,13,17,18-octaethyl-21H,23H-porphinato(2-)-N^{21},N^{22},N^{23},N^{24}]di- (Ru-Ru)	benzene	373(4.95),503(3.9), 527(3.89),545s(--), 650(3.66),624(3.72)	35-3500-84
$C_{73}H_{96}N_6O_{18}$ Urea, N-[(5S)-2,3-di-O-acetyl-5-C-[[(3β,20β)-11-oxo-20-[[[(phenyl-methyl)amino]carbonyl]amino]-30-norolean-12-en-3-yl]oxy]-4-O-	EtOH	205(4.77),248(4.36), 305(4.26),363(3.86)	65-2299-84

Compound	Solvent	$\lambda_{max}(\log \epsilon)$	Ref.
[(5S)-2,3,4-tri-O-acetyl-5-C-[[[(phenylmethyl)amino]carbonyl]-amino]-β-D-xylopyranosyl]-β-L-xylopyranosyl]-N'-(phenylmethyl)-(cont.)			65-2299-84
$C_{73}H_{114}N_6O_{24}$ L-Leucine, N-[[[(5S)-2,3-di-O-acetyl-5-C-[[[3β,20β(S)]]-20-[[[[(1-methoxycarbonyl)-3-methylbutyl]amino]carbonyl]amino]-11-oxo-30-norolean-12-en-3-yl]oxy]-4-O-[[5S-[5R*(R*)]]-2,3,4-tri-O-acetyl-5-C-[[[[1-(methoxycarbonyl)-3-methylbutyl]amino]carbonyl]-amino]-β-D-xylopyranosyl-β-L-xylopyranosyl]amino]carbonyl]-, methyl ester	EtOH	204(4.27),247(4.18), 292(3.89),314(3.87)	65-2299-84
$C_{74}H_{72}F_6N_4P_3Ru$ Ruthenium, octaethylporphinatobis-(triphenylphosphone)-, hexafluorophosphate	CH_2Cl_2	393(4.98),520s(4.15)	23-1238-84
$C_{74}H_{94}N_8O_3Ru_2$ Ruthenium, dimethoxybis[2,3,7,8,12,13-17,18-octaethyl-21H,23H-porphinato-(2-)-$N^{21},N^{22},N^{23},N^{24}$]-μ-oxodi-	CH_2Cl_2	376(5.29),511(3.10), 580(4.38)	35-5151-84
$C_{76}H_{46}N_4$ 21H,23H-Porphine, 5,10,15,20-tetra-9-anthracenyl-	CH_2Cl_2	249(5.61),255(5.56), 427(5.41),520(4.40), 552(3.74),593(3.89), 657(3.56)	64-1393-84B
$C_{78}H_{86}N_4O_6P_2Ru$ Ruthenium, [2,3,7,8,12,13,17,18-octaethyl-21H,23H-porphinato(2-)-$N^{21},N^{22},N^{23},N^{24}$]bis[tris(4-methoxyphenyl)phosphine-P]-, (OC-6-12)-	CH_2Cl_2	435(5.30),520(4.10), 555s(3.92)	23-0755-84
$C_{79}H_{132}N_6O_{18}$ Urea, N-[(5S)-2,3-di-O-acetyl-5-C-[[(3β,20β)-20-[[(nonylamino)carbonyl]amino]-11-oxo-30-norolean-12-en-3-yl]oxy]-4-O-[(5S)-2,3,4-tri-O-acetyl-5-C-[[(nonylamino)carbonyl]-amino]-β-D-xylopyranosyl-β-L-xylopyranosyl]-N'-nonyl-	EtOH	204(4.28),244(4.16), 313(3.67)	65-2299-84
$C_{80}H_{58}N_4P_2Ru$ Ruthenium, [5,10,15,20-tetraphenyl-21H,23H-porphinato(2-)-N^{21},N^{22}-N^{23},N^{24}]bis(triphenylphosphine)-	CH_2Cl_2 CH_2Cl_2-Ph_3P	433(5.32),515(4.02), 555s(3.84) 412(4.49),432(5.35), 520(3.99),548s(3.80)	23-0755-84 35-5151-84
$C_{80}H_{112}N_2O_{37}$ Arugomycin	MeOH	235(4.79),258(4.45), 292(4.01),476(4.25)	88-1937-84 +88-1941-84
$C_{80}H_{134}N_4$ 21H,23H-Porphine, 5,10,15,20-tetra-	benzene	418(5.77),520(4.21),	39-1483-84C

Compound	Solvent	$\lambda_{max}(\log \epsilon)$	Ref.
pentadecyl- (cont.)		554(4.11),600(3.57), 652(3.75)	39-1483-84C
	benzene-TFA	423(5.80),554(3.40), 586(3.98),634(4.49)	39-1483-84C
$C_{80}H_{114}N_2O_{37}$ Arugomycin methyl ester	MeOH	234(4.73),257(4.40), 289(4.22),478(4.15)	88-1941-84
$C_{82}H_{54}O_{52}$ Agrimoniin	MeOH	232(5.20),270s(5.05)	94-2165-84
$C_{84}H_{66}CrN_4O_2P_2$ Chromium(1+), [5,10,15,20-tetrakis(4-methylphenyl)-21H,23H-porphinato(2-)-$N^{21},N^{22},N^{23},N^{24}$]bis(triphenylphosphine oxide-O)-, hexafluorophosphate	CH_2Cl_2	361(4.27),396(4.53), 430s(4.38),449(5.48), 525(3.67),567(4.03), 606(4.07)	24-2261-84
$C_{88}H_{56}Cl_2N_8ORu_2$ Ruthenium, dichloro-μ-oxobis[5,10,15-20-tetraphenyl-21H,23H-porphinato-(2-)-$N^{21},N^{22},N^{23},N^{24}$]di-	CH_2Cl_2-EtOH-HCl	394(5.45),494(3.92), 534d(4.01),554(4.09), 604s(4.01),630(4.06)	35-5151-84
$C_{88}H_{150}N_4$ 21H,23H-Porphine, 5,10,15,20-tetra-heptadecyl-	benzene	402(4.86),421(5.61), 488(3.53),522(4.16), 556(4.01),603(3.60), 663(3.85)	64-1393-84B
$C_{88}H_{150}N_6O_{18}$ Urea, N-[(5S)-2,3-di-O-acetyl-5-C-[[(3β,20β)-20-[[(dodecylamino)-carbonyl]amino]-11-oxo-30-norolean-12-en-3-yl]oxy]-4-O-[(5S)-2,3,4-tri-O-acetyl-5-C-[[(dodecylamino)carbonyl]amino]-β-D-xylopyranosyl-β-L-xylopyranosyl]-N'-dodecyl-	EtOH	204(4.29),244(4.17), 315(3.78)	65-2299-84
$C_{90}H_{62}N_8O_3Ru_2$ Ruthenium, dimethoxy-μ-oxobis[5,10,15-20-tetraphenyl-21H,23H-porphinato-(2-)-$N^{21},N^{22},N^{23},N^{24}$]di-	CH_2Cl_2	398(5.26),529s(4.42), 549(4.51),584s(4.24)	35-5151-84
$C_{92}H_{56}F_6N_8O_5Ru_2$ Ruthenium, μ-oxobis[5,10,15,20-tetra-phenyl-21H,23H-porphinato(2-)-$N^{21},N^{22},N^{23},N^{24}$]-, bis(trifluoro-acetato-O)di]-	CH_2Cl_2-EtOH-CF_3COOH	389(5.49),530s(4.07), 549(4.12),599s(3.96), 640(4.10)	35-5151-84
$C_{96}H_{96}CeN_{16}$ Cerium, bis[2,9,16,23-tetrakis(1,1-di-methylethyl)-29H,31H-phthalocyanin-ato(2-)-$N^{29},N^{30},N^{31},N^{32}$]-	$o-C_6H_4Cl_2$	340(4.85),646(4.96), 686s(4.57)	65-1494-84
cation	$o-C_6H_4Cl_2$	322(4.87),344(4.85), 498(4.39),626(4.55), 695(4.82)	65-1494-84
dication	$o-C_6H_4Cl_2$	320(4.82),358(4.80), 520(4.55),748(4.21)	65-1494-84

$C_{98}H_{119}N_{15}O_{23}P_2S_2Si_4-C_{111}H_{112}O_{52}$

Compound	Solvent	$\lambda_{max}(\log \epsilon)$	Ref.
$C_{98}H_{119}N_{15}O_{23}P_2S_2Si_4$ Adenosine, (R)-N,N-dibenzoyl-2',3'-di-O-benzoyl-P(O)-methyl-P-thioadenyl-yl-(5'→2')-N-benzoyl-3'-O-[3-hydroxy-1,1,3,3-tetrakis(1-methylethyl)disil-oxanyl]-P(O)-methyl-P-thioadenyyl-(5'→2')-N-benzoyl-3',5'-O-[1,1,3,3-tetrakis(1-methylethyl)-1,2-disil-oxanediyl)]-	MeOH	278(4.77)	44-2314-84
$C_{100}H_{66}N_8O_5Ru_2$ Ruthenium, bis(1,2-benzenediolato-O)-μ-oxobis[5,10,15,20-tetraphenyl-21H,23H-porphinato(2-)-N^{21},N^{22},N^{23}-N^{24}]di-	CH_2Cl_2-catechol	392(5.38),532s(4.22), 554(4.32),564s(4.31), 587s(4.24)	35-5151-84
$C_{102}H_{70}N_8O_3Ru_2$ Ruthenium, bis(4-methylphenolato)-μ-oxobis[5,10,15,20-tetraphenyl-21H,23H-porphinato(2-)-N^{21},N^{22},N^{23}-N^{24}]di-	CH_2Cl_2-cresol	391(5.32),530s(4.31), 553(4.42),563s(4.41), 587s(4.29)	35-5151-84
$C_{108}H_{174}N_8O_4$ 21H,23H-Porphine-2,18-dipropanoic acid, 7,12-bis[[[4-(dihexadecylamino)phen-yl]amino]methyl]-3,8,13,17-tetra-methyl-	$CHCl_3$	402(5.12),501(4.11), 536(3.90),570(3.85), 623(3.56)	5-1057-84
$C_{110}H_{178}N_8O_4$ 21H,23H-Porphine-2,18-dipropanoic acid, 7,12-bis[[[4-(dihexadecylamino)phen-yl]amino]methyl]-3,8,13,17-tetra-methyl-, dimethyl ester	$CHCl_3$	402(5.26),500(4.20), 534(3.98),569(3.83), 622(3.65)	5-1057-84
$C_{111}H_{112}O_{52}$ Agrimoniin, nonacosa-O-methyl-	MeOH	218(5.39),255s(4.97), 295s(4.41)	94-2165-84

1-   -84, Acta Chem. Scand., B38 (1984)
0043   K. Aareskjold and S. Liaaen-Jensen
0109   B.A. Johnsen and K. Undheim
0117   J. Dale et al.
0141   B. Ringdahl
0293   B.A. Johnsen and K. Undheim
0309   E. Johansen et al.
0367   A. Grouiller and J. Chattopadhyaya
0607   O.H. Johansen et al.
0623   H. Priebe
0679   T. Simonen and R. Kivekas
0701   A. Norbbotten et al.
0709   C. Grøn and and C. Christophersen
0837   S. Brandange
0895   H. Priebe

2-   -84, Indian J. Chem., 23B (1984)
0049   P. Bhattacharyya et al.
0060   T.R. Juneja et al.
0101   B.S. Joshi et al.
0114   B.S. Joshi et al.
0168   A,C. Jain et al.
0177   C.B. Rao et al.
0203   S.B. Maiti et al.
0268   R.S. Mali and S.N. Yeola
0289   R.P. Iyer et al.
0316   P.B. Talukdar et al.
0331   H.R. Sonawane et al.
0339   H.R. Sonawane et al.
0363   K. Nagarajan et al.
0393   M. Gupta et al.
0395   D.K. NBanerjee et al.
0424   J.R. Merchant and R.B. Upasani
0435   A.H. Siddiqui et al.
0439   M. Ali and N.H. Khan
0445   D.R. Shridhar et al.
0478   P. Waykole and R.N. Usgaonkar
0498   B. Sundari and V. Ramamurthy
0502   K. Muthuramu and V. Ramamurthy
0509   R.R. Darii and A. Shah
0514   A.C. Ranale et al.
0543   K. Chakrabarty et al.
0603   A. Rao et al.
0611   K.R. Purushothaman et al.
0668   C.K. Ghosh and A. Bhattacharya
0677   S.N. Bose
0680   H.M. Chawla et al.
0712   V. Anjaneyulu and G.S. Rao
0736   A.N. Mandal and S. Bhattacharya
0766   D.N. Dhar and R. Raghunathan
0802   P.S.R. Anjaneyulu and A.K. Lala
0818   S.N. Yeloa and R.S. Mali
0821   P.K. Sen et al.
0863   J.R. Merchant and R.B. Upasani
0870   A.K. Sen and A.K. Mukhopadhyay
0885   A. Banerji and N.C. Goomer
0887   S.A. Khan and M. Krishnamurti
0889   D.I. Bruhmbhatt and B.H. Bhide
0904   A.K. Goswami et al.
0969   K.A. Kumar and G. Srimannaruyana
1001   A.C. Jain and P. Arya
1002   A.C. Jain et al.
1021   K.M. Biswas et al.
1028   A.C. Jain et al.
1030   A.C. Jain et al.
1036   A.C. Jain et al.

1040   P.L. Majumder and M. Joardar
1046   P.R. Kumar and K.T. Selvi
1048   C.K. Ghosh and C. Bundyopadhyay
1168   S.C. Roy et al.
1211   A.C. Jain et al.
1223   J. Banerji et al.
1289   M.S. Singh and K.N. Mehrotra

3-   -84, Anal. Chem., 56 (1984)
1935   J.H. Waite

4-   -84, J. Heterocyclic Chem., 21 (1984)
0001   S. Shtelzer et al.
0013   B. Chantegrel et al.
0017   B. Chantegrel et al.
0041   J. Becher et al.
0053   A. Katoh et al.
0103   A. Ohta et al.
0139   D. Lightner et al.
0219   A. Horvath et al.
0241   U. Slomczynska and G. Barany
0259   T. Kurihara et al.
0301   C. Deshayes et al.
0333   H. Griengl et al.
0337   H.A. Hammouda et al.
0353   A. Croisy et al.
0361   L. Mosti et al.
0389   C.K. Chu
0481   R.P. Gagnier et al.
0521   C. Parkanyi et al.
0525   E. Dominguez and E. Lete
0529   Y. Tarumi et al.
0551   M.-T. LeBrie
0569   G. Adembri et al.
0607   L. Della Ciana and A. Haim
0611   M.P. Georgiadis
0661   M.V. Jovanovic et al.
0697   W.L. Mitchell et al.
0753   A. Buschauer and W. Schunack
0759   M.S. Sinsky and R.G. Bass
0825   N.L. Agarwal et al.
0845   R. Sanduja et al.
0849   Y. Tarumi et al.
0861   P. Goya et al.
0865   R. Cassis et al.
0881   T. Biftu
0911   K.-Ch. Liu and H-Ho Chen
0917   L.A. Summers and S. Trotman
0937   R. Gatta and G. Settimj
0953   M. Muraoka et al.
0985   Y. Mesplie et al.
1013   H.M. Hassaneen et al.
1029   C. Parkanyi et al.
1041   S. Gronowitz and R. Kada
1049   A.O. Abdehamid et al.
1081   J. Gauthier and J.S. Duceppe
1097   E. Perrone et al.
1125   Z.T. Fomum et al.
1135   L.H. Klemm et al.
1149   A.H. Pedersen and K. Undheim
1157   S. Mataka et al.
1161   H. Gershon and A.T. Grefig
1189   M.C. Gomez-Gil et al.
1193   R.M. Srivastava et al.
1221   J.I. Andres et al.
1233   A.M. Fahmy et al.

1237	L.P. Schaub et al.
1245	J.-W. Chern et al.
1301	J. Kökösi et al.
1313	P.L. Desbène and N. Jehanno
1321	P.L. Desbène and N. Jehanno
1345	F. Clémence et al.
1437	G. Menozzi et al.
1441	G. Menozzi et al.
1445	M. Muraoka and T. Yamamoto
1473	N.R. El-Rayyes and F.H. Al-Hajjar
1499	V. Scartoni and T. Tognetti
1543	T.-L. Su and K.A. Watanabe
1557	D.T. Connor et al.
1561	D.T. Connor et al.
1565	N.S.K. Youssef et al.
1571	K.-C. Liu et al.
1585	B. Abarca et al.
1597	M. Weitzberg et al.
1615	L. Legrand and N. Lozac'h
1673	A.R. Katritzky et al.
1841	Y. Omote et al.
1865	B.G. Ugarkar et al.
1881	K. Faber and T. Kappe
1889	V.A. Kozinskij et al.

5-   -84, Ann. Chem. Liebigs (1984)

0031	J. Reisch et al.
0133	H. Quast and N. Jehön
0273	F. Seela and U. Liman
0306	K. Krohn et al.
0319	H. Laatsch
0340	B. Hagenbruch and S. Hünig
0381	H. Quast and N. Schön
0397	M. Cichy, V. Wray and G. Höfle
0426	J.-H. Fuhrhop et al.
0576	P. Welzel and J. Schmidt
0618	A. Aumüller and S. Hünig
0649	C. Reichardt et al.
0708	H.-D. Winkeler and F. Seela
0722	F. Seela and H. Driller
0773	J. Daub and T. Knochel
0820	I. Reichelt and H. Reissig
0867	W. Michels and E. Schlimme
0877	H. Quast and N. Schön
0920	M. Makowski et al.
1003	W. Ott et al.
1024	R. Mentlein and E. Vowinkel
1036	R. Aldag and G. Schröder
1046	W. Hain et al.
1057	J.H. Fuhrhop and T. Lehmann
1068	S. Antus et al.
1129	H. Frischkorn and E. Schinzel
1259	A. Botulinski et al.
1332	S. Backens et al.
1367	H. Laatsch
1386	J.H. Fuhrhop and T. Lehmann
1390	K. Gewald et al.
1399	G. Benz
1441	H. Böhm et al.
1454	K. Gottschall and H. Pleininger
1494	W. Plesch and M. Wiessler
1519	D. Schumann and A. Naumann
1605	G. Häfelinger and G. Ott
1616	T. Anke et al.
1696	H. Gnichtel et al.
1702	K. Gewald et al.

1711	W. Sucrow et al.
1719	F. Seela and and H. Steker
1746	G.G. Habermehl and L. Busam
1798	R. Baur et al.
1815	M. Kappel et al.
1873	C. Ibis
1883	G. Höfle et al.
1905	S. Hünig and B. Ort
1936	S. Hünig and B. Ort
1972	F. Seela and W. Bussmann

7-   -84, Ann. Chim. (Rome), 74 (1984)

0027	G. Sen and G. Alberti
0819	G. Sen et al.

9-   -84, Appl. Spectroscopy, 38 (1984)

0556	P.W. Albro et al.

10-   -84A, Chemica Scripta, 23 (1984)

0023	P. Finlander and E.B. Pedersen
0073	A. Jørgensen et al.
0098	G. Lindgren and G.H. Schmid
0170	O. Ceder and E. Andreasson

11-   -84B, Chemica Scripta, 24 (1984)

0038	G. Lindgren
0073	A. Jørgensen et al.
0208	F.E. Nielsen et al.

12-   -84, Australian J. Chem., 37 (1984)

0055	M.J. Gray et al.
0095	D.St.C. Black and L.M. Johnstone
0109	D.St.C. Black and L.M. Johnstone
0117	D.St.C. Black and L.M. Johnstone
0143	P.S. Clezy et al.
0221	O. Pancharoen et al.
0227	B.F. Bowden et al.
0341	K.J. Chapman et al.
0381	R.H. Prager et al.
0389	R.W. Parr and J.A. Reiss
0449	P. Tuntiwachwuttikul et al.
0545	B.F. Bowden et al.
0577	D.St.C. Black and L.M. Johnstone
0587	D.St.C. Black and L.M. Johnstone
0599	D.St.C. Black and L.M. Johnstone
0611	C.R. Tindale
0629	P.J. Babidge and R.A. Massy-West-rop
0635	E.L. Ghisalberti et al.
0777	D.ST.C. Black et al.
0885	J.A. Kennedy et al.
0893	D.B. Paul et al.
0985	R.W. Read et al.
1035	R.A. Russell et al.
1043	J.B. Bremner et al.
1081	P. Karuso et al.
1203	J.B. Bremner and K.N. Winzenberg
1217	R.D. Grant et al.
1231	R.D. Grant and J.T. Pinhey
1245	R.P. Kopinski et al.
1271	P. Karuso and W.C. Taylor
1283	I.J. Anthony and D. Wege
1329	H.-D. Becker et al.
1461	G.W. Breen et al.
1473	R.L.N. Harris and H.G. McFadden
1483	C. Copeland and R.V. Stick

1489	M.P. Hartshorn et al.
1511	I.K. Boddy et al.
1531	J.A. Elix et al.
1545	J.W. Blunt et al.
1659	J.B. Bremner and K.N. Winzenberg
1699	R.A. Russell et al.
1739	C. Mahidol et al.
1769	D.A. Burgess et al.
1775	R.C. Cambie et al.
2027	M.J. Gray et al.
2037	M.V. Baker et al.
2059	B. Raguse and D.D. Ridley
2073	R.C. Cambie et al.
2111	C.F. Carvalho et al.
2279	D.J. Collins et al.
2295	R.F.C. Brown et al.
2305	R.F.C. Brown et al.
2479	R.L.N. Harris and H.G. McFadden
2489	J. Rosevear and J.F.K. Wilshire
2525	P. Cacioli and J.A. Reiss
2599	P. Cacioli and J.A. Reiss

13-   -84A, Steroids, 43 (1984)
0179	E. Mincione et al.
0271	F. Felippone et al.
0283	L.S. Chagonda et al.

13-   -84B, Steroids, 44 (1984)
0095	M.J. Joyce et al.
0175	R.J. Park

18-   -84, Bull. Chem. Soc. Japan, 57 (1984)
0001	T. Nishimura and T. Motoyama
0022	T. Nogami et al.
0047	Y. Yamamoto and E. Toyota
0219	I. Matsuda et al.
0341	H. Uchimura et al.
0442	S. Yamaguchi et al.
0470	Y. Omote et al.
0591	T. Honjo
0605	I. Shibuya
0609	K. Imafuku et al.
0615	Y. Kanao and M. Oda
0621	H. Takeshita
0623	Y. Sudoh and K. Imafuku
0719	S. Kamiyama and A. Kasahara
0725	M. Minabe et al.
0791	Y. Miyagi et al.
0844	R.H. Abu-Eittah and M.M. Hammed
1128	T. Okuyama and T. Fueno
1138	H.A. El Sherief et al.
1147	Y. Jinnouchi et al.
1211	Y. Hara et al.
1567	K. Saito et al.
1653	M.Z.A. Badr et al.
1914	K. Kurosawa et al.
1961	E. Matsumura et al.
2009	K. Iio and A. Ichikawa
2013	M. Ishibashi et al.
2017	J. Kimura et al.
2121	N. Kishii et al.
2144	S. Shinkai et al.
2266	J. Kuniya et al.
2325	N. Kamigata et al.
2435	T. Izumi et al.
2484	T. Furuno et al.

2490	M. Tori et al.
2508	H. Nagamatsu et al.
2515	T. Sato et al.
2600	H. Nishida et al.
2879	S. Shinkai et al.
2885	M. Ishibashi et al.
2893	M. Ojika et al.
2905	M. Kobayashi et al.
2984	D.C. Holla and S. Seshadri
3019	H. Watanabe et al.
3343	J. Hasegawa et al.
3351	K. Abe et al.
3483	K. Saito et al.
3494	Y. Fujioka
3526	A. Izuoka et al.

19-   -84, Bull. Pol. Acad. Sci., 32 (1984)
0085	J. Wojcik et al.
0207	W. Sobotka and E. Chojecka-Koryn

20-   -84, Bull. Soc. Chim. Belges, 93 (1984)
0323	M.Z. Dimbi et al.
0459	L.I. Spiessens and M.J.O. Anteunis
1099	E. Dominguez et al.

22-   -84, Bull. soc. chim. France II (1984)
0071	V. Sanchez et al.
0077	A. Tambute and and A. Collet
0139	F. Roblot et al.
0145	A. Defoin et al.
0164	M.R. Mahmoud et al.
0187	J. Rigaudy et al.
0195	J. Rigaudy et al.
0241	S. Es-Seddiki et al.
0362	M. Franck-Neumann and M. Miesch
0435	G. Lewin and J. Poisson

23-   -84, Can. J. Chem., 62 (1984)
0016	K.K. Ogilvie
0121	S.J. Alward and A.G. Fallis
0258	R.E. Mitchell et al.
0320	J.P. Kutney et al.
0574	D.A. Holden
0586	I.W.J. Still et al.
0610	T.P. Ahern et al.
0741	C.M. Hemens et al.
0755	S. Ariel et al.
0763	H. Lee et al.
0967	T.-F. Ho et al.
1103	J. Das et al.
1238	B.R. James et al.
1407	J.P. Kutney and A.K. Singh
1414	R.A. Back
1622	K.K. Ogilvie et al.
1628	E. Buncel et al.
1709	G. Gallacher et al.
1767	P. Leblanc et al.
1780	A.K. Colter et al.
1785	R.M. Borg et al.
1922	L.M. Harwood et al.
1940	N. Jacobsen et al.
1954	J.R. Christensen and W. Reusch
1958	J.P. Guthrie et al.
1996	H.K. Al-Ehabi and G.A.W. Derwish
2006	G. Wenska and S. Poszyo

2054   A. Lecas-Nawrocka et al.
2101   G.W. Bushnell et al.
2429   S. Wolff and W.C. Agosta
2435   A. Carty et al.
2440   T. Tschamber et al.
2456   P. Umrigar et al.
2478   J. Pouliquen et al.
2506   A. Padwa and S.I. Goldstein
2546   R.A. Loutfy et al.
2583   M.D. Menachery and M.P. Cava
2592   H.E. Zimmerman and W.D. Ramsden
2612   P.E. Eaton et al.
2702   K.O. Ogilvie and H.R. Hanna
2769   C.O. Bender et al.
2813   J.P. Kutney and A.K. Singh
2818   N.N. Gerber and M.P. Lechevalier
2822   R.H. Burnell et al.
2841   S.A. Ibrahim and M.S.K. Youssef
2936   Y. Ueda et al.

24-   -84, Chem. Ber., 117 (1984)
0107   H. Meier et al.
0277   A. Krebs et al.
0310   A. Krebs et al.
0336   H. Scherübl et al.
0455   K.-L. Noble et al.
0474   K.-L. Noble et al.
0517   K. Beck et al.
0534   S. Hünig and F. Prokschy
0585   H. Wamhoff and G. Haffmanns
0633   K. Sarma and G. Schröder
0666   R.W. Saalfrank et al.
0702   H. Kessler et al.
0763   G. Märkl et al.
0809   W. Bauer et al.
0915   T.A. van der Knaap and F. Bickel-
        haupt
1077   S.A.L. Abdel-Hady et al.
1083   M.A. Badawy et al.
1424   W. Flitsch and W. Lubisch
1436   E. Langhals et al.
1455   W. Berning et al.
1476   J. Koros et al.
1523   H. Vorbruggen and K. Krolikiewicz
1542   M. Baudler et al.
1597   R. Allman et al.
1620   W. Sucrow et al.
1649   V. Batroff et al.
1726   G. Ege et al.
1994   E. Bayer et al.
1999   M. Herberhold et al.
2006   O. Schweikert et al.
2027   O. Schweikert et al.
2045   O. Schweikert et al.
2157   J. Brokatzky-Geiger and W. Eberbach
2205   E. Schaumann et al.
2261   J.W. Buchler et al.
2275   K. Fritzche and H, Langhals
2293   W. Tochtermann and M. Haase
2300   P. Lemmen and D. Lenoir
2337   G. Maier et al.
2351   G. Maier et al.
2409   W. Burgert and D. Rewicki
2422   H. Glombik and W. Tochtermann
2452   W. Winter et al.
2597   A. Buschauer et al.

2703   W. Tritschler et al.
2714   B. Gerecht et al.
2761   H. Quast and U. Nahr
2839   H.-G. Lohr et al.
2910   E. Vilsmaier and K. Joerg
3134   D. Kaufmann and A. de Meijere
3222   M. Al-Talib and J.C. Jochims
3473   P. Eilbracht et al.

25-   -84, Chem. and Ind.(London) (1984)
0301   P. Bhattacharyya and B.K. Chowd-
        hury
0352   P. Bhattacharyya and B.K. Chowd-
        hury
0632   F. Gomez et al.

27-   -84, Chimia, 38 (1984)
0157   U. Stampfli and M. Neuen-
        schwander

28-   -84A, Compt. rend., 298(1984)
0591   D. Cortes et al.
0627   J. Abaul et al.

28-   -84B, Compt. rend., 299 (1984)
0311   D. Cortes et al.

30-   -84, Doklady Akad Nauk S.S.S.R.,
        274-279 (1984)
0009   G.A. Zapevalov et al.
0094   N.S. Kozlov et al.
0192   V.N. Postnov et al.
0386   V.V. Bezuglov et al.

31-   -84, Experientia, 40 (1984)
0246   G. Cimino et al.
0339   G. Cimino et al.
0808   A.M. Rizk et al.
0930   J.M. Cassady et al.
0931   S.A. Look et al.
1248   D. Robeson et al.

32-   -84, Gazz. chim. ital., 114 (1984)
0015   P. Beltrame et al.
0041   D. Pitea et al.
0049   M. Nicoletti et al.
0055   F. Della Monache et al.
0107   C. Rossi et al.
0111   G. Tosi et al.
0169   V. Amico et al.
0189   G. Speranza et al.
0261   G. Tarzia et al.
0289   E.M. Beccalli et al.
0319   M. Alpegiani et al.
0431   G. Tarzia et al.

33-   -84, Helv. Chim. Acta, 67 (1984)
0054   H. Wagner et al.
0073   M. Karpf
0113   M. Hugentobler et al.
0120   A. Siewinski et al.
0129   A. Pascual et al.
0136   N. Bischofberger et al.
0160   I. Calis et al.
0170   W. Eschenmoser et al.
0175   M. Acemoglu et al.

0184	M. Acemoglu et al.		0355	D.E. Wellman and R. West
0201	F. Matloubi-Moghadam et al.		0440	W.E. Billups et al.
0237	K. Homberger and M. Hesse		0477	S.E. Biali and Z. Rappoport
0305	J. Wirz et al.		0715	C.H. LeeGo and W.H. Waddell
0332	P. Lupon et al.		0734	R.J. Xu et al.
0455	R.P. Borris et al.		0791	S.F. Nelsen et al.
0461	E. Marki-Fischer et al.		0821	M.J. Michalczyk et al.
0471	M. Acemoglu and C.H. Eugster		0934	D.A. Lightner et al.
0550	P.X. Iten et al.		0952	J. Diamond and G.A. Segal
0570	H. Marki-Danzig and C.H. Eugster		1065	A. Padwa et al.
0574	R. Neidlein et al.		1335	G.F. Mes et al.
0600	Z. Zhichen et al.		1351	J.P. Guthrie et al.
0690	G. Snatzke and C. Tamm		1446	L.A. Paquette et al.
0748	R. Kilger and P. Margaretha		1481	N. Baba et al.
0774	T.A. Lyle et al.		1508	R.D. Miller et al.
0804	R. Kyburz et al.		1518	L.A. Paquette et al.
0815	A. O'Sullivan et al.		1524	K.L. Rinehart, Jr. et al.
0885	M. Guex and C. Tamm		1779	R.S. Givens et al.
0916	B. Kojic-Prodic et al.		1819	R.J. Capon and D.J. Faulkner
0986	C.A. Barras et al.		1847	L.R. Chamberlain et al.
1003	A.C. Alder et al.		2645	R.W. Schoenleber et al.
1040	T. Darbre et al.		2672	H.C. Kang and V. Boekelheide
1116	P. Ruedi		2805	M. Shahbaz et al.
1154	W. Oppolzer and T. Godel		3140	R.K. Norris et al.
1175	K. Ishii et al.		3309	G.A. Eberlein and M.F. Powell
1208	D. Wuthier et al.		3366	S.P. Nelsen et al.
1298	W. Summermatter and H. Heimgartner		3370	V. Nair et al.
1348	H. Wyler et al.		3466	L. McElwee-White and D.A. Dough-
1379	H. Stadler, M. Rey and A.S. Dreid-			erty
	ing		3500	J.P. Collman et al.
1386	B. Szechner et al.		3551	A.S. Kende et al.
1402	E. Anklam et al.		3590	A.G. Schultz et al.
1493	R. Kilger et al.		3636	J.A. Leigh et al.
1503	A. Kocak and O. Bekaroglu		3699	S.W. Staley and T.D. Norden
1506	R. Huston et al.		3701	E.H. White and N. Egger
1515	X. Luo et al.		3943	B. Ward et al.
1523	A.C. Alder et al.		4211	M.H. Chang et al.
1535	M. Beyer et al.		4266	B.A. O'Brien et al.
1547	H. Hilpert and A.S. Dreiding		4336	R. Rossetti and L.E. Brus
1729	M. Hamburger et al.		4472	D. Lancon et al.
1734	M.E. Scheller and B. Frei		4511	R.A. McClelland et al.
1801	L. Walder et al.		4531	D.A. Holden et al.
1854	H. Stadler et al.		4539	J.S. Petersen et al.
1878	H.-J. Borschberg		4566	P.W. Raynolds and J.A. DeLoach
1897	M.E. Baumann et al.		4621	V. Buss et al.
1957	R. Ghaffari-Tabrizi et al.		4852	L.T. Scott et al.
2017	A. Katz and H. Rudin		4857	L.T. Scott and C.M. Adams
2037	W. Gessner et al.		5151	J.P. Collman et al.
2043	R. Buchecker et al.		5164	M.P. Suh et al.
2078	H. Schubel et al.		5217	M. Fontecave et al.
2111	S. Uesato et al.		5234	A.K. Schrock and G.B. Schuster
2192	R. Neidlein and G. Hartz		5271	R. Okazaki et al.
2226	M. Acemoglu and C.H. Eugster		5521	C.W. Carlson et al.
2270	M. Balogh et al.		5760	S. Ogawa et al.
			5879	J.M.J. Lamberts et al.

34-   -84, J. Chem. Eng. Data, 29 (1984)

0221	R.M. Srivastava and L.M. Mendes e		5997	J. Canceill et al.
	Silva		6015	K. Krageloh et al.
0225	N.R. El-Rayes et al.		6029	B.J. van Keulen and R.M. Kellogg
0237	B.N. Ghose		6084	M.V. Lakshmikantham et al.
0345	M.K. Das, P. Bose and N. Roy		6099	S. Firdous et al.
0358	A.A.H. Saeed		6101	T. Gozler et al.
0482	D.O. Ukponnwan et al.		6108	B. Halton et al.
			6155	S.J. Valenty et al.

35-   -84, J. Am. Chem. Soc., 106 (1984)

0265	M.J. Politi and J.H. Fendler		6379	Z. Kazimierczuk et al.
			6383	Z. Yoshida et al.
			6453	R. Breslow et al.

6456	R.E. Moore et al.
6457	K.M. Barkigia et al.
6726	E.B. Troughton et al.
6748	J.E. Hochlowski et al.
6768	G.R. Pettit et al.
6778	J.T. Slama et al.
7033	D.A. Hroval et al.
7082	G. Doddi and G. Ercolani
7131	S.D. Cox et al.
7175	P.N. Confalone and E.M. Huie
7195	C.F. Wilcox, Jr. and E.N. Farley
7261	M. Zandemeneghi et al.
7267	A.S. Nazran et al.
7268	Y. Sugihara et al.
7310	N.C. Yang et al.
7352	B.B. Lohray et al.
7367	M. Tomasz et al.
7540	M. Nassal
7624	J. Saltiel and D.W. Eaker
7776	R.H. Mitchell et al.
7916	G. Eberlein et al.
8024	F. Diederich and K. Dick
8301	R.F. Waldron et al.

36-   -84, J. Pharm. Sci., 73 (1984)

0411	M. Arisawa et al.
1241	O.W. Lever, Jr. et al.
1639	A.M. El-Fishawy et al.
1846	F.A.H. Rice and C.G. Chen

39-   -84B, J. Chem. Soc., Perkin II (1984)

0149	G. Hallas et al.
0231	P.B. de la Mare and P.A. Newman
0317	G. Consiglio et al.
0367	J.A. Barltrop and A.C. Jackson
0441	K. Maeda et al.
0537	R. Noto et al.
0641	T. Konakahara et al.
0781	G. Consiglio et al.
0785	V. Frenna et al.
0799	T. Keumi et al.
0833	R. Bonnett et al.
0841	A.R. Katritzky et al.
0849	A.R. Katritzky et al.
0867	A.R. Katritzky and D.E. Leahy
0879	A.R. Katritzky et al.
0891	M. Cavazza et al.
0943	A.K. Mishra and S.K. Dogra
0947	M. Swaminathan and S.K. Dogra
0965	M. Basato et al.
1093	I. Cano-Yelo and A. Deronzier
1099	J.R.L. Smith et al.
1187	J.C. Brand et al.
1227	Y. Yono and E. Ohya
1259	C.M. Evans and A.J. Kirby
1269	C.M. Evans and A.J. Kirby
1317	S.F. Tan et al.
1331	H. Laatsch et al.
1383	F. Cristiani et al.
1629	L. Szabo et al.
1693	H. Maskill et al.
1761	T.J. Cholerton et al.
1833	G.C.R. Ellis-Davies et al.
2031	T. Sakurai and H. Inoue

39-   -84C, J. Chem. Soc., Perkin I (1984)

0005	S. Bhattacharya et al.
0069	C.G. Newton et al.
0075	C.G. Newton et al.
0085	M. Yokoyama et al.
0101	N.J. Morrison
0135	S. Spyroudis and A. Varroglis
0165	R. McCague et al.
0175	R. McCague et al.
0183	Y. Nagao et al.
0203	A. Matsuo et al.
0215	A. Matsuo et al.
0229	L.J.S. Knutsen et al.
0239	T. Nishio and Y. Omote
0249	R. Purvis et al.
0261	A.K. Chakraborti et al.
0287	J.V. Greenhill et al.
0391	T. Nishio et al.
0397	J.R. Bull et al.
0487	I.D. Dicker et al.
0497	W. Claydon et al.
0503	M. Yamauchi et al.
0529	K. Yoshida et al.
0535	R.F.W. Jackson and R.A. Raphael
0567	M.A. O'Leary et al.
0651	J.E. Leet et al.
0687	M. Phialas et al.
0697	A.E. de Jesus et al.
0733	P.M. Collins et al.
0753	J.M. Mellor and R.N. Pathirina
0769	N. Hoshi et al.
0825	R.B. Herbert et al.
0859	K.R. Lawson et al.
0865	K.R. Lawson et al.
0879	R. Mornet et al.
0909	R. McCague et al.
0915	C.J. Moody et al.
0937	A.N. Phadnis et al.
0941	A.R. Katritzky et al.
0953	C.D. Ginger et al.
0959	R. Wrigglesworth et al.
1005	D.H.R. Barton et al.
1053	F.J. Leeper and J. Staunton
1069	M.F. Ansell et al.
1079	Z.T. Fomum et al.
1127	M.M.L. Crilley and R.J. Stoodley
1143	M. Lempert-Sreter et al.
1159	V. Van de Velde et al.
1213	J.V. Greenhill et al.
1219	J.F. Grove
1237	B. Danieli et al.
1263	C.B. Reese and P.A. Skow
1273	G.R. Lenz and F.J. Keszyk
1279	D.J. Mincher and G. Shaw
1297	M. Matsuoka et al.
1311	C.P. Gorst Allman et al.
1323	R. Ramage and I.A. Southwell
1395	M.J. Finn et al.
1413	M.R. Cann et al.
1441	M.S. Kemp and R.S. Burden
1445	R.S. Burden et al.
1465	J. Tsunetsugu et al.
1471	A.S. Jones et al.
1483	L.R. Milgrom

1507	B. Simoneau and P. Brassard
1515	G. Bhattacharjee et al.
1531	R. Ramage et al.
1539	R. Ramage et al.
1547	R. Ramage et al.
1555	R. Ramage and P.P. McCleery
1573	L.A.P. Anderson et al.
1577	F.R. Ahmed and T.P. Toube
1599	G. Brooks et al.
1605	C.F. Carvalho and M.V. Sargent
1613	C.F. Carvalho and M.V. Sargent
1631	S. Rajappa et al.
1635	S. Ohta et al.
1655	S. Hagishita et al.
1701	H.G. Theuns et al.
1755	F.M. Dean et al.
1769	H. Ishii et al.
1785	J.M. Brown et al.
1795	H. Hermecz et al.
1869	T. Katada et al.
1891	P. Molina et al.
1913	C.F. Carvalho and M.V. Sargent
1919	M.V. Sargent and E. Stanojevic
1933	M. Casey et al.
1957	J.L. Carey et al.
2013	K. Seno et al.
2023	S. Ogawa et al.
2049	P.J. Brown et al.
2069	H. Arzeno et al.
2077	D.H.R. Barton et al.
2089	M. Hetmanski et al.
2097	M. Yogo et al.
2117	P.G. Sammes et al.
2133	M. Devys et al.
2147	A.J. Davies et al.
2165	M. Tashiro and Y. Yamato
2173	J.R. Hanson and L. Yang-zhi
2177	S. Iwashima et al.
2221	V. Skaric et al.
2297	M.G. Hicks et al.
2305	J. Setsune et al.
2327	W.R. Bowman et al.
2367	J.G. Buchanan et al.
2375	T.R. Kasturi et al.
2421	S.N. Bose et al.
2447	F.J. Cuadrado et al.
2491	S. Chimichi et al.
2511	I. Kouno et al.
2529	M.B. Stringer et al.
2541	R.A. Jones et al.
2545	S. Araki and Y. Butsugan
2549	J. Lane et al.
2553	D. Farquhar et al.
2573	C.F. Carvalho and M.V. Sargent
2587	D.I. Patel and R.K. Smalley
2599	M.W. Anderson et al.
2655	F. Bigi et al.
2671	A. Padwa et al.
2677	S.R. Landor et al.
2767	B.C.R. Bezuidenhoudt et al.
2785	P.H. Crackett et al.
2815	P. Anastasis et al.
2859	M.K. Baynham and J.R. Hanson
2863	A. Gazit and Z. Rappoport
2871	H.G. Heller and G.A. Jenkins
2877	H.G. Heller and G.A. Jenkins

2881	M. Srebnik et al.
2903	C.J. Moody and J.G. Ward
2933	F. Stansfield
2941	J.R. Hanson et al.
2945	A.V. Vyas and N.B. Mulchandani

40- -84, Nippon Kagaku Kaishi (1984)

0001	T. Sakurai et al.
0022	M. Kuzuya et al.
0060	S. Yamada et al.
0090	K. Maruyama et al.
0145	K. Fukunaga and M. Kimura
0158	T. Kumagai et al.
0522	H. Nishi et al.
1158	M. Tamano et al.
1164	M. Tamano et al.

41- -84, J. chim. phys., 81(1984)

0021	C. Parkanyi et al.
0243	C. Lovain et al.
0629	J. Sotiropoulos and N. El Batouti

42- -84, J. Indian Chem. Soc., 61 (1984)

0142	P.L. Majumder and A. Kuendu
0146	J.S. Shukla and R. Misra
0430	A.I. Hashem et al.
0611	B.R. Sarkar et al.
0650	I. Bhattacharyya and A. Chakraborty
0728	J. Mathew and A.V. Subba Rao
0975	D.P. Chakraborty et al.
0985	S.K. Khanna et al.
1010	S.K. Talapatra et al.

44- -84, J. Org. Chem., 49 (1984)

0001	S.F. Nelsen and M.R. Willi
0026	C. Pac et al.
0034	J.P. McCormick et al.
0056	S.H. Rosenberg and H. Rapoport
0062	M. Nagarajan and H. Shechter
0080	Z. Rappoport et al.
0203	J.M. Muchowski and D.R. Solas
0220	J.W. Ulrich et al.
0228	F.-T. Chiu et al.
0241	F.J. Schmitz et al.
0276	A. Padwa et al.
0282	A. Padwa et al.
0289	V. Bakthavachalam et al.
0318	B.L. Chenard et al.
0363	R.C. Moschel et al.
0381	S. Amin et al.
0393	S. Singh and V. Ramamurthy
0396	M. Sakamoto et al.
0399	A. Padwa and L.A. Cohen
0406	P.J. Harris et al.
0413	E. Buncel et al.
0448	A.R. Katritzky and D. Rubio
0482	C.F. Bernasconi et al.
0491	J. Winterle et al.
0495	S. Nishida et al.
0528	E.M. Acton and K.J. Ryan
0536	V. Ramesh and V. Ramamurthy
0547	B. Danieli et al.
0559	C.M. Ireland et al.
0579	S.F. Nelsen et al.
0581	S. Smith, Jr. et al.

0587	T. Sheradsky and R. Moshenberg
0635	A.H. Hunt et al.
0649	A.H. Lewin et al.
0714	A. Couture et al.
0726	M.V. Lakshmikantham et al.
0742	M. Watanabe et al.
0770	M. Ballester et al.
0806	O. Lerman et al.
0813	T. Fukunaga and R.W. Begland
0827	T. Nishio
0856	A. Padwa et al.
0908	Y. Okamoto et al.
0942	J.M. Cassady et al.
0948	T. Takahashi et al.
0965	D.E. Wellman et al.
0997	D.G. Patil and M.R. Chedekel
1027	L.T.W. Lin and H. Hart
1030	T.K. Dobbs et al.
1033	S.F. Sellers et al.
1040	J.C. Martin and J.M. Muchowski
1091	S. Amin et al.
1130	P.L. Soutjwick et al.
1146	Y. Sugihara et al.
1177	Y. Miyahara et al.
1224	B.L. Chenard
1261	J.A. Moore et al.
1338	R. Wedinger et al.
1353	A. Padwa et al.
1412	C.W. Doecke et al.
1417	S.A. Look et al.
1439	G. Blasko et al.
1453	J. Alexander
1537	K. Nakayama et al.
1613	S.L. Goldstein and E. McNelis
1627	J. Tirado Rives et al.
1728	P.E. Eaton et al.
1772	W.R. Roush and T.A. Blizzard
1789	J. Fayos et al.
1793	V.V. Kane et al.
1832	E. Wenkert et al.
1853	V. Guay and P. Brassard
1898	R.K. Duke and R.W. Rickards
1937	N.F. Woolsey et al.
1941	R.W. Curley, Jr. and H.F. DeLuca
1944	R.W. Curley, Jr. and H.F. DeLuca
2039	J. Gharbi-Benarous et al.
2050	M.-C. Carre et al.
2057	J. Gramza et al.
2069	B.F. Plummer et al.
2121	J.P. Ferris and H. Yamagawa
2148	J.W. Morzycki et al.
2158	T.M. Stevenson and N.J. Leonard
2208	R.P. Thummel et al.
2212	C.S. Rooney et al.
2258	P.G. Gassman and C.K. Harrington
2273	G. Maas and W. Lorenz
2297	J.W. Wilt and C. George
2314	P.S. Nelson et al.
2355	A. Feigenbaum et al.
2368	S.J. Cristol et al.
2506	R.J. Capon and D.J. Faulkner
2546	A. Hassner et al.
2634	F. Sweet and B.R. Samant
2642	C.R. Dorn et al.
2708	S. Chao et al.
2714	R.F. Ruffauf et al.
2724	Y. Miura and M. Kinoshita
2738	V.V. Kaminski et al.
2803	P.F. King and R.F. O'Malley
2812	C.K. Ghosh et al.
2816	M.J. Wyvratt et al.
2884	M. Ballester et al.
2917	C.-C. Cheng et al.
2925	R.C. Mease and J.A. Hirsch
2954	A.J. Bridges and J.W. Fischer
2981	R.B. Bates et al.
2994	G. Delgado et al.
3174	A. Padwa et al.
3204	B. Sullivan and D.J. Faulkner
3275	G. Hugel and J. Levy
3314	A. Padwa et al.
3327	A.G.H. Wee et al.
3367	R.A. Olofson et al.
3398	J.H. Cardellina, II et al.
3424	A.T. Nakayama et al.
3436	W.E. Billups et al.
3489	P.G. Williard and S.E. de Laszlo
3534	S.G. Wood et al.
3579	S.B. Harmalker and D.T. Sawyer
3621	D.M. Jerina et al.
3672	K.T. Potts and W.R. Kuehnling
3733	E. Wenkert et al.
3762	M. Ganguli et al.
3766	K. Krohn and E. Broser
3791	J.G. Bauman et al.
3848	G. Mehta et al.
3869	S.W. Ayer et al.
3988	A.C. Jackson et al.
4021	V. Nair et al.
4029	M. Pomerantz et al.
4065	A. Annulli et al.
4128	W.D. Rohrbach et al.
4165	B.A.R.C. Murty et al.
4176	F. Terrier et al.
4192	J.R. Bodwell et al.
4209	T. Hogberg and B. Ulff
4258	G.R. Pettit et al.
4266	N. Haroda et al.
4332	W.R. Roush and T.A. Blizzard
4340	V. Nair and D.A. Young
4344	A. Padwa and G.D. Kennedy
4352	R.F. Childs et al.
4399	A.J. Castellino and H. Rapoport
4419	B. Chantegrel et al.
4429	A.G. Schultz et al.
4446	C.C. Duke et al.
4602	K.M. Smith et al.
4644	I. Kubo et al.
4647	B.B. Lohray et al.
4716	L.J. Valdes, III et al.
4733	J.A. Walker et al.
4736	T.J. Holmes, Jr. et al.
4752	R.C. Hartnedy and D.C. Dittmer
4755	T. Kitamura et al.
4761	L.I. Kruse and M.D. Meyer
4769	A.G. Anderson, Jr. and R.P. Ko
4773	T.C. Britton et al.
4860	E.H. White et al.
4872	E.H. White et al.
4923	C.V. Kumar et al.
4978	M.B. Sponsler and D.A. Dougherty
5000	R. Dolmazon and S. Gelin

5002	B.S. Ong and B. Keoshkerian
5035	R.B. Gammill et al.
5084	J. Nakayama et al.
5109	D.L. Coffen et al.
5116	M. Tracy and E.M. Acton
5124	R.A. Abramovitch et al.
5154	J.D. McChesney et al.
5157	R.M. Rosser and D.J. Faulkner
5160	R.P. Walker et al.
5164	J. Rebek, Jr. et al.
5202	A.G. Schultz and Y.S. Kulkarni
5243	R.I. Duclos, Jr. et al.
5265	M.I. Dawson et al.
5271	M. Ohtani et al.

46- -84, J. Phys. Chem., 88 (1984)

1157	J.R. Majer and Z.Y. Al-Saigh
1160	J. Waluk et al.
2297	P.V. Kamat and M.A. Fox
2714	Y. Maeda et al.
5544	G. Woessner et al.
6041	F. Zhao and M.J. Rosen
6185	A. Bachackashvilli et al.

47- -84, J. Polymer Sic., Polymer Chem. Ed., 22 (1984)

0069	S.P. Pappas et al.
0739	M. Tsunooka et al.
0813	N.M. Brahme and W.T. Smith, Jr.
2169	J.P.C. Bootsma et al.
2455	S.-B. Fang et al.
2789	T. Oishi and M. Fujimoto
2841	M.A. Aponte and G.B. Butler
3001	B. Roland et al.
3715	M.A. Aponte and G.B. Butler
3943	S.-B. Fang et al.

48- -84, J. prakt. Chem., 326 (1984)

0063	F. Walkow and J. Epperlein
0151	D. Simov and T. Deligeorgiev
0177	J. Fuhrmann et al.
0233	H.-D. Ilge et al.
0287	G.W. Fischer and M. Herrmann
0311	J. Liebscher and A. Rumler
0353	W. Freiberg et al.
0367	K. Kirschke et al.
0385	H. Heberer and H. Matschiner
0401	M. Augustin and M. Köhler
0457	W. Weissflog et al.
0467	G. Merkel et al.
0479	E. Fänghanel et al.
0579	M. Schulz et al.
0594	M. Augustin et al.
0647	G.W. Fischer et al.
0657	G.W. Fischer et al.
0695	A.I.M. Koraiem
0757	J. Wagner et al.
0799	M.F. Ismail et al.
0811	A.I.M. Koraiem
0829	H. Wilde et al.
0917	H. Schäfer et al.
0941	H. Kasch and K. Ponsold
0985	B. Schwarz et al.
1016	S. Schwarz et al.
1021	D.R. Gupta and R.K. Arora
1034	K. Behrmann et al.

49- -84, Monatsh. Chem., 115 (1984)

0101	H. Falk and U. Zrunek
0121	H. Falk et al.
0179	B. Pilarski et al.
0187	J. Moskal et al.
0197	E. Akgun and U. Pindur
0333	F. Kurzer et al.
0357	A. Daroca et al.
0477	O. Hofer and H. Greger
0793	F. Kurzer and J.N. Patel
0809	F. Kurzer and J.N. Patel
0837	J. Edinger et al.
0853	S. Li et al.
1071	H. Falk and U. Zrunek
1081	J. Edinger et al.
1207	O. Hofer et al.
1421	W. Fabian
1443	H. Falk et al.

51- -84, Naturwissenschaften, 71 (1984)

0425	A. Guerriero et al.

54- -84, Rec. trav. chim., 103 (1984)

0023	V.H.M. Elferink and H.J.T. Bos
0152	A.C. Brouwer and H.J.T. Bos

56- -84, Polish J. Chem., 58 (1984)

0157	Z.S. Arnold
0297	B. Osipowicz et al.
0411	W.E. Hahn and J. Osinski
0447	M.J. Korohoda and A.B. Bojarska
0585	K. Kiec-Kononowitz et al.
0705	R. Wydra and Z. Paryzek
0711	J.R. Jaszczynski et al.
0903	D. Maciejewska and L. Skulski
0917	W.E. Hahn and J. Lesiak
0935	I. Zadrosna and H. Ilcewick
1071	S. Mejer et al.
1077	J. Suwinski and K. Walczak

59- -84, Spectrochim. Acta, 40A (1984)

0075	J. Clark and G. Hitiris
0159	K. Jerwin and F. Wasgestian
0681	S.L. Srivastava et al.

60- -84, J. Chem. Sco., Faraday I (1984)

0341	H. Shizuka et al.
0383	H. Shizuka et al.
2323	P.E. Watkins and E. Whittle

61- -84, Ber. Bunsen Gesell. Phys. Chem., 88 (1984)

0042	B. Jahn and H. Dreeskamp
0759	O.S. Wolfbeis et al.

62- -84A, Z. phys. Chem. (Leipzig), 265 (1984)

0081	J. Wagner et al.
0154	M.S.A. Abdel-Mottaleb and A.M.K. Sherief
0302	M.R. Mahmoud et al.
0947	B. Vieth and W. Jugett

62- 84C, Z. phys. Chem. (Frankfurt), 140 (1984)

0167	J. Baro et al.

63-   -84, Z. physiol. Chem., 365 (1984)
1351   R. Herzog et al.

64-   -84B, Z. Naturforsch., 39B (1984)
0074   M.I. Attalla et al.
0095   K.H. Ongania and R. Pawlowski
0222   J.W. Buchler and C. Dreher
0238   O.S. Wolfbeis et al.
0244   H. Laatsch
0252   A. Bacher
0474   J.M.J. Lamberts and D.C. Neckers
0485   A. Gieren et al.
0524   Atta-ur-Rahman et al.
0683   H. Gotta et al.
0985   H. Hofmann and G. Low
1211   K. Gholivand et al.
1393   H. Volz and G. Herb
1399   B. Singer and G. Maas
1586   J. Daub et al.
1607   U. Hildebrand et al.
1781   H.A. Brune et al.
1790   H. Vogler and F.A. Neugebauer
1825   A. Morel et al.

64-   -84C, Z. Naturforsch., 39C (1984)
0010   W. Steglich et al.
0695   H. Anke et al.
0872   M. Kaouadji et al.

65-   -84, Zhur. Obshchei Khim., 54 (1984)
          (English translation pagination)
0134   N.I. Rtishchev et al.
0142   B.G. Rarog et al.
0148   V.D. Orlov et al.
0161   N.I. Shergina et al.
0289   P.P. Onys'ko et al.
0389   V.M. Ovrutskii et al.
0489   A.M. Panov et al.
0602   V.M. Tsentovskii et al.
0685   P.I. Yagodinets et al.
0836   E.I. Khabarova et al.
1067   V.I. Boev and A.V. Dombrovskii
1400   N.A. Bychkova et al.
1430   A.P. Apsitis and Y.T. Rotberg
1439   V.I. Boev and A.V. Dombrovskii
1494   L.G. Tomilova et al.
1597   V.S. Gamayurova et al.
1610   V.A. Makarov et al.
1657   D.B. Akopova et al.
1987   A.V. Moskvin et al.
2114   V.V. Alenin and V.D. Domkin
2299   L.A. Baltina et al.
2317   Y.V. Kolodyazhnyi et al.
2350   A.F. Mironov et al.
2359   P.S. Khokhlov et al.
2363   V.A. Timoshchuk
2414   P.S. Khokhlov et al.
2495   P.S. Khokhlov et al.

69-   -84, Biochemistry, 23 (1984)
0047   J.P. Albertini and A. Garnier-
          Suillerot
0589   G.J. Pielak et al.
1742   J.M. Andreu et al.
2779   V.E. Anderson et al.
3765   C.E. Seidman et al.

4589   S. Kobayashi et al.
5686   B.L. Gaffney et al.
5833   D.G. Brenner and J.R. Knowles
5839   D.G. Brenner and J.R. Knowles

70-   -84, Izvest. Akad. Nauk S.S.S.R.,
          33 (1984)
0134   A.V. Semenovskii et al.
0195   S.Z. Taits and V.N. Bulgakova
0310   N.N. Chipanina et al.
0414   V.A. Petukhov et al.
0442   G.A. Tolstikov et al.
0585   E.A. Chernyshev et al.
0630   E.I. Zhilyaeva et al.
0663   A.V. Kamernitskii et al.
0726   M.G. Voronkov et al.
0738   Y.N. Belokon' et al.
0842   V.A. Petukhov and V.V. Il'in
0936   B.Y. Ladygin
0948   A.I. Shienok et al.
0961   V.S. Bogdanov et al.
1044   N.A. Bumagin et al.
1268   N.I. Petrenko et al.
1285   A.S. Mikhailov et al.
1526   A.I. D'yachenko et al.
1603   P.F. Levin et al.
1668   V.A. Golubev and A.N. Rozenberg
1743   K.Y. Burshtein et al.
1901   L.A. Myshkina et al.
2005   Y.N. Malkin et al.
2056   G.I. Sarapulova et al.
2160   R.I. Khusnutdinov et al.
2328   G.V. Kryshtal' et al.
2345   N.V. Ivashkina et al.
2349   V.A. Reznikov and L.B. Volodarskii
2377   A.V. Kamernitskii et al.
2428   V.A. Nikanorov et al.
2437   A.I. D'yachenko et al.
2482   L.G. Kuz'mina et al.
2519   A.M. Moiseenkov et al.
2528   V.P. Litvinov et al.
2580   A.V. Kamernitskii et al.

73-   -84, Coll. Czech. Chem. Comm., 49
          (1984)
0065   A. Krutosikova et al.
0086   Z. Polivka et al.
0141   E. Klinotova et al.
0301   L. Kohout and J. Zajicek
0421   S. Marchalin et al.
0533   J. Stetinova et al.
0597   J. Kuthan et al.
0603   J. Jilek et al.
0621   Z. Polivka et al.
0704   J. Slavik and L. Slavikova
0840   M. Ledvina and J. Farkas
0984   R. Kada et al.
0992   J. Urban et al.
1009   I. Cervina and M. Protiva
1021   K. Sindelar et al.
1193   L. Fisera et al.
1311   M. Budesinsky et al.
1318   J. Slavik et al.
1360   J. Cacho et al.
1395   S. Marchalin and J. Kuthan
1412   J. Smidrkal

1421	P. Hrenciar and I. Culak
1529	A. Korcnova et al.
1536	S. Dvorackova et al.
1600	T. Gracza et al.
1617	R. Mickova
1622	J. Kopecki et al.
1699	L. Janda et al.
1713	P. Zalupsky and A. Martvon
1722	V. Kmonicek et al.
1764	V. Zvak et al.
1788	J. Prousek
1800	V. Bartl et al.
1816	V. Bartl et al.
1921	M. Lebl et al.
1949	L. Musil et al.
1990	L. Fisera and D. Pavlovic
2024	I. Valterova et al.
2141	R. Kada et al.
2148	A. Holy
2285	G.D. Krapivin et al.
2295	V. Bartl et al.
2309	S. Marchelin and J. Kuthan
2485	M. Kriz et al.
2496	R. Kada et al.
2502	V. Zvak et al.
2531	K. Sindelar et al.
2541	H. Pischel et al.
2635	I. Lukac et al.
2638	J. Jilek et al.
2689	H. Hrebabecky and J. Beranek
2713	V. Schwarz et al.
2751	J. Barek and I. Danhel
2776	J. Plesek et al.
2828	A. Cerny et al.
2912	G. Cik et al.

74-   -84A, Mikrochim. Acta I (1984)
| 0053 | W. Buchberger et al. |

74-   -84B, Mikrochim. Acta II (1984)
| 0407 | J.L. Gomez Ariza et al. |

74-   -84C, Mikrochim. Acta III (1984)
| 0103 | K. Ueda et al. |

77-   -84, J. Chem. Soc., Chem. Comm. (1984)
0055	S.T.A.K. Daley et al.
0067	D.T. Coxon et al.
0168	G.W. Gribble and M.G. Saunier
0200	M.M. Campbell et al.
0244	L. Crombie et al.
0266	B.S. Ong
0287	D. Barr et al.
0334	H. Suginome et al.
0405	J.R. Anderson et al.
0422	R.E. Bambury et al.
0423	H. Ogawa et al.
0430	Aziz-ur-Rahman et al.
0473	J.J. Kim Wright et al.
0509	J.B. Hansen et al.
0583	C. Leumann and A. Eschenmoser
0689	M. Yoshifuji et al.
0714	E. Wenkert and C.A. Broka
0733	L.W. Jenneskens et al.
0764	C.J. Rabie et al.
0821	D. Leaver and D. Skinner

0837	G.D. Meakins et al.
0917	J.R. Anderson et al.
0920	M.J. Crossley and L.G. King
0930	C. Iwata et al.
0999	A.K. Mahapatra et al.
1008	R. Ramage and A.M. MacLeod
1217	D.G. Gilheany et al.
1234	F. Toda et al.
1289	C.M. Pant et al.
1293	T. Nishio and Y. Omote
1319	C.K. Ghosh et al.
1365	A. Fassler et al.
1371	S. Firdous et al.
1450	J.P. Schaumberg et al.
1491	O. Hernandez and M.B. Gopinathan
1513	T. Yoshioka et al.
1525	M.J. Michalczyk et al.
1565	M. Yamauchi et al.
1691	M. Sako et al.
1693	G.R. Pettit et al.
1696	V. Buss et al.
1705	M. Hatanaka et al.

78-   -84, Tetrahedron, 40 (1984)
0003	B.L. Gaffney et al.
0059	F. Himmelsbach et al.
0079	H.P. Albrecht et al.
0119	J.G. Buchanan et al.
0125	F. Hansske et al.
0153	S. Honda et al.
0215	U. Berger et al.
0221	A. Previero et al.
0311	R. Beugelmans et al.
0369	Z. Bende et al.
0427	R. Cosstick et al.
0473	R.S.H. Lin et al.
0579	H. Ogoshi et al.
0613	J. Tortajada et al.
0665	A. Merz et al.
0695	J.P. Alazard et al.
0757	D.F. Maity and D. Mukherjee
0765	T.H.A. van der Knaap et al.
0793	E. Sonveaux
0799	C. Tringali et al.
0873	A. San Feliciano et al.
0879	P.H. Boyle and R.J. Lockhart
0919	H. Berner et al.
0931	F. Lucchesini et al.
1145	T. Gozler et al.
1391	W. Oppolzer et al.
1545	D. Babin and M. Julia
1585	D.N. Dhar and R. Ragunathan
1721	V. Amico et al.
1749	K.M. Smith and R.K. Pandey
1755	K. Mutai et al.
1835	A. Couture et al.
1907	J.E. Baldwin et al.
1919	D.E. Ames and A. Opalko
1971	G. Blasko et al.
1975	H. Guinaudeau et al.
2035	E. Armani et al.
2047	B. Yde et al.
2069	W.A. Ayer et al.
2133	H. Wagner et al.
2141	S.R. Landor et al.
2177	L. Toscano et al.

2189	J. de P. Teresa et al.	
2451	D.J. Chadwick et al.	
2461	M.M. Campbell et al.	
2513	S. Fazal Hussain et al.	
2519	Y. Shimojima et al.	
2541	S.K. Okwute et al.	
2651	A. Defoin et al.	
2677	S. Panichanun and I.R.C. Bick	
2685	S. Panichanun and I.R.C. Bick	
2773	R. Rossi et al.	
2795	P.D. Magnus and N.L. Sear	
2829	D.C. Lankin et al.	
2879	R. Ghirlando et al.	
2913	V.J. Paul and W. Fenical	
2951	A.M. Lamazouere and J. Sotiropoulos	
2959	J.K. Ray et al.	
2961	P. Bhan et al.	
3053	V.J. Paul and W. Fenical	
3117	L.W. Jenneskens et al.	
3129	R.D.H. Murray and Z.D. Jorge	
3339	S.J. Martinez et al.	
3455	J. Savard and P. Brassard	
3499	M. Augustin et al.	
3539	P.-L. Desbene et al.	
3559	P.-L. Desbene and J.-C. Cherton	
3567	P.-L. Desbene and J.-C. Cherton	
3617	H.A. Priestap	
3677	K. Krohn et al.	
3749	T.J. Eckersley et al.	
3823	S.J. Coval et al.	
3839	M. Srebnik and R. Mechoulam	
3957	E. Valencia et al.	
3979	S. Inel et al.	
3997	K.B. Sloan and K.G. Siver	
4003	G.R. Lenz	
4033	M.I. Chicarelli et al.	
4081	F. Bigi et al.	
4127	J. Bertrand et al.	
4245	A.S.R. Anjaneyulu et al.	
4253	D.A. Lightner et al.	
4351	K.M. Biswas et al.	
4359	M.A. Hai et al.	
4401	L.W. Jenneskens et al.	
4473	S. Ramakanth et al.	
4513	J.E. Baldwin et al.	
4549	J. Tamariz and P. Vogel	
4561	A. Echavarren et al.	
4579	A.E. Ashcroft et al.	
4609	K. Krohn and W. Priyono	
4633	J.S. Swenton et al.	
4649	M.J. Broadhurst et al.	
4657	R.C. Gupta et al.	
4677	S. Pencho et al.	
4685	J.M. McNanamara and Y. Kishi	
4701	T.-T. Li et al.	
4823	W.D. Rohrbach et al.	
4837	R. Alan Jones et al.	
4843	P. Flecker et al.	
4853	P. Flecker and E. Winterfeldt	
4897	C. Nallaiah	
4963	A. Haas and K.W. Kempf	
5039	V. Guay and P. Brassard	
5047	A. Banerji et al.	
5197	M. Miski et al.	
5229	R.D.H. Murray and Z.D. Jorge	
5249	N.P. Hacker and F.W. McOmie	

80-   -84, Revue Roumaine Chim., 29 (1984)
0193	D. Manoiu et al.
0345	H.A. Dessoki and A. Essawy
0447	S. Dumiterscu et al.
0755	R. Schwartz et al.
0817	V.T. Ciorba et al.
0857	D. Manoiu et al.

83-   -84, Arch. Pharm., 317 (1984)
0002	J. Dusemund and E. Kröger
0027	J. Dusemund and B. Gruschow
0042	R.W. Grauert
0092	J. Knabe and B. Itanke
0143	G. Dannhardt and R. Obergrusber-ger
0223	N. Takao et al.
0246	K.-A. Kovar and M. Abdel-Hunid
0329	H.J. Kallmayer and K. Sefang
0377	J. Dusemund and T. Schurreit
0379	G. Seitz and S. Dietrich
0381	J. Dusemund and E. Kröger
0443	K. Görlitzer and A. Dehne
0448	K. Görlitzer and P. Dehne
0472	K.C. Liu and B.-J. Shih
0488	U. Berger et al.
0526	K. Görlitzer and A. Dehne
0531	K. Görlitzer and A. Dehne
0547	A.A. El-Barbary and A. Hammouda
0624	P. Pachaly and K.S. Sin
0639	J. Reisch et al.
0680	G. Rücker et al.
0685	W. Löwe and G. Eggersmann
0724	K.A. Kovar and S. Keilwagen
0743	H.J. Kallmayer and K. Seyfang
0824	J. Reisch et al.
0855	H.J. Kallmayer and K. Seyfang
0867	K. Keppeler et al.
0906	E. Golpashin et al.
0970	F. Eiden and J. Schünemann
0984	R. Neidlein et al.
1029	K. Görlitzer and D. Buss

86-   -84, Talanta, 31 (1984)
0755	M. Grote et al.
1013	P. Verma and V.K. Gupta
1041	R.-Q. Yu et al.
1075	M. Gallego et al.
1121	R.-Q. Yu et al.
1125	D. Fompeydie and P. Levillain

87-   -84, J. Med. Chem., 27 (1984)
0052	G.A. Conway et al.
0160	D.T. Mao et al.
0175	G.E. Wright and L.W. Dudycz
0223	D.R. Buckle et al.
0232	J.B. Hynes et al.
0261	D.R. Friend and G.W. Chang
0266	P.C. Srivastava et al.
0270	D.C. Baker et al.
0274	I. Antonini et al.
0279	D.E. Bergstrom et al.
0306	T. Hogberg et al.
0376	J.I. DeGraw et al.
0410	R.A. Sharma et al.
0429	C.G. Wong and R.B. Meyer, Jr.
0440	R.N. Hunston et al.

0503	P.J. Machin et al.	1521	M. Arys et al.
0528	D.T. Connor et al.	1585	T. Kawashima et al.
0534	J.A. Secrist, III et al.	1613	J.W. Scheeren and J. Lange
0585	P. Krogsgaard-Larsen et al.	1665	C.L. Liotta et al.
0600	A. Rosowsky et al.	1693	H. Butenschon and A. de Meijere
0628	T.C. McKenzie et al.	1733	H. Satoh and T. Tsuji
0670	Y.F. Shealy et al.	1737	H. Satoh and T. Tsuji
0680	J.A. Montgomery et al.	1785	A. Beck et al.
0749	F.S. Sariaslani et al.	1809	M. Yoshifuji et al.
0792	M.-C. Carre et al.	1829	J.M. Boente et al.
0836	T. Fujioka et al.	1929	K. Mukai et al.
0870	D.E.V. Wilman et al.	1937	H. Kawai et al.
0875	J. Bagli et al.	1941	H. Kawai et al.
0888	A. Rosowsky et al.	1975	A.H. Beiter and W. Pfleiderer
0928	R.G. Christiansen et al.	1991	S. Akabori et al.
0981	F. Seela et al.	2241	Y. Ooi et al.
0990	K. Schaffner-Sabba et al.	2341	W. Steglich et al.
1026	B.G. Ugarkar et al.	2359	J. Polonsky et al.
1057	C.D. Jones et al.	2375	T. Sugawara et al.
1119	H.B. Cottam et al.	2431	G. Chang and M.P. Mertes
1161	C. Auclair et al.	2459	G. Fischer et al.
1225	C.U. Kim et al.	2471	Y. Takahashi et al.
1295	P. Strazzolini et al.	2531	D.O. Spry et al.
1300	A. Sugimoto et al.	2537	G. Bringmann and J.R. Jansen
1370	H. Edery et al.	2573	K. Saito et al.
1389	G.R. Revankar et al.	2731	L. Fisera et al.
1416	Y.F. Shealy et al.	2767	M.H. Brillanceau et al.
1486	M.J. Robins et al.	2783	A.L. Skaltsounis et al.
1516	M.I. Dawson et al.	2809	C. Nishino et al.
1536	M.-I. Lim et al.	2909	M. Franck-Neumann and M. Miesch
1559	W.K. Anderson and A.N. Jones	2923	G.I. Fray et al.
1613	R.E. Bambury et al.	2925	G. Cimino et al.
1639	R.D. Youssefyeh et al.	2953	R.B. Gammill and S.A. Nash
1677	K.A. Stevenson et al.	2989	H. Nakamura et al.
1710	A. Srinivasan et al.	3007	S. Bhattacharya et al.
1733	R.R. Chawla et al.	3025	K. Beautement and J.M. Clough
		3087	C. Beland and M. Guyot
88- -84, Tetrahedron Letters (1984)		3133	S. Greenfield et al.
0023	P.E. Eaton and W.H. Bunnelle	3163	E. Valencia et al.
0057	F. Bronberger and R. Huisgen	3195	A.F. Maggio et al.
0073	K. Takahashi et al.	3233	E. Okuyama et al.
0077	K. Takahashi et al.	3243	I. Saito et al.
0117	Z. Rappoport and C. Rav-Acha	3471	M. Olomucki et al.
0135	A.J. Paine	3483	S.H. Goh, C. Wei and R.R. Mali
0145	N. Le-Van et al.	3547	V. Nair and T.S. Jahnke
0247	V. Nair and G.A. Turner	3575	B. Fugmann et al.
0269	A.E. Mathew and C.C. Cheng	3603	M. Sato et al.
0325	T. Ohta et al.	3611	Y. Fugise et al.
0345	Z. Yoshida et al.	3719	H. Wu et al.
0371	C.D. Hufford et al.	3735	M. Niwa et al.
0419	S. Tsubotani et al.	3913	Atta-Ur-Rahman and S. Khanum
0523	S. Paphassarang et al.	4167	M. Alpegiani et al.
0561	M. Hasegawa et al.	4183	B. Chenera and W. Reusch
0889	J.M. Boente et al.	4191	J.T. Snow et al.
0905	T.-S. Lin and M.-C. Liu	4223	Z. Yoshida et al.
0959	M. Nitta and T. Kobayashi	4227	Z. Yoshida et al.
1007	R.A.S. Chandrarutna et al.	4267	L. Crombie and R. Denman
1031	W. Pfleiderer	4283	N.L. Leow and J.A.H. MacBride
1035	M.F. Zipplies and H.A. Staab	4285	F. Amat-Guerri et al.
1061	B. Maurer and A. Hauser	4429	Y. Yamashita et al.
1085	K. Mori et al.	4471	H.G. Thomas and H.-W. Schwager
1137	J. Brokatzky-Geiger and W. Eberbach	4525	Y. Ito et al.
1287	D.H.R. Barton et al.	4601	I. Kubo and T. Matsumoto
1373	M. Nakata et al.	4673	S. Takose et al.
1445	K. Hafner and G.F. Thiele	4689	M. Yamashita et al.
1505	Y. Fukazawa et al.	4697	M.B. Rubin

4707	M.K.W. Li and P.J. Scheuer		0497	T. Sano et al.
4733	M.S. Manhas et al.		0538	M. Ikeda et al.
4761	H. Saito et al.		0801	Y. Inamori et al.
4801	G.M. Gunaherath and A.A.L. Guna-		0808	T. Fukai et al.
	tilaka		0860	S. Kitanaka and M. Takido
4837	M. d'Ischia et al.		0930	K. Ogawa et al.
4917	K.A. Parker et al.		1170	M. Sakamoto et al.
4941	N. Fusetani et al.		1178	J. Maruyama and K. Ito
5127	P. Wipf and H. Heimgartner		1209	S. Kitagawa et al.
5143	S. Nakatsuji et al.		1256	S. Kadota et al.
5401	H. Nakamara et al.		1260	Y. Hano et al.
5419	N. Morita et al.		1313	M. Shimizu et al.
5599	A.G. Anastassiou and M.R. Saadein		1355	N. Tanaka et al.
5619	J. Zemlicka		1373	M. Nakagawa et al.
5669	T. Kumagai et al.		1423	K. Kaji, H. Nagashima and H. Oda
5809	L.B. Volodarsky et al.		1441	T. Yamaguchi and M. Saneyoshi
5847	B.S. Joshi et al.		1472	M. Iinuma et al.
5907	S. Tanaka et al.		1583	Y. Fujimoto et al.
5925	M. Kobayashi et al.		1597	T. Inaba et al.
5933	M.J. Campello et al.		1604	T. Ueda et al.
6051	Atta-Ur-Rahman and J. Fatima		1612	H. Tatematsu et al.
			1641	M. Tashiro et al.

89-   -84, Angew. Chem., Intl. Ed., 23 (1984)

0053B	R. Gompper and H. Glöckner		1647	H. Furukawa et al.
0059	N. Wiberg et al.		1665	Y. Tominaga et al.
0063	S. Yoneda et al.		1667	M. Kobayashi et al.
0074	J. Tamariz and P. Vogel		1709	M. Ito et al.
0081	E. Kölling et al.		1724	H. Asada et al.
0241	P. Spang and H. Durr		1761	M. Yoga et al.
0251	K. Bartels and H. Hopf		1808	T. Murakami et al.
0314	J. Bindl et al.		1815	T. Murakami et al.
0440	G. Quinkert et al.		1885	I. Yoshizawa et al.
0508	D. Kuck et al.		1998	T. Kuraishi et al.
0509	R. Schulz and A. Schweig		2111	T. Konishi et al.
0521	M. Herberhold and L. Haumaier		2126	K. Irie et al.
0622	T. Troll et al.		2146	S. Matsugo et al.
0631	H. Quast and J. Christ		2165	T. Okuda et al.
0717	E. Vogel et al.		2224	R. Yoda et al.
0719	E. Vogel et al.		2230	M. Hanaoka et al.
0736	H. Priebe et al.		2290	H. Kanamori et al.
0890	G. Seitz et al.		2296	M. Iinuma et al.
0894	G. Märkl and W. Burger		2406	A. Kusai and S. Ueda
0960	J. Daub et al.		2430	M. Miyashita et al.
			2456	T. Kurihara et al.
			2486	H. Takagi et al.

90-   -84, Polyhedron, 3 (1984)

0729	M.N. Bakola-Christianopoulou		2544	M. Taniguchi et al.
			2591	K. Hirota et al.
			2609	S. Ozaki et al.

93-   -84, J. Appl. Chem. U.S.S.R., 57 (1984)

1038	V.A. Gorchakova et al.		2617	T. Konoshima and T. Sawada
			2622	M. Fujita et al.
			2676	T. Tanaka et al.

94-   -84, Chem. Pharm. Bull., 32 (1984)

0011	K. Yakushijin et al.		2730	H. Tsukamoto et al.
0023	K. Yakushijin et al.		2745	K. Watanabe et al.
0094	K. Kawai et al.		2815	A. Numata et al.
0122	Y. Tominaga et al.		2832	I. Matsunaga et al.
0166	C.-C. Chen et al.		2910	Y. Tominaga et al.
0170	T. Ohmoto and K. Koike		2947	F. Abe et al.
0295	Y. Hirai et al.		3009	H. Kobayashi et al.
0325	A. Namata et al.		3043	T. Miyase and S. Fukushima
0336	E. Sato et al.		3053	M. Yamato et al.
0350	S. Ueda et al.		3093	Y. Asahi et al.
0354	S. Sakai et al.		3244	O. Nishikawa et al.
0362	G.-Q. Liu et al.		3267	T. Miyase et al.
0447	C. Iwata et al.		3354	M. Iinuma et al.
0457	S. Amemiya et al.		3378	Y. Chen et al.
0490	N. Tanaka et al.		3384	Y. Tominaga et al.
			3396	K. Yamakawa et al.

3417   H. Shibuya et al.
3436   S. Kitanaka and M. Takido
3452   T. Satoh et al.
3483   R. Yoda et al.
3493   Y. Kashiwada et al.
3501   Y. Kashiwada et al.
3518   E. Ohshima et al.
3525   E. Ohshima et al.
3579   T. Ohmoto and K. Koike
3674   R.A. Barnes et al.
3747   H. Tanaka et al.
3770   M. Kobayashi et al.
3774   Y. Hitotsuyanagi et al.
3787   T. Okuda et al.
3866   H. Sai et al.
3873   Y. Komoda et al.
3880   H. Kobayashi et al.
3926   Y. Kanaoka et al.
4003   Y. Hirai et al.
4137   S. Gopalakrishnan et al.
4217   M. Iinuma et al.
4220   M. Tashiro et al.
4233   Y. Hitotsuyanagi et al.
4252   T. Fujii et al.
4292   Y. Yamamoto et al.
4350   K. Ohta et al.
4388   M. Taguchi et al.
4410   T. Kurihara et al.
4437   K. Kaji et al.
4455   T. Noro et al.
4478   T. Kosuge et al.
4482   H. Tsukamoto et al.
4620   T. Satake et al.
4625   S. Kanatomo et al.
4653   S. Nishibe et al.
4662   T. Okuda et al.
4698   S. Yoshimura et al.
4707   T. Chiba et al.
4721   K. Matoba et al.
4842   T. Fujii et al.
4852   H. Takahashi et al.
4935   M. Iinuma et al.
5023   E. Shimizu et al.
5036   N. Fujii et al.

95-   -84, Yakugaku Zasshi, 104 (1984)
0037   T. Inoue, Y. Arai and M. Nagai
0127   Y. Tominaga et al.
0134   Y. Tominaga et al.
0142   T. Murakami et al.
0191   O. Tamemasa et al.
0222   S.I. Sakai et al.
0238   N. Ezaki and S. Sakai
0440   Y. Tominaga et al.
0839   H. Itokawa et al.
0848   I. Kitagawa et al.
0909   Y. Isomura et al.
0935   M. Nagai et al.
0946   M. Ju-ichi et al.
0951   H. Iwabuchi et al.
1122   E. Tomitori and T. Okamoto
1232   S. Uesato et al.

96-   -84, The Analyst, 109 (1984)
1003   T. Toyo'oka and K. Imai

97-   -84, Z. Chemie, 26 (1984)
0021   M. Reichenbacher and R. Paetzold
0146   M. Pulst et al.
0147   U. Hauer et al.
0183   M. Pulst et al.
0251   A.M. Richter et al.
0253   J. Konig et al.
0254   J. Wrubel and R. Mayer
0256   J. Wrubel and R. Mayer
0286   E. Fanghanel et al.
0287   M. Weissenfels et al.
0292   H.-D. Ilge et al.
0328   S. Leistner et al.
0331   M. Lischewski
0435   W. Schroth et al.
0438   G. Barnikow and A.A. Martin

98-   -84, J. Agr. Food Chem., 32 (1984)
0021   P.-W. Lu et al.
0596   J.J. McDonald et al.
0676   Y.S. Chauhan and B. Toth
0749   M.S. Blecke and J.E. Casida
0778   H.G. Cutler et al.
0873   P.K. Freeman and K.D. McCarthy
1134   G.B. Quistad and L.E. Staiger

99-   -84, Theor. Exptl. Chem., 20 (1984)
0415   A.A. Ishchenko et al.

100-   -84, J. Natural Products, 47 (1984)
0001   B.D. Krane et al.
0055   F.J.Q. Monte et al.
0076   S.H. Grode and J.H. Cardellina II
0106   M. Arisawa et al.
0117   X.Z. Feng et al.
0123   M. Parra et al.
0143   S. Funayama et al.
0179   R. Maurya et al.
0270   R.P. Borris and G.A. Cordell
0285   S. Funayama and G.A. Cordell
0325   S.C. Basa and R.N. Tripathy
0331   C.-T. Che et al.
0342   R. Hocquemiller et al.
0353   R. Hocquemiller et al.
0363   A.C. Monteiro de Farias et al.
0379   J. Bhattacharyya et al.
0391   S. Ahmad
0397   W. Wiegrebe et al.
0453   M. Arisawa et al.
0459   M. Abd El-kawi et al.
0465   M. Matsui and Y. Watanabe
0478   C. Kan et al.
0504   A. Jossang et al.
0525   A. Urzua and R. Torres
0527   M. Sahai and I. Kirson
0530   N. Shoji et al.
0539   R. Hocquemiller et al.
0547   W.S. Woo and S.S. Kang
0565   H. Guinaudeau et al.
0585   D.S. Bhakuni and R. Chaturredi
0615   C. Tringali et al.
0620   G. Delle Monache et al.
0626   A.G. Ober et al.
0648   A. Ali et al.
0652   T.C. Dabrah and A.T. Sneden

0687	E. Seguin et al.
0698	D.E. Nettleton, Jr.
0708	S.M. El-Sayyed et al.
0753	B. Gozler and M. Shamma
0775	M.S. Karawya et al.
0796	G.R. Pettit et al.
0809	O. Servettaz et al.
0828	R.H. McClanahan and L.W. Robertson
0835	P. Perera et al.
0858	M. Matsui et al.
0868	M. Parra et al.
0877	G. Cimino et al.
0882	F.M. Eckenrode
0913	G.R. Pettit et al.
0920	A.G. Ober et al.
0947	V. Amico et al.
0953	M. Nicoletti et al.
0988	X. Fang et al.
0994	J. Polonsky et al.
1003	A. Evidente et al.
1009	M.B. Ksebati et al.
1013	M.A. Al-Yahya et al.
1040	A. Rivera et al.
1048	G.I. Yakimov et al.
1050	E. Valencia et al.
1061	A. Evidente et al.

101-   -84A, J. Organomet. Chem., 260 (1984)
0199   M.E.E. Meijer-Veldman and H.J. de
        L. Meijer

101-   -84C, J. Organomet. Chem., 262 (1984)
0001   H. Ahlbrecht et al.
0201   V.N. Postnov et al.

101-   -84E, J. Organomet. Chem., 264 (1984)
0283   D.A. Lemenovskii et al.

101-   -84I, J. Organomet. Chem., 268 (1984)
0091   A.D. Ryabov

101-   -84L, J. Organomet. Chem., 271 (1984)
0353   A. Katti et al.

101-   -84N, J. Organomet. Chem., 273 (1984)
0047   S.K. Brahma et al.

101-   -84R, J. Organomet. Chem., 277 (1984)
0113   K.-P. Stahl et al.
0379   D. Sellmann and J. Muller

102-   -84, Phytochemistry, 23 (1984)
0125   L. Quijano et al.
0129   W. Adolf et al.
0133   A. Ulubelen and G. Topcu
0143   P.A. Ramaiah et al.
0151   G.M. Sheriha and K.M. Abou Amer
0167   D.T. Burns et al.
0181   D. Hofer and H. Greyev
0188   K. Ito et al.
0195   M. Pinar et al.
0200   D.J. Robins and M.A. Sefton
0301   H. Fukui et al.
0307   K. Inoue et al.
0313   K. Inoue et al.
0323   A.A.L. Gunatilaka et al.

0431	S. Huneck
0449	E. Dagne and W. Steglich
0464	F.C. Seaman et al.
0467	T. Iwagawa and T. Hase
0468	T. Iwagawa et al.
0471	P. Bhattacharyya and A. Chakra-borty
0533	D.W. Cameron et al.
0599	W. Herz et al.
0611	G. Javona et al.
0623	R.D.H. Murray et al.
0631	S.D. Fang et al.
0635	R. Osborne and K.H. Pegel
0639	D.D. Rowan and R.H. Newman
0661	L.M.V. Trevisan et al.
0671	P.L. Majumder and N. Datta
0675	G. Delgado et al.
0688	G.H. Dartayet et al.
0699	H. Tsukamoto et al.
0703	B. Achari et al.
0704	A. Chatterjee et al.
0709	Atta-Ur-Rahman and S. Khanum
0759	J.A. Schneider et al.
0765	K.A. Khan and A. Shoeb
0767	D.J. Robeson and G.A. Strobel
0813	E. Maldonado et al.
0817	F.C. Seaman et al.
0823	A.L. Perez C. et al.
0829	L. Quijano and N.H. Fischer
0833	L. Quijano and N.H. Fischer
0843	G. Savona et al.
0849	G. Savona et al.
0853	A. Bagchi et al.
0857	J.D. Connolly et al.
0859	J. Borges-del-Castillo et al.
0863	J. Lemmich and M. Shabana
0871	J.S. Dahiya et al.
0887	I. Murakoshi et al.
0897	A.V. Subba Rao et al.
0910	A.G. Ober et al.
0919	H.M.G. Al-Hazimi et al.
0929	A.A.L. Gunatilaka et al.
1003	K. Konno et al.
1029	D.E. Jackson and P.M. Dewick
1063	F.C. Seaman et al.
1125	C.M. Ojinnaka et al.
1129	M. Hirotani et al.
1143	J.-S. Liu et al.
1147	D.E. Jackson and P.M. Dewick
1167	S. Ghosal et al.
1198	H. Tanaka et al.
1199	C.A. Elliger
1204	N. Le Van
1206	N.B. Mulchandani and S.R. Venka-tachalam
1281	J. Gershenson et al.
1289	A.G. Ober et al.
1293	A.C. Pinto et al.
1323	W.H. Gerwick
1331	L.M.V. Tillekeratone and F.J. Schmitz
1431	R. Bernini et al.
1439	A.G. Ober et al.
1461	W. Adolf et al.
1465	M.-C. Rodriguez et al.
1485	M. Coskum et al.

1607	Y. Ohta et al.		2708	J.U. Oguakwa
1637	G.A. Miana and H.M.G. Al-Hazimi		2781	R. Capasso et al.
1647	G. Rucker et al.		2789	R.G. Powell et al.
1651	M. Martinez et al.		2839	H. Tsukamoto et al.
1665	R. Mata et al.		2885	C. Ino et al.
1677	A. Kelecom		2911	A.L. Perez et al.
1681	C. Tringali et al.		2931	N.K. Khare
1689	E.H. Seip and E. Hecker		2962	G.S.R. Subba Rao et al.
1717	W.C. Evans et al.			
1729	E. Dagne and W. Steglich	103-	-84,	Khim. Geterosikl. Soedin., 20
1733	P. Cheng and K.-H. Lee			(1984)
1737	R.M. Pinheiro et al.		0024	N.M. Turkevich and G.N. Sementsiv
1757	G. Delle Monache et al.		0029	E.K. Mikitenko and N.N. Romanov
1765	K.K. Bhutani, M. Ali and C.K. Atal		0047	G.P. Tokmakov et al.
1793	A. Ulubelen et al.		0052	V.N. Kopranenkov et al.
1805	B. Rodriguez et al.		0055	M.A. Yurovskaya et al.
1816	S. Walia and S.K. Mukerjee		0058	M.A. Yurovskaya et al.
1818	Z. Rozsa et al.		0062	M.G. Cheshmaritashvili et al.
1819	J. de Pascual Teresa et al.		0066	V.G. Kartsev et al.
1825	P. Bhattacharyya and B.K. Chowdhury		0077	V.M. Potapov et al.
1909	Y. Shoyama et al.		0081	V.M. Potapov et al.
1917	T. Tanahashi et al.		0097	A.Y. Tikhonov et al.
1947	G. Delle Monache et al.		0106	V.D. Orlov et al.
1951	J.W. Blunt et al.		0150	A.F. Pozharskii et al.
1967	Y.-L. Liu et al.		0183	Z.F. Solomko et al.
1971	L. Quijano et al.		0196	V.M. Dziomko et al.
2013	T.T. Dabrah and A.T. Sneden		0201	A.D. Shutalev et al.
2017	V. Amico et al.		0204	V.S. Mokrushin et al.
2021	L.M.X. Lopes et al.		0210	Y.N. Bulychev et al.
2033	W. Pushan et al.		0215	Y.N. Bulychev et al.
2039	A.V.B. Sankaram and V.V.N. Reddy		0263	E.V. Tsoi et al.
2043	M. Aritomi and T. Kawasaki		0275	E.I. Voronina et al.
2075	M. Pinar		0277	G.M. Ostapchuk et al.
2077	P.G. Waterman and I. Mohammad		0280	S.A. Samsoniya et al.
2094	L.R. Angeles et al.		0283	S.A. Samsoniya et al.
2095	J.B. del Castillo et al.		0295	N.S. Prostakov et al.
2101	P. Romoff et al.		0304	E.M. Peresleni et al.
2104	B.K. Rao et al.		0315	G.V. Grishina et al.
2108	S. Yahara and I. Nishioka		0343	V.G. Kartsev and T.S. Pokidova
2118	V.B. Pandey et al.		0359	M.A. Kudinova et al.
2193	S. Manulis et al.		0372	L.N. Chupina et al.
2293	G. Adam et al.		0376	S.A. Samsoniya et al.
2313	I. Calis et al.		0394	T.F. Ponasenkova et al.
2323	J. Banerji et al.		0399	T.F. Ponasenkova et al.
2329	F.S. El-Feraly		0416	V.G. Kartsev et al.
2359	A.R. Araujo et al.		0431	E.L. Khamina et al.
2379	K. Sachdev and D.K. Kulshreshtha		0439	L.V. Ershov et al.
2390	W.G. De Oliveira et al.		0443	E.E. Grinshtein et al.
2394	A. Guerriero and F. Pietra		0447	A.D. Shutalev et al.
2396	S. Huneck et al.		0455	V.L. Rusinov et al.
2407	D. Guillaume et al.		0505	V.G. Kartsev et al.
2409	P. Bhattacharyya et al.		0508	V.G. Kartsev et al.
2535	S. Uesato et al.		0515	A.I. Pavlyuchenko et al.
2539	I. Calis and O. Sticher		0519	N.S. Prostakov et al.
2545	G. Appendino et al.		0522	V.V. Ogle
2553	S.D. Fang and E. Rodriguez		0537	A.V. Ivashchenko et al.
2557	J. Gershenzon and T.J. Mabry		0542	V.D. Orlov et al.
2573	F.R. Melek et al.		0571	L.M. Gornostaev and G.F. Zeibert
2593	B.P. Pradhan et al.		0577	Y.A. Sharanin et al.
2617	I. Messana et al.		0607	V.M. Petrichenko and M.E. Konshin
2629	V. Lakshmi et al.		0665	M.S. Lyubich et al.
2647	L.M.X. Lopes et al.		0717	L.M. Gornostaev et al.
2659	P. Thepenier et al.		0734	S.M. Ramsh et al.
2672	Y. Shizuri et al.		0735	V.S. Fedenko et al.
2693	S.A. Sparace et al.		0748	A.M. Shul'gas and G.V. Ponomarev
2706	N. El-Sebakhy and P.G. Waterman		0761	A.M. Belostotskii and A.B. Shapiro

0767   A.A. Tolmachev et al.
0787   V.D. Orlov et al.
0791   K.V. Fedotov et al.
0838   V.K. Daukshas et al.
0918   Y.A. Manaev et al.
0921   E.K. Mikitenko and N.N. Romanov
0930   V.M. Dziomko et al.
0951   A.D. Dubonosov et al.
0957   V.P. Litvinov et al.
0962   Y.L. Gol'dfarb et al.
0976   M.S. Lyubich et al.
0978   V.A. Kovtunenko et al.
0988   L.M. Zorin and G.I. Zhungietu
0996   S.A. Samsoniya et al.
1033   I.Z. Lulle et al.
1035   Z.F. Solomko et al.
1038   Y.M. Shafran et al.
1088   G.A. Mironova et al.
1095   T.E. Khoshtariya et al.
1119   V.A. Glushkov et al.
1123   T.E. Khoshtariya et al.
1128   N.S. Prostakov et al.
1136   A.K. Sheinkman et al.
1152   V.D. Orlov et al.
1155   A.G. Drushlyak et al.
1162   G.V. Shishkin
1185   V.I. Gefenas and P.I. Vainila-
        vichyus
1222   M.A. Gal'bershtam et al.
1226   I.M. Gavrilyuk et al.
1241   A.A. Krauze et al.
1245   A.R. Katritzky et al.
1254   P.B. Kurapov et al.
1262   Y.R. Tymyanskii et al.
1266   V.P. Perevalov et al.
1270   O.A. Zagulyaeva et al.
1276   A.G. Drushlyak
1281   K.F. Turchin et al.
1287   I.D. Shingarova et al.
1342   Y.N. Kreitsberga et al.
1364   K.M. Dyumaev et al.
1370   V.D. Orlov et al.
1374   V.P. Sevodin et al.

104-   -84, Zhur. Organ. Khim., 20 (1984)
0061   V.A. Mironov et al.
0140   A.V. Moskvin et al.
0149   V.I. Letunov and T.G. Tkhov
0198   A.A. Gogachev et al.
0257   E.N. Manukov et al.
0278   G.A. Tolstikov et al.
0294   T.F. Titova et al.
0307   V.V. Bardin et al.
0370   Y.S. Andreichikov and N.V. Gel't
0381   L.S. Tul'chinskaya et al.
0385   N.A. Orlova et al.
0390   A.Y. Il'chenko et al.
0395   S.M. Makin et al.
0408   A.A. Suvorov et al.
0449   V.P. Ivshin et al.
0465   V.K. Daukshas et al.
0473   T.I. Bychkova et al.
0485   G.A. Tolstikov et al.
0523   R.G. Dubenko et al.
0534   Z.F. Solomko et al.
0565   I.M. Gershkovich et al.

0580   G.A. Lanovaya et al.
0599   V.V. Mel'nikov et al.
0604   M.V. Gorelik et al.
0733   L.V. Ektova and E.P. Fokin
0738   V.A. Shigalevskii et al.
0745   M.V. Gorelik and R.A. Alimova
0773   O.S. Povalyaeva et al.
0795   L.M. Gornostaev and T.I. Lavri-
        kova
0806   G.A. Tolstikov et al.
0809   I.E. Efremova et al.
0810   K.A. Balodis et al.
0861   V.A. Petukhov et al.
0967   A.A. Bogachev et al.
0976   V.P. Perevalov et al.
0980   Y.E. Gerasimenko et al.
1082   V.N. Drozd et al.
1088   T.D. Petrova et al.
1223   M.V. Povstyanoi et al.
1224   L.N. Markovskii et al.
1264   Z.F. Pavlova et al.
1353   V.A. Bren' et al.
1402   Y.K. Sharanin et al.
1415   M.V. Gorelik and R.A. Alimova
1423   D.S. Yufit et al.
1515   V.N. Baidin et al.
1525   T.A. Zhidkova et al.
1528   T.A. Zhidkova et al.
1574   I.A. Dorofeev et al.
1592   L.M. Gornostaev et al.
1595   Y.S. Andreichikov et al.
1609   V.N. Barinova et al.
1642   B.A. Trofimov et al.
1643   N.A. Keiko et al.
1669   E.P. Karpenko et al.
1732   T.F. Titova et al.
1766   M.V. Gorelik et al.
1771   L.L. Pushkina et al.
1780   S.A. Russkikh et al.
1828   Y.A. Sharanin et al.
1837   Y.A. Sharanin et al.
1844   N.G. Malyuta et al.
1884   G.G. Melikyan et al.
1911   N.N. Magdesieva et al.
1921   D.A. Oparin et al.
1930   A.P. Molchanov and R.R. Kostikov
2012   L.M. Gornostaev and T.I. Lavri-
        kova
2032   G.A. Tolstikov and E.E. Shul'ts
2045   L.M. Yagupol'skii et al.
2076   G.A. Tolstikov et al.
2172   V.M. Berestovitskaya et al.
2216   Y.A. Sharanin et al.
2225   Y.A. Sharanin et al.
2234   L.M. Gornostaev and V.A. Levdan-
        skii
2277   G.V. Nekrasova et al.
2285   N.Y. Kuz'mina et al.
2302   V.R. Denisov et al.
2335   A.A. Akhrem et al.
2375   O.B. Kun et al.
2395   T.A. Dashevskaya et al.

105-   -84, Khim. Prirodn. Soedin., 20 (1984)
0068   I.A. Bessonova et al.
0071   I.S. Israilov et al.

0076   I.A. Israilov et al.
0079   S.F. Aripova et al.
0113   V.Y. Bogirov et al.
0126   N.K. Utkina and S.A. Fedoreev
0182   N.D. Abdullaev et al.
0226   E.K. Batirov et al.
0233   S.S. Yusupova et al.
0300   U. Baltaev et al.
0354   G.G. Balezina et al.
0378   T.S. Tulyagonov et al.
0467   E.L. Kristallovich et al.
0486   V.N. Odinokov et al.
0509   E.N. Zhukovich and V.Y. Vachnadze
0592   M.E. Perel'son et al.
0599   D.M. Razakova et al.
0617   I.A. Kir'yanova and Y.E. Sklyar
0644   I.A. Israilov et al.
0700   Z. Saatov et al.

106-   -84, Die Pharmazie, 39 (1984)
0092   G. Schubert et al.
0117   L.B. Pruna et al.
0429   B. Proksa et al.
0687   B. Proksa
0711   M. Hauteville and J. Favre-Bonvin
0717   K. Kottke and H. Kuhmstedt

107-   -84, Synthetic Comm., 14 (1984)
0453   J.M. Muchowski and D.R. Solas
0725   F.E. Carlon and R.W. Draper
0737   R. Chenevert et al.
0925   S.A. Ferrino and L.A. Maldonado
0967   L. Rodriguez-Hahn et al.

108-   -84, Israel J. Chem., 24 (1984)
0302   S. Shinkai et al.

109-   -84, Doklady Phys. Chem., 274-279 (1984)
0338   V.V. Korshak et al.

110-   -84, Russian J. Phys. Chem., 58 (1984)
0057   G.K. Glushonok et al.
0556   V.V. Danilov et al.
0939   T.P. Filippova et al.

111-   -84, European J. Med. Chem., 19 (1984)
0089   F.G. de las Heras et al.
0223   A. Sugimoto et al.
0267   J. Sturtz and G. Guillamot
0365   M. Faulques et al.
0465   P. Dalmases et al.

112-   -84, Spectroscopy Letters, 17 (1984)
0765   C.A.L. Mahaffy and J. Rawlings

116-   -84, Macromolecules, 17 (1984)
1624   G.D. Andrews
1895   C. Jakus and G. Smets
2468   D.C. Neckers and I.I. Abu-Abdoun
2934   J.E. Fernandez and K. Al-Jumah

117-   -84, Org. Preps. and Procedures, 16 (1984)
0115   D. Loganathan et al.
0299   S. Julia et al.
0309   H.D. Hollis Showalter and J.M. Hoftiezer

118-   -84, Synthesis (1984)
0062   K. Gewald and U. Hain
0119   S. Seitz and H.-S. The
0128   J. Moskal and P. Milant
0132   A.R. Daniewski and W. Wojciech-owsky
0230   D. Villemin et al.
0262   T. Kampchen et al.
0275   H. Hartmann and J. Liebscher
0311   T.P. Johnston and G.S. McCaleb
0315   J.G. Schantl et al.
0401   V. Nair and S.D. Chamberlain
0410   D.E. Gibbs and C. Larsen
0418   P.K. Tripathy and A.K. Mikerjee
0424   V. Nair and T.S. Jahnke
0479   F.S. Guziec, Jr. and J.M. Russo
0485   P. Guerry and R. Neier
0582   M. Balogh et al.
0618   D.E. Orr
0686   A. Landau et al.
0728   N.A. Bumagin et al.
0732   I.A. Wolff et al.
0752   G. Seitz and and G. Offermann
0852   W.-D. Rudorf
0859   J.A.M. van den Goorbergh et al.
0938   N.R. Ayyangar et al.
0953   M. Matsuoka et al.
0961   N.B. Dyatkina and A.V. Azhayev
0965   R. Charubala et al.
0974   Z. Arnold et al.
1075   M. Igarashi et al.

119-   -84, S. African J. Chem., 37 (1984)
0171   J.L.M. Dillen and O. Meth-Cohn
0187   J.C.A. Boeyens et al.

120-   -84, Pakistan. J. Sci. Ind. Research, 27 (1984)
0062   S.M.M. El-Shafie
0066   A.M. El-Naggar et al.
0326   T.U. Qasi

121-   -84, J. Macromol. Sci. A, 21 (1984)
0021   U.G. Deshpande and J.R. Shah

123-   -84, Moscow U. Chem. Bull., 39 (1984)
0075   L.A. Kazitsyna

126-   -84, Makromol. Chem., 185 (1984)
0037   C. Matyjaszewski
0499   H. Kammerer and G. Hegemann
0687   H. Rosemeyer and F. Seela
1395   H. Shirai et al.
1727   M.A. Müller and G. Wagner
1795   H.A. Gaur et al.
1809   R.D. Grant et al.
2347   H. Hirai et al.
2497   F.X.W. Basset, Jr. and O. Vogl

126-   -84B, Makromol. Chem., Rapid Comm., 5 (1984)
0221   V. Böhmer et al.

128-   -84, Croatica Chem. Acta, 57 (1984)
0295   J. Kobe et al.
0415   V. Skaric et al.

131-   -84H, J. Mol. Structure, 114 (1984)
    0235   F. Fratev et al.
    0239   F. Fratev et al.
    0275   I. Timtcheva et al.
    0337   A. Mordzinski and A. Grabowska
    0417   L. Shishkova and V. Dryanska

131-   -84I, J. Mol. Structure, 115 (1984)
    0363   S. Stoyanov et al.

131-   -84J, J. Mol. Structure, 116 (1984)
    0377   D. Mirarchi and G.L.D. Ritchie

131-   -84K, J. Mol. Structure, 117 (1984)
    0019   L. Kania et al.

135-   -84, J. Appl. Spectroscopy, 40-41
            (1984)
    0312   T.A. Oleinikova et al.
    0577   A.I. Yalandina et al.
    0674   L.I. Loboda et al.
    1280   I.P. Zharkov et al.
    1330   B.M. Krasovitskii et al.

136-   -84B, Carbohydrate Research, 126 (1984)
    0081   L.M. Vazquez de Miguel et al.
    0091   J.A.G. Perez et al.

136-   -84C, Carbohydrate Research, 127 (1984)
    0141   L.M. Lerner
    0181   R.J. Clarke et al.
    0217   F.J.L. Herrera et al.

136-   -84E, Carbohydrate Research, 129 (1984)
    0131   J.A. Galbis Perez et al.

136-   -84F, Carbohydrate Research, 130 (1984)
    0179   H. Lutz and W. Pfleiderer

136-   -84G, Carbohydrate Research, 131 (1984)
    0071   J.A. Galbis Perez et al.

136-   -84H, Carbohydrate Research, 132 (1984)
    0019   F.J. Lopez Aparicio et al.
    0153   J.A. Galbis Perez
136-   -84J, Carbohydrate Research, 133 (1984)
    0033   R.A. Cadenas et al.
    0324   M.A. El Sekily et al.

136-   -84K, Carbohydrate Research, 134 (1984)
    0063   A.G. Sanchez et al.
    0075   A.G. Sanchez et al.

136-   -84L, Carbohydrate Research, 135 (1984)
    0101   A. Gomez-Sanchez et al.

137-   -84, Finnish Chem. Letters (1984)
    0097   E. Suokas et al.
    0102   B.O. Brown and B.C.L. Weedon

138-   -84, Chemistry Letters (1984)
    0053   T. Tamaki et al.
    0221   M. Okano et al.
    0231   J. Tanaka and T. Higa
    0317   M. Yoshifuji et al.
    0339   O. Soga et al.

    0393   Y. Nakadaira et al.
    0397   M. Ojika et al.
    0539   M. Niwa et al.
    0555   M. Ishibashi et al.
    0603   M. Yoshifuji et al.
    0627   T. Nozoe et al.
    0631   Y. Matsubara et al.
    0689   M. Takasugi et al.
    0693   M. Takasugi et al.
    0757   Y. Okamoto et al.
    0779   H. Tada and F. Yasuda
    1021   M. Arno et al.
    1025   R. Baur et al.
    1045   S. Araki et al.
    1145   T. Nozoe et al.
    1149   Y. Okamoto et al.
    1191   A. Yoneda and T. Azumi
    1379   H. Sakurai et al.
    1383   H. Sakurai et al.
    1397   H. Obara et al.
    1503   T. Shimo et al.
    1513   I. Kubo et al.
    1587   M. Niwa et al.
    1613   I. Kubo et al.
    1649   A.G. Gonzalez et al.
    1673   K. Tsujimoto et al.
    1687   H. Itokawa et al.
    1741   Y. Yamashita and T. Mukai
    1773   A.L. Weis and R. Vishkautsan
    1871   S. Igarashi and T. Yotsuyanagi
    1977   K. Yamamoto et al.
    2031   V.J. Freer and P. Yates
    2033   K. Satake et al.

139-   -84B, P and S and Related Elements,
            20 (1984)
    0121   F. Boberg and U. Puttins

139-   -84C, P and S and Related Elements,
            21 (1984)
    0005   B.S. Thyagarajan et al.
    0105   J. Navech et al.
    0253   G.P. Schiemenz et al.
    0259   G.P. Schiemenz and P. Nielsen
    0267   G.P. Schiemenz and P. Nielsen
    0315   F. Boberg et al.

140-   -84, J. Anal. Chem. S.S.S.R., 39
)1984)
    1028   G.N. Dudareva et al.
    1105   N.O. Mchedlov-Petrosyan et al.
    1598   A.T. Pilipenko et al.
    1797   A.V. Anisimov et al.

142-   -84, Heterocycles, 22 (1984)
    0009   S. Philip et al.
    0013   Y. Maki et al.
    0021   R. Huisgen and K. Niklas
    0059   M. Nakagawa et al.
    0063   W. Friedrichsen et al.
    0073   Z. Szulc et al.
    0097   J. Cossy and J.-P. Pete
    0101   M.L. Contreras et al.
    0107   H. Guinaudeau and D.P. Allais
    0233   V.S. Giri et al.
    0249   P.S. Khattri et al.

0261	M. Tsukayama et al.	149-	-84B, Photochem. Photobiol., 40 (184)
0333	S.C. Basa et al.	0545	M.W. Geiger et al.
0365	B. Chantegrel et al.	0641	Y. Kimura et al.
0379	G.W. Gribble et al.	0731	S. Sakura and D. Fujimoto
0473	T. Fukai et al.		
0505	A.R. Katritzky and M.T. Mokrosz	150-	-84M, J. Chem. Research (1984)
0675	K. Kaji et al.		(Microfiche)
0681	M. Yoshifuji et al.	0147	J.C. Carvetero et al.
0709	M. Tsukayama et al.	0701	P.J. Bowman et al.
0791	Y. Yamashita et al.	0965	S. Ghosal and K.S. Saini
0807	E. Belgodere et al.	1301	R.M. Acheson et al.
0993	A.D. El-Shabrawy et al.	1441	P. Uriac et al.
0997	T. Fujimoto and T. Nomura	1531	N. Khan and D.A. Wilson
1007	T. Fukai et al.	1901	J.P. Desvergne et al.
1021	M. Sako et al.	2060	M.T.M. Clements et al.
1031	S.T. Lu and I.L. Tsai	2656	W. Fatma et al.
1049	T. Sasaki and I. Shimizu	2738	I.K. Al-Khawaja et al.
1077	C. Parkanyi et al.	3001	J.A.H. MacBride and P.M. Wright
1179	M.A. Jimenez et al.	3201	A.S. Bailey et al.
1217	E. Dominguez et al.	3465	T. Nishiwaki and N. Kunishige
1319	T. Konakahara et al.	3501	N. Hoshi et al.
1369	K. Narasimhan and P.R. Kumar	3601	Y. Maki et al.
1513	J.V. Cooney and R.N. Hazlitt	3755	D.R. Buckle and D.J. Outred
1673	L.A. Mitscher et al.	3786	S.A. Patwardhan and A.S. Gupta
1701	M. Node et al.		
1709	A. Kascheres et al.	150-	-84S, J. Chem. Research (1984)
1747	S.H. Wilen et al.	0016	I. Yaveri and S. Abdi
1771	A.A. Haman and J.A. Joule	0018	P. Mencarelli and F. Stegel
1791	Y. Hano et al.	0168	C. Bank et al.
1943	T.S. Rao et al.	0232	S. Ghosal et al.
1965	S.K. Chattopadhyay et al.	0264	T. Nishiwaki and N. Abe
2047	S.W. Oliver and T.D. Smith	0266	T. Yamamoto and M. Muraoka
2203	H. Tada and F. Yasuda	0412	S. Ghosal and S. Razdan
2221	G. Baudouin et al.		
2231	I. Weiss et al.	151-	-84A, J. Photochem., 24 (1984)
2255	Y. Tsuda et al.	0249	D.J. Fife et al.
2259	K. Hirota et al.	0285	P.V. Kamat and M.A. Fox
2277	C. Herdeis and A. Dimmerling	0355	B. Savory and J.H. Turnbull
2317	A. Ohta et al.	0383	G.R. Seely and E.R. Shaw
2369	A. Ohta et al.		
2439	J.-W. Chern et al.	151-	-84C, J. Photochem., 26 (1984)
2453	N. Langlois et al.	0075	H.-D. Becker and K. Andersson
2467	Y. Miki et al.		
2509	M. Sako et al.	151-	-84D, J. Photochem., 27 (1984)
2529	A. Kascheres et al.	0017	H.-D. Brauer and R. Schmidt
2559	X.-D. Luo et al.	0085	I. Forsskahl and H. Tylli
2729	Y. Haus et al.	0101	L.L. Costanzo et al.
2739	M. Dezor-Mazur	0327	P.M. op den Brouw et al.
2805	T. Fukai et al.	0343	Y.N. Malkin et al.
2821	A. Patra et al.		
		152-	-84, Nouveau J. Chim., 8 (1984)
145-	-84, Arzneimittel Forsch., 34 (1984)	0223	J. Streith et al.
0233	H. Schaefer et al.	0475	W. Lindner et al.
0334	T. Shimizu et al.		
		154-	-84, J. Chem. Ecology, 10 (1984)
149-	-84A, Photochem. Photobiol., 39 (1984)	0547	I. Kubo et al.
0031	M.G. Neumann et al.	0641	D.W. Knight et al.
0057	D.E. Moore et al.	1435	R. Nishida et al.
0111	O.S. Wolfbeis and G. Uray		
0145	J. Blais et al.	155-	-84A, J. Fluorine Chem., 24 (1984)
0313	P.K. Das and R.S. Becker	0041	D.J. Dodsworth et al.
0391	P. Dupuis et al.	0263	J. Burdon et al.
0469	G.A. Epling et al.	0485	B.L. Booth et al.
0755	G. Vanderkooi		

155-   -84B, J. Fluorine Chem., 25 (1984)
0309   M.E. Allen et al.

156-   -84B, Photobiochem. Photobiophys.,
        8 (1984)
0051   P.C. Joshi et al.

157-   -84, Organometallics, 3 (1984)
0082   K.P.C. Vollhardt and T.W. Weidman
0141   H. Watanabe et al.
0440   J. Setsune and D. Dolphin
0495   A.J. Ashe, III and F.J. Drone
0653   P. Bickert et al.
0731   J.C. Stark et al.
1308   L. Engman

158-   -84, J. Antibiotics, 37 (1984)
0013   N. Onishi et al.
0096   L.A. Dolak et al.
0110   G. Werner et al.
0118   Y. Fukagawa et al.
0127   Y. Mutoh et al.
0143   J.M. Girodeau et al.
0187   S. Morimoto et al.
0191   M. Konishi et al.
0200   M. Konishi et al.
0211   T. Yoshioka et al.
0325   T.W. Flegel et al.
0354   S. Iwasaki et al.
0431   W.L. Parker et al.
0441   K.H. Michel et al.
0454   B. Kunze et al.
0610   H. Seto et al.
0670   K. Isono et al.
0685   G. Franceschi et al.
0700   S. Omura et al.
0706   I. Umezawa et al.
0718   T. Okuda et al.
0733   A.K. Samanta et al.
0738   H. Sano et al.
0750   H. Sano et al.
0760   H. Sano et al.
0823   T.A. Smitka et al.
0885   R.C. Mearman et al.
0931   T. Ito et al.
0941   K. Kato et al.
0943   M. Patel et al.
0984   K. Ogawa et al.
1091   H. Iwaki et al.
1149   M.D. Lee et al.
1153   K. Umehara et al.
1170   K. Takesako and T. Beppu
1187   K. Harada et al.
1260   L.A. Mitscher et al.
1264   S. Omura et al.
1273   P.F. Wiley et al.
1284   M. Yamashita et al.
1294   K. Kintaka et al.
1320   S. Takase et al.
1333   H. Kinashi et al.
1416   J. Shoji et al.
Also spectra numbered 84-1, 84-2, etc.

159-   -84, J. Carbohydrate Chem., 3 (1984)
0475   A. Calvo-matteo et al.

160-   -84, Analytical Letters, 17 (1984)
0329   J.J. Vallon et al.
1009   M. Blanco and S. Maspoch

161-   -84, Il Farmaco, 39 (1984)
0081   L. Mosti et al.
0162   M. Bianchi and A. Butti
0346   P. Roveri et al.